Reactions at Supports and Connections for a Three-Dimensional Stru

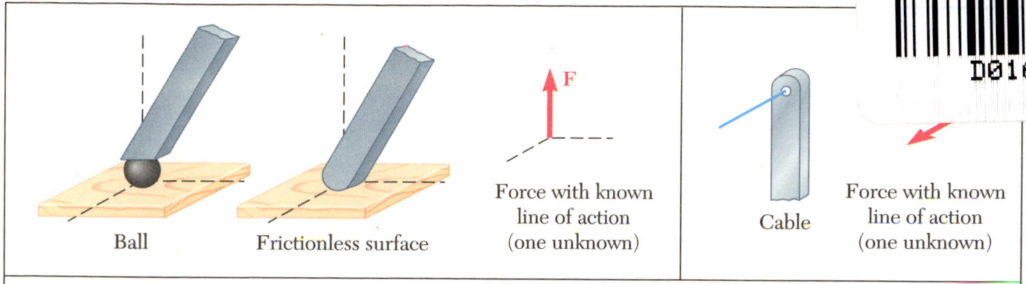

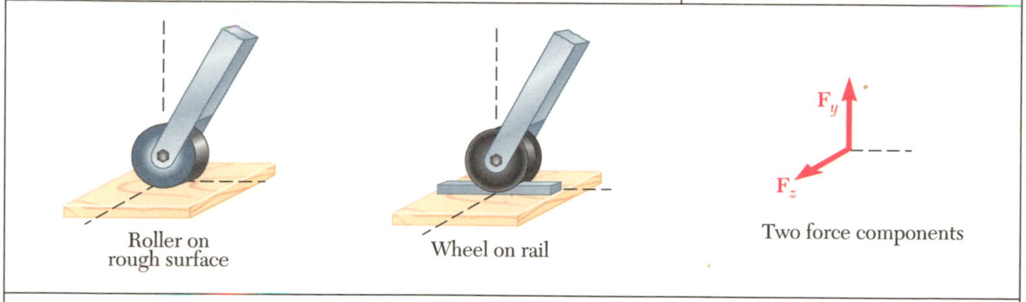

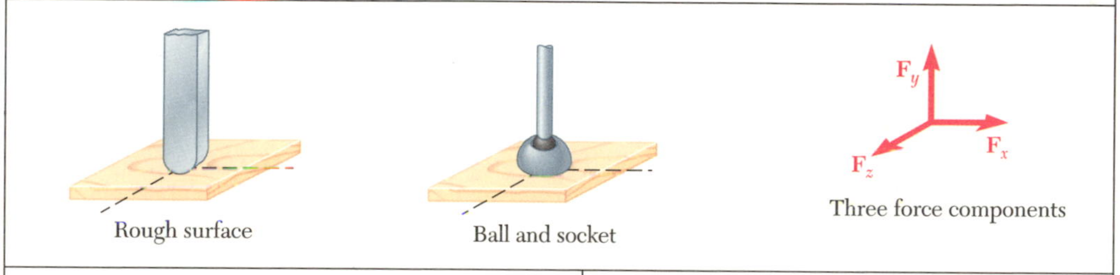

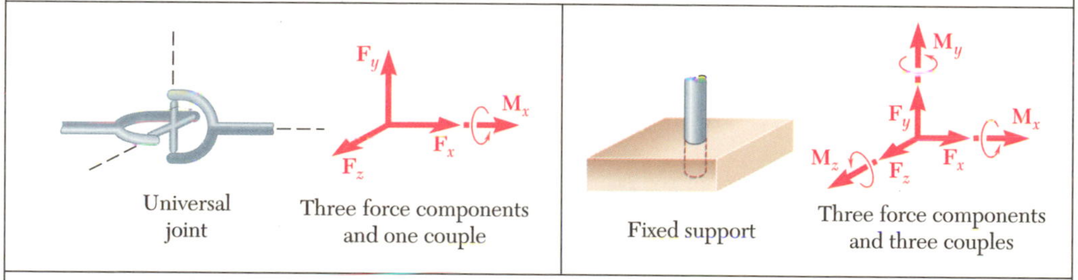

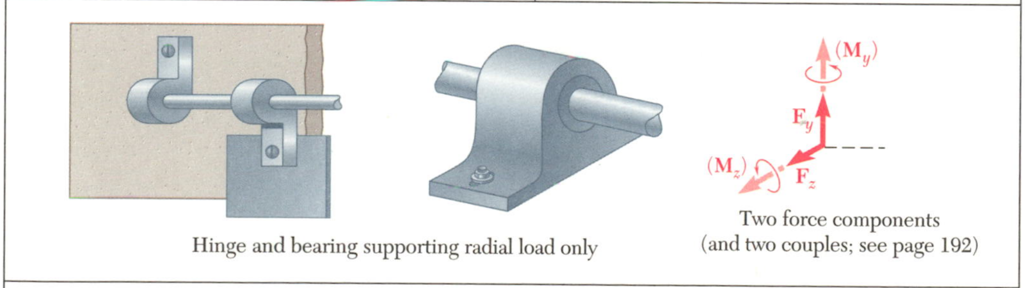

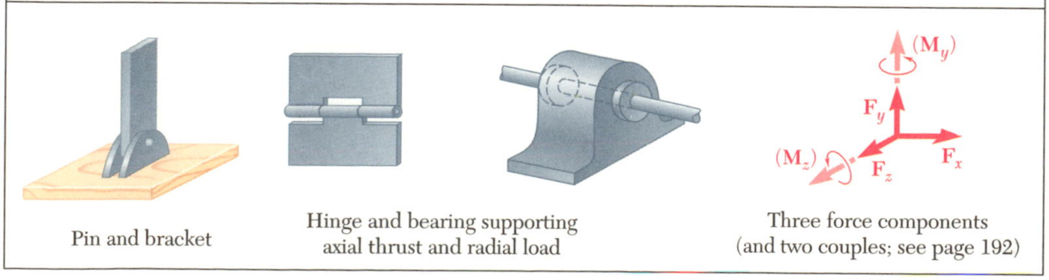

Student

Please enter my subscription for **Engineering News-Record**.

6 months ❑ $29.50 (Domestic)

Name

Address

City State Zip

❑ Payment enclosed ❑ Bill me later

McGraw_Hill CONSTRUCTION ENR

5EN2DMHE

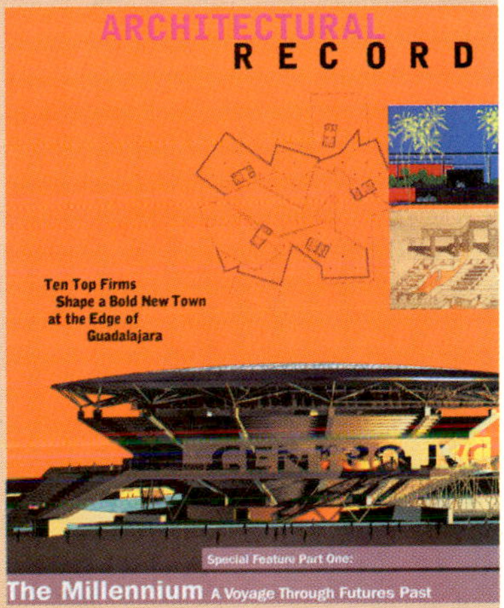

Friendly

Please enter my subscription for **Architectural Record**.

6 months ❑ $19.50 (Domestic)

Name

Address

City State Zip

❑ Payment enclosed ❑ Bill me later

McGraw_Hill CONSTRUCTION Architectural Record

5AR2DMHE

Savings

Please enter my subscription for **Aviation Week & Space Technology**.

6 months ❑ $29.95 (Domestic)

Name

Address

City State Zip

❑ Payment enclosed ❑ Bill me later

Aviation Week's AVIATION WEEK & SPACE TECHNOLOGY
www.AviationNow.com/awst

CAW34EDU

IMPORTANT:

HERE IS YOUR REGISTRATION CODE TO ACCESS YOUR PREMIUM McGRAW-HILL ONLINE RESOURCES.

For key premium online resources you need THIS CODE to gain access. Once the code is entered, you will be able to use the Web resources for the length of your course.

If your course is using **WebCT** or **Blackboard**, you'll be able to use this code to access the McGraw-Hill content within your instructor's online course.

Access is provided if you have purchased a new book. If the registration code is missing from this book, the registration screen on our Website, and within your WebCT or Blackboard course, will tell you how to obtain your new code.

Registering for McGraw-Hill Online Resources

To gain access to your McGraw-Hill web resources simply follow the steps below:

1. USE YOUR WEB BROWSER TO GO TO: **www.mhhe.com/beerjohnston7**
2. CLICK ON **FIRST TIME USER**.
3. ENTER THE REGISTRATION CODE* PRINTED ON THE TEAR-OFF BOOKMARK ON THE RIGHT.
4. AFTER YOU HAVE ENTERED YOUR REGISTRATION CODE, CLICK **REGISTER**.
5. FOLLOW THE INSTRUCTIONS TO SET-UP YOUR PERSONAL UserID AND PASSWORD.
6. WRITE YOUR UserID AND PASSWORD DOWN FOR FUTURE REFERENCE. KEEP IT IN A SAFE PLACE.

TO GAIN ACCESS to the McGraw-Hill content in your instructor's **WebCT** or **Blackboard** course simply log in to the course with the UserID and Password provided by your instructor. Enter the registration code exactly as it appears in the box to the right when prompted by the system. You will only need to use the code the first time you click on McGraw-Hill content.

Thank you, and welcome to your McGraw-Hill online resources!

Mc Graw Hill **Higher Education**

* YOUR REGISTRATION CODE CAN BE USED ONLY ONCE TO ESTABLISH ACCESS. IT IS NOT TRANSFERABLE.

0-07-292339-3 T/A BEER/JOHNSTON, VECTOR MECHANICS FOR ENGINEERS, 7/E

MCGRAW-HILL

ONLINE RESOURCES

REGISTRATION CODE

4WZU-4CRU-LF5D-WYQO-5S48

Mc Graw Hill Higher Education

Seventh Edition

VECTOR MECHANICS FOR ENGINEERS

Statics and Dynamics

FERDINAND P. BEER
Lehigh University

E. RUSSELL JOHNSTON, JR.
University of Connecticut

ELLIOT R. EISENBERG
The Pennsylvania State University

WILLIAM E. CLAUSEN
The Ohio State University

With the collaboration of
George H. Staab
The Ohio State University

McGraw Hill Higher Education

Boston Burr Ridge, IL Dubuque, IA Madison, WI New York San Francisco St. Louis
Bangkok Bogotá Caracas Kuala Lumpur Lisbon London Madrid Mexico City
Milan Montreal New Delhi Santiago Seoul Singapore Sydney Taipei Toronto

The McGraw·Hill Companies

Mc Graw Hill Higher Education

VECTOR MECHANICS FOR ENGINEERS: STATICS AND DYNAMICS, SEVENTH EDITION

Published by McGraw-Hill, a business unit of The McGraw-Hill Companies, Inc., 1221 Avenue of the Americas, New York, NY 10020.

This book is printed on acid-free paper.

3 4 5 6 7 8 9 0 VNH/VNH 0 9 8 7 6 5 4

ISBN 0–07–230491–X

Publisher: *Elizabeth A. Jones*
Associate sponsoring editor: *Debra D. Matteson*
Marketing manager: *Sarah Martin*
Project manager: *Jane Mohr*
Production supervisor: *Sherry L. Kane*
Senior media project manager: *Stacy A. Patch*
Senior media technology producer: *Phillip Meek*
Designer: *David W. Hash*
Lead photo research coordinator: *Carrie K. Burger*
Photo research: *Sabina Dowell*
Supplement producer: *Brenda A. Ernzen*
Compositor: *The GTS Companies*/York, PA Campus
Art house: *Fine Line Illustrations, Inc.*
Typeface: *10.5/12 New Caledonia*
Printer: *Von Hoffmann Corporation*

Cover Photograph: Courtesy of NASA. The Proteus aircraft was designed and built by Burt Rutan, president of Scaled Composites, Inc., to carry an 18-foot diameter telecommunications antenna system for relay of broadband data over major cities. The design allows for Proteus to be reconfigured at will for a variety of other missions such as atmospheric research, reconnaissance, commercial imaging, and launch of small space satellites. In October of 2000, Proteus set three world records for performance in its weight class, including a maximum altitude record of 62,786 feet. The flights were conducted under the sponsorship of the NASA Office of Earth Science with funding provided by the NOAA/DoD/NASA Integrated Program Office. This sponsorship includes evaluation of the Proteus aircraft as an atmospheric science and remote sensing airborne platform. The aircraft consists of an all composite airframe with graphite-epoxy sandwich construction. It has a wingspan of 77 feet 7 inches, expandable to 92 feet with removable wingtips installed. It is 56.3 feet long and 17.6 feet high and weighs 5,900 pounds, empty. Proteus is powered by two Williams-Rolls FJ44-2 turbofan engines developing 2,300 pounds of thrust each.

Cover Photograph: Courtesy Central Tunnel/Artery Project, Boston, MA. The Leonard P. Zakim Bunker Hill Bridge, shown illuminated at night, is at the north end of Boston. The Storrow Drive Connector Bridge is on the left. This shot is looking north and was taken on January 30, 2002. The bridge crosses the Charles River and connects Boston and Charlestown, Massachusetts. Conceived by Swiss bridge designer Christian Menn, the 270-foot inverted Y-shaped towers reflect the shape of the Bunker Hill Monument in Charlestown. The 1457-foot-long bridge is the widest cable-stayed bridge in the world, and the first in the United States to use an asymmetrical, hybrid design. Eight lanes of traffic pass through the legs of the twin towers, while an additional two lanes, supported by steel beams, are cantilevered from the east side of the bridge. The 745-foot-long, 183-foot-wide main span is supported by girders and two planes of cables; the back spans between the towers and the shores are supported by a single plane of cables.

The credits section for this book begins on page 1291 and is considered an extension of the copyright page.

Library of Congress Cataloging-in-Publication Data:
Vector mechanics for engineers : statics and dynamics / Ferdinand P. Beer . . . [et al.].—7th ed.
p. cm.
Includes index.
ISBN 0–07–230491–X (combined title)—ISBN 0–07–230492–8 (Dynamics)—ISBN 0–07–230493–6 (Statics).
1. Mechanics, Applied. 2. Vector analysis. 3. Statics. 4. Dynamics. I. Beer, Ferdinand Pierre, 1915–.
II. Beer, Ferdinand Pierre, 1915– Vector mechanics for engineers.

TA350.V34 2004
620.1′05—dc21 2003044560
CIP

www.mhhe.com

About the Authors

As one of the most successful author teams in engineering education, Ferd Beer and Russ Johnston are often asked how, with one author at Lehigh and the other at the University of Connecticut, they happened to write their books together and how they managed to keep collaborating on their many books with many successive revisions.

The answer to this question is simple. Russ Johnston's first teaching appointment was in the Department of Civil Engineering and Mechanics at Lehigh University. There he met Ferd Beer, who had joined that department two years earlier and was in charge of the courses in mechanics.

Ferd was delighted to discover that the young man who had been hired chiefly to teach graduate structural engineering courses was not only willing but eager to help him reorganize the mechanics courses. Both believed that these courses should be taught from a few basic principles and that the various concepts involved would be best understood and remembered by the students if they were presented to them in a graphic way. Together they wrote lecture notes in statics and dynamics, to which they later added problems they felt would appeal to future engineers, and soon they produced the manuscript of the first edition of *Mechanics for Engineers*.

The second edition of *Mechanics for Engineers* and the first edition of *Vector Mechanics for Engineers* found Russ Johnston at Worcester Polytechnic Institute and the next editions at the University of Connecticut. In the meantime, both Ferd and Russ assumed administrative responsibilities in their departments, and both were involved in research, consulting, and supervising graduate students—Ferd in the area of stochastic processes and random vibrations and Russ in the area of elastic stability and structural analysis design. However, their interest in improving the teaching of the basic mechanics courses had not subsided, and they both taught sections of these courses as they kept revising their texts and began writing the manuscript of the first edition of their *Mechanics of Materials* text.

Their collaboration has spanned many years and many successful revisions of all of their textbooks, and Ferd and Russ's contributions to engineering education have earned them a number of honors and awards. They were presented with the Western Electric Fund Award for excellence in the instruction of engineering students by their respective regional sections of the American Society for Engineering Education, and they both received the Distinguished Educator Award from the Mechanics Division of the same society. Starting in 2001, the New Mechanics Educator Award of the Mechanics Division has been named in honor of the Beer and Johnston author team.

Ferd and Russ are delighted to bring new co-authors and collaborators to the Beer and Johnston team for the seventh editions of *Vector Mechanics for Engineers.* Elliot Eisenberg worked as a collaborator on the sixth edition and is now a co-author for *Statics*. William Clausen, who studied with and was inspired by Ferd Beer, joins the seventh edition team as co-author for *Dynamics*. George Staab developed the interactive tutorial to accompany the sixth and seventh editions, and now he joins the seventh edition team as a collaborator on both the *Statics* and *Dynamics* texts.

Ferdinand P. Beer. Born in France and educated in France and Switzerland, Ferd holds an M.S. degree from the Sorbonne and an Sc.D. degree in theoretical mechanics from the University of Geneva. He came to the United States after serving in the French army during the early part of World War II and taught for four years at Williams College in the Williams-MIT joint arts and engineering program. Following his service at Williams College, Ferd joined the faculty of Lehigh University where he taught for thirty-seven years. He held several positions, including University Distinguished Professor and chairman of the Department of Mechanical Engineering and Mechanics, and in 1995 Ferd was awarded an honorary Doctor of Engineering degree by Lehigh University.

E. Russell Johnston, Jr. Born in Philadelphia, Russ holds a B.S. degree in civil engineering from the University of Delaware and an Sc.D. degree in the field of structural engineering from the Massachusetts Institute of Technology. He taught at Lehigh University and Worcester Polytechnic Institute before joining the faculty of the University of Connecticut where he held the position of Chairman of the Civil Engineering Department and taught for twenty-six years. In 1991 Russ received the Outstanding Civil Engineer Award from the Connecticut Section of the American Society of Civil Engineers.

Elliot R. Eisenberg. Elliot holds a B.S. degree in engineering and an M.E. degree, both from Cornell University. He has focused his scholarly activities on professional service and teaching, and he was recognized for this work in 1992 when the American Society of Mechanical Engineers awarded him the Ben C. Sparks Medal for his contributions to mechanical engineering and mechanical engineering technology education and for service to the American Society for Engineering Education. Elliot taught for thirty-two years, including twenty-nine years at Penn State where he was recognized with awards for both teaching and advising.

William E. Clausen. Bill holds a B.S. degree in engineering mechanics from Lehigh University and M.S. and Ph.D. degrees in engineering mechanics from the Ohio State University. Bill is a registered professional engineer specializing in structural dynamics and vibration measurements. He taught for thirty years and served as vice chairman in the Department of Engineering Mechanics at the Ohio State University and has also taught in the Department of Mechanical Engineering.

George H. Staab. George holds B.S., M.S., and Ph.D. degrees in Aeronautical Engineering from Purdue University. He worked at Sikorsky Aircraft for three years and is currently an Associate Professor in Mechanical Engineering at the Ohio State University where he has taught for twenty-two years. His scholarly activities are focused on teaching and service and he has received two college-wide awards: the Charles E. MacQuigg Outstanding Teaching Award in 1998 and the Boyer Teaching Award in 1999. George serves as the campus representative for the American Society for Engineering Education and is the faculty advisor for the student chapter of the American Society of Mechanical Engineering and several student project teams.

Contents

3
RIGID BODIES: EQUIVALENT SYSTEMS OF FORCES
73

4
EQUILIBRIUM OF RIGID BODIES
157

5
DISTRIBUTED FORCES: CENTROIDS AND CENTERS OF GRAVITY
219

6
ANALYSIS OF STRUCTURES
284

7
FORCES IN BEAMS AND CABLES
353

8
FRICTION
411

9
DISTRIBUTED FORCES: MOMENTS OF INERTIA
471

10
METHOD OF VIRTUAL WORK
557

11
KINEMATICS OF PARTICLES
601

12
KINETICS OF PARTICLES: NEWTON'S SECOND LAW

13
KINETICS OF PARTICLES: ENERGY AND MOMENTUM METHODS

14
SYSTEMS OF PARTICLES
855

15
KINEMATICS OF RIGID BODIES
915

16
PLANE MOTION OF RIGID BODIES: FORCES AND ACCELERATIONS
1025

17
PLANE MOTION OF RIGID BODIES: ENERGY AND MOMENTUM METHODS
1081

18
KINETICS OF RIGID BODIES IN THREE DIMENSIONS
1145

19
MECHANICAL VIBRATIONS

Appendix
FUNDAMENTALS OF ENGINEERING EXAMINATION

Preface

OBJECTIVES

The main objective of a first course in mechanics should be to develop in the engineering student the ability to analyze any problem in a simple and logical manner and to apply to its solution a few, well-understood, basic principles. This text is designed for the first courses in statics and dynamics offered in the sophomore or junior year, and it is hoped that it will help the instructor achieve this goal.†

GENERAL APPROACH

Vector analysis is introduced early in the text and used throughout the presentation of statics and dynamics. This approach leads to more concise derivations of the fundamental principles of mechanics. It also results in simpler solutions of three-dimensional problems in statics, and makes it possible to analyze many advanced problems in kinematics and kinetics, which could not be solved by scalar methods. The emphasis in this text, however, remains on the correct understanding of the principles of mechanics and on their application to the solution of engineering problems, and vector analysis is presented chiefly as a convenient tool.‡

Practical Applications Are Introduced Early. One of the characteristics of the approach used in this book is that mechanics of *particles* is clearly separated from the mechanics of *rigid bodies*. This approach makes it possible to consider simple practical applications at an early stage and to postpone the introduction of the more difficult concepts. For example:

- In *Statics,* the statics of particles is treated first (Chap. 2); after the rules of addition and subtraction of vectors are introduced, the principle of equilibrium of a particle is immediately applied

†This text is available in separate volumes, *Vector Mechanics for Engineers: Statics*, seventh edition, and *Vector Mechanics for Engineers: Dynamics*, seventh edition.

‡In a parallel text, *Mechanics for Engineers*, fourth edition, the use of vector algebra is limited to the addition and subtraction of vectors, and vector differentiation is omitted.

to practical situations involving only concurrent forces. The statics of rigid bodies is considered in Chaps. 3 and 4. In Chap. 3, the vector and scalar products of two vectors are introduced and used to define the moment of a force about a point and about an axis. The presentation of these new concepts is followed by a thorough and rigorous discussion of equivalent systems of forces leading, in Chap. 4, to many practical applications involving the equilibrium of rigid bodies under general force systems.

- In *Dynamics,* the same division is observed. The basic concepts of force, mass, and acceleration, of work and energy, and of impulse and momentum are introduced and first applied to problems involving only particles. Thus, students can familiarize themselves with the three basic methods used in dynamics and learn their respective advantages before facing the difficulties associated with the motion of rigid bodies.

New Concepts Are Introduced in Simple Terms. Since this text is designed for the first course in statics and dynamics new concepts are presented in simple terms and every step is explained in detail. On the other hand, by discussing the broader aspects of the problems considered, and by stressing methods of general applicability, a definite maturity of approach is achieved. For example:

- In *Statics,* the concepts of partial constraints and of statical indeterminacy are introduced early and are used throughout statics.
- In *Dynamics,* the concept of potential energy is discussed in the general case of a conservative force. Also, the study of the plane motion of rigid bodies is designed to lead naturally to the study of their general motion in space. This is true in kinematics as well as in kinetics, where the principle of equivalence of external and effective forces is applied directly to the analysis of plane motion, thus facilitating the transition to the study of three-dimensional motion.

Fundamental Principles Are Placed in the Context of Simple Applications. The fact that mechanics is essentially a *deductive* science based on a few fundamental principles is stressed. Derivations have been presented in their logical sequence and with all the rigor warranted at this level. However, the learning process being largely *inductive,* simple applications are considered first. For example:

- The statics of particles precedes the statics of rigid bodies, and problems involving internal forces are postponed until Chap. 6.
- In Chap. 4, equilibrium problems involving only coplanar forces are considered first and solved by ordinary algebra, while problems involving three-dimensional forces and requiring the full use of vector algebra are discussed in the second part of the chapter.
- The kinematics of particles (Chap. 11) precedes the kinematics of rigid bodies (Chap. 15).
- The fundamental principles of the kinetics rigid bodies are first applied to the solution of two-dimensional problems (Chaps. 16 and 17), which can be more easily visualized by the student, while three-dimensional problems are postponed until Chap. 18.

The Presentation of the Principles of Kinetics Is Unified. The seventh edition of *Vector Mechanics for Engineers* retains the unified presentation of the principles of kinetics which characterized the previous six editions. The concepts of linear and angular momentum are introduced in Chap. 12 so that Newton's second law of motion can be presented not only in its conventional form $\mathbf{F} = m\mathbf{a}$, but also as a law relating, respectively, the sum of the forces acting on a particle and the sum of their moments to the rates of change of the linear and angular momentum of the particle. This makes possible an earlier introduction of the principle of conservation of angular momentum and a more meaningful discussion of the motion of a particle under a central force (Sec. 12.9). More importantly, this approach can be readily extended to the study of the motion of a system of particles (Chap. 14) and leads to a more concise and unified treatment of the kinetics of rigid bodies in two and three dimensions (Chaps. 16 through 18).

Free-Body Diagrams Are Used Both to Solve Equilibrium Problems and to Express the Equivalence of Force Systems. Free-body diagrams are introduced early, and their importance is emphasized throughout the text. They are used not only to solve equilibrium problems but also to express the equivalence of two systems of forces or, more generally, of two systems of vectors. The advantage of this approach becomes apparent in the study of the dynamics of rigid bodies, where it is used to solve three-dimensional as well as two-dimensional problems. By placing the emphasis on "free-body-diagram equations" rather than on the standard algebraic equations of motion, a more intuitive and more complete understanding of the fundamental principles of dynamics can be achieved. This approach, which was first introduced in 1962 in the first edition of *Vector Mechanics for Engineers*, has now gained wide acceptance among mechanics teachers in this country. It is, therefore, used in preference to the method of dynamic equilibrium and to the equations of motion in the solution of all sample problems in this book.

A Four-Color Presentation Uses Color to Distinguish Vectors. Color has been used, not only to enhance the quality of the illustrations, but also to help students distinguish among the various types of vectors they will encounter. While there is no intention to "color code" this text, the same color was used in any given chapter to represent vectors of the same type. Throughout *Statics*, for example, red is used exclusively to represent forces and couples, while position vectors are shown in blue and dimensions in black. This makes it easier for the students to identify the forces acting on a given particle or rigid body and to follow the discussion of sample problems and other examples given in the text. In *Dynamics*, for the chapters on kinetics, red is used again for forces and couples, as well as for effective forces. Red is also used to represent impulses and momenta in free-body-diagram equations, while green is used for velocities, and blue for accelerations. In the two chapters on kinematics, which do not involve any forces, blue, green, and red are used, respectively, for displacements, velocities, and accelerations.

A Careful Balance Between SI and U.S. Customary Units Is Consistently Maintained. Because of the current trend in the American government and industry to adopt the international system of units (SI metric units), the SI units most frequently used in mechanics are introduced in Chap. 1 and are used throughout the text. Approximately half of the sample problems and 60 percent of the homework problems are stated in these units, while the remainder are in U.S. customary units. The authors believe that this approach will best serve the need of students, who, as engineers, will have to be conversant with both systems of units.

It also should be recognized that using both SI and U.S. customary units entails more than the use of conversion factors. Since the SI system of units is an absolute system based on the units of time, length, and mass, whereas the U.S. customary system is a gravitational system based on the units of time, length, and force, different approaches are required for the solution of many problems. For example, when SI units are used, a body is generally specified by its mass expressed in kilograms; in most problems of statics it will be necessary to determine the weight of the body in newtons, and an additional calculation will be required for this purpose. On the other hand, when U.S. customary units are used, a body is specified by its weight in pounds and, in dynamics problems, an additional calculation will be required to determine its mass in slugs (or $\text{lb} \cdot \text{s}^2/\text{ft}$). The authors, therefore, believe that problem assignments should include both systems of units.

The *Instructor's and Solutions Manual* provides six different lists of assignments so that an equal number of problems stated in SI units and in U.S. customary units can be selected. If so desired, two complete lists of assignments can also be selected with up to 75 percent of the problems stated in SI units.

Optional Sections Offer Advanced or Specialty Topics. A large number of optional sections have been included. These sections are indicated by asterisks and thus are easily distinguished from those which form the core of the basic mechanics course. They may be omitted without prejudice to the understanding of the rest of the text.

The topics covered in the optional sections in statics include the reduction of a system of forces to a wrench, applications to hydrostatics, shear and bending-moment diagrams for beams, equilibrium of cables, products of inertia and Mohr's circle, mass products of inertia and principal axes of inertia for three-dimensional bodies, and the method of virtual work. An optional section on the determination of the principal axes and moments of inertia of a body of arbitrary shape is included (Sec. 9.18). The sections on beams are especially useful when the course in statics is immediately followed by a course in mechanics of materials, while the sections on the inertia properties of three-dimensional bodies are primarily intended for the students who will later study in dynamics the three-dimensional motion of rigid bodies.

The topics covered in the optional sections in dynamics include graphical methods for the solution of rectilinear-motion problems, the trajectory of a particle under a central force, the deflection of fluid streams, problems involving jet and rocket propulsion, the kinematics and kinetics of rigid bodies in three dimensions, damped mechanical vibrations, and electrical analogues. These topics will be found of particular interest when dynamics is taught in the junior year.

The material presented in the text and most of the problems requires no previous mathematical knowledge beyond algebra, trigonometry, and elementary calculus; all the elements of vector algebra necessary to the understanding of the text are carefully presented in Chaps. 2 and 3. However, special problems are included, which make use of a more advanced knowledge of calculus, and certain sections, such as Secs. 19.8 and 19.9 on damped vibrations, should be assigned only if students possess the proper mathematical background. In portions of the text using elementary calculus, a greater emphasis is placed on the correct understanding and application of the concepts of differentiation and integration, than on the nimble manipulation of mathematical formulas. In this connection, it should be mentioned that the determination of the centroids of composite areas precedes the calculation of centroids by integration, thus making it possible to establish the concept of moment of area firmly before introducing the use of integration.

NEW TO THIS EDITION

While retaining the well-received approach and organization of previous editions, the *seventh edition* offers the following new features and improvements:

- Ninety percent of the homework problems for the seventh edition are new or revised. The emphasis on industry-related and discipline-specific questions of the new edition problems provides motivation for today's students.
- The computer problems have been revised to be used with popular computational software and the number of problems has been increased. The computer problems, many of which are relevant to the design process, are included in a special section at the end of each chapter. The problems focus on symbolic manipulation and plotting, as opposed to the programming-based computer problems in previous editions of the text.
- Numerous in-chapter photographs have been added to help students better visualize important concepts.
- Chapter outlines have been added to the introduction of each chapter to provide a preview of topics that will be covered in the chapter.
- A Fundamentals of Engineering Examination Appendix has been added for use when students prepare for the FE exam.

CHAPTER ORGANIZATION AND PEDAGOGICAL FEATURES

Chapter Introduction. Each chapter begins with an introductory section setting the purpose and goals of the chapter and describing in simple terms the material to be covered and its application to the solution of engineering problems. New chapter outlines provide students with a preview of chapter topics.

Chapter Lessons. The body of the text is divided into units, each consisting of one or several theory sections, one or several sample problems, and a large number of problems to be assigned. Each unit corresponds to a well-defined topic and generally can be covered in one lesson. In a number of cases, however, the instructor will find it desirable to devote more than one lesson to a given topic.

Sample Problems. The sample problems are set up in much the same form that students will use when solving the assigned problems. They thus serve the double purpose of amplifying the text and demonstrating the type of neat, orderly work that students should cultivate in their own solutions.

Solving Problems on Your Own. A section entitled *Solving Problems on Your Own* is included for each lesson, between the sample problems and the problems to be assigned. The purpose of these sections is to help students organize in their own minds the preceding theory of the text and the solution methods of the sample problems so that they can more successfully solve the homework problems. Also included in these sections are specific suggestions and strategies which will enable students to more efficiently attack any assigned problems.

Homework Problem Sets. Most of the problems are of a practical nature and should appeal to engineering students. They are primarily designed, however, to illustrate the material presented in the text and to help students understand the principles of mechanics. The problems are grouped according to the portions of material they illustrate and are arranged in order of increasing difficulty. Problems requiring special attention are indicated by asterisks. Answers to 70 percent of the problems are given at the end of the book. Problems for which the answers are given are set in straight type in the text, while problems for which no answer is given are set in italic.

Chapter Review and Summary. Each chapter ends with a review and summary of the material covered in that chapter. Marginal notes are used to help students organize their review work, and cross-references have been included to help them find the portions of material requiring their special attention.

Review Problems. A set of review problems is included at the end of each chapter. These problems provide students further opportunity to apply the most important concepts introduced in the chapter.

Computer Problems. Each chapter includes a set of problems designed to be solved with computational software. Many of these problems provide an introduction to the design process. In *Statics*, for example, they may involve the analysis of a structure for various configurations and loadings of the structure or the determination of the equilibrium positions of a mechanism which may require an iterative method of solution. In *Dynamics,* they may involve the determination of the motion of a particle under initial conditions, the kinematic or kinetic analysis of mechanisms in successive positions, or the numerical integration of various equations of motion. Developing the algorithm required to solve a given mechanics problem will benefit the students in two different ways: (1) it will help them gain a better understanding of the mechanics principles involved; (2) it will provide them with an opportunity to apply their computer skills to the solution of a meaningful engineering problem.

SUPPLEMENTS

An extensive supplements package for both instructors and students is available with the text. Instructor resources include: an instructor's solutions manual with complete solutions to all text problems; image sets with electronic files of all text art and photo images; PowerPoint lecture presentations for all text chapters; transparencies of additional solved problems; scripts in various computational software formats for all text computer problems; access to course management systems to accommodate your online course needs; and various other presentation and course organization resources.

Students have access to S.M.A.R.T. (Self-paced, Mechanics, Algorithmic, Review, and Tutorial), an online interactive tutorial with algorithmic quizzing which can also be used as a classroom presentation tool. Other student resources include: FE Exam-style multiple-choice quizzes with feedback; a guide for using computational software packages in mechanics courses; and many more internet-based content and learning tools.

Please visit our *Vector Mechanics for Engineers: Statics and Dynamics* seventh edition Online Learning Center (OLC) at www.mhhe.com/beerjohnston7 for more information on the supplements available with this text.

ACKNOWLEDGMENTS

The authors wish to acknowledge the collaboration of George Staab to this seventh edition of *Vector Mechanics for Engineers* and thank him especially for his crucial role in making the extensive problem set revision possible.

A special thanks go to our colleagues who thoroughly checked the solutions and answers of all problems in this edition and then prepared the solutions for the accompanying *Instructor's and Solution Manual:* Dean Updike of Lehigh University; Richard H. Lance of Cornell University; Daniel W. Yannitell of Louisiana State University; Kenneth Oster of University of Missouri, Rolla; Gerald Rehkugler of Cornell University; and Petru Petrina of Cornell University.

The authors thank the many companies that provided photographs for this edition. We also wish to recognize the determined efforts and patience of our photo researcher Sabina Dowell.

The authors also thank the members of the staff at McGraw-Hill for their support and dedication during the preparation of this new edition. We particularly wish to acknowledge the contributions of Sponsoring Editor Debra Matteson and Project Manager Jane Mohr.

The authors gratefully acknowledge the many helpful comments and suggestions offered by users of the previous editions of *Vector Mechanics for Engineers*.

Ferdinand P. Beer
E. Russell Johnston, Jr.
Elliot R. Eisenberg
William E. Clausen

Acknowledgments

The McGraw-Hill engineering mechanics editorial and marketing team would like to join the authors in gratefully acknowledging the individuals listed in the authors' preface for their contributions to the seventh editions of the texts and solutions manuals. In addition, we would also like to thank the following reviewers, focus group attendees, and symposium attendees whose comments and suggestions over the past few years have not only helped us to develop the supplemental and media resources for the seventh editions of *Vector Mechanics for Engineers* but have also helped us to establish standards and goals for developing other texts, supplements, and new media resources for statics and dynamics courses.

REVIEWERS FROM THE UNITED STATES

Shaaban Abdallah
University of Cincinnati

Charles A. Barnes
University of Massachusetts, Lowell

Liang-Wu Cai
Kansas State University

Brian Coller
University of Illinois, Chicago

Stephen C. Cowin
City College of New York

Ali El-Zeiny
California State University, Fresno

Christopher M. Foley
Marquette University

John H. Forrester
University of Tennessee

Orlando Hankins
North Carolina State University

Yohannes Ketema
University of Minnesota

Sang-Soo Kim
Ohio University

Roger L. Ludin
California State University, San Luis Obispo

Mohammad Mahinfalah
North Dakota State University

James R. Matthews
University of New Mexico

Wilfrid Nixon
University of Iowa

Karim Nohra
University of South Florida

David B. Oglesby
University of Missouri, Rolla

G.E. Ramey
Auburn University

William F. Reiter, Jr.
Oregon State University

Ali A. Selim
South Dakota State University

William Seto
San Jose State University

Teong E. Tan
The University of Memphis

Francis M. Thomas
University of Kansas

T. Calvin Tszeng
Illinois Institute of Technology

Christine Valle
University of Maine

INTERNATIONAL REVIEWERS

Yap Fook Fah
Nanyang Technical University

Y.A. Khulief
King Fahd University of Petroleum Minerals, Saudi Arabia

Moktar Ngah
Universiti Teknologi Malaysia, Malaysia

Ted Stathopoulos
Concordia University

Wen-Fang Wu
National Taiwan University, Taiwan

Chou Chu Yang
National Taiwan University, Taiwan

STATICS AND DYNAMICS FOCUS GROUP ATTENDEES

Makola M. Abdullah
Florida A&M University

Riad Alakkad
University of Dayton

Kurt Anderson
Rensselaer Polytechnic Institute

George Buzyna
Florida State University and Florida A&M University

Ravindar Chona
Texas A&M University

Ted Conway
University of Central Florida

Julie Coonrod
University of New Mexico

Phillip Cornwell
Rose-Hulman Institute

K.L. Devries
University of Utah

DeRome O. Dunn
North Carolina A&T State University

Gil Emmert
University of Wisconsin

Brian Fabian
University of Washington

Roy James Hartfield, Jr.
Auburn University

Lih-Min Hsia
California State University, Los Angeles

S. Graham Kelly
The University of Akron

Sam Kiger
University of Missouri, Columbia

Jeff Kuo
California State University, Fullerton

Marc Mignolet
Arizona State University

Masami Nakagawa
Colorado School of Mines

Steven O'Hare
Oklahoma State University

Charles K. Roby
Clemson University

Scott Schiff
Clemson University

Geoff Schiflett
University of Southern California

Ganesh Thiagarajan
Louisiana State University

Francis Thomas
University of Kansas

Erik Thompson
Colorado State University

Josef S. Torok
Rochester Institute of Technology

Robert Weber
Marquette University

STATICS AND DYNAMICS SYMPOSIUM ATTENDEES

Subhash Anand
Clemson University

Stephen Bechtel
The Ohio State University

Sherrill Biggers
Clemson University

Pasquale Cinnella
Mississippi State University

Adel El-Safty
University of Central Florida

Walt Haisler
Texas A&M University

Kimberly M. Hill
University of Illinois, Urbana–Champaign

James Jones
Purdue University

Yohannes Ketema
University of Minnesota

Chuck Krousgrill
Purdue University

Mohammad Mahinfalah
North Dakota State University

Christine Masters
Pennsylvania State University

Dan Mendelsohn
The Ohio State University

Mark Orwat
United States Military Academy

Art Peterson
University of Alberta

Tad Pfeffer
University of Colorado at Boulder

Sadanandan Pradeep
Clarkson University

Robert Rankin
Arizona State University

Michael Shelton
California State Polytechnic University

Larry Silverberg
North Carolina State University

Bob Witt
University of Wisconsin, Madison

Wan-Lee Yin
Georgia Institute of Technology

Musharraf Zaman
University of Oklahoma

List of Symbols

$\mathbf{a}$, a	Acceleration
a	Constant; radius; distance; semimajor axis of ellipse
$\overline{\mathbf{a}}$, $\overline{a}$	Acceleration of mass center
$\mathbf{a}_{B/A}$	Acceleration of B relative to frame in translation with A
$\mathbf{a}_{P/\mathcal{F}}$	Acceleration of P relative to rotating frame $\mathcal{F}$
$\mathbf{a}_c$	Coriolis acceleration
$\mathbf{A}$, $\mathbf{B}$, $\mathbf{C}$, . . .	Reactions at supports and connections
A, B, C, . . .	Points
A	Area
b	Width; distance; semiminor axis of ellipse
c	Constant; coefficient of viscous damping
C	Centroid; instantaneous center of rotation; capacitance
d	Distance
$\mathbf{e}_n$, $\mathbf{e}_t$	Unit vectors along normal and tangent
$\mathbf{e}_r$, $\mathbf{e}_\theta$	Unit vectors in radial and transverse directions
e	Coefficient of restitution; base of natural logarithms
E	Total mechanical energy; voltage
f	Scalar function
f_f	Frequency of forced vibration
f_n	Natural frequency
$\mathbf{F}$	Force; friction force
g	Acceleration of gravity
G	Center of gravity; mass center; constant of gravitation
h	Height; sag of cable; angular momentum per unit mass
$\mathbf{H}_O$	Angular momentum about point O
$\dot{\mathbf{H}}_G$	Rate of change of angular momentum $\mathbf{H}_G$ with respect to frame of fixed orientation
$(\dot{\mathbf{H}}_G)_{Gxyz}$	Rate of change of angular momentum $\mathbf{H}_G$ with respect to rotating frame $Gxyz$
$\mathbf{i}$, $\mathbf{j}$, $\mathbf{k}$	Unit vectors along coordinate axes
i	Current
I, I_x, . . .	Moments of inertia

$\bar{I}$	Centroidal moment of inertia
$I_{xy}, \ldots$	Products of inertia
J	Polar moment of inertia
k	Spring constant
k_x, k_y, k_O	Radii of gyration
$\bar{k}$	Centroidal radius of gyration
l	Length
$\mathbf{L}$	Linear momentum
L	Length; span; inductance
m	Mass
m'	Mass per unit length
$\mathbf{M}$	Couple; moment
$\mathbf{M}_O$	Moment about point O
$\mathbf{M}_O^R$	Moment resultant about point O
M	Magnitude of couple or moment; mass of earth
M_{OL}	Moment about axis OL
n	Normal direction
$\mathbf{N}$	Normal component of reaction
O	Origin of coordinates
p	Pressure
$\mathbf{P}$	Force; vector
$\dot{\mathbf{P}}$	Rate of change of vector **P** with respect to frame of fixed orientation
q	Mass rate of flow; electric charge
$\mathbf{Q}$	Force; vector
$\dot{\mathbf{Q}}$	Rate of change of vector **Q** with respect to frame of fixed orientation
$(\dot{\mathbf{Q}})_{Oxyz}$	Rate of change of vector **Q** with respect to frame *Oxyz*
$\mathbf{r}$	Position vector
$\mathbf{r}_{B/A}$	Position vector of B relative to A
r	Radius; distance; polar coordinate
$\mathbf{R}$	Resultant force; resultant vector; reaction
R	Radius of earth; resistance
$\mathbf{s}$	Position vector
s	Length of arc; length of cable
$\mathbf{S}$	Force; vector
t	Time; thickness; tangential direction
$\mathbf{T}$	Force
T	Tension; kinetic energy
$\mathbf{u}$	Velocity
u	Variable
U	Work
$\mathbf{v}, v$	Velocity
v	Speed
$\bar{\mathbf{v}}, \bar{v}$	Velocity of mass center
$\mathbf{v}_{B/A}$	Velocity of B relative to frame in translation with A
$\mathbf{v}_{P/\mathcal{F}}$	Velocity of P relative to rotating frame $\mathcal{F}$
$\mathbf{V}$	Vector product shearing force
V	Volume; potential energy; shear
w	Load per unit length

$\mathbf{W}$, W	Weight; load
x, y, z	Rectangular coordinates; distances
$\dot{x}$, $\dot{y}$, $\dot{z}$	Time derivatives of coordinates x, y, z
$\overline{x}$, $\overline{y}$, $\overline{z}$	Rectangular coordinates of centroid, center of gravity, or mass center
$\boldsymbol{\alpha}$, α	Angular acceleration
α, β, γ	Angles
γ	Specific weight
δ	Elongation
$\delta\mathbf{r}$	Virtual displacement
δU	Virtual work
ε	Eccentricity of conic section or of orbit
$\boldsymbol{\lambda}$	Unit vector along a line
η	Efficiency
θ	Angular coordinate; Eulerian angle; angle; polar coordinate
μ	Coefficient of friction
ρ	Density; radius of curvature
τ	Periodic time
τ_n	Period of free vibration
ϕ	Angle of friction; Eulerian angle; phase angle; angle
φ	Phase difference
ψ	Eulerian angle
$\boldsymbol{\omega}$, ω	Angular velocity
ω_f	Circular frequency of forced vibration
ω_n	Natural circular frequency
$\boldsymbol{\Omega}$	Angular velocity of frame of reference

C H A P T E R

1

Introduction

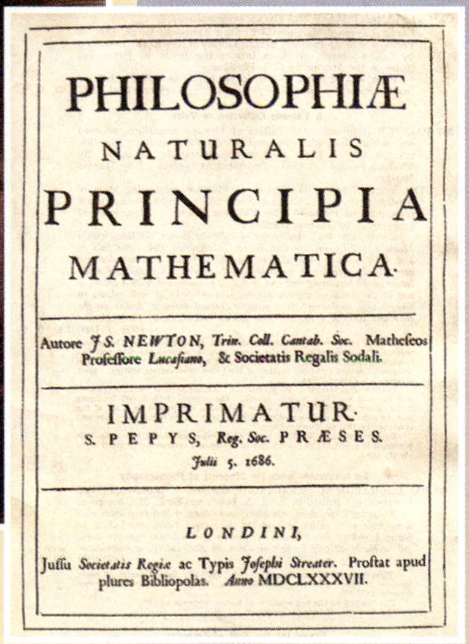

PHILOSOPHIÆ
NATURALIS
PRINCIPIA
MATHEMATICA.

Autore *J.S. NEWTON*, *Trin. Coll. Cantab. Soc.* Matheseos Professore *Lucasiano*, & Societatis Regalis Sodali.

IMPRIMATUR.
S. PEPYS, *Reg. Soc.* PRÆSES.
Julii 5. 1686.

LONDINI,

Jussu *Societatis Regiæ* ac Typis *Josephi Streater*. Prostat apud plures Bibliopolas. *Anno* MDCLXXXVII.

In the latter part of the seventeenth century, Sir Isaac Newton stated the fundamental principles of mechanics, which are the foundation of much of today's engineering. His *Principia,* published in 1687, summarized his work in terrestrial, celestial, and fluid mechanics.

INTRODUCTION

1.1. WHAT IS MECHANICS?

Mechanics can be defined as that science which describes and predicts the conditions of rest or motion of bodies under the action of forces. It is divided into three parts: mechanics of *rigid bodies,* mechanics of *deformable bodies,* and mechanics of *fluids.*

The mechanics of rigid bodies is subdivided into *statics* and *dynamics,* the former dealing with bodies at rest, the latter with bodies in motion. In this part of the study of mechanics, bodies are assumed to be perfectly rigid. Actual structures and machines, however, are never absolutely rigid and deform under the loads to which they are subjected. But these deformations are usually small and do not appreciably affect the conditions of equilibrium or motion of the structure under consideration. They are important, though, as far as the resistance of the structure to failure is concerned and are studied in mechanics of materials, which is a part of the mechanics of deformable bodies. The third division of mechanics, the mechanics of fluids, is subdivided into the study of *incompressible fluids* and of *compressible fluids.* An important subdivision of the study of incompressible fluids is *hydraulics,* which deals with problems involving water.

Mechanics is a physical science, since it deals with the study of physical phenomena. However, some associate mechanics with mathematics, while many consider it as an engineering subject. Both these views are justified in part. Mechanics is the foundation of most engineering sciences and is an indispensable prerequisite to their study. However, it does not have the *empiricism* found in some engineering sciences, that is, it does not rely on experience or observation alone; by its rigor and the emphasis it places on deductive reasoning it resembles mathematics. But, again, it is not an *abstract* or even a *pure* science; mechanics is an *applied* science. The purpose of mechanics is to explain and predict physical phenomena and thus to lay the foundations for engineering applications.

1.2. FUNDAMENTAL CONCEPTS AND PRINCIPLES

Although the study of mechanics goes back to the time of Aristotle (384–322 B.C.) and Archimedes (287–212 B.C.), one has to wait until Newton (1642–1727) to find a satisfactory formulation of its fundamental principles. These principles were later expressed in a modified form by d'Alembert, Lagrange, and Hamilton. Their validity remained unchallenged, however, until Einstein formulated his *theory of relativity* (1905). While its limitations have now been recognized, *newtonian mechanics* still remains the basis of today's engineering sciences.

The basic concepts used in mechanics are *space, time, mass,* and *force.* These concepts cannot be truly defined; they should be accepted on the basis of our intuition and experience and used as a mental frame of reference for our study of mechanics.

The concept of *space* is associated with the notion of the position of a point P. The position of P can be defined by three lengths measured from a certain reference point, or *origin,* in three given directions. These lengths are known as the *coordinates* of P.

To define an event, it is not sufficient to indicate its position in space. The *time* of the event should also be given.

The concept of *mass* is used to characterize and compare bodies on the basis of certain fundamental mechanical experiments. Two bodies of the same mass, for example, will be attracted by the earth in the same manner; they will also offer the same resistance to a change in translational motion.

A *force* represents the action of one body on another. It can be exerted by actual contact or at a distance, as in the case of gravitational forces and magnetic forces. A force is characterized by its *point of application,* its *magnitude,* and its *direction;* a force is represented by a *vector* (Sec. 2.3).

In newtonian mechanics, space, time, and mass are absolute concepts, independent of each other. (This is not true in *relativistic mechanics,* where the time of an event depends upon its position, and where the mass of a body varies with its velocity.) On the other hand, the concept of force is not independent of the other three. Indeed, one of the fundamental principles of newtonian mechanics listed below indicates that the resultant force acting on a body is related to the mass of the body and to the manner in which its velocity varies with time.

You will study the conditions of rest or motion of particles and rigid bodies in terms of the four basic concepts we have introduced. By *particle* we mean a very small amount of matter which may be assumed to occupy a single point in space. A *rigid body* is a combination of a large number of particles occupying fixed positions with respect to each other. The study of the mechanics of particles is obviously a prerequisite to that of rigid bodies. Besides, the results obtained for a particle can be used directly in a large number of problems dealing with the conditions of rest or motion of actual bodies.

The study of elementary mechanics rests on six fundamental principles based on experimental evidence.

The Parallelogram Law for the Addition of Forces. This states that two forces acting on a particle may be replaced by a single force, called their *resultant,* obtained by drawing the diagonal of the parallelogram which has sides equal to the given forces (Sec. 2.2).

The Principle of Transmissibility. This states that the conditions of equilibrium or of motion of a rigid body will remain unchanged if a force acting at a given point of the rigid body is replaced by a force of the same magnitude and same direction, but acting at a different point, provided that the two forces have the same line of action (Sec. 3.3).

Newton's Three Fundamental Laws. Formulated by Sir Isaac Newton in the latter part of the seventeenth century, these laws can be stated as follows:

FIRST LAW. If the resultant force acting on a particle is zero, the particle will remain at rest (if originally at rest) or will move with constant speed in a straight line (if originally in motion) (Sec. 2.10).

SECOND LAW. If the resultant force acting on a particle is not zero, the particle will have an acceleration proportional to the magnitude of the resultant and in the direction of this resultant force.

As you will see in Sec. 12.2, this law can be stated as

$$\mathbf{F} = m\mathbf{a} \tag{1.1}$$

where $\mathbf{F}$, m, and $\mathbf{a}$ represent, respectively, the resultant force acting on the particle, the mass of the particle, and the acceleration of the particle, expressed in a consistent system of units.

THIRD LAW. The forces of action and reaction between bodies in contact have the same magnitude, same line of action, and opposite sense (Sec. 6.1).

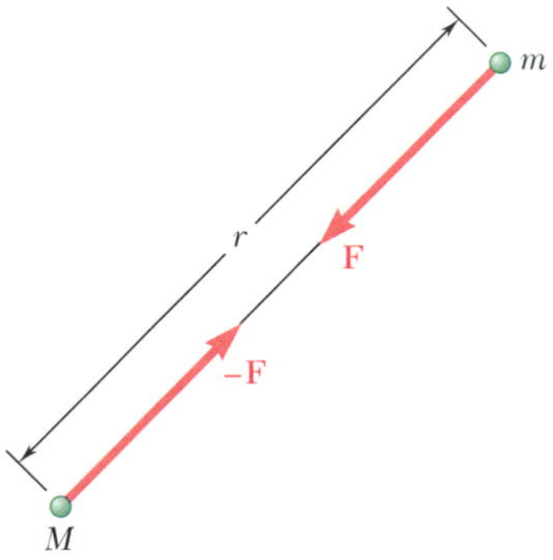

Fig. 1.1

Newton's Law of Gravitation. This states that two particles of mass M and m are mutually attracted with equal and opposite forces $\mathbf{F}$ and $-\mathbf{F}$ (Fig. 1.1) of magnitude F given by the formula

$$F = G\frac{Mm}{r^2} \tag{1.2}$$

where r = distance between the two particles
G = universal constant called the *constant of gravitation*

Newton's law of gravitation introduces the idea of an action exerted at a distance and extends the range of application of Newton's third law: the action $\mathbf{F}$ and the reaction $-\mathbf{F}$ in Fig. 1.1 are equal and opposite, and they have the same line of action.

A particular case of great importance is that of the attraction of the earth on a particle located on its surface. The force $\mathbf{F}$ exerted by the earth on the particle is then defined as the *weight* $\mathbf{W}$ of the particle. Taking M equal to the mass of the earth, m equal to the mass of the particle, and r equal to the radius R of the earth, and introducing the constant

$$g = \frac{GM}{R^2} \tag{1.3}$$

Photo 1.1 When in earth orbit, people and objects are said to be *weightless* even though the gravitational force acting is approximately 90% of that experienced on the surface of the earth. This apparent contradiction will be resolved in Chapter 12 when we apply Newton's second law to the motion of particles.

the magnitude W of the weight of a particle of mass m may be expressed as†

$$W = mg \tag{1.4}$$

The value of R in formula (1.3) depends upon the elevation of the point considered; it also depends upon its latitude, since the earth is not truly spherical. The value of g therefore varies with the position of the point considered. As long as the point actually remains on the surface of the earth, it is sufficiently accurate in most engineering computations to assume that g equals 9.81 m/s^2 or 32.2 ft/s^2.

†A more accurate definition of the weight $\mathbf{W}$ should take into account the rotation of the earth.

The principles we have just listed will be introduced in the course of our study of mechanics as they are needed. The study of the statics of particles carried out in Chap. 2 will be based on the parallelogram law of addition and on Newton's first law alone. The principle of transmissibility will be introduced in Chap. 3 as we begin the study of the statics of rigid bodies, and Newton's third law in Chap. 6 as we analyze the forces exerted on each other by the various members forming a structure. In the study of dynamics, Newton's second law and Newton's law of gravitation will be introduced. It will then be shown that Newton's first law is a particular case of Newton's second law (Sec. 12.2) and that the principle of transmissibility could be derived from the other principles and thus eliminated (Sec. 16.5). In the meantime, however, Newton's first and third laws, the parallelogram law of addition, and the principle of transmissibility will provide us with the necessary and sufficient foundation for the entire study of the statics of particles, rigid bodies, and systems of rigid bodies.

As noted earlier, the six fundamental principles listed above are based on experimental evidence. Except for Newton's first law and the principle of transmissibility, they are independent principles which cannot be derived mathematically from each other or from any other elementary physical principle. On these principles rests most of the intricate structure of newtonian mechanics. For more than two centuries a tremendous number of problems dealing with the conditions of rest and motion of rigid bodies, deformable bodies, and fluids were solved by applying these fundamental principles. Many of the solutions obtained could be checked experimentally, thus providing a further verification of the principles from which they were derived. It was only in the last century that Newton's mechanics was found at fault, in the study of the motion of atoms and in the study of the motion of certain planets, where it must be supplemented by the theory of relativity. But on the human or engineering scale, where velocities are small compared with the speed of light, Newton's mechanics has yet to be disproved.

1.3. SYSTEMS OF UNITS

Associated with the four fundamental concepts introduced in the preceding section are the so-called *kinetic units,* that is, the units of *length, time, mass,* and *force.* These units cannot be chosen independently if Eq. (1.1) is to be satisfied. Three of the units may be defined arbitrarily; they are then referred to as *base units.* The fourth unit, however, must be chosen in accordance with Eq. (1.1) and is referred to as a *derived unit.* Kinetic units selected in this way are said to form a *consistent system of units.*

International System of Units (SI Units†). In this system, which will be in universal use when the United States completes its conversion to SI units, the base units are the units of length, mass, and time, and they are called, respectively, the *meter* (m), the *kilogram* (kg), and the *second* (s). All three are arbitrarily defined. The second, which was originally chosen to represent 1/86 400 of the mean

†SI stands for *Système International d'Unités* (French).

solar day, is now defined as the duration of 9 192 631 770 cycles of the radiation corresponding to the transition between two levels of the fundamental state of the cesium-133 atom. The meter, originally defined as one ten-millionth of the distance from the equator to either pole, is now defined as 1 650 763.73 wavelengths of the orange-red light corresponding to a certain transition in an atom of krypton-86. The kilogram, which is approximately equal to the mass of 0.001 m^3 of water, is defined as the mass of a platinum-iridium standard kept at the International Bureau of Weights and Measures at Sèvres, near Paris, France. The unit of force is a derived unit. It is called the *newton* (N) and is defined as the force which gives an acceleration of 1 m/s^2 to a mass of 1 kg (Fig. 1.2). From Eq. (1.1) we write

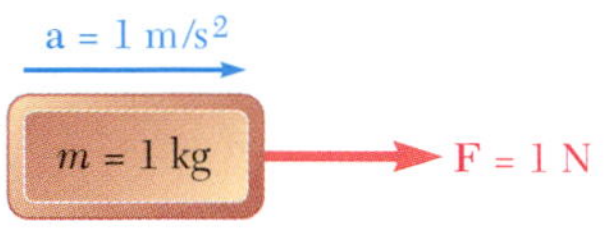

Fig. 1.2

$$1 \text{ N} = (1 \text{ kg})(1 \text{ m/s}^2) = 1 \text{ kg} \cdot \text{m/s}^2 \tag{1.5}$$

The SI units are said to form an *absolute* system of units. This means that the three base units chosen are independent of the location where measurements are made. The meter, the kilogram, and the second may be used anywhere on the earth; they may even be used on another planet. They will always have the same significance.

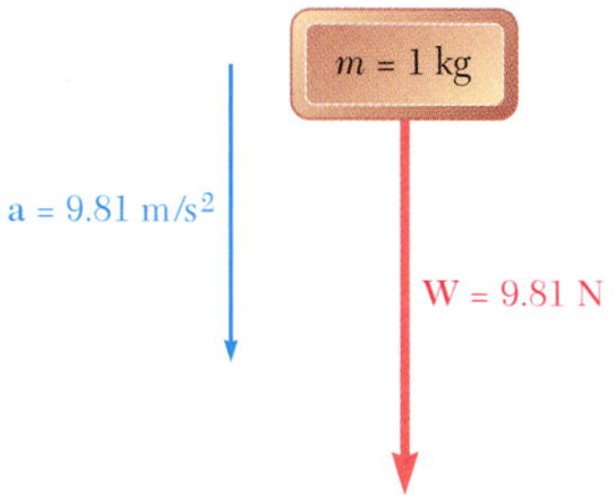

Fig. 1.3

The *weight* of a body, or the *force of gravity* exerted on that body, should, like any other force, be expressed in newtons. From Eq. (1.4) it follows that the weight of a body of mass 1 kg (Fig. 1.3) is

$$\begin{aligned} W &= mg \\ &= (1 \text{ kg})(9.81 \text{ m/s}^2) \\ &= 9.81 \text{ N} \end{aligned}$$

Multiples and submultiples of the fundamental SI units may be obtained through the use of the prefixes defined in Table 1.1. The multiples and submultiples of the units of length, mass, and force most frequently used in engineering are, respectively, the *kilometer* (km) and the *millimeter* (mm); the *megagram*† (Mg) and the *gram* (g); and the *kilonewton* (kN). According to Table 1.1, we have

$$1 \text{ km} = 1000 \text{ m} \qquad 1 \text{ mm} = 0.001 \text{ m}$$
$$1 \text{ Mg} = 1000 \text{ kg} \qquad 1 \text{ g} = 0.001 \text{ kg}$$
$$1 \text{ kN} = 1000 \text{ N}$$

The conversion of these units into meters, kilograms, and newtons, respectively, can be effected by simply moving the decimal point three places to the right or to the left. For example, to convert 3.82 km into meters, one moves the decimal point three places to the right:

$$3.82 \text{ km} = 3820 \text{ m}$$

Similarly, 47.2 mm is converted into meters by moving the decimal point three places to the left:

$$47.2 \text{ mm} = 0.0472 \text{ m}$$

†Also known as a *metric ton*.

Table 1.1. SI Prefixes

Multiplication Factor	Prefix†	Symbol
$1\ 000\ 000\ 000\ 000 = 10^{12}$	tera	T
$1\ 000\ 000\ 000 = 10^{9}$	giga	G
$1\ 000\ 000 = 10^{6}$	mega	M
$1\ 000 = 10^{3}$	kilo	k
$100 = 10^{2}$	hecto‡	h
$10 = 10^{1}$	deka‡	da
$0.1 = 10^{-1}$	deci‡	d
$0.01 = 10^{-2}$	centi‡	c
$0.001 = 10^{-3}$	milli	m
$0.000\ 001 = 10^{-6}$	micro	μ
$0.000\ 000\ 001 = 10^{-9}$	nano	n
$0.000\ 000\ 000\ 001 = 10^{-12}$	pico	p
$0.000\ 000\ 000\ 000\ 001 = 10^{-15}$	femto	f
$0.000\ 000\ 000\ 000\ 000\ 001 = 10^{-18}$	atto	a

†The first syllable of every prefix is accented so that the prefix will retain its identity. Thus, the preferred pronunciation of kilometer places the accent on the first syllable, not the second.

‡The use of these prefixes should be avoided, except for the measurement of areas and volumes and for the nontechnical use of centimeter, as for body and clothing measurements.

Using scientific notation, one may also write

$$3.82 \text{ km} = 3.82 \times 10^{3} \text{ m}$$
$$47.2 \text{ mm} = 47.2 \times 10^{-3} \text{ m}$$

The multiples of the unit of time are the *minute* (min) and the *hour* (h). Since 1 min = 60 s and 1 h = 60 min = 3600 s, these multiples cannot be converted as readily as the others.

By using the appropriate multiple or submultiple of a given unit, one can avoid writing very large or very small numbers. For example, one usually writes 427.2 km rather than 427 200 m, and 2.16 mm rather than 0.002 16 m.†

Units of Area and Volume. The unit of area is the *square meter* (m^2), which represents the area of a square of side 1 m; the unit of volume is the *cubic meter* (m^3), equal to the volume of a cube of side 1 m. In order to avoid exceedingly small or large numerical values in the computation of areas and volumes, one uses systems of subunits obtained by respectively squaring and cubing not only the millimeter but also two intermediate submultiples of the meter, namely, the *decimeter* (dm) and the *centimeter* (cm). Since, by definition,

$$1 \text{ dm} = 0.1 \text{ m} = 10^{-1} \text{ m}$$
$$1 \text{ cm} = 0.01 \text{ m} = 10^{-2} \text{ m}$$
$$1 \text{ mm} = 0.001 \text{ m} = 10^{-3} \text{ m}$$

†It should be noted that when more than four digits are used on either side of the decimal point to express a quantity in SI units—as in 427 200 m or 0.002 16 m—spaces, never commas, should be used to separate the digits into groups of three. This is to avoid confusion with the comma used in place of a decimal point, which is the convention in many countries.

the submultiples of the unit of area are

$$1\ \text{dm}^2 = (1\ \text{dm})^2 = (10^{-1}\ \text{m})^2 = 10^{-2}\ \text{m}^2$$
$$1\ \text{cm}^2 = (1\ \text{cm})^2 = (10^{-2}\ \text{m})^2 = 10^{-4}\ \text{m}^2$$
$$1\ \text{mm}^2 = (1\ \text{mm})^2 = (10^{-3}\ \text{m})^2 = 10^{-6}\ \text{m}^2$$

and the submultiples of the unit of volume are

$$1\ \text{dm}^3 = (1\ \text{dm})^3 = (10^{-1}\ \text{m})^3 = 10^{-3}\ \text{m}^3$$
$$1\ \text{cm}^3 = (1\ \text{cm})^3 = (10^{-2}\ \text{m})^3 = 10^{-6}\ \text{m}^3$$
$$1\ \text{mm}^3 = (1\ \text{mm})^3 = (10^{-3}\ \text{m})^3 = 10^{-9}\ \text{m}^3$$

It should be noted that when the volume of a liquid is being measured, the cubic decimeter (dm^3) is usually referred to as a *liter* (L).

Other derived SI units used to measure the moment of a force, the work of a force, etc., are shown in Table 1.2. While these units will be introduced in later chapters as they are needed, we should note an important rule at this time: When a derived unit is obtained by dividing a base unit by another base unit, a prefix may be used in the numerator of the derived unit but not in its denominator. For example, the constant k of a spring which stretches 20 mm under a load of 100 N will be expressed as

$$k = \frac{100\ \text{N}}{20\ \text{mm}} = \frac{100\ \text{N}}{0.020\ \text{m}} = 5000\ \text{N/m} \qquad \text{or} \qquad k = 5\ \text{kN/m}$$

but never as $k = 5\ \text{N/mm}$.

Table 1.2. Principal SI Units Used in Mechanics

Quantity	Unit	Symbol	Formula
Acceleration	Meter per second squared	. . .	m/s^2
Angle	Radian	rad	†
Angular acceleration	Radian per second squared	. . .	rad/s^2
Angular velocity	Radian per second	. . .	rad/s
Area	Square meter	. . .	m^2
Density	Kilogram per cubic meter	. . .	kg/m^3
Energy	Joule	J	$\text{N}\cdot\text{m}$
Force	Newton	N	$\text{kg}\cdot\text{m/s}^2$
Frequency	Hertz	Hz	s^{-1}
Impulse	Newton-second	. . .	$\text{kg}\cdot\text{m/s}$
Length	Meter	m	‡
Mass	Kilogram	kg	‡
Moment of a force	Newton-meter	. . .	$\text{N}\cdot\text{m}$
Power	Watt	W	J/s
Pressure	Pascal	Pa	N/m^2
Stress	Pascal	Pa	N/m^2
Time	Second	s	‡
Velocity	Meter per second	. . .	m/s
Volume			
Solids	Cubic meter	. . .	m^3
Liquids	Liter	L	$10^{-3}\ \text{m}^3$
Work	Joule	J	$\text{N}\cdot\text{m}$

†Supplementary unit (1 revolution = 2π rad = 360°).

‡Base unit.

U.S. Customary Units. Most practicing American engineers still commonly use a system in which the base units are the units of length, force, and time. These units are, respectively, the *foot* (ft), the *pound* (lb), and the *second* (s). The second is the same as the corresponding SI unit. The foot is defined as 0.3048 m. The pound is defined as the *weight* of a platinum standard, called the *standard pound,* which is kept at the National Institute of Standards and Technology outside Washington, the mass of which is 0.453 592 43 kg. Since the weight of a body depends upon the earth's gravitational attraction, which varies with location, it is specified that the standard pound should be placed at sea level and at a latitude of 45° to properly define a force of 1 lb. Clearly the U.S. customary units do not form an absolute system of units. Because of their dependence upon the gravitational attraction of the earth, they form a *gravitational* system of units.

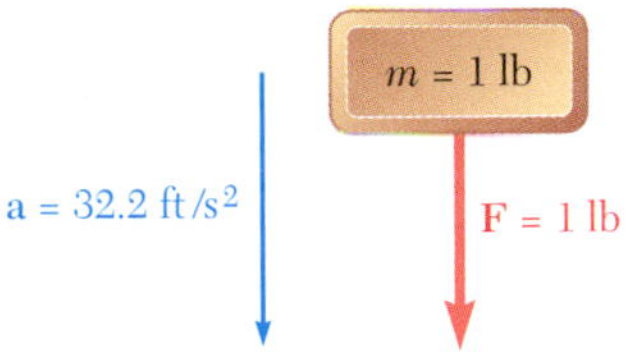

Fig. 1.4

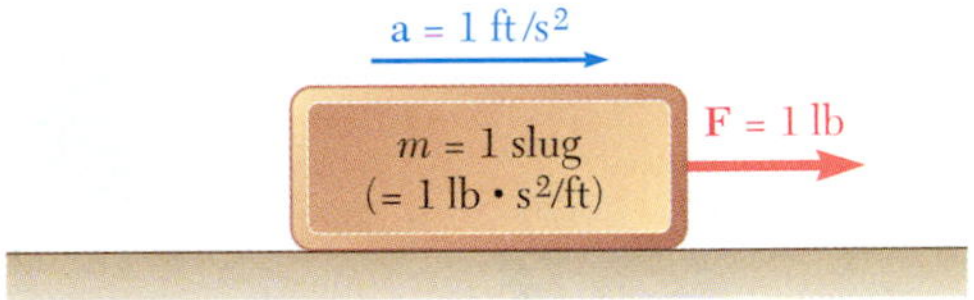

Fig. 1.5

While the standard pound also serves as the unit of mass in commercial transactions in the United States, it cannot be so used in engineering mechanics computations, since such a unit would not be consistent with the base units defined in the preceding paragraph. Indeed, when acted upon by a force of 1 lb, that is, when subjected to the force of gravity, the standard pound receives the acceleration of gravity, $g = 32.2 \text{ ft/s}^2$ (Fig. 1.4), not the unit acceleration required by Eq. (1.1). The unit of mass consistent with the foot, the pound, and the second is the mass which receives an acceleration of 1 ft/s^2 when a force of 1 lb is applied to it (Fig. 1.5). This unit, sometimes called a *slug,* can be derived from the equation $F = ma$ after substituting 1 lb and 1 ft/s^2 for F and a, respectively. We write

$$F = ma \qquad 1 \text{ lb} = (1 \text{ slug})(1 \text{ ft/s}^2)$$

and obtain

$$1 \text{ slug} = \frac{1 \text{ lb}}{1 \text{ ft/s}^2} = 1 \text{ lb} \cdot \text{s}^2/\text{ft} \tag{1.6}$$

Comparing Figs. 1.4 and 1.5, we conclude that the slug is a mass 32.2 times larger than the mass of the standard pound.

The fact that in the U.S. customary system of units bodies are characterized by their weight in pounds rather than by their mass in slugs will be a convenience in the study of statics, where one constantly deals with weights and other forces and only seldom with masses. However, in the study of dynamics, where forces, masses, and accelerations are involved, the mass m of a body will be expressed in slugs when its weight W is given in pounds. Recalling Eq. (1.4), we write

$$m = \frac{W}{g} \tag{1.7}$$

where g is the acceleration of gravity ($g = 32.2 \text{ ft/s}^2$).

Photo 1.2 The unit of mass is the only basic unit still based on a physical standard. Work is in progress to replace this standard with one based on unchanging natural phenomena.

Other U.S. customary units frequently encountered in engineering problems are the *mile* (mi), equal to 5280 ft; the *inch* (in.), equal to $\frac{1}{12}$ ft; and the *kilopound* (kip), equal to a force of 1000 lb. The *ton* is often used to represent a mass of 2000 lb but, like the pound, must be converted into slugs in engineering computations.

The conversion into feet, pounds, and seconds of quantities expressed in other U.S. customary units is generally more involved and requires greater attention than the corresponding operation in SI units. If, for example, the magnitude of a velocity is given as $v = 30$ mi/h, we convert it to ft/s as follows. First we write

$$v = 30\frac{\text{mi}}{\text{h}}$$

Since we want to get rid of the unit miles and introduce instead the unit feet, we should multiply the right-hand member of the equation by an expression containing miles in the denominator and feet in the numerator. But, since we do not want to change the value of the right-hand member, the expression used should have a value equal to unity. The quotient (5280 ft)/(1 mi) is such an expression. Operating in a similar way to transform the unit hour into seconds, we write

$$v = \left(30\frac{\text{mi}}{\text{h}}\right)\left(\frac{5280 \text{ ft}}{1 \text{ mi}}\right)\left(\frac{1 \text{ h}}{3600 \text{ s}}\right)$$

Carrying out the numerical computations and canceling out units which appear in both the numerator and the denominator, we obtain

$$v = 44\,\frac{\text{ft}}{\text{s}} = 44 \text{ ft/s}$$

Photo 1.3 The importance of including units in all calculations cannot be over emphasized. It was found that the $125 million Mars Climate Orbiter failed to go into orbit around Mars because the prime contractor had provided the navigation team with operating data based on U.S. units rather than the specified SI units.

1.4. CONVERSION FROM ONE SYSTEM OF UNITS TO ANOTHER

There are many instances when an engineer wishes to convert into SI units a numerical result obtained in U.S. customary units or vice versa. Because the unit of time is the same in both systems, only two kinetic base units need be converted. Thus, since all other kinetic units can be derived from these base units, only two conversion factors need be remembered.

Units of Length. By definition the U.S. customary unit of length is

$$1 \text{ ft} = 0.3048 \text{ m} \tag{1.8}$$

It follows that

$$1 \text{ mi} = 5280 \text{ ft} = 5280(0.3048 \text{ m}) = 1609 \text{ m}$$

or

$$1 \text{ mi} = 1.609 \text{ km} \tag{1.9}$$

Also

$$1 \text{ in.} = \tfrac{1}{12} \text{ ft} = \tfrac{1}{12}(0.3048 \text{ m}) = 0.0254 \text{ m}$$

or

$$1 \text{ in.} = 25.4 \text{ mm} \tag{1.10}$$

Units of Force. Recalling that the U.S. customary unit of force (pound) is defined as the weight of the standard pound (of mass 0.4536 kg) at sea level and at a latitude of 45° (where $g = 9.807 \text{ m/s}^2$) and using Eq. (1.4), we write

$$W = mg$$
$$1 \text{ lb} = (0.4536 \text{ kg})(9.807 \text{ m/s}^2) = 4.448 \text{ kg} \cdot \text{m/s}^2$$

or, recalling Eq. (1.5),

$$1 \text{ lb} = 4.448 \text{ N} \tag{1.11}$$

Units of Mass. The U.S. customary unit of mass (slug) is a derived unit. Thus, using Eqs. (1.6), (1.8), and (1.11), we write

$$1 \text{ slug} = 1 \text{ lb} \cdot \text{s}^2/\text{ft} = \frac{1 \text{ lb}}{1 \text{ ft/s}^2} = \frac{4.448 \text{ N}}{0.3048 \text{ m/s}^2} = 14.59 \text{ N} \cdot \text{s}^2/\text{m}$$

and, recalling Eq. (1.5),

$$1 \text{ slug} = 1 \text{ lb} \cdot \text{s}^2/\text{ft} = 14.59 \text{ kg} \tag{1.12}$$

Although it cannot be used as a consistent unit of mass, we recall that the mass of the standard pound is, by definition,

$$1 \text{ pound mass} = 0.4536 \text{ kg} \tag{1.13}$$

This constant may be used to determine the *mass* in SI units (kilograms) of a body which has been characterized by its *weight* in U.S. customary units (pounds).

To convert a derived U.S. customary unit into SI units, one simply multiplies or divides by the appropriate conversion factors. For example, to convert the moment of a force which was found to be $M = 47 \text{ lb} \cdot \text{in.}$ into SI units, we use formulas (1.10) and (1.11) and write

$$\begin{aligned} M = 47 \text{ lb} \cdot \text{in.} &= 47(4.448 \text{ N})(25.4 \text{ mm}) \\ &= 5310 \text{ N} \cdot \text{mm} = 5.31 \text{ N} \cdot \text{m} \end{aligned}$$

The conversion factors given in this section may also be used to convert a numerical result obtained in SI units into U.S. customary units. For example, if the moment of a force was found to be $M = 40 \text{ N} \cdot \text{m}$, we write, following the procedure used in the last paragraph of Sec. 1.3,

$$M = 40 \text{ N} \cdot \text{m} = (40 \text{ N} \cdot \text{m})\left(\frac{1 \text{ lb}}{4.448 \text{ N}}\right)\left(\frac{1 \text{ ft}}{0.3048 \text{ m}}\right)$$

Carrying out the numerical computations and canceling out units which appear in both the numerator and the denominator, we obtain

$$M = 29.5 \text{ lb} \cdot \text{ft}$$

The U.S. customary units most frequently used in mechanics are listed in Table 1.3 with their SI equivalents.

1.5. METHOD OF PROBLEM SOLUTION

You should approach a problem in mechanics as you would approach an actual engineering situation. By drawing on your own experience and intuition, you will find it easier to understand and formulate the problem. Once the problem has been clearly stated, however, there is

Table 1.3. U.S. Customary Units and Their SI Equivalents

Quantity	U.S. Customary Unit	SI Equivalent
Acceleration	ft/s^2	$0.3048 m/s^2$
	$in./s^2$	$0.0254 m/s^2$
Area	ft^2	$0.0929 m^2$
	in^2	$645.2 mm^2$
Energy	ft · lb	1.356 J
Force	kip	4.448 kN
	lb	4.448 N
	oz	0.2780 N
Impulse	lb · s	4.448 N · s
Length	ft	0.3048 m
	in.	25.40 mm
	mi	1.609 km
Mass	oz mass	28.35 g
	lb mass	0.4536 kg
	slug	14.59 kg
	ton	907.2 kg
Moment of a force	lb · ft	1.356 N · m
	lb · in.	0.1130 N · m
Moment of inertia		
Of an area	in^4	$0.4162 \times 10^6 mm^4$
Of a mass	$lb \cdot ft \cdot s^2$	$1.356 kg \cdot m^2$
Momentum	lb · s	4.448 kg · m/s
Power	ft · lb/s	1.356 W
	hp	745.7 W
Pressure or stress	lb/ft^2	47.88 Pa
	lb/in^2 (psi)	6.895 kPa
Velocity	ft/s	0.3048 m/s
	in./s	0.0254 m/s
	mi/h (mph)	0.4470 m/s
	mi/h (mph)	1.609 km/h
Volume	ft^3	$0.02832 m^3$
	in^3	$16.39 cm^3$
Liquids	gal	3.785 L
	qt	0.9464 L
Work	ft · lb	1.356 J

no place in its solution for your particular fancy. *The solution must be based on the six fundamental principles stated in Sec. 1.2 or on theorems derived from them.* Every step taken must be justified on that basis. Strict rules must be followed, which lead to the solution in an almost automatic fashion, leaving no room for your intuition or "feeling." After an answer has been obtained, it should be checked. Here again, you may call upon your common sense and personal experience. If not completely satisfied with the result obtained, you should carefully check your formulation of the problem, the validity of the methods used for its solution, and the accuracy of your computations.

The *statement* of a problem should be clear and precise. It should contain the given data and indicate what information is required. A neat drawing showing all quantities involved should be included. Separate diagrams should be drawn for all bodies involved, indicating clearly the forces acting on each body. These diagrams are known as *free-body diagrams* and are described in detail in Secs. 2.11 and 4.2.

The *fundamental principles* of mechanics listed in Sec. 1.2 *will be used to write equations* expressing the conditions of rest or motion of the bodies considered. Each equation should be clearly related to one of the free-body diagrams. You will then proceed to solve the problem, observing strictly the usual rules of algebra and recording neatly the various steps taken.

After the answer has been obtained, it should be *carefully checked*. Mistakes in *reasoning* can often be detected by checking the units. For example, to determine the moment of a force of 50 N about a point 0.60 m from its line of action, we would have written (Sec. 3.12)

$$M = Fd = (50 \text{ N})(0.60 \text{ m}) = 30 \text{ N} \cdot \text{m}$$

The unit N · m obtained by multiplying newtons by meters is the correct unit for the moment of a force; if another unit had been obtained, we would have known that some mistake had been made.

Errors in *computation* will usually be found by substituting the numerical values obtained into an equation which has not yet been used and verifying that the equation is satisfied. The importance of correct computations in engineering cannot be overemphasized.

1.6. NUMERICAL ACCURACY

The accuracy of the solution of a problem depends upon two items: (1) the accuracy of the given data and (2) the accuracy of the computations performed.

The solution cannot be more accurate than the less accurate of these two items. For example, if the loading of a bridge is known to be 75,000 lb with a possible error of 100 lb either way, the relative error which measures the degree of accuracy of the data is

$$\frac{100 \text{ lb}}{75{,}000 \text{ lb}} = 0.0013 = 0.13 \text{ percent}$$

In computing the reaction at one of the bridge supports, it would then be meaningless to record it as 14,322 lb. The accuracy of the solution cannot be greater than 0.13 percent, no matter how accurate the computations are, and the possible error in the answer may be as large as $(0.13/100)(14{,}322 \text{ lb}) \approx 20$ lb. The answer should be properly recorded as $14{,}320 \pm 20$ lb.

In engineering problems, the data are seldom known with an accuracy greater than 0.2 percent. It is therefore seldom justified to write the answers to such problems with an accuracy greater than 0.2 percent. A practical rule is to use 4 figures to record numbers beginning with a "1" and 3 figures in all other cases. Unless otherwise indicated, the data given in a problem should be assumed known with a comparable degree of accuracy. A force of 40 lb, for example, should be read 40.0 lb, and a force of 15 lb should be read 15.00 lb.

Pocket electronic calculators are widely used by practicing engineers and engineering students. The speed and accuracy of these calculators facilitate the numerical computations in the solution of many problems. However, students should not record more significant figures than can be justified merely because they are easily obtained. As noted above, an accuracy greater than 0.2 percent is seldom necessary or meaningful in the solution of practical engineering problems.

C H A P T E R

2

Statics of Particles

Shown above is the world's largest single-dish radio telescope, located in Arecibo, Puerto Rico. The 305-m-diameter spherical reflecting dish covers an area of approximately 20 acres. Suspended above the reflector is a 900-ton platform below which is the 328-ft-long azimuth arm. In this chapter, we will learn that by considering the platform as a *particle* in equilibrium, it will be possible to determine the tensions in the supporting cables.

STATICS OF PARTICLES

2.1. INTRODUCTION

In this chapter you will study the effect of forces acting on particles. First you will learn how to replace two or more forces acting on a given particle by a single force having the same effect as the original forces. This single equivalent force is the *resultant* of the original forces acting on the particle. Later the relations which exist among the various forces acting on a particle in a state of *equilibrium* will be derived and used to determine some of the forces acting on the particle.

The use of the word *particle* does not imply that our study will be limited to that of small corpuscles. What it means is that the size and shape of the bodies under consideration will not significantly affect the solution of the problems treated in this chapter and that all the forces acting on a given body will be assumed to be applied at the same point. Since such an assumption is verified in many practical applications, you will be able to solve a number of engineering problems in this chapter.

The first part of the chapter is devoted to the study of forces contained in a single plane, and the second part to the analysis of forces in three-dimensional space.

FORCES IN A PLANE

2.2. FORCE ON A PARTICLE. RESULTANT OF TWO FORCES

A force represents the action of one body on another and is generally characterized by its *point of application,* its *magnitude,* and its *direction.* Forces acting on a given particle, however, have the same point of application. Each force considered in this chapter will thus be completely defined by its magnitude and direction.

The magnitude of a force is characterized by a certain number of units. As indicated in Chap. 1, the SI units used by engineers to measure the magnitude of a force are the newton (N) and its multiple the kilonewton (kN), equal to 1000 N, while the U.S. customary units used for the same purpose are the pound (lb) and its multiple the kilopound (kip), equal to 1000 lb. The direction of a force is defined by the *line of action* and the *sense* of the force. The line of action is the infinite straight line along which the force acts; it is characterized by the angle it forms with some fixed axis (Fig. 2.1).

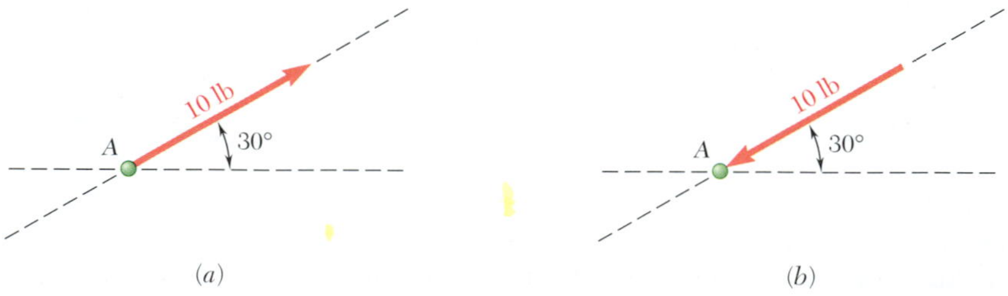

Fig. 2.1

The force itself is represented by a segment of that line; through the use of an appropriate scale, the length of this segment may be chosen to represent the magnitude of the force. Finally, the sense of the force should be indicated by an arrowhead. It is important in defining a force to indicate its sense. Two forces having the same magnitude and the same line of action but different sense, such as the forces shown in Fig. 2.1*a* and *b*, will have directly opposite effects on a particle.

Experimental evidence shows that two forces **P** and **Q** acting on a particle *A* (Fig. 2.2*a*) can be replaced by a single force **R** which has the same effect on the particle (Fig. 2.2*c*). This force is called the *resultant* of the forces **P** and **Q** and can be obtained, as shown in Fig. 2.2*b*, by constructing a parallelogram, using **P** and **Q** as two adjacent sides of the parallelogram. *The diagonal that passes through A represents the resultant.* This method for finding the resultant is known as the *parallelogram law* for the addition of two forces. This law is based on experimental evidence; it cannot be proved or derived mathematically.

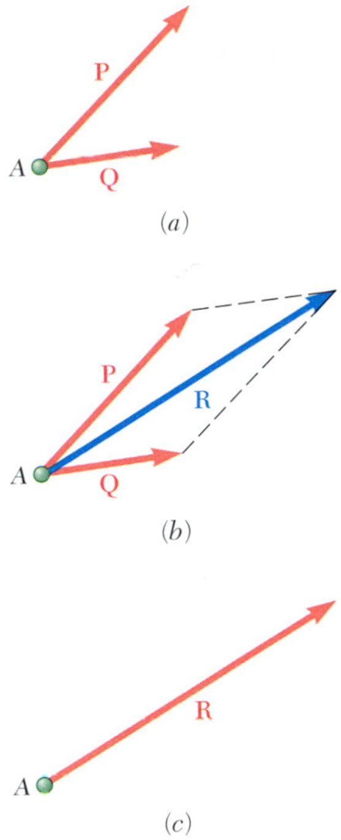

Fig. 2.2

2.3. VECTORS

It appears from the above that forces do not obey the rules of addition defined in ordinary arithmetic or algebra. For example, two forces acting at a right angle to each other, one of 4 lb and the other of 3 lb, add up to a force of 5 lb, *not* to a force of 7 lb. Forces are not the only quantities which follow the parallelogram law of addition. As you will see later, *displacements, velocities, accelerations,* and *momenta* are other examples of physical quantities possessing magnitude and direction that are added according to the parallelogram law. All these quantities can be represented mathematically by *vectors,* while those physical quantities which have magnitude but not direction, such as *volume, mass,* or *energy,* are represented by plain numbers or *scalars.*

Vectors are defined as *mathematical expressions possessing magnitude and direction, which add according to the parallelogram law.* Vectors are represented by arrows in the illustrations and will be distinguished from scalar quantities in this text through the use of boldface type (**P**). In longhand writing, a vector may be denoted by drawing a short arrow above the letter used to represent it ($\vec{P}$) or by underlining the letter ($\underline{P}$). The magnitude of a vector defines the length of the arrow used to represent the vector. In this text, italic type will be used to denote the magnitude of a vector. Thus, the magnitude of the vector **P** will be denoted by *P*.

A vector used to represent a force acting on a given particle has a well-defined point of application, namely, the particle itself. Such a vector is said to be a *fixed,* or *bound,* vector and cannot be moved without modifying the conditions of the problem. Other physical quantities, however, such as couples (see Chap. 3), are represented by vectors which may be freely moved in space; these vectors are called *free* vectors. Still other physical quantities, such as forces acting on a rigid body (see Chap. 3), are represented by vectors which can be moved, or slid, along their lines of action; they are known as *sliding* vectors.†

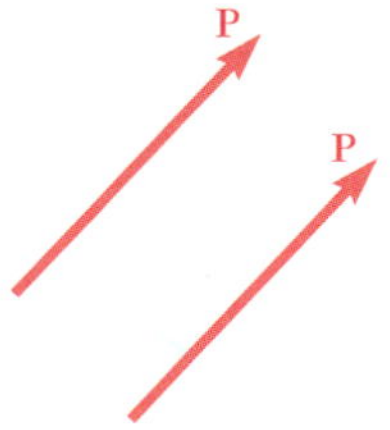

Fig. 2.4

Two vectors which have the same magnitude and the same direction are said to be *equal*, whether or not they also have the same point of application (Fig. 2.4); equal vectors may be denoted by the same letter.

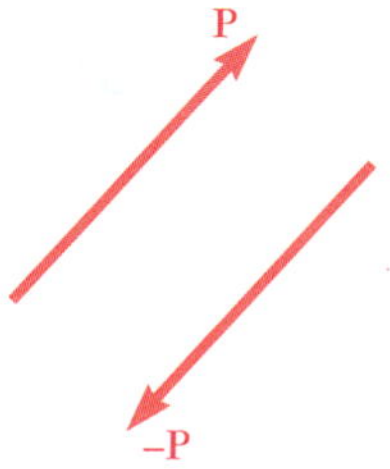

Fig. 2.5

The *negative vector* of a given vector **P** is defined as a vector having the same magnitude as **P** and a direction opposite to that of **P** (Fig. 2.5); the negative of the vector **P** is denoted by −**P**. The vectors **P** and −**P** are commonly referred to as *equal and opposite* vectors. Clearly, we have

$$\mathbf{P} + (-\mathbf{P}) = 0$$

2.4. ADDITION OF VECTORS

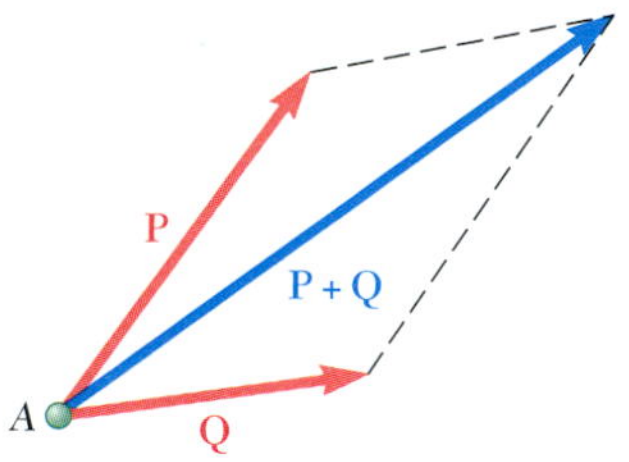

Fig. 2.6

We saw in the preceding section that, by definition, vectors add according to the parallelogram law. Thus, the sum of two vectors **P** and **Q** is obtained by attaching the two vectors to the same point A and constructing a parallelogram, using **P** and **Q** as two sides of the parallelogram (Fig. 2.6). The diagonal that passes through A represents the sum of the vectors **P** and **Q**, and this sum is denoted by **P** + **Q**. The fact that the sign + is used to denote both vector and scalar addition should not cause any confusion if vector and scalar quantities are always carefully distinguished. Thus, we should note that the magnitude of the vector **P** + **Q** is *not*, in general, equal to the sum $P + Q$ of the magnitudes of the vectors **P** and **Q**.

Since the parallelogram constructed on the vectors **P** and **Q** does not depend upon the order in which **P** and **Q** are selected, we conclude that the addition of two vectors is *commutative*, and we write

$$\mathbf{P} + \mathbf{Q} = \mathbf{Q} + \mathbf{P} \tag{2.1}$$

†Some expressions have magnitude and direction, but do not add according to the parallelogram law. While these expressions may be represented by arrows, they *cannot* be considered as vectors.

A group of such expressions is the finite rotations of a rigid body. Place a closed book on a table in front of you, so that it lies in the usual fashion, with its front cover up and its binding to the left. Now rotate it through 180° about an axis parallel to the binding (Fig. 2.3*a*); this rotation may be represented by an arrow of length equal to 180 units and oriented as shown. Picking up the book as it lies in its new position, rotate it now through

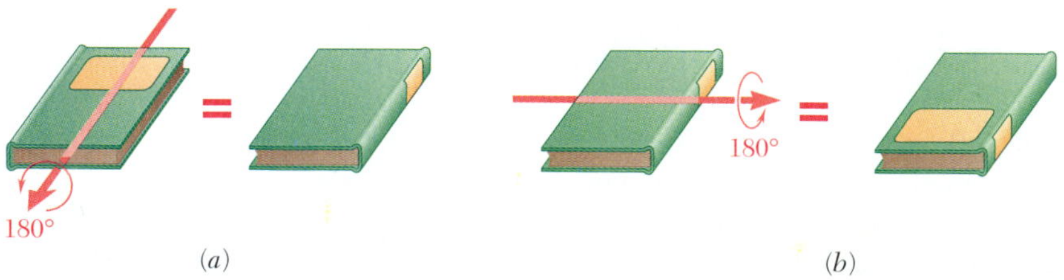

Fig. 2.3 Finite rotations of rigid body

From the parallelogram law, we can derive an alternative method for determining the sum of two vectors. This method, known as the *triangle rule,* is derived as follows. Consider Fig. 2.6, where the sum of the vectors **P** and **Q** has been determined by the parallelogram law. Since the side of the parallelogram opposite **Q** is equal to **Q** in magnitude and direction, we could draw only half of the parallelogram (Fig. 2.7*a*). The sum of the two vectors can thus be found by *arranging* **P** *and* **Q** *in tip-to-tail fashion and then connecting the tail of* **P** *with the tip of* **Q**. In Fig. 2.7*b*, the other half of the parallelogram is considered, and the same result is obtained. This confirms the fact that vector addition is commutative.

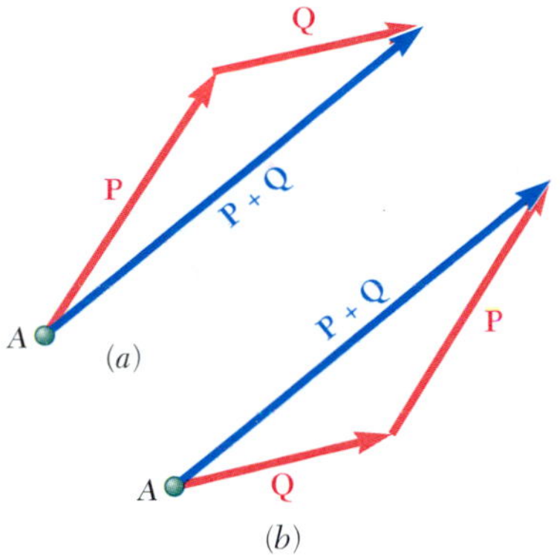

Fig. 2.7

The *subtraction* of a vector is defined as the addition of the corresponding negative vector. Thus, the vector **P** − **Q** representing the difference between the vectors **P** and **Q** is obtained by adding to **P** the negative vector −**Q** (Fig. 2.8). We write

$$\mathbf{P} - \mathbf{Q} = \mathbf{P} + (-\mathbf{Q}) \tag{2.2}$$

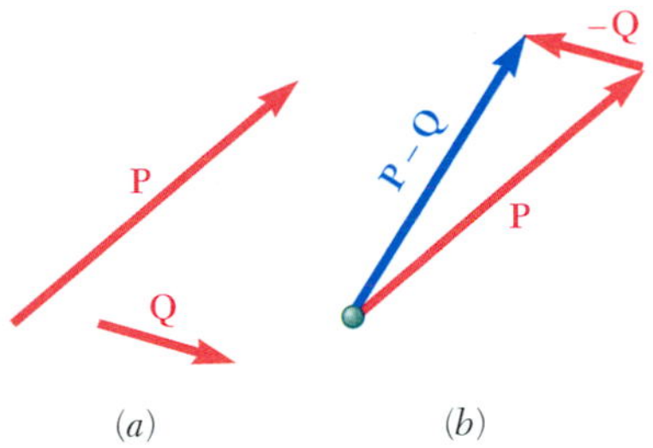

Fig. 2.8

Here again we should observe that, while the same sign is used to denote both vector and scalar subtraction, confusion will be avoided if care is taken to distinguish between vector and scalar quantities.

We will now consider the *sum of three or more vectors.* The sum of three vectors **P**, **Q**, and **S** will, *by definition,* be obtained by first adding the vectors **P** and **Q** and then adding the vector **S** to the vector **P** + **Q**. We thus write

$$\mathbf{P} + \mathbf{Q} + \mathbf{S} = (\mathbf{P} + \mathbf{Q}) + \mathbf{S} \tag{2.3}$$

Similarly, the sum of four vectors will be obtained by adding the fourth vector to the sum of the first three. It follows that the sum of any number of vectors can be obtained by applying repeatedly the parallelogram law to successive pairs of vectors until all the given vectors are replaced by a single vector.

180° about a horizontal axis perpendicular to the binding (Fig. 2.3*b*); this second rotation may be represented by an arrow 180 units long and oriented as shown. But the book could have been placed in this final position through a single 180° rotation about a vertical axis (Fig. 2.3*c*). We conclude that the sum of the two 180° rotations represented by arrows directed respectively along the *z* and *x* axes is a 180° rotation represented by an arrow directed along the *y* axis (Fig. 2.3*d*). Clearly, the finite rotations of a rigid body *do not* obey the parallelogram law of addition; therefore, they *cannot* be represented by vectors.

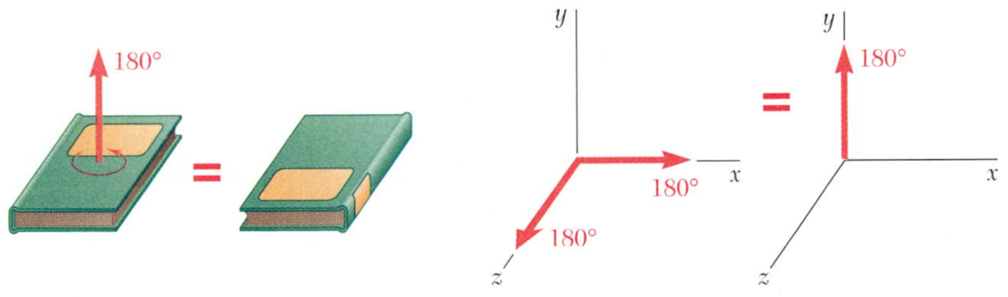

(*c*) (*d*)

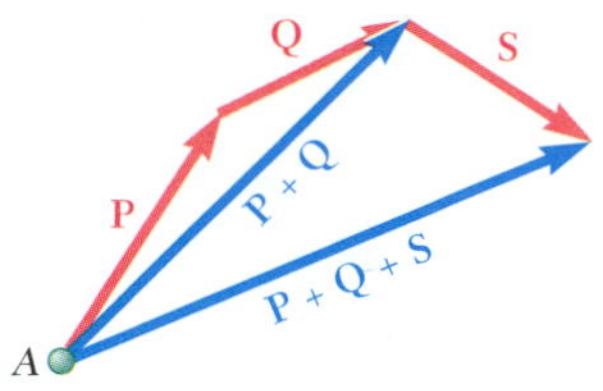

Fig. 2.9

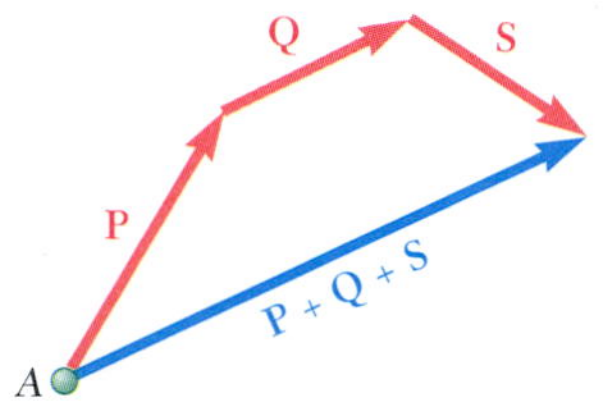

Fig. 2.10

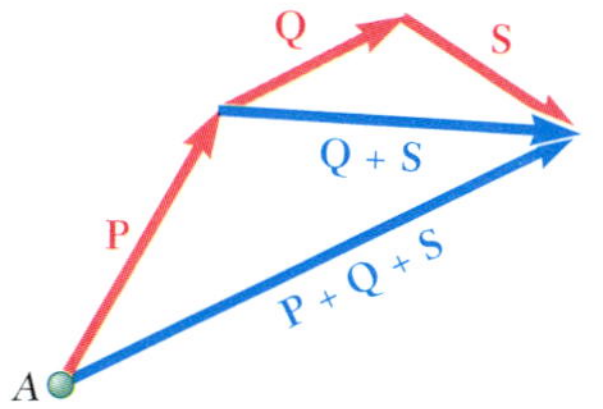

Fig. 2.11

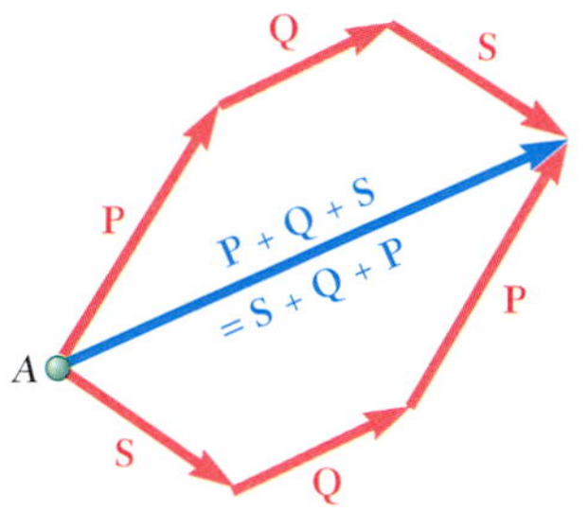

Fig. 2.12

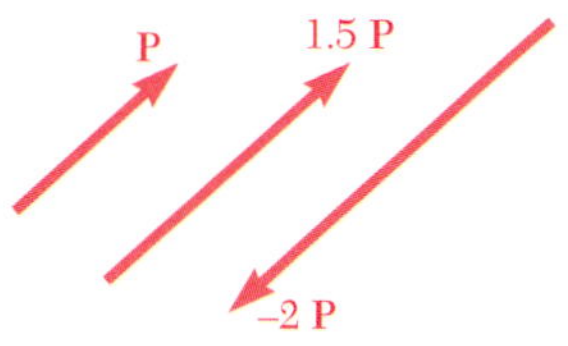

Fig. 2.13

If the given vectors are *coplanar,* that is, if they are contained in the same plane, their sum can be easily obtained graphically. For this case, the repeated application of the triangle rule is preferred to the application of the parallelogram law. In Fig. 2.9 the sum of three vectors **P**, **Q**, and **S** was obtained in that manner. The triangle rule was first applied to obtain the sum **P** + **Q** of the vectors **P** and **Q**; it was applied again to obtain the sum of the vectors **P** + **Q** and **S**. The determination of the vector **P** + **Q**, however, could have been omitted and the sum of the three vectors could have been obtained directly, as shown in Fig. 2.10, by *arranging the given vectors in tip-to-tail fashion and connecting the tail of the first vector with the tip of the last one.* This is known as the *polygon rule* for the addition of vectors.

We observe that the result obtained would have been unchanged if, as shown in Fig. 2.11, the vectors **Q** and **S** had been replaced by their sum **Q** + **S**. We may thus write

$$\mathbf{P} + \mathbf{Q} + \mathbf{S} = (\mathbf{P} + \mathbf{Q}) + \mathbf{S} = \mathbf{P} + (\mathbf{Q} + \mathbf{S}) \tag{2.4}$$

which expresses the fact that vector addition is *associative*. Recalling that vector addition has also been shown, in the case of two vectors, to be commutative, we write

$$\begin{aligned}\mathbf{P} + \mathbf{Q} + \mathbf{S} &= (\mathbf{P} + \mathbf{Q}) + \mathbf{S} = \mathbf{S} + (\mathbf{P} + \mathbf{Q}) \\ &= \mathbf{S} + (\mathbf{Q} + \mathbf{P}) = \mathbf{S} + \mathbf{Q} + \mathbf{P}\end{aligned} \tag{2.5}$$

This expression, as well as others which may be obtained in the same way, shows that the order in which several vectors are added together is immaterial (Fig. 2.12).

Product of a Scalar and a Vector. Since it is convenient to denote the sum **P** + **P** by 2**P**, the sum **P** + **P** + **P** by 3**P**, and, in general, the sum of n equal vectors **P** by the product n**P**, we will define the product n**P** of a positive integer n and a vector **P** as a vector having the same direction as **P** and the magnitude nP. Extending this definition to include all scalars, and recalling the definition of a negative vector given in Sec. 2.3, we define the product k**P** of a scalar k and a vector **P** as a vector having the same direction as **P** (if k is positive), or a direction opposite to that of **P** (if k is negative), and a magnitude equal to the product of P and of the absolute value of k (Fig. 2.13).

2.5. RESULTANT OF SEVERAL CONCURRENT FORCES

Consider a particle A acted upon by several coplanar forces, that is, by several forces contained in the same plane (Fig. 2.14*a*). Since the forces considered here all pass through A, they are also said to be *concurrent.* The vectors representing the forces acting on A may be added by the polygon rule (Fig. 2.14*b*). Since the use of the polygon rule is equivalent to the repeated application of the parallelogram law, the vector **R** thus obtained represents the resultant of the given concurrent forces, that is, the single force which has the same effect on the particle A as the given forces. As indicated above, the order in which the vectors **P**, **Q**, and **S** representing the given forces are added together is immaterial.

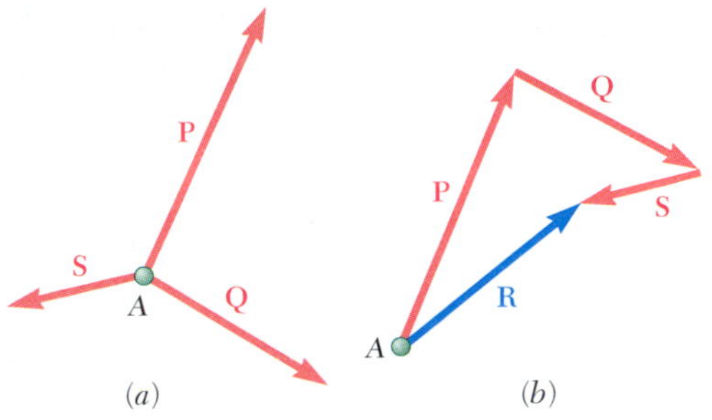

Fig. 2.14

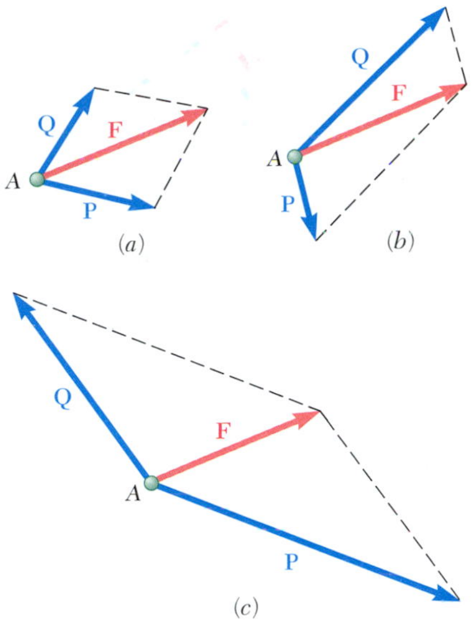

Fig. 2.15

2.6. RESOLUTION OF A FORCE INTO COMPONENTS

We have seen that two or more forces acting on a particle may be replaced by a single force which has the same effect on the particle. Conversely, a single force **F** acting on a particle may be replaced by two or more forces which, together, have the same effect on the particle. These forces are called the *components* of the original force **F**, and the process of substituting them for **F** is called *resolving the force* **F** *into components.*

Clearly, for each force **F** there exist an infinite number of possible sets of components. Sets of *two components* **P** *and* **Q** are the most important as far as practical applications are concerned. But, even then, the number of ways in which a given force **F** may be resolved into two components is unlimited (Fig. 2.15). Two cases are of particular interest:

1. *One of the Two Components,* **P**, *Is Known.* The second component, **Q**, is obtained by applying the triangle rule and joining the tip of **P** to the tip of **F** (Fig. 2.16); the magnitude and direction of **Q** are determined graphically or by trigonometry. Once **Q** has been determined, both components **P** and **Q** should be applied at *A*.
2. *The Line of Action of Each Component Is Known.* The magnitude and sense of the components are obtained by applying the parallelogram law and drawing lines, through the tip of **F**, parallel to the given lines of action (Fig. 2.17). This process leads to two well-defined components, **P** and **Q**, which can be determined graphically or computed trigonometrically by applying the law of sines.

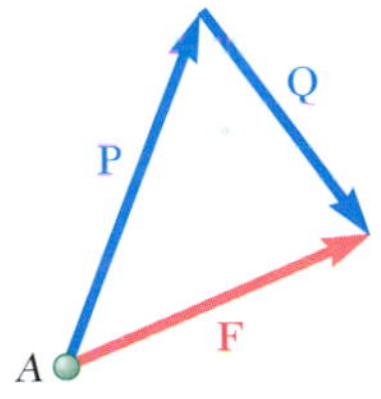

Fig. 2.16

Many other cases can be encountered; for example, the direction of one component may be known, while the magnitude of the other component is to be as small as possible (see Sample Prob. 2.2). In all cases the appropriate triangle or parallelogram which satisfies the given conditions is drawn.

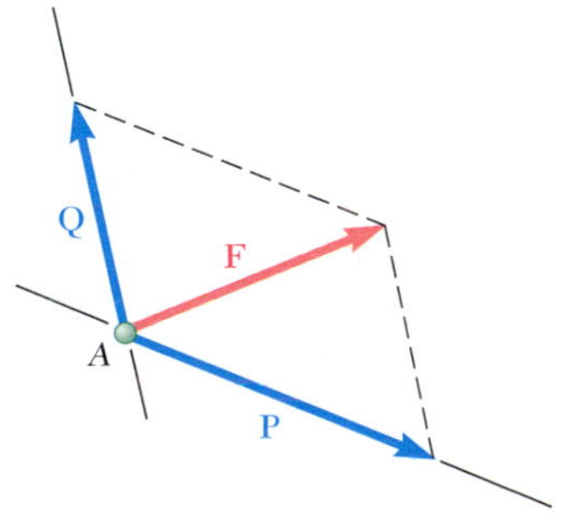

Fig. 2.17

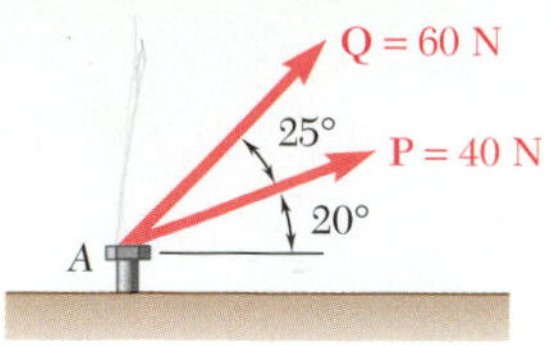

SAMPLE PROBLEM 2.1

The two forces **P** and **Q** act on a bolt A. Determine their resultant.

SOLUTION

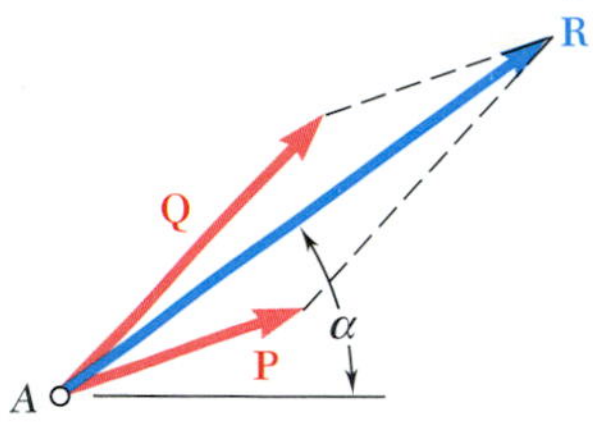

Graphical Solution. A parallelogram with sides equal to **P** and **Q** is drawn to scale. The magnitude and direction of the resultant are measured and found to be

$$R = 98\text{ N} \qquad \alpha = 35^\circ \qquad \mathbf{R} = 98\text{ N} \measuredangle 35^\circ \blacktriangleleft$$

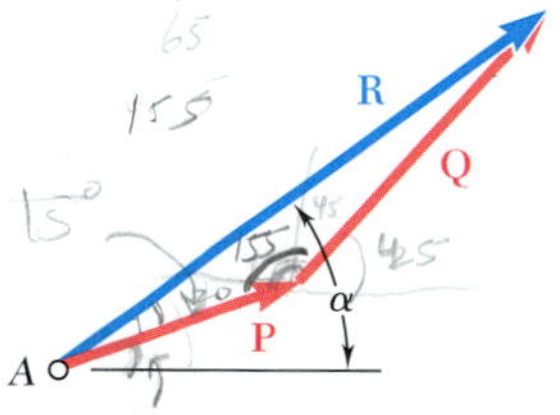

The triangle rule may also be used. Forces **P** and **Q** are drawn in tip-to-tail fashion. Again the magnitude and direction of the resultant are measured.

$$R = 98\text{ N} \qquad \alpha = 35^\circ \qquad \mathbf{R} = 98\text{ N} \measuredangle 35^\circ \blacktriangleleft$$

Trigonometric Solution. The triangle rule is again used; two sides and the included angle are known. We apply the law of cosines.

$$R^2 = P^2 + Q^2 - 2PQ \cos B$$
$$R^2 = (40\text{ N})^2 + (60\text{ N})^2 - 2(40\text{ N})(60\text{ N}) \cos 155^\circ$$
$$R = 97.73\text{ N}$$

Now, applying the law of sines, we write

$$\frac{\sin A}{Q} = \frac{\sin B}{R} \qquad \frac{\sin A}{60\text{ N}} = \frac{\sin 155^\circ}{97.73\text{ N}} \tag{1}$$

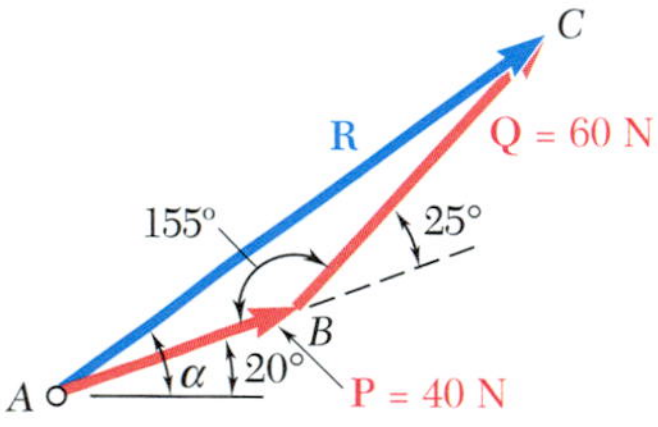

Solving Eq. (1) for sin A, we have

$$\sin A = \frac{(60\text{ N}) \sin 155^\circ}{97.73\text{ N}}$$

Using a calculator, we first compute the quotient, then its arc sine, and obtain

$$A = 15.04^\circ \qquad \alpha = 20^\circ + A = 35.04^\circ$$

We use 3 significant figures to record the answer (see Sec. 1.6):

$$\mathbf{R} = 97.7\text{ N} \measuredangle 35.0^\circ \blacktriangleleft$$

Alternative Trigonometric Solution. We construct the right triangle BCD and compute

$$CD = (60\text{ N}) \sin 25^\circ = 25.36\text{ N}$$
$$BD = (60\text{ N}) \cos 25^\circ = 54.38\text{ N}$$

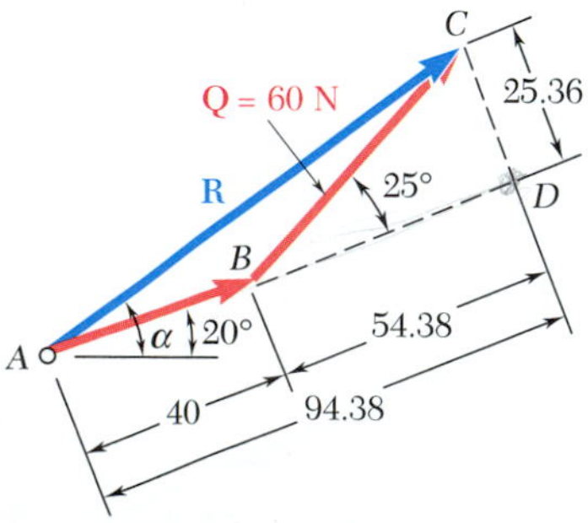

Then, using triangle ACD, we obtain

$$\tan A = \frac{25.36\text{ N}}{94.38\text{ N}} \qquad A = 15.04^\circ$$
$$R = \frac{25.36}{\sin A} \qquad R = 97.73\text{ N}$$

Again, $\alpha = 20^\circ + A = 35.04^\circ \qquad \mathbf{R} = 97.7\text{ N} \measuredangle 35.0^\circ \blacktriangleleft$

SAMPLE PROBLEM 2.2

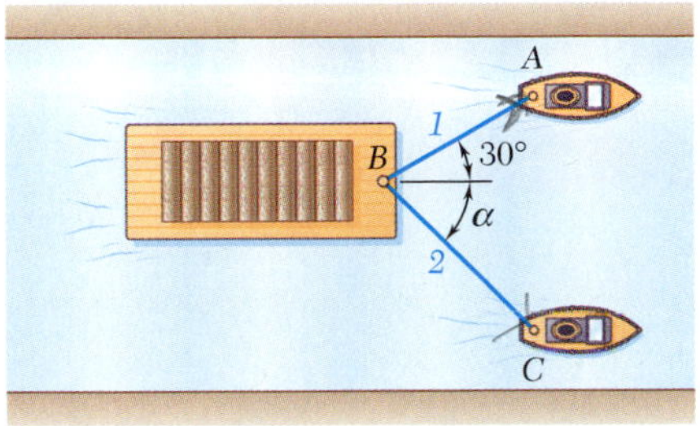

A barge is pulled by two tugboats. If the resultant of the forces exerted by the tugboats is a 5000-lb force directed along the axis of the barge, determine (*a*) the tension in each of the ropes knowing that $\alpha = 45°$, (*b*) the value of α for which the tension in rope 2 is minimum.

SOLUTION

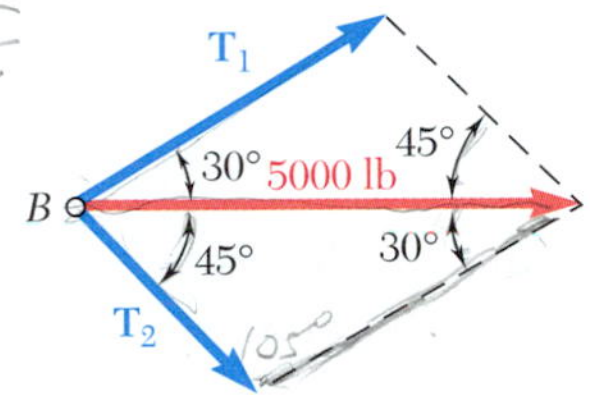

a*. Tension for $\alpha = 45°$.** ***Graphical Solution. The parallelogram law is used; the diagonal (resultant) is known to be equal to 5000 lb and to be directed to the right. The sides are drawn parallel to the ropes. If the drawing is done to scale, we measure

$T_1 = 3700$ lb $\quad T_2 = 2600$ lb ◀

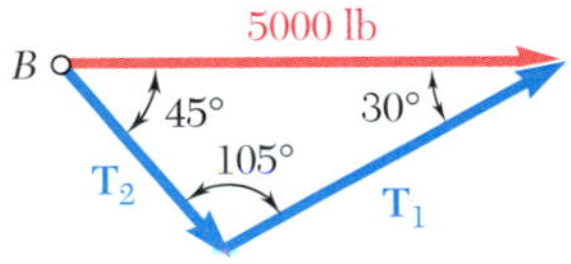

Trigonometric Solution. The triangle rule can be used. We note that the triangle shown represents half of the parallelogram shown above. Using the law of sines, we write

$$\frac{T_1}{\sin 45°} = \frac{T_2}{\sin 30°} = \frac{5000 \text{ lb}}{\sin 105°}$$

With a calculator, we first compute and store the value of the last quotient. Multiplying this value successively by sin 45° and sin 30°, we obtain

$T_1 = 3660$ lb $\quad T_2 = 2590$ lb ◀

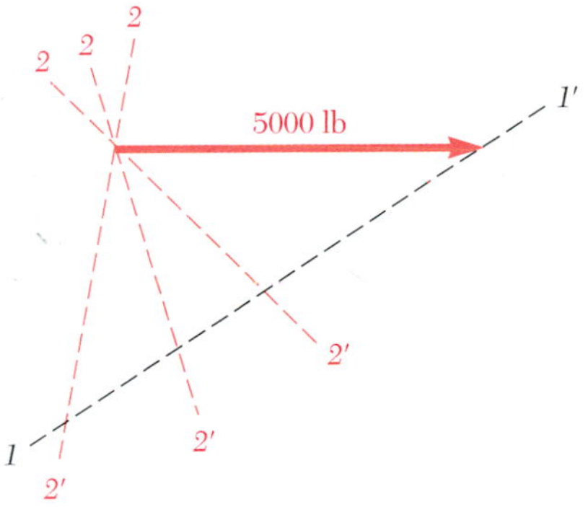

***b*. Value of α for Minimum T_2.** To determine the value of α for which the tension in rope 2 is minimum, the triangle rule is again used. In the sketch shown, line *1-1′* is the known direction of $\mathbf{T}_1$. Several possible directions of $\mathbf{T}_2$ are shown by the lines *2-2′*. We note that the minimum value of T_2 occurs when $\mathbf{T}_1$ and $\mathbf{T}_2$ are perpendicular. The minimum value of T_2 is

$$T_2 = (5000 \text{ lb}) \sin 30° = 2500 \text{ lb}$$

Corresponding values of T_1 and α are

$$T_1 = (5000 \text{ lb}) \cos 30° = 4330 \text{ lb}$$
$$\alpha = 90° - 30°$$

$\alpha = 60°$ ◀

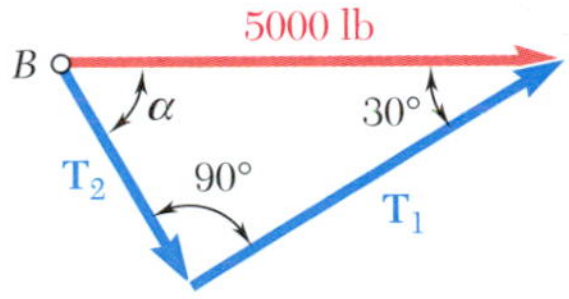

SOLVING PROBLEMS ON YOUR OWN

The preceding sections were devoted to introducing and applying the *parallelogram law* for the addition of vectors.

You will now be asked to solve problems on your own. Some may resemble one of the sample problems; others may not. What all problems and sample problems in this section have in common is that they can be solved by the direct application of the parallelogram law.

Your solution of a given problem should consist of the following steps:

1. Identify which of the forces are the applied forces and which is the resultant. It is often helpful to write the vector equation which shows how the forces are related. For example, in Sample Prob. 2.1 we would have

$$\mathbf{R} = \mathbf{P} + \mathbf{Q}$$

You should keep that relation in mind as you formulate the next part of your solution.

2. Draw a parallelogram with the applied forces as two adjacent sides and the resultant as the included diagonal (Fig. 2.2). Alternatively, you can *use the triangle rule,* with the applied forces drawn in tip-to-tail fashion and the resultant extending from the tail of the first vector to the tip of the second (Fig. 2.7).

3. Indicate all dimensions. Using one of the triangles of the parallelogram, or the triangle constructed according to the triangle rule, indicate all dimensions—whether sides or angles—and determine the unknown dimensions either graphically or by trigonometry. If you use trigonometry, remember that the law of cosines should be applied first if two sides and the included angle are known [Sample Prob. 2.1], and the law of sines should be applied first if one side and all angles are known [Sample Prob. 2.2].

As is evident from the figures of Sec. 2.6, the two components of a force need not be perpendicular. Thus, when asked to resolve a force into two components, it is essential that you align the two adjacent sides of your parallelogram with the specified lines of action of the components.

If you have had prior exposure to mechanics, you might be tempted to ignore the solution techniques of this lesson in favor of resolving the forces into rectangular components. While this latter method is important and will be considered in the next section, use of the parallelogram law simplifies the solution of many problems and should be mastered at this time.

Problems†

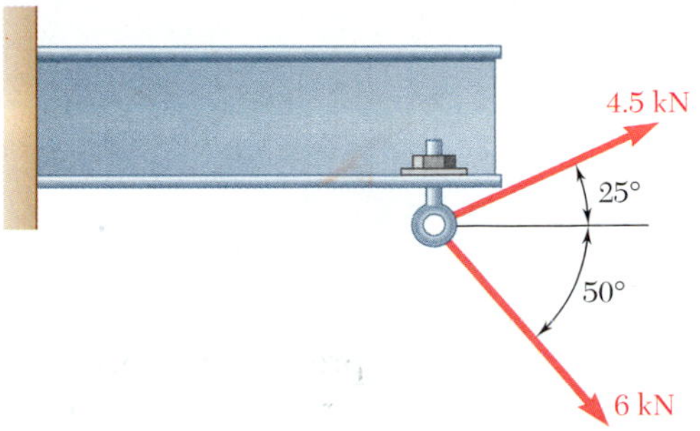

Fig. P2.1

2.1 Two forces are applied to an eye bolt fastened to a beam. Determine graphically the magnitude and direction of their resultant using (*a*) the parallelogram law, (*b*) the triangle rule.

2.2 The cable stays *AB* and *AD* help support pole *AC*. Knowing that the tension is 500 N in *AB* and 160 N in *AD*, determine graphically the magnitude and direction of the resultant of the forces exerted by the stays at *A* using (*a*) the parallelogram law, (*b*) the triangle rule.

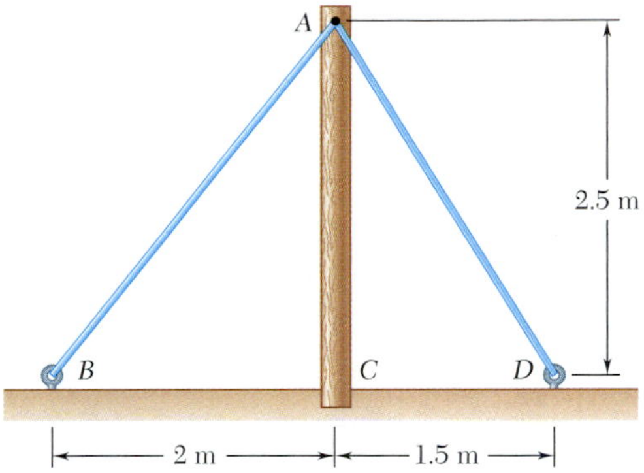

Fig. P2.2

2.3 Two forces **P** and **Q** are applied as shown at point *A* of a hook support. Knowing that $P = 15$ lb and $Q = 25$ lb, determine graphically the magnitude and direction of their resultant using (*a*) the parallelogram law, (*b*) the triangle rule.

2.4 Two forces **P** and **Q** are applied as shown at point *A* of a hook support. Knowing that $P = 45$ lb and $Q = 15$ lb, determine graphically the magnitude and direction of their resultant using (*a*) the parallelogram law, (*b*) the triangle rule.

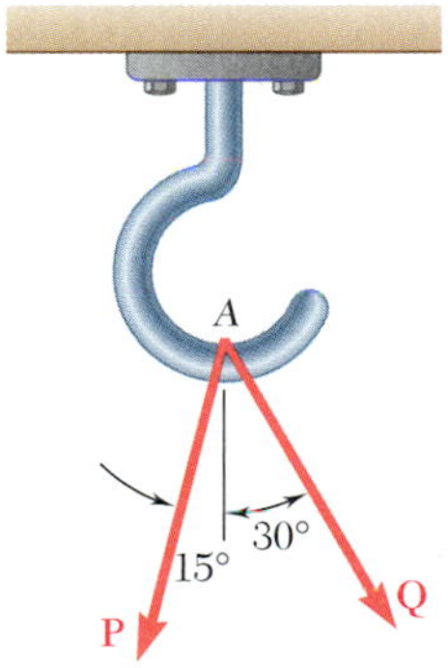

Fig. P2.3 and *P2.4*

2.5 Two control rods are attached at *A* to lever *AB*. Using trigonometry and knowing that the force in the left-hand rod is $F_1 = 120$ N, determine (*a*) the required force F_2 in the right-hand rod if the resultant **R** of the forces exerted by the rods on the lever is to be vertical, (*b*) the corresponding magnitude of **R**.

2.6 Two control rods are attached at *A* to lever *AB*. Using trigonometry and knowing that the force in the right-hand rod is $F_2 = 80$ N, determine (*a*) the required force F_1 in the left-hand rod if the resultant **R** of the forces exerted by the rods on the lever is to be vertical, (*b*) the corresponding magnitude of **R**.

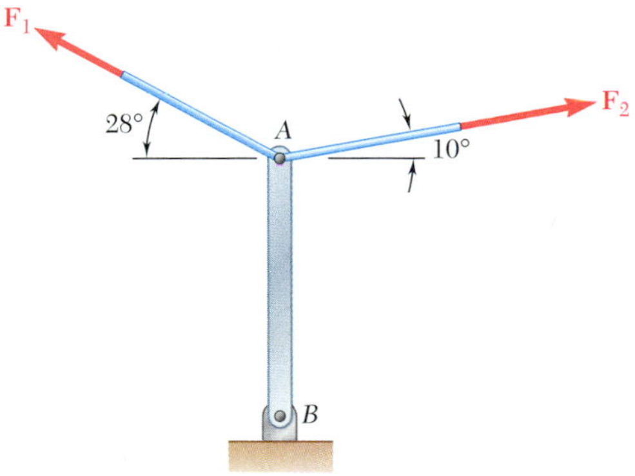

Fig. P2.5 and P2.6

†Answers to all problems set in straight type (such as **2.1**) are given at the end of the book. Answers to problems with a number set in italic type (such as ***2.2***) are not given.

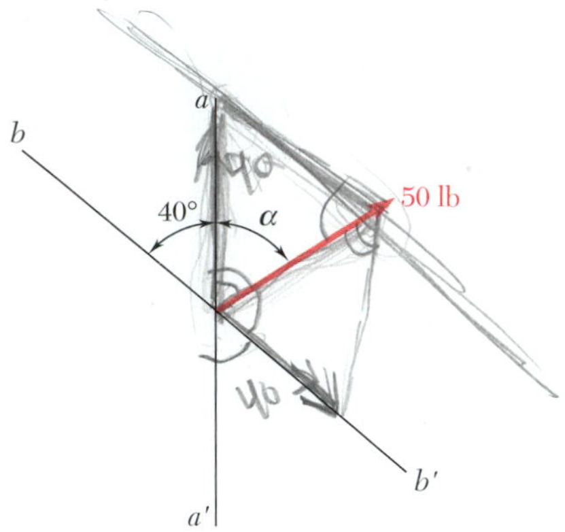

Fig. P2.7 and P2.8

2.7 The 50-lb force is to be resolved into components along lines $a-a'$ and $b-b'$. (*a*) Using trigonometry, determine the angle α knowing that the component along $a-a'$ is 35 lb. (*b*) What is the corresponding value of the component along $b-b'$?

2.8 The 50-lb force is to be resolved into components along lines $a-a'$ and $b-b'$. (*a*) Using trigonometry, determine the angle α knowing that the component along $b-b'$ is 30 lb. (*b*) What is the corresponding value of the component along $a-a'$?

2.9 To steady a sign as it is being lowered, two cables are attached to the sign at *A*. Using trigonometry and knowing that $\alpha = 25°$, determine (*a*) the required magnitude of the force **P** if the resultant **R** of the two forces applied at *A* is to be vertical, (*b*) the corresponding magnitude of **R**.

2.10 To steady a sign as it is being lowered, two cables are attached to the sign at *A*. Using trigonometry and knowing that the magnitude of **P** is 300 N, determine (*a*) the required angle α if the resultant **R** of the two forces applied at *A* is to be vertical, (*b*) the corresponding magnitude of **R**.

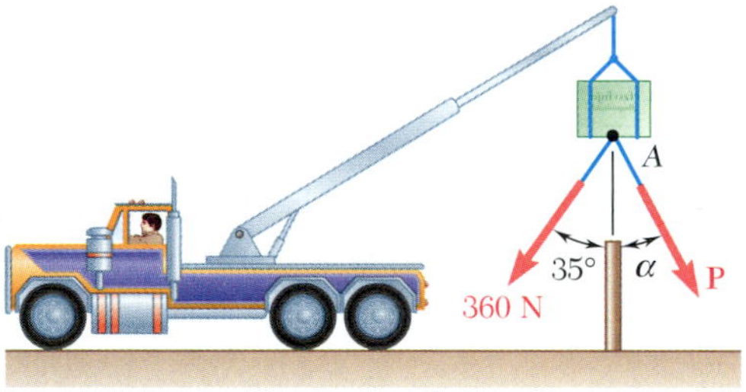

Fig. P2.9 and P2.10

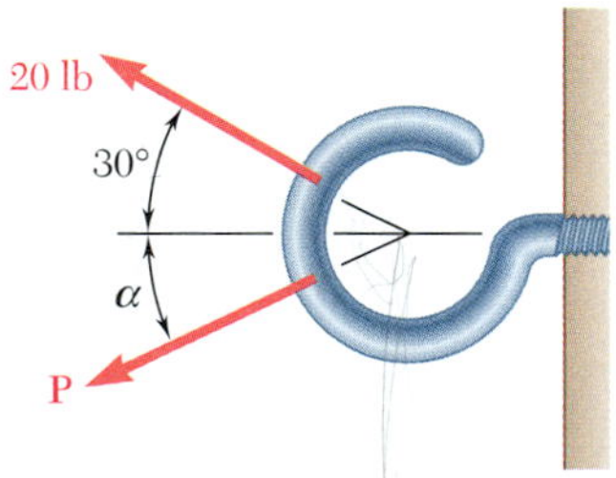

Fig. *P2.11*

2.11 Two forces are applied as shown to a hook support. Using trigonometry and knowing that the magnitude of **P** is 14 lb, determine (*a*) the required angle α if the resultant **R** of the two forces applied to the support is to be horizontal, (*b*) the corresponding magnitude of **R**.

2.12 For the hook support of Prob. 2.3, using trigonometry and knowing that the magnitude of **P** is 25 lb, determine (*a*) the required magnitude of the force **Q** if the resultant **R** of the two forces applied at *A* is to be vertical, (*b*) the corresponding magnitude of **R**.

2.13 For the hook support of Prob. 2.11, determine, using trigonometry, (*a*) the magnitude and direction of the smallest force **P** for which the resultant **R** of the two forces applied to the support is horizontal, (*b*) the corresponding magnitude of **R**.

2.14 As shown in Fig. P2.9, two cables are attached to a sign at *A* to steady the sign as it is being lowered. Using trigonometry, determine (*a*) the magnitude and direction of the smallest force **P** for which the resultant **R** of the two forces applied at *A* is vertical, (*b*) the corresponding magnitude of **R**.

2.15 For the hook support of Prob. 2.11, determine, using trigonometry, the magnitude and direction of the resultant of the two forces applied to the support knowing that $P = 10$ lb and $\alpha = 40°$.

2.16 Solve Prob. 2.1 using trigonometry.

2.17 Solve Prob. 2.2 using trigonometry.

2.18 Solve Prob. 2.3 using trigonometry.

2.19 Two structural members *A* and *B* are bolted to a bracket as shown. Knowing that both members are in compression and that the force is 30 kN in member *A* and 20 kN in member *B*, determine, using trigonometry, the magnitude and direction of the resultant of the forces applied to the bracket by members *A* and *B*.

2.20 Two structural members *A* and *B* are bolted to a bracket as shown. Knowing that both members are in compression and that the force is 20 kN in member *A* and 30 kN in member *B*, determine, using trigonometry, the magnitude and direction of the resultant of the forces applied to the bracket by members *A* and *B*.

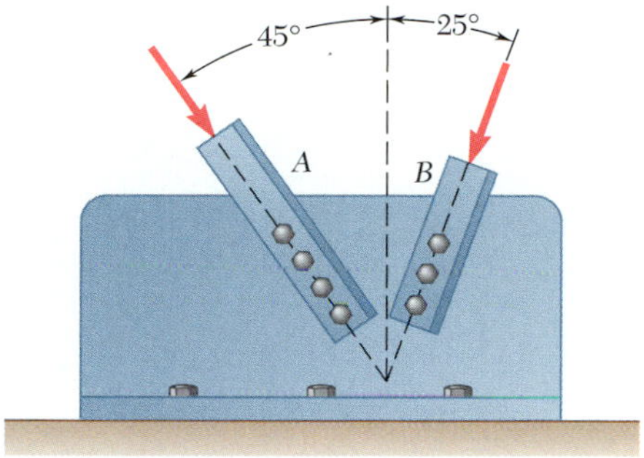

Fig. *P2.19* and *P2.20*

2.7. RECTANGULAR COMPONENTS OF A FORCE. UNIT VECTORS†

In many problems it will be found desirable to resolve a force into two components which are perpendicular to each other. In Fig. 2.18, the force **F** has been resolved into a component $\mathbf{F}_x$ along the x axis and a component $\mathbf{F}_y$ along the y axis. The parallelogram drawn to obtain the two components is a *rectangle*, and $\mathbf{F}_x$ and $\mathbf{F}_y$ are called *rectangular components*.

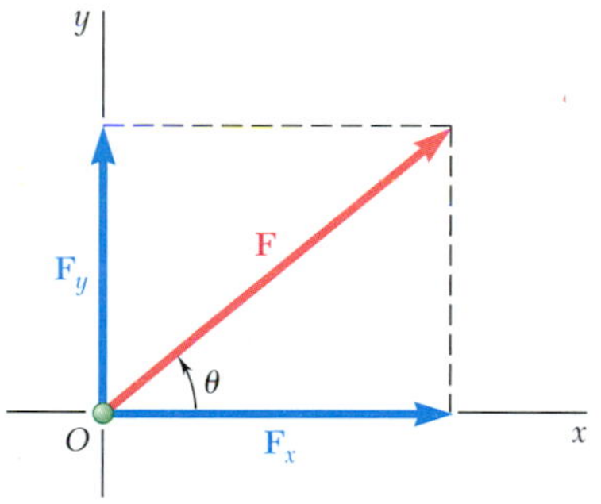

Fig. 2.18

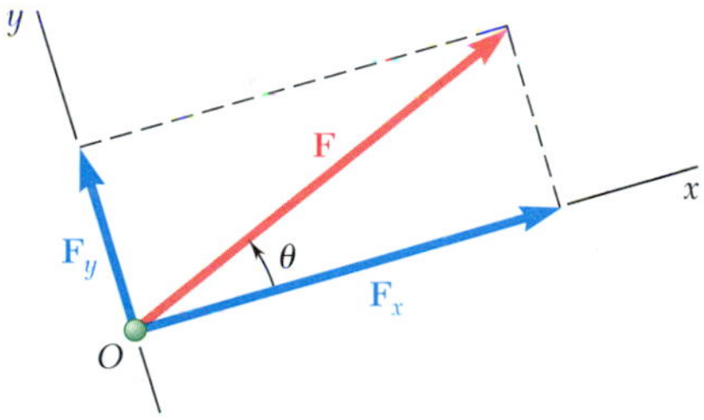

Fig. 2.19

The x and y axes are usually chosen horizontal and vertical, respectively, as in Fig. 2.18; they may, however, be chosen in any two perpendicular directions, as shown in Fig. 2.19. In determining the rectangular components of a force, the student should think of the construction lines shown in Figs. 2.18 and 2.19 as being *parallel* to the x and y axes, rather than *perpendicular* to these axes. This practice will help avoid mistakes in determining *oblique* components as in Sec. 2.6.

†The properties established in Secs. 2.7 and 2.8 may be readily extended to the rectangular components of any vector quantity.

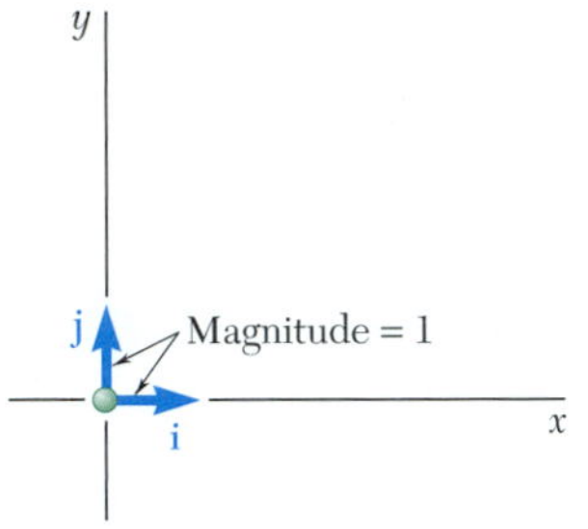

Fig. 2.20

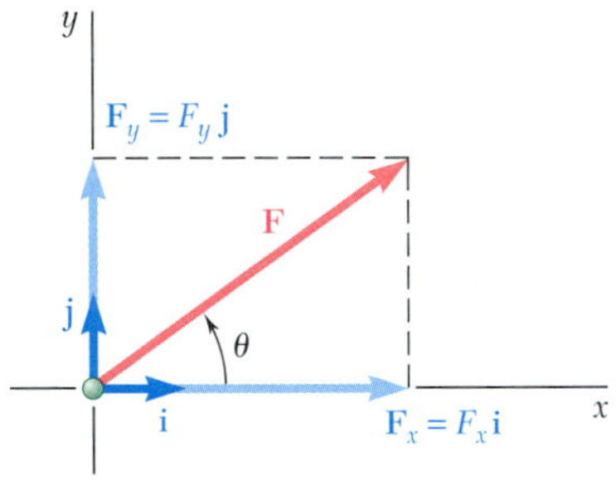

Fig. 2.21

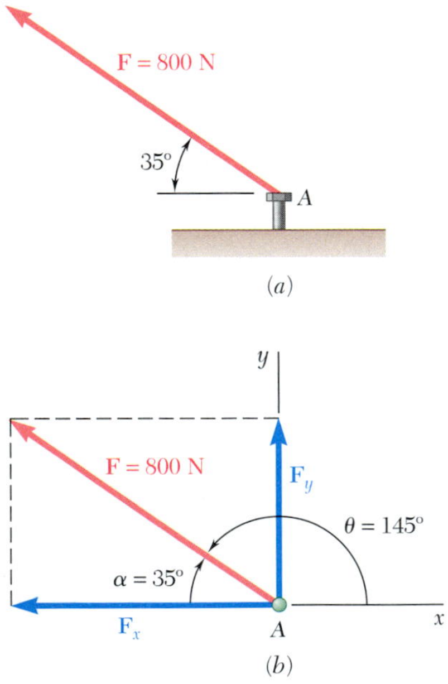

Fig. 2.22

Two vectors of unit magnitude, directed respectively along the positive x and y axes, will be introduced at this point. These vectors are called *unit vectors* and are denoted by **i** and **j**, respectively (Fig. 2.20). Recalling the definition of the product of a scalar and a vector given in Sec. 2.4, we note that the rectangular components $\mathbf{F}_x$ and $\mathbf{F}_y$ of a force **F** may be obtained by multiplying respectively the unit vectors **i** and **j** by appropriate scalars (Fig. 2.21). We write

$$\mathbf{F}_x = F_x\mathbf{i} \qquad \mathbf{F}_y = F_y\mathbf{j} \tag{2.6}$$

and

$$\mathbf{F} = F_x\mathbf{i} + F_y\mathbf{j} \tag{2.7}$$

While the scalars F_x and F_y may be positive or negative, depending upon the sense of $\mathbf{F}_x$ and of $\mathbf{F}_y$, their absolute values are respectively equal to the magnitudes of the component forces $\mathbf{F}_x$ and $\mathbf{F}_y$. The scalars F_x and F_y are called the *scalar components* of the force **F**, while the actual component forces $\mathbf{F}_x$ and $\mathbf{F}_y$ should be referred to as the *vector components* of **F**. However, when there exists no possibility of confusion, the vector as well as the scalar components of **F** may be referred to simply as the *components* of **F**. We note that the scalar component F_x is positive when the vector component $\mathbf{F}_x$ has the same sense as the unit vector **i** (that is, the same sense as the positive x axis) and is negative when $\mathbf{F}_x$ has the opposite sense. A similar conclusion may be drawn regarding the sign of the scalar component F_y.

Denoting by F the magnitude of the force **F** and by θ the angle between **F** and the x axis, measured counterclockwise from the positive x axis (Fig. 2.21), we may express the scalar components of **F** as follows:

$$F_x = F\cos\theta \qquad F_y = F\sin\theta \tag{2.8}$$

We note that the relations obtained hold for any value of the angle θ from 0° to 360° and that they define the signs as well as the absolute values of the scalar components F_x and F_y.

Example 1. A force of 800 N is exerted on a bolt A as shown in Fig. 2.22*a*. Determine the horizontal and vertical components of the force.

In order to obtain the correct sign for the scalar components F_x and F_y, the value $180° - 35° = 145°$ should be substituted for θ in Eqs. (2.8). However, it will be found more practical to determine by inspection the signs of F_x and F_y (Fig. 2.22*b*) and to use the trigonometric functions of the angle $\alpha = 35°$. We write, therefore,

$$F_x = -F\cos\alpha = -(800\text{ N})\cos 35° = -655\text{ N}$$
$$F_y = +F\sin\alpha = +(800\text{ N})\sin 35° = +459\text{ N}$$

The vector components of **F** are thus

$$\mathbf{F}_x = -(655\text{ N})\mathbf{i} \qquad \mathbf{F}_y = +(459\text{ N})\mathbf{j}$$

and we may write **F** in the form

$$\mathbf{F} = -(655\text{ N})\mathbf{i} + (459\text{ N})\mathbf{j}$$

Example 2. A man pulls with a force of 300 N on a rope attached to a building, as shown in Fig. 2.23*a*. What are the horizontal and vertical components of the force exerted by the rope at point *A*?

It is seen from Fig. 2.23*b* that

$$F_x = +(300\text{ N})\cos\alpha \qquad F_y = -(300\text{ N})\sin\alpha$$

Observing that $AB = 10$ m, we find from Fig. 2.23*a*

$$\cos\alpha = \frac{8\text{ m}}{AB} = \frac{8\text{ m}}{10\text{ m}} = \frac{4}{5} \qquad \sin\alpha = \frac{6\text{ m}}{AB} = \frac{6\text{ m}}{10\text{ m}} = \frac{3}{5}$$

We thus obtain

$$F_x = +(300\text{ N})\tfrac{4}{5} = +240\text{ N} \qquad F_y = -(300\text{ N})\tfrac{3}{5} = -180\text{ N}$$

and write

$$\mathbf{F} = (240\text{ N})\mathbf{i} - (180\text{ N})\mathbf{j}$$

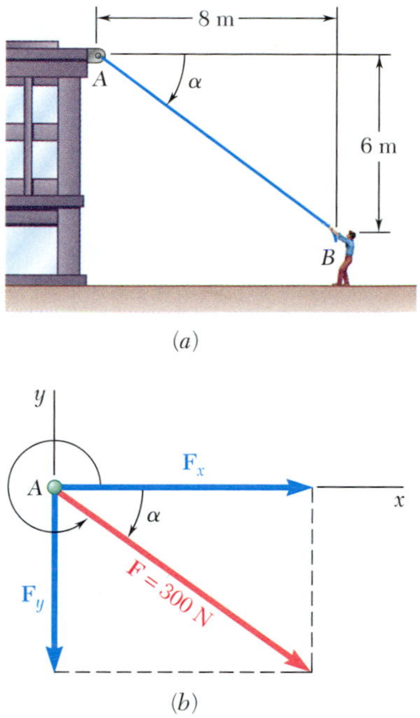

Fig. 2.23

When a force **F** is defined by its rectangular components F_x and F_y (see Fig. 2.21), the angle θ defining its direction can be obtained by writing

$$\tan\theta = \frac{F_y}{F_x} \tag{2.9}$$

The magnitude F of the force can be obtained by applying the Pythagorean theorem and writing

$$F = \sqrt{F_x^2 + F_y^2} \tag{2.10}$$

or by solving for F one of the Eqs. (2.8).

Example 3. A force $\mathbf{F} = (700\text{ lb})\mathbf{i} + (1500\text{ lb})\mathbf{j}$ is applied to a bolt *A*. Determine the magnitude of the force and the angle θ it forms with the horizontal.

First we draw a diagram showing the two rectangular components of the force and the angle θ (Fig. 2.24). From Eq. (2.9), we write

$$\tan\theta = \frac{F_y}{F_x} = \frac{1500\text{ lb}}{700\text{ lb}}$$

Using a calculator,† we enter 1500 lb and divide by 700 lb; computing the arc tangent of the quotient, we obtain $\theta = 65.0°$. Solving the second of Eqs. (2.8) for F, we have

$$F = \frac{F_y}{\sin\theta} = \frac{1500\text{ lb}}{\sin 65.0°} = 1655\text{ lb}$$

The last calculation is facilitated if the value of F_y is stored when originally entered; it may then be recalled to be divided by $\sin\theta$.

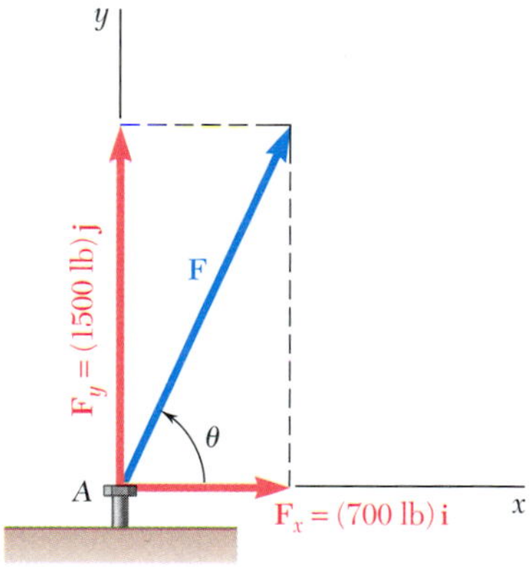

Fig. 2.24

†It is assumed that the calculator used has keys for the computation of trigonometric and inverse trigonometric functions. Some calculators also have keys for the direct conversion of rectangular coordinates into polar coordinates, and vice versa. Such calculators eliminate the need for the computation of trigonometric functions in Examples 1, 2, and 3 and in problems of the same type.

2.8. ADDITION OF FORCES BY SUMMING *X* AND *Y* COMPONENTS

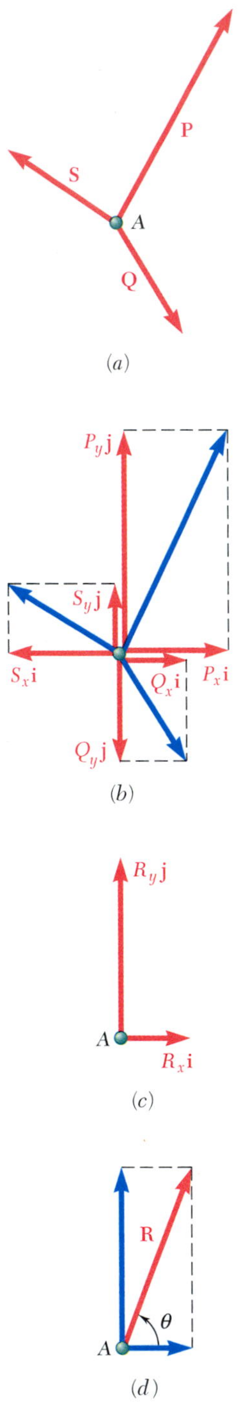

Fig. 2.25

It was seen in Sec. 2.2 that forces should be added according to the parallelogram law. From this law, two other methods, more readily applicable to the *graphical* solution of problems, were derived in Secs. 2.4 and 2.5: the triangle rule for the addition of two forces and the polygon rule for the addition of three or more forces. It was also seen that the force triangle used to define the resultant of two forces could be used to obtain a *trigonometric* solution.

When three or more forces are to be added, no practical trigonometric solution can be obtained from the force polygon which defines the resultant of the forces. In this case, an *analytic* solution of the problem can be obtained by resolving each force into two rectangular components. Consider, for instance, three forces **P**, **Q**, and **S** acting on a particle A (Fig. 2.25*a*). Their resultant **R** is defined by the relation

$$\mathbf{R} = \mathbf{P} + \mathbf{Q} + \mathbf{S} \tag{2.11}$$

Resolving each force into its rectangular components, we write

$$\begin{aligned} R_x\mathbf{i} + R_y\mathbf{j} &= P_x\mathbf{i} + P_y\mathbf{j} + Q_x\mathbf{i} + Q_y\mathbf{j} + S_x\mathbf{i} + S_y\mathbf{j} \\ &= (P_x + Q_x + S_x)\mathbf{i} + (P_y + Q_y + S_y)\mathbf{j} \end{aligned}$$

from which it follows that

$$R_x = P_x + Q_x + S_x \qquad R_y = P_y + Q_y + S_y \tag{2.12}$$

or, for short,

$$R_x = \Sigma F_x \qquad R_y = \Sigma F_y \tag{2.13}$$

We thus conclude that *the scalar components R_x and R_y of the resultant* **R** *of several forces acting on a particle are obtained by adding algebraically the corresponding scalar components of the given forces.*†

In practice, the determination of the resultant **R** is carried out in three steps as illustrated in Fig. 2.25. First the given forces shown in Fig. 2.25*a* are resolved into their x and y components (Fig. 2.25*b*). Adding these components, we obtain the x and y components of **R** (Fig. 2.25*c*). Finally, the resultant $\mathbf{R} = R_x\mathbf{i} + R_y\mathbf{j}$ is determined by applying the parallelogram law (Fig. 2.25*d*). The procedure just described will be carried out most efficiently if the computations are arranged in a table. While it is the only practical analytic method for adding three or more forces, it is also often preferred to the trigonometric solution in the case of the addition of two forces.

†Clearly, this result also applies to the addition of other vector quantities, such as velocities, accelerations, or momenta.

SAMPLE PROBLEM 2.3

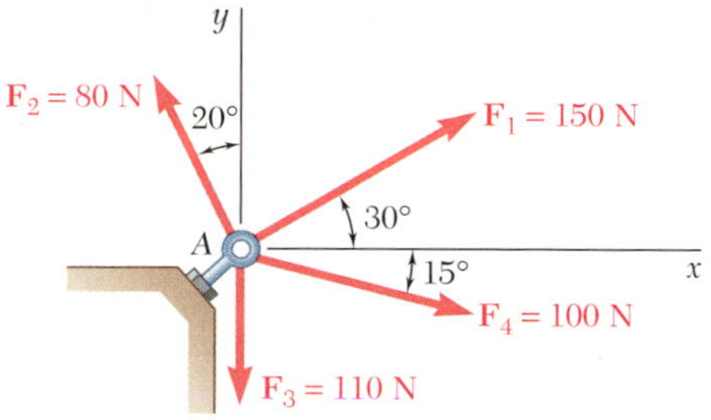

Four forces act on bolt A as shown. Determine the resultant of the forces on the bolt.

SOLUTION

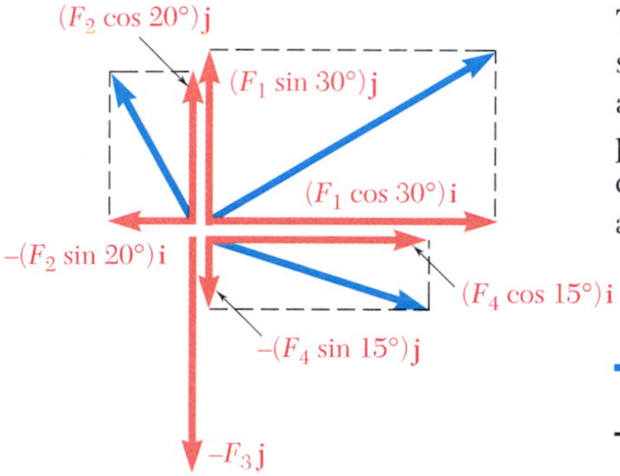

The x and y components of each force are determined by trigonometry as shown and are entered in the table below. According to the convention adopted in Sec. 2.7, the scalar number representing a force component is positive if the force component has the same sense as the corresponding coordinate axis. Thus, x components acting to the right and y components acting upward are represented by positive numbers.

Force	Magnitude, N	x Component, N	y Component, N
$\mathbf{F}_1$	150	+129.9	+75.0
$\mathbf{F}_2$	80	−27.4	+75.2
$\mathbf{F}_3$	110	0	−110.0
$\mathbf{F}_4$	100	+96.6	−25.9
		$R_x = +199.1$	$R_y = +14.3$

Thus, the resultant **R** of the four forces is

$$\mathbf{R} = R_x\mathbf{i} + R_y\mathbf{j} \qquad \mathbf{R} = (199.1\text{ N})\mathbf{i} + (14.3\text{ N})\mathbf{j} \quad ◀$$

The magnitude and direction of the resultant may now be determined. From the triangle shown, we have

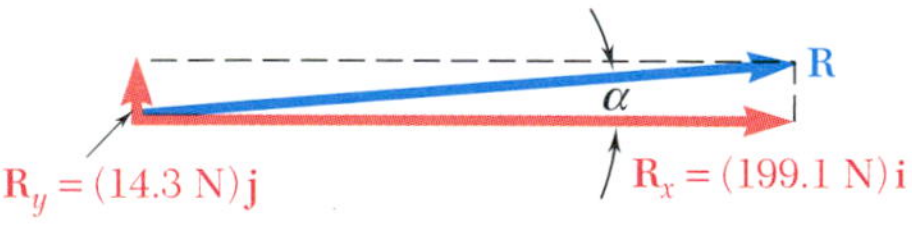

$$\tan \alpha = \frac{R_y}{R_x} = \frac{14.3\text{ N}}{199.1\text{ N}} \qquad \alpha = 4.1°$$

$$R = \frac{14.3\text{ N}}{\sin \alpha} = 199.6\text{ N} \qquad \mathbf{R} = 199.6\text{ N} \measuredangle 4.1° \quad ◀$$

With a calculator, the last computation may be facilitated if the value of R_y is stored when originally entered; it may then be recalled to be divided by sin α. (Also see the footnote on p. 29.)

SOLVING PROBLEMS ON YOUR OWN

You saw in the preceding lesson that the resultant of two forces can be determined either graphically or from the trigonometry of an oblique triangle.

A. When three or more forces are involved, the determination of their resultant **R** is best carried out by first resolving each force into *rectangular components.* Two cases may be encountered, depending upon the way in which each of the given forces is defined:

Case 1. The force* F *is defined by its magnitude F and the angle α it forms with the x axis. The x and y components of the force can be obtained by multiplying F by cos α and sin α, respectively [Example 1].

Case 2. The force* F *is defined by its magnitude F and the coordinates of two points A and B on its line of action (Fig. 2.23). The angle α that **F** forms with the x axis may first be determined by trigonometry. However, the components of **F** may also be obtained directly from proportions among the various dimensions involved, without actually determining α [Example 2].

B. Rectangular components of the resultant. The components R_x and R_y of the resultant can be obtained by adding algebraically the corresponding components of the given forces [Sample Prob. 2.3].

You can express the resultant in *vectorial form* using the unit vectors **i** and **j**, which are directed along the x and y axes, respectively:

$$\mathbf{R} = R_x\mathbf{i} + R_y\mathbf{j}$$

Alternatively, you can determine the *magnitude and direction* of the resultant by solving the right triangle of sides R_x and R_y for R and for the angle that **R** forms with the x axis.

In the examples and sample problem of this lesson, the x and y axes were horizontal and vertical, respectively. You should remember, however, that for some problems it will be more efficient to rotate the axes to align them with one or more of the applied forces.

Problems

2.21 Determine the x and y components of each of the forces shown.

2.22 Determine the x and y components of each of the forces shown.

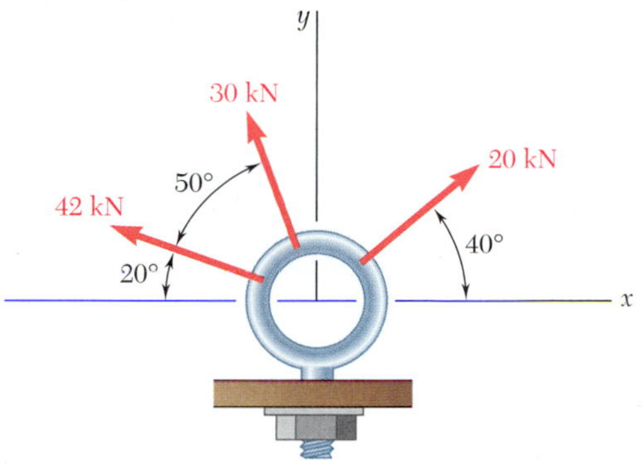

Fig. P2.21

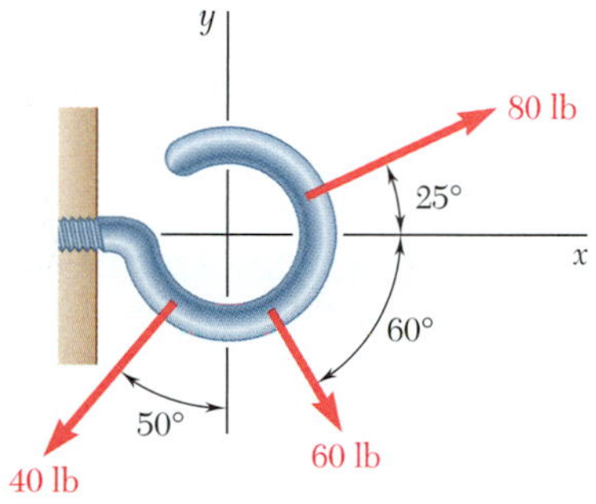

Fig. P2.22

2.23 and *2.24* Determine the x and y components of each of the forces shown.

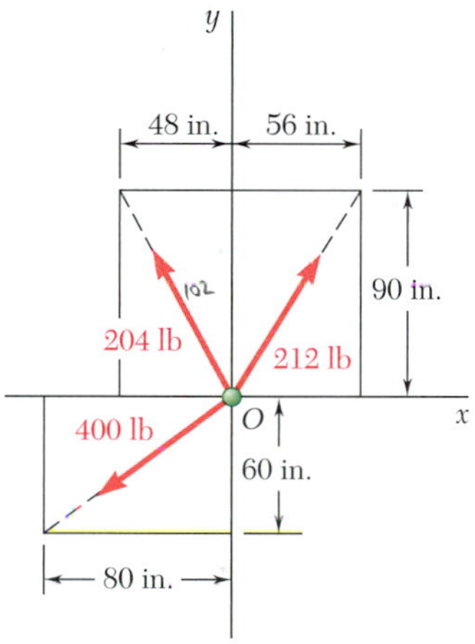

Fig. P2.23

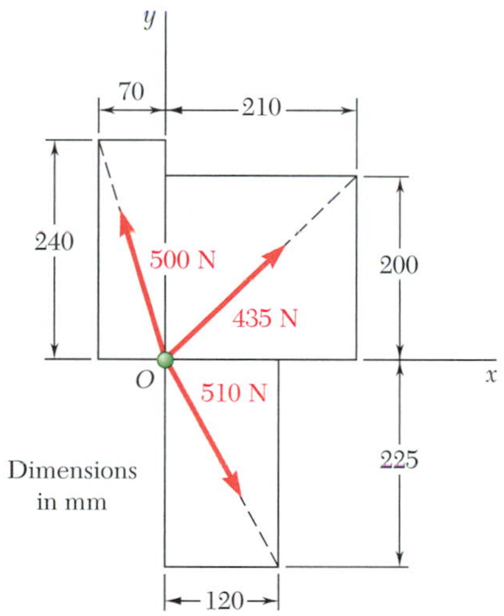

Fig. *P2.24*

2.25 While emptying a wheelbarrow, a gardener exerts on each handle AB a force $\mathbf{P}$ directed along line CD. Knowing that $\mathbf{P}$ must have a 135-N horizontal component, determine (*a*) the magnitude of the force $\mathbf{P}$, (*b*) its vertical component.

Fig. P2.25

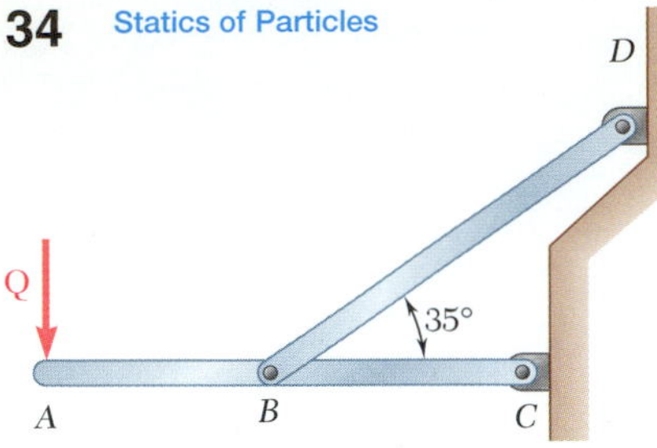

Fig. P2.26

2.26 Member BD exerts on member ABC a force **P** directed along line BD. Knowing that **P** must have a 960-N vertical component, determine (*a*) the magnitude of the force **P**, (*b*) its horizontal component.

2.27 Member CB of the vise shown exerts on block B a force **P** directed along line CB. Knowing that **P** must have a 260-lb horizontal component, determine (*a*) the magnitude of the force **P**, (*b*) its vertical component.

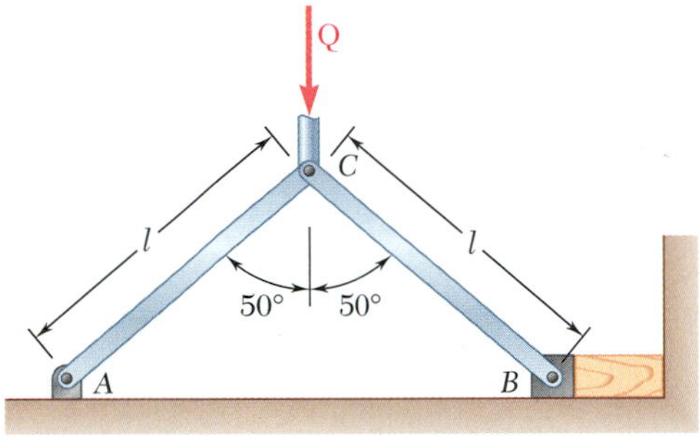

Fig. P2.27

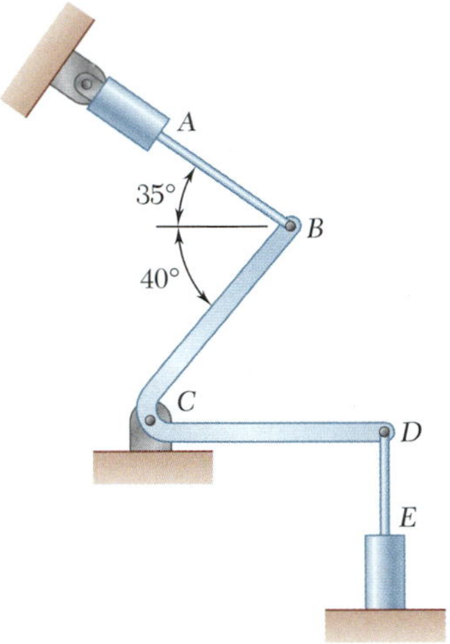

Fig. P2.28

2.28 Activator rod AB exerts on crank BCD a force **P** directed along line AB. Knowing that **P** must have a 25-lb component perpendicular to arm BC of the crank, determine (*a*) the magnitude of the force **P**, (*b*) its component along line BC.

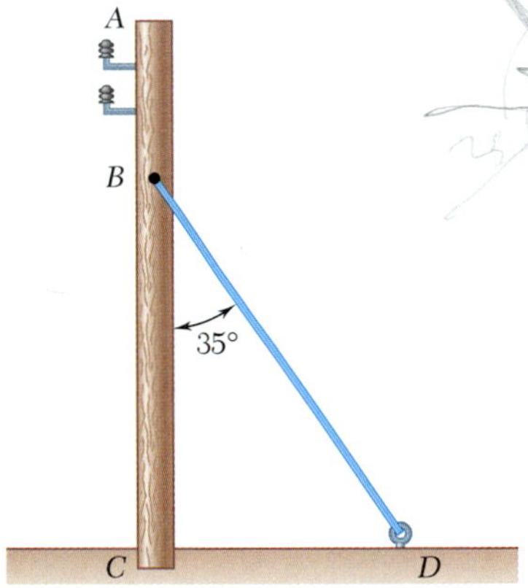

Fig. *P2.29* and *P2.30*

2.29 The guy wire BD exerts on the telephone pole AC a force **P** directed along BD. Knowing that **P** has a 450-N component along line AC, determine (*a*) the magnitude of the force **P**, (*b*) its component in a direction perpendicular to AC.

2.30 The guy wire BD exerts on the telephone pole AC a force **P** directed along BD. Knowing that **P** must have a 200-N component perpendicular to the pole AC, determine (*a*) the magnitude of the force **P**, (*b*) its component along line AC.

2.31 Determine the resultant of the three forces of Prob. 2.24.

2.32 Determine the resultant of the three forces of Prob. 2.21.

2.33 Determine the resultant of the three forces of Prob. 2.22.

2.34 Determine the resultant of the three forces of Prob. 2.23.

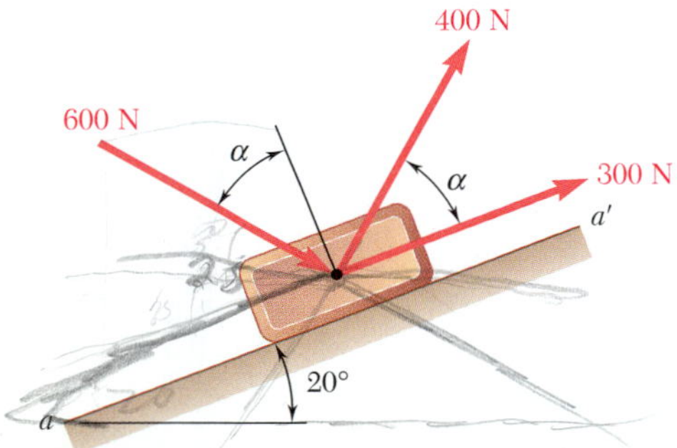

Fig. P2.35 and P2.36

2.35 Knowing that $\alpha = 35°$, determine the resultant of the three forces shown.

2.36 Knowing that $\alpha = 65°$, determine the resultant of the three forces shown.

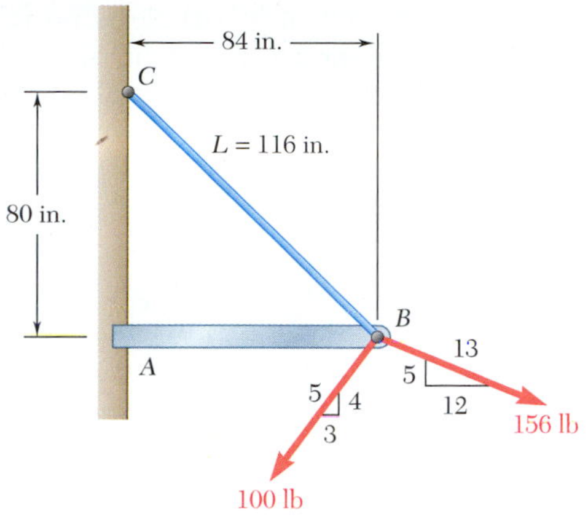

Fig. *P2.37*

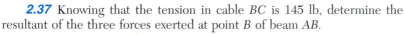

2.37 Knowing that the tension in cable BC is 145 lb, determine the resultant of the three forces exerted at point B of beam AB.

2.38 Knowing that $\alpha = 50°$, determine the resultant of the three forces shown.

2.39 Determine (a) the required value of α if the resultant of the three forces shown is to be vertical, (b) the corresponding magnitude of the resultant.

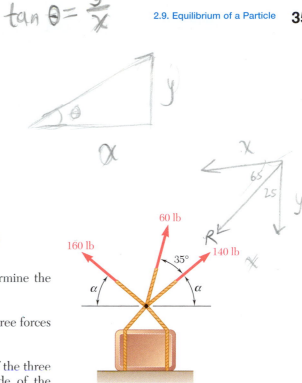

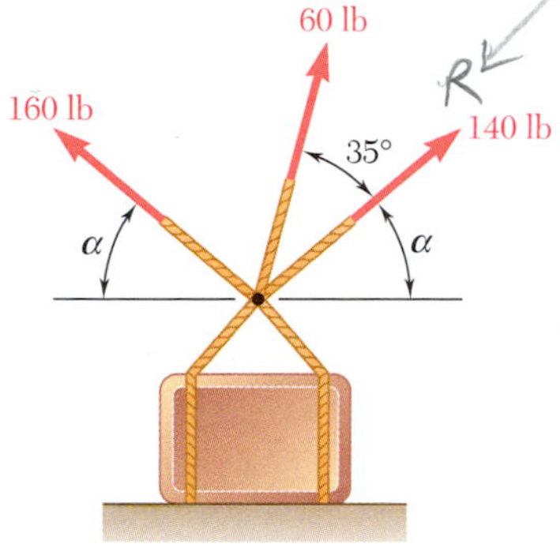

Fig. *P2.38* and P2.39

2.40 For the beam of Prob. 2.37, determine (a) the required tension in cable BC if the resultant of the three forces exerted at point B is to be vertical, (b) the corresponding magnitude of the resultant.

2.41 Boom AB is held in the position shown by three cables. Knowing that the tensions in cables AC and AD are 4 kN and 5.2 kN, respectively, determine (a) the tension in cable AE if the resultant of the tensions exerted at point A of the boom must be directed along AB, (b) the corresponding magnitude of the resultant.

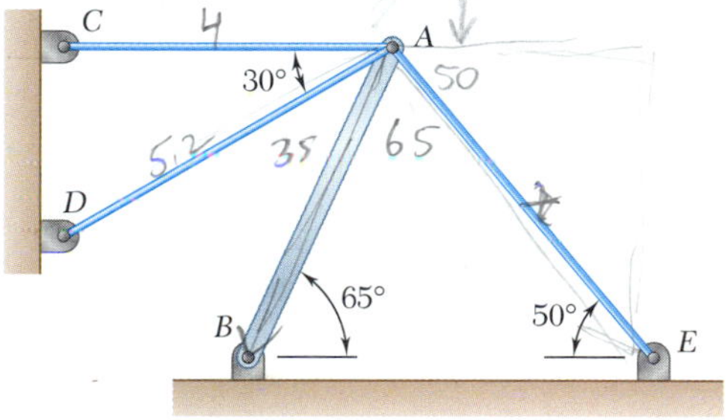

Fig. P2.41

2.42 For the block of Probs. 2.35 and 2.36, determine (a) the required value of α if the resultant of the three forces shown is to be parallel to the incline, (b) the corresponding magnitude of the resultant.

2.9. EQUILIBRIUM OF A PARTICLE

In the preceding sections, we discussed the methods for determining the resultant of several forces acting on a particle. Although it has not occurred in any of the problems considered so far, it is quite possible for the resultant to be zero. In such a case, the net effect of the given forces is zero, and the particle is said to be in equilibrium. We thus have the following definition: *When the resultant of all the forces acting on a particle is zero, the particle is in equilibrium.*

A particle which is acted upon by two forces will be in equilibrium if the two forces have the same magnitude and the same line of action but opposite sense. The resultant of the two forces is then zero. Such a case is shown in Fig. 2.26.

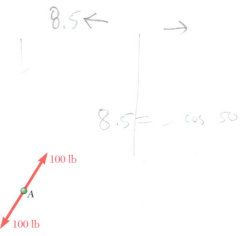

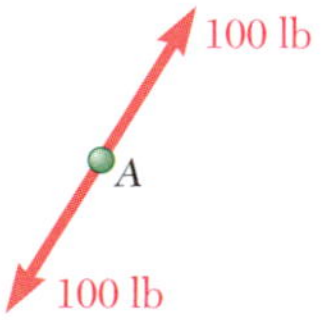

Fig. 2.26

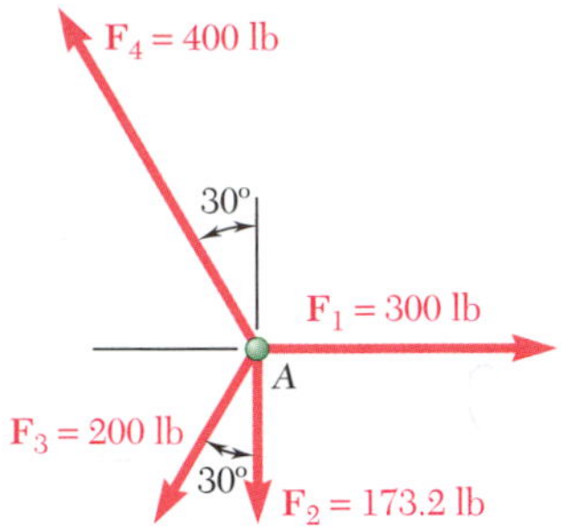

Fig. 2.27

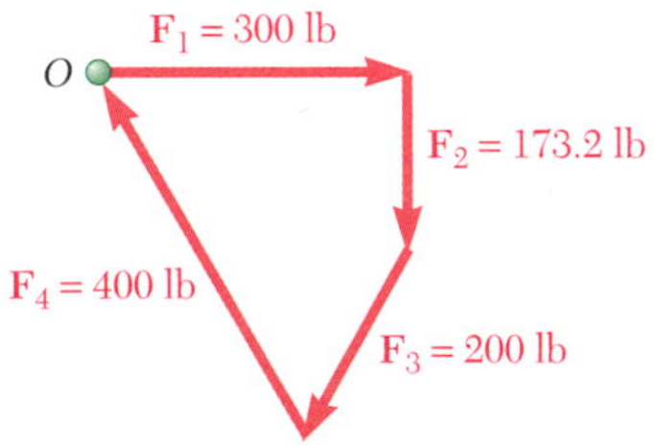

Fig. 2.28

Another case of equilibrium of a particle is represented in Fig. 2.27, where four forces are shown acting on A. In Fig. 2.28, the resultant of the given forces is determined by the polygon rule. Starting from point O with $\mathbf{F}_1$ and arranging the forces in tip-to-tail fashion, we find that the tip of $\mathbf{F}_4$ coincides with the starting point O. Thus the resultant $\mathbf{R}$ of the given system of forces is zero, and the particle is in equilibrium.

The closed polygon drawn in Fig. 2.28 provides a *graphical* expression of the equilibrium of A. To express *algebraically* the conditions for the equilibrium of a particle, we write

$$\mathbf{R} = \Sigma \mathbf{F} = 0 \tag{2.14}$$

Resolving each force $\mathbf{F}$ into rectangular components, we have

$$\Sigma(F_x \mathbf{i} + F_y \mathbf{j}) = 0 \qquad \text{or} \qquad (\Sigma F_x)\mathbf{i} + (\Sigma F_y)\mathbf{j} = 0$$

We conclude that the necessary and sufficient conditions for the equilibrium of a particle are

$$\Sigma F_x = 0 \qquad \Sigma F_y = 0 \tag{2.15}$$

Returning to the particle shown in Fig. 2.27, we check that the equilibrium conditions are satisfied. We write

$$\begin{aligned} \Sigma F_x &= 300 \text{ lb} - (200 \text{ lb}) \sin 30° - (400 \text{ lb}) \sin 30° \\ &= 300 \text{ lb} - 100 \text{ lb} - 200 \text{ lb} = 0 \\ \Sigma F_y &= -173.2 \text{ lb} - (200 \text{ lb}) \cos 30° + (400 \text{ lb}) \cos 30° \\ &= -173.2 \text{ lb} - 173.2 \text{ lb} + 346.4 \text{ lb} = 0 \end{aligned}$$

2.10. NEWTON'S FIRST LAW OF MOTION

In the latter part of the seventeenth century, Sir Isaac Newton formulated three fundamental laws upon which the science of mechanics is based. The first of these laws can be stated as follows:

If the resultant force acting on a particle is zero, the particle will remain at rest (if originally at rest) or will move with constant speed in a straight line (if originally in motion).

From this law and from the definition of equilibrium given in Sec. 2.9, it is seen that a particle in equilibrium either is at rest or is moving in a straight line with constant speed. In the following section, various problems concerning the equilibrium of a particle will be considered.

2.11. PROBLEMS INVOLVING THE EQUILIBRIUM OF A PARTICLE. FREE-BODY DIAGRAMS

In practice, a problem in engineering mechanics is derived from an actual physical situation. A sketch showing the physical conditions of the problem is known as a *space diagram.*

The methods of analysis discussed in the preceding sections apply to a system of forces acting on a particle. A large number of problems involving actual structures, however, can be reduced to problems concerning the equilibrium of a particle. This is done by choosing a significant particle and drawing a separate diagram showing this particle and all the forces acting on it. Such a diagram is called a *free-body diagram.*

As an example, consider the 75-kg crate shown in the space diagram of Fig. 2.29*a.* This crate was lying between two buildings, and it is now being lifted onto a truck, which will remove it. The crate is supported by a vertical cable, which is joined at *A* to two ropes which pass over pulleys attached to the buildings at *B* and *C*. It is desired to determine the tension in each of the ropes *AB* and *AC*.

In order to solve this problem, a free-body diagram showing a particle in equilibrium must be drawn. Since we are interested in the rope tensions, the free-body diagram should include at least one of these tensions or, if possible, both tensions. Point *A* is seen to be a good free body for this problem. The free-body diagram of point *A* is shown in Fig. 2.29*b*. It shows point *A* and the forces exerted on *A* by the vertical cable and the two ropes. The force exerted by the cable is directed downward, and its magnitude is equal to the weight *W* of the crate. Recalling Eq. (1.4), we write

$$W = mg = (75 \text{ kg})(9.81 \text{ m/s}^2) = 736 \text{ N}$$

and indicate this value in the free-body diagram. The forces exerted by the two ropes are not known. Since they are respectively equal in magnitude to the tensions in rope *AB* and rope *AC*, we denote them by $\mathbf{T}_{AB}$ and $\mathbf{T}_{AC}$ and draw them away from *A* in the directions shown in the space diagram. No other detail is included in the free-body diagram.

Since point *A* is in equilibrium, the three forces acting on it must form a closed triangle when drawn in tip-to-tail fashion. This *force triangle* has been drawn in Fig. 2.29*c*. The values T_{AB} and T_{AC} of the tension in the ropes may be found graphically if the triangle is drawn to scale, or they may be found by trigonometry. If the latter method of solution is chosen, we use the law of sines and write

$$\frac{T_{AB}}{\sin 60°} = \frac{T_{AC}}{\sin 40°} = \frac{736 \text{ N}}{\sin 80°}$$

$$T_{AB} = 647 \text{ N} \qquad T_{AC} = 480 \text{ N}$$

When a particle is in *equilibrium under three forces*, the problem can be solved by drawing a force triangle. When a particle is in *equilibrium under more than three forces,* the problem can be solved graphically by drawing a force polygon. If an analytic solution is desired, the *equations of equilibrium* given in Sec. 2.9 should be solved:

$$\Sigma F_x = 0 \qquad \Sigma F_y = 0 \tag{2.15}$$

These equations can be solved for no more than *two unknowns;* similarly, the force triangle used in the case of equilibrium under three forces can be solved for two unknowns.

The more common types of problems are those in which the two unknowns represent (1) the two components (or the magnitude and direction) of a single force, (2) the magnitudes of two forces, each of known direction. Problems involving the determination of the maximum or minimum value of the magnitude of a force are also encountered (see Probs. 2.59 through 2.64).

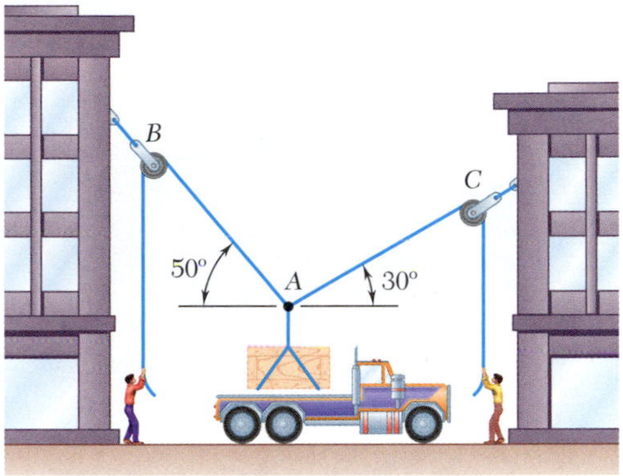

(*a*) Space diagram

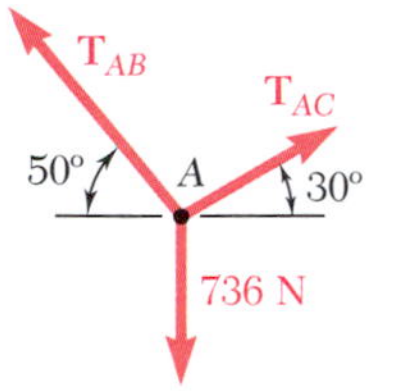

(*b*) Free-body diagram

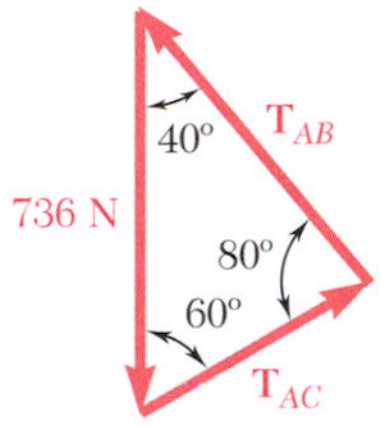

(*c*) Force triangle

Fig. 2.29

Photo 2.1 As illustrated in the above example, it is possible to determine the tensions in the cables supporting the shaft shown by treating the hook as a particle and then applying the equations of equilibrium to the forces acting on the hook.

SAMPLE PROBLEM 2.4

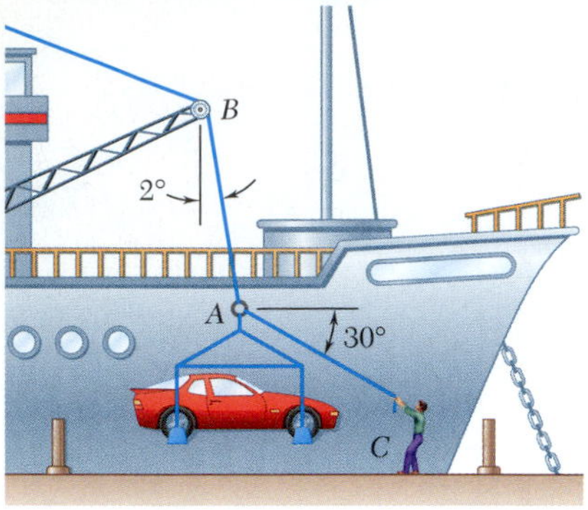

In a ship-unloading operation, a 3500-lb automobile is supported by a cable. A rope is tied to the cable at A and pulled in order to center the automobile over its intended position. The angle between the cable and the vertical is 2°, while the angle between the rope and the horizontal is 30°. What is the tension in the rope?

SOLUTION

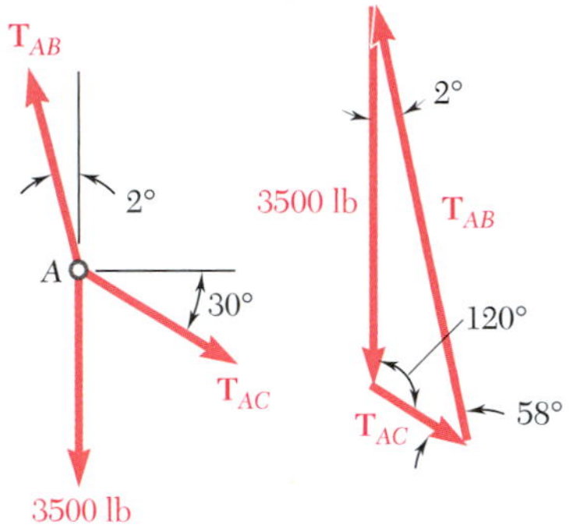

Free-Body Diagram. Point A is chosen as a free body, and the complete free-body diagram is drawn. T_{AB} is the tension in the cable AB, and T_{AC} is the tension in the rope.

Equilibrium Condition. Since only three forces act on the free body, we draw a force triangle to express that it is in equilibrium. Using the law of sines, we write

$$\frac{T_{AB}}{\sin 120°} = \frac{T_{AC}}{\sin 2°} = \frac{3500 \text{ lb}}{\sin 58°}$$

With a calculator, we first compute and store the value of the last quotient. Multiplying this value successively by sin 120° and sin 2°, we obtain

$$T_{AB} = 3570 \text{ lb} \qquad T_{AC} = 144 \text{ lb} \quad ◀$$

SAMPLE PROBLEM 2.5

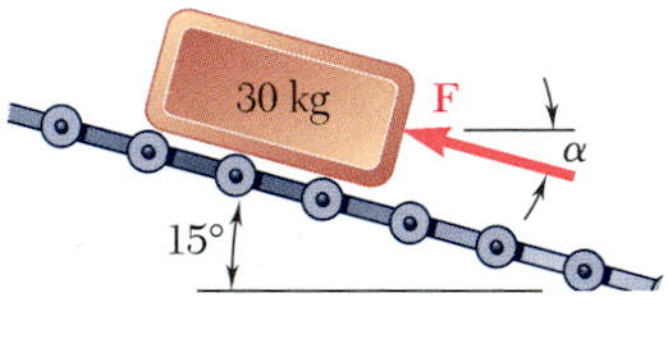

Determine the magnitude and direction of the smallest force **F** which will maintain the package shown in equilibrium. Note that the force exerted by the rollers on the package is perpendicular to the incline.

SOLUTION

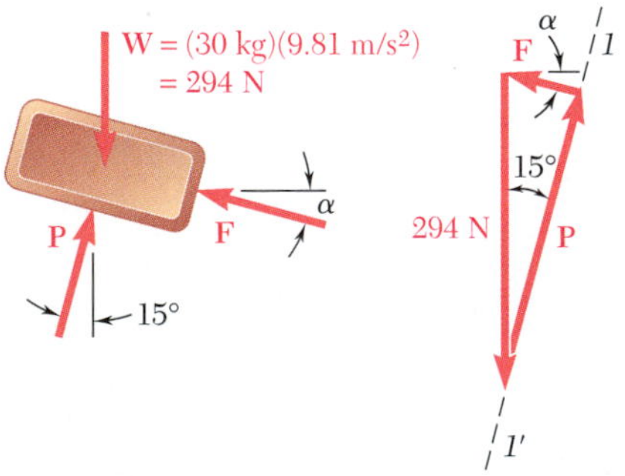

Free-Body Diagram. We choose the package as a free body, assuming that it can be treated as a particle. We draw the corresponding free-body diagram.

Equilibrium Condition. Since only three forces act on the free body, we draw a force triangle to express that it is in equilibrium. Line *1-1'* represents the known direction of **P**. In order to obtain the minimum value of the force **F**, we choose the direction of **F** perpendicular to that of **P**. From the geometry of the triangle obtained, we find

$$F = (294 \text{ N}) \sin 15° = 76.1 \text{ N} \qquad \alpha = 15°$$

$$\mathbf{F} = 76.1 \text{ N} \measuredangle 15° \quad ◀$$

SAMPLE PROBLEM 2.6

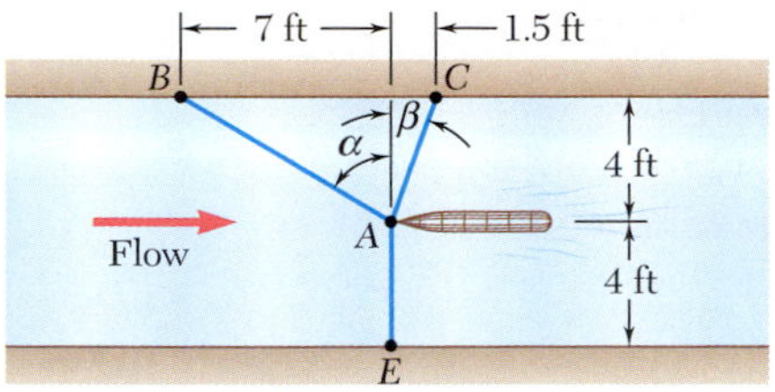

As part of the design of a new sailboat, it is desired to determine the drag force which may be expected at a given speed. To do so, a model of the proposed hull is placed in a test channel and three cables are used to keep its bow on the centerline of the channel. Dynamometer readings indicate that for a given speed, the tension is 40 lb in cable AB and 60 lb in cable AE. Determine the drag force exerted on the hull and the tension in cable AC.

SOLUTION

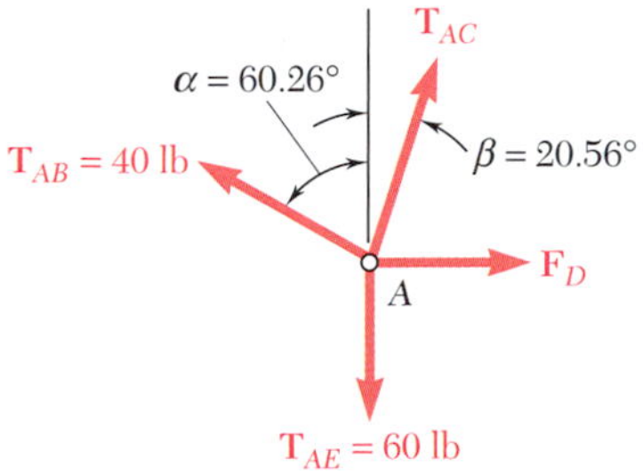

Determination of the Angles. First, the angles α and β defining the direction of cables AB and AC are determined. We write

$$\tan \alpha = \frac{7 \text{ ft}}{4 \text{ ft}} = 1.75 \qquad \tan \beta = \frac{1.5 \text{ ft}}{4 \text{ ft}} = 0.375$$

$$\alpha = 60.26° \qquad \beta = 20.56°$$

Free-Body Diagram. Choosing the hull as a free body, we draw the free-body diagram shown. It includes the forces exerted by the three cables on the hull, as well as the drag force $\mathbf{F}_D$ exerted by the flow.

Equilibrium Condition. We express that the hull is in equilibrium by writing that the resultant of all forces is zero:

$$\mathbf{R} = \mathbf{T}_{AB} + \mathbf{T}_{AC} + \mathbf{T}_{AE} + \mathbf{F}_D = 0 \tag{1}$$

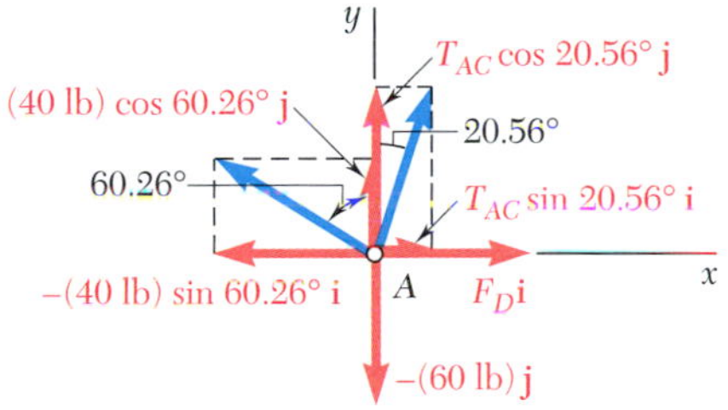

Since more than three forces are involved, we resolve the forces into x and y components:

$$\begin{aligned}
\mathbf{T}_{AB} &= -(40 \text{ lb}) \sin 60.26°\mathbf{i} + (40 \text{ lb}) \cos 60.26°\mathbf{j} \\
&= -(34.73 \text{ lb})\mathbf{i} + (19.84 \text{ lb})\mathbf{j} \\
\mathbf{T}_{AC} &= T_{AC} \sin 20.56°\mathbf{i} + T_{AC} \cos 20.56°\mathbf{j} \\
&= 0.3512T_{AC}\mathbf{i} + 0.9363T_{AC}\mathbf{j} \\
\mathbf{T}_{AE} &= -(60 \text{ lb})\mathbf{j} \\
\mathbf{F}_D &= F_D\mathbf{i}
\end{aligned}$$

Substituting the expressions obtained into Eq. (1) and factoring the unit vectors **i** and **j**, we have

$$(-34.73 \text{ lb} + 0.3512T_{AC} + F_D)\mathbf{i} + (19.84 \text{ lb} + 0.9363T_{AC} - 60 \text{ lb})\mathbf{j} = 0$$

This equation will be satisfied if, and only if, the coefficients of **i** and **j** are equal to zero. We thus obtain the following two equilibrium equations, which express, respectively, that the sum of the x components and the sum of the y components of the given forces must be zero.

$$\Sigma F_x = 0: \qquad -34.73 \text{ lb} + 0.3512T_{AC} + F_D = 0 \tag{2}$$

$$\Sigma F_y = 0: \qquad 19.84 \text{ lb} + 0.9363T_{AC} - 60 \text{ lb} = 0 \tag{3}$$

From Eq. (3) we find $\qquad T_{AC} = +42.9 \text{ lb}$ ◀

and, substituting this value into Eq. (2), $\qquad F_D = +19.66 \text{ lb}$ ◀

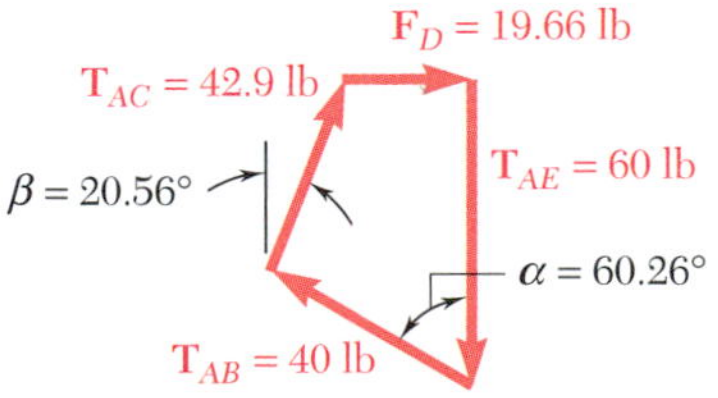

In drawing the free-body diagram, we assumed a sense for each unknown force. A positive sign in the answer indicates that the assumed sense is correct. The complete force polygon may be drawn to check the results.

SOLVING PROBLEMS ON YOUR OWN

When a particle is in *equilibrium,* the resultant of the forces acting on the particle must be zero. Expressing this fact in the case of a particle under *coplanar forces* will provide you with two relations among these forces. As you saw in the preceding sample problems, these relations can be used to determine two unknowns—such as the magnitude and direction of one force or the magnitudes of two forces.

Drawing a free-body diagram is the first step in the solution of a problem involving the equilibrium of a particle. This diagram shows the particle and all the forces acting on it. Indicate on your free-body diagram the magnitudes of known forces as well as any angle or dimensions that define the direction of a force. Any unknown magnitude or angle should be denoted by an appropriate symbol. Nothing else should be included on the free-body diagram.

Drawing a clear and accurate free-body diagram is a must in the solution of any equilibrium problem. Skipping this step might save you pencil and paper, but is very likely to lead you to a wrong solution.

Case 1. If only three forces are involved in the free-body diagram, the rest of the solution is best carried out by drawing these forces in tip-to-tail fashion to form a *force triangle.* This triangle can be solved graphically or using trigonometry for no more than two unknowns [Sample Probs. 2.4 and 2.5].

Case 2. If more than three forces are involved, it is to your advantage to use an *analytic solution.* Begin by selecting appropriate x and y axes and resolve each of the forces shown in the free-body diagram into x and y components. Expressing that the sum of the x components and the sum of the y components of all the forces are both zero, you will obtain two equations which you can solve for no more than two unknowns [Sample Prob. 2.6].

It is strongly recommended that when using an analytic solution the equations of equilibrium be written in the same form as Eqs. (2) and (3) of Sample Prob. 2.6. The practice adopted by some students of initially placing the unknowns on the left side of the equation and the known quantities on the right side may lead to confusion in assigning the appropriate sign to each term.

We have noted that regardless of the method used to solve a two-dimensional equilibrium problem we can determine at most two unknowns. If a two-dimensional problem involves more than two unknowns, one or more additional relations must be obtained from the information contained in the statement of the problem.

Some of the following problems contain small pulleys. We will assume that the pulleys are frictionless, so the tension in the rope or cable passing over a pulley is the same on each side of the pulley. In Chap. 4 we will discuss why the tension is the same.

Problems

2.43 Two cables are tied together at C and are loaded as shown. Determine the tension (*a*) in cable AC, (*b*) in cable BC.

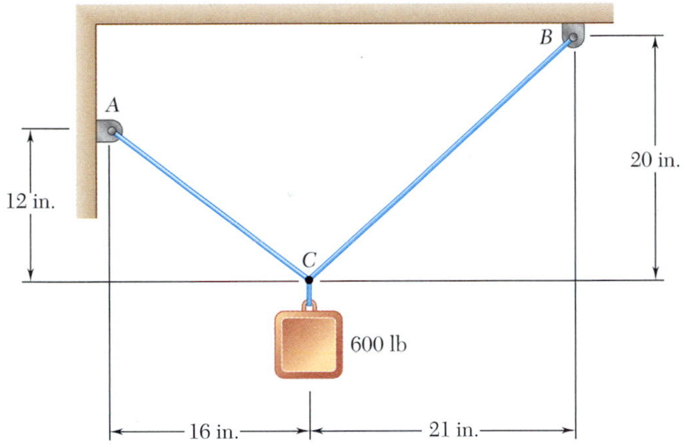

Fig. P2.43

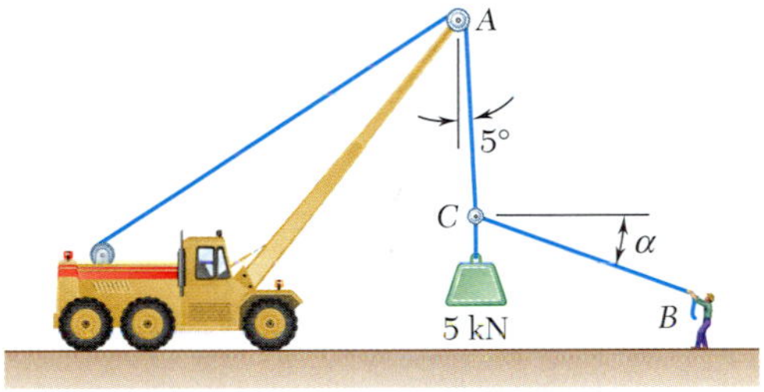

Fig. P2.44

2.44 Knowing that $\alpha = 25°$, determine the tension (*a*) in cable AC, (*b*) in rope BC.

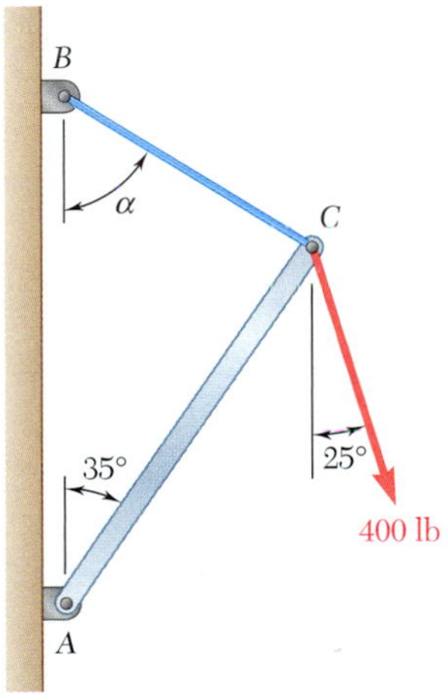

Fig. P2.45

2.45 Knowing that $\alpha = 50°$ and that boom AC exerts on pin C a force directed along line AC, determine (*a*) the magnitude of that force, (*b*) the tension in cable BC.

2.46 Two cables are tied together at C and are loaded as shown. Knowing that $\alpha = 30°$, determine the tension (*a*) in cable AC, (*b*) in cable BC.

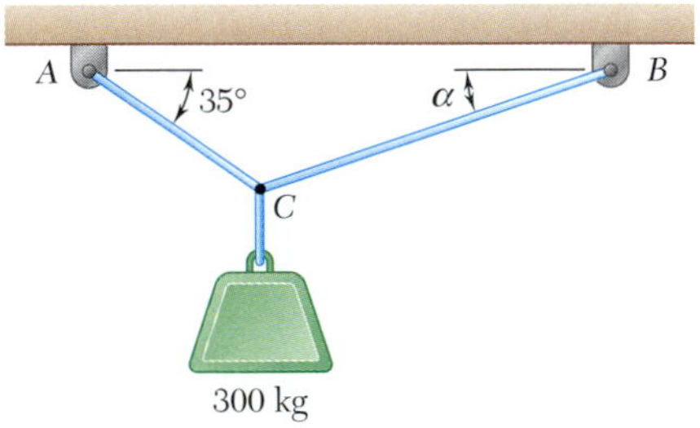

Fig. P2.46

2.47 A chairlift has been stopped in the position shown. Knowing that each chair weighs 300 N and that the skier in chair *E* weighs 890 N, determine the weight of the skier in chair *F*.

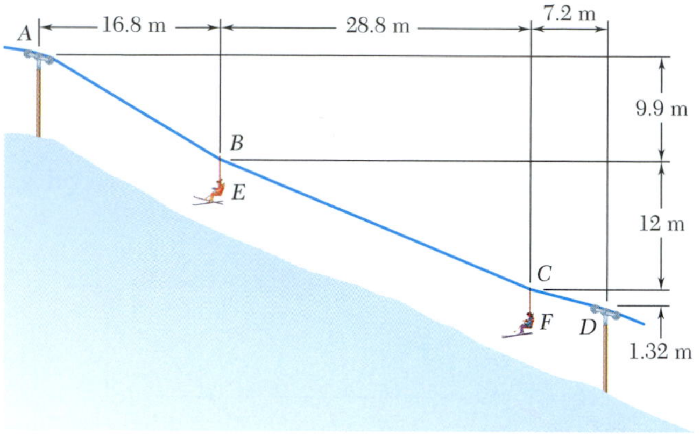

Fig. *P2.47* and *P2.48*

2.48 A chairlift has been stopped in the position shown. Knowing that each chair weighs 300 N and that the skier in chair *F* weighs 800 N, determine the weight of the skier in chair *E*.

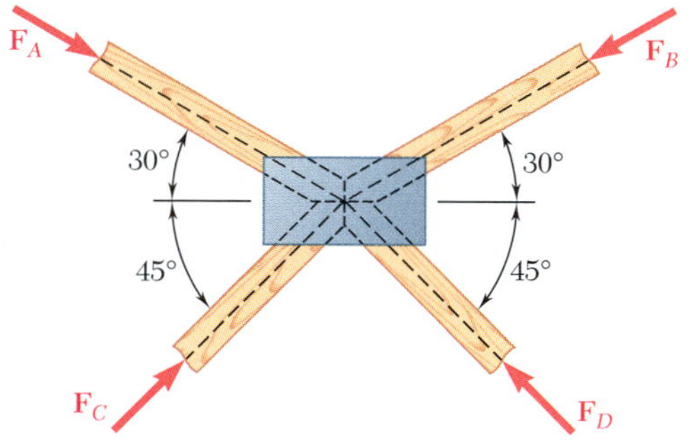

Fig. P2.49 and P2.50

2.49 Four wooden members are joined with metal plate connectors and are in equilibrium under the action of the four forces shown. Knowing that $F_A = 510$ lb and $F_B = 480$ lb, determine the magnitudes of the other two forces.

2.50 Four wooden members are joined with metal plate connectors and are in equilibrium under the action of the four forces shown. Knowing that $F_A = 420$ lb and $F_C = 540$ lb, determine the magnitudes of the other two forces.

2.51 Two forces **P** and **Q** are applied as shown to an aircraft connection. Knowing that the connection is in equilibrium and that $P = 400$ lb and $Q = 520$ lb, determine the magnitudes of the forces exerted on the rods *A* and *B*.

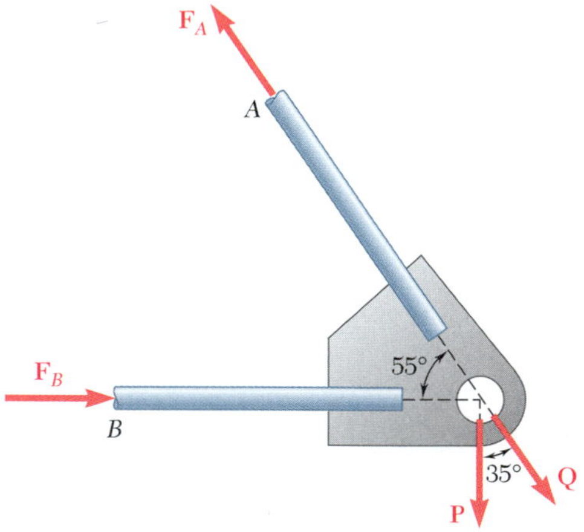

Fig. *P2.51* and *P2.52*

2.52 Two forces **P** and **Q** are applied as shown to an aircraft connection. Knowing that the connection is in equilibrium and that the magnitudes of the forces exerted on rods *A* and *B* are $F_A = 600$ lb and $F_B = 320$ lb, determine the magnitudes of **P** and **Q**.

2.53 Two cables tied together at C are loaded as shown. Knowing that $W = 840$ N, determine the tension (*a*) in cable AC, (*b*) in cable BC.

2.54 Two cables tied together at C are loaded as shown. Determine the range of values of W for which the tension will not exceed 1050 N in either cable.

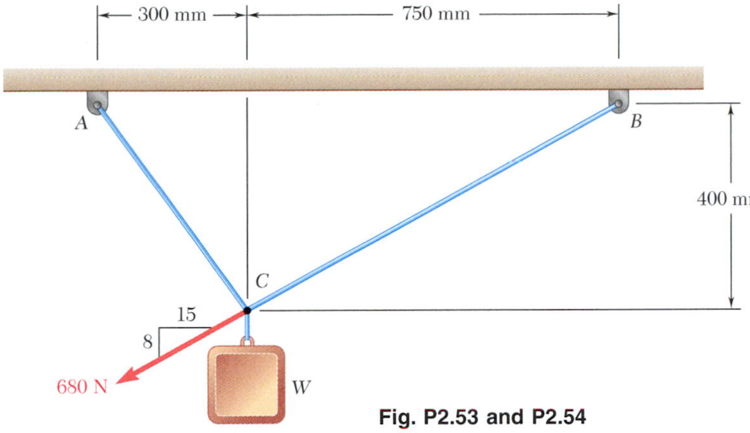

Fig. P2.53 and P2.54

2.55 The cabin of an aerial tramway is suspended from a set of wheels that can roll freely on the support cable ACB and is being pulled at a constant speed by cable DE. Knowing that $\alpha = 40°$ and $\beta = 35°$, that the combined weight of the cabin, its support system, and its passengers is 24.8 kN, and assuming the tension in cable DF to be negligible, determine the tension (*a*) in the support cable ACB, (*b*) in the traction cable DE.

2.56 The cabin of an aerial tramway is suspended from a set of wheels that can roll freely on the support cable ACB and is being pulled at a constant speed by cable DE. Knowing that $\alpha = 42°$ and $\beta = 32°$, that the tension in cable DE is 20 kN, and assuming the tension in cable DF to be negligible, determine (*a*) the combined weight of the cabin, its support system, and its passengers, (*b*) the tension in the support cable ACB.

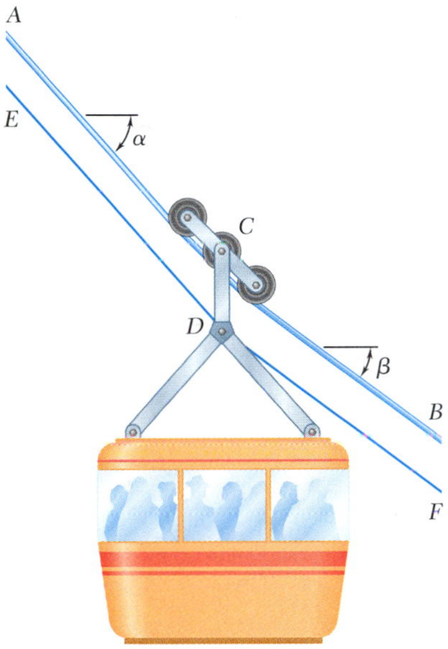

Fig. P2.55 and P2.56

2.57 A block of weight W is suspended from a 500-mm long cord and two springs of which the unstretched lengths are 450 mm. Knowing that the constants of the springs are $k_{AB} = 1500$ N/m and $k_{AD} = 500$ N/m, determine (*a*) the tension in the cord, (*b*) the weight of the block.

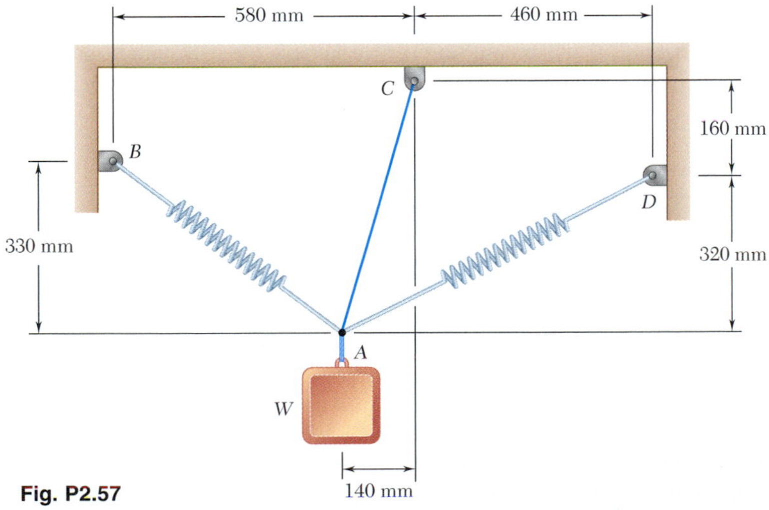

Fig. P2.57

2.58 A load of weight 400 N is suspended from a spring and two cords which are attached to blocks of weights $3W$ and W as shown. Knowing that the constant of the spring is 800 N/m, determine (*a*) the value of W, (*b*) the unstretched length of the spring.

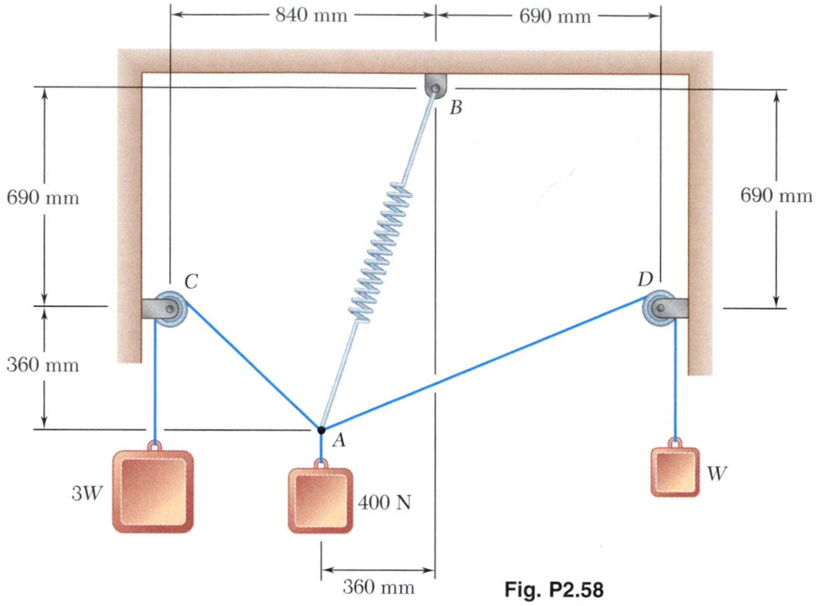

Fig. P2.58

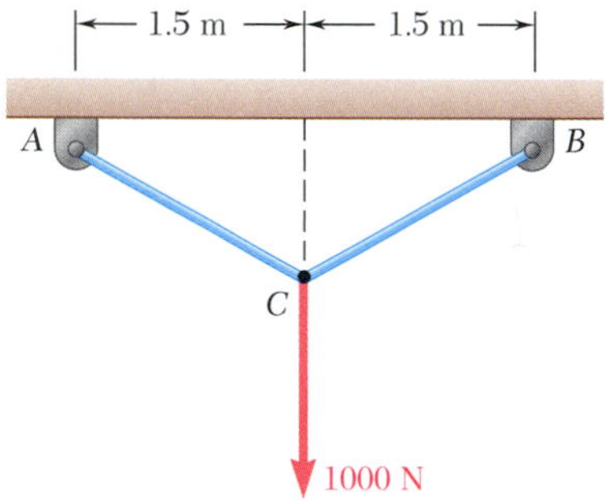

Fig. P2.60

2.59 For the cables and loading of Prob. 2.46, determine (*a*) the value of α for which the tension in cable BC is as small as possible, (*b*) the corresponding value of the tension.

2.60 Knowing that portions AC and BC of cable ACB must be equal, determine the shortest length of cable which can be used to support the load shown if the tension in the cable is not to exceed 725 N.

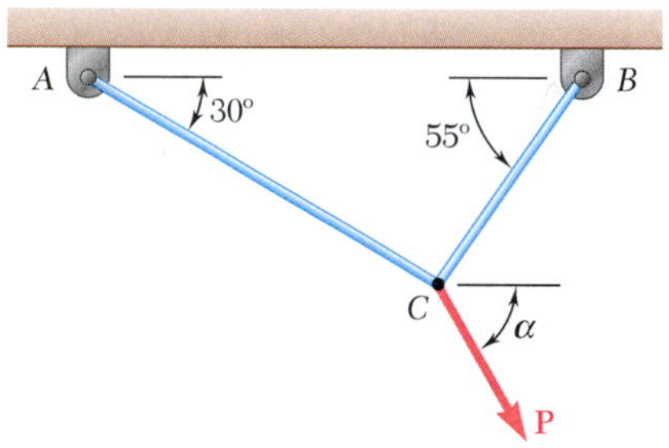

Fig. P2.61 and *P2.62*

2.61 Two cables tied together at C are loaded as shown. Knowing that the maximum allowable tension in each cable is 200 lb, determine (*a*) the magnitude of the largest force **P** which may be applied at C, (*b*) the corresponding value of α.

2.62 Two cables tied together at C are loaded as shown. Knowing that the maximum allowable tension is 300 lb in cable AC and 150 lb in cable BC, determine (*a*) the magnitude of the largest force **P** which may be applied at C, (*b*) the corresponding value of α.

2.63 For the structure and loading of Prob. 2.45, determine (*a*) the value of α for which the tension in cable BC is as small as possible, (*b*) the corresponding value of the tension.

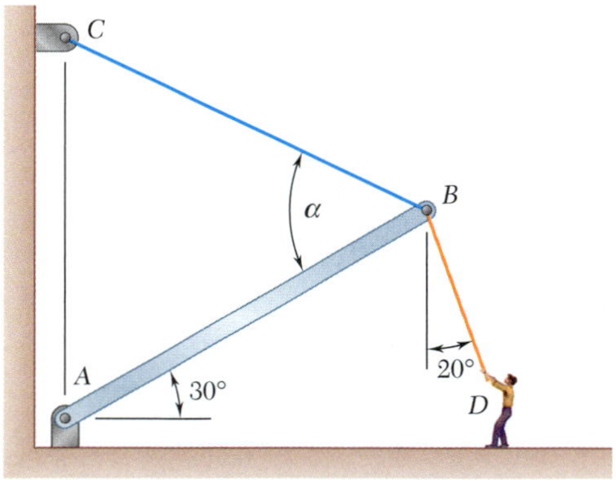

Fig. P2.64

2.64 Boom AB is supported by cable BC and a hinge at A. Knowing that the boom exerts on pin B a force directed along the boom and that the tension in rope BD is 70 lb, determine (*a*) the value of α for which the tension in cable BC is as small as possible, (*b*) the corresponding value of the tension.

2.65 Collar A shown in Fig. P2.65 and P2.66 can slide on a frictionless vertical rod and is attached as shown to a spring. The constant of the spring is 660 N/m, and the spring is unstretched when $h = 300$ mm. Knowing that the system is in equilibrium when $h = 400$ mm, determine the weight of the collar.

2.66 The 40-N collar A can slide on a frictionless vertical rod and is attached as shown to a spring. The spring is unstretched when $h = 300$ mm. Knowing that the constant of the spring is 560 N/m, determine the value of h for which the system is in equilibrium.

2.67 A 280-kg crate is supported by several rope-and-pulley arrangements as shown. Determine for each arrangement the tension in the rope. (*Hint:* The tension in the rope is the same on each side of a simple pulley. This can be proved by the methods of Chap. 4.)

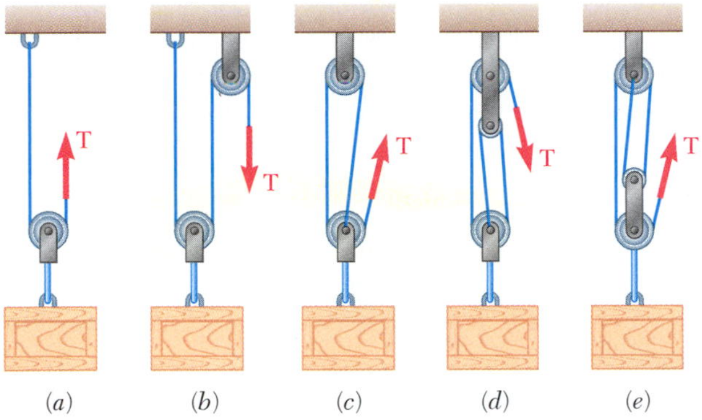

Fig. ***P2.67***

2.68 Solve parts b and d of Prob. 2.67 assuming that the free end of the rope is attached to the crate.

2.69 A 350-lb load is supported by the rope-and-pulley arrangement shown. Knowing that $\beta = 25°$, determine the magnitude and direction of the force **P** which should be exerted on the free end of the rope to maintain equilibrium. (See the hint of Prob. 2.67.)

2.70 A 350-lb load is supported by the rope-and-pulley arrangement shown. Knowing that $\alpha = 35°$, determine (a) the angle β, (b) the magnitude of the force **P** which should be exerted on the free end of the rope to maintain equilibrium. (See the hint of Prob. 2.67.)

2.71 A load **Q** is applied to the pulley C, which can roll on the cable ACB. The pulley is held in the position shown by a second cable CAD, which passes over the pulley A and supports a load **P**. Knowing that $P = 800$ N, determine (a) the tension in cable ACB, (b) the magnitude of load **Q**.

2.72 A 2000-N load **Q** is applied to the pulley C, which can roll on the cable ACB. The pulley is held in the position shown by a second cable CAD, which passes over the pulley A and supports a load **P**. Determine (a) the tension in cable ACB, (b) the magnitude of load **P**.

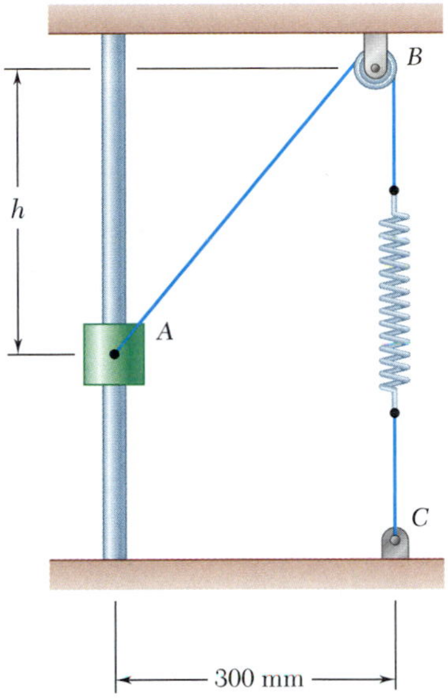

Fig. P2.65 and P2.66

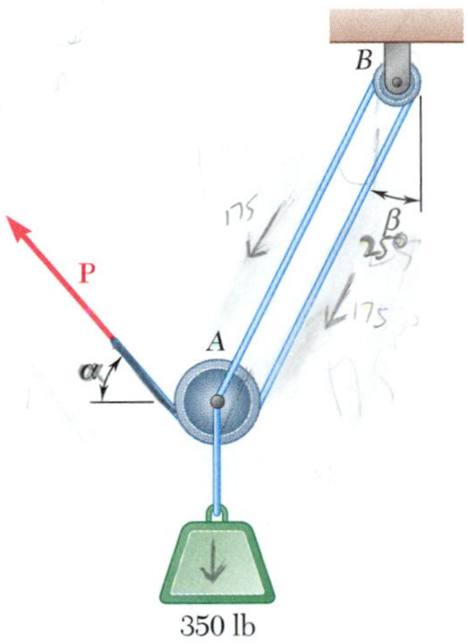

Fig. P2.69 and P2.70

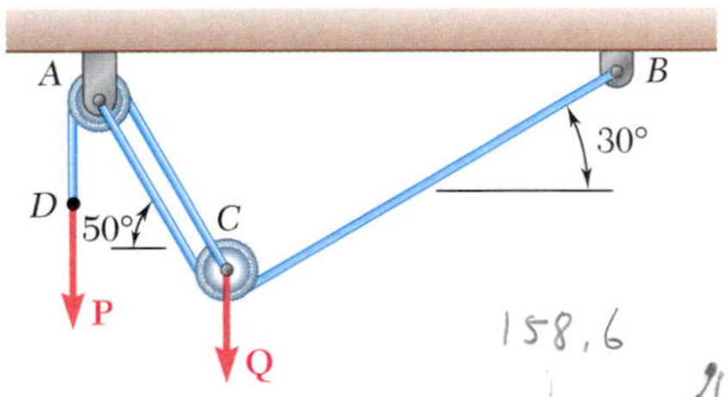

Fig. P2.71 and *P2.72*

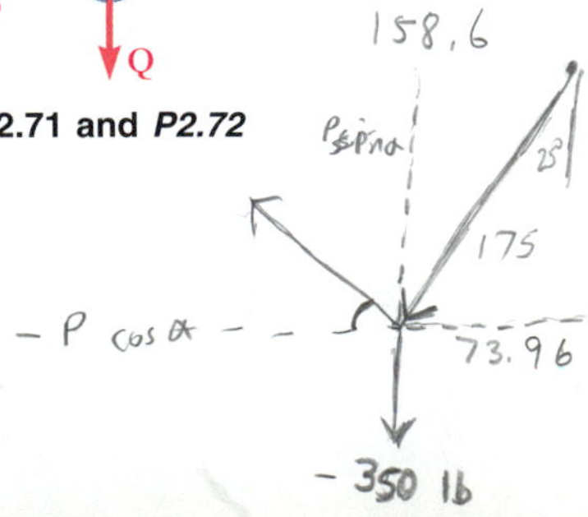

FORCES IN SPACE

2.12. RECTANGULAR COMPONENTS OF A FORCE IN SPACE

The problems considered in the first part of this chapter involved only two dimensions; they could be formulated and solved in a single plane. In this section and in the remaining sections of the chapter, we will discuss problems involving the three dimensions of space.

Consider a force **F** acting at the origin O of the system of rectangular coordinates x, y, z. To define the direction of **F**, we draw the

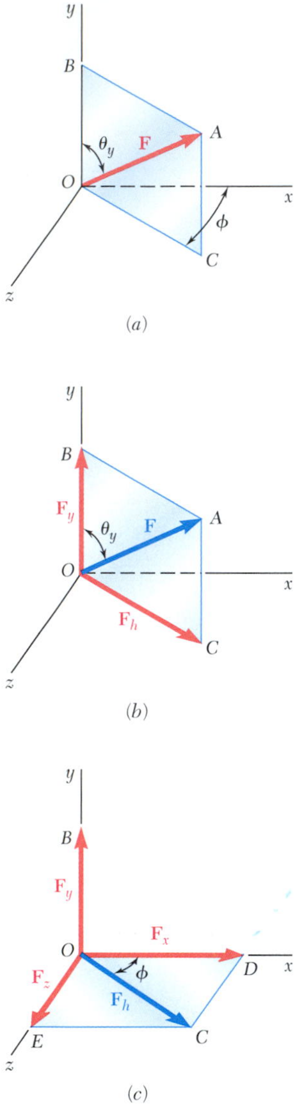

Fig. 2.30

vertical plane *OBAC* containing **F** (Fig. 2.30*a*). This plane passes through the vertical y axis; its orientation is defined by the angle ϕ it forms with the xy plane. The direction of **F** within the plane is defined by the angle θ_y that **F** forms with the y axis. The force **F** may be resolved into a vertical component $\mathbf{F}_y$ and a horizontal component $\mathbf{F}_h$; this operation, shown in Fig. 2.30*b*, is carried out in plane *OBAC* according to the rules developed in the first part of the chapter. The corresponding scalar components are

$$F_y = F \cos \theta_y \qquad F_h = F \sin \theta_y \tag{2.16}$$

But $\mathbf{F}_h$ may be resolved into two rectangular components $\mathbf{F}_x$ and $\mathbf{F}_z$ along the x and z axes, respectively. This operation, shown in Fig. 2.30*c*, is carried out in the xz plane. We obtain the following expressions for the corresponding scalar components:

$$\begin{aligned} F_x &= F_h \cos \phi = F \sin \theta_y \cos \phi \\ F_z &= F_h \sin \phi = F \sin \theta_y \sin \phi \end{aligned} \tag{2.17}$$

The given force **F** has thus been resolved into three rectangular vector components $\mathbf{F}_x$, $\mathbf{F}_y$, $\mathbf{F}_z$, which are directed along the three coordinate axes.

Applying the Pythagorean theorem to the triangles *OAB* and *OCD* of Fig. 2.30, we write

$$F^2 = (OA)^2 = (OB)^2 + (BA)^2 = F_y^2 + F_h^2$$
$$F_h^2 = (OC)^2 = (OD)^2 + (DC)^2 = F_x^2 + F_z^2$$

Eliminating F_h^2 from these two equations and solving for F, we obtain the following relation between the magnitude of **F** and its rectangular scalar components:

$$F = \sqrt{F_x^2 + F_y^2 + F_z^2} \tag{2.18}$$

The relationship existing between the force **F** and its three components $\mathbf{F}_x$, $\mathbf{F}_y$, $\mathbf{F}_z$ is more easily visualized if a "box" having $\mathbf{F}_x$, $\mathbf{F}_y$, $\mathbf{F}_z$ for edges is drawn as shown in Fig. 2.31. The force **F** is then represented by the diagonal *OA* of this box. Figure 2.31*b* shows the right triangle *OAB* used to derive the first of the formulas (2.16): $F_y = F \cos \theta_y$. In Fig. 2.31*a* and *c*, two other right triangles have also been drawn: *OAD* and *OAE*. These triangles are seen to occupy in the box positions comparable with that of triangle *OAB*. Denoting by θ_x and θ_z, respectively, the angles that **F** forms with the x and z axes, we can derive two formulas similar to $F_y = F \cos \theta_y$. We thus write

$$F_x = F \cos \theta_x \qquad F_y = F \cos \theta_y \qquad F_z = F \cos \theta_z \tag{2.19}$$

The three angles θ_x, θ_y, θ_z define the direction of the force **F**; they are more commonly used for this purpose than the angles θ_y and ϕ introduced at the beginning of this section. The cosines of θ_x, θ_y, θ_z are known as the *direction cosines* of the force **F**.

Introducing the unit vectors **i**, **j**, and **k**, directed respectively along the x, y, and z axes (Fig. 2.32), we can express **F** in the form

$$\mathbf{F} = F_x\mathbf{i} + F_y\mathbf{j} + F_z\mathbf{k} \tag{2.20}$$

where the scalar components F_x, F_y, F_z are defined by the relations (2.19).

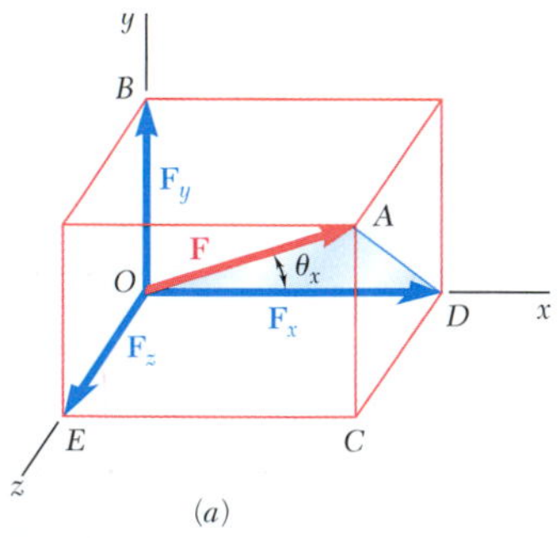

Fig. 2.31

Example 1. A force of 500 N forms angles of 60°, 45°, and 120°, respectively, with the x, y, and z axes. Find the components F_x, F_y, and F_z of the force.

Substituting $F = 500$ N, $\theta_x = 60°$, $\theta_y = 45°$, $\theta_z = 120°$ into formulas (2.19), we write

$$F_x = (500 \text{ N}) \cos 60° = +250 \text{ N}$$
$$F_y = (500 \text{ N}) \cos 45° = +354 \text{ N}$$
$$F_z = (500 \text{ N}) \cos 120° = -250 \text{ N}$$

Carrying into Eq. (2.20) the values obtained for the scalar components of **F**, we have

$$\mathbf{F} = (250 \text{ N})\mathbf{i} + (354 \text{ N})\mathbf{j} - (250 \text{ N})\mathbf{k}$$

As in the case of two-dimensional problems, a plus sign indicates that the component has the same sense as the corresponding axis, and a minus sign indicates that it has the opposite sense.

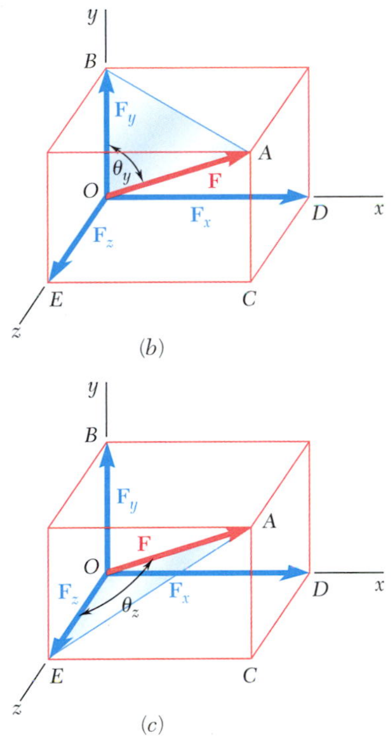

Fig. 2.31

The angle a force **F** forms with an axis should be measured from the positive side of the axis and will always be between 0 and 180°. An angle θ_x smaller than 90° (acute) indicates that **F** (assumed attached to O) is on the same side of the yz plane as the positive x axis; $\cos \theta_x$ and F_x will then be positive. An angle θ_x larger than 90° (obtuse) indicates that **F** is on the other side of the yz plane; $\cos \theta_x$ and F_x will then be negative. In Example 1 the angles θ_x and θ_y are acute, while θ_z is obtuse; consequently, F_x and F_y are positive, while F_z is negative.

Substituting into (2.20) the expressions obtained for F_x, F_y, F_z in (2.19), we write

$$\mathbf{F} = F(\cos \theta_x \mathbf{i} + \cos \theta_y \mathbf{j} + \cos \theta_z \mathbf{k}) \tag{2.21}$$

which shows that the force **F** can be expressed as the product of the scalar F and the vector

$$\boldsymbol{\lambda} = \cos \theta_x \mathbf{i} + \cos \theta_y \mathbf{j} + \cos \theta_z \mathbf{k} \tag{2.22}$$

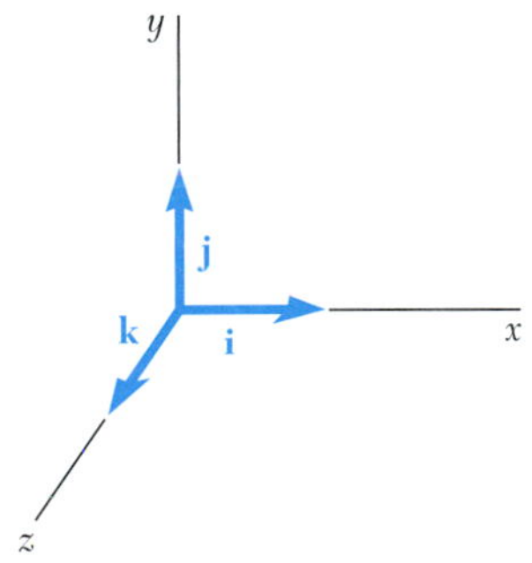

Fig. 2.32

Clearly, the vector $\boldsymbol{\lambda}$ is a vector whose magnitude is equal to 1 and whose direction is the same as that of **F** (Fig. 2.33). The vector $\boldsymbol{\lambda}$ is referred to as the *unit vector* along the line of action of **F**. It follows from (2.22) that the components of the unit vector $\boldsymbol{\lambda}$ are respectively equal to the direction cosines of the line of action of **F**:

$$\lambda_x = \cos \theta_x \qquad \lambda_y = \cos \theta_y \qquad \lambda_z = \cos \theta_z \tag{2.23}$$

We should observe that the values of the three angles θ_x, θ_y, θ_z are not independent. Recalling that the sum of the squares of the components of a vector is equal to the square of its magnitude, we write

$$\lambda_x^2 + \lambda_y^2 + \lambda_z^2 = 1$$

or, substituting for λ_x, λ_y, λ_z from (2.23),

$$\cos^2 \theta_x + \cos^2 \theta_y + \cos^2 \theta_z = 1 \tag{2.24}$$

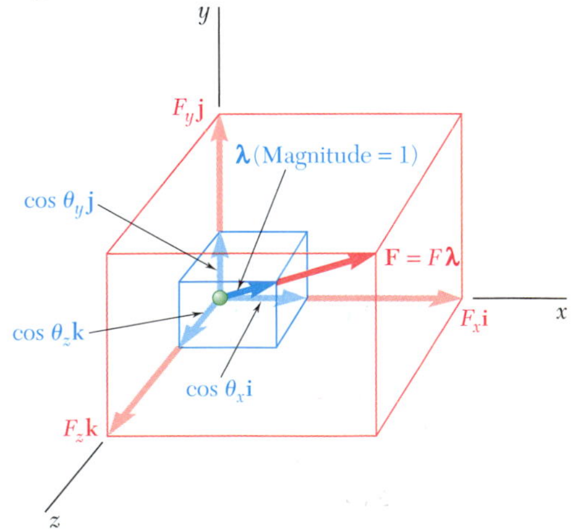

Fig. 2.33

In Example 1, for instance, once the values $\theta_x = 60°$ and $\theta_y = 45°$ have been selected, the value of θ_z *must* be equal to 60° or 120° in order to satisfy identity (2.24).

When the components F_x, F_y, F_z of a force **F** are given, the magnitude F of the force is obtained from (2.18). The relations (2.19) can then be solved for the direction cosines,

$$\cos \theta_x = \frac{F_x}{F} \qquad \cos \theta_y = \frac{F_y}{F} \qquad \cos \theta_z = \frac{F_z}{F} \tag{2.25}$$

and the angles θ_x, θ_y, θ_z characterizing the direction of **F** can be found.

Example 2. A force **F** has the components $F_x = 20$ lb, $F_y = -30$ lb, $F_z = 60$ lb. Determine its magnitude F and the angles θ_x, θ_y, θ_z it forms with the coordinate axes.

From formula (2.18) we obtain†

$$\begin{aligned} F &= \sqrt{F_x^2 + F_y^2 + F_z^2} \\ &= \sqrt{(20 \text{ lb})^2 + (-30 \text{ lb})^2 + (60 \text{ lb})^2} \\ &= \sqrt{4900} \text{ lb} = 70 \text{ lb} \end{aligned}$$

Substituting the values of the components and magnitude of **F** into Eqs. (2.25), we write

$$\cos\theta_x = \frac{F_x}{F} = \frac{20 \text{ lb}}{70 \text{ lb}} \qquad \cos\theta_y = \frac{F_y}{F} = \frac{-30 \text{ lb}}{70 \text{ lb}} \qquad \cos\theta_z = \frac{F_z}{F} = \frac{60 \text{ lb}}{70 \text{ lb}}$$

Calculating successively each quotient and its arc cosine, we obtain

$$\theta_x = 73.4° \qquad \theta_y = 115.4° \qquad \theta_z = 31.0°$$

These computations can be carried out easily with a calculator.

2.13. FORCE DEFINED BY ITS MAGNITUDE AND TWO POINTS ON ITS LINE OF ACTION

In many applications, the direction of a force **F** is defined by the coordinates of two points, $M(x_1, y_1, z_1)$ and $N(x_2, y_2, z_2)$, located on its line of action (Fig. 2.34). Consider the vector $\overrightarrow{MN}$ joining M and N

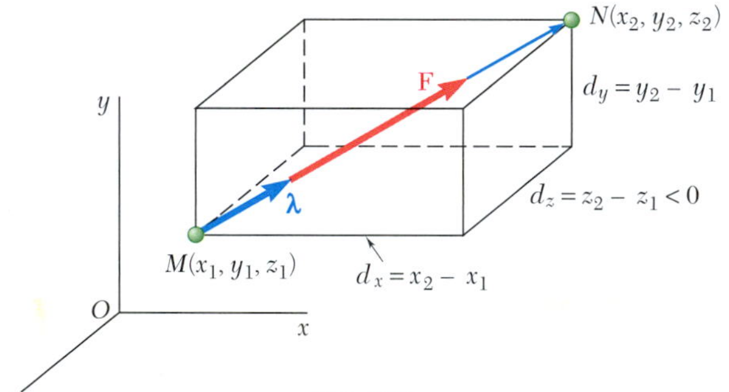

Fig. 2.34

and of the same sense as **F**. Denoting its scalar components by d_x, d_y, d_z, respectively, we write

$$\overrightarrow{MN} = d_x\mathbf{i} + d_y\mathbf{j} + d_z\mathbf{k} \tag{2.26}$$

The unit vector $\boldsymbol{\lambda}$ along the line of action of **F** (i.e., along the line MN) may be obtained by dividing the vector $\overrightarrow{MN}$ by its magnitude MN. Substituting for $\overrightarrow{MN}$ from (2.26) and observing that MN is equal to the distance d from M to N, we write

$$\boldsymbol{\lambda} = \frac{\overrightarrow{MN}}{MN} = \frac{1}{d}(d_x\mathbf{i} + d_y\mathbf{j} + d_z\mathbf{k}) \tag{2.27}$$

†With a calculator programmed to convert rectangular coordinates into polar coordinates, the following procedure will be found more expeditious for computing F: First determine F_h from its two rectangular components F_x and F_z (Fig. 2.30*c*), then determine F from its two rectangular components F_h and F_y (Fig. 2.30*b*). The actual order in which the three components F_x, F_y, F_z are entered is immaterial.

Recalling that **F** is equal to the product of F and λ, we have

$$\mathbf{F} = F\boldsymbol{\lambda} = \frac{F}{d}(d_x\mathbf{i} + d_y\mathbf{j} + d_z\mathbf{k}) \tag{2.28}$$

from which it follows that the scalar components of **F** are, respectively,

$$F_x = \frac{Fd_x}{d} \qquad F_y = \frac{Fd_y}{d} \qquad F_z = \frac{Fd_z}{d} \tag{2.29}$$

The relations (2.29) considerably simplify the determination of the components of a force **F** of given magnitude F when the line of action of **F** is defined by two points M and N. Subtracting the coordinates of M from those of N, we first determine the components of the vector $\overrightarrow{MN}$ and the distance d from M to N:

$$d_x = x_2 - x_1 \qquad d_y = y_2 - y_1 \qquad d_z = z_2 - z_1$$
$$d = \sqrt{d_x^2 + d_y^2 + d_z^2}$$

Substituting for F and for d_x, d_y, d_z, and d into the relations (2.29), we obtain the components F_x, F_y, F_z of the force.

The angles θ_x, θ_y, θ_z that **F** forms with the coordinate axes can then be obtained from Eqs. (2.25). Comparing Eqs. (2.22) and (2.27), we can also write

$$\cos\theta_x = \frac{d_x}{d} \qquad \cos\theta_y = \frac{d_y}{d} \qquad \cos\theta_z = \frac{d_z}{d} \tag{2.30}$$

and determine the angles θ_x, θ_y, θ_z directly from the components and magnitude of the vector $\overrightarrow{MN}$.

2.14. ADDITION OF CONCURRENT FORCES IN SPACE

The resultant **R** of two or more forces in space will be determined by summing their rectangular components. Graphical or trigonometric methods are generally not practical in the case of forces in space.

The method followed here is similar to that used in Sec. 2.8 with coplanar forces. Setting

$$\mathbf{R} = \Sigma\mathbf{F}$$

we resolve each force into its rectangular components and write

$$\begin{aligned} R_x\mathbf{i} + R_y\mathbf{j} + R_z\mathbf{k} &= \Sigma(F_x\mathbf{i} + F_y\mathbf{j} + F_z\mathbf{k}) \\ &= (\Sigma F_x)\mathbf{i} + (\Sigma F_y)\mathbf{j} + (\Sigma F_z)\mathbf{k} \end{aligned}$$

from which it follows that

$$R_x = \Sigma F_x \qquad R_y = \Sigma F_y \qquad R_z = \Sigma F_z \tag{2.31}$$

The magnitude of the resultant and the angles θ_x, θ_y, θ_z that the resultant forms with the coordinate axes are obtained using the method discussed in Sec. 2.12. We write

$$R = \sqrt{R_x^2 + R_y^2 + R_z^2} \tag{2.32}$$

$$\cos\theta_x = \frac{R_x}{R} \qquad \cos\theta_y = \frac{R_y}{R} \qquad \cos\theta_z = \frac{R_z}{R} \tag{2.33}$$

SAMPLE PROBLEM 2.7

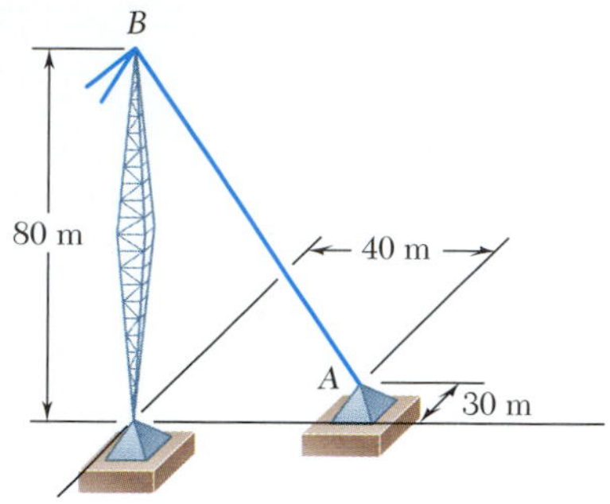

A tower guy wire is anchored by means of a bolt at A. The tension in the wire is 2500 N. Determine (*a*) the components F_x, F_y, F_z of the force acting on the bolt, (*b*) the angles θ_x, θ_y, θ_z defining the direction of the force.

SOLUTION

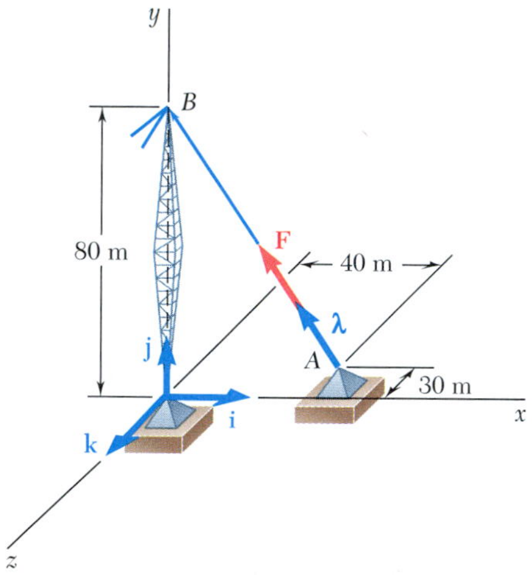

***a.* Components of the Force.** The line of action of the force acting on the bolt passes through A and B, and the force is directed from A to B. The components of the vector $\overrightarrow{AB}$, which has the same direction as the force, are

$$d_x = -40 \text{ m} \qquad d_y = +80 \text{ m} \qquad d_z = +30 \text{ m}$$

The total distance from A to B is

$$AB = d = \sqrt{d_x^2 + d_y^2 + d_z^2} = 94.3 \text{ m}$$

Denoting by $\mathbf{i}$, $\mathbf{j}$, $\mathbf{k}$ the unit vectors along the coordinate axes, we have

$$\overrightarrow{AB} = -(40 \text{ m})\mathbf{i} + (80 \text{ m})\mathbf{j} + (30 \text{ m})\mathbf{k}$$

Introducing the unit vector $\boldsymbol{\lambda} = \overrightarrow{AB}/AB$, we write

$$\mathbf{F} = F\boldsymbol{\lambda} = F\frac{\overrightarrow{AB}}{AB} = \frac{2500 \text{ N}}{94.3 \text{ m}}\overrightarrow{AB}$$

Substituting the expression found for $\overrightarrow{AB}$, we obtain

$$\mathbf{F} = \frac{2500 \text{ N}}{94.3 \text{ m}}[-(40 \text{ m})\mathbf{i} + (80 \text{ m})\mathbf{j} + (30 \text{ m})\mathbf{k}]$$

$$\mathbf{F} = -(1060 \text{ N})\mathbf{i} + (2120 \text{ N})\mathbf{j} + (795 \text{ N})\mathbf{k}$$

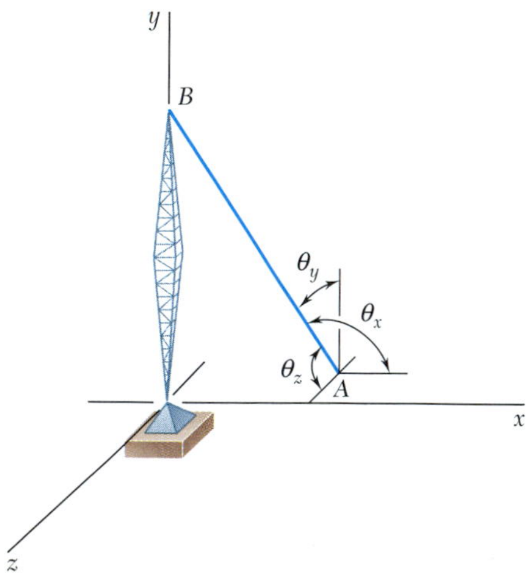

The components of $\mathbf{F}$, therefore, are

$$F_x = -1060 \text{ N} \qquad F_y = +2120 \text{ N} \qquad F_z = +795 \text{ N} \quad ◀$$

***b.* Direction of the Force.** Using Eqs. (2.25), we write

$$\cos\theta_x = \frac{F_x}{F} = \frac{-1060 \text{ N}}{2500 \text{ N}} \qquad \cos\theta_y = \frac{F_y}{F} = \frac{+2120 \text{ N}}{2500 \text{ N}}$$

$$\cos\theta_z = \frac{F_z}{F} = \frac{+795 \text{ N}}{2500 \text{ N}}$$

Calculating successively each quotient and its arc cosine, we obtain

$$\theta_x = 115.1° \qquad \theta_y = 32.0° \qquad \theta_z = 71.5° \quad ◀$$

(*Note:* This result could have been obtained by using the components and magnitude of the vector $\overrightarrow{AB}$ rather than those of the force $\mathbf{F}$.)

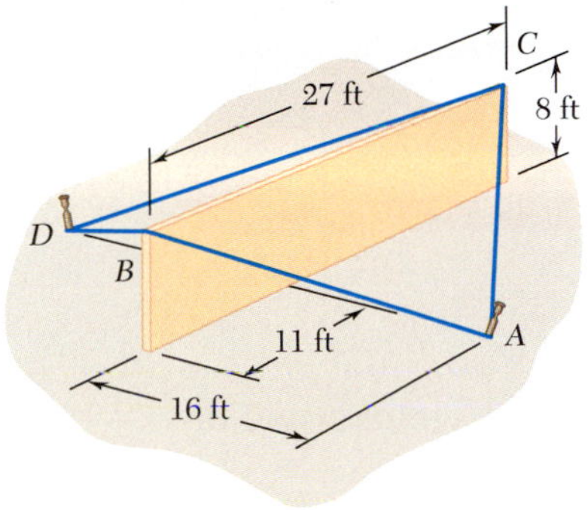

SAMPLE PROBLEM 2.8

A wall section of precast concrete is temporarily held by the cables shown. Knowing that the tension is 840 lb in cable *AB* and 1200 lb in cable *AC*, determine the magnitude and direction of the resultant of the forces exerted by cables *AB* and *AC* on stake *A*.

SOLUTION

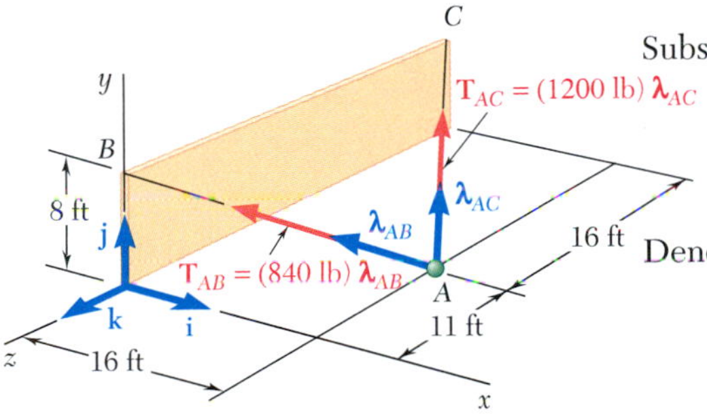

Components of the Forces. The force exerted by each cable on stake *A* will be resolved into *x*, *y*, and *z* components. We first determine the components and magnitude of the vectors $\overrightarrow{AB}$ and $\overrightarrow{AC}$, measuring them from *A* toward the wall section. Denoting by **i**, **j**, **k** the unit vectors along the coordinate axes, we write

$$\overrightarrow{AB} = -(16\text{ ft})\mathbf{i} + (8\text{ ft})\mathbf{j} + (11\text{ ft})\mathbf{k} \qquad AB = 21\text{ ft}$$
$$\overrightarrow{AC} = -(16\text{ ft})\mathbf{i} + (8\text{ ft})\mathbf{j} - (16\text{ ft})\mathbf{k} \qquad AC = 24\text{ ft}$$

Denoting by $\boldsymbol{\lambda}_{AB}$ the unit vector along *AB*, we have

$$\mathbf{T}_{AB} = T_{AB}\boldsymbol{\lambda}_{AB} = T_{AB}\frac{\overrightarrow{AB}}{AB} = \frac{840\text{ lb}}{21\text{ ft}}\overrightarrow{AB}$$

Substituting the expression found for $\overrightarrow{AB}$, we obtain

$$\mathbf{T}_{AB} = \frac{840\text{ lb}}{21\text{ ft}}[-(16\text{ ft})\mathbf{i} + (8\text{ ft})\mathbf{j} + (11\text{ ft})\mathbf{k}]$$
$$\mathbf{T}_{AB} = -(640\text{ lb})\mathbf{i} + (320\text{ lb})\mathbf{j} + (440\text{ lb})\mathbf{k}$$

Denoting by $\boldsymbol{\lambda}_{AC}$ the unit vector along *AC*, we obtain in a similar way

$$\mathbf{T}_{AC} = T_{AC}\boldsymbol{\lambda}_{AC} = T_{AC}\frac{\overrightarrow{AC}}{AC} = \frac{1200\text{ lb}}{24\text{ ft}}\overrightarrow{AC}$$
$$\mathbf{T}_{AC} = -(800\text{ lb})\mathbf{i} + (400\text{ lb})\mathbf{j} - (800\text{ lb})\mathbf{k}$$

Resultant of the Forces. The resultant **R** of the forces exerted by the two cables is

$$\mathbf{R} = \mathbf{T}_{AB} + \mathbf{T}_{AC} = -(1440\text{ lb})\mathbf{i} + (720\text{ lb})\mathbf{j} - (360\text{ lb})\mathbf{k}$$

The magnitude and direction of the resultant are now determined:

$$R = \sqrt{R_x^2 + R_y^2 + R_z^2} = \sqrt{(-1440)^2 + (720)^2 + (-360)^2}$$

$$R = 1650\text{ lb} \quad ◀$$

From Eqs. (2.33) we obtain

$$\cos\theta_x = \frac{R_x}{R} = \frac{-1440\text{ lb}}{1650\text{ lb}} \qquad \cos\theta_y = \frac{R_y}{R} = \frac{+720\text{ lb}}{1650\text{ lb}}$$

$$\cos\theta_z = \frac{R_z}{R} = \frac{-360\text{ lb}}{1650\text{ lb}}$$

Calculating successively each quotient and its arc cosine, we have

$$\theta_x = 150.8° \qquad \theta_y = 64.1° \qquad \theta_z = 102.6° \quad ◀$$

SOLVING PROBLEMS ON YOUR OWN

In this lesson we saw that a *force in space* can be defined by its magnitude and direction or by its three rectangular components F_x, F_y, and F_z.

A. When a force is defined by its magnitude and direction, its rectangular components F_x, F_y, and F_z can be found as follows:

Case 1. If the direction of the force **F** is defined by the angles θ_y and ϕ shown in Fig. 2.30, projections of **F** through these angles or their complements will yield the components of **F** [Eqs. (2.17)]. Note that the x and z components of **F** are found by first projecting **F** onto the horizontal plane; the projection $\mathbf{F}_h$ obtained in this way is then resolved into the components $\mathbf{F}_x$ and $\mathbf{F}_z$ (Fig. 2.30*c*).

When solving problems of this type, we strongly encourage you first to sketch the force **F** and then its projection $\mathbf{F}_h$ and components F_x, F_y, and F_z before beginning the mathematical part of the solution.

Case 2. If the direction of the force **F** is defined by the angles θ_x, θ_y, θ_z that **F** forms with the coordinate axes, each component can be obtained by multiplying the magnitude F of the force by the cosine of the corresponding angle [Example 1]:

$$F_x = F \cos \theta_x \qquad F_y = F \cos \theta_y \qquad F_z = F \cos \theta_z$$

Case 3. If the direction of the force **F** is defined by two points M and N located on its line of action (Fig. 2.34), you will first express the vector $\overrightarrow{MN}$ drawn from M to N in terms of its components d_x, d_y, d_z and the unit vectors **i**, **j**, **k**:

$$\overrightarrow{MN} = d_x\mathbf{i} + d_y\mathbf{j} + d_z\mathbf{k}$$

Next, you will determine the unit vector $\boldsymbol{\lambda}$ along the line of action of **F** by dividing the vector $\overrightarrow{MN}$ by its magnitude MN. Multiplying $\boldsymbol{\lambda}$ by the magnitude of **F**, you will obtain the desired expression for **F** in terms of its rectangular components [Sample Prob. 2.7]:

$$\mathbf{F} = F\boldsymbol{\lambda} = \frac{F}{d}(d_x\mathbf{i} + d_y\mathbf{j} + d_z\mathbf{k})$$

It is advantageous to use a consistent and meaningful system of notation when determining the rectangular components of a force. The method used in this text is illustrated in Sample Prob. 2.8 where, for example, the force $\mathbf{T}_{AB}$ acts from stake A toward point B. Note that the subscripts have been ordered to agree with the direction of the force. It is recommended that you adopt the same notation, as it will help you identify point 1 (the first subscript) and point 2 (the second subscript).

When forming the vector defining the line of action of a force, you may think of its scalar components as the number of steps you must take in each coordinate direction to go from point 1 to point 2. It is essential that you always remember to assign the correct sign to each of the components.

B. *When a force is defined by its rectangular components F_x, F_y, F_z,* you can obtain its magnitude F by writing

$$F = \sqrt{F_x^2 + F_y^2 + F_z^2}$$

You can determine the direction cosines of the line of action of **F** by dividing the components of the force by F:

$$\cos \theta_x = \frac{F_x}{F} \qquad \cos \theta_y = \frac{F_y}{F} \qquad \cos \theta_z = \frac{F_z}{F}$$

From the direction cosines you can obtain the angles θ_x, θ_y, θ_z that **F** forms with the coordinate axes [Example 2].

C. *To determine the resultant* R *of two or more forces* in three-dimensional space, first determine the rectangular components of each force by one of the procedures described above. Adding these components will yield the components R_x, R_y, R_z of the resultant. The magnitude and direction of the resultant can then be obtained as indicated above for the force **F** [Sample Prob. 2.8].

Problems

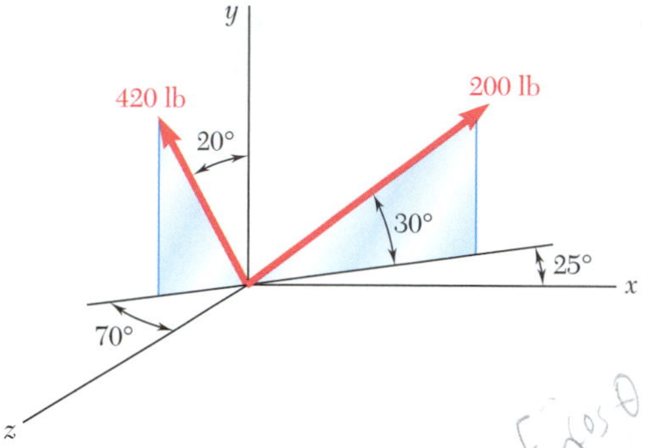

Fig. P2.73 and *P2.74*

2.73 Determine (*a*) the *x*, *y*, and *z* components of the 200-lb force, (*b*) the angles θ_x, θ_y, and θ_z that the force forms with the coordinate axes.

2.74 Determine (*a*) the *x*, *y*, and *z* components of the 420-lb force, (*b*) the angles θ_x, θ_y, and θ_z that the force forms with the coordinate axes.

2.75 To stabilize a tree partially uprooted in a storm, cables *AB* and *AC* are attached to the upper trunk of the tree and then are fastened to steel rods anchored in the ground. Knowing that the tension in cable *AB* is 4.2 kN, determine (*a*) the components of the force exerted by this cable on the tree, (*b*) the angles θ_x, θ_y, and θ_z that the force forms with axes at *A* which are parallel to the coordinate axes.

2.76 To stabilize a tree partially uprooted in a storm, cables *AB* and *AC* are attached to the upper trunk of the tree and then are fastened to steel rods anchored in the ground. Knowing that the tension in cable *AC* is 3.6 kN, determine (*a*) the components of the force exerted by this cable on the tree, (*b*) the angles θ_x, θ_y, and θ_z that the force forms with axes at *A* which are parallel to the coordinate axes.

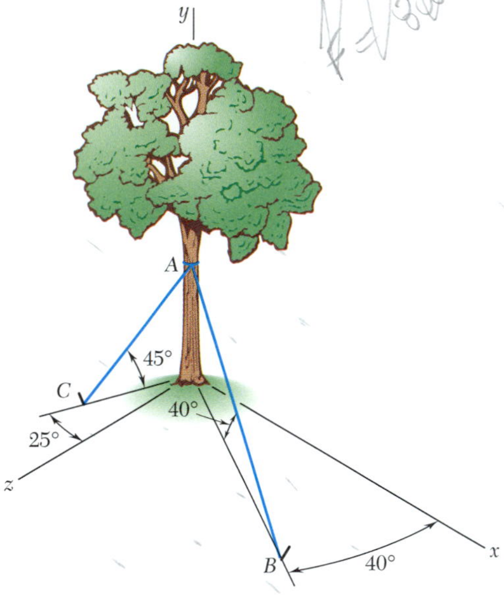

Fig. P2.75 and P2.76

2.77 A horizontal circular plate is suspended as shown from three wires which are attached to a support at *D* and form 30° angles with the vertical. Knowing that the *x* component of the force exerted by wire *AD* on the plate is 220.6 N, determine (*a*) the tension in wire *AD*, (*b*) the angles θ_x, θ_y, and θ_z that the force exerted at *A* forms with the coordinate axes.

2.78 A horizontal circular plate is suspended as shown from three wires which are attached to a support at *D* and form 30° angles with the vertical. Knowing that the *z* component of the force exerted by wire *BD* on the plate is −64.28 N, determine (*a*) the tension in wire *BD*, (*b*) the angles θ_x, θ_y, and θ_z that the force exerted at *B* forms with the coordinate axes.

2.79 A horizontal circular plate is suspended as shown from three wires which are attached to a support at *D* and form 30° angles with the vertical. Knowing that the tension in wire *CD* is 120 lb, determine (*a*) the components of the force exerted by this wire on the plate, (*b*) the angles θ_x, θ_y, and θ_z that the force forms with the coordinate axes.

2.80 A horizontal circular plate is suspended as shown from three wires which are attached to a support at *D* and form 30° angles with the vertical. Knowing that the *x* component of the force exerted by wire *CD* on the plate is −40 lb, determine (*a*) the tension in wire *CD*, (*b*) the angles θ_x, θ_y, and θ_z that the force exerted at *C* forms with the coordinate axes.

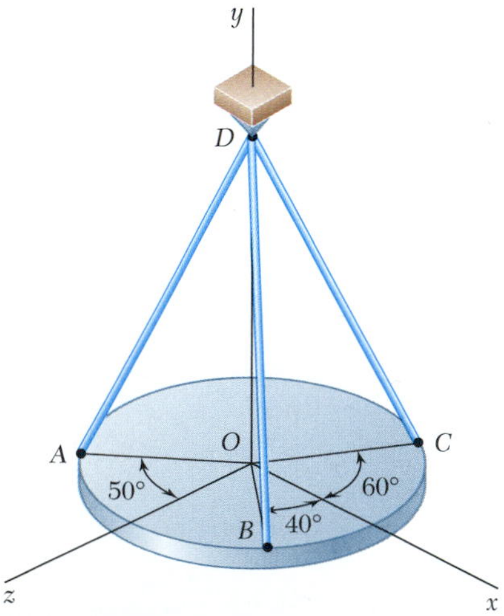

Fig. P2.77, P2.78, P2.79, and *P2.80*

2.81 Determine the magnitude and direction of the force $\mathbf{F} = (800 \text{ lb})\mathbf{i} + (260 \text{ lb})\mathbf{j} - (320 \text{ lb})\mathbf{k}$.

2.82 Determine the magnitude and direction of the force $\mathbf{F} = (400 \text{ N})\mathbf{i} - (1200 \text{ N})\mathbf{j} + (300 \text{ N})\mathbf{k}$.

2.83 A force acts at the origin of a coordinate system in a direction defined by the angles $\theta_x = 64.5°$ and $\theta_z = 55.9°$. Knowing that the y component of the force is -200 N, determine (*a*) the angle θ_y, (*b*) the other components and the magnitude of the force.

2.84 A force acts at the origin of a coordinate system in a direction defined by the angles $\theta_x = 75.4°$ and $\theta_y = 132.6°$. Knowing that the z component of the force is -60 N, determine (*a*) the angle θ_z, (*b*) the other components and the magnitude of the force.

2.85 A force **F** of magnitude 400 N acts at the origin of a coordinate system. Knowing that $\theta_x = 28.5°$, $F_y = -80$ N, and $F_z > 0$, determine (*a*) the components F_x and F_z, (*b*) the angles θ_y and θ_z.

2.86 A force **F** of magnitude 600 lb acts at the origin of a coordinate system. Knowing that $F_x = 200$ lb, $\theta_z = 136.8°$, and $F_y < 0$, determine (*a*) the components F_y and F_z, (*b*) the angles θ_x and θ_y.

2.87 A transmission tower is held by three guy wires anchored by bolts at *B*, *C*, and *D*. If the tension in wire *AB* is 2100 N, determine the components of the force exerted by the wire on the bolt at *B*.

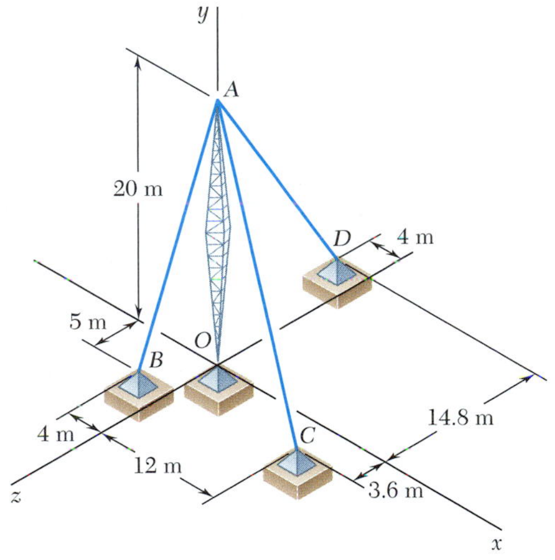

Fig. P2.87 and *P2.88*

2.88 A transmission tower is held by three guy wires anchored by bolts at *B*, *C*, and *D*. If the tension in wire *AD* is 1260 N, determine the components of the force exerted by the wire on the bolt at *D*.

2.89 A rectangular plate is supported by three cables as shown. Knowing that the tension in cable *AB* is 204 lb, determine the components of the force exerted on the plate at *B*.

2.90 A rectangular plate is supported by three cables as shown. Knowing that the tension in cable *AD* is 195 lb, determine the components of the forces exerted on the plate at *D*.

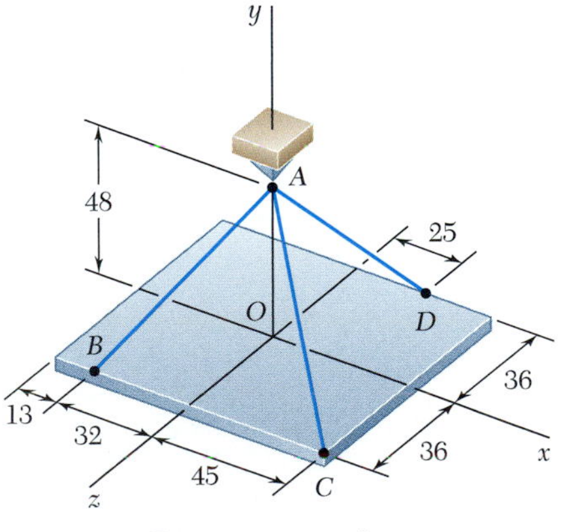

Fig. P2.89 and P2.90

2.91 A steel rod is bent into a semicircular ring of radius 0.96 m and is supported in part by cables *BD* and *BE* which are attached to the ring at *B*. Knowing that the tension in cable *BD* is 220 N, determine the components of this force exerted by the cable on the support at *D*.

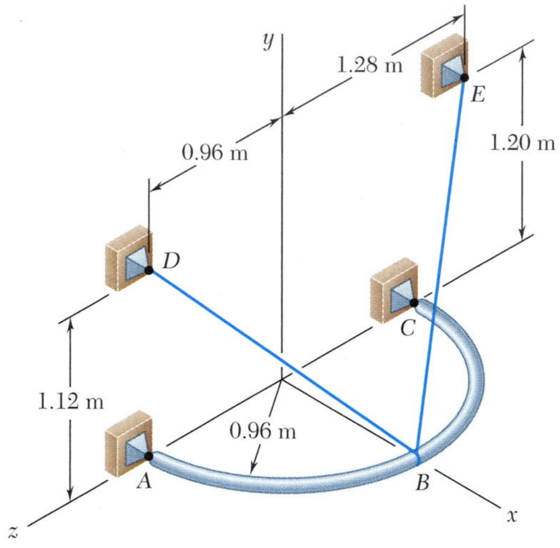

Fig. P2.91 and *P2.92*

2.92 A steel rod is bent into a semicircular ring of radius 0.96 m and is supported in part by cables *BD* and *BE* which are attached to the ring at *B*. Knowing that the tension in cable *BE* is 250 N, determine the components of this force exerted by the cable on the support at *E*.

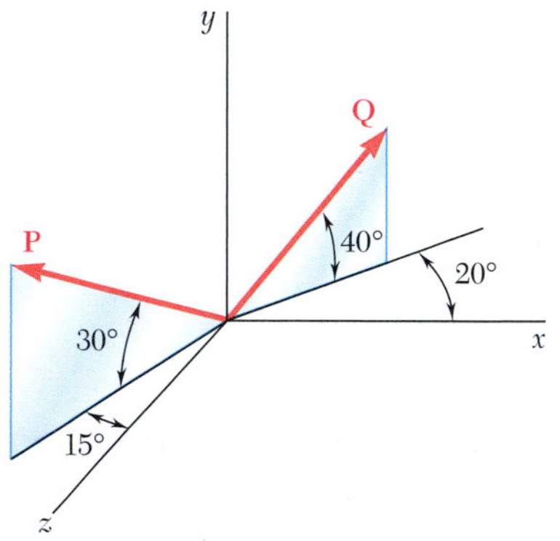

Fig. *P2.93* and P2.94

2.93 Find the magnitude and direction of the resultant of the two forces shown knowing that $P = 500$ lb and $Q = 600$ lb.

2.94 Find the magnitude and direction of the resultant of the two forces shown knowing that $P = 600$ N and $Q = 400$ N.

2.95 Knowing that the tension is 850 N in cable *AB* and 1020 N in cable *AC*, determine the magnitude and direction of the resultant of the forces exerted at *A* by the two cables.

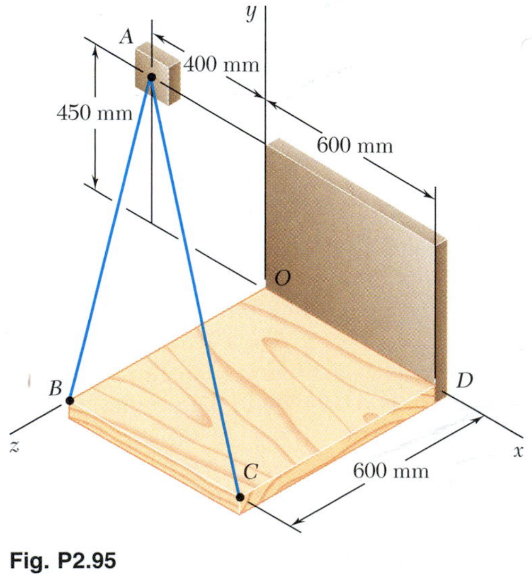

Fig. P2.95

2.96 Assuming that in Prob. 2.95 the tension is 1020 N in cable *AB* and 850 N in cable *AC*, determine the magnitude and direction of the resultant of the forces exerted at *A* by the two cables.

2.97 For the semicircular ring of Prob. 2.91, determine the magnitude and direction of the resultant of the forces exerted by the cables at *B* knowing that the tensions in cables *BD* and *BE* are 220 N and 250 N, respectively.

2.98 To stabilize a tree partially uprooted in a storm, cables *AB* and *AC* are attached to the upper trunk of the tree and then are fastened to steel rods anchored in the ground. Knowing that the tension in *AB* is 920 lb and that the resultant of the forces exerted at *A* by cables *AB* and *AC* lies in the *yz* plane, determine (*a*) the tension in *AC*, (*b*) the magnitude and direction of the resultant of the two forces.

2.99 To stabilize a tree partially uprooted in a storm, cables *AB* and *AC* are attached to the upper trunk of the tree and then are fastened to steel rods anchored in the ground. Knowing that the tension in *AC* is 850 lb and that the resultant of the forces exerted at *A* by cables *AB* and *AC* lies in the *yz* plane, determine (*a*) the tension in *AB*, (*b*) the magnitude and direction of the resultant of the two forces.

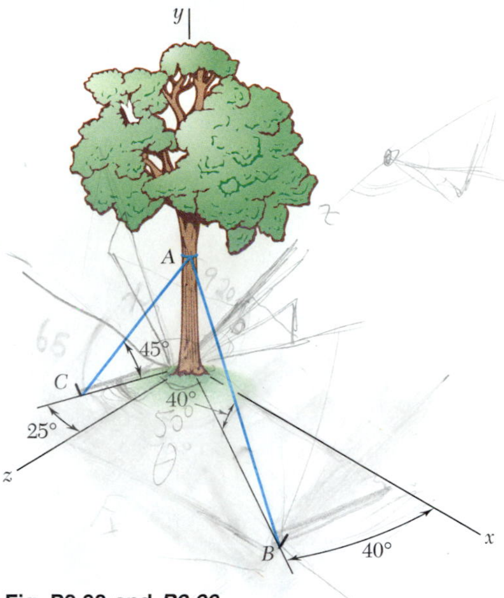

Fig. P2.98 and *P2.99*

2.100 For the plate of Prob. 2.89, determine the tension in cables *AB* and *AD* knowing that the tension in cable *AC* is 27 lb and that the resultant of the forces exerted by the three cables at *A* must be vertical.

2.15. EQUILIBRIUM OF A PARTICLE IN SPACE

According to the definition given in Sec. 2.9, a particle *A* is in equilibrium if the resultant of all the forces acting on *A* is zero. The components R_x, R_y, R_z of the resultant are given by the relations (2.31); expressing that the components of the resultant are zero, we write

$$\Sigma F_x = 0 \qquad \Sigma F_y = 0 \qquad \Sigma F_z = 0 \tag{2.34}$$

Photo 2.2 While the tension in the *four* cables supporting the shipping container cannot be found using the *three* equations of (2.34), a relation between the tensions can be obtained by considering the equilibrium of the hook.

Equations (2.34) represent the necessary and sufficient conditions for the equilibrium of a particle in space. They can be used to solve problems dealing with the equilibrium of a particle involving no more than three unknowns.

To solve such problems, you first should draw a free-body diagram showing the particle in equilibrium and *all* the forces acting on it. You can then write the equations of equilibrium (2.34) and solve them for three unknowns. In the more common types of problems, these unknowns will represent (1) the three components of a single force or (2) the magnitude of three forces, each of known direction.

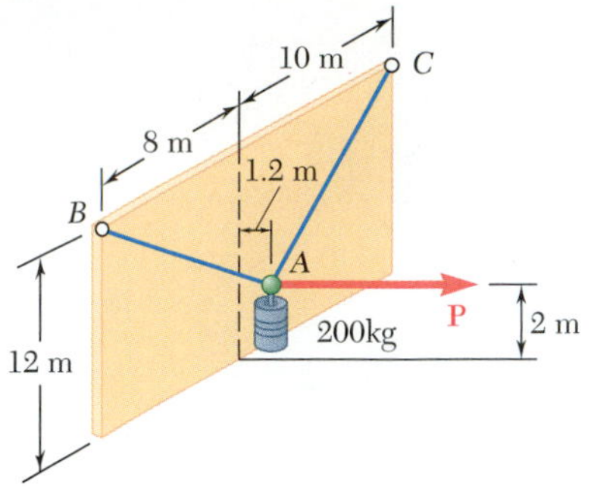

SAMPLE PROBLEM 2.9

A 200-kg cylinder is hung by means of two cables AB and AC, which are attached to the top of a vertical wall. A horizontal force $\mathbf{P}$ perpendicular to the wall holds the cylinder in the position shown. Determine the magnitude of $\mathbf{P}$ and the tension in each cable.

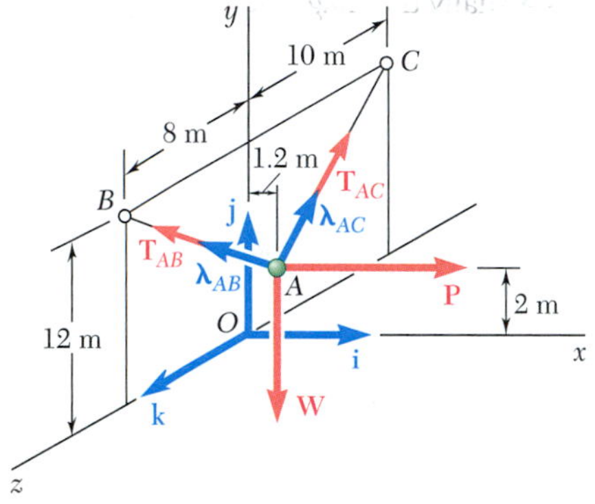

SOLUTION

Free-body Diagram. Point A is chosen as a free body; this point is subjected to four forces, three of which are of unknown magnitude.

Introducing the unit vectors $\mathbf{i}$, $\mathbf{j}$, $\mathbf{k}$, we resolve each force into rectangular components.

$$\mathbf{P} = P\mathbf{i}$$
$$\mathbf{W} = -mg\mathbf{j} = -(200 \text{ kg})(9.81 \text{ m/s}^2)\mathbf{j} = -(1962 \text{ N})\mathbf{j} \qquad (1)$$

In the case of $\mathbf{T}_{AB}$ and $\mathbf{T}_{AC}$, it is necessary first to determine the components and magnitudes of the vectors $\overrightarrow{AB}$ and $\overrightarrow{AC}$. Denoting by $\boldsymbol{\lambda}_{AB}$ the unit vector along AB, we write

$$\overrightarrow{AB} = -(1.2 \text{ m})\mathbf{i} + (10 \text{ m})\mathbf{j} + (8 \text{ m})\mathbf{k} \qquad AB = 12.862 \text{ m}$$
$$\boldsymbol{\lambda}_{AB} = \frac{\overrightarrow{AB}}{12.862 \text{ m}} = -0.09330\mathbf{i} + 0.7775\mathbf{j} + 0.6220\mathbf{k}$$
$$\mathbf{T}_{AB} = T_{AB}\boldsymbol{\lambda}_{AB} = -0.09330T_{AB}\mathbf{i} + 0.7775T_{AB}\mathbf{j} + 0.6220T_{AB}\mathbf{k} \qquad (2)$$

Denoting by $\boldsymbol{\lambda}_{AC}$ the unit vector along AC, we write in a similar way

$$\overrightarrow{AC} = -(1.2 \text{ m})\mathbf{i} + (10 \text{ m})\mathbf{j} - (10 \text{ m})\mathbf{k} \qquad AC = 14.193 \text{ m}$$
$$\boldsymbol{\lambda}_{AC} = \frac{\overrightarrow{AC}}{14.193 \text{ m}} = -0.08455\mathbf{i} + 0.7046\mathbf{j} - 0.7046\mathbf{k}$$
$$\mathbf{T}_{AC} = T_{AC}\boldsymbol{\lambda}_{AC} = -0.08455T_{AC}\mathbf{i} + 0.7046T_{AC}\mathbf{j} - 0.7046T_{AC}\mathbf{k} \qquad (3)$$

Equilibrium Condition. Since A is in equilibrium, we must have

$$\Sigma\mathbf{F} = 0: \qquad \mathbf{T}_{AB} + \mathbf{T}_{AC} + \mathbf{P} + \mathbf{W} = 0$$

or, substituting from (1), (2), (3) for the forces and factoring $\mathbf{i}$, $\mathbf{j}$, $\mathbf{k}$,

$$(-0.09330T_{AB} - 0.08455T_{AC} + P)\mathbf{i} + (0.7775T_{AB} + 0.7046T_{AC} - 1962 \text{ N})\mathbf{j} + (0.6220T_{AB} - 0.7046T_{AC})\mathbf{k} = 0$$

Setting the coefficients of $\mathbf{i}$, $\mathbf{j}$, $\mathbf{k}$ equal to zero, we write three scalar equations, which express that the sums of the x, y, and z components of the forces are respectively equal to zero.

$$\Sigma F_x = 0: \qquad -0.09330T_{AB} - 0.08455T_{AC} + P = 0$$
$$\Sigma F_y = 0: \qquad +0.7775T_{AB} + 0.7046T_{AC} - 1962 \text{ N} = 0$$
$$\Sigma F_z = 0: \qquad +0.6220T_{AB} - 0.7046T_{AC} = 0$$

Solving these equations, we obtain

$$P = 235 \text{ N} \qquad T_{AB} = 1402 \text{ N} \qquad T_{AC} = 1238 \text{ N} \quad \blacktriangleleft$$

SOLVING PROBLEMS ON YOUR OWN

We saw earlier that when a particle is in *equilibrium*, the resultant of the forces acting on the particle must be zero. Expressing this fact in the case of the equilibrium of a *particle in three-dimensional space* will provide you with three relations among the forces acting on the particle. These relations can be used to determine three unknowns—usually the magnitudes of three forces.

Your solution will consist of the following steps:

1. ***Draw a free-body diagram of the particle.*** This diagram shows the particle and all the forces acting on it. Indicate on the diagram the magnitudes of known forces, as well as any angles or dimensions that define the direction of a force. Any unknown magnitude or angle should be denoted by an appropriate symbol. Nothing else should be included on your free-body diagram.

2. ***Resolve each of the forces into rectangular components.*** Following the method used in the preceding lesson, you will determine for each force $\mathbf{F}$ the unit vector $\boldsymbol{\lambda}$ defining the direction of that force and express $\mathbf{F}$ as the product of its magnitude F and the unit vector $\boldsymbol{\lambda}$. When two points on the line of action of $\mathbf{F}$ are known, you will obtain an expression of the form

$$\mathbf{F} = F\boldsymbol{\lambda} = \frac{F}{d}(d_x\mathbf{i} + d_y\mathbf{j} + d_z\mathbf{k})$$

where d, d_x, d_y, and d_z are dimensions obtained from the free-body diagram of the particle. We also showed the direction of $\mathbf{F}$ can be defined in terms of the angles θ_y and ϕ. If a force is known in magnitude as well as in direction, then F is known and the expression obtained for $\mathbf{F}$ is completely defined; otherwise F is one of the three unknowns that should be determined.

3. ***Set the resultant, or sum, of the forces exerted on the particle equal to zero.*** You will obtain a vectorial equation consisting of terms containing the unit vectors $\mathbf{i}$, $\mathbf{j}$, or $\mathbf{k}$. You will group the terms containing the same unit vector and factor that vector. For the vectorial equation to be satisfied, the coefficient of each of the unit vectors must be equal to zero. Thus, setting each coefficient equal to zero will yield three scalar equations that you can solve for no more than three unknowns [Sample Prob. 2.9].

Problems

2.101 The support assembly shown is bolted in place at B, C, and D and supports a downward force **P** at A. Knowing that the forces in members AB, AC, and AD are directed along the respective members and that the force in member AB is 146 N, determine the magnitude of **P**.

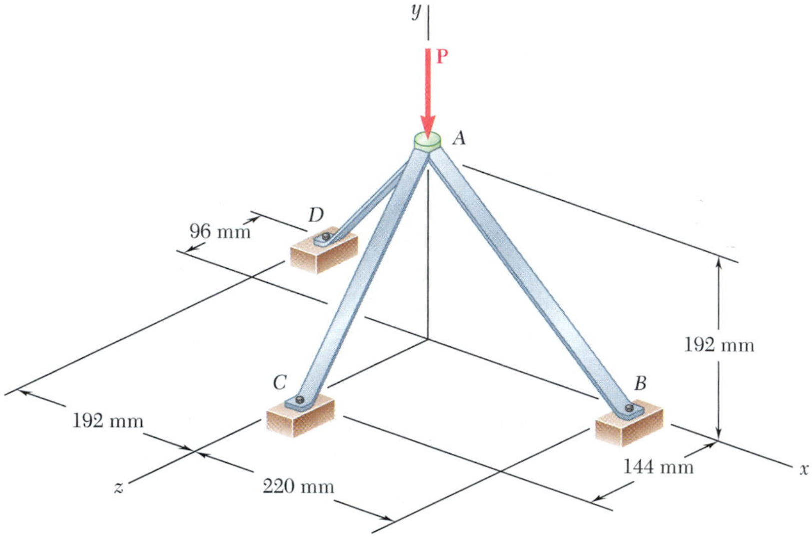

Fig. P2.101 and *P2.102*

2.102 The support assembly shown is bolted in place at B, C, and D and supports a downward force **P** at A. Knowing that the forces in members AB, AC, and AD are directed along the respective members and that $P = 200$ N, determine the forces in the members.

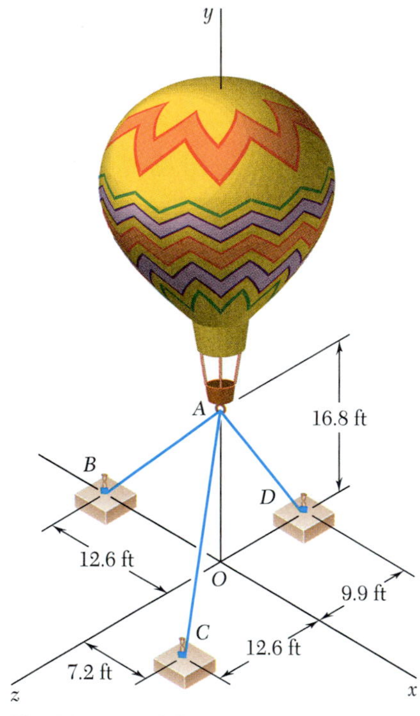

Fig. P2.103 and P2.104

2.103 Three cables are used to tether a balloon as shown. Determine the vertical force **P** exerted by the balloon at A knowing that the tension in cable AB is 60 lb.

2.104 Three cables are used to tether a balloon as shown. Determine the vertical force **P** exerted by the balloon at A knowing that the tension in cable AC is 100 lb.

2.105 The crate shown in Fig. P2.105 and P2.108 is supported by three cables. Determine the weight of the crate knowing that the tension in cable AB is 3 kN.

2.106 For the crate of Prob. 2.105, determine the weight of the crate knowing that the tension in cable AD is 2.8 kN.

2.107 For the crate of Prob. 2.105, determine the weight of the crate knowing that the tension in cable AC is 2.4 kN.

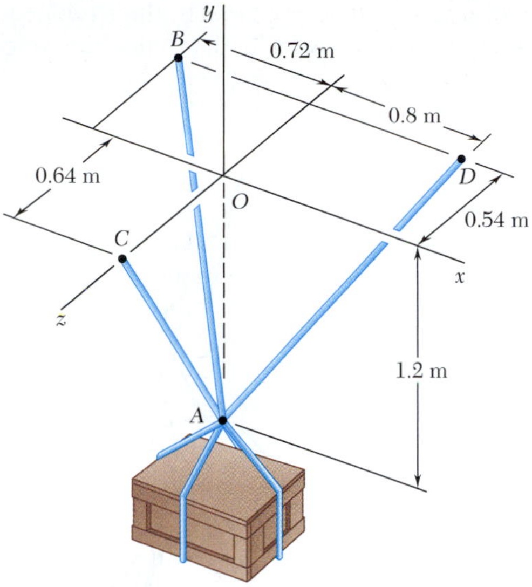

Fig. P2.105 and P2.108

2.108 A 750-kg crate is supported by three cables as shown. Determine the tension in each cable.

2.109 A force **P** is applied as shown to a uniform cone which is supported by three cords, where the lines of action of the cords pass through the vertex A of the cone. Knowing that $P = 0$ and that the tension in cord BE is 0.2 lb, determine the weight W of the cone.

2.110 A force **P** is applied as shown to a uniform cone which is supported by three cords, where the lines of action of the cords pass through the vertex A of the cone. Knowing that the cone weighs 1.6 lb, determine the range of values of P for which cord CF is taut.

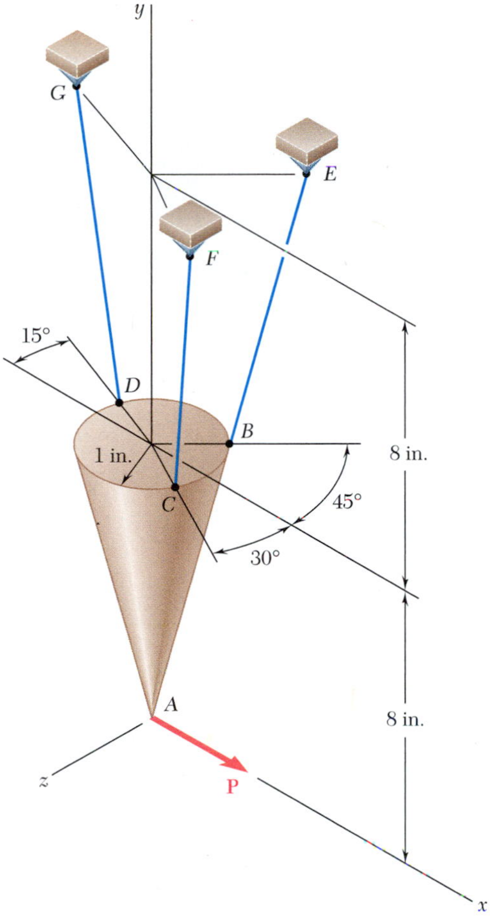

Fig. P2.109 and *P2.110*

2.111 A transmission tower is held by three guy wires attached to a pin at A and anchored by bolts at B, C, and D. If the tension in wire AB is 3.6 kN, determine the vertical force **P** exerted by the tower on the pin at A.

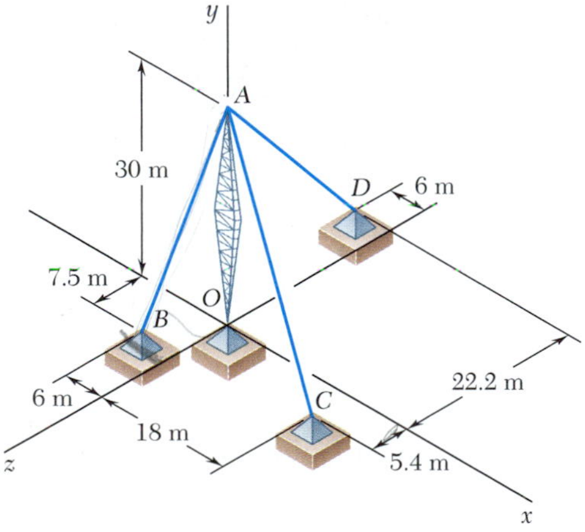

Fig. P2.111 and P2.112

2.112 A transmission tower is held by three guy wires attached to a pin at A and anchored by bolts at B, C, and D. If the tension in wire AC is 2.6 kN, determine the vertical force **P** exerted by the tower on the pin at A.

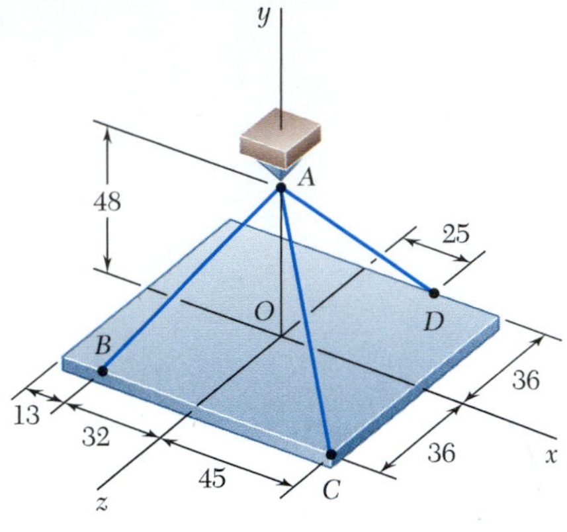

Fig. P2.113 and *P2.114*

2.113 A rectangular plate is supported by three cables as shown. Knowing that the tension in cable *AC* is 15 lb, determine the weight of the plate.

2.114 A rectangular plate is supported by three cables as shown. Knowing that the tension in cable *AD* is 120 lb, determine the weight of the plate.

2.115 A horizontal circular plate having a mass of 28 kg is suspended as shown from three wires which are attached to a support *D* and form 30° angles with the vertical. Determine the tension in each wire.

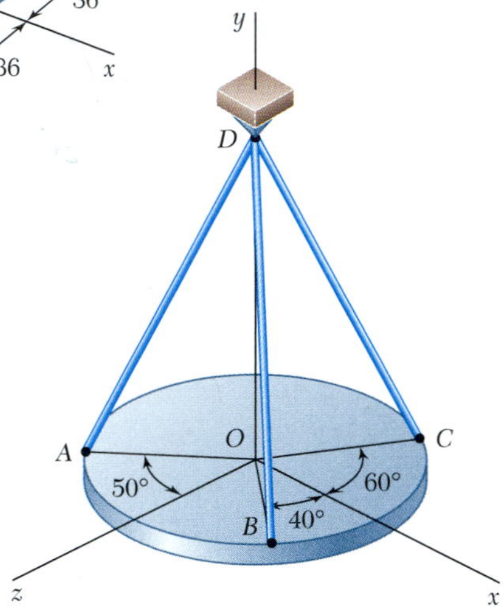

Fig. P2.115

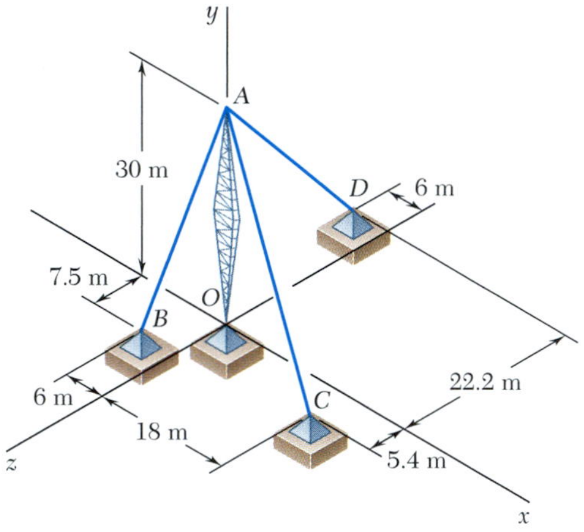

Fig. P2.116

2.116 A transmission tower is held by three guy wires attached to a pin at *A* and anchored by bolts at *B*, *C*, and *D*. Knowing that the tower exerts on the pin at *A* an upward vertical force of 8 kN, determine the tension in each wire.

2.117 For the rectangular plate of Probs. 2.113 and 2.114, determine the tension in each of the three cables knowing that the weight of the plate is 180 lb.

2.118 For the cone of Prob. 2.110, determine the range of values of *P* for which cord *DG* is taut if **P** is directed in the $-x$ direction.

2.119 A force **P** is applied as shown to a uniform cone which is supported by three cords, where the lines of action of the cords pass through the vertex *A* of the cone. Knowing that the cone weighs 2.4 lb and that $P = 0$, determine the tension in each cord.

2.120 A force **P** is applied as shown to a uniform cone which is supported by three cords, where the lines of action of the cords pass through the vertex *A* of the cone. Knowing that the cone weighs 2.4 lb and that $P = 0.1$ lb, determine the tension in each cord.

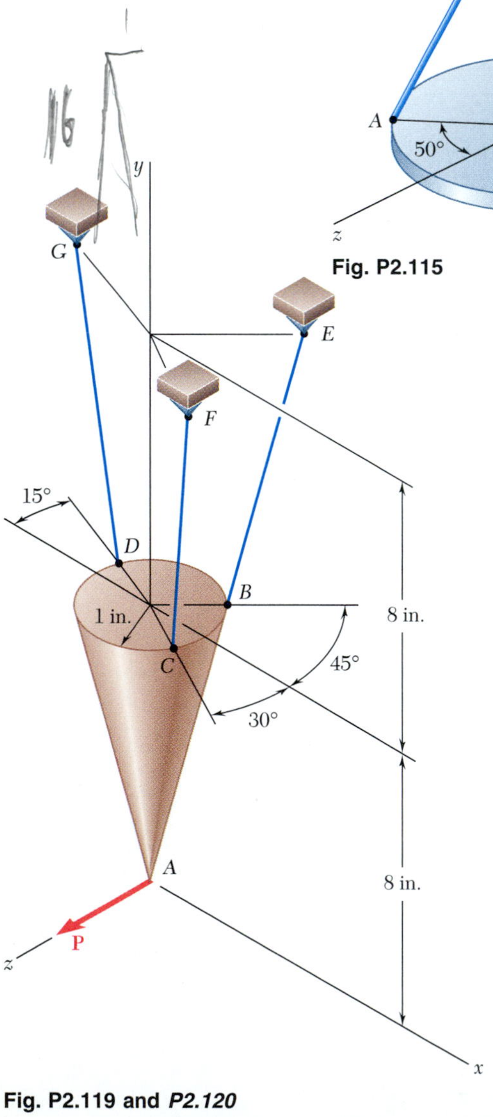

Fig. P2.119 and *P2.120*

2.121 Using two ropes and a roller chute, two workers are unloading a 200-kg cast-iron counterweight from a truck. Knowing that at the instant shown the counterweight is kept from moving and that the positions of points A, B, and C are, respectively, $A(0, -0.5 \text{ m}, 1 \text{ m})$, $B(-0.6 \text{ m}, 0.8 \text{ m}, 0)$, and $C(0.7 \text{ m}, 0.9 \text{ m}, 0)$, and assuming that no friction exists between the counterweight and the chute, determine the tension in each rope. (*Hint:* Since there is no friction, the force exerted by the chute on the counterweight must be perpendicular to the chute.)

2.122 Solve Prob. 2.121 assuming that a third worker is exerting a force $\mathbf{P} = -(180 \text{ N})\mathbf{i}$ on the counterweight.

2.123 A piece of machinery of weight W is temporarily supported by cables AB, AC, and ADE. Cable ADE is attached to the ring at A, passes over the pulley at D and back through the ring, and is attached to the support at E. Knowing that $W = 320$ lb, determine the tension in each cable. (*Hint:* The tension is the same in all portions of cable ADE.)

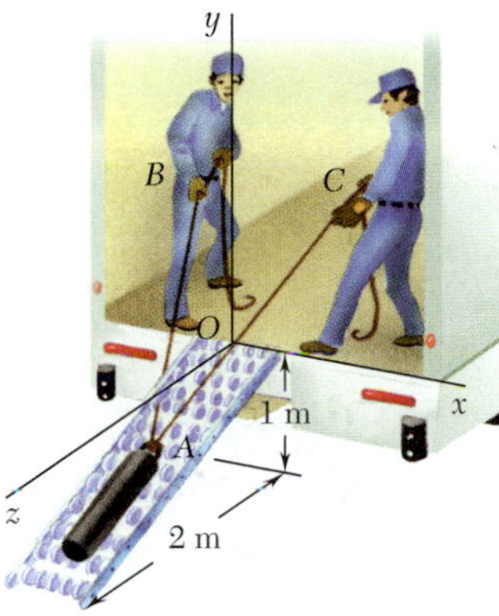

Fig. P2.121

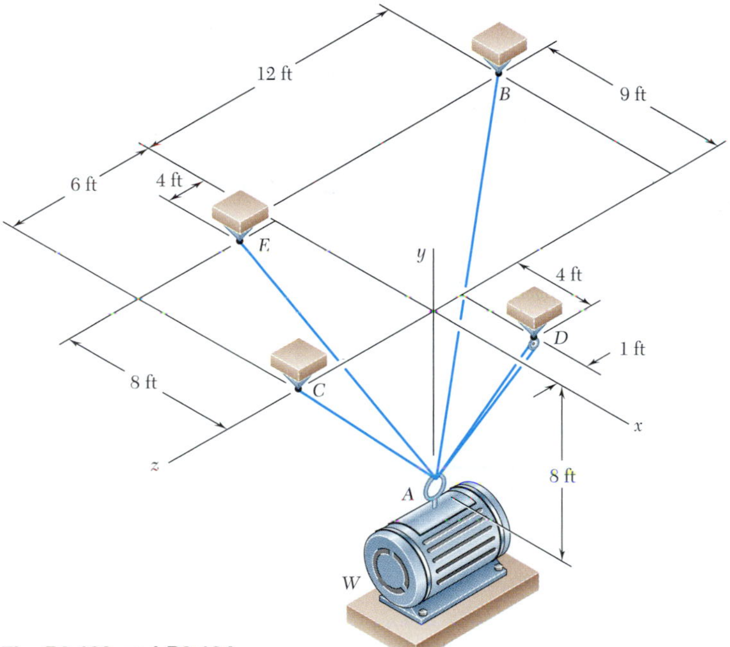

Fig. P2.123 and P2.124

2.124 A piece of machinery of weight W is temporarily supported by cables AB, AC, and ADE. Cable ADE is attached to the ring at A, passes over the pulley at D and back through the ring, and is attached to the support at E. Knowing that the tension in cable AB is 68 lb, determine (*a*) the tension in AC, (*b*) the tension in ADE, (*c*) the weight W. (*Hint:* The tension is the same in all portions of cable ADE.)

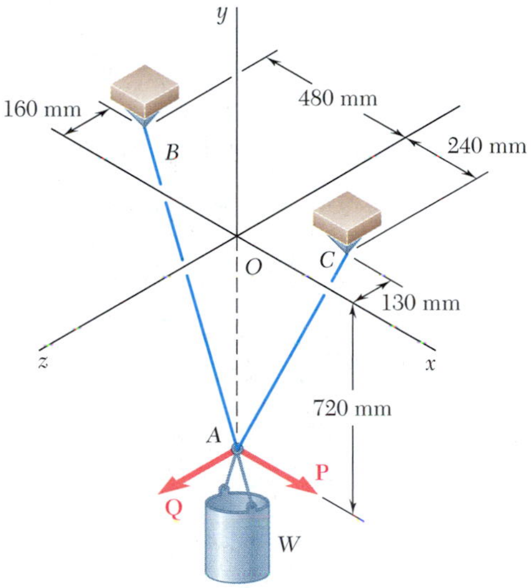

Fig. P2.125

2.125 A container of weight W is suspended from ring A. Cable BAC passes through the ring and is attached to fixed supports at B and C. Two forces $\mathbf{P} = P\mathbf{i}$ and $\mathbf{Q} = Q\mathbf{k}$ are applied to the ring to maintain the container in the position shown. Knowing that $W = 1200$ N, determine P and Q. (*Hint:* The tension is the same in both portions of cable BAC.)

2.126 For the system of Prob. 2.125, determine W and P knowing that $Q = 160$ N.

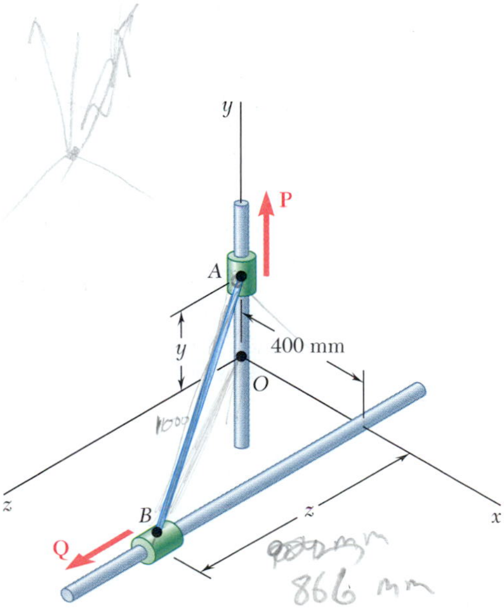

Fig. P2.127

2.127 Collars A and B are connected by a 1-m-long wire and can slide freely on frictionless rods. If a force $\mathbf{P} = (680 \text{ N})\mathbf{j}$ is applied at A, determine (*a*) the tension in the wire when $y = 300$ mm, (*b*) the magnitude of the force $\mathbf{Q}$ required to maintain the equilibrium of the system.

2.128 Solve Prob. 2.127 assuming $y = 550$ mm.

REVIEW AND SUMMARY FOR CHAPTER 2

In this chapter we have studied the effect of forces on particles, that is, on bodies of such shape and size that all forces acting on them may be assumed applied at the same point.

Resultant of two forces

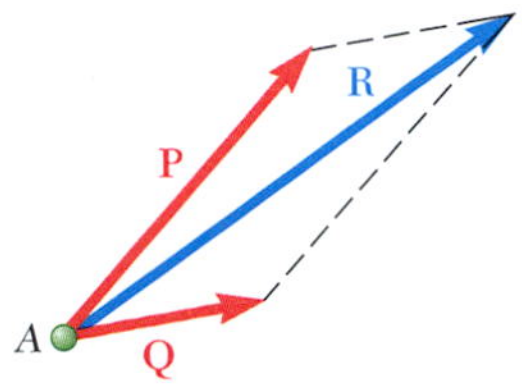

Fig. 2.35

Forces are *vector quantities*; they are characterized by a *point of application*, a *magnitude*, and a *direction*, and they add according to the *parallelogram law* (Fig. 2.35). The magnitude and direction of the resultant **R** of two forces **P** and **Q** can be determined either graphically or by trigonometry, using successively the law of cosines and the law of sines [Sample Prob. 2.1].

Components of a force

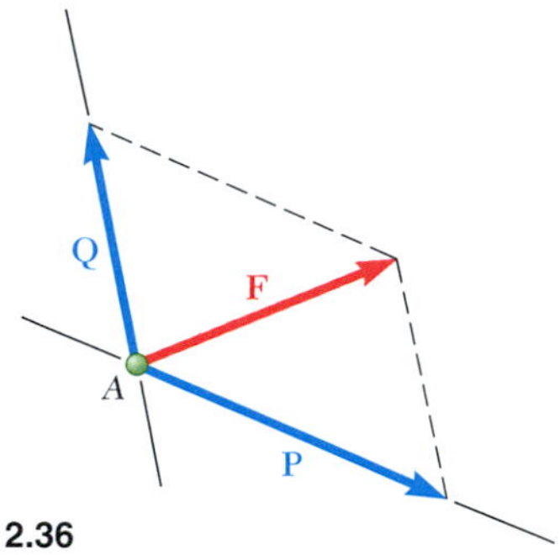

Fig. 2.36

Any given force acting on a particle can be resolved into two or more *components*, that is, it can be replaced by two or more forces which have the same effect on the particle. A force **F** can be resolved into two components **P** and **Q** by drawing a parallelogram which has **F** for its diagonal; the components **P** and **Q** are then represented by the two adjacent sides of the parallelogram (Fig. 2.36) and can be determined either graphically or by trigonometry [Sec. 2.6].

Rectangular components
Unit vectors

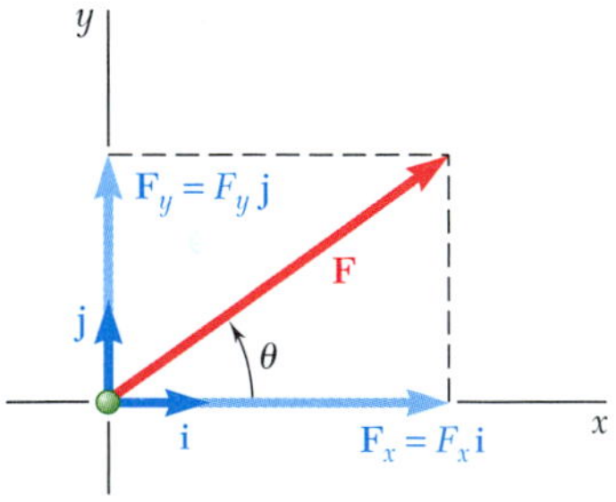

Fig. 2.37

A force **F** is said to have been resolved into two *rectangular components* if its components $\mathbf{F}_x$ and $\mathbf{F}_y$ are perpendicular to each other and are directed along the coordinate axes (Fig. 2.37). Introducing the *unit vectors* **i** and **j** along the x and y axes, respectively, we write [Sec. 2.7]

$$\mathbf{F}_x = F_x\mathbf{i} \qquad \mathbf{F}_y = F_y\mathbf{j} \tag{2.6}$$

and

$$\mathbf{F} = F_x\mathbf{i} + F_y\mathbf{j} \tag{2.7}$$

where F_x and F_y are the *scalar components* of **F**. These components, which can be positive or negative, are defined by the relations

$$F_x = F\cos\theta \qquad F_y = F\sin\theta \tag{2.8}$$

When the rectangular components F_x and F_y of a force **F** are given, the angle θ defining the direction of the force can be obtained by writing

$$\tan\theta = \frac{F_y}{F_x} \tag{2.9}$$

The magnitude F of the force can then be obtained by solving one of the equations (2.8) for F or by applying the Pythagorean theorem and writing

$$F = \sqrt{F_x^2 + F_y^2} \tag{2.10}$$

Resultant of several coplanar forces

When *three or more coplanar forces* act on a particle, the rectangular components of their resultant **R** can be obtained by adding algebraically the corresponding components of the given forces [Sec. 2.8]. We have

$$R_x = \Sigma F_x \qquad R_y = \Sigma F_y \tag{2.13}$$

The magnitude and direction of **R** can then be determined from relations similar to Eqs. (2.9) and (2.10) [Sample Prob. 2.3].

Forces in space

A force **F** in *three-dimensional space* can be resolved into rectangular components $\mathbf{F}_x$, $\mathbf{F}_y$, and $\mathbf{F}_z$ [Sec. 2.12]. Denoting by θ_x, θ_y, and θ_z, respectively, the angles that **F** forms with the x, y, and z axes (Fig. 2.38), we have

$$F_x = F\cos\theta_x \qquad F_y = F\cos\theta_y \qquad F_z = F\cos\theta_z \tag{2.19}$$

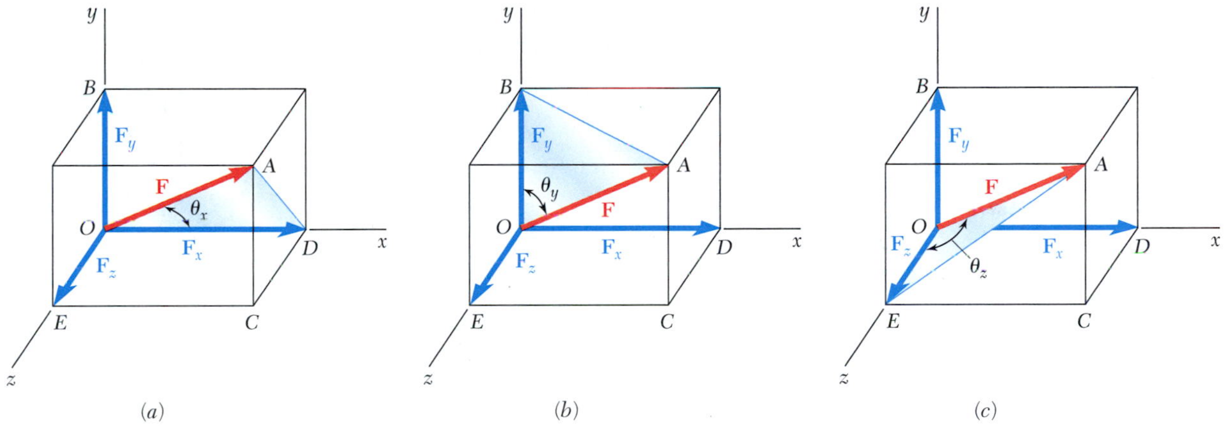

Fig. 2.38

Direction cosines

The cosines of θ_x, θ_y, θ_z are known as the *direction cosines* of the force **F**. Introducing the unit vectors **i**, **j**, **k** along the coordinate axes, we write

$$\mathbf{F} = F_x\mathbf{i} + F_y\mathbf{j} + F_z\mathbf{k} \tag{2.20}$$

or

$$\mathbf{F} = F(\cos\theta_x\mathbf{i} + \cos\theta_y\mathbf{j} + \cos\theta_z\mathbf{k}) \tag{2.21}$$

which shows (Fig. 2.39) that **F** is the product of its magnitude F and the unit vector

$$\boldsymbol{\lambda} = \cos\theta_x\mathbf{i} + \cos\theta_y\mathbf{j} + \cos\theta_z\mathbf{k}$$

Since the magnitude of $\boldsymbol{\lambda}$ is equal to unity, we must have

$$\cos^2\theta_x + \cos^2\theta_y + \cos^2\theta_z = 1 \tag{2.24}$$

When the rectangular components F_x, F_y, F_z of a force **F** are given, the magnitude F of the force is found by writing

$$F = \sqrt{F_x^2 + F_y^2 + F_z^2} \tag{2.18}$$

and the direction cosines of **F** are obtained from Eqs. (2.19). We have

$$\cos\theta_x = \frac{F_x}{F} \qquad \cos\theta_y = \frac{F_y}{F} \qquad \cos\theta_z = \frac{F_z}{F} \tag{2.25}$$

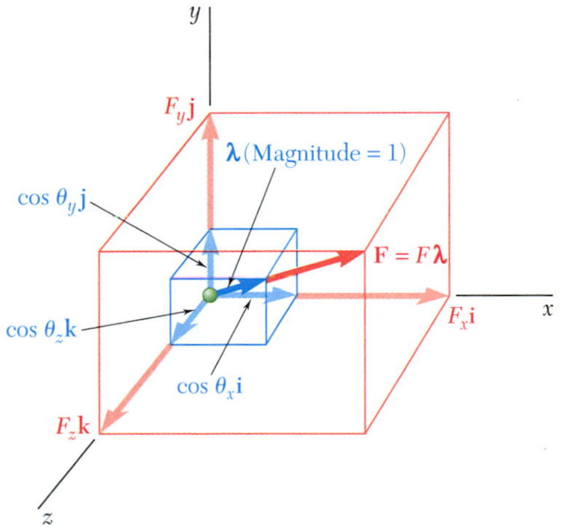

Fig. 2.39

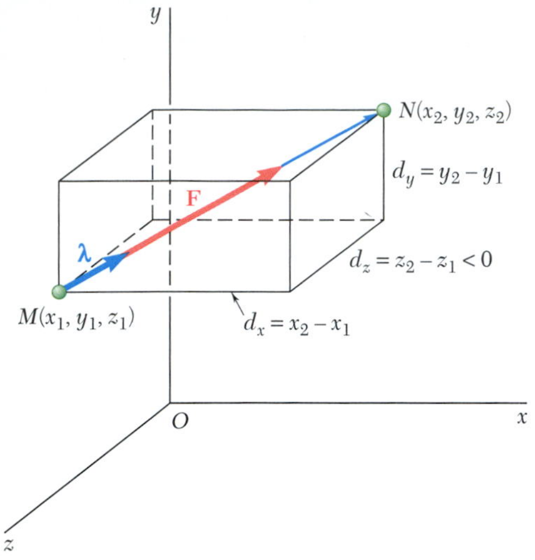

Fig. 2.40

When a force **F** is defined in three-dimensional space by its magnitude F and two points M and N on its line of action [Sec. 2.13], its rectangular components can be obtained as follows. We first express the vector $\overrightarrow{MN}$ joining points M and N in terms of its components d_x, d_y, and d_z (Fig. 2.40); we write

$$\overrightarrow{MN} = d_x\mathbf{i} + d_y\mathbf{j} + d_z\mathbf{k} \quad (2.26)$$

We next determine the unit vector $\boldsymbol{\lambda}$ along the line of action of **F** by dividing $\overrightarrow{MN}$ by its magnitude $MN = d$:

$$\boldsymbol{\lambda} = \frac{\overrightarrow{MN}}{MN} = \frac{1}{d}(d_x\mathbf{i} + d_y\mathbf{j} + d_z\mathbf{k}) \quad (2.27)$$

Recalling that **F** is equal to the product of F and $\boldsymbol{\lambda}$, we have

$$\mathbf{F} = F\boldsymbol{\lambda} = \frac{F}{d}(d_x\mathbf{i} + d_y\mathbf{j} + d_z\mathbf{k}) \quad (2.28)$$

from which it follows [Sample Probs. 2.7 and 2.8] that the scalar components of **F** are, respectively,

$$F_x = \frac{Fd_x}{d} \qquad F_y = \frac{Fd_y}{d} \qquad F_z = \frac{Fd_z}{d} \quad (2.29)$$

Resultant of forces in space

When *two or more forces* act on a particle in *three-dimensional space*, the rectangular components of their resultant **R** can be obtained by adding algebraically the corresponding components of the given forces [Sec. 2.14]. We have

$$R_x = \Sigma F_x \qquad R_y = \Sigma F_y \qquad R_z = \Sigma F_z \quad (2.31)$$

The magnitude and direction of **R** can then be determined from relations similar to Eqs. (2.18) and (2.25) [Sample Prob. 2.8].

Equilibrium of a particle

A particle is said to be in *equilibrium* when the resultant of all the forces acting on it is zero [Sec. 2.9]. The particle will then remain at rest (if originally at rest) or move with constant speed in a straight line (if originally in motion) [Sec. 2.10].

Free-body diagram

To solve a problem involving a particle in equilibrium, one first should draw a *free-body diagram* of the particle showing all the forces acting on it [Sec. 2.11]. If *only three coplanar forces* act on the particle, a *force triangle* may be drawn to express that the particle is in equilibrium. Using graphical methods or trigonometry, this triangle can be solved for no more than two unknowns [Sample Prob. 2.4]. If *more than three coplanar forces* are involved, the equations of equilibrium

$$\Sigma F_x = 0 \qquad \Sigma F_y = 0 \quad (2.15)$$

should be used. These equations can be solved for no more than two unknowns [Sample Prob. 2.6].

Equilibrium in space

When a particle is in *equilibrium in three-dimensional space* [Sec. 2.15], the three equations of equilibrium

$$\Sigma F_x = 0 \qquad \Sigma F_y = 0 \qquad \Sigma F_z = 0 \quad (2.34)$$

should be used. These equations can be solved for no more than three unknowns [Sample Prob. 2.9].

Review Problems

2.129 Member BD exerts on member ABC a force **P** directed along line BD. Knowing that **P** must have a 300-lb horizontal component, determine (*a*) the magnitude of the force **P**, (*b*) its vertical component.

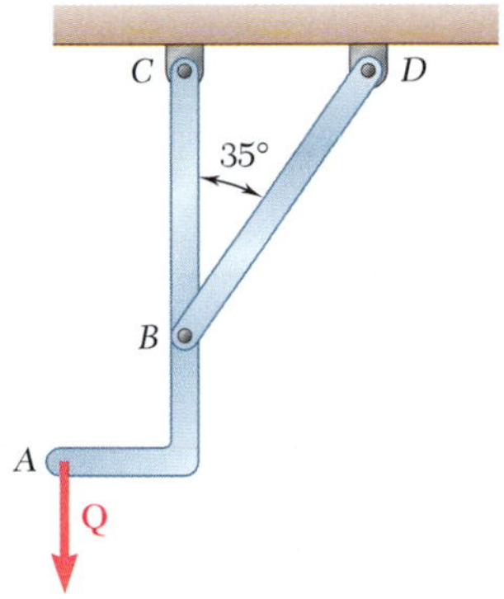

Fig. P2.129

2.130 A container of weight W is suspended from ring A, to which cables AC and AE are attached. A force **P** is applied to the end F of a third cable which passes over a pulley at B and through ring A and which is attached to a support at D. Knowing that $W = 1000$ N, determine the magnitude of **P**. (*Hint:* The tension is the same in all portions of cable $FBAD$.)

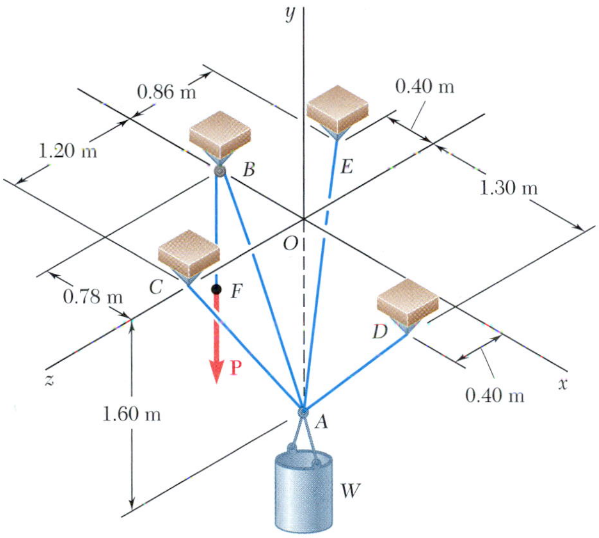

Fig. *P2.130* and P2.131

2.131 A container of weight W is suspended from ring A, to which cables AC and AE are attached. A force **P** is applied to the end F of a third cable which passes over a pulley at B and through ring A and which is attached to a support at D. Knowing that the tension in cable AC is 150 N, determine (*a*) the magnitude of the force **P**, (*b*) the weight W of the container. (See the hint for Prob. 2.130.)

2.132 Two cables tied together at C are loaded as shown. Knowing that $Q = 60$ lb, determine the tension (*a*) in cable AC, (*b*) in cable BC.

2.133 Two cables tied together at C are loaded as shown. Determine the range of values of Q for which the tension will not exceed 60 lb in either cable.

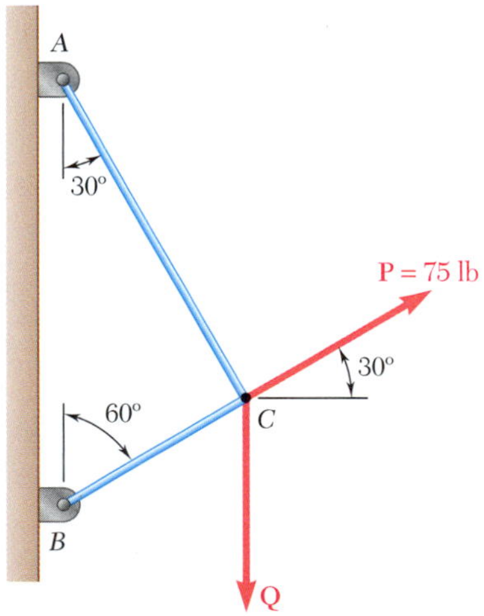

Fig. P2.132 and *P2.133*

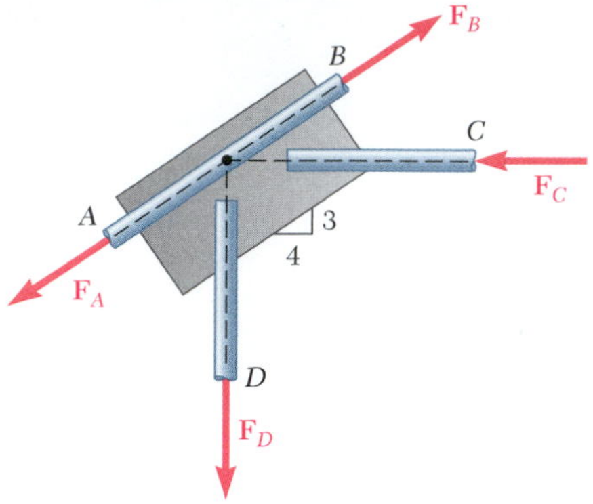

Fig. P2.134 and P2.135

2.134 A welded connection is in equilibrium under the action of the four forces shown. Knowing that $F_A = 8$ kN and $F_B = 16$ kN, determine the magnitudes of the other two forces.

2.135 A welded connection is in equilibrium under the action of the four forces shown. Knowing that $F_A = 5$ kN and $F_D = 6$ kN, determine the magnitudes of the other two forces.

2.136 Collar A is connected as shown to a 50-lb load and can slide on a frictionless horizontal rod. Determine the magnitude of the force **P** required to maintain the equilibrium of the collar when (*a*) $x = 4.5$ in., (*b*) $x = 15$ in.

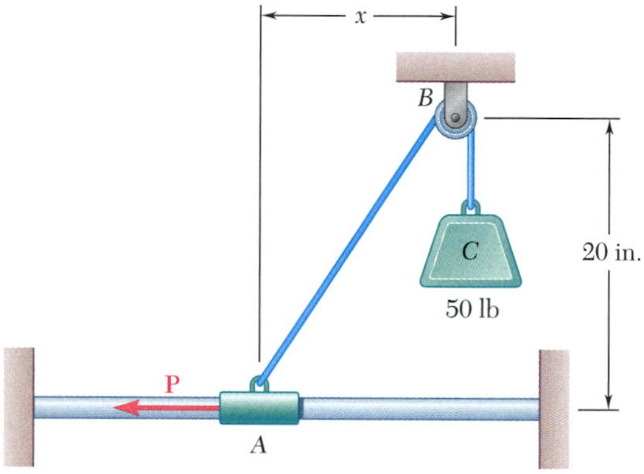

Fig. *P2.136* and P2.137

2.137 Collar A is connected as shown to a 50-lb load and can slide on a frictionless horizontal rod. Determine the distance x for which the collar is in equilibrium when $P = 48$ lb.

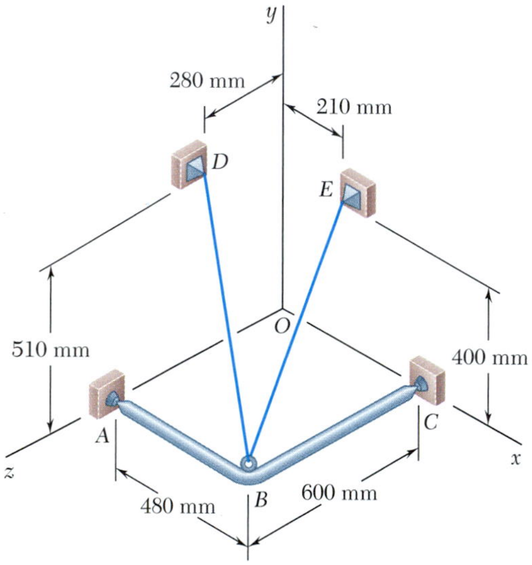

Fig. *P2.138* and P2.139

2.138 A frame ABC is supported in part by cable DBE which passes through a frictionless ring at B. Knowing that the tension in the cable is 385 N, determine the components of the force exerted by the cable on the support at D.

2.139 A frame ABC is supported in part by cable DBE which passes through a frictionless ring at B. Determine the magnitude and direction of the resultant of the forces exerted by the cable at B knowing that the tension in the cable is 385 N.

2.140 A steel tank is to be positioned in an excavation. Using trigonometry, determine (*a*) the magnitude and direction of the smallest force **P** for which the resultant **R** of the two forces applied at A is vertical, (*b*) the corresponding magnitude of **R**.

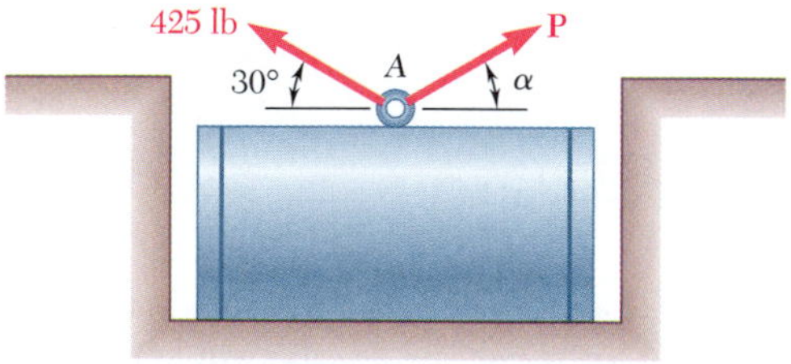

Fig. P2.140

Computer Problems

2.C1 Using computational software, determine the magnitude and direction of the resultant of n coplanar forces applied at a point A. Use this software to solve Probs. 2.32, 2.33, 2.34, and 2.36.

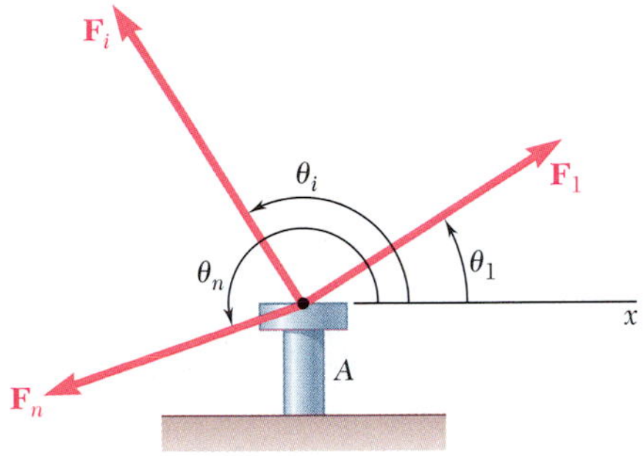

Fig. P2.C1

2.C2 A worker plans to lift a 60-lb 5-gallon bucket of paint by tying a rope to the scaffold at A and then passing the rope through the bail of the bucket at B and finally over the pulley at C. (*a*) Plot the tension in the rope as a function of the height y for 2 ft $\leq y \leq$ 18 ft. (*b*) Evaluate the worker's plan.

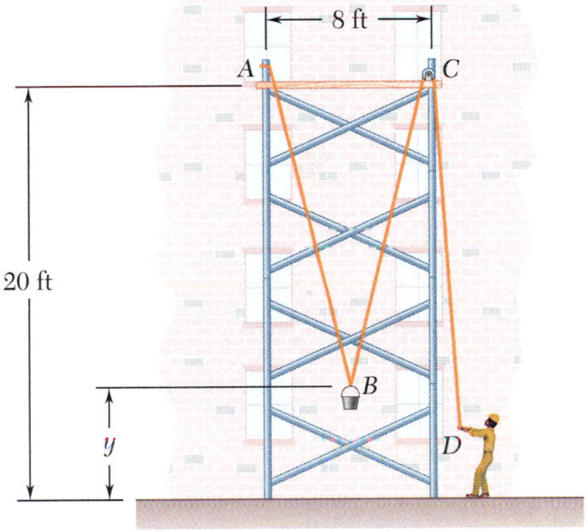

Fig. P2.C2

2.C3 The collar A can slide freely on the horizontal frictionless rod shown. The spring attached to the collar has a spring constant k and is undeformed when the collar is directly below support B. Express in terms of k and the distance c the magnitude of the force **P** required to maintain equilibrium of the system. Plot P as a function of c for values of c from 0 to 600 mm when (*a*) k = 2 N/mm, (*b*) k = 3 N/mm, (*c*) k = 4 N/mm.

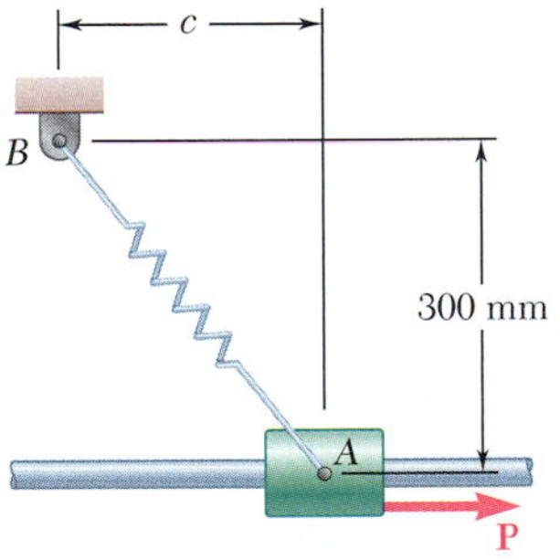

Fig. P2.C3

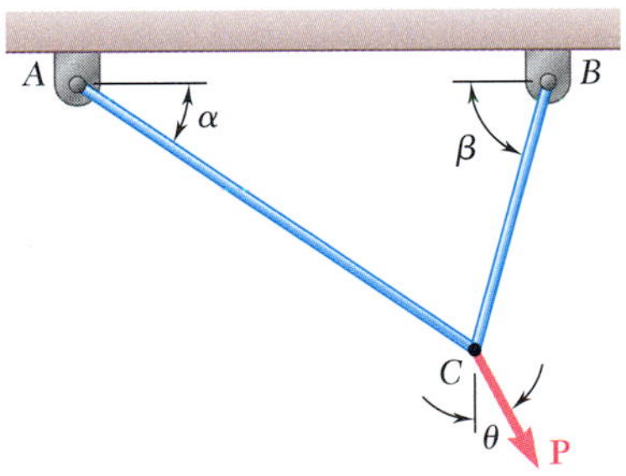

Fig. P2.C4

2.C4 A load **P** is supported by two cables as shown. Using computational software, determine the tension in each cable as a function of P and θ. For the following three sets of numerical values, plot the tensions for values of θ ranging from $\theta_1 = \beta - 90°$ to $\theta_2 = 90°$ and then determine from the graphs (*a*) the value of θ for which the tension in the two cables is as small as possible, (*b*) the corresponding value of the tension.

(1) $\alpha = 35°$, $\beta = 75°$, $P = 1.6$ kN
(2) $\alpha = 50°$, $\beta = 30°$, $P = 2.4$ kN
(3) $\alpha = 40°$, $\beta = 60°$, $P = 1.0$ kN

2.C5 Cables *AC* and *BC* are tied together at *C* and are loaded as shown. Knowing that $P = 100$ lb, (*a*) express the tension in each cable as a function of θ. (*b*) Plot the tension in each cable for $0 \le \theta \le 90°$. (*c*) From the graph obtained in part *a*, determine the smallest value of θ for which both cables are in tension.

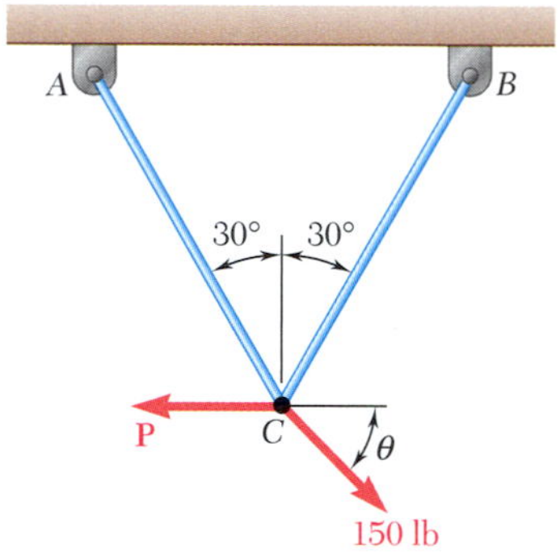

Fig. P2.C5

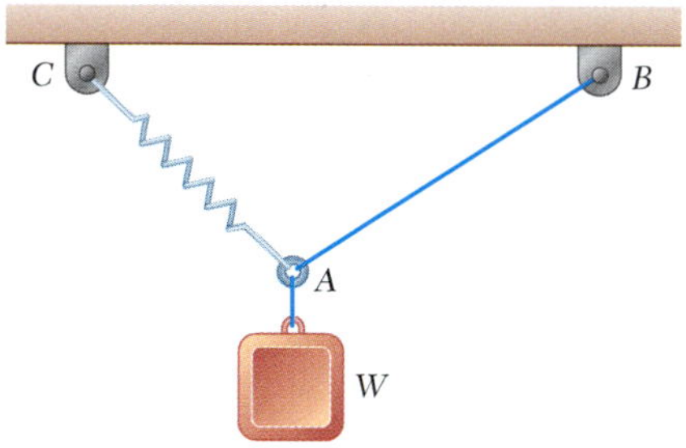

Fig. P2.C6

2.C6 A container of weight *W* is suspended from ring *A* to which cable *AB* of length 5 m and spring *AC* are attached. The constant of the spring is 100 N/m, and its unstretched length is 3 m. Determine the tension in the cable when (*a*) $W = 120$ N, (*b*) $W = 160$ N.

2.C7 An acrobat is walking on a tightrope of length $L = 80.3$ ft attached to supports *A* and *B* at a distance of 80.0 ft from each other. The combined weight of the acrobat and his balancing pole is 200 lb, and the friction between his shoes and the rope is large enough to prevent him from slipping. Neglecting the weight of the rope and any elastic deformation, use computational software to determine the deflection *y* and the tension in portions *AC* and *BC* of the rope for values of *x* from 0.5 ft to 40 ft using 0.5-ft increments. From the results obtained, determine (*a*) the maximum deflection of the rope, (*b*) the maximum tension in the rope, (*c*) the minimum values of the tension in portions *AC* and *BC* of the rope.

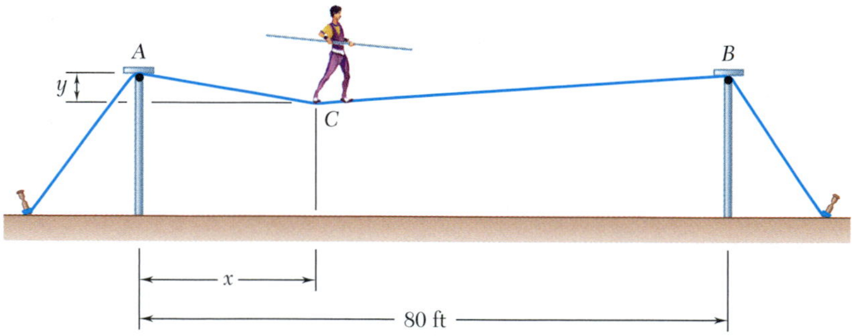

Fig. P2.C7

2.C8 The transmission tower shown is guyed by three cables attached to a pin at A and anchored at points B, C, and D. Cable AD is 21 m long and the tension in that cable is 20 kN. (*a*) Express the x, y, and z components of the force exerted by cable AD on the anchor at D and the corresponding angles θ_x, θ_y, and θ_z in terms of α. (*b*) Plot the components of the force and the angles θ_x, θ_y, and θ_z for $0 \leq \alpha \leq 60°$.

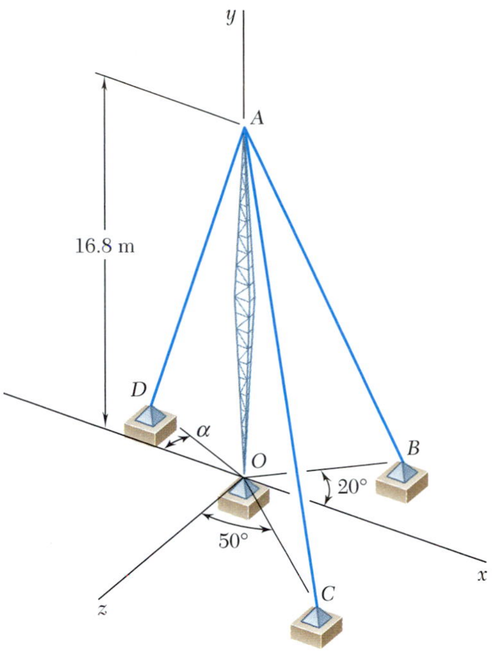

Fig. P2.C8

2.C9 A tower is guyed by cables AB and AC. A worker ties a rope of length 12 m to the tower at A and exerts a constant force of 160 N on the rope. (*a*) Express the tension in each cable as a function of θ knowing that the resultant of the tensions in the cables and the rope is directed downward. (*b*) Plot the tension in each cable as a function of θ for $0 \leq \theta \leq 180°$, and determine from the graph the range of values of θ for which both cables are taut.

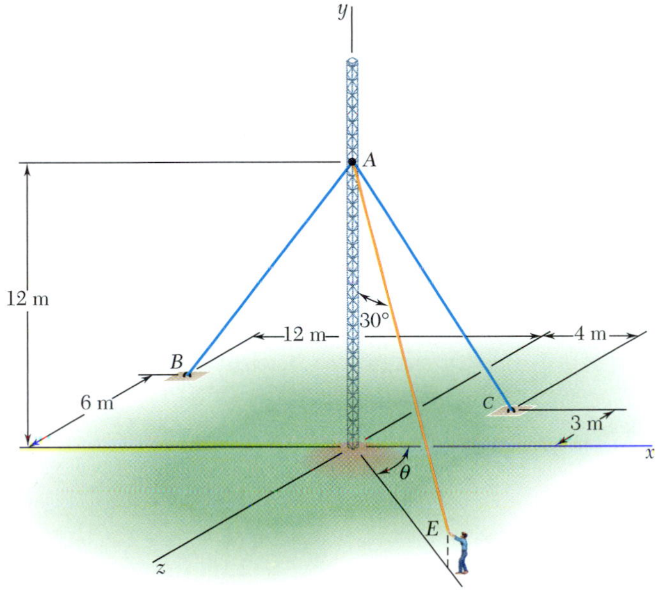

Fig. P2.C9

2.C10 Collars A and B are connected by a 10-in.-long wire and can slide freely on frictionless rods. If a force **Q** of magnitude 25 lb is applied to collar B as shown, determine the tension in the wire and the corresponding magnitude of the force **P** required to maintain equilibrium. Plot the tension in the wire and the magnitude of the force **P** for $0 \leq x \leq 5$ in.

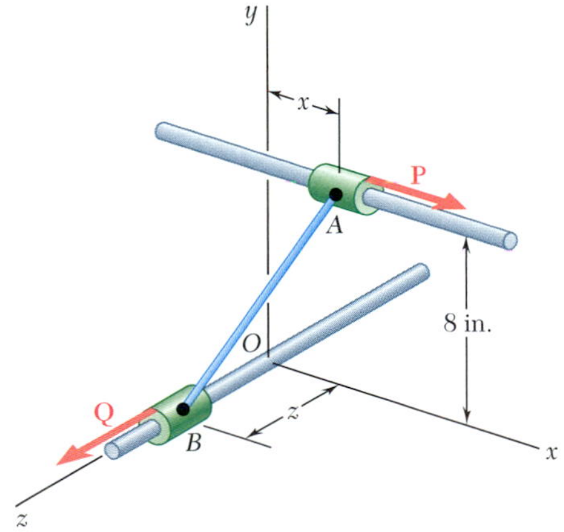

Fig. P2.C10

C H A P T E R

Rigid Bodies: Equivalent Systems of Forces

3

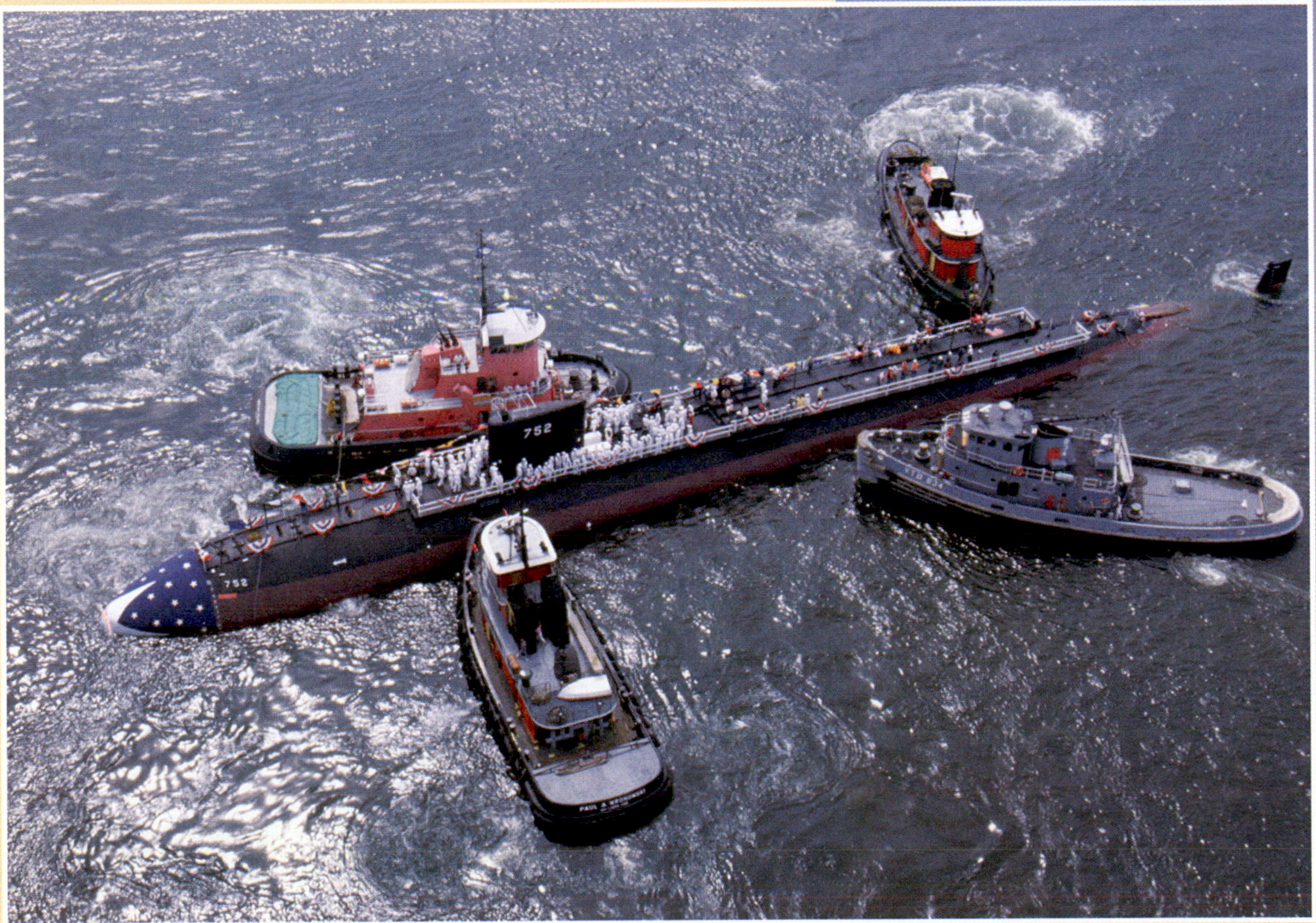

It will be shown in this chapter that the forces exerted by the four tugboats on the USS *Pasadena* could be replaced with a single equivalent force exerted by one tugboat.

RIGID BODIES: EQUIVALENT SYSTEMS OF FORCES

3.1. INTRODUCTION

In the preceding chapter it was assumed that each of the bodies considered could be treated as a single particle. Such a view, however, is not always possible, and a body, in general, should be treated as a combination of a large number of particles. The size of the body will have to be taken into consideration, as well as the fact that forces will act on different particles and thus will have different points of application.

Most of the bodies considered in elementary mechanics are assumed to be *rigid*, a *rigid body* being defined as one which does not deform. Actual structures and machines, however, are never absolutely rigid and deform under the loads to which they are subjected. But these deformations are usually small and do not appreciably affect the conditions of equilibrium or motion of the structure under consideration. They are important, though, as far as the resistance of the structure to failure is concerned and are considered in the study of mechanics of materials.

In this chapter you will study the effect of forces exerted on a rigid body, and you will learn how to replace a given system of forces by a simpler equivalent system. This analysis will rest on the fundamental assumption that the effect of a given force on a rigid body remains unchanged if that force is moved along its line of action (*principle of transmissibility*). It follows that forces acting on a rigid body can be represented by *sliding vectors,* as indicated earlier in Sec. 2.3.

Two important concepts associated with the effect of a force on a rigid body are the *moment of a force about a point* (Sec. 3.6) and the *moment of a force about an axis* (Sec. 3.11). Since the determination of these quantities involves the computation of vector products and scalar products of two vectors, the fundamentals of vector algebra will be introduced in this chapter and applied to the solution of problems involving forces acting on rigid bodies.

Another concept introduced in this chapter is that of a *couple,* that is, the combination of two forces which have the same magnitude, parallel lines of action, and opposite sense (Sec. 3.12). As you will see, any system of forces acting on a rigid body can be replaced by an equivalent system consisting of one force acting at a given point and one couple. This basic system is called a *force-couple system*. In the case of concurrent, coplanar, or parallel forces, the equivalent force-couple system can be further reduced to a single force, called the *resultant* of the system, or to a single couple, called the *resultant couple* of the system.

3.2. EXTERNAL AND INTERNAL FORCES

Forces acting on rigid bodies can be separated into two groups: (1) *external forces* and (2) *internal forces*.

1. The *external forces* represent the action of other bodies on the rigid body under consideration. They are entirely responsible for the external behavior of the rigid body. They will either cause it to move or ensure that it remains at rest. We shall be concerned only with external forces in this chapter and in Chaps. 4 and 5.

2. The *internal forces* are the forces which hold together the particles forming the rigid body. If the rigid body is structurally composed of several parts, the forces holding the component parts together are also defined as internal forces. Internal forces will be considered in Chaps. 6 and 7.

As an example of external forces, let us consider the forces acting on a disabled truck that three people are pulling forward by means of a rope attached to the front bumper (Fig. 3.1). The external forces acting on the truck are shown in a *free-body diagram* (Fig. 3.2). Let us first consider the *weight* of the truck. Although it embodies the effect of the earth's pull on each of the particles forming the truck, the weight can be represented by the single force **W**. The *point of application* of this force, that is, the point at which the force acts, is defined as the *center of gravity* of the truck. It will be seen in Chap. 5 how centers of gravity can be determined. The weight **W** tends to make the truck move vertically downward. In fact, it would actually cause the truck to move downward, that is, to fall, if it were not for the presence of the ground. The ground opposes the downward motion of the truck by means of the reactions $\mathbf{R}_1$ and $\mathbf{R}_2$. These forces are exerted *by* the ground *on* the truck and must therefore be included among the external forces acting on the truck.

Fig. 3.1

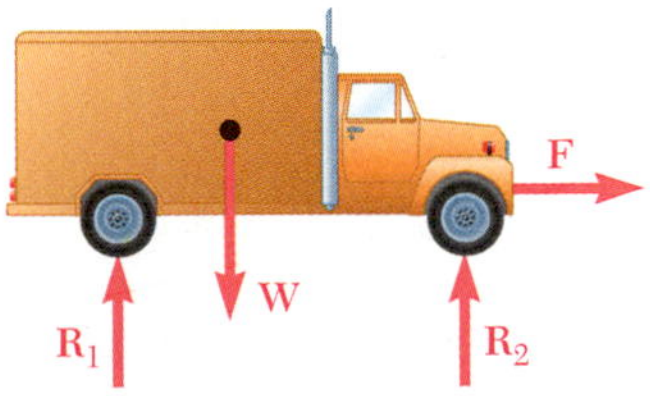

Fig. 3.2

The people pulling on the rope exert the force **F**. The point of application of **F** is on the front bumper. The force **F** tends to make the truck move forward in a straight line and does actually make it move, since no external force opposes this motion. (Rolling resistance has been neglected here for simplicity.) This forward motion of the truck, during which each straight line keeps its original orientation (the floor of the truck remains horizontal, and the walls remain vertical), is known as a *translation*. Other forces might cause the truck to move differently. For example, the force exerted by a jack placed under the front axle would cause the truck to pivot about its rear axle. Such a motion is a *rotation*. It can be concluded, therefore, that each of the *external forces* acting on a *rigid body* can, if unopposed, impart to the rigid body a motion of translation or rotation, or both.

3.3. PRINCIPLE OF TRANSMISSIBILITY. EQUIVALENT FORCES

The *principle of transmissibility* states that the conditions of equilibrium or motion of a rigid body will remain unchanged if a force **F** acting at a given point of the rigid body is replaced by a force **F′** of the same magnitude and same direction, but acting at a different point, *provided that the two forces have the same line of action* (Fig. 3.3). The two forces **F** and **F′** have the same effect on the rigid body and are said to be *equivalent*. This principle, which states that the action of a force may be *transmitted* along its line of action, is based on experimental evidence. It *cannot* be derived from the properties established so far in this text and must therefore be accepted as an experimental law. However, as you will see in Sec. 16.5, the principle of transmissibility can be derived from the study of the dynamics of rigid bodies, but this study requires the introduction of Newton's

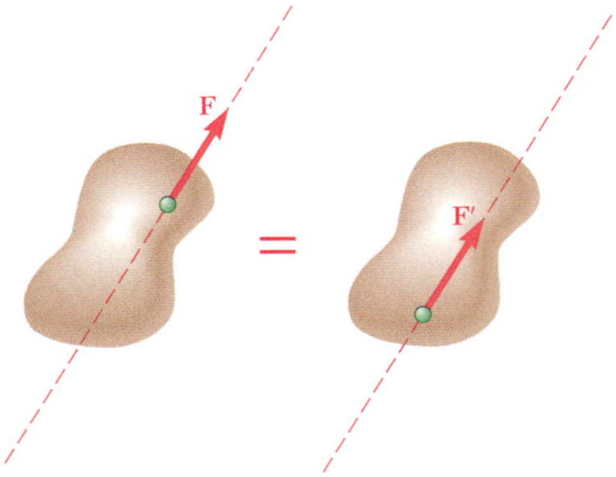

Fig. 3.3

second and third laws and of a number of other concepts as well. Therefore, our study of the statics of rigid bodies will be based on the three principles introduced so far, that is, the parallelogram law of addition, Newton's first law, and the principle of transmissibility.

It was indicated in Chap. 2 that the forces acting on a particle could be represented by vectors. These vectors had a well-defined point of application, namely, the particle itself, and were therefore fixed, or bound, vectors. In the case of forces acting on a rigid body, however, the point of application of the force does not matter, as long as the line of action remains unchanged. Thus, forces acting on a rigid body must be represented by a different kind of vector, known as a *sliding vector,* since forces may be allowed to slide along their lines of action. We should note that all the properties which will be derived in the following sections for the forces acting on a rigid body will be valid more generally for any system of sliding vectors. In order to keep our presentation more intuitive, however, we will carry it out in terms of physical forces rather than in terms of mathematical sliding vectors.

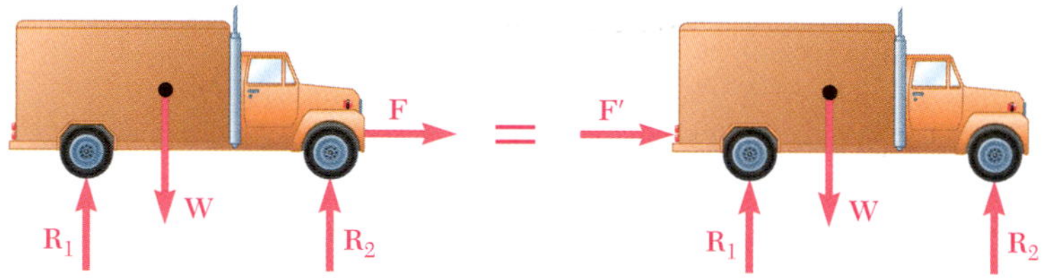

Fig. 3.4

Returning to the example of the truck, we first observe that the line of action of the force **F** is a horizontal line passing through both the front and the rear bumpers of the truck (Fig. 3.4). Using the principle of transmissibility, we can therefore replace **F** by an *equivalent force* **F**′ acting on the rear bumper. In other words, the conditions of motion are unaffected, and all the other external forces acting on the truck (**W**, $\mathbf{R}_1$, $\mathbf{R}_2$) remain unchanged if the people push on the rear bumper instead of pulling on the front bumper.

The principle of transmissibility and the concept of equivalent forces have limitations, however. Consider, for example, a short bar AB acted upon by equal and opposite axial forces $\mathbf{P}_1$ and $\mathbf{P}_2$, as shown in Fig. 3.5*a*. According to the principle of transmissibility, the force $\mathbf{P}_2$ can be replaced by a force $\mathbf{P}'_2$ having the same magnitude, the same direction, and the same line of action but acting at A instead of B (Fig. 3.5*b*). The forces $\mathbf{P}_1$ and $\mathbf{P}'_2$ acting on the same particle can

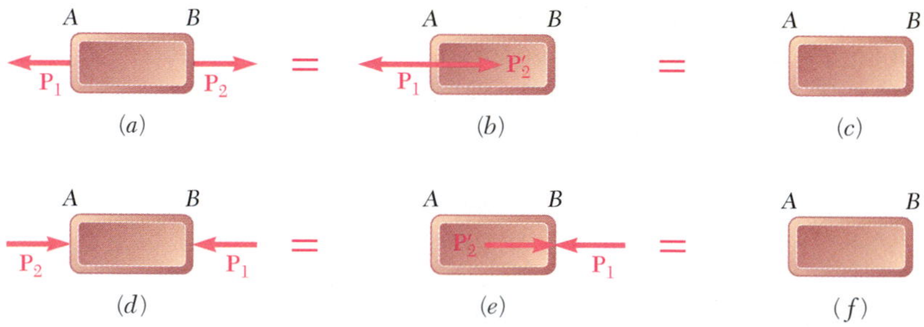

Fig. 3.5

be added according to the rules of Chap. 2, and, as these forces are equal and opposite, their sum is equal to zero. Thus, in terms of the external behavior of the bar, the original system of forces shown in Fig. 3.5*a* is equivalent to no force at all (Fig. 3.5*c*).

Consider now the two equal and opposite forces $\mathbf{P}_1$ and $\mathbf{P}_2$ acting on the bar *AB* as shown in Fig. 3.5*d*. The force $\mathbf{P}_2$ can be replaced by a force $\mathbf{P}'_2$ having the same magnitude, the same direction, and the same line of action but acting at *B* instead of at *A* (Fig. 3.5*e*). The forces $\mathbf{P}_1$ and $\mathbf{P}'_2$ can then be added, and their sum is again zero (Fig. 3.5*f*). From the point of view of the mechanics of rigid bodies, the systems shown in Fig. 3.5*a* and *d* are thus equivalent. But the *internal forces* and *deformations* produced by the two systems are clearly different. The bar of Fig. 3.5*a* is in *tension* and, if not absolutely rigid, will increase in length slightly; the bar of Fig. 3.5*d* is in *compression* and, if not absolutely rigid, will decrease in length slightly. Thus, while the principle of transmissibility may be used freely to determine the conditions of motion or equilibrium of rigid bodies and to compute the external forces acting on these bodies, it should be avoided, or at least used with care, in determining internal forces and deformations.

3.4. VECTOR PRODUCT OF TWO VECTORS

In order to gain a better understanding of the effect of a force on a rigid body, a new concept, the concept of *a moment of a force about a point,* will be introduced at this time. This concept will be more clearly understood, and applied more effectively, if we first add to the mathematical tools at our disposal the *vector product* of two vectors.

The vector product of two vectors **P** and **Q** is defined as the vector **V** which satisfies the following conditions.

1. The line of action of **V** is perpendicular to the plane containing **P** and **Q** (Fig. 3.6*a*).
2. The magnitude of **V** is the product of the magnitudes of **P** and **Q** and of the sine of the angle θ formed by **P** and **Q** (the measure of which will always be 180° or less); we thus have

$$V = PQ \sin \theta \tag{3.1}$$

3. The direction of **V** is obtained from the *right-hand rule*. Close your right hand and hold it so that your fingers are curled in the same sense as the rotation through θ which brings the vector **P** in line with the vector **Q**; your thumb will then indicate the direction of the vector **V** (Fig. 3.6*b*). Note that if **P** and **Q** do not have a common point of application, they should first be redrawn from the same point. The three vectors **P**, **Q**, and **V**—taken in that order—are said to form a *right-handed triad*.†

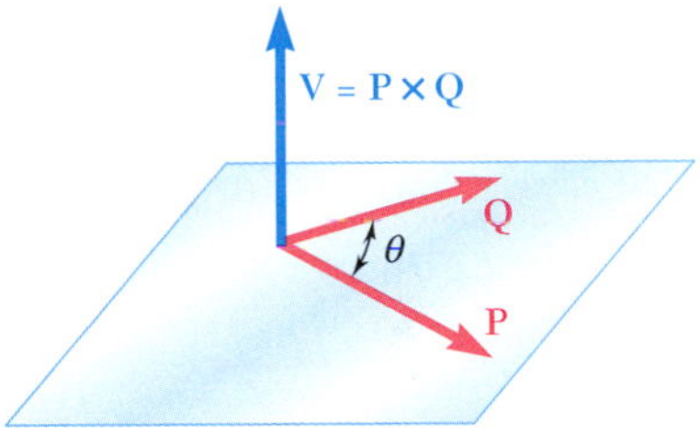

(*a*)

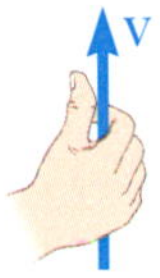

(*b*)

Fig. 3.6

†We should note that the *x*, *y*, and *z* axes used in Chap. 2 form a right-handed system of orthogonal axes and that the unit vectors **i**, **j**, **k** defined in Sec. 2.12 form a right-handed orthogonal triad.

As stated above, the vector **V** satisfying these three conditions (which define it uniquely) is referred to as the vector product of **P** and **Q**; it is represented by the mathematical expression

$$\mathbf{V} = \mathbf{P} \times \mathbf{Q} \tag{3.2}$$

Because of the notation used, the vector product of two vectors **P** and **Q** is also referred to as the *cross product* of **P** and **Q**.

It follows from Eq. (3.1) that, when two vectors **P** and **Q** have either the same direction or opposite directions, their vector product is zero. In the general case when the angle θ formed by the two vectors is neither 0° nor 180°, Eq. (3.1) can be given a simple geometric interpretation: The magnitude V of the vector product of **P** and **Q** is equal to the area of the parallelogram which has **P** and **Q** for sides (Fig. 3.7). The vector product **P** × **Q** will therefore remain unchanged if we replace **Q** by a vector **Q**′ which is coplanar with **P** and **Q** and such that the line joining the tips of **Q** and **Q**′ is parallel to **P**. We write

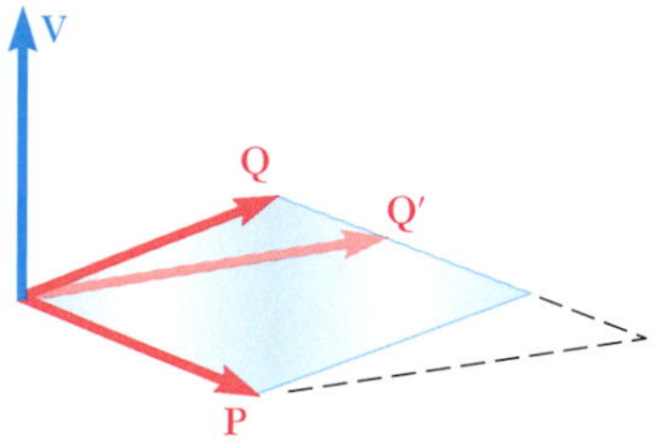

Fig. 3.7

$$\mathbf{V} = \mathbf{P} \times \mathbf{Q} = \mathbf{P} \times \mathbf{Q}' \tag{3.3}$$

From the third condition used to define the vector product **V** of **P** and **Q**, namely, the condition stating that **P**, **Q**, and **V** must form a right-handed triad, it follows that vector products *are not commutative*, that is, **Q** × **P** is not equal to **P** × **Q**. Indeed, we can easily check that **Q** × **P** is represented by the vector −**V**, which is equal and opposite to **V**. We thus write

$$\mathbf{Q} \times \mathbf{P} = -(\mathbf{P} \times \mathbf{Q}) \tag{3.4}$$

Example. Let us compute the vector product **V** = **P** × **Q** where the vector **P** is of magnitude 6 and lies in the zx plane at an angle of 30° with the x axis, and where the vector **Q** is of magnitude 4 and lies along the x axis (Fig. 3.8).

It follows immediately from the definition of the vector product that the vector **V** must lie along the y axis, have the magnitude

$$V = PQ \sin \theta = (6)(4) \sin 30° = 12$$

and be directed upward.

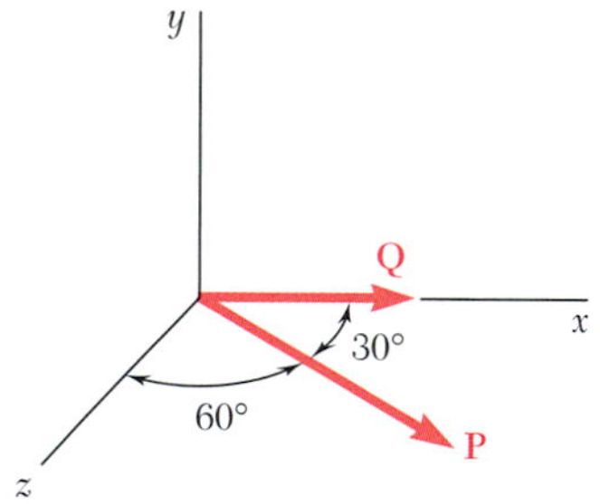

Fig. 3.8

We saw that the commutative property does not apply to vector products. We may wonder whether the *distributive* property holds, that is, whether the relation

$$\mathbf{P} \times (\mathbf{Q}_1 + \mathbf{Q}_2) = \mathbf{P} \times \mathbf{Q}_1 + \mathbf{P} \times \mathbf{Q}_2 \tag{3.5}$$

is valid. The answer is *yes*. Many readers are probably willing to accept without formal proof an answer which they intuitively feel is correct. However, since the entire structure of both vector algebra and statics depends upon the relation (3.5), we should take time out to derive it.

We can, without any loss of generality, assume that **P** is directed along the y axis (Fig. 3.9*a*). Denoting by **Q** the sum of $\mathbf{Q}_1$ and $\mathbf{Q}_2$, we drop perpendiculars from the tips of **Q**, $\mathbf{Q}_1$, and $\mathbf{Q}_2$ onto the zx plane, defining in this way the vectors **Q**′, $\mathbf{Q}'_1$, and $\mathbf{Q}'_2$. These vectors will be referred to, respectively, as the *projections* of **Q**, $\mathbf{Q}_1$, and $\mathbf{Q}_2$ on the zx plane. Recalling the property expressed by Eq. (3.3), we note that the left-hand member of Eq. (3.5) can be replaced by $\mathbf{P} \times \mathbf{Q}'$ and that, similarly, the vector products $\mathbf{P} \times \mathbf{Q}_1$ and $\mathbf{P} \times \mathbf{Q}_2$ can respectively be replaced by $\mathbf{P} \times \mathbf{Q}'_1$ and $\mathbf{P} \times \mathbf{Q}'_2$. Thus, the relation to be proved can be written in the form

$$\mathbf{P} \times \mathbf{Q}' = \mathbf{P} \times \mathbf{Q}'_1 + \mathbf{P} \times \mathbf{Q}'_2 \qquad (3.5')$$

We now observe that $\mathbf{P} \times \mathbf{Q}'$ can be obtained from $\mathbf{Q}'$ by multiplying this vector by the scalar P and rotating it counterclockwise through 90° in the zx plane (Fig. 3.9b); the other two vector products

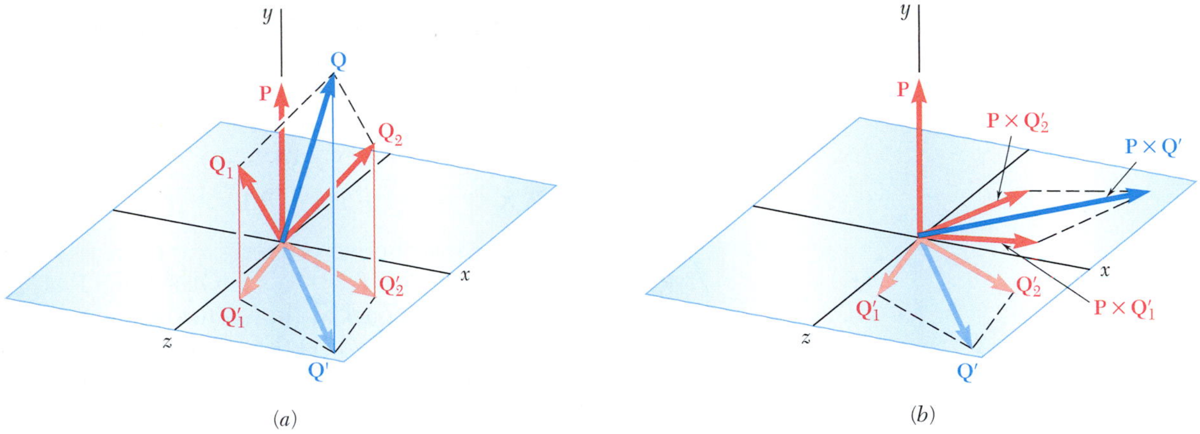

Fig. 3.9

in (3.5′) can be obtained in the same manner from $\mathbf{Q}'_1$ and $\mathbf{Q}'_2$, respectively. Now, since the projection of a parallelogram onto an arbitrary plane is a parallelogram, the projection $\mathbf{Q}'$ of the sum $\mathbf{Q}$ of $\mathbf{Q}_1$ and $\mathbf{Q}_2$ must be the sum of the projections $\mathbf{Q}'_1$ and $\mathbf{Q}'_2$ of $\mathbf{Q}_1$ and $\mathbf{Q}_2$ on the same plane (Fig. 3.9a). This relation between the vectors $\mathbf{Q}'$, $\mathbf{Q}'_1$, and $\mathbf{Q}'_2$ will still hold after the three vectors have been multiplied by the scalar P and rotated through 90° (Fig. 3.9b). Thus, the relation (3.5′) has been proved, and we can now be sure that the distributive property holds for vector products.

A third property, the associative property, does not apply to vector products; we have in general

$$(\mathbf{P} \times \mathbf{Q}) \times \mathbf{S} \neq \mathbf{P} \times (\mathbf{Q} \times \mathbf{S}) \qquad (3.6)$$

3.5. VECTOR PRODUCTS EXPRESSED IN TERMS OF RECTANGULAR COMPONENTS

Let us now determine the vector product of any two of the unit vectors $\mathbf{i}$, $\mathbf{j}$, and $\mathbf{k}$, which were defined in Chap. 2. Consider first the product $\mathbf{i} \times \mathbf{j}$ (Fig. 3.10a). Since both vectors have a magnitude equal to 1 and since they are at a right angle to each other, their vector product will also be a unit vector. This unit vector must be $\mathbf{k}$, since the vectors $\mathbf{i}$, $\mathbf{j}$, and $\mathbf{k}$ are mutually perpendicular and form a right-handed triad. On the other hand, it follows from the right-hand rule given on page 77 that the product $\mathbf{j} \times \mathbf{i}$ will be equal to $-\mathbf{k}$ (Fig. 3.10b). Finally, it should be observed that the vector product of a unit

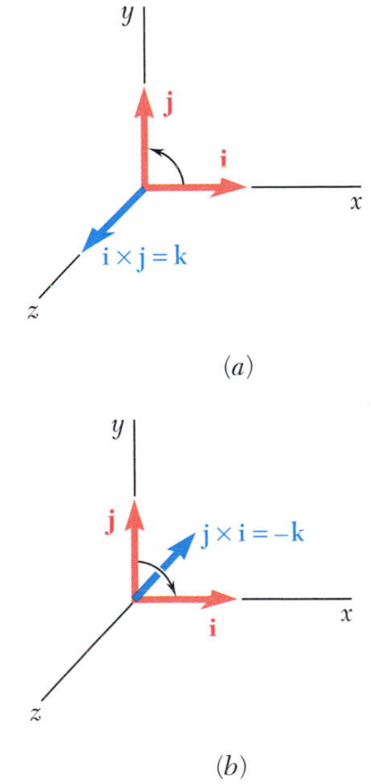

Fig. 3.10

vector with itself, such as $\mathbf{i} \times \mathbf{i}$, is equal to zero, since both vectors have the same direction. The vector products of the various possible pairs of unit vectors are

$$\begin{array}{lll} \mathbf{i} \times \mathbf{i} = 0 & \mathbf{j} \times \mathbf{i} = -\mathbf{k} & \mathbf{k} \times \mathbf{i} = \mathbf{j} \\ \mathbf{i} \times \mathbf{j} = \mathbf{k} & \mathbf{j} \times \mathbf{j} = 0 & \mathbf{k} \times \mathbf{j} = -\mathbf{i} \\ \mathbf{i} \times \mathbf{k} = -\mathbf{j} & \mathbf{j} \times \mathbf{k} = \mathbf{i} & \mathbf{k} \times \mathbf{k} = 0 \end{array} \tag{3.7}$$

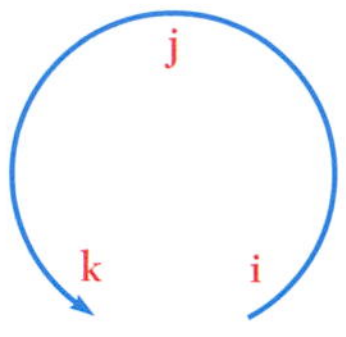

Fig. 3.11

By arranging in a circle and in counterclockwise order the three letters representing the unit vectors (Fig. 3.11), we can simplify the determination of the sign of the vector product of two unit vectors: The product of two unit vectors will be positive if they follow each other in counterclockwise order and will be negative if they follow each other in clockwise order.

We can now easily express the vector product $\mathbf{V}$ of two given vectors $\mathbf{P}$ and $\mathbf{Q}$ in terms of the rectangular components of these vectors. Resolving $\mathbf{P}$ and $\mathbf{Q}$ into components, we first write

$$\mathbf{V} = \mathbf{P} \times \mathbf{Q} = (P_x\mathbf{i} + P_y\mathbf{j} + P_z\mathbf{k}) \times (Q_x\mathbf{i} + Q_y\mathbf{j} + Q_z\mathbf{k})$$

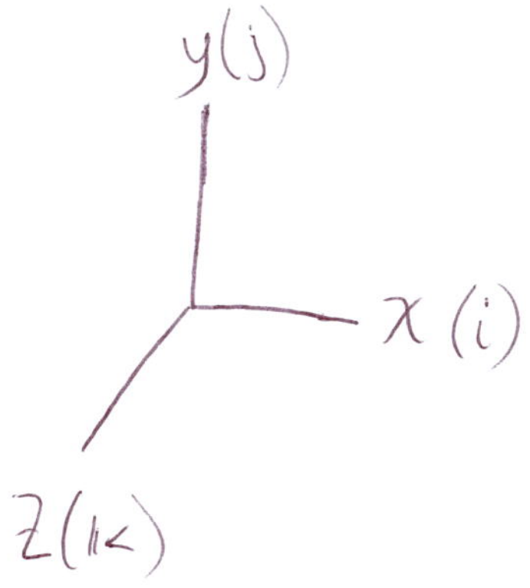

Making use of the distributive property, we express $\mathbf{V}$ as the sum of vector products, such as $P_x\mathbf{i} \times Q_y\mathbf{j}$. Observing that each of the expressions obtained is equal to the vector product of two unit vectors, such as $\mathbf{i} \times \mathbf{j}$, multiplied by the product of two scalars, such as P_xQ_y, and recalling the identities (3.7), we obtain, after factoring out $\mathbf{i}$, $\mathbf{j}$, and $\mathbf{k}$,

$$\mathbf{V} = (P_yQ_z - P_zQ_y)\mathbf{i} + (P_zQ_x - P_xQ_z)\mathbf{j} + (P_xQ_y - P_yQ_x)\mathbf{k} \tag{3.8}$$

The rectangular components of the vector product $\mathbf{V}$ are thus found to be

$$\begin{aligned} V_x &= P_yQ_z - P_zQ_y \\ V_y &= P_zQ_x - P_xQ_z \\ V_z &= P_xQ_y - P_yQ_x \end{aligned} \tag{3.9}$$

Returning to Eq. (3.8), we observe that its right-hand member represents the expansion of a determinant. The vector product $\mathbf{V}$ can thus be expressed in the following form, which is more easily memorized:†

$$\mathbf{V} = \begin{vmatrix} \mathbf{i} & \mathbf{j} & \mathbf{k} \\ P_x & P_y & P_z \\ Q_x & Q_y & Q_z \end{vmatrix} \tag{3.10}$$

†Any determinant consisting of three rows and three columns can be evaluated by repeating the first and second columns and forming products along each diagonal line. The sum of the products obtained along the red lines is then subtracted from the sum of the products obtained along the black lines.

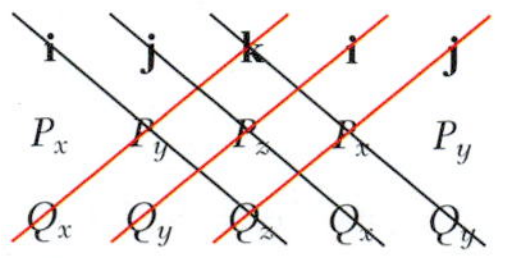

3.6. MOMENT OF A FORCE ABOUT A POINT

Let us now consider a force **F** acting on a rigid body (Fig. 3.12*a*). As we know, the force **F** is represented by a vector which defines its magnitude and direction. However, the effect of the force on the rigid body depends also upon its point of application *A*. The position of *A* can be conveniently defined by the vector **r** which joins the fixed reference point *O* with *A*; this vector is known as the *position vector* of *A*.† The position vector **r** and the force **F** define the plane shown in Fig. 3.12*a*.

We will define the *moment of* **F** *about O* as the vector product of **r** and **F**:

$$\mathbf{M}_O = \mathbf{r} \times \mathbf{F} \tag{3.11}$$

According to the definition of the vector product given in Sec. 3.4, the moment $\mathbf{M}_O$ must be perpendicular to the plane containing *O* and the force **F**. The sense of $\mathbf{M}_O$ is defined by the sense of the rotation which will bring the vector **r** in line with the vector **F**; this rotation will be observed as *counterclockwise* by an observer located at the tip of $\mathbf{M}_O$. Another way of defining the sense of $\mathbf{M}_O$ is furnished by a variation of the right-hand rule: Close your right hand and hold it so that your fingers are curled in the sense of the rotation that **F** would impart to the rigid body about a fixed axis directed along the line of action of $\mathbf{M}_O$; your thumb will indicate the sense of the moment $\mathbf{M}_O$ (Fig. 3.12*b*).

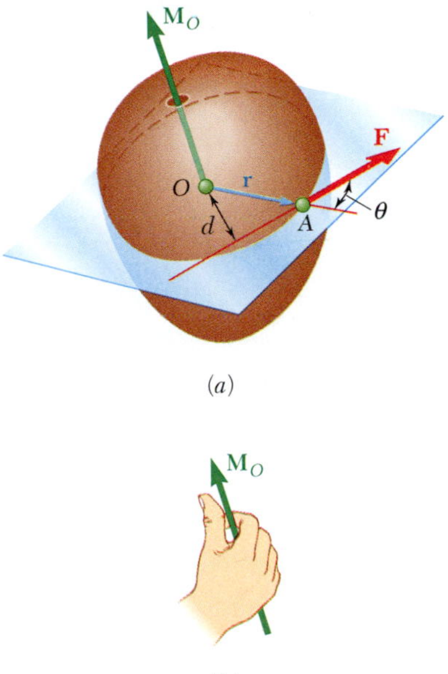

(*a*)

(*b*)

Fig. 3.12

Finally, denoting by θ the angle between the lines of action of the position vector **r** and the force **F**, we find that the magnitude of the moment of **F** about *O* is

$$M_O = rF \sin \theta = Fd \tag{3.12}$$

where *d* represents the perpendicular distance from *O* to the line of action of **F**. Since the tendency of a force **F** to make a rigid body rotate about a fixed axis perpendicular to the force depends upon the distance of **F** from that axis as well as upon the magnitude of **F**, we note that *the magnitude of* $\mathbf{M}_O$ *measures the tendency of the force* **F** *to make the rigid body rotate about a fixed axis directed along* $\mathbf{M}_O$.

In the SI system of units, where a force is expressed in newtons (N) and a distance in meters (m), the moment of a force is expressed in newton-meters (N · m). In the U.S. customary system of units, where a force is expressed in pounds and a distance in feet or inches, the moment of a force is expressed in lb · ft or lb · in.

We can observe that although the moment $\mathbf{M}_O$ of a force about a point depends upon the magnitude, the line of action, and the sense of the force, it does *not* depend upon the actual position of the point of application of the force along its line of action. Conversely, the moment $\mathbf{M}_O$ of a force **F** does not characterize the position of the point of application of **F**.

†We can easily verify that position vectors obey the law of vector addition and, thus, are truly vectors. Consider, for example, the position vectors **r** and **r**′ of *A* with respect to two reference points *O* and *O*′ and the position vector **s** of *O* with respect to *O*′ (Fig. 3.40*a*, Sec. 3.16). We verify that the position vector $\mathbf{r}' = \overrightarrow{O'A}$ can be obtained from the position vectors $\mathbf{s} = \overrightarrow{O'O}$ and $\mathbf{r} = \overrightarrow{OA}$ by applying the triangle rule for the addition of vectors.

However, as it will be seen presently, the moment $\mathbf{M}_O$ of a force $\mathbf{F}$ of given magnitude and direction *completely defines the line of action of* $\mathbf{F}$. Indeed, the line of action of $\mathbf{F}$ must lie in a plane through O perpendicular to the moment $\mathbf{M}_O$; its distance d from O must be equal to the quotient M_O/F of the magnitudes of $\mathbf{M}_O$ and $\mathbf{F}$; and the sense of $\mathbf{M}_O$ determines whether the line of action of $\mathbf{F}$ is to be drawn on one side or the other of the point O.

We recall from Sec. 3.3 that the principle of transmissibility states that two forces $\mathbf{F}$ and $\mathbf{F}'$ are equivalent (that is, have the same effect on a rigid body) if they have the same magnitude, same direction, and same line of action. This principle can now be restated as follows: *Two forces* $\mathbf{F}$ *and* $\mathbf{F}'$ *are equivalent if, and only if, they are equal* (that is, have the same magnitude and same direction) *and have equal moments about a given point O*. The necessary and sufficient conditions for two forces $\mathbf{F}$ and $\mathbf{F}'$ to be equivalent are thus

$$\mathbf{F} = \mathbf{F}' \qquad \text{and} \qquad \mathbf{M}_O = \mathbf{M}'_O \tag{3.13}$$

We should observe that it follows from this statement that if the relations (3.13) hold for a given point O, they will hold for any other point.

Problems Involving Only Two Dimensions. Many applications deal with two-dimensional structures, that is, structures which have length and breadth but only negligible depth and which are subjected to forces contained in the plane of the structure. Two-dimensional structures and the forces acting on them can be readily represented on a sheet of paper or on a blackboard. Their analysis is therefore considerably simpler than that of three-dimensional structures and forces.

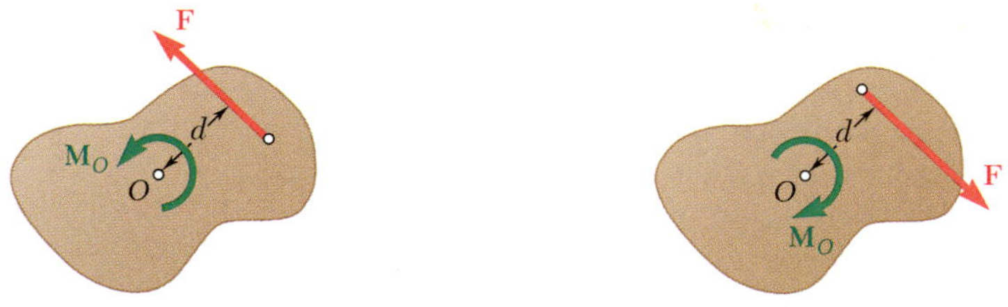

(a) $\mathbf{M}_O = +Fd$ (b) $\mathbf{M}_O = -Fd$

Fig. 3.13

Consider, for example, a rigid slab acted upon by a force $\mathbf{F}$ (Fig. 3.13). The moment of $\mathbf{F}$ about a point O chosen in the plane of the figure is represented by a vector $\mathbf{M}_O$ perpendicular to that plane and of magnitude Fd. In the case of Fig. 3.13*a* the vector $\mathbf{M}_O$ points *out of* the paper, while in the case of Fig. 3.13*b* it points *into* the paper. As we look at the figure, we observe in the first case that $\mathbf{F}$ tends to rotate the slab counterclockwise and in the second case that it tends to rotate the slab clockwise. Therefore, it is natural to refer to the sense of the moment of $\mathbf{F}$ about O in Fig. 3.13*a* as counterclockwise ↺, and in Fig. 3.13*b* as clockwise ↻.

Since the moment of a force $\mathbf{F}$ acting in the plane of the figure must be perpendicular to that plane, we need only specify the *magnitude* and the *sense* of the moment of $\mathbf{F}$ about O. This can be done by assigning to the magnitude M_O of the moment a positive or negative sign according to whether the vector $\mathbf{M}_O$ points out of or into the paper.

3.7. VARIGNON'S THEOREM

The distributive property of vector products can be used to determine the moment of the resultant of several *concurrent forces*. If several forces $\mathbf{F}_1$, $\mathbf{F}_2$, . . . are applied at the same point A (Fig. 3.14), and if we denote by $\mathbf{r}$ the position vector of A, it follows immediately from Eq. (3.5) of Sec. 3.4 that

$$\mathbf{r} \times (\mathbf{F}_1 + \mathbf{F}_2 + \cdots) = \mathbf{r} \times \mathbf{F}_1 + \mathbf{r} \times \mathbf{F}_2 + \cdots \qquad (3.14)$$

In words, *the moment about a given point O of the resultant of several concurrent forces is equal to the sum of the moments of the various forces about the same point O.* This property, which was originally established by the French mathematician Pierre Varignon (1654–1722) long before the introduction of vector algebra, is known as *Varignon's theorem.*

The relation (3.14) makes it possible to replace the direct determination of the moment of a force $\mathbf{F}$ by the determination of the moments of two or more component forces. As you will see in the next section, $\mathbf{F}$ will generally be resolved into components parallel to the coordinate axes. However, it may be more expeditious in some instances to resolve $\mathbf{F}$ into components which are not parallel to the coordinate axes (see Sample Prob. 3.3).

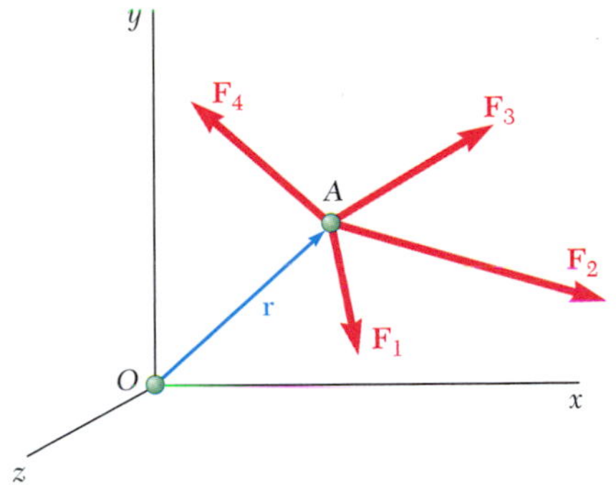

Fig. 3.14

3.8. RECTANGULAR COMPONENTS OF THE MOMENT OF A FORCE

In general, the determination of the moment of a force in space will be considerably simplified if the force and the position vector of its point of application are resolved into rectangular x, y, and z components. Consider, for example, the moment $\mathbf{M}_O$ about O of a force $\mathbf{F}$ whose components are F_x, F_y, and F_z and which is applied at a point A of coordinates x, y, and z (Fig. 3.15). Observing that the components of the position vector $\mathbf{r}$ are respectively equal to the coordinates x, y, and z of the point A, we write

$$\mathbf{r} = x\mathbf{i} + y\mathbf{j} + z\mathbf{k} \qquad (3.15)$$
$$\mathbf{F} = F_x\mathbf{i} + F_y\mathbf{j} + F_z\mathbf{k} \qquad (3.16)$$

Substituting for $\mathbf{r}$ and $\mathbf{F}$ from (3.15) and (3.16) into

$$\mathbf{M}_O = \mathbf{r} \times \mathbf{F} \qquad (3.11)$$

and recalling the results obtained in Sec. 3.5, we write the moment $\mathbf{M}_O$ of $\mathbf{F}$ about O in the form

$$\mathbf{M}_O = M_x\mathbf{i} + M_y\mathbf{j} + M_z\mathbf{k} \qquad (3.17)$$

where the components M_x, M_y, and M_z are defined by the relations

$$\begin{aligned} M_x &= yF_z - zF_y \\ M_y &= zF_x - xF_z \\ M_z &= xF_y - yF_x \end{aligned} \qquad (3.18)$$

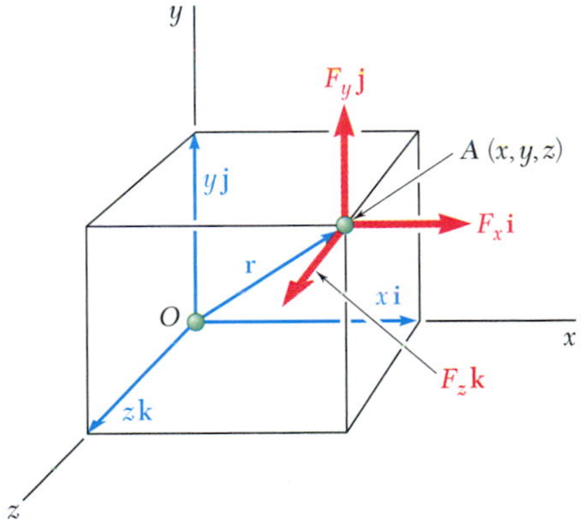

Fig. 3.15

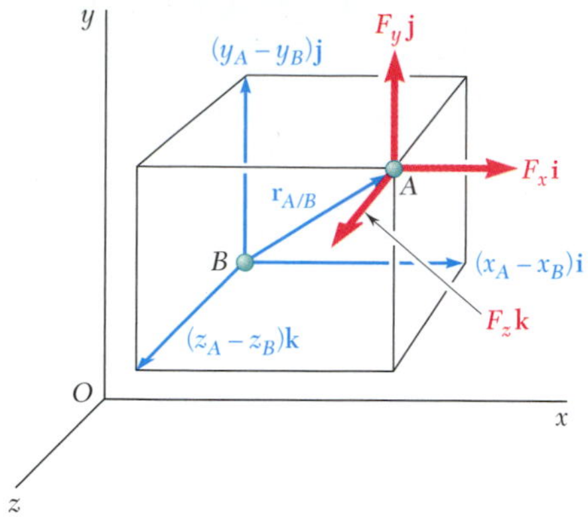

Fig 3.16

As you will see in Sec. 3.11, the scalar components M_x, M_y, and M_z of the moment $\mathbf{M}_O$ measure the tendency of the force $\mathbf{F}$ to impart to a rigid body a motion of rotation about the x, y, and z axes, respectively. Substituting from (3.18) into (3.17), we can also write $\mathbf{M}_O$ in the form of the determinant

$$\mathbf{M}_O = \begin{vmatrix} \mathbf{i} & \mathbf{j} & \mathbf{k} \\ x & y & z \\ F_x & F_y & F_z \end{vmatrix} \tag{3.19}$$

To compute the moment $\mathbf{M}_B$ about an arbitrary point B of a force $\mathbf{F}$ applied at A (Fig. 3.16), we must replace the position vector $\mathbf{r}$ in Eq. (3.11) by a vector drawn from B to A. This vector is the *position vector of A relative to B* and will be denoted by $\mathbf{r}_{A/B}$. Observing that $\mathbf{r}_{A/B}$ can be obtained by subtracting $\mathbf{r}_B$ from $\mathbf{r}_A$, we write

$$\mathbf{M}_B = \mathbf{r}_{A/B} \times \mathbf{F} = (\mathbf{r}_A - \mathbf{r}_B) \times \mathbf{F} \tag{3.20}$$

or, using the determinant form,

$$\mathbf{M}_B = \begin{vmatrix} \mathbf{i} & \mathbf{j} & \mathbf{k} \\ x_{A/B} & y_{A/B} & z_{A/B} \\ F_x & F_y & F_z \end{vmatrix} \tag{3.21}$$

where $x_{A/B}$, $y_{A/B}$, and $z_{A/B}$ denote the components of the vector $\mathbf{r}_{A/B}$:

$$x_{A/B} = x_A - x_B \qquad y_{A/B} = y_A - y_B \qquad z_{A/B} = z_A - z_B$$

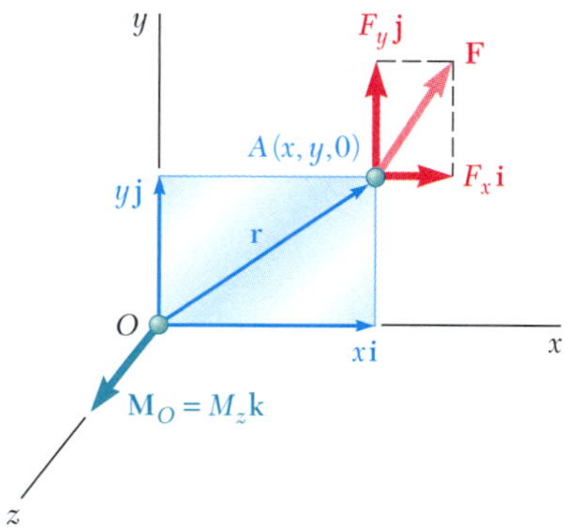

Fig. 3.17

In the case of *problems involving only two dimensions,* the force $\mathbf{F}$ can be assumed to lie in the xy plane (Fig. 3.17). Setting $z = 0$ and $F_z = 0$ in Eq. (3.19), we obtain

$$\mathbf{M}_O = (xF_y - yF_x)\mathbf{k}$$

We verify that the moment of $\mathbf{F}$ about O is perpendicular to the plane of the figure and that it is completely defined by the scalar

$$M_O = M_z = xF_y - yF_x \tag{3.22}$$

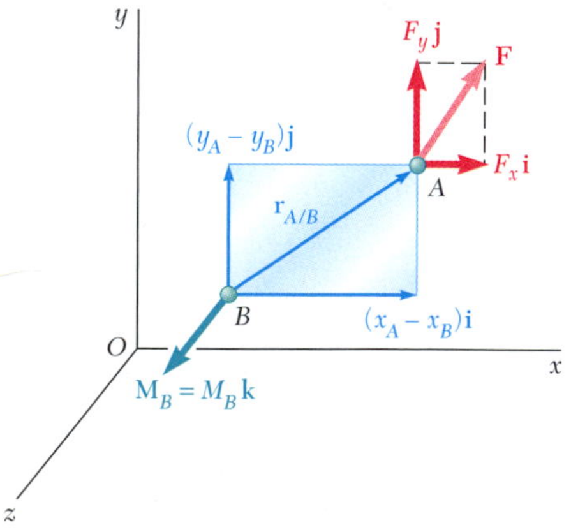

Fig. 3.18

As noted earlier, a positive value for M_O indicates that the vector $\mathbf{M}_O$ points out of the paper (the force $\mathbf{F}$ tends to rotate the body counterclockwise about O), and a negative value indicates that the vector $\mathbf{M}_O$ points into the paper (the force $\mathbf{F}$ tends to rotate the body clockwise about O).

To compute the moment about $B(x_B, y_B)$ of a force lying in the xy plane and applied at $A(x_A, y_A)$ (Fig. 3.18), we set $z_{A/B} = 0$ and $F_z = 0$ in the relations (3.21) and note that the vector $\mathbf{M}_B$ is perpendicular to the xy plane and is defined in magnitude and sense by the scalar

$$M_B = (x_A - x_B)F_y - (y_A - y_B)F_x \tag{3.23}$$

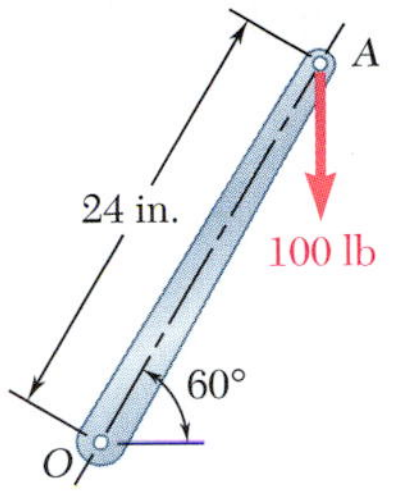

SAMPLE PROBLEM 3.1

A 100-lb vertical force is applied to the end of a lever which is attached to a shaft at O. Determine (a) the moment of the 100-lb force about O; (b) the horizontal force applied at A which creates the same moment about O; (c) the smallest force applied at A which creates the same moment about O; (d) how far from the shaft a 240-lb vertical force must act to create the same moment about O; (e) whether any one of the forces obtained in parts b, c, and d is equivalent to the original force.

SOLUTION

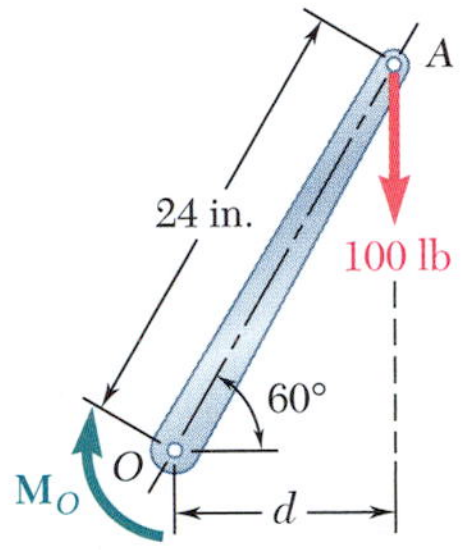

***a.* Moment about *O*.** The perpendicular distance from O to the line of action of the 100-lb force is

$$d = (24 \text{ in.}) \cos 60° = 12 \text{ in.}$$

The magnitude of the moment about O of the 100-lb force is

$$M_O = Fd = (100 \text{ lb})(12 \text{ in.}) = 1200 \text{ lb} \cdot \text{in.}$$

Since the force tends to rotate the lever clockwise about O, the moment will be represented by a vector $\mathbf{M}_O$ perpendicular to the plane of the figure and pointing *into* the paper. We express this fact by writing

$$\mathbf{M}_O = 1200 \text{ lb} \cdot \text{in. } \circlearrowright \quad ◀$$

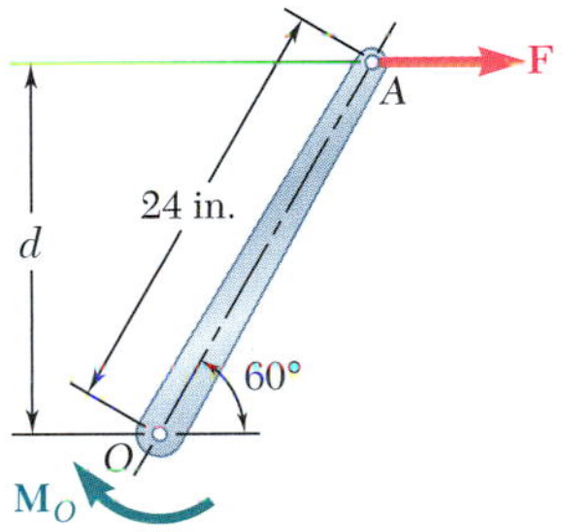

***b.* Horizontal Force.** In this case, we have

$$d = (24 \text{ in.}) \sin 60° = 20.8 \text{ in.}$$

Since the moment about O must be 1200 lb · in., we write

$$M_O = Fd$$
$$1200 \text{ lb} \cdot \text{in.} = F(20.8 \text{ in.})$$
$$F = 57.7 \text{ lb} \qquad \mathbf{F} = 57.7 \text{ lb} \rightarrow \quad ◀$$

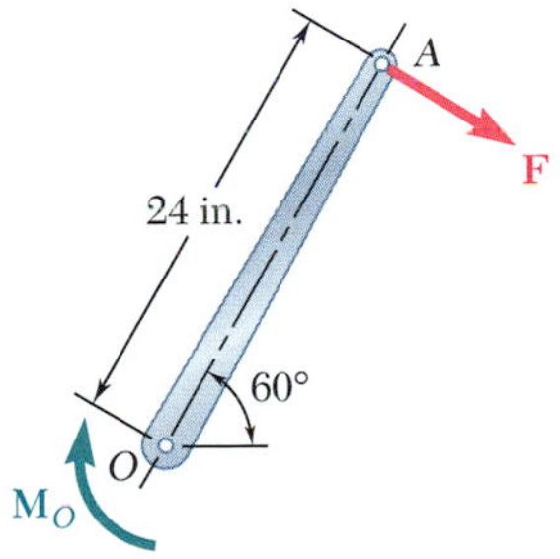

***c.* Smallest Force.** Since $M_O = Fd$, the smallest value of F occurs when d is maximum. We choose the force perpendicular to OA and note that $d = 24$ in.; thus

$$M_O = Fd$$
$$1200 \text{ lb} \cdot \text{in.} = F(24 \text{ in.})$$
$$F = 50 \text{ lb} \qquad \mathbf{F} = 50 \text{ lb} \measuredangle 30° \quad ◀$$

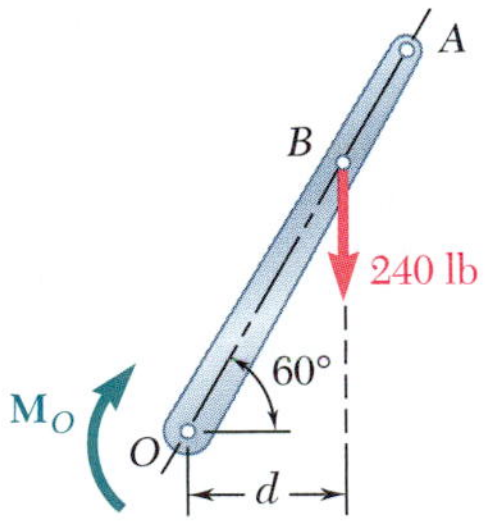

***d.* 240-lb Vertical Force.** In this case $M_O = Fd$ yields

$$1200 \text{ lb} \cdot \text{in.} = (240 \text{ lb})d \qquad d = 5 \text{ in.}$$

but

$$OB \cos 60° = d \qquad OB = 10 \text{ in.} \quad ◀$$

e. None of the forces considered in parts b, c, and d is equivalent to the original 100-lb force. Although they have the same moment about O, they have different x and y components. In other words, although each force tends to rotate the shaft in the same manner, each causes the lever to pull on the shaft in a different way.

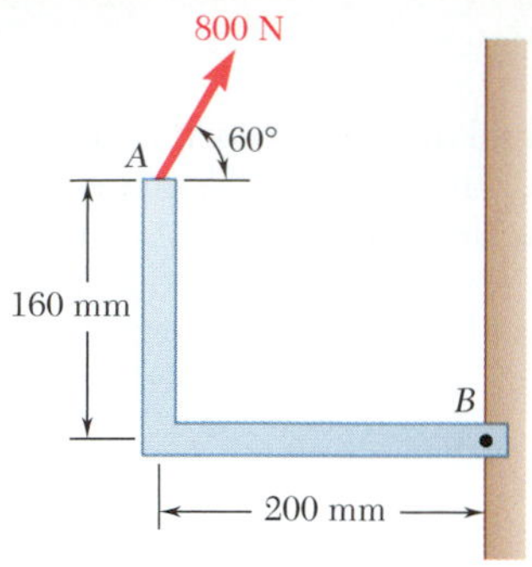

SAMPLE PROBLEM 3.2

A force of 800 N acts on a bracket as shown. Determine the moment of the force about B.

SOLUTION

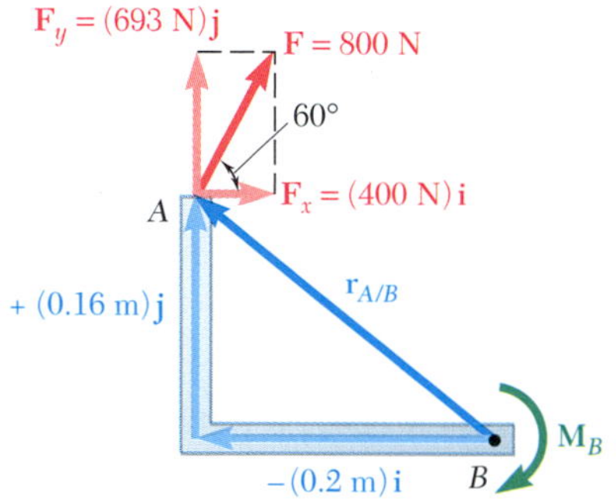

The moment $\mathbf{M}_B$ of the force $\mathbf{F}$ about B is obtained by forming the vector product

$$\mathbf{M}_B = \mathbf{r}_{A/B} \times \mathbf{F}$$

where $\mathbf{r}_{A/B}$ is the vector drawn from B to A. Resolving $\mathbf{r}_{A/B}$ and $\mathbf{F}$ into rectangular components, we have

$$\begin{aligned}\mathbf{r}_{A/B} &= -(0.2 \text{ m})\mathbf{i} + (0.16 \text{ m})\mathbf{j} \\ \mathbf{F} &= (800 \text{ N}) \cos 60°\mathbf{i} + (800 \text{ N}) \sin 60°\mathbf{j} \\ &= (400 \text{ N})\mathbf{i} + (693 \text{ N})\mathbf{j}\end{aligned}$$

Recalling the relations (3.7) for the vector products of unit vectors (Sec. 3.5), we obtain

$$\begin{aligned}\mathbf{M}_B = \mathbf{r}_{A/B} \times \mathbf{F} &= [-(0.2 \text{ m})\mathbf{i} + (0.16 \text{ m})\mathbf{j}] \times [(400 \text{ N})\mathbf{i} + (693 \text{ N})\mathbf{j}] \\ &= -(138.6 \text{ N} \cdot \text{m})\mathbf{k} - (64.0 \text{ N} \cdot \text{m})\mathbf{k} \\ &= -(202.6 \text{ N} \cdot \text{m})\mathbf{k}\end{aligned}$$

$$\mathbf{M}_B = 203 \text{ N} \cdot \text{m} \curvearrowright \quad ◀$$

The moment $\mathbf{M}_B$ is a vector perpendicular to the plane of the figure and pointing *into* the paper.

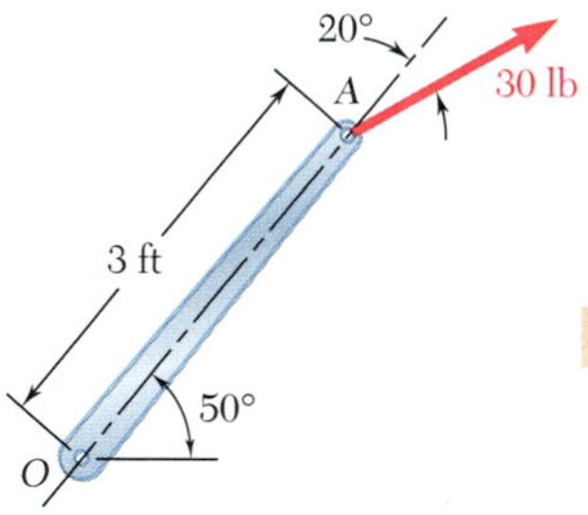

SAMPLE PROBLEM 3.3

A 30-lb force acts on the end of the 3-ft lever as shown. Determine the moment of the force about O.

SOLUTION

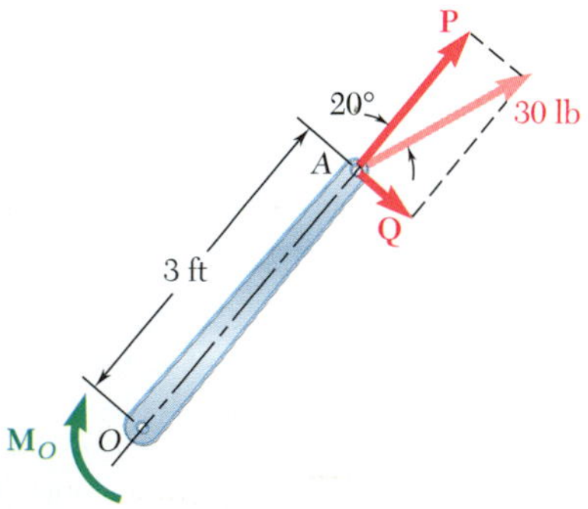

The force is replaced by two components, one component $\mathbf{P}$ in the direction of OA and one component $\mathbf{Q}$ perpendicular to OA. Since O is on the line of action of $\mathbf{P}$, the moment of $\mathbf{P}$ about O is zero and the moment of the 30-lb force reduces to the moment of $\mathbf{Q}$, which is clockwise and, thus, is represented by a negative scalar.

$$\begin{aligned}Q &= (30 \text{ lb}) \sin 20° = 10.26 \text{ lb} \\ M_O &= -Q(3 \text{ ft}) = -(10.26 \text{ lb})(3 \text{ ft}) = -30.8 \text{ lb} \cdot \text{ft}\end{aligned}$$

Since the value obtained for the scalar M_O is negative, the moment $\mathbf{M}_O$ points *into* the paper. We write

$$\mathbf{M}_O = 30.8 \text{ lb} \cdot \text{ft} \curvearrowright \quad ◀$$

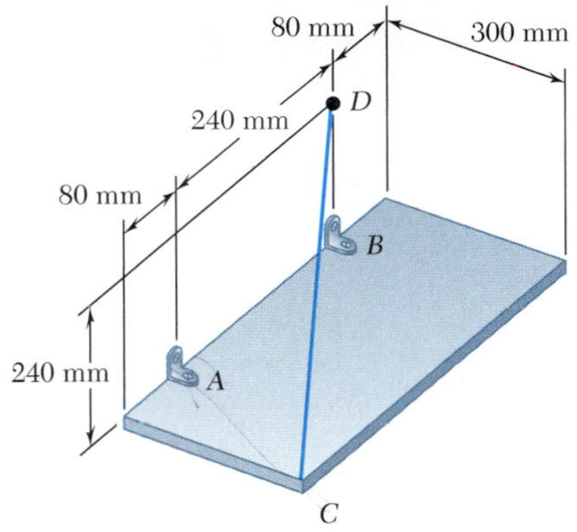

SAMPLE PROBLEM 3.4

A rectangular plate is supported by brackets at A and B and by a wire CD. Knowing that the tension in the wire is 200 N, determine the moment about A of the force exerted by the wire on point C.

SOLUTION

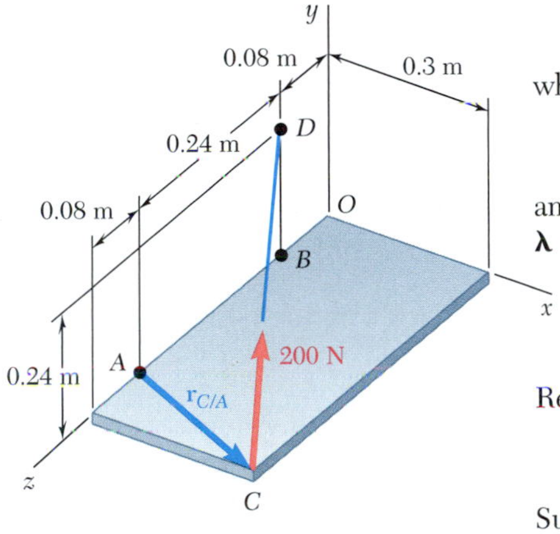

The moment $\mathbf{M}_A$ about A of the force $\mathbf{F}$ exerted by the wire on point C is obtained by forming the vector product

$$\mathbf{M}_A = \mathbf{r}_{C/A} \times \mathbf{F} \tag{1}$$

where $\mathbf{r}_{C/A}$ is the vector drawn from A to C,

$$\mathbf{r}_{C/A} = \overrightarrow{AC} = (0.3\text{ m})\mathbf{i} + (0.08\text{ m})\mathbf{k} \tag{2}$$

and $\mathbf{F}$ is the 200-N force directed along CD. Introducing the unit vector $\boldsymbol{\lambda} = \overrightarrow{CD}/CD$, we write

$$\mathbf{F} = F\boldsymbol{\lambda} = (200\text{ N})\frac{\overrightarrow{CD}}{CD} \tag{3}$$

Resolving the vector $\overrightarrow{CD}$ into rectangular components, we have

$$\overrightarrow{CD} = -(0.3\text{ m})\mathbf{i} + (0.24\text{ m})\mathbf{j} - (0.32\text{ m})\mathbf{k} \qquad CD = 0.50\text{ m}$$

Substituting into (3), we obtain

$$\begin{aligned} \mathbf{F} &= \frac{200\text{ N}}{0.50\text{ m}}[-(0.3\text{ m})\mathbf{i} + (0.24\text{ m})\mathbf{j} - (0.32\text{ m})\mathbf{k}] \\ &= -(120\text{ N})\mathbf{i} + (96\text{ N})\mathbf{j} - (128\text{ N})\mathbf{k} \end{aligned} \tag{4}$$

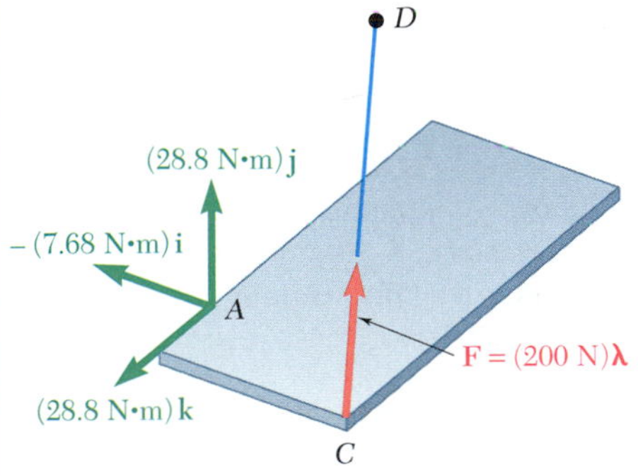

Substituting for $\mathbf{r}_{C/A}$ and $\mathbf{F}$ from (2) and (4) into (1) and recalling the relations (3.7) of Sec. 3.5, we obtain

$$\begin{aligned} \mathbf{M}_A = \mathbf{r}_{C/A} \times \mathbf{F} &= (0.3\mathbf{i} + 0.08\mathbf{k}) \times (-120\mathbf{i} + 96\mathbf{j} - 128\mathbf{k}) \\ &= (0.3)(96)\mathbf{k} + (0.3)(-128)(-\mathbf{j}) + (0.08)(-120)\mathbf{j} + (0.08)(96)(-\mathbf{i}) \end{aligned}$$

$$\mathbf{M}_A = -(7.68\text{ N}\cdot\text{m})\mathbf{i} + (28.8\text{ N}\cdot\text{m})\mathbf{j} + (28.8\text{ N}\cdot\text{m})\mathbf{k} \quad ◀$$

Alternative Solution. As indicated in Sec. 3.8, the moment $\mathbf{M}_A$ can be expressed in the form of a determinant:

$$\mathbf{M}_A = \begin{vmatrix} \mathbf{i} & \mathbf{j} & \mathbf{k} \\ x_C - x_A & y_C - y_A & z_C - z_A \\ F_x & F_y & F_z \end{vmatrix} = \begin{vmatrix} \mathbf{i} & \mathbf{j} & \mathbf{k} \\ 0.3 & 0 & 0.08 \\ -120 & 96 & -128 \end{vmatrix}$$

$$\mathbf{M}_A = -(7.68\text{ N}\cdot\text{m})\mathbf{i} + (28.8\text{ N}\cdot\text{m})\mathbf{j} + (28.8\text{ N}\cdot\text{m})\mathbf{k} \quad ◀$$

SOLVING PROBLEMS ON YOUR OWN

In this lesson we introduced the *vector product* or *cross product* of two vectors. In the following problems, you will use the vector product to compute the *moment of a force about a point* and also to determine the *perpendicular distance* from a point to a line.

We defined the moment of the force **F** about the point O of a rigid body as

$$\mathbf{M}_O = \mathbf{r} \times \mathbf{F} \tag{3.11}$$

where **r** is the position vector *from O to any point* on the line of action of **F**. Since the vector product is not commutative, it is absolutely necessary when computing such a product that you place the vectors in the proper order and that each vector have the correct sense. The moment $\mathbf{M}_O$ is important because its magnitude is a measure of the tendency of the force **F** to cause the rigid body to rotate about an axis directed along $\mathbf{M}_O$.

1. Computing the moment $\mathbf{M}_O$ of a force in two dimensions. You can use one of the following procedures:

a. Use Eq. (3.12), $M_O = Fd$, which expresses the magnitude of the moment as the product of the magnitude of **F** and the *perpendicular distance* d from O to the line of action of **F** (Sample Prob. 3.1).

b. Express **r** and **F** in component form and formally evaluate the vector product $\mathbf{M}_O = \mathbf{r} \times \mathbf{F}$ [Sample Prob. 3.2].

c. Resolve **F** into components respectively parallel and perpendicular to the position vector **r**. Only the perpendicular component contributes to the moment of **F** [Sample Prob. 3.3].

d. Use Eq. (3.22), $M_O = M_z = xF_y - yF_x$. When applying this method, the simplest approach is to treat the scalar components of **r** and **F** as positive and then to assign, by observation, the proper sign to the moment produced by each force component. For example, applying this method to solve Sample Prob. 3.2, we observe that both force components tend to produce a clockwise rotation about B. Therefore, the moment of each force about B should be represented by a negative scalar. We then have for the total moment

$$M_B = -(0.16 \text{ m})(400 \text{ N}) - (0.20 \text{ m})(693 \text{ N}) = -202.6 \text{ N} \cdot \text{m}$$

2. Computing the moment $\mathbf{M}_O$ of a force F in three dimensions. Following the method of Sample Prob. 3.4, the first step in the process is to select the most convenient (simplest) position vector **r**. You should next express **F** in terms of its rectangular components. The final step is to evaluate the vector product $\mathbf{r} \times \mathbf{F}$ to determine the moment. In most three-dimensional problems you will find it easiest to calculate the vector product using a determinant.

3. Determining the perpendicular distance *d* from a point *A* to a given line. First assume that a force **F** of known magnitude F lies along the given line. Next determine its moment about A by forming the vector product $\mathbf{M}_A = \mathbf{r} \times \mathbf{F}$, and calculate this product as indicated above. Then compute its magnitude M_A. Finally, substitute the values of F and M_A into the equation $M_A = Fd$ and solve for d.

Problems

3.1 A 13.2-N force **P** is applied to the lever which controls the auger of a snowblower. Determine the moment of **P** about A when α is equal to 30°.

3.2 The force **P** is applied to the lever which controls the auger of a snowblower. Determine the magnitude and the direction of the smallest force **P** which has a 2.20-N · m counterclockwise moment about A.

3.3 A 13.1-N force **P** is applied to the lever which controls the auger of a snowblower. Determine the value of α knowing that the moment of **P** about A is counterclockwise and has a magnitude of 1.95 N · m.

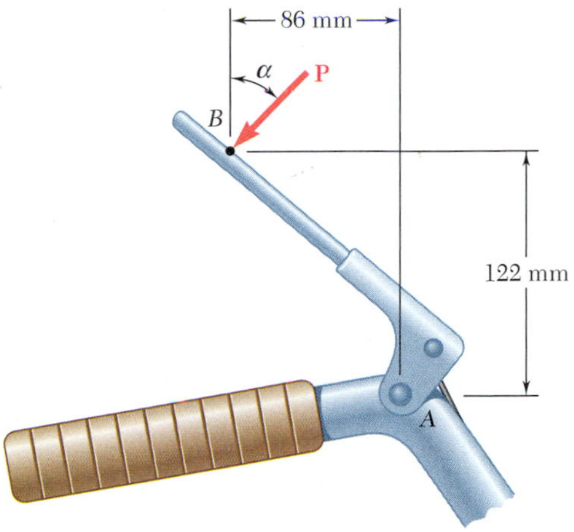

Fig. P3.1, P3.2, and P3.3

3.4 A foot valve for a pneumatic system is hinged at B. Knowing that $\alpha = 28°$, determine the moment of the 4-lb force about point B by resolving the force into horizontal and vertical components.

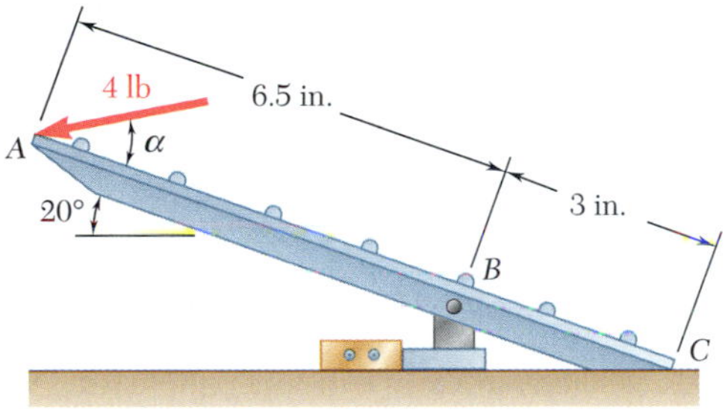

Fig. P3.4 and *P3.5*

3.5 A foot valve for a pneumatic system is hinged at B. Knowing that $\alpha = 28°$, determine the moment of the 4-lb force about point B by resolving the force into components along ABC and in a direction perpendicular to ABC.

3.6 It is known that a vertical force of 800 N is required to remove the nail at C from the board. As the nail first starts moving, determine (*a*) the moment about B of the force exerted on the nail, (*b*) the magnitude of the force **P** which creates the same moment about B if $\alpha = 10°$, (*c*) the smallest force **P** which creates the same moment about B.

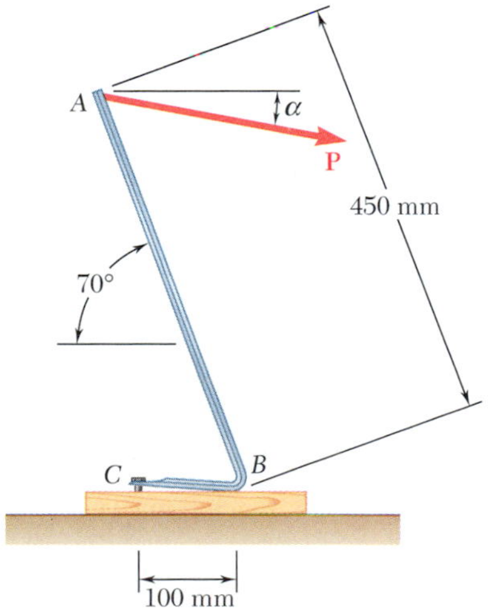

Fig. P3.6

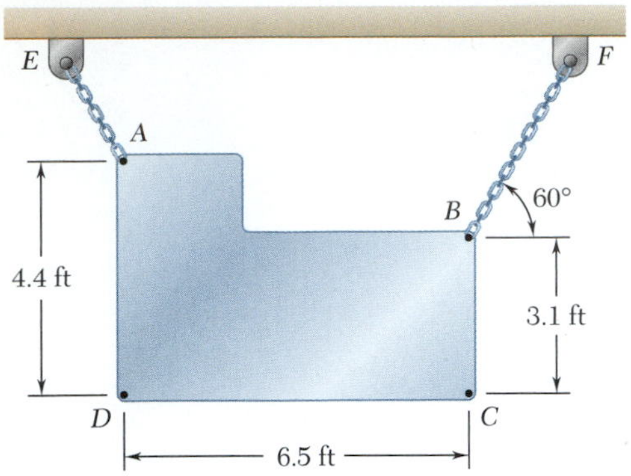

Fig. P3.7 and *P3.8*

3.7 A sign is suspended from two chains *AE* and *BF*. Knowing that the tension in *BF* is 45 lb, determine (*a*) the moment about *A* of the force exerted by the chain at *B*, (*b*) the smallest force applied at *C* which creates the same moment about *A*.

3.8 A sign is suspended from two chains *AE* and *BF*. Knowing that the tension in *BF* is 45 lb, determine (*a*) the moment about *A* of the force exerted by the chain at *B*, (*b*) the magnitude and sense of the vertical force applied at *C* which creates the same moment about *A*, (*c*) the smallest force applied at *B* which creates the same moment about *A*.

3.9 and 3.10 The tailgate of a car is supported by the hydraulic lift *BC*. If the lift exerts a 125-N force directed along its center line on the ball and socket at *B*, determine the moment of the force about *A*.

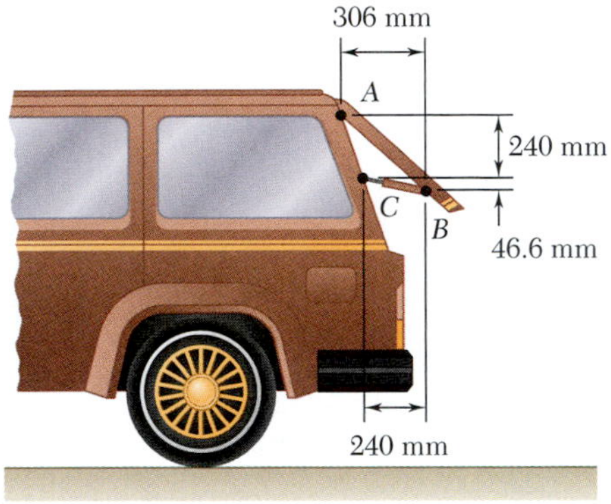

Fig. P3.9

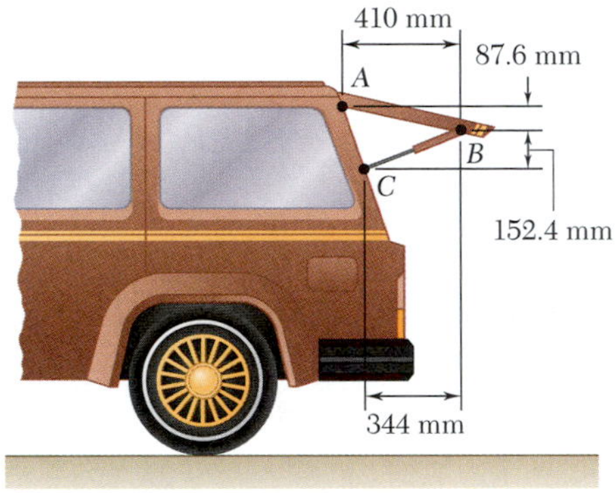

Fig. P3.10

3.11 A winch puller *AB* is used to straighten a fence post. Knowing that the tension in cable *BC* is 260 lb, length *a* is 8 in., length *b* is 35 in., and length *d* is 76 in., determine the moment about *D* of the force exerted by the cable at *C* by resolving that force into horizontal and vertical components applied (*a*) at point *C*, (*b*) at point *E*.

3.12 It is known that a force with a moment of 7840 lb · in. about *D* is required to straighten the fence post *CD*. If $a = 8$ in., $b = 35$ in., and $d = 112$ in., determine the tension that must be developed in the cable of winch puller *AB* to create the required moment about point *D*.

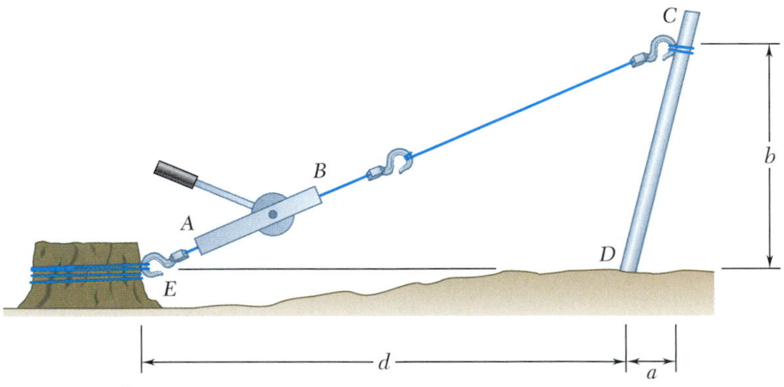

Fig. P3.11, P3.12, and *P3.13*

3.13 It is known that a force with a moment of 1152 N · m about *D* is required to straighten the fence post *CD*. If the capacity of the winch puller *AB* is 2880 N, determine the minimum value of distance *d* to create the specified moment about point *D* knowing that $a = 0.24$ m and $b = 1.05$ m.

3.14 A mechanic uses a piece of pipe AB as a lever when tightening an alternator belt. When he pushes down at A, a force of 580 N is exerted on the alternator B. Determine the moment of that force about bolt C if its line of action passes through O.

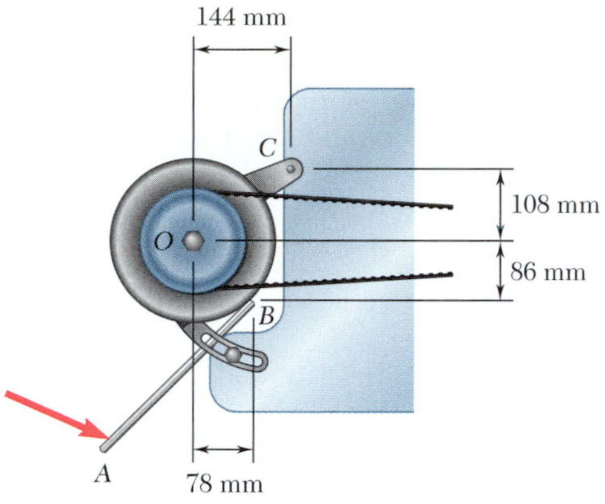

Fig. *P3.14*

3.15 Form the vector products $\mathbf{B} \times \mathbf{C}$ and $\mathbf{B}' \times \mathbf{C}$, where $B = B'$, and use the results obtained to prove the identity

$$\sin \alpha \cos \beta = \tfrac{1}{2} \sin (\alpha + \beta) + \tfrac{1}{2} \sin (\alpha - \beta).$$

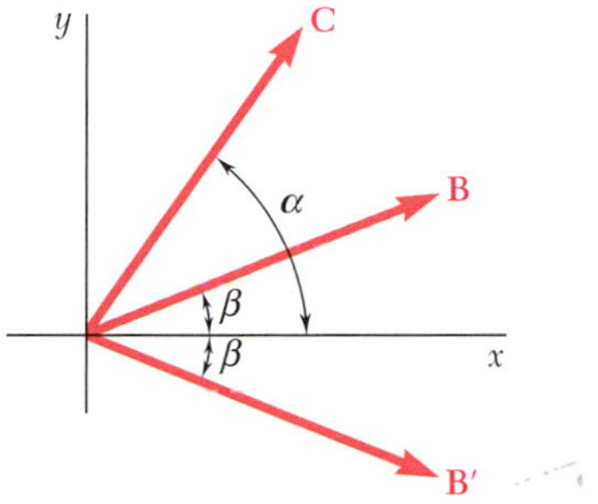

Fig. P3.15

3.16 A line passes through the points (420 mm, −150 mm) and (−140 mm, 180 mm). Determine the perpendicular distance d from the line to the origin O of the system of coordinates.

3.17 A plane contains the vectors **A** and **B**. Determine the unit vector normal to the plane when **A** and **B** are equal to, respectively, (*a*) $4\mathbf{i} - 2\mathbf{j} + 3\mathbf{k}$ and $-2\mathbf{i} + 6\mathbf{j} - 5\mathbf{k}$, (*b*) $7\mathbf{i} + \mathbf{j} - 4\mathbf{k}$ and $-6\mathbf{i} - 3\mathbf{j} + 2\mathbf{k}$.

3.18 The vectors **P** and **Q** are two adjacent sides of a parallelogram. Determine the area of the parallelogram when (*a*) $\mathbf{P} = (8 \text{ in.})\mathbf{i} + (2 \text{ in.})\mathbf{j} - (1 \text{ in.})\mathbf{k}$ and $\mathbf{Q} = -(3 \text{ in.})\mathbf{i} + (4 \text{ in.})\mathbf{j} + (2 \text{ in.})\mathbf{k}$, (*b*) $\mathbf{P} = -(3 \text{ in.})\mathbf{i} + (6 \text{ in.})\mathbf{j} + (4 \text{ in.})\mathbf{k}$ and $\mathbf{Q} = (2 \text{ in.})\mathbf{i} + (5 \text{ in.})\mathbf{j} - (3 \text{ in.})\mathbf{k}$.

3.19 Determine the moment about the origin O of the force $\mathbf{F} = -(5 \text{ N})\mathbf{i} - (2 \text{ N})\mathbf{j} + (3 \text{ N})\mathbf{k}$ which acts at a point A. Assume that the position vector of A is (*a*) $\mathbf{r} = (4 \text{ m})\mathbf{i} - (2 \text{ m})\mathbf{j} - (1 \text{ m})\mathbf{k}$, (*b*) $\mathbf{r} = -(8 \text{ m})\mathbf{i} + (3 \text{ m})\mathbf{j} + (4 \text{ m})\mathbf{k}$, (*c*) $\mathbf{r} = (7.5 \text{ m})\mathbf{i} + (3 \text{ m})\mathbf{j} - (4.5 \text{ m})\mathbf{k}$.

3.20 Determine the moment about the origin O of the force $\mathbf{F} = -(1.5 \text{ lb})\mathbf{i} + (3 \text{ lb})\mathbf{j} - (2 \text{ lb})\mathbf{k}$ which acts at a point A. Assume that the position vector of A is (*a*) $\mathbf{r} = (2.5 \text{ ft})\mathbf{i} - (1 \text{ ft})\mathbf{j} + (2 \text{ ft})\mathbf{k}$, (*b*) $\mathbf{r} = (4.5 \text{ ft})\mathbf{i} - (9 \text{ ft})\mathbf{j} + (6 \text{ ft})\mathbf{k}$, (*c*) $\mathbf{r} = (4 \text{ ft})\mathbf{i} - (1 \text{ ft})\mathbf{j} + (7 \text{ ft})\mathbf{k}$.

3.21 Before the trunk of a large tree is felled, cables AB and BC are attached as shown. Knowing that the tension in cables AB and BC are 777 N and 990 N, respectively, determine the moment about O of the resultant force exerted on the tree by the cables at B.

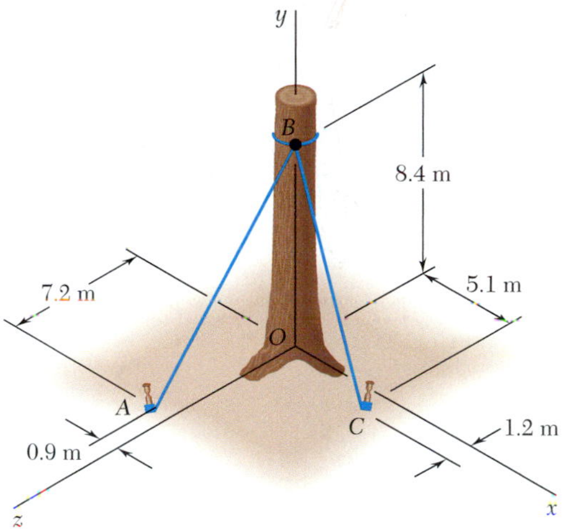

Fig. P3.21

3.22 Before a telephone cable is strung, rope BAC is tied to a stake at B and is passed over a pulley at A. Knowing that portion AC of the rope lies in a plane parallel to the xy plane and that the tension **T** in the rope is 124 N, determine the moment about O of the resultant force exerted on the pulley by the rope.

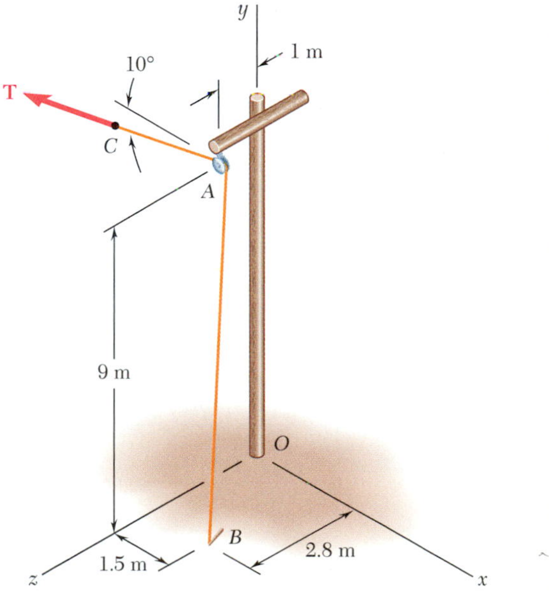

Fig. *P3.22*

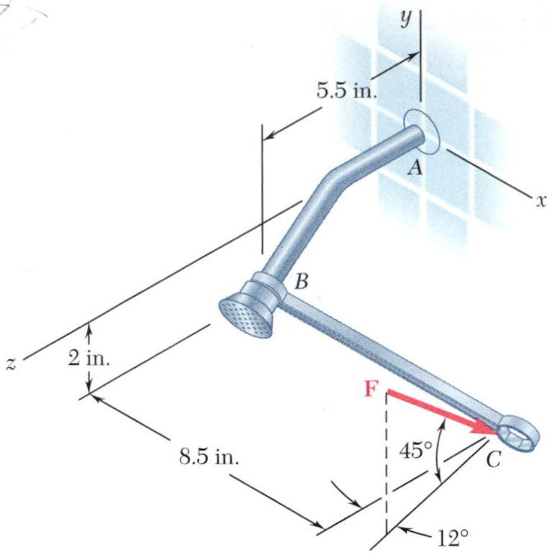

Fig. P3.23

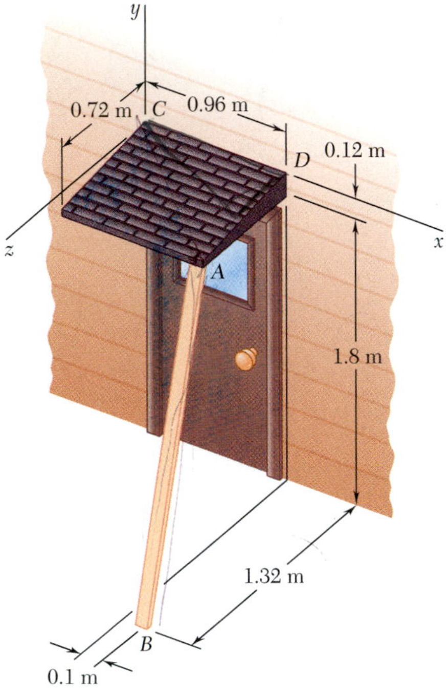

Fig. P3.24

3.23 An 8-lb force is applied to a wrench to tighten a showerhead. Knowing that the centerline of the wrench is parallel to the x axis, determine the moment of the force about A.

3.24 A wooden board AB, which is used as a temporary prop to support a small roof, exerts at point A of the roof a 228 N force directed along BA. Determine the moment about C of that force.

3.25 The ramp $ABCD$ is supported by cables at corners C and D. The tension in each of the cables is 360 lb. Determine the moment about A of the force exerted by (a) the cable at D, (b) the cable at C.

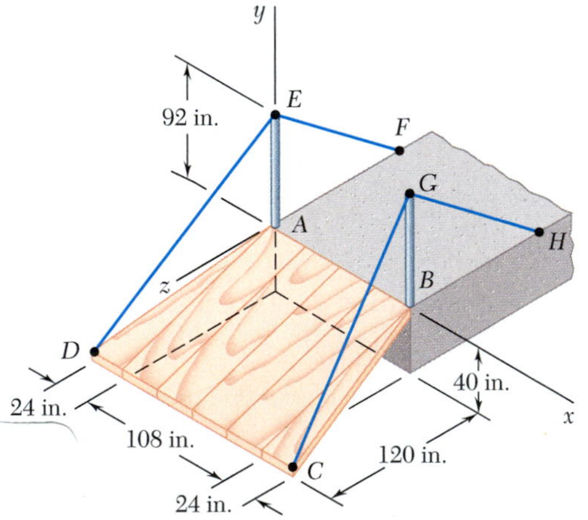

Fig. P3.25

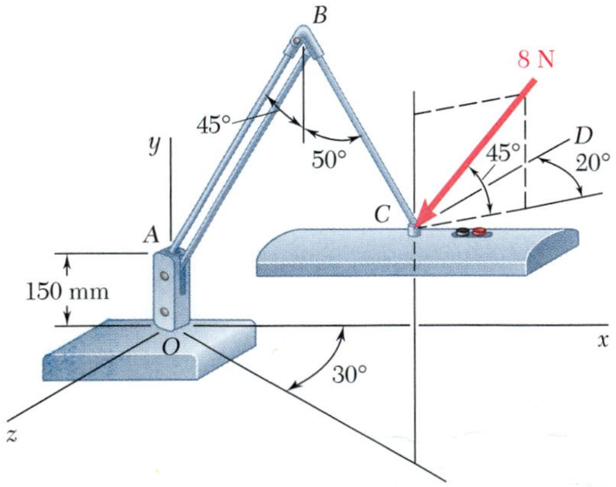

Fig. P3.26

3.26 The arms AB and BC of a desk lamp lie in a vertical plane that forms an angle of 30° with the xy plane. To reposition the light, a force of magnitude 8 N is applied at C as shown. Determine the moment of the force about O knowing that $AB = 450$ mm, $BC = 325$ mm, and line CD is parallel to the z axis.

3.27 In Prob. 3.21, determine the perpendicular distance from point O to cable AB.

3.28 In Prob. 3.21, determine the perpendicular distance from point O to cable BC.

3.29 In Prob. 3.24, determine the perpendicular distance from point D to a line drawn through points A and B.

3.30 In Prob. 3.24, determine the perpendicular distance from point C to a line drawn through points A and B.

3.31 In Prob. 3.25, determine the perpendicular distance from point A to portion DE of cable DEF.

3.32 In Prob. 3.25, determine the perpendicular distance from point A to a line drawn through points C and G.

3.33 In Prob. 3.25, determine the perpendicular distance from point B to a line drawn through points D and E.

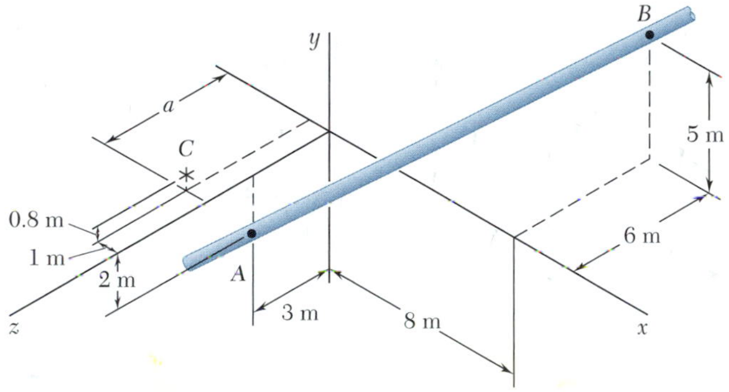

Fig. P3.34

3.34 Determine the value of a which minimizes the perpendicular distance from point C to a section of pipeline that passes through points A and B.

3.9. SCALAR PRODUCT OF TWO VECTORS

The *scalar product* of two vectors **P** and **Q** is defined as the product of the magnitudes of **P** and **Q** and of the cosine of the angle θ formed by **P** and **Q** (Fig. 3.19). The scalar product of **P** and **Q** is denoted by $\mathbf{P} \cdot \mathbf{Q}$. We write therefore

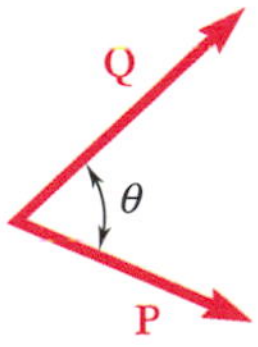

Fig. 3.19

$$\mathbf{P} \cdot \mathbf{Q} = PQ \cos \theta \tag{3.24}$$

Note that the expression just defined is not a vector but a *scalar,* which explains the name *scalar product*; because of the notation used, $\mathbf{P} \cdot \mathbf{Q}$ is also referred to as the *dot product* of the vectors **P** and **Q**.

It follows from its very definition that the scalar product of two vectors is *commutative,* that is, that

$$\mathbf{P} \cdot \mathbf{Q} = \mathbf{Q} \cdot \mathbf{P} \tag{3.25}$$

To prove that the scalar product is also *distributive,* we must prove the relation

$$\mathbf{P} \cdot (\mathbf{Q}_1 + \mathbf{Q}_2) = \mathbf{P} \cdot \mathbf{Q}_1 + \mathbf{P} \cdot \mathbf{Q}_2 \tag{3.26}$$

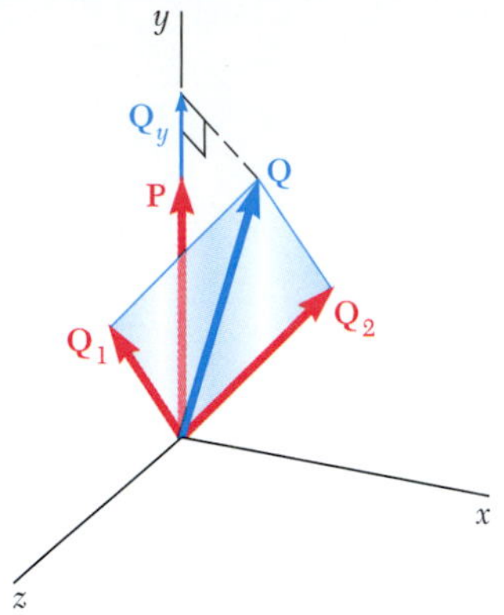

Fig. 3.20

We can, without any loss of generality, assume that **P** is directed along the y axis (Fig. 3.20). Denoting by **Q** the sum of $\mathbf{Q}_1$ and $\mathbf{Q}_2$ and by θ_y the angle **Q** forms with the y axis, we express the left-hand member of (3.26) as follows:

$$\mathbf{P} \cdot (\mathbf{Q}_1 + \mathbf{Q}_2) = \mathbf{P} \cdot \mathbf{Q} = PQ \cos \theta_y = PQ_y \tag{3.27}$$

where Q_y is the y component of **Q**. We can, in a similar way, express the right-hand member of (3.26) as

$$\mathbf{P} \cdot \mathbf{Q}_1 + \mathbf{P} \cdot \mathbf{Q}_2 = P(Q_1)_y + P(Q_2)_y \tag{3.28}$$

Since **Q** is the sum of $\mathbf{Q}_1$ and $\mathbf{Q}_2$, its y component must be equal to the sum of the y components of $\mathbf{Q}_1$ and $\mathbf{Q}_2$. Thus, the expressions obtained in (3.27) and (3.28) are equal, and the relation (3.26) has been proved.

As far as the third property—the associative property—is concerned, we note that this property cannot apply to scalar products. Indeed, $(\mathbf{P} \cdot \mathbf{Q}) \cdot \mathbf{S}$ has no meaning, since $\mathbf{P} \cdot \mathbf{Q}$ is not a vector but a scalar.

The scalar product of two vectors **P** and **Q** can be expressed in terms of their rectangular components. Resolving **P** and **Q** into components, we first write

$$\mathbf{P} \cdot \mathbf{Q} = (P_x\mathbf{i} + P_y\mathbf{j} + P_z\mathbf{k}) \cdot (Q_x\mathbf{i} + Q_y\mathbf{j} + Q_z\mathbf{k})$$

Making use of the distributive property, we express $\mathbf{P} \cdot \mathbf{Q}$ as the sum of scalar products, such as $P_x\mathbf{i} \cdot Q_x\mathbf{i}$ and $P_x\mathbf{i} \cdot Q_y\mathbf{j}$. However, from the definition of the scalar product it follows that the scalar products of the unit vectors are either zero or one.

$$\begin{array}{lll} \mathbf{i} \cdot \mathbf{i} = 1 & \mathbf{j} \cdot \mathbf{j} = 1 & \mathbf{k} \cdot \mathbf{k} = 1 \\ \mathbf{i} \cdot \mathbf{j} = 0 & \mathbf{j} \cdot \mathbf{k} = 0 & \mathbf{k} \cdot \mathbf{i} = 0 \end{array} \tag{3.29}$$

Thus, the expression obtained for $\mathbf{P} \cdot \mathbf{Q}$ reduces to

$$\mathbf{P} \cdot \mathbf{Q} = P_xQ_x + P_yQ_y + P_zQ_z \tag{3.30}$$

In the particular case when **P** and **Q** are equal, we note that

$$\mathbf{P} \cdot \mathbf{P} = P_x^2 + P_y^2 + P_z^2 = P^2 \tag{3.31}$$

Applications

1. *Angle formed by two given vectors.* Let two vectors be given in terms of their components:

$$\mathbf{P} = P_x\mathbf{i} + P_y\mathbf{j} + P_z\mathbf{k}$$
$$\mathbf{Q} = Q_x\mathbf{i} + Q_y\mathbf{j} + Q_z\mathbf{k}$$

To determine the angle formed by the two vectors, we equate the expressions obtained in (3.24) and (3.30) for their scalar product and write

$$PQ \cos \theta = P_xQ_x + P_yQ_y + P_zQ_z$$

Solving for $\cos \theta$, we have

$$\cos \theta = \frac{P_xQ_x + P_yQ_y + P_zQ_z}{PQ} \tag{3.32}$$

2. *Projection of a vector on a given axis.* Consider a vector **P** forming an angle θ with an axis, or directed line, OL (Fig. 3.21). The *projection of* **P** *on the axis* OL is defined as the scalar

$$P_{OL} = P \cos \theta \tag{3.33}$$

We note that the projection P_{OL} is equal in absolute value to the length of the segment OA; it will be positive if OA has the same sense as the axis OL, that is, if θ is acute, and negative otherwise. If **P** and OL are at a right angle, the projection of **P** on OL is zero.

Consider now a vector **Q** directed along OL and of the same sense as OL (Fig. 3.22). The scalar product of **P** and **Q** can be expressed as

$$\mathbf{P} \cdot \mathbf{Q} = PQ \cos \theta = P_{OL}Q \tag{3.34}$$

from which it follows that

$$P_{OL} = \frac{\mathbf{P} \cdot \mathbf{Q}}{Q} = \frac{P_xQ_x + P_yQ_y + P_zQ_z}{Q} \tag{3.35}$$

In the particular case when the vector selected along OL is the unit vector $\boldsymbol{\lambda}$ (Fig. 3.23), we write

$$P_{OL} = \mathbf{P} \cdot \boldsymbol{\lambda} \tag{3.36}$$

Resolving **P** and $\boldsymbol{\lambda}$ into rectangular components and recalling from Sec. 2.12 that the components of $\boldsymbol{\lambda}$ along the coordinate axes are respectively equal to the direction cosines of OL, we express the projection of **P** on OL as

$$P_{OL} = P_x \cos \theta_x + P_y \cos \theta_y + P_z \cos \theta_z \tag{3.37}$$

where θ_x, θ_y, and θ_z denote the angles that the axis OL forms with the coordinate axes.

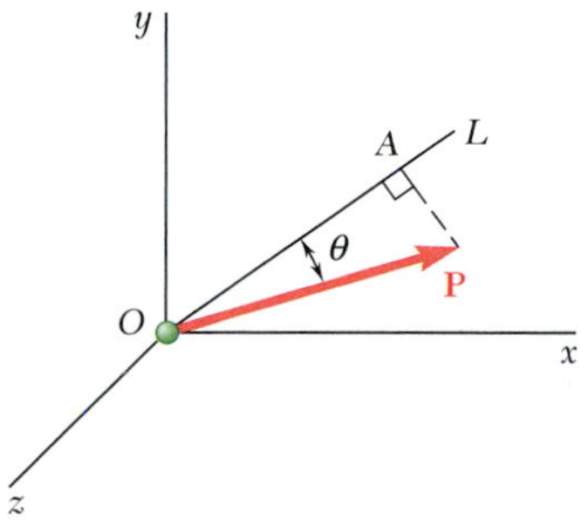

Fig. 3.21

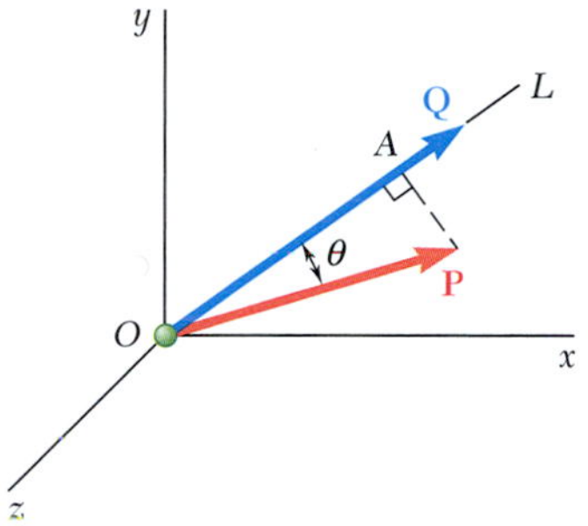

Fig. 3.22

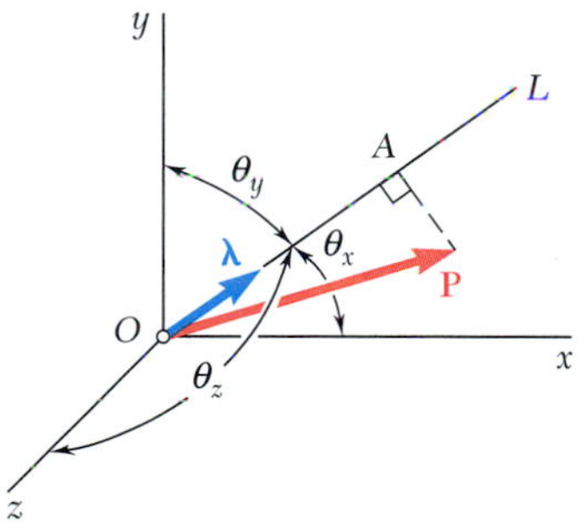

Fig. 3.23

3.10. MIXED TRIPLE PRODUCT OF THREE VECTORS

We define the *mixed triple product* of the three vectors **S**, **P**, and **Q** as the scalar expression

$$\mathbf{S} \cdot (\mathbf{P} \times \mathbf{Q}) \tag{3.38}$$

obtained by forming the scalar product of **S** with the vector product of **P** and **Q**.†

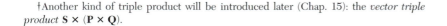

†Another kind of triple product will be introduced later (Chap. 15): the *vector triple product* $\mathbf{S} \times (\mathbf{P} \times \mathbf{Q})$.

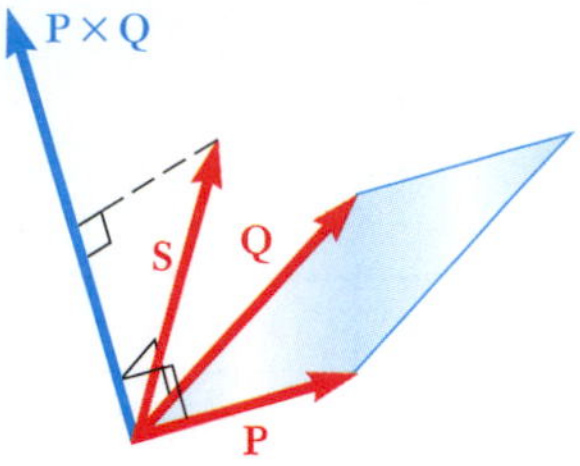

Fig. 3.24

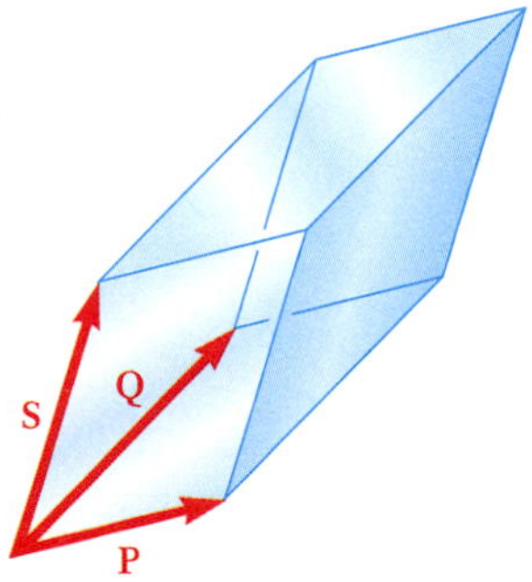

Fig. 3.25

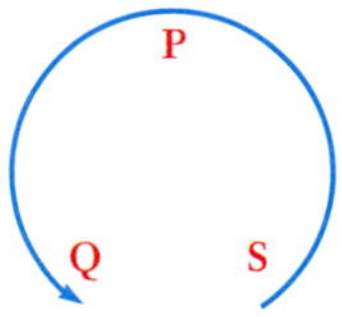

Fig. 3.26

A simple geometrical interpretation can be given for the mixed triple product of **S**, **P**, and **Q** (Fig. 3.24). We first recall from Sec. 3.4 that the vector **P** × **Q** is perpendicular to the plane containing **P** and **Q** and that its magnitude is equal to the area of the parallelogram which has **P** and **Q** for sides. On the other hand, Eq. (3.34) indicates that the scalar product of **S** and **P** × **Q** can be obtained by multiplying the magnitude of **P** × **Q** (that is, the area of the parallelogram defined by **P** and **Q**) by the projection of **S** on the vector **P** × **Q** (that is, by the projection of **S** on the normal to the plane containing the parallelogram). The mixed triple product is thus equal, in absolute value, to the volume of the parallelepiped having the vectors **S**, **P**, and **Q** for sides (Fig. 3.25). We note that the sign of the mixed triple product will be positive if **S**, **P**, and **Q** form a right-handed triad and negative if they form a left-handed triad [that is, $\mathbf{S} \cdot (\mathbf{P} \times \mathbf{Q})$ will be negative if the rotation which brings **P** into line with **Q** is observed as clockwise from the tip of **S**]. The mixed triple product will be zero if **S**, **P**, and **Q** are coplanar.

Since the parallelepiped defined in the preceding paragraph is independent of the order in which the three vectors are taken, the six mixed triple products which can be formed with **S**, **P**, and **Q** will all have the same absolute value, although not the same sign. It is easily shown that

$$\begin{aligned} \mathbf{S} \cdot (\mathbf{P} \times \mathbf{Q}) &= \mathbf{P} \cdot (\mathbf{Q} \times \mathbf{S}) = \mathbf{Q} \cdot (\mathbf{S} \times \mathbf{P}) \\ &= -\mathbf{S} \cdot (\mathbf{Q} \times \mathbf{P}) = -\mathbf{P} \cdot (\mathbf{S} \times \mathbf{Q}) = -\mathbf{Q} \cdot (\mathbf{P} \times \mathbf{S}) \end{aligned} \tag{3.39}$$

Arranging in a circle and in counterclockwise order the letters representing the three vectors (Fig. 3.26), we observe that the sign of the mixed triple product remains unchanged if the vectors are permuted in such a way that they are still read in counterclockwise order. Such a permutation is said to be a *circular permutation*. It also follows from Eq. (3.39) and from the commutative property of scalar products that the mixed triple product of **S**, **P**, and **Q** can be defined equally well as $\mathbf{S} \cdot (\mathbf{P} \times \mathbf{Q})$ or $(\mathbf{S} \times \mathbf{P}) \cdot \mathbf{Q}$.

The mixed triple product of the vectors **S**, **P**, and **Q** can be expressed in terms of the rectangular components of these vectors. Denoting **P** × **Q** by **V** and using formula (3.30) to express the scalar product of **S** and **V**, we write

$$\mathbf{S} \cdot (\mathbf{P} \times \mathbf{Q}) = \mathbf{S} \cdot \mathbf{V} = S_x V_x + S_y V_y + S_z V_z$$

Substituting from the relations (3.9) for the components of **V**, we obtain

$$\begin{aligned} \mathbf{S} \cdot (\mathbf{P} \times \mathbf{Q}) = S_x(P_y Q_z - P_z Q_y) &+ S_y(P_z Q_x - P_x Q_z) \\ &+ S_z(P_x Q_y - P_y Q_x) \end{aligned} \tag{3.40}$$

This expression can be written in a more compact form if we observe that it represents the expansion of a determinant:

$$\mathbf{S} \cdot (\mathbf{P} \times \mathbf{Q}) = \begin{vmatrix} S_x & S_y & S_z \\ P_x & P_y & P_z \\ Q_x & Q_y & Q_z \end{vmatrix} \tag{3.41}$$

By applying the rules governing the permutation of rows in a determinant, we could easily verify the relations (3.39) which were derived earlier from geometrical considerations.

3.11. MOMENT OF A FORCE ABOUT A GIVEN AXIS

Now that we have further increased our knowledge of vector algebra, we can introduce a new concept, the concept of *moment of a force about an axis.* Consider again a force $\mathbf{F}$ acting on a rigid body and the moment $\mathbf{M}_O$ of that force about O (Fig. 3.27). Let OL be an axis through O; *we define the moment M_{OL} of $\mathbf{F}$ about OL as the projection OC of the moment $\mathbf{M}_O$ onto the axis OL.* Denoting by $\boldsymbol{\lambda}$ the unit vector along OL and recalling from Secs. 3.9 and 3.6, respectively, the expressions (3.36) and (3.11) obtained for the projection of a vector on a given axis and for the moment $\mathbf{M}_O$ of a force $\mathbf{F}$, we write

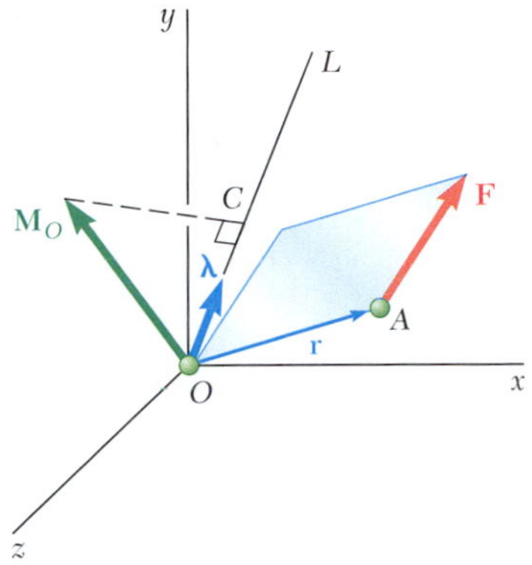

Fig. 3.27

$$M_{OL} = \boldsymbol{\lambda} \cdot \mathbf{M}_O = \boldsymbol{\lambda} \cdot (\mathbf{r} \times \mathbf{F}) \tag{3.42}$$

which shows that the moment M_{OL} of $\mathbf{F}$ about the axis OL is the scalar obtained by forming the mixed triple product of $\boldsymbol{\lambda}$, $\mathbf{r}$, and $\mathbf{F}$. Expressing M_{OL} in the form of a determinant, we write

$$M_{OL} = \begin{vmatrix} \lambda_x & \lambda_y & \lambda_z \\ x & y & z \\ F_x & F_y & F_z \end{vmatrix} \tag{3.43}$$

where $\lambda_x, \lambda_y, \lambda_z$ = direction cosines of axis OL
x, y, z = coordinates of point of application of $\mathbf{F}$
F_x, F_y, F_z = components of force $\mathbf{F}$

The physical significance of the moment M_{OL} of a force $\mathbf{F}$ about a fixed axis OL becomes more apparent if we resolve $\mathbf{F}$ into two rectangular components $\mathbf{F}_1$ and $\mathbf{F}_2$, with $\mathbf{F}_1$ parallel to OL and $\mathbf{F}_2$ lying in a plane P perpendicular to OL (Fig. 3.28). Resolving $\mathbf{r}$ similarly into two components $\mathbf{r}_1$ and $\mathbf{r}_2$ and substituting for $\mathbf{F}$ and $\mathbf{r}$ into (3.42), we write

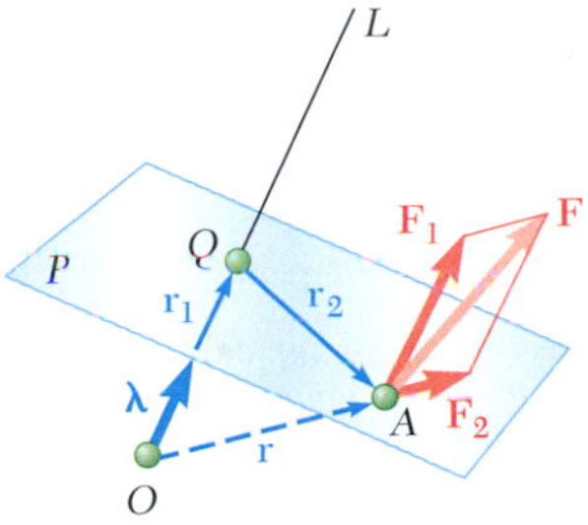

Fig. 3.28

$$\begin{aligned} M_{OL} &= \boldsymbol{\lambda} \cdot [(\mathbf{r}_1 + \mathbf{r}_2) \times (\mathbf{F}_1 + \mathbf{F}_2)] \\ &= \boldsymbol{\lambda} \cdot (\mathbf{r}_1 \times \mathbf{F}_1) + \boldsymbol{\lambda} \cdot (\mathbf{r}_1 \times \mathbf{F}_2) + \boldsymbol{\lambda} \cdot (\mathbf{r}_2 \times \mathbf{F}_1) + \boldsymbol{\lambda} \cdot (\mathbf{r}_2 \times \mathbf{F}_2) \end{aligned}$$

Noting that all of the mixed triple products except the last one are equal to zero, since they involve vectors which are coplanar when drawn from a common origin (Sec. 3.10), we have

$$M_{OL} = \boldsymbol{\lambda} \cdot (\mathbf{r}_2 \times \mathbf{F}_2) \tag{3.44}$$

The vector product $\mathbf{r}_2 \times \mathbf{F}_2$ is perpendicular to the plane P and represents the moment of the component $\mathbf{F}_2$ of $\mathbf{F}$ about the point Q where OL intersects P. Therefore, the scalar M_{OL}, which will be positive if $\mathbf{r}_2 \times \mathbf{F}_2$ and OL have the same sense and negative otherwise, measures the tendency of $\mathbf{F}_2$ to make the rigid body rotate about the fixed axis OL. Since the other component $\mathbf{F}_1$ of $\mathbf{F}$ does not tend to make the body rotate about OL, we conclude that *the moment M_{OL} of $\mathbf{F}$ about OL measures the tendency of the force $\mathbf{F}$ to impart to the rigid body a motion of rotation about the fixed axis OL.*

It follows from the definition of the moment of a force about an axis that the moment of **F** about a coordinate axis is equal to the component of $\mathbf{M}_O$ along that axis. Substituting successively each of the unit vectors **i**, **j**, and **k** for **λ** in (3.42), we observe that the expressions thus obtained for the *moments of* **F** *about the coordinate axes* are respectively equal to the expressions obtained in Sec. 3.8 for the components of the moment $\mathbf{M}_O$ of **F** about O:

$$\begin{aligned} M_x &= yF_z - zF_y \\ M_y &= zF_x - xF_z \\ M_z &= xF_y - yF_x \end{aligned} \tag{3.18}$$

We observe that just as the components F_x, F_y, and F_z of a force **F** acting on a rigid body measure, respectively, the tendency of **F** to move the rigid body in the x, y, and z directions, the moments M_x, M_y, and M_z of **F** about the coordinate axes measure the tendency of **F** to impart to the rigid body a motion of rotation about the x, y, and z axes, respectively.

More generally, the moment of a force **F** applied at A about an axis which does not pass through the origin is obtained by choosing an arbitrary point B on the axis (Fig. 3.29) and determining the projection on the axis BL of the moment $\mathbf{M}_B$ of **F** about B. We write

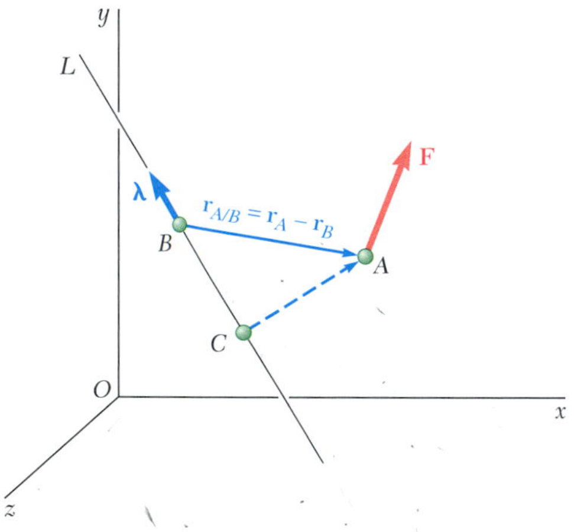

Fig. 3.29

$$M_{BL} = \boldsymbol{\lambda} \cdot \mathbf{M}_B = \boldsymbol{\lambda} \cdot (\mathbf{r}_{A/B} \times \mathbf{F}) \tag{3.45}$$

where $\mathbf{r}_{A/B} = \mathbf{r}_A - \mathbf{r}_B$ represents the vector drawn from B to A. Expressing M_{BL} in the form of a determinant, we have

$$M_{BL} = \begin{vmatrix} \lambda_x & \lambda_y & \lambda_z \\ x_{A/B} & y_{A/B} & z_{A/B} \\ F_x & F_y & F_z \end{vmatrix} \tag{3.46}$$

where $\lambda_x, \lambda_y, \lambda_z$ = direction cosines of axis BL

$x_{A/B} = x_A - x_B \qquad y_{A/B} = y_A - y_B \qquad z_{A/B} = z_A - z_B$

F_x, F_y, F_z = components of force **F**

It should be noted that the result obtained is independent of the choice of the point B on the given axis. Indeed, denoting by M_{CL} the result obtained with a different point C, we have

$$\begin{aligned} M_{CL} &= \boldsymbol{\lambda} \cdot [(\mathbf{r}_A - \mathbf{r}_C) \times \mathbf{F}] \\ &= \boldsymbol{\lambda} \cdot [(\mathbf{r}_A - \mathbf{r}_B) \times \mathbf{F}] + \boldsymbol{\lambda} \cdot [(\mathbf{r}_B - \mathbf{r}_C) \times \mathbf{F}] \end{aligned}$$

But, since the vectors **λ** and $\mathbf{r}_B - \mathbf{r}_C$ lie in the same line, the volume of the parallelepiped having the vectors **λ**, $\mathbf{r}_B - \mathbf{r}_C$, and **F** for sides is zero, as is the mixed triple product of these three vectors (Sec. 3.10). The expression obtained for M_{CL} thus reduces to its first term, which is the expression used earlier to define M_{BL}. In addition, it follows from Sec. 3.6 that, when computing the moment of **F** about the given axis, A can be any point on the line of action of **F**.

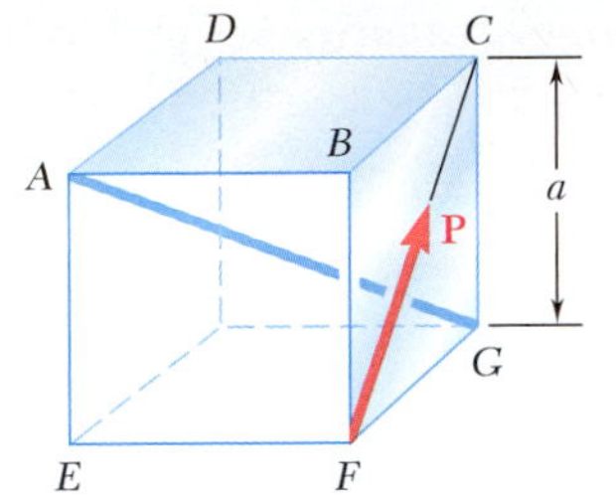

SAMPLE PROBLEM 3.5

A cube of side a is acted upon by a force $\mathbf{P}$ as shown. Determine the moment of $\mathbf{P}$ (*a*) about A, (*b*) about the edge AB, (*c*) about the diagonal AG of the cube. (*d*) Using the result of part *c*, determine the perpendicular distance between AG and FC.

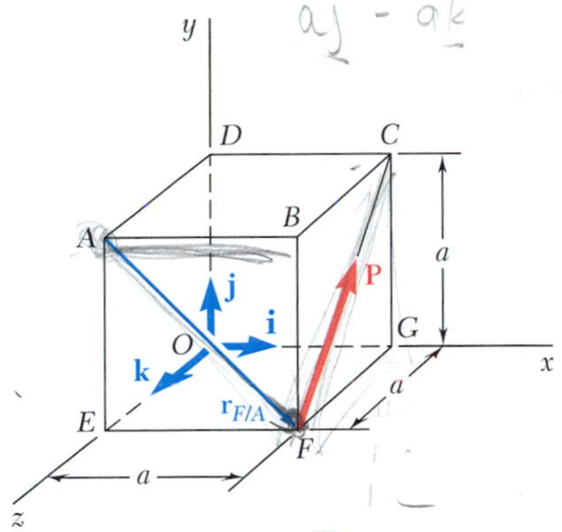

SOLUTION

***a.* Moment about *A*.** Choosing x, y, and z axes as shown, we resolve into rectangular components the force $\mathbf{P}$ and the vector $\mathbf{r}_{F/A} = \overrightarrow{AF}$ drawn from A to the point of application F of $\mathbf{P}$.

$$\mathbf{r}_{F/A} = a\mathbf{i} - a\mathbf{j} = a(\mathbf{i} - \mathbf{j})$$

$$\mathbf{P} = (P/\sqrt{2})\mathbf{j} - (P/\sqrt{2})\mathbf{k} = (P/\sqrt{2})(\mathbf{j} - \mathbf{k})$$

The moment of $\mathbf{P}$ about A is

$$\mathbf{M}_A = \mathbf{r}_{F/A} \times \mathbf{P} = a(\mathbf{i} - \mathbf{j}) \times (P/\sqrt{2})(\mathbf{j} - \mathbf{k})$$

$$\mathbf{M}_A = (aP/\sqrt{2})(\mathbf{i} + \mathbf{j} + \mathbf{k}) \quad ◀$$

***b.* Moment about *AB*.** Projecting $\mathbf{M}_A$ on AB, we write

$$M_{AB} = \mathbf{i} \cdot \mathbf{M}_A = \mathbf{i} \cdot (aP/\sqrt{2})(\mathbf{i} + \mathbf{j} + \mathbf{k})$$

$$M_{AB} = aP/\sqrt{2} \quad ◀$$

We verify that, since AB is parallel to the x axis, M_{AB} is also the x component of the moment $\mathbf{M}_A$.

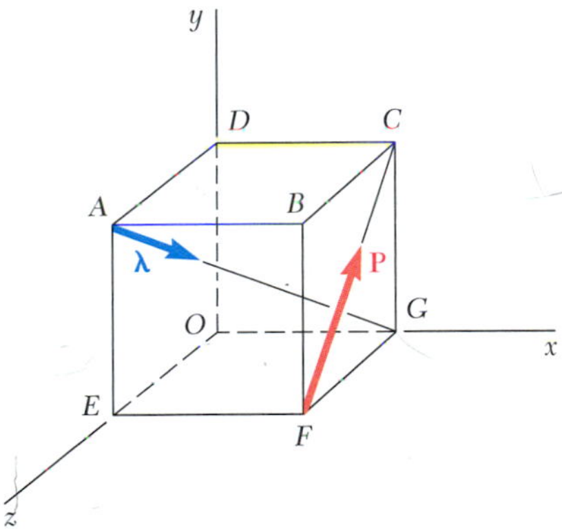

***c.* Moment about Diagonal *AG*.** The moment of $\mathbf{P}$ about AG is obtained by projecting $\mathbf{M}_A$ on AG. Denoting by $\boldsymbol{\lambda}$ the unit vector along AG, we have

$$\boldsymbol{\lambda} = \frac{\overrightarrow{AG}}{AG} = \frac{a\mathbf{i} - a\mathbf{j} - a\mathbf{k}}{a\sqrt{3}} = (1/\sqrt{3})(\mathbf{i} - \mathbf{j} - \mathbf{k})$$

$$M_{AG} = \boldsymbol{\lambda} \cdot \mathbf{M}_A = (1/\sqrt{3})(\mathbf{i} - \mathbf{j} - \mathbf{k}) \cdot (aP/\sqrt{2})(\mathbf{i} + \mathbf{j} + \mathbf{k})$$

$$M_{AG} = (aP/\sqrt{6})(1 - 1 - 1) \qquad M_{AG} = -aP/\sqrt{6} \quad ◀$$

Alternative Method. The moment of $\mathbf{P}$ about AG can also be expressed in the form of a determinant:

$$M_{AG} = \begin{vmatrix} \lambda_x & \lambda_y & \lambda_z \\ x_{F/A} & y_{F/A} & z_{F/A} \\ F_x & F_y & F_z \end{vmatrix} = \begin{vmatrix} 1/\sqrt{3} & -1/\sqrt{3} & -1/\sqrt{3} \\ a & -a & 0 \\ 0 & P/\sqrt{2} & -P/\sqrt{2} \end{vmatrix} = -aP/\sqrt{6}$$

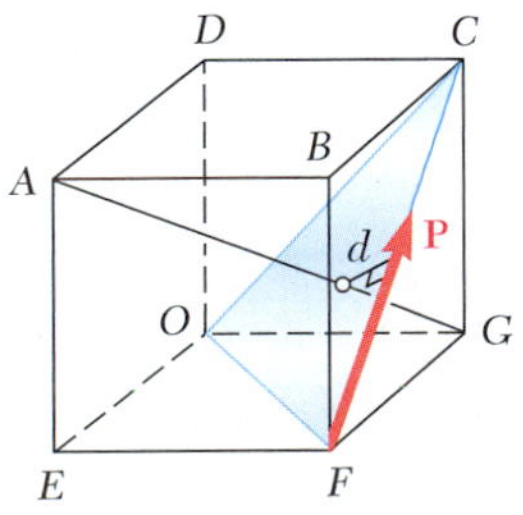

***d.* Perpendicular Distance between *AG* and *FC*.** We first observe that $\mathbf{P}$ is perpendicular to the diagonal AG. This can be checked by forming the scalar product $\mathbf{P} \cdot \boldsymbol{\lambda}$ and verifying that it is zero:

$$\mathbf{P} \cdot \boldsymbol{\lambda} = (P/\sqrt{2})(\mathbf{j} - \mathbf{k}) \cdot (1/\sqrt{3})(\mathbf{i} - \mathbf{j} - \mathbf{k}) = (P\sqrt{6})(0 - 1 + 1) = 0$$

The moment M_{AG} can then be expressed as $-Pd$, where d is the perpendicular distance from AG to FC. (The negative sign is used since the rotation imparted to the cube by $\mathbf{P}$ appears as clockwise to an observer at G.) Recalling the value found for M_{AG} in part *c*,

$$M_{AG} = -Pd = -aP/\sqrt{6} \qquad d = a/\sqrt{6} \quad ◀$$

SOLVING PROBLEMS ON YOUR OWN

In the problems for this lesson you will apply the *scalar product* or *dot product* of two vectors to determine the *angle formed by two given vectors* and the *projection of a force on a given axis.* You will also use the *mixed triple product* of three vectors to find the *moment of a force about a given axis* and the *perpendicular distance between two lines.*

1. *Calculating the angle formed by two given vectors.* First express the vectors in terms of their components and determine the magnitudes of the two vectors. The cosine of the desired angle is then obtained by dividing the scalar product of the two vectors by the product of their magnitudes [Eq. (3.32)].

2. *Computing the projection of a vector* P *on a given axis OL.* In general, begin by expressing $\mathbf{P}$ and the unit vector $\boldsymbol{\lambda}$, that defines the direction of the axis, in component form. Take care that $\boldsymbol{\lambda}$ has the correct sense (that is, $\boldsymbol{\lambda}$ is directed from O to L). The required projection is then equal to the scalar product $\mathbf{P} \cdot \boldsymbol{\lambda}$. However, if you know the angle θ formed by $\mathbf{P}$ and $\boldsymbol{\lambda}$, the projection is also given by $P \cos \theta$.

3. *Determining the moment* M_{OL} *of a force about a given axis OL.* We defined M_{OL} as

$$M_{OL} = \boldsymbol{\lambda} \cdot \mathbf{M}_O = \boldsymbol{\lambda} \cdot (\mathbf{r} \times \mathbf{F}) \tag{3.42}$$

where $\boldsymbol{\lambda}$ is the unit vector along OL and $\mathbf{r}$ is a position vector *from any point* on the line OL *to any point* on the line of action of $\mathbf{F}$. As was the case for the moment of a force about a point, choosing the most convenient position vector will simplify your calculations. Also, recall the warning of the previous lesson: the vectors $\mathbf{r}$ and $\mathbf{F}$ must have the correct sense, and they must be placed in the proper order. The procedure you should follow when computing the moment of a force about an axis is illustrated in part *c* of Sample Prob. 3.5. The two essential steps in this procedure are to first express $\boldsymbol{\lambda}$, $\mathbf{r}$, and $\mathbf{F}$ in terms of their rectangular components and to then evaluate the mixed triple product $\boldsymbol{\lambda} \cdot (\mathbf{r} \times \mathbf{F})$ to determine the moment about the axis. In most three-dimensional problems the most convenient way to compute the mixed triple product is by using a determinant.

As noted in the text, when $\boldsymbol{\lambda}$ is directed along one of the coordinate axes, M_{OL} is equal to the scalar component of $\mathbf{M}_O$ along that axis.

4. Determining the perpendicular distance between two lines. You should remember that it is the perpendicular component $\mathbf{F}_2$ of the force $\mathbf{F}$ that tends to make a body rotate about a given axis OL (Fig. 3.28). It then follows that

$$M_{OL} = F_2 d$$

where M_{OL} is the moment of $\mathbf{F}$ about axis OL and d is the perpendicular distance between OL and the line of action of $\mathbf{F}$. This last equation gives us a simple technique for determining d. First assume that a force $\mathbf{F}$ of known magnitude F lies along one of the given lines and that the unit vector $\boldsymbol{\lambda}$ lies along the other line. Next compute the moment M_{OL} of the force $\mathbf{F}$ about the second line using the method discussed above. The magnitude of the parallel component, F_1, of $\mathbf{F}$ is obtained using the scalar product:

$$F_1 = \mathbf{F} \cdot \boldsymbol{\lambda}$$

The value of F_2 is then determined from

$$F_2 = \sqrt{F^2 - F_1^2}$$

Finally, substitute the values of M_{OL} and F_2 into the equation $M_{OL} = F_2 d$ and solve for d.

You should now realize that the calculation of the perpendicular distance in part d of Sample Prob. 3.5 was simplified by $\mathbf{P}$ being perpendicular to the diagonal AG. In general, the two given lines will not be perpendicular, so that the technique just outlined will have to be used when determining the perpendicular distance between them.

Problems

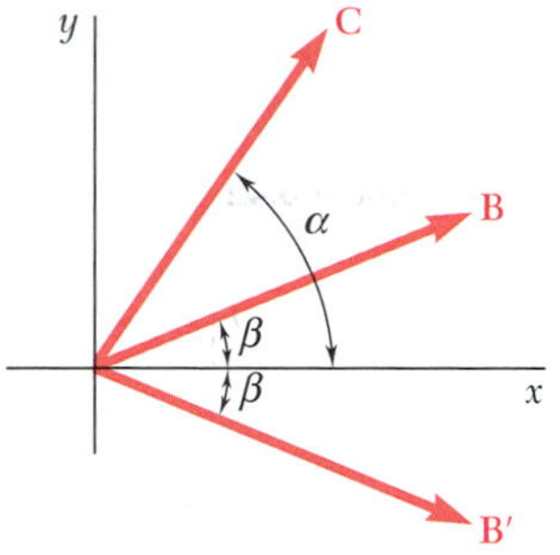

Fig. P3.36

3.35 Given the vectors $\mathbf{P} = 7\mathbf{i} - 2\mathbf{j} + 5\mathbf{k}$, $\mathbf{Q} = -3\mathbf{i} - 4\mathbf{j} + 6\mathbf{k}$, and $\mathbf{S} = 8\mathbf{i} + \mathbf{j} - 9\mathbf{k}$, compute the scalar products $\mathbf{P} \cdot \mathbf{Q}$, $\mathbf{P} \cdot \mathbf{S}$, and $\mathbf{Q} \cdot \mathbf{S}$.

3.36 Form the scalar products $\mathbf{B} \cdot \mathbf{C}$ and $\mathbf{B}' \cdot \mathbf{C}$, where $B = B'$, and use the results obtained to prove the identity

$$\cos\alpha\cos\beta = \tfrac{1}{2}\cos(\alpha + \beta) + \tfrac{1}{2}\cos(\alpha - \beta).$$

3.37 Consider the volleyball net shown. Determine the angle formed by guy wires AB and AC.

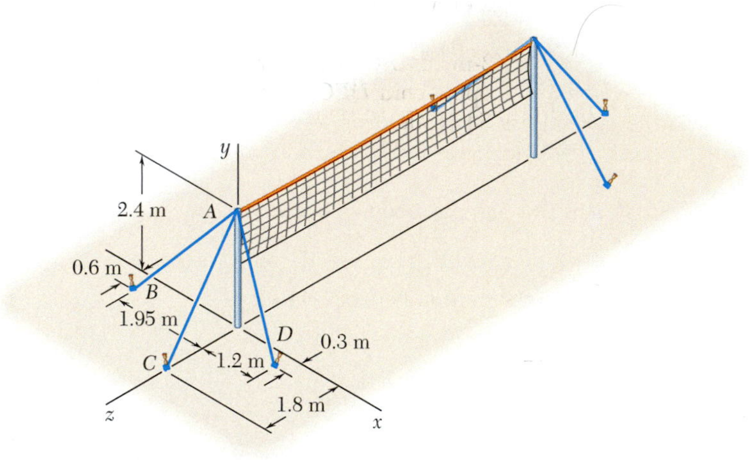

Fig. P3.37 and *P3.38*

3.38 Consider the volleyball net shown. Determine the angle formed by guy wires AC and AD.

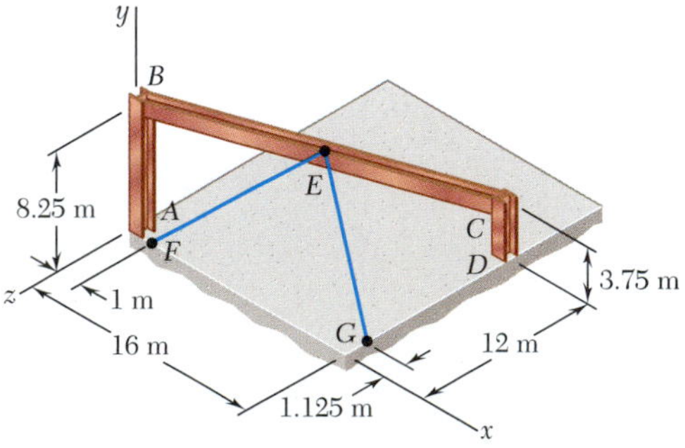

Fig. P3.39 and P3.40

3.39 Steel framing members AB, BC, and CD are joined at B and C and are braced using cables EF and EG. Knowing that E is at the midpoint of BC and that the tension in cable EF is 330 N, determine (*a*) the angle between EF and member BC, (*b*) the projection on BC of the force exerted by cable EF at point E.

3.40 Steel framing members AB, BC, and CD are joined at B and C and are braced using cables EF and EG. Knowing that E is at the midpoint of BC and that the tension in cable EG is 445 N, determine (*a*) the angle between EG and member BC, (*b*) the projection on BC of the force exerted by cable EG at point E.

3.41 Slider P can move along rod OA. An elastic cord PC is attached to the slider and to the vertical member BC. Knowing that the distance from O to P is 0.12 m and the tension in the cord is 30 N, determine (*a*) the angle between the elastic cord and the rod OA, (*b*) the projection on OA of the force exerted by cord PC at point P.

3.42 Slider P can move along rod OA. An elastic cord PC is attached to the slider and to the vertical member BC. Determine the distance from O to P for which cord PC and rod OA are perpendicular.

3.43 Determine the volume of the parallelepiped of Fig. 3.25 when (*a*) $\mathbf{P} = -(7\text{ in.})\mathbf{i} - (1\text{ in.})\mathbf{j} + (2\text{ in.})\mathbf{k}$, $\mathbf{Q} = (3\text{ in.})\mathbf{i} - (2\text{ in.})\mathbf{j} + (4\text{ in.})\mathbf{k}$, and $\mathbf{S} = -(5\text{ in.})\mathbf{i} + (6\text{ in.})\mathbf{j} - (1\text{ in.})\mathbf{k}$, (*b*) $\mathbf{P} = (1\text{ in.})\mathbf{i} + (2\text{ in.})\mathbf{j} - (1\text{ in.})\mathbf{k}$, $\mathbf{Q} = -(8\text{ in.})\mathbf{i} - (1\text{ in.})\mathbf{j} + (9\text{ in.})\mathbf{k}$, and $\mathbf{S} = (2\text{ in.})\mathbf{i} + (3\text{ in.})\mathbf{j} + (1\text{ in.})\mathbf{k}$.

3.44 Given the vectors $\mathbf{P} = 4\mathbf{i} - 2\mathbf{j} + P_z\mathbf{k}$, $\mathbf{Q} = \mathbf{i} + 3\mathbf{j} - 5\mathbf{k}$, and $\mathbf{S} = -6\mathbf{i} + 2\mathbf{j} - \mathbf{k}$, determine the value of P_z for which the three vectors are coplanar.

3.45 The 0.732×1.2-m lid $ABCD$ of a storage bin is hinged along side AB and is held open by looping cord DEC over a frictionless hook at E. If the tension in the cord is 54 N, determine the moment about each of the coordinate axes of the force exerted by the cord at D.

3.46 The 0.732×1.2-m lid $ABCD$ of a storage bin is hinged along side AB and is held open by looping cord DEC over a frictionless hook at E. If the tension in the cord is 54 N, determine the moment about each of the coordinate axes of the force exerted by the cord at C.

3.47 A fence consists of wooden posts and a steel cable fastened to each post and anchored in the ground at A and D. Knowing that the sum of the moments about the z axis of the forces exerted by the cable on the posts at B and C is -66 N · m, determine the magnitude of $\mathbf{T}_{CD}$ when $T_{BA} = 56$ N.

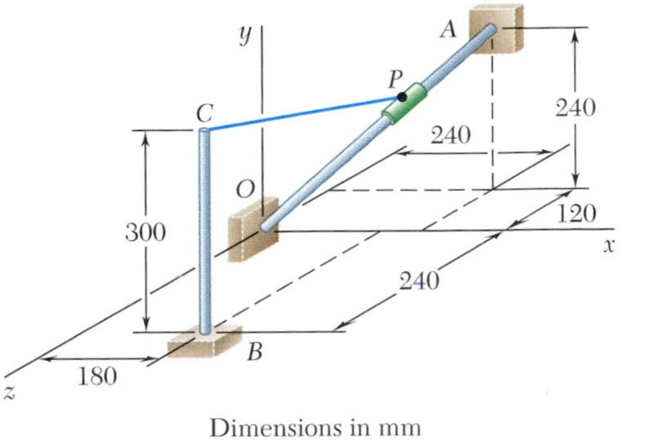

Fig. P3.41 and *P3.42*

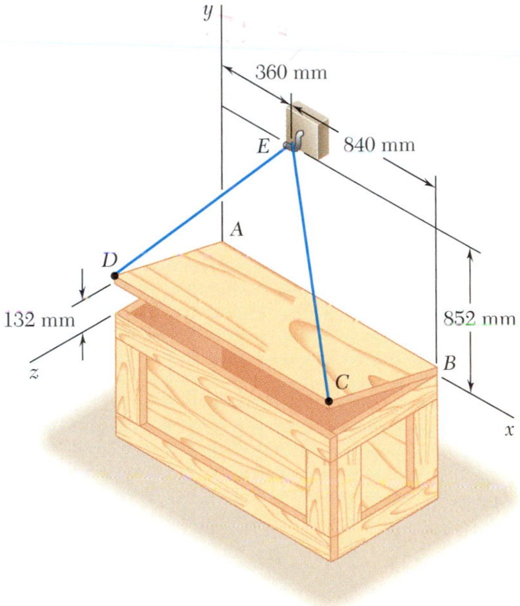

Fig. P3.45 and P3.46

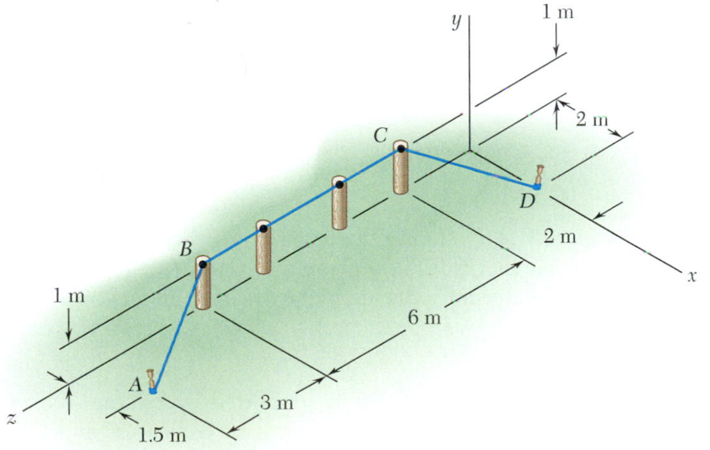

Fig. P3.47 and P3.48

3.48 A fence consists of wooden posts and a steel cable fastened to each post and anchored in the ground at A and D. Knowing that the sum of the moments about the y axis of the forces exerted by the cable on the posts at B and C is 212 N · m, determine the magnitude of $\mathbf{T}_{BA}$ when $T_{CD} = 33$ N.

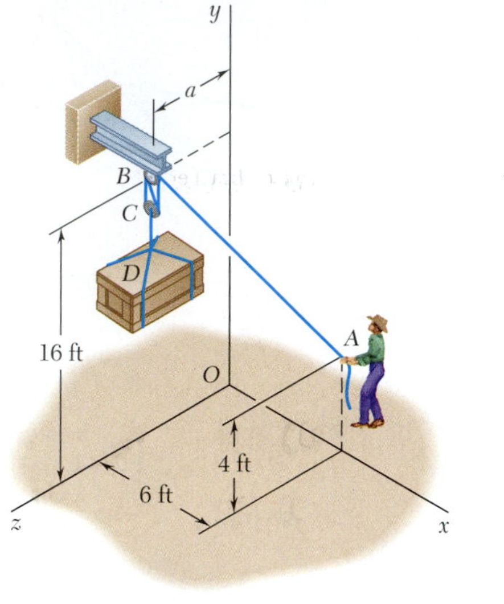

Fig. P3.49 and *P3.50*

3.49 To lift a heavy crate, a man uses a block and tackle attached to the bottom of an I-beam at hook B. Knowing that the moments about the y and z axes of the force exerted at B by portion AB of the rope are, respectively, 100 lb · ft and −400 lb · ft, determine the distance a.

3.50 To lift a heavy crate, a man uses a block and tackle attached to the bottom of an I-beam at hook B. Knowing that the man applies a 200-lb force to end A of the rope and that the moment of that force about the y axis is 175 lb · ft, determine the distance a.

3.51 A force **P** is applied to the lever of an arbor press. Knowing that **P** lies in a plane parallel to the yz plane and that $M_x = 230$ lb · in., $M_y = -200$ lb · in., and $M_z = -35$ lb · in., determine the magnitude of **P** and the values of ϕ and θ.

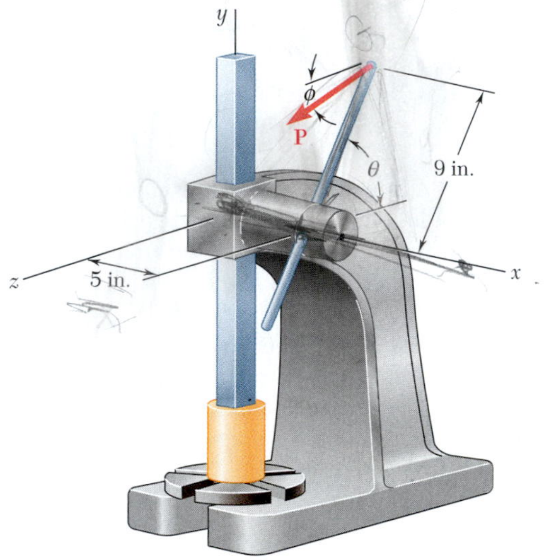

Fig. P3.51 and *P3.52*

3.52 A force **P** is applied to the lever of an arbor press. Knowing that **P** lies in a plane parallel to the yz plane and that $M_y = -180$ lb · in. and $M_z = -30$ lb · in., determine the moment M_x of **P** about the x axis when $\theta = 60°$.

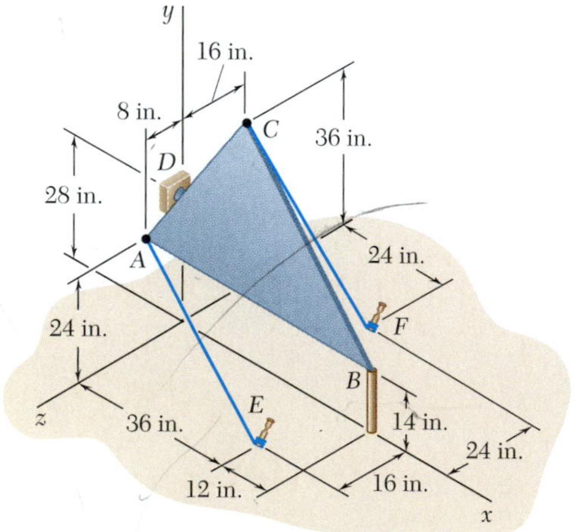

Fig. P3.53 and *P3.54*

3.53 The triangular plate ABC is supported by ball-and-socket joints at B and D and is held in the position shown by cables AE and CF. If the force exerted by cable AE at A is 220 lb, determine the moment of that force about the line joining points D and B.

3.54 The triangular plate ABC is supported by ball-and-socket joints at B and D and is held in the position shown by cables AE and CF. If the force exerted by cable CF at C is 132 lb, determine the moment of that force about the line joining points D and B.

3.55 A mast is mounted on the roof of a house using bracket *ABCD* and is guyed by cables *EF*, *EG*, and *EH*. Knowing that the force exerted by cable *EF* at *E* is 66 N, determine the moment of that force about the line joining points *D* and *I*.

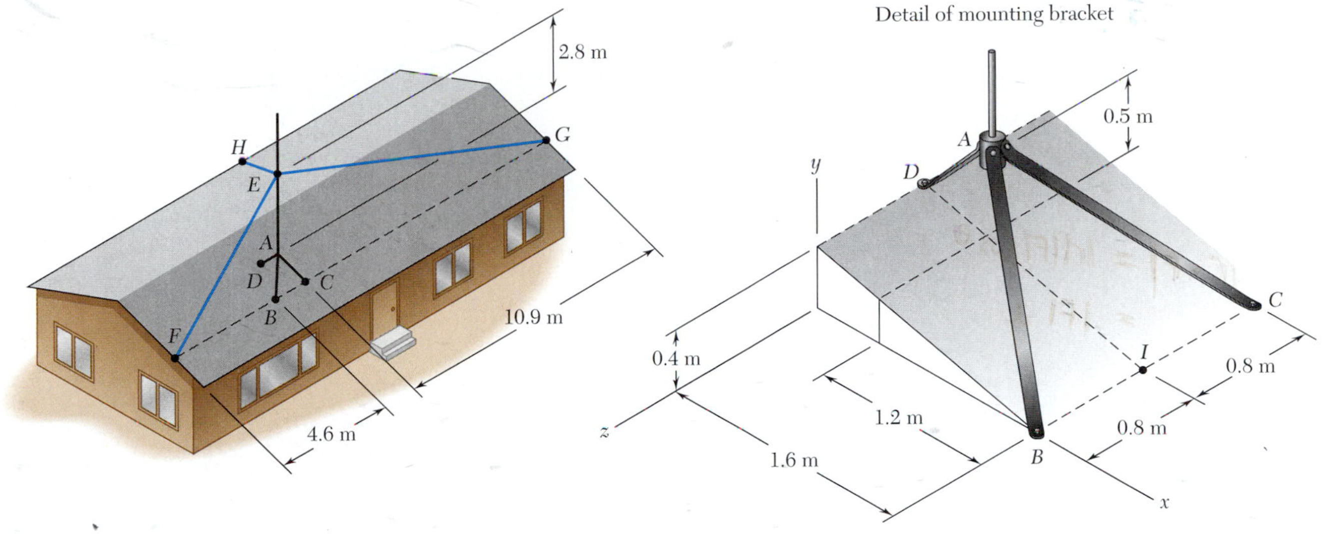

Fig. P3.55 and P3.56

3.56 A mast is mounted on the roof of a house using bracket *ABCD* and is guyed by cables *EF*, *EG*, and *EH*. Knowing that the force exerted by cable *EG* at *E* is 61.5 N, determine the moment of that force about the line joining points *D* and *I*.

3.57 A rectangular tetrahedron has six edges of length *a*. A force **P** is directed as shown along edge *BC*. Determine the moment of **P** about edge *OA*.

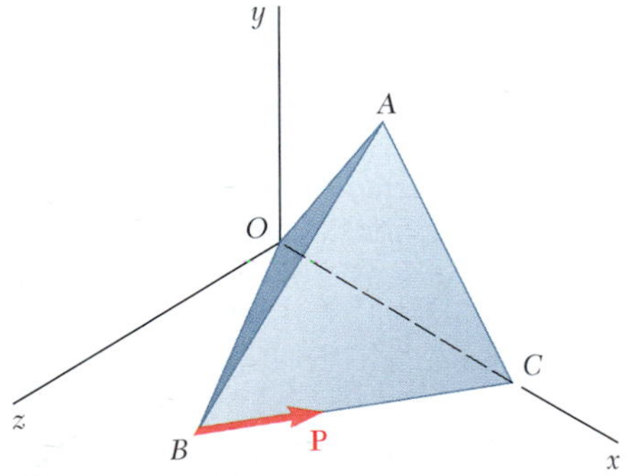

Fig. P3.57 and P3.58

3.58 A rectangular tetrahedron has six edges of length *a*. (*a*) Show that two opposite edges, such as *OA* and *BC*, are perpendicular to each other. (*b*) Use this property and the result obtained in Prob. 3.57 to determine the perpendicular distance between edges *OA* and *BC*.

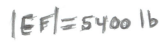

3.59 The 8-ft-wide portion *ABCD* of an inclined, cantilevered walkway is partially supported by members *EF* and *GH*. Knowing that the compressive force exerted by member *EF* on the walkway at *F* is 5400 lb, determine the moment of that force about edge *AD*.

$|EF| = 5400$ lb

$|r \times F| = |r||F| \cos\theta$
$= |F|\, d$

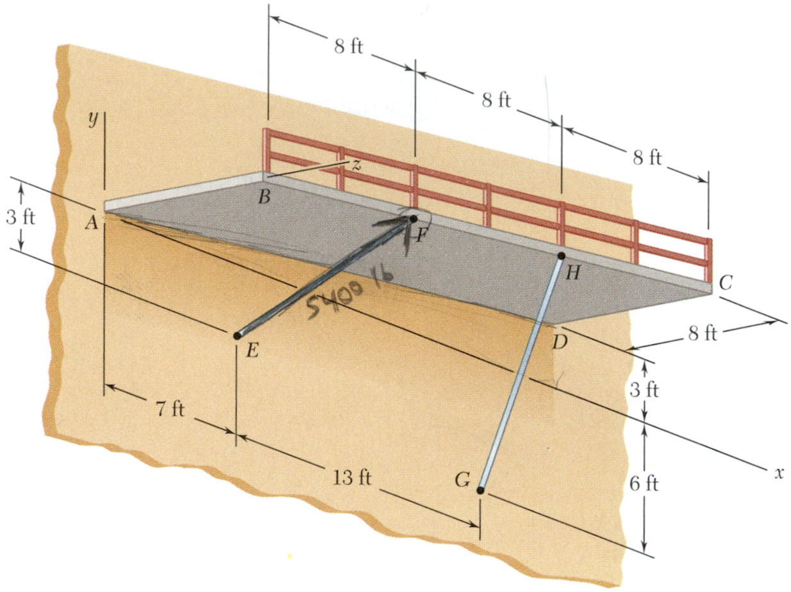

Fig. P3.59 and *P3.60*

3.60 The 8-ft-wide portion *ABCD* of an inclined, cantilevered walkway is partially supported by members *EF* and *GH*. Knowing that the compressive force exerted by member *GH* on the walkway at *H* is 4800 lb, determine the moment of that force about edge *AD*.

3.61 Two forces $\mathbf{F}_1$ and $\mathbf{F}_2$ in space have the same magnitude *F*. Prove that the moment of $\mathbf{F}_1$ about the line of action of $\mathbf{F}_2$ is equal to the moment of $\mathbf{F}_2$ about the line of action of $\mathbf{F}_1$.

***3.62** In Prob. 3.53, determine the perpendicular distance between cable *AE* and the line joining points *D* and *B*.

***3.63** In Prob. 3.54, determine the perpendicular distance between cable *CF* and the line joining points *D* and *B*.

***3.64** In Prob. 3.55, determine the perpendicular distance between cable *EF* and the line joining points *D* and *I*.

****3.65*** In Prob. 3.56, determine the perpendicular distance between cable *EG* and the line joining points *D* and *I*.

***3.66** In Prob. 3.41, determine the perpendicular distance between post *BC* and the line connecting points *O* and *A*.

****3.67*** In Prob. 3.45, determine the perpendicular distance between cord *DE* and the *y* axis.

3.12. MOMENT OF A COUPLE

Two forces **F** *and* −**F** *having the same magnitude, parallel lines of action, and opposite sense are said to form a couple.* (Fig. 3.30). Clearly, the sum of the components of the two forces in any direction is zero. The sum of the moments of the two forces about a given point, however, is not zero. While the two forces will not translate the body on which they act, they will tend to make it rotate.

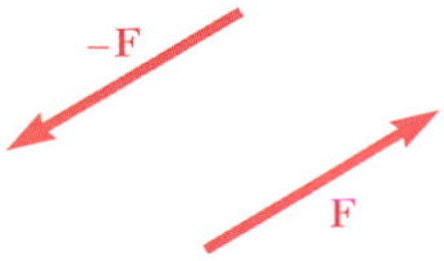

Fig. 3.30

Denoting by $\mathbf{r}_A$ and $\mathbf{r}_B$, respectively, the position vectors of the points of application of **F** and −**F** (Fig. 3.31), we find that the sum of the moments of the two forces about O is

$$\mathbf{r}_A \times \mathbf{F} + \mathbf{r}_B \times (-\mathbf{F}) = (\mathbf{r}_A - \mathbf{r}_B) \times \mathbf{F}$$

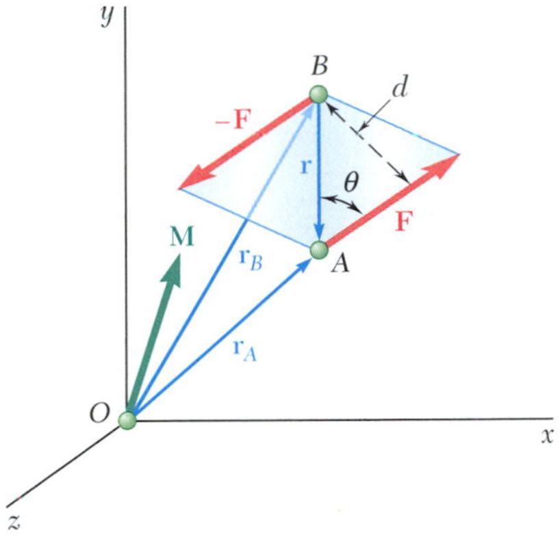

Fig. 3.31

Setting $\mathbf{r}_A - \mathbf{r}_B = \mathbf{r}$, where **r** is the vector joining the points of application of the two forces, we conclude that the sum of the moments of **F** and −**F** about O is represented by the vector

$$\mathbf{M} = \mathbf{r} \times \mathbf{F} \tag{3.47}$$

The vector **M** is called the *moment of the couple;* it is a vector perpendicular to the plane containing the two forces, and its magnitude is

$$M = rF \sin \theta = Fd \tag{3.48}$$

where d is the perpendicular distance between the lines of action of **F** and −**F**. The sense of **M** is defined by the right-hand rule.

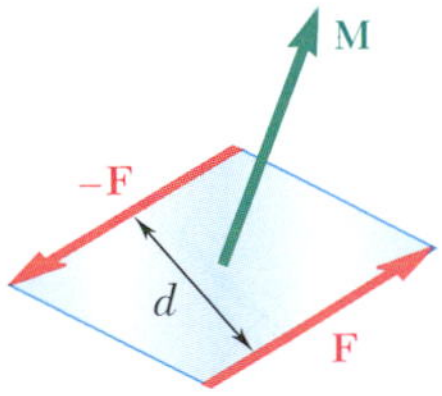

Fig. 3.32

Since the vector **r** in (3.47) is independent of the choice of the origin O of the coordinate axes, we note that the same result would have been obtained if the moments of **F** and −**F** had been computed about a different point O'. Thus, the moment **M** of a couple is a *free vector* (Sec. 2.3) which can be applied at any point (Fig. 3.32).

From the definition of the moment of a couple, it also follows that two couples, one consisting of the forces $\mathbf{F}_1$ and $-\mathbf{F}_1$, the other of the forces $\mathbf{F}_2$ and $-\mathbf{F}_2$ (Fig. 3.33), will have equal moments if

$$F_1 d_1 = F_2 d_2 \tag{3.49}$$

and if the two couples lie in parallel planes (or in the same plane) and have the same sense.

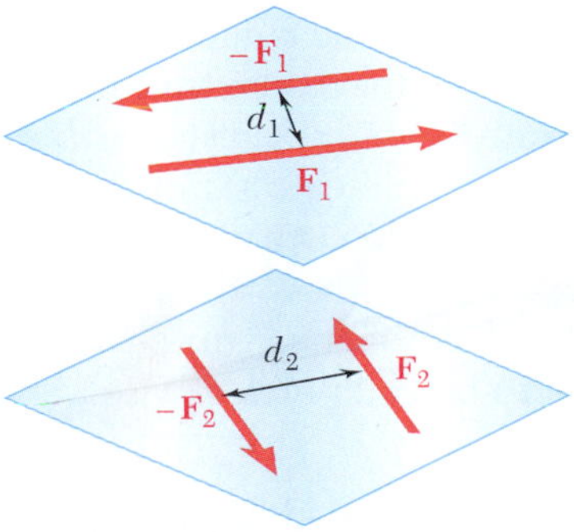

Fig. 3.33

Photo 3.1 The parallel upward and downward forces of equal magnitude exerted on the arms of the lug nut wrench are an example of a couple.

3.13. EQUIVALENT COUPLES

Figure 3.34 shows three couples which act successively on the same rectangular box. As seen in the preceding section, the only motion a couple can impart to a rigid body is a rotation. Since each of the three couples shown has the same moment **M** (same direction and same magnitude $M = 120$ lb · in.), we can expect the three couples to have the same effect on the box.

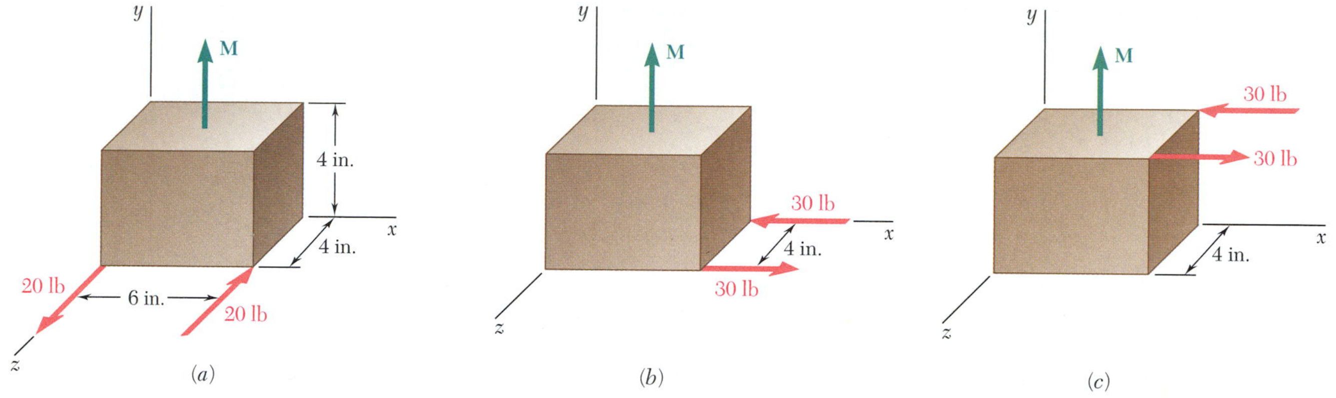

Fig. 3.34

As reasonable as this conclusion appears, we should not accept it hastily. While intuitive feeling is of great help in the study of mechanics, it should not be accepted as a substitute for logical reasoning. Before stating that two systems (or groups) of forces have the same effect on a rigid body, we should prove that fact on the basis of the experimental evidence introduced so far. This evidence consists of the parallelogram law for the addition of two forces (Sec. 2.2) and the principle of transmissibility (Sec. 3.3). Therefore, we will state that *two systems of forces are equivalent* (that is, they have the same effect on a rigid body) *if we can transform one of them into the other by means of one or several of the following operations:* (1) replacing two forces acting on the same particle by their resultant; (2) resolving a force into two components; (3) canceling two equal and opposite forces acting on the same particle; (4) attaching to the same particle two equal and opposite forces; (5) moving a force along its line of action. Each of these operations is easily justified on the basis of the parallelogram law or the principle of transmissibility.

Let us now prove that *two couples having the same moment* **M** *are equivalent.* First consider two couples contained in the same plane, and assume that this plane coincides with the plane of the figure (Fig. 3.35). The first couple consists of the forces $\mathbf{F}_1$ and $-\mathbf{F}_1$ of magnitude F_1, which are located at a distance d_1 from each other (Fig. 3.35*a*), and the second couple consists of the forces $\mathbf{F}_2$ and $-\mathbf{F}_2$ of magnitude F_2, which are located at a distance d_2 from each other (Fig. 3.35*d*). Since the two couples have the same moment **M**, which is perpendicular to the plane of the figure, they must have the same sense (assumed here to be counterclockwise), and the relation

$$F_1 d_1 = F_2 d_2 \tag{3.49}$$

must be satisfied. To prove that they are equivalent, we shall show that the first couple can be transformed into the second by means of the operations listed above.

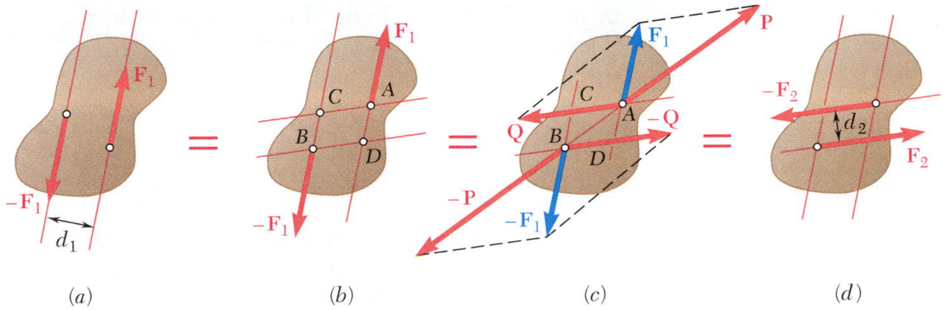

Fig. 3.35

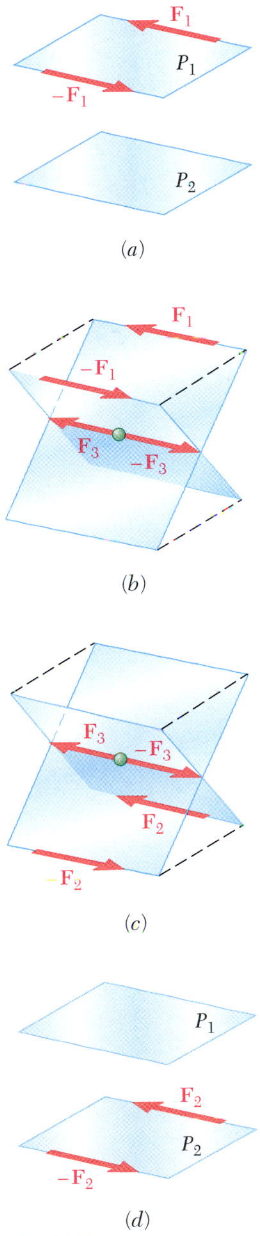

Fig. 3.36

Denoting by *A*, *B*, *C*, *D* the points of intersection of the lines of action of the two couples, we first slide the forces $\mathbf{F}_1$ and $-\mathbf{F}_1$ until they are attached, respectively, at *A* and *B*, as shown in Fig. 3.35*b*. The force $\mathbf{F}_1$ is then resolved into a component **P** along line *AB* and a component **Q** along *AC* (Fig. 3.35*c*); similarly, the force $-\mathbf{F}_1$ is resolved into $-\mathbf{P}$ along *AB* and $-\mathbf{Q}$ along *BD*. The forces **P** and $-\mathbf{P}$ have the same magnitude, the same line of action, and opposite sense; they can be moved along their common line of action until they are applied at the same point and may then be canceled. Thus the couple formed by $\mathbf{F}_1$ and $-\mathbf{F}_1$ reduces to a couple consisting of **Q** and $-\mathbf{Q}$.

We will now show that the forces **Q** and $-\mathbf{Q}$ are respectively equal to the forces $-\mathbf{F}_2$ and $\mathbf{F}_2$. The moment of the couple formed by **Q** and $-\mathbf{Q}$ can be obtained by computing the moment of **Q** about *B*; similarly, the moment of the couple formed by $\mathbf{F}_1$ and $-\mathbf{F}_1$ is the moment of $\mathbf{F}_1$ about *B*. But, by Varignon's theorem, the moment of $\mathbf{F}_1$ is equal to the sum of the moments of its components **P** and **Q**. Since the moment of **P** about *B* is zero, the moment of the couple formed by **Q** and $-\mathbf{Q}$ must be equal to the moment of the couple formed by $\mathbf{F}_1$ and $-\mathbf{F}_1$. Recalling (3.49), we write

$$Qd_2 = F_1d_1 = F_2d_2 \qquad \text{and} \qquad Q = F_2$$

Thus the forces **Q** and $-\mathbf{Q}$ are respectively equal to the forces $-\mathbf{F}_2$ and $\mathbf{F}_2$, and the couple of Fig. 3.35*a* is equivalent to the couple of Fig. 3.35*d*.

Next consider two couples contained in parallel planes P_1 and P_2; we will prove that they are equivalent if they have the same moment. In view of the foregoing, we can assume that the couples consist of forces of the same magnitude *F* acting along parallel lines (Fig. 3.36*a* and *d*). We propose to show that the couple contained in plane P_1 can be transformed into the couple contained in plane P_2 by means of the standard operations listed above.

Let us consider the two planes defined respectively by the lines of action of $\mathbf{F}_1$ and $-\mathbf{F}_2$ and by those of $-\mathbf{F}_1$ and $\mathbf{F}_2$ (Fig. 3.36*b*). At a point on their line of intersection we attach two forces $\mathbf{F}_3$ and $-\mathbf{F}_3$, respectively equal to $\mathbf{F}_1$ and $-\mathbf{F}_1$. The couple formed by $\mathbf{F}_1$ and $-\mathbf{F}_3$ can be replaced by a couple consisting of $\mathbf{F}_3$ and $-\mathbf{F}_2$ (Fig. 3.36*c*), since both couples clearly have the same moment and are contained in the same plane. Similarly, the couple formed by $-\mathbf{F}_1$ and $\mathbf{F}_3$ can be replaced by a couple consisting of $-\mathbf{F}_3$ and $\mathbf{F}_2$. Canceling the two equal and opposite forces $\mathbf{F}_3$ and $-\mathbf{F}_3$, we obtain the desired couple in plane P_2 (Fig. 3.36*d*). Thus, we conclude that two couples having the same moment $\mathbf{M}$ are equivalent, whether they are contained in the same plane or in parallel planes.

The property we have just established is very important for the correct understanding of the mechanics of rigid bodies. It indicates that when a couple acts on a rigid body, it does not matter where the two forces forming the couple act or what magnitude and direction they have. The only thing which counts is the *moment* of the couple (magnitude and direction). Couples with the same moment will have the same effect on the rigid body.

3.14. ADDITION OF COUPLES

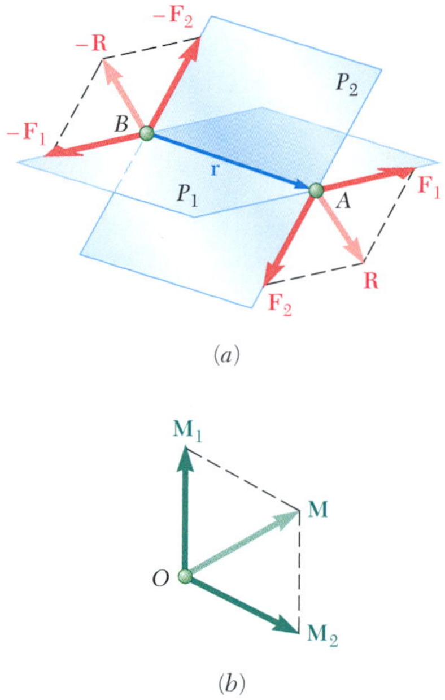

Fig. 3.37

Consider two intersecting planes P_1 and P_2 and two couples acting respectively in P_1 and P_2. We can, without any loss of generality, assume that the couple in P_1 consists of two forces $\mathbf{F}_1$ and $-\mathbf{F}_1$ perpendicular to the line of intersection of the two planes and acting respectively at A and B (Fig. 3.37*a*). Similarly, we assume that the couple in P_2 consists of two forces $\mathbf{F}_2$ and $-\mathbf{F}_2$ perpendicular to AB and acting respectively at A and B. It is clear that the resultant $\mathbf{R}$ of $\mathbf{F}_1$ and $\mathbf{F}_2$ and the resultant $-\mathbf{R}$ of $-\mathbf{F}_1$ and $-\mathbf{F}_2$ form a couple. Denoting by $\mathbf{r}$ the vector joining B to A and recalling the definition of the moment of a couple (Sec. 3.12), we express the moment $\mathbf{M}$ of the resulting couple as follows:

$$\mathbf{M} = \mathbf{r} \times \mathbf{R} = \mathbf{r} \times (\mathbf{F}_1 + \mathbf{F}_2)$$

and, by Varignon's theorem,

$$\mathbf{M} = \mathbf{r} \times \mathbf{F}_1 + \mathbf{r} \times \mathbf{F}_2$$

But the first term in the expression obtained represents the moment $\mathbf{M}_1$ of the couple in P_1, and the second term represents the moment $\mathbf{M}_2$ of the couple in P_2. We have

$$\mathbf{M} = \mathbf{M}_1 + \mathbf{M}_2 \tag{3.50}$$

and we conclude that the sum of two couples of moments $\mathbf{M}_1$ and $\mathbf{M}_2$ is a couple of moment $\mathbf{M}$ equal to the vector sum of $\mathbf{M}_1$ and $\mathbf{M}_2$ (Fig. 3.37*b*).

3.15. COUPLES CAN BE REPRESENTED BY VECTORS

As we saw in Sec. 3.13, couples which have the same moment, whether they act in the same plane or in parallel planes, are equivalent. There is therefore no need to draw the actual forces forming a given couple in order to define its effect on a rigid body (Fig. 3.38*a*). It is sufficient to draw an arrow equal in magnitude and direction to the moment $\mathbf{M}$ of the couple (Fig. 3.38*b*). On the other hand, we saw in Sec. 3.14 that the sum of two couples is itself a couple and that the moment $\mathbf{M}$ of the resultant couple can be obtained by forming the vector sum of the moments $\mathbf{M}_1$ and $\mathbf{M}_2$ of the given couples. Thus, couples obey the law of addition of vectors, and the arrow used in Fig. 3.38*b* to represent the couple defined in Fig. 3.38*a* can truly be considered a vector.

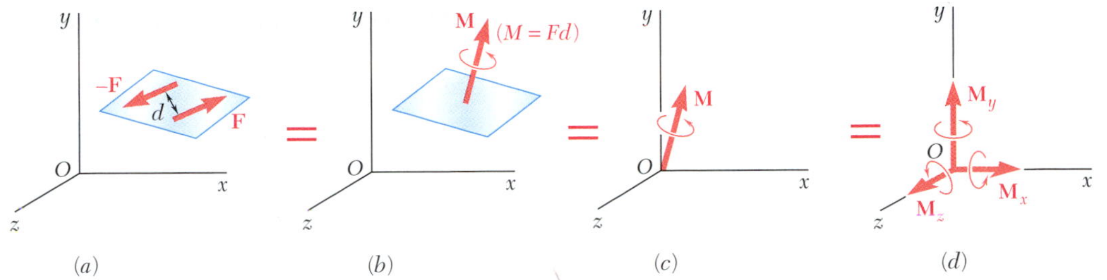

Fig. 3.38

The vector representing a couple is called a *couple vector.* Note that, in Fig. 3.38, a red arrow is used to distinguish the couple vector, *which represents the couple itself,* from the *moment* of the couple, which was represented by a green arrow in earlier figures. Also note that the symbol ↰ is added to this red arrow to avoid any confusion with vectors representing forces. A couple vector, like the moment of a couple, is a free vector. Its point of application, therefore, can be chosen at the origin of the system of coordinates, if so desired (Fig. 3.38*c*). Furthermore, the couple vector **M** can be resolved into component vectors $\mathbf{M}_x$, $\mathbf{M}_y$, and $\mathbf{M}_z$, which are directed along the coordinate axes (Fig. 3.38*d*). These component vectors represent couples acting, respectively, in the *yz*, *zx*, and *xy* planes.

3.16. RESOLUTION OF A GIVEN FORCE INTO A FORCE AT *O* AND A COUPLE

Consider a force **F** acting on a rigid body at a point *A* defined by the position vector **r** (Fig. 3.39*a*). Suppose that for some reason we would rather have the force act at point *O*. While we can move **F** along its line of action (principle of transmissibility), we cannot move it to a point *O* which does not lie on the original line of action without modifying the action of **F** on the rigid body.

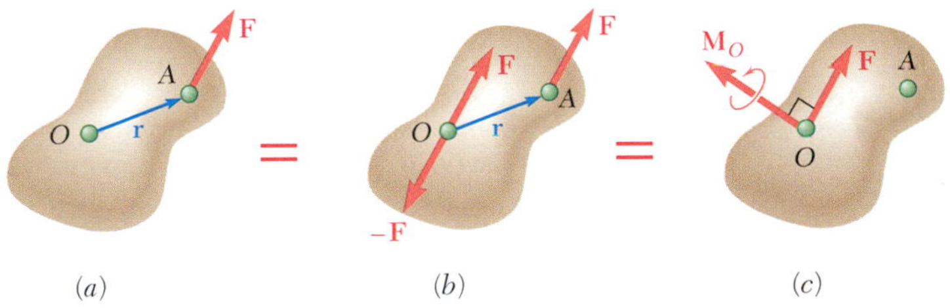

Fig. 3.39

We can, however, attach two forces at point *O*, one equal to **F** and the other equal to −**F**, without modifying the action of the original force on the rigid body (Fig. 3.39*b*). As a result of this transformation, a force **F** is now applied at *O*; the other two forces form a

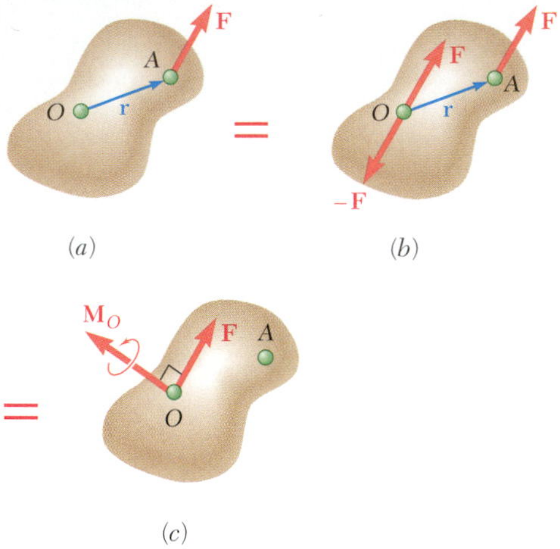

Fig. 3.39 (*repeated*)

couple of moment $\mathbf{M}_O = \mathbf{r} \times \mathbf{F}$. Thus, *any force* **F** *acting on a rigid body can be moved to an arbitrary point O provided that a couple is added whose moment is equal to the moment of* **F** *about O.* The couple tends to impart to the rigid body the same rotational motion about O that the force **F** tended to produce before it was transferred to O. The couple is represented by a couple vector $\mathbf{M}_O$ perpendicular to the plane containing **r** and **F**. Since $\mathbf{M}_O$ is a free vector, it may be applied anywhere; for convenience, however, the couple vector is usually attached at O, together with **F**, and the combination obtained is referred to as a *force-couple system* (Fig. 3.39*c*).

If the force **F** had been moved from A to a different point O' (Fig. 3.40*a* and *c*), the moment $\mathbf{M}_{O'} = \mathbf{r}' \times \mathbf{F}$ of **F** about O' should have been computed, and a new force-couple system, consisting of **F** and of the couple vector $\mathbf{M}_{O'}$, would have been attached at O'. The relation existing between the moments of **F** about O and O' is obtained by writing

$$\mathbf{M}_{O'} = \mathbf{r}' \times \mathbf{F} = (\mathbf{r} + \mathbf{s}) \times \mathbf{F} = \mathbf{r} \times \mathbf{F} + \mathbf{s} \times \mathbf{F}$$

$$\mathbf{M}_{O'} = \mathbf{M}_O + \mathbf{s} \times \mathbf{F} \tag{3.51}$$

where **s** is the vector joining O' to O. Thus, the moment $\mathbf{M}_{O'}$ of **F** about O' is obtained by adding to the moment $\mathbf{M}_O$ of **F** about O the vector product $\mathbf{s} \times \mathbf{F}$ representing the moment about O' of the force **F** applied at O.

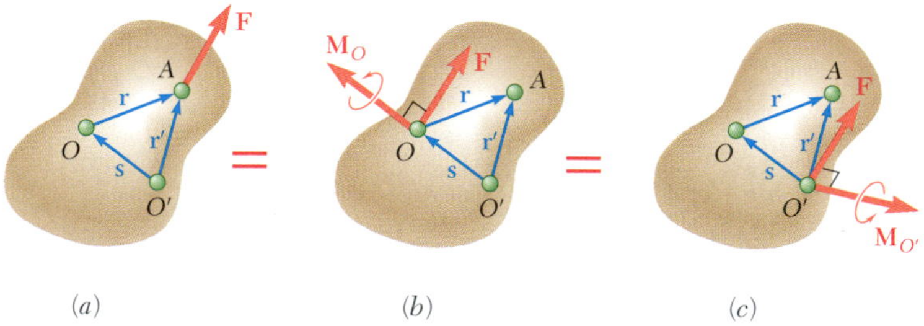

Fig. 3.40

This result could also have been established by observing that, in order to transfer to O' the force-couple system attached at O (Fig. 3.40*b* and *c*), the couple vector $\mathbf{M}_O$ can be freely moved to O'; to move the force **F** from O to O', however, it is necessary to add to **F** a couple vector whose moment is equal to the moment about O' of the force **F** applied at O. Thus, the couple vector $\mathbf{M}_{O'}$ must be the sum of $\mathbf{M}_O$ and the vector $\mathbf{s} \times \mathbf{F}$.

As noted above, the force-couple system obtained by transferring a force **F** from a point A to a point O consists of **F** and a couple vector $\mathbf{M}_O$ perpendicular to **F**. Conversely, any force-couple system consisting of a force **F** and a couple vector $\mathbf{M}_O$ which are *mutually perpendicular* can be replaced by a single equivalent force. This is done by moving the force **F** in the plane perpendicular to $\mathbf{M}_O$ until its moment about O is equal to the moment of the couple to be eliminated.

Photo 3.2 The force exerted by each hand on the wrench could be replaced with an equivalent force-couple system acting on the nut.

SAMPLE PROBLEM 3.6

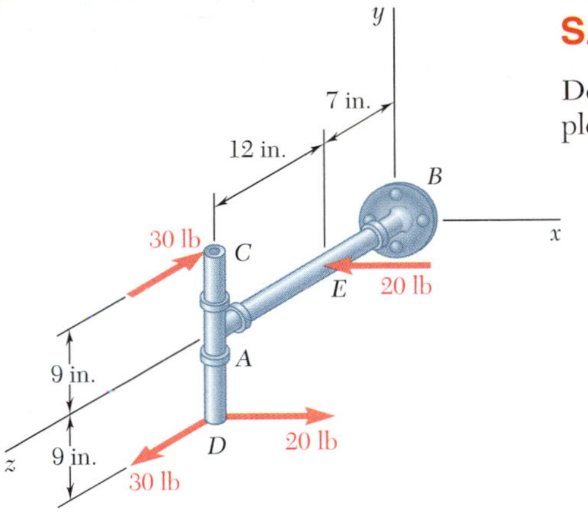

Determine the components of the single couple equivalent to the two couples shown.

SOLUTION

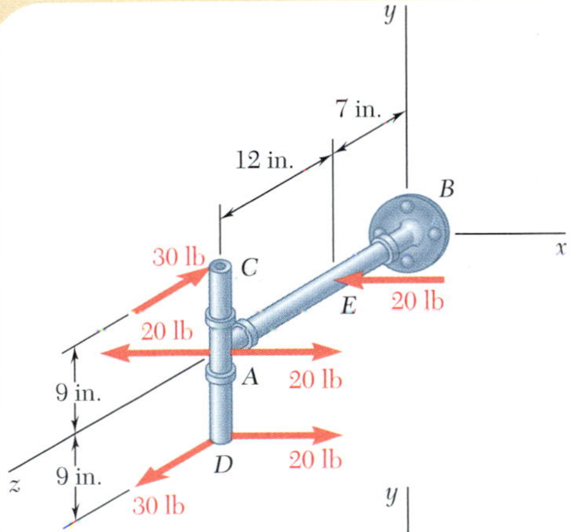

Our computations will be simplified if we attach two equal and opposite 20-lb forces at A. This enables us to replace the original 20-lb-force couple by two new 20-lb-force couples, one of which lies in the zx plane and the other in a plane parallel to the xy plane. The three couples shown in the adjoining sketch can be represented by three couple vectors $\mathbf{M}_x$, $\mathbf{M}_y$, and $\mathbf{M}_z$ directed along the coordinate axes. The corresponding moments are

$$M_x = -(30 \text{ lb})(18 \text{ in.}) = -540 \text{ lb} \cdot \text{in.}$$
$$M_y = +(20 \text{ lb})(12 \text{ in.}) = +240 \text{ lb} \cdot \text{in.}$$
$$M_z = +(20 \text{ lb})(9 \text{ in.}) = +180 \text{ lb} \cdot \text{in.}$$

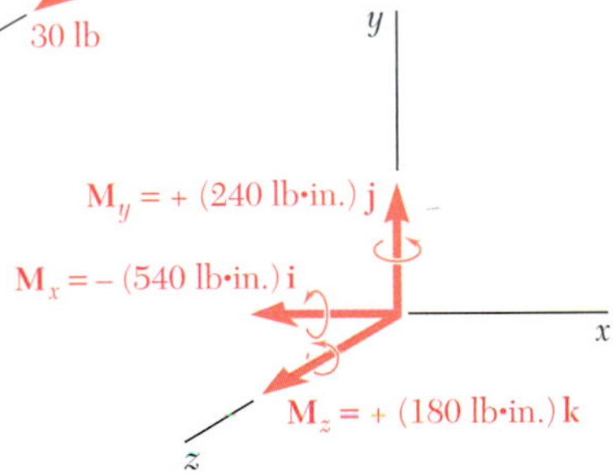

These three moments represent the components of the single couple $\mathbf{M}$ equivalent to the two given couples. We write

$$\mathbf{M} = -(540 \text{ lb} \cdot \text{in.})\mathbf{i} + (240 \text{ lb} \cdot \text{in.})\mathbf{j} + (180 \text{ lb} \cdot \text{in.})\mathbf{k} \quad ◀$$

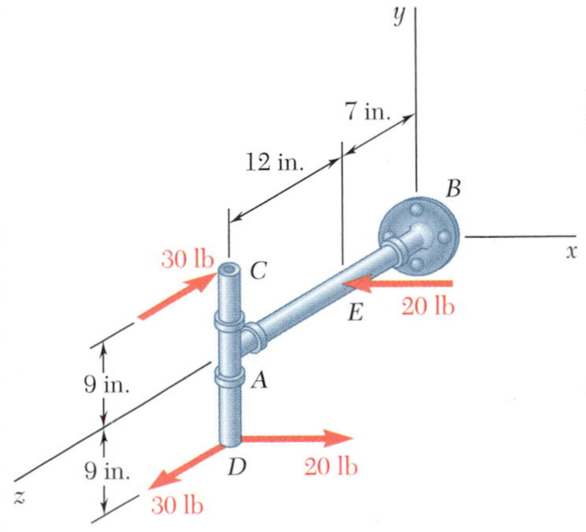

Alternative Solution. The components of the equivalent single couple $\mathbf{M}$ can also be obtained by computing the sum of the moments of the four given forces about an arbitrary point. Selecting point D, we write

$$\mathbf{M} = \mathbf{M}_D = (18 \text{ in.})\mathbf{j} \times (-30 \text{ lb})\mathbf{k} + [(9 \text{ in.})\mathbf{j} - (12 \text{ in.})\mathbf{k}] \times (-20 \text{ lb})\mathbf{i}$$

and, after computing the various vector products,

$$\mathbf{M} = -(540 \text{ lb} \cdot \text{in.})\mathbf{i} + (240 \text{ lb} \cdot \text{in.})\mathbf{j} + (180 \text{ lb} \cdot \text{in.})\mathbf{k} \quad ◀$$

SAMPLE PROBLEM 3.7

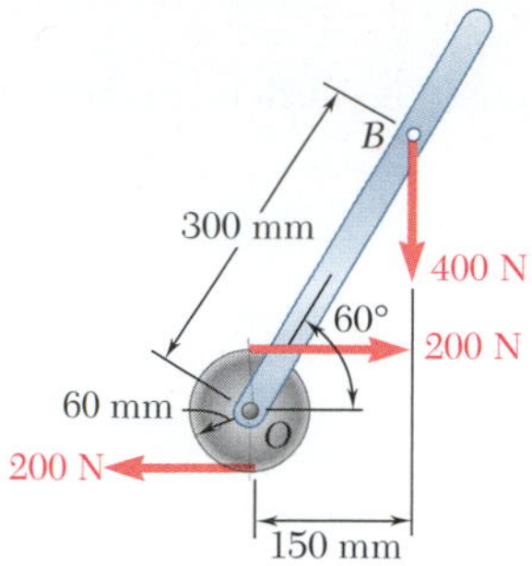

Replace the couple and force shown by an equivalent single force applied to the lever. Determine the distance from the shaft to the point of application of this equivalent force.

SOLUTION

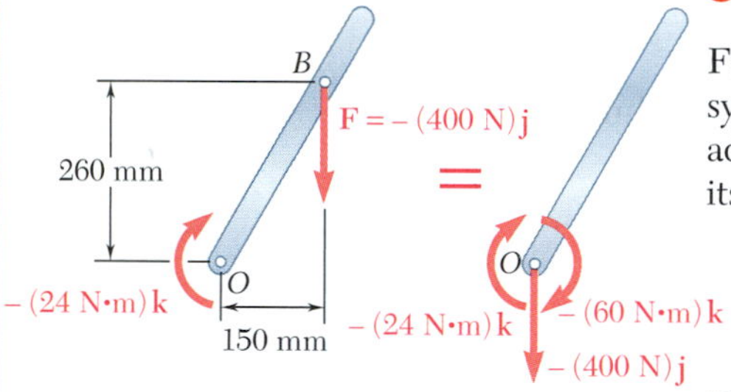

First the given force and couple are replaced by an equivalent force-couple system at O. We move the force $\mathbf{F} = -(400 \text{ N})\mathbf{j}$ to O and at the same time add a couple of moment $\mathbf{M}_O$ equal to the moment about O of the force in its original position.

$$\mathbf{M}_O = \overrightarrow{OB} \times \mathbf{F} = [(0.150 \text{ m})\mathbf{i} + (0.260 \text{ m})\mathbf{j}] \times (-400 \text{ N})\mathbf{j}$$
$$= -(60 \text{ N} \cdot \text{m})\mathbf{k}$$

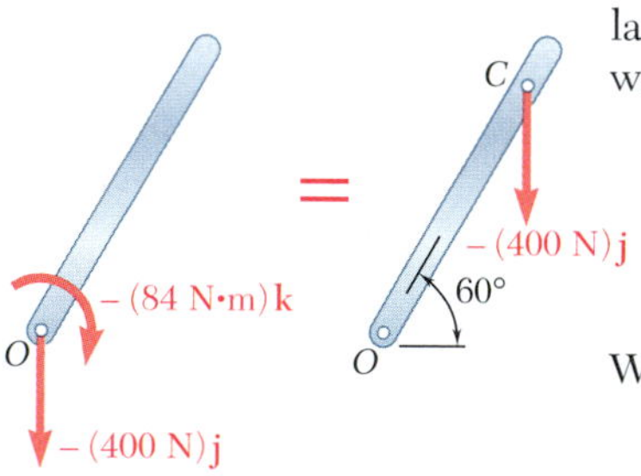

This couple is added to the couple of moment $-(24 \text{ N} \cdot \text{m})\mathbf{k}$ formed by the two 200-N forces, and a couple of moment $-(84 \text{ N} \cdot \text{m})\mathbf{k}$ is obtained. This last couple can be eliminated by applying $\mathbf{F}$ at a point C chosen in such a way that

$$-(84 \text{ N} \cdot \text{m})\mathbf{k} = \overrightarrow{OC} \times \mathbf{F}$$
$$= [(OC) \cos 60°\mathbf{i} + (OC) \sin 60°\mathbf{j}] \times (-400 \text{ N})\mathbf{j}$$
$$= -(OC) \cos 60°(400 \text{ N})\mathbf{k}$$

We conclude that

$$(OC) \cos 60° = 0.210 \text{ m} = 210 \text{ mm} \qquad OC = 420 \text{ mm} \blacktriangleleft$$

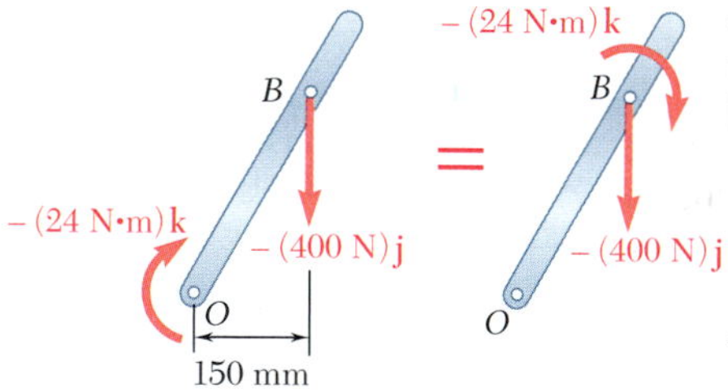

Alternative Solution. Since the effect of a couple does not depend on its location, the couple of moment $-(24 \text{ N} \cdot \text{m})\mathbf{k}$ can be moved to B; we thus obtain a force-couple system at B. The couple can now be eliminated by applying $\mathbf{F}$ at a point C chosen in such a way that

$$-(24 \text{ N} \cdot \text{m})\mathbf{k} = \overrightarrow{BC} \times \mathbf{F}$$
$$= -(BC) \cos 60°(400 \text{ N})\mathbf{k}$$

We conclude that

$$(BC) \cos 60° = 0.060 \text{ m} = 60 \text{ mm} \qquad BC = 120 \text{ mm}$$
$$OC = OB + BC = 300 \text{ mm} + 120 \text{ mm} \qquad OC = 420 \text{ mm} \blacktriangleleft$$

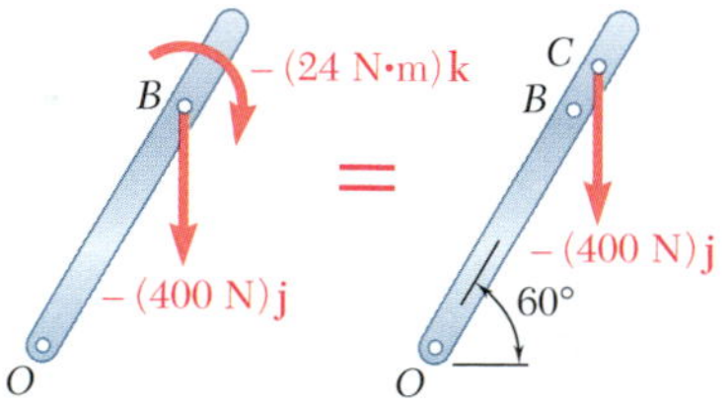

SOLVING PROBLEMS ON YOUR OWN

In this lesson we discussed the properties of *couples*. To solve the problems which follow, you will need to remember that the net effect of a couple is to produce a moment **M**. Since this moment is independent of the point about which it is computed, **M** is a *free vector* and thus remains unchanged as it is moved from point to point. Also, two couples are *equivalent* (that is, they have the same effect on a given rigid body) if they produce the same moment.

When determining the moment of a couple, all previous techniques for computing moments apply. Also, since the moment of a couple is a free vector, it should be computed relative to the most convenient point.

Because the only effect of a couple is to produce a moment, it is possible to represent a couple with a vector, the *couple vector,* which is equal to the moment of the couple. The couple vector is a free vector and will be represented by a special symbol, ↗↻, to distinguish it from force vectors.

In solving the problems in this lesson, you will be called upon to perform the following operations:

1. Adding two or more couples. This results in a new couple, the moment of which is obtained by adding vectorially the moments of the given couples [Sample Prob. 3.6].

2. Replacing a force with an equivalent force-couple system at a specified point. As explained in Sec. 3.16, the force of the force-couple system is equal to the original force, while the required couple vector is equal to the moment of the original force about the given point. In addition, it is important to observe that the force and the couple vector are perpendicular to each other. Conversely, it follows that a force-couple system can be reduced to a single force only if the force and couple vector are mutually perpendicular (see the next paragraph).

3. Replacing a force-couple system (with F perpendicular to M) with a single equivalent force. Note that the requirement that **F** and **M** be mutually perpendicular will be satisfied in all two-dimensional problems. The single equivalent force is equal to **F** and is applied in such a way that its moment about the original point of application is equal to **M** [Sample Prob. 3.7].

Problems

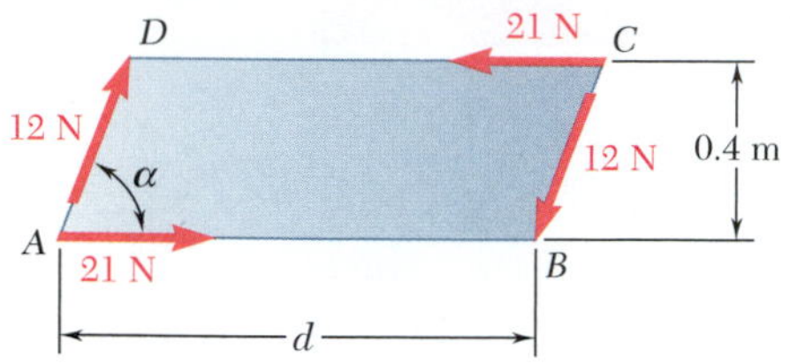

Fig. P3.68

3.68 A plate in the shape of a parallelogram is acted upon by two couples. Determine (*a*) the moment of the couple formed by the two 21-N forces, (*b*) the perpendicular distance between the 12-N forces if the resultant of the two couples is zero, (*c*) the value of α if the resultant couple is 1.8 N · m clockwise and d is 1.05 m.

3.69 A couple **M** of magnitude 10 lb · ft is applied to the handle of a screwdriver to tighten a screw into a block of wood. Determine the magnitudes of the two smallest horizontal forces that are equivalent to **M** if they are applied (*a*) at corners A and D, (*b*) at corners B and C, (*c*) anywhere on the block.

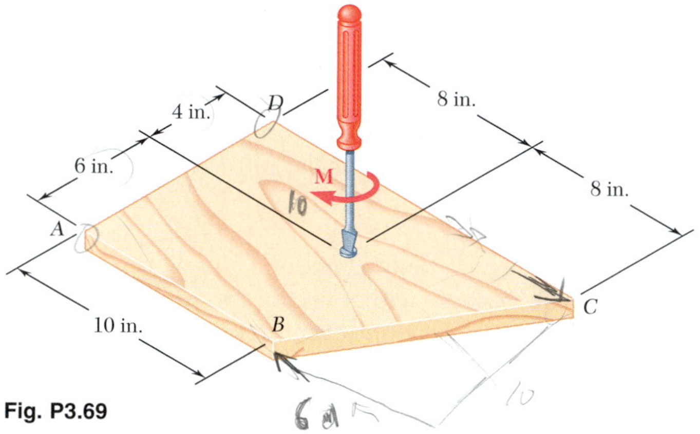

Fig. P3.69

3.70 Two 60-mm-diameter pegs are mounted on a steel plate at A and C, and two rods are attached to the plate at B and D. A cord is passed around the pegs and pulled as shown, while the rods exert on the plate 10-N forces as indicated. (*a*) Determine the resulting couple acting on the plate when $T = 36$ N. (*b*) If only the cord is used, in what direction should it be pulled to create the same couple with the minimum tension in the cord? (*c*) Determine the value of that minimum tension.

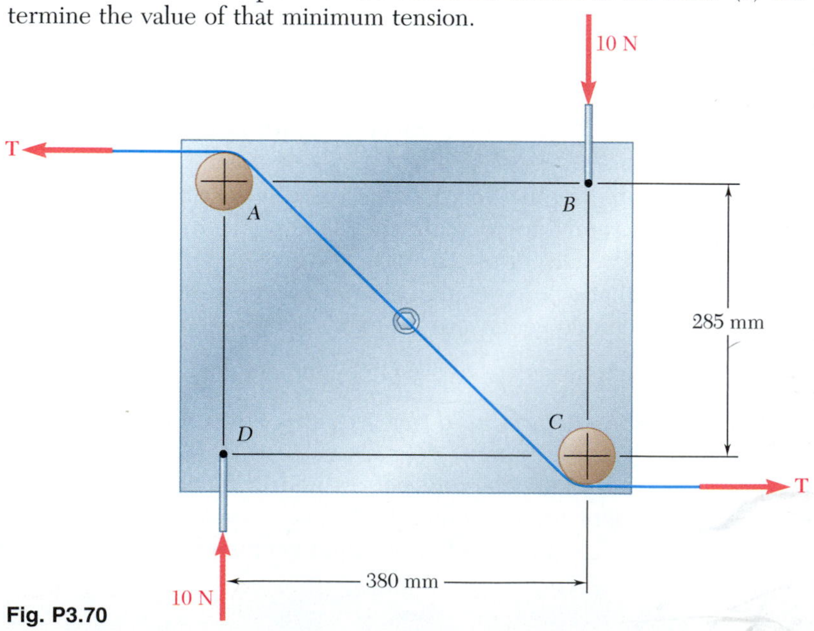

Fig. P3.70

3.71 The steel plate shown will support six 50-mm-diameter idler rollers mounted on the plate as shown. Two flat belts pass around the rollers, and rollers A and D will be adjusted so that the tension in each belt is 45 N. Determine (*a*) the resultant couple acting on the plate if $a = 0.2$ m, (*b*) the value of a so that the resultant couple acting on the plate is 54 N · m clockwise.

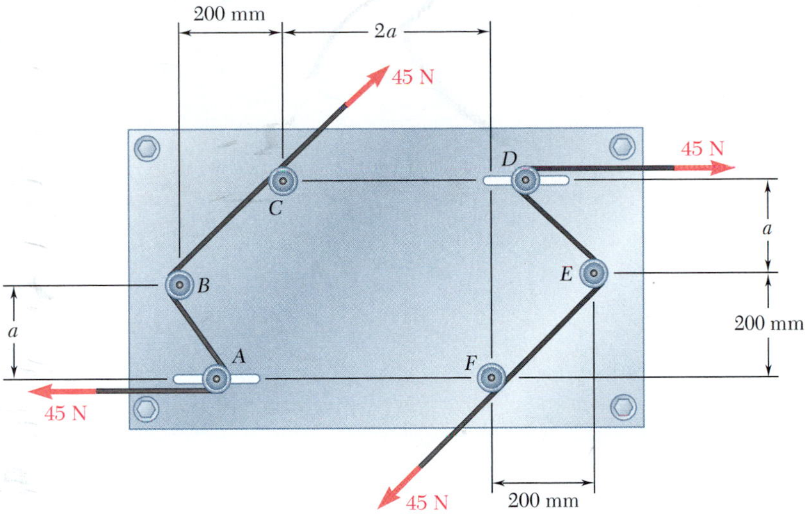

Fig. P3.71

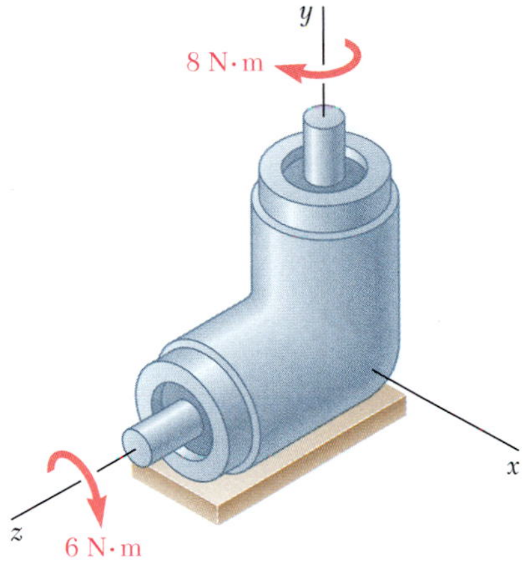

Fig. P3.72

3.72 The shafts of an angle drive are acted upon by the two couples shown. Replace the two couples with a single equivalent couple, specifying its magnitude and the direction of its axis.

3.73 and 3.74 Knowing that $P = 0$, replace the two remaining couples with a single equivalent couple, specifying its magnitude and the direction of its axis.

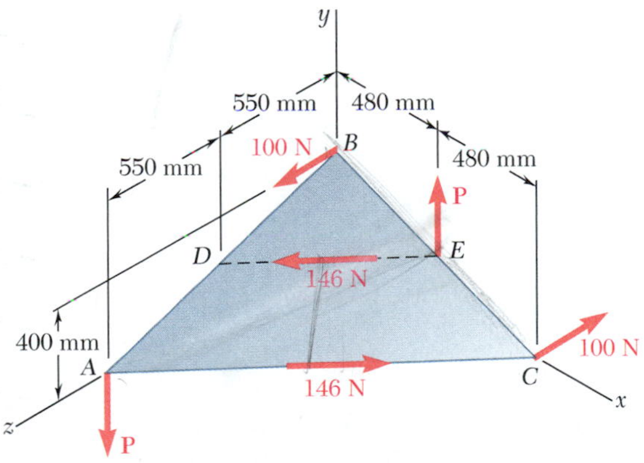

Fig. P3.73 and P3.76

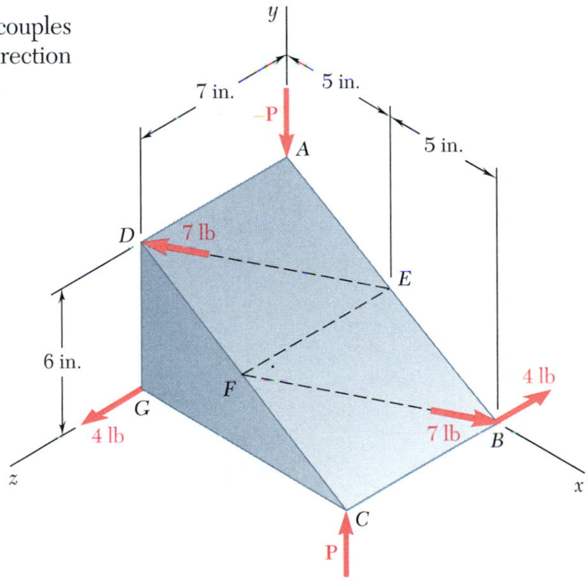

Fig. P3.74 and *P3.75*

3.75 Knowing that $P = 5$ lb, replace the three couples with a single equivalent couple, specifying its magnitude and the direction of its axis.

3.76 Knowing that $P = 210$ N, replace the three couples with a single equivalent couple, specifying its magnitude and the direction of its axis.

3.77 In a manufacturing operation, three holes are drilled simultaneously in a workpiece. Knowing that the holes are perpendicular to the surfaces of the workpiece, replace the couples applied to the drills with a single equivalent couple, specifying its magnitude and the direction of its axis.

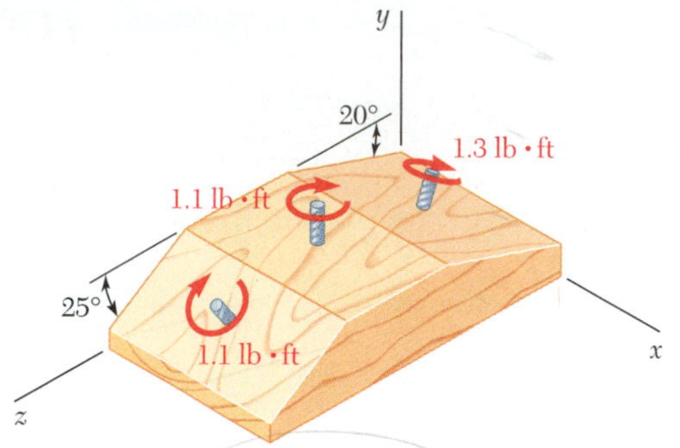

Fig. *P3.77*

3.78 The tension in the cable attached to the end C of an adjustable boom ABC is 1000 N. Replace the force exerted by the cable at C with an equivalent force-couple system (*a*) at A, (*b*) at B.

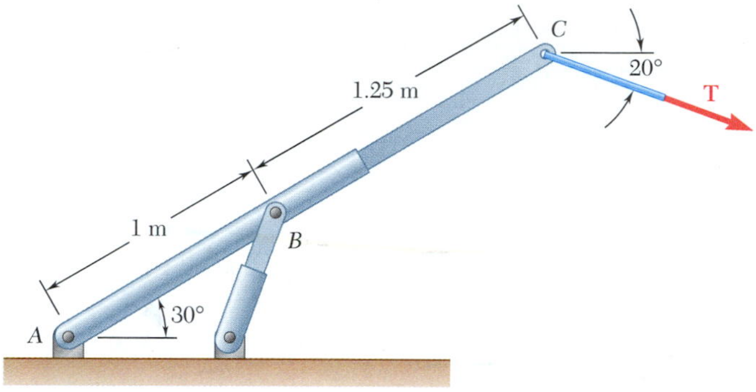

Fig. P3.78

3.79 The 20-lb horizontal force **P** acts on a bell crank as shown. (*a*) Replace **P** with an equivalent force-couple system at B. (*b*) Find the two vertical forces at C and D which are equivalent to the couple found in part *a*.

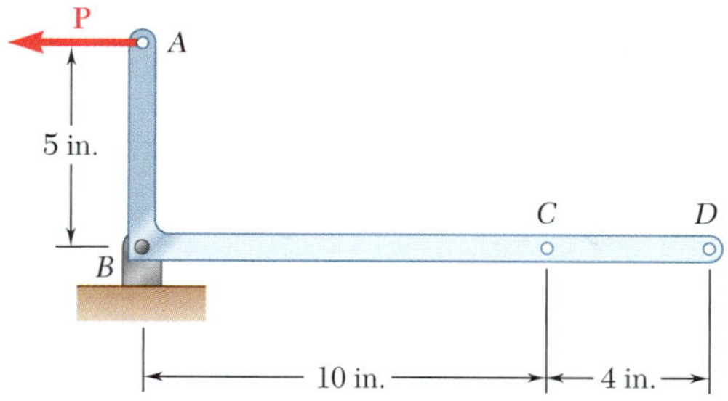

Fig. P3.79

3.80 A 700-N force **P** is applied at point A of a structural member. Replace **P** with (*a*) an equivalent force-couple system at C, (*b*) an equivalent system consisting of a vertical force at B and a second force at D.

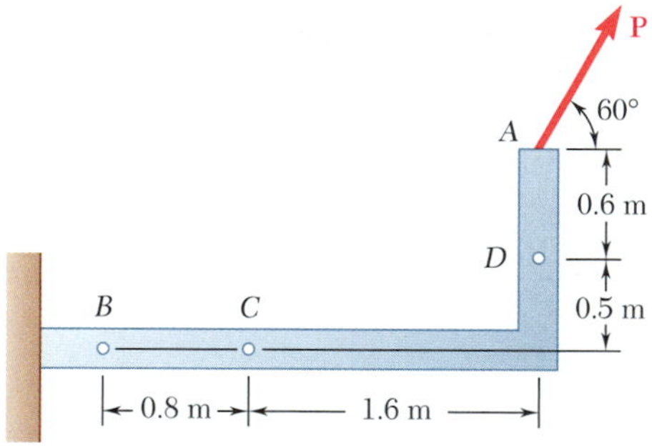

Fig. *P3.80*

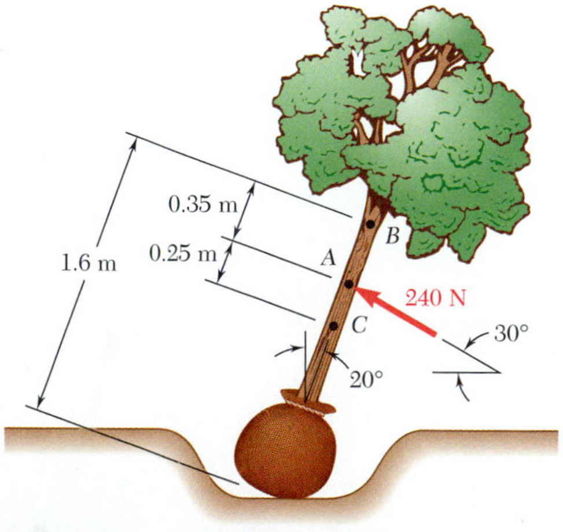

Fig. P3.81 and *P3.82*

3.81 A landscaper tries to plumb a tree by applying a 240-N force as shown. Two helpers then attempt to plumb the same tree, with one pulling at B and the other pushing with a parallel force at C. Determine these two forces so that they are equivalent to the single 240-N force shown in the figure.

3.82 A landscaper tries to plumb a tree by applying a 240-N force as shown. (*a*) Replace that force with an equivalent force-couple system at C. (*b*) Two helpers attempt to plumb the same tree, with one applying a horizontal force at C and the other pulling at B. Determine these two forces if they are to be equivalent to the single force of part *a*.

3.83 A dirigible is tethered by a cable attached to its cabin at B. If the tension in the cable is 250 lb, replace the force exerted by the cable at B with an equivalent system formed by two parallel forces applied at A and C.

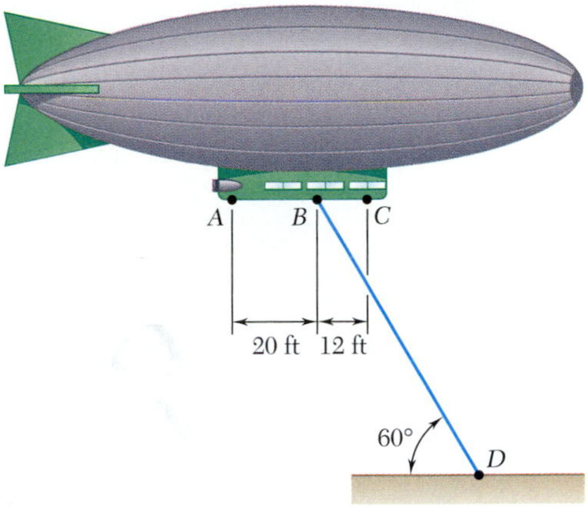

Fig. P3.83

3.84 Three workers trying to move a $3 \times 3 \times 4$-ft crate apply to the crate the three horizontal forces shown. (a) If $P = 60$ lb, replace the three forces with an equivalent force-couple system at A. (b) Replace the force-couple system of part a with a single force, and determine where it should be applied to side AB. (c) Determine the magnitude of $\mathbf{P}$ so that the three forces can be replaced with a single equivalent force applied at B.

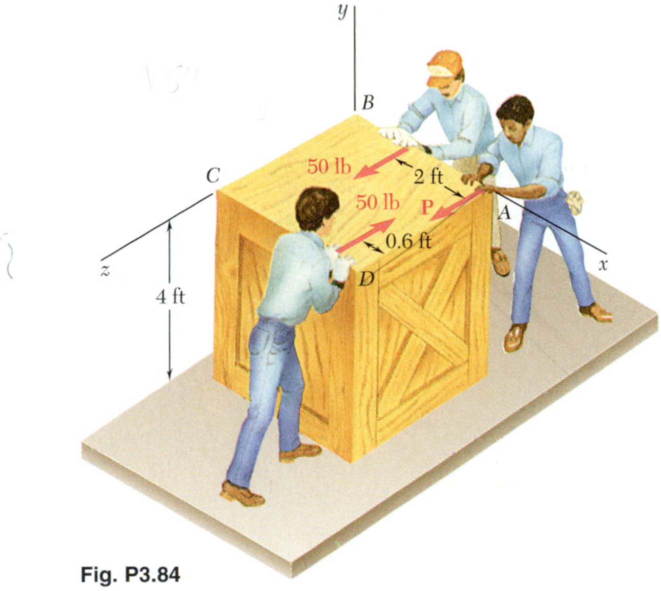

Fig. P3.84

3.85 A force and a couple are applied to a beam. (a) Replace this system with a single force $\mathbf{F}$ applied at point G, and determine the distance d. (b) Solve part a assuming that the directions of the two 600-N forces are reversed.

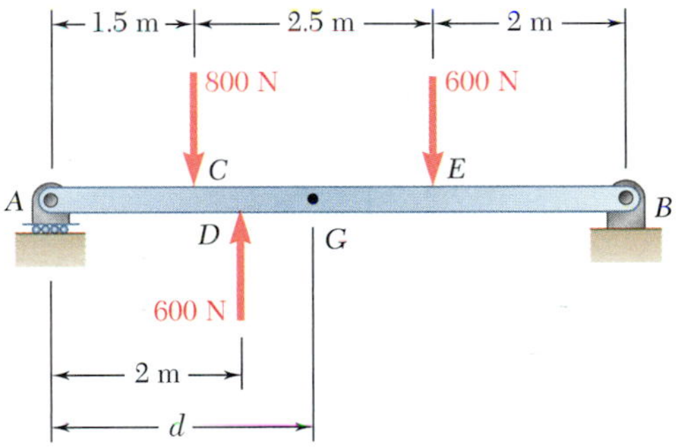

Fig. P3.85

3.86 Three cables attached to a disk exert on it the forces shown. (a) Replace the three forces with an equivalent force-couple system at A. (b) Determine the single force which is equivalent to the force-couple system obtained in part a, and specify its point of application on a line drawn through points A and D.

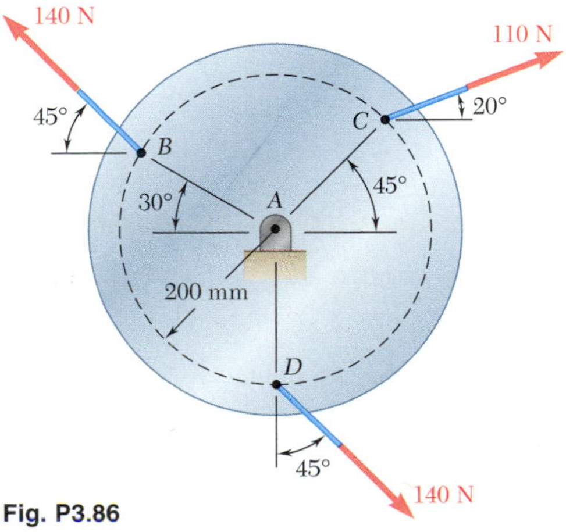

Fig. P3.86

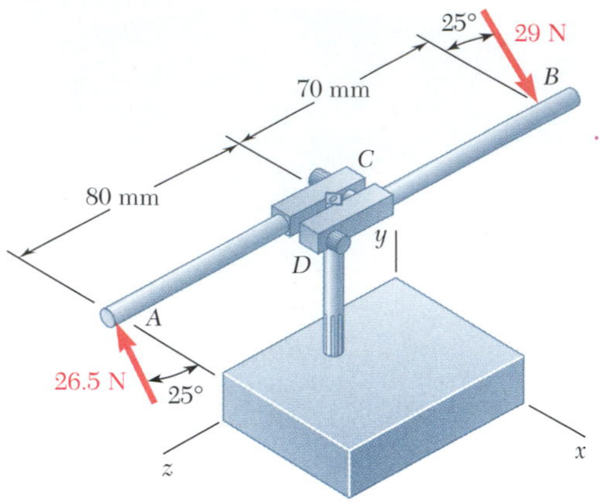

Fig. *P3.87*

3.87 While tapping a hole, a machinist applies the horizontal forces shown to the handle of the tap wrench. Show that these forces are equivalent to a single force, and specify, if possible, the point of application of the single force on the handle.

3.88 A rectangular plate is acted upon by the force and couple shown. This system is to be replaced with a single equivalent force. (*a*) For $\alpha = 40°$, specify the magnitude and the line of action of the equivalent force. (*b*) Specify the value of α if the line of action of the equivalent force is to intersect line *CD* 12 in. to the right of *D*.

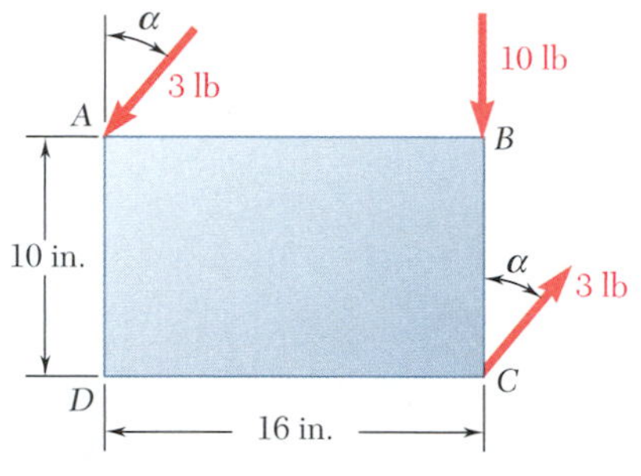

Fig. P3.88

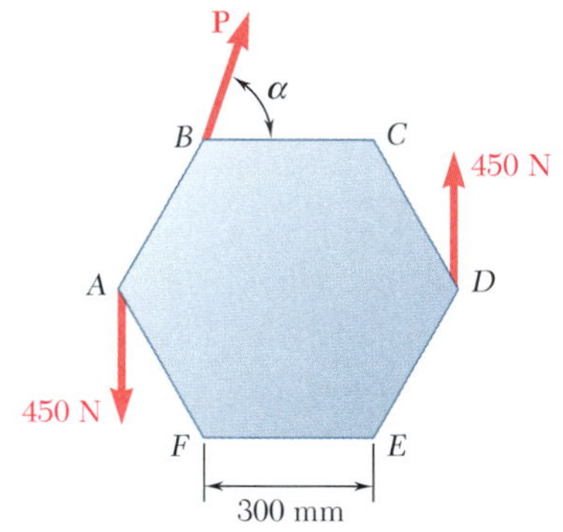

Fig. *P3.89*

3.89 A hexagonal plate is acted upon by the force **P** and the couple shown. Determine the magnitude and the direction of the smallest force **P** for which this system can be replaced with a single force at *E*.

3.90 An eccentric, compressive 270-lb force **P** is applied to the end of a cantilever beam. Replace **P** with an equivalent force-couple system at *G*.

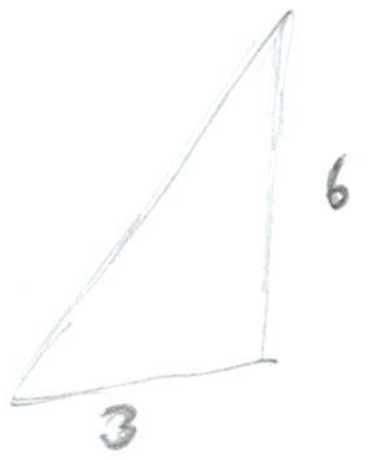

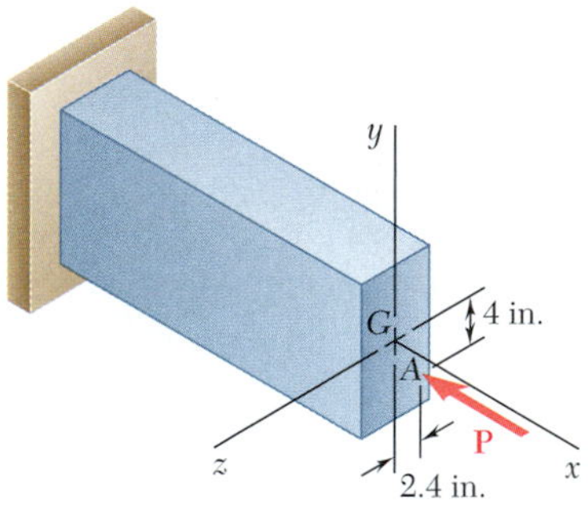

Fig. P3.90

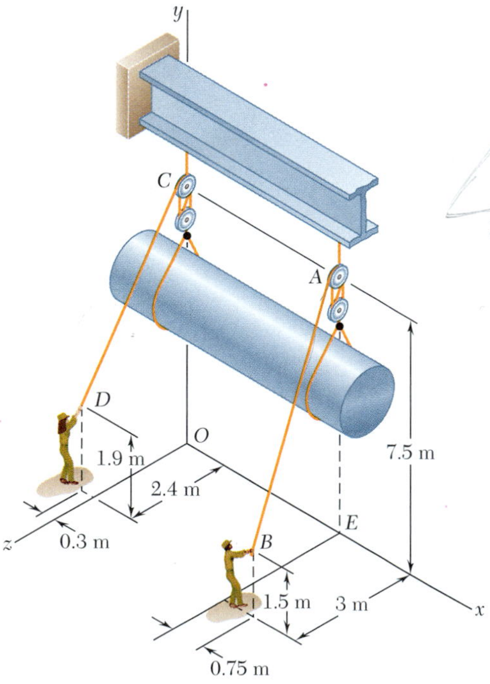

Fig. P3.91 and P3.92

3.91 Two workers use blocks and tackles attached to the bottom of an I-beam to lift a large cylindrical tank. Knowing that the tension in rope *AB* is 324 N, replace the force exerted at *A* by rope *AB* with an equivalent force-couple system at *E*.

3.92 Two workers use blocks and tackles attached to the bottom of an I-beam to lift a large cylindrical tank. Knowing that the tension in rope *CD* is 366 N, replace the force exerted at *C* by rope *CD* with an equivalent force-couple system at *O*.

3.93 To keep a door closed, a wooden stick is wedged between the floor and the doorknob. The stick exerts at B a 45-lb force directed along line AB. Replace that force with an equivalent force-couple system at C.

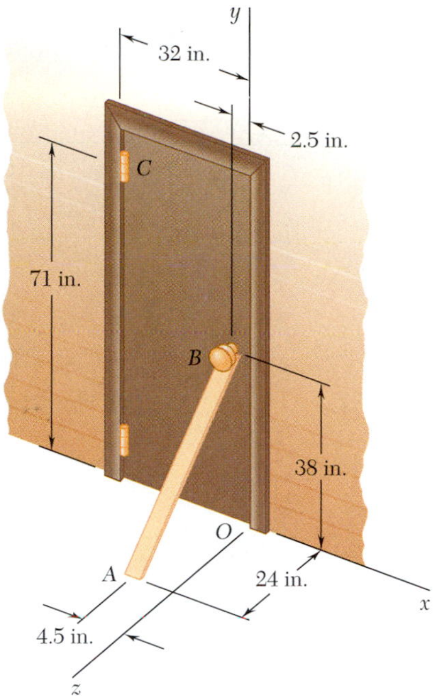

Fig. P3.93

3.94 A 25-lb force acting in a vertical plane parallel to the yz plane is applied to the 8-in.-long horizontal handle AB of a socket wrench. Replace the force with an equivalent force-couple system at the origin O of the coordinate system.

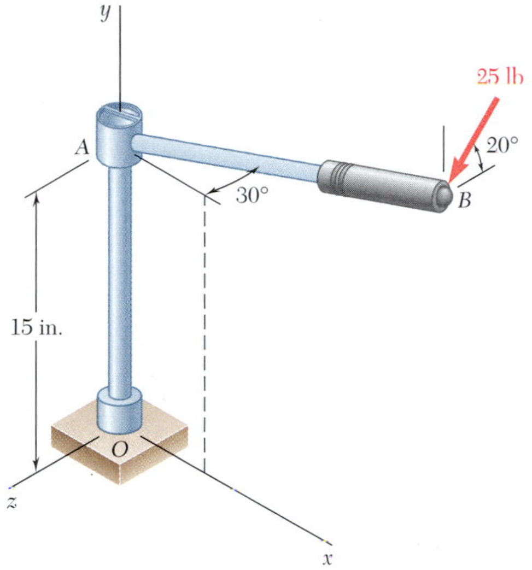

Fig. *P3.94*

3.95 A 315-N force **F** and 70-N · m couple **M** are applied to corner A of the block shown. Replace the given force-couple system with an equivalent force-couple system at corner D.

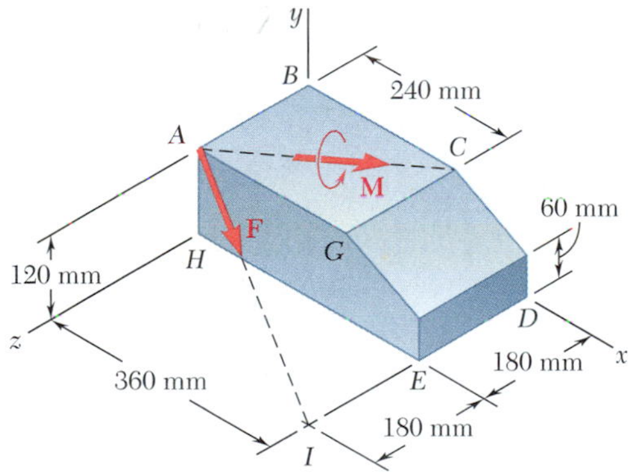

Fig. P3.95

3.96 The handpiece of a miniature industrial grinder weighs 2.4 N, and its center of gravity is located on the y axis. The head of the handpiece is offset in the xz plane in such a way that line BC forms an angle of 25° with the x direction. Show that the weight of the handpiece and the two couples $\mathbf{M}_1$ and $\mathbf{M}_2$ can be replaced with a single equivalent force. Further assuming that $M_1 = 0.068$ N · m and $M_2 = 0.065$ N · m, determine (*a*) the magnitude and the direction of the equivalent force, (*b*) the point where its line of action intersects the xz plane.

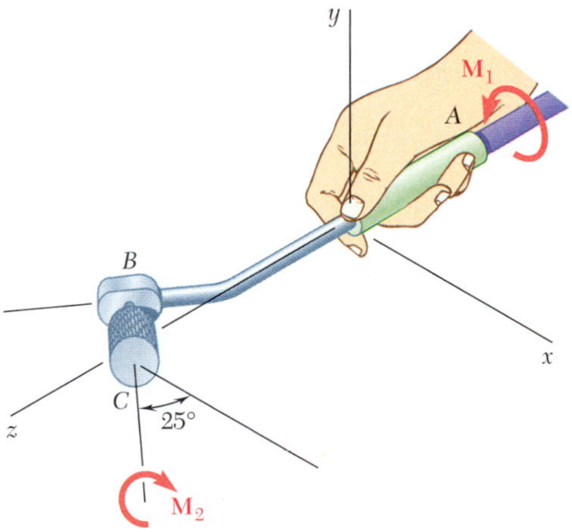

Fig. P3.96

3.97 A 20-lb force $\mathbf{F}_1$ and a 40-lb · ft couple $\mathbf{M}_1$ are applied to corner E of the bent plate shown. If $\mathbf{F}_1$ and $\mathbf{M}_1$ are to be replaced with an equivalent force-couple system ($\mathbf{F}_2$, $\mathbf{M}_2$) at corner B and if $(M_2)_z = 0$, determine (*a*) the distance d, (*b*) $\mathbf{F}_2$ and $\mathbf{M}_2$.

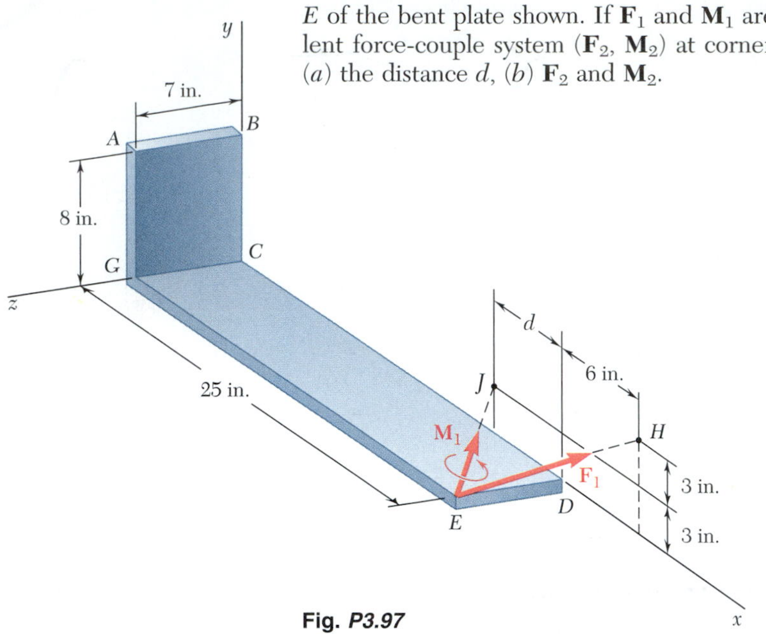

Fig. ***P3.97***

3.17. REDUCTION OF A SYSTEM OF FORCES TO ONE FORCE AND ONE COUPLE

Consider a system of forces $\mathbf{F}_1$, $\mathbf{F}_2$, $\mathbf{F}_3$, . . . , acting on a rigid body at the points A_1, A_2, A_3, . . . , *defined by the position vectors* $\mathbf{r}_1$, $\mathbf{r}_2$, $\mathbf{r}_3$, *etc.* (Fig. 3.41*a*). As seen in the preceding section, $\mathbf{F}_1$ can be moved from A_1 to a given point O if a couple of moment $\mathbf{M}_1$ equal to the moment $\mathbf{r}_1 \times \mathbf{F}_1$ of $\mathbf{F}_1$ about O is added to the original system of forces. Repeating this procedure with $\mathbf{F}_2$, $\mathbf{F}_3$, . . . , we obtain the system shown in Fig. 3.41*b*, which consists of the original forces, now acting at O, and the added couple vectors. Since the forces are now concurrent, they can be added vectorially and replaced by their resultant $\mathbf{R}$. Similarly, the couple vectors $\mathbf{M}_1$, $\mathbf{M}_2$, $\mathbf{M}_3$, . . . , can be added vectorially and replaced by a single couple vector $\mathbf{M}_O^R$. Any system of forces, however complex, can thus be reduced to an *equivalent force-couple system acting at a given point O* (Fig. 3.41*c*). We should note that while each of the couple vectors $\mathbf{M}_1$, $\mathbf{M}_2$, $\mathbf{M}_3$, . . . , in Fig. 3.41*b* is perpendicular to its corresponding force, the resultant force $\mathbf{R}$ and the resultant couple vector $\mathbf{M}_O^R$ in Fig. 3.41*c* will not, in general, be perpendicular to each other.

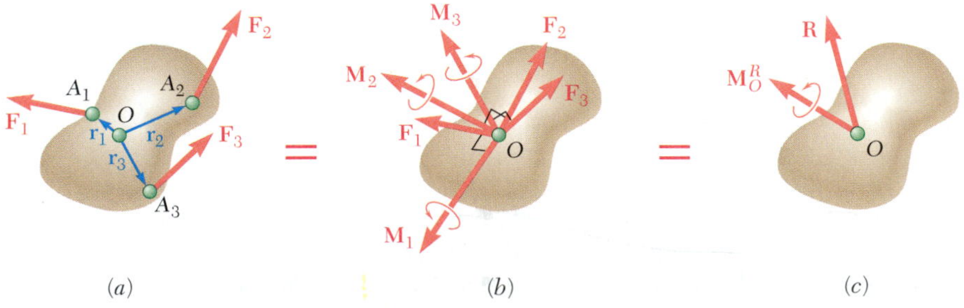

Fig. 3.41

The equivalent force-couple system is defined by the equations

$$\mathbf{R} = \Sigma\mathbf{F} \qquad \mathbf{M}_O^R = \Sigma\mathbf{M}_O = \Sigma(\mathbf{r} \times \mathbf{F}) \tag{3.52}$$

which express that the force **R** is obtained by adding all the forces of the system, while the moment of the resultant couple vector $\mathbf{M}_O^R$, called the *moment resultant* of the system, is obtained by adding the moments about O of all the forces of the system.

Once a given system of forces has been reduced to a force and a couple at a point O, it can easily be reduced to a force and a couple at another point O'. While the resultant force **R** will remain unchanged, the new moment resultant $\mathbf{M}_{O'}^R$ will be equal to the sum of $\mathbf{M}_O^R$ and the moment about O' of the force **R** attached at O (Fig. 3.42). We have

$$\mathbf{M}_{O'}^R = \mathbf{M}_O^R + \mathbf{s} \times \mathbf{R} \tag{3.53}$$

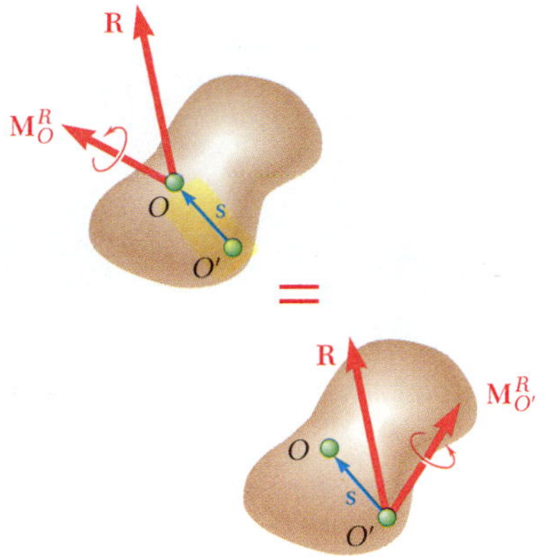

Fig. 3.42

In practice, the reduction of a given system of forces to a single force **R** at O and a couple vector $\mathbf{M}_O^R$ will be carried out in terms of components. Resolving each position vector **r** and each force **F** of the system into rectangular components, we write

$$\mathbf{r} = x\mathbf{i} + y\mathbf{j} + z\mathbf{k} \tag{3.54}$$
$$\mathbf{F} = F_x\mathbf{i} + F_y\mathbf{j} + F_z\mathbf{k} \tag{3.55}$$

Substituting for **r** and **F** in (3.52) and factoring out the unit vectors **i**, **j**, **k**, we obtain **R** and $\mathbf{M}_O^R$ in the form

$$\mathbf{R} = R_x\mathbf{i} + R_y\mathbf{j} + R_z\mathbf{k} \qquad \mathbf{M}_O^R = M_x^R\mathbf{i} + M_y^R\mathbf{j} + M_z^R\mathbf{k} \tag{3.56}$$

The components R_x, R_y, R_z represent, respectively, the sums of the x, y, and z components of the given forces and measure the tendency of the system to impart to the rigid body a motion of translation in the x, y, or z direction. Similarly, the components M_x^R, M_y^R, M_z^R represent, respectively, the sum of the moments of the given forces about the x, y, and z axes and measure the tendency of the system to impart to the rigid body a motion of rotation about the x, y, or z axis.

If the magnitude and direction of the force **R** are desired, they can be obtained from the components R_x, R_y, R_z by means of the relations (2.18) and (2.19) of Sec. 2.12; similar computations will yield the magnitude and direction of the couple vector $\mathbf{M}_O^R$.

3.18. EQUIVALENT SYSTEMS OF FORCES

We saw in the preceding section that any system of forces acting on a rigid body can be reduced to a force-couple system at a given point O. This equivalent force-couple system characterizes completely the effect of the given force system on the rigid body. *Two systems of forces are equivalent, therefore, if they can be reduced to the same*

Photo 3.3 The forces exerted by the tow rope and the rear tugboat on the ship can be replaced with an equivalent force-couple system when analyzing the motion of the ship.

force-couple system at a given point O. Recalling that the force-couple system at O is defined by the relations (3.52), we state that *two systems of forces*, $\mathbf{F}_1$, $\mathbf{F}_2$, $\mathbf{F}_3$, . . . , and $\mathbf{F}'_1$, $\mathbf{F}'_2$, $\mathbf{F}'_3$, . . . , *which act on the same rigid body are equivalent if, and only if, the sums of the forces and the sums of the moments about a given point O of the forces of the two systems are, respectively, equal*. Expressed mathematically, the necessary and sufficient conditions for the two systems of forces to be equivalent are

$$\Sigma\mathbf{F} = \Sigma\mathbf{F}' \quad \text{and} \quad \Sigma\mathbf{M}_O = \Sigma\mathbf{M}'_O \tag{3.57}$$

Note that to prove that two systems of forces are equivalent, the second of the relations (3.57) must be established with respect to *only one point O*. It will hold, however, with respect to *any point* if the two systems are equivalent.

Resolving the forces and moments in (3.57) into their rectangular components, we can express the necessary and sufficient conditions for the equivalence of two systems of forces acting on a rigid body as follows:

$$\begin{array}{lll} \Sigma F_x = \Sigma F'_x & \Sigma F_y = \Sigma F'_y & \Sigma F_z = \Sigma F'_z \\ \Sigma M_x = \Sigma M'_x & \Sigma M_y = \Sigma M'_y & \Sigma M_z = \Sigma M'_z \end{array} \tag{3.58}$$

These equations have a simple physical significance. They express that two systems of forces are equivalent if they tend to impart to the rigid body (1) the same translation in the x, y, and z directions, respectively, and (2) the same rotation about the x, y, and z axes, respectively.

3.19. EQUIPOLLENT SYSTEMS OF VECTORS

In general, when two systems of vectors satisfy Eqs. (3.57) or (3.58), that is, when their resultants and their moment resultants about an arbitrary point O are respectively equal, the two systems are said to be *equipollent*. The result established in the preceding section can thus be restated as follows: *If two systems of forces acting on a rigid body are equipollent, they are also equivalent*.

It is important to note that this statement does not apply to *any* system of vectors. Consider, for example, a system of forces acting on a set of independent particles which do *not* form a rigid body. A different system of forces acting on the same particles may happen to be equipollent to the first one; that is, it may have the same resultant and the same moment resultant. Yet, since different forces will now act on the various particles, their effects on these particles will be different; the two systems of forces, while equipollent, are *not equivalent*.

3.20. FURTHER REDUCTION OF A SYSTEM OF FORCES

We saw in Sec. 3.17 that any given system of forces acting on a rigid body can be reduced to an equivalent force-couple system at O consisting of a force $\mathbf{R}$ equal to the sum of the forces of the system and a couple vector $\mathbf{M}_O^R$ of moment equal to the moment resultant of the system.

When $\mathbf{R} = 0$, the force-couple system reduces to the couple vector $\mathbf{M}_O^R$. The given system of forces can then be reduced to a single couple, called the *resultant couple* of the system.

Let us now investigate the conditions under which a given system of forces can be reduced to a single force. It follows from Sec. 3.16 that the force-couple system at O can be replaced by a single force $\mathbf{R}$ acting along a new line of action if $\mathbf{R}$ and $\mathbf{M}_O^R$ are mutually perpendicular. The systems of forces which can be reduced to a single force, or *resultant,* are therefore the systems for which the force $\mathbf{R}$ and the couple vector $\mathbf{M}_O^R$ are mutually perpendicular. While this condition *is generally not satisfied* by systems of forces in space, it *will be satisfied* by systems consisting of (1) concurrent forces, (2) coplanar forces, or (3) parallel forces. These three cases will be discussed separately.

1. *Concurrent forces* are applied at the same point and can therefore be added directly to obtain their resultant $\mathbf{R}$. Thus, they always reduce to a single force. Concurrent forces were discussed in detail in Chap. 2.
2. *Coplanar forces* act in the same plane, which may be assumed to be the plane of the figure (Fig. 3.43*a*). The sum $\mathbf{R}$ of the forces of the system will also lie in the plane of the figure, while the moment of each force about O, and thus the moment resultant $\mathbf{M}_O^R$, will be perpendicular to that plane. The force-couple system at O consists, therefore, of a force $\mathbf{R}$ and a couple vector $\mathbf{M}_O^R$ which are mutually perpendicular (Fig. 3.43*b*).† They can be reduced to a single force $\mathbf{R}$ by moving $\mathbf{R}$ in the plane of the figure until its moment about O becomes equal to $\mathbf{M}_O^R$. The distance from O to the line of action of $\mathbf{R}$ is $d = M_O^R/R$ (Fig. 3.43*c*).

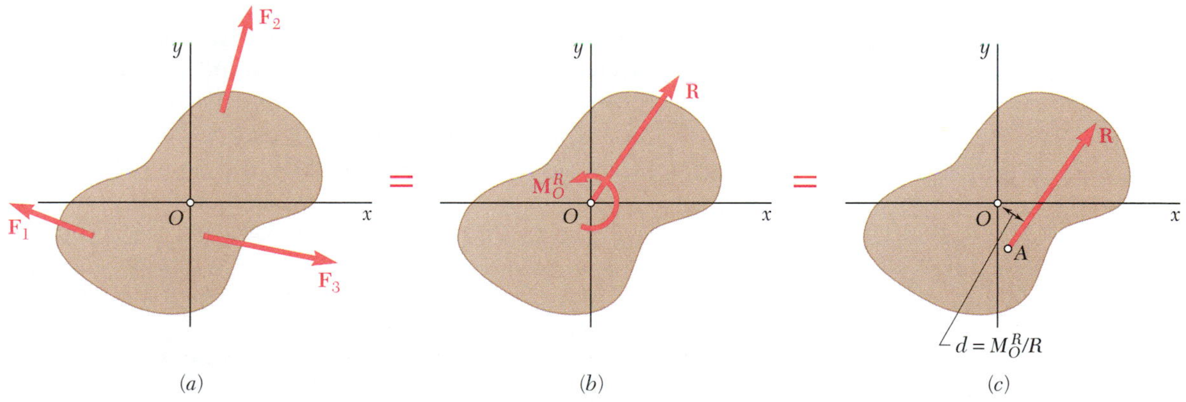

Fig. 3.43

†Since the couple vector $\mathbf{M}_O^R$ is perpendicular to the plane of the figure, it has been represented by the symbol ↺. A counterclockwise couple ↺ represents a vector pointing out of the paper, and a clockwise couple ↻ represents a vector pointing into the paper.

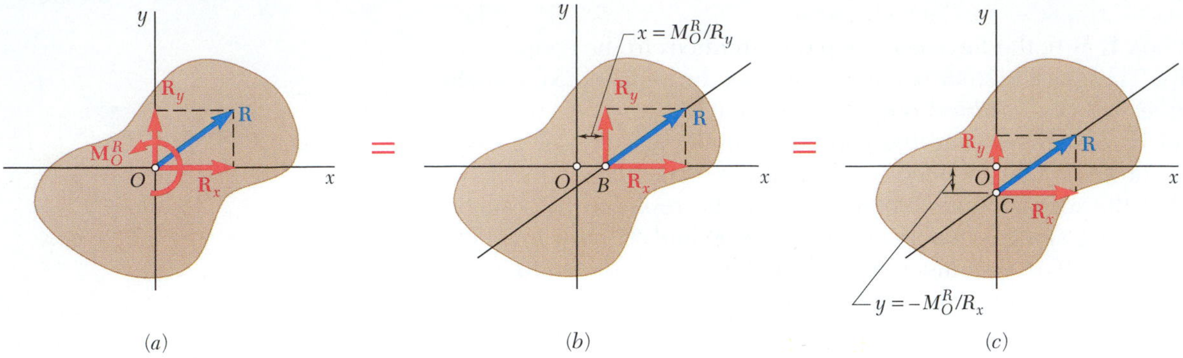

Fig. 3.44

As noted in Sec. 3.17, the reduction of a system of forces is considerably simplified if the forces are resolved into rectangular components. The force-couple system at O is then characterized by the components (Fig. 3.44*a*)

$$R_x = \Sigma F_x \qquad R_y = \Sigma F_y \qquad M_z^R = M_O^R = \Sigma M_O \tag{3.59}$$

To reduce the system to a single force **R**, we express that the moment of **R** about O must be equal to $\mathbf{M}_O^R$. Denoting by x and y the coordinates of the point of application of the resultant and recalling formula (3.22) of Sec. 3.8, we write

$$xR_y - yR_x = M_O^R$$

which represents the equation of the line of action of **R**. We can also determine directly the x and y intercepts of the line of action of the resultant by noting that $\mathbf{M}_O^R$ must be equal to the moment about O of the y component of **R** when **R** is attached at B (Fig. 3.44*b*) and to the moment of its x component when **R** is attached at C (Fig. 3.44*c*).

3. *Parallel forces* have parallel lines of action and may or may not have the same sense. Assuming here that the forces are parallel to the y axis (Fig. 3.45*a*), we note that their sum **R** will also be parallel to the y axis. On the other hand, since the moment of a given force must be perpendicular to that force, the moment about O of each force of the system, and thus the moment resultant $\mathbf{M}_O^R$, will lie in the zx plane. The force-couple system at O consists, therefore, of a force **R** and

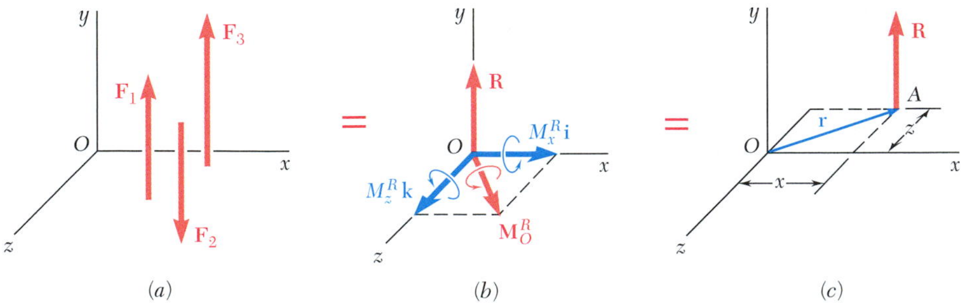

Fig. 3.45

a couple vector $\mathbf{M}_O^R$ which are mutually perpendicular (Fig. 3.45*b*). They can be reduced to a single force $\mathbf{R}$ (Fig. 3.45*c*) or, if $\mathbf{R} = 0$, to a single couple of moment $\mathbf{M}_O^R$.

In practice, the force-couple system at O will be characterized by the components

$$R_y = \Sigma F_y \qquad M_x^R = \Sigma M_x \qquad M_z^R = \Sigma M_z \tag{3.60}$$

The reduction of the system to a single force can be carried out by moving $\mathbf{R}$ to a new point of application $A(x, 0, z)$ chosen so that the moment of $\mathbf{R}$ about O is equal to $\mathbf{M}_O^R$. We write

$$\mathbf{r} \times \mathbf{R} = \mathbf{M}_O^R$$
$$(x\mathbf{i} + z\mathbf{k}) \times R_y\mathbf{j} = M_x^R\mathbf{i} + M_z^R\mathbf{k}$$

By computing the vector products and equating the coefficients of the corresponding unit vectors in both members of the equation, we obtain two scalar equations which define the coordinates of A:

$$-zR_y = M_x^R \qquad xR_y = M_z^R$$

These equations express that the moments of $\mathbf{R}$ about the x and z axes must, respectively, be equal to M_x^R and M_z^R.

Photo 3.4 The parallel wind forces acting on the highway signs can be reduced to a single equivalent force. Determining this force can simplify the calculation of the forces acting on the supports of the frame to which the signs are attached.

*3.21. REDUCTION OF A SYSTEM OF FORCES TO A WRENCH

In the general case of a system of forces in space, the equivalent force-couple system at O consists of a force $\mathbf{R}$ and a couple vector $\mathbf{M}_O^R$ which are not perpendicular, and neither of which is zero (Fig. 3.46*a*). Thus, the system of forces *cannot* be reduced to a single force or to a single couple. The couple vector, however, can be replaced by two other couple vectors obtained by resolving $\mathbf{M}_O^R$ into a component $\mathbf{M}_1$ along $\mathbf{R}$ and a component $\mathbf{M}_2$ in a plane perpendicular to $\mathbf{R}$ (Fig. 3.46*b*). The couple vector $\mathbf{M}_2$ and the force $\mathbf{R}$ can then be replaced by a single force $\mathbf{R}$ acting along a new line of action. The original system of forces thus reduces to $\mathbf{R}$ and to the couple vector $\mathbf{M}_1$ (Fig. 3.46*c*), that is, to $\mathbf{R}$ and a couple acting in the plane perpendicular to $\mathbf{R}$. This particular force-couple system is called a *wrench,* and the resulting combination of push and twist is found in drilling and tapping operations

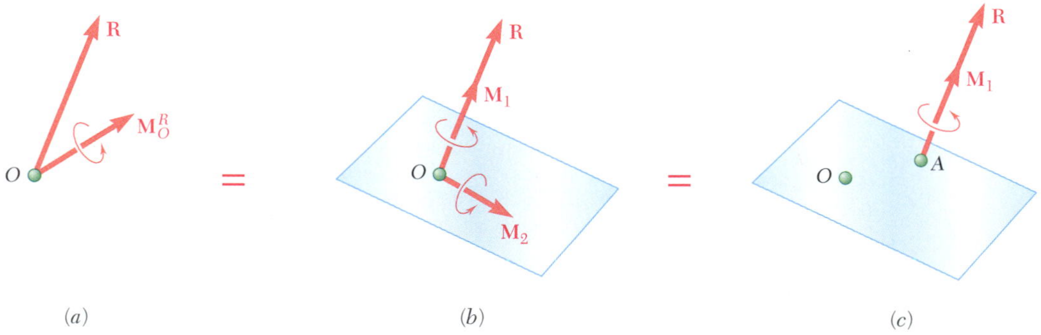

Fig. 3.46

and in the tightening and the loosening of screws. The line of action of $\mathbf{R}$ is known as the *axis of the wrench*, and the ratio $p = M_1/R$ is called the *pitch* of the wrench. A wrench, therefore, consists of two collinear vectors, namely, a force $\mathbf{R}$ and a couple vector

$$\mathbf{M}_1 = p\mathbf{R} \tag{3.61}$$

Recalling the expression (3.35) obtained in Sec. 3.9 for the projection of a vector on the line of action of another vector, we note that the projection of $\mathbf{M}_O^R$ on the line of action of $\mathbf{R}$ is

$$M_1 = \frac{\mathbf{R} \cdot \mathbf{M}_O^R}{R}$$

Thus, the pitch of the wrench can be expressed as†

$$p = \frac{M_1}{R} = \frac{\mathbf{R} \cdot \mathbf{M}_O^R}{R^2} \tag{3.62}$$

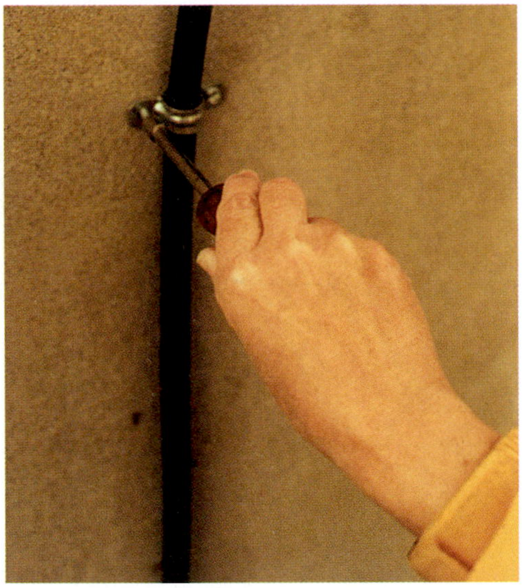

Photo 3.5 The pushing-turning action associated with the tightening of a screw illustrates the collinear lines of action of the force and couple vector that constitute a wrench.

To define the axis of the wrench, we can write a relation involving the position vector $\mathbf{r}$ of an arbitrary point P located on that axis. Attaching the resultant force $\mathbf{R}$ and couple vector $\mathbf{M}_1$ at P (Fig. 3.47) and expressing that the moment about O of this force-couple system is equal to the moment resultant $\mathbf{M}_O^R$ of the original force system, we write

$$\mathbf{M}_1 + \mathbf{r} \times \mathbf{R} = \mathbf{M}_O^R \tag{3.63}$$

or, recalling Eq. (3.61),

$$p\mathbf{R} + \mathbf{r} \times \mathbf{R} = \mathbf{M}_O^R \tag{3.64}$$

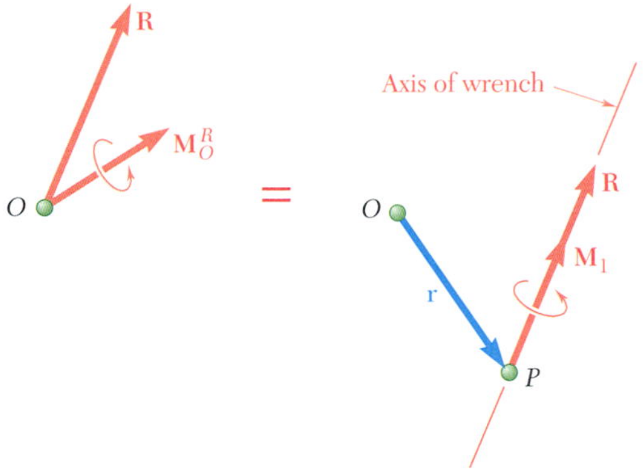

Fig. 3.47

†The expressions obtained for the projection of the couple vector on the line of action of $\mathbf{R}$ and for the pitch of the wrench are independent of the choice of point O. Using the relation (3.53) of Sec. 3.17, we note that if a different point O' had been used, the numerator in (3.62) would have been

$$\mathbf{R} \cdot \mathbf{M}_{O'}^R = \mathbf{R} \cdot (\mathbf{M}_O^R + \mathbf{s} \times \mathbf{R}) = \mathbf{R} \cdot \mathbf{M}_O^R + \mathbf{R} \cdot (\mathbf{s} \times \mathbf{R})$$

Since the mixed triple product $\mathbf{R} \cdot (\mathbf{s} \times \mathbf{R})$ is identically equal to zero, we have

$$\mathbf{R} \cdot \mathbf{M}_{O'}^R = \mathbf{R} \cdot \mathbf{M}_O^R$$

Thus, the scalar product $\mathbf{R} \cdot \mathbf{M}_O^R$ is independent of the choice of point O.

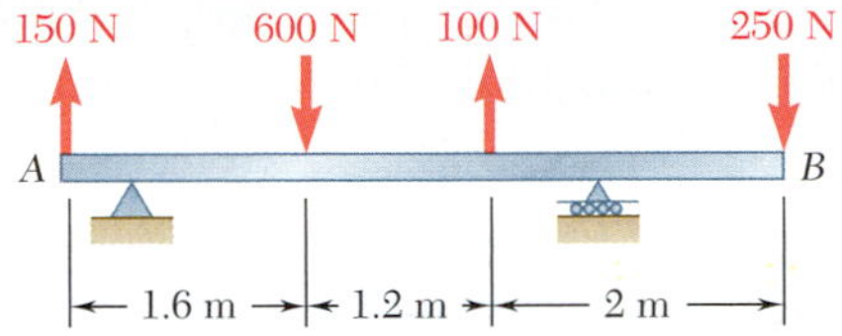

SAMPLE PROBLEM 3.8

A 4.80-m-long beam is subjected to the forces shown. Reduce the given system of forces to (*a*) an equivalent force-couple system at *A*, (*b*) an equivalent force-couple system at *B*, (*c*) a single force or resultant.

Note. Since the reactions at the supports are not included in the given system of forces, the given system will not maintain the beam in equilibrium.

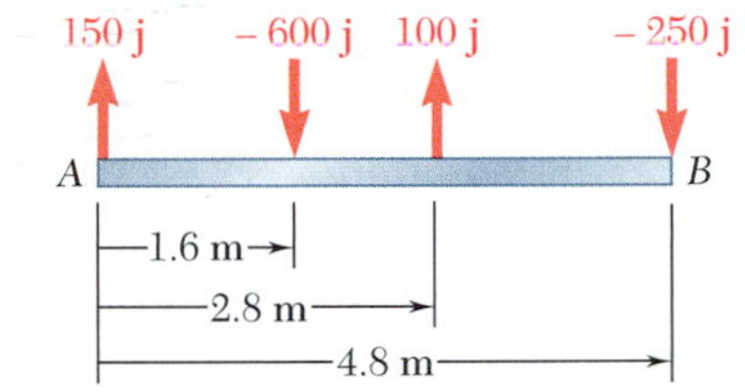

SOLUTION

***a*. Force-Couple System at *A*.** The force-couple system at *A* equivalent to the given system of forces consists of a force **R** and a couple $\mathbf{M}_A^R$ defined as follows:

$$\begin{aligned}\mathbf{R} &= \Sigma\mathbf{F}\\ &= (150\text{ N})\mathbf{j} - (600\text{ N})\mathbf{j} + (100\text{ N})\mathbf{j} - (250\text{ N})\mathbf{j} = -(600\text{ N})\mathbf{j}\\ \mathbf{M}_A^R &= \Sigma(\mathbf{r}\times\mathbf{F})\\ &= (1.6\mathbf{i})\times(-600\mathbf{j}) + (2.8\mathbf{i})\times(100\mathbf{j}) + (4.8\mathbf{i})\times(-250\mathbf{j})\\ &= -(1880\text{ N}\cdot\text{m})\mathbf{k}\end{aligned}$$

The equivalent force-couple system at *A* is thus

$\mathbf{R} = 600\text{ N}\downarrow \qquad \mathbf{M}_A^R = 1880\text{ N}\cdot\text{m}\ \circlearrowright$ ◀

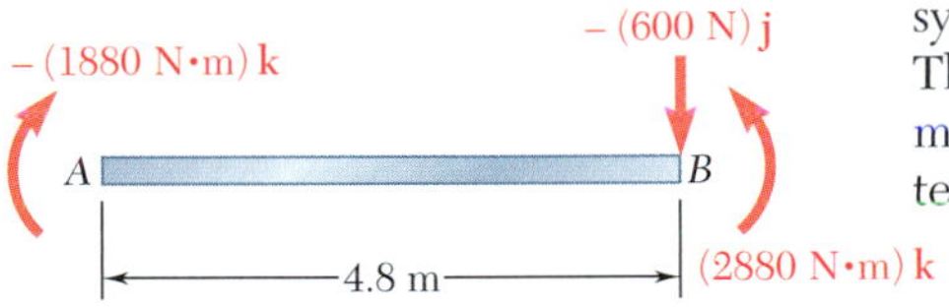

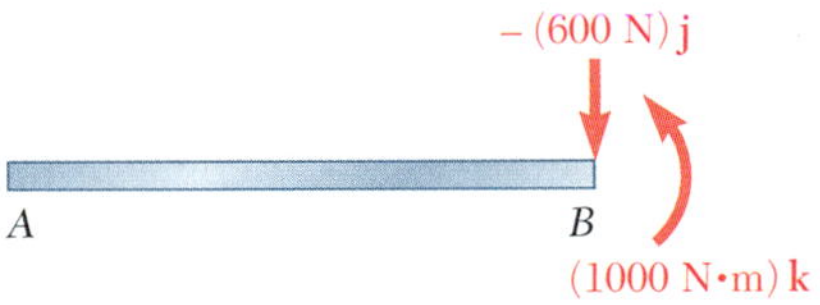

***b*. Force-Couple System at *B*.** We propose to find a force-couple system at *B* equivalent to the force-couple system at *A* determined in part *a*. The force **R** is unchanged, but a new couple $\mathbf{M}_B^R$ must be determined, the moment of which is equal to the moment about *B* of the force-couple system determined in part *a*. Thus, we have

$$\begin{aligned}\mathbf{M}_B^R &= \mathbf{M}_A^R + \overrightarrow{BA}\times\mathbf{R}\\ &= -(1880\text{ N}\cdot\text{m})\mathbf{k} + (-4.8\text{ m})\mathbf{i}\times(-600\text{ N})\mathbf{j}\\ &= -(1880\text{ N}\cdot\text{m})\mathbf{k} + (2880\text{ N}\cdot\text{m})\mathbf{k} = +(1000\text{ N}\cdot\text{m})\mathbf{k}\end{aligned}$$

The equivalent force-couple system at *B* is thus

$\mathbf{R} = 600\text{ N}\downarrow \qquad \mathbf{M}_B^R = 1000\text{ N}\cdot\text{m}\ \circlearrowleft$ ◀

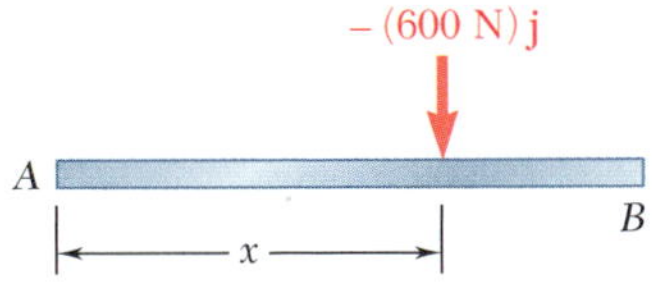

***c*. Single Force or Resultant.** The resultant of the given system of forces is equal to **R**, and its point of application must be such that the moment of **R** about *A* is equal to $\mathbf{M}_A^R$. We write

$$\begin{aligned}\mathbf{r}\times\mathbf{R} &= \mathbf{M}_A^R\\ x\mathbf{i}\times(-600\text{ N})\mathbf{j} &= -(1880\text{ N}\cdot\text{m})\mathbf{k}\\ -x(600\text{ N})\mathbf{k} &= -(1880\text{ N}\cdot\text{m})\mathbf{k}\end{aligned}$$

and conclude that $x = 3.13$ m. Thus, the single force equivalent to the given system is defined as

$\mathbf{R} = 600\text{ N}\downarrow \qquad x = 3.13\text{ m}$ ◀

SAMPLE PROBLEM 3.9

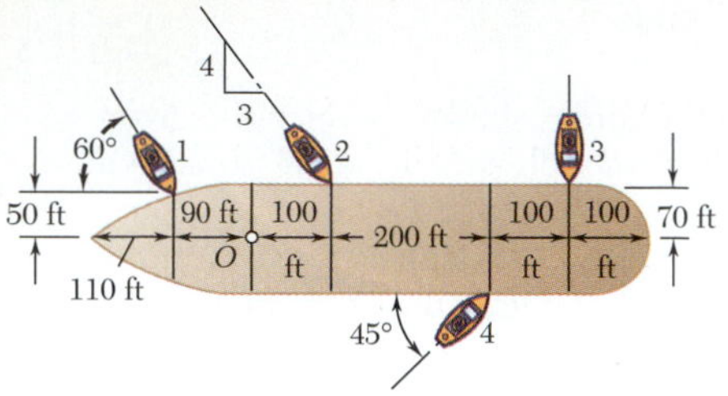

Four tugboats are used to bring an ocean liner to its pier. Each tugboat exerts a 5000-lb force in the direction shown. Determine (*a*) the equivalent force-couple system at the foremast *O*, (*b*) the point on the hull where a single, more powerful tugboat should push to produce the same effect as the original four tugboats.

SOLUTION

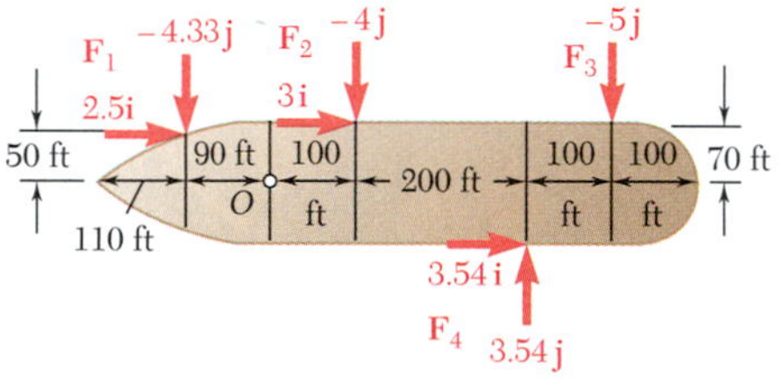

***a.* Force-Couple System at *O*.** Each of the given forces is resolved into components in the diagram shown (kip units are used). The force-couple system at *O* equivalent to the given system of forces consists of a force **R** and a couple $\mathbf{M}_O^R$ defined as follows:

$$\mathbf{R} = \Sigma\mathbf{F}$$
$$= (2.50\mathbf{i} - 4.33\mathbf{j}) + (3.00\mathbf{i} - 4.00\mathbf{j}) + (-5.00\mathbf{j}) + (3.54\mathbf{i} + 3.54\mathbf{j})$$
$$= 9.04\mathbf{i} - 9.79\mathbf{j}$$

$$\mathbf{M}_O^R = \Sigma(\mathbf{r} \times \mathbf{F})$$
$$= (-90\mathbf{i} + 50\mathbf{j}) \times (2.50\mathbf{i} - 4.33\mathbf{j})$$
$$+ (100\mathbf{i} + 70\mathbf{j}) \times (3.00\mathbf{i} - 4.00\mathbf{j})$$
$$+ (400\mathbf{i} + 70\mathbf{j}) \times (-5.00\mathbf{j})$$
$$+ (300\mathbf{i} - 70\mathbf{j}) \times (3.54\mathbf{i} + 3.54\mathbf{j})$$
$$= (390 - 125 - 400 - 210 - 2000 + 1062 + 248)\mathbf{k}$$
$$= -1035\mathbf{k}$$

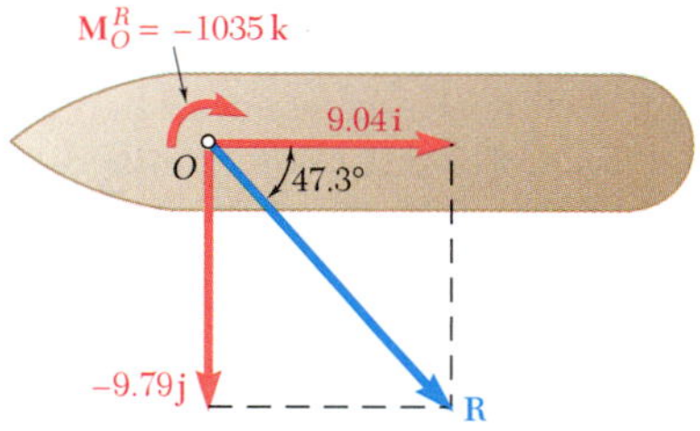

The equivalent force-couple system at *O* is thus

$$\mathbf{R} = (9.04 \text{ kips})\mathbf{i} - (9.79 \text{ kips})\mathbf{j} \qquad \mathbf{M}_O^R = -(1035 \text{ kip} \cdot \text{ft})\mathbf{k}$$

or $\qquad \mathbf{R} = 13.33 \text{ kips} \measuredangle 47.3° \qquad \mathbf{M}_O^R = 1035 \text{ kip} \cdot \text{ft} \circlearrowright$ ◀

Remark. Since all the forces are contained in the plane of the figure, we could have expected the sum of their moments to be perpendicular to that plane. Note that the moment of each force component could have been obtained directly from the diagram by first forming the product of its magnitude and perpendicular distance to *O* and then assigning to this product a positive or a negative sign depending upon the sense of the moment.

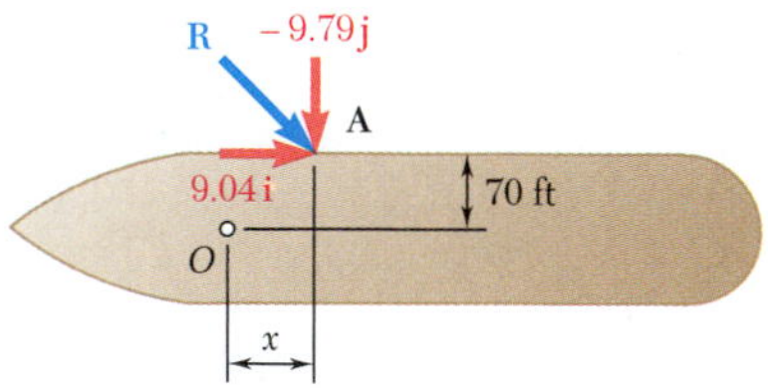

***b.* Single Tugboat.** The force exerted by a single tugboat must be equal to **R**, and its point of application *A* must be such that the moment of **R** about *O* is equal to $\mathbf{M}_O^R$. Observing that the position vector of *A* is

$$\mathbf{r} = x\mathbf{i} + 70\mathbf{j}$$

we write

$$\mathbf{r} \times \mathbf{R} = \mathbf{M}_O^R$$
$$(x\mathbf{i} + 70\mathbf{j}) \times (9.04\mathbf{i} - 9.79\mathbf{j}) = -1035\mathbf{k}$$
$$-x(9.79)\mathbf{k} - 633\mathbf{k} = -1035\mathbf{k} \qquad x = 41.1 \text{ ft}$$ ◀

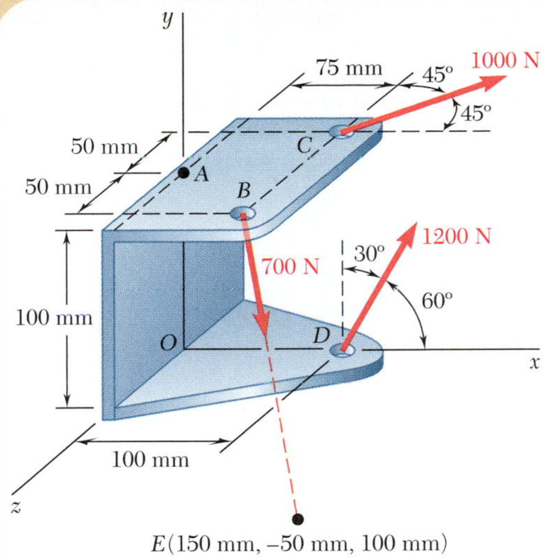

SAMPLE PROBLEM 3.10

Three cables are attached to a bracket as shown. Replace the forces exerted by the cables with an equivalent force-couple system at A.

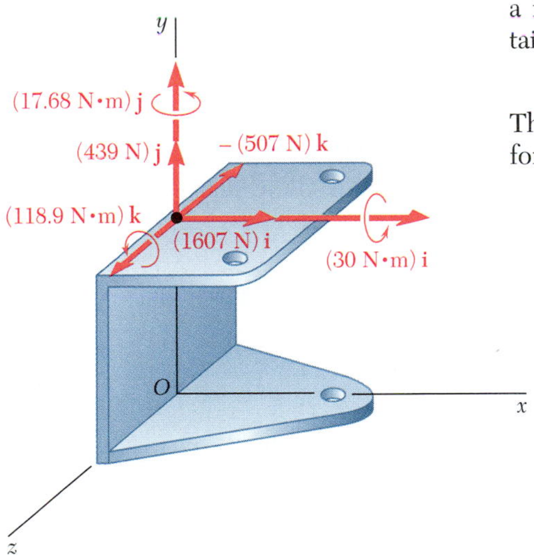

SOLUTION

We first determine the relative position vectors drawn from point A to the points of application of the various forces and resolve the forces into rectangular components. Observing that $\mathbf{F}_B = (700\ \text{N})\boldsymbol{\lambda}_{BE}$ where

$$\boldsymbol{\lambda}_{BE} = \frac{\overrightarrow{BE}}{BE} = \frac{75\mathbf{i} - 150\mathbf{j} + 50\mathbf{k}}{175}$$

we have, using meters and newtons,

$$\begin{aligned}
\mathbf{r}_{B/A} &= \overrightarrow{AB} = 0.075\mathbf{i} + 0.050\mathbf{k} & \mathbf{F}_B &= 300\mathbf{i} - 600\mathbf{j} + 200\mathbf{k}\\
\mathbf{r}_{C/A} &= \overrightarrow{AC} = 0.075\mathbf{i} - 0.050\mathbf{k} & \mathbf{F}_C &= 707\mathbf{i} - 707\mathbf{k}\\
\mathbf{r}_{D/A} &= \overrightarrow{AD} = 0.100\mathbf{i} - 0.100\mathbf{j} & \mathbf{F}_D &= 600\mathbf{i} + 1039\mathbf{j}
\end{aligned}$$

The force-couple system at A equivalent to the given forces consists of a force $\mathbf{R} = \Sigma\mathbf{F}$ and a couple $\mathbf{M}_A^R = \Sigma(\mathbf{r} \times \mathbf{F})$. The force $\mathbf{R}$ is readily obtained by adding respectively the x, y, and z components of the forces:

$$\mathbf{R} = \Sigma\mathbf{F} = (1607\ \text{N})\mathbf{i} + (439\ \text{N})\mathbf{j} - (507\ \text{N})\mathbf{k} \quad ◀$$

The computation of $\mathbf{M}_A^R$ will be facilitated if we express the moments of the forces in the form of determinants (Sec. 3.8):

$$\mathbf{r}_{B/A} \times \mathbf{F}_B = \begin{vmatrix} \mathbf{i} & \mathbf{j} & \mathbf{k} \\ 0.075 & 0 & 0.050 \\ 300 & -600 & 200 \end{vmatrix} = 30\mathbf{i} \qquad -45\mathbf{k}$$

$$\mathbf{r}_{C/A} \times \mathbf{F}_C = \begin{vmatrix} \mathbf{i} & \mathbf{j} & \mathbf{k} \\ 0.075 & 0 & -0.050 \\ 707 & 0 & -707 \end{vmatrix} = \qquad 17.68\mathbf{j}$$

$$\mathbf{r}_{D/A} \times \mathbf{F}_D = \begin{vmatrix} \mathbf{i} & \mathbf{j} & \mathbf{k} \\ 0.100 & -0.100 & 0 \\ 600 & 1039 & 0 \end{vmatrix} = \qquad 163.9\mathbf{k}$$

Adding the expressions obtained, we have

$$\mathbf{M}_A^R = \Sigma(\mathbf{r} \times \mathbf{F}) = (30\ \text{N}\cdot\text{m})\mathbf{i} + (17.68\ \text{N}\cdot\text{m})\mathbf{j} + (118.9\ \text{N}\cdot\text{m})\mathbf{k} \quad ◀$$

The rectangular components of the force $\mathbf{R}$ and the couple $\mathbf{M}_A^R$ are shown in the adjoining sketch.

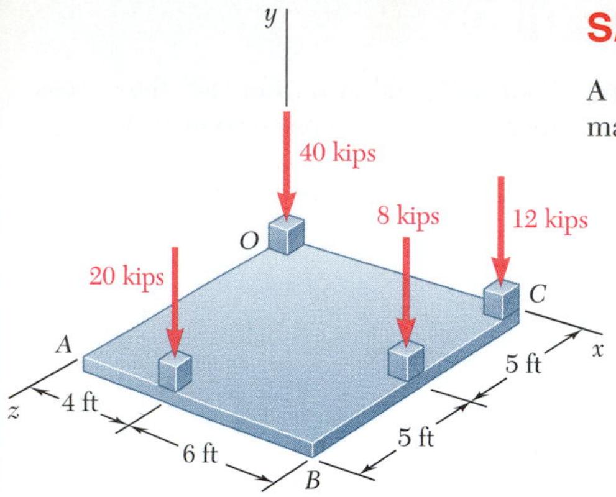

SAMPLE PROBLEM 3.11

A square foundation mat supports the four columns shown. Determine the magnitude and point of application of the resultant of the four loads.

SOLUTION

We first reduce the given system of forces to a force-couple system at the origin O of the coordinate system. This force-couple system consists of a force $\mathbf{R}$ and a couple vector $\mathbf{M}_O^R$ defined as follows:

$$\mathbf{R} = \Sigma\mathbf{F} \qquad \mathbf{M}_O^R = \Sigma(\mathbf{r} \times \mathbf{F})$$

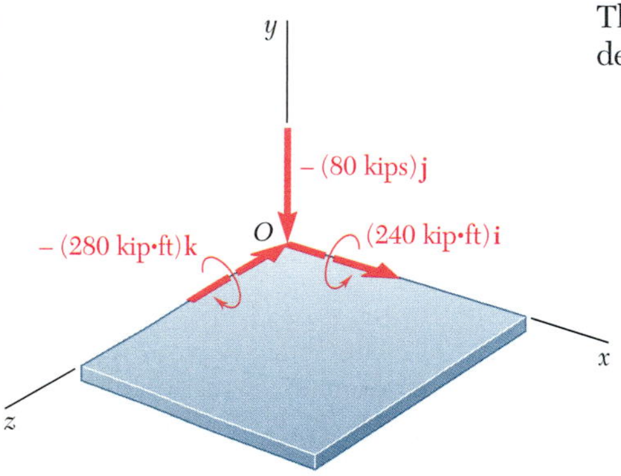

The position vectors of the points of application of the various forces are determined, and the computations are arranged in tabular form.

r, ft	**F**, kips	**r** × **F**, kip · ft
0	$-40\mathbf{j}$	0
$10\mathbf{i}$	$-12\mathbf{j}$	$-120\mathbf{k}$
$10\mathbf{i} + 5\mathbf{k}$	$-8\mathbf{j}$	$40\mathbf{i} - 80\mathbf{k}$
$4\mathbf{i} + 10\mathbf{k}$	$-20\mathbf{j}$	$200\mathbf{i} - 80\mathbf{k}$
	$\mathbf{R} = -80\mathbf{j}$	$\mathbf{M}_O^R = 240\mathbf{i} - 280\mathbf{k}$

Since the force $\mathbf{R}$ and the couple vector $\mathbf{M}_O^R$ are mutually perpendicular, the force-couple system obtained can be reduced further to a single force $\mathbf{R}$. The new point of application of $\mathbf{R}$ will be selected in the plane of the mat and in such a way that the moment of $\mathbf{R}$ about O will be equal to $\mathbf{M}_O^R$. Denoting by $\mathbf{r}$ the position vector of the desired point of application, and by x and z its coordinates, we write

$$\mathbf{r} \times \mathbf{R} = \mathbf{M}_O^R$$
$$(x\mathbf{i} + z\mathbf{k}) \times (-80\mathbf{j}) = 240\mathbf{i} - 280\mathbf{k}$$
$$-80x\mathbf{k} + 80z\mathbf{i} = 240\mathbf{i} - 280\mathbf{k}$$

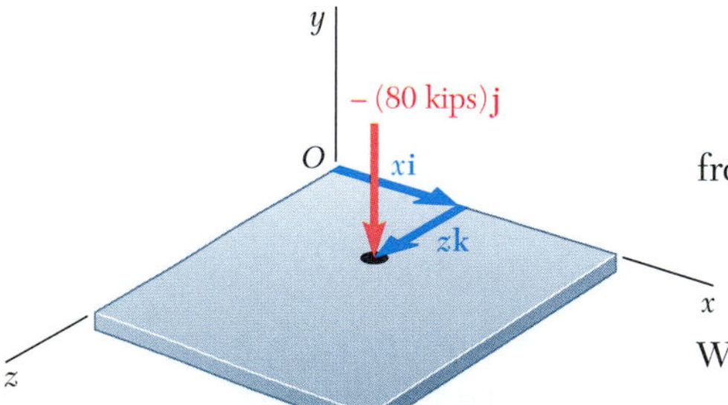

from which it follows that

$$-80x = -280 \qquad 80z = 240$$
$$x = 3.50 \text{ ft} \qquad z = 3.00 \text{ ft}$$

We conclude that the resultant of the given system of forces is

$$\mathbf{R} = 80 \text{ kips} \downarrow \qquad \text{at } x = 3.50 \text{ ft}, z = 3.00 \text{ ft}$$ ◀

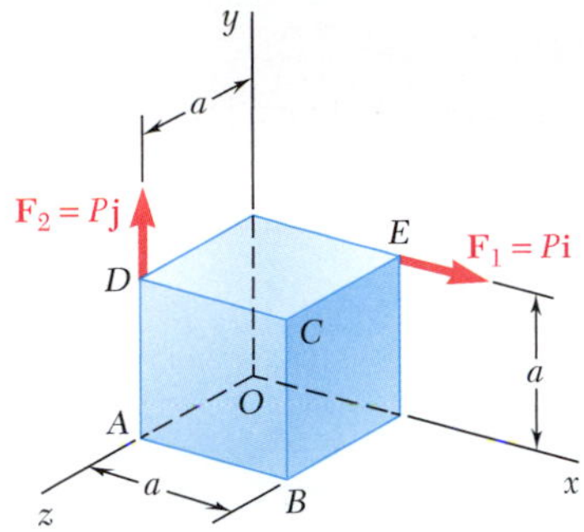

SAMPLE PROBLEM 3.12

Two forces of the same magnitude P act on a cube of side a as shown. Replace the two forces by an equivalent wrench, and determine (a) the magnitude and direction of the resultant force $\mathbf{R}$, (b) the pitch of the wrench, (c) the point where the axis of the wrench intersects the yz plane.

SOLUTION

Equivalent Force-Couple System at *O*. We first determine the equivalent force-couple system at the origin O. We observe that the position vectors of the points of application E and D of the two given forces are $\mathbf{r}_E = a\mathbf{i} + a\mathbf{j}$ and $\mathbf{r}_D = a\mathbf{j} + a\mathbf{k}$. The resultant $\mathbf{R}$ of the two forces and their moment resultant $\mathbf{M}_O^R$ about O are

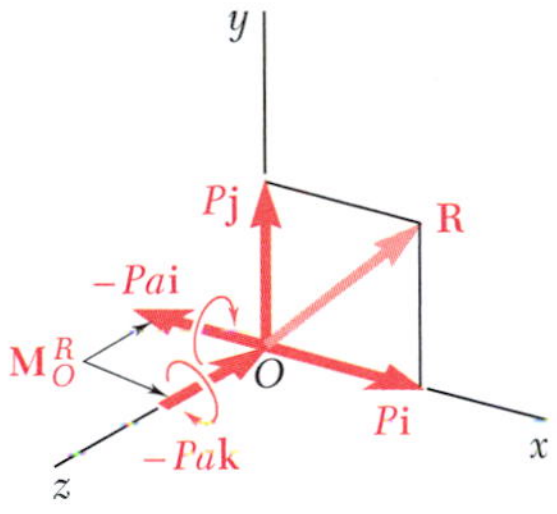

$$\mathbf{R} = \mathbf{F}_1 + \mathbf{F}_2 = P\mathbf{i} + P\mathbf{j} = P(\mathbf{i} + \mathbf{j}) \qquad (1)$$

$$\mathbf{M}_O^R = \mathbf{r}_E \times \mathbf{F}_1 + \mathbf{r}_D \times \mathbf{F}_2 = (a\mathbf{i} + a\mathbf{j}) \times P\mathbf{i} + (a\mathbf{j} + a\mathbf{k}) \times P\mathbf{j}$$

$$= -Pa\mathbf{k} - Pa\mathbf{i} = -Pa(\mathbf{i} + \mathbf{k}) \qquad (2)$$

***a.* Resultant Force R.** It follows from Eq. (1) and the adjoining sketch that the resultant force $\mathbf{R}$ has the magnitude $R = P\sqrt{2}$, lies in the xy plane, and forms angles of 45° with the x and y axes. Thus

$$R = P\sqrt{2} \qquad \theta_x = \theta_y = 45° \qquad \theta_z = 90° \quad ◀$$

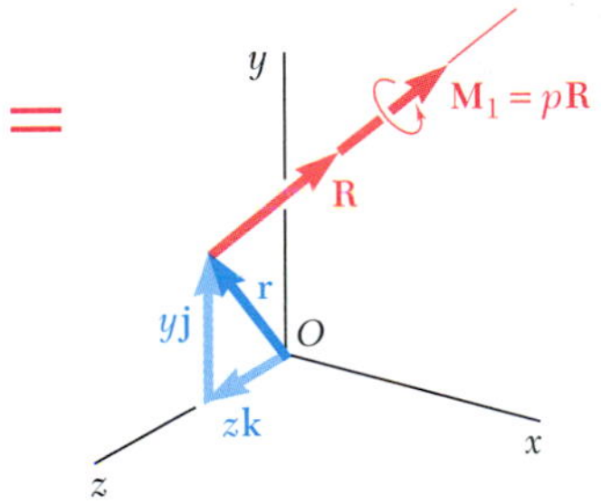

***b.* Pitch of Wrench.** Recalling formula (3.62) of Sec. 3.21 and Eqs. (1) and (2) above, we write

$$p = \frac{\mathbf{R} \cdot \mathbf{M}_O^R}{R^2} = \frac{P(\mathbf{i} + \mathbf{j}) \cdot (-Pa)(\mathbf{i} + \mathbf{k})}{(P\sqrt{2})^2} = \frac{-P^2a(1 + 0 + 0)}{2P^2} \qquad p = -\frac{a}{2} \quad ◀$$

***c.* Axis of Wrench.** It follows from the above and from Eq. (3.61) that the wrench consists of the force $\mathbf{R}$ found in (1) and the couple vector

$$\mathbf{M}_1 = p\mathbf{R} = -\frac{a}{2}P(\mathbf{i} + \mathbf{j}) = -\frac{Pa}{2}(\mathbf{i} + \mathbf{j}) \qquad (3)$$

To find the point where the axis of the wrench intersects the yz plane, we express that the moment of the wrench about O is equal to the moment resultant $\mathbf{M}_O^R$ of the original system:

$$\mathbf{M}_1 + \mathbf{r} \times \mathbf{R} = \mathbf{M}_O^R$$

or, noting that $\mathbf{r} = y\mathbf{j} + z\mathbf{k}$ and substituting for $\mathbf{R}$, $\mathbf{M}_O^R$, and $\mathbf{M}_1$ from Eqs. (1), (2), and (3),

$$-\frac{Pa}{2}(\mathbf{i} + \mathbf{j}) + (y\mathbf{j} + z\mathbf{k}) \times P(\mathbf{i} + \mathbf{j}) = -Pa(\mathbf{i} + \mathbf{k})$$

$$-\frac{Pa}{2}\mathbf{i} - \frac{Pa}{2}\mathbf{j} - Py\mathbf{k} + Pz\mathbf{j} - Pz\mathbf{i} = -Pa\mathbf{i} - Pa\mathbf{k}$$

Equating the coefficients of $\mathbf{k}$, and then the coefficients of $\mathbf{j}$, we find

$$y = a \qquad z = a/2 \quad ◀$$

SOLVING PROBLEMS ON YOUR OWN

This lesson was devoted to the reduction and simplification of force systems. In solving the problems which follow, you will be asked to perform the operations discussed below.

1. *Reducing a force system to a force and a couple at a given point A.* The force is the *resultant* **R** of the system and is obtained by adding the various forces; the moment of the couple is the *moment resultant* of the system and is obtained by adding the moments about A of the various forces. We have

$$\mathbf{R} = \Sigma\mathbf{F} \qquad \mathbf{M}_A^R = \Sigma(\mathbf{r} \times \mathbf{F})$$

where the position vector **r** is drawn from A to *any point* on the line of action of **F**.

2. *Moving a force-couple system from point A to point B.* If you wish to reduce a given force system to a force-couple system at point B after you have reduced it to a force-couple system at point A, you need not recompute the moments of the forces about B. The resultant **R** remains unchanged, and the new moment resultant $\mathbf{M}_B^R$ can be obtained by adding to $\mathbf{M}_A^R$ the moment about B of the force **R** applied at A [Sample Prob. 3.8]. Denoting by **s** the vector drawn from B to A, you can write

$$\mathbf{M}_B^R = \mathbf{M}_A^R + \mathbf{s} \times \mathbf{R}$$

3. *Checking whether two force systems are equivalent.* First reduce each force system to a force-couple system *at the same, but arbitrary, point* A (as explained in paragraph 1). The two systems are equivalent (that is, they have the same effect on the given rigid body) if the two force-couple systems you have obtained are identical, that is, if

$$\Sigma\mathbf{F} = \Sigma\mathbf{F}' \qquad \text{and} \qquad \Sigma\mathbf{M}_A = \Sigma\mathbf{M}'_A$$

You should recognize that if the first of these equations is not satisfied, that is, if the two systems do not have the same resultant **R**, the two systems cannot be equivalent and there is then no need to check whether or not the second equation is satisfied.

4. *Reducing a given force system to a single force.* First reduce the given system to a force-couple system consisting of the resultant **R** and the couple vector $\mathbf{M}_A^R$ at some convenient point A (as explained in paragraph 1). You will recall from the previous lesson that further reduction to a single force is possible *only if the force* **R** *and the couple vector* $\mathbf{M}_A^R$ *are mutually perpendicular*. This will certainly be the case for systems of forces which are either *concurrent, coplanar,* or *parallel*. The required single force can then be obtained by moving **R** until its moment about A is equal to $\mathbf{M}_A^R$, as you did in several problems of the preceding lesson. More formally, you can write that the position vector **r** drawn from A to any point on the line of action of the single force **R** must satisfy the equation

$$\mathbf{r} \times \mathbf{R} = \mathbf{M}_A^R$$

This procedure was used in Sample Probs. 3.8, 3.9, and 3.11.

5. *Reducing a given force system to a wrench.* If the given system is comprised of forces which are not concurrent, coplanar, or parallel, the equivalent force-couple system at a point A will consist of a force **R** and a couple vector $\mathbf{M}_A^R$ which, in general, *are not mutually perpendicular*. (To check whether **R** and $\mathbf{M}_A^R$ are mutually perpendicular, form their scalar product. If this product is zero, they are mutually perpendicular; otherwise, they are not.) If **R** and $\mathbf{M}_A^R$ are not mutually perpendicular, the force-couple system (and thus the given system of forces) *cannot be reduced to a single force*. However, the system can be reduced to a *wrench*—the combination of a force **R** and a couple vector $\mathbf{M}_1$ directed along a common line of action called the *axis of the wrench* (Fig. 3.47). The ratio $p = M_1/R$ is called the *pitch* of the wrench.

To reduce a given force system to a wrench, you should follow these steps:

a. Reduce the given system to an equivalent force-couple system $(\mathbf{R}, \mathbf{M}_O^R)$, typically located at the origin O.

b. Determine the pitch p from Eq. (3.62)

$$p = \frac{M_1}{R} = \frac{\mathbf{R} \cdot \mathbf{M}_O^R}{R^2} \tag{3.62}$$

and the couple vector from $\mathbf{M}_1 = p\mathbf{R}$.

c. Express that the moment about O of the wrench is equal to the moment resultant $\mathbf{M}_O^R$ of the force-couple system at O:

$$\mathbf{M}_1 + \mathbf{r} \times \mathbf{R} = \mathbf{M}_O^R \tag{3.63}$$

This equation allows you to determine the point where the line of action of the wrench intersects a specified plane, since the position vector **r** is directed from O to that point.

These steps are illustrated in Sample Prob. 3.12. Although the determination of a wrench and the point where its axis intersects a plane may appear difficult, the process is simply the application of several of the ideas and techniques developed in this chapter. Thus, once you have mastered the wrench, you can feel confident that you understand much of Chap. 3.

Problems

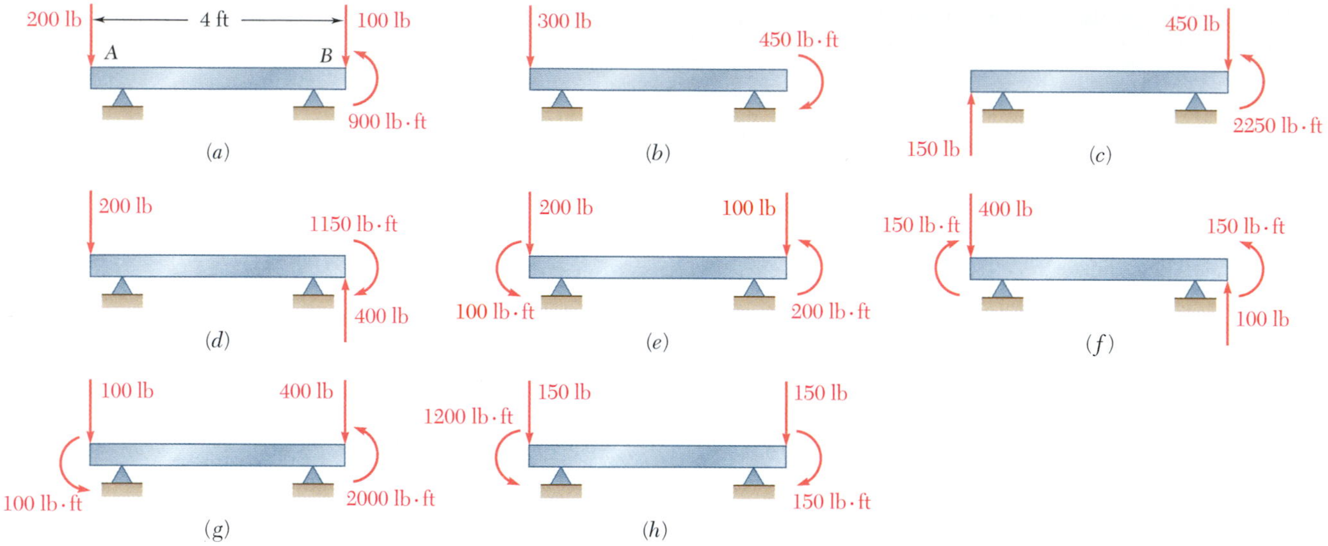

Fig. P3.98

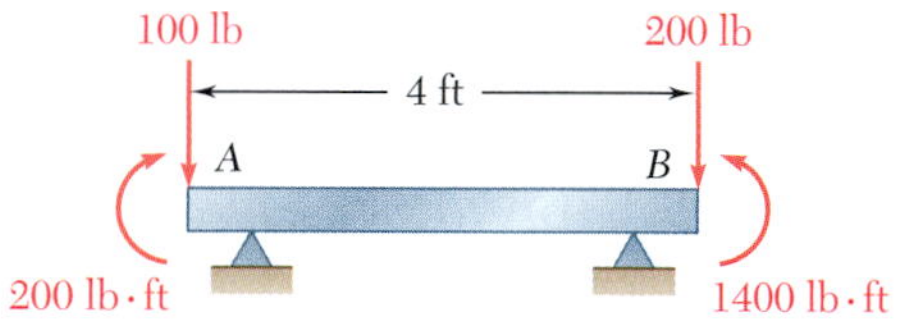

Fig. P3.99

3.98 A 4-ft-long beam is subjected to a variety of loadings. (*a*) Replace each loading with an equivalent force-couple system at end *A* of the beam. (*b*) Which of the loadings are equivalent?

3.99 A 4-ft-long beam is loaded as shown. Determine the loading of Prob. 3.98 which is equivalent to this loading.

3.100 Determine the single equivalent force and the distance from point *A* to its line of action for the beam and loading of (*a*) Prob. 3.98*b*, (*b*) Prob. 3.98*d*, (*c*) Prob. 3.98*e*.

3.101 Five separate force-couple systems act at the corners of a metal block, which has been machined into the shape shown. Determine which of these systems is equivalent to a force $\mathbf{F} = (10\ \text{N})\mathbf{j}$ and a couple of moment $\mathbf{M} = (6\ \text{N} \cdot \text{m})\mathbf{i} + (4\ \text{N} \cdot \text{m})\mathbf{k}$ located at point *A*.

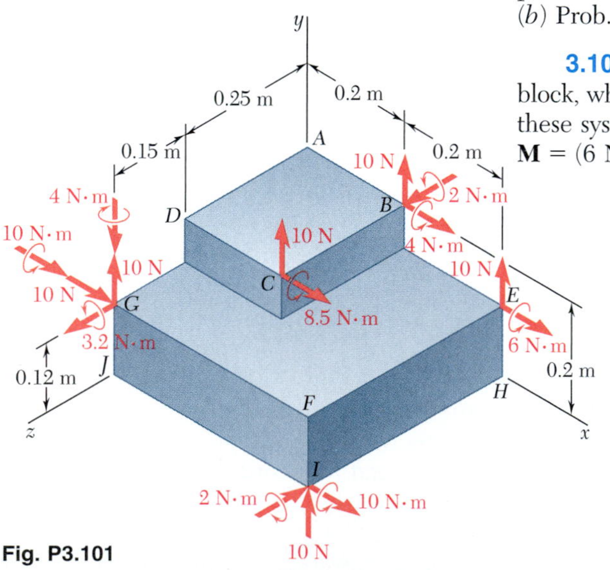

Fig. P3.101

3.102 The masses of two children sitting at ends A and B of a seesaw are 38 kg and 29 kg, respectively. Where should a third child sit so that the resultant of the weights of the three children will pass through C if she has a mass of (*a*) 27 kg, (*b*) 24 kg.

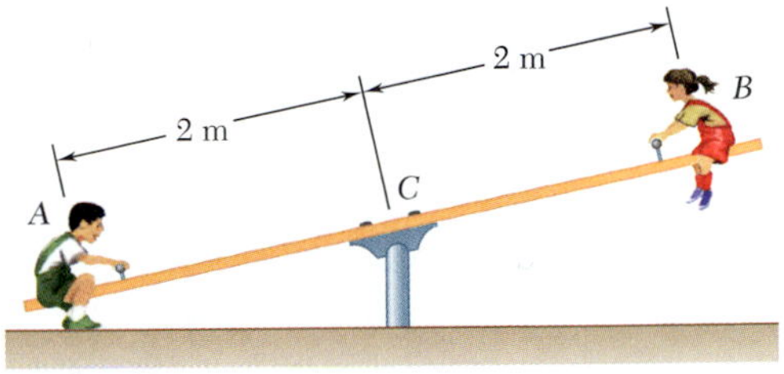

Fig. *P3.102*

3.103 Three stage lights are mounted on a pipe as shown. The mass of each light is $m_A = m_B = 1.8$ kg and $m_C = 1.6$ kg. (*a*) If $d = 0.75$ m, determine the distance from D to the line of action of the resultant of the weights of the three lights. (*b*) Determine the value of d so that the resultant of the weights passes through the midpoint of the pipe.

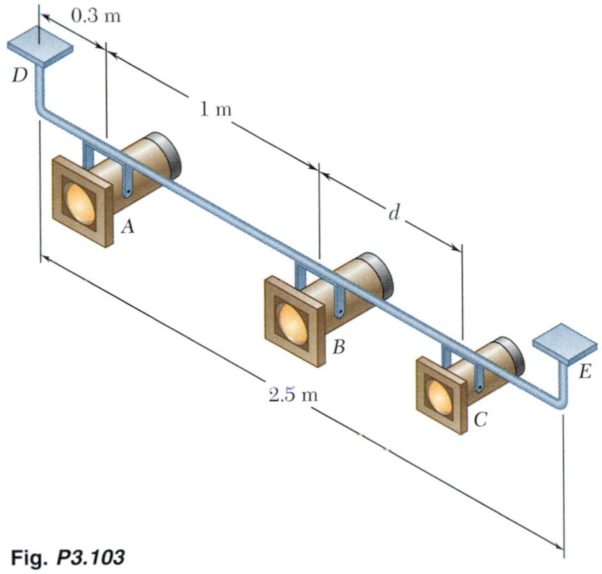

Fig. *P3.103*

3.104 Three hikers are shown crossing a footbridge. Knowing that the weights of the hikers at points C, D, and E are 800 N, 700 N, and 540 N, respectively, determine (*a*) the horizontal distance from A to the line of action of the resultant of the three weights when $a = 1.1$ m, (*b*) the value of a so that the loads on the bridge supports at A and B are equal.

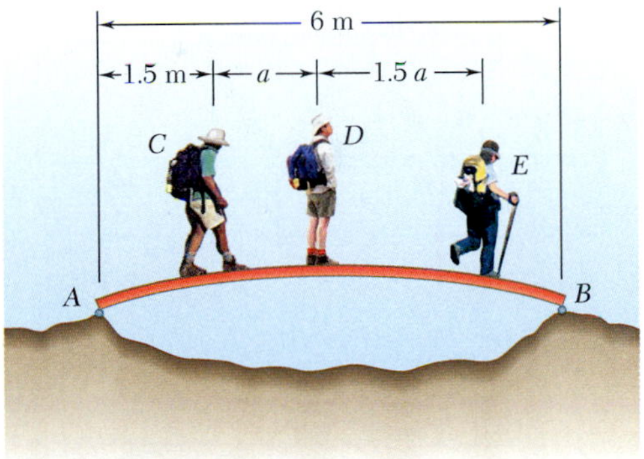

Fig. P3.104

3.105 Gear C is rigidly attached to arm AB. If the forces and couple shown can be reduced to a single equivalent force at A, determine the equivalent force and the magnitude of the couple **M**.

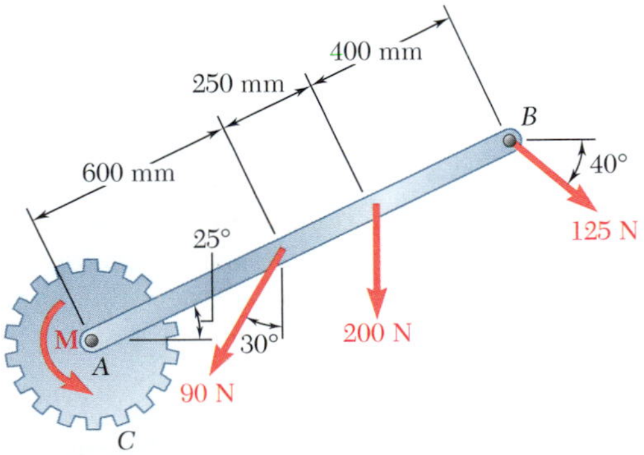

Fig. P3.105

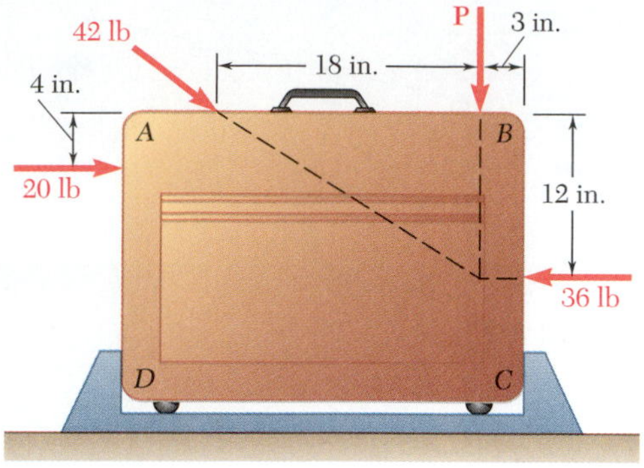

Fig. P3.106

3.106 To test the strength of a 25 × 20-in. suitcase, forces are applied as shown. Knowing that $P = 18$ lb, (*a*) determine the resultant of the applied forces, (*b*) locate the two points where the line of action of the resultant intersects the edge of the suitcase.

3.107 Solve Prob. 3.106 assuming that $P = 28$ lb.

3.108 As four holes are punched simultaneously in a piece of aluminum sheet metal, the punches exert on the piece the forces shown. Knowing that the forces are perpendicular to the surfaces of the piece, determine (*a*) the resultant of the applied forces when $\alpha = 45°$ and the point of intersection of the line of action of that resultant with a line drawn through points *A* and *B*, (*b*) the value of α so that the line of action of the resultant passes through fold *EF*.

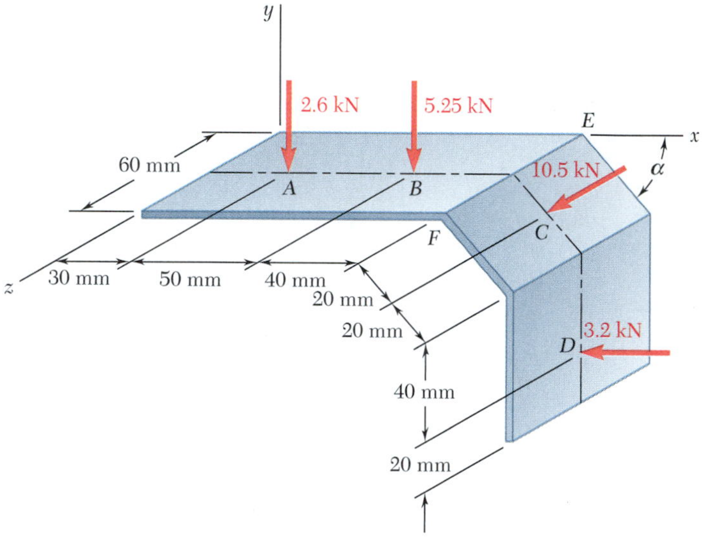

Fig. P3.108 and P3.109

3.109 As four holes are punched simultaneously in a piece of aluminum sheet metal, the punches exert on the piece the forces shown. Knowing that the forces are perpendicular to the surfaces of the piece, determine (*a*) the value of α so that the resultant of the applied forces is parallel to the 10.5 N force, (*b*) the corresponding resultant of the applied forces and the point of intersection of its line of action with a line drawn through points *A* and *B*.

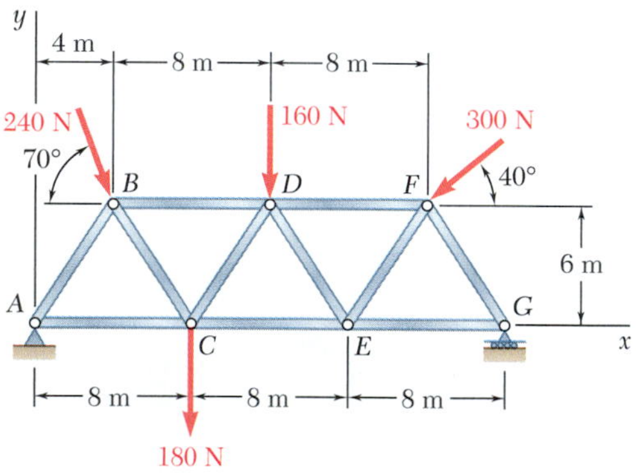

Fig. P3.110

3.110 A truss supports the loading shown. Determine the equivalent force acting on the truss and the point of intersection of its line of action with a line through points *A* and *G*.

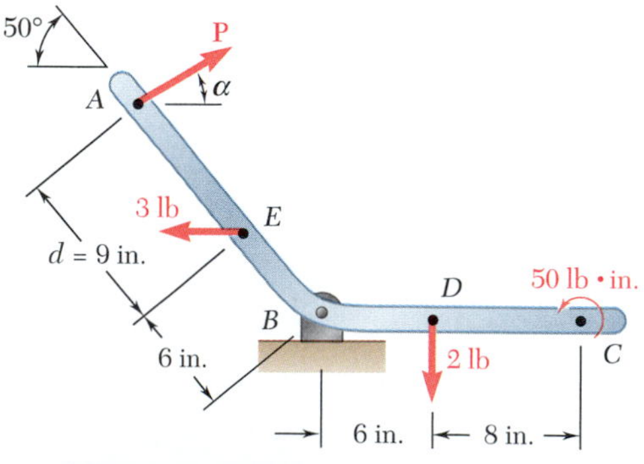

Fig. P3.111 and *P3.112*

3.111 Three forces and a couple act on crank *ABC*. For $P = 5$ lb and $\alpha = 40°$, (*a*) determine the resultant of the given system of forces, (*b*) locate the point where the line of action of the resultant intersects a line drawn through points *B* and *C*, (*c*) locate the point where the line of action of the resultant intersects a line drawn through points *A* and *B*.

3.112 Three forces and a couple act on crank *ABC*. Determine the value of *d* so that the given system of forces is equivalent to zero at (*a*) point *B*, (*b*) point *D*.

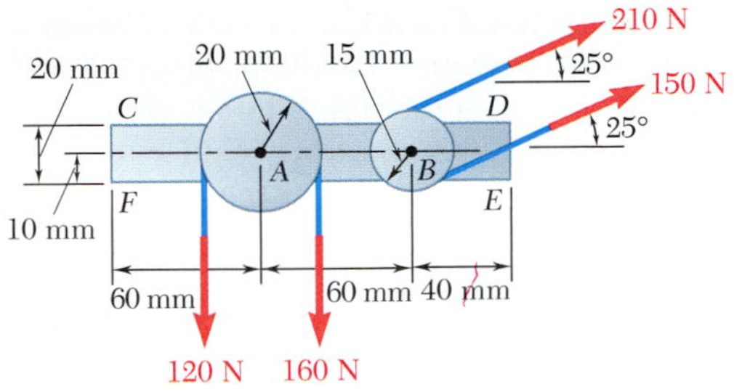

Fig. P3.113

3.113 Pulleys A and B are mounted on bracket $CDEF$. The tension on each side of the two belts is as shown. Replace the four forces with a single equivalent force, and determine where its line of action intersects the bottom edge of the bracket.

3.114 As follower AB rolls along the surface of member C, it exerts a constant force **F** perpendicular to the surface. (*a*) Replace **F** with an equivalent force-couple system at the point D obtained by drawing the perpendicular from the point of contact to the x axis. (*b*) For $a = 1$ m and $b = 2$ m, determine the value of x for which the moment of the equivalent force-couple system at D is maximum.

3.115 As plastic bushings are inserted into a 3-in.-diameter cylindrical sheet metal container, the insertion tool exerts the forces shown on the enclosure. Each of the forces is parallel to one of the coordinate axes. Replace these forces with an equivalent force-couple system at C.

3.116 Two 300-mm-diameter pulleys are mounted on line shaft AD. The belts B and C lie in vertical planes parallel to the yz plane. Replace the belt forces shown with an equivalent force-couple system at A.

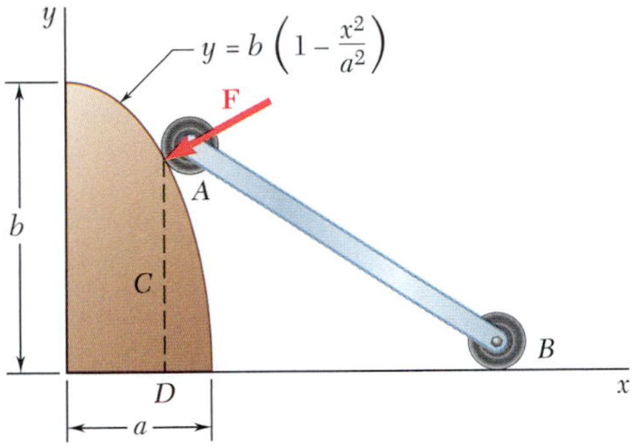

Fig. P3.114

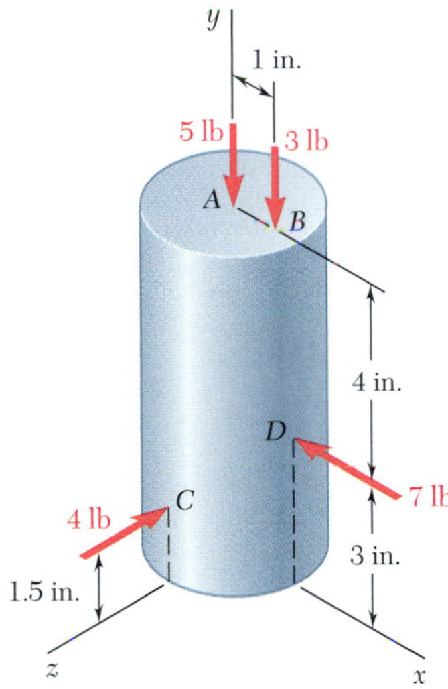

Fig. P3.115

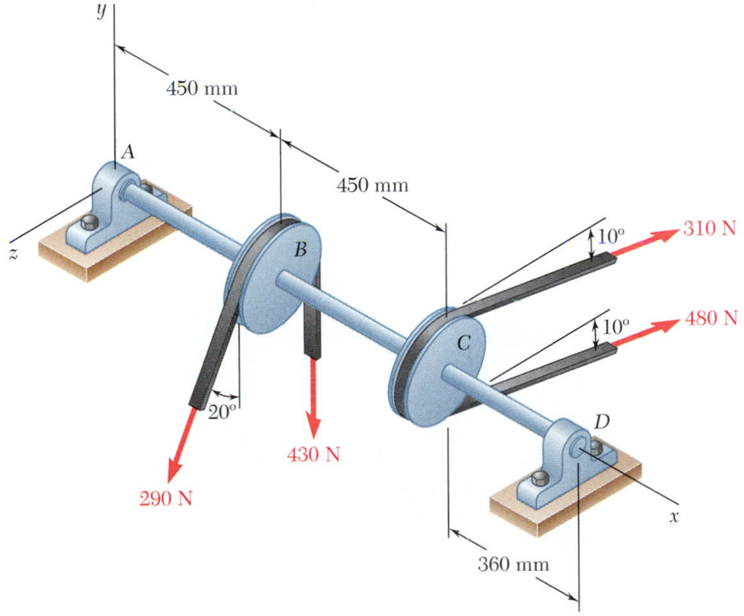

Fig. P3.116

3.117 A mechanic uses a crowfoot wrench to loosen a bolt at C. The mechanic holds the socket wrench handle at points A and B and applies forces at these points. Knowing that these forces are equivalent to a force-couple system at C consisting of the force $\mathbf{C} = -(40\text{ N})\mathbf{i} + (20\text{ N})\mathbf{k}$ and the couple $\mathbf{M}_C = (40\text{ N}\cdot\text{m})\mathbf{i}$, determine the forces applied at A and B when $A_z = 10$ N.

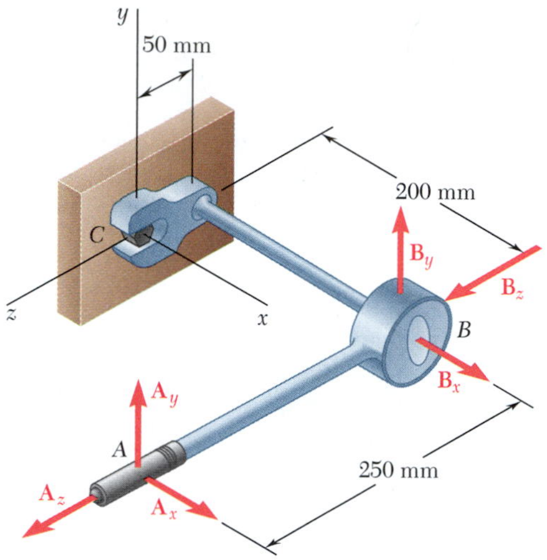

Fig. P3.117

3.118 While using a pencil sharpener, a student applies the forces and couple shown. (*a*) Determine the forces exerted at B and C knowing that these forces and the couple are equivalent to a force-couple system at A consisting of the force $\mathbf{R} = (3.9\text{ lb})\mathbf{i} + R_y\mathbf{j} - (1.1\text{ lb})\mathbf{k}$ and the couple $\mathbf{M}_A^R = M_x\mathbf{i} + (1.5\text{ lb}\cdot\text{ft})\mathbf{j} - (1.1\text{ lb}\cdot\text{ft})\mathbf{k}$. (*b*) Find the corresponding values of R_y and M_x.

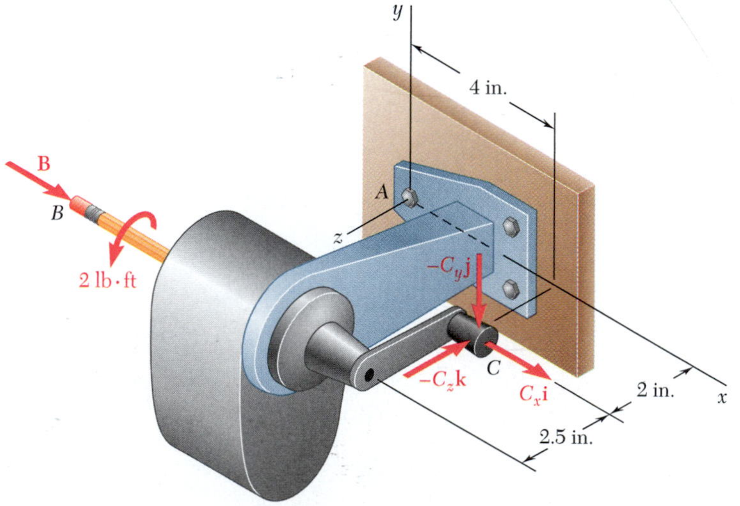

Fig. *P3.118*

3.119 A portion of the flue for a furnace is attached to the ceiling at *A*. While supporting the free end of the flue at *F*, a worker pushes in at *E* and pulls out at *F* to align end *E* with the furnace. Knowing that the 10-lb force at *F* lies in a plane parallel to the *yz* plane, determine (*a*) the angle α the force at *F* should form with the horizontal if duct *AB* is not to tend to rotate about the vertical, (*b*) the force-couple system at *B* equivalent to the given force system when this condition is satisfied.

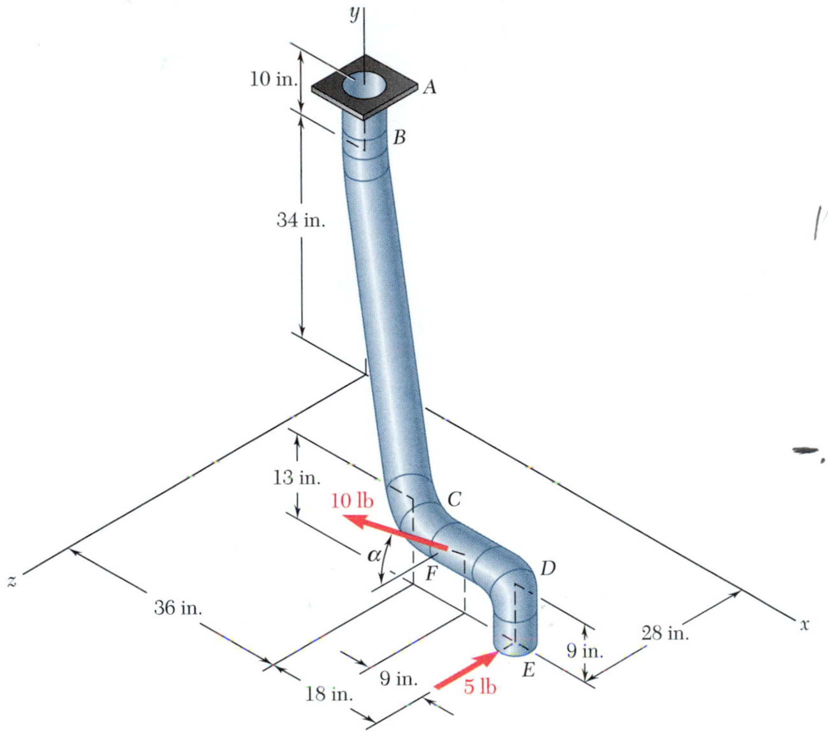

Fig. P3.119 and P3.120

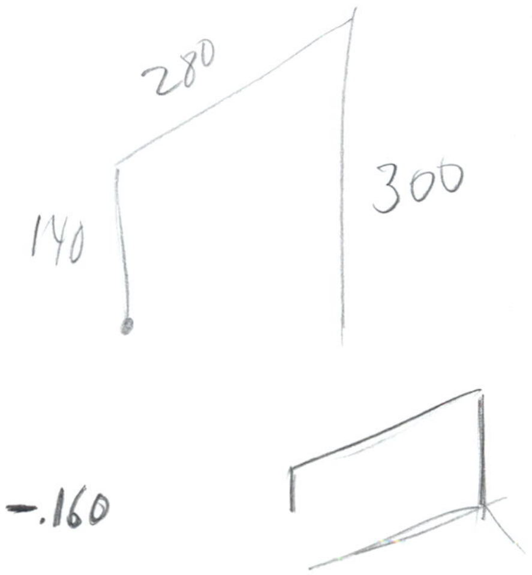

3.120 A portion of the flue for a furnace is attached to the ceiling at *A*. While supporting the free end of the flue at *F*, a worker pushes in at *E* and pulls out at *F* to align end *E* with the furnace. Knowing that the 10-lb force at *F* lies in a plane parallel to the *yz* plane and that $\alpha = 60°$, (*a*) replace the given force system with an equivalent force-couple system at *C*, (*b*) determine whether duct *CD* will tend to rotate clockwise or counterclockwise relative to elbow *C*, as viewed from *D* to *C*.

3.121 The head-and-motor assembly of a radial drill press was originally positioned with arm *AB* parallel to the *z* axis and the axis of the chuck and bit parallel to the *y* axis. The assembly was then rotated 25° about the *y* axis and 20° about the centerline of the horizontal arm *AB*, bringing it into the position shown. The drilling process was started by switching on the motor and rotating the handle to bring the bit into contact with the workpiece. Replace the force and couple exerted by the drill press with an equivalent force-couple system at the center *O* of the base of the vertical column.

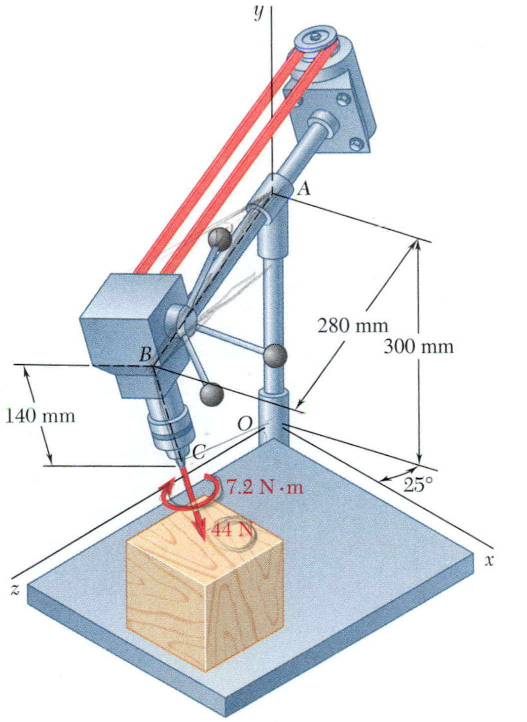

Fig. P3.121

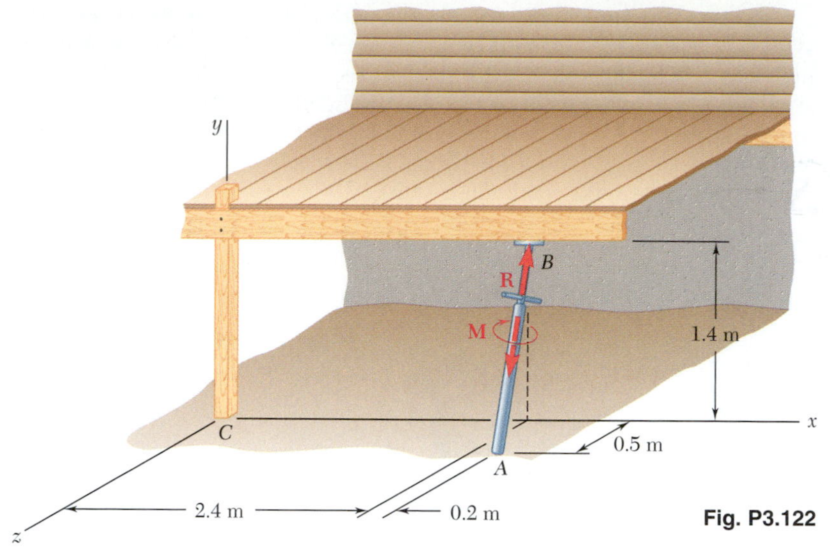

Fig. P3.122

3.122 While a sagging porch is leveled and repaired, a screw jack is used to support the front of the porch. As the jack is expanded, it exerts on the porch the force-couple system shown, where $R = 300$ N and $M = 37.5$ N · m. Replace this force-couple system with an equivalent force-couple system at C.

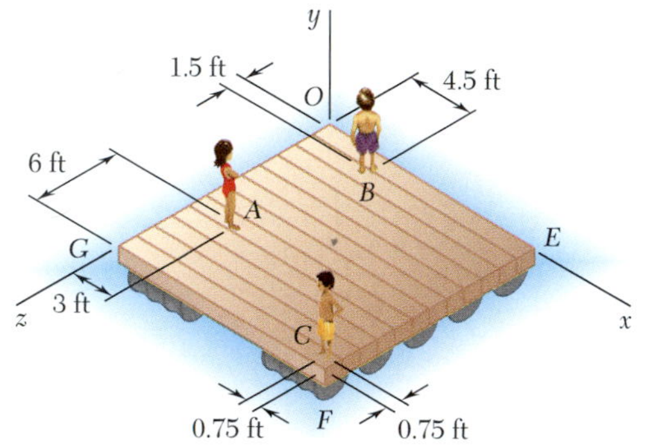

Fig. P3.123 and P3.124

3.123 Three children are standing on a 15 × 15-ft raft. If the weights of the children at points A, B, and C are 85 lb, 60 lb, and 90 lb, respectively, determine the magnitude and the point of application of the resultant of the three weights.

3.124 Three children are standing on a 15 × 15-ft raft. The weights of the children at points A, B, and C are 85 lb, 60 lb, and 90 lb, respectively. If a fourth child of weight 95 lb climbs onto the raft, determine where she should stand if the other children remain in the positions shown and the line of action of the resultant of the four weights is to pass through the center of the raft.

3.125 The forces shown are the resultant downward loads on sections of the flat roof of a building because of accumulated snow. Determine the magnitude and the point of application of the resultant of these four loads.

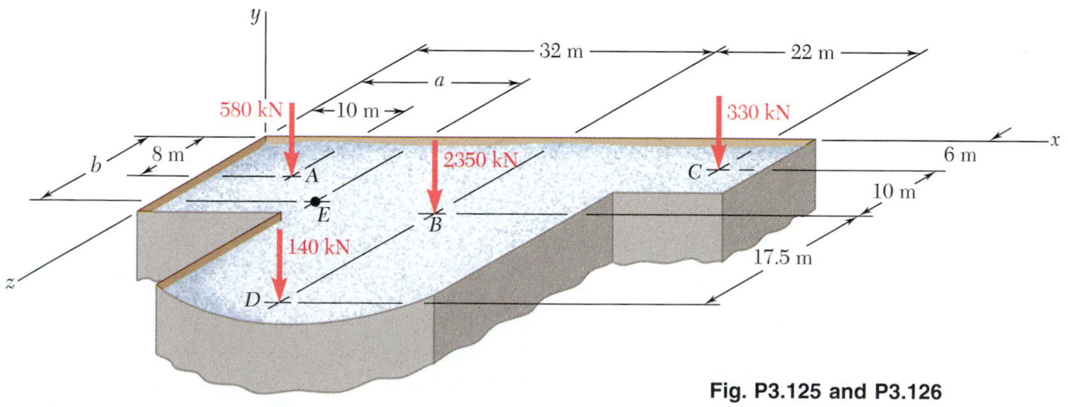

Fig. P3.125 and P3.126

3.126 The forces shown are the resultant downward loads on sections of the flat roof of a building because of accumulated snow. If the snow represented by the 580-kN force is shoveled so that this load acts at E, determine a and b knowing that the point of application of the resultant of the four loads is then at B.

***3.127** A group of students loads a 2 × 4-m flatbed trailer with two 0.6 × 0.6 × 0.6-m boxes and one 0.6 × 0.6 × 1.2-m box. Each of the boxes at the rear of the trailer is positioned so that it is aligned with both the back and a side of the trailer. Determine the smallest load the students should place in a second 0.6 × 0.6 × 1.2-m box and where on the trailer they should secure it, without any part of the box overhanging the sides of the trailer, if each box is uniformly loaded and the line of action of the resultant of the weights of the four boxes is to pass through the point of intersection of the centerlines of the trailer and the axle. (*Hint:* Keep in mind that the box may be placed either on its side or on its end.)

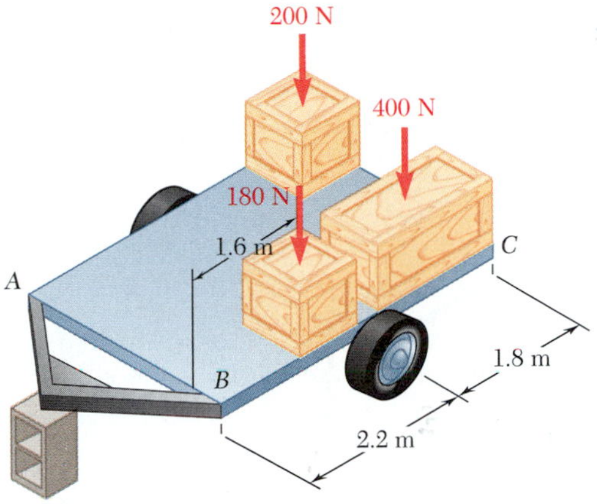

Fig. *P3.127*

***3.128** Solve Prob. 3.127 if the students want to place as much weight as possible in the fourth box and that at least one side of the box must coincide with a side of the trailer.

***3.129** A block of wood is acted upon by three forces of the same magnitude P and having the directions shown. Replace the three forces with an equivalent wrench and determine (*a*) the magnitude and direction of the resultant **R**, (*b*) the pitch of the wrench, (*c*) the point where the axis of the wrench intersects the xy plane.

***3.130** A piece of sheet metal is bent into the shape shown and is acted upon by three forces. Replace the three forces with an equivalent wrench and determine (*a*) the magnitude and direction of the resultant **R**, (*b*) the pitch of the wrench, (*c*) the point where the axis of the wrench intersects the yz plane.

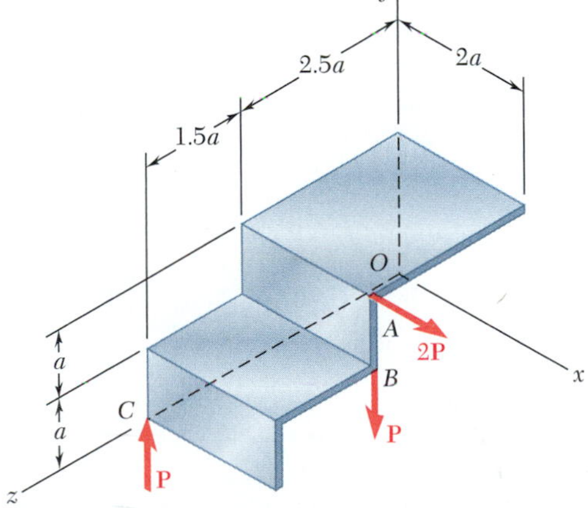

Fig. *P3.130*

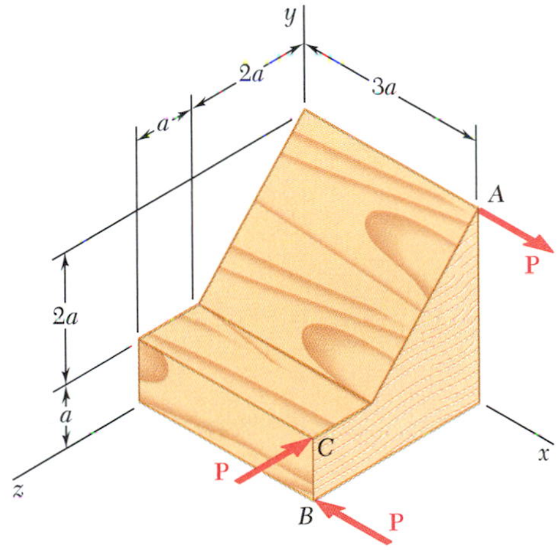

Fig. P3.129

***3.131 and *3.132** The forces and couples shown are applied to two screws as a piece of sheet metal is fastened to a block of wood. Reduce the forces and the couples to an equivalent wrench and determine (*a*) the resultant force **R**, (*b*) the pitch of the wrench, (*c*) the point where the axis of the wrench intersects the *xz* plane.

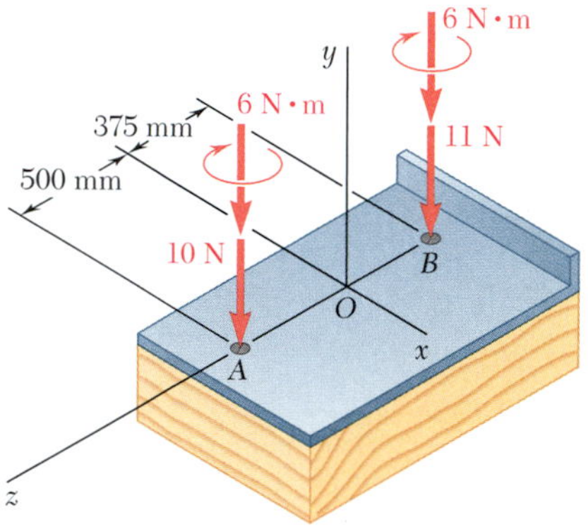

Fig. P3.131

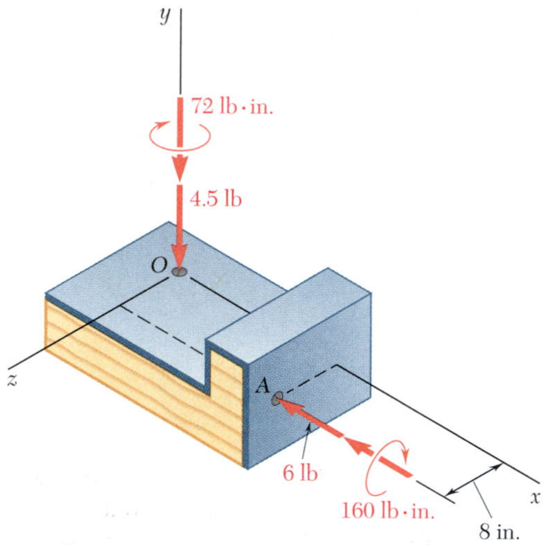

Fig. *P3.132*

***3.133 and *3.134** Two bolts *A* and *B* are tightened by applying the forces and couple shown. Replace the two wrenches with a single equivalent wrench and determine (*a*) the resultant **R**, (*b*) the pitch of the single equivalent wrench, (*c*) the point where the axis of the wrench intersects the *xz* plane.

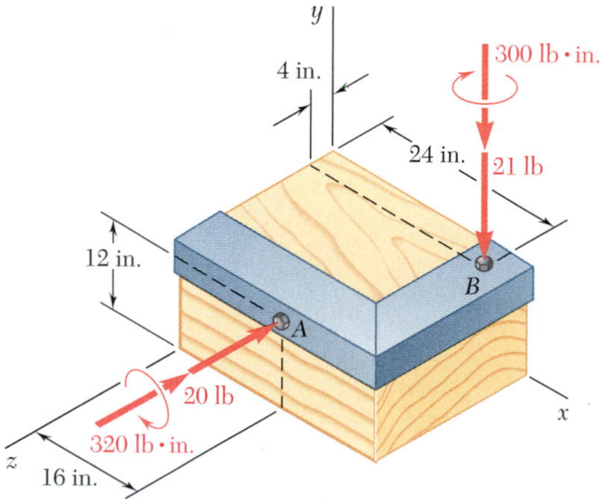

Fig. P3.133

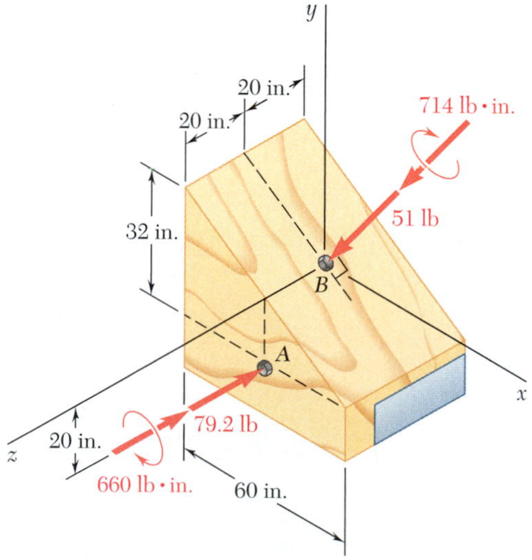

Fig. *P3.134*

***3.135** A flagpole is guyed by three cables. If the tensions in the cables have the same magnitude P, replace the forces exerted on the pole with an equivalent wrench and determine (*a*) the resultant force **R**, (*b*) the pitch of the wrench, (*c*) the point where the axis of the wrench intersects the xz plane.

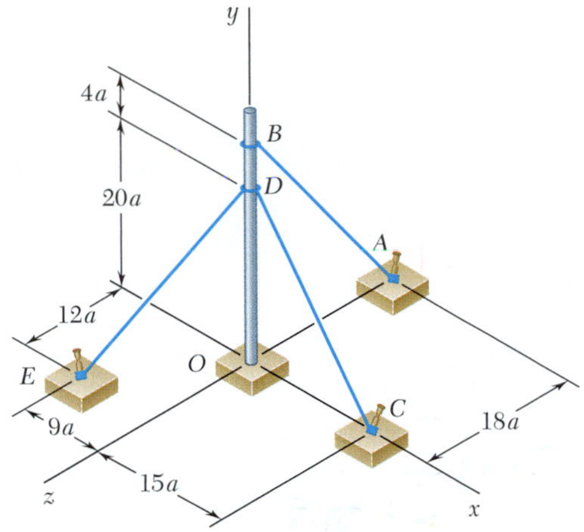

Fig. P3.135

***3.136 and *3.137** Determine whether the force-and-couple system shown can be reduced to a single equivalent force **R**. If it can, determine **R** and the point where the line of action of **R** intersects the yz plane. If it cannot be so reduced, replace the given system with an equivalent wrench and determine its resultant, its pitch, and the point where its axis intersects the yz plane.

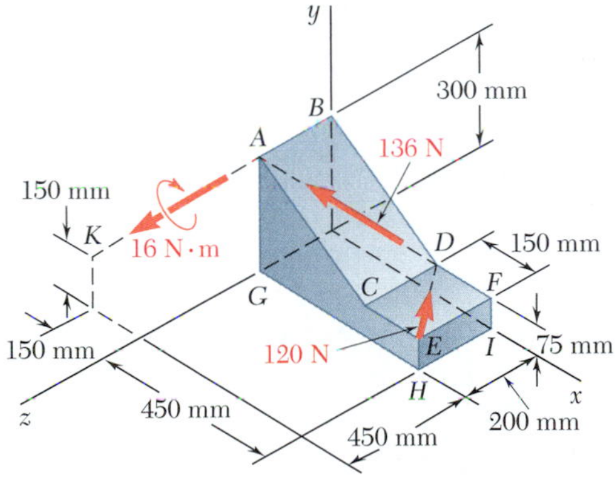

Fig. P3.136

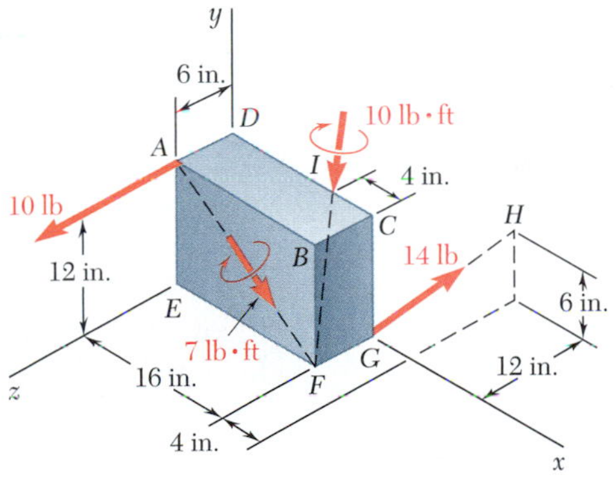

Fig. P3.137

***3.138** Replace the wrench shown with an equivalent system consisting of two forces perpendicular to the y axis and applied respectively at A and B.

***3.139** Show that, in general, a wrench can be replaced with two forces chosen in such a way that one force passes through a given point while the other force lies in a given plane.

***3.140** Show that a wrench can be replaced with two perpendicular forces, one of which is applied at a given point.

***3.141** Show that a wrench can be replaced with two forces, one of which has a prescribed line of action.

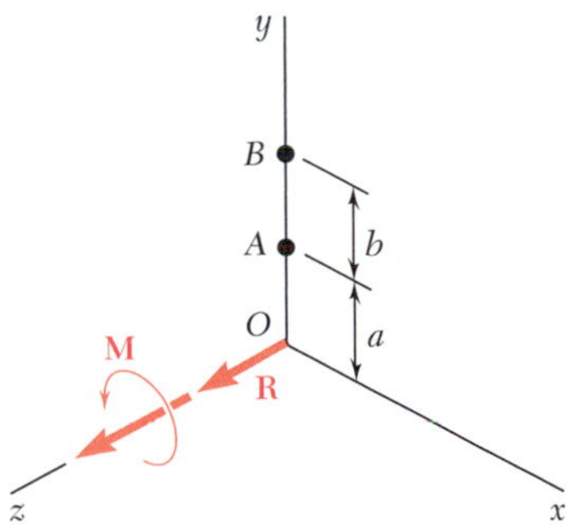

Fig. P3.138

REVIEW AND SUMMARY FOR CHAPTER 3

Principle of transmissibility

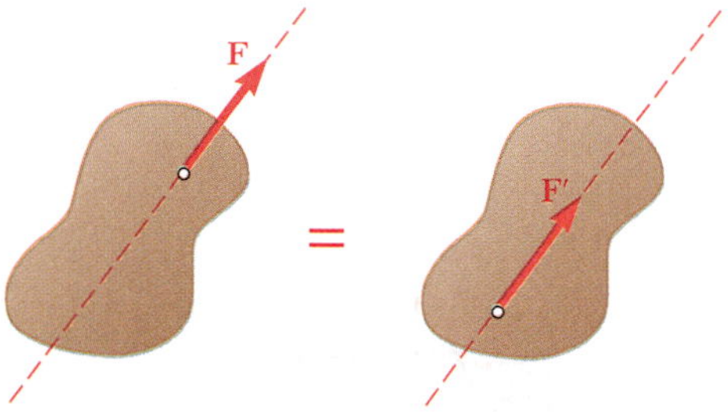

Fig. 3.48

In this chapter we studied the effect of forces exerted on a rigid body. We first learned to distinguish between *external* and *internal* forces [Sec. 3.2] and saw that, according to the *principle of transmissibility,* the effect of an external force on a rigid body remains unchanged if that force is moved along its line of action [Sec. 3.3]. In other words, two forces **F** and **F′** acting on a rigid body at two different points have the same effect on that body if they have the same magnitude, same direction, and same line of action (Fig. 3.48). Two such forces are said to be *equivalent*.

Vector product of two vectors

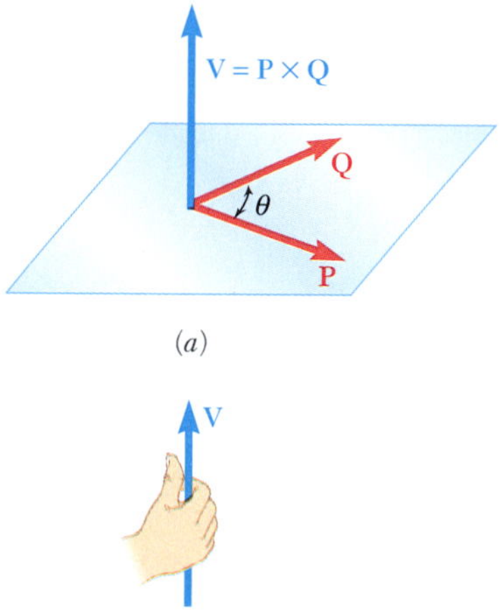

Fig. 3.49

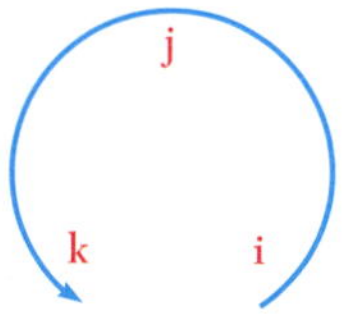

Fig. 3.50

Before proceeding with the discussion of *equivalent systems of forces,* we introduced the concept of the *vector product of two vectors* [Sec. 3.4]. The vector product

$$\mathbf{V} = \mathbf{P} \times \mathbf{Q}$$

of the vectors **P** and **Q** was defined as a vector perpendicular to the plane containing **P** and **Q** (Fig. 3.49), of magnitude

$$V = PQ \sin \theta \tag{3.1}$$

and directed in such a way that a person located at the tip of **V** will observe as counterclockwise the rotation through θ which brings the vector **P** in line with the vector **Q**. The three vectors **P**, **Q**, and **V**—taken in that order—are said to form a *right-handed triad*. It follows that the vector products $\mathbf{Q} \times \mathbf{P}$ and $\mathbf{P} \times \mathbf{Q}$ are represented by equal and opposite vectors. We have

$$\mathbf{Q} \times \mathbf{P} = -(\mathbf{P} \times \mathbf{Q}) \tag{3.4}$$

It also follows from the definition of the vector product of two vectors that the vector products of the unit vectors **i**, **j**, and **k** are

$$\mathbf{i} \times \mathbf{i} = 0 \qquad \mathbf{i} \times \mathbf{j} = \mathbf{k} \qquad \mathbf{j} \times \mathbf{i} = -\mathbf{k}$$

and so on. The sign of the vector product of two unit vectors can be obtained by arranging in a circle and in counterclockwise order the three letters representing the unit vectors (Fig. 3.50): The vector product of two unit vectors will be positive if they follow each other in counterclockwise order and negative if they follow each other in clockwise order.

Rectangular components of vector product

The *rectangular components of the vector product* **V** of two vectors **P** and **Q** were expressed [Sec. 3.5] as

$$\begin{aligned} V_x &= P_yQ_z - P_zQ_y \\ V_y &= P_zQ_x - P_xQ_z \\ V_z &= P_xQ_y - P_yQ_x \end{aligned} \quad (3.9)$$

Using a determinant, we also wrote

$$\mathbf{V} = \begin{vmatrix} \mathbf{i} & \mathbf{j} & \mathbf{k} \\ P_x & P_y & P_z \\ Q_x & Q_y & Q_z \end{vmatrix} \quad (3.10)$$

Moment of a force about a point

The *moment of a force* **F** *about a point O* was defined [Sec. 3.6] as the vector product

$$\mathbf{M}_O = \mathbf{r} \times \mathbf{F} \quad (3.11)$$

where **r** is the *position vector* drawn from O to the point of application A of the force **F** (Fig. 3.51). Denoting by θ the angle between the lines of action of **r** and **F**, we found that the magnitude of the moment of **F** about O can be expressed as

$$M_O = rF \sin\theta = Fd \quad (3.12)$$

where d represents the perpendicular distance from O to the line of action of **F**.

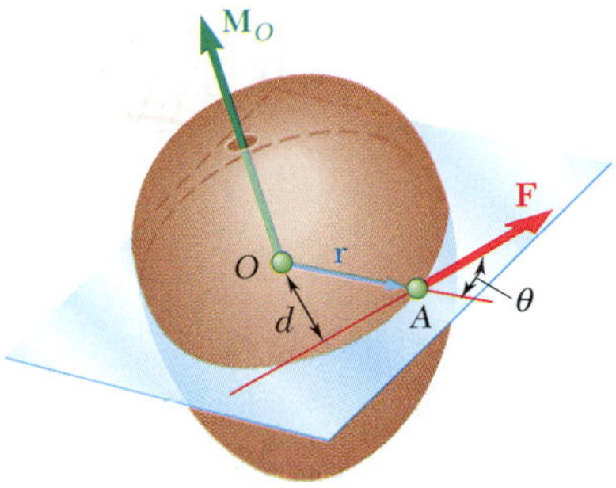

Fig. 3.51

Rectangular components of moment

The *rectangular components of the moment* $\mathbf{M}_O$ *of a force* **F** were expressed [Sec. 3.8] as

$$\begin{aligned} M_x &= yF_z - zF_y \\ M_y &= zF_x - xF_z \\ M_z &= xF_y - yF_x \end{aligned} \quad (3.18)$$

where x, y, z are the components of the position vector **r** (Fig. 3.52). Using a determinant form, we also wrote

$$\mathbf{M}_O = \begin{vmatrix} \mathbf{i} & \mathbf{j} & \mathbf{k} \\ x & y & z \\ F_x & F_y & F_z \end{vmatrix} \quad (3.19)$$

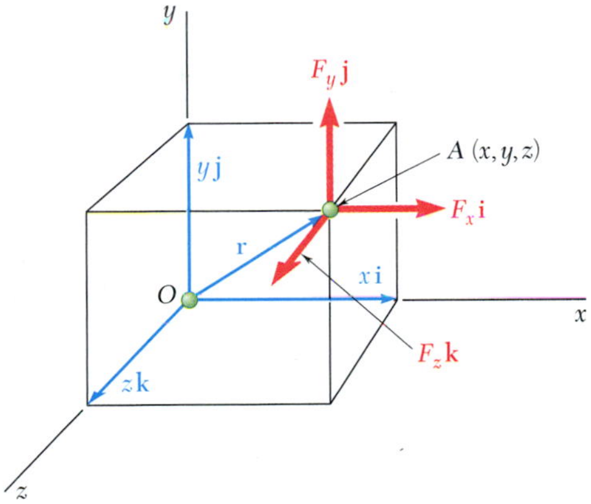

Fig. 3.52

In the more general case of the moment about an arbitrary point B of a force **F** applied at A, we had

$$\mathbf{M}_B = \begin{vmatrix} \mathbf{i} & \mathbf{j} & \mathbf{k} \\ x_{A/B} & y_{A/B} & z_{A/B} \\ F_x & F_y & F_z \end{vmatrix} \quad (3.21)$$

where $x_{A/B}$, $y_{A/B}$, and $z_{A/B}$ denote the components of the vector $\mathbf{r}_{A/B}$:

$$x_{A/B} = x_A - x_B \qquad y_{A/B} = y_A - y_B \qquad z_{A/B} = z_A - z_B$$

In the case of *problems involving only two dimensions,* the force **F** can be assumed to lie in the xy plane. Its moment $\mathbf{M}_B$ about a point B in the same plane is perpendicular to that plane (Fig. 3.53) and is completely defined by the scalar

$$M_B = (x_A - x_B)F_y - (y_A - y_B)F_x \qquad (3.23)$$

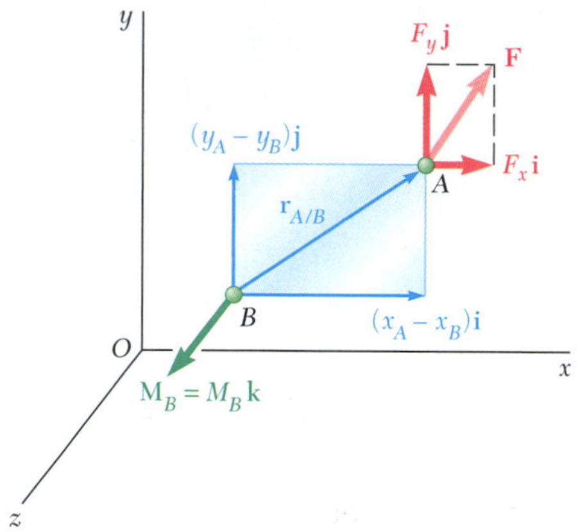

Fig. 3.53

Various methods for the computation of the moment of a force about a point were illustrated in Sample Probs. 3.1 through 3.4.

Scalar product of two vectors

The *scalar product* of two vectors **P** and **Q** [Sec. 3.9] was denoted by $\mathbf{P} \cdot \mathbf{Q}$ and was defined as the scalar quantity

$$\mathbf{P} \cdot \mathbf{Q} = PQ \cos \theta \qquad (3.24)$$

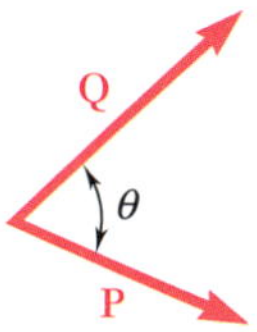

Fig. 3.54

where θ is the angle between **P** and **Q** (Fig. 3.54). By expressing the scalar product of **P** and **Q** in terms of the rectangular components of the two vectors, we determined that

$$\mathbf{P} \cdot \mathbf{Q} = P_xQ_x + P_yQ_y + P_zQ_z \qquad (3.30)$$

Projection of a vector on an axis

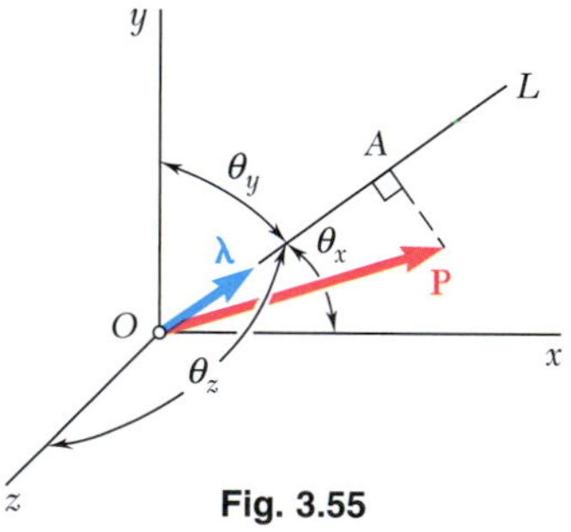

Fig. 3.55

The *projection of a vector* **P** *on an axis* OL (Fig. 3.55) can be obtained by forming the scalar product of **P** and the unit vector $\boldsymbol{\lambda}$ along OL. We have

$$P_{OL} = \mathbf{P} \cdot \boldsymbol{\lambda} \qquad (3.36)$$

or, using rectangular components,

$$P_{OL} = P_x \cos \theta_x + P_y \cos \theta_y + P_z \cos \theta_z \qquad (3.37)$$

where θ_x, θ_y, and θ_z denote the angles that the axis OL forms with the coordinate axes.

Mixed triple product of three vectors

The *mixed triple product* of the three vectors **S**, **P**, and **Q** was defined as the scalar expression

$$\mathbf{S} \cdot (\mathbf{P} \times \mathbf{Q}) \qquad (3.38)$$

obtained by forming the scalar product of **S** with the vector product of **P** and **Q** [Sec. 3.10]. It was shown that

$$\mathbf{S} \cdot (\mathbf{P} \times \mathbf{Q}) = \begin{vmatrix} S_x & S_y & S_z \\ P_x & P_y & P_z \\ Q_x & Q_y & Q_z \end{vmatrix} \tag{3.41}$$

where the elements of the determinant are the rectangular components of the three vectors.

Moment of a force about an axis

The *moment of a force* **F** *about an axis OL* [Sec. 3.11] was defined as the projection OC on OL of the moment $\mathbf{M}_O$ of the force **F** (Fig. 3.56), that is, as the mixed triple product of the unit vector $\boldsymbol{\lambda}$, the position vector **r**, and the force **F**:

$$M_{OL} = \boldsymbol{\lambda} \cdot \mathbf{M}_O = \boldsymbol{\lambda} \cdot (\mathbf{r} \times \mathbf{F}) \tag{3.42}$$

Using the determinant form for the mixed triple product, we have

$$M_{OL} = \begin{vmatrix} \lambda_x & \lambda_y & \lambda_z \\ x & y & z \\ F_x & F_y & F_z \end{vmatrix} \tag{3.43}$$

where $\lambda_x, \lambda_y, \lambda_z$ = direction cosines of axis OL
x, y, z = components of **r**
F_x, F_y, F_z = components of **F**

An example of the determination of the moment of a force about a skew axis was given in Sample Prob. 3.5.

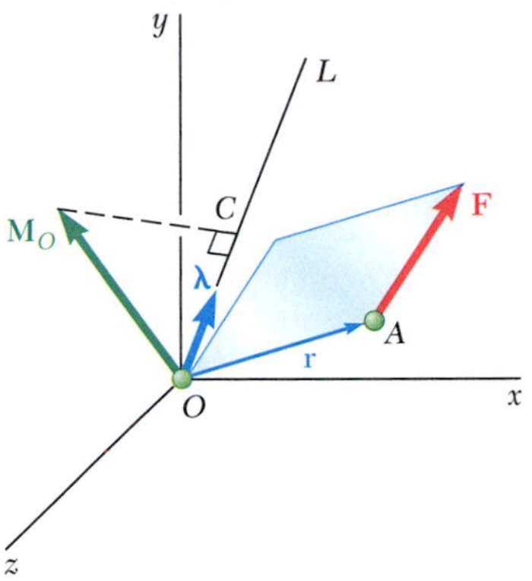

Fig. 3.56

Couples

Two forces **F** *and* −**F** *having the same magnitude, parallel lines of action, and opposite sense are said to form a couple* [Sec. 3.12]. It was shown that the moment of a couple is independent of the point about which it is computed; it is a vector **M** perpendicular to the plane of the couple and equal in magnitude to the product of the common magnitude F of the forces and the perpendicular distance d between their lines of action (Fig. 3.57).

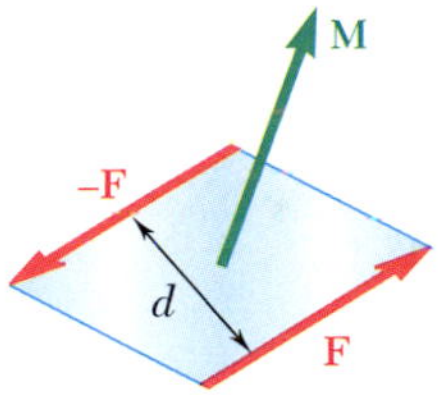

Fig. 3.57

Two couples having the same moment **M** are *equivalent,* that is, they have the same effect on a given rigid body [Sec. 3.13]. The sum of two couples is itself a couple [Sec. 3.14], and the moment **M** of the resultant couple can be obtained by adding vectorially the moments $\mathbf{M}_1$ and $\mathbf{M}_2$ of the original couples [Sample Prob. 3.6]. It follows that a couple can be represented by a vector, called a *couple vector,* equal in magnitude and direction to the moment **M** of the couple [Sec. 3.15]. A couple vector is a *free vector* which can be attached to the origin O if so desired and resolved into components (Fig. 3.58).

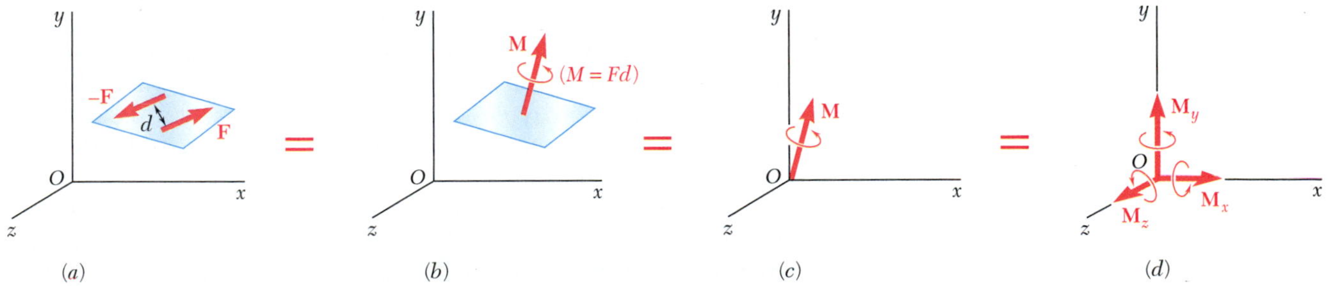

Fig. 3.58

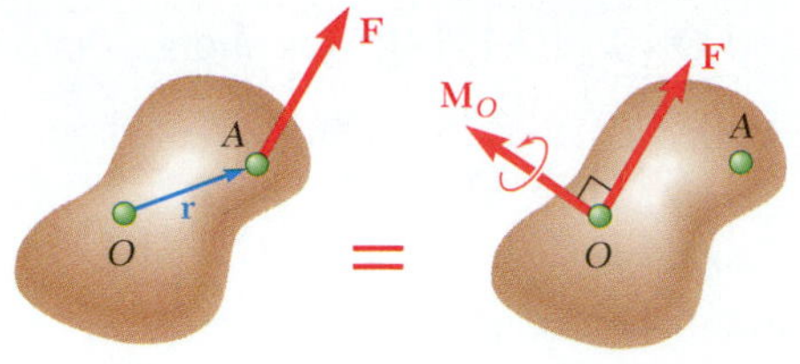

Fig. 3.59

Force-couple system

Any force **F** acting at a point *A* of a rigid body can be replaced by a *force-couple system* at an arbitrary point *O*, consisting of the force **F** applied at *O* and a couple of moment $\mathbf{M}_O$ equal to the moment about *O* of the force **F** in its original position [Sec. 3.16]; it should be noted that the force **F** and the couple vector $\mathbf{M}_O$ are always perpendicular to each other (Fig. 3.59).

Reduction of a system of forces to a force-couple system

It follows [Sec. 3.17] that *any system of forces can be reduced to a force-couple system at a given point O* by first replacing each of the forces of the system by an equivalent force-couple system at *O* (Fig. 3.60) and then adding all the forces and all the couples determined in this manner to obtain a resultant force **R** and a resultant couple vector $\mathbf{M}_O^R$ [Sample Probs. 3.8 through 3.11]. Note that, in general, the resultant **R** and the couple vector $\mathbf{M}_O^R$ will not be perpendicular to each other.

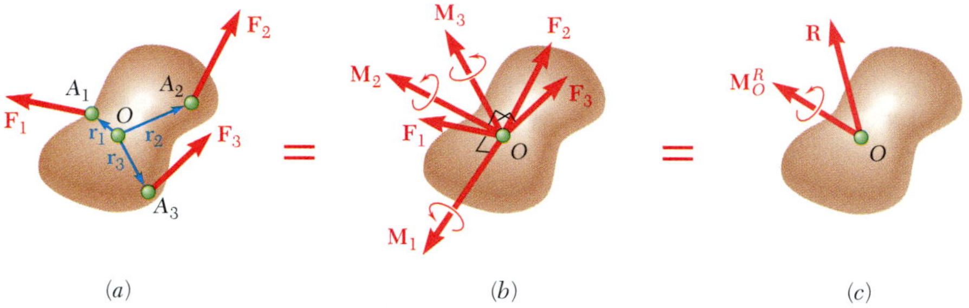

Fig. 3.60

Equivalent systems of forces

We concluded from the above [Sec. 3.18] that, as far as a rigid body is concerned, *two systems of forces,* $\mathbf{F}_1, \mathbf{F}_2, \mathbf{F}_3, \ldots$ *and* $\mathbf{F}'_1, \mathbf{F}'_2, \mathbf{F}'_3, \ldots$, *are equivalent if, and only if,*

$$\Sigma\mathbf{F} = \Sigma\mathbf{F}' \qquad \text{and} \qquad \Sigma\mathbf{M}_O = \Sigma\mathbf{M}'_O \tag{3.57}$$

Further reduction of a system of forces

If the resultant force **R** and the resultant couple vector $\mathbf{M}_O^R$ are perpendicular to each other, the force-couple system at *O* can be further reduced to a single resultant force [Sec. 3.20]. This will be the case for systems consisting either of (*a*) concurrent forces (cf. Chap. 2), (*b*) coplanar forces [Sample Probs. 3.8 and 3.9], or (*c*) parallel forces [Sample Prob. 3.11]. If the resultant **R** and the couple vector $\mathbf{M}_O^R$ are *not* perpendicular to each other, the system *cannot* be reduced to a single force. It can, however, be reduced to a special type of force-couple system called a *wrench,* consisting of the resultant **R** and a couple vector $\mathbf{M}_1$ directed along **R** [Sec. 3.21 and Sample Prob. 3.12].

Review Problems

3.142 A worker tries to move a rock by applying a 360-N force to a steel bar as shown. (*a*) Replace that force with an equivalent force-couple system at *D*. (*b*) Two workers attempt to move the same rock by applying a vertical force at *A* and another force at *D*. Determine these two forces if they are to be equivalent to the single force of part *a*.

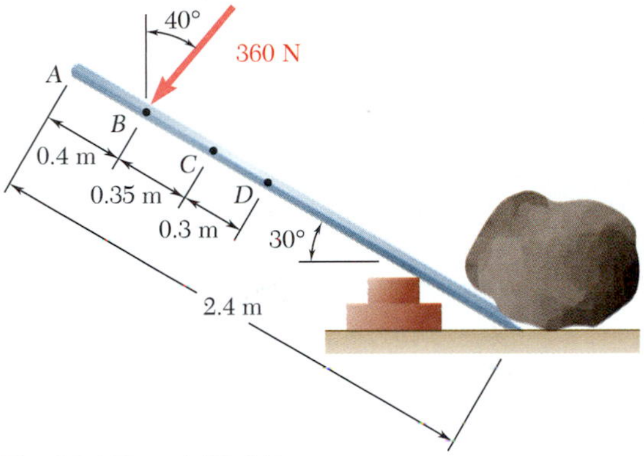

Fig. P3.142 and *P3.143*

3.143 A worker tries to move a rock by applying a 360-N force to a steel bar as shown. If two workers attempt to move the same rock by applying a force at *A* and a parallel force at *C*, determine these two forces so that they will be equivalent to the single 360-N force shown in the figure.

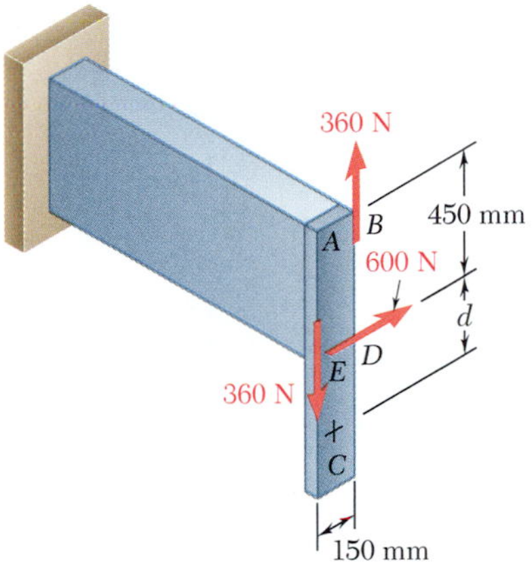

Fig. P3.144

3.144 A force and a couple are applied as shown to the end of a cantilever beam. (*a*) Replace this system with a single force **F** applied at point *C*, and determine the distance *d* from *C* to a line drawn through points *D* and *E*. (*b*) Solve part *a* if the directions of the two 360-N forces are reversed.

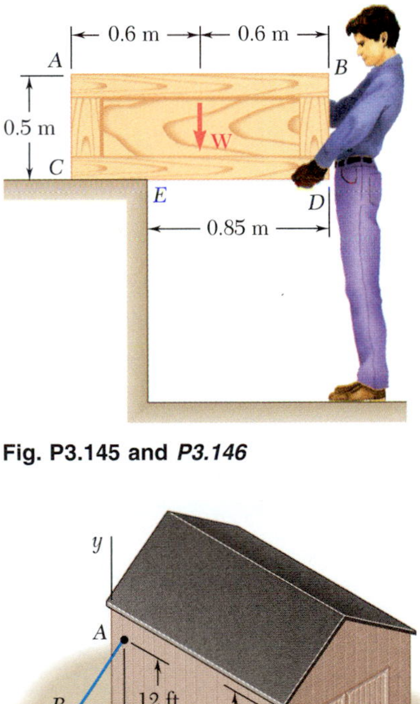

Fig. P3.145 and *P3.146*

3.145 A crate of mass 80 kg is held in the position shown. Determine (*a*) the moment produced by the weight **W** of the crate about *E*, (*b*) the smallest force applied at *B* which creates a moment of equal magnitude and opposite sense about *E*.

3.146 A crate of mass 80 kg is held in the position shown. Determine (*a*) the moment produced by the weight **W** of the crate about *E*, (*b*) the smallest force applied at *A* which creates a moment of equal magnitude and opposite sense about *E*, (*c*) the magnitude, sense, and point of application on the bottom of the crate of the smallest vertical force which creates a moment of equal magnitude and opposite sense about *E*.

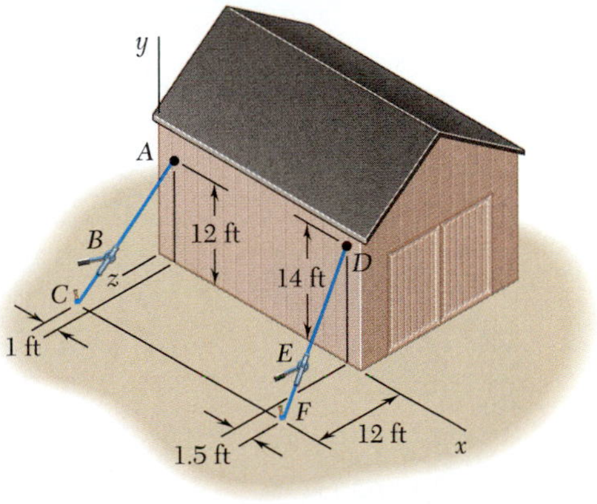

Fig. P3.147

3.147 A farmer uses cables and winch pullers *B* and *E* to plumb one side of a small barn. Knowing that the sum of the moments about the *x* axis of the forces exerted by the cables on the barn at points *A* and *D* is equal to 4728 lb · ft, determine the magnitude of $\mathbf{T}_{DE}$ when $T_{AB} = 255$ lb.

3.148 Solve Prob. 3.147 when the tension in cable *AB* is 306 lb.

3.149 As an adjustable brace BC is used to bring a wall into plumb, the force-couple system shown is exerted on the wall. Replace this force-couple system with an equivalent force-couple system at A knowing that $R = 21.2$ lb and $M = 13.25$ lb · ft.

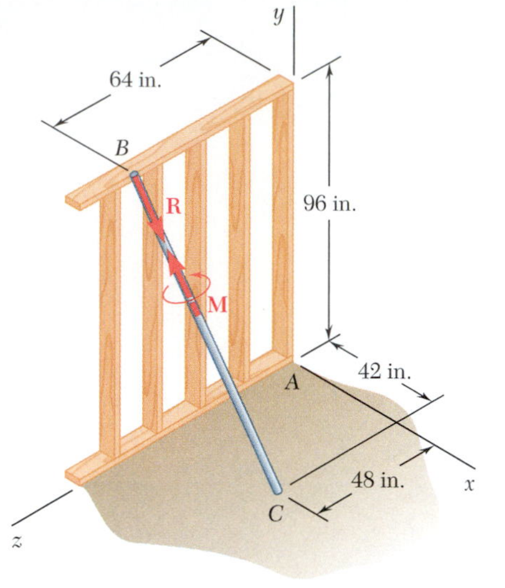

Fig. P3.149

3.150 Two parallel 60-N forces are applied to a lever as shown. Determine the moment of the couple formed by the two forces (*a*) by resolving each force into horizontal and vertical components and adding the moments of the two resulting couples, (*b*) by using the perpendicular distance between the two forces, (*c*) by summing the moments of the two forces about point A.

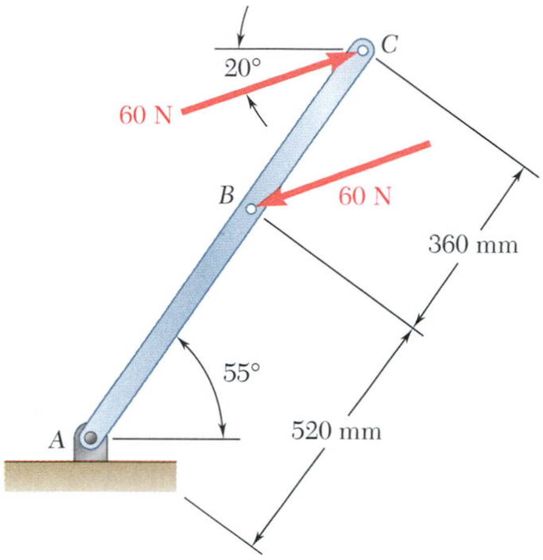

Fig. P3.150

3.151 A 32-lb motor is mounted on the floor. Find the resultant of the weight and the forces exerted on the belt, and determine where the line of action of the resultant intersects the floor.

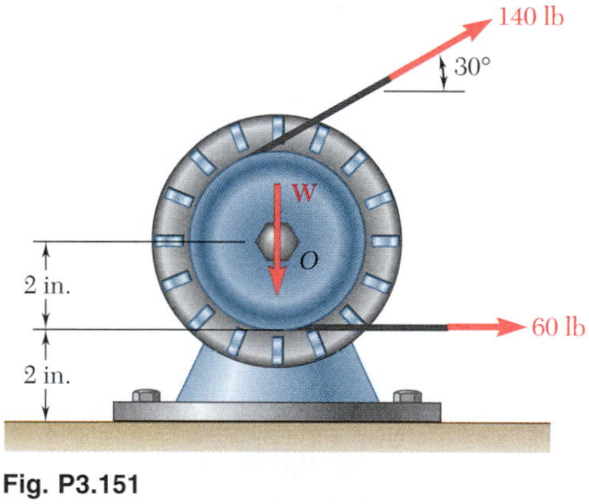

Fig. P3.151

3.152 To loosen a frozen valve, a force **F** of magnitude 70 lb is applied to the handle of the valve. Knowing that $\theta = 25°$, $M_x = -61$ lb · ft, and $M_z = -43$ lb · ft, determine ϕ and d.

3.153 When a force **F** is applied to the handle of the valve shown, its moments about the x and z axes are, respectively, $M_x = -77$ lb · ft and $M_z = -81$ lb · ft. For $d = 27$ in., determine the moment M_y of **F** about the y axis.

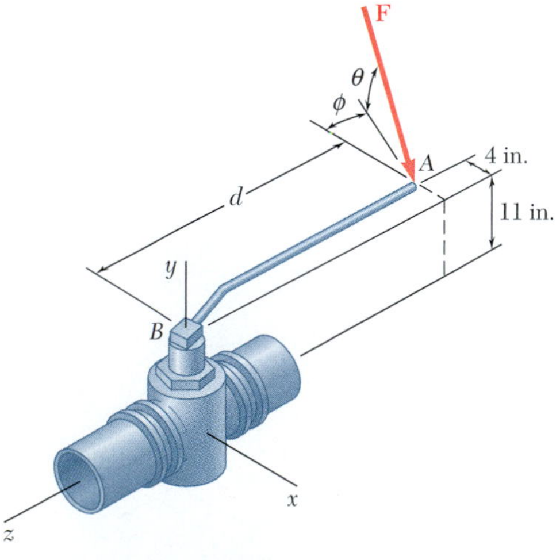

Fig. P3.152 and *P3.153*

Computer Problems

3.C1 Rod AB is held in place by a cord AC that has a tension T. Using computational software, determine the moment about B of the force exerted by the cord at point A in terms of the tension T and the distance c. Plot the moment about B for 320 mm $\leq c \leq$ 960 mm when (*a*) $T = 50$ N, (*b*) $T = 75$ N, (*c*) $T = 100$ N.

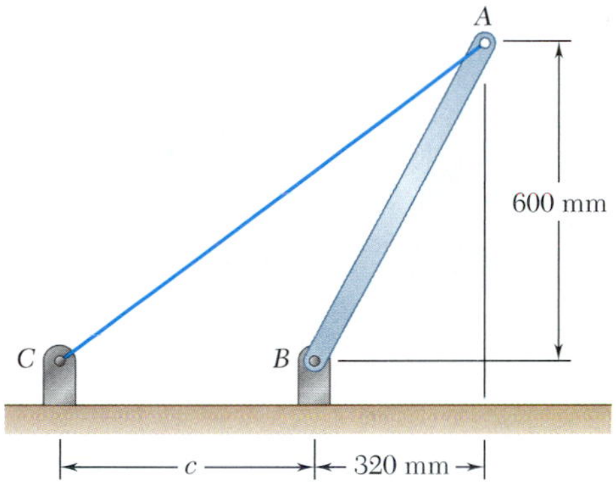

Fig. P3.C1

3.C2 A 400-mm tube AB can slide freely along a horizontal rod. The ends A and B of the tube are connected by elastic cords fixed to point C. (*a*) Determine the angle θ between the two cords AC and BC as a function of x. (*b*) Plot the angle θ between the cords AC and BC as a function of x for -400 mm $\leq x \leq$ 400 mm.

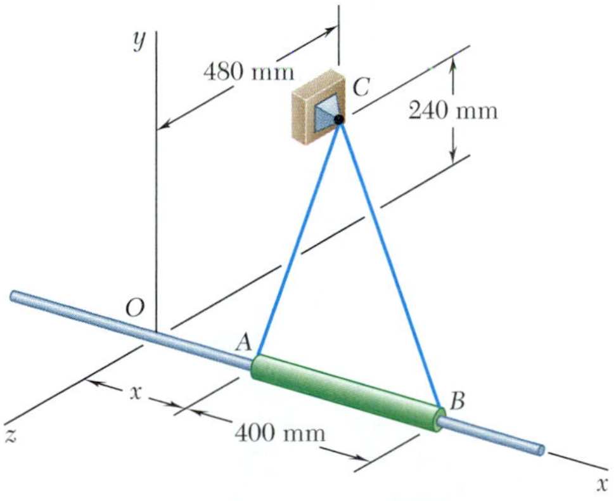

Fig. P3.C2

3.C3 Using computational software, determine the perpendicular distance between the line of action of a force **F** and the line OL. Use this software to solve (*a*) Prob. 3.62, (*b*) Prob. 3.63, (*c*) Prob. 3.65.

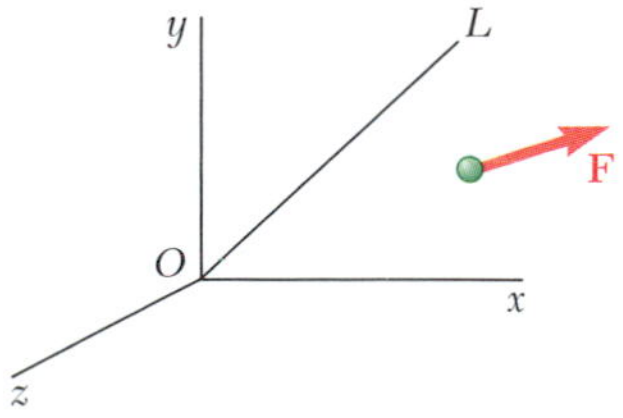

Fig. P3.C3

3.C4 A friend asks for your help in designing flower planter boxes. The boxes are to have 4, 5, 6, or 8 sides, which are to tilt outward at 10°, 20°, or 30°. Using computational software, determine the bevel angle α for each of the twelve planter designs. (*Hint:* The bevel angle α is equal to one-half of the angle formed by the inward normals of two adjacent sides.)

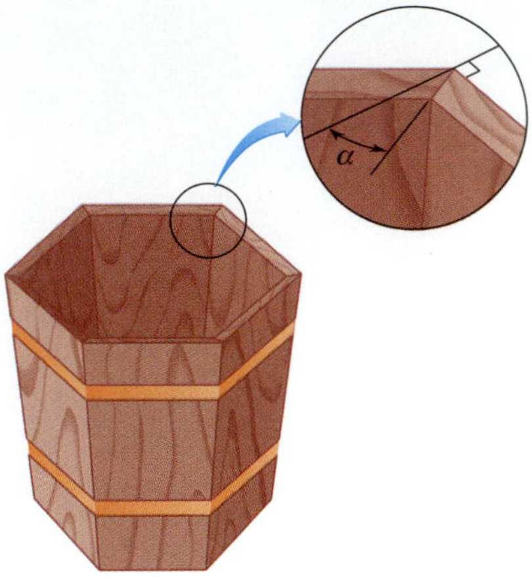

Fig. P3.C4

3.C5 A crane is oriented so that the end of the 50-ft boom OA lies in the yz plane as shown. The tension in cable AB is 875 lb. Determine the moment about each coordinate axes of the force exerted on A by cable AB as a function of x. Plot each moment as a function of x for $-15 \text{ ft} \leq x \leq 15 \text{ ft}$.

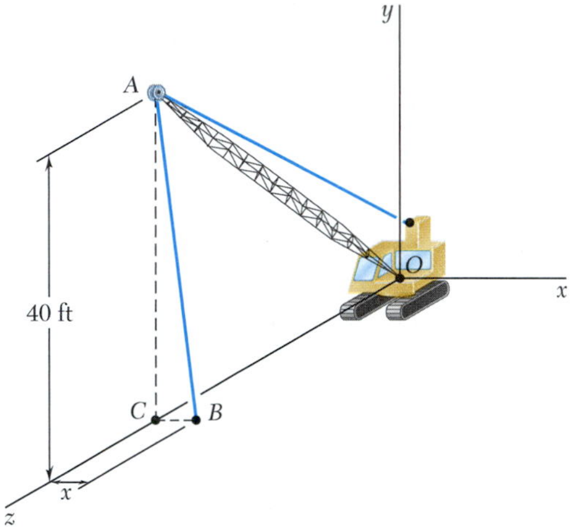

Fig. P3.C5

3.C6 A physical therapy patient lying on her right side holds a 1.5-kg dumbbell A in her left hand and slowly raises it along a circular path. Knowing that the center of the path is at B and that the path lies in the xy plane, plot the scalar components of the moment about C of the weight of the dumbbell as functions of θ for $-70° \leq \theta \leq 40°$.

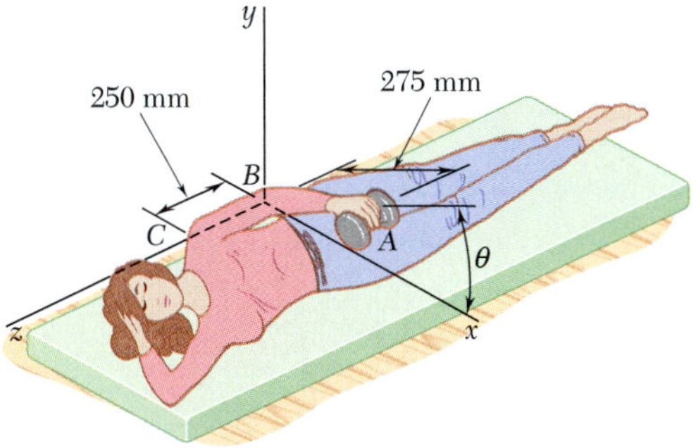

Fig. P3.C6

3.C7 The right angle horizontal speed-reducer shown weighs 40 lb, and its center of gravity is located on the y axis. Knowing that $M_1 = An^2$ and $M_2 = An^3$, determine the single force that is equivalent to the weight of the unit and the two couples acting on it and the point of coordinates $(x, 0, z)$ where the line of action of the single force intersects the floor. Plot x and z as functions of n for $2 \leq n \leq 6$ when (*a*) $A = 0.75 \text{ lb} \cdot \text{ft}$, (*b*) $A = 1.5 \text{ lb} \cdot \text{ft}$, (*c*) $A = 2 \text{ lb} \cdot \text{ft}$.

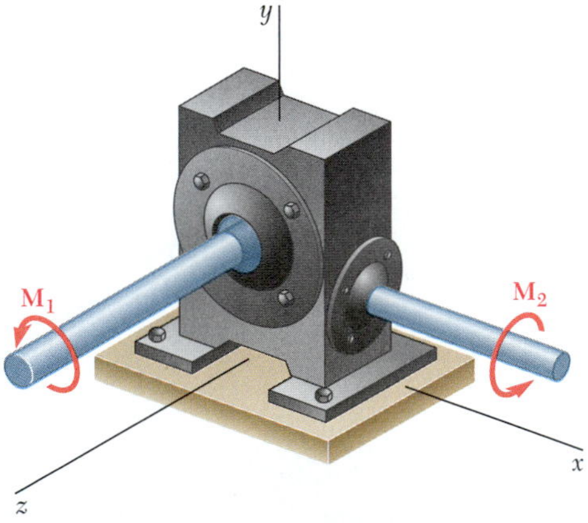

Fig. P3.C7

3.C8 A beam *AB* is subjected to several vertical forces as shown. Using computational software, determine the magnitude of the resultant of the forces and the distance x_C to point *C*, the point where the line of action of the resultant intersects *AB*. Use this software to solve (*a*) Prob. 3.103*a*, (*b*) Prob. 3.104*a*.

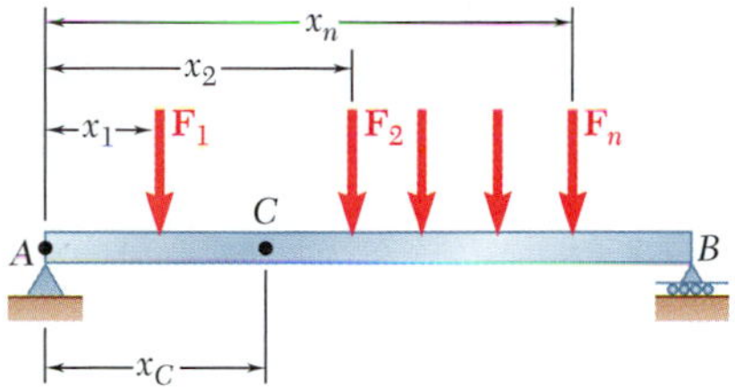

Fig. P3.C8

3.C9 Using computational software, determine the magnitude and the point of application of the resultant of the vertical forces $\mathbf{P}_1$, $\mathbf{P}_2$, . . . , $\mathbf{P}_n$ which act at points $\mathbf{A}_1$, $\mathbf{A}_2$, . . . , $\mathbf{A}_n$ that are located in the *xz* plane. Use this software to solve (*a*) Prob. 3.123, (*b*) Prob. 3.125.

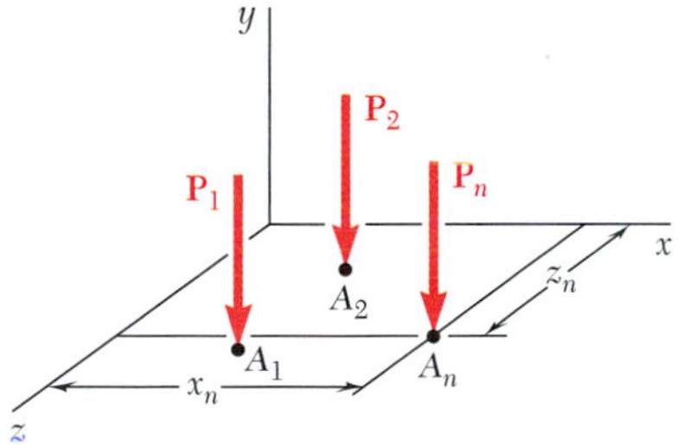

Fig. P3.C9

3.C10 Using computational software, determine whether a system of forces and couples can be reduced to a single equivalent force **R**, and, if it can, determine **R** and the point where the line of action of **R** intersects the *yz* plane. Use this software to solve (*a*) Prob. 3.136, (*b*) Prob. 3.137.

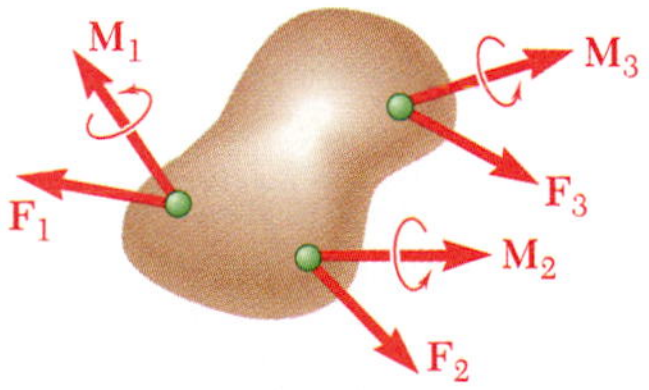

Fig. P3.C10

C H A P T E R

4

Equilibrium of Rigid Bodies

On many construction projects, large cranes such as those shown are used. The sum of the moments of the loads and the counterweights about the base of a crane must be adjusted so that the couple of the reaction at the base does not cause failure of the tower.

EQUILIBRIUM OF RIGID BODIES

4.1. INTRODUCTION

We saw in the preceding chapter that the external forces acting on a rigid body can be reduced to a force-couple system at some arbitrary point O. When the force and the couple are both equal to zero, the external forces form a system equivalent to zero, and the rigid body is said to be in *equilibrium.*

The necessary and sufficient conditions for the equilibrium of a rigid body, therefore, can be obtained by setting $\mathbf{R}$ and $\mathbf{M}_O^R$ equal to zero in the relations (3.52) of Sec. 3.17:

$$\Sigma \mathbf{F} = 0 \qquad \Sigma \mathbf{M}_O = \Sigma(\mathbf{r} \times \mathbf{F}) = 0 \tag{4.1}$$

Resolving each force and each moment into its rectangular components, we can express the necessary and sufficient conditions for the equilibrium of a rigid body with the following six scalar equations:

$$\Sigma F_x = 0 \qquad \Sigma F_y = 0 \qquad \Sigma F_z = 0 \tag{4.2}$$
$$\Sigma M_x = 0 \qquad \Sigma M_y = 0 \qquad \Sigma M_z = 0 \tag{4.3}$$

The equations obtained can be used to determine unknown forces applied to the rigid body or unknown reactions exerted on it by its supports. We note that Eqs. (4.2) express the fact that the components of the external forces in the x, y, and z directions are balanced; Eqs. (4.3) express the fact that the moments of the external forces about the x, y, and z axes are balanced. Therefore, for a rigid body in equilibrium, the system of the external forces will impart no translational or rotational motion to the body considered.

In order to write the equations of equilibrium for a rigid body, it is essential to first identify all of the forces acting on that body and then to draw the corresponding *free-body diagram.* In this chapter we first consider the equilibrium of *two-dimensional structures* subjected to forces contained in their planes and learn how to draw their free-body diagrams. In addition to the forces *applied* to a structure, the *reactions* exerted on the structure by its supports will be considered. A specific reaction will be associated with each type of support. You will learn how to determine whether the structure is properly supported, so that you can know in advance whether the equations of equilibrium can be solved for the unknown forces and reactions.

Later in the chapter, the equilibrium of three-dimensional structures will be considered, and the same kind of analysis will be given to these structures and their supports.

4.2. FREE-BODY DIAGRAM

In solving a problem concerning the equilibrium of a rigid body, it is essential to consider *all* of the forces acting on the body; it is equally important to exclude any force which is not directly applied to the body. Omitting a force or adding an extraneous one would destroy the conditions of equilibrium. Therefore, the first step in the solution of the problem should be to draw a *free-body diagram* of the rigid body under consideration. Free-body diagrams have already been used on many occasions in Chap. 2. However, in view of their importance to the solution of equilibrium problems, we summarize here the various steps which must be followed in drawing a free-body diagram.

Photo 4.1 A free-body diagram of the tractor shown would include all of the external forces acting on the tractor: the weight of the tractor, the weight of the load in the bucket, and the forces exerted by the ground on the tires.

Photo 4.2 In Chap. 6, we will discuss how to determine the internal forces in structures made of several connected pieces, such as the forces in the members that support the bucket of the tractor of Photo 4.1.

1. A clear decision should be made regarding the choice of the free body to be used. This body is then detached from the ground and is separated from all other bodies. The contour of the body thus isolated is sketched.

2. All external forces should be indicated on the free-body diagram. These forces represent the actions exerted *on* the free body *by* the ground and *by* the bodies which have been detached; they should be applied at the various points where the free body was supported by the ground or was connected to the other bodies. The *weight* of the free body should also be included among the external forces, since it represents the attraction exerted by the earth on the various particles forming the free body. As will be seen in Chap. 5, the weight should be applied at the center of gravity of the body. When the free body is made of several parts, the forces the various parts exert on each other should *not* be included among the external forces. These forces are internal forces as far as the free body is concerned.

3. The magnitudes and directions of the *known external forces* should be clearly marked on the free-body diagram. When indicating the directions of these forces, it must be remembered that the forces shown on the free-body diagram must be those which are exerted *on,* and not *by,* the free body. Known external forces generally include the *weight* of the free body and *forces applied* for a given purpose.

4. *Unknown external forces* usually consist of the *reactions,* through which the ground and other bodies oppose a possible motion of the free body. The reactions constrain the free body to remain in the same position and, for that reason, are sometimes called *constraining forces.* Reactions are exerted at the points where the free body is *supported by* or *connected to* other bodies and should be clearly indicated. Reactions are discussed in detail in Secs. 4.3 and 4.8.

5. The free-body diagram should also include dimensions, since these may be needed in the computation of moments of forces. Any other detail, however, should be omitted.

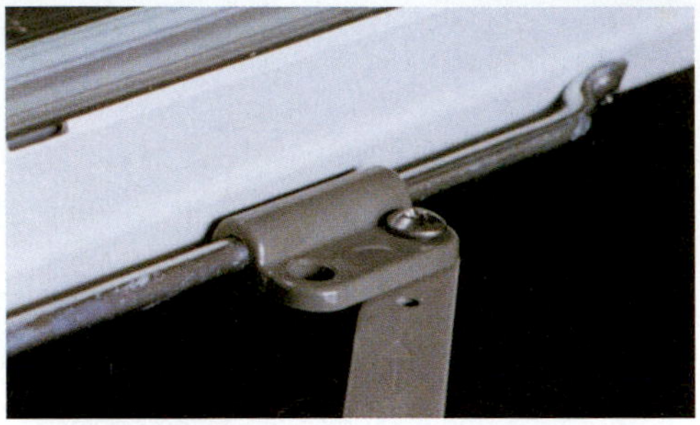

Photo 4.3 As the link of the awning window opening mechanism is extended, the force it exerts on the slider results in a normal force being applied to the rod, which causes the window to open.

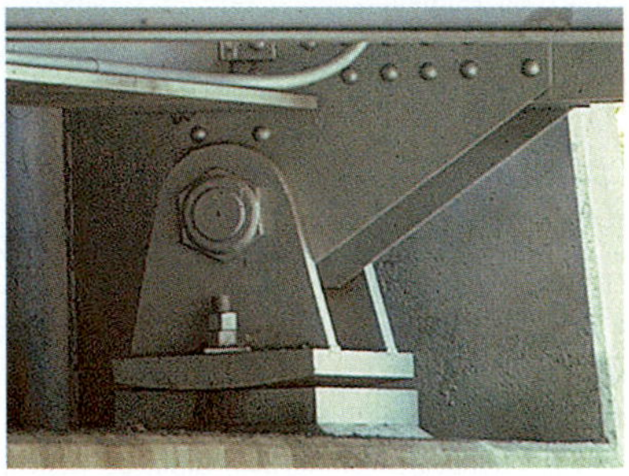

Photo 4.4 The abutment-mounted rocker bearing shown is used to support the roadway of a bridge.

Photo 4.5 Shown is the rocker expansion bearing of a plate girder bridge. The convex surface of the rocker allows the support of the girder to move horizontally.

EQUILIBRIUM IN TWO DIMENSIONS

4.3. REACTIONS AT SUPPORTS AND CONNECTIONS FOR A TWO-DIMENSIONAL STRUCTURE

In the first part of this chapter, the equilibrium of a two-dimensional structure is considered; that is, it is assumed that the structure being analyzed and the forces applied to it are contained in the same plane. Clearly, the reactions needed to maintain the structure in the same position will also be contained in this plane.

The reactions exerted on a two-dimensional structure can be divided into three groups corresponding to three types of *supports*, or *connections:*

1. *Reactions Equivalent to a Force with Known Line of Action.* Supports and connections causing reactions of this type include *rollers, rockers, frictionless surfaces, short links and cables, collars on frictionless rods,* and *frictionless pins in slots.* Each of these supports and connections can prevent motion in one direction only. They are shown in Fig. 4.1, together with the reactions they produce. Each of these reactions involves *one unknown,* namely, the magnitude of the reaction; this magnitude should be denoted by an appropriate letter. The line of action of the reaction is known and should be indicated clearly in the free-body diagram. The sense of the reaction must be as shown in Fig. 4.1 for the cases of a frictionless surface (toward the free body) or a cable (away from the free body). The reaction can be directed either way in the case of double-track rollers, links, collars on rods, and pins in slots. Single-track rollers and rockers are generally assumed to be reversible, and thus the corresponding reactions can also be directed either way.

2. *Reactions Equivalent to a Force of Unknown Direction and Magnitude.* Supports and connections causing reactions of this type include *frictionless pins in fitted holes, hinges,* and *rough surfaces.* They can prevent translation of the free body in all directions, but they cannot prevent the body from rotating about the connection. Reactions of this group involve *two unknowns* and are usually represented by their x and y components. In the case of a rough surface, the component normal to the surface must be directed away from the surface, and thus is directed toward the free body.

3. *Reactions Equivalent to a Force and a Couple.* These reactions are caused by *fixed supports,* which oppose any motion of the free body and thus constrain it completely. Fixed supports actually produce forces over the entire surface of contact; these forces, however, form a system which can be reduced to a force and a couple. Reactions of this group involve *three unknowns,* consisting usually of the two components of the force and the moment of the couple.

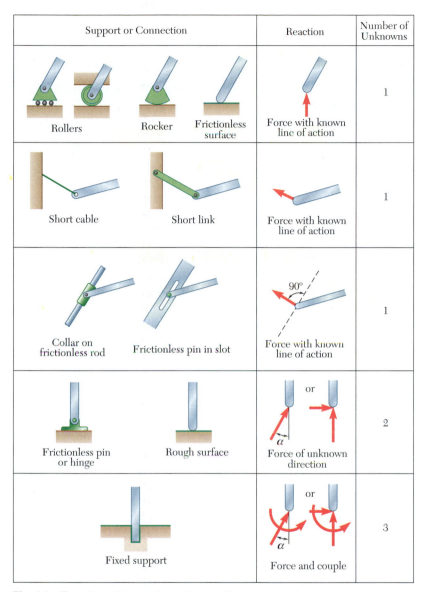

Support or Connection	Reaction	Number of Unknowns
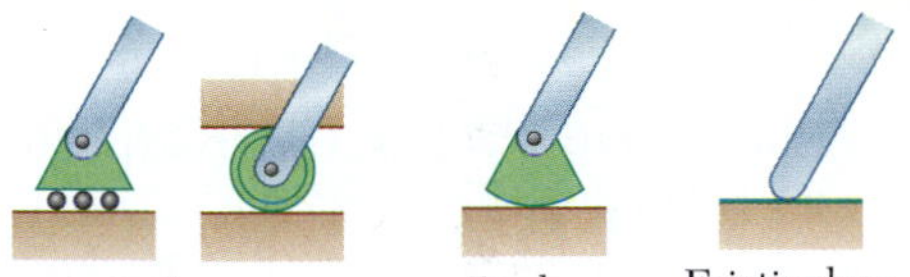 Rollers; Rocker; Frictionless surface	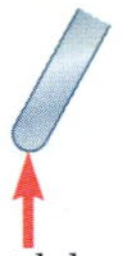 Force with known line of action	1
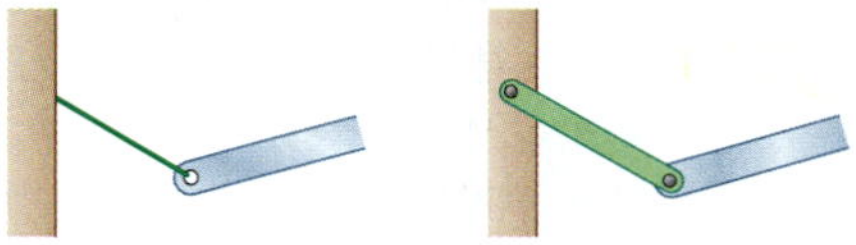 Short cable; Short link	 Force with known line of action	1
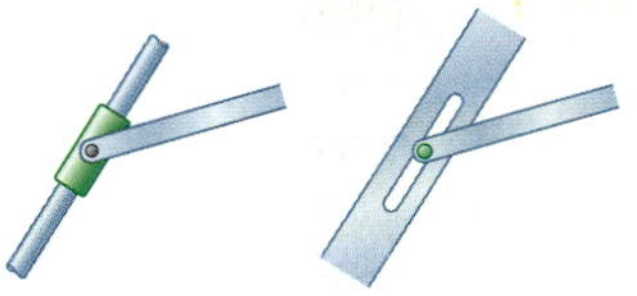 Collar on frictionless rod; Frictionless pin in slot	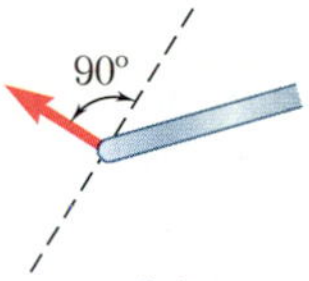 Force with known line of action	1
 Frictionless pin or hinge; Rough surface	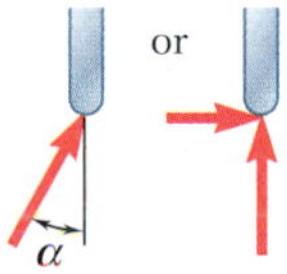 Force of unknown direction	2
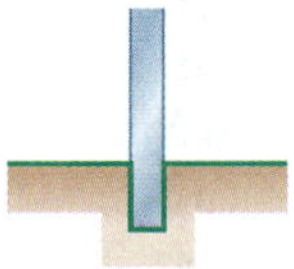 Fixed support	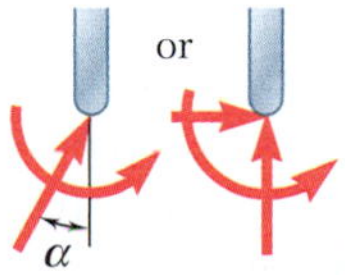 Force and couple	3

Fig. 4.1 Reactions at supports and connections.

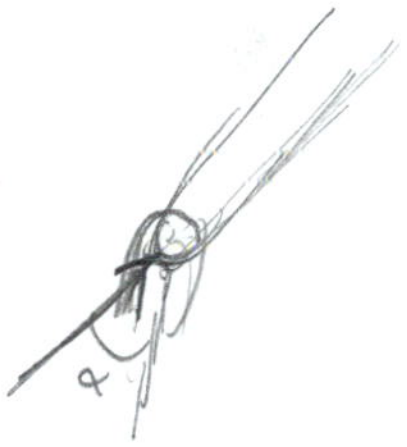

When the sense of an unknown force or couple is not readily apparent, no attempt should be made to determine it. Instead, the sense of the force or couple should be arbitrarily assumed; the sign of the answer obtained will indicate whether the assumption is correct or not.

4.4. EQUILIBRIUM OF A RIGID BODY IN TWO DIMENSIONS

The conditions stated in Sec. 4.1 for the equilibrium of a rigid body become considerably simpler for the case of a two-dimensional structure. Choosing the x and y axes to be in the plane of the structure, we have

$$F_z = 0 \qquad M_x = M_y = 0 \qquad M_z = M_O$$

for each of the forces applied to the structure. Thus, the six equations of equilibrium derived in Sec. 4.1 reduce to

$$\Sigma F_x = 0 \qquad \Sigma F_y = 0 \qquad \Sigma M_O = 0 \tag{4.4}$$

and to three trivial identities, $0 = 0$. Since $\Sigma M_O = 0$ must be satisfied regardless of the choice of the origin O, we can write the equations of equilibrium for a two-dimensional structure in the more general form

$$\Sigma F_x = 0 \qquad \Sigma F_y = 0 \qquad \Sigma M_A = 0 \tag{4.5}$$

where A is any point in the plane of the structure. The three equations obtained can be solved for no more than *three unknowns.*

We saw in the preceding section that unknown forces include reactions and that the number of unknowns corresponding to a given reaction depends upon the type of support or connection causing that reaction. Referring to Sec. 4.3, we observe that the equilibrium equations (4.5) can be used to determine the reactions associated with two rollers and one cable, one fixed support, or one roller and one pin in a fitted hole, etc.

Consider Fig. 4.2*a*, in which the truss shown is subjected to the given forces **P**, **Q**, and **S**. The truss is held in place by a pin at A and a roller at B. The pin prevents point A from moving by exerting on the truss a force which can be resolved into the components $\mathbf{A}_x$ and $\mathbf{A}_y$; the roller keeps the truss from rotating about A by exerting the vertical force **B**. The free-body diagram of the truss is shown in Fig. 4.2*b*; it includes the reactions $\mathbf{A}_x$, $\mathbf{A}_y$, and **B** as well as the applied forces **P**, **Q**, **S** and the weight **W** of the truss. Expressing that the sum of the moments about A of all of the forces shown in Fig. 4.2*b* is zero, we write the equation $\Sigma M_A = 0$, which can be used to determine the magnitude B since it does not contain A_x or A_y. Next, expressing that the sum of the x components and the sum of the y components of the forces are zero, we write the equations $\Sigma F_x = 0$ and $\Sigma F_y = 0$, from which we can obtain the components A_x and A_y, respectively.

An additional equation could be obtained by expressing that the sum of the moments of the external forces about a point other than A is zero. We could write, for instance, $\Sigma M_B = 0$. Such a statement, however, does not contain any new information, since it has already been established that the system of the forces shown in Fig. 4.2*b* is equivalent to zero. The additional equation *is not independent* and cannot be used to determine a fourth unknown. It will be useful,

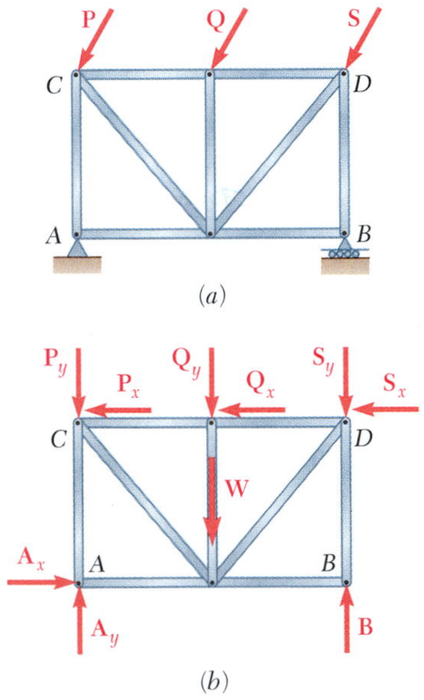

Fig. 4.2

however, for checking the solution obtained from the original three equations of equilibrium.

While the three equations of equilibrium cannot be *augmented* by additional equations, any of them can be *replaced* by another equation. Therefore, an alternative system of equations of equilibrium is

$$\Sigma F_x = 0 \qquad \Sigma M_A = 0 \qquad \Sigma M_B = 0 \tag{4.6}$$

where the second point about which the moments are summed (in this case, point B) cannot lie on the line parallel to the y axis that passes through point A (Fig. 4.2*b*). These equations are sufficient conditions for the equilibrium of the truss. The first two equations indicate that the external forces must reduce to a single vertical force at A. Since the third equation requires that the moment of this force be zero about a point B which is not on its line of action, the force must be zero, and the rigid body is in equilibrium.

A third possible set of equations of equilibrium is

$$\Sigma M_A = 0 \qquad \Sigma M_B = 0 \qquad \Sigma M_C = 0 \tag{4.7}$$

where the points A, B, and C do not lie in a straight line (Fig. 4.2*b*). The first equation requires that the external forces reduce to a single force at A; the second equation requires that this force pass through B; and the third equation requires that it pass through C. Since the points A, B, C do not lie in a straight line, the force must be zero, and the rigid body is in equilibrium.

The equation $\Sigma M_A = 0$, which expresses that the sum of the moments of the forces about pin A is zero, possesses a more definite physical meaning than either of the other two equations (4.7). These two equations express a similar idea of balance, but with respect to points about which the rigid body is not actually hinged. They are, however, as useful as the first equation, and our choice of equilibrium equations should not be unduly influenced by the physical meaning of these equations. Indeed, it will be desirable in practice to choose equations of equilibrium containing only one unknown, since this eliminates the necessity of solving simultaneous equations. Equations containing only one unknown can be obtained by summing moments about the point of intersection of the lines of action of two unknown forces or, if these forces are parallel, by summing components in a direction perpendicular to their common direction. For example, in Fig. 4.3, in which the truss shown is held by rollers at A and B and a short link at D, the reactions at A and B can be eliminated by summing x components. The reactions at A and D will be eliminated by summing moments about C, and the reactions at B and D by summing moments about D. The equations obtained are

$$\Sigma F_x = 0 \qquad \Sigma M_C = 0 \qquad \Sigma M_D = 0$$

Each of these equations contains only one unknown.

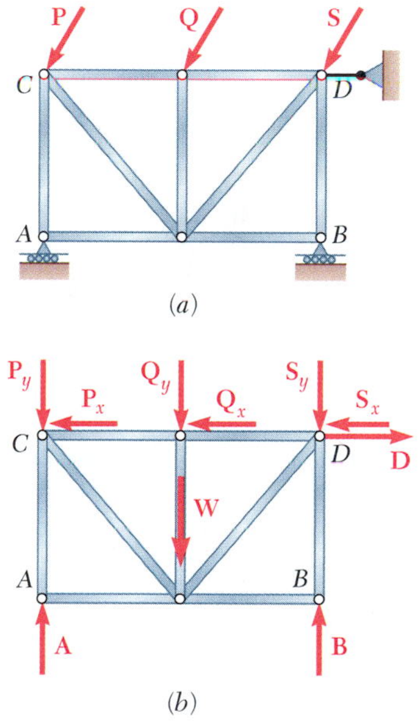

Fig. 4.3

4.5. STATICALLY INDETERMINATE REACTIONS. PARTIAL CONSTRAINTS

In the two examples considered in the preceding section (Figs. 4.2 and 4.3), the types of supports used were such that the rigid body could not possibly move under the given loads or under any other loading conditions. In such cases, the rigid body is said to be *completely constrained.* We also recall that the reactions corresponding to these supports involved *three unknowns* and could be determined by solving the three equations of equilibrium. When such a situation exists, the reactions are said to be *statically determinate.*

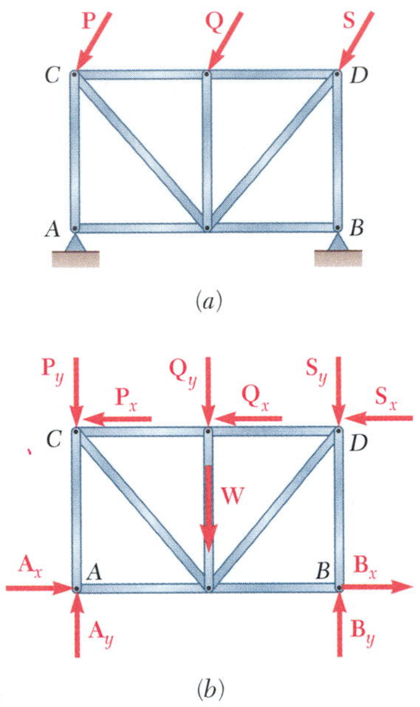

Fig. 4.4 Statically indeterminate reactions.

Consider Fig. 4.4*a*, in which the truss shown is held by pins at *A* and *B*. These supports provide more constraints than are necessary to keep the truss from moving under the given loads or under any other loading conditions. We also note from the free-body diagram of Fig. 4.4*b* that the corresponding reactions involve *four unknowns.* Since, as was pointed out in Sec. 4.4, only three independent equilibrium equations are available, there are *more unknowns than equations;* thus, all of the unknowns cannot be determined. While the equations $\Sigma M_A = 0$ and $\Sigma M_B = 0$ yield the vertical components B_y and A_y, respectively, the equation $\Sigma F_x = 0$ gives only the sum $A_x + B_x$ of the horizontal components of the reactions at *A* and *B*. The components A_x and B_x are said to be *statically indeterminate.* They could be determined by considering the deformations produced in the truss by the given loading, but this method is beyond the scope of statics and belongs to the study of mechanics of materials.

The supports used to hold the truss shown in Fig. 4.5*a* consist of rollers at *A* and *B*. Clearly, the constraints provided by these supports are not sufficient to keep the truss from moving. While any vertical motion is prevented, the truss is free to move horizontally. The truss is said to be *partially constrained.*† Turning our attention to Fig. 4.5*b*, we note that the reactions at *A* and *B* involve only *two unknowns.* Since three equations of equilibrium must still be satisfied, there are *fewer unknowns than equations,* and, in general, one of the equilibrium equations will not be satisfied. While the equations $\Sigma M_A = 0$ and $\Sigma M_B = 0$ can be satisfied by a proper choice of reactions at *A* and *B*, the equation $\Sigma F_x = 0$ will not be satisfied unless the sum of the horizontal components of the applied forces happens to be zero. We thus observe that the equlibrium of the truss of Fig. 4.5 cannot be maintained under general loading conditions.

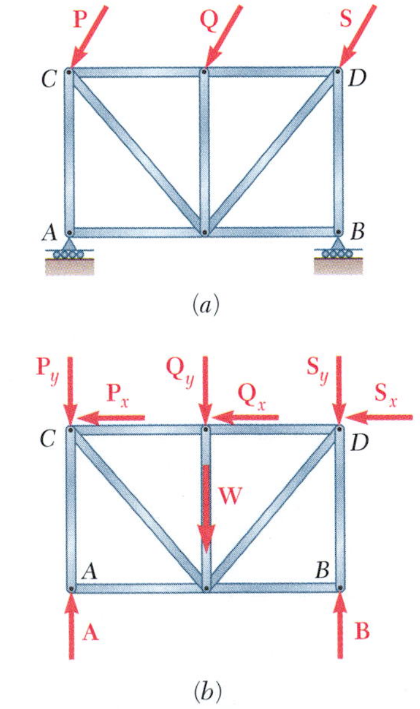

Fig. 4.5 Partial constraints.

It appears from the above that if a rigid body is to be completely constrained and if the reactions at its supports are to be statically determinate, *there must be as many unknowns as there are equations of equilibrium.* When this condition is *not* satisfied, we can be certain either that the rigid body is not completely constrained or that the reactions at its supports are not statically determinate; it is also possible that the rigid body is not completely constrained *and* that the reactions are statically indeterminate.

We should note, however, that, while *necessary,* the above condition is *not sufficient.* In other words, the fact that the number of

†Partially constrained bodies are often referred to as *unstable.* However, to avoid confusion between this type of instability, due to insufficient constraints, and the type of instability considered in Chap. 10, which relates to the behavior of a rigid body when its equilibrium is disturbed, we will restrict the use of the words *stable* and *unstable* to the latter case.

unknowns is equal to the number of equations is no guarantee that the body is completely constrained or that the reactions at its supports are statically determinate. Consider Fig. 4.6*a*, in which the truss shown is held by rollers at *A*, *B*, and *E*. While there are three unknown reactions, **A**, **B**, and **E** (Fig. 4.6*b*), the equation $\Sigma F_x = 0$ will not be satisfied unless the sum of the horizontal components of the applied forces happens to be zero. Although there are a sufficient number of constraints, these constraints are not properly arranged, and the truss is free to move horizontally. We say that the truss is *improperly constrained.* Since only two equilibrium equations are left for determining three unknowns, the reactions will be statically indeterminate. Thus, improper constraints also produce static indeterminacy.

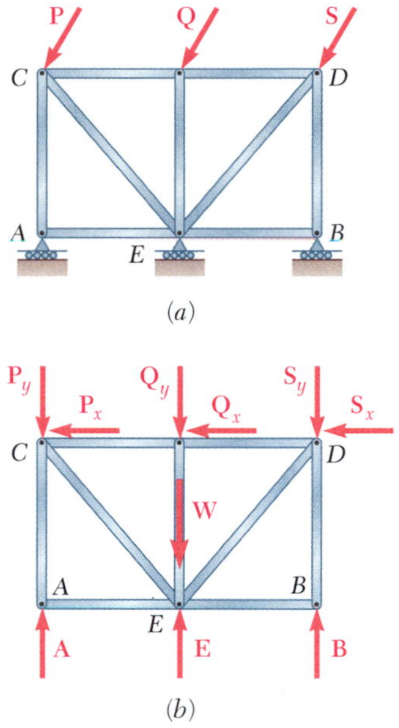

Fig. 4.6 Improper constraints.

Another example of improper constraints—and of static indeterminacy—is provided by the truss shown in Fig. 4.7. This truss is held by a pin at *A* and by rollers at *B* and *C*, which altogether involve four unknowns. Since only three independent equilibrium equations are available, the reactions at the supports are statically indeterminate. On the other hand, we note that the equation $\Sigma M_A = 0$ cannot be satisfied under general loading conditions, since the lines of action of the reactions **B** and **C** pass through *A*. We conclude that the truss can rotate about *A* and that it is improperly constrained.†

The examples of Figs. 4.6 and 4.7 lead us to conclude that *a rigid body is improperly constrained whenever the supports,* even though they may provide a sufficient number of reactions, *are arranged in such a way that the reactions must be either concurrent or parallel.*‡

In summary, to be sure that a two-dimensional rigid body is completely constrained and that the reactions at its supports are statically determinate, we should verify that the reactions involve three—and only three—unknowns and that the supports are arranged in such a way that they do not require the reactions to be either concurrent or parallel.

Supports involving statically indeterminate reactions should be used with care in the *design* of structures and only with a full knowledge of the problems they may cause. On the other hand, the *analysis* of structures possessing statically indeterminate reactions often can be partially carried out by the methods of statics. In the case of the truss of Fig. 4.4, for example, the vertical components of the reactions at *A* and *B* were obtained from the equilibrium equations.

For obvious reasons, supports producing partial or improper constraints should be avoided in the design of stationary structures. However, a partially or improperly constrained structure will not necessarily collapse; under particular loading conditions, equilibrium can be maintained. For example, the trusses of Figs. 4.5 and 4.6 will be in equilibrium if the applied forces **P**, **Q**, and **S** are vertical. Besides, structures which are designed to move *should* be only partially constrained. A railroad car, for instance, would be of little use if it were completely constrained by having its brakes applied permanently.

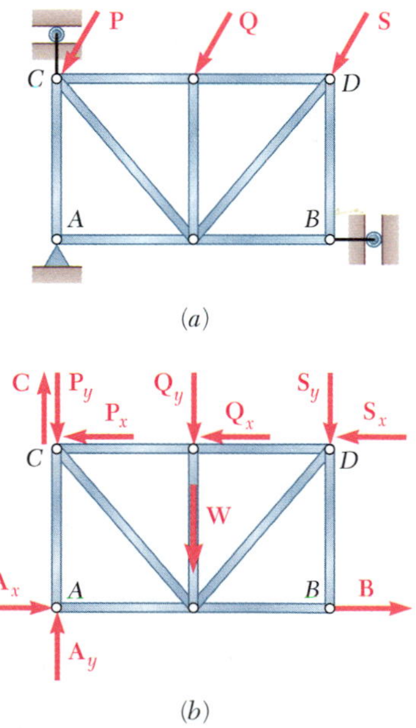

Fig. 4.7 Improper constraints.

†Rotation of the truss about *A* requires some "play" in the supports at *B* and *C*. In practice such play will always exist. In addition, we note that if the play is kept small, the displacements of the rollers *B* and *C* and, thus, the distances from *A* to the lines of action of the reactions **B** and **C** will also be small. The equation $\Sigma M_A = 0$ then requires that **B** and **C** be very large, a situation which can result in the failure of the supports at *B* and *C*.

‡Because this situation arises from an inadequate arrangement or *geometry* of the supports, it is often referred to as *geometric instability.*

SAMPLE PROBLEM 4.1

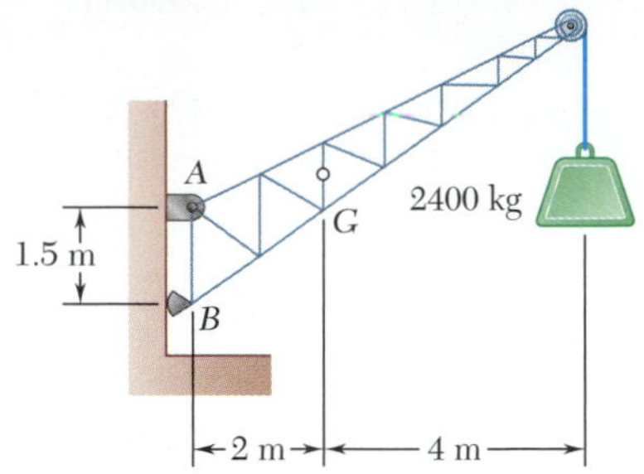

A fixed crane has a mass of 1000 kg and is used to lift a 2400-kg crate. It is held in place by a pin at A and a rocker at B. The center of gravity of the crane is located at G. Determine the components of the reactions at A and B.

SOLUTION

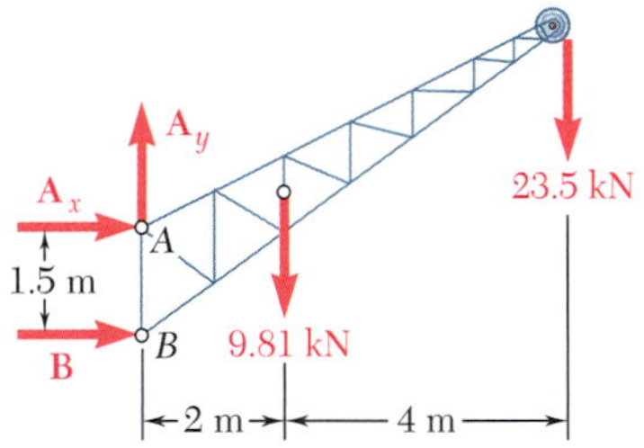

Free-Body Diagram. A free-body diagram of the crane is drawn. By multiplying the masses of the crane and of the crate by $g = 9.81\ \text{m/s}^2$, we obtain the corresponding weights, that is, 9810 N or 9.81 kN, and 23 500 N or 23.5 kN. The reaction at pin A is a force of unknown direction; it is represented by its components $\mathbf{A}_x$ and $\mathbf{A}_y$. The reaction at the rocker B is perpendicular to the rocker surface; thus, it is horizontal. We assume that $\mathbf{A}_x$, $\mathbf{A}_y$, and $\mathbf{B}$ act in the directions shown.

Determination of *B*. We express that the sum of the moments of all external forces about point A is zero. The equation obtained will contain neither A_x nor A_y, since the moments of $\mathbf{A}_x$ and $\mathbf{A}_y$ about A are zero. Multiplying the magnitude of each force by its perpendicular distance from A, we write

$$+\circlearrowleft \Sigma M_A = 0: \qquad +B(1.5\text{ m}) - (9.81\text{ kN})(2\text{ m}) - (23.5\text{ kN})(6\text{ m}) = 0$$
$$B = +107.1\text{ kN} \qquad \mathbf{B} = 107.1\text{ kN} \rightarrow \blacktriangleleft$$

Since the result is positive, the reaction is directed as assumed.

Determination of $\mathbf{A}_x$. The magnitude of $\mathbf{A}_x$ is determined by expressing that the sum of the horizontal components of all external forces is zero.

$$\overset{+}{\rightarrow}\Sigma F_x = 0: \qquad A_x + B = 0$$
$$A_x + 107.1\text{ kN} = 0$$
$$A_x = -107.1\text{ kN} \qquad \mathbf{A}_x = 107.1\text{ kN} \leftarrow \blacktriangleleft$$

Since the result is negative, the sense of $\mathbf{A}_x$ is opposite to that assumed originally.

Determination of A_y. The sum of the vertical components must also equal zero.

$$+\uparrow\Sigma F_y = 0: \qquad A_y - 9.81\text{ kN} - 23.5\text{ kN} = 0$$
$$A_y = +33.3\text{ kN} \qquad \mathbf{A}_y = 33.3\text{ kN} \uparrow \blacktriangleleft$$

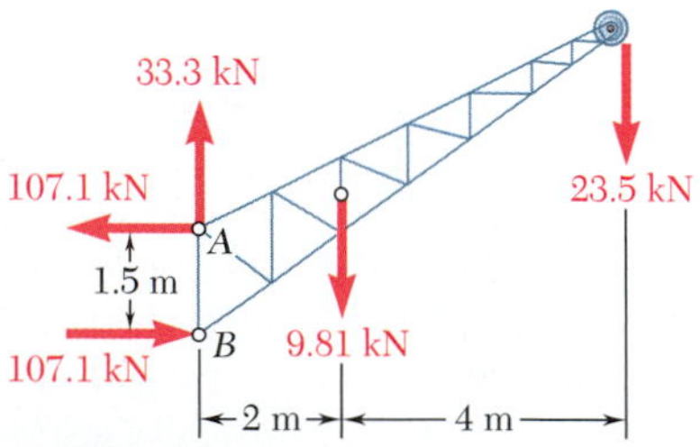

Adding vectorially the components $\mathbf{A}_x$ and $\mathbf{A}_y$, we find that the reaction at A is 112.2 kN ⦪17.3°.

Check. The values obtained for the reactions can be checked by recalling that the sum of the moments of all of the external forces about any point must be zero. For example, considering point B, we write

$$+\circlearrowleft\Sigma M_B = -(9.81\text{ kN})(2\text{ m}) - (23.5\text{ kN})(6\text{ m}) + (107.1\text{ kN})(1.5\text{ m}) = 0$$

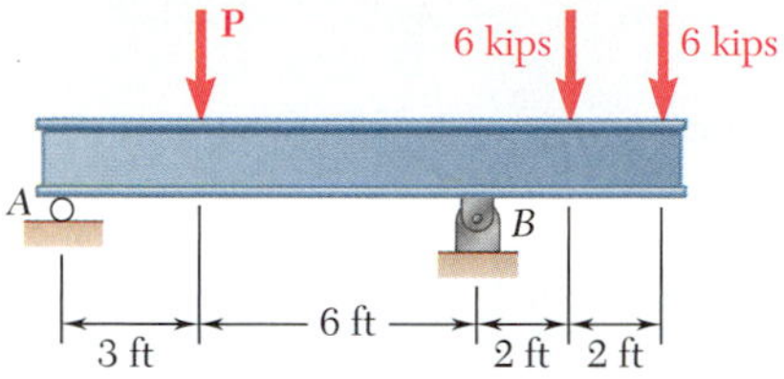

SAMPLE PROBLEM 4.2

Three loads are applied to a beam as shown. The beam is supported by a roller at A and by a pin at B. Neglecting the weight of the beam, determine the reactions at A and B when $P = 15$ kips.

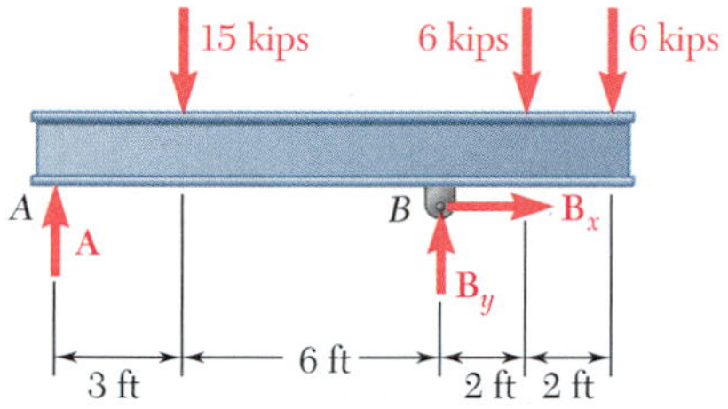

SOLUTION

Free-Body Diagram. A free-body diagram of the beam is drawn. The reaction at A is vertical and is denoted by $\mathbf{A}$. The reaction at B is represented by components $\mathbf{B}_x$ and $\mathbf{B}_y$. Each component is assumed to act in the direction shown.

Equilibrium Equations. We write the following three equilibrium equations and solve for the reactions indicated:

$$\overset{+}{\rightarrow}\Sigma F_x = 0: \qquad B_x = 0 \qquad \mathbf{B}_x = 0 \blacktriangleleft$$

$$+\circlearrowleft\Sigma M_A = 0:$$
$$-(15 \text{ kips})(3 \text{ ft}) + B_y(9 \text{ ft}) - (6 \text{ kips})(11 \text{ ft}) - (6 \text{ kips})(13 \text{ ft}) = 0$$
$$B_y = +21.0 \text{ kips} \qquad \mathbf{B}_y = 21.0 \text{ kips} \uparrow \blacktriangleleft$$

$$+\circlearrowleft\Sigma M_B = 0:$$
$$-A(9 \text{ ft}) + (15 \text{ kips})(6 \text{ ft}) - (6 \text{ kips})(2 \text{ ft}) - (6 \text{ kips})(4 \text{ ft}) = 0$$
$$A = +6.00 \text{ kips} \qquad \mathbf{A} = 6.00 \text{ kips} \uparrow \blacktriangleleft$$

Check. The results are checked by adding the vertical components of all of the external forces:

$$+\uparrow\Sigma F_y = +6.00 \text{ kips} - 15 \text{ kips} + 21.0 \text{ kips} - 6 \text{ kips} - 6 \text{ kips} = 0$$

Remark. In this problem the reactions at both A and B are vertical; however, these reactions are vertical for different reasons. At A, the beam is supported by a roller; hence the reaction cannot have a horizontal component. At B, the horizontal component of the reaction is zero because it must satisfy the equilibrium equation $\Sigma F_x = 0$ and because none of the other forces acting on the beam has a horizontal component.

We could have noticed at first glance that the reaction at B was vertical and dispensed with the horizontal component $\mathbf{B}_x$. This, however, is a bad practice. In following it, we would run the risk of forgetting the component $\mathbf{B}_x$ when the loading conditions require such a component (that is, when a horizontal load is included). Also, the component $\mathbf{B}_x$ was found to be zero by using and solving an equilibrium equation, $\Sigma F_x = 0$. By setting $\mathbf{B}_x$ equal to zero immediately, we might not realize that we actually made use of this equation and thus might lose track of the number of equations available for solving the problem.

SAMPLE PROBLEM 4.3

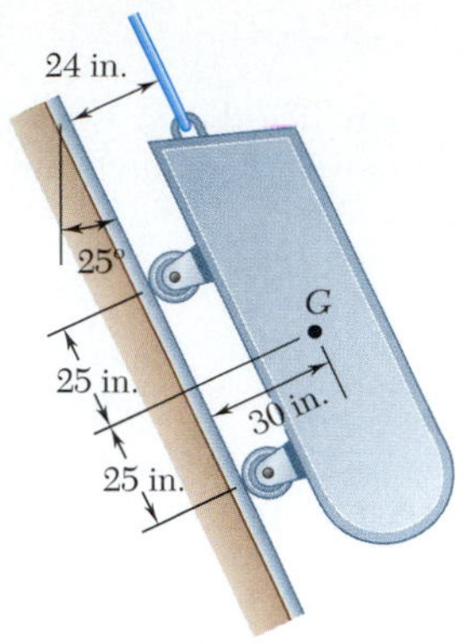

A loading car is at rest on a track forming an angle of 25° with the vertical. The gross weight of the car and its load is 5500 lb, and it is applied at a point 30 in. from the track, halfway between the two axles. The car is held by a cable attached 24 in. from the track. Determine the tension in the cable and the reaction at each pair of wheels.

SOLUTION

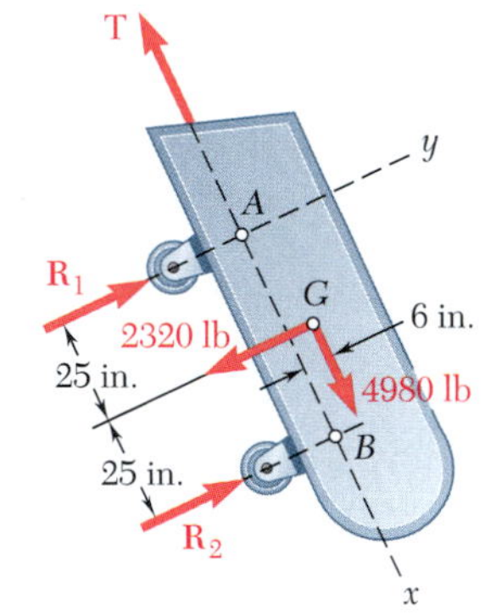

Free-Body Diagram. A free-body diagram of the car is drawn. The reaction at each wheel is perpendicular to the track, and the tension force **T** is parallel to the track. For convenience, we choose the x axis parallel to the track and the y axis perpendicular to the track. The 5500-lb weight is then resolved into x and y components.

$$W_x = +(5500 \text{ lb}) \cos 25° = +4980 \text{ lb}$$
$$W_y = -(5500 \text{ lb}) \sin 25° = -2320 \text{ lb}$$

Equilibrium Equations. We take moments about A to eliminate **T** and $\mathbf{R}_1$ from the computation.

$$+\circlearrowleft \Sigma M_A = 0\text{:} \quad -(2320 \text{ lb})(25 \text{ in.}) - (4980 \text{ lb})(6 \text{ in.}) + R_2(50 \text{ in.}) = 0$$
$$R_2 = +1758 \text{ lb} \qquad \mathbf{R}_2 = 1758 \text{ lb} \nearrow \blacktriangleleft$$

Now, taking moments about B to eliminate **T** and $\mathbf{R}_2$ from the computation, we write

$$+\circlearrowleft \Sigma M_B = 0\text{:} \quad (2320 \text{ lb})(25 \text{ in.}) - (4980 \text{ lb})(6 \text{ in.}) - R_1(50 \text{ in.}) = 0$$
$$R_1 = +562 \text{ lb} \qquad \mathbf{R}_1 = 562 \text{ lb} \nearrow \blacktriangleleft$$

The value of T is found by writing

$$\searrow + \Sigma F_x = 0\text{:} \quad +4980 \text{ lb} - T = 0$$
$$T = +4980 \text{ lb} \qquad \mathbf{T} = 4980 \text{ lb} \nwarrow \blacktriangleleft$$

The computed values of the reactions are shown in the adjacent sketch.

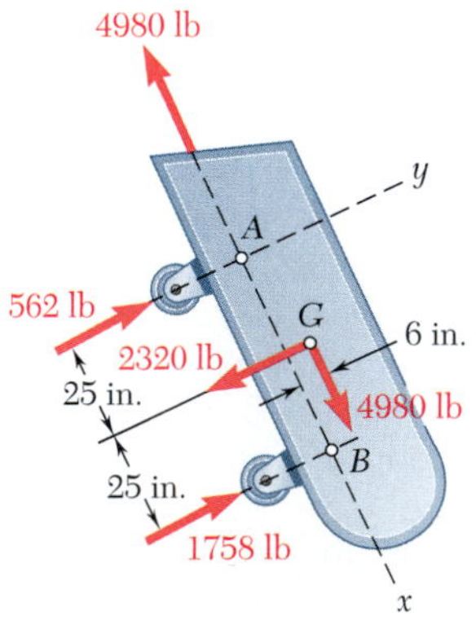

Check. The computations are verified by writing

$$\nearrow + \Sigma F_y = +562 \text{ lb} + 1758 \text{ lb} - 2320 \text{ lb} = 0$$

The solution could also have been checked by computing moments about any point other than A or B.

SAMPLE PROBLEM 4.4

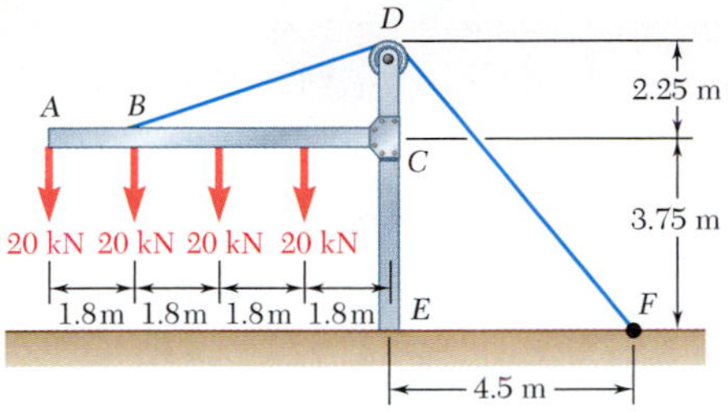

The frame shown supports part of the roof of a small building. Knowing that the tension in the cable is 150 kN, determine the reaction at the fixed end E.

SOLUTION

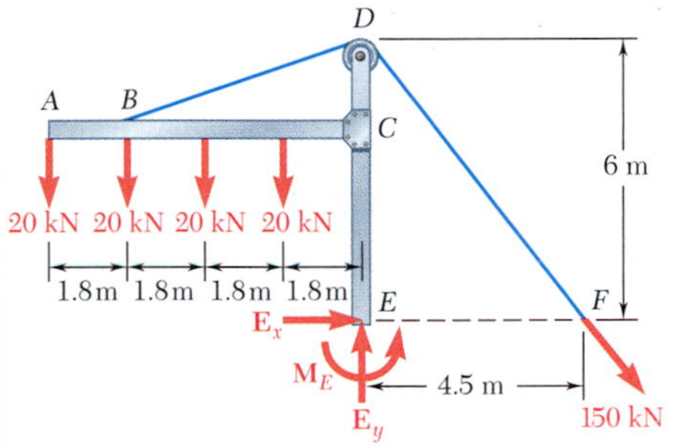

Free-Body Diagram. A free-body diagram of the frame and of the cable BDF is drawn. The reaction at the fixed end E is represented by the force components $\mathbf{E}_x$ and $\mathbf{E}_y$ and the couple $\mathbf{M}_E$. The other forces acting on the free body are the four 20-kN loads and the 150-kN force exerted at end F of the cable.

Equilibrium Equations. Noting that $DF = \sqrt{(4.5\text{ m})^2 + (6\text{ m})^2} = 7.5$ m, we write

$$\overset{+}{\rightarrow}\Sigma F_x = 0: \qquad E_x + \frac{4.5}{7.5}(150\text{ kN}) = 0$$

$$E_x = -90.0\text{ kN} \qquad \mathbf{E}_x = 90.0\text{ kN} \leftarrow \blacktriangleleft$$

$$+\uparrow\Sigma F_y = 0: \qquad E_y - 4(20\text{ kN}) - \frac{6}{7.5}(150\text{ kN}) = 0$$

$$E_y = +200\text{ kN} \qquad \mathbf{E}_y = 200\text{ kN}\uparrow \blacktriangleleft$$

$$+\circlearrowleft\Sigma M_E = 0: \qquad (20\text{ kN})(7.2\text{ m}) + (20\text{ kN})(5.4\text{ m}) + (20\text{ kN})(3.6\text{ m})$$
$$+ (20\text{ kN})(1.8\text{ m}) - \frac{6}{7.5}(150\text{ kN})(4.5\text{ m}) + M_E = 0$$

$$M_E = +180.0\text{ kN}\cdot\text{m} \qquad \mathbf{M}_E = 180.0\text{ kN}\cdot\text{m}\circlearrowleft \blacktriangleleft$$

SAMPLE PROBLEM 4.5

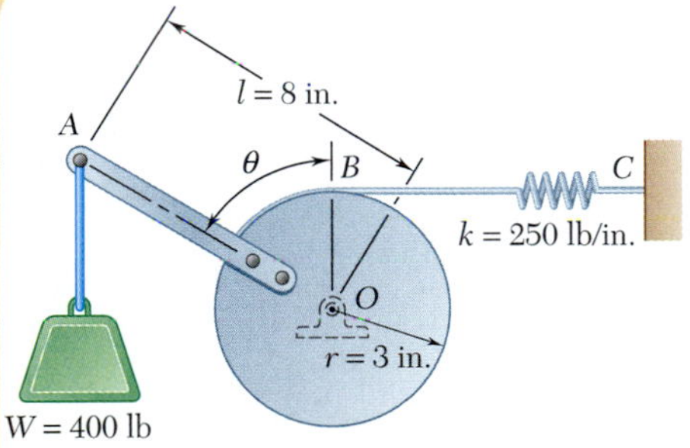

A 400-lb weight is attached at A to the lever shown. The constant of the spring BC is $k = 250$ lb/in., and the spring is unstretched when $\theta = 0$. Determine the position of equilibrium.

SOLUTION

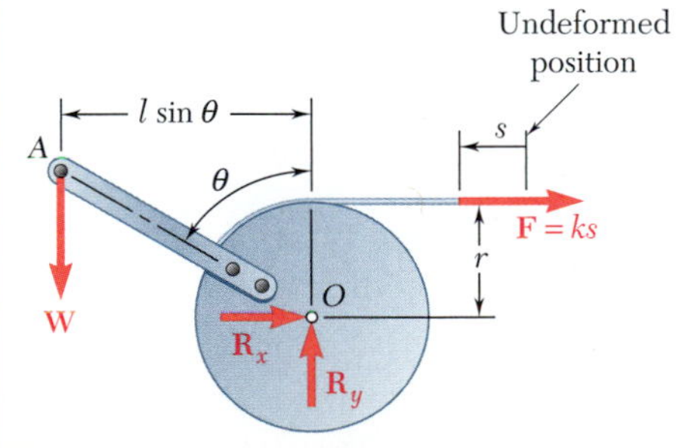

Free-Body Diagram. We draw a free-body diagram of the lever and cylinder. Denoting by s the deflection of the spring from its undeformed position, and noting that $s = r\theta$, we have $F = ks = kr\theta$.

Equilibrium Equation. Summing the moments of $\mathbf{W}$ and $\mathbf{F}$ about O, we write

$$+\circlearrowleft\Sigma M_O = 0: \qquad Wl\sin\theta - r(kr\theta) = 0 \qquad \sin\theta = \frac{kr^2}{Wl}\theta$$

Substituting the given data, we obtain

$$\sin\theta = \frac{(250\text{ lb/in.})(3\text{ in.})^2}{(400\text{ lb})(8\text{ in.})}\theta \qquad \sin\theta = 0.703\theta$$

Solving numerically, we find $\qquad \theta = 0 \qquad \theta = 80.3° \ \blacktriangleleft$

SOLVING PROBLEMS ON YOUR OWN

You saw that the external forces acting on a rigid body in equilibrium form a system equivalent to zero. To solve an equilibrium problem your first task is to draw a neat, reasonably large *free-body diagram* on which you will show all external forces and relevant dimensions. Both known and unknown forces must be included.

For a two-dimensional rigid body, the reactions at the supports can involve one, two, or three unknowns depending on the type of support (Fig. 4.1). For the successful solution of a problem, a correct free-body diagram is essential. Never proceed with the solution of a problem until you are sure that your free-body diagram includes all loads, all reactions, and the weight of the body (if appropriate).

As you construct your free-body diagrams, it will be necessary to assign directions to the unknown reactions. We suggest you always assume these forces act in a positive direction, so that positive answers always imply forces acting in a positive direction, while negative answers always imply forces acting in a negative direction. Similarly, we recommend you always assume the unknown force in a rod or cable is tensile, so that a positive result always means a tensile reaction. While a negative or comprehensive reaction is possible for a rod, a negative answer for a cable is impossible and, therefore, implies that there is an error in your solution.

1. ***You can write three equilibrium equations*** and solve them for *three unknowns.* The three equations might be

$$\Sigma F_x = 0 \qquad \Sigma F_y = 0 \qquad \Sigma M_O = 0$$

However, there are usually several sets of equations that you can write, such as

$$\Sigma F_x = 0 \qquad \Sigma M_A = 0 \qquad \Sigma M_B = 0$$

where point B is chosen in such a way that the line AB is not parallel to the y axis, or

$$\Sigma M_A = 0 \qquad \Sigma M_B = 0 \qquad \Sigma M_C = 0$$

where the points A, B, and C do not lie in a straight line.

2. ***To simplify your solution,*** it may be helpful to use one of the following solution techniques if applicable.

a. By summing moments about the point of intersection of the lines of action of two unknown forces, you will obtain an equation in a single unknown.

b. By summing components in a direction perpendicular to two unknown parallel forces, you will obtain an equation in a single unknown.

In some of the following problems you will be asked to determine the *allowable range of values* of the applied load for a given set of constraints, such as the maximum reaction at a support or the maximum force in one or more cables or rods. For problems of this type, you first assume a *maximum* loading situation (for example, the maximum allowed force in a rod), and then apply the equations of equilibrium to determine the corresponding unknown reactions and applied load. If the reactions satisfy the constraints, then the applied load is either the maximum or minimum value of the allowable range. However, if the solution violates a constraint (for example, the force in a cable is compressive), the initial assumption is wrong and another loading condition must be assumed (for the previous example, you would assume the force in the cable is zero, the *minimum* allowed reaction). The solution process is then repeated for another possible *maximum* loading to complete the determination of the allowable range of values of the applied load.

As in Chap. 2, we strongly recommend you always write the equations of equilibrium in the same form that we have used in the preceding sample problems. That is, both the known and unknown quantities are placed on the left side of the equation, and their sum is set equal to zero.

Problems

4.1 The boom on a 4300-kg truck is used to unload a pallet of shingles of mass 1600 kg. Determine the reaction at each of the two (*a*) rear wheels *B*, (*b*) front wheels *C*.

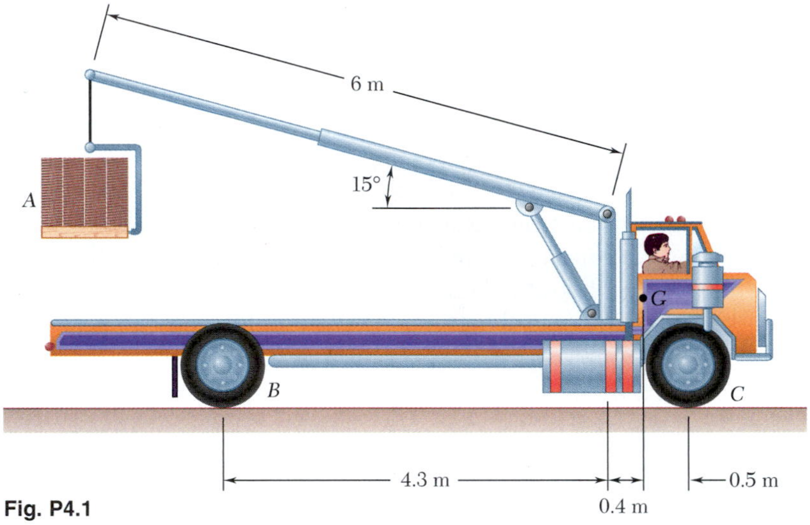

Fig. P4.1

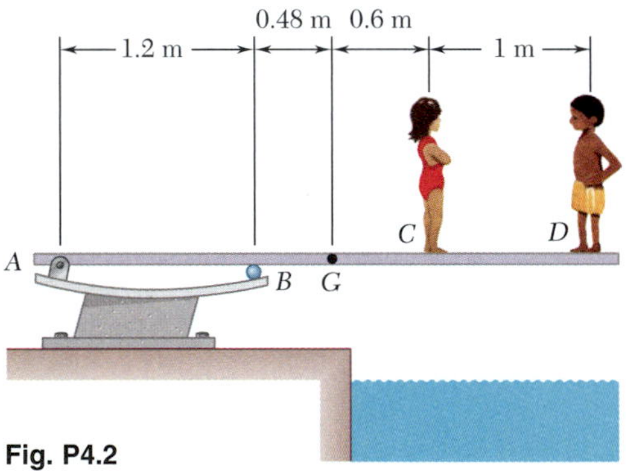

Fig. P4.2

4.2 Two children are standing on a diving board of mass 65 kg. Knowing that the masses of the children at *C* and *D* are 28 kg and 40 kg, respectively, determine (*a*) the reaction at *A*, (*b*) the reaction at *B*.

4.3 Two crates, each weighing 250 lb, are placed as shown in the bed of a 3000-lb pickup truck. Determine the reactions at each of the two (*a*) rear wheels *A*, (*b*) front wheels *B*.

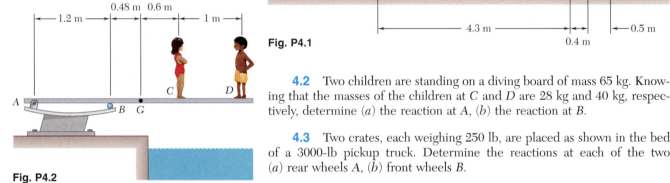

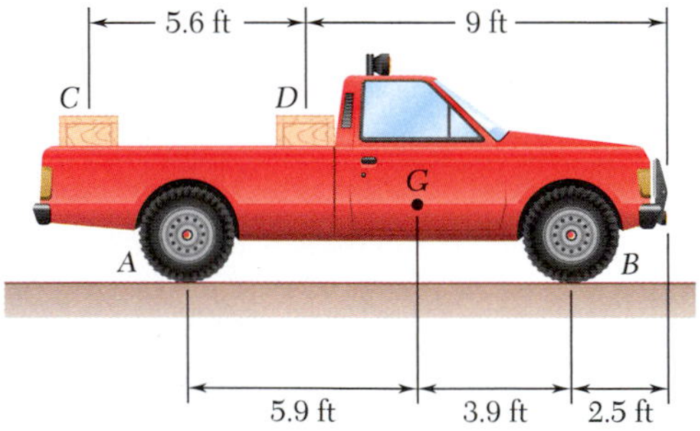

Fig. P4.3

4.4 Solve Prob. 4.3 assuming that crate *D* is removed and that the position of crate *C* is unchanged.

4.5 A T-shaped bracket supports the four loads shown. Determine the reactions at A and B if (*a*) $a = 100$ mm, (*b*) $a = 70$ mm.

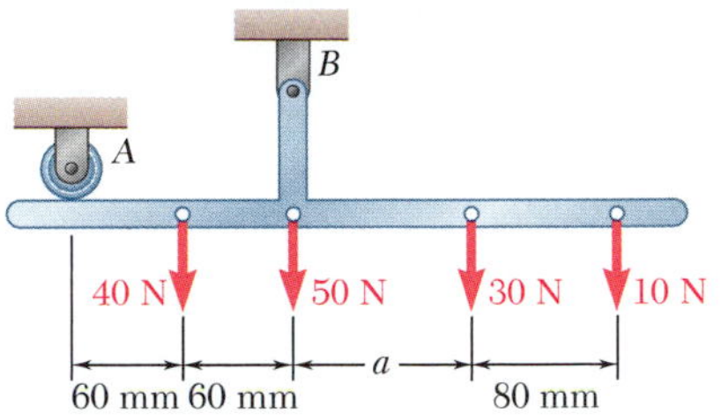

Fig. P4.5

4.6 For the bracket and loading of Prob. 4.5, determine the smallest distance a if the bracket is not to move.

4.7 A hand truck is used to move two barrels, each weighing 80 lb. Neglecting the weight of the hand truck, determine (*a*) the vertical force **P** which should be applied to the handle to maintain equilibrium when $\alpha = 35°$, (*b*) the corresponding reaction at each of the two wheels.

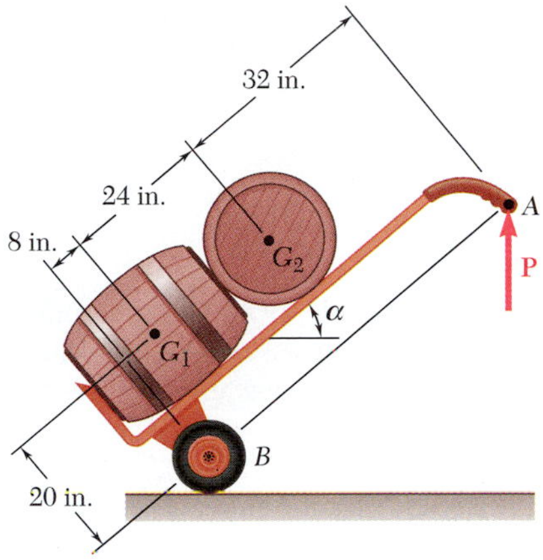

Fig. *P4.7*

4.8 Solve Prob. 4.7 when $\alpha = 40°$.

4.9 Four boxes are placed on a uniform 14-kg wooden plank which rests on two sawhorses. Knowing that the masses of boxes B and D are 4.5 kg and 45 kg, respectively, determine the range of values of the mass of box A so that the plank remains in equilibrium when box C is removed.

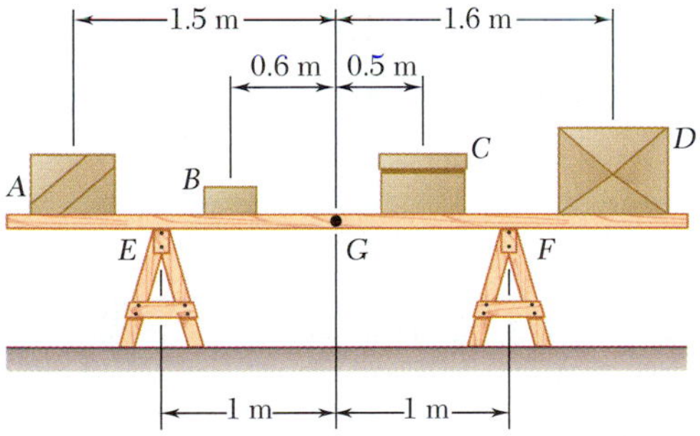

Fig. P4.9

4.10 A control rod is attached to a crank at A and cords are attached at B and C. For the given force in the rod, determine the range of values of the tension in the cord at C knowing that the cords must remain taut and that the maximum allowed tension in a cord is 180 N.

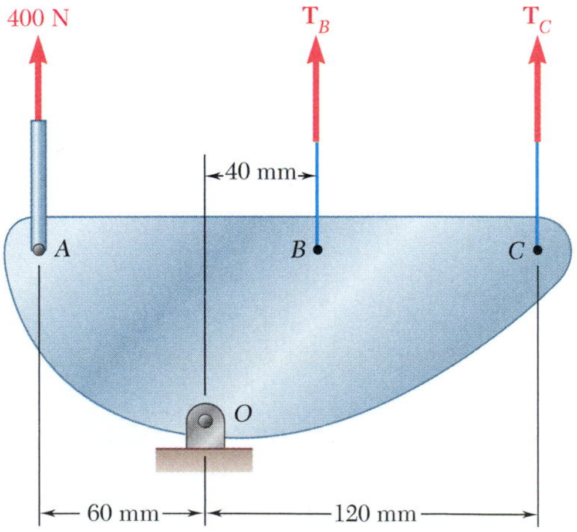

Fig. P4.10

4.11 The maximum allowable value of each of the reactions is 360 N. Neglecting the weight of the beam, determine the range of values of the distance d for which the beam is safe.

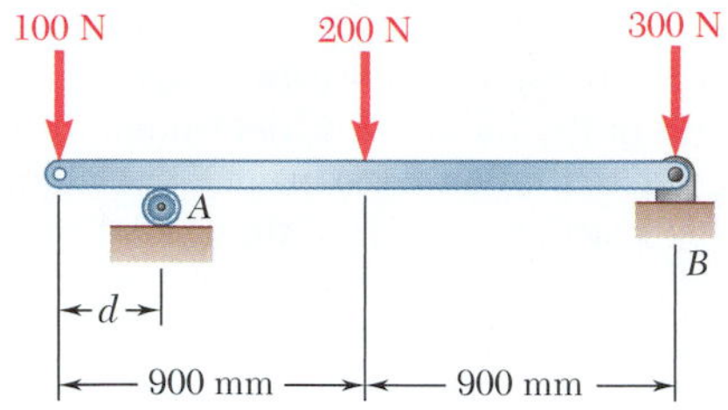

Fig. *P4.11*

4.12 Solve Prob. 4.11 assuming that the 100-N load is replaced by a 160-N load.

4.13 For the beam of Sample Prob. 4.2, determine the range of values of P for which the beam will be safe knowing that the maximum allowable value of each of the reactions is 45 kips and that the reaction at A must be directed upward.

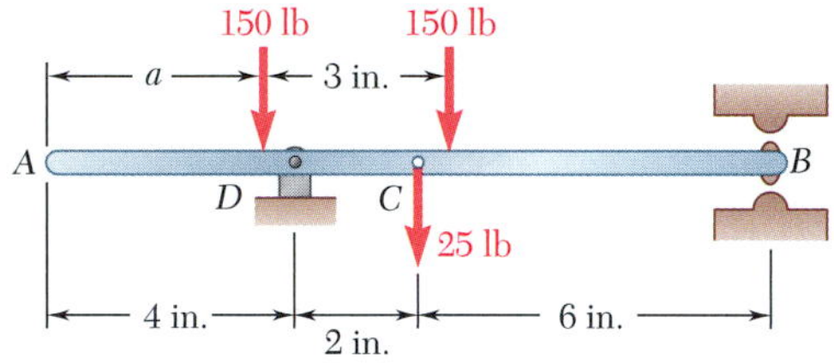

Fig. P4.14

4.14 For the beam and loading shown, determine the range of values of the distance a for which the reaction at B does not exceed 50 lb downward or 100 lb upward.

4.15 A follower $ABCD$ is held against a circular cam by a stretched spring, which exerts a force of 21 N for the position shown. Knowing that the tension in rod BE is 14 N, determine (*a*) the force exerted on the roller at A, (*b*) the reaction at bearing C.

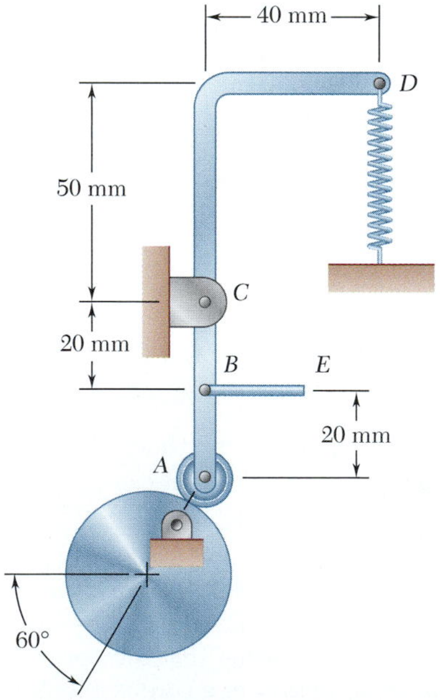

Fig. P4.15

4.16 A 6-m-long pole AB is placed in a hole and is guyed by three cables. Knowing that the tensions in cables BD and BE are 442 N and 322 N, respectively, determine (*a*) the tension in cable CD, (*b*) the reaction at A.

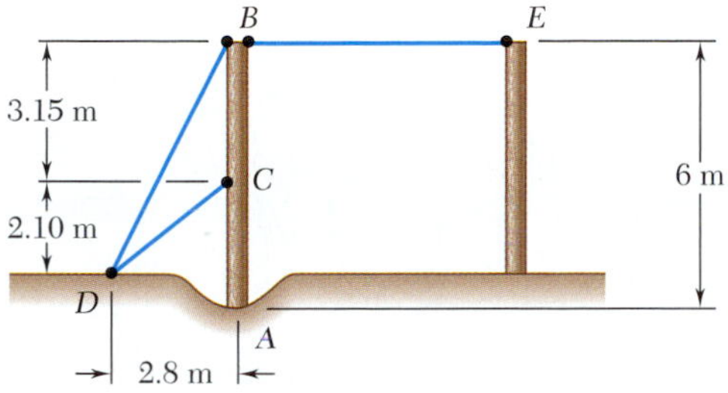

Fig. P4.16

4.17 Determine the reactions at A and C when (*a*) $\alpha = 0$, (*b*) $\alpha = 30°$.

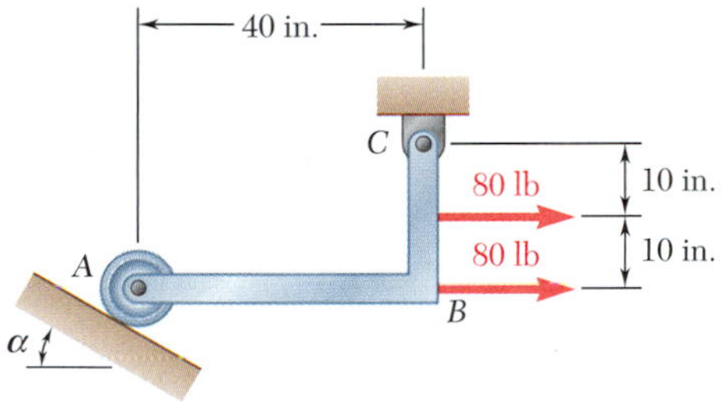

Fig. P4.17

4.18 Determine the reactions at A and B when (*a*) $h = 0$, (*b*) $h = 8$ in.

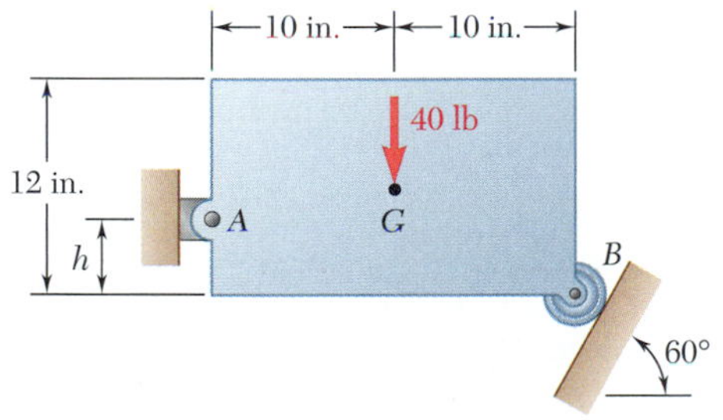

Fig. P4.18

4.19 The lever BCD is hinged at C and is attached to a control rod at B. If $P = 200$ N, determine (*a*) the tension in rod AB, (*b*) the reaction at C.

4.20 The lever BCD is hinged at C and is attached to a control rod at B. Determine the maximum force **P** which can be safely applied at D if the maximum allowable value of the reaction at C is 500 N.

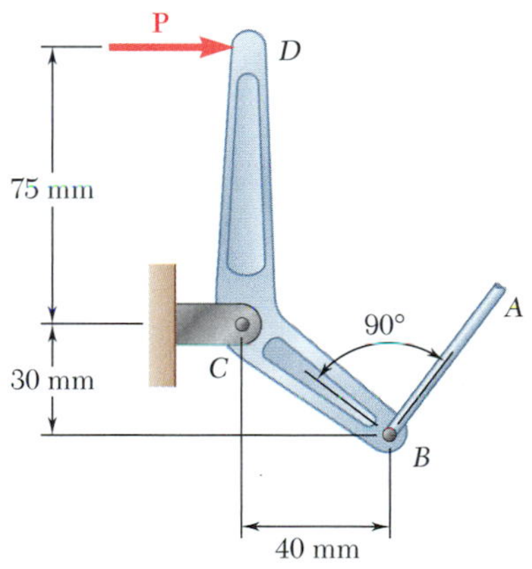

Fig. *P4.19* and *P4.20*

4.21 The required tension in cable AB is 800 N. Determine (*a*) the vertical force **P** which must be applied to the pedal, (*b*) the corresponding reaction at C.

4.22 Determine the maximum tension which can be developed in cable AB if the maximum allowable value of the reaction at C is 1000 N.

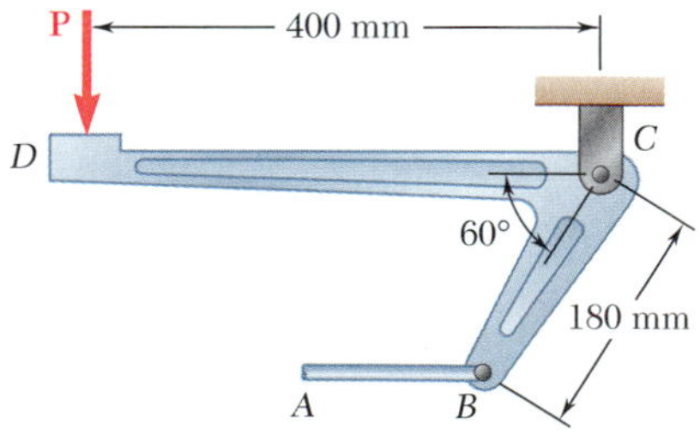

Fig. P4.21 and P4.22

4.23 and 4.24 A steel rod is bent to form a mounting bracket. For each of the mounting brackets and loadings shown, determine the reactions at A and B.

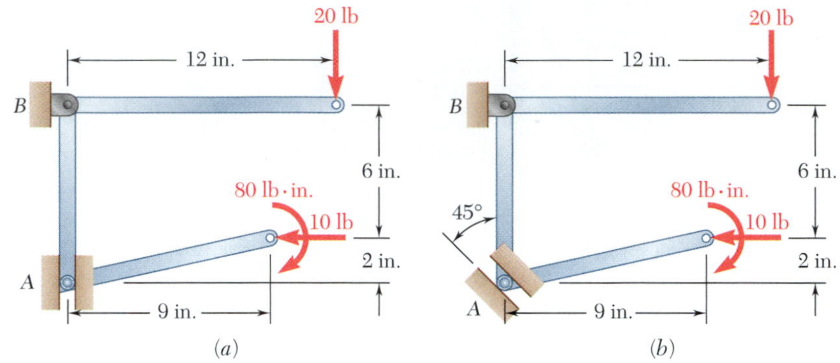

Fig. P4.23

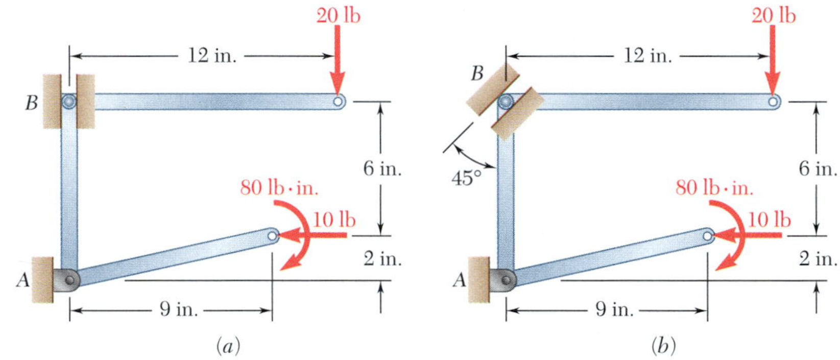

Fig. P4.24

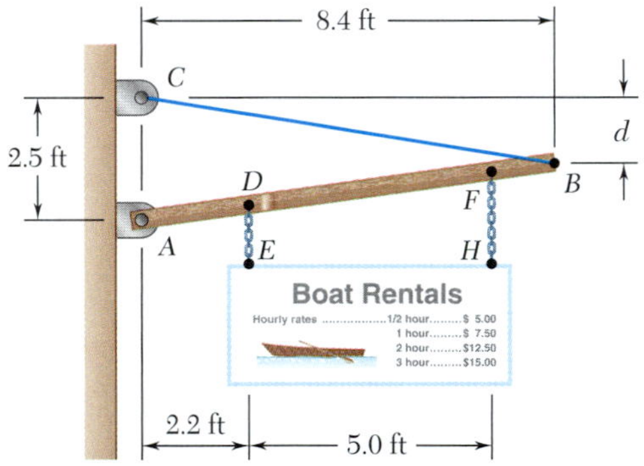

Fig. *P4.25* and *P4.26*

4.25 A sign is hung by two chains from mast AB. The mast is hinged at A and is supported by cable BC. Knowing that the tensions in chains DE and FH are 50 lb and 30 lb, respectively, and that $d = 1.3$ ft, determine (*a*) the tension in cable BC, (*b*) the reaction at A.

4.26 A sign is hung by two chains from mast AB. The mast is hinged at A and is supported by cable BC. Knowing that the tensions in chains DE and FH are 30 lb and 20 lb, respectively, and that $d = 1.54$ ft, determine (*a*) the tension in cable BC, (*b*) the reaction at A.

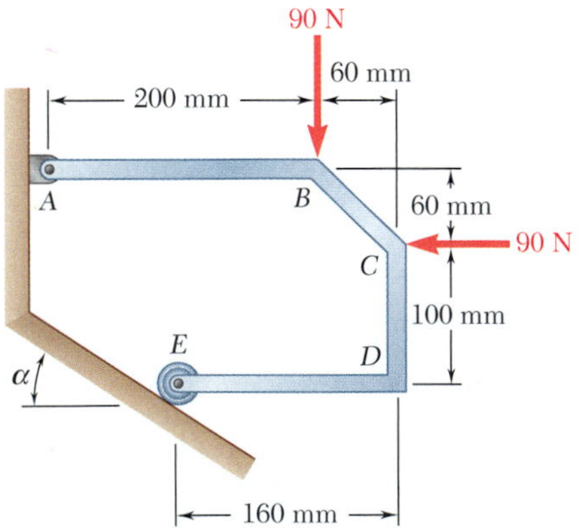

Fig. P4.27

4.27 For the frame and loading shown, determine the reactions at A and E when (*a*) $\alpha = 30°$, (*b*) $\alpha = 45°$.

4.28 A lever AB is hinged at C and is attached to a control cable at A. If the lever is subjected to a 300-N vertical force at B, determine (a) the tension in the cable, (b) the reaction at C.

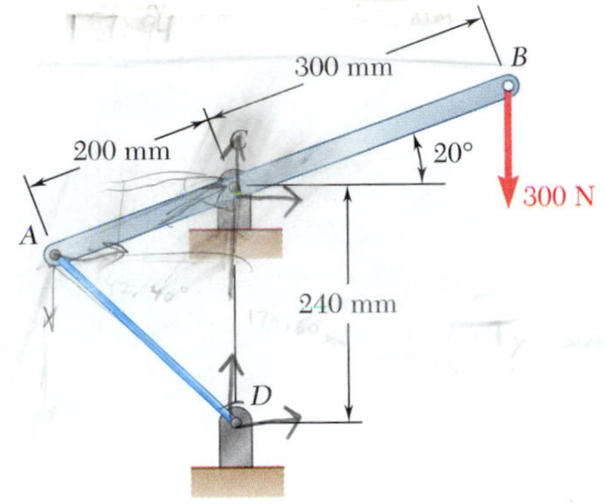

Fig. P4.28

4.29 Neglecting friction and the radius of the pulley, determine the tension in cable BCD and the reaction at support A when $d = 80$ mm.

4.30 Neglecting friction and the radius of the pulley, determine the tension in cable BCD and the reaction at support A when $d = 144$ mm.

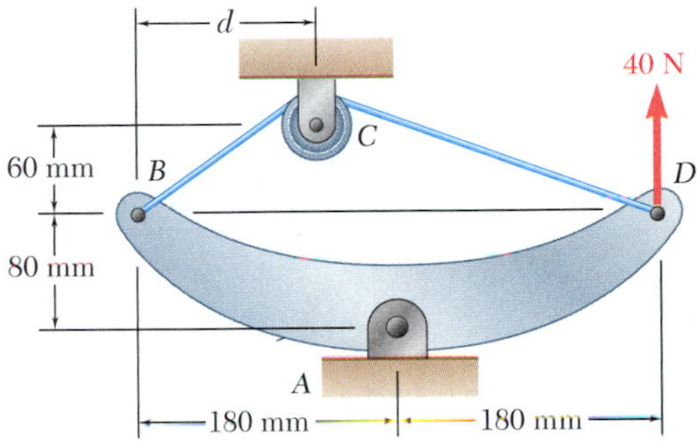

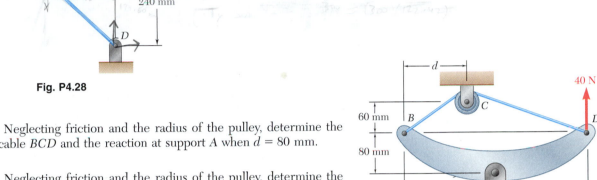

Fig. P4.29 and P4.30

4.31 Neglecting friction, determine the tension in cable ABD and the reaction at support C.

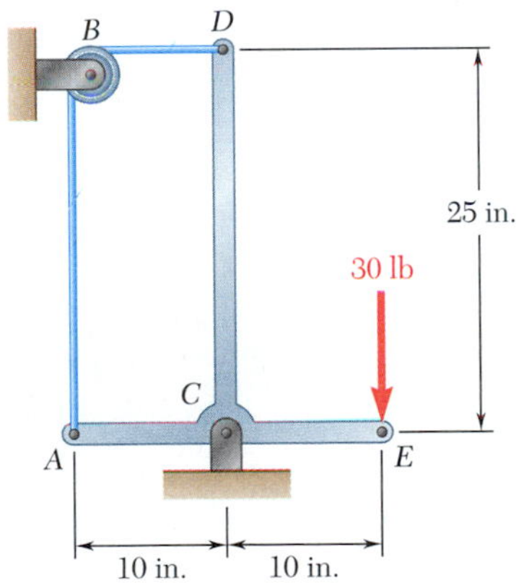

Fig. P4.31

4.32 Rod ABC is bent in the shape of a circular arc of radius R. Knowing that $\theta = 35°$, determine the reaction (a) at B, (b) at C.

4.33 Rod ABC is bent in the shape of a circular arc of radius R. Knowing that $\theta = 50°$, determine the reaction (a) at B, (b) at C.

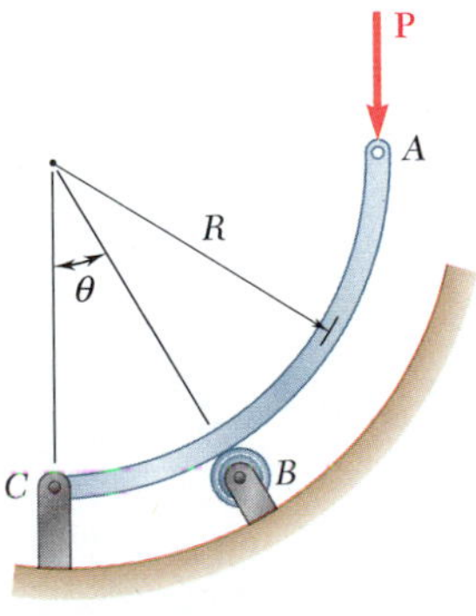

Fig. *P4.32* and P4.33

4.34 Neglecting friction and the radius of the pulley, determine (*a*) the tension in cable *ADB*, (*b*) the reaction at *C*.

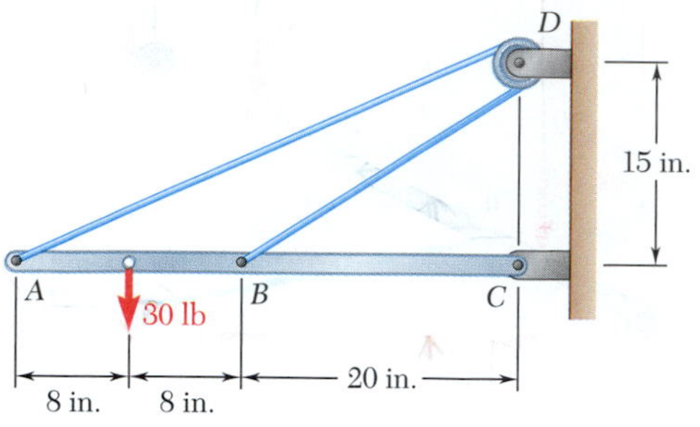

Fig. P4.34

4.35 Neglecting friction, determine the tension in cable *ABD* and the reaction at *C* when $\theta = 60°$.

4.36 Neglecting friction, determine the tension in cable *ABD* and the reaction at *C* when $\theta = 30°$.

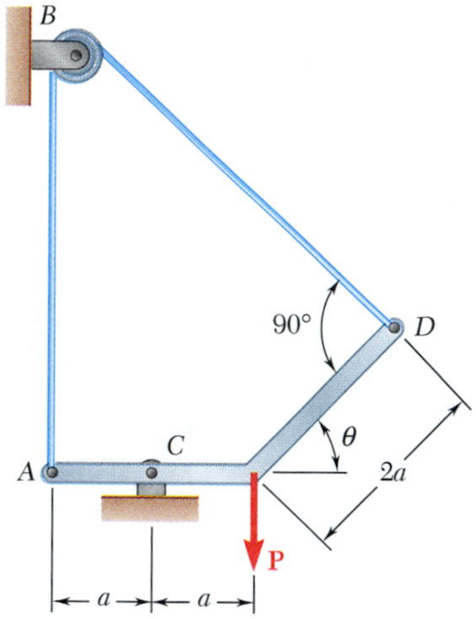

Fig. P4.35 and *P4.36*

4.37 Determine the tension in each cable and the reaction at *D*.

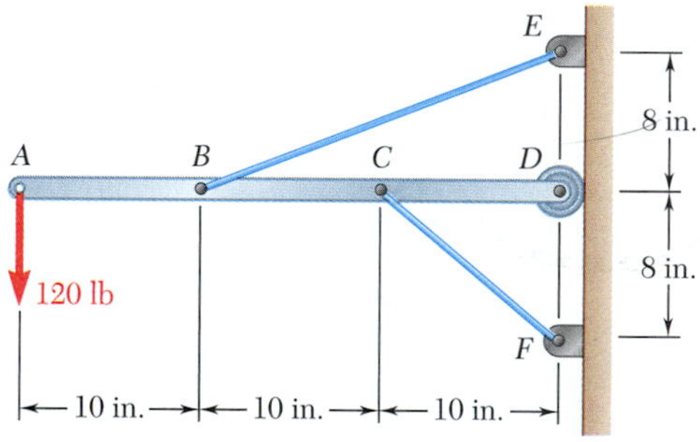

Fig. P4.37

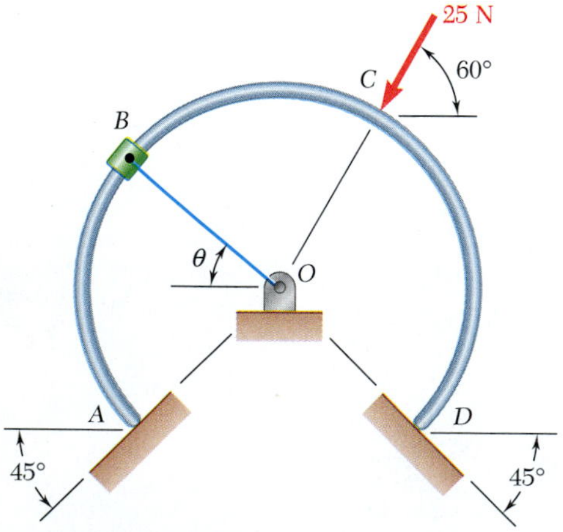

Fig. P4.38 and P4.39

4.38 Rod *ABCD* is bent in the shape of a circular arc of radius 80 mm and rests against frictionless surfaces at *A* and *D*. Knowing that the collar at *B* can move freely on the rod and that $\theta = 45°$, determine (*a*) the tension in cord *OB*, (*b*) the reactions at *A* and *D*.

4.39 Rod *ABCD* is bent in the shape of a circular arc of radius 80 mm and rests against frictionless surfaces at *A* and *D*. Knowing that the collar at *B* can move freely on the rod, determine (*a*) the value of θ for which the tension in cord *OB* is as small as possible, (*b*) the corresponding value of the tension, (*c*) the reactions at *A* and *D*.

4.40 Bar AC supports two 100-lb loads as shown. Rollers A and C rest against frictionless surfaces and a cable BD is attached at B. Determine (*a*) the tension in cable BD, (*b*) the reaction at A, (*c*) the reaction at C.

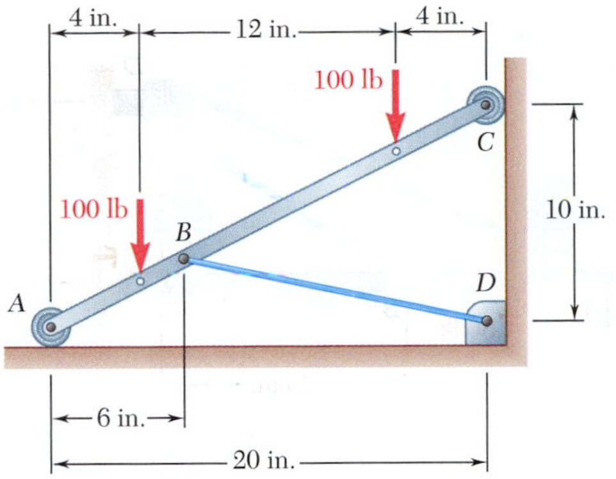

Fig. *P4.40*

4.41 A parabolic slot has been cut in plate AD, and the plate has been placed so that the slot fits two fixed, frictionless pins B and C. The equation of the slot is $y = x^2/100$, where x and y are expressed in mm. Knowing that the input force $P = 4$ N, determine (*a*) the force each pin exerts on the plate, (*b*) the output force $\mathbf{Q}$.

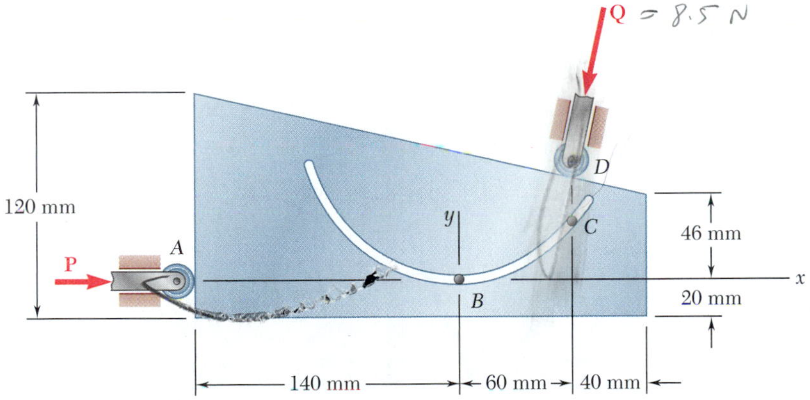

Fig. P4.41 and P4.42

4.42 A parabolic slot has been cut in plate AD, and the plate has been placed so that the slot fits two fixed, frictionless pins B and C. The equation of the slot is $y = x^2/100$, where x and y are expressed in mm. Knowing that the maximum allowable force exerted on the roller at D is 8.5 N, determine (*a*) the corresponding magnitude of the input force $\mathbf{P}$, (*b*) the force each pin exerts on the plate.

4.43 A movable bracket is held at rest by a cable attached at E and by frictionless rollers. Knowing that the width of post FG is slightly less than the distance between the rollers, determine the force exerted on the post by each roller when $\alpha = 20°$.

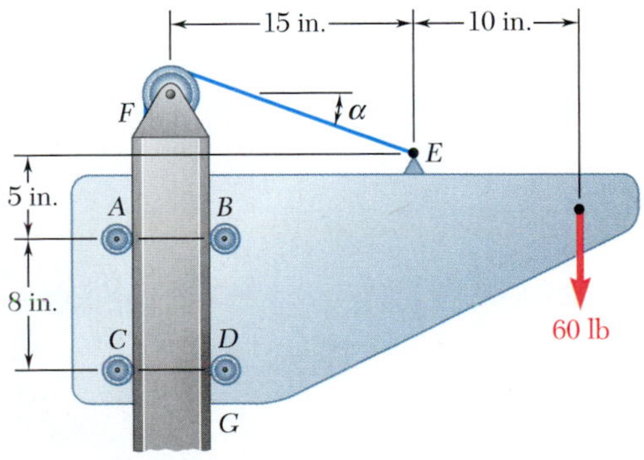

Fig. P4.43

4.44 Solve Prob. 4.43 when $\alpha = 30°$.

4.45 A 20-lb weight can be supported in the three different ways shown. Knowing that the pulleys have a 4-in. radius, determine the reaction at A in each case.

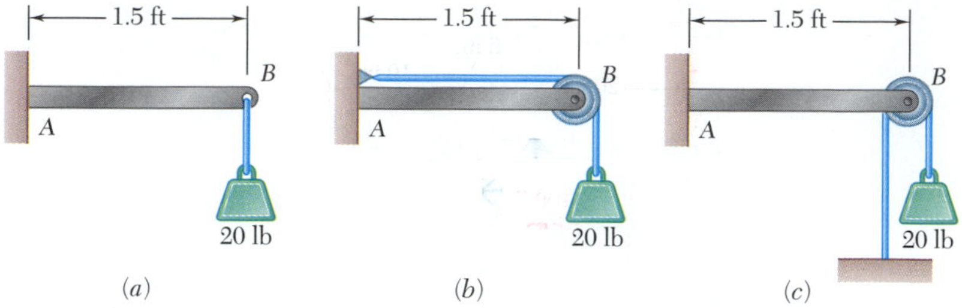

Fig. P4.45

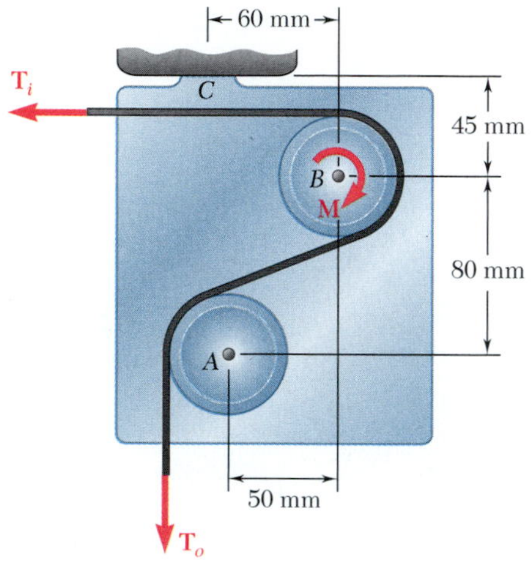

Fig. P4.46 and P4.47

4.46 A belt passes over two 50-mm-diameter pulleys which are mounted on a bracket as shown. Knowing that $M = 0$ and $T_i = T_o = 24$ N, determine the reaction at C.

4.47 A belt passes over two 50-mm-diameter pulleys which are mounted on a bracket as shown. Knowing that $M = 0.40$ N · m and that T_i and T_o are equal to 32 N and 16 N, respectively, determine the reaction at C.

4.48 A 350-lb utility pole is used to support at C the end of an electric wire. The tension in the wire is 120 lb, and the wire forms an angle of 15° with the horizontal at C. Determine the largest and smallest allowable tensions in the guy cable BD if the magnitude of the couple at A may not exceed 200 lb · ft.

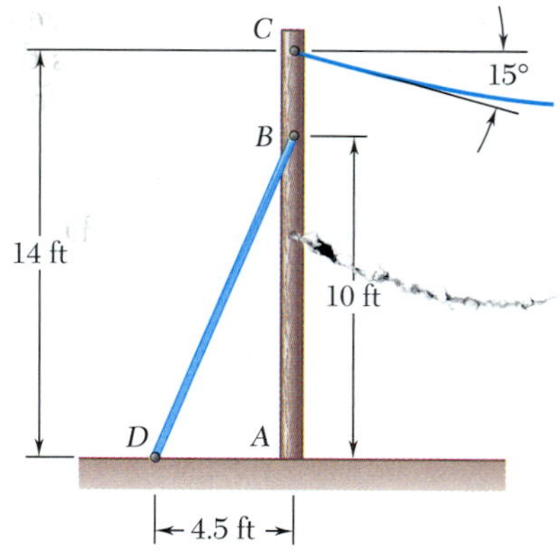

Fig. P4.48

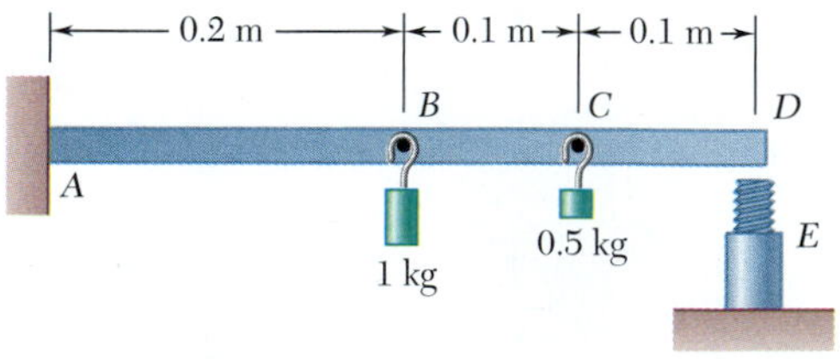

Fig. P4.49 and P4.50

4.49 In a laboratory experiment, students hang the masses shown from a beam of negligible mass. (a) Determine the reaction at the fixed support A knowing that end D of the beam does not touch support E. (b) Determine the reaction at the fixed support A knowing that the adjustable support E exerts an upward force of 6 N on the beam.

4.50 In a laboratory experiment, students hang the masses shown from a beam of negligible mass. Determine the range of values of the force exerted on the beam by the adjustable support E for which the magnitude of the couple at A does not exceed 2.5 N · m.

4.51 Knowing that the tension in wire BD is 300 lb, determine the reaction at fixed support C for the frame shown.

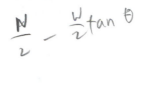

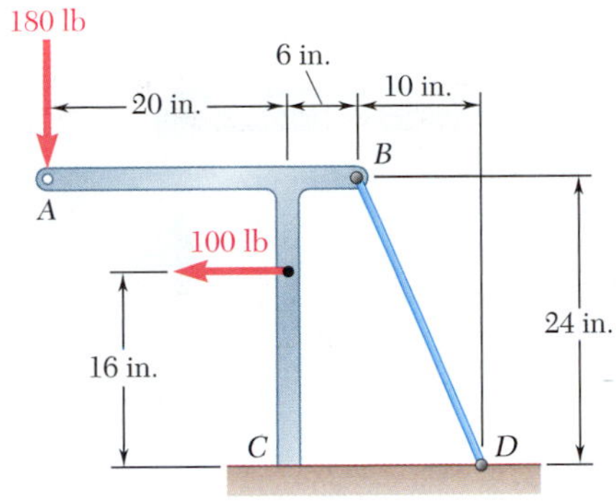

Fig. *P4.51* and *P4.52*

4.52 Determine the range of allowable values of the tension in wire BD if the magnitude of the couple at the fixed support C is not to exceed 75 lb · ft.

4.53 Uniform rod AB of length l and weight W lies in a vertical plane and is acted upon by a couple $\mathbf{M}$. The ends of the rod are connected to small rollers which rest against frictionless surfaces. (a) Express the angle θ corresponding to equilibrium in terms of M, W, and l. (b) Determine the value of θ corresponding to equilibrium when $M = 1.5$ lb · ft, $W = 4$ lb, and $l = 2$ ft.

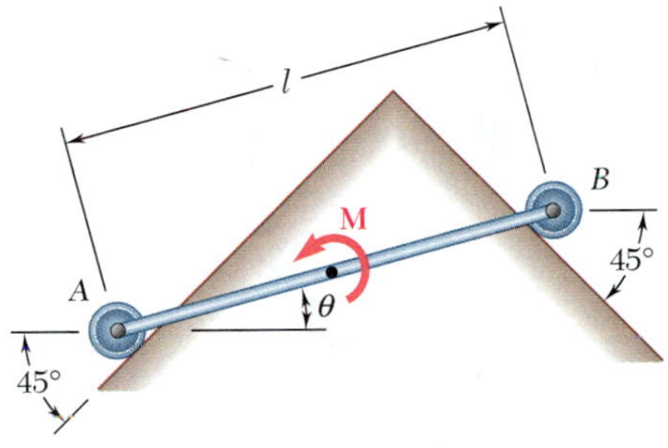

Fig. P4.53

4.54 A slender rod AB, of weight W, is attached to blocks A and B, which move freely in the guides shown. The blocks are connected by an elastic cord which passes over a pulley at C. (a) Express the tension in the cord in terms of W and θ. (b) Determine the value of θ for which the tension in the cord is equal to $3W$.

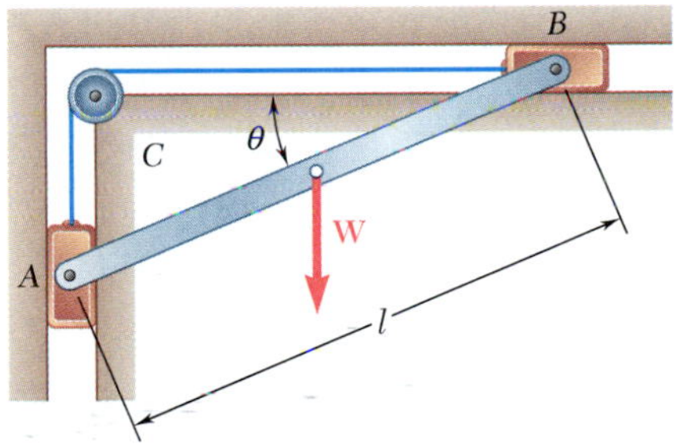

Fig. P4.54

4.55 A thin, uniform ring of mass m and radius R is attached by a frictionless pin to a collar at A and rests against a small roller at B. The ring lies in a vertical plane, and the collar can move freely on a horizontal rod and is acted upon by a horizontal force $\mathbf{P}$. (a) Express the angle θ corresponding to equilibrium in terms of m and P. (b) Determine the value of θ corresponding to equilibrium when $m = 500$ g and $P = 5$ N.

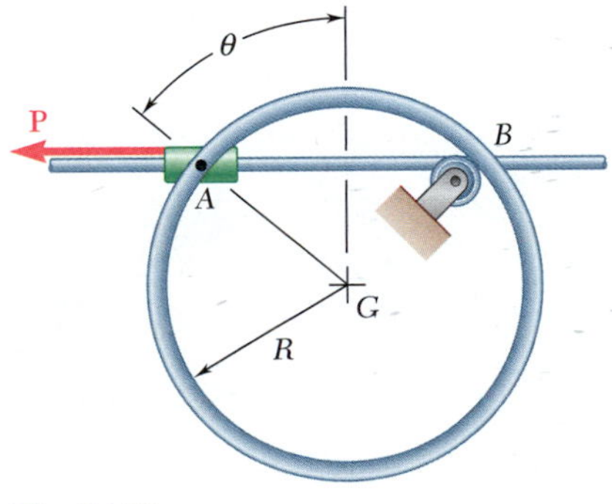

Fig. P4.55

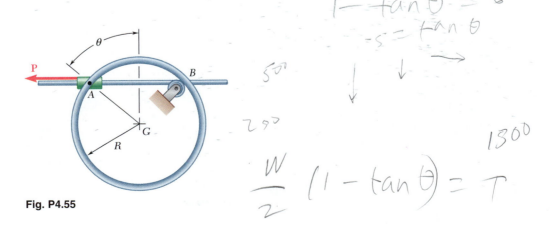

4.56 Rod AB is acted upon by a couple **M** and two forces, each of magnitude P. (*a*) Derive an equation in θ, P, M, and l which must be satisfied when the rod is in equilibrium. (*b*) Determine the value of θ corresponding to equilibrium when $M = 150$ lb · in., $P = 20$ lb, and $l = 6$ in.

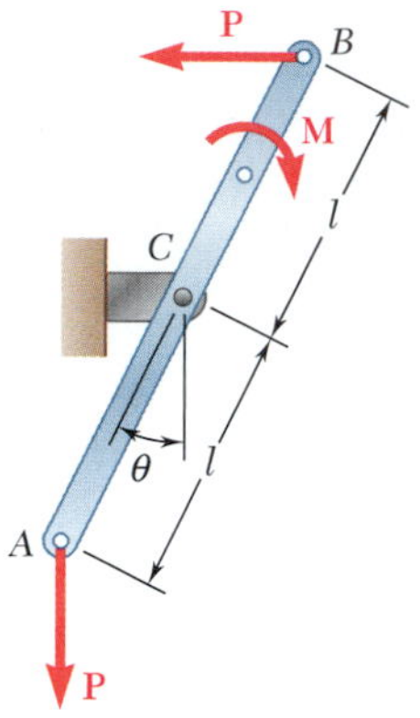

Fig. P4.56

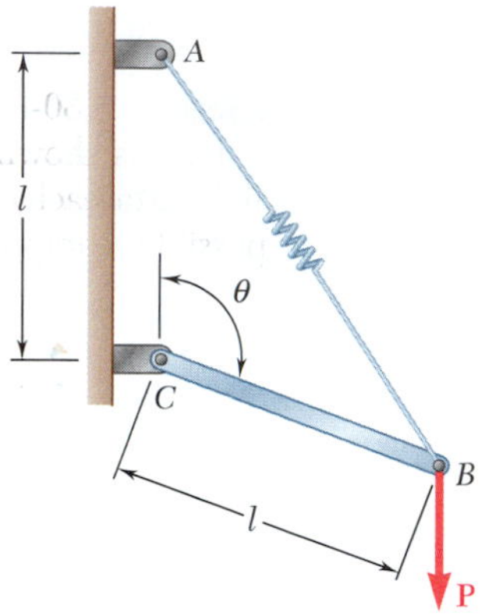

Fig. P4.57

4.57 A vertical load **P** is applied at end B of rod BC. The constant of the spring is k, and the spring is unstretched when $\theta = 90°$. (*a*) Neglecting the weight of the rod, express the angle θ corresponding to equilibrium in terms of P, k, and l. (*b*) Determine the value of θ corresponding to equilibrium when $P = \frac{1}{4}kl$.

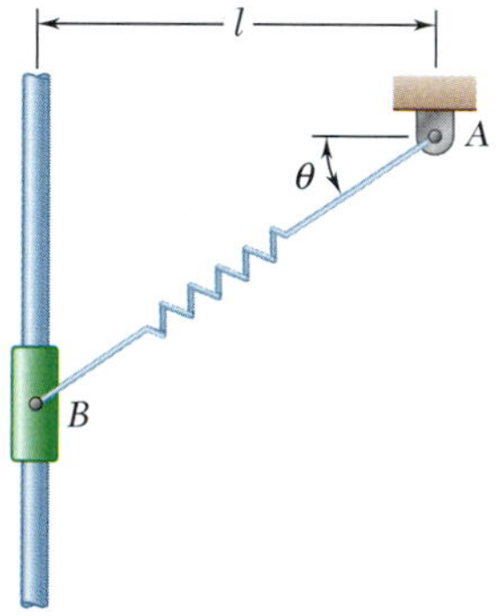

Fig. P4.59

4.58 Solve Sample Prob. 4.5 assuming that the spring is unstretched when $\theta = 90°$.

4.59 A collar B of weight W can move freely along the vertical rod shown. The constant of the spring is k, and the spring is unstretched when $\theta = 0$. (*a*) Derive an equation in θ, W, k, and l which must be satisfied when the collar is in equilibrium. (*b*) Knowing that $W = 3$ lb, $l = 6$ in., and $k = 8$ lb/ft, determine the value of θ corresponding to equilibrium.

4.60 A slender rod AB, of mass m, is attached to blocks A and B which move freely in the guides shown. The constant of the spring is k, and the spring is unstretched when $\theta = 0$. (*a*) Neglecting the mass of the blocks, derive an equation in m, g, k, l, and θ which must be satisfied when the rod is in equilibrium. (*b*) Determine the value of θ when $m = 2$ kg, $l = 750$ mm, and $k = 30$ N/m.

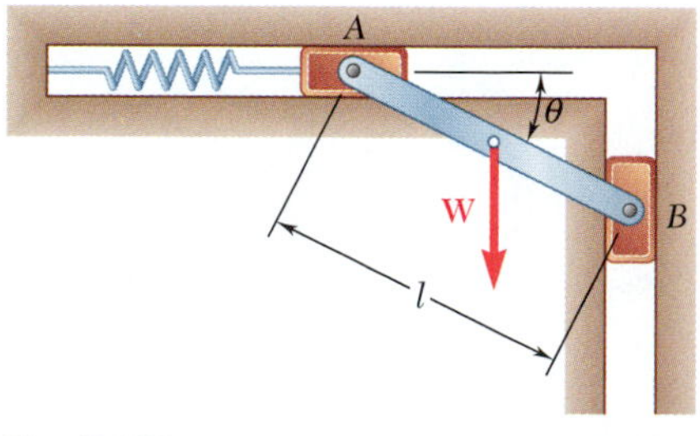

Fig. *P4.60*

4.61 The bracket ABC can be supported in the eight different ways shown. All connections consist of smooth pins, rollers, or short links. In each case, determine whether (a) the plate is completely, partially, or improperly constrained, (b) the reactions are statically determinate or indeterminate, (c) the equilibrium of the plate is maintained in the position shown. Also, wherever possible, compute the reactions assuming that the magnitude of the force **P** is 100 N.

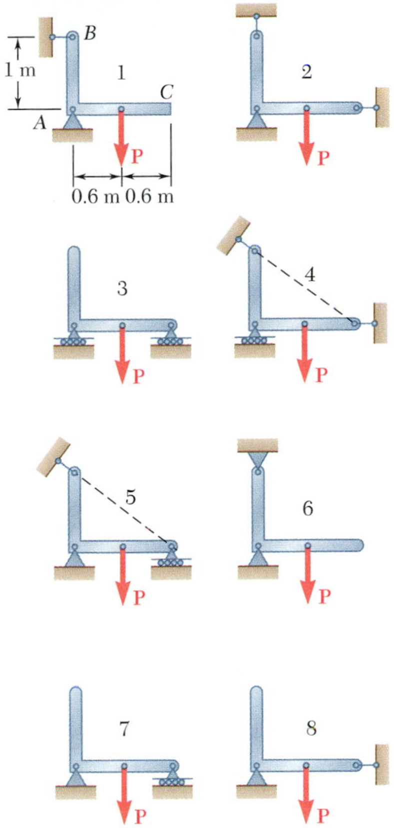

Fig. P4.61

4.62 Eight identical 20 × 30-in. rectangular plates, each weighing 50 lb, are held in a vertical plane as shown. All connections consist of frictionless pins, rollers, or short links. For each case, answer the questions listed in Prob. 4.61, and, wherever possible, compute the reactions.

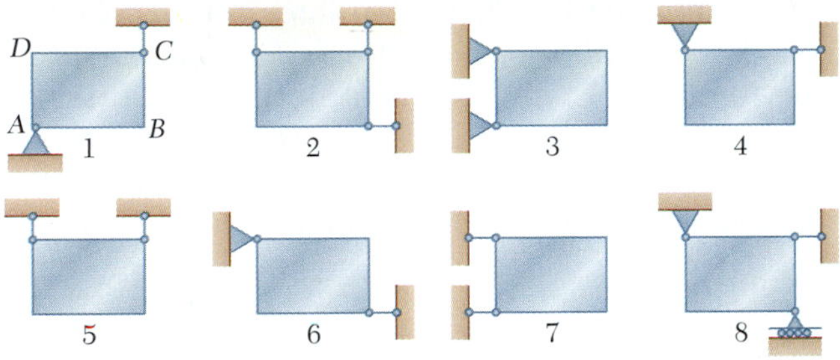

Fig. *P4.62*

4.6. EQUILIBRIUM OF A TWO-FORCE BODY

A particular case of equilibrium which is of considerable interest is that of a rigid body subjected to two forces. Such a body is commonly called a *two-force body.* It will be shown that *if a two-force body is in equilibrium, the two forces must have the same magnitude, the same line of action, and opposite sense.*

Consider a corner plate subjected to two forces $\mathbf{F}_1$ and $\mathbf{F}_2$ acting at A and B, respectively (Fig. 4.8a). If the plate is to be in equilibrium, the sum of the moments of $\mathbf{F}_1$ and $\mathbf{F}_2$ about any point must be zero. First, we sum moments about A. Since the moment of $\mathbf{F}_1$ is obviously zero, the moment of $\mathbf{F}_2$ must also be zero and the line of action of $\mathbf{F}_2$ must pass through A (Fig. 4.8b). Summing moments about B, we prove similarly that the line of action of $\mathbf{F}_1$ must pass through B (Fig. 4.8c). Therefore, both forces have the same line of action (line AB). From either of the equations $\Sigma F_x = 0$ and $\Sigma F_y = 0$ it is seen that they must also have the same magnitude but opposite sense.

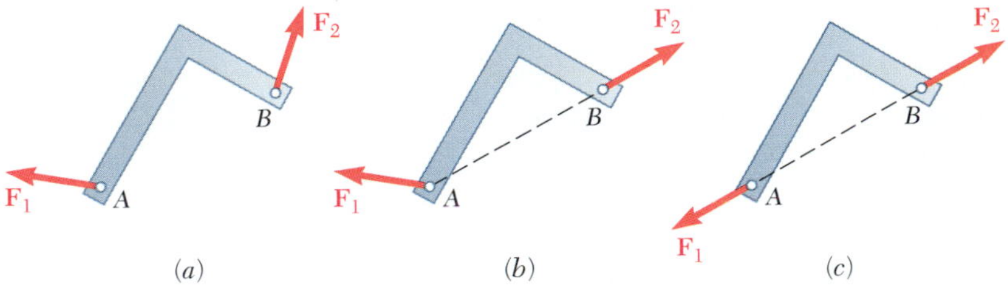

Fig. 4.8

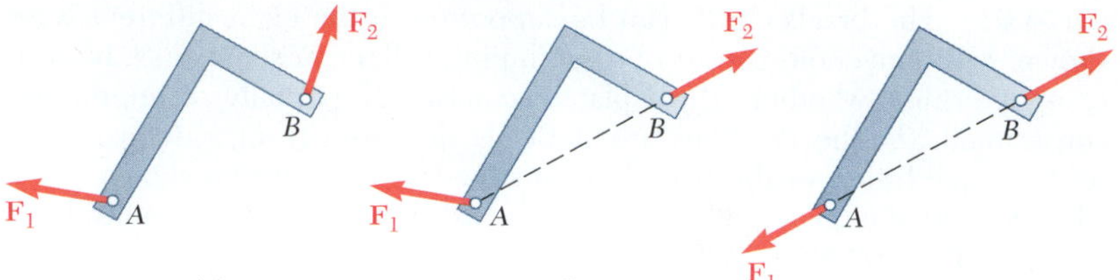

(a) (b) (c)

Fig. 4.8 *(repeated)*

If several forces act at two points A and B, the forces acting at A can be replaced by their resultant $\mathbf{F}_1$ and those acting at B can be replaced by their resultant $\mathbf{F}_2$. Thus a two-force body can be more generally defined as *a rigid body subjected to forces acting at only two points*. The resultants $\mathbf{F}_1$ and $\mathbf{F}_2$ then must have the same line of action, the same magnitude, and opposite sense (Fig. 4.8).

In the study of structures, frames, and machines, you will see how the recognition of two-force bodies simplifies the solution of certain problems.

4.7. EQUILIBRIUM OF A THREE-FORCE BODY

Another case of equilibrium that is of great interest is that of a *three-force body*, that is, a rigid body subjected to three forces or, more generally, *a rigid body subjected to forces acting at only three points*. Consider a rigid body subjected to a system of forces which can be reduced to three forces $\mathbf{F}_1$, $\mathbf{F}_2$, and $\mathbf{F}_3$ acting at A, B, and C, respectively (Fig. 4.9*a*). It will be shown that if the body is in equilibrium, *the lines of action of the three forces must be either concurrent or parallel*.

Since the rigid body is in equilibrium, the sum of the moments of $\mathbf{F}_1$, $\mathbf{F}_2$, and $\mathbf{F}_3$ about any point must be zero. Assuming that the lines of action of $\mathbf{F}_1$ and $\mathbf{F}_2$ intersect and denoting their point of intersection by D, we sum moments about D (Fig. 4.9*b*). Since the moments of $\mathbf{F}_1$ and $\mathbf{F}_2$ about D are zero, the moment of $\mathbf{F}_3$ about D must also be zero, and the line of action of $\mathbf{F}_3$ must pass through D (Fig. 4.9*c*). Therefore, the three lines of action are concurrent. The only exception occurs when none of the lines intersect; the lines of action are then parallel.

Although problems concerning three-force bodies can be solved by the general methods of Secs. 4.3 to 4.5, the property just established can be used to solve them either graphically or mathematically from simple trigonometric or geometric relations.

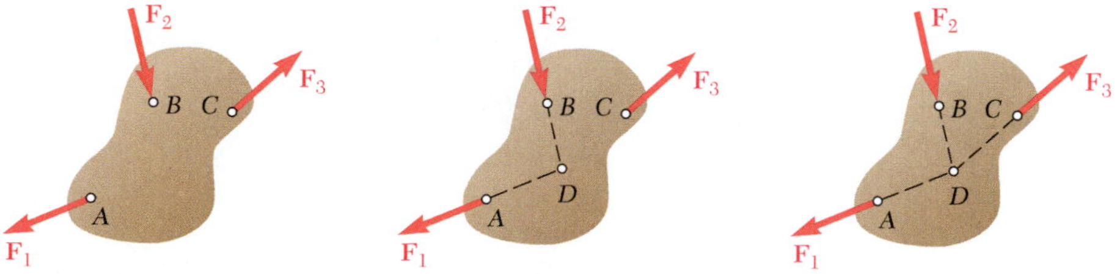

(a) (b) (c)

Fig. 4.9

SAMPLE PROBLEM 4.6

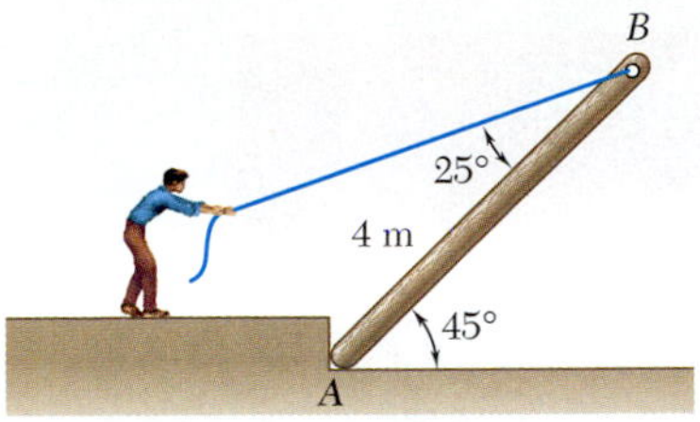

A man raises a 10-kg joist, of length 4 m, by pulling on a rope. Find the tension T in the rope and the reaction at A.

SOLUTION

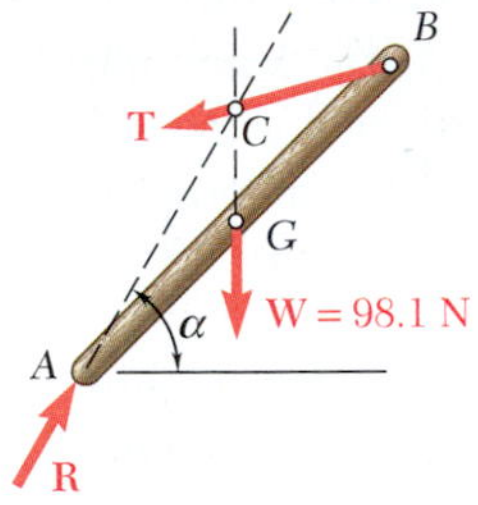

Free-Body Diagram. The joist is a three-force body, since it is acted upon by three forces: its weight **W**, the force **T** exerted by the rope, and the reaction **R** of the ground at A. We note that

$$W = mg = (10 \text{ kg})(9.81 \text{ m/s}^2) = 98.1 \text{ N}$$

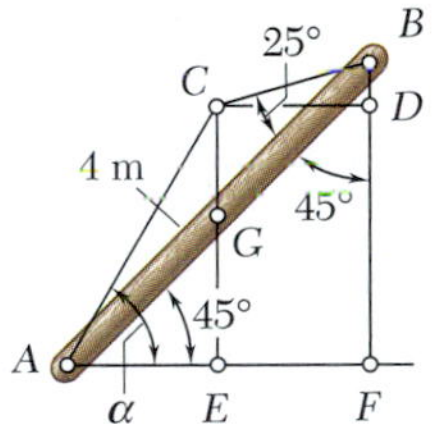

Three-Force Body. Since the joist is a three-force body, the forces acting on it must be concurrent. The reaction **R**, therefore, will pass through the point of intersection C of the lines of action of the weight **W** and the tension force **T**. This fact will be used to determine the angle α that **R** forms with the horizontal.

Drawing the vertical BF through B and the horizontal CD through C, we note that

$$AF = BF = (AB) \cos 45° = (4 \text{ m}) \cos 45° = 2.828 \text{ m}$$
$$CD = EF = AE = \tfrac{1}{2}(AF) = 1.414 \text{ m}$$
$$BD = (CD) \cot (45° + 25°) = (1.414 \text{ m}) \tan 20° = 0.515 \text{ m}$$
$$CE = DF = BF - BD = 2.828 \text{ m} - 0.515 \text{ m} = 2.313 \text{ m}$$

We write

$$\tan \alpha = \frac{CE}{AE} = \frac{2.313 \text{ m}}{1.414 \text{ m}} = 1.636$$

$\alpha = 58.6°$ ◀

We now know the direction of all the forces acting on the joist.

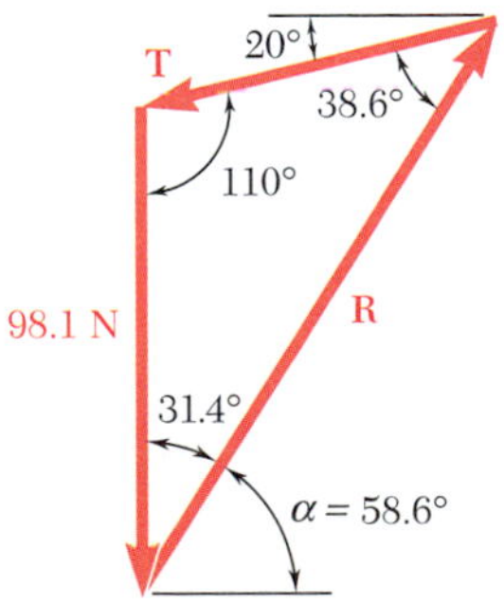

Force Triangle. A force triangle is drawn as shown, and its interior angles are computed from the known directions of the forces. Using the law of sines, we write

$$\frac{T}{\sin 31.4°} = \frac{R}{\sin 110°} = \frac{98.1 \text{ N}}{\sin 38.6°}$$

$T = 81.9 \text{ N}$ ◀

$\mathbf{R} = 147.8 \text{ N} \measuredangle 58.6°$ ◀

SOLVING PROBLEMS ON YOUR OWN

The preceding sections covered two particular cases of equilibrium of a rigid body.

1. *A two-force body is a body subjected to forces at only two points.* The resultants of the forces acting at each of these points must have the *same magnitude, the same line of action, and opposite sense.* This property will allow you to simplify the solutions of some problems by replacing the two unknown components of a reaction by a single force of unknown magnitude but of *known direction*.

2. *A three-force body is subjected to forces at only three points.* The resultants of the forces acting at each of these points must be *concurrent or parallel.* To solve a problem involving a three-force body with concurrent forces, draw your free-body diagram showing that the lines of action of these three forces pass through the same point. The use of geometry will then allow you to complete the solution using a force triangle [Sample Prob. 4.6].

Although the principle noted above for the solution of problems involving three-force bodies is easily understood, it can be difficult to determine the needed geometric constructions. If you encounter difficulty, first draw a reasonably large free-body diagram and then seek a relation between known or easily calculated lengths and a dimension that involves an unknown. This was done in Sample Prob. 4.6, where the easily calculated dimensions AE and CE were used to determine the angle α.

Problems

4.63 Horizontal and vertical links are hinged to a wheel, and forces are applied to the links as shown. Knowing that $a = 3.0$ in., determine the value of P and the reaction at A.

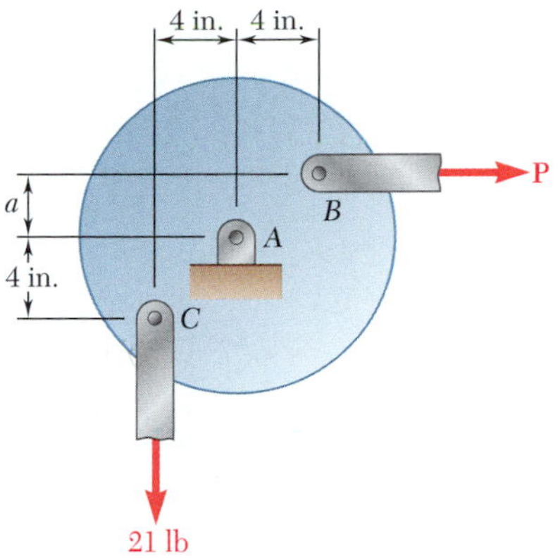

Fig. P4.63 and P4.64

4.64 Horizontal and vertical links are hinged to a wheel, and forces are applied to the links as shown. Determine the range of values of the distance a for which the magnitude of the reaction at A does not exceed 42 lb.

4.65 Using the method of Sec. 4.7, solve Prob. 4.21.

4.66 Using the method of Sec. 4.7, solve Prob. 4.22.

4.67 To remove a nail, a small block of wood is placed under a crowbar, and a horizontal force **P** is applied as shown. Knowing that $l = 3.5$ in. and $P = 30$ lb, determine the vertical force exerted on the nail and the reaction at B.

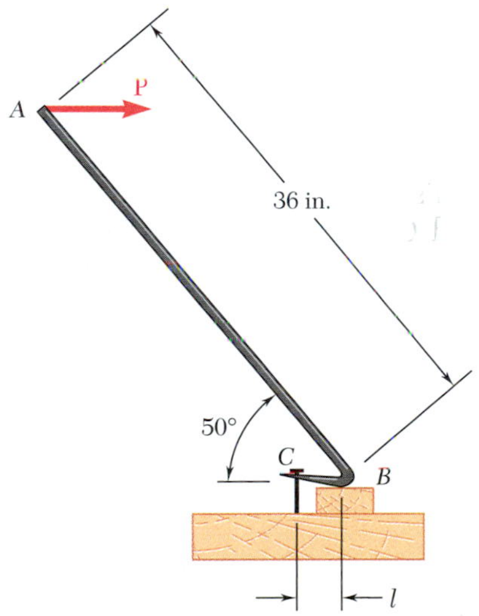

Fig. *P4.67* and *P4.68*

4.68 To remove a nail, a small block of wood is placed under a crowbar, and a horizontal force **P** is applied as shown. Knowing that the maximum vertical force needed to extract the nail is 600 lb and that the horizontal force **P** is not to exceed 65 lb, determine the largest acceptable value of distance l.

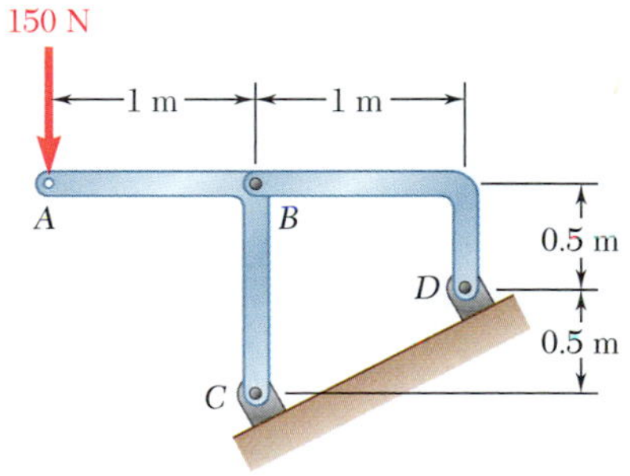

Fig. P4.69

4.69 For the frame and loading shown, determine the reactions at C and D.

4.70 For the frame and loading shown, determine the reactions at A and C.

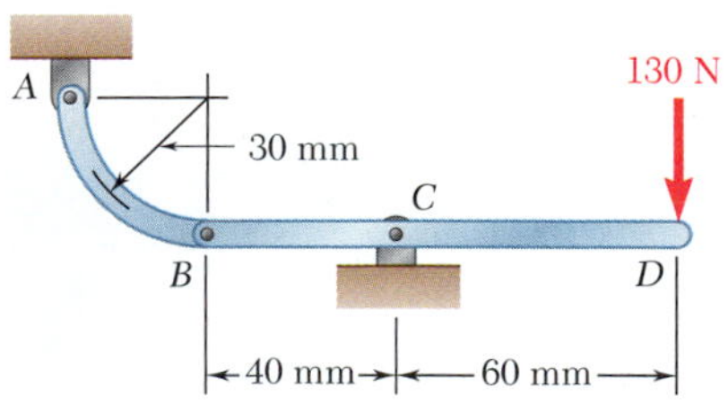

Fig. P4.70

4.71 To remove the lid from a 5-gallon pail, the tool shown is used to apply an upward and radially outward force to the bottom inside rim of the lid. Assuming that the rim rests against the tool at *A* and that a 100-N force is applied as indicated to the handle, determine the force acting on the rim.

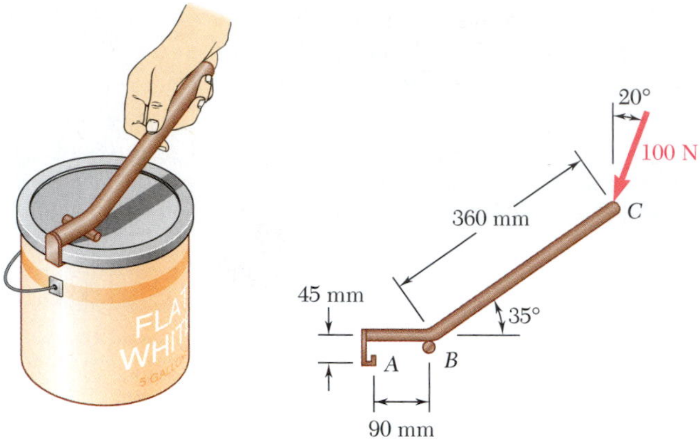

Fig. P4.71

4.72 To remove the lid from a 5-gallon pail, the tool shown is used to apply an upward and radially outward force to the bottom inside rim of the lid. Assuming that the top and the rim of the lid rest against the tool at *A* and *B*, respectively, and that a 60-N force is applied as indicated to the handle, determine the force acting on the rim.

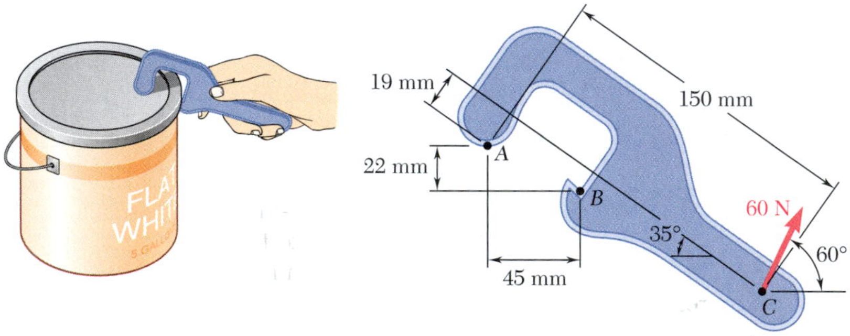

Fig. P4.72

4.73 A 200-lb crate is attached to the trolley-beam system shown. Knowing that $a = 1.5$ ft, determine (*a*) the tension in cable *CD*, (*b*) the reaction at *B*.

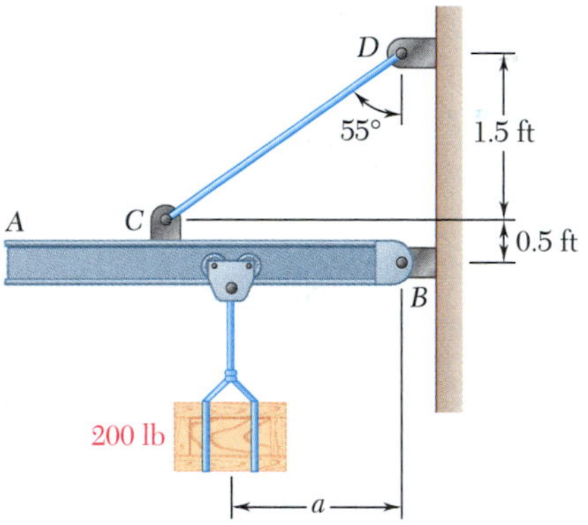

Fig. P4.73

4.74 Solve Prob. 4.73 assuming that $a = 3$ ft.

4.75 A 20-kg roller, of diameter 200 mm, which is to be used on a tile floor, is resting directly on the subflooring as shown. Knowing that the thickness of each tile is 8 mm, determine the force **P** required to move the roller onto the tiles if the roller is pushed to the left.

4.76 A 20-kg roller, of diameter 200 mm, which is to be used on a tile floor, is resting directly on the subflooring as shown. Knowing that the thickness of each tile is 8 mm, determine the force **P** required to move the roller onto the tiles if the roller is pulled to the right.

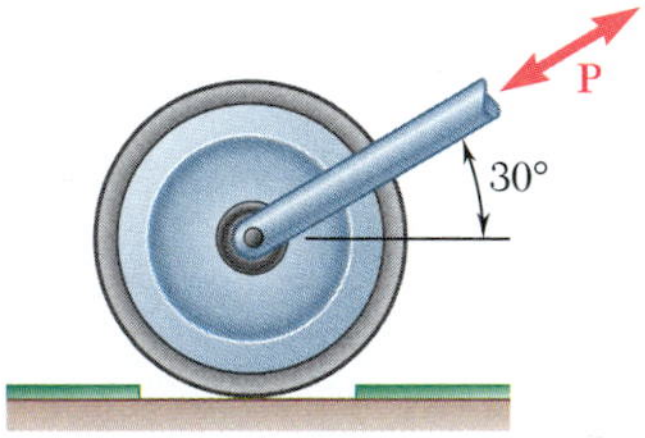

Fig. *P4.75* and P4.76

4.77 A small hoist is mounted on the back of a pickup truck and is used to lift a 120-kg crate. Determine (*a*) the force exerted on the hoist by the hydraulic cylinder *BC*, (*b*) the reaction at *A*.

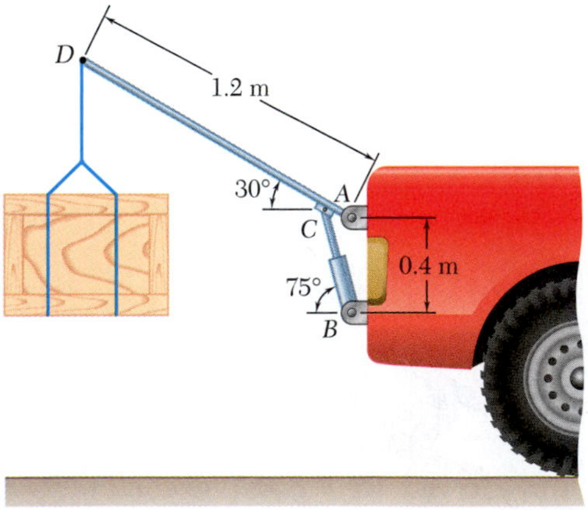

Fig. P4.77

4.78 The clamp shown is used to hold the rough workpiece C. Knowing that the maximum allowable compressive force on the workpiece is 200 N and neglecting the effect of friction at A, determine the corresponding (a) reaction at B, (b) reaction at A, (c) tension in the bolt.

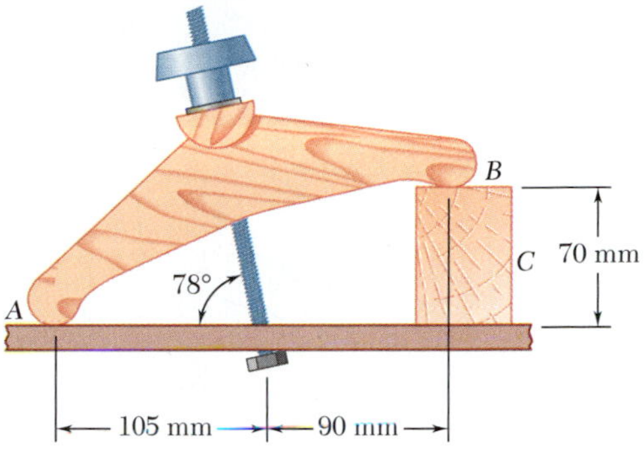

Fig. P4.78

4.79 A modified peavey is used to lift a 0.2-m-diameter log of mass 36 kg. Knowing that $\theta = 45°$ and that the force exerted at C by the worker is perpendicular to the handle of the peavey, determine (a) the force exerted at C, (b) the reaction at A.

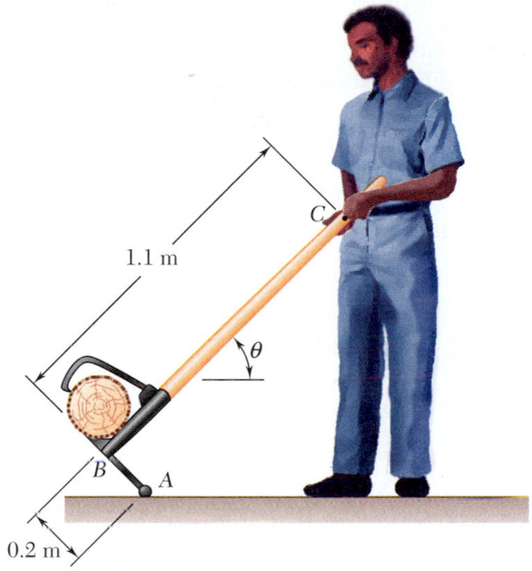

Fig. P4.79 and P4.80

***4.80** A modified peavey is used to lift a 0.2-m-diameter log of mass 36 kg. Knowing that $\theta = 60°$ and that the force exerted at C by the worker is perpendicular to the handle of the peavey, determine (a) the force exerted at C, (b) the reaction at A.

4.81 Member ABC is supported by a pin and bracket at B and by an inextensible cord attached at A and C and passing over a frictionless pulley at D. The tension may be assumed to be the same in portion AD and CD of the cord. For the loading shown and neglecting the size of the pulley, determine the tension in the cord and the reaction at B.

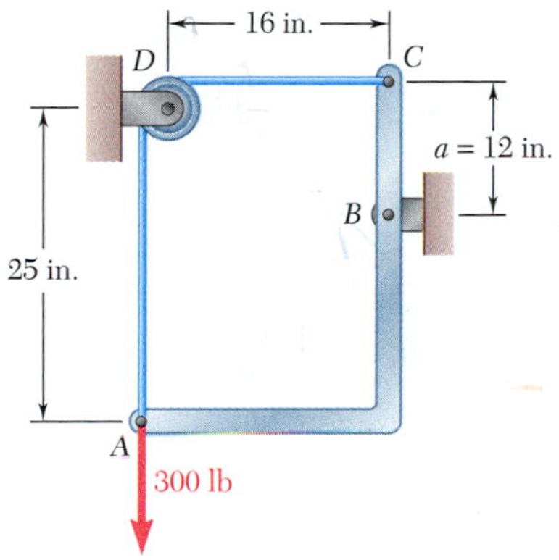

Fig. P4.81

4.82 Member $ABCD$ is supported by a pin and bracket at C and by an inextensible cord attached at A and D and passing over frictionless pulleys at B and E. Neglecting the size of the pulleys, determine the tension in the cord and the reaction at C.

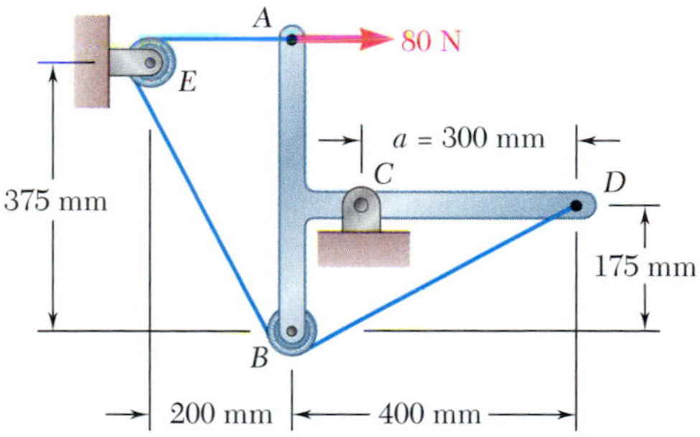

Fig. P4.82

4.83 Using the method of Sec. 4.7, solve Prob. 4.18.

4.84 Using the method of Sec. 4.7, solve Prob. 4.28.

4.85 Knowing that $\theta = 35°$, determine the reaction (a) at B, (b) at C.

4.86 Knowing that $\theta = 50°$, determine the reaction (a) at B, (b) at C.

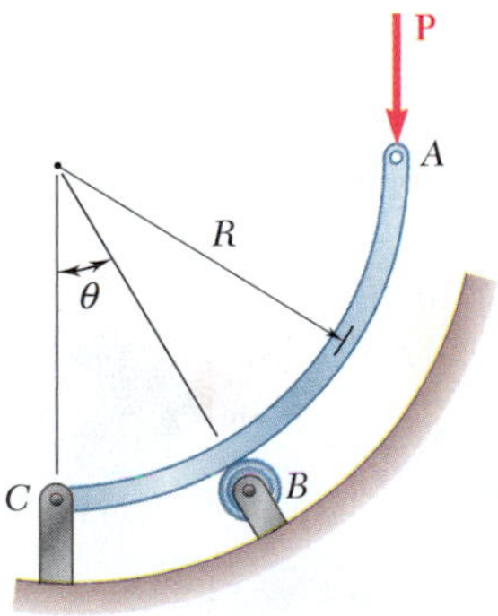

Fig. P4.85 and P4.86

4.87 A slender rod of length L and weight W is held in equilibrium as shown, with one end against a frictionless wall and the other end attached to a cord of length S. Derive an expression for the distance h in terms of L and S. Show that this position of equilibrium does not exist if $S > 2L$.

4.88 A slender rod of length $L = 200$ mm is held in equilibrium as shown, with one end against a frictionless wall and the other end attached to a cord of length $S = 300$ mm. Knowing that the mass of the rod is 1.5 kg, determine (*a*) the distance h, (*b*) the tension in the cord, (*c*) the reaction at B.

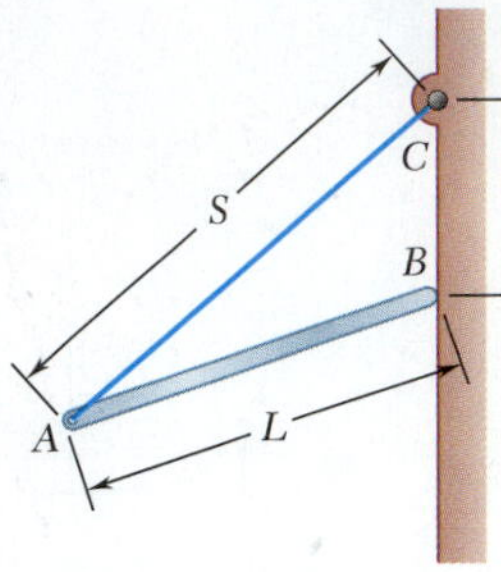

Fig. P4.87 and P4.88

4.89 A slender rod of length L and weight W is attached to collars which can slide freely along the guides shown. Knowing that the rod is in equilibrium, derive an expression for the angle θ in terms of the angle β.

4.90 A 10-kg slender rod of length L is attached to collars which can slide freely along the guides shown. Knowing that the rod is in equilibrium and that $\beta = 25°$, determine (*a*) the angle θ that the rod forms with the vertical, (*b*) the reactions at A and B.

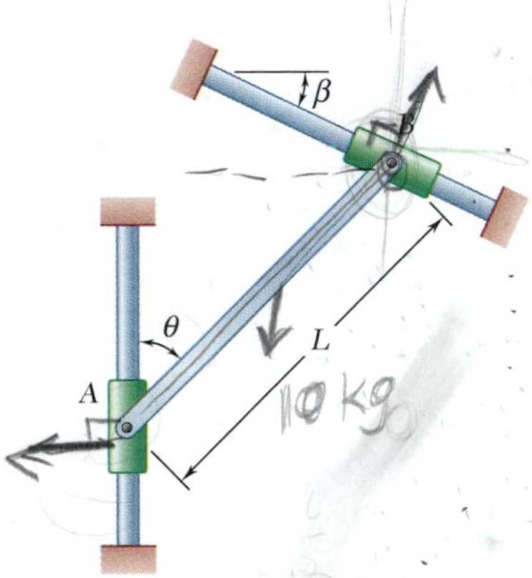

Fig. *P4.89* and P4.90

4.91 A uniform slender rod of mass 5 g and length 250 mm is balanced on a glass of inner diameter 70 mm. Neglecting friction, determine the angle θ corresponding to equilibrium.

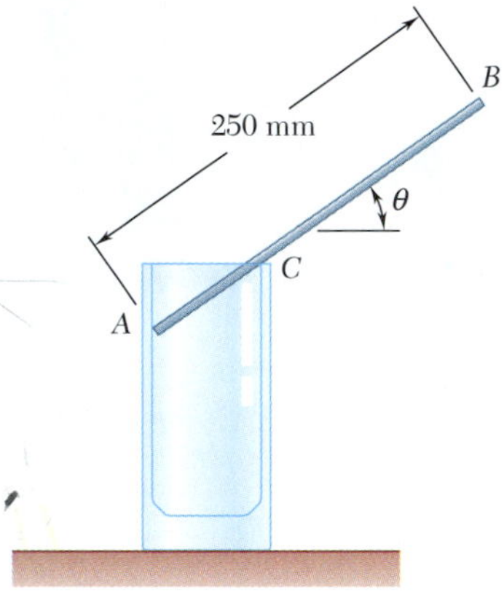

Fig. P4.91

4.92 Rod AB is bent into the shape of a circular arc and is lodged between two pegs D and E. It supports a load **P** at end B. Neglecting friction and the weight of the rod, determine the distance c corresponding to equilibrium when $a = 1$ in. and $R = 5$ in.

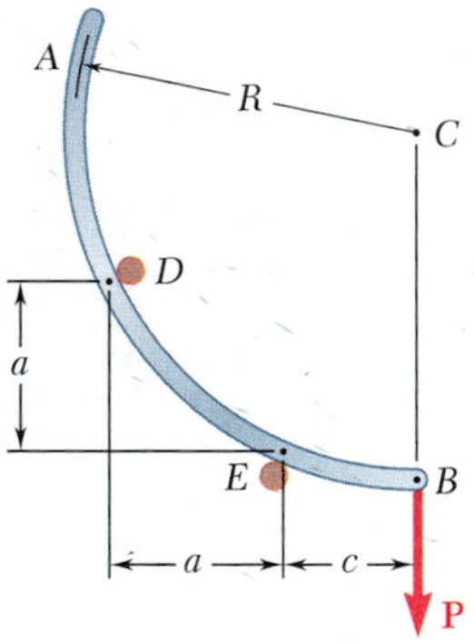

Fig. *P4.92*

4.93 A uniform rod AB of weight W and length $2R$ rests inside a hemispherical bowl of radius R as shown. Neglecting friction, determine the angle θ corresponding to equilibrium.

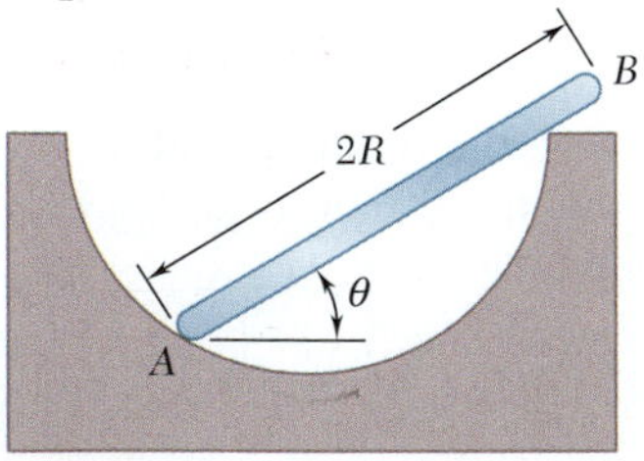

Fig. *P4.93*

4.94 A uniform slender rod of mass m and length $4r$ rests on the surface shown and is held in the given equilibrium position by the force **P**. Neglecting the effect of friction at A and C, (*a*) determine the angle θ, (*b*) derive an expression for P in terms of m.

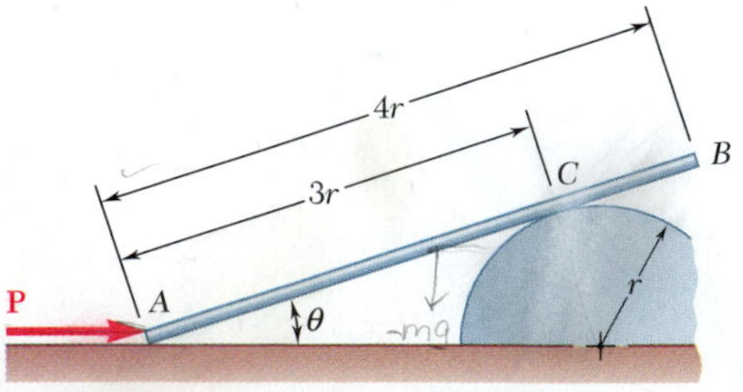

Fig. P4.94

4.95 A uniform slender rod of length $2L$ and mass m rests against a roller at D and is held in the equilibrium position shown by a cord of length a. Knowing that $L = 200$ mm, determine (*a*) the angle θ, (*b*) the length a.

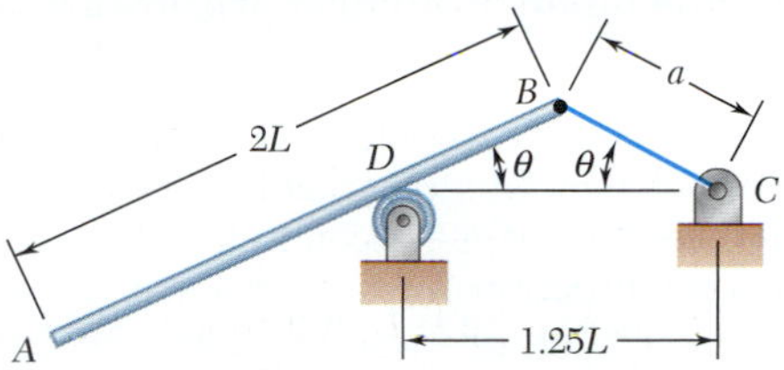

Fig. P4.95

EQUILIBRIUM IN THREE DIMENSIONS

4.8. EQUILIBRIUM OF A RIGID BODY IN THREE DIMENSIONS

We saw in Sec. 4.1 that six scalar equations are required to express the conditions for the equilibrium of a rigid body in the general three-dimensional case:

$$\Sigma F_x = 0 \qquad \Sigma F_y = 0 \qquad \Sigma F_z = 0 \tag{4.2}$$
$$\Sigma M_x = 0 \qquad \Sigma M_y = 0 \qquad \Sigma M_z = 0 \tag{4.3}$$

These equations can be solved for no more than *six unknowns*, which generally will represent reactions at supports or connections.

In most problems the scalar equations (4.2) and (4.3) will be more conveniently obtained if we first express in vector form the conditions for the equilibrium of the rigid body considered. We write

$$\Sigma \mathbf{F} = 0 \qquad \Sigma \mathbf{M}_O = \Sigma(\mathbf{r} \times \mathbf{F}) = 0 \tag{4.1}$$

and express the forces **F** and position vectors **r** in terms of scalar components and unit vectors. Next, we compute all vector products, either by direct calculation or by means of determinants (see Sec. 3.8). We observe that as many as three unknown reaction components may be eliminated from these computations through a judicious choice of the point O. By equating to zero the coefficients of the unit vectors in each of the two relations (4.1), we obtain the desired scalar equations.†

4.9. REACTIONS AT SUPPORTS AND CONNECTIONS FOR A THREE-DIMENSIONAL STRUCTURE

The reactions on a three-dimensional structure range from the single force of known direction exerted by a frictionless surface to the force-couple system exerted by a fixed support. Consequently, in problems involving the equilibrium of a three-dimensional structure, there can be between one and six unknowns associated with the reaction at each

†In some problems, it will be found convenient to eliminate the reactions at two points A and B from the solution by writing the equilibrium equation $\Sigma M_{AB} = 0$, which involves the determination of the moments of the forces about the axis AB joining points A and B (see Sample Prob. 4.10).

support or connection. Various types of supports and connections are shown in Fig. 4.10 with their corresponding reactions. A simple way of determining the type of reaction corresponding to a given support or connection and the number of unknowns involved is to find which of the six fundamental motions (translation in the x, y, and z directions, rotation about the x, y, and z axes) are allowed and which motions are prevented.

Photo 4.6 Universal joints, easily seen on the drive shafts of rear-wheel-drive cars and trucks, allow rotational motion to be transferred between two non-collinear shafts.

Ball supports, frictionless surfaces, and cables, for example, prevent translation in one direction only and thus exert a single force whose line of action is known; each of these supports involves one unknown, namely, the magnitude of the reaction. Rollers on rough surfaces and wheels on rails prevent translation in two directions; the corresponding reactions consist of two unknown force components. Rough surfaces in direct contact and ball-and-socket supports prevent translation in three directions; these supports involve three unknown force components.

Some supports and connections can prevent rotation as well as translation; the corresponding reactions include couples as well as forces. For example, the reaction at a fixed support, which prevents any motion (rotation as well as translation), consists of three unknown forces and three unknown couples. A universal joint, which is designed to allow rotation about two axes, will exert a reaction consisting of three unknown force components and one unknown couple.

Other supports and connections are primarily intended to prevent translation; their design, however, is such that they also prevent some rotations. The corresponding reactions consist essentially of force components but *may* also include couples. One group of supports of this type includes hinges and bearings designed to support radial loads only (for example, journal bearings, roller bearings). The corresponding reactions consist of two force components but may also include two couples. Another group includes pin-and-bracket supports, hinges, and bearings designed to support an axial thrust as well as a radial load (for example, ball bearings). The corresponding reactions consist of three force components but may include two couples. However, these supports will not exert any appreciable couples under normal conditions of use. Therefore, *only* force components should be included in their analysis *unless* it is found that couples are necessary to maintain the equilibrium of the rigid body, or unless the support is known to have been specifically designed to exert a couple (see Probs. 4.128 through 4.131).

Photo 4.7 The pillow block bearing shown supports the shaft of a fan used to ventilate a foundry.

If the reactions involve more than six unknowns, there are more unknowns than equations, and some of the reactions are *statically indeterminate*. If the reactions involve fewer than six unknowns, there are more equations than unknowns, and some of the equations of equilibrium cannot be satisfied under general loading conditions; the rigid body is only *partially constrained*. Under the particular loading conditions corresponding to a given problem, however, the extra equations often reduce to trivial identities, such as $0 = 0$, and can be disregarded; although only partially constrained, the rigid body remains in equilibrium (see Sample Probs. 4.7 and 4.8). Even with six or more unknowns, it is possible that some equations of equilibrium will not be satisfied. This can occur when the reactions associated with the given supports either are parallel or intersect the same line; the rigid body is then *improperly constrained*.

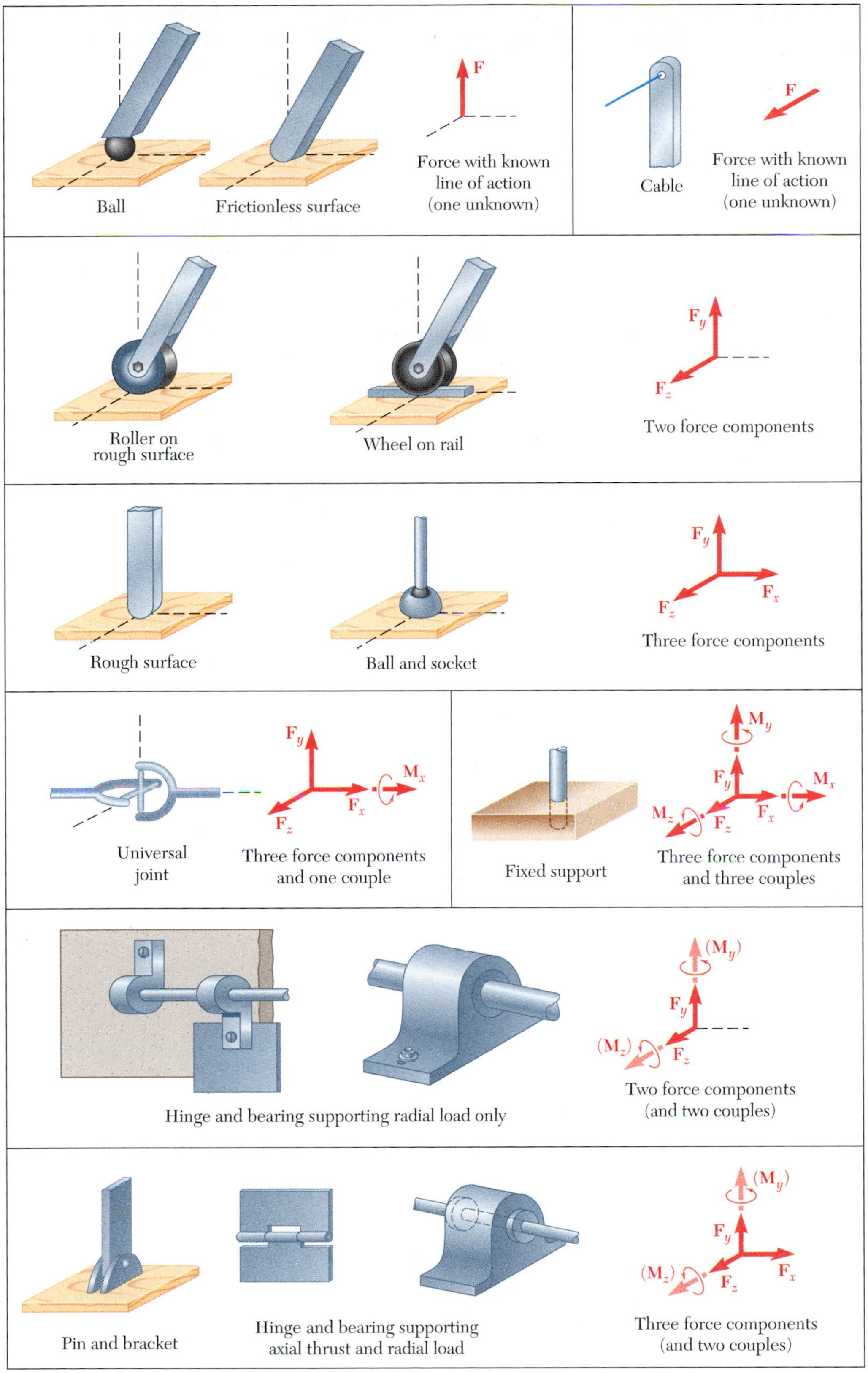

Fig. 4.10 Reactions at supports and connections.

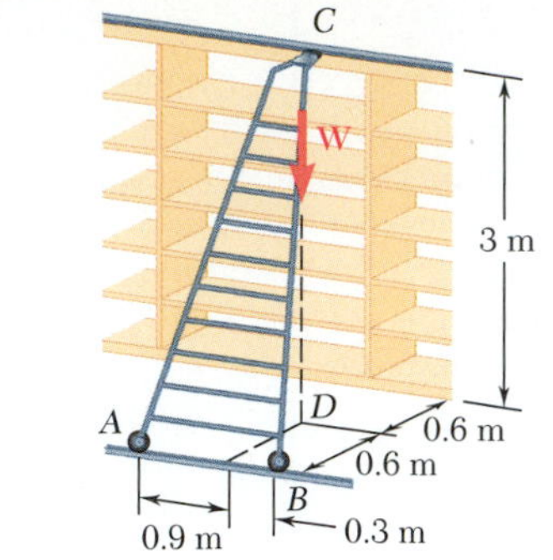

SAMPLE PROBLEM 4.7

A 20-kg ladder used to reach high shelves in a storeroom is supported by two flanged wheels A and B mounted on a rail and by an unflanged wheel C resting against a rail fixed to the wall. An 80-kg man stands on the ladder and leans to the right. The line of action of the combined weight $\mathbf{W}$ of the man and ladder intersects the floor at point D. Determine the reactions at A, B, and C.

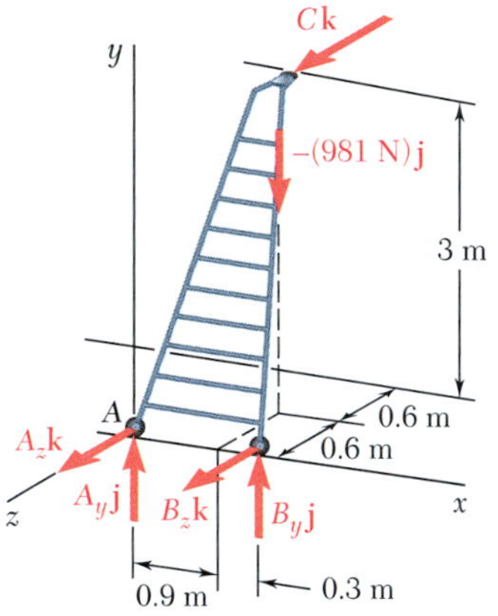

SOLUTION

Free-Body Diagram. A free-body diagram of the ladder is drawn. The forces involved are the combined weight of the man and ladder.

$$\mathbf{W} = -mg\mathbf{j} = -(80 \text{ kg} + 20 \text{ kg})(9.81 \text{ m/s}^2)\mathbf{j} = -(981 \text{ N})\mathbf{j}$$

and five unknown reaction components, two at each flanged wheel and one at the unflanged wheel. The ladder is thus only partially constrained; it is free to roll along the rails. It is, however, in equilibrium under the given load since the equation $\Sigma F_x = 0$ is satisfied.

Equilibrium Equations. We express that the forces acting on the ladder form a system equivalent to zero:

$$\Sigma\mathbf{F} = 0: \quad A_y\mathbf{j} + A_z\mathbf{k} + B_y\mathbf{j} + B_z\mathbf{k} - (981 \text{ N})\mathbf{j} + C\mathbf{k} = 0$$

$$(A_y + B_y - 981 \text{ N})\mathbf{j} + (A_z + B_z + C)\mathbf{k} = 0 \qquad (1)$$

$$\Sigma\mathbf{M}_A = \Sigma(\mathbf{r} \times \mathbf{F}) = 0: \quad 1.2\mathbf{i} \times (B_y\mathbf{j} + B_z\mathbf{k}) + (0.9\mathbf{i} - 0.6\mathbf{k}) \times (-981\mathbf{j}) + (0.6\mathbf{i} + 3\mathbf{j} - 1.2\mathbf{k}) \times C\mathbf{k} = 0$$

Computing the vector product, we have†

$$1.2B_y\mathbf{k} - 1.2B_z\mathbf{j} - 882.9\mathbf{k} - 588.6\mathbf{i} - 0.6C\mathbf{j} + 3C\mathbf{i} = 0$$

$$(3C - 588.6)\mathbf{i} - (1.2B_z + 0.6C)\mathbf{j} + (1.2B_y - 882.9)\mathbf{k} = 0 \qquad (2)$$

Setting the coefficients of **i**, **j**, **k** equal to zero in Eq. (2), we obtain the following three scalar equations, which express that the sum of the moments about each coordinate axis must be zero:

$$3C - 588.6 = 0 \qquad C = +196.2 \text{ N}$$

$$1.2B_z + 0.6C = 0 \qquad B_z = -98.1 \text{ N}$$

$$1.2B_y - 882.9 = 0 \qquad B_y = +736 \text{ N}$$

The reactions at B and C are therefore

$$\mathbf{B} = +(736 \text{ N})\mathbf{j} - (98.1 \text{ N})\mathbf{k} \qquad \mathbf{C} = +(196.2 \text{ N})\mathbf{k} \quad ◀$$

Setting the coefficients of **j** and **k** equal to zero in Eq. (1), we obtain two scalar equations expressing that the sums of the components in the y and z directions are zero. Substituting for B_y, B_z, and C the values obtained above, we write

$$A_y + B_y - 981 = 0 \qquad A_y + 736 - 981 = 0 \qquad A_y = +245 \text{ N}$$

$$A_z + B_z + C = 0 \qquad A_z - 98.1 + 196.2 = 0 \qquad A_z = -98.1 \text{ N}$$

We conclude that the reaction at A is $\quad \mathbf{A} = +(245 \text{ N})\mathbf{j} - (98.1 \text{ N})\mathbf{k} \quad ◀$

†The moments in this sample problem and in Sample Probs. 4.8 and 4.9 can also be expressed in the form of determinants (see Sample Prob. 3.10).

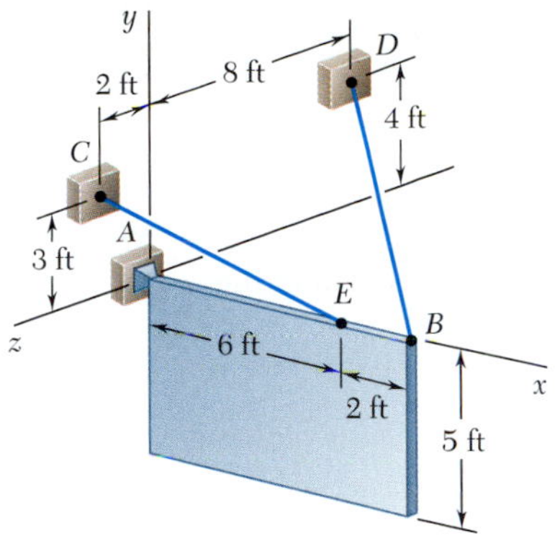

SAMPLE PROBLEM 4.8

A 5 × 8-ft sign of uniform density weighs 270 lb and is supported by a ball-and-socket joint at A and by two cables. Determine the tension in each cable and the reaction at A.

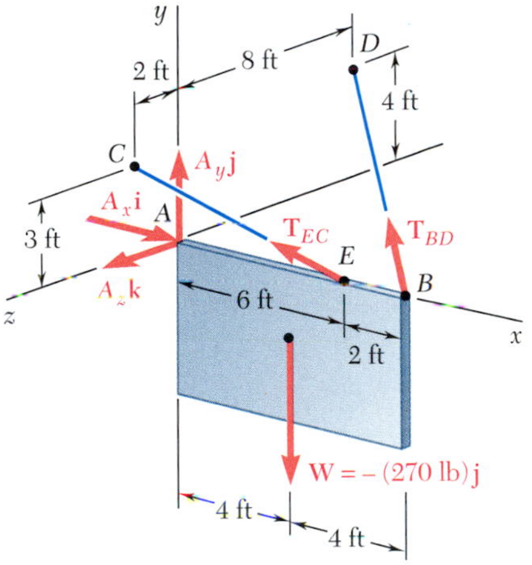

SOLUTION

Free-Body Diagram. A free-body diagram of the sign is drawn. The forces acting on the free body are the weight $\mathbf{W} = -(270 \text{ lb})\mathbf{j}$ and the reactions at A, B, and E. The reaction at A is a force of unknown direction and is represented by three unknown components. Since the directions of the forces exerted by the cables are known, these forces involve only one unknown each, namely, the magnitudes T_{BD} and T_{EC}. Since there are only five unknowns, the sign is partially constrained. It can rotate freely about the x axis; it is, however, in equilibrium under the given loading, since the equation $\Sigma M_x = 0$ is satisfied.

The components of the forces $\mathbf{T}_{BD}$ and $\mathbf{T}_{EC}$ can be expressed in terms of the unknown magnitudes T_{BD} and T_{EC} by writing

$$\overrightarrow{BD} = -(8 \text{ ft})\mathbf{i} + (4 \text{ ft})\mathbf{j} - (8 \text{ ft})\mathbf{k} \qquad BD = 12 \text{ ft}$$
$$\overrightarrow{EC} = -(6 \text{ ft})\mathbf{i} + (3 \text{ ft})\mathbf{j} + (2 \text{ ft})\mathbf{k} \qquad EC = 7 \text{ ft}$$
$$\mathbf{T}_{BD} = T_{BD}\left(\frac{\overrightarrow{BD}}{BD}\right) = T_{BD}\left(-\tfrac{2}{3}\mathbf{i} + \tfrac{1}{3}\mathbf{j} - \tfrac{2}{3}\mathbf{k}\right)$$
$$\mathbf{T}_{EC} = T_{EC}\left(\frac{\overrightarrow{EC}}{EC}\right) = T_{EC}\left(-\tfrac{6}{7}\mathbf{i} + \tfrac{3}{7}\mathbf{j} + \tfrac{2}{7}\mathbf{k}\right)$$

Equilibrium Equations. We express that the forces acting on the sign form a system equivalent to zero:

$$\Sigma \mathbf{F} = 0: \qquad A_x\mathbf{i} + A_y\mathbf{j} + A_z\mathbf{k} + \mathbf{T}_{BD} + \mathbf{T}_{EC} - (270 \text{ lb})\mathbf{j} = 0$$
$$(A_x - \tfrac{2}{3}T_{BD} - \tfrac{6}{7}T_{EC})\mathbf{i} + (A_y + \tfrac{1}{3}T_{BD} + \tfrac{3}{7}T_{EC} - 270 \text{ lb})\mathbf{j}$$
$$+ (A_z - \tfrac{2}{3}T_{BD} + \tfrac{2}{7}T_{EC})\mathbf{k} = 0 \qquad (1)$$

$$\Sigma \mathbf{M}_A = \Sigma(\mathbf{r} \times \mathbf{F}) = 0:$$
$$(8 \text{ ft})\mathbf{i} \times T_{BD}(-\tfrac{2}{3}\mathbf{i} + \tfrac{1}{3}\mathbf{j} - \tfrac{2}{3}\mathbf{k}) + (6 \text{ ft})\mathbf{i} \times T_{EC}(-\tfrac{6}{7}\mathbf{i} + \tfrac{3}{7}\mathbf{j} + \tfrac{2}{7}\mathbf{k})$$
$$+ (4 \text{ ft})\mathbf{i} \times (-270 \text{ lb})\mathbf{j} = 0$$
$$(2.667T_{BD} + 2.571T_{EC} - 1080 \text{ lb})\mathbf{k} + (5.333T_{BD} - 1.714T_{EC})\mathbf{j} = 0 \qquad (2)$$

Setting the coefficients of $\mathbf{j}$ and $\mathbf{k}$ equal to zero in Eq. (2), we obtain two scalar equations which can be solved for T_{BD} and T_{EC}:

$$T_{BD} = 101.3 \text{ lb} \qquad T_{EC} = 315 \text{ lb} \quad ◀$$

Setting the coefficients of $\mathbf{i}$, $\mathbf{j}$, and $\mathbf{k}$ equal to zero in Eq. (1), we obtain three more equations, which yield the components of $\mathbf{A}$. We have

$$\mathbf{A} = +(338 \text{ lb})\mathbf{i} + (101.2 \text{ lb})\mathbf{j} - (22.5 \text{ lb})\mathbf{k} \quad ◀$$

SAMPLE PROBLEM 4.9

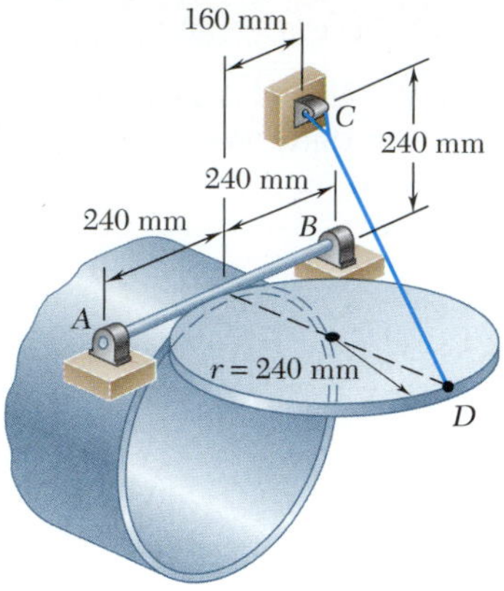

A uniform pipe cover of radius $r = 240$ mm and mass 30 kg is held in a horizontal position by the cable CD. Assuming that the bearing at B does not exert any axial thrust, determine the tension in the cable and the reactions at A and B.

SOLUTION

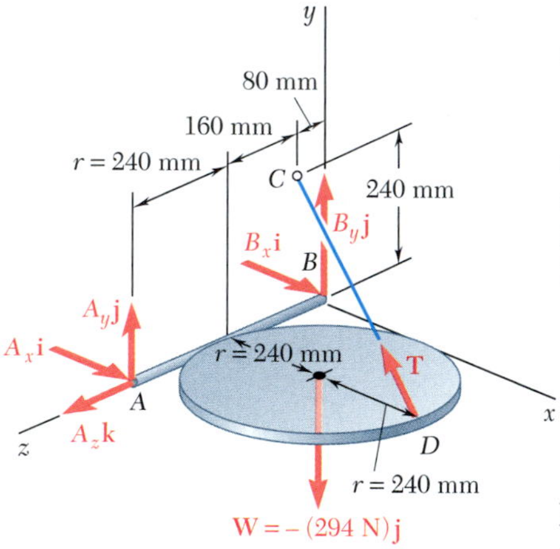

Free-Body Diagram. A free-body diagram is drawn with the coordinate axes shown. The forces acting on the free body are the weight of the cover,

$$\mathbf{W} = -mg\mathbf{j} = -(30 \text{ kg})(9.81 \text{ m/s}^2)\mathbf{j} = -(294 \text{ N})\mathbf{j}$$

and reactions involving six unknowns, namely, the magnitude of the force **T** exerted by the cable, three force components at hinge A, and two at hinge B. The components of **T** are expressed in terms of the unknown magnitude T by resolving the vector $\overrightarrow{DC}$ into rectangular components and writing

$$\overrightarrow{DC} = -(480 \text{ mm})\mathbf{i} + (240 \text{ mm})\mathbf{j} - (160 \text{ mm})\mathbf{k} \qquad DC = 560 \text{ mm}$$

$$\mathbf{T} = T\frac{\overrightarrow{DC}}{DC} = -\tfrac{6}{7}T\mathbf{i} + \tfrac{3}{7}T\mathbf{j} - \tfrac{2}{7}T\mathbf{k}$$

Equilibrium Equations. We express that the forces acting on the pipe cover form a system equivalent to zero:

$\Sigma\mathbf{F} = 0$: $\quad A_x\mathbf{i} + A_y\mathbf{j} + A_z\mathbf{k} + B_x\mathbf{i} + B_y\mathbf{j} + \mathbf{T} - (294 \text{ N})\mathbf{j} = 0$

$$(A_x + B_x - \tfrac{6}{7}T)\mathbf{i} + (A_y + B_y + \tfrac{3}{7}T - 294 \text{ N})\mathbf{j} + (A_z - \tfrac{2}{7}T)\mathbf{k} = 0 \qquad (1)$$

$\Sigma\mathbf{M}_B = \Sigma(\mathbf{r} \times \mathbf{F}) = 0$:

$$2r\mathbf{k} \times (A_x\mathbf{i} + A_y\mathbf{j} + A_z\mathbf{k}) + (2r\mathbf{i} + r\mathbf{k}) \times (-\tfrac{6}{7}T\mathbf{i} + \tfrac{3}{7}T\mathbf{j} - \tfrac{2}{7}T\mathbf{k}) + (r\mathbf{i} + r\mathbf{k}) \times (-294 \text{ N})\mathbf{j} = 0$$

$$(-2A_y - \tfrac{3}{7}T + 294 \text{ N})r\mathbf{i} + (2A_x - \tfrac{2}{7}T)r\mathbf{j} + (\tfrac{6}{7}T - 294 \text{ N})r\mathbf{k} = 0 \qquad (2)$$

Setting the coefficients of the unit vectors equal to zero in Eq. (2), we write three scalar equations, which yield

$$A_x = +49.0 \text{ N} \qquad A_y = +73.5 \text{ N} \qquad T = 343 \text{ N} \quad ◀$$

Setting the coefficients of the unit vectors equal to zero in Eq. (1), we obtain three more scalar equations. After substituting the values of T, A_x, and A_y into these equations, we obtain

$$A_z = +98.0 \text{ N} \qquad B_x = +245 \text{ N} \qquad B_y = +73.5 \text{ N}$$

The reactions at A and B are therefore

$$\mathbf{A} = +(49.0 \text{ N})\mathbf{i} + (73.5 \text{ N})\mathbf{j} + (98.0 \text{ N})\mathbf{k} \quad ◀$$

$$\mathbf{B} = +(245 \text{ N})\mathbf{i} + (73.5 \text{ N})\mathbf{j} \quad ◀$$

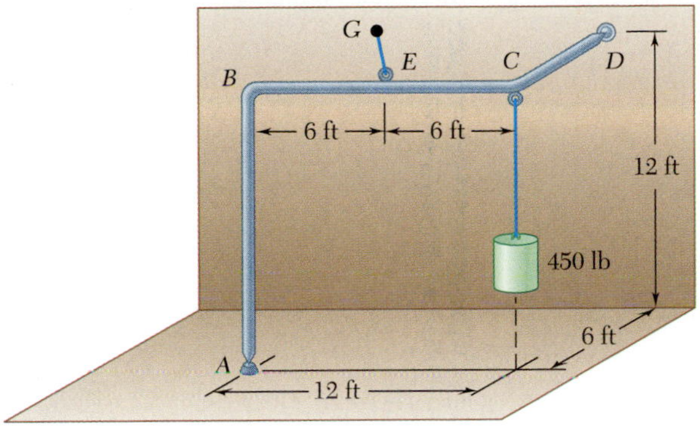

SAMPLE PROBLEM 4.10

A 450-lb load hangs from the corner C of a rigid piece of pipe $ABCD$ which has been bent as shown. The pipe is supported by the ball-and-socket joints A and D, which are fastened, respectively, to the floor and to a vertical wall, and by a cable attached at the midpoint E of the portion BC of the pipe and at a point G on the wall. Determine (*a*) where G should be located if the tension in the cable is to be minimum, (*b*) the corresponding minimum value of the tension.

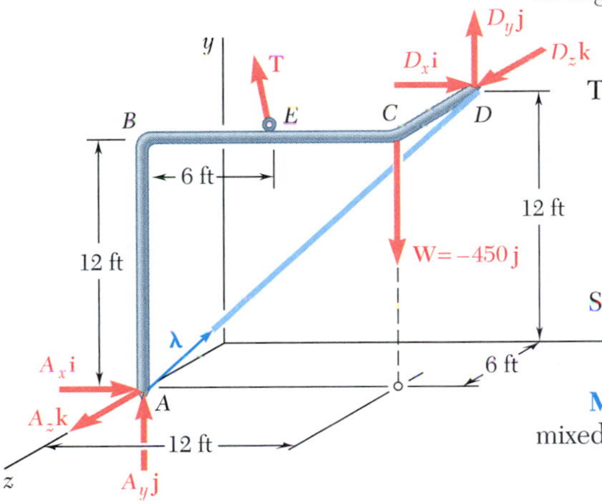

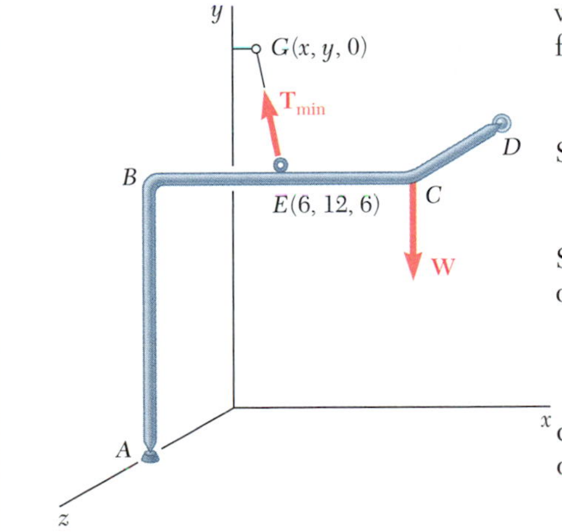

SOLUTION

Free-Body Diagram. The free-body diagram of the pipe includes the load $\mathbf{W} = (-450 \text{ lb})\mathbf{j}$, the reactions at A and D, and the force $\mathbf{T}$ exerted by the cable. To eliminate the reactions at A and D from the computations, we express that the sum of the moments of the forces about AD is zero. Denoting by $\boldsymbol{\lambda}$ the unit vector along AD, we write

$$\Sigma M_{AD} = 0: \qquad \boldsymbol{\lambda} \cdot (\overrightarrow{AE} \times \mathbf{T}) + \boldsymbol{\lambda} \cdot (\overrightarrow{AC} \times \mathbf{W}) = 0 \tag{1}$$

The second term in Eq. (1) can be computed as follows:

$$\overrightarrow{AC} \times \mathbf{W} = (12\mathbf{i} + 12\mathbf{j}) \times (-450\mathbf{j}) = -5400\mathbf{k}$$

$$\boldsymbol{\lambda} = \frac{\overrightarrow{AD}}{AD} = \frac{12\mathbf{i} + 12\mathbf{j} - 6\mathbf{k}}{18} = \tfrac{2}{3}\mathbf{i} + \tfrac{2}{3}\mathbf{j} - \tfrac{1}{3}\mathbf{k}$$

$$\boldsymbol{\lambda} \cdot (\overrightarrow{AC} \times \mathbf{W}) = (\tfrac{2}{3}\mathbf{i} + \tfrac{2}{3}\mathbf{j} - \tfrac{1}{3}\mathbf{k}) \cdot (-5400\mathbf{k}) = +1800$$

Substituting the value obtained into Eq. (1), we write

$$\boldsymbol{\lambda} \cdot (\overrightarrow{AE} \times \mathbf{T}) = -1800 \text{ lb} \cdot \text{ft} \tag{2}$$

Minimum Value of Tension. Recalling the commutative property for mixed triple products, we rewrite Eq. (2) in the form

$$\mathbf{T} \cdot (\boldsymbol{\lambda} \times \overrightarrow{AE}) = -1800 \text{ lb} \cdot \text{ft} \tag{3}$$

which shows that the projection of $\mathbf{T}$ on the vector $\boldsymbol{\lambda} \times \overrightarrow{AE}$ is a constant. It follows that $\mathbf{T}$ is minimum when parallel to the vector

$$\boldsymbol{\lambda} \times \overrightarrow{AE} = (\tfrac{2}{3}\mathbf{i} + \tfrac{2}{3}\mathbf{j} - \tfrac{1}{3}\mathbf{k}) \times (6\mathbf{i} + 12\mathbf{j}) = 4\mathbf{i} - 2\mathbf{j} + 4\mathbf{k}$$

Since the corresponding unit vector is $\tfrac{2}{3}\mathbf{i} - \tfrac{1}{3}\mathbf{j} + \tfrac{2}{3}\mathbf{k}$, we write

$$\mathbf{T}_{\min} = T(\tfrac{2}{3}\mathbf{i} - \tfrac{1}{3}\mathbf{j} + \tfrac{2}{3}\mathbf{k}) \tag{4}$$

Substituting for $\mathbf{T}$ and $\boldsymbol{\lambda} \times \overrightarrow{AE}$ in Eq. (3) and computing the dot products, we obtain $6T = -1800$ and, thus, $T = -300$. Carrying this value into (4), we obtain

$$\mathbf{T}_{\min} = -200\mathbf{i} + 100\mathbf{j} - 200\mathbf{k} \qquad T_{\min} = 300 \text{ lb} \blacktriangleleft$$

Location of G. Since the vector $\overrightarrow{EG}$ and the force $\mathbf{T}_{\min}$ have the same direction, their components must be proportional. Denoting the coordinates of G by x, y, 0, we write

$$\frac{x - 6}{-200} = \frac{y - 12}{+100} = \frac{0 - 6}{-200} \qquad x = 0 \qquad y = 15 \text{ ft} \blacktriangleleft$$

SOLVING PROBLEMS ON YOUR OWN

The equilibrium of a *three-dimensional body* was considered in the sections you just completed. It is again most important that you draw a complete *free-body diagram* as the first step of your solution.

1. As you draw the free-body diagram, pay particular attention to the reactions at the supports. The number of unknowns at a support can range from one to six (Fig. 4.10). To decide whether an unknown reaction or reaction component exists at a support, ask yourself whether the support prevents motion of the body in a certain direction or about a certain axis.

a. If motion is prevented in a certain direction, include in your free-body diagram an unknown reaction or *reaction component* that acts in the *same direction.*

b. If a support prevents rotation about a certain axis, include in your free-body diagram a *couple* of unknown magnitude that acts about the *same* axis.

2. The external forces acting on a three-dimensional body form a system equivalent to zero. Writing $\Sigma\mathbf{F} = 0$ and $\Sigma\mathbf{M}_A = 0$ about an appropriate point A, and setting the coefficients of **i**, **j**, **k** in both equations equal to zero will provide you with six scalar equations. In general, these equations will contain six unknowns and can be solved for these unknowns.

3. After completing your free-body diagram, you may want to seek equations involving as few unknowns as possible. The following strategies may help you.

a. By summing moments about a ball-and-socket support or a hinge, you will obtain equations from which three unknown reaction components have been eliminated [Sample Probs. 4.8 and 4.9].

b. If you can draw an axis through the points of application of all but one of the unknown reactions, summing moments about that axis will yield an equation in a single unknown [Sample Prob. 4.10].

4. After drawing your free-body diagram, we encourage you to compare the number of unknowns to the number of nontrivial, scalar equations of equilibrium for the given problem. Doing so will tell you if the body is properly or partially constrained and whether the problem is statically determinate or indeterminate. Further, as we consider more complex problems in later chapters, keeping track of the numbers of unknowns and equations will help you to develop correct solutions.

Problems

4.96 Gears A and B are attached to a shaft supported by bearings at C and D. The diameters of gears A and B are 150 mm and 75 mm, respectively, and the tangential and radial forces acting on the gears are as shown. Knowing that the system rotates at a constant rate, determine the reactions at C and D. Assume that the bearing at C does not exert any axial force, and neglect the weights of the gears and the shaft.

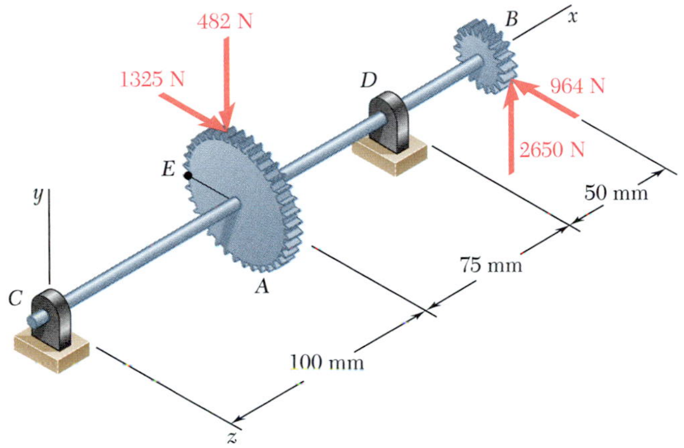

Fig. P4.96

4.97 Solve Prob. 4.96 assuming that for gear A the tangential and radial forces are acting at E, so that $\mathbf{F}_A = (1325\ \text{N})\mathbf{j} + (482\ \text{N})\mathbf{k}$.

4.98 Two transmission belts pass over sheaves welded to an axle supported by bearings at B and D. The sheave at A has a radius of 50 mm, and the sheave at C has a radius of 40 mm. Knowing that the system rotates with a constant rate, determine (*a*) the tension T, (*b*) the reactions at B and D. Assume that the bearing at D does not exert any axial thrust and neglect the weights of the sheaves and the axle.

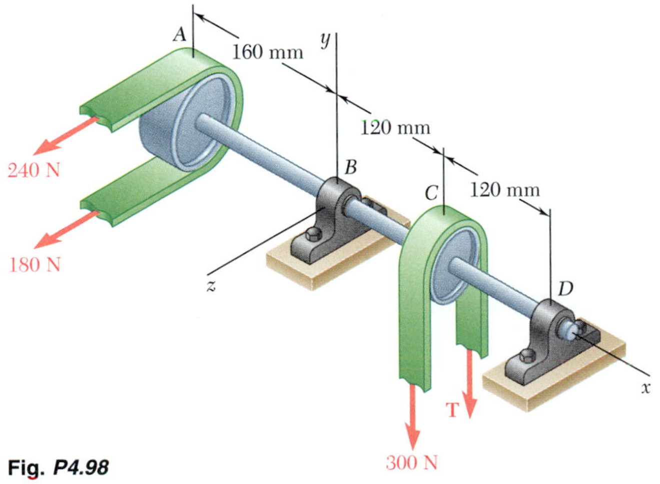

Fig. *P4.98*

4.99 For the portion of a machine shown, the 4-in.-diameter pulley *A* and wheel *B* are fixed to a shaft supported by bearings at *C* and *D*. The spring of constant 2 lb/in. is unstretched when $\theta = 0$, and the bearing at *C* does not exert any axial force. Knowing that $\theta = 180°$ and that the machine is at rest and in equilibrium, determine (*a*) the tension *T*, (*b*) the reactions at *C* and *D*. Neglect the weights of the shaft, pulley, and wheel.

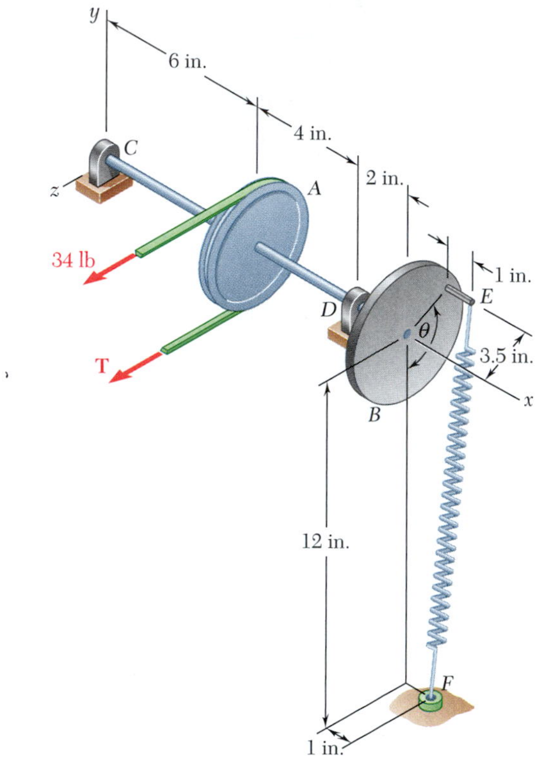

Fig. P4.99

4.100 Solve Prob. 4.99 for $\theta = 90°$.

4.101 A 1.2 × 2.4-m sheet of plywood having a mass of 17 kg has been temporarily placed among three pipe supports. The lower edge of the sheet rests on small collars *A* and *B* and its upper edge leans against pipe *C*. Neglecting friction at all surfaces, determine the reactions at *A*, *B*, and *C*.

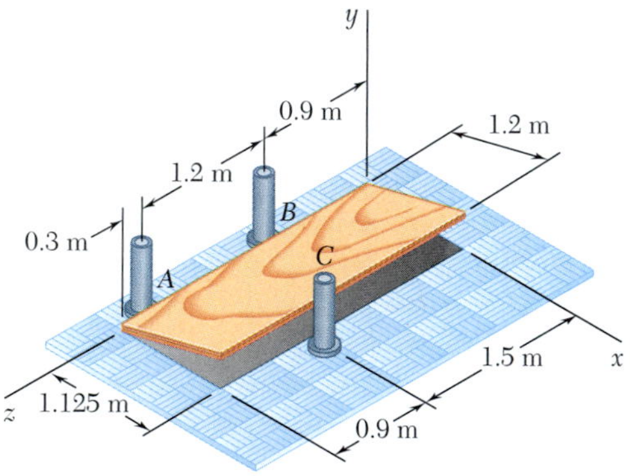

Fig. *P4.101*

4.102 The 200 × 200-mm square plate shown has a mass of 25 kg and is supported by three vertical wires. Determine the tension in each wire.

4.103 The 200 × 200-mm square plate shown has a mass of 25 kg and is supported by three vertical wires. Determine the mass and location of the lightest block which should be placed on the plate if the tensions in the three cables are to be equal.

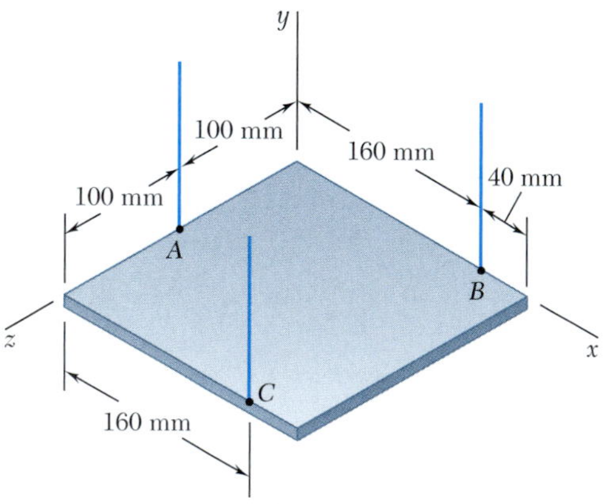

Fig. P4.102 and P4.103

4.104 A camera of mass 240 g is mounted on a small tripod of mass 200 g. Assuming that the mass of the camera is uniformly distributed and that the line of action of the weight of the tripod passes through D, determine (a) the vertical components of the reactions at A, B, and C when $\theta = 0$, (b) the maximum value of θ if the tripod is not to tip over.

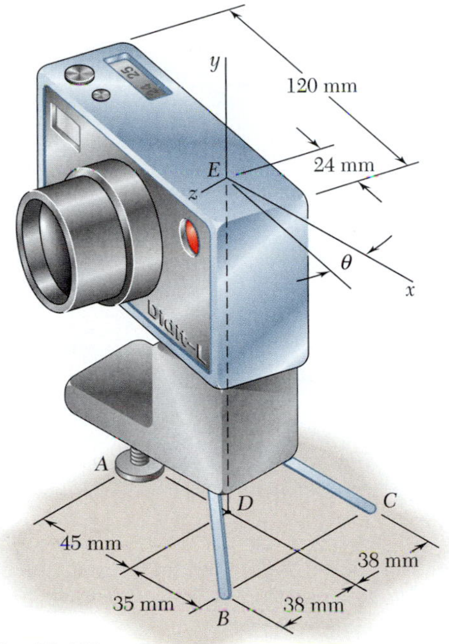

Fig. P4.104

4.105 Two steel pipes AB and BC, each having a weight per unit length of 5 lb/ft, are welded together at B and are supported by three wires. Knowing that $a = 1.25$ ft, determine the tension in each wire.

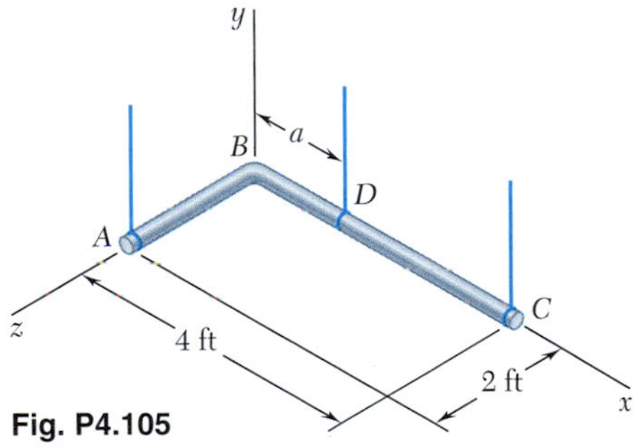

Fig. P4.105

4.106 For the pipe assembly of Prob. 4.105, determine (a) the largest permissible value of a if the assembly is not to tip, (b) the corresponding tension in each wire.

4.107 A uniform aluminum rod of weight W is bent into a circular ring of radius R and is supported by three wires as shown. Determine the tension in each wire.

4.108 A uniform aluminum rod of weight W is bent into a circular ring of radius R and is supported by three wires as shown. A small collar of weight W' is then placed on the ring and positioned so that the tensions in the three wires are equal. Determine (a) the position of the collar, (b) the value of W', (c) the tension in the wires.

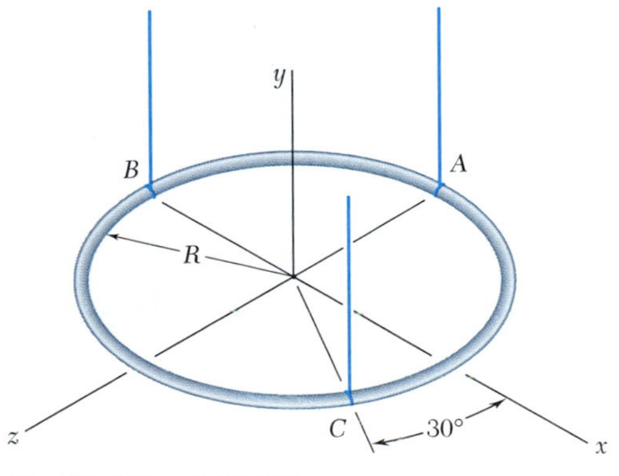

Fig. P4.107 and *P4.108*

4.109 An opening in a floor is covered by a 3×4-ft sheet of plywood weighing 12 lb. The sheet is hinged at A and B and is maintained in a position slightly above the floor by a small block C. Determine the vertical component of the reaction (*a*) at A, (*b*) at B, (*c*) at C.

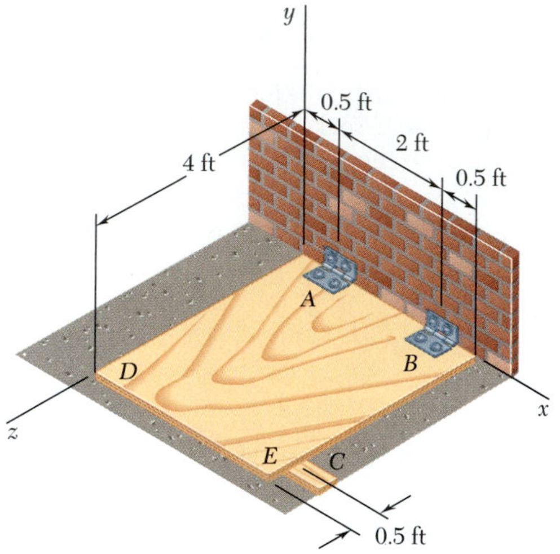

Fig. P4.109

4.110 Solve Prob. 4.109 assuming that the small block C is moved and placed under edge DE at a point 0.5 ft from corner E.

4.111 The 10-kg square plate shown is supported by three vertical wires. Determine (*a*) the tension in each wire when $a = 100$ mm, (*b*) the value of a for which tensions in the three wires are equal.

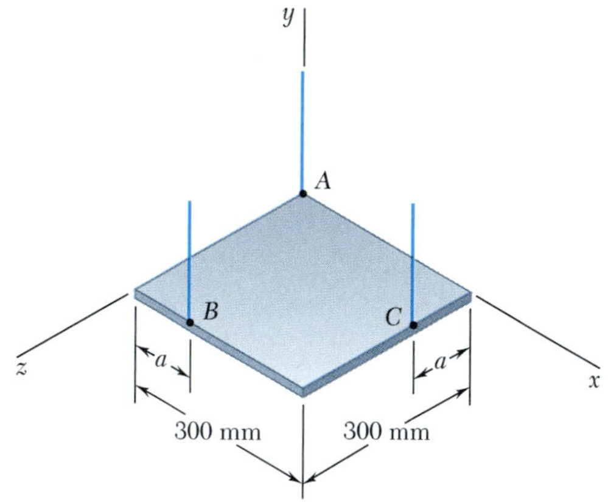

Fig. P4.111

4.112 The 3-m flagpole AC forms an angle of 30° with the z axis. It is held by a ball-and-socket joint at C and by two thin braces BD and BE. Knowing that the distance BC is 0.9 m, determine the tension in each brace and the reaction at C.

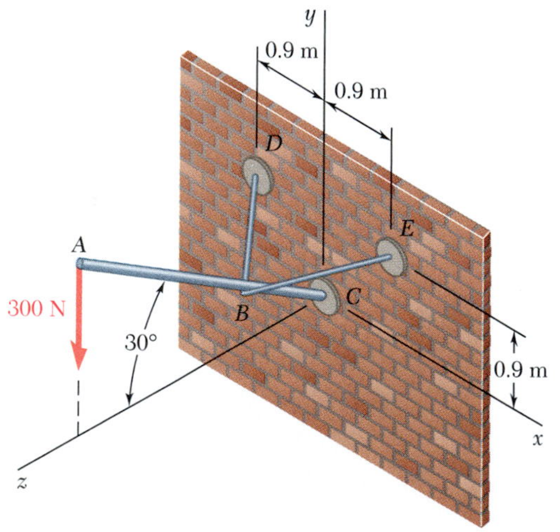

Fig. P4.112

4.113 A 3-m boom is acted upon by the 4-kN force shown. Determine the tension in each cable and the reaction at the ball-and-socket joint at A.

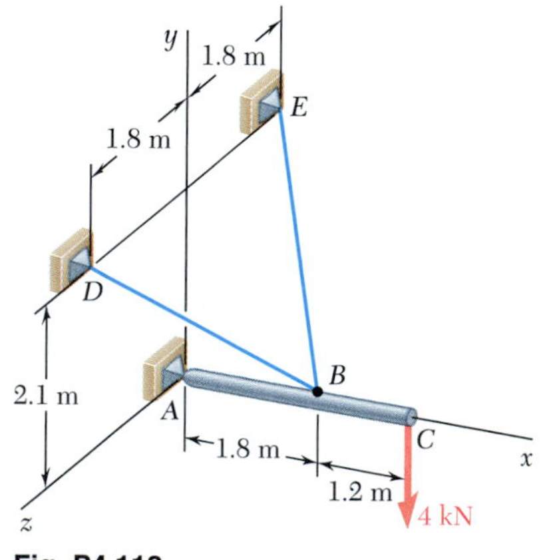

Fig. P4.113

4.114 An 8-ft-long boom is held by a ball-and-socket joint at C and by two cables AD and BE. Determine the tension in each cable and the reaction at C.

4.115 Solve Prob. 4.114 assuming that the given 198-lb load is replaced with two 99-lb loads applied at A and B.

4.116 The 18-ft pole ABC is acted upon by a 210-lb force as shown. The pole is held by a ball-and-socket joint at A and by two cables BD and BE. For $a = 9$ ft, determine the tension in each cable and the reaction at A.

4.117 Solve Prob. 4.116 for $a = 4.5$ ft.

4.118 Two steel pipes $ABCD$ and EBF are welded together at B to form the boom shown. The boom is held by a ball-and-socket joint at D and by two cables EG and $ICFH$; cable $ICFH$ passes around frictionless pulleys at C and F. For the loading shown, determine the tension in each cable and the reaction at D.

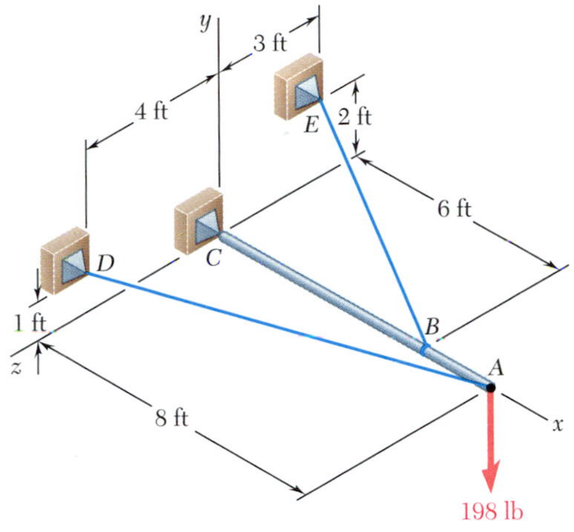

Fig. P4.114

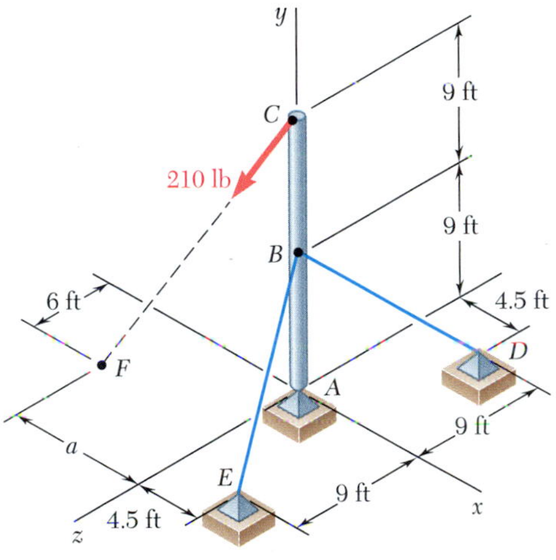

Fig. P4.116

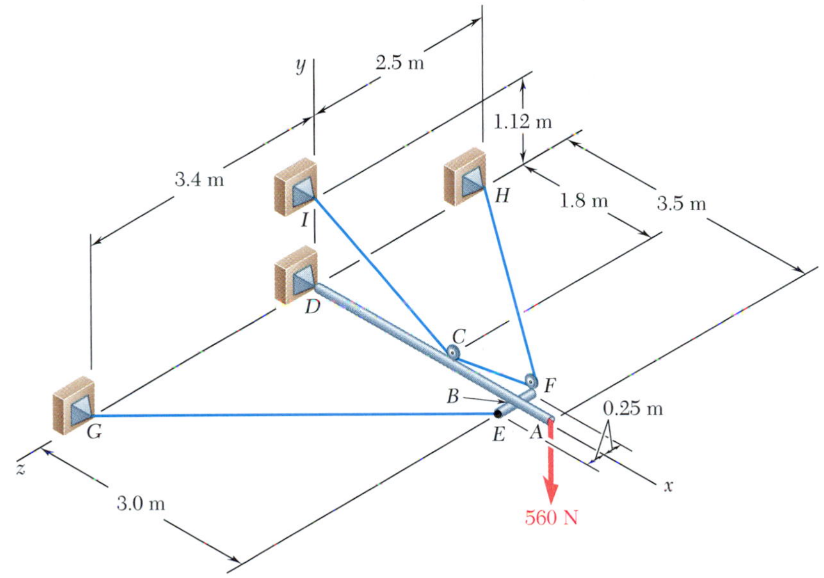

Fig. P4.118

4.119 Solve Prob. 4.118 assuming that the 560-N load is applied at B.

4.120 The lever AB is welded to the bent rod BCD which is supported by bearings at E and F and by cable DG. Knowing that the bearing at E does not exert any axial thrust, determine (*a*) the tension in cable DG, (*b*) the reactions at E and F.

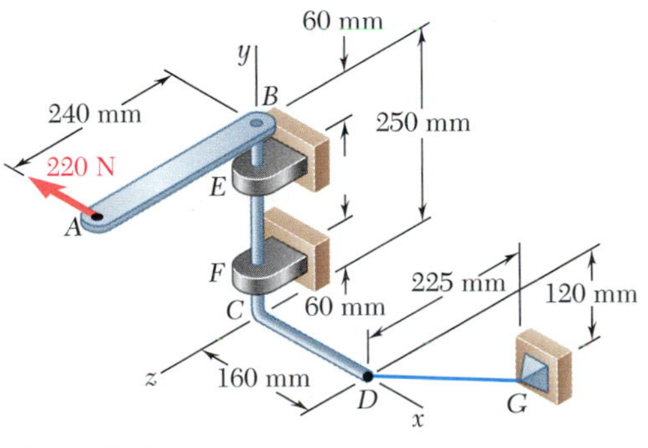

Fig. P4.120

4.121 A 30-kg cover for a roof opening is hinged at corners *A* and *B*. The roof forms an angle of 30° with the horizontal, and the cover is maintained in a horizontal position by the brace *CE*. Determine (*a*) the magnitude of the force exerted by the brace, (*b*) the reactions at the hinges. Assume that the hinge at *A* does not exert any axial thrust.

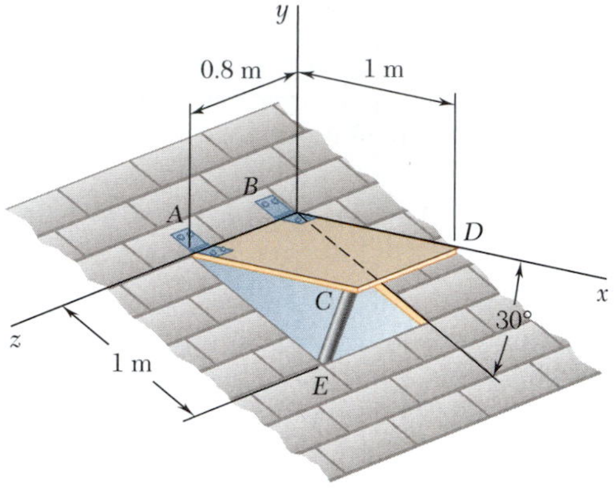

Fig. *P4.121*

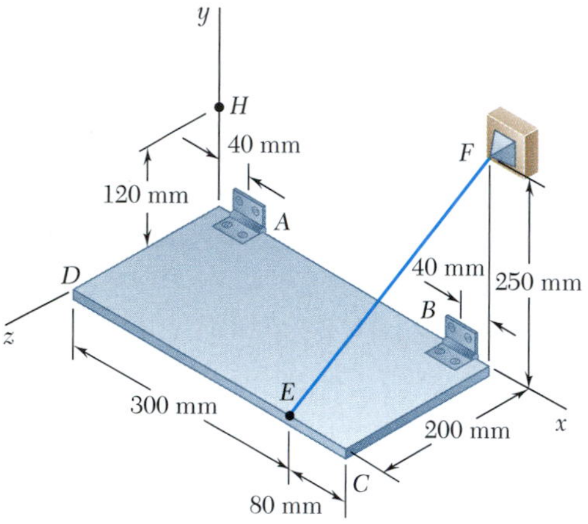

Fig. P4.122

4.122 The rectangular plate shown has a mass of 15 kg and is held in the position shown by hinges *A* and *B* and cable *EF*. Assuming that the hinge at *B* does not exert any axial thrust, determine (*a*) the tension in the cable, (*b*) the reactions at *A* and *B*.

4.123 Solve Prob. 4.122 assuming that cable *EF* is replaced by a cable attached at points *E* and *H*.

4.124 A small door weighing 16 lb is attached by hinges *A* and *B* to a wall and is held in the horizontal position shown by rope *EFH*. The rope passes around a small, frictionless pulley at *F* and is tied to a fixed cleat at *H*. Assuming that the hinge at *A* does not exert any axial thrust, determine (*a*) the tension in the rope, (*b*) the reactions at *A* and *B*.

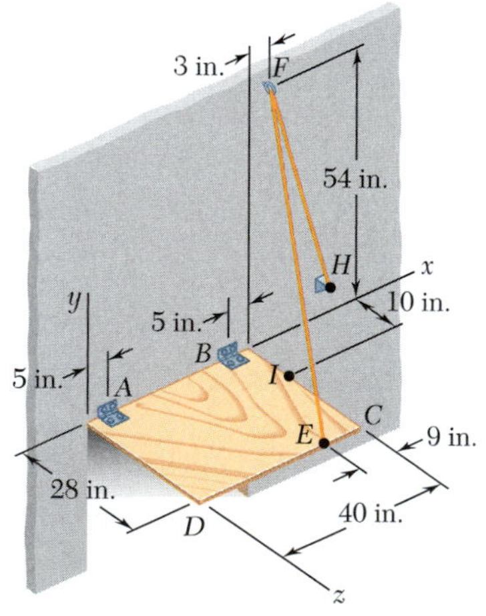

Fig. P4.124

4.125 Solve Prob. 4.124 assuming that the rope is attached to the door at *I*.

4.126 A 285-lb uniform rectangular plate is supported in the position shown by hinges *A* and *B* and by cable *DCE*, which passes over a frictionless hook at *C*. Assuming that the tension is the same in both parts of the cable, determine (*a*) the tension in the cable, (*b*) the reactions at *A* and *B*. Assume that the hinge at *B* does not exert any axial thrust.

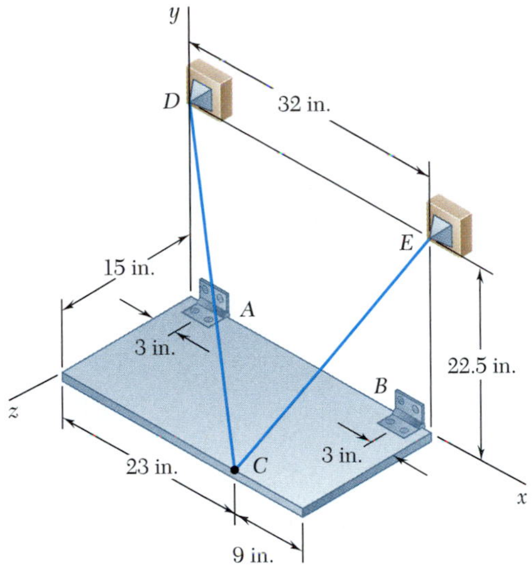

Fig. P4.126

4.127 Solve Prob. 4.126 assuming that cable *DCE* is replaced by a cable attached to point *E* and hook *C*.

4.128 The tensioning mechanism of a belt drive consists of frictionless pulley *A*, mounting plate *B*, and spring *C*. Attached below the mounting plate is slider block *D* which is free to move in the frictionless slot of bracket *E*. Knowing that the pulley and the belt lie in a horizontal plane, with portion *F* of the belt parallel to the *x* axis and portion *G* forming an angle of 30° with the *x* axis, determine (*a*) the force in the spring, (*b*) the reaction at *D*.

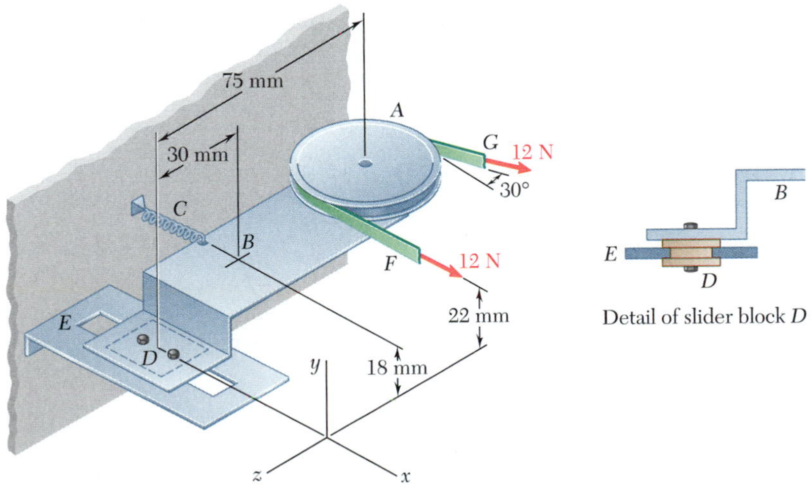

Fig. P4.128

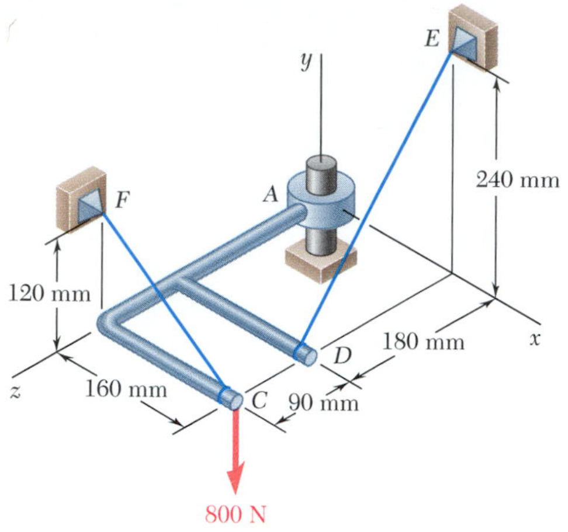

Fig. P4.129

4.129 The assembly shown is welded to collar A which fits on the vertical pin shown. The pin can exert couples about the x and z axes but does not prevent motion about or along the y axis. For the loading shown, determine the tension in each cable and the reaction at A.

4.130 The lever AB is welded to the bent rod BCD which is supported by bearing E and by cable DG. Assuming that the bearing can exert an axial thrust and couples about axes parallel to the x and z axes, determine (a) the tension in cable DG, (b) the reaction at E.

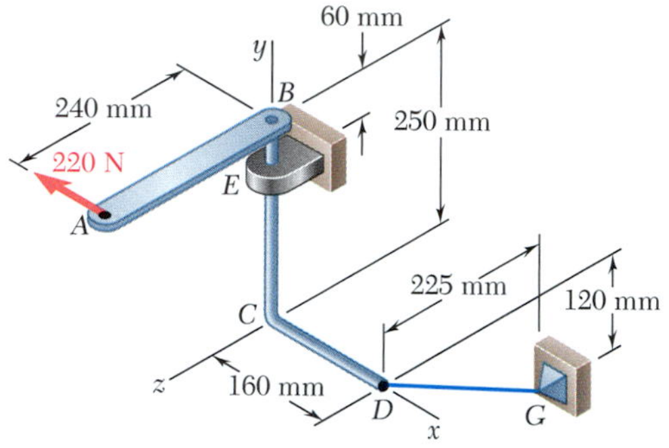

Fig. P4.130

4.131 Solve Prob. 4.124 assuming that the hinge at A is removed and that the hinge at B can exert couples about the y and z axes.

4.132 The frame shown is supported by three cables and a ball-and-socket joint at A. For $\mathbf{P} = 0$, determine the tension in each cable and the reaction at A.

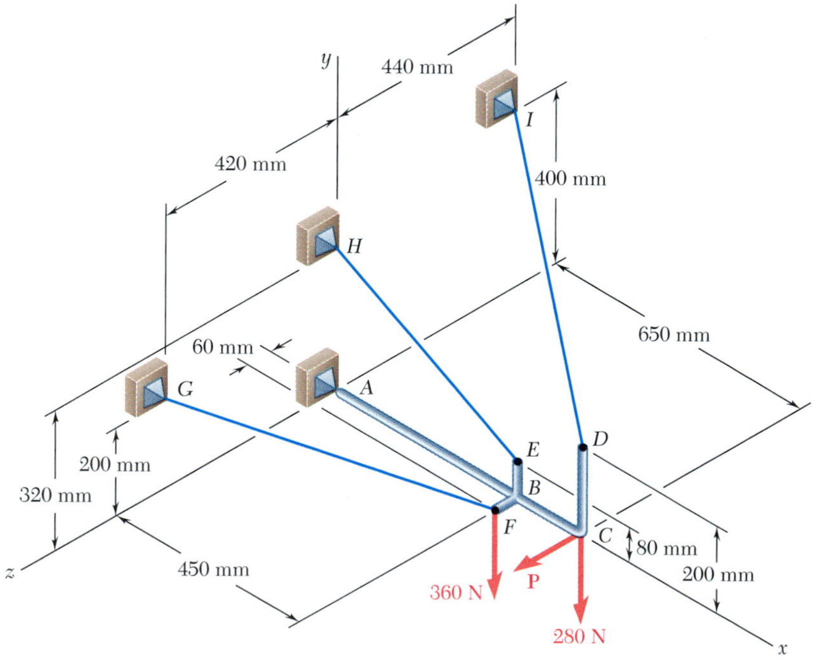

Fig. P4.132 and P4.133

4.133 The frame shown is supported by three cables and a ball-and-socket joint at A. For $P = 50$ N, determine the tension in each cable and the reaction at A.

4.134 The rigid L-shaped member *ABF* is supported by a ball-and-socket joint at *A* and by three cables. For the loading shown, determine the tension in each cable and the reaction at *A*.

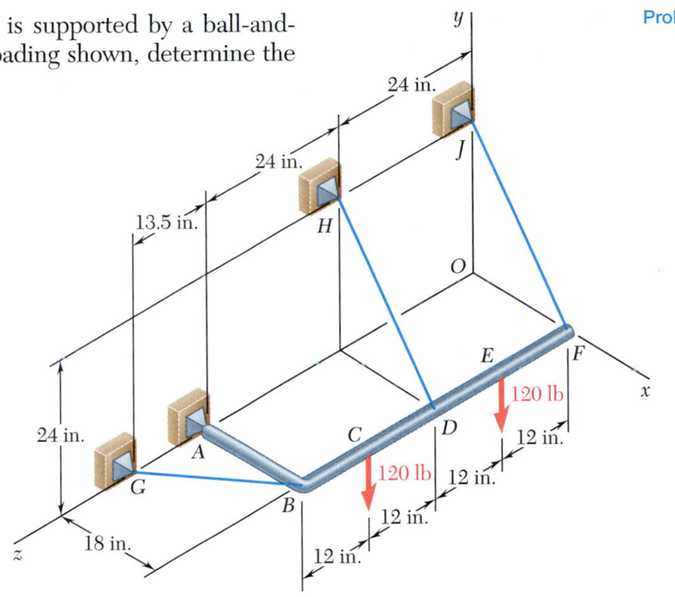

Fig. *P4.134*

4.135 Solve Prob. 4.134 assuming that the load at *C* has been removed.

4.136 In order to clean the clogged drainpipe *AE*, a plumber has disconnected both ends of the pipe and inserted a power snake through the opening at *A*. The cutting head of the snake is connected by a heavy cable to an electric motor which rotates at a constant speed as the plumber forces the cable into the pipe. The forces exerted by the plumber and the motor on the end of the cable can be represented by the wrench $\mathbf{F} = -(60\text{ N})\mathbf{k}$, $\mathbf{M} = -(108\text{ N}\cdot\text{m})\mathbf{k}$. Determine the additional reactions at *B*, *C*, and *D* caused by the cleaning operation. Assume that the reaction at each support consists of two force components perpendicular to the pipe.

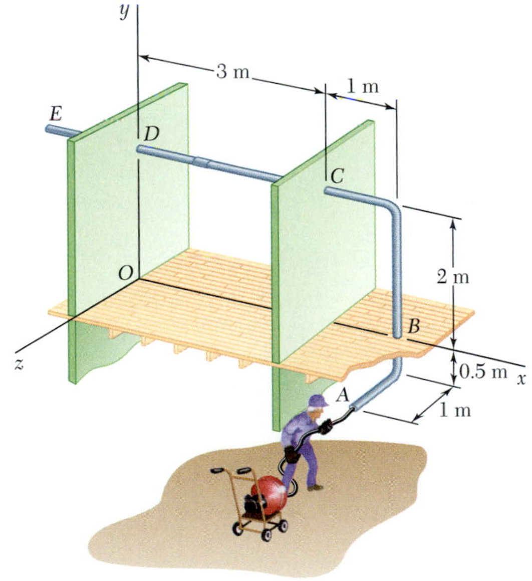

Fig. P4.136

4.137 Solve Prob. 4.136 assuming that the plumber exerts a force $\mathbf{F} = -(60\text{ N})\mathbf{k}$ and that the motor is turned off ($\mathbf{M} = 0$).

4.138 Three rods are welded together to form a "corner" which is supported by three eyebolts. Neglecting friction, determine the reactions at A, B, and C when $P = 240$ N, $a = 120$ mm, $b = 80$ mm, and $c = 100$ mm.

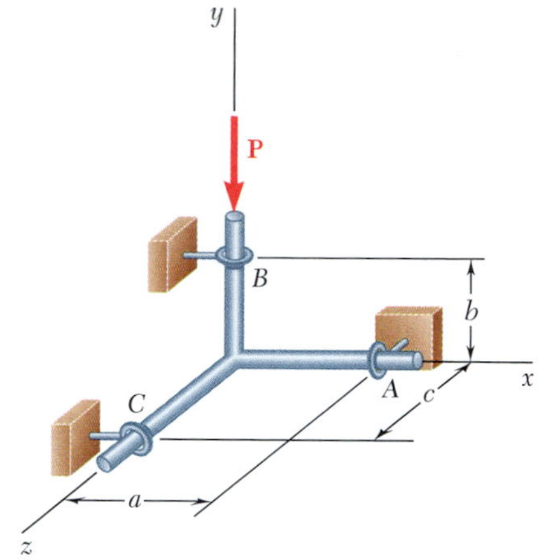

Fig. P4.138

4.139 Solve Prob. 4.138 assuming that the force **P** is removed and is replaced by a couple $\mathbf{M} = +(6 \text{ N} \cdot \text{m})\mathbf{j}$ acting at B.

4.140 The uniform 10-lb rod AB is supported by a ball-and-socket joint at A and leans against both the rod CD and the vertical wall. Neglecting the effects of friction, determine (*a*) the force which rod CD exerts on AB, (*b*) the reactions at A and B. (*Hint:* The force exerted by CD on AB must be perpendicular to both rods.)

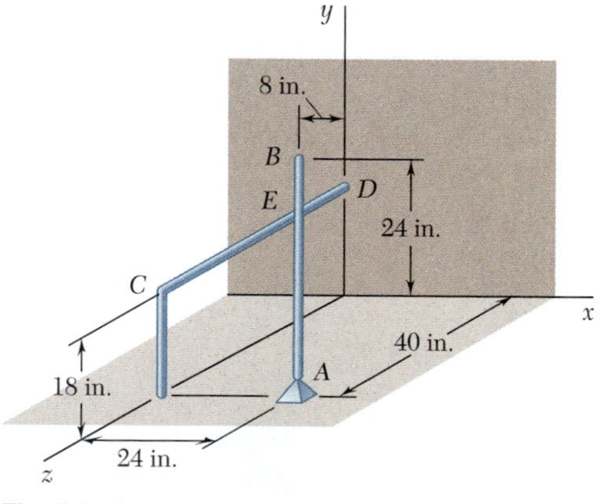

Fig. P4.140

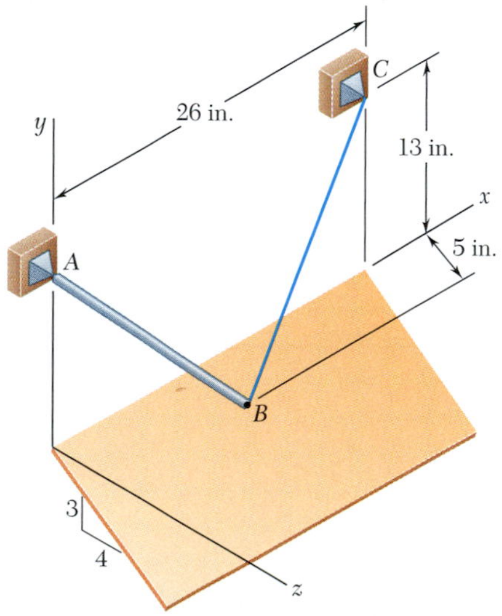

Fig. P4.141

4.141 A 21-in.-long uniform rod AB weighs 6.4 lb and is attached to a ball-and-socket joint at A. The rod rests against an inclined frictionless surface and is held in the position shown by cord BC. Knowing that the cord is 21 in. long, determine (*a*) the tension in the cord, (*b*) the reactions at A and B.

4.142 While being installed, the 56-lb chute *ABCD* is attached to a wall with brackets *E* and *F* and is braced with props *GH* and *IJ*. Assuming that the weight of the chute is uniformly distributed, determine the magnitude of the force exerted on the chute by prop *GH* if prop *IJ* is removed.

4.143 While being installed, the 56-lb chute *ABCD* is attached to a wall with brackets *E* and *F* and is braced with props *GH* and *IJ*. Assuming that the weight of the chute is uniformly distributed, determine the magnitude of the force exerted on the chute by prop *IJ* if prop *GH* is removed.

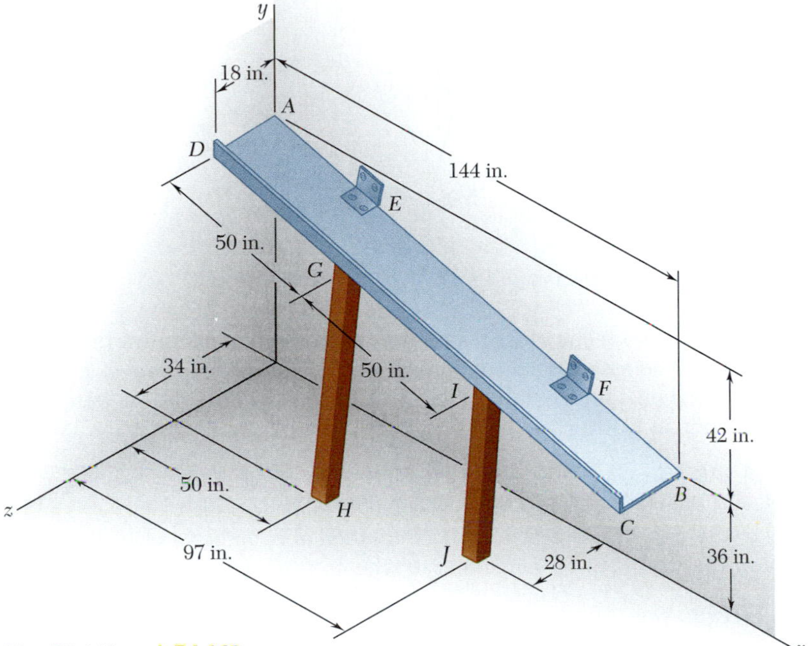

Fig. P4.142 and *P4.143*

4.144 To water seedlings, a gardener joins three lengths of pipe, *AB*, *BC*, and *CD*, fitted with spray nozzles and suspends the assembly using hinged supports at *A* and *D* and cable *EF*. Knowing that the pipe weighs 0.85 lb/ft, determine the tension in the cable.

4.145 Solve Prob. 4.144 assuming that cable *EF* is replaced by a cable connecting *E* and *C*.

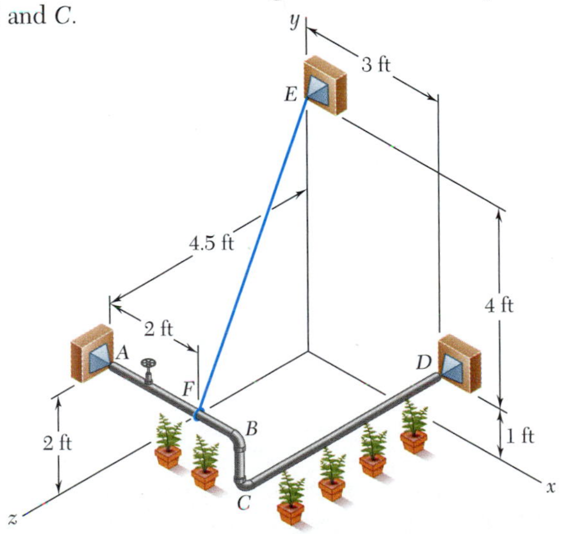

Fig. P4.144

4.146 The bent rod $ABDE$ is supported by ball-and-socket joints at A and E and by the cable DF. If a 600-N load is applied at C as shown, determine the tension in the cable.

4.147 Solve Prob. 4.146 assuming that cable DF is replaced by a cable connecting B and F.

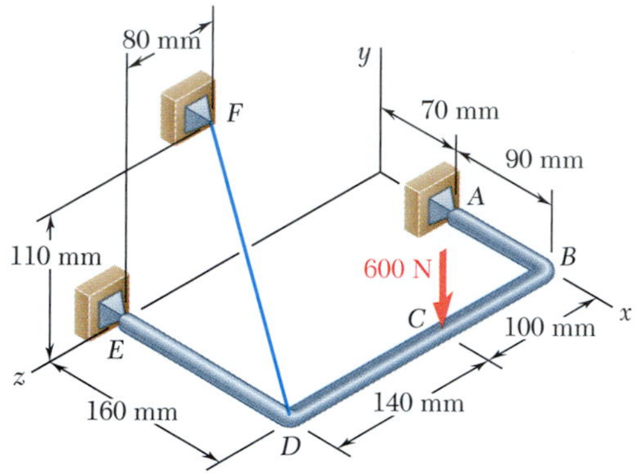

Fig. P4.146

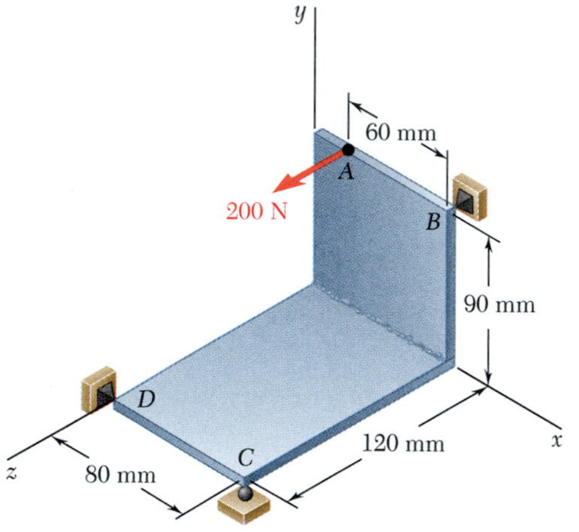

Fig. P4.148

4.148 Two rectangular plates are welded together to form the assembly shown. The assembly is supported by ball-and-socket joints at B and D and by a ball on a horizontal surface at C. For the loading shown, determine the reaction at C.

4.149 Two 1 × 2-m plywood panels, each of mass 15 kg, are nailed together as shown. The panels are supported by ball-and-socket joints at A and F and by the wire BH. Determine (*a*) the location of H in the xy plane if the tension in the wire is to be minimum, (*b*) the corresponding minimum tension.

4.150 Solve Prob. 4.149 subject to the restriction that H must lie on the y axis.

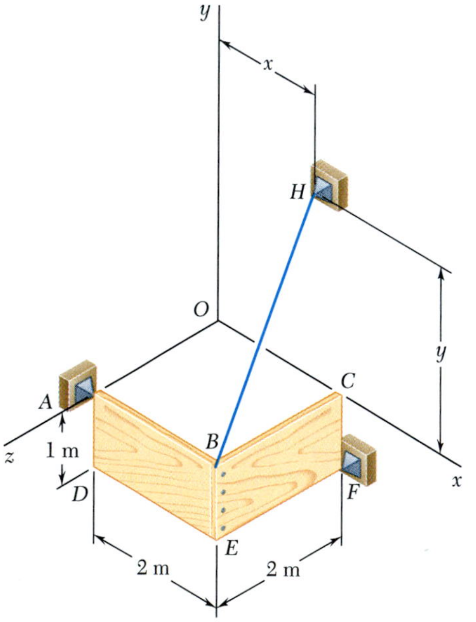

Fig. P4.149

4.151 A uniform 20 × 30-in. steel plate $ABCD$ weighs 85 lb and is attached to ball-and-socket joints at A and B. Knowing that the plate leans against a frictionless vertical wall at D, determine (*a*) the location of D, (*b*) the reaction at D.

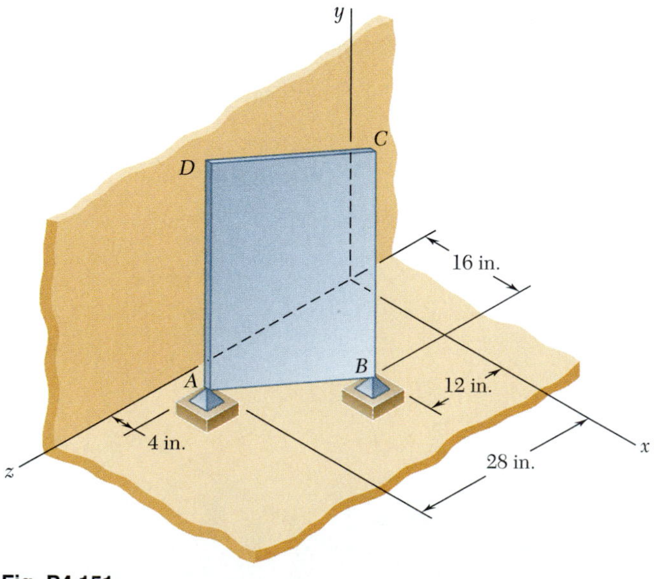

Fig. P4.151

REVIEW AND SUMMARY FOR CHAPTER 4

Equilibrium equations

This chapter was devoted to the study of the *equilibrium of rigid bodies,* that is, to the situation when the external forces acting on a rigid body *form a system equivalent to zero* [Sec. 4.1]. We then have

$$\Sigma \mathbf{F} = 0 \qquad \Sigma \mathbf{M}_O = \Sigma(\mathbf{r} \times \mathbf{F}) = 0 \tag{4.1}$$

Resolving each force and each moment into its rectangular components, we can express the necessary and sufficient conditions for the equilibrium of a rigid body with the following six scalar equations:

$$\Sigma F_x = 0 \qquad \Sigma F_y = 0 \qquad \Sigma F_z = 0 \tag{4.2}$$

$$\Sigma M_x = 0 \qquad \Sigma M_y = 0 \qquad \Sigma M_z = 0 \tag{4.3}$$

These equations can be used to determine unknown forces applied to the rigid body or unknown reactions exerted by its supports.

Free-body diagram

When solving a problem involving the equilibrium of a rigid body, it is essential to consider *all* of the forces acting on the body. Therefore, the first step in the solution of the problem should be to draw a *free-body diagram* showing the body under consideration and all of the unknown as well as known forces acting on it [Sec. 4.2].

Equilibrium of a two-dimensional structure

In the first part of the chapter, we considered the *equilibrium of a two-dimensional structure;* that is, we assumed that the structure considered and the forces applied to it were contained in the same plane. We saw that each of the reactions exerted on the structure by its supports could involve one, two, or three unknowns, depending upon the type of support [Sec. 4.3].

In the case of a two-dimensional structure, Eqs. (4.1), or Eqs. (4.2) and (4.3), reduce to *three equilibrium equations,* namely

$$\Sigma F_x = 0 \qquad \Sigma F_y = 0 \qquad \Sigma M_A = 0 \tag{4.5}$$

where A is an arbitrary point in the plane of the structure [Sec. 4.4]. These equations can be used to solve for three unknowns. While the three equilibrium equations (4.5) cannot be *augmented* with additional equations, any of them can be *replaced* by another equation. Therefore, we can write alternative sets of equilibrium equations, such as

$$\Sigma F_x = 0 \qquad \Sigma M_A = 0 \qquad \Sigma M_B = 0 \tag{4.6}$$

where point B is chosen in such a way that the line AB is not parallel to the y axis, or

$$\Sigma M_A = 0 \qquad \Sigma M_B = 0 \qquad \Sigma M_C = 0 \tag{4.7}$$

where the points A, B, and C do not lie in a straight line.

Statical indeterminacy

Partial constraints

Improper constraints

Since any set of equilibrium equations can be solved for only three unknowns, the reactions at the supports of a rigid two-dimensional structure cannot be completely determined if they involve *more than three unknowns;* they are said to be *statically indeterminate* [Sec. 4.5]. On the other hand, if the reactions involve *fewer than three unknowns,* equilibrium will not be maintained under general loading conditions; the structure is said to be *partially constrained.* The fact that the reactions involve exactly three unknowns is no guarantee that the equilibrium equations can be solved for all three unknowns. If the supports are arranged in such a way that the reactions are *either concurrent or parallel,* the reactions are statically indeterminate, and the structure is said to be *improperly constrained.*

Two-force body

Three-force body

Two particular cases of equilibrium of a rigid body were given special attention. In Sec. 4.6, a *two-force body* was defined as a rigid body subjected to forces at only two points, and it was shown that the resultants $\mathbf{F}_1$ and $\mathbf{F}_2$ of these forces must have the *same magnitude, the same line of action, and opposite sense* (Fig. 4.11), a property which will simplify the solution of certain problems in later chapters. In Sec. 4.7, a *three-force body* was defined as a rigid body subjected to forces at only three points, and it was shown that the resultants $\mathbf{F}_1$, $\mathbf{F}_2$, and $\mathbf{F}_3$ of these forces must be *either concurrent* (Fig. 4.12) *or parallel.* This properly provides us with an alternative approach to the solution of problems involving a three-force body [Sample Prob. 4.6].

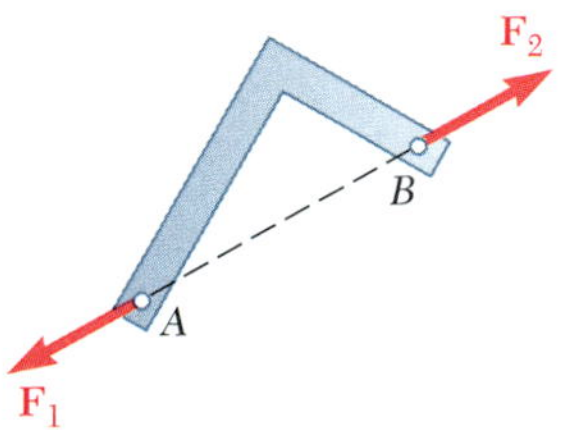

Fig. 4.11

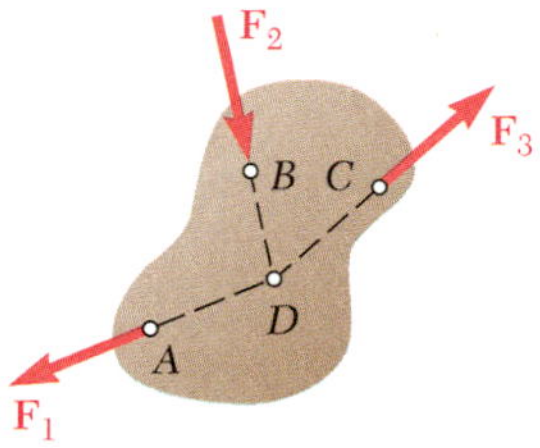

Fig. 4.12

Equilibrium of a three-dimensional body

In the second part of the chapter, we considered the *equilibrium of a three-dimensional body* and saw that each of the reactions exerted on the body by its supports could involve between one and six unknowns, depending upon the type of support [Sec. 4.8].

In the general case of the equilibrium of a three-dimensional body, all of the six scalar equilibrium equations (4.2) and (4.3) listed at the beginning of this review should be used and solved for *six unknowns* [Sec. 4.9]. In most problems, however, these equations will be more conveniently obtained if we first write

$$\Sigma \mathbf{F} = 0 \qquad \Sigma \mathbf{M}_O = \Sigma(\mathbf{r} \times \mathbf{F}) = 0 \tag{4.1}$$

and express the forces $\mathbf{F}$ and position vectors $\mathbf{r}$ in terms of scalar components and unit vectors. The vector products can then be computed either directly or by means of determinants, and the desired scalar equations obtained by equating to zero the coefficients of the unit vectors [Sample Probs. 4.7 through 4.9].

We noted that as many as three unknown reaction components may be eliminated from the computation of $\Sigma \mathbf{M}_O$ in the second of the relations (4.1) through a judicious choice of point O. Also, the reactions at two points A and B can be eliminated from the solution of some problems by writing the equation $\Sigma M_{AB} = 0$, which involves the computation of the moments of the forces about an axis AB joining points A and B [Sample Prob. 4.10].

If the reactions involve more than six unknowns, some of the reactions are *statically indeterminate;* if they involve fewer than six unknowns, the rigid body is only *partially constrained.* Even with six or more unknowns, the rigid body will be *improperly constrained* if the reactions associated with the given supports either are parallel or intersect the same line.

Review Problems

4.152 Beam AD carries the two 40-lb loads shown. The beam is held by a fixed support at D and by the cable BE which is attached to the counterweight W. Determine the reaction at D when (*a*) $W = 100$ lb, (*b*) $W = 90$ lb.

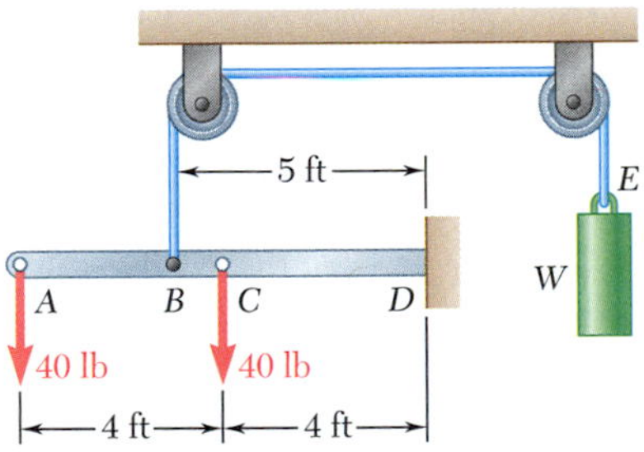

Fig. P4.152 and *P4.153*

4.153 For the beam and loading shown, determine the range of values of W for which the magnitude of the couple at D does not exceed 40 lb · ft.

4.154 Determine the reactions at A and D when $\beta = 30°$.

4.155 Determine the reactions at A and D when $\beta = 60°$.

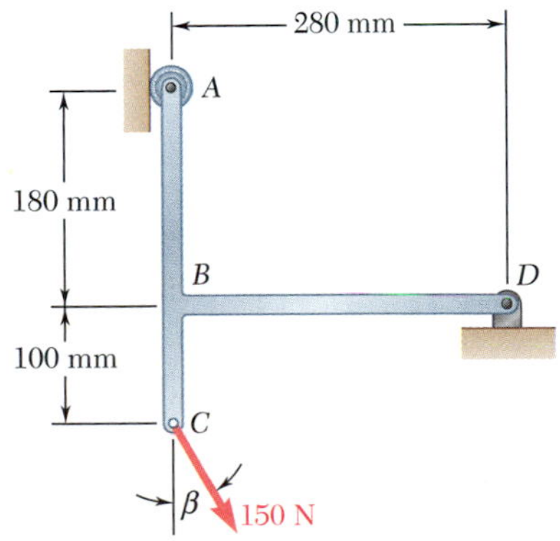

Fig. P4.154 and *P4.155*

4.156 A 2100-lb tractor is used to lift 900 lb of gravel. Determine the reaction at each of the two (*a*) rear wheels *A*, (*b*) front wheels *B*.

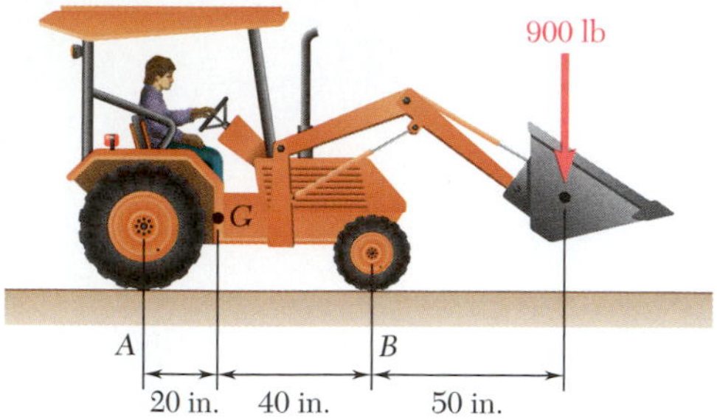

Fig. P4.156

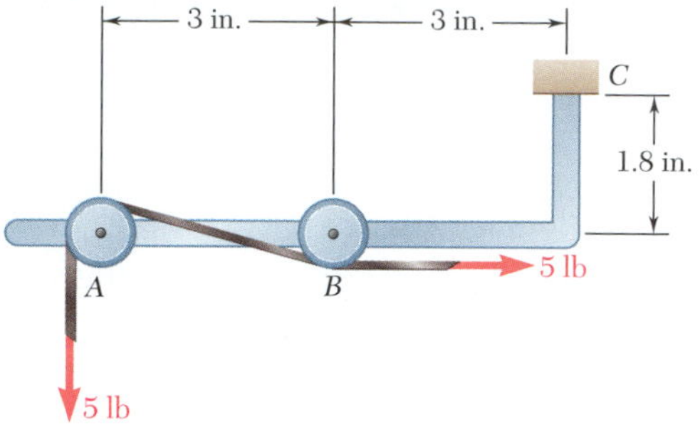

Fig. P4.157

4.157 A tension of 5 lb is maintained in a tape as it passes the support system shown. Knowing that the radius of each pulley is 0.4 in., determine the reaction at *C*.

4.158 Solve Prob. 4.157 assuming that 0.6-in.-radius pulleys are used.

4.159 The bent rod *ABEF* is supported by bearings at *C* and *D* and by wire *AH*. Knowing that portion *AB* of the rod is 250 mm long, determine (*a*) the tension in wire *AH*, (*b*) the reactions at *C* and *D*. Assume that the bearing at *D* does not exert any axial thrust.

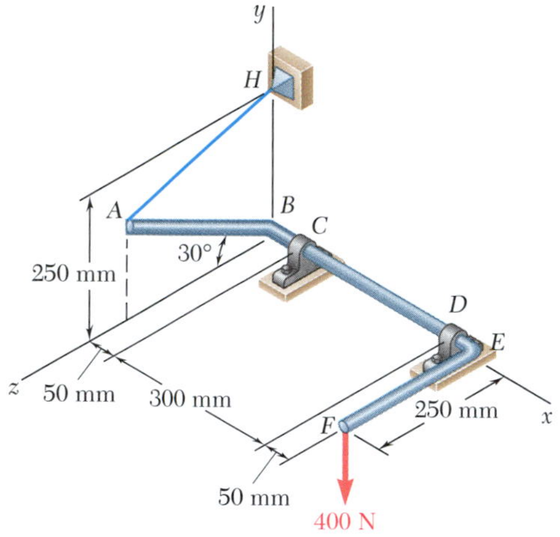

Fig. P4.159

4.160 For the beam and loading shown, determine (*a*) the reaction at *A*, (*b*) the tension in cable *BC*.

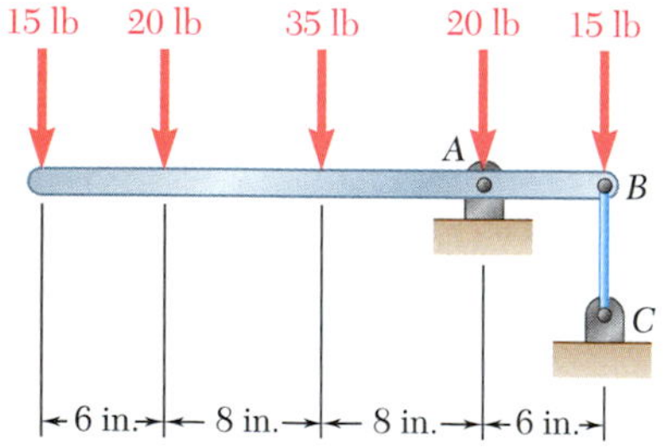

Fig. *P4.160*

4.161 Frame *ABCD* is supported by a ball-and-socket joint at *A* and by three cables. For $a = 150$ mm, determine the tension in each cable and the reaction at *A*.

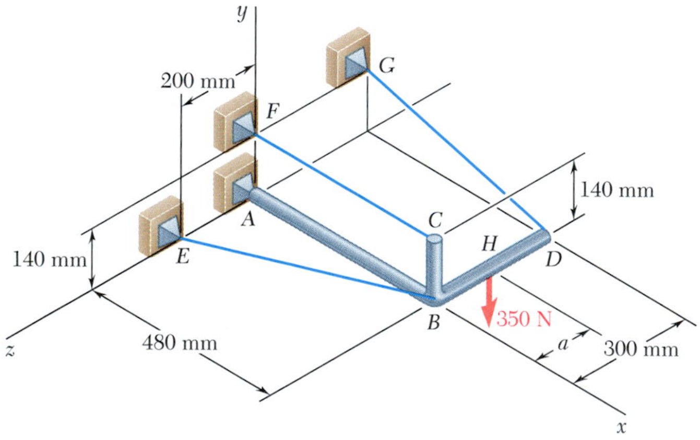

Fig. P4.161 and *P4.162*

4.162 Frame *ABCD* is supported by a ball-and-socket joint at *A* and by three cables. Knowing that the 350-N load is applied at *D* ($a = 300$ mm), determine the tension in each cable and the reaction at *A*.

***4.163** In the problems listed below, the rigid bodies considered were completely constrained and the reactions were statically determinate. For each of these rigid bodies it is possible to create an improper set of constraints by changing a dimension of the body. In each of the following problems determine the value of *a* which results in improper constraints. (*a*) Prob. 4.81, (*b*) Prob. 4.82.

Computer Problems

4.C1 A slender rod AB of weight W is attached to blocks at A and B which can move freely in the guides shown. The constant of the spring is k, and the spring is unstretched when the rod is horizontal. Neglecting the weight of the blocks, derive an equation in terms of θ, W, l, and k which must be satisfied when the rod is in equilibrium. Knowing that $W = 10$ lb and $l = 40$ in., (*a*) calculate and plot the value of the spring constant k as a function of the angle θ for $15° \leq \theta \leq 40°$, (*b*) determine the two values of the angle θ corresponding to equilibrium when $k = 0.7$ lb/in.

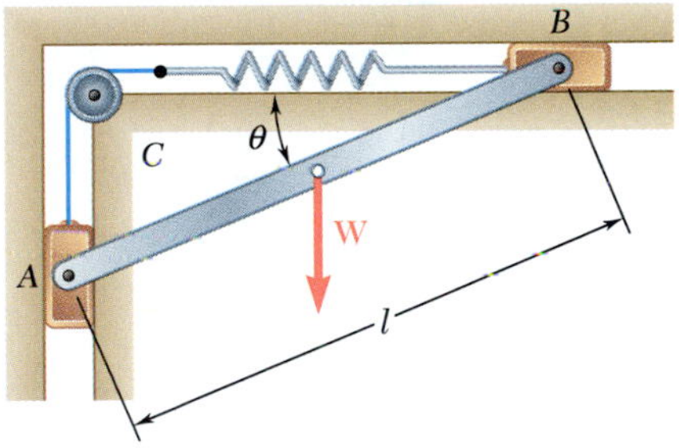

Fig. P4.C1

4.C2 The position of the L-shaped rod shown is controlled by a cable attached at point B. Knowing that the rod supports a load of magnitude $P = 200$ N, use computational software to calculate and plot the tension T in the cable as a function of θ for values of θ from 0 to 120°. Determine the maximum tension T_{max} and the corresponding value of θ.

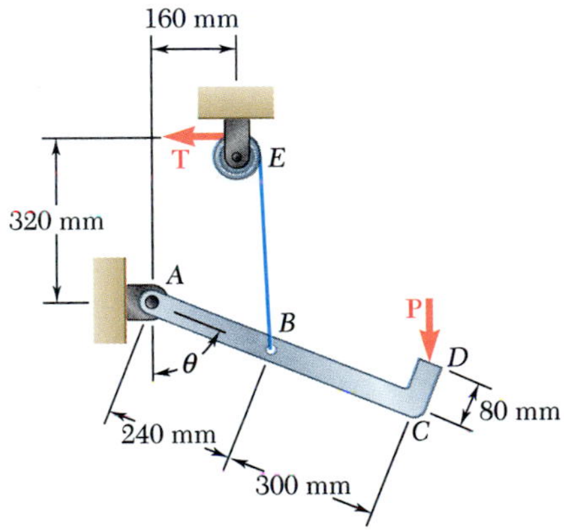

Fig. P4.C2

4.C3 The position of the 20-lb rod AB is controlled by the block shown, which is slowly moved to the left by the force **P**. Neglecting the effect of friction, use computational software to calculate and plot the magnitude P of the force as a function of x for values of x decreasing from 30 in. to 0. Determine the maximum value of P and the corresponding value of x.

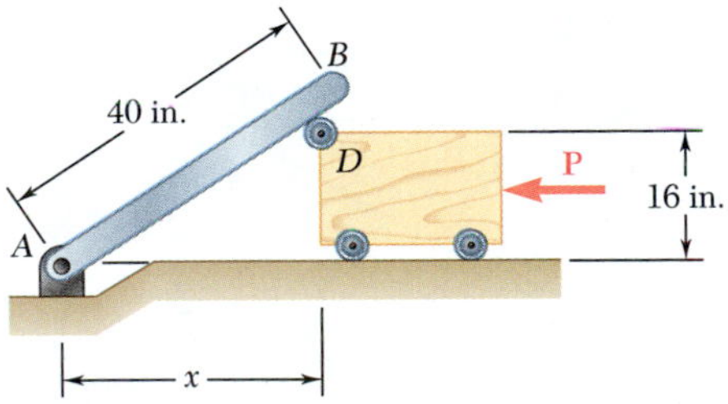

Fig. P4.C3

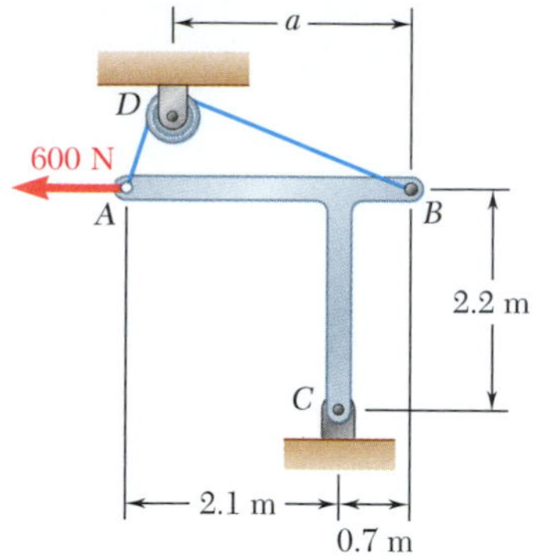

Fig. P4.C4

***4.C4** Member ABC is supported by a pin and bracket at C and by an inextensible cable of length 3.5 m that is attached at A and B and passes over a frictionless pulley at D. Neglecting the mass of ABC and the radius of the pulley, (a) plot the tension in the cable as a function of a for $0 \le a \le 2.4$ m, (b) determine the largest value of a for which equilibrium can be maintained.

4.C5 and 4.C6 The constant of spring AB is k, and the spring is unstretched when $\theta = 0$. Knowing that $R = 200$ mm, $a = 400$ mm, and $k = 1$ kN/m, use computational software to calculate and plot the mass m corresponding to equilibrium as a function of θ for values of θ from 0 to 90°. Determine the value of θ corresponding to equilibrium when $m = 2$ kg.

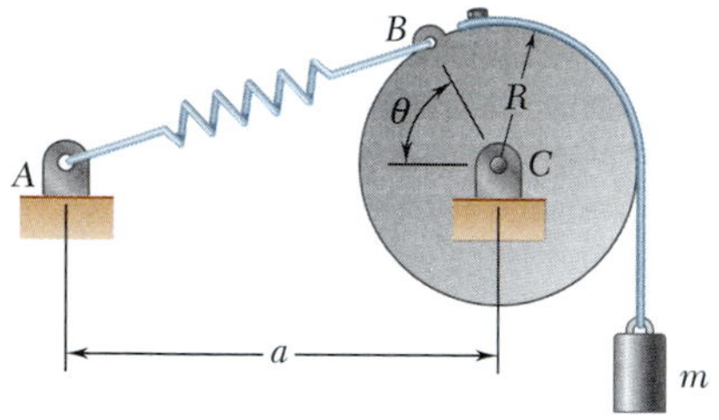

Fig. P4.C5

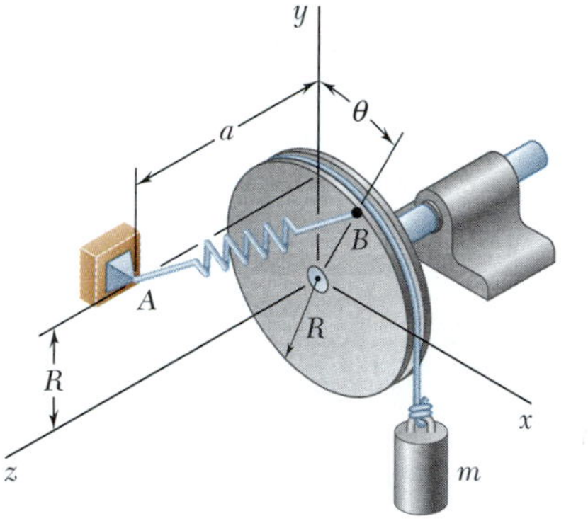

Fig. P4.C6

4.C7 An 8 × 10-in. panel of weight $W = 40$ lb is supported by hinges along edge AB. Cable CDE is attached to the panel at point C, passes over a small pulley at D, and supports a cylinder of weight W. Neglecting the effect of friction, use computational software to calculate and plot the weight of the cylinder corresponding to equilibrium as a function of θ for values of θ from 0 to 90°. Determine the value of θ corresponding to equilibrium when $W = 20$ lb.

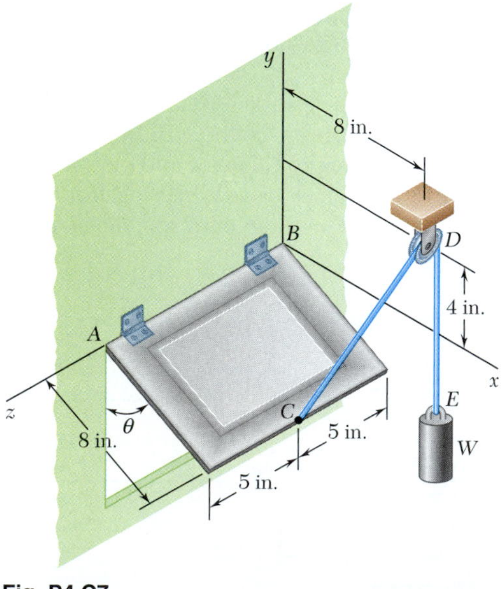

Fig. P4.C7

4.C8 A uniform circular plate of radius 300 mm and mass 26 kg is supported by three vertical wires that are equally spaced around its edge. A small 3-kg block E is placed on the plate at D and is then slowly moved along diameter CD until it reaches C. (*a*) Plot the tension in wires A and C as functions of a, where a is the distance of the block from D. (*b*) Determine the value of a for which the tension in wires A and C is minimum.

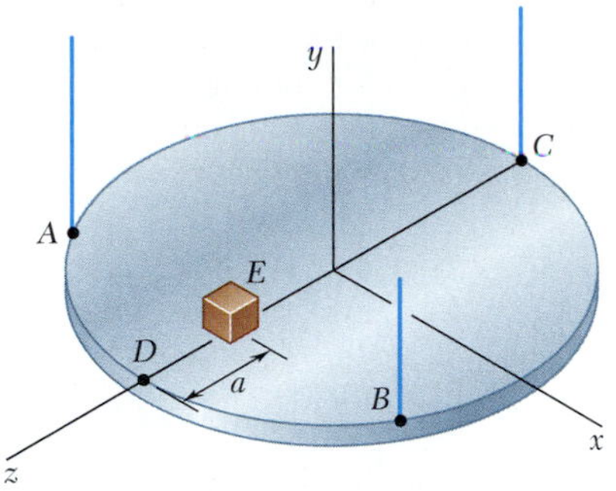

Fig. P4.C8

4.C9 The derrick shown supports a 4000-lb crate. It is held by a ball-and-socket joint at point A and by two cables attached at points D and E. Knowing that the derrick lies in a vertical plane forming an angle ϕ with the xy plane, use computational software to calculate and plot the tension in each cable as a function of ϕ for values of ϕ from 0 to 40°. Determine the value of ϕ for which the tension in cable BE is maximum.

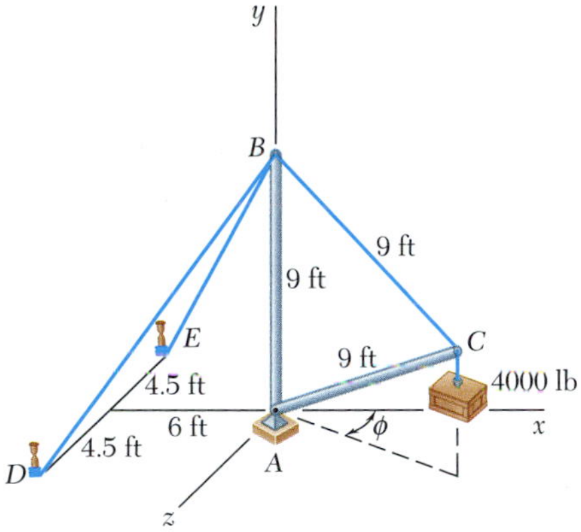

Fig. P4.C9

4.C10 The 140-lb uniform steel plate $ABCD$ is welded to shaft EF and is maintained in the position shown by the couple **M**. Knowing that collars prevent the shaft from sliding in the bearings and that the shaft lies in the yz plane, plot the magnitude M of the couple as a function of θ for $0 \leq \theta \leq 90°$.

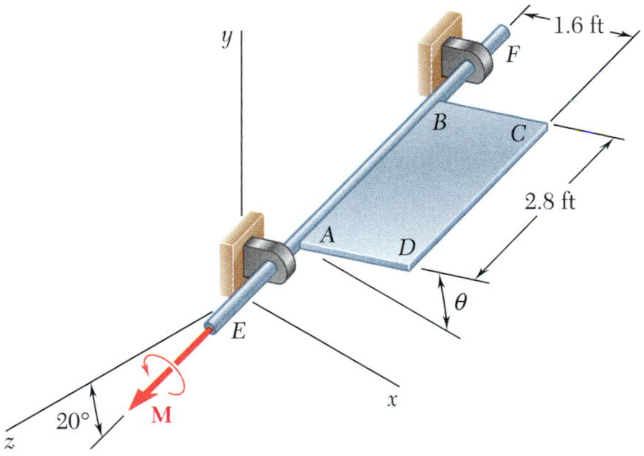

Fig. P4.C10

CHAPTER

Distributed Forces: Centroids and Centers of Gravity

5

A precast section of roadway for a new interchange on Interstate 93 is shown being lowered from a gantry crane. In this chapter we will introduce the concept of the centroid of an area; in subsequent courses the relation between the location of the centroid and the behavior of the roadway under loading will be established.

DISTRIBUTED FORCES: CENTROIDS AND CENTERS OF GRAVITY

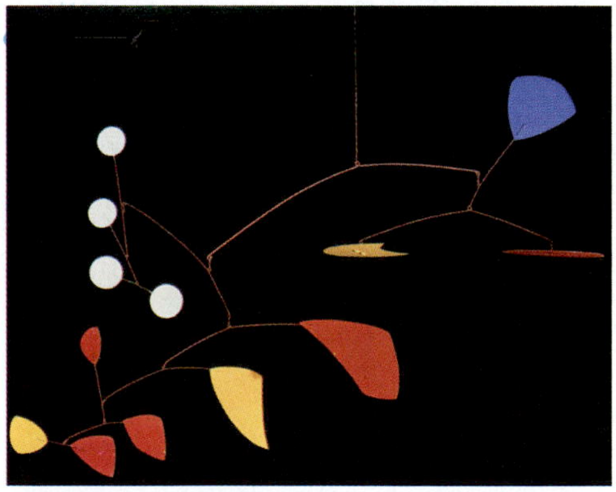

Photo 5.1 The precise balancing of the components of a mobile requires an understanding of centers of gravity and centroids, the main topics of this chapter.

5.1. INTRODUCTION

We have assumed so far that the attraction exerted by the earth on a rigid body could be represented by a single force **W**. This force, called the force of gravity or the weight of the body, was to be applied at the *center of gravity* of the body (Sec. 3.2). Actually, the earth exerts a force on each of the particles forming the body. The action of the earth on a rigid body should thus be represented by a large number of small forces distributed over the entire body. You will learn in this chapter, however, that all of these small forces can be replaced by a single equivalent force **W**. You will also learn how to determine the center of gravity, that is, the point of application of the resultant **W**, for bodies of various shapes.

In the first part of the chapter, two-dimensional bodies, such as flat plates and wires contained in a given plane, are considered. Two concepts closely associated with the determination of the center of gravity of a plate or a wire are introduced: the concept of the *centroid* of an area or a line and the concept of the *first moment* of an area or a line with respect to a given axis.

You will also learn that the computation of the area of a surface of revolution or of the volume of a body of revolution is directly related to the determination of the centroid of the line or area used to generate that surface or body of revolution (Theorems of Pappus-Guldinus). And, as is shown in Secs. 5.8 and 5.9, the determination of the centroid of an area simplifies the analysis of beams subjected to distributed loads and the computation of the forces exerted on submerged rectangular surfaces, such as hydraulic gates and portions of dams.

In the last part of the chapter, you will learn how to determine the center of gravity of a three-dimensional body as well as the centroid of a volume and the first moments of that volume with respect to the coordinate planes.

AREAS AND LINES

5.2. CENTER OF GRAVITY OF A TWO-DIMENSIONAL BODY

Let us first consider a flat horizontal plate (Fig. 5.1). We can divide the plate into n small elements. The coordinates of the first element

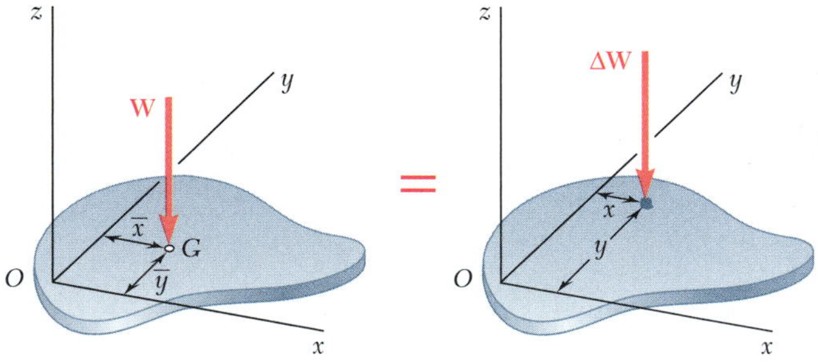

$$\Sigma M_y: \quad \bar{x}\,W = \Sigma x\,\Delta W$$

$$\Sigma M_x: \quad \bar{y}\,W = \Sigma y\,\Delta W$$

Fig. 5.1 Center of gravity of a plate.

are denoted by x_1 and y_1, those of the second element by x_2 and y_2, etc. The forces exerted by the earth on the elements of plate will be denoted, respectively, by $\Delta\mathbf{W}_1$, $\Delta\mathbf{W}_2$, . . . , $\Delta\mathbf{W}_n$. These forces or weights are directed toward the center of the earth; however, for all practical purposes they can be assumed to be parallel. Their resultant is therefore a single force in the same direction. The magnitude W of this force is obtained by adding the magnitudes of the elemental weights:

$$\Sigma F_z: \qquad W = \Delta W_1 + \Delta W_2 + \cdots + \Delta W_n$$

To obtain the coordinates $\bar{x}$ and $\bar{y}$ of the point G where the resultant **W** should be applied, we write that the moments of **W** about the y and x axes are equal to the sum of the corresponding moments of the elemental weights,

$$\begin{aligned} \Sigma M_y: \qquad \bar{x}W &= x_1\,\Delta W_1 + x_2\,\Delta W_2 + \cdots + x_n\,\Delta W_n \\ \Sigma M_x: \qquad \bar{y}W &= y_1\,\Delta W_1 + y_2\,\Delta W_2 + \cdots + y_n\,\Delta W_n \end{aligned} \tag{5.1}$$

If we now increase the number of elements into which the plate is divided and simultaneously decrease the size of each element, we obtain in the limit the following expressions:

$$W = \int dW \qquad \bar{x}W = \int x\,dW \qquad \bar{y}W = \int y\,dW \tag{5.2}$$

These equations define the weight **W** and the coordinates $\bar{x}$ and $\bar{y}$ of the center of gravity G of a flat plate. The same equations can be derived for a wire lying in the xy plane (Fig. 5.2). We note that the center of gravity G of a wire is usually not located on the wire.

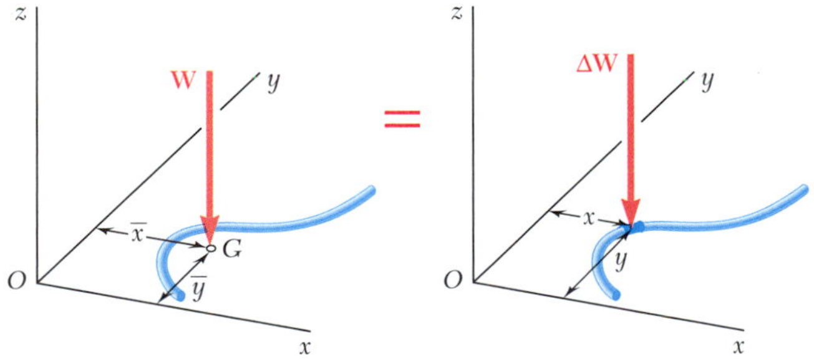

$$\Sigma M_y: \quad \bar{x}W = \Sigma x\,\Delta W$$
$$\Sigma M_x: \quad \bar{y}W = \Sigma y\,\Delta W$$

Fig. 5.2 Center of gravity of a wire.

5.3. CENTROIDS OF AREAS AND LINES

In the case of a flat homogeneous plate of uniform thickness, the magnitude ΔW of the weight of an element of the plate can be expressed as

$$\Delta W = \gamma t \, \Delta A$$

where γ = specific weight (weight per unit volume) of the material
t = thickness of the plate
ΔA = area of the element

Similarly, we can express the magnitude W of the weight of the entire plate as

$$W = \gamma t A$$

where A is the total area of the plate.

If U.S. customary units are used, the specific weight γ should be expressed in lb/ft^3, the thickness t in feet, and the areas ΔA and A in square feet. We observe that ΔW and W will then be expressed in pounds. If SI units are used, γ should be expressed in N/m^3, t in meters, and the areas ΔA and A in square meters; the weights ΔW and W will then be expressed in newtons.†

Substituting for ΔW and W in the moment equations (5.1) and dividing throughout by γt, we obtain

$$\Sigma M_y: \quad \bar{x}A = x_1 \, \Delta A_1 + x_2 \, \Delta A_2 + \cdots + x_n \, \Delta A_n$$
$$\Sigma M_x: \quad \bar{y}A = y_1 \, \Delta A_1 + y_2 \, \Delta A_2 + \cdots + y_n \, \Delta A_n$$

If we increase the number of elements into which the area A is divided and simultaneously decrease the size of each element, we obtain in the limit

$$\bar{x}A = \int x \, dA \qquad \bar{y}A = \int y \, dA \tag{5.3}$$

These equations define the coordinates $\bar{x}$ and $\bar{y}$ of the center of gravity of a homogeneous plate. The point whose coordinates are $\bar{x}$ and $\bar{y}$ is also known as the *centroid C of the area A* of the plate (Fig. 5.3). If the plate is not homogeneous, these equations cannot be used to determine the center of gravity of the plate; they still define, however, the centroid of the area.

In the case of a homogeneous wire of uniform cross section, the magnitude ΔW of the weight of an element of wire can be expressed as

$$\Delta W = \gamma a \, \Delta L$$

where γ = specific weight of the material
a = cross-sectional area of the wire
ΔL = length of the element

†It should be noted that in the SI system of units a given material is generally characterized by its density ρ (mass per unit volume) rather than by its specific weight γ. The specific weight of the material can then be obtained from the relation

$$\gamma = \rho g$$

where $g = 9.81$ m/s^2. Since ρ is expressed in kg/m^3, we observe that γ will be expressed in (kg/m^3)(m/s^2), that is, in N/m^3.

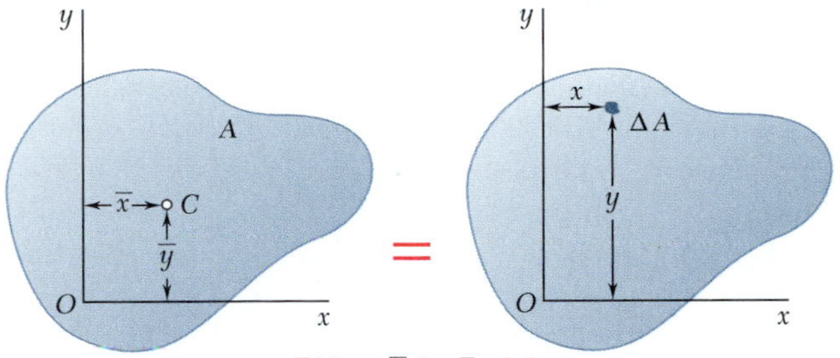

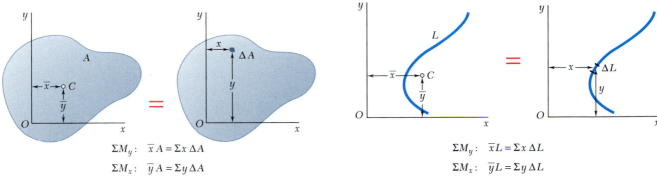

$\Sigma M_y: \quad \bar{x}A = \Sigma x\,\Delta A$

$\Sigma M_x: \quad \bar{y}A = \Sigma y\,\Delta A$

Fig. 5.3 Centroid of an area.

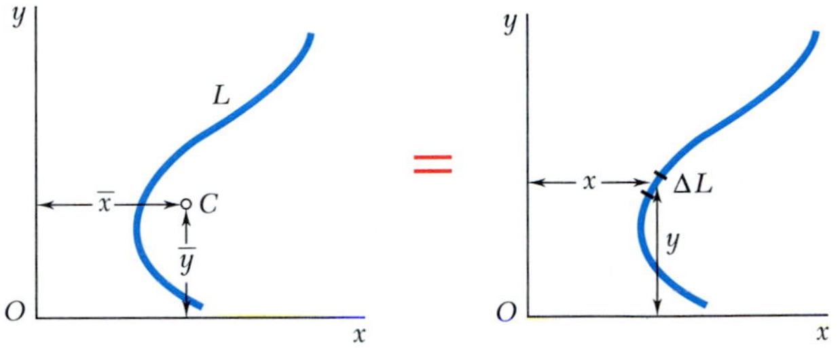

$\Sigma M_y: \quad \bar{x}L = \Sigma x\,\Delta L$

$\Sigma M_x: \quad \bar{y}L = \Sigma y\,\Delta L$

Fig. 5.4 Centroid of a line.

The center of gravity of the wire then coincides with the *centroid C of the line L* defining the shape of the wire (Fig. 5.4). The coordinates $\bar{x}$ and $\bar{y}$ of the centroid of the line L are obtained from the equations

$$\bar{x}L = \int x\,dL \qquad \bar{y}L = \int y\,dL \tag{5.4}$$

5.4. FIRST MOMENTS OF AREAS AND LINES

The integral $\int x\,dA$ in Eqs. (5.3) of the preceding section is known as the *first moment of the area A with respect to the y axis* and is denoted by Q_y. Similarly, the integral $\int y\,dA$ defines the *first moment of A with respect to the x axis* and is denoted by Q_x. We write

$$Q_y = \int x\,dA \qquad Q_x = \int y\,dA \tag{5.5}$$

Comparing Eqs. (5.3) with Eqs. (5.5), we note that the first moments of the area A can be expressed as the products of the area and the coordinates of its centroid:

$$Q_y = \bar{x}A \qquad Q_x = \bar{y}A \tag{5.6}$$

It follows from Eqs. (5.6) that the coordinates of the centroid of an area can be obtained by dividing the first moments of that area by the area itself. The first moments of the area are also useful in mechanics of materials for determining the shearing stresses in beams under transverse loadings. Finally, we observe from Eqs. (5.6) that if the centroid of an area is located on a coordinate axis, the first moment of the area with respect to that axis is zero. Conversely, if the first moment of an area with respect to a coordinate axis is zero, then the centroid of the area is located on that axis.

Relations similar to Eqs. (5.5) and (5.6) can be used to define the first moments of a line with respect to the coordinate axes and to express these moments as the products of the length L of the line and the coordinates $\bar{x}$ and $\bar{y}$ of its centroid.

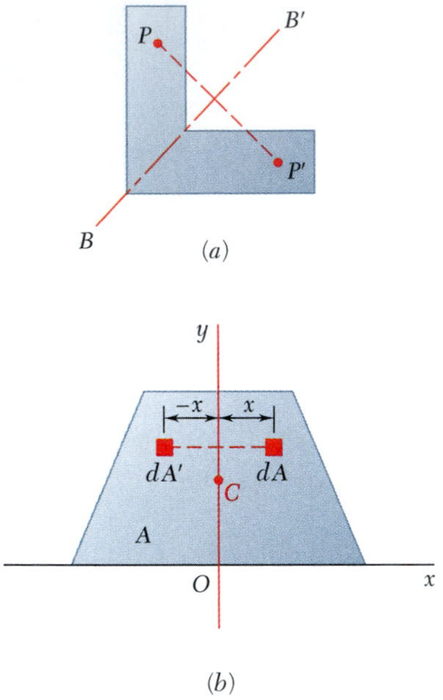

Fig. 5.5

An area A is said to be *symmetric with respect to an axis BB'* if for every point P of the area there exists a point P' of the same area such that the line PP' is perpendicular to BB' and is divided into two equal parts by that axis (Fig. 5.5a). A line L is said to be symmetric with respect to an axis BB' if it satisfies similar conditions. When an area A or a line L possesses an axis of symmetry BB', its first moment with respect to BB' is zero, and its centroid is located on that axis. For example, in the case of the area A of Fig. 5.5b, which is symmetric with respect to the y axis, we observe that for every element of area dA of abscissa x there exists an element dA' of equal area and with abscissa $-x$. It follows that the integral in the first of Eqs. (5.5) is zero and, thus, that $Q_y = 0$. It also follows from the first of the relations (5.3) that $\bar{x} = 0$. Thus, if an area A or a line L possesses an axis of symmetry, its centroid C is located on that axis.

We further note that if an area or line possesses two axes of symmetry, its centroid C must be located at the intersection of the two axes (Fig. 5.6). This property enables us to determine immediately the centroid of areas such as circles, ellipses, squares, rectangles, equilateral triangles, or other symmetric figures as well as the centroid of lines in the shape of the circumference of a circle, the perimeter of a square, etc.

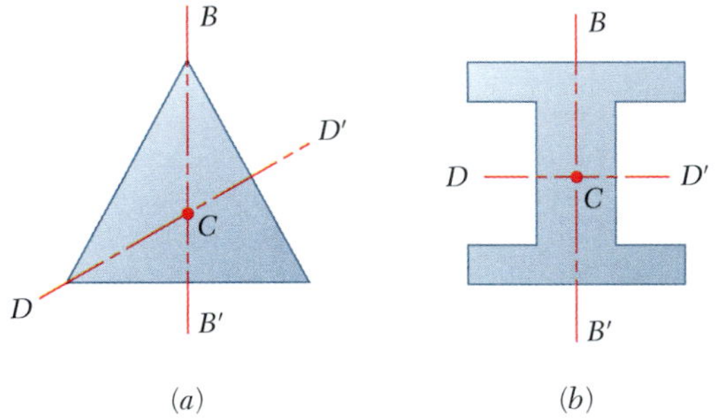

Fig. 5.6

An area A is said to be *symmetric with respect to a center O* if for every element of area dA of coordinates x and y there exists an element dA' of equal area with coordinates $-x$ and $-y$ (Fig. 5.7). It then follows that the integrals in Eqs. (5.5) are both zero and that $Q_x = Q_y = 0$. It also follows from Eqs. (5.3) that $\bar{x} = \bar{y} = 0$, that is, that the centroid of the area coincides with its center of symmetry O. Similarly, if a line possesses a center of symmetry O, the centroid of the line will coincide with the center O.

It should be noted that a figure possessing a center of symmetry does not necessarily possess an axis of symmetry (Fig. 5.7), while a figure possessing two axes of symmetry does not necessarily possess a center of symmetry (Fig. 5.6a). However, if a figure possesses two axes of symmetry at a right angle to each other, the point of intersection of these axes is a center of symmetry (Fig. 5.6b).

Determining the centroids of unsymmetrical areas and lines and of areas and lines possessing only one axis of symmetry will be discussed in Secs. 5.6 and 5.7. Centroids of common shapes of areas and lines are shown in Fig. 5.8A and B.

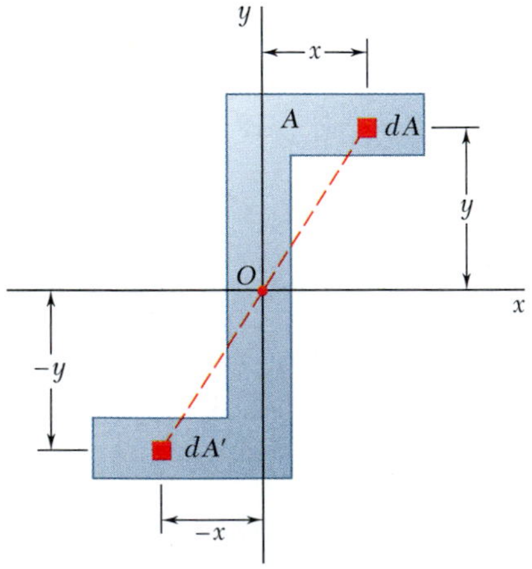

Fig. 5.7

Shape		$\bar{x}$	$\bar{y}$	Area
Triangular area	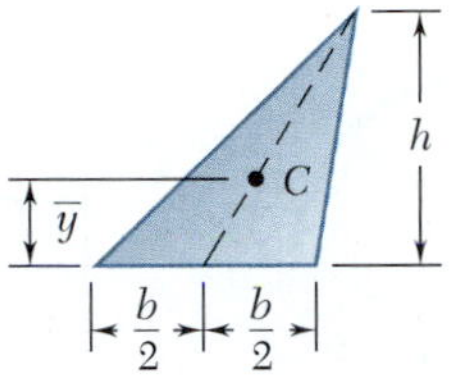		$\frac{h}{3}$	$\frac{bh}{2}$
Quarter-circular area	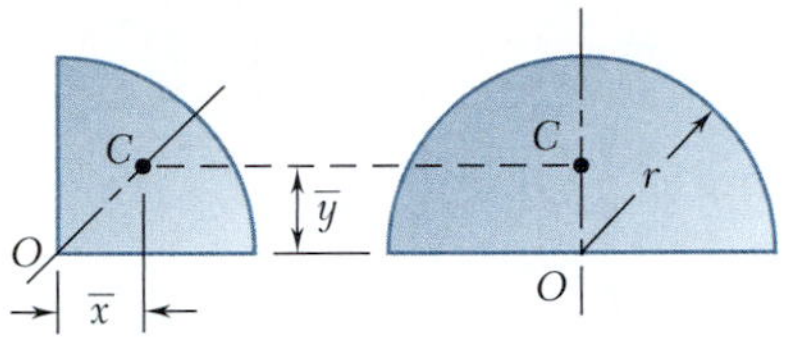	$\frac{4r}{3\pi}$	$\frac{4r}{3\pi}$	$\frac{\pi r^2}{4}$
Semicircular area		0	$\frac{4r}{3\pi}$	$\frac{\pi r^2}{2}$
Quarter-elliptical area	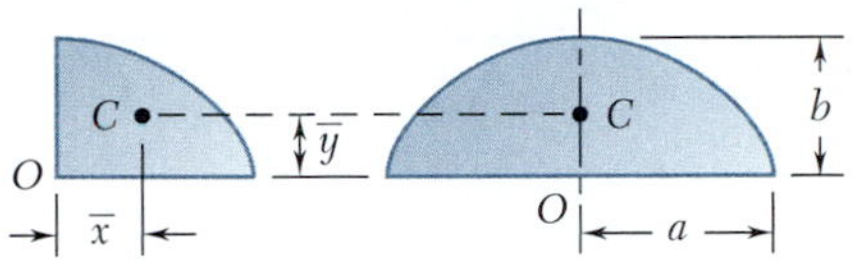	$\frac{4a}{3\pi}$	$\frac{4b}{3\pi}$	$\frac{\pi ab}{4}$
Semielliptical area		0	$\frac{4b}{3\pi}$	$\frac{\pi ab}{2}$
Semiparabolic area	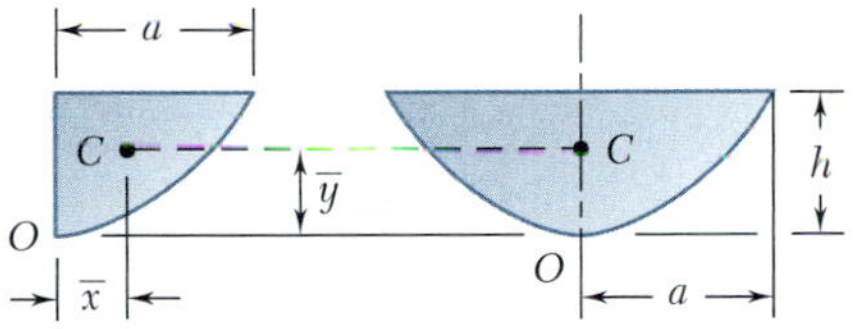	$\frac{3a}{8}$	$\frac{3h}{5}$	$\frac{2ah}{3}$
Parabolic area		0	$\frac{3h}{5}$	$\frac{4ah}{3}$
Parabolic spandrel	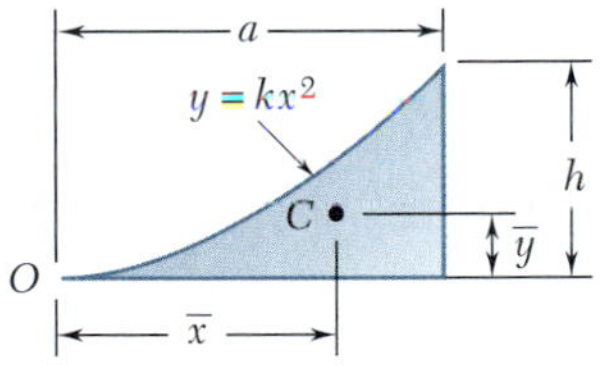	$\frac{3a}{4}$	$\frac{3h}{10}$	$\frac{ah}{3}$
General spandrel	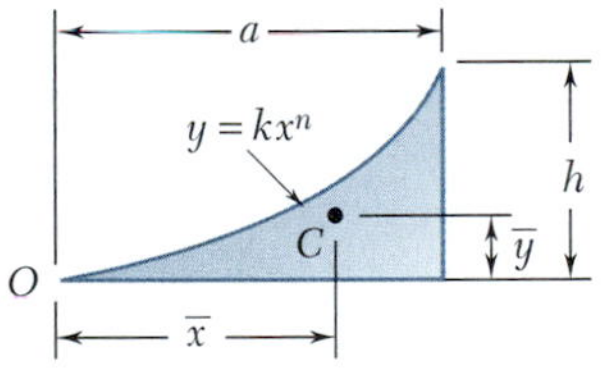	$\frac{n+1}{n+2}a$	$\frac{n+1}{4n+2}h$	$\frac{ah}{n+1}$
Circular sector	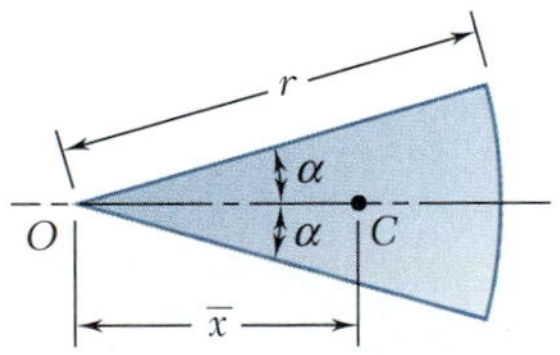	$\frac{2r \sin \alpha}{3\alpha}$	0	αr^2

Fig. 5.8A Centroids of common shapes of areas.

Shape		$\bar{x}$	$\bar{y}$	Length
Quarter-circular arc	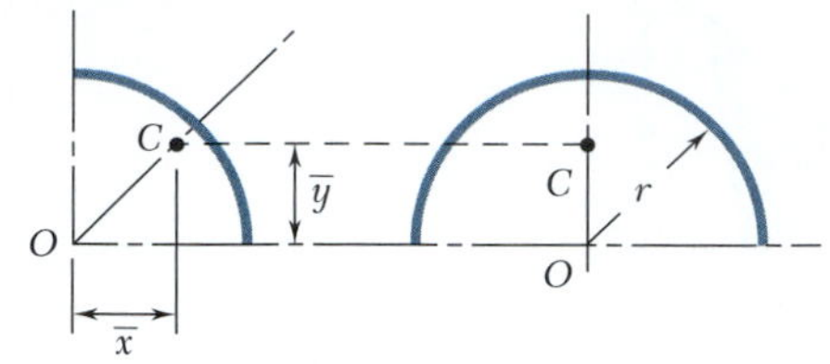	$\frac{2r}{\pi}$	$\frac{2r}{\pi}$	$\frac{\pi r}{2}$
Semicircular arc		0	$\frac{2r}{\pi}$	πr
Arc of circle	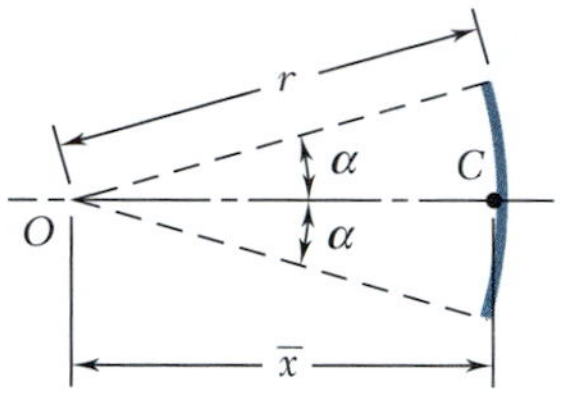	$\frac{r \sin \alpha}{\alpha}$	0	$2\alpha r$

Fig. 5.8B Centroids of common shapes of lines.

5.5. COMPOSITE PLATES AND WIRES

In many instances, a flat plate can be divided into rectangles, triangles, or the other common shapes shown in Fig. 5.8A. The abscissa $\bar{X}$ of its center of gravity G can be determined from the abscissas $\bar{x}_1, \bar{x}_2, \ldots, \bar{x}_n$ of the centers of gravity of the various parts by expressing that the moment of the weight of the whole plate about the y axis is equal to the sum of the moments of the weights of the various parts about the same axis (Fig. 5.9). The ordinate $\bar{Y}$ of the center of gravity of the plate is found in a similar way by equating moments about the x axis. We write

$$\Sigma M_y: \quad \bar{X}(W_1 + W_2 + \cdots + W_n) = \bar{x}_1 W_1 + \bar{x}_2 W_2 + \cdots + \bar{x}_n W_n$$

$$\Sigma M_x: \quad \bar{Y}(W_1 + W_2 + \cdots + W_n) = \bar{y}_1 W_1 + \bar{y}_2 W_2 + \cdots + \bar{y}_n W_n$$

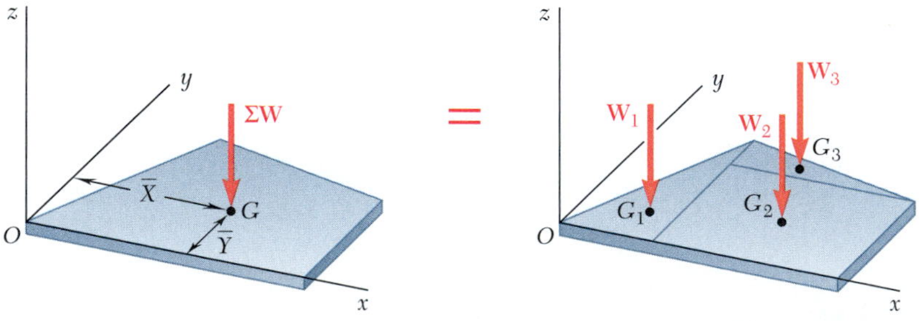

$$\Sigma M_y: \quad \bar{X}\,\Sigma W = \Sigma \bar{x} W$$

$$\Sigma M_x: \quad \bar{Y}\,\Sigma W = \Sigma \bar{y} W$$

Fig. 5.9 Center of gravity of a composite plate.

or, for short,

$$\bar{X}\Sigma W = \Sigma \bar{x} W \qquad \bar{Y}\Sigma W = \Sigma \bar{y} W \tag{5.7}$$

These equations can be solved for the coordinates $\bar{X}$ and $\bar{Y}$ of the center of gravity of the plate.

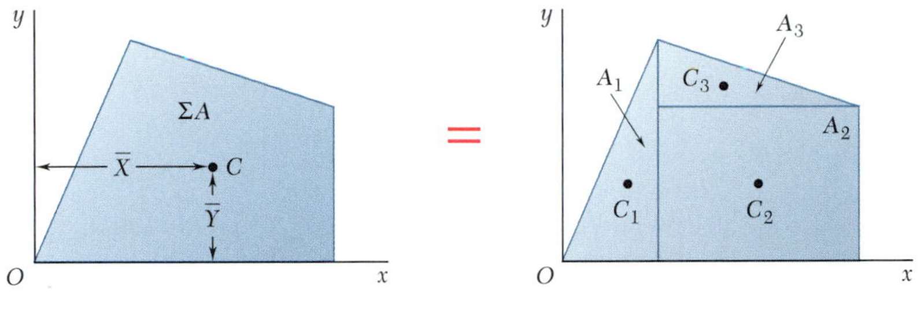

$$Q_y = \bar{X}\Sigma A = \Sigma \bar{x} A$$
$$Q_x = \bar{Y}\Sigma A = \Sigma \bar{y} A$$

Fig. 5.10 Centroid of a composite area.

If the plate is homogeneous and of uniform thickness, the center of gravity coincides with the centroid C of its area. The abscissa $\bar{X}$ of the centroid of the area can be determined by noting that the first moment Q_y of the composite area with respect to the y axis can be expressed both as the product of $\bar{X}$ and the total area and as the sum of the first moments of the elementary areas with respect to the y axis (Fig. 5.10). The ordinate $\bar{Y}$ of the centroid is found in a similar way by considering the first moment Q_x of the composite area. We have

$$Q_y = \bar{X}(A_1 + A_2 + \cdots + A_n) = \bar{x}_1 A_1 + \bar{x}_2 A_2 + \cdots + \bar{x}_n A_n$$
$$Q_x = \bar{Y}(A_1 + A_2 + \cdots + A_n) = \bar{y}_1 A_1 + \bar{y}_2 A_2 + \cdots + \bar{y}_n A_n$$

or, for short,

$$Q_y = \bar{X}\Sigma A = \Sigma \bar{x} A \qquad Q_x = \bar{Y}\Sigma A = \Sigma \bar{y} A \tag{5.8}$$

These equations yield the first moments of the composite area, or they can be used to obtain the coordinates $\bar{X}$ and $\bar{Y}$ of its centroid.

Care should be taken to assign the appropriate sign to the moment of each area. First moments of areas, like moments of forces, can be positive or negative. For example, an area whose centroid is located to the left of the y axis will have a negative first moment with respect to that axis. Also, the area of a hole should be assigned a negative sign (Fig. 5.11).

Similarly, it is possible in many cases to determine the center of gravity of a composite wire or the centroid of a composite line by dividing the wire or line into simpler elements (see Sample Prob. 5.2).

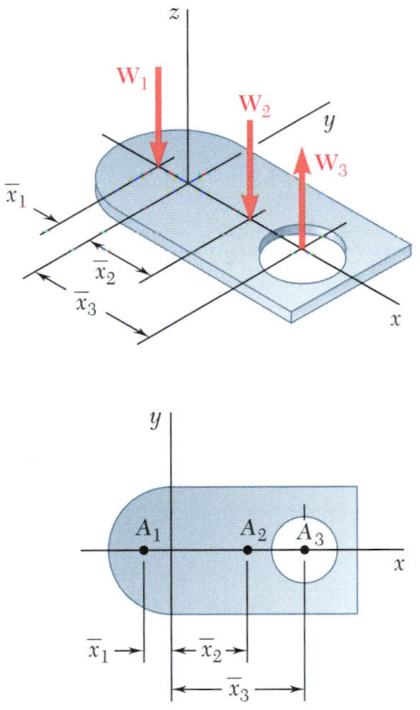

	$\bar{x}$	A	$\bar{x}A$
A_1 Semicircle	−	+	−
A_2 Full rectangle	+	+	+
A_3 Circular hole	+	−	−

Fig. 5.11

SAMPLE PROBLEM 5.1

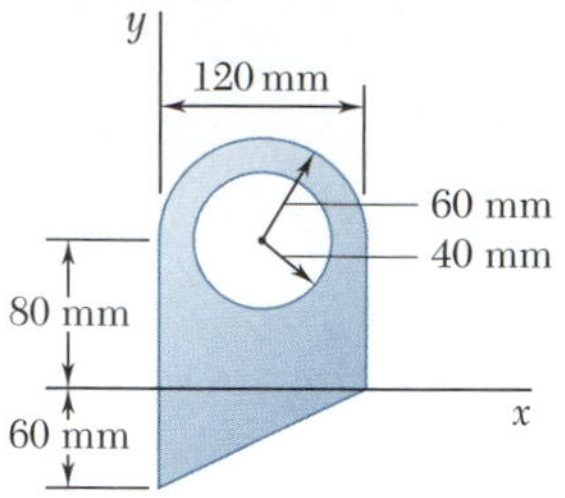

For the plane area shown, determine (*a*) the first moments with respect to the x and y axes, (*b*) the location of the centroid.

SOLUTION

Components of Area. The area is obtained by adding a rectangle, a triangle, and a semicircle and by then subtracting a circle. Using the coordinate axes shown, the area and the coordinates of the centroid of each of the component areas are determined and entered in the table below. The area of the circle is indicated as negative, since it is to be subtracted from the other areas. We note that the coordinate $\bar{y}$ of the centroid of the triangle is negative for the axes shown. The first moments of the component areas with respect to the coordinate axes are computed and entered in the table.

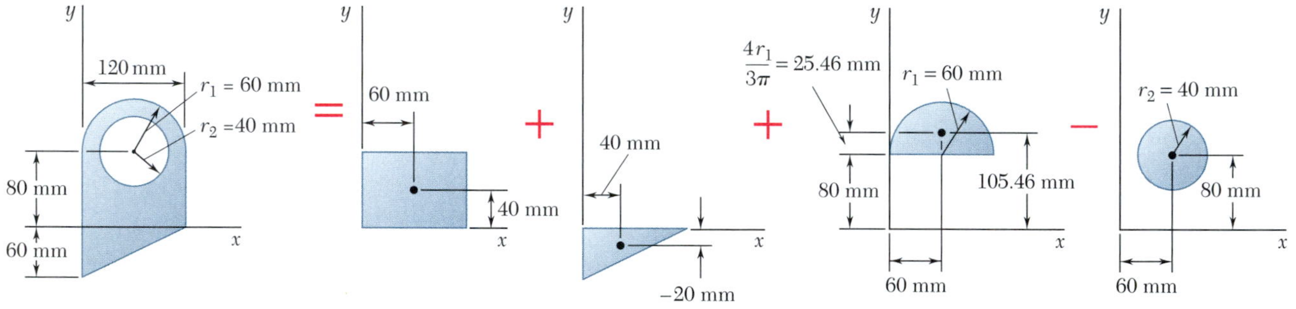

Component	A, mm²	$\bar{x}$, mm	$\bar{y}$, mm	$\bar{x}A$, mm³	$\bar{y}A$, mm³
Rectangle	$(120)(80) = 9.6 \times 10^3$	60	40	$+576 \times 10^3$	$+384 \times 10^3$
Triangle	$\frac{1}{2}(120)(60) = 3.6 \times 10^3$	40	−20	$+144 \times 10^3$	-72×10^3
Semicircle	$\frac{1}{2}\pi(60)^2 = 5.655 \times 10^3$	60	105.46	$+339.3 \times 10^3$	$+596.4 \times 10^3$
Circle	$-\pi(40)^2 = -5.027 \times 10^3$	60	80	-301.6×10^3	-402.2×10^3
	$\Sigma A = 13.828 \times 10^3$			$\Sigma\bar{x}A = +757.7 \times 10^3$	$\Sigma\bar{y}A = +506.2 \times 10^3$

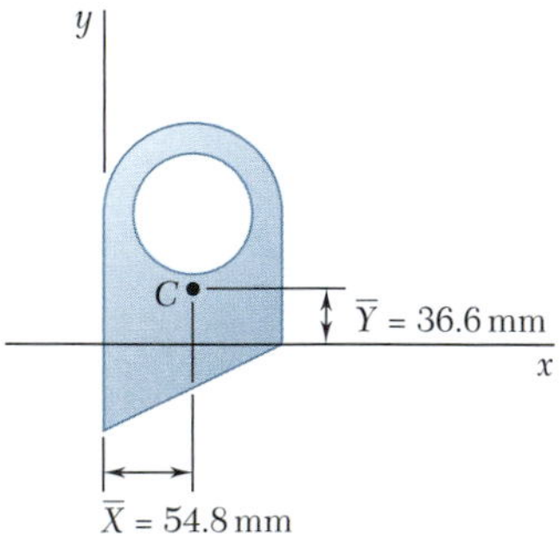

***a.* First Moments of the Area.** Using Eqs. (5.8), we write

$$Q_x = \Sigma\bar{y}A = 506.2 \times 10^3 \text{ mm}^3 \qquad Q_x = 506 \times 10^3 \text{ mm}^3 \blacktriangleleft$$
$$Q_y = \Sigma\bar{x}A = 757.7 \times 10^3 \text{ mm}^3 \qquad Q_y = 758 \times 10^3 \text{ mm}^3 \blacktriangleleft$$

***b.* Location of Centroid.** Substituting the values given in the table into the equations defining the centroid of a composite area, we obtain

$$\bar{X}\Sigma A = \Sigma\bar{x}A: \qquad \bar{X}(13.828 \times 10^3 \text{ mm}^2) = 757.7 \times 10^3 \text{ mm}^3$$
$$\bar{X} = 54.8 \text{ mm} \blacktriangleleft$$

$$\bar{Y}\Sigma A = \Sigma\bar{y}A: \qquad \bar{Y}(13.828 \times 10^3 \text{ mm}^2) = 506.2 \times 10^3 \text{ mm}^3$$
$$\bar{Y} = 36.6 \text{ mm} \blacktriangleleft$$

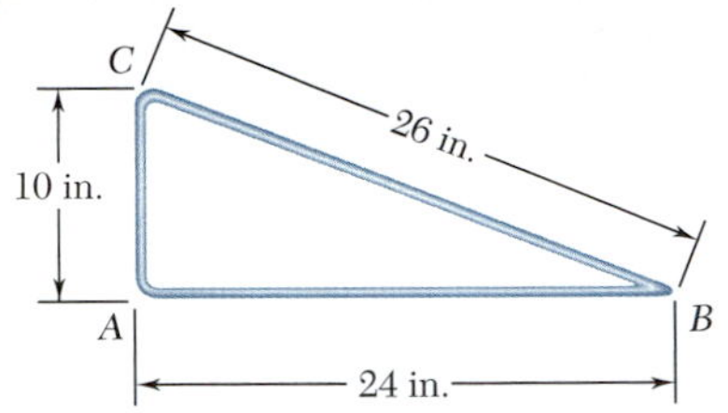

SAMPLE PROBLEM 5.2

The figure shown is made from a piece of thin, homogeneous wire. Determine the location of its center of gravity.

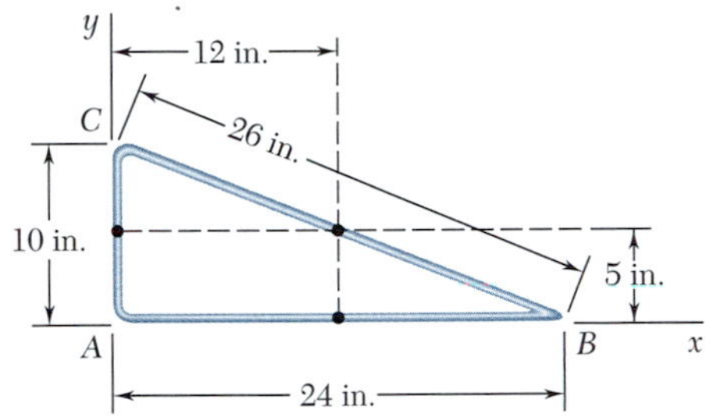

SOLUTION

Since the figure is formed of homogeneous wire, its center of gravity coincides with the centroid of the corresponding line. Therefore, that centroid will be determined. Choosing the coordinate axes shown, with origin at A, we determine the coordinates of the centroid of each line segment and compute the first moments with respect to the coordinate axes.

Segment	L, in.	$\bar{x}$, in.	$\bar{y}$, in.	$\bar{x}L$, in^2	$\bar{y}L$, in^2
AB	24	12	0	288	0
BC	26	12	5	312	130
CA	10	0	5	0	50
	$\Sigma L = 60$			$\Sigma \bar{x}L = 600$	$\Sigma \bar{y}L = 180$

Substituting the values obtained from the table into the equations defining the centroid of a composite line, we obtain

$\bar{X}\Sigma L = \Sigma \bar{x}L$: $\quad \bar{X}(60 \text{ in.}) = 600 \text{ in}^2$ $\quad \bar{X} = 10 \text{ in.}$ ◀

$\bar{Y}\Sigma L = \Sigma \bar{y}L$: $\quad \bar{Y}(60 \text{ in.}) = 180 \text{ in}^2$ $\quad \bar{Y} = 3 \text{ in.}$ ◀

SAMPLE PROBLEM 5.3

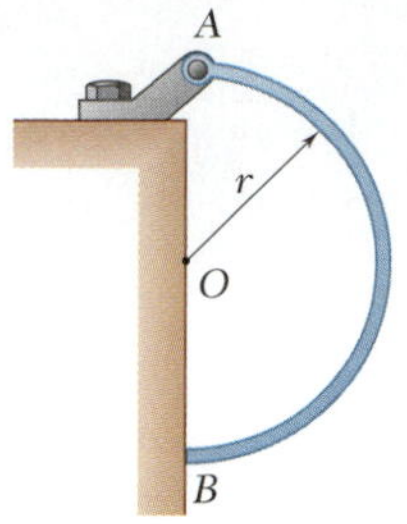

A uniform semicircular rod of weight W and radius r is attached to a pin at A and rests against a frictionless surface at B. Determine the reactions at A and B.

SOLUTION

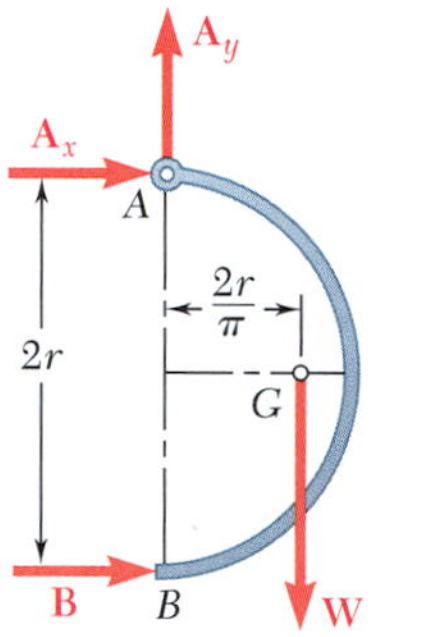

Free-Body Diagram. A free-body diagram of the rod is drawn. The forces acting on the rod are its weight $\mathbf{W}$, which is applied at the center of gravity G (whose position is obtained from Fig. 5.8B); a reaction at A, represented by its components $\mathbf{A}_x$ and $\mathbf{A}_y$; and a horizontal reaction at B.

Equilibrium Equations

$+\circlearrowleft \Sigma M_A = 0$: $\quad B(2r) - W\left(\frac{2r}{\pi}\right) = 0$

$$B = +\frac{W}{\pi} \qquad \mathbf{B} = \frac{W}{\pi} \rightarrow \blacktriangleleft$$

$\xrightarrow{+} \Sigma F_x = 0$: $\quad A_x + B = 0$

$$A_x = -B = -\frac{W}{\pi} \qquad \mathbf{A}_x = \frac{W}{\pi} \leftarrow$$

$+\uparrow \Sigma F_y = 0$: $\quad A_y - W = 0 \qquad \mathbf{A}_y = W\uparrow$

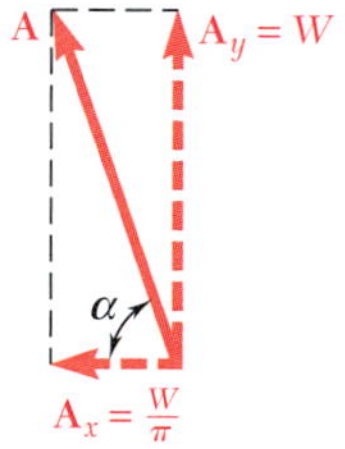

Adding the two components of the reaction at A:

$$A = \left[W^2 + \left(\frac{W}{\pi}\right)^2\right]^{1/2} \qquad A = W\left(1 + \frac{1}{\pi^2}\right)^{1/2} \blacktriangleleft$$

$$\tan \alpha = \frac{W}{W/\pi} = \pi \qquad \alpha = \tan^{-1} \pi \blacktriangleleft$$

The answers can also be expressed as follows:

$$\mathbf{A} = 1.049W \measuredangle 72.3° \qquad \mathbf{B} = 0.318W \rightarrow \blacktriangleleft$$

SOLVING PROBLEMS ON YOUR OWN

In this lesson we developed the general equations for locating the centers of gravity of two-dimensional bodies and wires [Eqs. (5.2)] and the centroids of plane areas [Eqs. (5.3)] and lines [Eqs. (5.4)]. In the following problems, you will have to locate the centroids of composite areas and lines or determine the first moments of the area for composite plates [Eqs. (5.8)].

1. ***Locating the centroids of composite areas and lines.*** Sample Problems 5.1 and 5.2 illustrate the procedure you should follow when solving problems of this type. There are, however, several points that should be emphasized.

a. The first step in your solution should be to decide how to construct the given area or line from the common shapes of Fig. 5.8. You should recognize that for plane areas it is often possible to construct a particular shape in more than one way. Also, showing the different components (as is done in Sample Prob. 5.1) will help you to correctly establish their centroids and areas or lengths. Do not forget that you can subtract areas as well as add them to obtain a desired shape.

b. We strongly recommend that for each problem you construct a table containing the areas or lengths and the respective coordinates of the centroids. It is essential for you to remember that areas which are "removed" (for example, holes) are treated as negative. Also, the sign of negative coordinates must be included. Therefore, you should always carefully note the location of the origin of the coordinate axes.

c. When possible, use symmetry [Sec. 5.4] to help you determine the location of a centroid.

d. In the formulas for the circular sector and for the arc of a circle in Fig. 5.8, the angle α must always be expressed in radians.

2. ***Calculating the first moments of an area.*** The procedures for locating the centroid of an area and for determining the first moments of an area are similar; however, for the latter it is not necessary to compute the total area. Also, as noted in Sec. 5.4, you should recognize that the first moment of an area relative to a centroidal axis is zero.

3. ***Solving problems involving the center of gravity.*** The bodies considered in the following problems are homogeneous; thus, their centers of gravity and centroids coincide. In addition, when a body that is suspended from a single pin is in equilibrium, the pin and the body's center of gravity must lie on the same vertical line.

It may appear that many of the problems in this lesson have little to do with the study of mechanics. However, being able to locate the centroid of composite shapes will be essential in several topics that you will soon encounter.

Problems

5.1 through 5.8 Locate the centroid of the plane area shown.

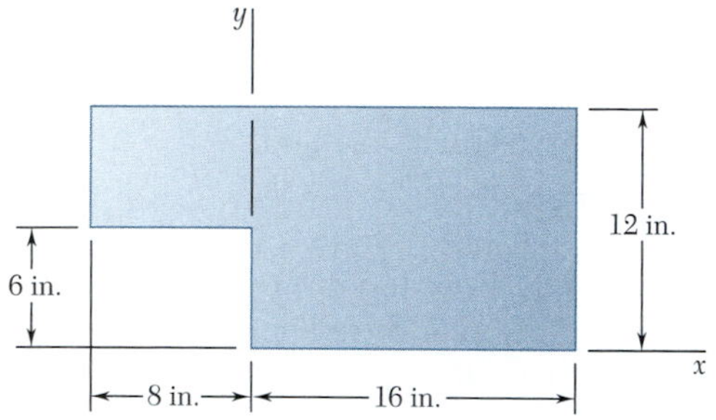

Fig. P5.1

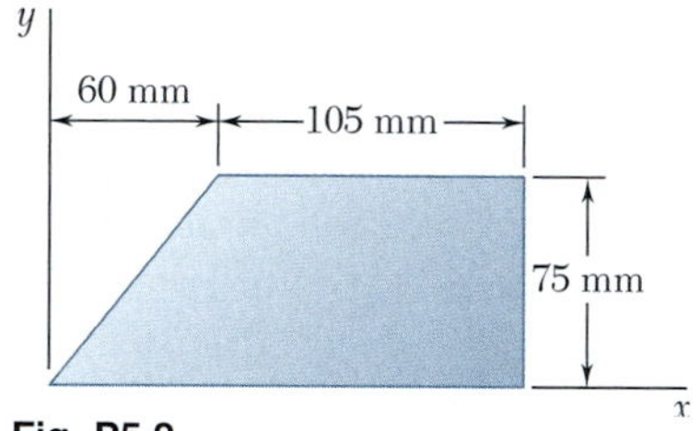

Fig. P5.2

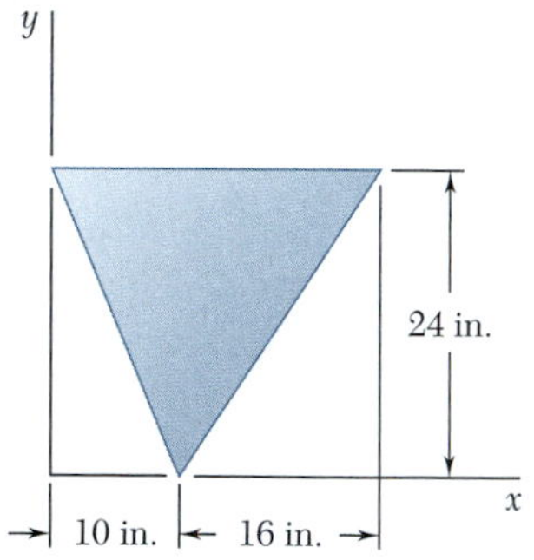

Fig. *P5.3*

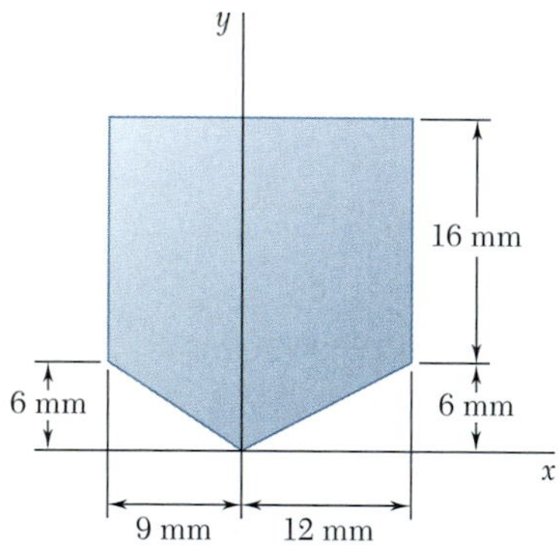

Fig. P5.4

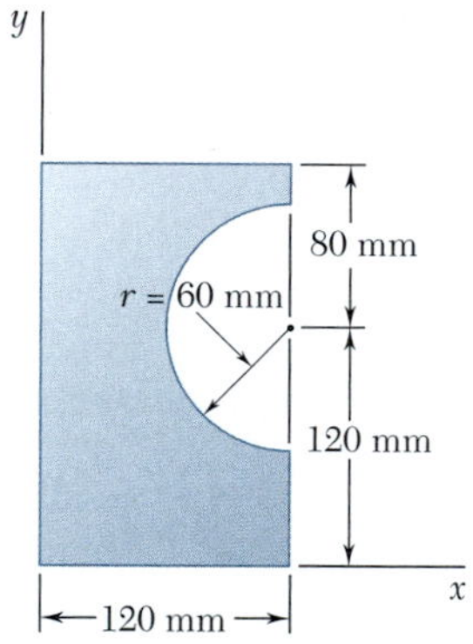

Fig. P5.5

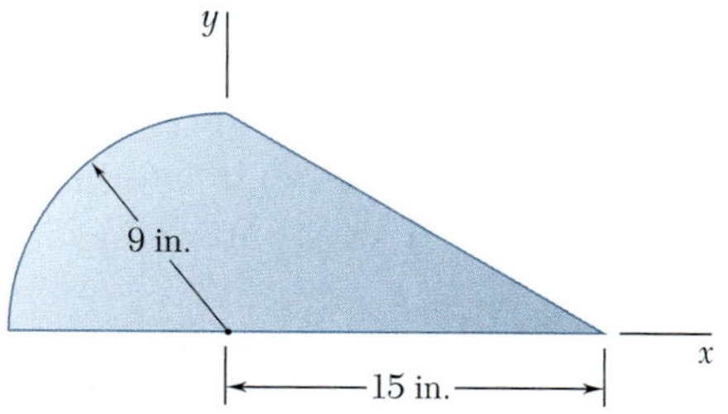

Fig. P5.6

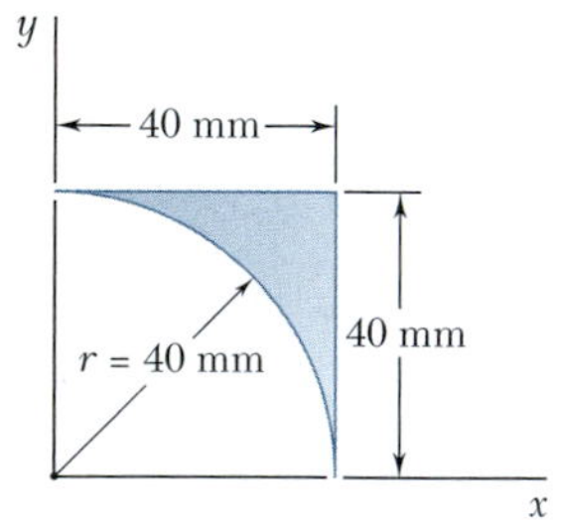

Fig. P5.7

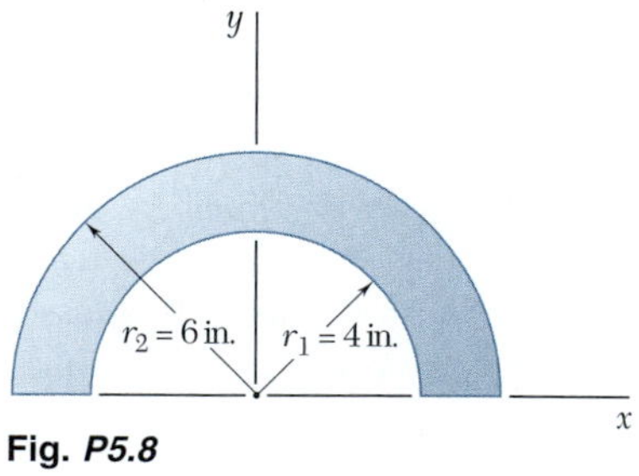

Fig. *P5.8*

5.9 For the area of Prob. 5.8, determine the ratio r_2/r_1 so that $\bar{y} = 3r_1/4$.

5.10 Show that as r_1 approaches r_2, the location of the centroid approaches that of a circular arc of radius $(r_1 + r_2)/2$.

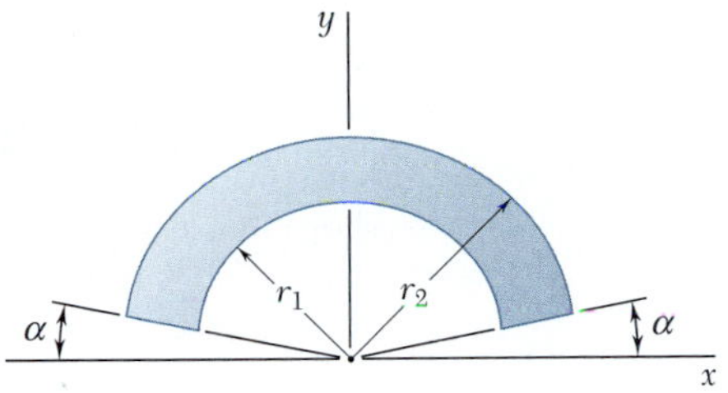

Fig. P5.10

5.11 through 5.16 Locate the centroid of the plane area shown.

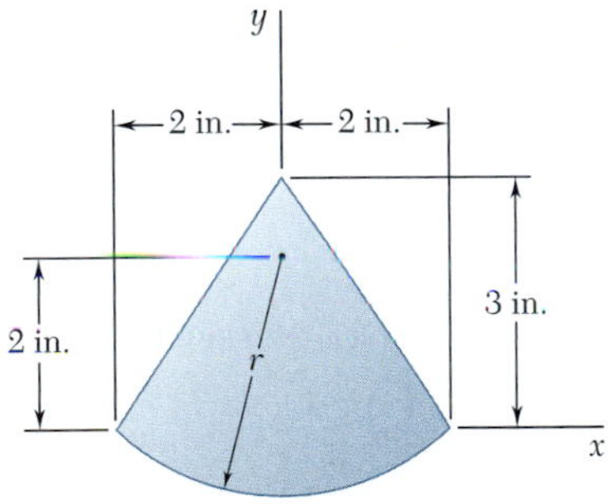

Fig. P5.11

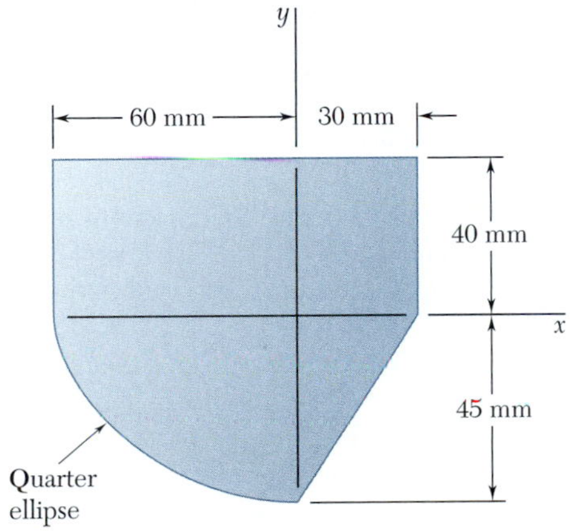

Fig. P5.12

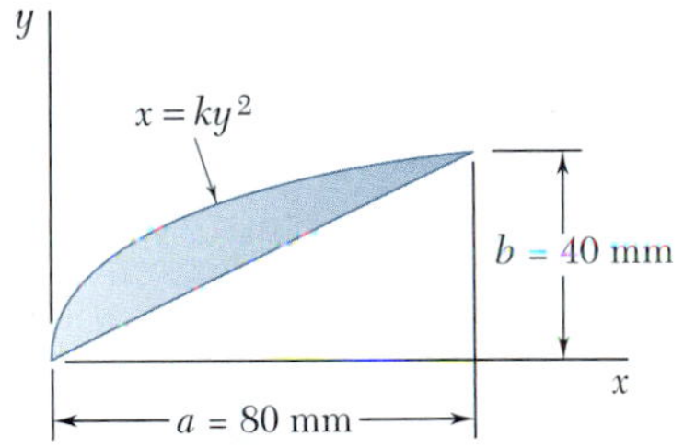

Fig. *P5.13*

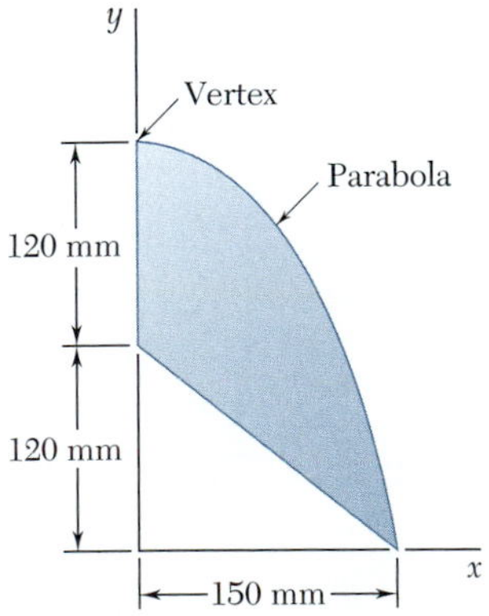

Fig. *P5.14*

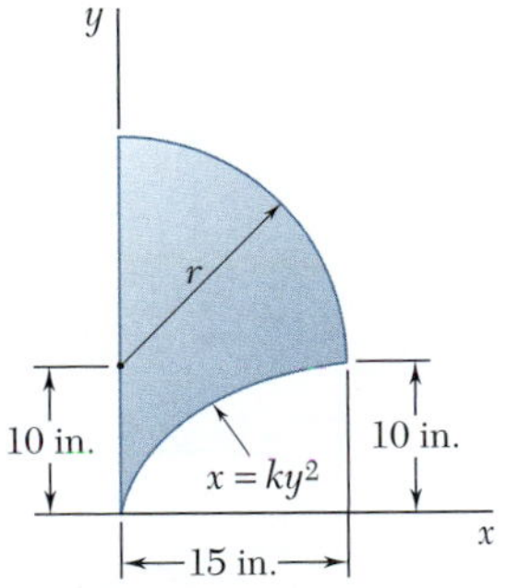

Fig. P5.15

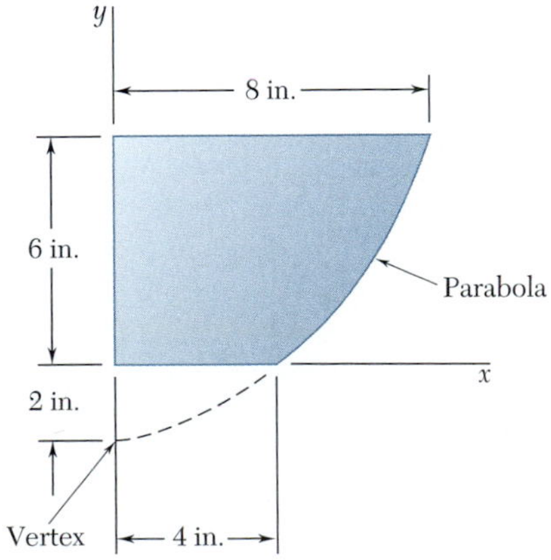

Fig. P5.16

5.17 and 5.18 The horizontal x axis is drawn through the centroid C of the area shown and divides the area into two component areas A_1 and A_2. Determine the first moment of each component area with respect to the x axis, and explain the results obtained.

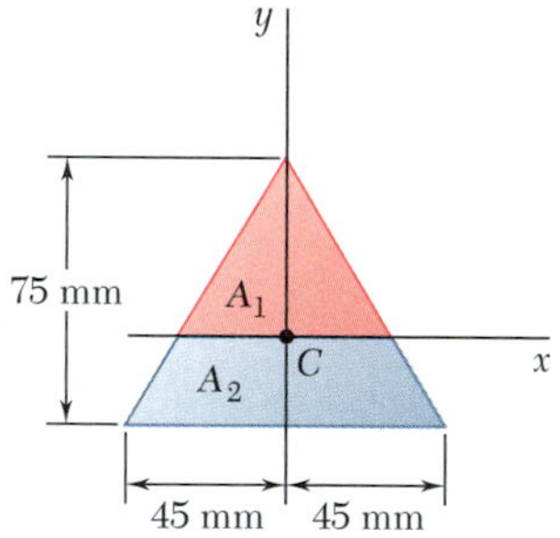

Fig. P5.17

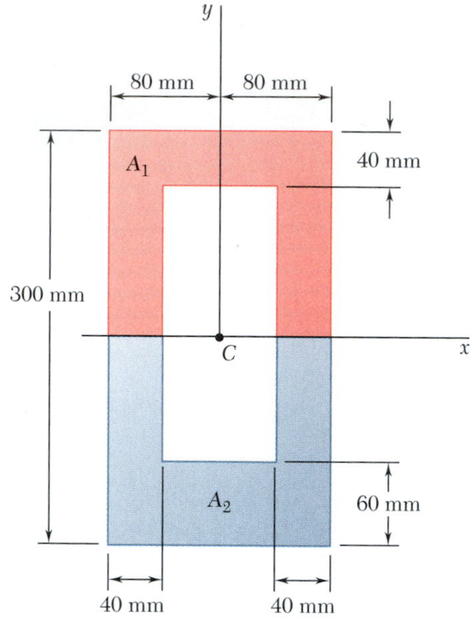

Fig. P5.18

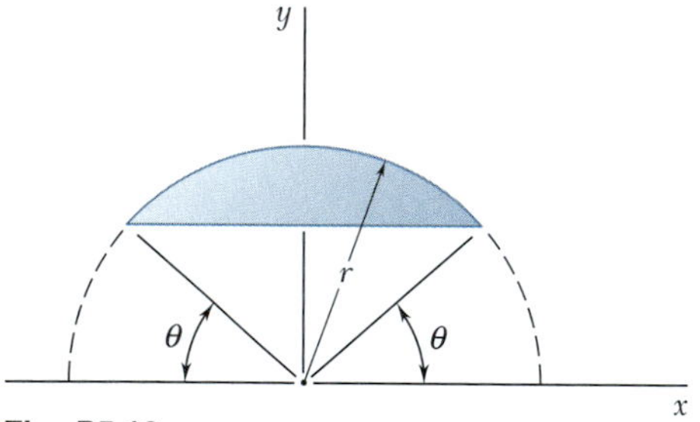

Fig. *P5.19*

5.19 The first moment of the shaded area with respect to the x axis is denoted by Q_x. (*a*) Express Q_x in terms of r and θ. (*b*) For what value of θ is Q_x maximum, and what is the maximum value?

5.20 A composite beam is constructed by bolting four plates to four $2 \times 2 \times 3/8$-in. angles as shown. The bolts are equally spaced along the beam, and the beam supports a vertical load. As proved in mechanics of materials, the shearing forces exerted on the bolts at A and B are proportional to the first moments with respect to the centroidal x axis of the red shaded areas shown, respectively, in parts a and b of the figure. Knowing that the force exerted on the bolt at A is 70 lb, determine the force exerted on the bolt at B.

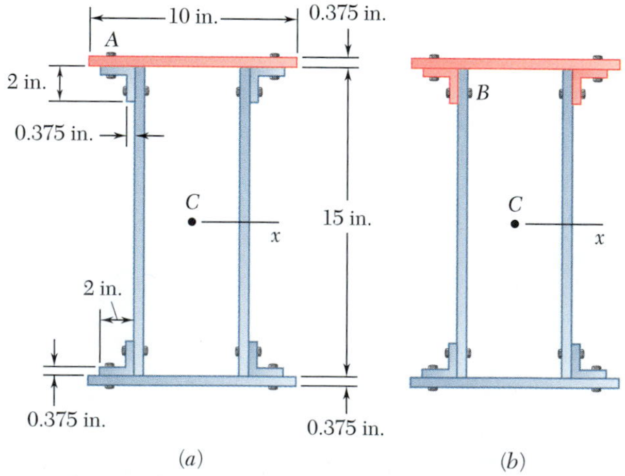

Fig. P5.20

5.21 through 5.24 A thin, homogeneous wire is bent to form the perimeter of the figure indicated. Locate the center of gravity of the wire figure thus formed.

5.21 Fig. P5.1.
5.22 Fig. P5.2.
5.23 Fig. P5.4.
5.24 Fig. P5.8.

5.25 A 750-g uniform steel rod is bent into a circular arc of radius 500 mm as shown. The rod is supported by a pin at A and the cord BC. Determine (*a*) the tension in the cord, (*b*) the reaction at A.

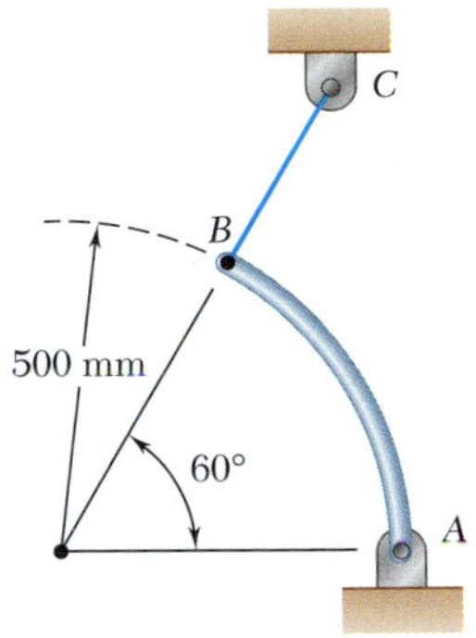

Fig. P5.25

5.26 The homogeneous wire $ABCD$ is bent as shown and is supported by a pin at B. Knowing that $l = 8$ in., determine the angle θ for which portion BC of the wire is horizontal.

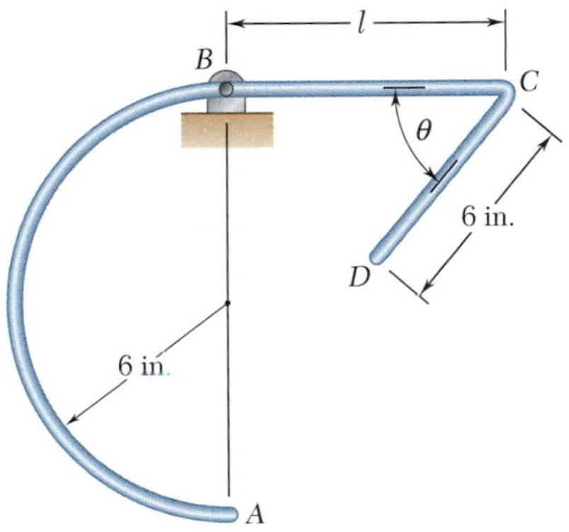

Fig. P5.26 and *P5.27*

5.27 The homogeneous wire $ABCD$ is bent as shown and is supported by a pin at B. Knowing that $\theta = 30°$, determine the length l for which portion CD of the wire is horizontal.

5.28 The homogeneous wire $ABCD$ is bent as shown and is attached to a hinge at C. Determine the length L for which the portion BCD of the wire is horizontal.

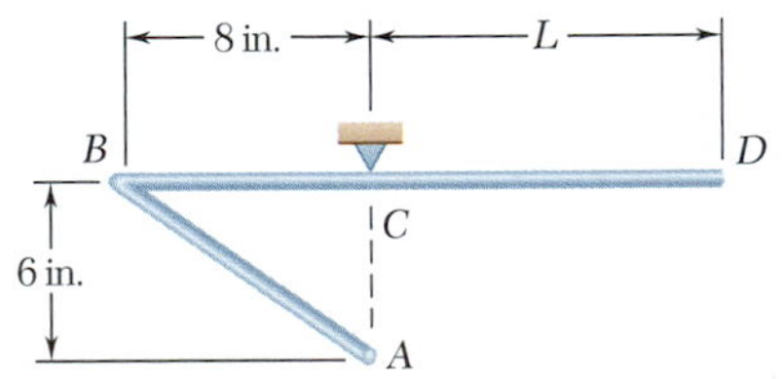

Fig. P5.28

5.29 Determine the distance h so that the centroid of the shaded area is as close to line BB' as possible when (*a*) $k = 0.2$, (*b*) $k = 0.6$.

5.30 Show when the distance h is selected to minimize the distance $\overline{y}$ from line BB' to the centroid of the shaded area that $\overline{y} = h$.

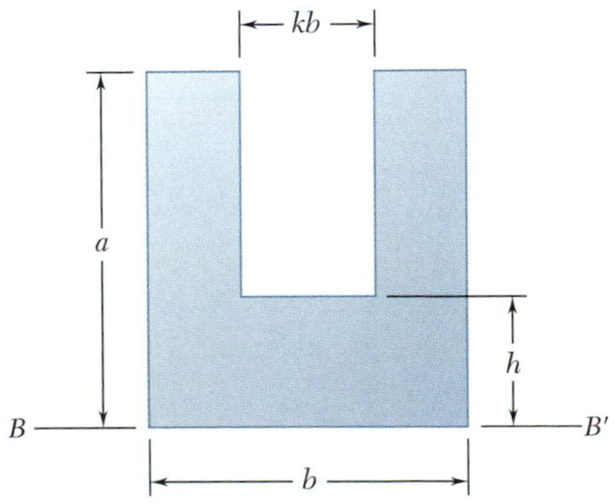

Fig. P5.29 and P5.30

5.6. DETERMINATION OF CENTROIDS BY INTEGRATION

The centroid of an area bounded by analytical curves (that is, curves defined by algebraic equations) is usually determined by evaluating the integrals in Eqs. (5.3) of Sec. 5.3:

$$\bar{x}A = \int x\,dA \qquad \bar{y}A = \int y\,dA \tag{5.3}$$

If the element of area dA is a small rectangle of sides dx and dy, the evaluation of each of these integrals requires a *double integration* with respect to x and y. A double integration is also necessary if polar coordinates are used for which dA is a small element of sides dr and $r\,d\theta$.

In most cases, however, it is possible to determine the coordinates of the centroid of an area by performing a single integration. This is achieved by choosing dA to be a thin rectangle or strip or a thin sector or pie-shaped element (Fig. 5.12A); the centroid of the thin rectangle is located at its center, and the centroid of the thin sector is located at a distance $\frac{2}{3}r$ from its vertex (as it is for a triangle). The coordinates of the centroid of the area under consideration are then obtained by expressing that the first moment of the entire area with respect to each of the coordinate axes is equal to the sum (or integral) of the corresponding moments of the elements of area. Denoting by $\bar{x}_{el}$ and $\bar{y}_{el}$ the coordinates of the centroid of the element dA, we write

$$\begin{aligned} Q_y &= \bar{x}A = \int \bar{x}_{el}\,dA \\ Q_x &= \bar{y}A = \int \bar{y}_{el}\,dA \end{aligned} \tag{5.9}$$

If the area A is not already known, it can also be computed from these elements.

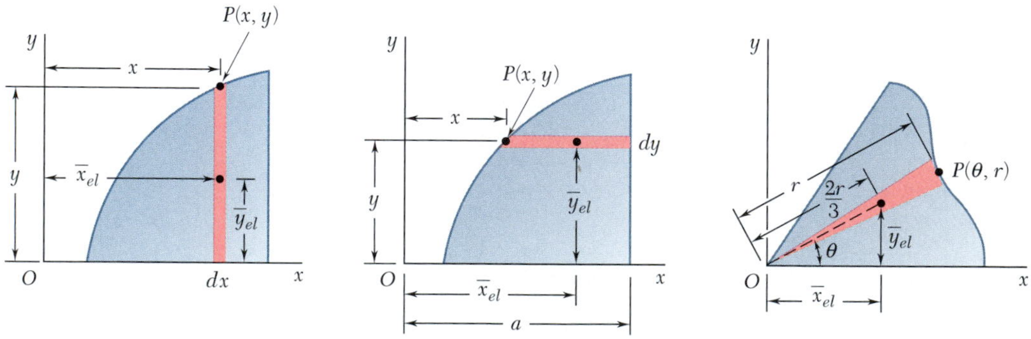

Fig. 5.12A Centroids and areas of differential elements.

The coordinates $\bar{x}_{el}$ and $\bar{y}_{el}$ of the centroid of the element of area dA should be expressed in terms of the coordinates of a point located on the curve bounding the area under consideration. Also, the area of the element dA should be expressed in terms of the coordinates of that point and the appropriate differentials. This has been done in Fig. 5.12B for three common types of elements; the pie-shaped element of part c should be used when the equation of the curve bounding the area is given in polar coordinates. The appropriate expressions

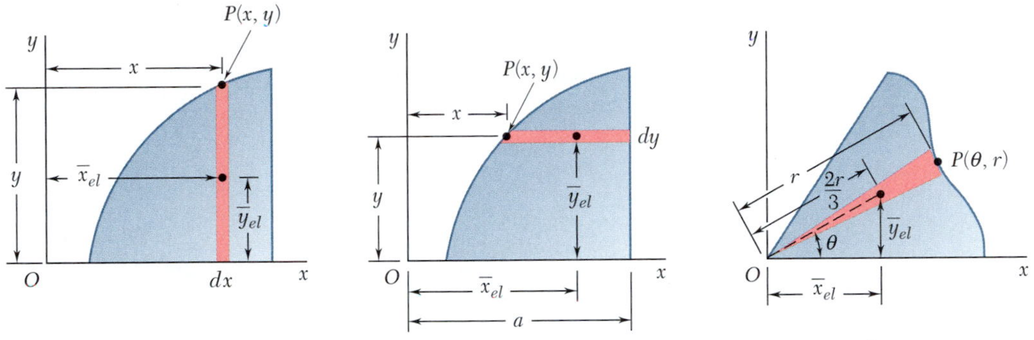

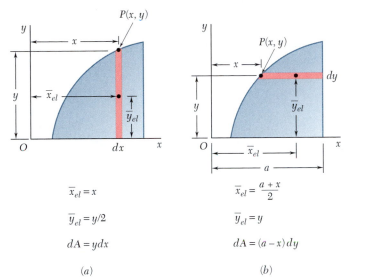

$\bar{x}_{el} = x$ | $\bar{y}_{el} = y/2$ | $dA = y\,dx$

(*a*)

$\bar{x}_{el} = \dfrac{a + x}{2}$ | $\bar{y}_{el} = y$ | $dA = (a - x)\,dy$

(*b*)

$\bar{x}_{el} = \dfrac{2r}{3}\cos\theta$ | $\bar{y}_{el} = \dfrac{2r}{3}\sin\theta$ | $dA = \dfrac{1}{2}r^2\,d\theta$

(*c*)

Fig. 5.12B Centroids and areas of differential elements.

should be substituted into formulas (5.9), and the equation of the bounding curve should be used to express one of the coordinates in terms of the other. The integration is thus reduced to a single integration. Once the area has been determined and the integrals in Eqs. (5.9) have been evaluated, these equations can be solved for the coordinates $\bar{x}$ and $\bar{y}$ of the centroid of the area.

When a line is defined by an algebraic equation, its centroid can be determined by evaluating the integrals in Eqs. (5.4) of Sec. 5.3:

$$\bar{x}L = \int x\,dL \qquad \bar{y}L = \int y\,dL \tag{5.4}$$

The differential length dL should be replaced by one of the following expressions, depending upon which coordinate, x, y, or θ, is chosen as the independent variable in the equation used to define the line (these expressions can be derived using the Pythagorean theorem):

$$dL = \sqrt{1 + \left(\frac{dy}{dx}\right)^2}\,dx \qquad dL = \sqrt{1 + \left(\frac{dx}{dy}\right)^2}\,dy$$

$$dL = \sqrt{r^2 + \left(\frac{dr}{d\theta}\right)^2}\,d\theta$$

After the equation of the line has been used to express one of the coordinates in terms of the other, the integration can be performed, and Eqs. (5.4) can be solved for the coordinates $\bar{x}$ and $\bar{y}$ of the centroid of the line.

5.7. THEOREMS OF PAPPUS-GULDINUS

These theorems, which were first formulated by the Greek geometer Pappus during the third century A.D. and later restated by the Swiss mathematician Guldinus, or Guldin, (1577–1643) deal with surfaces and bodies of revolution.

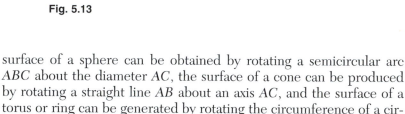

A *surface of revolution* is a surface which can be generated by rotating a plane curve about a fixed axis. For example (Fig. 5.13), the surface of a sphere can be obtained by rotating a semicircular arc ABC about the diameter AC, the surface of a cone can be produced by rotating a straight line AB about an axis AC, and the surface of a torus or ring can be generated by rotating the circumference of a circle about a nonintersecting axis. A *body of revolution* is a body which can be generated by rotating a plane area about a fixed axis. As shown in Fig. 5.14, a sphere, a cone, and a torus can each be generated by rotating the appropriate shape about the indicated axis.

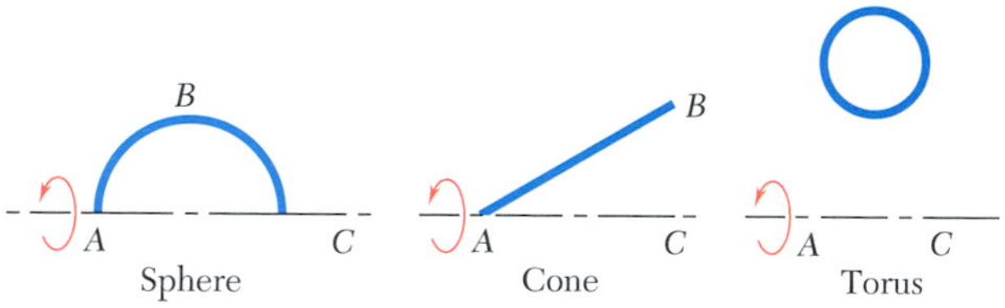

Fig. 5.13

Photo 5.2 The storage tanks shown are all bodies of revolution. Thus, their surface areas and volumes can be determined using the theorems of Pappus-Guldinus.

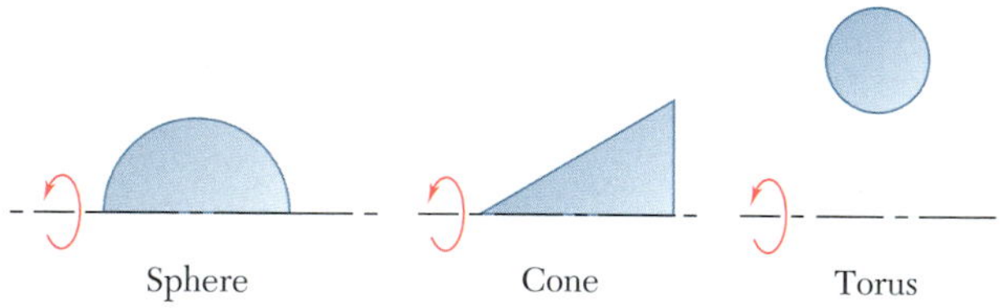

Fig. 5.14

THEOREM I. *The area of a surface of revolution is equal to the length of the generating curve times the distance traveled by the centroid of the curve while the surface is being generated.*

Proof. Consider an element dL of the line L (Fig. 5.15), which is revolved about the x axis. The area dA generated by the element dL is equal to $2\pi y\, dL$. Thus, the entire area generated by L is $A = \int 2\pi y\, dL$. Recalling that we found in Sec. 5.3 that the integral $\int y\, dL$

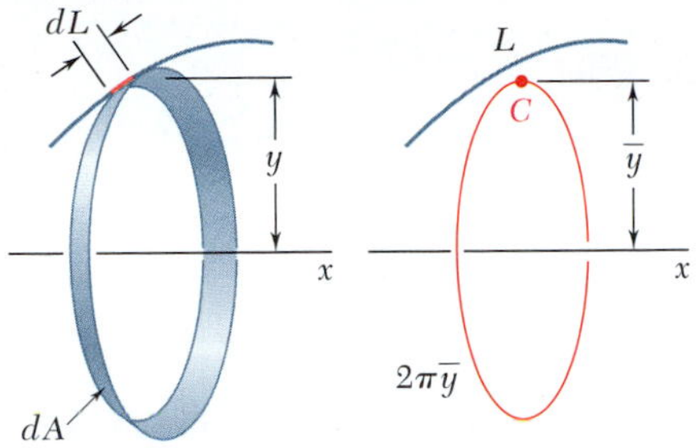

Fig. 5.15

is equal to $\overline{y}L$, we therefore have

$$A = 2\pi\overline{y}L \tag{5.10}$$

where $2\pi\overline{y}$ is the distance traveled by the centroid of L (Fig. 5.15). It should be noted that the generating curve must not cross the axis about which it is rotated; if it did, the two sections on either side of the axis would generate areas having opposite signs, and the theorem would not apply.

THEOREM II. *The volume of a body of revolution is equal to the generating area times the distance traveled by the centroid of the area while the body is being generated.*

Proof. Consider an element dA of the area A which is revolved about the x axis (Fig. 5.16). The volume dV generated by the element

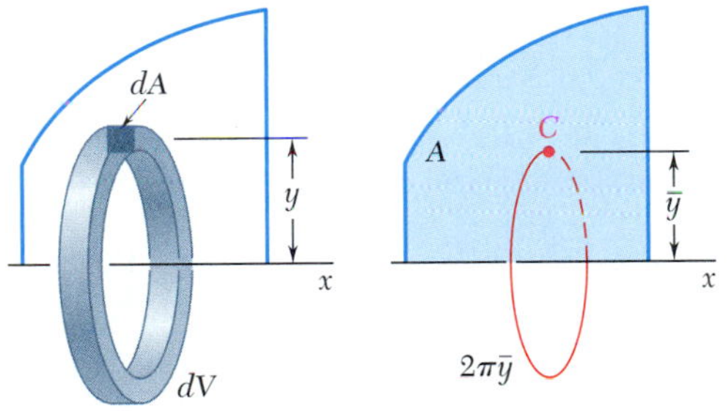

Fig. 5.16

dA is equal to $2\pi y\, dA$. Thus, the entire volume generated by A is $V = \int 2\pi y\, dA$, and since the integral $\int y\, dA$ is equal to $\overline{y}A$ (Sec. 5.3), we have

$$V = 2\pi\overline{y}A \tag{5.11}$$

where $2\pi\overline{y}$ is the distance traveled by the centroid of A. Again, it should be noted that the theorem does not apply if the axis of rotation intersects the generating area.

The theorems of Pappus-Guldinus offer a simple way to compute the areas of surfaces of revolution and the volumes of bodies of revolution. Conversely, they can also be used to determine the centroid of a plane curve when the area of the surface generated by the curve is known or to determine the centroid of a plane area when the volume of the body generated by the area is known (see Sample Prob. 5.8).

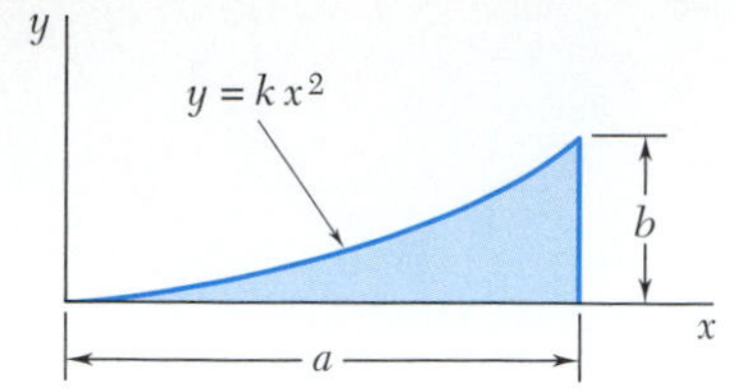

SAMPLE PROBLEM 5.4

Determine by direct integration the location of the centroid of a parabolic spandrel.

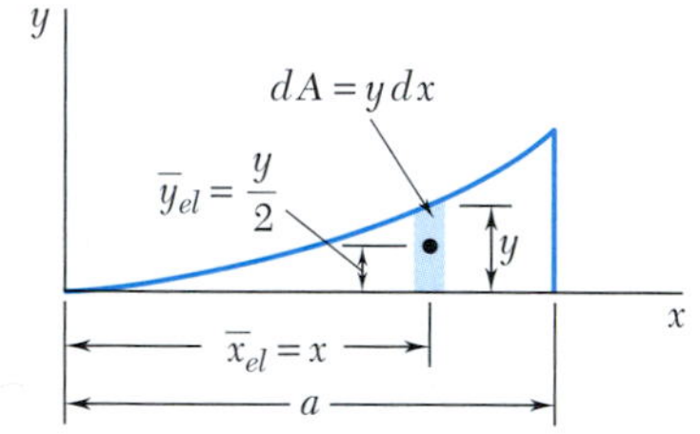

SOLUTION

Determination of the Constant k. The value of k is determined by substituting $x = a$ and $y = b$ into the given equation. We have $b = ka^2$ or $k = b/a^2$. The equation of the curve is thus

$$y = \frac{b}{a^2}x^2 \qquad \text{or} \qquad x = \frac{a}{b^{1/2}}y^{1/2}$$

Vertical Differential Element. We choose the differential element shown and find the total area of the figure.

$$A = \int dA = \int y\,dx = \int_0^a \frac{b}{a^2}x^2\,dx = \left[\frac{b}{a^2}\frac{x^3}{3}\right]_0^a = \frac{ab}{3}$$

The first moment of the differential element with respect to the y axis is $\bar{x}_{el}\,dA$; hence, the first moment of the entire area with respect to this axis is

$$Q_y = \int \bar{x}_{el}\,dA = \int xy\,dx = \int_0^a x\left(\frac{b}{a^2}x^2\right)dx = \left[\frac{b}{a^2}\frac{x^4}{4}\right]_0^a = \frac{a^2b}{4}$$

Since $Q_y = \bar{x}A$, we have

$$\bar{x}A = \int \bar{x}_{el}\,dA \qquad \bar{x}\frac{ab}{3} = \frac{a^2b}{4} \qquad \bar{x} = \tfrac{3}{4}a \;\blacktriangleleft$$

Likewise, the first moment of the differential element with respect to the x axis is $\bar{y}_{el}\,dA$, and the first moment of the entire area is

$$Q_x = \int \bar{y}_{el}\,dA = \int \frac{y}{2}y\,dx = \int_0^a \frac{1}{2}\left(\frac{b}{a^2}x^2\right)^2 dx = \left[\frac{b^2}{2a^4}\frac{x^5}{5}\right]_0^a = \frac{ab^2}{10}$$

Since $Q_x = \bar{y}A$, we have

$$\bar{y}A = \int \bar{y}_{el}\,dA \qquad \bar{y}\frac{ab}{3} = \frac{ab^2}{10} \qquad \bar{y} = \tfrac{3}{10}b \;\blacktriangleleft$$

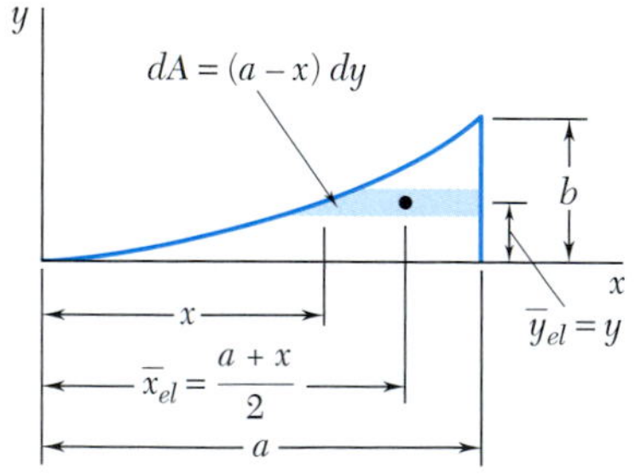

Horizontal Differential Element. The same results can be obtained by considering a horizontal element. The first moments of the area are

$$Q_y = \int \bar{x}_{el}\,dA = \int \frac{a+x}{2}(a-x)\,dy = \int_0^b \frac{a^2-x^2}{2}\,dy$$

$$= \frac{1}{2}\int_0^b \left(a^2 - \frac{a^2}{b}y\right)dy = \frac{a^2b}{4}$$

$$Q_x = \int \bar{y}_{el}\,dA = \int y(a-x)\,dy = \int y\left(a - \frac{a}{b^{1/2}}y^{1/2}\right)dy$$

$$= \int_0^b \left(ay - \frac{a}{b^{1/2}}y^{3/2}\right)dy = \frac{ab^2}{10}$$

To determine $\bar{x}$ and $\bar{y}$, the expressions obtained are again substituted into the equations defining the centroid of the area.

SAMPLE PROBLEM 5.5

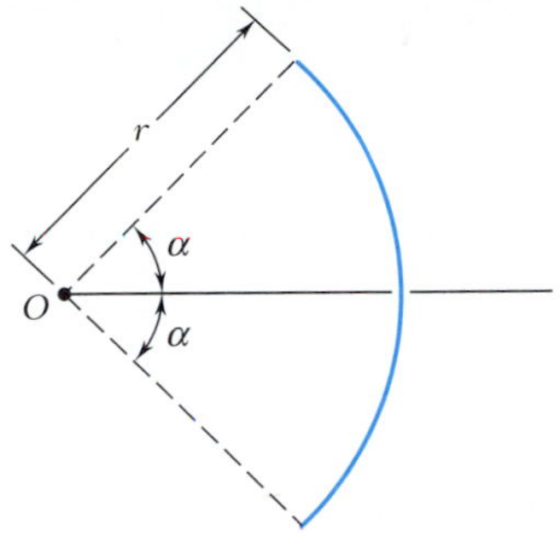

Determine the location of the centroid of the circular arc shown.

SOLUTION

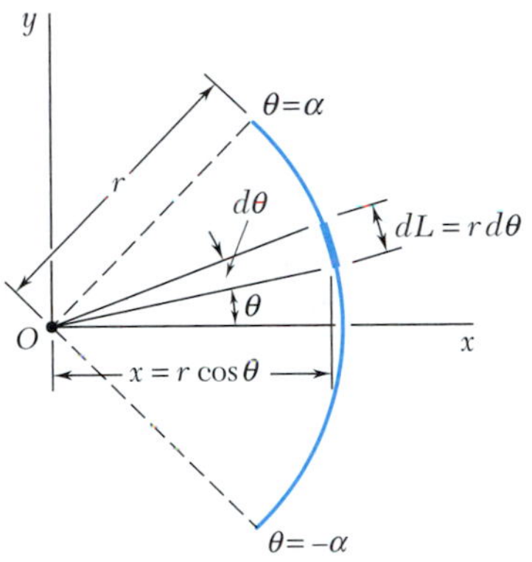

Since the arc is symmetrical with respect to the x axis, $\bar{y} = 0$. A differential element is chosen as shown, and the length of the arc is determined by integration.

$$L = \int dL = \int_{-\alpha}^{\alpha} r\,d\theta = r\int_{-\alpha}^{\alpha} d\theta = 2r\alpha$$

The first moment of the arc with respect to the y axis is

$$Q_y = \int x\,dL = \int_{-\alpha}^{\alpha} (r\cos\theta)(r\,d\theta) = r^2\int_{-\alpha}^{\alpha} \cos\theta\,d\theta$$
$$= r^2[\sin\theta]_{-\alpha}^{\alpha} = 2r^2\sin\alpha$$

Since $Q_y = \bar{x}L$, we write

$$\bar{x}(2r\alpha) = 2r^2\sin\alpha \qquad \bar{x} = \frac{r\sin\alpha}{\alpha} \blacktriangleleft$$

SAMPLE PROBLEM 5.6

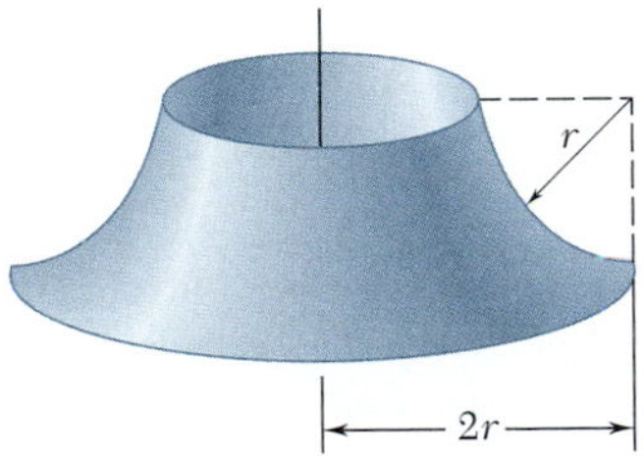

Determine the area of the surface of revolution shown, which is obtained by rotating a quarter-circular arc about a vertical axis.

SOLUTION

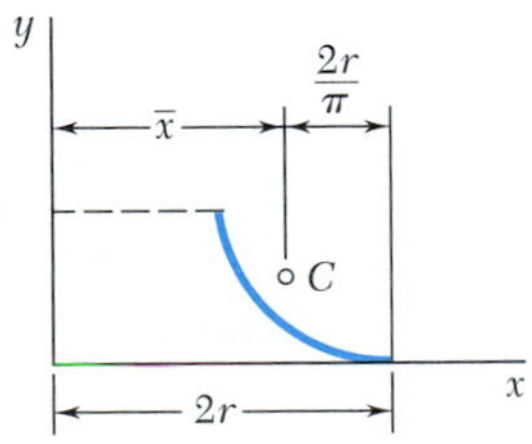

According to Theorem I of Pappus-Guldinus, the area generated is equal to the product of the length of the arc and the distance traveled by its centroid. Referring to Fig. 5.8B, we have

$$\bar{x} = 2r - \frac{2r}{\pi} = 2r\left(1 - \frac{1}{\pi}\right)$$

$$A = 2\pi\bar{x}L = 2\pi\left[2r\left(1 - \frac{1}{\pi}\right)\right]\left(\frac{\pi r}{2}\right)$$

$$A = 2\pi r^2(\pi - 1) \blacktriangleleft$$

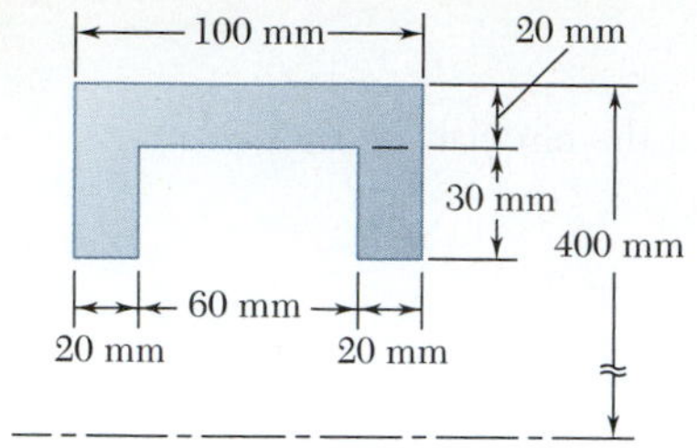

SAMPLE PROBLEM 5.7

The outside diameter of a pulley is 0.8 m, and the cross section of its rim is as shown. Knowing that the pulley is made of steel and that the density of steel is $\rho = 7.85 \times 10^3$ kg/m^3, determine the mass and the weight of the rim.

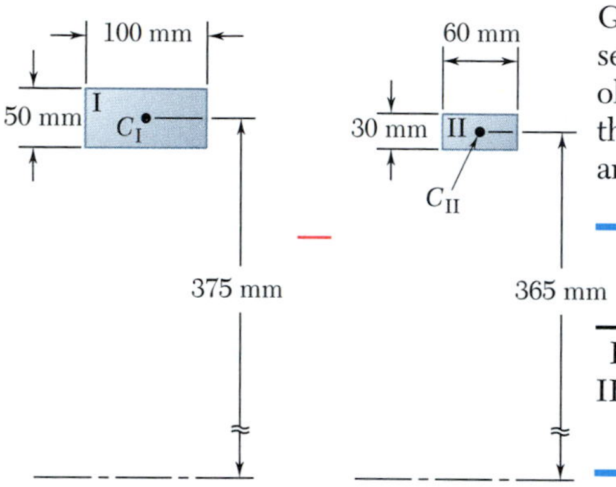

SOLUTION

The volume of the rim can be found by applying Theorem II of Pappus-Guldinus, which states that the volume equals the product of the given cross-sectional area and the distance traveled by its centroid in one complete revolution. However, the volume can be more easily determined if we observe that the cross section can be formed from rectangle I, whose area is positive, and rectangle II, whose area is negative.

	Area, mm^2	$\bar{y}$, mm	Distance Traveled by C, mm	Volume, mm^3
I	+5000	375	$2\pi(375) = 2356$	$(5000)(2356) = 11.78 \times 10^6$
II	−1800	365	$2\pi(365) = 2293$	$(-1800)(2293) = -4.13 \times 10^6$
				Volume of rim $= 7.65 \times 10^6$

Since 1 mm $= 10^{-3}$ m, we have 1 mm$^3 = (10^{-3}\text{ m})^3 = 10^{-9}$ m^3, and we obtain $V = 7.65 \times 10^6 \text{ mm}^3 = (7.65 \times 10^6)(10^{-9}\text{ m}^3) = 7.65 \times 10^{-3}\text{ m}^3$.

$$m = \rho V = (7.85 \times 10^3 \text{ kg/m}^3)(7.65 \times 10^{-3} \text{ m}^3) \qquad m = 60.0 \text{ kg} \blacktriangleleft$$

$$W = mg = (60.0 \text{ kg})(9.81 \text{ m/s}^2) = 589 \text{ kg} \cdot \text{m/s}^2 \qquad W = 589 \text{ N} \blacktriangleleft$$

SAMPLE PROBLEM 5.8

Using the theorems of Pappus-Guldinus, determine (*a*) the centroid of a semicircular area, (*b*) the centroid of a semicircular arc. We recall that the volume and the surface area of a sphere are $\frac{4}{3}\pi r^3$ and $4\pi r^2$, respectively.

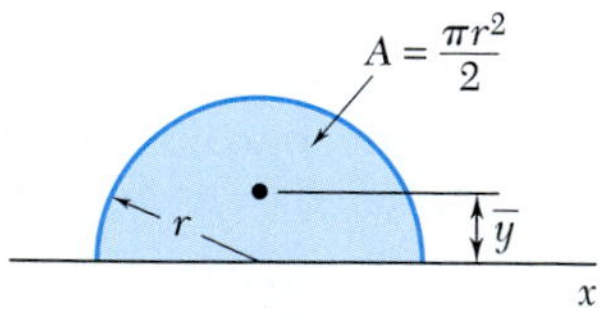

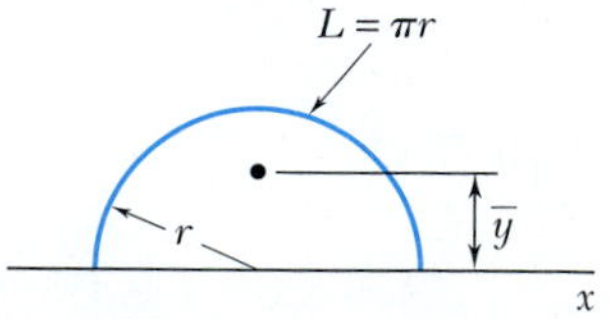

SOLUTION

The volume of a sphere is equal to the product of the area of a semicircle and the distance traveled by the centroid of the semicircle in one revolution about the x axis.

$$V = 2\pi\bar{y}A \qquad \tfrac{4}{3}\pi r^3 = 2\pi\bar{y}(\tfrac{1}{2}\pi r^2) \qquad \bar{y} = \frac{4r}{3\pi} \blacktriangleleft$$

Likewise, the area of a sphere is equal to the product of the length of the generating semicircle and the distance traveled by its centroid in one revolution.

$$A = 2\pi\bar{y}L \qquad 4\pi r^2 = 2\pi\bar{y}(\pi r) \qquad \bar{y} = \frac{2r}{\pi} \blacktriangleleft$$

SOLVING PROBLEMS ON YOUR OWN

In the problems for this lesson, you will use the equations

$$\bar{x}A = \int x\,dA \qquad \bar{y}A = \int y\,dA \tag{5.3}$$

$$\bar{x}L = \int x\,dL \qquad \bar{y}L = \int y\,dL \tag{5.4}$$

to locate the centroids of plane areas and lines, respectively. You will also apply the theorems of Pappus-Guldinus (Sec. 5.7) to determine the areas of surfaces of revolution and the volumes of bodies of revolution.

1. *Determining by direct integration the centroids of areas and lines.* When solving problems of this type, you should follow the method of solution shown in Sample Probs. 5.4 and 5.5: compute A or L, determine the first moments of the area or the line, and solve Eqs. (5.3) or (5.4) for the coordinates of the centroid. In addition, you should pay particular attention to the following points.

a. Begin your solution by carefully defining or determining each term in the applicable integral formulas. We strongly encourage you to show on your sketch of the given area or line your choice for dA or dL and the distances to its centroid.

b. As explained in Sec. 5.6, the x and the y in the above equations represent the *coordinates of the centroid* of the differential elements dA and dL. It is important to recognize that the coordinates of the centroid of dA are not equal to the coordinates of a point located on the curve bounding the area under consideration. You should carefully study Fig. 5.12 until you fully understand this important point.

c. To possibly simplify or minimize your computations, always examine the shape of the given area or line before defining the differential element that you will use. For example, sometimes it may be preferable to use horizontal rectangular elements instead of vertical ones. Also, it will usually be advantageous to use polar coordinates when a line or an area has circular symmetry.

d. Although most of the integrations in this lesson are straightforward, at times it may be necessary to use more advanced techniques, such as trigonometric substitution or integration by parts. Of course, using a table of integrals is the fastest method to evaluate difficult integrals.

2. *Applying the theorems of Pappus-Guldinus.* As shown in Sample Probs. 5.6 through 5.8, these simple, yet very useful theorems allow you to apply your knowledge of centroids to the computation of areas and volumes. Although the theorems refer to the distance traveled by the centroid and to the length of the generating curve or to the generating area, the resulting equations [Eqs. (5.10) and (5.11)] contain the products of these quantities, which are simply the first moments of a line ($\bar{y}L$) and an area ($\bar{y}A$), respectively. Thus, for those problems for which the generating line or area consists of more than one common shape, you need only determine $\bar{y}L$ or $\bar{y}A$; you do not have to calculate the length of the generating curve or the generating area.

Problems

5.31 through 5.33 Determine by direct integration the centroid of the area shown. Express your answer in terms of a and h.

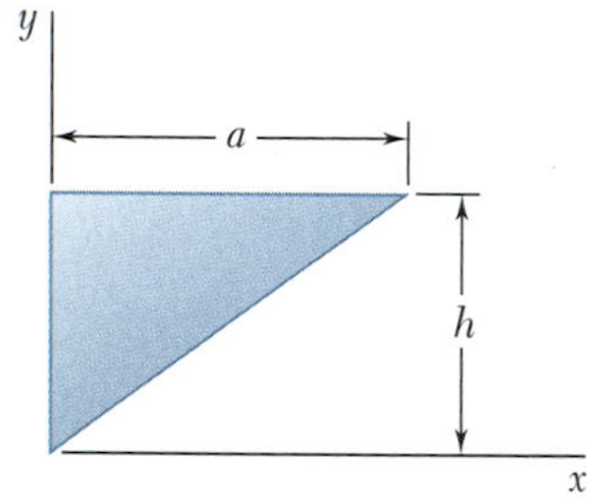

Fig. P5.31

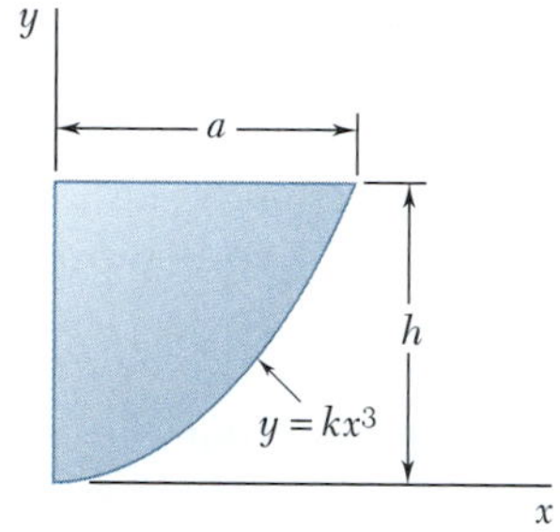

Fig. P5.32

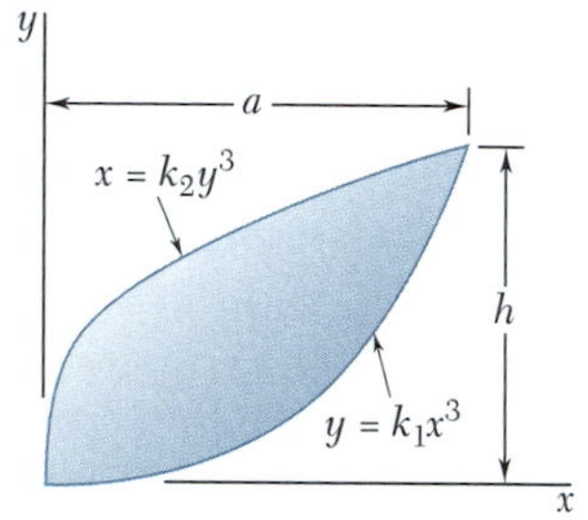

Fig. P5.33

5.34 through 5.36 Determine by direct integration the centroid of the area shown.

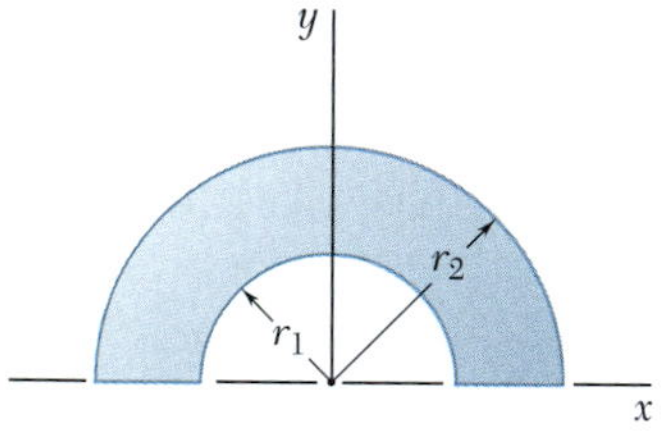

Fig. *P5.34*

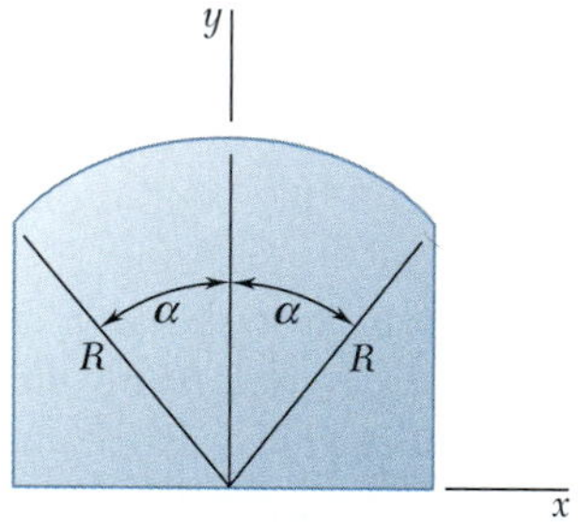

Fig. P5.35

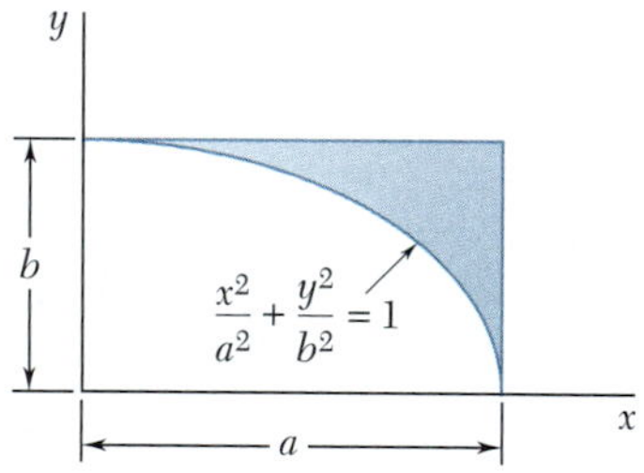

Fig. *P5.36*

5.37 and 5.38 Determine by direct integration the centroid of the area shown. Express your answer in terms of a and b.

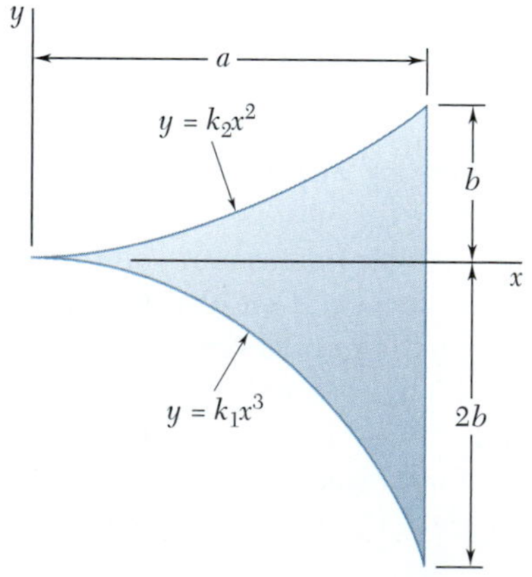

Fig. P5.37

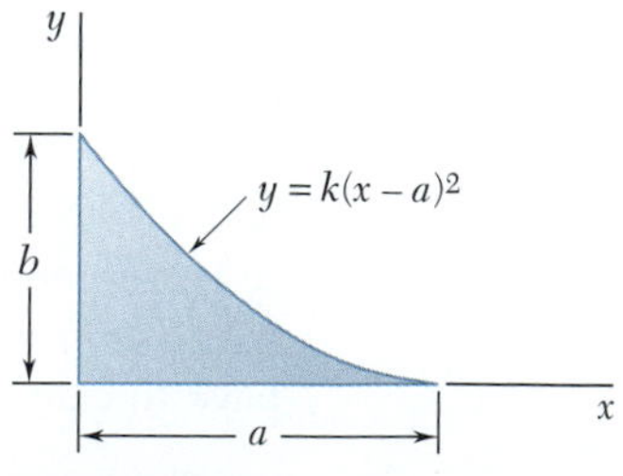

Fig. P5.38

5.39 Determine by direct integration the centroid of the area shown.

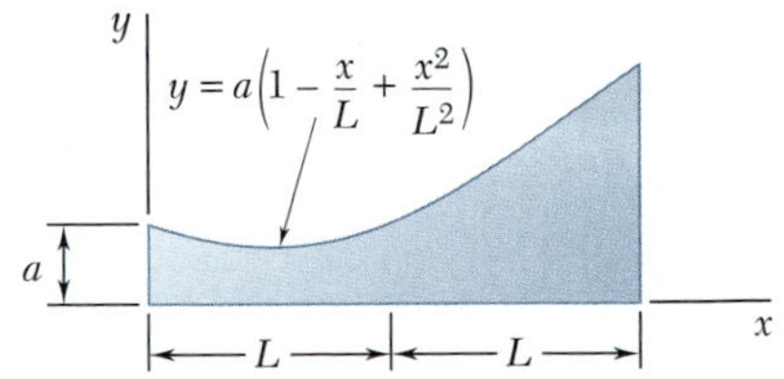

Fig. P5.39

***5.40* and 5.41** Determine by direct integration the centroid of the area shown. Express your answer in terms of a and b.

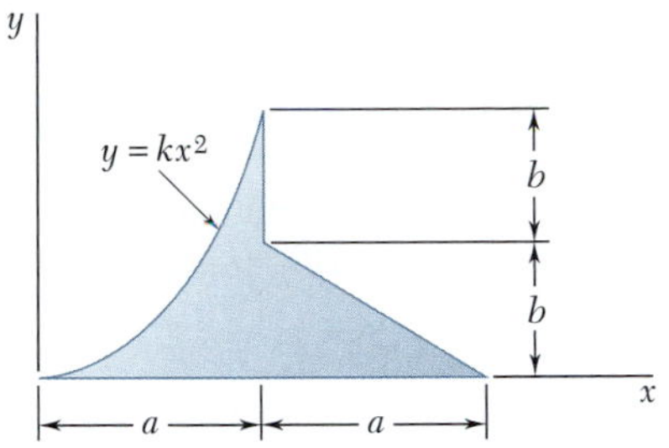

Fig. *P5.40*

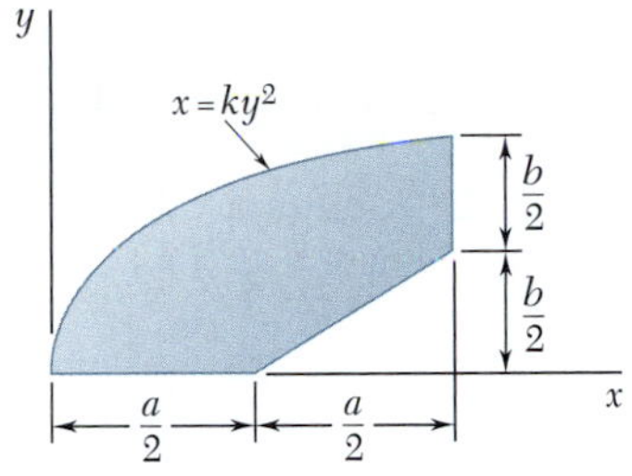

Fig. P5.41

5.42 A homogeneous wire is bent into the shape shown. Determine by direct integration the x coordinate of its centroid. Express your answer in terms of a.

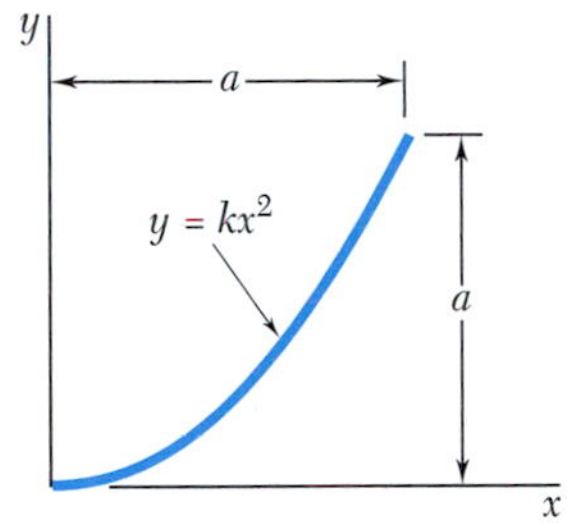

Fig. *P5.42*

5.43 and 5.44 A homogeneous wire is bent into the shape shown. Determine by direct integration the x coordinate of its centroid.

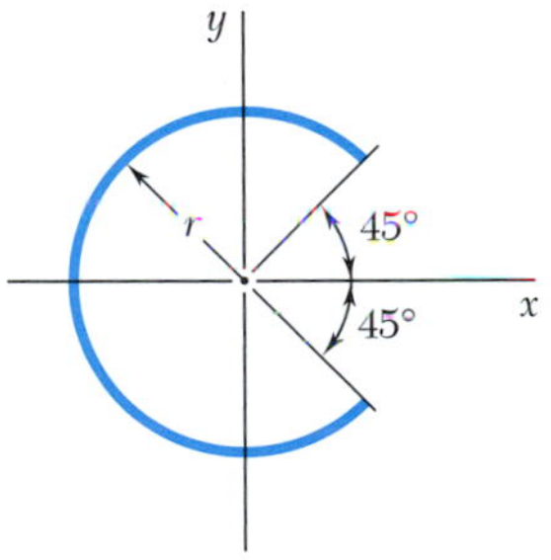

Fig. P5.43

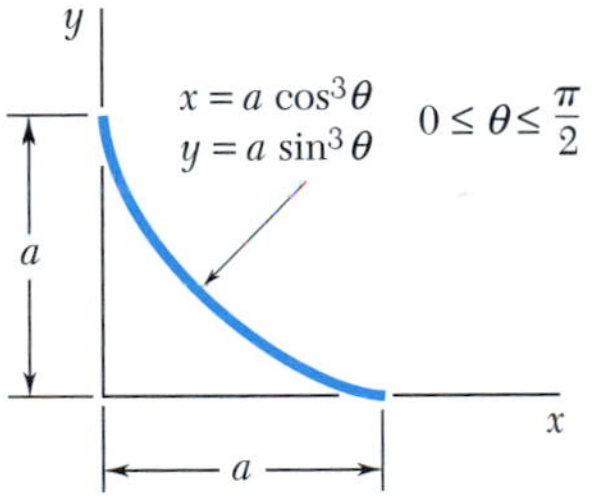

Fig. P5.44

***5.45 and *5.46** Determine by direct integration the centroid of the area shown.

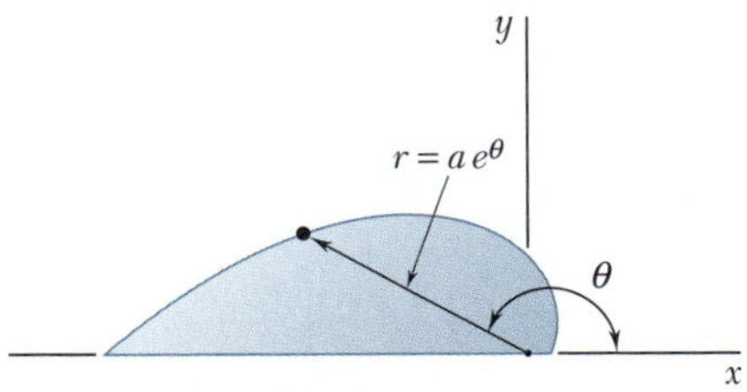

Fig. P5.45

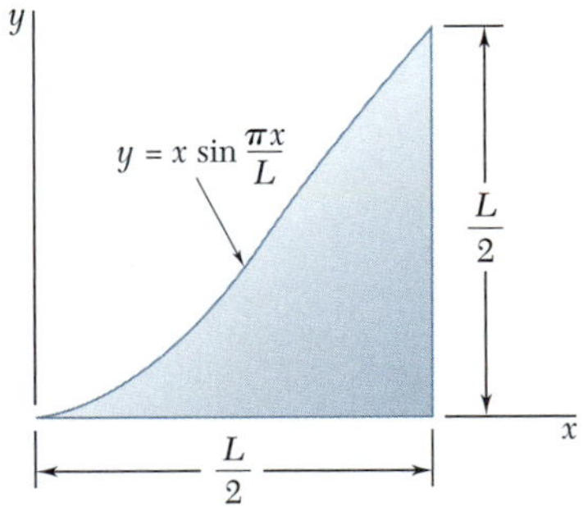

Fig. P5.46

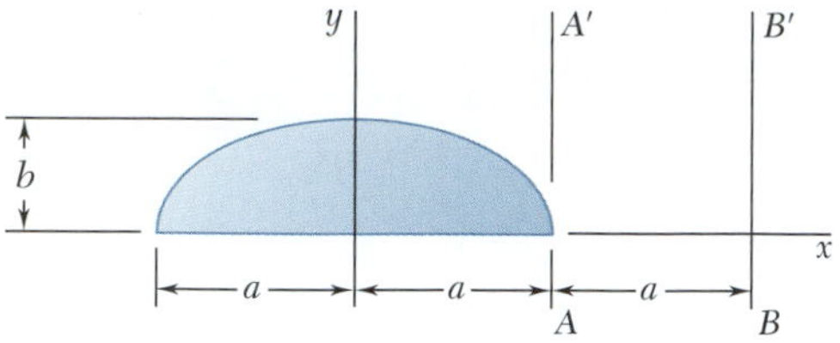

Fig. P5.50

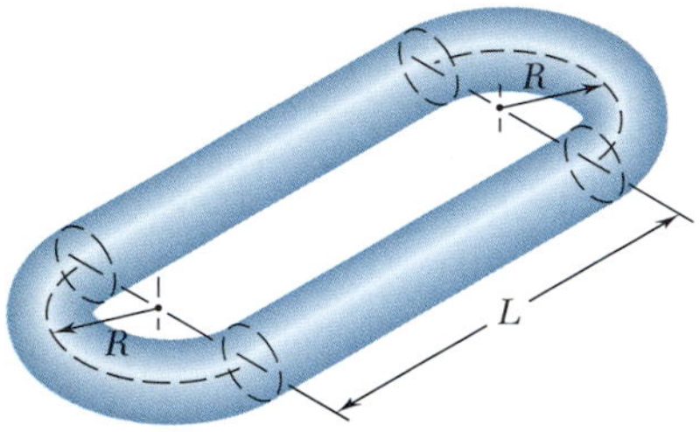

Fig. P5.51

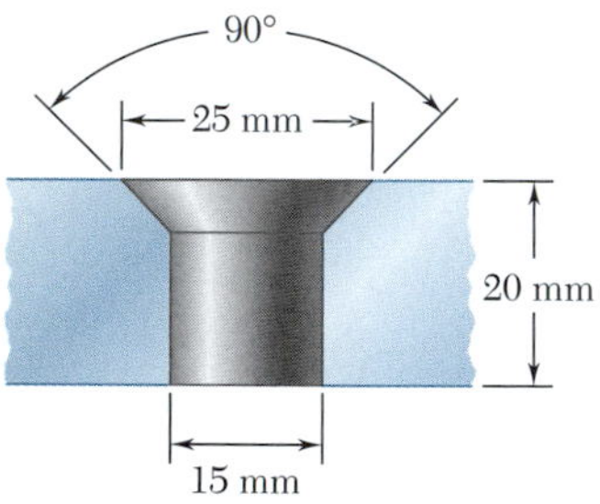

Fig. P5.53

5.47 Determine the volume and the surface area of the solid obtained by rotating the area of Prob. 5.2 about (*a*) the x axis, (*b*) the line $x = 165$ mm.

5.48 Determine the volume and the surface area of the solid obtained by rotating the area of Prob. 5.4 about (*a*) the line $y = 22$ mm, (*b*) the line $x = 12$ mm.

5.49 Determine the volume and the surface area of the solid obtained by rotating the area of Prob. 5.1 about (*a*) the x axis, (*b*) the line $x = 16$ in.

5.50 Determine the volume of the solid generated by rotating the semi-elliptical area shown about (*a*) the axis AA', (*b*) the axis BB', (*c*) the y axis.

5.51 Determine the volume and the surface area of the chain link shown, which is made from a 2-in.-diameter bar, if $R = 3$ in. and $L = 10$ in.

5.52 Verify that the expressions for the volumes of the first four shapes in Fig. 5.21 on page 261 are correct.

5.53 A 15-mm-diameter hole is drilled in a piece of 20-mm-thick steel; the hole is then countersunk as shown. Determine the volume of steel removed during the countersinking process.

5.54 Three different drive belt profiles are to be studied. If at any given time each belt makes contact with one-half of the circumference of its pulley, determine the *contact area* between the belt and the pulley for each design.

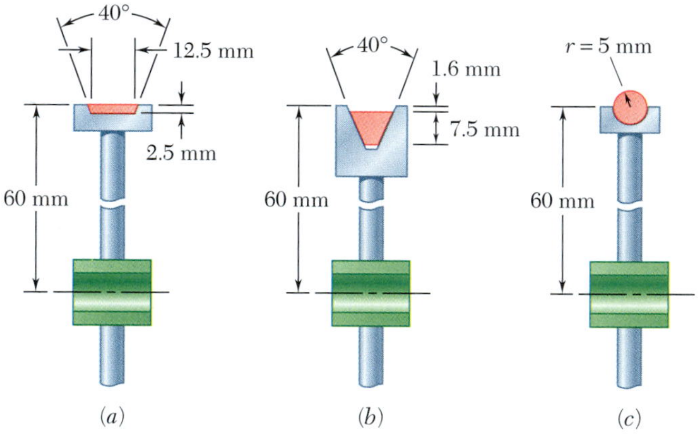

Fig. P5.54

5.55 Determine the capacity, in gallons, of the punch bowl shown if $R = 12$ in.

Fig. *P5.55*

5.56 The aluminum shade for a small high-intensity lamp has a uniform thickness of 3/32 in. Knowing that the specific weight of aluminum is 0.101 lb/in^3, determine the weight of the shade.

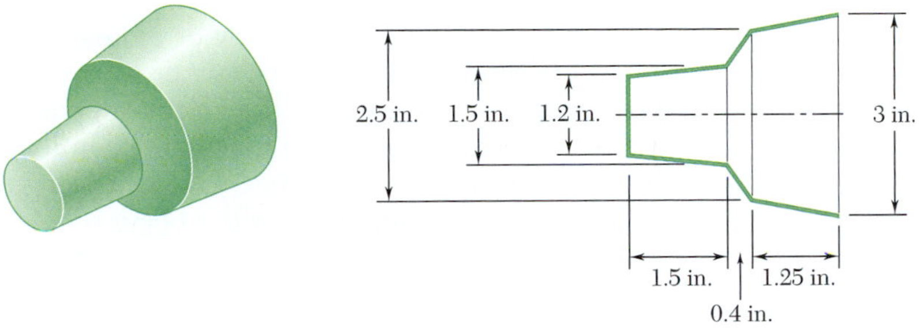

Fig. P5.56

5.57 The top of a round wooden table has the edge profile shown. Knowing that the diameter of the top is 1100 mm before shaping and that the density of the wood is 690 kg/m^3, determine the weight of the waste wood resulting from the production of 5000 tops.

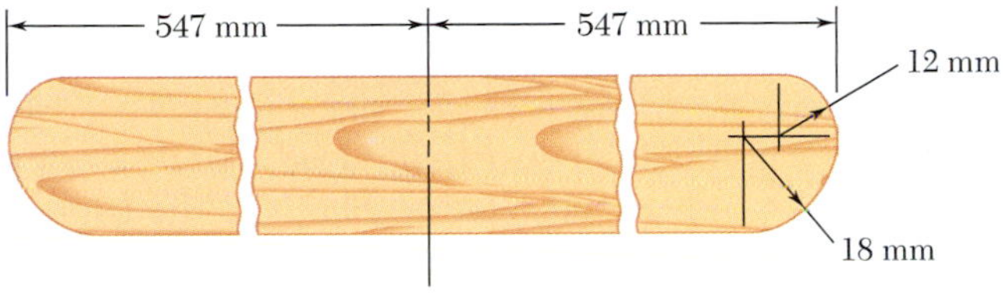

Fig. P5.57 and *P5.58*

5.58 The top of a round wooden table has the shape shown. Determine how many liters of lacquer are required to finish 5000 tops knowing that each top is given three coats of lacquer and that 1 liter of lacquer covers 12 m^2.

5.59 The escutcheon (a decorative plate placed on a pipe where the pipe exits from a wall) shown is cast from yellow brass. Knowing that the specific weight of yellow brass is 0.306 lb/in^3, determine the weight of the escutcheon.

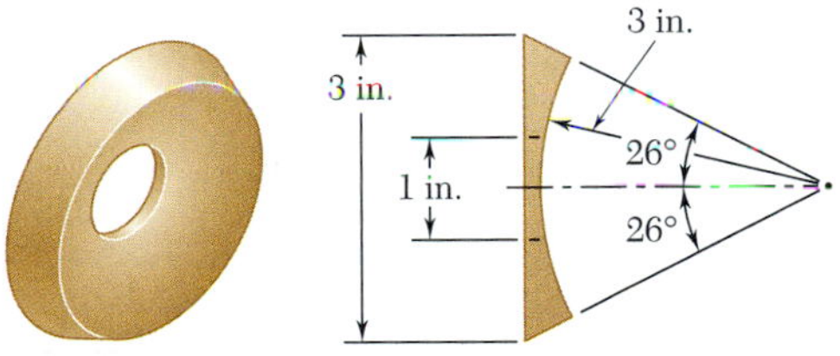

Fig. P5.59

***5.60** The reflector of a small flashlight has the parabolic shape shown. Determine the surface area of the inside of the reflector.

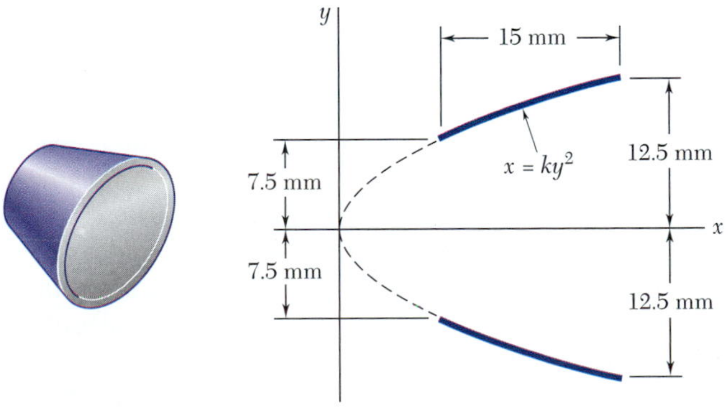

Fig. P5.60

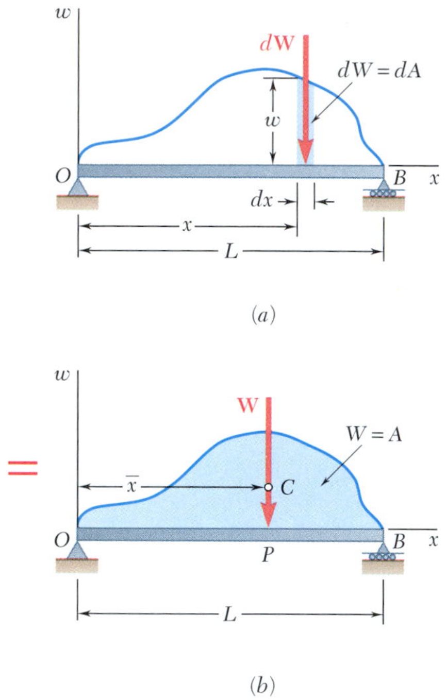

Fig. 5.17

Photo 5.3 The roofs of the buildings shown must be able to support not only the total weight of the snow but also the nonsymmetric distributed loads resulting from drifting of the snow.

*5.8. DISTRIBUTED LOADS ON BEAMS

The concept of the centroid of an area can be used to solve other problems besides those dealing with the weights of flat plates. Consider, for example, a beam supporting a *distributed load;* this load may consist of the weight of materials supported directly or indirectly by the beam, or it may be caused by wind or hydrostatic pressure. The distributed load can be represented by plotting the load w supported per unit length (Fig. 5.17); this load is expressed in N/m or in lb/ft. The magnitude of the force exerted on an element of beam of length dx is $dW = w\,dx$, and the total load supported by the beam is

$$W = \int_0^L w\,dx$$

We observe that the product $w\,dx$ is equal in magnitude to the element of area dA shown in Fig. 5.17*a*. The load W is thus equal in magnitude to the total area A under the load curve:

$$W = \int dA = A$$

We now determine where a *single concentrated load* $\mathbf{W}$, of the same magnitude W as the total distributed load, should be applied on the beam if it is to produce the same reactions at the supports (Fig. 5.17*b*). However, this concentrated load $\mathbf{W}$, which represents the resultant of the given distributed loading, is equivalent to the loading only when considering the free-body diagram of the entire beam. The point of application P of the equivalent concentrated load $\mathbf{W}$ is obtained by expressing that the moment of $\mathbf{W}$ about point O is equal to the sum of the moments of the elemental loads $d\mathbf{W}$ about O:

$$(OP)W = \int x\,dW$$

or, since $dW = w\,dx = dA$ and $W = A$,

$$(OP)A = \int_0^L x\,dA \qquad (5.12)$$

Since the integral represents the first moment with respect to the w axis of the area under the load curve, it can be replaced by the product $\bar{x}A$. We therefore have $OP = \bar{x}$, where $\bar{x}$ is the distance from the w axis to the centroid C of the area A (this is *not* the centroid of the beam).

A distributed load on a beam can thus be replaced by a concentrated load; the magnitude of this single load is equal to the area under the load curve, and its line of action passes through the centroid of that area. It should be noted, however, that the concentrated load is equivalent to the given loading only as far as external forces are concerned. It can be used to determine reactions but should not be used to compute internal forces and deflections.

*5.9. FORCES ON SUBMERGED SURFACES

The approach used in the preceding section can be used to determine the resultant of the hydrostatic pressure forces exerted on a *rectangular surface* submerged in a liquid. Consider the rectangular plate shown in Fig. 5.18, which is of length L and width b, where b is measured perpendicular to the plane of the figure. As noted in Sec. 5.8, the load exerted on an element of the plate of length dx is $w\,dx$, where w is the load per unit length. However, this load can also be expressed as $p\,dA = pb\,dx$, where p is the gage pressure in the liquid† and b is the width of the plate; thus, $w = bp$. Since the gage pressure in a liquid is $p = \gamma h$, where γ is the specific weight of the liquid and h is the vertical distance from the free surface, it follows that

$$w = bp = b\gamma h \tag{5.13}$$

which shows that the load per unit length w is proportional to h and, thus, varies linearly with x.

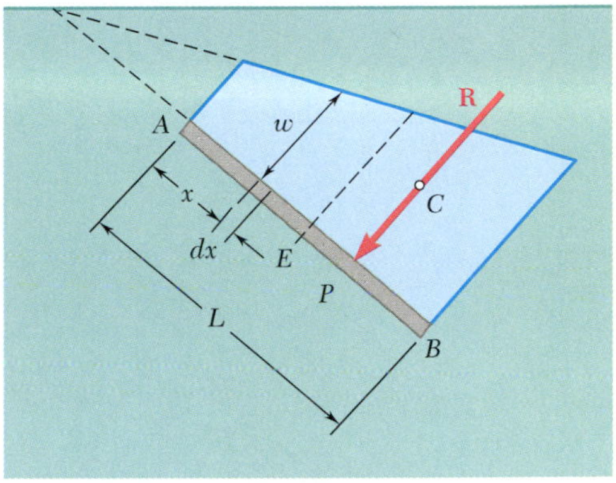

Fig. 5.18

Recalling the results of Sec. 5.8, we observe that the resultant **R** of the hydrostatic forces exerted on one side of the plate is equal in magnitude to the trapezoidal area under the load curve and that its line of action passes through the centroid C of that area. The point P of the plate where **R** is applied is known as the *center of pressure.*‡

Next, we consider the forces exerted by a liquid on a curved surface of constant width (Fig. 5.19*a*). Since the determination of the resultant **R** of these forces by direct integration would not be easy, we consider the free body obtained by detaching the volume of liquid ABD bounded by the curved surface AB and by the two plane surfaces AD and DB shown in Fig. 5.19*b*. The forces acting on the free body ABD are the weight **W** of the detached volume of liquid, the resultant $\mathbf{R}_1$ of the forces exerted on AD, the resultant $\mathbf{R}_2$ of the forces exerted on BD, and the resultant $-\mathbf{R}$ of the forces exerted *by the curved surface on the liquid.* The resultant $-\mathbf{R}$ is equal and opposite to, and has the same line of action as, the resultant **R** of the forces exerted *by the liquid on the curved surface.* The forces **W**, $\mathbf{R}_1$, and $\mathbf{R}_2$ can be determined by standard methods; after their values have been found, the force $-\mathbf{R}$ is obtained by solving the equations of equilibrium for the free body of Fig. 5.19*b*. The resultant **R** of the hydrostatic forces exerted on the curved surface is then obtained by reversing the sense of $-\mathbf{R}$.

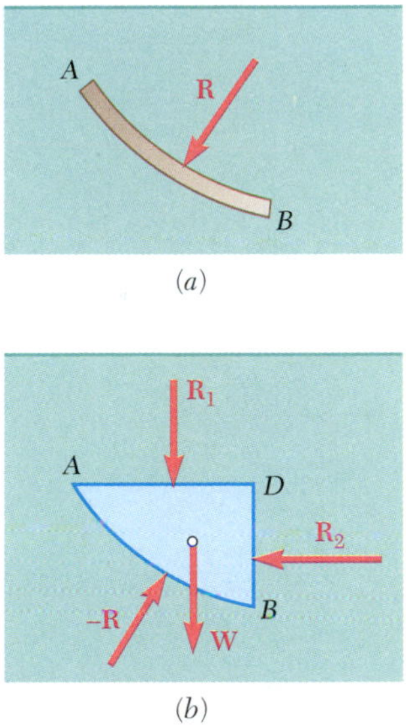

Fig. 5.19

The methods outlined in this section can be used to determine the resultant of the hydrostatic forces exerted on the surfaces of dams and rectangular gates and vanes. The resultants of forces on submerged surfaces of variable width will be determined in Chap. 9.

Photo 5.4 As discussed in this section, the Grand Coulee Dam supports three different kinds of distributed forces: the weights of its constitutive elements, the pressure forces exerted by the water on its submerged face, and the pressure forces exerted by the ground on its base.

†The pressure p, which represents a load per unit area, is expressed in N/m^2 or in lb/ft^2. The derived SI unit N/m^2 is called a *pascal* (Pa).

‡Noting that the area under the load curve is equal to $w_E L$, where w_E is the load per unit length at the center E of the plate, and recalling Eq. (5.13), we can write

$$R = w_E L = (bp_E)L = p_E(bL) = p_E A$$

where A denotes the area of the *plate.* Thus, the magnitude of **R** can be obtained by multiplying the area of the plate by the pressure at its center E. The resultant **R**, however, *should be applied at P, not at E.*

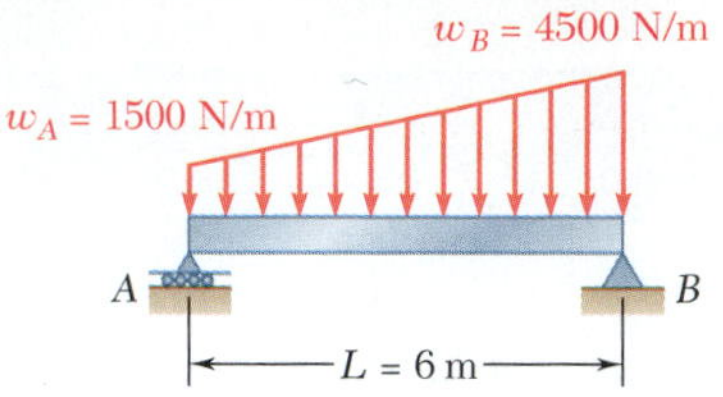

SAMPLE PROBLEM 5.9

A beam supports a distributed load as shown. (*a*) Determine the equivalent concentrated load. (*b*) Determine the reactions at the supports.

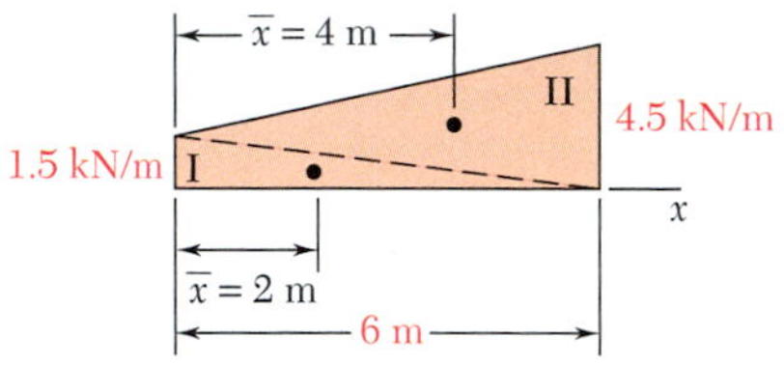

SOLUTION

***a.* Equivalent Concentrated Load.** The magnitude of the resultant of the load is equal to the area under the load curve, and the line of action of the resultant passes through the centroid of the same area. We divide the area under the load curve into two triangles and construct the table below. To simplify the computations and tabulation, the given loads per unit length have been converted into kN/m.

Component	A, kN	$\bar{x}$, m	$\bar{x}A$, kN · m
Triangle I	4.5	2	9
Triangle II	13.5	4	54
	$\Sigma A = 18.0$		$\Sigma \bar{x}A = 63$

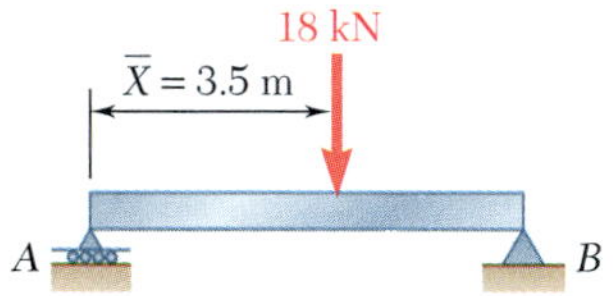

Thus, $\bar{X}\Sigma A = \Sigma \bar{x}A$: $\quad \bar{X}(18 \text{ kN}) = 63 \text{ kN} \cdot \text{m} \quad \bar{X} = 3.5 \text{ m}$

The equivalent concentrated load is

$$\mathbf{W} = 18 \text{ kN} \downarrow \quad ◀$$

and its line of action is located at a distance

$$\bar{X} = 3.5 \text{ m to the right of } A \quad ◀$$

***b.* Reactions.** The reaction at A is vertical and is denoted by **A**; the reaction at B is represented by its components $\mathbf{B}_x$ and $\mathbf{B}_y$. The given load can be considered to be the sum of two triangular loads as shown. The resultant of each triangular load is equal to the area of the triangle and acts at its centroid. We write the following equilibrium equations for the free body shown:

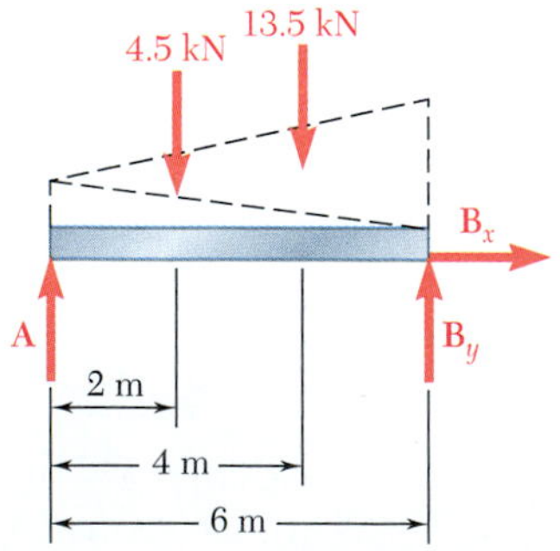

$\xrightarrow{+} \Sigma F_x = 0$: $\qquad \mathbf{B}_x = 0 \quad ◀$

$+\circlearrowleft \Sigma M_A = 0$: $\qquad -(4.5 \text{ kN})(2 \text{ m}) - (13.5 \text{ kN})(4 \text{ m}) + B_y(6 \text{ m}) = 0$

$$\mathbf{B}_y = 10.5 \text{ kN} \uparrow \quad ◀$$

$+\circlearrowleft \Sigma M_B = 0$: $\qquad +(4.5 \text{ kN})(4 \text{ m}) + (13.5 \text{ kN})(2 \text{ m}) - A(6 \text{ m}) = 0$

$$\mathbf{A} = 7.5 \text{ kN} \uparrow \quad ◀$$

Alternative Solution. The given distributed load can be replaced by its resultant, which was found in part *a*. The reactions can be determined by writing the equilibrium equations $\Sigma F_x = 0$, $\Sigma M_A = 0$, and $\Sigma M_B = 0$. We again obtain

$$\mathbf{B}_x = 0 \qquad \mathbf{B}_y = 10.5 \text{ kN} \uparrow \qquad \mathbf{A} = 7.5 \text{ kN} \uparrow \quad ◀$$

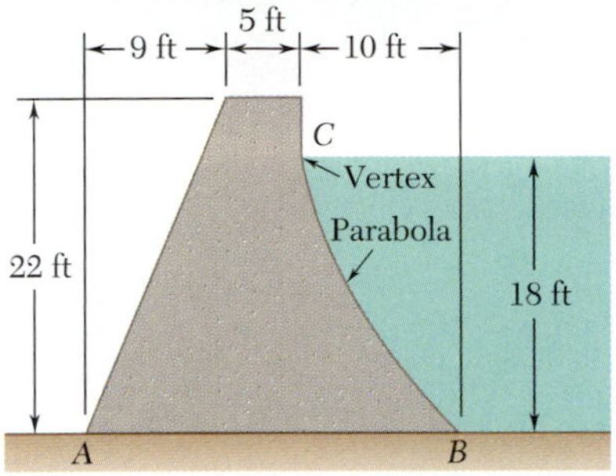

SAMPLE PROBLEM 5.10

The cross section of a concrete dam is as shown. Consider a 1-ft-thick section of the dam, and determine (*a*) the resultant of the reaction forces exerted by the ground on the base *AB* of the dam, (*b*) the resultant of the pressure forces exerted by the water on the face *BC* of the dam. The specific weights of concrete and water are 150 lb/ft^3 and 62.4 lb/ft^3, respectively.

SOLUTION

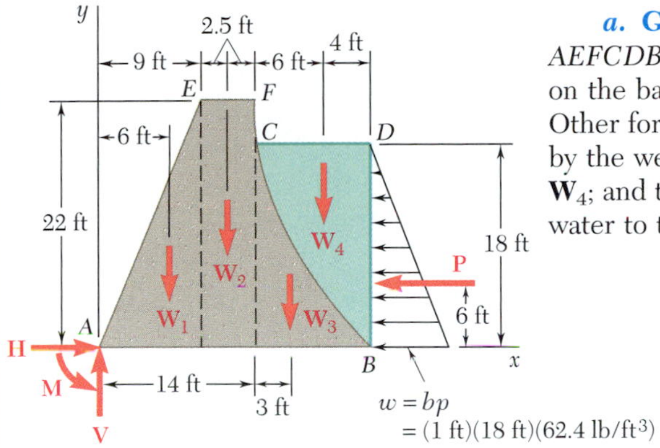

a. Ground Reaction. We choose as a free body the 1-ft-thick section *AEFCDB* of the dam and water. The reaction forces exerted by the ground on the base *AB* are represented by an equivalent force-couple system at *A*. Other forces acting on the free body are the weight of the dam, represented by the weights of its components $\mathbf{W}_1$, $\mathbf{W}_2$, and $\mathbf{W}_3$; the weight of the water $\mathbf{W}_4$; and the resultant $\mathbf{P}$ of the pressure forces exerted on section *BD* by the water to the right of section *BD*. We have

$$W_1 = \tfrac{1}{2}(9\text{ ft})(22\text{ ft})(1\text{ ft})(150\text{ lb/ft}^3) = 14{,}850\text{ lb}$$
$$W_2 = (5\text{ ft})(22\text{ ft})(1\text{ ft})(150\text{ lb/ft}^3) = 16{,}500\text{ lb}$$
$$W_3 = \tfrac{1}{3}(10\text{ ft})(18\text{ ft})(1\text{ ft})(150\text{ lb/ft}^3) = 9000\text{ lb}$$
$$W_4 = \tfrac{2}{3}(10\text{ ft})(18\text{ ft})(1\text{ ft})(62.4\text{ lb/ft}^3) = 7488\text{ lb}$$
$$P = \tfrac{1}{2}(18\text{ ft})(1\text{ ft})(18\text{ ft})(62.4\text{ lb/ft}^3) = 10{,}109\text{ lb}$$

Equilibrium Equations

$\overset{+}{\rightarrow}\Sigma F_x = 0$: $\quad H - 10{,}109\text{ lb} = 0 \qquad$ $\mathbf{H} = 10{,}110\text{ lb} \rightarrow$ ◀

$+\uparrow\Sigma F_y = 0$: $\quad V - 14{,}850\text{ lb} - 16{,}500\text{ lb} - 9000\text{ lb} - 7488\text{ lb} = 0$

$\mathbf{V} = 47{,}840\text{ lb}\uparrow$ ◀

$+\circlearrowleft\Sigma M_A = 0$: $\quad -(14{,}850\text{ lb})(6\text{ ft}) - (16{,}500\text{ lb})(11.5\text{ ft}) - (9000\text{ lb})(17\text{ ft}) - (7488\text{ lb})(20\text{ ft}) + (10{,}109\text{ lb})(6\text{ ft}) + M = 0$

$\mathbf{M} = 520{,}960\text{ lb}\cdot\text{ft}\circlearrowleft$ ◀

We can replace the force-couple system obtained by a single force acting at a distance *d* to the right of *A*, where

$$d = \frac{520{,}960\text{ lb}\cdot\text{ft}}{47{,}840\text{ lb}} = 10.89\text{ ft}$$

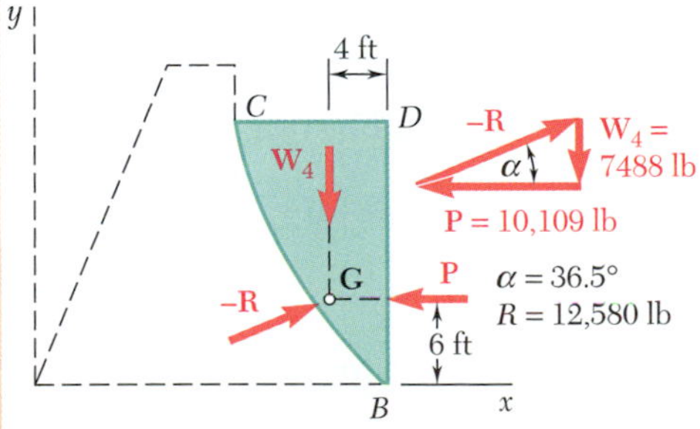

b. Resultant R of Water Forces. The parabolic section of water *BCD* is chosen as a free body. The forces involved are the resultant $-\mathbf{R}$ of the forces exerted by the dam on the water, the weight $\mathbf{W}_4$, and the force $\mathbf{P}$. Since these forces must be concurrent, $-\mathbf{R}$ passes through the point of intersection *G* of $\mathbf{W}_4$ and $\mathbf{P}$. A force triangle is drawn from which the magnitude and direction of $-\mathbf{R}$ are determined. The resultant $\mathbf{R}$ of the forces exerted by the water on the face *BC* is equal and opposite:

$\mathbf{R} = 12{,}580\text{ lb} \measuredangle 36.5°$ ◀

SOLVING PROBLEMS ON YOUR OWN

The problems in this lesson involve two common and very important types of loading: distributed loads on beams and forces on submerged surfaces of constant width. As we discussed in Secs. 5.8 and 5.9 and illustrated in Sample Probs. 5.9 and 5.10, determining the single equivalent force for each of these loadings requires a knowledge of centroids.

1. *Analyzing beams subjected to distributed loads.* In Sec. 5.8, we showed that a distributed load on a beam can be replaced by a single equivalent force. The magnitude of this force is equal to the area under the distributed load curve and its line of action passes through the centroid of that area. Thus, you should begin your solution by replacing the various distributed loads on a given beam by their respective single equivalent forces. The reactions at the supports of the beam can then be determined by using the methods of Chap. 4.

When possible, complex distributed loads should be divided into the common-shape areas shown in Fig. 5.8A [Sample Prob. 5.9]. Each of these areas can then be replaced by a single equivalent force. If required, the system of equivalent forces can be reduced further to a single equivalent force. As you study Sample Prob. 5.9, note how we have used the analogy between force and area and the techniques for locating the centroid of a composite area to analyze a beam subjected to a distributed load.

2. *Solving problems involving forces on submerged bodies.* The following points and techniques should be remembered when solving problems of this type.

a. The pressure p at a depth h below the free surface of a liquid is equal to γh or $\rho g h$, where γ and ρ are the specific weight and the density of the liquid, respectively. The load per unit length w acting on a submerged surface of constant width b is then

$$w = bp = b\gamma h = b\rho g h$$

b. The line of action of the resultant force $\mathbf{R}$ acting on a submerged plane surface is perpendicular to the surface.

c. For a vertical or inclined plane rectangular surface of width b, the loading on the surface can be represented by a linearly distributed load which is trapezoidal in shape (Fig. 5.18). Further, the magnitude of $\mathbf{R}$ is given by

$$R = \gamma h_E A$$

where h_E is the vertical distance to the center of the surface and A is the area of the surface.

d. The load curve will be triangular (rather than trapezoidal) when the top edge of a plane rectangular surface coincides with the free surface of the liquid, since the pressure of the liquid at the free surface is zero. For this case, the line of action of **R** is easily determined, for it passes through the centroid of a *triangular* distributed load.

e. For the general case, rather than analyzing a trapezoid, we suggest that you use the method indicated in part *b* of Sample Prob. 5.9. First divide the trapezoidal distributed load into two triangles, and then compute the magnitude of the resultant of each triangular load. (The magnitude is equal to the area of the triangle times the width of the plate.) Note that the line of action of each resultant force passes through the centroid of the corresponding triangle and that the sum of these forces is equivalent to **R**. Thus, rather than using **R**, you can use the two equivalent resultant forces, whose points of application are easily calculated. Of course, the equation given for R in paragraph c should be used when only the magnitude of **R** is needed.

f. When the submerged surface of constant width is curved, the resultant force acting on the surface is obtained by considering the equilibrium of the volume of liquid bounded by the curved surface and by horizontal and vertical planes (Fig. 5.19). Observe that the force $\mathbf{R}_1$ of Fig. 5.19 is equal to the weight of the liquid lying above the plane AD. The method of solution for problems involving curved surfaces is shown in part *b* of Sample Prob. 5.10.

In subsequent mechanics courses (in particular, mechanics of materials and fluid mechanics), you will have ample opportunity to use the ideas introduced in this lesson.

Problems

5.61 and 5.62 For the beam and loading shown, determine (*a*) the magnitude and location of the resultant of the distributed load, (*b*) the reactions at the beam supports.

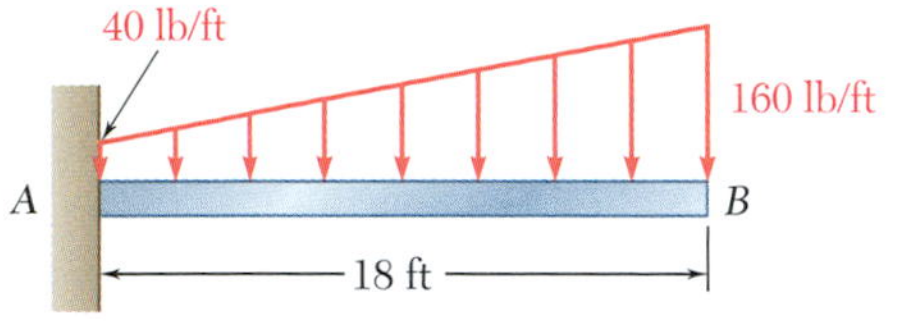

Fig. P5.61

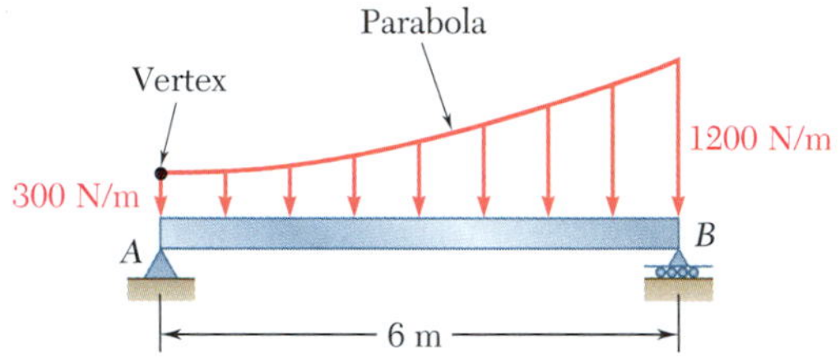

Fig. P5.62

5.63 through 5.68 Determine the reactions at the beam supports for the given loading.

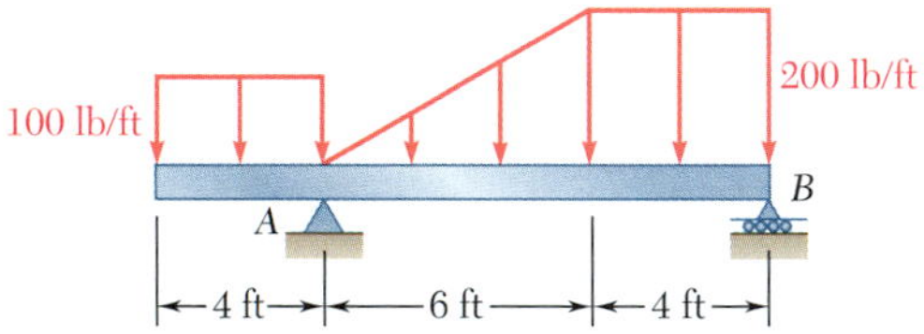

Fig. P5.63

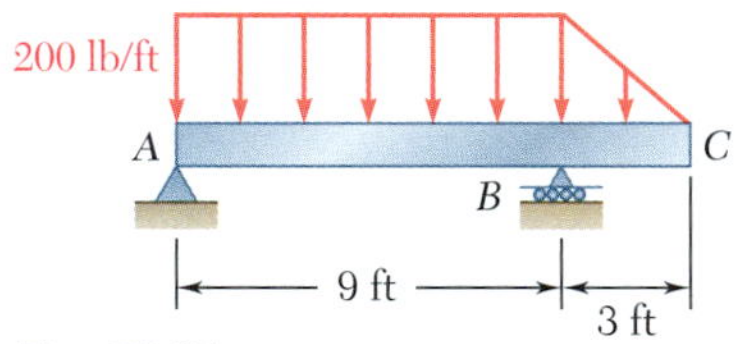

Fig. *P5.64*

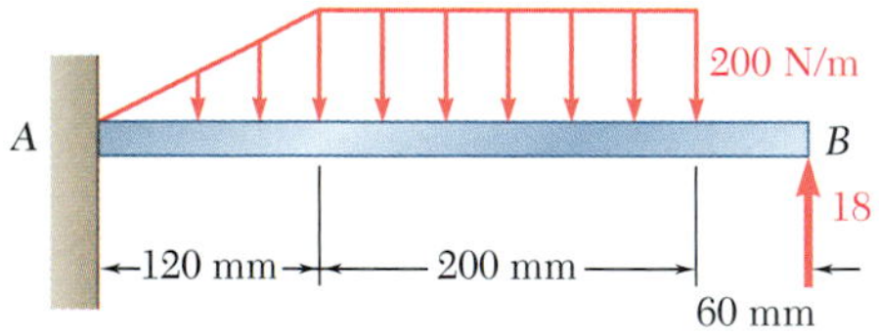

Fig. P5.65

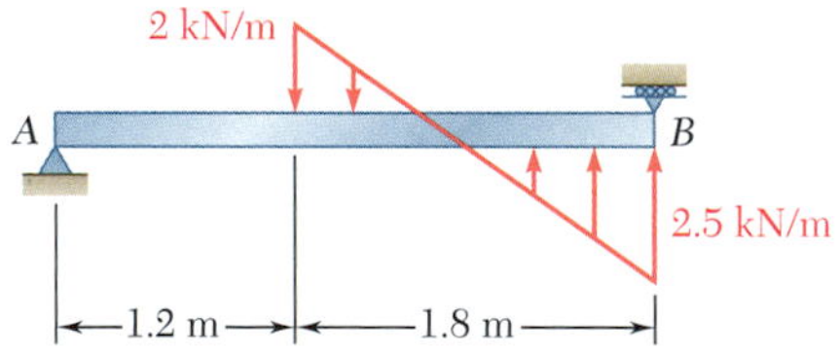

Fig. P5.66

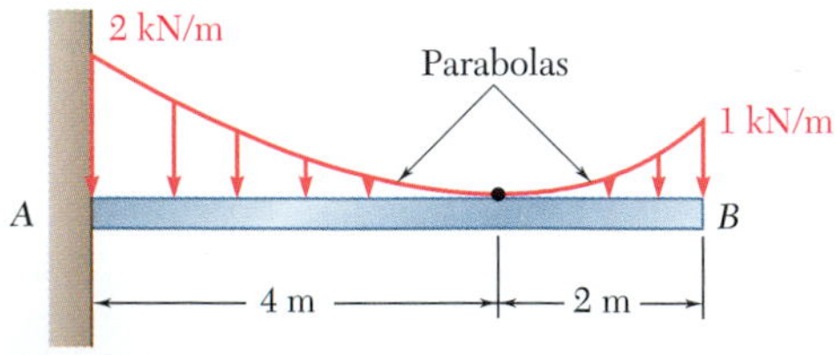

Fig. P5.67

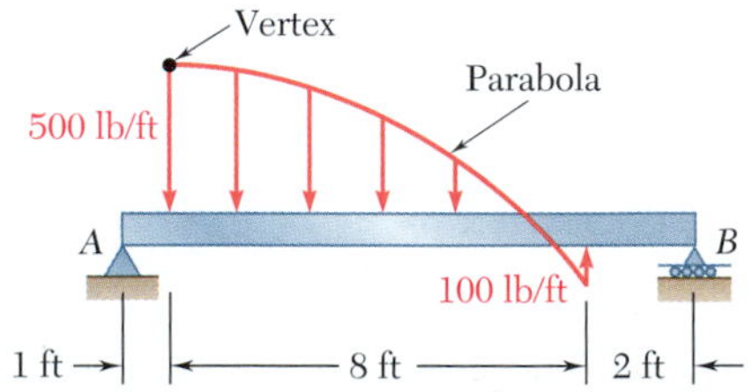

Fig. *P5.68*

5.69 Determine (*a*) the distance a so that the vertical reactions at supports A and B are equal, (*b*) the corresponding reactions at the supports.

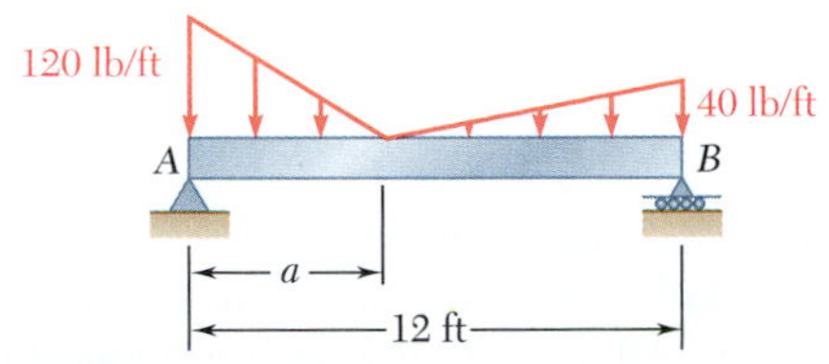

Fig. P5.69 and *P5.70*

5.70 Determine (*a*) the distance a so that the vertical reaction at support B is minimum, (*b*) the corresponding reactions at the supports.

5.71 Determine the reactions at the beam supports for the given loading when $w_0 = 1.5$ kN/m.

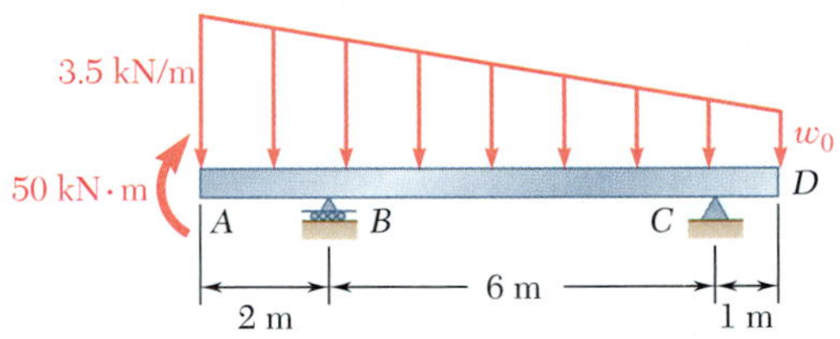

Fig. P5.71 and P5.72

5.72 Determine (*a*) the distributed load w_0 at the end D of the beam $ABCD$ for which the reaction at B is zero, (*b*) the corresponding reactions at C.

5.73 A grade beam AB supports three concentrated loads and rests on soil and the top of a large rock. The soil exerts an upward distributed load, and the rock exerts a concentrated load $\mathbf{R}_R$ as shown. Knowing that $P = 4$ kN and $w_B = \frac{1}{2} w_A$, determine the values of w_A and R_R corresponding to equilibrium.

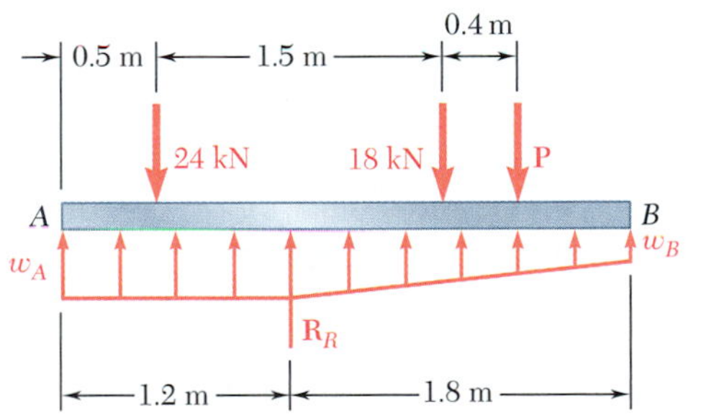

Fig. P5.73 and *P5.74*

5.74 A grade beam AB supports three concentrated loads and rests on soil and the top of a large rock. The soil exerts an upward distributed load, and the rock exerts a concentrated load $\mathbf{R}_R$ as shown. Knowing that $w_B = 0.4w_A$, determine (*a*) the largest value of $\mathbf{P}$ for which the beam is in equilibrium, (*b*) the corresponding value of w_A.

In the following problems, use $\gamma = 62.4$ lb/ft^3 for the specific weight of fresh water and $\gamma = 150$ lb/ft^3 for the specific weight of concrete if U.S. customary units are used. With SI units, use $\rho = 10^3$ kg/m^3 for the density of fresh water and $\rho_c = 2.40 \times 10^3$ kg/m^3 for the density of concrete. (See the footnote on page 222 for how to determine the specific weight of a material given its density.)

5.75 and 5.76 The cross section of a concrete dam is as shown. For a dam section of unit width, determine (*a*) the reaction forces exerted by the ground on the base *AB* of the dam, (*b*) the point of application of the resultant of the reaction forces of part *a*, (*c*) the resultant of the pressure forces exerted by the water on the face *BC* of the dam.

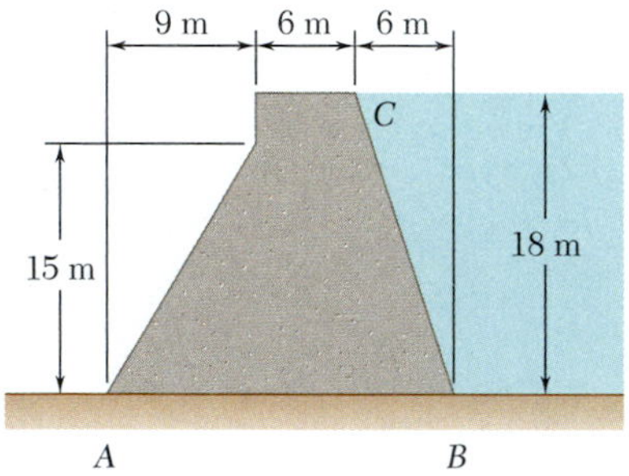

Fig. P5.75

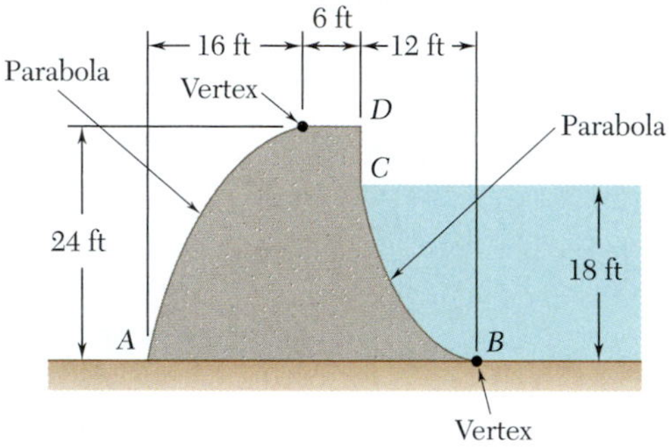

Fig. P5.76

5.77 The 9 × 12-ft side *AB* of a tank is hinged at its bottom *A* and is held in place by a thin rod *BC*. The maximum tensile force the rod can withstand without breaking is 40 kips, and the design specifications require the force in the rod not to exceed 20 percent of this value. If the tank is slowly filled with water, determine the maximum allowable depth of water *d* in the tank.

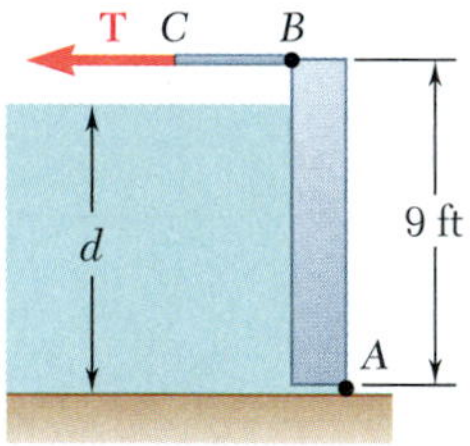

Fig. P5.77 and P5.78

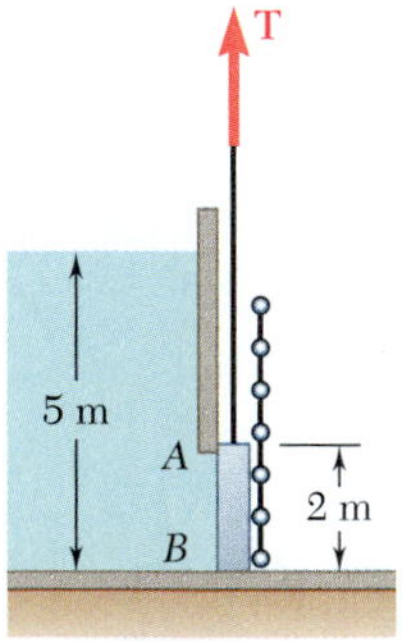

Fig. *P5.79*

5.78 The 9 × 12-ft side of an open tank is hinged at its bottom *A* and is held in place by a thin rod. The tank is filled with glycerine, whose specific weight is 80 lb/ft^3. Determine the force **T** in the rod and the reactions at the hinge after the tank is filled to a depth of 8 ft.

5.79 The friction force between a 2 × 2-m square sluice gate *AB* and its guides is equal to 10 percent of the resultant of the pressure forces exerted by the water on the face of the gate. Determine the initial force needed to lift the gate knowing that its mass is 500 kg.

5.80 The dam for a lake is designed to withstand the additional force caused by silt which has settled on the lake bottom. Assuming that silt is equivalent to a liquid of density $\rho_s = 1.76 \times 10^3$ kg/m^3 and considering a 1-m-wide section of dam, determine the percentage increase in the force acting on the dam face for a silt accumulation of depth 1.5 m.

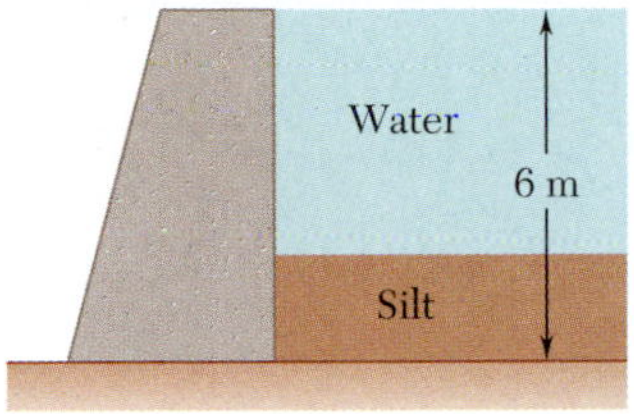

Fig. P5.80 and *P5.81*

5.81 The base of a dam for a lake is designed to resist up to 150 percent of the horizontal force of the water. After construction, it is found that silt (which is equivalent to a liquid of density $\rho_s = 1.76 \times 10^3$ kg/m^3) is settling on the lake bottom at a rate of 20 mm/yr. Considering a 1-m-wide section of dam, determine the number of years until the dam becomes unsafe.

5.82 The square gate *AB* is held in the position shown by hinges along its top edge *A* and by a shear pin at *B*. For a depth of water $d = 3.5$ m, determine the force exerted on the gate by the shear pin.

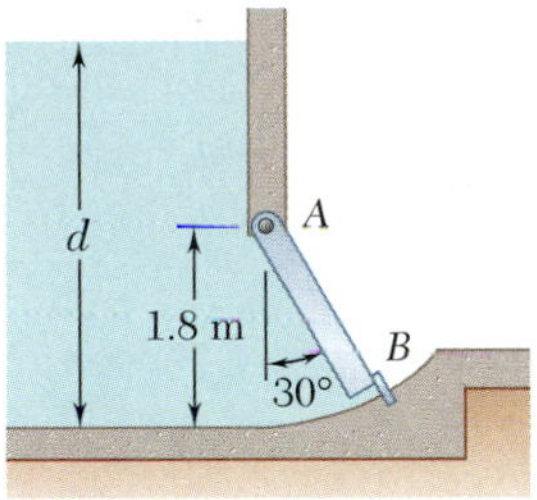

Fig. P5.82

5.83 A temporary dam is constructed in a 5-ft-wide fresh water channel by nailing two boards to pilings located at the sides of the channel and propping a third board *AB* against the pilings and the floor of the channel. Neglecting friction, determine the reactions at *A* and *B* when rope *BC* is slack.

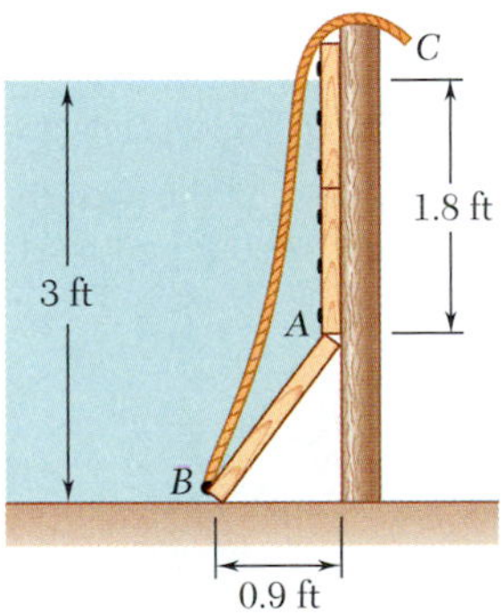

Fig. P5.83 and *P5.84*

5.84 A temporary dam is constructed in a 5-ft-wide fresh water channel by nailing two boards to pilings located at the sides of the channel and propping a third board *AB* against the pilings and the floor of the channel. Neglecting friction, determine the magnitude and direction of the minimum tension required in rope *BC* to move board *AB*.

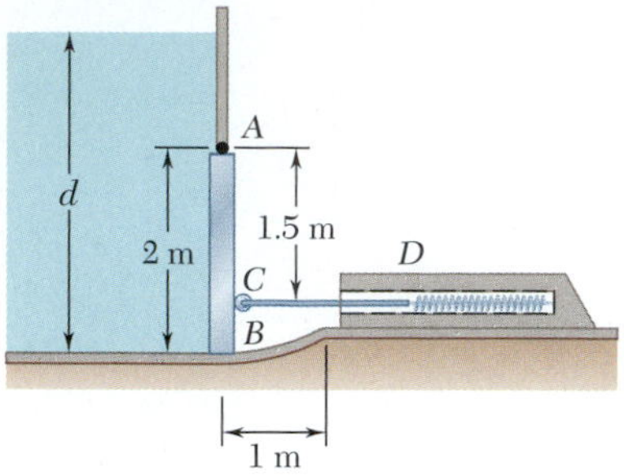

Fig. P5.85

5.85 A 2×3-m gate is hinged at A and is held in position by rod CD. End D rests against a spring whose constant is 12 kN/m. The spring is undeformed when the gate is vertical. Assuming that the force exerted by rod CD on the gate remains horizontal, determine the minimum depth of water d for which the bottom B of the gate will move to the end of the cylindrical portion of the floor.

5.86 Solve Prob. 5.85 if the mass of the gate is 500 kg.

5.87 The gate at the end of a 3-ft-wide fresh water channel is fabricated from three 240-lb, rectangular steel plates. The gate is hinged at A and rests against a frictionless support at D. Knowing that $d = 2.5$ ft, determine the reactions at A and D.

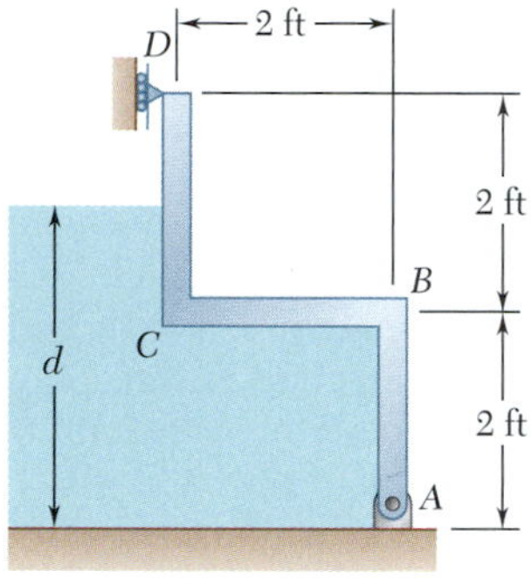

Fig. P5.87 and *P5.88*

5.88 The gate at the end of a 3-ft-wide fresh water channel is fabricated from three 240-lb, rectangular steel plates. The gate is hinged at A and rests against a frictionless support at D. Determine the depth of water d for which the gate will open.

5.89 A rain gutter is supported from the roof of a house by hangers that are spaced 0.6 m apart. After leaves clog the gutter's drain, the gutter slowly fills with rainwater. When the gutter is completely filled with water, determine (a) the resultant of the pressure force exerted by the water on the 0.6-m section of the curved surface of the gutter, (b) the force-couple system exerted on a hanger where it is attached to the gutter.

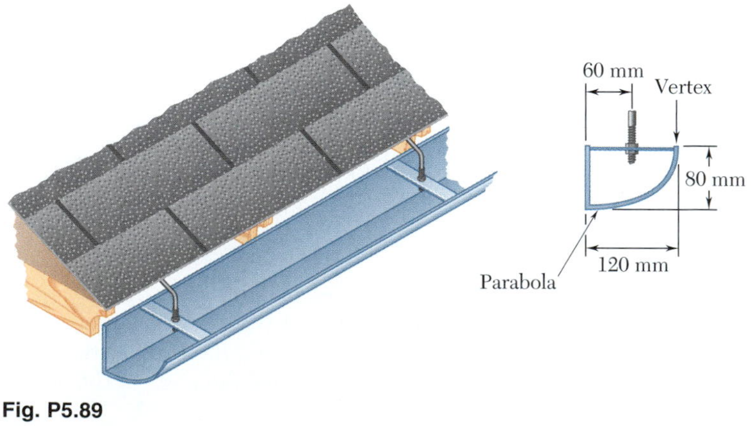

Fig. P5.89

5.10. CENTER OF GRAVITY OF A THREE-DIMENSIONAL BODY. CENTROID OF A VOLUME

The *center of gravity* G of a three-dimensional body is obtained by dividing the body into small elements and by then expressing that the weight $\mathbf{W}$ of the body acting at G is equivalent to the system of distributed forces $\Delta\mathbf{W}$ representing the weights of the small elements. Choosing the y axis to be vertical with positive sense upward (Fig. 5.20) and denoting by $\bar{\mathbf{r}}$ the position vector of G, we write that

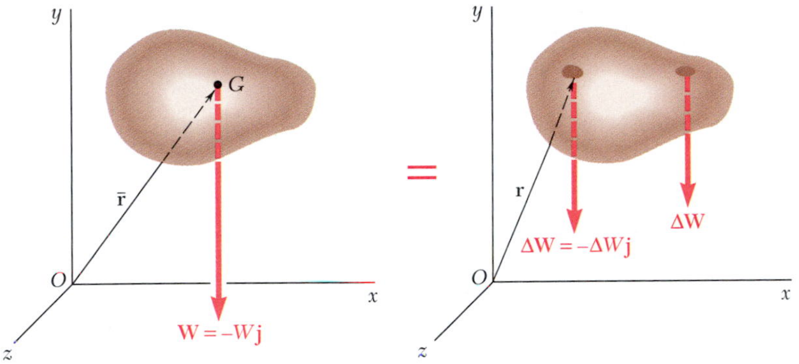

Fig. 5.20

$\mathbf{W}$ is equal to the sum of the elemental weights $\Delta\mathbf{W}$ and that its moment about O is equal to the sum of the moments about O of the elemental weights:

$$\Sigma\mathbf{F}: \qquad -W\mathbf{j} = \Sigma(-\Delta W\mathbf{j})$$
$$\Sigma\mathbf{M}_O: \qquad \bar{\mathbf{r}} \times (-W\mathbf{j}) = \Sigma[\mathbf{r} \times (-\Delta W\mathbf{j})] \qquad (5.13)$$

Rewriting the last equation in the form

$$\bar{\mathbf{r}}W \times (-\mathbf{j}) = (\Sigma\mathbf{r}\,\Delta W) \times (-\mathbf{j}) \qquad (5.14)$$

we observe that the weight $\mathbf{W}$ of the body is equivalent to the system of the elemental weights $\Delta\mathbf{W}$ if the following conditions are satisfied:

$$W = \Sigma\,\Delta W \qquad \bar{\mathbf{r}}W = \Sigma\mathbf{r}\,\Delta W$$

Increasing the number of elements and simultaneously decreasing the size of each element, we obtain in the limit

$$W = \int dW \qquad \bar{\mathbf{r}}W = \int \mathbf{r}\,dW \qquad (5.15)$$

We note that the relations obtained are independent of the orientation of the body. For example, if the body and the coordinate axes were rotated so that the z axis pointed upward, the unit vector $-\mathbf{j}$ would be replaced by $-\mathbf{k}$ in Eqs. (5.13) and (5.14), but the relations (5.15) would remain unchanged. Resolving the vectors $\bar{\mathbf{r}}$ and $\mathbf{r}$ into rectangular components, we note that the second of the relations (5.15) is equivalent to the three scalar equations

$$\bar{x}W = \int x\,dW \qquad \bar{y}W = \int y\,dW \qquad \bar{z}W = \int z\,dW \qquad (5.16)$$

Photo 5.5 To predict the flight characteristics of the modified Boeing 747 when used to transport a *Space Shuttle Orbiter*, the center of gravity of each craft had to be determined.

If the body is made of a homogeneous material of specific weight γ, the magnitude dW of the weight of an infinitesimal element can be expressed in terms of the volume dV of the element, and the magnitude W of the total weight can be expressed in terms of the total volume V. We write

$$dW = \gamma\, dV \qquad W = \gamma V$$

Substituting for dW and W in the second of the relations (5.15), we write

$$\bar{\mathbf{r}}V = \int \mathbf{r}\, dV \tag{5.17}$$

or, in scalar form,

$$\bar{x}V = \int x\, dV \qquad \bar{y}V = \int y\, dV \qquad \bar{z}V = \int z\, dV \tag{5.18}$$

The point whose coordinates are $\bar{x}$, $\bar{y}$, $\bar{z}$ is also known as the *centroid C of the volume* V of the body. If the body is not homogeneous, Eqs. (5.18) cannot be used to determine the center of gravity of the body; however, Eqs. (5.18) still define the centroid of the volume.

The integral $\int x\, dV$ is known as the *first moment of the volume with respect to the yz plane.* Similarly, the integrals $\int y\, dV$ and $\int z\, dV$ define the first moments of the volume with respect to the zx plane and the xy plane, respectively. It is seen from Eqs. (5.18) that if the centroid of a volume is located in a coordinate plane, the first moment of the volume with respect to that plane is zero.

A volume is said to be symmetrical with respect to a given plane if for every point P of the volume there exists a point P' of the same volume, such that the line PP' is perpendicular to the given plane and is bisected by that plane. The plane is said to be a *plane of symmetry* for the given volume. When a volume V possesses a plane of symmetry, the first moment of V with respect to that plane is zero, and the centroid of the volume is located in the plane of symmetry. When a volume possesses two planes of symmetry, the centroid of the volume is located on the line of intersection of the two planes. Finally, when a volume possesses three planes of symmetry which intersect at a well-defined point (that is, not along a common line), the point of intersection of the three planes coincides with the centroid of the volume. This property enables us to determine immediately the locations of the centroids of spheres, ellipsoids, cubes, rectangular parallelepipeds, etc.

The centroids of unsymmetrical volumes or of volumes possessing only one or two planes of symmetry should be determined by integration (Sec. 5.12). The centroids of several common volumes are shown in Fig. 5.21. It should be observed that in general the centroid of a volume of revolution *does not coincide* with the centroid of its cross section. Thus, the centroid of a hemisphere is different from that of a semicircular area, and the centroid of a cone is different from that of a triangle.

Shape		$\bar{x}$	Volume
Hemisphere	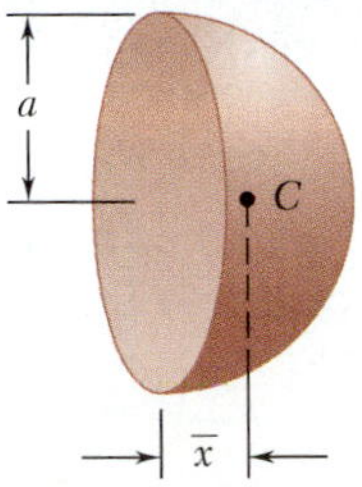	$\frac{3a}{8}$	$\frac{2}{3}\pi a^3$
Semiellipsoid of revolution	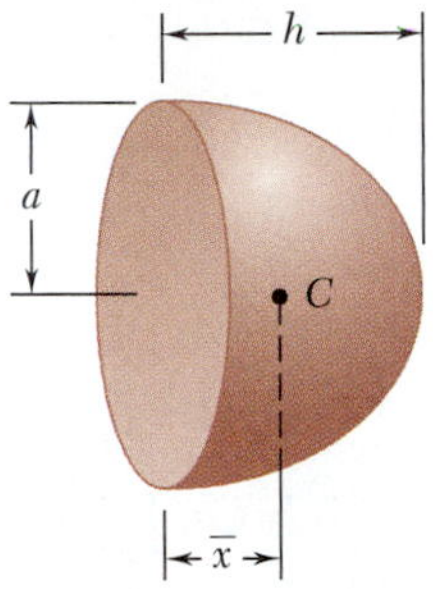	$\frac{3h}{8}$	$\frac{2}{3}\pi a^2 h$
Paraboloid of revolution	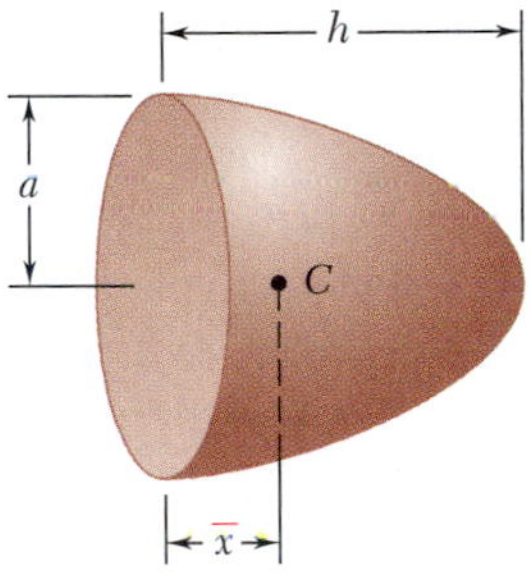	$\frac{h}{3}$	$\frac{1}{2}\pi a^2 h$
Cone	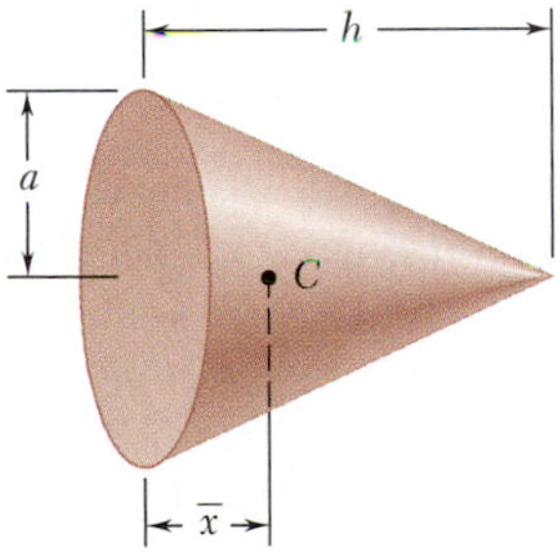	$\frac{h}{4}$	$\frac{1}{3}\pi a^2 h$
Pyramid	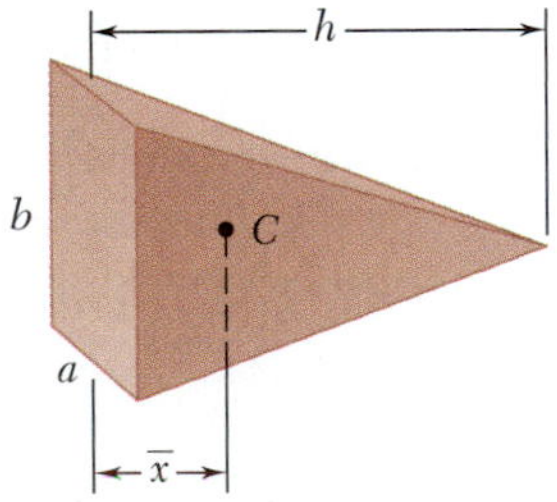	$\frac{h}{4}$	$\frac{1}{3}abh$

Fig. 5.21 Centroids of common shapes and volumes.

5.11. COMPOSITE BODIES

If a body can be divided into several of the common shapes shown in Fig. 5.21, its center of gravity G can be determined by expressing that the moment about O of its total weight is equal to the sum of the moments about O of the weights of the various component parts. Proceeding as in Sec. 5.10, we obtain the following equations defining the coordinates $\bar{X}$, $\bar{Y}$, $\bar{Z}$ of the center of gravity G:

$$\bar{X}\Sigma W = \Sigma \bar{x} W \qquad \bar{Y}\Sigma W = \Sigma \bar{y} W \qquad \bar{Z}\Sigma W = \Sigma \bar{z} W \tag{5.19}$$

If the body is made of a homogeneous material, its center of gravity coincides with the centroid of its volume, and we obtain

$$\bar{X}\Sigma V = \Sigma \bar{x} V \qquad \bar{Y}\Sigma V = \Sigma \bar{y} V \qquad \bar{Z}\Sigma V = \Sigma \bar{z} V \tag{5.20}$$

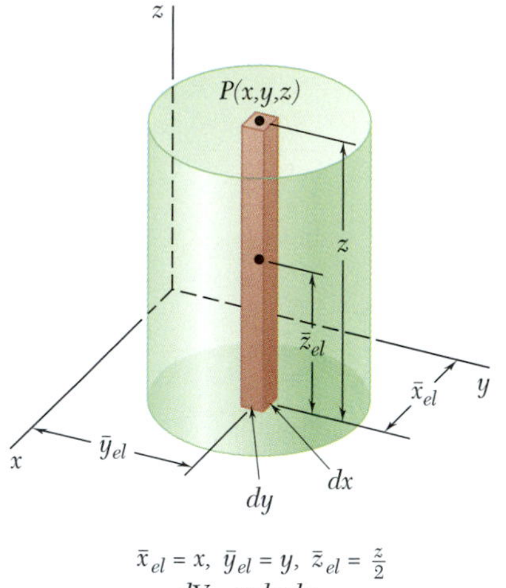

Fig. 5.22 Determination of the centroid of a volume by double integration.

5.12. DETERMINATION OF CENTROIDS OF VOLUMES BY INTEGRATION

The centroid of a volume bounded by analytical surfaces can be determined by evaluating the integrals given in Sec. 5.10:

$$\bar{x}V = \int x\,dV \qquad \bar{y}V = \int y\,dV \qquad \bar{z}V = \int z\,dV \tag{5.21}$$

If the element of volume dV is chosen to be equal to a small cube of sides dx, dy, and dz, the evaluation of each of these integrals requires a *triple integration*. However, it is possible to determine the coordinates of the centroid of most volumes by *double integration* if dV is chosen to be equal to the volume of a thin filament (Fig. 5.22). The coordinates of the centroid of the volume are then obtained by rewriting Eqs. (5.21) as

$$\bar{x}V = \int \bar{x}_{el}\,dV \qquad \bar{y}V = \int \bar{y}_{el}\,dV \qquad \bar{z}V = \int \bar{z}_{el}\,dV \tag{5.22}$$

and by then substituting the expressions given in Fig. 5.22 for the volume dV and the coordinates $\bar{x}_{el}$, $\bar{y}_{el}$, $\bar{z}_{el}$. By using the equation of the surface to express z in terms of x and y, the integration is reduced to a double integration in x and y.

If the volume under consideration possesses *two planes of symmetry*, its centroid must be located on the line of intersection of the two planes. Choosing the x axis to lie along this line, we have

$$\bar{y} = \bar{z} = 0$$

and the only coordinate to determine is $\bar{x}$. This can be done with a *single integration* by dividing the given volume into thin slabs parallel to the yz plane and expressing dV in terms of x and dx in the equation

$$\bar{x}V = \int \bar{x}_{el}\,dV \tag{5.23}$$

For a body of revolution, the slabs are circular and their volume is given in Fig. 5.23.

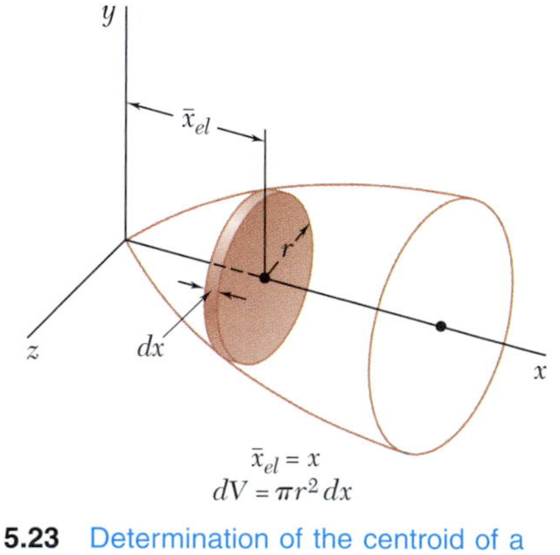

Fig. 5.23 Determination of the centroid of a body of revolution.

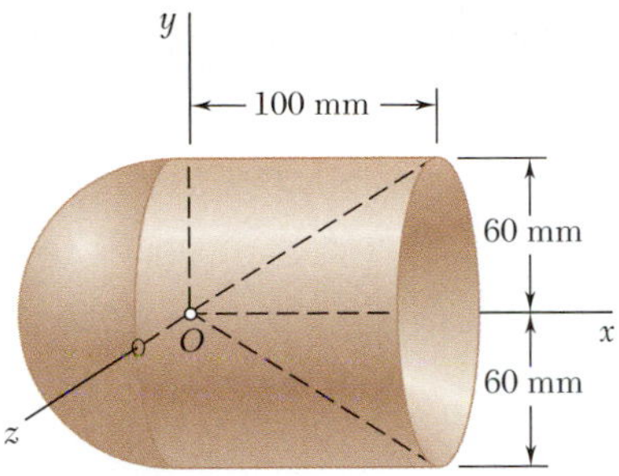

SAMPLE PROBLEM 5.11

Determine the location of the center of gravity of the homogeneous body of revolution shown, which was obtained by joining a hemisphere and a cylinder and carving out a cone.

SOLUTION

Because of symmetry, the center of gravity lies on the x axis. As shown in the figure below, the body can be obtained by adding a hemisphere to a cylinder and then subtracting a cone. The volume and the abscissa of the centroid of each of these components are obtained from Fig. 5.21 and are entered in the table below. The total volume of the body and the first moment of its volume with respect to the yz plane are then determined.

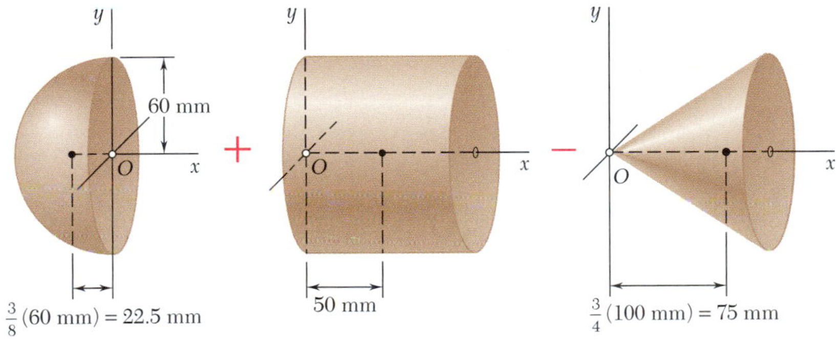

Component	Volume, mm^3	$\bar{x}$, mm	$\bar{x}V$, mm^4
Hemisphere	$\frac{1}{2}\frac{4\pi}{3}(60)^3 = 0.4524 \times 10^6$	-22.5	-10.18×10^6
Cylinder	$\pi(60)^2(100) = 1.1310 \times 10^6$	$+50$	$+56.55 \times 10^6$
Cone	$-\frac{\pi}{3}(60)^2(100) = -0.3770 \times 10^6$	$+75$	-28.28×10^6
	$\Sigma V = 1.206 \times 10^6$		$\Sigma \bar{x} V = +18.09 \times 10^6$

Thus,

$\bar{X}\Sigma V = \Sigma \bar{x} V$: $\quad \bar{X}(1.206 \times 10^6 \text{ mm}^3) = 18.09 \times 10^6 \text{ mm}^4$

$\bar{X} = 15 \text{ mm}$ ◀

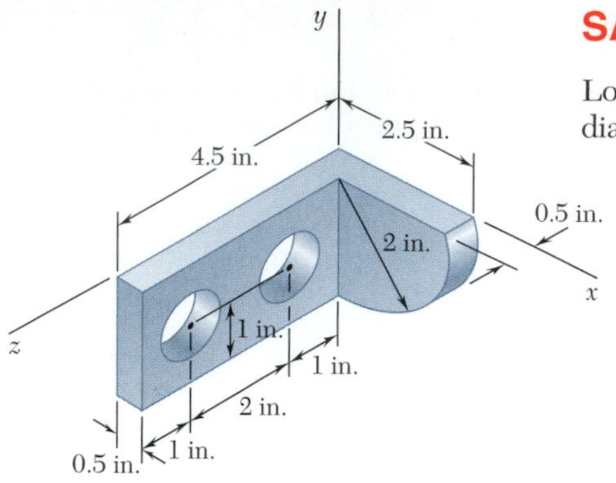

SAMPLE PROBLEM 5.12

Locate the center of gravity of the steel machine element shown. The diameter of each hole is 1 in.

SOLUTION

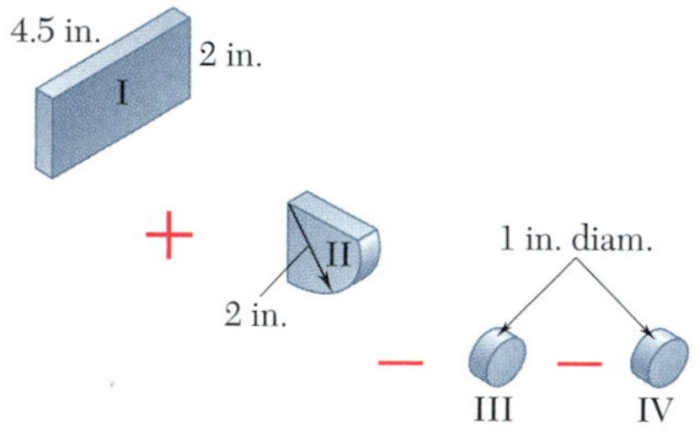

The machine element can be obtained by adding a rectangular parallelepiped (I) to a quarter cylinder (II) and then subtracting two 1-in.-diameter cylinders (III and IV). The volume and the coordinates of the centroid of each component are determined and are entered in the table below. Using the data in the table, we then determine the total volume and the moments of the volume with respect to each of the coordinate planes.

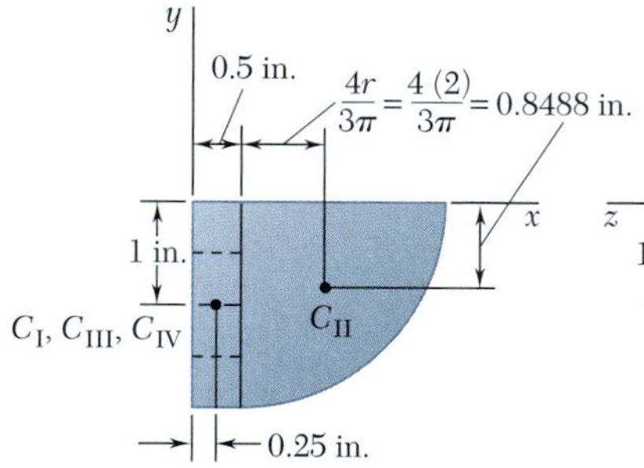

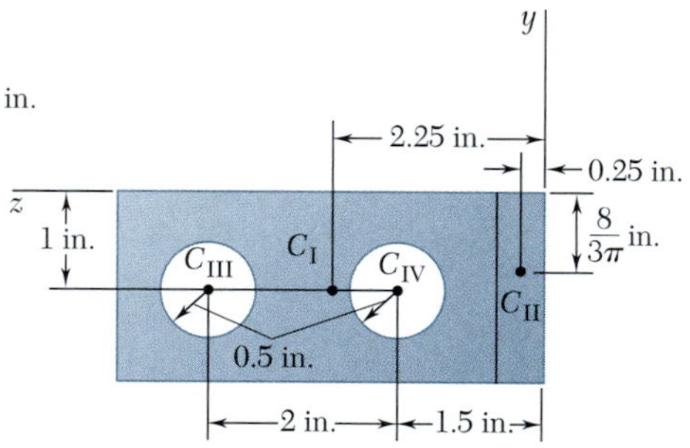

	V, in^3	$\bar{x}$, in.	$\bar{y}$, in.	$\bar{z}$, in.	$\bar{x}V$, in^4	$\bar{y}V$, in^4	$\bar{z}V$, in^4
I	$(4.5)(2)(0.5) = 4.5$	0.25	−1	2.25	1.125	−4.5	10.125
II	$\frac{1}{4}\pi(2)^2(0.5) = 1.571$	1.3488	−0.8488	0.25	2.119	−1.333	0.393
III	$-\pi(0.5)^2(0.5) = -0.3927$	0.25	−1	3.5	−0.098	0.393	−1.374
IV	$-\pi(0.5)^2(0.5) = -0.3927$	0.25	−1	1.5	−0.098	0.393	−0.589
	$\Sigma V = 5.286$				$\Sigma\bar{x}V = 3.048$	$\Sigma\bar{y}V = -5.047$	$\Sigma\bar{z}V = 8.555$

Thus,

$$\bar{X}\Sigma V = \Sigma\bar{x}V: \quad \bar{X}(5.286 \text{ in}^3) = 3.048 \text{ in}^4 \qquad \bar{X} = 0.577 \text{ in.} \blacktriangleleft$$
$$\bar{Y}\Sigma V = \Sigma\bar{y}V: \quad \bar{Y}(5.286 \text{ in}^3) = -5.047 \text{ in}^4 \qquad \bar{Y} = -0.955 \text{ in.} \blacktriangleleft$$
$$\bar{Z}\Sigma V = \Sigma\bar{z}V: \quad \bar{Z}(5.286 \text{ in}^3) = 8.555 \text{ in}^4 \qquad \bar{Z} = 1.618 \text{ in.} \blacktriangleleft$$

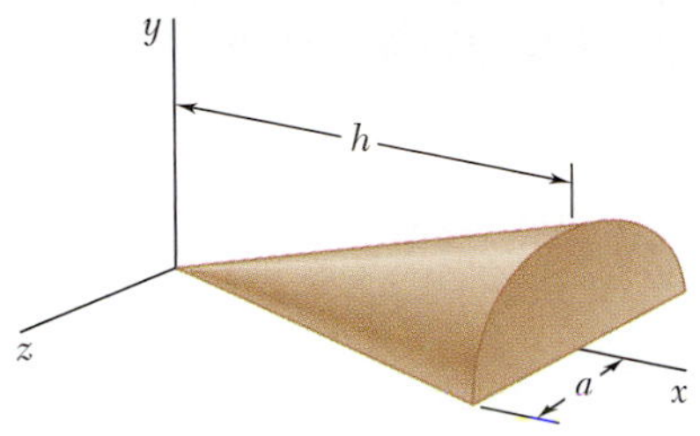

SAMPLE PROBLEM 5.13

Determine the location of the centroid of the half right circular cone shown.

SOLUTION

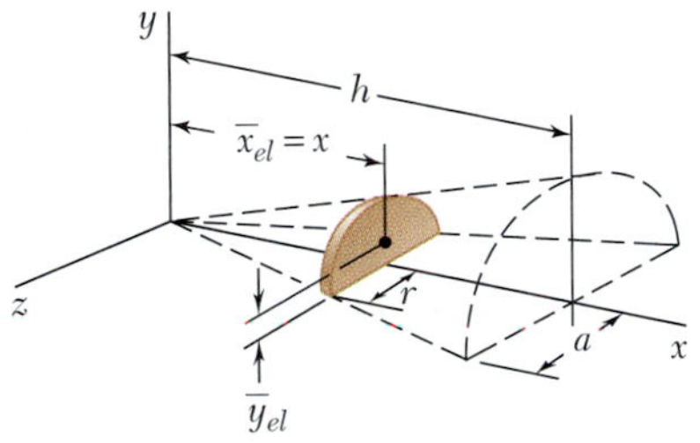

Since the xy plane is a plane of symmetry, the centroid lies in this plane and $\bar{z} = 0$. A slab of thickness dx is chosen as a differential element. The volume of this element is

$$dV = \tfrac{1}{2}\pi r^2\, dx$$

The coordinates $\bar{x}_{el}$ and $\bar{y}_{el}$ of the centroid of the element are

$$\bar{x}_{el} = x \qquad \bar{y}_{el} = \frac{4r}{3\pi}$$

where $\bar{y}_{el}$ is obtained from Fig. 5.8 (semicircular area).

We observe that r is proportional to x and write

$$\frac{r}{x} = \frac{a}{h} \qquad r = \frac{a}{h}x$$

The volume of the body is

$$V = \int dV = \int_0^h \tfrac{1}{2}\pi r^2\, dx = \int_0^h \tfrac{1}{2}\pi\left(\frac{a}{h}x\right)^2 dx = \frac{\pi a^2 h}{6}$$

The moment of the differential element with respect to the yz plane is $\bar{x}_{el}\,dV$; the total moment of the body with respect to this plane is

$$\int \bar{x}_{el}\, dV = \int_0^h x(\tfrac{1}{2}\pi r^2)\, dx = \int_0^h x(\tfrac{1}{2}\pi)\left(\frac{a}{h}x\right)^2 dx = \frac{\pi a^2 h^2}{8}$$

Thus,

$$\bar{x}V = \int \bar{x}_{el}\, dV \qquad \bar{x}\frac{\pi a^2 h}{6} = \frac{\pi a^2 h^2}{8} \qquad \bar{x} = \tfrac{3}{4}h \blacktriangleleft$$

Likewise, the moment of the differential element with respect to the zx plane is $\bar{y}_{el}\, dV$; the total moment is

$$\int \bar{y}_{el}\, dV = \int_0^h \frac{4r}{3\pi}(\tfrac{1}{2}\pi r^2)\, dx = \frac{2}{3}\int_0^h \left(\frac{a}{h}x\right)^3 dx = \frac{a^3 h}{6}$$

Thus,

$$\bar{y}V = \int \bar{y}_{el}\, dV \qquad \bar{y}\frac{\pi a^2 h}{6} = \frac{a^3 h}{6} \qquad \bar{y} = \frac{a}{\pi} \blacktriangleleft$$

SOLVING PROBLEMS ON YOUR OWN

In the problems for this lesson, you will be asked to locate the centers of gravity of three-dimensional bodies or the centroids of their volumes. All of the techniques we previously discussed for two-dimensional bodies—using symmetry, dividing the body into common shapes, choosing the most efficient differential element, etc.—may also be applied to the general three-dimensional case.

1. *Locating the centers of gravity of composite bodies.* In general, Eqs. (5.19) must be used:

$$\bar{X}\Sigma W = \Sigma \bar{x} W \qquad \bar{Y}\Sigma W = \Sigma \bar{y} W \qquad \bar{Z}\Sigma W = \Sigma \bar{z} W \tag{5.19}$$

However, for the case of a *homogeneous body,* the center of gravity of the body coincides with the *centroid of its volume*. Therefore, for this special case, the center of gravity of the body can also be located using Eqs. (5.20):

$$\bar{X}\Sigma V = \Sigma \bar{x} V \qquad \bar{Y}\Sigma V = \Sigma \bar{y} V \qquad \bar{Z}\Sigma V = \Sigma \bar{z} V \tag{5.20}$$

You should realize that these equations are simply an extension of the equations used for the two-dimensional problems considered earlier in the chapter. As the solutions of Sample Probs. 5.11 and 5.12 illustrate, the methods of solution for two- and three-dimensional problems are identical. Thus, we once again strongly encourage you to construct appropriate diagrams and tables when analyzing composite bodies. Also, as you study Sample Prob. 5.12, observe how the x and y coordinates of the centroid of the quarter cylinder were obtained using the equations for the centroid of a quarter circle.

We note that *two special cases* of interest occur when the given body consists of either uniform wires or uniform plates made of the same material.

a. For a body made of *several wire elements* of the *same uniform cross section,* the cross-sectional area A of the wire elements will factor out of Eqs. (5.20) when V is replaced with the product AL, where L is the length of a given element. Equations (5.20) thus reduce in this case to

$$\bar{X}\Sigma L = \Sigma \bar{x} L \qquad \bar{Y}\Sigma L = \Sigma \bar{y} L \qquad \bar{Z}\Sigma L = \Sigma \bar{z} L$$

b. For a body made of *several plates* of the *same uniform thickness,* the thickness t of the plates will factor out of Eqs. (5.20) when V is replaced with the product tA, where A is the area of a given plate. Equations (5.20) thus reduce in this case to

$$\bar{X}\Sigma A = \Sigma \bar{x} A \qquad \bar{Y}\Sigma A = \Sigma \bar{y} A \qquad \bar{Z}\Sigma A = \Sigma \bar{z} A$$

2. *Locating the centroids of volumes by direct integration.* As explained in Sec. 5.11, evaluating the integrals of Eqs. (5.21) can be simplified by choosing either a thin filament (Fig. 5.22) or a thin slab (Fig. 5.23) for the element of volume dV. Thus, you should begin your solution by identifying, if possible, the dV which produces the single or double integrals that are the easiest to compute. For bodies of revolution, this may be a thin slab (as in Sample Prob. 5.13) or a thin cylindrical shell. However, it is important to remember that the relationship that you establish among the variables (like the relationship between r and x in Sample Prob. 5.13) will directly affect the complexity of the integrals you will have to compute. Finally, we again remind you that $\bar{x}_{el}$, $\bar{y}_{el}$, and $\bar{z}_{el}$ in Eqs. (5.22) are the coordinates of the centroid of dV.

Problems

5.90 The composite body shown is formed by removing a hemisphere of radius r from a cylinder of radius R and height $2R$. Determine (a) the y coordinate of the centroid when $r = 3R/4$, (b) the ratio r/R for which $\overline{y} = -1.2R$.

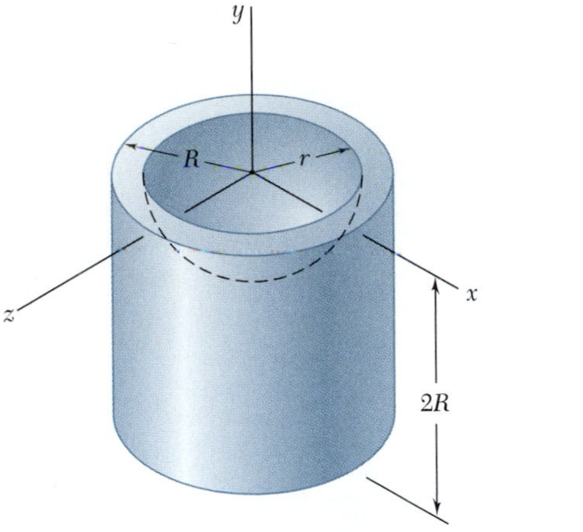

Fig. P5.90

5.91 Determine the y coordinate of the centroid of the body shown.

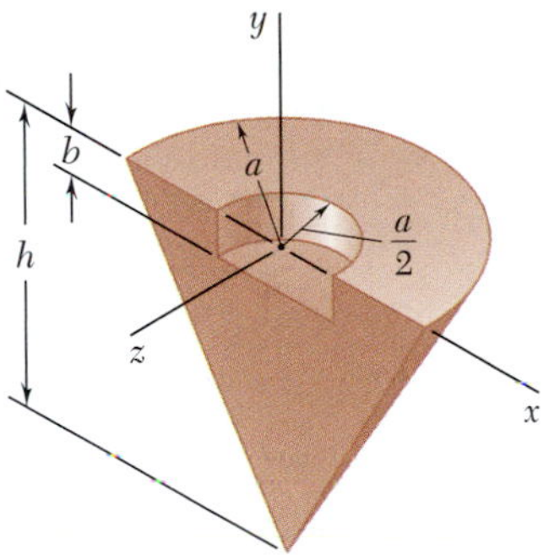

Fig. P5.91 and P5.92

5.92 Determine the z coordinate of the centroid of the body shown. (*Hint:* Use the result of Sample Prob. 5.13.)

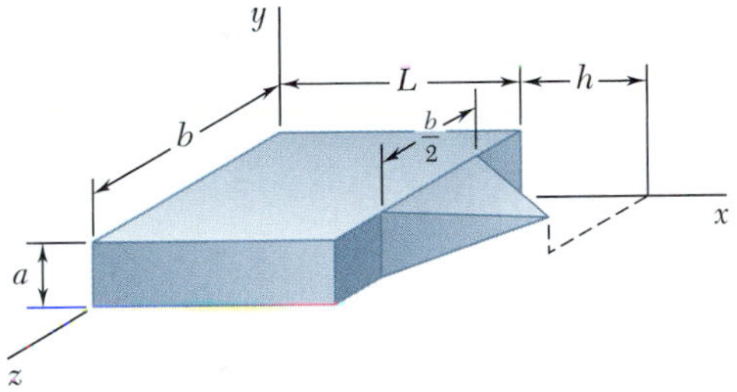

Fig. P5.93

5.93 Consider the composite body shown. Determine (a) the value of $\overline{x}$ when $h = L/2$, (b) the ratio h/L for which $\overline{x} = L$.

5.94 For the machine element shown, determine the x coordinate of the center of gravity.

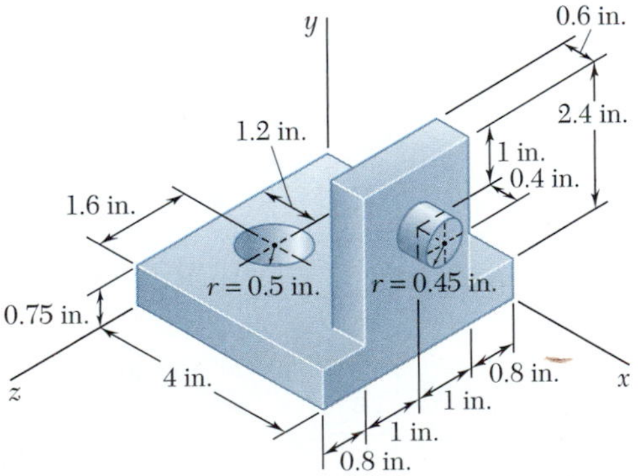

Fig. P5.94 and P5.95

5.95 For the machine element shown, determine the y coordinate of the center of gravity.

5.96 For the machine element shown, locate the x coordinate of the center of gravity.

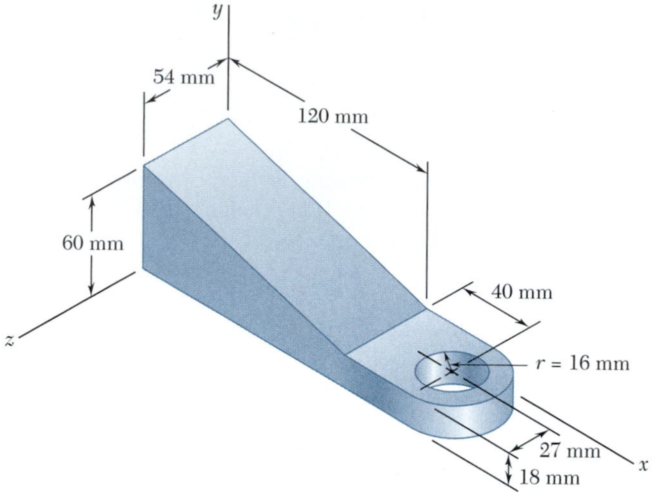

Fig. P5.96 and P5.97

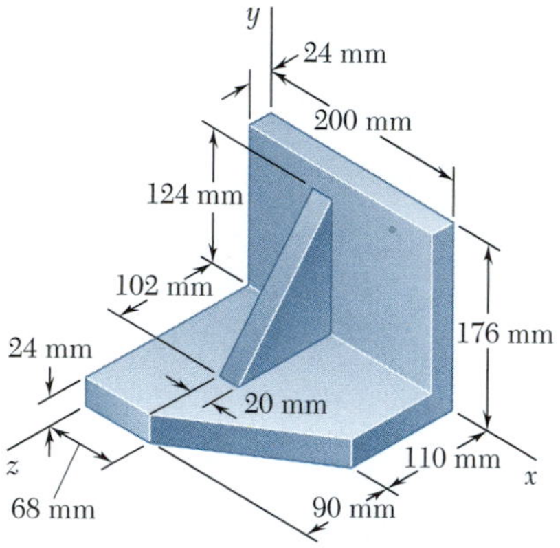

Fig. *P5.98* and *P5.99*

5.97 For the machine element shown, locate the y coordinate of the center of gravity.

5.98 For the stop bracket shown, locate the x coordinate of the center of gravity.

5.99 For the stop bracket shown, locate the z coordinate of the center of gravity.

5.100 Locate the center of gravity of the sheet-metal form shown.

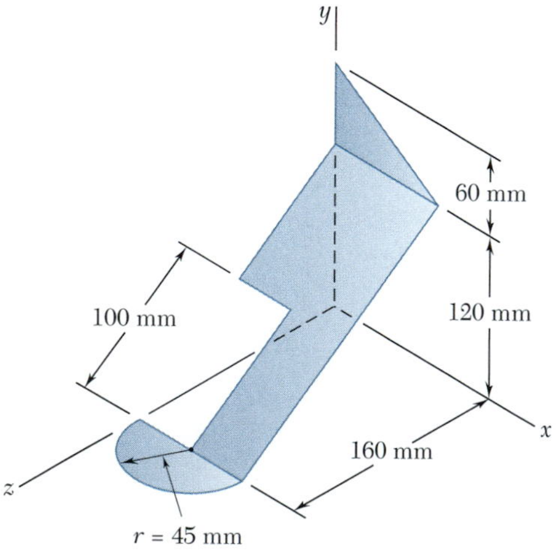

Fig. P5.100

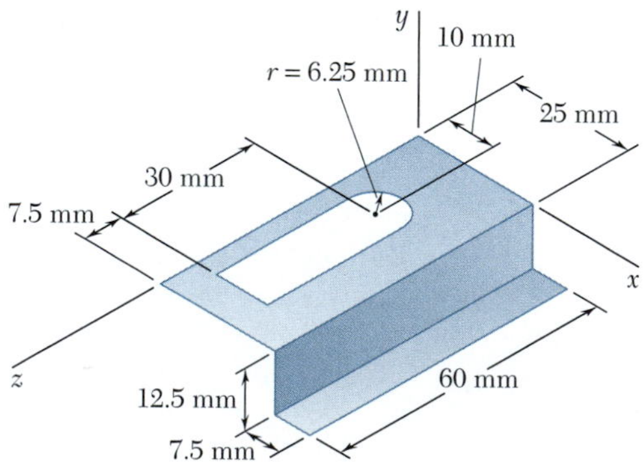

Fig. P5.101

5.101 A mounting bracket for electronic components is formed from sheet metal of uniform thickness. Locate the center of gravity of the bracket.

5.102 Locate the center of gravity of the sheet-metal form shown.

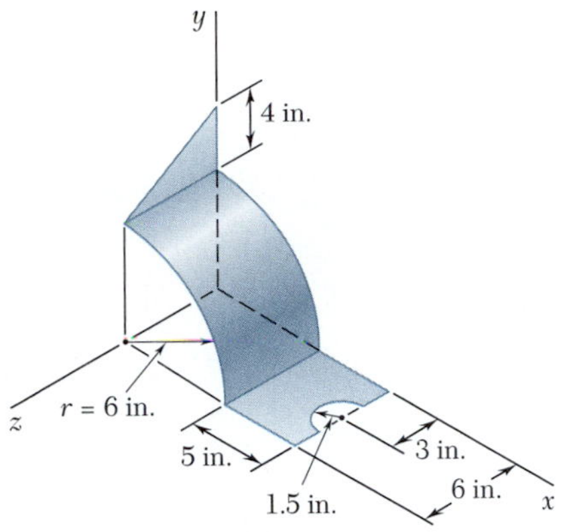

Fig. *P5.102*

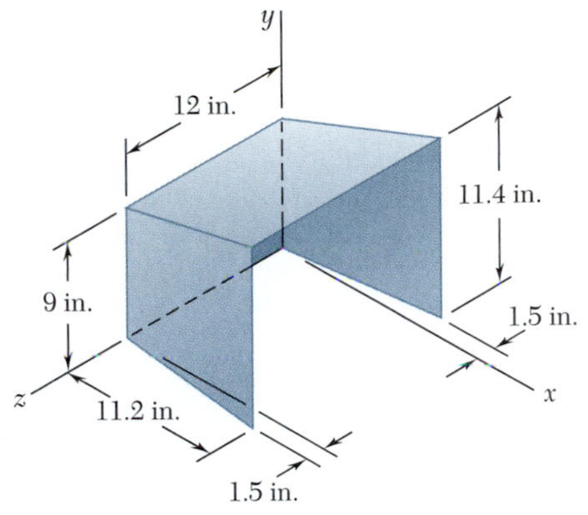

Fig. P5.103

***5.103** An enclosure for an electronic device is formed from sheet metal of uniform thickness. Locate the center of gravity of the enclosure.

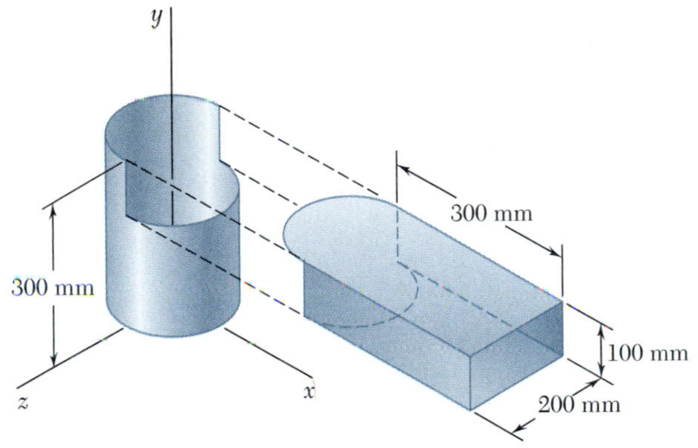

Fig. P5.104

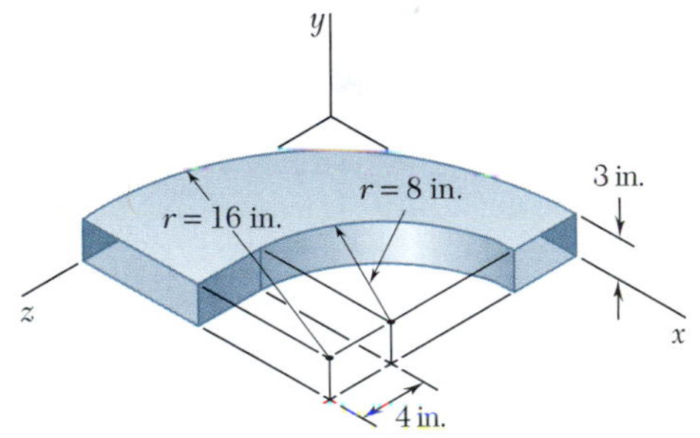

Fig. *P5.105*

5.104 A 200-mm-diameter cylindrical duct and a 100 × 200-mm rectangular duct are to be joined as indicated. Knowing that the ducts are fabricated from the same sheet metal, which is of uniform thickness, locate the center of gravity of the assembly.

5.105 An elbow for the duct of a ventilating system is made of sheet metal of uniform thickness. Locate the center of gravity of the elbow.

5.106 A window awning is fabricated from sheet metal of uniform thickness. Locate the center of gravity of the awning.

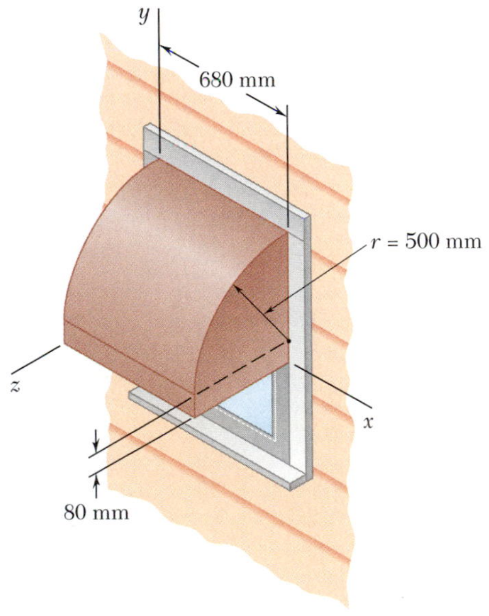

Fig. P5.106

5.107 The thin, plastic front cover of a wall clock is of uniform thickness. Locate the center of gravity of the cover.

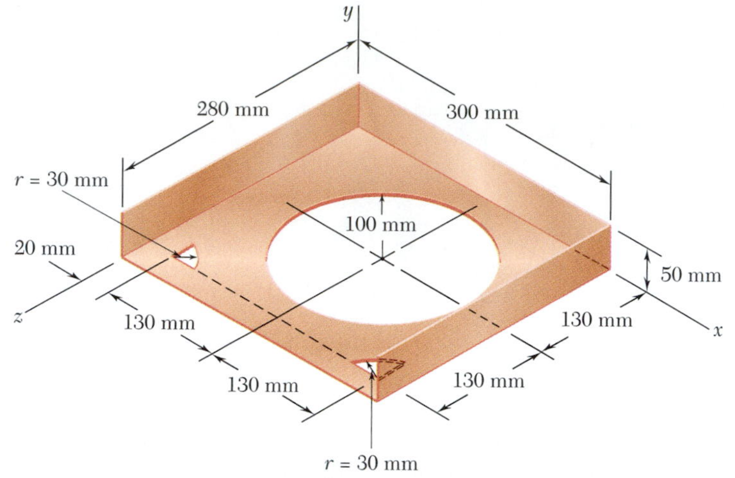

Fig. P5.107

5.108 and 5.109 A thin steel wire of uniform cross section is bent into the shape shown, where arc BC is a quarter circle of radius R. Locate its center of gravity.

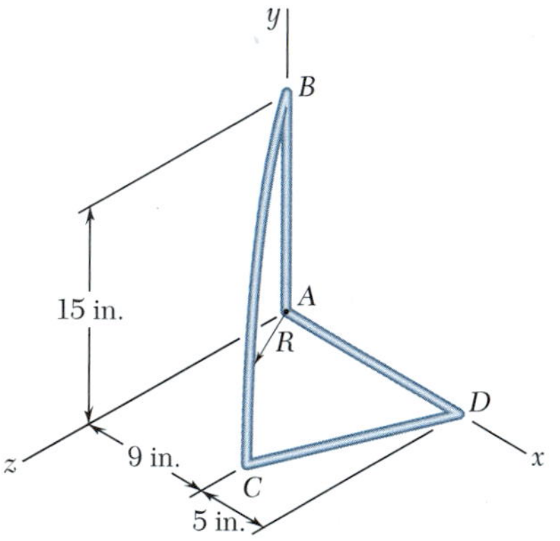

Fig. P5.108

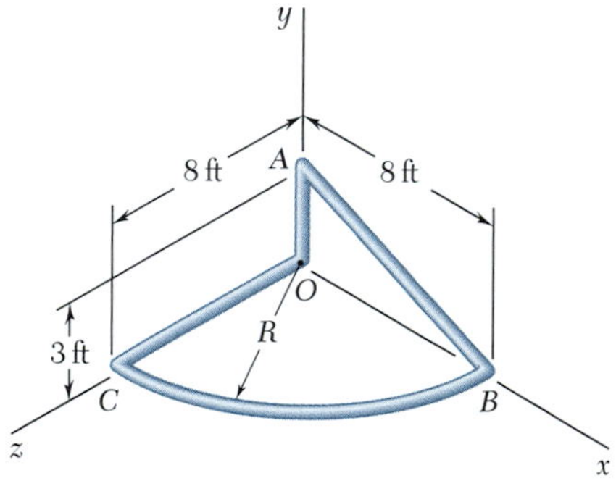

Fig. P5.109

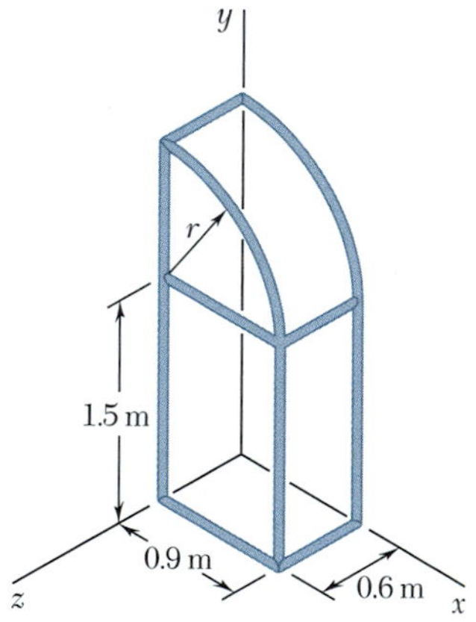

Fig. P5.110

5.110 The frame of a greenhouse is constructed from uniform aluminum channels. Locate the center of gravity of the portion of the frame shown.

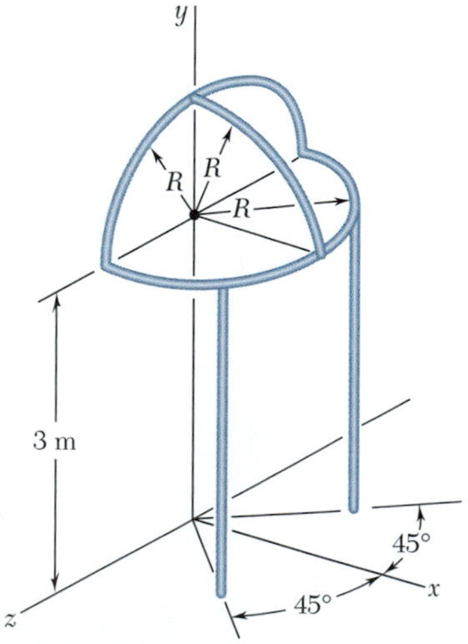

Fig. P5.111

5.111 The decorative metalwork at the entrance of a store is fabricated from uniform steel structural tubing. Knowing that $R = 1.2$ m, locate the center of gravity of the metalwork.

5.112 A scratch awl has a plastic handle and a steel blade and shank. Knowing that the specific weight of plastic is 0.0374 lb/in^3 and of steel is 0.284 lb/in^3, locate the center of gravity of the awl.

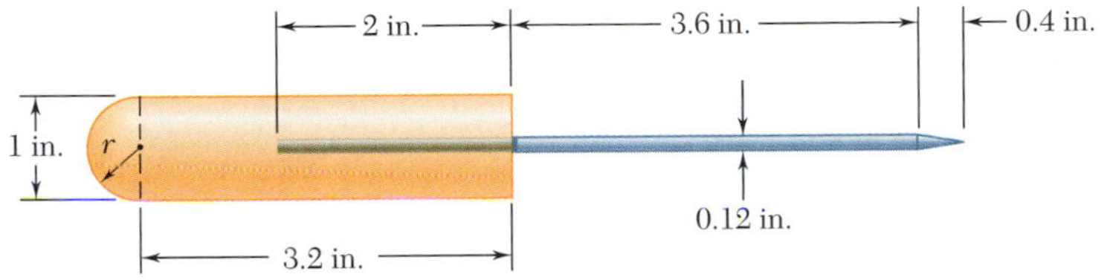

Fig. *P5.112*

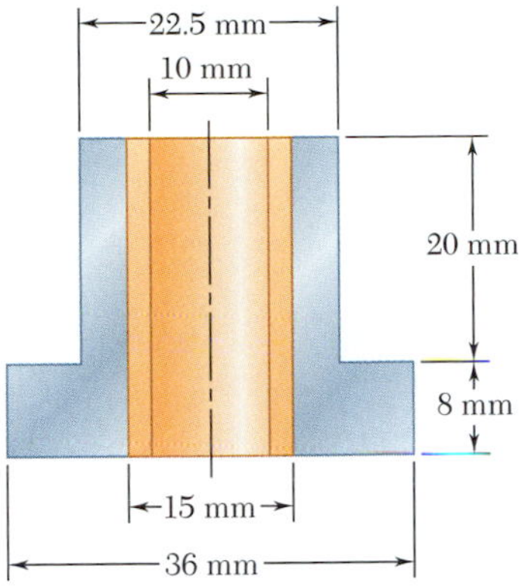

Fig. P5.113

5.113 A bronze bushing is mounted inside a steel sleeve. Knowing that the density of bronze is 8800 kg/m^3 and of steel is 7860 kg/m^3, determine the center of gravity of the assembly.

5.114 A marker for a garden path consists of a truncated regular pyramid carved from stone of specific weight 160 lb/ft^3. The pyramid is mounted on a steel base of thickness h. Knowing that the specific weight of steel is 490 lb/ft^3 and that steel plate is available in ¼-in. increments, specify the minimum thickness h for which the center of gravity of the marker is approximately 12 in. above the top of the base.

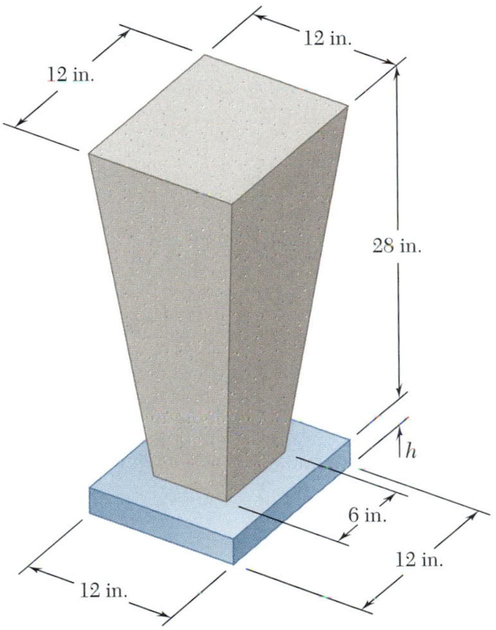

Fig. *P5.114*

***5.115** The ends of the park bench shown are made of concrete, while the seat and back are wooden boards. Each piece of wood is $36 \times 120 \times 1180$ mm. Knowing that the density of concrete is 2320 kg/m^3 and of wood is 470 kg/m^3, determine the x and y coordinates of the center of gravity of the bench.

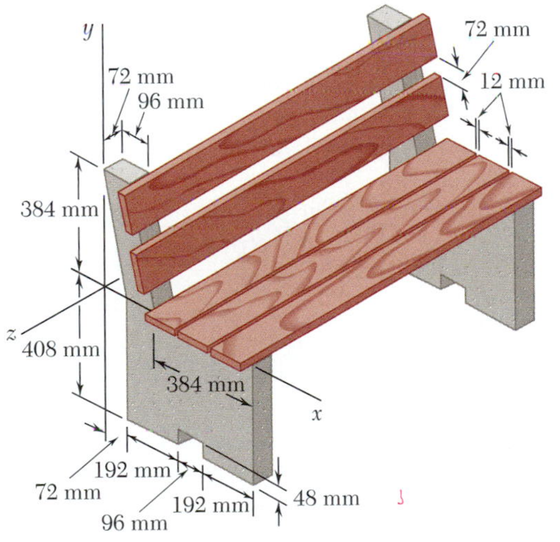

Fig. P5.115

5.116 through 5.118 Determine by direct integration the values of $\bar{x}$ for the two volumes obtained by passing a vertical cutting plane through the given shape of Fig. 5.21. The cutting plane is parallel to the base of the given shape and divides the shape into two volumes of equal height.

5.116 A hemisphere.
5.117 A semiellipsoid of revolution.
5.118 A paraboloid of revolution.

5.119 and *5.120* Locate the centroid of the volume obtained by rotating the shaded area about the x axis.

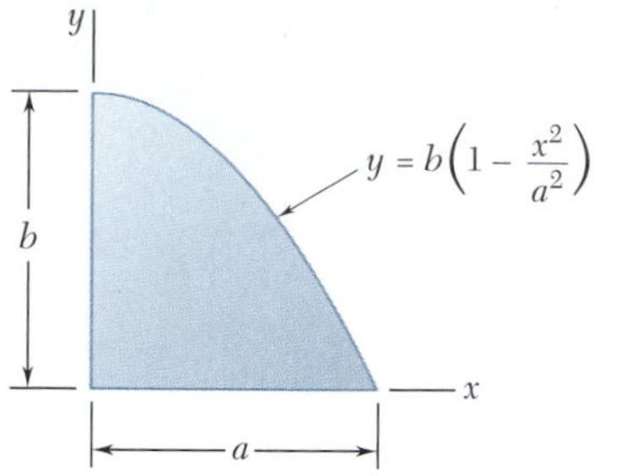

Fig. P5.119

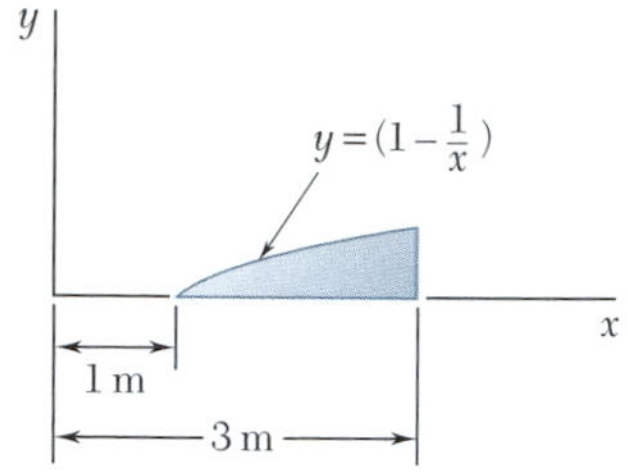

Fig. *P5.120*

5.121 Locate the centroid of the volume obtained by rotating the shaded area about the line $x = h$.

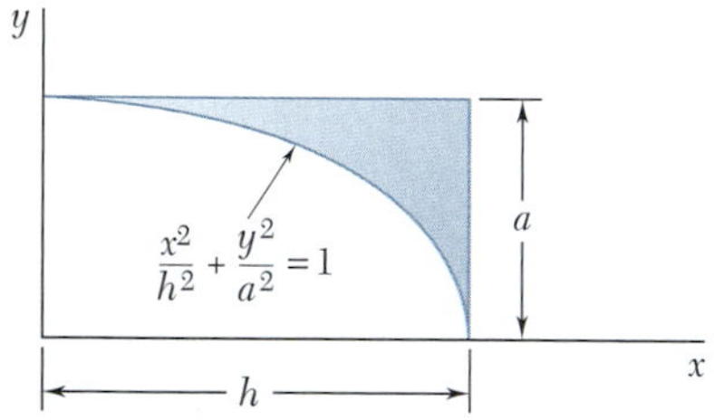

Fig. P5.121

***5.122** Locate the centroid of the volume generated by revolving the portion of the sine curve shown about the x axis.

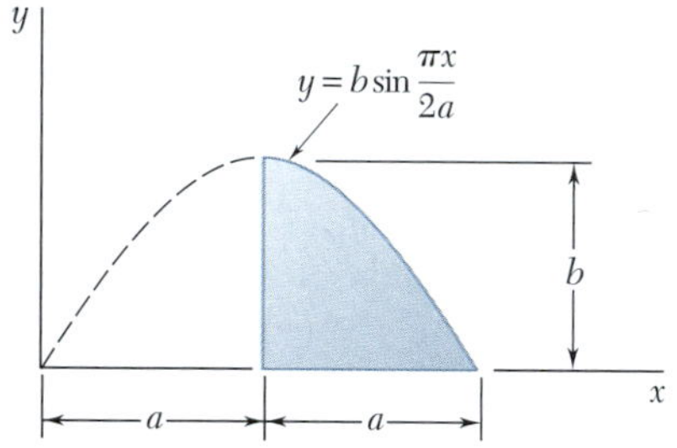

Fig. P5.122 and P5.123

***5.123** Locate the centroid of the volume generated by revolving the portion of the sine curve shown about the y axis. (*Hint:* Use a thin cylindrical shell of radius r and thickness dr as the element of volume.)

***5.124** Show that for a regular pyramid of height h and n sides ($n = 3, 4, \ldots$) the centroid of the volume of the pyramid is located at a distance $h/4$ above the base.

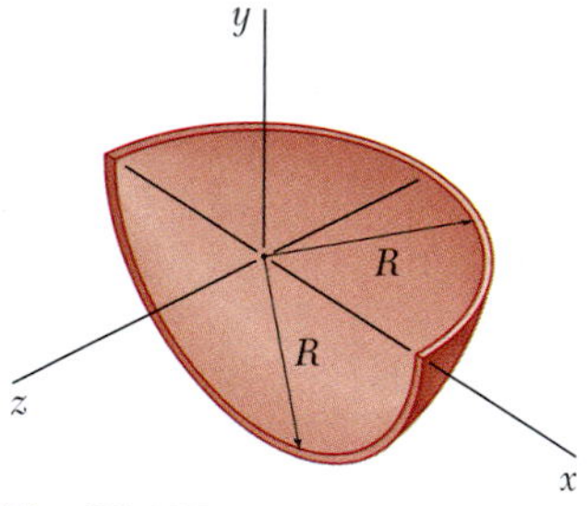

Fig. P5.125

5.125 Determine by direct integration the location of the centroid of one-half of a thin, uniform hemispherical shell of radius R.

5.126 The sides and the base of a punch bowl are of uniform thickness t. If $t << R$ and $R = 350$ mm, determine the location of the center of gravity of (a) the bowl, (b) the punch.

Fig. *P5.126*

5.127 After grading a lot, a builder places four stakes to designate the corners of the slab for a house. To provide a firm, level base for the slab, the builder places a minimum of 60 mm of gravel beneath the slab. Determine the volume of gravel needed and the x coordinate of the centroid of the volume of the gravel. (*Hint:* The bottom of the gravel is an oblique plane, which can be represented by the equation $y = a + bx + cz$.)

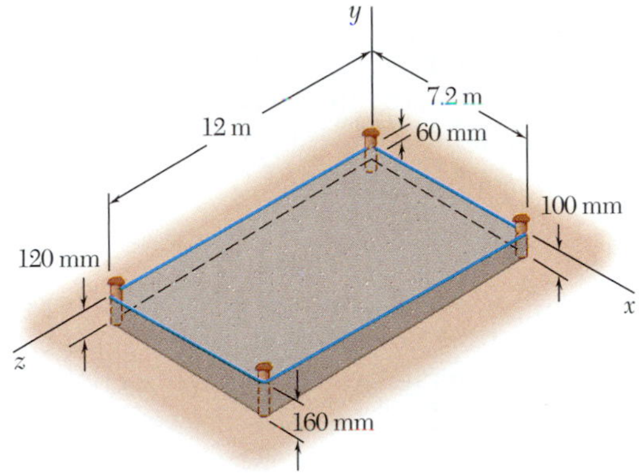

Fig. P5.127

5.128 Determine by direct integration the location of the centroid of the volume between the xz plane and the portion shown of the surface $y = 16h(ax - x^2)(bz - z^2)/a^2b^2$.

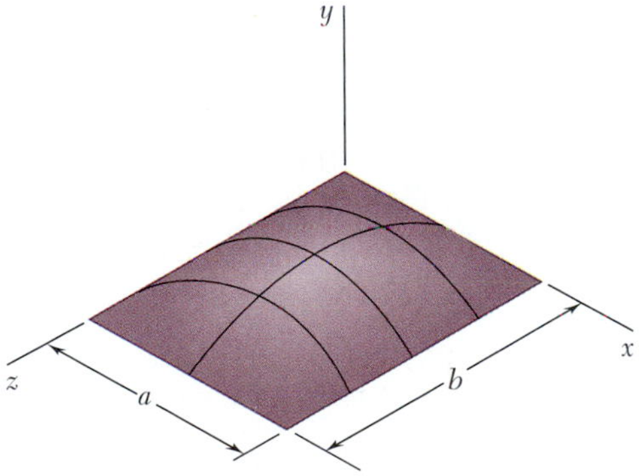

Fig. *P5.128*

5.129 Locate the centroid of the section shown, which was cut from a circular cylinder by an inclined plane.

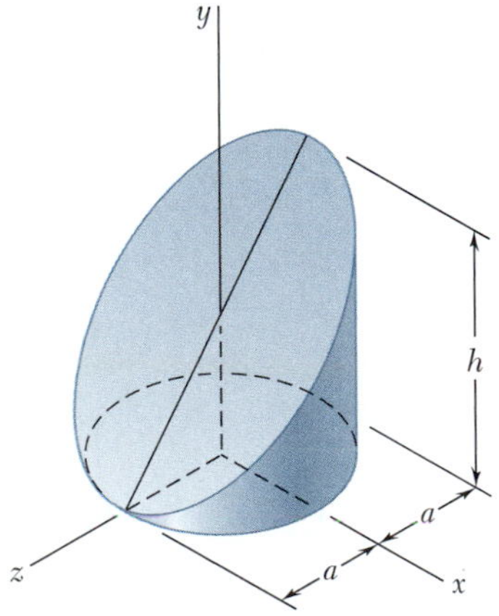

Fig. P5.129

REVIEW AND SUMMARY FOR CHAPTER 5

This chapter was devoted chiefly to the determination of the *center of gravity* of a rigid body, that is, to the determination of the point G where a single force $\mathbf{W}$, called the *weight* of the body, can be applied to represent the effect of the earth's attraction on the body.

Center of gravity of a two-dimensional body

In the first part of the chapter, we considered *two-dimensional bodies,* such as flat plates and wires contained in the xy plane. By adding force components in the vertical z direction and moments about the horizontal y and x axes [Sec. 5.2], we derived the relations

$$W = \int dW \qquad \bar{x}W = \int x\, dW \qquad \bar{y}W = \int y\, dW \tag{5.2}$$

which define the weight of the body and the coordinates $\bar{x}$ and $\bar{y}$ of its center of gravity.

Centroid of an area or line

In the case of a *homogeneous flat plate of uniform thickness* [Sec. 5.3], the center of gravity G of the plate coincides with the *centroid C of the area A* of the plate, the coordinates of which are defined by the relations

$$\bar{x}A = \int x\, dA \qquad \bar{y}A = \int y\, dA \tag{5.3}$$

Similarly, the determination of the center of gravity of a *homogeneous wire of uniform cross section* contained in a plane reduces to the determination of the *centroid C of the line L* representing the wire; we have

$$\bar{x}L = \int x\, dL \qquad \bar{y}L = \int y\, dL \tag{5.4}$$

First moments

The integrals in Eqs. (5.3) are referred to as the *first moments* of the area A with respect to the y and x axes and are denoted by Q_y and Q_x, respectively [Sec. 5.4]. We have

$$Q_y = \bar{x}A \qquad Q_x = \bar{y}A \tag{5.6}$$

The first moments of a line can be defined in a similar way.

Properties of symmetry

The determination of the centroid C of an area or line is simplified when the area or line possesses certain *properties of symmetry.* If the area or line is symmetric with respect to an axis, its centroid C lies on that axis; if it is symmetric with respect to two axes, C is located at the intersection of the two axes; if it is symmetric with respect to a center O, C coincides with O.

Center of gravity of a composite body

The *areas and the centroids of various common shapes* are tabulated in Fig. 5.8. When a flat plate can be divided into several of these shapes, the coordinates $\bar{X}$ and $\bar{Y}$ of its center of gravity G can be determined from the coordinates $\bar{x}_1, \bar{x}_2, \ldots$ and $\bar{y}_1, \bar{y}_2, \ldots$ of the centers of gravity $G_1, G_2, \ldots$ of the various parts [Sec. 5.5]. Equating moments about the y and x axes, respectively (Fig. 5.24), we have

$$\bar{X}\Sigma W = \Sigma\bar{x}W \qquad \bar{Y}\Sigma W = \Sigma\bar{y}W \tag{5.7}$$

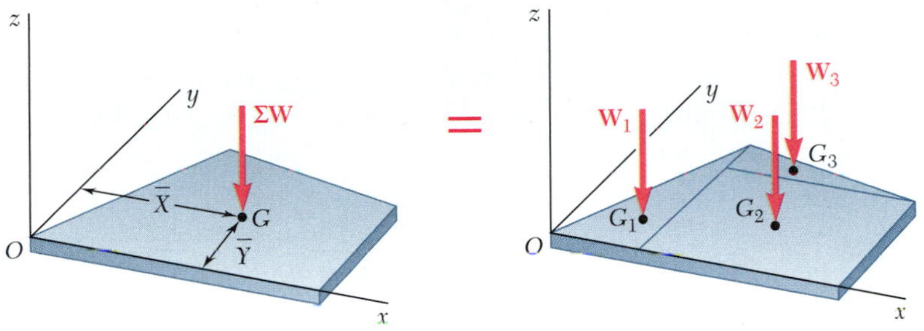

Fig. 5.24

If the plate is homogeneous and of uniform thickness, its center of gravity coincides with the centroid C of the area of the plate, and Eqs. (5.7) reduce to

$$Q_y = \bar{X}\Sigma A = \Sigma\bar{x}A \qquad Q_x = \bar{Y}\Sigma A = \Sigma\bar{y}A \tag{5.8}$$

These equations yield the first moments of the composite area, or they can be solved for the coordinates $\bar{X}$ and $\bar{Y}$ of its centroid [Sample Prob. 5.1]. The determination of the center of gravity of a composite wire is carried out in a similar fashion [Sample Prob. 5.2].

Determination of centroid by integration

When an area is bounded by analytical curves, the coordinates of its centroid can be determined by *integration* [Sec. 5.6]. This can be done by evaluating either the double integrals in Eqs. (5.3) or a *single integral* which uses one of the thin rectangular or pie-shaped elements of area shown in Fig. 5.12. Denoting by $\bar{x}_{el}$ and $\bar{y}_{el}$ the coordinates of the centroid of the element dA, we have

$$Q_y = \bar{x}A = \int \bar{x}_{el}\, dA \qquad Q_x = \bar{y}A = \int \bar{y}_{el}\, dA \tag{5.9}$$

It is advantageous to use the same element of area to compute both of the first moments Q_y and Q_x; the same element can also be used to determine the area A [Sample Prob. 5.4].

Theorems of Pappus-Guldinus

The *theorems of Pappus-Guldinus* relate the determination of the area of a surface of revolution or the volume of a body of revolution to the determination of the centroid of the generating curve or area [Sec. 5.7]. The area A of the surface generated by rotating a curve of length L about a fixed axis (Fig. 5.25*a*) is

$$A = 2\pi\bar{y}L \qquad (5.10)$$

where $\bar{y}$ represents the distance from the centroid C of the curve to the fixed axis. Similarly, the volume V of the body generated by rotating an area A about a fixed axis (Fig. 5.25*b*) is

$$V = 2\pi\bar{y}A \qquad (5.11)$$

where $\bar{y}$ represents the distance from the centroid C of the area to the fixed axis.

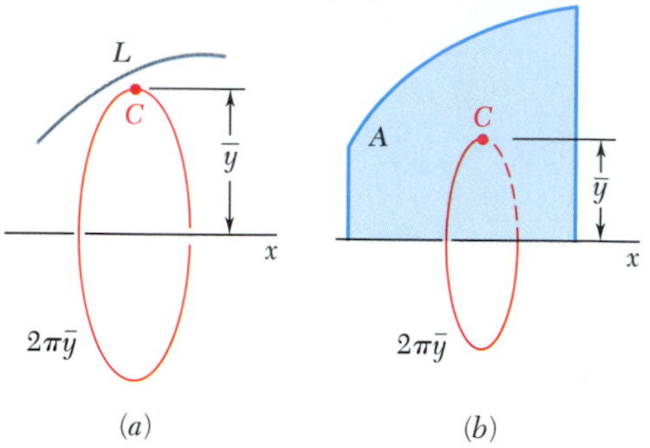

Fig. 5.25

Distributed loads

The concept of centroid of an area can also be used to solve problems other than those dealing with the weight of flat plates. For example, to determine the reactions at the supports of a beam [Sec. 5.8], we can replace a *distributed load* w by a concentrated load $\mathbf{W}$ equal in magnitude to the area A under the load curve and passing through the centroid C of that area (Fig. 5.26). The same approach can be used to determine the resultant of the hydrostatic forces exerted on a *rectangular plate submerged in a liquid* [Sec. 5.9].

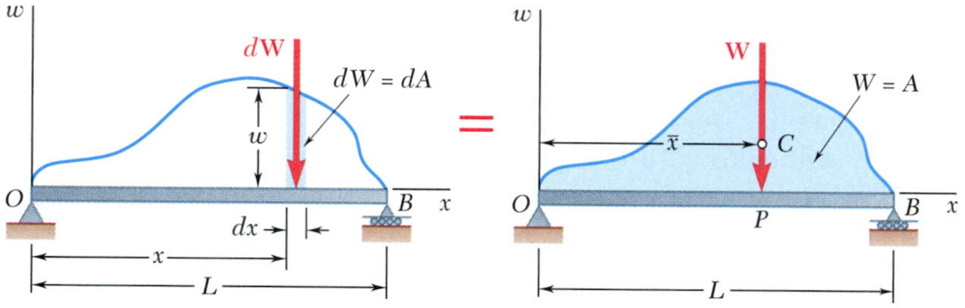

Fig. 5.26

Center of gravity of a three-dimensional body

The last part of the chapter was devoted to the determination of the *center of gravity G of a three-dimensional body.* The coordinates $\bar{x}$, $\bar{y}$, $\bar{z}$ of G were defined by the relations

$$\bar{x}W = \int x\,dW \qquad \bar{y}W = \int y\,dW \qquad \bar{z}W = \int z\,dW \qquad (5.16)$$

Centroid of a volume

In the case of a *homogeneous body,* the center of gravity G coincides with the *centroid C of the volume V* of the body; the coordinates of C are defined by the relations

$$\bar{x}V = \int x\,dV \qquad \bar{y}V = \int y\,dV \qquad \bar{z}V = \int z\,dV \qquad (5.18)$$

If the volume possesses a *plane of symmetry,* its centroid C will lie in that plane; if it possesses two planes of symmetry, C will be located on the line of intersection of the two planes; if it possesses three planes of symmetry which intersect at only one point, C will coincide with that point [Sec. 5.10].

Center of gravity of a composite body

The *volumes and centroids of various common three-dimensional shapes* are tabulated in Fig. 5.21. When a body can be divided into several of these shapes, the coordinates $\bar{X}, \bar{Y}, \bar{Z}$ of its center of gravity G can be determined from the corresponding coordinates of the centers of gravity of its various parts [Sec. 5.11]. We have

$$\bar{X}\Sigma W = \Sigma \bar{x} W \qquad \bar{Y}\Sigma W = \Sigma \bar{y} W \qquad \bar{Z}\Sigma W = \Sigma \bar{z} W \quad (5.19)$$

If the body is made of a homogeneous material, its center of gravity coincides with the centroid C of its volume, and we write [Sample Probs. 5.11 and 5.12]

$$\bar{X}\Sigma V = \Sigma \bar{x} V \qquad \bar{Y}\Sigma V = \Sigma \bar{y} V \qquad \bar{Z}\Sigma V = \Sigma \bar{z} V \quad (5.20)$$

Determination of centroid by integration

When a volume is bounded by analytical surfaces, the coordinates of its centroid can be determined by *integration* [Sec. 5.12]. To avoid the computation of the triple integrals in Eqs. (5.18), we

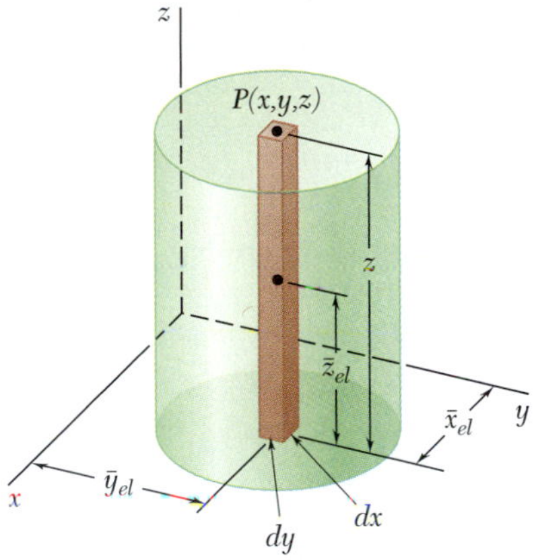

$\bar{x}_{el} = x, \; \bar{y}_{el} = y, \; \bar{z}_{el} = \frac{z}{2}$
$dV = z\,dx\,dy$

Fig. 5.27

can use elements of volume in the shape of thin filaments, as shown in Fig. 5.27. Denoting by $\bar{x}_{el}$, $\bar{y}_{el}$, and $\bar{z}_{el}$ the coordinates of the centroid of the element dV, we rewrite Eqs. (5.18) as

$$\bar{x}V = \int \bar{x}_{el}\,dV \qquad \bar{y}V = \int \bar{y}_{el}\,dV \qquad \bar{z}V = \int \bar{z}_{el}\,dV \qquad (5.22)$$

which involve only double integrals. If the volume possesses *two planes of symmetry*, its centroid C is located on their line of intersection. Choosing the x axis to lie along that line and dividing the volume into thin slabs parallel to the yz plane, we can determine C from the relation

$$\bar{x}V = \int \bar{x}_{el}\,dV \qquad (5.23)$$

with a *single integration* [Sample Prob. 5.13]. For a body of revolution, these slabs are circular and their volume is given in Fig. 5.28.

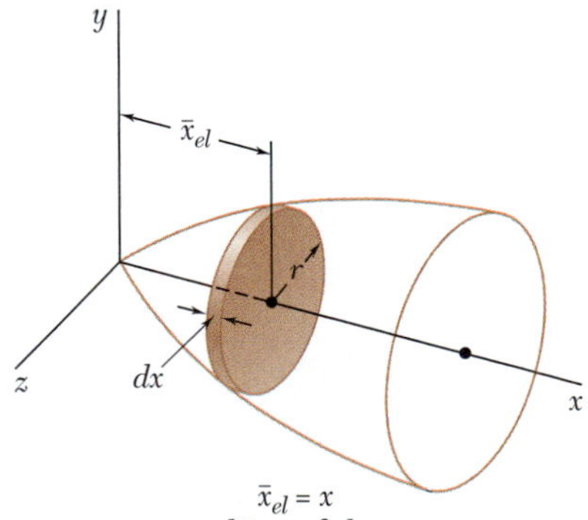

$\bar{x}_{el} = x$
$dV = \pi r^2\,dx$

Fig. 5.28

Review Problems

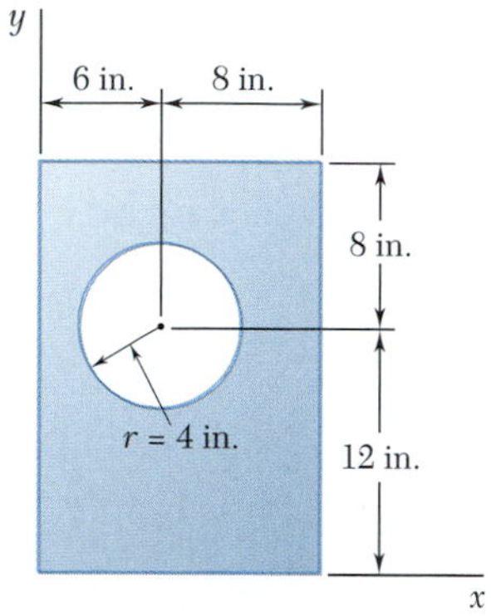

Fig. P5.130

5.130 Locate the centroid of the plane area shown.

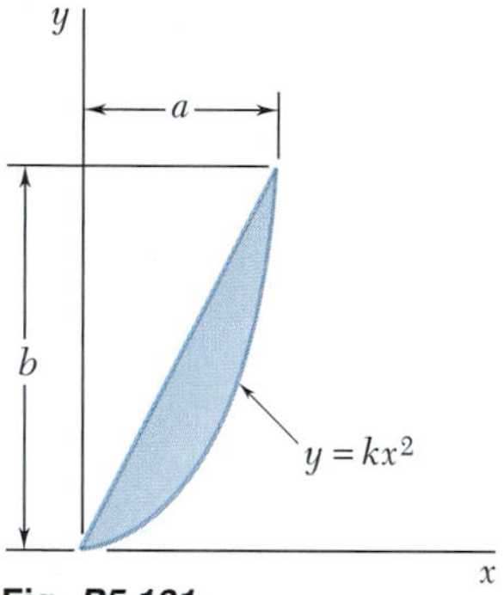

Fig. *P5.131*

5.131 For the area shown, determine the ratio a/b for which $\bar{x} = \bar{y}$.

5.132 Locate the centroid of the plane area shown.

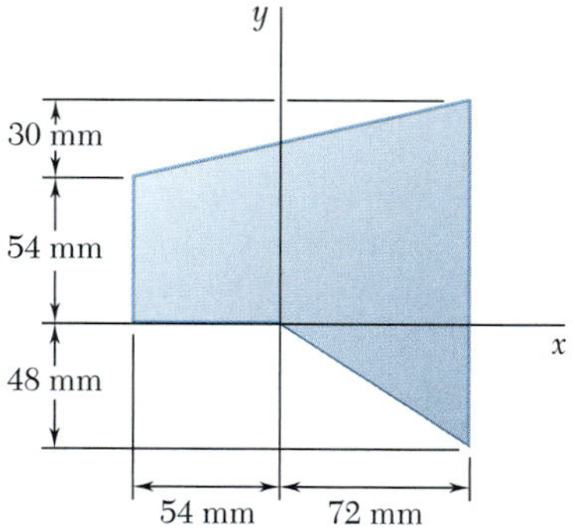

Fig. P5.132

5.133 Determine by direct integration the centroid of the area shown. Express your answer in terms of a and h.

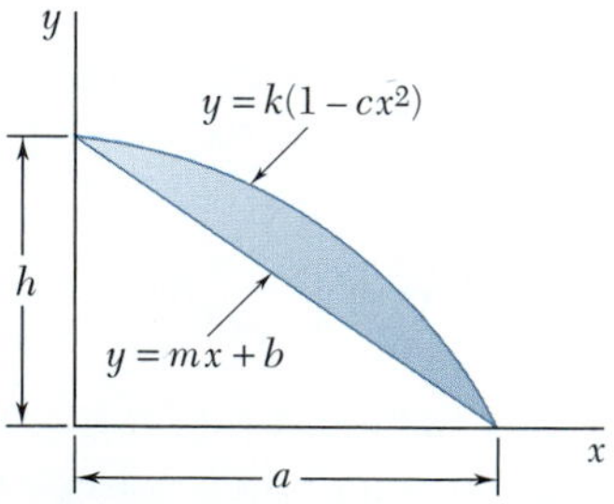

Fig. *P5.133*

5.134 Member *ABCDE* is a component of a mobile and is formed from a single piece of aluminum tubing. Knowing that the member is supported at *C* and that $l = 2$ m, determine the distance *d* so that portion *BCD* of the member is horizontal.

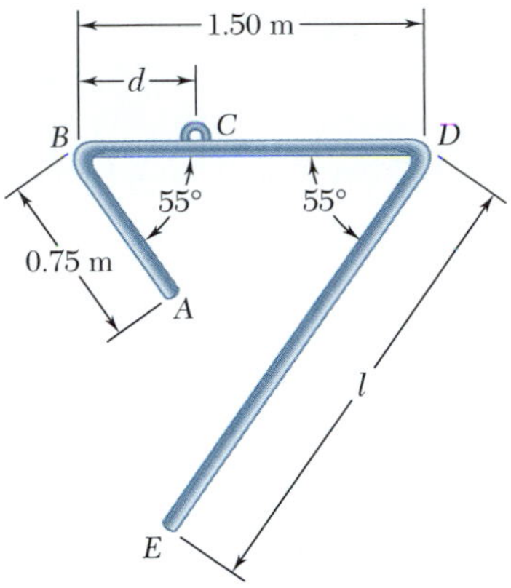

Fig. P5.134

5.135 A cylindrical hole is drilled through the center of a steel ball bearing shown here in cross section. Denoting the length of the *hole* by *L*, show that the volume of the steel remaining is equal to the volume of a sphere of diameter *L*.

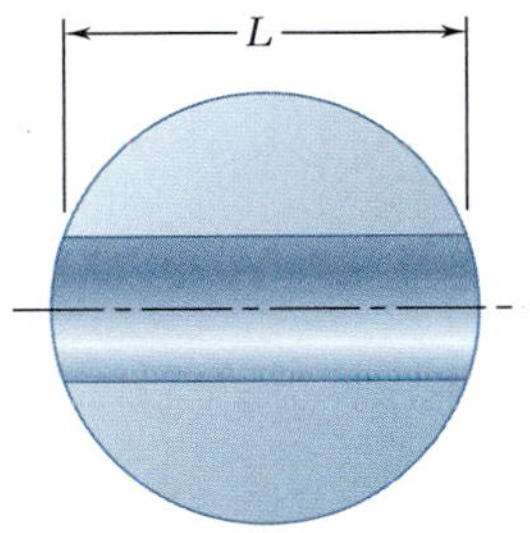

Fig. P5.135

5.136 For the beam and loading shown, determine (*a*) the magnitude and location of the resultant of the distributed load, (*b*) the reactions at the beam supports.

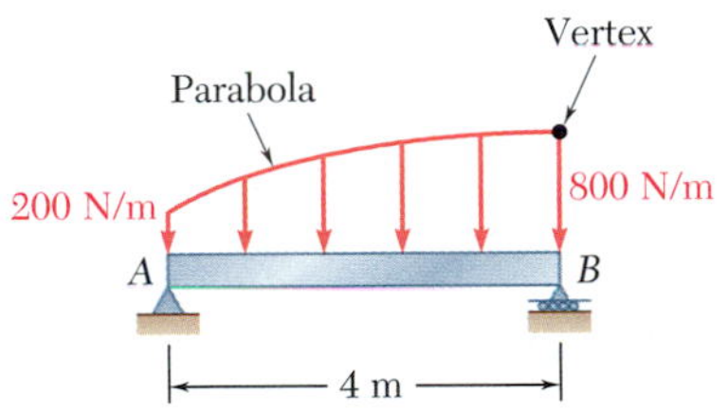

Fig. *P5.136*

5.137 Determine the reactions at the beam supports for the given loading.

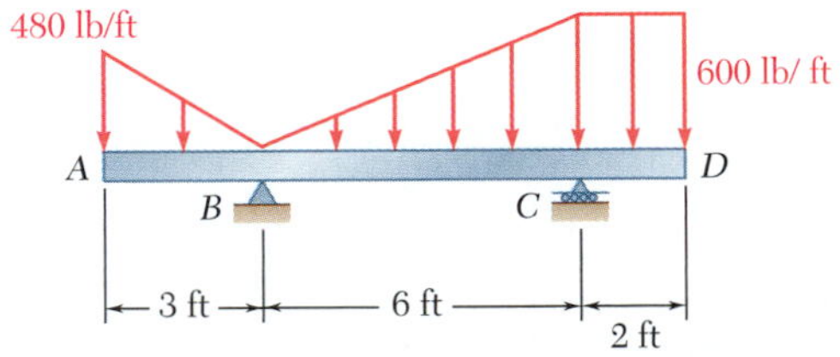

Fig. P5.137

5.138 Locate the center of gravity of the sheet-metal form shown.

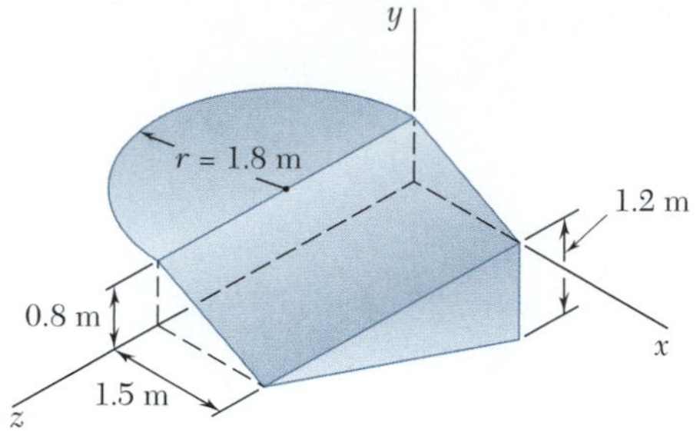

Fig. *P5.138*

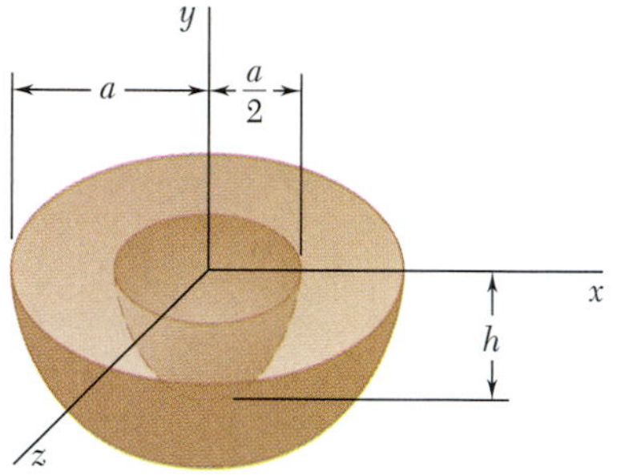

Fig. P5.139

5.139 The composite body shown is formed by removing a semiellipsoid of revolution of semimajor axis h and semiminor axis $a/2$ from a hemisphere of radius a. Determine (a) the y coordinate of the centroid when $h = a/2$, (b) the ratio h/a for which $\bar{y} = -0.4a$.

5.140 A thin steel wire of uniform cross section is bent into the shape shown. Locate its center of gravity.

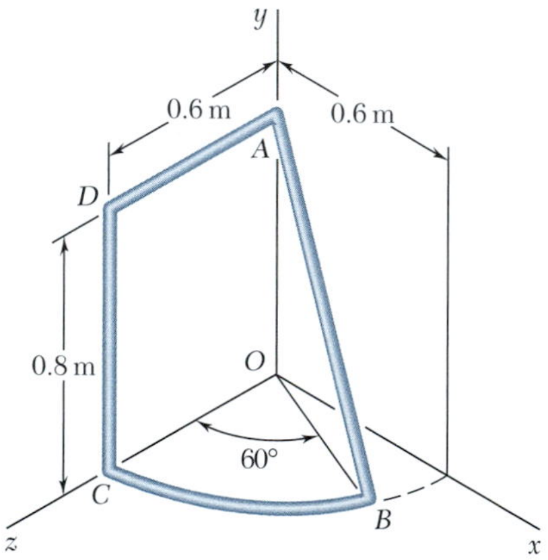

Fig. P5.140

5.141 Locate the centroid of the volume obtained by rotating the shaded area about the x axis.

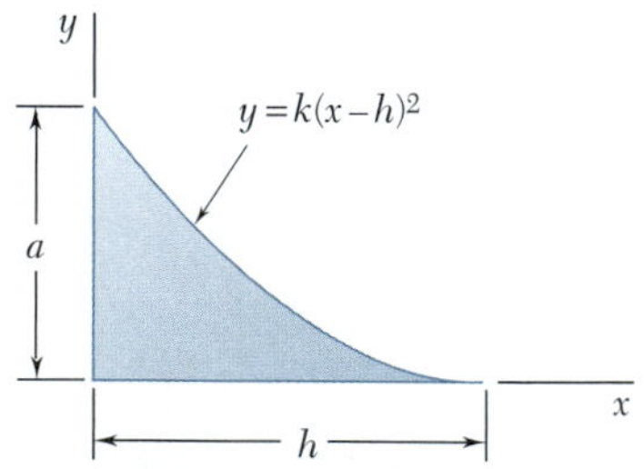

Fig. P5.141

Computer Problems

5.C1 For the trapezoid shown, determine the x and y coordinates of the centroid knowing that $h_1 = a/n^2$ and $h_2 = a/n$. Plot the values of $\bar{x}$ and $\bar{y}$ for $1 \leq n \leq 4$ and $a = 6$ in.

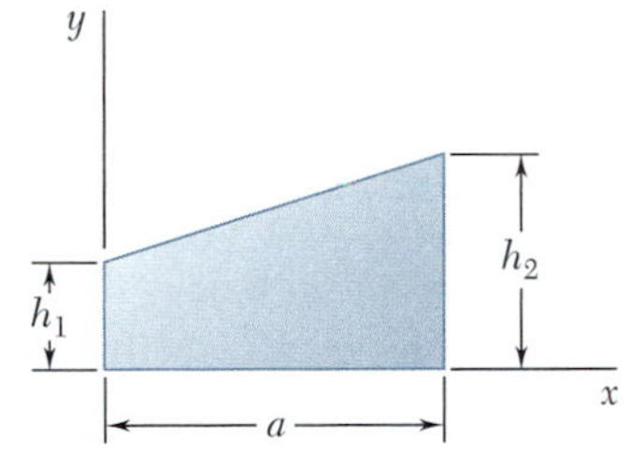

Fig. P5.C1

5.C2 Determine the distance h for which the centroid of the shaded area is as high above BB' as possible. Plot the ratio h/b as a function of k for $0.125 \leq k \leq 0.875$.

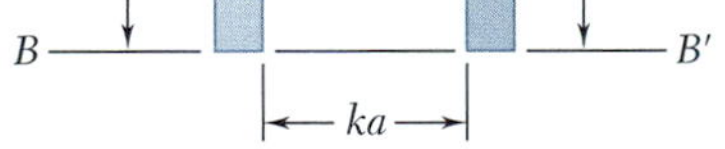

Fig. P5.C2

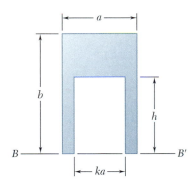

5.C3 Approximate the general spandrel shown using a series of n rectangles, each of width Δa and of the form $bcc'b'$, and then use computational software to calculate the coordinates of the centroid of the area. Locate the centroid when (*a*) $m = 2$, $a = 4$ in., $h = 4$ in.; (*b*) $m = 2$, $a = 4$ in., $h = 25$ in.; (*c*) $m = 5$, $a = 4$ in., $h = 4$ in.; (*d*) $m = 5$, $a = 4$ in., $h = 25$ in. In each case compare the answers obtained to the exact values of $\bar{x}$ and $\bar{y}$ computed from the formulas given in Fig. 5.8A and determine the percentage error.

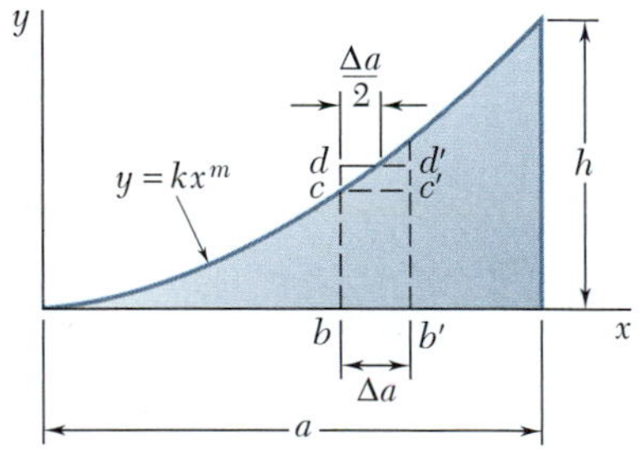

Fig. P5.C3

5.C4 Determine the volume and the surface area of the solid obtained by rotating the area shown about the y axis when $a = 80$ mm and (*a*) $n = 1$, (*b*) $n = 2$, (*c*) $n = 3$.

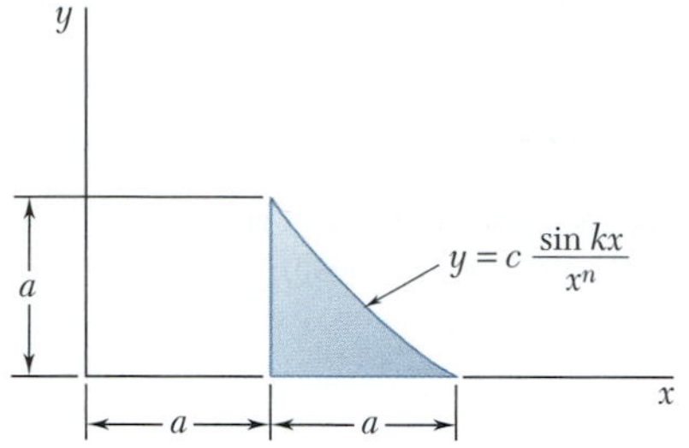

Fig. P5.C4

5.C5 Gate AB of uniform width w is held in the position shown by a hinge along its top edge A and by a single shear pin located at the center of its lower edge B. Determine the force on shear pin B as a function of the water depth d. Plot the force on pin B as a function of water depth for $0.54 \text{ m} \leq d \leq 1.5 \text{ m}$ when (*a*) $w = 0.25$ m, (*b*) $w = 0.50$ m, (*c*) $w = 0.75$ m, (*d*) $w = 1$ m. The density of water is $\rho = 10^3 \text{ kg/m}^3$.

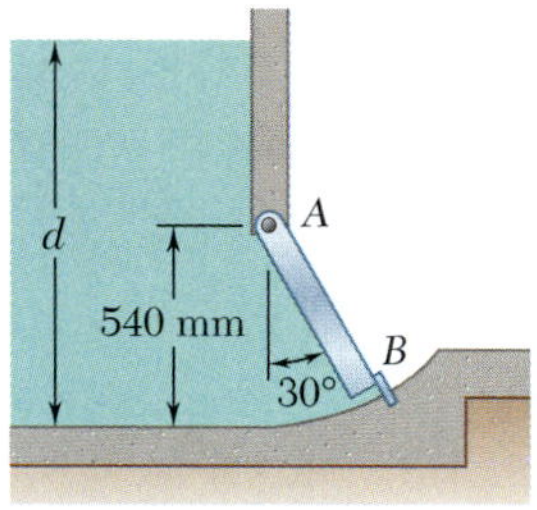

Fig. P5.C5

5.C6 An open tank is to be slowly filled with water. Determine the resultant and direction of the pressure force exerted by the water on a 4-ft-wide section of side ABC of the tank as a function of the depth d. Using computational software, plot the pressure force as a function of d for $0 \leq d \leq 10$ ft. The specific weight of water is 62.4 lb/ft^3.

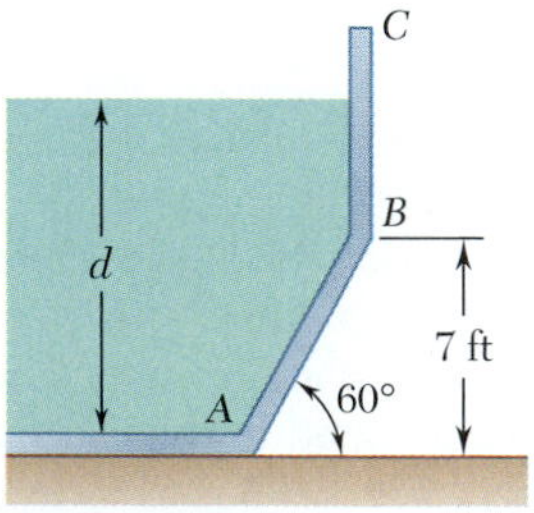

Fig. P5.C6

5.C7 The cross section of a concrete dam is as shown. For a dam section of 1-m width, plot the magnitude of the resultant of the reaction forces exerted by the ground on the base *AB* of the dam and the resultant of the pressure forces exerted by the water on the face *BC* of the dam as functions of the depth *d* of the water for $0 \leq d \leq 16$ m. The densities of concrete and water are 2.40×10^3 kg/m^3 and 10^3 kg/m^3, respectively.

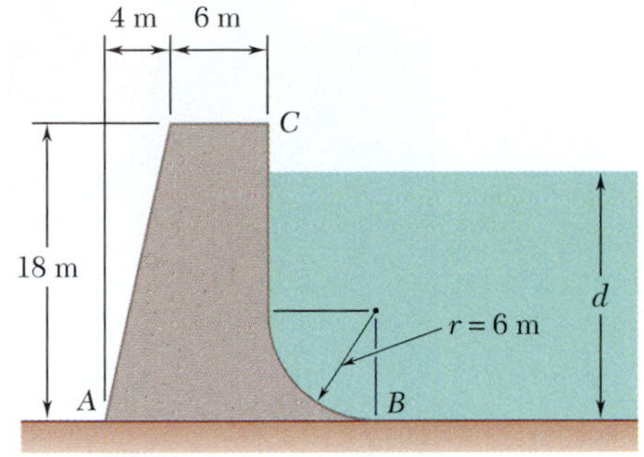

Fig. P5.C7

5.C8 A beam is subjected to the loading shown. Plot the magnitude of the vertical reactions at supports *A* and *B* as functions of distance *a* for $0 \leq a \leq 3$ m.

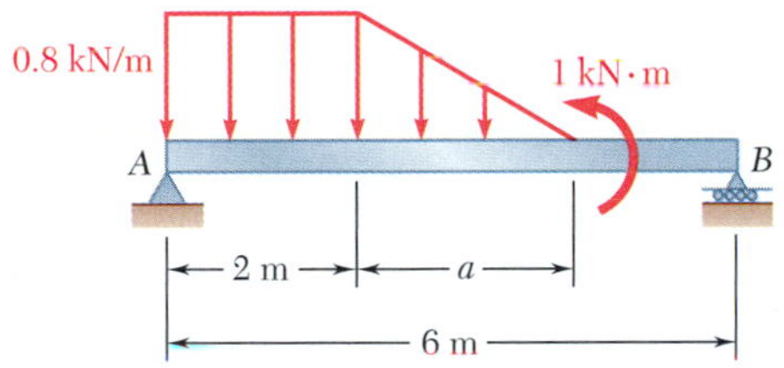

Fig. P5.C8

5.C9 The three-dimensional structure shown is fabricated from five thin steel rods of equal diameter. Using computational software, determine the coordinates of the center of gravity of the structure. Locate the coordinates of the center of gravity when (*a*) $h = 40$ ft, $R = 15$ ft, $\alpha = 90°$; (*b*) $h = 22$ in., $R = 30$ in., $\alpha = 30°$; (*c*) $h = 70$ ft, $R = 65$ ft, $\alpha = 135°$.

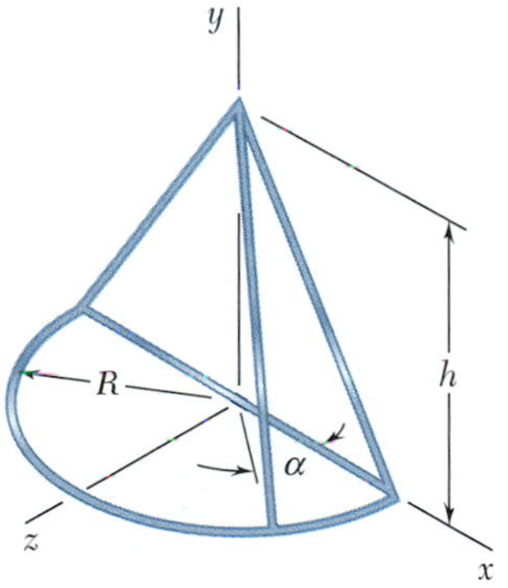

Fig. P5.C9

5.C10 Determine the *y* coordinate of the centroid of the body shown when $h = nb$ and $h = n^2b$. Plot $\overline{y}$ as a function of *n* for both cases using $1 \leq n \leq 10$ and (*a*) $b = 4$ in., (*b*) $b = 6$ in., (*c*) $b = 8$ in.

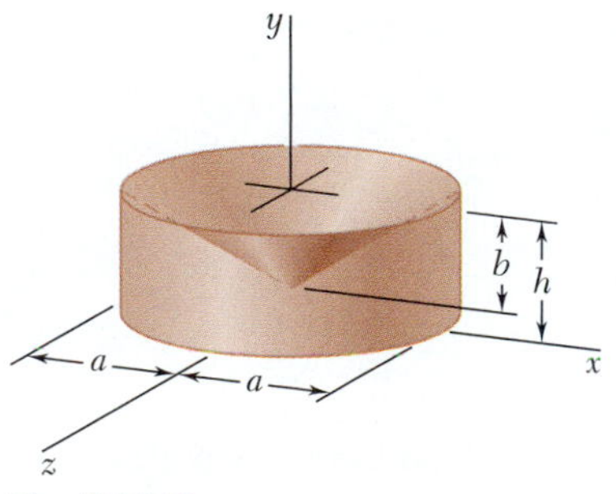

Fig. P5.C10

C H A P T E R

6

Analysis of Structures

Trusses, such as those used for the roof of Safeco Field in Seattle, provide both a practical and an economical solution to many engineering problems.

6.1. INTRODUCTION

The problems considered in the preceding chapters concerned the equilibrium of a single rigid body, and all forces involved were external to the rigid body. We now consider problems dealing with the equilibrium of structures made of several connected parts. These problems call for the determination not only of the external forces acting on the structure but also of the forces which hold together the various parts of the structure. From the point of view of the structure as a whole, these forces are *internal forces.*

Consider, for example, the crane shown in Fig. 6.1*a*, which carries a load W. The crane consists of three beams AD, CF, and BE connected by frictionless pins; it is supported by a pin at A and by a cable DG. The free-body diagram of the crane has been drawn in Fig. 6.1*b*. The external forces, which are shown in the diagram, include the weight $\mathbf{W}$, the two components $\mathbf{A}_x$ and $\mathbf{A}_y$ of the reaction at A, and the force $\mathbf{T}$ exerted by the cable at D. The internal forces holding the various parts of the crane together do not appear in the diagram. If, however, the crane is dismembered and if a free-body diagram is drawn for each of its component parts, the forces holding the three beams together will also be represented, since these forces are external forces from the point of view of each component part (Fig. 6.1*c*).

ANALYSIS OF STRUCTURES

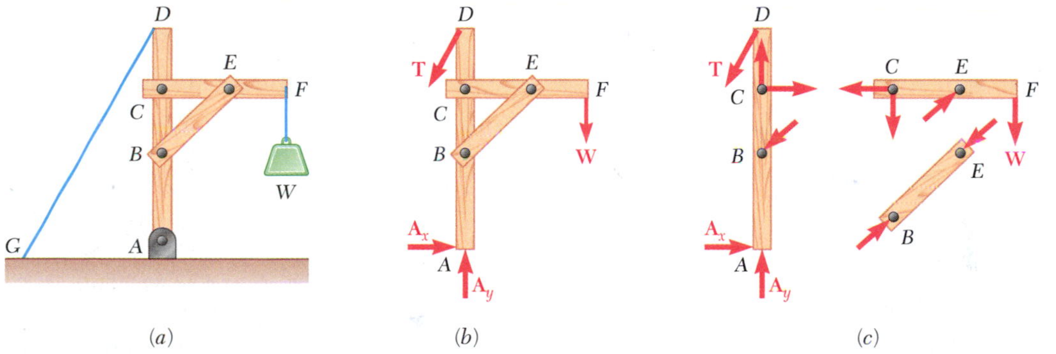

Fig. 6.1

It will be noted that the force exerted at B by member BE on member AD has been represented as equal and opposite to the force exerted at the same point by member AD on member BE; the force exerted at E by BE on CF is shown equal and opposite to the force exerted by CF on BE; and the components of the force exerted at C by CF on AD are shown equal and opposite to the components of the force exerted by AD on CF. This is in conformity with Newton's third law, which states that *the forces of action and reaction between bodies in contact have the same magnitude, same line of action, and opposite sense.* As pointed out in Chap. 1, this law, which is based on experimental evidence, is one of the six fundamental principles of elementary mechanics, and its application is essential to the solution of problems involving connected bodies.

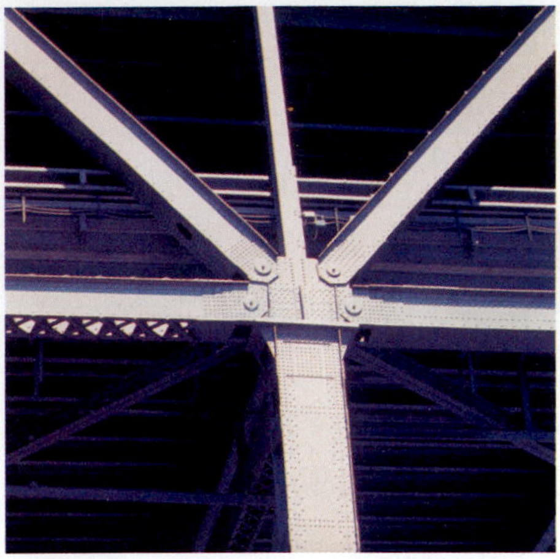

Photo 6.1 Shown is a pin-jointed connection on the approach span to the San Francisco–Oakland Bay Bridge.

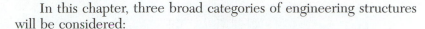

In this chapter, three broad categories of engineering structures will be considered:

1. *Trusses,* which are designed to support loads and are usually stationary, fully constrained structures. Trusses consist exclusively of straight members connected at joints located at the ends of each member. Members of a truss, therefore, are *two-force members,* that is, members each acted upon by two equal and opposite forces directed along the member.
2. *Frames,* which are also designed to support loads and are also usually stationary, fully constrained structures. However, like the crane of Fig. 6.1, frames always contain at least one *multiforce member,* that is, a member acted upon by three or more forces which, in general, are not directed along the member.
3. *Machines,* which are designed to transmit and modify forces and are structures containing moving parts. Machines, like frames, always contain at least one multiforce member.

TRUSSES

6.2. DEFINITION OF A TRUSS

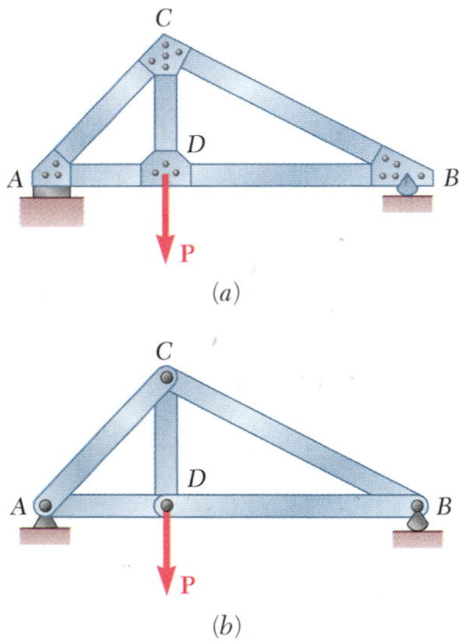

Fig. 6.2

The truss is one of the major types of engineering structures. It provides both a practical and an economical solution to many engineering situations, especially in the design of bridges and buildings. A typical truss is shown in Fig. 6.2*a*. A truss consists of straight members connected at joints. Truss members are connected at their extremities only; thus no member is continuous through a joint. In Fig. 6.2*a*, for example, there is no member AB; there are instead two distinct members AD and DB. Most actual structures are made of several trusses joined together to form a space framework. Each truss is designed to carry those loads which act in its plane and thus may be treated as a two-dimensional structure.

In general, the members of a truss are slender and can support little lateral load; all loads, therefore, must be applied to the various joints, and not to the members themselves. When a concentrated load is to be applied between two joints, or when a distributed load is to be supported by the truss, as in the case of a bridge truss, a floor system must be provided which, through the use of stringers and floor beams, transmits the load to the joints (Fig. 6.3).

The weights of the members of the truss are also assumed to be applied to the joints, half of the weight of each member being applied to each of the two joints the member connects. Although the members are actually joined together by means of bolted or welded connections, it is customary to assume that the members are pinned together; therefore, the forces acting at each end of a member reduce to a single force and no couple. Thus, the only forces assumed to be applied to a truss member are a single force at each end of the

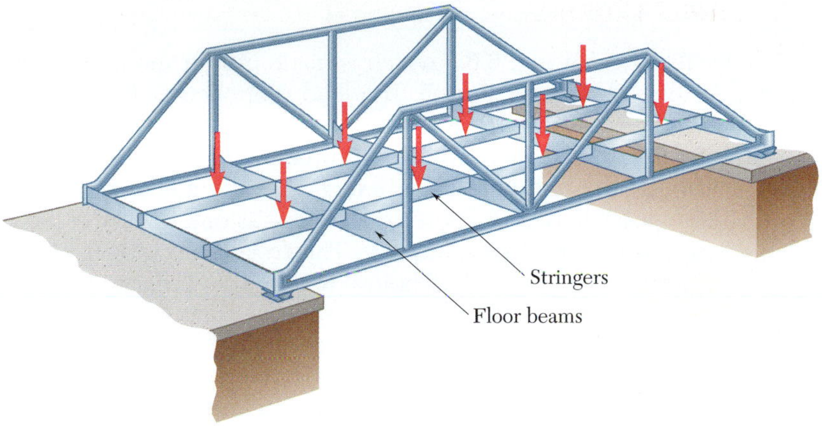

Fig. 6.3

member. Each member can then be treated as a two-force member, and the entire truss can be considered as a group of pins and two-force members (Fig. 6.2*b*). An individual member can be acted upon as shown in either of the two sketches of Fig. 6.4. In Fig. 6.4*a*, the forces tend to pull the member apart, and the member is in tension; in Fig. 6.4*b*, the forces tend to compress the member, and the member is in compression. A number of typical trusses are shown in Fig. 6.5.

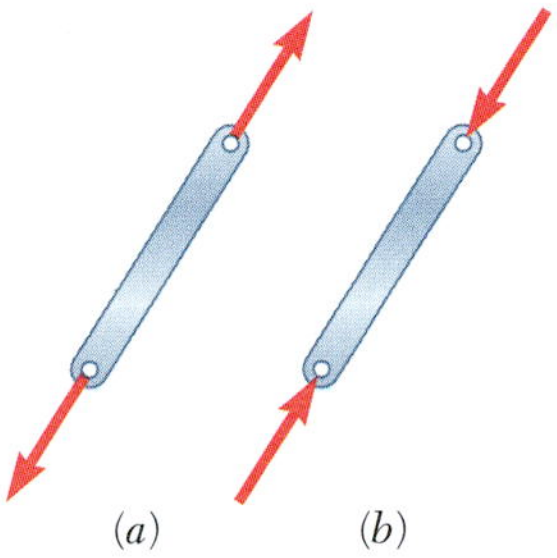

Fig. 6.4

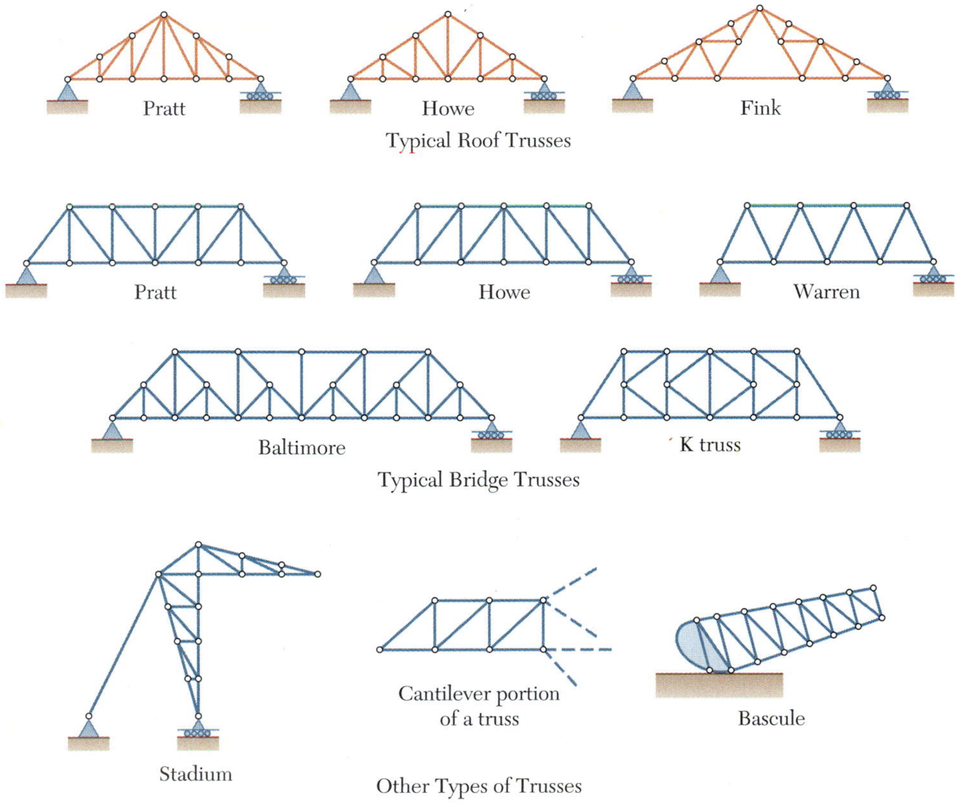

Fig. 6.5

6.3. SIMPLE TRUSSES

Consider the truss of Fig. 6.6*a*, which is made of four members connected by pins at *A*, *B*, *C*, and *D*. If a load is applied at *B*, the truss will greatly deform, completely losing its original shape. In contrast, the truss of Fig. 6.6*b*, which is made of three members connected by pins at *A*, *B*, and *C*, will deform only slightly under a load applied at *B*. The only possible deformation for this truss is one involving small changes in the length of its members. The truss of Fig. 6.6*b* is said to be a *rigid truss*, the term rigid being used here to indicate that the truss *will not collapse*.

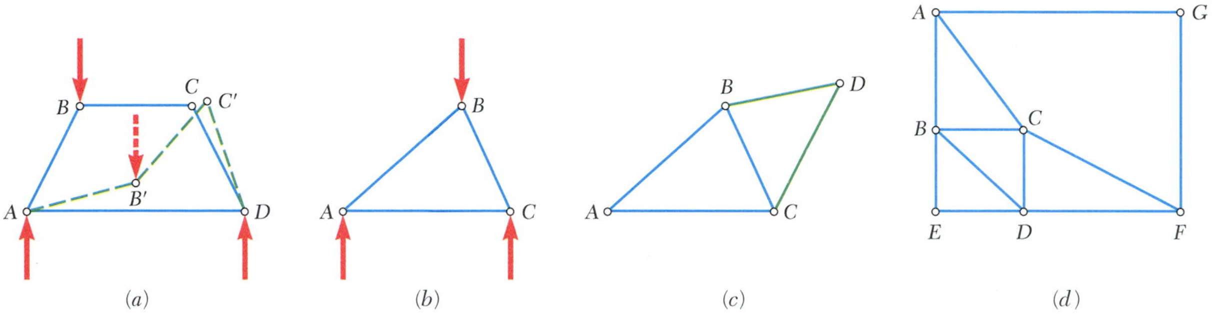

Fig. 6.6

As shown in Fig. 6.6*c*, a larger rigid truss can be obtained by adding two members *BD* and *CD* to the basic triangular truss of Fig. 6.6*b*. This procedure can be repeated as many times as desired, and the resulting truss will be rigid if each time two new members are added, they are attached to two existing joints and are connected at a new joint.† A truss which can be constructed in this manner is called a *simple truss*.

Photo 6.2 Two K-trusses were used as the main components of the movable bridge shown which moved above a large stockpile of ore. The bucket below the trusses picked up ore and redeposited it until the ore was thoroughly mixed. The ore was then sent to the mill for processing into steel.

It should be noted that a simple truss is not necessarily made only of triangles. The truss of Fig. 6.6*d*, for example, is a simple truss which was constructed from triangle *ABC* by adding successively the joints *D*, *E*, *F*, and *G*. On the other hand, rigid trusses are not always simple trusses, even when they appear to be made of triangles. The Fink and Baltimore trusses shown in Fig. 6.5, for instance, are not simple trusses, since they cannot be constructed from a single triangle in the manner described above. All the other trusses shown in Fig. 6.5 are simple trusses, as may be easily checked. (For the K truss, start with one of the central triangles.)

Returning to Fig. 6.6, we note that the basic triangular truss of Fig. 6.6*b* has three members and three joints. The truss of Fig. 6.6*c* has two more members and one more joint, that is, five members and four joints altogether. Observing that every time two new members are added, the number of joints is increased by one, we find that in a simple truss the total number of members is $m = 2n - 3$, where n is the total number of joints.

†The three joints must not be in a straight line.

6.4. ANALYSIS OF TRUSSES BY THE METHOD OF JOINTS

We saw in Sec. 6.2 that a truss can be considered as a group of pins and two-force members. The truss of Fig. 6.2, whose free-body diagram is shown in Fig. 6.7*a*, can thus be dismembered, and a free-body diagram can be drawn for each pin and each member (Fig. 6.7*b*). Each member is acted upon by two forces, one at each end; these forces have the same magnitude, same line of action, and opposite sense (Sec. 4.6). Furthermore, Newton's third law indicates that the forces of action and reaction between a member and a pin are equal and opposite. Therefore, the forces exerted by a member on the two pins it connects must be directed along that member and be equal and opposite. The common magnitude of the forces exerted by a member on the two pins it connects is commonly referred to as the *force in the member* considered, even though this quantity is actually a scalar. Since the lines of action of all the internal forces in a truss are known, the analysis of a truss reduces to computing the forces in its various members and to determining whether each of its members is in tension or in compression.

Since the entire truss is in equilibrium, each pin must be in equilibrium. The fact that a pin is in equilibrium can be expressed by drawing its free-body diagram and writing two equilibrium equations (Sec. 2.9). If the truss contains n pins, there will, therefore, be $2n$ equations available, which can be solved for $2n$ unknowns. In the case of a simple truss, we have $m = 2n - 3$, that is, $2n = m + 3$, and the number of unknowns which can be determined from the free-body diagrams of the pins is thus $m + 3$. This means that the forces in all the members, the two components of the reaction $\mathbf{R}_A$, and the reaction $\mathbf{R}_B$ can be found by considering the free-body diagrams of the pins.

The fact that the entire truss is a rigid body in equilibrium can be used to write three more equations involving the forces shown in the free-body diagram of Fig. 6.7*a*. Since they do not contain any new information, these equations are not independent of the equations associated with the free-body diagrams of the pins. Nevertheless, they can be used to determine the components of the reactions at the supports. The arrangement of pins and members in a simple truss is such that it will then always be possible to find a joint involving only two unknown forces. These forces can be determined by the methods of Sec. 2.11 and their values transferred to the adjacent joints and treated as known quantities at these joints. This procedure can be repeated until all unknown forces have been determined.

As an example, the truss of Fig. 6.7 will be analyzed by considering the equilibrium of each pin successively, starting with a joint at which only two forces are unknown. In the truss considered, all pins are subjected to at least three unknown forces. Therefore, the reactions at the supports must first be determined by considering the entire truss as a free body and using the equations of equilibrium of a rigid body. We find in this way that $\mathbf{R}_A$ is vertical and determine the magnitudes of $\mathbf{R}_A$ and $\mathbf{R}_B$.

The number of unknown forces at joint A is thus reduced to two, and these forces can be determined by considering the equilibrium of pin A. The reaction $\mathbf{R}_A$ and the forces $\mathbf{F}_{AC}$ and $\mathbf{F}_{AD}$ exerted on pin

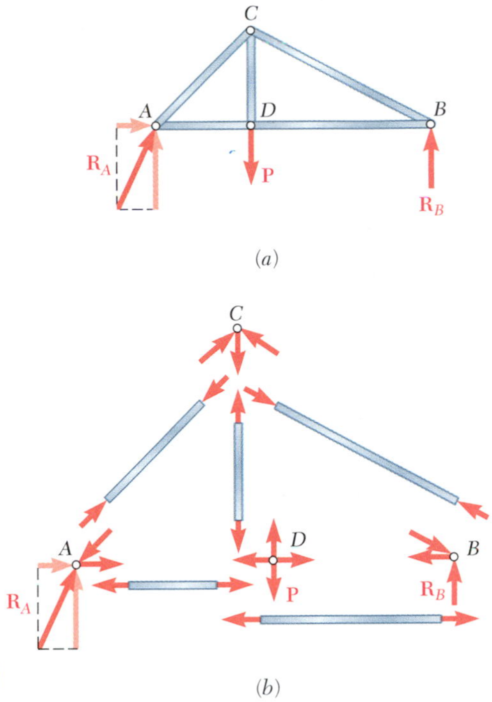

Fig. 6.7

Photo 6.3 Because roof trusses, such as those shown, require support only at their ends, it is possible to construct buildings with large, unobstructed floor areas.

A by members *AC* and *AD*, respectively, must form a force triangle. First we draw $\mathbf{R}_A$ (Fig. 6.8); noting that $\mathbf{F}_{AC}$ and $\mathbf{F}_{AD}$ are directed along *AC* and *AD*, respectively, we complete the triangle and determine the magnitude and sense of $\mathbf{F}_{AC}$ and $\mathbf{F}_{AD}$. The magnitudes F_{AC} and F_{AD} represent the forces in members *AC* and *AD*. Since $\mathbf{F}_{AC}$ is directed down and to the left, that is, *toward* joint *A*, member *AC* pushes on pin *A* and is in compression. Since $\mathbf{F}_{AD}$ is directed *away* from joint *A*, member *AD* pulls on pin *A* and is in tension.

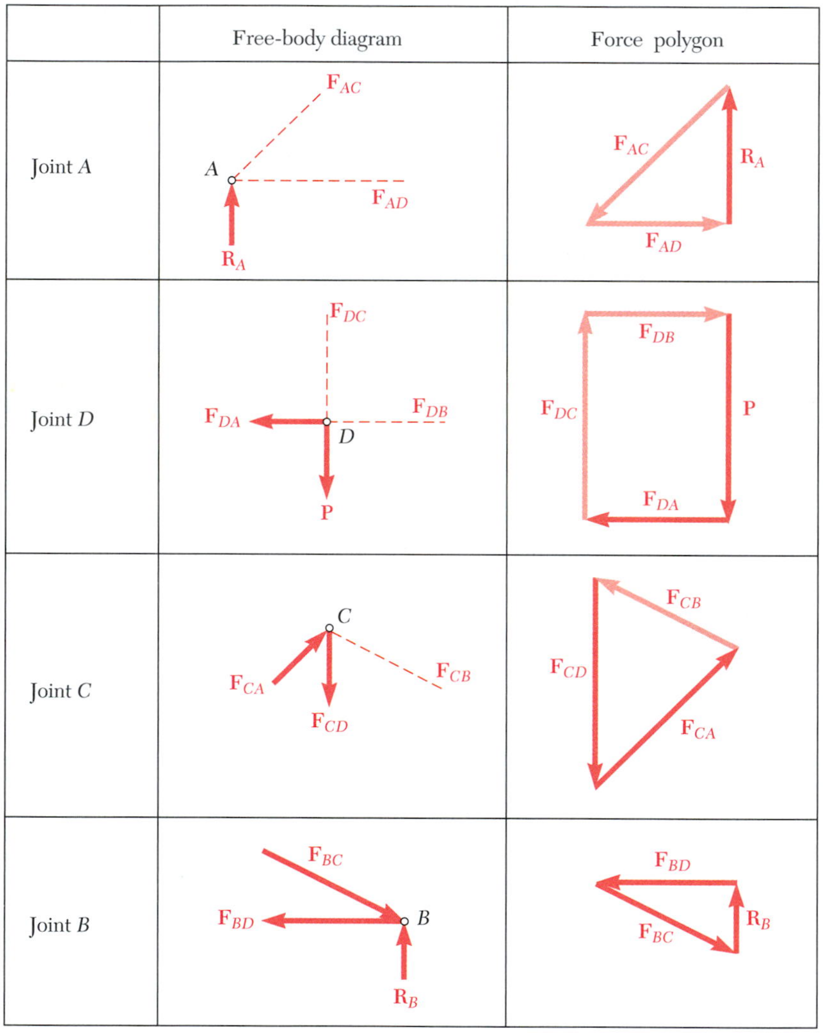

Fig. 6.8

We can now proceed to joint *D*, where only two forces, $\mathbf{F}_{DC}$ and $\mathbf{F}_{DB}$, are still unknown. The other forces are the load **P**, which is given, and the force $\mathbf{F}_{DA}$ exerted on the pin by member *AD*. As indicated above, this force is equal and opposite to the force $\mathbf{F}_{AD}$ exerted by the same member on pin *A*. We can draw the force polygon corresponding to joint *D*, as shown in Fig. 6.8, and determine the forces

$\mathbf{F}_{DC}$ and $\mathbf{F}_{DB}$ from that polygon. However, when more than three forces are involved, it is usually more convenient to solve the equations of equilibrium $\Sigma F_x = 0$ and $\Sigma F_y = 0$ for the two unknown forces. Since both of these forces are found to be directed away from joint D, members DC and DB pull on the pin and are in tension.

Next, joint C is considered; its free-body diagram is shown in Fig. 6.8. It is noted that both $\mathbf{F}_{CD}$ and $\mathbf{F}_{CA}$ are known from the analysis of the preceding joints and that only $\mathbf{F}_{CB}$ is unknown. Since the equilibrium of each pin provides sufficient information to determine two unknowns, a check of our analysis is obtained at this joint. The force triangle is drawn, and the magnitude and sense of $\mathbf{F}_{CB}$ are determined. Since $\mathbf{F}_{CB}$ is directed toward joint C, member CB pushes on pin C and is in compression. The check is obtained by verifying that the force $\mathbf{F}_{CB}$ and member CB are parallel.

At joint B, all of the forces are known. Since the corresponding pin is in equilibrium, the force triangle must close and an additional check of the analysis is obtained.

It should be noted that the force polygons shown in Fig. 6.8 are not unique. Each of them could be replaced by an alternative configuration. For example, the force triangle corresponding to joint A could be drawn as shown in Fig. 6.9. The triangle actually shown in Fig. 6.8 was obtained by drawing the three forces $\mathbf{R}_A$, $\mathbf{F}_{AC}$, and $\mathbf{F}_{AD}$ in tip-to-tail fashion in the order in which their lines of action are encountered when moving clockwise around joint A. The other force polygons in Fig. 6.8, having been drawn in the same way, can be made to fit into a single diagram, as shown in Fig. 6.10. Such a diagram, known as *Maxwell's diagram,* greatly facilitates the *graphical analysis* of truss problems.

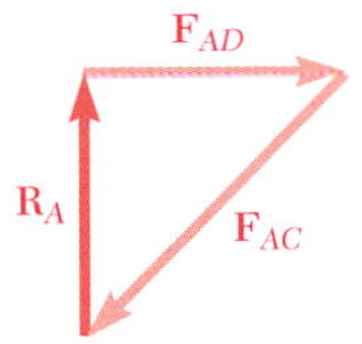

Fig. 6.9

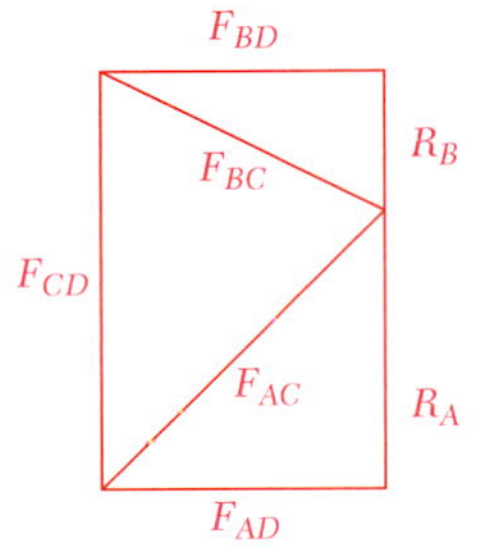

Fig. 6.10

*6.5. JOINTS UNDER SPECIAL LOADING CONDITIONS

Consider Fig. 6.11*a*, in which the joint shown connects four members lying in two intersecting straight lines. The free-body diagram of Fig. 6.11*b* shows that pin A is subjected to two pairs of directly opposite forces. The corresponding force polygon, therefore, must be a parallelogram (Fig. 6.11*c*), and *the forces in opposite members must be equal.*

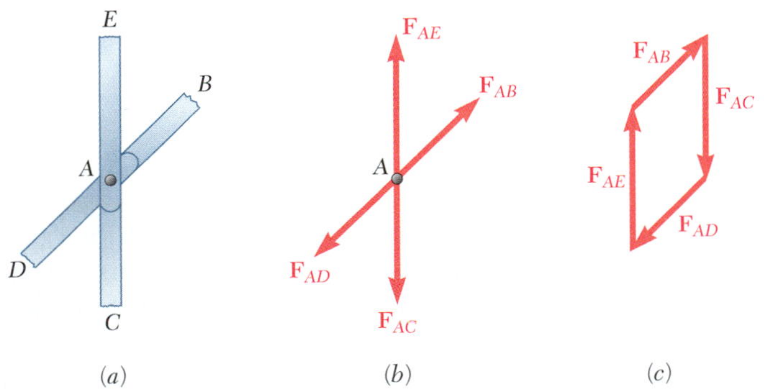

Fig. 6.11

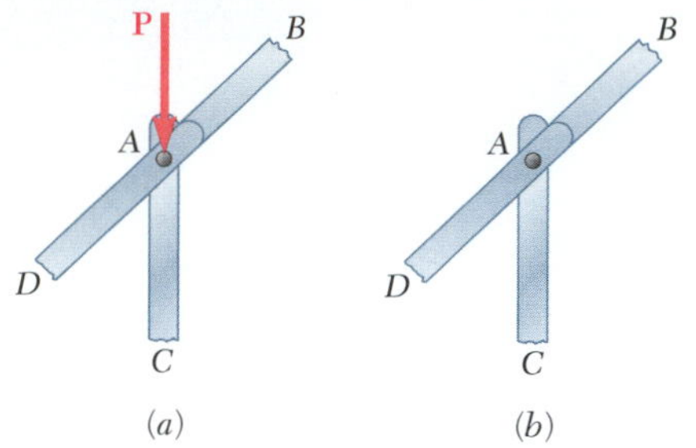

Fig. 6.12

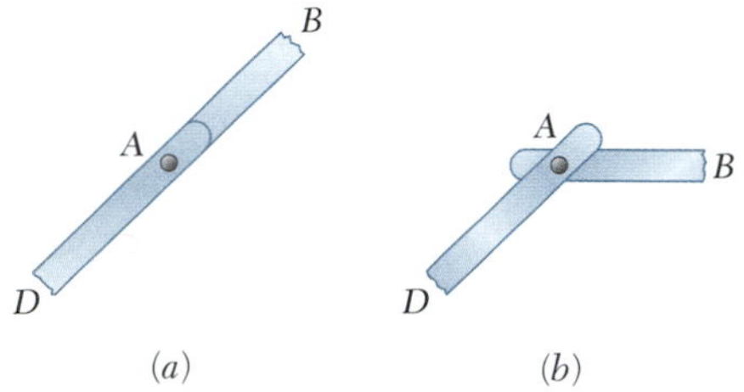

Fig. 6.13

Consider next Fig. 6.12*a*, in which the joint shown connects three members and supports a load **P**. Two of the members lie in the same line, and the load **P** acts along the third member. The free-body diagram of pin *A* and the corresponding force polygon will be as shown in Fig. 6.11*b* and *c*, with $\mathbf{F}_{AE}$ replaced by the load **P**. Thus, *the forces in the two opposite members must be equal, and the force in the other member must equal P.* A particular case of special interest is shown in Fig. 6.12*b*. Since, in this case, no external load is applied to the joint, we have $P = 0$, and the force in member *AC* is zero. Member *AC* is said to be a *zero-force member.*

Consider now a joint connecting two members only. From Sec. 2.9, we know that a particle which is acted upon by two forces will be in equilibrium if the two forces have the same magnitude, same line of action, and opposite sense. In the case of the joint of Fig. 6.13*a*, which connects two members *AB* and *AD* lying in the same line, *the forces in the two members must be equal* for pin *A* to be in equilibrium. In the case of the joint of Fig. 6.13*b*, pin *A* cannot be in equilibrium unless the forces in both members are zero. Members connected as shown in Fig. 6.13*b*, therefore, must be *zero-force members.*

Spotting the joints which are under the special loading conditions listed above will expedite the analysis of a truss. Consider, for example, a Howe truss loaded as shown in Fig. 6.14. All of the members represented by green lines will be recognized as zero-force members. Joint *C* connects three members, two of which lie in the same line, and is not subjected to any external load; member *BC* is thus a zero-force member. Applying the same reasoning to joint *K*, we find that member *JK* is also a zero-force member. But joint *J* is now in the same situation as joints *C* and *K*, and member *IJ* must be a zero-force member. The examination of joints *C*, *J*, and *K* also shows that the forces in members *AC* and *CE* are equal, that the forces in members *HJ* and *JL* are equal, and that the forces in members *IK* and *KL* are equal. Turning our attention to joint *I*, where the 20-kN load and member *HI* are collinear, we note that the force in member *HI* is 20 kN (tension) and that the forces in members *GI* and *IK* are equal. Hence, the forces in members *GI*, *IK*, and *KL* are equal.

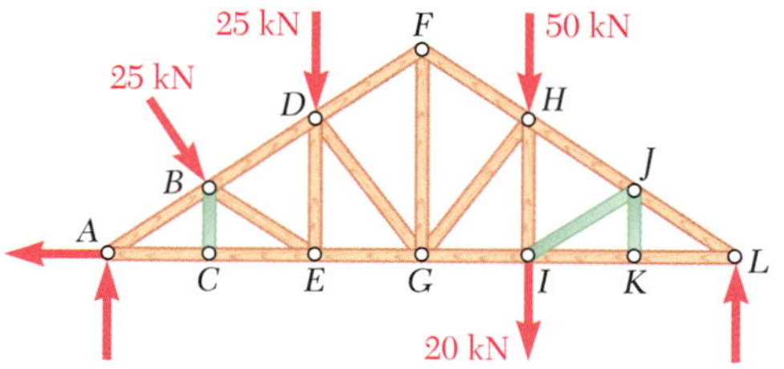

Fig. 6.14

Note that the conditions described above do not apply to joints *B* and *D* in Fig. 6.14, and it would be wrong to assume that the force in member *DE* is 25 kN or that the forces in members *AB* and *BD* are equal. The forces in these members and in all remaining members should be found by carrying out the analysis of joints *A*, *B*, *D*, *E*, *F*, *G*, *H*, and *L* in the usual manner. Thus, until you have become thoroughly familiar with the conditions under which the rules established in this section can be applied, you should draw the free-body diagrams of all pins and write the corresponding equilibrium equations (or draw the corresponding force polygons) whether or not the joints being considered are under one of the special loading conditions described above.

A final remark concerning zero-force members: These members are not useless. For example, although the zero-force members of Fig. 6.14 do not carry any loads under the loading conditions shown, the same members would probably carry loads if the loading conditions were changed. Besides, even in the case considered, these members are needed to support the weight of the truss and to maintain the truss in the desired shape.

*6.6. SPACE TRUSSES

When several straight members are joined together at their extremities to form a three-dimensional configuration, the structure obtained is called a *space truss*.

We recall from Sec. 6.3 that the most elementary two-dimensional rigid truss consisted of three members joined at their extremities to form the sides of a triangle; by adding two members at a time to this basic configuration, and connecting them at a new joint, it was possible to obtain a larger rigid structure which was defined as a simple truss. Similarly, the most elementary rigid space truss consists of six members joined at their extremities to form the edges of a tetrahedron $ABCD$ (Fig. 6.15*a*). By adding three members at a time to this basic configuration, such as AE, BE, and CE, attaching them to three existing joints, and connecting them at a new joint,† we can obtain a larger rigid structure which is defined as a *simple space truss* (Fig. 6.15*b*). Observing that the basic tetrahedron has six members and four joints and that every time three members are added, the number of joints is increased by one, we conclude that in a simple space truss the total number of members is $m = 3n - 6$, where n is the total number of joints.

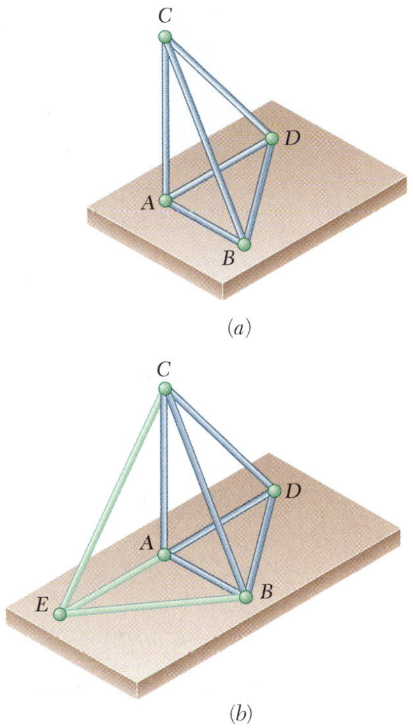

Fig. 6.15

If a space truss is to be completely constrained and if the reactions at its supports are to be statically determinate, the supports should consist of a combination of balls, rollers, and balls and sockets which provides six unknown reactions (see Sec. 4.9). These unknown reactions may be readily determined by solving the six equations expressing that the three-dimensional truss is in equilibrium.

Photo 6.4 Three-dimensional or space trusses are used for broadcast and power transmission line towers, roof framing, and spacecraft applications, such as components of the *International Space Station*.

Although the members of a space truss are actually joined together by means of bolted or welded connections, it is assumed that each joint consists of a ball-and-socket connection. Thus, no couple will be applied to the members of the truss, and each member can be treated as a two-force member. The conditions of equilibrium for each joint will be expressed by the three equations $\Sigma F_x = 0$, $\Sigma F_y = 0$, and $\Sigma F_z = 0$. In the case of a simple space truss containing n joints, writing the conditions of equilibrium for each joint will thus yield $3n$ equations. Since $m = 3n - 6$, these equations suffice to determine all unknown forces (forces in m members and six reactions at the supports). However, to avoid the necessity of solving simultaneous equations, care should be taken to select joints in such an order that no selected joint will involve more than three unknown forces.

†The four joints must not lie in a plane.

SAMPLE PROBLEM 6.1

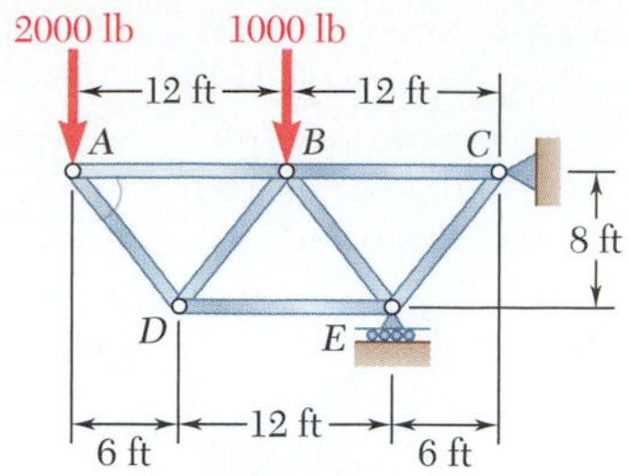

Using the method of joints, determine the force in each member of the truss shown.

SOLUTION

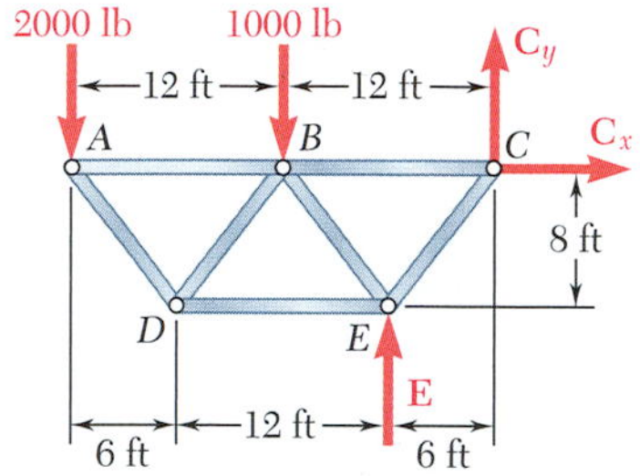

Free-Body: Entire Truss. A free-body diagram of the entire truss is drawn; external forces acting on this free body consist of the applied loads and the reactions at C and E. We write the following equilibrium equations.

$$+\circlearrowleft \Sigma M_C = 0: \quad (2000 \text{ lb})(24 \text{ ft}) + (1000 \text{ lb})(12 \text{ ft}) - E(6 \text{ ft}) = 0$$

$$E = +10{,}000 \text{ lb} \qquad \mathbf{E} = 10{,}000 \text{ lb}\uparrow$$

$$\xrightarrow{+} \Sigma F_x = 0: \qquad \mathbf{C}_x = 0$$

$$+\uparrow \Sigma F_y = 0: \quad -2000 \text{ lb} - 1000 \text{ lb} + 10{,}000 \text{ lb} + C_y = 0$$

$$C_y = -7000 \text{ lb} \qquad \mathbf{C}_y = 7000 \text{ lb}\downarrow$$

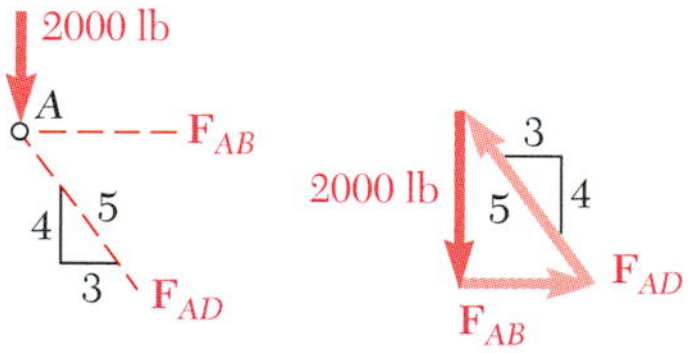

Free-Body: Joint *A*. This joint is subjected to only two unknown forces, namely, the forces exerted by members AB and AD. A force triangle is used to determine $\mathbf{F}_{AB}$ and $\mathbf{F}_{AD}$. We note that member AB pulls on the joint and thus is in tension and that member AD pushes on the joint and thus is in compression. The magnitudes of the two forces are obtained from the proportion

$$\frac{2000 \text{ lb}}{4} = \frac{F_{AB}}{3} = \frac{F_{AD}}{5}$$

$$F_{AB} = 1500 \text{ lb } T \quad ◀$$
$$F_{AD} = 2500 \text{ lb } C \quad ◀$$

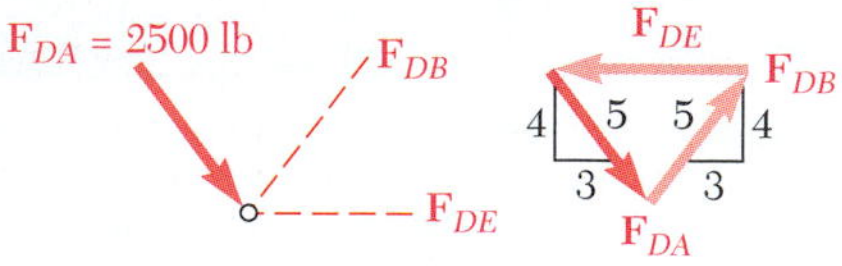

Free-Body: Joint *D*. Since the force exerted by member AD has been determined, only two unknown forces are now involved at this joint. Again, a force triangle is used to determine the unknown forces in members DB and DE.

$$F_{DB} = F_{DA} \qquad F_{DB} = 2500 \text{ lb } T \quad ◀$$
$$F_{DE} = 2(\tfrac{3}{5})F_{DA} \qquad F_{DE} = 3000 \text{ lb } C \quad ◀$$

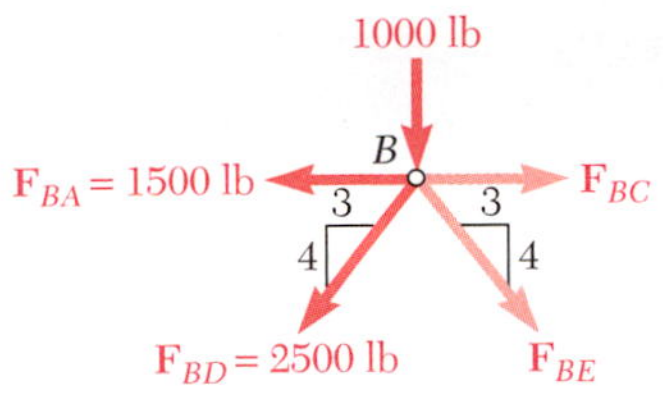

Free-Body: Joint *B*. Since more than three forces act at this joint, we determine the two unknown forces $\mathbf{F}_{BC}$ and $\mathbf{F}_{BE}$ by solving the equilibrium equations $\Sigma F_x = 0$ and $\Sigma F_y = 0$. We arbitrarily assume that both unknown forces act away from the joint, that is, that the members are in tension. The positive value obtained for F_{BC} indicates that our assumption was correct; member *BC* is in tension. The negative value of F_{BE} indicates that our assumption was wrong; member *BE* is in compression.

$$+\uparrow\Sigma F_y = 0: \quad -1000 - \tfrac{4}{5}(2500) - \tfrac{4}{5}F_{BE} = 0$$
$$F_{BE} = -3750 \text{ lb} \qquad F_{BE} = 3750 \text{ lb } C \blacktriangleleft$$

$$\overset{+}{\rightarrow}\Sigma F_x = 0: \quad F_{BC} - 1500 - \tfrac{3}{5}(2500) - \tfrac{3}{5}(3750) = 0$$
$$F_{BC} = +5250 \text{ lb} \qquad F_{BC} = 5250 \text{ lb } T \blacktriangleleft$$

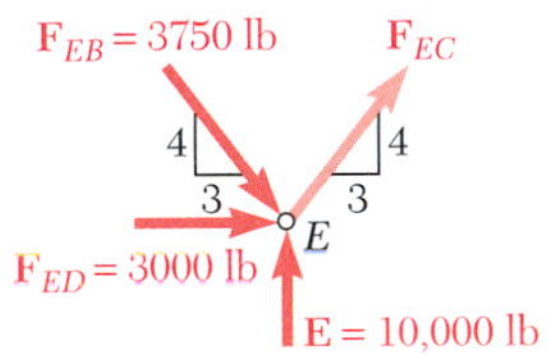

Free-Body: Joint *E*. The unknown force $\mathbf{F}_{EC}$ is assumed to act away from the joint. Summing *x* components, we write

$$\overset{+}{\rightarrow}\Sigma F_x = 0: \quad \tfrac{3}{5}F_{EC} + 3000 + \tfrac{3}{5}(3750) = 0$$
$$F_{EC} = -8750 \text{ lb} \qquad F_{EC} = 8750 \text{ lb } C \blacktriangleleft$$

Summing *y* components, we obtain a check of our computations:

$$+\uparrow\Sigma F_y = 10{,}000 - \tfrac{4}{5}(3750) - \tfrac{4}{5}(8750)$$
$$= 10{,}000 - 3000 - 7000 = 0 \qquad \text{(checks)}$$

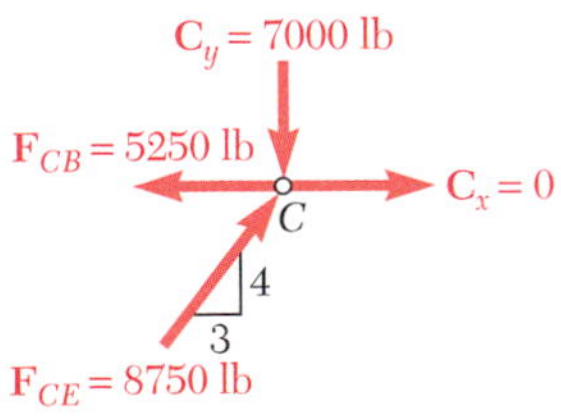

Free-Body: Joint *C*. Using the computed values of $\mathbf{F}_{CB}$ and $\mathbf{F}_{CE}$, we can determine the reactions $\mathbf{C}_x$ and $\mathbf{C}_y$ by considering the equilibrium of this joint. Since these reactions have already been determined from the equilibrium of the entire truss, we will obtain two checks of our computations. We can also simply use the computed values of all forces acting on the joint (forces in members and reactions) and check that the joint is in equilibrium:

$$\overset{+}{\rightarrow}\Sigma F_x = -5250 + \tfrac{3}{5}(8750) = -5250 + 5250 = 0 \qquad \text{(checks)}$$
$$+\uparrow\Sigma F_y = -7000 + \tfrac{4}{5}(8750) = -7000 + 7000 = 0 \qquad \text{(checks)}$$

SOLVING PROBLEMS ON YOUR OWN

In this lesson you learned to use the *method of joints* to determine the forces in the members of a *simple truss*, that is, a truss that can be constructed from a basic triangular truss by adding to it two new members at a time and connecting them at a new joint. We note that the analysis of a *simple* truss can always be carried out by the method of joints.

Your solution will consist of the following steps:

1. *Draw a free-body diagram of the entire truss,* and use this diagram to determine the reactions at the supports.

2. *Note that the choice of the first joint is not unique.* Once you have determined the reactions at the supports of the truss, you can choose either of two joints as a starting point for your analysis. In Sample Prob. 6.1, we started at joint A and proceeded through joints D, B, E, and C, but we could also have started at joint C and proceeded through joints E, B, D, and A. On the other hand, having selected a first joint, you may in some cases reach a point in your analysis beyond which you cannot proceed. You must then start again from another joint to complete your solution. Because there is usually more than one way to analyze a truss, we encourage you to outline your solution *before* starting any computations.

3. *Locate a joint connecting only two members, and draw the free-body diagram of its pin.* Use this free-body diagram to determine the unknown force in each of the two members. If only three forces are involved (the two unknown forces and a known one), you will probably find it more convenient to draw and solve the corresponding force triangle. If more than three forces are involved, you should write and solve the equilibrium equations for the pin, $\Sigma F_x = 0$ and $\Sigma F_y = 0$, assuming that the members are in tension. A positive answer means that the member is in tension, a negative answer that the member is in compression. Once the forces have been found, enter their values on a sketch of the truss, with T for tension and C for compression.

In Sec. 6.4, we discussed how to use a force triangle to determine the unknown forces at a joint, and in Sample Prob. 6.1 we applied this technique to joints A and D. In each of these cases, the directions of the unknown forces at a joint were not specified in the associated free-body diagram—only their lines of action were shown. Thus, when you construct a force triangle, draw the known force first and then add the lines of action of the unknown forces, remembering that the forces must be arranged tip to tail to form a triangle. It is in this last step that the directions of the unknown forces are established.

4. *Next, locate a joint where the forces in only two of the connected members are still unknown.* Draw the free-body diagram of the pin and use it as indicated above to determine the two unknown forces.

5. *Repeat this procedure until the forces in all the members of the truss have been found.* Since you previously used the three equilibrium equations associated with the free-body diagram of the entire truss to determine the reactions at the supports, you will end up with three extra equations. These equations can be used to check your computations.

Problems

6.1 through 6.8 Using the method of joints, determine the force in each member of the truss shown. State whether each member is in tension or compression.

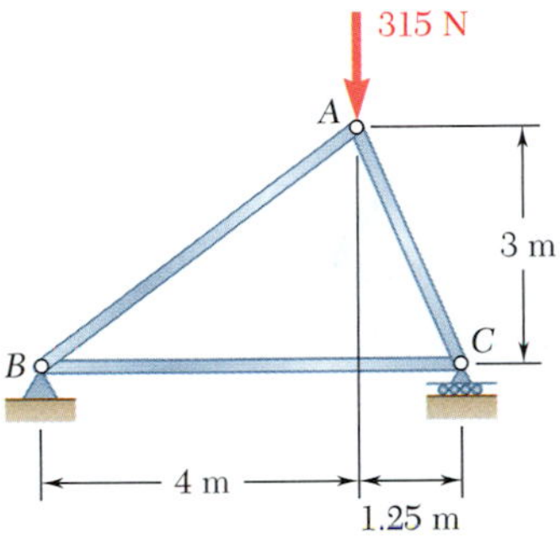

Fig. P6.1

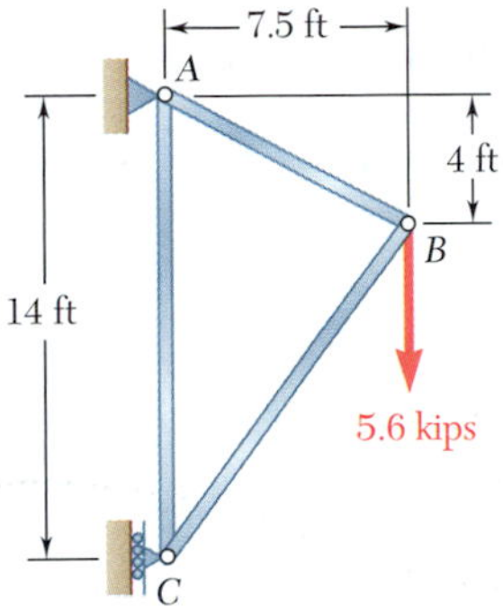

Fig. P6.2

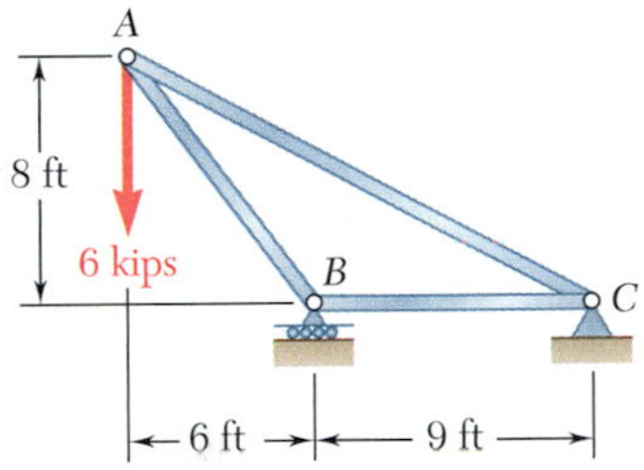

Fig. *P6.3*

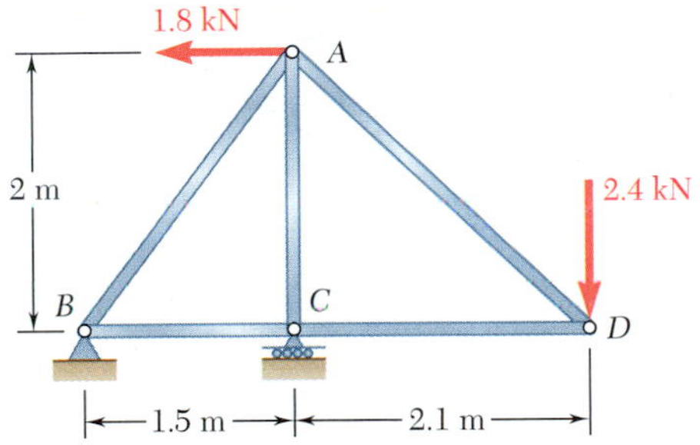

Fig. P6.4

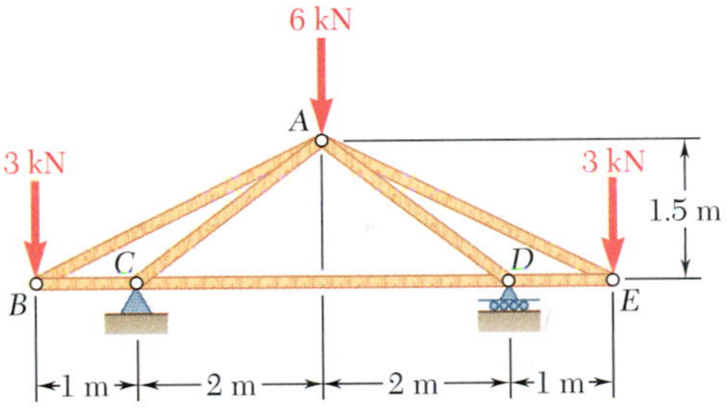

Fig. P6.5

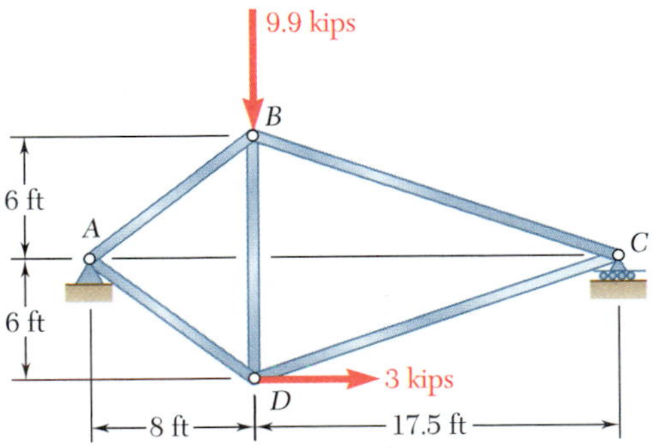

Fig. P6.6

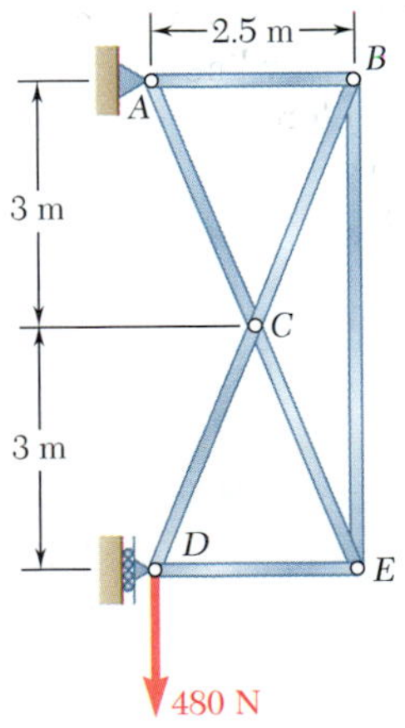

Fig. P6.7

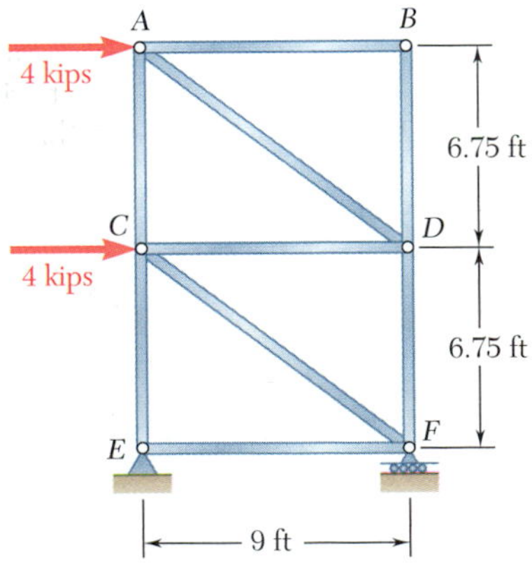

Fig. *P6.8*

6.9 Determine the force in each member of the Gambrel roof truss shown. State whether each member is in tension or compression.

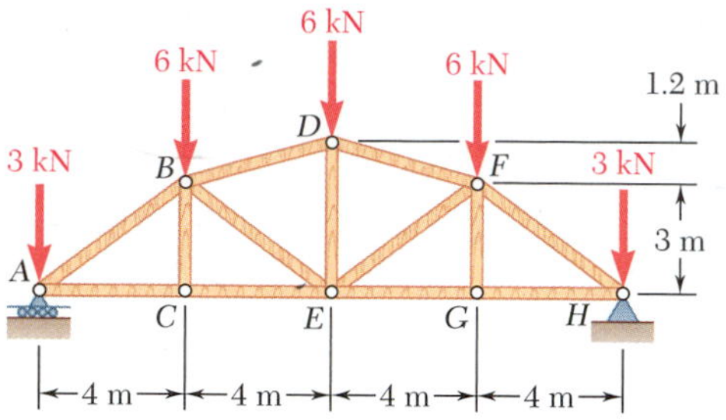

Fig. P6.9

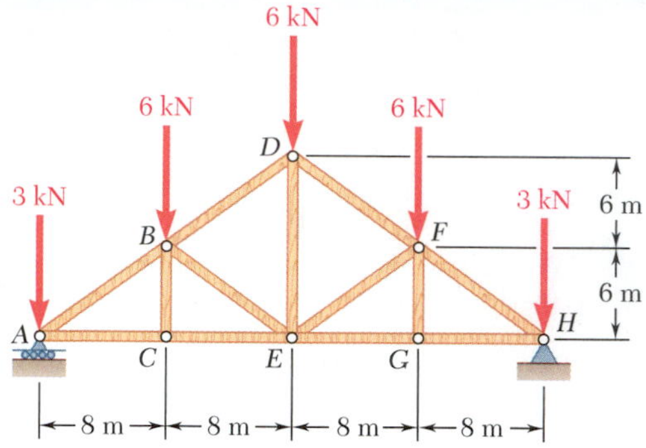

Fig. P6.10

6.10 Determine the force in each member of the Howe roof truss shown. State whether each member is in tension or compression.

6.11 Determine the force in each member of the Fink roof truss shown. State whether each member is in tension or compression.

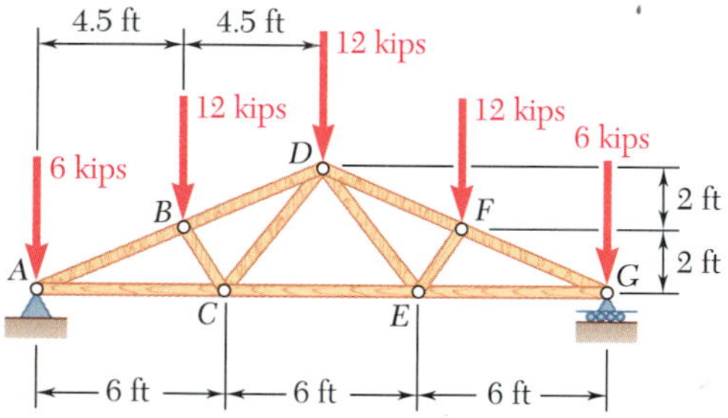

Fig. P6.11

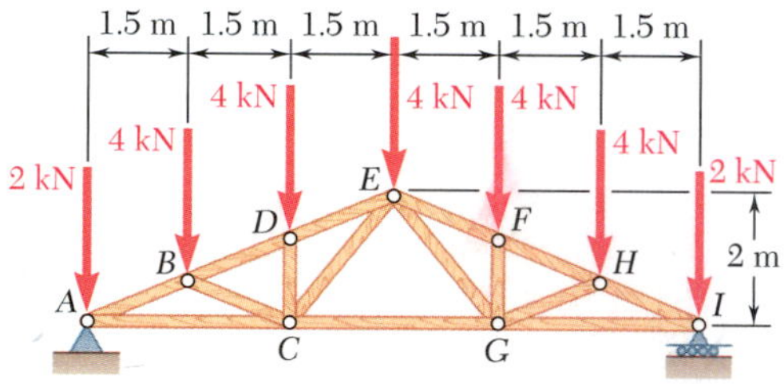

Fig. *P6.12*

6.12 Determine the force in each member of the fan roof truss shown. State whether each member is in tension or compression.

6.13 Determine the force in each member of the roof truss shown. State whether each member is in tension or compression.

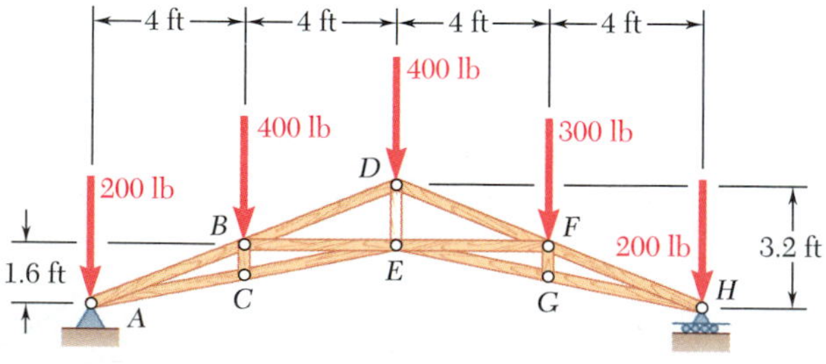

Fig. P6.13

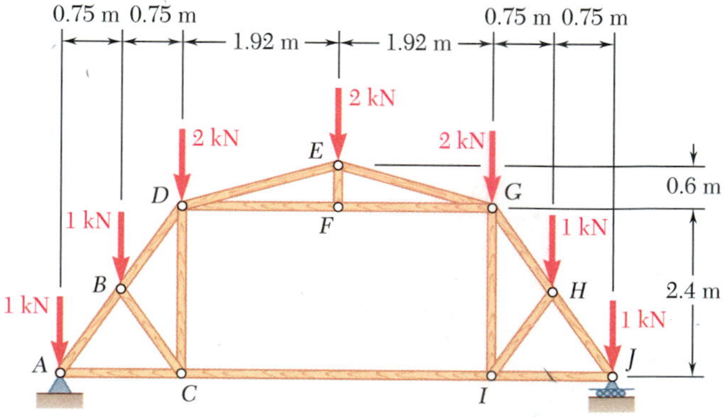

Fig. *P6.14*

6.14 Determine the force in each member of the Gambrel roof truss shown. State whether each member is in tension or compression.

6.15 Determine the force in each member of the Pratt bridge truss shown. State whether each member is in tension or compression.

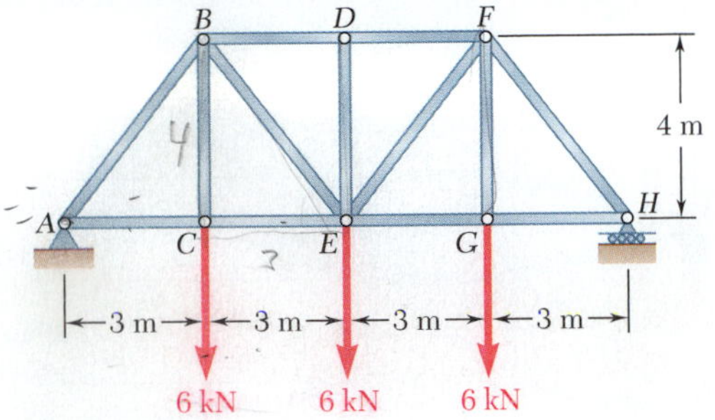

Fig. P6.15

6.16 Solve Prob. 6.15 assuming that the load at G has been removed.

6.17 Determine the force in member DE and in each of the members located to the left of DE for the inverted Howe roof truss shown. State whether each member is in tension or compression.

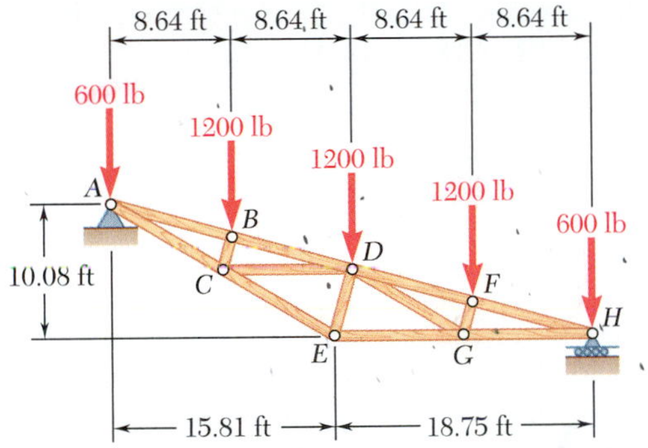

Fig. P6.17 and P6.18

6.18 Determine the force in each of the members located to the right of DE for the inverted Howe roof truss shown. State whether each member is in tension or compression.

6.19 Determine the force in each of the members located to the left of member FG for the scissor roof truss shown. State whether each member is in tension or compression.

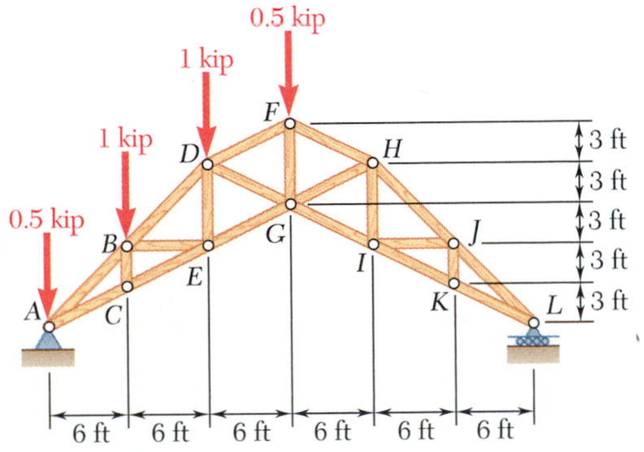

Fig. P6.19 and P6.20

6.20 Determine the force in member FG and in each of the members located to the right of member FG for the scissor roof truss shown. State whether each member is in tension or compression.

6.21 The portion of truss shown represents the upper part of a power transmission line tower. For the given loading, determine the force in each of the members located above HJ. State whether each member is in tension or compression.

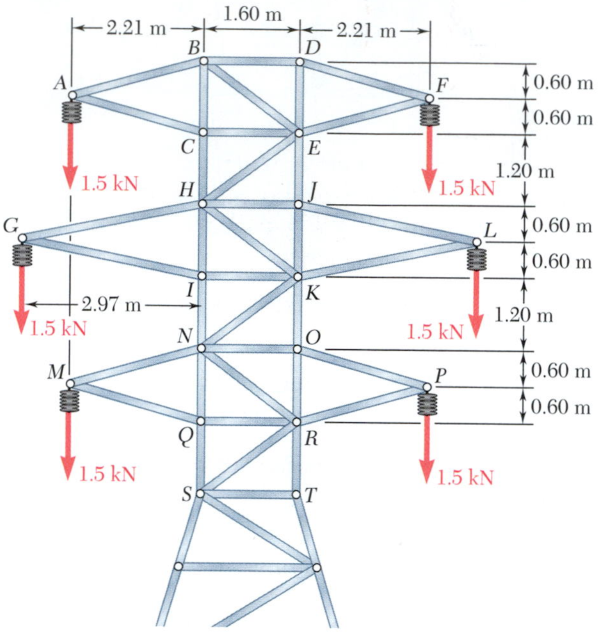

Fig. P6.21

6.22 For the tower and loading of Prob. 6.21 and knowing that $F_{CH} = F_{EJ} = 1.5$ kN C and $F_{EH} = 0$, determine the force in member HJ and in each of the members located between HJ and NO. State whether each member is in tension or compression.

6.23 Determine the force in each member of the truss shown. State whether each member is in tension or compression.

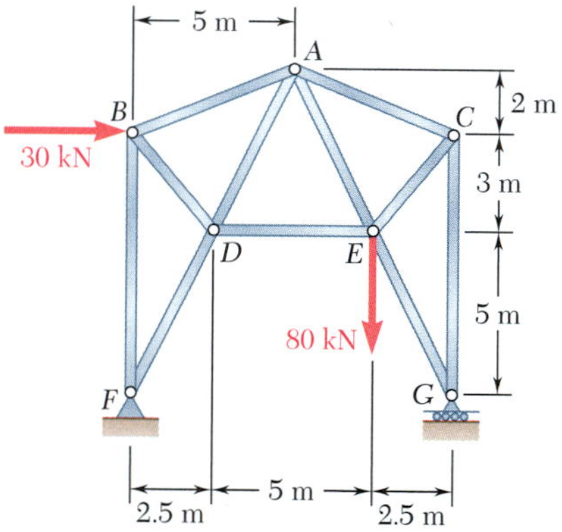

Fig. *P6.23*

6.24 Determine the force in each member of the truss shown. State whether each member is in tension or compression.

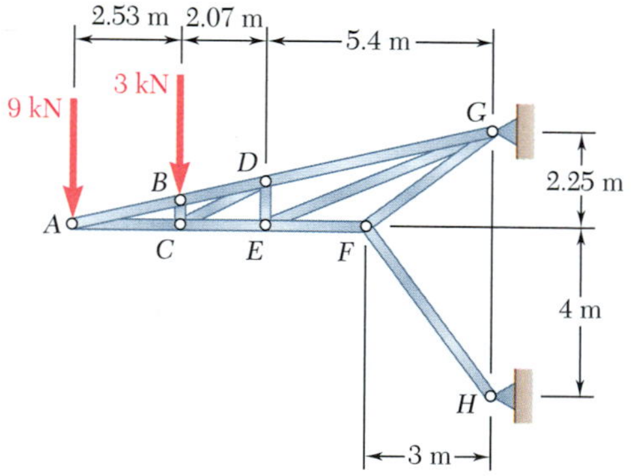

Fig. P6.24

6.25 For the roof truss shown in Fig. *P6.25* and *P6.26*, determine the force in each of the members located to the left of member GH. State whether each member is in tension or compression.

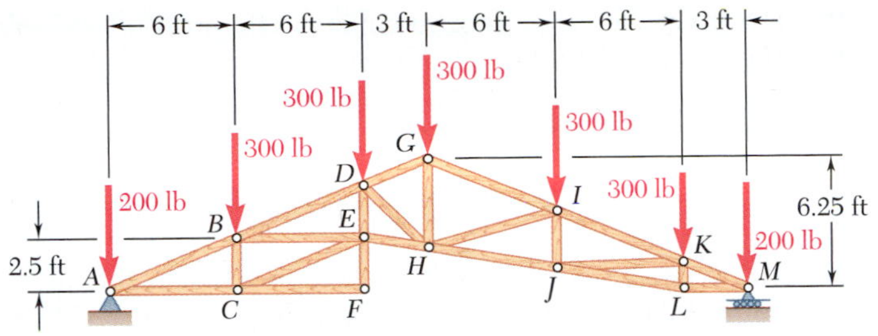

Fig. *P6.25* and *P6.26*

6.26 Determine the force in member *GH* and in each of the members located to the right of *GH* for the roof truss shown. State whether each member is in tension or compression.

6.27 Determine whether the trusses of Probs. 6.13, 6.14, and 6.25 are simple trusses.

6.28 Determine whether the trusses of Probs. 6.19, 6.21, and 6.23 are simple trusses.

***6.29* and 6.30** For the given loading, determine the zero-force members in the truss shown.

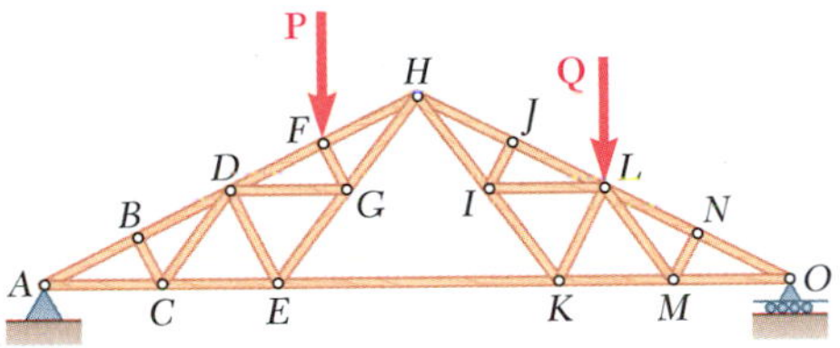

Fig. *P6.29*

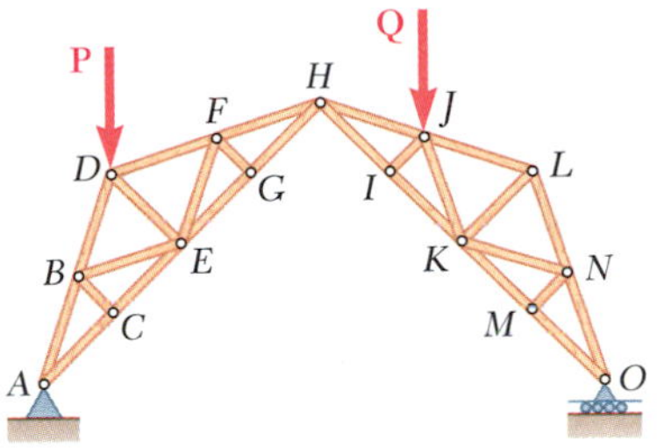

Fig. P6.30

***6.31* and 6.32** For the given loading, determine the zero-force members in the truss shown.

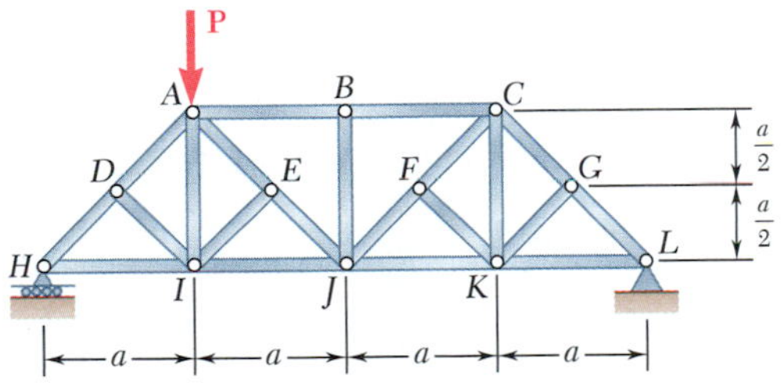

Fig. *P6.31*

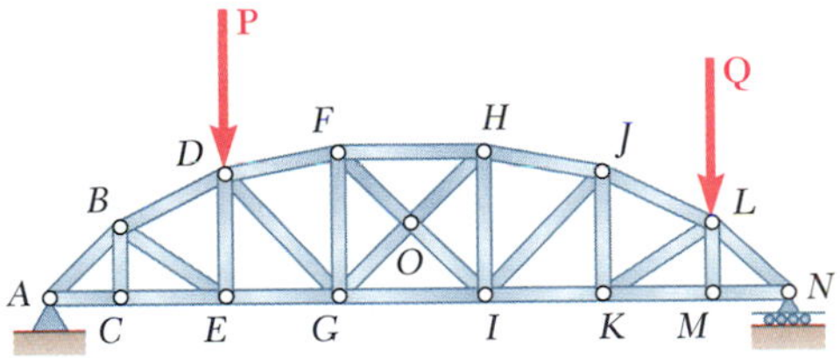

Fig. P6.32

6.33 and 6.34 For the given loading, determine the zero-force members in the truss shown.

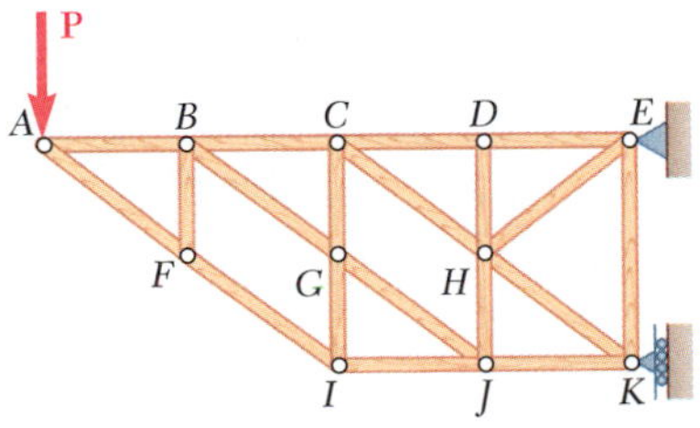

Fig. P6.33

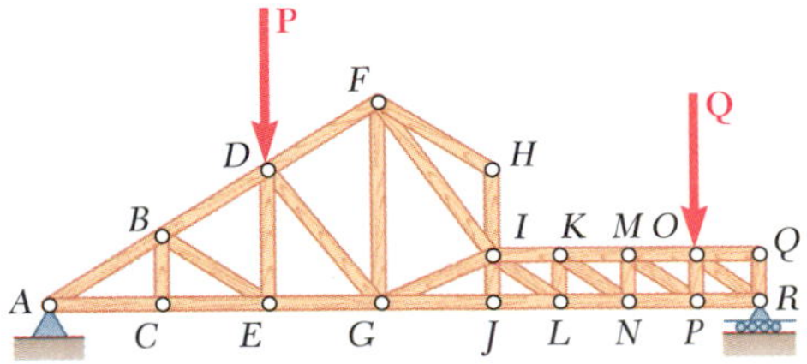

Fig. P6.34

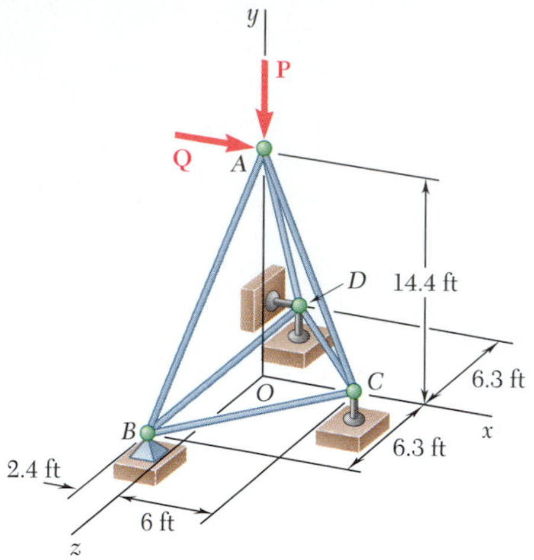

Fig. P6.36 and P6.37

6.35 Determine the zero-force members in the truss of (*a*) Prob. 6.9, (*b*) Prob. 6.19.

***6.36** The truss shown consists of six members and is supported by a ball and socket at *B*, a short link at *C*, and two short links at *D*. Determine the force in each of the members for $\mathbf{P} = (-5670 \text{ lb})\mathbf{j}$ and $\mathbf{Q} = 0$.

***6.37** The truss shown consists of six members and is supported by a ball and socket at *B*, a short link at *C*, and two short links at *D*. Determine the force in each of the members for $\mathbf{P} = 0$ and $\mathbf{Q} = (7420 \text{ lb})\mathbf{i}$.

***6.38** The truss shown consists of six members and is supported by a short link at *A*, two short links at *B*, and a ball and socket at *D*. Determine the force in each of the members for the given loading.

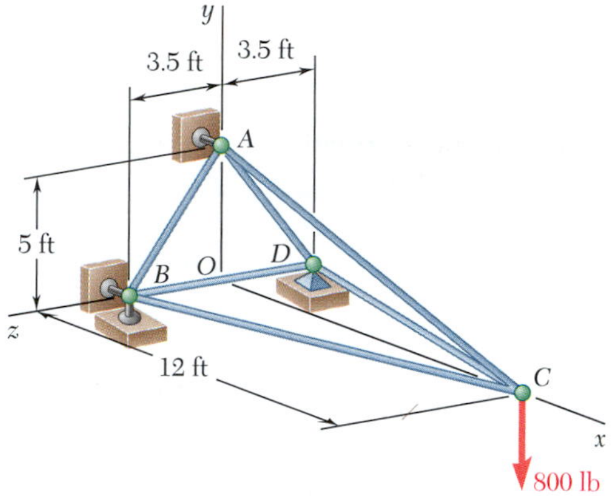

Fig. P6.38

***6.39** The portion of a power line transmission tower shown consists of nine members and is supported by a ball and socket at *B* and short links at *C*, *D*, and *E*. Determine the force in each of the members for the given loading.

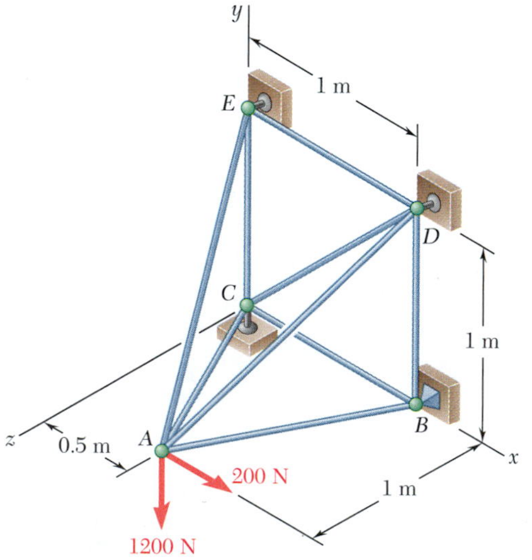

Fig. P6.39

***6.40** The truss shown consists of 18 members and is supported by a ball and socket at A, two short links at B, and one short link at G. (a) Check that this truss is a simple truss, that it is completely constrained, and that the reactions at its supports are statically determinate. (b) For the given loading, determine the force in each of the six members joined at E.

***6.41** The truss shown consists of 18 members and is supported by a ball and socket at A, two short links at B, and one short link at G. (a) Check that this truss is a simple truss, that it is completely constrained, and that the reactions at its supports are statically determinate. (b) For the given loading, determine the force in each of the six members joined at G.

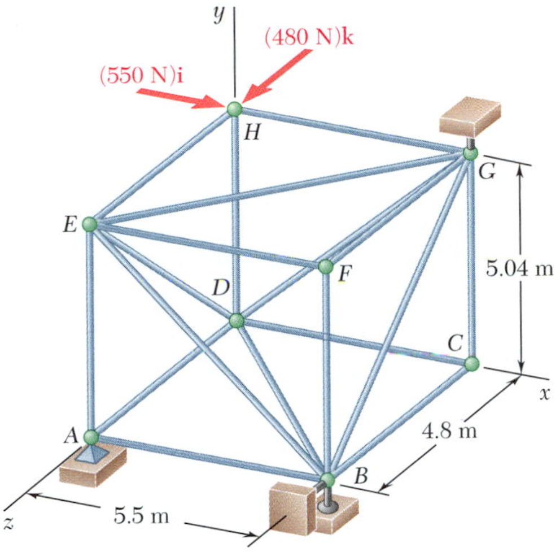

Fig. P6.40 and P6.41

6.7. ANALYSIS OF TRUSSES BY THE METHOD OF SECTIONS

The method of joints is most effective when the forces in all the members of a truss are to be determined. If, however, the force in only one member or the forces in a very few members are desired, another method, the method of sections, is more efficient.

Assume, for example, that we want to determine the force in member BD of the truss shown in Fig. 6.16a. To do this, we must determine the force with which member BD acts on either joint B or joint D. If we were to use the method of joints, we would choose either joint B or joint D as a free body. However, we can also choose as a free body a larger portion of the truss, composed of several joints and members, provided that the desired force is one of the external forces acting on that portion. If, in addition, the portion of the truss is chosen so that there is a total of only three unknown forces acting upon it, the desired force can be obtained by solving the equations of equilibrium for this portion of the truss. In practice, the portion of the truss to be utilized is obtained by *passing a section* through three members of the truss, one of which is the desired member, that is, by drawing a line which divides the truss into two completely separate parts but does not intersect more than three members. Either of the two portions of the truss obtained after the intersected members have been removed can then be used as a free body.†

In Fig. 6.16a, the section nn has been passed through members BD, BE, and CE, and the portion ABC of the truss is chosen as the free body (Fig. 6.16b). The forces acting on the free body are the loads $\mathbf{P}_1$ and $\mathbf{P}_2$ at points A and B and the three unknown forces $\mathbf{F}_{BD}$, $\mathbf{F}_{BE}$, and $\mathbf{F}_{CE}$. Since it is not known whether the members removed were in tension or compression, the three forces have been arbitrarily drawn away from the free body as if the members were in tension.

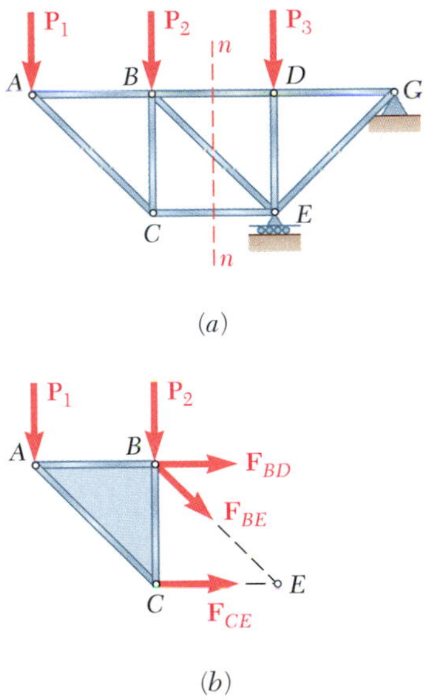

Fig. 6.16

†In the analysis of certain trusses, sections are passed which intersect more than three members; the forces in one, or possibly two, of the intersected members may be obtained if equilibrium equations can be found, each of which involves only one unknown (see Probs. 6.60 through 6.63).

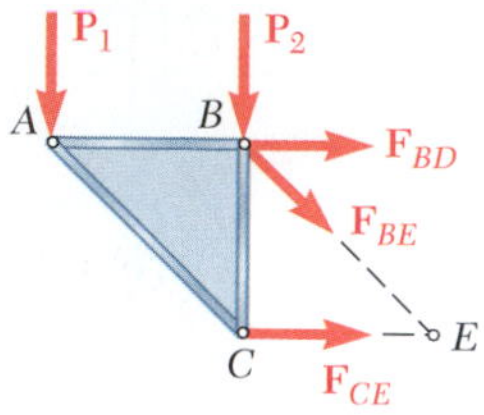

Fig. 6.16b (*repeated*)

The fact that the rigid body *ABC* is in equilibrium can be expressed by writing three equations which can be solved for the three unknown forces. If only the force $\mathbf{F}_{BD}$ is desired, we need write only one equation, provided that the equation does not contain the other unknowns. Thus the equation $\Sigma M_E = 0$ yields the value of the magnitude F_{BD} of the force $\mathbf{F}_{BD}$ (Fig. 6.16*b*). A positive sign in the answer will indicate that our original assumption regarding the sense of $\mathbf{F}_{BD}$ was correct and that member *BD* is in tension; a negative sign will indicate that our assumption was incorrect and that *BD* is in compression.

On the other hand, if only the force $\mathbf{F}_{CE}$ is desired, an equation which does not involve $\mathbf{F}_{BD}$ or $\mathbf{F}_{BE}$ should be written; the appropriate equation is $\Sigma M_B = 0$. Again a positive sign for the magnitude F_{CE} of the desired force indicates a correct assumption, that is, tension; and a negative sign indicates an incorrect assumption, that is, compression.

If only the force $\mathbf{F}_{BE}$ is desired, the appropriate equation is $\Sigma F_y = 0$. Whether the member is in tension or compression is again determined from the sign of the answer.

When the force in only one member is determined, no independent check of the computation is available. However, when all the unknown forces acting on the free body are determined, the computations can be checked by writing an additional equation. For instance, if $\mathbf{F}_{BD}$, $\mathbf{F}_{BE}$, and $\mathbf{F}_{CE}$ are determined as indicated above, the computation can be checked by verifying that $\Sigma F_x = 0$.

*6.8. TRUSSES MADE OF SEVERAL SIMPLE TRUSSES

Consider two simple trusses *ABC* and *DEF*. If they are connected by three bars *BD*, *BE*, and *CE* as shown in Fig. 6.17*a*, they will form together a rigid truss *ABDF*. The trusses *ABC* and *DEF* can also be combined into a single rigid truss by joining joints *B* and *D* into a single joint *B* and by connecting joints *C* and *E* by a bar *CE* (Fig. 6.17*b*). The truss thus obtained is known as a *Fink truss*. It should be noted that the trusses of Fig. 6.17*a* and *b* are *not* simple trusses; they cannot be constructed from a triangular truss by adding successive pairs of members as prescribed in Sec. 6.3. They are rigid trusses, however, as we can check by comparing the systems of connections used to hold the simple trusses *ABC* and *DEF* together (three bars in Fig. 6.17*a*, one pin and one bar in Fig. 6.17*b*) with the systems of supports discussed in Secs. 4.4 and 4.5. Trusses made of several simple trusses rigidly connected are known as *compound trusses*.

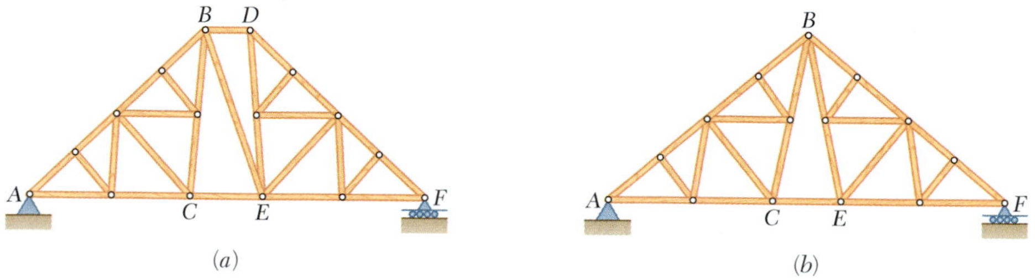

Fig. 6.17

In a compound truss the number of members m and the number of joints n are still related by the formula $m = 2n - 3$. This can be verified by observing that, if a compound truss is supported by a frictionless pin and a roller (involving three unknown reactions), the total number of unknowns is $m + 3$, and this number must be equal to the number $2n$ of equations obtained by expressing that the n pins are in equilibrium; it follows that $m = 2n - 3$. Compound trusses supported by a pin and a roller, or by an equivalent system of supports, are *statically determinate, rigid,* and *completely constrained.* This means that all of the unknown reactions and the forces in all the members can be determined by the methods of statics, and that the truss will neither collapse nor move. The forces in the members, however, cannot all be determined by the method of joints, except by solving a large number of simultaneous equations. In the case of the compound truss of Fig. 6.17*a*, for example, it is more efficient to pass a section through members *BD*, *BE*, and *CE* to determine the forces in these members.

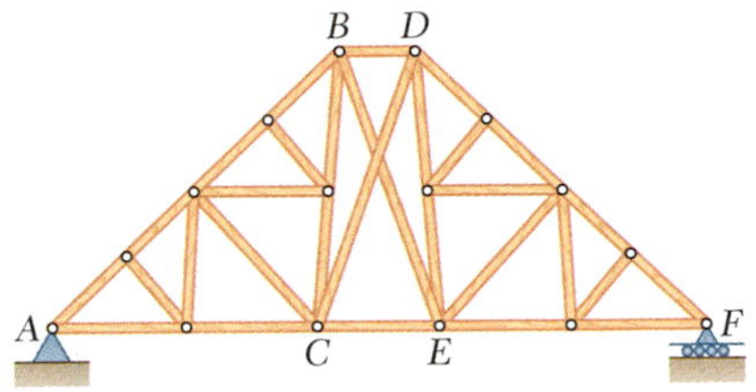

Fig. 6.18

Suppose, now, that the simple trusses *ABC* and *DEF* are connected by *four* bars *BD*, *BE*, *CD*, and *CE* (Fig. 6.18). The number of members m is now larger than $2n - 3$; the truss obtained is *overrigid*, and one of the four members *BD*, *BE*, *CD*, or *CE* is said to be *redundant.* If the truss is supported by a pin at *A* and a roller at *F*, the total number of unknowns is $m + 3$. Since $m > 2n - 3$, the number $m + 3$ of unknowns is now larger than the number $2n$ of available independent equations; the truss is *statically indeterminate.*

Finally, let us assume that the two simple trusses *ABC* and *DEF* are joined by a pin as shown in Fig. 6.19*a*. The number of members m is smaller than $2n - 3$. If the truss is supported by a pin at *A* and a roller at *F*, the total number of unknowns is $m + 3$. Since $m < 2n - 3$, the number $m + 3$ of unknowns is now smaller than the number $2n$ of equilibrium equations which should be satisfied; the truss is *nonrigid* and will collapse under its own weight. However, if two pins are used to support it, the truss becomes *rigid* and will not collapse (Fig. 6.19*b*). We note that the total number of unknowns is now $m + 4$ and is equal to the number $2n$ of equations. More generally, if the reactions at the supports involve r unknowns, the condition for a compound truss to be statically determinate, rigid, and completely constrained is $m + r = 2n$. However, while necessary, this condition is not sufficient for the equilibrium of a structure which ceases to be rigid when detached from its supports (see Sec. 6.11).

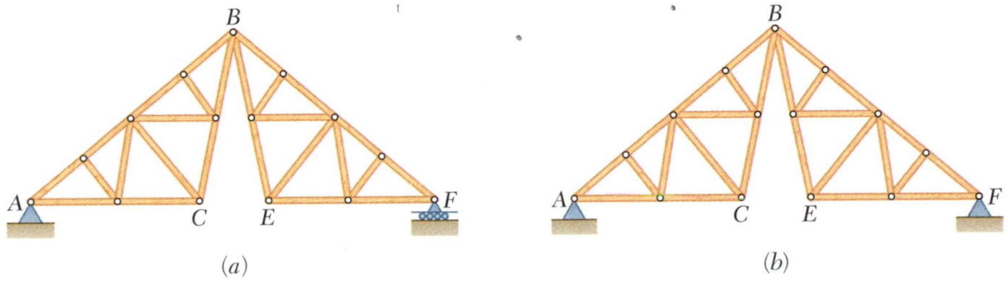

Fig. 6.19

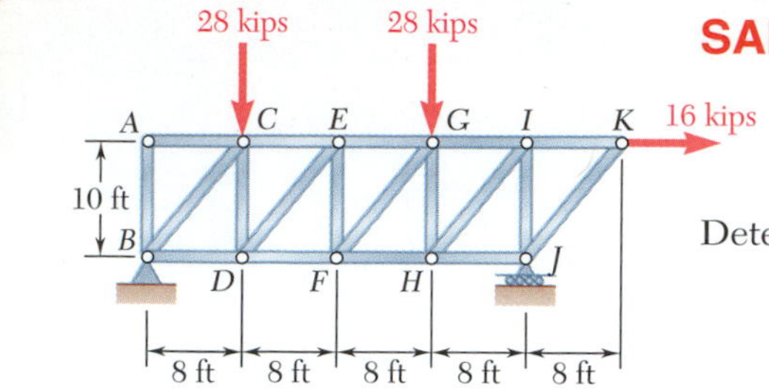

SAMPLE PROBLEM 6.2

Determine the force in members EF and GI of the truss shown.

SOLUTION

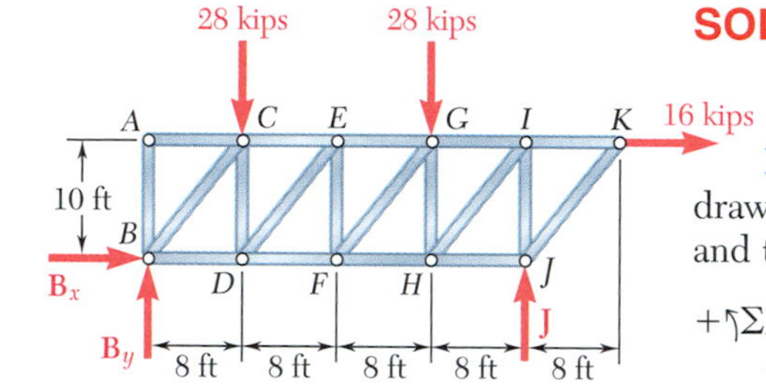

Free-Body: Entire Truss. A free-body diagram of the entire truss is drawn; external forces acting on this free body consist of the applied loads and the reactions at B and J. We write the following equilibrium equations.

$+\circlearrowleft \Sigma M_B = 0$:

$$-(28 \text{ kips})(8 \text{ ft}) - (28 \text{ kips})(24 \text{ ft}) - (16 \text{ kips})(10 \text{ ft}) + J(32 \text{ ft}) = 0$$

$$J = +33 \text{ kips} \qquad \mathbf{J} = 33 \text{ kips}\uparrow$$

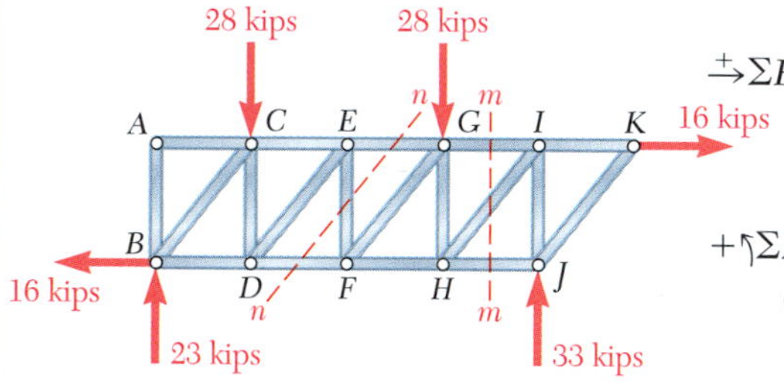

$\overset{+}{\rightarrow}\Sigma F_x = 0$: $\qquad B_x + 16 \text{ kips} = 0$

$$B_x = -16 \text{ kips} \qquad \mathbf{B}_x = 16 \text{ kips}\leftarrow$$

$+\circlearrowleft \Sigma M_J = 0$:

$$(28 \text{ kips})(24 \text{ ft}) + (28 \text{ kips})(8 \text{ ft}) - (16 \text{ kips})(10 \text{ ft}) - B_y(32 \text{ ft}) = 0$$

$$B_y = +23 \text{ kips} \qquad \mathbf{B}_y = 23 \text{ kips}\uparrow$$

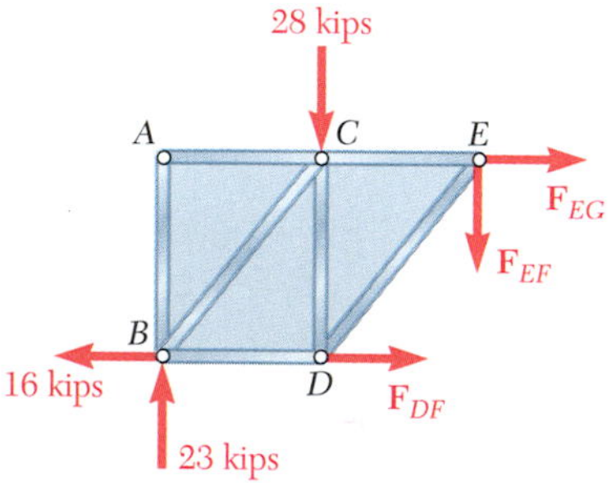

Force in Member *EF*. Section nn is passed through the truss so that it intersects member EF and only two additional members. After the intersected members have been removed, the left-hand portion of the truss is chosen as a free body. Three unknowns are involved; to eliminate the two horizontal forces, we write

$+\uparrow\Sigma F_y = 0$: $\qquad +23 \text{ kips} - 28 \text{ kips} - F_{EF} = 0$

$$F_{EF} = -5 \text{ kips}$$

The sense of $\mathbf{F}_{EF}$ was chosen assuming member EF to be in tension; the negative sign obtained indicates that the member is in compression.

$$F_{EF} = 5 \text{ kips } C \quad ◀$$

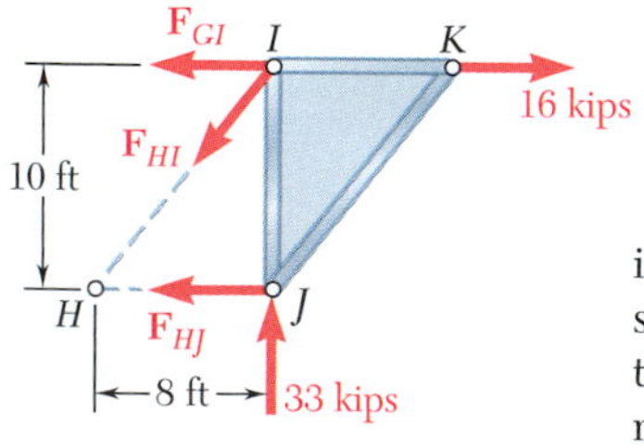

Force in Member *GI*. Section mm is passed through the truss so that it intersects member GI and only two additional members. After the intersected members have been removed, we choose the right-hand portion of the truss as a free body. Three unknown forces are again involved; to eliminate the two forces passing through point H, we write

$+\circlearrowleft \Sigma M_H = 0$: $\qquad (33 \text{ kips})(8 \text{ ft}) - (16 \text{ kips})(10 \text{ ft}) + F_{GI}(10 \text{ ft}) = 0$

$$F_{GI} = -10.4 \text{ kips} \qquad F_{GI} = 10.4 \text{ kips } C \quad ◀$$

SAMPLE PROBLEM 6.3

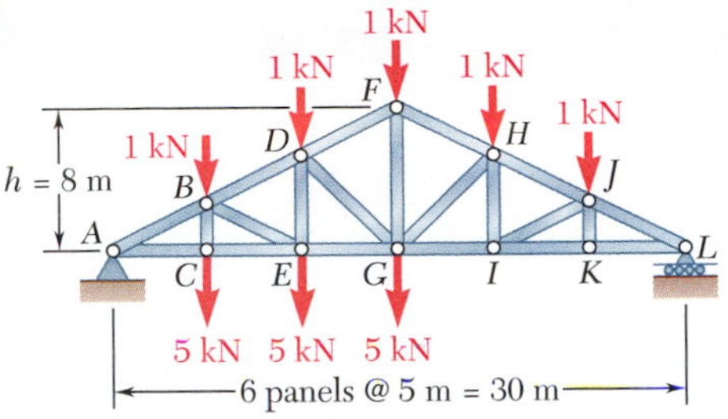

Determine the force in members *FH*, *GH*, and *GI* of the roof truss shown.

SOLUTION

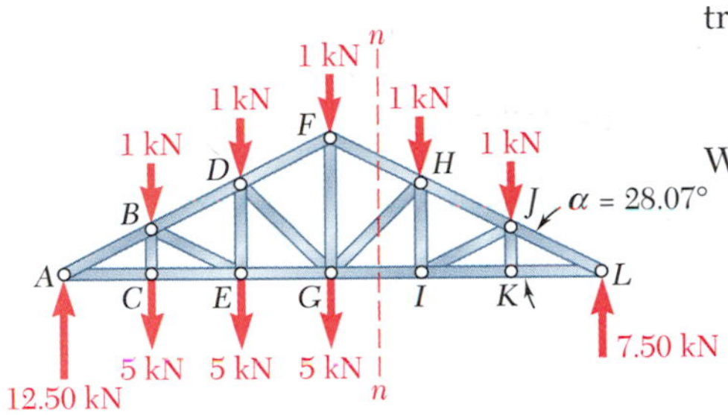

Free Body: Entire Truss. From the free-body diagram of the entire truss, we find the reactions at A and L:

$$\mathbf{A} = 12.50 \text{ kN}\uparrow \qquad \mathbf{L} = 7.50 \text{ kN}\uparrow$$

We note that

$$\tan \alpha = \frac{FG}{GL} = \frac{8 \text{ m}}{15 \text{ m}} = 0.5333 \qquad \alpha = 28.07°$$

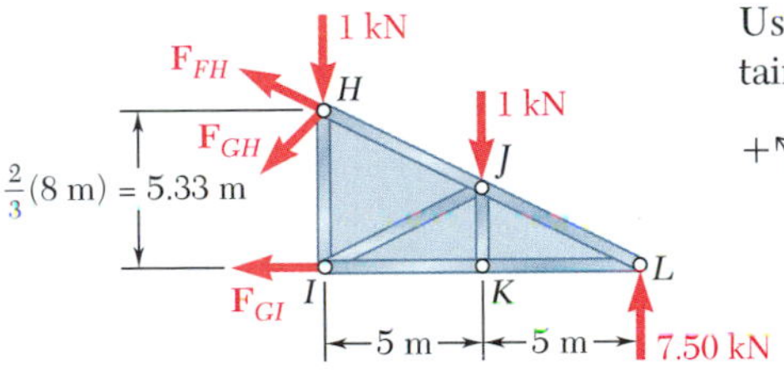

Force in Member *GI*. Section *nn* is passed through the truss as shown. Using the portion *HLI* of the truss as a free body, the value of F_{GI} is obtained by writing

$$+\circlearrowleft\Sigma M_H = 0: \qquad (7.50 \text{ kN})(10 \text{ m}) - (1 \text{ kN})(5 \text{ m}) - F_{GI}(5.33 \text{ m}) = 0$$

$$F_{GI} = +13.13 \text{ kN} \qquad F_{GI} = 13.13 \text{ kN } T \blacktriangleleft$$

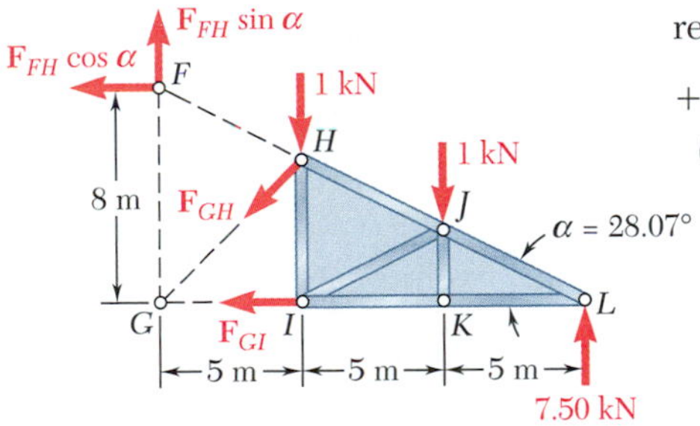

Force in Member *FH*. The value of F_{FH} is obtained from the equation $\Sigma M_G = 0$. We move $\mathbf{F}_{FH}$ along its line of action until it acts at point F, where it is resolved into its x and y components. The moment of $\mathbf{F}_{FH}$ with respect to point G is now equal to $(F_{FH} \cos \alpha)(8 \text{ m})$.

$$+\circlearrowleft\Sigma M_G = 0:$$

$$(7.50 \text{ kN})(15 \text{ m}) - (1 \text{ kN})(10 \text{ m}) - (1 \text{ kN})(5 \text{ m}) + (F_{FH} \cos \alpha)(8 \text{ m}) = 0$$

$$F_{FH} = -13.81 \text{ kN} \qquad F_{FH} = 13.81 \text{ kN } C \blacktriangleleft$$

Force in Member *GH*. We first note that

$$\tan \beta = \frac{GI}{HI} = \frac{5 \text{ m}}{\frac{2}{3}(8 \text{ m})} = 0.9375 \qquad \beta = 43.15°$$

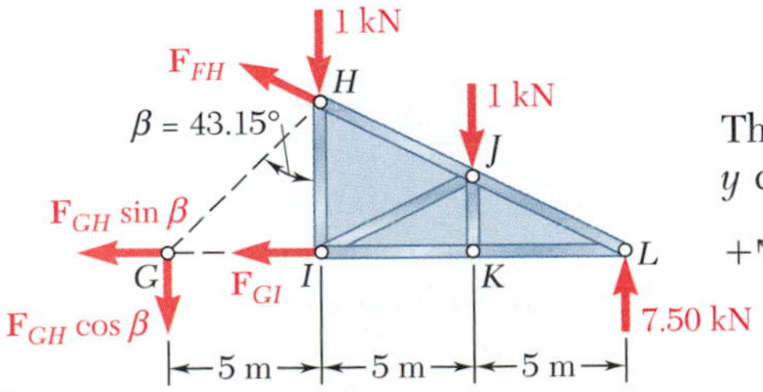

The value of F_{GH} is then determined by resolving the force $\mathbf{F}_{GH}$ into x and y components at point G and solving the equation $\Sigma M_L = 0$.

$$+\circlearrowleft\Sigma M_L = 0: \qquad (1 \text{ kN})(10 \text{ m}) + (1 \text{ kN})(5 \text{ m}) + (F_{GH} \cos \beta)(15 \text{ m}) = 0$$

$$F_{GH} = -1.371 \text{ kN} \qquad F_{GH} = 1.371 \text{ kN } C \blacktriangleleft$$

SOLVING PROBLEMS ON YOUR OWN

The *method of joints* that you studied earlier is usually the best method to use when the forces *in all the members* of a simple truss are to be found. However, the method of sections, which was covered in this lesson, is more effective when the force *in only one member* or the forces *in a very few members* of a simple truss are desired. The method of sections must also be used when the truss *is not a simple truss.*

A. To determine the force in a given truss member by the method of sections, you should follow these steps.

1. Draw a free-body diagram of the entire truss, and use this diagram to determine the reactions at the supports.

2. Pass a section through three members of the truss, one of which is the desired member. After you have removed these members, you will obtain two separate portions of the truss.

3. Select one of the two portions of the truss you have obtained, and draw its free-body diagram. This diagram should include the external forces applied to the selected portion as well as the forces exerted on it by the intersected members before these members were removed.

4. You can now write three equilibrium equations which can be solved for the forces in the three intersected members.

5. An alternative approach is to write a single equation, which can be solved for the force in the desired member. To do so, first observe whether the forces exerted by the other two members on the free body are parallel or whether their lines of action intersect.

a. If these forces are parallel, they can be eliminated by writing an equilibrium equation involving *components in a direction perpendicular* to these two forces.

b. If their lines of action intersect at a point H, they can be eliminated by writing an equilibrium equation involving *moments about H.*

6. Keep in mind that the section you use must intersect three members only. This is because the equilibrium equations in step 4 can be solved for three unknowns only. However, you can pass a section through more than three members to find the force in one of those members if you can write an equilibrium equation containing only that force as an unknown. Examples of such special situations are found in Probs. 6.60 through 6.63. Also, although we have considered separately the method of joints and the method of sections, you should recognize that these two techniques can be used sequentially to determine the force in a given member of a truss when the force cannot be found using a single section.

B. In Probs. 6.64–6.67 you will be asked to determine the tensile forces in the active counters of trusses (the forces in the other counters are zero). To determine the forces in each pair of counters, begin by determining the reactions at the supports and then pass a section through the truss which passes through the pair of counters. After selecting the portion of the truss to be analyzed and drawing its free-body diagram, next *assume* which counter is acting and use an equilibrium equation to determine the force in that counter. If the force is tensile, your assumption was correct; if not, you must repeat the analysis using the other counter of the pair as the active counter. However, you can often deduce which is the active counter by examining the free-body diagram, as a mental summation of forces of the external forces (loads and reactions at the supports) will imply which counter must act for equilibrium to exist (remembering that the force in a counter can only be tensile).

C. *About completely constrained and determinate trusses:*

1. ***First note that any simple truss which is simply supported*** is a completely constrained and determinate truss.

2. ***To determine whether any other truss is or is not completely constrained and determinate,*** you first count the number m of its members, the number n of its joints, and the number r of the reaction components at its supports. You then compare the sum $m + r$ representing the number of unknowns and the product $2n$ representing the number of available independent equilibrium equations.

   a. If $m + r < 2n$**,** there are fewer unknowns than equations. Thus, some of the equations cannot be satisfied; the truss is only *partially constrained.*

   b. If $m + r > 2n$**,** there are more unknowns than equations. Thus, some of the unknowns cannot be determined; the truss is *indeterminate.*

   c. If $m + r = 2n$**,** there are as many unknowns as there are equations. This, however, does not mean that all the unknowns can be determined and that all the equations can be satisfied. To find out whether the truss is *completely* or *improperly constrained,* you should try to determine the reactions at its supports and the forces in its members. If all can be found, the truss is *completely constrained and determinate.*

Problems

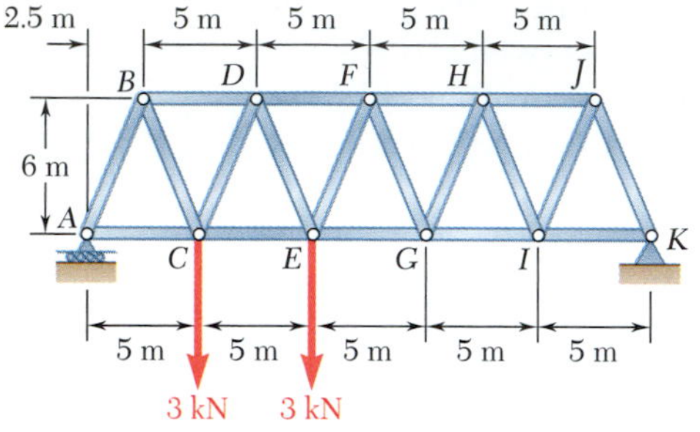

Fig. P6.42 and *P6.43*

6.42 A Warren bridge truss is loaded as shown. Determine the force in members *CE*, *DE*, and *DF*.

6.43 A Warren bridge truss is loaded as shown. Determine the force in members *EG*, *FG*, and *FH*.

6.44 A parallel chord Howe truss is loaded as shown. Determine the force in members *CE*, *DE*, and *DF*.

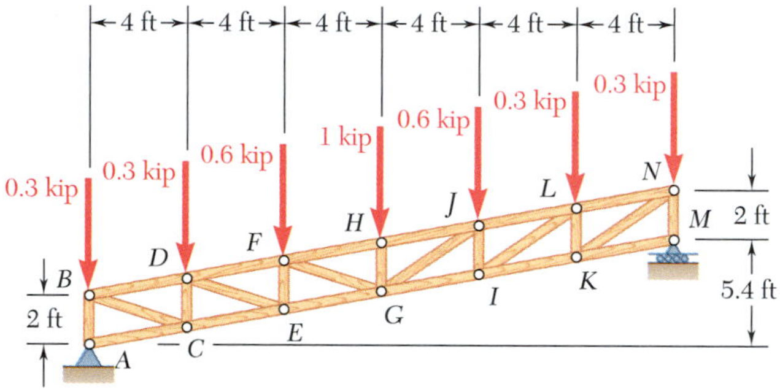

Fig. P6.44 and P6.45

6.45 A parallel chord Howe truss is loaded as shown. Determine the force in members *GI*, *GJ*, and *HJ*.

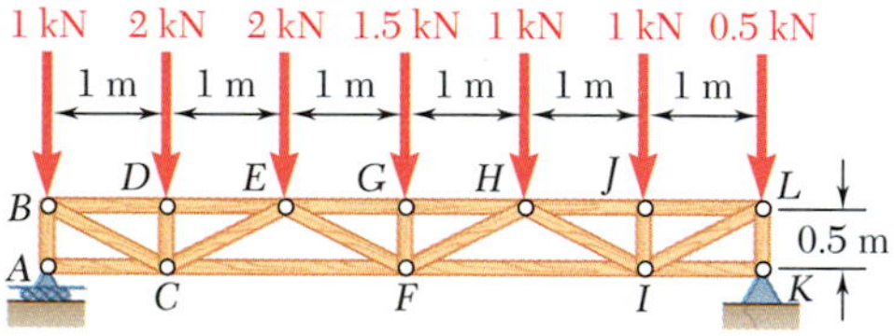

Fig. P6.46 and *P6.47*

6.46 A floor truss is loaded as shown. Determine the force in members *CF*, *EF*, and *EG*.

6.47 A floor truss is loaded as shown. Determine the force in members *FI*, *HI*, and *HJ*.

6.48 A pitched flat roof truss is loaded as shown. Determine the force in members *CE*, *DE*, and *DF*.

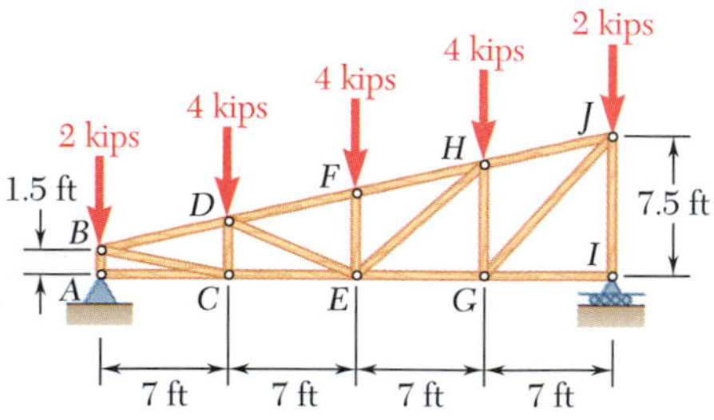

Fig. P6.48 and *P6.49*

6.49 A pitched flat roof truss is loaded as shown. Determine the force in members *EG*, *GH*, and *HJ*.

6.50 A Howe scissors roof truss is loaded as shown. Determine the force in members *DF*, *DG*, and *EG*.

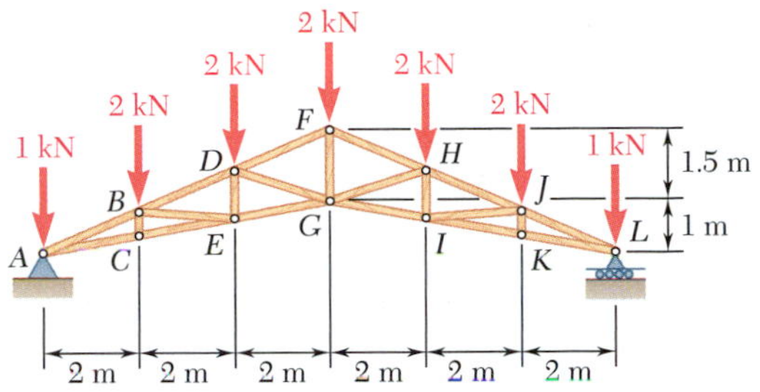

Fig. P6.50 and P6.51

6.51 A Howe scissors roof truss is loaded as shown. Determine the force in members *GI*, *HI*, and *HJ*.

6.52 A Fink roof truss is loaded as shown. Determine the force in members *BD*, *CD*, and *CE*.

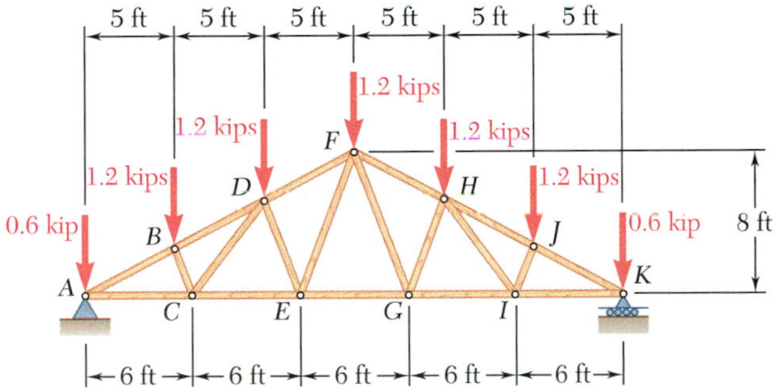

Fig. P6.52 and P6.53

6.53 A Fink roof truss is loaded as shown. Determine the force in members *FH*, *FG*, and *EG*.

6.54 A Fink roof truss is loaded as shown. Determine the force in members *DF*, *DG*, and *EG*. (*Hint:* First determine the force in member *EK*.)

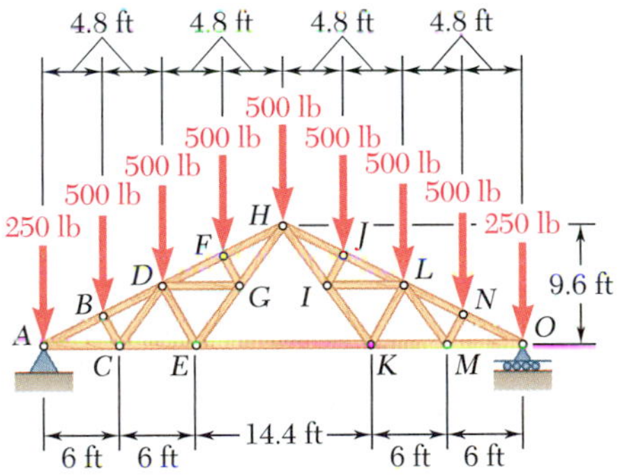

Fig. *P6.54*

6.55 A roof truss is loaded as shown. Determine the force in members *FH*, *GJ*, and *GI*.

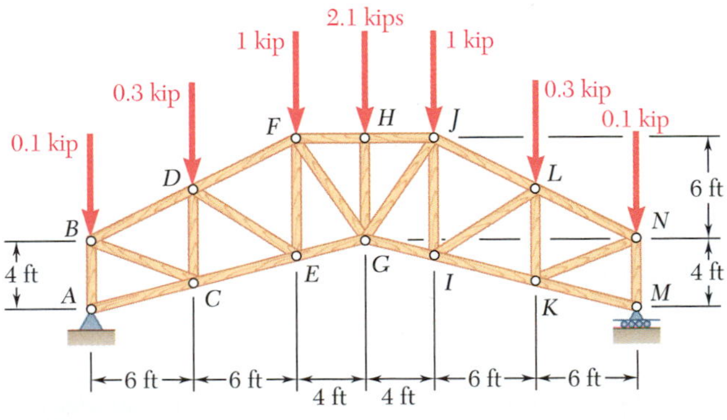

Fig. P6.55

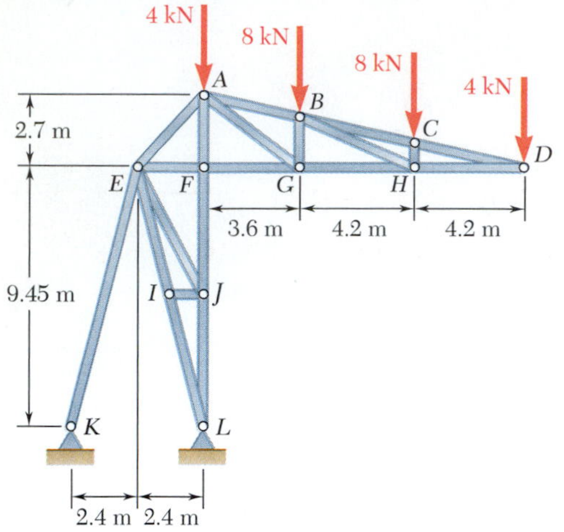

Fig. P6.56 and P6.57

6.56 A stadium roof truss is loaded as shown. Determine the force in members *AB*, *AG*, and *FG*.

6.57 A stadium roof truss is loaded as shown. Determine the force in members *AE*, *EF*, and *FJ*.

6.58 A vaulted roof truss is loaded as shown. Determine the force in members *BE*, *CE*, and *DF*.

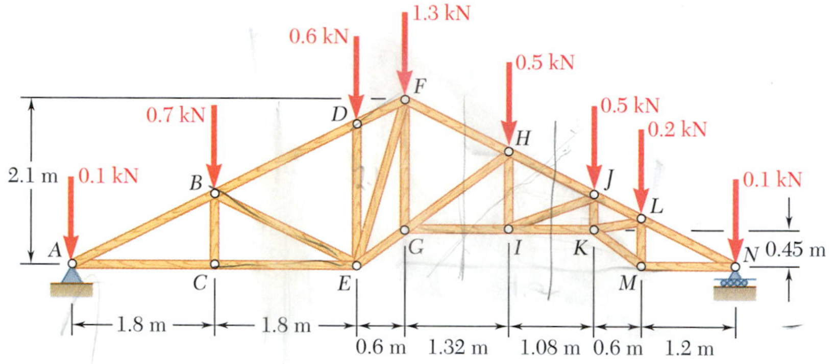

Fig. P6.58 and P6.59

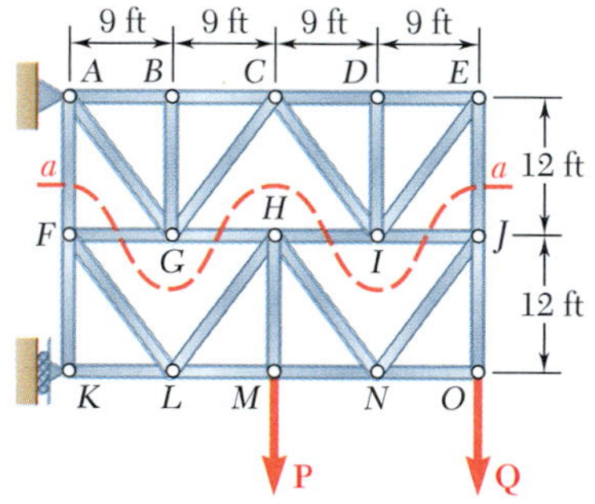

Fig. P6.60 and *P6.61*

6.59 A vaulted roof truss is loaded as shown. Determine the force in members *HJ*, *IJ*, and *GI*.

6.60 Determine the force in members *AF* and *EJ* of the truss shown when $P = Q = 2$ kips. (*Hint:* Use section *aa*.)

6.61 Determine the force in members *AF* and *EJ* of the truss shown when $P = 2$ kips and $Q = 0$. (*Hint:* Use section *aa*.)

6.62 Determine the force in members *DG* and *FH* of the truss shown. (*Hint:* Use section *aa*.)

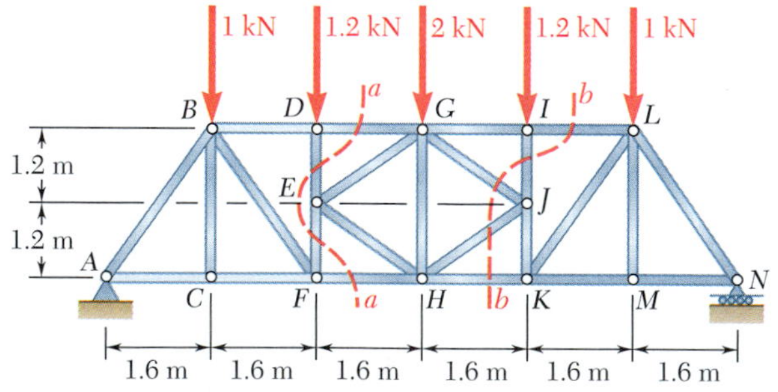

Fig. P6.62 and P6.63

6.63 Determine the force in members *IL*, *GJ*, and *HK* of the truss shown. (*Hint:* Begin with pins *I* and *J* and then use section *bb*.)

6.64 and 6.65 The diagonal members in the center panels of the truss shown are very slender and can act only in tension; such members are known as *counters*. Determine the forces in the counters which are acting under the given loading.

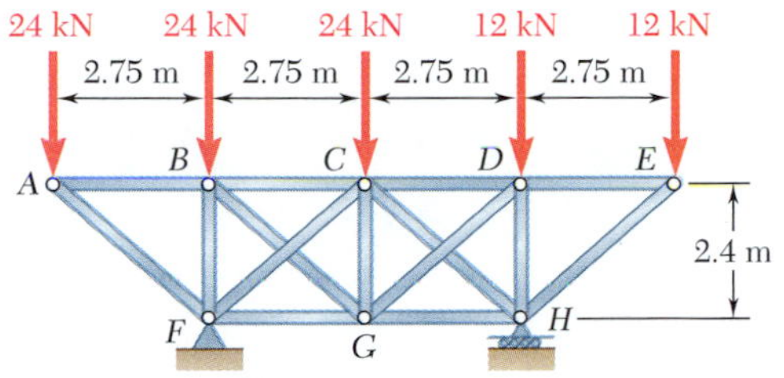

Fig. P6.64

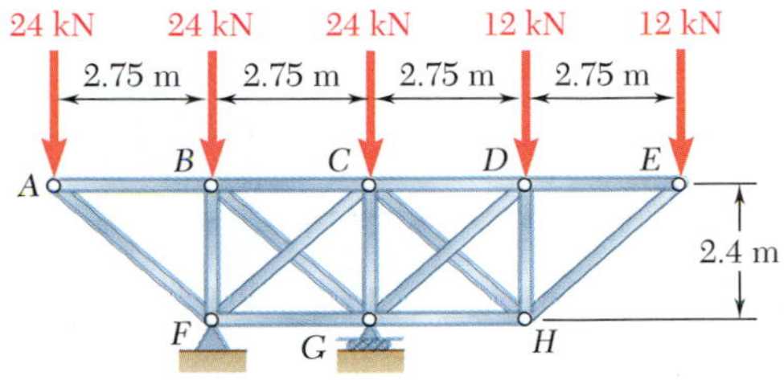

Fig. P6.65

6.66 The diagonal members in the center panel of the truss shown are very slender and can act only in tension; such members are known as *counters*. Determine the force in members *BD* and *CE* and in the counter which is acting when $P = 3$ kips.

6.67 Solve Prob. 6.66 when $P = 1.5$ kips.

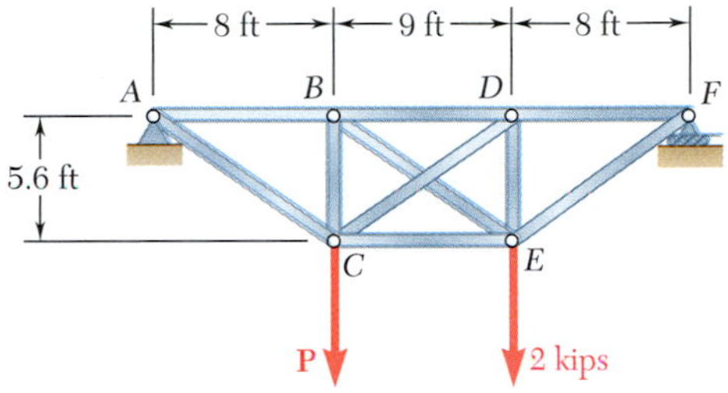

Fig. P6.66

6.68 The diagonal members *CF* and *DE* of the truss shown are very slender and can act only in tension; such members are known as *counters*. Determine the force in members *CE* and *DF* and in the counter which is acting.

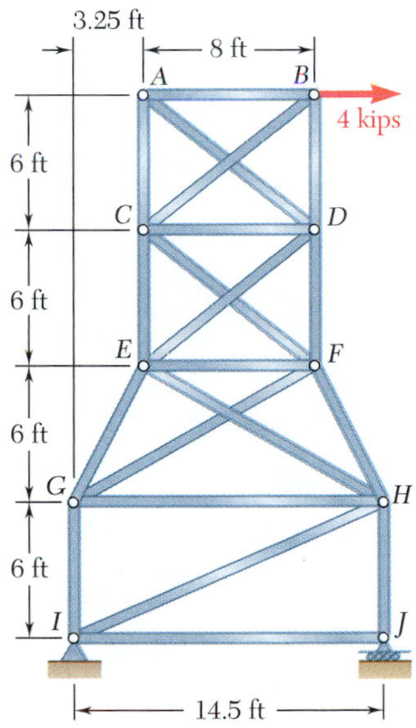

Fig. P6.68 and *P6.69*

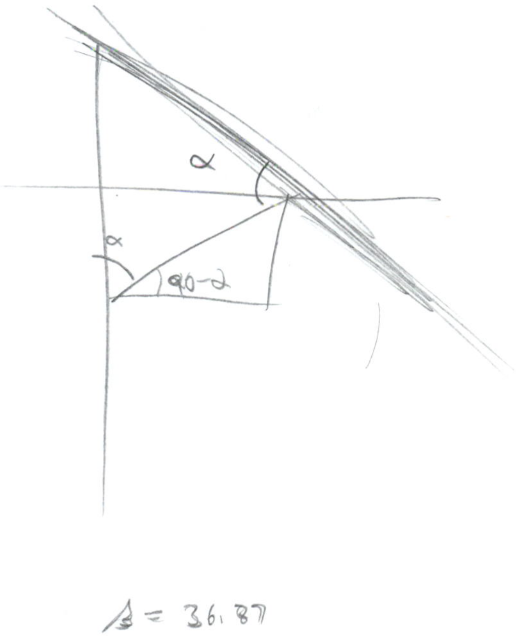

6.69 The diagonal members *EH* and *FG* of the truss shown are very slender and can act only in tension; such members are known as *counters*. Determine the force in members *EG* and *FH* and in the counter which is acting.

6.70 through 6.75 Classify each of the structures shown as completely, partially, or improperly constrained; if completely constrained, further classify it as determinate or indeterminate. (All members can act both in tension and in compression.)

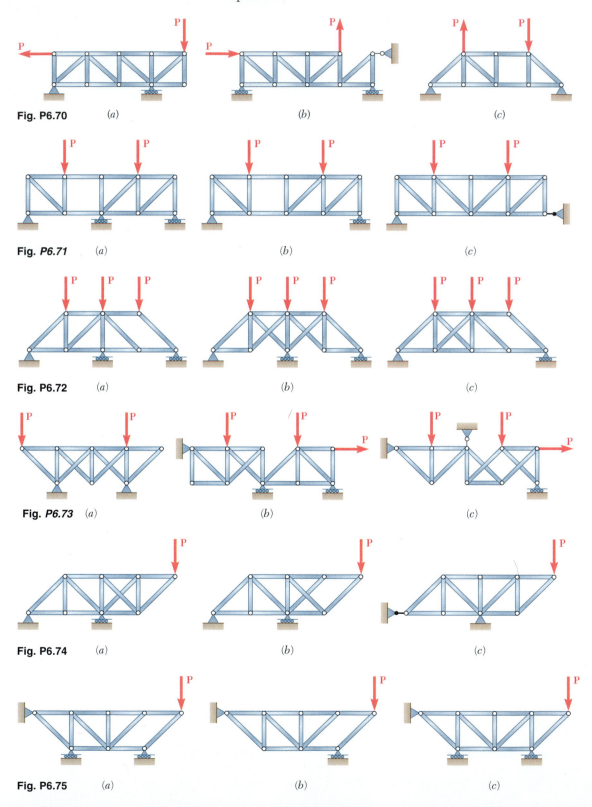

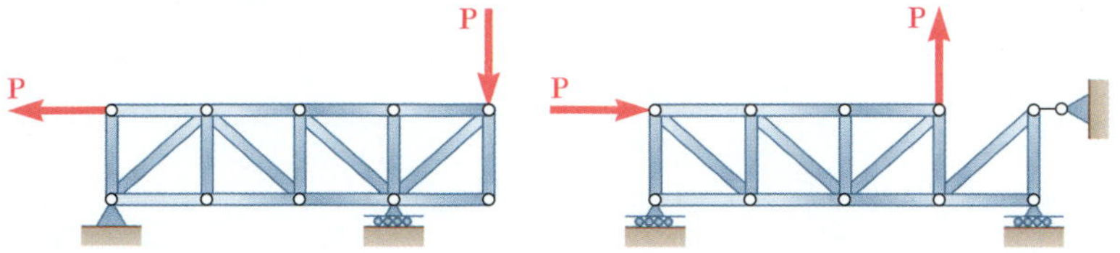

Fig. P6.70 (*a*) (*b*) (*c*)

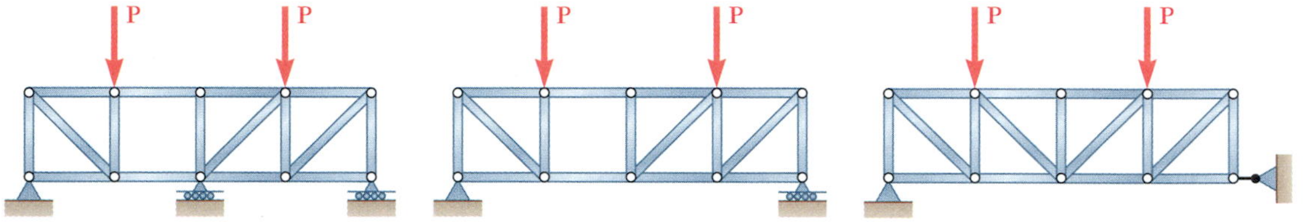

Fig. *P6.71* (*a*) (*b*) (*c*)

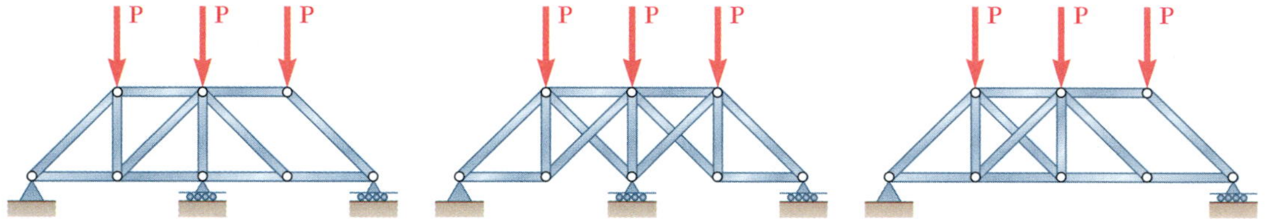

Fig. P6.72 (*a*) (*b*) (*c*)

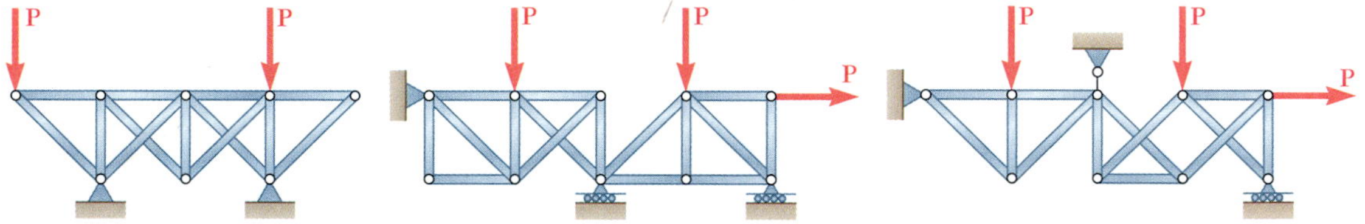

Fig. *P6.73* (*a*) (*b*) (*c*)

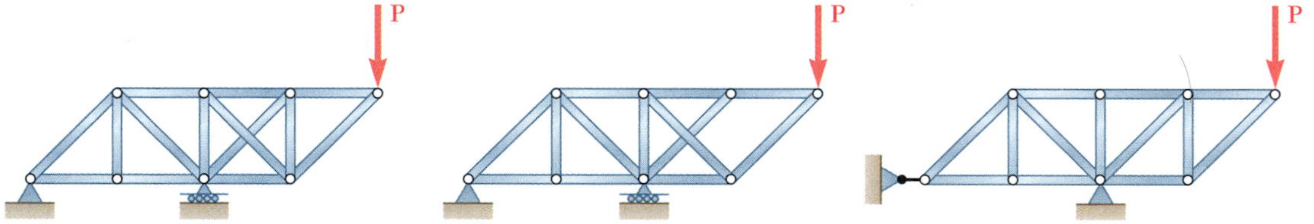

Fig. P6.74 (*a*) (*b*) (*c*)

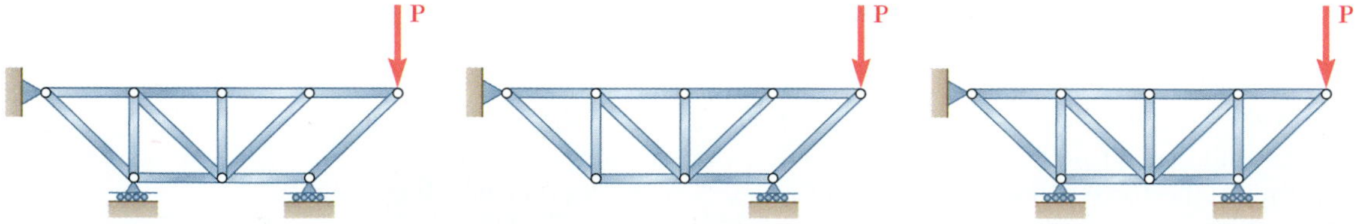

Fig. P6.75 (*a*) (*b*) (*c*)

FRAMES AND MACHINES

6.9. STRUCTURES CONTAINING MULTIFORCE MEMBERS

Under trusses, we have considered structures consisting entirely of pins and straight two-force members. The forces acting on the two-force members were known to be directed along the members themselves. We now consider structures in which at least one of the members is a *multiforce* member, that is, a member acted upon by three or more forces. These forces will generally not be directed along the members on which they act; their direction is unknown, and they should be represented therefore by two unknown components.

Frames and machines are structures containing multiforce members. *Frames* are designed to support loads and are usually stationary, fully constrained structures. *Machines* are designed to transmit and modify forces; they may or may not be stationary and will always contain moving parts.

6.10. ANALYSIS OF A FRAME

As a first example of analysis of a frame, the crane described in Sec. 6.1, which carries a given load W (Fig. 6.20*a*), will again be considered. The free-body diagram of the entire frame is shown in Fig. 6.20*b*. This diagram can be used to determine the external forces acting on the frame. Summing moments about A, we first determine the force $\mathbf{T}$ exerted by the cable; summing x and y components, we then determine the components $\mathbf{A}_x$ and $\mathbf{A}_y$ of the reaction at the pin A.

In order to determine the internal forces holding the various parts of a frame together, we must dismember the frame and draw a free-body diagram for each of its component parts (Fig. 6.20*c*). First, the two-force members should be considered. In this frame, member BE is the only two-force member. The forces acting at each end of this member must have the same magnitude, same line of action, and opposite sense (Sec. 4.6). They are therefore directed along BE and will be denoted, respectively, by $\mathbf{F}_{BE}$ and $-\mathbf{F}_{BE}$. Their sense will be arbitrarily assumed as shown in Fig. 6.20*c*; later the sign obtained for the common magnitude F_{BE} of the two forces will confirm or deny this assumption.

Next, we consider the multiforce members, that is, the members which are acted upon by three or more forces. According to Newton's third law, the force exerted at B by member BE on member AD must be equal and opposite to the force $\mathbf{F}_{BE}$ exerted by AD on BE. Similarly, the force exerted at E by member BE on member CF must be equal and opposite to the force $-\mathbf{F}_{BE}$ exerted by CF on BE. Thus the forces that the two-force member BE exerts on AD and CF are, respectively, equal to $-\mathbf{F}_{BE}$ and $\mathbf{F}_{BE}$; they have the same magnitude F_{BE} and opposite sense, and should be directed as shown in Fig. 6.20*c*.

At C two multiforce members are connected. Since neither the direction nor the magnitude of the forces acting at C is known, these forces will be represented by their x and y components. The components $\mathbf{C}_x$ and $\mathbf{C}_y$ of the force acting on member AD will be arbitrarily directed to the right and upward. Since, according to Newton's third law, the forces exerted by member CF on AD and by member AD on

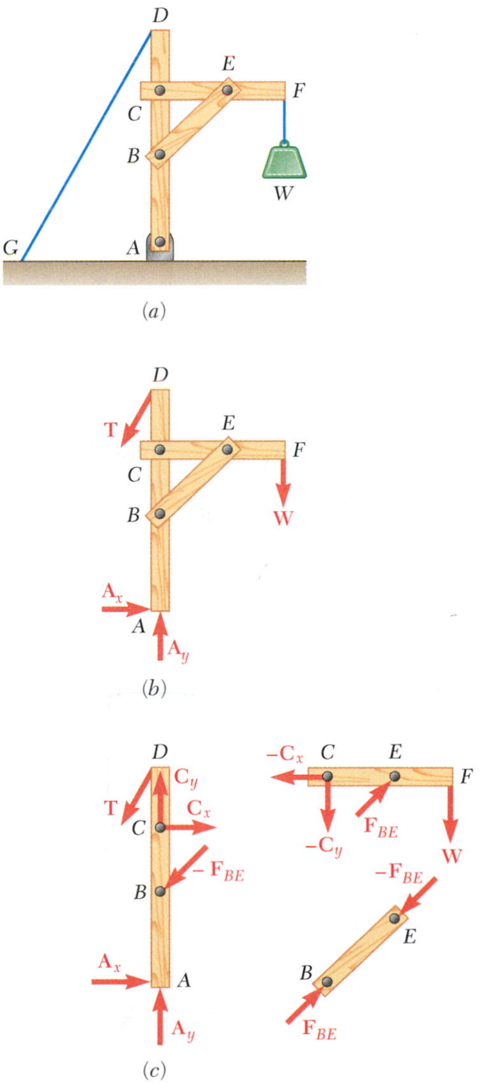

Fig. 6.20

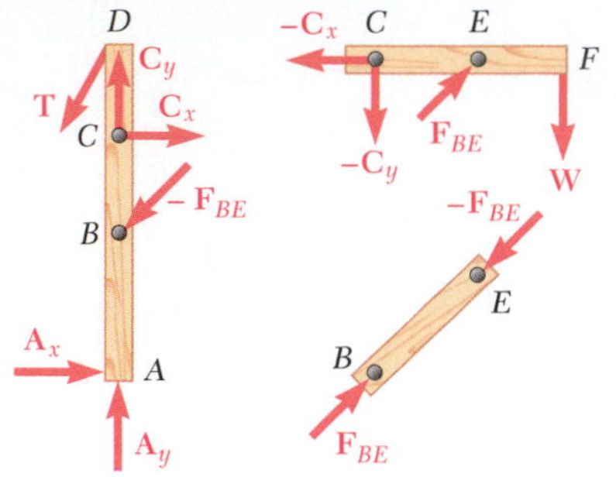

Fig. 6.20*c* (*repeated*)

CF are equal and opposite, the components of the force acting on member CF *must* be directed to the left and downward; they will be denoted, respectively, by $-\mathbf{C}_x$ and $-\mathbf{C}_y$. Whether the force $\mathbf{C}_x$ is actually directed to the right and the force $-\mathbf{C}_x$ is actually directed to the left will be determined later from the sign of their common magnitude C_x, a plus sign indicating that the assumption made was correct, and a minus sign that it was wrong. The free-body diagrams of the multiforce members are completed by showing the external forces acting at A, D, and F.†

The internal forces can now be determined by considering the free-body diagram of either of the two multiforce members. Choosing the free-body diagram of CF, for example, we write the equations $\Sigma M_C = 0$, $\Sigma M_E = 0$, and $\Sigma F_x = 0$, which yield the values of the magnitudes F_{BE}, C_y, and C_x, respectively. These values can be checked by verifying that member AD is also in equilibrium.

It should be noted that the pins in Fig. 6.20 were assumed to form an integral part of one of the two members they connected and so it was not necessary to show their free-body diagrams. This assumption can always be used to simplify the analysis of frames and machines. When a pin connects three or more members, however, or when a pin connects a support and two or more members, or when a load is applied to a pin, a clear decision must be made in choosing the member to which the pin will be assumed to belong. (If multiforce members are involved, the pin should be attached to one of these members.) The various forces exerted on the pin should then be clearly identified. This is illustrated in Sample Prob. 6.6.

6.11. FRAMES WHICH CEASE TO BE RIGID WHEN DETACHED FROM THEIR SUPPORTS

The crane analyzed in Sec. 6.10 was so constructed that it could keep the same shape without the help of its supports; it was therefore considered as a rigid body. Many frames, however, will collapse if detached from their supports; such frames cannot be considered as rigid bodies. Consider, for example, the frame shown in Fig. 6.21*a*, which consists of two members AC and CB carrying loads $\mathbf{P}$ and $\mathbf{Q}$ at their midpoints; the members are supported by pins at A and B and are connected by a pin at C. If detached from its supports, this frame will not maintain its shape; it should therefore be considered as made of *two distinct rigid parts* AC and CB.

†It is not strictly necessary to use a minus sign to distinguish the force exerted by one member on another from the equal and opposite force exerted by the second member on the first, since the two forces belong to different free-body diagrams and thus cannot easily be confused. In the Sample Problems, the same symbol is used to represent equal and opposite forces which are applied to different free bodies. It should be noted that, under these conditions, the sign obtained for a given force component will not directly relate the sense of that component to the sense of the corresponding coordinate axis. Rather, a positive sign will indicate that *the sense assumed for that component in the free-body diagram* is correct, and a negative sign will indicate that it is wrong.

The equations $\Sigma F_x = 0, \Sigma F_y = 0, \Sigma M = 0$ (about any given point) express the conditions for the *equilibrium of a rigid body* (Chap. 4); we should use them, therefore, in connection with the free-body diagrams of rigid bodies, namely, the free-body diagrams of members AC and CB (Fig. 6.21*b*). Since these members are multiforce members, and since pins are used at the supports and at the connection, the reactions at A and B and the forces at C will each be represented by two components. In accordance with Newton's third law, the components of the force exerted by CB on AC and the components of the force exerted by AC on CB will be represented by vectors of the same magnitude and opposite sense; thus, if the first pair of components consists of $\mathbf{C}_x$ and $\mathbf{C}_y$, the second pair will be represented by $-\mathbf{C}_x$ and $-\mathbf{C}_y$. We note that four unknown force components act on free body AC, while only three independent equations can be used to express that the body is in equilibrium; similarly, four unknowns, but only three equations, are associated with CB. However, only six different unknowns are involved in the analysis of the two members, and altogether six equations are available to express that the members are in equilibrium. Writing $\Sigma M_A = 0$ for free body AC and $\Sigma M_B = 0$ for CB, we obtain two simultaneous equations which may be solved for the common magnitude C_x of the components $\mathbf{C}_x$ and $-\mathbf{C}_x$ and for the common magnitude C_y of the components $\mathbf{C}_y$ and $-\mathbf{C}_y$. We then write $\Sigma F_x = 0$ and $\Sigma F_y = 0$ for each of the two free bodies, obtaining, successively, the magnitudes A_x, A_y, B_x, and B_y.

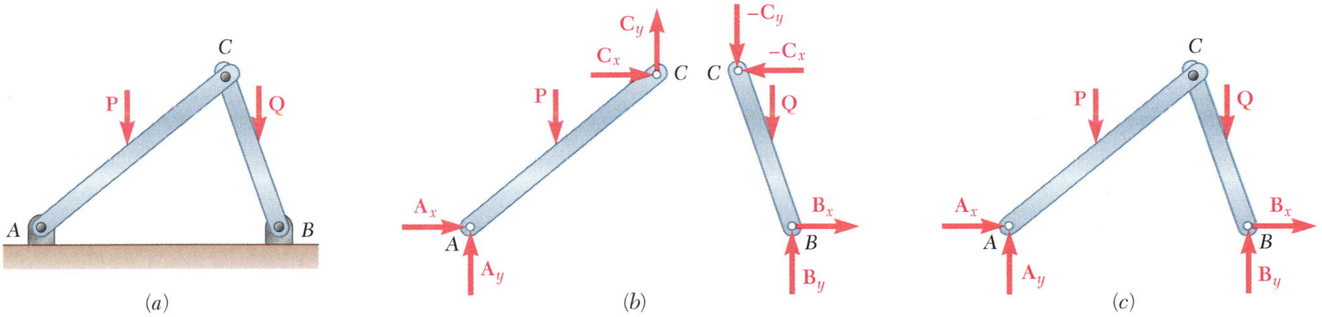

Fig. 6.21

It can now be observed that since the equations of equilibrium $\Sigma F_x = 0$, $\Sigma F_y = 0$, and $\Sigma M = 0$ (about any given point) are satisfied by the forces acting on free body AC, and since they are also satisfied by the forces acting on free body CB, they must be satisfied when the forces acting on the two free bodies are considered simultaneously. Since the internal forces at C cancel each other, we find that the equations of equilibrium must be satisfied by the external forces shown on the free-body diagram of the frame ACB itself (Fig. 6.21*c*), although the frame is not a rigid body. These equations can be used to determine some of the components of the reactions at A and B. We will also find, however, that *the reactions cannot be completely determined from the free-body diagram of the whole frame*. It is thus

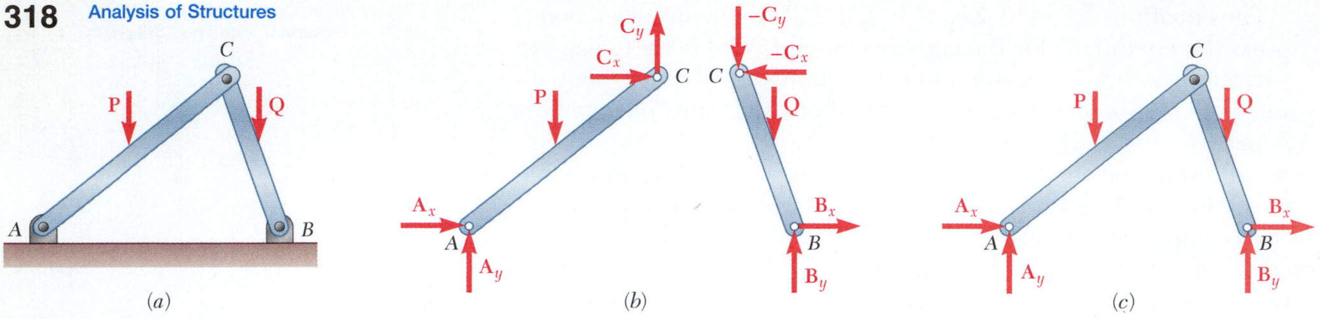

Fig. 6.21 (*repeated*)

necessary to dismember the frame and to consider the free-body diagrams of its component parts (Fig. 6.21*b*), even when we are interested in determining external reactions only. This is because the equilibrium equations obtained for free body *ACB are necessary conditions* for the equilibrium of a nonrigid structure, *but are not sufficient conditions*.

The method of solution outlined in the second paragraph of this section involved simultaneous equations. A more efficient method is now presented, which utilizes the free body *ACB* as well as the free bodies *AC* and *CB*. Writing $\Sigma M_A = 0$ and $\Sigma M_B = 0$ for free body *ACB*, we obtain B_y and A_y. Writing $\Sigma M_C = 0$, $\Sigma F_x = 0$, and $\Sigma F_y = 0$ for free body *AC*, we obtain, successively, A_x, C_x, and C_y. Finally, writing $\Sigma F_x = 0$ for *ACB*, we obtain B_x.

We noted above that the analysis of the frame of Fig. 6.21 involves six unknown force components and six independent equilibrium equations. (The equilibrium equations for the whole frame were obtained from the original six equations and, therefore, are not independent.) Moreover, we checked that all unknowns could be actually determined and that all equations could be satisfied. The frame considered is *statically determinate and rigid*.† In general, to determine whether a structure is statically determinate and rigid, we should draw a free-body diagram for each of its component parts and count the reactions and internal forces involved. We should also determine the number of independent equilibrium equations (excluding equations expressing the equilibrium of the whole structure or of groups of component parts already analyzed). If there are more unknowns than equations, the structure is *statically indeterminate*. If there are fewer unknowns than equations, the structure is *nonrigid*. If there are as many unknowns as equations, *and if all unknowns can be determined and all equations satisfied* under general loading conditions, the structure is *statically determinate and rigid*. If, however, due to an *improper arrangement* of members and supports, all unknowns cannot be determined and all equations cannot be satisfied, the structure is *statically indeterminate and nonrigid*.

†The word *rigid* is used here to indicate that the frame will maintain its shape as long as it remains attached to its supports.

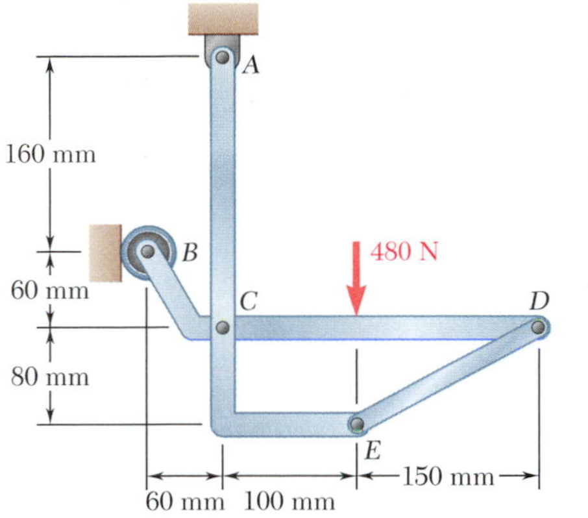

SAMPLE PROBLEM 6.4

In the frame shown, members *ACE* and *BCD* are connected by a pin at *C* and by the link *DE*. For the loading shown, determine the force in link *DE* and the components of the force exerted at *C* on member *BCD*.

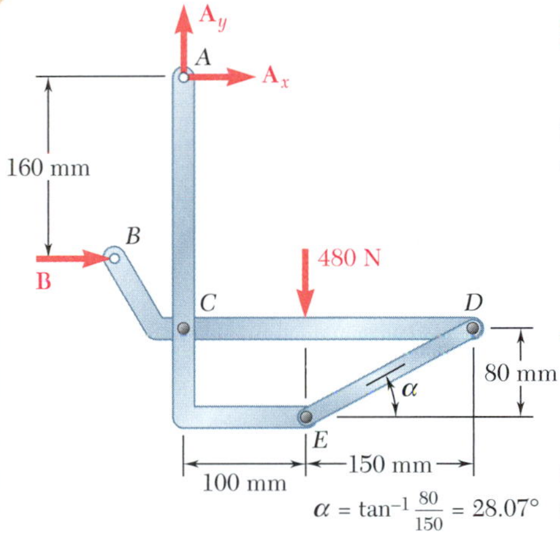

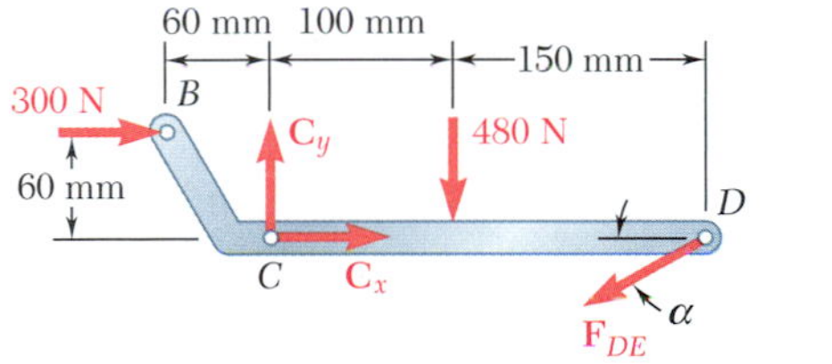

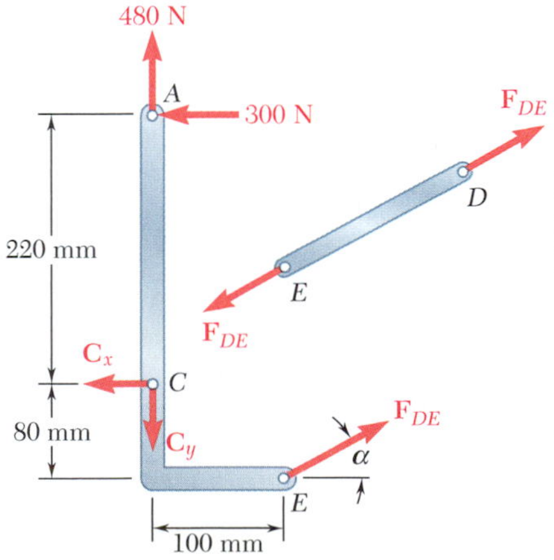

SOLUTION

Free Body: Entire Frame. Since the external reactions involve only three unknowns, we compute the reactions by considering the free-body diagram of the entire frame.

$+\uparrow\Sigma F_y = 0$: $\quad A_y - 480\text{ N} = 0 \qquad A_y = +480\text{ N} \qquad \mathbf{A}_y = 480\text{ N}\uparrow$

$+\circlearrowleft\Sigma M_A = 0$: $\quad -(480\text{ N})(100\text{ mm}) + B(160\text{ mm}) = 0$

$B = +300\text{ N} \qquad \mathbf{B} = 300\text{ N}\rightarrow$

$\xrightarrow{+}\Sigma F_x = 0$: $\quad B + A_x = 0$

$300\text{ N} + A_x = 0 \qquad A_x = -300\text{ N} \qquad \mathbf{A}_x = 300\text{ N}\leftarrow$

Members. We now dismember the frame. Since only two members are connected at *C*, the components of the unknown forces acting on *ACE* and *BCD* are, respectively, equal and opposite and are assumed directed as shown. We assume that link *DE* is in tension and exerts equal and opposite forces at *D* and *E*, directed as shown.

Free Body: Member BCD. Using the free body *BCD*, we write

$+\circlearrowleft\Sigma M_C = 0$:

$$-(F_{DE}\sin\alpha)(250\text{ mm}) - (300\text{ N})(80\text{ mm}) - (480\text{ N})(100\text{ mm}) = 0$$

$F_{DE} = -561\text{ N} \qquad F_{DE} = 561\text{ N } C$ ◀

$\xrightarrow{+}\Sigma F_x = 0$: $\quad C_x - F_{DE}\cos\alpha + 300\text{ N} = 0$

$C_x - (-561\text{ N})\cos 28.07° + 300\text{ N} = 0 \qquad C_x = -795\text{ N}$

$+\uparrow\Sigma F_y = 0$: $\quad C_y - F_{DE}\sin\alpha - 480\text{ N} = 0$

$C_y - (-561\text{ N})\sin 28.07° - 480\text{ N} = 0 \qquad C_y = +216\text{ N}$

From the signs obtained for C_x and C_y we conclude that the force components $\mathbf{C}_x$ and $\mathbf{C}_y$ exerted on member *BCD* are directed, respectively, to the left and up. We have

$\mathbf{C}_x = 795\text{ N}\leftarrow, \ \mathbf{C}_y = 216\text{ N}\uparrow$ ◀

Free Body: Member ACE (Check). The computations are checked by considering the free body *ACE*. For example,

$$+\circlearrowleft\Sigma M_A = (F_{DE}\cos\alpha)(300\text{ mm}) + (F_{DE}\sin\alpha)(100\text{ mm}) - C_x(220\text{ mm})$$
$$= (-561\cos\alpha)(300) + (-561\sin\alpha)(100) - (-795)(220) = 0$$

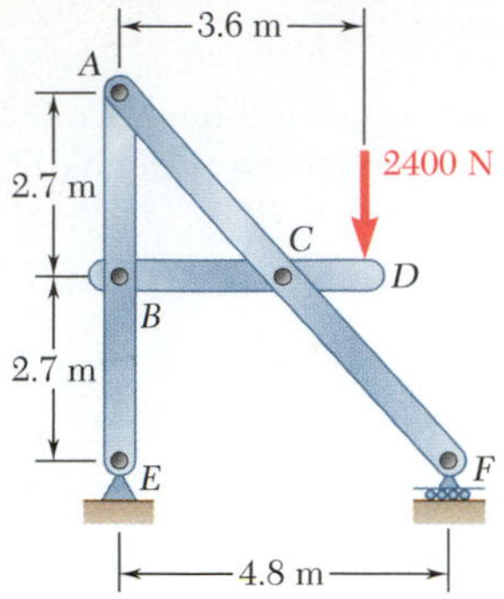

SAMPLE PROBLEM 6.5

Determine the components of the forces acting on each member of the frame shown.

SOLUTION

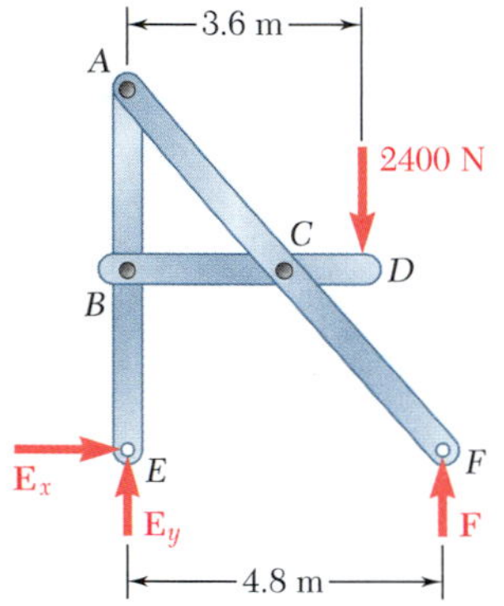

Free Body: Entire Frame. Since the external reactions involve only three unknowns, we compute the reactions by considering the free-body diagram of the entire frame.

$$+\circlearrowleft\Sigma M_E = 0: \quad -(2400\text{ N})(3.6\text{ m}) + F(4.8\text{ m}) = 0$$
$$F = +1800\text{ N} \qquad \mathbf{F} = 1800\text{ N}\uparrow \blacktriangleleft$$
$$+\uparrow\Sigma F_y = 0: \quad -2400\text{ N} + 1800\text{ N} + E_y = 0$$
$$E_y = +600\text{ N} \qquad \mathbf{E}_y = 600\text{ N}\uparrow \blacktriangleleft$$
$$\overset{+}{\rightarrow}\Sigma F_x = 0: \quad E_x = 0 \qquad \mathbf{E}_x = 0 \blacktriangleleft$$

Members. The frame is now dismembered; since only two members are connected at each joint, equal and opposite components are shown on each member at each joint.

Free Body: Member BCD

$$+\circlearrowleft\Sigma M_B = 0: \quad -(2400\text{ N})(3.6\text{ m}) + C_y(2.4\text{ m}) = 0 \qquad C_y = +3600\text{ N} \blacktriangleleft$$
$$+\circlearrowleft\Sigma M_C = 0: \quad -(2400\text{ N})(1.2\text{ m}) + B_y(2.4\text{ m}) = 0 \qquad B_y = +1200\text{ N} \blacktriangleleft$$
$$\overset{+}{\rightarrow}\Sigma F_x = 0: \quad -B_x + C_x = 0$$

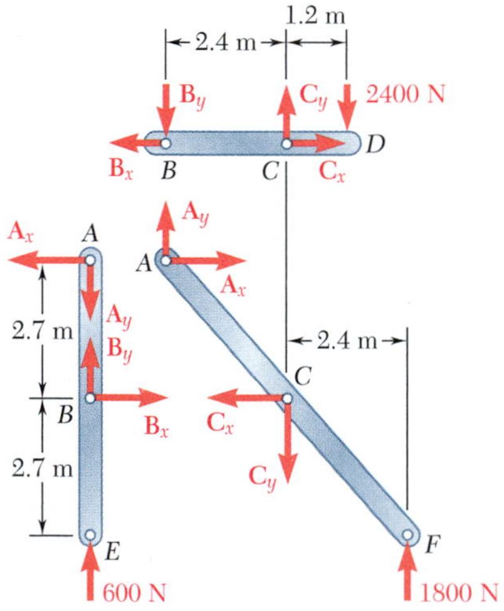

We note that neither B_x nor C_x can be obtained by considering only member *BCD*. The positive values obtained for B_y and C_y indicate that the force components $\mathbf{B}_y$ and $\mathbf{C}_y$ are directed as assumed.

Free Body: Member ABE

$$+\circlearrowleft\Sigma M_A = 0: \quad B_x(2.7\text{ m}) = 0 \qquad B_x = 0 \blacktriangleleft$$
$$\overset{+}{\rightarrow}\Sigma F_x = 0: \quad +B_x - A_x = 0 \qquad A_x = 0 \blacktriangleleft$$
$$+\uparrow\Sigma F_y = 0: \quad -A_y + B_y + 600\text{ N} = 0$$
$$-A_y + 1200\text{ N} + 600\text{ N} = 0 \qquad A_y = +1800\text{ N} \blacktriangleleft$$

Free Body: Member BCD. Returning now to member *BCD*, we write

$$\overset{+}{\rightarrow}\Sigma F_x = 0: \quad -B_x + C_x = 0 \qquad 0 + C_x = 0 \qquad C_x = 0 \blacktriangleleft$$

Free Body: Member ACF (Check). All unknown components have now been found; to check the results, we verify that member *ACF* is in equilibrium.

$$+\circlearrowleft\Sigma M_C = (1800\text{ N})(2.4\text{ m}) - A_y(2.4\text{ m}) - A_x(2.7\text{ m})$$
$$= (1800\text{ N})(2.4\text{ m}) - (1800\text{ N})(2.4\text{ m}) - 0 = 0 \qquad \text{(checks)}$$

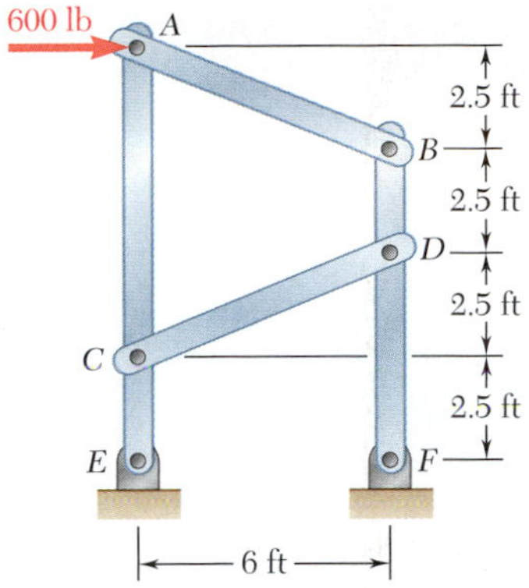

SAMPLE PROBLEM 6.6

A 600-lb horizontal force is applied to pin A of the frame shown. Determine the forces acting on the two vertical members of the frame.

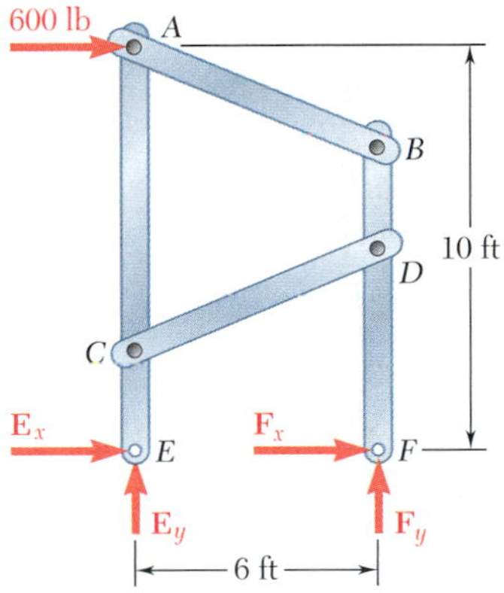

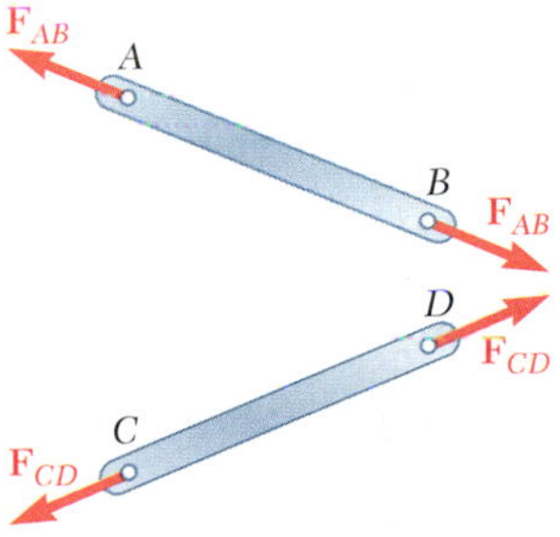

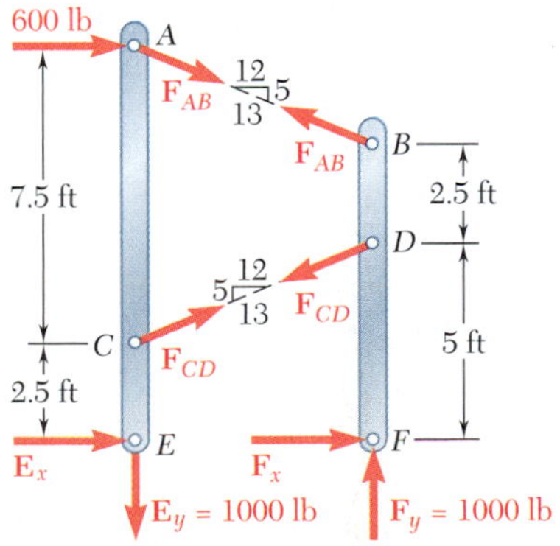

SOLUTION

Free Body: Entire Frame. The entire frame is chosen as a free body; although the reactions involve four unknowns, $\mathbf{E}_y$ and $\mathbf{F}_y$ may be determined by writing

$$+\circlearrowleft\Sigma M_E = 0: \quad -(600 \text{ lb})(10 \text{ ft}) + F_y(6 \text{ ft}) = 0$$
$$F_y = +1000 \text{ lb} \qquad \mathbf{F}_y = 1000 \text{ lb}\uparrow \blacktriangleleft$$
$$+\uparrow\Sigma F_y = 0: \quad E_y + F_y = 0$$
$$E_y = -1000 \text{ lb} \qquad \mathbf{E}_y = 1000 \text{ lb}\downarrow \blacktriangleleft$$

Members. The equations of equilibrium of the entire frame are not sufficient to determine $\mathbf{E}_x$ and $\mathbf{F}_x$. The free-body diagrams of the various members must now be considered in order to proceed with the solution. In dismembering the frame we will assume that pin A is attached to the multiforce member ACE and, thus, that the 600-lb force is applied to that member. We also note that AB and CD are two-force members.

Free Body: Member ACE

$$+\uparrow\Sigma F_y = 0: \quad -\tfrac{5}{13}F_{AB} + \tfrac{5}{13}F_{CD} - 1000 \text{ lb} = 0$$
$$+\circlearrowleft\Sigma M_E = 0: \quad -(600 \text{ lb})(10 \text{ ft}) - (\tfrac{12}{13}F_{AB})(10 \text{ ft}) - (\tfrac{12}{13}F_{CD})(2.5 \text{ ft}) = 0$$

Solving these equations simultaneously, we find

$$F_{AB} = -1040 \text{ lb} \qquad F_{CD} = +1560 \text{ lb} \blacktriangleleft$$

The signs obtained indicate that the sense assumed for F_{CD} was correct and the sense for F_{AB} was incorrect. Summing now x components,

$$\overset{+}{\rightarrow}\Sigma F_x = 0: \quad 600 \text{ lb} + \tfrac{12}{13}(-1040 \text{ lb}) + \tfrac{12}{13}(+1560 \text{ lb}) + E_x = 0$$
$$E_x = -1080 \text{ lb} \qquad \mathbf{E}_x = 1080 \text{ lb}\leftarrow \blacktriangleleft$$

Free Body: Entire Frame. Since $\mathbf{E}_x$ has been determined, we can return to the free-body diagram of the entire frame and write

$$\overset{+}{\rightarrow}\Sigma F_x = 0: \quad 600 \text{ lb} - 1080 \text{ lb} + F_x = 0$$
$$F_x = +480 \text{ lb} \qquad \mathbf{F}_x = 480 \text{ lb}\rightarrow \blacktriangleleft$$

***Free Body: Member BDF* (Check).** We can check our computations by verifying that the equation $\Sigma M_B = 0$ is satisfied by the forces acting on member BDF.

$$+\circlearrowleft\Sigma M_B = -\left(\tfrac{12}{13}F_{CD}\right)(2.5 \text{ ft}) + (F_x)(7.5 \text{ ft})$$
$$= -\tfrac{12}{13}(1560 \text{ lb})(2.5 \text{ ft}) + (480 \text{ lb})(7.5 \text{ ft})$$
$$= -3600 \text{ lb} \cdot \text{ft} + 3600 \text{ lb} \cdot \text{ft} = 0 \qquad \text{(checks)}$$

SOLVING PROBLEMS ON YOUR OWN

In this lesson you learned to analyze *frames containing one or more multiforce members*. In the problems that follow you will be asked to determine the external reactions exerted on the frame and the internal forces that hold together the members of the frame.

In solving problems involving frames containing one or more multiforce members, follow these steps:

1. Draw a free-body diagram of the entire frame. Use this free-body diagram to calculate, to the extent possible, the reactions at the supports. (In Sample Prob. 6.6 only two of the four reaction components could be found from the free body of the entire frame.)

2. Dismember the frame, and draw a free-body diagram of each member.

3. Considering first the two-force members, apply equal and opposite forces to each two-force member at the points where it is connected to another member. If the two-force member is a straight member, these forces will be directed along the axis of the member. If you cannot tell at this point whether the member is in tension or compression, just *assume* that the member is in tension and *direct both of the forces away from the member*. Since these forces have the same unknown magnitude, give them both the *same name* and, to avoid any confusion later, *do not use a plus sign or a minus sign*.

4. Next, consider the multiforce members. For each of these members, show all the forces acting on the member, including *applied loads, reactions, and internal forces at connections*. The magnitude and direction of any reaction or reaction component found earlier from the free-body diagram of the entire frame should be clearly indicated.

a. Where a multiforce member is connected to a two-force member, apply to the multiforce member a force *equal and opposite* to the force drawn on the free-body diagram of the two-force member, *giving it the same name*.

b. Where a multiforce member is connected to another multiforce member, use *horizontal and vertical components* to represent the internal forces at that point, since neither the direction nor the magnitude of these forces is known. The direction you choose for each of the two force components exerted on the first multiforce member is arbitrary, but *you must apply equal and opposite force components of the same name* to the other multiforce member. Again, *do not use a plus sign or a minus sign*.

5. *The internal forces may now be determined,* as well as any *reactions* that you have not already found.

a. *The free-body diagram* of each of the multiforce members can provide you with *three equilibrium equations*.

b. *To simplify your solution,* you should seek a way to write an equation involving a single unknown. If you can locate *a point where all but one of the unknown force components intersect,* you will obtain an equation in a single unknown by summing moments about that point. *If all unknown forces except one are parallel,* you will obtain an equation in a single unknown by summing force components in a direction perpendicular to the parallel forces.

c. *Since you arbitrarily chose the direction of each of the unknown forces,* you cannot determine until the solution is completed whether your guess was correct. To do that, consider the *sign* of the value found for each of the unknowns: a *positive* sign means that the direction you selected was *correct;* a *negative* sign means that the direction is *opposite* to the direction you assumed.

6. *To be more effective and efficient* as you proceed through your solution, observe the following rules:

a. *If an equation involving only one unknown can be found,* write that equation and *solve it for that unknown*. Immediately *replace* that unknown wherever it appears on other free-body diagrams *by the value you have found.* Repeat this process by seeking equilibrium equations involving only one unknown until you have found all of the internal forces and unknown reactions.

b. *If an equation involving only one unknown cannot be found,* you may have to *solve a pair of simultaneous equations*. Before doing so, check that you have shown the values of all of the reactions that were obtained from the free-body diagram of the entire frame.

c. *The total number of equations* of equilibrium for the entire frame and for the individual members *will be larger than the number of unknown forces and reactions*. After you have found all the reactions and all the internal forces, you can use the remaining dependent equations to check the accuracy of your solution.

Problems

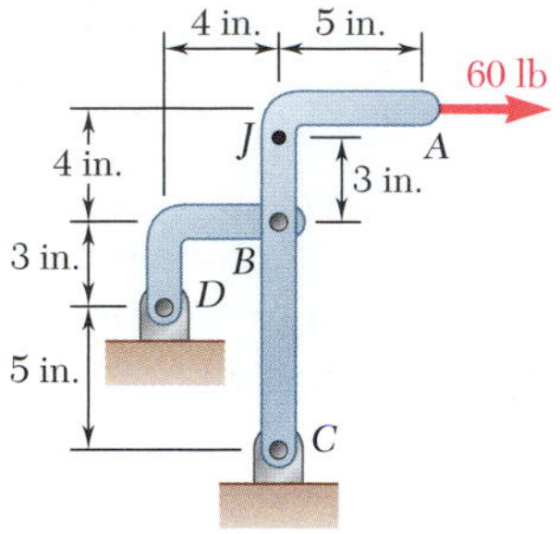

Fig. P6.76

6.76 For the frame and loading shown, determine the force acting on member ABC (*a*) at B, (*b*) at C.

6.77 Determine the force in member AC and the reaction at B when (*a*) $\theta = 30°$, (*b*) $\theta = 60°$.

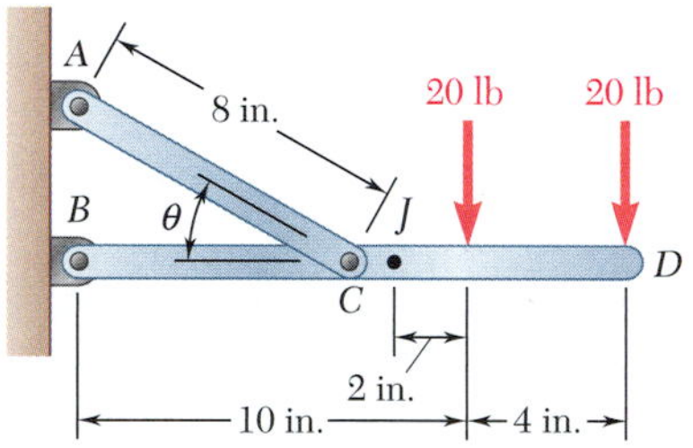

Fig. *P6.77*

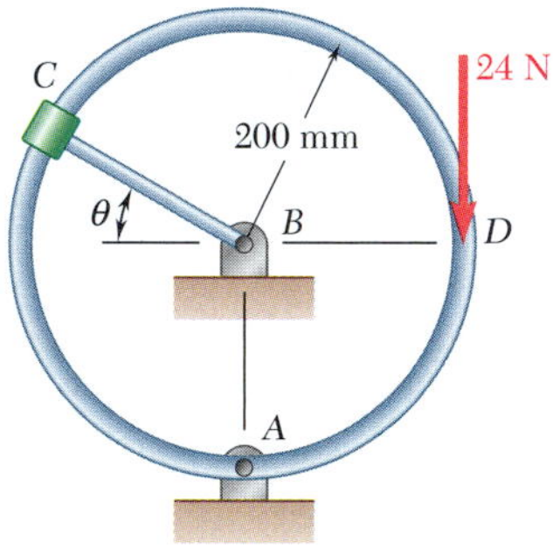

Fig. P6.78

6.78 A circular ring of radius 200 mm is pinned at A and is supported by rod BC, which is fitted with a collar at C that can be moved along the ring. For the position when $\theta = 35°$, determine (*a*) the force in rod BC, (*b*) the reaction at A.

6.79 Solve Prob. 6.78 when $\theta = -20°$.

6.80 For the frame and loading shown, determine the components of all forces acting on member ABC.

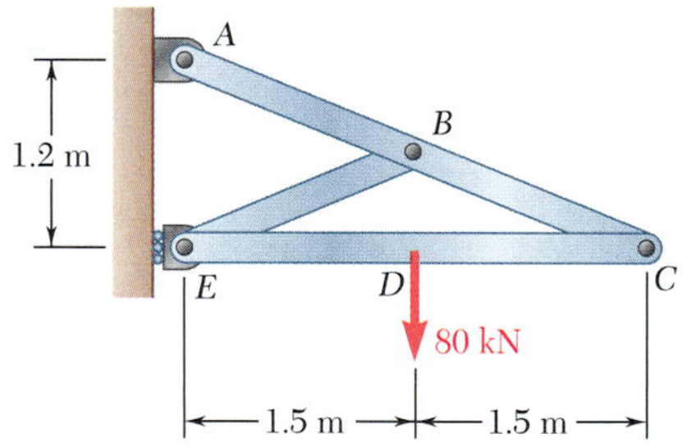

Fig. P6.80

6.81 Solve Prob. 6.80 assuming that the 80-kN load is replaced by a clockwise couple of magnitude 120 kN · m applied to member EDC at point D.

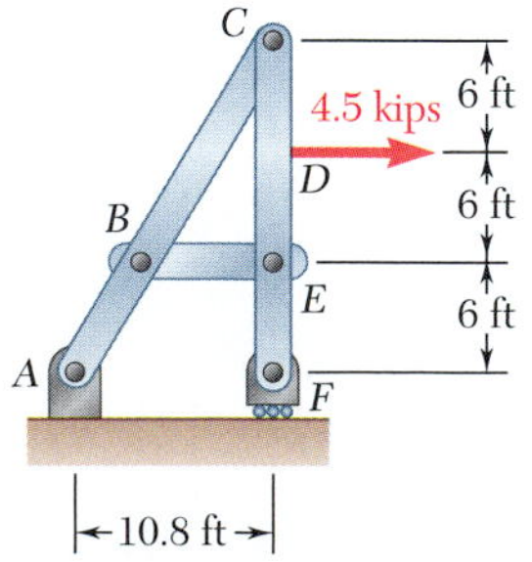

Fig. P6.82

6.82 For the frame and loading shown, determine the components of all forces acting on member ABC.

6.83 Solve Prob. 6.82 assuming that the 4.5-kip load is replaced by a clockwise couple of magnitude 54 kip · ft applied to member $CDEF$ at point D.

6.84 Determine the components of the reactions at A and E when a 160-lb force directed vertically downward is applied (*a*) at B, (*b*) at D.

6.85 Determine the components of the reactions at A and E when a 120-N force directed vertically downward is applied (*a*) at B, (*b*) at D.

6.86 Determine the components of the reactions at A and E when the frame is loaded by a clockwise couple of magnitude 360 lb · in. applied (*a*) at B, (*b*) at D.

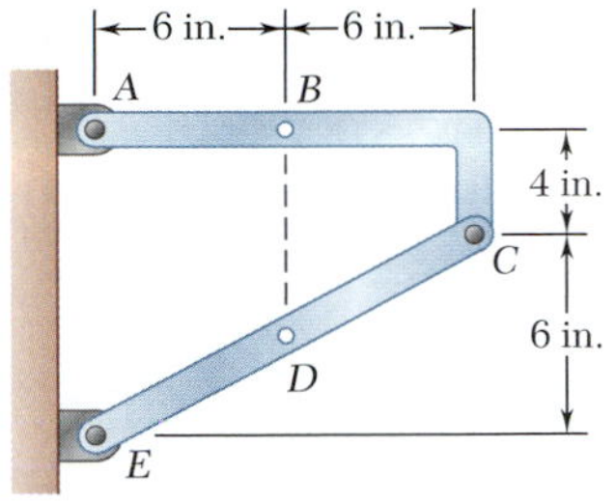

Fig. P6.84 and P6.86

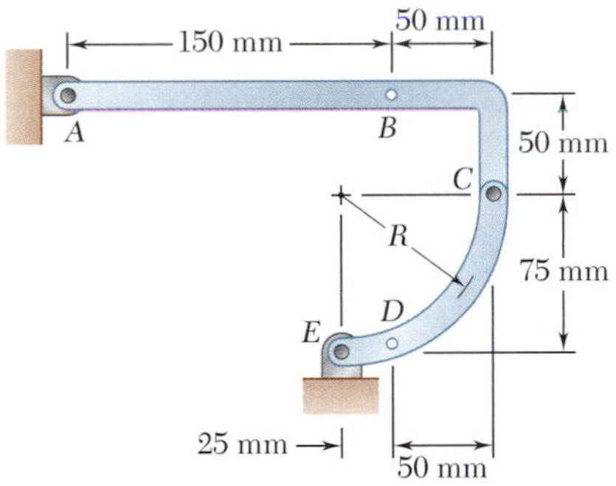

Fig. P6.85 and *P6.87*

6.87 Determine the components of the reactions at A and E when the frame is loaded by a counterclockwise couple of magnitude 24 N · m applied (*a*) at B, (*b*) at D.

6.88 Determine the components of the reactions at A and B when (*a*) the 240-N load is applied as shown, (*b*) the 240-N load is moved along its line of action and is applied at E.

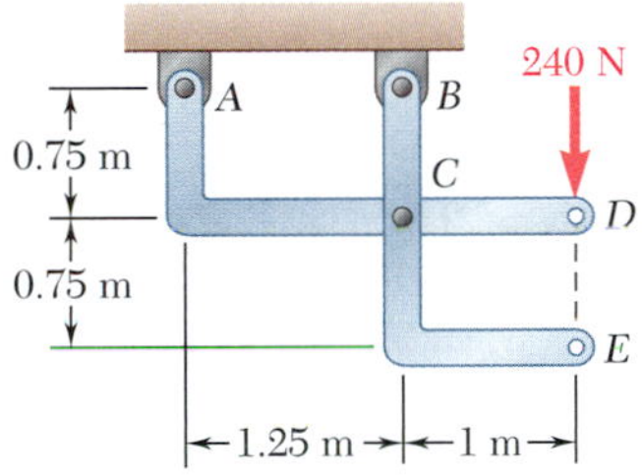

Fig. P6.88

6.89 The 192-N load can be moved along the line of action shown and applied at A, D, or E. Determine the components of the reactions at B and F when the 192-N load is applied (*a*) at A, (*b*) at D, (*c*) at E.

6.90 The 192-N load is removed and a 288 N · m clockwise couple is applied successively at A, D, and E. Determine the components of the reactions at B and F when the couple is applied (*a*) at A, (*b*) at B, (*c*) at E.

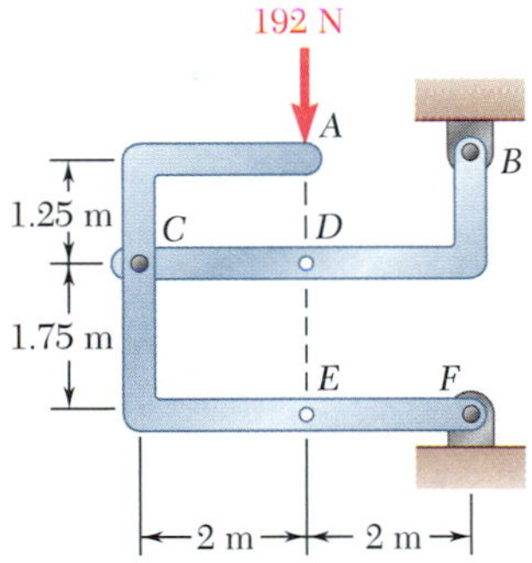

Fig. P6.89 and *P6.90*

6.91 (*a*) Show that when a frame supports a pulley at A, an equivalent loading of the frame and of each of its component parts can be obtained by removing the pulley and applying at A two forces equal and parallel to the forces that the cable exerted on the pulley. (*b*) Show that if one end of the cable is attached to the frame at point B, a force of magnitude equal to the tension in the cable should also be applied at B.

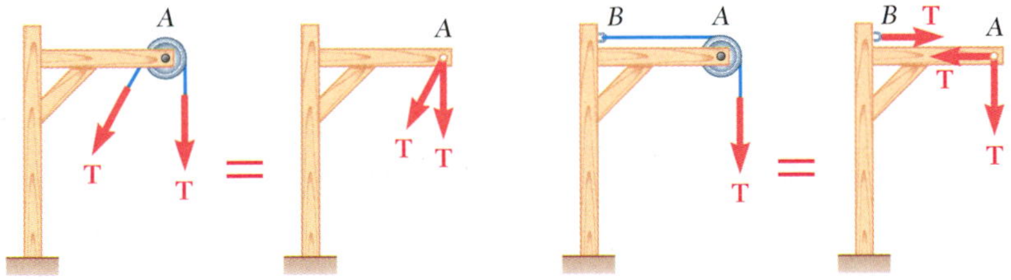

(*a*) (*b*)

Fig. P6.91

6.92 Knowing that the pulley has a radius of 1.5 ft, determine the components of the reactions at A and E.

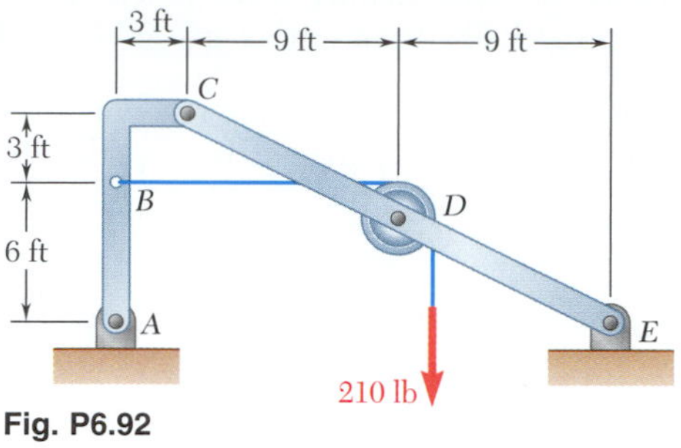

Fig. P6.92

6.93 Knowing that the pulley has a radius of 1.25 in., determine the components of the reactions at B and E.

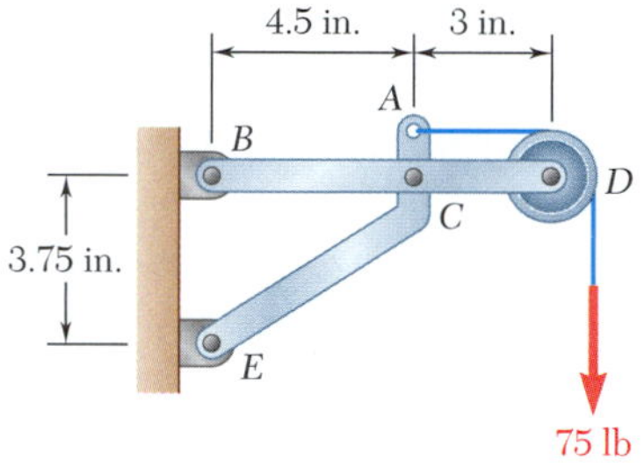

Fig. P6.93

6.94 Two 200-mm-diameter pipes (pipe *1* and pipe *2*) are supported every 3 m by a small frame like the one shown. Knowing that the combined mass per unit length of each pipe and its contents is 32 kg/m and assuming frictionless surfaces, determine the components of the reactions at A and E when $a = 0$.

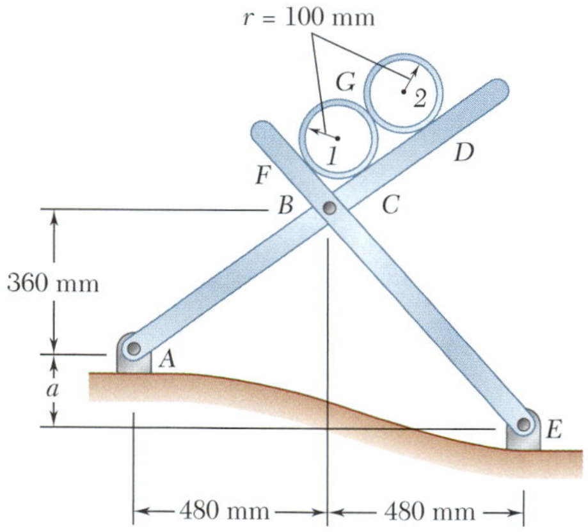

Fig. P6.94

6.95 Solve Prob. 6.94 when $a = 280$ mm.

6.96 The tractor and scraper units shown are connected by a vertical pin located 1.8 ft behind the tractor wheels. The distance from C to D is 2.25 ft. The center of gravity of the 20-kip tractor unit is located at G_t, while the centers of gravity of the 16-kip scraper unit and the 90-kip load are located at G_s and G_l, respectively. Knowing that the tractor is at rest with its brakes released, determine (*a*) the reactions at each of the four wheels, (*b*) the forces exerted on the tractor unit at C and D.

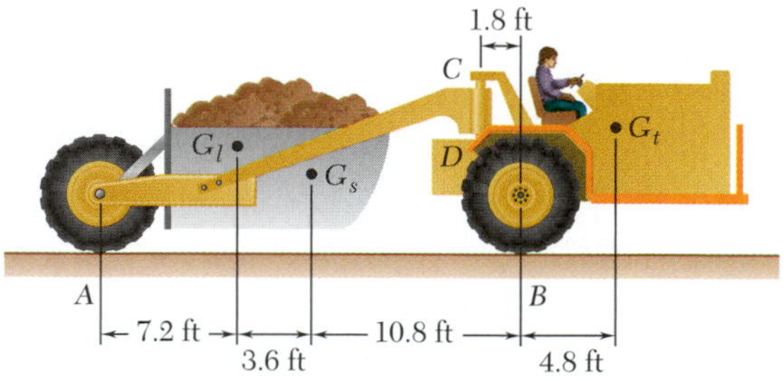

Fig. P6.96

6.97 Solve Prob. 6.96 assuming that the 90-kip load has been removed.

6.98 and 6.99 For the frame and loading shown, determine the components of all forces acting on member *ABE*.

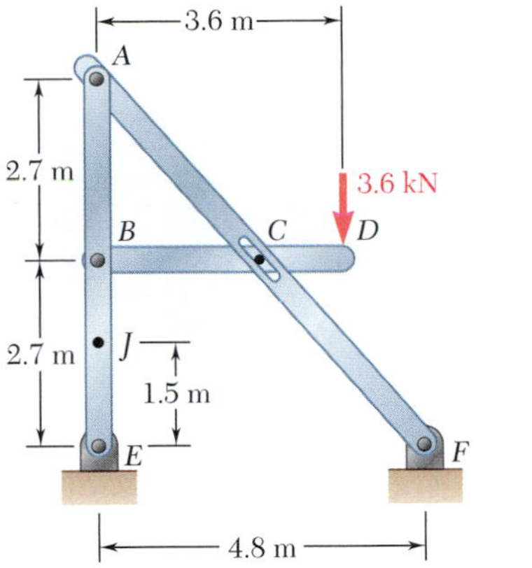

Fig. P6.98

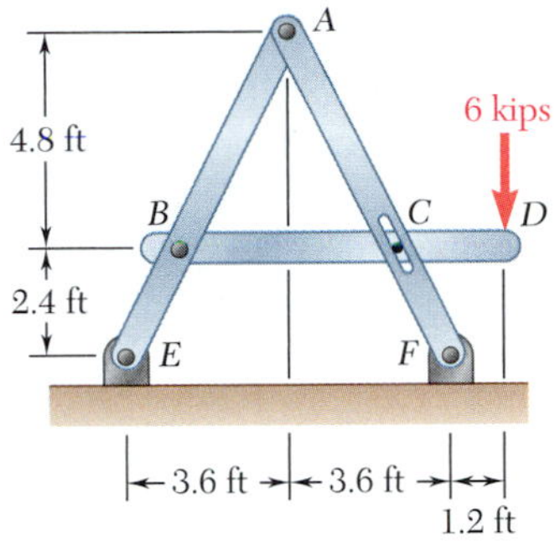

Fig. P6.99

6.100 For the frame and loading shown, determine the components of the forces acting on member *ABC* at *B* and *C*.

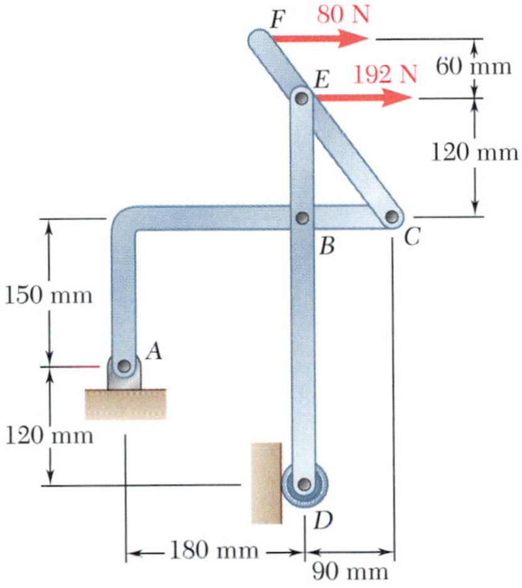

Fig. P6.100

6.101 For the frame and loading shown, determine the components of the forces acting on member *CDE* at *C* and *D*.

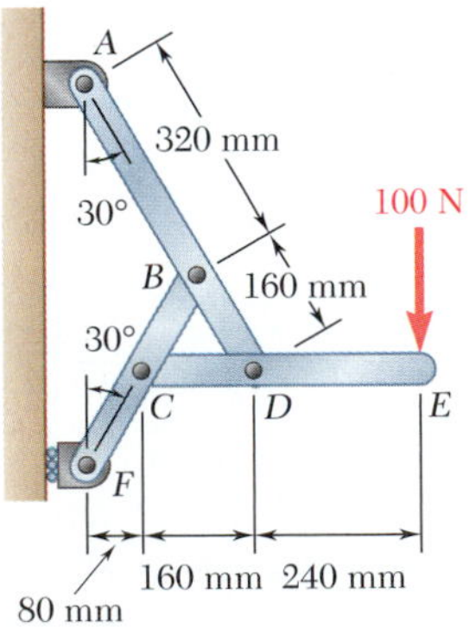

Fig. *P6.101*

6.102 Knowing that $P = 15$ N and $Q = 65$ N, determine the components of the forces exerted (*a*) on member *BCDF* at *C* and *D*, (*b*) on member *ACEG* at *E*.

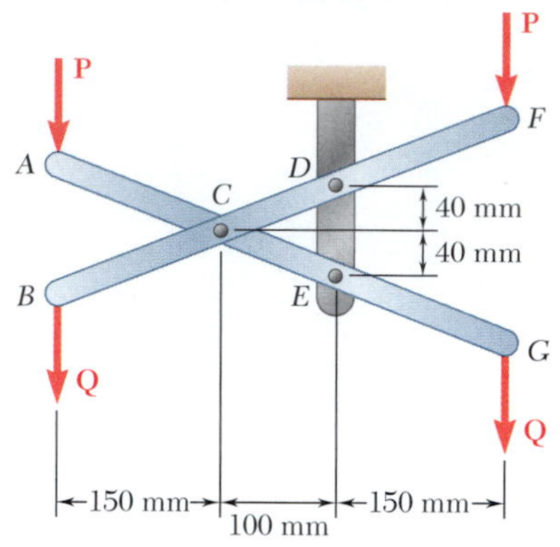

Fig. P6.102 and *P6.103*

6.103 Knowing that $P = 25$ N and $Q = 55$ N, determine the components of the forces exerted (*a*) on member *BCDF* at *C* and *D*, (*b*) on member *ACEG* at *E*.

6.104 Knowing that $P = 822$ N and $Q = 0$, determine for the frame and loading shown (*a*) the reaction at *D*, (*b*) the force in member *BF*.

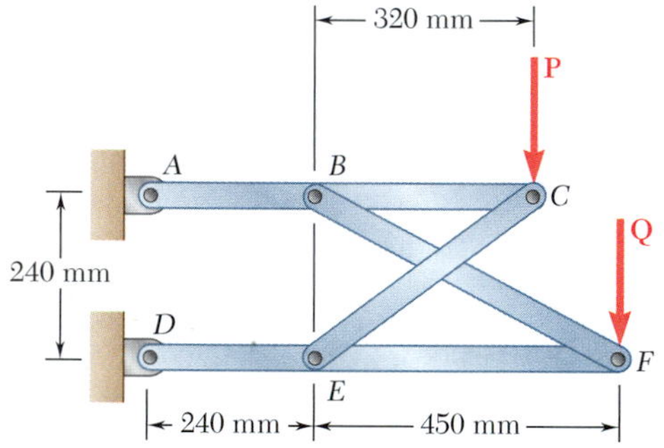

Fig. P6.104 and P6.105

6.105 Knowing that $P = 0$ and $Q = 1096$ N, determine for the frame and loading shown (*a*) the reaction at *D*, (*b*) the force in member *BF*.

6.106 The axis of the three-hinge arch *ABC* is a parabola with vertex at *B*. Knowing that $P = 28$ kips and $Q = 35$ kips, determine (*a*) the components of the reaction at *A*, (*b*) the components of the force exerted at *B* on segment *AB*.

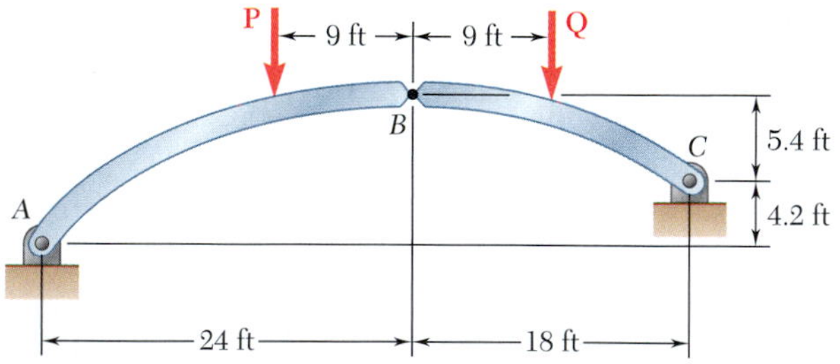

Fig. P6.106 and *P6.107*

6.107 The axis of the three-hinge arch *ABC* is a parabola with vertex at *B*. Knowing that $P = 35$ kips and $Q = 28$ kips, determine (*a*) the components of the reaction at *A*, (*b*) the components of the force exerted at *B* on segment *AB*.

6.108 For the frame and loading shown, determine the reactions at A, B, D, and E. Assume that the surface at each support is frictionless.

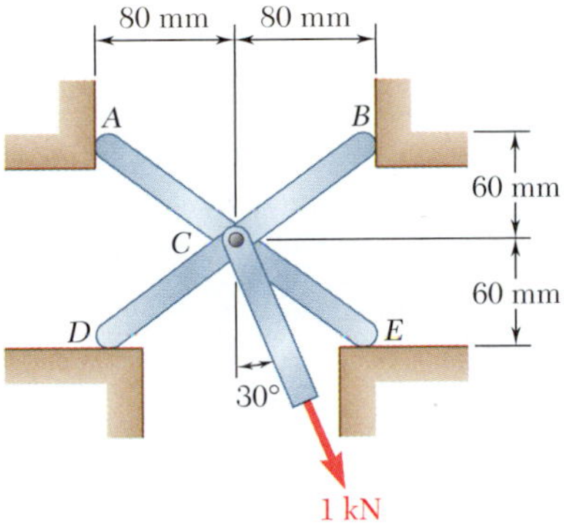

Fig. P6.108

6.109 through 6.111 Members ABC and CDE are pin-connected at C and are supported by four links. For the loading shown, determine the force in each link.

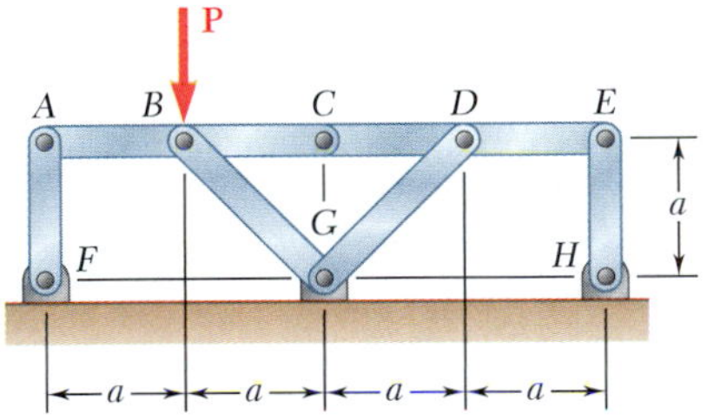

Fig. P6.109

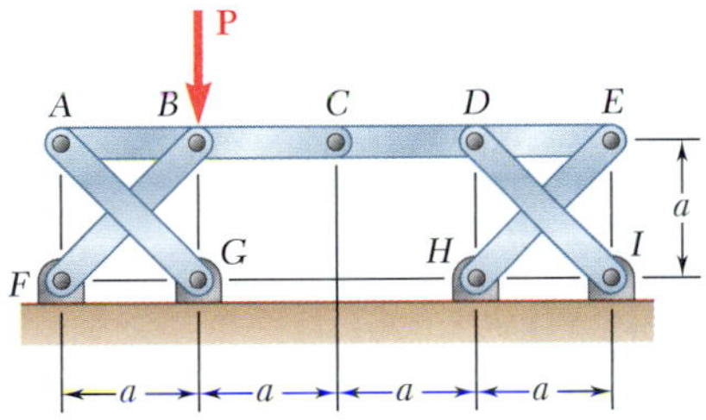

Fig. P6.110

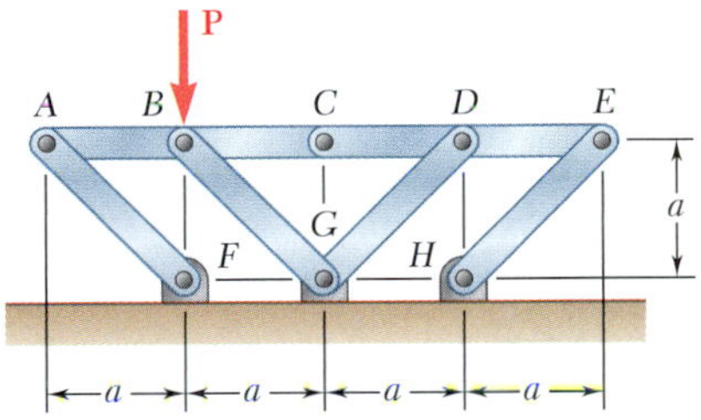

Fig. P6.111

6.112 Solve Prob. 6.109 assuming that the force **P** is replaced by a clockwise couple of moment $\mathbf{M}_0$ applied to member CDE at D.

6.113 Four wooden beams, each of length $2a$, are nailed together at their midpoints to form the support system shown. Assuming that only vertical forces are exerted at the connections, determine the vertical reactions at A, D, E, and H.

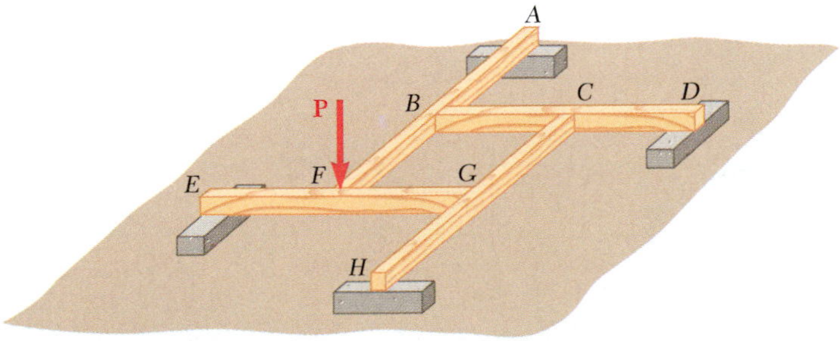

Fig. P6.113

6.114 Three wooden beams, each of length of $3a$, are nailed together to form the support system shown. Assuming that only vertical forces are exerted at the connections, determine the vertical reactions at A, D, and F.

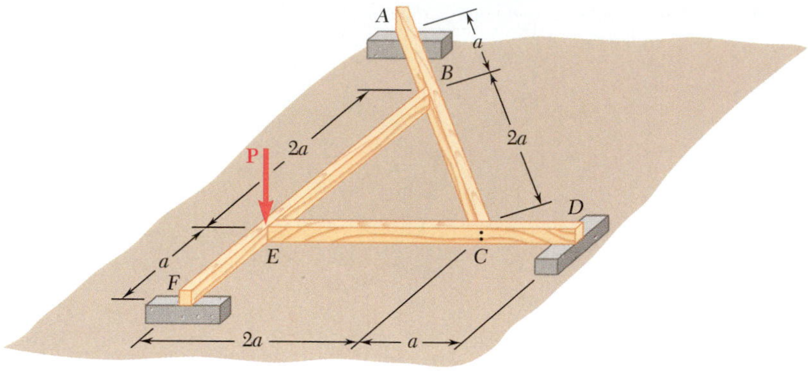

Fig. P6.114

6.115 through 6.117 Each of the frames shown consists of two L-shaped members connected by two rigid links. For each frame, determine the reactions at the supports and indicate whether the frame is rigid.

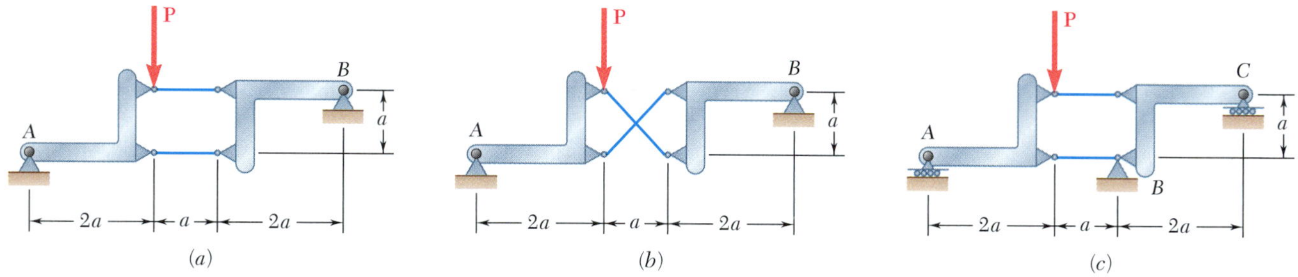

Fig. P6.115

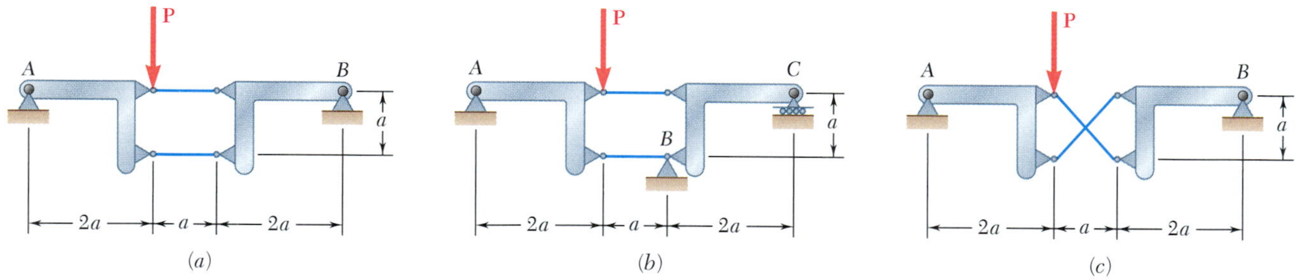

Fig. P6.116

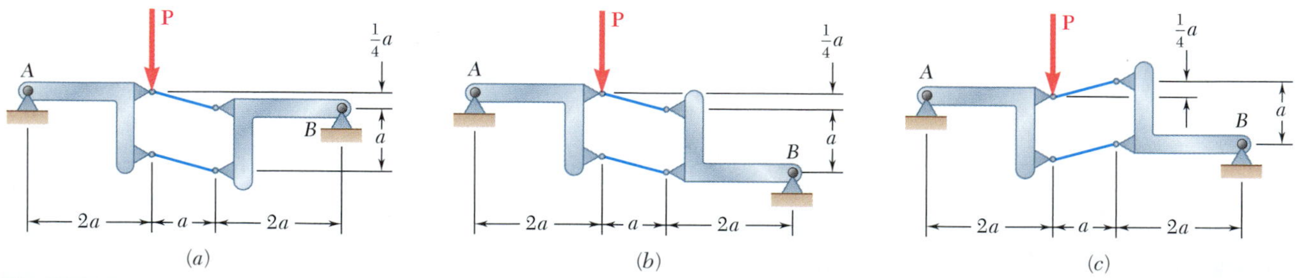

Fig. P6.117

6.12. MACHINES

Machines are structures designed to transmit and modify forces. Whether they are simple tools or include complicated mechanisms, their main purpose is to transform *input forces* into *output forces*. Consider, for example, a pair of cutting pliers used to cut a wire (Fig. 6.22*a*). If we apply two equal and opposite forces **P** and $-$**P** on their handles, they will exert two equal and opposite forces **Q** and $-$**Q** on the wire (Fig. 6.22*b*).

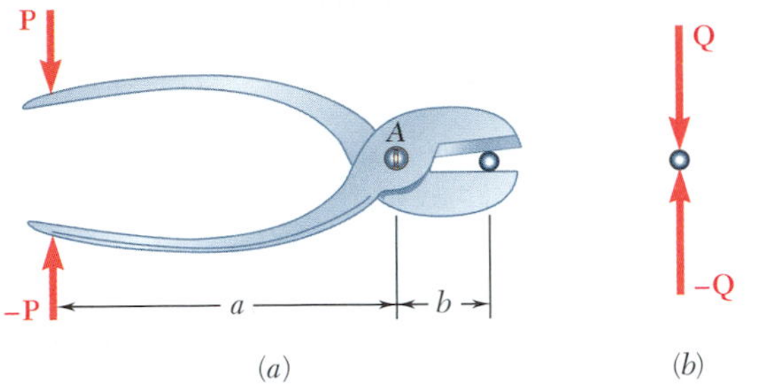

Fig. 6.22

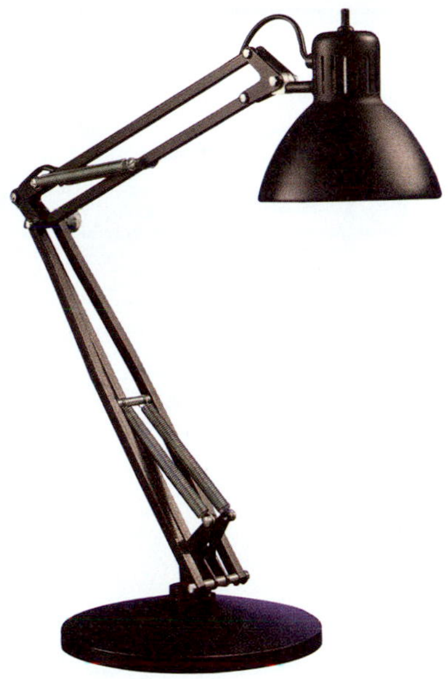

Photo 6.5 The lamp shown can be placed in many positions. By considering various free bodies, the force in the springs and the internal forces at the joints can be determined.

To determine the magnitude Q of the output forces when the magnitude P of the input forces is known (or, conversely, to determine P when Q is known), we draw a free-body diagram of the pliers *alone*, showing the input forces **P** and $-$**P** and the *reactions* $-$**Q** and **Q** that the wire exerts on the pliers (Fig. 6.23). However, since

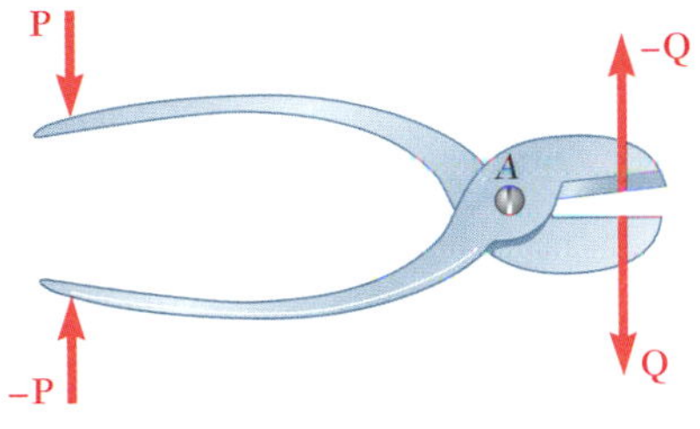

Fig. 6.23

a pair of pliers forms a nonrigid structure, we must use one of the component parts as a free body in order to determine the unknown forces. Considering Fig. 6.24*a*, for example, and taking moments about A, we obtain the relation $Pa = Qb$, which defines the magnitude Q in terms of P or P in terms of Q. The same free-body diagram can be used to determine the components of the internal force at A; we find $A_x = 0$ and $A_y = P + Q$.

In the case of more complicated machines, it generally will be necessary to use several free-body diagrams and, possibly, to solve simultaneous equations involving various internal forces. The free bodies should be chosen to include the input forces and the reactions to the output forces, and the total number of unknown force components involved should not exceed the number of available independent equations. It is advisable, before attempting to solve a problem, to determine whether the structure considered is determinate. There is no point, however, in discussing the rigidity of a machine, since a machine includes moving parts and thus *must* be nonrigid.

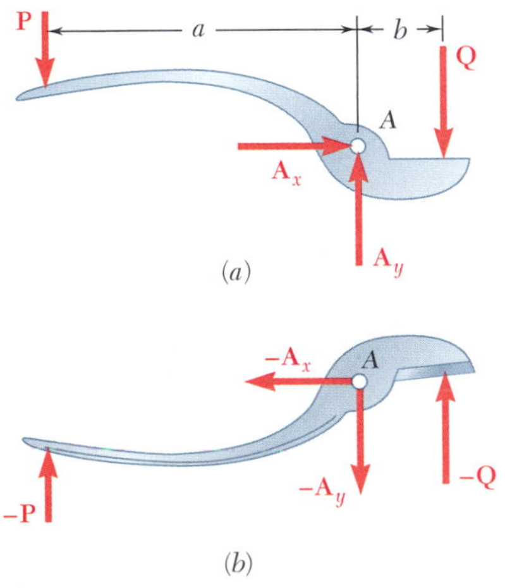

Fig. 6.24

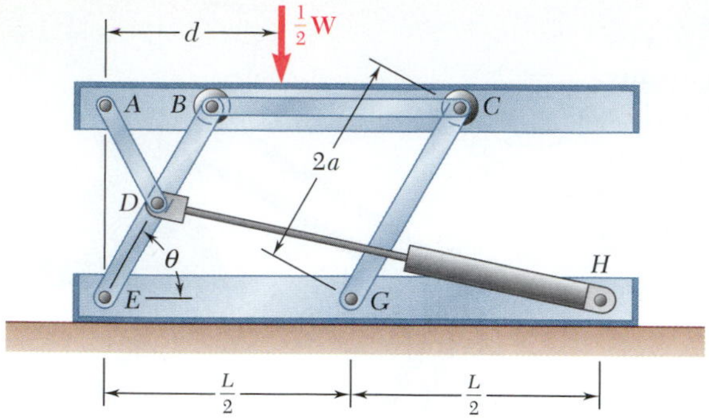

SAMPLE PROBLEM 6.7

A hydraulic-lift table is used to raise a 1000-kg crate. It consists of a platform and two identical linkages on which hydraulic cylinders exert equal forces. (Only one linkage and one cylinder are shown.) Members *EDB* and *CG* are each of length $2a$, and member *AD* is pinned to the midpoint of *EDB*. If the crate is placed on the table, so that half of its weight is supported by the system shown, determine the force exerted by each cylinder in raising the crate for $\theta = 60°$, $a = 0.70$ m, and $L = 3.20$ m. Show that the result obtained is independent of the distance d.

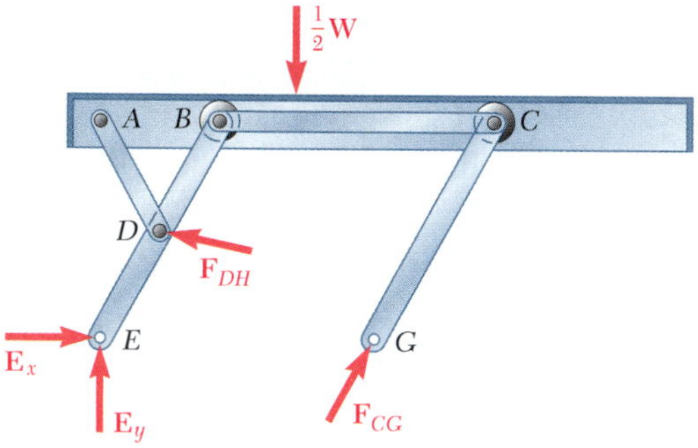

SOLUTION

The machine considered consists of the platform and the linkage. Its free-body diagram includes a force $\mathbf{F}_{DH}$ exerted by the cylinder, the weight $\frac{1}{2}\mathbf{W}$, and reactions at *E* and *G* that we assume to be directed as shown. Since more than three unknowns are involved, this diagram will not be used. The mechanism is dismembered and a free-body diagram is drawn for each of its component parts. We note that *AD*, *BC*, and *CG* are two-force members. We already assumed member *CG* to be in compression. We now assume that *AD* and *BC* are in tension; the forces exerted on them are then directed as shown. Equal and opposite vectors will be used to represent the forces exerted by the two-force members on the platform, on member *BDE*, and on roller *C*.

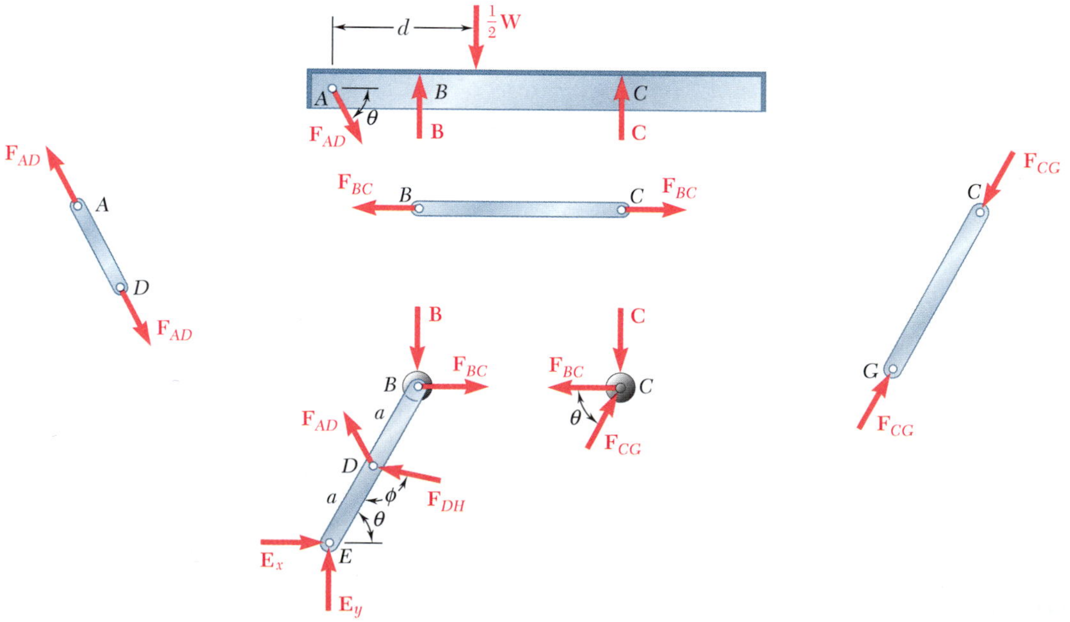

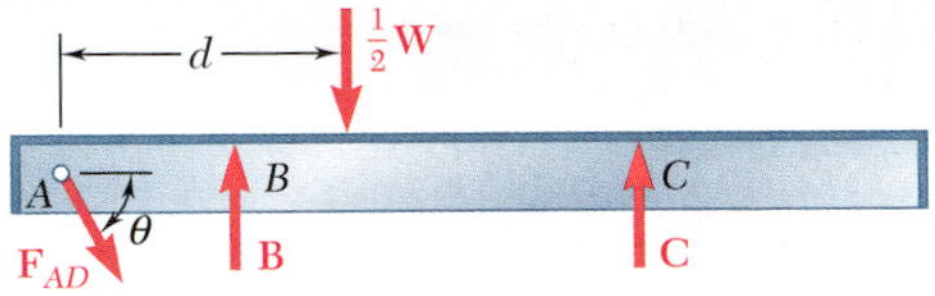

Free Body: Platform *ABC*.

$$\overset{+}{\rightarrow}\Sigma F_x = 0: \quad F_{AD}\cos\theta = 0 \qquad F_{AD} = 0$$
$$+\uparrow\Sigma F_y = 0: \quad B + C - \tfrac{1}{2}W = 0 \qquad B + C = \tfrac{1}{2}W \tag{1}$$

Free Body: Roller *C*. We draw a force triangle and obtain $F_{BC} = C\cot\theta$.

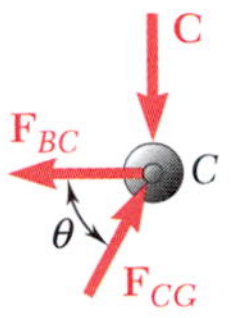

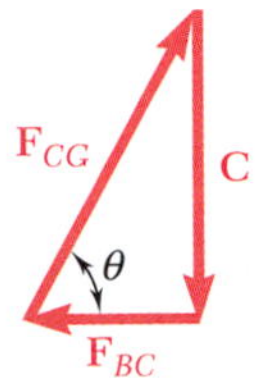

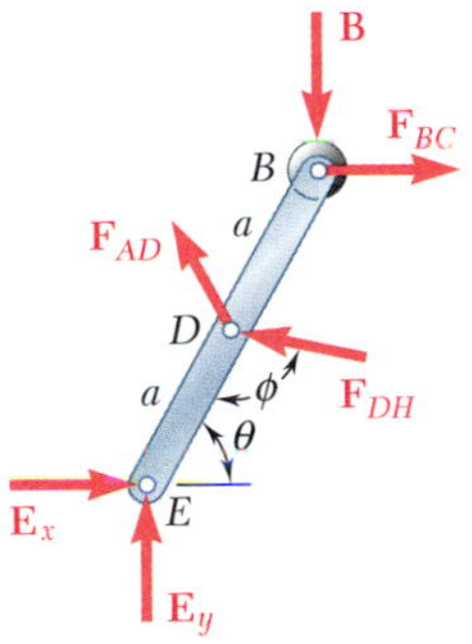

Free Body: Member *BDE*. Recalling that $F_{AD} = 0$,

$$+\circlearrowleft\Sigma M_E = 0: \quad F_{DH}\cos(\phi - 90°)a - B(2a\cos\theta) - F_{BC}(2a\sin\theta) = 0$$
$$F_{DH}a\sin\phi - B(2a\cos\theta) - (C\cot\theta)(2a\sin\theta) = 0$$
$$F_{DH}\sin\phi - 2(B + C)\cos\theta = 0$$

Recalling Eq. (1), we have

$$F_{DH} = W\frac{\cos\theta}{\sin\phi} \tag{2}$$

and we observe that *the result obtained is independent of d.* ◀

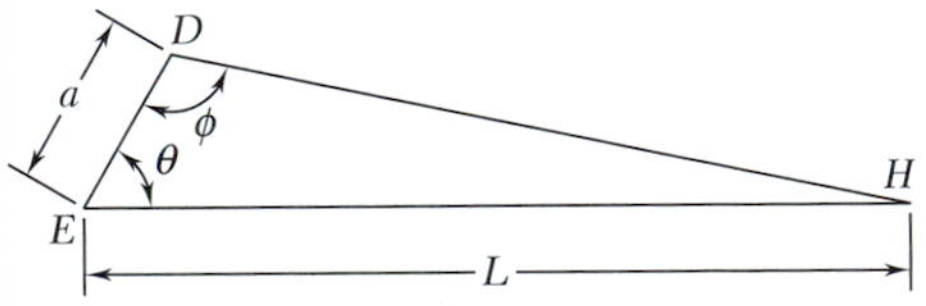

Applying first the law of sines to triangle EDH, we write

$$\frac{\sin\phi}{EH} = \frac{\sin\theta}{DH} \qquad \sin\phi = \frac{EH}{DH}\sin\theta \tag{3}$$

Using now the law of cosines, we have

$$(DH)^2 = a^2 + L^2 - 2aL\cos\theta$$
$$= (0.70)^2 + (3.20)^2 - 2(0.70)(3.20)\cos 60°$$
$$(DH)^2 = 8.49 \qquad DH = 2.91 \text{ m}$$

We also note that

$$W = mg = (1000 \text{ kg})(9.81 \text{ m/s}^2) = 9810 \text{ N} = 9.81 \text{ kN}$$

Substituting for $\sin\phi$ from (3) into (2) and using the numerical data, we write

$$F_{DH} = W\frac{DH}{EH}\cot\theta = (9.81 \text{ kN})\frac{2.91 \text{ m}}{3.20 \text{ m}}\cot 60°$$

$F_{DH} = 5.15$ kN ◀

SOLVING PROBLEMS ON YOUR OWN

This lesson was devoted to the analysis of *machines*. Since machines are designed to transmit or modify forces, they always contain moving parts. However, the machines considered here will always be at rest, and you will be working with the set of *forces required to maintain the equilibrium of the machine*.

Known forces that act on a machine are called *input forces. A machine transforms the input forces into output forces,* such as the cutting forces applied by the pliers of Fig. 6.22. You will determine the output forces by finding the forces equal and opposite to the output forces that should be applied to the machine to maintain its equilibrium.

In the preceding lesson you analyzed frames; you will now use almost the same procedure to analyze machines:

1. *Draw a free-body diagram of the whole machine,* and use it to determine as many as possible of the unknown forces exerted on the machine.

2. *Dismember the machine, and draw a free-body diagram of each member.*

3. *Considering first the two-force members,* apply equal and opposite forces to each two-force member at the points where it is connected to another member. If you cannot tell at this point whether the member is in tension or in compression just *assume* that the member is in tension and *direct both of the forces away from the member*. Since these forces have the same unknown magnitude, *give them both the same name*.

4. *Next consider the multiforce members.* For each of these members, show all the forces acting on the member, including applied loads and forces, reactions, and internal forces at connections.

a. *Where a multiforce member is connected to a two-force member,* apply to the multiforce member a force *equal and opposite* to the force drawn on the free-body diagram of the two-force member, *giving it the same name*.

b. *Where a multiforce member is connected to another multiforce member,* use *horizontal and vertical components* to represent the internal forces at that point. The directions you choose for each of the two force components exerted on the first multiforce member are arbitrary, but *you must apply equal and opposite force components of the same name* to the other multiforce member.

5. *Equilibrium equations can be written* after you have completed the various free-body diagrams.

a. *To simplify your solution,* you should, whenever possible, write and solve equilibrium equations involving single unknowns.

b. *Since you arbitrarily chose the direction of each of the unknown forces,* you must determine at the end of the solution whether your guess was correct. To that effect, *consider the sign* of the value found for each of the unknowns. A *positive* sign indicates that your guess was correct, and a *negative* sign indicates that it was not.

6. *Finally, you should check your solution* by substituting the results obtained into an equilibrium equation that you have not previously used.

Problems

6.118 A 50-N force directed vertically downward is applied to the toggle vise at C. Knowing that link BD is 150 mm long and that $a = 100$ mm, determine the horizontal force exerted on block E.

6.119 A 50-N force directed vertically downward is applied to the toggle vise at C. Knowing that link BD is 150 mm long and that $a = 200$ mm, determine the horizontal force exerted on block E.

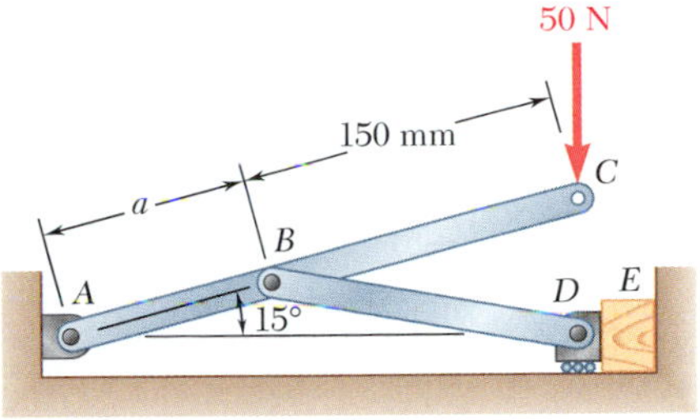

Fig. P6.118 and *P6.119*

6.120 The press shown is used to emboss a small seal at E. Knowing that $P = 60$ lb, determine (a) the vertical component of the force exerted on the seal, (b) the reaction at A.

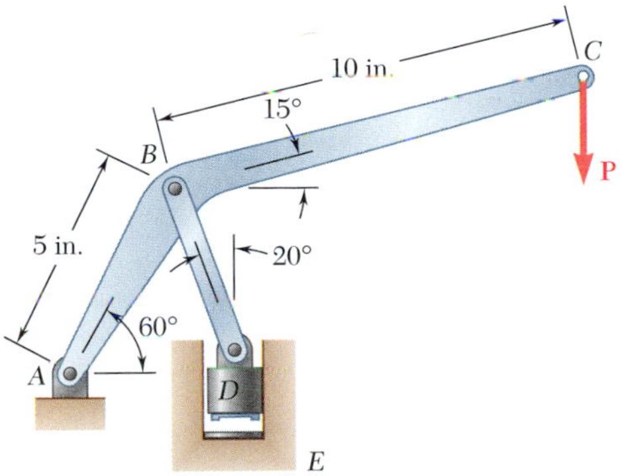

Figs. P6.120 and P6.121

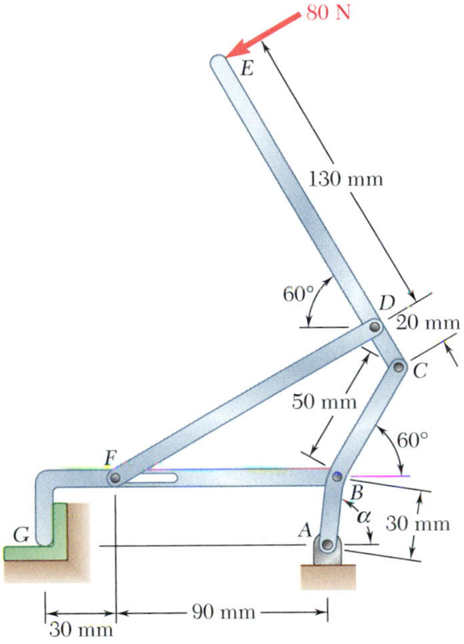

Fig. P6.122 and *P6.123*

6.121 The press shown is used to emboss a small seal at E. Knowing that the vertical component of the force exerted on the seal must be 240 lb, determine (a) the required vertical force **P**, (b) the corresponding reaction at A.

6.122 The double toggle latching mechanism shown is used to hold member G against the support. Knowing that $\alpha = 60°$, determine the force exerted on G.

6.123 The double toggle latching mechanism shown is used to hold member G against the support. Knowing that $\alpha = 75°$, determine the force exerted on G.

6.124 For the system and loading shown, determine (a) the force **P** required for equilibrium, (b) the corresponding force in member BD, (c) the corresponding reaction at C.

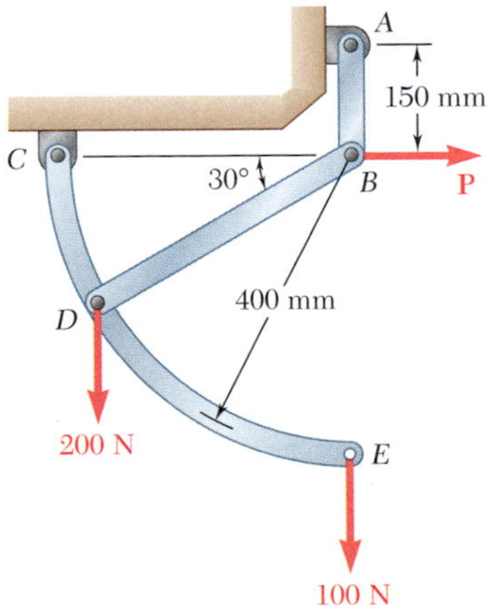

Fig. *P6.124*

6.125 A couple **M** of magnitude 15 kip · in. is applied to the crank of the engine system shown. For each of the two positions shown, determine the force **P** required to hold the system in equilibrium.

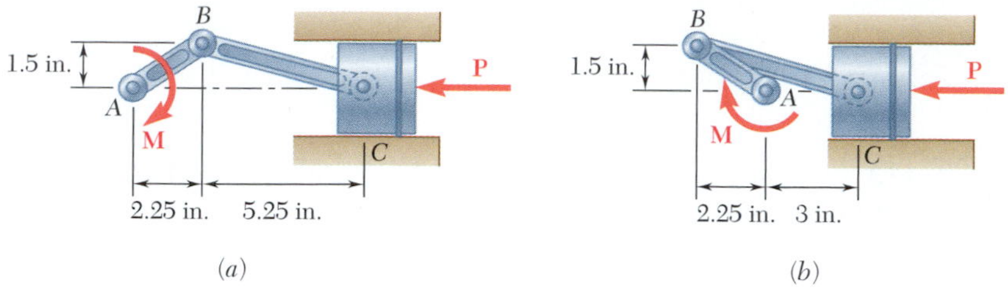

Fig. P6.125 and *P6.126*

6.126 A force **P** of magnitude 4 kips is applied to the piston of the engine system shown. For each of the two positions shown, determine the couple **M** required to hold the system in equilibrium.

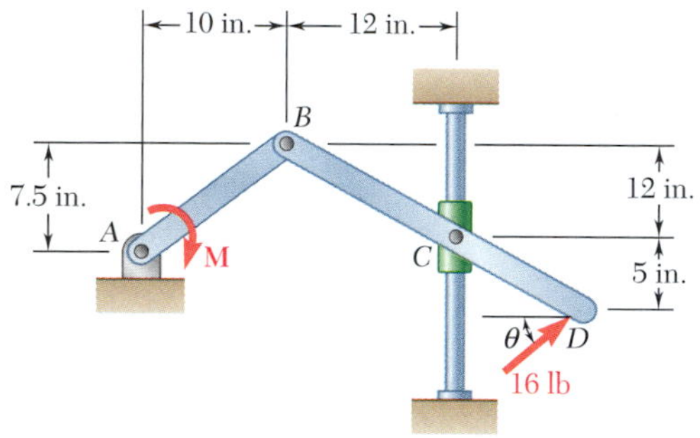

Fig. P6.127 and P6.128

6.127 Arm *BCD* is connected by pins to crank *AB* at *B* and to a collar at *C*. Neglecting the effect of friction, determine the couple **M** required to hold the system in equilibrium when $\theta = 0$.

6.128 Arm *BCD* is connected by pins to crank *AB* at *B* and to a collar at *C*. Neglecting the effect of friction, determine the couple **M** required to hold the system in equilibrium when $\theta = 45°$.

6.129 The double toggle mechanism shown is used in a punching machine. Knowing that links *AB* and *BC* are each of length 150 mm, determine the couple **M** required to hold the system in equilibrium when $\phi = 20°$.

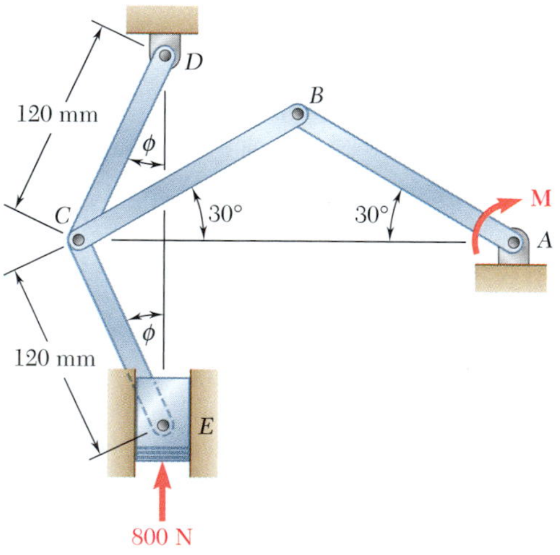

Fig. P6.129 and P6.130

6.130 The double toggle mechanism shown is used in a punching machine. Knowing that links *AB* and *BC* are each of length 150 mm and that $M = 75$ N · m, determine the angle ϕ if the system is in equilibrium.

6.131 and 6.132 Two rods are connected by a slider block as shown. Neglecting the effect of friction, determine the couple $\mathbf{M}_A$ required to hold the system in equilibrium.

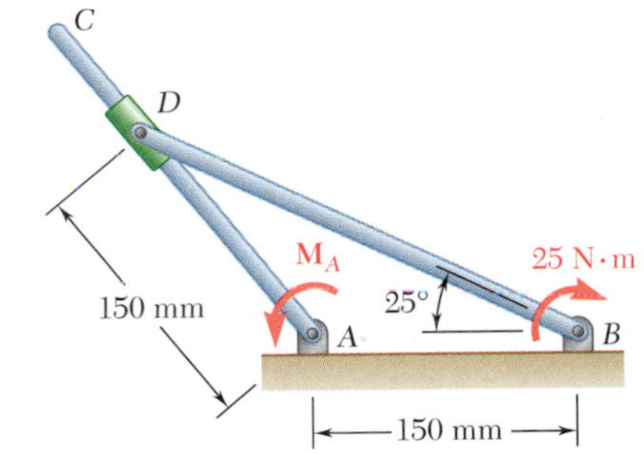

Fig. P6.131

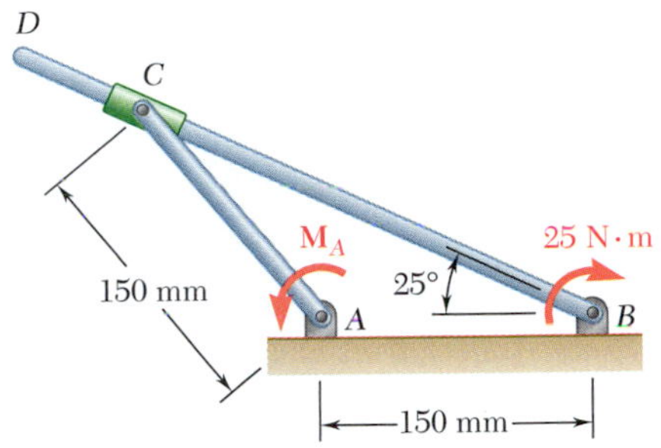

Fig. P6.132

6.133 and *6.134* Rod *CD* is attached to the collar *D* and passes through a collar welded to end *B* of lever *AB*. Neglecting the effect of friction, determine the couple **M** required to hold the system in equilibrium when $\theta = 30°$.

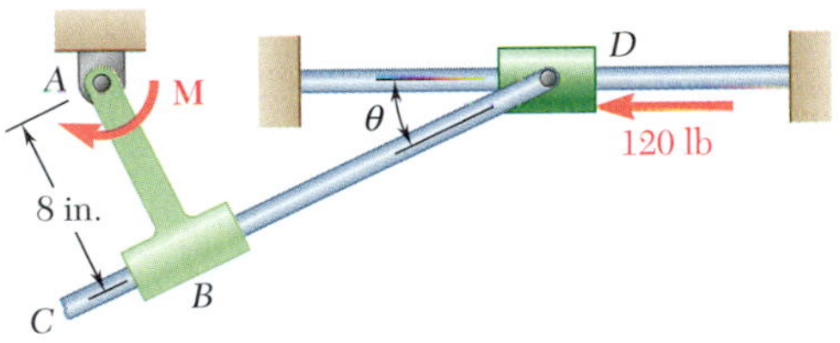

Fig. P6.133

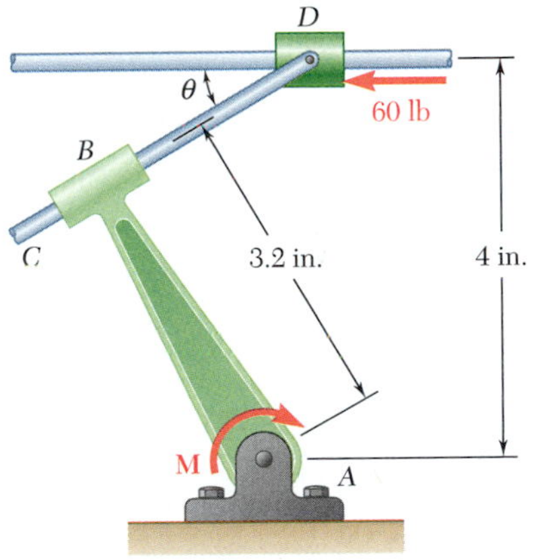

Fig. *P6.134*

6.135 The slab tongs shown are used to lift a 960-lb block of granite. Determine the forces exerted at *D* and *F* on tong *BDF*.

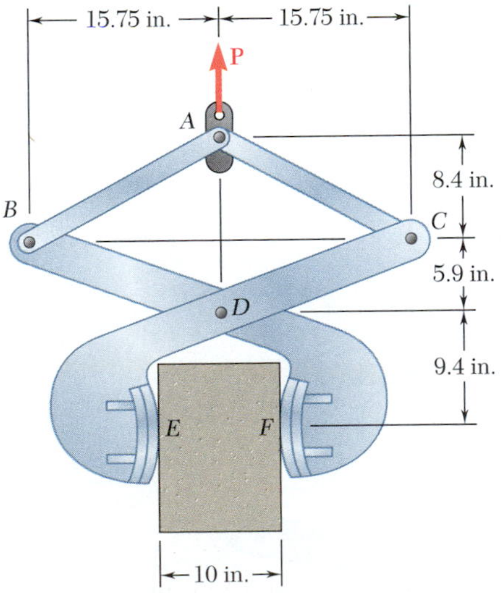

Fig. P6.135

6.136 The pallet puller shown is used to pull a loaded pallet to the rear of a truck. Knowing that $P = 420$ lb, determine the forces exerted at G and H on tong FGH.

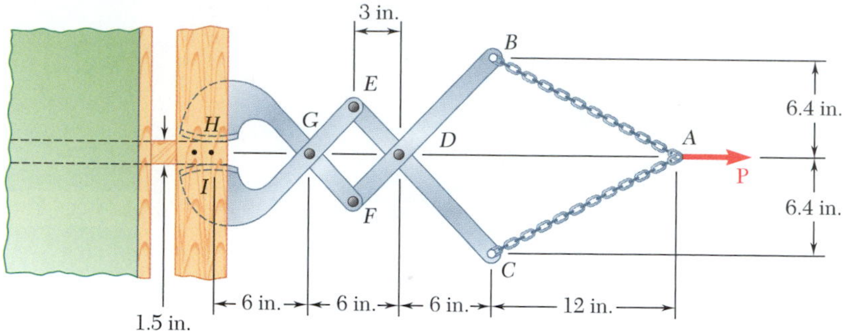

Fig. *P6.136*

6.137 The drum lifter shown is used to lift a steel drum. Knowing that the mass of the drum and its contents is 240 kg, determine the forces exerted at F and H on member DFH.

6.138 A small barrel having a mass of 72 kg is lifted by a pair of tongs as shown. Knowing that $a = 100$ mm, determine the forces exerted at B and D on tong ABD.

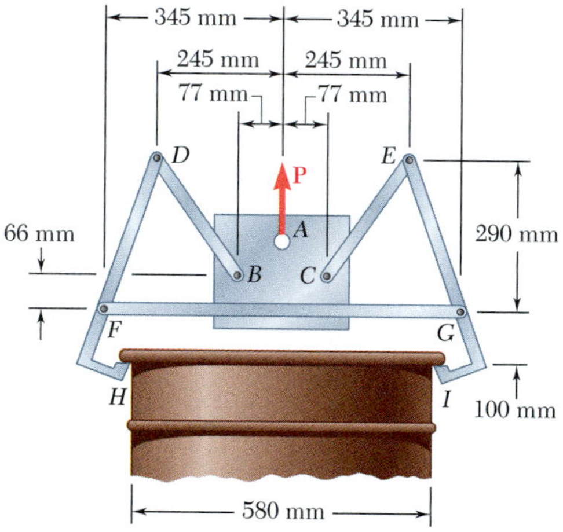

Fig. P6.137

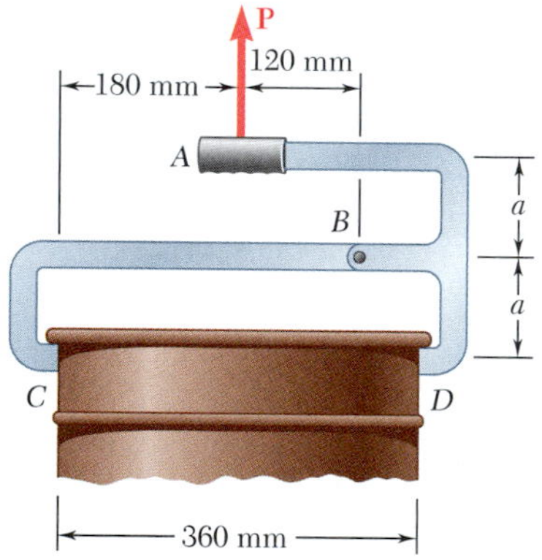

Fig. P6.138

6.139 Determine the magnitude of the gripping forces exerted along line aa on the nut when two 240-N forces are applied to the handles as shown. Assume that pins A and D slide freely in slots cut in the jaws.

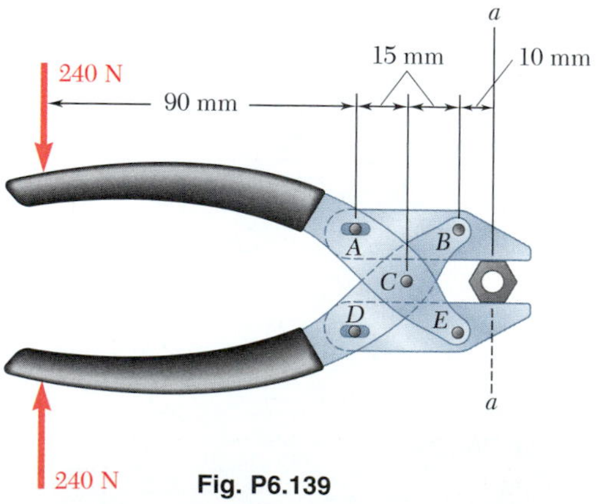

Fig. P6.139

6.140 In using the bolt cutter shown, a worker applies two 75-lb forces to the handles. Determine the magnitude of the forces exerted by the cutter on the bolt.

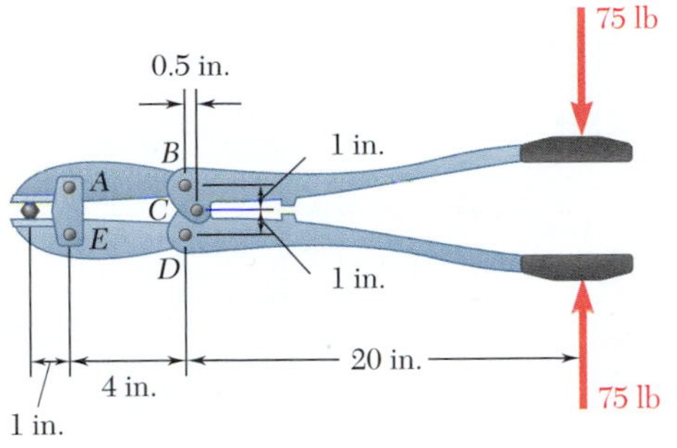

Fig. P6.140

6.141 Determine the magnitude of the gripping forces produced when two 50-lb forces are applied as shown.

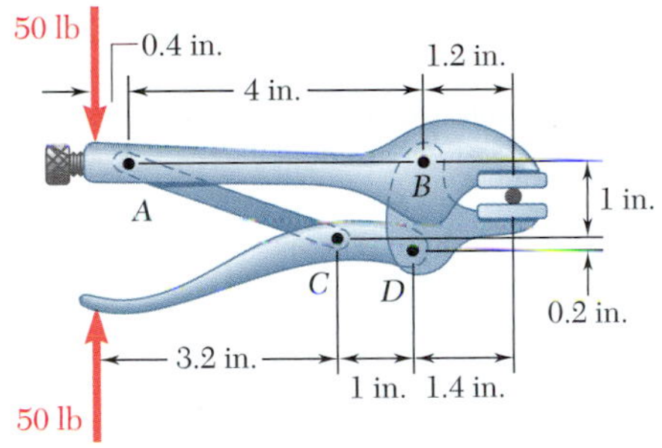

Fig. P6.141

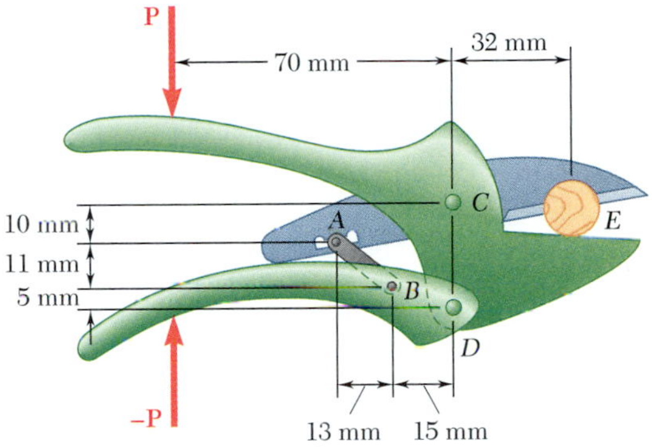

Fig. P6.142

6.142 The compound-lever pruning shears shown can be adjusted by placing pin A at various ratchet positions on blade ACE. Knowing that 1.5-kN vertical forces are required to complete the pruning of a small branch, determine the magnitude P of the forces that must be applied to the handles when the shears are adjusted as shown.

6.143 Determine the force **P** which must be applied to the toggle BCD to maintain equilibrium in the position shown.

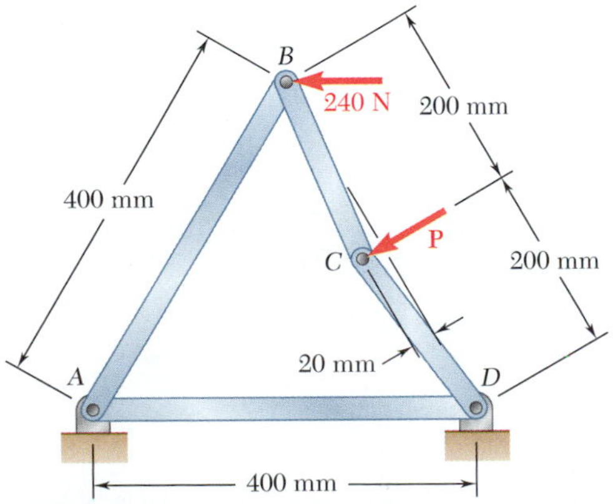

Fig. P6.143

6.144 In the locked position shown, the toggle clamp exerts at *A* a vertical 1.2-kN force on the wooden block, and handle *CF* rests against the stop at *G*. Determine the force **P** required to release the clamp. (*Hint:* To release the clamp, the forces of contact at *G* must be zero.)

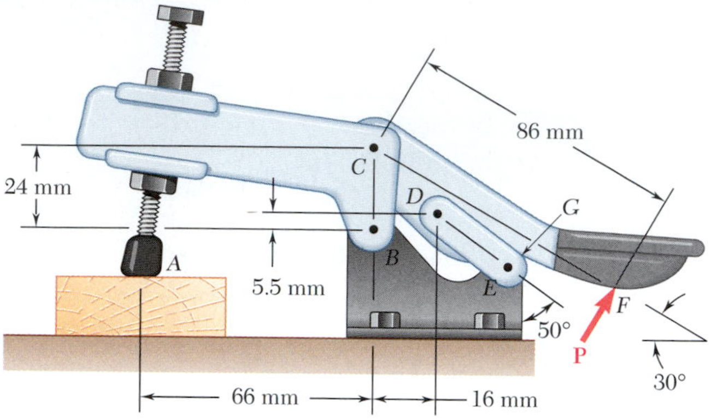

Fig. *P6.144*

6.145 The garden shears shown consist of two blades and two handles. The two handles are connected by pin *C* and the two blades are connected by pin *D*. The left blade and the right handle are connected by pin *A*; the right blade and the left handle are connected by pin *B*. Determine the magnitude of the forces exerted on the small branch *E* when two 20-lb forces are applied to the handles as shown.

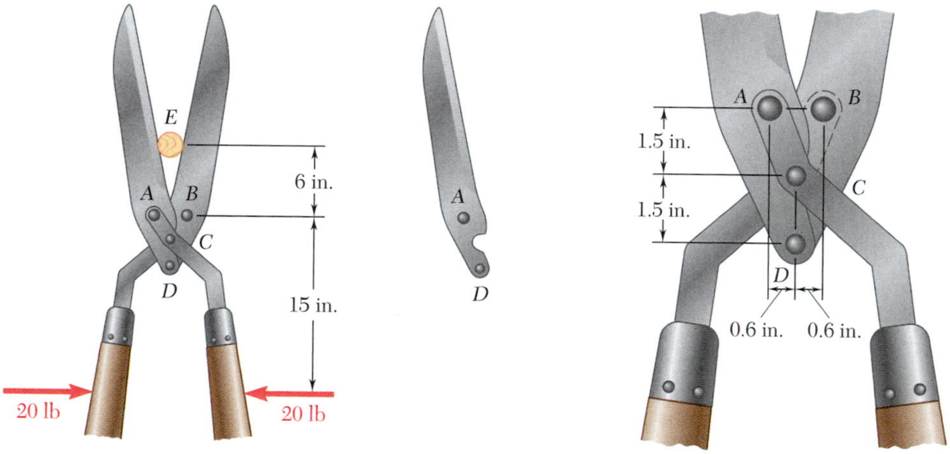

Fig. P6.145

6.146 The bone rongeur shown is used in surgical procedures to cut small bones. Determine the magnitude of the forces exerted on the bone at *E* when two 25-lb forces are applied as shown.

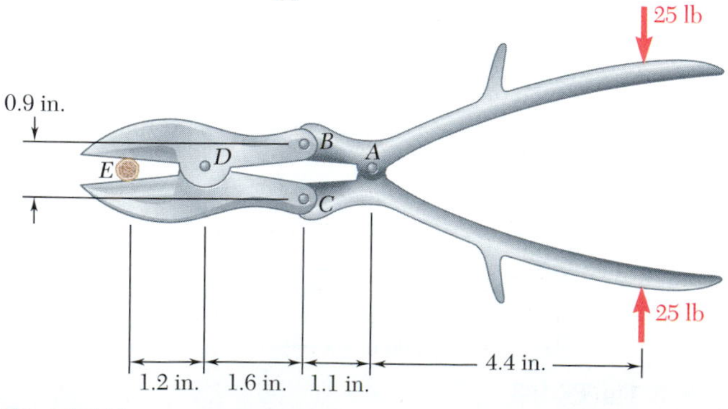

Fig. P6.146

6.147 The telescoping arm ABC is used to provide an elevated platform for construction workers. The workers and the platform together have a mass of 240 kg and have a combined center of gravity located directly above C. For the position when $\theta = 24°$, determine (*a*) the force exerted at B by the single hydraulic cylinder BD, (*b*) the force exerted on the supporting carriage at A.

6.148 The telescoping arm ABC can be lowered until end C is close to the ground, so that workers can easily board the platform. For the position when $\theta = -18°$, determine (*a*) the force exerted at B by the single hydraulic cylinder BD, (*b*) the force exerted on the supporting carriage at A.

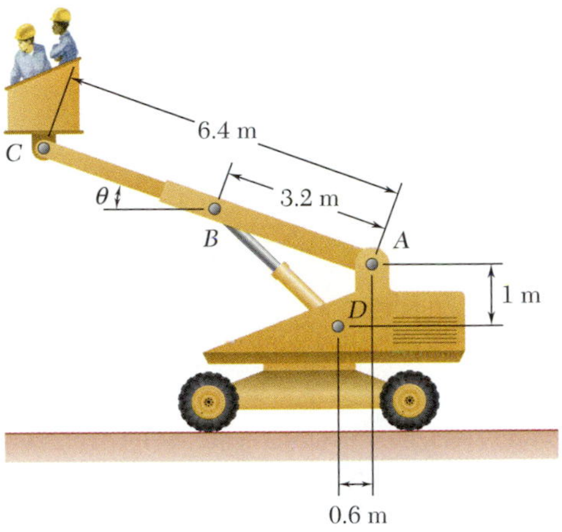

Fig. P6.147 and P6.148

6.149 The bucket of the front-end loader shown carries a 12-kN load. The motion of the bucket is controlled by two identical mechanisms, only one of which is shown. Knowing that the mechanism shown supports one-half of the 12-kN load, determine the force exerted (*a*) by cylinder CD, (*b*) by cylinder FH.

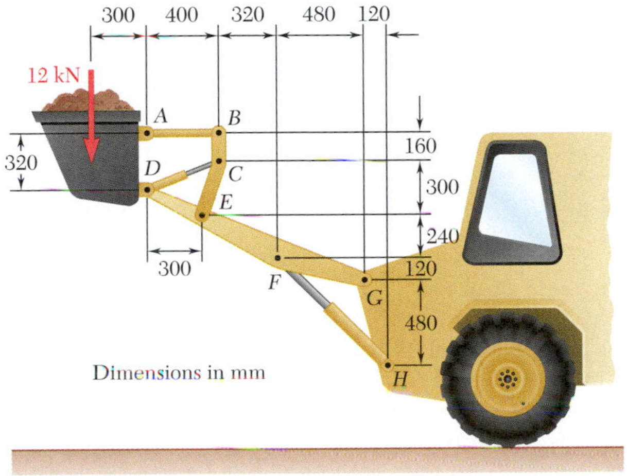

Fig. P6.149

6.150 The motion of the bucket of the front-end loader shown is controlled by two arms and a linkage which are pin-connected at D. The arms are located symmetrically with respect to the central, vertical, and longitudinal plane of the loader; one arm AFJ and its control cylinder EF are shown. The single linkage $GHDB$ and its control cylinder BC are located in the plane of symmetry. For the position shown, determine the force exerted (*a*) by cylinder BC, (*b*) by cylinder EF.

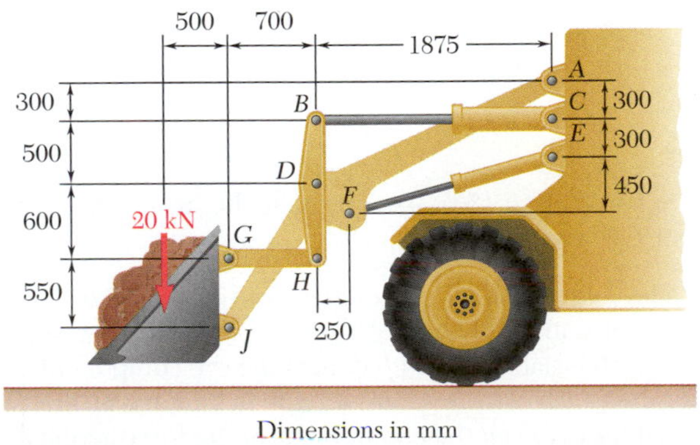

Fig. *P6.150*

6.151 In the planetary gear system shown, the radius of the central gear A is $a = 20$ mm, the radius of the planetary gears is b, and the radius of the outer gear E is $(a + 2b)$. A clockwise couple of magnitude $M_A = 20$ N · m is applied to the central gear A, and a counterclockwise couple of magnitude $M_S = 100$ N · m is applied to the spider BCD. If the system is to be in equilibrium, determine (*a*) the required radius b of the planetary gears, (*b*) the couple $\mathbf{M}_E$ that must be applied to the outer gear E.

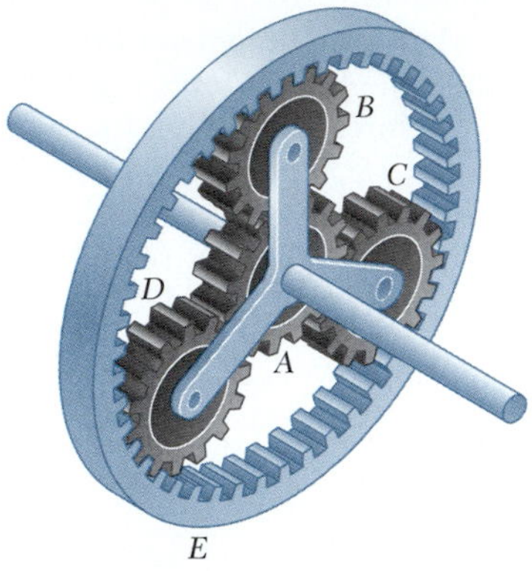

Fig. P6.151

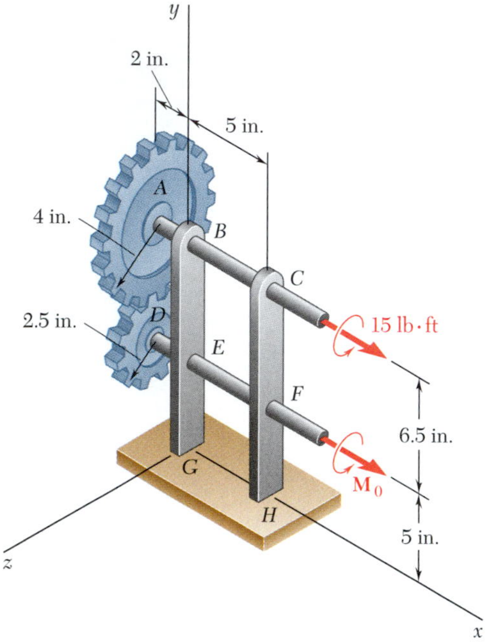

Fig. *P6.152*

6.152 Gears A and D are rigidly attached to horizontal shafts that are held by frictionless bearings. Determine (*a*) the couple $\mathbf{M}_0$ that must be applied to shaft DEF to maintain equilibrium, (*b*) the reactions at G and H.

***6.153** Two shafts AC and CF, which lie in the vertical xy plane, are connected by a universal joint at C. The bearings at B and D do not exert any axial force. A couple of magnitude 50 N · m (clockwise when viewed from the positive x axis) is applied to shaft CF at F. At a time when the arm of the crosspiece attached to shaft CF is horizontal, determine (*a*) the magnitude of the couple which must be applied to shaft AC at A to maintain equilibrium, (*b*) the reactions at B, D, and E. (*Hint:* The sum of the couples exerted on the crosspiece must be zero.)

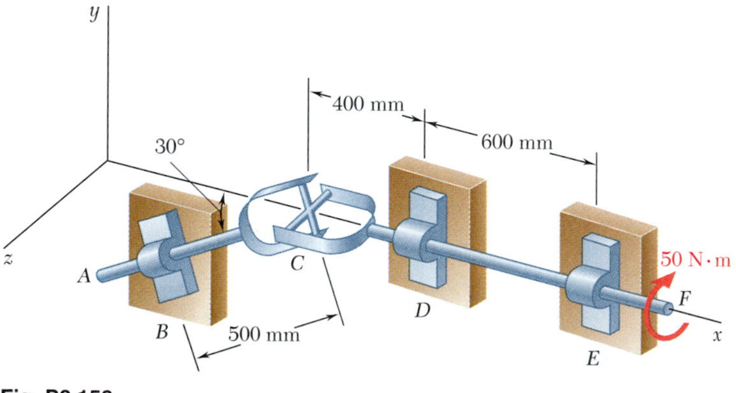

Fig. P6.153

***6.154** Solve Prob. 6.153 assuming that the arm of the crosspiece attached to shaft CF is vertical.

***6.155** The large mechanical tongs shown are used to grab and lift a thick 1800-lb steel slab HJ. Knowing that slipping does not occur between the tong grips and the slab at H and J, determine the components of all forces acting on member EFH. (*Hint:* Consider the symmetry of the tongs to establish relationships between the components of the force acting at E on EFH and the components of the force acting at D on CDF.)

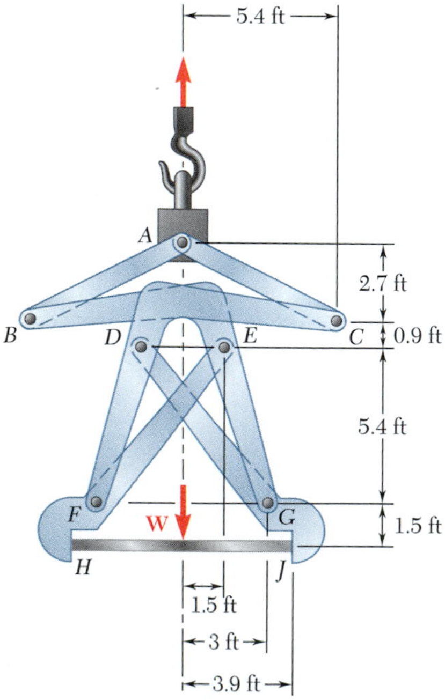

Fig. P6.155

REVIEW AND SUMMARY FOR CHAPTER 6

In this chapter you learned to determine the *internal forces* holding together the various parts of a structure.

Analysis of trusses

The first half of the chapter was devoted to the analysis of *trusses,* that is, to the analysis of structures consisting of *straight members connected at their extremities only*. The members being slender and unable to support lateral loads, all the loads must be applied at the joints; a truss may thus be assumed to consist of *pins and two-force members* [Sec. 6.2].

Simple trusses

A truss is said to be *rigid* if it is designed in such a way that it will not greatly deform or collapse under a small load. A triangular truss consisting of three members connected at three joints is clearly a rigid truss (Fig. 6.25*a*) and so will be the truss obtained by adding two new members to the first one and connecting them at a new joint (Fig. 6.25*b*). Trusses obtained by repeating this procedure are called *simple trusses*. We may check that in a simple truss the total number of members is $m = 2n - 3$, where n is the total number of joints [Sec. 6.3].

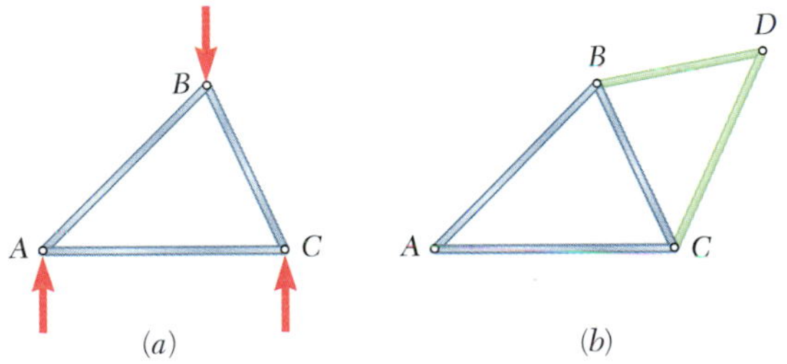

Fig. 6.25

Method of joints

The forces in the various members of a simple truss can be determined by the *method of joints* [Sec. 6.4]. First, the reactions at the supports can be obtained by considering the entire truss as a free body. The free-body diagram of each pin is then drawn, showing the forces exerted on the pin by the members or supports it connects. Since the members are straight two-force members, the force exerted by a member on the pin is directed along that member, and only the magnitude of the force is unknown. It is always possible in the case of a simple truss to draw the free-body diagrams of the pins in such an order that only two unknown forces are included in each diagram. These forces can be obtained from the corresponding two equilibrium equations or—if only three forces are involved—from the corresponding force triangle. If the force exerted by a member on a pin is directed toward that pin, the member is in *compression;* if it is directed away from the pin, the member is in *tension* [Sample Prob. 6.1]. The analysis of a truss is sometimes expedited by first recognizing *joints under special loading conditions* [Sec. 6.5]. The method of joints can also be extended to the analysis of three-dimensional or *space trusses* [Sec. 6.6].

Method of sections

The *method of sections* is usually preferred to the method of joints when the force in only one member—or very few members—of a truss is desired [Sec. 6.7]. To determine the force in member BD of the truss of Fig. 6.26*a*, for example, we *pass a section* through members BD, BE, and CE, remove these members, and use the portion ABC of the truss as a free body (Fig. 6.26*b*). Writing $\Sigma M_E = 0$, we determine the magnitude of the force $\mathbf{F}_{BD}$, which represents the force in member BD. A positive sign indicates that the member is in *tension;* a negative sign indicates that it is in *compression* [Sample Probs. 6.2 and 6.3].

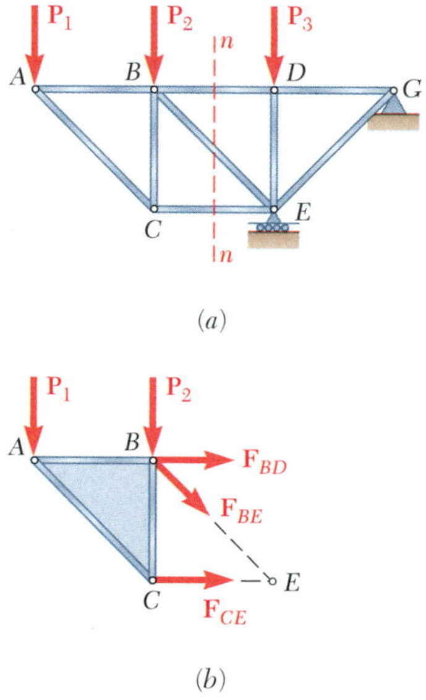

Fig. 6.26

Compound trusses

The method of sections is particularly useful in the analysis of *compound trusses,* that is, trusses which cannot be constructed from the basic triangular truss of Fig. 6.25*a* but which can be obtained by rigidly connecting several simple trusses [Sec. 6.8]. If the component trusses have been properly connected (for example, one pin and one link, or three nonconcurrent and nonparallel links) and if the resulting structure is properly supported (for example, one pin and one roller), the compound truss is *statically determinate, rigid, and completely constrained*. The following necessary—but not sufficient—condition is then satisfied: $m + r = 2n$, where m is the number of members, r is the number of unknowns representing the reactions at the supports, and n is the number of joints.

Frames and machines

The second part of the chapter was devoted to the analysis of *frames and machines*. Frames and machines are structures which contain *multiforce members,* that is, members acted upon by three or more forces. Frames are designed to support loads and are usually stationary, fully constrained structures. Machines are designed to transmit or modify forces and always contain moving parts [Sec. 6.9].

Analysis of a frame

To *analyze a frame,* we first consider the *entire frame as a free body* and write three equilibrium equations [Sec. 6.10]. If the frame remains rigid when detached from its supports, the reactions involve only three unknowns and may be determined from these equations [Sample Probs. 6.4 and 6.5]. On the other hand, if the frame ceases to be rigid when detached from its supports, the reactions involve more than three unknowns and cannot be completely determined from the equilibrium equations of the frame [Sec. 6.11; Sample Prob. 6.6].

Multiforce members

We then *dismember the frame* and identify the various members as either two-force members or multiforce members; pins are assumed to form an integral part of one of the members they connect. We draw the free-body diagram of each of the multiforce members, noting that when two multiforce members are connected to the same two-force member, they are acted upon by that member with *equal and opposite forces of unknown magnitude but known direction*. When two multiforce members are connected by a pin, they exert on each other *equal and opposite forces of unknown direction,* which should be represented by *two unknown components*. The equilibrium equations obtained from the free-body diagrams of the multiforce members can then be solved for the various internal forces [Sample Probs. 6.4 and 6.5]. The equilibrium equations can also be used to complete the determination of the reactions at the supports [Sample Prob. 6.6]. Actually, if the frame is *statically determinate and rigid,* the free-body diagrams of the multiforce members could provide as many equations as there are unknown forces (including the reactions) [Sec. 6.11]. However, as suggested above, it is advisable to first consider the free-body diagram of the entire frame to minimize the number of equations that must be solved simultaneously.

Analysis of a machine

To *analyze a machine,* we dismember it and, following the same procedure as for a frame, draw the free-body diagram of each of the multiforce members. The corresponding equilibrium equations yield the *output forces* exerted by the machine in terms of the *input forces* applied to it, as well as the *internal forces* at the various connections [Sec. 6.12; Sample Prob. 6.7].

Review Problems

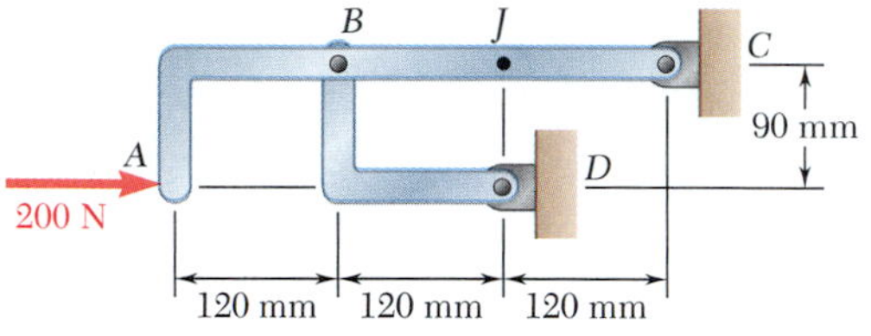

Fig. ***P6.156***

6.156 For the frame and loading shown, determine the force acting on member *ABC* (*a*) at *B*, (*b*) at *C*.

6.157 Determine the force in each member of the truss shown. State whether each member is in tension or compression.

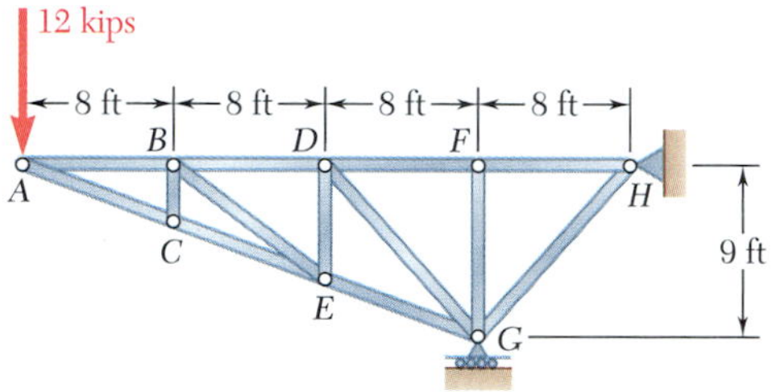

Fig. P6.157

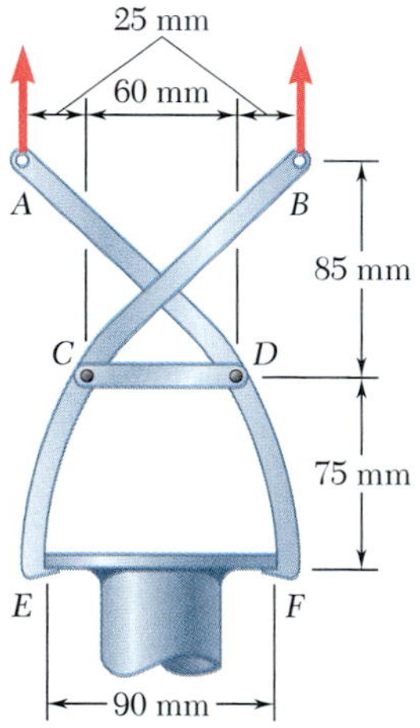

Fig. ***P6.158***

6.158 The tongs shown are used to apply a total upward force of 45 kN on a pipe cap. Determine the forces exerted at *D* and *F* on tong *ADF*.

6.159 A stadium roof truss is loaded as shown. Determine the force in members *BC*, *BH*, and *GH*.

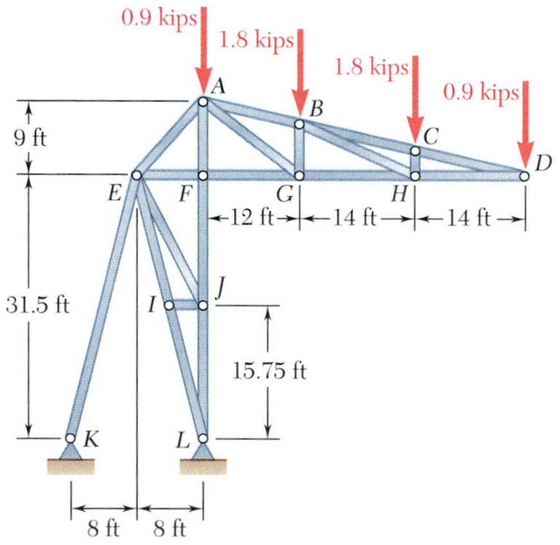

Fig. P6.159 and ***P6.160***

6.160 A stadium roof truss is loaded as shown. Determine the force in members *EJ*, *FJ*, and *EI*.

6.161 For the frame and loading shown, determine the components of the forces acting on member *DABC* at *B* and at *D*.

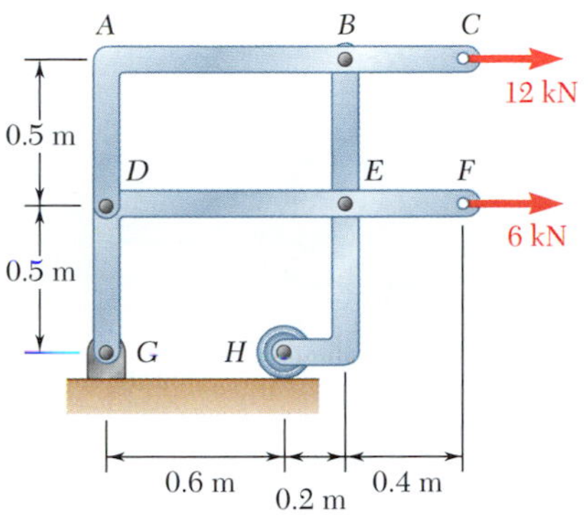

Fig. P6.161

6.162 Using the method of joints, determine the force in each member of the truss shown. State whether each member is in tension or compression.

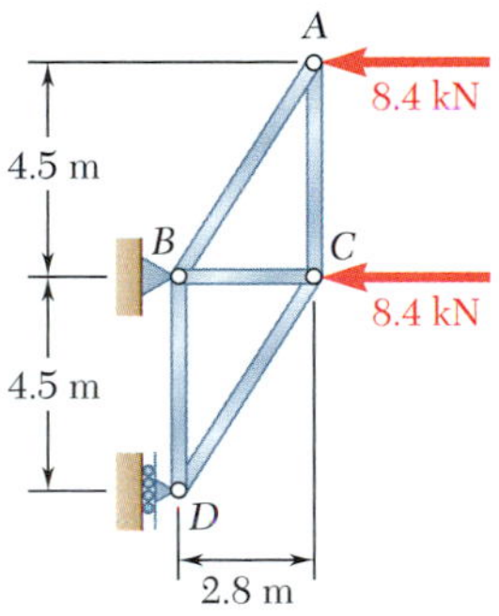

Fig. P6.162

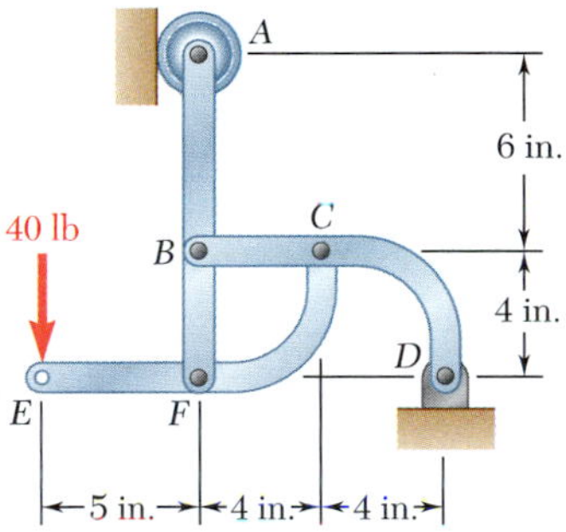

Fig. *P6.163*

6.163 For the frame and loading shown, determine the components of the forces acting on member *CFE* at *C* and at *F*.

6.164 A Mansard roof truss is loaded as shown. Determine the force in members *DF*, *DG*, and *EG*.

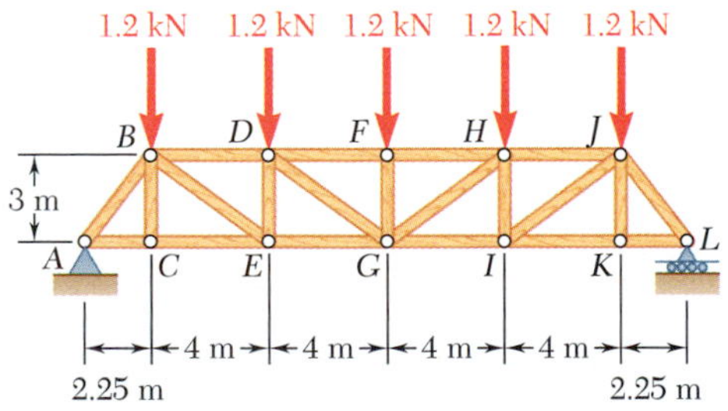

Fig. P6.164 and P6.165

6.165 A Mansard roof truss is loaded as shown. Determine the force in members *GI*, *HI*, and *HJ*.

6.166 Rod CD is fitted with a collar at D that can be moved along rod AB, which is bent in the shape of a circular arc. For the position when $\theta = 30°$, determine (*a*) the force in rod CD, (*b*) the reaction at B.

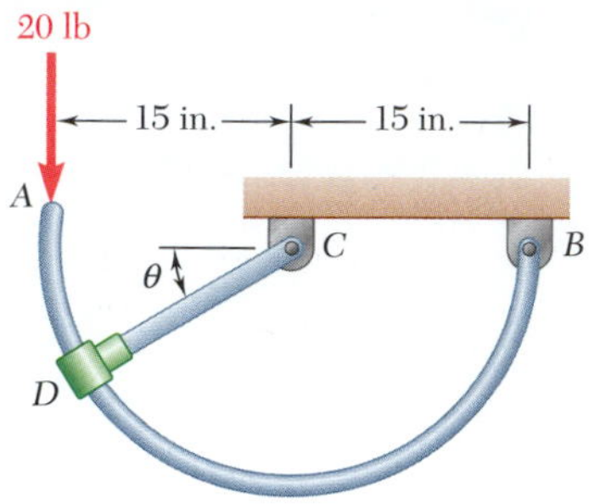

Fig. P6.166

6.167 A log weighing 800 lb is lifted by a pair of tongs as shown. Determine the forces exerted at E and at F on tong DEF.

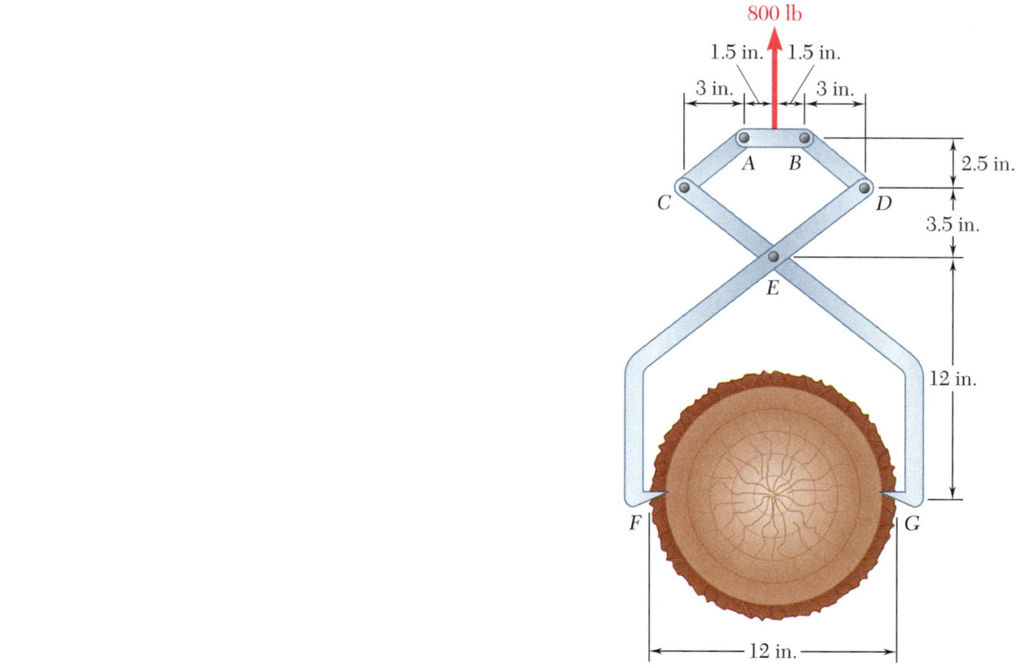

Fig. P6.167

Computer Problems

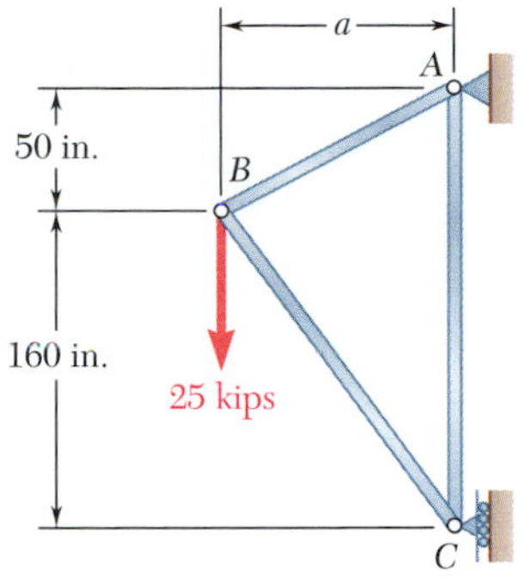

Fig. P6.C1

6.C1 For the loading shown, determine the force in each member of the truss as a function of the dimension a. Plot the force in each member for 40 in. $\leq a \leq$ 240 in., with tensile forces plotted as positive and compressive forces as negative.

6.C2 In the Fink truss shown, a single 3.75-kip load is to be applied to the top chord of the truss at one of the numbered joints. Calculate the force in member CD as the load is successively applied at joints 1, 2, . . . , 9.

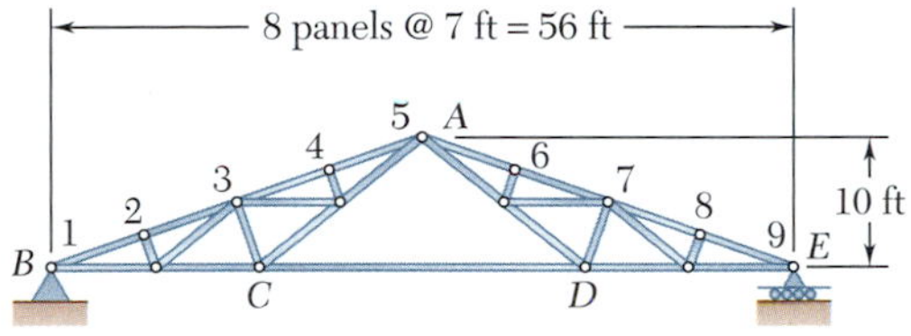

Fig. P6.C2

6.C3 The floor of a bridge will rest on stringers that will be simply supported by transverse floor beams, as in Fig 6.3. The ends of the beams will be connected to the upper joints of two trusses, one of which is shown in Fig. P6.C3. As part of the design of the bridge, it is desired to simulate the effect on this truss of driving a 3-kip truck over the bridge. Knowing that the distance between the truck's axles is b = 7.5 ft and assuming that the weight of the truck is equally distributed over its four wheels, use computational software to calculate and plot the forces created by the truck in members BH and GH as a function of x for $0 \leq x \leq 57.5$ ft. From the results obtained, determine (a) the maximum tensile force in BH, (b) the maximum compressive force in BH, (c) the maximum tensile force in GH. Indicate in each case the corresponding value of x.

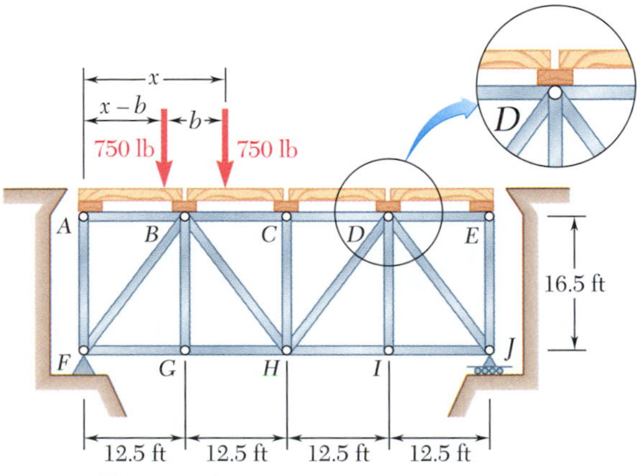

Fig. P6.C3

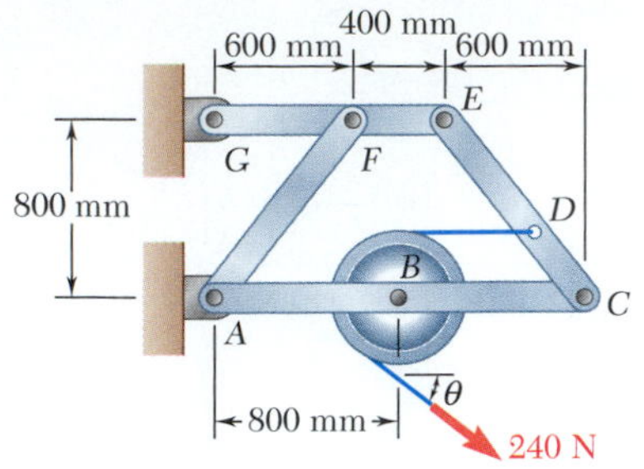

Fig. P6.C4

6.C4 Knowing that the radius of the pulley is 300 mm, plot the magnitude of the reactions at A and G as functions of θ for $0 \leq \theta \leq 90°$.

6.C5 The design of a robotic system calls for the two-rod mechanism shown. Rods AC and BD are connected by a slider block D as shown. Neglecting the effect of friction, calculate and plot the couple $\mathbf{M}_A$ required to hold the rods in equilibrium for values of θ from 0 to 120°. For the same values of θ, calculate and plot the magnitude of the force $\mathbf{F}$ exerted by rod AC on the slider block.

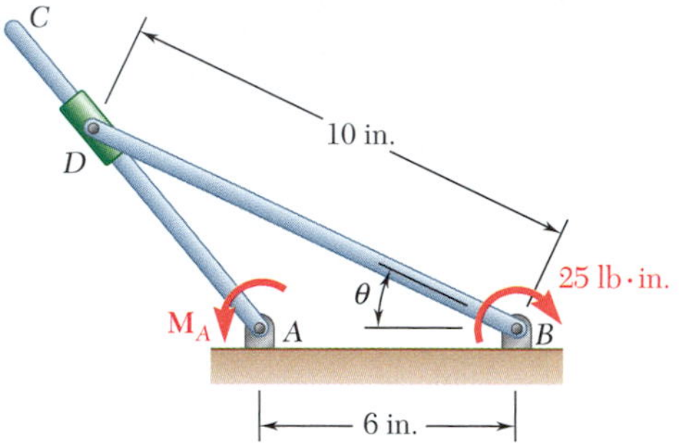

Fig. P6.C5

6.C6 The magnitude of the force $\mathbf{P}$ applied to the piston of an engine system during one revolution of crank AB is shown in the figure. Plot the magnitude of the couple $\mathbf{M}$ required to hold the system in equilibrium as a function of θ for $0 \leq \theta \leq 2\pi$.

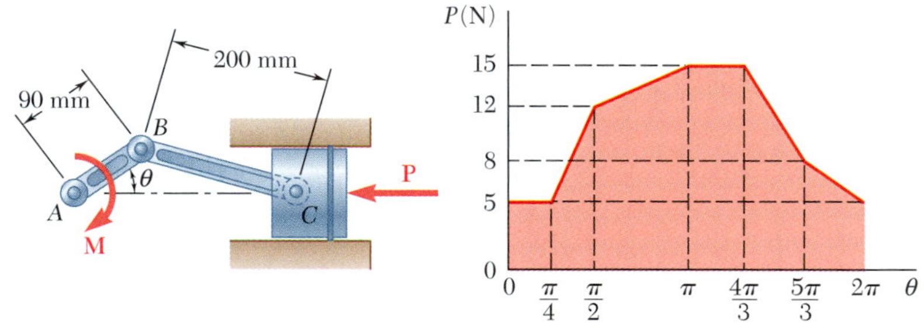

Fig. P6.C6

6.C7 The shelf ABC is held horizontally by a self-locking brace that consists of two parts BDE and EDF hinged at E and bearing against each other at D. Determine the force $\mathbf{P}$ required to release the brace as a function of the angle θ. Knowing that the mass of the shelf is 18 kg, plot the magnitude of $\mathbf{P}$ as a function of θ for $0° \leq \theta \leq 90°$.

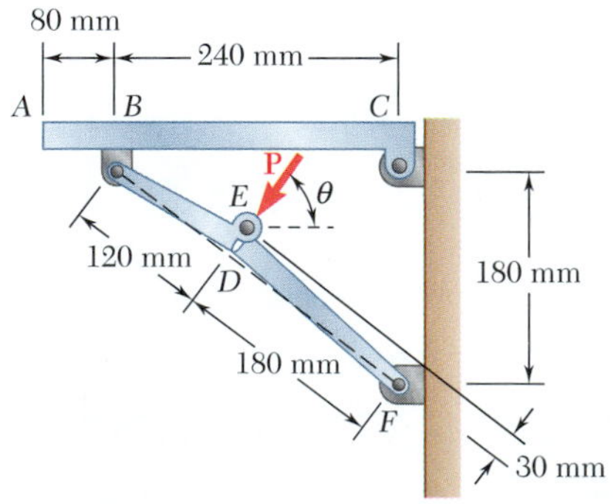

Fig. P6.C7

6.C8 In the mechanism shown, the position of boom AC is controlled by arm BD. For the loading shown, calculate and plot the reaction at A and the couple **M** required to hold the system in equilibrium as functions of θ for $-30° \leq \theta \leq 90°$. As part of the design process of the mechanism, determine (*a*) the value of θ for which M is maximum and the corresponding value of M, (*b*) the value of θ for which the reaction at A is maximum and the corresponding magnitude of this reaction.

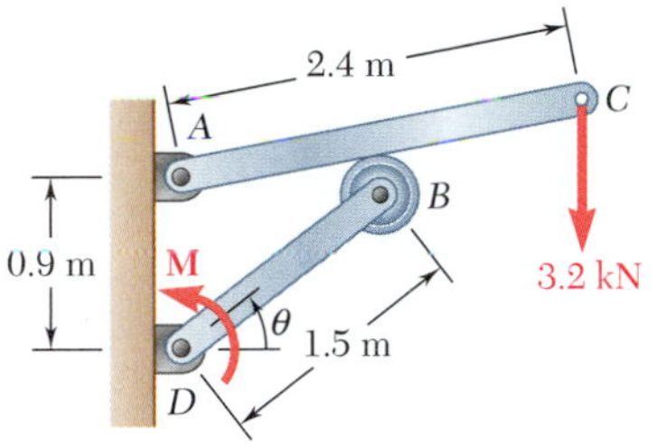

Fig. P6.C8

6.C9 Rod CD is attached to collar D and passes through a collar welded to end B of lever AB. As an initial step in the design of lever AB, use computational software to calculate and plot the magnitude M of the couple required to hold the system in equilibrium as a function of θ for $15° \leq \theta \leq 90°$. Determine the value of θ for which M is minimum and the corresponding value of M.

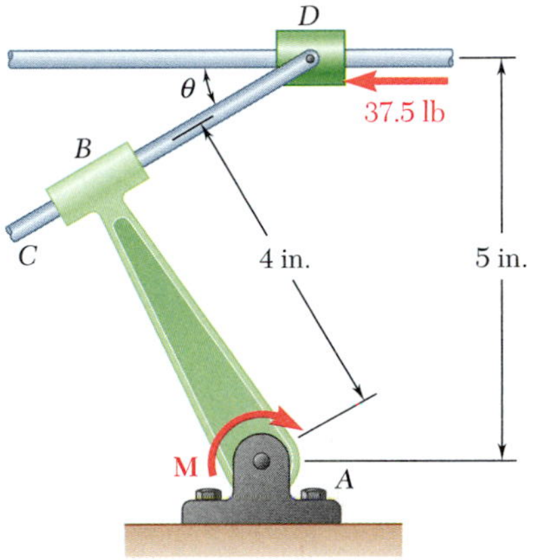

Fig. P6.C9

***6.C10** The tree trimmer shown is used to prune a 25-mm-diameter branch. Knowing that the tension in the rope is 70 N and neglecting the radii of the pulleys, plot the magnitude of the force exerted on the branch by blade BC as a function of θ for $0 \leq \theta \leq 90°$, where the cutting edge of the blade is vertical and just contacts the branch when $\theta = 0$. (*Hint:* Assume that the line of action of the force acting on the branch passes through the center of the branch and is perpendicular to the blade and that portions AD and AE of the rope are parallel to a line drawn through points A and E.)

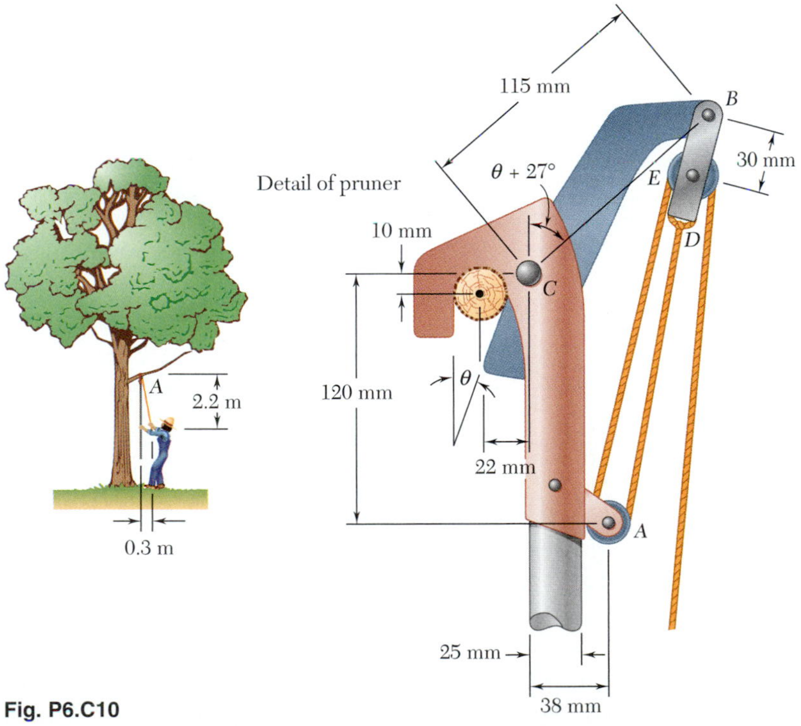

Fig. P6.C10

C H A P T E R

7

Forces in Beams and Cables

The Gateshead Millennium Bridge for pedestrians in Newcastle-upon-Tyne, England is unique in both appearance and how it rotates to permit the passage of taller ships. The determination of the internal forces in a member, such as those in the curved footpath and supporting arch of the bridge, is the focus of this chapter.

FORCES IN BEAMS AND CABLES

*7.1. INTRODUCTION

In preceding chapters, two basic problems involving structures were considered: (1) determining the external forces acting on a structure (Chap. 4) and (2) determining the forces which hold together the various members forming a structure (Chap. 6). The problem of determining the internal forces which hold together the various parts of a given member will now be considered.

We will first analyze the internal forces in the members of a frame, such as the crane considered in Secs. 6.1 and 6.10, noting that whereas the internal forces in a straight two-force member can produce only *tension* or *compression* in that member, the internal forces in any other type of member usually produce *shear* and *bending* as well.

Most of this chapter will be devoted to the analysis of the internal forces in two important types of engineering structures, namely,

1. *Beams,* which are usually long, straight prismatic members designed to support loads applied at various points along the member.
2. *Cables,* which are flexible members capable of withstanding only tension and are designed to support either concentrated or distributed loads. Cables are used in many engineering applications, such as suspension bridges and transmission lines.

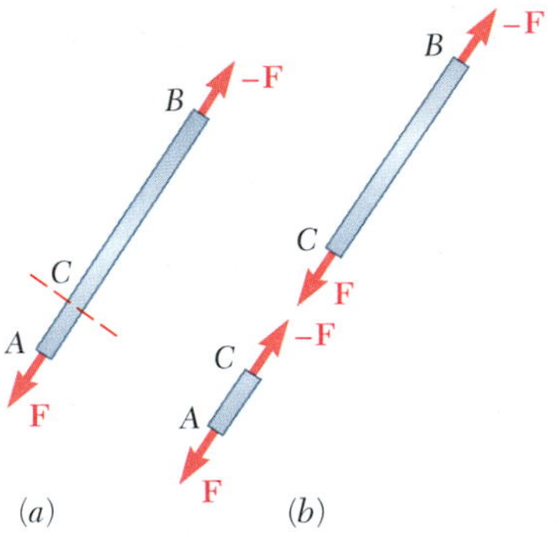

Fig. 7.1

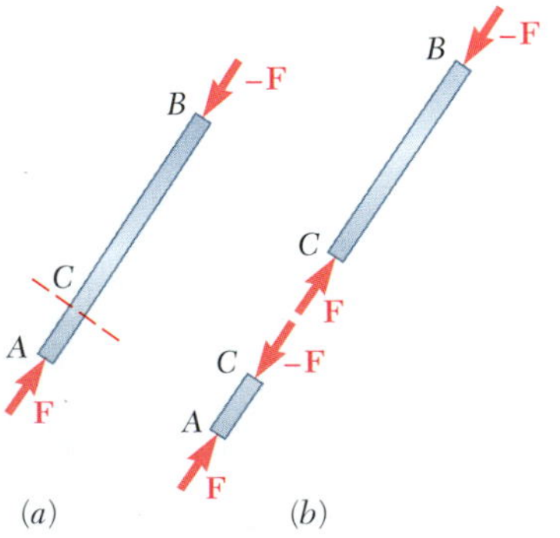

Fig. 7.2

*7.2. INTERNAL FORCES IN MEMBERS

Let us first consider a *straight two-force member AB* (Fig. 7.1*a*). From Sec. 4.6, we know that the forces **F** and −**F** acting at *A* and *B*, respectively, must be directed along *AB* in opposite sense and have the same magnitude *F*. Now, let us cut the member at *C*. To maintain the equilibrium of the free bodies *AC* and *CB* thus obtained, we must apply to *AC* a force −**F** equal and opposite to **F**, and to *CB* a force **F** equal and opposite to −**F** (Fig. 7.1*b*). These new forces are directed along *AB* in opposite sense and have the same magnitude *F*. Since the two parts *AC* and *CB* were in equilibrium before the member was cut, *internal forces* equivalent to these new forces must have existed in the member itself. We conclude that in the case of a straight two-force member, the internal forces that the two portions of the member exert on each other are equivalent to *axial forces.* The common magnitude *F* of these forces does not depend upon the location of the section *C* and is referred to as the *force in member AB.* In the case considered, the member is in tension and will elongate under the action of the internal forces. In the case represented in Fig. 7.2, the member is in compression and will decrease in length under the action of the internal forces.

Next, let us consider a *multiforce member.* Take, for instance, member *AD* of the crane analyzed in Sec. 6.10. This crane is shown again in Fig. 7.3*a*, and the free-body diagram of member *AD* is drawn in Fig. 7.3*b*. We now cut member *AD* at *J* and draw a free-body diagram for each of the portions *JD* and *AJ* of the member (Fig. 7.3*c* and *d*). Considering the free body *JD*, we find that its equilibrium will

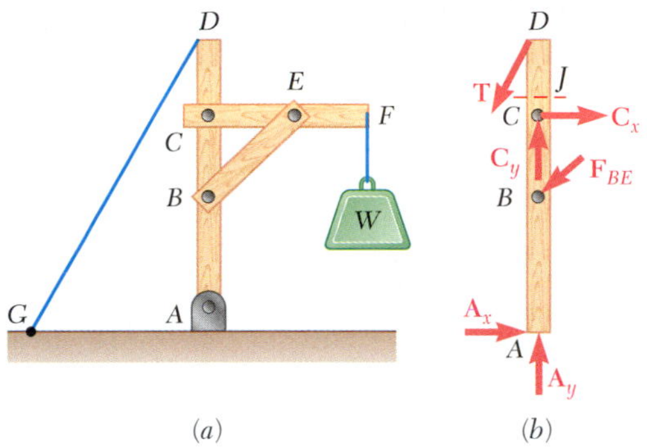

Fig. 7.3

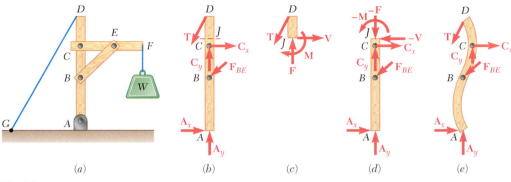

be maintained if we apply at *J* a force **F** to balance the vertical component of **T**, a force **V** to balance the horizontal component of **T**, and a couple **M** to balance the moment of **T** about *J*. Again we conclude that internal forces must have existed at *J* before member *AD* was cut. The internal forces acting on the portion *JD* of member *AD* are equivalent to the force-couple system shown in Fig. 7.3*c*. According to Newton's third law, the internal forces acting on *AJ* must be equivalent to an equal and opposite force-couple system, as shown in Fig. 7.3*d*. It is clear that the action of the internal forces in member *AD* *is not limited to producing tension or compression* as in the case of straight two-force members; the internal forces *also produce shear and bending*. The force **F** is an *axial force;* the force **V** is called a *shearing force;* and the moment **M** of the couple is known as the *bending moment at J*. We note that when determining internal forces in a member, we should clearly indicate on which portion of the member the forces are supposed to act. The deformation which will occur in member *AD* is sketched in Fig. 7.3*e*. The actual analysis of such a deformation is part of the study of mechanics of materials.

It should be noted that in a *two-force member which is not straight,* the internal forces are also equivalent to a force-couple system. This is shown in Fig. 7.4, where the two-force member *ABC* has been cut at *D*.

Photo 7.1 The design of the shaft of a circular saw must account for the internal forces resulting from the forces applied to the teeth of the blade. At a given point in the shaft, these internal forces are equivalent to a force-couple system consisting of axial and shearing forces and a couple representing the bending and torsional moments.

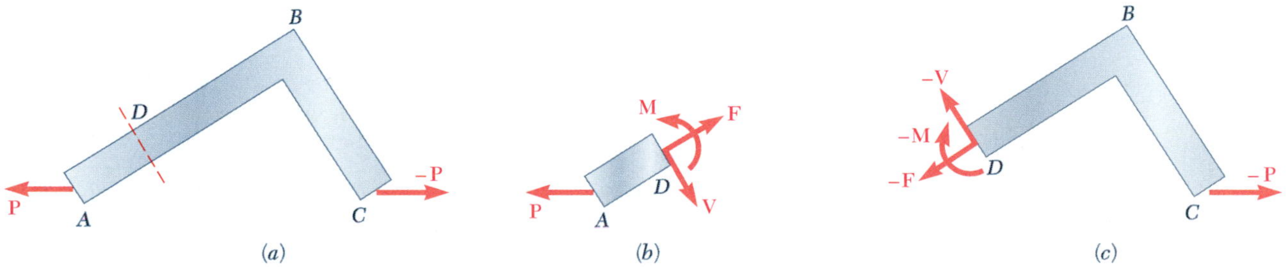

Fig. 7.4

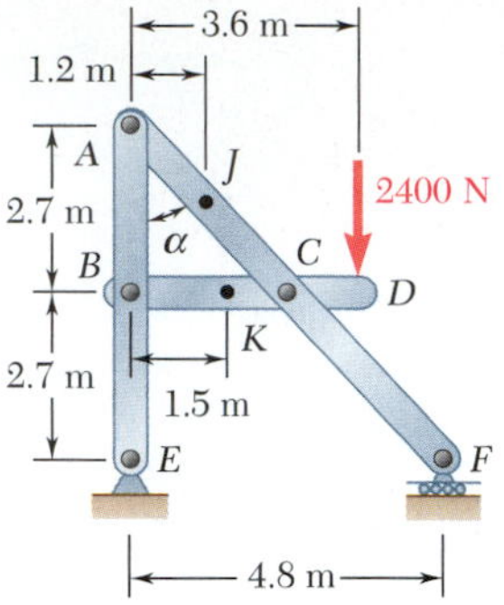

SAMPLE PROBLEM 7.1

In the frame shown, determine the internal forces (*a*) in member *ACF* at point *J*, (*b*) in member *BCD* at point *K*. This frame has been previously considered in Sample Prob. 6.5.

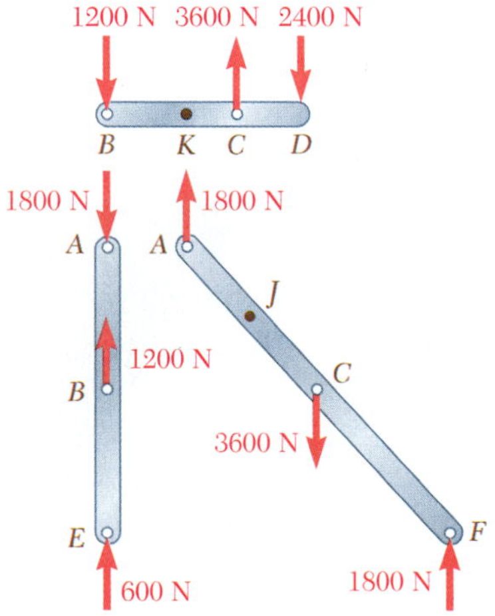

SOLUTION

Reactions and Forces at Connections. The reactions and the forces acting on each member of the frame are determined; this has been previously done in Sample Prob. 6.5, and the results are repeated here.

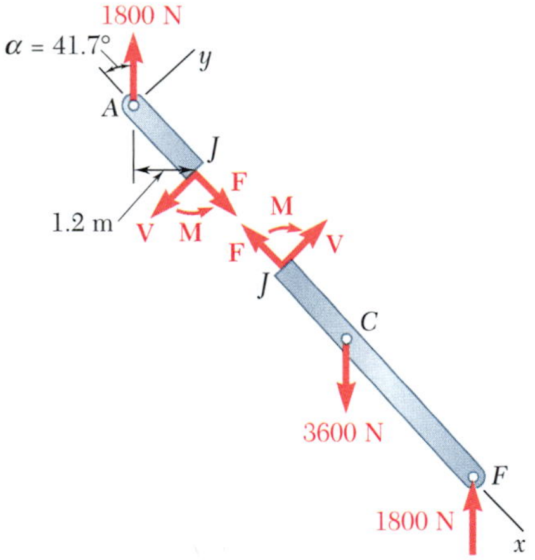

***a.* Internal Forces at *J*.** Member *ACF* is cut at point *J*, and the two parts shown are obtained. The internal forces at *J* are represented by an equivalent force-couple system and can be determined by considering the equilibrium of either part. Considering the *free body AJ*, we write

$+\circlearrowleft\Sigma M_J = 0$: $\quad -(1800\text{ N})(1.2\text{ m}) + M = 0$

$M = +2160\text{ N}\cdot\text{m}$ $\qquad$ $\mathbf{M} = 2160\text{ N}\cdot\text{m}\circlearrowleft$ ◀

$+\searrow\Sigma F_x = 0$: $\quad F - (1800\text{ N})\cos 41.7° = 0$

$F = +1344\text{ N}$ $\qquad$ $\mathbf{F} = 1344\text{ N}\searrow$ ◀

$+\nearrow\Sigma F_y = 0$: $\quad -V + (1800\text{ N})\sin 41.7° = 0$

$V = +1197\text{ N}$ $\qquad$ $\mathbf{V} = 1197\text{ N}\swarrow$ ◀

The internal forces at *J* are therefore equivalent to a couple **M**, an axial force **F**, and a shearing force **V**. The internal force-couple system acting on part *JCF* is equal and opposite.

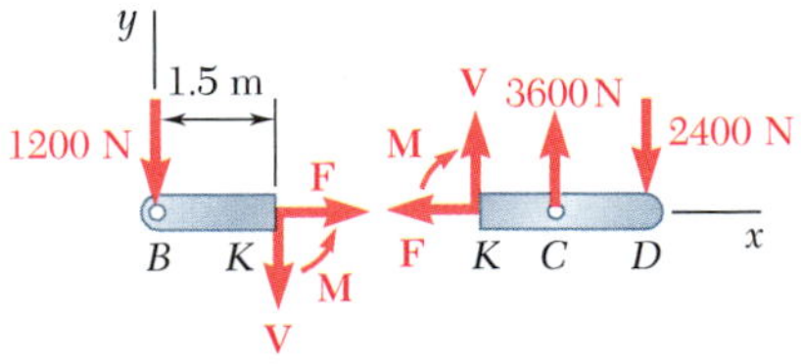

***b.* Internal Forces at *K*.** We cut member *BCD* at *K* and obtain the two parts shown. Considering the *free body BK*, we write

$+\circlearrowleft\Sigma M_K = 0$: $\quad (1200\text{ N})(1.5\text{ m}) + M = 0$

$M = -1800\text{ N}\cdot\text{m}$ $\qquad$ $\mathbf{M} = 1800\text{ N}\cdot\text{m}\circlearrowright$ ◀

$\xrightarrow{+}\Sigma F_x = 0$: $\quad F = 0$ $\qquad$ $\mathbf{F} = 0$ ◀

$+\uparrow\Sigma F_y = 0$: $\quad -1200\text{ N} - V = 0$

$V = -1200\text{ N}$ $\qquad$ $\mathbf{V} = 1200\text{ N}\uparrow$ ◀

SOLVING PROBLEMS ON YOUR OWN

In this lesson you learned to determine the internal forces in the member of a frame. The internal forces at a given point in a *straight two-force member* reduce to an axial force, but in all other cases, they are equivalent to a *force-couple system* consisting of an *axial force* **F**, a *shearing force* **V**, and a couple **M** representing the *bending moment* at that point.

To determine the internal forces at a given point *J* of the member of a frame, you should take the following steps.

1. Draw a free-body diagram of the entire frame, and use it to determine as many of the reactions at the supports as you can.

2. Dismember the frame, and draw a free-body diagram of each of its members. Write as many equilibrium equations as are necessary to find all the forces acting on the member on which point *J* is located.

3. Cut the member at point J, and draw a free-body diagram of each of the two portions of the member that you have obtained, applying to each portion at point *J* the force components and couple representing the internal forces exerted by the other portion. Note that these force components and couples are equal in magnitude and opposite in sense.

4. Select one of the two free-body diagrams you have drawn, and use it to write three equilibrium equations for the corresponding portion of the member.

a. Summing moments about J and equating them to zero will yield the bending moment at point *J*.

b. Summing components in directions parallel and perpendicular to the member at *J* and equating them to zero will yield, respectively, the axial and shearing force.

5. When recording your answers, be sure to specify the portion of the member you have used, since the forces and couples acting on the two portions have opposite senses.

Since the solutions of the problems in this lesson require the determination of the forces exerted on each other by the various members of a frame, be sure to review the methods used in Chap. 6 to solve this type of problem. When frames involve pulleys and cables, for instance, remember that the forces exerted by a pulley on the member of the frame to which it is attached have the same magnitude and direction as the forces exerted by the cable on the pulley [Prob. 6.91].

Problems

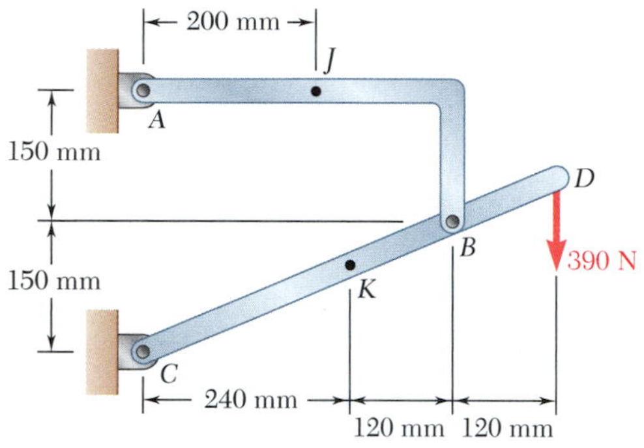

Fig. P7.5 and *P7.6*

7.1 and 7.2 Determine the internal forces (axial force, shearing force, and bending moment) at point J of the structure indicated:
7.1 Frame and loading of Prob. 6.77.
7.2 Frame and loading of Prob. 6.76.

7.3 For the frame and loading of Prob. 6.80, determine the internal forces at a point J located halfway between points A and B.

7.4 For the frame and loading of Prob. 6.101, determine the internal forces at a point J located halfway between points A and B.

7.5 Determine the internal forces at point J of the structure shown.

7.6 Determine the internal forces at point K of the structure shown.

7.7 A semicircular rod is loaded as shown. Determine the internal forces at point J.

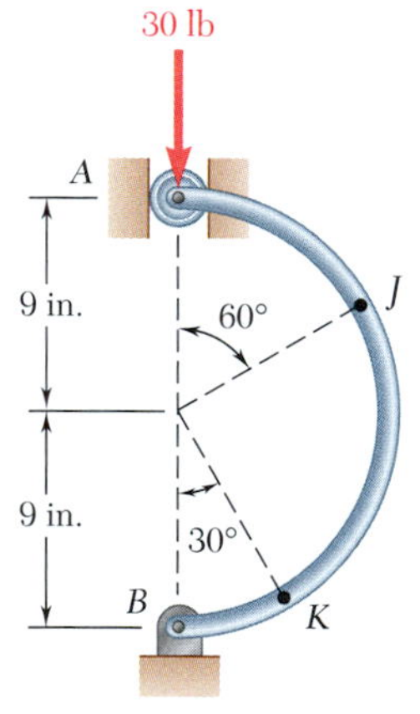

Fig. P7.7 and *P7.8*

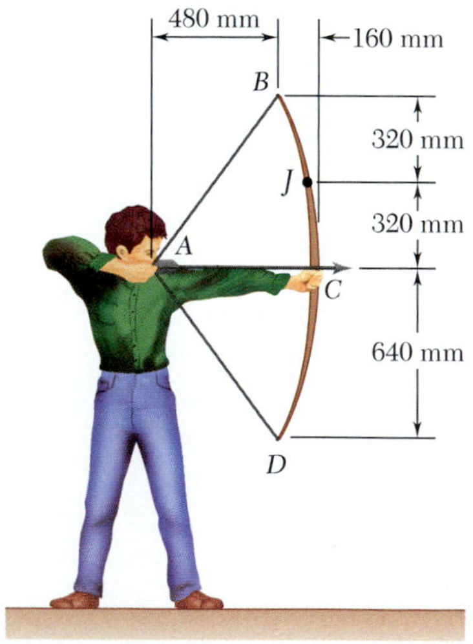

Fig. P7.9

7.8 A semicircular rod is loaded as shown. Determine the internal forces at point K.

7.9 An archer aiming at a target is pulling with a 210-N force on the bowstring. Assuming that the shape of the bow can be approximated by a parabola, determine the internal forces at point J.

7.10 For the bow of Prob. 7.9, determine the magnitude and location of the maximum (*a*) axial force, (*b*) shearing force, (*c*) bending moment.

7.11 A semicircular rod is loaded as shown. Determine the internal forces at point J knowing that $\theta = 30°$.

7.12 A semicircular rod is loaded as shown. Determine the magnitude and location of the maximum bending moment in the rod.

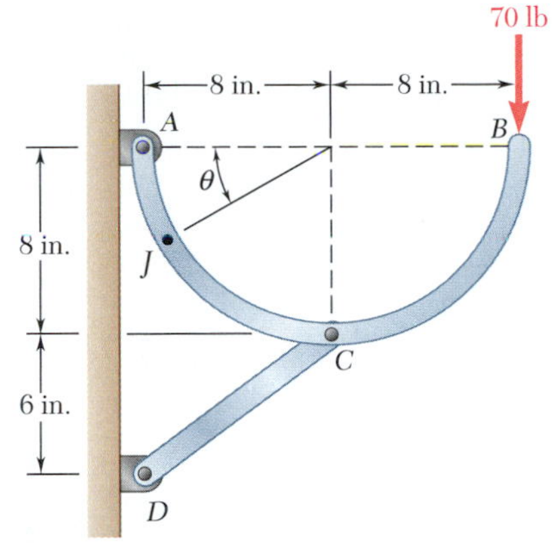

Fig. P7.11 and *P7.12*

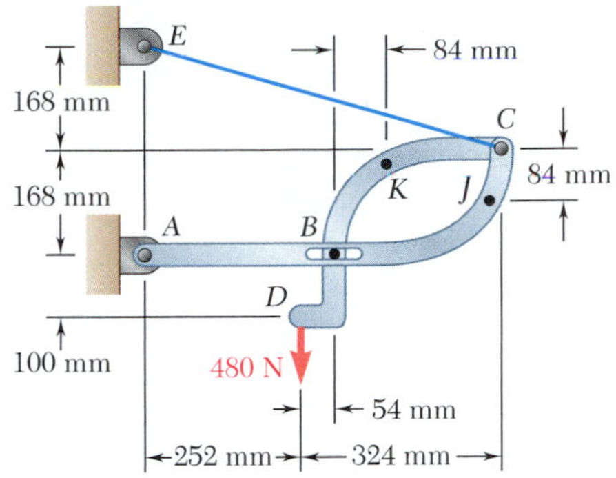

Fig. P7.13 and P7.14

7.13 Two members, each consisting of straight and 168-mm-radius quarter-circle portions, are connected as shown and support a 480-N load at D. Determine the internal forces at point J.

7.14 Two members, each consisting of straight and 168-mm-radius quarter-circle portions, are connected as shown and support a 480-N load at D. Determine the internal forces at point K.

7.15 Knowing that the radius of each pulley is 7.2 in. and neglecting friction, determine the internal forces at point J of the frame shown.

7.16 Knowing that the radius of each pulley is 7.2 in. and neglecting friction, determine the internal forces at point K of the frame shown.

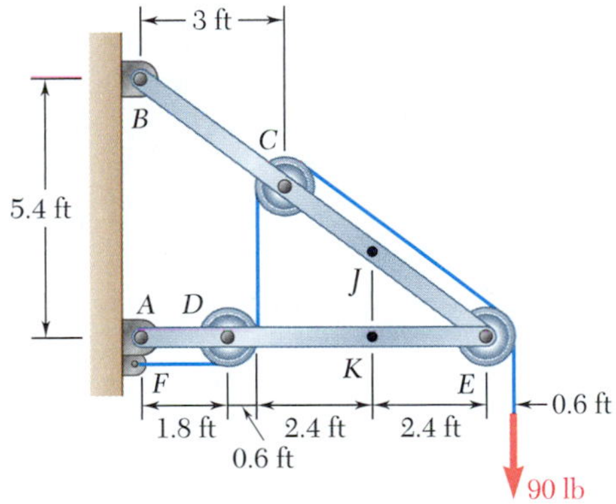

Fig. P7.15 and P7.16

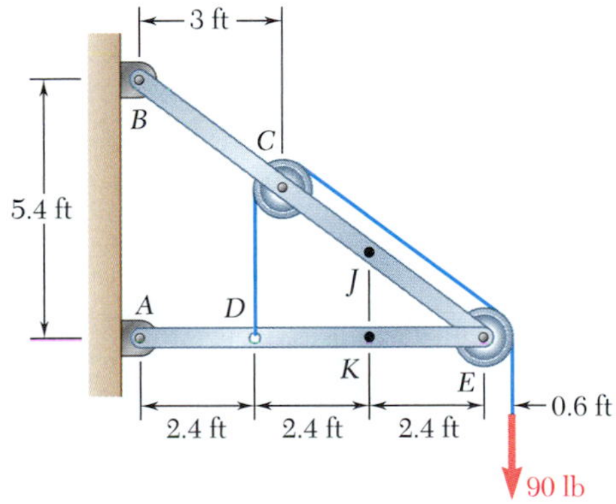

Fig. *P7.17* and *P7.18*

7.17 Knowing that the radius of each pulley is 7.2 in. and neglecting friction, determine the internal forces at point J of the frame shown.

7.18 Knowing that the radius of each pulley is 7.2 in. and neglecting friction, determine the internal forces at point K of the frame shown.

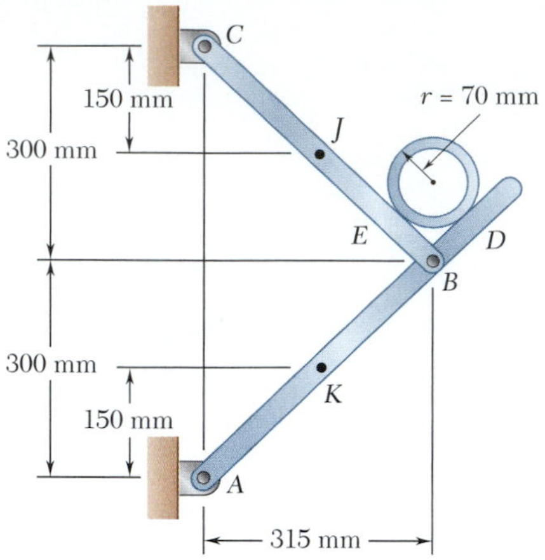

Fig. P7.19 and P7.20

7.19 A 140-mm-diameter pipe is supported every 3 m by a small frame consisting of two members as shown. Knowing that the combined mass per unit length of the pipe and its contents is 28 kg/m and neglecting the effect of friction, determine the internal forces at point *J*.

7.20 A 140-mm-diameter pipe is supported every 3 m by a small frame consisting of two members as shown. Knowing that the combined mass per unit length of the pipe and its contents is 28 kg/m and neglecting the effect of friction, determine the internal forces at point *K*.

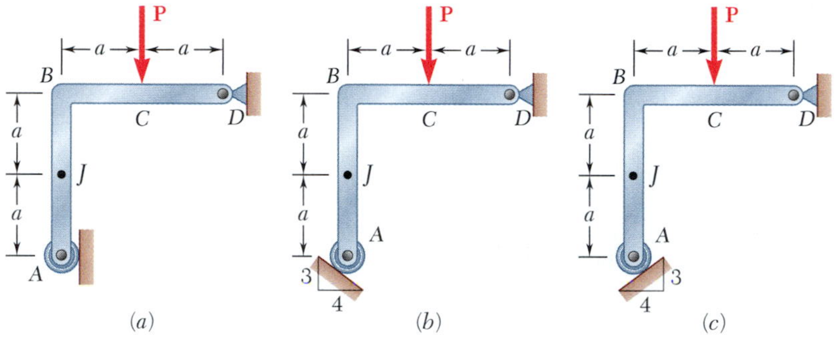

Fig. P7.21

7.21 and 7.22 A force **P** is applied to a bent rod which is supported by a roller and a pin and bracket. For each of the three cases shown, determine the internal forces at point *J*.

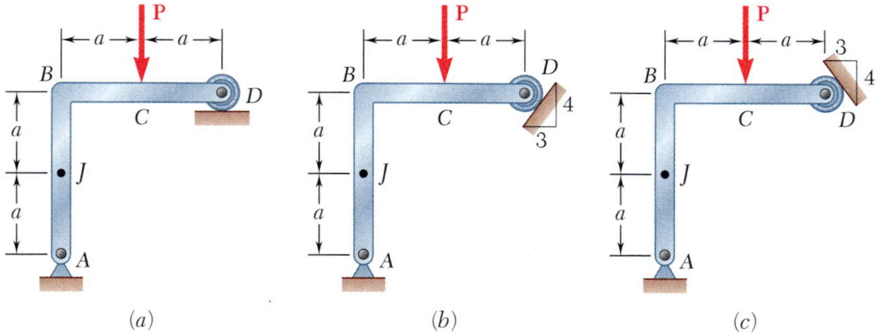

Fig. P7.22

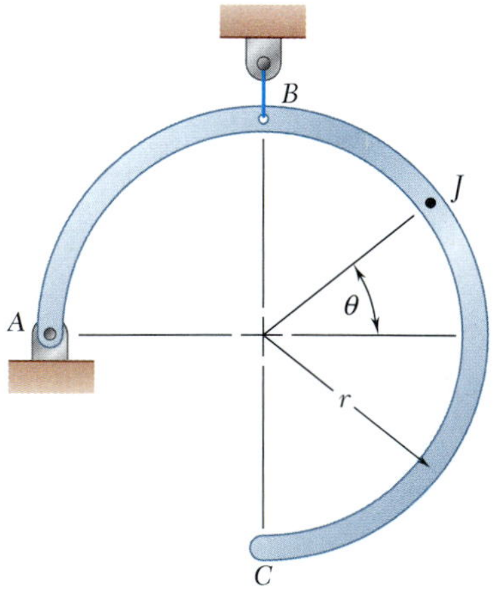

Fig. P7.23 and P7.24

7.23 A rod of weight *W* and uniform cross section is bent into the circular arc of radius *r* shown. Determine the bending moment at point *J* when $\theta = 30°$.

7.24 A rod of weight *W* and uniform cross section is bent into the circular arc of radius *r* shown. Determine the bending moment at point *J* when $\theta = 120°$.

7.25 and *7.26* A quarter-circular rod of weight W and uniform cross section is supported as shown. Determine the bending moment at point J when $\theta = 30°$.

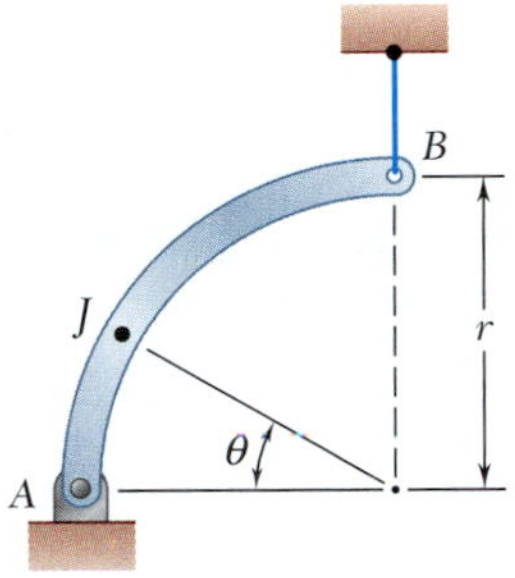

Fig. P7.25

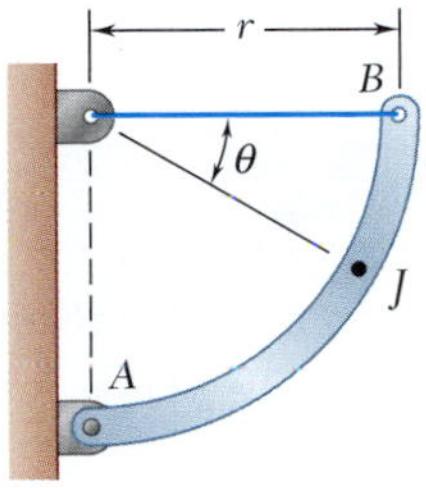

Fig. *P7.26*

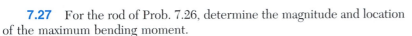

7.27 For the rod of Prob. 7.26, determine the magnitude and location of the maximum bending moment.

7.28 For the rod of Prob. 7.25, determine the magnitude and location of the maximum bending moment.

BEAMS

*7.3. VARIOUS TYPES OF LOADING AND SUPPORT

A structural member designed to support loads applied at various points along the member is known as a *beam*. In most cases, the loads are perpendicular to the axis of the beam and will cause only shear and bending in the beam. When the loads are not at a right angle to the beam, they will also produce axial forces in the beam.

Beams are usually long, straight prismatic bars. Designing a beam for the most effective support of the applied loads is a two-part process: (1) determining the shearing forces and bending moments produced by the loads and (2) selecting the cross section best suited to resist the shearing forces and bending moments determined in the first part. Here we are concerned with the first part of the problem of beam design. The second part belongs to the study of mechanics of materials.

A beam can be subjected to *concentrated loads* $\mathbf{P}_1$, $\mathbf{P}_2$, . . . , expressed in newtons, pounds, or their multiples kilonewtons and kips (Fig. 7.5*a*), to a *distributed load* w, expressed in N/m, kN/m, lb/ft, or kips/ft (Fig. 7.5*b*), or to a combination of both. When the load w per unit length has a constant value over part of the beam (as between A and B in Fig. 7.5*b*), the load is said to be *uniformly distributed* over that part of the beam. The determination of the reactions at the supports is considerably simplified if distributed loads are replaced by equivalent concentrated loads, as explained in Sec. 5.8. This substitution, however, should not be performed, or at least should be performed with care, when internal forces are being computed (see Sample Prob. 7.3).

Beams are classified according to the way in which they are supported. Several types of beams frequently used are shown in

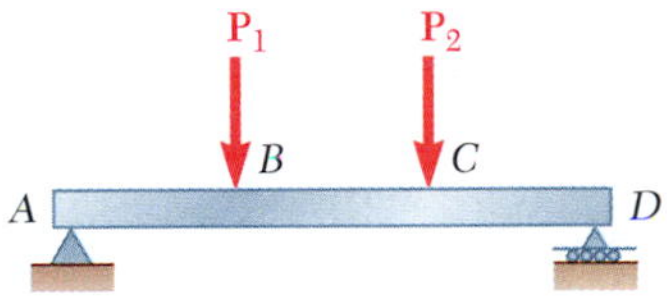

(*a*) Concentrated loads

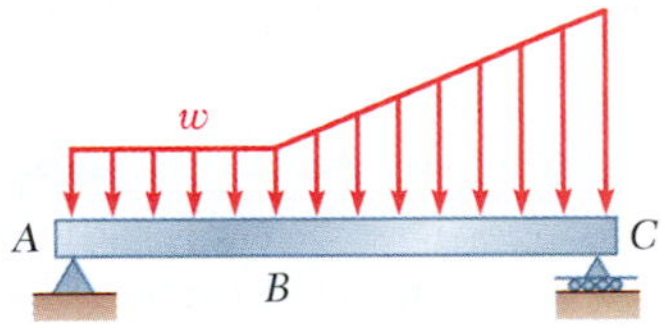

(*b*) Distributed load

Fig. 7.5

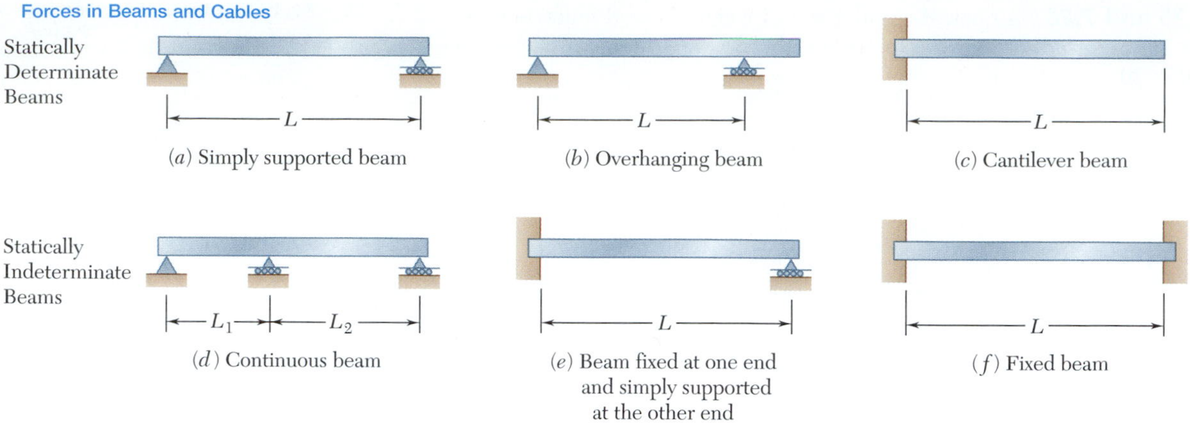

Fig. 7.6

Fig. 7.6. The distance L between supports is called the *span*. It should be noted that the reactions will be determinate if the supports involve only three unknowns. If more unknowns are involved, the reactions will be statically indeterminate and the methods of statics will not be sufficient to determine the reactions; the properties of the beam with regard to its resistance to bending must then be taken into consideration. Beams supported by two rollers are not shown here; they are only partially constrained and will move under certain loadings.

Sometimes two or more beams are connected by hinges to form a single continuous structure. Two examples of beams hinged at a point H are shown in Fig. 7.7. It will be noted that the reactions at the supports involve four unknowns and cannot be determined from the free-body diagram of the two-beam system. They can be determined, however, by considering the free-body diagram of each beam separately; six unknowns are involved (including two force components at the hinge), and six equations are available.

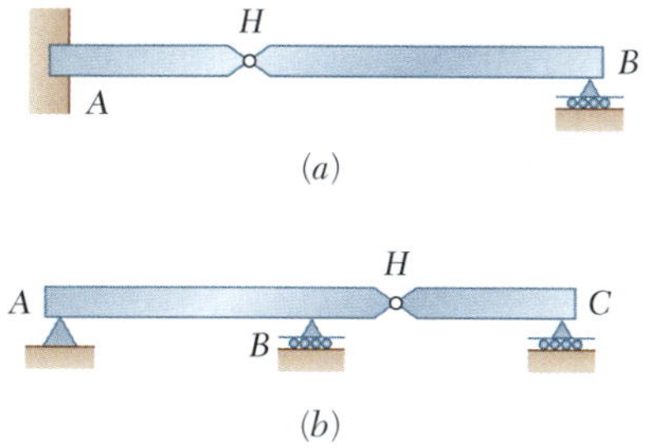

Fig. 7.7

*7.4. SHEAR AND BENDING MOMENT IN A BEAM

Consider a beam AB subjected to various concentrated and distributed loads (Fig. 7.8*a*). We propose to determine the shearing force and bending moment at any point of the beam. In the example considered here, the beam is simply supported, but the method used could be applied to any type of statically determinate beam.

First we determine the reactions at A and B by choosing the entire beam as a free body (Fig. 7.8*b*); writing $\Sigma M_A = 0$ and $\Sigma M_B = 0$, we obtain, respectively, $\mathbf{R}_B$ and $\mathbf{R}_A$.

To determine the internal forces at C, we cut the beam at C and draw the free-body diagrams of the portions AC and CB of the beam (Fig. 7.8*c*). Using the free-body diagram of AC, we can determine the shearing force $\mathbf{V}$ at C by equating to zero the sum of the vertical components of all forces acting on AC. Similarly, the bending moment $\mathbf{M}$ at C can be found by equating to zero the sum of the moments about C of all forces and couples acting on AC. Alternatively, we could use the free-body diagram of CB† and determine the shearing force $\mathbf{V}'$ and the bending moment $\mathbf{M}'$ by equating to zero the sum of the

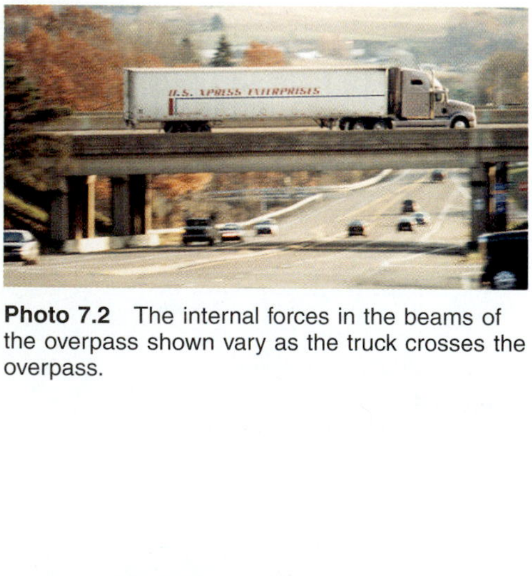

Photo 7.2 The internal forces in the beams of the overpass shown vary as the truck crosses the overpass.

†The force and couple representing the internal forces acting on CB will now be denoted by $\mathbf{V}'$ and $\mathbf{M}'$, rather than by $-\mathbf{V}$ and $-\mathbf{M}$ as done earlier, in order to avoid confusion when applying the sign convention which we are about to introduce.

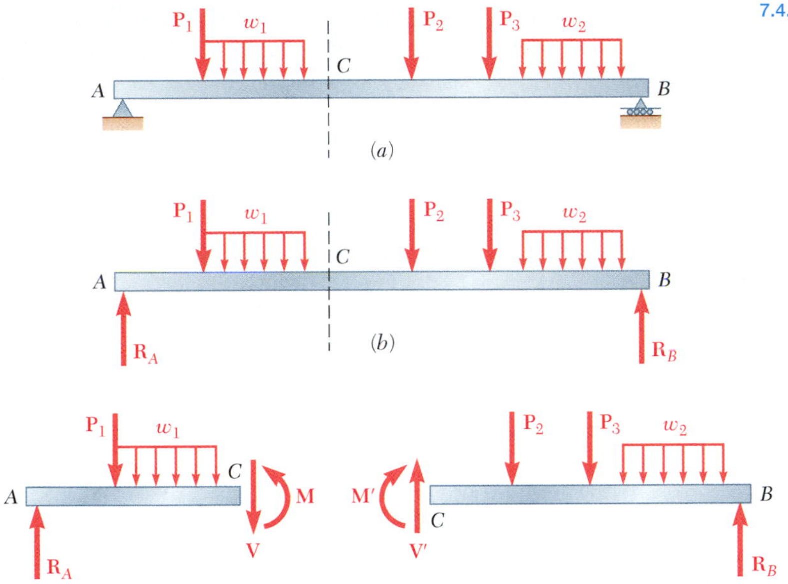

Fig. 7.8

vertical components and the sum of the moments about C of all forces and couples acting on CB. While this choice of free bodies may facilitate the computation of the numerical values of the shearing force and bending moment, it makes it necessary to indicate on which portion of the beam the internal forces considered are acting. If the shearing force and bending moment are to be computed at every point of the beam and efficiently recorded, we must find a way to avoid having to specify every time which portion of the beam is used as a free body. We shall adopt, therefore, the following conventions:

In determining the shearing force in a beam, *it will always be assumed* that the internal forces **V** and **V′** are directed as shown in Fig. 7.8*c*. A positive value obtained for their common magnitude V will indicate that this assumption was correct and that the shearing forces are actually directed as shown. A negative value obtained for V will indicate that the assumption was wrong and that the shearing forces are directed in the opposite way. Thus, only the magnitude V, together with a plus or minus sign, needs to be recorded to define completely the shearing forces at a given point of the beam. The scalar V is commonly referred to as the *shear* at the given point of the beam.

Similarly, *it will always be assumed* that the internal couples **M** and **M′** are directed as shown in Fig. 7.8*c*. A positive value obtained for their magnitude M, commonly referred to as the bending moment, will indicate that this assumption was correct, and a negative value will indicate that it was wrong. Summarizing the sign conventions we have presented, we state:

The shear V and the bending moment M at a given point of a beam are said to be positive when the internal forces and couples acting on each portion of the beam are directed as shown in Fig. 7.9a.

These conventions can be more easily remembered if we note that:

1. *The shear at C is positive when the* **external** *forces (loads and reactions) acting on the beam tend to shear off the beam at C as indicated in Fig. 7.9b.*

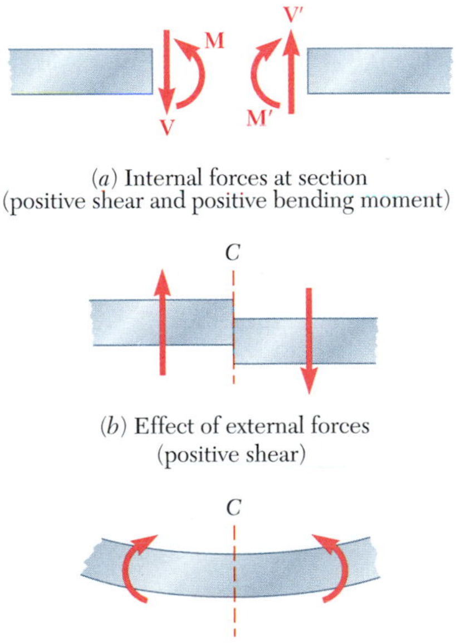

(*a*) Internal forces at section (positive shear and positive bending moment)

(*b*) Effect of external forces (positive shear)

(*c*) Effect of external forces (positive bending moment)

Fig. 7.9

2. *The bending moment at C is positive when the* **external** *forces acting on the beam tend to bend the beam at C as indicated in Fig. 7.9c.*

It may also help to note that the situation described in Fig. 7.9, in which the values of the shear and of the bending moment are positive, is precisely the situation which occurs in the left half of a simply supported beam carrying a single concentrated load at its midpoint. This particular example is fully discussed in the following section.

*7.5. SHEAR AND BENDING-MOMENT DIAGRAMS

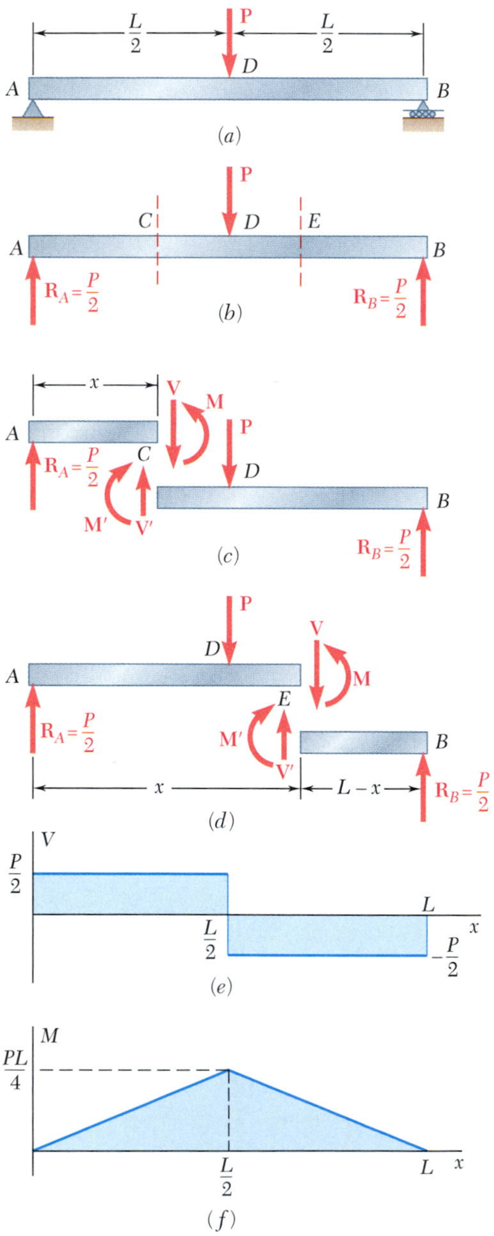

Fig. 7.10

Now that shear and bending moment have been clearly defined in sense as well as in magnitude, we can easily record their values at any point of a beam by plotting these values against the distance x measured from one end of the beam. The graphs obtained in this way are called, respectively, the *shear diagram* and the *bending-moment diagram*. As an example, consider a simply supported beam AB of span L subjected to a single concentrated load $\mathbf{P}$ applied at its midpoint D (Fig. 7.10*a*). We first determine the reactions at the supports from the free-body diagram of the entire beam (Fig. 7.10*b*); we find that the magnitude of each reaction is equal to $P/2$.

Next we cut the beam at a point C between A and D and draw the free-body diagrams of AC and CB (Fig. 7.10*c*). *Assuming that shear and bending moment are positive,* we direct the internal forces $\mathbf{V}$ and $\mathbf{V}'$ and the internal couples $\mathbf{M}$ and $\mathbf{M}'$ as indicated in Fig. 7.9*a*. Considering the free body AC and writing that the sum of the vertical components and the sum of the moments about C of the forces acting on the free body are zero, we find $V = +P/2$ and $M = +Px/2$. Both shear and bending moment are therefore positive; this can be checked by observing that the reaction at A tends to shear off and to bend the beam at C as indicated in Fig. 7.9*b* and *c*. We can plot V and M between A and D (Fig. 7.10*e* and *f*); the shear has a constant value $V = P/2$, while the bending moment increases linearly from $M = 0$ at $x = 0$ to $M = PL/4$ at $x = L/2$.

Cutting, now, the beam at a point E between D and B and considering the free body EB (Fig. 7.10*d*), we write that the sum of the vertical components and the sum of the moments about E of the forces acting on the free body are zero. We obtain $V = -P/2$ and $M = P(L - x)/2$. The shear is therefore negative and the bending moment positive; this can be checked by observing that the reaction at B bends the beam at E as indicated in Fig. 7.9*c* but tends to shear it off in a manner opposite to that shown in Fig. 7.9*b*. We can complete, now, the shear and bending-moment diagrams of Fig. 7.10*e* and *f*; the shear has a constant value $V = -P/2$ between D and B, while the bending moment decreases linearly from $M = PL/4$ at $x = L/2$ to $M = 0$ at $x = L$.

It should be noted that when a beam is subjected to concentrated loads only, the shear is of constant value between loads and the bending moment varies linearly between loads, but when a beam is subjected to distributed loads, the shear and bending moment vary quite differently (see Sample Prob. 7.3).

SAMPLE PROBLEM 7.2

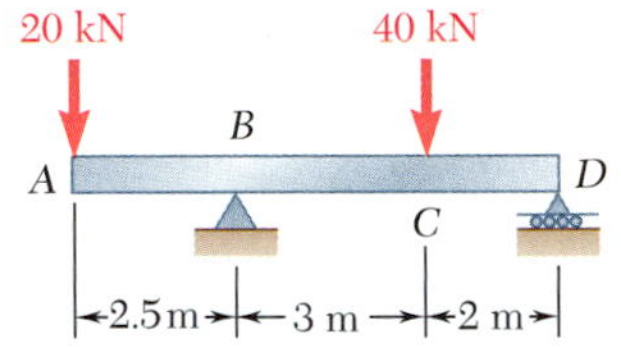

Draw the shear and bending-moment diagrams for the beam and loading shown.

SOLUTION

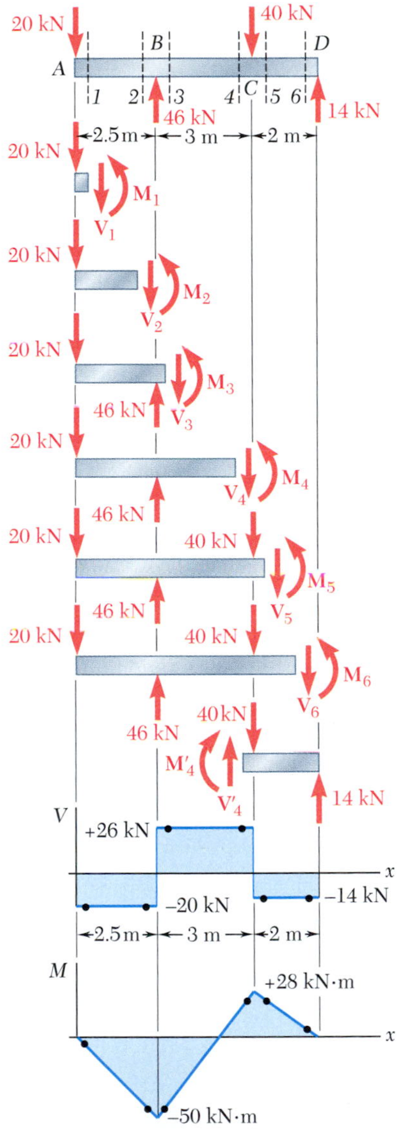

Free-Body: Entire Beam. From the free-body diagram of the entire beam, we find the reactions at B and D:

$$\mathbf{R}_B = 46 \text{ kN} \uparrow \qquad \mathbf{R}_D = 14 \text{ kN} \uparrow$$

Shear and Bending Moment. We first determine the internal forces just to the right of the 20-kN load at A. Considering the stub of the beam to the left of section *1* as a free body and assuming V and M to be positive (according to the standard convention), we write

$$+\uparrow \Sigma F_y = 0: \quad -20 \text{ kN} - V_1 = 0 \qquad V_1 = -20 \text{ kN}$$
$$+\circlearrowleft \Sigma M_1 = 0: \quad (20 \text{ kN})(0 \text{ m}) + M_1 = 0 \qquad M_1 = 0$$

We next consider as a free body the portion of the beam to the left of section *2* and write

$$+\uparrow \Sigma F_y = 0: \quad -20 \text{ kN} - V_2 = 0 \qquad V_2 = -20 \text{ kN}$$
$$+\circlearrowleft \Sigma M_2 = 0: \quad (20 \text{ kN})(2.5 \text{ m}) + M_2 = 0 \qquad M_2 = -50 \text{ kN}\cdot\text{m}$$

The shear and bending moment at sections *3*, *4*, *5*, and *6* are determined in a similar way from the free-body diagrams shown. We obtain

$$V_3 = +26 \text{ kN} \qquad M_3 = -50 \text{ kN}\cdot\text{m}$$
$$V_4 = +26 \text{ kN} \qquad M_4 = +28 \text{ kN}\cdot\text{m}$$
$$V_5 = -14 \text{ kN} \qquad M_5 = +28 \text{ kN}\cdot\text{m}$$
$$V_6 = -14 \text{ kN} \qquad M_6 = 0$$

For several of the latter sections, the results are more easily obtained by considering as a free body the portion of the beam to the right of the section. For example, considering the portion of the beam to the right of section *4*, we write

$$+\uparrow \Sigma F_y = 0: \quad V_4 - 40 \text{ kN} + 14 \text{ kN} = 0 \qquad V_4 = +26 \text{ kN}$$
$$+\circlearrowleft \Sigma M_4 = 0: \quad -M_4 + (14 \text{ kN})(2 \text{ m}) = 0 \qquad M_4 = +28 \text{ kN}\cdot\text{m}$$

Shear and Bending-Moment Diagrams. We can now plot the six points shown on the shear and bending-moment diagrams. As indicated in Sec. 7.5, the shear is of constant value between concentrated loads, and the bending moment varies linearly; we therefore obtain the shear and bending-moment diagrams shown.

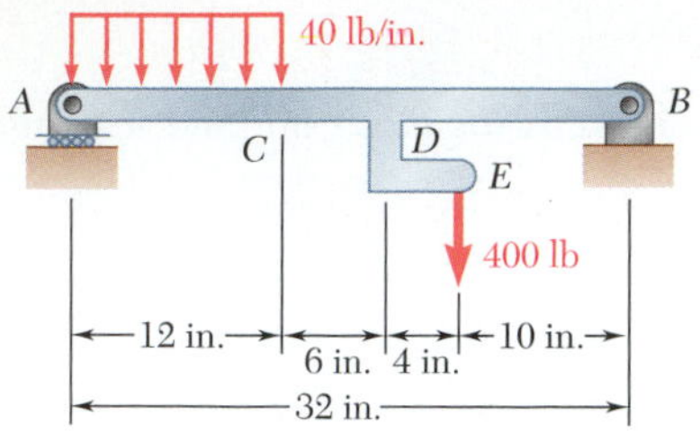

SAMPLE PROBLEM 7.3

Draw the shear and bending-moment diagrams for the beam AB. The distributed load of 40 lb/in. extends over 12 in. of the beam, from A to C, and the 400-lb load is applied at E.

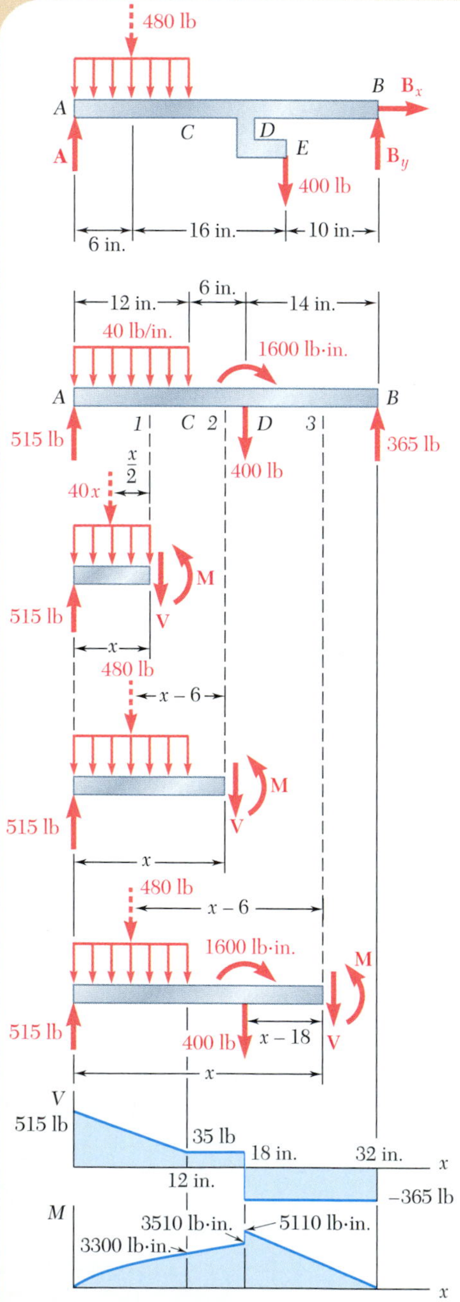

SOLUTION

Free-Body: Entire Beam. The reactions are determined by considering the entire beam as a free body.

$+\circlearrowleft\Sigma M_A = 0$: $B_y(32 \text{ in.}) - (480 \text{ lb})(6 \text{ in.}) - (400 \text{ lb})(22 \text{ in.}) = 0$

$B_y = +365 \text{ lb}$ $\qquad\mathbf{B}_y = 365 \text{ lb}\uparrow$

$+\circlearrowleft\Sigma M_B = 0$: $(480 \text{ lb})(26 \text{ in.}) + (400 \text{ lb})(10 \text{ in.}) - A(32 \text{ in.}) = 0$

$A = +515 \text{ lb}$ $\qquad\mathbf{A} = 515 \text{ lb}\uparrow$

$\overset{+}{\rightarrow}\Sigma F_x = 0$: $B_x = 0$ $\qquad\mathbf{B}_x = 0$

The 400-lb load is now replaced by an equivalent force-couple system acting on the beam at point D.

Shear and Bending Moment. ***From A to C.*** We determine the internal forces at a distance x from point A by considering the portion of the beam to the left of section *1*. That part of the distributed load acting on the free body is replaced by its resultant, and we write

$+\uparrow\Sigma F_y = 0$: $515 - 40x - V = 0$ $\qquad V = 515 - 40x$

$+\circlearrowleft\Sigma M_1 = 0$: $-515x + 40x(\frac{1}{2}x) + M = 0$ $\qquad M = 515x - 20x^2$

Since the free-body diagram shown can be used for all values of x smaller than 12 in., the expressions obtained for V and M are valid throughout the region $0 < x < 12$ in.

From C to D. Considering the portion of the beam to the left of section *2* and again replacing the distributed load by its resultant, we obtain

$+\uparrow\Sigma F_y = 0$: $515 - 480 - V = 0$ $\qquad V = 35 \text{ lb}$

$+\circlearrowleft\Sigma M_2 = 0$: $-515x + 480(x - 6) + M = 0$ $\qquad M = (2880 + 35x) \text{ lb}\cdot\text{in.}$

These expressions are valid in the region 12 in. $< x <$ 18 in.

From D to B. Using the portion of the beam to the left of section *3*, we obtain for the region 18 in. $< x <$ 32 in.

$+\uparrow\Sigma F_y = 0$: $515 - 480 - 400 - V = 0$ $\qquad V = -365 \text{ lb}$

$+\circlearrowleft\Sigma M_3 = 0$: $-515x + 480(x - 6) - 1600 + 400(x - 18) + M = 0$

$M = (11{,}680 - 365x) \text{ lb}\cdot\text{in.}$

Shear and Bending-Moment Diagrams. The shear and bending-moment diagrams for the entire beam can be plotted. We note that the couple of moment 1600 lb · in. applied at point D introduces a discontinuity into the bending-moment diagram.

SOLVING PROBLEMS ON YOUR OWN

In this lesson you learned to determine the *shear* V and the *bending moment* M at any point in a beam. You also learned to draw the *shear diagram* and the *bending-moment diagram* for the beam by plotting, respectively, V and M against the distance x measured along the beam.

A. *Determining the shear and bending moment in a beam.* To determine the shear V and the bending moment M at a given point C of a beam, you should take the following steps.

1. *Draw a free-body diagram of the entire beam,* and use it to determine the reactions at the beam supports.

2. *Cut the beam at point C,* and, using the original loading, select one of the two portions of the beam you have obtained.

3. *Draw the free-body diagram of the portion of the beam you have selected,* showing:

a. *The loads and the reaction* exerted on that portion of the beam, replacing each distributed load by an equivalent concentrated load as explained earlier in Sec. 5.8.

b. *The shearing force and the bending couple representing the internal forces at C.* To facilitate recording the shear V and the bending moment M after they have been determined, follow the convention indicated in Figs. 7.8 and 7.9. Thus, if you are using the portion of the beam located to the *left of C*, apply at C a *shearing force* **V** *directed downward* and a *bending moment* **M** *directed counterclockwise*. If you are using the portion of the beam located to the *right of C*, apply at C a *shearing force* **V**′ *directed upward* and a *bending moment* **M**′ *directed clockwise* [Sample Prob. 7.2].

4. *Write the equilibrium equations for the portion of the beam you have selected.* Solve the equation $\Sigma F_y = 0$ for V and the equation $\Sigma M_C = 0$ for M.

5. *Record the values of V and M with the sign obtained for each of them.* A positive sign for V means that the shearing forces exerted at C on each of the two portions of the beam are directed as shown in Figs. 7.8 and 7.9; a negative sign means that they have the opposite sense. Similarly, a positive sign for M means that the bending couples at C are directed as shown in these figures, and a negative sign means that they have the opposite sense. In addition, a positive sign for M means that the concavity of the beam at C is directed upward, and a negative sign means that it is directed downward.

(continued)

B. Drawing the shear and bending-moment diagrams for a beam. These diagrams are obtained by plotting, respectively, V and M against the distance x measured along the beam. However, in most cases the values of V and M need to be computed only at a few points.

1. For a beam supporting only concentrated loads, we note [Sample Prob. 7.2] that

a. The shear diagram consists of segments of horizontal lines. Thus, to draw the shear diagram of the beam you will need to compute V only just to the left or just to the right of the points where the loads or the reactions are applied.

b. The bending-moment diagram consists of segments of oblique straight lines. Thus, to draw the bending-moment diagram of the beam you will need to compute M only at the points where the loads or the reactions are applied.

2. For a beam supporting uniformly distributed loads, we note [Sample Prob. 7.3] that under each of the distributed loads:

a. The shear diagram consists of a segment of an oblique straight line. Thus, you will need to compute V only where the distributed load begins and where it ends.

b. The bending-moment diagram consists of a parabolic arc. In most cases you will need to compute M only where the distributed load begins and where it ends.

3. For a beam with a more complicated loading, it is necessary to consider the free-body diagram of a portion of the beam of arbitrary length x and determine V and M as functions of x. This procedure may have to be repeated several times, since V and M are often represented by different functions in various parts of the beam [Sample Prob. 7.3].

4. When a couple is applied to a beam, the shear has the same value on both sides of the point of application of the couple, but the bending-moment diagram will show a discontinuity at that point, rising or falling by an amount equal to the magnitude of the couple. Note that a couple can either be applied directly to the beam, or result from the application of a load to a member rigidly attached to the beam [Sample Prob. 7.3].

Problems

7.29 through 7.32 For the beam and loading shown, (*a*) draw the shear and bending-moment diagrams, (*b*) determine the maximum absolute values of the shear and bending moment.

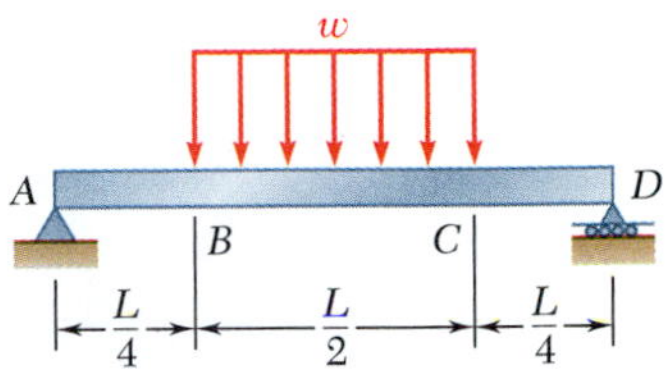

Fig. P7.29

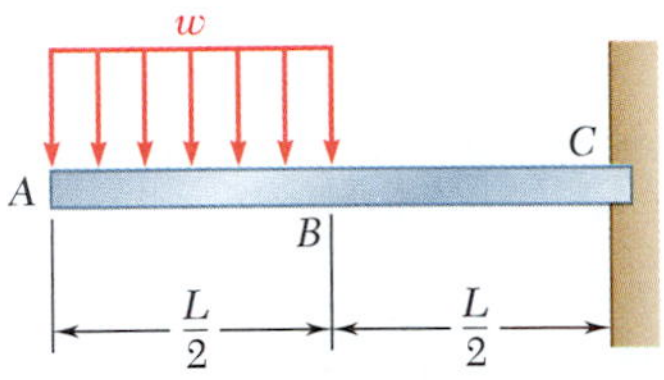

Fig. P7.30

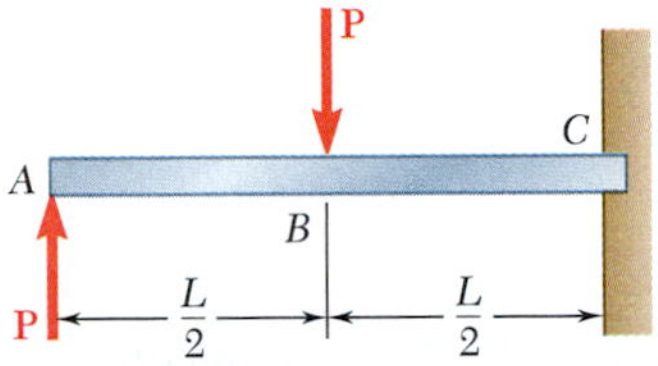

Fig. *P7.31*

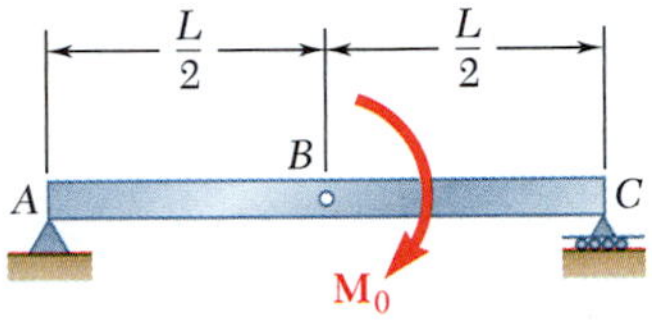

Fig. *P7.32*

7.33 and 7.34 For the beam and loading shown, (*a*) draw the shear and bending-moment diagrams, (*b*) determine the maximum absolute values of the shear and bending moment.

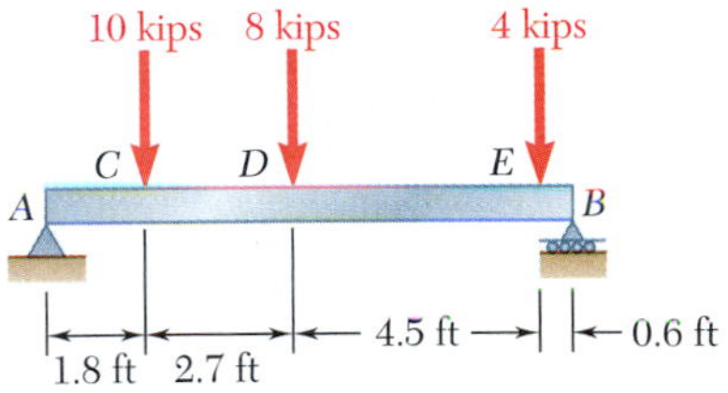

Fig. P7.33

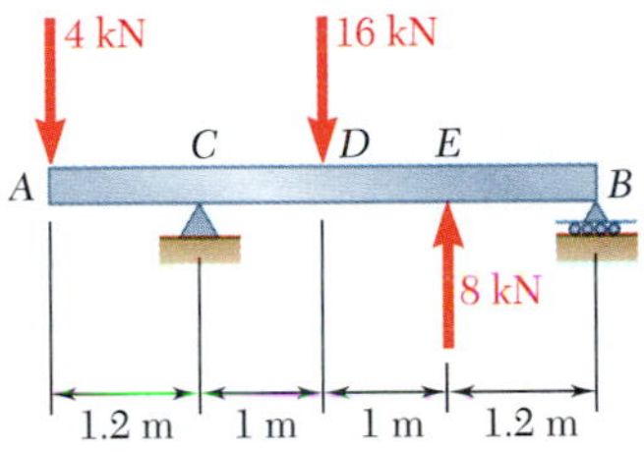

Fig. P7.34

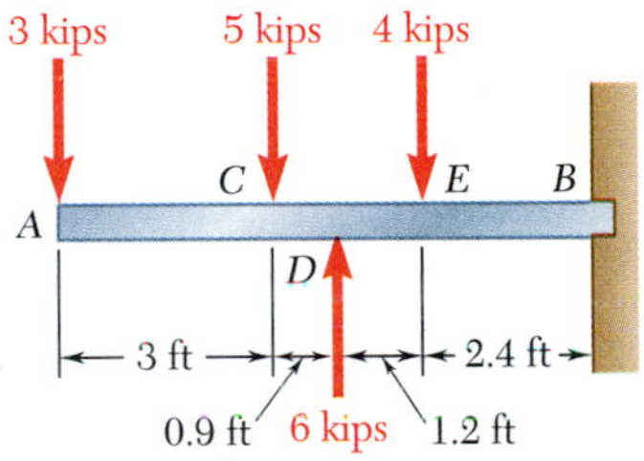

Fig. P7.35

7.35 and 7.36 For the beam and loading shown, (*a*) draw the shear and bending-moment diagrams, (*b*) determine the maximum absolute values of the shear and bending moment.

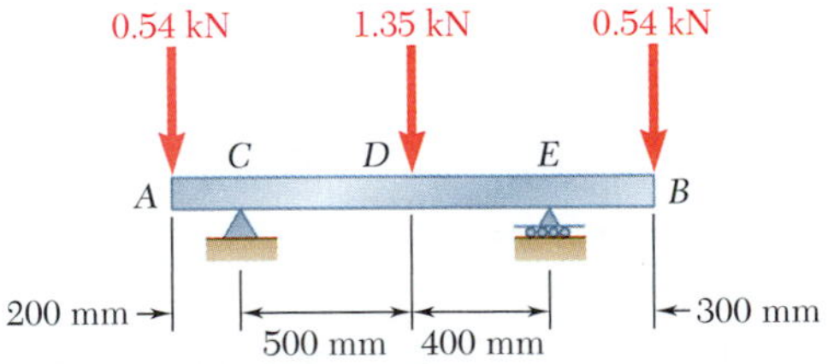

Fig. P7.36

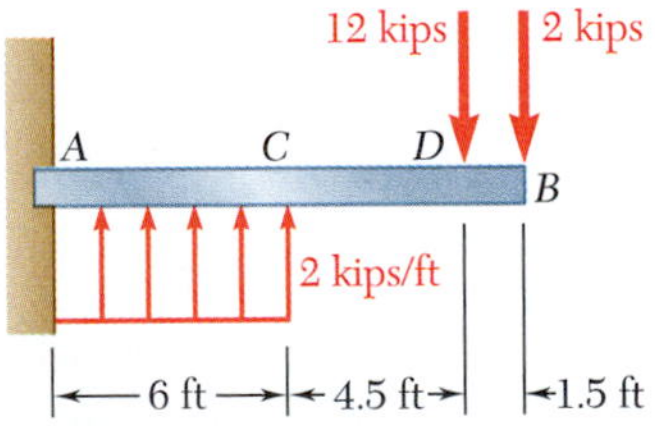

Fig. P7.37

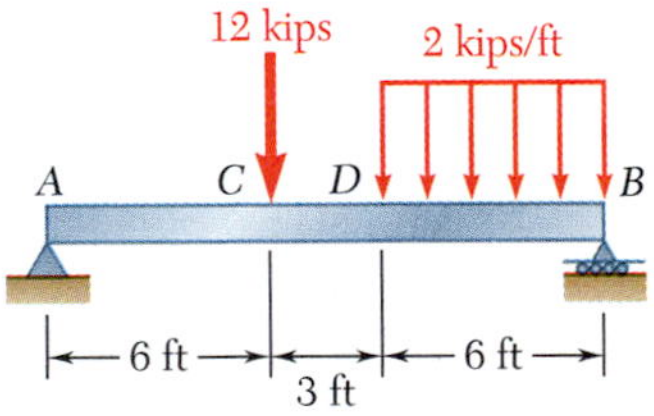

Fig. P7.38

7.37 and 7.38 For the beam and loading shown, (*a*) draw the shear and bending-moment diagrams, (*b*) determine the maximum absolute values of the shear and bending moment.

7.39 and 7.40 For the beam and loading shown, (*a*) draw the shear and bending-moment diagrams, (*b*) determine the maximum absolute values of the shear and bending moment.

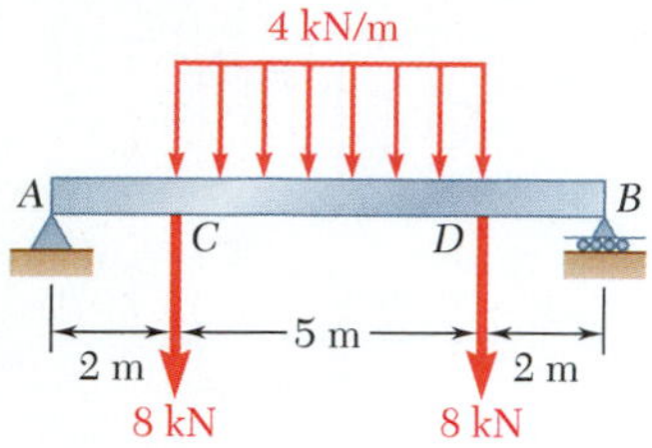

Fig. P7.39

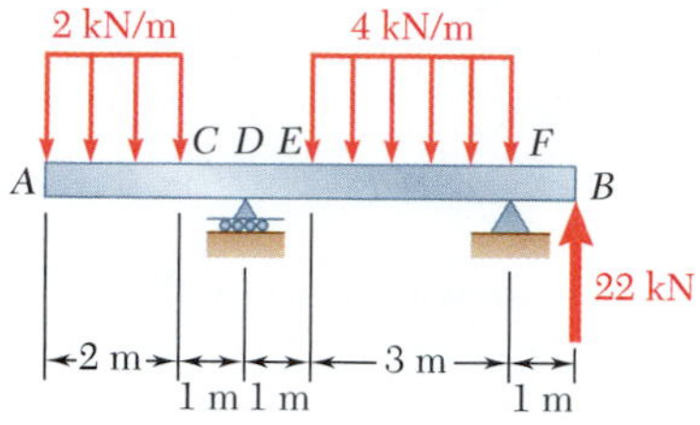

Fig. P7.40

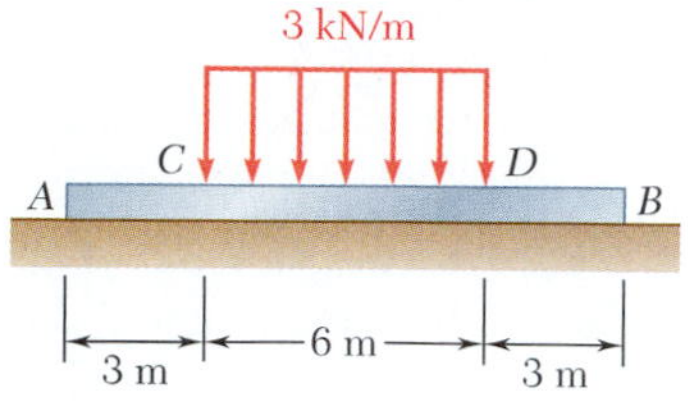

Fig. P7.41

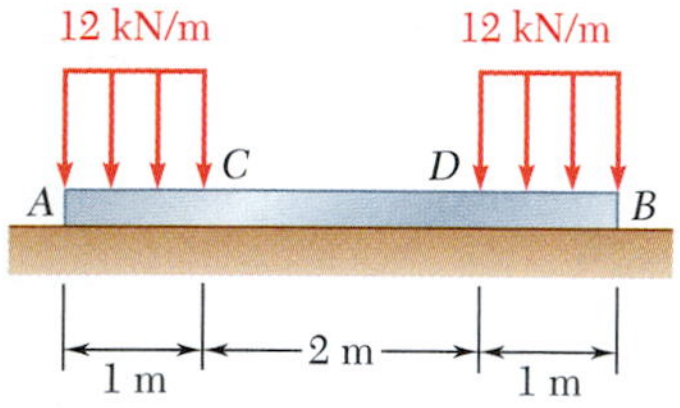

Fig. P7.42

7.41 and 7.42 Assuming the upward reaction of the ground on beam *AB* to be uniformly distributed, (*a*) draw the shear and bending-moment diagrams, (*b*) determine the maximum absolute values of the shear and bending moment.

7.43 Assuming the upward reaction of the ground on beam *AB* to be uniformly distributed and knowing that $a = 0.9$ ft, (*a*) draw the shear and bending-moment diagrams, (*b*) determine the maximum absolute values of the shear and bending moment.

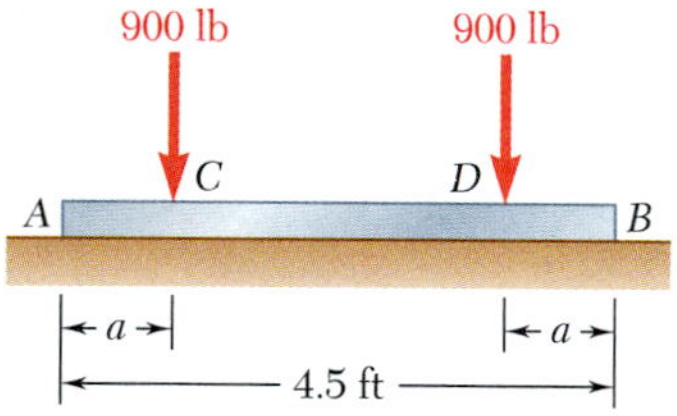

Fig. P7.43

7.44 Solve Prob. 7.43 assuming that $a = 1.5$ ft.

7.45 Two short angle sections *CE* and *DF* are bolted to the uniform beam *AB* of weight 3.33 kN, and the assembly is temporarily supported by the vertical cables *EG* and *FH* as shown. A second beam resting on beam *AB* at *I* exerts a downward force of 3 kN on *AB*. Knowing that $a = 0.3$ m and neglecting the weight of the angle sections, (*a*) draw the shear and bending-moment diagrams for beam *AB*, (*b*) determine the maximum absolute values of the shear and bending moment in the beam.

7.46 Solve Prob. 7.45 when $a = 0.6$ m.

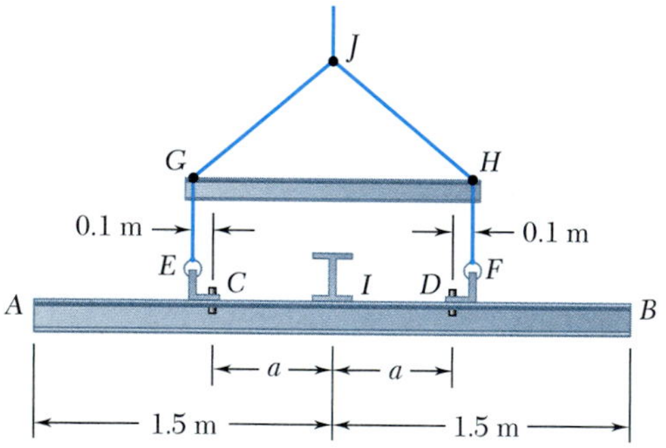

Fig. P7.45

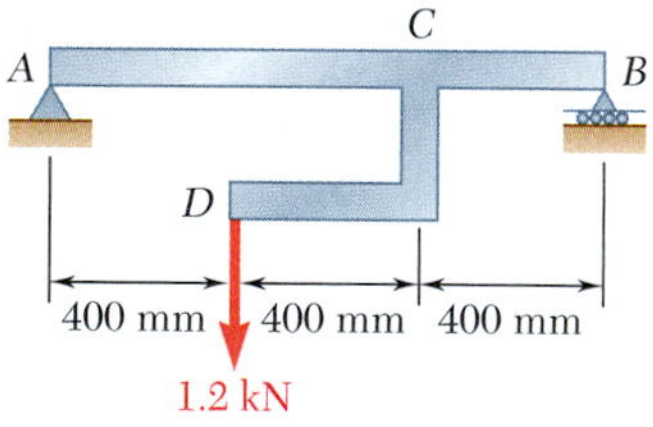

Fig. P7.47

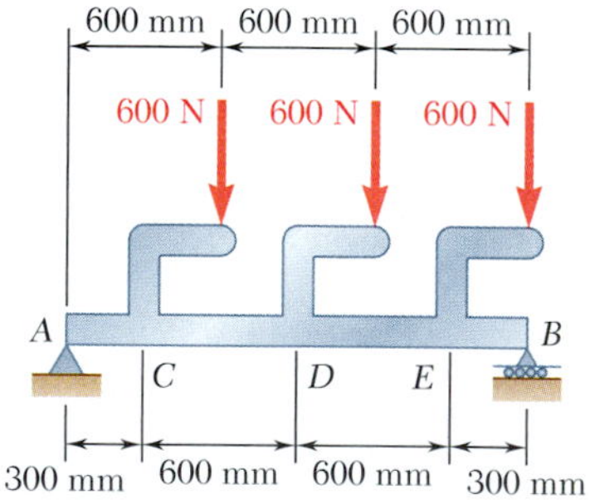

Fig. P7.48

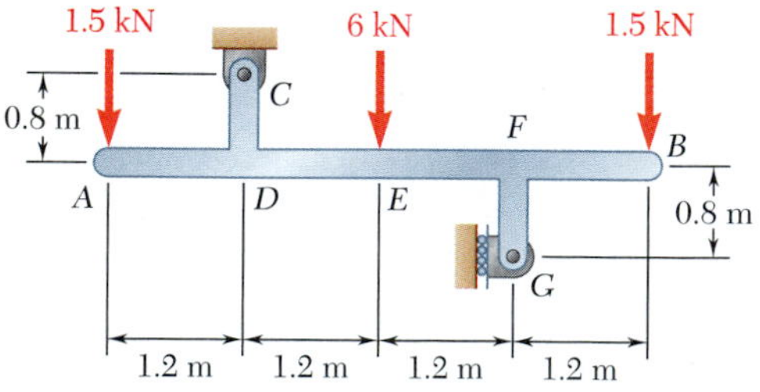

Fig. P7.49

7.47 Draw the shear and bending-moment diagrams for the beam *AB*, and determine the shear and bending moment (*a*) just to the left of *C*, (*b*) just to the right of *C*.

7.48 and 7.49 Draw the shear and bending-moment diagrams for the beam *AB*, and determine the maximum absolute values of the shear and bending moment.

7.50 Neglecting the size of the pulley at G, (*a*) draw the shear and bending-moment diagrams for the beam AB, (*b*) determine the maximum absolute values of the shear and bending moment.

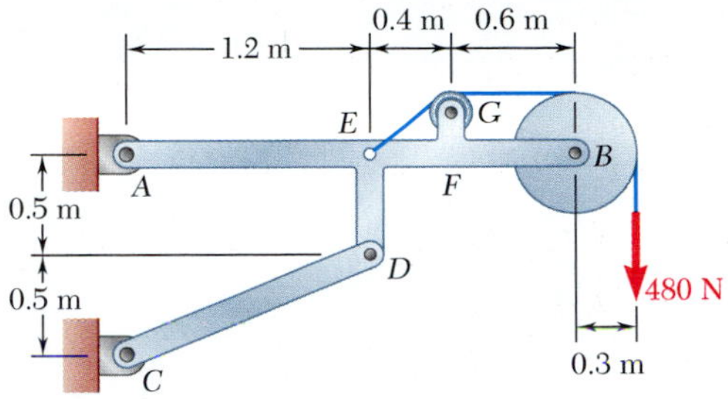

Fig. P7.50

7.51 For the beam of Prob. 7.43, determine (*a*) the distance a for which the maximum absolute value of the bending moment in the beam is as small as possible, (*b*) the corresponding value of $|M|_{max}$. (*Hint:* Draw the bending-moment diagram and then equate the absolute values of the largest positive and negative bending moments obtained.)

7.52 For the assembly of Prob. 7.45, determine (*a*) the distance a for which the maximum absolute value of the bending moment in beam AB is as small as possible, (*b*) the corresponding value of $|M|_{max}$. (See hint for Prob. 7.51.)

7.53 For the beam shown, determine (*a*) the magnitude P of the two upward forces for which the maximum value of the bending moment is as small as possible, (*b*) the corresponding value of $|M|_{max}$. (See hint for Prob. 7.51.)

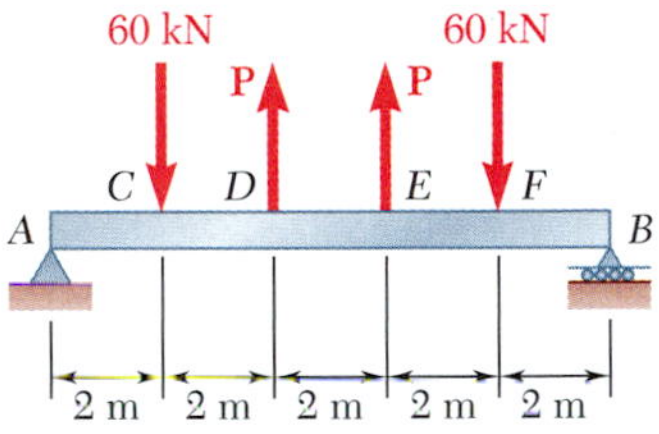

Fig. P7.53

7.54 For the beam and loading shown, determine (*a*) the distance a for which the maximum absolute value of the bending moment in the beam is as small as possible, (*b*) the corresponding value of $|M|_{max}$. (See hint for Prob. 7.51.)

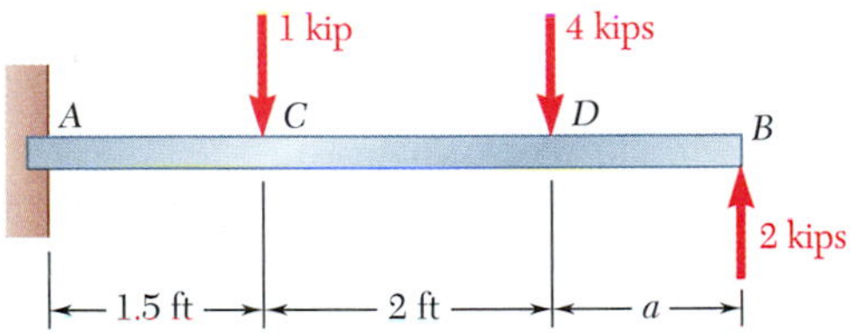

Fig. P7.54

7.55 Knowing that $P = Q = 375$ lb, determine (*a*) the distance a for which the maximum absolute value of the bending moment in beam AB is as small as possible, (*b*) the corresponding value of $|M|_{max}$. (See hint for Prob. 7.51.)

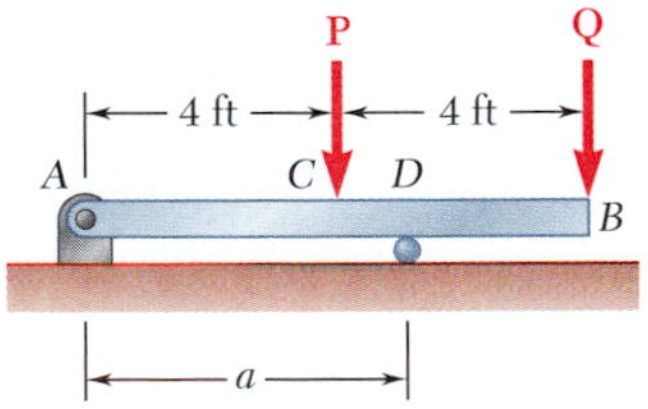

Fig. P7.55

7.56 Solve Prob. 7.55 assuming that $P = 750$ lb and $Q = 375$ lb.

7.57 In order to reduce the bending moment in the cantilever beam AB, a cable and counterweight are permanently attached at end B. Determine the magnitude of the counterweight for which the maximum absolute value of the bending moment in the beam is as small as possible and the corresponding value of $|M|_{max}$. Consider (*a*) the case when the distributed load is permanently applied to the beam, (*b*) the more general case when the distributed load may either be applied or removed.

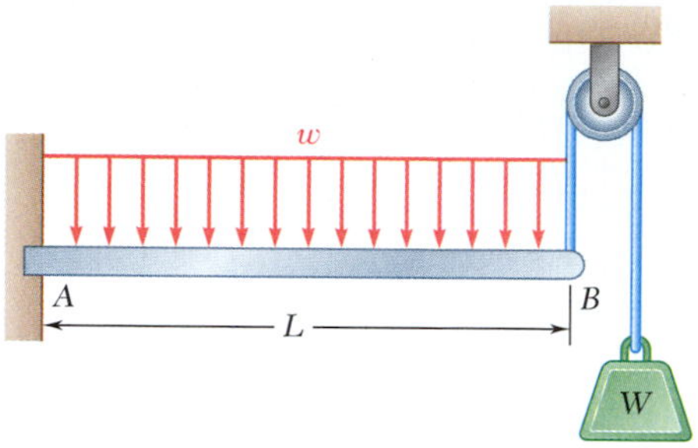

Fig. *P7.57*

*7.6. RELATIONS AMONG LOAD, SHEAR, AND BENDING MOMENT

When a beam carries more than two or three concentrated loads, or when it carries distributed loads, the method outlined in Sec. 7.5 for plotting shear and bending moment is likely to be quite cumbersome. The construction of the shear diagram and, especially, of the bending-moment diagram will be greatly facilitated if certain relations existing among load, shear, and bending moment are taken into consideration.

Let us consider a simply supported beam AB carrying a distributed load w per unit length (Fig. 7.11*a*), and let C and C' be two points of the beam at a distance Δx from each other. The shear and bending moment at C will be denoted by V and M, respectively, and will be assumed positive; the shear and bending moment at C' will be denoted by $V + \Delta V$ and $M + \Delta M$.

Let us now detach the portion of beam CC' and draw its free-body diagram (Fig. 7.11*b*). The forces exerted on the free body include a load of magnitude $w\,\Delta x$ and internal forces and couples at C and C'. Since shear and bending moment have been assumed positive, the forces and couples will be directed as shown in the figure.

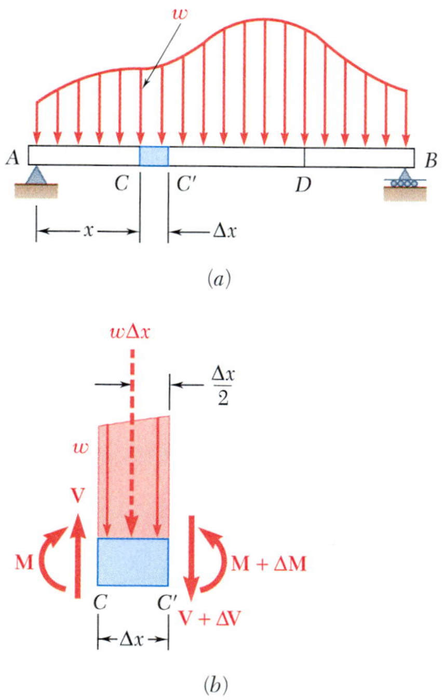

Fig. 7.11

Relations between Load and Shear. We write that the sum of the vertical components of the forces acting on the free body CC' is zero:

$$V - (V + \Delta V) - w\,\Delta x = 0$$
$$\Delta V = -w\,\Delta x$$

Dividing both members of the equation by Δx and then letting Δx approach zero, we obtain

$$\frac{dV}{dx} = -w \tag{7.1}$$

Formula (7.1) indicates that for a beam loaded as shown in Fig. 7.11*a*, the slope dV/dx of the shear curve is negative; the absolute value of the slope at any point is equal to the load per unit length at that point. Integrating (7.1) between points C and D, we obtain

$$V_D - V_C = -\int_{x_C}^{x_D} w\,dx \tag{7.2}$$

$$V_D - V_C = -(\text{area under load curve between } C \text{ and } D) \tag{7.2'}$$

Note that this result could also have been obtained by considering the equilibrium of the portion of beam CD, since the area under the load curve represents the total load applied between C and D.

It should be observed that formula (7.1) *is not valid* at a point where a concentrated load is applied; the shear curve is discontinuous at such a point, as seen in Sec. 7.5. Similarly, formulas (7.2) and (7.2′) cease to be valid when concentrated loads are applied between C and D, since they do not take into account the sudden change in shear caused by a concentrated load. Formulas (7.2) and (7.2′), therefore, should be applied only between successive concentrated loads.

Relations between Shear and Bending Moment. Returning to the free-body diagram of Fig. 7.11*b*, and writing now that the sum of the moments about C' is zero, we obtain

$$(M + \Delta M) - M - V\,\Delta x + w\,\Delta x\frac{\Delta x}{2} = 0$$

$$\Delta M = V\,\Delta x - \tfrac{1}{2}w(\Delta x)^2$$

Dividing both members of the equation by Δx and then letting Δx approach zero, we obtain

$$\frac{dM}{dx} = V \tag{7.3}$$

Formula (7.3) indicates that the slope dM/dx of the bending-moment curve is equal to the value of the shear. This is true at any point where the shear has a well-defined value, that is, at any point where no concentrated load is applied. Formula (7.3) also shows that the shear is zero at points where the bending moment is maximum. This property facilitates the determination of the points where the beam is likely to fail under bending.

Integrating (7.3) between points C and D, we obtain

$$M_D - M_C = \int_{x_C}^{x_D} V\,dx \tag{7.4}$$

$$M_D - M_C = \text{area under shear curve between } C \text{ and } D \tag{7.4'}$$

Note that the area under the shear curve should be considered positive where the shear is positive and negative where the shear is negative. Formulas (7.4) and (7.4′) are valid even when concentrated loads are applied between C and D, as long as the shear curve has been correctly drawn. The formulas cease to be valid, however, if a *couple* is applied at a point between C and D, since they do not take into account the sudden change in bending moment caused by a couple (see Sample Prob. 7.7).

Example. Let us consider a simply supported beam AB of span L carrying a uniformly distributed load w (Fig. 7.12a). From the free-body diagram of the entire beam we determine the magnitude of the reactions at the supports: $R_A = R_B = wL/2$ (Fig. 7.12b). Next, we draw the shear diagram. Close to the end A of the beam, the shear is equal to R_A, that is, to $wL/2$, as we can check by considering a very small portion of the beam as a free body. Using formula (7.2), we can then determine the shear V at any distance x from A. We write

$$V - V_A = -\int_0^x w\,dx = -wx$$

$$V = V_A - wx = \frac{wL}{2} - wx = w\left(\frac{L}{2} - x\right)$$

The shear curve is thus an oblique straight line which crosses the x axis at $x = L/2$ (Fig. 7.12c). Considering, now, the bending moment, we first observe that $M_A = 0$. The value M of the bending moment at any distance x from A can then be obtained from formula (7.4); we have

$$M - M_A = \int_0^x V dx$$

$$M = \int_0^x w\left(\frac{L}{2} - x\right) dx = \frac{w}{2}(Lx - x^2)$$

The bending-moment curve is a parabola. The maximum value of the bending moment occurs when $x = L/2$, since V (and thus dM/dx) is zero for that value of x. Substituting $x = L/2$ in the last equation, we obtain $M_{max} = wL^2/8$.

In most engineering applications, the value of the bending moment needs to be known only at a few specific points. Once the shear diagram has been drawn, and after M has been determined at one of the ends of the beam, the value of the bending moment can then be obtained at any given point by computing the area under the shear curve and using formula (7.4′). For instance, since $M_A = 0$ for the beam of Fig. 7.12, the maximum value of the bending moment for that beam can be obtained simply by calculating the area of the shaded triangle in the shear diagram:

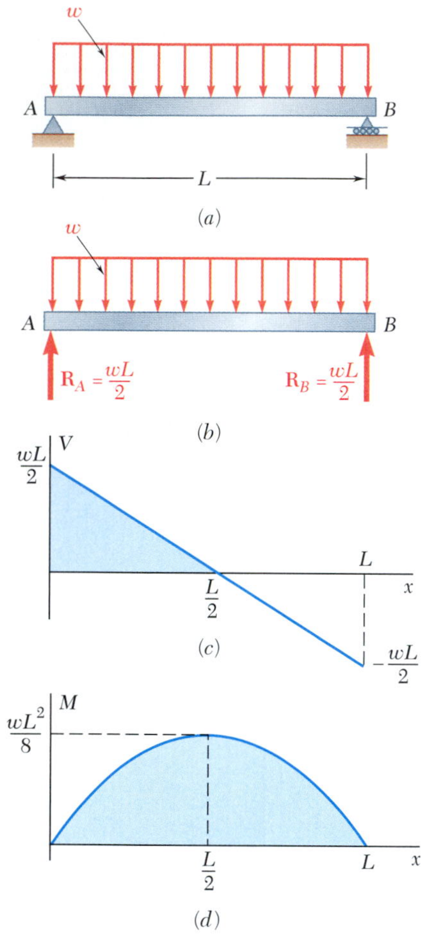

Fig. 7.12

$$M_{max} = \frac{1}{2}\frac{L}{2}\frac{wL}{2} = \frac{wL^2}{8}$$

In this example, the load curve is a horizontal straight line, the shear curve is an oblique straight line, and the bending-moment curve is a parabola. If the load curve had been an oblique straight line (first degree), the shear curve would have been a parabola (second degree), and the bending-moment curve would have been a cubic (third degree). The shear and bending-moment curves will always be, respectively, one and two degrees higher than the load curve. Thus, once a few values of the shear and bending moment have been computed, we should be able to sketch the shear and bending-moment diagrams without actually determining the functions $V(x)$ and $M(x)$. The sketches obtained will be more accurate if we make use of the fact that at any point where the curves are continuous, the slope of the shear curve is equal to $-w$ and the slope of the bending-moment curve is equal to V.

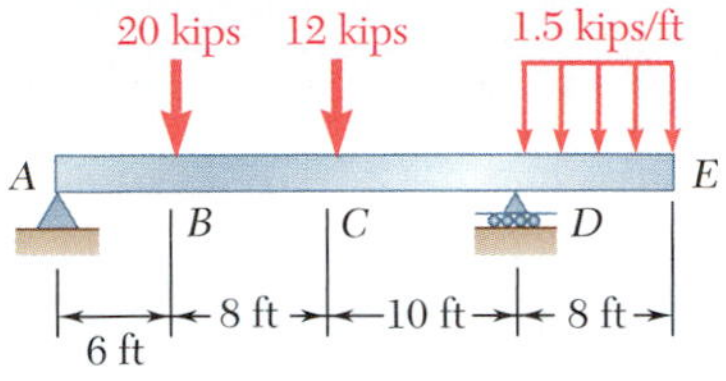

SAMPLE PROBLEM 7.4

Draw the shear and bending-moment diagrams for the beam and loading shown.

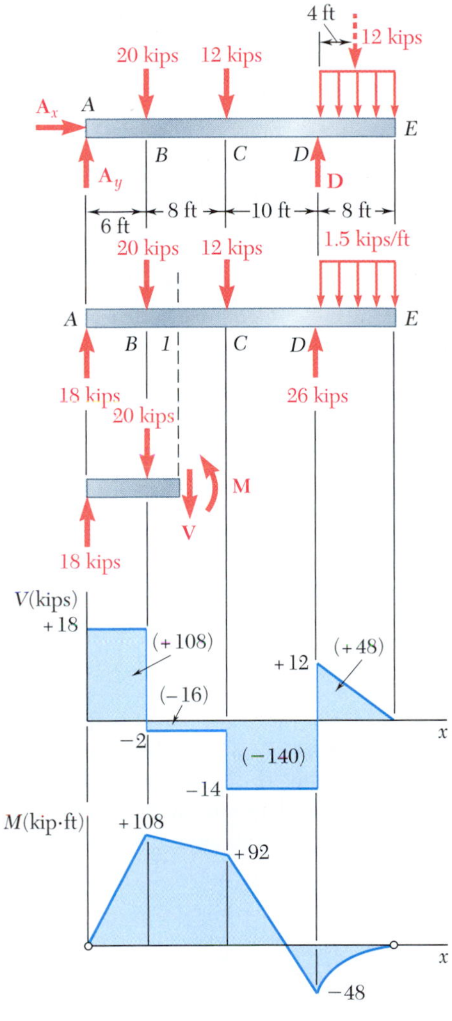

SOLUTION

Free-Body: Entire Beam. Considering the entire beam as a free body, we determine the reactions:

$+\circlearrowleft\Sigma M_A = 0$:

$D(24 \text{ ft}) - (20 \text{ kips})(6 \text{ ft}) - (12 \text{ kips})(14 \text{ ft}) - (12 \text{ kips})(28 \text{ ft}) = 0$

$D = +26 \text{ kips}$ $\qquad$ $\mathbf{D} = 26 \text{ kips} \uparrow$

$+\uparrow\Sigma F_y = 0$: $\quad A_y - 20 \text{ kips} - 12 \text{ kips} + 26 \text{ kips} - 12 \text{ kips} = 0$

$A_y = +18 \text{ kips}$ $\qquad$ $\mathbf{A}_y = 18 \text{ kips} \uparrow$

$\xrightarrow{+}\Sigma F_x = 0$: $\quad A_x = 0$ $\qquad$ $\mathbf{A}_x = 0$

We also note that at both A and E the bending moment is zero; thus two points (indicated by small circles) are obtained on the bending-moment diagram.

Shear Diagram. Since $dV/dx = -w$, we find that between concentrated loads and reactions the slope of the shear diagram is zero (that is, the shear is constant). The shear at any point is determined by dividing the beam into two parts and considering either part as a free body. For example, using the portion of beam to the left of section *1*, we obtain the shear between B and C:

$+\uparrow\Sigma F_y = 0$: $\quad +18 \text{ kips} - 20 \text{ kips} - V = 0 \qquad V = -2 \text{ kips}$

We also find that the shear is +12 kips just to the right of D and zero at end E. Since the slope $dV/dx = -w$ is constant between D and E, the shear diagram between these two points is a straight line.

Bending-Moment Diagram. We recall that the area under the shear curve between two points is equal to the change in bending moment between the same two points. For convenience, the area of each portion of the shear diagram is computed and is indicated on the diagram. Since the bending moment M_A at the left end is known to be zero, we write

$$\begin{aligned} M_B - M_A &= +108 & M_B &= +108 \text{ kip}\cdot\text{ft} \\ M_C - M_B &= -\ 16 & M_C &= +\ 92 \text{ kip}\cdot\text{ft} \\ M_D - M_C &= -140 & M_D &= -\ 48 \text{ kip}\cdot\text{ft} \\ M_E - M_D &= +\ 48 & M_E &= 0 \end{aligned}$$

Since M_E is known to be zero, a check of the computations is obtained.

Between the concentrated loads and reactions the shear is constant; thus the slope dM/dx is constant, and the bending-moment diagram is drawn by connecting the known points with straight lines. Between D and E, where the shear diagram is an oblique straight line, the bending-moment diagram is a parabola.

From the V and M diagrams we note that $V_{max} = 18$ kips and $M_{max} = 108$ kip · ft.

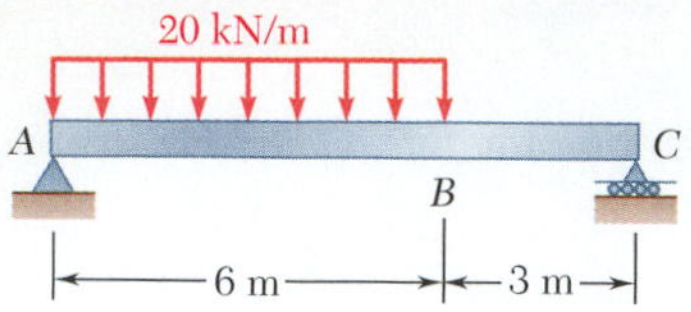

SAMPLE PROBLEM 7.5

Draw the shear and bending-moment diagrams for the beam and loading shown, and determine the location and magnitude of the maximum bending moment.

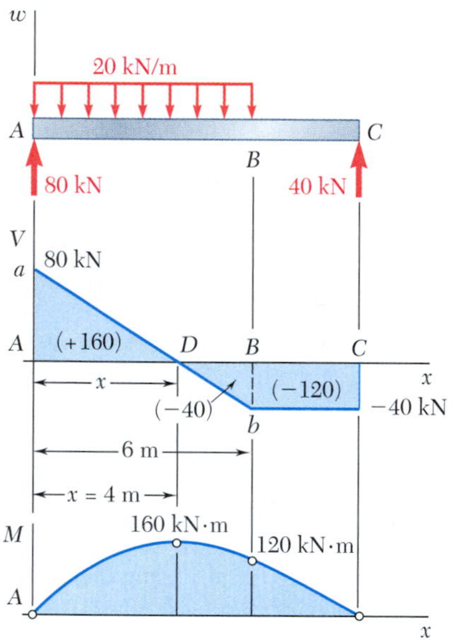

SOLUTION

Free-Body: Entire Beam. Considering the entire beam as a free body, we obtain the reactions

$$\mathbf{R}_A = 80 \text{ kN} \uparrow \qquad \mathbf{R}_C = 40 \text{ kN} \uparrow$$

Shear Diagram. The shear just to the right of A is $V_A = +80$ kN. Since the change in shear between two points is equal to *minus* the area under the load curve between the same two points, we obtain V_B by writing

$$V_B - V_A = -(20 \text{ kN/m})(6 \text{ m}) = -120 \text{ kN}$$
$$V_B = -120 + V_A = -120 + 80 = -40 \text{ kN}$$

Since the slope $dV/dx = -w$ is constant between A and B, the shear diagram between these two points is represented by a straight line. Between B and C, the area under the load curve is zero; therefore,

$$V_C - V_B = 0 \qquad V_C = V_B = -40 \text{ kN}$$

and the shear is constant between B and C.

Bending-Moment Diagram. We note that the bending moment at each end of the beam is zero. In order to determine the maximum bending moment, we locate the section D of the beam where $V = 0$. We write

$$V_D - V_A = -wx$$
$$0 - 80 \text{ kN} = -(20 \text{ kN/m})x$$

and, solving for x: $x = 4$ m ◀

The maximum bending moment occurs at point D, where we have $dM/dx = V = 0$. The areas of the various portions of the shear diagram are computed and are given (in parentheses) on the diagram. Since the area of the shear diagram between two points is equal to the change in bending moment between the same two points, we write

$$M_D - M_A = +160 \text{ kN}\cdot\text{m} \qquad M_D = +160 \text{ kN}\cdot\text{m}$$
$$M_B - M_D = -\ 40 \text{ kN}\cdot\text{m} \qquad M_B = +120 \text{ kN}\cdot\text{m}$$
$$M_C - M_B = -120 \text{ kN}\cdot\text{m} \qquad M_C = 0$$

The bending-moment diagram consists of a parabolic arc followed by a segment of a straight line; the slope of the parabola at A is equal to the value of V at that point.

The maximum bending moment is

$$M_{max} = M_D = +160 \text{ kN}\cdot\text{m} \quad ◀$$

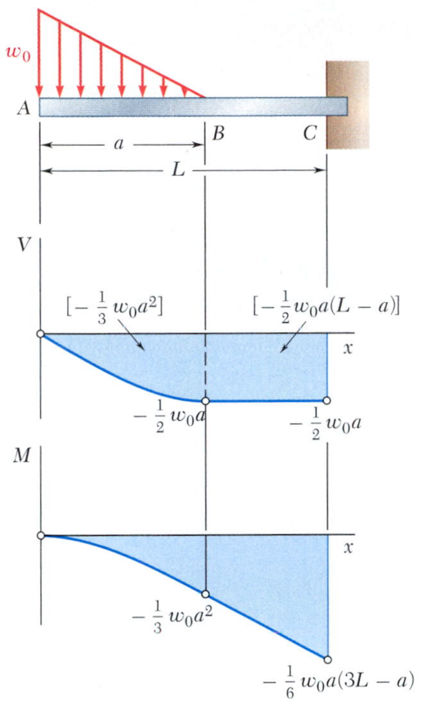

SAMPLE PROBLEM 7.6

Sketch the shear and bending-moment diagrams for the cantilever beam shown.

SOLUTION

Shear Diagram. At the free end of the beam, we find $V_A = 0$. Between A and B, the area under the load curve is $\frac{1}{2}w_0a$; we find V_B by writing

$$V_B - V_A = -\tfrac{1}{2}w_0a \qquad V_B = -\tfrac{1}{2}w_0a$$

Between B and C, the beam is not loaded; thus $V_C = V_B$. At A, we have $w = w_0$, and, according to Eq. (7.1), the slope of the shear curve is $dV/dx = -w_0$, while at B the slope is $dV/dx = 0$. Between A and B, the loading decreases linearly, and the shear diagram is parabolic. Between B and C, $w = 0$, and the shear diagram is a horizontal line.

Bending-Moment Diagram. We note that $M_A = 0$ at the free end of the beam. We compute the area under the shear curve and write

$$M_B - M_A = -\tfrac{1}{3}w_0a^2 \qquad M_B = -\tfrac{1}{3}w_0a^2$$
$$M_C - M_B = -\tfrac{1}{2}w_0a(L-a)$$
$$M_C = -\tfrac{1}{6}w_0a(3L-a)$$

The sketch of the bending-moment diagram is completed by recalling that $dM/dx = V$. We find that between A and B the diagram is represented by a cubic curve with zero slope at A, and between B and C the diagram is represented by a straight line.

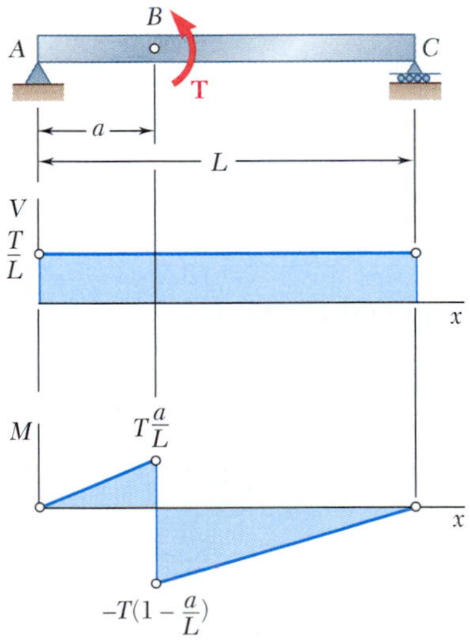

SAMPLE PROBLEM 7.7

The simple beam AC is loaded by a couple of magnitude T applied at point B. Draw the shear and bending-moment diagrams for the beam.

SOLUTION

Free Body: Entire Beam. The entire beam is taken as a free body, and we obtain

$$\mathbf{R}_A = \frac{T}{L}\uparrow \qquad \mathbf{R}_C = \frac{T}{L}\downarrow$$

Shear and Bending-Moment Diagrams. The shear at any section is constant and equal to T/L. Since a couple is applied at B, the bending-moment diagram is discontinuous at B; the bending moment decreases suddenly by an amount equal to T.

SOLVING PROBLEMS ON YOUR OWN

In this lesson you learned how to use the relations existing among load, shear, and bending moment to simplify the drawing of the shear and bending-moment diagrams. These relations are

$$\frac{dV}{dx} = -w \qquad (7.1)$$

$$\frac{dM}{dx} = V \qquad (7.3)$$

$$V_D - V_C = -(\text{area under load curve between } C \text{ and } D) \qquad (7.2')$$

$$M_D - M_C = (\text{area under shear curve between } C \text{ and } D) \qquad (7.4')$$

Taking into account these relations, you can use the following procedure to draw the shear and bending-moment diagrams for a beam.

1. *Draw a free-body diagram of the entire beam,* and use it to determine the reactions at the beam supports.

2. *Draw the shear diagram.* This can be done as in the preceding lesson by cutting the beam at various points and considering the free-body diagram of one of the two portions of the beam that you have obtained [Sample Prob. 7.3]. You can, however, consider one of the following alternative procedures.

a. *The shear V at any point of the beam is the sum of the reactions and loads to the left of that point;* an upward force is counted as positive, and a downward force is counted as negative.

b. *For a beam carrying a distributed load,* you can start from a point where you know V and use Eq. (7.2′) repeatedly to find V at all the other points of interest.

3. *Draw the bending-moment diagram,* using the following procedure.

a. *Compute the area under each portion of the shear curve,* assigning a positive sign to areas located above the x axis and a negative sign to areas located below the x axis.

b. *Apply Eq. (7.4′) repeatedly* [Sample Probs. 7.4 and 7.5], starting from the left end of the beam, where $M = 0$ (except if a couple is applied at that end, or if the beam is a cantilever beam with a fixed left end).

c. *Where a couple is applied to the beam,* be careful to show a discontinuity in the bending-moment diagram by *increasing* the value of M at that point by an amount equal to the magnitude of the couple if the couple is *clockwise,* or *decreasing* the value of M by that amount if the couple is *counterclockwise* [Sample Prob. 7.7].

4. *Determine the location and magnitude of $|M|_{max}$.* The maximum absolute value of the bending moment occurs at one of the points where $dM/dx = 0$, that is, according to Eq. (7.3), at a point where V is equal to zero or changes sign. You should, therefore:

a. Determine the values of $|M|$, using the area under the shear curve, at the points where V changes sign; these points will occur under concentrated loads [Sample Prob. 7.4].

b. Determine the points where $V = 0$ and the corresponding values of $|M|$; these points will occur under a distributed load. To find the distance x between point C, where the distributed load starts, and point D, where the shear is zero, use Eq. (7.2′); for V_C use the known value of the shear at point C, for V_D use zero, and express the area under the load curve as a function of x [Sample Prob. 7.5].

5. *You can improve the quality of your drawings* by keeping in mind that at any given point, according to Eqs. (7.1) and (7.3), the slope of the V curve is equal to $-w$ and the slope of the M curve is equal to V. In particular, your drawings should clearly show where the slope of the curves is zero.

6. *Finally, for beams supporting a distributed load expressed as a function $w(x)$,* remember that the shear V can be obtained by integrating the function $-w(x)$, and the bending moment M can be obtained by integrating $V(x)$ [Eqs. (7.3) and (7.4)].

Problems

7.58 Using the method of Sec. 7.6, solve Prob. 7.29.

7.59 Using the method of Sec. 7.6, solve Prob. 7.30.

7.60 Using the method of Sec. 7.6, solve Prob. 7.31.

7.61 Using the method of Sec. 7.6, solve Prob. 7.32.

7.62 Using the method of Sec. 7.6, solve Prob. 7.33.

7.63 Using the method of Sec. 7.6, solve Prob. 7.36.

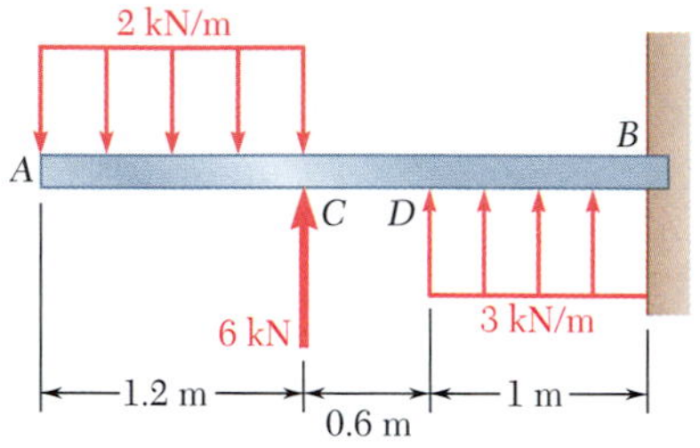

Fig. P7.64

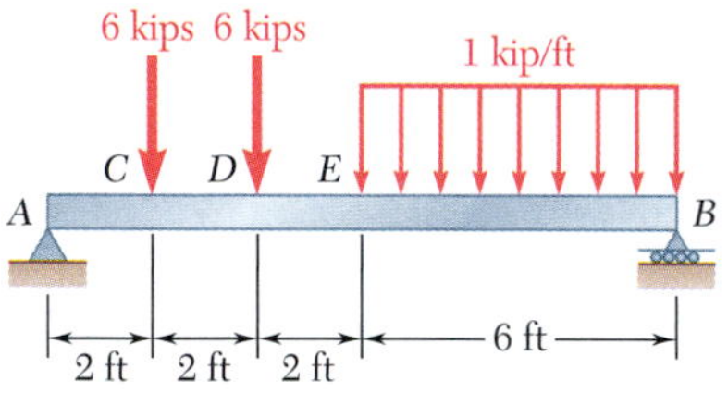

Fig. P7.65

7.64 and 7.65 For the beam and loading shown, (*a*) draw the shear and bending-moment diagrams, (*b*) determine the maximum absolute values of the shear and bending moment.

7.66 Using the method of Sec. 7.6, solve Prob. 7.37.

7.67 Using the method of Sec. 7.6, solve Prob. 7.38.

7.68 Using the method of Sec. 7.6, solve Prob. 7.39.

7.69 Using the method of Sec. 7.6, solve Prob. 7.40.

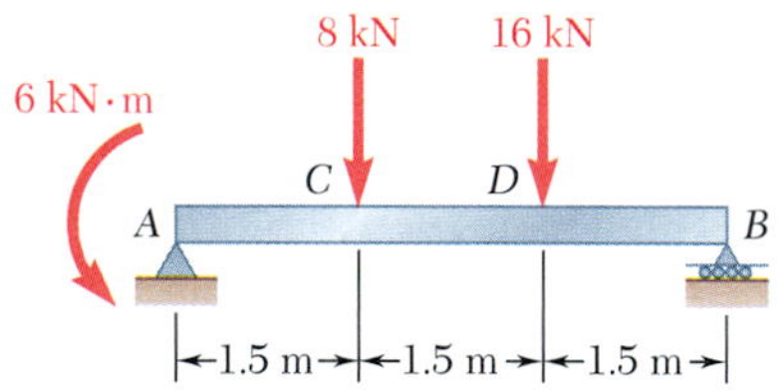

Fig. *P7.70*

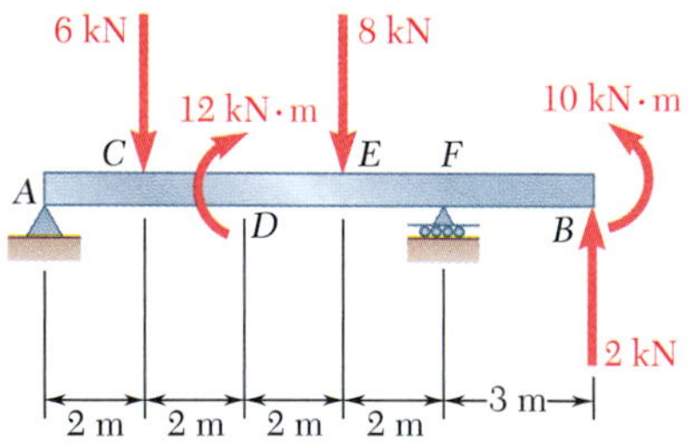

Fig. P7.71

***7.70* and 7.71** For the beam and loading shown, (*a*) draw the shear and bending-moment diagrams, (*b*) determine the maximum absolute values of the shear and bending moment.

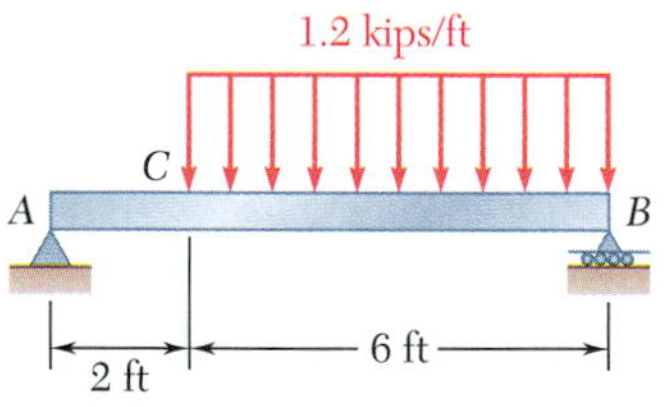

Fig. P7.72

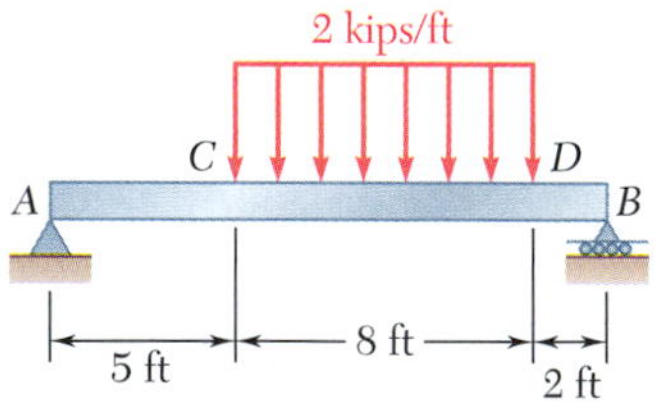

Fig. P7.73

7.72 and 7.73 For the beam and loading shown, (*a*) draw the shear and bending-moment diagrams, (*b*) determine the magnitude and location of the maximum bending moment.

7.74 For the beam shown, draw the shear and bending-moment diagrams, and determine the maximum absolute value of the bending moment knowing that (*a*) $P = 14$ kN, (*b*) $P = 20$ kN.

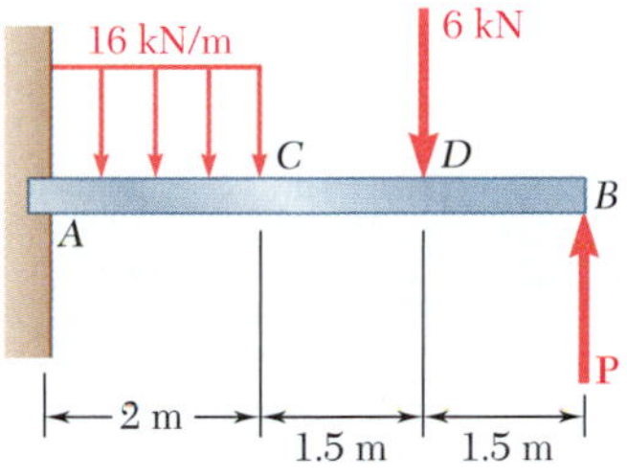

Fig. P7.74

7.75 For the beam shown, draw the shear and bending-moment diagrams, and determine the magnitude and location of the maximum absolute value of the bending moment knowing that (*a*) $M = 0$, (*b*) $M = 12$ kN · m.

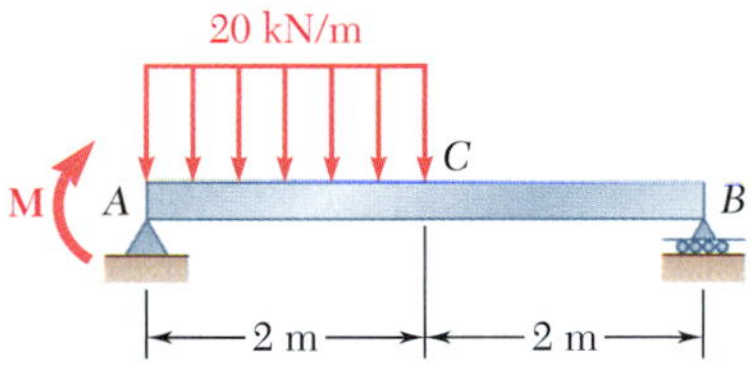

Fig. P7.75

7.76 For the beam and loading shown, (*a*) draw the shear and bending-moment diagrams, (*b*) determine the magnitude and location of the maximum absolute value of the bending moment.

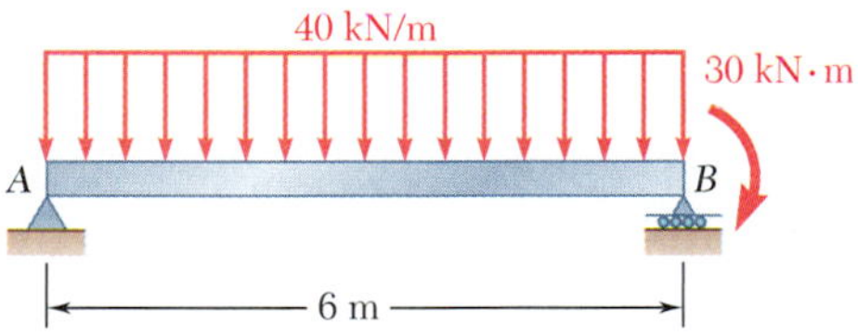

Fig. *P7.76*

7.77 Solve Prob. 7.76 assuming that the 30 kN · m couple applied at *B* is counterclockwise.

7.78 For beam *AB*, (*a*) draw the shear and bending-moment diagrams, (*b*) determine the magnitude and location of the maximum absolute value of the bending moment.

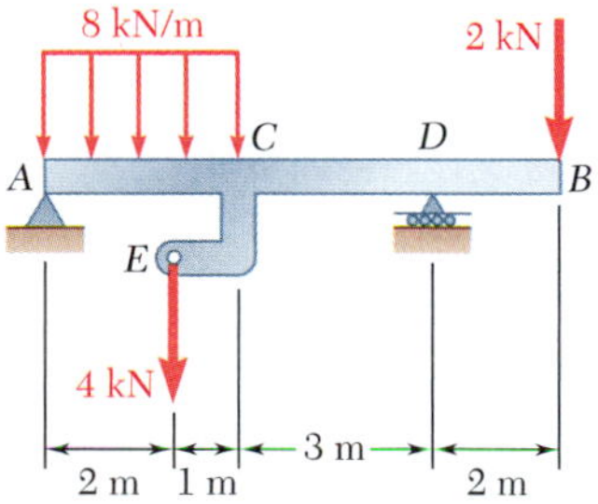

Fig. P7.78

7.79 Solve Prob. 7.78 assuming that the 4-kN force applied at *E* is directed upward.

7.80 and 7.81 For the beam and loading shown, (*a*) derive the equations of the shear and bending-moment curves, (*b*) draw the shear and bending-moment diagrams, (*c*) determine the magnitude and location of the maximum bending moment.

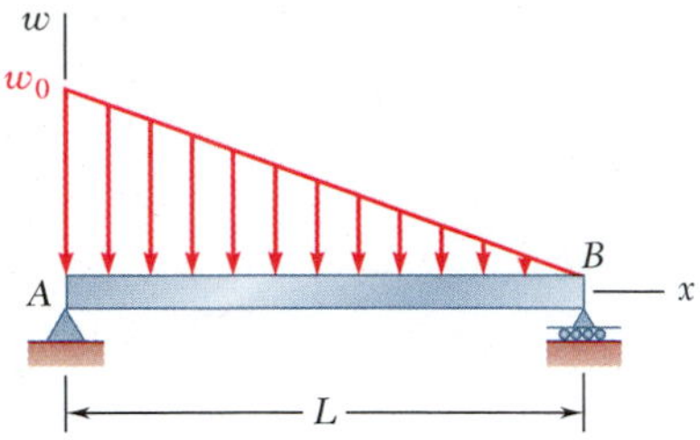

Fig. P7.80

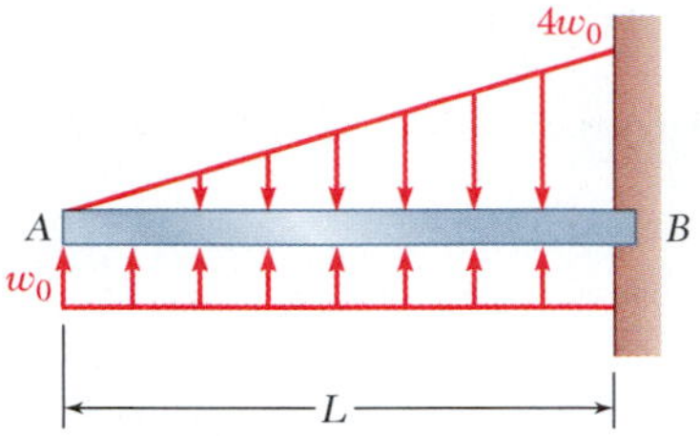

Fig. P7.81

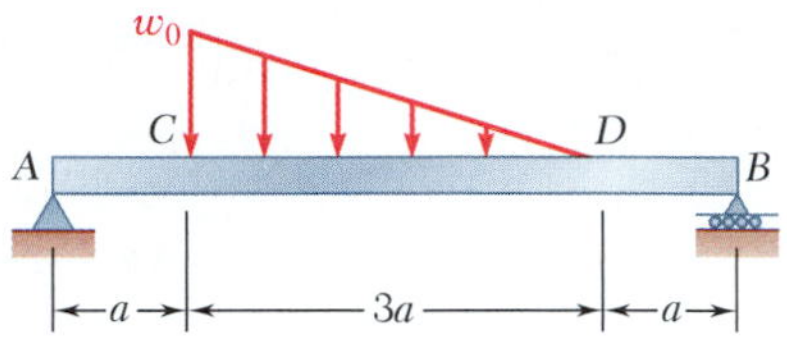

Fig. P7.82

7.82 For the beam shown, (*a*) draw the shear and bending-moment diagrams, (*b*) determine the magnitude and location of the maximum bending moment. (*Hint:* Derive the equations of the shear and bending-moment curves for portion *CD* of the beam.)

7.83 Beam *AB*, which lies on the ground, supports the parabolic load shown. Assuming the upward reaction of the ground to be uniformly distributed, (*a*) write the equations of the shear and bending-moment curves, (*b*) determine the maximum bending moment.

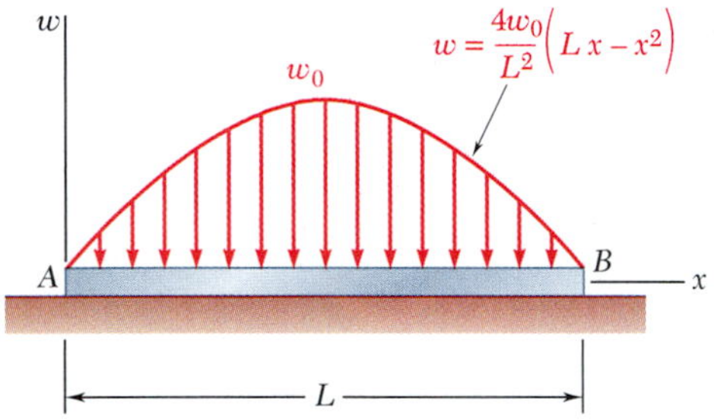

Fig. *P7.83*

7.84 The beam *AB* is subjected to the uniformly distributed load shown and to two unknown forces **P** and **Q**. Knowing that it has been experimentally determined that the bending moment is +325 lb · ft at *D* and +800 lb · ft at *E*, (*a*) determine **P** and **Q**, (*b*) draw the shear and bending-moment diagrams for the beam.

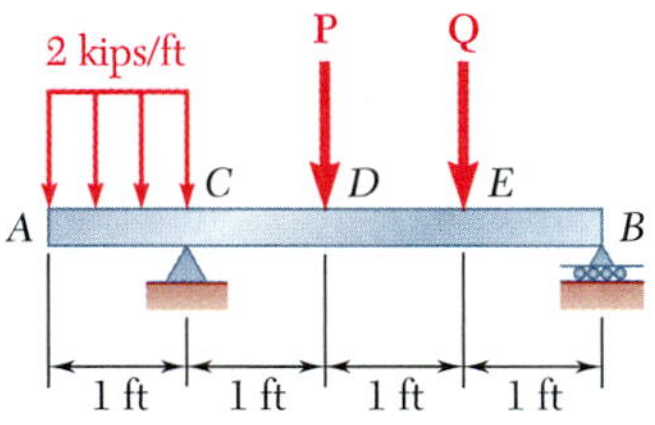

Fig. P7.84

7.85 Solve Prob. 7.84 assuming that the bending moment was found to be +260 lb · ft at *D* and +860 lb · ft at *E*.

***7.86** The beam *AB* is subjected to the uniformly distributed load shown and to two unknown forces **P** and **Q**. Knowing that it has been experimentally determined that the bending moment is +7 kN · m at *D* and +5 kN · m at *E*, (*a*) determine **P** and **Q**, (*b*) draw the shear and bending-moment diagrams for the beam.

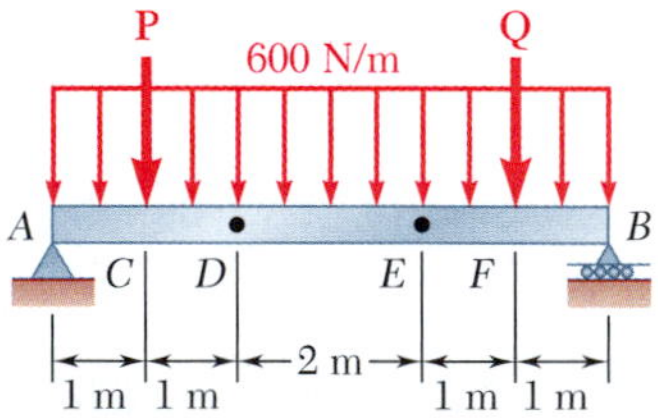

Fig. P7.86

***7.87** Solve Prob. 7.86 assuming that the bending moment was found to be +3.6 kN · m at *D* and +4.14 kN · m at *E*.

*7.7. CABLES WITH CONCENTRATED LOADS

Cables are used in many engineering applications, such as suspension bridges, transmission lines, aerial tramways, guy wires for high towers, etc. Cables may be divided into two categories according to their loading: (1) cables supporting concentrated loads, (2) cables supporting distributed loads. In this section, cables of the first category are examined.

Consider a cable attached to two fixed points A and B and supporting n vertical concentrated loads $\mathbf{P}_1, \mathbf{P}_2, \ldots, \mathbf{P}_n$ (Fig. 7.13*a*). We assume that the cable is *flexible*, that is, that its resistance to bending is small and can be neglected. We further assume that the *weight of the cable is negligible* compared with the loads supported by the cable. Any portion of cable between successive loads can therefore be considered as a two-force member, and the internal forces at any point in the cable reduce to a *force of tension directed along the cable*.

We assume that each of the loads lies in a given vertical line, that is, that the horizontal distance from support A to each of the loads is known; we also assume that the horizontal and vertical distances between the supports are known. We propose to determine the shape of the cable, that is, the vertical distance from support A to each of the points $C_1, C_2, \ldots, C_n$, and also the tension T in each portion of the cable.

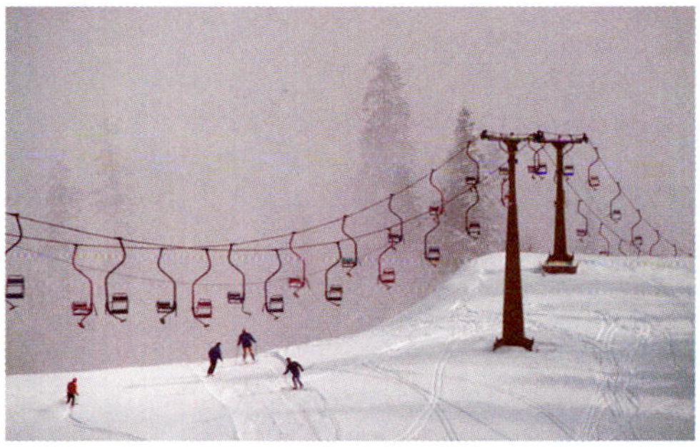

Photo 7.3 Since the weight of the cable of the chairlift shown is negligible compared to the weights of the chairs and skiers, the methods of this section can be used to determine the force at any point in the cable.

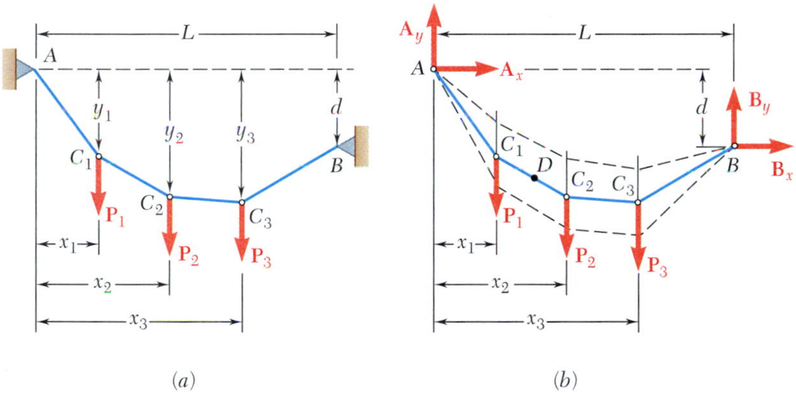

Fig. 7.13

We first draw the free-body diagram of the entire cable (Fig. 7.13*b*). Since the slope of the portions of cable attached at A and B is not known, the reactions at A and B must be represented by two components each. Thus, four unknowns are involved, and the three equations of equilibrium are not sufficient to determine the reactions at A and B.† We must therefore obtain an additional equation by considering the equilibrium of a portion of the cable. This is possible if

†Clearly, the cable is not a rigid body; the equilibrium equations represent, therefore, *necessary but not sufficient conditions* (see Sec. 6.11).

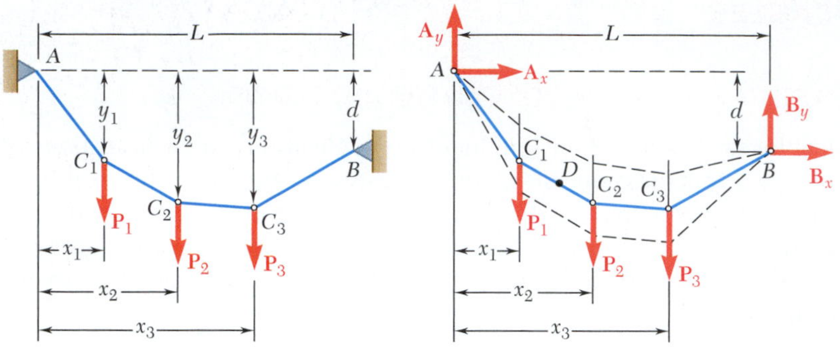

(a) (b)

Fig. 7.13 *(repeated)*

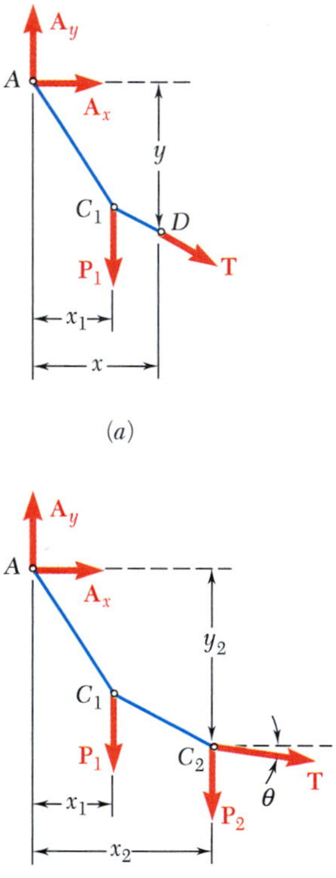

(a)

(b)

Fig. 7.14

we know the coordinates x and y of a point D of the cable. Drawing the free-body diagram of the portion of cable AD (Fig. 7.14*a*) and writing $\Sigma M_D = 0$, we obtain an additional relation between the scalar components A_x and A_y and can determine the reactions at A and B. The problem would remain indeterminate, however, if we did not know the coordinates of D, unless some other relation between A_x and A_y (or between B_x and B_y) were given. The cable might hang in any of various possible ways, as indicated by the dashed lines in Fig. 7.13*b*.

Once A_x and A_y have been determined, the vertical distance from A to any point of the cable can easily be found. Considering point C_2, for example, we draw the free-body diagram of the portion of cable AC_2 (Fig. 7.14*b*). Writing $\Sigma M_{C_2} = 0$, we obtain an equation which can be solved for y_2. Writing $\Sigma F_x = 0$ and $\Sigma F_y = 0$, we obtain the components of the force $\mathbf{T}$ representing the tension in the portion of cable to the right of C_2. We observe that $T \cos \theta = -A_x$; *the horizontal component of the tension force is the same at any point of the cable.* It follows that the tension T is maximum when $\cos \theta$ is minimum, that is, in the portion of cable which has the largest angle of inclination θ. Clearly, this portion of cable must be adjacent to one of the two supports of the cable.

*7.8. CABLES WITH DISTRIBUTED LOADS

Consider a cable attached to two fixed points A and B and carrying a *distributed load* (Fig. 7.15*a*). We saw in the preceding section that for a cable supporting concentrated loads, the internal force at any point is a force of tension directed along the cable. In the case of a cable carrying a distributed load, the cable hangs in the shape of a curve, and the internal force at a point D is a force of tension $\mathbf{T}$ *directed along the tangent to the curve.* In this section, you will learn to determine the tension at any point of a cable supporting a given distributed load. In the following sections, the shape of the cable will be determined for two particular types of distributed loads.

Considering the most general case of distributed load, we draw the free-body diagram of the portion of cable extending from the lowest point C to a given point D of the cable (Fig. 7.15*b*). The forces

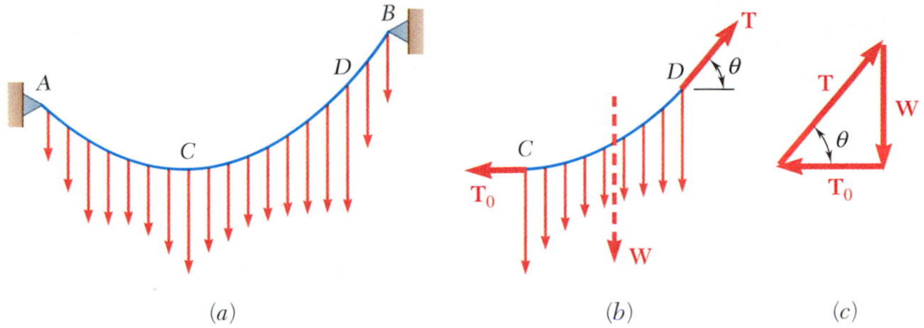

Fig. 7.15

acting on the free body are the tension force $\mathbf{T}_0$ at C, which is horizontal, the tension force $\mathbf{T}$ at D, directed along the tangent to the cable at D, and the resultant $\mathbf{W}$ of the distributed load supported by the portion of cable CD. Drawing the corresponding force triangle (Fig. 7.15c), we obtain the following relations:

$$T \cos \theta = T_0 \qquad T \sin \theta = W \tag{7.5}$$

$$T = \sqrt{T_0^2 + W^2} \qquad \tan \theta = \frac{W}{T_0} \tag{7.6}$$

From the relations (7.5), it appears that the horizontal component of the tension force $\mathbf{T}$ is the same at any point and that the vertical component of $\mathbf{T}$ is equal to the magnitude W of the load measured from the lowest point. Relations (7.6) show that the tension T is minimum at the lowest point and maximum at one of the two points of support.

*7.9. PARABOLIC CABLE

Let us assume, now, that the cable AB carries a load *uniformly distributed along the horizontal* (Fig. 7.16a). Cables of suspension bridges may be assumed loaded in this way, since the weight of the cables is small compared with the weight of the roadway. We denote by w the load per unit length (*measured horizontally*) and express it in N/m or in lb/ft. Choosing coordinate axes with origin at the lowest point C of the cable, we find that the magnitude W of the total load carried by the portion of cable extending from C to the point D of coordinates x and y is $W = wx$. The relations (7.6) defining the magnitude and direction of the tension force at D become

$$T = \sqrt{T_0^2 + w^2x^2} \qquad \tan \theta = \frac{wx}{T_0} \tag{7.7}$$

Moreover, the distance from D to the line of action of the resultant $\mathbf{W}$ is equal to half the horizontal distance from C to D (Fig. 7.16b). Summing moments about D, we write

$$+\circlearrowleft \Sigma M_D = 0: \qquad wx\frac{x}{2} - T_0 y = 0$$

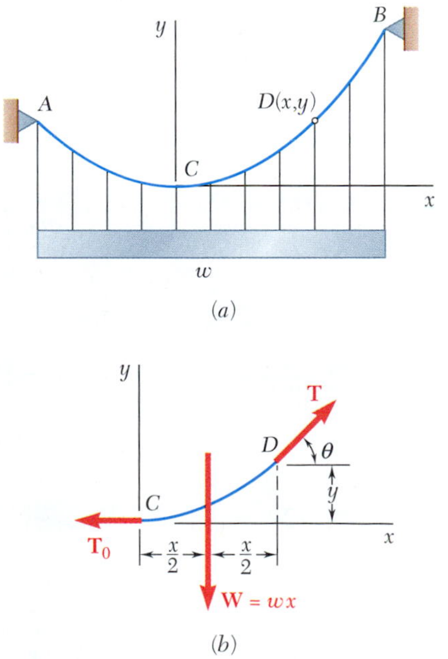

Fig. 7.16

and, solving for y,

$$y = \frac{wx^2}{2T_0} \tag{7.8}$$

This is the equation of a *parabola* with a vertical axis and its vertex at the origin of coordinates. The curve formed by cables loaded uniformly along the horizontal is thus a parabola.†

When the supports A and B of the cable have the same elevation, the distance L between the supports is called the *span* of the cable and the vertical distance h from the supports to the lowest point is called the *sag* of the cable (Fig. 7.17*a*). If the span and sag of a cable are known, and if the load w per unit horizontal length is given, the minimum tension T_0 may be found by substituting $x = L/2$ and $y = h$ in Eq. (7.8). Equations (7.7) will then yield the tension and the slope at any point of the cable, and Eq. (7.8) will define the shape of the cable.

When the supports have different elevations, the position of the lowest point of the cable is not known and the coordinates x_A, y_A and x_B, y_B of the supports must be determined. To this effect, we express that the coordinates of A and B satisfy Eq. (7.8) and that $x_B - x_A = L$ and $y_B - y_A = d$, where L and d denote, respectively, the horizontal and vertical distances between the two supports (Fig. 7.17*b* and *c*).

The length of the cable from its lowest point C to its support B can be obtained from the formula

$$s_B = \int_0^{x_B} \sqrt{1 + \left(\frac{dy}{dx}\right)^2}\, dx \tag{7.9}$$

Differentiating (7.8), we obtain the derivative $dy/dx = wx/T_0$; substituting into (7.9) and using the binomial theorem to expand the radical in an infinite series, we have

$$s_B = \int_0^{x_B} \sqrt{1 + \frac{w^2x^2}{T_0^2}}\, dx = \int_0^{x_B} \left(1 + \frac{w^2x^2}{2T_0^2} - \frac{w^4x^4}{8T_0^4} + \cdots\right) dx$$

$$s_B = x_B\left(1 + \frac{w^2x_B^2}{6T_0^2} - \frac{w^4x_B^4}{40T_0^4} + \cdots\right)$$

and, since $wx_B^2/2T_0 = y_B$,

$$s_B = x_B\left[1 + \frac{2}{3}\left(\frac{y_B}{x_B}\right)^2 - \frac{2}{5}\left(\frac{y_B}{x_B}\right)^4 + \cdots\right] \tag{7.10}$$

The series converges for values of the ratio y_B/x_B less than 0.5; in most cases, this ratio is much smaller, and only the first two terms of the series need be computed.

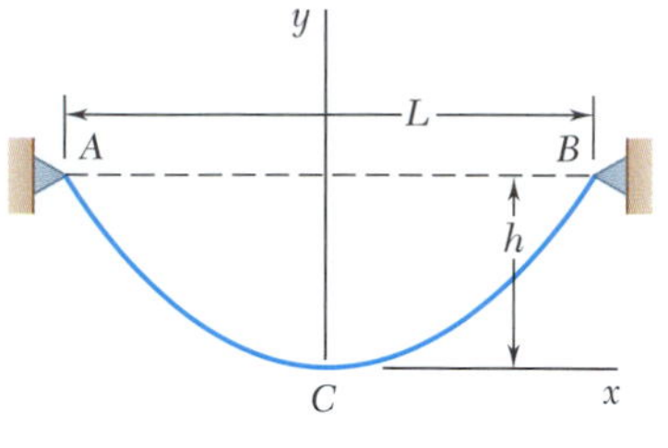

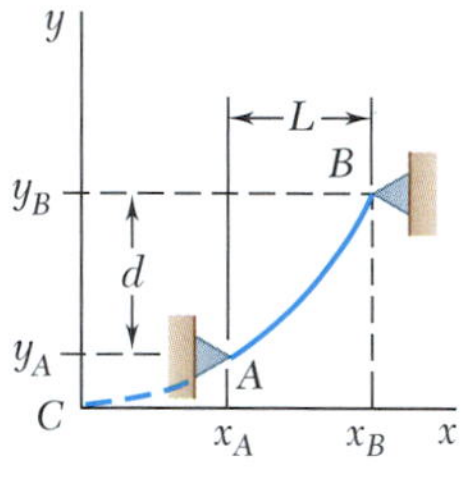

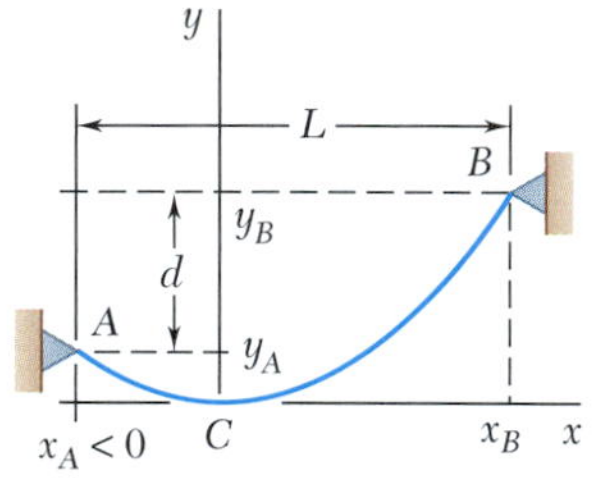

Fig. 7.17

Photo 7.4 Suspension bridges, in which cables support the roadway, are used to span rivers and estuaries. The Verrazano-Narrows Bridge, which connects Staten Island and Brooklyn in New York City, has the longest span of all bridges in the United States.

†Cables hanging under their own weight are not loaded uniformly along the horizontal, and they do not form a parabola. The error introduced by assuming a parabolic shape for cables hanging under their weight, however, is small when the cable is sufficiently taut. A complete discussion of cables hanging under their own weight is given in the next section.

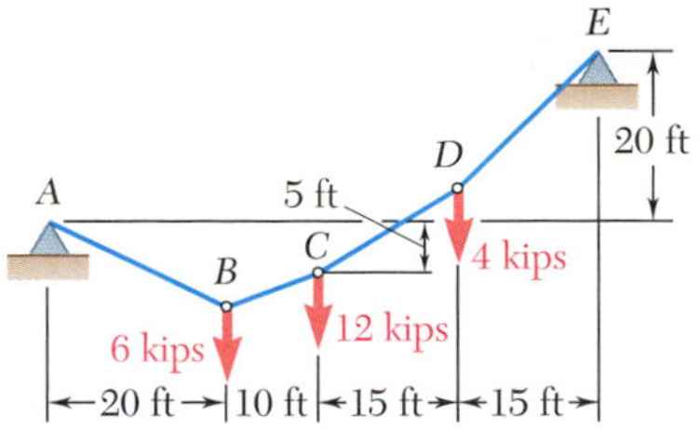

SAMPLE PROBLEM 7.8

The cable AE supports three vertical loads from the points indicated. If point C is 5 ft below the left support, determine (*a*) the elevation of points B and D, (*b*) the maximum slope and the maximum tension in the cable.

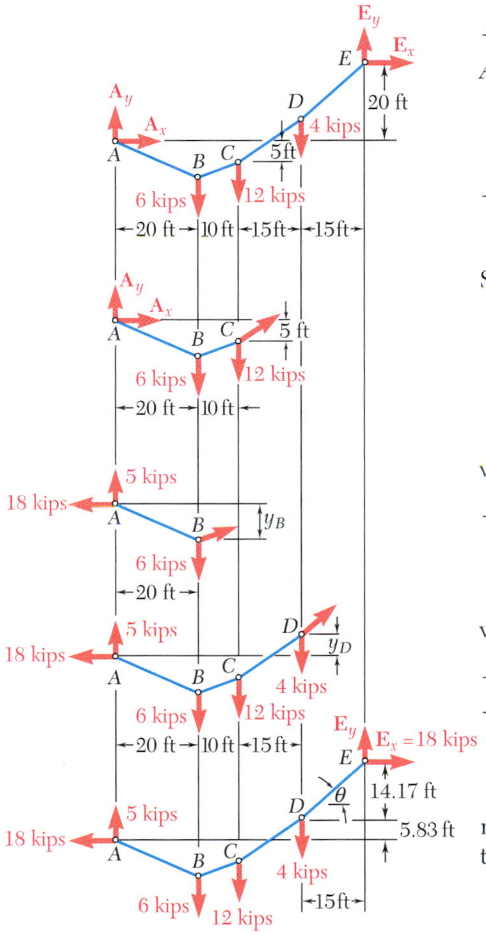

SOLUTION

Reactions at Supports. The reaction components $\mathbf{A}_x$ and $\mathbf{A}_y$ are determined as follows:

Free Body: Entire Cable

$+\circlearrowleft\Sigma M_E = 0$:

$$A_x(20 \text{ ft}) - A_y(60 \text{ ft}) + (6 \text{ kips})(40 \text{ ft}) + (12 \text{ kips})(30 \text{ ft}) + (4 \text{ kips})(15 \text{ ft}) = 0$$

$$20A_x - 60A_y + 660 = 0$$

Free Body: ABC

$$+\circlearrowleft\Sigma M_C = 0: \qquad -A_x(5 \text{ ft}) - A_y(30 \text{ ft}) + (6 \text{ kips})(10 \text{ ft}) = 0$$

$$-5A_x - 30A_y + 60 = 0$$

Solving the two equations simultaneously, we obtain

$$A_x = -18 \text{ kips} \qquad \mathbf{A}_x = 18 \text{ kips} \leftarrow$$
$$A_y = +5 \text{ kips} \qquad \mathbf{A}_y = 5 \text{ kips} \uparrow$$

a. Elevation of Points B and D.

Free Body: AB Considering the portion of cable AB as a free body, we write

$$+\circlearrowleft\Sigma M_B = 0: \qquad (18 \text{ kips})y_B - (5 \text{ kips})(20 \text{ ft}) = 0$$

$y_B = 5.56$ ft below A ◀

Free Body: ABCD. Using the portion of cable $ABCD$ as a free body, we write

$+\circlearrowleft\Sigma M_D = 0$:

$$-(18 \text{ kips})y_D - (5 \text{ kips})(45 \text{ ft}) + (6 \text{ kips})(25 \text{ ft}) + (12 \text{ kips})(15 \text{ ft}) = 0$$

$y_D = 5.83$ ft above A ◀

b. Maximum Slope and Maximum Tension. We observe that the maximum slope occurs in portion DE. Since the horizontal component of the tension is constant and equal to 18 kips, we write

$$\tan\theta = \frac{14.17}{15 \text{ ft}} \qquad \theta = 43.4° \blacktriangleleft$$

$$T_{max} = \frac{18 \text{ kips}}{\cos\theta} \qquad T_{max} = 24.8 \text{ kips} \blacktriangleleft$$

SAMPLE PROBLEM 7.9

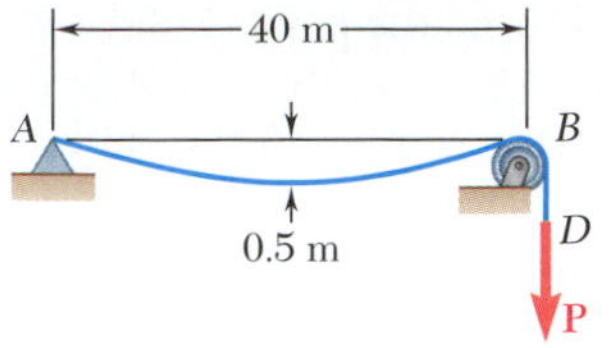

A light cable is attached to a support at A, passes over a small pulley at B, and supports a load **P**. Knowing that the sag of the cable is 0.5 m and that the mass per unit length of the cable is 0.75 kg/m, determine (a) the magnitude of the load **P**, (b) the slope of the cable at B, (c) the total length of the cable from A to B. Since the ratio of the sag to the span is small, assume the cable to be parabolic. Also, neglect the weight of the portion of cable from B to D.

SOLUTION

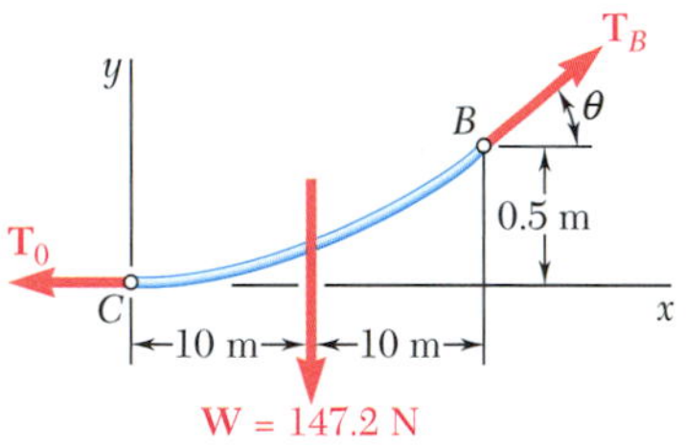

***a.* Load P.** We denote by C the lowest point of the cable and draw the free-body diagram of the portion CB of cable. Assuming the load to be uniformly distributed along the horizontal, we write

$$w = (0.75 \text{ kg/m})(9.81 \text{ m/s}^2) = 7.36 \text{ N/m}$$

The total load for the portion CB of cable is

$$W = wx_B = (7.36 \text{ N/m})(20 \text{ m}) = 147.2 \text{ N}$$

and is applied halfway between C and B. Summing moments about B, we write

$$+\circlearrowleft \Sigma M_B = 0: \qquad (147.2 \text{ N})(10 \text{ m}) - T_0(0.5 \text{ m}) = 0 \qquad T_0 = 2944 \text{ N}$$

From the force triangle we obtain

$$T_B = \sqrt{T_0^2 + W^2}$$
$$= \sqrt{(2944 \text{ N})^2 + (147.2 \text{ N})^2} = 2948 \text{ N}$$

Since the tension on each side of the pulley is the same, we find

$$P = T_B = 2948 \text{ N} \quad ◀$$

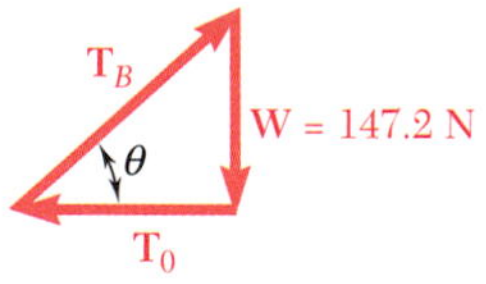

***b.* Slope of Cable at *B*.** We also obtain from the force triangle

$$\tan \theta = \frac{W}{T_0} = \frac{147.2 \text{ N}}{2944 \text{ N}} = 0.05$$

$$\theta = 2.9° \quad ◀$$

***c.* Length of Cable.** Applying Eq. (7.10) between C and B, we write

$$s_B = x_B\left[1 + \frac{2}{3}\left(\frac{y_B}{x_B}\right)^2 + \cdots\right]$$

$$= (20 \text{ m})\left[1 + \frac{2}{3}\left(\frac{0.5 \text{ m}}{20 \text{ m}}\right)^2 + \cdots\right] = 20.00833 \text{ m}$$

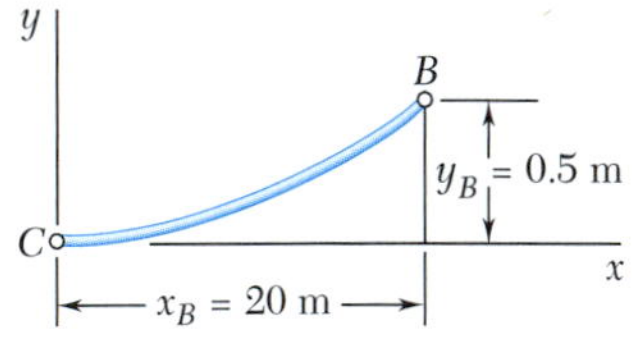

The total length of the cable between A and B is twice this value,

$$\text{Length} = 2s_B = 40.0167 \text{ m} \quad ◀$$

SOLVING PROBLEMS ON YOUR OWN

In the problems of this section you will apply the equations of equilibrium to *cables that lie in a vertical plane.* We assume that a cable cannot resist bending, so that the force of tension in the cable is always directed along the cable.

A. In the first part of this lesson we considered cables subjected to concentrated loads. Since the weight of the cable is neglected, the cable is straight between loads.

Your solution will consist of the following steps:

1. Draw a free-body diagram of the entire cable showing the loads and the horizontal and vertical components of the reaction at each support. Use this free-body diagram to write the corresponding equilibrium equations.

2. You will be confronted with four unknown components and only three equations of equilibrium (see Fig. 7.13). You must therefore find an additional piece of information, such as the *position* of a point on the cable or the *slope* of the cable at a given point.

3. After you have identified the point of the cable where the additional information exists, cut the cable at that point, and draw a free-body diagram of one of the two portions of the cable you have obtained.

a. If you know the position of the point where you have cut the cable, writing $\Sigma M = 0$ about that point for the new free body will yield the additional equation required to solve for the four unknown components of the reactions. [Sample Prob. 7.8].

b. If you know the slope of the portion of the cable you have cut, writing $\Sigma F_x = 0$ and $\Sigma F_y = 0$ for the new free body will yield two equilibrium equations which, together with the original three, can be solved for the four reaction components and for the tension in the cable where it has been cut.

4. To find the elevation of a given point of the cable and the slope and tension at that point once the reactions at the supports have been found, you should cut the cable at that point and draw a free-body diagram of one of the two portions of the cable you have obtained. Writing $\Sigma M = 0$ about the given point yields its elevation. Writing $\Sigma F_x = 0$ and $\Sigma F_y = 0$ yields the components of the tension force, from which its magnitude and direction can easily be found.

(continued)

5. For a cable supporting vertical loads only, you will observe that *the horizontal component of the tension force is the same at any point.* It follows that, for such a cable, the *maximum tension occurs in the steepest portion of the cable.*

B. In the second portion of this lesson we considered cables carrying a load uniformly distributed along the horizontal. The shape of the cable is then parabolic.

Your solution will use one or more of the following concepts:

1. Placing the origin of coordinates at the lowest point of the cable and directing the x and y axes to the right and upward, respectively, we find that *the equation of the parabola* is

$$y = \frac{wx^2}{2T_0} \tag{7.8}$$

The minimum cable tension occurs at the origin, where the cable is horizontal, and the maximum tension is at the support where the slope is maximum.

2. If the supports of the cable have the same elevation, the sag h of the cable is the vertical distance from the lowest point of the cable to the horizontal line joining the supports. To solve a problem involving such a parabolic cable, you should write Eq. (7.8) for one of the supports; this equation can be solved for one unknown.

3. If the supports of the cable have different elevations, you will have to write Eq. (7.8) for each of the supports (see Fig. 7.17).

4. To find the length of the cable from the lowest point to one of the supports, you can use Eq. (7.10). In most cases, you will need to compute only the first two terms of the series.

In Sample Prob. 7.9 we considered a cable hanging under its own weight. We observe that for taut cables their weights per unit length in the horizontal direction are approximately constant. This was the external loading condition of the cables analyzed in Sec. 7.9. Thus, we are able to apply the results obtained for this type of external loading to the special case of *taut* cables hanging under their own weights if w is now taken as the weight per unit length of the cables.

Problems

7.88 Two loads are suspended as shown from cable *ABCD*. Knowing that $d_C = 1.5$ ft, determine (*a*) the distance d_B, (*b*) the components of the reaction at *A*, (*c*) the maximum tension in the cable.

7.89 Two loads are suspended as shown from cable *ABCD*. Knowing that the maximum tension in the cable is 720 lb, determine (*a*) the distance d_B, (*b*) the distance d_C.

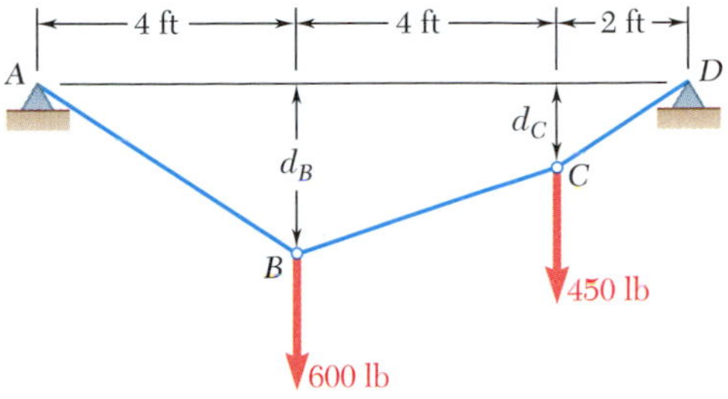

Fig. P7.88 and *P7.89*

7.90 Knowing that $d_C = 4$ m, determine (*a*) the reaction at *A*, (*b*) the reaction at *E*.

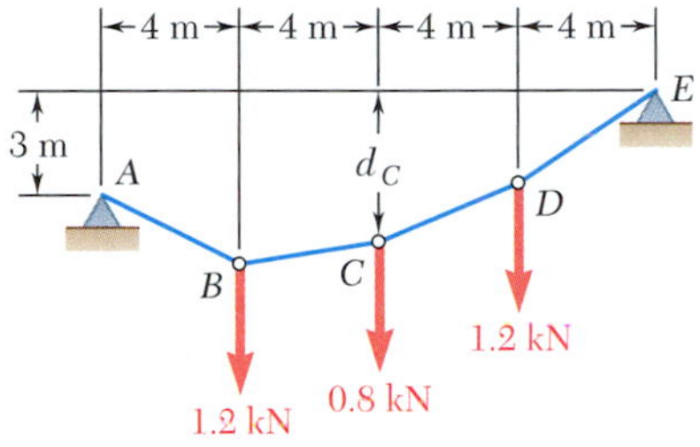

Fig. P7.90 and P7.91

7.91 Knowing that $d_C = 2.25$ m, determine (*a*) the reaction at *A*, (*b*) the reaction at *E*.

7.92 Cable *ABCDE* supports three loads as shown. Knowing that $d_C = 3.6$ ft, determine (*a*) the reaction at *E*, (*b*) the distances d_B and d_D.

7.93 Cable *ABCDE* supports three loads as shown. Determine (*a*) the distance d_C for which portion *CD* of the cable is horizontal, (*b*) the corresponding reactions at the supports.

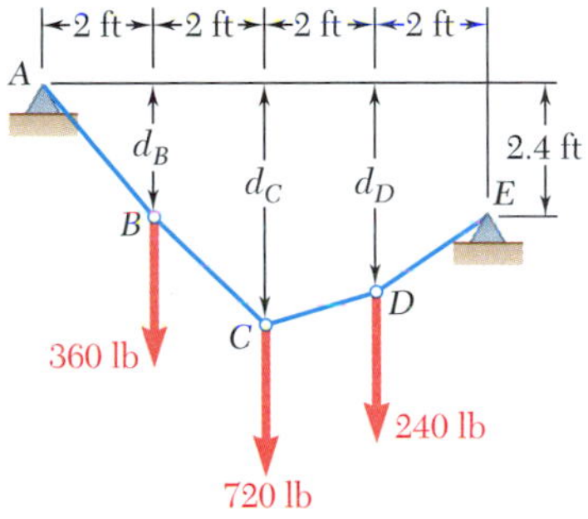

Fig. P7.92 and *P7.93*

7.94 An oil pipeline is supported at 6-m intervals by vertical hangers attached to the cable shown. Due to the combined weight of the pipe and its contents, the tension in each hanger is 4 kN. Knowing that $d_C = 12$ m, determine (*a*) the maximum tension in the cable, (*b*) the distance d_D.

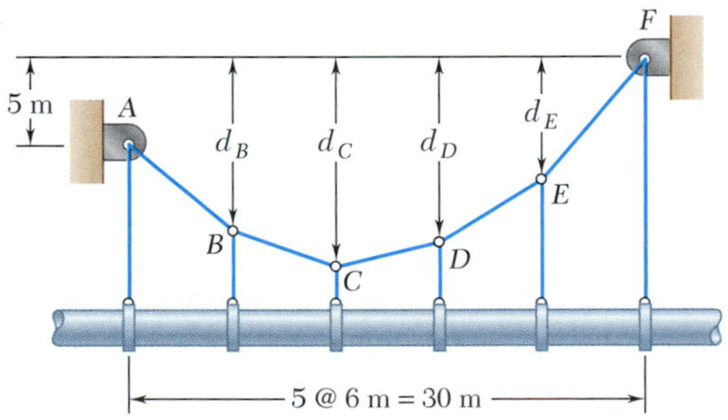

Fig. P7.94

7.95 Solve Prob. 7.94 assuming that $d_C = 9$ m.

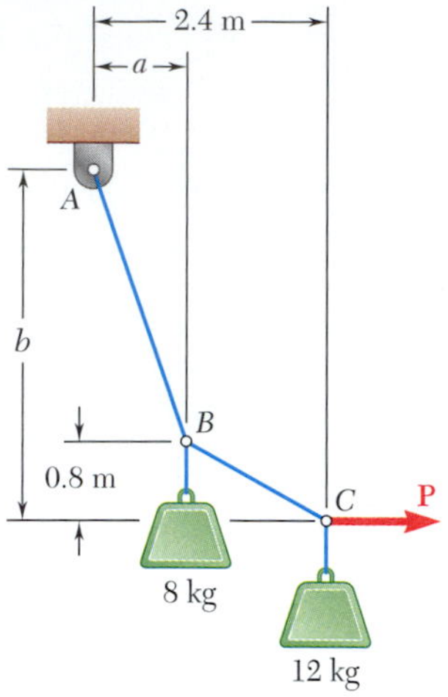

Fig. P7.96 and *P7.97*

7.96 Cable ABC supports two boxes as shown. Knowing that b = 3.6 m, determine (*a*) the required magnitude of the horizontal force **P**, (*b*) the corresponding distance a.

7.97 Cable ABC supports two boxes as shown. Determine the distances a and b when a horizontal force **P** of magnitude 100 N is applied at C.

7.98 Knowing that W_B = 150 lb and W_C = 50 lb, determine the magnitude of the force **P** required to maintain equilibrium.

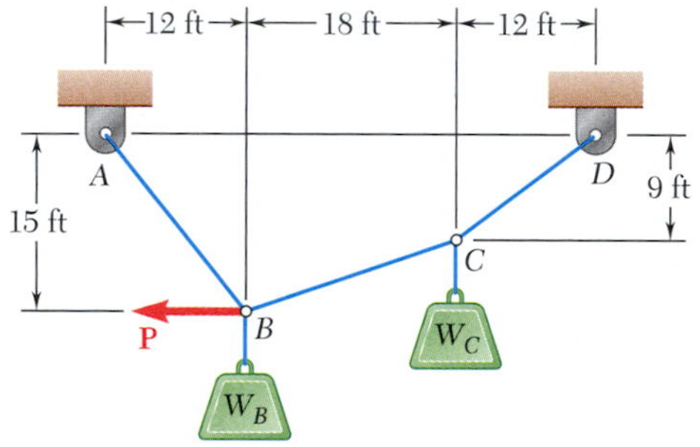

Fig. P7.98 and P7.99

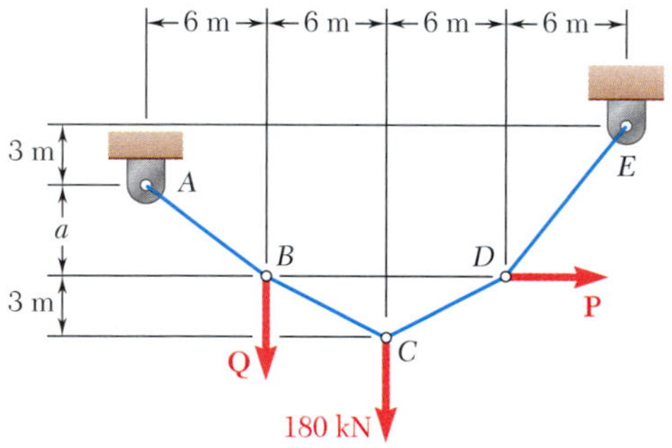

Fig. P7.100 and *P7.101*

7.99 Knowing that W_B = 40 lb and W_C = 22 lb, determine the magnitude of the force **P** required to maintain equilibrium.

7.100 If a = 4.5 m, determine the magnitudes of **P** and **Q** required to maintain the cable in the shape shown.

7.101 If a = 6 m, determine the magnitudes of **P** and **Q** required to maintain the cable in the shape shown.

7.102 A transmission cable having a mass per unit length of 1 kg/m is strung between two insulators at the same elevation that are 60 m apart. Knowing that the sag of the cable is 1.2 m, determine (*a*) the maximum tension in the cable, (*b*) the length of the cable.

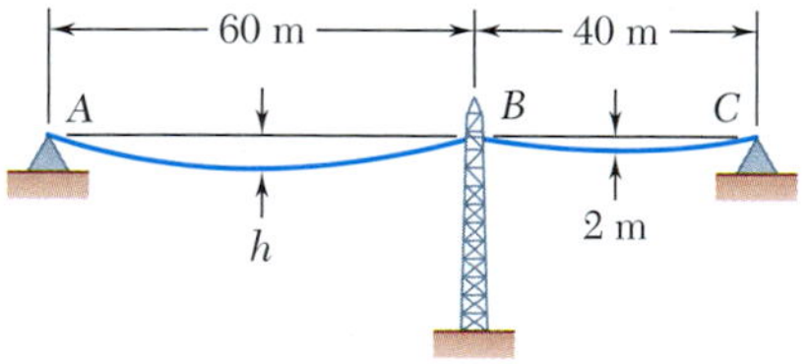

Fig. *P7.103*

7.103 Two cables of the same gauge are attached to a transmission tower at B. Since the tower is slender, the horizontal component of the resultant of the forces exerted by the cables at B is to be zero. Knowing that the mass per unit length of the cables is 0.4 kg/m, determine (*a*) the required sag h, (*b*) the maximum tension in each cable.

7.104 The center span of the George Washington Bridge, as originally constructed, consisted of a uniform roadway suspended from four cables. The uniform load supported by each cable was w = 9.75 kips/ft along the horizontal. Knowing that the span L is 3500 ft and that the sag h is 316 ft, determine for the original configuration (*a*) the maximum tension in each cable, (*b*) the length of each cable.

7.105 Each cable of the Golden Gate Bridge supports a load w = 11.1 kips/ft along the horizontal. Knowing that the span L is 4150 ft and that the sag h is 464 ft, determine (*a*) the maximum tension in each cable, (*b*) the length of each cable.

7.106 To mark the positions of the rails on the posts of a fence, a homeowner ties a cord to the post at A, passes the cord over a short piece of pipe attached to the post at B, and ties the free end of the cord to a bucket filled with bricks having a total mass of 20 kg. Knowing that the mass per unit length of the rope is 0.02 kg/m and assuming that A and B are at the same elevation, determine (*a*) the sag h, (*b*) the slope of the cable at B. Neglect the effect of friction.

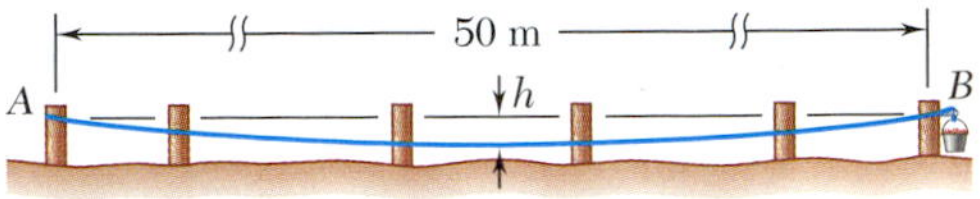

Fig. P7.106

7.107 A small ship is tied to a pier with a 5-m length of rope as shown. Knowing that the current exerts on the hull of the ship a 300-N force directed from the bow to the stern and that the mass per unit length of the rope is 2.2 kg/m, determine (*a*) the maximum tension in the rope, (*b*) the sag h. [*Hint:* Use only the first two terms of Eq. (7.10).]

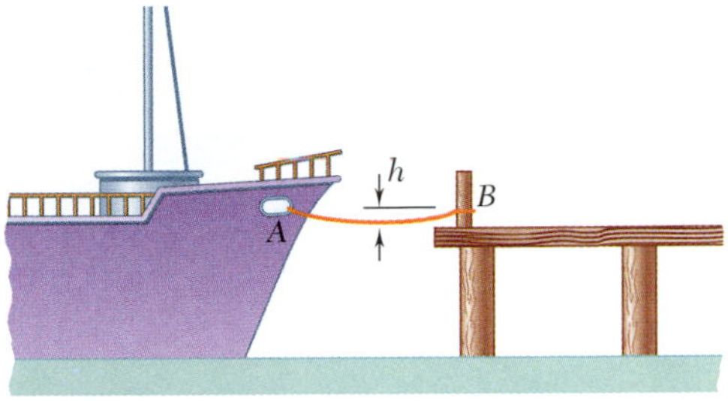

Fig. *P7.107*

7.108 The center span of the Verrazano-Narrows Bridge consists of two uniform roadways suspended from four cables. The design of the bridge allowed for the effect of extreme temperature changes which cause the sag of the center span to vary from $h_w = 386$ ft in winter to $h_s = 394$ ft in summer. Knowing that the span is $L = 4260$ ft, determine the change in length of the cables due to extreme temperature changes.

7.109 A cable of length $L + \Delta$ is suspended between two points which are at the same elevation and a distance L apart. (*a*) Assuming that Δ is small compared to L and that the cable is parabolic, determine the approximate sag in terms of L and Δ. (*b*) If $L = 30$ m and $\Delta = 1.2$ m, determine the approximate sag. [*Hint:* Use only the first two terms of Eq. (7.10).]

7.110 Each cable of the side spans of the Golden Gate Bridge supports a load $w = 10.2$ kips/ft along the horizontal. Knowing that for the side spans the maximum vertical distance h from each cable to the chord AB is 30 ft and occurs at midspan, determine (*a*) the maximum tension in each cable, (*b*) the slope at B.

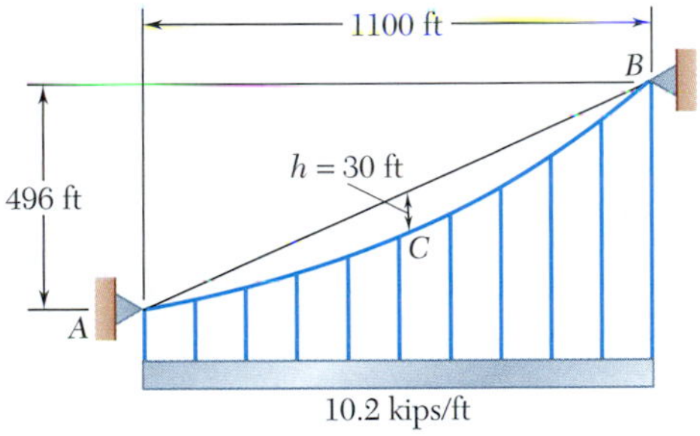

Fig. P7.110

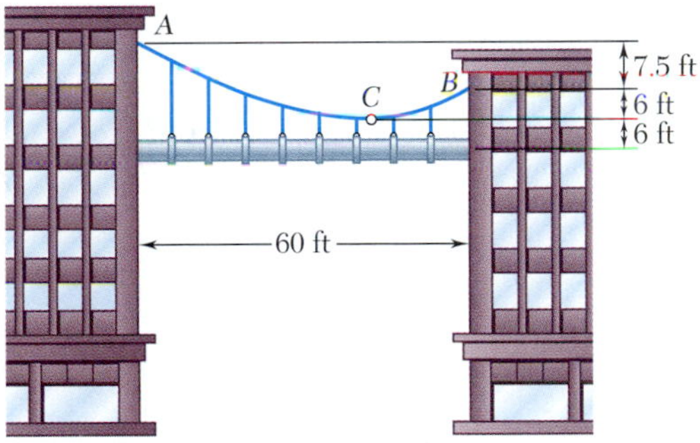

Fig. P7.111

7.111 A steam pipe weighing 50 lb/ft that passes between two buildings 60 ft apart is supported by a system of cables as shown. Assuming that the weight of the cable is equivalent to a uniformly distributed loading of 7.5 lb/ft, determine (*a*) the location of the lowest point C of the cable, (*b*) the maximum tension in the cable.

7.112 Chain AB supports a horizontal, uniform steel beam having a mass per unit length of 85 kg/m. If the maximum tension in the cable is not to exceed 8 kN, determine (*a*) the horizontal distance a from A to the lowest point C of the chain, (*b*) the approximate length of the chain.

7.113 Chain AB of length 6.4 m supports a horizontal, uniform steel beam having a mass per unit length of 85 kg/m. Determine (*a*) the horizontal distance a from A to the lowest point C of the chain, (*b*) the maximum tension in the chain.

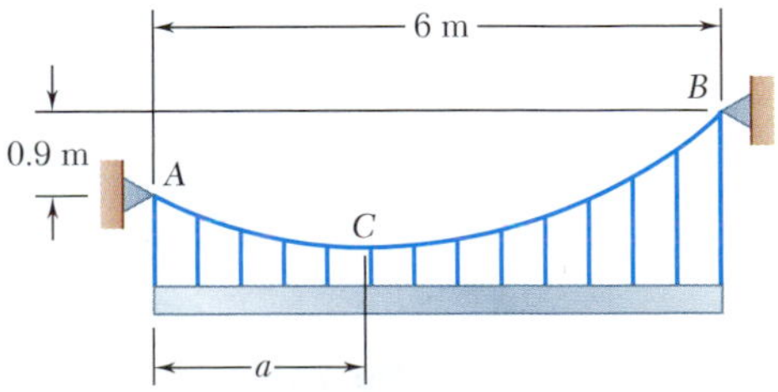

Fig. P7.112 and P7.113

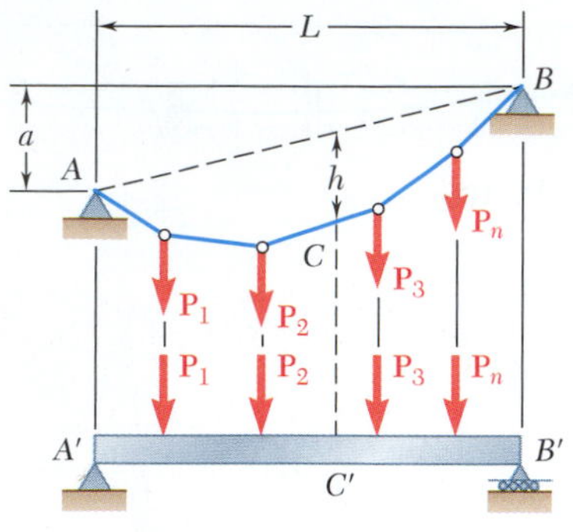

Fig. P7.114

***7.114** A cable AB of span L and a simple beam $A'B'$ of the same span are subjected to identical vertical loadings as shown. Show that the magnitude of the bending moment at a point C' in the beam is equal to the product T_0h, where T_0 is the magnitude of the horizontal component of the tension force in the cable and h is the vertical distance between point C and the chord joining the points of support A and B.

7.115 through 7.118 Making use of the property established in Prob. 7.114, solve the problem indicated by first solving the corresponding beam problem.

7.115 Prob. 7.89*a*.
7.116 Prob. 7.92*b*.
7.117 Prob. 7.94*b*.
7.118 Prob. 7.95*b*.

***7.119** Show that the curve assumed by a cable that carries a distributed load $w(x)$ is defined by the differential equation $d^2y/dx^2 = w(x)/T_0$, where T_0 is the tension at the lowest point.

***7.120** Using the property indicated in Prob. 7.119, determine the curve assumed by a cable of span L and sag h carrying a distributed load $w = w_0 \cos(\pi x/L)$, where x is measured from mid-span. Also determine the maximum and minimum values of the tension in the cable.

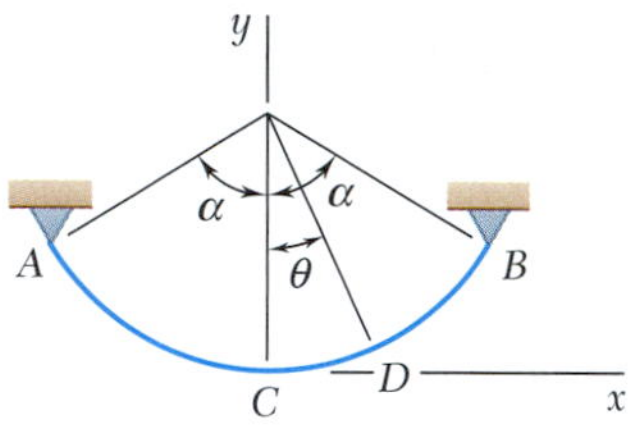

Fig. P7.121

***7.121** If the weight per unit length of the cable AB is $w_0/\cos^2\theta$, prove that the curve formed by the cable is a circular arc. (*Hint:* Use the property indicated in Prob. 7.119.)

*7.10. CATENARY

Let us now consider a cable AB carrying a load *uniformly distributed along the cable itself* (Fig. 7.18*a*). Cables hanging under their own weight are loaded in this way. We denote by w the load per unit length (*measured along the cable*) and express it in N/m or in lb/ft. The magnitude W of the total load carried by a portion of cable of length s extending from the lowest point C to a point D is $W = ws$. Substituting this value for W in formula (7.6), we obtain the tension at D:

$$T = \sqrt{T_0^2 + w^2s^2}$$

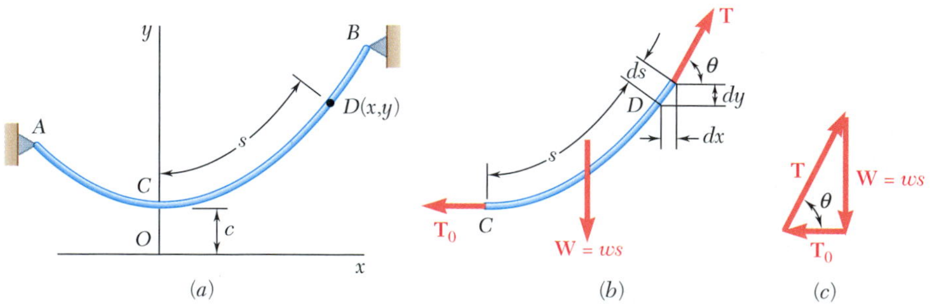

Fig. 7.18

In order to simplify the subsequent computations, we introduce the constant $c = T_0/w$. We thus write

$$T_0 = wc \qquad W = ws \qquad T = w\sqrt{c^2 + s^2} \tag{7.11}$$

The free-body diagram of the portion of cable CD is shown in Fig. 7.18*b*. This diagram, however, cannot be used to obtain directly the equation of the curve assumed by the cable, since we do not know the horizontal distance from D to the line of action of the resultant $\mathbf{W}$ of the load. To obtain this equation, we first write that the horizontal projection of a small element of cable of length ds is $dx = ds \cos \theta$. Observing from Fig. 7.18*c* that $\cos \theta = T_0/T$ and using (7.11), we write

$$dx = ds \cos \theta = \frac{T_0}{T} ds = \frac{wc\, ds}{w\sqrt{c^2 + s^2}} = \frac{ds}{\sqrt{1 + s^2/c^2}}$$

Photo 7.5 The forces on the supports and the internal forces in the cables of the power line shown are discussed in this section.

Selecting the origin O of the coordinates at a distance c directly below C (Fig. 7.18*a*) and integrating from $C(0, c)$ to $D(x, y)$, we obtain†

$$x = \int_0^s \frac{ds}{\sqrt{1 + s^2/c^2}} = c\left[\sinh^{-1} \frac{s}{c}\right]_0^s = c \sinh^{-1} \frac{s}{c}$$

This equation, which relates the length s of the portion of cable CD and the horizontal distance x, can be written in the form

$$s = c \sinh \frac{x}{c} \tag{7.15}$$

The relation between the coordinates x and y can now be obtained by writing $dy = dx \tan \theta$. Observing from Fig. 7.18*c* that $\tan \theta = W/T_0$ and using (7.11) and (7.15), we write

$$dy = dx \tan \theta = \frac{W}{T_0} dx = \frac{s}{c} dx = \sinh \frac{x}{c} dx$$

†This integral can be found in all standard integral tables. The function

$$z = \sinh^{-1} u$$

(read "arc hyperbolic sine u") is the *inverse* of the function $u = \sinh z$ (read "hyperbolic sine z"). This function and the function $v = \cosh z$ (read "hyperbolic cosine z") are defined as follows:

$$u = \sinh z = \tfrac{1}{2}(e^z - e^{-z}) \qquad v = \cosh z = \tfrac{1}{2}(e^z + e^{-z})$$

Numerical values of the functions sinh z and cosh z are found in *tables of hyperbolic functions.* They may also be computed on most calculators either directly or from the above definitions. The student is referred to any calculus text for a complete description of the properties of these functions. In this section, we use only the following properties, which are easily derived from the above definitions:

$$\frac{d \sinh z}{dz} = \cosh z \qquad \frac{d \cosh z}{dz} = \sinh z \tag{7.12}$$

$$\sinh 0 = 0 \qquad \cosh 0 = 1 \tag{7.13}$$

$$\cosh^2 z - \sinh^2 z = 1 \tag{7.14}$$

Integrating from $C(0, c)$ to $D(x, y)$ and using (7.12) and (7.13), we obtain

$$y - c = \int_0^x \sinh\frac{x}{c}\,dx = c\left[\cosh\frac{x}{c}\right]_0^x = c\left(\cosh\frac{x}{c} - 1\right)$$
$$y - c = c\cosh\frac{x}{c} - c$$

which reduces to

$$y = c\cosh\frac{x}{c} \tag{7.16}$$

This is the equation of a *catenary* with a vertical axis. The ordinate c of the lowest point C is called the *parameter* of the catenary. Squaring both sides of Eqs. (7.15) and (7.16), subtracting, and taking (7.14) into account, we obtain the following relation between y and s:

$$y^2 - s^2 = c^2 \tag{7.17}$$

Solving (7.17) for s^2 and carrying into the last of the relations (7.11), we write these relations as follows:

$$T_0 = wc \qquad W = ws \qquad T = wy \tag{7.18}$$

The last relation indicates that the tension at any point D of the cable is proportional to the vertical distance from D to the horizontal line representing the x axis.

When the supports A and B of the cable have the same elevation, the distance L between the supports is called the *span* of the cable and the vertical distance h from the supports to the lowest point C is called the *sag* of the cable. These definitions are the same as those given in the case of parabolic cables, but it should be noted that because of our choice of coordinate axes, the sag h is now

$$h = y_A - c \tag{7.19}$$

It should also be observed that certain catenary problems involve transcendental equations which must be solved by successive approximations (see Sample Prob. 7.10). When the cable is fairly taut, however, the load can be assumed uniformly distributed *along the horizontal* and the catenary can be replaced by a parabola. This greatly simplifies the solution of the problem, and the error introduced is small.

When the supports A and B have different elevations, the position of the lowest point of the cable is not known. The problem can then be solved in a manner similar to that indicated for parabolic cables, by expressing that the cable must pass through the supports and that $x_B - x_A = L$ and $y_B - y_A = d$, where L and d denote, respectively, the horizontal and vertical distances between the two supports.

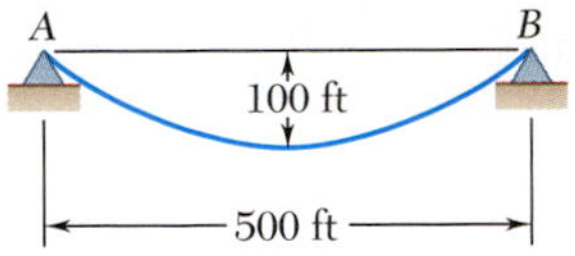

SAMPLE PROBLEM 7.10

A uniform cable weighing 3 lb/ft is suspended between two points A and B as shown. Determine (*a*) the maximum and minimum values of the tension in the cable, (*b*) the length of the cable.

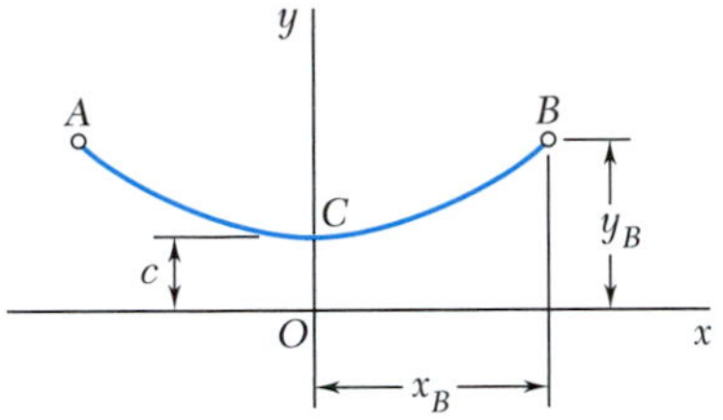

SOLUTION

Equation of Cable. The origin of coordinates is placed at a distance c below the lowest point of the cable. The equation of the cable is given by Eq. (7.16),

$$y = c \cosh \frac{x}{c}$$

The coordinates of point B are

$$x_B = 250 \text{ ft} \qquad y_B = 100 + c$$

Substituting these coordinates into the equation of the cable, we obtain

$$100 + c = c \cosh \frac{250}{c}$$

$$\frac{100}{c} + 1 = \cosh \frac{250}{c}$$

The value of c is determined by assuming successive trial values, as shown in the following table:

c	$\frac{250}{c}$	$\frac{100}{c}$	$\frac{100}{c} + 1$	$\cosh \frac{250}{c}$
300	0.833	0.333	1.333	1.367
350	0.714	0.286	1.286	1.266
330	0.758	0.303	1.303	1.301
328	0.762	0.305	1.305	1.305

Taking $c = 328$, we have

$$y_B = 100 + c = 428 \text{ ft}$$

***a.* Maximum and Minimum Values of the Tension.** Using Eqs. (7.18), we obtain

$$T_{\text{min}} = T_0 = wc = (3 \text{ lb/ft})(328 \text{ ft}) \qquad T_{\text{min}} = 984 \text{ lb} \blacktriangleleft$$

$$T_{\text{max}} = T_B = wy_B = (3 \text{ lb/ft})(428 \text{ ft}) \qquad T_{\text{max}} = 1284 \text{ lb} \blacktriangleleft$$

***b.* Length of Cable.** One-half the length of the cable is found by solving Eq. (7.17):

$$y_B^2 - s_{CB}^2 = c^2 \qquad s_{CB}^2 = y_B^2 - c^2 = (428)^2 - (328)^2 \qquad s_{CB} = 275 \text{ ft}$$

The total length of the cable is therefore

$$s_{AB} = 2s_{CB} = 2(275 \text{ ft}) \qquad s_{AB} = 550 \text{ ft} \blacktriangleleft$$

SOLVING PROBLEMS ON YOUR OWN

In the last section of this chapter you learned to solve problems involving a *cable carrying a load uniformly distributed along the cable*. The shape assumed by the cable is a catenary and is defined by the equation:

$$y = c \cosh \frac{x}{c} \tag{7.16}$$

1. *You should keep in mind that the origin of coordinates for a catenary is located at a distance c directly below the lowest point of the catenary.* The length of the cable from the origin to any point is expressed as

$$s = c \sinh \frac{x}{c} \tag{7.15}$$

2. *You should first identify all of the known and unknown quantities.* Then consider each of the equations listed in the text (Eqs. 7.15 through 7.19), and solve an equation that contains only one unknown. Substitute the value found into another equation, and solve that equation for another unknown.

3. *If the sag h is given,* use Eq. (7.19) to replace y by $h + c$ in Eq. (7.16) if x is known [Sample Prob. 7.10], or in Eq. (7.17) if s is known, and solve the equation obtained for the constant c.

4. *Many of the problems that you will encounter will involve the solution by trial and error* of an equation involving a hyperbolic sine or cosine. You can make your work easier by keeping track of your calculations in a table, as in Sample Prob. 7.10. Of course, you also can use a computer- or calculator-based numerical solution technique.

5. *It is important to recognize that the defining equation may have more than one solution.* To help you visualize this possibility, we encourage you first to graph the equation and then to solve for the root(s). If there is more than one root, you must determine which of the roots correspond to physically possible situations.

Problems

7.122 Two hikers are standing 30-ft apart and are holding the ends of a 35-ft length of rope as shown. Knowing that the weight per unit length of the rope is 0.05 lb/ft, determine (*a*) the sag h, (*b*) the magnitude of the force exerted on the hand of a hiker.

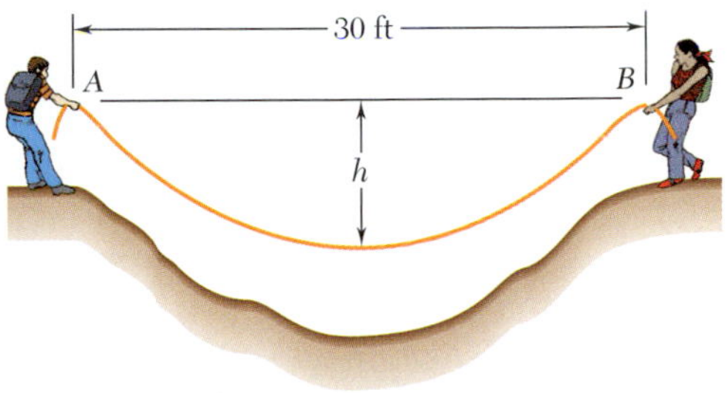

Fig. P7.122

7.123 A 60-ft chain weighing 120 lb is suspended between two points at the same elevation. Knowing that the sag is 24 ft, determine (*a*) the distance between the supports, (*b*) the maximum tension in the chain.

7.124 A 200-ft steel surveying tape weighs 4 lb. If the tape is stretched between two points at the same elevation and pulled until the tension at each end is 16 lb, determine the horizontal distance between the ends of the tape. Neglect the elongation of the tape due to the tension.

7.125 An electric transmission cable of length 130 m and mass per unit length of 3.4 kg/m is suspended between two points at the same elevation. Knowing that the sag is 30 m, determine the horizontal distance between the supports and the maximum tension.

7.126 A 30-m length of wire having a mass per unit length of 0.3 kg/m is attached to a fixed support at A and to a collar at B. Neglecting the effect of friction, determine (*a*) the force **P** for which $h = 12$ m, (*b*) the corresponding span L.

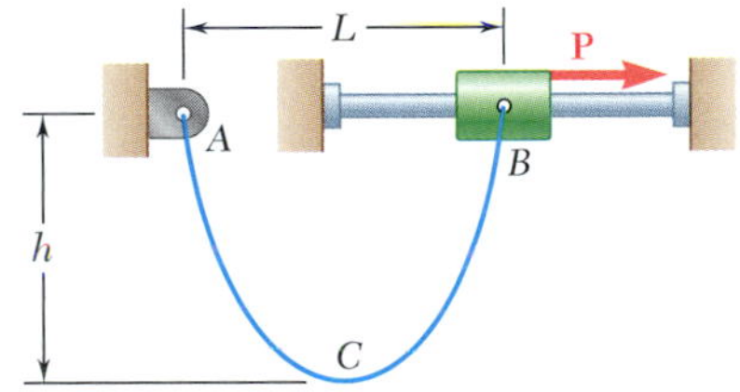

Fig. P7.126, *P7.127,* and P7.128

7.127 A 30-m length of wire having a mass per unit length of 0.3 kg/m is attached to a fixed support at A and to a collar at B. Knowing that the magnitude of the horizontal force applied to the collar is $P = 30$ N, determine (*a*) the sag h, (*b*) the corresponding span L.

7.128 A 30-m length of wire having a mass per unit length of 0.3 kg/m is attached to a fixed support at A and to a collar at B. Neglecting the effect of friction, determine (*a*) the sag h for which $L = 22.5$ m, (*b*) the corresponding force **P**.

7.129 A 30-ft wire is suspended from two points at the same elevation that are 20 ft apart. Knowing that the maximum tension is 80 lb, determine (*a*) the sag of the wire, (*b*) the total weight of the wire.

7.130 Determine the sag of a 45-ft chain which is attached to two points at the same elevation that are 20 ft apart.

7.131 A 10-m rope is attached to two supports A and B as shown. Determine (*a*) the span of the rope for which the span is equal to the sag, (*b*) the corresponding angle θ_B.

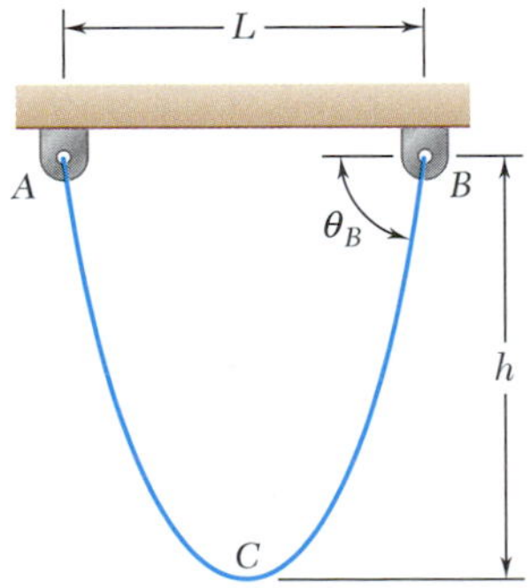

Fig. P7.131

7.132 A cable having a mass per unit length of 3 kg/m is suspended between two points at the same elevation that are 48 m apart. Determine the smallest allowable sag of the cable if the maximum tension is not to exceed 1800 N.

7.133 An 8-m length of chain having a mass per unit length of 3.72 kg/m is attached to a beam at A and passes over a small pulley at B as shown. Neglecting the effect of friction, determine the values of distance a for which the chain is in equilibrium.

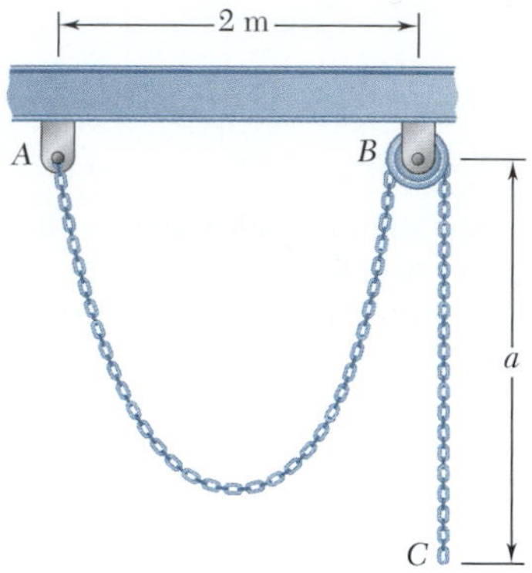

Fig. P7.133

7.134 A motor M is used to slowly reel in the cable shown. Knowing that the weight per unit length of the cable is 0.5 lb/ft, determine the maximum tension in the cable when $h = 15$ ft.

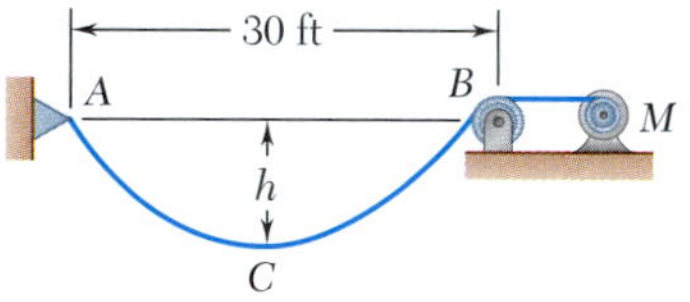

Fig. P7.134 and *P7.135*

7.135 A motor M is used to slowly reel in the cable shown. Knowing that the weight per unit length of the cable is 0.5 lb/ft, determine the maximum tension in the cable when $h = 9$ ft.

7.136 To the left of point B the long cable $ABDE$ rests on the rough horizontal surface shown. Knowing that the weight per unit length of the cable is 1.5 lb/ft, determine the force $\mathbf{F}$ when $a = 10.8$ ft.

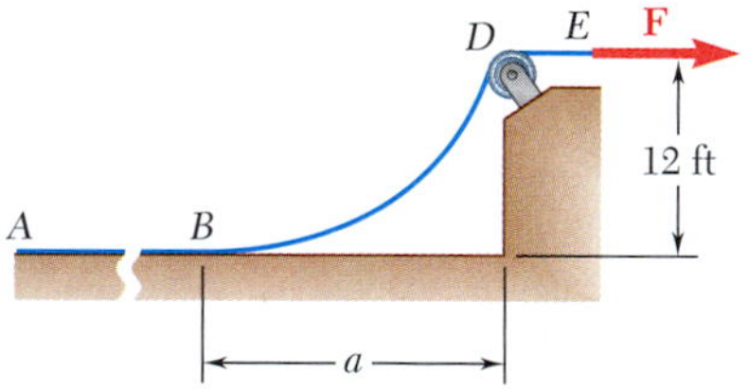

Fig. P7.136 and P7.137

7.137 To the left of point B the long cable $ABDE$ rests on the rough horizontal surface shown. Knowing that the weight per unit length of the cable is 1.5 lb/ft, determine the force $\mathbf{F}$ when $a = 18$ ft.

7.138 The uniform cable shown in Fig. P7.138 and P7.139 has a mass per unit length of 4 kg/m and is held in the position shown by a horizontal force $\mathbf{P}$ applied at B. Knowing that $P = 800$ N and $\theta_A = 60°$, determine (*a*) the location of point B, (*b*) the length of the cable.

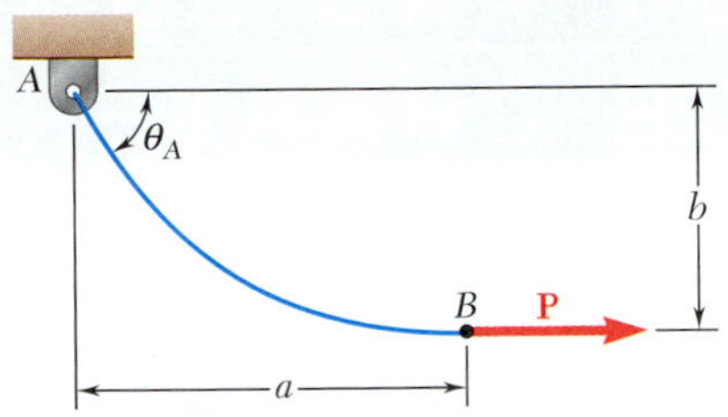

Fig. P7.138 and P7.139

7.139 A uniform cable having a mass per unit length of 4 kg/m is held in the position shown by a horizontal force **P** applied at B. Knowing that $P =$ 600 N and $\theta_A = 60°$, determine (*a*) the location of point B, (*b*) the length of the cable.

7.140 The cable ACB weighs 0.3 lb/ft. Knowing that the lowest point of the cable is located at a distance $a = 1.8$ ft below the support A, determine (*a*) the location of the lowest point C, (*b*) the maximum tension in the cable.

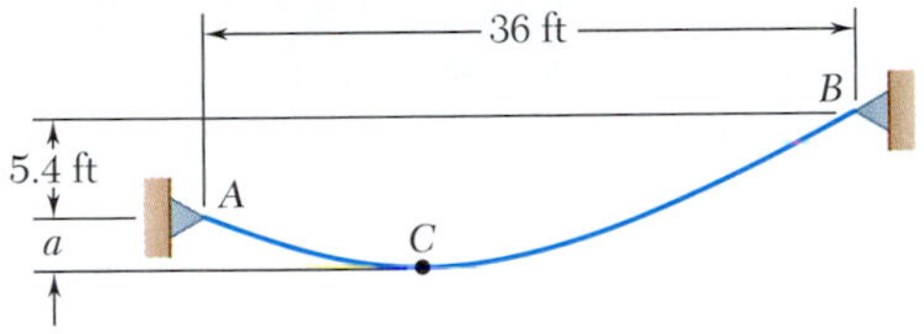

Fig. P7.140 and P7.141

7.141 The cable ACB weighs 0.3 lb/ft. Knowing that the lowest point of the cable is located at a distance $a = 6$ ft below the support A, determine (*a*) the location of the lowest point C, (*b*) the maximum tension in the cable.

7.142 Denoting by θ the angle formed by a uniform cable and the horizontal, show that at any point (*a*) $s = c \tan \theta$, (*b*) $y = c \sec \theta$.

7.143 (*a*) Determine the maximum allowable horizontal span for a uniform cable of mass per unit length m' if the tension in the cable is not to exceed a given value T_m. (*b*) Using the result of part *a*, determine the maximum span of a steel wire for which $m' = 0.34$ kg/m and $T_m = 32$ kN.

***7.144** A cable has a weight per unit length of 2 lb/ft and is supported as shown. Knowing that the span L is 18 ft, determine the two values of the sag h for which the maximum tension is 80 lb.

***7.145** Determine the sag-to-span ratio for which the maximum tension in the cable is equal to the total weight of the entire cable AB.

***7.146** A cable of weight w per unit length is suspended between two points at the same elevation that are a distance L apart. Determine (*a*) the sag-to-span ratio for which the maximum tension is as small as possible, (*b*) the corresponding values of θ_B and T_m.

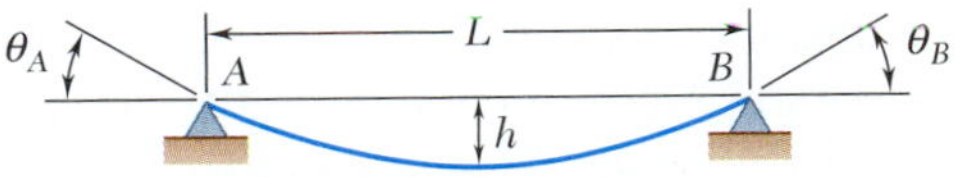

Fig. *P7.144, P7.145,* and P7.146

REVIEW AND SUMMARY FOR CHAPTER 7

In this chapter you learned to determine the internal forces which hold together the various parts of a given member in a structure.

Forces in straight two-force members

Considering first a *straight two-force member AB* [Sec. 7.2], we recall that such a member is subjected at A and B to equal and opposite forces $\mathbf{F}$ and $-\mathbf{F}$ directed along AB (Fig. 7.19a). Cutting member AB at C and drawing the free-body diagram of portion AC, we conclude that the internal forces which existed at C in member AB are equivalent to an *axial force* $-\mathbf{F}$ equal and opposite to $\mathbf{F}$ (Fig. 7.19b). We note that in the case of a two-force member which is not straight, the internal forces reduce to a force-couple system and not to a single force.

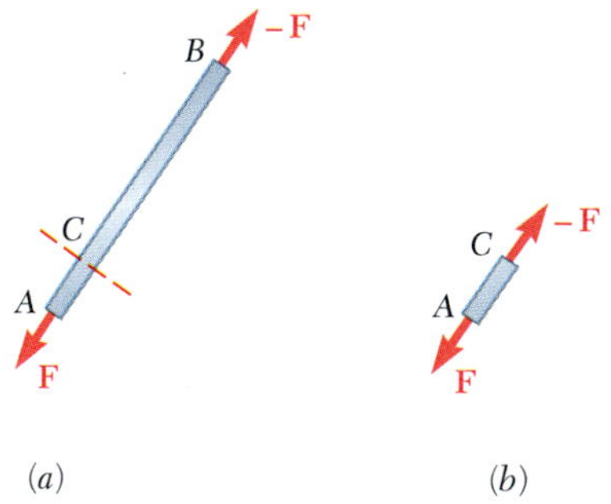

Fig. 7.19

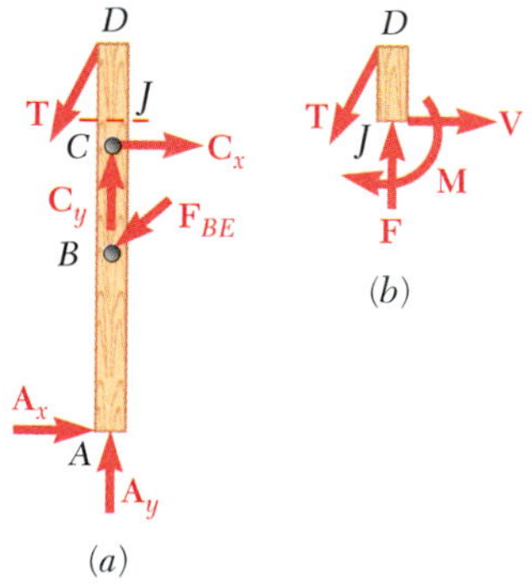

Fig. 7.20

Forces in multiforce members

Considering next a *multiforce member AD* (Fig. 7.20a), cutting it at J, and drawing the free-body diagram of portion JD, we conclude that the internal forces at J are equivalent to a force-couple system consisting of the *axial force* $\mathbf{F}$, the *shearing force* $\mathbf{V}$, and a couple $\mathbf{M}$ (Fig. 7.20b). The magnitude of the shearing force measures the *shear* at point J, and the moment of the couple is referred to as the *bending moment* at J. Since an equal and opposite force-couple system would have been obtained by considering the free-body diagram of portion AJ, it is necessary to specify which portion of member AD was used when recording the answers [Sample Prob. 7.1].

Forces in beams

Most of the chapter was devoted to the analysis of the internal forces in two important types of engineering structures: *beams* and *cables*. *Beams* are usually long, straight prismatic members designed to support loads applied at various points along the member. In general the loads are perpendicular to the axis of the beam and produce only *shear and bending* in the beam. The loads may be either *concentrated* at specific points, or *distributed* along the entire length or a portion of the beam. The beam itself may be supported in various ways; since only statically determinate beams are considered in this text, we limited our analysis to that of *simply supported beams, overhanging beams*, and *cantilever beams* [Sec. 7.3].

Shear and bending moment in a beam

To obtain the *shear* V and *bending moment* M at a given point C of a beam, we first determine the reactions at the supports by considering the entire beam as a free body. We then cut the beam at C and use the free-body diagram of one of the two portions obtained in this fashion to determine V and M. In order to avoid any confusion regarding the sense of the shearing force $\mathbf{V}$ and couple $\mathbf{M}$ (which act in opposite directions on the two portions of the beam), the sign convention illustrated in Fig. 7.21 was adopted [Sec. 7.4]. Once the values of the shear and bending moment have been determined at a few selected points of the beam, it is usually possible to draw a *shear diagram* and a *bending-moment diagram* representing, respectively, the shear and bending moment at any point of the beam [Sec. 7.5]. When a beam is subjected to concentrated loads only, the shear is of constant value between loads and the bending moment varies linearly between loads [Sample Prob. 7.2]. On the other hand, when a beam is subjected to distributed loads, the shear and bending moment vary quite differently [Sample Prob. 7.3].

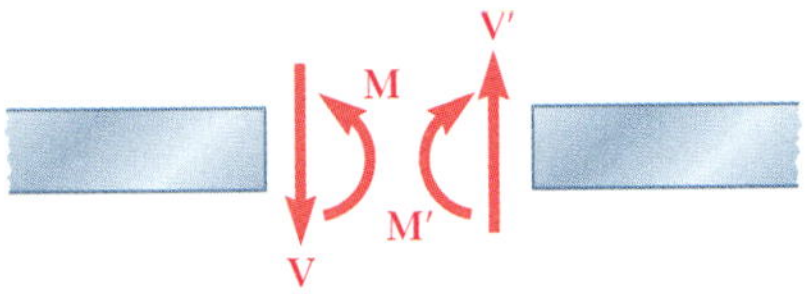

Internal forces at section
(positive shear and positive bending moment)

Fig. 7.21

Relations among load, shear, and bending moment

The construction of the shear and bending-moment diagrams is facilitated if the following relations are taken into account. Denoting by w the distributed load per unit length (assumed positive if directed downward), we have [Sec. 7.5]:

$$\frac{dV}{dx} = -w \tag{7.1}$$

$$\frac{dM}{dx} = V \tag{7.3}$$

or, in integrated form,

$$V_D - V_C = -(\text{area under load curve between } C \text{ and } D) \tag{7.2'}$$
$$M_D - M_C = \text{area under shear curve between } C \text{ and } D \tag{7.4'}$$

Equation (7.2′) makes it possible to draw the shear diagram of a beam from the curve representing the distributed load on that beam and the value of V at one end of the beam. Similarly, Eq. (7.4′) makes it possible to draw the bending-moment diagram from the shear diagram and the value of M at one end of the beam. However, concentrated loads introduce discontinuities in the shear diagram and concentrated couples introduce discontinuities in the bending-moment diagram, none of which are accounted for in these equations [Sample Probs. 7.4 and 7.7]. Finally, we note from Eq. (7.3) that the points of the beam where the bending moment is maximum or minimum are also the points where the shear is zero [Sample Prob. 7.5].

Cables with concentrated loads

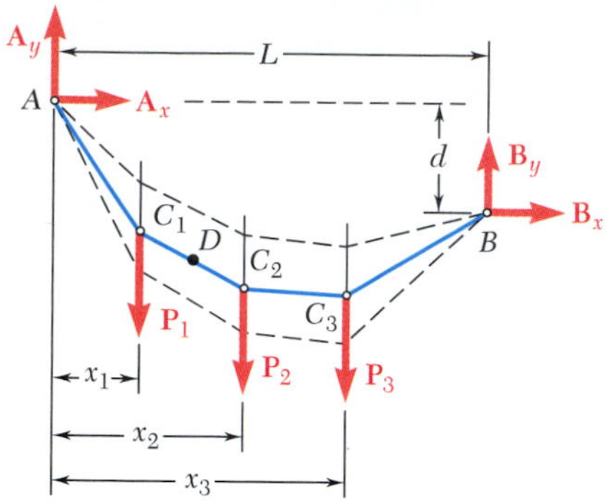

Fig. 7.22

The second half of the chapter was devoted to the analysis of *flexible cables*. We first considered a cable of negligible weight supporting *concentrated loads* [Sec. 7.7]. Using the entire cable AB as a free body (Fig. 7.22), we noted that the three available equilibrium equations were not sufficient to determine the four unknowns representing the reactions at the supports A and B. However, if the coordinates of a point D of the cable are known, an additional equation can be obtained by considering the free-body diagram of the portion AD or DB of the cable. Once the reactions at the supports have been determined, the elevation of any point of the cable and the tension in any portion of the cable can be found from the appropriate free-body diagram [Sample Prob. 7.8]. It was noted that the horizontal component of the force **T** representing the tension is the same at any point of the cable.

Cables with distributed loads

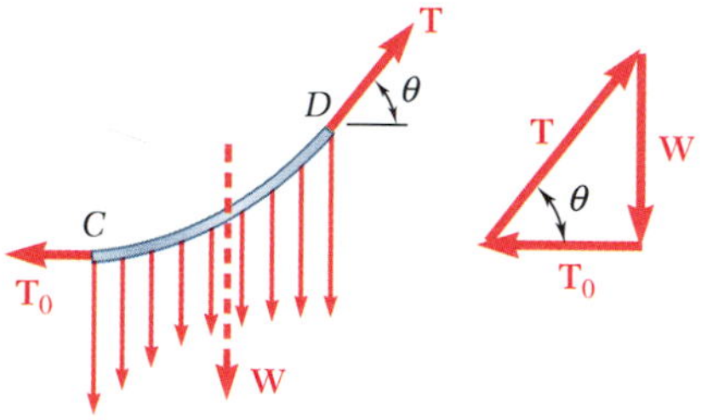

Fig. 7.23

We next considered cables carrying *distributed loads* [Sec. 7.8]. Using as a free body a portion of cable CD extending from the lowest point C to an arbitrary point D of the cable (Fig. 7.23), we observed that the horizontal component of the tension force **T** at D is constant and equal to the tension T_0 at C, while its vertical component is equal to the weight W of the portion of cable CD. The magnitude and direction of **T** were obtained from the force triangle:

$$T = \sqrt{T_0^2 + W^2} \qquad \tan\theta = \frac{W}{T_0} \tag{7.6}$$

Parabolic cable

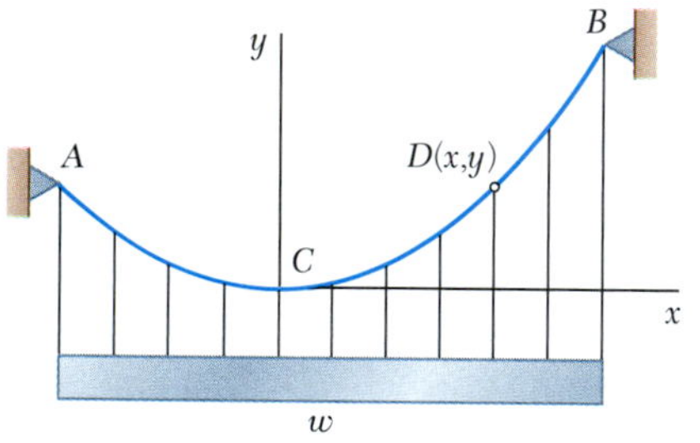

Fig. 7.24

In the case of a load *uniformly distributed along the horizontal*—as in a suspension bridge (Fig. 7.24)—the load supported by portion CD is $W = wx$, where w is the constant load per unit horizontal length [Sec. 7.9]. We also found that the curve formed by the cable is a *parabola* of equation

$$y = \frac{wx^2}{2T_0} \tag{7.8}$$

and that the length of the cable can be found by using the expansion in series given in Eq. (7.10) [Sample Prob. 7.9].

Catenary

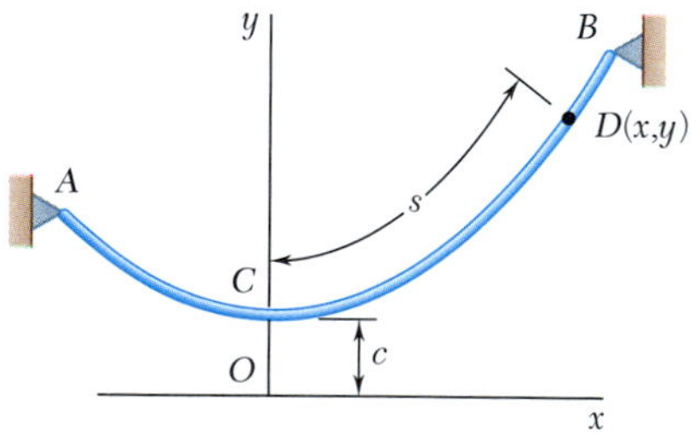

Fig. 7.25

In the case of a load *uniformly distributed along the cable itself* [for example, a cable hanging under its own weight (Fig. 7.25)] the load supported by portion CD is $W = ws$, where s is the length measured along the cable and w is the constant load per unit length [Sec. 7.10]. Choosing the origin O of the coordinate axes at a distance $c = T_0/w$ below C, we derived the relations

$$s = c \sinh \frac{x}{c} \tag{7.15}$$

$$y = c \cosh \frac{x}{c} \tag{7.16}$$

$$y^2 - s^2 = c^2 \tag{7.17}$$

$$T_0 = wc \qquad W = ws \qquad T = wy \tag{7.18}$$

which can be used to solve problems involving cables hanging under their own weight [Sample Prob. 7.10]. Equation (7.16), which defines the shape of the cable, is the equation of a *catenary*.

Review Problems

7.147 For the beam and loading shown, (*a*) draw the shear and bending moment diagrams, (*b*) determine the maximum absolute values of the shear and bending moment.

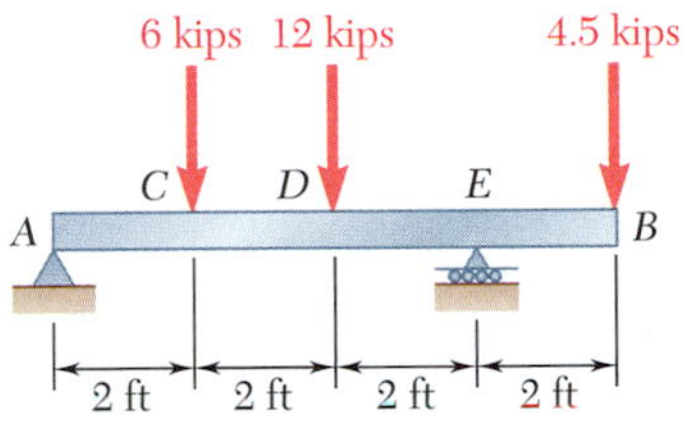

Fig. P7.147

7.148 For the beam and loading shown, (*a*) draw the shear and bending moment diagrams, (*b*) determine the maximum absolute values of the shear and bending moment.

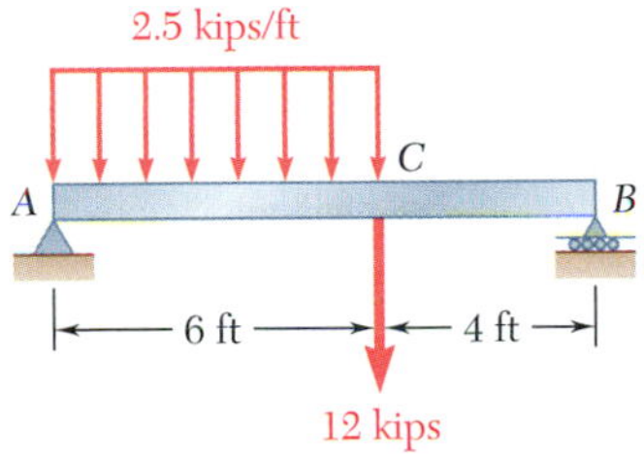

Fig. *P7.148*

7.149 Two loads are suspended as shown from the cable *ABCD*. Knowing that $h_B = 1.8$ m, determine (*a*) the distance h_C, (*b*) the components of the reaction at *D*, (*c*) the maximum tension in the cable.

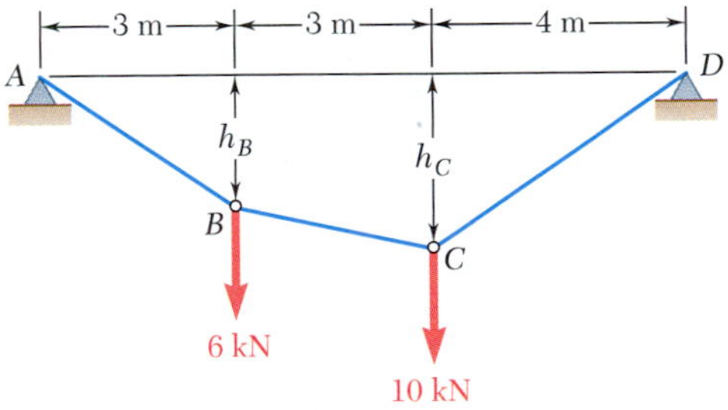

Fig. P7.149 and *P7.150*

7.150 Knowing that the maximum tension in cable *ABCD* is 15 kN, determine (*a*) the distance h_B, (*b*) the distance h_C.

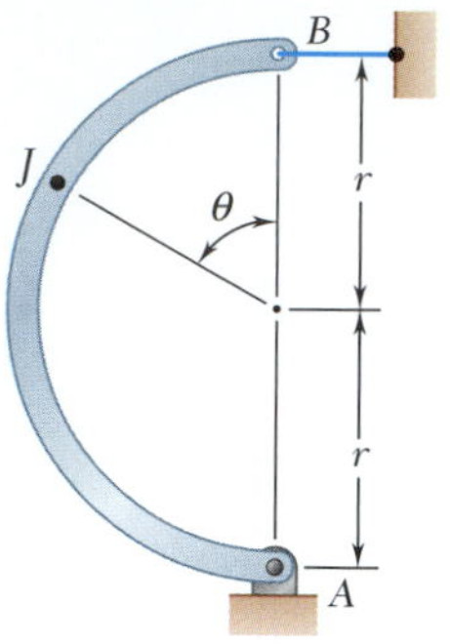

Fig. P7.151 and P7.152

7.151 A semicircular rod of weight W and uniform cross section is supported as shown. Determine the bending moment at point J when $\theta = 60°$.

7.152 A semicircular rod of weight W and uniform cross section is supported as shown. Determine the bending moment at point J when $\theta = 150°$.

7.153 Determine the internal forces at point J of the structure shown.

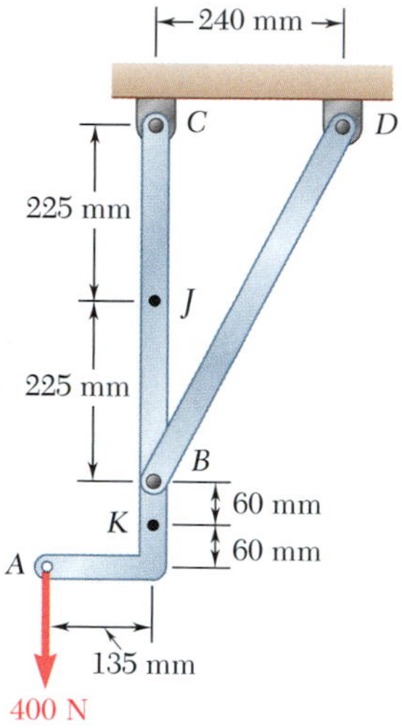

Fig. P7.153 and *P7.154*

7.154 Determine the internal forces at point K of the structure shown.

7.155 Two small channel sections DF and EH have been welded to the uniform beam AB of weight $W = 3$ kN to form the rigid structural member shown. This member is being lifted by two cables attached at D and E. Knowing that $\theta = 30°$ and neglecting the weight of the channel sections, (*a*) draw the shear and bending-moment diagrams for beam AB, (*b*) determine the maximum absolute values of the shear and bending moment in the beam.

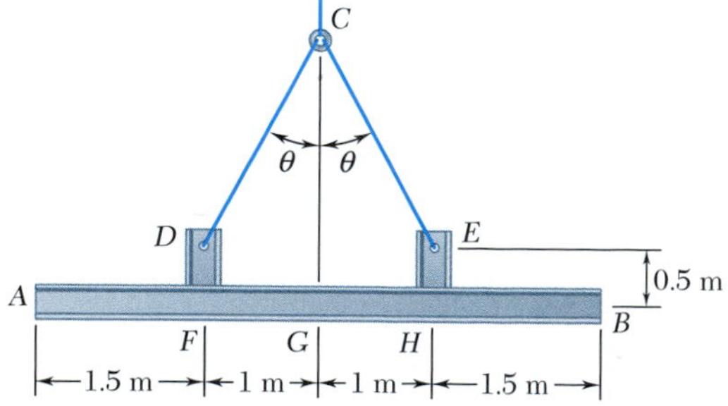

Fig. P7.155

7.156 (*a*) Draw the shear and bending moment diagrams for beam *AB*, (*b*) determine the magnitude and location of the maximum absolute value of the bending momkent.

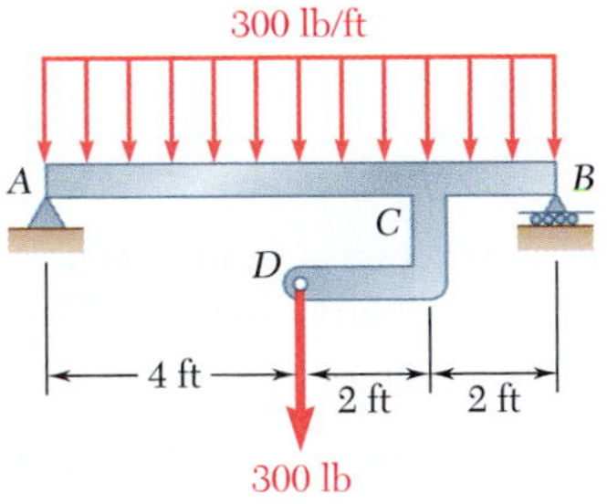

Fig. P7.156

7.157 Cable *ABC* supports two loads as shown. Knowing that $b = 4$ ft, determine (*a*) the required magnitude of the horizontal force **P**, (*b*) the corresponding distance *a*.

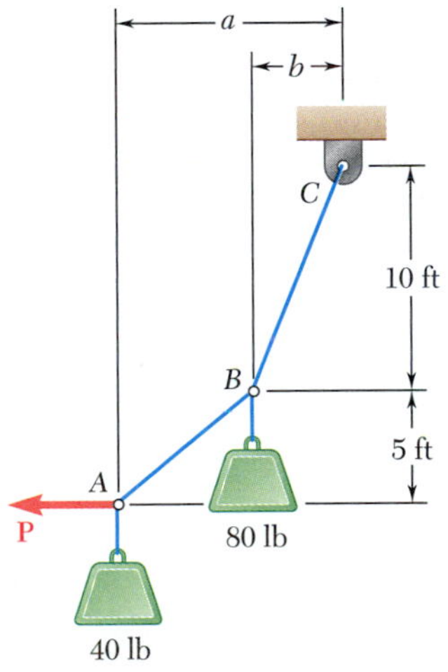

Fig. P7.157 and *P7.158*

7.158 Cable *ABC* supports two loads as shown. Determine the distances *a* and *b* when a horizontal force **P** of magnitude 60 lb is applied at *A*.

Computer Problems

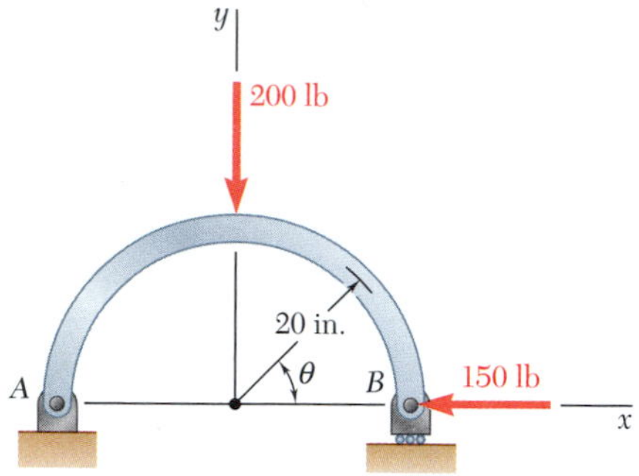

Fig. P7.C1

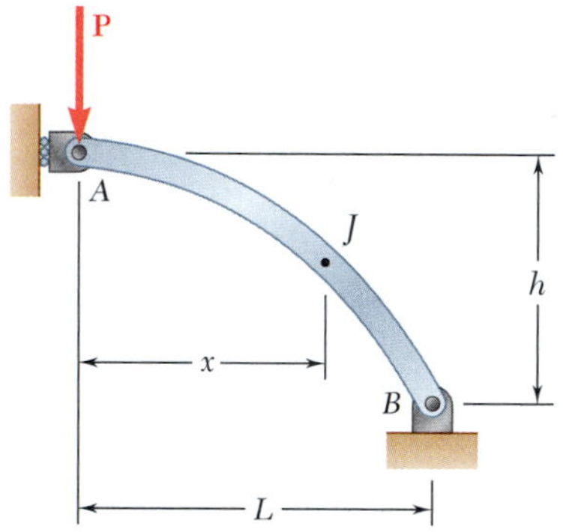

Fig. P7.C2

7.C1 A semicircular member of radius 20 in. is loaded as shown. Plot the internal forces (axial force, shearing force, and bending moment) as functions of θ for $0 < \theta < 90°$.

7.C2 The axis of the curved member AB is a parabola for which the vertex is at A and $L = nh$. Knowing that a vertical load **P** of magnitude 1.2 kN is applied at A, determine the internal forces and bending moment at an arbitrary point J. Plot the internal forces and bending moment as functions of x for $0.25L \leq x \leq 0.75L$ when $L = 400$ mm and (*a*) $n = 2$, (*b*) $n = 3$, (*c*) $n = 4$.

7.C3 The floor of a bridge will consist of narrow planks resting on two simply supported beams, one of which is shown in the figure. As part of the design of the bridge, it is desired to simulate the effect that driving a 12-kN truck over the bridge will have on this beam. The distance between the truck's axles is 1.8 m, and it assumed that the weight of the truck is equally distributed over its four wheels. (*a*) Using computational software, calculate, using 0.15-m increments, and plot the magnitude and location of the maximum bending moment in the beam as a function of x for $-0.9 \text{ m} \leq x \leq 3 \text{ m}$. (*b*) Using smaller increments if necessary, determine the largest value of the bending moment that occurs in the beam as the truck is driven over the bridge, and determine the corresponding value of x.

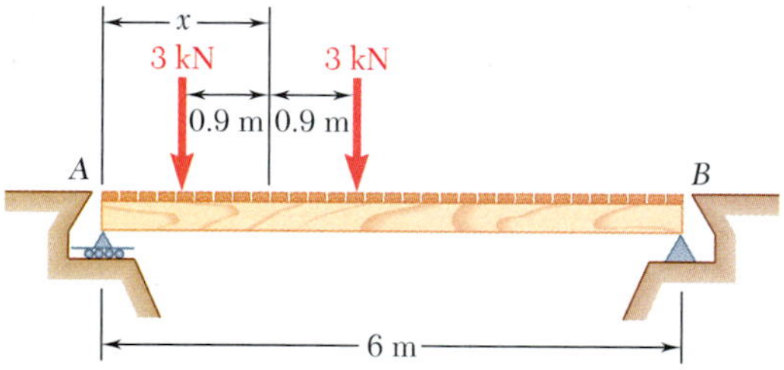

Fig. P7.C3

7.C4 Determine the equations for the shear and bending moment curves for beams 1 and 2 shown assuming that $w_0 = 15$ kN/m and $L = 3$ m. Plot the shear and bending moment diagrams for each beam.

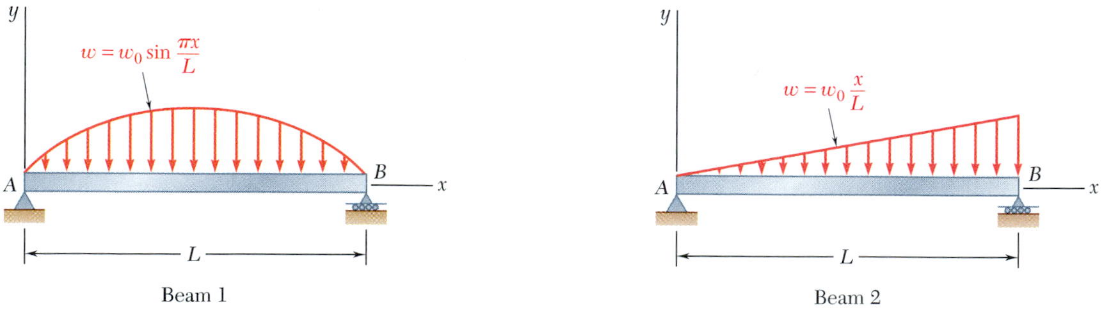

Fig. P7.C4

7.C5 A beam AB hinged at B and supported by a roller at D is to be designed to carry a load uniformly distributed from its end A to its midpoint C with maximum efficiency. As part of the design process, use computational software to determine the distance a from end A to the point D where the roller should be placed to minimize the absolute value of the bending moment M in the beam. (*Hint:* A short preliminary analysis will show that the roller should be placed under the load and that the largest negative value of M will occur at D, while its largest positive value will occur somewhere between D and C. Thus, you should draw the bending-moment diagram and then equate the absolute values of the largest positive and negative bending moments obtained.)

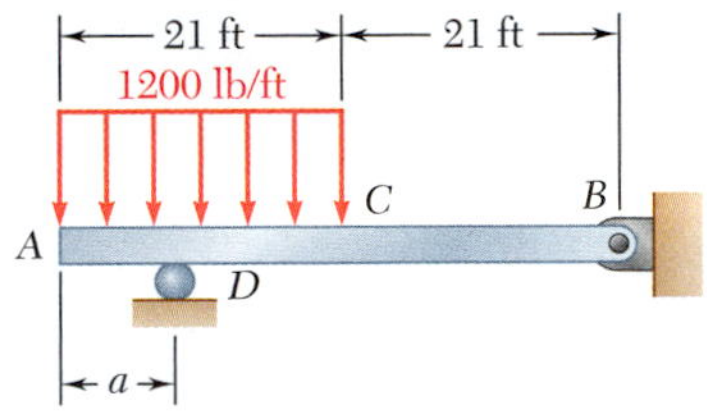

Fig. P7.C5

7.C6 Cable AB supports a load distributed uniformly along the horizontal as shown. The lowest portion of the cable is located at a distance a = 3.6 m below support A, and support B is located a distance $b = na$ above A. (*a*) Determine the maximum tension in the cable as a function of n. (*b*) Plot the maximum tension in the cable for $2 \leq n \leq 6$.

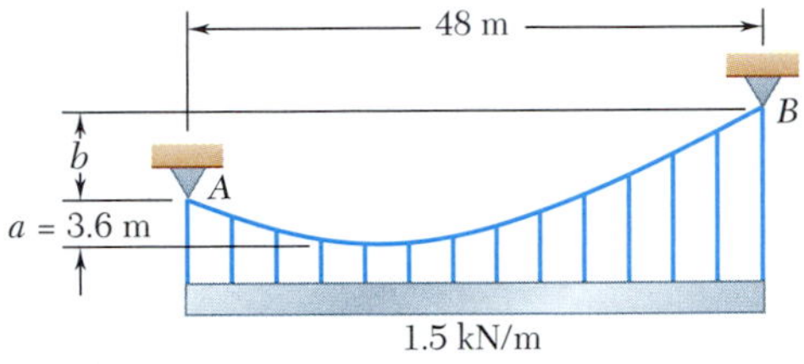

Fig. P7.C6

7.C7 Using computational software, solve Prob. 7.127 for values of P from 0 to 75 N using 5-N increments.

7.C8 A typical transmission-line installation consists of a cable of length s_{AB} and mass per unit length m' suspended as shown between two points at the same elevation. Using computational software, construct a table that can be used in the design of future installations. The table should present the dimensionless quantities h/L, s_{AB}/L, $T_0/m'gL$ and $T_{max}/m'gL$ for values of c/L from 0.2 to 0.5 using 0.025 increments and from 1 to 4 using 0.5 increments.

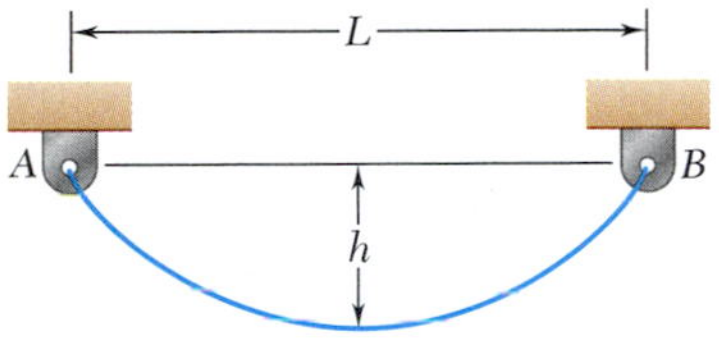

Fig. P7.C8

7.C9 The electrician shown is installing an electric cable weighing 0.2 lb/ft. As the electrician slowly climbs the ladder, the cable unwinds from the spool so that the tension T_B in the cable at B is 1.2 lb. Knowing that h = 4 ft when the electrician is standing on the ground, plot the magnitude of the force exerted on the hand of the electrician by the cable as a function of h for 4 ft $\leq h \leq$ 16 ft.

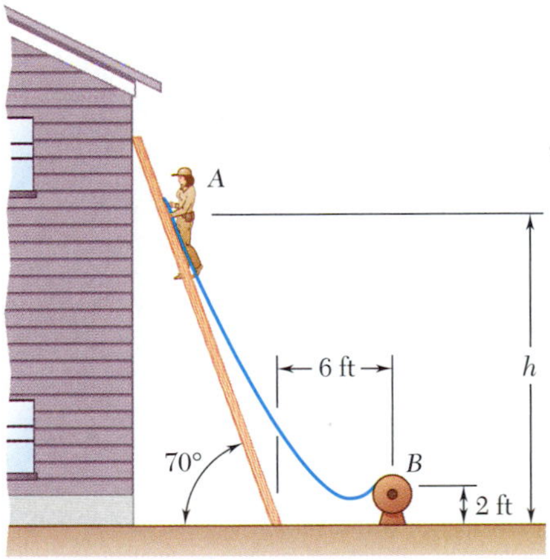

Fig. P7.C9

CHAPTER

8

Friction

The operation of the bicycle brake shown depends upon the normal and friction forces exerted by the brake pads on the rim of the wheel as the pads are pressed against the rim.

FRICTION

8.1. INTRODUCTION

In the preceding chapters, it was assumed that surfaces in contact were either *frictionless* or *rough*. If they were frictionless, the force each surface exerted on the other was normal to the surfaces and the two surfaces could move freely with respect to each other. If they were rough, it was assumed that tangential forces could develop to prevent the motion of one surface with respect to the other.

This view was a simplified one. Actually, no perfectly frictionless surface exists. When two surfaces are in contact, tangential forces, called *friction forces*, will always develop if one attempts to move one surface with respect to the other. On the other hand, these friction forces are limited in magnitude and will not prevent motion if sufficiently large forces are applied. The distinction between frictionless and rough surfaces is thus a matter of degree. This will be seen more clearly in the present chapter, which is devoted to the study of friction and of its applications to common engineering situations.

There are two types of friction: *dry friction*, sometimes called *Coulomb friction*, and *fluid friction*. Fluid friction develops between layers of fluid moving at different velocities. Fluid friction is of great importance in problems involving the flow of fluids through pipes and orifices or dealing with bodies immersed in moving fluids. It is also basic in the analysis of the motion of *lubricated mechanisms*. Such problems are considered in texts on fluid mechanics. The present study is limited to dry friction, that is, to problems involving rigid bodies which are in contact along *nonlubricated* surfaces.

In the first part of this chapter, the equilibrium of various rigid bodies and structures, assuming dry friction at the surfaces of contact, is analyzed. Later a number of specific engineering applications where dry friction plays an important role are considered: wedges, square-threaded screws, journal bearings, thrust bearings, rolling resistance, and belt friction.

8.2. THE LAWS OF DRY FRICTION. COEFFICIENTS OF FRICTION

The laws of dry friction are exemplified by the following experiment. A block of weight **W** is placed on a horizontal plane surface (Fig. 8.1*a*). The forces acting on the block are its weight **W** and the reaction of the surface. Since the weight has no horizontal component, the reaction of the surface also has no horizontal component; the reaction is therefore *normal* to the surface and is represented by **N** in Fig. 8.1*a*. Suppose, now, that a horizontal force **P** is applied to the block (Fig. 8.1*b*). If **P** is small, the block will not move; some other horizontal force must therefore exist, which balances **P**. This other force is the *static-friction force* **F**, which is actually the resultant of a great number of forces acting over the entire surface of contact between the block and the plane. The nature of these forces is not known exactly, but it is generally assumed that these forces are due to the irregularities of the surfaces in contact and, to a certain extent, to molecular attraction.

If the force **P** is increased, the friction force **F** also increases, continuing to oppose **P**, until its magnitude reaches a certain *maximum value* F_m (Fig. 8.1*c*). If **P** is further increased, the friction force cannot balance it any more and the block starts sliding.† As soon as the block has been set in motion, the magnitude of **F** drops from F_m to a lower value F_k. This is because there is less interpenetration between the irregularities of the surfaces in contact when these surfaces move with respect to each other. From then on, the block keeps sliding with increasing velocity while the friction force, denoted by $\mathbf{F}_k$ and called the *kinetic-friction force,* remains approximately constant.

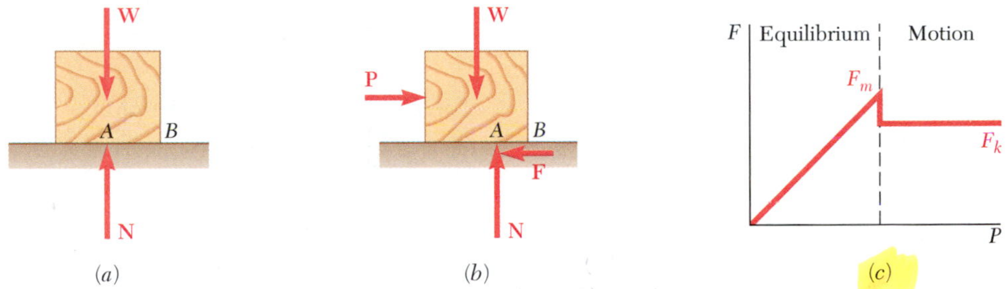

Fig. 8.1

Experimental evidence shows that the maximum value F_m of the static-friction force is proportional to the normal component N of the reaction of the surface. We have

$$F_m = \mu_s N \tag{8.1}$$

where μ_s is a constant called the *coefficient of static friction*. Similarly, the magnitude F_k of the kinetic-friction force may be put in the form

$$F_k = \mu_k N \tag{8.2}$$

where μ_k is a constant called the *coefficient of kinetic friction*. The coefficients of friction μ_s and μ_k do not depend upon the area of the surfaces in contact. Both coefficients, however, depend strongly on the *nature* of the surfaces in contact. Since they also depend upon the exact condition of the surfaces, their value is seldom known with an accuracy greater than 5 percent. Approximate values of coefficients

†It should be noted that, as the magnitude F of the friction force increases from 0 to F_m, the point of application A of the resultant **N** of the normal forces of contact moves to the right, so that the couples formed, respectively, by **P** and **F** and by **W** and **N** remain balanced. If **N** reaches B before F reaches its maximum value F_m, the block will tip about B before it can start sliding (see Probs. 8.15 and 8.16).

of static friction for various dry surfaces are given in Table 8.1. The corresponding values of the coefficient of kinetic friction would be about 25 percent smaller. Since coefficients of friction are dimensionless quantities, the values given in Table 8.1 can be used with both SI units and U.S. customary units.

Table 8.1. Approximate Values of Coefficient of Static Friction for Dry Surfaces

Metal on metal	0.15–0.60
Metal on wood	0.20–0.60
Metal on stone	0.30–0.70
Metal on leather	0.30–0.60
Wood on wood	0.25–0.50
Wood on leather	0.25–0.50
Stone on stone	0.40–0.70
Earth on earth	0.20–1.00
Rubber on concrete	0.60–0.90

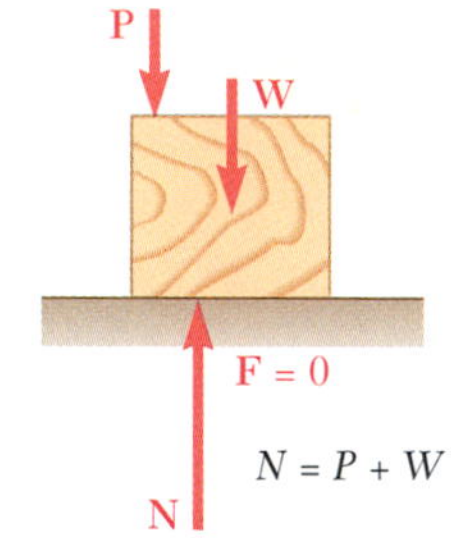

(a) No friction ($P_x = 0$)

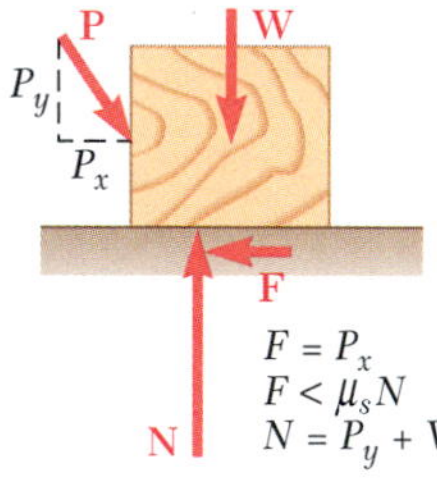

(b) No motion ($P_x < F_m$)

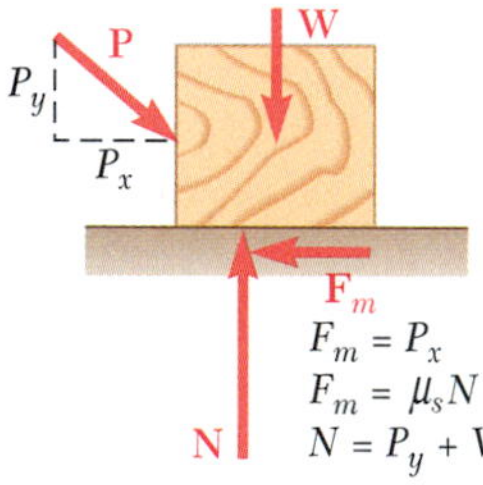

(c) Motion impending ⟶ ($P_x = F_m$)

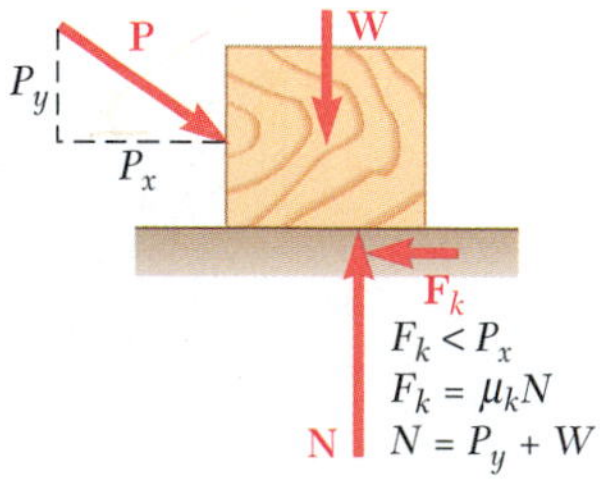

(d) Motion ⟶ ($P_x > F_m$)

Fig. 8.2

From the description given above, it appears that four different situations can occur when a rigid body is in contact with a horizontal surface:

1. The forces applied to the body do not tend to move it along the surface of contact; there is no friction force (Fig. 8.2a).

2. The applied forces tend to move the body along the surface of contact but are not large enough to set it in motion. The friction force **F** which has developed can be found by solving the equations of equilibrium for the body. Since there is no evidence that **F** has reached its maximum value, the equation $F_m = \mu_s N$ *cannot be used* to determine the friction force (Fig. 8.2b).

3. The applied forces are such that the body is just about to slide. We say that *motion is impending*. The friction force **F** has reached its maximum value F_m and, together with the normal force **N**, balances the applied forces. Both the equations of equilibrium and the equation $F_m = \mu_s N$ *can be used*. We also note that the friction force has a sense opposite to the sense of impending motion (Fig. 8.2c).

4. The body is sliding under the action of the applied forces, and the equations of equilibrium do not apply any more. However, **F** is now equal to $\mathbf{F}_k$ and the equation $F_k = \mu_k N$ may be used. The sense of $\mathbf{F}_k$ is opposite to the sense of motion (Fig. 8.2d).

8.3. ANGLES OF FRICTION

It is sometimes convenient to replace the normal force **N** and the friction force **F** by their resultant **R**. Let us consider again a block of weight **W** resting on a horizontal plane surface. If no horizontal force is applied to the block, the resultant **R** reduces to the normal force **N** (Fig. 8.3*a*). However, if the applied force **P** has a horizontal component $\mathbf{P}_x$ which tends to move the block, the force **R** will have a horizontal component **F** and, thus, will form an angle ϕ with the normal to the surface (Fig. 8.3*b*). If $\mathbf{P}_x$ is increased until motion becomes impending, the angle between **R** and the vertical grows and reaches a maximum value (Fig. 8.3*c*). This value is called the *angle of static friction* and is denoted by ϕ_s. From the geometry of Fig. 8.3*c*, we note that

$$\tan \phi_s = \frac{F_m}{N} = \frac{\mu_s N}{N}$$

$$\tan \phi_s = \mu_s \tag{8.3}$$

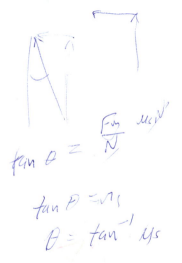

If motion actually takes place, the magnitude of the friction force drops to F_k; similarly, the angle ϕ between **R** and **N** drops to a lower value ϕ_k, called the *angle of kinetic friction* (Fig. 8.3*d*). From the geometry of Fig. 8.3*d*, we write

$$\tan \phi_k = \frac{F_k}{N} = \frac{\mu_k N}{N}$$

$$\tan \phi_k = \mu_k \tag{8.4}$$

Another example will show how the angle of friction can be used to advantage in the analysis of certain types of problems. Consider a block resting on a board and subjected to no other force than its weight **W** and the reaction **R** of the board. The board can be given any desired inclination. If the board is horizontal, the force **R** exerted by the board on the block is perpendicular to the board and balances the weight **W** (Fig. 8.4*a*). If the board is given a small angle of inclina-

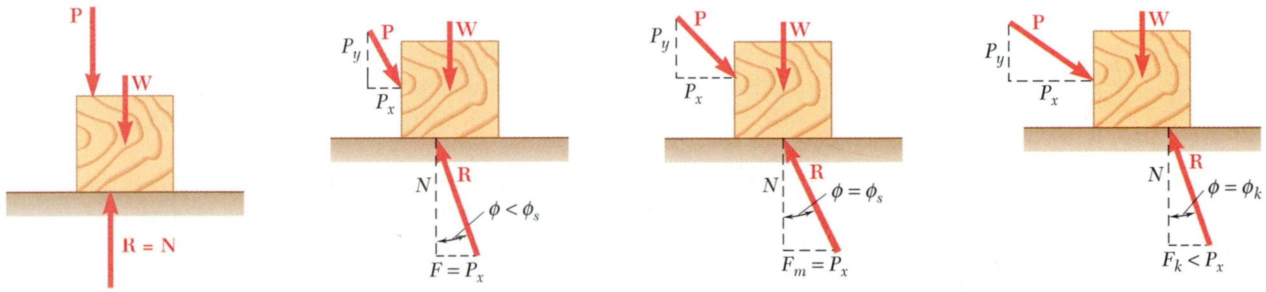

(*a*) No friction (*b*) No motion (*c*) Motion impending ⟶ (*d*) Motion ⟶

Fig. 8.3

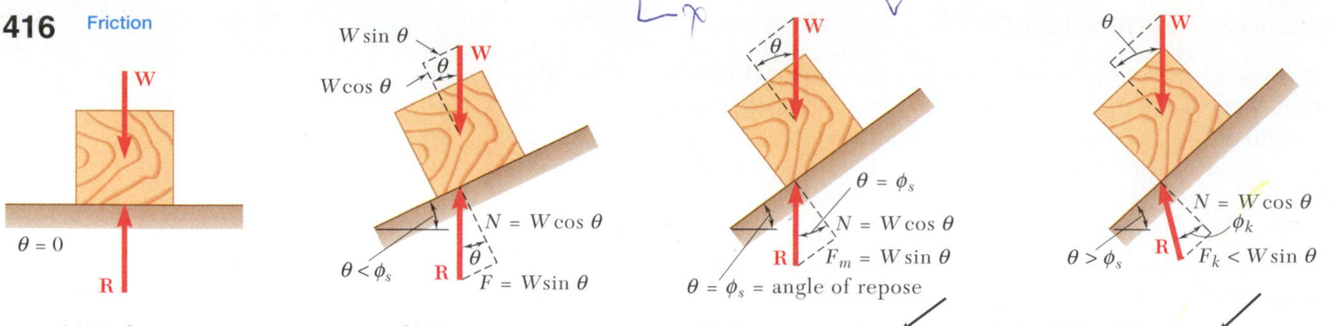

(*a*) No friction (*b*) No motion (*c*) Motion impending (*d*) Motion

Fig. 8.4

Photo 8.1 The coefficient of static friction between a package and the inclined conveyer belt must be sufficiently large to enable the package to be transported without slipping.

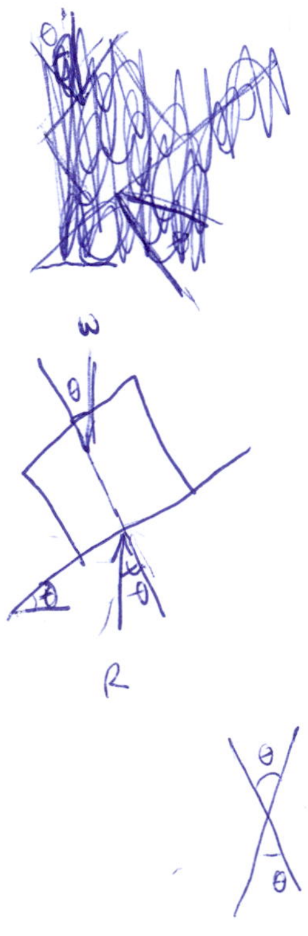

tion θ, the force **R** will deviate from the perpendicular to the board by the angle θ and will keep balancing **W** (Fig. 8.4*b*); it will then have a normal component **N** of magnitude $N = W \cos \theta$ and a tangential component **F** of magnitude $F = W \sin \theta$.

If we keep increasing the angle of inclination, motion will soon become impending. At that time, the angle between **R** and the normal will have reached its maximum value ϕ_s (Fig. 8.4*c*). The value of the angle of inclination corresponding to impending motion is called the *angle of repose*. Clearly, the angle of repose is equal to the angle of static friction ϕ_s. If the angle of inclination θ is further increased, motion starts and the angle between **R** and the normal drops to the lower value ϕ_k (Fig. 8.4*d*). The reaction **R** is not vertical any more, and the forces acting on the block are unbalanced.

8.4. PROBLEMS INVOLVING DRY FRICTION

Problems involving dry friction are found in many engineering applications. Some deal with simple situations such as the block sliding on a plane described in the preceding sections. Others involve more complicated situations as in Sample Prob. 8.3; many deal with the stability of rigid bodies in accelerated motion and will be studied in dynamics. Also, a number of common machines and mechanisms can be analyzed by applying the laws of dry friction. These include wedges, screws, journal and thrust bearings, and belt transmissions. They will be studied in the following sections.

The *methods* which should be used to solve problems involving dry friction are the same that were used in the preceding chapters. If a problem involves only a motion of translation, with no possible rotation, the body under consideration can usually be treated as a particle, and the methods of Chap. 2 can be used. If the problem involves a possible rotation, the body must be considered as a rigid body, and the methods of Chap. 4 should be used. If the structure considered is made of several parts, the principle of action and reaction must be used as was done in Chap. 6.

If the body considered is acted upon by more than three forces (including the reactions at the surfaces of contact), the reaction at each surface will be represented by its components **N** and **F** and the problem will be solved from the equations of equilibrium. If only three forces act on the body under consideration, it may be more convenient to represent each reaction by a single force **R** and to solve the problem by drawing a force triangle.

Most problems involving friction fall into one of the following *three groups:* In the *first group* of problems, all applied forces are given and the coefficients of friction are known; we are to determine whether the body considered will remain at rest or slide. The friction force **F** *required to maintain equilibrium* is unknown (its magnitude is *not* equal to $\mu_s N$) and should be determined, together with the normal force **N**, by drawing a free-body diagram and *solving the equations of equilibrium* (Fig. 8.5*a*). The value found for the magnitude F of the friction force is then compared with the maximum value $F_m = \mu_s N$. If F is smaller than or equal to F_m, the body remains at rest. If the value found for F is larger than F_m, equilibrium cannot be maintained and motion takes place; the actual magnitude of the friction force is then $F_k = \mu_k N$.

In problems of the *second group*, all applied forces are given and the motion is known to be impending; we are to determine the value of the coefficient of static friction. Here again, we determine the friction force and the normal force by drawing a free-body diagram and solving the equations of equilibrium (Fig. 8.5*b*). Since we know that the value found for F is the maximum value F_m, the coefficient of friction may be found by writing and solving the equation $F_m = \mu_s N$.

In problems of the *third group*, the coefficient of static friction is given, and it is known that the motion is impending in a given direction; we are to determine the magnitude or the direction of one of the applied forces. The friction force should be shown in the free-body diagram with a *sense opposite to that of the impending motion* and with a magnitude $F_m = \mu_s N$ (Fig. 8.5*c*). The equations of equilibrium can then be written, and the desired force determined.

As noted above, when only three forces are involved it may be more convenient to represent the reaction of the surface by a single force **R** and to solve the problem by drawing a force triangle. Such a solution is used in Sample Prob. 8.2.

When two bodies A and B are in contact (Fig. 8.6*a*), the forces of friction exerted, respectively, by A on B and by B on A are equal and opposite (Newton's third law). In drawing the free-body diagram of one of the bodies, it is important to include the appropriate friction force with its correct sense. The following rule should then be observed: *The sense of the friction force acting on A is opposite to that of the motion (or impending motion) of A as observed from B* (Fig. 8.6*b*).† The sense of the friction force acting on B is determined in a similar way (Fig. 8.6*c*). Note that the motion of A as observed from B is a *relative motion*. For example, if body A is fixed and body B moves, body A will have a relative motion with respect to B. Also, if both B and A are moving down but B is moving faster than A, body A will be observed, from B, to be moving up.

†It is therefore *the same as that of the motion of B as observed from A.*

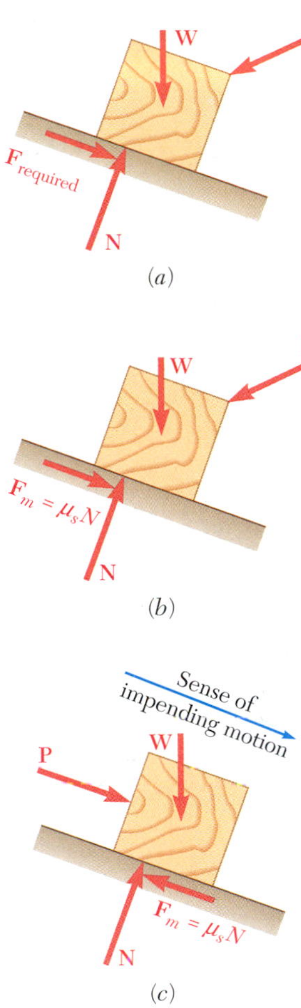

Fig. 8.5

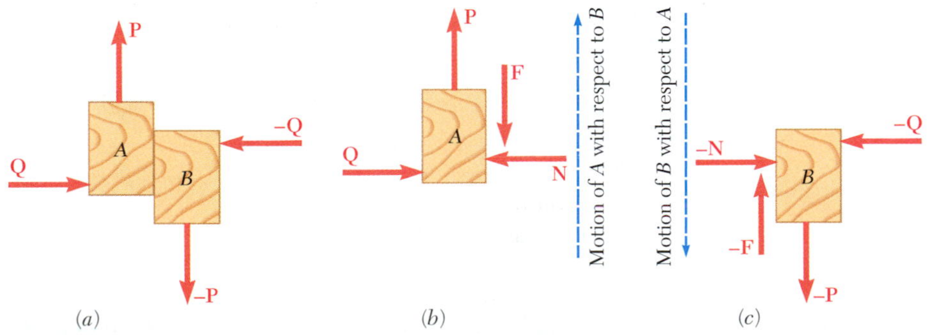

Fig. 8.6

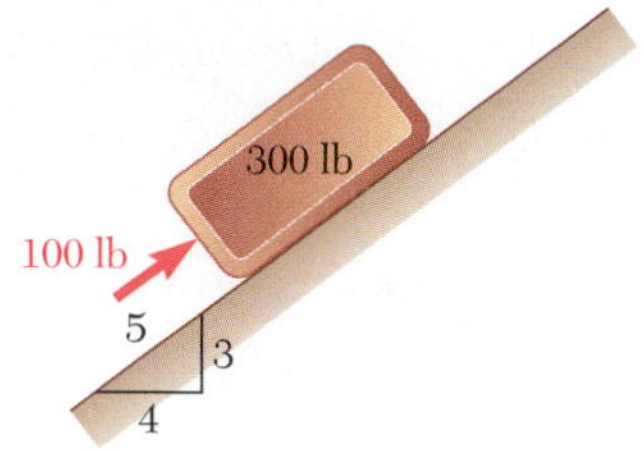

SAMPLE PROBLEM 8.1

A 100-lb force acts as shown on a 300-lb block placed on an inclined plane. The coefficients of friction between the block and the plane are $\mu_s = 0.25$ and $\mu_k = 0.20$. Determine whether the block is in equilibrium, and find the value of the friction force.

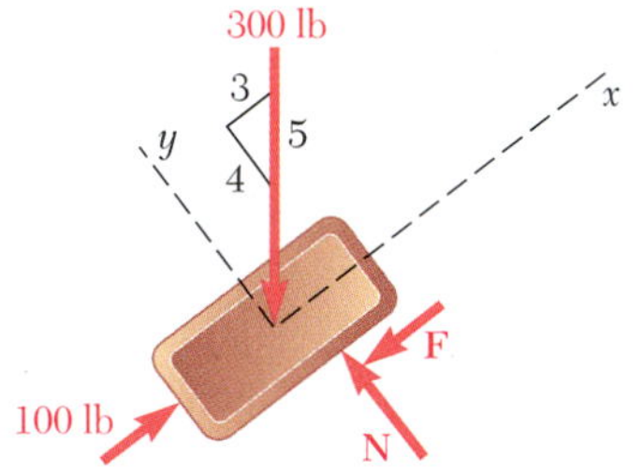

SOLUTION

Force Required for Equilibrium. We first determine the value of the friction force *required to maintain equilibrium.* Assuming that **F** is directed down and to the left, we draw the free-body diagram of the block and write

$$+\nearrow\Sigma F_x = 0: \qquad 100 \text{ lb} - \tfrac{3}{5}(300 \text{ lb}) - F = 0$$
$$F = -80 \text{ lb} \qquad \mathbf{F} = 80 \text{ lb} \nearrow$$

$$+\nwarrow\Sigma F_y = 0: \qquad N - \tfrac{4}{5}(300 \text{ lb}) = 0$$
$$N = +240 \text{ lb} \qquad \mathbf{N} = 240 \text{ lb} \nwarrow$$

The force **F** required to maintain equilibrium is an 80-lb force directed up and to the right; the tendency of the block is thus to move down the plane.

Maximum Friction Force. The magnitude of the maximum friction force which can be developed is

$$F_m = \mu_s N \qquad F_m = 0.25(240 \text{ lb}) = 60 \text{ lb}$$

Since the value of the force required to maintain equilibrium (80 lb) is larger than the maximum value which can be obtained (60 lb), equilibrium will not be maintained and *the block will slide down the plane.*

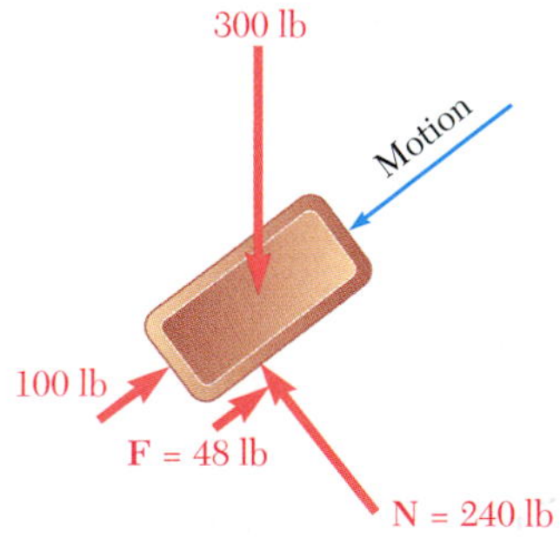

Actual Value of Friction Force. The magnitude of the actual friction force is obtained as follows:

$$F_{\text{actual}} = F_k = \mu_k N$$
$$= 0.20(240 \text{ lb}) = 48 \text{ lb}$$

The sense of this force is opposite to the sense of motion; the force is thus directed up and to the right:

$$\mathbf{F}_{\text{actual}} = 48 \text{ lb} \nearrow \quad ◀$$

It should be noted that the forces acting on the block are not balanced; the resultant is

$$\tfrac{3}{5}(300 \text{ lb}) - 100 \text{ lb} - 48 \text{ lb} = 32 \text{ lb} \swarrow$$

SAMPLE PROBLEM 8.2

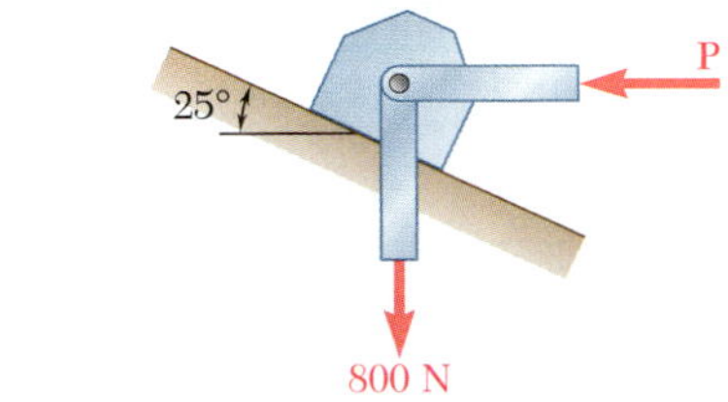

A support block is acted upon by two forces as shown. Knowing that the coefficients of friction between the block and the incline are $\mu_s = 0.35$ and $\mu_k = 0.25$, determine (*a*) the force **P** for which motion of the block up the incline is impending, (*b*) the friction force when the block is moving up, (*c*) the smallest force **P** required to prevent the block from sliding down.

SOLUTION

Free-Body Diagram. For each part of the problem we draw a free-body diagram of the block and a force triangle including the 800-N vertical force, the horizontal force **P**, and the force **R** exerted on the block by the incline. The direction of **R** must be determined in each separate case. We note that since **P** is perpendicular to the 800-N force, the force triangle is a right triangle, which can easily be solved for **P**. In most other problems, however, the force triangle will be an oblique triangle and should be solved by applying the law of sines.

a. Force P for Impending Motion of the Block Up the Incline

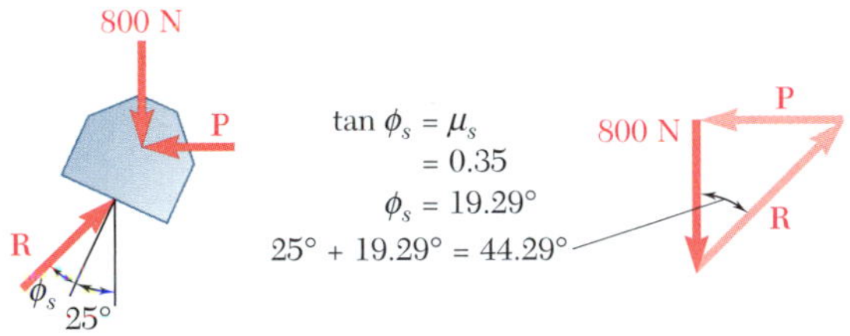

$P = (800 \text{ N}) \tan 44.29°$ **P = 780 N←** ◀

b. Friction Force F when the Block Moves Up the Incline

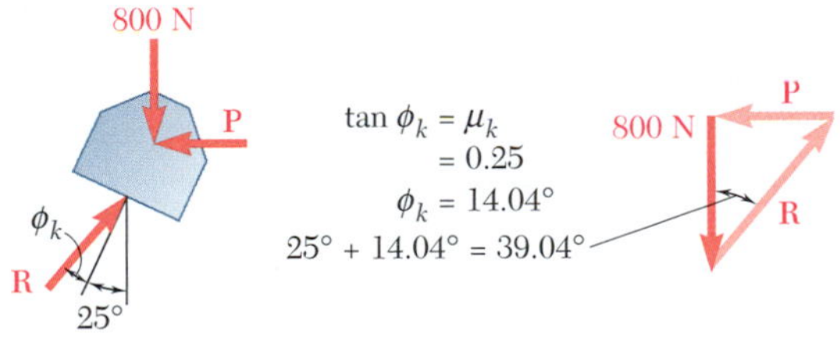

$$R = (800 \text{ N})/\cos 39.04° = 1029.9 \text{ N}$$

$$F = R \sin \phi_k = (1029.9 \text{ N}) \sin 14.04°$$

F = 250 N↘ ◀

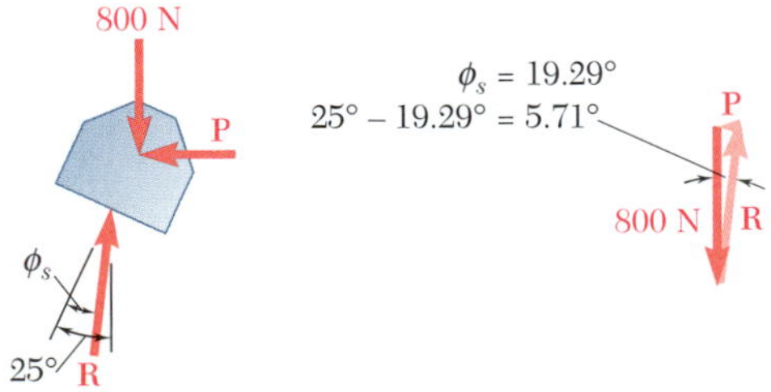

c. Force P for Impending Motion of the Block Down the Incline

$P = (800 \text{ N}) \tan 5.71°$ **P = 80.0 N←** ◀

SAMPLE PROBLEM 8.3

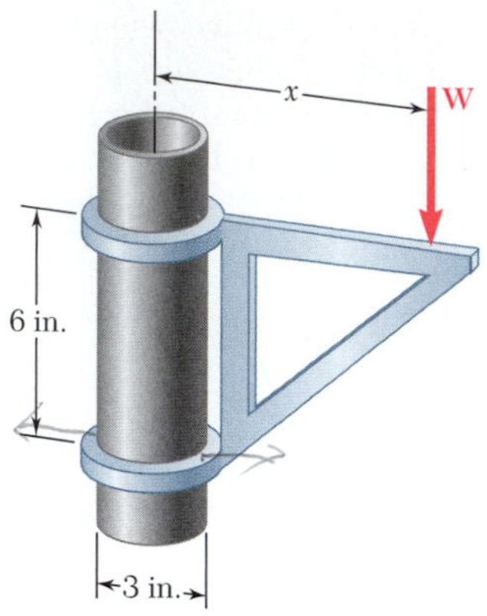

The movable bracket shown may be placed at any height on the 3-in.-diameter pipe. If the coefficient of static friction between the pipe and bracket is 0.25, determine the minimum distance x at which the load **W** can be supported. Neglect the weight of the bracket.

SOLUTION

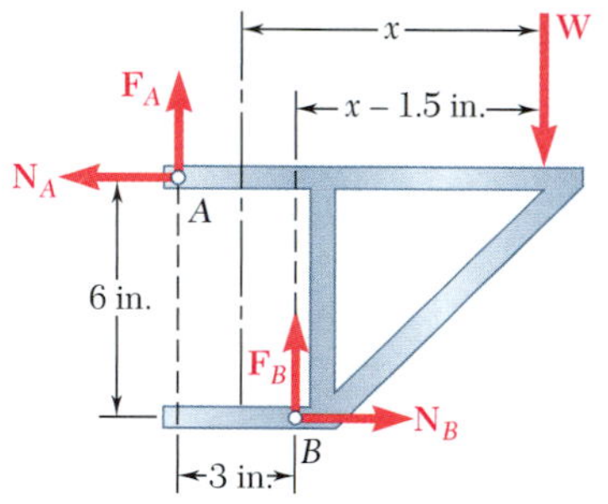

Free-Body Diagram. We draw the free-body diagram of the bracket. When **W** is placed at the minimum distance x from the axis of the pipe, the bracket is just about to slip, and the forces of friction at A and B have reached their maximum values:

$$F_A = \mu_s N_A = 0.25 N_A$$
$$F_B = \mu_s N_B = 0.25 N_B$$

Equilibrium Equations

$\xrightarrow{+} \Sigma F_x = 0$: $\quad N_B - N_A = 0$

$$N_B = N_A$$

$+\uparrow \Sigma F_y = 0$: $\quad F_A + F_B - W = 0$

$$0.25 N_A + 0.25 N_B = W$$

And, since N_B has been found equal to N_A,

$$0.50 N_A = W$$
$$N_A = 2W$$

$+\circlearrowleft \Sigma M_B = 0$: $\quad N_A(6 \text{ in.}) - F_A(3 \text{ in.}) - W(x - 1.5 \text{ in.}) = 0$

$$6N_A - 3(0.25N_A) - Wx + 1.5W = 0$$
$$6(2W) - 0.75(2W) - Wx + 1.5W = 0$$

Dividing through by W and solving for x,

$$x = 12 \text{ in.} \quad ◀$$

SOLVING PROBLEMS ON YOUR OWN

In this lesson you studied and applied the *laws of dry friction.* Previously you had encountered only (*a*) frictionless surfaces that could move freely with respect to each other, (*b*) rough surfaces that allowed no motion relative to each other.

A. *In solving problems involving dry friction,* you should keep the following in mind.

1. *The reaction R exerted by a surface on a free body* can be resolved into a normal component **N** and a tangential component **F**. The tangential component is known as the *friction force.* When a body is in contact with a fixed surface the direction of the friction force **F** is opposite to that of the actual or impending motion of the body.

a. No motion will occur as long as F does not exceed the maximum value $F_m = \mu_s N$, where μ_s is the *coefficient of static friction.*

b. Motion will occur if a value of F larger than F_m is required to maintain equilibrium. As motion takes place, the actual value of F drops to $F_k = \mu_k N$, where μ_k is the *coefficient of kinetic friction* [Sample Prob. 8.1].

2. *When only three forces are involved* an alternative approach to the analysis of friction may be preferred [Sample Prob. 8.2]. The reaction **R** is defined by its magnitude R and the angle ϕ it forms with the normal to the surface. No motion will occur as long as ϕ does not exceed the maximum value ϕ_s, where $\tan \phi_s = \mu_s$. Motion will occur if a value of ϕ larger than ϕ_s is required to maintain equilibrium, and the actual value of ϕ will drop to ϕ_k, where $\tan \phi_k = \mu_k$.

3. *When two bodies are in contact* the sense of the actual or impending relative motion at the point of contact must be determined. On each of the two bodies a friction force **F** should be shown in a direction opposite to that of the actual or impending motion of the body as seen from the other body.

B. *Methods of solution.* The first step in your solution is to *draw a free-body diagram* of the body under consideration, resolving the force exerted on each surface where friction exists into a normal component **N** and a friction force **F**. If several bodies are involved, draw a free-body diagram of each of them, labeling and directing the forces at each surface of contact as you learned to do when analyzing frames in Chap. 6.

The problem you have to solve may fall into one of the following five categories:

1. *All the applied forces and the coefficients of friction are known, and you must determine whether equilibrium is maintained.* Note that in this situation the friction force is unknown and *cannot be assumed to be equal* to $\mu_s N$.

a. Write the equations of equilibrium to determine N and F.

b. Calculate the maximum allowable friction force, $F_m = \mu_s N$. If $F \leq F_m$, equilibrium is maintained. If $F > F_m$, motion occurs, and the magnitude of the friction force is $F_k = \mu_k N$ [Sample Prob. 8.1].

2. *All the applied forces are known, and you must find the smallest allowable value of μ_s for which equilibrium is maintained.* You will assume that motion is impending and determine the corresponding value of μ_s.

a. Write the equations of equilibrium to determine N and F.

b. Since motion is impending, $F = F_m$. Substitute the values found for N and F into the equation $F_m = \mu_s N$ and solve for μ_s.

3. ***The motion of the body is impending and μ_s is known; you must find some unknown quantity,*** such as a distance, an angle, the magnitude of a force, or the direction of a force.

a. Assume a possible impending motion of the body and, on the free-body diagram, draw the friction force in a direction opposite to that of the assumed motion.

b. Since motion is impending, $F = F_m = \mu_s N$. Substituting for μ_s its known value, you can express F in terms of N on the free-body diagram, thus eliminating one unknown.

c. Write and solve the equilibrium equations for the unknown you seek [Sample Prob. 8.3].

4. ***The system consists of several bodies and more than one state of impending motion is possible; you must determine a force, a distance, or a coefficient of friction for which motion is impending.***

a. Count the number of unknowns (forces, angles, etc.) and the total number of independent equations of equilibrium; the difference is equal to the number of surfaces at which motion can be impending.

b. Assume where motion is impending, and on the free-body diagrams of the bodies, draw the friction forces in directions opposite to the assumed motion.

c. At those surfaces where motion is assumed to be impending, $F = F_m = \mu_s N$.

d. Write and solve the equations of equilibrium for all unknowns.

e. Check your assumption by computing F and F_m for the surfaces where motion was assumed not to be impending. If $F < F_m$, your assumption was correct; however, if $F > F_m$, your assumption was incorrect, and it is then necessary to make a new assumption as to where motion impends and to repeat the solution process. You should always include a statement of your assumption and the associated check.

5. ***The impending motion of the body can be either sliding or tipping about a given point; you must find a force, a distance, or a coefficient of friction.***

a. Assume a possible impending sliding motion of the body and, on the free-body diagram, draw the friction force or forces in the direction opposite to that of the assumed motion.

b. Separately consider impending sliding and impending tipping.

i. For impending sliding, $F = F_m = \mu_s N$. Write the equations of equilibrium, and solve for the required unknown.

ii. For impending tipping, assume that the external reactions go to zero at all supports except for the reactions at the point about which tipping is assumed to be impending. Note for this case that $F \neq F_m$ and that the required unknown is determined from the equations of equilibrium.

The problems in this section require careful analysis, for they are some of the most difficult problems you will encounter in your statics course. You must always carefully draw free-body diagrams, paying particular attention to the directions of the friction forces. When it is necessary to assume the direction of the friction force, a positive answer implies a correct assumption, while a negative answer implies the friction force is directed opposite to the assumed direction. Also, the solutions of some problems are simplified using either the property of three-force bodies [Sec. 4.7] or the geometry of the system (for example, Prob. 8.23).

Problems

8.1 Determine whether the block shown is in equilibrium, and find the magnitude and direction of the friction force when $\theta = 30°$ and $P = 200$ N.

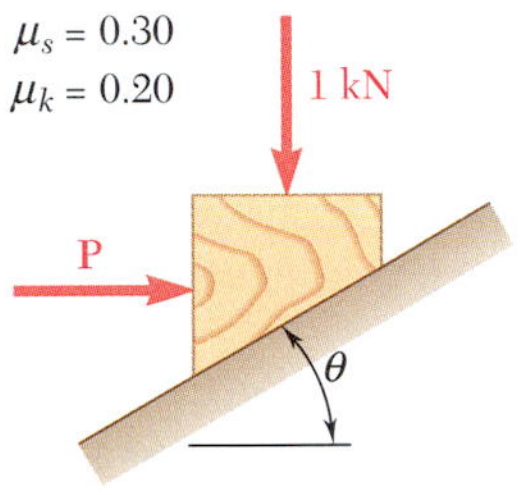

Fig. P8.1 and P8.2

8.2 Determine whether the block shown is in equilibrium, and find the magnitude and direction of the friction force when $\theta = 35°$ and $P = 400$ N.

8.3 Determine whether the 20-lb block shown is in equilibrium, and find the magnitude and direction of the friction force when $P = 8$ lb and $\theta = 20°$.

8.4 Determine whether the 20-lb block shown is in equilibrium, and find the magnitude and direction of the friction force when $P = 12.5$ lb and $\theta = 15°$.

8.5 Knowing that $\theta = 25°$, determine the range of values of P for which equilibrium is maintained.

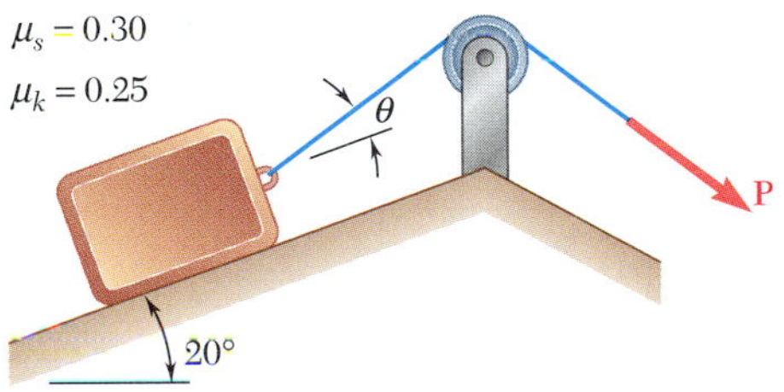

Fig. P8.3, *P8.4*, and P8.5

8.6 Knowing that the coefficient of friction between the 60-lb block and the incline is $\mu_s = 0.25$, determine (*a*) the smallest value of P for which motion of the block up the incline is impending, (*b*) the corresponding value of β.

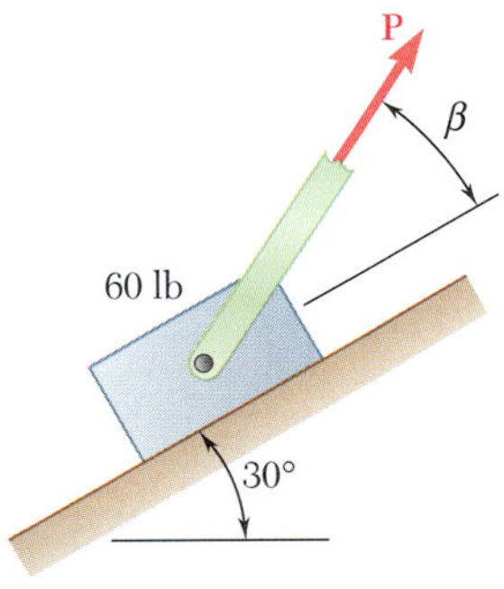

Fig. P8.6

8.7 Considering only values of θ less than 90°, determine the smallest value of θ for which motion of the block to the right is impending when (*a*) $m = 30$ kg, (*b*) $m = 40$ kg.

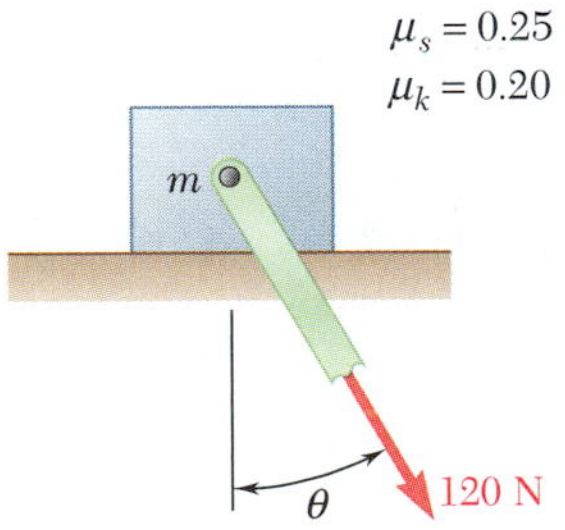

Fig. P8.7

8.8 Knowing that the coefficient of friction between the 30-lb block and the incline is $\mu_s = 0.25$, determine (*a*) the smallest value of P required to maintain the block in equilibrium, (*b*) the corresponding value of β.

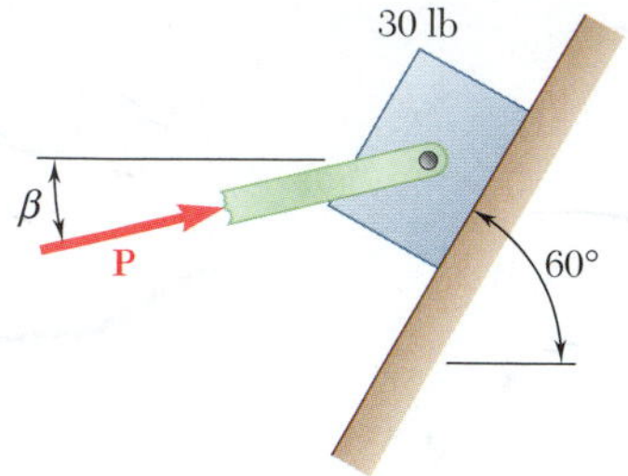

Fig. *P8.8*

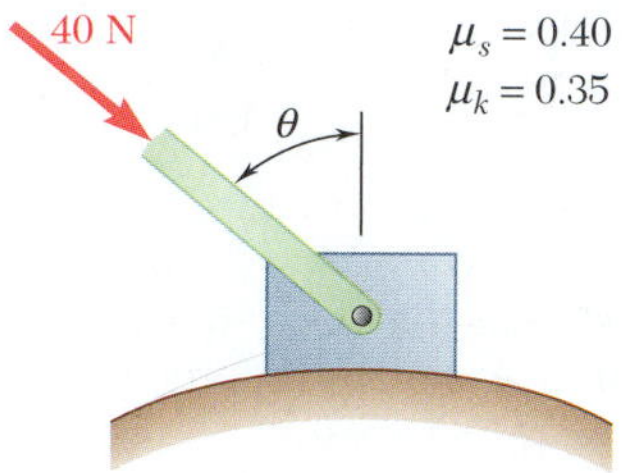

Fig. P8.9

8.9 A 6-kg block is at rest as shown. Determine the positive range of values of θ for which the block is in equilibrium if (*a*) θ is less than 90°, (*b*) θ is between 90° and 180°.

8.10 Knowing that $P = 25$ lb, determine the range of values of θ for which equilibrium of the 18-lb block is maintained.

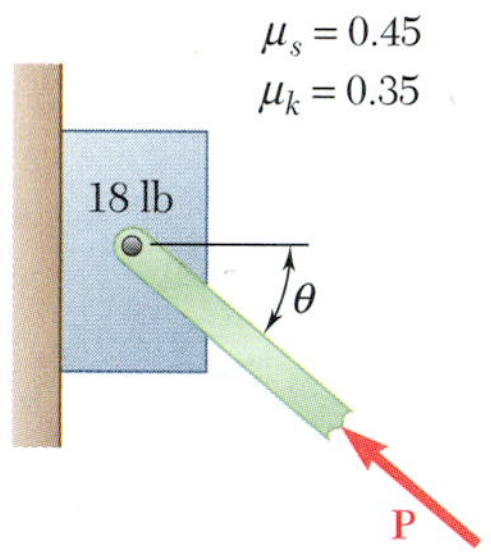

Fig. *P8.10*

8.11 and 8.12 The coefficients of friction are $\mu_s = 0.40$ and $\mu_k = 0.30$ between all surfaces of contact. Determine the force **P** for which motion of the 60-lb block is impending if cable AB (*a*) is attached as shown, (*b*) is removed.

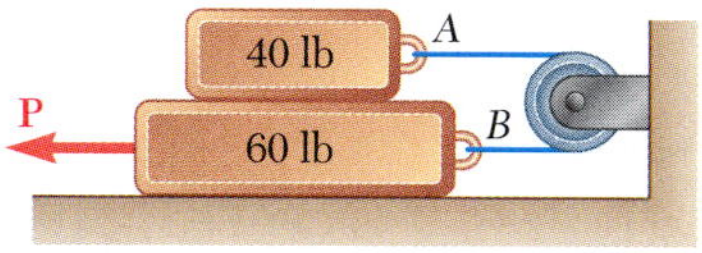

Fig. P8.11

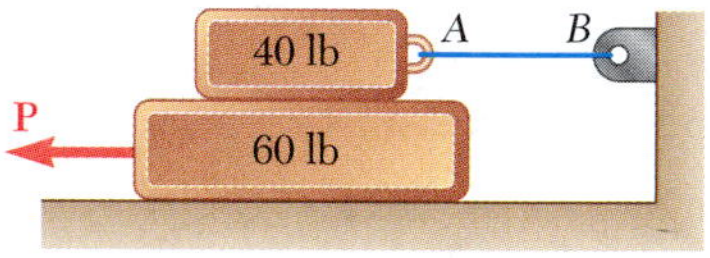

Fig. P8.12

8.13 The 8-kg block A is attached to link AC and rests on the 12-kg block B. Knowing that the coefficient of static friction is 0.20 between all surfaces of contact and neglecting the mass of the link, determine the value of θ for which motion of block B is impending.

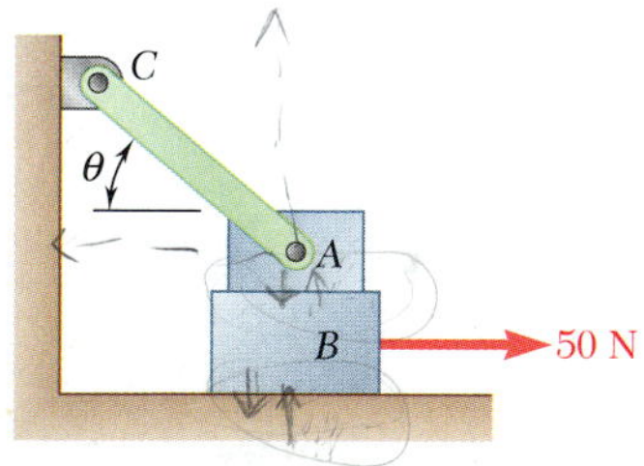

Fig. P8.13

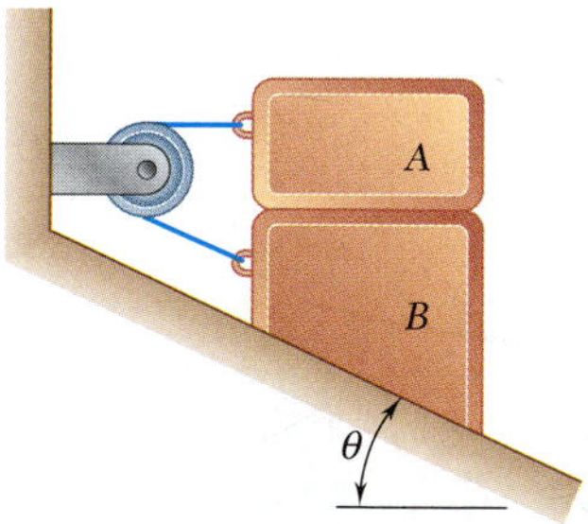

Fig. *P8.14*

8.14 The 8-kg block A and the 16-kg block B are at rest on an incline as shown. Knowing that the coefficient of static friction is 0.25 between all surfaces of contact, determine the value of θ for which motion is impending.

8.15 A 48-kg cabinet is mounted on casters which can be locked to prevent their rotation. The coefficient of static friction between the floor and each caster is 0.30. Knowing that $h = 640$ mm, determine the magnitude of the force **P** required for impending motion of the cabinet to the right (*a*) if all casters are locked, (*b*) if the casters at *B* are locked and the casters at *A* are free to rotate, (*c*) if the casters at *A* are locked and the casters at *B* are free to rotate.

8.16 A 48-kg cabinet is mounted on casters which can be locked to prevent their rotation. The coefficient of static friction between the floor and each caster is 0.30. Assuming that the casters at *A* and *B* are locked, determine (*a*) the force **P** required for impending motion of the cabinet to the right, (*b*) the largest allowable height *h* if the cabinet is not to tip over.

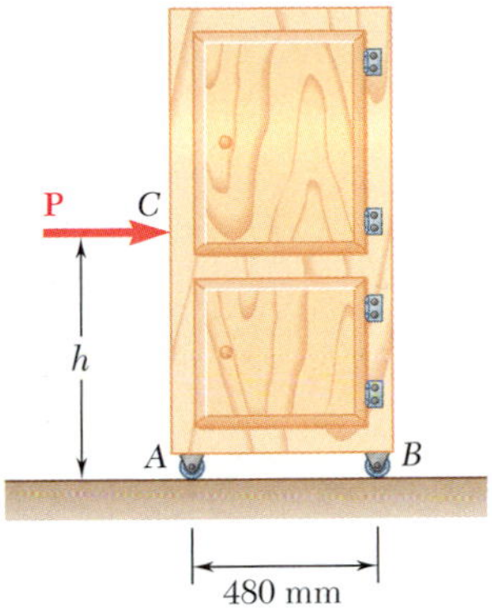

Fig. P8.15 and *P8.16*

8.17 The cylinder shown is of weight *W* and radius *r*, and the coefficient of static friction μ_s is the same at *A* and *B*. Determine the magnitude of the largest couple **M** which can be applied to the cylinder if it is not to rotate.

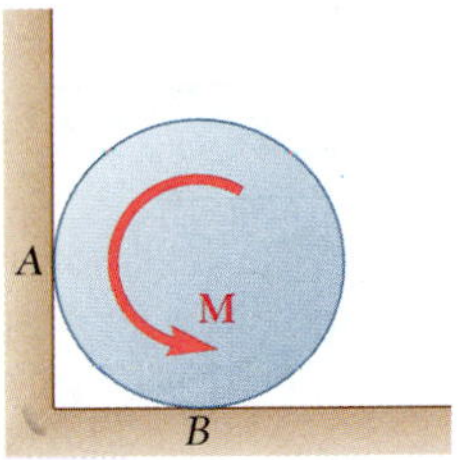

Fig. P8.17 and P8.18

8.18 The cylinder shown is of weight *W* and radius *r*. Express in terms of *W* and *r* the magnitude of the largest couple **M** which can be applied to the cylinder if it is not to rotate assuming that the coefficient of static friction is (*a*) zero at *A* and 0.36 at *B*, (*b*) 0.30 at *A* and 0.36 at *B*.

8.19 The hydraulic cylinder shown exerts a force of 680 lb directed to the right on point *B* and to the left on point *E*. Determine the magnitude of the couple **M** required to rotate the drum clockwise at a constant speed.

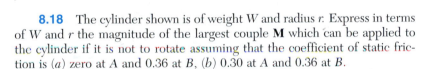

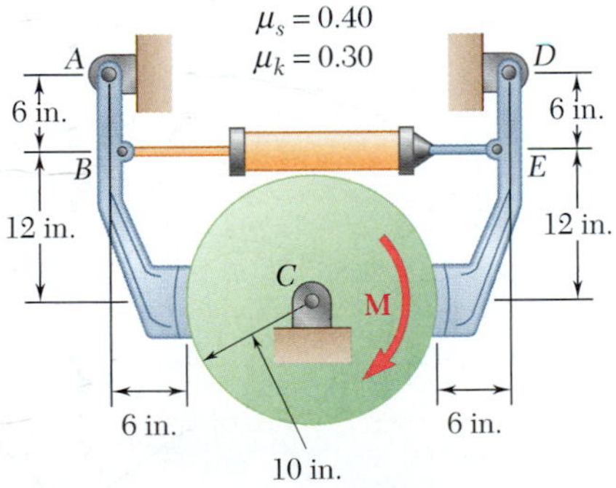

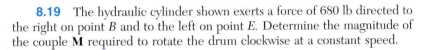

Fig. P8.19 and *P8.20*

8.20 A couple **M** of magnitude 70 lb · ft is applied to the drum as shown. Determine the smallest force which must be exerted by the hydraulic cylinder on joints *B* and *E* if the drum is not to rotate.

8.21 and 8.22 A 19.5-ft ladder AB leans against a wall as shown. Assuming that the coefficient of static friction μ_s is the same at A and B, determine the smallest value of μ_s for which equilibrium is maintained.

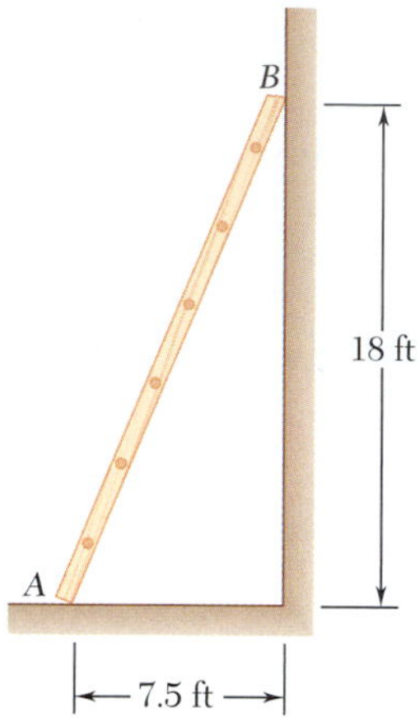

Fig. P8.21

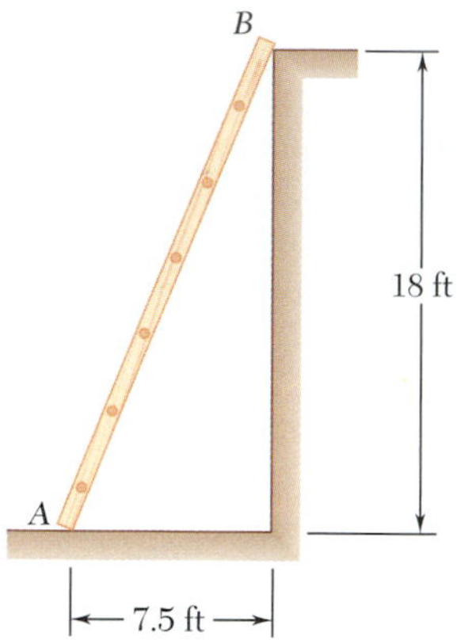

Fig. P8.22

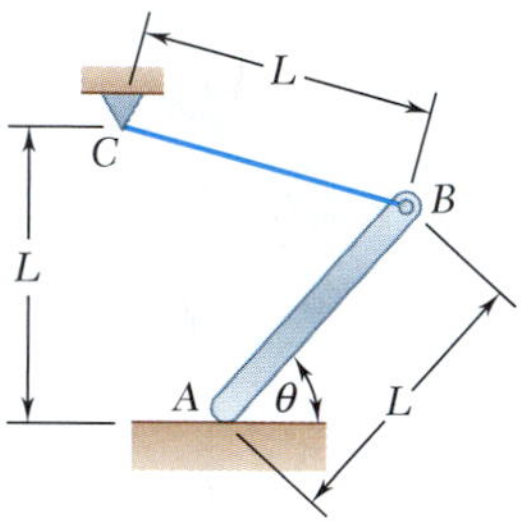

Fig. P8.23

8.23 End A of a slender, uniform rod of weight W and length L bears on a horizontal surface as shown, while end B is supported by a cord BC of length L. Knowing that the coefficient of static friction is 0.40, determine (*a*) the value of θ for which motion is impending, (*b*) the corresponding value of the tension in the cord.

8.24 A slender rod of length L is lodged between peg C and the vertical wall and supports a load **P** at end A. Knowing that the coefficient of static friction between the peg and the rod is 0.25 and neglecting friction at the roller, determine the range of values of the ratio L/a for which equilibrium is maintained.

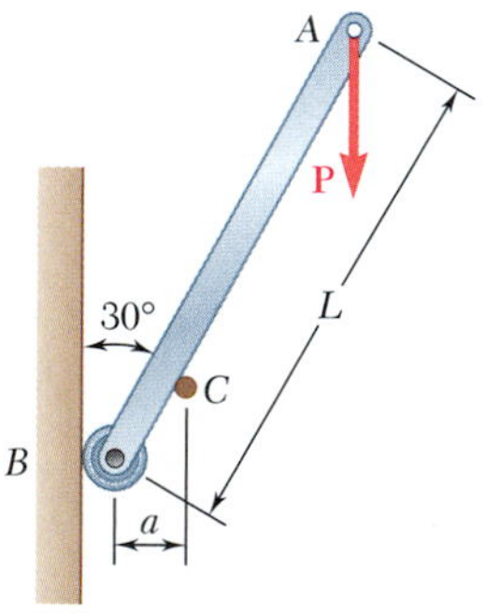

Fig. *P8.24*

8.25 The basic components of a clamping device are bar AB, locking plate CD, and lever EFG; the dimensions of the slot in CD are slightly larger than those of the cross section of AB. To engage the clamp, AB is pushed against the workpiece, and then force **P** is applied. Knowing that $P = 160$ N and neglecting the friction force between the lever and the plate, determine the smallest allowable value of the static coefficient of friction between the bar and the plate.

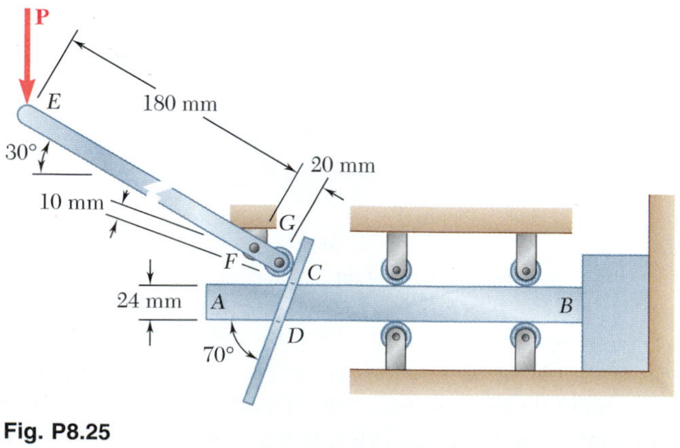

Fig. P8.25

8.26 A window sash having a mass of 4 kg is normally supported by two 2-kg sash weights. Knowing that the window remains open after one sash cord has broken, determine the smallest possible value of the coefficient of static friction. (Assume that the sash is slightly smaller than the frame and will bind only at points *A* and *D*.)

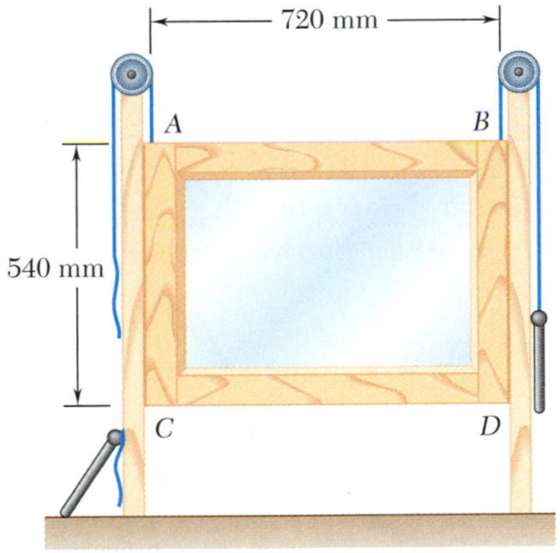

Fig. *P8.26*

8.27 The steel-plate clamp shown is used to lift a steel plate *H* of mass 250 kg. Knowing that the normal force exerted on steel cam *EG* by pin *D* forms an angle of 40° with the horizontal and neglecting the friction force between the cam and the pin, determine the smallest allowable value of the coefficient of static friction.

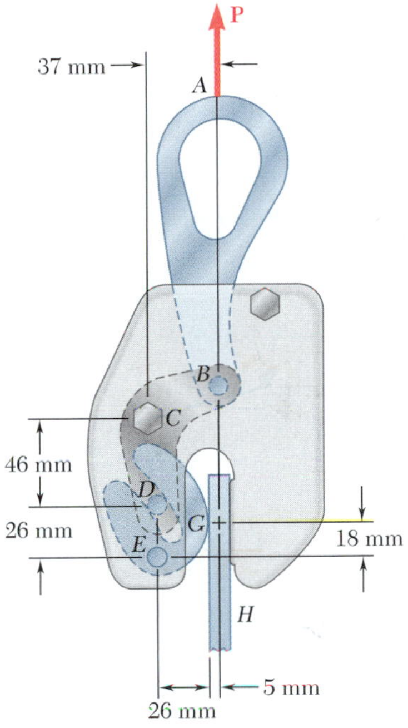

Fig. P8.27

8.28 The 5-in.-radius cam shown is used to control the motion of the plate *CD*. Knowing that the coefficient of static friction between the cam and the plate is 0.45 and neglecting friction at the roller supports, determine (*a*) the force **P** for which motion of the plate is impending knowing that the plate is 1 in. thick, (*b*) the largest thickness of the plate for which the mechanism is self-locking (that is, for which the plate cannot be moved however large the force **P** may be).

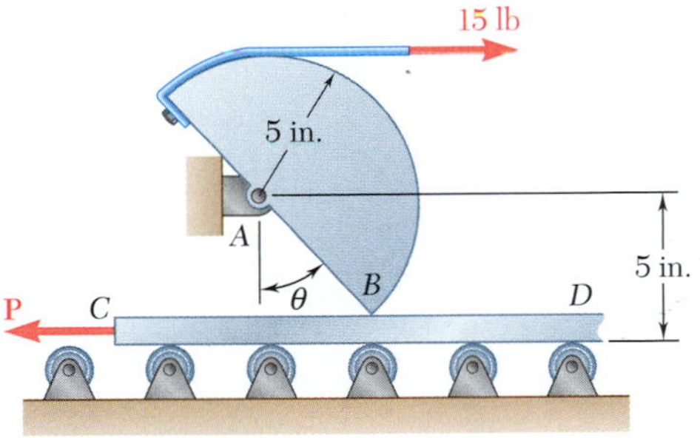

Fig. P8.28

8.29 A child having a mass of 18 kg is seated halfway between the ends of a small, 16-kg table as shown. The coefficient of static friction is 0.20 between the ends of the table and the floor. If a second child pushes on edge *B* of the table top at a point directly opposite to the first child with a force **P** lying in a vertical plane parallel to the ends of the table and having a magnitude of 66 N, determine the range of values of θ for which the table will (*a*) tip, (*b*) slide.

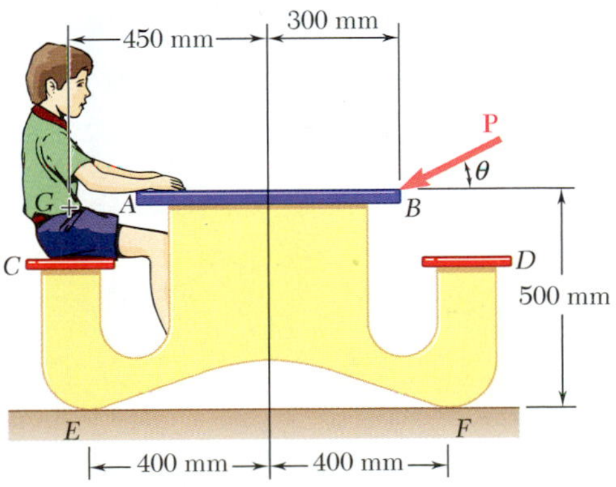

Fig. *P8.29*

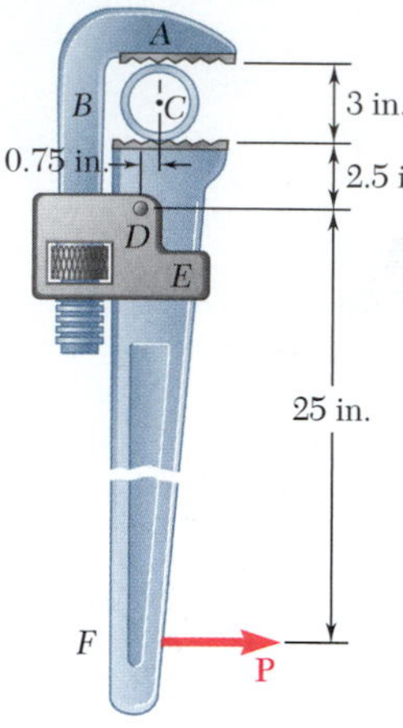

Fig. P8.30

8.30 A pipe of diameter 3 in. is gripped by the stillson wrench shown. Portions AB and DE of the wrench are rigidly attached to each other, and portion CF is connected by a pin at D. If the wrench is to grip the pipe and be self-locking, determine the required minimum coefficients of friction at A and C.

8.31 Solve Prob. 8.30 assuming that the diameter of the pipe is 1.5 in.

8.32 The 25-kg plate $ABCD$ is attached at A and D to collars which can slide on the vertical rod. Knowing that the coefficient of static friction is 0.40 between both collars and the rod, determine whether the plate is in equilibrium in the position shown when the magnitude of the vertical force applied at E is (*a*) $P = 0$, (*b*) $P = 80$ N.

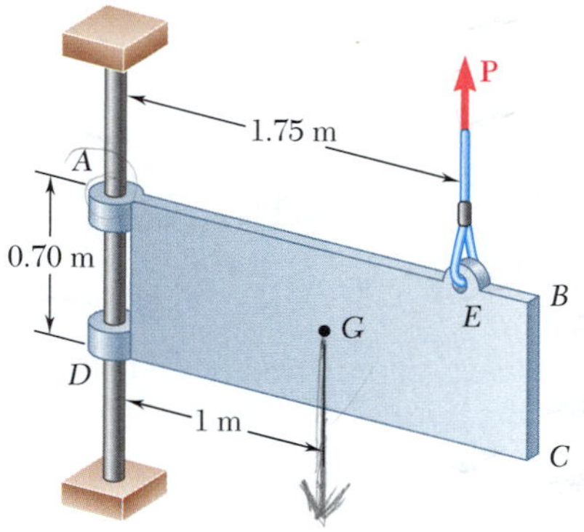

Fig. P8.32

8.33 In Prob. 8.32, determine the range of values of the magnitude P of the vertical force applied at E for which the plate will move downward.

8.34 and 8.35 A collar B of weight W is attached to the spring AB and can move along the rod shown. The constant of the spring is 1.5 kN/m and the spring is unstretched when $\theta = 0$. Knowing that the coefficient of static friction between the collar and the rod is 0.40, determine the range of values of W for which equilibrium is maintained when (*a*) $\theta = 20°$, (*b*) $\theta = 30°$.

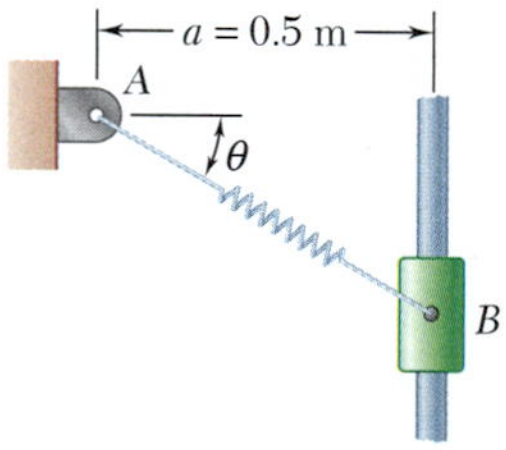

Fig. P8.34

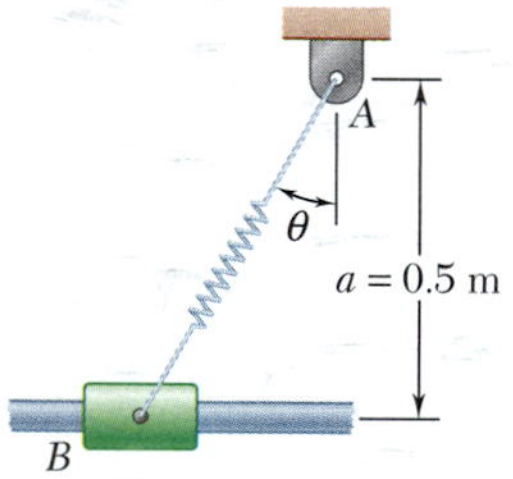

Fig. P8.35

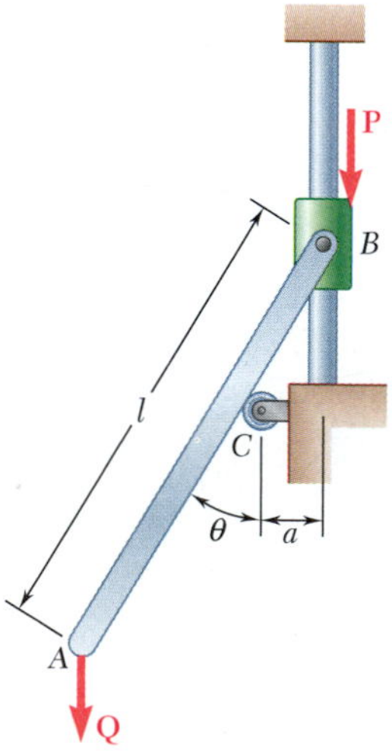

Fig. P8.36

8.36 The slender rod AB of length $l = 30$ in. is attached to a collar at B and rests on a small wheel located at a horizontal distance $a = 4$ in. from the vertical rod on which the collar slides. Knowing that the coefficient of static friction between the collar and the vertical rod is 0.25 and neglecting the radius of the wheel, determine the range of values of P for which equilibrium is maintained when $Q = 25$ lb and $\theta = 30°$.

8.37 The 4.5-kg block A and the 3-kg block B are connected by a slender rod of negligible mass. The coefficient of static friction is 0.40 between all surfaces of contact. Knowing that for the position shown the rod is horizontal, determine the range of values of P for which equilibrium is maintained.

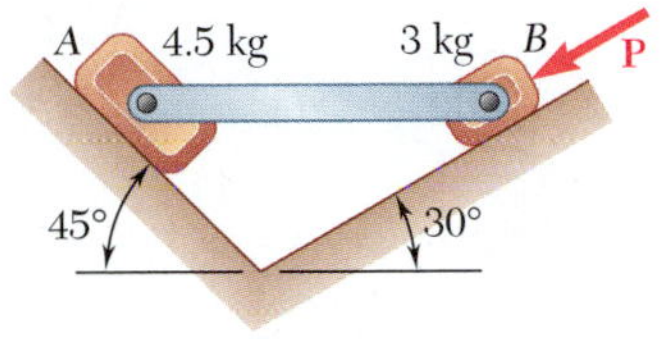

Fig. P8.37

8.38 Bar AB is attached to collars which can slide on the inclined rods shown. A force **P** is applied at point D located at a distance a from end A. Knowing that the coefficient of static friction μ_s between each collar and the rod upon which it slides is 0.30 and neglecting the weights of the bar and of the collars, determine the smallest value of the ratio a/L for which equilibrium is maintained.

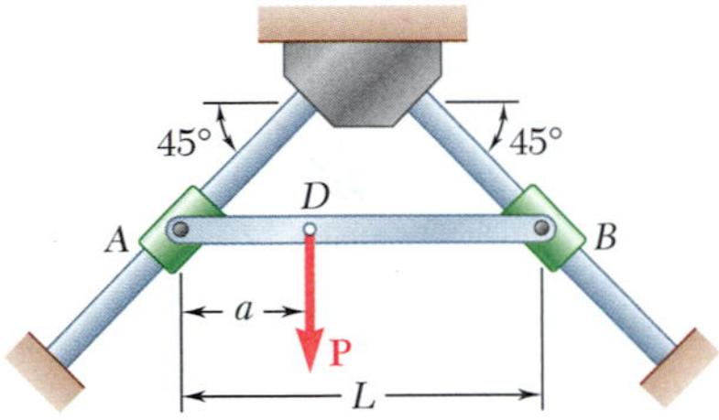

Fig. *P8.38*

8.39 The 6-kg slender rod AB is pinned at A and rests on the 18-kg cylinder C. Knowing that the diameter of the cylinder is 250 mm and that the coefficient of static friction is 0.35 between all surfaces of contact, determine the largest magnitude of the force **P** for which equilibrium is maintained.

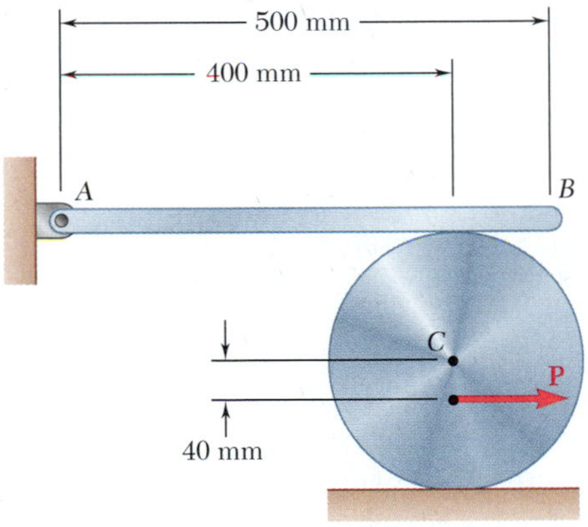

Fig. P8.39

8.40 Two rods are connected by a collar at B. A couple $\mathbf{M}_A$ of magnitude 12 lb · ft is applied to rod AB. Knowing that $\mu_s = 0.30$ between the collar and rod AB, determine the largest couple $\mathbf{M}_C$ for which equilibrium will be maintained.

8.41 In Prob. 8.40, determine the smallest couple $\mathbf{M}_C$ for which equilibrium will be maintained.

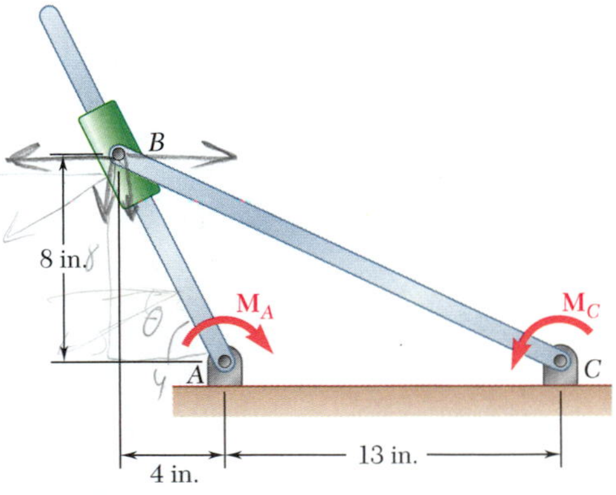

Fig. P8.40

8.42 Blocks A, B, and C having the masses shown are at rest on an incline. Denoting by μ_s the coefficient of static friction between all surfaces of contact, determine the smallest value of μ_s for which equilibrium is maintained.

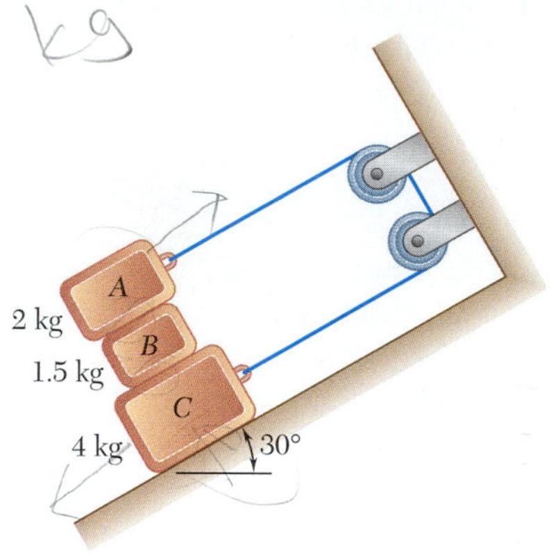

Fig. P8.42

8.43 A slender steel rod of length 9 in. is placed inside a pipe as shown. Knowing that the coefficient of static friction between the rod and the pipe is 0.20, determine the largest value of θ for which the rod will not fall into the pipe.

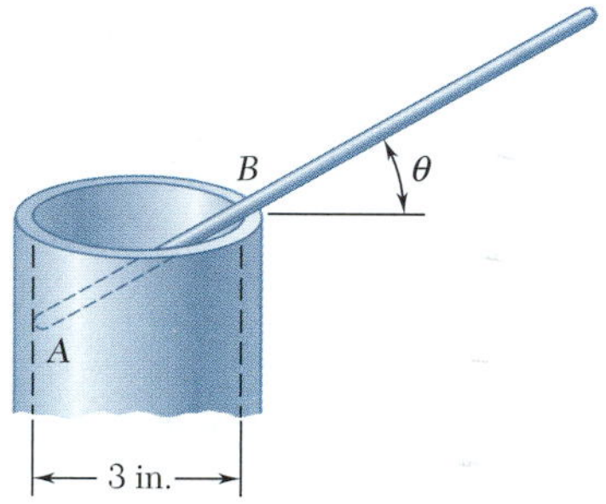

Fig. P8.43

8.44 In Prob. 8.43, determine the smallest value of θ for which the rod will not fall out of the pipe.

8.45 Two slender rods of negligible weight are pin-connected at C and attached to blocks A and B, each of weight W. Knowing that $\theta = 70°$ and that the coefficient of static friction between the blocks and the horizontal surface is 0.30, determine the largest value of P for which equilibrium is maintained.

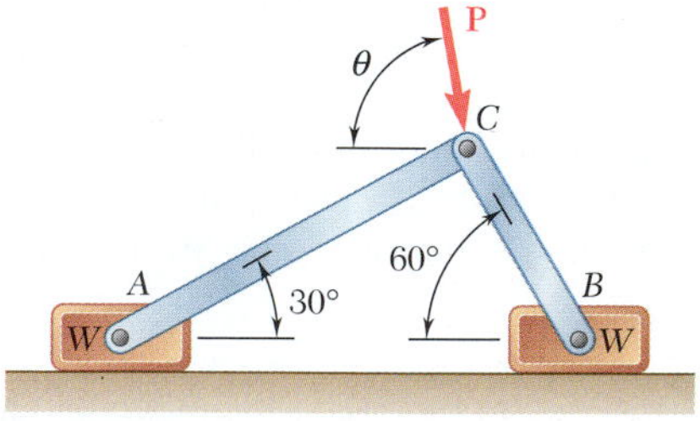

Fig. P8.45

8.5. WEDGES

Wedges are simple machines used to raise large stone blocks and other heavy loads. These loads can be raised by applying to the wedge a force usually considerably smaller than the weight of the load. In addition, because of the friction between the surfaces in contact, a properly shaped wedge will remain in place after being forced under the load. Wedges can thus be used advantageously to make small adjustments in the position of heavy pieces of machinery.

Consider the block A shown in Fig. 8.7*a*. This block rests against a vertical wall B and is to be raised slightly by forcing a wedge C between block A and a second wedge D. We want to find the minimum value of the force **P** which must be applied to the wedge C to move the block. It will be assumed that the weight **W** of the block is known, either given in pounds or determined in newtons from the mass of the block expressed in kilograms.

The free-body diagrams of block A and of wedge C have been drawn in Fig. 8.7*b* and *c*. The forces acting on the block include its weight and the normal and friction forces at the surfaces of contact with wall B and wedge C. The magnitudes of the friction forces $\mathbf{F}_1$ and $\mathbf{F}_2$ are equal, respectively, to $\mu_s N_1$ and $\mu_s N_2$ since the motion of the block is impending. It is important to show the friction forces with their correct sense. Since the block will move upward, the force $\mathbf{F}_1$ exerted by the wall on the block must be directed downward. On the other hand, since the wedge C will move to the right, the relative motion of A with respect to C is to the left and the force $\mathbf{F}_2$ exerted by C on A must be directed to the right.

Considering now the free body C in Fig. 8.7*c*, we note that the forces acting on C include the applied force **P** and the normal and friction forces at the surfaces of contact with A and D. The weight of the wedge is small compared with the other forces involved and can be neglected. The forces exerted by A on C are equal and opposite to the forces $\mathbf{N}_2$ and $\mathbf{F}_2$ exerted by C on A and are denoted, respectively, by $-\mathbf{N}_2$ and $-\mathbf{F}_2$; the friction force $-\mathbf{F}_2$ must therefore be directed to the left. We check that the force $\mathbf{F}_3$ exerted by D is also directed to the left.

The total number of unknowns involved in the two free-body diagrams can be reduced to four if the friction forces are expressed in terms of the normal forces. Expressing that block A and wedge C are in equilibrium will provide four equations which can be solved to obtain the magnitude of **P**. It should be noted that in the example considered here, it will be more convenient to replace each pair of normal and friction forces by their resultant. Each free body is then subjected to only three forces, and the problem can be solved by drawing the corresponding force triangles (see Sample Prob. 8.4).

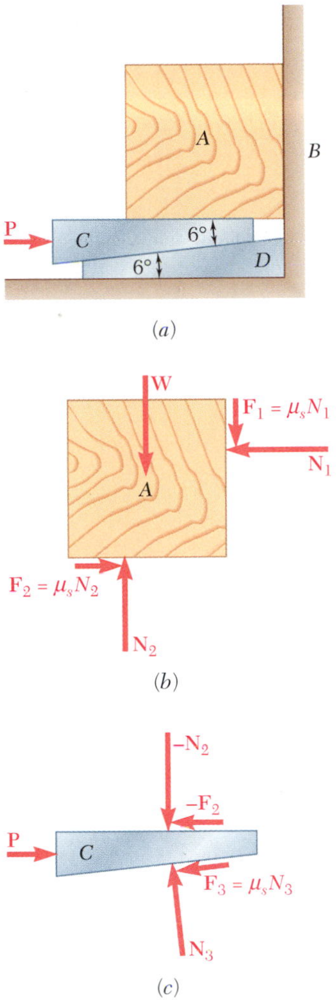

Fig. 8.7

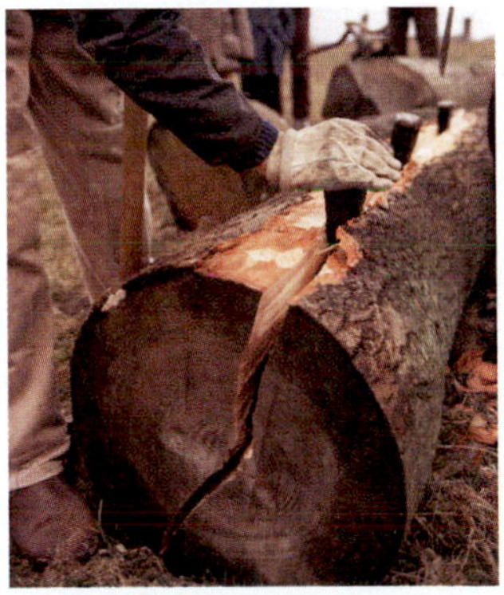

Photo 8.2 Wedges are used as shown to split tree trunks because the normal forces exerted by the wedges on the wood are much larger than the forces required to insert the wedges.

8.6. SQUARE-THREADED SCREWS

Square-threaded screws are frequently used in jacks, presses, and other mechanisms. Their analysis is similar to the analysis of a block sliding along an inclined plane.

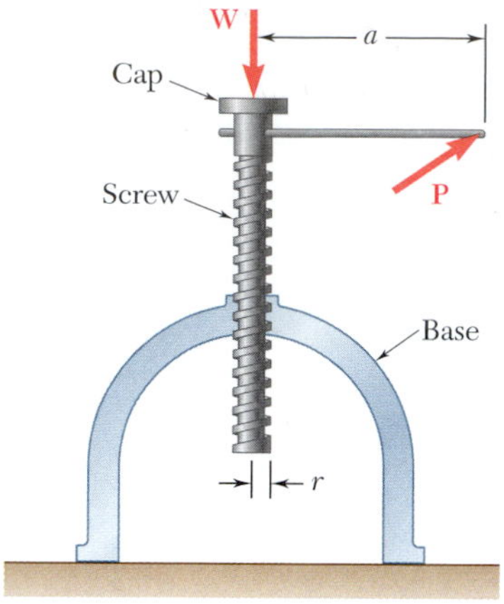

Fig. 8.8

Consider the jack shown in Fig. 8.8. The screw carries a load **W** and is supported by the base of the jack. Contact between the screw and the base takes place along a portion of their threads. By applying a force **P** on the handle, the screw can be made to turn and to raise the load **W**.

The thread of the base has been unwrapped and shown as a straight line in Fig. 8.9*a*. The correct slope was obtained by plotting horizontally the product $2\pi r$, where r is the mean radius of the thread, and vertically the *lead* L of the screw, that is, the distance through which the screw advances in one turn. The angle θ this line forms with the horizontal is the *lead angle*. Since the force of friction between two surfaces in contact does not depend upon the area of contact, a much smaller than actual area of contact between the two threads can be assumed, and the screw can be represented by the block shown in Fig. 8.9*a*. It should be noted, however, that in this analysis of the jack, the friction between the cap and the screw is neglected.

The free-body diagram of the block should include the load **W**, the reaction **R** of the base thread, and a horizontal force **Q** having the same effect as the force **P** exerted on the handle. The force **Q** should have the same moment as **P** about the axis of the screw and its magnitude should thus be $Q = Pa/r$. The force **Q**, and thus the force **P** required to raise the load **W**, can be obtained from the free-body diagram shown in Fig. 8.9*a*. The friction angle is taken equal to ϕ_s since the load will presumably be raised through a succession of short strokes. In mechanisms providing for the continuous rotation of a screw, it may be desirable to distinguish between the force required for impending motion (using ϕ_s) and that required to maintain motion (using ϕ_k).

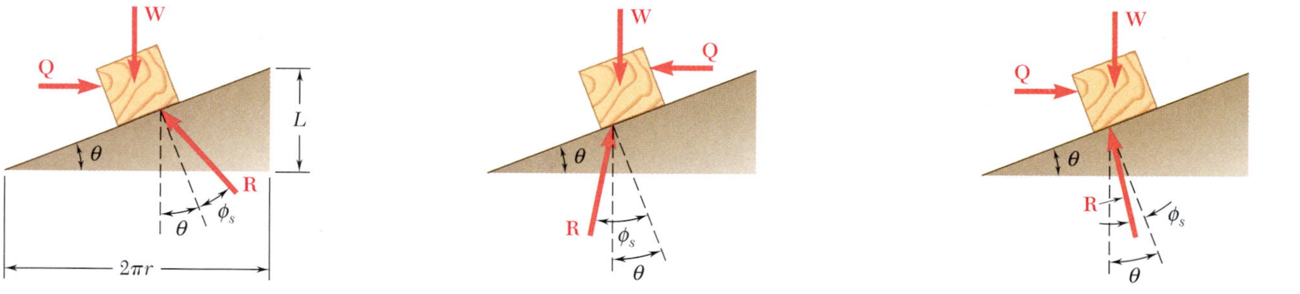

(*a*) Impending motion upward (*b*) Impending motion downward with $\phi_s > \theta$ (*c*) Impending motion downward with $\phi_s < \theta$

Fig. 8.9 Block-and-incline analysis of a screw.

If the friction angle ϕ_s is larger than the lead angle θ, the screw is said to be *self-locking;* it will remain in place under the load. To lower the load, we must then apply the force shown in Fig. 8.9*b*. If ϕ_s is smaller than θ, the screw will unwind under the load; it is then necessary to apply the force shown in Fig. 8.9*c* to maintain equilibrium.

The lead of a screw should not be confused with its *pitch*. The lead was defined as the distance through which the screw advances in one turn; the pitch is the distance measured between two consecutive threads. While lead and pitch are equal in the case of *single-threaded* screws, they are different in the case of *multiple-threaded* screws, that is, screws having several independent threads. It is easily verified that for double-threaded screws, the lead is twice as large as the pitch; for triple-threaded screws, it is three times as large as the pitch; etc.

SAMPLE PROBLEM 8.4

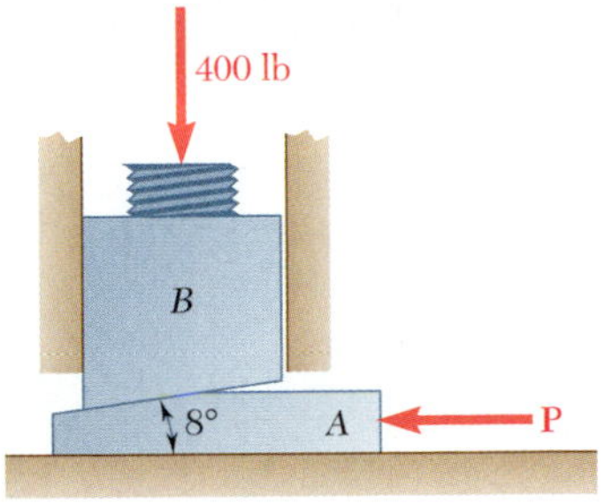

The position of the machine block B is adjusted by moving the wedge A. Knowing that the coefficient of static friction is 0.35 between all surfaces of contact, determine the force **P** for which motion of block B (*a*) is impending upward, (*b*) is impending downward.

SOLUTION

For each part, the free-body diagrams of block B and wedge A are drawn, together with the corresponding force triangles, and the law of sines is used to find the desired forces. We note that since $\mu_s = 0.35$, the angle of friction is

$$\phi_s = \tan^{-1} 0.35 = 19.3°$$

a. Force P for Impending Motion of Block Upward

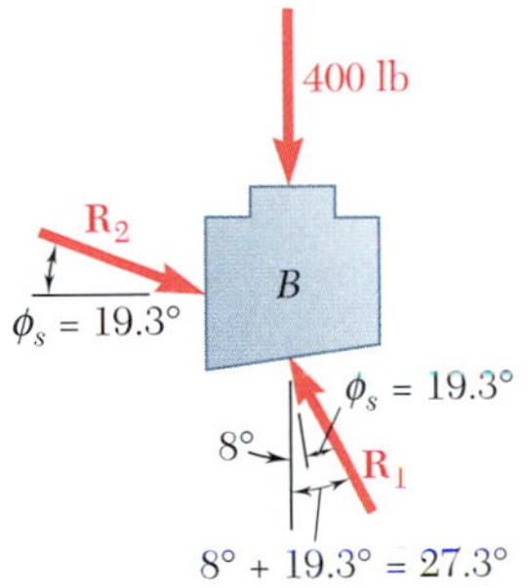

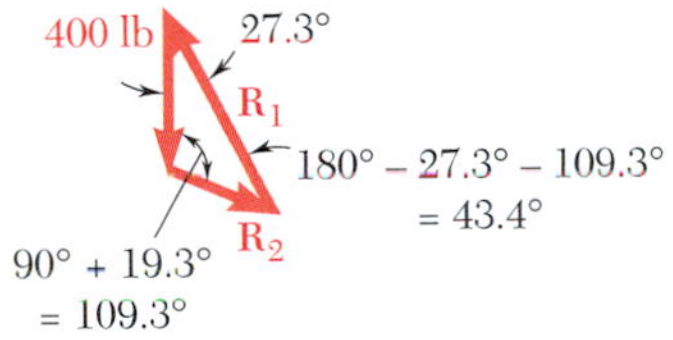

Free Body: Block B

$$\frac{R_1}{\sin 109.3°} = \frac{400 \text{ lb}}{\sin 43.4°}$$

$$R_1 = 549 \text{ lb}$$

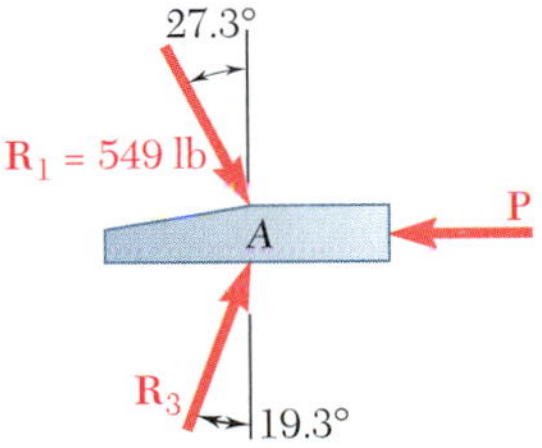

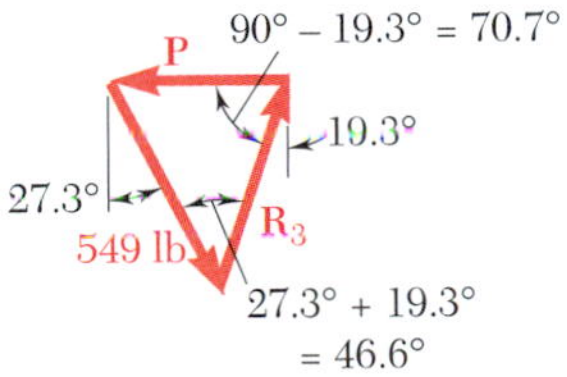

Free Body: Wedge A

$$\frac{P}{\sin 46.6°} = \frac{549 \text{ lb}}{\sin 70.7°}$$

$$P = 423 \text{ lb} \qquad \mathbf{P} = 423 \text{ lb} \leftarrow \blacktriangleleft$$

b. Force P for Impending Motion of Block Downward

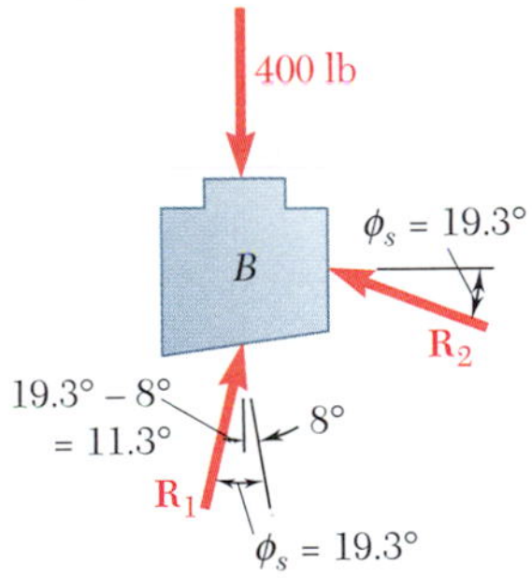

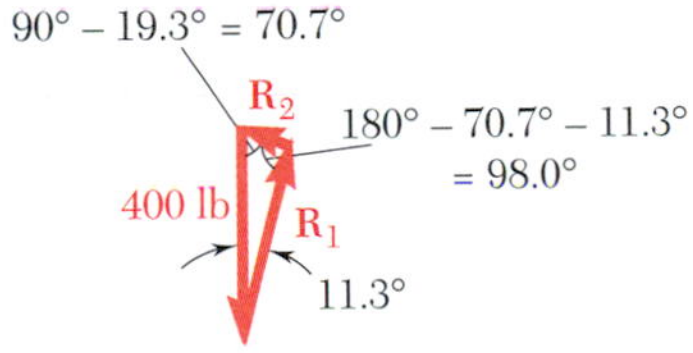

Free Body: Block B

$$\frac{R_1}{\sin 70.7°} = \frac{400 \text{ lb}}{\sin 98.0°}$$

$$R_1 = 381 \text{ lb}$$

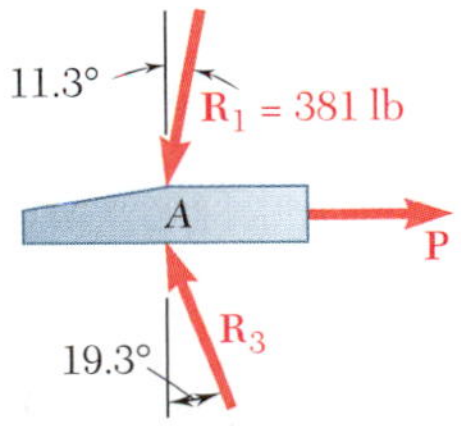

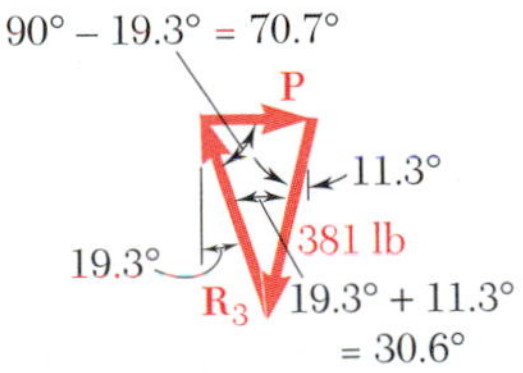

Free Body: Wedge A

$$\frac{P}{\sin 30.6°} = \frac{381 \text{ lb}}{\sin 70.7°}$$

$$P = 206 \text{ lb} \qquad \mathbf{P} = 206 \text{ lb} \rightarrow \blacktriangleleft$$

SAMPLE PROBLEM 8.5

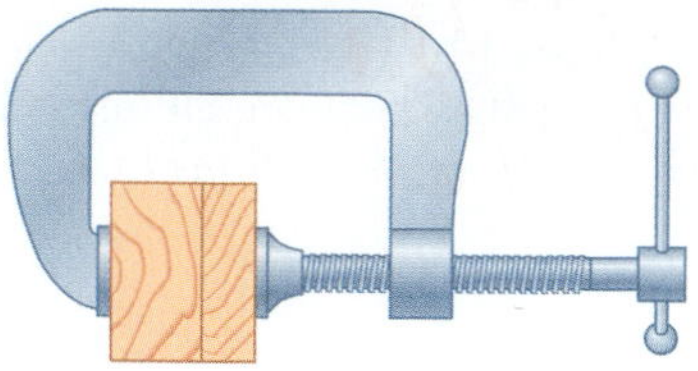

A clamp is used to hold two pieces of wood together as shown. The clamp has a double square thread of mean diameter equal to 10 mm with a pitch of 2 mm. The coefficient of friction between threads is $\mu_s = 0.30$. If a maximum torque of 40 N · m is applied in tightening the clamp, determine (*a*) the force exerted on the pieces of wood, (*b*) the torque required to loosen the clamp.

SOLUTION

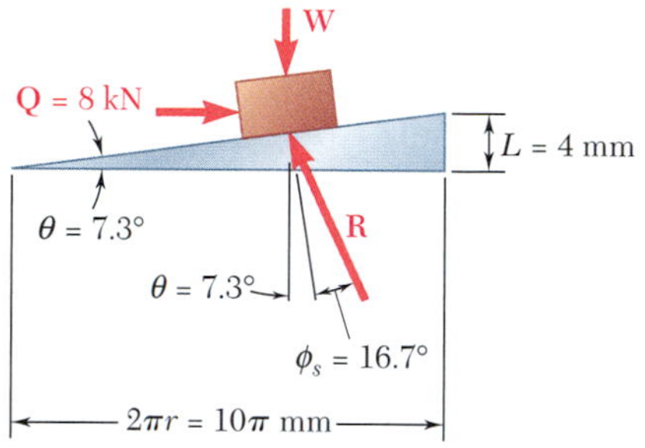

a. Force Exerted by Clamp. The mean radius of the screw is r = 5 mm. Since the screw is double-threaded, the lead L is equal to twice the pitch: $L = 2(2 \text{ mm}) = 4$ mm. The lead angle θ and the friction angle ϕ_s are obtained by writing

$$\tan\theta = \frac{L}{2\pi r} = \frac{4 \text{ mm}}{10\pi \text{ mm}} = 0.1273 \qquad \theta = 7.3^\circ$$

$$\tan\phi_s = \mu_s = 0.30 \qquad \phi_s = 16.7^\circ$$

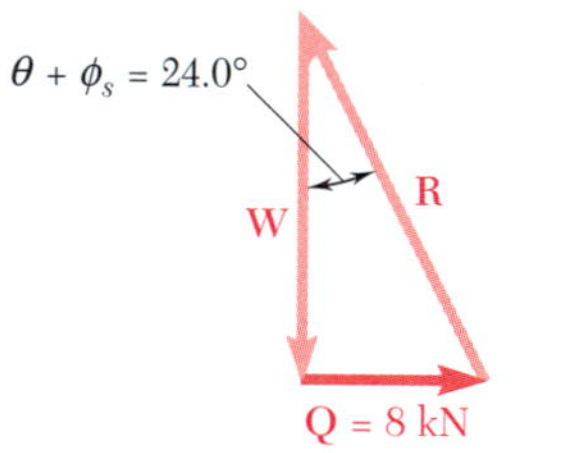

The force **Q** which should be applied to the block representing the screw is obtained by expressing that its moment Qr about the axis of the screw is equal to the applied torque.

$$Q(5 \text{ mm}) = 40 \text{ N} \cdot \text{m}$$

$$Q = \frac{40 \text{ N} \cdot \text{m}}{5 \text{ mm}} = \frac{40 \text{ N} \cdot \text{m}}{5 \times 10^{-3} \text{ m}} = 8000 \text{ N} = 8 \text{ kN}$$

The free-body diagram and the corresponding force triangle can now be drawn for the block; the magnitude of the force **W** exerted on the pieces of wood is obtained by solving the triangle.

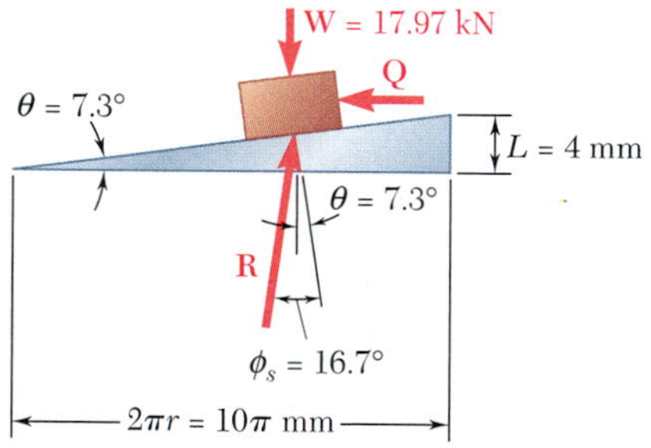

$$W = \frac{Q}{\tan(\theta + \phi_s)} = \frac{8 \text{ kN}}{\tan 24.0^\circ}$$

$W = 17.97$ kN ◀

b. Torque Required to Loosen Clamp. The force **Q** required to loosen the clamp and the corresponding torque are obtained from the free-body diagram and force triangle shown.

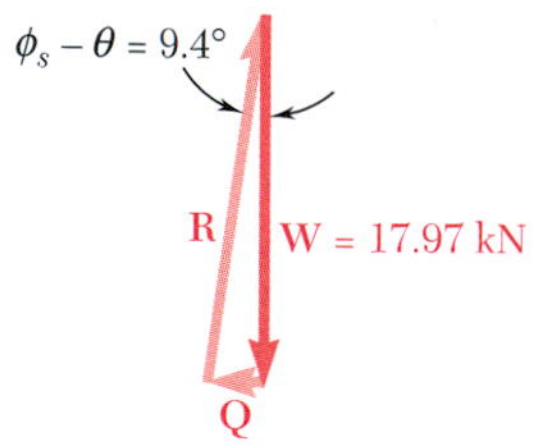

$$Q = W \tan(\phi_s - \theta) = (17.97 \text{ kN}) \tan 9.4^\circ$$
$$= 2.975 \text{ kN}$$

$$\text{Torque} = Qr = (2.975 \text{ kN})(5 \text{ mm})$$
$$= (2.975 \times 10^3 \text{ N})(5 \times 10^{-3} \text{ m}) = 14.87 \text{ N} \cdot \text{m}$$

Torque = 14.87 N · m ◀

SOLVING PROBLEMS ON YOUR OWN

In this lesson you learned to apply the laws of friction to the solution of problems involving *wedges* and *square-threaded screws*.

1. Wedges. Keep the following in mind when solving a problem involving a wedge:

a. First draw a free-body diagram of the wedge and of all the other bodies involved. Carefully note the sense of the relative motion of all surfaces of contact and show each friction force acting in *a direction opposite* to the direction of that relative motion.

b. Show the maximum static friction force $\mathbf{F}_m$ at each surface if the wedge is to be inserted or removed, *since motion will be impending in each of these cases*.

c. The reaction R and the angle of friction, rather than the normal force and the friction force, can be used in many applications. You can then draw one or more force triangles and determine the unknown quantities either graphically or by trigonometry [Sample Prob. 8.4].

2. Square-Threaded Screws. The analysis of a square-threaded screw is equivalent to the analysis of a block sliding on an incline. To draw the appropriate incline, you should unwrap the thread of the screw and represent it by a straight line [Sample Prob. 8.5]. When solving a problem involving a square-threaded screw, keep the following in mind:

a. Do not confuse the pitch of a screw with the lead of a screw. The *pitch* of a screw is the distance between two consecutive threads, while the *lead* of a screw is the distance the screw advances in one full turn. The lead and the pitch are equal only in single-threaded screws. In a double-threaded screw, the lead is twice the pitch.

b. The torque required to tighten a screw is different from the torque required to loosen it. Also, screws used in jacks and clamps are usually *self-locking;* that is, the screw will remain stationary as long as no torque is applied to it, and a torque must be applied to the screw to loosen it [Sample Prob. 8.5].

Problems

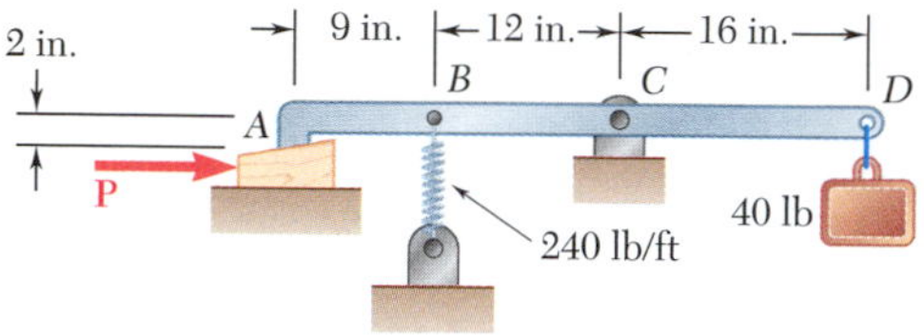

Fig. P8.46

8.46 A 40-lb weight is hung from a lever which rests against a 10° wedge at A and is supported by a frictionless hinge at C. Knowing that the coefficient of static friction is 0.25 at both surfaces of the wedge and that for the position shown the spring is stretched 4 in., determine (*a*) the magnitude of the force **P** for which motion of the wedge is impending, (*b*) the components of the corresponding reaction at C.

8.47 Solve Prob. 8.46 assuming that force **P** is directed to the left.

8.48 and 8.49 Two 8° wedges of negligible mass are used to move and position a 240-kg block. Knowing that the coefficient of static friction is 0.40 at all surfaces of contact, determine the magnitude of the force **P** for which motion of the block is impending.

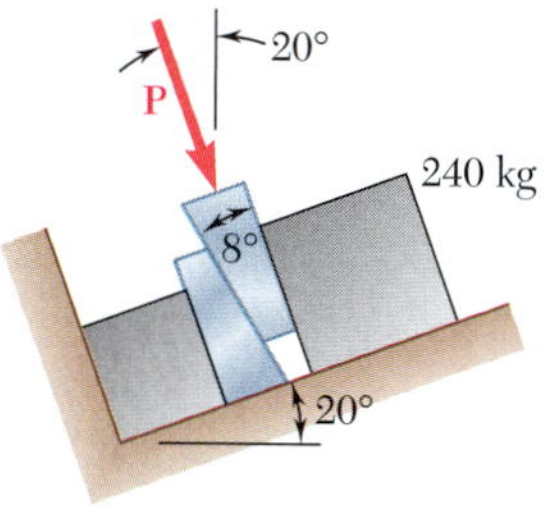

Fig. P8.48

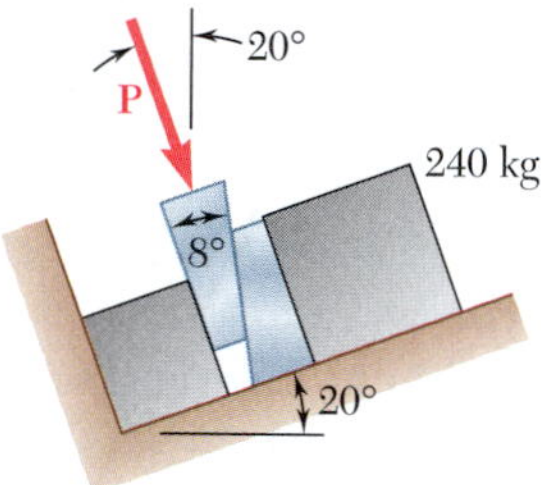

Fig. P8.49

8.50* and *8.51 The elevation of the end of the steel beam supported by a concrete floor is adjusted by means of the steel wedges E and F. The base plate CD has been welded to the lower flange of the beam, and the end reaction of the beam is known to be 150 kN. The coefficient of static friction is 0.30 between the two steel surfaces and 0.60 between the steel and the concrete. If the horizontal motion of the beam is prevented by the force **Q**, determine (*a*) the force **P** required to raise the beam, (*b*) the corresponding force **Q**.

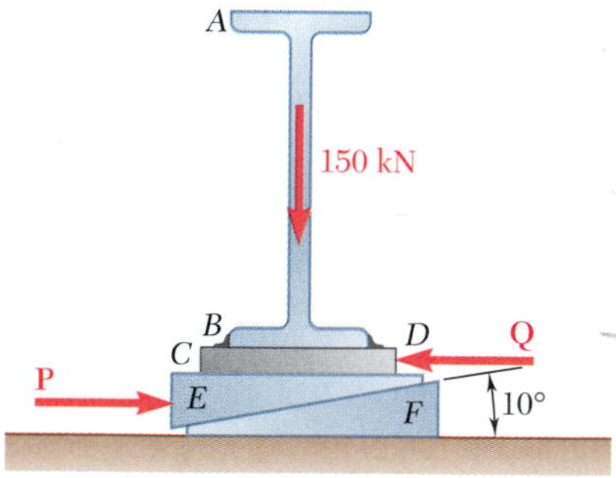

Fig. *P8.50*

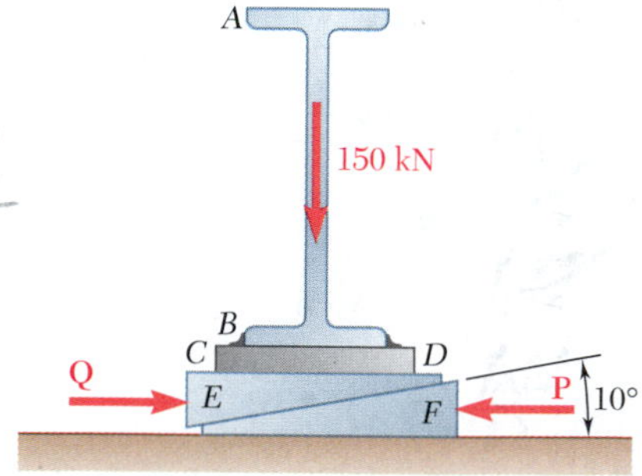

Fig. *P8.51*

8.52 Block A supports a pipe column and rests as shown on wedge B. Knowing that the coefficient of static friction at all surfaces of contact is 0.25 and that $\theta = 45°$, determine the smallest force **P** required to raise block A.

8.53 Block A supports a pipe column and rests as shown on wedge B. Knowing that the coefficient of static friction at all surfaces of contact is 0.25 and that $\theta = 45°$, determine the smallest force **P** for which equilibrium is maintained.

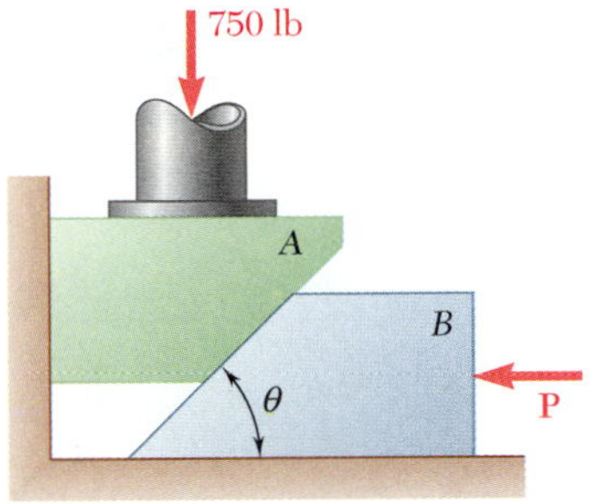

Fig. P8.52 and *P8.53*

8.54 and 8.55 A 16° wedge A of negligible mass is placed between two 80-kg blocks B and C which are at rest on inclined surfaces as shown. The coefficient of static friction is 0.40 between both the wedge and the blocks and block C and the incline. Determine the magnitude of the force **P** for which motion of the wedge is impending when the coefficient of static friction between block B and the incline is (a) 0.40, (b) 0.60.

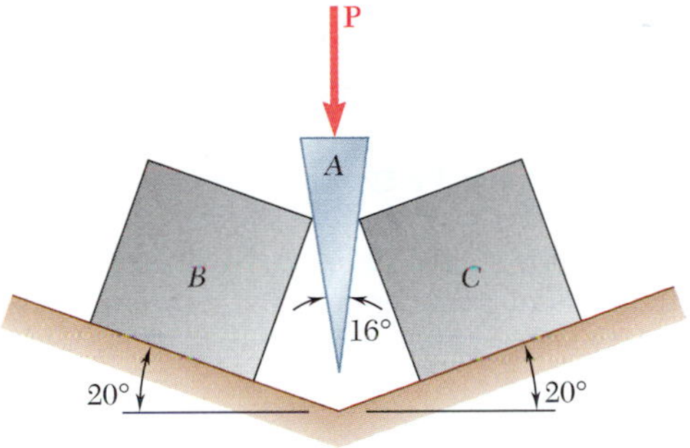

Fig. P8.54

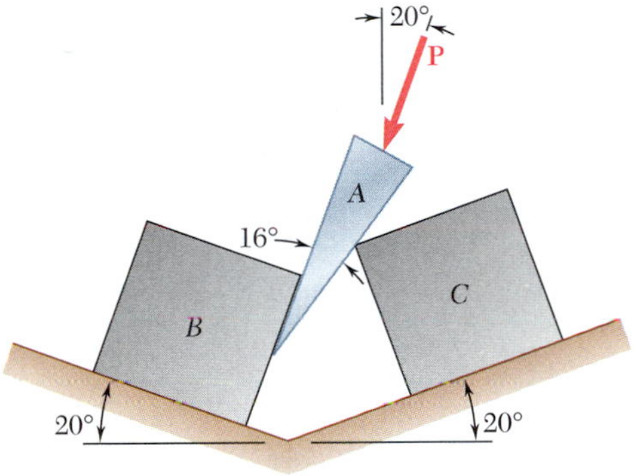

Fig. P8.55

8.56 A 10° wedge is to be forced under end B of the 12-lb rod AB. Knowing that the coefficient of static friction is 0.45 between the wedge and the rod and 0.25 between the wedge and the floor, determine the smallest force **P** required to raise end B of the rod.

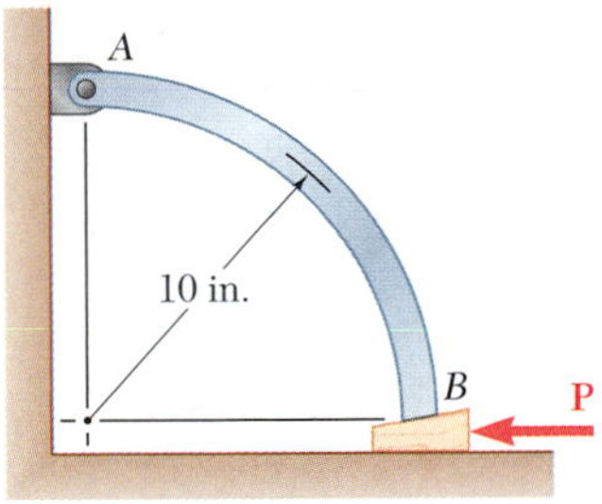

Fig. *P8.56*

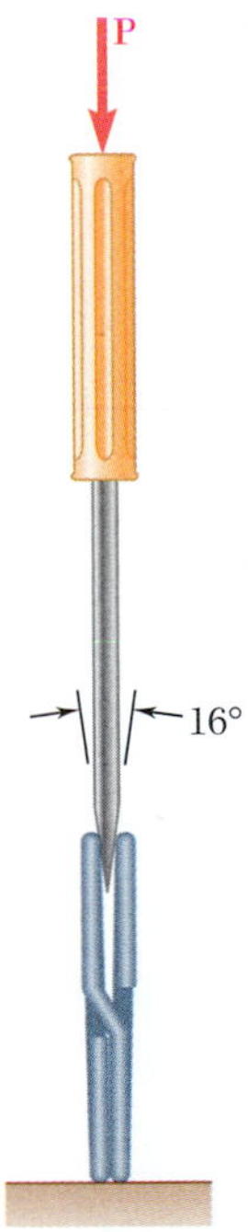

Fig. P8.57

8.57 A small screwdriver is used to pry apart the two coils of a circular key ring. The wedge angle of the screwdriver blade is 16° and the coefficient of static friction is 0.12 between the coils and the blade. Knowing that a force **P** of magnitude 0.8 lb was required to insert the screwdriver to the equilibrium position shown, determine the magnitude of the forces exerted on the ring by the screwdriver immediately after force **P** is removed.

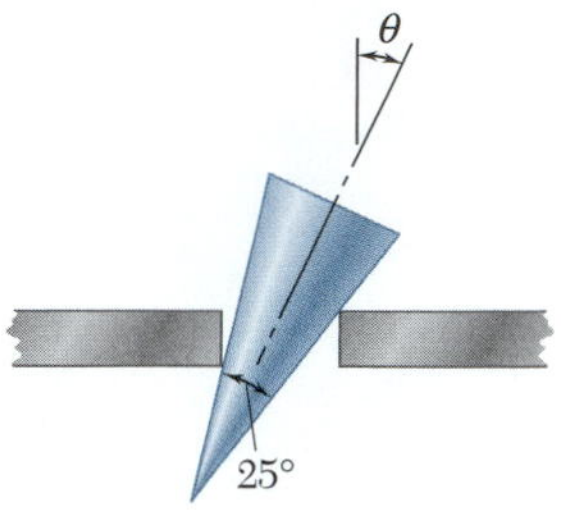

Fig. P8.58

8.58 A conical wedge is placed between two horizontal plates that are then slowly moved toward each other. Indicate what will happen to the wedge (*a*) if $\mu_s = 0.20$, (*b*) if $\mu_s = 0.30$.

8.59 A 6° steel wedge is driven into the end of an ax handle to lock the handle to the ax head. The coefficient of static friction between the wedge and the handle is 0.35. Knowing that a force **P** of magnitude 250 N was required to insert the wedge to the equilibrium position shown, determine the magnitude of the forces exerted on the handle by the wedge after force **P** is removed.

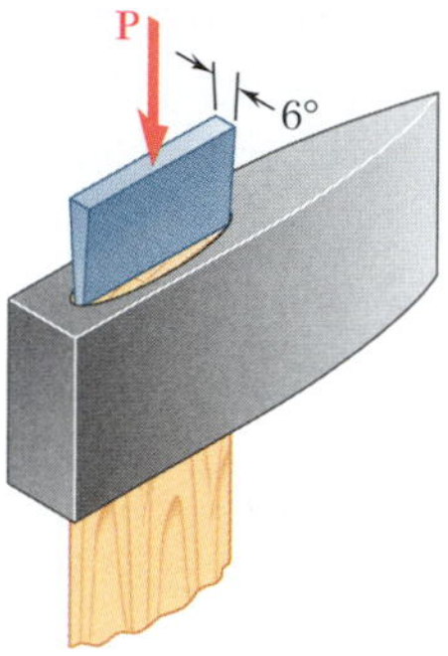

Fig. P8.59

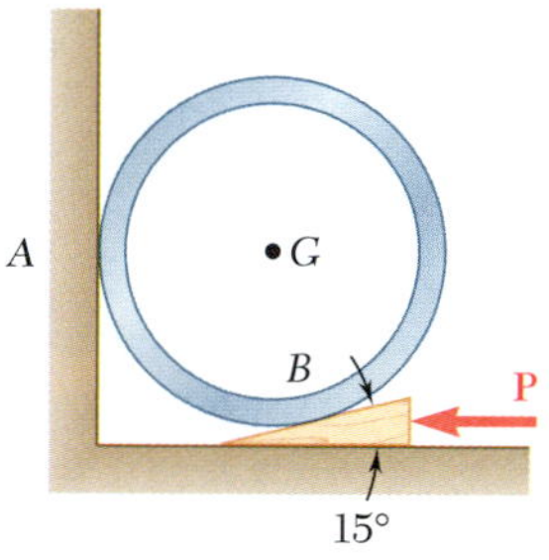

Fig. P8.60 and *P8.61*

8.60 A 15° wedge is forced under a 100-lb pipe as shown. The coefficient of static friction at all surfaces is 0.20. Determine (*a*) at which surface slipping of the pipe will first occur, (*b*) the force **P** for which motion of the wedge is impending.

8.61 A 15° wedge is forced under a 100-lb pipe as shown. Knowing that the coefficient of static friction at both surfaces of the wedge is 0.20, determine the largest coefficient of static friction between the pipe and the vertical wall for which slipping is impending at *A*.

8.62 Bags of grass seed are stored on a wooden plank as shown. To move the plank, a 9° wedge is driven under end *A*. Knowing that the weight of the grass seed can be represented by the distributed load shown and that the coefficient of static friction is 0.45 between all surfaces of contact, (*a*) determine the force **P** for which motion of the wedge is impending, (*b*) indicate whether the plank will slide on the floor.

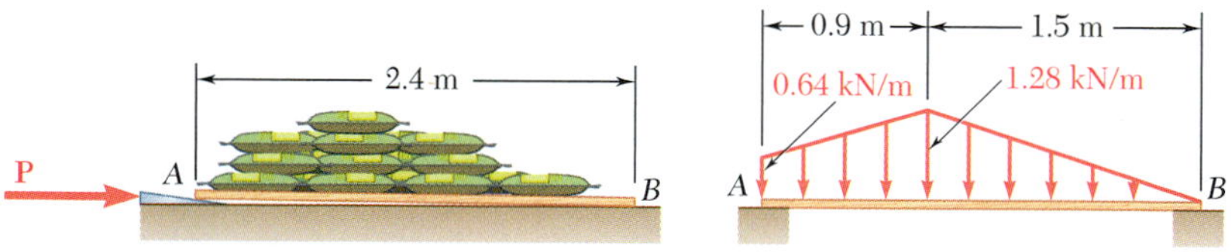

Fig. P8.62

8.63 Solve Prob. 8.62 assuming that the wedge is driven under the plank at *B* instead of at *A*.

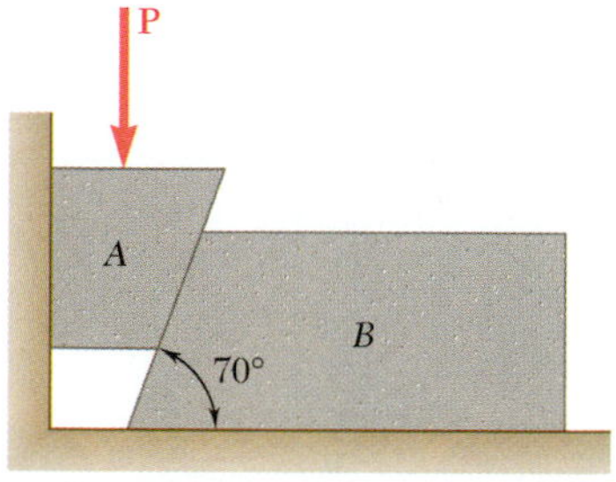

Fig. P8.64

***8.64** The 20-lb block *A* is at rest against the 100-lb block *B* as shown. The coefficient of static friction μ_s is the same between blocks *A* and *B* and between block *B* and the floor, while friction between block *A* and the wall can be neglected. Knowing that $P = 30$ lb, determine the value of μ_s for which motion is impending.

***8.65** Solve Prob. 8.64 assuming that μ_s is the coefficient of static friction between all surfaces of contact.

8.66 Derive the following formulas relating the load **W** and the force **P** exerted on the handle of the jack discussed in Sec. 8.6. (*a*) $P = (Wr/a) \tan (\theta + \phi_s)$, to raise the load; (*b*) $P = (Wr/a) \tan (\phi_s - \theta)$, to lower the load if the screw is self-locking; (*c*) $P = (Wr/a) \tan (\theta - \phi_s)$, to hold the load if the screw is not self-locking.

8.67 The square-threaded worm gear shown has a mean radius of 30 mm and a lead of 7.5 mm. The larger gear is subjected to a constant clockwise torque of 720 N · m. Knowing that the coefficient of static friction between the two gears is 0.12, determine the torque that must be applied to shaft *AB* in order to rotate the large gear counterclockwise. Neglect friction in the bearings at *A*, *B*, and *C*.

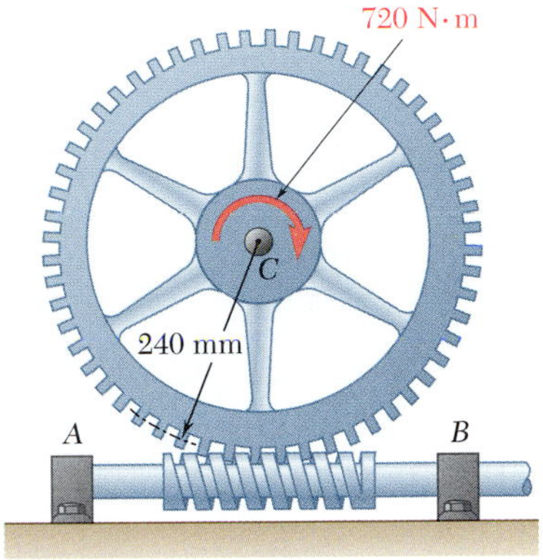

Fig. P8.67

8.68 In Prob. 8.67, determine the torque that must be applied to shaft *AB* in order to rotate the gear clockwise.

8.69 High-strength bolts are used in the construction of many steel structures. For a 24-mm-nominal-diameter bolt the required minimum bolt tension is 210 kN. Assuming the coefficient of friction to be 0.40, determine the required torque that should be applied to the bolt and nut. The mean diameter of the thread is 22.6 mm, and the lead is 3 mm. Neglect friction between the nut and washer, and assume the bolt to be square-threaded.

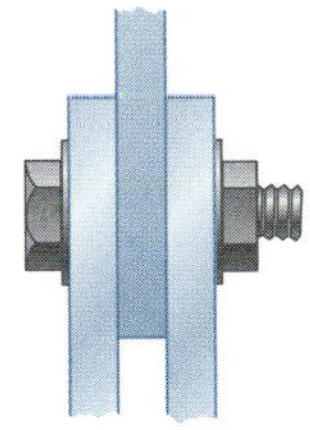

Fig. *P8.69*

8.70 The ends of two fixed rods *A* and *B* are each made in the form of a single-threaded screw of mean radius 0.3 in. and pitch 0.1 in. Rod *A* has a right-handed thread and rod *B* a left-handed thread. The coefficient of static friction between the rods and the threaded sleeve is 0.12. Determine the magnitude of the couple that must be applied to the sleeve in order to draw the rods closer together.

Fig. P8.70

8.71 Assuming that in Prob. 8.70 a right-handed thread is used on *both* rods *A* and *B*, determine the magnitude of the couple that must be applied to the sleeve in order to rotate it.

8.72 The position of the automobile jack shown is controlled by a screw *ABC* that is single-threaded at each end (right-handed thread at *A*, left-handed thread at *C*). Each thread has a pitch of 2 mm and a mean diameter of 7.5 mm. If the coefficient of static friction is 0.15, determine the magnitude of the couple **M** that must be applied to raise the automobile.

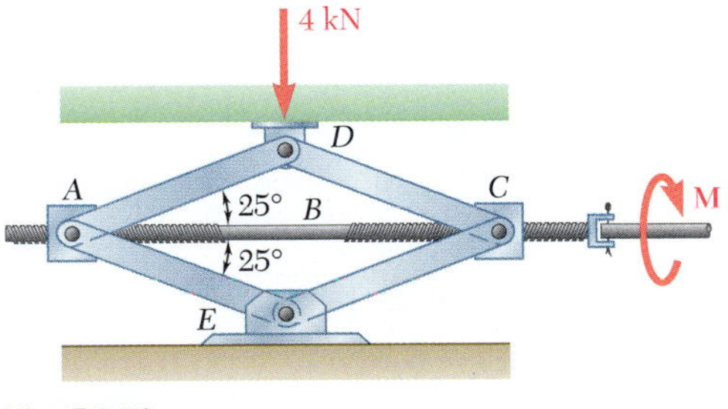

Fig. P8.72

8.73 For the jack of Prob. 8.72, determine the magnitude of the couple **M** that must be applied to lower the automobile.

8.74 In the gear-pulling assembly shown, the square-threaded screw *AB* has a mean radius of 22.5 mm and a lead of 6 mm. Knowing that the coefficient of static friction is 0.10, determine the torque which must be applied to the screw in order to produce a force of 4.5 kN on the gear. Neglect friction at end *A* of the screw.

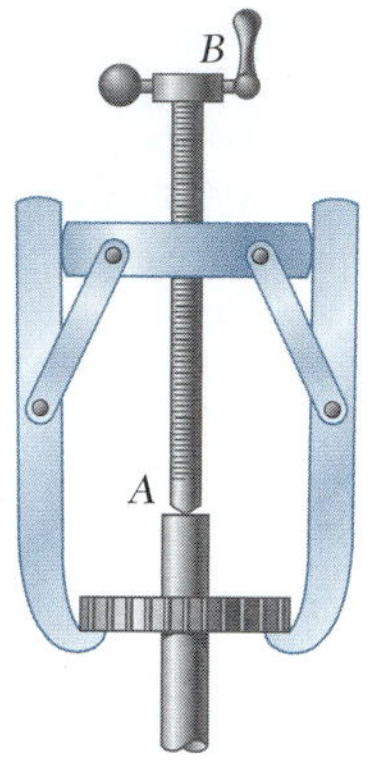

Fig. P8.74

*8.7. JOURNAL BEARINGS. AXLE FRICTION

Journal bearings are used to provide lateral support to rotating shafts and axles. Thrust bearings, which will be studied in the next section, are used to provide axial support to shafts and axles. If the journal bearing is fully lubricated, the frictional resistance depends upon the speed of rotation, the clearance between the axle and the bearing, and the viscosity of the lubricant. As indicated in Sec. 8.1, such problems are studied in fluid mechanics. The methods of this chapter, however, can be applied to the study of axle friction when the bearing is not lubricated or is only partially lubricated. It can then be assumed that the axle and the bearing are in direct contact along a single straight line.

Consider two wheels, each of weight **W**, rigidly mounted on an axle supported symmetrically by two journal bearings (Fig. 8.10*a*). If the wheels rotate, we find that to keep them rotating at constant speed, it is necessary to apply to each of them a couple **M**. The free-body diagram in Fig. 8.10*c* represents one of the wheels and the corresponding half axle in projection on a plane perpendicular to the axle. The forces acting on the free body include the weight **W** of the wheel, the couple **M** required to maintain its motion, and a force **R** representing the reaction of the bearing. This force is vertical, equal, and opposite to **W** but does not pass through the center *O* of the axle; **R** is located to the right of *O* at a distance such that its moment about *O* balances the moment **M** of the couple. Therefore, contact between the axle and bearing does not take place at the lowest point *A* when the axle rotates. It takes place at point *B* (Fig. 8.10*b*) or, rather, along a straight line intersecting the plane of the figure at *B*. Physically, this is explained by the fact that when the wheels are set in motion, the axle "climbs" in the bearings until slippage occurs. After sliding back slightly, the axle settles more or less in the position shown. This position is such that the angle between the reaction **R** and the normal to the surface of the bearing is equal to the angle of kinetic friction ϕ_k.

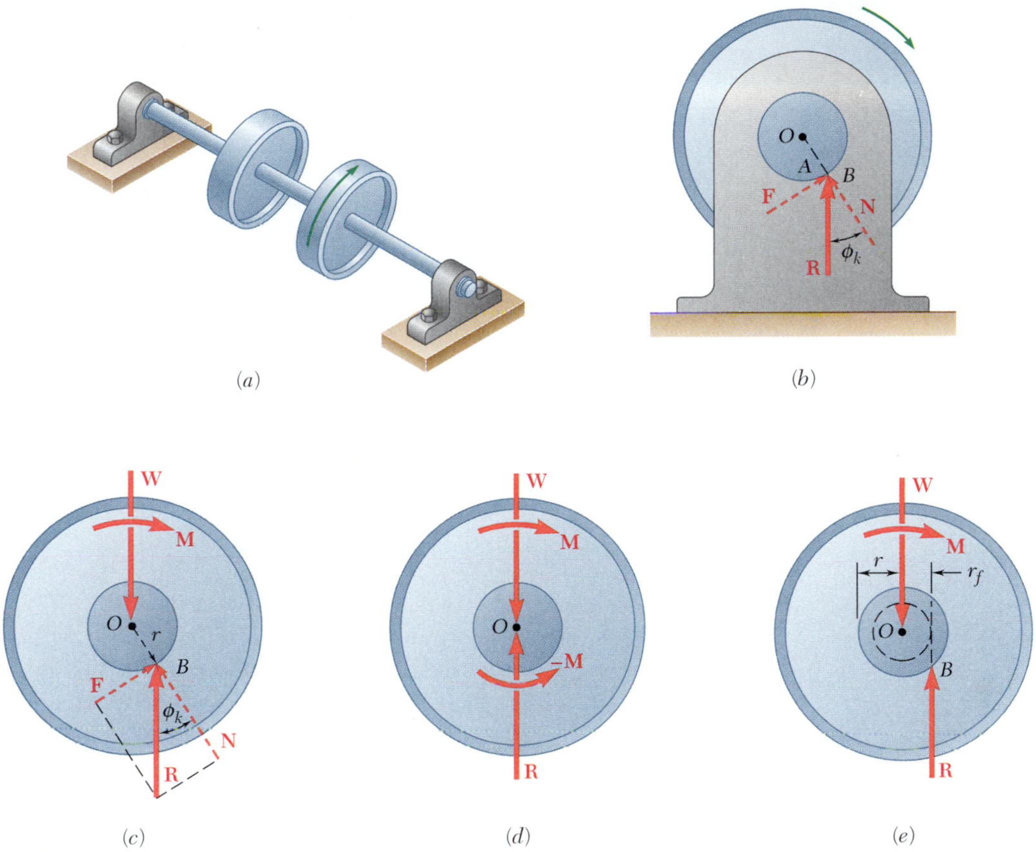

Fig. 8.10

The distance from O to the line of action of $\mathbf{R}$ is thus $r \sin \phi_k$, where r is the radius of the axle. Writing that $\Sigma M_O = 0$ for the forces acting on the free body considered, we obtain the magnitude of the couple $\mathbf{M}$ required to overcome the frictional resistance of one of the bearings:

$$M = Rr \sin \phi_k \tag{8.5}$$

Observing that, for small values of the angle of friction, $\sin \phi_k$ can be replaced by $\tan \phi_k$, that is, by μ_k, we write the approximate formula

$$M \approx Rr\mu_k \tag{8.6}$$

In the solution of certain problems, it may be more convenient to let the line of action of $\mathbf{R}$ pass through O, as it does when the axle does not rotate. A couple $-\mathbf{M}$ of the same magnitude as the couple $\mathbf{M}$ but of opposite sense must then be added to the reaction $\mathbf{R}$ (Fig. 8.10*d*). This couple represents the frictional resistance of the bearing.

In case a graphical solution is preferred, the line of action of $\mathbf{R}$ can be readily drawn (Fig. 8.10*e*) if we note that it must be tangent to a circle centered at O and of radius

$$r_f = r \sin \phi_k \approx r\mu_k \tag{8.7}$$

This circle is called the *circle of friction* of the axle and bearing and is independent of the loading conditions of the axle.

*8.8. THRUST BEARINGS. DISK FRICTION

Two types of thrust bearings are used to provide axial support to rotating shafts and axles: (1) *end bearings* and (2) *collar bearings* (Fig. 8.11). In the case of collar bearings, friction forces develop between the two ring-shaped areas which are in contact. In the case of end bearings, friction takes place over full circular areas, or over ring-shaped areas when the end of the shaft is hollow. Friction between circular areas, called *disk friction,* also occurs in other mechanisms, such as *disk clutches*.

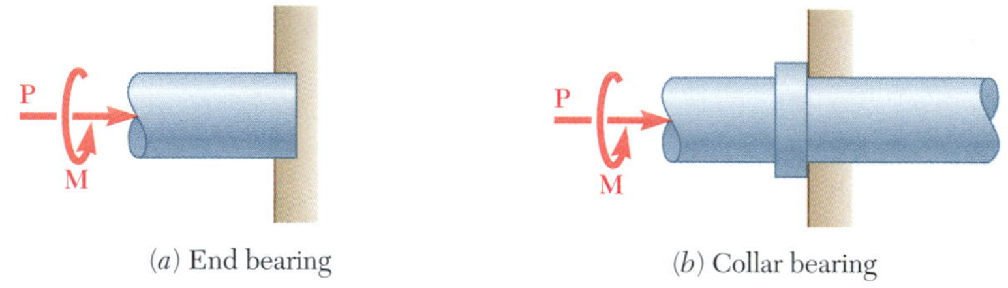

Fig. 8.11 Thrust bearings.

To obtain a formula which is valid in the most general case of disk friction, let us consider a rotating hollow shaft. A couple **M** keeps the shaft rotating at constant speed while a force **P** maintains it in contact with a fixed bearing (Fig. 8.12). Contact between the shaft and

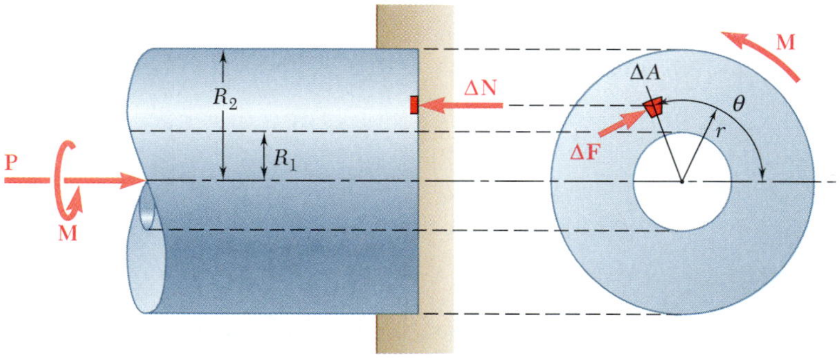

Fig. 8.12

the bearing takes place over a ring-shaped area of inner radius R_1 and outer radius R_2. Assuming that the pressure between the two surfaces in contact is uniform, we find that the magnitude of the normal force $\Delta\mathbf{N}$ exerted on an element of area ΔA is $\Delta N = P\,\Delta A/A$, where $A = \pi(R_2^2 - R_1^2)$, and that the magnitude of the friction force $\Delta\mathbf{F}$ acting on ΔA is $\Delta F = \mu_k\,\Delta N$. Denoting by r the distance from the axis of the shaft to the element of area ΔA, we express the magnitude ΔM of the moment of $\Delta\mathbf{F}$ about the axis of the shaft as follows:

$$\Delta M = r\,\Delta F = \frac{r\mu_k P\,\Delta A}{\pi(R_2^2 - R_1^2)}$$

The equilibrium of the shaft requires that the moment **M** of the couple applied to the shaft be equal in magnitude to the sum of the moments of the friction forces $\Delta\mathbf{F}$. Replacing ΔA by the infinitesimal element $dA = r\,d\theta\,dr$ used with polar coordinates, and integrating over the area of contact, we thus obtain the following expression for the magnitude of the couple **M** required to overcome the frictional resistance of the bearing:

$$M = \frac{\mu_k P}{\pi(R_2^2 - R_1^2)} \int_0^{2\pi} \int_{R_1}^{R_2} r^2\,dr\,d\theta$$

$$= \frac{\mu_k P}{\pi(R_2^2 - R_1^2)} \int_0^{2\pi} \tfrac{1}{3}(R_2^3 - R_1^3)\,d\theta$$

$$M = \tfrac{2}{3}\mu_k P \frac{R_2^3 - R_1^3}{R_2^2 - R_1^2} \qquad (8.8)$$

When contact takes place over a full circle of radius R, formula (8.8) reduces to

$$M = \tfrac{2}{3}\mu_k PR \qquad (8.9)$$

The value of M is then the same as would be obtained if contact between the shaft and the bearing took place at a single point located at a distance $2R/3$ from the axis of the shaft.

The largest torque which can be transmitted by a disk clutch without causing slippage is given by a formula similar to (8.9), where μ_k is replaced by the coefficient of static friction μ_s.

*8.9. WHEEL FRICTION. ROLLING RESISTANCE

The wheel is one of the most important inventions of our civilization. Its use makes it possible to move heavy loads with relatively little effort. Because the point of the wheel in contact with the ground at any given instant has no relative motion with respect to the ground, the wheel eliminates the large friction forces which would arise if the load were in direct contact with the ground. However, some resistance to the wheel's motion exists. This resistance has two distinct causes. It is due (1) to a combined effect of axle friction and friction at the rim and (2) to the fact that the wheel and the ground deform, with the result that contact between the wheel and the ground takes place over a certain area, rather than at a single point.

To understand better the first cause of resistance to the motion of a wheel, let us consider a railroad car supported by eight wheels mounted on axles and bearings. The car is assumed to be moving to the right at constant speed along a straight horizontal track. The

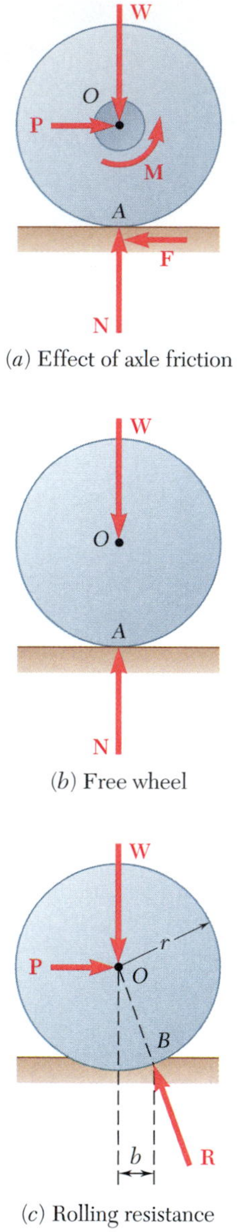

Fig. 8.13

free-body diagram of one of the wheels is shown in Fig. 8.13*a*. The forces acting on the free body include the load **W** supported by the wheel and the normal reaction **N** of the track. Since **W** is drawn through the center *O* of the axle, the frictional resistance of the bearing should be represented by a counterclockwise couple **M** (see Sec. 8.7). To keep the free body in equilibrium, we must add two equal and opposite forces **P** and **F**, forming a clockwise couple of moment −**M**. The force **F** is the friction force exerted by the track on the wheel, and **P** represents the force which should be applied to the wheel to keep it rolling at constant speed. Note that the forces **P** and **F** would not exist if there were no friction between the wheel and the track. The couple **M** representing the axle friction would then be zero; thus, the wheel would slide on the track without turning in its bearing.

The couple **M** and the forces **P** and **F** also reduce to zero when there is no axle friction. For example, a wheel which is not held in bearings and rolls freely and at constant speed on horizontal ground (Fig. 8.13*b*) will be subjected to only two forces: its own weight **W** and the normal reaction **N** of the ground. Regardless of the value of the coefficient of friction between the wheel and the ground, no friction force will act on the wheel. A wheel rolling freely on horizontal ground should thus keep rolling indefinitely.

Experience, however, indicates that the wheel will slow down and eventually come to rest. This is due to the second type of resistance mentioned at the beginning of this section, known as the *rolling resistance.* Under the load **W**, both the wheel and the ground deform slightly, causing the contact between the wheel and the ground to take place over a certain area. Experimental evidence shows that the resultant of the forces exerted by the ground on the wheel over this area is a force **R** applied at a point *B*, which is not located directly under the center *O* of the wheel but slightly in front of it (Fig. 8.13*c*). To balance the moment of **W** about *B* and to keep the wheel rolling at constant speed, it is necessary to apply a horizontal force **P** at the center of the wheel. Writing $\Sigma M_B = 0$, we obtain

$$Pr = Wb \tag{8.10}$$

where r = radius of wheel
b = horizontal distance between O and B

The distance b is commonly called the *coefficient of rolling resistance.* It should be noted that b is not a dimensionless coefficient since it represents a length; b is usually expressed in inches or in millimeters. The value of b depends upon several parameters in a manner which has not yet been clearly established. Values of the coefficient of rolling resistance vary from about 0.01 in. or 0.25 mm for a steel wheel on a steel rail to 5.0 in. or 125 mm for the same wheel on soft ground.

SAMPLE PROBLEM 8.6

A pulley of diameter 4 in. can rotate about a fixed shaft of diameter 2 in. The coefficient of static friction between the pulley and shaft is 0.20. Determine (*a*) the smallest vertical force **P** required to start raising a 500-lb load, (*b*) the smallest vertical force **P** required to hold the load, (*c*) the smallest horizontal force **P** required to start raising the same load.

SOLUTION

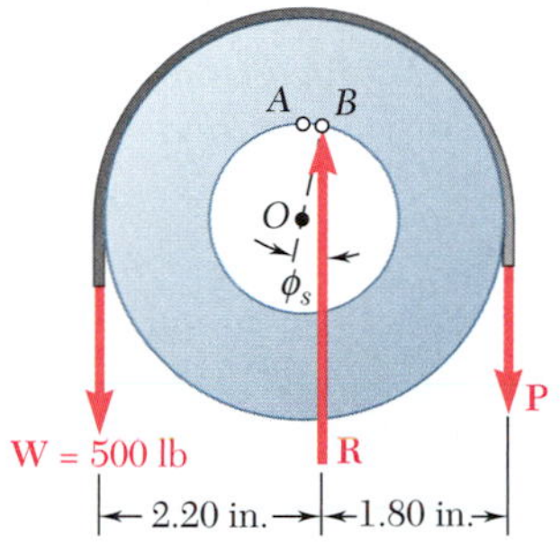

***a.* Vertical Force P Required to Start Raising the Load.** When the forces in both parts of the rope are equal, contact between the pulley and shaft takes place at *A*. When **P** is increased, the pulley rolls around the shaft slightly and contact takes place at *B*. The free-body diagram of the pulley when motion is impending is drawn. The perpendicular distance from the center *O* of the pulley to the line of action of **R** is

$$r_f = r \sin \phi_s \approx r\mu_s \qquad r_f \approx (1 \text{ in.})0.20 = 0.20 \text{ in.}$$

Summing moments about *B*, we write

$+\circlearrowleft \Sigma M_B = 0$: $\quad (2.20 \text{ in.})(500 \text{ lb}) - (1.80 \text{ in.})P = 0$

$P = 611$ lb $\qquad$ **P** = 611 lb ↓ ◀

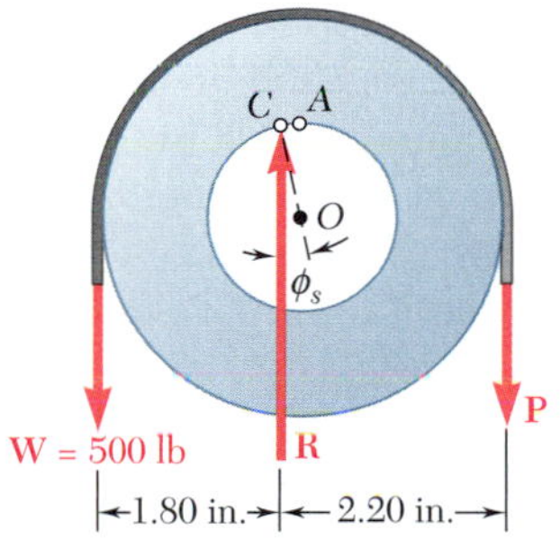

***b.* Vertical Force P to Hold the Load.** As the force **P** is decreased, the pulley rolls around the shaft and contact takes place at *C*. Considering the pulley as a free body and summing moments about *C*, we write

$+\circlearrowleft \Sigma M_C = 0$: $\quad (1.80 \text{ in.})(500 \text{ lb}) - (2.20 \text{ in.})P = 0$

$P = 409$ lb $\qquad$ **P** = 409 lb ↓ ◀

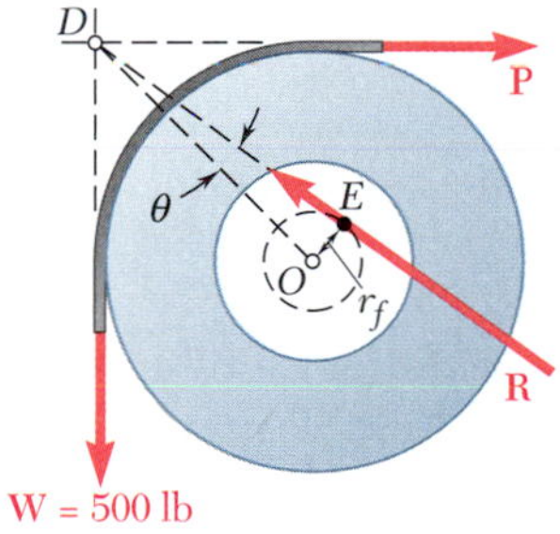

***c.* Horizontal Force P to Start Raising the Load.** Since the three forces **W**, **P**, and **R** are not parallel, they must be concurrent. The direction of **R** is thus determined from the fact that its line of action must pass through the point of intersection *D* of **W** and **P** and must be tangent to the circle of friction. Recalling that the radius of the circle of friction is $r_f = 0.20$ in., we write

$$\sin \theta = \frac{OE}{OD} = \frac{0.20 \text{ in.}}{(2 \text{ in.})\sqrt{2}} = 0.0707 \qquad \theta = 4.1°$$

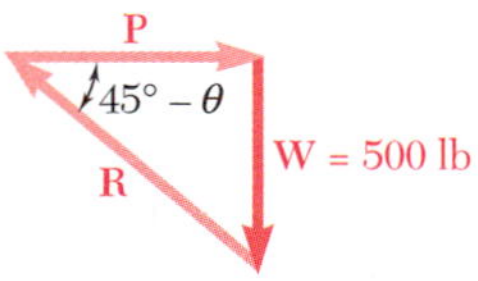

From the force triangle, we obtain

$$P = W \cot (45° - \theta) = (500 \text{ lb}) \cot 40.9°$$
$$= 577 \text{ lb}$$

P = 577 lb → ◀

SOLVING PROBLEMS ON YOUR OWN

In this lesson you learned about several additional engineering applications of the laws of friction.

1. *Journal bearings and axle friction.* In journal bearings, the *reaction does not pass through the center of the shaft or axle* which is being supported. The distance from the center of the shaft or axle to the line of action of the reaction (Fig. 8.10) is defined by the equation

$$r_f = r \sin \phi_k \approx r\mu_k$$

if motion is actually taking place, and by the equation

$$r_f = r \sin \phi_s \approx r\mu_s$$

if the motion is impending.

Once you have determined the line of action of the reaction, you can draw a *free-body diagram* and use the corresponding equations of equilibrium to complete your solution [Sample Prob. 8.6]. In some problems, it is useful to observe that the line of action of the reaction must be tangent to a circle of radius $r_f \approx r\mu_k$, or $r_f \approx r\mu_s$, known as the *circle of friction* [Sample Prob. 8.6, part *c*]. It is important to remember that the reaction is always directed to oppose the actual or impending rotation of the shaft or axle.

2. *Thrust bearings and disk friction.* In a *thrust bearing* the magnitude of the couple required to overcome frictional resistance is equal to the sum of the moments of the *kinetic friction forces* exerted on the elements of the end of the shaft [Eqs. (8.8) and (8.9)].

An example of disk friction is the *disk clutch.* It is analyzed in the same way as a thrust bearing, except that to determine the largest torque that can be transmitted, you must compute the sum of the moments of the *maximum static friction forces* exerted on the disk.

3. *Wheel friction and rolling resistance.* You saw that the rolling resistance of a wheel is caused by deformations of both the wheel and the ground. The line of action of the reaction **R** of the ground on the wheel intersects the ground at a horizontal distance b from the center of the wheel. The distance b is known as the *coefficient of rolling resistance* and is expressed in inches or millimeters.

4. *In problems involving both rolling resistance and axle friction,* your free-body diagram should show that the line of action of the reaction **R** of the ground on the wheel is tangent to the friction circle of the axle and intersects the ground at a horizontal distance from the center of the wheel equal to the coefficient of rolling resistance.

Problems

8.75 A 120-mm-radius pulley of mass 5 kg is attached to a 30-mm-radius shaft which fits loosely in a fixed bearing. It is observed that the pulley will just start rotating if a 0.5-kg mass is added to block A. Determine the coefficient of static friction between the shaft and the bearing.

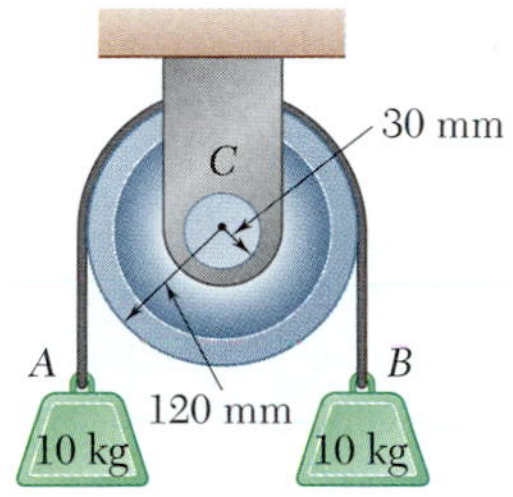

Fig. P8.75

8.76 and 8.77 The double pulley shown is attached to a 0.5-in.-radius shaft which fits loosely in a fixed bearing. Knowing that the coefficient of static friction between the shaft and the poorly lubricated bearing is 0.40, determine the magnitude of the force **P** required to start raising the load.

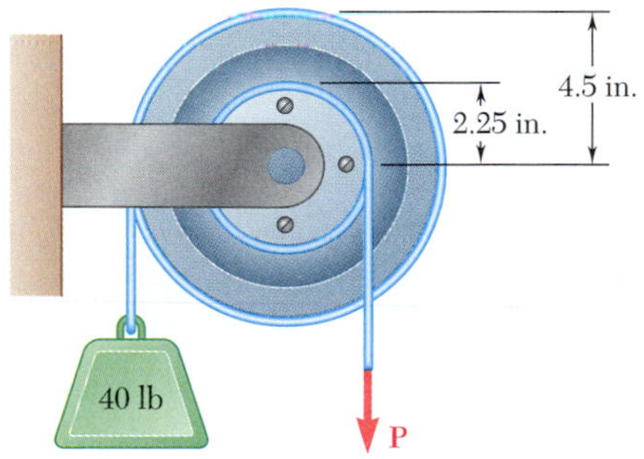

Fig. P8.76 and *P8.78*

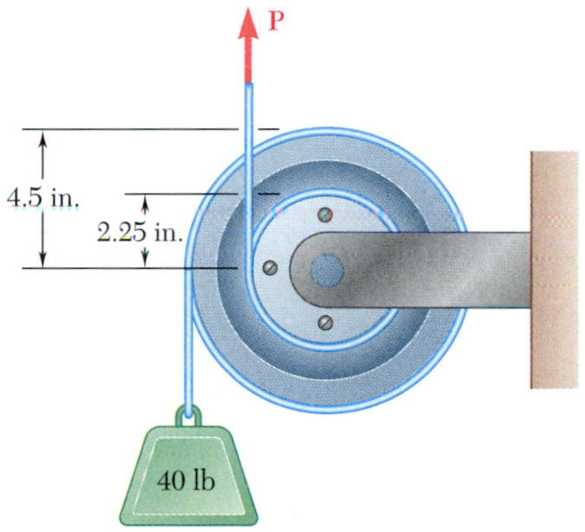

Fig. P8.77 and *P8.79*

8.78* and *8.79 The double pulley shown is attached to a 0.5-in.-radius shaft which fits loosely in a fixed bearing. Knowing that the coefficient of static friction between the shaft and the poorly lubricated bearing is 0.40, determine the magnitude of the force **P** required to maintain equilibrium.

8.80 Control lever ABC fits loosely on a 18-mm-diameter shaft at support B. Knowing that $P = 130$ N for impending clockwise rotation of the lever, determine (*a*) the coefficient of static friction between the pin and the lever, (*b*) the magnitude of the force **P** for which counterclockwise rotation of the lever is impending.

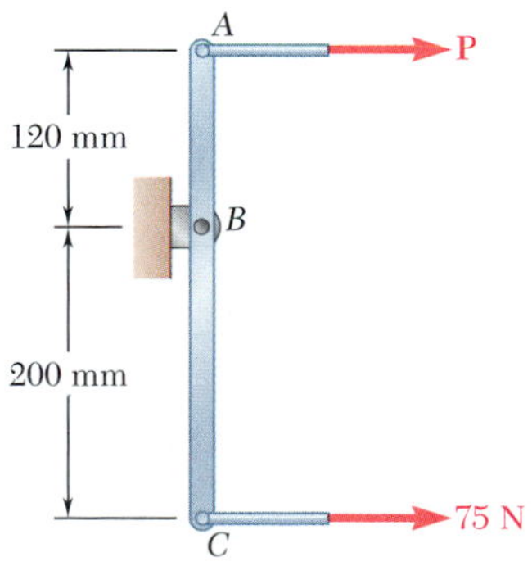

Fig. P8.80

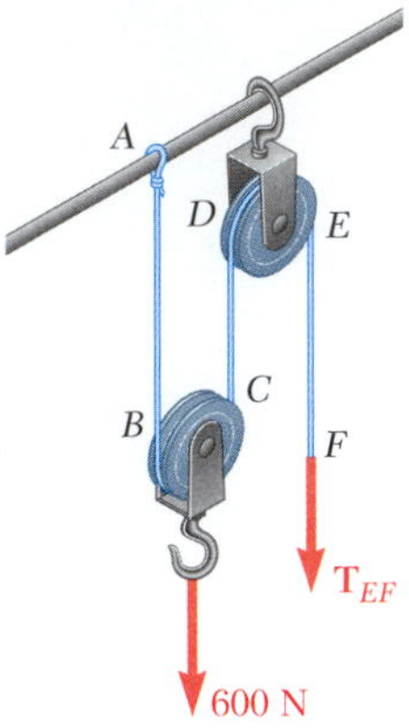

Fig. P8.81 and P8.82

8.81 The block and tackle shown are used to raise a 600-N load. Each of the 60-mm-diameter pulleys rotates on a 10-mm-diameter axle. Knowing that the coefficient of kinetic friction is 0.20, determine the tension in each portion of the rope as the load is slowly raised.

8.82 The block and tackle shown are used to lower a 600-N load. Each of the 60-mm-diameter pulleys rotates on a 10-mm-diameter axle. Knowing that the coefficient of kinetic friction is 0.20, determine the tension in each portion of the rope as the load is slowly lowered.

8.83 The link arrangement shown is frequently used in highway bridge construction to allow for expansion due to changes in temperature. At each of the 3-in.-diameter pins *A* and *B* the coefficient of static friction is 0.20. Knowing that the vertical component of the force exerted by *BC* on the link is 50 kips, determine (*a*) the horizontal force which should be exerted on beam *BC* to just move the link, (*b*) the angle that the resulting force exerted by beam *BC* on the link will form with the vertical.

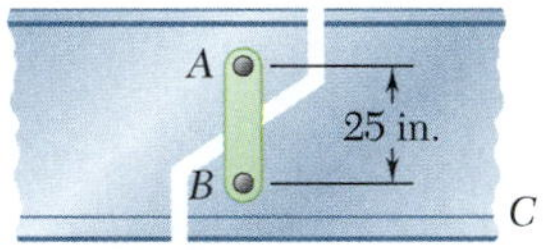

Fig. P8.83

8.84 and 8.85 A gate assembly consisting of a 24-kg gate *ABC* and a 66-kg counterweight *D* is attached to a 24-mm-diameter shaft *B* which fits loosely in a fixed bearing. Knowing that the coefficient of static friction is 0.20 between the shaft and the bearing, determine the magnitude of the force **P** for which counterclockwise rotation of the gate is impending.

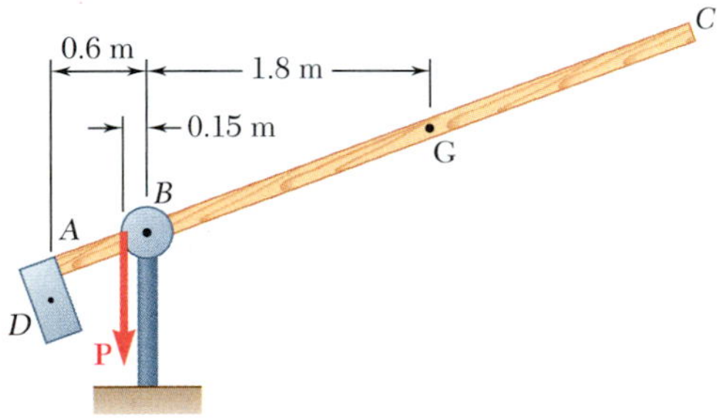

Fig. P8.84 and *P8.86*

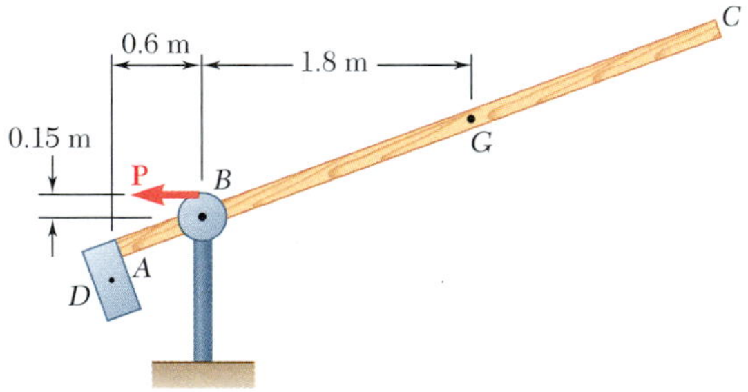

Fig. P8.85 and *P8.87*

8.86* and *8.87 A gate assembly consisting of a 24-kg gate *ABC* and a 66-kg counterweight *D* is attached to a 24-mm-diameter shaft *B* which fits loosely in a fixed bearing. Knowing that the coefficient of static friction is 0.20 between the shaft and the bearing, determine the magnitude of the force **P** for which clockwise rotation of the gate is impending.

8.88 A loaded railroad car has a weight of 35 tons and is supported by eight 32-in.-diameter wheels with 5-in.-diameter axles. Knowing that the coefficients of friction are $\mu_s = 0.020$ and $\mu_k = 0.015$, determine the horizontal force required (*a*) for impending motion of the car, (*b*) to keep the car moving at a constant speed. Neglect rolling resistance between the wheels and the track.

8.89 A scooter is designed to roll down a 2 percent slope at a constant speed. Assuming that the coefficient of kinetic friction between the 1-in.-diameter axles and the bearing is 0.10, determine the required diameter of the wheels. Neglect the rolling resistance between the wheels and the ground.

8.90 A 25-kg electric floor polisher is operated on a surface for which the coefficient of kinetic friction is 0.25. Assuming that the normal force per unit area between the disk and the floor is uniformly distributed, determine the magnitude Q of the horizontal forces required to prevent motion of the machine.

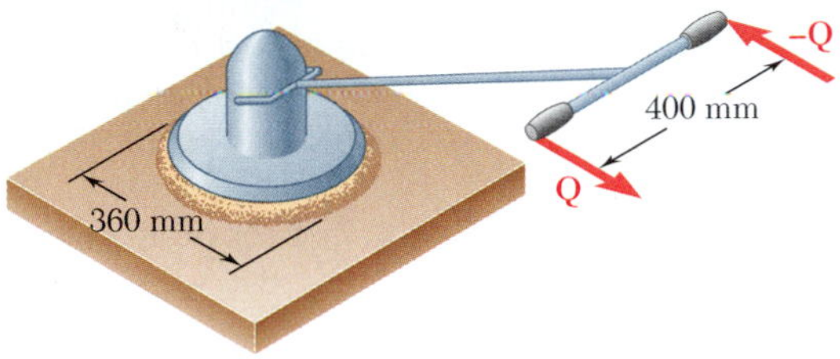

Fig. P8.90

8.91 The pivot for the seat of a desk chair consists of the steel plate A, which supports the seat, the solid steel shaft B which is welded to A and which turns freely in the tubular member C, and the nylon bearing D. If a person of weight $W = 180$ lb is seated directly above the pivot, determine the magnitude of the couple **M** for which rotation of the seat is impending knowing that the coefficient of static friction is 0.15 between the tubular member and the bearing.

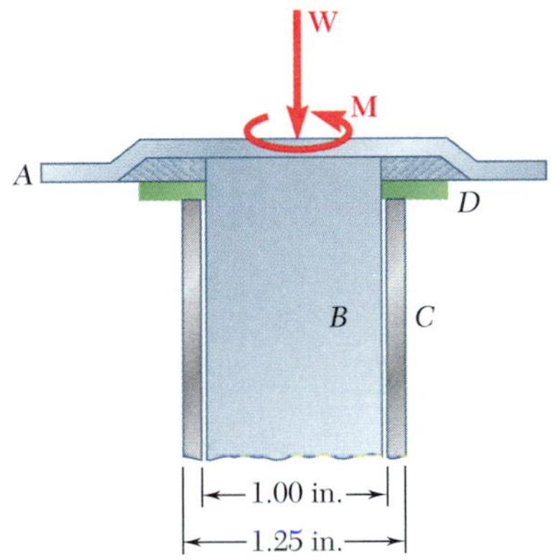

Fig. P8.91

***8.92** As the surfaces of a shaft and a bearing wear out, the frictional resistance of a thrust bearing decreases. It is generally assumed that the wear is directly proportional to the distance traveled by any given point of the shaft and thus to the distance r from the point to the axis of the shaft. Assuming, then, that the normal force per unit area is inversely proportional to r, show that the magnitude M of the couple required to overcome the frictional resistance of a worn-out end bearing (with contact over the full circular area) is equal to 75 percent of the value given by formula (8.9) for a new bearing.

***8.93** Assuming that bearings wear out as indicated in Prob. 8.92, show that the magnitude M of the couple required to overcome the frictional resistance of a worn-out collar bearing is

$$M = \frac{1}{2}\mu_k P(R_1 + R_2)$$

where P = magnitude of the total axial force
R_1, R_2 = inner and outer radii of collar

***8.94** Assuming that the pressure between the surfaces of contact is uniform, show that the magnitude M of the couple required to overcome frictional resistance for the conical bearing shown is

$$M = \frac{2}{3}\frac{\mu_k P}{\sin\theta}\frac{R_2^3 - R_1^3}{R_2^2 - R_1^2}$$

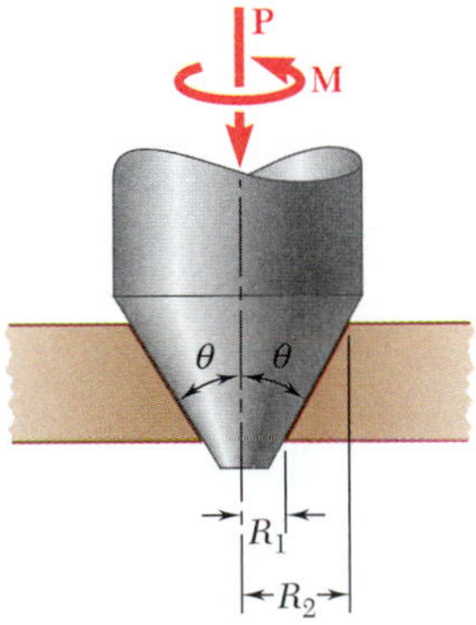

Fig. *P8.94*

8.95 Solve Prob. 8.90 assuming that the normal force per unit area between the disk and the floor varies linearly from a maximum at the center to zero at the circumference of the disk.

Fig. P8.96

8.96 A 1-ton machine base is rolled along a concrete floor using a series of steel pipes with outside diameters of 5 in. Knowing that the coefficient of rolling resistance is 0.025 in. between the pipes and the base and 0.0625 in. between the pipes and the concrete floor, determine the magnitude of the force **P** required to slowly move the base along the floor.

8.97 Knowing that a 120-mm-diameter disk rolls at a constant velocity down a 2 percent incline, determine the coefficient of rolling resistance between the disk and the incline.

8.98 Determine the horizontal force required to move a 1-Mg automobile with 460-mm-diameter tires along a horizontal road at a constant speed. Neglect all forms of friction except rolling resistance, and assume the coefficient of rolling resistance to be 1 mm.

8.99 Solve Prob. 8.88 including the effect of a coefficient of rolling resistance of 0.02 in.

8.100 Solve Prob. 8.89 including the effect of a coefficient of rolling resistance of 0.07 in.

8.10. BELT FRICTION

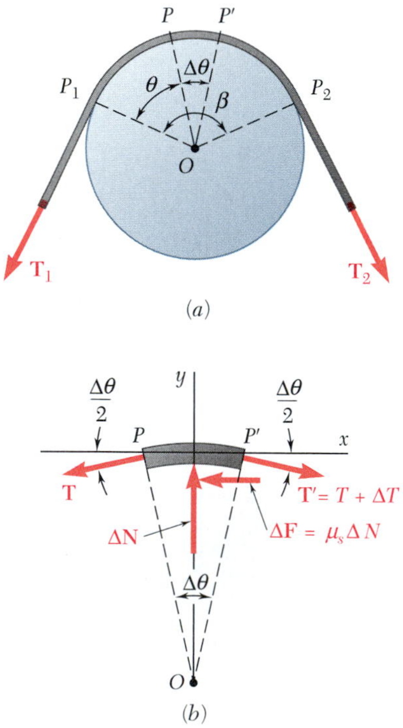

Fig. 8.14

Consider a flat belt passing over a fixed cylindrical drum (Fig. 8.14*a*). We propose to determine the relation existing between the values T_1 and T_2 of the tension in the two parts of the belt when the belt is just about to slide toward the right.

Let us detach from the belt a small element PP' subtending an angle $\Delta\theta$. Denoting by T the tension at P and by $T + \Delta T$ the tension at P', we draw the free-body diagram of the element of the belt (Fig. 8.14*b*). Besides the two forces of tension, the forces acting on the free body are the normal component $\Delta\mathbf{N}$ of the reaction of the drum and the friction force $\Delta\mathbf{F}$. Since motion is assumed to be impending, we have $\Delta F = \mu_s \Delta N$. It should be noted that if $\Delta\theta$ is made to approach zero, the magnitudes ΔN and ΔF, and the *difference* ΔT between the tension at P and the tension at P', will also approach zero; the value T of the tension at P, however, will remain unchanged. This observation helps in understanding our choice of notation.

Choosing the coordinate axes shown in Fig. 8.14*b*, we write the equations of equilibrium for the element PP':

$$\Sigma F_x = 0: \qquad (T + \Delta T)\cos\frac{\Delta\theta}{2} - T\cos\frac{\Delta\theta}{2} - \mu_s \Delta N = 0 \qquad (8.11)$$

$$\Sigma F_y = 0: \qquad \Delta N - (T + \Delta T)\sin\frac{\Delta\theta}{2} - T\sin\frac{\Delta\theta}{2} = 0 \qquad (8.12)$$

Solving Eq. (8.12) for ΔN and substituting into (8.11), we obtain after reductions

$$\Delta T \cos \frac{\Delta\theta}{2} - \mu_s(2T + \Delta T) \sin \frac{\Delta\theta}{2} = 0$$

Both terms are now divided by $\Delta\theta$. For the first term, this is done simply by dividing ΔT by $\Delta\theta$. The division of the second term is carried out by dividing the terms in the parentheses by 2 and the sine by $\Delta\theta/2$. We write

$$\frac{\Delta T}{\Delta\theta} \cos \frac{\Delta\theta}{2} - \mu_s\left(T + \frac{\Delta T}{2}\right) \frac{\sin(\Delta\theta/2)}{\Delta\theta/2} = 0$$

If we now let $\Delta\theta$ approach 0, the cosine approaches 1 and $\Delta T/2$ approaches zero, as noted above. The quotient of $\sin(\Delta\theta/2)$ over $\Delta\theta/2$ approaches 1, according to a lemma derived in all calculus textbooks. Since the limit of $\Delta T/\Delta\theta$ is by definition equal to the derivative $dT/d\theta$, we write

$$\frac{dT}{d\theta} - \mu_s T = 0 \qquad \frac{dT}{T} = \mu_s d\theta$$

Both members of the last equation (Fig. 8.14*a*) will now be integrated from P_1 to P_2. At P_1, we have $\theta = 0$ and $T = T_1$; at P_2, we have $\theta = \beta$ and $T = T_2$. Integrating between these limits, we write

$$\int_{T_1}^{T_2} \frac{dT}{T} = \int_0^{\beta} \mu_s \, d\theta$$

$$\ln T_2 - \ln T_1 = \mu_s\beta$$

or, noting that the left-hand member is equal to the natural logarithm of the quotient of T_2 and T_1,

$$\ln \frac{T_2}{T_1} = \mu_s\beta \qquad (8.13)$$

This relation can also be written in the form

$$\frac{T_2}{T_1} = e^{\mu_s\beta} \qquad (8.14)$$

The formulas we have derived apply equally well to problems involving flat belts passing over fixed cylindrical drums and to problems involving ropes wrapped around a post or capstan. They can also be used to solve problems involving band brakes. In such problems, it is the drum which is about to rotate, while the band remains fixed. The formulas can also be applied to problems involving belt drives. In

Photo 8.3 By wrapping the rope around the bollard, the force exerted by the worker to control the rope is much smaller than the tension in the taut portion of the rope.

these problems, both the pulley and the belt rotate; our concern is then to find whether the belt will slip, that is, whether it will move *with respect* to the pulley.

Formulas (8.13) and (8.14) should be used only if the belt, rope, or brake is *about to slip*. Formula (8.14) will be used if T_1 or T_2 is desired; formula (8.13) will be preferred if either μ_s or the angle of contact β is desired. We should note that T_2 is always larger than T_1; T_2 therefore represents the tension in that part of the belt or rope which *pulls*, while T_1 is the tension in the part which *resists*. We should also observe that the angle of contact β must be expressed in *radians*. The angle β may be larger than 2π; for example, if a rope is wrapped n times around a post, β is equal to $2\pi n$.

If the belt, rope, or brake is actually slipping, formulas similar to (8.13) and (8.14), but involving the coefficient of kinetic friction μ_k, should be used. If the belt, rope, or brake is not slipping and is not about to slip, none of these formulas can be used.

The belts used in belt drives are often V-shaped. In the V belt shown in Fig. 8.15*a* contact between the belt and the pulley takes

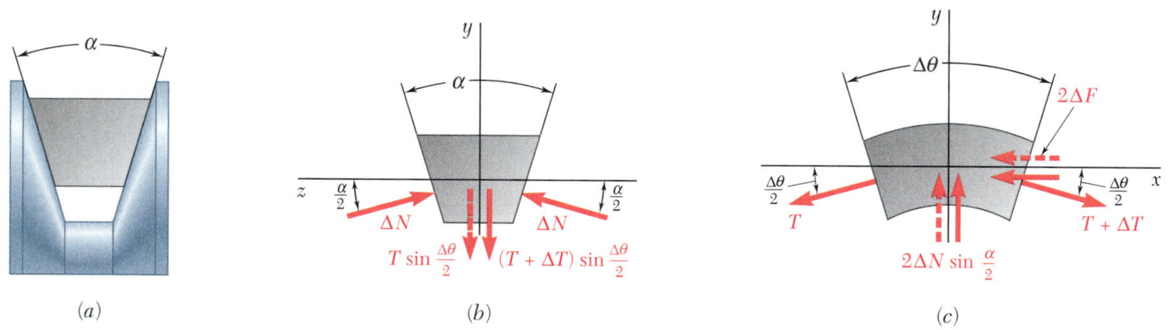

Fig. 8.15

place along the sides of the groove. The relation existing between the values T_1 and T_2 of the tension in the two parts of the belt when the belt is just about to slip can again be obtained by drawing the free-body diagram of an element of belt (Fig. 8.15*b* and *c*). Equations similar to (8.11) and (8.12) are derived, but the magnitude of the total friction force acting on the element is now $2\ \Delta F$, and the sum of the y components of the normal forces is $2\ \Delta N \sin(\alpha/2)$. Proceeding as above, we obtain

$$\ln \frac{T_2}{T_1} = \frac{\mu_s \beta}{\sin(\alpha/2)} \tag{8.15}$$

or,

$$\frac{T_2}{T_1} = e^{\mu_s \beta / \sin(\alpha/2)} \tag{8.16}$$

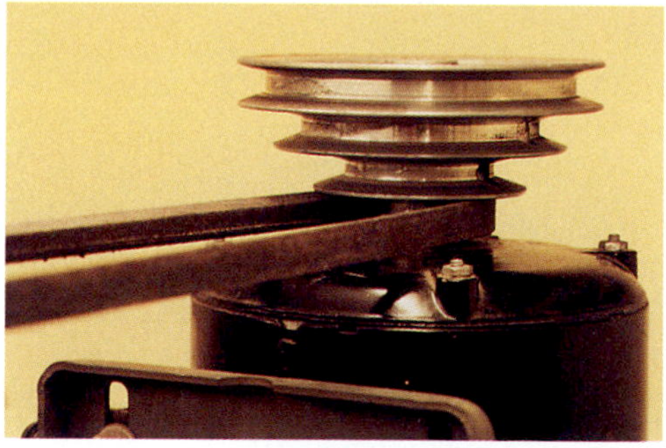

Photo 8.4 Large normal and friction forces on the sloping sides of the V belt increase the power that can be transmitted before slipping occurs.

SAMPLE PROBLEM 8.7

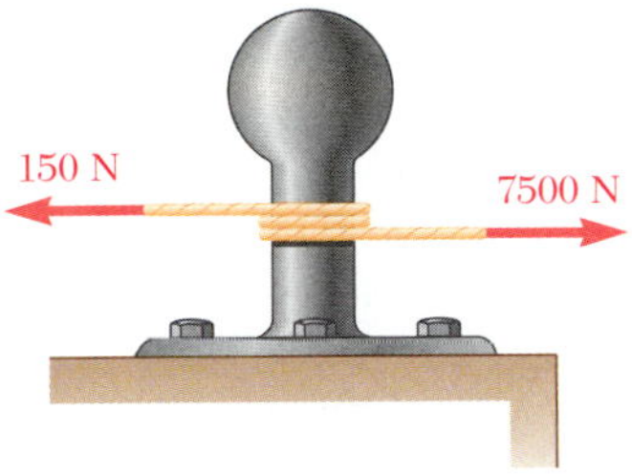

A hawser thrown from a ship to a pier is wrapped two full turns around a bollard. The tension in the hawser is 7500 N; by exerting a force of 150 N on its free end, a dockworker can just keep the hawser from slipping. (*a*) Determine the coefficient of friction between the hawser and the bollard. (*b*) Determine the tension in the hawser that could be resisted by the 150-N force if the hawser were wrapped three full turns around the bollard.

SOLUTION

***a*. Coefficient of Friction.** Since slipping of the hawser is impending, we use Eq. (8.13):

$$\ln \frac{T_2}{T_1} = \mu_s \beta$$

Since the hawser is wrapped two full turns around the bollard, we have

$$\beta = 2(2\pi \text{ rad}) = 12.57 \text{ rad}$$
$$T_1 = 150 \text{ N} \qquad T_2 = 7500 \text{ N}$$

Therefore,

$$\mu_s \beta = \ln \frac{T_2}{T_1}$$

$$\mu_s(12.57 \text{ rad}) = \ln \frac{7500 \text{ N}}{150 \text{ N}} = \ln 50 = 3.91$$

$$\mu_s = 0.311 \qquad\qquad \mu_s = 0.31 \blacktriangleleft$$

***b*. Hawser Wrapped Three Turns around Bollard.** Using the value of μ_s obtained in part *a*, we now have

$$\beta = 3(2\pi \text{ rad}) = 18.85 \text{ rad}$$
$$T_1 = 150 \text{ N} \qquad \mu_s = 0.311$$

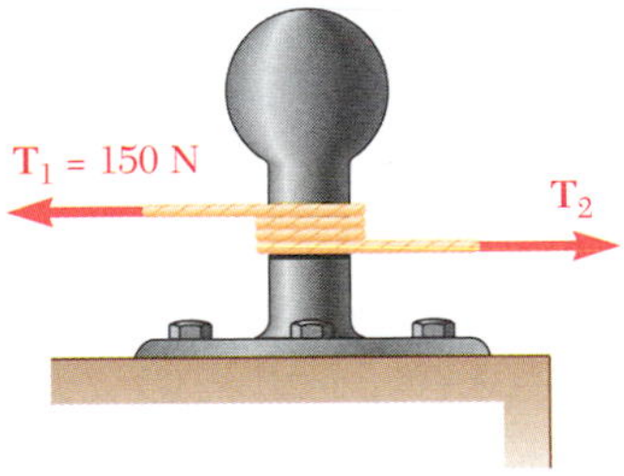

Substituting these values into Eq. (8.14), we obtain

$$\frac{T_2}{T_1} = e^{\mu_s \beta}$$

$$\frac{T_2}{150 \text{ N}} = e^{(0.311)(18.85)} = e^{5.862} = 351.5$$

$$T_2 = 52\,725 \text{ N}$$

$$T_2 = 52.7 \text{ kN} \blacktriangleleft$$

SAMPLE PROBLEM 8.8

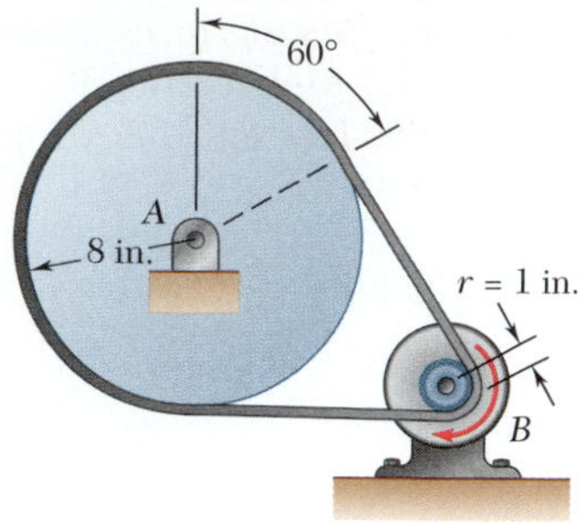

A flat belt connects pulley A, which drives a machine tool, to pulley B, which is attached to the shaft of an electric motor. The coefficients of friction are $\mu_s = 0.25$ and $\mu_k = 0.20$ between both pulleys and the belt. Knowing that the maximum allowable tension in the belt is 600 lb, determine the largest torque which can be exerted by the belt on pulley A.

SOLUTION

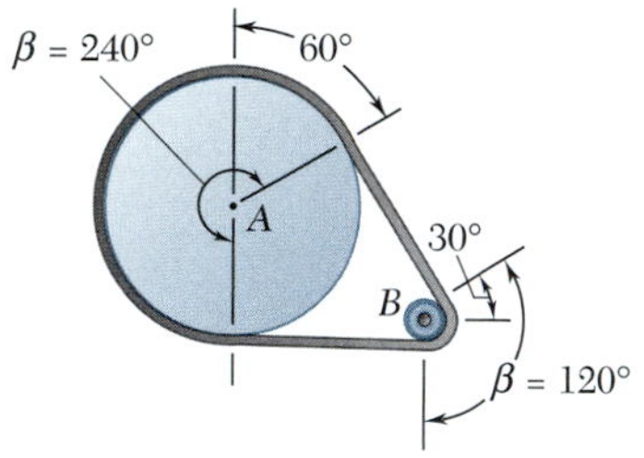

Since the resistance to slippage depends upon the angle of contact β between pulley and belt, as well as upon the coefficient of static friction μ_s, and since μ_s is the same for both pulleys, slippage will occur first on pulley B, for which β is smaller.

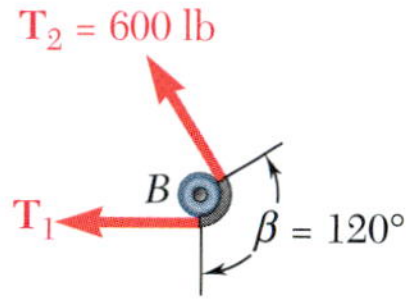

Pulley *B*. Using Eq. (8.14) with $T_2 = 600$ lb, $\mu_s = 0.25$, and $\beta = 120° = 2\pi/3$ rad, we write

$$\frac{T_2}{T_1} = e^{\mu_s \beta} \qquad \frac{600 \text{ lb}}{T_1} = e^{0.25(2\pi/3)} = 1.688$$

$$T_1 = \frac{600 \text{ lb}}{1.688} = 355.4 \text{ lb}$$

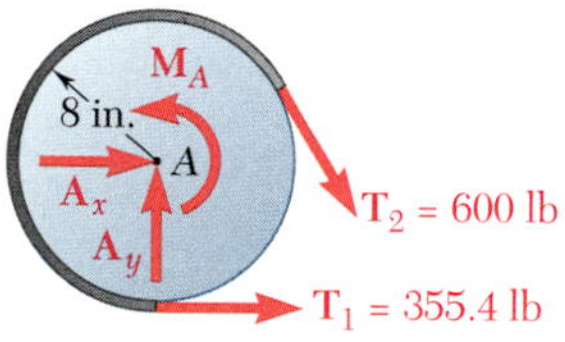

Pulley *A*. We draw the free-body diagram of pulley A. The couple $\mathbf{M}_A$ is applied to the pulley by the machine tool to which it is attached and is equal and opposite to the torque exerted by the belt. We write

$$+\circlearrowleft \Sigma M_A = 0: \quad M_A - (600 \text{ lb})(8 \text{ in.}) + (355.4 \text{ lb})(8 \text{ in.}) = 0$$

$$M_A = 1957 \text{ lb} \cdot \text{in.} \qquad M_A = 163.1 \text{ lb} \cdot \text{ft} \quad ◀$$

Note. We may check that the belt does not slip on pulley A by computing the value of μ_s required to prevent slipping at A and verifying that it is smaller than the actual value of μ_s. From Eq. (8.13) we have

$$\mu_s \beta = \ln \frac{T_2}{T_1} = \ln \frac{600 \text{ lb}}{355.4 \text{ lb}} = 0.524$$

and, since $\beta = 240° = 4\pi/3$ rad,

$$\frac{4\pi}{3} \mu_s = 0.524 \qquad \mu_s = 0.125 < 0.25$$

SOLVING PROBLEMS ON YOUR OWN

In the preceding section you learned about *belt friction*. The problems you will have to solve include belts passing over fixed drums, band brakes in which the drum rotates while the band remains fixed, and belt drives.

1. ***Problems involving belt friction*** fall into one of the following two categories:

a. Problems in which slipping is impending. One of the following formulas, involving the *coefficient of static friction* μ_s, may then be used:

$$\ln \frac{T_2}{T_1} = \mu_s \beta \tag{8.13}$$

or

$$\frac{T_2}{T_1} = e^{\mu_s \beta} \tag{8.14}$$

b. Problems in which slipping is occurring. The formulas to be used can be obtained from Eqs. (8.13) and (8.14) by replacing μ_s with the *coefficient of kinetic friction* μ_k.

2. ***As you start solving a belt-friction problem,*** be sure to remember the following:

a. The angle β must be expressed in radians. In a belt-and-drum problem, this is the angle subtending the arc of the drum on which the belt is wrapped.

b. The larger tension is always denoted by T_2, and the smaller tension is denoted by T_1.

c. The larger tension occurs at the end of the belt which is in the direction of the motion, or impending motion, of the belt relative to the drum. You can also determine the direction of the larger tension by remembering that the friction force acting on the belt and the larger tension are always opposed.

3. ***In each of the problems you will be asked to solve,*** three of the four quantities T_1, T_2, β, and μ_s (or μ_k) will either be given or readily found, and you will then solve the appropriate equation for the fourth quantity. There are two kinds of problems that you will encounter:

a. Find μ_s between the belt and the drum knowing that slipping is impending. From the given data, determine T_1, T_2, and β; substitute these values into Eq. (8.13) and solve for μ_s [Sample Prob. 8.7, part *a*]. Follow the same procedure to find the *smallest value* of μ_s for which slipping will not occur.

b. Find the magnitude of a force or couple applied to the belt or drum knowing that slipping is impending. The given data should include μ_s and β. If it also includes T_1 or T_2, use Eq. (8.14) to find the other tension. If neither T_1 nor T_2 is known but some other data is given, use the free-body diagram of the belt-drum system to write an equilibrium equation that you will solve simultaneously with Eq. (8.14) for T_1 and T_2. You will then be able to find the magnitude of the specified force or couple from the free-body diagram of the system. Follow the same procedure to determine the *largest value* of a force or couple which can be applied to the belt or drum if no slipping is to occur [Sample Prob. 8.8].

Problems

8.101 A hawser is wrapped two full turns around a bollard. By exerting a 320-N force on the free end of the hawser, a dockworker can resist a force of 20 kN on the other end of the hawser. Determine (*a*) the coefficient of static friction between the hawser and the bollard, (*b*) the number of times the hawser should be wrapped around the bollard if a 80-kN force is to be resisted by the same 320-N force.

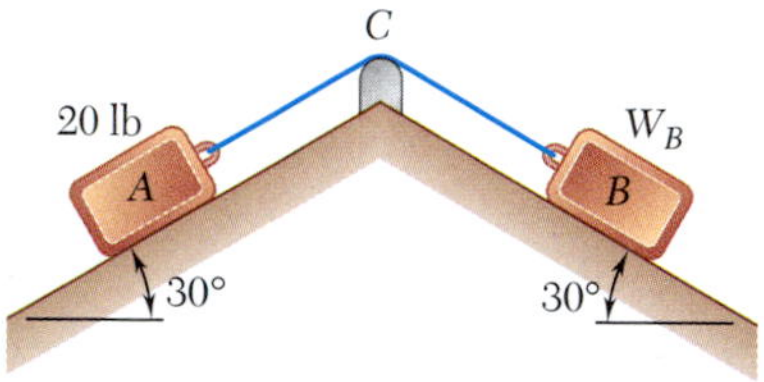

Fig. P8.102 and P8.103

8.102 Blocks A and B are connected by a cable that passes over support C. Friction between the blocks and the inclined surfaces can be neglected. Knowing that motion of block B up the incline is impending when $W_B = 16$ lb, determine (*a*) the coefficient of static friction between the rope and the support, (*b*) the largest value of W_B for which equilibrium is maintained. (*Hint:* See Prob. 8.128.)

8.103 Blocks A and B are connected by a cable that passes over support C. Friction between the blocks and the inclined surfaces can be neglected. Knowing that the coefficient of static friction between the rope and the support is 0.50, determine the range of values of W_B for which equilibrium is maintained. (*Hint:* See Prob. 8.128.)

8.104 A 120-kg block is supported by a rope which is wrapped $1\frac{1}{2}$ times around a horizontal rod. Knowing that the coefficient of static friction between the rope and the rod is 0.15, determine the range of values of P for which equilibrium is maintained.

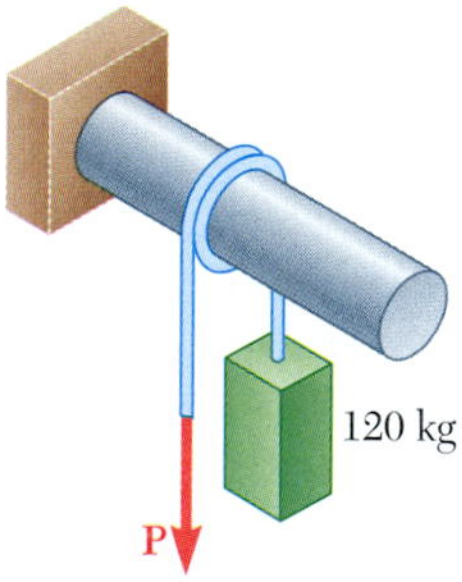

Fig. P8.104

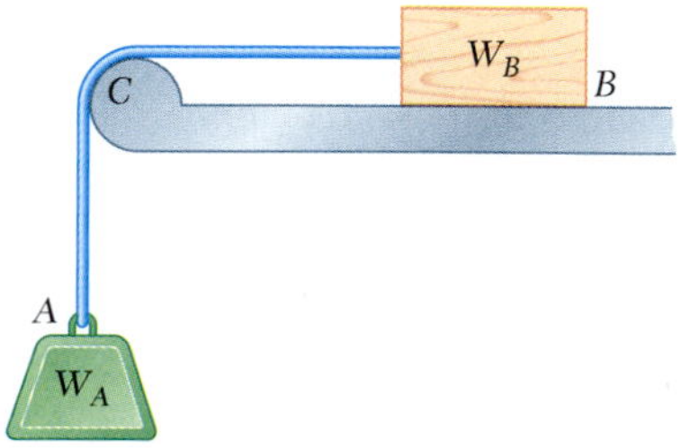

Fig. *P8.105* and *P8.106*

8.105 The coefficient of static friction between block B and the horizontal surface and between the rope and support C is 0.40. Knowing that $W_A = 30$ lb, determine the smallest weight of block B for which equilibrium is maintained.

8.106 The coefficient of static friction μ_s is the same between block B and the horizontal surface and between the rope and support C. Knowing that $W_A = W_B$, determine the smallest value of μ_s for which equilibrium is maintained.

8.107 In the pivoted motor mount shown, the weight **W** of the 85-kg motor is used to maintain tension in the drive belt. Knowing that the coefficient of static friction between the flat belt and drums *A* and *B* is 0.40, and neglecting the weight of platform *CD*, determine the largest torque which can be transmitted to drum *B* when the drive drum *A* is rotating clockwise.

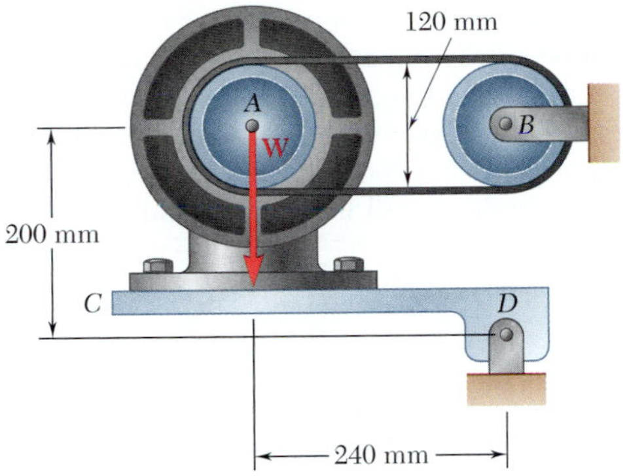

Fig. P8.107

8.108 Solve Prob. 8.107 assuming that the drive drum *A* is rotating counterclockwise.

8.109 A flat belt is used to transmit a torque from pulley *A* to pulley *B*. The radius of each pulley is 3 in., and a force of magnitude $P = 225$ lb is applied as shown to the axle of pulley *A*. Knowing that the coefficient of static friction is 0.35, determine (*a*) the largest torque which can be transmitted, (*b*) the corresponding maximum value of the tension in the belt.

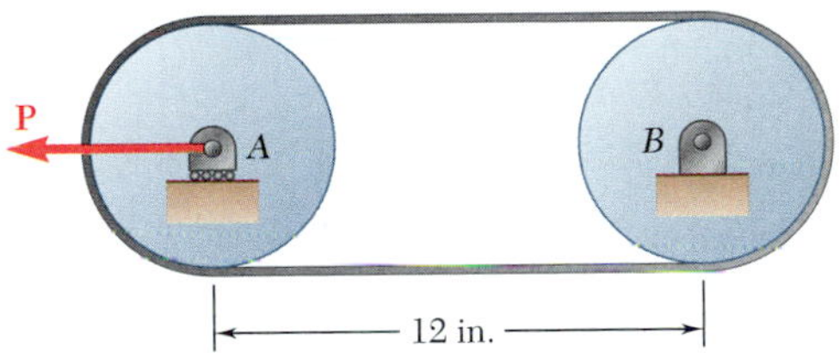

Fig. P8.109

8.110 Solve Prob. 8.109 assuming that the belt is looped around the pulleys in a figure eight.

8.111 A couple $\mathbf{M}_B$ of magnitude 2 lb · ft is applied to the drive drum *B* of a portable belt sander to maintain the sanding belt *C* at a constant speed. The total downward force exerted on the wooden workpiece *E* is 12 lb, and $\mu_k = 0.10$ between the belt and the sanding platen *D*. Knowing that $\mu_s = 0.35$ between the belt and the drive drum and that the radii of drums *A* and *B* are 1.00 in., determine (*a*) the minimum tension in the lower portion of the belt if no slipping is to occur between the belt and the drive drum, (*b*) the value of the coefficient of kinetic friction between the belt and the workpiece.

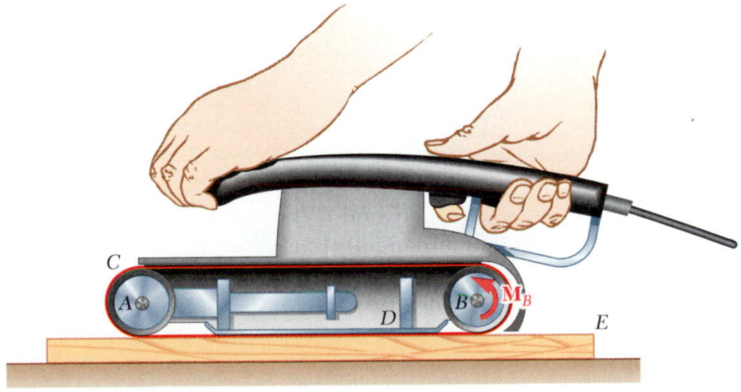

Fig. P8.111

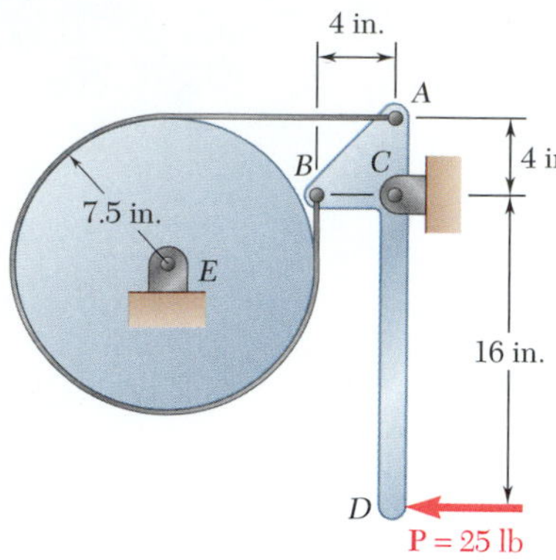

Fig. *P8.112*

8.112 A band belt is used to control the speed of a flywheel as shown. Determine the magnitude of the couple being applied to the flywheel knowing that the coefficient of kinetic friction between the belt and the flywheel is 0.25 and that the flywheel is rotating clockwise at a constant speed. Show that the same result is obtained if the flywheel rotates counterclockwise.

8.113 The drum brake shown permits clockwise rotation of the drum but prevents rotation in the counterclockwise direction. Knowing that the maximum allowed tension in the belt is 7.2 kN, determine (*a*) the magnitude of the largest counterclockwise couple that can be applied to the drum, (*b*) the smallest value of the coefficient of static friction between the belt and the drum for which the drum will not rotate counterclockwise.

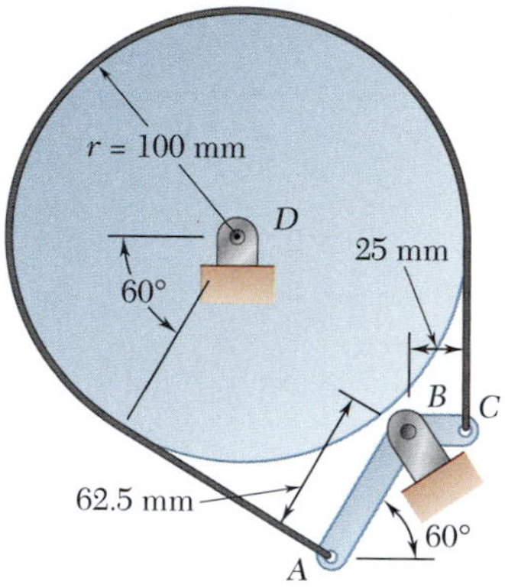

Fig. P8.113

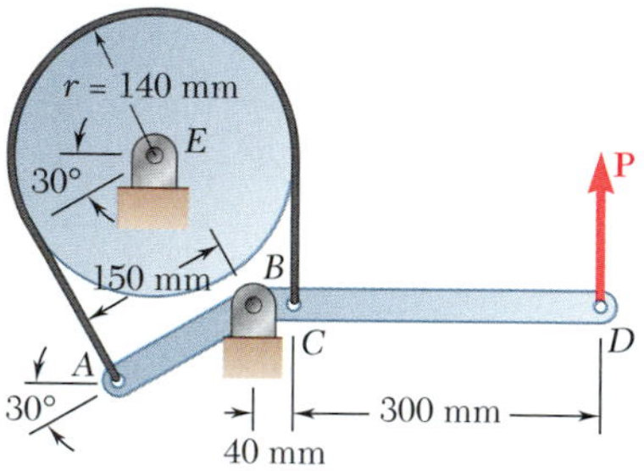

Fig. P8.114 and P8.115

8.114 A differential band brake is used to control the speed of a drum which rotates at a constant speed. Knowing that the coefficient of kinetic friction between the belt and the drum is 0.30 and that a couple of magnitude 150 N · m is applied to the drum, determine the corresponding magnitude of the force **P** that is exerted on end *D* of the lever when the drum is rotating (*a*) clockwise, (*b*) counterclockwise.

8.115 A differential band brake is used to control the speed of a drum. Determine the minimum value of the coefficient of static friction for which the brake is self-locking when the drum rotates counterclockwise.

8.116 Bucket *A* and block *C* are connected by a cable that passes over drum *B*. Knowing that drum *B* rotates slowly counterclockwise and that the coefficients of friction at all surfaces are $\mu_s = 0.35$ and $\mu_k = 0.25$, determine the smallest combined weight *W* of the bucket and its contents for which block *C* will (*a*) remain at rest, (*b*) be about to move up the incline, (*c*) continue moving up the incline at a constant speed.

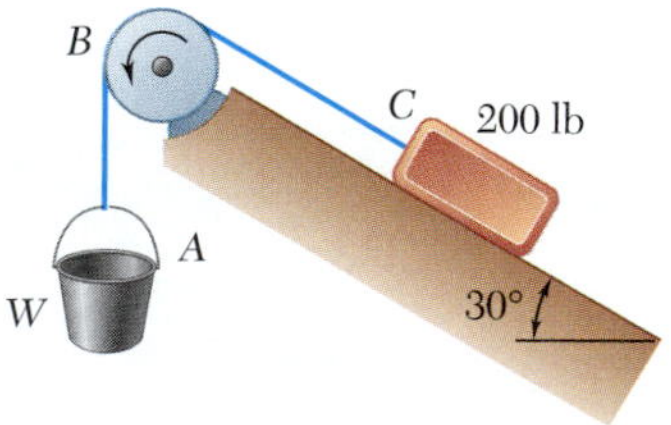

Fig. P8.116

8.117 Solve Prob. 8.116 assuming that drum *B* is frozen and cannot rotate.

8.118 and 8.120 A cable passes around three 30-mm-radius pulleys and supports two blocks as shown. Pulleys C and E are locked to prevent rotation, and the coefficients of friction between the cable and the pulleys are $\mu_s = 0.20$ and $\mu_k = 0.15$. Determine the range of values of the mass of block A for which equilibrium is maintained (a) if pulley D is locked, (b) if pulley D is free to rotate.

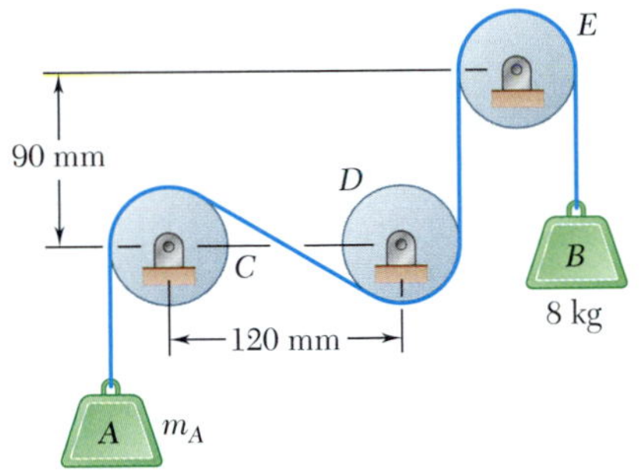

Fig. P8.118 and P8.119

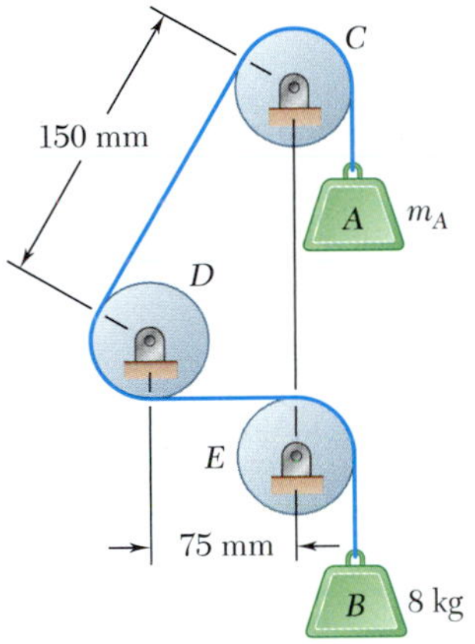

Fig. P8.120 and P8.121

8.119 and 8.121 A cable passes around three 30-mm-radius pulleys and supports two blocks as shown. Two of the pulleys are locked to prevent rotation, while the third pulley is rotated slowly at a constant speed. Knowing that the coefficients of friction between the cable and the pulleys are $\mu_s = 0.20$ and $\mu_k = 0.15$, determine the largest mass m_A which can be raised (a) if pulley C is rotated, (b) if pulley E is rotated.

8.122 A recording tape passes over the 1-in.-radius drive drum B and under the idler drum C. Knowing that the coefficients of friction between the tape and the drums are $\mu_s = 0.40$ and $\mu_k = 0.30$ and that drum C is free to rotate, determine the smallest allowable value of P if slipping of the tape on drum B is not to occur.

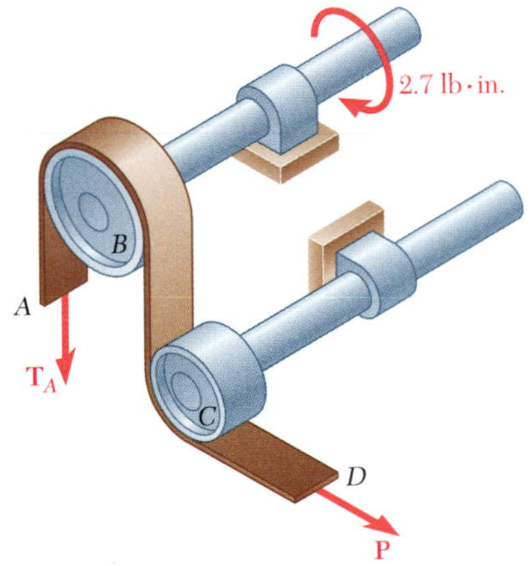

Fig. P8.122

8.123 Solve Prob. 8.122 assuming that the idler drum C is frozen and cannot rotate.

8.124 For the band brake shown, the maximum allowed tension in either belt is 5.6 kN. Knowing that the coefficient of static friction between the belt and the 160-mm-radius drum is 0.25, determine (*a*) the largest clockwise moment $\mathbf{M}_0$ that can be applied to the drum if slipping is not to occur, (*b*) the corresponding force $\mathbf{P}$ exerted on end *E* of the lever.

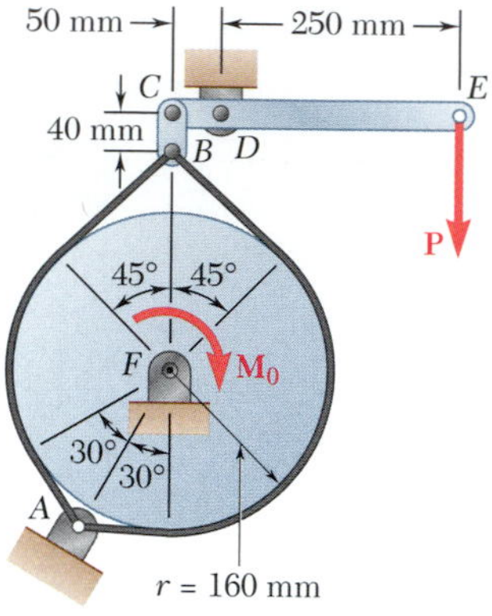

Fig. P8.124

8.125 Solve Prob. 8.124 assuming that a counterclockwise moment is applied to the drum.

8.126 The strap wrench shown is used to grip the pipe firmly without marring the surface of the pipe. Knowing that the coefficient of static friction is the same for all surfaces of contact, determine the smallest value of μ_s for which the wrench will be self-locking when $a = 10$ in., $r = 1.5$ in., and $\theta = 65°$.

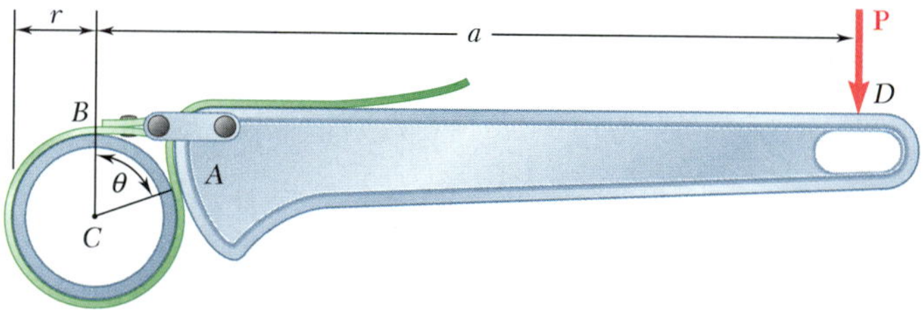

Fig. P8.126

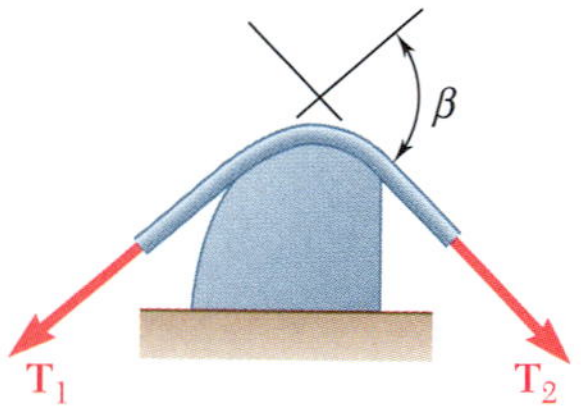

Fig. *P8.128*

8.127 Solve Prob. 8.126 assuming that $\theta = 75°$.

8.128 Prove that Eqs. (8.13) and (8.14) are valid for any shape of surface provided that the coefficient of friction is the same at all points of contact.

8.129 Complete the derivation of Eq. (8.15), which relates the tension in both parts of a V belt.

8.130 Solve Prob. 8.107 assuming that the flat belt and drums are replaced by a V belt and V pulleys with $\alpha = 36°$. (The angle α is as shown in Fig. 8.15*a*.)

8.131 Solve Prob. 8.109 assuming that the flat belt and drums are replaced by a V belt and V pulleys with $\alpha = 36°$. (The angle α is as shown in Fig. 8.15*a*.)

REVIEW AND SUMMARY FOR CHAPTER 8

This chapter was devoted to the study of *dry friction*, that is, to problems involving rigid bodies which are in contact along *nonlubricated surfaces*.

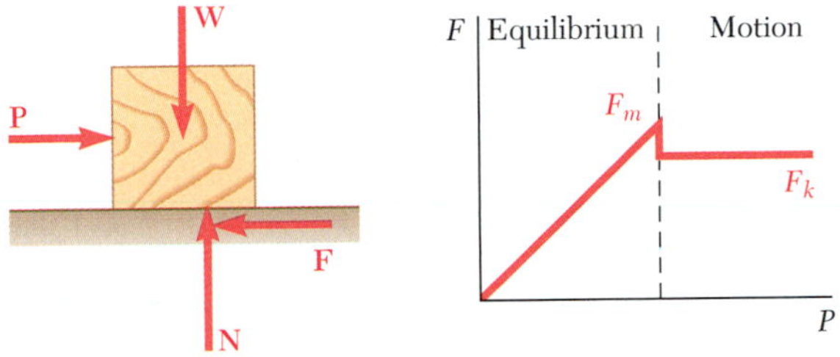

Fig. 8.16

Static and kinetic friction

Applying a horizontal force **P** to a block resting on a horizontal surface [Sec. 8.2], we note that the block at first does not move. This shows that a *friction force* **F** must have developed to balance **P** (Fig. 8.16). As the magnitude of **P** is increased, the magnitude of **F** also increases until it reaches a maximum value F_m. If **P** is further increased, the block starts sliding and the magnitude of **F** drops from F_m to a lower value F_k. Experimental evidence shows that F_m and F_k are proportional to the normal component N of the reaction of the surface. We have

$$F_m = \mu_s N \qquad F_k = \mu_k N \qquad (8.1, 8.2)$$

where μ_s and μ_k are called, respectively, the *coefficient of static friction* and the *coefficient of kinetic friction*. These coefficients depend on the nature and the condition of the surfaces in contact. Approximate values of the coefficients of static friction were given in Table 8.1.

Angles of friction

It is sometimes convenient to replace the normal force **N** and the friction force **F** by their resultant **R** (Fig. 8.17). As the friction force increases and reaches its maximum value $F_m = \mu_s N$, the angle ϕ that **R** forms with the normal to the surface increases and reaches a maximum value ϕ_s, called the *angle of static friction*. If motion actually takes place, the magnitude of **F** drops to F_k; similarly the angle ϕ drops to a lower value ϕ_k, called the *angle of kinetic friction*. As shown in Sec. 8.3, we have

$$\tan \phi_s = \mu_s \qquad \tan \phi_k = \mu_k \qquad (8.3, 8.4)$$

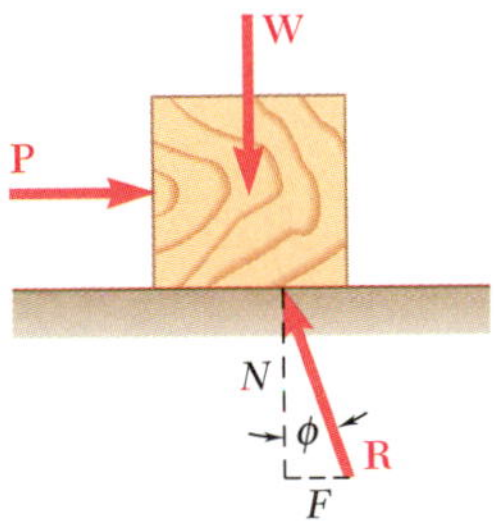

Fig. 8.17

Problems involving friction

When solving equilibrium problems involving friction, we should keep in mind that the magnitude F of the friction force is equal to $F_m = \mu_s N$ *only if the body is about to slide* [Sec. 8.4]. *If motion is not impending,* F and N should be considered as *independent unknowns* to be determined from the equilibrium equations (Fig. 8.18*a*). We should also check that the value of F required to maintain equilibrium is not larger than F_m; if it were, the body would move and the magnitude of the friction force would be $F_k = \mu_k N$ [Sample Prob. 8.1]. On the other hand, *if motion is known to be impending,* F has reached its maximum value $F_m = \mu_s N$ (Fig. 8.18*b*), and this expression may be substituted for F in the equilibrium equations [Sample Prob. 8.3]. When only three forces are involved in a free-body diagram, including the reaction **R** of the surface in contact with the body, it is usually more convenient to solve the problem by drawing a force triangle [Sample Prob. 8.2].

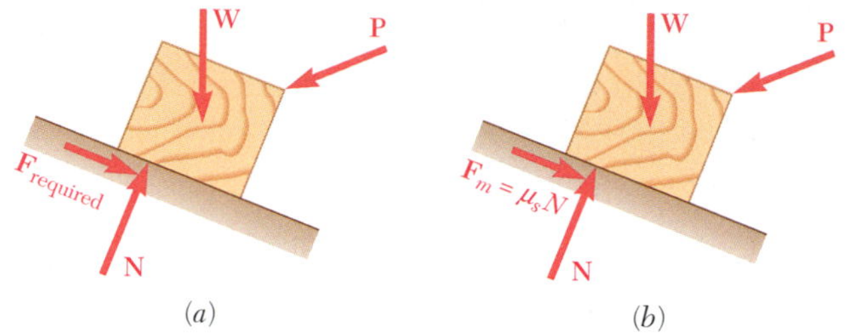

Fig. 8.18

When a problem involves the analysis of the forces exerted on each other by *two bodies A and B,* it is important to show the friction forces with their correct sense. The correct sense for the friction force exerted by B on A, for instance, is opposite to that of the *relative motion* (or impending motion) of A with respect to B [Fig. 8.6].

Wedges and screws

In the second part of the chapter we considered a number of specific engineering applications where dry friction plays an important role. In the case of *wedges,* which are simple machines used to raise heavy loads [Sec. 8.5], two or more free-body diagrams were drawn and care was taken to show each friction force with its correct sense [Sample Prob. 8.4]. The analysis of *square-threaded screws,* which are frequently used in jacks, presses, and other mechanisms, was reduced to the analysis of a block sliding on an incline by unwrapping the thread of the screw and showing it as a straight line [Sec. 8.6]. This is done again in Fig. 8.19, where r denotes the *mean radius* of the thread, L is the *lead* of the screw, that is, the distance through which the screw advances in one turn, **W** is the load, and Qr is equal to the torque exerted on the screw. It was noted that in the case of multiple-threaded screws the lead L of the screw is *not* equal to its pitch, which is the distance measured between two consecutive threads.

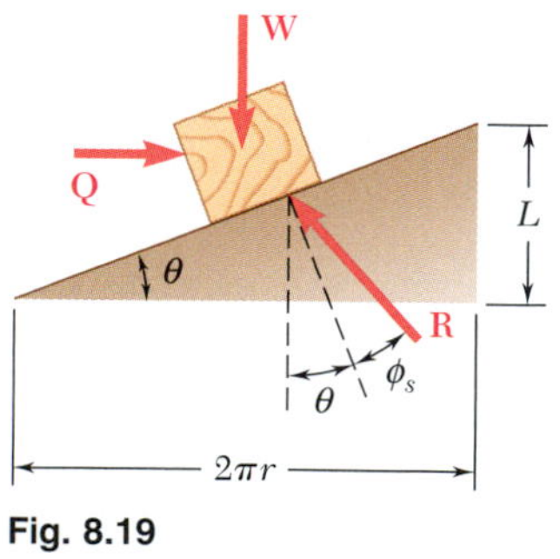

Fig. 8.19

Other engineering applications considered in this chapter were *journal bearings* and *axle friction* [Sec. 8.7], *thrust bearings* and *disk friction* [Sec. 8.8], *wheel friction* and *rolling resistance* [Sec. 8.9], and *belt friction* [Sec. 8.10].

Belt friction

In solving a problem involving a *flat belt* passing over a fixed cylinder, it is important to first determine the direction in which the belt slips or is about to slip. If the drum is rotating, the motion or impending motion of the belt should be determined *relative* to the rotating drum. For instance, if the belt shown in Fig. 8.20 is about to slip to the right relative to the drum, the friction forces exerted by the drum on the belt will be directed to the left and the tension will be larger in the right-hand portion of the belt than in the left-hand portion. Denoting the larger tension by T_2, the smaller tension by T_1, the coefficient of static friction by μ_s, and the angle (in radians) subtended by the belt by β, we derived in Sec. 8.10 the formulas

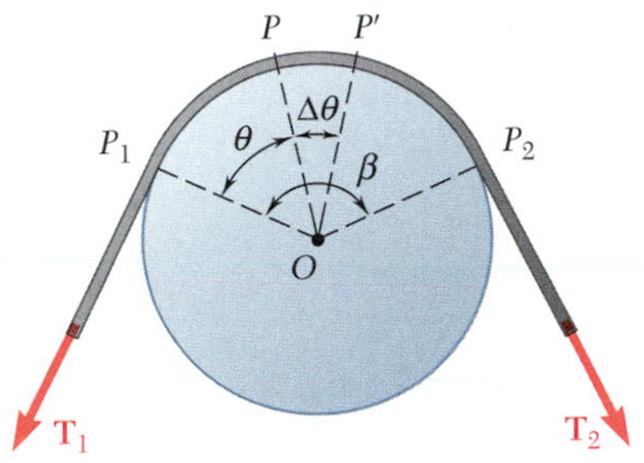

Fig. 8.20

$$\ln \frac{T_2}{T_1} = \mu_s \beta \qquad (8.13)$$

$$\frac{T_2}{T_1} = e^{\mu_s \beta} \qquad (8.14)$$

which were used in solving Sample Probs. 8.7 and 8.8. If the belt actually slips on the drum, the coefficient of static friction μ_s should be replaced by the coefficient of kinetic friction μ_k in both of these formulas.

Review Problems

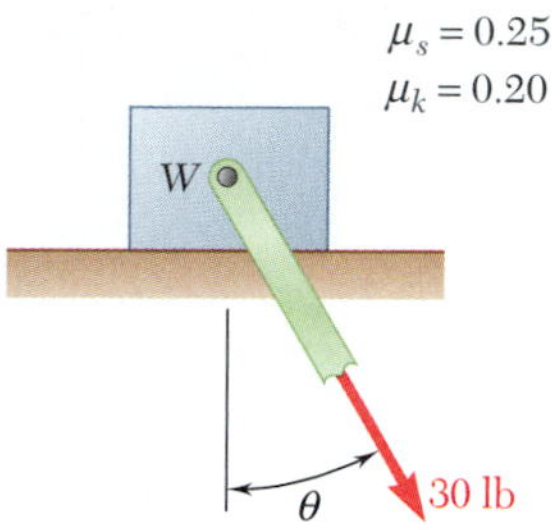

Fig. *P8.132*

8.132 Considering only values of θ less than 90°, determine the smallest value of θ required to start the block moving to the right when (*a*) $W = 75$ lb, (*b*) $W = 100$ lb.

8.133 The machine base shown has a mass of 75 kg and is fitted with skids at A and B. The coefficient of static friction between the skids and the floor is 0.30. If a force **P** of magnitude 500 N is applied at corner C, determine the range of values of θ for which the base will not move.

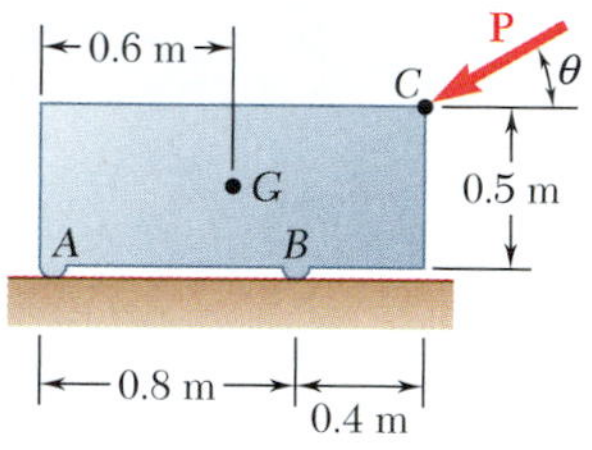

Fig. P8.133

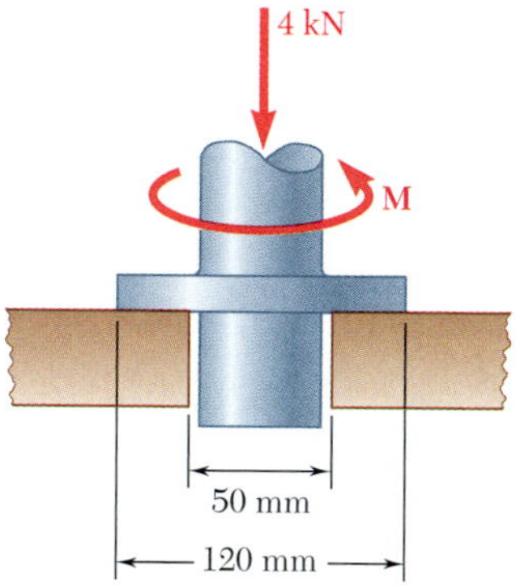

Fig. *P8.134*

8.134 Knowing that a couple of magnitude 30 N · m is required to start the vertical shaft rotating, determine the coefficient of static friction between the annular surfaces of contact.

8.135 The 20-lb block A and the 30-lb block B are supported by an incline which is held in the position shown. Knowing that the coefficient of static friction is 0.15 between the two blocks and zero between block B and the incline, determine the value of θ for which motion is impending.

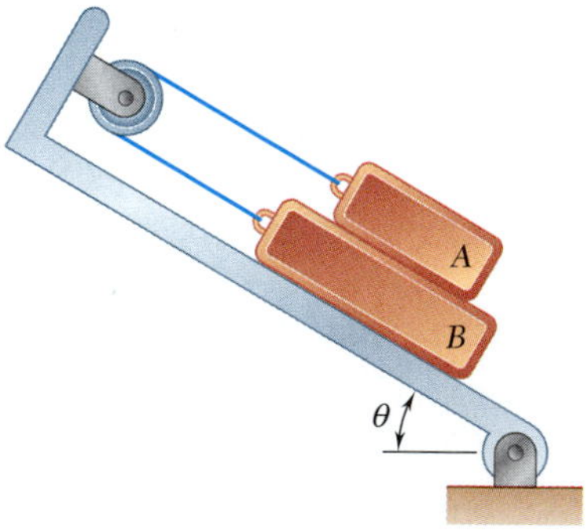

Fig. P8.135 and *P8.136*

8.136 The 20-lb block A and the 30-lb block B are supported by an incline which is held in the position shown. Knowing that the coefficient of static friction is 0.15 between all surfaces of contact, determine the value of θ for which motion is impending.

8.137 Two cylinders are connected by a rope that passes over two fixed rods as shown. Knowing that the coefficient of static friction between the rope and the rods is 0.40, determine the range of values of the mass m of cylinder D for which equilibrium is maintained.

8.138 Two cylinders are connected by a rope that passes over two fixed rods as shown. Knowing that for cylinder D upward motion impends when $m = 20$ kg, determine (a) the coefficient of static friction between the rope and the rods, (b) the corresponding tension in portion BC of the rope.

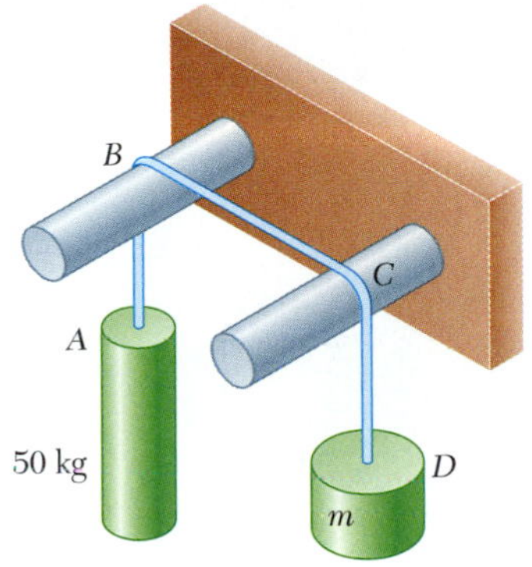

Fig. P8.137 and *P8.138*

8.139 A 10° wedge is used to split a section of a log. The coefficient of static friction between the wedge and the log is 0.35. Knowing that a force **P** of magnitude 600 lb was required to insert the wedge, determine the magnitude of the forces exerted on the wood by the wedge after insertion.

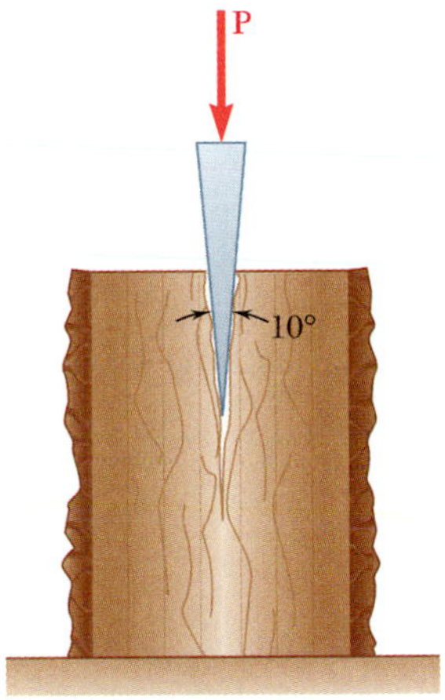

Fig. P8.139

8.140 A flat belt is used to transmit a torque from drum B to drum A. Knowing that the coefficient of static friction is 0.40 and that the allowable belt tension is 450 N, determine the largest torque that can be exerted on drum A.

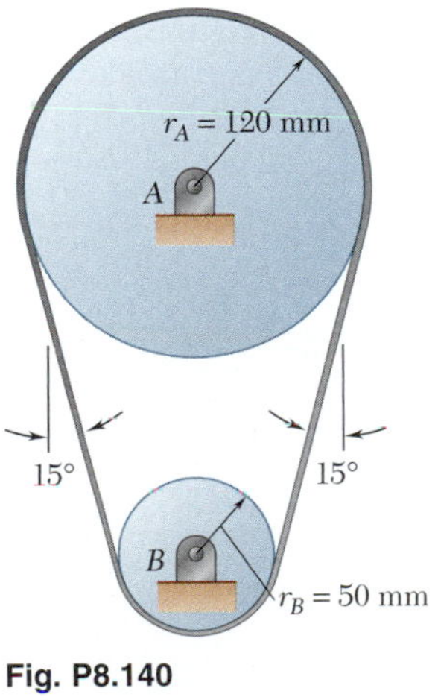

Fig. P8.140

8.141 Two 10-lb blocks A and B are connected by a slender rod of negligible weight. The coefficient of static friction is 0.30 between all surfaces of contact, and the rod forms an angle $\theta = 30°$ with the vertical. (*a*) Show that the system is in equilibrium when $P = 0$. (*b*) Determine the largest value of P for which equilibrium is maintained.

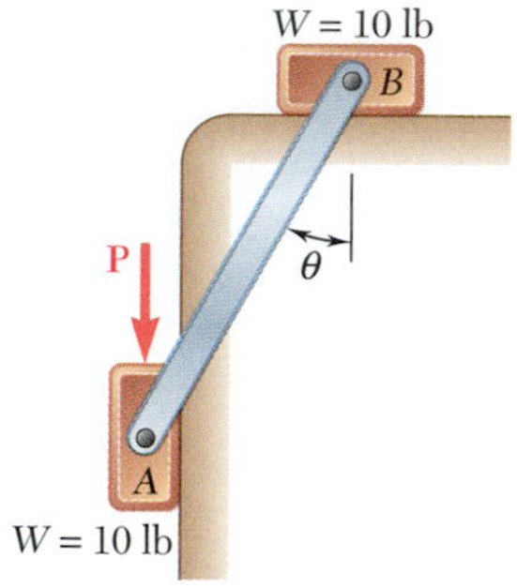

Fig. P8.141

8.142 Determine the range of values of P for which equilibrium of the block shown is maintained.

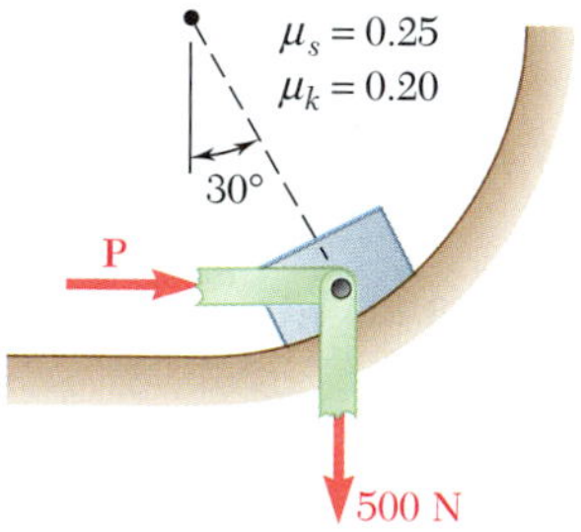

Fig. P8.142

8.143 Two identical uniform boards, each of weight 40 lb, are temporarily leaned against each other as shown. Knowing that the coefficient of static friction between all surfaces is 0.40, determine (*a*) the largest magnitude of the force **P** for which equilibrium will be maintained, (*b*) the surface at which motion will impend.

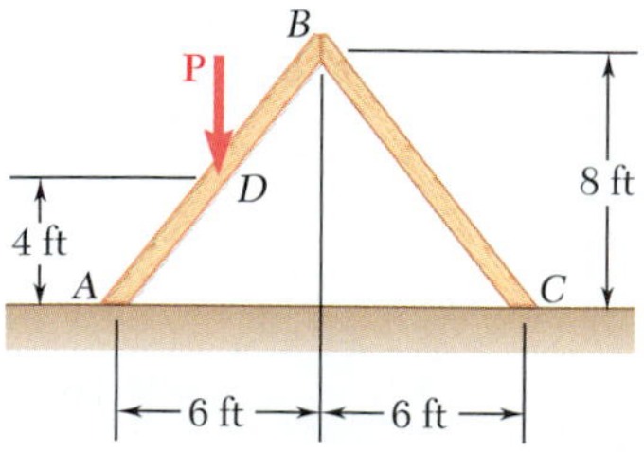

Fig. P8.143

Computer Problems

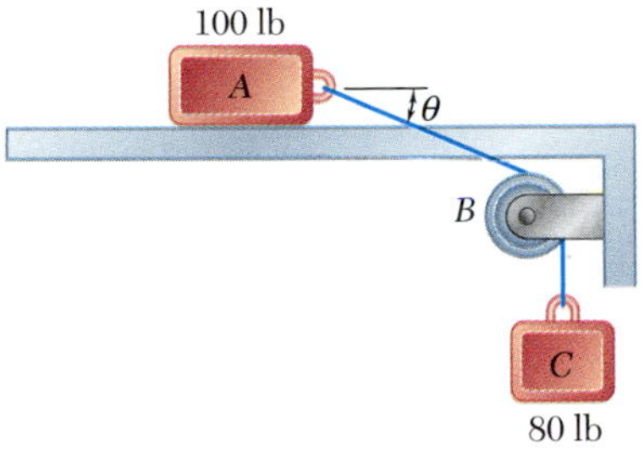

Fig. P8.C1

8.C1 Two blocks are connected by a cable as shown. Knowing that the coefficient of static friction between block A and the horizontal surface varies between 0 and 0.6, determine the values of θ for which motion is impending and plot these values as a function of the coefficient of static friction.

8.C2 The slender rod AB of length $l = 15$ in. is attached to a collar A and rests on a wheel located at a vertical distance $a = 2.5$ in. from the horizontal rod on which the collar slides. Neglecting friction at C, use computational software to determine the range of values of the magnitude of **Q** for which equilibrium is maintained. Plot the maximum and minimum values of **Q** as functions of the coefficient of static friction for $0.10 \leq \mu_s \leq 0.55$ when $P = 15$ lb and $\theta = 20°$.

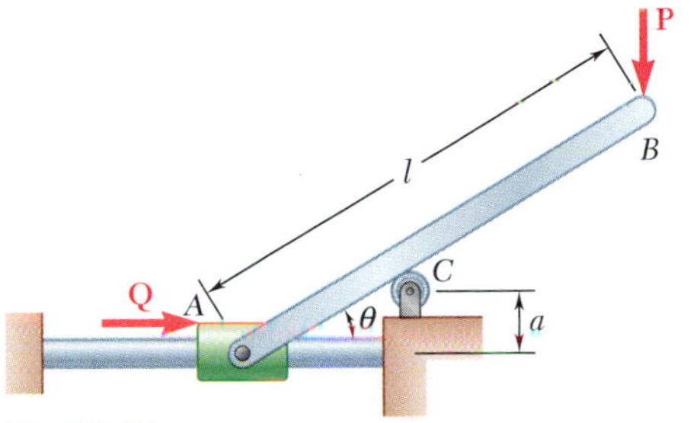

Fig. P8.C2

8.C3 Blocks A and B are supported by an incline which is held in the position shown. Knowing that the mass of block A is 10 kg and that the coefficient of static friction between all surfaces of contact is 0.15, use computational software to determine the value of θ for which motion is impending for values of the mass of block B from 0 to 50 kg using 5-kg increments.

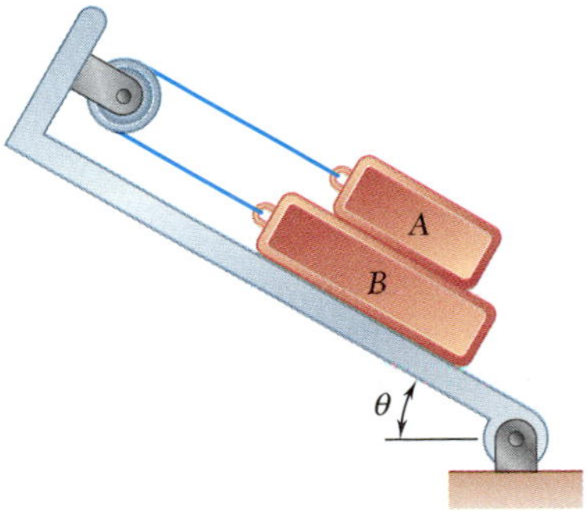

Fig. P8.C3

8.C4 The position of the 20-lb rod AB is controlled by the 4-lb block shown, which is slowly moved to the left by the force **P**. Knowing that the coefficient of kinetic friction between all surfaces of contact is 0.25, use computational software to plot the magnitude P of the force as a function of x for values of x from 45 in. to 5 in. Determine the maximum value of P and the corresponding value of x.

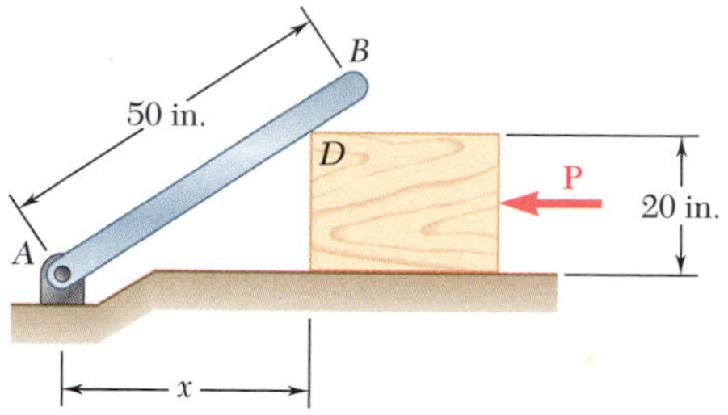

Fig. P8.C4

8.C5 A 0.6-lb cylinder C rests on cylinder D as shown. Knowing that the coefficient of static friction μ_s is the same at A and B, use computational software to determine, for values of μ_s from 0 to 0.40 and using 0.05 increments, the largest counterclockwise couple **M** which can be applied to cylinder D if it is not to rotate.

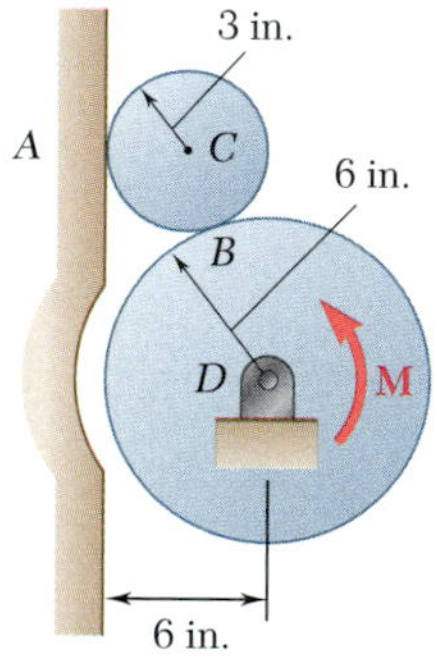

Fig. P8.C5

8.C6 Two identical wedges of negligible mass are used to move and position a 200-kg block. Knowing that the coefficient of static friction μ_s is the same between all surfaces of contact, plot the magnitude of the force **P** for which motion of the block is impending as a function of μ_s for $0.2 \leq \mu_s \leq 0.8$ when (*a*) $\theta = 8°$, (*b*) $\theta = 10°$, (*c*) $\theta = 12°$.

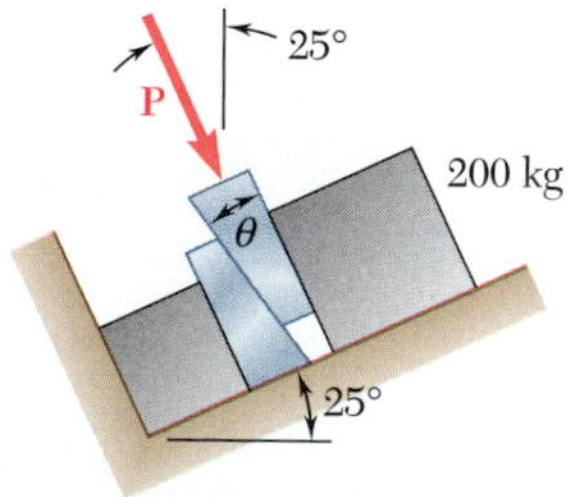

Fig. P8.C6

8.C7 The axle of the pulley shown is frozen and cannot rotate with respect to the block. Knowing that the coefficient of static friction between cable *ABCD* and the pulley varies between 0 and 0.55, determine (*a*) the corresponding values of θ for the system to remain in equilibrium, (*b*) the corresponding reactions at *A* and *D*, and (*c*) plot the values of θ as a function of the coefficient of friction.

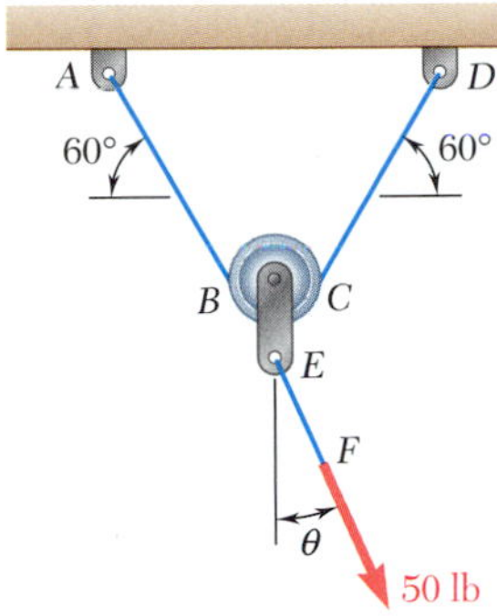

Fig. P8.C7

8.C8 A cable passes around two 50-mm-radius pulleys and supports two blocks as shown. Pulley *B* can only rotate clockwise and pulley *C* can only rotate counterclockwise. Knowing that the coefficient of static friction between the cable and the pulleys is 0.25, plot the minimum and maximum values of the mass of block *D* for which equilibrium is maintained as functions of θ for $15° \leq \theta \leq 40°$.

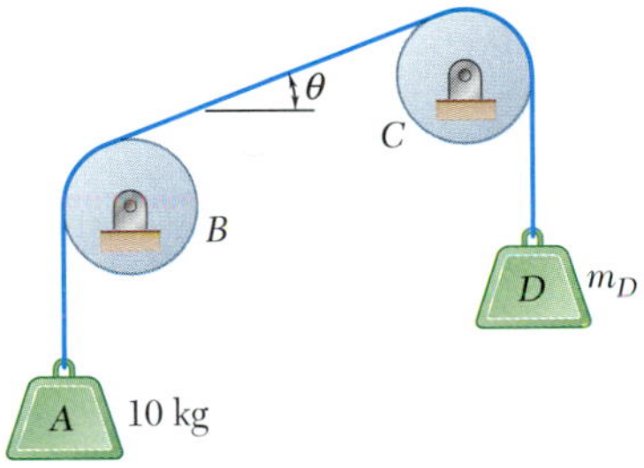

Fig. P8.C8

8.C9 A flat belt is used to transmit a torque from drum *A* to drum *B*. The radius of each drum is 4 in., and the system is fitted with an idler wheel *C* that is used to increase the contact between the belt and the drums. The allowable belt tension is 50 lb, and the coefficient of static friction between the belt and the drums is 0.30. Using computational software, calculate and plot the largest torque that can be transmitted as a function of θ for $0 \leq \theta \leq 30°$.

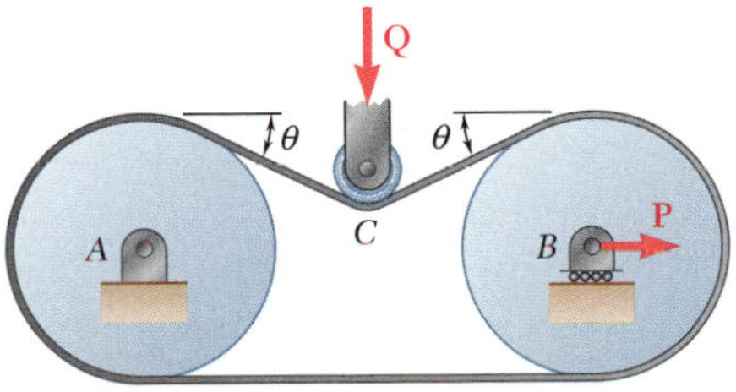

Fig. P8.C9

CHAPTER

9

Distributed Forces: Moments of Inertia

The strength of structural members depends to a large extent on the properties of their cross sections, particularly on the second moments, or moments of inertia, of their areas.

DISTRIBUTED FORCES: MOMENTS OF INERTIA

9.1. INTRODUCTION

In Chap. 5, we analyzed various systems of forces distributed over an area or volume. The three main types of forces considered were (1) weights of homogeneous plates of uniform thickness (Secs. 5.3 through 5.6), (2) distributed loads on beams (Sec. 5.8) and hydrostatic forces (Sec. 5.9), and (3) weights of homogeneous three-dimensional bodies (Secs. 5.10 and 5.11). In the case of homogeneous plates, the magnitude ΔW of the weight of an element of a plate was proportional to the area ΔA of the element. For distributed loads on beams, the magnitude ΔW of each elemental weight was represented by an element of area $\Delta A = \Delta W$ under the load curve; in the case of hydrostatic forces on submerged rectangular surfaces, a similar procedure was followed. In the case of homogeneous three-dimensional bodies, the magnitude ΔW of the weight of an element of the body was proportional to the volume ΔV of the element. Thus, in all cases considered in Chap. 5, the distributed forces were proportional to the elemental areas or volumes associated with them. The resultant of these forces, therefore, could be obtained by summing the corresponding areas or volumes, and the moment of the resultant about any given axis could be determined by computing the first moments of the areas or volumes about that axis.

In the first part of this chapter, we consider distributed forces $\Delta\mathbf{F}$ whose magnitudes depend not only upon the elements of area ΔA on which these forces act but also upon the distance from ΔA to some given axis. More precisely, the magnitude of the force per unit area $\Delta F/\Delta A$ is assumed to vary linearly with the distance to the axis. As indicated in the next section, forces of this type are found in the study of the bending of beams and in problems involving submerged nonrectangular surfaces. Assuming that the elemental forces involved are distributed over an area A and vary linearly with the distance y to the x axis, it will be shown that while the magnitude of their resultant $\mathbf{R}$ depends upon the first moment $Q_x = \int y\,dA$ of the area A, the location of the point where $\mathbf{R}$ is applied depends upon the *second moment*, or *moment of inertia*, $I_x = \int y^2\,dA$ of the same area with respect to the x axis. You will learn to compute the moments of inertia of various areas with respect to given x and y axes. Also introduced in the first part of this chapter is the *polar moment of inertia* $J_O = \int r^2\,dA$ of an area, where r is the distance from the element of area dA to the point O. To facilitate your computations, a relation will be established between the moment of inertia I_x of an area A with respect to a given x axis and the moment of inertia $I_{x'}$ of the same area with respect to the parallel centroidal x' axis (parallel-axis theorem). You will also study the transformation of the moments of inertia of a given area when the coordinate axes are rotated (Secs. 9.9 and 9.10).

In the second part of the chapter, you will learn how to determine the moments of inertia of various *masses* with respect to a given axis. As you will see in Sec. 9.11, the moment of inertia of a given mass about an axis AA' is defined as $I = \int r^2\,dm$, where r is the distance from the axis AA' to the element of mass dm. Moments of inertia of masses are encountered in dynamics in problems involving the rotation of a rigid body about an axis. To facilitate the computation of mass moments of inertia, the parallel-axis theorem will be introduced (Sec. 9.12). Finally, you will learn to analyze the transformation of moments of inertia of masses when the coordinate axes are rotated (Secs. 9.16 through 9.18).

9.2. SECOND MOMENT, OR MOMENT OF INERTIA, OF AN AREA

In the first part of this chapter, we consider distributed forces $\mathbf{\Delta F}$ whose magnitudes ΔF are proportional to the elements of area ΔA on which the forces act and at the same time vary linearly with the distance from ΔA to a given axis.

Consider, for example, a beam of uniform cross section which is subjected to two equal and opposite couples applied at each end of the beam. Such a beam is said to be in *pure bending*, and it is shown in mechanics of materials that the internal forces in any section of the beam are distributed forces whose magnitudes $\Delta F = ky\,\Delta A$ vary linearly with the distance y between the element of area ΔA and an axis passing through the centroid of the section. This axis, represented by the x axis in Fig. 9.1, is known as the *neutral axis* of the section. The forces on one side of the neutral axis are forces of compression, while those on the other side are forces of tension; on the neutral axis itself the forces are zero.

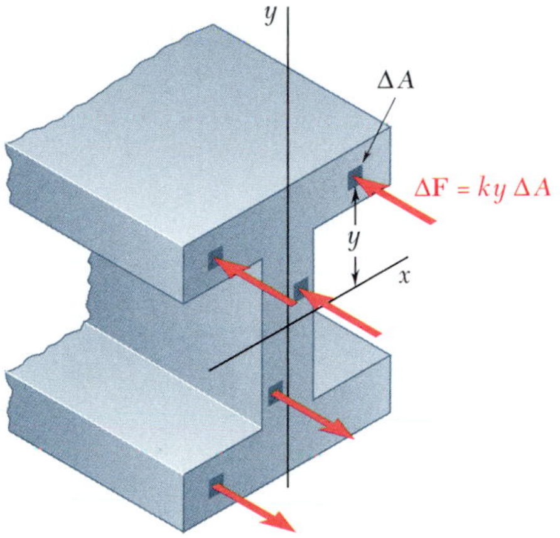

Fig. 9.1

The magnitude of the resultant $\mathbf{R}$ of the elemental forces $\mathbf{\Delta F}$ which act over the entire section is

$$R = \int ky\,dA = k\int y\,dA$$

The last integral obtained is recognized as the *first moment* Q_x of the section about the x axis; it is equal to $\bar{y}A$ and is thus equal to zero, since the centroid of the section is located on the x axis. The system of the forces $\mathbf{\Delta F}$ thus reduces to a couple. The magnitude M of this couple (bending moment) must be equal to the sum of the moments $\Delta M_x = y\,\Delta F = ky^2\,\Delta A$ of the elemental forces. Integrating over the entire section, we obtain

$$M = \int ky^2\,dA = k\int y^2\,dA$$

The last integral is known as the *second moment*, or *moment of inertia*,† of the beam section with respect to the x axis and is denoted by I_x. It is obtained by multiplying each element of area dA by the *square of its distance* from the x axis and integrating over the beam section. Since each product $y^2\,dA$ is positive, regardless of the sign of y, or zero (if y is zero), the integral I_x will always be positive.

Another example of a second moment, or moment of inertia, of an area is provided by the following problem from hydrostatics: A vertical circular gate used to close the outlet of a large reservoir is submerged under water as shown in Fig. 9.2. What is the resultant of the forces exerted by the water on the gate, and what is the moment of the resultant about the line of intersection of the plane of the gate and the water surface (x axis)?

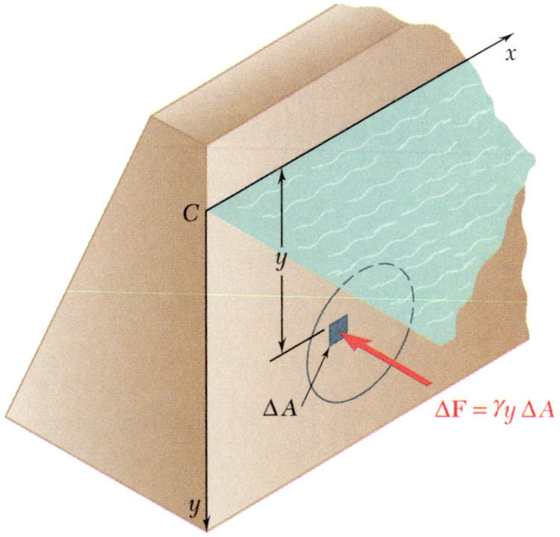

Fig. 9.2

†The term *second moment* is more proper than the term *moment of inertia*, since, logically, the latter should be used only to denote integrals of mass (see Sec. 9.11). In engineering practice, however, moment of inertia is used in connection with areas as well as masses.

If the gate were rectangular, the resultant of the forces of pressure could be determined from the pressure curve, as was done in Sec. 5.9. Since the gate is circular, however, a more general method must be used. Denoting by y the depth of an element of area ΔA and by γ the specific weight of water, the pressure at the element is $p = \gamma y$, and the magnitude of the elemental force exerted on ΔA is $\Delta F = p\,\Delta A = \gamma y\,\Delta A$. The magnitude of the resultant of the elemental forces is thus

$$R = \int \gamma y\,dA = \gamma \int y\,dA$$

and can be obtained by computing the first moment $Q_x = \int y\,dA$ of the area of the gate with respect to the x axis. The moment M_x of the resultant must be equal to the sum of the moments $\Delta M_x = y\,\Delta F = \gamma y^2\,\Delta A$ of the elemental forces. Integrating over the area of the gate, we have

$$M_x = \int \gamma y^2\,dA = \gamma \int y^2\,dA$$

Here again, the integral obtained represents the second moment, or moment of inertia, I_x of the area with respect to the x axis.

9.3. DETERMINATION OF THE MOMENT OF INERTIA OF AN AREA BY INTEGRATION

We defined in the preceding section the second moment, or moment of inertia, of an area A with respect to the x axis. Defining in a similar way the moment of inertia I_y of the area A with respect to the y axis, we write (Fig. 9.3*a*)

$$I_x = \int y^2\,dA \qquad I_y = \int x^2\,dA \tag{9.1}$$

These integrals, known as the *rectangular moments of inertia* of the area A, can be more easily evaluated if we choose dA to be a thin strip parallel to one of the coordinate axes. To compute I_x, the strip is chosen parallel to the x axis, so that all of the points of the strip are at the same distance y from the x axis (Fig. 9.3*b*); the moment of inertia dI_x of the strip is then obtained by multiplying the area dA of the strip by y^2. To compute I_y, the strip is chosen parallel to the y axis so that all of the points of the strip are at the same distance x from the y axis (Fig. 9.3*c*); the moment of inertia dI_y of the strip is $x^2\,dA$.

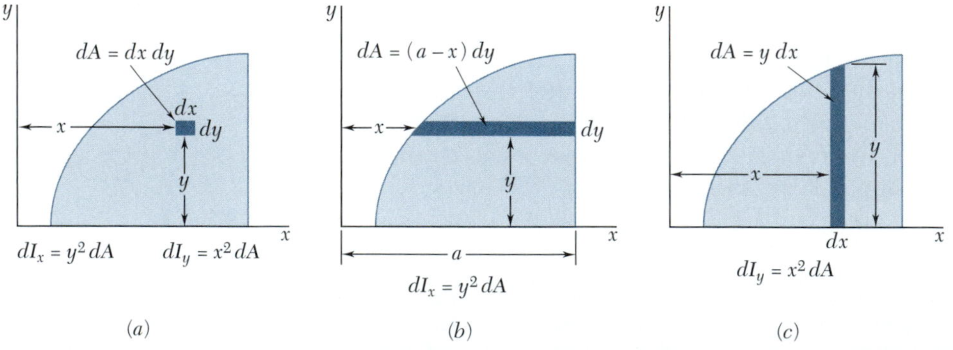

Fig. 9.3

Moment of Inertia of a Rectangular Area. As an example, let us determine the moment of inertia of a rectangle with respect to its base (Fig. 9.4). Dividing the rectangle into strips parallel to the x axis, we obtain

$$dA = b\,dy \qquad dI_x = y^2 b\,dy$$

$$I_x = \int_0^h by^2\,dy = \tfrac{1}{3}bh^3 \tag{9.2}$$

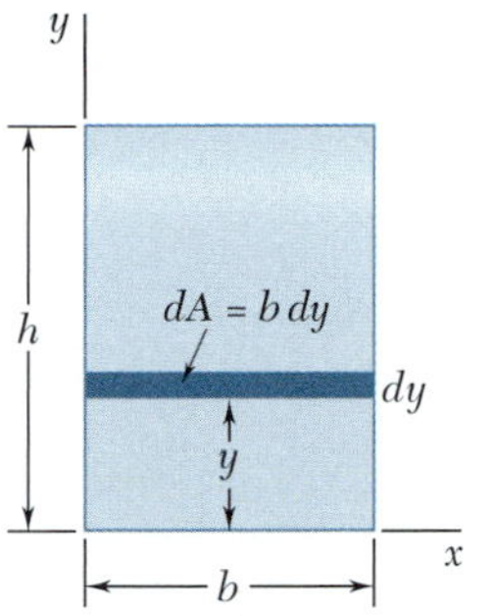

Fig. 9.4

Computing I_x and I_y Using the Same Elemental Strip. The formula just derived can be used to determine the moment of inertia dI_x with respect to the x axis of a rectangular strip which is parallel to the y axis, such as the strip shown in Fig. 9.3c. Setting $b = dx$ and $h = y$ in formula (9.2), we write

$$dI_x = \tfrac{1}{3}y^3\,dx$$

On the other hand, we have

$$dI_y = x^2\,dA = x^2 y\,dx$$

The same element can thus be used to compute the moments of inertia I_x and I_y of a given area (Fig. 9.5a). The analogous results for the area of Fig. 9.3b are shown in Fig. 9.5b.

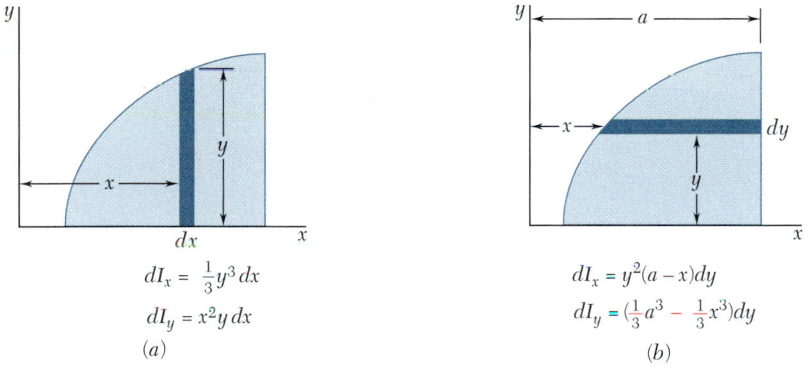

Fig. 9.5

9.4. POLAR MOMENT OF INERTIA

An integral of great importance in problems concerning the torsion of cylindrical shafts and in problems dealing with the rotation of slabs is

$$J_O = \int r^2\,dA \tag{9.3}$$

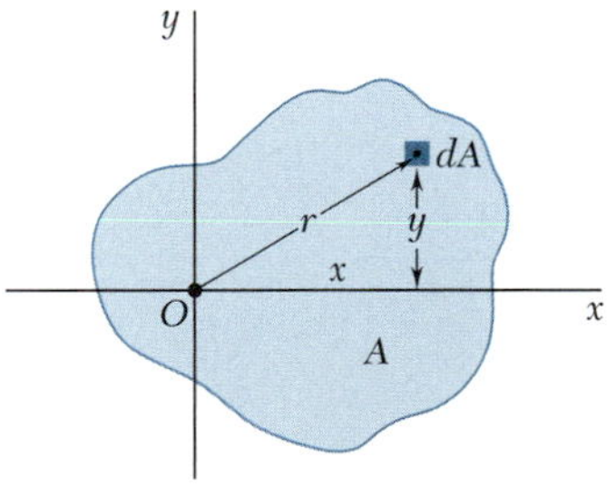

Fig. 9.6

where r is the distance from O to the element of area dA (Fig. 9.6). This integral is the *polar moment of inertia* of the area A with respect to *the "pole"* O.

The polar moment of inertia of a given area can be computed from the rectangular moments of inertia I_x and I_y of the area if these quantities are already known. Indeed, noting that $r^2 = x^2 + y^2$, we write

$$J_O = \int r^2\,dA = \int (x^2 + y^2)\,dA = \int y^2\,dA + \int x^2\,dA$$

that is,

$$J_O = I_x + I_y \tag{9.4}$$

9.5. RADIUS OF GYRATION OF AN AREA

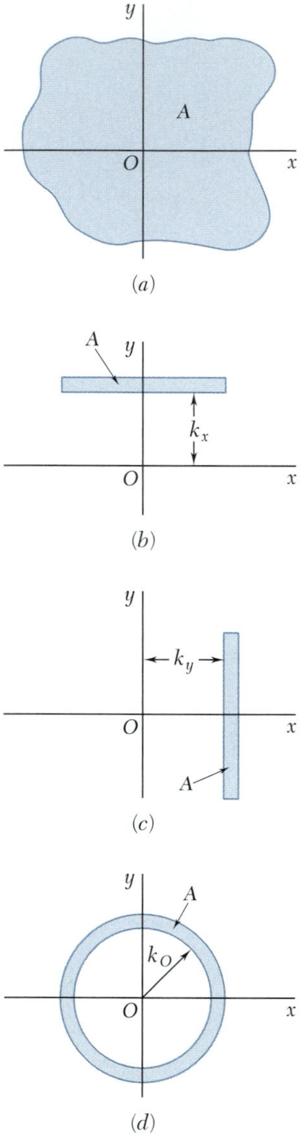

Fig. 9.7

Consider an area A which has a moment of inertia I_x with respect to the x axis (Fig. 9.7*a*). Let us imagine that we concentrate this area into a thin strip parallel to the x axis (Fig. 9.7*b*). If the area A, thus concentrated, is to have the same moment of inertia with respect to the x axis, the strip should be placed at a distance k_x from the x axis, where k_x is defined by the relation

$$I_x = k_x^2 A$$

Solving for k_x, we write

$$k_x = \sqrt{\frac{I_x}{A}} \tag{9.5}$$

The distance k_x is referred to as the *radius of gyration* of the area with respect to the x axis. In a similar way, we can define the radii of gyration k_y and k_O (Fig. 9.7*c* and *d*); we write

$$I_y = k_y^2 A \qquad k_y = \sqrt{\frac{I_y}{A}} \tag{9.6}$$

$$J_O = k_O^2 A \qquad k_O = \sqrt{\frac{J_O}{A}} \tag{9.7}$$

If we rewrite Eq. (9.4) in terms of the radii of gyration, we find that

$$k_O^2 = k_x^2 + k_y^2 \tag{9.8}$$

Example. For the rectangle shown in Fig. 9.8, let us compute the radius of gyration k_x with respect to its base. Using formulas (9.5) and (9.2), we write

$$k_x^2 = \frac{I_x}{A} = \frac{\frac{1}{3}bh^3}{bh} = \frac{h^2}{3} \qquad k_x = \frac{h}{\sqrt{3}}$$

The radius of gyration k_x of the rectangle is shown in Fig. 9.8. It should not be confused with the ordinate $\bar{y} = h/2$ of the centroid of the area. While k_x depends upon the *second moment,* or moment of inertia, of the area, the ordinate $\bar{y}$ is related to the *first moment* of the area.

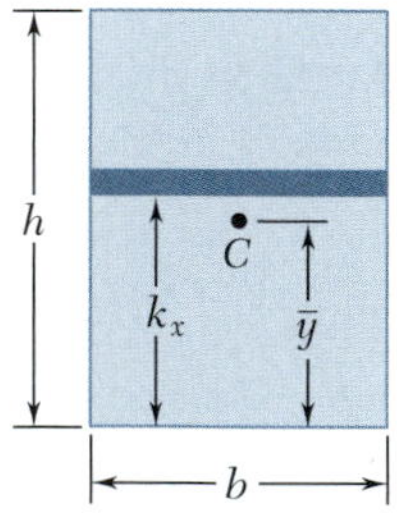

Fig. 9.8

SAMPLE PROBLEM 9.1

Determine the moment of inertia of a triangle with respect to its base.

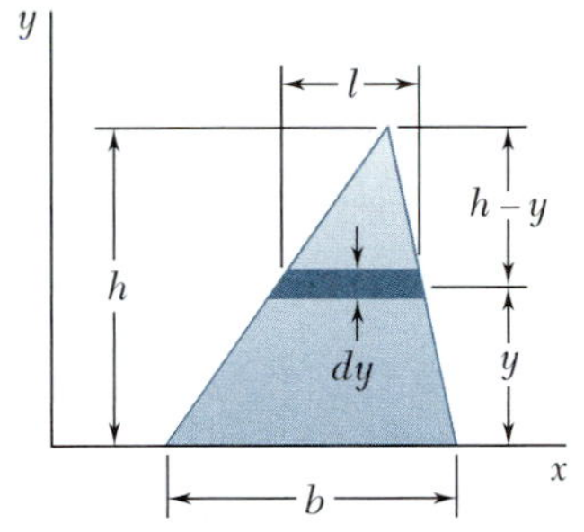

SOLUTION

A triangle of base b and height h is drawn; the x axis is chosen to coincide with the base. A differential strip parallel to the x axis is chosen to be dA. Since all portions of the strip are at the same distance from the x axis, we write

$$dI_x = y^2\,dA \qquad dA = l\,dy$$

Using similar triangles, we have

$$\frac{l}{b} = \frac{h-y}{h} \qquad l = b\frac{h-y}{h} \qquad dA = b\frac{h-y}{h}\,dy$$

Integrating dI_x from $y = 0$ to $y = h$, we obtain

$$I_x = \int y^2\,dA = \int_0^h y^2 b\frac{h-y}{h}\,dy = \frac{b}{h}\int_0^h (hy^2 - y^3)\,dy$$

$$= \frac{b}{h}\left[h\frac{y^3}{3} - \frac{y^4}{4}\right]_0^h \qquad I_x = \frac{bh^3}{12} \blacktriangleleft$$

SAMPLE PROBLEM 9.2

(a) Determine the centroidal polar moment of inertia of a circular area by direct integration. (b) Using the result of part a, determine the moment of inertia of a circular area with respect to a diameter.

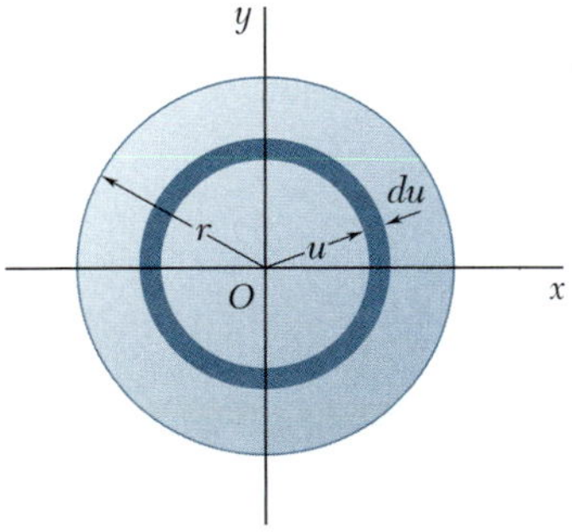

SOLUTION

a. Polar Moment of Inertia. An annular differential element of area is chosen to be dA. Since all portions of the differential area are at the same distance from the origin, we write

$$dJ_O = u^2\,dA \qquad dA = 2\pi u\,du$$

$$J_O = \int dJ_O = \int_0^r u^2(2\pi u\,du) = 2\pi\int_0^r u^3\,du$$

$$J_O = \frac{\pi}{2}r^4 \blacktriangleleft$$

b. Moment of Inertia with Respect to a Diameter. Because of the symmetry of the circular area, we have $I_x = I_y$. We then write

$$J_O = I_x + I_y = 2I_x \qquad \frac{\pi}{2}r^4 = 2I_x \qquad I_{\text{diameter}} = I_x = \frac{\pi}{4}r^4 \blacktriangleleft$$

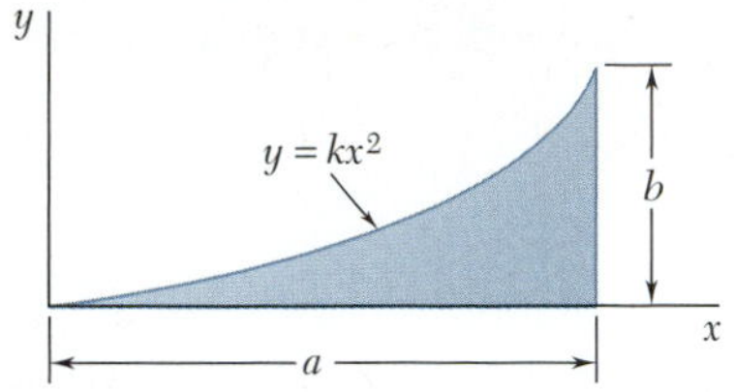

SAMPLE PROBLEM 9.3

(*a*) Determine the moment of inertia of the shaded area shown with respect to each of the coordinate axes. (Properties of this area were considered in Sample Prob. 5.4.) (*b*) Using the results of part *a*, determine the radius of gyration of the shaded area with respect to each of the coordinate axes.

SOLUTION

Referring to Sample Prob. 5.4, we obtain the following expressions for the equation of the curve and the total area:

$$y = \frac{b}{a^2}x^2 \qquad A = \tfrac{1}{3}ab$$

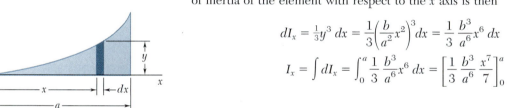

Moment of Inertia I_x. A vertical differential element of area is chosen to be dA. Since all portions of this element are *not* at the same distance from the x axis, we must treat the element as a thin rectangle. The moment of inertia of the element with respect to the x axis is then

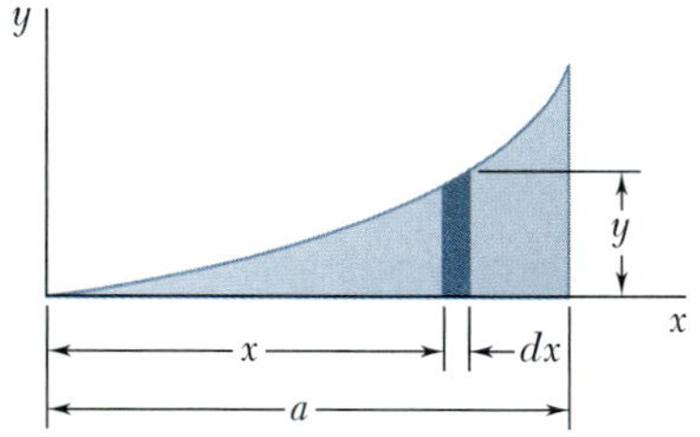

$$dI_x = \tfrac{1}{3}y^3\,dx = \frac{1}{3}\left(\frac{b}{a^2}x^2\right)^3 dx = \frac{1}{3}\frac{b^3}{a^6}x^6\,dx$$

$$I_x = \int dI_x = \int_0^a \frac{1}{3}\frac{b^3}{a^6}x^6\,dx = \left[\frac{1}{3}\frac{b^3}{a^6}\frac{x^7}{7}\right]_0^a$$

$$I_x = \frac{ab^3}{21} \blacktriangleleft$$

Moment of Inertia I_y. The same vertical differential element of area is used. Since all portions of the element are at the same distance from the y axis, we write

$$dI_y = x^2\,dA = x^2(y\,dx) = x^2\left(\frac{b}{a^2}x^2\right)dx = \frac{b}{a^2}x^4\,dx$$

$$I_y = \int dI_y = \int_0^a \frac{b}{a^2}x^4\,dx = \left[\frac{b}{a^2}\frac{x^5}{5}\right]_0^a$$

$$I_y = \frac{a^3b}{5} \blacktriangleleft$$

Radii of Gyration k_x and k_y. We have, by definition,

$$k_x^2 = \frac{I_x}{A} = \frac{ab^3/21}{ab/3} = \frac{b^2}{7} \qquad k_x = \sqrt{\tfrac{1}{7}}\,b \blacktriangleleft$$

and

$$k_y^2 = \frac{I_y}{A} = \frac{a^3b/5}{ab/3} = \tfrac{3}{5}a^2 \qquad k_y = \sqrt{\tfrac{3}{5}}\,a \blacktriangleleft$$

SOLVING PROBLEMS ON YOUR OWN

The purpose of this lesson was to introduce the *rectangular and polar moments of inertia of areas* and the corresponding *radii of gyration*. Although the problems you are about to solve may appear to be more appropriate for a calculus class than for one in mechanics, we hope that our introductory comments have convinced you of the relevance of the moments of inertia to your study of a variety of engineering topics.

1. *Calculating the rectangular moments of inertia I_x and I_y.* We defined these quantities as

$$I_x = \int y^2\, dA \qquad I_y = \int x^2\, dA \tag{9.1}$$

where dA is a differential element of area $dx\, dy$. The moments of inertia are *the second moments of the area;* it is for that reason that I_x, for example, depends on the perpendicular distance y to the area dA. As you study Sec. 9.3, you should recognize the importance of carefully defining the shape and the orientation of dA. Further, you should note the following points.

a. The moments of inertia of most areas can be obtained by means of a single integration. The expressions given in Figs. 9.3*b* and *c* and Fig. 9.5 can be used to calculate I_x and I_y. Regardless of whether you use a single or a double integration, be sure to show on your sketch the element dA that you have chosen.

b. The moment of inertia of an area is always positive, regardless of the location of the area with respect to the coordinate axes. This is because it is obtained by integrating the product of dA and the *square* of a distance. (Note how this differs from the results for the first moment of the area.) Only when an area is *removed* (as in the case for a hole) will its moment of inertia be entered in your computations with a minus sign.

c. As a partial check of your work, observe that the moments of inertia are equal to an area times the square of a length. Thus, every term in an expression for a moment of inertia must be a *length to the fourth power.*

2. *Computing the polar moment of inertia J_O.* We defined J_O as

$$J_O = \int r^2\, dA \tag{9.3}$$

where $r^2 = x^2 + y^2$. If the given area has circular symmetry (as in Sample Prob. 9.2), it is possible to express dA as a function of r and to compute J_O with a single integration. When the area lacks circular symmetry, it is usually easier first to calculate I_x and I_y and then to determine J_O from

$$J_O = I_x + I_y \tag{9.4}$$

Lastly, if the equation of the curve that bounds the given area is expressed in polar coordinates, then $dA = r\, dr\, d\theta$ and a double integration is required to compute the integral for J_O [see Prob. 9.27].

3. *Determining the radii of gyration k_x and k_y and the polar radius of gyration k_O.* These quantities were defined in Sec. 9.5, and you should realize that they can be determined only after the area and the appropriate moments of inertia have been computed. It is important to remember that k_x is measured in the y direction, while k_y is measured in the x direction; you should carefully study Sec. 9.5 until you understand this point.

Problems

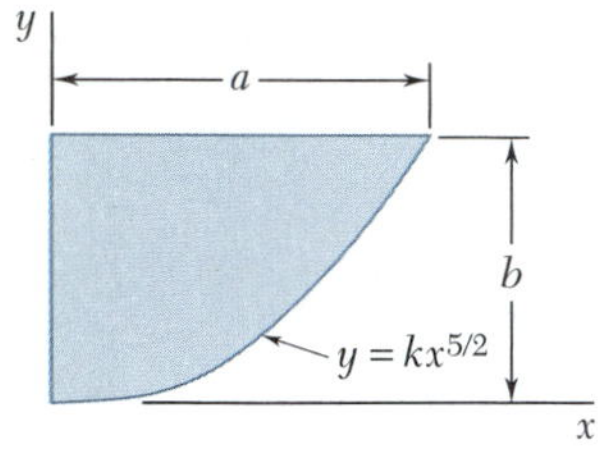

Fig. P9.1 and P9.5

9.1 through 9.4 Determine by direct integration the moment of inertia of the shaded area with respect to the y axis.

9.5 through 9.8 Determine by direct integration the moment of inertia of the shaded area with respect to the x axis.

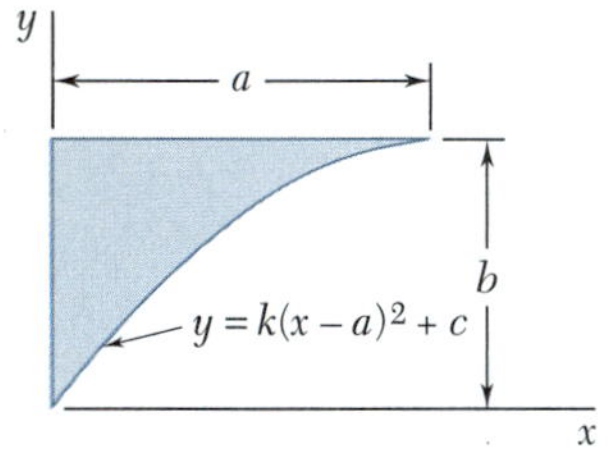

Fig. P9.2 and *P9.6*

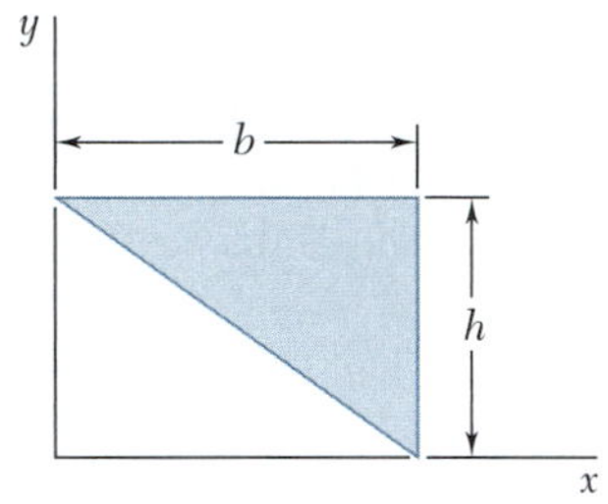

Fig. P9.3 and *P9.7*

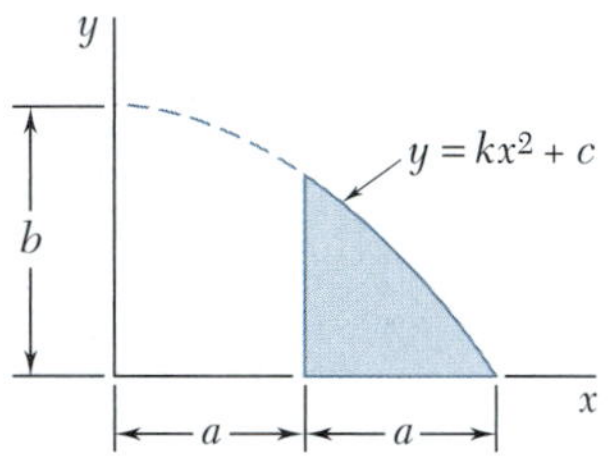

Fig. P9.4 and P9.8

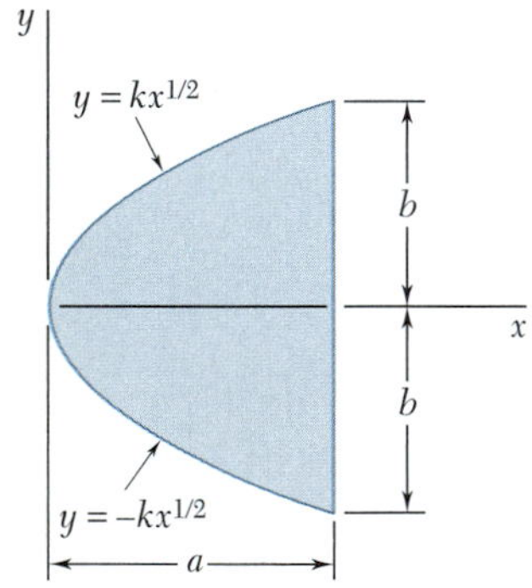

Fig. P9.9 and *P9.12*

9.9 through 9.11 Determine by direct integration the moment of inertia of the shaded area with respect to the x axis.

9.12 through 9.14 Determine by direct integration the moment of inertia of the shaded area with respect to the y axis.

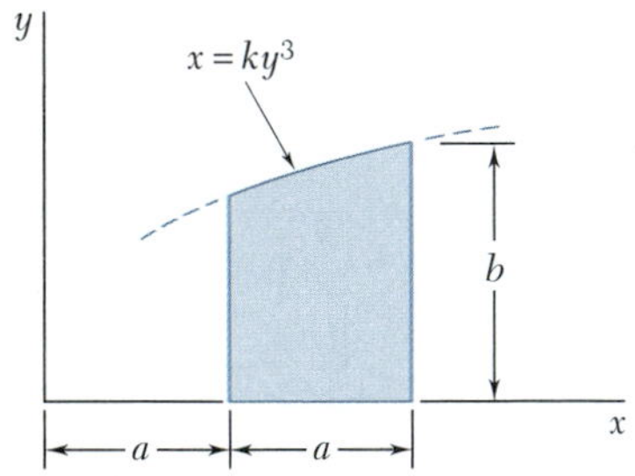

Fig. P9.10 and P9.13

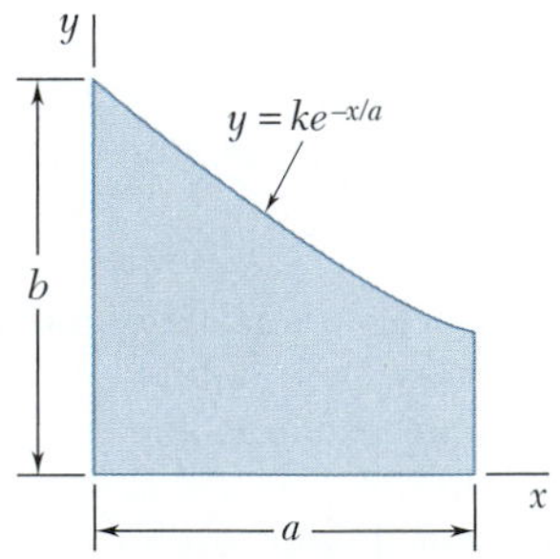

Fig. *P9.11* and P9.14

9.15 and *9.16* Determine the moment of inertia and the radius of gyration of the shaded area shown with respect to the x axis.

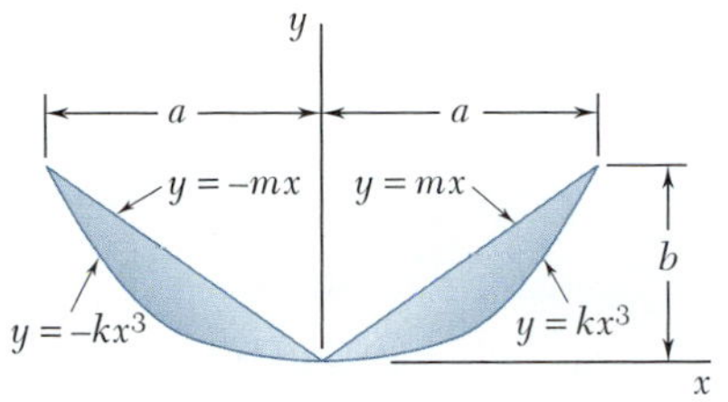

Fig. P9.15 and P9.17

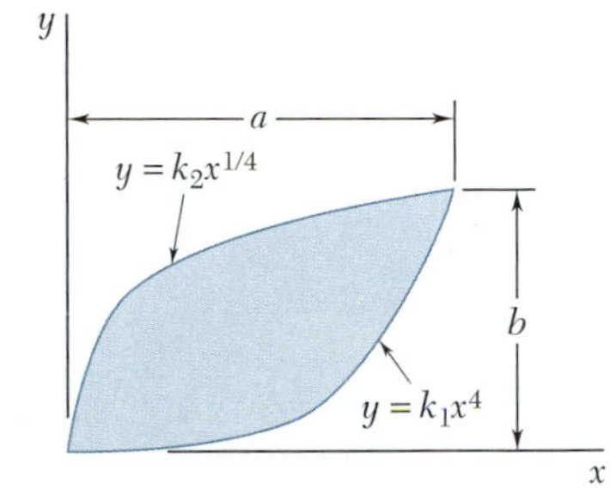

Fig. *P9.16* and P9.18

9.17 and 9.18 Determine the moment of inertia and the radius of gyration of the shaded area shown with respect to the y axis.

9.19 Determine the moment of inertia and the radius of gyration of the shaded area shown with respect to the x axis.

9.20 Determine the moment of inertia and the radius of gyration of the shaded area shown with respect to the y axis.

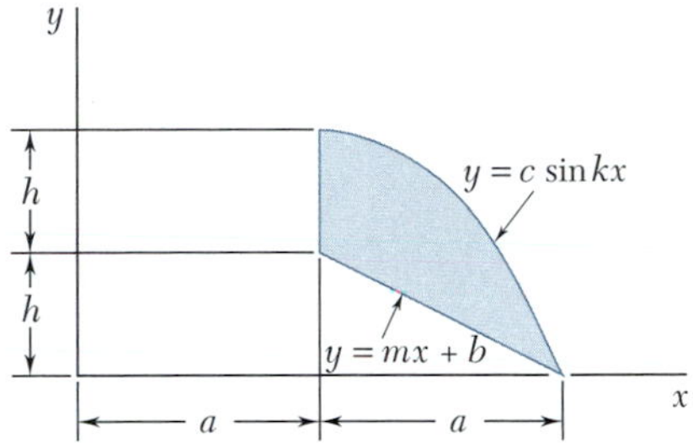

Fig. *P9.19* and P9.20

9.21 and 9.22 Determine the polar moment of inertia and the polar radius of gyration of the shaded area shown with respect to point P.

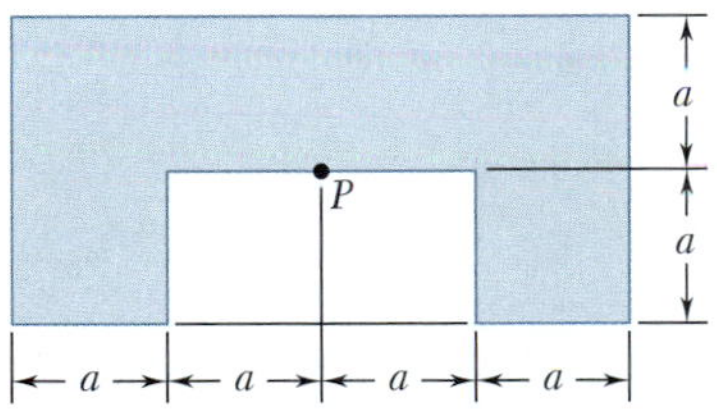

Fig. P9.21

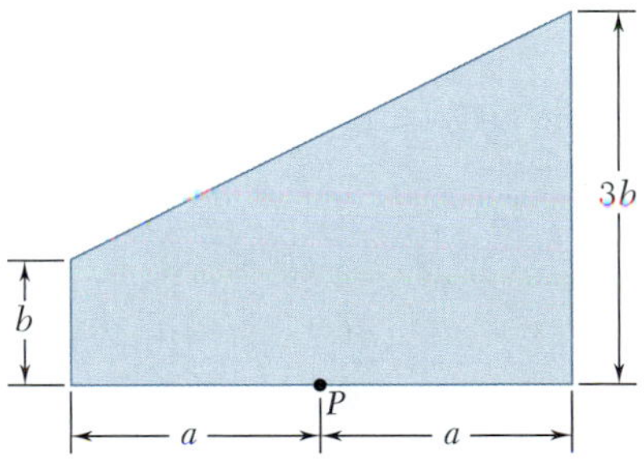

Fig. P9.22

9.23 and *9.24* Determine the polar moment of inertia and the polar radius of gyration of the shaded area shown with respect to point P.

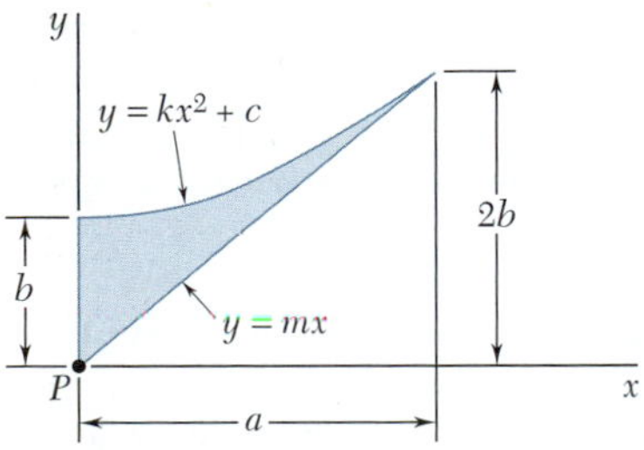

Fig. P9.23

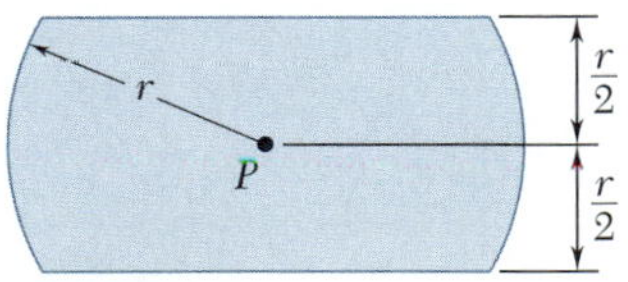

Fig. *P9.24*

9.25 (*a*) Determine by direct integration the polar moment of inertia of the area shown with respect to point O. (*b*) Using the result of part a, determine the moments of inertia of the given area with respect to the x and y axes.

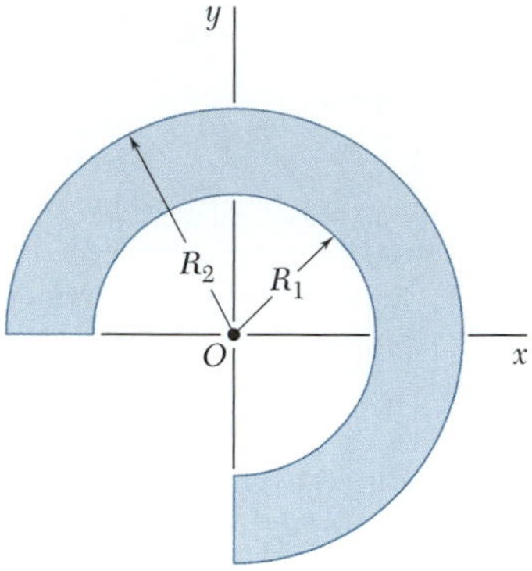

Fig. *P9.25* and P9.26

9.26 (*a*) Show that the polar radius of gyration k_O of the area shown is approximately equal to the mean radius $R_m = (R_1 + R_2)/2$ for small values of the thickness $t = R_2 - R_1$. (*b*) Determine the percentage error introduced by using R_m in place of k_O for the following values of t/R_m: 1, $\frac{1}{4}$, $\frac{1}{16}$.

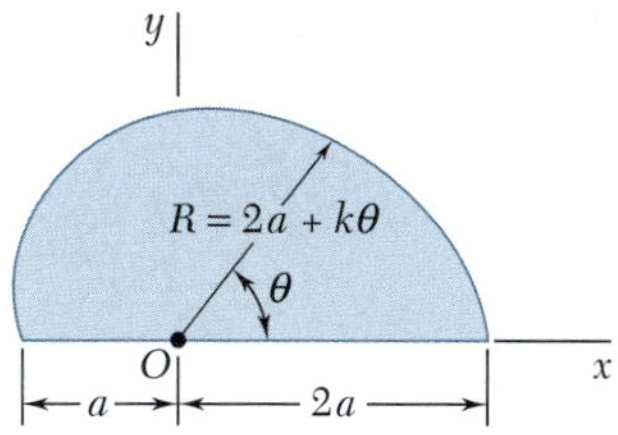

Fig. P9.27

9.27 Determine the polar moment of inertia and the polar radius of gyration of the shaded area shown with respect to point O.

9.28 Determine the polar moment of inertia and the radius of gyration of the isosceles triangle shown with respect to point O.

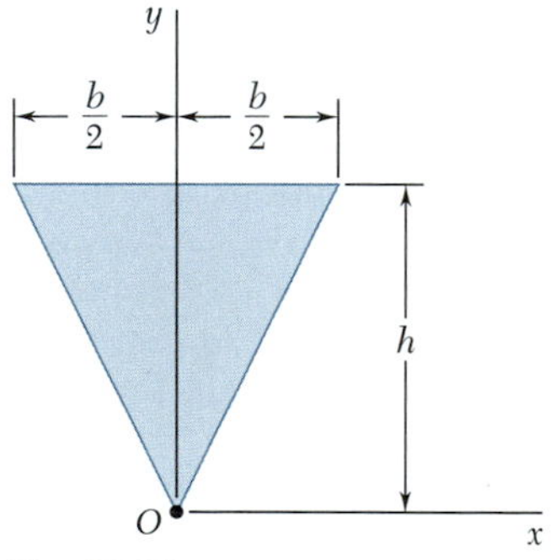

Fig. P9.28

***9.29** Using the polar moment of inertia of the isosceles triangle of Prob. 9.28, show that the centroidal polar moment of inertia of a circular area of radius r is $\pi r^4/2$. (*Hint:* As a circular area is divided into an increasing number of equal circular sectors, what is the approximate shape of each circular sector?)

***9.30** Prove that the centroidal polar moment of inertia of a given area A cannot be smaller than $A^2/2\pi$. (*Hint:* Compare the moment of inertia of the given area with the moment of inertia of a circle which has the same area and the same centroid.)

9.6. PARALLEL-AXIS THEOREM

Consider the moment of inertia I of an area A with respect to an axis AA' (Fig. 9.9). Denoting by y the distance from an element of area dA to AA', we write

$$I = \int y^2 \, dA$$

Let us now draw through the centroid C of the area an axis BB' parallel to AA'; this axis is called a *centroidal axis*. Denoting by y' the

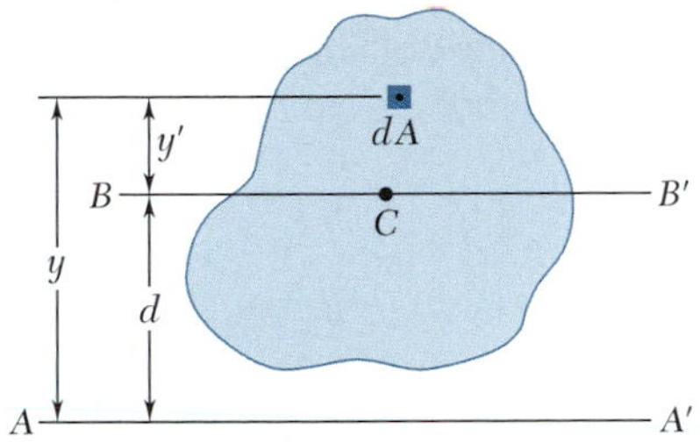

Fig. 9.9

distance from the element dA to BB', we write $y = y' + d$, where d is the distance between the axes AA' and BB'. Substituting for y in the above integral, we write

$$\begin{aligned} I &= \int y^2 \, dA = \int (y' + d)^2 \, dA \\ &= \int y'^2 \, dA + 2d \int y' \, dA + d^2 \int dA \end{aligned}$$

The first integral represents the moment of inertia $\bar{I}$ of the area with respect to the centroidal axis BB'. The second integral represents the first moment of the area with respect to BB'; since the centroid C of the area is located on that axis, the second integral must be zero. Finally, we observe that the last integral is equal to the total area A. Therefore, we have

$$I = \bar{I} + Ad^2 \qquad (9.9)$$

This formula expresses that the moment of inertia I of an area with respect to any given axis AA' is equal to the moment of inertia $\bar{I}$ of the area with respect to the centroidal axis BB' parallel to AA' *plus* the product of the area A and the square of the distance d between the two axes. This theorem is known as the *parallel-axis theorem*. Substituting k^2A for I and $\bar{k}^2A$ for $\bar{I}$, the theorem can also be expressed as

$$k^2 = \bar{k}^2 + d^2 \qquad (9.10)$$

A similar theorem can be used to relate the polar moment of inertia J_O of an area about a point O to the polar moment of inertia $\bar{J}_C$ of the same area about its centroid C. Denoting by d the distance between O and C, we write

$$J_O = \bar{J}_C + Ad^2 \qquad \text{or} \qquad k_O^2 = \bar{k}_C^2 + d^2 \qquad (9.11)$$

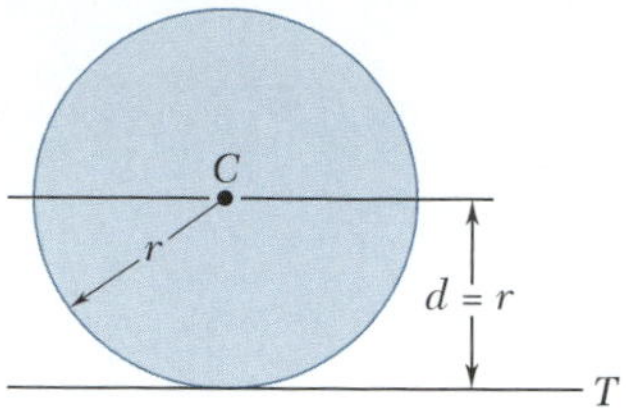

Fig. 9.10

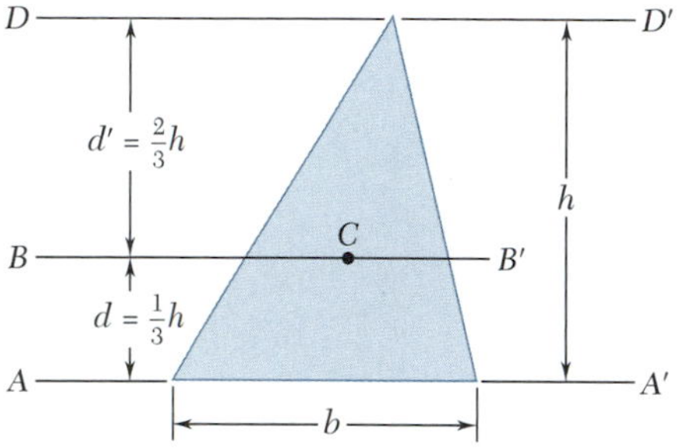

Fig. 9.11

Photo 9.1 Figure 9.13 tabulates data for a small sample of the rolled-steel shapes that are readily available. Shown above are two examples of wide-flange shapes that are commonly used in the construction of buildings.

Example 1. As an application of the parallel-axis theorem, let us determine the moment of inertia I_T of a circular area with respect to a line tangent to the circle (Fig. 9.10). We found in Sample Prob. 9.2 that the moment of inertia of a circular area about a centroidal axis is $\bar{I} = \frac{1}{4}\pi r^4$. We can write, therefore,

$$I_T = \bar{I} + Ad^2 = \tfrac{1}{4}\pi r^4 + (\pi r^2)r^2 = \tfrac{5}{4}\pi r^4$$

Example 2. The parallel-axis theorem can also be used to determine the centroidal moment of inertia of an area when the moment of inertia of the area with respect to a parallel axis is known. Consider, for instance, a triangular area (Fig. 9.11). We found in Sample Prob. 9.1 that the moment of inertia of a triangle with respect to its base AA' is equal to $\frac{1}{12}bh^3$. Using the parallel-axis theorem, we write

$$I_{AA'} = \bar{I}_{BB'} + Ad^2$$
$$\bar{I}_{BB'} = I_{AA'} - Ad^2 = \tfrac{1}{12}bh^3 - \tfrac{1}{2}bh(\tfrac{1}{3}h)^2 = \tfrac{1}{36}bh^3$$

It should be observed that the product Ad^2 was *subtracted* from the given moment of inertia in order to obtain the centroidal moment of inertia of the triangle. Note that this product is *added* when transferring *from* a centroidal axis to a parallel axis, but it should be *subtracted* when transferring *to* a centroidal axis. In other words, the moment of inertia of an area is always smaller with respect to a centroidal axis than with respect to any parallel axis.

Returning to Fig. 9.11, we observe that the moment of inertia of the triangle with respect to the line DD' (which is drawn through a vertex) can be obtained by writing

$$I_{DD'} = \bar{I}_{BB'} + Ad'^2 = \tfrac{1}{36}bh^3 + \tfrac{1}{2}bh(\tfrac{2}{3}h)^2 = \tfrac{1}{4}bh^3$$

Note that $I_{DD'}$ could not have been obtained directly from $I_{AA'}$. The parallel-axis theorem can be applied only if one of the two parallel axes passes through the centroid of the area.

9.7. MOMENTS OF INERTIA OF COMPOSITE AREAS

Consider a composite area A made of several component areas A_1, A_2, A_3, . . . Since the integral representing the moment of inertia of A can be subdivided into integrals evaluated over A_1, A_2, A_3, . . . , the moment of inertia of A with respect to a given axis is obtained by adding the moments of inertia of the areas A_1, A_2, A_3, . . . with respect to the same axis. The moment of inertia of an area consisting of several of the common shapes shown in Fig. 9.12 can thus be obtained by using the formulas given in that figure. Before adding the moments of inertia of the component areas, however, the parallel-axis theorem may have to be used to transfer each moment of inertia to the desired axis. This is shown in Sample Probs. 9.4 and 9.5.

The properties of the cross sections of various structural shapes are given in Fig. 9.13. As noted in Sec. 9.2, the moment of inertia of a beam section about its neutral axis is closely related to the computation of the bending moment in that section of the beam. The determination of moments of inertia is thus a prerequisite to the analysis and design of structural members.

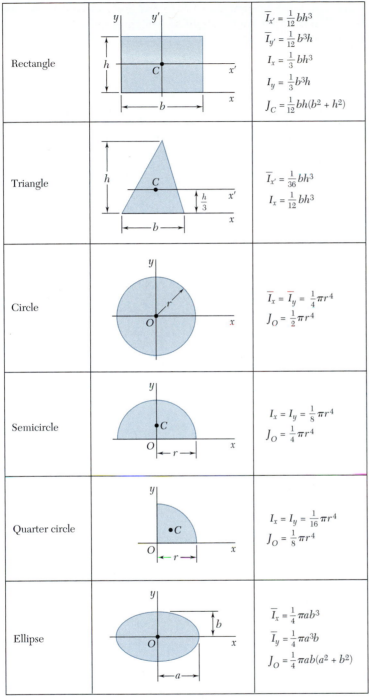

Rectangle	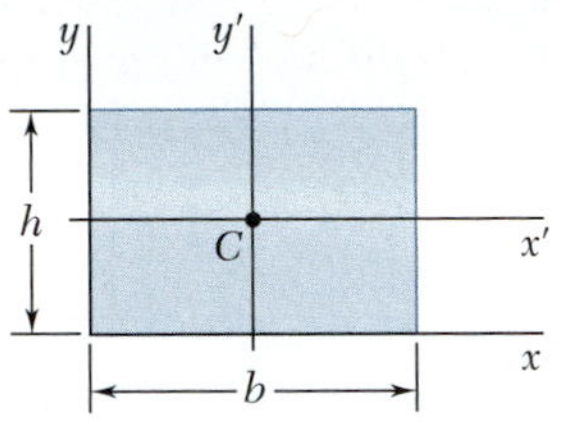	$\bar{I}_{x'} = \frac{1}{12}bh^3$ $\bar{I}_{y'} = \frac{1}{12}b^3h$ $I_x = \frac{1}{3}bh^3$ $I_y = \frac{1}{3}b^3h$ $J_C = \frac{1}{12}bh(b^2 + h^2)$
Triangle	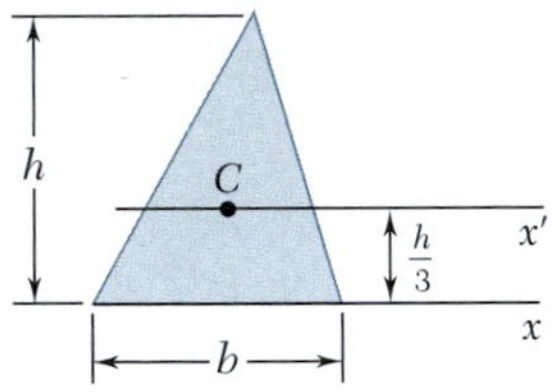	$\bar{I}_{x'} = \frac{1}{36}bh^3$ $I_x = \frac{1}{12}bh^3$
Circle	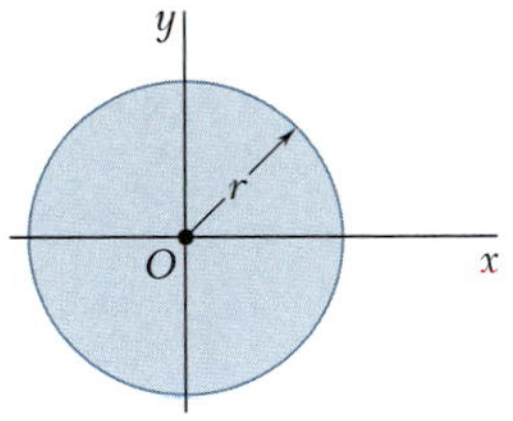	$\bar{I}_x = \bar{I}_y = \frac{1}{4}\pi r^4$ $J_O = \frac{1}{2}\pi r^4$
Semicircle	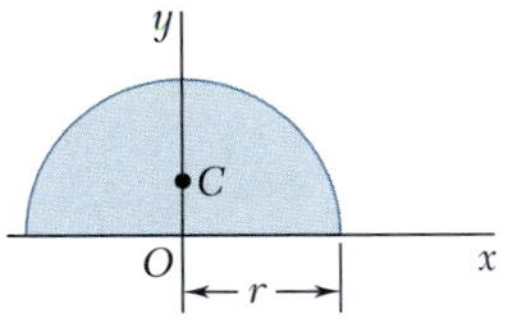	$I_x = I_y = \frac{1}{8}\pi r^4$ $J_O = \frac{1}{4}\pi r^4$
Quarter circle	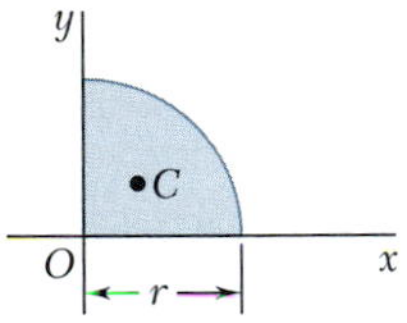	$I_x = I_y = \frac{1}{16}\pi r^4$ $J_O = \frac{1}{8}\pi r^4$
Ellipse	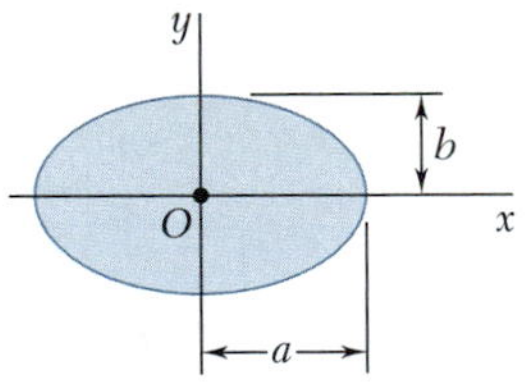	$\bar{I}_x = \frac{1}{4}\pi ab^3$ $\bar{I}_y = \frac{1}{4}\pi a^3b$ $J_O = \frac{1}{4}\pi ab(a^2 + b^2)$

Fig. 9.12 Moments of inertia of common geometric shapes.

It should be noted that the radius of gyration of a composite area is *not* equal to the sum of the radii of gyration of the component areas. In order to determine the radius of gyration of a composite area, it is first necessary to compute the moment of inertia of the area.

	Designation	Area in^2	Depth in.	Width in.	Axis X-X $\bar{I}_x$, in^4	$\bar{k}_x$, in.	$\bar{y}$, in.	Axis Y-Y $\bar{I}_y$, in^4	$\bar{k}_y$, in.	$\bar{x}$, in.
W Shapes (Wide-Flange Shapes) 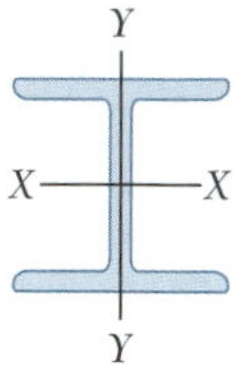	W18 × 50†	14.7	17.99	7.495	800	7.38		40.1	1.65	
	W16 × 40	11.8	16.01	6.995	518	6.63		28.9	1.57	
	W14 × 30	8.85	13.84	6.730	291	5.73		19.6	1.49	
	W8 × 24	7.08	7.93	6.495	82.8	3.42		18.3	1.61	
S Shapes (American Standard Shapes) 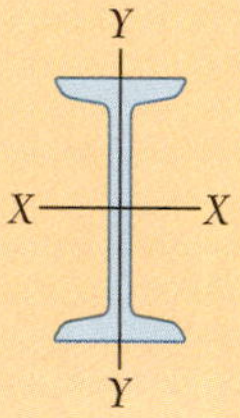	S18 × 70†	20.6	18.00	6.251	926	6.71		24.1	1.08	
	S12 × 50	14.7	12.00	5.477	305	4.55		15.7	1.03	
	S10 × 35	10.3	10.00	4.944	147	3.78		8.36	0.901	
	S6 × 17.25	5.07	6.00	3.565	26.3	2.28		2.31	0.675	
C Shapes (American Standard Channels) 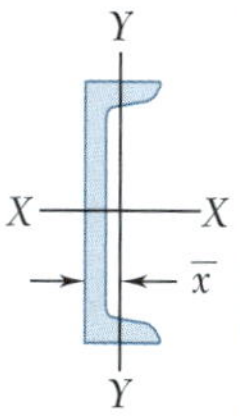	C12 × 25†	7.35	12.00	3.047	144	4.43		4.47	0.780	0.674
	C10 × 20	5.88	10.00	2.739	78.9	3.66		2.81	0.692	0.606
	C8 × 13.75	4.04	8.00	2.343	36.1	2.99		1.53	0.615	0.553
	C6 × 10.5	3.09	6.00	2.034	15.2	2.22		0.866	0.529	0.499
Angles 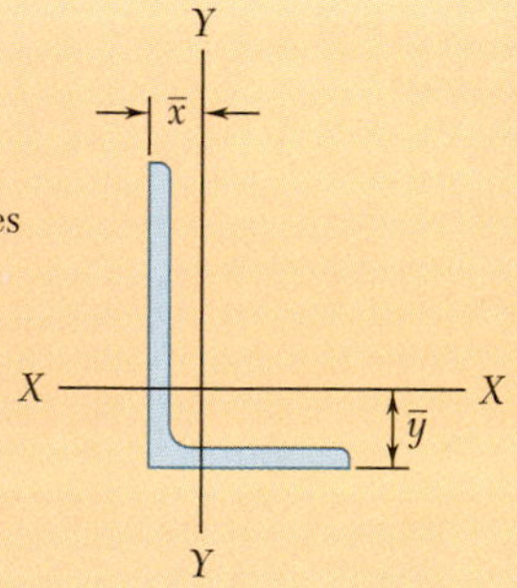	L6 × 6 × $\frac{3}{4}$‡	8.44			28.2	1.83	1.78	28.2	1.83	1.78
	L4 × 4 × $\frac{1}{2}$	3.75			5.56	1.22	1.18	5.56	1.22	1.18
	L3 × 3 × $\frac{1}{4}$	1.44			1.24	0.930	0.842	1.24	0.930	0.842
	L6 × 4 × $\frac{1}{2}$	4.75			17.4	1.91	1.99	6.27	1.15	0.987
	L5 × 3 × $\frac{1}{2}$	3.75			9.45	1.59	1.75	2.58	0.829	0.750
	L3 × 2 × $\frac{1}{4}$	1.19			1.09	0.957	0.993	0.392	0.574	0.493

Fig. 9.13A Properties of Rolled-Steel Shapes (U.S. Customary Units).*

*Courtesy of the American Institute of Steel Construction, Chicago, Illinois

†Nominal depth in inches and weight in pounds per foot

‡Depth, width, and thickness in inches

	Designation	Area mm^2	Depth mm	Width mm	Axis X-X: $\bar{I}_x$ 10^6 mm^4	Axis X-X: $\bar{k}_x$ mm	Axis X-X: $\bar{y}$ mm	Axis Y-Y: $\bar{I}_y$ 10^6 mm^4	Axis Y-Y: $\bar{k}_y$ mm	Axis Y-Y: $\bar{x}$ mm
W Shapes (Wide-Flange Shapes) 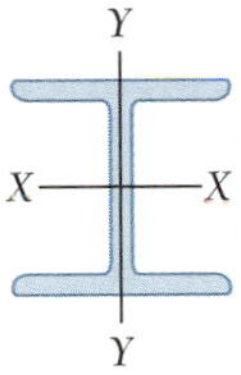	W460 × 74†	9450	457	190	333	188		16.6	41.9	
	W410 × 60	7580	407	178	216	169		12.1	40.0	
	W360 × 44	5730	352	171	122	146		8.18	37.8	
	W200 × 35.9	4580	201	165	34.4	86.7		7.64	40.8	
S Shapes (American Standard Shapes) 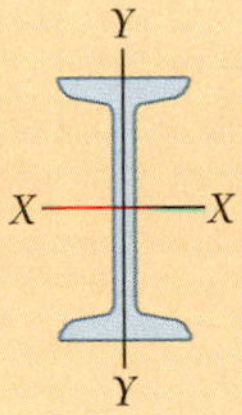	S460 × 104†	13300	457	159	385	170		10.4	27.5	
	S310 × 74	9480	305	139	126	115		6.69	26.1	
	S250 × 52	6670	254	126	61.2	95.8		3.59	22.9	
	S150 × 25.7	3270	152	91	10.8	57.5		1.00	17.2	
C Shapes (American Standard Channels) 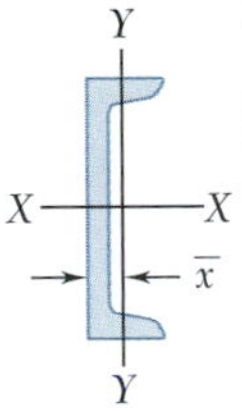	C310 × 37†	5690	305	77	59.7	112		1.83	19.7	17.0
	C250 × 30	3780	254	69	32.6	92.9		1.14	17.4	15.3
	C200 × 27.9	3560	203	64	18.2	71.5		0.817	15.1	14.3
	C150 × 15.6	1980	152	51	6.21	56.0		0.347	13.2	12.5
Angles 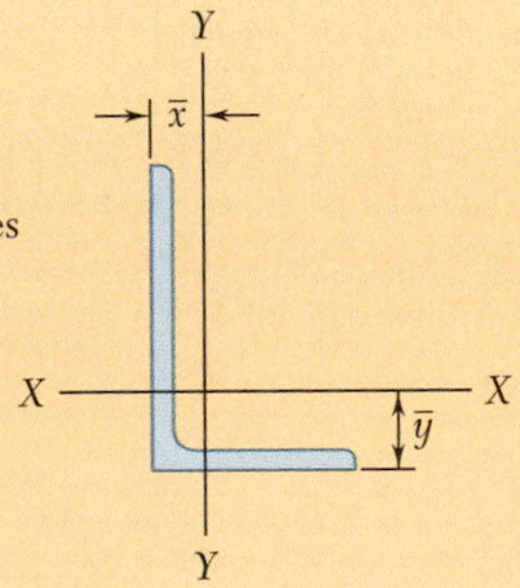	L152 × 152 × 19.0‡	5420			11.6	46.3	44.9	11.6	46.3	44.9
	L102 × 102 × 12.7	2430			2.34	31.0	30.2	2.34	31.0	30.2
	L76 × 76 × 6.4	932			0.517	23.6	21.4	0.517	23.6	21.4
	L152 × 102 × 12.7	3060			7.20	48.5	50.3	2.64	29.4	25.3
	L127 × 76 × 12.7	2420			3.93	40.3	44.4	1.06	20.9	19.0
	L76 × 51 × 6.4	772			0.453	24.2	25.1	0.166	14.7	12.6

Fig. 9.13B Properties of Rolled-Steel Shapes (SI Units).

†Nominal depth in millimeters and mass in kilograms per meter

‡Depth, width, and thickness in millimeters

SAMPLE PROBLEM 9.4

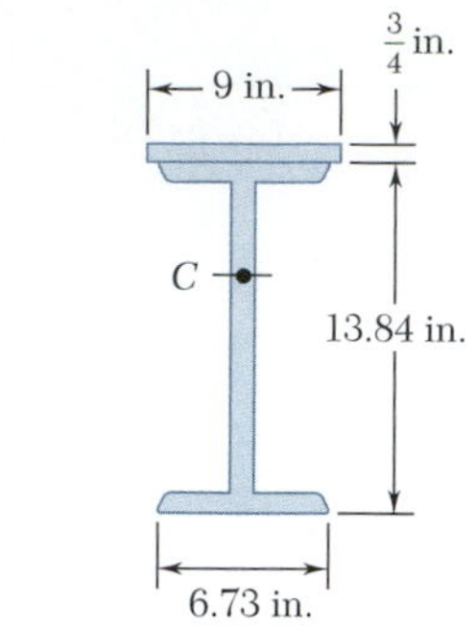

The strength of a W14 × 30 rolled-steel beam is increased by attaching a 9 × $\frac{3}{4}$-in. plate to its upper flange as shown. Determine the moment of inertia and the radius of gyration of the composite section with respect to an axis which is parallel to the plate and passes through the centroid C of the section.

SOLUTION

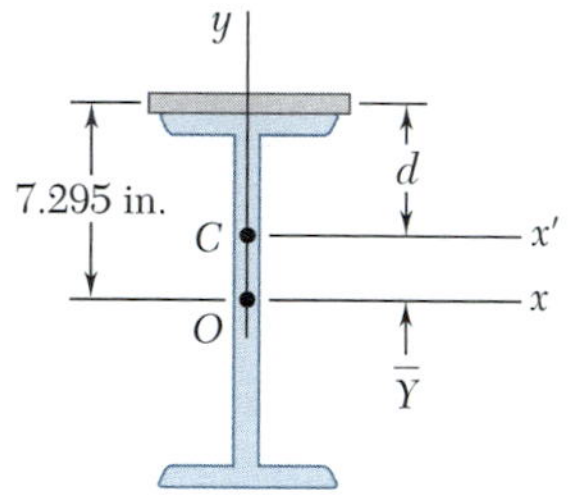

The origin O of the coordinates is placed at the centroid of the wide-flange shape, and the distance $\overline{Y}$ to the centroid of the composite section is computed using the methods of Chap. 5. The area of the wide-flange shape is found by referring to Fig. 9.13A. The area and the y coordinate of the centroid of the plate are

$$A = (9 \text{ in.})(0.75 \text{ in.}) = 6.75 \text{ in}^2$$
$$\overline{y} = \tfrac{1}{2}(13.84 \text{ in.}) + \tfrac{1}{2}(0.75 \text{ in.}) = 7.295 \text{ in.}$$

Section	Area, in^2	$\overline{y}$, in.	$\overline{y}A$, in^3
Plate	6.75	7.295	49.24
Wide-flange shape	8.85	0	0
	$\Sigma A = 15.60$		$\Sigma\overline{y}A = 49.24$

$$\overline{Y}\Sigma A = \Sigma\overline{y}A \qquad \overline{Y}(15.60) = 49.24 \qquad \overline{Y} = 3.156 \text{ in.}$$

Moment of Inertia. The parallel-axis theorem is used to determine the moments of inertia of the wide-flange shape and the plate with respect to the x' axis. This axis is a centroidal axis for the composite section but *not* for either of the elements considered separately. The value of $\overline{I}_x$ for the wide-flange shape is obtained from Fig. 9.13A.

For the wide-flange shape,

$$I_{x'} = \overline{I}_x + A\overline{Y}^2 = 291 + (8.85)(3.156)^2 = 379.1 \text{ in}^4$$

For the plate,

$$I_{x'} = \overline{I}_x + Ad^2 = (\tfrac{1}{12})(9)(\tfrac{3}{4})^3 + (6.75)(7.295 - 3.156)^2 = 116.0 \text{ in}^4$$

For the composite area,

$$I_{x'} = 379.1 + 116.0 = 495.1 \text{ in}^4 \qquad I_{x'} = 495 \text{ in}^4 \blacktriangleleft$$

Radius of Gyration. We have

$$k_{x'}^2 = \frac{I_{x'}}{A} = \frac{495.1 \text{ in}^4}{15.60 \text{ in}^2} \qquad k_{x'} = 5.63 \text{ in.} \blacktriangleleft$$

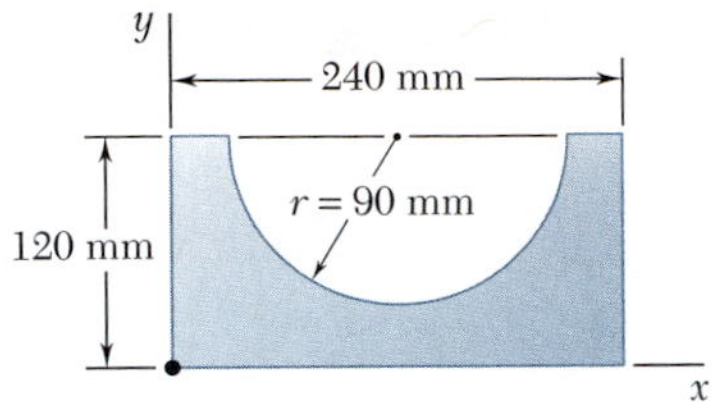

SAMPLE PROBLEM 9.5

Determine the moment of inertia of the shaded area with respect to the x axis.

SOLUTION

The given area can be obtained by subtracting a half circle from a rectangle. The moments of inertia of the rectangle and the half circle will be computed separately.

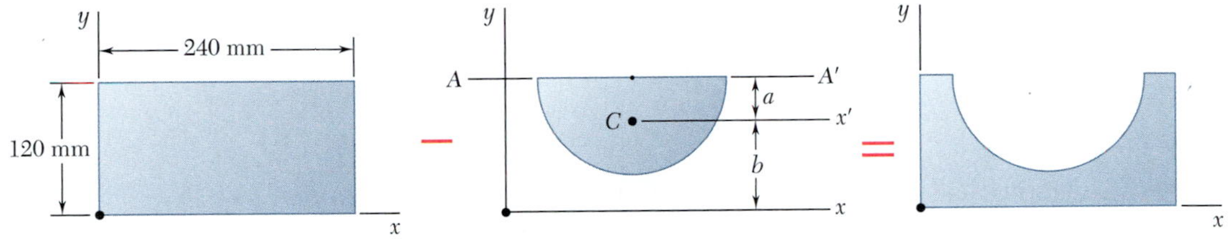

Moment of Inertia of Rectangle. Referring to Fig. 9.12, we obtain

$$I_x = \tfrac{1}{3}bh^3 = \tfrac{1}{3}(240 \text{ mm})(120 \text{ mm})^3 = 138.2 \times 10^6 \text{ mm}^4$$

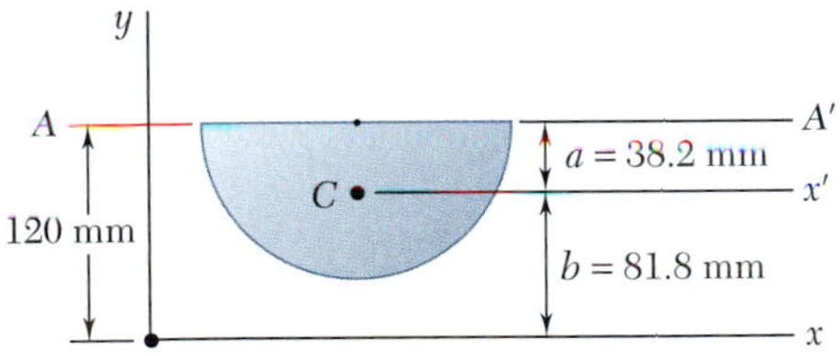

Moment of Inertia of Half Circle. Referring to Fig. 5.8, we determine the location of the centroid C of the half circle with respect to diameter AA'.

$$a = \frac{4r}{3\pi} = \frac{(4)(90 \text{ mm})}{3\pi} = 38.2 \text{ mm}$$

The distance b from the centroid C to the x axis is

$$b = 120 \text{ mm} - a = 120 \text{ mm} - 38.2 \text{ mm} = 81.8 \text{ mm}$$

Referring now to Fig. 9.12, we compute the moment of inertia of the half circle with respect to diameter AA'; we also compute the area of the half circle.

$$I_{AA'} = \tfrac{1}{8}\pi r^4 = \tfrac{1}{8}\pi(90 \text{ mm})^4 = 25.76 \times 10^6 \text{ mm}^4$$
$$A = \tfrac{1}{2}\pi r^2 = \tfrac{1}{2}\pi(90 \text{ mm})^2 = 12.72 \times 10^3 \text{ mm}^2$$

Using the parallel-axis theorem, we obtain the value of $\bar{I}_{x'}$:

$$I_{AA'} = \bar{I}_{x'} + Aa^2$$
$$25.76 \times 10^6 \text{ mm}^4 = \bar{I}_{x'} + (12.72 \times 10^3 \text{ mm}^2)(38.2 \text{ mm})^2$$
$$\bar{I}_{x'} = 7.20 \times 10^6 \text{ mm}^4$$

Again using the parallel-axis theorem, we obtain the value of I_x:

$$I_x = \bar{I}_{x'} + Ab^2 = 7.20 \times 10^6 \text{ mm}^4 + (12.72 \times 10^3 \text{ mm}^2)(81.8 \text{ mm})^2$$
$$= 92.3 \times 10^6 \text{ mm}^4$$

Moment of Inertia of Given Area. Subtracting the moment of inertia of the half circle from that of the rectangle, we obtain

$$I_x = 138.2 \times 10^6 \text{ mm}^4 - 92.3 \times 10^6 \text{ mm}^4$$

$$I_x = 45.9 \times 10^6 \text{ mm}^4 \quad ◀$$

SOLVING PROBLEMS ON YOUR OWN

In this lesson we introduced the *parallel-axis theorem* and illustrated how it can be used to simplify the computation of moments and polar moments of inertia of composite areas. The areas that you will consider in the following problems will consist of common shapes and rolled-steel shapes. You will also use the parallel-axis theorem to locate the point of application (the center of pressure) of the resultant of the hydrostatic forces acting on a submerged plane area.

1. *Applying the parallel-axis theorem.* In Sec. 9.6 we derived the parallel-axis theorem

$$I = \bar{I} + Ad^2 \qquad (9.9)$$

which states that the moment of inertia I of an area A with respect to a given axis is equal to the sum of the moment of inertia $\bar{I}$ of that area with respect to the *parallel centroidal axis* and the product Ad^2, where d is the distance between the two axes. It is important that you remember the following points as you use the parallel-axis theorem.

a. *The centroidal moment of inertia $\bar{I}$ of an area A can be obtained by subtracting the product Ad^2* from the moment of inertia I of the area with respect to a parallel axis. Therefore, as we noted in Example 2 and illustrated in Sample Prob. 9.5, it follows that the moment of inertia $\bar{I}$ is *smaller* than the moment of inertia I of the same area with respect to any parallel axis.

b. *The parallel-axis theorem can be applied only if one of the two axes involved is a centroidal axis.* Therefore, as we noted in Example 2, to compute the moment of inertia of an area with respect to a *noncentroidal axis* when the moment of inertia of the area is known with respect to *another noncentroidal axis*, it is necessary to *first compute* the moment of inertia of the area with respect to the *centroidal axis parallel to the two given axes*.

2. *Computing the moments and polar moments of inertia of composite areas.* Sample Probs. 9.4 and 9.5 illustrate the steps you should follow to solve problems of this type. As with all composite-area problems, you should show on your sketch the common shapes or rolled-steel shapes that constitute the various elements of the given area, as well as the distances between the centroidal axes of the elements and the axes about which the moments of inertia are to be computed. In addition, it is important that the following points be noted.

a. The moment of inertia of an area is always positive, regardless of the location of the axis with respect to which it is computed. As pointed out in the comments for the preceding lesson, it is only when an area is *removed* (as in the case of a hole) that its moment of inertia should be entered in your computations with a minus sign.

b. The moments of inertia of a semiellipse and a quarter ellipse can be determined by dividing the moment of inertia of an ellipse by 2 and 4, respectively. It should be noted, however, that the moments of inertia obtained in this manner are *with respect to the axes of symmetry of the ellipse*. To obtain the *centroidal* moments of inertia of these shapes, the parallel-axis theorem should be used. Note that this remark also applies to a semicircle and to a quarter circle and that the expressions given for these shapes in Fig. 9.12 are *not* centroidal moments of inertia.

c. To calculate the polar moment of inertia of a composite area, you can use either the expressions given in Fig. 9.12 for J_O or the relationship

$$J_O = I_x + I_y \tag{9.4}$$

depending on the shape of the given area.

d. Before computing the centroidal moments of inertia of a given area, you may find it necessary to first locate the centroid of the area using the methods of Chap. 5.

3. ***Locating the point of application of the resultant of a system of hydrostatic forces.*** In Sec. 9.2 we found that

$$R = \gamma \int y \, dA = \gamma \bar{y} A$$
$$M_x = \gamma \int y^2 \, dA = \gamma I_x$$

where $\bar{y}$ is the distance from the x axis to the centroid of the submerged plane area. Since $\mathbf{R}$ is equivalent to the system of elemental hydrostatic forces, it follows that

$$\Sigma M_x: \qquad y_P R = M_x$$

where y_P is the depth of the point of application of $\mathbf{R}$. Then

$$y_P(\gamma \bar{y} A) = \gamma I_x \qquad \text{or} \qquad y_P = \frac{I_x}{\bar{y} A}$$

In closing, we encourage you to carefully study the notation used in Fig. 9.13 for the rolled-steel shapes, as you will likely encounter it again in subsequent engineering courses.

Problems

9.31 and 9.32 Determine the moment of inertia and the radius of gyration of the shaded area with respect to the x axis.

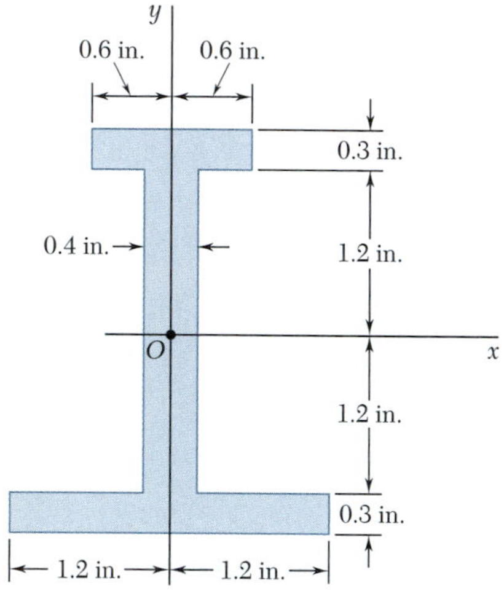

Fig. P9.31 and P9.33

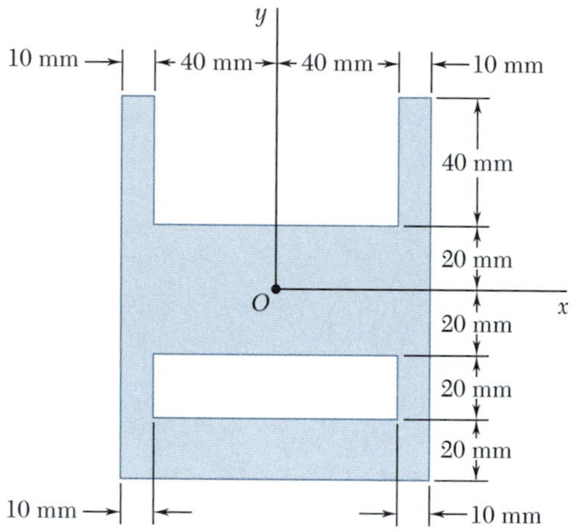

Fig. P9.32 and *P9.34*

9.33 and *9.34* Determine the moment of inertia and the radius of gyration of the shaded area with respect to the y axis.

9.35 and *9.36* Determine the moments of inertia of the shaded area shown with respect to the x and y axes.

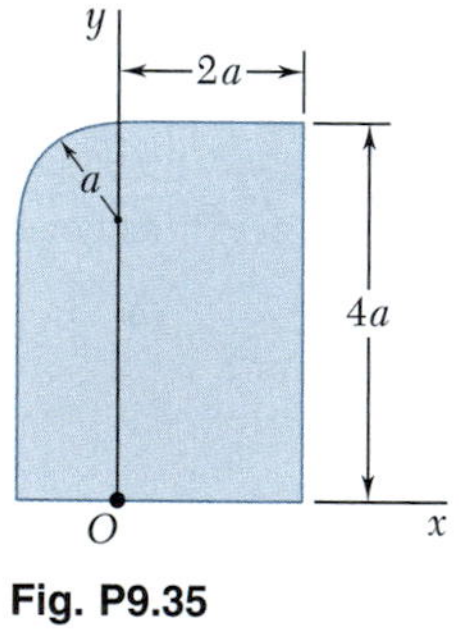

Fig. P9.35

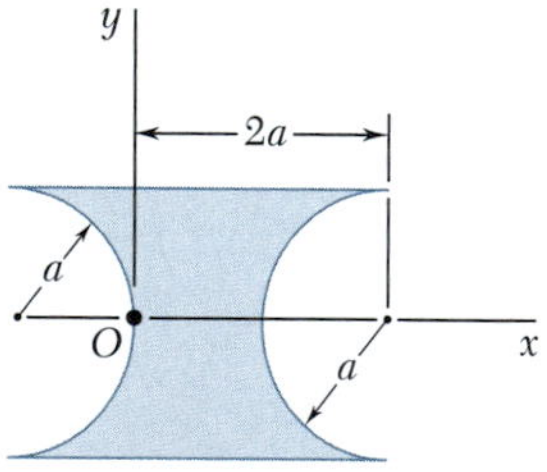

Fig. *P9.36*

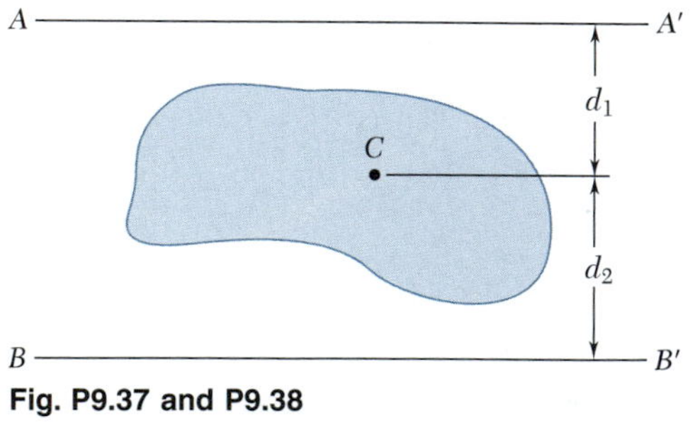

Fig. P9.37 and P9.38

9.37 For the 6-in^2 shaded area shown, determine the distance d_2 and the moment of inertia with respect to the centroidal axis parallel to AA' knowing that the moments of inertia with respect to AA' and BB' are 30 in^4 and 58 in^4, respectively, and that $d_1 = 1.25$ in.

9.38 Determine for the shaded region the area and the moment of inertia with respect to the centroidal axis parallel to BB' knowing that $d_1 = 1.25$ in. and $d_2 = 0.75$ in. and that the moments of inertia with respect to AA' and BB' are 20 in^4 and 15 in^4, respectively.

9.39 The centroidal polar moment of inertia $\bar{J}_C$ of the 15.5×10^3 mm^2 shaded region is 250×10^6 mm^4. Determine the polar moments of inertia J_B and J_D of the shaded region knowing that $J_D = 2J_B$ and $d = 100$ mm.

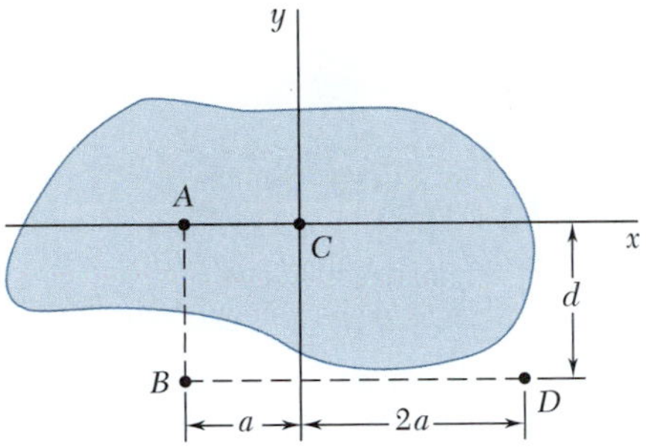

Fig. P9.39 and P9.40

9.40 Determine the centroidal polar moment of inertia $\bar{J}_C$ of the 10×10^3 mm^2 shaded area knowing that the polar moments of inertia of the area with respect to points A, B, and D are $J_A = 45 \times 10^6$ mm^4, $J_B = 130 \times 10^6$ mm^4, and $J_D = 252 \times 10^6$ mm^4, respectively.

9.41 through 9.44 Determine the moments of inertia $\bar{I}_x$ and $\bar{I}_y$ of the area shown with respect to centroidal axes respectively parallel and perpendicular to side AB.

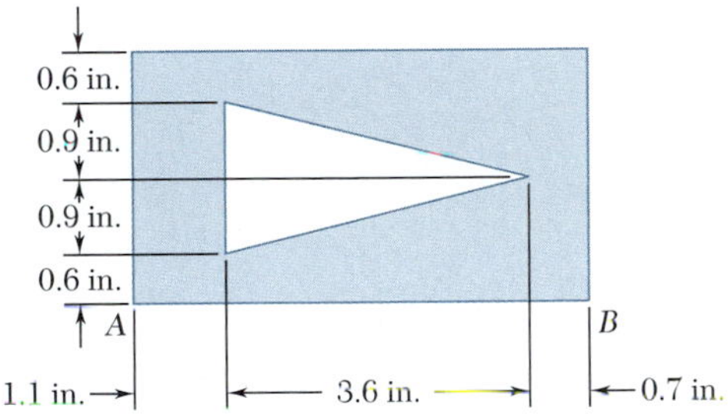

Fig. P9.41

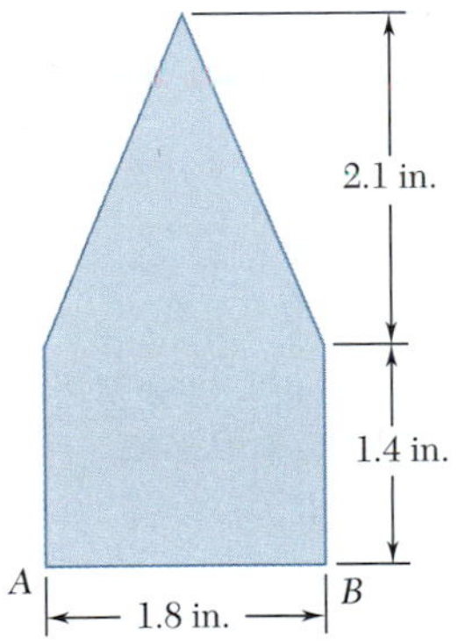

Fig. *P9.42*

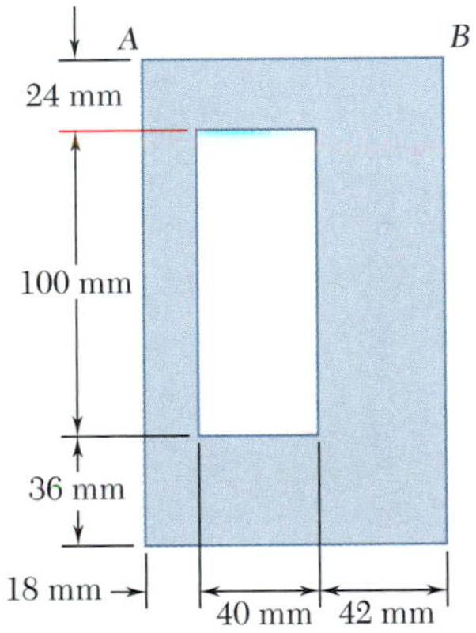

Fig. P9.43

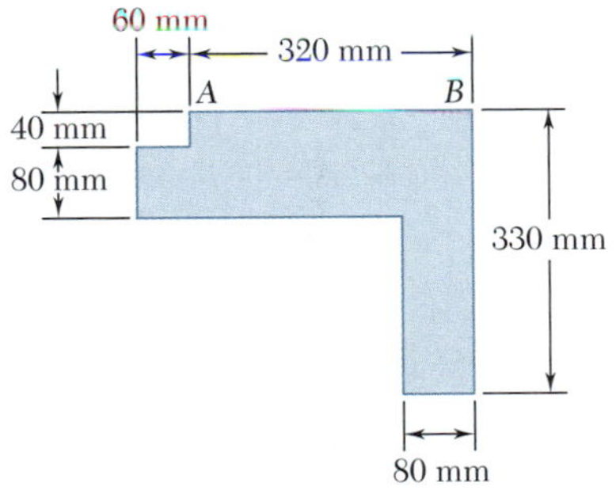

Fig. P9.44

***9.45* and 9.46** Determine the polar moment of inertia of the area shown with respect to (a) point O, (b) the centroid of the area.

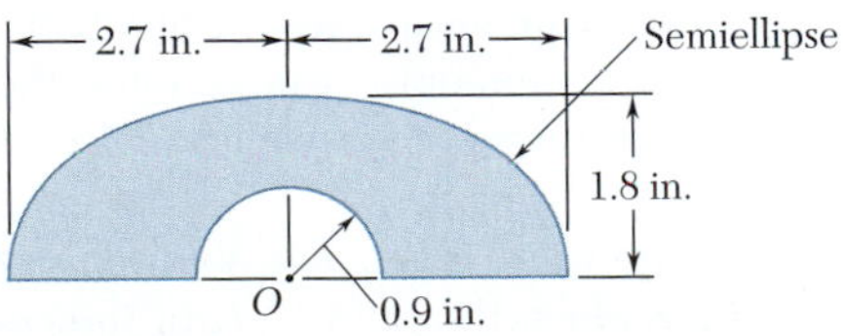

Fig. *P9.45*

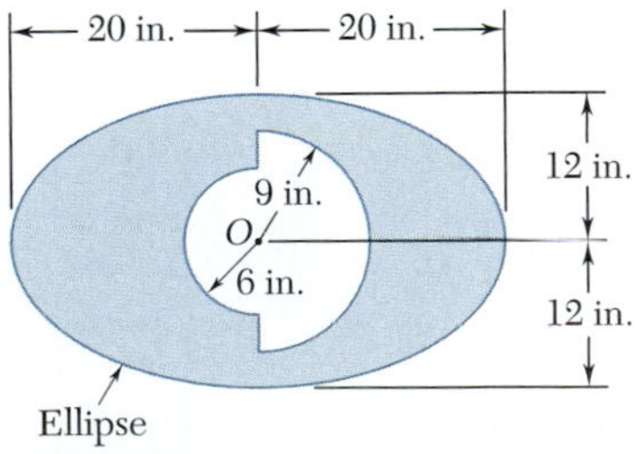

Fig. P9.46

9.47 and 9.48 Determine the polar moment of inertia of the area shown with respect to (*a*) point *O*, (*b*) the centroid of the area.

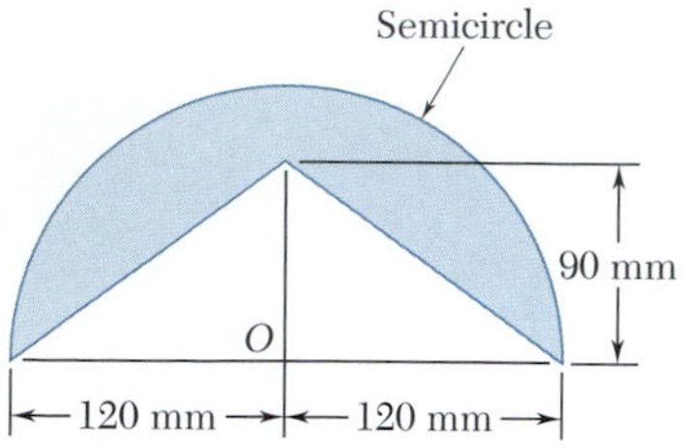

Fig. P9.47

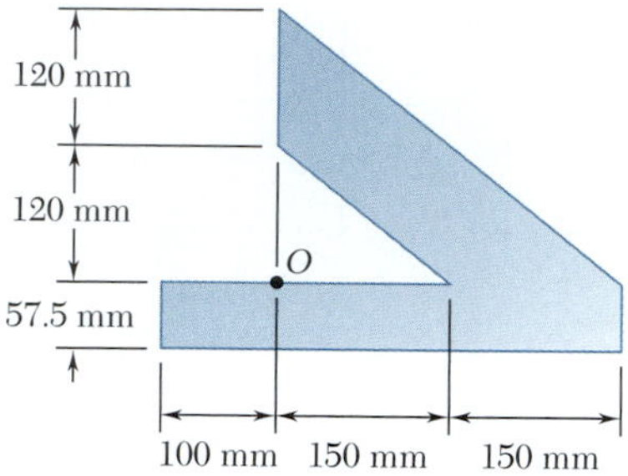

Fig. P9.48

9.49 Two 1-in. steel plates are welded to a rolled S section as shown. Determine the moments of inertia and the radii of gyration of the section with respect to the centroidal *x* and *y* axes.

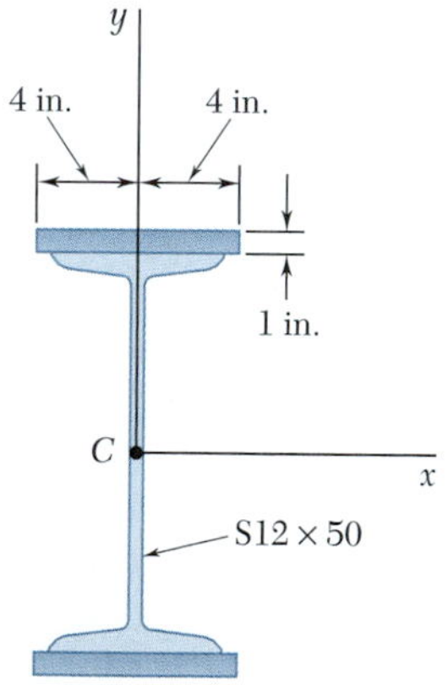

Fig. P9.49

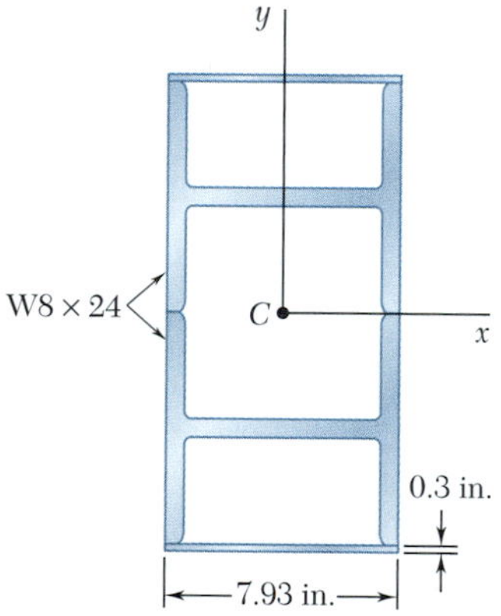

Fig. *P9.50*

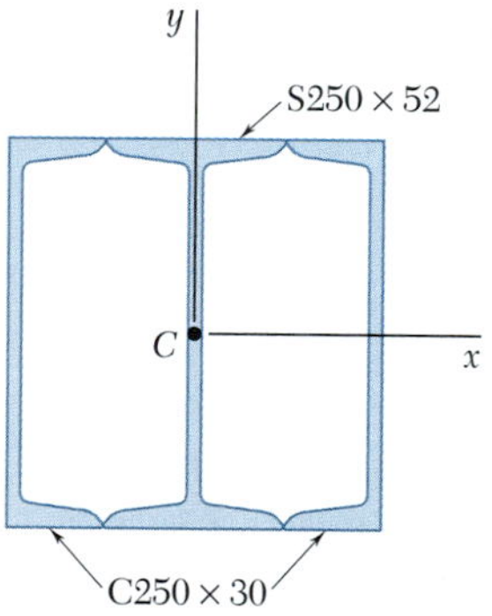

Fig. P9.51

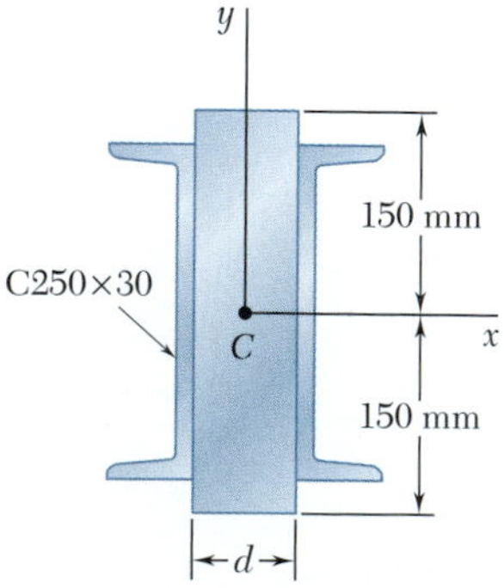

Fig. P9.52

9.50 To form a reinforced box section, two rolled W sections and two plates are welded together. Determine the moments of inertia and the radii of gyration of the combined section with respect to the centroidal axes shown.

9.51 Two C250 × 30 channels are welded to a 250 × 52 rolled S section as shown. Determine the moments of inertia and the radii of gyration of the combined section with respect to its centroidal *x* and *y* axes.

9.52 Two channels are welded to a $d \times 300$-mm steel plate as shown. Determine the width d for which the ratio $\bar{I}_x/\bar{I}_y$ of the centroidal moments of inertia of the section is 16.

9.53 Two L3 × 3 × $\frac{1}{4}$-in. angles are welded to a C10 × 20 channel. Determine the moments of inertia of the combined section with respect to centroidal axes respectively parallel and perpendicular to the web of the channel.

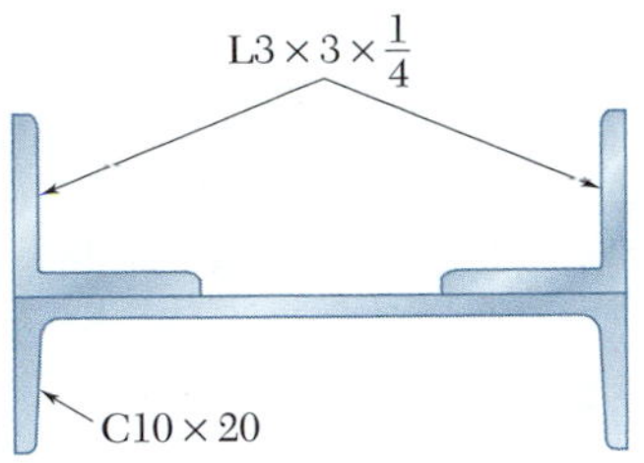

Fig. P9.53

9.54 To form an unsymmetrical girder, two L3 × 3 × $\frac{1}{4}$-in. angles and two L6 × 4 × $\frac{1}{2}$-in. angles are welded to a 0.8-in. steel plate as shown. Determine the moments of inertia of the combined section with respect to its centroidal x and y axes.

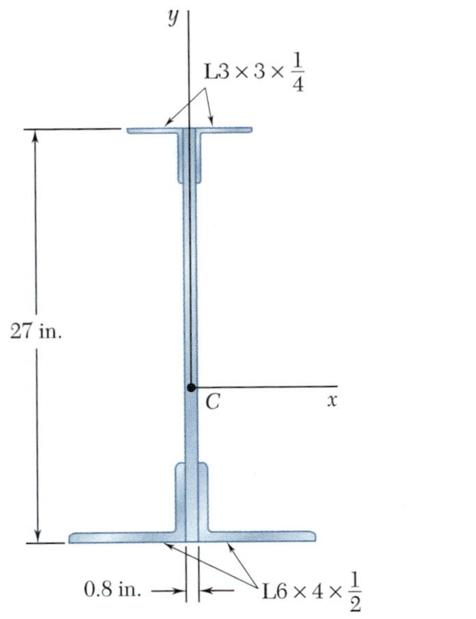

Fig. *P9.54*

9.55 Two L127 × 76 × 12.7-mm angles are welded to a 10-mm steel plate. Determine the distance b and the centroidal moments of inertia $\bar{I}_x$ and $\bar{I}_y$ of the combined section knowing that $\bar{I}_y = 3\bar{I}_x$.

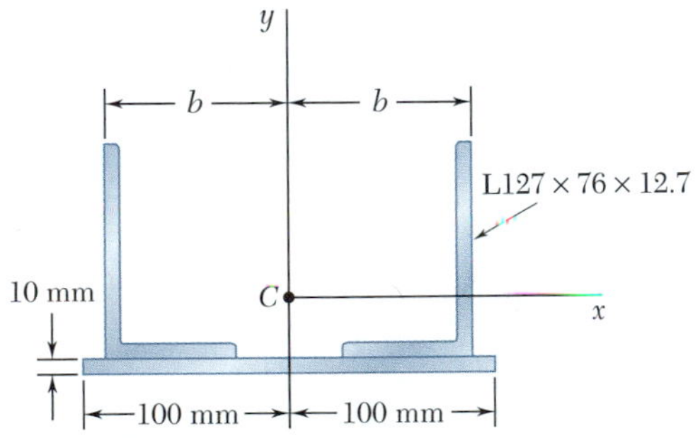

Fig. P9.55

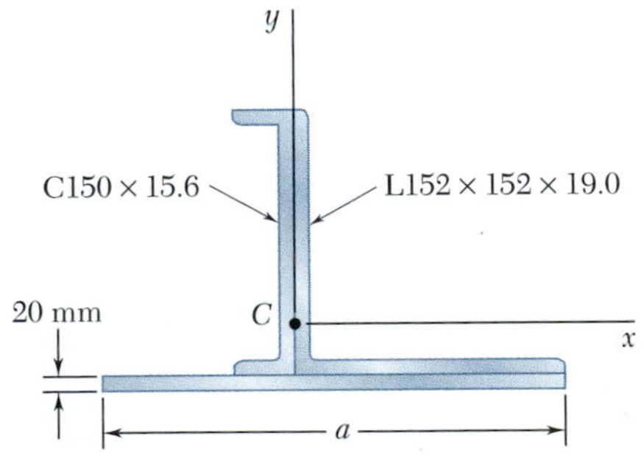

Fig. P9.56

9.56 A channel and an angle are welded to an a × 20-mm steel plate. Knowing that the centroidal y axis is located as shown, determine (a) the width a, (b) the moments of inertia with respect to the centroidal x and y axes.

9.57 and *9.58* The panel shown forms the end of a trough which is filled with water to the line AA'. Referring to Sec. 9.2, determine the depth of the point of application of the resultant of the hydrostatic forces acting on the panel (the center of pressure).

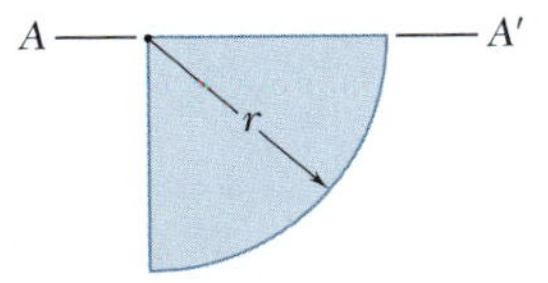

Fig. P9.57

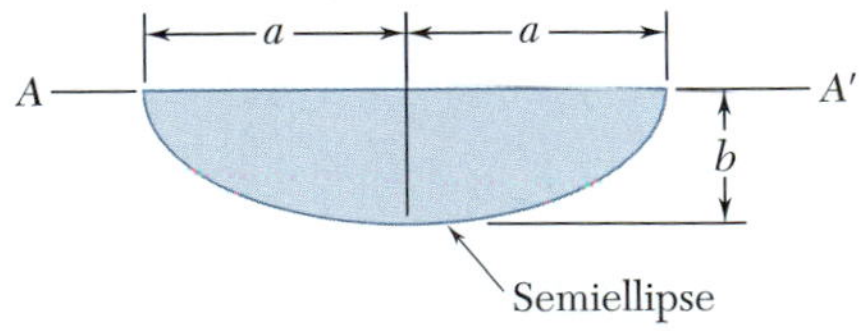

Fig. *P9.58*

9.59 and 9.60 The panel shown forms the end of a trough which is filled with water to the line AA'. Referring to Sec. 9.2, determine the depth of the point of application of the resultant of the hydrostatic forces acting on the panel (the center of pressure).

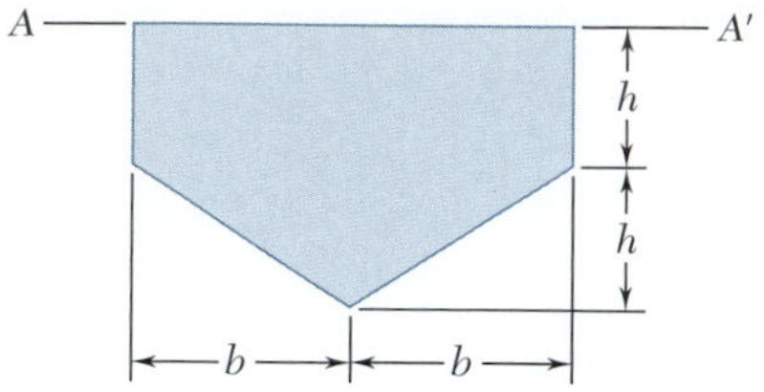

Fig. *P9.59*

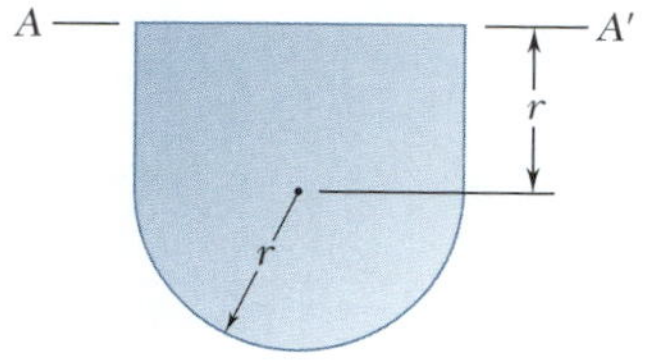

Fig. P9.60

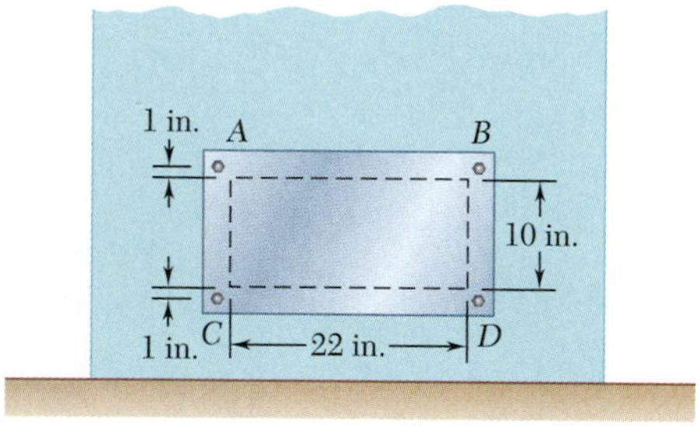

Fig. P9.61

9.61 The cover for a 10 × 22-in. access hole in an oil storage tank is attached to the outside of the tank with four bolts as shown. Knowing that the specific weight of the oil is 57.4 lb/ft^3 and that the center of the cover is located 10 ft below the surface of the oil, determine the additional force on each bolt because of the pressure of the oil.

9.62 A vertical trapezoidal gate that is used as an automatic valve is held shut by two springs attached to hinges located along edge AB. Knowing that each spring exerts a couple of magnitude 8.50 kN · m, determine the depth d of water for which the gate will open.

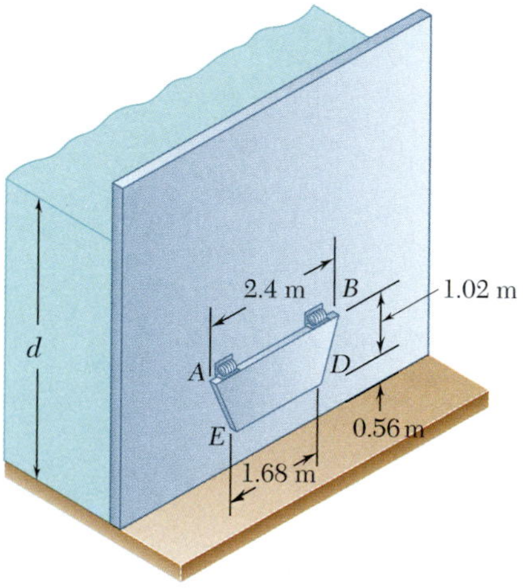

Fig. P9.62

***9.63** Determine the x coordinate of the centroid of the volume shown. (*Hint:* The height y of the volume is proportional to the x coordinate; consider an analogy between this height and the water pressure on a submerged surface.)

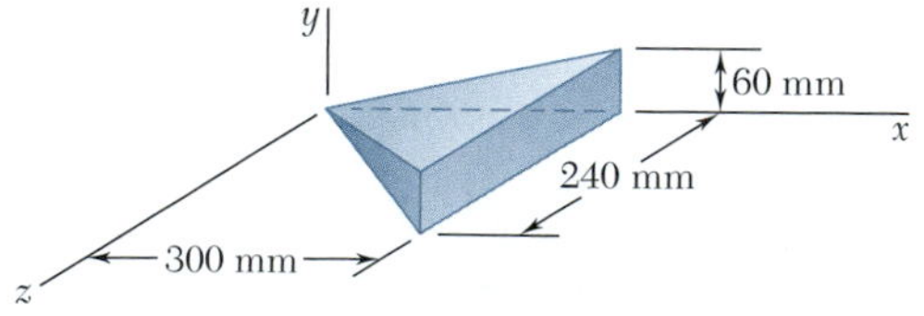

Fig. P9.63

***9.64** Determine the x coordinate of the centroid of the volume shown; this volume was obtained by intersecting an elliptic cylinder with an oblique plane. (*Hint:* The height y of the volume is proportional to the x coordinate; consider an analogy between this height and the water pressure on a submerged surface.)

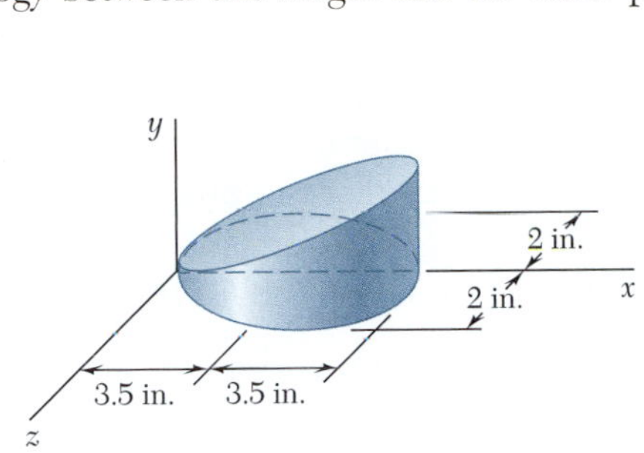

Fig. P9.64

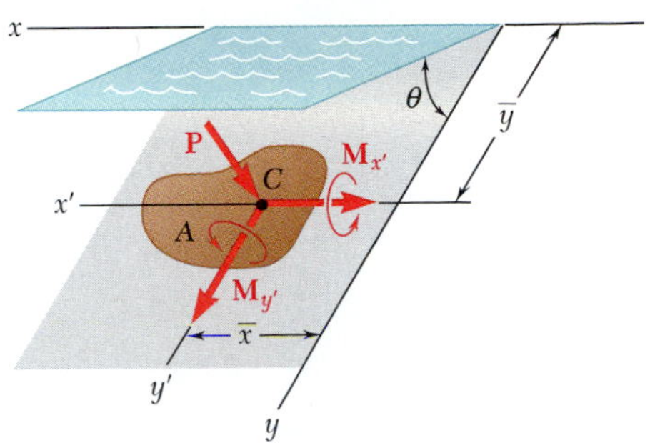

Fig. P9.65

***9.65** Show that the system of hydrostatic forces acting on a submerged plane area A can be reduced to a force $\mathbf{P}$ at the centroid C of the area and two couples. The force $\mathbf{P}$ is perpendicular to the area and is of magnitude $P = \gamma A\bar{y} \sin \theta$, where γ is the specific weight of the liquid, and the couples are $\mathbf{M}_{x'} = (\gamma \bar{I}_{x'} \sin \theta)\mathbf{i}$ and $\mathbf{M}_{y'} = (\gamma \bar{I}_{x'y'} \sin \theta)\mathbf{j}$, where $\bar{I}_{x'y'} = \int x'y'dA$ (see Sec. 9.8). Note that the couples are independent of the depth at which the area is submerged.

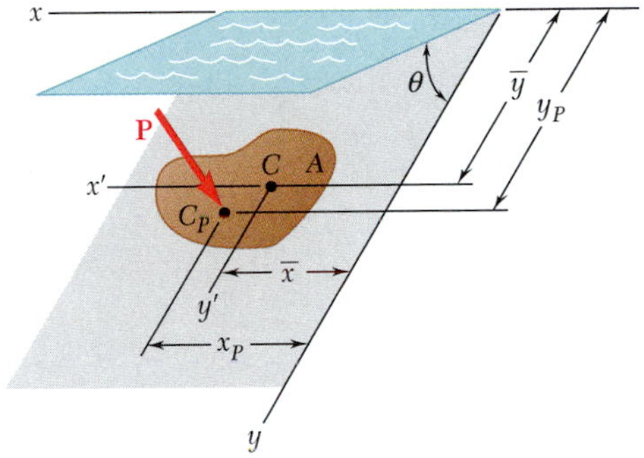

Fig. P9.66

***9.66** Show that the resultant of the hydrostatic forces acting on a submerged plane area A is a force $\mathbf{P}$ perpendicular to the area and of magnitude $P = \gamma A\bar{y} \sin \theta = \bar{p}A$, where γ is the specific weight of the liquid and $\bar{p}$ is the pressure at the centroid C of the area. Show that $\mathbf{P}$ is applied at a point C_p, called the center of pressure, whose coordinates are $x_p = I_{xy}/A\bar{y}$ and $y_p = I_x/A\bar{y}$, where $\bar{I}_{xy} = \int xy\, dA$ (see Sec. 9.8). Show also that the difference of ordinates $y_p - \bar{y}$ is equal to $\bar{k}_{x'}^2/\bar{y}$ and thus depends upon the depth at which the area is submerged.

*9.8. PRODUCT OF INERTIA

The integral

$$I_{xy} = \int xy\, dA \tag{9.12}$$

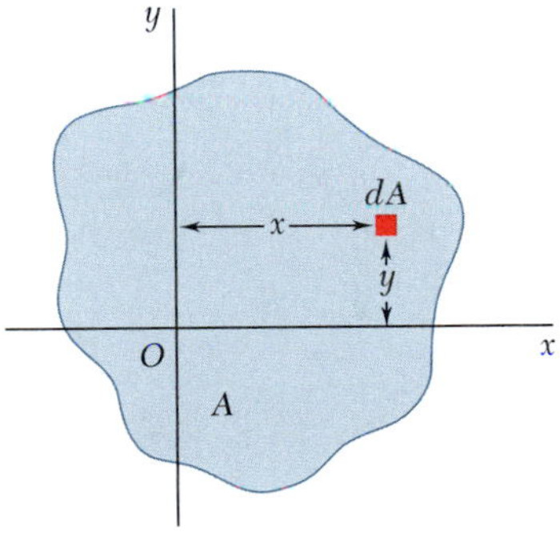

Fig. 9.14

which is obtained by multiplying each element dA of an area A by its coordinates x and y and integrating over the area (Fig. 9.14), is known as the *product of inertia* of the area A with respect to the x and y axes. Unlike the moments of inertia I_x and I_y, the product of inertia I_{xy} can be positive, negative, or zero.

When one or both of the x and y axes are axes of symmetry for the area A, the product of inertia I_{xy} is zero. Consider, for example, the channel section shown in Fig. 9.15. Since this section is symmetrical with respect to the x axis, we can associate with each element dA of coordinates x and y an element dA' of coordinates x and $-y$. Clearly, the contributions to I_{xy} of any pair of elements chosen in this way cancel out, and the integral (9.12) reduces to zero.

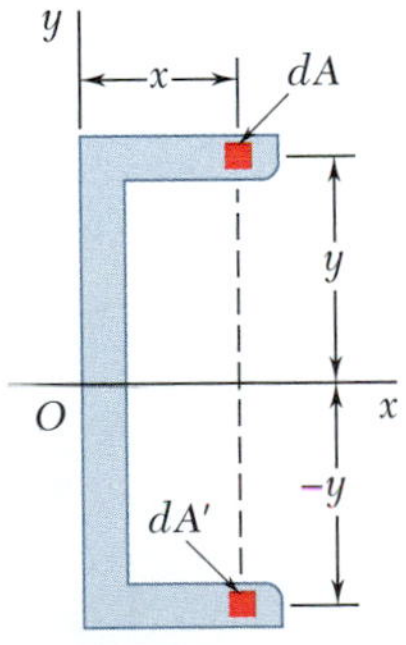

Fig. 9.15

A parallel-axis theorem similar to the one established in Sec. 9.6 for moments of inertia can be derived for products of inertia.

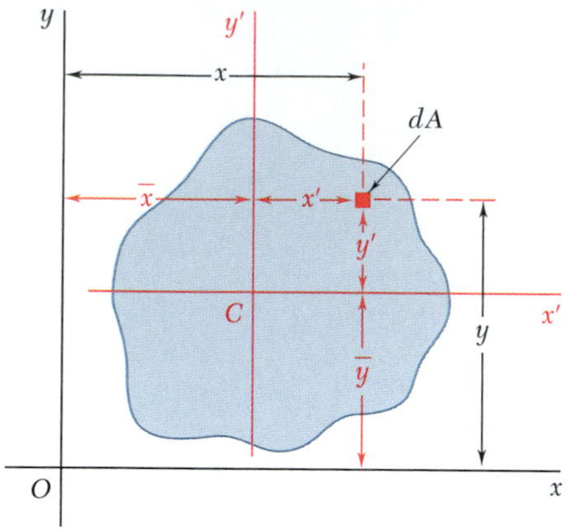

Fig. 9.16

Consider an area A and a system of rectangular coordinates x and y (Fig. 9.16). Through the centroid C of the area, of coordinates $\bar{x}$ and $\bar{y}$, we draw two *centroidal axes* x' and y' which are parallel, respectively, to the x and y axes. Denoting by x and y the coordinates of an element of area dA with respect to the original axes, and by x' and y' the coordinates of the same element with respect to the centroidal axes, we write $x = x' + \bar{x}$ and $y = y' + \bar{y}$. Substituting into (9.12), we obtain the following expression for the product of inertia I_{xy}:

$$I_{xy} = \int xy\, dA = \int (x' + \bar{x})(y' + \bar{y})\, dA$$
$$= \int x'y'\, dA + \bar{y} \int x'\, dA + \bar{x} \int y'\, dA + \bar{x}\bar{y} \int dA$$

The first integral represents the product of inertia $\bar{I}_{x'y'}$ of the area A with respect to the centroidal axes x' and y'. The next two integrals represent first moments of the area with respect to the centroidal axes; they reduce to zero, since the centroid C is located on these axes. Finally, we observe that the last integral is equal to the total area A. Therefore, we have

$$I_{xy} = \bar{I}_{x'y'} + \bar{x}\bar{y}A \qquad (9.13)$$

*9.9. PRINCIPAL AXES AND PRINCIPAL MOMENTS OF INERTIA

Consider the area A and the coordinate axes x and y (Fig. 9.17). Assuming that the moments and product of inertia

$$I_x = \int y^2\, dA \qquad I_y = \int x^2\, dA \qquad I_{xy} = \int xy\, dA \qquad (9.14)$$

of the area A are known, we propose to determine the moments and product of inertia $I_{x'}$, $I_{y'}$, and $I_{x'y'}$ of A with respect to new axes x' and y' which are obtained by rotating the original axes about the origin through an angle θ.

We first note the following relations between the coordinates x', y' and x, y of an element of area dA:

$$x' = x \cos\theta + y \sin\theta \qquad y' = y \cos\theta - x \sin\theta$$

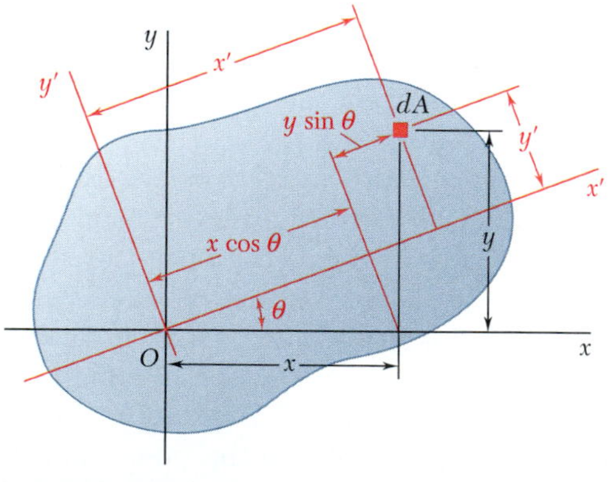

Fig. 9.17

Substituting for y' in the expression for $I_{x'}$, we write

$$I_{x'} = \int (y')^2\, dA = \int (y \cos \theta - x \sin \theta)^2\, dA$$
$$= \cos^2 \theta \int y^2\, dA - 2 \sin \theta \cos \theta \int xy\, dA + \sin^2 \theta \int x^2\, dA$$

Using the relations (9.14), we write

$$I_{x'} = I_x \cos^2 \theta - 2I_{xy} \sin \theta \cos \theta + I_y \sin^2 \theta \qquad (9.15)$$

Similarly, we obtain for $I_{y'}$ and $I_{x'y'}$ the expressions

$$I_{y'} = I_x \sin^2 \theta + 2I_{xy} \sin \theta \cos \theta + I_y \cos^2 \theta \qquad (9.16)$$
$$I_{x'y'} = (I_x - I_y) \sin \theta \cos \theta + I_{xy}(\cos^2 \theta - \sin^2 \theta) \qquad (9.17)$$

Recalling the trigonometric relations

$$\sin 2\theta = 2 \sin \theta \cos \theta \qquad \cos 2\theta = \cos^2 \theta - \sin^2 \theta$$

and

$$\cos^2 \theta = \frac{1 + \cos 2\theta}{2} \qquad \sin^2 \theta = \frac{1 - \cos 2\theta}{2}$$

we can write (9.15), (9.16), and (9.17) as follows:

$$I_{x'} = \frac{I_x + I_y}{2} + \frac{I_x - I_y}{2} \cos 2\theta - I_{xy} \sin 2\theta \qquad (9.18)$$

$$I_{y'} = \frac{I_x + I_y}{2} - \frac{I_x - I_y}{2} \cos 2\theta + I_{xy} \sin 2\theta \qquad (9.19)$$

$$I_{x'y'} = \frac{I_x - I_y}{2} \sin 2\theta + I_{xy} \cos 2\theta \qquad (9.20)$$

Adding (9.18) and (9.19) we observe that

$$I_{x'} + I_{y'} = I_x + I_y \qquad (9.21)$$

This result could have been anticipated, since both members of (9.21) are equal to the polar moment of inertia J_O.

Equations (9.18) and (9.20) are the parametric equations of a circle. This means that if we choose a set of rectangular axes and plot a point M of abscissa $I_{x'}$ and ordinate $I_{x'y'}$ for any given value of the parameter θ, all of the points thus obtained will lie on a circle. To establish this property, we eliminate θ from Eqs. (9.18) and (9.20); this is done by transposing $(I_x + I_y)/2$ in Eq. (9.18), squaring both members of Eqs. (9.18) and (9.20), and adding. We write

$$\left(I_{x'} - \frac{I_x + I_y}{2}\right)^2 + I_{x'y'}^2 = \left(\frac{I_x - I_y}{2}\right)^2 + I_{xy}^2 \qquad (9.22)$$

Setting

$$I_{\text{ave}} = \frac{I_x + I_y}{2} \qquad \text{and} \qquad R = \sqrt{\left(\frac{I_x - I_y}{2}\right)^2 + I_{xy}^2} \qquad (9.23)$$

we write the identity (9.22) in the form

$$(I_{x'} - I_{\text{ave}})^2 + I_{x'y'}^2 = R^2 \qquad (9.24)$$

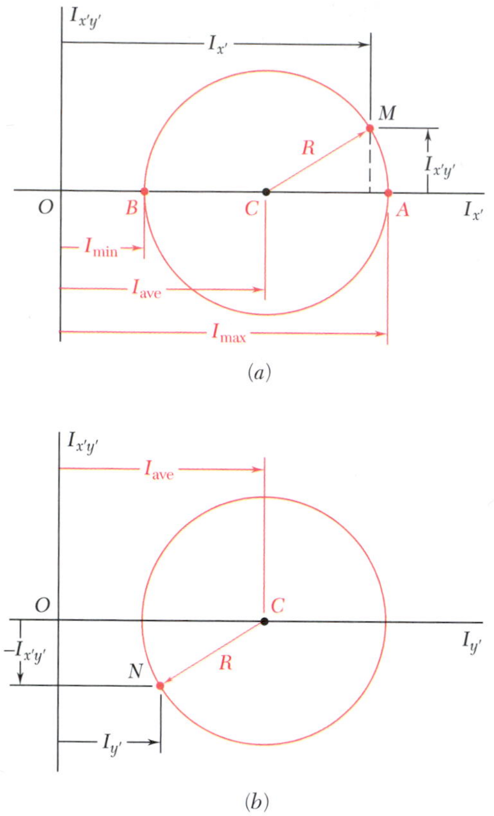

Fig. 9.18

which is the equation of a circle of radius R centered at the point C whose x and y coordinates are I_{ave} and 0, respectively (Fig. 9.18*a*). We observe that Eqs. (9.19) and (9.20) are the parametric equations of the same circle. Furthermore, because of the symmetry of the circle about the horizontal axis, the same result would have been obtained if instead of plotting M, we had plotted a point N of coordinates $I_{y'}$ and $-I_{x'y'}$ (Fig. 9.18*b*). This property will be used in Sec. 9.10.

The two points A and B where the above circle intersects the horizontal axis (Fig. 9.18*a*) are of special interest: Point A corresponds to the maximum value of the moment of inertia $I_{x'}$, while point B corresponds to its minimum value. In addition, both points correspond to a zero value of the product of inertia $I_{x'y'}$. Thus, the values θ_m of the parameter θ which correspond to the points A and B can be obtained by setting $I_{x'y'} = 0$ in Eq. (9.20). We obtain†

$$\tan 2\theta_m = -\frac{2I_{xy}}{I_x - I_y} \tag{9.25}$$

This equation defines two values $2\theta_m$ which are 180° apart and thus two values θ_m which are 90° apart. One of these values corresponds to point A in Fig. 9.18*a* and to an axis through O in Fig. 9.17 with respect to which the moment of inertia of the given area is maximum; the other value corresponds to point B and to an axis through O with respect to which the moment of inertia of the area is minimum. The two axes thus defined, which are perpendicular to each other, are called the *principal axes of the area about O*, and the corresponding values I_{max} and I_{min} of the moment of inertia are called the *principal moments of inertia of the area about O*. Since the two values θ_m defined by Eq. (9.25) were obtained by setting $I_{x'y'} = 0$ in Eq. (9.20), it is clear that the product of inertia of the given area with respect to its principal axes is zero.

We observe from Fig. 9.18*a* that

$$I_{max} = I_{ave} + R \qquad I_{min} = I_{ave} - R \tag{9.26}$$

Using the values for I_{ave} and R from formulas (9.23), we write

$$I_{max,min} = \frac{I_x + I_y}{2} \pm \sqrt{\left(\frac{I_x - I_y}{2}\right)^2 + I_{xy}^2} \tag{9.27}$$

Unless it is possible to tell by inspection which of the two principal axes corresponds to I_{max} and which corresponds to I_{min}, it is necessary to substitute one of the values of θ_m into Eq. (9.18) in order to determine which of the two corresponds to the maximum value of the moment of inertia of the area about O.

Referring to Sec. 9.8, we note that if an area possesses an axis of symmetry through a point O, this axis must be a principal axis of the area about O. On the other hand, a principal axis does not need to be an axis of symmetry; whether or not an area possesses any axes of symmetry, it will have two principal axes of inertia about any point O.

The properties we have established hold for any point O located inside or outside the given area. If the point O is chosen to coincide with the centroid of the area, any axis through O is a centroidal axis; the two principal axes of the area about its centroid are referred to as the *principal centroidal axes of the area*.

†This relation can also be obtained by differentiating $I_{x'}$ in Eq. (9.18) and setting $dI_{x'}/d\theta = 0$.

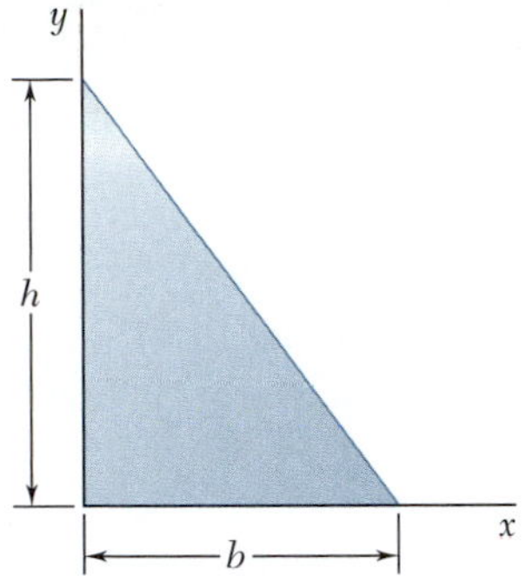

SAMPLE PROBLEM 9.6

Determine the product of inertia of the right triangle shown (*a*) with respect to the x and y axes and (*b*) with respect to centroidal axes parallel to the x and y axes.

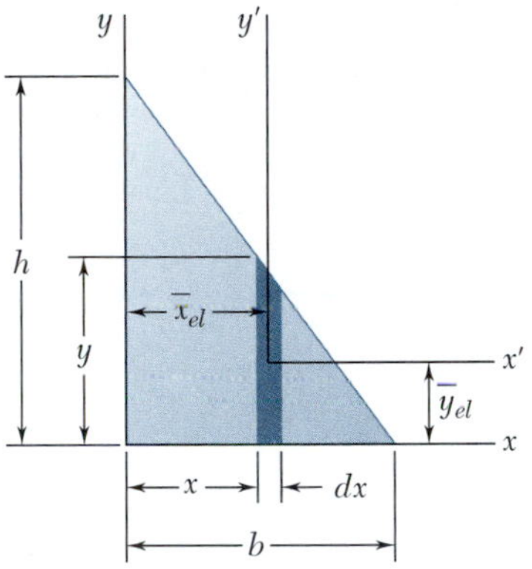

SOLUTION

a. Product of Inertia I_{xy}. A vertical rectangular strip is chosen as the differential element of area. Using the parallel-axis theorem, we write

$$dI_{xy} = dI_{x'y'} + \bar{x}_{el}\bar{y}_{el}\,dA$$

Since the element is symmetrical with respect to the x' and y' axes, we note that $dI_{x'y'} = 0$. From the geometry of the triangle, we obtain

$$y = h\left(1 - \frac{x}{b}\right) \qquad dA = y\,dx = h\left(1 - \frac{x}{b}\right)dx$$

$$\bar{x}_{el} = x \qquad \bar{y}_{el} = \tfrac{1}{2}y = \tfrac{1}{2}h\left(1 - \frac{x}{b}\right)$$

Integrating dI_{xy} from $x = 0$ to $x = b$, we obtain

$$I_{xy} = \int dI_{xy} = \int \bar{x}_{el}\bar{y}_{el}\,dA = \int_0^b x(\tfrac{1}{2})h^2\left(1 - \frac{x}{b}\right)^2 dx$$

$$= h^2\int_0^b \left(\frac{x}{2} - \frac{x^2}{b} + \frac{x^3}{2b^2}\right)dx = h^2\left[\frac{x^2}{4} - \frac{x^3}{3b} + \frac{x^4}{8b^2}\right]_0^b$$

$$I_{xy} = \tfrac{1}{24}b^2h^2 \quad ◀$$

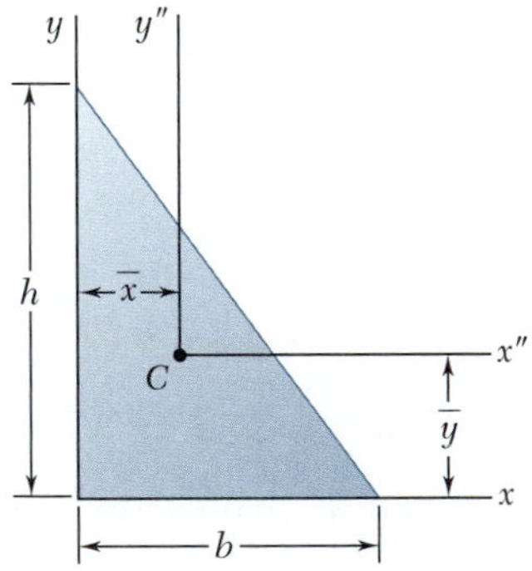

b. Product of Inertia $\bar{I}_{x''y''}$. The coordinates of the centroid of the triangle relative to the x and y axes are

$$\bar{x} = \tfrac{1}{3}b \qquad \bar{y} = \tfrac{1}{3}h$$

Using the expression for I_{xy} obtained in part *a*, we apply the parallel-axis theorem and write

$$I_{xy} = \bar{I}_{x''y''} + \bar{x}\bar{y}A$$

$$\tfrac{1}{24}b^2h^2 = \bar{I}_{x''y''} + (\tfrac{1}{3}b)(\tfrac{1}{3}h)(\tfrac{1}{2}bh)$$

$$\bar{I}_{x''y''} = \tfrac{1}{24}b^2h^2 - \tfrac{1}{18}b^2h^2$$

$$\bar{I}_{x''y''} = -\tfrac{1}{72}b^2h^2 \quad ◀$$

SAMPLE PROBLEM 9.7

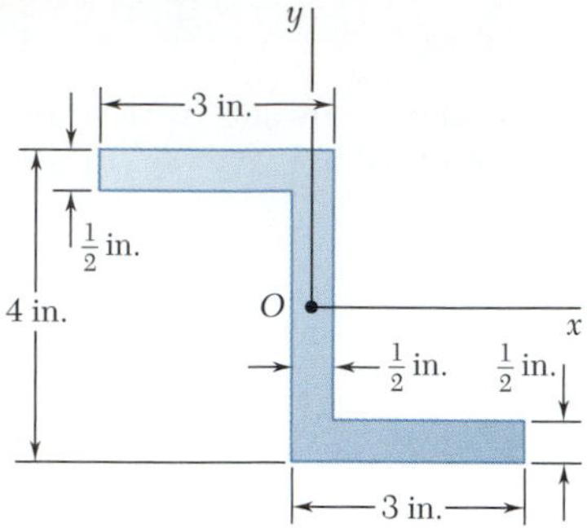

For the section shown, the moments of inertia with respect to the x and y axes have been computed and are known to be

$$I_x = 10.38 \text{ in}^4 \qquad I_y = 6.97 \text{ in}^4$$

Determine (*a*) the orientation of the principal axes of the section about O, (*b*) the values of the principal moments of inertia of the section about O.

SOLUTION

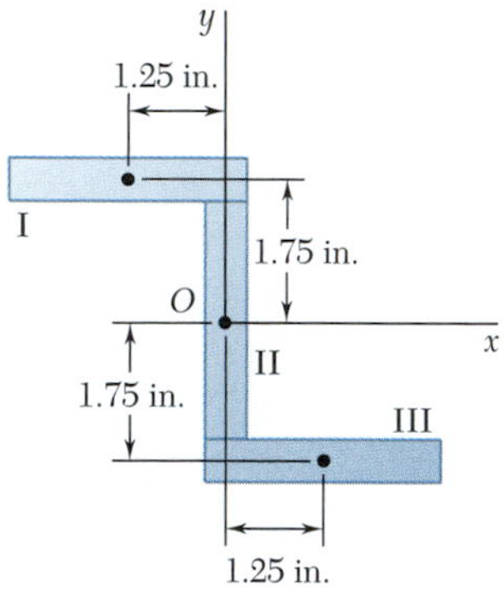

We first compute the product of inertia with respect to the x and y axes. The area is divided into three rectangles as shown. We note that the product of inertia $\bar{I}_{x'y'}$ with respect to centroidal axes parallel to the x and y axes is zero for each rectangle. Using the parallel-axis theorem $I_{xy} = \bar{I}_{x'y'} + \bar{x}\bar{y}A$, we find that I_{xy} reduces to $\bar{x}\bar{y}A$ for each rectangle.

Rectangle	Area, in^2	$\bar{x}$, in.	$\bar{y}$, in.	$\bar{x}\bar{y}A$, in^4
I	1.5	−1.25	+1.75	−3.28
II	1.5	0	0	0
III	1.5	+1.25	−1.75	−3.28
				$\Sigma\bar{x}\bar{y}A = -6.56$

$$I_{xy} = \Sigma\bar{x}\bar{y}A = -6.56 \text{ in}^4$$

***a.* Principal Axes.** Since the magnitudes of I_x, I_y, and I_{xy} are known, Eq. (9.25) is used to determine the values of θ_m:

$$\tan 2\theta_m = -\frac{2I_{xy}}{I_x - I_y} = -\frac{2(-6.56)}{10.38 - 6.97} = +3.85$$

$$2\theta_m = 75.4° \text{ and } 255.4°$$

$$\theta_m = 37.7° \qquad \text{and} \qquad \theta_m = 127.7° \quad ◀$$

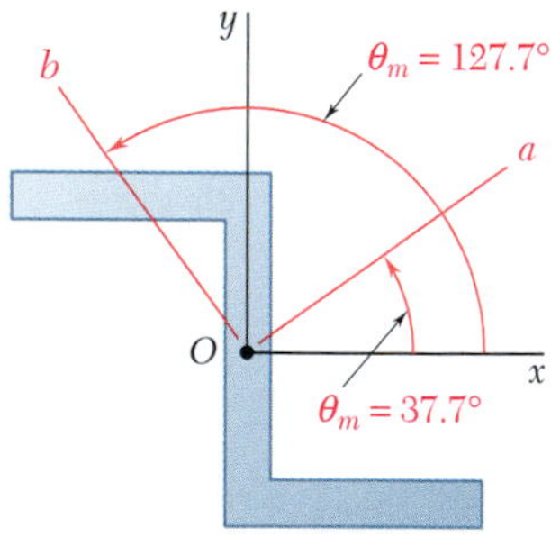

***b.* Principal Moments of Inertia.** Using Eq. (9.27), we write

$$I_{\max,\min} = \frac{I_x + I_y}{2} \pm \sqrt{\left(\frac{I_x - I_y}{2}\right)^2 + I_{xy}^2}$$

$$= \frac{10.38 + 6.97}{2} \pm \sqrt{\left(\frac{10.38 - 6.97}{2}\right)^2 + (-6.56)^2}$$

$$I_{\max} = 15.45 \text{ in}^4 \qquad I_{\min} = 1.897 \text{ in}^4 \quad ◀$$

Noting that the elements of the area of the section are more closely distributed about the b axis than about the a axis, we conclude that $I_a = I_{\max} = 15.45 \text{ in}^4$ and $I_b = I_{\min} = 1.897 \text{ in}^4$. This conclusion can be verified by substituting $\theta = 37.7°$ into Eqs. (9.18) and (9.19).

SOLVING PROBLEMS ON YOUR OWN

In the problems for this lesson, you will continue your work with *moments of inertia* and will utilize various techniques for computing *products of inertia*. Although the problems are generally straightforward, several items are worth noting.

1. *Calculating the product of inertia I_{xy} by integration.* We defined this quantity as

$$I_{xy} = \int xy\, dA \qquad (9.12)$$

and stated that its value can be positive, negative, or zero. The product of inertia can be computed directly from the above equation using double integration, or it can be determined using single integration as shown in Sample Prob. 9.6. When applying the latter technique and using the parallel-axis theorem, it is important to remember that $\bar{x}_{el}$ and $\bar{y}_{el}$ in the equation

$$dI_{xy} = dI_{x'y'} + \bar{x}_{el}\bar{y}_{el}\, dA$$

are the coordinates of the centroid of the element of area dA. Thus, if dA is not in the first quadrant, one or both of these coordinates will be negative.

2. *Calculating the products of inertia of composite areas.* They can easily be computed from the products of inertia of their component parts by using the parallel-axis theorem

$$I_{xy} = \bar{I}_{x'y'} + \bar{x}\bar{y}A \qquad (9.13)$$

The proper technique to use for problems of this type is illustrated in Sample Probs. 9.6 and 9.7. In addition to the usual rules for composite-area problems, it is essential that you remember the following points.

a. If either of the centroidal axes of a component area is an axis of symmetry for that area, the product of inertia $\bar{I}_{x'y'}$ for that area is zero. Thus, $\bar{I}_{x'y'}$ is zero for component areas such as circles, semicircles, rectangles, and isosceles triangles which possess an axis of symmetry parallel to one of the coordinate axes.

b. Pay careful attention to the signs of the coordinates $\bar{x}$ and $\bar{y}$ of each component area when you use the parallel-axis theorem [Sample Prob. 9.7].

3. *Determining the moments of inertia and the product of inertia for rotated coordinate axes.* In Sec. 9.9 we derived Eqs. (9.18), (9.19), and (9.20), from which the moments of inertia and the product of inertia can be computed for coordinate axes which have been rotated about the origin O. To apply these equations, you must know a set of values I_x, I_y, and I_{xy} for a given orientation of the axes, and you must remember that θ is positive for counterclockwise rotations of the axes and negative for clockwise rotations of the axes.

4. *Computing the principal moments of inertia.* We showed in Sec. 9.9 that there is a particular orientation of the coordinate axes for which the moments of inertia attain their maximum and minimum values, I_{max} and I_{min}, and for which the product of inertia is zero. Equation (9.27) can be used to compute these values, known as the *principal moments of inertia* of the area about O. The corresponding axes are referred to as the *principal axes* of the area about O, and their orientation is defined by Eq. (9.25). *To determine which of the principal axes corresponds to I_{max} and which corresponds to I_{min}*, you can either follow the procedure outlined in the text after Eq. (9.27) or observe about which of the two principal axes the area is more closely distributed; that axis corresponds to I_{min} [Sample Prob. 9.7].

Problems

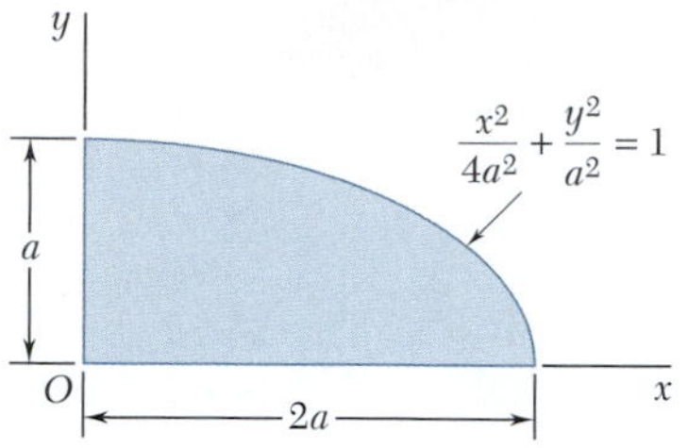

Fig. P9.67

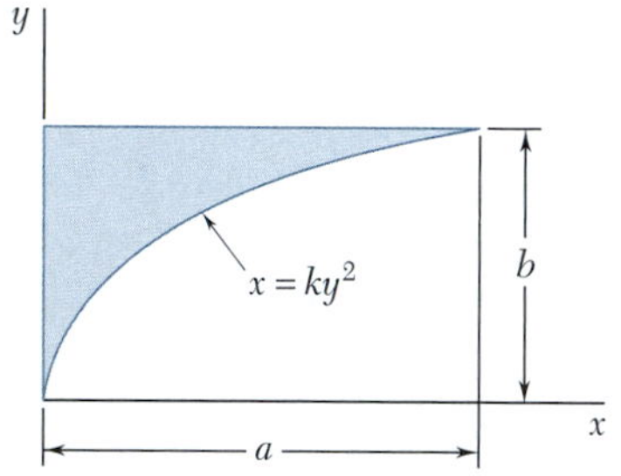

Fig. P9.68

9.67 through 9.70 Determine by direct integration the product of inertia of the given area with respect to the x and y axes.

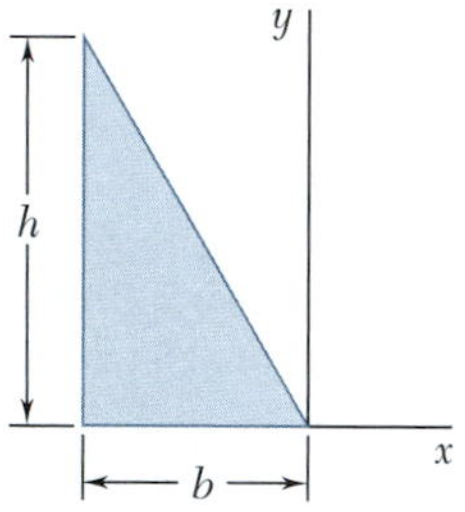

Fig. P9.69

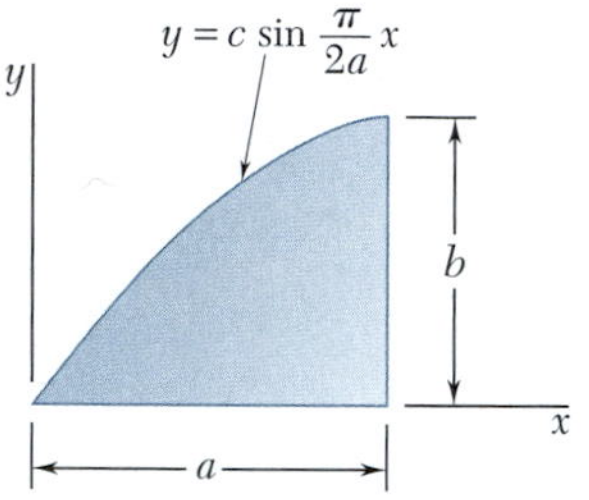

Fig. P9.70

9.71 through 9.74 Using the parallel-axis theorem, determine the product of inertia of the area shown with respect to the centroidal x and y axes.

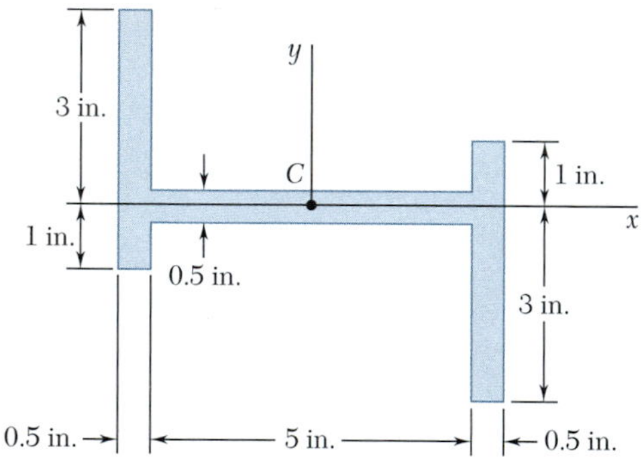

Fig. P9.71

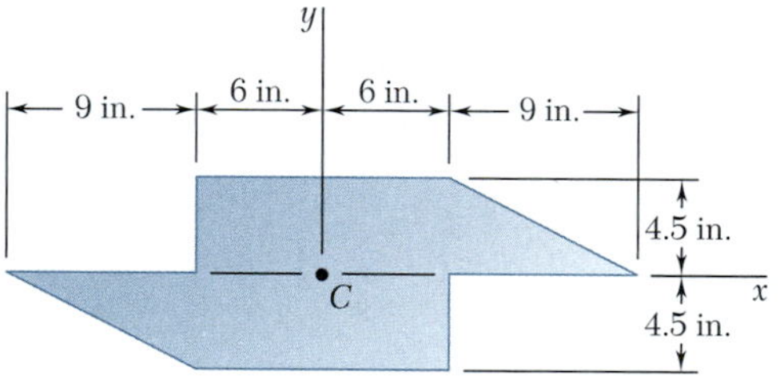

Fig. P9.72

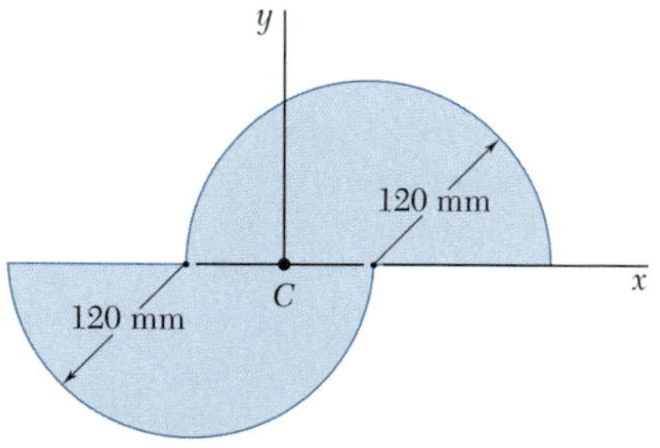

Fig. P9.73

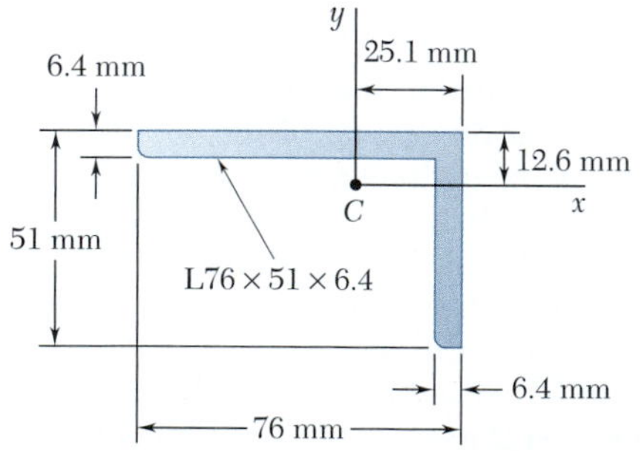

Fig. P9.74

9.75 through 9.78 Using the parallel-axis theorem, determine the product of inertia of the area shown with respect to the centroidal x and y axes.

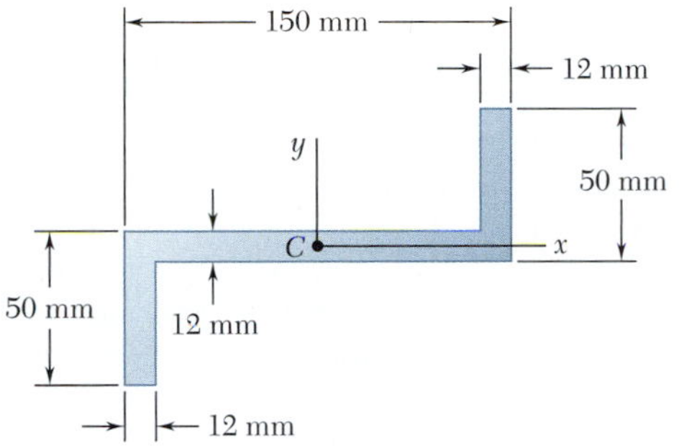

Fig. P9.75

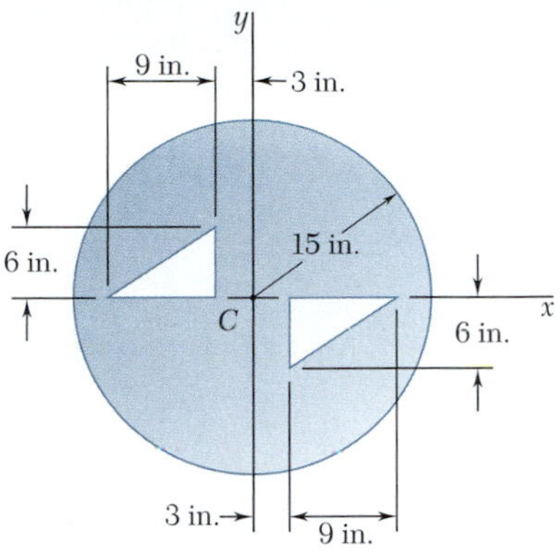

Fig. *P9.76*

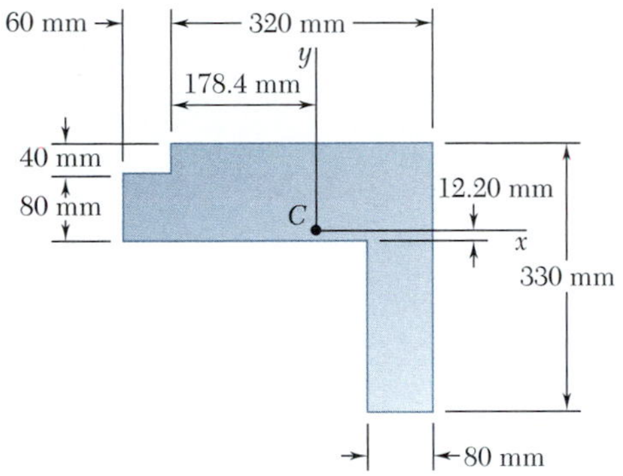

Fig. P9.77

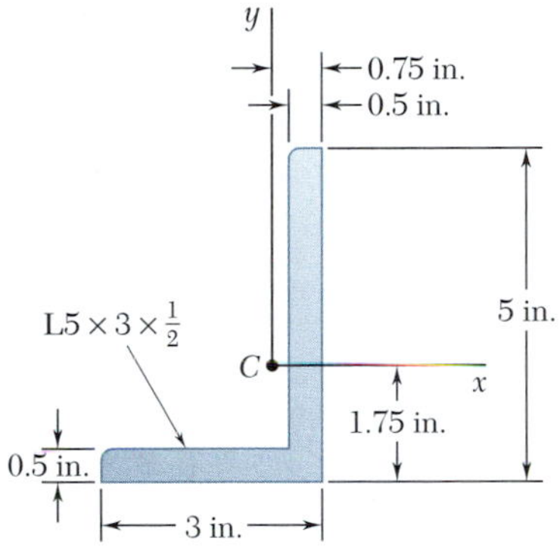

Fig. P9.78

9.79 Determine for the quarter ellipse of Prob. 9.67 the moments of inertia and the product of inertia with respect to new axes obtained by rotating the x and y axes about O (*a*) through 45° counterclockwise, (*b*) through 30° clockwise.

9.80 Determine the moments of inertia and the product of inertia of the area of Prob. 9.72 with respect to new centroidal axes obtained by rotating the x and y axes 45° clockwise.

9.81 Determine the moments of inertia and the product of inertia of the area of Prob. 9.73 with respect to new centroidal axes obtained by rotating the x and y axes through 30° clockwise.

9.82 Determine the moments of inertia and the product of inertia of the area of Prob. 9.75 with respect to new centroidal axes obtained by rotating the x and y axes through 60° counterclockwise.

9.83 Determine the moments of inertia and the product of inertia of the L76 × 51 × 6.4-mm angle cross section of Prob. 9.74 with respect to new centroidal axes obtained by rotating the x and y axes through 45° clockwise.

9.84 Determine the moments of inertia and the product of inertia of the L5 × 3 × $\frac{1}{2}$-in. angle cross section of Prob. 9.78 with respect to new centroidal axes obtained by rotating the x and y axes through 30° counterclockwise.

9.85 For the quarter ellipse of Prob. 9.67, determine the orientation of the principal axes at the origin and the corresponding values of the moments of inertia.

9.86 through 9.88 For the area indicated, determine the orientation of the principal axes at the origin and the corresponding values of the moments of inertia.

9.86 Area of Prob. 9.72
9.87 Area of Prob. 9.73
9.88 Area of Prob. 9.75

9.89 and 9.90 For the angle cross section indicated, determine the orientation of the principal axes at the origin and the corresponding values of the moments of inertia.

9.89 The L76 × 51 × 6.4-mm angle cross section of Prob. 9.74
9.90 The L5 × 3 × $\frac{1}{2}$-in. angle cross section of Prob. 9.78

*9.10. MOHR'S CIRCLE FOR MOMENTS AND PRODUCTS OF INERTIA

The circle used in the preceding section to illustrate the relations existing between the moments and products of inertia of a given area with respect to axes passing through a fixed point O was first introduced by the German engineer Otto Mohr (1835–1918) and is known as *Mohr's circle*. It will be shown that if the moments and product of inertia of an area A are known with respect to two rectangular x and y axes which pass through a point O, Mohr's circle can be used to graphically determine (*a*) the principal axes and principal moments of inertia of the area about O and (*b*) the moments and product of inertia of the area with respect to any other pair of rectangular axes x' and y' through O.

Consider a given area A and two rectangular coordinate axes x and y (Fig. 9.19*a*). Assuming that the moments of inertia I_x and I_y and the product of inertia I_{xy} are known, we will represent them on a diagram by plotting a point X of coordinates I_x and I_{xy} and a point Y of coordinates I_y and $-I_{xy}$ (Fig. 9.19*b*). If I_{xy} is positive, as assumed in Fig. 9.19*a*, point X is located above the horizontal axis and point Y is located below, as shown in Fig. 9.19*b*. If I_{xy} is negative, X is located below the horizontal axis and Y is located above. Joining X and Y with a straight line, we denote by C the point of intersection of line XY with the horizontal axis and draw the circle of center C and diameter XY. Noting that the abscissa of C and the radius of the circle are respectively equal to the quantities I_{ave} and R defined by the formula (9.23), we conclude that the circle obtained is Mohr's circle for the given area about point O. Thus, the abscissas of the points A and B where the circle intersects the horizontal axis represent respectively the principal moments of inertia $I_{\max}$ and $I_{\min}$ of the area.

We also note that, since $\tan(XCA) = 2I_{xy}/(I_x - I_y)$, the angle XCA is equal in magnitude to one of the angles $2\theta_m$ which satisfy Eq. (9.25);

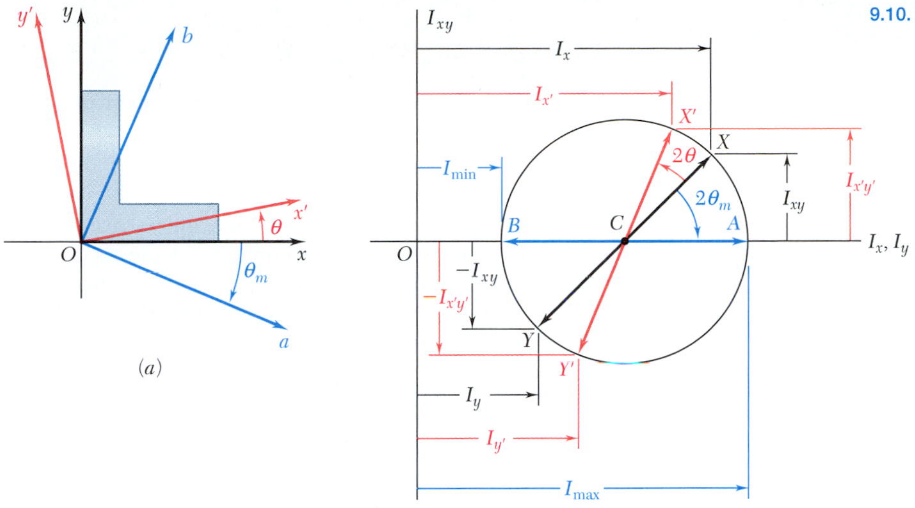

Fig. 9.19

thus, the angle θ_m, which defines in Fig. 9.19*a* the principal axis Oa corresponding to point A in Fig. 9.19*b*, is equal to half of the angle XCA of Mohr's circle. We further observe that if $I_x > I_y$ and $I_{xy} > 0$, as in the case considered here, the rotation which brings CX into CA is clockwise. Also, under these conditions, the angle θ_m obtained from Eq. (9.25), which defines the principal axis Oa in Fig. 9.19*a*, is negative; thus, the rotation which brings Ox into Oa is also clockwise. We conclude that the senses of rotation in both parts of Fig. 9.19 are the same. If a clockwise rotation through $2\theta_m$ is required to bring CX into CA on Mohr's circle, a clockwise rotation through θ_m will bring Ox into the corresponding principal axis Oa in Fig. 9.19*a*.

Since Mohr's circle is uniquely defined, the same circle can be obtained by considering the moments and product of inertia of the area A with respect to the rectangular axes x' and y' (Fig. 9.19*a*). The point X' of coordinates $I_{x'}$ and $I_{x'y'}$ and the point Y' of coordinates $I_{y'}$ and $-I_{x'y'}$ are thus located on Mohr's circle, and the angle $X'CA$ in Fig. 9.19*b* must be equal to twice the angle $x'Oa$ in Fig. 9.19*a*. Since, as noted above, the angle XCA is twice the angle xOa, it follows that the angle XCX' in Fig. 9.19*b* is twice the angle xOx' in Fig. 9.19*a*. The diameter $X'Y'$, which defines the moments and product of inertia $I_{x'}$, $I_{y'}$, and $I_{x'y'}$ of the given area with respect to rectangular axes x' and y' forming an angle θ with the x and y axes, can be obtained by rotating through an angle 2θ the diameter XY, which corresponds to the moments and product of inertia I_x, I_y, and I_{xy}. We note that the rotation which brings the diameter XY into the diameter $X'Y'$ in Fig. 9.19*b* has the same sense as the rotation which brings the x and y axes into the x' and y' axes in Fig. 9.19*a*.

It should be noted that the use of Mohr's circle is not limited to graphical solutions, that is, to solutions based on the careful drawing and measuring of the various parameters involved. By merely sketching Mohr's circle and using trigonometry, one can easily derive the various relations required for a numerical solution of a given problem (see Sample Prob. 9.8).

SAMPLE PROBLEM 9.8

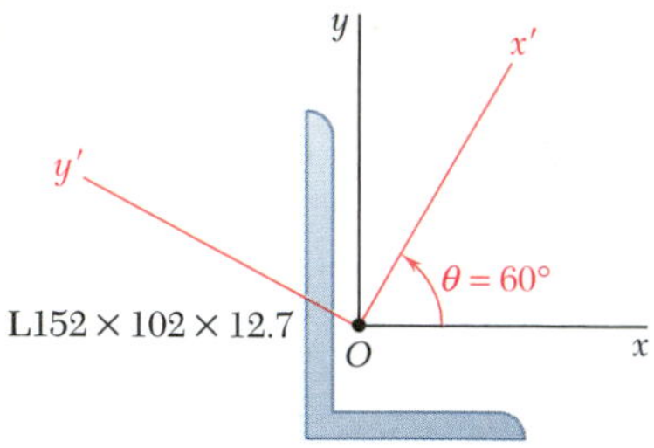

For the section shown, the moments and product of inertia with respect to the x and y axes are known to be

$$I_x = 7.24 \times 10^6 \text{ mm}^4 \qquad I_y = 2.61 \times 10^6 \text{ mm}^4 \qquad I_{xy} = -2.54 \times 10^6 \text{ mm}^4$$

Using Mohr's circle, determine (*a*) the principal axes of the section about O, (*b*) the values of the principal moments of inertia of the section about O, (*c*) the moments and product of inertia of the section with respect to the x' and y' axes which form an angle of 60° with the x and y axes.

SOLUTION

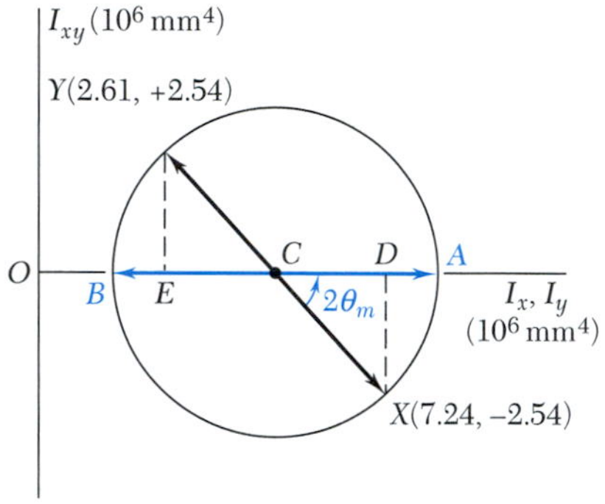

Drawing Mohr's Circle. We first plot point X of coordinates $I_x = 7.24$, $I_{xy} = -2.54$, and point Y of coordinates $I_y = 2.61$, $-I_{xy} = +2.54$. Joining X and Y with a straight line, we define the center C of Mohr's circle. The abscissa of C, which represents I_{ave}, and the radius R of the circle can be measured directly or calculated as follows:

$$I_{ave} = OC = \tfrac{1}{2}(I_x + I_y) = \tfrac{1}{2}(7.24 \times 10^6 + 2.61 \times 10^6) = 4.925 \times 10^6 \text{ mm}^4$$
$$CD = \tfrac{1}{2}(I_x - I_y) = \tfrac{1}{2}(7.24 \times 10^6 - 2.61 \times 10^6) = 2.315 \times 10^6 \text{ mm}^4$$
$$R = \sqrt{(CD)^2 + (DX)^2} = \sqrt{(2.315 \times 10^6)^2 + (2.54 \times 10^6)^2} = 3.437 \times 10^6 \text{ mm}^4$$

***a.* Principal Axes.** The principal axes of the section correspond to points A and B on Mohr's circle, and the angle through which we should rotate CX to bring it into CA defines $2\theta_m$. We have

$$\tan 2\theta_m = \frac{DX}{CD} = \frac{2.54}{2.315} = 1.097 \qquad 2\theta_m = 47.6° \circlearrowleft \qquad \theta_m = 23.8° \circlearrowleft \quad ◀$$

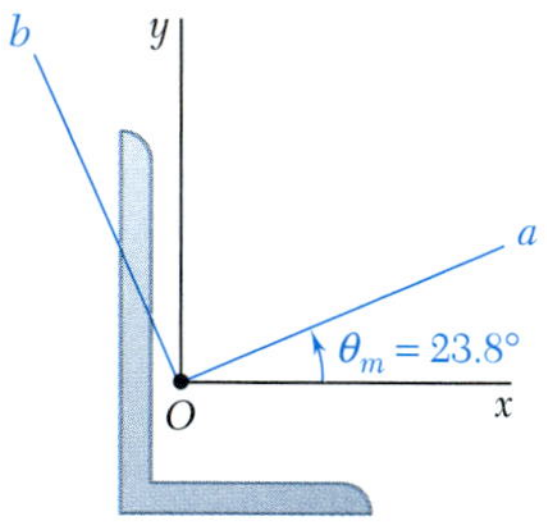

Thus, the principal axis Oa corresponding to the maximum value of the moment of inertia is obtained by rotating the x axis through 23.8° counterclockwise; the principal axis Ob corresponding to the minimum value of the moment of inertia is obtained by rotating the y axis through the same angle.

***b.* Principal Moments of Inertia.** The principal moments of inertia are represented by the abscissas of A and B. We have

$$I_{max} = OA = OC + CA = I_{ave} + R = (4.925 + 3.437)10^6 \text{ mm}^4$$
$$I_{max} = 8.36 \times 10^6 \text{ mm}^4 \quad ◀$$
$$I_{min} = OB = OC - BC = I_{ave} - R = (4.925 - 3.437)10^6 \text{ mm}^4$$
$$I_{min} = 1.49 \times 10^6 \text{ mm}^4 \quad ◀$$

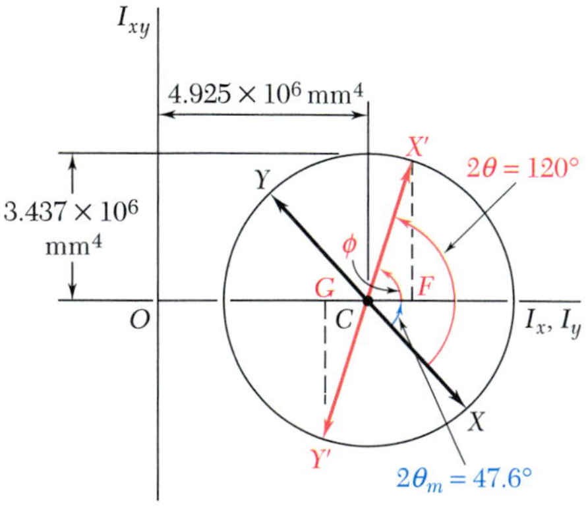

***c.* Moments and Product of Inertia with Respect to the x' and y' Axes.** On Mohr's circle, the points X' and Y', which correspond to the x' and y' axes, are obtained by rotating CX and CY through an angle $2\theta = 2(60°) = 120°$ counterclockwise. The coordinates of X' and Y' yield the desired moments and product of inertia. Noting that the angle that CX' forms with the horizontal axis is $\phi = 120° - 47.6° = 72.4°$, we write

$$I_{x'} = OF = OC + CF = 4.925 \times 10^6 \text{ mm}^4 + (3.437 \times 10^6 \text{ mm}^4) \cos 72.4°$$
$$I_{x'} = 5.96 \times 10^6 \text{ mm}^4 \quad ◀$$
$$I_{y'} = OG = OC - GC = 4.925 \times 10^6 \text{ mm}^4 - (3.437 \times 10^6 \text{ mm}^4) \cos 72.4°$$
$$I_{y'} = 3.89 \times 10^6 \text{ mm}^4 \quad ◀$$
$$I_{x'y'} = FX' = (3.437 \times 10^6 \text{ mm}^4) \sin 72.4°$$
$$I_{x'y'} = 3.28 \times 10^6 \text{ mm}^4 \quad ◀$$

SOLVING PROBLEMS ON YOUR OWN

In the problems for this lesson, you will use *Mohr's circle* to determine the moments and products of inertia of a given area for different orientations of the coordinate axes. Although in some cases using Mohr's circle may not be as direct as substituting into the appropriate equations [Eqs. (9.18) through (9.20)], this method of solution has the advantage of providing a visual representation of the relationships among the various variables. Further, Mohr's circle shows all of the values of the moments and products of inertia which are possible for a given problem.

Using Mohr's circle. The underlying theory was presented in Sec. 9.9, and we discussed the application of this method in Sec. 9.10 and in Sample Prob. 9.8. In the same problem, we presented the steps you should follow to determine the *principal axes*, the *principal moments of inertia*, and the *moments and product of inertia with respect to a specified orientation of the coordinate axes*. When you use Mohr's circle to solve problems, it is important that you remember the following points.

a. Mohr's circle is completely defined by the quantities R and I_{ave}, which represent, respectively, the radius of the circle and the distance from the origin O to the center C of the circle. These quantities can be obtained from Eqs. (9.23) if the moments and product of inertia are known for a given orientation of the axes. However, Mohr's circle can be defined by other combinations of known values [Probs. 9.105, 9.108, and 9.109]. For these cases, it may be necessary to first make one or more assumptions, such as choosing an arbitrary location for the center when I_{ave} is unknown, assigning relative magnitudes to the moments of inertia (for example, $I_x > I_y$), or selecting the sign of the product of inertia.

b. Point X of coordinates (I_x, I_{xy}) and point Y of coordinates (I_y, $-I_{xy}$) are both located on Mohr's circle and are diametrically opposite.

c. Since moments of inertia must be positive, the entire Mohr's circle must lie to the right of the I_{xy} axis; it follows that $I_{ave} > R$ for all cases.

d. As the coordinate axes are rotated through an angle θ, the associated rotation of the diameter of Mohr's circle is equal to 2θ and is in the same sense (clockwise or counterclockwise). We strongly suggest that the known points on the circumference of the circle be labeled with the appropriate capital letter, as was done in Fig. 9.19*b* and for the Mohr's circles of Sample Prob. 9.8. This will enable you to determine, for each value of θ, the sign of the corresponding product of inertia and to determine which moment of inertia is associated with each of the coordinate axes [Sample Prob. 9.8, parts *a* and *c*].

Although we have introduced Mohr's circle within the specific context of the study of moments and products of inertia, the Mohr's circle technique is also applicable to the solution of analogous but physically different problems in mechanics of materials. This multiple use of a specific technique is not unique, and as you pursue your engineering studies, you will encounter several methods of solution which can be applied to a variety of problems.

Problems

9.91 Using Mohr's circle, determine for the quarter ellipse of Prob. 9.67 the moments of inertia and the product of inertia with respect to new axes obtained by rotating the x and y axes about O (a) through 45° counterclockwise, (b) through 30° clockwise.

9.92 Using Mohr's circle, determine the moments of inertia and the product of inertia of the area of Prob. 9.72 with respect to new centroidal axes obtained by rotating the x and y axes 45° clockwise.

9.93 Using Mohr's circle, determine the moments of inertia and the product of inertia of the area of Prob. 9.73 with respect to new centroidal axes obtained by rotating the x and y axes through 30° clockwise.

9.94 Using Mohr's circle, determine the moments of inertia and the product of inertia of the area of Prob. 9.75 with respect to new centroidal axes obtained by rotating the x and y axes through 60° counterclockwise.

9.95 Using Mohr's circle, determine the moments of inertia and the product of inertia of the L76 × 51 × 6.4-mm angle cross section of Prob. 9.74 with respect to new centroidal axes obtained by rotating the x and y axes through 45° clockwise.

9.96 Using Mohr's circle, determine the moments of inertia and the product of inertia of the L5 × 3 × $\frac{1}{2}$-in. angle cross section of Prob. 9.78 with respect to new centroidal axes obtained by rotating the x and y axes through 30° counterclockwise.

9.97 For the quarter ellipse of Prob. 9.67, use Mohr's circle to determine the orientation of the principal axes at the origin and the corresponding values of the moments of inertia.

9.98 through 9.104 Using Mohr's circle, determine for the area indicated the orientation of the principal centroidal axes and the corresponding values of the moments of inertia.

9.98 Area of Prob. 9.72
9.99 Area of Prob. 9.76
9.100 Area of Prob. 9.73
9.101 Area of Prob. 9.74
9.102 Area of Prob. 9.75
9.103 Area of Prob. 9.71
9.104 Area of Prob. 9.77

(The moments of inertia $\bar{I}_x$ and $\bar{I}_y$ of the area of Prob. 9.104 were determined in Prob. 9.44.)

9.105 The moments and product of inertia for an L102 × 76 × 6.4-mm angle cross section with respect to two rectangular axes x and y through C are, respectively, $\bar{I}_x = 0.166 \times 10^6$ mm^4, $\bar{I}_y = 0.453 \times 10^6$ mm^4, and $\bar{I}_{xy} < 0$, with the minimum value of the moment of inertia of the area with respect to any axis through C being $\bar{I}_{\min} = 0.051 \times 10^6$ mm^4. Using Mohr's circle, determine (a) the product of inertia $\bar{I}_{xy}$ of the area, (b) the orientation of the principal axes, (c) the value of $\bar{I}_{\max}$.

9.106 and 9.107 Using Mohr's circle, determine for the cross section of the rolled-steel angle shown the orientation of the principal centroidal axes and the corresponding values of the moments of inertia. (Properties of the cross sections are given in Fig. 9.13.)

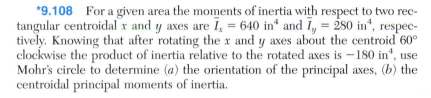

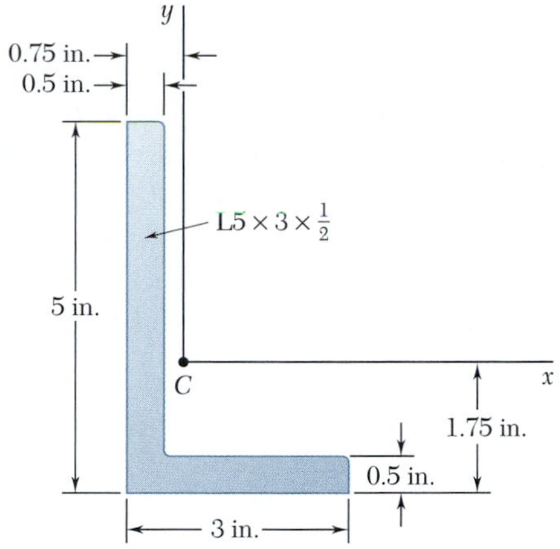

Fig. P9.106

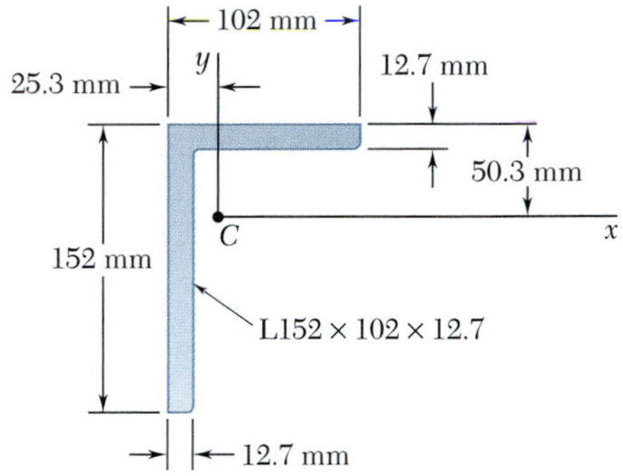

Fig. P9.107

***9.108** For a given area the moments of inertia with respect to two rectangular centroidal x and y axes are $\bar{I}_x = 640 \text{ in}^4$ and $\bar{I}_y = 280 \text{ in}^4$, respectively. Knowing that after rotating the x and y axes about the centroid 60° clockwise the product of inertia relative to the rotated axes is -180 in^4, use Mohr's circle to determine (*a*) the orientation of the principal axes, (*b*) the centroidal principal moments of inertia.

9.109 It is known that for a given area $\bar{I}_y = 300 \text{ in}^4$ and $\bar{I}_{xy} = -125 \text{ in}^4$, where the x and y axes are rectangular centroidal axes. If the axis corresponding to the maximum product of inertia is obtained by rotating the x axis 67.5° counterclockwise about C, use Mohr's circle to determine (*a*) the moment of inertia $\bar{I}_x$ of the area, (*b*) the principal centroidal moments of inertia.

9.110 Using Mohr's circle, show that for any regular polygon (such as a pentagon) (*a*) the moment of inertia with respect to every axis through the centroid is the same, (*b*) the product of inertia with respect to every pair of rectangular axes through the centroid is zero.

9.111 Using Mohr's circle, prove that the expression $I_{x'}I_{y'} - I^2_{x'y'}$ is independent of the orientation of the x' and y' axes, where $I_{x'}$, $I_{y'}$, and $I_{x'y'}$ represent the moments and product of inertia, respectively, of a given area with respect to a pair of rectangular axes x' and y' through a given point O. Also show that the given expression is equal to the square of the length of the tangent drawn from the origin of the coordinate system to Mohr's circle.

9.112 Using the invariance property established in the preceding problem, express the product of inertia I_{xy} of an area A with respect to a pair of rectangular axes through O in terms of the moments of inertia I_x and I_y of A and the principal moments of inertia $I_{\min}$ and $I_{\max}$ of A about O. Use the formula obtained to calculate the product of inertia $\bar{I}_{xy}$ of the $76 \times 51 \times 6.4$-mm angle cross section shown in Fig. 9.13B knowing that its maximum moment of inertia is $524 \times 10^3 \text{ mm}^4$.

MOMENTS OF INERTIA OF MASSES

9.11. MOMENT OF INERTIA OF A MASS

Consider a small mass Δm mounted on a rod of negligible mass which can rotate freely about an axis AA' (Fig. 9.20*a*). If a couple is applied to the system, the rod and mass, assumed to be initially at rest, will start rotating about AA'. The details of this motion will be studied later in dynamics. At present, we wish only to indicate that the time required for the system to reach a given speed of rotation is proportional to the mass Δm and to the square of the distance r. The product $r^2\,\Delta m$ provides, therefore, a measure of the *inertia* of the system, that is, a measure of the resistance the system offers when we try to set it in motion. For this reason, the product $r^2\,\Delta m$ is called the *moment of inertia* of the mass Δm with respect to the axis AA'.

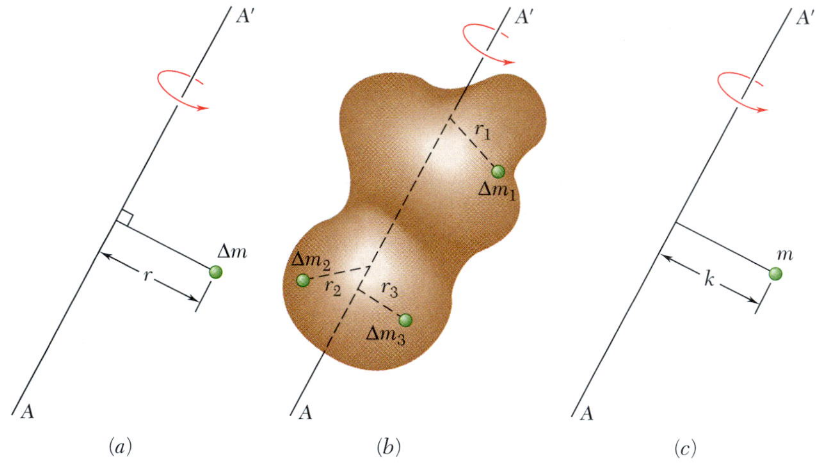

Fig. 9.20

Consider now a body of mass m which is to be rotated about an axis AA' (Fig. 9.20*b*). Dividing the body into elements of mass Δm_1, Δm_2, etc., we find that the body's resistance to being rotated is measured by the sum $r_1^2\,\Delta m_1 + r_2^2\,\Delta m_2 + \cdots$. This sum defines, therefore, the moment of inertia of the body with respect to the axis AA'. Increasing the number of elements, we find that the moment of inertia is equal, in the limit, to the integral

$$I = \int r^2\,dm \qquad (9.28)$$

The *radius of gyration* k of the body with respect to the axis AA' is defined by the relation

$$I = k^2 m \qquad \text{or} \qquad k = \sqrt{\frac{I}{m}} \tag{9.29}$$

The radius of gyration k represents, therefore, the distance at which the entire mass of the body should be concentrated if its moment of inertia with respect to AA' is to remain unchanged (Fig. 9.20*c*). Whether it is kept in its original shape (Fig. 9.20*b*) or whether it is concentrated as shown in Fig. 9.20*c*, the mass m will react in the same way to a rotation, or *gyration*, about AA'.

If SI units are used, the radius of gyration k is expressed in meters and the mass m in kilograms, and thus the unit used for the moment of inertia of a mass is $\text{kg} \cdot \text{m}^2$. If U.S. customary units are used, the radius of gyration is expressed in feet and the mass in slugs (that is, in $\text{lb} \cdot \text{s}^2/\text{ft}$), and thus the derived unit used for the moment of inertia of a mass is $\text{lb} \cdot \text{ft} \cdot \text{s}^2$.†

The moment of inertia of a body with respect to a coordinate axis can easily be expressed in terms of the coordinates x, y, z of the element of mass dm (Fig. 9.21). Noting, for example, that the square of the distance r from the element dm to the y axis is $z^2 + x^2$, we express the moment of inertia of the body with respect to the y axis as

$$I_y = \int r^2\, dm = \int (z^2 + x^2)\, dm$$

Similar expressions can be obtained for the moments of inertia with respect to the x and z axes. We write

$$I_x = \int (y^2 + z^2)\, dm$$
$$I_y = \int (z^2 + x^2)\, dm \tag{9.30}$$
$$I_z = \int (x^2 + y^2)\, dm$$

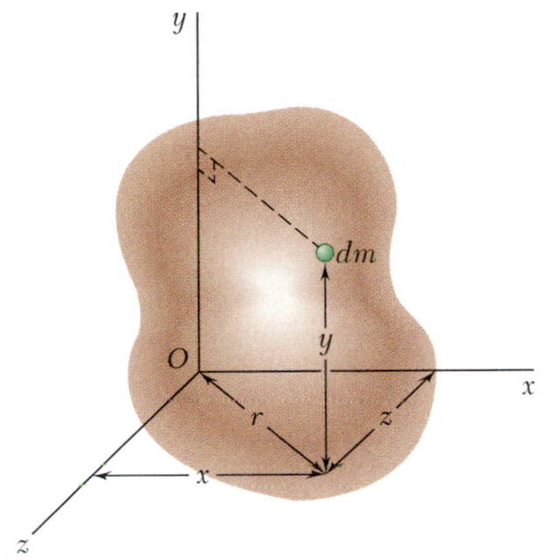

Fig. 9.21

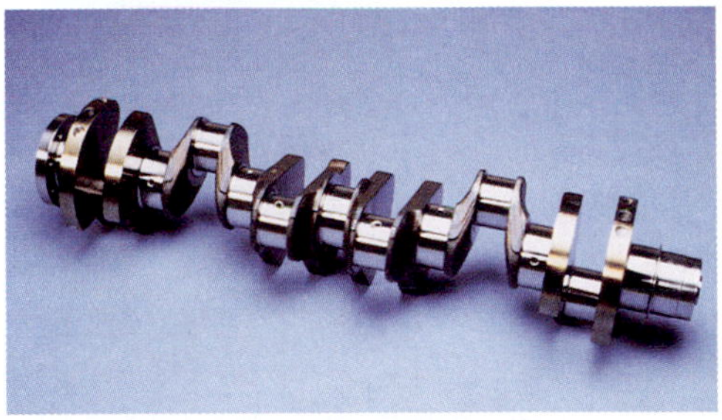
Photo 9.2 As you will discuss in your dynamics course, the rotational behavior of the camshaft shown is dependent upon the mass moment of inertia of the camshaft with respect to its axis of rotation.

†It should be kept in mind when converting the moment of inertia of a mass from U.S. customary units to SI units that the base unit *pound* used in the derived unit $\text{lb} \cdot \text{ft} \cdot \text{s}^2$ is a unit of force (*not* of mass) and should therefore be converted into newtons. We have

$$1\ \text{lb} \cdot \text{ft} \cdot \text{s}^2 = (4.45\ \text{N})(0.3048\ \text{m})(1\ \text{s})^2 = 1.356\ \text{N} \cdot \text{m} \cdot \text{s}^2$$

or, since $1\ \text{N} = 1\ \text{kg} \cdot \text{m/s}^2$,

$$1\ \text{lb} \cdot \text{ft} \cdot \text{s}^2 = 1.356\ \text{kg} \cdot \text{m}^2$$

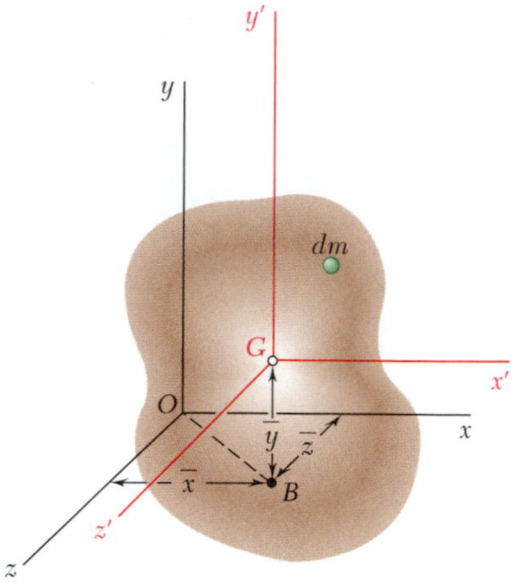

Fig. 9.22

9.12. PARALLEL-AXIS THEOREM

Consider a body of mass m. Let $Oxyz$ be a system of rectangular coordinates whose origin is at the arbitrary point O, and $Gx'y'z'$ a system of parallel *centroidal axes,* that is, a system whose origin is at the center of gravity G of the body† and whose axes x', y', z' are parallel to the x, y, and z axes, respectively (Fig. 9.22). Denoting by $\bar{x}$, $\bar{y}$, $\bar{z}$ the coordinates of G with respect to $Oxyz$, we write the following relations between the coordinates x, y, z of the element dm with respect to $Oxyz$ and its coordinates x', y', z' with respect to the centroidal axes $Gx'y'z'$:

$$x = x' + \bar{x} \qquad y = y' + \bar{y} \qquad z = z' + \bar{z} \tag{9.31}$$

Referring to Eqs. (9.30), we can express the moment of inertia of the body with respect to the x axis as follows:

$$\begin{aligned} I_x &= \int (y^2 + z^2)\, dm = \int [(y' + \bar{y})^2 + (z' + \bar{z})^2]\, dm \\ &= \int (y'^2 + z'^2)\, dm + 2\bar{y} \int y'\, dm + 2\bar{z} \int z'\, dm + (\bar{y}^2 + \bar{z}^2) \int dm \end{aligned}$$

The first integral in this expression represents the moment of inertia $\bar{I}_{x'}$ of the body with respect to the centroidal axis x'; the second and third integrals represent the first moment of the body with respect to the $z'x'$ and $x'y'$ planes, respectively, and, since both planes contain G, the two integrals are *zero;* the last integral is equal to the total mass m of the body. We write, therefore,

$$I_x = \bar{I}_{x'} + m(\bar{y}^2 + \bar{z}^2) \tag{9.32}$$

and, similarly,

$$I_y = \bar{I}_{y'} + m(\bar{z}^2 + \bar{x}^2) \qquad I_z = \bar{I}_{z'} + m(\bar{x}^2 + \bar{y}^2) \tag{9.32$'$}$$

We easily verify from Fig. 9.22 that the sum $\bar{z}^2 + \bar{x}^2$ represents the square of the distance OB, between the y and y' axes. Similarly, $\bar{y}^2 + \bar{z}^2$ and $\bar{x}^2 + \bar{y}^2$ represent the squares of the distance between the x and x' axes and the z and z' axes, respectively. Denoting by d the distance between an arbitrary axis AA' and the parallel centroidal axis BB' (Fig. 9.23), we can, therefore, write the following general relation between the moment of inertia I of the body with respect to AA' and its moment of inertia $\bar{I}$ with respect to BB':

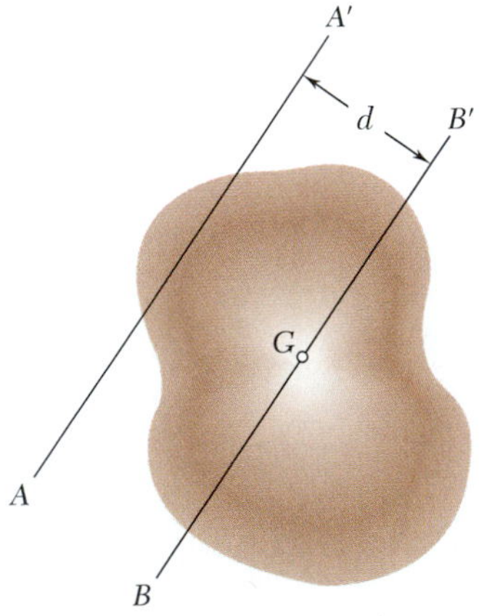

Fig. 9.23

$$I = \bar{I} + md^2 \tag{9.33}$$

Expressing the moments of inertia in terms of the corresponding radii of gyration, we can also write

$$k^2 = \bar{k}^2 + d^2 \tag{9.34}$$

where k and $\bar{k}$ represent the radii of gyration of the body about AA' and BB', respectively.

†Note that the term *centroidal* is used here to define an axis passing through the center of gravity G of the body, whether or not G coincides with the centroid of the volume of the body.

9.13. MOMENTS OF INERTIA OF THIN PLATES

Consider a thin plate of uniform thickness t, which is made of a homogeneous material of density ρ (density = mass per unit volume). The mass moment of inertia of the plate with respect to an axis AA' *contained in the plane* of the plate (Fig. 9.24*a*) is

$$I_{AA',\text{ mass}} = \int r^2\, dm$$

Since $dm = \rho t\, dA$, we write

$$I_{AA',\text{ mass}} = \rho t \int r^2\, dA$$

But r represents the distance of the element of area dA to the axis

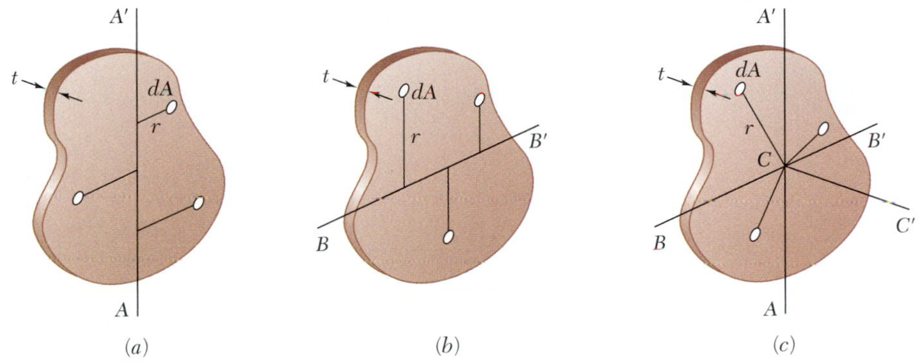

Fig. 9.24

AA'; the integral is therefore equal to the moment of inertia of the area of the plate with respect to AA'. We have

$$I_{AA',\text{ mass}} = \rho t I_{AA',\text{ area}} \tag{9.35}$$

Similarly, for an axis BB' which is contained in the plane of the plate and is perpendicular to AA' (Fig. 9.24*b*), we have

$$I_{BB',\text{ mass}} = \rho t I_{BB',\text{ area}} \tag{9.36}$$

Considering now the axis CC' which is *perpendicular* to the plate and passes through the point of intersection C of AA' and BB' (Fig. 9.24*c*), we write

$$I_{CC',\text{ mass}} = \rho t J_{C,\text{ area}} \tag{9.37}$$

where J_C is the *polar* moment of inertia of the area of the plate with respect to point C.

Recalling the relation $J_C = I_{AA'} + I_{BB'}$ which exists between polar and rectangular moments of inertia of an area, we write the following relation between the mass moments of inertia of a thin plate:

$$I_{CC'} = I_{AA'} + I_{BB'} \tag{9.38}$$

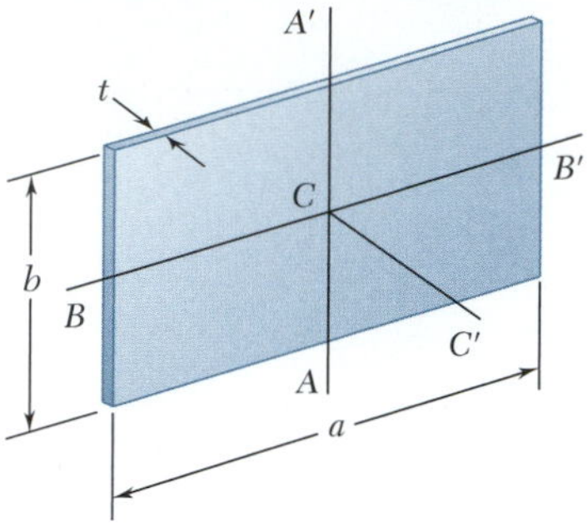

Fig. 9.25

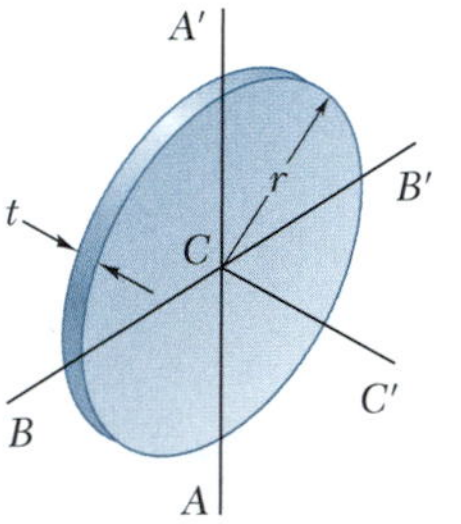

Fig. 9.26

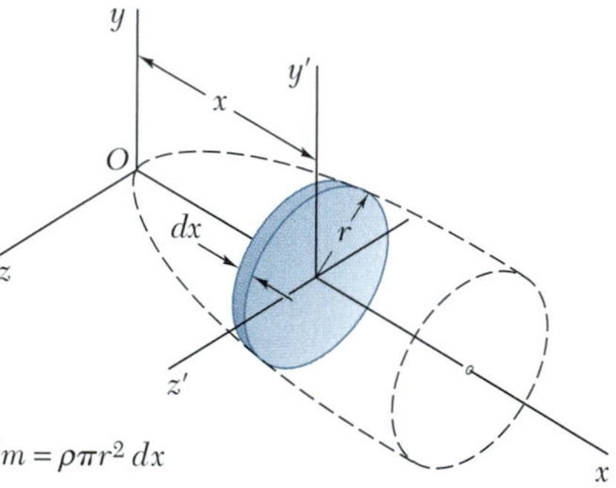

$dm = \rho\pi r^2\,dx$

$dI_x = \frac{1}{2}r^2\,dm$

$dI_y = dI_{y'} + x^2\,dm = \left(\frac{1}{4}r^2 + x^2\right)dm$

$dI_z = dI_{z'} + x^2\,dm = \left(\frac{1}{4}r^2 + x^2\right)dm$

Fig. 9.27 Determination of the moment of inertia of a body of revolution.

Rectangular Plate. In the case of a rectangular plate of sides a and b (Fig. 9.25), we obtain the following mass moments of inertia with respect to axes through the center of gravity of the plate:

$$I_{AA',\,\text{mass}} = \rho t I_{AA',\,\text{area}} = \rho t(\tfrac{1}{12}a^3b)$$
$$I_{BB',\,\text{mass}} = \rho t I_{BB',\,\text{area}} = \rho t(\tfrac{1}{12}ab^3)$$

Observing that the product ρabt is equal to the mass m of the plate, we write the mass moments of inertia of a thin rectangular plate as follows:

$$I_{AA'} = \tfrac{1}{12}ma^2 \qquad I_{BB'} = \tfrac{1}{12}mb^2 \tag{9.39}$$
$$I_{CC'} = I_{AA'} + I_{BB'} = \tfrac{1}{12}m(a^2 + b^2) \tag{9.40}$$

Circular Plate. In the case of a circular plate, or disk, of radius r (Fig. 9.26), we write

$$I_{AA',\,\text{mass}} = \rho t I_{AA',\,\text{area}} = \rho t(\tfrac{1}{4}\pi r^4)$$

Observing that the product $\rho\pi r^2 t$ is equal to the mass m of the plate and that $I_{AA'} = I_{BB'}$, we write the mass moments of inertia of a circular plate as follows:

$$I_{AA'} = I_{BB'} = \tfrac{1}{4}mr^2 \tag{9.41}$$
$$I_{CC'} = I_{AA'} + I_{BB'} = \tfrac{1}{2}mr^2 \tag{9.42}$$

9.14. DETERMINATION OF THE MOMENT OF INERTIA OF A THREE-DIMENSIONAL BODY BY INTEGRATION

The moment of inertia of a three-dimensional body is obtained by evaluating the integral $I = \int r^2\,dm$. If the body is made of a homogeneous material of density ρ, the element of mass dm is equal to ρdV and we can write $I = \rho \int r^2\,dV$. This integral depends only upon the shape of the body. Thus, in order to compute the moment of inertia of a three-dimensional body, it will generally be necessary to perform a triple, or at least a double, integration.

However, if the body possesses two planes of symmetry, it is usually possible to determine the body's moment of inertia with a single integration by choosing as the element of mass dm a thin slab which is perpendicular to the planes of symmetry. In the case of bodies of revolution, for example, the element of mass would be a thin disk (Fig. 9.27). Using formula (9.42), the moment of inertia of the disk with respect to the axis of revolution can be expressed as indicated in Fig. 9.27. Its moment of inertia with respect to each of the other two coordinate axes is obtained by using formula (9.41) and the parallel-axis theorem. Integration of the expression obtained yields the desired moment of inertia of the body.

9.15. MOMENTS OF INERTIA OF COMPOSITE BODIES

The moments of inertia of a few common shapes are shown in Fig. 9.28. For a body consisting of several of these simple shapes, the moment of inertia of the body with respect to a given axis can be obtained by first computing the moments of inertia of its component parts about the desired axis and then adding them together. As was the case for areas, the radius of gyration of a composite body *cannot* be obtained by adding the radii of gyration of its component parts.

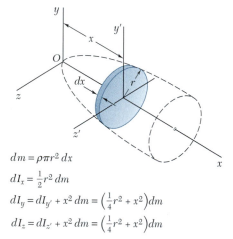

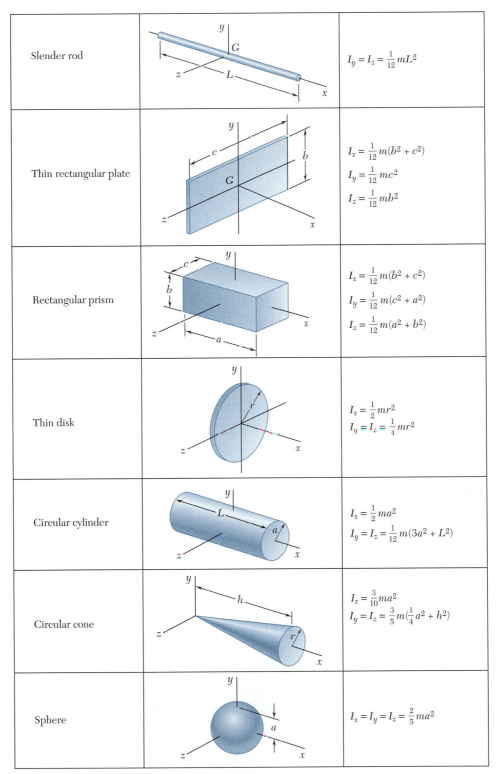

Slender rod	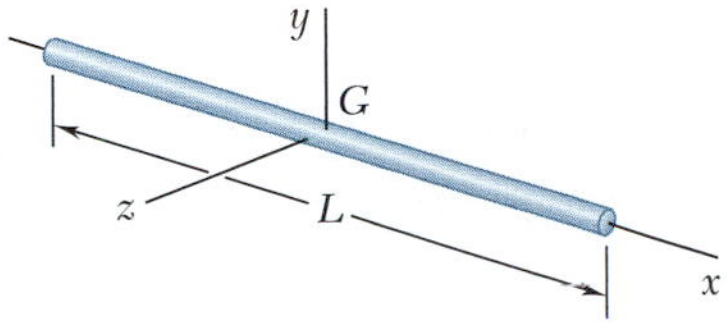	$I_y = I_z = \frac{1}{12}mL^2$
Thin rectangular plate	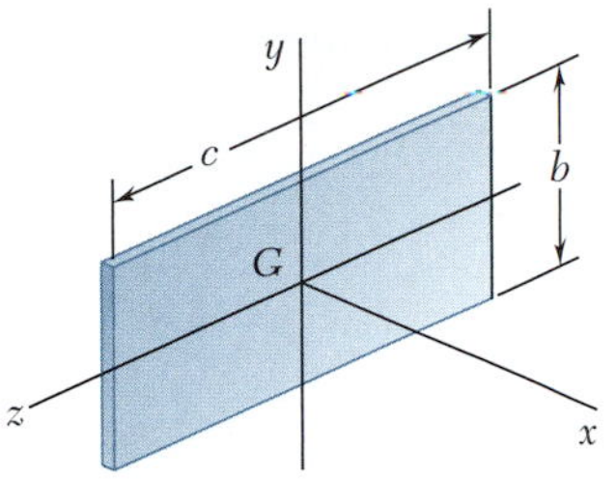	$I_x = \frac{1}{12}m(b^2 + c^2)$ $I_y = \frac{1}{12}mc^2$ $I_z = \frac{1}{12}mb^2$
Rectangular prism	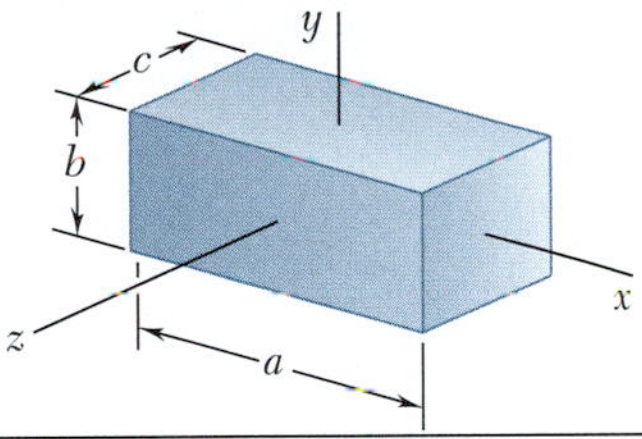	$I_x = \frac{1}{12}m(b^2 + c^2)$ $I_y = \frac{1}{12}m(c^2 + a^2)$ $I_z = \frac{1}{12}m(a^2 + b^2)$
Thin disk	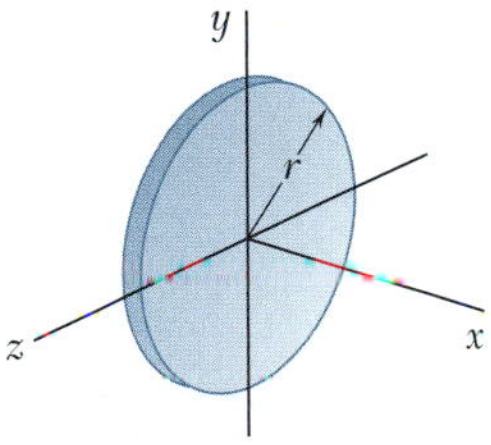	$I_x = \frac{1}{2}mr^2$ $I_y = I_z = \frac{1}{4}mr^2$
Circular cylinder	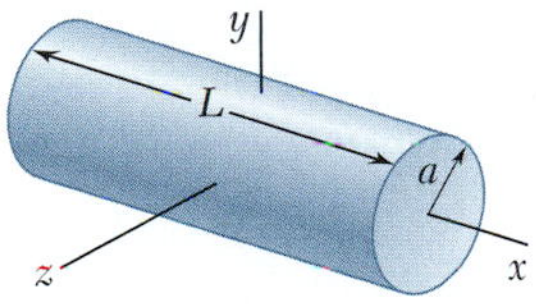	$I_x = \frac{1}{2}ma^2$ $I_y = I_z = \frac{1}{12}m(3a^2 + L^2)$
Circular cone	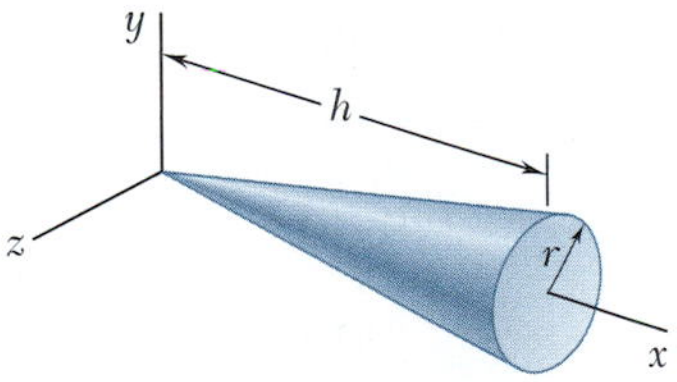	$I_x = \frac{3}{10}ma^2$ $I_y = I_z = \frac{3}{5}m(\frac{1}{4}a^2 + h^2)$
Sphere	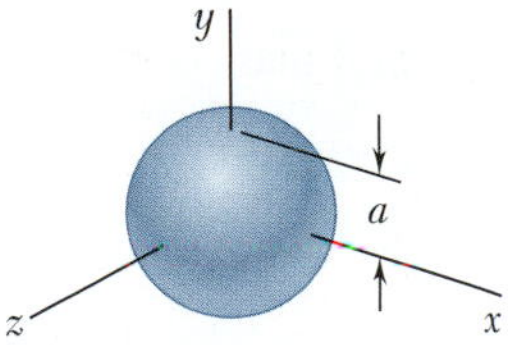	$I_x = I_y = I_z = \frac{2}{5}ma^2$

Fig. 9.28 Mass moments of inertia of common geometric shapes.

SAMPLE PROBLEM 9.9

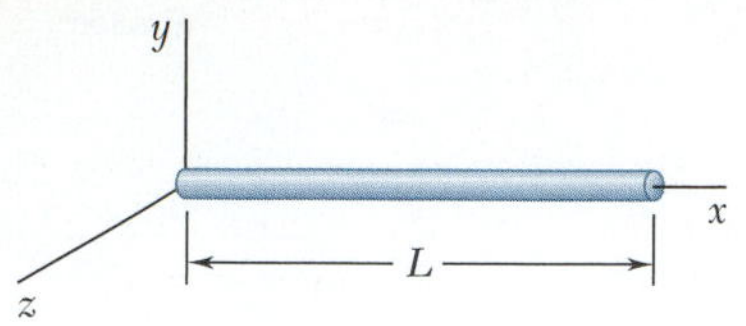

Determine the moment of inertia of a slender rod of length L and mass m with respect to an axis which is perpendicular to the rod and passes through one end of the rod.

SOLUTION

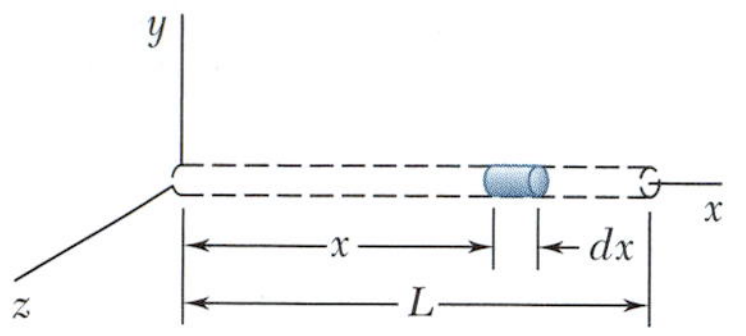

Choosing the differential element of mass shown, we write

$$dm = \frac{m}{L}\,dx$$

$$I_y = \int x^2\,dm = \int_0^L x^2 \frac{m}{L}\,dx = \left[\frac{m}{L}\frac{x^3}{3}\right]_0^L \qquad I_y = \tfrac{1}{3}mL^2 \blacktriangleleft$$

SAMPLE PROBLEM 9.10

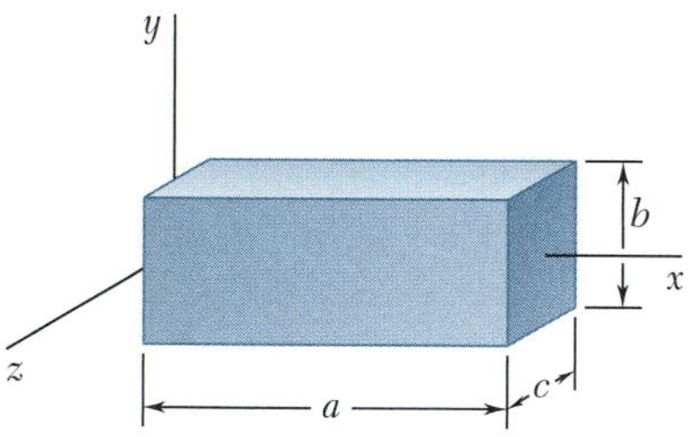

For the homogeneous rectangular prism shown, determine the moment of inertia with respect to the z axis.

SOLUTION

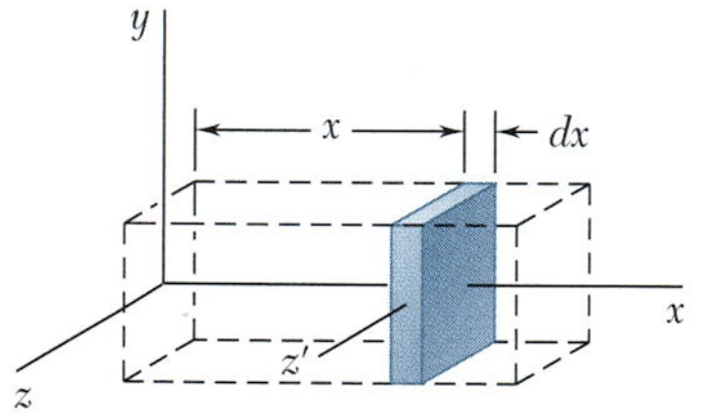

We choose as the differential element of mass the thin slab shown; thus

$$dm = \rho bc\,dx$$

Referring to Sec. 9.13, we find that the moment of inertia of the element with respect to the z' axis is

$$dI_{z'} = \tfrac{1}{12}b^2\,dm$$

Applying the parallel-axis theorem, we obtain the mass moment of inertia of the slab with respect to the z axis.

$$dI_z = dI_{z'} + x^2\,dm = \tfrac{1}{12}b^2\,dm + x^2\,dm = (\tfrac{1}{12}b^2 + x^2)\,\rho bc\,dx$$

Integrating from $x = 0$ to $x = a$, we obtain

$$I_z = \int dI_z = \int_0^a (\tfrac{1}{12}b^2 + x^2)\,\rho bc\,dx = \rho abc(\tfrac{1}{12}b^2 + \tfrac{1}{3}a^2)$$

Since the total mass of the prism is $m = \rho abc$, we can write

$$I_z = m(\tfrac{1}{12}b^2 + \tfrac{1}{3}a^2) \qquad I_z = \tfrac{1}{12}m(4a^2 + b^2) \blacktriangleleft$$

We note that if the prism is thin, b is small compared to a, and the expression for I_z reduces to $\frac{1}{3}ma^2$, which is the result obtained in Sample Prob. 9.9 when $L = a$.

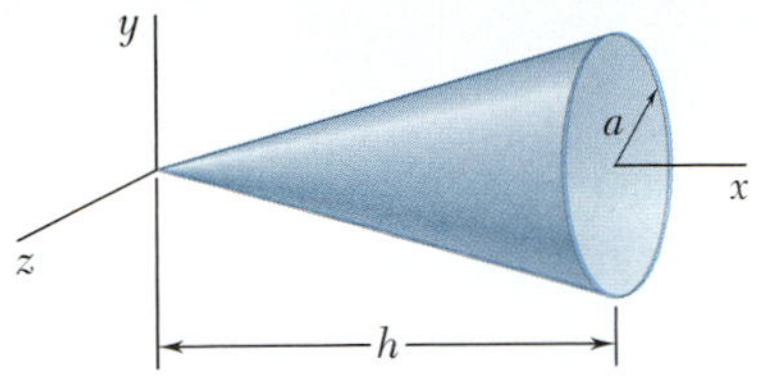

SAMPLE PROBLEM 9.11

Determine the moment of inertia of a right circular cone with respect to (*a*) its longitudinal axis, (*b*) an axis through the apex of the cone and perpendicular to its longitudinal axis, (*c*) an axis through the centroid of the cone and perpendicular to its longitudinal axis.

SOLUTION

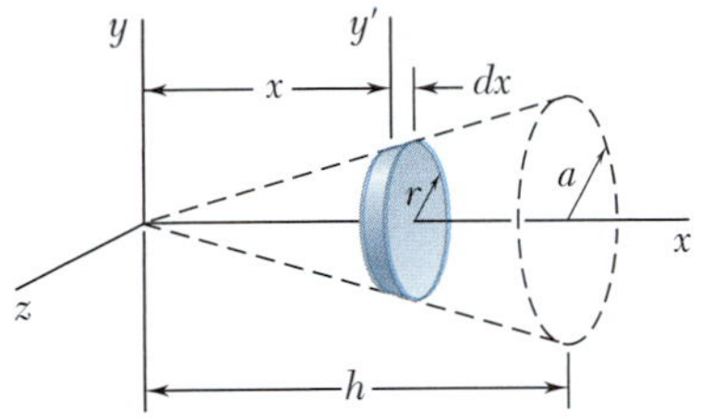

We choose the differential element of mass shown.

$$r = a\frac{x}{h} \qquad dm = \rho\pi r^2\, dx = \rho\pi\frac{a^2}{h^2}x^2\, dx$$

a. Moment of Inertia I_x. Using the expression derived in Sec. 9.13 for a thin disk, we compute the mass moment of inertia of the differential element with respect to the x axis.

$$dI_x = \tfrac{1}{2}r^2\, dm = \tfrac{1}{2}\left(a\frac{x}{h}\right)^2\left(\rho\pi\frac{a^2}{h^2}x^2\, dx\right) = \tfrac{1}{2}\rho\pi\frac{a^4}{h^4}x^4\, dx$$

Integrating from $x = 0$ to $x = h$, we obtain

$$I_x = \int dI_x = \int_0^h \tfrac{1}{2}\rho\pi\frac{a^4}{h^4}x^4\, dx = \tfrac{1}{2}\rho\pi\frac{a^4}{h^4}\frac{h^5}{5} = \tfrac{1}{10}\rho\pi a^4 h$$

Since the total mass of the cone is $m = \tfrac{1}{3}\rho\pi a^2 h$, we can write

$$I_x = \tfrac{1}{10}\rho\pi a^4 h = \tfrac{3}{10}a^2(\tfrac{1}{3}\rho\pi a^2 h) = \tfrac{3}{10}ma^2 \qquad I_x = \tfrac{3}{10}ma^2 \blacktriangleleft$$

b. Moment of Inertia I_y. The same differential element is used. Applying the parallel-axis theorem and using the expression derived in Sec. 9.13 for a thin disk, we write

$$dI_y = dI_{y'} + x^2\, dm = \tfrac{1}{4}r^2\, dm + x^2\, dm = (\tfrac{1}{4}r^2 + x^2)\, dm$$

Substituting the expressions for r and dm into the equation, we obtain

$$dI_y = \left(\frac{1}{4}\frac{a^2}{h^2}x^2 + x^2\right)\left(\rho\pi\frac{a^2}{h^2}x^2\, dx\right) = \rho\pi\frac{a^2}{h^2}\left(\frac{a^2}{4h^2} + 1\right)x^4\, dx$$

$$I_y = \int dI_y = \int_0^h \rho\pi\frac{a^2}{h^2}\left(\frac{a^2}{4h^2} + 1\right)x^4\, dx = \rho\pi\frac{a^2}{h^2}\left(\frac{a^2}{4h^2} + 1\right)\frac{h^5}{5}$$

Introducing the total mass of the cone m, we rewrite I_y as follows:

$$I_y = \tfrac{3}{5}(\tfrac{1}{4}a^2 + h^2)\tfrac{1}{3}\rho\pi a^2 h \qquad I_y = \tfrac{3}{5}m(\tfrac{1}{4}a^2 + h^2) \blacktriangleleft$$

c. Moment of Inertia $\bar{I}_{y''}$. We apply the parallel-axis theorem and write

$$I_y = \bar{I}_{y''} + m\bar{x}^2$$

Solving for $\bar{I}_{y''}$ and recalling from Fig. 5.21 that $\bar{x} = \tfrac{3}{4}h$, we have

$$\bar{I}_{y''} = I_y - m\bar{x}^2 = \tfrac{3}{5}m(\tfrac{1}{4}a^2 + h^2) - m(\tfrac{3}{4}h)^2$$

$$\bar{I}_{y''} = \tfrac{3}{20}m(a^2 + \tfrac{1}{4}h^2) \blacktriangleleft$$

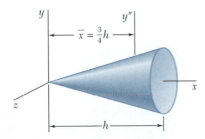

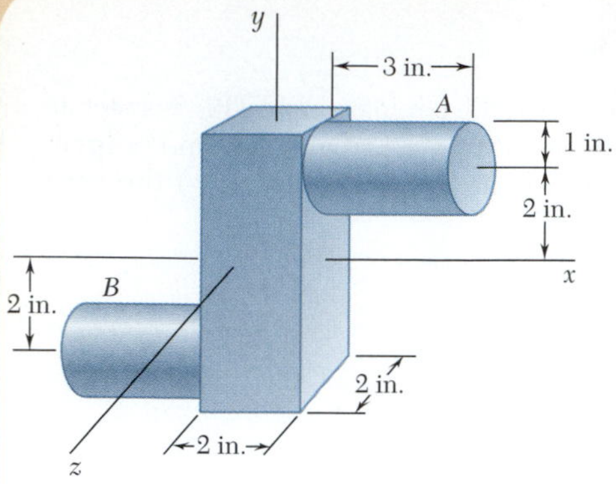

SAMPLE PROBLEM 9.12

A steel forging consists of a 6 × 2 × 2-in. rectangular prism and two cylinders of diameter 2 in. and length 3 in. as shown. Determine the moments of inertia of the forging with respect to the coordinate axes knowing that the specific weight of steel is 490 lb/ft^3.

SOLUTION

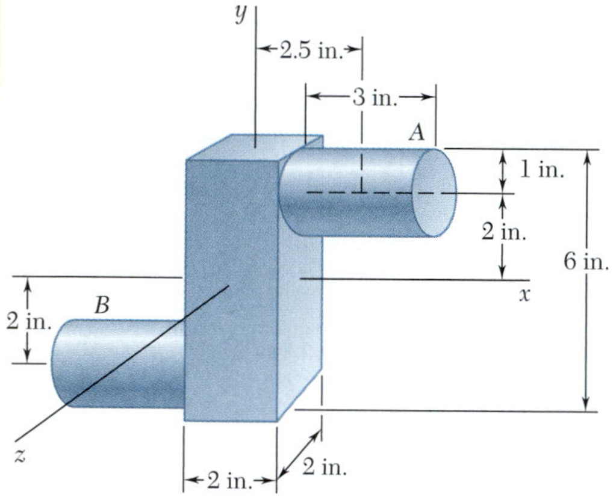

Computation of Masses

Prism

$$V = (2 \text{ in.})(2 \text{ in.})(6 \text{ in.}) = 24 \text{ in}^3$$

$$W = (24 \text{ in}^3)(490 \text{ lb/ft}^3)\left(\frac{1 \text{ ft}}{12 \text{ in.}}\right)^3 = 6.81 \text{ lb}$$

$$m = \frac{6.81 \text{ lb}}{32.2 \text{ ft/s}^2} = 0.211 \text{ lb}\cdot\text{s}^2\text{/ft}$$

Each Cylinder

$$V = \pi(1 \text{ in.})^2(3 \text{ in.}) = 9.42 \text{ in}^3$$

$$W = (9.42 \text{ in}^3)(490 \text{ lb/ft}^3)\left(\frac{1 \text{ ft}}{12 \text{ in.}}\right)^3 = 2.67 \text{ lb}$$

$$m = \frac{2.67 \text{ lb}}{32.2 \text{ ft/s}^2} = 0.0829 \text{ lb}\cdot\text{s}^2\text{/ft}$$

Moments of Inertia. The moments of inertia of each component are computed from Fig. 9.28, using the parallel-axis theorem when necessary. Note that all lengths are expressed in feet.

Prism

$$I_x = I_z = \tfrac{1}{12}(0.211 \text{ lb}\cdot\text{s}^2\text{/ft})[(\tfrac{6}{12}\text{ ft})^2 + (\tfrac{2}{12}\text{ ft})^2] = 4.88 \times 10^{-3} \text{ lb}\cdot\text{ft}\cdot\text{s}^2$$

$$I_y = \tfrac{1}{12}(0.211 \text{ lb}\cdot\text{s}^2\text{/ft})[(\tfrac{2}{12}\text{ ft})^2 + (\tfrac{2}{12}\text{ ft})^2] = 0.977 \times 10^{-3} \text{ lb}\cdot\text{ft}\cdot\text{s}^2$$

Each Cylinder

$$I_x = \tfrac{1}{2}ma^2 + m\bar{y}^2 = \tfrac{1}{2}(0.0829 \text{ lb}\cdot\text{s}^2\text{/ft})(\tfrac{1}{12}\text{ ft})^2 + (0.0829 \text{ lb}\cdot\text{s}^2\text{/ft})(\tfrac{2}{12}\text{ ft})^2$$
$$= 2.59 \times 10^{-3} \text{ lb}\cdot\text{ft}\cdot\text{s}^2$$

$$I_y = \tfrac{1}{12}m(3a^2 + L^2) + m\bar{x}^2 = \tfrac{1}{12}(0.0829 \text{ lb}\cdot\text{s}^2\text{/ft})[3(\tfrac{1}{12}\text{ ft})^2 + (\tfrac{3}{12}\text{ ft})^2]$$
$$+ (0.0829 \text{ lb}\cdot\text{s}^2\text{/ft})(\tfrac{2.5}{12}\text{ ft})^2 = 4.17 \times 10^{-3} \text{ lb}\cdot\text{ft}\cdot\text{s}^2$$

$$I_z = \tfrac{1}{12}m(3a^2 + L^2) + m(\bar{x}^2 + \bar{y}^2) = \tfrac{1}{12}(0.0829 \text{ lb}\cdot\text{s}^2\text{/ft})[3(\tfrac{1}{12}\text{ ft})^2 + (\tfrac{3}{12}\text{ ft})^2]$$
$$+ (0.0829 \text{ lb}\cdot\text{s}^2\text{/ft})[(\tfrac{2.5}{12}\text{ ft})^2 + (\tfrac{2}{12}\text{ ft})^2] = 6.48 \times 10^{-3} \text{ lb}\cdot\text{ft}\cdot\text{s}^2$$

Entire Body. Adding the values obtained,

$$I_x = 4.88 \times 10^{-3} + 2(2.59 \times 10^{-3}) \qquad I_x = 10.06 \times 10^{-3} \text{ lb}\cdot\text{ft}\cdot\text{s}^2 \quad \blacktriangleleft$$

$$I_y = 0.977 \times 10^{-3} + 2(4.17 \times 10^{-3}) \qquad I_y = 9.32 \times 10^{-3} \text{ lb}\cdot\text{ft}\cdot\text{s}^2 \quad \blacktriangleleft$$

$$I_z = 4.88 \times 10^{-3} + 2(6.48 \times 10^{-3}) \qquad I_z = 17.84 \times 10^{-3} \text{ lb}\cdot\text{ft}\cdot\text{s}^2 \quad \blacktriangleleft$$

SAMPLE PROBLEM 9.13

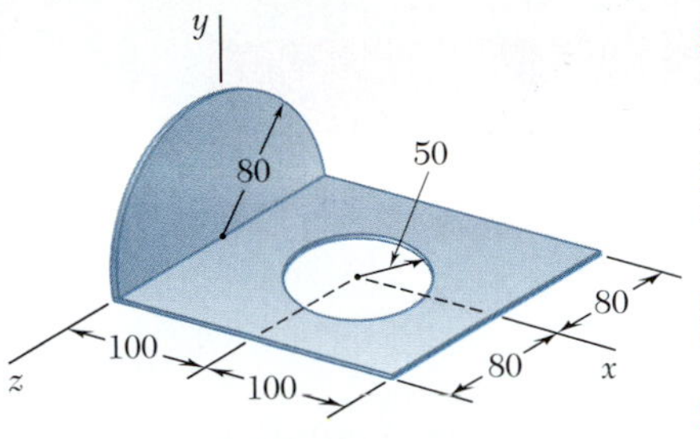

Dimensions in mm

A thin steel plate which is 4 mm thick is cut and bent to form the machine part shown. Knowing that the density of steel is 7850 kg/m^3, determine the moments of inertia of the machine part with respect to the coordinate axes.

SOLUTION

We observe that the machine part consists of a semicircular plate and a rectangular plate from which a circular plate has been removed.

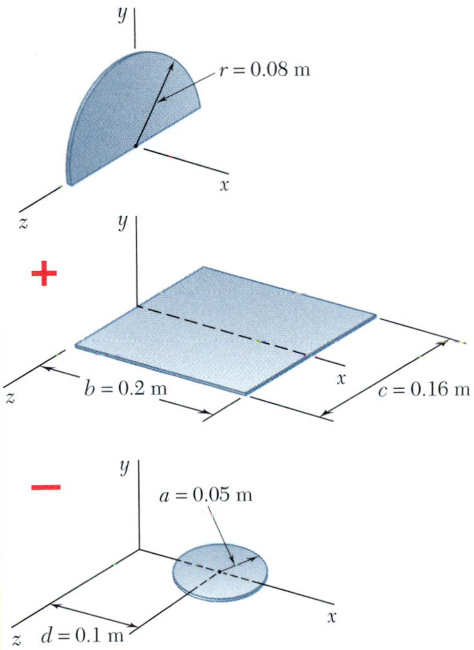

Computation of Masses. ***Semicircular Plate***

$$V_1 = \tfrac{1}{2}\pi r^2 t = \tfrac{1}{2}\pi(0.08\text{ m})^2(0.004\text{ m}) = 40.21 \times 10^{-6}\text{ m}^3$$
$$m_1 = \rho V_1 = (7.85 \times 10^3\text{ kg/m}^3)(40.21 \times 10^{-6}\text{ m}^3) = 0.3156\text{ kg}$$

Rectangular Plate

$$V_2 = (0.200\text{ m})(0.160\text{ m})(0.004\text{ m}) = 128 \times 10^{-6}\text{ m}^3$$
$$m_2 = \rho V_2 = (7.85 \times 10^3\text{ kg/m}^3)(128 \times 10^{-6}\text{ m}^3) = 1.005\text{ kg}$$

Circular Plate

$$V_3 = \pi a^2 t = \pi(0.050\text{ m})^2(0.004\text{ m}) = 31.42 \times 10^{-6}\text{ m}^3$$
$$m_3 = \rho V_3 = (7.85 \times 10^3\text{ kg/m}^3)(31.42 \times 10^{-6}\text{ m}^3) = 0.2466\text{ kg}$$

Moments of Inertia. Using the method presented in Sec. 9.13, we compute the moments of inertia of each component.

Semicircular Plate. From Fig. 9.28, we observe that for a circular plate of mass m and radius r

$$I_x = \tfrac{1}{2}mr^2 \qquad I_y = I_z = \tfrac{1}{4}mr^2$$

Because of symmetry, we note that for a semicircular plate

$$I_x = \tfrac{1}{2}(\tfrac{1}{2}mr^2) \qquad I_y = I_z = \tfrac{1}{2}(\tfrac{1}{4}mr^2)$$

Since the mass of the semicircular plate is $m_1 = \tfrac{1}{2}m$, we have

$$I_x = \tfrac{1}{2}m_1r^2 = \tfrac{1}{2}(0.3156\text{ kg})(0.08\text{ m})^2 = 1.010 \times 10^{-3}\text{ kg}\cdot\text{m}^2$$
$$I_y = I_z = \tfrac{1}{4}(\tfrac{1}{2}mr^2) = \tfrac{1}{4}m_1r^2 = \tfrac{1}{4}(0.3156\text{ kg})(0.08\text{ m})^2 = 0.505 \times 10^{-3}\text{ kg}\cdot\text{m}^2$$

Rectangular Plate

$$I_x = \tfrac{1}{12}m_2c^2 = \tfrac{1}{12}(1.005\text{ kg})(0.16\text{ m})^2 = 2.144 \times 10^{-3}\text{ kg}\cdot\text{m}^2$$
$$I_z = \tfrac{1}{3}m_2b^2 = \tfrac{1}{3}(1.005\text{ kg})(0.2\text{ m})^2 = 13.400 \times 10^{-3}\text{ kg}\cdot\text{m}^2$$
$$I_y = I_x + I_z = (2.144 + 13.400)(10^{-3}) = 15.544 \times 10^{-3}\text{ kg}\cdot\text{m}^2$$

Circular Plate

$$I_x = \tfrac{1}{4}m_3a^2 = \tfrac{1}{4}(0.2466\text{ kg})(0.05\text{ m})^2 = 0.154 \times 10^{-3}\text{ kg}\cdot\text{m}^2$$
$$I_y = \tfrac{1}{2}m_3a^2 + m_3d^2$$
$$= \tfrac{1}{2}(0.2466\text{ kg})(0.05\text{ m})^2 + (0.2466\text{ kg})(0.1\text{ m})^2 = 2.774 \times 10^{-3}\text{ kg}\cdot\text{m}^2$$
$$I_z = \tfrac{1}{4}m_3a^2 + m_3d^2 = \tfrac{1}{4}(0.2466\text{ kg})(0.05\text{ m})^2 + (0.2466\text{ kg})(0.1\text{ m})^2$$
$$= 2.620 \times 10^{-3}\text{ kg}\cdot\text{m}^2$$

Entire Machine Part

$$I_x = (1.010 + 2.144 - 0.154)(10^{-3})\text{ kg}\cdot\text{m}^2 \qquad I_x = 3.00 \times 10^{-3}\text{ kg}\cdot\text{m}^2 \blacktriangleleft$$
$$I_y = (0.505 + 15.544 - 2.774)(10^{-3})\text{ kg}\cdot\text{m}^2 \qquad I_y = 13.28 \times 10^{-3}\text{ kg}\cdot\text{m}^2 \blacktriangleleft$$
$$I_z = (0.505 + 13.400 - 2.620)(10^{-3})\text{ kg}\cdot\text{m}^2 \qquad I_z = 11.29 \times 10^{-3}\text{ kg}\cdot\text{m}^2 \blacktriangleleft$$

SOLVING PROBLEMS ON YOUR OWN

In this lesson we introduced the *mass moment of inertia* and the *radius of gyration* of a three-dimensional body with respect to a given axis [Eqs. (9.28) and (9.29)]. We also derived a *parallel-axis theorem* for use with mass moments of inertia and discussed the computation of the mass moments of inertia of thin plates and three-dimensional bodies.

1. *Computing mass moments of inertia.* The mass moment of inertia I of a body with respect to a given axis can be calculated directly from the definition given in Eq. (9.28) for simple shapes [Sample Prob. 9.9]. In most cases, however, it is necessary to divide the body into thin slabs, compute the moment of inertia of a typical slab with respect to the given axis—using the parallel-axis theorem if necessary—and integrate the expression obtained.

2. *Applying the parallel-axis theorem.* In Sec. 9.12 we derived the parallel-axis theorem for mass moments of inertia

$$I = \bar{I} + md^2 \qquad (9.33)$$

which states that the moment of inertia I of a body of mass m with respect to a given axis is equal to the sum of the moment of inertia $\bar{I}$ of that body with respect to the *parallel centroidal axis* and the product md^2, where d is the distance between the two axes. When the moment of inertia of a three-dimensional body is calculated with respect to one of the coordinate axes, d^2 can be replaced by the sum of the squares of distances measured along the other two coordinate axes [Eqs. (9.32) and (9.32′)].

3. *Avoiding unit-related errors.* To avoid errors, it is essential that you be consistent in your use of units. Thus, all lengths should be expressed in meters or feet, as appropriate, and for problems using U.S. customary units, masses should be given in $\text{lb} \cdot \text{s}^2/\text{ft}$. In addition, we strongly recommend that you include units as you perform your calculations [Sample Probs. 9.12 and 9.13].

4. *Calculating the mass moment of inertia of thin plates.* We showed in Sec. 9.13 that the mass moment of inertia of a thin plate with respect to a given axis can be obtained by multiplying the corresponding moment of inertia of the area of the plate by the density ρ and the thickness t of the plate [Eqs. (9.35) through (9.37)]. Note that since the axis CC' in Fig. 9.24*c* is *perpendicular to the plate*, $I_{CC', \text{mass}}$ is associated with the *polar* moment of inertia $J_{C, \text{area}}$.

Instead of calculating directly the moment of inertia of a thin plate with respect to a specified axis, you may sometimes find it convenient to first compute its moment of inertia with respect to an axis parallel to the specified axis and then apply the parallel-axis theorem. Further, to determine the moment of inertia of a thin plate with respect to an axis perpendicular to the plate, you may wish to first determine its moments of inertia with respect to two perpendicular in-plane axes and then use Eq. (9.38). Finally, remember that the mass of a plate of area A, thickness t, and density ρ is $m = \rho t A$.

5. *Determining the moment of inertia of a body by direct single integration.* We discussed in Sec. 9.14 and illustrated in Sample Probs. 9.10 and 9.11 how single integration can be used to compute the moment of inertia of a body that can be divided into a series of thin, parallel slabs. For such cases, you will often need to express the mass of the body in terms of the body's density and dimensions. Assuming that the body has been divided, as in the sample problems, into thin slabs perpendicular to the x axis, you will need to express the dimensions of each slab as functions of the variable x.

a. *In the special case of a body of revolution,* the elemental slab is a thin disk, and the equations given in Fig. 9.27 should be used to determine the moments of inertia of the body [Sample Prob. 9.11].

b. *In the general case, when the body is not of revolution,* the differential element is not a disk, but a thin slab of a different shape, and the equations of Fig. 9.27 cannot be used. See, for example, Sample Prob. 9.10, where the element was a thin, rectangular slab. For more complex configurations, you may want to use one or more of the following equations, which are based on Eqs. (9.32) and (9.32′) of Sec. 9.12.

$$dI_x = dI_{x'} + (\overline{y}^2_{el} + \overline{z}^2_{el})\, dm$$
$$dI_y = dI_{y'} + (\overline{z}^2_{el} + \overline{x}^2_{el})\, dm$$
$$dI_z = dI_{z'} + (\overline{x}^2_{el} + \overline{y}^2_{el})\, dm$$

where the primes denote the centroidal axes of each elemental slab, and where $\overline{x}_{el}$, $\overline{y}_{el}$, and $\overline{z}_{el}$ represent the coordinates of its centroid. The centroidal moments of inertia of the slab are determined in the manner described earlier for a thin plate: Referring to Fig. 9.12 on page 485, calculate the corresponding moments of inertia of the area of the slab and multiply the result by the density ρ and the thickness t of the slab. Also, assuming that the body has been divided into thin slabs perpendicular to the x axis, remember that you can obtain $dI_{x'}$ by adding $dI_{y'}$ and $dI_{z'}$ instead of computing it directly. Finally, using the geometry of the body, express the result obtained in terms of the single variable x and integrate in x.

6. *Computing the moment of inertia of a composite body.* As stated in Sec. 9.15, the moment of inertia of a composite body with respect to a specified axis is equal to the sum of the moments of its components with respect to that axis. Sample Probs. 9.12 and 9.13 illustrate the appropriate method of solution. You must also remember that the moment of inertia of a component will be negative only if the component is *removed* (as in the case of a hole).

Although the composite-body problems in this lesson are relatively straightforward, you will have to work carefully to avoid computational errors. In addition, if some of the moments of inertia that you need are not given in Fig. 9.28, you will have to derive your own formulas using the techniques of this lesson.

Problems

9.113 The quarter ring shown has a mass m and was cut from a thin, uniform plate. Knowing that $r_1 = \frac{1}{2}r_2$, determine the mass moment of inertia of the quarter ring with respect to (*a*) axis AA', (*b*) the centroidal axis CC' that is perpendicular to the plane of the quarter ring.

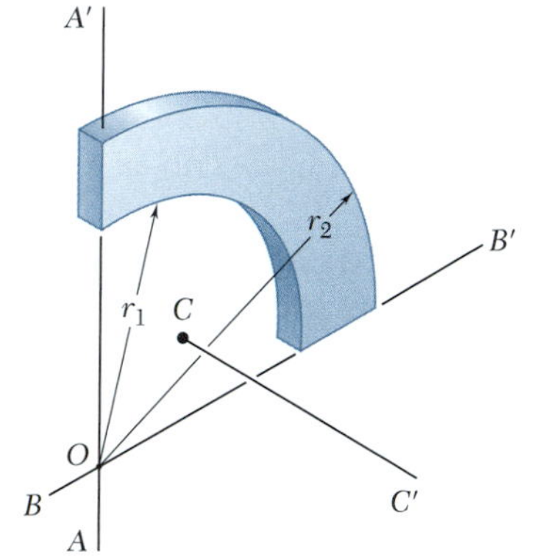

Fig. P9.113

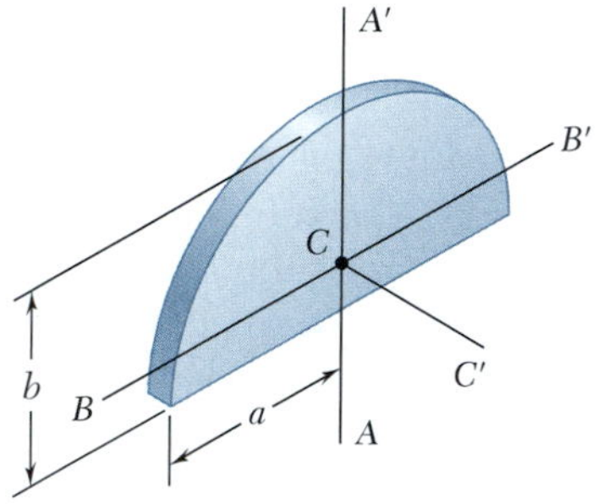

Fig. P9.114

9.114 A thin, semielliptical plate has a mass m. Determine the mass moment of inertia of the plate with respect to (*a*) the centroidal axis BB', (*b*) the centroidal axis CC' that is perpendicular to the plate.

9.115 The elliptical ring shown was cut from a thin, uniform plate. Denoting the mass of the ring by m, determine its moment of inertia with respect to (*a*) the centroidal axis BB', (*b*) the centroidal axis CC' that is perpendicular to the plane of the ring.

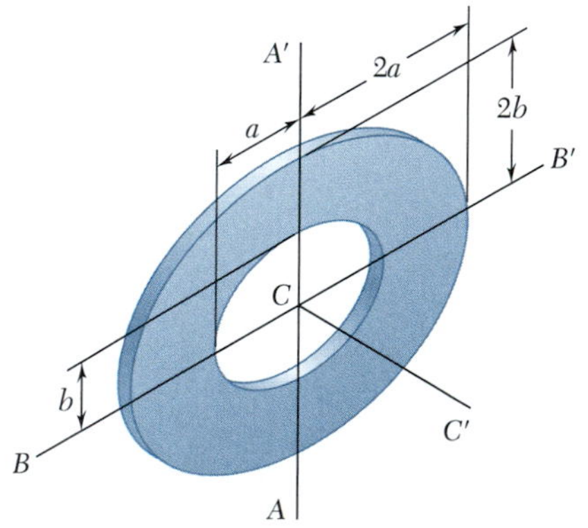

Fig. *P9.115*

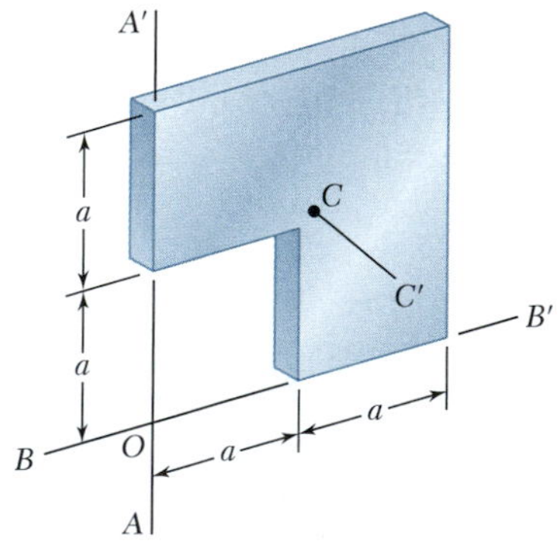

Fig. P9.116

9.116 The machine component shown was cut from a thin, uniform plate. Denoting the mass of the component by m, determine its mass moment of inertia with respect to (*a*) the axis BB', (*b*) the centroidal axis CC' that is perpendicular to the plane of the component.

9.117 The rhombus shown has a mass m and was cut from a thin, uniform plate. Determine the mass moment of inertia of the rhombus with respect to (*a*) the x axis, (*b*) the y axis.

9.118 The rhombus shown has a mass m and was cut from a thin, uniform plate. Knowing that the AA' and BB' axes are parallel to the z axis and lie in a plane parallel to and at a distance a above the zx plane, determine the mass moment of inertia of the rhombus with respect to (*a*) the axis AA', (*b*) the axis BB'.

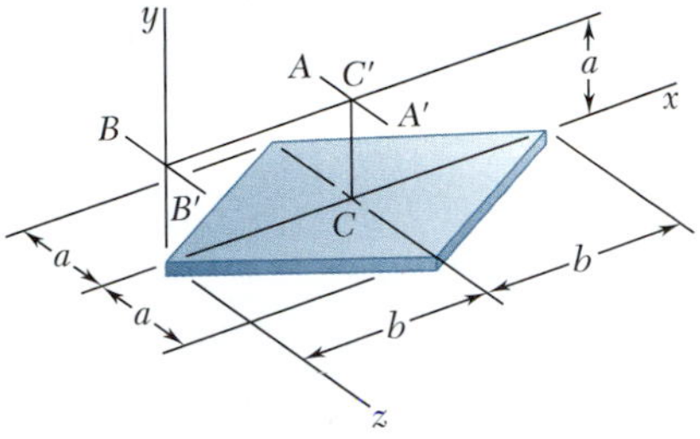

Fig. P9.117 and *P9.118*

9.119 A thin plate of mass m has the trapezoidal shape shown. Determine the mass moment of inertia of the plate with respect to (*a*) the x axis, (*b*) the y axis.

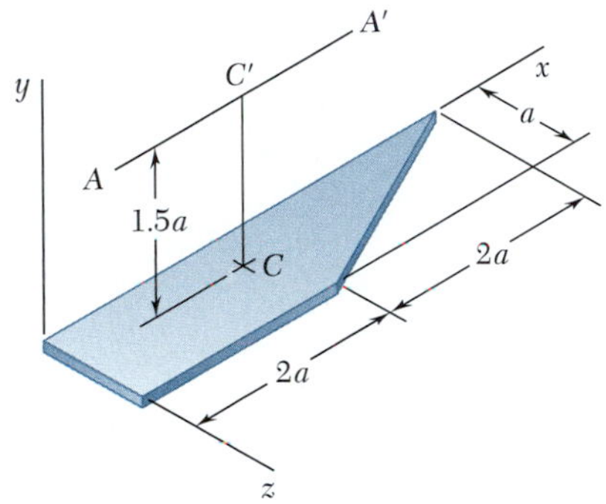

Fig. P9.119 and P9.120

9.120 A thin plate of mass m has the trapezoidal shape shown. Determine the mass moment of inertia of the plate with respect to (*a*) the centroidal axis CC' that is perpendicular to the plate, (*b*) the axis AA' which is parallel to the x axis and is located at a distance $1.5a$ from the plate.

9.121 The parabolic spandrel shown is revolved about the x axis to form a homogeneous solid of revolution of mass m. Using direct integration, express the moment of inertia of the solid with respect to the x axis in terms of m and b.

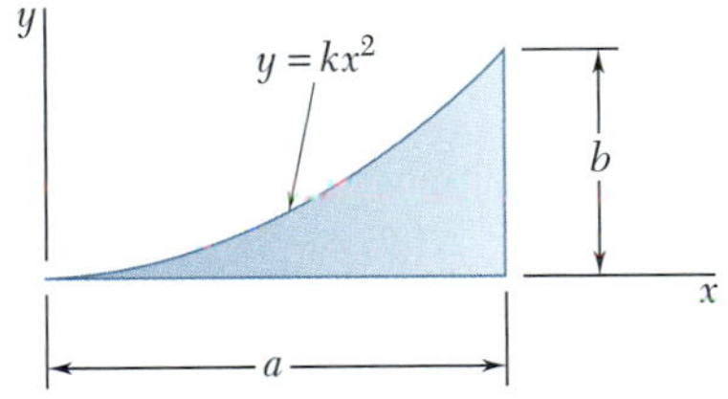

Fig. P9.121

9.122 Determine by direct integration the moment of inertia with respect to the z axis of the right circular cylinder shown assuming that it has a uniform density and a mass m.

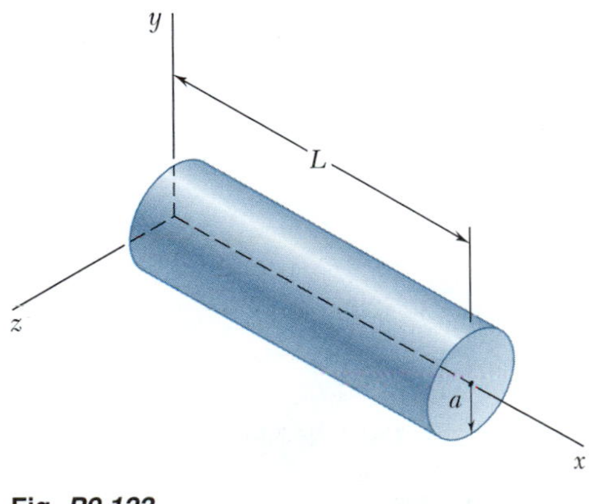

Fig. *P9.122*

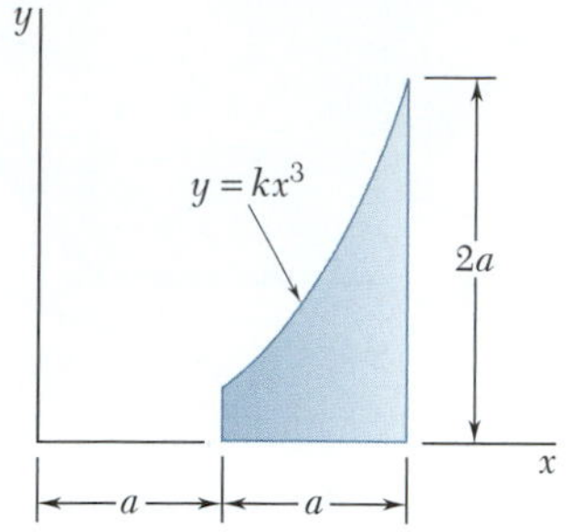

Fig. P9.123

9.123 The area shown is revolved about the x axis to form a homogeneous solid of revolution of mass m. Determine by direct integration the moment of inertia of the solid with respect to (a) the x axis, (b) the y axis. Express your answers in terms of m and a.

9.124 Determine by direct integration the moment of inertia with respect to the x axis of the tetrahedron shown assuming that it has a uniform density and a mass m.

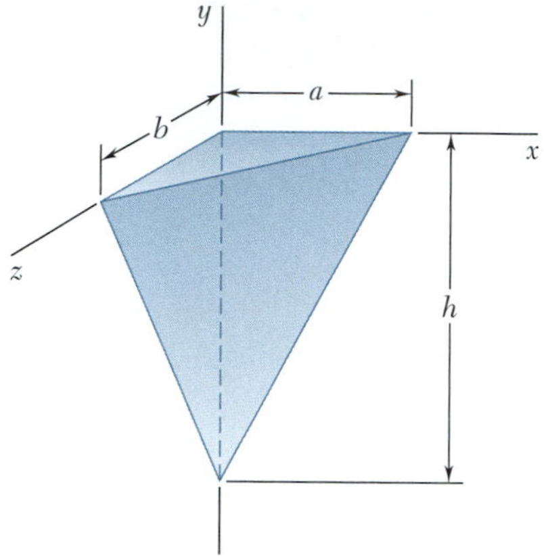

Fig. P9.124 and *P9.125*

9.125 Determine by direct integration the moment of inertia with respect to the y axis of the tetrahedron shown assuming that it has a uniform density and a mass m.

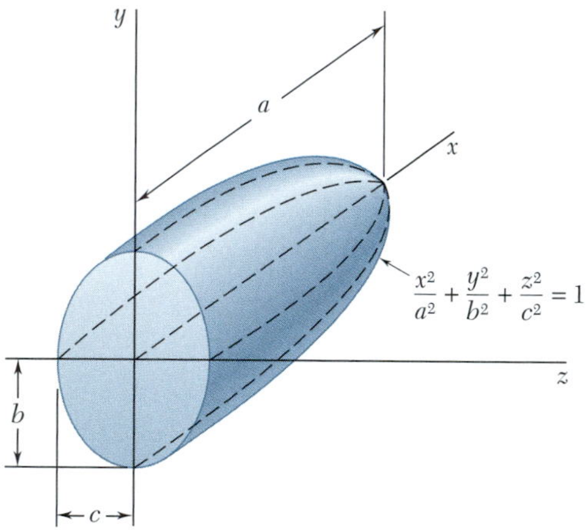

Fig. P9.126

***9.126** Determine by direct integration the moment of inertia with respect to the z axis of the semiellipsoid shown assuming that it has a uniform density and a mass m.

***9.127** A thin steel wire is bent into the shape shown. Denoting the mass per unit length of the wire by m', determine by direct integration the moment of inertia of the wire with respect to each of the coordinate axes.

9.128 A thin triangular plate of mass m is welded along its base AB to a block as shown. Knowing that the plate forms an angle θ with the y axis, determine by direct integration the mass moment of inertia of the plate with respect to (a) the x axis, (b) the y axis, (c) the z axis.

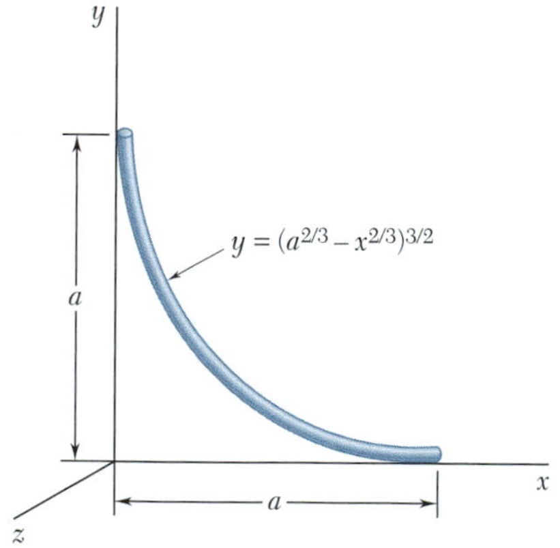

Fig. P9.127

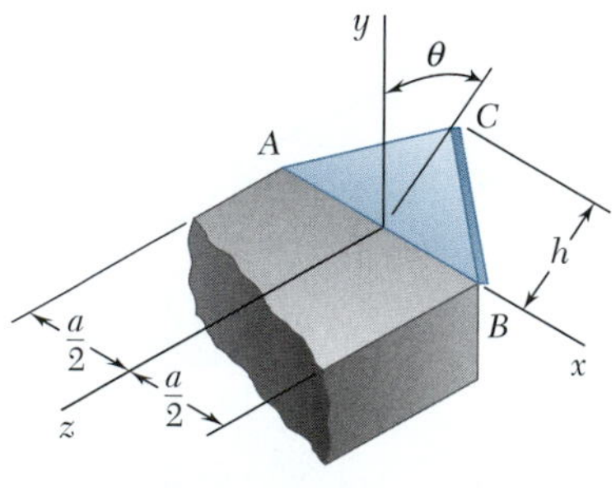

Fig. P9.128

9.129 Shown is the cross section of a molded flat-belt pulley. Determine its mass moment of inertia and its radius of gyration with respect to the axis AA'. (The specific weight of brass is 0.306 lb/in^3 and the specific weight of the fiber-reinforced polycarbonate used is 0.0433 lb/in^3.)

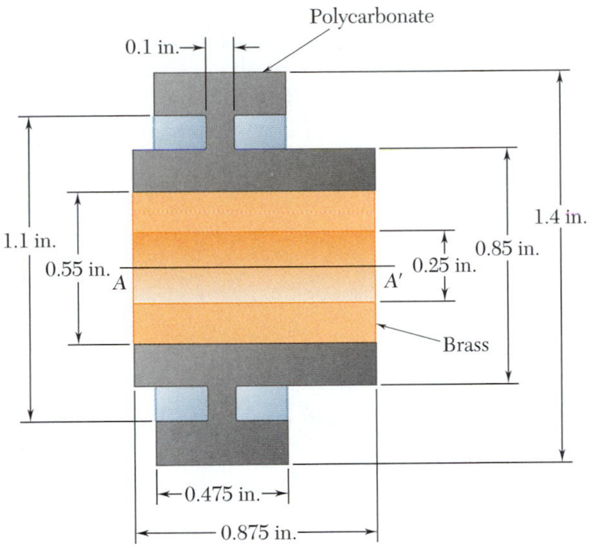

Fig. P9.129

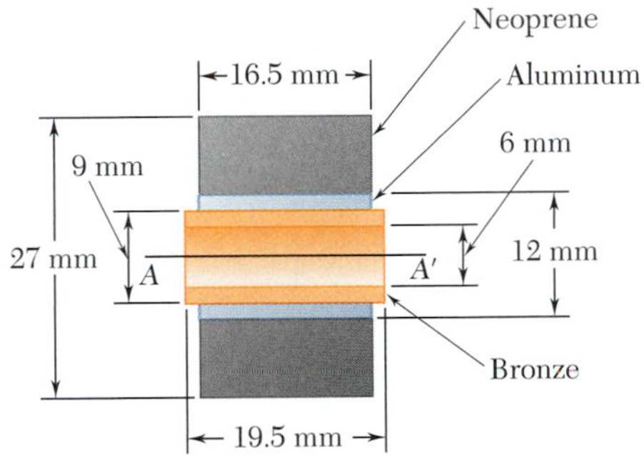

Fig. P9.130

9.130 Shown is the cross section of an idler roller. Determine its moment of inertia and its radius of gyration with respect to the axis AA'. (The density of bronze is 8580 kg/m^3; of aluminum, 2770 kg/m^3; and of neoprene, 1250 kg/m^3.)

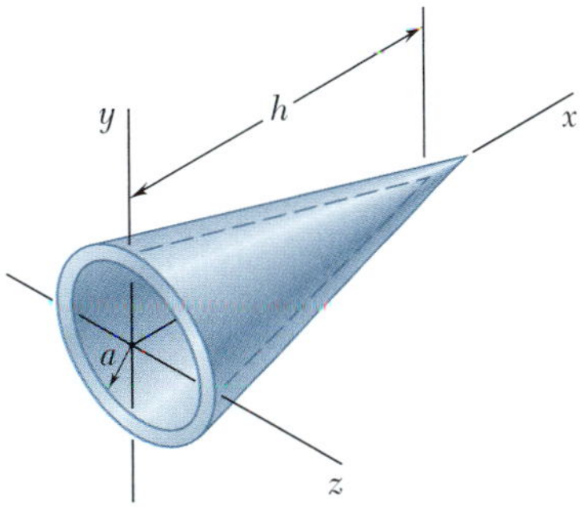

Fig. P9.131

9.131 Given the dimensions and the mass m of the thin conical shell shown, determine the moment of inertia and the radius of gyration of the shell with respect to the x axis. (*Hint:* Assume that the shell was formed by removing a cone with a circular base of radius a from a cone with a circular base of radius $a + t$. In the resulting expressions, neglect terms containing t^2, t^3, etc. Do not forget to account for the difference in the heights of the two cones.)

9.132 A portion of an 8-in.-long steel rod of diameter 1.50 in. is turned to form the conical section shown. Knowing that the turning process reduces the moment of inertia of the rod with respect to the x axis by 20 percent, determine the height h of the cone.

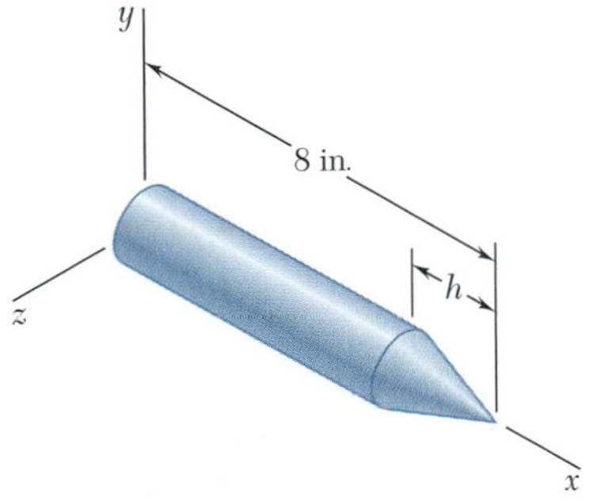

Fig. P9.132

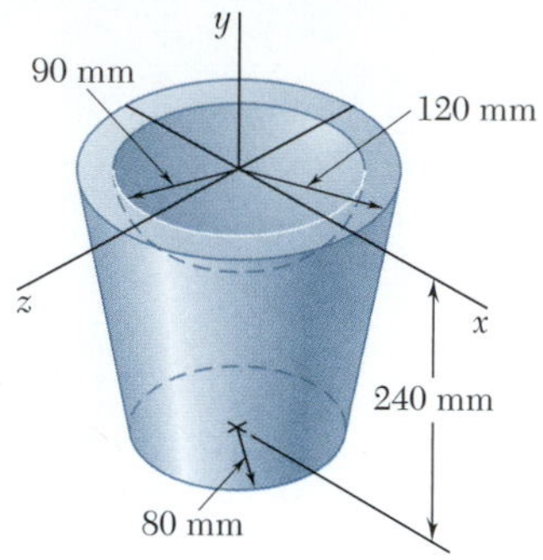

Fig. P9.133

9.133 The steel machine component shown is formed by machining a hemisphere into the base of a truncated cone. Knowing that the density of steel is 7850 kg/m^3, determine the mass moment of inertia of the component with respect to the y axis.

9.134 After a period of use, one of the blades of a shredder has been worn to the shape shown and is of weight 0.4 lb. Knowing that the moments of inertia of the blade with respect to the AA' and BB' axes are 0.6×10^{-3} lb · ft · s^2 and 1.26×10^{-3} lb · ft · s^2, respectively, determine (*a*) the location of the centroidal axis GG', (*b*) the radius of gyration with respect to axis GG'.

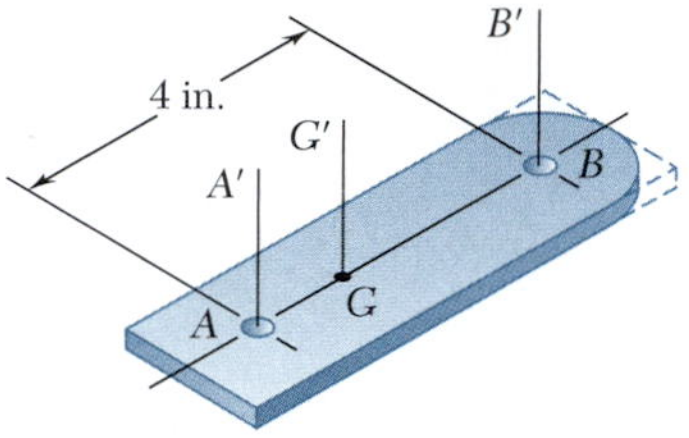

Fig. *P9.134*

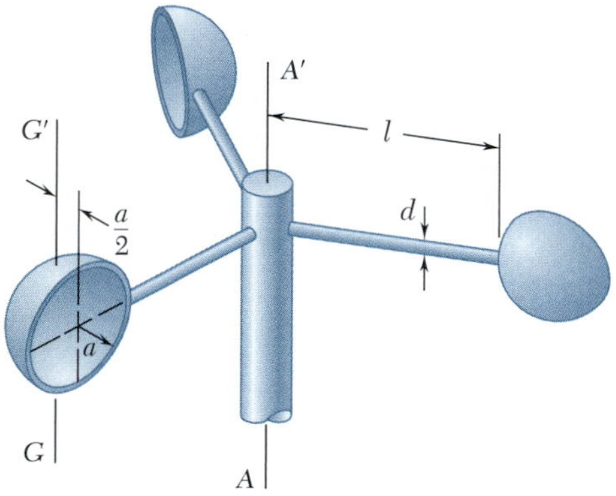

Fig. *P9.135*

9.135 The cups and the arms of an anemometer are fabricated from a material of density ρ. Knowing that the moment of inertia of a thin, hemispherical shell of mass m and thickness t with respect to its centroidal axis GG' is $5ma^2/12$, determine (*a*) the moment of inertia of the anemometer with respect to the axis AA', (*b*) the ratio of a to l for which the centroidal moment of inertia of the cups is equal to 1 percent of the moment of inertia of the cups with respect to the axis AA'.

9.136 A square hole is centered in and extends through the aluminum machine component shown. Determine (*a*) the value of a for which the mass moment of inertia of the component with respect to the axis AA', which bisects the top surface of the hole, is maximum, (*b*) the corresponding values of the mass moment of inertia and the radius of gyration with respect to the axis AA'. (The density of aluminum is 2800 kg/m^3.)

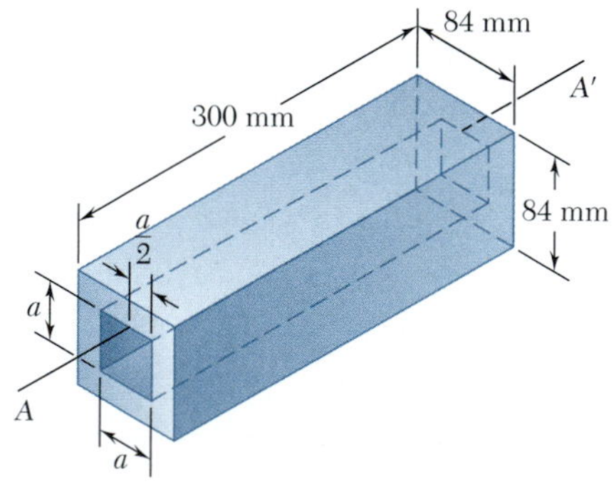

Fig. P9.136

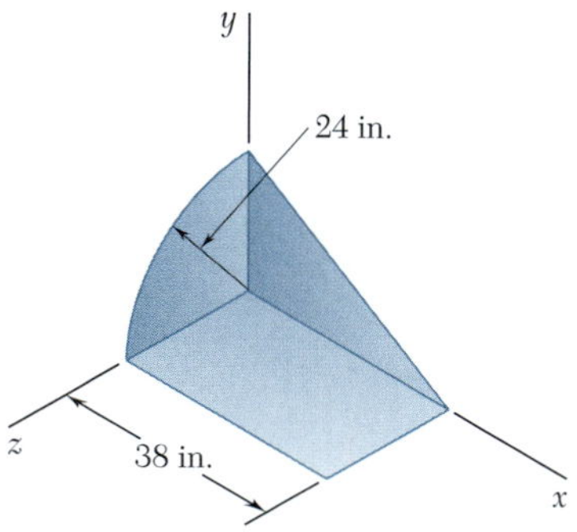

Fig. P9.137

9.137 A 0.1-in.-thick piece of sheet metal is cut and bent into the machine component shown. Knowing that the specific weight of steel is 0.284 lb/in^3, determine the moment of inertia of the component with respect to each of the coordinate axes.

9.138 A 3-mm-thick piece of sheet metal is cut and bent into the machine component shown. Knowing that the density of steel is 7850 kg/m^3, determine the moment of inertia of the component with respect to each of the coordinate axes.

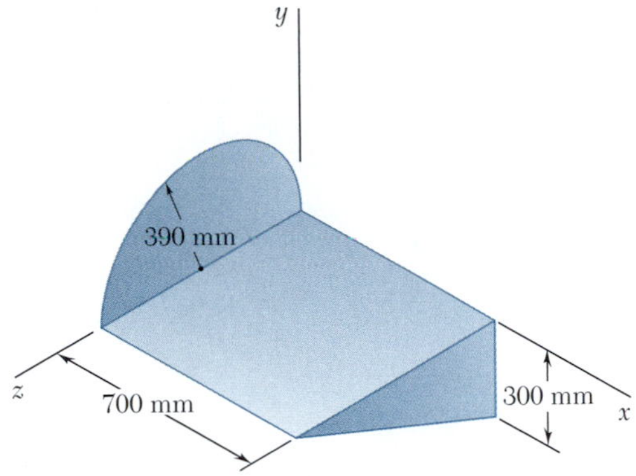

Fig. *P9.138*

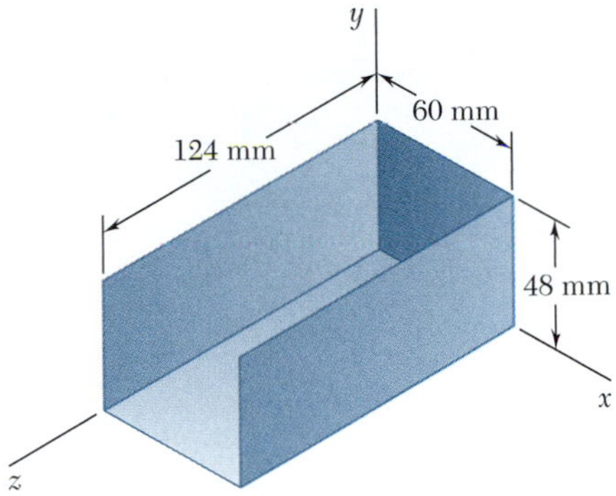

Fig. P9.139

9.139 The cover of an electronic device is formed from sheet aluminum that is 2 mm thick. Determine the mass moment of inertia of the cover with respect to each of the coordinate axes. (The density of aluminum is 2770 kg/m^3.)

9.140 A framing anchor is formed of 2-mm-thick galvanized steel. Determine the mass moment of inertia of the anchor with respect to each of the coordinate axes. (The density of galvanized steel is 7530 kg/m^3.)

9.141 A 2-mm-thick piece of sheet steel is cut and bent into the machine component shown. Knowing that the density of steel is 7850 kg/m^3, determine the moment of inertia of the component with respect to each of the coordinate axes.

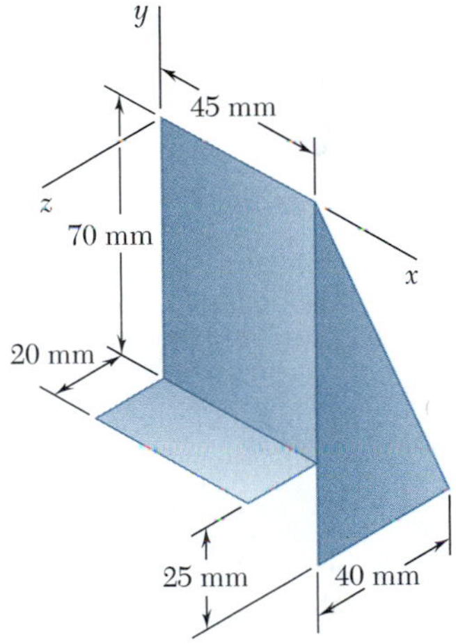

Fig. *P9.140*

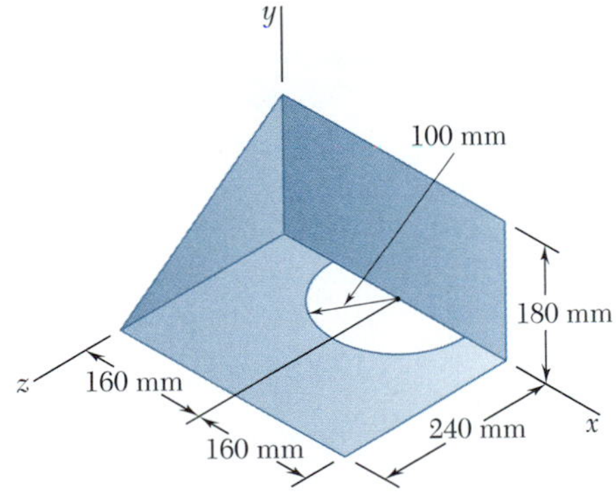

Fig. P9.141

***9.142** The piece of roof flashing shown is formed from sheet copper that is 0.032 in. thick. Knowing that the specific weight of copper is 558 lb/ft^3, determine the mass moment of inertia of the flashing with respect to each of the coordinate axes.

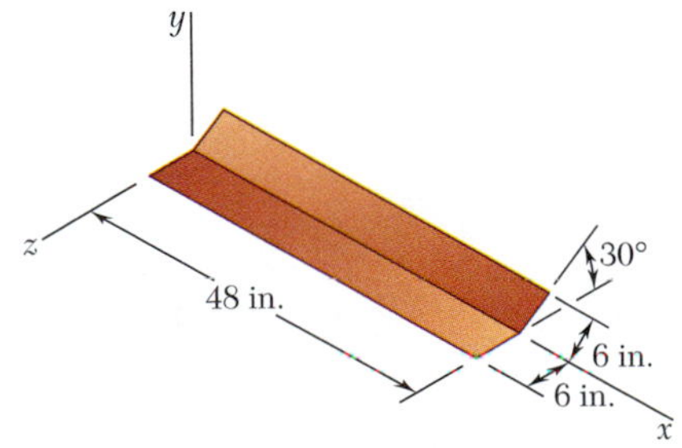

Fig. P9.142

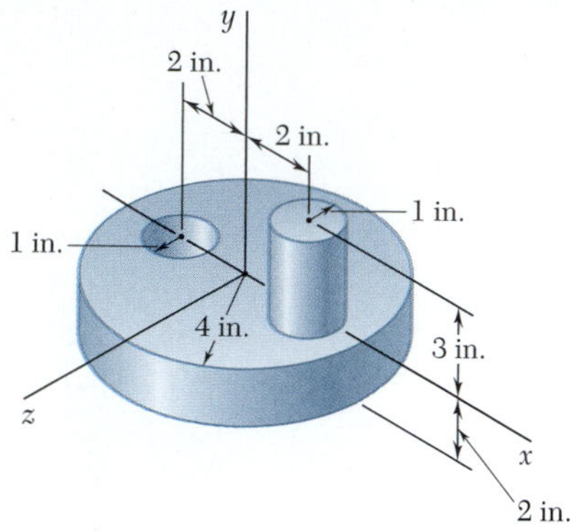

Fig. P9.143

9.143 The machine element shown is fabricated from steel. Determine the mass moment of inertia of the assembly with respect to (*a*) the x axis, (*b*) the y axis, (*c*) the z axis. (The specific weight of steel is 0.284 lb/in^3.)

9.144 Determine the mass moment of inertia of the steel machine element shown with respect to the y axis. (The density of steel is 7850 kg/m^3.)

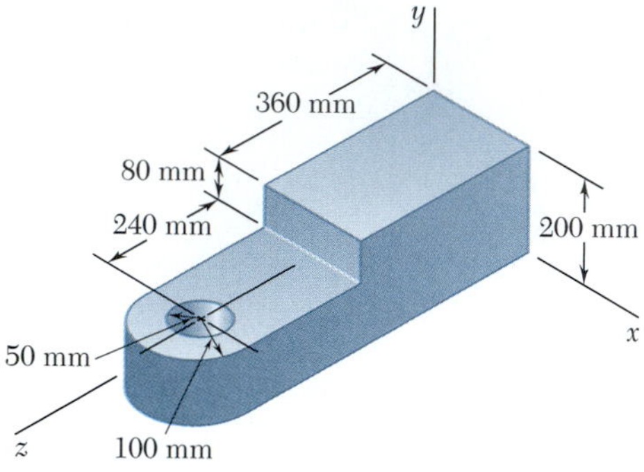

Fig. P9.144 and P9.145

9.145 Determine the mass moment of inertia of the steel machine element shown with respect to the z axis. (The density of steel is 7850 kg/m^3.)

9.146 An aluminum casting has the shape shown. Knowing that the specific weight of aluminum is 0.100 lb/in^3, determine the moment of inertia of the casting with respect to the z axis.

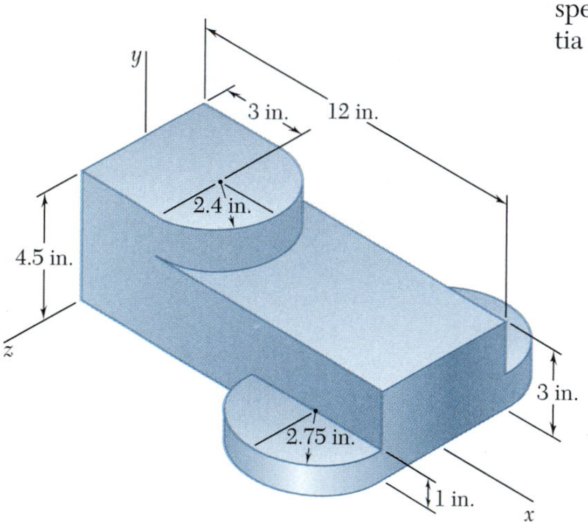

Fig. *P9.146*

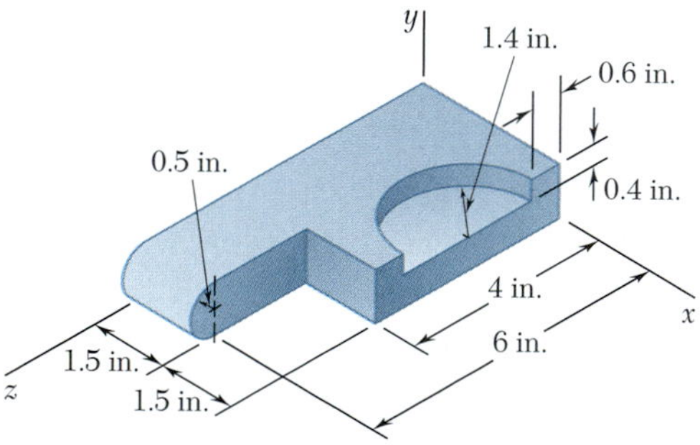

Fig. P9.147

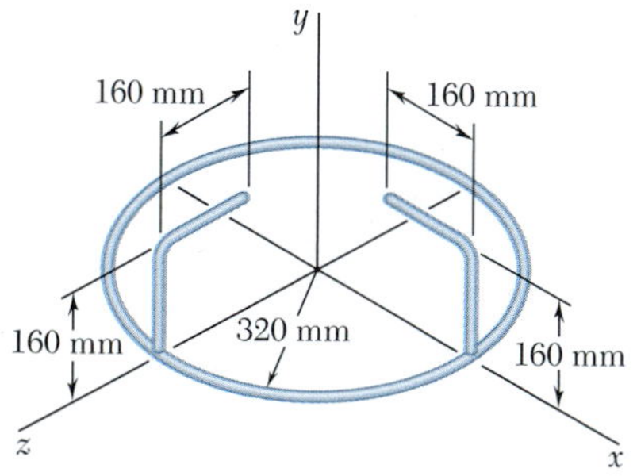

Fig. P9.148

9.147 Determine the moment of inertia of the steel machine element shown with respect to (*a*) the x axis, (*b*) the y axis, (*c*) the z axis. (The specific weight of steel is 490 lb/ft^3.)

9.148 Aluminum wire with a mass per unit length of 0.049 kg/m is used to form the circle and the straight members of the figure shown. Determine the mass moment of inertia of the assembly with respect to each of the coordinate axes.

9.149 The figure shown is formed of 3-mm-diameter steel wire. Knowing that the density of the steel is 7850 kg/m^3, determine the mass moment of inertia of the wire with respect to each of the coordinate axes.

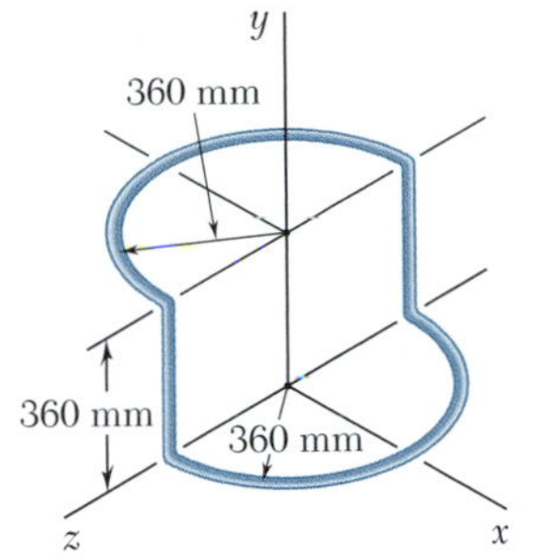

Fig. P9.149

9.150 A homogeneous wire with a weight per unit length of 0.041 lb/ft is used to form the figure shown. Determine the moment of inertia of the wire with respect to each of the coordinate axes.

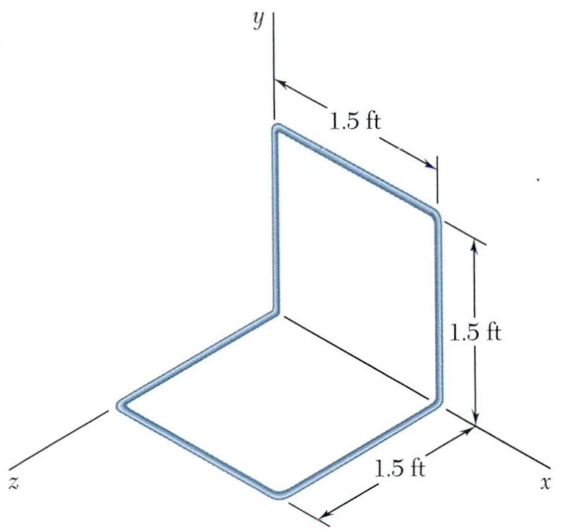

Fig. *P9.150*

*9.16. MOMENT OF INERTIA OF A BODY WITH RESPECT TO AN ARBITRARY AXIS THROUGH *O*. MASS PRODUCTS OF INERTIA

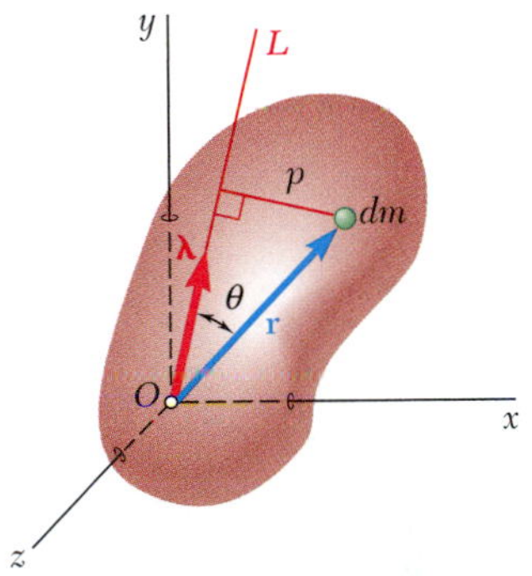

Fig. 9.29

In this section you will see how the moment of inertia of a body can be determined with respect to an arbitrary axis OL through the origin (Fig. 9.29) if its moments of inertia with respect to the three coordinate axes, as well as certain other quantities to be defined below, have already been determined.

The moment of inertia I_{OL} of the body with respect to OL is equal to $\int p^2\,dm$, where p denotes the perpendicular distance from the element of mass dm to the axis OL. If we denote by $\boldsymbol{\lambda}$ the unit vector along OL and by $\mathbf{r}$ the position vector of the element dm, we observe that the perpendicular distance p is equal to $r \sin\theta$, which is the magnitude of the vector product $\boldsymbol{\lambda} \times \mathbf{r}$. We therefore write

$$I_{OL} = \int p^2\,dm = \int |\boldsymbol{\lambda} \times \mathbf{r}|^2\,dm \tag{9.43}$$

Expressing $|\boldsymbol{\lambda} \times \mathbf{r}|^2$ in terms of the rectangular components of the vector product, we have

$$I_{OL} = \int [(\lambda_x y - \lambda_y x)^2 + (\lambda_y z - \lambda_z y)^2 + (\lambda_z x - \lambda_x z)^2]\,dm$$

where the components λ_x, λ_y, λ_z of the unit vector $\boldsymbol{\lambda}$ represent the direction cosines of the axis OL and the components x, y, z of $\mathbf{r}$ represent the coordinates of the element of mass dm. Expanding the squares and rearranging the terms, we write

$$I_{OL} = \lambda_x^2 \int (y^2 + z^2)\,dm + \lambda_y^2 \int (z^2 + x^2)\,dm + \lambda_z^2 \int (x^2 + y^2)\,dm$$
$$- 2\lambda_x\lambda_y \int xy\,dm - 2\lambda_y\lambda_z \int yz\,dm - 2\lambda_z\lambda_x \int zx\,dm \tag{9.44}$$

Referring to Eqs. (9.30), we note that the first three integrals in (9.44) represent, respectively, the moments of inertia I_x, I_y, and I_z of the body with respect to the coordinate axes. The last three integrals in (9.44), which involve products of coordinates, are called the *products of inertia* of the body with respect to the x and y axes, the y and z axes, and the z and x axes, respectively. We write

$$I_{xy} = \int xy\,dm \qquad I_{yz} = \int yz\,dm \qquad I_{zx} = \int zx\,dm \tag{9.45}$$

Rewriting Eq. (9.44) in terms of the integrals defined in Eqs. (9.30) and (9.45), we have

$$I_{OL} = I_x\lambda_x^2 + I_y\lambda_y^2 + I_z\lambda_z^2 - 2I_{xy}\lambda_x\lambda_y - 2I_{yz}\lambda_y\lambda_z - 2I_{zx}\lambda_z\lambda_x \tag{9.46}$$

We note that the definition of the products of inertia of a mass given in Eqs. (9.45) is an extension of the definition of the product of inertia of an area (Sec. 9.8). Mass products of inertia reduce to zero under the same conditions of symmetry as do products of inertia of areas, and the parallel-axis theorem for mass products of inertia is expressed by relations similar to the formula derived for the product of inertia of an area. Substituting the expressions for x, y, and z given in Eqs. (9.31) into Eqs. (9.45), we find that

$$\begin{aligned} I_{xy} &= \bar{I}_{x'y'} + m\bar{x}\bar{y} \\ I_{yz} &= \bar{I}_{y'z'} + m\bar{y}\bar{z} \\ I_{zx} &= \bar{I}_{z'x'} + m\bar{z}\bar{x} \end{aligned} \tag{9.47}$$

where $\bar{x}$, $\bar{y}$, $\bar{z}$ are the coordinates of the center of gravity G of the body and $\bar{I}_{x'y'}$, $\bar{I}_{y'z'}$, $\bar{I}_{z'x'}$ denote the products of inertia of the body with respect to the centroidal axes x', y', z' (Fig. 9.22).

*9.17. ELLIPSOID OF INERTIA. PRINCIPAL AXES OF INERTIA

Let us assume that the moment of inertia of the body considered in the preceding section has been determined with respect to a large number of axes OL through the fixed point O and that a point Q has been plotted on each axis OL at a distance $OQ = 1/\sqrt{I_{OL}}$ from O. The locus of the points Q thus obtained forms a surface (Fig. 9.30). The equation of that surface can be obtained by substituting $1/(OQ)^2$ for I_{OL} in (9.46) and then multiplying both sides of the equation by $(OQ)^2$. Observing that

$$(OQ)\lambda_x = x \qquad (OQ)\lambda_y = y \qquad (OQ)\lambda_z = z$$

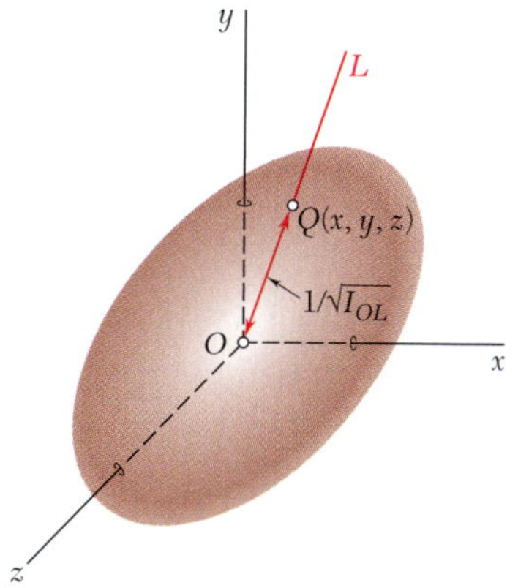

Fig. 9.30

where x, y, z denote the rectangular coordinates of Q, we write

$$I_x x^2 + I_y y^2 + I_z z^2 - 2I_{xy}xy - 2I_{yz}yz - 2I_{zx}zx = 1 \tag{9.48}$$

The equation obtained is the equation of a *quadric surface*. Since the moment of inertia I_{OL} is different from zero for every axis OL, no point Q can be at an infinite distance from O. Thus, the quadric surface obtained is an *ellipsoid*. This ellipsoid, which defines the moment of inertia of the body with respect to any axis through O, is known as the *ellipsoid of inertia* of the body at O.

We observe that if the axes in Fig. 9.30 are rotated, the coefficients of the equation defining the ellipsoid change, since they are equal to the moments and products of inertia of the body with respect to the rotated coordinate axes. However, the *ellipsoid itself remains unaffected,* since its shape depends only upon the distribution of mass in the given body. Suppose that we choose as coordinate axes the principal axes x', y', z' of the ellipsoid of inertia (Fig. 9.31). The equation of the ellipsoid with respect to these coordinate axes is known to be of the form

$$I_{x'}x'^2 + I_{y'}y'^2 + I_{z'}z'^2 = 1 \qquad (9.49)$$

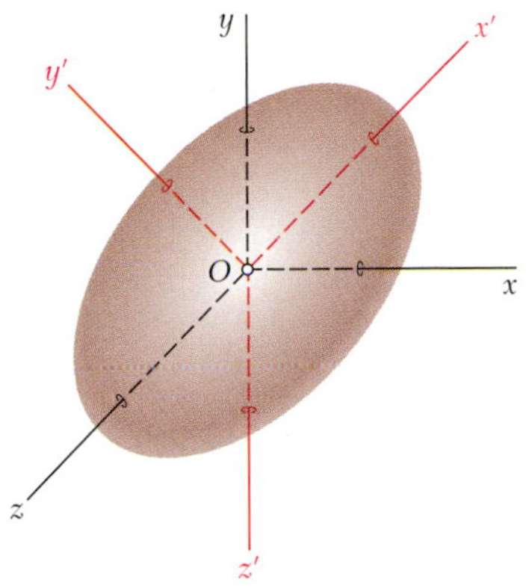

Fig. 9.31

which does not contain any products of the coordinates. Comparing Eqs. (9.48) and (9.49), we observe that the products of inertia of the body with respect to the x', y', z' axes must be zero. The x', y', z' axes are known as the *principal axes of inertia* of the body at O, and the coefficients $I_{x'}$, $I_{y'}$, $I_{z'}$ are referred to as the *principal moments of inertia* of the body at O. Note that, given a body of arbitrary shape and a point O, it is always possible to find axes which are the principal axes of inertia of the body at O, that is, axes with respect to which the products of inertia of the body are zero. Indeed, whatever the shape of the body, the moments and products of inertia of the body with respect to x, y, and z axes through O will define an ellipsoid, and this ellipsoid will have principal axes which, by definition, are the principal axes of inertia of the body at O.

If the principal axes of inertia x', y', z' are used as coordinate axes, the expression obtained in Eq. (9.46) for the moment of inertia of a body with respect to an arbitrary axis through O reduces to

$$I_{OL} = I_{x'}\lambda_{x'}^2 + I_{y'}\lambda_{y'}^2 + I_{z'}\lambda_{z'}^2 \qquad (9.50)$$

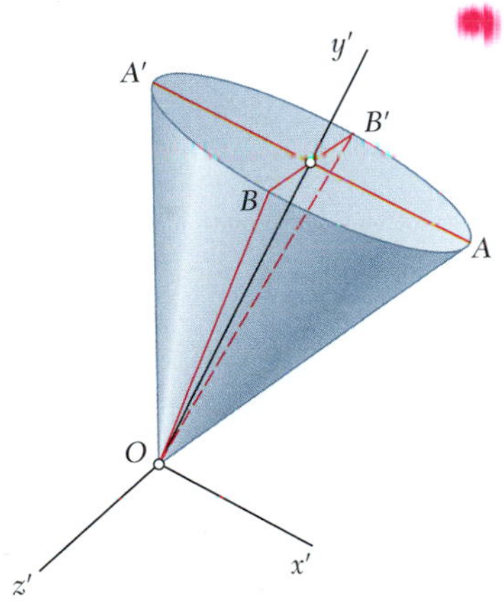

Fig. 9.32

The determination of the principal axes of inertia of a body of arbitrary shape is somewhat involved and will be discussed in the next section. There are many cases, however, where these axes can be spotted immediately. Consider, for instance, the homogeneous cone of elliptical base shown in Fig. 9.32; this cone possesses two mutually perpendicular planes of symmetry OAA' and OBB'. From the definition (9.45), we observe that if the $x'y'$ and $y'z'$ planes are chosen to coincide with the two planes of symmetry, all of the products of inertia are zero. The x', y', and z' axes thus selected are therefore the principal axes of inertia of the cone at O. In the case of the homogeneous regular tetrahedron $OABC$ shown in Fig. 9.33, the line joining the corner O to the center D of the opposite face is a principal axis of inertia at O, and any line through O perpendicular to OD is also a principal axis of inertia at O. This property is apparent if we observe that rotating the tetrahedron through 120° about OD leaves its shape and mass distribution unchanged. It follows that the ellipsoid of inertia at O also remains unchanged under this rotation. The ellipsoid, therefore, is a body of revolution whose axis of revolution is OD, and the line OD, as well as any perpendicular line through O, must be a principal axis of the ellipsoid.

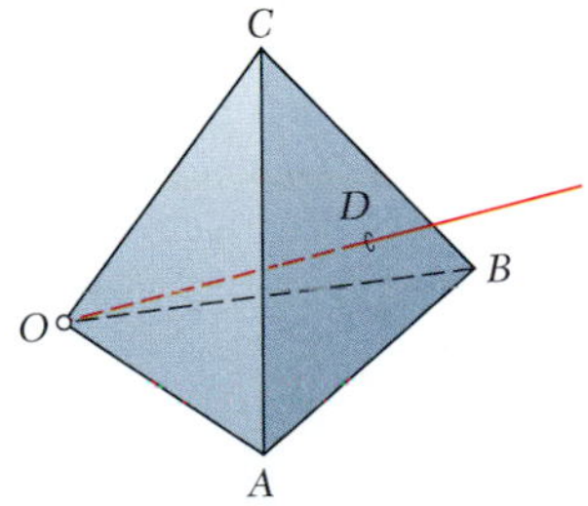

Fig. 9.33

*9.18. DETERMINATION OF THE PRINCIPAL AXES AND PRINCIPAL MOMENTS OF INERTIA OF A BODY OF ARBITRARY SHAPE

The method of analysis described in this section should be used when the body under consideration has no obvious property of symmetry.

Consider the ellipsoid of inertia of the body at a given point O (Fig. 9.34); let $\mathbf{r}$ be the radius vector of a point P on the surface of the ellipsoid and let $\mathbf{n}$ be the unit vector along the normal to that surface at P. We observe that the only points where $\mathbf{r}$ and $\mathbf{n}$ are collinear are the points P_1, P_2, and P_3, where the principal axes intersect the visible portion of the surface of the ellipsoid, and the corresponding points on the other side of the ellipsoid.

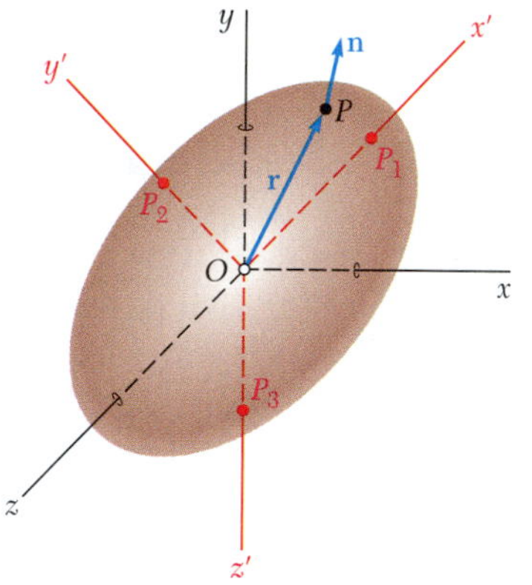

Fig. 9.34

We now recall from calculus that the direction of the normal to a surface of equation $f(x, y, z) = 0$ at a point $P(x, y, z)$ is defined by the gradient ∇f of the function f at that point. To obtain the points where the principal axes intersect the surface of the ellipsoid of inertia, we must therefore write that $\mathbf{r}$ and ∇f are collinear,

$$\nabla f = (2K)\mathbf{r} \tag{9.51}$$

where K is a constant, $\mathbf{r} = x\mathbf{i} + y\mathbf{j} + z\mathbf{k}$, and

$$\nabla f = \frac{\partial f}{\partial x}\mathbf{i} + \frac{\partial f}{\partial y}\mathbf{j} + \frac{\partial f}{\partial x}\mathbf{k}$$

Recalling Eq. (9.48), we note that the function $f(x, y, z)$ corresponding to the ellipsoid of inertia is

$$f(x, y, z) = I_x x^2 + I_y y^2 + I_z z^2 - 2I_{xy}xy - 2I_{yz}yz - 2I_{zx}zx - 1$$

Substituting for $\mathbf{r}$ and ∇f into Eq. (9.51) and equating the coefficients of the unit vectors, we write

$$\begin{aligned} I_x x - I_{xy}y - I_{zx}z &= Kx \\ -I_{xy}x + I_y y - I_{yz}z &= Ky \\ -I_{zx}x - I_{yz}y + I_z z &= Kz \end{aligned} \tag{9.52}$$

Dividing each term by the distance r from O to P, we obtain similar equations involving the direction cosines λ_x, λ_y, and λ_z:

$$\begin{aligned} I_x\lambda_x - I_{xy}\lambda_y - I_{zx}\lambda_z &= K\lambda_x \\ -I_{xy}\lambda_x + I_y\lambda_y - I_{yz}\lambda_z &= K\lambda_y \\ -I_{zx}\lambda_x - I_{yz}\lambda_y + I_z\lambda_z &= K\lambda_z \end{aligned} \tag{9.53}$$

Transposing the right-hand members leads to the following homogeneous linear equations:

$$\begin{aligned} (I_x - K)\lambda_x - I_{xy}\lambda_y - I_{zx}\lambda_z &= 0 \\ -I_{xy}\lambda_x + (I_y - K)\lambda_y - I_{yz}\lambda_z &= 0 \\ -I_{zx}\lambda_x - I_{yz}\lambda_y + (I_z - K)\lambda_z &= 0 \end{aligned} \tag{9.54}$$

For this system of equations to have a solution different from $\lambda_x = \lambda_y = \lambda_z = 0$, its discriminant must be zero:

$$\begin{vmatrix} I_x - K & -I_{xy} & -I_{zx} \\ -I_{xy} & I_y - K & -I_{yz} \\ -I_{zx} & -I_{yz} & I_z - K \end{vmatrix} = 0 \tag{9.55}$$

Expanding this determinant and changing signs, we write

$$\begin{aligned} K^3 - (I_x + I_y + I_z)K^2 + (I_xI_y + I_yI_z + I_zI_x - I_{xy}^2 - I_{yz}^2 - I_{zx}^2)K \\ - (I_xI_yI_z - I_xI_{yz}^2 - I_yI_{zx}^2 - I_zI_{xy}^2 - 2I_{xy}I_{yz}I_{zx}) = 0 \end{aligned} \tag{9.56}$$

This is a cubic equation in K, which yields three real, positive roots K_1, K_2, and K_3.

To obtain the direction cosines of the principal axis corresponding to the root K_1, we substitute K_1 for K in Eqs. (9.54). Since these equations are now linearly dependent, only two of them may be used to determine λ_x, λ_y, and λ_z. An additional equation may be obtained, however, by recalling from Sec. 2.12 that the direction cosines must satisfy the relation

$$\lambda_x^2 + \lambda_y^2 + \lambda_z^2 = 1 \tag{9.57}$$

Repeating this procedure with K_2 and K_3, we obtain the direction cosines of the other two principal axes.

We will now show that *the roots K_1, K_2, and K_3 of Eq. (9.56) are the principal moments of inertia of the given body*. Let us substitute for K in Eqs. (9.53) the root K_1, and for λ_x, λ_y, and λ_z the corresponding values $(\lambda_x)_1$, $(\lambda_y)_1$, and $(\lambda_z)_1$ of the direction cosines; the three equations will be satisfied. We now multiply by $(\lambda_x)_1$, $(\lambda_y)_1$, and $(\lambda_z)_1$, respectively, each term in the first, second, and third equation and add the equations obtained in this way. We write

$$\begin{aligned} I_x^2(\lambda_x)_1^2 + I_y^2(\lambda_y)_1^2 + I_z^2(\lambda_z)_1^2 - 2I_{xy}(\lambda_x)_1(\lambda_y)_1 \\ - 2I_{yz}(\lambda_y)_1(\lambda_z)_1 - 2I_{zx}(\lambda_z)_1(\lambda_x)_1 = K_1[(\lambda_x)_1^2 + (\lambda_y)_1^2 + (\lambda_z)_1^2] \end{aligned}$$

Recalling Eq. (9.46), we observe that the left-hand member of this equation represents the moment of inertia of the body with respect to the principal axis corresponding to K_1; it is thus the principal moment of inertia corresponding to that root. On the other hand, recalling Eq. (9.57), we note that the right-hand member reduces to K_1. Thus K_1 itself is the principal moment of inertia. We can show in the same fashion that K_2 and K_3 are the other two principal moments of inertia of the body.

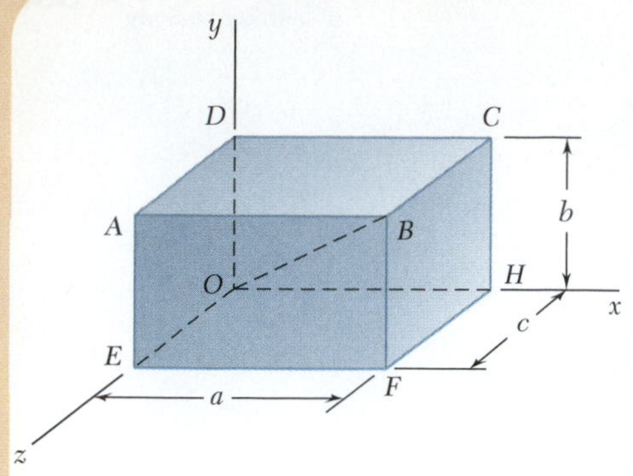

SAMPLE PROBLEM 9.14

Consider a rectangular prism of mass m and sides a, b, c. Determine (*a*) the moments and products of inertia of the prism with respect to the coordinate axes shown, (*b*) its moment of inertia with respect to the diagonal OB.

SOLUTION

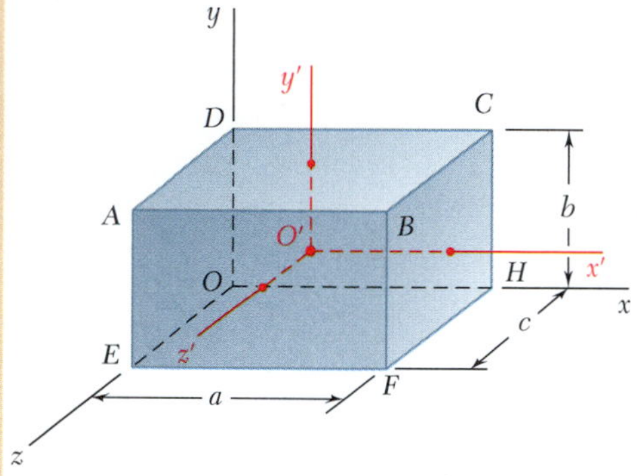

a. Moments and Products of Inertia with Respect to the Coordinate Axes. ***Moments of Inertia.*** Introducing the centroidal axes x', y', z', with respect to which the moments of inertia are given in Fig. 9.28, we apply the parallel-axis theorem:

$$I_x = \bar{I}_{x'} + m(\bar{y}^2 + \bar{z}^2) = \tfrac{1}{12}m(b^2 + c^2) + m(\tfrac{1}{4}b^2 + \tfrac{1}{4}c^2)$$

$$I_x = \tfrac{1}{3}m(b^2 + c^2) \quad ◀$$

Similarly, $\qquad I_y = \tfrac{1}{3}m(c^2 + a^2) \qquad I_z = \tfrac{1}{3}m(a^2 + b^2)$ ◀

Products of Inertia. Because of symmetry, the products of inertia with respect to the centroidal axes x', y', z' are zero, and these axes are principal axes of inertia. Using the parallel-axis theorem, we have

$$I_{xy} = \bar{I}_{x'y'} + m\bar{x}\bar{y} = 0 + m(\tfrac{1}{2}a)(\tfrac{1}{2}b) \qquad I_{xy} = \tfrac{1}{4}mab \quad ◀$$

Similarly, $\qquad I_{yz} = \tfrac{1}{4}mbc \qquad I_{zx} = \tfrac{1}{4}mca$ ◀

b. Moment of Inertia with Respect to OB. We recall Eq. (9.46):

$$I_{OB} = I_x\lambda_x^2 + I_y\lambda_y^2 + I_z\lambda_z^2 - 2I_{xy}\lambda_x\lambda_y - 2I_{yz}\lambda_y\lambda_z - 2I_{zx}\lambda_z\lambda_x$$

where the direction cosines of OB are

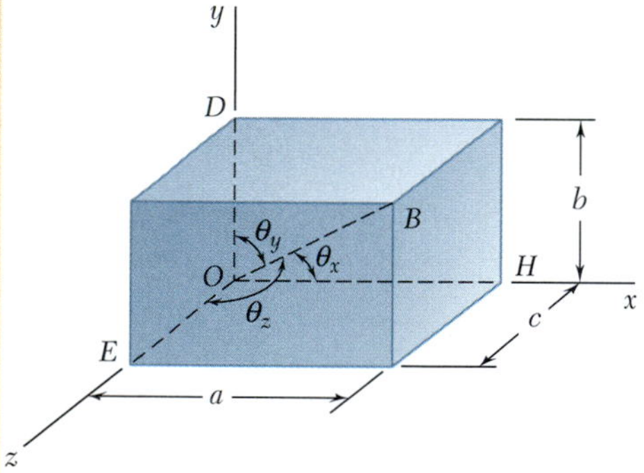

$$\lambda_x = \cos\theta_x = \frac{OH}{OB} = \frac{a}{(a^2 + b^2 + c^2)^{1/2}}$$

$$\lambda_y = \frac{b}{(a^2 + b^2 + c^2)^{1/2}} \qquad \lambda_z = \frac{c}{(a^2 + b^2 + c^2)^{1/2}}$$

Substituting the values obtained for the moments and products of inertia and for the direction cosines into the equation for I_{OB}, we have

$$I_{OB} = \frac{1}{a^2 + b^2 + c^2}[\tfrac{1}{3}m(b^2 + c^2)a^2 + \tfrac{1}{3}m(c^2 + a^2)b^2 + \tfrac{1}{3}m(a^2 + b^2)c^2 - \tfrac{1}{2}ma^2b^2 - \tfrac{1}{2}mb^2c^2 - \tfrac{1}{2}mc^2a^2]$$

$$I_{OB} = \frac{m}{6}\,\frac{a^2b^2 + b^2c^2 + c^2a^2}{a^2 + b^2 + c^2} \quad ◀$$

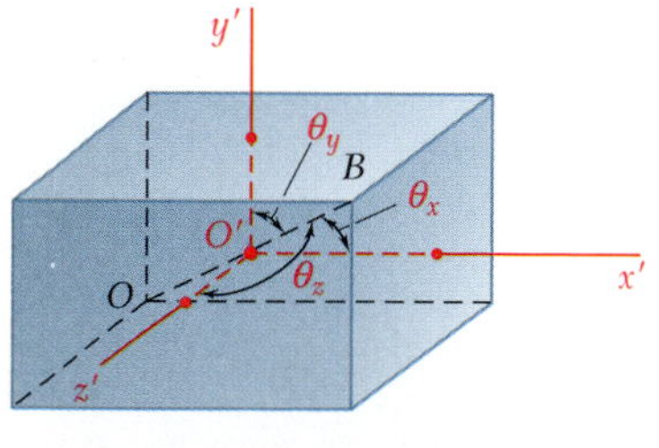

Alternative Solution. The moment of inertia I_{OB} can be obtained directly from the principal moments of inertia $\bar{I}_{x'}$, $\bar{I}_{y'}$, $\bar{I}_{z'}$, since the line OB passes through the centroid O'. Since the x', y', z' axes are principal axes of inertia, we use Eq. (9.50) to write

$$I_{OB} = \bar{I}_{x'}\lambda_x^2 + \bar{I}_{y'}\lambda_y^2 + \bar{I}_{z'}\lambda_z^2$$

$$= \frac{1}{a^2 + b^2 + c^2}\left[\frac{m}{12}(b^2 + c^2)a^2 + \frac{m}{12}(c^2 + a^2)b^2 + \frac{m}{12}(a^2 + b^2)c^2\right]$$

$$I_{OB} = \frac{m}{6}\,\frac{a^2b^2 + b^2c^2 + c^2a^2}{a^2 + b^2 + c^2} \quad ◀$$

SAMPLE PROBLEM 9.15

If $a = 3c$ and $b = 2c$ for the rectangular prism of Sample Prob. 9.14, determine (*a*) the principal moments of inertia at the origin *O*, (*b*) the principal axes of inertia at *O*.

SOLUTION

a. Principal Moments of Inertia at the Origin O. Substituting $a = 3c$ and $b = 2c$ into the solution to Sample Prob. 9.14, we have

$$I_x = \tfrac{5}{3}mc^2 \qquad I_y = \tfrac{10}{3}mc^2 \qquad I_z = \tfrac{13}{3}mc^2$$
$$I_{xy} = \tfrac{3}{2}mc^2 \qquad I_{yz} = \tfrac{1}{2}mc^2 \qquad I_{zx} = \tfrac{3}{4}mc^2$$

Substituting the values of the moments and products of inertia into Eq. (9.56) and collecting terms yields

$$K^3 - (\tfrac{28}{3}mc^2)K^2 + (\tfrac{3479}{144}m^2c^4)K - \tfrac{589}{54}m^3c^6 = 0$$

We then solve for the roots of this equation; from the discussion in Sec. 9.18, it follows that these roots are the principal moments of inertia of the body at the origin.

$$K_1 = 0.568867mc^2 \qquad K_2 = 4.20885mc^2 \qquad K_3 = 4.55562mc^2$$
$$K_1 = 0.569mc^2 \qquad K_2 = 4.21mc^2 \qquad K_3 = 4.56mc^2 \quad ◀$$

b. Principal Axes of Inertia at O. To determine the direction of a principal axis of inertia, we first substitute the corresponding value of *K* into two of the equations (9.54); the resulting equations together with Eq. (9.57) constitute a system of three equations from which the direction cosines of the corresponding principal axis can be determined. Thus, we have for the first principal moment of inertia K_1:

$$(\tfrac{5}{3} - 0.568867)\,mc^2(\lambda_x)_1 - \tfrac{3}{2}mc^2(\lambda_y)_1 - \tfrac{3}{4}mc^2(\lambda_z)_1 = 0$$
$$-\tfrac{3}{2}mc^2(\lambda_x)_1 + (\tfrac{10}{3} - 0.568867)\,mc^2(\lambda_y)_1 - \tfrac{1}{2}mc^2(\lambda_z)_1 = 0$$
$$(\lambda_x)_1^2 + (\lambda_y)_1^2 + (\lambda_z)_1^2 = 1$$

Solving yields

$$(\lambda_x)_1 = 0.836600 \qquad (\lambda_y)_1 = 0.496001 \qquad (\lambda_z)_1 = 0.232557$$

The angles that the first principal axis of inertia forms with the coordinate axes are then

$$(\theta_x)_1 = 33.2° \qquad (\theta_y)_1 = 60.3° \qquad (\theta_z)_1 = 76.6° \quad ◀$$

Using the same set of equations successively with K_2 and K_3, we find that the angles associated with the second and third principal moments of inertia at the origin are, respectively,

$$(\theta_x)_2 = 57.8° \qquad (\theta_y)_2 = 146.6° \qquad (\theta_z)_2 = 98.0° \quad ◀$$

and

$$(\theta_x)_3 = 82.8° \qquad (\theta_y)_3 = 76.1° \qquad (\theta_z)_3 = 164.3° \quad ◀$$

SOLVING PROBLEMS ON YOUR OWN

In this lesson we defined the *mass products of inertia* I_{xy}, I_{yz}, and I_{zx} of a body and showed you how to determine the moments of inertia of that body with respect to an arbitrary axis passing through the origin O. You also learned how to determine at the origin O the *principal axes of inertia* of a body and the corresponding *principal moments of inertia.*

1. Determining the mass products of inertia of a composite body. The mass products of inertia of a composite body with respect to the coordinate axes can be expressed as the sums of the products of inertia of its component parts with respect to those axes. For each component part, we can use the parallel-axis theorem and write Eqs. (9.47)

$$I_{xy} = \bar{I}_{x'y'} + m\bar{x}\bar{y} \qquad I_{yz} = \bar{I}_{y'z'} + m\bar{y}\bar{z} \qquad I_{zx} = \bar{I}_{z'x'} + m\bar{z}\bar{x}$$

where the primes denote the centroidal axes of each component part and where $\bar{x}$, $\bar{y}$, and $\bar{z}$ represent the coordinates of its center of gravity. Keep in mind that the mass products of inertia can be positive, negative, or zero, and be sure to take into account the signs of $\bar{x}$, $\bar{y}$, and $\bar{z}$.

a. From the properties of symmetry of a component part, you can deduce that two or all three of its centroidal mass products of inertia are zero. For instance, you can verify that for a thin plate parallel to the xy plane; a wire lying in a plane parallel to the xy plane; a body with a plane of symmetry parallel to the xy plane; and a body with an axis of symmetry parallel to the z axis, *the products of inertia* $\bar{I}_{y'z'}$ *and* $\bar{I}_{z'x'}$ *are zero.*

For rectangular, circular, or semicircular plates with axes of symmetry parallel to the coordinate axes; straight wires parallel to a coordinate axis; circular and semicircular wires with axes of symmetry parallel to the coordinate axes; and rectangular prisms with axes of symmetry parallel to the coordinate axes, *the products of inertia* $\bar{I}_{x'y'}$, $\bar{I}_{y'z'}$, *and* $\bar{I}_{z'x'}$ *are all zero.*

b. Mass products of inertia which are different from zero can be computed from Eqs. (9.45). Although, in general, a triple integration is required to determine a mass product of inertia, a single integration can be used if the given body can be divided into a series of thin, parallel slabs. The computations are then similar to those discussed in the previous lesson for moments of inertia.

2. *Computing the moment of inertia of a body with respect to an arbitrary axis OL.* An expression for the moment of inertia I_{OL} was derived in Sec. 9.16 and is given in Eq. (9.46). Before computing I_{OL}, you must first determine the mass moments and products of inertia of the body with respect to the given coordinate axes as well as the direction cosines of the unit vector $\boldsymbol{\lambda}$ along OL.

3. *Calculating the principal moments of inertia of a body and determining its principal axes of inertia.* You saw in Sec. 9.17 that it is always possible to find an orientation of the coordinate axes for which the mass products of inertia are zero. These axes are referred to as the *principal axes of inertia* and the corresponding moments of inertia are known as the *principal moments of inertia* of the body. In many cases, the principal axes of inertia of a body can be determined from its properties of symmetry. The procedure required to determine the principal moments and principal axes of inertia of a body with no obvious property of symmetry was discussed in Sec. 9.18 and was illustrated in Sample Prob. 9.15. It consists of the following steps.

a. Expand the determinant in Eq. (9.55) and solve the resulting cubic equation. The solution can be obtained by trial and error or, preferably, with an advanced scientific calculator or with the appropriate computer software. The roots K_1, K_2, and K_3 of this equation are the principal moments of inertia of the body.

b. To determine the direction of the principal axis corresponding to K_1, substitute this value for K in two of the equations (9.54) and solve these equations together with Eq. (9.57) for the direction cosines of the principal axis corresponding to K_1.

c. Repeat this procedure with K_2 and K_3 to determine the directions of the other two principal axes. As a check of your computations, you may wish to verify that the scalar product of any two of the unit vectors along the three axes you have obtained is zero and, thus, that these axes are perpendicular to each other.

d. When a principal moment of inertia is approximately equal to a moment of inertia with respect to a coordinate axis, the calculated values of the corresponding direction cosines will be very sensitive to the number of significant figures used in your computations. Thus, for this case we suggest you express your intermediate answers in terms of six or seven significant figures to avoid possible errors.

Problems

9.151 Determine the products of inertia I_{xy}, I_{yz}, and I_{zx} of the steel machine element shown. (The specific weight of steel is 490 lb/ft^3.)

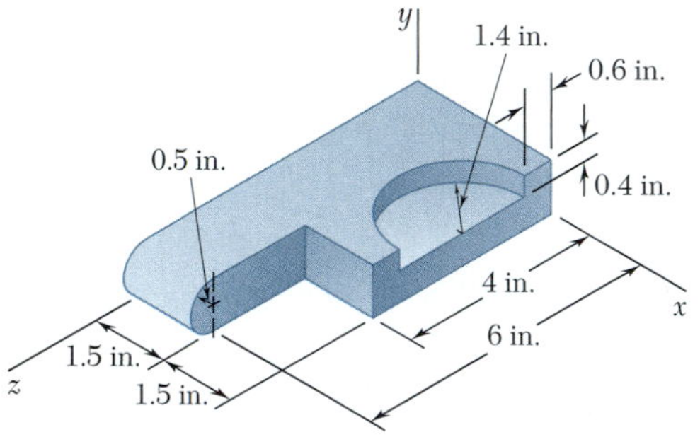

Fig. P9.151

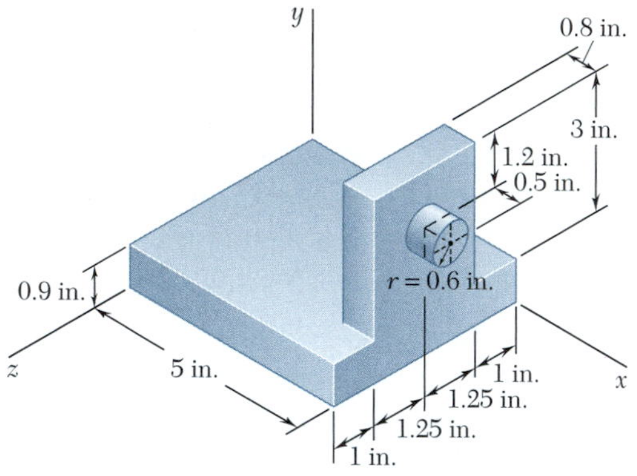

Fig. P9.152

9.152 Determine the products of inertia I_{xy}, I_{yz}, and I_{zx} of the steel machine element shown. (The specific weight of steel is 0.284 lb/in^3.)

9.153 and 9.154 Determine the mass products of inertia I_{xy}, I_{yz}, and I_{zx} of the cast aluminum machine component shown. (The density of aluminum is 2700 kg/m^3.)

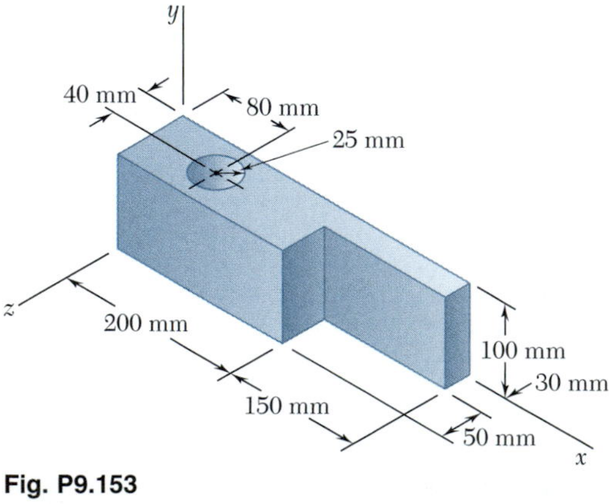

Fig. P9.153

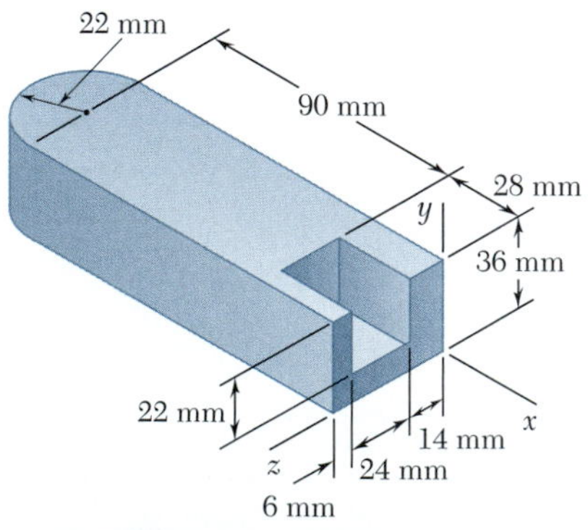

Fig. P9.154

9.155 through 9.157 A section of sheet steel 3 mm thick is cut and bent into the machine component shown. Knowing that the density of the steel is 7860 kg/m^3, determine the mass products of inertia I_{xy}, I_{yz}, and I_{zx} of the component.

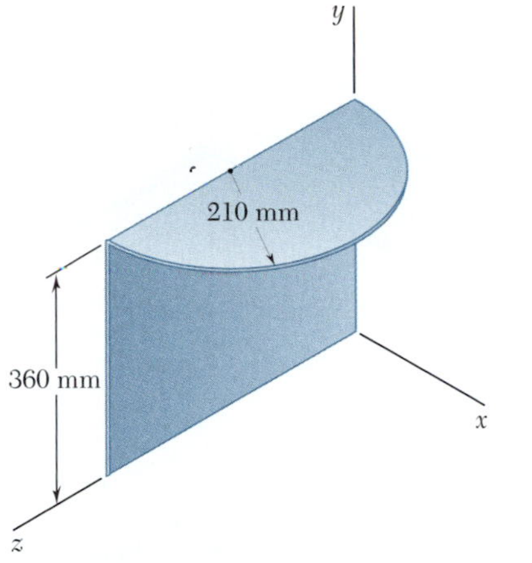

Fig. *P9.155*

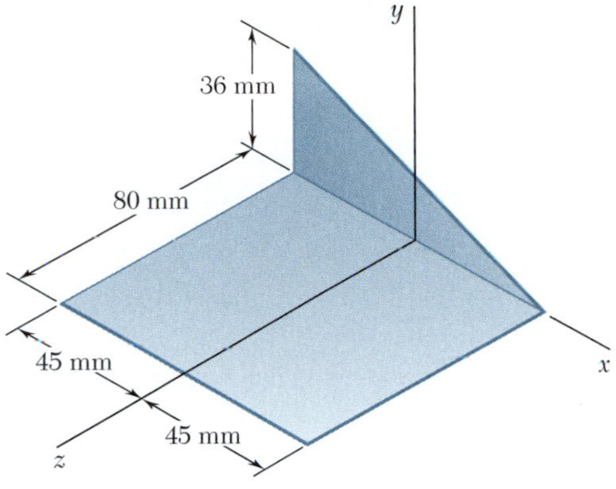

Fig. *P9.156*

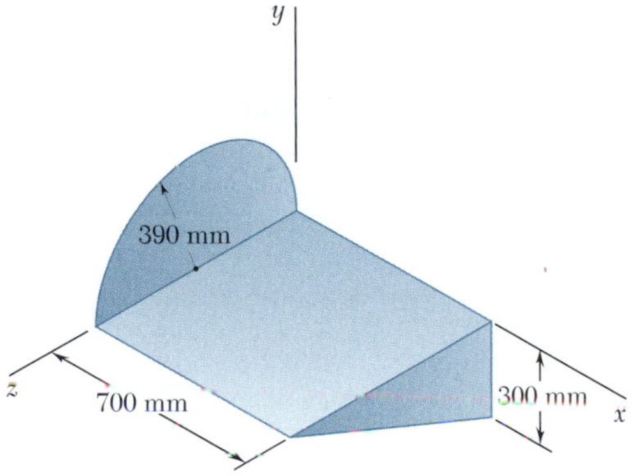

Fig. P9.157

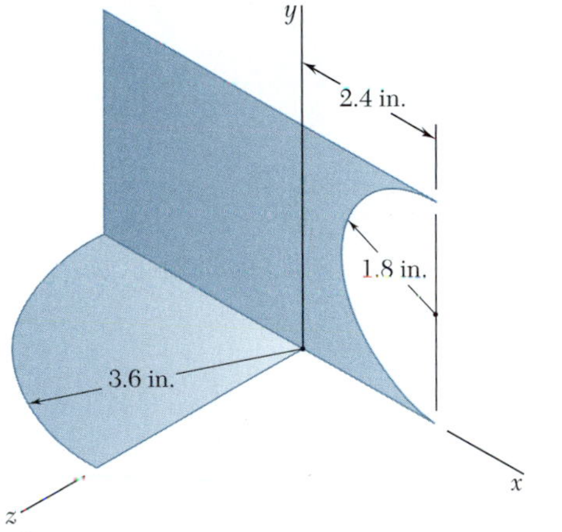

Fig. P9.158

9.158 A section of sheet steel 0.08 in. thick is cut and bent into the machine component shown. Knowing that the specific weight of steel is 490 lb/ft^3, determine the mass products of inertia I_{xy}, I_{yz}, and I_{zx} of the component.

9.159 and *9.160* Brass wire with a weight per unit length w is used to form the figure shown. Determine the products of inertia I_{xy}, I_{yz}, and I_{zx} of the wire figure.

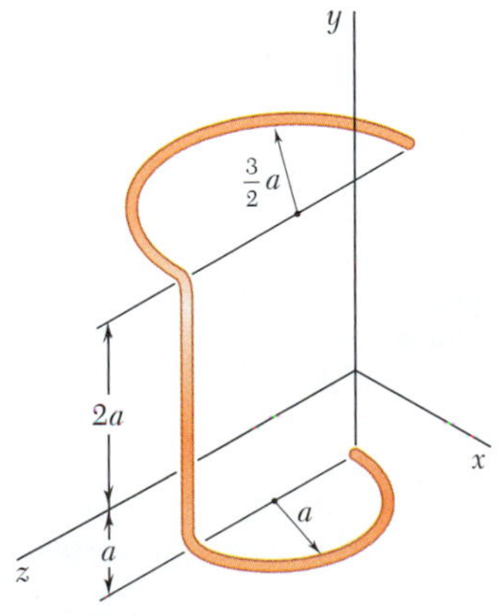

Fig. P9.159

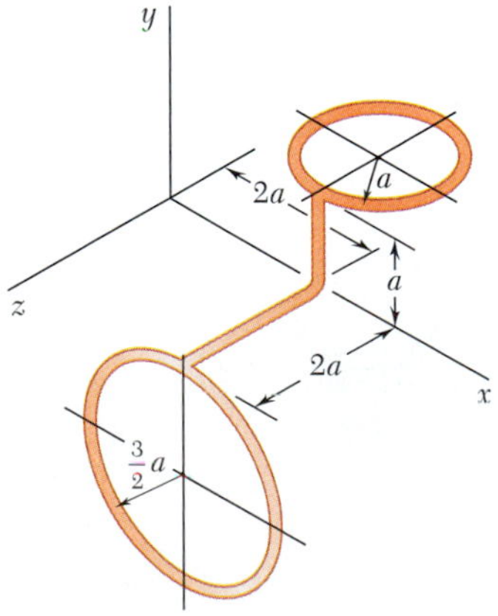

Fig. *P9.160*

9.161 The figure shown is formed of 0.075-in.-diameter aluminum wire. Knowing that the specific weight of aluminum is 0.10 lb/in^3, determine the products of inertia I_{xy}, I_{yz}, and I_{zx} of the wire figure.

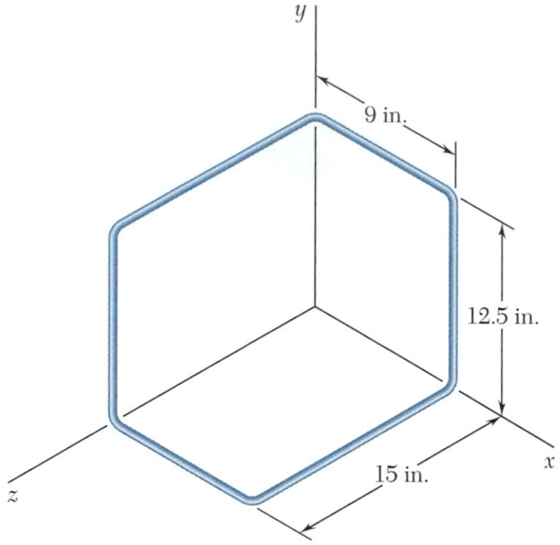

Fig. P9.161

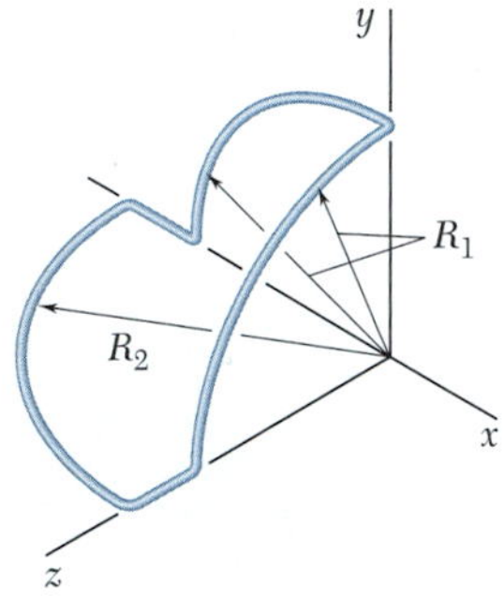

Fig. P9.162

9.162 Thin aluminum wire of uniform diameter is used to form the figure shown. Denoting by m' the mass per unit length of the wire, determine the products of inertia I_{xy}, I_{yz}, and I_{zx} of the wire figure.

9.163 Complete the derivation of Eqs. (9.47), which express the parallel-axis theorem for mass products of inertia.

9.164 For the homogeneous tetrahedron of mass m shown, (a) determine by direct integration the product of inertia I_{zx}, (b) deduce I_{yz} and I_{xy} from the results obtained in part a.

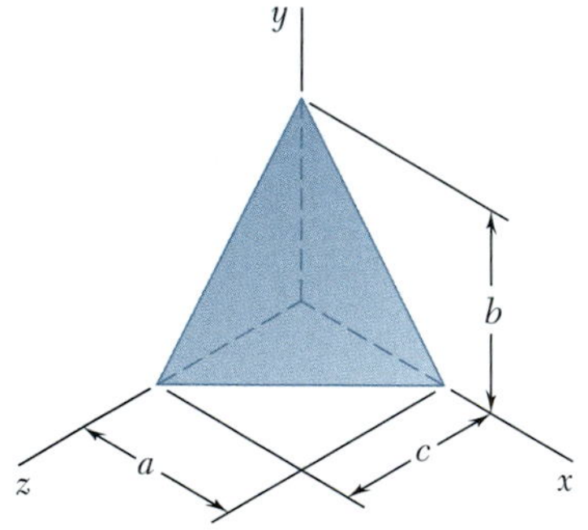

Fig. P9.164

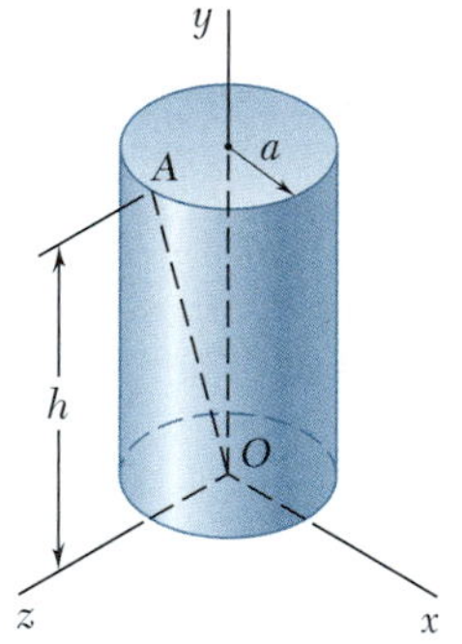

Fig. P9.165

9.165 The homogeneous circular cylinder shown has a mass m. Determine the moment of inertia of the cylinder with respect to the line joining the origin O and point A which is located on the perimeter of the top surface of the cylinder.

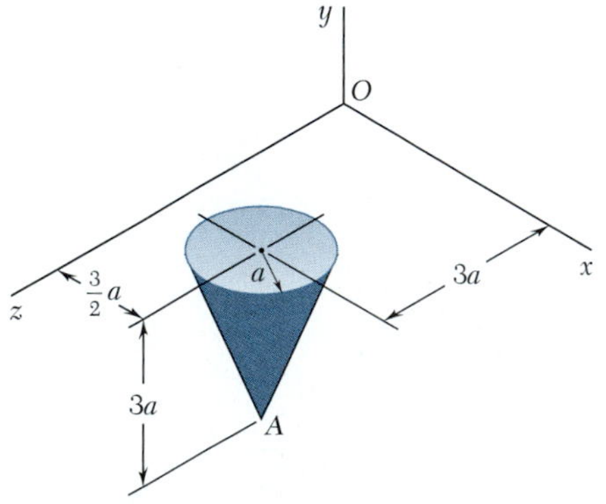

Fig. P9.166

9.166 The homogeneous circular cone shown has a mass m. Determine the moment of inertia of the cone with respect to the line joining the origin O and point A.

9.167 Shown is the machine element of Prob. 9.143. Determine its moment of inertia with respect to the line joining the origin O and point A.

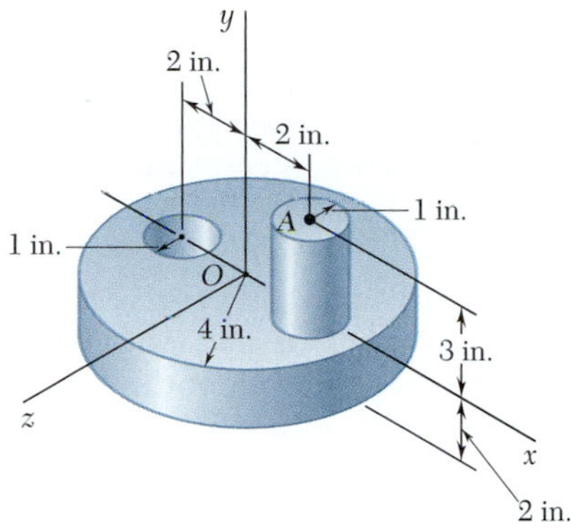

Fig. P9.167

9.168 Determine the moment of inertia of the steel machine element of Probs. 9.147 and 9.151 with respect to the axis through the origin which forms equal angles with the x, y, and z axes.

9.169 The thin bent plate shown is of uniform density and weight W. Determine its mass moment of inertia with respect to the line joining the origin O and point A.

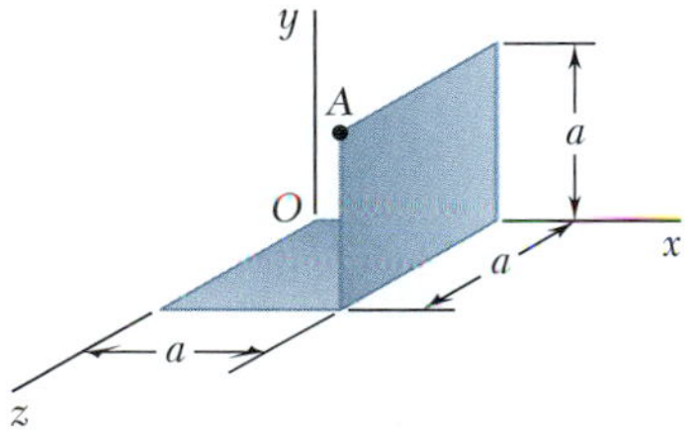

Fig. P9.169

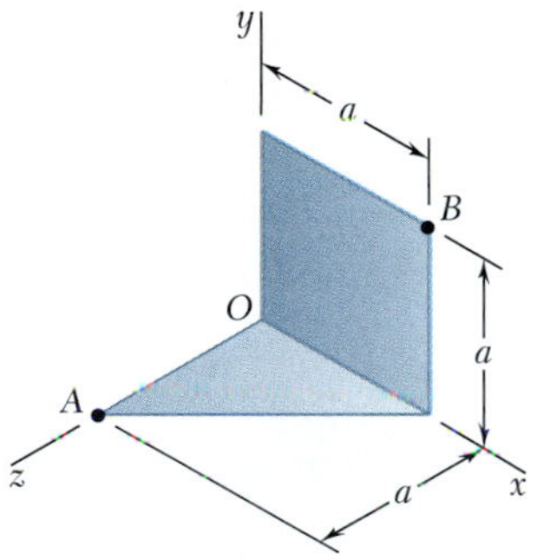

Fig. P9.170

9.170 A piece of sheet metal of thickness t and density ρ is cut and bent into the shape shown. Determine its mass moment of inertia with respect to a line joining points A and B.

9.171 Determine the mass moment of inertia of the machine component of Probs. 9.138 and 9.157 with respect to the axis through the origin characterized by the unit vector $\boldsymbol{\lambda} = (-4\mathbf{i} + 8\mathbf{j} + \mathbf{k})/9$.

9.172 through 9.174 For the wire figure of the problem indicated, determine the mass moment of inertia of the figure with respect to the axis through the origin characterized by the unit vector $\boldsymbol{\lambda} = (-3\mathbf{i} - 6\mathbf{j} + 2\mathbf{k})/7$.

9.172 Prob. 9.150
9.173 Prob. 9.149
9.174 Prob. 9.148

9.175 For the rectangular prism shown, determine the values of the ratios b/a and c/a so that the ellipsoid of inertia of the prism is a sphere when computed (a) at point A, (b) at point B.

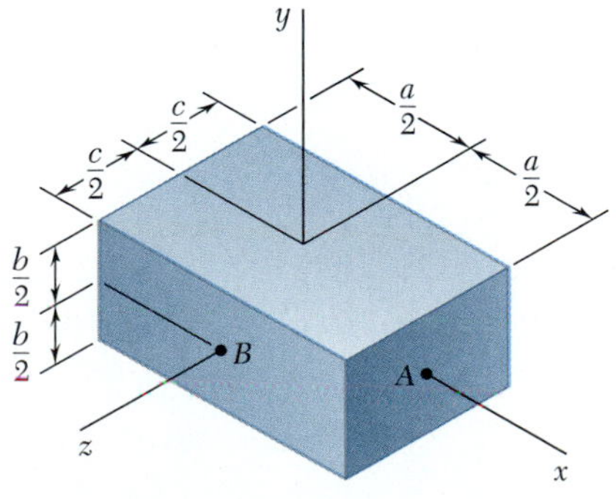

Fig. P9.175

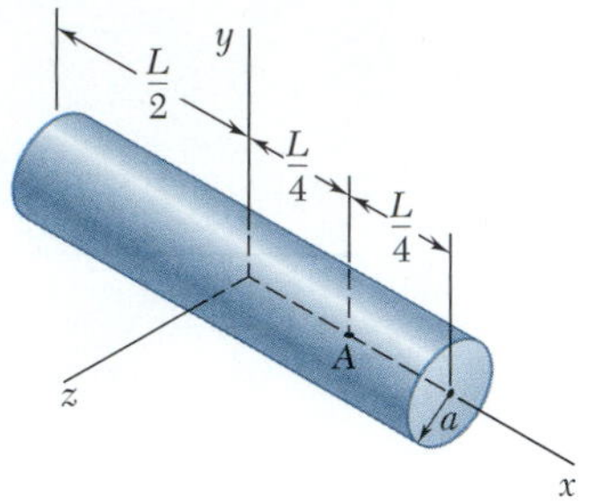

Fig. P9.177

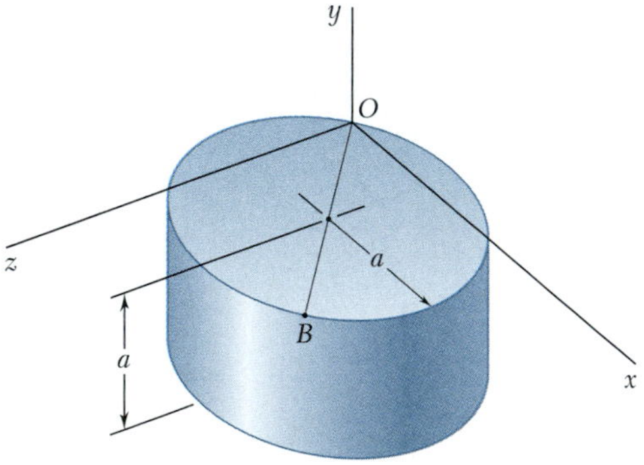

Fig. P9.181

9.176 For the right circular cone of Sample Prob. 9.11, determine the value of the ratio a/h for which the ellipsoid of inertia of the cone is a sphere when computed (*a*) at the apex of the cone, (*b*) at the center of the base of the cone.

9.177 For the homogeneous circular cylinder shown, of radius a and length L, determine the value of the ratio a/L for which the ellipsoid of inertia of the cylinder is a sphere when computed (*a*) at the centroid of the cylinder, (*b*) at point A.

9.178 Given an arbitrary body and three rectangular axes x, y, and z, prove that the moment of inertia of the body with respect to any one of the three axes cannot be larger than the sum of the moments of inertia of the body with respect to the other two axes. That is, prove that the inequality $I_x \leq I_y + I_z$ and the two similar inequalities are satisfied. Further, prove that $I_y \geq \frac{1}{2}I_x$ if the body is a homogeneous solid of revolution, where x is the axis of revolution and y is a transverse axis.

9.179 Consider a cube of mass m and side a. (*a*) Show that the ellipsoid of inertia at the center of the cube is a sphere, and use this property to determine the moment of inertia of the cube with respect to one of its diagonals. (*b*) Show that the ellipsoid of inertia at one of the corners of the cube is an ellipsoid of revolution, and determine the principal moments of inertia of the cube at that point.

9.180 Given a homogeneous body of mass m and of arbitrary shape and three rectangular axes x, y, and z with origin at O, prove that the sum $I_x + I_y + I_z$ of the moments of inertia of the body cannot be smaller than the similar sum computed for a sphere of the same mass and the same material centered at O. Further, using the results of Prob. 9.178, prove that if the body is a solid of revolution, where x is the axis of revolution, its moment of inertia I_y about a transverse axis y cannot be smaller than $3ma^2/10$, where a is the radius of the sphere of the same mass and the same material.

***9.181** The homogeneous circular cylinder shown has a mass m, and the diameter OB of its top surface forms 45° angles with the x and z axes. (*a*) Determine the principal moments of inertia of the cylinder at the origin O. (*b*) Compute the angles that the principal axes of inertia at O form with the coordinate axes. (*c*) Sketch the cylinder, and show the orientation of the principal axes of inertia relative to the x, y, and z axes.

9.182 through 9.186 For the component described in the problem indicated, determine (*a*) the principal moments of inertia at the origin, (*b*) the principal axes of inertia at the origin. Sketch the body and show the orientation of the principal axes of inertia relative to the x, y, and z axes.

***9.182** Prob. 9.167
***9.183** Probs. 9.147 and 9.151
***9.184** Prob. 9.169
***9.185** Prob. 9.170
***9.186** Probs. 9.150 and 9.172

REVIEW AND SUMMARY FOR CHAPTER 9

In the first half of this chapter, we discussed the determination of the resultant **R** of forces $\Delta\mathbf{F}$ distributed over a plane area A when the magnitudes of these forces are proportional to both the areas ΔA of the elements on which they act and the distances y from these elements to a given x axis; we thus had $\Delta F = ky\,\Delta A$. We found that the magnitude of the resultant **R** is proportional to the first moment $Q_x = \int y\,dA$ of the area A, while the moment of **R** about the x axis is proportional to the *second moment*, or *moment of inertia*, $I_x = \int y^2\,dA$ of A with respect to the same axis [Sec. 9.2].

Rectangular moments of inertia

The *rectangular moments of inertia I_x and I_y of an area* [Sec. 9.3] were obtained by evaluating the integrals

$$I_x = \int y^2\,dA \qquad I_y = \int x^2\,dA \tag{9.1}$$

These computations can be reduced to single integrations by choosing dA to be a thin strip parallel to one of the coordinate axes. We also recall that it is possible to compute I_x and I_y from the same elemental strip (Fig. 9.35) using the formula for the moment of inertia of a rectangular area [Sample Prob. 9.3].

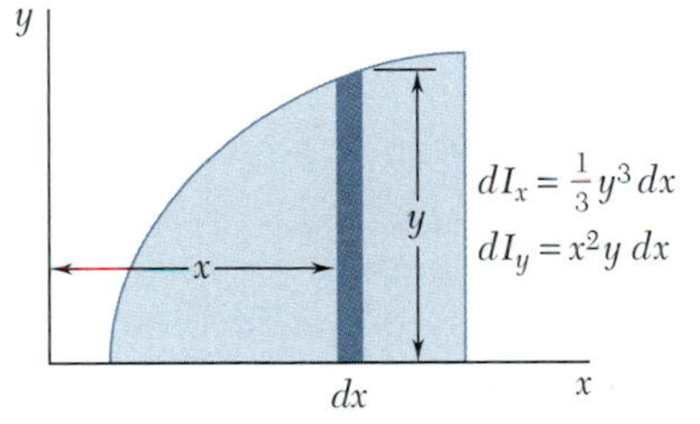

Fig. 9.35

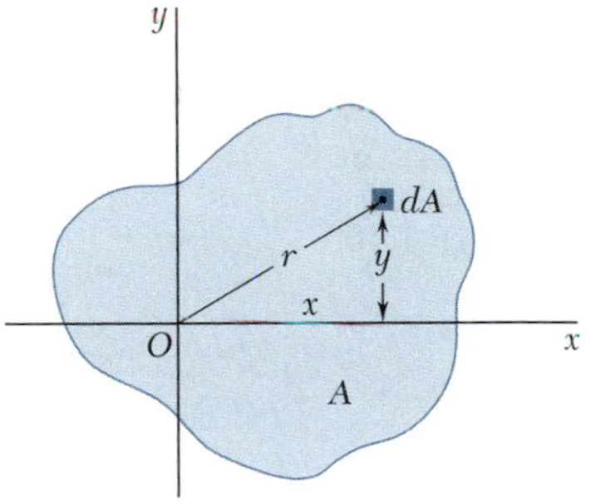

Fig. 9.36

Polar moment of inertia

The *polar moment of inertia of an area A* with respect to the pole O [Sec. 9.4] was defined as

$$J_O = \int r^2\,dA \tag{9.3}$$

where r is the distance from O to the element of area dA (Fig. 9.36). Observing that $r^2 = x^2 + y^2$, we established the relation

$$J_O = I_x + I_y \tag{9.4}$$

Radius of gyration

The *radius of gyration of an area A* with respect to the x axis [Sec. 9.5] was defined as the distance k_x, where $I_x = k_x^2 A$. With similar definitions for the radii of gyration of A with respect to the y axis and with respect to O, we had

$$k_x = \sqrt{\frac{I_x}{A}} \qquad k_y = \sqrt{\frac{I_y}{A}} \qquad k_O = \sqrt{\frac{J_O}{A}} \tag{9.5–9.7}$$

Parallel-axis theorem

The parallel-axis theorem was presented in Sec. 9.6. It states that the moment of inertia I of an area with respect to any given axis AA' (Fig. 9.37) is equal to the moment of inertia $\bar{I}$ of the area with respect to the centroidal axis BB' that is parallel to AA' *plus* the product of the area A and the square of the distance d between the two axes:

$$I = \bar{I} + Ad^2 \tag{9.9}$$

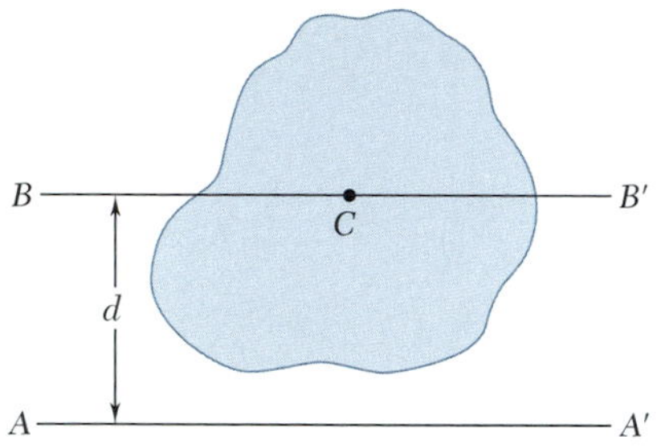

Fig. 9.37

This formula can also be used to determine the moment of inertia $\bar{I}$ of an area with respect to a centroidal axis BB' when its moment of inertia I with respect to a parallel axis AA' is known. In this case, however, the product Ad^2 should be *subtracted* from the known moment of inertia I.

A similar relation holds between the polar moment of inertia J_O of an area about a point O and the polar moment of inertia J_C of the same area about its centroid C. Letting d be the distance between O and C, we have

$$J_O = \bar{J}_C + Ad^2 \tag{9.11}$$

Composite areas

The parallel-axis theorem can be used very effectively to compute the *moment of inertia of a composite area* with respect to a given axis [Sec. 9.7]. Considering each component area separately, we first compute the moment of inertia of each area with respect to its centroidal axis, using the data provided in Figs. 9.12 and 9.13 whenever possible. The parallel-axis theorem is then applied to determine the moment of inertia of each component area with respect to the desired axis, and the various values obtained are added [Sample Probs. 9.4 and 9.5].

Product of inertia

Sections 9.8 through 9.10 were devoted to the transformation of the moments of inertia of an area *under a rotation of the coordinate axes.* First, we defined the *product of inertia of an area A* as

$$I_{xy} = \int xy\,dA \tag{9.12}$$

and showed that $I_{xy} = 0$ if the area A is symmetrical with respect to either or both of the coordinate axes. We also derived the *parallel-axis theorem for products of inertia.* We had

$$I_{xy} = \bar{I}_{x'y'} + \bar{x}\bar{y}A \tag{9.13}$$

where $\bar{I}_{x'y'}$ is the product of inertia of the area with respect to the centroidal axes x' and y' which are parallel to the x and y axes, respectively, and $\bar{x}$ and $\bar{y}$ are the coordinates of the centroid of the area [Sec. 9.8].

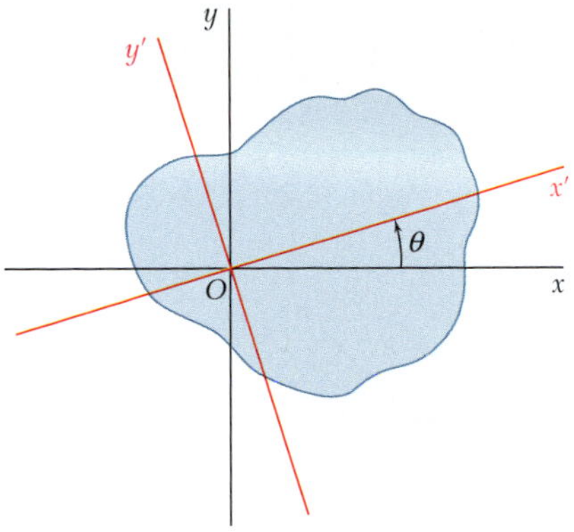

Fig. 9.38

Rotation of axes

In Sec. 9.9 we determined the moments and product of inertia $I_{x'}$, $I_{y'}$, and $I_{x'y'}$ of an area with respect to x' and y' axes obtained by rotating the original x and y coordinate axes through an angle θ counterclockwise (Fig. 9.38). We expressed $I_{x'}$, $I_{y'}$, and $I_{x'y'}$ in terms of the moments and product of inertia I_x, I_y, and I_{xy} computed with respect to the original x and y axes. We had

$$I_{x'} = \frac{I_x + I_y}{2} + \frac{I_x - I_y}{2}\cos 2\theta - I_{xy}\sin 2\theta \qquad (9.18)$$

$$I_{y'} = \frac{I_x + I_y}{2} - \frac{I_x - I_y}{2}\cos 2\theta + I_{xy}\sin 2\theta \qquad (9.19)$$

$$I_{x'y'} = \frac{I_x - I_y}{2}\sin 2\theta + I_{xy}\cos 2\theta \qquad (9.20)$$

Principal axes

The *principal axes of the area about O* were defined as the two axes perpendicular to each other and with respect to which the moments of inertia of the area are maximum and minimum. The corresponding values of θ, denoted by θ_m, were obtained from the formula

$$\tan 2\theta_m = -\frac{2I_{xy}}{I_x - I_y} \qquad (9.25)$$

Principal moments of inertia

The corresponding maximum and minimum values of I are called the *principal moments of inertia* of the area about O; we had

$$I_{\max,\min} = \frac{I_x + I_y}{2} \pm \sqrt{\left(\frac{I_x - I_y}{2}\right)^2 + I_{xy}^2} \qquad (9.27)$$

We also noted that the corresponding value of the product of inertia is zero.

Mohr's circle

The transformation of the moments and product of inertia of an area under a rotation of axes can be represented graphically by drawing *Mohr's circle* [Sec. 9.10]. Given the moments and product of inertia I_x, I_y, and I_{xy} of the area with respect to the x and y

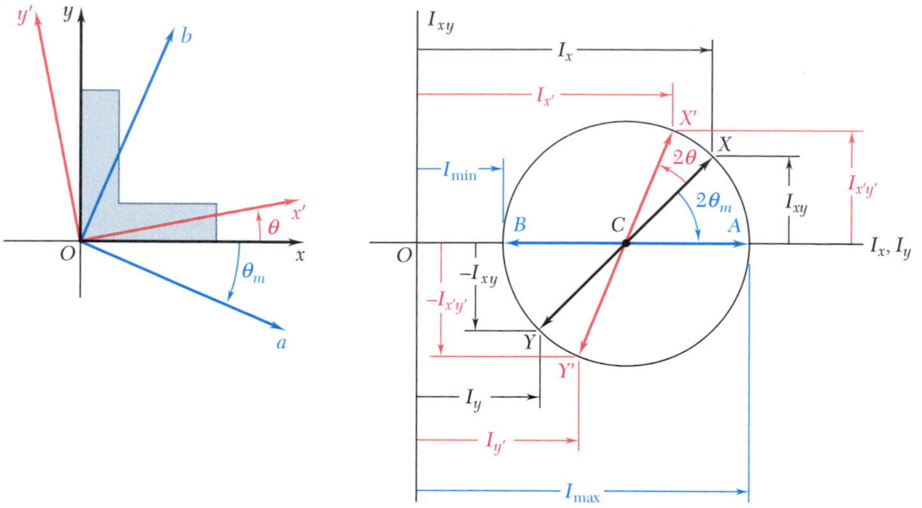

Fig. 9.39

coordinate axes, we plot points $X\ (I_x, I_{xy})$ and $Y\ (I_y, -I_{xy})$ and draw the line joining these two points (Fig. 9.39). This line is a diameter of Mohr's circle and thus defines this circle. As the coordinate axes are rotated through θ, the diameter rotates through *twice that angle,* and the coordinates of X' and Y' yield the new values $I_{x'}$, $I_{y'}$, and $I_{x'y'}$ of the moments and product of inertia of the area. Also, the angle θ_m and the coordinates of points A and B define the principal axes a and b and the principal moments of inertia of the area [Sample Prob. 9.8].

Moments of inertia of masses

The second half of the chapter was devoted to the determination of *moments of inertia of masses,* which are encountered in dynamics in problems involving the rotation of a rigid body about an axis. The mass moment of inertia of a body with respect to an axis AA' (Fig. 9.40) was defined as

$$I = \int r^2\, dm \tag{9.28}$$

where r is the distance from AA' to the element of mass [Sec. 9.11]. The *radius of gyration* of the body was defined as

$$k = \sqrt{\frac{I}{m}} \tag{9.29}$$

The moments of inertia of a body with respect to the coordinate axes were expressed as

$$\begin{aligned} I_x &= \int (y^2 + z^2)\, dm \\ I_y &= \int (z^2 + x^2)\, dm \qquad (9.30) \\ I_z &= \int (x^2 + y^2)\, dm \end{aligned}$$

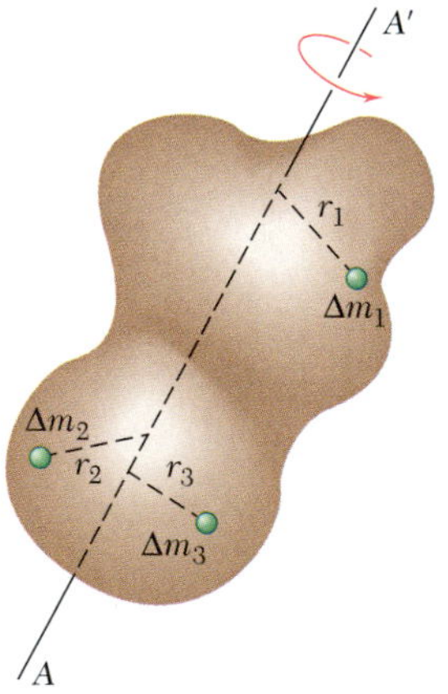

Fig. 9.40

Parallel-axis theorem

We saw that the *parallel-axis theorem* also applies to mass moments of inertia [Sec. 9.12]. Thus, the moment of inertia I of a body with respect to an arbitrary axis AA' (Fig. 9.41) can be expressed as

$$I = \bar{I} + md^2 \tag{9.33}$$

where $\bar{I}$ is the moment of inertia of the body with respect to the centroidal axis BB' which is parallel to the axis AA', m is the mass of the body, and d is the distance between the two axes.

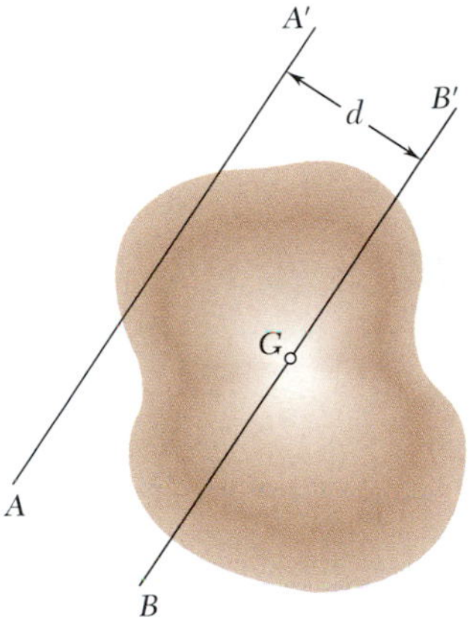

Fig. 9.41

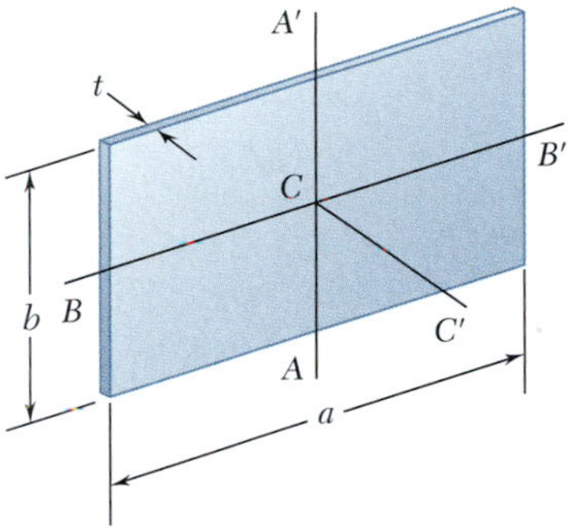

Fig. 9.42

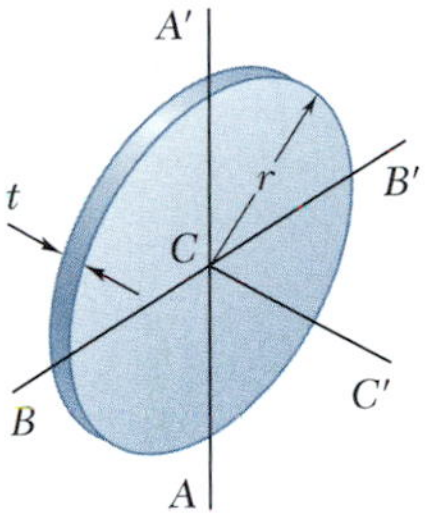

Fig. 9.43

Moments of inertia of thin plates

The moments of inertia of *thin plates* can be readily obtained from the moments of inertia of their areas [Sec. 9.13]. We found that for a *rectangular plate* the moments of inertia with respect to the axes shown (Fig. 9.42) are

$$I_{AA'} = \tfrac{1}{12}ma^2 \qquad I_{BB'} = \tfrac{1}{12}mb^2 \tag{9.39}$$

$$I_{CC'} = I_{AA'} + I_{BB'} = \tfrac{1}{12}m(a^2 + b^2) \tag{9.40}$$

while for a *circular plate* (Fig. 9.43) they are

$$I_{AA'} = I_{BB'} = \tfrac{1}{4}mr^2 \tag{9.41}$$

$$I_{CC'} = I_{AA'} + I_{BB'} = \tfrac{1}{2}mr^2 \tag{9.42}$$

Composite bodies

When a body possesses *two planes of symmetry*, it is usually possible to use a single integration to determine its moment of inertia with respect to a given axis by selecting the element of mass dm to be a thin plate [Sample Probs. 9.10 and 9.11]. On the other hand, when a body consists of *several common geometric shapes*, its moment of inertia with respect to a given axis can be obtained by using the formulas given in Fig. 9.28 together with the parallel-axis theorem [Sample Probs. 9.12 and 9.13].

Moment of inertia with respect to an arbitrary axis

In the last portion of the chapter, we learned to determine the moment of inertia of a body *with respect to an arbitrary axis OL* which is drawn through the origin O [Sec. 9.16]. Denoting by λ_x,

λ_y, λ_z the components of the unit vector $\boldsymbol{\lambda}$ along OL (Fig. 9.44) and introducing the *products of inertia*

$$I_{xy} = \int xy\, dm \qquad I_{yz} = \int yz\, dm \qquad I_{zx} = \int zx\, dm \quad (9.45)$$

we found that the moment of inertia of the body with respect to OL could be expressed as

$$I_{OL} = I_x\lambda_x^2 + I_y\lambda_y^2 + I_z\lambda_z^2 - 2I_{xy}\lambda_x\lambda_y - 2I_{yz}\lambda_y\lambda_z - 2I_{zx}\lambda_z\lambda_x \quad (9.46)$$

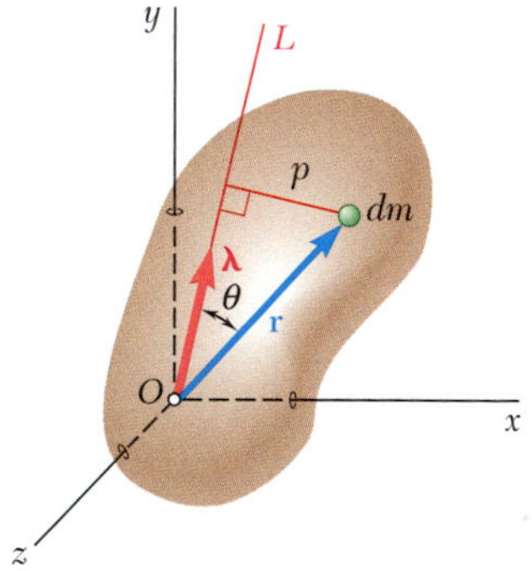

Fig. 9.44

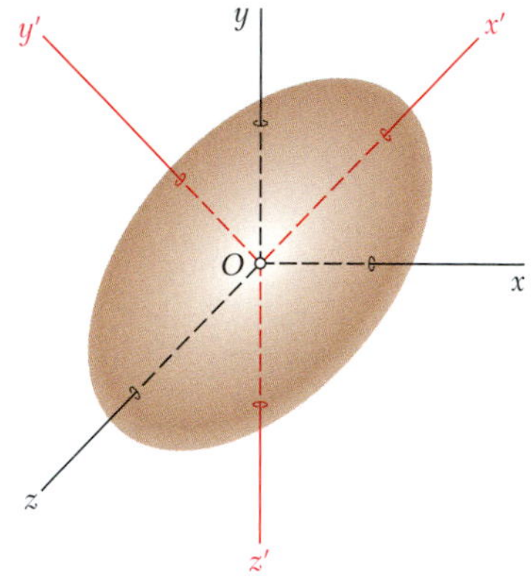

Fig. 9.45

Ellipsoid of inertia

By plotting a point Q along each axis OL at a distance $OQ = 1/\sqrt{I_{OL}}$ from O [Sec. 9.17], we obtained the surface of an ellipsoid, known as the *ellipsoid of inertia* of the body at point O. The principal axes x', y', z' of this ellipsoid (Fig. 9.45) are the *principal axes of inertia* of the body; that is, the products of inertia $I_{x'y'}$, $I_{y'z'}$, $I_{z'x'}$ of the body with respect to these axes are all zero. There are many situations when the principal axes of inertia of a body can be deduced from properties of symmetry of the body. Choosing these axes to be the coordinate axes, we can then express I_{OL} as

Principal axes of inertia

Principal moments of inertia

$$I_{OL} = I_{x'}\lambda_{x'}^2 + I_{y'}\lambda_{y'}^2 + I_{z'}\lambda_{z'}^2 \quad (9.50)$$

where $I_{x'}$, $I_{y'}$, $I_{z'}$ are the *principal moments of inertia* of the body at O.

When the principal axes of inertia cannot be obtained by observation [Sec. 9.17], it is necessary to solve the cubic equation

$$K^3 - (I_x + I_y + I_z)K^2 + (I_xI_y + I_yI_z + I_zI_x - I_{xy}^2 - I_{yz}^2 - I_{zx}^2)K$$
$$- (I_xI_yI_z - I_xI_{yz}^2 - I_yI_{zx}^2 - I_zI_{xy}^2 - 2I_{xy}I_{yz}I_{zx}) = 0 \quad (9.56)$$

We found [Sec. 9.18] that the roots K_1, K_2, and K_3 of this equation are the principal moments of inertia of the given body. The direction cosines $(\lambda_x)_1$, $(\lambda_y)_1$, and $(\lambda_z)_1$ of the principal axis corresponding to the principal moment of inertia K_1 are then determined by substituting K_1 into Eqs. (9.54) and solving two of these equations and Eq. (9.57) simultaneously. The same procedure is then repeated using K_2 and K_3 to determine the direction cosines of the other two principal axes [Sample Prob. 9.15].

Review Problems

9.187 Determine by direct integration the moment of inertia of the shaded area with respect to the y axis.

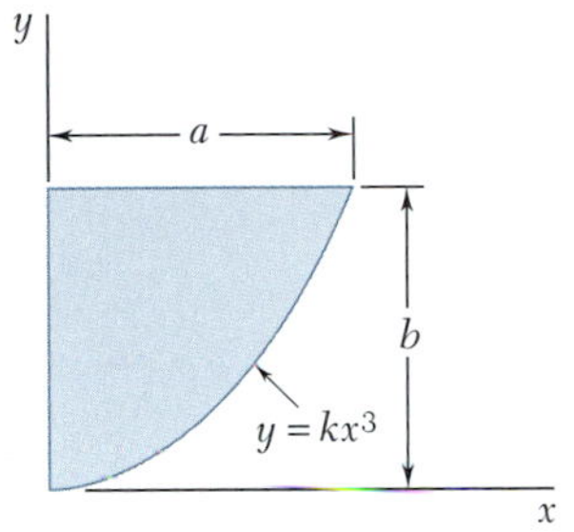

Fig. P9.187

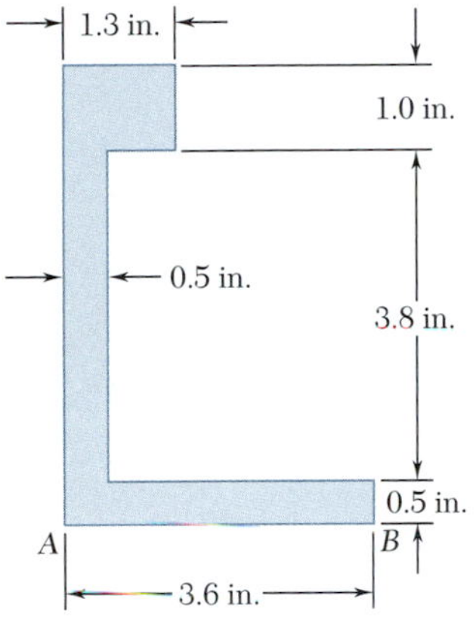

Fig. P9.188

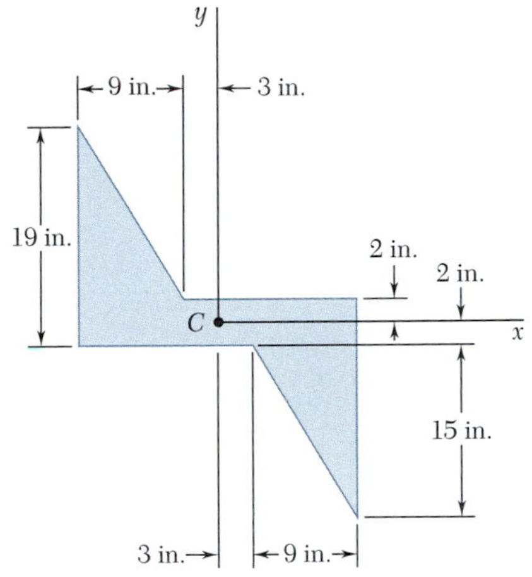

Fig. *P9.189*

9.188 Determine the moments of inertia $\bar{I}_x$ and $\bar{I}_y$ of the area shown with respect to centroidal axes respectively parallel and perpendicular to side AB.

9.189 Using the parallel-axis theorem, determine the product of inertia of the area shown with respect to the centroidal x and y axes.

9.190 A piece of thin, uniform sheet metal is cut to form the machine component shown. Denoting the mass of the component by m, determine its moment of inertia with respect to (a) the x axis, (b) the y axis.

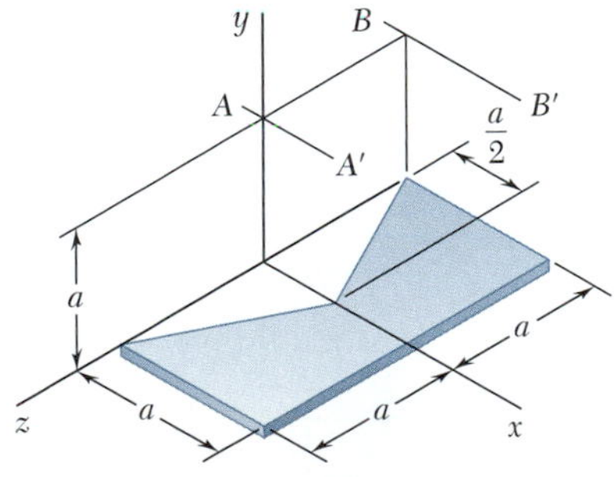

Fig. P9.190 and *P9.191*

9.191 A piece of thin, uniform sheet metal is cut to form the machine component shown. Denoting the mass of the component by m, determine its moment of inertia with respect to (a) the axis AA', (b) the axis BB', where the AA' and BB' axes are parallel to the x axis and lie in a plane parallel to and at a distance a above the zx plane.

9.192 For the $5 \times 3 \times \frac{1}{2}$-in. angle cross section shown, use Mohr's circle to determine (*a*) the moments of inertia and the product of inertia with respect to new centroidal axes obtained by rotating the x and y axes 45° counterclockwise, (*b*) the orientation of new centroidal axes for which $\bar{I}_{x'} = 2 \text{ in}^4$ and $\bar{I}_{x'y'} < 0$.

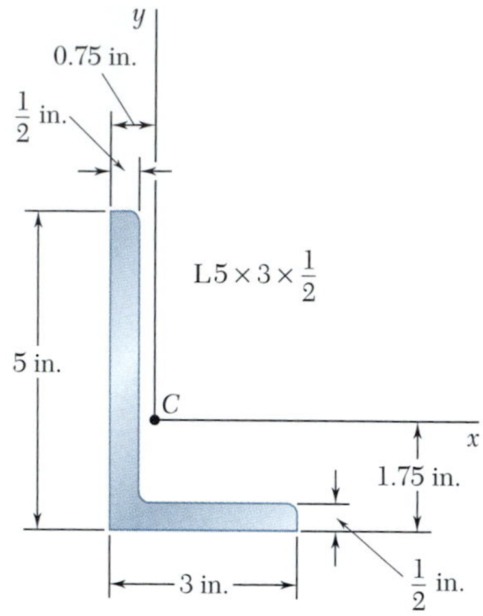

Fig. P9.192

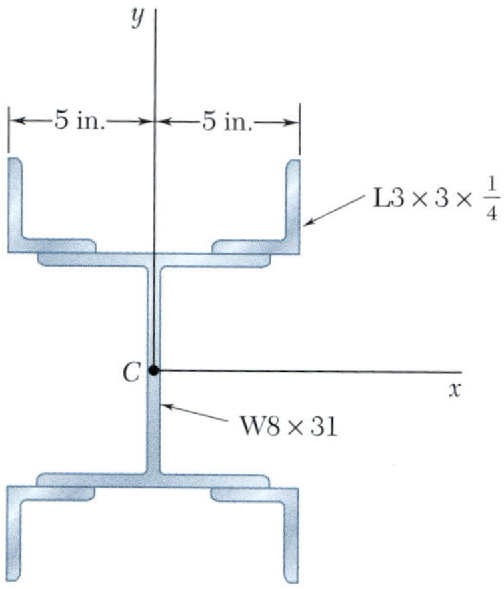

Fig. *P9.193*

9.193 Four $3 \times 3 \times \frac{1}{4}$-in. angles are welded to a rolled W section as shown. Determine the moments of inertia and the radii of gyration of the combined section with respect to its centroidal x and y axes.

9.194 For the 2-kg connecting rod shown, it has been experimentally determined that the mass moments of inertia of the rod with respect to the center-line axes of the bearings AA' and BB' are, respectively, $I_{AA'} = 78 \text{ g} \cdot \text{m}^2$ and $I_{BB'} = 41 \text{ g} \cdot \text{m}^2$. Knowing that $r_a + r_b = 290$ mm, determine (*a*) the location of the centroidal axis GG', (*b*) the radius of gyration with respect to axis GG'.

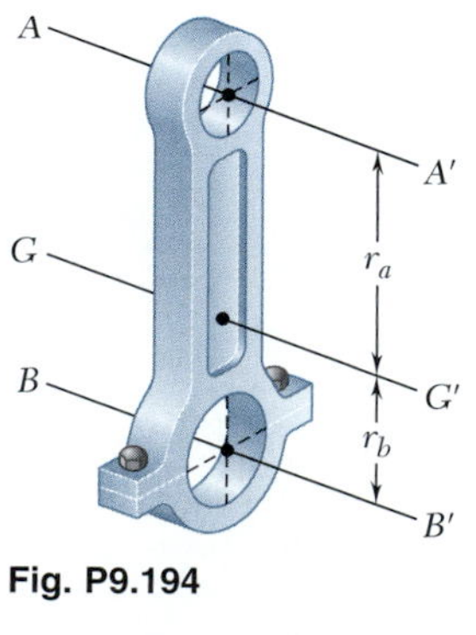

Fig. P9.194

9.195 Determine the mass moment of inertia of the 0.9-lb machine component shown with respect to the axis AA'.

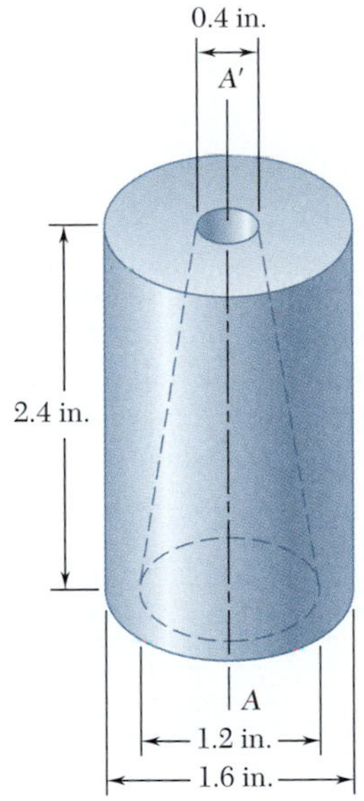

Fig. *P9.195*

9.196 Determine the moments of inertia and the radii of gyration of the steel machine element shown with respect to the x and y axes. (The density of steel is 7850 kg/m^3.)

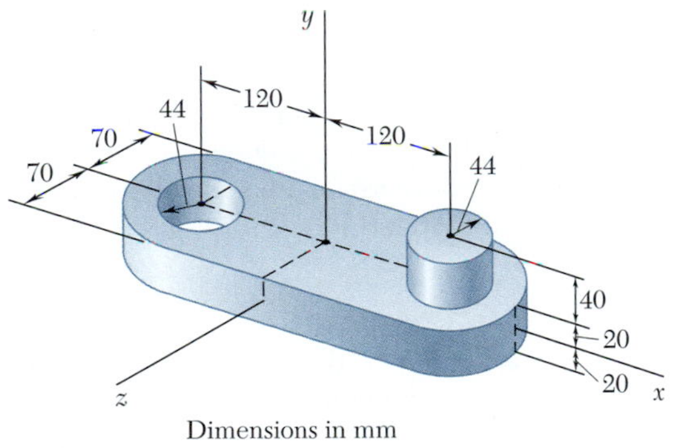

Fig. P9.196

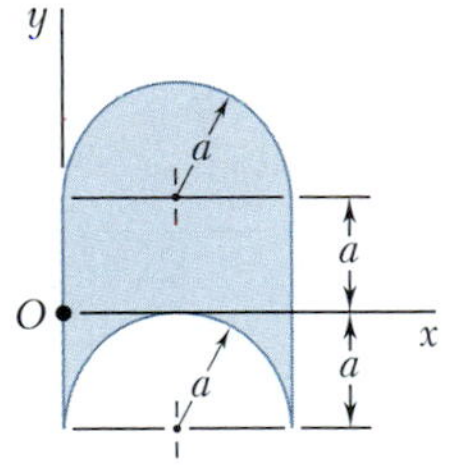

Fig. P9.197

9.197 Determine the moments of inertia of the shaded area shown with respect to the x and y axes.

9.198 A 20-mm-diameter hole is bored in a 32-mm-diameter rod as shown. Determine the depth d of the hole so that the ratio of the moments of inertia of the rod with and without the hole with respect to the axis AA' is 0.96.

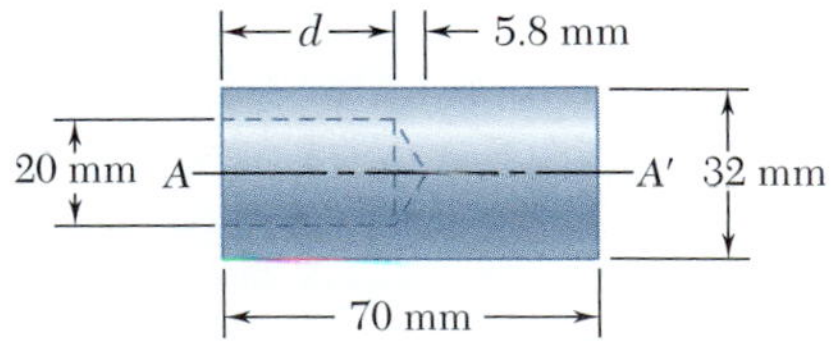

Fig. P9.198

Computer Problems

9.C1 A pattern of 6-mm-diameter holes is drilled in a circular steel plate of diameter d to form the drain cover shown. Knowing that the center-to-center distance between adjacent holes is 14 mm and that the holes lie within a circle of diameter $(d - 30)$ mm, determine the moment of inertia of the cover with respect to the x axis when (*a*) $d = 200$ mm, (*b*) $d = 450$ mm, (*c*) $d = 720$ mm.

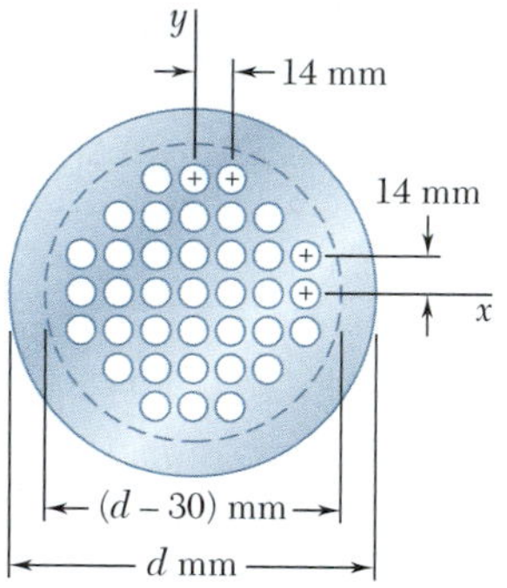

Fig. P9.C1

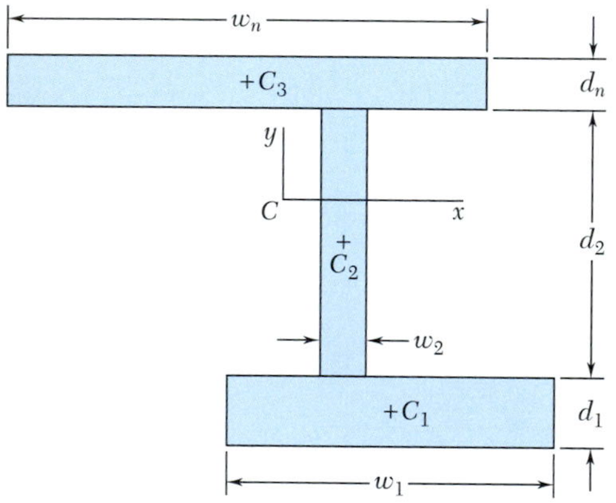

Fig. P9.C2

9.C2 Many cross sections can be approximated by a series of rectangles as shown. Using computational software, calculate the moments of inertia and the radii of gyration of cross sections of this type with respect to horizontal and vertical centroidal axes. Apply the above software to the cross section shown in (*a*) Fig. P9.31 and P9.33, (*b*) Fig. P9.32 and P9.34, (*c*) Fig. P9.43, (*d*) Fig. P9.44.

9.C3 The panel shown forms the end of a trough which is filled with water to line AA'. Determine the depth d of the point of application of the resultant of the hydrostatic forces acting on the panel (center of pressure). Knowing that $h = 400$ mm, $b = 200$ mm, and $a = nb$, where n varies from 1 to 8, plot the depth of the center of pressure d as a function of n.

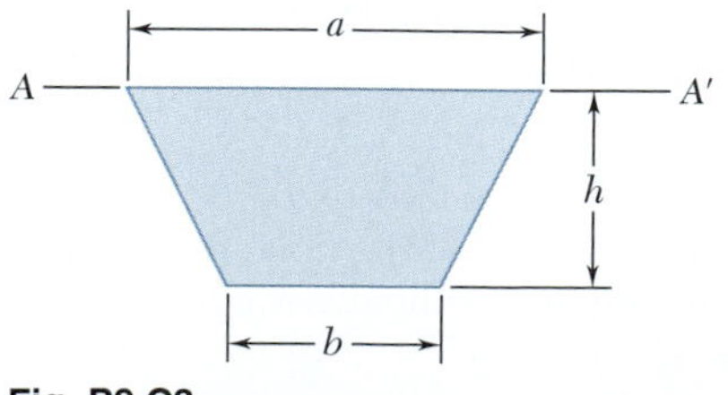

Fig. P9.C3

9.C4 For the area shown, plot I_{xy} as a function of n for values of n from 1 to 10 knowing that $a = 80$ mm and $b = 60$ mm.

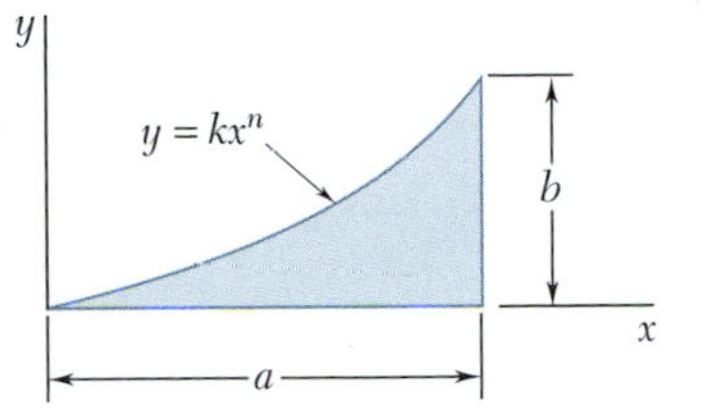

Fig. P9.C4

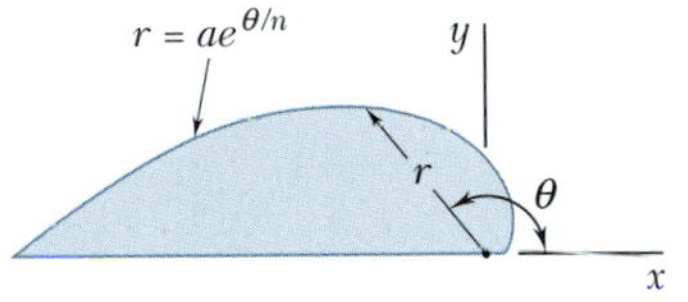

Fig. P9.C5

***9.C5** The area shown is bounded by the logarithmic spiral $r = ae^{\theta/n}$ with $0 \leq \theta \leq \pi$. Using computational software, determine the product of inertia of the area with respect to the x and y axes when $a = 3$ in. and (*a*) $n = 2$, (*b*) $n = 2.5$, (*c*) $n = 4$.

9.C6 An area with known moments and product of inertia I_x, I_y, and I_{xy} can be used to calculate the moments and product of inertia of the area with respect to axes x' and y' obtained by rotating the original axes counterclockwise through an angle θ. Using computational software, compute and plot $I_{x'}$, $I_{y'}$, and $I_{x'y'}$, as functions of θ for $0 \leq \theta \leq 90°$ for the area of Prob. 9.74.

9.C7 A homogeneous wire with a mass per unit length of 0.08 kg/m is used to form the figure shown. Approximating the figure using 10 straight line segments, use computational software to determine the mass moment of inertia I_x of the wire with respect to the x-axis. Using the software, determine I_x when (*a*) $a = 20$ mm, $L = 220$ mm, $h = 80$ mm, (*b*) $a = 40$ mm, $L = 340$ mm, $h = 200$ mm, (*c*) $a = 100$ mm, $L = 500$ mm, $h = 120$ mm.

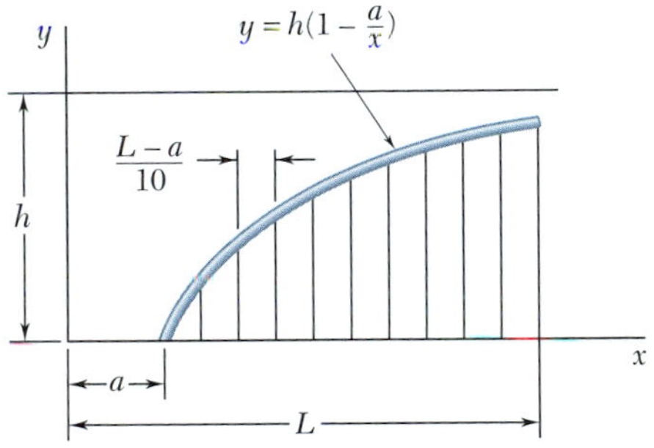

Fig. P9.C7

9.C8 The area shown is revolved about the x-axis to form a homogeneous solid of weight W. Approximating the area using a series of 400 rectangles of the form $bcc'b'$, each of width Δl, use computational software to determine the mass moment of inertia of the solid with respect to the x-axis. Assuming that $W = 4$ lb, $a = 5$ in., and $b = 20$ in., use the software to solve (*a*) Prob. 9.121, (*b*) Prob. 9.123.

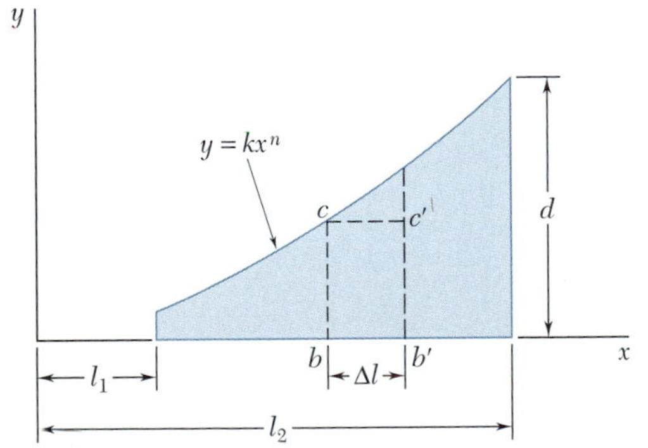

Fig. P9.C8

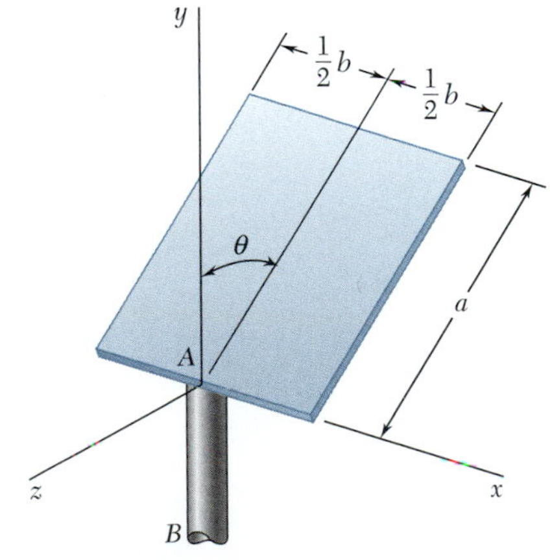

Fig. P9.C9

9.C9 A thin rectangular plate of weight W is welded to a vertical shaft AB with which it forms an angle θ. Knowing that $W = 1$ lb, $a = 10$ in., and $b = 8$ in., plot for values of θ from 0 to 90° the mass moment of inertia of the plate with respect to (*a*) the y axis, (*b*) the z axis.

CHAPTER

10

Method of Virtual Work

Using the analytical methods discussed in this chapter, it is easily shown that the force exerted by the hydraulic cylinder to raise the scissor lift shown is independent of the worker's location on the platform.

METHOD OF VIRTUAL WORK

*10.1. INTRODUCTION

In the preceding chapters, problems involving the equilibrium of rigid bodies were solved by expressing that the external forces acting on the bodies were balanced. The equations of equilibrium $\Sigma F_x = 0$, $\Sigma F_y = 0$, $\Sigma M_A = 0$ were written and solved for the desired unknowns. A different method, which will prove more effective for solving certain types of equilibrium problems, will now be considered. This method is based on the *principle of virtual work* and was first formally used by the Swiss mathematician Jean Bernoulli in the eighteenth century.

As you will see in Sec. 10.3, the principle of virtual work states that if a particle or rigid body, or, more generally, a system of connected rigid bodies, which is in equilibrium under various external forces, is given an arbitrary displacement from that position of equilibrium, the total work done by the external forces during the displacement is zero. This principle is particularly effective when applied to the solution of problems involving the equilibrium of machines or mechanisms consisting of several connected members.

In the second part of the chapter, the method of virtual work will be applied in an alternative form based on the concept of *potential energy*. It will be shown in Sec. 10.8 that if a particle, rigid body, or system of rigid bodies is in equilibrium, then the derivative of its potential energy with respect to a variable defining its position must be zero.

In this chapter, you will also learn to evaluate the mechanical efficiency of a machine (Sec. 10.5) and to determine whether a given position of equilibrium is stable, unstable, or neutral (Sec. 10.9).

*10.2. WORK OF A FORCE

Let us first define the terms *displacement* and *work* as they are used in mechanics. Consider a particle which moves from a point A to a neighboring point A' (Fig. 10.1). If $\mathbf{r}$ denotes the position vector corresponding to point A, the small vector joining A and A' may be denoted by the differential $d\mathbf{r}$; the vector $d\mathbf{r}$ is called the *displacement* of the particle. Now let us assume that a force $\mathbf{F}$ is acting on the particle. The *work of the force* $\mathbf{F}$ *corresponding to the displacement* $d\mathbf{r}$ is defined as the quantity

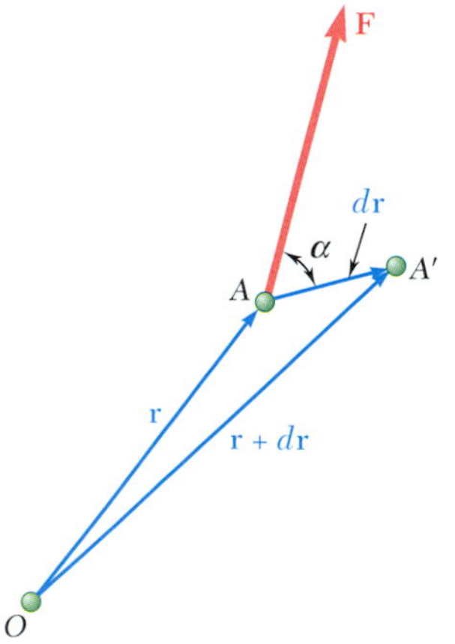

Fig. 10.1

$$dU = \mathbf{F} \cdot d\mathbf{r} \tag{10.1}$$

obtained by forming the scalar product of the force $\mathbf{F}$ and the displacement $d\mathbf{r}$. Denoting respectively by F and ds the magnitudes of the force and of the displacement, and by α the angle formed by $\mathbf{F}$ and $d\mathbf{r}$, and recalling the definition of the scalar product of two vectors (Sec. 3.9), we write

$$dU = F\, ds \cos \alpha \tag{10.1$'$}$$

Being a *scalar quantity*, work has a magnitude and a sign, but no direction. We also note that work should be expressed in units obtained by multiplying units of length by units of force. Thus, if U.S. customary units are used, work should be expressed in ft · lb or in · lb. If SI units are used, work should be expressed in N · m. The unit of work N · m is called a *joule* (J).†

It follows from (10.1′) that the work dU is positive if the angle α is acute and negative if α is obtuse. Three particular cases are of special interest. If the force **F** has the same direction as $d\mathbf{r}$, the work dU reduces to $F\,ds$. If **F** has a direction opposite to that of $d\mathbf{r}$, the work is $dU = -F\,ds$. Finally, if **F** is perpendicular to $d\mathbf{r}$, the work dU is zero.

The work dU of a force **F** during a displacement $d\mathbf{r}$ can also be considered as the product of F and the component $ds \cos \alpha$ of the displacement $d\mathbf{r}$ along **F** (Fig. 10.2*a*). This view is particularly useful

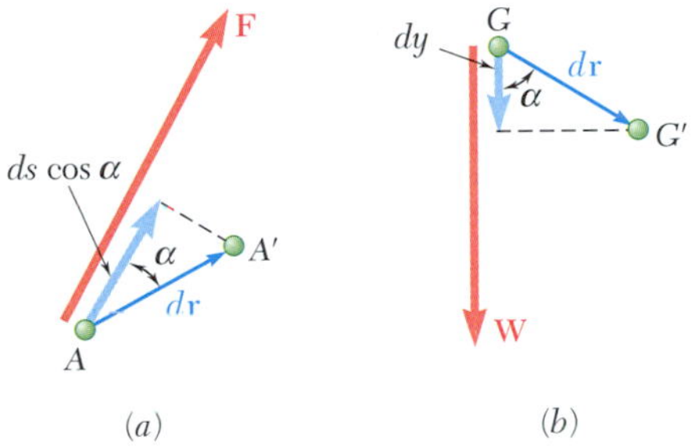

Fig. 10.2

Photo 10.1 The forces exerted by the hydraulic cylinders to position the bucket lift shown can be effectively determined using the method of virtual work since a simple relation exists among the displacements of the points of application of the forces acting on the members of the lift.

in the computation of the work done by the weight **W** of a body (Fig. 10.2*b*). The work of **W** is equal to the product of W and the vertical displacement dy of the center of gravity G of the body. If the displacement is downward, the work is positive; if it is upward, the work is negative.

A number of forces frequently encountered in statics *do no work:* forces applied to fixed points ($ds = 0$) or acting in a direction perpendicular to the displacement ($\cos \alpha = 0$). Among these forces are the reaction at a frictionless pin when the body supported rotates about the pin; the reaction at a frictionless surface when the body in contact moves along the surface; the reaction at a roller moving along its track; the weight of a body when its center of gravity moves horizontally; and the friction force acting on a wheel rolling without slipping (since at any instant the point of contact does not move). Examples of forces which *do work* are the weight of a body (except in the case considered above), the friction force acting on a body sliding on a rough surface, and most forces applied to a moving body.

†The joule is the SI unit of *energy*, whether in mechanical form (work, potential energy, kinetic energy) or in chemical, electrical, or thermal form. We should note that even though N · m = J, the moment of a force must be expressed in N · m, and not in joules, since the moment of a force is not a form of energy.

In certain cases, the sum of the work done by several forces is zero. Consider, for example, two rigid bodies AC and BC connected at C by a *frictionless pin* (Fig. 10.3a). Among the forces acting on AC is the force $\mathbf{F}$ exerted at C by BC. In general, the work of this force

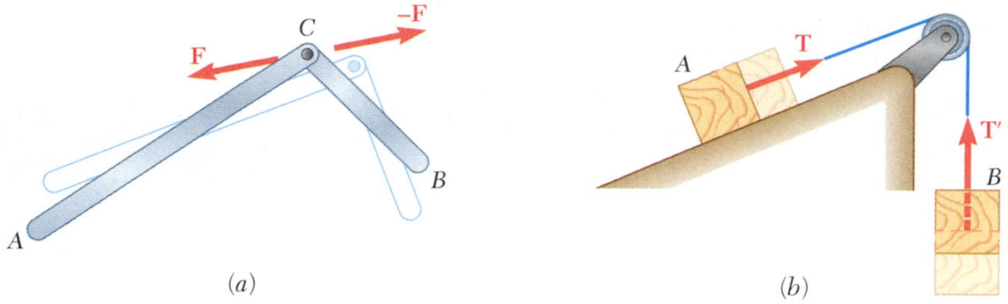

Fig. 10.3

will not be zero, but it will be equal in magnitude and opposite in sign to the work of the force $-\mathbf{F}$ exerted by AC on BC, since these forces are equal and opposite and are applied to the same particle. Thus, when the total work done by all the forces acting on AB and BC is considered, the work of the two internal forces at C cancels out. A similar result is obtained if we consider a system consisting of two blocks connected by an *inextensible cord AB* (Fig. 10.3b). The work of the tension force $\mathbf{T}$ at A is equal in magnitude to the work of the tension force $\mathbf{T}'$ at B, since these forces have the same magnitude and the points A and B move through the same distance; but in one case the work is positive, and in the other it is negative. Thus, the work of the internal forces again cancels out.

It can be shown that the total work of the internal forces holding together the particles of a rigid body is zero. Consider two particles A and B of a rigid body and the two equal and opposite forces $\mathbf{F}$ and $-\mathbf{F}$ they exert on each other (Fig. 10.4). While, in general, small dis-

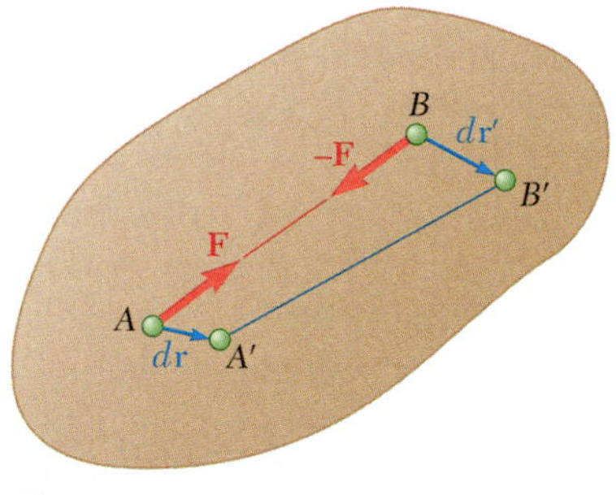

Fig. 10.4

placements $d\mathbf{r}$ and $d\mathbf{r}'$ of the two particles are different, the components of these displacements along AB must be equal; otherwise, the particles would not remain at the same distance from each other, and the body would not be rigid. Therefore, the work of $\mathbf{F}$ is equal in magnitude and opposite in sign to the work of $-\mathbf{F}$, and their sum is zero.

In computing the work of the external forces acting on a rigid body, it is often convenient to determine the work of a couple without considering separately the work of each of the two forces forming the couple. Consider the two forces **F** and $-\mathbf{F}$ forming a couple of moment **M** and acting on a rigid body (Fig. 10.5). Any small displacement of the rigid body bringing A and B, respectively, into A' and B'' can be divided into two parts, one in which points A and B undergo equal displacements $d\mathbf{r}_1$, the other in which A' remains fixed while B' moves into B'' through a displacement $d\mathbf{r}_2$ of magnitude $ds_2 = r\,d\theta$. In the first part of the motion, the work of **F** is equal in magnitude and opposite in sign to the work of $-\mathbf{F}$, and their sum is zero. In the second part of the motion, only force **F** does work, and its work is $dU = F\,ds_2 = Fr\,d\theta$. But the product Fr is equal to the magnitude M of the moment of the couple. Thus, the work of a couple of moment **M** acting on a rigid body is

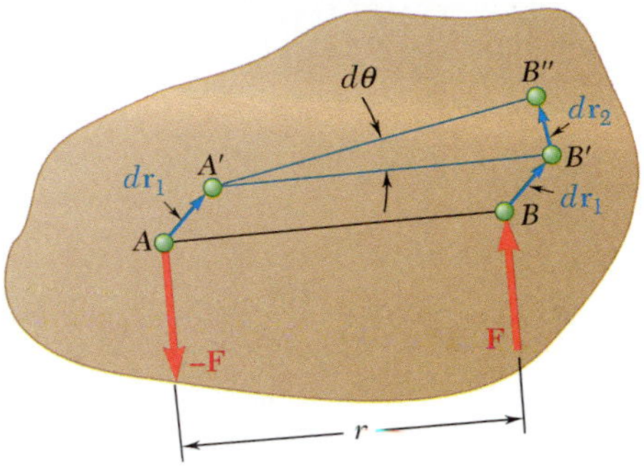

Fig. 10.5

$$dU = M\,d\theta \tag{10.2}$$

where $d\theta$ is the small angle expressed in radians through which the body rotates. We again note that work should be expressed in units obtained by multiplying units of force by units of length.

*10.3. PRINCIPLE OF VIRTUAL WORK

Consider a particle acted upon by several forces $\mathbf{F}_1, \mathbf{F}_2, \ldots, \mathbf{F}_n$ (Fig. 10.6). We can imagine that the particle undergoes a small displacement from A to A'. This displacement is possible, but it will not necessarily take place. The forces may be balanced and the particle at rest, or the particle may move under the action of the given forces in a direction different from that of AA'. Since the displacement considered does not actually occur, it is called a *virtual displacement* and is denoted by $\delta\mathbf{r}$. The symbol $\delta\mathbf{r}$ represents a differential of the first order; it is used to distinguish the virtual displacement from the displacement $d\mathbf{r}$ which would take place under actual motion. As you will see, virtual displacements can be used to determine whether the conditions of equilibrium of a particle are satisfied.

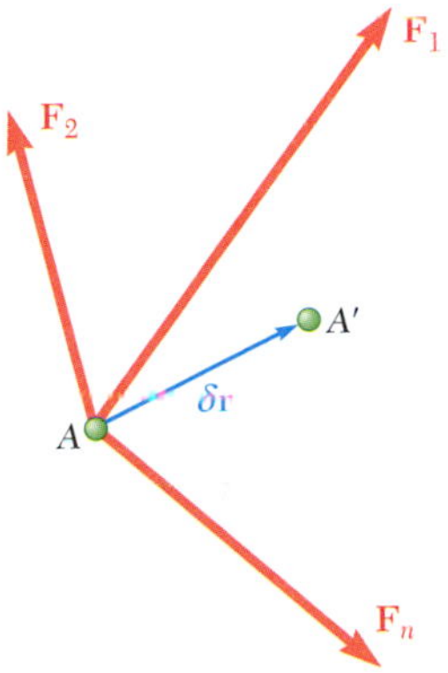

Fig. 10.6

The work of each of the forces $\mathbf{F}_1, \mathbf{F}_2, \ldots, \mathbf{F}_n$ during the virtual displacement $\delta\mathbf{r}$ is called *virtual work*. The virtual work of all the forces acting on the particle of Fig. 10.6 is

$$\begin{aligned}\delta U &= \mathbf{F}_1 \cdot \delta\mathbf{r} + \mathbf{F}_2 \cdot \delta\mathbf{r} + \cdots + \mathbf{F}_n \cdot \delta\mathbf{r} \\ &= (\mathbf{F}_1 + \mathbf{F}_2 + \cdots + \mathbf{F}_n) \cdot \delta\mathbf{r}\end{aligned}$$

or

$$\delta U = \mathbf{R} \cdot \delta\mathbf{r} \tag{10.3}$$

where **R** is the resultant of the given forces. Thus, the total virtual work of the forces $\mathbf{F}_1, \mathbf{F}_2, \ldots, \mathbf{F}_n$ is equal to the virtual work of their resultant **R**.

The principle of virtual work for a particle states that *if a particle is in equilibrium, the total virtual work of the forces acting on the particle is zero for any virtual displacement of the particle*. This condition is necessary: if the particle is in equilibrium, the resultant **R** of the forces is zero, and it follows from (10.3) that the total virtual work δU is zero. The condition is also sufficient: if the total virtual work δU is zero for any virtual displacement, the scalar product $\mathbf{R} \cdot \delta\mathbf{r}$ is zero for any $\delta\mathbf{r}$, and the resultant **R** must be zero.

In the case of a rigid body, the principle of virtual work states that *if a rigid body is in equilibrium, the total virtual work of the external forces acting on the rigid body is zero for any virtual displacement of the body*. The condition is necessary: if the body is in equilibrium, all the particles forming the body are in equilibrium and the total virtual work of the forces acting on all the particles must be zero; but we have seen in the preceding section that the total work of the internal forces is zero; the total work of the external forces must therefore also be zero. The condition can also be proved to be sufficient.

The principle of virtual work can be extended to the case of a *system of connected rigid bodies*. If the system remains connected during the virtual displacement, *only the work of the forces external to the system need be considered*, since the total work of the internal forces at the various connections is zero.

*10.4. APPLICATIONS OF THE PRINCIPLE OF VIRTUAL WORK

The principle of virtual work is particularly effective when applied to the solution of problems involving machines or mechanisms consisting of several connected rigid bodies. Consider, for instance, the toggle vise ACB of Fig. 10.7a, used to compress a wooden block. We

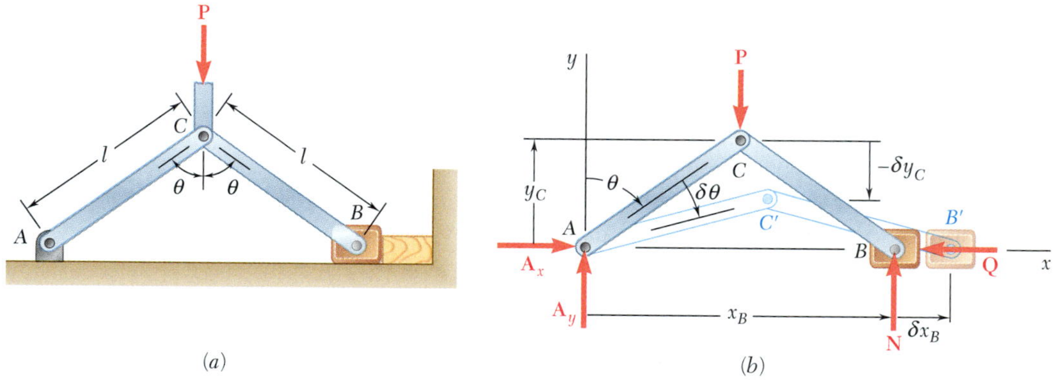

Fig. 10.7

wish to determine the force exerted by the vise on the block when a given force **P** is applied at C assuming that there is no friction. Denoting by **Q** the reaction of the block on the vise, we draw the free-body diagram of the vise and consider the virtual displacement obtained by giving a positive increment $\delta\theta$ to the angle θ (Fig. 10.7b). Choosing a system of coordinate axes with origin at A, we note that x_B increases while y_C decreases. This is indicated in the figure, where a positive increment δx_B and a negative increment $-\delta y_C$ are shown. The reactions $\mathbf{A}_x$, $\mathbf{A}_y$, and **N** will do no work during the virtual displacement considered, and we need only compute the work of **P** and **Q**. Since **Q** and δx_B have opposite senses, the virtual work of **Q** is $\delta U_Q = -Q\,\delta x_B$. Since **P** and the increment shown $(-\delta y_C)$ have the same sense, the virtual work of **P** is $\delta U_P = +P(-\delta y_C) = -P\,\delta y_C$. The minus signs obtained could have been predicted by simply noting that the forces **Q** and **P** are directed opposite to the positive x and y axes, respectively. Expressing the coordinates x_B and y_C in terms of the angle θ and differentiating, we obtain

$$\begin{aligned} x_B &= 2l \sin \theta & y_C &= l \cos \theta \\ \delta x_B &= 2l \cos \theta \, \delta\theta & \delta y_C &= -l \sin \theta \, \delta\theta \end{aligned} \tag{10.4}$$

The total virtual work of the forces **Q** and **P** is thus

$$\begin{aligned} \delta U = \delta U_Q + \delta U_P &= -Q \, \delta x_B - P \, \delta y_C \\ &= -2Ql \cos \theta \, \delta\theta + Pl \sin \theta \, \delta\theta \end{aligned}$$

Setting $\delta U = 0$, we obtain

$$2Ql \cos \theta \, \delta\theta = Pl \sin \theta \, \delta\theta \tag{10.5}$$

$$Q = \tfrac{1}{2}P \tan \theta \tag{10.6}$$

The superiority of the method of virtual work over the conventional equilibrium equations in the problem considered here is clear: by using the method of virtual work, we were able to eliminate all unknown reactions, while the equation $\Sigma M_A = 0$ would have eliminated only two of the unknown reactions. This property of the method of virtual work can be used in solving many problems involving machines and mechanisms. *If the virtual displacement considered is consistent with the constraints imposed by the supports and connections, all reactions and internal forces are eliminated and only the work of the loads, applied forces, and friction forces need be considered.*

The method of virtual work can also be used to solve problems involving completely constrained structures, although the virtual displacements considered will never actually take place. Consider, for example, the frame *ACB* shown in Fig. 10.8*a*. If point *A* is kept fixed, while *B* is given a horizontal virtual displacement (Fig. 10.8*b*), we need consider only the work of **P** and $\mathbf{B}_x$. We can thus determine the

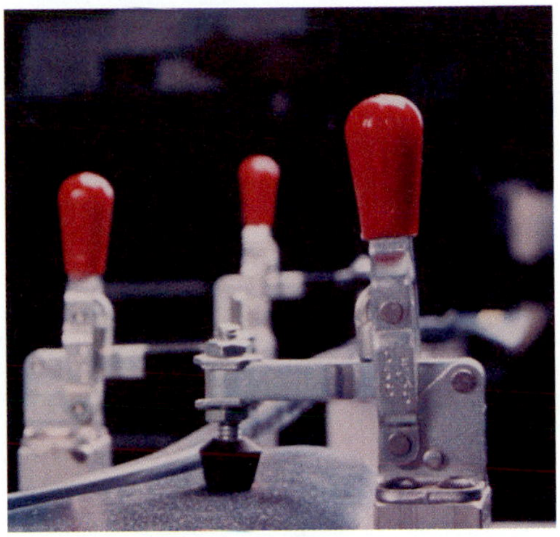

Photo 10.2 The clamping force of the toggle clamp shown can be expressed as a function of the force applied to the handle by first establishing the geometric relations among the members of the clamp and then applying the method of virtual work.

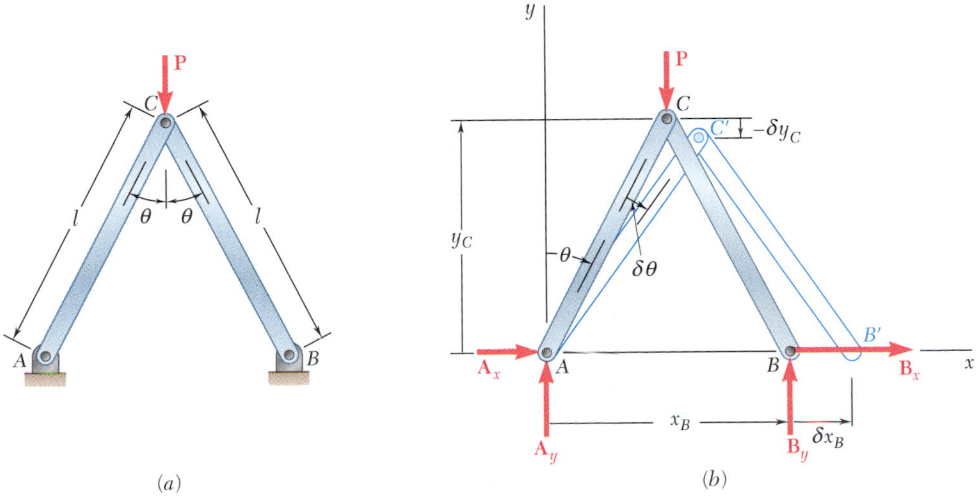

Fig. 10.8

reaction component $\mathbf{B}_x$ in the same way as the force $\mathbf{Q}$ of the preceding example (Fig. 10.7*b*); we have

$$B_x = -\tfrac{1}{2}P \tan \theta$$

Keeping B fixed and giving to A a horizontal virtual displacement, we can similarly determine the reaction component $\mathbf{A}_x$. The components $\mathbf{A}_y$ and $\mathbf{B}_y$ can be determined by rotating the frame ACB as a rigid body about B and A, respectively.

The method of virtual work can also be used to determine the configuration of a system in equilibrium under given forces. For example, the value of the angle θ for which the linkage of Fig. 10.7 is in equilibrium under two given forces $\mathbf{P}$ and $\mathbf{Q}$ can be obtained by solving Eq. (10.6) for $\tan \theta$.

It should be noted, however, that the attractiveness of the method of virtual work depends to a large extent upon the existence of simple geometric relations among the various virtual displacements involved in the solution of a given problem. When no such simple relations exist, it is usually advisable to revert to the conventional method of Chap. 6.

*10.5. REAL MACHINES. MECHANICAL EFFICIENCY

In analyzing the toggle vise in the preceding section, we assumed that no friction forces were involved. Thus, the virtual work consisted only of the work of the applied force $\mathbf{P}$ and of the reaction $\mathbf{Q}$. But the work of the reaction $\mathbf{Q}$ is equal in magnitude and opposite in sign to the work of the force exerted by the vise on the block. Equation (10.5), therefore, expresses that the *output work* $2Ql \cos \theta\, \delta\theta$ is equal to the *input work* $Pl \sin \theta\, \delta\theta$. A machine in which input and output work are equal is said to be an "ideal" machine. In a "real" machine, friction forces will always do some work, and the output work will be smaller than the input work.

Consider, for example, the toggle vise of Fig. 10.7*a*, and assume now that a friction force $\mathbf{F}$ develops between the sliding block B and the horizontal plane (Fig. 10.9). Using the conventional methods of statics and summing moments about A, we find $N = P/2$. Denoting by μ the coefficient of friction between block B and the horizontal

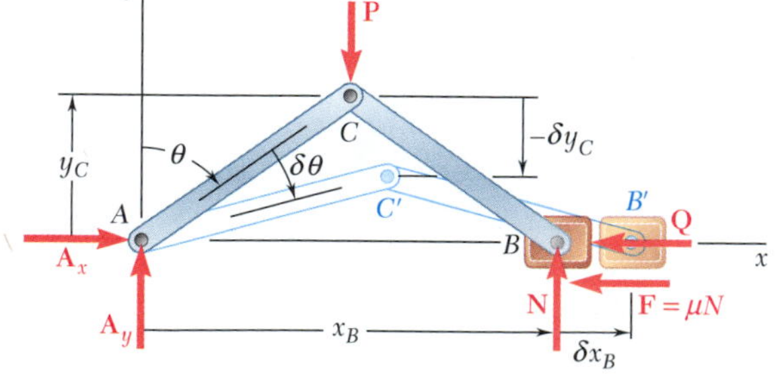

Fig. 10.9

plane, we have $F = \mu N = \mu P/2$. Recalling formulas (10.4), we find that the total virtual work of the forces **Q**, **P**, and **F** during the virtual displacement shown in Fig. 10.9 is

$$\begin{aligned}\delta U &= -Q\,\delta x_B - P\,\delta y_C - F\,\delta x_B \\ &= -2Ql\cos\theta\,\delta\theta + Pl\sin\theta\,\delta\theta - \mu Pl\cos\theta\,\delta\theta\end{aligned}$$

Setting $\delta U = 0$, we obtain

$$2Ql\cos\theta\,\delta\theta = Pl\sin\theta\,\delta\theta - \mu Pl\cos\theta\,\delta\theta \tag{10.7}$$

which expresses that the output work is equal to the input work minus the work of the friction force. Solving for Q, we have

$$Q = \tfrac{1}{2}P(\tan\theta - \mu) \tag{10.8}$$

We note that $Q = 0$ when $\tan\theta = \mu$, that is, when θ is equal to the angle of friction ϕ, and that $Q < 0$ when $\theta < \phi$. The toggle vise may thus be used only for values of θ larger than the angle of friction.

The *mechanical efficiency* of a machine is defined as the ratio

$$\eta = \frac{\text{output work}}{\text{input work}} \tag{10.9}$$

Clearly, the mechanical efficiency of an ideal machine is $\eta = 1$, since input and output work are then equal, while the mechanical efficiency of a real machine will always be less than 1.

In the case of the toggle vise we have just analyzed, we write

$$\eta = \frac{\text{output work}}{\text{input work}} = \frac{2Ql\cos\theta\,\delta\theta}{Pl\sin\theta\,\delta\theta}$$

Substituting from (10.8) for Q, we obtain

$$\eta = \frac{P(\tan\theta - \mu)l\cos\theta\,\delta\theta}{Pl\sin\theta\,\delta\theta} = 1 - \mu\cot\theta \tag{10.10}$$

We check that in the absence of friction forces, we would have $\mu = 0$ and $\eta = 1$. In the general case, when μ is different from zero, the efficiency η becomes zero for $\mu\cot\theta = 1$, that is, for $\tan\theta = \mu$, or $\theta = \tan^{-1}\mu = \phi$. We note again that the toggle vise can be used only for values of θ larger than the angle of friction ϕ.

SAMPLE PROBLEM 10.1

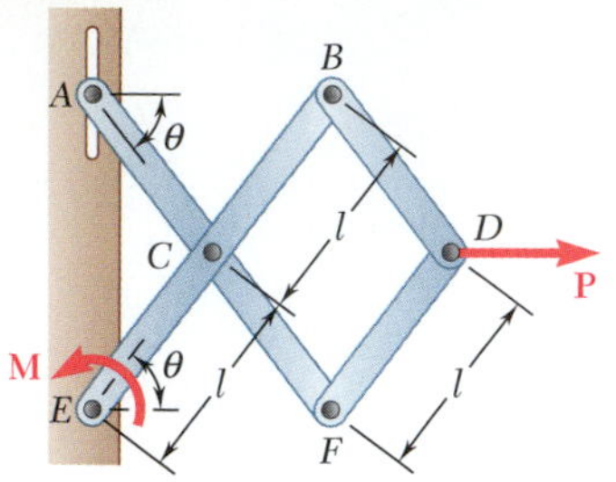

Using the method of virtual work, determine the magnitude of the couple **M** required to maintain the equilibrium of the mechanism shown.

SOLUTION

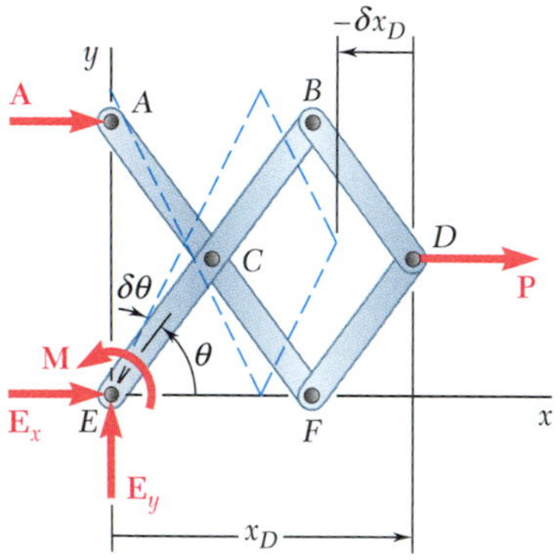

Choosing a coordinate system with origin at E, we write

$$x_D = 3l \cos\theta \qquad \delta x_D = -3l \sin\theta\,\delta\theta$$

Principle of Virtual Work. Since the reactions **A**, $\mathbf{E}_x$, and $\mathbf{E}_y$ will do no work during the virtual displacement, the total virtual work done by **M** and **P** must be zero. Noting that **P** acts in the positive x direction and **M** acts in the positive θ direction, we write

$\delta U = 0$:
$$+M\,\delta\theta + P\,\delta x_D = 0$$
$$+M\,\delta\theta + P(-3l \sin\theta\,\delta\theta) = 0$$
$$M = 3Pl \sin\theta \quad ◀$$

SAMPLE PROBLEM 10.2

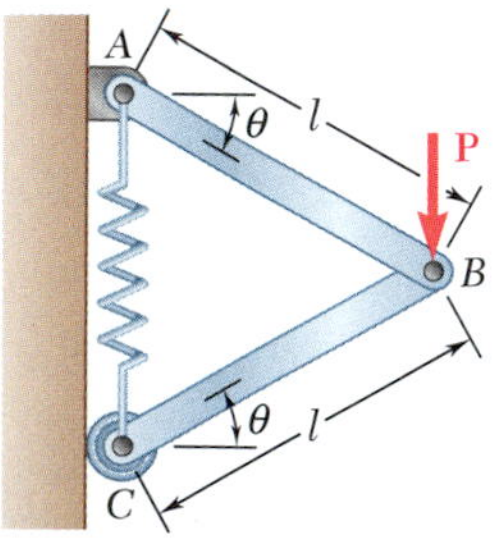

Determine the expressions for θ and for the tension in the spring which correspond to the equilibrium position of the mechanism. The unstretched length of the spring is h, and the constant of the spring is k. Neglect the weight of the mechanism.

SOLUTION

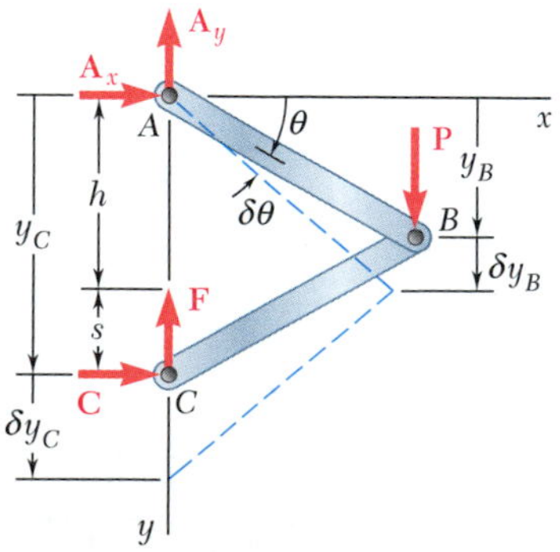

With the coordinate system shown

$$y_B = l \sin\theta \qquad y_C = 2l \sin\theta$$
$$\delta y_B = l \cos\theta\,\delta\theta \qquad \delta y_C = 2l \cos\theta\,\delta\theta$$

The elongation of the spring is

$$s = y_C - h = 2l \sin\theta - h$$

The magnitude of the force exerted at C by the spring is

$$F = ks = k(2l \sin\theta - h) \qquad (1)$$

Principle of Virtual Work. Since the reactions $\mathbf{A}_x$, $\mathbf{A}_y$, and **C** do no work, the total virtual work done by **P** and **F** must be zero.

$\delta U = 0$:
$$P\,\delta y_B - F\,\delta y_C = 0$$
$$P(l \cos\theta\,\delta\theta) - k(2l \sin\theta - h)(2l \cos\theta\,\delta\theta) = 0$$
$$\sin\theta = \frac{P + 2kh}{4kl} \quad ◀$$

Substituting this expression into (1), we obtain $\qquad F = \tfrac{1}{2}P \quad ◀$

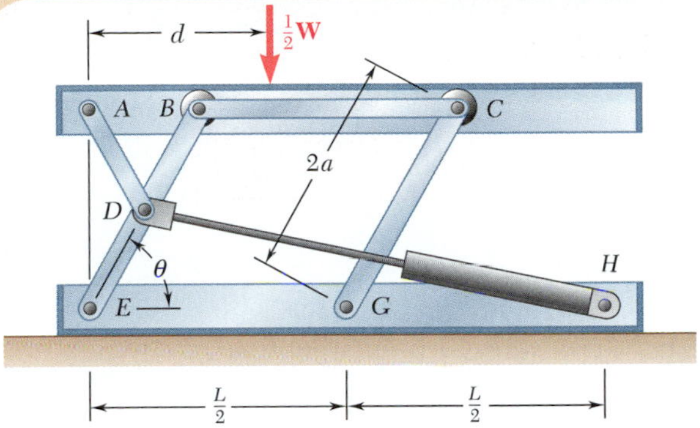

SAMPLE PROBLEM 10.3

A hydraulic-lift table is used to raise a 1000-kg crate. It consists of a platform and of two identical linkages on which hydraulic cylinders exert equal forces. (Only one linkage and one cylinder are shown.) Members *EDB* and *CG* are each of length $2a$, and member *AD* is pinned to the midpoint of *EDB*. If the crate is placed on the table, so that half of its weight is supported by the system shown, determine the force exerted by each cylinder in raising the crate for $\theta = 60°$, $a = 0.70$ m, and $L = 3.20$ m. This mechanism has been previously considered in Sample Prob. 6.7.

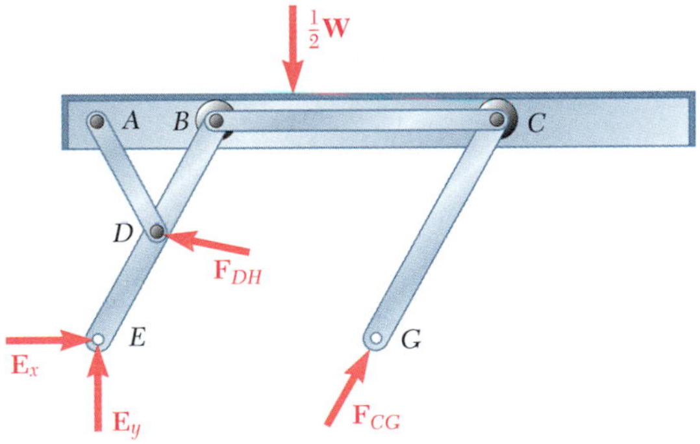

SOLUTION

The machine considered consists of the platform and of the linkage, with an input force $\mathbf{F}_{DH}$ exerted by the cylinder and an output force equal and opposite to $\frac{1}{2}\mathbf{W}$.

Principle of Virtual Work. We first observe that the reactions at *E* and *G* do no work. Denoting by y the elevation of the platform above the base, and by s the length *DH* of the cylinder-and-piston assembly, we write

$$\delta U = 0: \qquad -\tfrac{1}{2}W\,\delta y + F_{DH}\,\delta s = 0 \tag{1}$$

The vertical displacement δy of the platform is expressed in terms of the angular displacement $\delta\theta$ of *EDB* as follows:

$$y = (EB)\sin\theta = 2a\sin\theta$$
$$\delta y = 2a\cos\theta\,\delta\theta$$

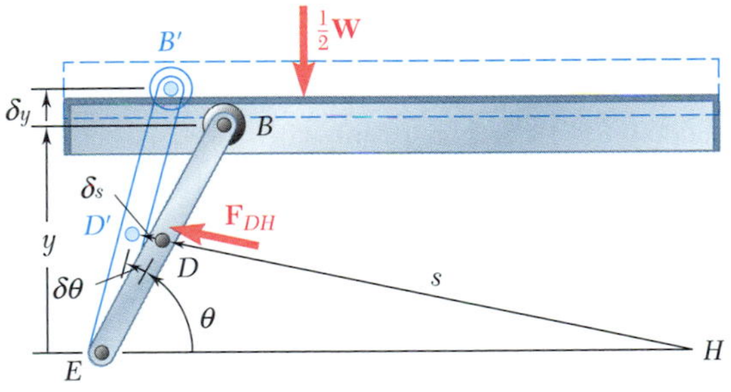

To express δs similarly in terms of $\delta\theta$, we first note that by the law of cosines,

$$s^2 = a^2 + L^2 - 2aL\cos\theta$$

Differentiating,

$$2s\,\delta s = -2aL(-\sin\theta)\,\delta\theta$$
$$\delta s = \frac{aL\sin\theta}{s}\,\delta\theta$$

Substituting for δy and δs into (1), we write

$$(-\tfrac{1}{2}W)2a\cos\theta\,\delta\theta + F_{DH}\frac{aL\sin\theta}{s}\,\delta\theta = 0$$
$$F_{DH} = W\frac{s}{L}\cot\theta$$

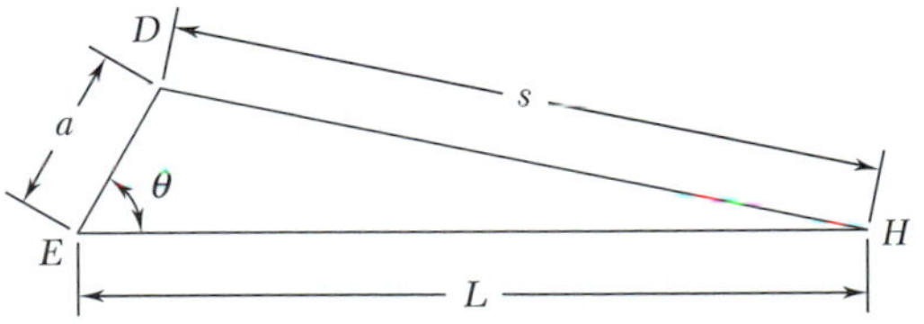

With the given numerical data, we have

$$W = mg = (1000\text{ kg})(9.81\text{ m/s}^2) = 9810\text{ N} = 9.81\text{ kN}$$
$$s^2 = a^2 + L^2 - 2aL\cos\theta$$
$$= (0.70)^2 + (3.20)^2 - 2(0.70)(3.20)\cos 60° = 8.49$$
$$s = 2.91\text{ m}$$

$$F_{DH} = W\frac{s}{L}\cot\theta = (9.81\text{ kN})\frac{2.91\text{ m}}{3.20\text{ m}}\cot 60°$$

$F_{DH} = 5.15$ kN ◀

SOLVING PROBLEMS ON YOUR OWN

In this lesson you learned to use the *method of virtual work,* which is a different way of solving problems involving the equilibrium of rigid bodies.

The work done by a force during a displacement of its point of application or by a couple during a rotation is found by using Eqs. (10.1) and (10.2), respectively:

$$dU = F\,ds\cos\alpha \tag{10.1}$$
$$dU = M\,d\theta \tag{10.2}$$

Principle of virtual work. In its more general and more useful form, this principle can be stated as follows: *If a system of connected rigid bodies is in equilibrium, the total virtual work of the external forces applied to the system is zero for any virtual displacement of the system.*

As you apply the principle of virtual work, keep in mind the following:

1. *Virtual displacement.* A machine or mechanism in equilibrium has no tendency to move. However, *we can cause, or imagine, a small displacement*. Since it does not actually occur, such a displacement is called a *virtual displacement*.

2. *Virtual work.* The work done by a force or couple during a virtual displacement is called *virtual work*.

3. *You need consider only the forces which do work* during the virtual displacement. It is important to remember that only the component of a virtual displacement parallel to a given force need be considered when computing the virtual work. This is why, for example, in Sample Prob. 10.2 only the vertical displacement of pin B was determined.

4. *Forces which do no work* during a virtual displacement that is consistent with the constraints imposed on the system include the following:
- ***a.*** Reactions at supports
- ***b.*** Internal forces at connections
- ***c.*** Forces exerted by inextensible cords and cables

None of these forces need be considered when you use the method of virtual work.

5. *Be sure to express the various virtual displacements* involved in your computations in terms of a *single virtual displacement*. This is done in each of the three preceding sample problems, where the virtual displacements are all expressed in terms of $\delta\theta$. In addition, virtual displacements containing second and higher order terms [for example, $(\delta\theta)^2$ and $(\delta x)^2$] can be ignored when computing the virtual work.

6. *Remember that the method of virtual work is effective only in those cases* where the geometry of the system makes it relatively easy to relate the displacements involved.

7. *The work of the force* F *exerted by a spring during a virtual displacement* δx is given by $\delta U = F\delta x$. In the next lesson we will determine the work done during a finite displacement of the force exerted by a spring.

Problems

10.1 and 10.2 Determine the vertical force **P** which must be applied at G to maintain the equilibrium of the linkage.

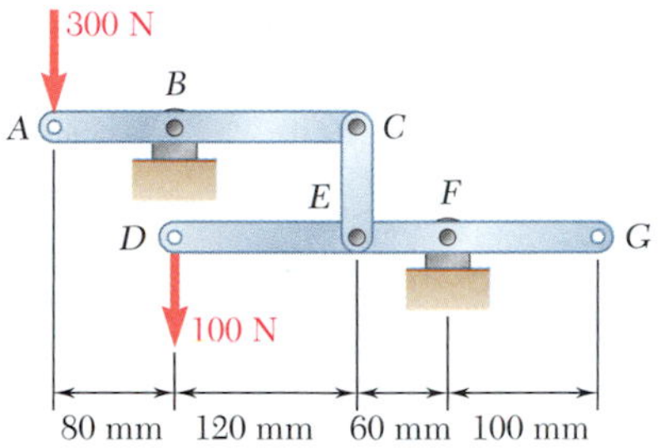

Fig. P10.1 and P10.4

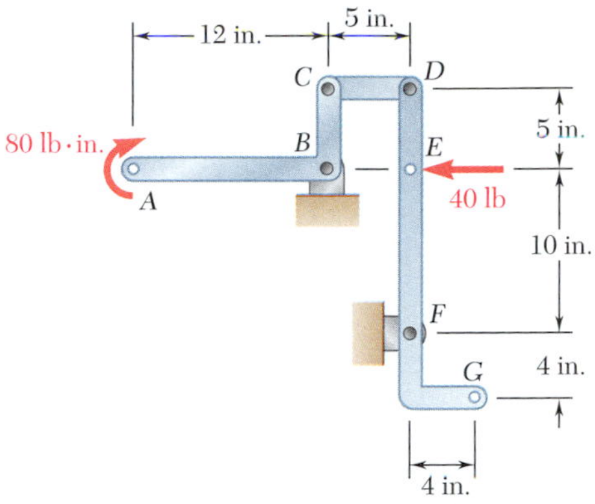

Fig. P10.2 and *P10.3*

***10.3* and 10.4** Determine the couple **M** which must be applied to member $DEFG$ to maintain the equilibrium of the linkage.

10.5 An unstretched spring of constant 4 lb/in. is attached to pins at points C and I as shown. The pin at B is attached to member BDE and can slide freely along the slot in the fixed plate. Determine the force in the spring and the horizontal displacement of point H when a 20-lb horizontal force directed to the right is applied (*a*) at point G, (*b*) at points G and H.

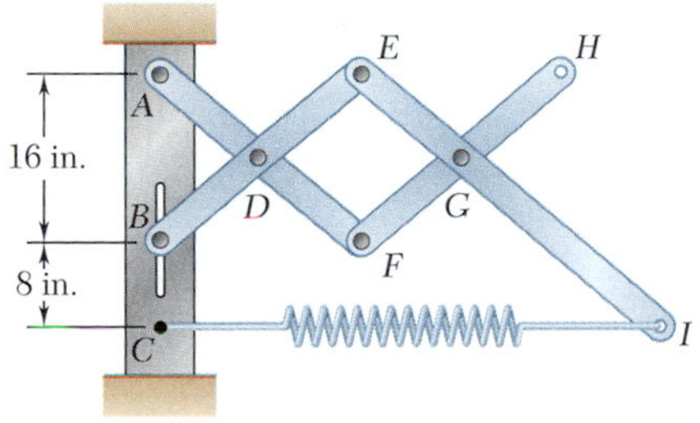

Fig. P10.5 and *P10.6*

10.6 An unstretched spring of constant 4 lb/in. is attached to pins at points C and I as shown. The pin at B is attached to member BDE and can slide freely along the slot in the fixed plate. Determine the force in the spring and the horizontal displacement of point H when a 20-lb horizontal force directed to the right is applied (*a*) at point E, (*b*) at points D and E.

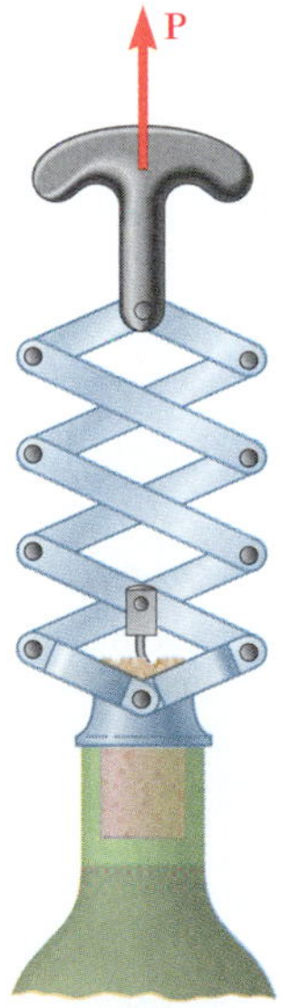

Fig. P10.7

10.7 Knowing that the maximum friction force exerted by the bottle on the cork is 300 N, determine (*a*) the force **P** which must be applied to the corkscrew to open the bottle, (*b*) the maximum force exerted by the base of the corkscrew on the top of the bottle.

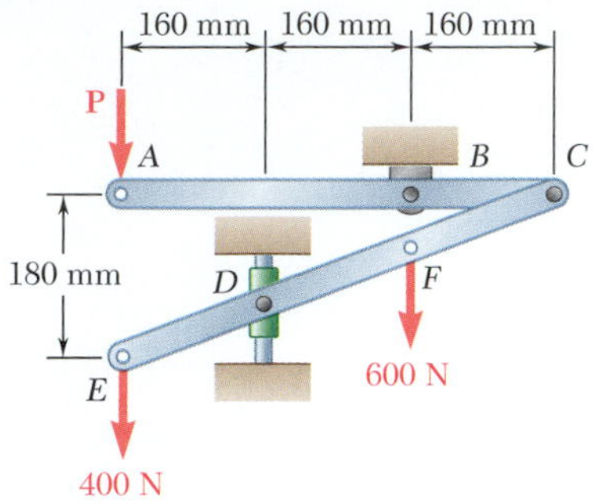

Fig. P10.8

10.8 The two-bar linkage shown is supported by a pin and bracket at B and a collar at D that slides freely on a vertical rod. Determine the force **P** required to maintain the equilibrium of the linkage.

10.9 The mechanism shown is acted upon by the force **P**; derive an expression for the magnitude of the force **Q** required for equilibrium.

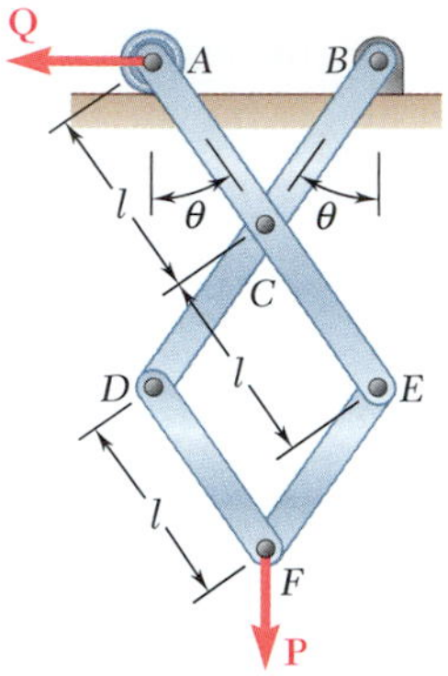

Fig. P10.9

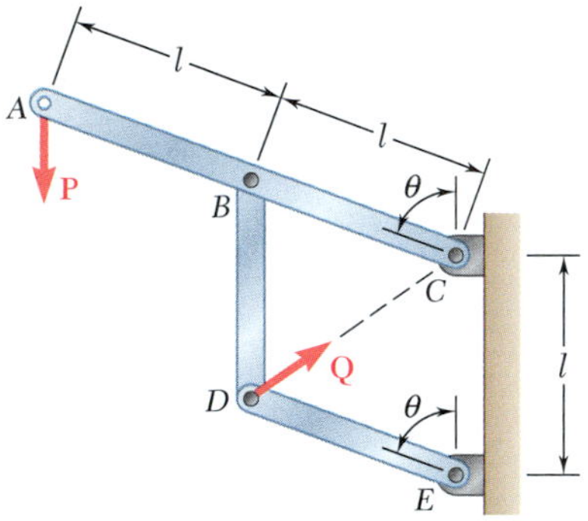

Fig. *P10.10*

10.10 Knowing that the line of action of the force **Q** passes through point C, derive an expression for the magnitude of **Q** required to maintain equilibrium.

10.11 Solve Prob. 10.10 assuming that the force **P** applied at point A acts horizontally to the left.

10.12 The slender rod AB is attached to a collar A and rests on a small wheel at C. Neglecting the radius of the wheel and the effect of friction, derive an expression for the magnitude of the force **Q** required to maintain the equilibrium of the rod.

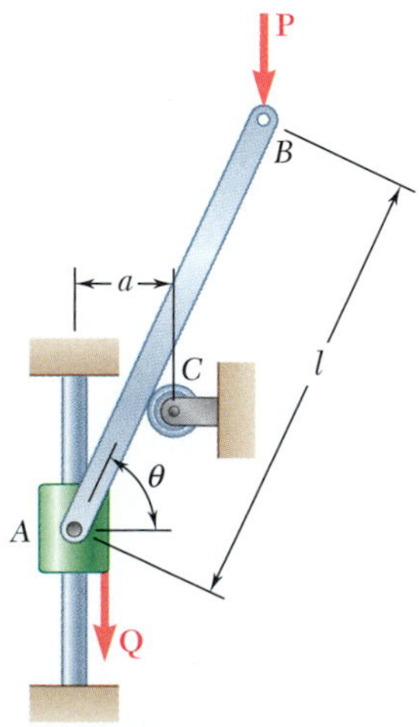

Fig. P10.12

10.13 A double scissor lift table is used to raise a 1000-lb machine component. The table consists of a platform and two identical linkages on which hydraulic cylinders exert equal forces. (Only one linkage and one cylinder are shown.) Each member of the linkage is of length 24 in., and pins C and G are at the midpoints of their respective members. The hydraulic cylinder is pinned at A to the base of the table and at F which is 6 in. from E. If the component is placed on the table so that half of its weight is supported by the system shown, determine the force exerted by each cylinder when $\theta = 30°$.

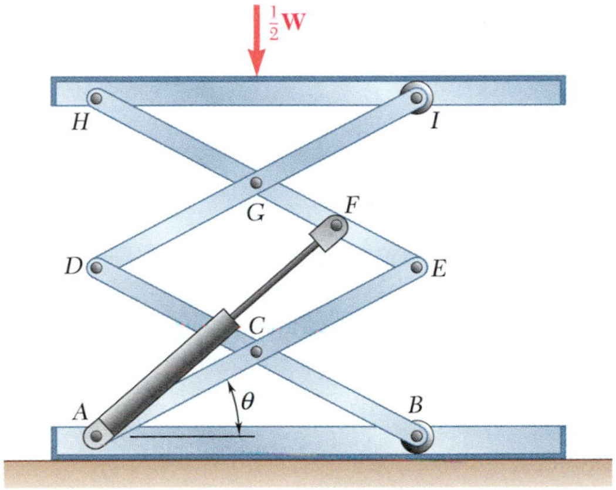

Fig. P10.13 and P10.14

10.14 A double scissor lift table is used to raise a 1000-lb machine component. The table consists of a platform and two identical linkages on which hydraulic cylinders exert equal forces. (Only one linkage and one cylinder are shown.) Each member of the linkage is of length 24 in., and pins C and G are at the midpoints of their respective members. The hydraulic cylinder is pinned at A to the base of the table and at F which is 6 in. from E. If the component is placed on the table so that half of its weight is supported by the system shown, determine the smallest allowable value of θ knowing that the maximum force each cylinder can exert is 8 kips.

10.15 through 10.17 Derive an expression for the magnitude of the couple **M** required to maintain the equilibrium of the linkage shown.

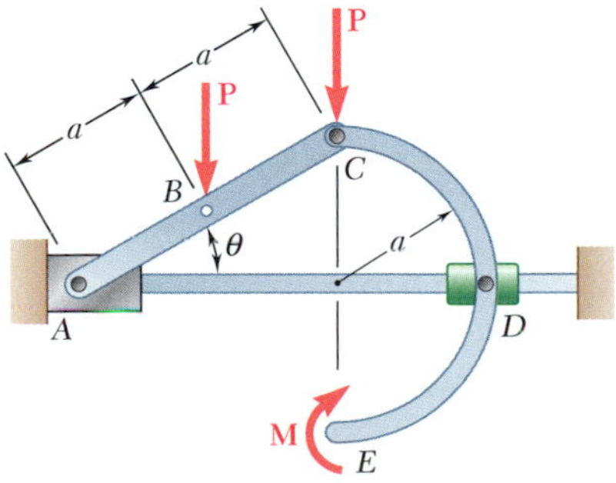

Fig. P10.15

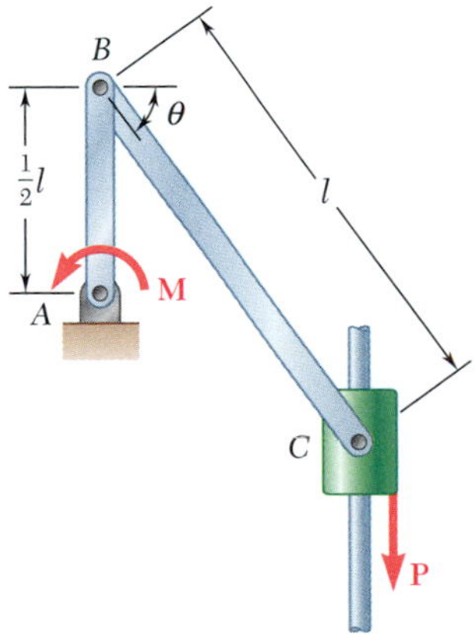

Fig. *P10.16*

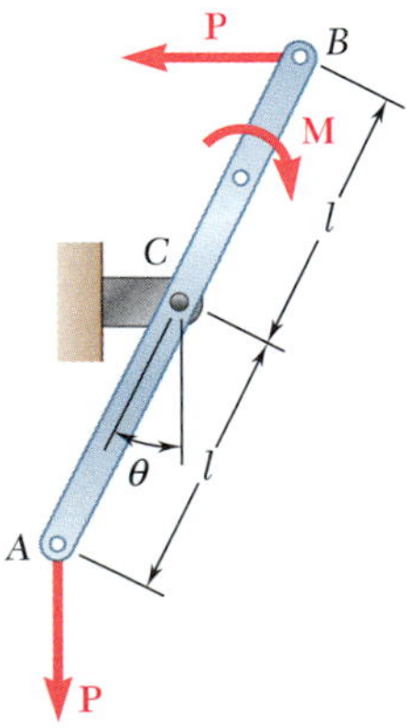

Fig. *P10.17*

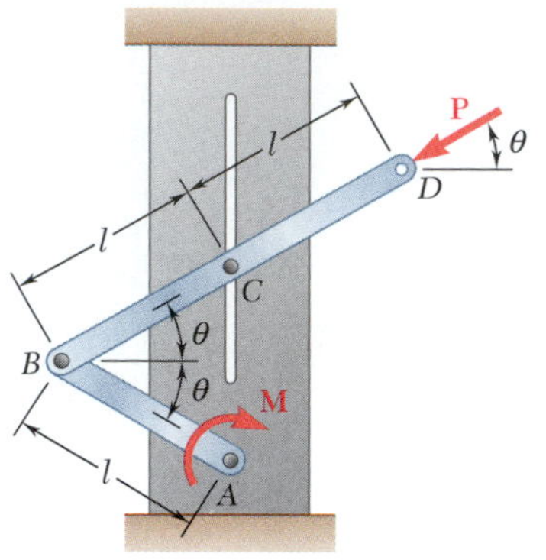

Fig. P10.18

10.18 The pin at C is attached to member BCD and can slide along a slot cut in the fixed plate shown. Neglecting the effect of friction, derive an expression for the magnitude of the couple **M** required to maintain equilibrium when the force **P** which acts at D is directed (*a*) as shown, (*b*) vertically downward, (*c*) horizontally to the right.

10.19 A 1-kip force **P** is applied as shown to the piston of the engine system. Knowing that $AB = 2.5$ in. and $BC = 10$ in., determine the couple **M** required to maintain the equilibrium of the system when (*a*) $\theta = 30°$, (*b*) $\theta = 150°$.

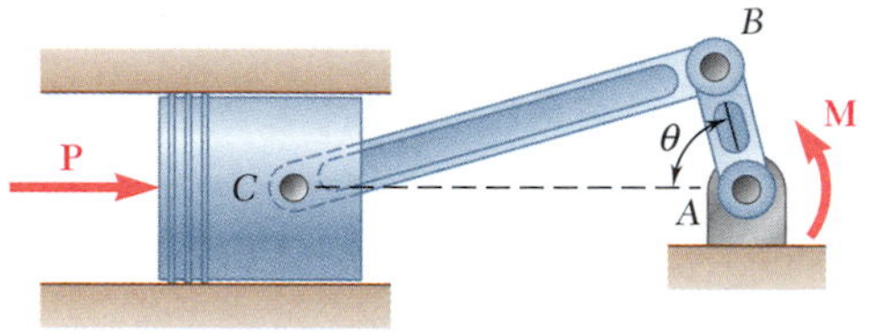

Fig. P10.19 and P10.20

10.20 A couple **M** of magnitude 75 lb · ft is applied as shown to the crank of the engine system. Knowing that $AB = 2.5$ in. and $BC = 10$ in., determine the force **P** required to maintain the equilibrium of the system when (*a*) $\theta = 60°$, (*b*) $\theta = 120°$.

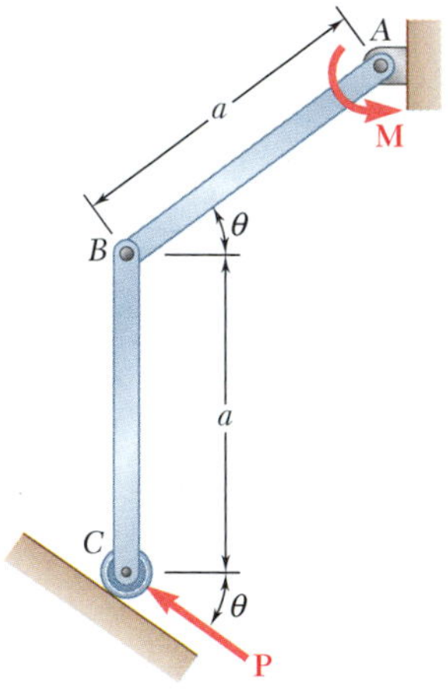

Fig. P10.21 and P10.22

10.21 For the linkage shown, determine the force **P** required for equilibrium when $a = 18$ in., $M = 240$ lb · in., and $\theta = 30°$.

10.22 For the linkage shown, determine the couple **M** required for equilibrium when $a = 2$ ft, $P = 30$ lb, and $\theta = 40°$.

10.23 Determine the value of θ corresponding to the equilibrium position of the mechanism of Prob. 10.10 when $P = 60$ lb and $Q = 75$ lb.

10.24 Determine the value of θ corresponding to the equilibrium position of the mechanism of Prob. 10.11 when $P = 20$ lb and $Q = 25$ lb.

10.25 A slender rod of length l is attached to a collar at B and rests on a portion of a circular cylinder of radius r. Neglecting the effect of friction, determine the value of θ corresponding to the equilibrium position of the mechanism when $l = 300$ mm, $r = 90$ mm, $P = 60$ N, and $Q = 120$ N.

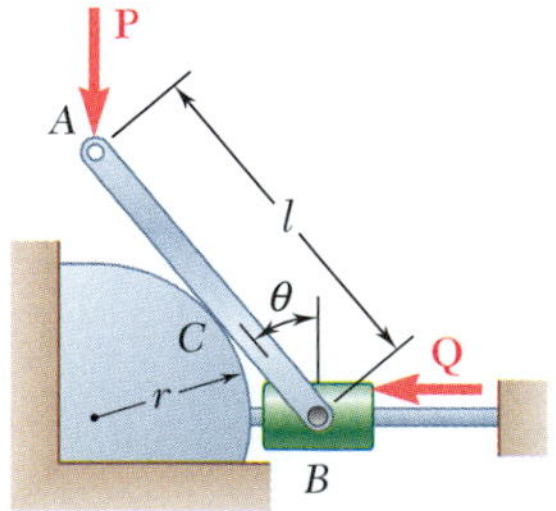

Fig. P10.25 and P10.26

10.26 A slender rod of length l is attached to a collar at B and rests on a portion of a circular cylinder of radius r. Neglecting the effect of friction, determine the value of θ corresponding to the equilibrium position of the mechanism when $l = 280$ mm, $r = 100$ mm, $P = 300$ N, and $Q = 600$ N.

10.27 Determine the value of θ corresponding to the equilibrium position of the mechanism of Prob. 10.12 when $l = 600$ mm, $a = 100$ mm, $P = 100$ N, and $Q = 160$ N.

10.28 Determine the value of θ corresponding to the equilibrium position of the mechanism of Prob. 10.13 when $l = 900$ mm, $a = 150$ mm, $P = 75$ N, and $Q = 135$ N.

10.29 Two rods AC and CE are connected by a pin at C and by a spring AE. The constant of the spring is k, and the spring is unstretched when $\theta = 30°$. For the loading shown, derive an equation in P, θ, l, and k that must be satisfied when the system is in equilibrium.

10.30 Two rods AC and CE are connected by a pin at C and by a spring AE. The constant of the spring is 300 N/m, and the spring is unstretched when $\theta = 30°$. Knowing that $l = 200$ mm and neglecting the mass of the rods, determine the value of θ corresponding to equilibrium when $P = 160$ N.

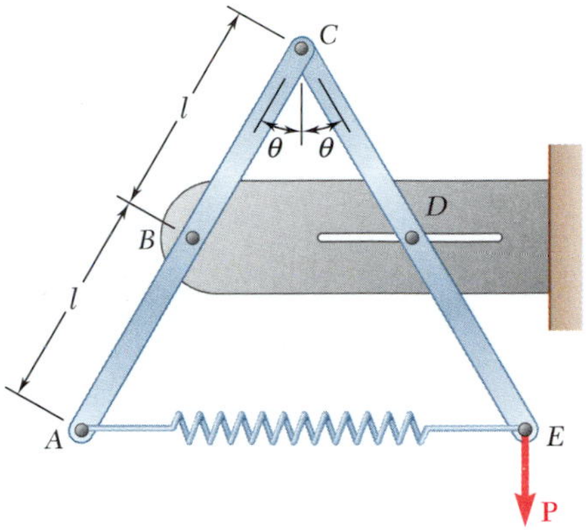

Fig. *P10.29* and P10.30

10.31 Solve Prob. 10.30 assuming that force **P** is moved to C and acts vertically downward.

10.32 For the mechanism shown, block A can move freely in its guide and rests against a spring of constant 15 lb/in. that is undeformed when $\theta = 45°$. For the loading shown, determine the value of θ corresponding to equilibrium.

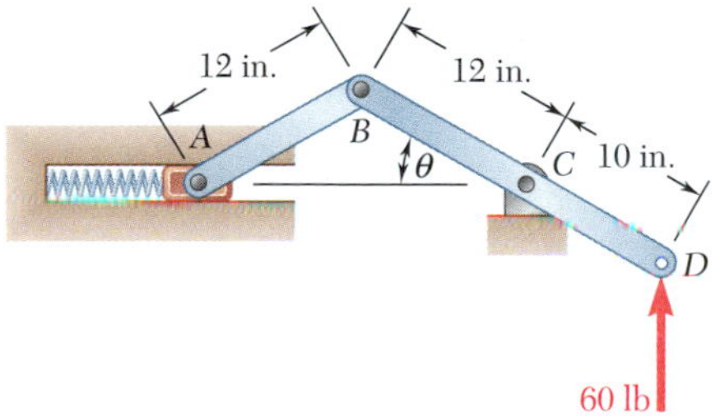

Fig. P10.32

10.33 A force **P** of magnitude 150 lb is applied to the linkage at B. The constant of the spring is 12.5 lb/in., and the spring is unstretched when AB and BC are horizontal. Neglecting the weight of the linkage and knowing that $l = 15$ in., determine the value of θ corresponding to equilibrium.

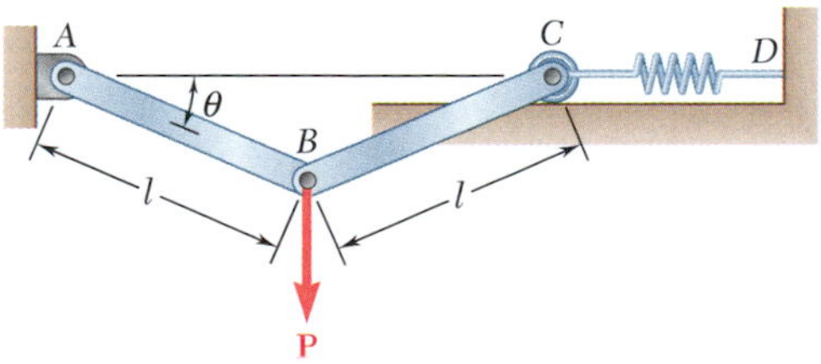

Fig. P10.33 and *P10.34*

10.34 A vertical force **P** is applied to the linkage at B. The constant of the spring is k, and the spring is unstretched when AB and BC are horizontal. Neglecting the weight of the linkage, derive an equation in θ, P, l, and k that must be satisfied when the linkage is in equilibrium.

10.35 and 10.36 Knowing that the constant of spring CD is k and that the spring is unstretched when rod ABC is horizontal, determine the value of θ corresponding to equilibrium for the data indicated.

10.35 $P = 150$ lb, $l = 30$ in., $k = 40$ lb/in.

10.36 $P = 600$ N, $l = 800$ mm, $k = 4$ kN/m.

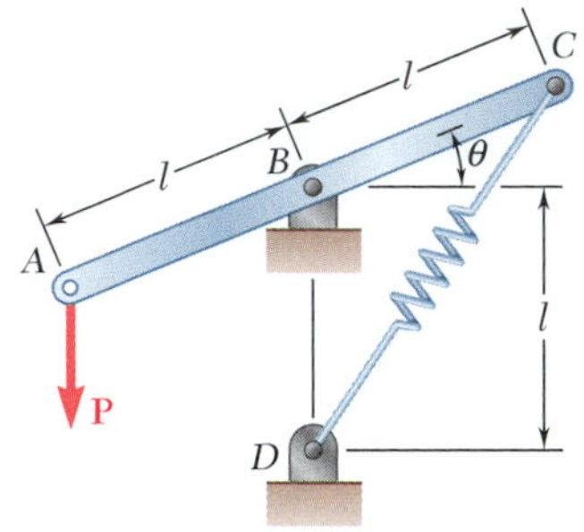

Fig. *P10.35* and P10.36

10.37 A horizontal force **P** of magnitude 160 N is applied to the mechanism at C. The constant of the spring is $k = 1.8$ kN/m, and the spring is unstretched when $\theta = 0$. Neglecting the mass of the mechanism, determine the value of θ corresponding to equilibrium.

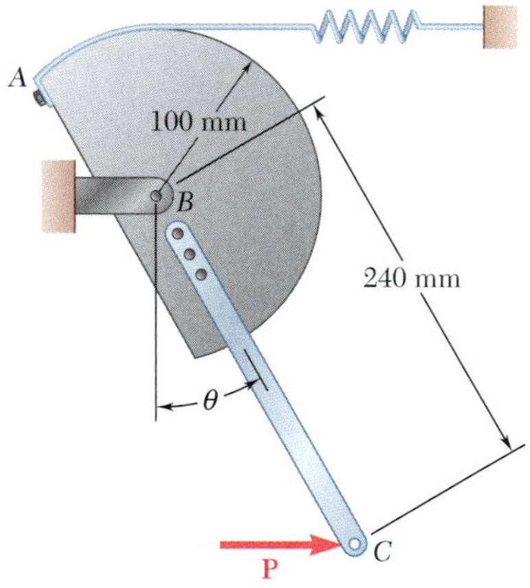

Fig. P10.37

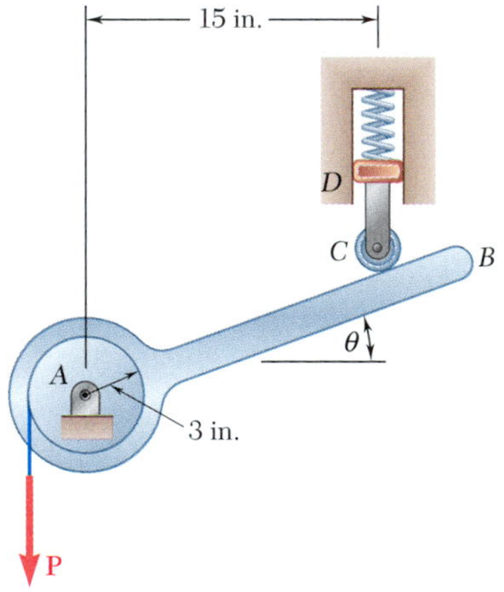

Fig. *P10.38*

10.38 A cord is wrapped around drum A which is attached to member AB. Block D can move freely in its guide and is fastened to link CD. Neglecting the weight of AB and knowing that the spring is of constant 4 lb/in. and is undeformed when $\theta = 0$, determine the value of θ corresponding to equilibrium when a downward force **P** of magnitude 96 lb is applied to the end of the cord.

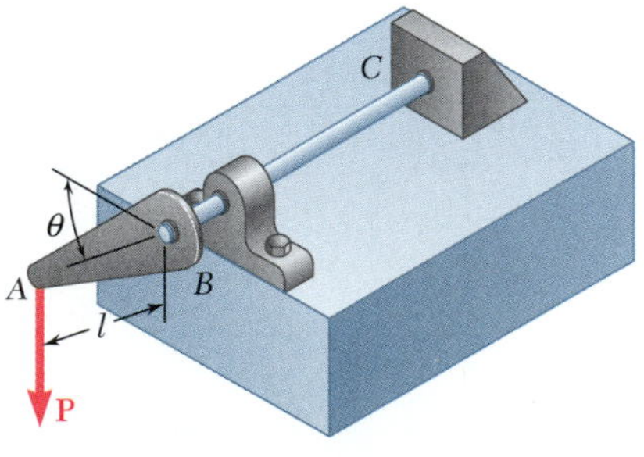

Fig. P10.39

10.39 The lever AB is attached to the horizontal shaft BC which passes through a bearing and is welded to a fixed support at C. The torsional spring constant of the shaft BC is K; that is, a couple of magnitude K is required to rotate end B through 1 rad. Knowing that the shaft is untwisted when AB is horizontal, determine the value of θ corresponding to the position of equilibrium when $P = 400$ lb, $l = 10$ in., and $K = 150$ lb · ft/rad.

10.40 Solve Prob. 10.39 assuming that $P = 1.26$ kips, $l = 10$ in., and $K = 150$ lb · ft/rad. Obtain answers in each of the following quadrants: $0 < \theta < 90°$, $270° < \theta < 360°$, and $360° < \theta < 450°$.

10.41 The position of crank BCD is controlled by the hydraulic cylinder AB. For the loading shown, determine the force exerted by the hydraulic cylinder on pin B knowing that $\theta = 60°$.

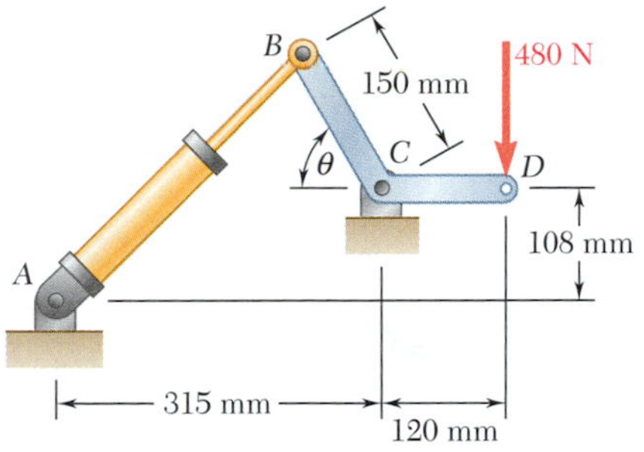

Fig. P10.41 and P10.42

10.42 The position of crank BCD is controlled by the hydraulic cylinder AB. Determine the angle θ knowing that the hydraulic cylinder exerts a 420-N force on pin B when the crank is in the position shown.

10.43 For the linkage shown, determine the force **P** required for equilibrium when $M = 40$ N · m.

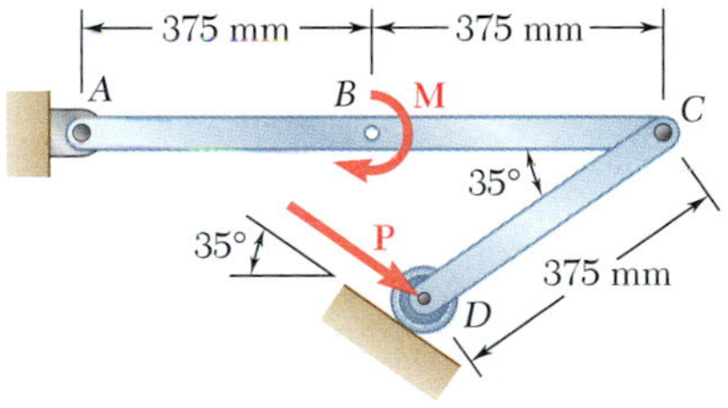

Fig. P10.43

10.44 A cord is wrapped around a drum of radius a that is pinned at A. The constant of the spring is 3 kN/m, and the spring is unstretched when $\theta = 0$. Knowing that $a = 150$ mm and neglecting the mass of the drum, determine the value of θ corresponding to equilibrium when a downward force **P** of magnitude 48 N is applied to the end of the cord.

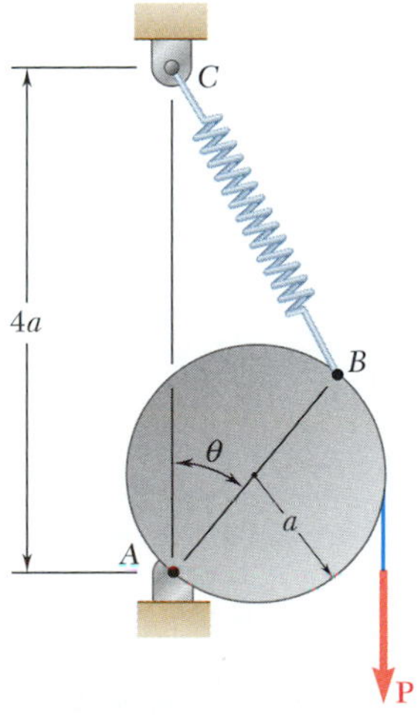

Fig. P10.44

10.45 The telescoping arm ABC is used to provide an elevated platform for construction workers. The workers and the platform together have a mass of 204 kg, and their combined center of gravity is located directly above C. For the position when $\theta = 20°$, determine the force exerted on pin B by the single hydraulic cylinder BD.

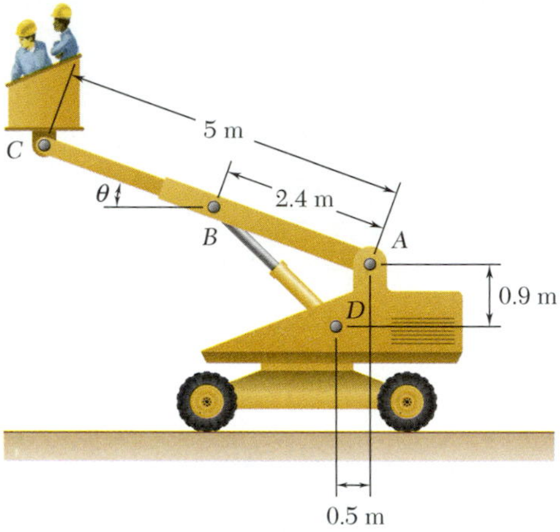

Fig. P10.45

10.46 Solve Prob. 10.45 assuming that the workers are lowered to a point near the ground so that $\theta = -20°$.

10.47 A block of weight W is pulled up a plane forming an angle α with the horizontal by a force $\mathbf{P}$ directed along the plane. If μ is the coefficient of friction between the block and the plane, derive an expression for the mechanical efficiency of the system. Show that the mechanical efficiency cannot exceed $\frac{1}{2}$ if the block is to remain in place when the force $\mathbf{P}$ is removed.

10.48 Denoting by μ_s the coefficient of static friction between collar C and the vertical rod, derive an expression for the magnitude of the largest couple $\mathbf{M}$ for which equilibrium is maintained in the position shown. Explain what happens if $\mu_s \geq \tan\theta$.

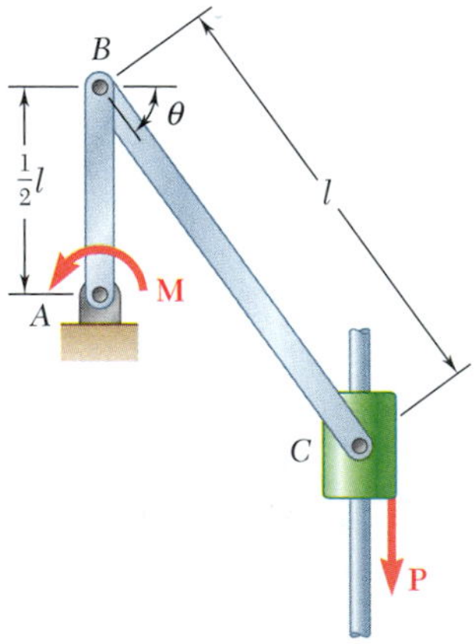

Fig. *P10.48* and P10.49

10.49 Knowing that the coefficient of static friction between collar C and the vertical rod is 0.40, determine the magnitude of the largest and smallest couple $\mathbf{M}$ for which equilibrium is maintained in the position shown when $\theta = 35°$, $l = 30$ in., and $P = 1.2$ kips.

10.50 Derive an expression for the mechanical efficiency of the jack discussed in Sec. 8.6. Show that if the jack is to be self-locking, the mechanical efficiency cannot exceed $\frac{1}{2}$.

10.51 Denoting by μ_s the coefficient of static friction between the block attached to rod ACE and the horizontal surface, derive expressions in terms of P, μ_s, and θ for the largest and smallest magnitudes of the force **Q** for which equilibrium is maintained.

10.52 Knowing that the coefficient of static friction between the block attached to rod ACE and the horizontal surface is 0.15, determine the magnitudes of the largest and smallest force **Q** for which equilibrium is maintained when $\theta = 30°$, $l = 8$ in., and $P = 160$ lb.

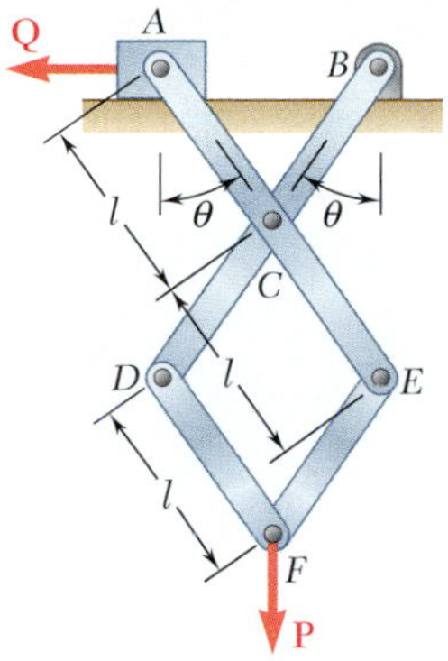

Fig. *P10.51* and P10.52

10.53 Using the method of virtual work, determine separately the force and the couple representing the reaction at A.

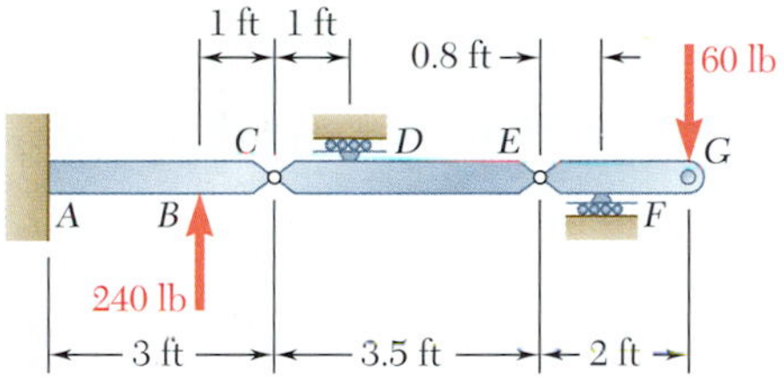

Fig. P10.53 and P10.54

10.54 Using the method of virtual work, determine the reaction at D.

10.55 Referring to Prob. 10.41 and using the value found for the force exerted by the hydraulic cylinder AB, determine the change in the length of AB required to raise the 480-N load 18 mm.

10.56 Referring to Prob. 10.45 and using the value found for the force exerted by the hydraulic cylinder BD, determine the change in the length of BD required to raise the platform attached at C by 50 mm.

10.57 Determine the vertical movement of joint D if the length of member BF is increased by 75 mm. (*Hint:* Apply a vertical load at joint D, and, using the methods of Chap. 6, compute the force exerted by member BF on joints B and F. Then apply the method of virtual work for a virtual displacement resulting in the specified increase in length of member BF. This method should be used only for small changes in the lengths of members.)

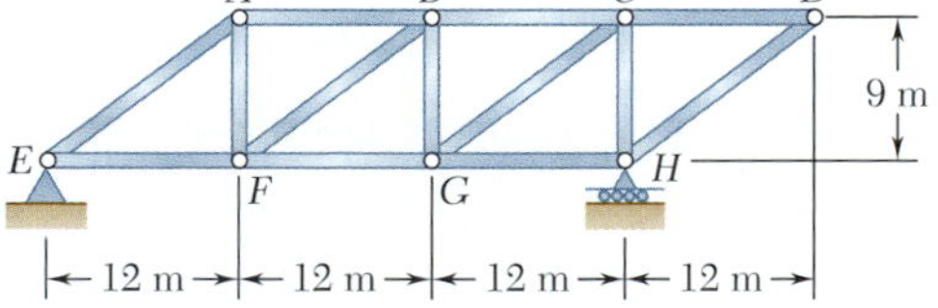

Fig. P10.57 and *P10.58*

10.58 Determine the horizontal movement of joint D if the length of member BF is increased by 75 mm. (See the hint for Prob. 10.57.)

*10.6. WORK OF A FORCE DURING A FINITE DISPLACEMENT

Consider a force **F** acting on a particle. The work of **F** corresponding to an infinitesimal displacement $d\mathbf{r}$ of the particle was defined in Sec. 10.2 as

$$dU = \mathbf{F} \cdot d\mathbf{r} \tag{10.1}$$

The work of **F** corresponding to a finite displacement of the particle from A_1 to A_2 (Fig. 10.10*a*) is denoted by $U_{1\rightarrow 2}$ and is obtained by integrating (10.1) along the curve described by the particle:

$$U_{1\rightarrow 2} = \int_{A_1}^{A_2} \mathbf{F} \cdot d\mathbf{r} \tag{10.11}$$

Using the alternative expression

$$dU = F\, ds \cos \alpha \tag{10.1'}$$

given in Sec. 10.2 for the elementary work dU, we can also express the work $U_{1\rightarrow 2}$ as

$$U_{1\rightarrow 2} = \int_{s_1}^{s_2} (F \cos \alpha)\, ds \tag{10.11'}$$

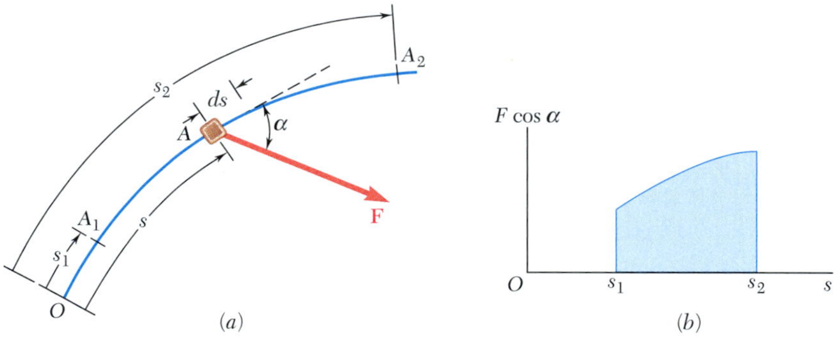

Fig. 10.10

where the variable of integration s measures the distance along the path traveled by the particle. The work $U_{1\rightarrow 2}$ is represented by the area under the curve obtained by plotting $F \cos \alpha$ against s (Fig. 10.10*b*). In the case of a force **F** of constant magnitude acting in the direction of motion, formula (10.11′) yields $U_{1\rightarrow 2} = F(s_2 - s_1)$.

Recalling from Sec. 10.2 that the work of a couple of moment **M** during an infinitesimal rotation $d\theta$ of a rigid body is

$$dU = M\, d\theta \tag{10.2}$$

we express as follows the work of the couple during a finite rotation of the body:

$$U_{1\rightarrow 2} = \int_{\theta_1}^{\theta_2} M\, d\theta \tag{10.12}$$

In the case of a constant couple, formula (10.12) yields

$$U_{1\rightarrow 2} = M(\theta_2 - \theta_1)$$

Work of a Weight. It was stated in Sec. 10.2 that the work of the weight **W** of a body during an infinitesimal displacement of the body is equal to the product of W and the vertical displacement of the center of gravity of the body. With the y axis pointing upward, the work of **W** during a finite displacement of the body (Fig. 10.11) is obtained by writing

$$dU = -W\,dy$$

Integrating from A_1 to A_2, we have

$$U_{1\rightarrow 2} = -\int_{y_1}^{y_2} W\,dy = Wy_1 - Wy_2 \tag{10.13}$$

or

$$U_{1\rightarrow 2} = -W(y_2 - y_1) = -W\,\Delta y \tag{10.13'}$$

where Δy is the vertical displacement from A_1 to A_2. The work of the weight **W** is thus equal to *the product of W and the vertical displacement of the center of gravity of the body*. The work is *positive* when $\Delta y < 0$, that is, *when the body moves down*.

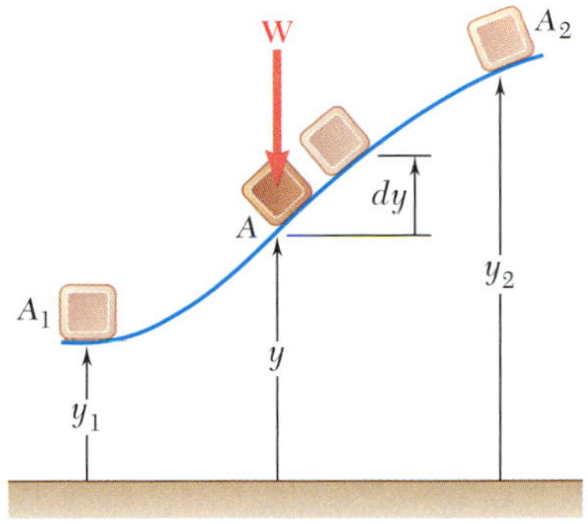

Fig. 10.11

Work of the Force Exerted by a Spring. Consider a body A attached to a fixed point B by a spring; it is assumed that the spring is undeformed when the body is at A_0 (Fig. 10.12*a*). Experimental evidence shows that the magnitude of the force **F** exerted by the spring on a body A is proportional to the deflection x of the spring measured from the position A_0. We have

$$F = kx \tag{10.14}$$

where k is the *spring constant,* expressed in N/m if SI units are used and expressed in lb/ft or lb/in. if U.S. customary units are used. The work of the force **F** exerted by the spring during a finite displacement of the body from $A_1(x = x_1)$ to $A_2(x = x_2)$ is obtained by writing

$$dU = -F\,dx = -kx\,dx$$

$$U_{1\rightarrow 2} = -\int_{x_1}^{x_2} kx\,dx = \tfrac{1}{2}kx_1^2 - \tfrac{1}{2}kx_2^2 \tag{10.15}$$

Care should be taken to express k and x in consistent units. For example, if U.S. customary units are used, k should be expressed in lb/ft and x expressed in feet, or k in lb/in. and x in inches; in the first case, the work is obtained in ft · lb; in the second case, in in · lb. We note that the work of the force **F** exerted by the spring on the body is *positive* when $x_2 < x_1$, that is, *when the spring is returning to its undeformed position*.

Since Eq. (10.14) is the equation of a straight line of slope k passing through the origin, the work $U_{1\rightarrow 2}$ of **F** during the displacement from A_1 to A_2 can be obtained by evaluating the area of the trapezoid shown in Fig. 10.12*b*. This is done by computing the values F_1 and F_2 and multiplying the base Δx of the trapezoid by its mean height $\frac{1}{2}(F_1 + F_2)$. Since the work of the force **F** exerted by the spring is positive for a negative value of Δx, we write

$$U_{1\rightarrow 2} = -\tfrac{1}{2}(F_1 + F_2)\,\Delta x \tag{10.16}$$

Formula (10.16) is usually more convenient to use than (10.15) and affords fewer chances of confusing the units involved.

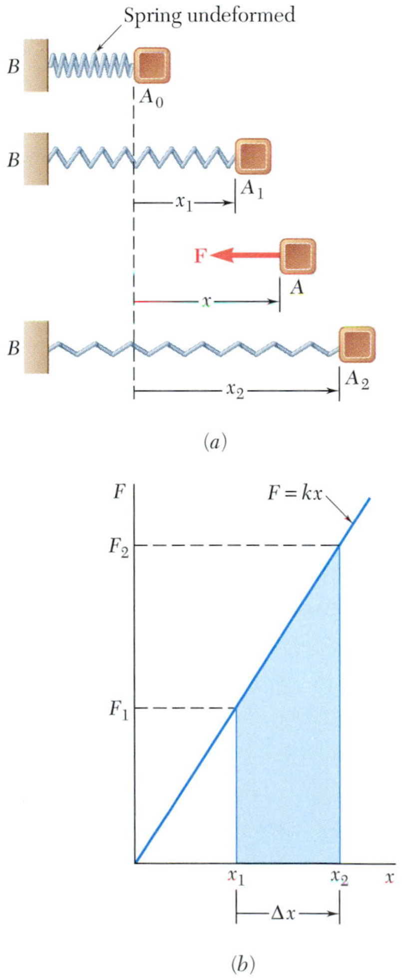

Fig. 10.12

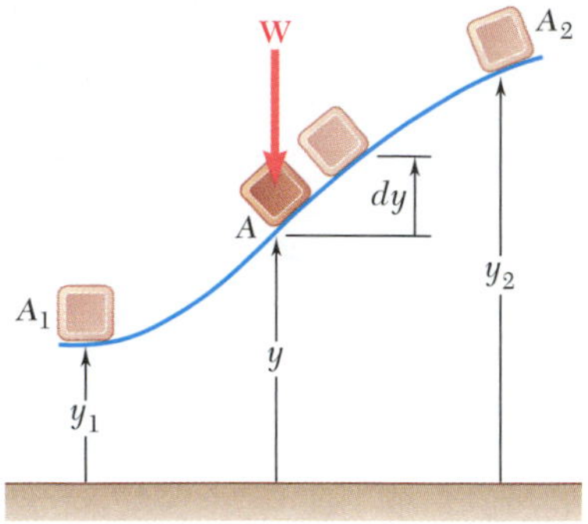

Fig. 10.11 *(repeated)*

*10.7. POTENTIAL ENERGY

Considering again the body of Fig. 10.11, we note from Eq. (10.13) that the work of the weight **W** during a finite displacement is obtained by subtracting the value of the function Wy corresponding to the second position of the body from its value corresponding to the first position. The work of **W** is thus independent of the actual path followed; it depends only upon the initial and final values of the function Wy. This function is called the *potential energy* of the body with respect to the *force of gravity* **W** and is denoted by V_g. We write

$$U_{1\rightarrow 2} = (V_g)_1 - (V_g)_2 \qquad \text{with } V_g = Wy \tag{10.17}$$

We note that if $(V_g)_2 > (V_g)_1$, that is, *if the potential energy increases* during the displacement (as in the case considered here), *the work $U_{1\rightarrow 2}$ is negative*. If, on the other hand, the work of **W** is positive, the potential energy decreases. Therefore, the potential energy V_g of the body provides a measure of *the work which can be done* by its weight **W**. Since only the *change* in potential energy, and not the actual value of V_g, is involved in formula (10.17), an arbitrary constant can be added to the expression obtained for V_g. In other words, the level from which the elevation y is measured can be chosen arbitrarily. Note that potential energy is expressed in the same units as work, that is, in joules (J) if SI units are used† and in ft · lb or in · lb if U.S. customary units are used.

Considering now the body of Fig. 10.12*a*, we note from Eq. (10.15) that the work of the elastic force **F** is obtained by subtracting the value of the function $\frac{1}{2}kx^2$ corresponding to the second position of the body from its value corresponding to the first position. This function is denoted by V_e and is called the *potential energy* of the body with respect to the *elastic force* **F**. We write

$$U_{1\rightarrow 2} = (V_e)_1 - (V_e)_2 \qquad \text{with } V_e = \tfrac{1}{2}kx^2 \tag{10.18}$$

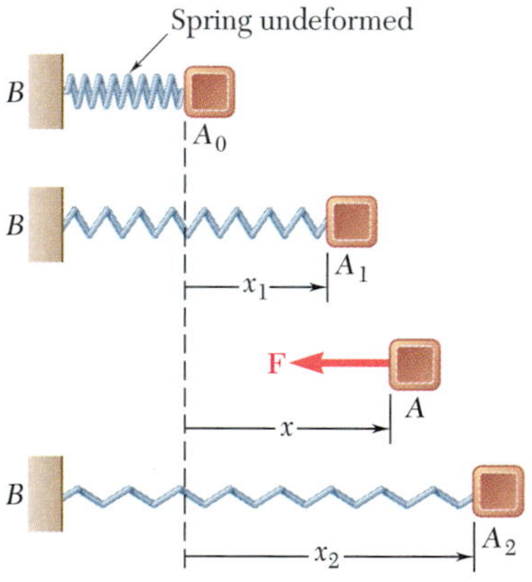

Fig. 10.12a *(repeated)*

and observe that during the displacement considered, the work of the force **F** exerted by the spring on the body is negative and the potential energy V_e increases. We should note that the expression obtained for V_e is valid only if the deflection of the spring is measured from its undeformed position.

The concept of potential energy can be used when forces other than gravity forces and elastic forces are involved. It remains valid as long as the elementary work dU of the force considered is an *exact differential*. It is then possible to find a function V, called potential energy, such that

$$dU = -dV \tag{10.19}$$

Integrating (10.19) over a finite displacement, we obtain the general formula

$$U_{1\rightarrow 2} = V_1 - V_2 \tag{10.20}$$

which expresses that *the work of the force is independent of the path followed and is equal to minus the change in potential energy*. A force which satisfies Eq. (10.20) is said to be a *conservative force*.‡

†See footnote, page 559.

‡A detailed discussion of conservative forces is given in Sec. 13.7 of *Dynamics*.

*10.8. POTENTIAL ENERGY AND EQUILIBRIUM

The application of the principle of virtual work is considerably simplified when the potential energy of a system is known. In the case of a virtual displacement, formula (10.19) becomes $\delta U = -\delta V$. Moreover, if the position of the system is defined by a single independent variable θ, we can write $\delta V = (dV/d\theta)\,\delta\theta$. Since $\delta\theta$ must be different from zero, the condition $\delta U = 0$ for the equilibrium of the system becomes

$$\frac{dV}{d\theta} = 0 \qquad (10.21)$$

In terms of potential energy, therefore, the principle of virtual work states that *if a system is in equilibrium, the derivative of its total potential energy is zero*. If the position of the system depends upon several independent variables (the system is then said to possess *several degrees of freedom*), the partial derivatives of V with respect to each of the independent variables must be zero.

Consider, for example, a structure made of two members AC and CB and carrying a load W at C. The structure is supported by a pin at A and a roller at B, and a spring BD connects B to a fixed point D (Fig. 10.13*a*). The constant of the spring is k, and it is assumed that the natural length of the spring is equal to AD and thus that the spring is undeformed when B coincides with A. Neglecting the friction forces and the weight of the members, we find that the only forces which do work during a displacement of the structure are the weight $\mathbf{W}$ and the force $\mathbf{F}$ exerted by the spring at point B (Fig. 10.13*b*). The total potential energy of the system will thus be obtained by adding the potential energy V_g corresponding to the gravity force $\mathbf{W}$ and the potential energy V_e corresponding to the elastic force $\mathbf{F}$.

Choosing a coordinate system with origin at A and noting that the deflection of the spring, measured from its undeformed position, is $AB = x_B$, we write

$$V_e = \tfrac{1}{2}kx_B^2 \qquad V_g = Wy_C$$

Expressing the coordinates x_B and y_C in terms of the angle θ, we have

$$x_B = 2l \sin\theta \qquad y_C = l\cos\theta$$
$$V_e = \tfrac{1}{2}k(2l\sin\theta)^2 \qquad V_g = W(l\cos\theta)$$
$$V = V_e + V_g = 2kl^2\sin^2\theta + Wl\cos\theta \qquad (10.22)$$

The positions of equilibrium of the system are obtained by equating to zero the derivative of the potential energy V. We write

$$\frac{dV}{d\theta} = 4kl^2 \sin\theta\cos\theta - Wl\sin\theta = 0$$

or, factoring $l\sin\theta$,

$$\frac{dV}{d\theta} = l\sin\theta(4kl\cos\theta - W) = 0$$

There are therefore two positions of equilibrium, corresponding to the values $\theta = 0$ and $\theta = \cos^{-1}(W/4kl)$, respectively.†

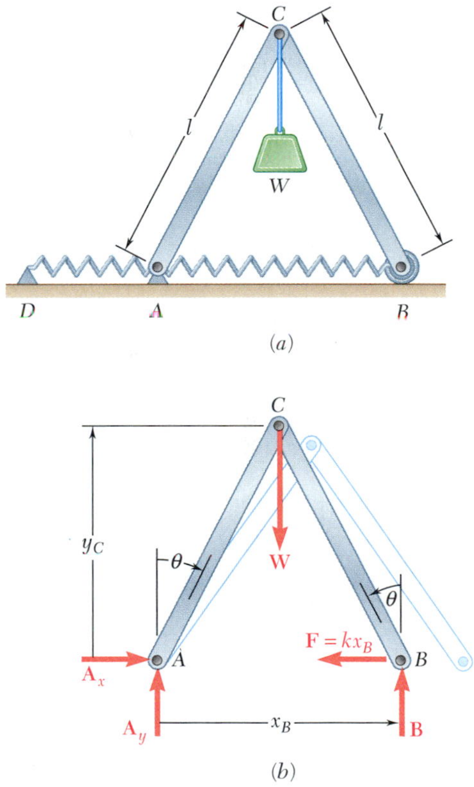

Fig. 10.13

†The second position does not exist if $W > 4kl$.

*10.9. STABILITY OF EQUILIBRIUM

Consider the three uniform rods of length $2a$ and weight $\mathbf{W}$ shown in Fig. 10.14. While each rod is in equilibrium, there is an important difference between the three cases considered. Suppose that each rod is slightly disturbed from its position of equilibrium and then released: rod a will move back toward its original position, rod b will keep moving away from its original position, and rod c will remain in its new position. In case a, the equilibrium of the rod is said to be *stable;* in case b, it is said to be *unstable;* and, in case c, it is said to be *neutral*.

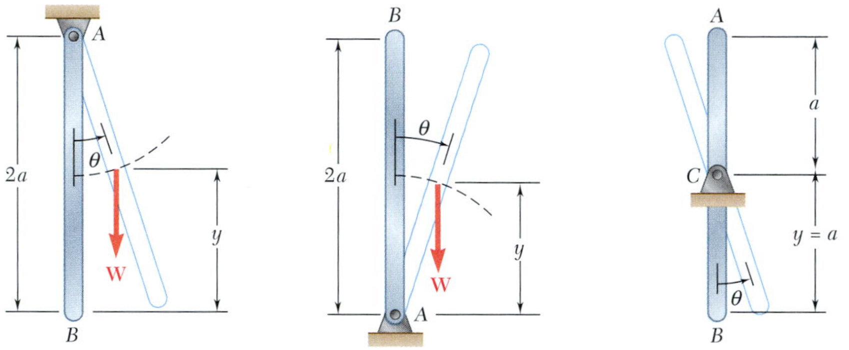

(*a*) Stable equilibrium (*b*) Unstable equilibrium (*c*) Neutral equilibrium

Fig. 10.14

Recalling from Sec. 10.7 that the potential energy V_g with respect to gravity is equal to Wy, where y is the elevation of the point of application of $\mathbf{W}$ measured from an arbitrary level, we observe that the potential energy of rod a is minimum in the position of equilibrium considered, that the potential energy of rod b is maximum, and that the potential energy of rod c is constant. Equilibrium is thus *stable, unstable,* or *neutral* according to whether the potential energy is *minimum, maximum,* or *constant* (Fig. 10.15).

That the result obtained is quite general can be seen as follows: We first observe that a force always tends to do positive work and thus to decrease the potential energy of the system on which it is applied. Therefore, when a system is disturbed from its position of equilibrium, the forces acting on the system will tend to bring it back to its original position if V is minimum (Fig. 10.15a) and to move it farther away if V is maximum (Fig. 10.15b). If V is constant (Fig. 10.15c), the forces will not tend to move the system either way.

Recalling from calculus that a function is minimum or maximum according to whether its second derivative is positive or negative, we can summarize the conditions for the equilibrium of a system with one degree of freedom (that is, a system the position of which is defined by a single independent variable θ) as follows:

$$\begin{aligned} \frac{dV}{d\theta} &= 0 \qquad \frac{d^2V}{d\theta^2} > 0\text{: stable equilibrium} \\ \frac{dV}{d\theta} &= 0 \qquad \frac{d^2V}{d\theta^2} < 0\text{: unstable equilibrium} \end{aligned} \tag{10.23}$$

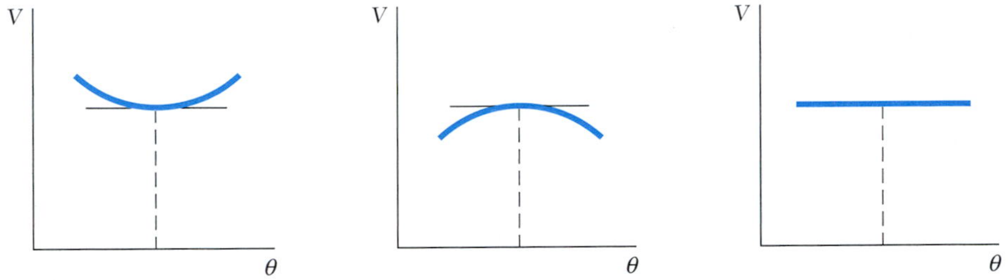

(a) Stable equilibrium (b) Unstable equilibrium (c) Neutral equilibrium

Fig. 10.15

If both the first and the second derivatives of V are zero, it is necessary to examine derivatives of a higher order to determine whether the equilibrium is stable, unstable, or neutral. The equilibrium will be neutral if all derivatives are zero, since the potential energy V is then a constant. The equilibrium will be stable if the first derivative found to be different from zero is of even order and positive. In all other cases the equilibrium will be unstable.

If the system considered possesses *several degrees of freedom,* the potential energy V depends upon several variables, and it is thus necessary to apply the theory of functions of several variables to determine whether V is minimum. It can be verified that a system with 2 degrees of freedom will be stable, and the corresponding potential energy $V(\theta_1, \theta_2)$ will be minimum, if the following relations are satisfied simultaneously:

$$\begin{gathered} \frac{\partial V}{\partial \theta_1} = \frac{\partial V}{\partial \theta_2} = 0 \\ \left(\frac{\partial^2 V}{\partial \theta_1\, \partial \theta_2}\right)^2 - \frac{\partial^2 V}{\partial \theta_1^2}\frac{\partial^2 V}{\partial \theta_2^2} < 0 \\ \frac{\partial^2 V}{\partial \theta_1^2} > 0 \qquad \text{or} \qquad \frac{\partial^2 V}{\partial \theta_2^2} > 0 \end{gathered} \tag{10.24}$$

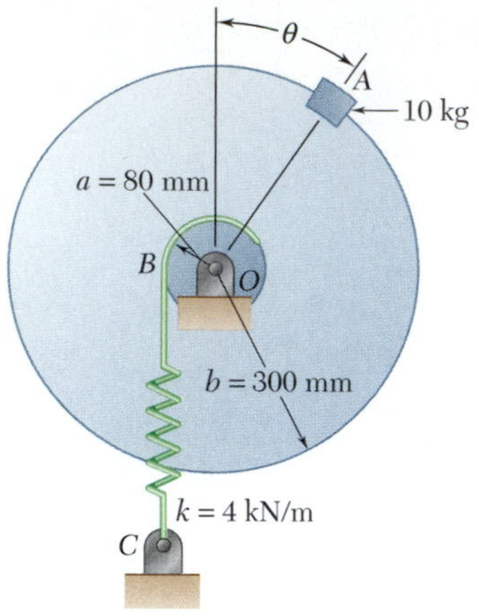

SAMPLE PROBLEM 10.4

A 10-kg block is attached to the rim of a 300-mm-radius disk as shown. Knowing that spring BC is unstretched when $\theta = 0$, determine the position or positions of equilibrium, and state in each case whether the equilibrium is stable, unstable, or neutral.

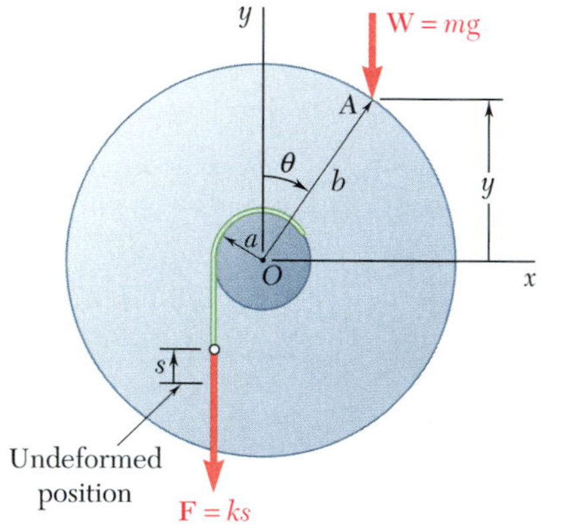

SOLUTION

Potential Energy. Denoting by s the deflection of the spring from its undeformed position and placing the origin of coordinates at O, we obtain

$$V_e = \tfrac{1}{2}ks^2 \qquad V_g = Wy = mgy$$

Measuring θ in radians, we have

$$s = a\theta \qquad y = b\cos\theta$$

Substituting for s and y in the expressions for V_e and V_g, we write

$$V_e = \tfrac{1}{2}ka^2\theta^2 \qquad V_g = mgb\cos\theta$$
$$V = V_e + V_g = \tfrac{1}{2}ka^2\theta^2 + mgb\cos\theta$$

Positions of Equilibrium. Setting $dV/d\theta = 0$, we write

$$\frac{dV}{d\theta} = ka^2\theta - mgb\sin\theta = 0$$
$$\sin\theta = \frac{ka^2}{mgb}\theta$$

Substituting $a = 0.08$ m, $b = 0.3$ m, $k = 4$ kN/m, and $m = 10$ kg, we obtain

$$\sin\theta = \frac{(4\text{ kN/m})(0.08\text{ m})^2}{(10\text{ kg})(9.81\text{ m/s}^2)(0.3\text{ m})}\theta,$$
$$\sin\theta = 0.8699\,\theta$$

where θ is expressed in radians. Solving numerically for θ, we find

$$\theta = 0 \quad\text{and}\quad \theta = 0.902\text{ rad}$$
$$\theta = 0 \quad\text{and}\quad \theta = 51.7° \quad ◀$$

Stability of Equilibrium. The second derivative of the potential energy V with respect to θ is

$$\frac{d^2V}{d\theta^2} = ka^2 - mgb\cos\theta$$
$$= (4\text{ kN/m})(0.08\text{ m})^2 - (10\text{ kg})(9.81\text{ m/s}^2)(0.3\text{ m})\cos\theta$$
$$= 25.6 - 29.43\cos\theta$$

For $\theta = 0$: $$\frac{d^2V}{d\theta^2} = 25.6 - 29.43\cos 0 = -3.83 < 0$$

The equilibrium is unstable for $\theta = 0$ ◀

For $\theta = 51.7°$: $$\frac{d^2V}{d\theta^2} = 25.6 - 29.43\cos 51.7° = +7.36 > 0$$

The equilibrium is stable for $\theta = 51.7°$ ◀

SOLVING PROBLEMS ON YOUR OWN

In this lesson we defined the *work of a force during a finite displacement* and the *potential energy* of a rigid body or a system of rigid bodies. You learned to use the concept of potential energy to determine the *equilibrium position* of a rigid body or a system of rigid bodies.

1. *The potential energy V of a system* is the sum of the potential energies associated with the various forces acting on the system that *do work* as the system moves. In the problems of this lesson you will determine the following:

a. Potential energy of a weight. This is the potential energy due to *gravity*, $V_g = Wy$, where y is the elevation of the weight W measured from some arbitrary reference level. Note that the potential energy V_g can be used with any vertical force **P** of constant magnitude directed downward; we write $V_g = Py$.

b. Potential energy of a spring. The potential energy, $V_e = \frac{1}{2}kx^2$, is due to the *elastic* force exerted by a spring, where k is the constant of the spring and x is the deformation of the spring *measured from its unstretched position*.

Reactions at fixed supports, internal forces at connections, forces exerted by inextensible cords and cables, and other forces which do no work do not contribute to the potential energy of the system.

2. *Express all distances and angles in terms of a single variable,* such as an angle θ, when computing the potential energy V of a system. This is necessary, since the determination of the equilibrium position of the system requires the computation of the derivative $dV/d\theta$.

3. *When a system is in equilibrium, the first derivative of its potential energy is zero.* Therefore:

a. To determine a position of equilibrium of a system, once its potential energy V has been expressed in terms of the single variable θ, compute its derivative and solve the equation $dV/d\theta = 0$ for θ.

b. To determine the force or couple required to maintain a system in a given position of equilibrium, substitute the known value of θ in the equation $dV/d\theta = 0$ and solve this equation for the desired force or couple.

4. *Stability of equilibrium.* The following rules generally apply:

a. Stable equilibrium occurs when the potential energy of the system is *minimum*, that is, when $dV/d\theta = 0$ and $d^2V/d\theta^2 > 0$ (Figs. 10.14*a* and 10.15*a*).

b. Unstable equilibrium occurs when the potential energy of the system is *maximum*, that is, when $dV/d\theta = 0$ and $d^2V/d\theta^2 < 0$ (Figs. 10.14*b* and 10.15*b*).

c. Neutral equilibrium occurs when the potential energy of the system is *constant;* $dV/d\theta$, $dV^2/d\theta^2$, and all the successive derivatives of V are then equal to zero (Figs. 10.14*c* and 10.15*c*).

See page 583 for a discussion of the case when $dV/d\theta$, $dV^2/d\theta^2$ but *not all* of the successive derivatives of V are equal to zero.

Problems

10.59 Using the method of Sec. 10.8, solve Prob. 10.29.

10.60 Using the method of Sec. 10.8, solve Prob. 10.30.

10.61 Using the method of Sec. 10.8, solve Prob. 10.31.

10.62 Using the method of Sec. 10.8, solve Prob. 10.33.

10.63 Using the method of Sec. 10.8, solve Prob. 10.34.

10.64 Using the method of Sec. 10.8, solve Prob. 10.35.

10.65 Using the method of Sec. 10.8, solve Prob. 10.36.

10.66 Using the method of Sec. 10.8, solve Prob. 10.38.

10.67 Show that the equilibrium is neutral in Prob. 10.1.

10.68 Show that the equilibrium is neutral in Prob. 10.2.

10.69 Two identical uniform rods, each of weight W and length L, are attached to pulleys that are connected by a belt as shown. Assuming that no slipping occurs between the belt and the pulleys, determine the positions of equilibrium of the system and state in each case whether the equilibrium is stable, unstable, or neutral.

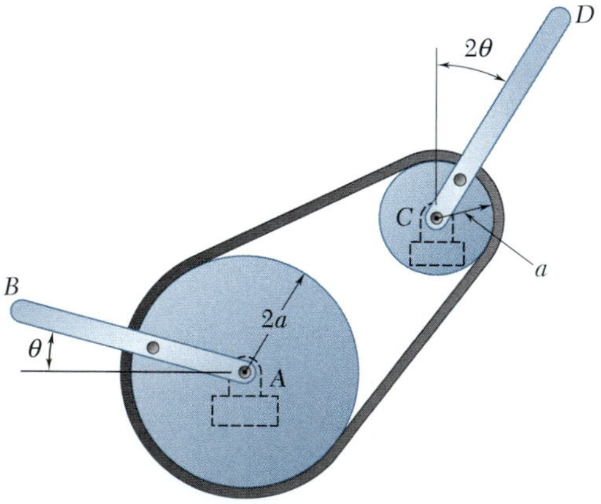

Fig. P10.69

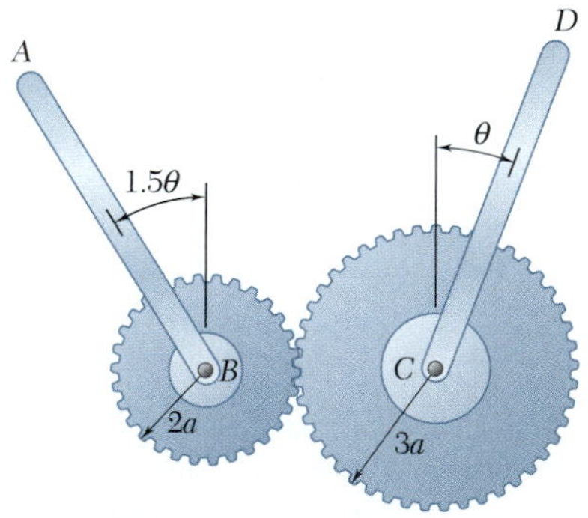

Fig. P10.70

10.70 Two uniform rods, each of mass m and length l, are attached to gears as shown. For the range $0 \leq \theta \leq 180°$, determine the positions of equilibrium of the system and state in each case whether the equilibrium is stable, unstable, or neutral.

10.71 Two uniform rods, each of mass m, are attached to gears of equal radii as shown. Determine the positions of equilibrium of the system and state in each case whether the equilibrium is stable, unstable, or neutral.

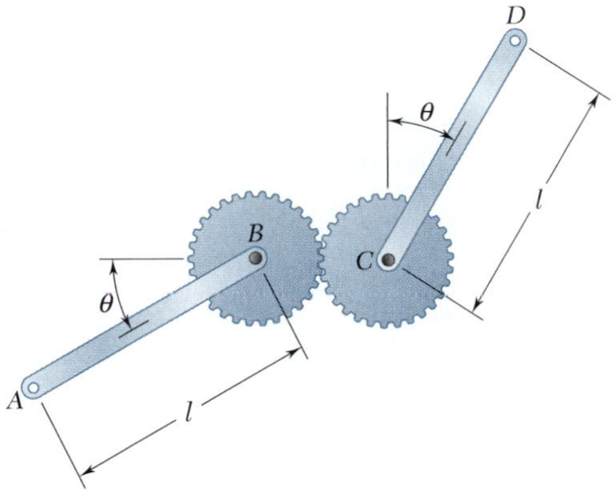

Fig. *P10.71* and *P10.72*

10.72 Two uniform rods, AB and CD, are attached to gears of equal radii as shown. Knowing that $m_{AB} = 3.5$ kg and $m_{CD} = 1.75$ kg, determine the positions of equilibrium of the system and state in each case whether the equilibrium is stable, unstable, or neutral.

10.73 Using the method of Sec. 10.8, solve Prob. 10.39. Determine whether the equilibrium is stable, unstable, or neutral. (*Hint:* The potential energy corresponding to the couple exerted by a torsional spring is $\frac{1}{2}K\theta^2$, where K is the torsional spring constant and θ is the angle of twist.)

10.74 In Prob. 10.40, determine whether each of the positions of equilibrium is stable, unstable, or neutral. (See the hint for Prob. 10.73.)

10.75 Angle θ is equal to 45° after a block of mass m is hung from member AB as shown. Neglecting the mass of AB and knowing that the spring is unstretched when $\theta = 20°$, determine the value of m and state whether the equilibrium is stable, unstable, or neutral.

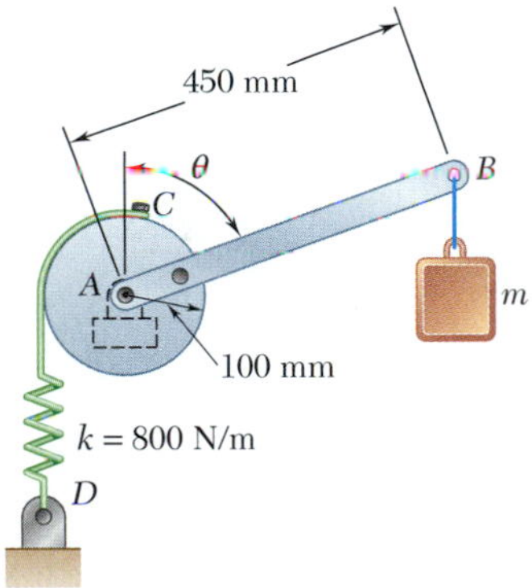

Fig. P10.75 and P10.76

10.76 A block of mass m is hung from member AB as shown. Neglecting the mass of AB and knowing that the spring is unstretched when $\theta = 20°$, determine the value of θ corresponding to equilibrium when $m = 3$ kg. State whether the equilibrium is stable, unstable, or neutral.

10.77 A slender rod AB, of mass m, is attached to two blocks A and B which can move freely in the guides shown. Knowing that the spring is unstretched when $y = 0$, determine the value of y corresponding to equilibrium when $m = 12$ kg, $l = 750$ mm, and $k = 900$ N/m.

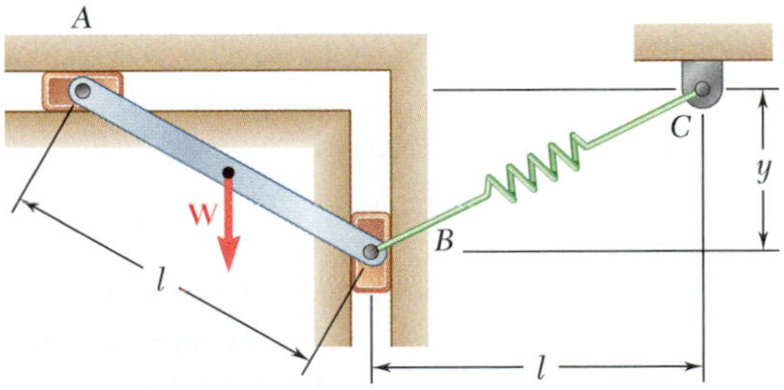

Fig. *P10.77*

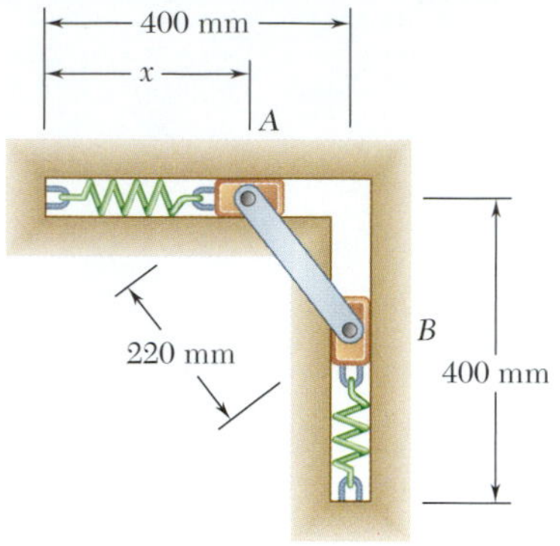

Fig. P10.78

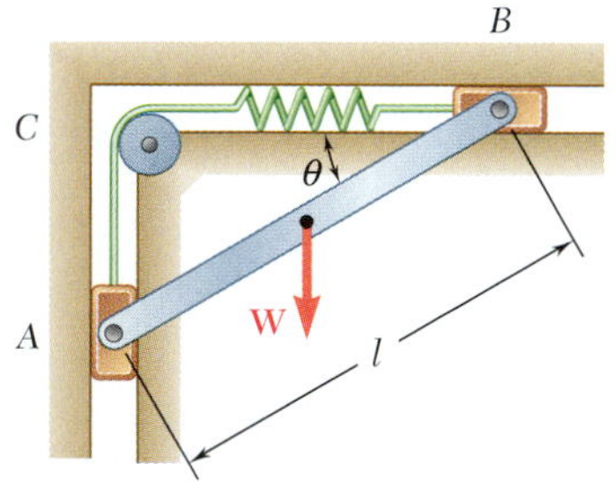

Fig. *P10.79* and P10.80

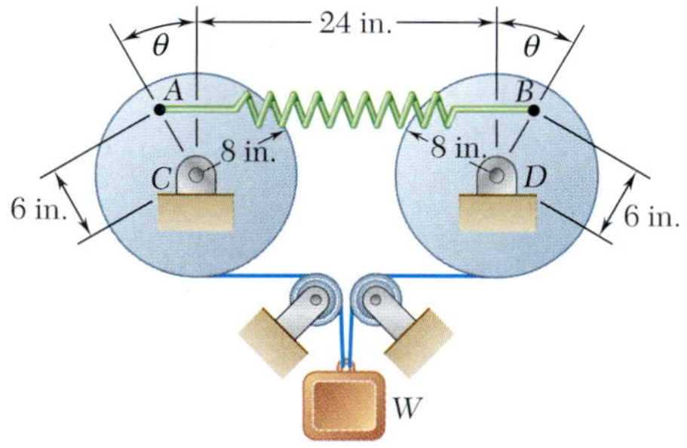

Fig. P10.81 and P10.82

10.78 The slender rod AB of negligible mass is attached to two 4-kg blocks A and B that can move freely in the guides shown. Knowing that the constant of the springs is 160 N/m and that the unstretched length of each spring is 150 mm, determine the value of x corresponding to equilibrium.

10.79 A slender rod AB, of mass m, is attached to two blocks A and B that can move freely in the guides shown. The constant of the spring is k, and the spring is unstretched when AB is horizontal. Neglecting the weight of the blocks, derive an equation in θ, m, l, and k that must be satisfied when the rod is in equilibrium.

10.80 A slender rod AB, of mass m, is attached to two blocks A and B that can move freely in the guides shown. Knowing that the spring is unstretched when AB is horizontal, determine three values of θ corresponding to equilibrium when $m = 125$ kg, $l = 320$ mm, and $k = 15$ kN/mm. State in each case whether the equilibrium is stable, unstable, or neutral.

10.81 Spring AB of constant 10 lb/in. is attached to two identical drums as shown. Knowing that the spring is unstretched when $\theta = 0$, determine (*a*) the range of values of the weight W of the block for which a position of equilibrium exists, (*b*) the range of values of θ for which the equilibrium is stable.

10.82 Spring AB of constant 10 lb/in. is attached to two identical drums as shown. Knowing that the spring is unstretched when $\theta = 0$ and that $W =$ 40 lb, determine the values of θ less than 180° corresponding to equilibrium. State in each case whether the equilibrium is stable, unstable, or neutral.

10.83 A slender rod AB of negligible weight is attached to two collars A and B that can move freely along the guide rods shown. Knowing that $\beta = 30°$ and $P = Q = 100$ lb, determine the value of the angle θ corresponding to equilibrium.

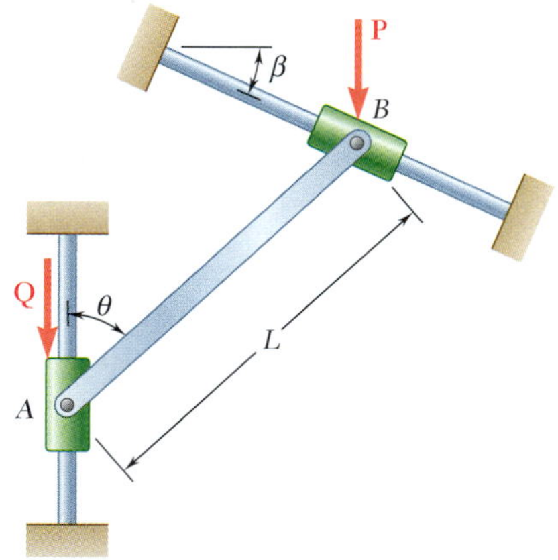

Fig. P10.83 and P10.84

10.84 A slender rod AB of negligible weight is attached to two collars A and B that can move freely along the guide rods shown. Knowing that $\beta = 30°$, $P = 40$ lb, and $Q = 10$ lb, determine the value of the angle θ corresponding to equilibrium.

10.85 and *10.86* Collar A can slide freely on the semicircular rod shown. Knowing that the constant of the spring is k and that the unstretched length of the spring is equal to the radius r, determine the value of θ corresponding to equilibrium when $m = 20$ kg, $r = 180$ mm, and $k = 3$ kN/m.

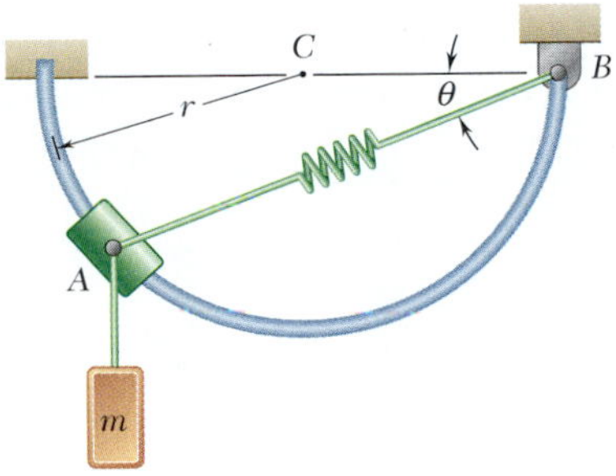

Fig. P10.85

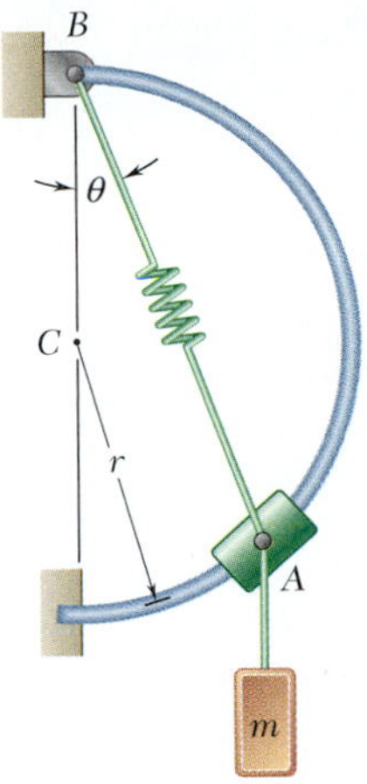

Fig. *P10.86*

10.87 The 12-kg block D can slide freely on the inclined surface. Knowing that the constant of the spring is 480 N/m and that the spring is unstretched when $\theta = 0$, determine the value of θ corresponding to equilibrium.

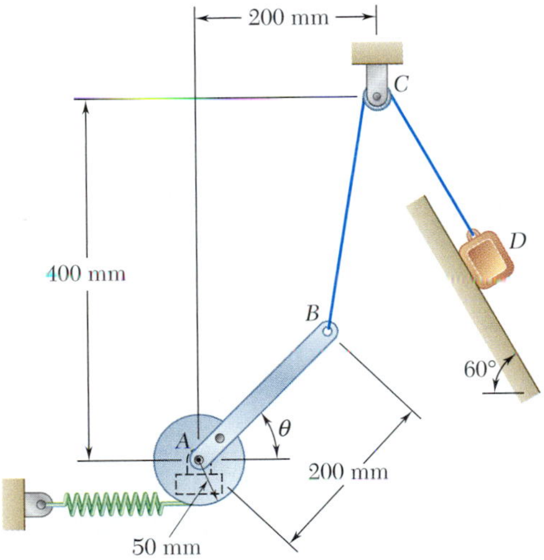

Fig. P10.87

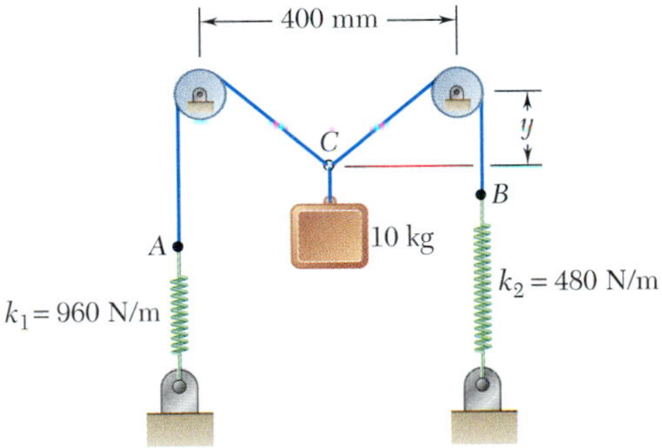

Fig. *P10.88*

10.88 Cable AB is attached to two springs and passes through a ring at C. Knowing that the springs are unstretched when $y = 0$, determine the distance y corresponding to equilibrium.

10.89 Rod AB is attached to a hinge at A and to two springs, each of constant k. If $h = 50$ in., $d = 24$ in., and $W = 160$ lb, determine the range of values of k for which the equilibrium of the rod is stable in the position shown. Each spring can act in either tension or compression.

10.90 Rod AB is attached to a hinge at A and to two springs, each of constant k. If $h = 30$ in., $k = 4$ lb/in., and $W = 40$ lb, determine the smallest distance d for which the equilibrium of the rod is stable in the position shown. Each spring can act in either tension or compression.

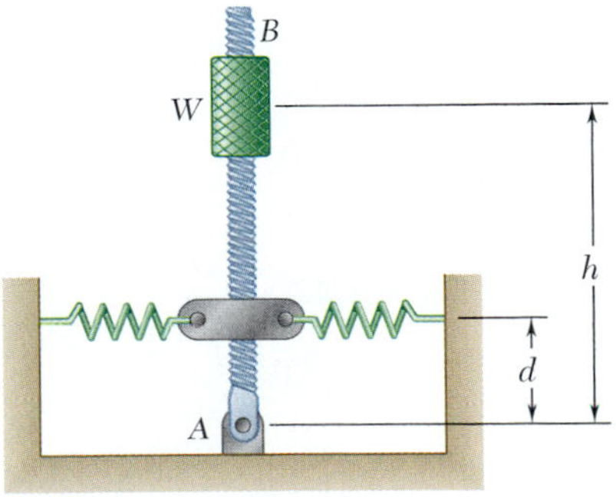

Fig. P10.89 and P10.90

10.91 The uniform plate $ABCD$ of negligible mass is attached to four springs of constant k and is in equilibrium in the position shown. Knowing that the springs can act in either tension or compression and are undeformed in the given position, determine the range of values of the magnitude P of two equal and opposite horizontal forces **P** and −**P** for which the equilibrium position is stable.

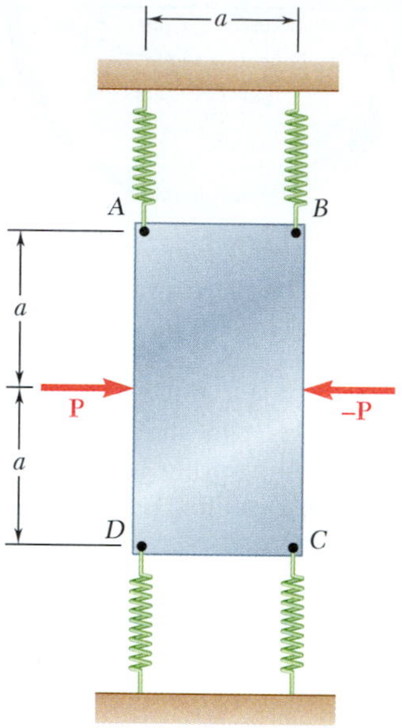

Fig. P10.91

10.92 and 10.93 Two bars are attached to a single spring of constant k that is unstretched when the bars are vertical. Determine the range of values of P for which the equilibrium of the system is stable in the position shown.

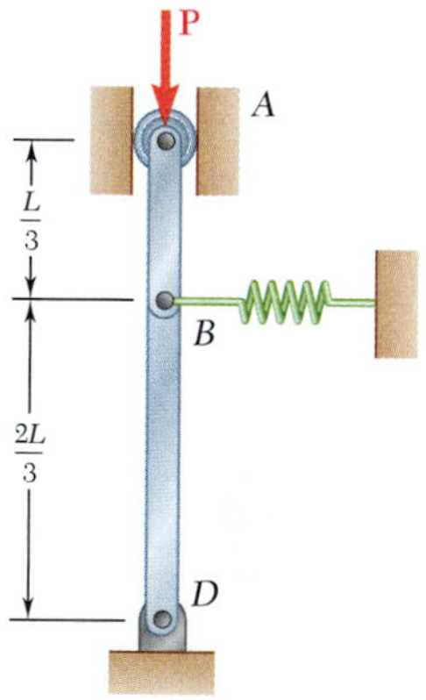

Fig. *P10.92*

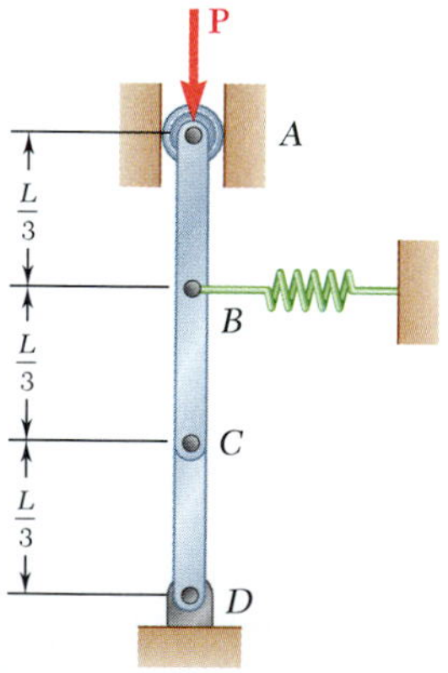

Fig. *P10.93*

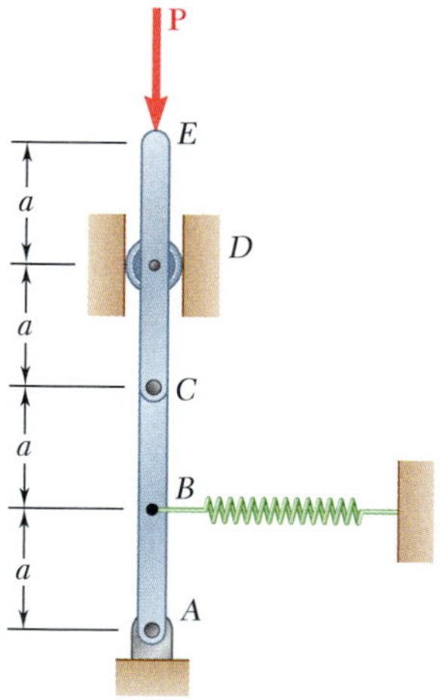

Fig. P10.94

10.94 Bar AC is attached to a hinge at A and to a spring of constant k that is undeformed when the bar is vertical. Knowing that the spring can act in either tension or compression, determine the range of values of P for which the equilibrium of the system is stable in the position shown.

10.95 The horizontal bar BEH is connected to three vertical bars. The collar at E can slide freely on bar DF. Determine the range of values of P for which the equilibrium of the system is stable in the position shown when $a = 300$ mm, $b = 400$ mm, and $Q = 90$ N.

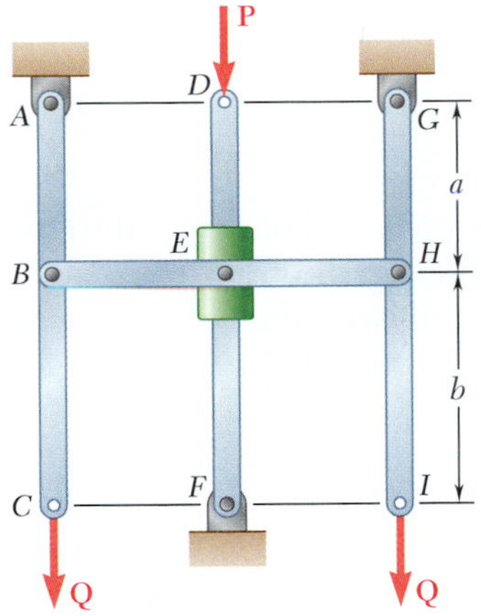

Fig. P10.95 and P10.96

10.96 The horizontal bar BEH is connected to three vertical bars. The collar at E can slide freely on bar DF. Determine the range of values of Q for which the equilibrium of the system is stable in the position shown when $a = 480$ mm, $b = 400$ mm, and $P = 600$ N.

10.97 Bars AB and BC, each of length l and of negligible weight, are attached to two springs, each of constant k. The springs are undeformed and the system is in equilibrium when $\theta_1 = \theta_2 = 0$. Determine the range of values of P for which the equilibrium position is stable.

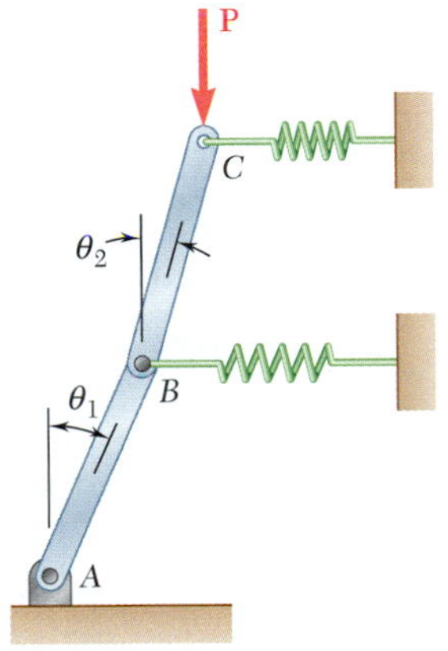

Fig. *P10.97*

10.98 Solve Prob. 10.97 knowing that $l = 400$ mm and $k = 1.25$ kN/m.

***10.99** Bar ABC of length $2a$ and negligible weight is hinged at C to a drum of radius a as shown. Knowing that the constant of each spring is k and that the springs are undeformed when $\theta_1 = \theta_2 = 0$, determine the range of values of P for which the equilibrium position $\theta_1 = \theta_2 = 0$ is stable.

***10.100** Solve Prob. 10.99 knowing that $k = 10$ lb/in. and $a = 14$ in.

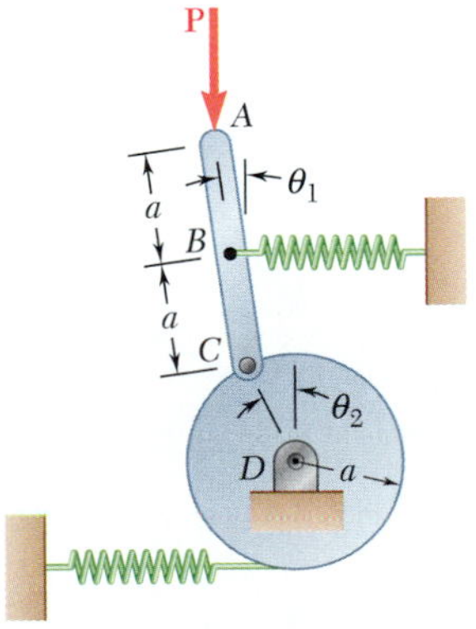

Fig. P10.99

REVIEW AND SUMMARY FOR CHAPTER 10

Work of a force

The first part of this chapter was devoted to the *principle of virtual work* and to its direct application to the solution of equilibrium problems. We first defined the *work of a force* **F** *corresponding to the small displacement* $d\mathbf{r}$ [Sec. 10.2] as the quantity

$$dU = \mathbf{F} \cdot d\mathbf{r} \tag{10.1}$$

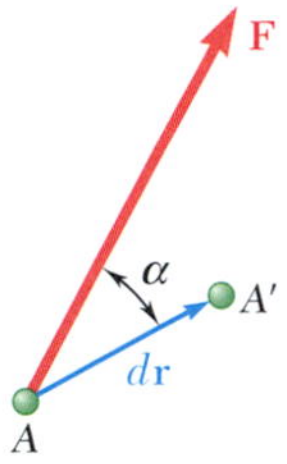

Fig. 10.16

obtained by forming the scalar product of the force **F** and the displacement $d\mathbf{r}$ (Fig. 10.16). Denoting respectively by F and ds the magnitudes of the force and of the displacement, and by α the angle formed by **F** and $d\mathbf{r}$, we wrote

$$dU = F\, ds \cos \alpha \tag{10.1$'$}$$

The work dU is positive if $\alpha < 90°$, zero if $\alpha = 90°$, and negative if $\alpha > 90°$. We also found that the *work of a couple of moment* **M** acting on a rigid body is

$$dU = M\, d\theta \tag{10.2}$$

where $d\theta$ is the small angle expressed in radians through which the body rotates.

Virtual displacement

Considering a particle located at A and acted upon by several forces $\mathbf{F}_1, \mathbf{F}_2, \ldots, \mathbf{F}_n$ [Sec. 10.3], we imagined that the particle moved to a new position A' (Fig. 10.17). Since this displacement did not actually take place, it was referred to as a *virtual displacement* and denoted by $\delta\mathbf{r}$, while the corresponding work of the forces was called *virtual work* and denoted by δU. We had

$$\delta U = \mathbf{F}_1 \cdot \delta\mathbf{r} + \mathbf{F}_2 \cdot \delta\mathbf{r} + \cdots + \mathbf{F}_n \cdot \delta\mathbf{r}$$

Principle of virtual work

The *principle of virtual work* states that *if a particle is in equilibrium, the total virtual work* δU *of the forces acting on the particle is zero for any virtual displacement of the particle.*

The principle of virtual work can be extended to the case of rigid bodies and systems of rigid bodies. Since it involves *only forces which do work,* its application provides a useful alternative to the use of the equilibrium equations in the solution of many engineering problems. It is particularly effective in the case of machines and mechanisms consisting of connected rigid bodies, since the work of the reactions at the supports is zero and the work of the internal forces at the pin connections cancels out [Sec. 10.4; Sample Probs. 10.1, 10.2, and 10.3].

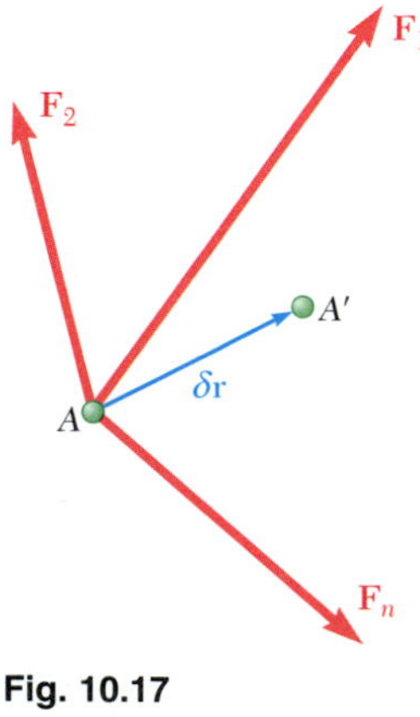

Fig. 10.17

In the case of *real machines* [Sec. 10.5], however, the work of the friction forces should be taken into account, with the result that the *output work will be less than the input work*. Defining the *mechanical efficiency* of a machine as the ratio

Mechanical efficiency

$$\eta = \frac{\text{output work}}{\text{input work}} \tag{10.9}$$

we also noted that for an ideal machine (no friction) $\eta = 1$, while for a real machine $\eta < 1$.

Work of a force over a finite displacement

In the second part of the chapter we considered the *work of forces corresponding to finite displacements* of their points of application. The work $U_{1\to2}$ of the force **F** corresponding to a displacement of the particle A from A_1 to A_2 (Fig. 10.18) was obtained by integrating the right-hand member of Eq. (10.1) or (10.1′) along the curve described by the particle [Sec. 10.6]:

$$U_{1\to2} = \int_{A_1}^{A_2} \mathbf{F} \cdot d\mathbf{r} \tag{10.11}$$

or

$$U_{1\to2} = \int_{s_1}^{s_2} (F \cos \alpha)\, ds \tag{10.11'}$$

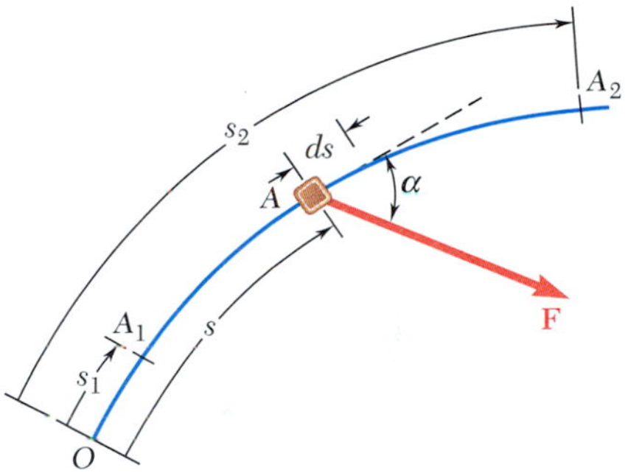

Fig. 10.18

Similarly, the work of a couple of moment **M** corresponding to a finite rotation from θ_1 to θ_2 of a rigid body was expressed as

$$U_{1\to2} = \int_{\theta_1}^{\theta_2} M\, d\theta \tag{10.12}$$

Work of a weight

The *work of the weight* **W** *of a body* as its center of gravity moves from the elevation y_1 to y_2 (Fig. 10.19) can be obtained by setting $F = W$ and $\alpha = 180°$ in Eq. (10.11′):

$$U_{1\to2} = -\int_{y_1}^{y_2} W\, dy = Wy_1 - Wy_2 \tag{10.13}$$

The work of **W** is therefore positive *when the elevation y decreases.*

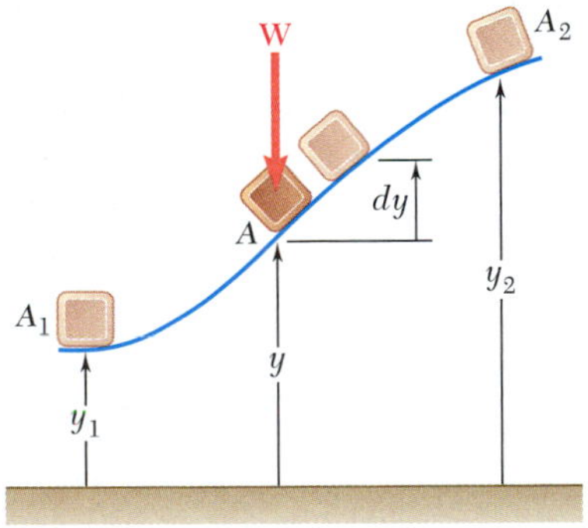

Fig. 10.19

Work of the force exerted by a spring

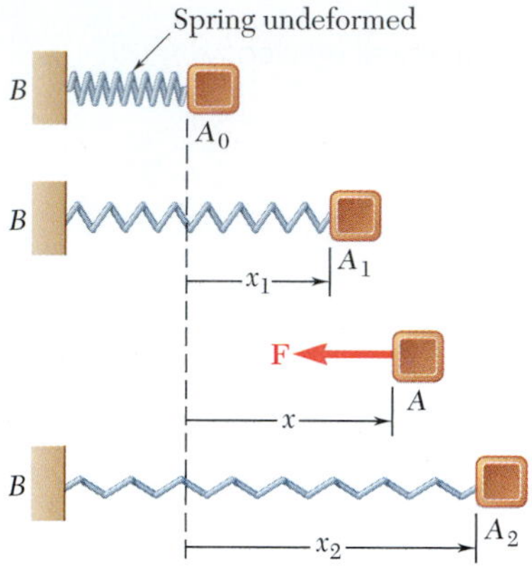

Fig. 10.20

The *work of the force* **F** *exerted by a spring* on a body *A* as the spring is stretched from x_1 to x_2 (Fig. 10.20) can be obtained by setting $F = kx$, where *k* is the constant of the spring, and $\alpha = 180°$ in Eq. (10.11′):

$$U_{1\rightarrow 2} = -\int_{x_1}^{x_2} kx\,dx = \tfrac{1}{2}kx_1^2 - \tfrac{1}{2}kx_2^2 \qquad (10.15)$$

The work of **F** is therefore positive *when the spring is returning to its undeformed position.*

Potential energy

When the work of a force **F** is independent of the path actually followed between A_1 and A_2, the force is said to be a *conservative force,* and its work can be expressed as

$$U_{1\rightarrow 2} = V_1 - V_2 \qquad (10.20)$$

where *V* is the *potential energy* associated with **F**, and V_1 and V_2 represent the values of *V* at A_1 and A_2, respectively [Sec. 10.7]. The potential energies associated, respectively, with the *force of gravity* **W** and the *elastic force* **F** exerted by a spring were found to be

$$V_g = Wy \qquad \text{and} \qquad V_e = \tfrac{1}{2}kx^2 \qquad (10.17, 10.18)$$

Alternative expression for the principle of virtual work

When the position of a mechanical system depends upon a single independent variable θ, the potential energy of the system is a function $V(\theta)$ of that variable, and it follows from Eq. (10.20) that $\delta U = -\delta V = -(dV/d\theta)\,\delta\theta$. The condition $\delta U = 0$ required by the principle of virtual work for the equilibrium of the system can thus be replaced by the condition

$$\frac{dV}{d\theta} = 0 \qquad (10.21)$$

When all the forces involved are conservative, it may be preferable to use Eq. (10.21) rather than to apply the principle of virtual work directly [Sec. 10.8; Sample Prob. 10.4].

Stability of equilibrium

This approach presents another advantage, since it is possible to determine from the sign of the second derivative of *V* whether the equilibrium of the system is *stable, unstable,* or *neutral* [Sec. 10.9]. If $d^2V/d\theta^2 > 0$, *V* is *minimum* and the equilibrium is *stable;* if $d^2V/d\theta^2 < 0$, *V* is *maximum* and the equilibrium is *unstable;* if $d^2V/d\theta^2 = 0$, it is necessary to examine derivatives of a higher order.

Review Problems

10.101 Derive an expression for the magnitude of the force **Q** required to maintain the equilibrium of the mechanism shown.

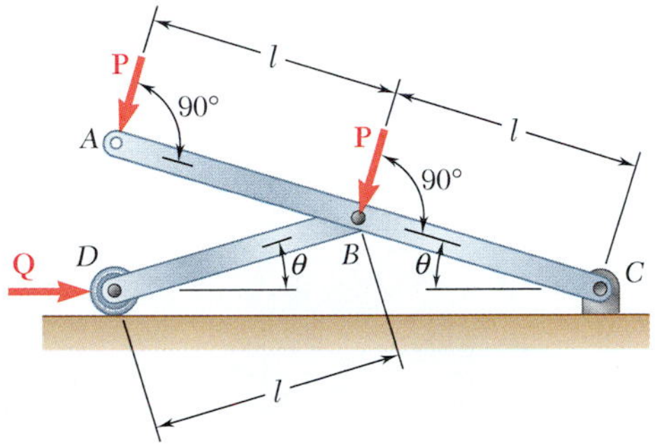

Fig. *P10.101*

10.102 The position of boom ABC is controlled by the hydraulic cylinder BD. For the loading shown, determine the force exerted by the hydraulic cylinder on pin B when $\theta = 70°$.

10.103 The position of boom ABC is controlled by the hydraulic cylinder BD. For the loading shown, determine the largest allowable value of the angle θ if the maximum force that the cylinder can exert on pin B is 25 kips.

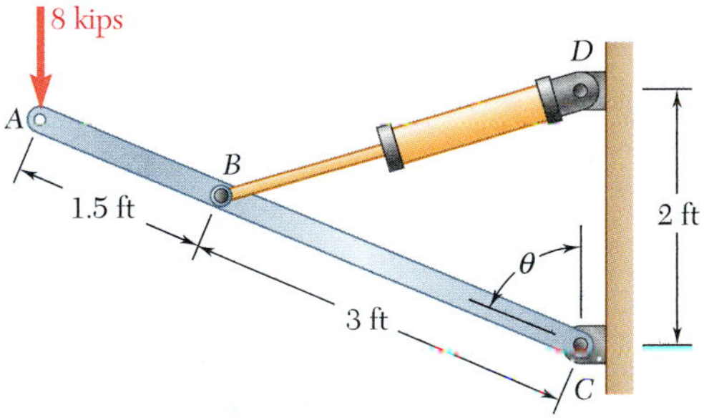

Fig. P10.102 and *P10.103*

10.104 A vertical bar AD is attached to two springs of constant k and is in equilibrium in the position shown. Determine the range of values of the magnitude P of two equal and opposite vertical forces **P** and $-$**P** for which the equilibrium position is stable if (a) $AB = CD$, (b) $AB = 2CD$.

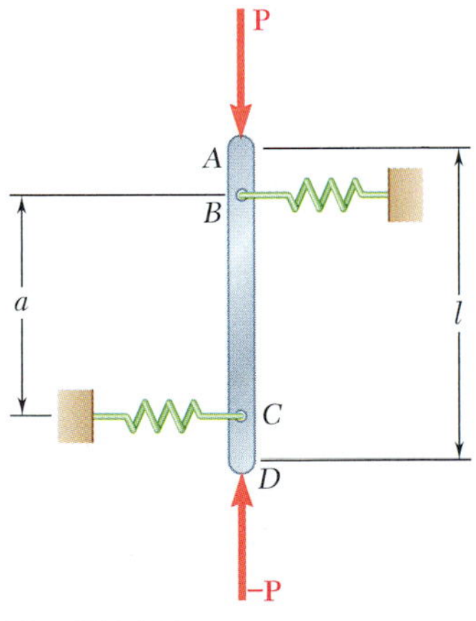

Fig. P10.104

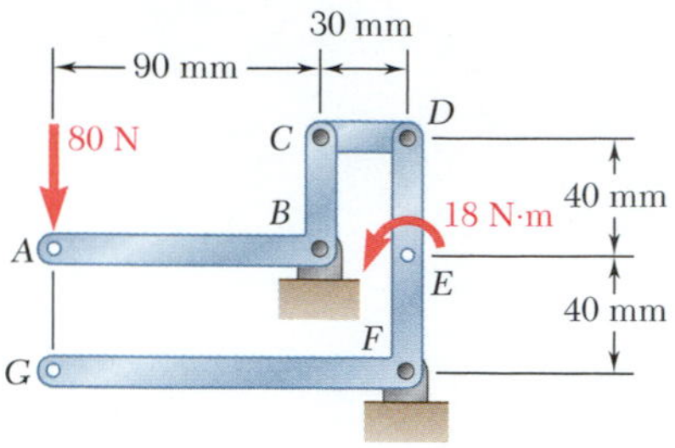

Fig. P10.105 and *P10.106*

10.105 Determine the vertical force **P** which must be applied at G to maintain the equilibrium of the linkage.

10.106 Determine the couple **M** which must be applied to member $DEFG$ to maintain the equilibrium of the linkage.

10.107 Two uniform rods, each of mass m and length l, are attached to drums that are connected by a belt as shown. Assuming that no slipping occurs between the belt and the drums, determine the positions of equilibrium of the system and state in each case whether the equilibrium is stable, unstable, or neutral.

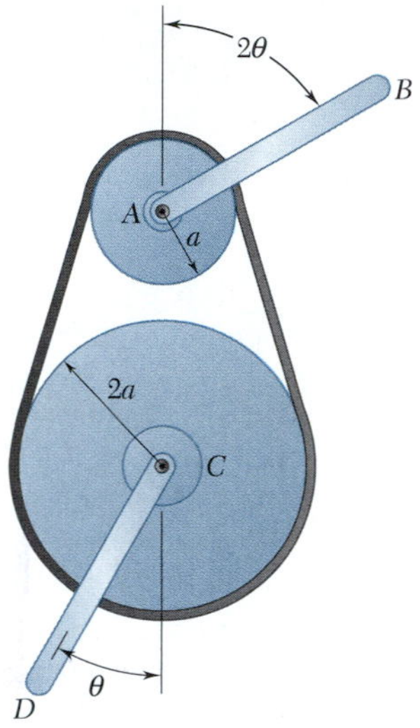

Fig. P10.107

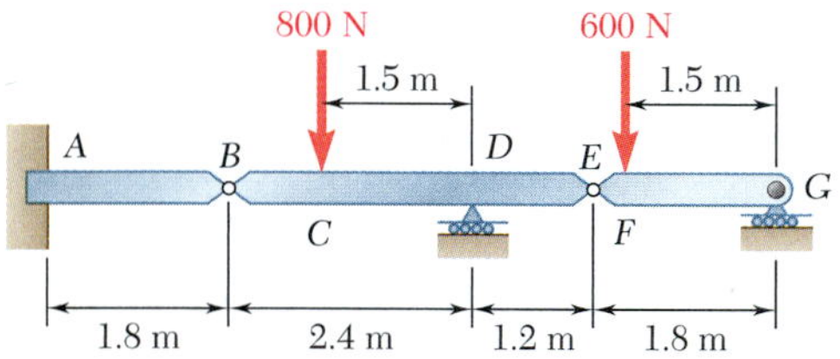

Fig. P10.108 and P10.109

10.108 Using the method of virtual work, determine separately the force and the couple representing the reaction at A.

10.109 Using the method of virtual work, determine the reaction at D.

10.110 The slender rod AB is attached to a collar A and rests on a small wheel at C. Neglecting the radius of the wheel and the effect of friction, derive an expression for the magnitude of the force **Q** required to maintain the equilibrium of the rod.

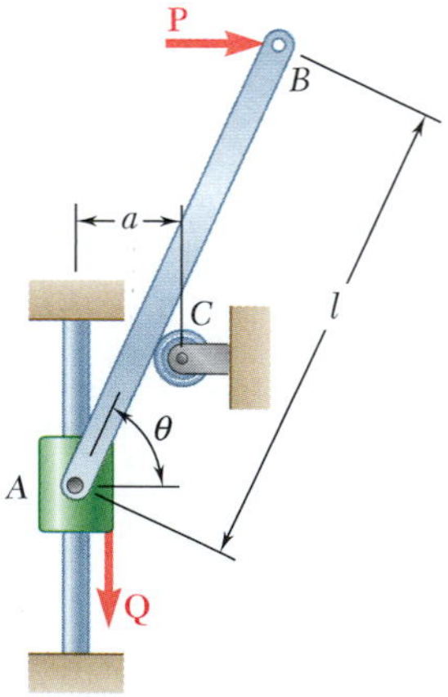

Fig. P10.110

10.111 A load **W** of magnitude 100 lb is applied to the mechanism at C. Knowing that the spring is unstretched when $\theta = 15°$, determine the value of θ corresponding to equilibrium and check that the equilibrium is stable.

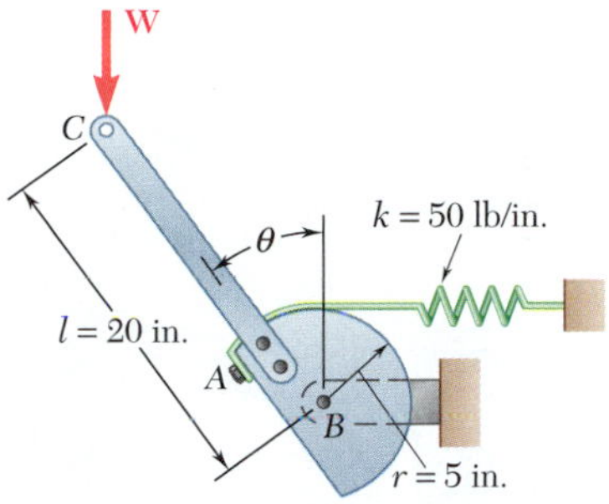

Fig. *P10.111* and P10.112

10.112 A load **W** of magnitude 100 lb is applied to the mechanism at C. Knowing that the spring is unstretched when $\theta = 30°$, determine the value of θ corresponding to equilibrium and check that the equilibrium is stable.

Computer Problems

10.C1 Knowing that $a = 20$ in., $b = 6$ in., $L = 20$ in., and $P = 25$ lb, use computational software to plot the force in member BD as a function of θ for values of θ from 30° to 150°. Determine the range of values of θ for which the absolute value of the force in member BD is less than 100 lb.

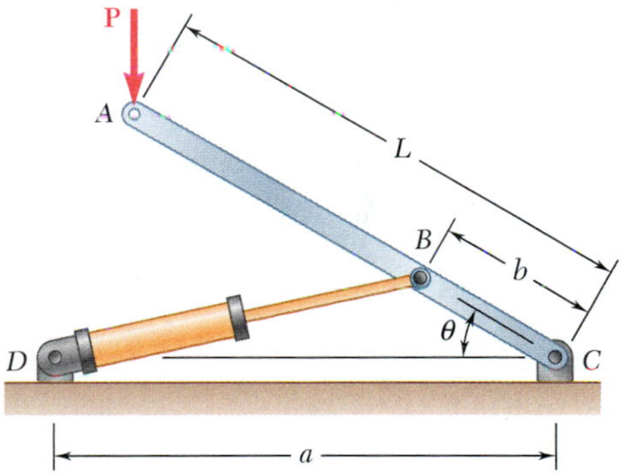

Fig. P10.C1

10.C2 Collars A and B are connected by the wire AB and can slide freely on the rods shown, where rod DE lies in the xy plane. Knowing that the length of the wire is 500 mm and that the mass of each collar is 2.5 kg, (a) use the principle of virtual work to express in terms of the distance x the magnitude of the force **P** required to maintain the system in equilibrium, (b) plot P as a function of x for 100 mm $\leq x \leq$ 400 mm.

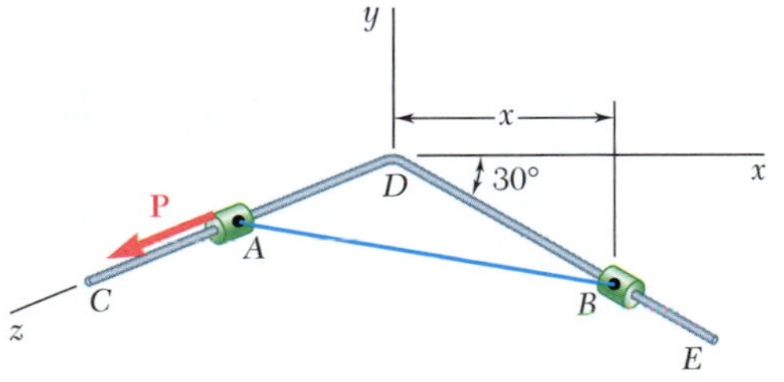

Fig. P10.C2

10.C3 A couple **M** is applied to crank AB to maintain the equilibrium of the engine system shown when a force **P** is applied to the piston. (*a*) Knowing that $b = 48$ mm and $l = 150$ mm, use computational software to plot the ratio M/P as a function of θ for values of θ from 0 to 180°. (*b*) Determine the value of θ for which the ratio M/P is maximum and the corresponding value of M/P.

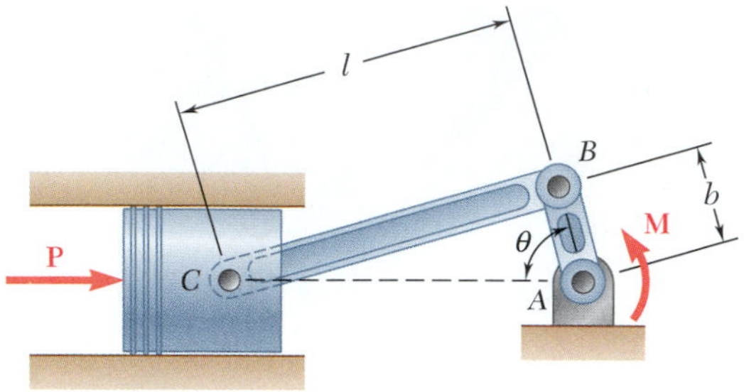

Fig. P10.C3

10.C4 The constant of spring AB is k, and the spring is unstretched when $\theta = 0$. (*a*) Neglecting the weight of member BCD, express the potential energy and its derivative $dV/d\theta$ in terms of a, k, W, and θ. (*b*) Using computational software and knowing that $W = 600$ N, $a = 200$ mm, and $k = 15$ kN/m, plot the potential energy and $dV/d\theta$ as functions of θ for values of θ from 0 to 165°. (*c*) Determine the values of θ for which the system is in equilibrium, and state in each case whether the equilibrium is stable, unstable, or neutral.

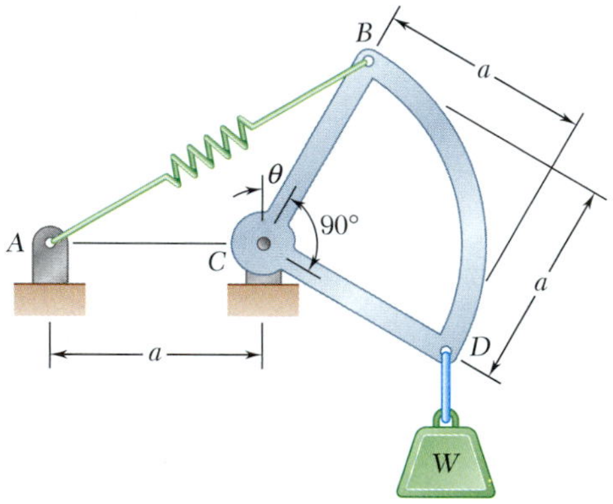

Fig. P10.C4

10.C5 A vertical load **W** is applied to the linkage at B. The constant of the spring is k, and the spring is unstretched when AB and BC are horizontal. For $l = 11$ in. and $k = 12.5$ lb/in., plot W as a function of θ for values of θ from 0 to 80°.

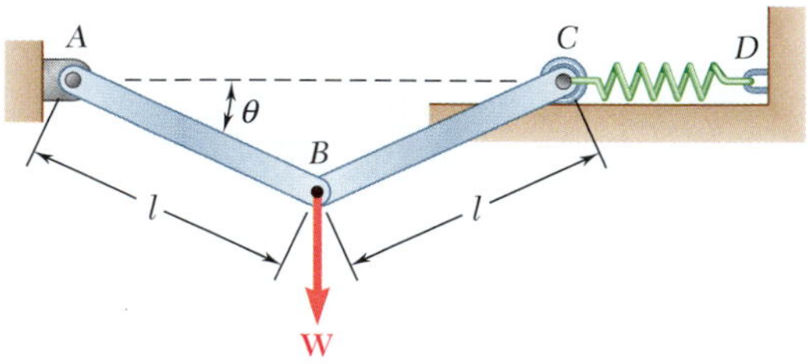

Fig. P10.C5

10.C6 The piston of the boom ABC is controlled by the hydraulic cylinder BD. Using computational software, plot the magnitude of the force exerted by the hydraulic cylinder on pin B as a function of θ for values of θ from 5° to 120°.

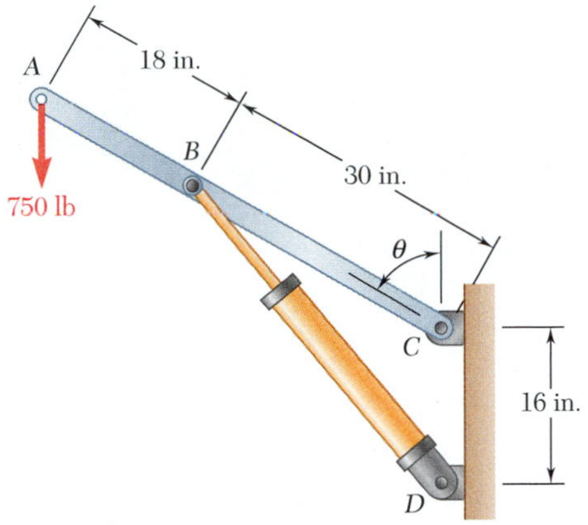

Fig. P10.C6

***10.C7** An 8-kg block is hung from the midpoint C of cable AB that is attached to two springs as shown. Knowing that the springs are unstretched when $y = 0$, plot the distance y corresponding to equilibrium as a function of the spring constant k_2 for $400 \text{ N/m} \leq k_2 \leq 600 \text{ N/m}$.

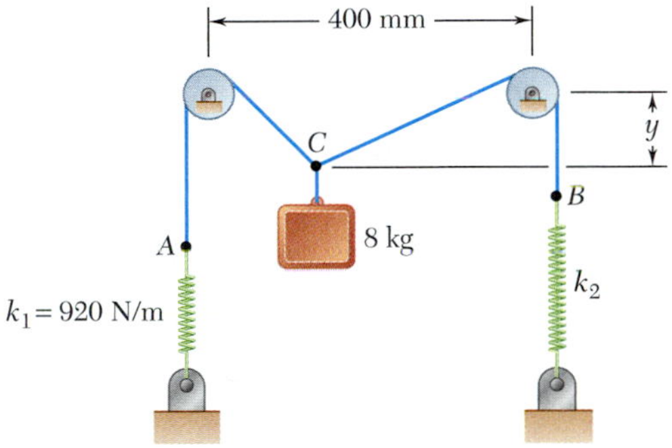

Fig. P10.C7

10.C8 A slender rod AB of weight W is attached to blocks A and B which can move freely in the guides shown. The spring constant is k and the spring is unstretched when AB is horizontal. For $l = 20$ in. and $k = 1$ lb/in., use computational software to determine the three values of the angle θ corresponding to equilibrium for values of W from 1 lb to 12 lb in 1 lb increments.

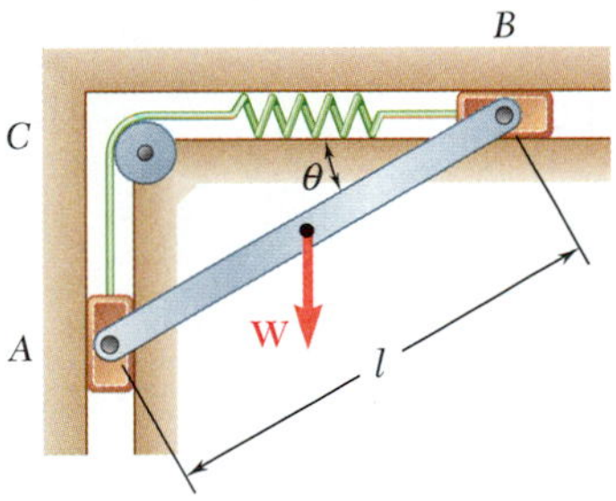

Fig. P10.C8

10.C9 A vertical load $\mathbf{W}$ is applied to the mechanism at C. Neglecting the weight of the mechanism, use computational software to plot θ as a function of W for values of W from 40 N to 480 N knowing that $k = 5$ kN/m and that the spring is unstretched when $\theta = 0$.

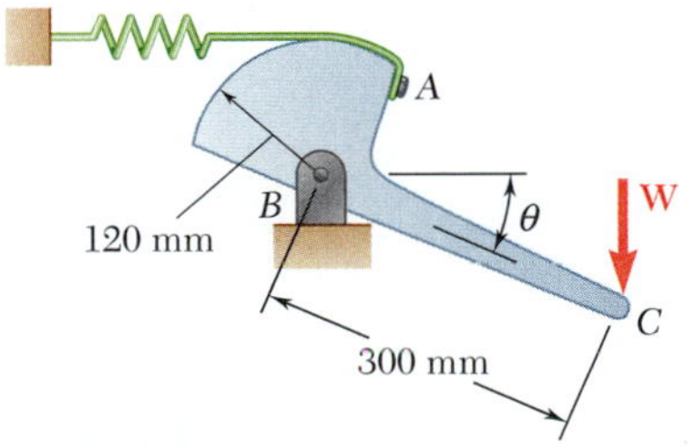

Fig. P10.C9

C H A P T E R

11

Kinematics of Particles

The motion of the aircraft shown is characterized at any instant by its *position, velocity,* and *acceleration.* Also of interest is the motion of the aircraft relative to the moving air. During the approach to a crosswind landing, the velocity of the aircraft relative to the air is a vector in the direction of the longitudinal axis of the aircraft, and the velocity of the aircraft relative to the ground is a vector in the direction of the runway.

KINEMATICS OF PARTICLES

11.1. INTRODUCTION TO DYNAMICS

Chapters 1 to 10 were devoted to *statics*, that is, to the analysis of bodies at rest. We now begin the study of *dynamics*, the part of mechanics that deals with the analysis of bodies in motion.

While the study of statics goes back to the time of the Greek philosophers, the first significant contribution to dynamics was made by Galileo (1564–1642). Galileo's experiments on uniformly accelerated bodies led Newton (1642–1727) to formulate his fundamental laws of motion.

Dynamics includes:

1. *Kinematics*, which is the study of the geometry of motion. Kinematics is used to relate displacement, velocity, acceleration, and time, without reference to the cause of the motion.
2. *Kinetics*, which is the study of the relation existing between the forces acting on a body, the mass of the body, and the motion of the body. Kinetics is used to predict the motion caused by given forces or to determine the forces required to produce a given motion.

Chapters 11 to 14 are devoted to the *dynamics of particles;* in Chap. 11 the *kinematics of particles* will be considered. The use of the word *particles* does not mean that our study will be restricted to small corpuscles; rather, it indicates that in these first chapters the motion of bodies—possibly as large as cars, rockets, or airplanes—will be considered without regard to their size. By saying that the bodies are analyzed as particles, we mean that only their motion as an entire unit will be considered; any rotation about their own mass center will be neglected. There are cases, however, when such a rotation is not negligible; the bodies cannot then be considered as particles. Such motions will be analyzed in later chapters, dealing with the *dynamics of rigid bodies.*

In the first part of Chap. 11, the rectilinear motion of a particle will be analyzed; that is, the position, velocity, and acceleration of a particle will be determined at every instant as it moves along a straight line. First, general methods of analysis will be used to study the motion of a particle; then two important particular cases will be considered, namely, the uniform motion and the uniformly accelerated motion of a particle (Secs. 11.4 and 11.5). In Sec. 11.6 the simultaneous motion of several particles will be considered, and the concept of the relative motion of one particle with respect to another will be introduced. The first part of this chapter concludes with a study of graphical methods of analysis and their application to the solution of various problems involving the rectilinear motion of particles (Secs. 11.7 and 11.8).

In the second part of this chapter, the motion of a particle as it moves along a curved path will be analyzed. Since the position, velocity, and acceleration of a particle will be defined as vector quantities, the concept of the derivative of a vector function will be introduced in Sec. 11.10 and added to our mathematical tools. Applications in which the motion of a particle is defined by the rectangular components of its velocity and acceleration will then be considered; at this point, the motion of a projectile will be analyzed (Sec. 11.11). In Sec. 11.12, the motion of a particle relative to a reference frame in translation will be considered. Finally, the curvilinear motion of a particle will be analyzed in terms of components other than rectangular. The tangential and normal components of the velocity and acceleration of a particle will be introduced in Sec. 11.13 and the radial and transverse components of its velocity and acceleration in Sec. 11.14.

RECTILINEAR MOTION OF PARTICLES

11.2. POSITION, VELOCITY, AND ACCELERATION

A particle moving along a straight line is said to be in *rectilinear motion.* At any given instant t, the particle will occupy a certain position on the straight line. To define the position P of the particle, we choose a fixed origin O on the straight line and a positive direction along the line. We measure the distance x from O to P and record it with a plus or minus sign, according to whether P is reached from O by moving along the line in the positive or the negative direction. The distance x, with the appropriate sign, completely defines the position of the particle; it is called the *position coordinate* of the particle considered. For example, the position coordinate corresponding to P in Fig. 11.1*a* is $x = +5$ m; the coordinate corresponding to P' in Fig. 11.1*b* is $x' = -2$ m.

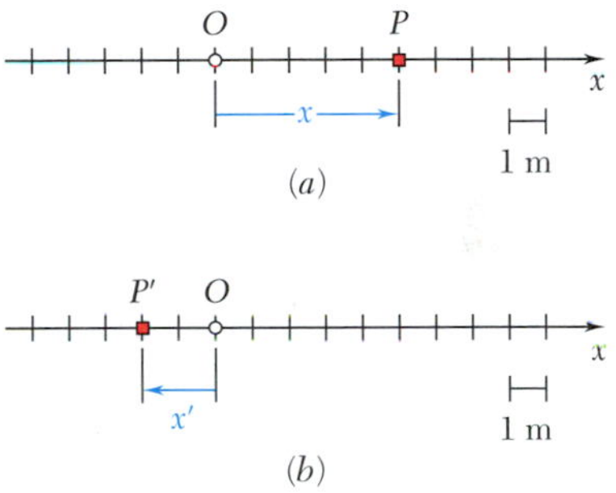

Fig. 11.1

When the position coordinate x of a particle is known for every value of time t, we say that the motion of the particle is known. The "timetable" of the motion can be given in the form of an equation in x and t, such as $x = 6t^2 - t^3$, or in the form of a graph of x versus t as shown in Fig. 11.6 on page 606. The units most often used to measure the position coordinate x are the meter (m) in the SI system of units† and the foot (ft) in the U.S. customary system of units. Time t is usually measured in seconds (s).

Consider the position P occupied by the particle at time t and the corresponding coordinate x (Fig. 11.2). Consider also the position P' occupied by the particle at a later time $t + \Delta t$; the position coordinate of P' can be obtained by adding to the coordinate x of P the small displacement Δx, which will be positive or negative according to whether P' is to the right or to the left of P. The *average velocity* of the particle over the time interval Δt is defined as the quotient of the displacement Δx and the time interval Δt:

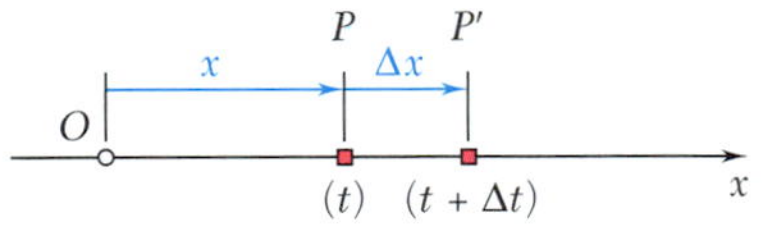

Fig. 11.2

$$\text{Average velocity} = \frac{\Delta x}{\Delta t}$$

†Cf. Sec. 1.3.

If SI units are used, Δx is expressed in meters and Δt in seconds; the average velocity will thus be expressed in meters per second (m/s). If U.S. customary units are used, Δx is expressed in feet and Δt in seconds; the average velocity will then be expressed in feet per second (ft/s).

The *instantaneous velocity* v of the particle at the instant t is obtained from the average velocity by choosing shorter and shorter time intervals Δt and displacements Δx:

$$\text{Instantaneous velocity} = v = \lim_{\Delta t \to 0} \frac{\Delta x}{\Delta t}$$

The instantaneous velocity will also be expressed in m/s or ft/s. Observing that the limit of the quotient is equal, by definition, to the derivative of x with respect to t, we write

$$v = \frac{dx}{dt} \tag{11.1}$$

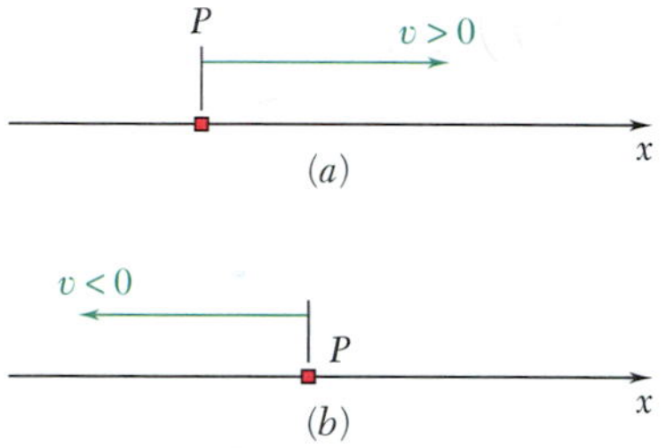

Fig. 11.3

The velocity v is represented by an algebraic number which can be positive or negative.† A positive value of v indicates that x increases, that is, that the particle moves in the positive direction (Fig. 11.3*a*); a negative value of v indicates that x decreases, that is, that the particle moves in the negative direction (Fig. 11.3*b*). The magnitude of v is known as the *speed* of the particle.

Consider the velocity v of the particle at time t and also its velocity $v + \Delta v$ at a later time $t + \Delta t$ (Fig. 11.4). The *average acceleration* of the particle over the time interval Δt is defined as the quotient of Δv and Δt:

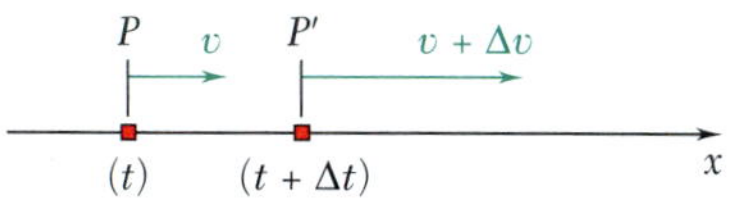

Fig. 11.4

$$\text{Average acceleration} = \frac{\Delta v}{\Delta t}$$

If SI units are used, Δv is expressed in m/s and Δt in seconds; the average acceleration will thus be expressed in m/s^2. If U.S. customary units are used, Δv is expressed in ft/s and Δt in seconds; the average acceleration will then be expressed in ft/s^2.

The *instantaneous acceleration* a of the particle at the instant t is obtained from the average acceleration by choosing smaller and smaller values for Δt and Δv:

$$\text{Instantaneous acceleration} = a = \lim_{\Delta t \to 0} \frac{\Delta v}{\Delta t}$$

The instantaneous acceleration will also be expressed in m/s^2 or ft/s^2. The limit of the quotient, which is by definition the derivative of v

†As you will see in Sec. 11.9, the velocity is actually a vector quantity. However, since we are considering here the rectilinear motion of a particle, where the velocity of the particle has a known and fixed direction, we need only specify the sense and magnitude of the velocity; this can be conveniently done by using a scalar quantity with a plus or minus sign. The same is true of the acceleration of a particle in rectilinear motion.

with respect to t, measures the rate of change of the velocity. We write

$$a = \frac{dv}{dt} \tag{11.2}$$

or, substituting for v from (11.1),

$$a = \frac{d^2x}{dt^2} \tag{11.3}$$

The acceleration a is represented by an algebraic number which can be positive or negative.† A positive value of a indicates that the velocity (that is, the algebraic number v) increases. This may mean that the particle is moving faster in the positive direction (Fig. 11.5a) or that it is moving more slowly in the negative direction (Fig. 11.5b); in both cases, Δv is positive. A negative value of a indicates that the velocity decreases; either the particle is moving more slowly in the positive direction (Fig. 11.5c) or it is moving faster in the negative direction (Fig. 11.5d).

Photo 11.1 As the truck moves along a straight path, its motion is characterized at any instant by its position coordinate *x*, its velocity *v*, and its acceleration *a*.

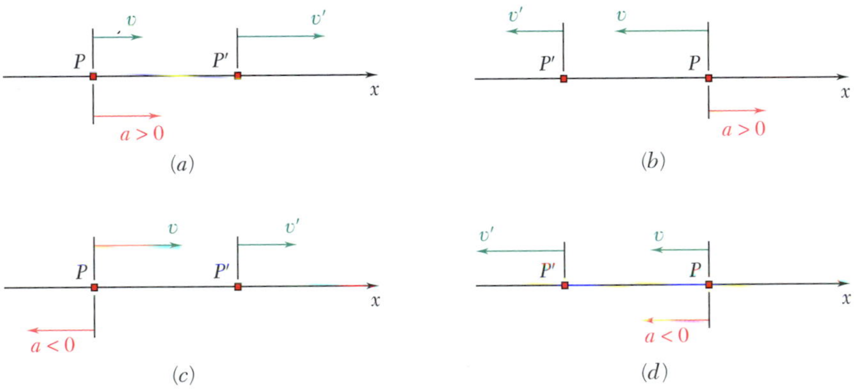

Fig. 11.5

The term *deceleration* is sometimes used to refer to a when the speed of the particle (that is, the magnitude of v) decreases; the particle is then moving more slowly. For example, the particle of Fig. 11.5 is decelerated in parts b and c; it is truly accelerated (that is, moves faster) in parts a and d.

Another expression for the acceleration can be obtained by eliminating the differential dt in Eqs. (11.1) and (11.2). Solving (11.1) for dt, we obtain $dt = dx/v$; substituting into (11.2), we write

$$a = v\frac{dv}{dx} \tag{11.4}$$

†See footnote, page 604.

Example. Consider a particle moving in a straight line, and assume that its position is defined by the equation

$$x = 6t^2 - t^3$$

where t is expressed in seconds and x in meters. The velocity v at any time t is obtained by differentiating x with respect to t:

$$v = \frac{dx}{dt} = 12t - 3t^2$$

The acceleration a is obtained by differentiating again with respect to t:

$$a = \frac{dv}{dt} = 12 - 6t$$

The position coordinate, the velocity, and the acceleration have been plotted against t in Fig. 11.6. The curves obtained are known as *motion curves*. Keep in mind, however, that the particle does not move along any of these curves; the particle moves in a straight line. Since the derivative of a function measures the slope of the corresponding curve, the slope of the x–t curve at any given time is equal to the value of v at that time and the slope of the v–t curve is equal to the value of a. Since $a = 0$ at $t = 2$ s, the slope of the v–t curve must be zero at $t = 2$ s; the velocity reaches a maximum at this instant. Also, since $v = 0$ at $t = 0$ and at $t = 4$ s, the tangent to the x–t curve must be horizontal for both of these values of t.

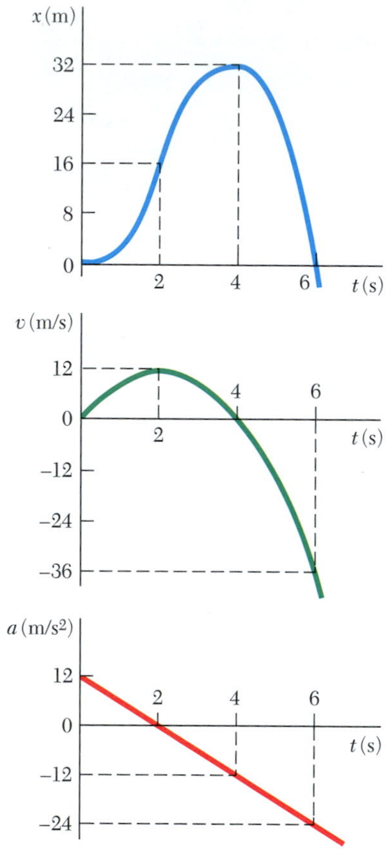

Fig. 11.6

A study of the three motion curves of Fig. 11.6 shows that the motion of the particle from $t = 0$ to $t = \infty$ can be divided into four phases:

1. The particle starts from the origin, $x = 0$, with no velocity but with a positive acceleration. Under this acceleration, the particle gains a positive velocity and moves in the positive direction. From $t = 0$ to $t = 2$ s, x, v, and a are all positive.

2. At $t = 2$ s, the acceleration is zero; the velocity has reached its maximum value. From $t = 2$ s to $t = 4$ s, v is positive, but a is negative; the particle still moves in the positive direction but more and more slowly; the particle is decelerating.

3. At $t = 4$ s, the velocity is zero; the position coordinate x has reached its maximum value. From then on, both v and a are negative; the particle is accelerating and moves in the negative direction with increasing speed.

4. At $t = 6$ s, the particle passes through the origin; its coordinate x is then zero, while the total distance traveled since the beginning of the motion is 64 m. For values of t larger than 6 s, x, v, and a will all be negative. The particle keeps moving in the negative direction, away from O, faster and faster.

11.3. DETERMINATION OF THE MOTION OF A PARTICLE

We saw in the preceding section that the motion of a particle is said to be known if the position of the particle is known for every value of the time t. In practice, however, a motion is seldom defined by a relation between x and t. More often, the conditions of the motion will be specified by the type of acceleration that the particle possesses. For example, a freely falling body will have a constant acceleration, directed downward and equal to 9.81 m/s^2, or 32.2 ft/s^2, a mass attached to a spring which has been stretched will have an acceleration proportional to the instantaneous elongation of the spring measured from the equilibrium position; etc. In general, the acceleration of the particle can be expressed as a function of one or more of the variables x, v, and t. In order to determine the position coordinate x in terms of t, it will thus be necessary to perform two successive integrations.

Let us consider three common classes of motion:

1. $a = f(t)$. *The Acceleration Is a Given Function of* t. Solving (11.2) for dv and substituting $f(t)$ for a, we write

$$dv = a\,dt$$
$$dv = f(t)\,dt$$

Integrating both members, we obtain the equation

$$\int dv = \int f(t)\,dt$$

which defines v in terms of t. It should be noted, however, that an arbitrary constant will be introduced as a result of the integration. This is due to the fact that there are many motions which correspond to the given acceleration $a = f(t)$. In order to uniquely define the motion of the particle, it is necessary to specify the *initial conditions* of the motion, that is, the value v_0 of the velocity and the value x_0 of the position coordinate at $t = 0$. Replacing the indefinite integrals by *definite integrals* with lower limits corresponding to the initial conditions $t = 0$ and $v = v_0$ and upper limits corresponding to $t = t$ and $v = v$, we write

$$\int_{v_0}^{v} dv = \int_0^t f(t)\,dt$$

$$v - v_0 = \int_0^t f(t)\,dt$$

which yields v in terms of t.

Equation (11.1) can now be solved for dx,

$$dx = v\,dt$$

and the expression just obtained substituted for v. Both members are then integrated, the left-hand member with respect to x from $x = x_0$ to $x = x$, and the right-hand member with respect to t from $t = 0$ to $t = t$. The position coordinate x is thus obtained in terms of t; the motion is completely determined.

Two important particular cases will be studied in greater detail in Secs. 11.4 and 11.5: the case when $a = 0$, corresponding to a *uniform motion,* and the case when $a =$ constant, corresponding to a *uniformly accelerated motion.*

2. *$a = f(x)$. The Acceleration Is a Given Function of x.* Rearranging Eq. (11.4) and substituting $f(x)$ for a, we write

$$v\,dv = a\,dx$$
$$v\,dv = f(x)\,dx$$

Since each member contains only one variable, we can integrate the equation. Denoting again by v_0 and x_0, respectively, the initial values of the velocity and of the position coordinate, we obtain

$$\int_{v_0}^{v} v\,dv = \int_{x_0}^{x} f(x)\,dx$$

$$\tfrac{1}{2}v^2 - \tfrac{1}{2}v_0^2 = \int_{x_0}^{x} f(x)\,dx$$

which yields v in terms of x. We now solve (11.1) for dt,

$$dt = \frac{dx}{v}$$

and substitute for v the expression just obtained. Both members can then be integrated to obtain the desired relation between x and t. However, in most cases this last integration cannot be performed analytically and one must resort to a numerical method of integration.

3. *$a = f(v)$. The Acceleration Is a Given Function of v.* We can now substitute $f(v)$ for a in either (11.2) or (11.4) to obtain either of the following relations:

$$f(v) = \frac{dv}{dt} \qquad f(v) = v\frac{dv}{dx}$$

$$dt = \frac{dv}{f(v)} \qquad dx = \frac{v\,dv}{f(v)}$$

Integration of the first equation will yield a relation between v and t; integration of the second equation will yield a relation between v and x. Either of these relations can be used in conjunction with Eq. (11.1) to obtain the relation between x and t which characterizes the motion of the particle.

SAMPLE PROBLEM 11.1

The position of a particle which moves along a straight line is defined by the relation $x = t^3 - 6t^2 - 15t + 40$, where x is expressed in feet and t in seconds. Determine (*a*) the time at which the velocity will be zero, (*b*) the position and distance traveled by the particle at that time, (*c*) the acceleration of the particle at that time, (*d*) the distance traveled by the particle from $t = 4$ s to $t = 6$ s.

SOLUTION

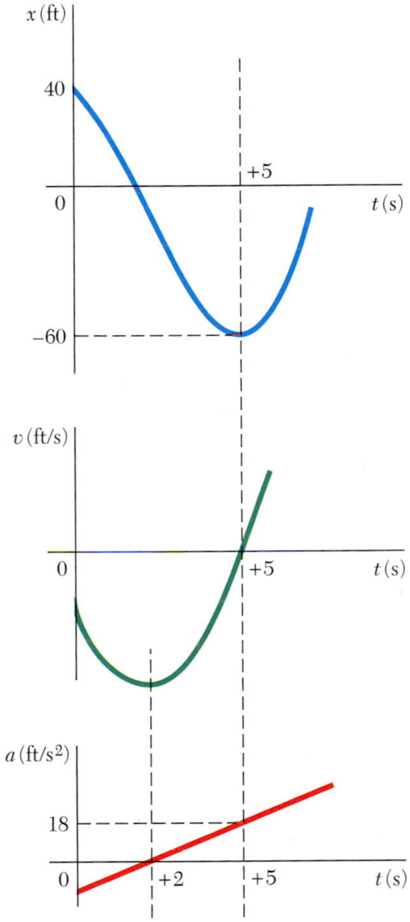

The equations of motion are

$$x = t^3 - 6t^2 - 15t + 40 \qquad (1)$$

$$v = \frac{dx}{dt} = 3t^2 - 12t - 15 \qquad (2)$$

$$a = \frac{dv}{dt} = 6t - 12 \qquad (3)$$

***a.* Time at Which $v = 0$.** We set $v = 0$ in (2):

$$3t^2 - 12t - 15 = 0 \qquad t = -1 \text{ s} \qquad \text{and} \qquad t = +5 \text{ s} \quad ◀$$

Only the root $t = +5$ s corresponds to a time after the motion has begun: for $t < 5$ s, $v < 0$, the particle moves in the negative direction; for $t > 5$ s, $v > 0$, the particle moves in the positive direction.

***b.* Position and Distance Traveled When $v = 0$.** Carrying $t = +5$ s into (1), we have

$$x_5 = (5)^3 - 6(5)^2 - 15(5) + 40 \qquad x_5 = -60 \text{ ft} \quad ◀$$

The initial position at $t = 0$ was $x_0 = +40$ ft. Since $v \neq 0$ during the interval $t = 0$ to $t = 5$ s, we have

$$\text{Distance traveled} = x_5 - x_0 = -60 \text{ ft} - 40 \text{ ft} = -100 \text{ ft}$$

Distance traveled = 100 ft in the negative direction ◀

***c.* Acceleration When $v = 0$.** We substitute $t = +5$ s into (3):

$$a_5 = 6(5) - 12 \qquad a_5 = +18 \text{ ft/s}^2 \quad ◀$$

***d.* Distance Traveled from $t = 4$ s to $t = 6$ s.** The particle moves in the negative direction from $t = 4$ s to $t = 5$ s and in the positive direction from $t = 5$ s to $t = 6$ s; therefore, the distance traveled during each of these time intervals will be computed separately.

From $t = 4$ s to $t = 5$ s: $\quad x_5 = -60$ ft

$$x_4 = (4)^3 - 6(4)^2 - 15(4) + 40 = -52 \text{ ft}$$

$$\text{Distance traveled} = x_5 - x_4 = -60 \text{ ft} - (-52 \text{ ft}) = -8 \text{ ft}$$
$$= 8 \text{ ft in the negative direction}$$

From $t = 5$ s to $t = 6$ s: $\quad x_5 = -60$ ft

$$x_6 = (6)^3 - 6(6)^2 - 15(6) + 40 = -50 \text{ ft}$$

$$\text{Distance traveled} = x_6 - x_5 = -50 \text{ ft} - (-60 \text{ ft}) = +10 \text{ ft}$$
$$= 10 \text{ ft in the positive direction}$$

Total distance traveled from $t = 4$ s to $t = 6$ s is 8 ft + 10 ft = 18 ft ◀

SAMPLE PROBLEM 11.2

A ball is tossed with a velocity of 10 m/s directed vertically upward from a window located 20 m above the ground. Knowing that the acceleration of the ball is constant and equal to 9.81 m/s^2 downward, determine (*a*) the velocity v and elevation y of the ball above the ground at any time t, (*b*) the highest elevation reached by the ball and the corresponding value of t, (*c*) the time when the ball will hit the ground and the corresponding velocity. Draw the v–t and y–t curves.

SOLUTION

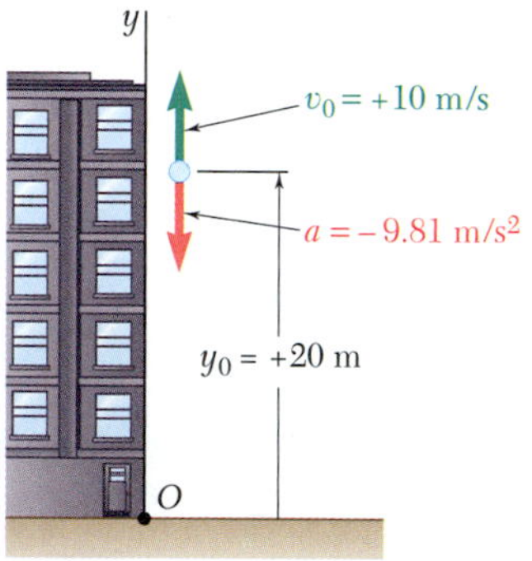

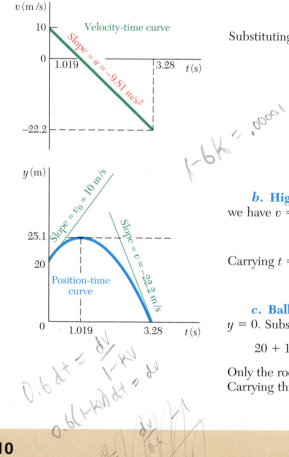

a. Velocity and Elevation. The y axis measuring the position coordinate (or elevation) is chosen with its origin O on the ground and its positive sense upward. The value of the acceleration and the initial values of v and y are as indicated. Substituting for a in $a = dv/dt$ and noting that at $t = 0$, $v_0 = +10$ m/s, we have

$$\frac{dv}{dt} = a = -9.81 \text{ m/s}^2$$

$$\int_{v_0=10}^{v} dv = -\int_0^t 9.81\, dt$$

$$[v]_{10}^{v} = -[9.81t]_0^t$$

$$v - 10 = -9.81t$$

$$v = 10 - 9.81t \qquad (1) \quad ◀$$

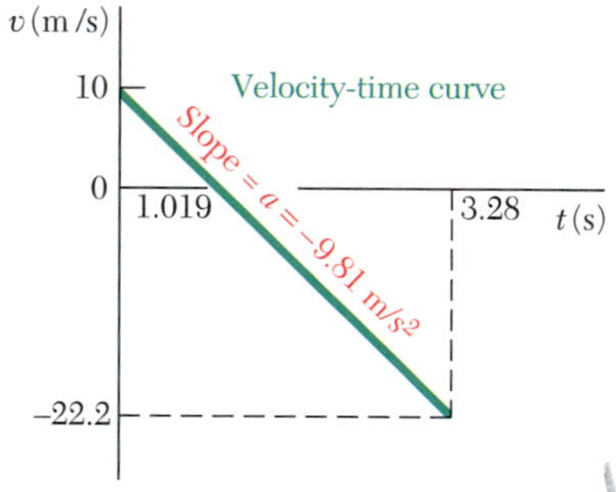

Substituting for v in $v = dy/dt$ and noting that at $t = 0$, $y_0 = 20$ m, we have

$$\frac{dy}{dt} = v = 10 - 9.81t$$

$$\int_{y_0=20}^{y} dy = \int_0^t (10 - 9.81t)\, dt$$

$$[y]_{20}^{y} = [10t - 4.905t^2]_0^t$$

$$y - 20 = 10t - 4.905t^2$$

$$y = 20 + 10t - 4.905t^2 \qquad (2) \quad ◀$$

b. Highest Elevation. When the ball reaches its highest elevation, we have $v = 0$. Substituting into (1), we obtain

$$10 - 9.81t = 0 \qquad t = 1.019 \text{ s} \quad ◀$$

Carrying $t = 1.019$ s into (2), we have

$$y = 20 + 10(1.019) - 4.905(1.019)^2 \qquad y = 25.1 \text{ m} \quad ◀$$

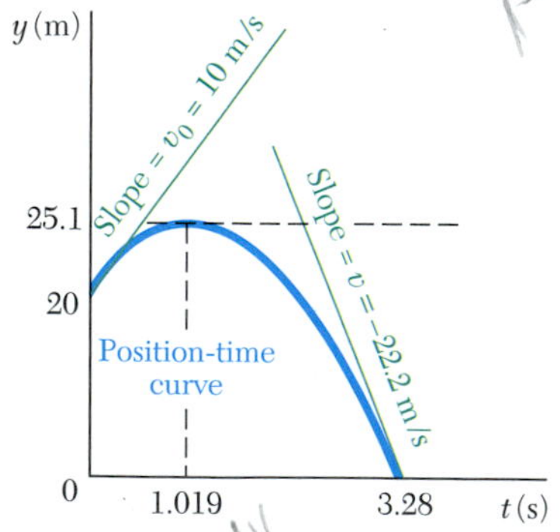

c. Ball Hits the Ground. When the ball hits the ground, we have $y = 0$. Substituting into (2), we obtain

$$20 + 10t - 4.905t^2 = 0 \qquad t = -1.243 \text{ s} \quad \text{and} \quad t = +3.28 \text{ s} \quad ◀$$

Only the root $t = +3.28$ s corresponds to a time after the motion has begun. Carrying this value of t into (1), we have

$$v = 10 - 9.81(3.28) = -22.2 \text{ m/s} \qquad v = 22.2 \text{ m/s} \downarrow \quad ◀$$

SAMPLE PROBLEM 11.3

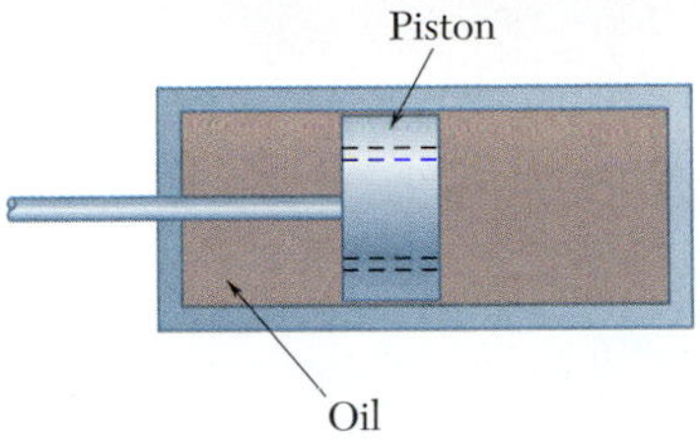

The brake mechanism used to reduce recoil in certain types of guns consists essentially of a piston attached to the barrel and moving in a fixed cylinder filled with oil. As the barrel recoils with an initial velocity v_0, the piston moves and oil is forced through orifices in the piston, causing the piston and the barrel to decelerate at a rate proportional to their velocity; that is, $a = -kv$. Express (*a*) v in terms of t, (*b*) x in terms of t, (*c*) v in terms of x. Draw the corresponding motion curves.

SOLUTION

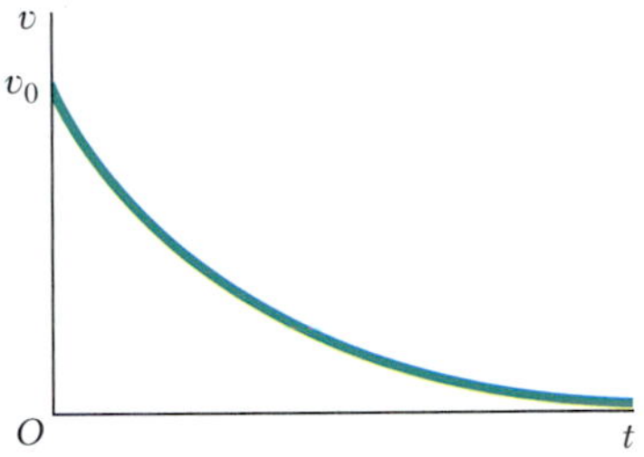

***a. v* in Terms of *t*.** Substituting $-kv$ for a in the fundamental formula defining acceleration, $a = dv/dt$, we write

$$-kv = \frac{dv}{dt} \qquad \frac{dv}{v} = -k\,dt \qquad \int_{v_0}^{v} \frac{dv}{v} = -k\int_0^t dt$$

$$\ln\frac{v}{v_0} = -kt \qquad\qquad v = v_0 e^{-kt} \blacktriangleleft$$

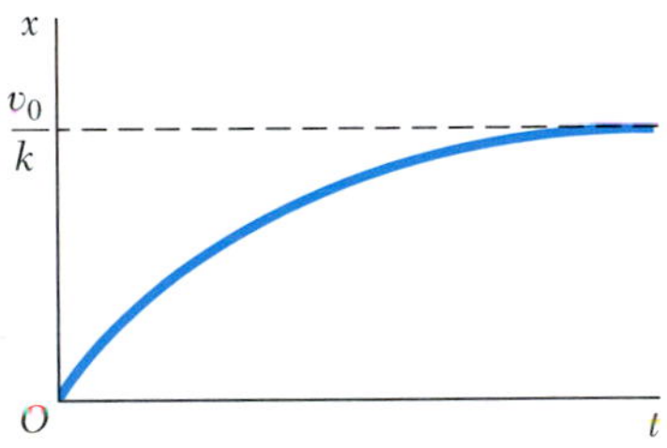

***b. x* in Terms of *t*.** Substituting the expression just obtained for v into $v = dx/dt$, we write

$$v_0 e^{-kt} = \frac{dx}{dt}$$

$$\int_0^x dx = v_0 \int_0^t e^{-kt}\,dt$$

$$x = -\frac{v_0}{k}[e^{-kt}]_0^t = -\frac{v_0}{k}(e^{-kt} - 1)$$

$$x = \frac{v_0}{k}(1 - e^{-kt}) \blacktriangleleft$$

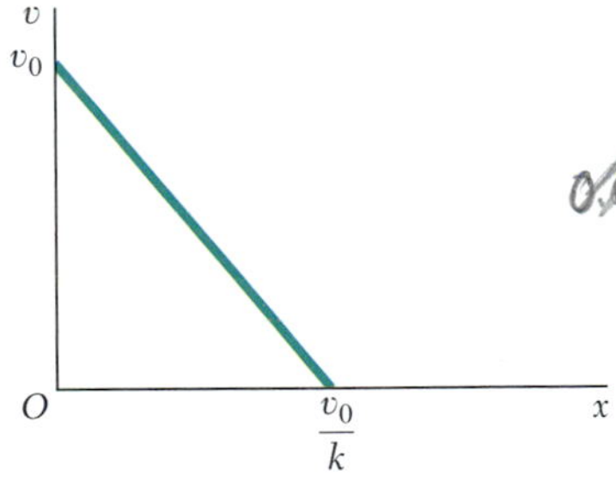

***c. v* in Terms of *x*.** Substituting $-kv$ for a in $a = v\,dv/dx$, we write

$$-kv = v\frac{dv}{dx}$$

$$dv = -k\,dx$$

$$\int_{v_0}^{v} dv = -k\int_0^x dx$$

$$v - v_0 = -kx \qquad\qquad v = v_0 - kx \blacktriangleleft$$

Check. Part *c* could have been solved by eliminating t from the answers obtained for parts *a* and *b*. This alternative method can be used as a check. From part *a* we obtain $e^{-kt} = v/v_0$; substituting into the answer of part *b*, we obtain

$$x = \frac{v_0}{k}(1 - e^{-kt}) = \frac{v_0}{k}\left(1 - \frac{v}{v_0}\right) \qquad v = v_0 - kx \qquad \text{(checks)}$$

SOLVING PROBLEMS ON YOUR OWN

In the problems for this lesson, you will be asked to determine the *position,* the *velocity,* or the *acceleration* of a particle in *rectilinear motion.* As you read each problem, it is important that you identify both the independent variable (typically t or x) and what is required (for example, the need to express v as a function of x). You may find it helpful to start each problem by writing down both the given information and a simple statement of what is to be determined.

1. *Determining $v(t)$ and $a(t)$ for a given $x(t)$.* As explained in Sec. 11.2, the first and the second derivatives of x with respect to t are respectively equal to the velocity and the acceleration of the particle [Eqs. (11.1) and (11.2)]. If the velocity and the acceleration have opposite signs, the particle can come to rest and then move in the opposite direction [Sample Prob. 11.1]. Thus, when computing the total distance traveled by a particle, you should first determine if the particle will come to rest during the specified interval of time. Constructing a diagram similar to that of Sample Prob. 11.1 that shows the position and the velocity of the particle at each critical instant ($v = v_{max}$, $v = 0$, etc.) will help you to visualize the motion.

2. *Determining $v(t)$ and $x(t)$ for a given $a(t)$.* The solution of problems of this type was discussed in the first part of Sec. 11.3. We used the initial conditions, $t = 0$ and $v = v_0$, for the lower limits of the integrals in t and v, but any other known state (for example, $t = t_1$, $v = v_1$) could have been used instead. Also, if the given function $a(t)$ contains an unknown constant (for example, the constant k if $a = kt$), you will first have to determine that constant by substituting a set of known values of t and a in the equation defining $a(t)$.

3. *Determining $v(x)$ and $x(t)$ for a given $a(x)$.* This is the second case considered in Sec. 11.3. We again note that the lower limits of integration can be any known state (for example, $x = x_1$, $v = v_1$). In addition, since $v = v_{max}$ when $a = 0$, the positions where the maximum values of the velocity occur are easily determined by writing $a(x) = 0$ and solving for x.

4. *Determining $v(x)$, $v(t)$, and $x(t)$ for a given $a(v)$.* This is the last case treated in Sec. 11.3; the appropriate solution techniques for problems of this type are illustrated in Sample Prob. 11.3. All of the general comments for the preceding cases once again apply. Note that Sample Prob. 11.3 provides a summary of how and when to use the equations $v = dx/dt$, $a = dv/dt$, and $a = v\,dv/dx$.

Problems†

11.1 The motion of a particle is defined by the relation $x = 8t^3 - 8 + 30 \sin \pi t$, where x and t are expressed in millimeters and seconds, respectively. Determine the position, the velocity, and the acceleration of the particle when $t = 5$ s.

11.2 The motion of a particle is defined by the relation $x = \frac{10}{3}t^3 - \frac{5}{2}t^2 - 20t + 10$, where x and t are expressed in meters and seconds, respectively. Determine the time, the position, and the acceleration of the particle when $v = 0$.

11.3 The motion of the slider A is defined by the relation $x = 20 \sin kt$, where x and t are expressed in inches and seconds, respectively, and k is a constant. Knowing that $k = 10$ rad/s, determine the position, the velocity, and the acceleration of slider A when $t = 0.05$ s.

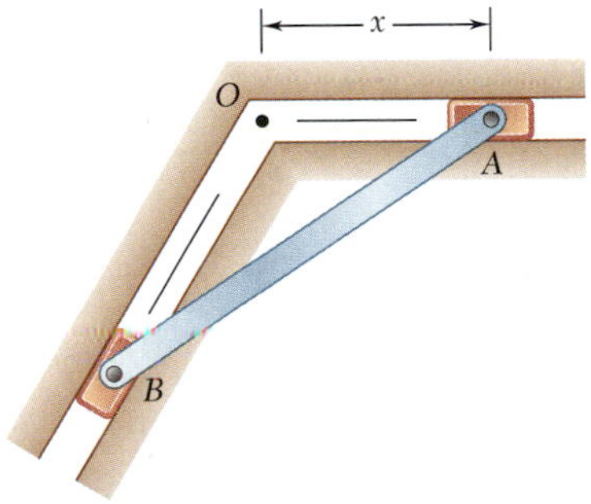

Fig. P11.3 and P11.4

11.4 The motion of the slider A is defined by the relation $x = 20 \sin(k_1 t - k_2 t^2)$, where x and t are expressed in inches and seconds, respectively. The constants k_1 and k_2 are known to be 1 rad/s and 0.5 rad/s^2, respectively. Consider the range $0 < t < 2$ s and determine the position and acceleration of slider A when $v = 0$.

11.5 The motion of a particle is defined by the relation $x = 5t^4 - 4t^3 + 3t - 2$, where x and t are expressed in feet and seconds, respectively. Determine the position, the velocity, and the acceleration of the particle when $t = 2$ s.

11.6 The motion of a particle is defined by the relation $x = 6t^4 + 8t^3 - 14t^2 - 10t + 16$, where x and t are expressed in inches and seconds, respectively. Determine the position, the velocity, and the acceleration of the particle when $t = 3$ s.

†Answers to all problems set in straight type (such as **11.1**) are given at the end of the book. Answers to problems with a number set in italic type (such as ***11.6***) are not given.

11.7 The motion of a particle is defined by the relation $x = 2t^3 - 12t^2 - 72t - 80$, where x and t are expressed in meters and seconds, respectively. Determine (*a*) when the velocity is zero, (*b*) the velocity, the acceleration, and the total distance traveled when $x = 0$.

11.8 The motion of a particle is defined by the relation $x = 2t^3 - 18t^2 + 48t - 16$, where x and t are expressed in millimeters and seconds, respectively. Determine (*a*) when the velocity is zero, (*b*) the position and the total distance traveled when the acceleration is zero.

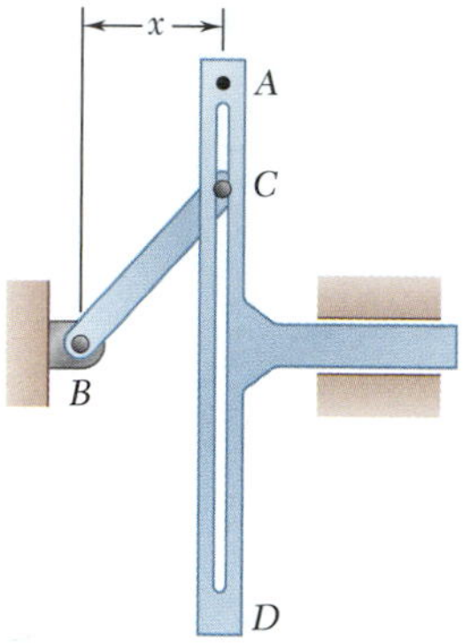

Fig. P11.9 and P11.10

11.9 The acceleration of point A is defined by the relation $a = -1.8 \sin kt$, where a and t are expressed in m/s^2 and seconds, respectively, and $k = 3$ rad/s. Knowing that $x = 0$ and $v = 0.6$ m/s when $t = 0$, determine the velocity and position of point A when $t = 0.5$ s.

11.10 The acceleration of point A is defined by the relation $a = -1.08 \sin kt - 1.44 \cos kt$, where a and t are expressed in m/s^2 and seconds, respectively, and $k = 3$ rad/s. Knowing that $x = 0.16$ m and $v = 0.36$ m/s when $t = 0$, determine the velocity and position of point A when $t = 0.5$ s.

11.11 The acceleration of a particle is directly proportional to the square of the time t. When $t = 0$, the particle is at $x = 36$ ft. Knowing that at $t = 9$ s, $x = 144$ ft and $v = 27$ ft/s, express x and v in terms of t.

11.12 It is known that from $t = 2$ s to $t = 10$ s the acceleration of a particle is inversely proportional to the cube of the time t. When $t = 2$ s, $v = -15$ ft/s, and when $t = 10$ s, $v = 0.36$ ft/s. Knowing that the particle is twice as far from the origin when $t = 2$ s as it is when $t = 10$ s, determine (*a*) the position of the particle when $t = 2$ s and when $t = 10$ s, (*b*) the total distance traveled by the particle from $t = 2$ s to $t = 10$ s.

11.13 The acceleration of a particle is directly proportional to the time t. At $t = 0$, the velocity of the particle is 400 mm/s. Knowing that $v = 370$ mm/s and $x = 500$ mm when $t = 1$ s, determine the velocity, the position, and the total distance traveled when $t = 7$ s.

11.14 The acceleration of a particle is defined by the relation $a = 0.15\ m/s^2$. Knowing that $x = -10$ m when $t = 0$ and $v = -0.15$ m/s when $t = 2$ s, determine the velocity, the position, and the total distance traveled when $t = 5$ s.

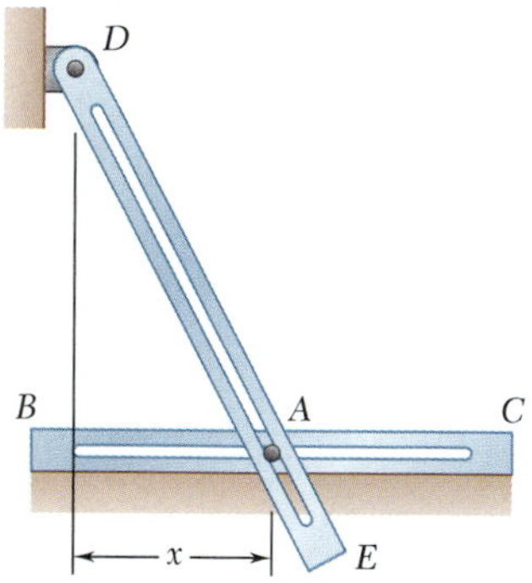

Fig. P11.15 and P11.16

11.15 The acceleration of point A is defined by the relation $a = 200x(1 + kx^2)$, where a and x are expressed in m/s^2 and meters, respectively, and k is a constant. Knowing that the velocity of A is 2.5 m/s when $x = 0$ and 5 m/s when $x = 0.15$ m, determine the value of k.

11.16 The acceleration of point A is defined by the relation $a = 200x + 3200x^3$, where a and x are expressed in m/s^2 and meters, respectively. Knowing that the velocity of A is 2.5 m/s and $x = 0$ when $t = 0$, determine the velocity and position of A when $t = 0.05$ s.

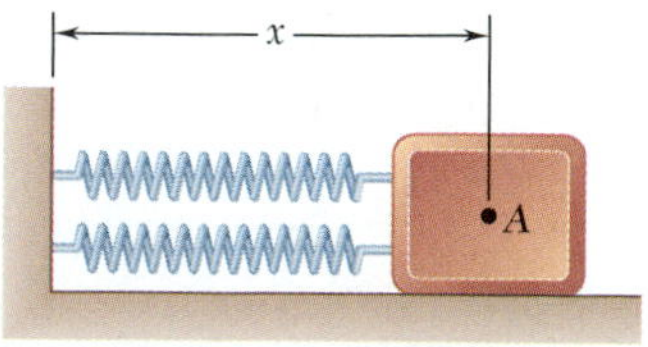

Fig. P11.17 and P11.18

11.17 Point A oscillates with an acceleration $a = 2880 - 144x$, where a and x are expressed in in./s^2 and inches, respectively. The magnitude of the velocity is 11 in./s when $x = 20.4$ in. Determine (*a*) the maximum velocity of A, (*b*) the two positions at which the velocity of A is zero.

11.18 Point A oscillates with an acceleration $a = 144(20 - x)$, where a and x are expressed in in./s^2 and inches, respectively. Knowing that the system starts at time $t = 0$ with $v = 0$ and $x = 19$ in., determine the position and the velocity of A when $t = 0.2$ s.

11.19 The acceleration of a particle is defined by the relation $a = 12x - 28$, where a and x are expressed in m/s^2 and meters, respectively. Knowing that $v = 8$ m/s when $x = 0$, determine (*a*) the maximum value of x, (*b*) the velocity when the particle has traveled a total distance of 3 m.

11.20 The acceleration of a particle is defined by the relation $a = k(1 - e^{-x})$, where k is a constant. Knowing that the velocity of the particle is $v = +9$ m/s when $x = -3$ m and that the particle comes to rest at the origin, determine (*a*) the value of k, (*b*) the velocity of the particle when $x = -2$ m.

11.21 The acceleration of a particle is defined by the relation $a = -k\sqrt{v}$, where k is a constant. Knowing that $x = 0$ and $v = 25$ ft/s at $t = 0$, and that $v = 12$ ft/s when $x = 6$ ft, determine (*a*) the velocity of the particle when $x = 8$ ft, (*b*) the time required for the particle to come to rest.

11.22 Starting from $x = 0$ with no initial velocity, a particle is given an acceleration $a = 0.8\sqrt{v^2 + 49}$, where a and v are expressed in ft/s^2 and ft/s, respectively. Determine (*a*) the position of the particle when $v = 24$ ft/s, (*b*) the speed of the particle when $x = 40$ ft.

11.23 The acceleration of slider A is defined by the relation $a = -2k\sqrt{k^2 - v^2}$, where a and v are expressed in m/s^2 and m/s, respectively, and k is a constant. The system starts at time $t = 0$ with $x = 0.5$ m and $v = 0$. Knowing that $x = 0.4$ m when $t = 0.2$ s, determine the value of k.

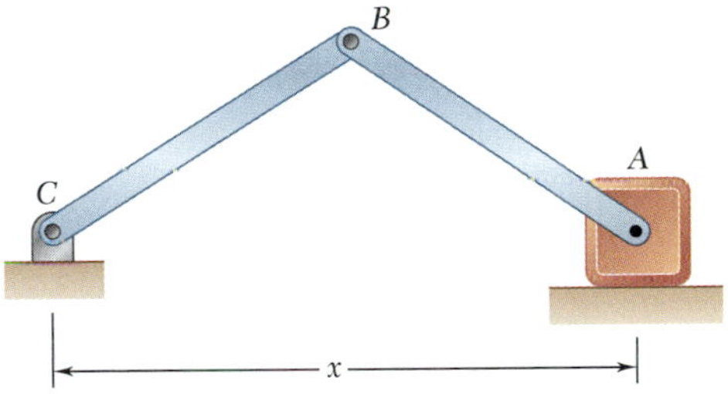

Fig. P11.23 and P11.24

11.24 The acceleration of slider A is defined by the relation $a = -2\sqrt{1 - v^2}$, where a and v are expressed in m/s^2 and m/s, respectively. The system starts at time $t = 0$ with $x = 0.5$ m and $v = 0$. Determine (*a*) the position of A when $v = -0.8$ m/s, (*b*) the position of A when $t = 0.2$ s.

11.25 The acceleration of a particle is defined by the relation $a = -kv^{2.5}$, where k is a constant. The particle starts at $x = 0$ with a velocity of 16 mm/s, and when $x = 6$ mm the velocity is observed to be 4 mm/s. Determine (*a*) the velocity of the particle when $x = 5$ mm, (*b*) the time at which the velocity of the particle is 9 mm/s.

11.26 The acceleration of a particle is defined by the relation $a = 0.6(1 - kv)$, where k is a constant. Knowing that at $t = 0$ the particle starts from rest at $x = 6$ m and that $v = 6$ m/s when $t = 20$ s, determine (*a*) the constant k, (*b*) the position of the particle when $v = 7.5$ m/s, (*c*) the maximum velocity of the particle.

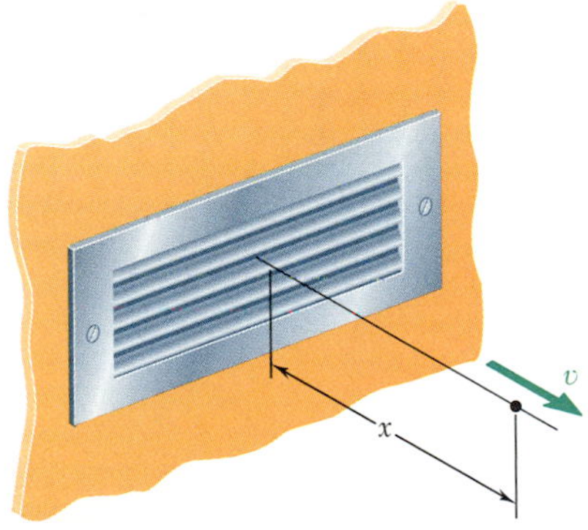

Fig. P11.27

11.27 Experimental data indicate that in a region downstream of a given louvered supply vent the velocity of the emitted air is defined by $v = 0.18v_0/x$, where v and x are expressed in ft/s and feet, respectively, and v_0 is the initial discharge velocity of the air. For $v_0 = 12$ ft/s, determine (*a*) the acceleration of the air at $x = 6$ ft, (*b*) the time required for the air to flow from $x = 3$ ft to $x = 10$ ft.

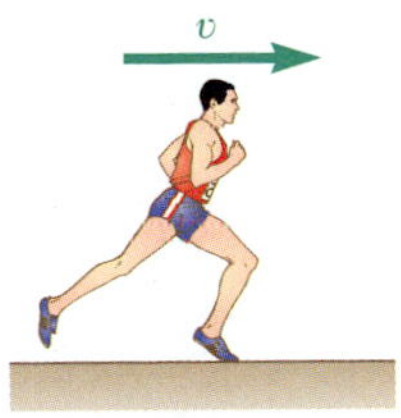

Fig. P11.28

11.28 Based on observations, the speed of a jogger can be approximated by the relation $v = 7.5(1 - 0.04x)^{0.3}$, where v and x are expressed in km/h and kilometers, respectively. Knowing that $x = 0$ at $t = 0$, determine (*a*) the distance the jogger has run when $t = 1$ h, (*b*) the jogger's acceleration in m/s^2 at $t = 0$, (*c*) the time required for the jogger to run 6 km.

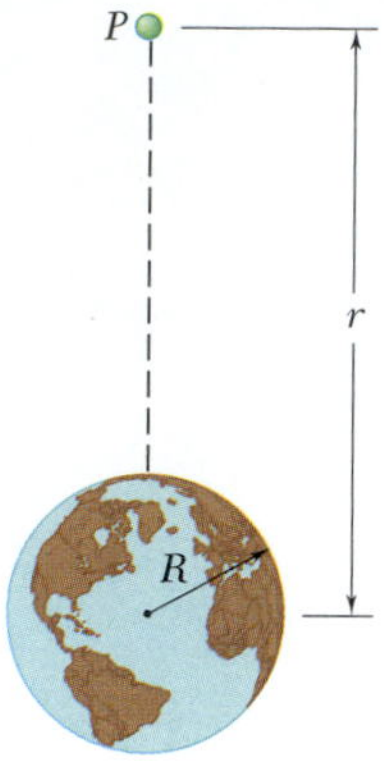

Fig. *P11.29*

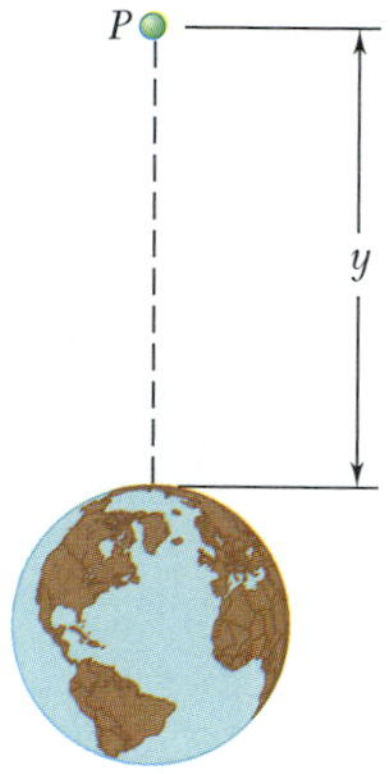

Fig. *P11.30*

11.29 The acceleration due to gravity of a particle falling toward the earth is $a = -gR^2/r^2$, where r is the distance from the *center* of the earth to the particle, R is the radius of the earth, and g is the acceleration due to gravity at the surface of the earth. If $R = 3960$ mi, calculate the *escape velocity,* that is, the minimum velocity with which a particle must be projected upward from the surface of the earth if it is not to return to earth. (*Hint:* $v = 0$ for $r = \infty$.)

11.30 The acceleration due to gravity at an altitude y above the surface of the earth can be expressed as

$$a = \frac{-32.2}{[1 + (y/20.9 \times 10^6)]^2}$$

where a and y are expressed in ft/s^2 and feet, respectively. Using this expression, compute the height reached by a projectile fired vertically upward from the surface of the earth if its initial velocity is (*a*) $v = 2400$ ft/s, (*b*) $v = 4000$ ft/s, (*c*) $v = 40{,}000$ ft/s.

11.31 The velocity of a slider is defined by the relation $v = v' \sin(\omega_n t + \phi)$. Denoting the velocity and the position of the slider at $t = 0$ by v_0 and x_0, respectively, and knowing that the maximum displacement of the slider is $2x_0$, show that (*a*) $v' = (v_0^2 + x_0^2\omega_n^2)/2x_0\omega_n$, (*b*) the maximum value of the velocity occurs when $x = x_0[3 - (v_0/x_0\omega_n)^2]/2$.

11.32 The velocity of a particle is $v = v_0[1 - \sin(\pi t/T)]$. Knowing that the particle starts from the origin with an initial velocity v_0, determine (*a*) its position and its acceleration at $t = 3T$, (*b*) its average velocity during the interval $t = 0$ to $t = T$.

11.4. UNIFORM RECTILINEAR MOTION

Uniform rectilinear motion is a type of straight-line motion which is frequently encountered in practical applications. In this motion, the acceleration a of the particle is zero for every value of t. The velocity v is therefore constant, and Eq. (11.1) becomes

$$\frac{dx}{dt} = v = \text{constant}$$

The position coordinate x is obtained by integrating this equation. Denoting by x_0 the initial value of x, we write

$$\int_{x_0}^{x} dx = v \int_0^t dt$$

$$x - x_0 = vt$$

$$x = x_0 + vt \qquad (11.5)$$

This equation can be used *only if the velocity of the particle is known to be constant.*

11.5. UNIFORMLY ACCELERATED RECTILINEAR MOTION

Uniformly accelerated rectilinear motion is another common type of motion. In this motion, the acceleration a of the particle is constant, and Eq. (11.2) becomes

$$\frac{dv}{dt} = a = \text{constant}$$

The velocity v of the particle is obtained by integrating this equation:

$$\int_{v_0}^{v} dv = a \int_0^t dt$$

$$v - v_0 = at$$

$$v = v_0 + at \qquad (11.6)$$

where v_0 is the initial velocity. Substituting for v in (11.1), we write

$$\frac{dx}{dt} = v_0 + at$$

Denoting by x_0 the initial value of x and integrating, we have

$$\int_{x_0}^{x} dx = \int_0^t (v_0 + at)\, dt$$

$$x - x_0 = v_0 t + \tfrac{1}{2}at^2$$

$$x = x_0 + v_0 t + \tfrac{1}{2}at^2 \qquad (11.7)$$

We can also use Eq. (11.4) and write

$$v\frac{dv}{dx} = a = \text{constant}$$

$$v\, dv = a\, dx$$

Integrating both sides, we obtain

$$\int_{v_0}^{v} v\, dv = a \int_{x_0}^{x} dx$$

$$\tfrac{1}{2}(v^2 - v_0^2) = a(x - x_0)$$

$$v^2 = v_0^2 + 2a(x - x_0) \qquad (11.8)$$

The three equations we have derived provide useful relations among position coordinate, velocity, and time in the case of a uniformly accelerated motion, as soon as appropriate values have been substituted for a, v_0, and x_0. The origin O of the x axis should first be defined and a positive direction chosen along the axis; this direction will be used to determine the signs of a, v_0, and x_0. Equation (11.6) relates v and t and should be used when the value of v corresponding to a given value of t is desired, or inversely. Equation (11.7) relates x and t; Eq. (11.8) relates v and x. An important application of uniformly accelerated motion is the motion of a *freely falling body*. The acceleration of a freely falling body (usually denoted by g) is equal to 9.81 m/s^2 or 32.2 ft/s^2.

It is important to keep in mind that the three equations above can be used *only when the acceleration of the particle is known to be constant*. If the acceleration of the particle is variable, its motion should be determined from the fundamental equations (11.1) to (11.4) according to the methods outlined in Sec. 11.3.

11.6. MOTION OF SEVERAL PARTICLES

When several particles move independently along the same line, independent equations of motion can be written for each particle. Whenever possible, time should be recorded from the same initial instant for all particles, and displacements should be measured from the same origin and in the same direction. In other words, a single clock and a single measuring tape should be used.

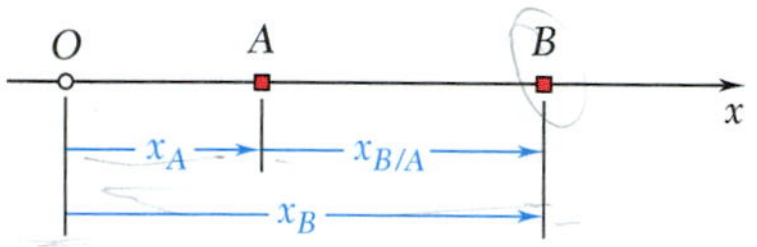

Fig. 11.7

Relative Motion of Two Particles. Consider two particles A and B moving along the same straight line (Fig. 11.7). If the position coordinates x_A and x_B are measured from the same origin, the difference $x_B - x_A$ defines the *relative position coordinate of B with respect to A* and is denoted by $x_{B/A}$. We write

$$x_{B/A} = x_B - x_A \qquad \text{or} \qquad x_B = x_A + x_{B/A} \qquad (11.9)$$

Regardless of the positions of A and B with respect to the origin, a positive sign for $x_{B/A}$ means that B is to the right of A, and a negative sign means that B is to the left of A.

The rate of change of $x_{B/A}$ is known as the *relative velocity of B with respect to A* and is denoted by $v_{B/A}$. Differentiating (11.9), we write

$$v_{B/A} = v_B - v_A \qquad \text{or} \qquad v_B = v_A + v_{B/A} \qquad (11.10)$$

A positive sign for $v_{B/A}$ means that B is *observed from A* to move in the positive direction; a negative sign means that it is observed to move in the negative direction.

The rate of change of $v_{B/A}$ is known as the *relative acceleration of B with respect to A* and is denoted by $a_{B/A}$. Differentiating (11.10), we obtain†

$$a_{B/A} = a_B - a_A \qquad \text{or} \qquad a_B = a_A + a_{B/A} \qquad (11.11)$$

†Note that the product of the subscripts A and B/A used in the right-hand member of Eqs. (11.9), (11.10), and (11.11) is equal to the subscript B used in their left-hand member.

Dependent Motions. Sometimes, the position of a particle will depend upon the position of another particle or of several other particles. The motions are then said to be *dependent*. For example, the position of block B in Fig. 11.8 depends upon the position of block A. Since the rope $ACDEFG$ is of constant length, and since the lengths of the portions of rope CD and EF wrapped around the pulleys remain constant, it follows that the sum of the lengths of the segments AC, DE, and FG is constant. Observing that the length of the segment AC differs from x_A only by a constant and that, similarly, the lengths of the segments DE and FG differ from x_B only by a constant, we write

$$x_A + 2x_B = \text{constant}$$

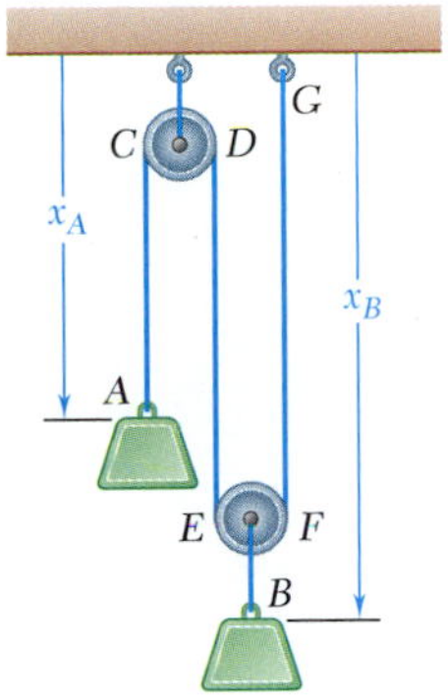

Fig. 11.8

Since only one of the two coordinates x_A and x_B can be chosen arbitrarily, we say that the system shown in Fig. 11.8 has *one degree of freedom*. From the relation between the position coordinates x_A and x_B, it follows that if x_A is given an increment Δx_A, that is, if block A is lowered by an amount Δx_A, the coordinate x_B will receive an increment $\Delta x_B = -\frac{1}{2}\Delta x_A$. In other words, block B will rise by half the same amount; this can easily be checked directly from Fig. 11.8.

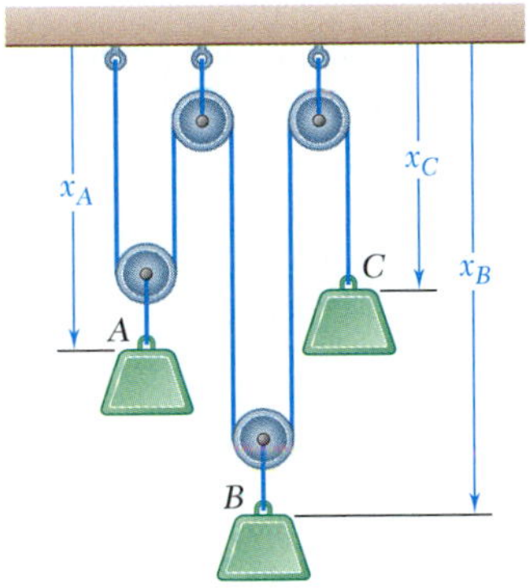

Fig. 11.9

In the case of the three blocks of Fig. 11.9, we can again observe that the length of the rope which passes over the pulleys is constant, and thus the following relation must be satisfied by the position coordinates of the three blocks:

$$2x_A + 2x_B + x_C = \text{constant}$$

Since two of the coordinates can be chosen arbitrarily, we say that the system shown in Fig. 11.9 has *two degrees of freedom*.

When the relation existing between the position coordinates of several particles is *linear*, a similar relation holds between the velocities and between the accelerations of the particles. In the case of the blocks of Fig. 11.9, for instance, we differentiate twice the equation obtained and write

$$2\frac{dx_A}{dt} + 2\frac{dx_B}{dt} + \frac{dx_C}{dt} = 0 \qquad \text{or} \qquad 2v_A + 2v_B + v_C = 0$$

$$2\frac{dv_A}{dt} + 2\frac{dv_B}{dt} + \frac{dv_C}{dt} = 0 \qquad \text{or} \qquad 2a_A + 2a_B + a_C = 0$$

SAMPLE PROBLEM 11.4

A ball is thrown vertically upward from the 12-m level in an elevator shaft with an initial velocity of 18 m/s. At the same instant an open-platform elevator passes the 5-m level, moving upward with a constant velocity of 2 m/s. Determine (*a*) when and where the ball will hit the elevator, (*b*) the relative velocity of the ball with respect to the elevator when the ball hits the elevator.

SOLUTION

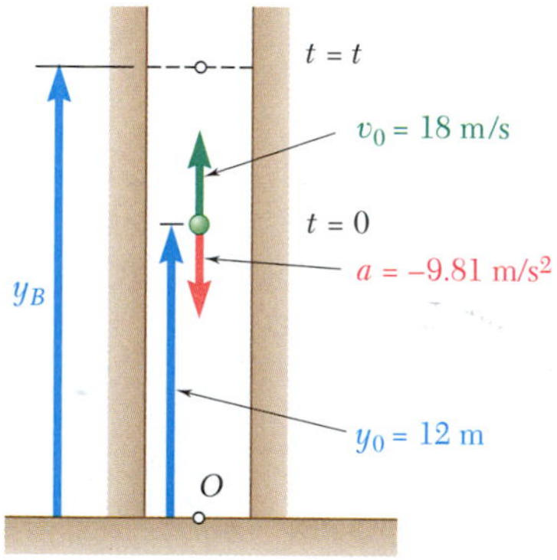

Motion of Ball. Since the ball has a constant acceleration, its motion is *uniformly accelerated.* Placing the origin O of the y axis at ground level and choosing its positive direction upward, we find that the initial position is $y_0 = +12$ m, the initial velocity is $v_0 = +18$ m/s, and the acceleration is $a = -9.81\text{ m/s}^2$. Substituting these values in the equations for uniformly accelerated motion, we write

$$v_B = v_0 + at \qquad v_B = 18 - 9.81t \qquad (1)$$

$$y_B = y_0 + v_0 t + \tfrac{1}{2}at^2 \qquad y_B = 12 + 18t - 4.905t^2 \qquad (2)$$

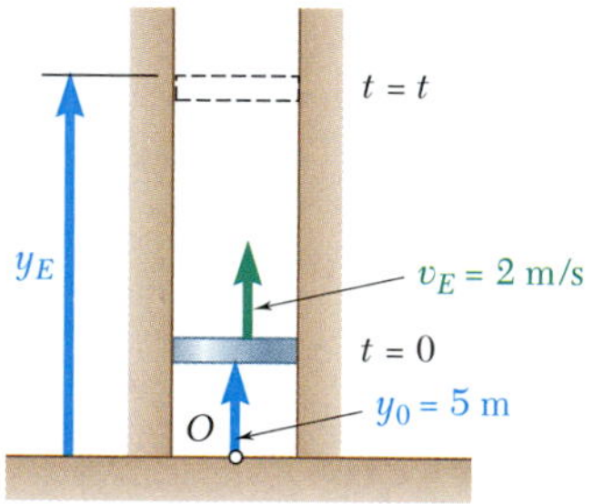

Motion of Elevator. Since the elevator has a constant velocity, its motion is *uniform.* Again placing the origin O at the ground level and choosing the positive direction upward, we note that $y_0 = +5$ m and write

$$v_E = +2\text{ m/s} \qquad (3)$$

$$y_E = y_0 + v_E t \qquad y_E = 5 + 2t \qquad (4)$$

Ball Hits Elevator. We first note that the same time t and the same origin O were used in writing the equations of motion of both the ball and the elevator. We see from the figure that when the ball hits the elevator,

$$y_E = y_B \qquad (5)$$

Substituting for y_E and y_B from (2) and (4) into (5), we have

$$5 + 2t = 12 + 18t - 4.905t^2$$

$$t = -0.39\text{ s} \qquad \text{and} \qquad t = 3.65\text{ s} \blacktriangleleft$$

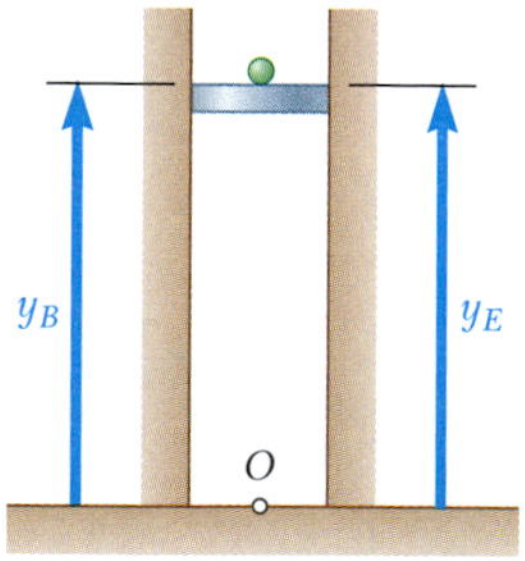

Only the root $t = 3.65$ s corresponds to a time after the motion has begun. Substituting this value into (4), we have

$$y_E = 5 + 2(3.65) = 12.30\text{ m}$$

$$\text{Elevation from ground} = 12.30\text{ m} \blacktriangleleft$$

The relative velocity of the ball with respect to the elevator is

$$v_{B/E} = v_B - v_E = (18 - 9.81t) - 2 = 16 - 9.81t$$

When the ball hits the elevator at time $t = 3.65$ s, we have

$$v_{B/E} = 16 - 9.81(3.65) \qquad v_{B/E} = -19.81\text{ m/s} \blacktriangleleft$$

The negative sign means that the ball is observed from the elevator to be moving in the negative sense (downward).

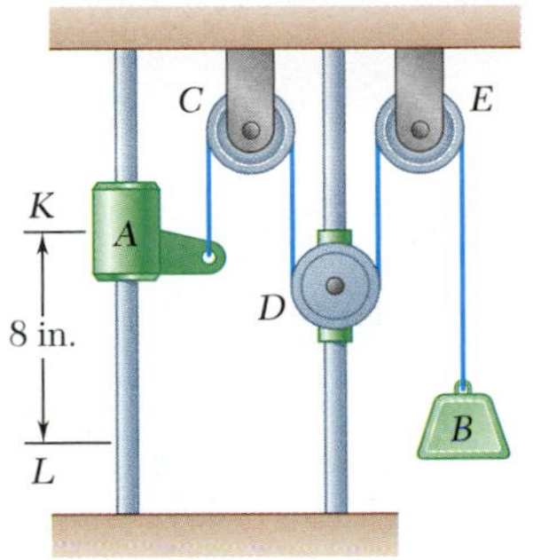

SAMPLE PROBLEM 11.5

Collar A and block B are connected by a cable passing over three pulleys C, D, and E as shown. Pulleys C and E are fixed, while D is attached to a collar which is pulled downward with a constant velocity of 3 in./s. At $t = 0$, collar A starts moving downward from position K with a constant acceleration and no initial velocity. Knowing that the velocity of collar A is 12 in./s as it passes through point L, determine the change in elevation, the velocity, and the acceleration of block B when collar A passes through L.

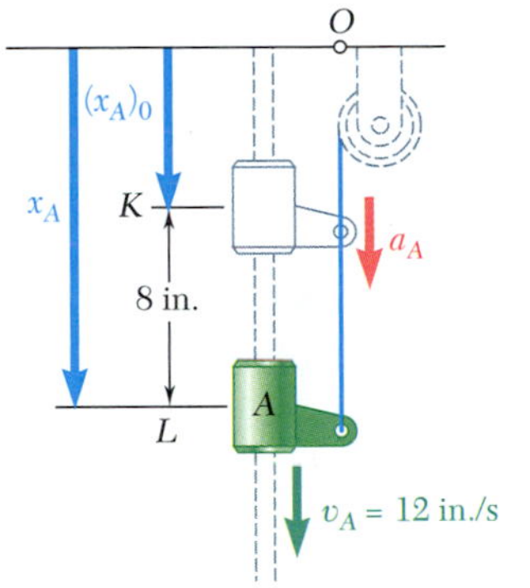

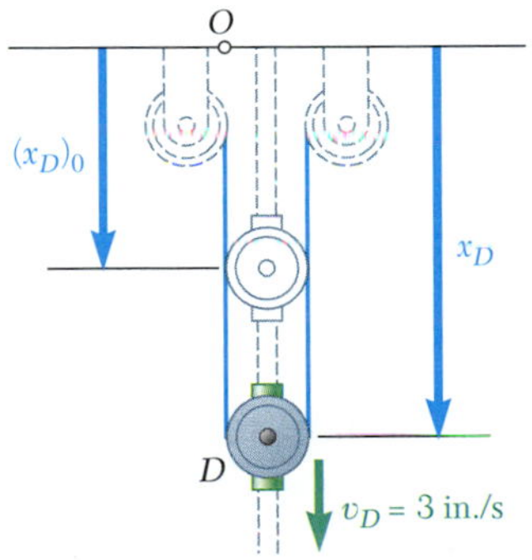

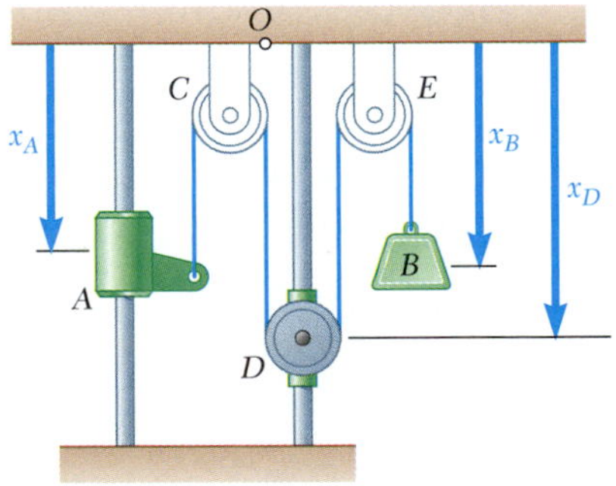

SOLUTION

Motion of Collar *A*. We place the origin O at the upper horizontal surface and choose the positive direction downward. We observe that when $t = 0$, collar A is at the position K and $(v_A)_0 = 0$. Since $v_A = 12$ in./s and $x_A - (x_A)_0 = 8$ in. when the collar passes through L, we write

$$v_A^2 = (v_A)_0^2 + 2a_A[x_A - (x_A)_0] \qquad (12)^2 = 0 + 2a_A(8)$$
$$a_A = 9 \text{ in./s}^2$$

The time at which collar A reaches point L is obtained by writing

$$v_A = (v_A)_0 + a_A t \qquad 12 = 0 + 9t \qquad t = 1.333 \text{ s}$$

Motion of Pulley *D*. Recalling that the positive direction is downward, we write

$$a_D = 0 \qquad v_D = 3 \text{ in./s} \qquad x_D = (x_D)_0 + v_D t = (x_D)_0 + 3t$$

When collar A reaches L, at $t = 1.333$ s, we have

$$x_D = (x_D)_0 + 3(1.333) = (x_D)_0 + 4$$

Thus,
$$x_D - (x_D)_0 = 4 \text{ in.}$$

Motion of Block *B*. We note that the total length of cable $ACDEB$ differs from the quantity $(x_A + 2x_D + x_B)$ only by a constant. Since the cable length is constant during the motion, this quantity must also remain constant. Thus, considering the times $t = 0$ and $t = 1.333$ s, we write

$$x_A + 2x_D + x_B = (x_A)_0 + 2(x_D)_0 + (x_B)_0 \qquad (1)$$
$$[x_A - (x_A)_0] + 2[x_D - (x_D)_0] + [x_B - (x_B)_0] = 0 \qquad (2)$$

But we know that $x_A - (x_A)_0 = 8$ in. and $x_D - (x_D)_0 = 4$ in.; substituting these values in (2), we find

$$8 + 2(4) + [x_B - (x_B)_0] = 0 \qquad x_B - (x_B)_0 = -16 \text{ in.}$$

Thus: Change in elevation of B = 16 in.↑ ◀

Differentiating (1) twice, we obtain equations relating the velocities and the accelerations of A, B, and D. Substituting for the velocities and accelerations of A and D at $t = 1.333$ s, we have

$$v_A + 2v_D + v_B = 0: \qquad 12 + 2(3) + v_B = 0$$
$$v_B = -18 \text{ in./s} \qquad v_B = 18 \text{ in./s}\uparrow \quad ◀$$

$$a_A + 2a_D + a_B = 0: \qquad 9 + 2(0) + a_B = 0$$
$$a_B = -9 \text{ in./s}^2 \qquad a_B = 9 \text{ in./s}^2\uparrow \quad ◀$$

SOLVING PROBLEMS ON YOUR OWN

In this lesson we derived the equations that describe *uniform rectilinear motion* (constant velocity) and *uniformly accelerated rectilinear motion* (constant acceleration). We also introduced the concept of *relative motion.* The equations for relative motion [Eqs. (11.9) to (11.11)] can be applied to the independent or dependent motions of any two particles moving along the same straight line.

A. *Independent motion of one or more particles.* The solution of problems of this type should be organized as follows:

1. *Begin your solution* by listing the given information, sketching the system, and selecting the origin and the positive direction of the coordinate axis [Sample Prob. 11.4]. It is always advantageous to have a visual representation of problems of this type.

2. *Write the equations* that describe the motions of the various particles as well as those that describe how these motions are related [Eq. (5) of Sample Prob. 11.4].

3. *Define the initial conditions,* that is, specify the state of the system corresponding to $t = 0$. This is especially important if the motions of the particles begin at different times. In such cases, either of two approaches can be used.

a. Let $t = 0$ be the time when the last particle begins to move. You must then determine the initial position x_0 and the initial velocity v_0 of each of the other particles.

b. Let $t = 0$ be the time when the first particle begins to move. You must then, in each of the equations describing the motion of another particle, replace t with $t - t_0$, where t_0 is the time at which that specific particle begins to move. It is important to recognize that the equations obtained in this way are valid only for $t \geq t_0$.

B. Dependent motion of two or more particles. In problems of this type the particles of the system are connected to each other, typically by ropes or by cables. The method of solution of these problems is similar to that of the preceding group of problems, except that it will now be necessary to describe the *physical connections* between the particles. In the following problems, the connection is provided by one or more cables. For each cable, you will have to write equations similar to the last three equations of Sec. 11.6. We suggest that you use the following procedure:

1. Draw a sketch of the system and select a coordinate system, indicating clearly a positive sense for each of the coordinate axes. For example, in Sample Prob. 11.5 lengths are measured downward from the upper horizontal support. It thus follows that those displacements, velocities, and accelerations which have positive values are directed downward.

2. Write the equation describing the constraint imposed by each cable on the motion of the particles involved. Differentiating this equation twice, you will obtain the corresponding relations among velocities and accelerations.

3. If several directions of motion are involved, you must select a coordinate axis and a positive sense for each of these directions. You should also try to locate the origins of your coordinate axes so that the equations of constraints will be as simple as possible. For example, in Sample Prob. 11.5 it is easier to define the various coordinates by measuring them downward from the upper support than by measuring them upward from the bottom support.

Finally, keep in mind that the method of analysis described in this lesson and the corresponding equations can be used only for particles moving with *uniform or uniformly accelerated rectilinear motion.*

Problems

11.33 An airplane begins its take-off run at A with zero velocity and a constant acceleration a. Knowing that it becomes airborne 30 s later at B and that the distance AB is 900 m, determine (*a*) the acceleration a, (*b*) the take-off velocity v_B.

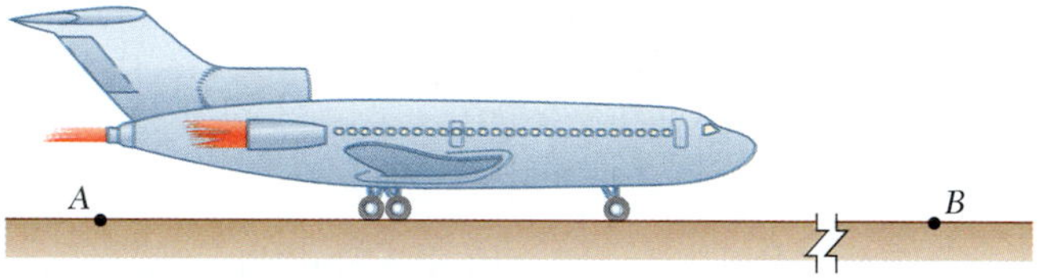

Fig. P11.33

11.34 Steep safety ramps are built beside mountain highways to enable vehicles with defective brakes to stop safely. A truck enters a 250-m ramp at a high speed v_0 and travels 180 m in 6 s at constant deceleration before its speed is reduced to $v_0/2$. Assuming the same constant deceleration, determine (*a*) the additional time required for the truck to stop, (*b*) the additional distance traveled by the truck.

Fig. P11.34

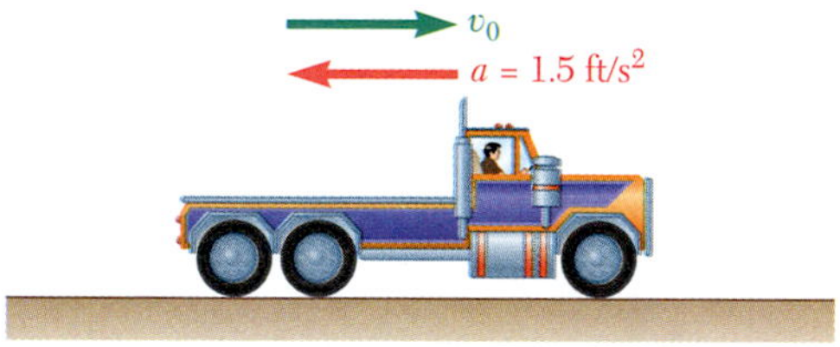

Fig. P11.35

11.35 A truck travels 540 ft in 8 s while being decelerated at a constant rate of 1.5 ft/s^2. Determine (*a*) its initial velocity, (*b*) its final velocity, (*c*) the distance traveled during the first 0.6 s.

11.36 A motorist enters a freeway at 25 mi/h and accelerates uniformly to 65 mi/h. From the odometer in the car, the motorist knows that she traveled 0.1 mi while accelerating. Determine (*a*) the acceleration of the car, (*b*) the time required to reach 65 mi/h.

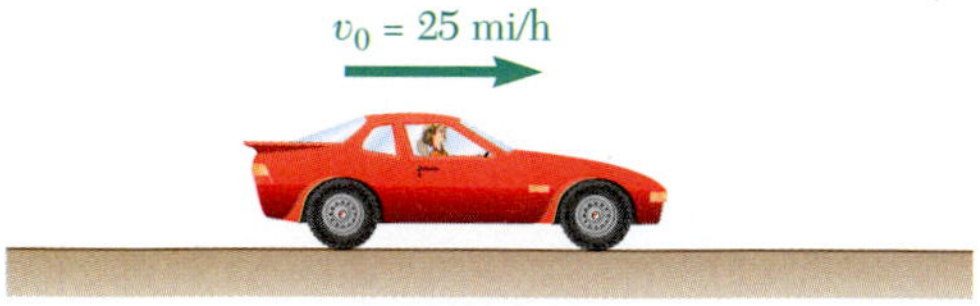

Fig. P11.36

11.37 A group of students launches a model rocket in the vertical direction. Based on tracking data, they determine that the altitude of the rocket was 27.5 m at the end of the powered portion of the flight and that the rocket landed 16 s later. Knowing that the descent parachute failed to deploy so that the rocket fell freely to the ground after reaching its maximum altitude and assuming that $g = 9.81 \text{ m/s}^2$, determine (*a*) the speed v_1 of the rocket at the end of powered flight, (*b*) the maximum altitude reached by the rocket.

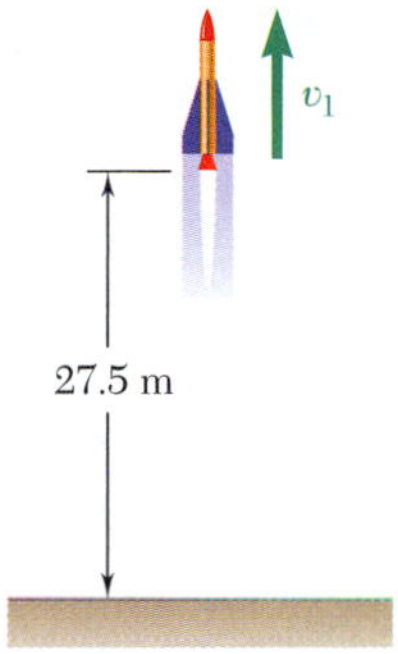

Fig. *P11.37*

11.38 A sprinter in a 400-m race accelerates uniformly for the first 130 m and then runs with constant velocity. If the sprinter's time for the first 130 m is 25 s, determine (*a*) his acceleration, (*b*) his final velocity, (*c*) his time for the race.

11.39 In a close harness race, horse 2 passes horse 1 at point *A*, where the two velocities are $v_2 = 7$ m/s and $v_1 = 6.8$ m/s. Horse 1 later passes horse 2 at point *B* and goes on to win the race at point *C*, 400 m from *A*. The elapsed times from *A* to *C* for horse 1 and horse 2 are $t_1 = 61.5$ s and $t_2 = 62.0$ s, respectively. Assuming uniform accelerations for both horses between *A* and *C*, determine (*a*) the distance from *A* to *B*, (*b*) the position of horse 1 relative to horse 2 when horse 1 reaches the finish line *C*.

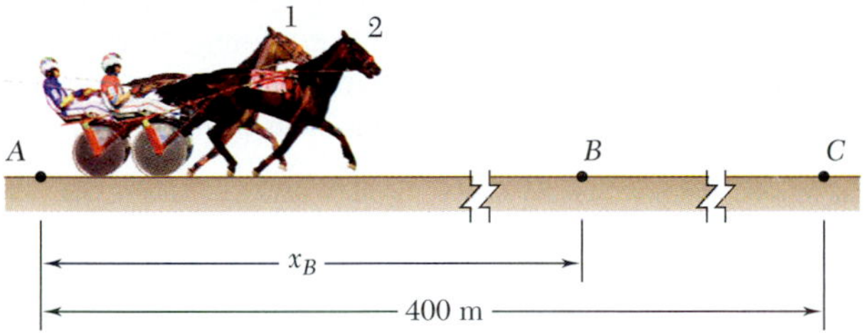

Fig. P11.39

11.40 Two rockets are launched at a fireworks performance. Rocket *A* is launched with an initial velocity v_0 and rocket *B* is launched 4 s later with the same initial velocity. The two rockets are timed to explode simultaneously at a height of 80 m, as *A* is falling and *B* is rising. Assuming a constant acceleration $g = 9.81 \text{ m/s}^2$, determine (*a*) the initial velocity v_0, (*b*) the velocity of *B* relative to *A* at the time of the explosion.

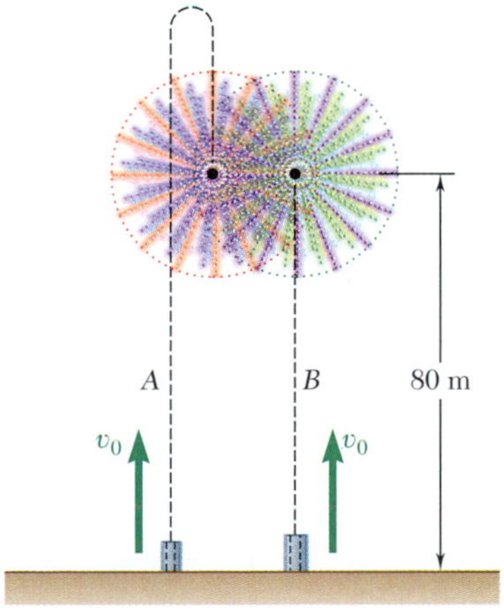

Fig. P11.40

11.41 A police officer in a patrol car parked in a 65 mi/h speed zone observes a passing automobile traveling at a slow, constant speed. Believing that the driver of the automobile might be intoxicated, the officer starts his car, accelerates uniformly to 85 mi/h in 8 s, and, maintaining a constant velocity of 85 mi/h, overtakes the motorist 42 s after the automobile passed him. Knowing that 18 s elapsed before the officer began pursuing the motorist, determine (*a*) the distance the officer traveled before overtaking the motorist, (*b*) the motorist's speed.

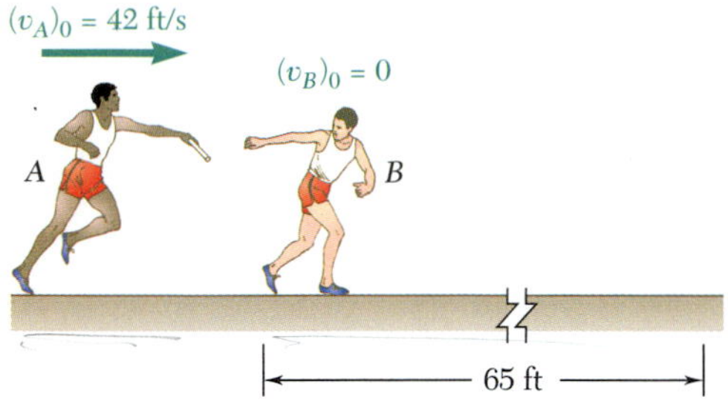

Fig. P11.42

11.42 As relay runner *A* enters the 65-ft-long exchange zone with a speed of 42 ft/s, he begins to slow down. He hands the baton to runner *B* 1.82 s later as they leave the exchange zone with the same velocity. Determine (*a*) the uniform acceleration of each of the runners, (*b*) when runner *B* should begin to run.

11.43 In a boat race, boat *A* is leading boat *B* by 38 m and both boats are traveling at a constant speed of 168 km/h. At $t = 0$, the boats accelerate at constant rates. Knowing that when *B* passes *A*, $t = 8$ s and $v_A = 228$ km/h, determine (*a*) the acceleration of *A*, (*b*) the acceleration of *B*.

11.44 Car *A* is parked along the northbound lane of a highway, and car *B* is traveling in the southbound lane at a constant speed of 96 km/h. At $t = 0$, *A* starts and accelerates at a constant rate a_A, while at $t = 5$ s, *B* begins to slow down with a constant deceleration of magnitude $a_A/6$. Knowing that when the cars pass each other $x = 90$ m and $v_A = v_B$, determine (*a*) the acceleration a_A, (*b*) when the vehicles pass each other, (*c*) the distance between the vehicles at $t = 0$.

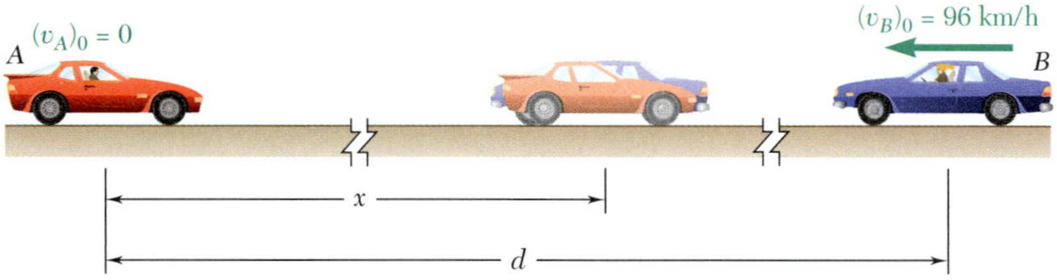

Fig. P11.44

11.45 Two automobiles *A* and *B* traveling in the same direction in adjacent lanes are stopped at a traffic signal. As the signal turns green, automobile *A* accelerates at a constant rate of 6.5 ft/s^2. Two seconds later, automobile *B* starts and accelerates at a constant rate of 11.7 ft/s^2. Determine (*a*) when and where *B* will overtake *A*, (*b*) the speed of each automobile at that time.

11.46 Two automobiles *A* and *B* are approaching each other in adjacent highway lanes. At $t = 0$, *A* and *B* are 0.62 mi apart, their speeds are $v_A = 68$ mi/h and $v_B = 39$ mi/h, and they are at points *P* and *Q*, respectively. Knowing that *A* passes point *Q* 40 s after *B* was there and that *B* passes point *P* 42 s after *A* was there, determine (*a*) the uniform accelerations of *A* and *B*, (*b*) when the vehicles pass each other, (*c*) the speed of *B* at that time.

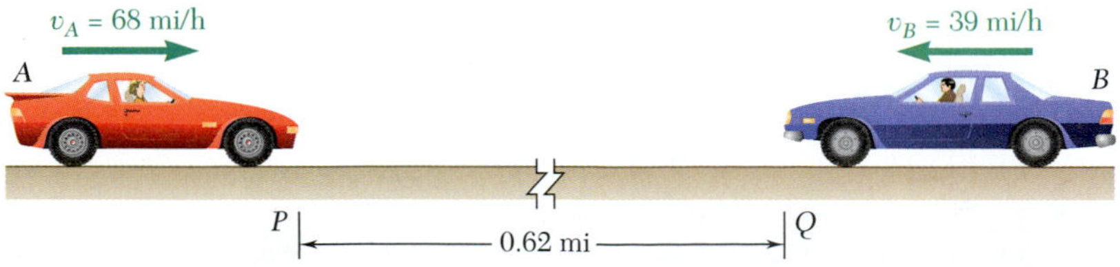

Fig. *P11.46*

11.47 Block A moves down with a constant velocity of 2 ft/s. Determine (*a*) the velocity of block C, (*b*) the velocity of collar B relative to block A, (*c*) the relative velocity of portion D of the cable with respect to block A.

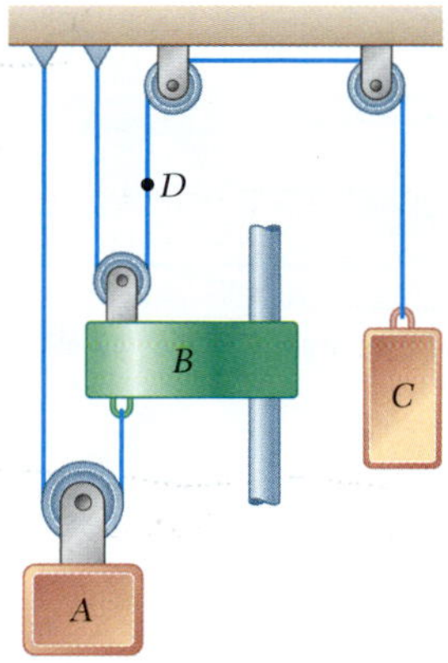

Fig. P11.47 and P11.48

11.48 Block C starts from rest and moves down with a constant acceleration. Knowing that after block A has moved 1.5 ft its velocity is 0.6 ft/s, determine (*a*) the accelerations of A and C, (*b*) the velocity and the change in position of block B after 2 s.

11.49 Block C starts from rest and moves downward with a constant acceleration. Knowing that after 12 s the velocity of block A is 456 mm/s, determine (*a*) the accelerations of A, B, and C, (*b*) the velocity and the change in position of block B after 8 s.

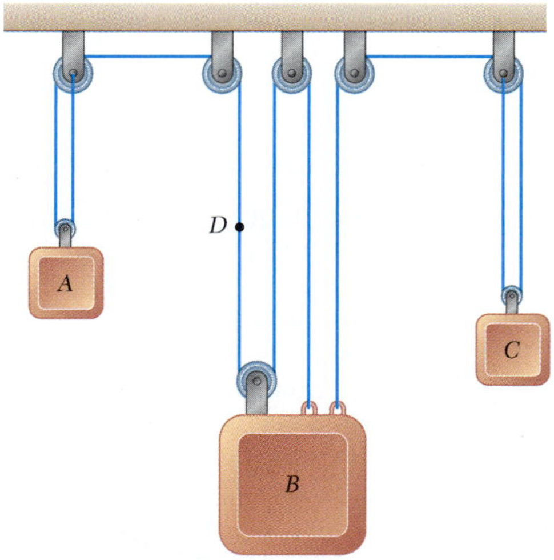

Fig. P11.49 and P11.50

11.50 Block B moves downward with a constant velocity of 610 mm/s. Determine (*a*) the velocity of block A, (*b*) the velocity of block C, (*c*) the velocity of portion D of the cable, (*d*) the relative velocity of portion D of the cable with respect to block B.

11.51 In the position shown, collar B moves to the left with a constant velocity of 300 mm/s. Determine (a) the velocity of collar A, (b) the velocity of portion C of the cable, (c) the relative velocity of portion C of the cable with respect to collar B.

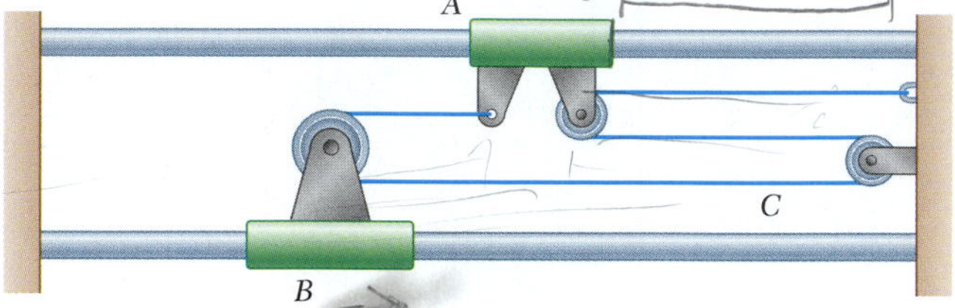

Fig. P11.51 and P11.52

11.52 Collar A starts from rest and moves to the right with a constant acceleration. Knowing that after 8 s the relative velocity of collar B with respect to collar A is 610 mm/s, determine (a) the accelerations of A and B, (b) the velocity and the change in position of B after 6 s.

11.53 At the instant shown, slider block B is moving to the right with a constant acceleration, and its speed is 6 in./s. Knowing that after slider block A has moved 10 in. to the right its velocity is 2.4 in./s, determine (a) the accelerations of A and B, (b) the acceleration of portion D of the cable, (c) the velocity and the change in position of slider block B after 4 s.

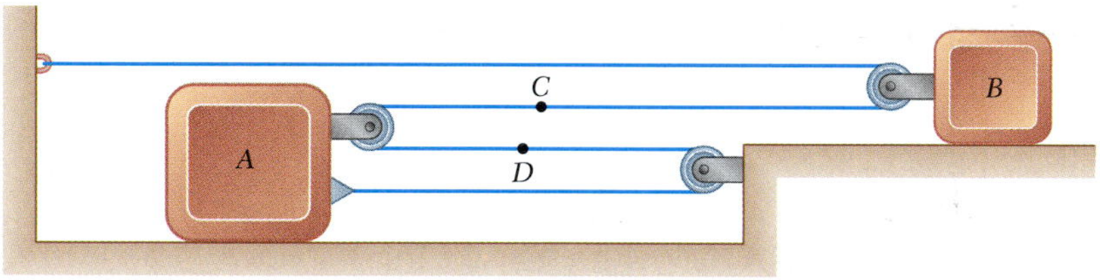

Fig. *P11.53* and *P11.54*

11.54 Slider block B moves to the right with a constant velocity of 12 in./s. Determine (a) the velocity of slider block A, (b) the velocity of portion C of the cable, (c) the velocity of portion D of the cable, (d) the relative velocity of portion C of the cable with respect to slider block A.

11.55 Collars A and B start from rest, and collar A moves upward with an acceleration of $3t^2$ mm/s^2. Knowing that collar B moves downward with a constant acceleration and that its velocity is 200 mm/s after moving 800 mm, determine (a) the acceleration of block C, (b) the distance through which block C will have moved after 3 s.

11.56 Collar A starts from rest at $t = 0$ and moves downward with a constant acceleration of 180 mm/s^2. Collar B moves upward with a constant acceleration, and its initial velocity is 200 mm/s. Knowing that collar B moves through 500 mm between $t = 0$ and $t = 2$ s, determine (a) the accelerations of collar B and block C, (b) the time at which the velocity of block C is zero, (c) the distance through which block C will have moved at that time.

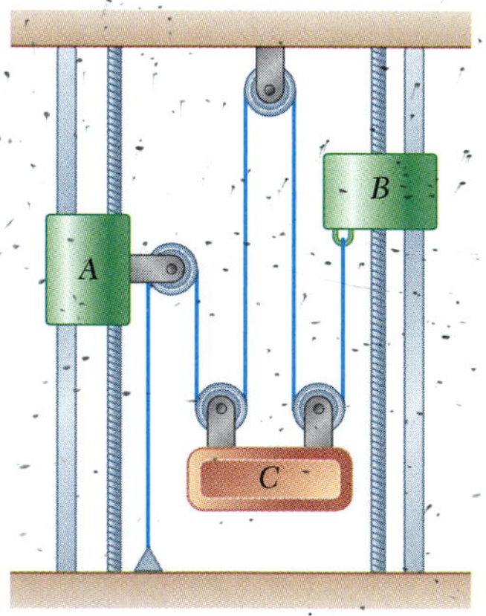

Fig. P11.55 and P11.56

11.57 Slider block B moves to the left with a constant velocity of 2 in./s. At $t = 0$, slider block A is moving to the right with a constant acceleration and a velocity of 4 in./s. Knowing that at $t = 2$ s slider block C has moved 1.5 in. to the right, determine (a) the velocity of slider block C at $t = 0$, (b) the velocity of portion D of the cable at $t = 0$, (c) the accelerations of A and C.

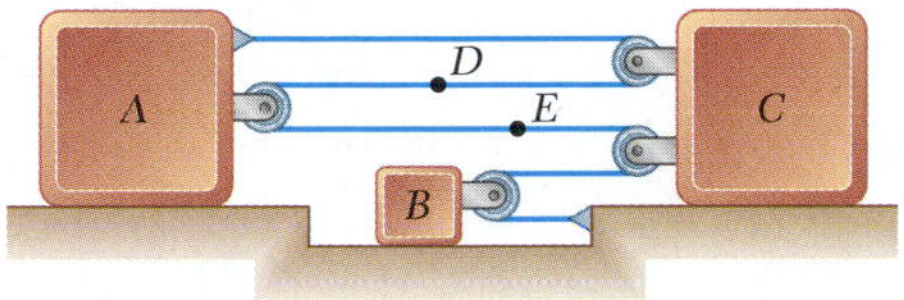

Fig. P11.57 and P11.58

11.58 Slider block A starts with an initial velocity at $t = 0$ and a constant acceleration of 9 in./s^2 to the right. Slider block C starts from rest at $t = 0$ and moves to the right with constant acceleration. Knowing that at $t = 2$ s, the velocities of A and B are 14 in./s to the right and 1 in./s to the left, respectively, determine (a) the accelerations of B and C, (b) the initial velocities of A and B, (c) the initial velocity of portion E of the cable.

***11.59** The system shown starts from rest, and the length of the upper cord is adjusted so that A, B, and C are initially at the same level. Each component moves with a constant acceleration. Knowing that when the relative velocity of collar B with respect to block A is 40 mm/s downward, the displacements of A and B are 80 mm downward and 160 mm downward, respectively, determine (a) the accelerations of A and B, (b) the change in position of block D when the velocity of block C is 300 mm/s upward.

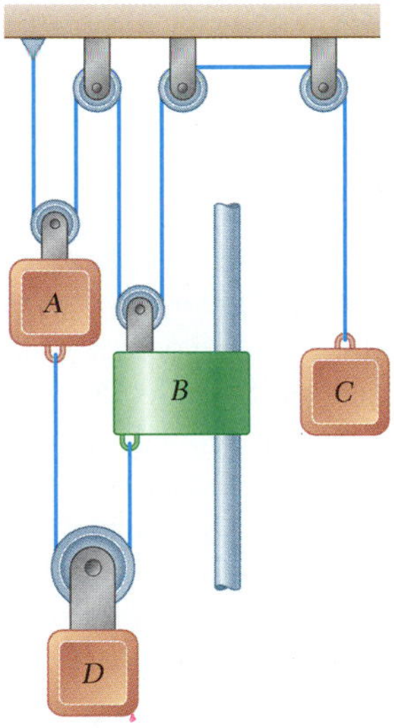

Fig. *P11.59* and *P11.60*

11.60 The system shown starts from rest, and each component moves with a constant acceleration. If the relative acceleration of block C with respect to collar B is 120 mm/s^2 upward and the relative acceleration of block D with respect to block A is 220 mm/s^2 downward, determine (a) the velocity of block C after 6 s, (b) the change in position of block D after 10 s.

*11.7. GRAPHICAL SOLUTION OF RECTILINEAR-MOTION PROBLEMS

It was observed in Sec. 11.2 that the fundamental formulas

$$v = \frac{dx}{dt} \qquad \text{and} \qquad a = \frac{dv}{dt}$$

have a geometrical significance. The first formula expresses that the velocity at any instant is equal to the slope of the x–t curve at the same instant (Fig. 11.10). The second formula expresses that the acceleration is equal to the slope of the v–t curve. These two properties can be used to determine graphically the v–t and a–t curves of a motion when the x–t curve is known.

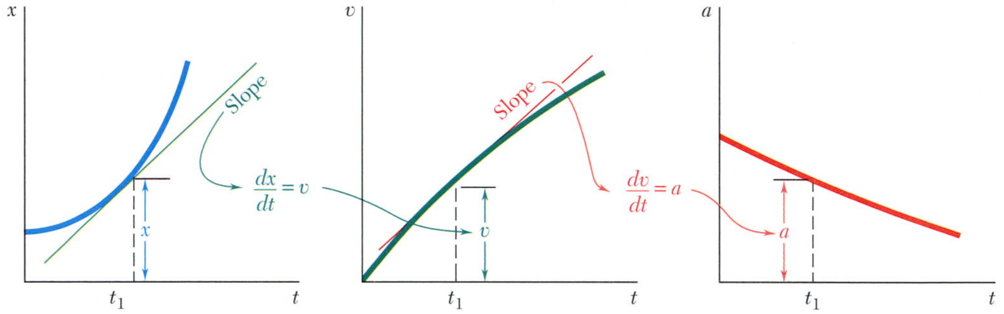

Fig. 11.10

Integrating the two fundamental formulas from a time t_1 to a time t_2, we write

$$x_2 - x_1 = \int_{t_1}^{t_2} v\,dt \qquad \text{and} \qquad v_2 - v_1 = \int_{t_1}^{t_2} a\,dt \qquad (11.12)$$

The first formula expresses that the area measured under the v–t curve from t_1 to t_2 is equal to the change in x during that time interval (Fig. 11.11). Similarly, the second formula expresses that the area measured under the a–t curve from t_1 to t_2 is equal to the change in v during that time interval. These two properties can be used to determine graphically the x–t curve of a motion when its v–t curve or its a–t curve is known (see Sample Prob. 11.6).

Graphical solutions are particularly useful when the motion considered is defined from experimental data and when x, v, and a are not analytical functions of t. They can also be used to advantage when the motion consists of distinct parts and when its analysis requires writing a different equation for each of its parts. When using a graphical solution, however, one should be careful to note that (1) the area under the v–t curve measures the *change in* x, not x itself, and similarly, that the area under the a–t curve measures the change in v; (2) an area above the t axis corresponds to an *increase* in x or v, while an area located below the t axis measures a *decrease* in x or v.

It will be useful to remember in drawing motion curves that if the velocity is constant, it will be represented by a horizontal straight line; the position coordinate x will then be a linear function of t and will be represented by an oblique straight line. If the acceleration is

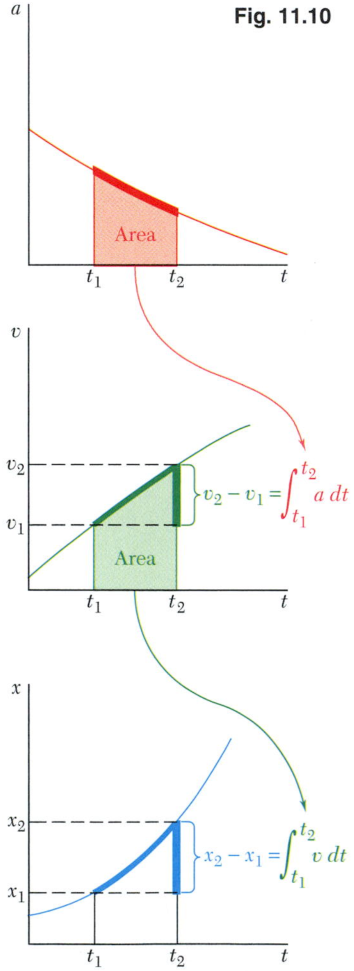

Fig. 11.11

constant and different from zero, it will be represented by a horizontal straight line; v will then be a linear function of t, represented by an oblique straight line, and x will be expressed as a second-degree polynomial in t, represented by a parabola. If the acceleration is a linear function of t, the velocity and the position coordinate will be equal, respectively, to second-degree and third-degree polynomials; a will then be represented by an oblique straight line, v by a parabola, and x by a cubic. In general, if the acceleration is a polynomial of degree n in t, the velocity will be a polynomial of degree $n + 1$ and the position coordinate a polynomial of degree $n + 2$; these polynomials are represented by motion curves of a corresponding degree.

*11.8. OTHER GRAPHICAL METHODS

An alternative graphical solution can be used to determine the position of a particle at a given instant directly from the a–t curve. Denoting the values of x and v at $t = 0$ as x_0 and v_0 and their values at $t = t_1$ as x_1 and v_1, and observing that the area under the v–t curve can be divided into a rectangle of area $v_0 t_1$ and horizontal differential elements of area $(t_1 - t)\,dv$ (Fig. 11.12*a*), we write

$$x_1 - x_0 = \text{area under } v\text{–}t \text{ curve} = v_0 t_1 + \int_{v_0}^{v_1} (t_1 - t)\,dv$$

Substituting $dv = a\,dt$ in the integral, we obtain

$$x_1 - x_0 = v_0 t_1 + \int_0^{t_1} (t_1 - t)\,a\,dt$$

Referring to Fig. 11.12*b*, we note that the integral represents the first moment of the area under the a–t curve with respect to the line $t = t_1$ bounding the area on the right. This method of solution is known, therefore, as the *moment-area method*. If the abscissa $\bar{t}$ of the centroid C of the area is known, the position coordinate x_1 can be obtained by writing

$$x_1 = x_0 + v_0 t_1 + (\text{area under } a\text{–}t \text{ curve})(t_1 - \bar{t}) \qquad (11.13)$$

If the area under the a–t curve is a composite area, the last term in (11.13) can be obtained by multiplying each component area by the distance from its centroid to the line $t = t_1$. Areas above the t axis should be considered as positive and areas below the t axis as negative.

Another type of motion curve, the v–x curve, is sometimes used. If such a curve has been plotted (Fig. 11.13), the acceleration a can be obtained at any time by drawing the normal AC to the curve and *measuring the subnormal BC*. Indeed, observing that the angle between AC and AB is equal to the angle θ between the horizontal and the tangent at A (the slope of which is $\tan\theta = dv/dx$), we write

$$BC = AB \tan\theta = v\frac{dv}{dx}$$

and thus, recalling formula (11.4),

$$BC = a$$

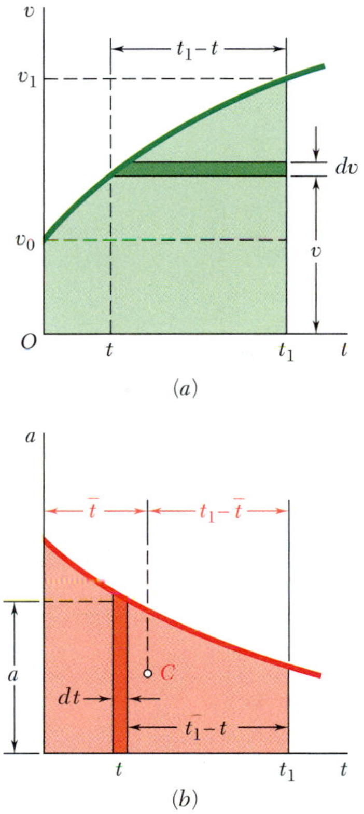

Fig. 11.12

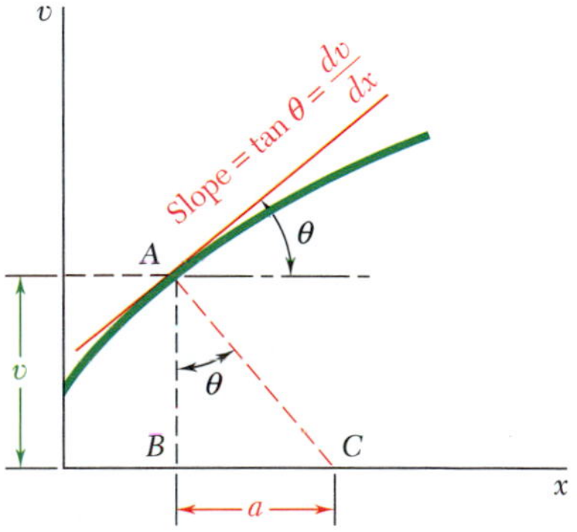

Fig. 11.13

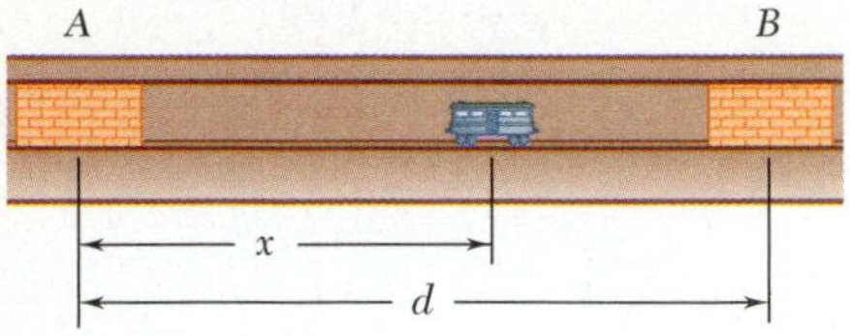

SAMPLE PROBLEM 11.6

A subway car leaves station A; it gains speed at the rate of 4 ft/s^2 for 6 s and then at the rate of 6 ft/s^2 until it has reached the speed of 48 ft/s. The car maintains the same speed until it approaches station B; brakes are then applied, giving the car a constant deceleration and bringing it to a stop in 6 s. The total running time from A to B is 40 s. Draw the a–t, v–t, and x–t curves, and determine the distance between stations A and B.

SOLUTION

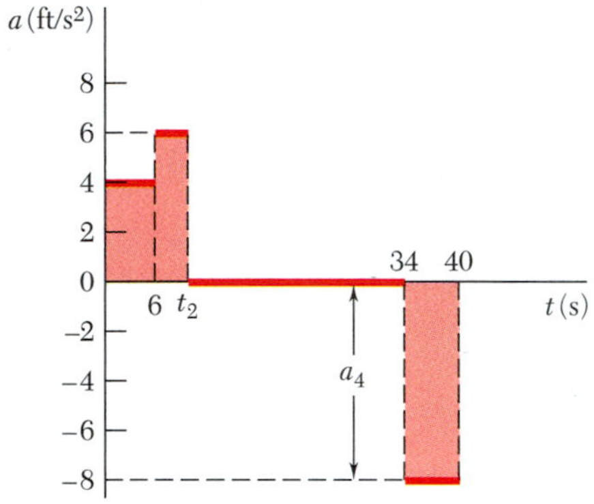

Acceleration-Time Curve. Since the acceleration is either constant or zero, the a–t curve is made of horizontal straight-line segments. The values of t_2 and a_4 are determined as follows:

$0 < t < 6$: Change in v = area under a–t curve

$$v_6 - 0 = (6\text{ s})(4\text{ ft/s}^2) = 24\text{ ft/s}$$

$6 < t < t_2$: Since the velocity increases from 24 to 48 ft/s,

$$\text{Change in } v = \text{area under } a\text{–}t \text{ curve}$$

$$48\text{ ft/s} - 24\text{ ft/s} = (t_2 - 6)(6\text{ ft/s}^2) \qquad t_2 = 10\text{ s}$$

$t_2 < t < 34$: Since the velocity is constant, the acceleration is zero.

$34 < t < 40$: Change in v = area under a–t curve

$$0 - 48\text{ ft/s} = (6\text{ s})a_4 \qquad a_4 = -8\text{ ft/s}^2$$

The acceleration being negative, the corresponding area is below the t axis; this area represents a decrease in velocity.

Velocity-Time Curve. Since the acceleration is either constant or zero, the v–t curve is made of straight-line segments connecting the points determined above.

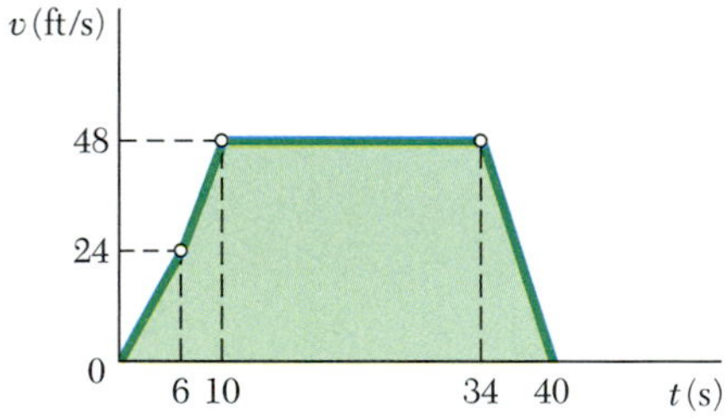

Change in x = area under v–t curve

$0 < t < 6$: $\quad x_6 - 0 = \frac{1}{2}(6)(24) = 72\text{ ft}$

$6 < t < 10$: $\quad x_{10} - x_6 = \frac{1}{2}(4)(24 + 48) = 144\text{ ft}$

$10 < t < 34$: $\quad x_{34} - x_{10} = (24)(48) = 1152\text{ ft}$

$34 < t < 40$: $\quad x_{40} - x_{34} = \frac{1}{2}(6)(48) = 144\text{ ft}$

Adding the changes in x, we obtain the distance from A to B:

$$d = x_{40} - 0 = 1512\text{ ft}$$

$d = 1512\text{ ft}$ ◀

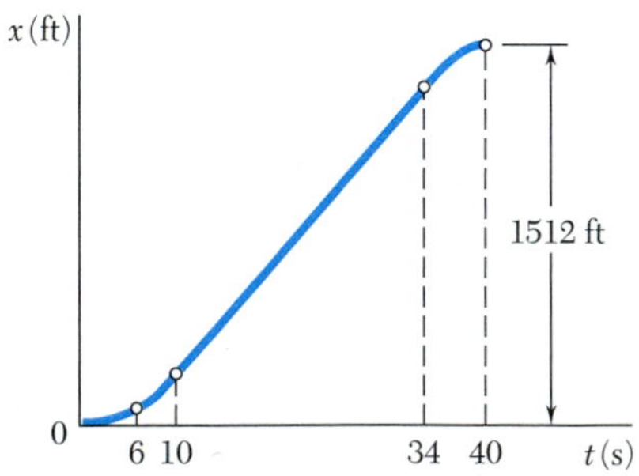

Position-Time Curve. The points determined above should be joined by three arcs of parabola and one straight-line segment. In constructing the x–t curve, keep in mind that for any value of t the slope of the tangent to the x–t curve is equal to the value of v at that instant.

SOLVING PROBLEMS ON YOUR OWN

In this lesson (Secs. 11.7 and 11.8), we reviewed and developed several *graphical techniques for the solution of problems involving rectilinear motion*. These techniques can be used to solve problems directly or to complement analytical methods of solution by providing a visual description, and thus a better understanding, of the motion of a given body. We suggest that you sketch one or more motion curves for several of the problems in this lesson, even if these problems are not part of your homework assignment.

1. *Drawing x–t, v–t, and a–t curves and applying graphical methods.* The following properties were indicated in Sec. 11.7 and should be kept in mind as you use a graphical method of solution.

a. *The slopes of the x–t and v–t curves* at a time t_1 are respectively equal to the *velocity* and the *acceleration* at time t_1.

b. *The areas under the a–t and v–t curves* between the times t_1 and t_2 are respectively equal to the change Δv in the velocity and to the change Δx in the position coordinate during that time interval.

c. *If one of the motion curves is known,* the fundamental properties we have summarized in paragraphs *a* and *b* will enable you to construct the other two curves. However, when using the properties of paragraph *b*, the velocity and the position coordinate at time t_1 must be known in order to determine the velocity and the position coordinate at time t_2. Thus, in Sample Prob. 11.6, knowing that the initial value of the velocity was zero allowed us to find the velocity at $t = 6$ s: $v_6 = v_0 + \Delta v = 0 + 24 \text{ ft/s} = 24 \text{ ft/s}$.

If you have previously studied the shear and bending-moment diagrams for a beam, you should recognize the analogy that exists between the three motion curves and the three diagrams representing respectively the distributed load, the shear, and the bending moment in the beam. Thus, any techniques that you learned regarding the construction of these diagrams can be applied when drawing the motion curves.

2. *Using approximate methods.* When the a–t and v–t curves are not represented by analytical functions or when they are based on experimental data, it is often necessary to use approximate methods to calculate the areas under these curves. In those cases, the given area is approximated by a series of rectangles of width Δt. The smaller the value of Δt, the smaller the error introduced by the approximation. The velocity and the position coordinate are obtained by writing

$$v = v_0 + \Sigma a_{\text{ave}}\, \Delta t \qquad x = x_0 + \Sigma v_{\text{ave}}\, \Delta t$$

where a_{ave} and v_{ave} are the heights of an acceleration rectangle and a velocity rectangle, respectively.

(continued)

3. *Applying the moment-area method.* This graphical technique is used when the a–t curve is given and the change in the position coordinate is to be determined. We found in Sec. 11.8 that the position coordinate x_1 can be expressed as

$$x_1 = x_0 + v_0 t_1 + (\text{area under } a\text{–}t \text{ curve})(t_1 - \bar{t}) \qquad (11.13)$$

Keep in mind that when the area under the a–t curve is a composite area, the same value of t_1 should be used for computing the contribution of each of the component areas.

4. *Determining the acceleration from a v–x curve.* You saw in Sec. 11.8 that it is possible to determine the acceleration from a v–x curve by direct measurement. It is important to note, however, that this method is applicable only if the same linear scale is used for the v and x axes (for example, 1 in. = 10 ft and 1 in. = 10 ft/s). When this condition is not satisfied, the acceleration can still be determined from the equation

$$a = v\frac{dv}{dx}$$

where the slope dv/dx is obtained as follows: First, draw the tangent to the curve at the point of interest. Next, using appropriate scales, measure along that tangent corresponding increments Δx and Δv. The desired slope is equal to the ratio $\Delta v/\Delta x$.

Problems

11.61 A particle moves in a straight line with a constant acceleration of -4 ft/s^2 for 6 s, zero acceleration for the next 4 s, and a constant acceleration of $+4$ ft/s^2 for the next 4 s. Knowing that the particle starts from the origin and that its velocity is -8 ft/s during the zero acceleration time interval, (*a*) construct the v–t and x–t curves for $0 \leq t \leq 14$ s, (*b*) determine the position and the velocity of the particle and the total distance traveled when $t = 14$ s.

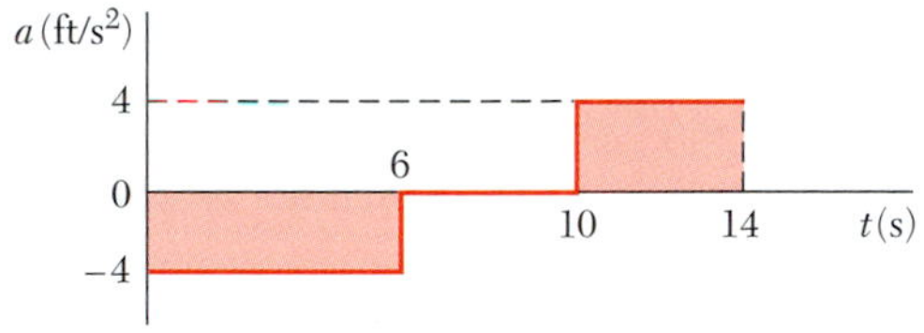

Fig. P11.61 and P11.62

11.62 A particle moves in a straight line with a constant acceleration of -4 ft/s^2 for 6 s, zero acceleration for the next 4 s, and a constant acceleration of $+4$ ft/s^2 for the next 4 s. Knowing that the particle starts from the origin with $v_0 = 16$ ft/s, (*a*) construct the v–t and x–t curves for $0 \leq t \leq 14$ s, (*b*) determine the amount of time during which the particle is further than 16 ft from the origin.

11.63 A particle moves in a straight line with the velocity shown in the figure. Knowing that $x = -540$ m at $t = 0$, (*a*) construct the a–t and x–t curves for $0 < t < 50$ s, and determine (*b*) the total distance traveled by the particle when $t = 50$ s, (*c*) the two times at which $x = 0$.

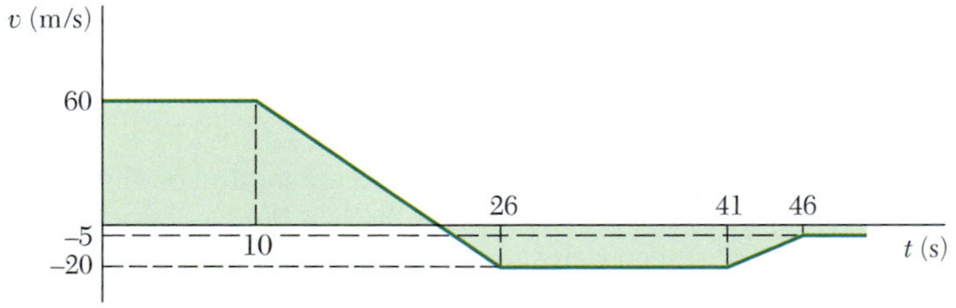

Fig. P11.63 and P11.64

11.64 A particle moves in a straight line with the velocity shown in the figure. Knowing that $x = -540$ m at $t = 0$, (*a*) construct the a–t and x–t curves for $0 < t < 50$ s, and determine (*b*) the maximum value of the position coordinate of the particle, (*c*) the values of t for which the particle is at $x = 100$ m.

11.65 A parachutist is in free fall at a rate of 180 ft/s when he opens his parachute at an altitude of 1900 ft. Following a rapid and constant deceleration, he then descends at a constant rate of 44 ft/s from 1800 ft to 100 ft, where he maneuvers the parachute into the wind to further slow his descent. Knowing that the parachutist lands with a negligible downward velocity, determine (*a*) the time required for the parachutist to land after opening his parachute, (*b*) the initial deceleration.

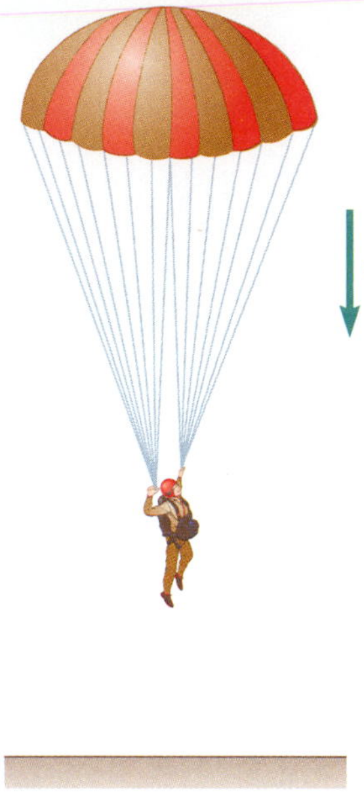

Fig. P11.65

11.66 A machine component is spray-painted while it is mounted on a pallet that travels 12 ft in 20 s. The pallet has an initial velocity of 3 in./s and can be accelerated at a maximum rate of 2 in./s^2. Knowing that the painting process requires 15 s to complete and is performed as the pallet moves with a constant speed, determine the smallest possible value of the maximum speed of the pallet.

11.67 In a 400-m race, runner A reaches her maximum velocity v_A in 4 s with constant acceleration and maintains that velocity until she reaches the halfway point with a split time of 25 s. Runner B reaches her maximum velocity v_B in 5 s with constant acceleration and maintains that velocity until she reaches the halfway point with a split time of 25.2 s. Both runners then run the second half of the race with the same constant deceleration of 0.1 m/s^2. Determine (*a*) the race times for both runners, (*b*) the position of the winner relative to the loser when the winner reaches the finish line.

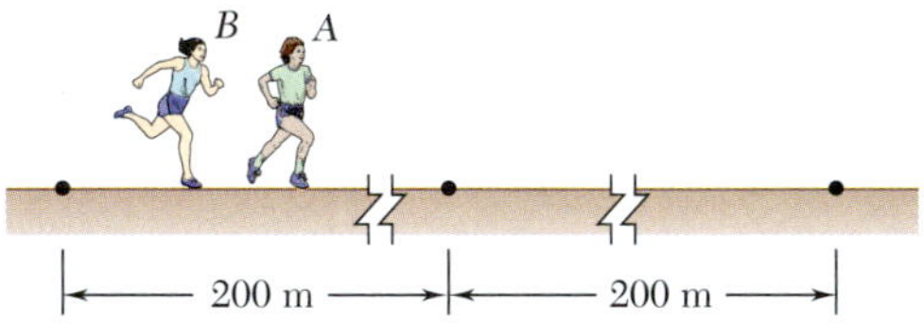

Fig. *P11.67*

11.68 A commuter train traveling at 64 km/h is 4.8 km from a station. The train then decelerates so that its speed is 32 km/h when it is 800 m from the station. Knowing that the train arrives at the station 7.5 min after beginning to decelerate and assuming constant decelerations, determine (*a*) the time required to travel the first 4 km, (*b*) the speed of the train as it arrives at the station, (*c*) the final constant deceleration of the train.

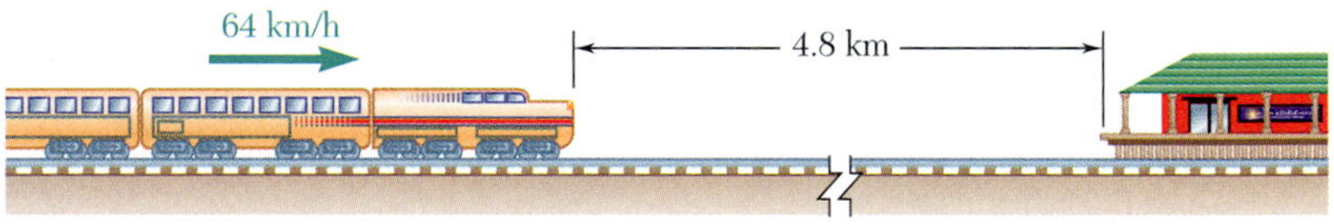

Fig. *P11.68*

11.69 Two road rally checkpoints A and B are located on the same highway and are 8 mi apart. The speed limits for the first 5 mi and the last 3 mi are 60 mi/h and 35 mi/h, respectively. Drivers must stop at each checkpoint, and the specified time between points A and B is 10 min 20 s. Knowing that a driver accelerates and decelerates at the same constant rate, determine the magnitude of her acceleration if she travels at the speed limit as much as possible.

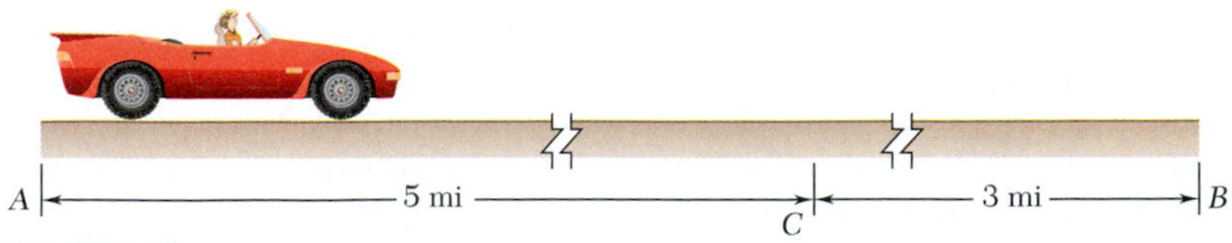

Fig. P11.69

11.70 In a water-tank test involving the launching of a small model boat, the model's initial horizontal velocity is 20 ft/s and its horizontal acceleration varies linearly from -40 ft/s^2 at $t = 0$ to -6 ft/s^2 at $t = t_1$ and then remains equal to -6 ft/s^2 until $t = 1.4$ s. Knowing that $v = 6$ ft/s when $t = t_1$, determine (*a*) the value of t_1, (*b*) the velocity and position of the model at $t = 1.4$ s.

11.71 A bus is parked along the side of a highway when it is passed by a truck traveling at a constant speed of 70 km/h. Two minutes later, the bus starts and accelerates until it reaches a speed of 100 km/h, which it then maintains. Knowing that 12 min after the truck passed the bus, the bus is 1.2 km ahead of the truck, determine (*a*) when and where the bus passed the truck, (*b*) the uniform acceleration of the bus.

11.72 Cars *A* and *B* are $d = 60$ m apart and are traveling respectively at the constant speeds of $(v_A)_0 = 32$ km/h and $(v_B)_0 = 24$ km/h on an ice-covered road. Knowing that 45 s after driver *A* applies his brakes to avoid overtaking car *B* the two cars collide, determine (*a*) the uniform deceleration of car *A*, (*b*) the relative velocity of car *A* with respect to car *B* when they collide.

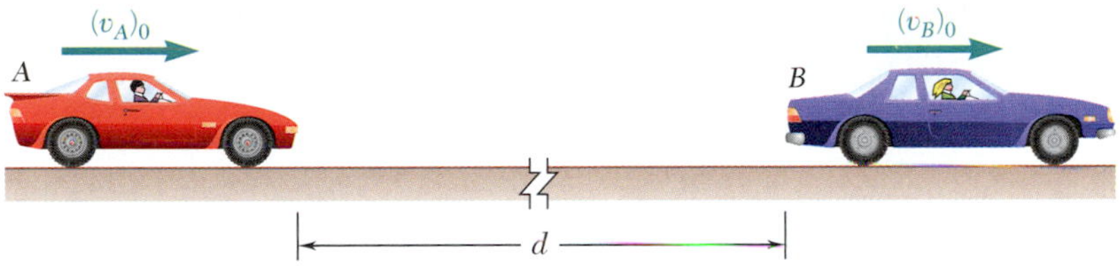

Fig. P11.72 and *P11.73*

11.73 Cars *A* and *B* are traveling respectively at the constant speeds of $(v_A)_0 = 22$ mi/h and $(v_B)_0 = 13$ mi/h on an ice-covered road. To avoid overtaking car *B*, the driver of car *A* applies his brakes so that his car decelerates at a constant rate of 0.14 ft/s^2. Determine the distance *d* between the cars at which the driver of car *A* must apply his brakes to just avoid colliding with car *B*.

11.74 An elevator starts from rest and moves upward, accelerating at a rate of 4 ft/s^2 until it reaches a speed of 24 ft/s, which it then maintains. Two seconds after the elevator begins to move, a man standing 40 ft above the initial position of the top of the elevator throws a ball upward with an initial velocity of 64 ft/s. Determine when the ball will hit the elevator.

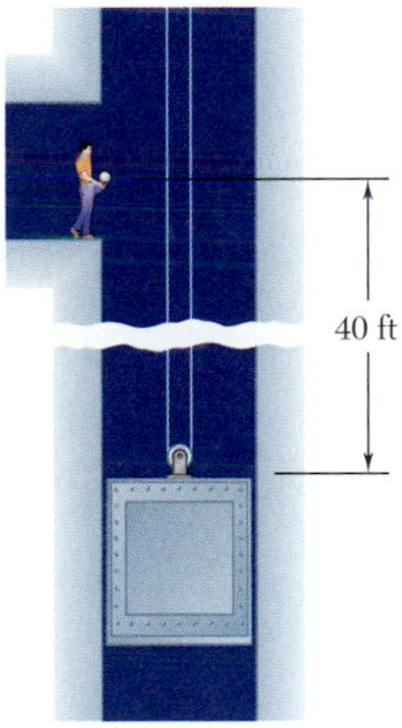

Fig. *P11.74*

11.75 Car *A* is traveling at 70 km/h when it enters a 50 km/h speed zone. The driver of car *A* decelerates at a rate of 5 m/s^2 until reaching a speed of 50 km/h, which she then maintains. When car *B*, which was initially 19 m behind car *A* and traveling with a constant speed of 72 km/h, enters the speed zone, its driver decelerates at a rate of 6 m/s^2 until reaching a speed of 48 km/h. Knowing that the driver of car *B* maintains a speed of 48 km/h, determine (*a*) the closest that car *B* comes to car *A*, (*b*) the time at which car *A* is 22 m in front of car *B*.

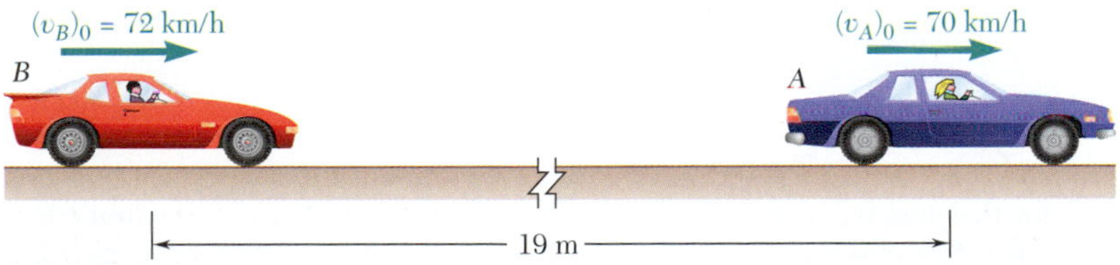

Fig. P11.75

11.76 Car A is traveling on a highway at a constant speed $(v_A)_0 =$ 100 km/h and is 120 m from the entrance of an access ramp when car B enters the acceleration lane at that point at a speed $(v_B)_0 = 25$ km/h. Car B accelerates uniformly and enters the main traffic lane after traveling 70 m in 5 s. It then continues to accelerate at the same rate until it reaches a speed of 100 km/h, which it then maintains. Determine the final distance between the two cars.

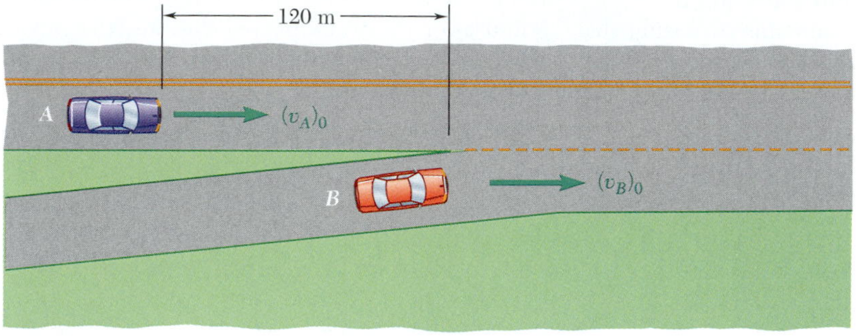

Fig. P11.76

11.77 During a manufacturing process, a conveyor belt starts from rest and travels a total of 0.36 m before temporarily coming to rest. Knowing that the jerk, or rate of change of acceleration, is limited to ± 1.5 m/s^2 per second, determine (*a*) the shortest time required for the belt to move 0.36 m, (*b*) the maximum and average values of the velocity of the belt during that time.

11.78 An airport shuttle train travels between two terminals that are 5 km apart. To maintain passenger comfort, the acceleration of the train is limited to ± 1.25 m/s^2, and the jerk, or rate of change of acceleration, is limited to ± 0.25 m/s^2 per second. If the shuttle has a maximum speed of 32 km/h, determine (*a*) the shortest time for the shuttle to travel between the two terminals, (*b*) the corresponding average velocity of the shuttle.

11.79 An elevator starts from rest and rises 125 ft to its maximum velocity in T s with the acceleration record shown in the figure. Determine (*a*) the required time T, (*b*) the maximum velocity, (*c*) the velocity and position of the elevator at $t = T/2$.

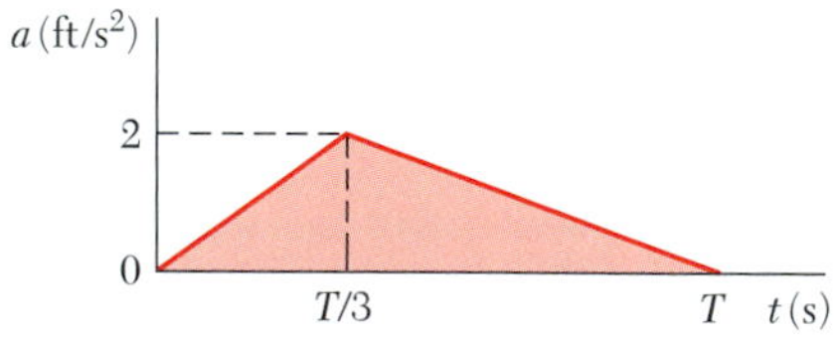

Fig. *P11.79*

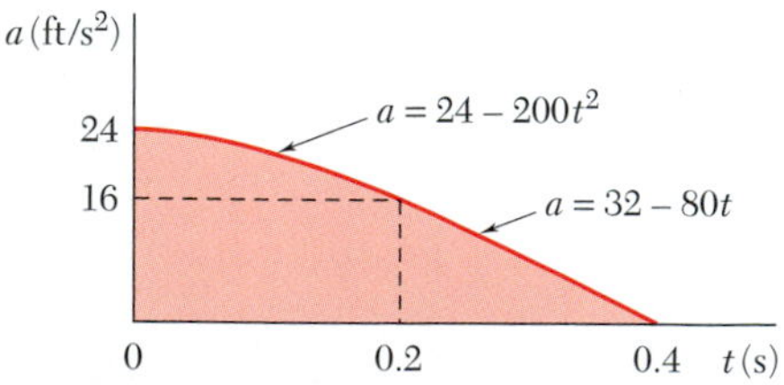

Fig. *P11.80*

11.80 An accelerometer record for the motion of a given part of a mechanism is approximated by an arc of a parabola for 0.2 s and a straight line for the next 0.2 s as shown in the figure. Knowing that $v = 0$ when $t = 0$ and $x = 0.8$ ft when $t = 0.4$ s, (*a*) construct the v–t curve for $0 \le t \le 0.4$ s, (*b*) determine the position of the part at $t = 0.3$ s and $t = 0.2$ s.

11.81 Two seconds are required to bring the piston rod of an air cylinder to rest; the acceleration record of the piston rod during the 2 s is as shown. Determine by approximate means (*a*) the initial velocity of the piston rod, (*b*) the distance traveled by the piston rod as it is brought to rest.

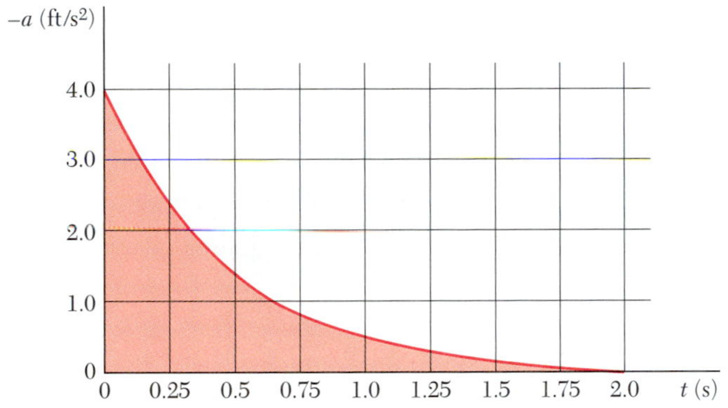

Fig. P11.81

11.82 The acceleration record shown was obtained during the speed trials of a sports car. Knowing that the car starts from rest, determine by approximate means (*a*) the velocity of the car at $t = 8$ s, (*b*) the distance the car traveled at $t = 20$ s.

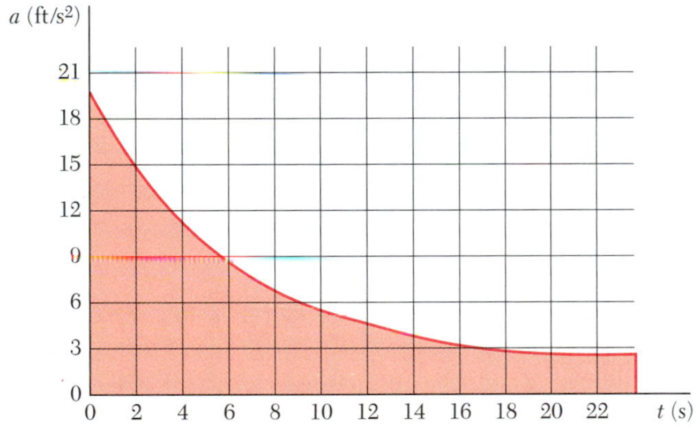

Fig. P11.82

11.83 A training airplane has a velocity of 32 m/s when it lands on an aircraft carrier. As the arresting gear of the carrier brings the airplane to rest, the velocity and the acceleration of the airplane are recorded; the results are shown (solid curve) in the figure. Determine by approximate means (*a*) the time required for the airplane to come to rest, (*b*) the distance traveled in that time.

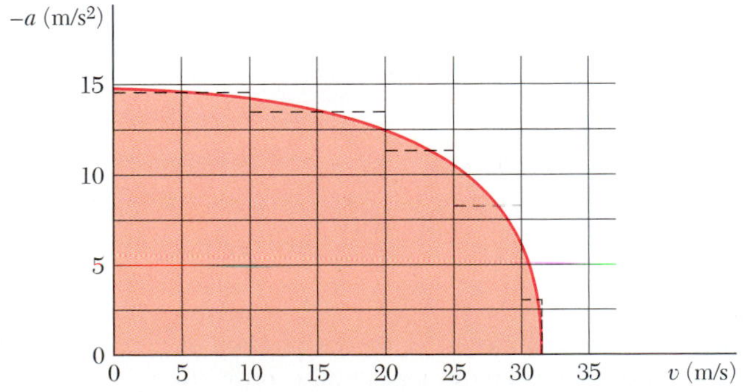

Fig. P11.83

11.84 Shown in the figure is a portion of the experimentally determined v–x curve for a shuttle cart. Determine by approximate means the acceleration of the cart (*a*) when $x = 0.25$ m, (*b*) when $v = 2$ m/s.

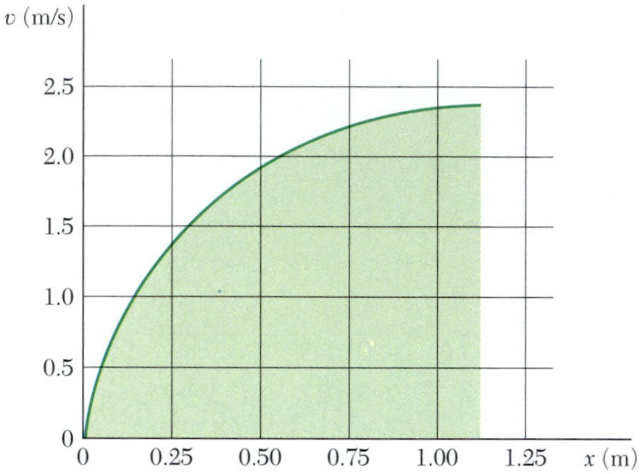

Fig. P11.84

11.85 Using the method of Sec. 11.8, derive the formula $x = x_0 + v_0t + \frac{1}{2}at^2$ for the position coordinate of a particle in uniformly accelerated rectilinear motion.

11.86 Using the method of Sec. 11.8, determine the position of the particle of Prob. 11.61 when $t = 12$ s.

11.87 An automobile begins a braking test with a velocity of 30 m/s at $t = 0$ and comes to a stop at $t = t_1$ with the acceleration record shown. Knowing that the area under the a–t curve from $t = 0$ to $t = T$ is a semiparabolic area, use the method of Sec. 11.8 to determine the distance traveled by the automobile before coming to a stop if (*a*) $T = 0.2$ s, (*b*) $T = 0.8$ s.

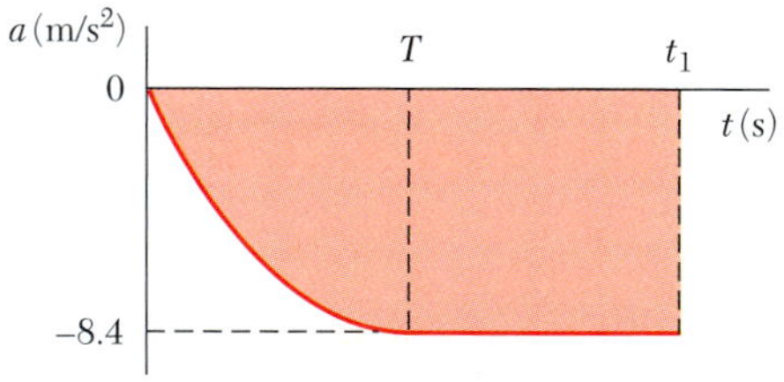

Fig. *P11.87*

11.88 For the particle of Prob. 11.63, draw the a–t curve and determine, using the method of Sec 11.8, (*a*) the position of the particle when $t = 52$ s, (*b*) the maximum value of its position coordinate.

11.9. POSITION VECTOR, VELOCITY, AND ACCELERATION

When a particle moves along a curve other than a straight line, we say that the particle is in *curvilinear motion*. To define the position P occupied by the particle at a given time t, we select a fixed reference system, such as the x, y, z axes shown in Fig. 11.14*a*, and draw the vector **r** joining the origin O and point P. Since the vector **r** is characterized by its magnitude r and its direction with respect to the reference axes, it completely defines the position of the particle with respect to those axes; the vector **r** is referred to as the *position vector* of the particle at time t.

Consider now the vector **r**′ defining the position P' occupied by the same particle at a later time $t + \Delta t$. The vector $\Delta\mathbf{r}$ joining P and P' represents the change in the position vector during the time interval Δt since, as we can easily check from Fig. 11.14*a*, the vector **r**′ is obtained by adding the vectors **r** and $\Delta\mathbf{r}$ according to the triangle rule. We note that $\Delta\mathbf{r}$ represents a change in *direction* as well as a change in *magnitude* of the position vector **r**. The *average velocity* of the particle over the time interval Δt is defined as the quotient of $\Delta\mathbf{r}$ and Δt. Since $\Delta\mathbf{r}$ is a vector and Δt is a scalar, the quotient $\Delta\mathbf{r}/\Delta t$ is a vector attached at P, of the same direction as $\Delta\mathbf{r}$ and of magnitude equal to the magnitude of $\Delta\mathbf{r}$ divided by Δt (Fig. 11.14*b*).

The *instantaneous velocity* of the particle at time t is obtained by choosing shorter and shorter time intervals Δt and, correspondingly, shorter and shorter vector increments $\Delta\mathbf{r}$. The instantaneous velocity is thus represented by the vector

$$\mathbf{v} = \lim_{\Delta t \to 0} \frac{\Delta\mathbf{r}}{\Delta t} \tag{11.14}$$

As Δt and $\Delta\mathbf{r}$ become shorter, the points P and P' get closer; the vector **v** obtained in the limit must therefore be tangent to the path of the particle (Fig. 11.14*c*).

Since the position vector **r** depends upon the time t, we can refer to it as a *vector function* of the scalar variable t and denote it by $\mathbf{r}(t)$. Extending the concept of derivative of a scalar function introduced in elementary calculus, we will refer to the limit of the quotient $\Delta\mathbf{r}/\Delta t$ as the *derivative* of the vector function $\mathbf{r}(t)$. We write

$$\mathbf{v} = \frac{d\mathbf{r}}{dt} \tag{11.15}$$

The magnitude v of the vector **v** is called the *speed* of the particle. It can be obtained by substituting for the vector $\Delta\mathbf{r}$ in formula (11.14) the magnitude of this vector represented by the straight-line segment PP'. But the length of the segment PP' approaches the length Δs of the arc PP' as Δt decreases (Fig. 11.14*a*), and we can write

$$v = \lim_{\Delta t \to 0} \frac{PP'}{\Delta t} = \lim_{\Delta t \to 0} \frac{\Delta s}{\Delta t} \qquad v = \frac{ds}{dt} \tag{11.16}$$

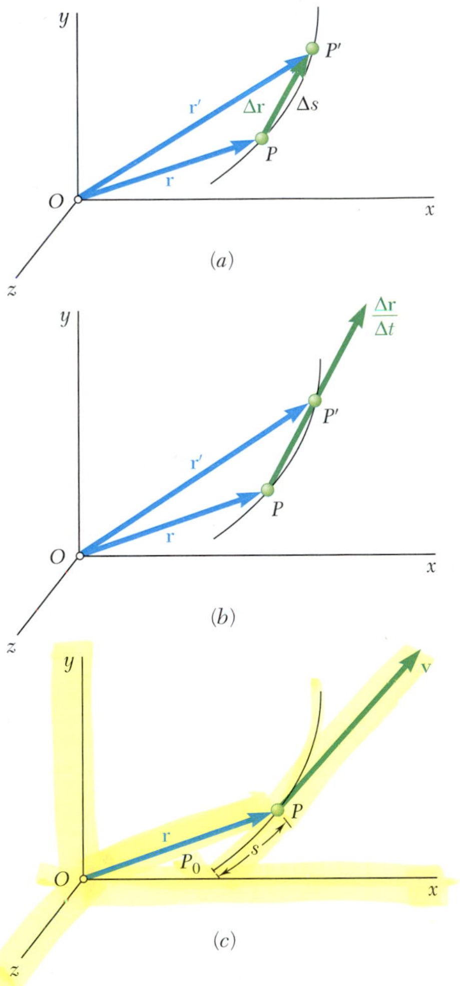

Fig. 11.14

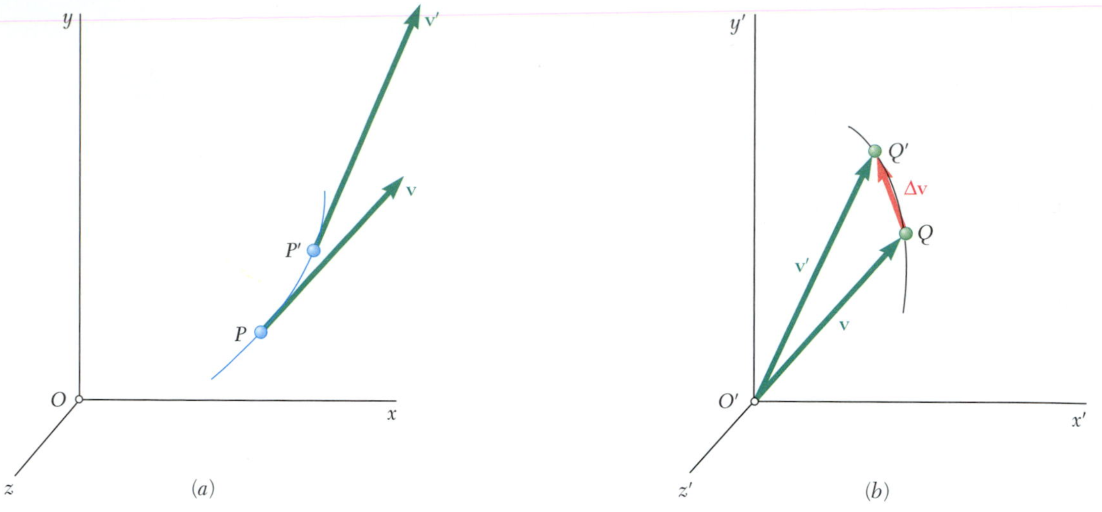

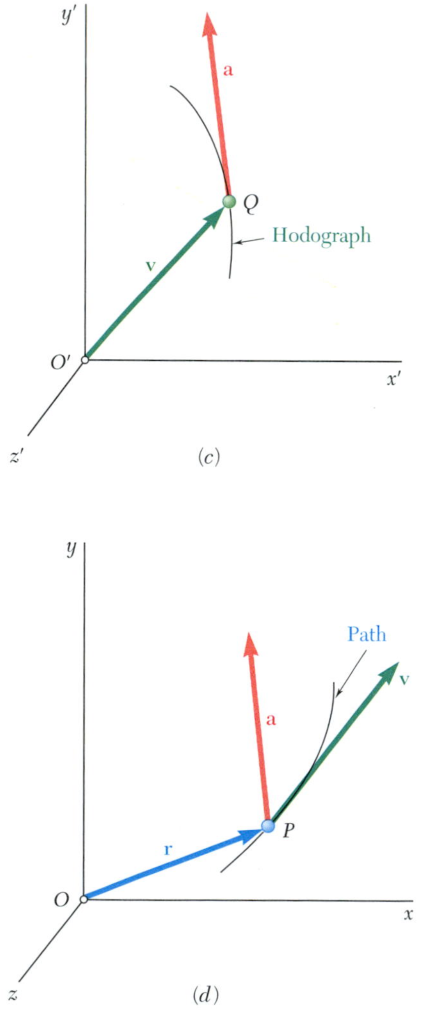

Fig. 11.15

The speed v can thus be obtained by differentiating with respect to t the length s of the arc described by the particle.

Consider the velocity $\mathbf{v}$ of the particle at time t and its velocity $\mathbf{v}'$ at a later time $t + \Delta t$ (Fig. 11.15*a*). Let us draw both vectors $\mathbf{v}$ and $\mathbf{v}'$ from the same origin O' (Fig. 11.15*b*). The vector $\Delta\mathbf{v}$ joining Q and Q' represents the change in the velocity of the particle during the time interval Δt, since the vector $\mathbf{v}'$ can be obtained by adding the vectors $\mathbf{v}$ and $\Delta\mathbf{v}$. We should note that $\Delta\mathbf{v}$ represents a change in the *direction* of the velocity as well as a change in *speed*. The *average acceleration* of the particle over the time interval Δt is defined as the quotient of $\Delta\mathbf{v}$ and Δt. Since $\Delta\mathbf{v}$ is a vector and Δt a scalar, the quotient $\Delta\mathbf{v}/\Delta t$ is a vector of the same direction as $\Delta\mathbf{v}$.

The *instantaneous acceleration* of the particle at time t is obtained by choosing smaller and smaller values for Δt and $\Delta\mathbf{v}$. The instantaneous acceleration is thus represented by the vector

$$\mathbf{a} = \lim_{\Delta t \to 0} \frac{\Delta\mathbf{v}}{\Delta t} \tag{11.17}$$

Noting that the velocity $\mathbf{v}$ is a vector function $\mathbf{v}(t)$ of the time t, we can refer to the limit of the quotient $\Delta\mathbf{v}/\Delta t$ as the derivative of $\mathbf{v}$ with respect to t. We write

$$\mathbf{a} = \frac{d\mathbf{v}}{dt} \tag{11.18}$$

We observe that the acceleration $\mathbf{a}$ is tangent to the curve described by the tip Q of the vector $\mathbf{v}$ when the latter is drawn from a fixed origin O' (Fig. 11.15*c*) and that, in general, the acceleration is *not* tangent to the path of the particle (Fig. 11.15*d*). The curve described by the tip of $\mathbf{v}$ and shown in Fig. 11.15*c* is called the *hodograph* of the motion.

11.10. DERIVATIVES OF VECTOR FUNCTIONS

We saw in the preceding section that the velocity $\mathbf{v}$ of a particle in curvilinear motion can be represented by the derivative of the vector function $\mathbf{r}(t)$ characterizing the position of the particle. Similarly, the acceleration $\mathbf{a}$ of the particle can be represented by the derivative of the vector function $\mathbf{v}(t)$. In this section, we will give a formal definition of the derivative of a vector function and establish a few rules governing the differentiation of sums and products of vector functions.

Let $\mathbf{P}(u)$ be a vector function of the scalar variable u. By that we mean that the scalar u completely defines the magnitude and direction of the vector $\mathbf{P}$. If the vector $\mathbf{P}$ is drawn from a fixed origin O and the scalar u is allowed to vary, the tip of $\mathbf{P}$ will describe a given curve in space. Consider the vectors $\mathbf{P}$ corresponding, respectively, to the values u and $u + \Delta u$ of the scalar variable (Fig. 11.16*a*). Let $\Delta\mathbf{P}$ be the vector joining the tips of the two given vectors; we write

$$\Delta\mathbf{P} = \mathbf{P}(u + \Delta u) - \mathbf{P}(u)$$

Dividing through by Δu and letting Δu approach zero, *we define the derivative of the vector function* $\mathbf{P}(u)$:

$$\frac{d\mathbf{P}}{du} = \lim_{\Delta u \to 0} \frac{\Delta\mathbf{P}}{\Delta u} = \lim_{\Delta u \to 0} \frac{\mathbf{P}(u + \Delta u) - \mathbf{P}(u)}{\Delta u} \tag{11.19}$$

As Δu approaches zero, the line of action of $\Delta\mathbf{P}$ becomes tangent to the curve of Fig. 11.16*a*. Thus, the derivative $d\mathbf{P}/du$ of the vector function $\mathbf{P}(u)$ *is tangent to the curve described by the tip of* $\mathbf{P}(u)$ (Fig. 11.16*b*).

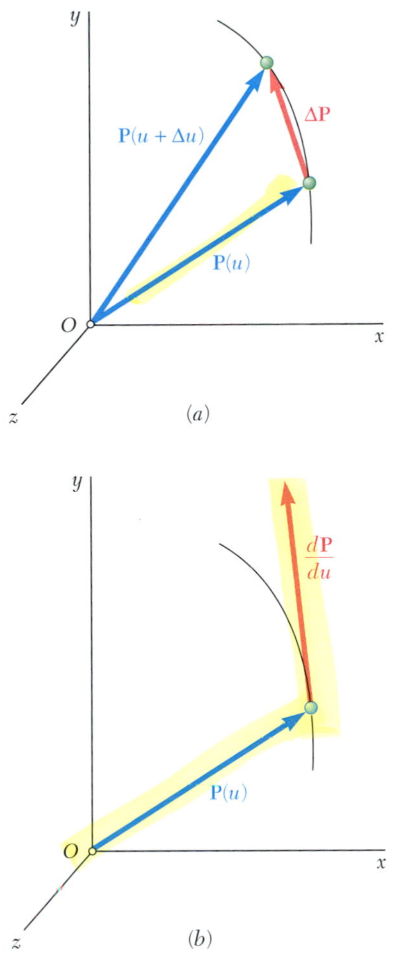

Fig. 11.16

The standard rules for the differentiation of the sums and products of scalar functions can be extended to vector functions. Consider first the *sum of two vector functions* $\mathbf{P}(u)$ and $\mathbf{Q}(u)$ of the same scalar variable u. According to the definition given in (11.19), the derivative of the vector $\mathbf{P} + \mathbf{Q}$ is

$$\frac{d(\mathbf{P} + \mathbf{Q})}{du} = \lim_{\Delta u \to 0} \frac{\Delta(\mathbf{P} + \mathbf{Q})}{\Delta u} = \lim_{\Delta u \to 0} \left(\frac{\Delta\mathbf{P}}{\Delta u} + \frac{\Delta\mathbf{Q}}{\Delta u} \right)$$

or since the limit of a sum is equal to the sum of the limits of its terms,

$$\frac{d(\mathbf{P} + \mathbf{Q})}{du} = \lim_{\Delta u \to 0} \frac{\Delta\mathbf{P}}{\Delta u} + \lim_{\Delta u \to 0} \frac{\Delta\mathbf{Q}}{\Delta u}$$

$$\frac{d(\mathbf{P} + \mathbf{Q})}{du} = \frac{d\mathbf{P}}{du} + \frac{d\mathbf{Q}}{du} \tag{11.20}$$

The *product of a scalar function* $f(u)$ *and a vector function* $\mathbf{P}(u)$ of the same scalar variable u will now be considered. The derivative of the vector $f\mathbf{P}$ is

$$\frac{d(f\mathbf{P})}{du} = \lim_{\Delta u \to 0} \frac{(f + \Delta f)(\mathbf{P} + \Delta\mathbf{P}) - f\mathbf{P}}{\Delta u} = \lim_{\Delta u \to 0} \left(\frac{\Delta f}{\Delta u}\mathbf{P} + f\frac{\Delta\mathbf{P}}{\Delta u} \right)$$

or recalling the properties of the limits of sums and products,

$$\frac{d(f\mathbf{P})}{du} = \frac{df}{du}\mathbf{P} + f\frac{d\mathbf{P}}{du} \tag{11.21}$$

The derivatives of the *scalar product* and the *vector product* of two vector functions $\mathbf{P}(u)$ and $\mathbf{Q}(u)$ can be obtained in a similar way. We have

$$\frac{d(\mathbf{P}\cdot\mathbf{Q})}{du} = \frac{d\mathbf{P}}{du}\cdot\mathbf{Q} + \mathbf{P}\cdot\frac{d\mathbf{Q}}{du} \tag{11.22}$$

$$\frac{d(\mathbf{P}\times\mathbf{Q})}{du} = \frac{d\mathbf{P}}{du}\times\mathbf{Q} + \mathbf{P}\times\frac{d\mathbf{Q}}{du} \qquad (11.23)\dagger$$

The properties established above can be used to determine the *rectangular components of the derivative of a vector function* $\mathbf{P}(u)$. Resolving $\mathbf{P}$ into components along fixed rectangular axes x, y, z, we write

$$\mathbf{P} = P_x\mathbf{i} + P_y\mathbf{j} + P_z\mathbf{k} \tag{11.24}$$

where P_x, P_y, P_z are the rectangular scalar components of the vector $\mathbf{P}$, and $\mathbf{i}$, $\mathbf{j}$, $\mathbf{k}$ the unit vectors corresponding, respectively, to the x, y, and z axes (Sec. 2.12). By (11.20), the derivative of $\mathbf{P}$ is equal to the sum of the derivatives of the terms in the right-hand member. Since each of these terms is the product of a scalar and a vector function, we should use (11.21). But the unit vectors $\mathbf{i}$, $\mathbf{j}$, $\mathbf{k}$ have a constant magnitude (equal to 1) and fixed directions. Their derivatives are therefore zero, and we write

$$\frac{d\mathbf{P}}{du} = \frac{dP_x}{du}\mathbf{i} + \frac{dP_y}{du}\mathbf{j} + \frac{dP_z}{du}\mathbf{k} \tag{11.25}$$

Noting that the coefficients of the unit vectors are, by definition, the scalar components of the vector $d\mathbf{P}/du$, we conclude that *the rectangular scalar components of the derivative* $d\mathbf{P}/du$ *of the vector function* $\mathbf{P}(u)$ are obtained by differentiating the corresponding scalar components of $\mathbf{P}$.

Rate of Change of a Vector. When the vector $\mathbf{P}$ is a function of the time t, its derivative $d\mathbf{P}/dt$ represents the *rate of change* of $\mathbf{P}$ with respect to the frame $Oxyz$. Resolving $\mathbf{P}$ into rectangular components, we have, by (11.25),

$$\frac{d\mathbf{P}}{dt} = \frac{dP_x}{dt}\mathbf{i} + \frac{dP_y}{dt}\mathbf{j} + \frac{dP_z}{dt}\mathbf{k}$$

or, using dots to indicate differentiation with respect to t,

$$\dot{\mathbf{P}} = \dot{P}_x\mathbf{i} + \dot{P}_y\mathbf{j} + \dot{P}_z\mathbf{k} \tag{11.25'}$$

†Since the vector product is not commutative (Sec. 3.4), the order of the factors in Eq. (11.23) must be maintained.

As you will see in Sec. 15.10, the rate of change of a vector as observed from a *moving frame of reference* is, in general, different from its rate of change as observed from a fixed frame of reference. However, if the moving frame $O'x'y'z'$ is in *translation*, that is, if its axes remain parallel to the corresponding axes of the fixed frame $Oxyz$ (Fig. 11.17), the same unit vectors **i**, **j**, **k** are used in both frames, and at any given instant the vector **P** has the same components P_x, P_y, P_z in both frames. It follows from (11.25′) that the rate of change $\dot{\mathbf{P}}$ is the same with respect to the frames $Oxyz$ and $O'x'y'z'$. We state, therefore: *The rate of change of a vector is the same with respect to a fixed frame and with respect to a frame in translation.* This property will greatly simplify our work, since we will be concerned mainly with frames in translation.

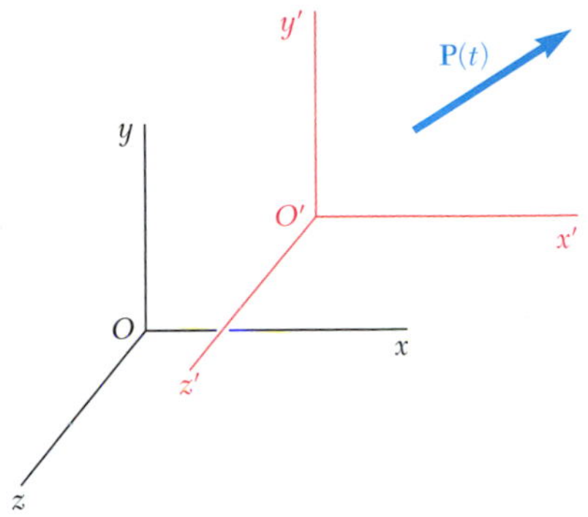

Fig. 11.17

11.11. RECTANGULAR COMPONENTS OF VELOCITY AND ACCELERATION

When the position of a particle P is defined at any instant by its rectangular coordinates x, y, and z, it is convenient to resolve the velocity **v** and the acceleration **a** of the particle into rectangular components (Fig. 11.18).

Resolving the position vector **r** of the particle into rectangular components, we write

$$\mathbf{r} = x\mathbf{i} + y\mathbf{j} + z\mathbf{k} \qquad (11.26)$$

where the coordinates x, y, z are functions of t. Differentiating twice, we obtain

$$\mathbf{v} = \frac{d\mathbf{r}}{dt} = \dot{x}\mathbf{i} + \dot{y}\mathbf{j} + \dot{z}\mathbf{k} \qquad (11.27)$$

$$\mathbf{a} = \frac{d\mathbf{v}}{dt} = \ddot{x}\mathbf{i} + \ddot{y}\mathbf{j} + \ddot{z}\mathbf{k} \qquad (11.28)$$

where $\dot{x}$, $\dot{y}$, $\dot{z}$ and $\ddot{x}$, $\ddot{y}$, $\ddot{z}$ represent, respectively, the first and second derivatives of x, y, and z with respect to t. It follows from (11.27) and (11.28) that the scalar components of the velocity and acceleration are

$$v_x = \dot{x} \qquad v_y = \dot{y} \qquad v_z = \dot{z} \qquad (11.29)$$

$$a_x = \ddot{x} \qquad a_y = \ddot{y} \qquad a_z = \ddot{z} \qquad (11.30)$$

A positive value for v_x indicates that the vector component $\mathbf{v}_x$ is directed to the right, and a negative value indicates that it is directed to the left. The sense of each of the other vector components can be determined in a similar way from the sign of the corresponding scalar component. If desired, the magnitudes and directions of the velocity and acceleration can be obtained from their scalar components by the methods of Secs. 2.7 and 2.12.

The use of rectangular components to describe the position, the velocity, and the acceleration of a particle is particularly effective when the component a_x of the acceleration depends only upon t, x, and/or v_x, and when, similarly, a_y depends only upon t, y, and/or v_y,

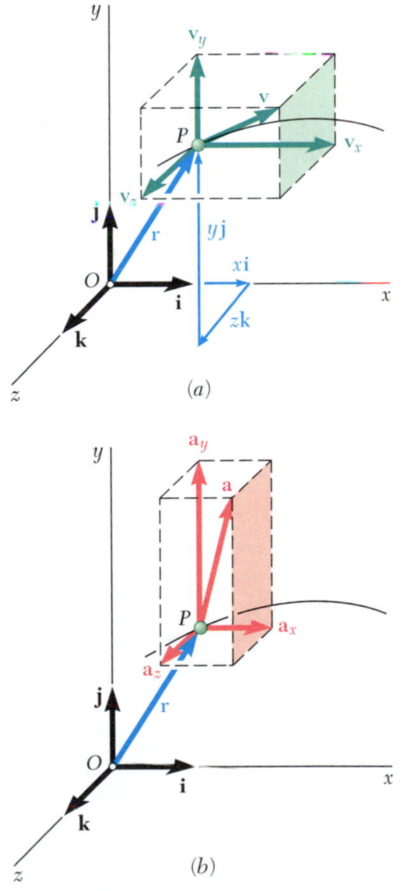

Fig. 11.18

and a_z upon t, z, and/or v_z. Equations (11.30) can then be integrated independently, and so can Eqs. (11.29). In other words, the motion of the particle in the x direction, its motion in the y direction, and its motion in the z direction can be considered separately.

In the case of the *motion of a projectile,* for example, it can be shown (see Sec. 12.5) that the components of the acceleration are

$$a_x = \ddot{x} = 0 \qquad a_y = \ddot{y} = -g \qquad a_z = \ddot{z} = 0$$

if the resistance of the air is neglected. Denoting by x_0, y_0, and z_0 the coordinates of a gun, and by $(v_x)_0$, $(v_y)_0$, and $(v_z)_0$ the components of the initial velocity $\mathbf{v}_0$ of the projectile (a bullet), we integrate twice in t and obtain

$$\begin{aligned} v_x &= \dot{x} = (v_x)_0 & v_y &= \dot{y} = (v_y)_0 - gt & v_z &= \dot{z} = (v_z)_0 \\ x &= x_0 + (v_x)_0 t & y &= y_0 + (v_y)_0 t - \tfrac{1}{2}gt^2 & z &= z_0 + (v_z)_0 t \end{aligned}$$

If the projectile is fired in the xy plane from the origin O, we have $x_0 = y_0 = z_0 = 0$ and $(v_z)_0 = 0$, and the equations of motion reduce to

$$\begin{aligned} v_x &= (v_x)_0 & v_y &= (v_y)_0 - gt & v_z &= 0 \\ x &= (v_x)_0 t & y &= (v_y)_0 t - \tfrac{1}{2}gt^2 & z &= 0 \end{aligned}$$

These equations show that the projectile remains in the xy plane, that its motion in the horizontal direction is uniform, and that its motion in the vertical direction is uniformly accelerated. The motion of a projectile can thus be replaced by two independent rectilinear motions, which are easily visualized if we assume that the projectile is fired vertically with an initial velocity $(\mathbf{v}_y)_0$ from a platform moving with a constant horizontal velocity $(\mathbf{v}_x)_0$ (Fig. 11.19). The coordinate x of the projectile is equal at any instant to the distance traveled by the platform, and its coordinate y can be computed as if the projectile were moving along a vertical line.

It can be observed that the equations defining the coordinates x and y of a projectile at any instant are the parametric equations of a parabola. Thus, the trajectory of a projectile is *parabolic.* This result, however, ceases to be valid when the resistance of the air or the variation with altitude of the acceleration of gravity is taken into account.

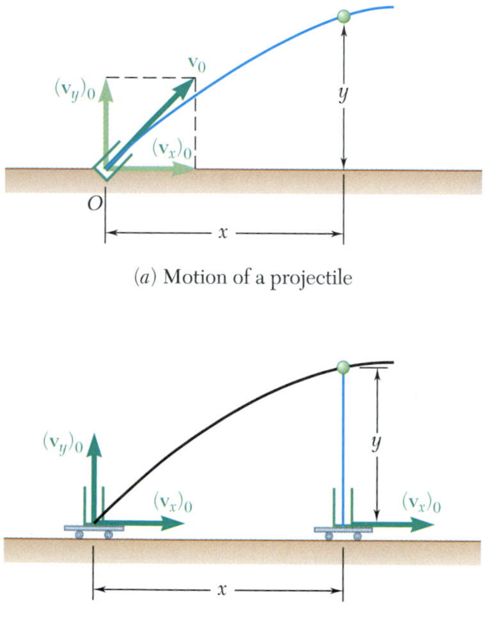

(*a*) Motion of a projectile

(*b*) Equivalent rectilinear motions

Fig. 11.19

Photo 11.2 When air resistance is neglected, the ski jumper may be considered as a projectile having a parabolic trajectory.

11.12. MOTION RELATIVE TO A FRAME IN TRANSLATION

In the preceding section, a single frame of reference was used to describe the motion of a particle. In most cases this frame was attached to the earth and was considered as fixed. Situations in which it is convenient to use several frames of reference simultaneously will now be analyzed. If one of the frames is attached to the earth, it will be called a *fixed frame of reference,* and the other frames will be referred to as *moving frames of reference.* It should be understood, however, that the selection of a fixed frame of reference is purely arbitrary. Any frame can be designated as "fixed"; all other frames not rigidly attached to this frame will then be described as "moving."

Consider two particles A and B moving in space (Fig. 11.20); the vectors $\mathbf{r}_A$ and $\mathbf{r}_B$ define their positions at any given instant with respect to the fixed frame of reference $Oxyz$. Consider now a system of axes x', y', z' centered at A and parallel to the x, y, z axes. While the origin of these axes moves, their orientation remains the same; the frame of reference $Ax'y'z'$ is in *translation* with respect to $Oxyz$. The vector $\mathbf{r}_{B/A}$ joining A and B defines *the position of B relative to the moving frame* $Ax'y'z'$ (or, for short, *the position of B relative to A*).

We note from Fig. 11.20 that the position vector $\mathbf{r}_B$ of particle B is the sum of the position vector $\mathbf{r}_A$ of particle A and of the position vector $\mathbf{r}_{B/A}$ of B relative to A; we write

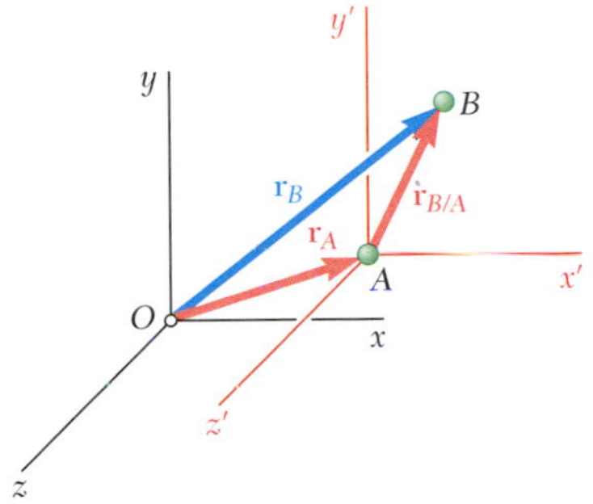

Fig. 11.20

$$\mathbf{r}_B = \mathbf{r}_A + \mathbf{r}_{B/A} \qquad (11.31)$$

Differentiating (11.31) with respect to t within the fixed frame of reference, and using dots to indicate time derivatives, we have

$$\dot{\mathbf{r}}_B = \dot{\mathbf{r}}_A + \dot{\mathbf{r}}_{B/A} \qquad (11.32)$$

The derivatives $\dot{\mathbf{r}}_A$ and $\dot{\mathbf{r}}_B$ represent, respectively, the velocities $\mathbf{v}_A$ and $\mathbf{v}_B$ of the particles A and B. Since $Ax'y'z'$ is in translation, the derivative $\dot{\mathbf{r}}_{B/A}$ represents the rate of change of $\mathbf{r}_{B/A}$ with respect to the frame $Ax'y'z'$ as well as with respect to the fixed frame (Sec. 11.10). This derivative, therefore, defines *the velocity* $\mathbf{v}_{B/A}$ *of B relative to the frame* $Ax'y'z'$ (or, for short, *the velocity* $\mathbf{v}_{B/A}$ *of B relative to A*). We write

$$\mathbf{v}_B = \mathbf{v}_A + \mathbf{v}_{B/A} \qquad (11.33)$$

Differentiating Eq. (11.33) with respect to t, and using the derivative $\dot{\mathbf{v}}_{B/A}$ to define *the acceleration* $\mathbf{a}_{B/A}$ *of B relative to the frame* $Ax'y'z'$ (or, for short, *the acceleration* $\mathbf{a}_{B/A}$ *of B relative to A*), we write

$$\mathbf{a}_B = \mathbf{a}_A + \mathbf{a}_{B/A} \qquad (11.34)$$

Photo 11.3 When a frame of reference attached to the ship moves in translation, the relative motion of the helicopter with respect to the ship is characterized by Eqs. (11.31) through (11.34).

The motion of B with respect to the fixed frame $Oxyz$ is referred to as the *absolute motion of B*. The equations derived in this section show that *the absolute motion of B can be obtained by combining the motion of A and the relative motion of B with respect to the moving frame attached to A*. Equation (11.33), for example, expresses that the absolute velocity $\mathbf{v}_B$ of particle B can be obtained by adding vectorially the velocity of A and the velocity of B relative to the frame $Ax'y'z'$. Equation (11.34) expresses a similar property in terms of the accelerations.† We should keep in mind, however, that *the frame* $Ax'y'z'$ *is in translation;* that is, while it moves with A, it maintains the same orientation. As you will see later (Sec. 15.14), different relations must be used in the case of a rotating frame of reference.

$V_A = -80$

V_B

†Note that the product of the subscripts A and B/A used in the right-hand member of Eqs. (11.31) through (11.34) is equal to the subscript B used in their left-hand member.

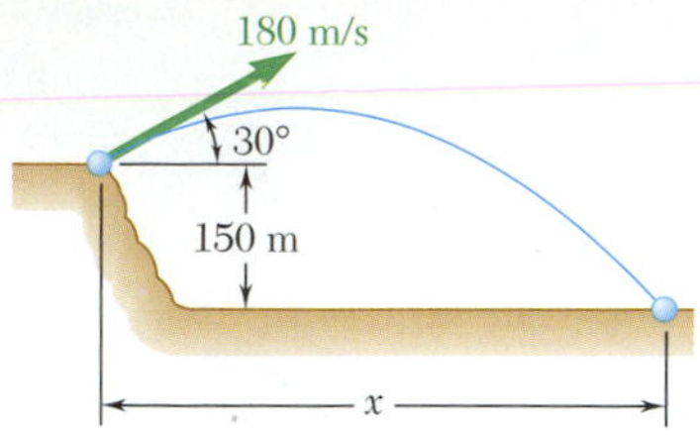

SAMPLE PROBLEM 11.7

A projectile is fired from the edge of a 150-m cliff with an initial velocity of 180 m/s at an angle of 30° with the horizontal. Neglecting air resistance, find (*a*) the horizontal distance from the gun to the point where the projectile strikes the ground, (*b*) the greatest elevation above the ground reached by the projectile.

SOLUTION

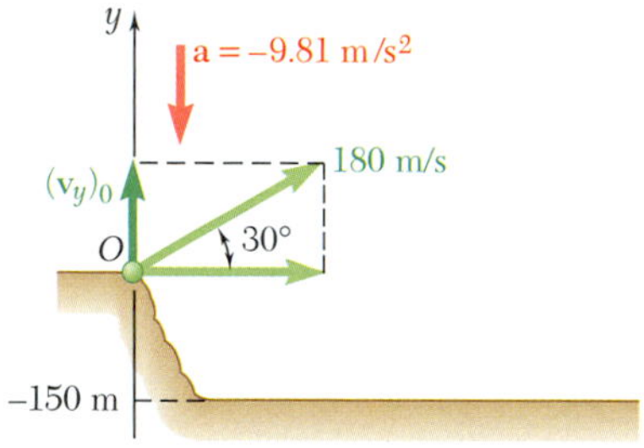

The vertical and the horizontal motion will be considered separately.

Vertical Motion. ***Uniformly Accelerated Motion.*** Choosing the positive sense of the y axis upward and placing the origin O at the gun, we have

$$(v_y)_0 = (180 \text{ m/s}) \sin 30° = +90 \text{ m/s}$$
$$a = -9.81 \text{ m/s}^2$$

Substituting into the equations of uniformly accelerated motion, we have

$$v_y = (v_y)_0 + at \qquad v_y = 90 - 9.81t \qquad (1)$$
$$y = (v_y)_0 t + \tfrac{1}{2}at^2 \qquad y = 90t - 4.90t^2 \qquad (2)$$
$$v_y^2 = (v_y)_0^2 + 2ay \qquad v_y^2 = 8100 - 19.62y \qquad (3)$$

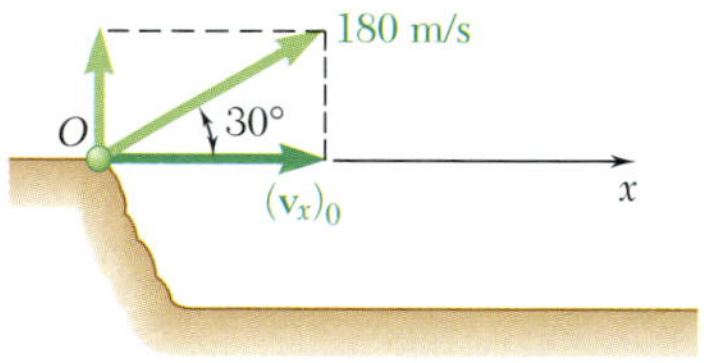

Horizontal Motion. ***Uniform Motion.*** Choosing the positive sense of the x axis to the right, we have

$$(v_x)_0 = (180 \text{ m/s}) \cos 30° = +155.9 \text{ m/s}$$

Substituting into the equation of uniform motion, we obtain

$$x = (v_x)_0 t \qquad x = 155.9t \qquad (4)$$

***a.* Horizontal Distance.** When the projectile strikes the ground, we have

$$y = -150 \text{ m}$$

Carrying this value into Eq. (2) for the vertical motion, we write

$$-150 = 90t - 4.90t^2 \qquad t^2 - 18.37t - 30.6 = 0 \qquad t = 19.91 \text{ s}$$

Carrying $t = 19.91$ s into Eq. (4) for the horizontal motion, we obtain

$$x = 155.9(19.91) \qquad x = 3100 \text{ m} \blacktriangleleft$$

***b.* Greatest Elevation.** When the projectile reaches its greatest elevation, we have $v_y = 0$; carrying this value into Eq. (3) for the vertical motion, we write

$$0 = 8100 - 19.62y \qquad y = 413 \text{ m}$$

Greatest elevation above ground = 150 m + 413 m = 563 m ◀

SAMPLE PROBLEM 11.8

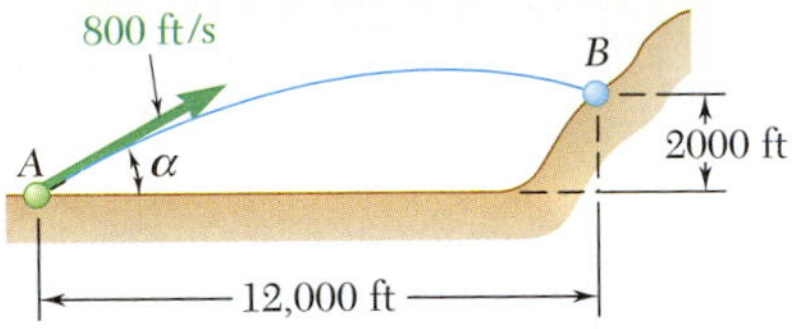

A projectile is fired with an initial velocity of 800 ft/s at a target B located 2000 ft above the gun A and at a horizontal distance of 12,000 ft. Neglecting air resistance, determine the value of the firing angle α.

SOLUTION

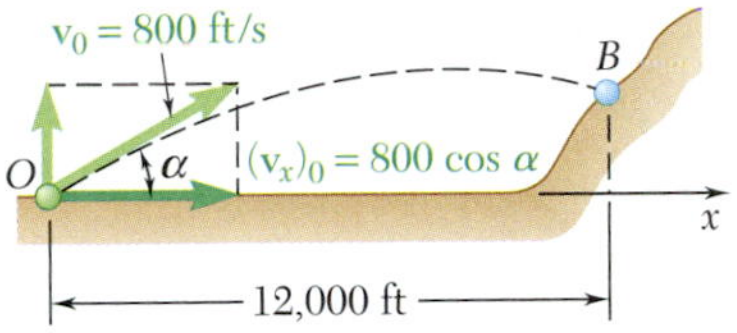

The horizontal and the vertical motion will be considered separately.

Horizontal Motion. Placing the origin of the coordinate axes at the gun, we have

$$(v_x)_0 = 800 \cos \alpha$$

Substituting into the equation of uniform horizontal motion, we obtain

$$x = (v_x)_0 t \qquad x = (800 \cos \alpha)t$$

The time required for the projectile to move through a horizontal distance of 12,000 ft is obtained by setting x equal to 12,000 ft.

$$12{,}000 = (800 \cos \alpha)t$$

$$t = \frac{12{,}000}{800 \cos \alpha} = \frac{15}{\cos \alpha}$$

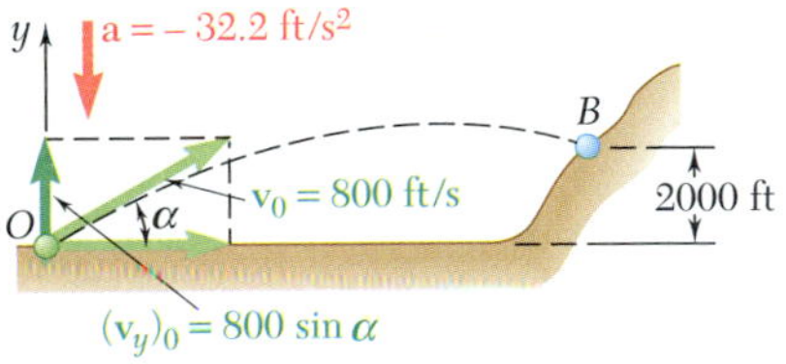

Vertical Motion

$$(v_y)_0 = 800 \sin \alpha \qquad a = -32.2 \text{ ft/s}^2$$

Substituting into the equation of uniformly accelerated vertical motion, we obtain

$$y = (v_y)_0 t + \tfrac{1}{2}at^2 \qquad y = (800 \sin \alpha)t - 16.1t^2$$

Projectile Hits Target. When $x = 12{,}000$ ft, we must have $y = 2000$ ft. Substituting for y and setting t equal to the value found above, we write

$$2000 = 800 \sin \alpha \frac{15}{\cos \alpha} - 16.1 \left(\frac{15}{\cos \alpha}\right)^2$$

Since $1/\cos^2 \alpha = \sec^2 \alpha = 1 + \tan^2 \alpha$, we have

$$2000 = 800(15) \tan \alpha - 16.1(15^2)(1 + \tan^2 \alpha)$$
$$3622 \tan^2 \alpha - 12{,}000 \tan \alpha + 5622 = 0$$

Solving this quadratic equation for $\tan \alpha$, we have

$$\tan \alpha = 0.565 \qquad \text{and} \qquad \tan \alpha = 2.75$$

$$\alpha = 29.5° \qquad \text{and} \qquad \alpha = 70.0° \quad ◀$$

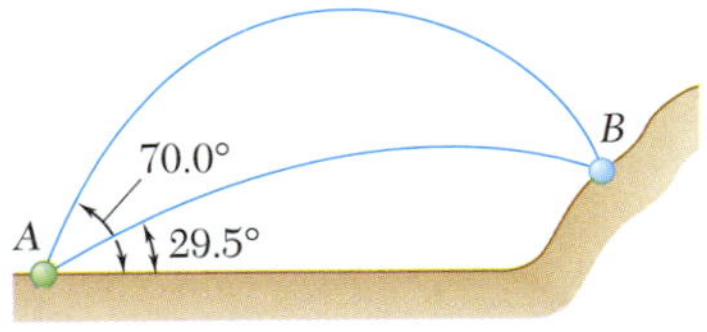

The target will be hit if either of these two firing angles is used (see figure).

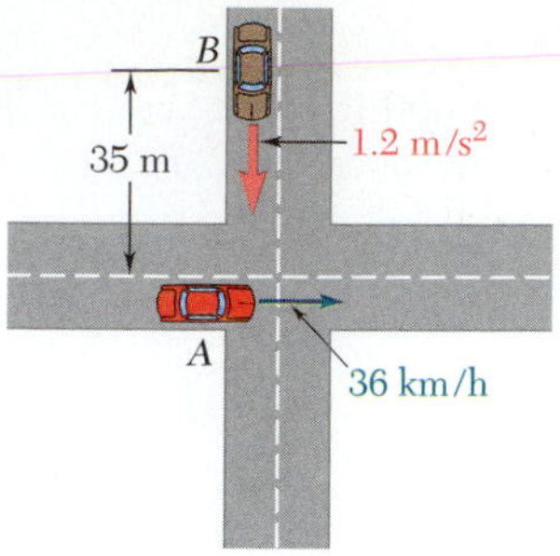

SAMPLE PROBLEM 11.9

Automobile A is traveling east at the constant speed of 36 km/h. As automobile A crosses the intersection shown, automobile B starts from rest 35 m north of the intersection and moves south with a constant acceleration of 1.2 m/s^2. Determine the position, velocity, and acceleration of B relative to A 5 s after A crosses the intersection.

SOLUTION

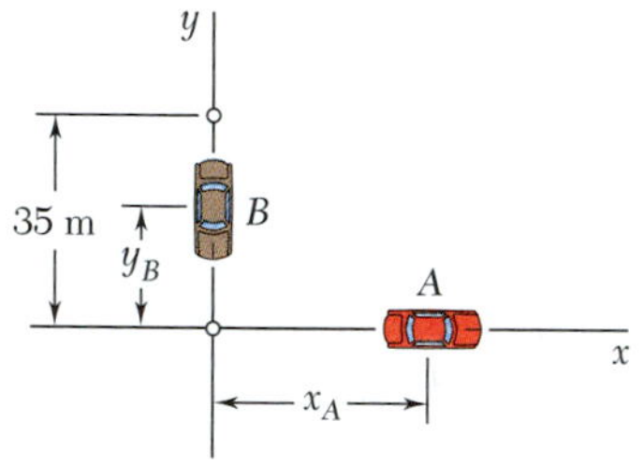

We choose x and y axes with origin at the intersection of the two streets and with positive senses directed respectively east and north.

Motion of Automobile A. First the speed is expressed in m/s:

$$v_A = \left(36\frac{\text{km}}{\text{h}}\right)\left(\frac{1000 \text{ m}}{1 \text{ km}}\right)\left(\frac{1 \text{ h}}{3600 \text{ s}}\right) = 10 \text{ m/s}$$

Noting that the motion of A is uniform, we write, for any time t,

$$a_A = 0$$
$$v_A = +10 \text{ m/s}$$
$$x_A = (x_A)_0 + v_A t = 0 + 10t$$

For $t = 5$ s, we have

$$a_A = 0 \qquad \mathbf{a}_A = 0$$
$$v_A = +10 \text{ m/s} \qquad \mathbf{v}_A = 10 \text{ m/s} \rightarrow$$
$$x_A = +(10 \text{ m/s})(5 \text{ s}) = +50 \text{ m} \qquad \mathbf{r}_A = 50 \text{ m} \rightarrow$$

Motion of Automobile B. We note that the motion of B is uniformly accelerated and write

$$a_B = -1.2 \text{ m/s}^2$$
$$v_B = (v_B)_0 + at = 0 - 1.2\,t$$
$$y_B = (y_B)_0 + (v_B)_0 t + \tfrac{1}{2}a_B t^2 = 35 + 0 - \tfrac{1}{2}(1.2)t^2$$

For $t = 5$ s, we have

$$a_B = -1.2 \text{ m/s}^2 \qquad \mathbf{a}_B = 1.2 \text{ m/s}^2 \downarrow$$
$$v_B = -(1.2 \text{ m/s}^2)(5 \text{ s}) = -6 \text{ m/s} \qquad \mathbf{v}_B = 6 \text{ m/s} \downarrow$$
$$y_B = 35 - \tfrac{1}{2}(1.2 \text{ m/s}^2)(5 \text{ s})^2 = +20 \text{ m} \qquad \mathbf{r}_B = 20 \text{ m} \uparrow$$

Motion of B Relative to A. We draw the triangle corresponding to the vector equation $\mathbf{r}_B = \mathbf{r}_A + \mathbf{r}_{B/A}$ and obtain the magnitude and direction of the position vector of B relative to A.

$$r_{B/A} = 53.9 \text{ m} \qquad \alpha = 21.8° \qquad \mathbf{r}_{B/A} = 53.9 \text{ m} \nwarrow 21.8° \quad ◀$$

Proceeding in a similar fashion, we find the velocity and acceleration of B relative to A.

$$\mathbf{v}_B = \mathbf{v}_A + \mathbf{v}_{B/A}$$
$$v_{B/A} = 11.66 \text{ m/s} \qquad \beta = 31.0° \qquad \mathbf{v}_{B/A} = 11.66 \text{ m/s} \swarrow 31.0° \quad ◀$$
$$\mathbf{a}_B = \mathbf{a}_A + \mathbf{a}_{B/A} \qquad \mathbf{a}_{B/A} = 1.2 \text{ m/s}^2 \downarrow \quad ◀$$

SOLVING PROBLEMS ON YOUR OWN

In the problems for this lesson, you will analyze the *two- and three-dimensional motion* of a particle. While the physical interpretations of the velocity and acceleration are the same as in the first lessons of the chapter, you should remember that these quantities are vectors. In addition, you should understand from your experience with vectors in statics that it will often be advantageous to express position vectors, velocities, and accelerations in terms of their rectangular scalar components [Eqs. (11.27) and (11.28)]. Furthermore, given two vectors **A** and **B**, recall that $\mathbf{A} \cdot \mathbf{B} = 0$ if **A** and **B** are perpendicular to each other, while $\mathbf{A} \times \mathbf{B} = 0$ if **A** and **B** are parallel.

A. *Analyzing the motion of a projectile.* Many of the following problems deal with the two-dimensional motion of a projectile, where the resistance of the air can be neglected. In Sec. 11.11, we developed the equations which describe this type of motion, and we observed that the horizontal component of the velocity remained constant (uniform motion) while the vertical component of the acceleration was constant (uniformly accelerated motion). We were able to consider separately the horizontal and the vertical motions of the particle. Assuming that the projectile is fired from the origin, we can write the two equations

$$x = (v_x)_0 t \qquad y = (v_y)_0 t - \tfrac{1}{2} g t^2$$

1. *If the initial velocity and firing angle are known,* the value of y corresponding to any given value of x (or the value of x for any value of y) can be obtained by solving one of the above equations for t and substituting for t into the other equation [Sample Prob. 11.7].

2. *If the initial velocity and the coordinates of a point of the trajectory are known,* and you wish to *determine the firing angle* α, begin your solution by expressing the components $(v_x)_0$ and $(v_y)_0$ of the initial velocity as functions of the angle α. These expressions and the known values of x and y are then substituted into the above equations. Finally, solve the first equation for t and substitute that value of t into the second equation to obtain a trigonometric equation in α, which you can solve for that unknown [Sample Prob. 11.8].

(continued)

B. Solving translational two-dimensional relative-motion problems. You saw in Sec. 11.12 that the absolute motion of a particle B can be obtained by combining the motion of a particle A and the *relative motion* of B with respect to a frame attached to A which is in *translation.* The velocity and acceleration of B can then be expressed as shown in Eqs. (11.33) and (11.34), respectively.

1. ***To visualize the relative motion of B with respect to A,*** imagine that you are attached to particle A as you observe the motion of particle B. For example, to a passenger in automobile A of Sample Prob. 11.9, automobile B appears to be heading in a southwesterly direction (*south* should be obvious; and *west* is due to the fact that automobile A is moving to the east—automobile B then appears to travel to the west). Note that this conclusion is consistent with the direction of $\mathbf{v}_{B/A}$.

2. ***To solve a relative-motion problem,*** first write the vector equations (11.31), (11.33), and (11.34), which relate the motions of particles A and B. You may then use either of the following methods:

a. Construct the corresponding vector triangles and solve them for the desired position vector, velocity, and acceleration [Sample Prob. 11.9].

b. Express all vectors in terms of their rectangular components and solve the two independent sets of scalar equations obtained in that way. If you choose this approach, be sure to select the same positive direction for the displacement, velocity, and acceleration of each particle.

Problems

11.89 The motion of a particle is defined by the equations $x = 17.6 - 0.8t(t^2 - 9t + 24)$ and $y = -13.8 + 0.6t(t^2 - 9t + 24)$, where x and y are expressed in feet and t is expressed in seconds. Show that the path of the particle is a portion of a straight line, and determine the velocity and acceleration when (*a*) $t = 2$ s, (*b*) $t = 3$ s, (*c*) $t = 4$ s.

11.90 The motion of a particle is defined by the equations $x = (4 \cos \pi t - 2)/(2 - \cos \pi t)$ and $y = 3 \sin \pi t/(2 - \cos \pi t)$, where x and y are expressed in feet and t is expressed in seconds. Show that the path of the particle is part of the ellipse shown, and determine the velocity when (*a*) $t = 0$, (*b*) $t = 1/3$ s, (*c*) $t = 1$ s.

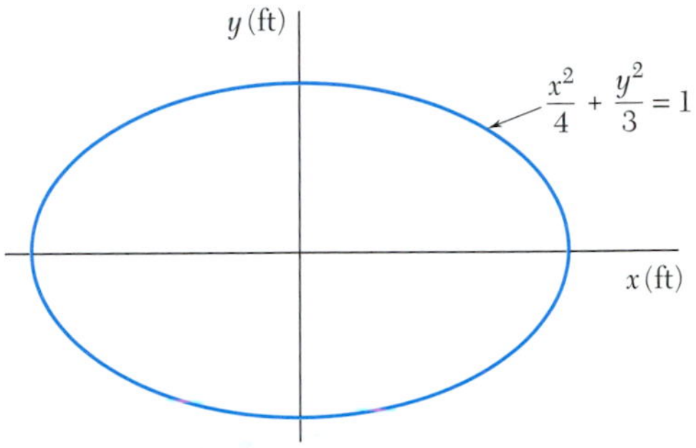

Fig. P11.90

11.91 The motion of a particle is defined by the equations $x = [(t - 4)^3/6] + t^2$ and $y = (t^3/6) - (t - 1)^2/4$, where x and y are expressed in meters and t is expressed in seconds. Determine (*a*) the magnitude of the smallest velocity reached by the particle, (*b*) the corresponding time, position, and direction of the velocity.

11.92 The motion of a particle is defined by the equations $x = 6t - 3 \sin t$ and $y = 6 - 3 \cos t$, where x and y are expressed in meters and t is expressed in seconds. Sketch the path of the particle for the time interval $0 \leq t \leq 2\pi$, and determine (*a*) the magnitudes of the smallest and largest velocities reached by the particle, (*b*) the corresponding times, positions, and directions of the velocities.

11.93 The motion of a particle is defined by the position vector $\mathbf{r} = A(\cos t + t \sin t)\mathbf{i} + A(\sin t - t \cos t)\mathbf{j}$, where t is expressed in seconds. Determine the values of t for which the position vector and the acceleration vector are (*a*) perpendicular, (*b*) parallel.

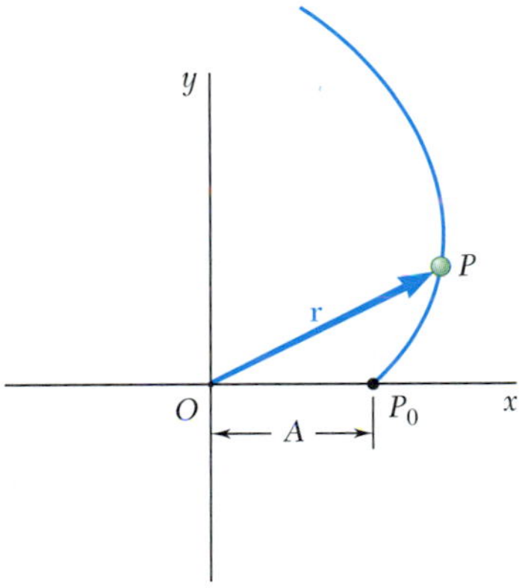

Fig. *P11.93*

11.94 The damped motion of a vibrating particle is defined by the position vector $\mathbf{r} = x_1[1 - 1/(t + 1)]\mathbf{i} + (y_1 e^{-\pi t/2} \cos 2\pi t)\mathbf{j}$, where t is expressed in seconds. For $x_1 = 30$ in. and $y_1 = 20$ in., determine the position, the velocity, and the acceleration of the particle when (*a*) $t = 0$, (*b*) $t = 1.5$ s.

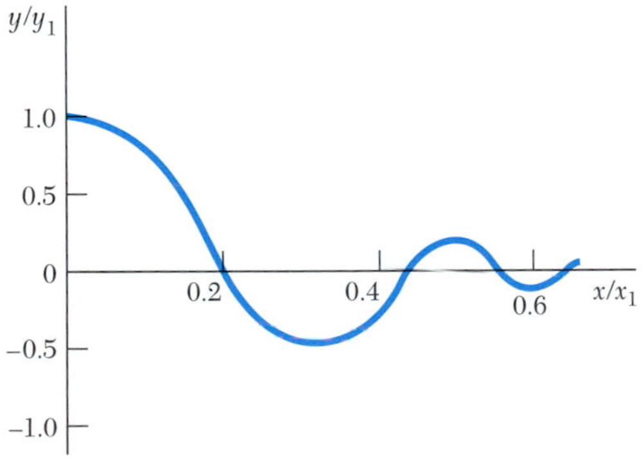

Fig. P11.94

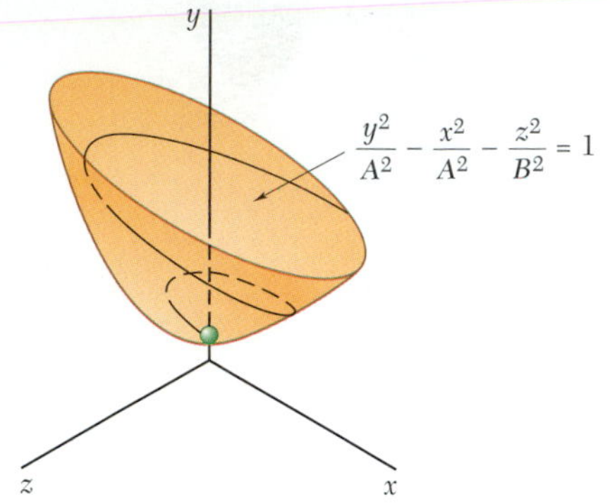

Fig. P11.96

11.95 The three-dimensional motion of a particle is defined by the position vector $\mathbf{r} = (Rt \cos \omega_n t)\mathbf{i} + ct\mathbf{j} + (Rt \sin \omega_n t)\mathbf{k}$. Determine the magnitudes of the velocity and acceleration of the particle. (The space curve described by the particle is a conic helix.)

***11.96** The three-dimensional motion of a particle is defined by the position vector $\mathbf{r} = (At \cos t)\mathbf{i} + (A\sqrt{t^2 + 1})\mathbf{j} + (Bt \sin t)\mathbf{k}$, where r and t are expressed in feet and seconds, respectively. Show that the curve described by the particle lies on the hyperboloid $(y/A)^2 - (x/A)^2 - (z/B)^2 = 1$. For $A = 3$ and $B = 1$, determine (*a*) the magnitudes of the velocity and acceleration when $t = 0$, (*b*) the smallest nonzero value of t for which the position vector and the velocity vector are perpendicular to each other.

11.97 A baseball pitching machine "throws" baseballs with a horizontal velocity $\mathbf{v}_0$. Knowing that height h varies between 788 mm and 1068 mm, determine (*a*) the range of values of v_0, (*b*) the values of α corresponding to $h = 788$ mm and $h = 1068$ mm.

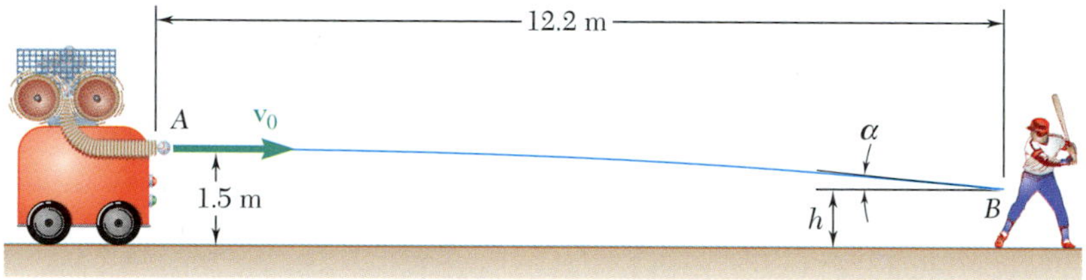

Fig. P11.97

11.98 While delivering newspapers, a girl throws a paper with a horizontal velocity $\mathbf{v}_0$. Determine the range of values of v_0 if the newspaper is to land between points B and C.

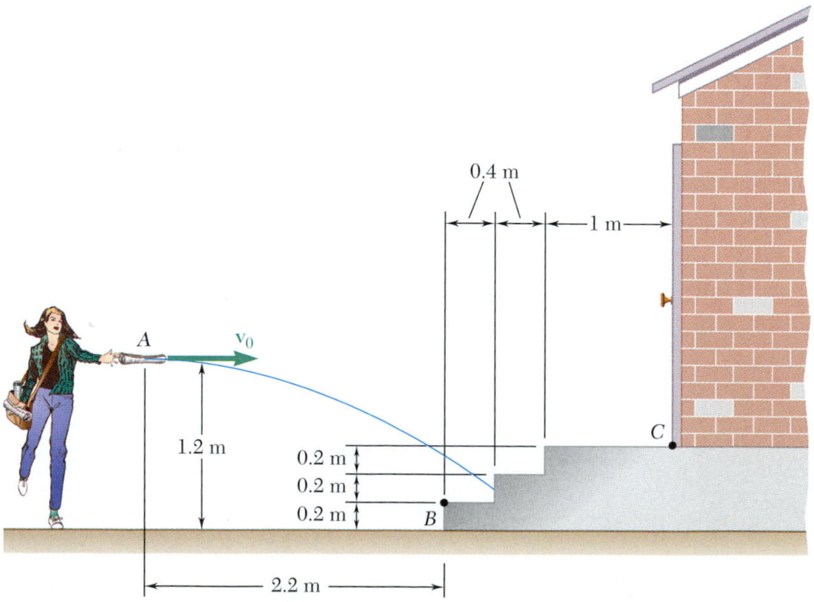

Fig. P11.98

11.99 A ski jumper starts with a horizontal take-off velocity of 80 ft/s and lands on a straight landing hill inclined at 30°. Determine (*a*) the time between take-off and landing, (*b*) the length d of the jump, (*c*) the maximum vertical distance between the jumper and the landing hill.

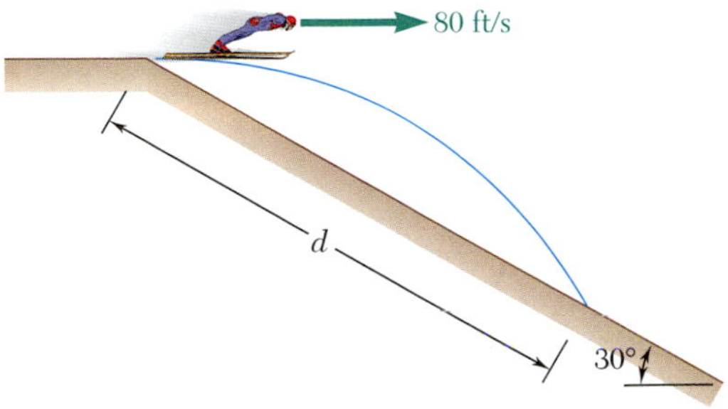

Fig. P11.99

11.100 A golfer aims his shot to clear the top of a tree by a distance h at the peak of the trajectory and to miss the pond on the opposite side. Knowing that the magnitude of $\mathbf{v}_0$ is 85 ft/s, determine the range of values of h which must be avoided.

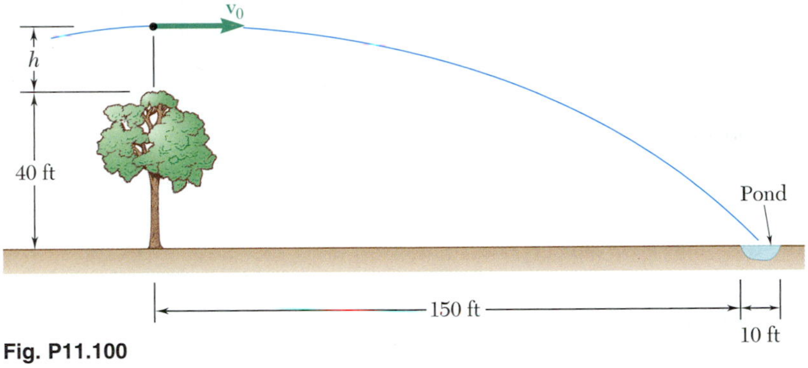

Fig. P11.100

11.101 A golfer hits a golf ball with an initial velocity of 48 m/s at an angle of 25° with the horizontal. Knowing that the fairway slopes downward at an average angle of 5°, determine the distance d between the golfer and point B where the ball first lands.

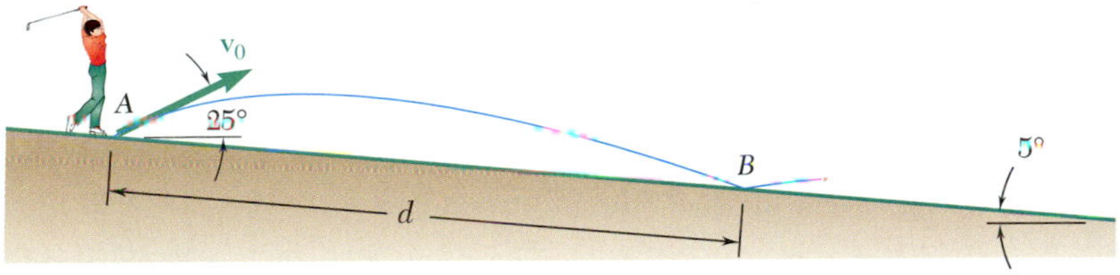

Fig. P11.101

11.102 Water flows from a drain spout with an initial velocity of 0.76 m/s at an angle of 15° with the horizontal. Determine the range of values of the distance d for which the water will enter the trough BC.

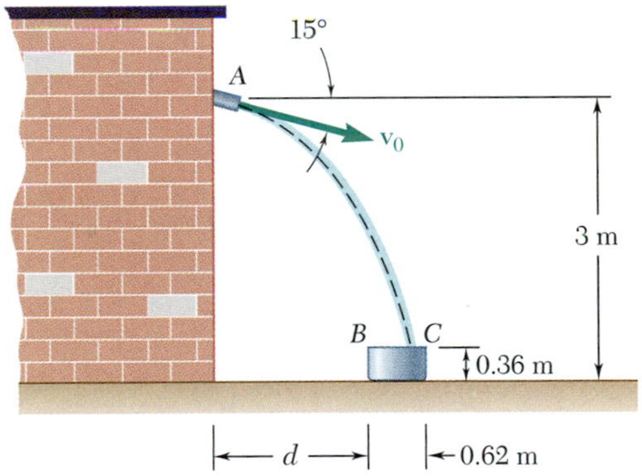

Fig. P11.102

11.103 In slow pitch softball the underhand pitch must reach a maximum height of between 6 ft and 12 ft above the ground. A pitch is made with an initial velocity $\mathbf{v}_0$ of magnitude 43 ft/s at an angle of 33° with the horizontal. Determine (*a*) if the pitch meets the maximum height requirement, (*b*) the height of the ball as it reaches the batter.

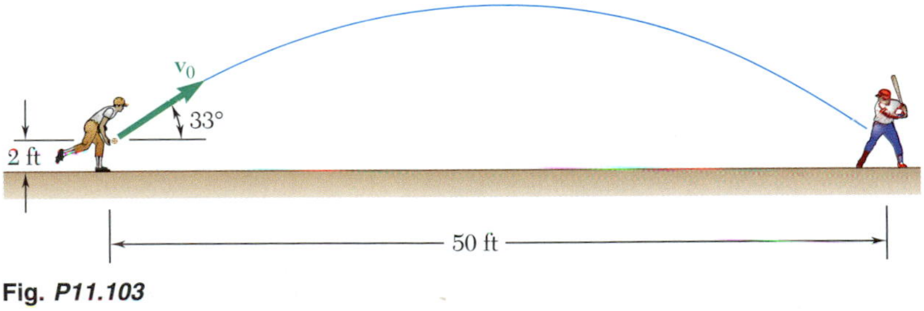

Fig. *P11.103*

11.104 A tennis player serves the ball at a height h with an initial velocity of 120 ft/s at an angle of 4° with the horizontal. Knowing that the ball clears the 3-ft net by 6 in., determine (*a*) the height h, (*b*) the distance d from the net to where the ball will land.

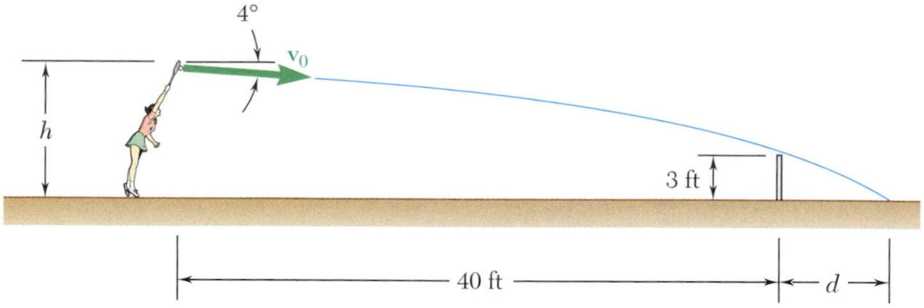

Fig. P11.104

11.105 A homeowner uses a snow blower to clear his driveway. Knowing that the snow is discharged at an average angle of 40° with the horizontal, determine the initial speed v_0 of the snow.

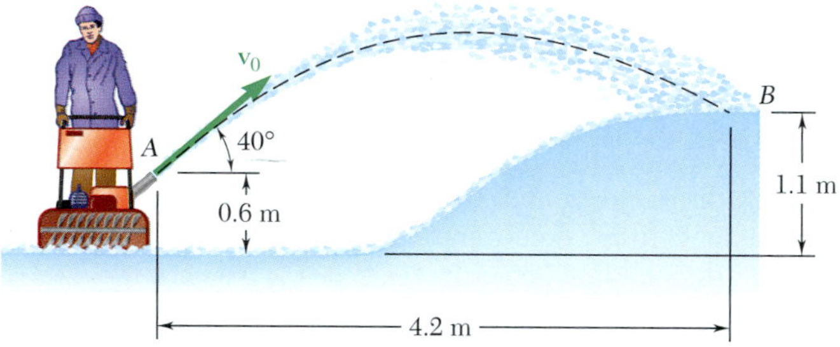

Fig. P11.105

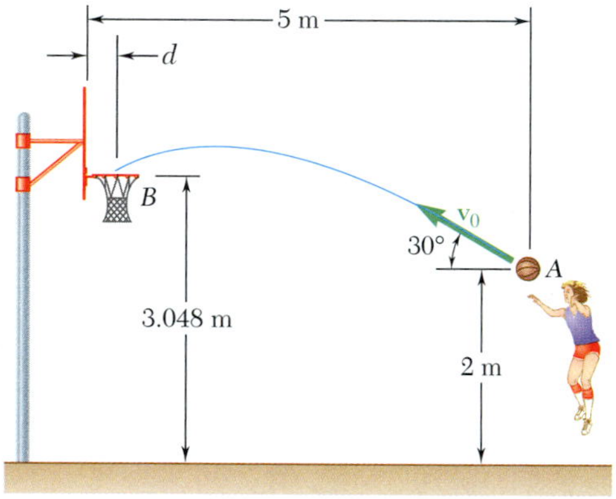

Fig. P11.106

11.106 A basketball player shoots when she is 5 m from the backboard. Knowing that the ball has an initial velocity $\mathbf{v}_0$ at an angle of 30° with the horizontal, determine the value of v_0 when d is equal to (*a*) 228 mm, (*b*) 430 mm.

11.107 The nozzle at A discharges cooling water with an initial velocity $\mathbf{v}_0$ at an angle of 6° with the horizontal onto a grinding wheel 13.8 in. in diameter. Determine the range of values of the initial velocity for which the water will land on the grinding wheel between points B and C.

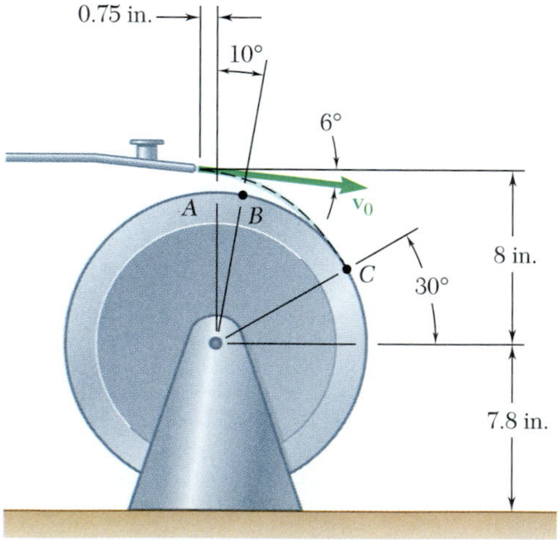

Fig. P11.107

11.108 A ball is dropped onto a step at point A and rebounds with a velocity $\mathbf{v}_0$ at an angle of 15° with the vertical. Determine the value of v_0 knowing that just before the ball bounces at point B its velocity $\mathbf{v}_B$ forms an angle of 12° with the vertical.

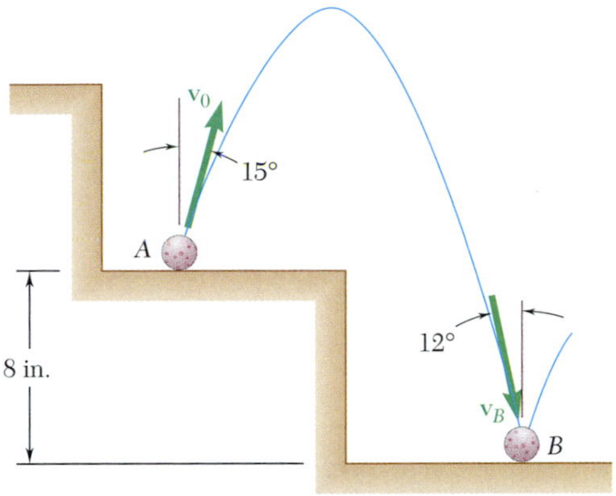

Fig. P11.108

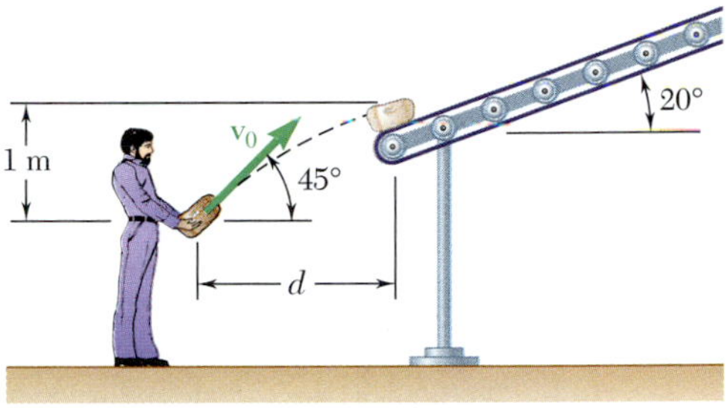

Fig. *P11.109*

11.109 The conveyor belt, which forms an angle of 20° with the horizontal, is used to load an airplane. Knowing that a worker tosses a package with an initial velocity $\mathbf{v}_0$ at an angle of 45° so that its velocity is parallel to the belt as it lands 1 m above the release point, determine (*a*) the magnitude of v_0, (*b*) the horizontal distance d.

11.110 A golfer hits a ball with an initial velocity of magnitude v_0 at an angle α with the horizontal. Knowing that the ball must clear the tops of two trees and land as close as possible to the flag, determine v_0 and the distance d when the golfer uses (*a*) a six-iron with $\alpha = 31°$, (*b*) a five-iron with $\alpha = 27°$.

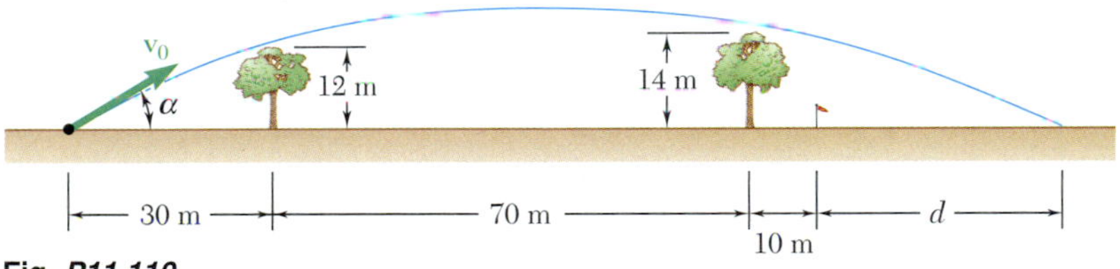

Fig. *P11.110*

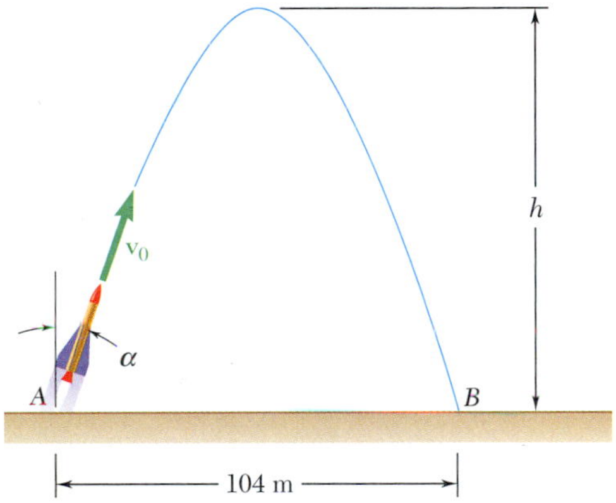

Fig. P11.111

11.111 A model rocket is launched from point A with an initial velocity $\mathbf{v}_0$ of 86 m/s. If the rocket's descent parachute does not deploy and the rocket lands 104 m from A, determine (*a*) the angle α that $\mathbf{v}_0$ forms with the vertical, (*b*) the maximum height h reached by the rocket, (*c*) the duration of the flight.

11.112 The initial velocity $\mathbf{v}_0$ of a hockey puck is 170 km/h. Determine (*a*) the largest value (less than 45°) of the angle α for which the puck will enter the net, (*b*) the corresponding time required for the puck to reach the net.

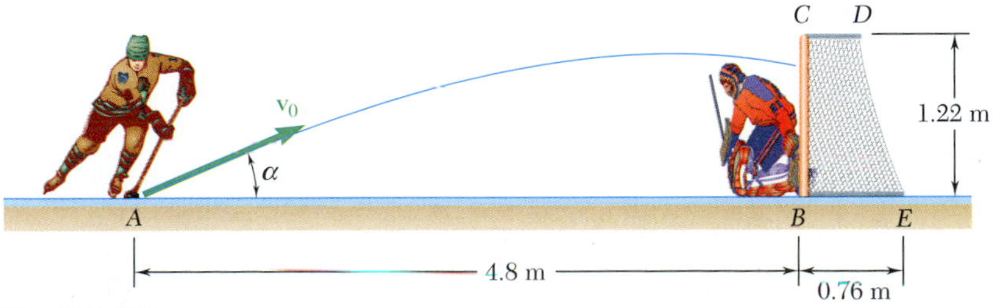

Fig. P11.112

11.113 The pitcher in a softball game throws a ball with an initial velocity $\mathbf{v}_0$ of 40 mi/h at an angle α with the horizontal. If the height of the ball at point B is 2.2 ft, determine (*a*) the angle α, (*b*) the angle θ that the velocity of the ball at point B forms with the horizontal.

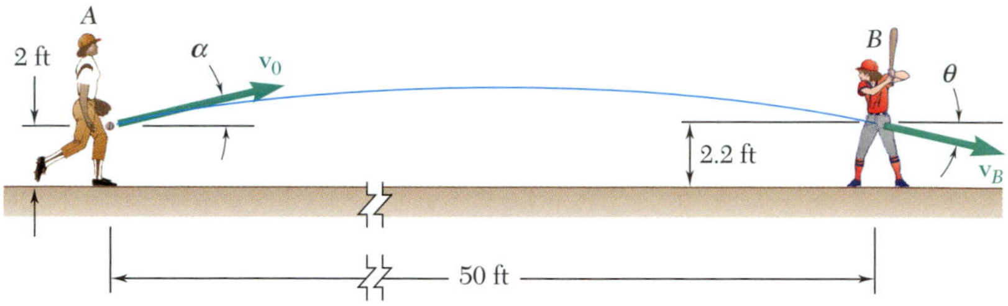

Fig. *P11.113*

11.114 A worker uses high-pressure water to clean the inside of a long drainpipe. If the water is discharged with an initial velocity $\mathbf{v}_0$ of 35 ft/s, determine (*a*) the distance d to the farthest point B on the top of the pipe that the water can wash from his position at A, (*b*) the corresponding angle α.

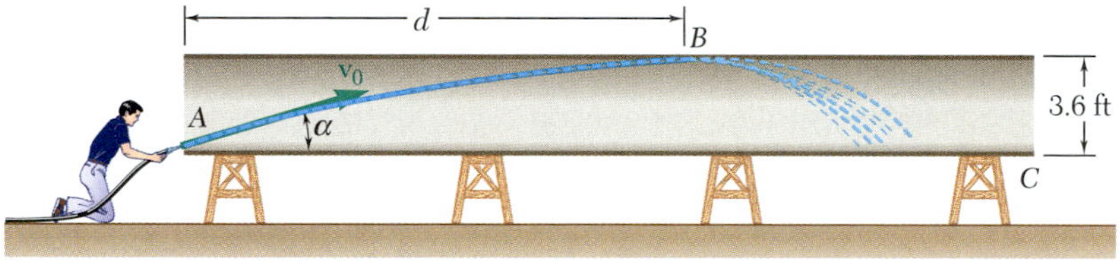

Fig. *P11.114*

11.115 A nozzle at A discharges water with an initial velocity of 12 m/s at an angle α with the horizontal. Determine (*a*) the distance d to the farthest point B on the roof that the water can reach, (*b*) the corresponding angle α. Check that the stream will clear the edge of the roof.

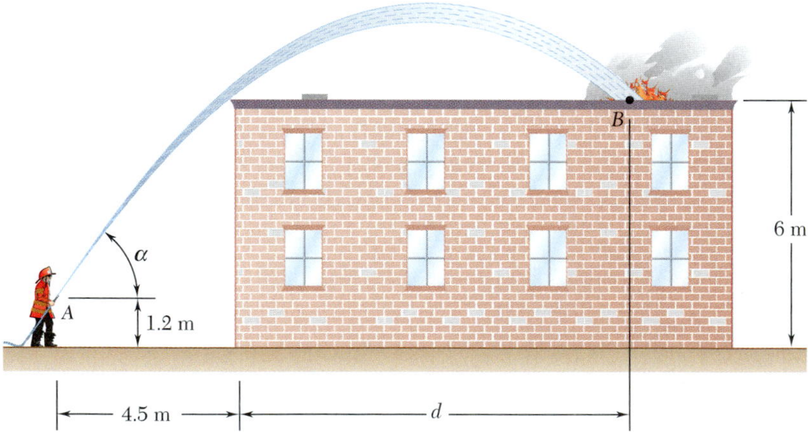

Fig. P11.115

11.116 A projectile is launched from point A with an initial velocity $\mathbf{v}_0$ of 40 m/s at an angle α with the vertical. Determine (*a*) the distance d to the farthest point B on the hill that the projectile can reach, (*b*) the corresponding angle α, (*c*) the maximum height of the projectile above the surface.

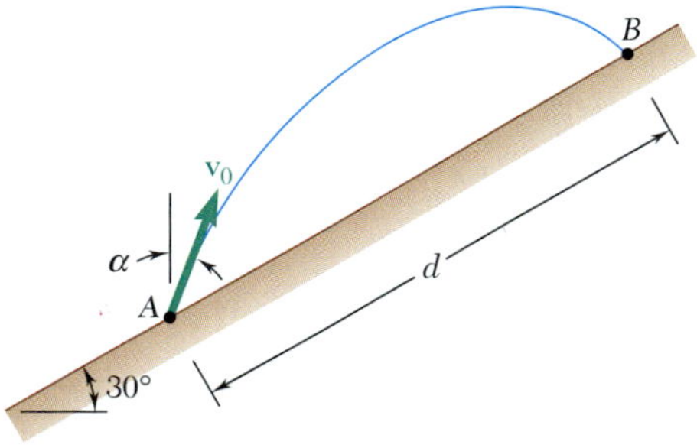

Fig. P11.116

11.117 At an intersection car A is traveling south with a velocity of 40 km/h when it is struck by car B traveling 30° north of east with a velocity of 50 km/h. Determine the relative velocity of car B with respect to car A.

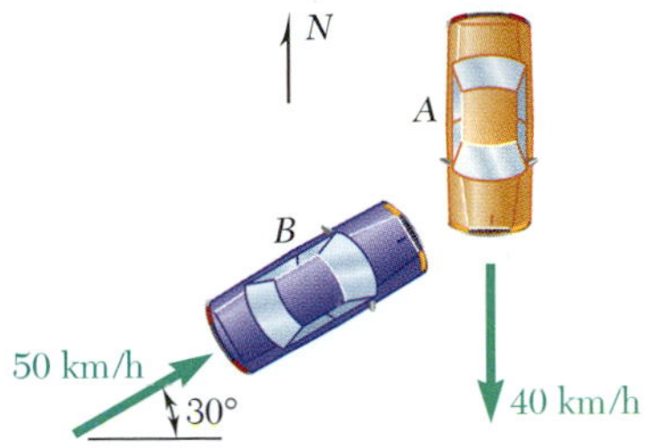

Fig. P11.117

11.118 Small wheels attached to the ends of rod AB roll along two surfaces. Knowing that at the instant shown the velocity $\mathbf{v}_A$ of wheel A is 1.5 m/s to the right and the relative velocity $\mathbf{v}_{B/A}$ of wheel B with respect to wheel A is perpendicular to rod AB, determine (a) the relative velocity $\mathbf{v}_{B/A}$, (b) the velocity $\mathbf{v}_B$ of wheel B.

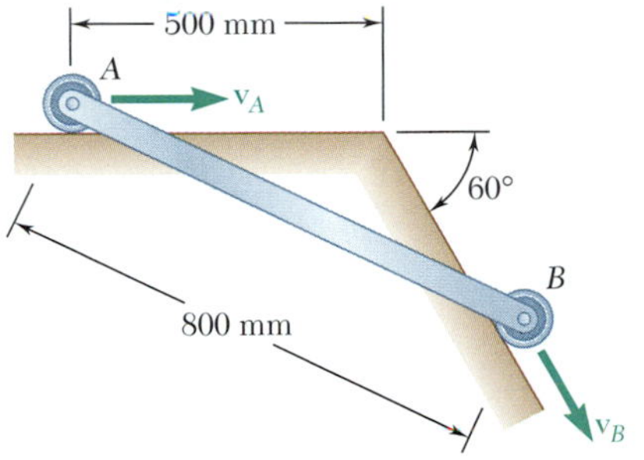

Fig. P11.118

11.119 As slider block A moves downward at a speed of 1.6 ft/s, the velocity with respect to A of the portion of belt B between the idler pulleys C and D is $\mathbf{v}_{CD/A} = 6.4$ ft/s $\measuredangle\,\theta$. Determine the velocity of portion CD of the belt when (a) $\theta = 45°$, (b) $\theta = 60°$.

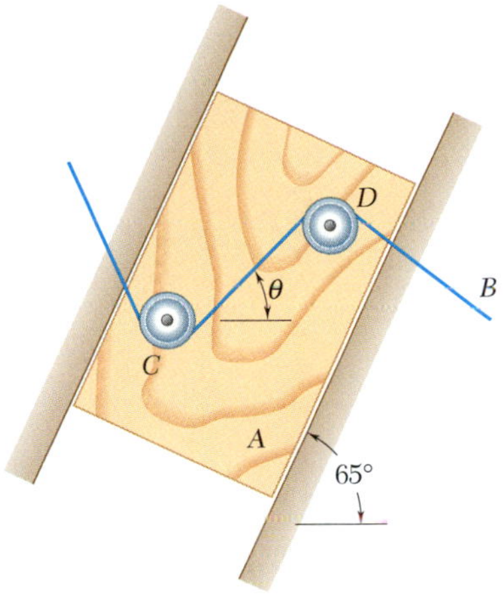

Fig. P11.119

11.120 Shore-based radar indicates that a ferry leaves its slip with a velocity $\mathbf{v} = 10$ knots $\measuredangle$ 65°, while instruments aboard the ferry indicate a speed of 10.4 knots and a heading of 35° west of south relative to the river. Determine the velocity of the river.

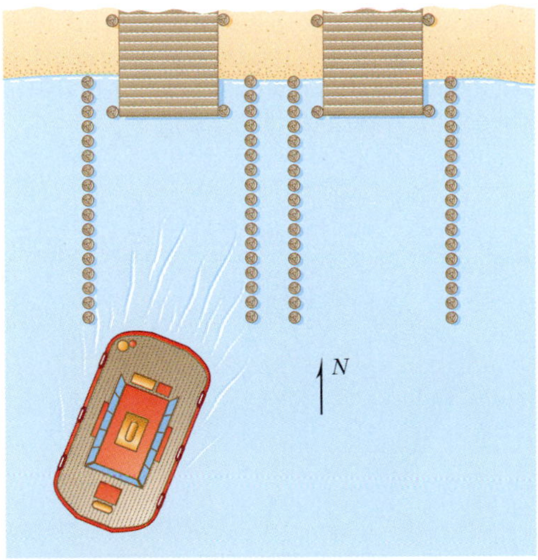

Fig. P11.120

11.121 The velocities of commuter trains A and B are as shown. Knowing that the speed of each train is constant and that B reaches the crossing 10 min after A passed through the same crossing, determine (*a*) the relative velocity of B with respect to A, (*b*) the distance between the fronts of the engines 3 min after A passed through the crossing.

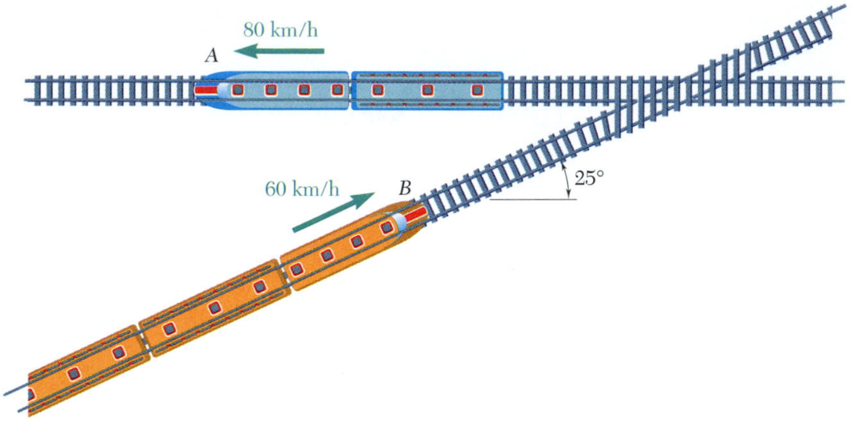

Fig. ***P11.121***

Fig. ***P11.122***

11.122 Airplanes A and B are flying at the same altitude and are tracking the eye of hurricane C. The relative velocity of C with respect to A is $\mathbf{v}_{C/A} = 470$ km/h ∠ 75°, and the relative velocity of C with respect to B is $\mathbf{v}_{C/B} = 520$ km/h ∠ 40°. Determine (*a*) the relative velocity of B with respect to A, (*b*) the velocity of A if ground-based radar indicates that the hurricane is moving at a speed of 48 km/h due north, (*c*) the change in position of C with respect to B during a 15-min interval.

11.123 Slider block B starts from rest and moves to the right with a constant acceleration of 300 mm/s^2. Determine (*a*) the relative acceleration of portion C of the cable with respect to slider block A, (*b*) the velocity of portion C of the cable after 2 s.

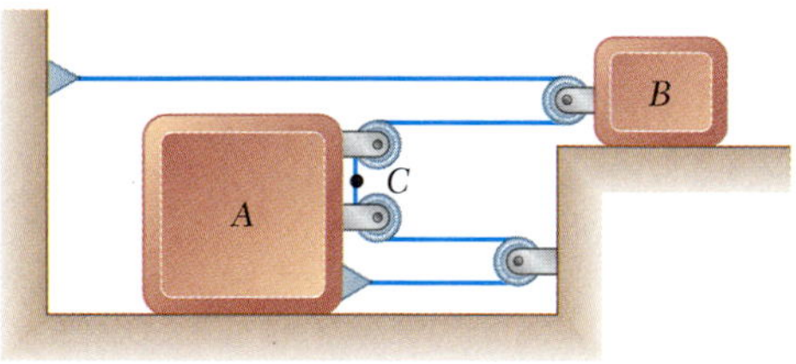

Fig. P11.123

11.124 Pin P moves at a constant speed of 8 in./s in a counterclockwise sense along a circular slot which has been milled in the block A as shown. Knowing that the block moves up the incline at a constant speed of 4.8 in./s, determine the magnitude and the direction relative to the xy axes of the velocity of pin P when (*a*) $\theta = 30°$, (*b*) $\theta = 135°$.

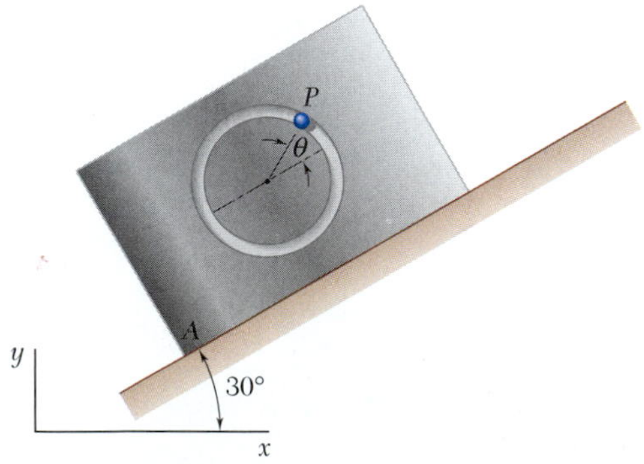

Fig. P11.124

11.125 As the truck shown begins to back up with a constant acceleration of 4 ft/s^2, the outer section B of its boom starts to retract with a constant acceleration of 1.6 ft/s^2 relative to the truck. Determine (*a*) the acceleration of section B, (*b*) the velocity of section B when $t = 2$ s.

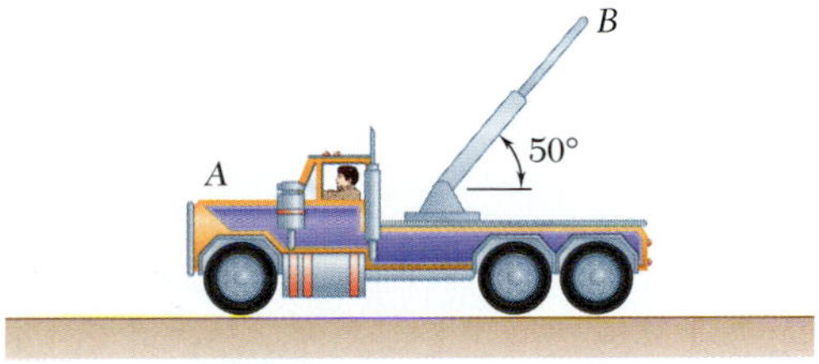

Fig. P11.125

11.126 A boat is moving to the right with a constant deceleration of 1 ft/s^2 when a boy standing on the deck D throws a ball with an initial velocity relative to the deck which is vertical. The ball rises to a maximum height of 25 ft above the release point and the boy must step forward a distance d to catch it at the same height as the release point. Determine (*a*) the distance d, (*b*) the relative velocity of the ball with respect to the deck when the ball is caught.

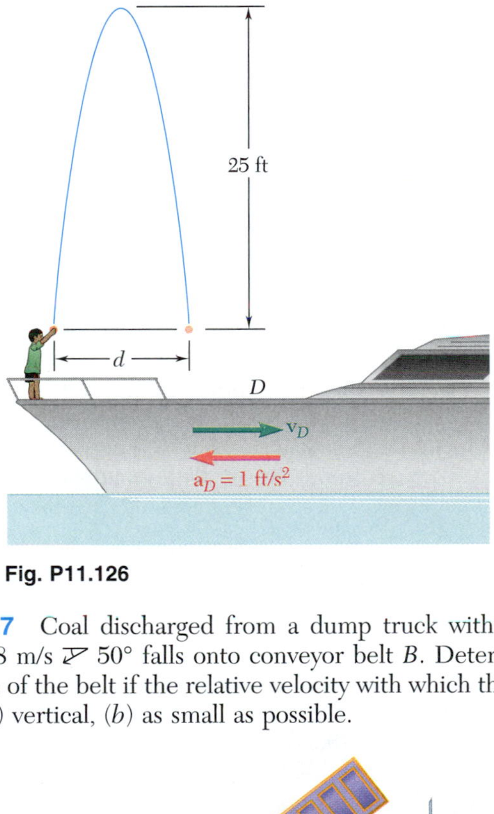

Fig. P11.126

11.127 Coal discharged from a dump truck with an initial velocity $(\mathbf{v}_C)_0 = 1.8$ m/s $\measuredangle$ 50° falls onto conveyor belt B. Determine the required velocity $\mathbf{v}_B$ of the belt if the relative velocity with which the coal hits the belt is to be (*a*) vertical, (*b*) as small as possible.

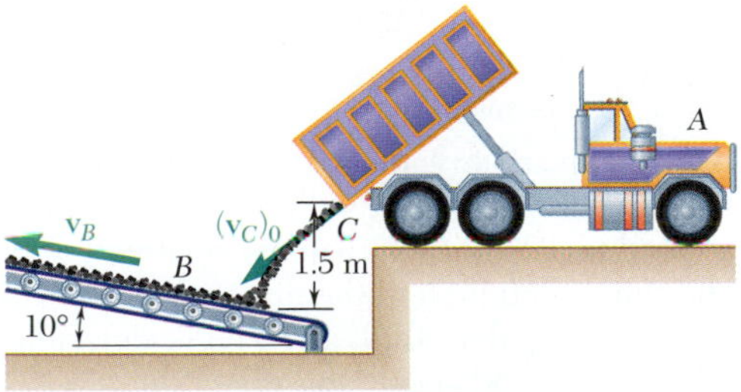

Fig. P11.127

11.128 Conveyor belt A, which forms a 20° angle with the horizontal, moves at a constant speed of 1.2 m/s and is used to load an airplane. Knowing that a worker tosses duffel bag B with an initial velocity of 0.76 m/s at an angle of 30° with the horizontal, determine the velocity of the bag relative to the belt as it lands on the belt.

Fig. P11.128

11.129 As the driver of an automobile travels north at 20 km/h in a parking lot, he observes a truck approaching from the northwest. After he reduces his speed to 12 km/h and turns so that he is traveling in a northwest direction, the truck appears to be approaching from the west. Assuming that the velocity of the truck is constant during that period of observation, determine the magnitude and the direction of the velocity of the truck.

11.130 The paths of raindrops during a storm appear to form an angle of 75° with the vertical and to be directed to the left when observed through a left-side window of an automobile traveling north at a speed of 64 km/h. When observed through the right-side window of an automobile traveling south at a speed of 48 km/h, the raindrops appear to form an angle of 60° with the vertical. If the driver of the automobile traveling north were to stop, at what angle and with what speed would she observe the drops to fall?

Fig. *P11.131*

11.131 When a small boat travels north at 3 mi/h, a flag mounted on its stern forms an angle $\theta = 50°$ with the centerline of the boat as shown. A short time later, when the boat travels east at 12 mi/h, angle θ is again 50°. Determine the speed and the direction of the wind.

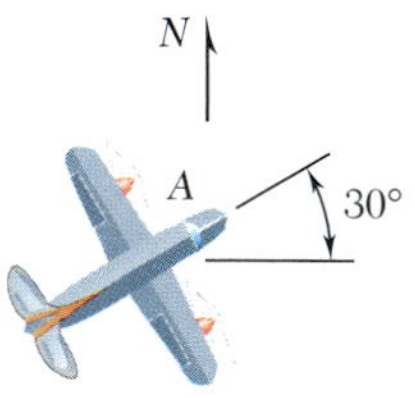

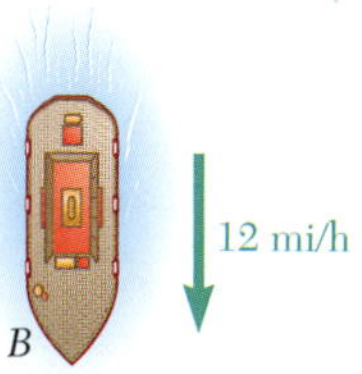

Fig. *P11.132*

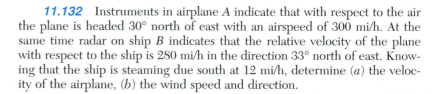

11.132 Instruments in airplane A indicate that with respect to the air the plane is headed 30° north of east with an airspeed of 300 mi/h. At the same time radar on ship B indicates that the relative velocity of the plane with respect to the ship is 280 mi/h in the direction 33° north of east. Knowing that the ship is steaming due south at 12 mi/h, determine (*a*) the velocity of the airplane, (*b*) the wind speed and direction.

11.13. TANGENTIAL AND NORMAL COMPONENTS

We saw in Sec. 11.9 that the velocity of a particle is a vector tangent to the path of the particle but that, in general, the acceleration is not tangent to the path. It is sometimes convenient to resolve the acceleration into components directed, respectively, along the tangent and the normal to the path of the particle.

Plane Motion of a Particle. First, let us consider a particle which moves along a curve contained in the plane of the figure. Let P be the position of the particle at a given instant. We attach at P a unit vector $\mathbf{e}_t$ tangent to the path of the particle and pointing in the direction of motion (Fig. 11.21*a*). Let $\mathbf{e}_t'$ be the unit vector corresponding to the position P' of the particle at a later instant. Drawing both vectors from the same origin O', we define the vector $\Delta\mathbf{e}_t = \mathbf{e}_t' - \mathbf{e}_t$ (Fig. 11.21*b*). Since $\mathbf{e}_t$ and $\mathbf{e}_t'$ are of unit length, their tips lie on a circle of radius 1. Denoting by $\Delta\theta$ the angle formed by $\mathbf{e}_t$ and $\mathbf{e}_t'$, we find that the magnitude of $\Delta\mathbf{e}_t$ is $2 \sin (\Delta\theta/2)$. Considering now the vector $\Delta\mathbf{e}_t/\Delta\theta$, we note that as $\Delta\theta$ approaches zero, this vector becomes tangent to the unit circle of Fig. 11.21*b*, that is, perpendicular to $\mathbf{e}_t$, and that its magnitude approaches

$$\lim_{\Delta\theta \to 0} \frac{2 \sin (\Delta\theta/2)}{\Delta\theta} = \lim_{\Delta\theta \to 0} \frac{\sin (\Delta\theta/2)}{\Delta\theta/2} = 1$$

Thus, the vector obtained in the limit is a unit vector along the normal to the path of the particle, in the direction toward which $\mathbf{e}_t$ turns. Denoting this vector by $\mathbf{e}_n$, we write

$$\mathbf{e}_n = \lim_{\Delta\theta \to 0} \frac{\Delta\mathbf{e}_t}{\Delta\theta}$$

$$\mathbf{e}_n = \frac{d\mathbf{e}_t}{d\theta} \tag{11.35}$$

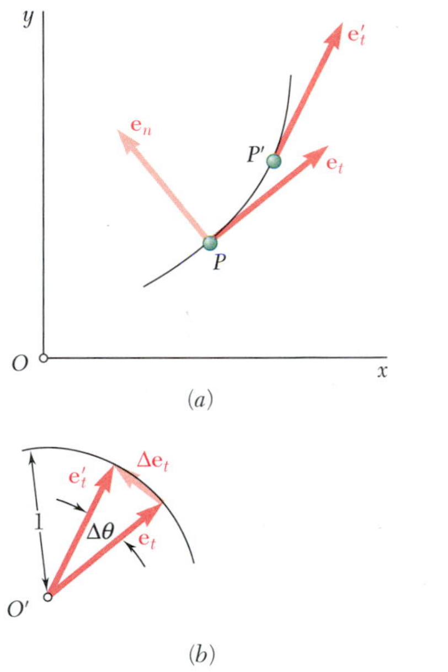

Fig. 11.21

Since the velocity $\mathbf{v}$ of the particle is tangent to the path, it can be expressed as the product of the scalar v and the unit vector $\mathbf{e}_t$. We have

$$\mathbf{v} = v\mathbf{e}_t \tag{11.36}$$

To obtain the acceleration of the particle, (11.36) will be differentiated with respect to t. Applying the rule for the differentiation of the product of a scalar and a vector function (Sec. 11.10), we write

$$\mathbf{a} = \frac{d\mathbf{v}}{dt} = \frac{dv}{dt}\mathbf{e}_t + v\frac{d\mathbf{e}_t}{dt} \tag{11.37}$$

But

$$\frac{d\mathbf{e}_t}{dt} = \frac{d\mathbf{e}_t}{d\theta}\frac{d\theta}{ds}\frac{ds}{dt}$$

Recalling from (11.16) that $ds/dt = v$, from (11.35) that $d\mathbf{e}_t/d\theta = \mathbf{e}_n$, and from elementary calculus that $d\theta/ds$ is equal to $1/\rho$, where ρ is the radius of curvature of the path at P (Fig. 11.22), we have

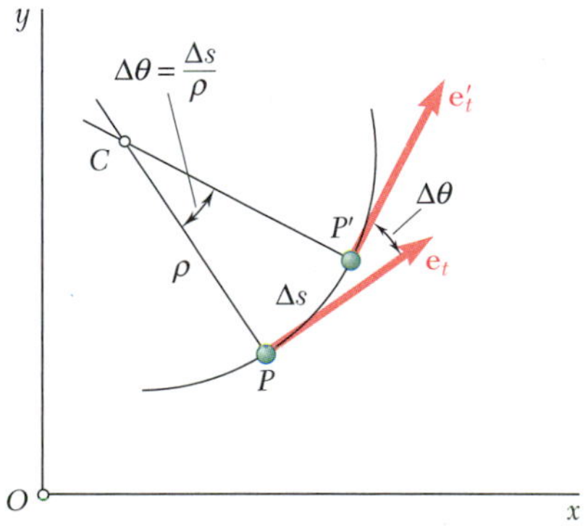

Fig. 11.22

$$\frac{d\mathbf{e}_t}{dt} = \frac{v}{\rho}\mathbf{e}_n \tag{11.38}$$

Substituting into (11.37), we obtain

$$\mathbf{a} = \frac{dv}{dt}\mathbf{e}_t + \frac{v^2}{\rho}\mathbf{e}_n \tag{11.39}$$

Thus, the scalar components of the acceleration are

$$a_t = \frac{dv}{dt} \qquad a_n = \frac{v^2}{\rho} \tag{11.40}$$

Photo 11.4 As he travels on the curved portion of a racetrack, each cyclist is subjected to a normal component of acceleration directed toward the center of curvature of his path.

The relations obtained express that the *tangential component* of the acceleration is equal to the *rate of change of the speed of the particle*, while the *normal component* is equal to the *square of the speed divided by the radius of curvature of the path at P.* If the speed of the particle increases, a_t is positive and the vector component $\mathbf{a}_t$ points in the direction of motion. If the speed of the particle decreases, a_t is negative and $\mathbf{a}_t$ points against the direction of motion. The vector component $\mathbf{a}_n$, on the other hand, *is always directed toward the center of curvature C of the path* (Fig. 11.23).

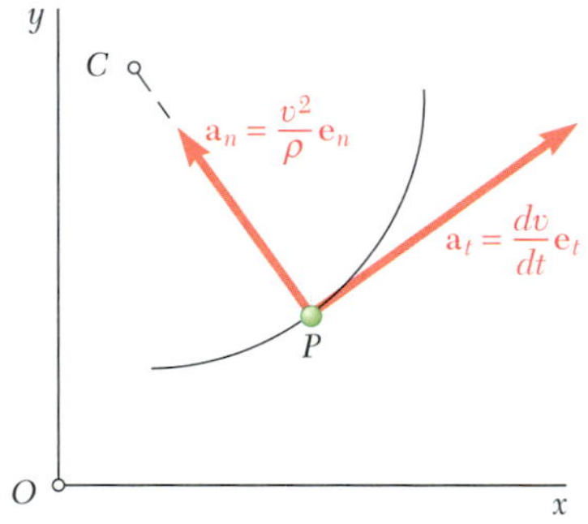

Fig 11.23

We conclude from the above that the tangential component of the acceleration reflects a change in the speed of the particle, while its normal component reflects a change in the direction of motion of the particle. The acceleration of a particle will be zero only if both its components are zero. Thus, the acceleration of a particle moving with constant speed along a curve will not be zero unless the particle happens to pass through a point of inflection of the curve (where the radius of curvature is infinite) or unless the curve is a straight line.

The fact that the normal component of the acceleration depends upon the radius of curvature of the path followed by the particle is taken into account in the design of structures or mechanisms as widely different as airplane wings, railroad tracks, and cams. In order to avoid sudden changes in the acceleration of the air particles flowing past a wing, wing profiles are designed without any sudden change in curvature. Similar care is taken in designing railroad curves, to avoid sudden changes in the acceleration of the cars (which would be hard on the equipment and unpleasant for the passengers). A straight section of track, for instance, is never directly followed by a circular section. Special transition sections are used to help pass smoothly from the infinite radius of curvature of the straight section to the finite radius of the circular track. Likewise, in the design of high-speed cams, abrupt changes in acceleration are avoided by using transition curves which produce a continuous change in acceleration.

Motion of a Particle in Space. The relations (11.39) and (11.40) still hold in the case of a particle moving along a space curve. However, since there are an infinite number of straight lines which are perpendicular to the tangent at a given point *P* of a space curve, it is necessary to define more precisely the direction of the unit vector $\mathbf{e}_n$.

Let us consider again the unit vectors $\mathbf{e}_t$ and $\mathbf{e}_t'$ tangent to the path of the particle at two neighboring points *P* and *P'* (Fig. 11.24*a*) and the vector $\Delta\mathbf{e}_t$ representing the difference between $\mathbf{e}_t$ and $\mathbf{e}_t'$

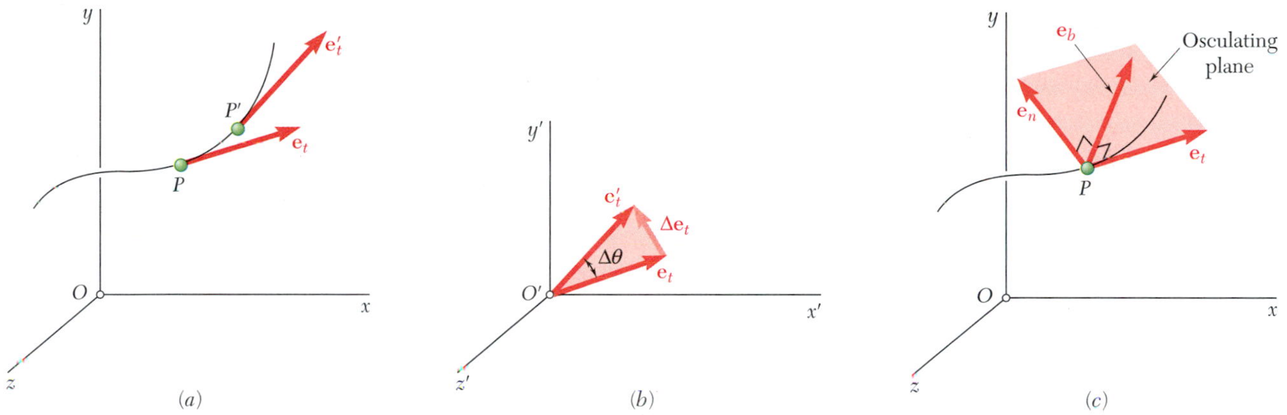

Fig 11.24

(Fig. 11.24*b*). Let us now imagine a plane through *P* (Fig. 11.24*a*) parallel to the plane defined by the vectors $\mathbf{e}_t$, $\mathbf{e}_t'$, and $\Delta\mathbf{e}_t$ (Fig. 11.24*b*). This plane contains the tangent to the curve at *P* and is parallel to the tangent at *P'*. If we let *P'* approach *P*, we obtain in the limit the plane which fits the curve most closely in the neighborhood of *P*. This plane is called the *osculating plane* at *P*.† It follows from this definition that the osculating plane contains the unit vector $\mathbf{e}_n$, since this vector represents the limit of the vector $\Delta\mathbf{e}_t/\Delta\theta$. The normal defined by $\mathbf{e}_n$ is thus contained in the osculating plane; it is called the *principal normal* at *P*. The unit vector $\mathbf{e}_b = \mathbf{e}_t \times \mathbf{e}_n$ which completes the right-handed triad $\mathbf{e}_t$, $\mathbf{e}_n$, $\mathbf{e}_b$ (Fig. 11.24*c*) defines the *binormal* at *P*. The binormal is thus perpendicular to the osculating plane. We conclude that the acceleration of the particle at *P* can be resolved into two components, one along the tangent, the other along the principal normal at *P*, as indicated in Eq. (11.39). Note that the acceleration has no component along the binormal.

†From the Latin *osculari*, to kiss.

11.14. RADIAL AND TRANSVERSE COMPONENTS

In certain problems of plane motion, the position of the particle P is defined by its polar coordinates r and θ (Fig. 11.25*a*). It is then convenient to resolve the velocity and acceleration of the particle into components parallel and perpendicular, respectively, to the line OP. These components are called *radial* and *transverse components*.

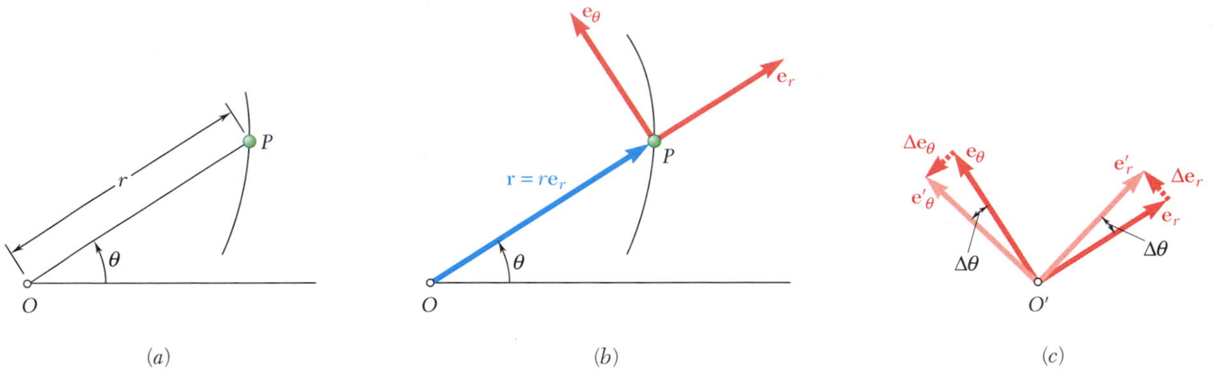

Fig. 11.25

We attach at P two unit vectors, $\mathbf{e}_r$ and $\mathbf{e}_\theta$ (Fig. 11.25*b*). The vector $\mathbf{e}_r$ is directed along OP and the vector $\mathbf{e}_\theta$ is obtained by rotating $\mathbf{e}_r$ through 90° counterclockwise. The unit vector $\mathbf{e}_r$ defines the *radial* direction, that is, the direction in which P would move if r were increased and θ were kept constant; the unit vector $\mathbf{e}_\theta$ defines the *transverse* direction, that is, the direction in which P would move if θ were increased and r were kept constant. A derivation similar to the one we used in Sec. 11.13 to determine the derivative of the unit vector $\mathbf{e}_t$ leads to the relations

$$\frac{d\mathbf{e}_r}{d\theta} = \mathbf{e}_\theta \qquad \frac{d\mathbf{e}_\theta}{d\theta} = -\mathbf{e}_r \tag{11.41}$$

where $-\mathbf{e}_r$ denotes a unit vector of sense opposite to that of $\mathbf{e}_r$ (Fig. 11.25*c*). Using the chain rule of differentiation, we express the time derivatives of the unit vectors $\mathbf{e}_r$ and $\mathbf{e}_\theta$ as follows:

$$\frac{d\mathbf{e}_r}{dt} = \frac{d\mathbf{e}_r}{d\theta}\frac{d\theta}{dt} = \mathbf{e}_\theta\frac{d\theta}{dt} \qquad \frac{d\mathbf{e}_\theta}{dt} = \frac{d\mathbf{e}_\theta}{d\theta}\frac{d\theta}{dt} = -\mathbf{e}_r\frac{d\theta}{dt}$$

or, using dots to indicate differentiation with respect to t,

$$\dot{\mathbf{e}}_r = \dot\theta\mathbf{e}_\theta \qquad \dot{\mathbf{e}}_\theta = -\dot\theta\mathbf{e}_r \tag{11.42}$$

To obtain the velocity $\mathbf{v}$ of the particle P, we express the position vector $\mathbf{r}$ of P as the product of the scalar r and the unit vector $\mathbf{e}_r$ and differentiate with respect to t:

$$\mathbf{v} = \frac{d}{dt}(r\mathbf{e}_r) = \dot r\mathbf{e}_r + r\dot{\mathbf{e}}_r$$

or, recalling the first of the relations (11.42),

$$\mathbf{v} = \dot r\mathbf{e}_r + r\dot\theta\mathbf{e}_\theta \tag{11.43}$$

Differentiating again with respect to t to obtain the acceleration, we write

$$\mathbf{a} = \frac{d\mathbf{v}}{dt} = \ddot{r}\mathbf{e}_r + \dot{r}\dot{\mathbf{e}}_r + \dot{r}\dot{\theta}\mathbf{e}_\theta + r\ddot{\theta}\mathbf{e}_\theta + r\dot{\theta}\dot{\mathbf{e}}_\theta$$

or, substituting for $\dot{\mathbf{e}}_r$ and $\dot{\mathbf{e}}_\theta$ from (11.42) and factoring $\mathbf{e}_r$ and $\mathbf{e}_\theta$,

$$\mathbf{a} = (\ddot{r} - r\dot{\theta}^2)\mathbf{e}_r + (r\ddot{\theta} + 2\dot{r}\dot{\theta})\mathbf{e}_\theta \qquad (11.44)$$

The scalar components of the velocity and the acceleration in the radial and transverse directions are, therefore,

$$v_r = \dot{r} \qquad v_\theta = r\dot{\theta} \qquad (11.45)$$

$$a_r = \ddot{r} - r\dot{\theta}^2 \qquad a_\theta = r\ddot{\theta} + 2\dot{r}\dot{\theta} \qquad (11.46)$$

It is important to note that a_r is *not* equal to the time derivative of v_r and that a_θ is *not* equal to the time derivative of v_θ.

In the case of a particle moving along a circle of center O, we have $r =$ constant and $\dot{r} = \ddot{r} = 0$, and the formulas (11.43) and (11.44) reduce, respectively, to

$$\mathbf{v} = r\dot{\theta}\mathbf{e}_\theta \qquad \mathbf{a} = -r\dot{\theta}^2\mathbf{e}_r + r\ddot{\theta}\mathbf{e}_\theta \qquad (11.47)$$

Extension to the Motion of a Particle in Space: Cylindrical Coordinates. The position of a particle P in space is sometimes defined by its cylindrical coordinates R, θ, and z (Fig. 11.26*a*). It is then convenient to use the unit vectors $\mathbf{e}_R$, $\mathbf{e}_\theta$, and $\mathbf{k}$ shown in Fig. 11.26*b*. Resolving the position vector $\mathbf{r}$ of the particle P into components along the unit vectors, we write

$$\mathbf{r} = R\mathbf{e}_R + z\mathbf{k} \qquad (11.48)$$

Observing that $\mathbf{e}_R$ and $\mathbf{e}_\theta$ define, respectively, the radial and transverse directions in the horizontal xy plane, and that the vector $\mathbf{k}$, which defines the *axial* direction, is constant in direction as well as in magnitude, we easily verify that

$$\mathbf{v} = \frac{d\mathbf{r}}{dt} = \dot{R}\mathbf{e}_R + R\dot{\theta}\mathbf{e}_\theta + \dot{z}\mathbf{k} \qquad (11.49)$$

$$\mathbf{a} = \frac{d\mathbf{v}}{dt} = (\ddot{R} - R\dot{\theta}^2)\mathbf{e}_R + (R\ddot{\theta} + 2\dot{R}\dot{\theta})\mathbf{e}_\theta + \ddot{z}\mathbf{k} \qquad (11.50)$$

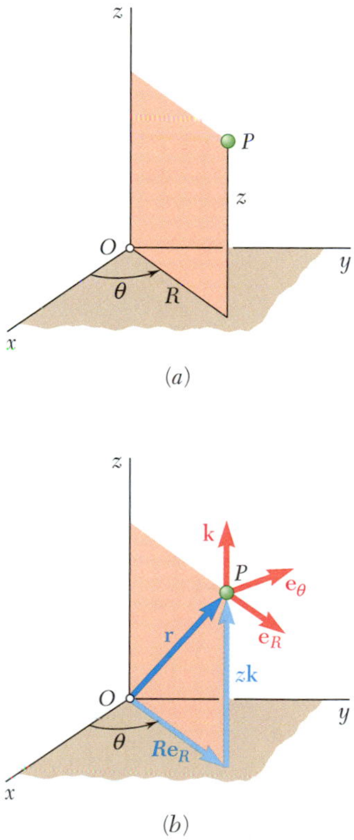

Fig. 11.26

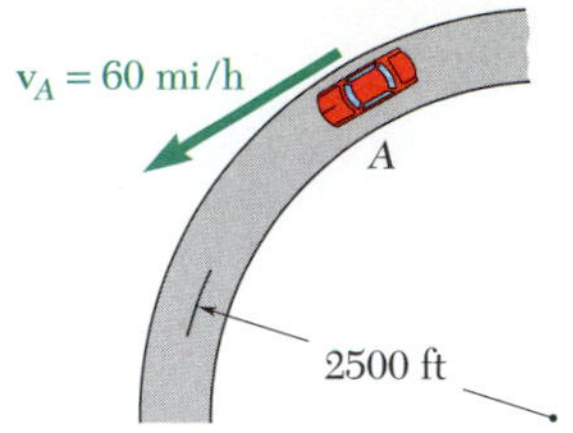

SAMPLE PROBLEM 11.10

A motorist is traveling on a curved section of highway of radius 2500 ft at the speed of 60 mi/h. The motorist suddenly applies the brakes, causing the automobile to slow down at a constant rate. Knowing that after 8 s the speed has been reduced to 45 mi/h, determine the acceleration of the automobile immediately after the brakes have been applied.

SOLUTION

Tangential Component of Acceleration. First the speeds are expressed in ft/s.

$$60 \text{ mi/h} = \left(60\frac{\text{mi}}{\text{h}}\right)\left(\frac{5280 \text{ ft}}{1 \text{ mi}}\right)\left(\frac{1 \text{ h}}{3600 \text{ s}}\right) = 88 \text{ ft/s}$$

$$45 \text{ mi/h} = 66 \text{ ft/s}$$

Since the automobile slows down at a constant rate, we have

$$a_t = \text{average } a_t = \frac{\Delta v}{\Delta t} = \frac{66 \text{ ft/s} - 88 \text{ ft/s}}{8 \text{ s}} = -2.75 \text{ ft/s}^2$$

Normal Component of Acceleration. Immediately after the brakes have been applied, the speed is still 88 ft/s, and we have

$$a_n = \frac{v^2}{\rho} = \frac{(88 \text{ ft/s})^2}{2500 \text{ ft}} = 3.10 \text{ ft/s}^2$$

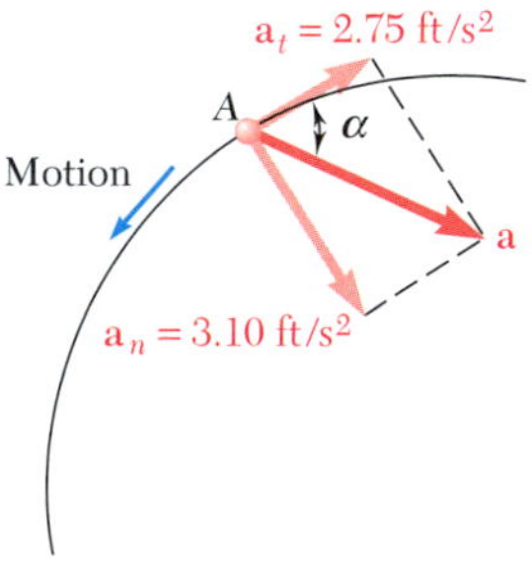

Magnitude and Direction of Acceleration. The magnitude and direction of the resultant **a** of the components $\mathbf{a}_n$ and $\mathbf{a}_t$ are

$$\tan \alpha = \frac{a_n}{a_t} = \frac{3.10 \text{ ft/s}^2}{2.75 \text{ ft/s}^2} \qquad \alpha = 48.4° \blacktriangleleft$$

$$a = \frac{a_n}{\sin \alpha} = \frac{3.10 \text{ ft/s}^2}{\sin 48.4°} \qquad \mathbf{a} = 4.14 \text{ ft/s}^2 \blacktriangleleft$$

SAMPLE PROBLEM 11.11

Determine the minimum radius of curvature of the trajectory described by the projectile considered in Sample Prob. 11.7.

SOLUTION

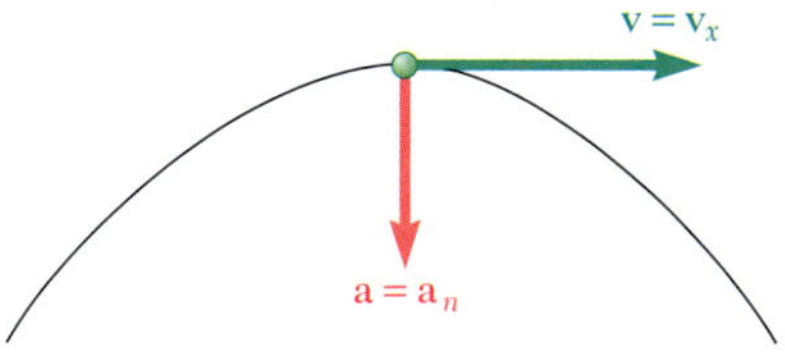

Since $a_n = v^2/\rho$, we have $\rho = v^2/a_n$. The radius will be small when v is small or when a_n is large. The speed v is minimum at the top of the trajectory since $v_y = 0$ at that point; a_n is maximum at that same point, since the direction of the vertical coincides with the direction of the normal. Therefore, the minimum radius of curvature occurs at the top of the trajectory. At this point, we have

$$v = v_x = 155.9 \text{ m/s} \qquad a_n = a = 9.81 \text{ m/s}^2$$

$$\rho = \frac{v^2}{a_n} = \frac{(155.9 \text{ m/s})^2}{9.81 \text{ m/s}^2} \qquad \rho = 2480 \text{ m} \blacktriangleleft$$

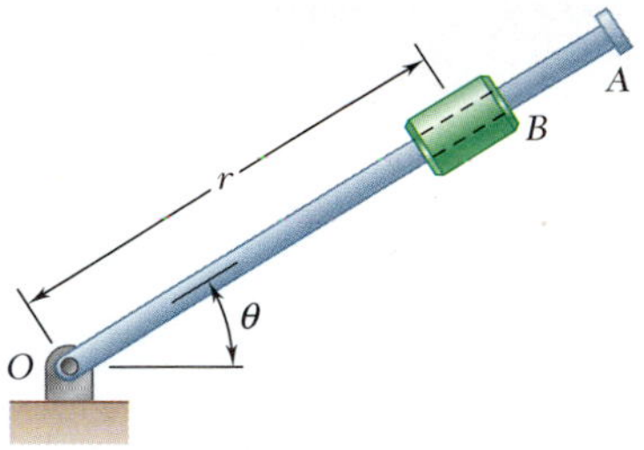

SAMPLE PROBLEM 11.12

The rotation of the 0.9-m arm OA about O is defined by the relation $\theta = 0.15t^2$, where θ is expressed in radians and t in seconds. Collar B slides along the arm in such a way that its distance from O is $r = 0.9 - 0.12t^2$, where r is expressed in meters and t in seconds. After the arm OA has rotated through 30°, determine (*a*) the total velocity of the collar, (*b*) the total acceleration of the collar, (*c*) the relative acceleration of the collar with respect to the arm.

SOLUTION

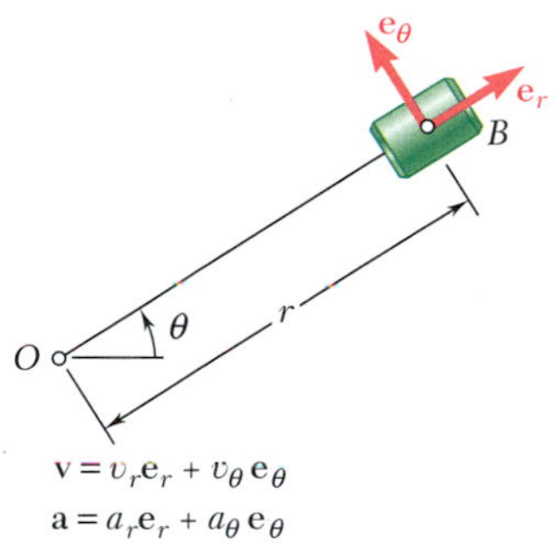

Time *t* at which *θ* = 30°. Substituting $\theta = 30° = 0.524$ rad into the expression for θ, we obtain

$$\theta = 0.15t^2 \qquad 0.524 = 0.15t^2 \qquad t = 1.869 \text{ s}$$

Equations of Motion. Substituting $t = 1.869$ s in the expressions for r, θ, and their first and second derivatives, we have

$$\begin{aligned} r &= 0.9 - 0.12t^2 = 0.481 \text{ m} & \theta &= 0.15t^2 = 0.524 \text{ rad} \\ \dot{r} &= -0.24t = -0.449 \text{ m/s} & \dot{\theta} &= 0.30t = 0.561 \text{ rad/s} \\ \ddot{r} &= -0.24 = -0.240 \text{ m/s}^2 & \ddot{\theta} &= 0.30 = 0.300 \text{ rad/s}^2 \end{aligned}$$

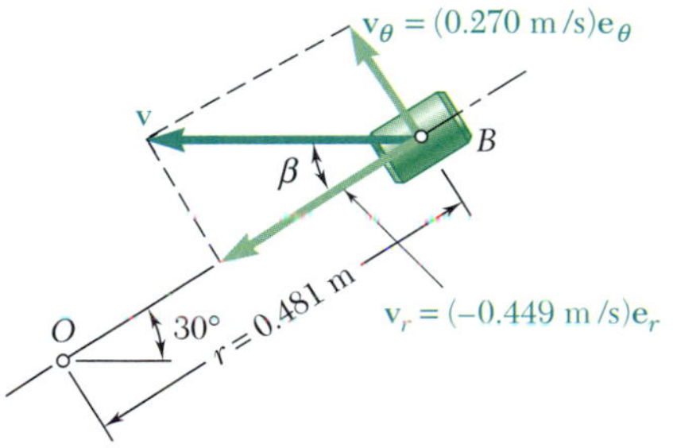

***a.* Velocity of *B*.** Using Eqs. (11.45), we obtain the values of v_r and v_θ when $t = 1.869$ s.

$$\begin{aligned} v_r &= \dot{r} = -0.449 \text{ m/s} \\ v_\theta &= r\dot{\theta} = 0.481(0.561) = 0.270 \text{ m/s} \end{aligned}$$

Solving the right triangle shown, we obtain the magnitude and direction of the velocity,

$$v = 0.524 \text{ m/s} \qquad \beta = 31.0° \quad ◀$$

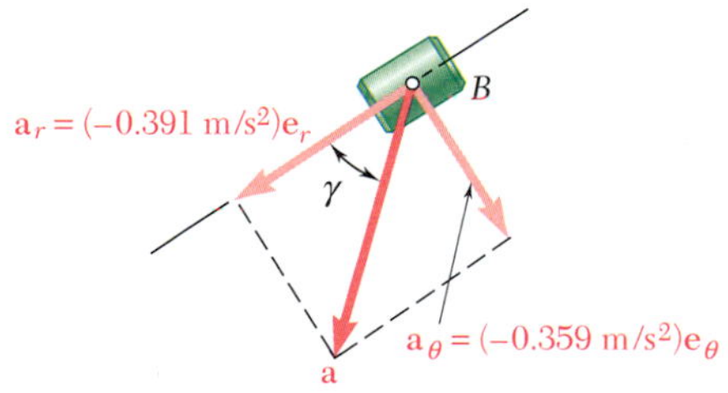

***b.* Acceleration of *B*.** Using Eqs. (11.46), we obtain

$$\begin{aligned} a_r &= \ddot{r} - r\dot{\theta}^2 \\ &= -0.240 - 0.481(0.561)^2 = -0.391 \text{ m/s}^2 \\ a_\theta &= r\ddot{\theta} + 2\dot{r}\dot{\theta} \\ &= 0.481(0.300) + 2(-0.449)(0.561) = -0.359 \text{ m/s}^2 \end{aligned}$$

$$a = 0.531 \text{ m/s}^2 \qquad \gamma = 42.6° \quad ◀$$

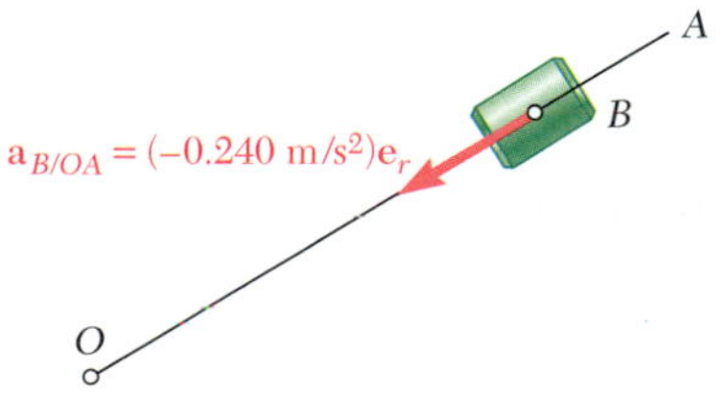

***c.* Acceleration of *B* with Respect to Arm *OA*.** We note that the motion of the collar with respect to the arm is rectilinear and defined by the coordinate r. We write

$$a_{B/OA} = \ddot{r} = -0.240 \text{ m/s}^2$$

$$a_{B/OA} = 0.240 \text{ m/s}^2 \text{ toward } O. \quad ◀$$

SOLVING PROBLEMS ON YOUR OWN

You will be asked in the following problems to express the velocity and the acceleration of particles in terms of either their *tangential and normal components* or their *radial and transverse components*. Although those components may not be as familiar to you as the rectangular components, you will find that they can simplify the solution of many problems and that certain types of motion are more easily described when they are used.

1. *Using tangential and normal components.* These components are most often used when the particle of interest travels along a circular path or when the radius of curvature of the path is to be determined. Remember that the unit vector $\mathbf{e}_t$ is tangent to the path of the particle (and thus aligned with the velocity) while the unit vector $\mathbf{e}_n$ is directed along the normal to the path and always points toward its center of curvature. It follows that, as the particle moves, the directions of the two unit vectors are constantly changing.

2. *Expressing the acceleration in terms of its tangential and normal components.* We derived in Sec. 11.13 the following equation, applicable to both the two-dimensional and the three-dimensional motion of a particle:

$$\mathbf{a} = \frac{dv}{dt}\mathbf{e}_t + \frac{v^2}{\rho}\mathbf{e}_n \tag{11.39}$$

The following observations may help you in solving the problems of this lesson.

a. *The tangential component* of the acceleration measures the rate of change of the speed: $a_t = dv/dt$. It follows that when a_t is constant, the equations for uniformly accelerated motion can be used with the acceleration equal to a_t. Furthermore, when a particle moves at a constant speed, we have $a_t = 0$ and the acceleration of the particle reduces to its normal component.

b. *The normal component* of the acceleration is always directed toward the center of curvature of the path of the particle, and its magnitude is $a_n = v^2/\rho$. Thus, the normal component can be easily determined if the speed of the particle and the radius of curvature ρ of the path are known. Conversely, when the speed and normal acceleration of the particle are known, the radius of curvature of the path can be obtained by solving this equation for ρ [Sample Prob. 11.11].

c. *In three-dimensional motion,* a third unit vector is used, $\mathbf{e}_b = \mathbf{e}_t \times \mathbf{e}_n$, which defines the direction of the *binormal*. Since this vector is perpendicular to both the velocity and the acceleration, it can be obtained by writing

$$\mathbf{e}_b = \frac{\mathbf{v} \times \mathbf{a}}{|\mathbf{v} \times \mathbf{a}|}$$

3. *Using radial and transverse components.* These components are used to analyze the plane motion of a particle P, when the position of P is defined by its polar coordinates r and θ. As shown in Fig. 11.25, the unit vector $\mathbf{e}_r$, which defines the *radial* direction, is attached to P and points away from the fixed point O, while the unit vector $\mathbf{e}_\theta$, which defines the *transverse* direction, is obtained by rotating $\mathbf{e}_r$ *counterclockwise* through 90°. The velocity and the acceleration of a particle were expressed in terms of their radial and transverse components in Eqs. (11.43) and (11.44), respectively. You will note that the expressions obtained contain the first and second derivatives with respect to t of both coordinates r and θ.

In the problems of this lesson, you will encounter the following types of problems involving radial and transverse components:

a. Both r and θ are known functions of t. In this case, you will compute the first and second derivatives of r and θ and substitute the expressions obtained into Eqs. (11.43) and (11.44).

b. A certain relationship exists between r and θ. First, you should determine this relationship from the geometry of the given system and use it to express r as a function of θ. Once the function $r = f(\theta)$ is known, you can apply the chain rule to determine $\dot{r}$ in terms of θ and $\dot{\theta}$, and $\ddot{r}$ in terms of θ, $\dot{\theta}$ and $\ddot{\theta}$:

$$\dot{r} = f'(\theta)\dot{\theta}$$
$$\ddot{r} = f''(\theta)\dot{\theta}^2 + f'(\theta)\ddot{\theta}$$

The expressions obtained can then be substituted into Eqs. (11.43) and (11.44).

c. The three-dimensional motion of a particle, as indicated at the end of Sec. 11.14, can often be effectively described in terms of the *cylindrical coordinates* R, θ, and z (Fig. 11.26). The unit vectors then should consist of $\mathbf{e}_R$, $\mathbf{e}_\theta$, and $\mathbf{k}$. The corresponding components of the velocity and the acceleration are given in Eqs. (11.49) and (11.50). Please note that the radial distance R is always measured in a plane parallel to the xy plane, and be careful not to confuse the position vector $\mathbf{r}$ with its radial component $R\mathbf{e}_R$.

Problems

11.133 The diameter of the eye of a stationary hurricane is 20 mi and the maximum wind speed is 100 mi/h at the eye wall, $r = 10$ mi. Assuming that the wind speed is constant for constant r and decreases uniformly with increasing r to 40 mi/h at $r = 110$ mi, determine the magnitude of the acceleration of the air at (*a*) $r = 10$ mi, (*b*) $r = 60$ mi, (*c*) $r = 110$ mi.

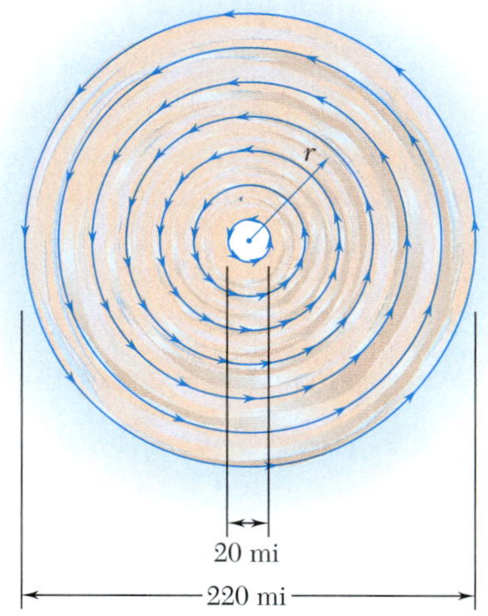

Fig. P11.133

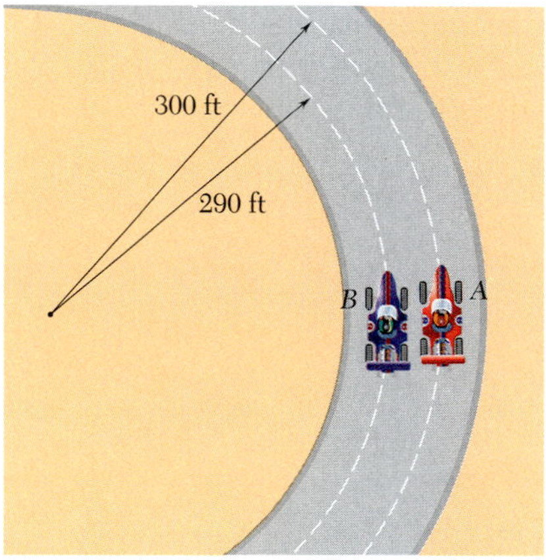

Fig. P11.134

11.134 At the instant shown, race car A is passing race car B with a relative velocity of 3 ft/s. Knowing that the speeds of both cars are constant and that the relative acceleration of car A with respect to car B is 0.9 ft/s^2 directed toward the center of curvature, determine (*a*) the speed of car A, (*b*) the speed of car B.

11.135 Determine the maximum speed that the cars of the roller-coaster can reach along the circular portion AB of the track if the normal component of their acceleration cannot exceed $3g$.

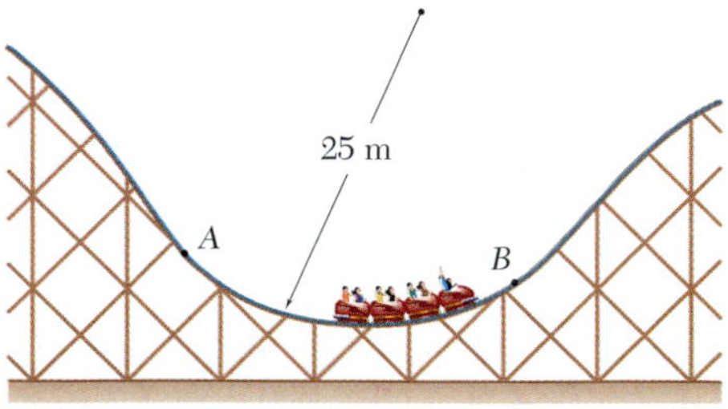

Fig. *P11.135*

11.136 As cam A rotates, follower wheel B rolls without slipping on the face of the cam. Knowing that the normal components of the acceleration of the points of contact at C of the cam A and the wheel B are 0.66 m/s^2 and 6.8 m/s^2, respectively, determine the diameter of the follower wheel.

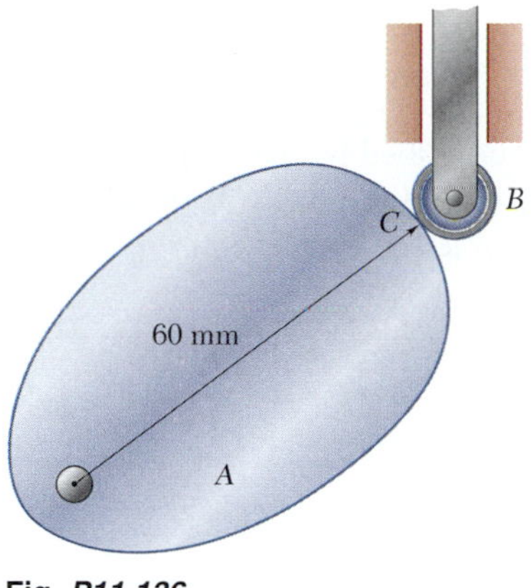

Fig. P11.136

11.137 Pin A, which is attached to link AB, is constrained to move in the circular slot CD. Knowing that at $t = 0$ the pin starts from rest and moves so that its speed increases at a constant rate of 0.8 in./s^2, determine the magnitude of its total acceleration when (*a*) $t = 0$, (*b*) $t = 2$ s.

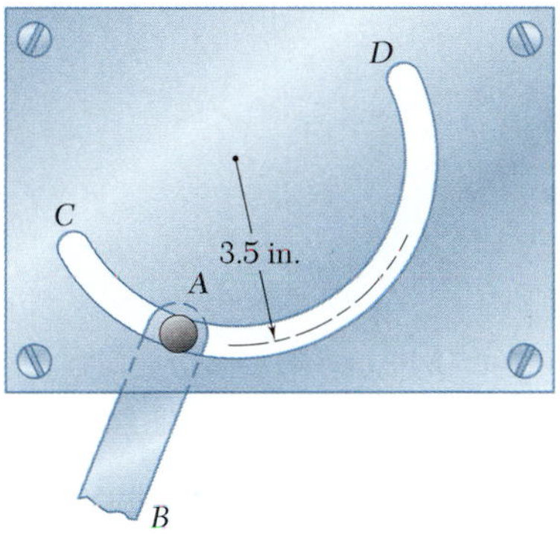

Fig. P11.137

11.138 The peripheral speed of the tooth of a 10-in.-diameter circular saw blade is 150 ft/s when the power to the saw is turned off. The speed of the tooth decreases at a constant rate, and the blade comes to rest in 9 s. Determine the time at which the total acceleration of the tooth is 130 ft/s^2.

11.139 An outdoor track is 130 m in diameter. A runner increases her speed at a constant rate from 4 to 7 m/s over a distance of 30 m. Determine the total acceleration of the runner 2 s after she begins to increase her speed.

11.140 A motorist starts from rest at point A on a circular entrance ramp when $t = 0$, increases the speed of her automobile at a constant rate and enters the highway at point B. Knowing that her speed continues to increase at the same rate until it reaches 100 km/h at point C, determine (*a*) the speed at point B, (*b*) the magnitude of the total acceleration when $t = 20$ s.

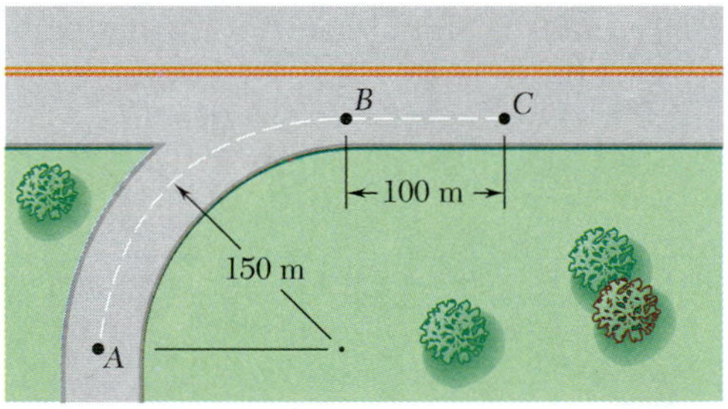

Fig. P11.140

11.141 At a given instant in an airplane race, airplane A is flying horizontally in a straight line, and its speed is being increased at a rate of 6 m/s^2. Airplane B is flying at the same altitude as airplane A and, as it rounds a pylon, is following a circular path of 200-m radius. Knowing that at the given instant the speed of B is being decreased at the rate of 2 m/s^2, determine, for the positions shown, (*a*) the velocity of B relative to A, (*b*) the acceleration of B relative to A.

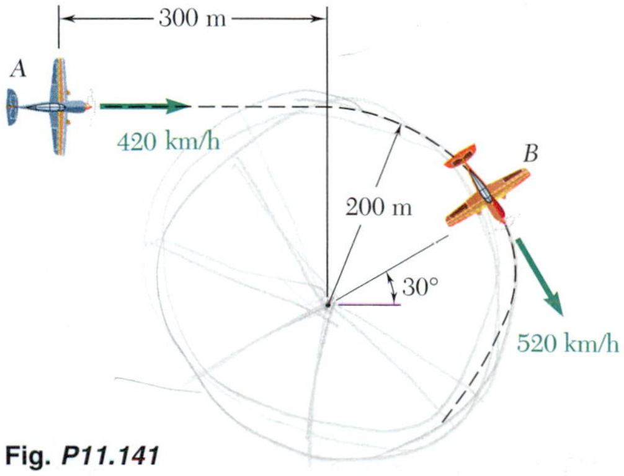

Fig. P11.141

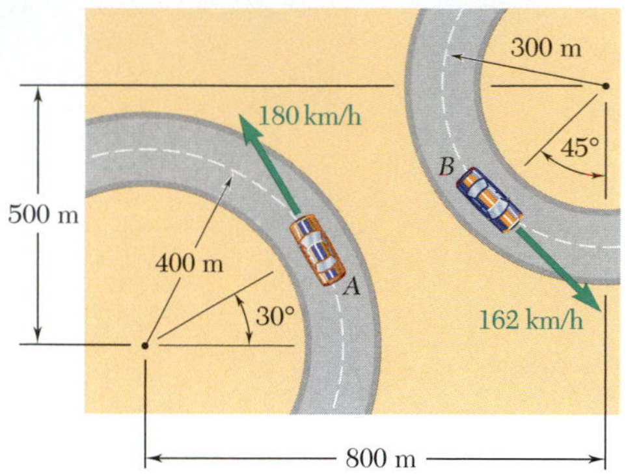

Fig. *P11.142*

11.142 Racing cars A and B are traveling on circular portions of a race track. At the instant shown, the speed of A is decreasing at the rate of 8 m/s^2, and the speed of B is increasing at the rate of 3 m/s^2. For the positions shown, determine (*a*) the velocity of B relative to A, (*b*) the acceleration of B relative to A.

11.143 A projectile is launched from point A with an initial velocity $\mathbf{v}_0$ of 40 m/s at an angle of 30° with the vertical. Determine the radius of curvature of the trajectory described by the projectile (*a*) at point A, (*b*) at the point on the trajectory where the velocity is parallel to the incline.

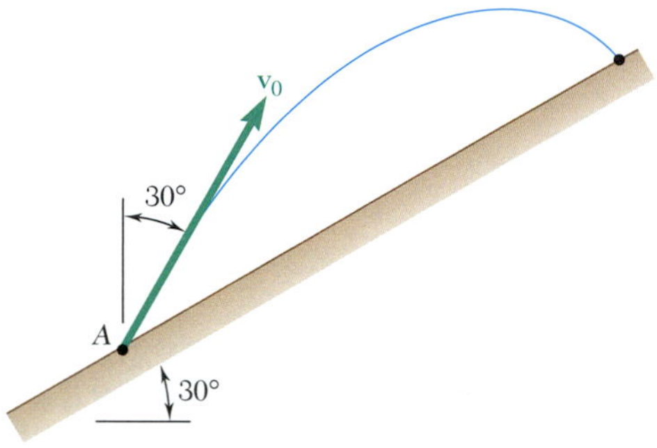

Fig. P11.143

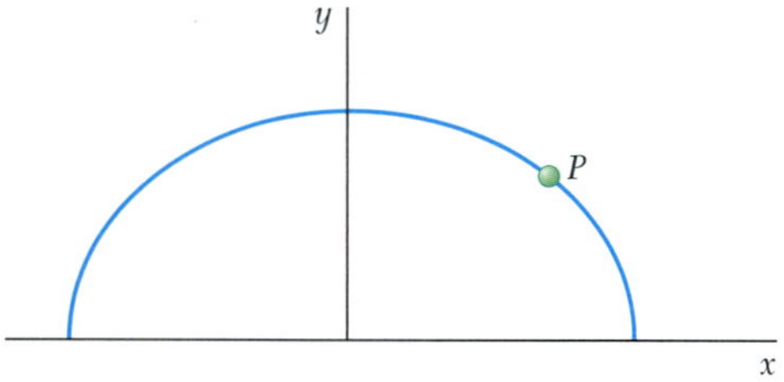

Fig. P11.144

11.144 The motion of particle P on the elliptical path shown is defined by the equations $x = (2\cos \pi t - 1)/(2 - \cos \pi t)$ and $y = 1.5 \sin \pi t/(2 - \cos \pi t)$, where x and y are expressed in meters and t is expressed in seconds. Determine the radius of curvature of the elliptical path when (*a*) $t = 0$, (*b*) $t = 1/3$ s, (*c*) $t = 1$ s.

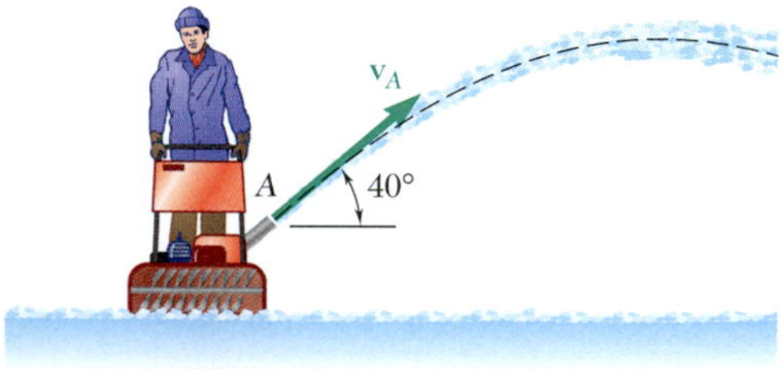

Fig. P11.145

11.145 From a photograph of a homeowner using a snowblower, it is determined that the radius of curvature of the trajectory of the snow was 30 ft as the snow left the discharge chute at A. Determine (*a*) the discharge velocity $\mathbf{v}_A$ of the snow, (*b*) the radius of curvature of the trajectory at its maximum height.

11.146 A child throws a ball from point A with an initial velocity $\mathbf{v}_A$ of 60 ft/s at an angle of 25° with the horizontal. Determine the velocity of the ball at the points of the trajectory described by the ball where the radius of curvature is equal to three-quarters of its value at A.

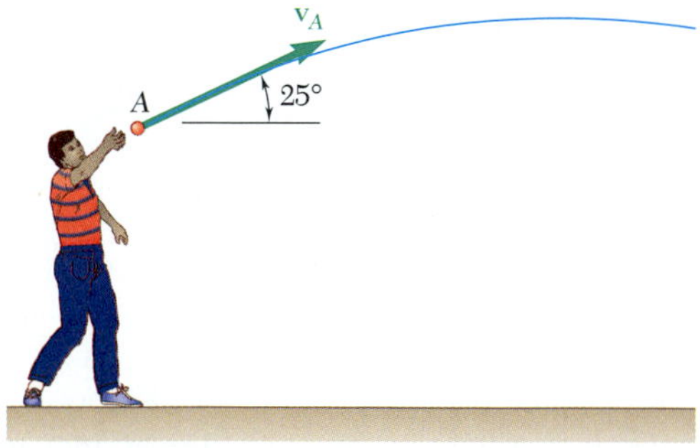

Fig. P11.146

11.147 Coal is discharged from the tailgate of a dump truck with an initial velocity $\mathbf{v}_A = 2$ m/s ⦨ 50°. Determine the radius of curvature of the trajectory described by the coal (*a*) at point *A*, (*b*) at the point of the trajectory 1 m below point *A*.

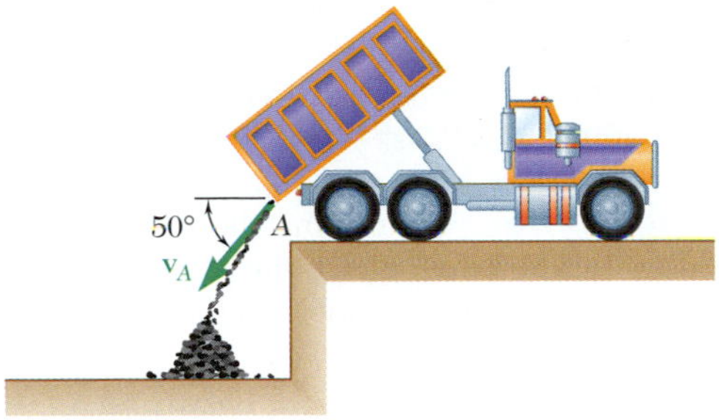

Fig. P11.147

11.148 A horizontal pipe discharges at point *A* a stream of water into a reservoir. Express the radius of curvature of the stream at point *B* in terms of the magnitudes of the velocities $\mathbf{v}_A$ and $\mathbf{v}_B$.

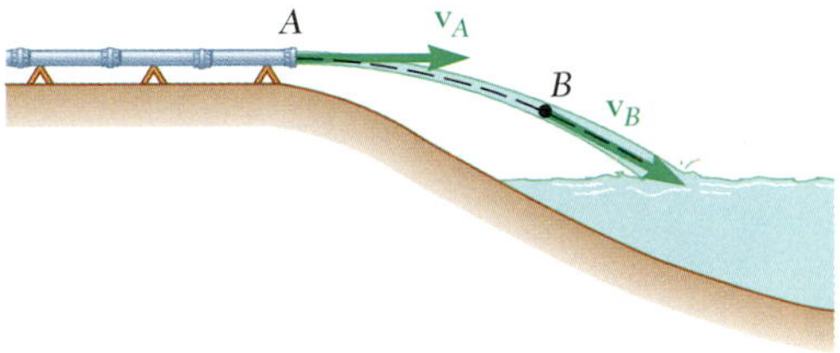

Fig. P11.148

11.149 A projectile is fired from point *A* with an initial velocity $\mathbf{v}_0$. (*a*) Show that the radius of curvature of the trajectory of the projectile reaches its minimum value at the highest point *B* of the trajectory. (*b*) Denoting by θ the angle formed by the trajectory and the horizontal at a given point *C*, show that the radius of curvature of the trajectory at *C* is $\rho = \rho_{min}/\cos^3 \theta$.

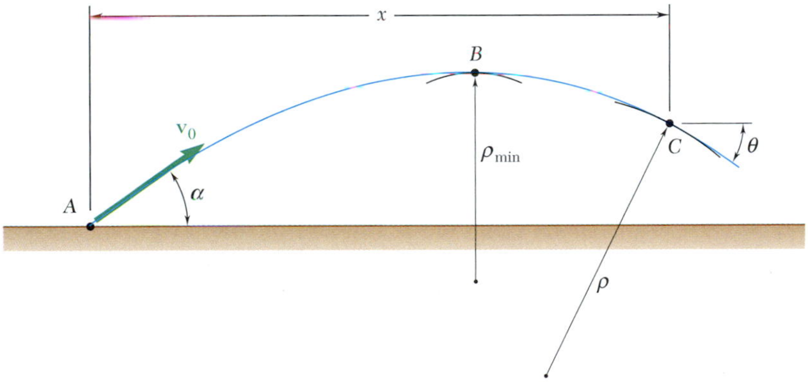

Fig. *P11.149* and *P11.150*

11.150 A projectile is fired from point *A* with an initial velocity $\mathbf{v}_0$ which forms an angle α with the horizontal. Express the radius of curvature of the trajectory of the projectile at point *C* in terms of x, v_0, α, and g.

***11.151** Determine the radius of curvature of the path described by the particle of Prob. 11.95 when $t = 0$.

***11.152** Determine the radius of curvature of the path described by the particle of Prob. 11.96 when $t = 0$, $A = 3$, and $B = 1$.

11.153 and 11.154 A satellite will travel indefinitely in a circular orbit around a planet if the normal component of the acceleration of the satellite is equal to $g(R/r)^2$, where g is the acceleration of gravity at the surface of the planet, R is the radius of the planet, and r is the distance from the center of the planet to the satellite. Knowing that the diameter of the sun is 1.39 Gm and that the acceleration of gravity at its surface is 274 m/s^2, determine the radius of the orbit of the indicated planet around the sun assuming that the orbit is circular.

11.153 Earth: $(v_{mean})_{orbit} = 107$ Mm/h
11.154 Saturn: $(v_{mean})_{orbit} = 34.7$ Mm/h

11.155 through 11.157 Determine the speed of a satellite relative to the indicated planet if the satellite is to travel indefinitely in a circular orbit 100 mi above the surface of the planet. (See information given in Probs. 11.153–11.154.)

11.155 Venus: $g = 29.20$ ft/s^2, $R = 3761$ mi.
11.156 Mars: $g = 12.24$ ft/s^2, $R = 2070$ mi.
11.157 Jupiter: $g = 75.35$ ft/s^2, $R = 44{,}432$ mi.

11.158 Knowing that the radius of the earth is 3960 mi, determine the time of one orbit of the Hubble Space Telescope if the telescope travels in a circular orbit 370 mi above the surface of the earth. (See information given in Probs. 11.153–11.154.)

11.159 A satellite is traveling in a circular orbit around Mars at an altitude of 290 km. After the altitude of the satellite is adjusted, it is found that the time of one orbit has increased by 10 percent. Knowing that the radius of Mars is 3333 km, determine the new altitude of the satellite. (See information given in Probs. 11.153–11.154.)

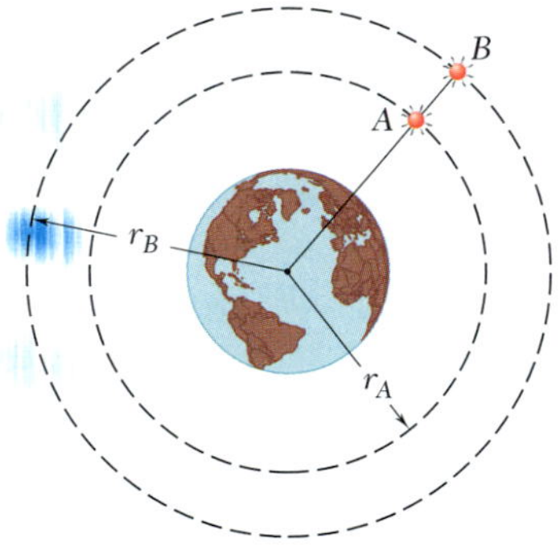

Fig. P11.160

11.160 Satellites A and B are traveling in the same plane in circular orbits around the earth at altitudes of 190 and 320 km, respectively. If at $t = 0$ the satellites are aligned as shown and knowing that the radius of the earth is $R = 6370$ km, determine when the satellites will next be radially aligned. (See information given in Probs. 11.153–11.154.)

11.161 The rotation of rod OA about O is defined by the relation $\theta = 0.5e^{-0.8t} \sin 3\pi t$, where θ and t are expressed in radians and seconds, respectively. Collar B slides along the rod so that its distance from O is $r = 0.2 + 1.92t - 6.72t^2 + 6.4t^3$, where r and t are expressed in meters and seconds, respectively. When $t = 0.5$ s, determine (*a*) the velocity of the collar, (*b*) the acceleration of the collar, (*c*) the acceleration of the collar relative to the rod.

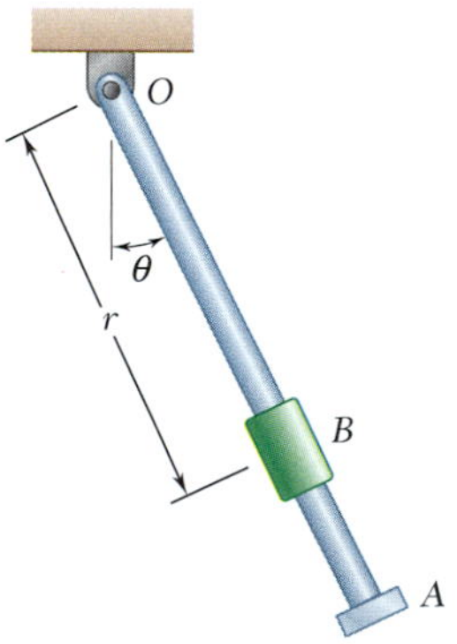

Fig. *P11.161* and *P11.162*

11.162 The oscillation of rod OA about O is defined by the relation $\theta = (4/\pi)(\sin \pi t)$, where θ and t are expressed in radians and seconds, respectively. Collar B slides along the rod so that its distance from O is $r = 10/(t + 6)$, where r and t are expressed in mm and seconds, respectively. When $t = 1$ s, determine (*a*) the velocity of the collar, (*b*) the total acceleration of the collar, (*c*) the acceleration of the collar relative to the rod.

11.163 The two-dimensional motion of a particle is defined by the relations $r = 6(4 - 2e^{-t})$ and $\theta = 2(2t + 4e^{-2t})$, where r is expressed in feet, t in seconds, and θ in radians. Determine the velocity and the acceleration of the particle (*a*) when $t = 0$, (*b*) as t approaches infinity. What conclusions can you draw regarding the final path of the particle?

11.164 The path of a particle P is a limaçon. The motion of the particle is defined by the relations $r = b(2 + \cos \pi t)$ and $\theta = \pi t$, where t and θ are expressed in seconds and radians, respectively. Determine (*a*) the velocity and the acceleration of the particle when $t = 2$ s, (*b*) the value of θ for which the magnitude of the velocity is maximum.

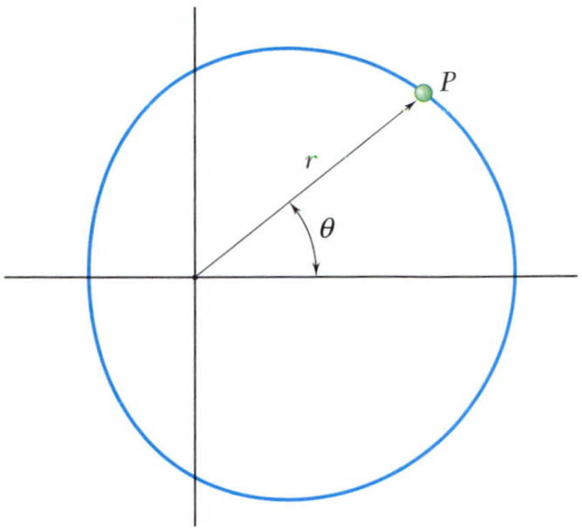

Fig. P11.164

11.165 The motion of particle P on the parabolic path shown is defined by the equations $r = 2t\sqrt{1 + 4t^2}$ and $\theta = \tan^{-1} 2t$, where r is expressed in meters, θ in radians, and t in seconds. Determine the velocity and acceleration of the particle when (*a*) $t = 0$, (*b*) $t = 0.5$ s.

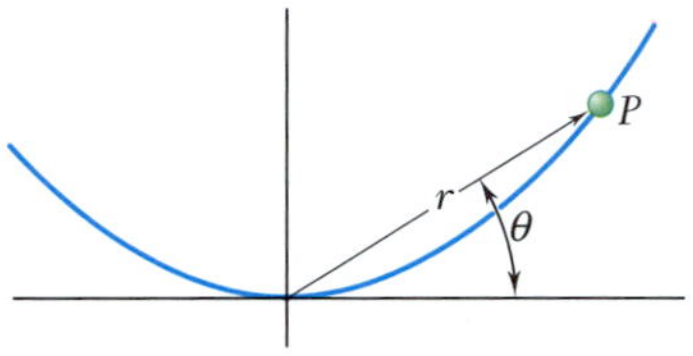

Fig. P11.165

11.166 The two-dimensional motion of a particle is defined by the relations $r = \dfrac{1}{\sin \theta - \cos \theta}$ and $\tan \theta = 1 + \dfrac{1}{t^2}$, where r and θ are expressed in meters and radians, respectively, and t is expressed in seconds. Determine (*a*) the magnitudes of the velocity and acceleration at any instant, (*b*) the radius of curvature of the path. What conclusion can you draw regarding the path of the particle?

11.167 To study the performance of a race car, a high-speed motion-picture camera is positioned at point A. The camera is mounted on a mechanism which permits it to record the motion of the car as the car travels on straightaway BC. Determine the speed of the car in terms of b, θ, and $\dot{\theta}$.

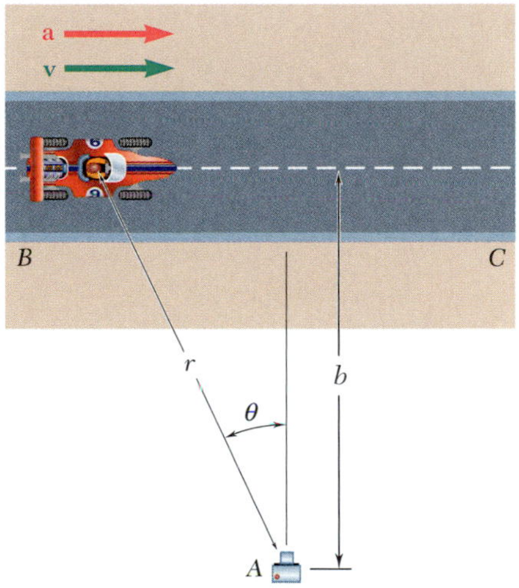

Fig. *P11.167*

11.168 Determine the magnitude of the acceleration of the race car of Prob. 11.167 in terms of b, θ, $\dot{\theta}$, and $\ddot{\theta}$.

11.169 After taking off, a helicopter climbs in a straight line at a constant angle β. Its flight is tracked by radar from point A. Determine the speed of the helicopter in terms of d, β, θ, and $\dot{\theta}$.

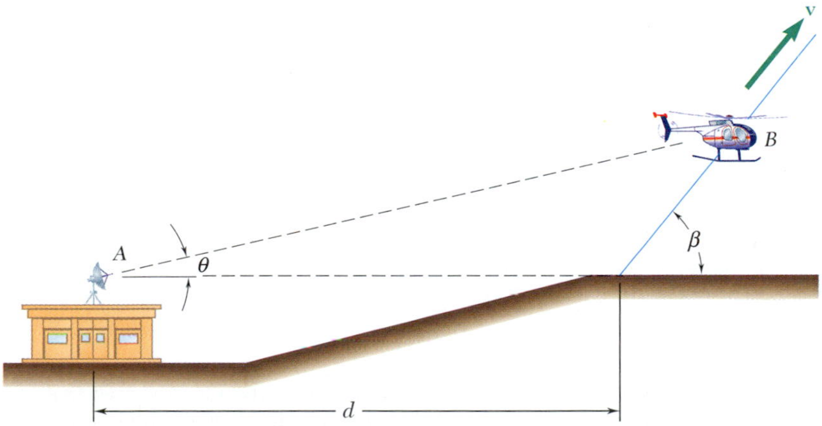

Fig. P11.169

11.170 Pin C is attached to rod BC and slides freely in the slot of rod OA which rotates at the constant rate ω. At the instant when $\beta = 60°$, determine (a) $\dot{r}$ and $\dot{\theta}$, (b) $\ddot{r}$ and $\ddot{\theta}$. Express your answers in terms of d and ω.

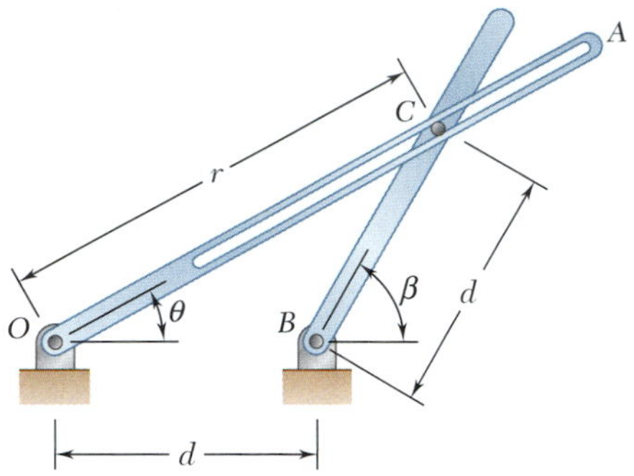

Fig. P11.170

11.171 For the race car of Prob. 11.167, it was found that it took 0.4 s for the car to travel from the position $\theta = 60°$ to the position $\theta = 35°$. Knowing that $b = 80$ ft, determine the average speed of the car during the 0.4-s interval.

11.172 For the helicopter of Prob. 11.169, it was found that when the helicopter was at B, the distance and the angle of elevation of the helicopter were $r = 900$ m and $\theta = 20°$, respectively. Five seconds later, the radar station sighted the helicopter at $r = 996$ m and $\theta = 23.1°$. Determine the average speed and the angle of climb β of the helicopter during the 4-s interval.

11.173* and *11.174 A particle moves along the spiral shown. Determine the magnitude of the velocity of the particle in terms of b, θ, and $\dot{\theta}$.

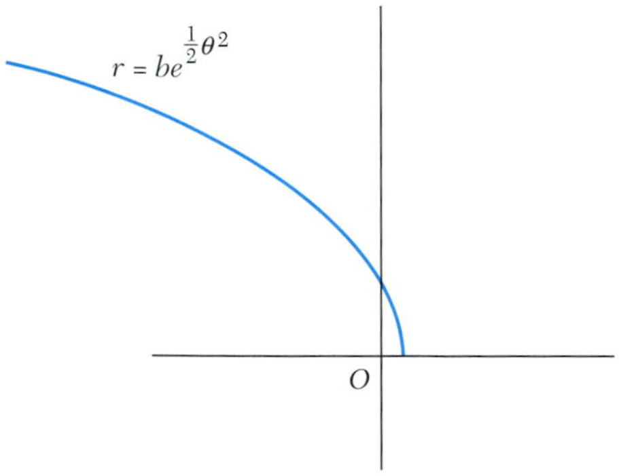

Fig. *P11.173* and P11.175

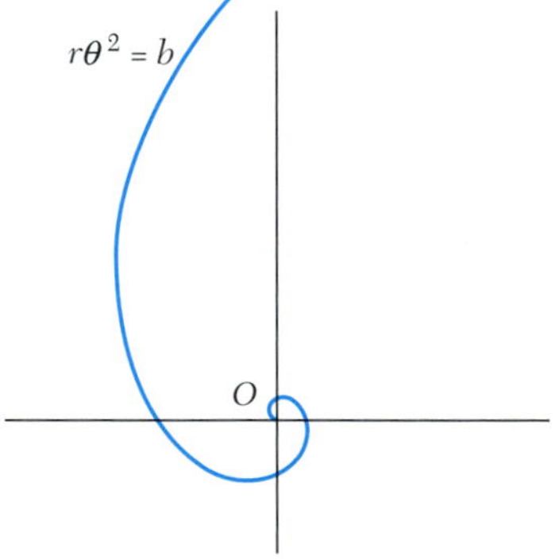

Fig. *P11.174* and P11.176

11.175 and 11.176 A particle moves along the spiral shown. Knowing that $\dot{\theta}$ is constant and denoting this constant by ω, determine the magnitude of the acceleration of the particle in terms of b, θ, and ω.

11.177 Show that $\dot{r} = h\dot{\phi} \sin \theta$ knowing that at the instant shown, step AB of the step exerciser is rotating counterclockwise at a constant rate $\dot{\phi}$.

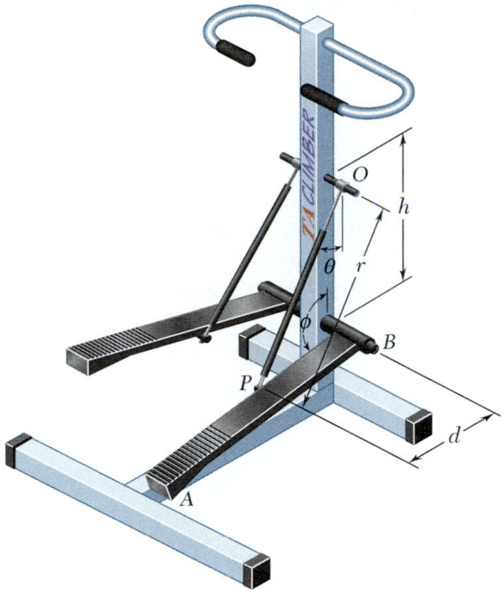

Fig. P11.177

11.178 The three-dimensional motion of a particle is defined by the cylindrical coordinates (see Fig. 11.26) $R = A/(t + 1)$, $\theta = Bt$, and $z = Ct/(t + 1)$. Determine the magnitudes of the velocity and acceleration when (*a*) $t = 0$, (*b*) $t = \infty$.

11.179 The motion of a particle on the surface of a right circular cylinder is defined by the relations $R = A$, $\theta = 2\pi t$, and $z = At^2/4$, where A is a constant. Determine the magnitudes of the velocity and acceleration of the particle at any time t.

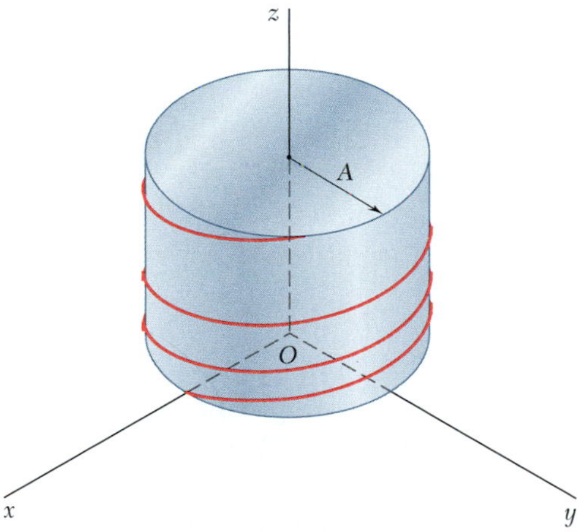

Fig. *P11.179*

***11.180** For the conic helix of Prob. 11.95, determine the angle that the oscillating plane forms with the y axis.

***11.181** Determine the direction of the binormal of the path described by the particle of Prob. 11.96 when (*a*) $t = 0$, (*b*) $t = \pi/2$ s.

REVIEW AND SUMMARY FOR CHAPTER 11

Position coordinate of a particle in rectilinear motion

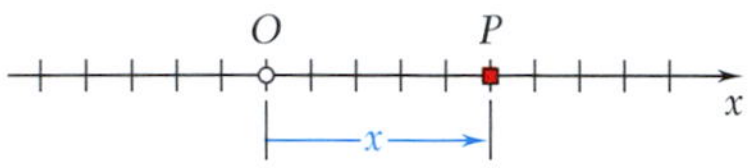

Fig. 11.27

In the first half of the chapter, we analyzed the *rectilinear motion of a particle,* that is, the motion of a particle along a straight line. To define the position P of the particle on that line, we chose a fixed origin O and a positive direction (Fig. 11.27). The distance x from O to P, with the appropriate sign, completely defines the position of the particle on the line and is called the *position coordinate* of the particle [Sec. 11.2].

Velocity and acceleration in rectilinear motion

The *velocity* v of the particle was shown to be equal to the time derivative of the position coordinate x,

$$v = \frac{dx}{dt} \tag{11.1}$$

and the *acceleration* a was obtained by differentiating v with respect to t,

$$a = \frac{dv}{dt} \tag{11.2}$$

or

$$a = \frac{d^2x}{dt^2} \tag{11.3}$$

We also noted that a could be expressed as

$$a = v\frac{dv}{dx} \tag{11.4}$$

We observed that the velocity v and the acceleration a were represented by algebraic numbers which can be positive or negative. A positive value for v indicates that the particle moves in the positive direction, and a negative value that it moves in the negative direction. A positive value for a, however, may mean that the particle is truly accelerated (that is, moves faster) in the positive direction, or that it is decelerated (that is, moves more slowly) in the negative direction. A negative value for a is subject to a similar interpretation [Sample Prob. 11.1].

Determination of the velocity and acceleration by integration

In most problems, the conditions of motion of a particle are defined by the type of acceleration that the particle possesses and by the initial conditions [Sec. 11.3]. The velocity and position of the particle can then be obtained by integrating two of the equations (11.1) to (11.4). Which of these equations should be selected depends upon the type of acceleration involved [Sample Probs. 11.2 and 11.3].

Uniform rectilinear motion

Uniformly accelerated rectilinear motion

Two types of motion are frequently encountered: the *uniform rectilinear motion* [Sec. 11.4], in which the velocity v of the particle is constant and

$$x = x_0 + vt \tag{11.5}$$

and the *uniformly accelerated rectilinear motion* [Sec. 11.5], in which the acceleration a of the particle is constant and we have

$$v = v_0 + at \tag{11.6}$$

$$x = x_0 + v_0 t + \tfrac{1}{2}at^2 \tag{11.7}$$

$$v^2 = v_0^2 + 2a(x - x_0) \tag{11.8}$$

Relative motion of two particles

When two particles A and B move along the same straight line, we may wish to consider the *relative motion* of B with respect to A

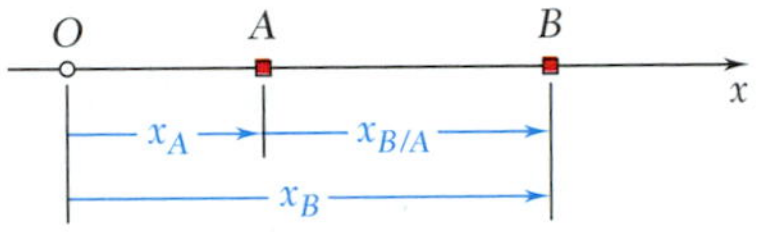

Fig. 11.28

[Sec. 11.6]. Denoting by $x_{B/A}$ the *relative position coordinate* of B with respect to A (Fig. 11.28), we had

$$x_B = x_A + x_{B/A} \tag{11.9}$$

Differentiating Eq. (11.9) twice with respect to t, we obtained successively

$$v_B = v_A + v_{B/A} \tag{11.10}$$

$$a_B = a_A + a_{B/A} \tag{11.11}$$

where $v_{B/A}$ and $a_{B/A}$ represent, respectively, the *relative velocity* and the *relative acceleration* of B with respect to A.

Blocks connected by inextensible cords

When several blocks are *connected by inextensible cords,* it is possible to write a *linear relation* between their position coordinates. Similar relations can then be written between their velocities and between their accelerations and can be used to analyze their motion [Sample Prob. 11.5].

Graphical solutions

It is sometimes convenient to use a *graphical solution* for problems involving the rectilinear motion of a particle [Secs. 11.7 and 11.8]. The graphical solution most commonly used involves the x–t, v–t, and a–t curves [Sec. 11.7; Sample Prob. 11.6]. It was shown that, at any given time t,

$$v = \text{slope of } x\text{–}t \text{ curve}$$
$$a = \text{slope of } v\text{–}t \text{ curve}$$

while, over any given time interval from t_1 to t_2,

$$v_2 - v_1 = \text{area under } a\text{–}t \text{ curve}$$
$$x_2 - x_1 = \text{area under } v\text{–}t \text{ curve}$$

Position vector and velocity in curvilinear motion

In the second half of the chapter, we analyzed the *curvilinear motion of a particle,* that is, the motion of a particle along a curved path. The position P of the particle at a given time [Sec. 11.9] was

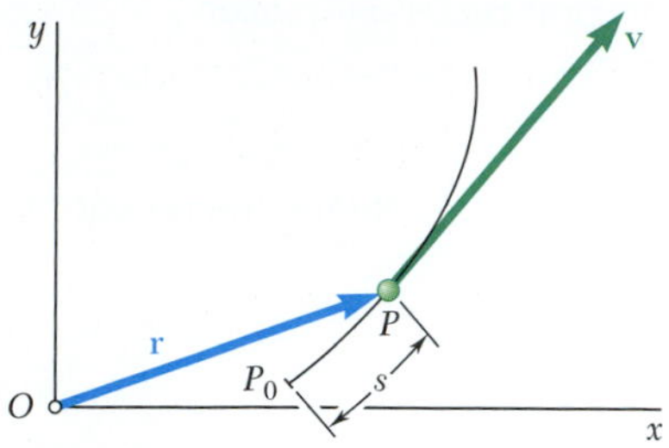

Fig. 11.29

defined by the *position vector* $\mathbf{r}$ joining the origin O of the coordinates and point P (Fig. 11.29). The *velocity* $\mathbf{v}$ of the particle was defined by the relation

$$\mathbf{v} = \frac{d\mathbf{r}}{dt} \tag{11.15}$$

and was found to be a *vector tangent to the path of the particle* and of magnitude v (called the *speed* of the particle) equal to the time derivative of the length s of the arc described by the particle:

$$v = \frac{ds}{dt} \tag{11.16}$$

Acceleration in curvilinear motion

The *acceleration* $\mathbf{a}$ of the particle was defined by the relation

$$\mathbf{a} = \frac{d\mathbf{v}}{dt} \tag{11.18}$$

and we noted that, in general, *the acceleration is not tangent to the path of the particle.*

Derivative of a vector function

Before proceeding to the consideration of the components of velocity and acceleration, we reviewed the formal definition of the derivative of a vector function and established a few rules governing the differentiation of sums and products of vector functions. We then showed that the rate of change of a vector is the same with respect to a fixed frame and with respect to a frame in translation [Sec. 11.10].

Rectangular components of velocity and acceleration

Denoting by x, y, and z the rectangular coordinates of a particle P, we found that the rectangular components of the velocity and acceleration of P equal, respectively, the first and second derivatives with respect to t of the corresponding coordinates:

$$v_x = \dot{x} \qquad v_y = \dot{y} \qquad v_z = \dot{z} \tag{11.29}$$
$$a_x = \ddot{x} \qquad a_y = \ddot{y} \qquad a_z = \ddot{z} \tag{11.30}$$

Component motions

When the component a_x of the acceleration depends only upon t, x, and/or v_x, and when similarly a_y depends only upon t, y, and/or v_y, and a_z upon t, z, and/or v_z, Eqs. (11.30) can be integrated independently. The analysis of the given curvilinear motion can thus be reduced to the analysis of three independent rectilinear component motions [Sec. 11.11]. This approach is particularly effective in the study of the motion of projectiles [Sample Probs. 11.7 and 11.8].

Relative motion of two particles

For two particles A and B moving in space (Fig. 11.30), we considered the relative motion of B with respect to A, or more precisely, with respect to a moving frame attached to A and in translation with A [Sec. 11.12]. Denoting by $\mathbf{r}_{B/A}$ the *relative position vector* of B with respect to A (Fig. 11.30), we had

$$\mathbf{r}_B = \mathbf{r}_A + \mathbf{r}_{B/A} \tag{11.31}$$

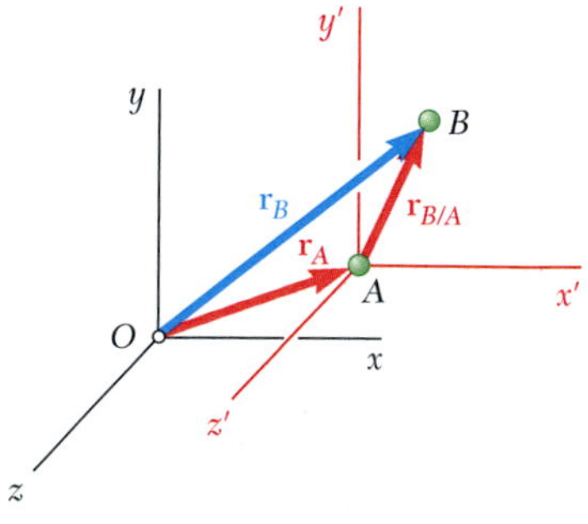

Fig. 11.30

Denoting by $\mathbf{v}_{B/A}$ and $\mathbf{a}_{B/A}$, respectively, the *relative velocity* and the *relative acceleration* of B with respect to A, we also showed that

$$\mathbf{v}_B = \mathbf{v}_A + \mathbf{v}_{B/A} \tag{11.33}$$

and

$$\mathbf{a}_B = \mathbf{a}_A + \mathbf{a}_{B/A} \tag{11.34}$$

Tangential and normal components

It is sometimes convenient to resolve the velocity and acceleration of a particle P into components other than the rectangular x, y, and z components. For a particle P moving along a path contained in a plane, we attached to P unit vectors $\mathbf{e}_t$ tangent to the path and $\mathbf{e}_n$ normal to the path and directed toward the center of curvature of the path [Sec. 11.13]. We then expressed the velocity and acceleration of the particle in terms of tangential and normal components. We wrote

$$\mathbf{v} = v\mathbf{e}_t \tag{11.36}$$

and

$$\mathbf{a} = \frac{dv}{dt}\mathbf{e}_t + \frac{v^2}{\rho}\mathbf{e}_n \tag{11.39}$$

where v is the speed of the particle and ρ the radius of curvature of its path [Sample Probs. 11.10 and 11.11]. We observed that while the velocity $\mathbf{v}$ is directed along the tangent to the path, the acceleration $\mathbf{a}$ consists of a component $\mathbf{a}_t$ directed along the tangent to the path and a component $\mathbf{a}_n$ directed toward the center of curvature of the path (Fig. 11.31).

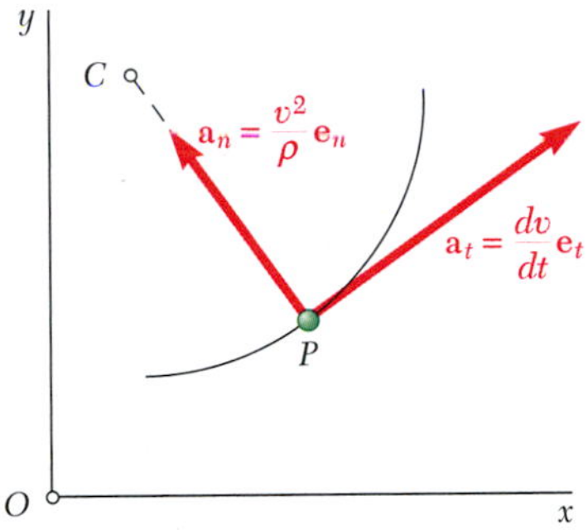

Fig. 11.31

Motion along a space curve

For a particle P moving along a space curve, we defined the plane which most closely fits the curve in the neighborhood of P as the *osculating plane*. This plane contains the unit vectors $\mathbf{e}_t$ and $\mathbf{e}_n$ which define, respectively, the tangent and principal normal to the curve. The unit vector $\mathbf{e}_b$ which is perpendicular to the osculating plane defines the *binormal*.

Radial and transverse components

When the position of a particle P moving in a plane is defined by its polar coordinates r and θ, it is convenient to use radial and transverse components directed, respectively, along the position vector $\mathbf{r}$ of the particle and in the direction obtained by rotating $\mathbf{r}$ through 90° counterclockwise [Sec. 11.14]. We attached to P unit vectors $\mathbf{e}_r$ and $\mathbf{e}_\theta$ directed, respectively, in the radial and transverse directions (Fig. 11.32). We then expressed the velocity and acceleration of the particle in terms of radial and transverse components

$$\mathbf{v} = \dot{r}\mathbf{e}_r + r\dot{\theta}\mathbf{e}_\theta \tag{11.43}$$

$$\mathbf{a} = (\ddot{r} - r\dot{\theta}^2)\mathbf{e}_r + (r\ddot{\theta} + 2\dot{r}\dot{\theta})\mathbf{e}_\theta \tag{11.44}$$

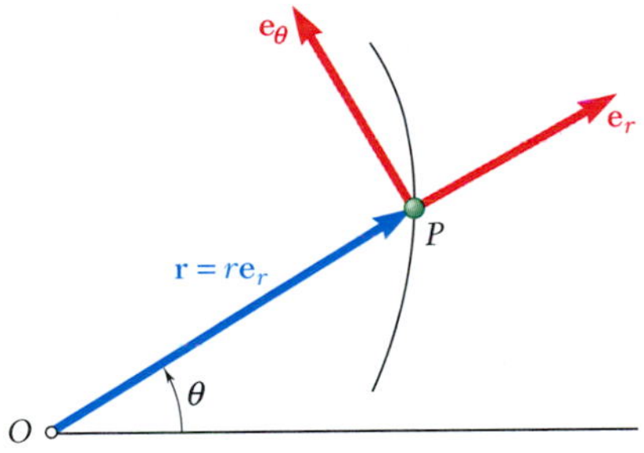

Fig. 11.32

where dots are used to indicate differentiation with respect to time. The scalar components of the velocity and acceleration in the radial and transverse directions are therefore

$$v_r = \dot{r} \qquad v_\theta = r\dot{\theta} \tag{11.45}$$

$$a_r = \ddot{r} - r\dot{\theta}^2 \qquad a_\theta = r\ddot{\theta} + 2\dot{r}\dot{\theta} \tag{11.46}$$

It is important to note that a_r is *not* equal to the time derivative of v_r, and that a_θ is *not* equal to the time derivative of v_θ [Sample Prob. 11.12].

The chapter ended with a discussion of the use of cylindrical coordinates to define the position and motion of a particle in space.

Review Problems

11.182 The acceleration of a particle is defined by the relation $a = A - 6t^2$, where A is a constant. At $t = 0$, the particle starts at $x = 8$ m with $v = 0$. Knowing that at $t = 1$ s, $v = 30$ m/s, determine (*a*) the times at which the velocity is zero, (*b*) the total distance traveled by the particle when $t = 5$ s.

11.183 At $t = 0$, a particle starts from $x = 0$ with a velocity v_0 and an acceleration defined by the relation $a = -5/(2v_0 - v)$, where a and v are expressed in ft/s^2 and ft/s, respectively. Knowing that $v = 0.5v_0$ at $t = 2$ s, determine (*a*) the initial velocity of the particle, (*b*) the time required for the particle to come to rest, (*c*) the position where the velocity is 1 ft/s.

11.184 Assuming a uniform acceleration of 11 ft/s^2 and knowing that the speed of a car as it passes A is 30 mi/h, determine (*a*) the time required for the car to reach B, (*b*) the speed of the car as it passes B.

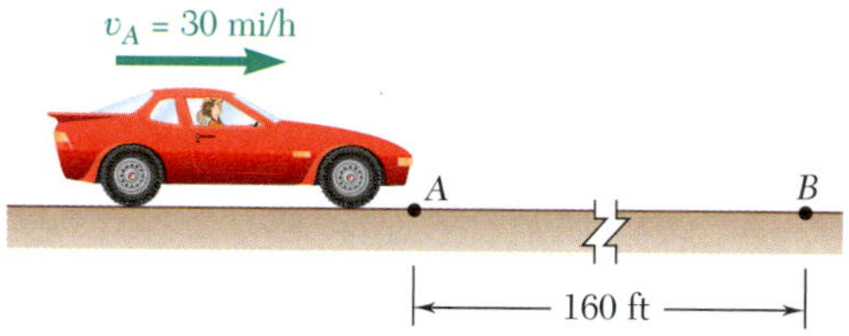

Fig. *P11.184*

11.185 Block B starts from rest and moves downward with a constant acceleration. Knowing that after slider block A has moved 400 mm its velocity is 4 m/s, determine (*a*) the accelerations of A and B, (*b*) the velocity and the change in position of B after 2 s.

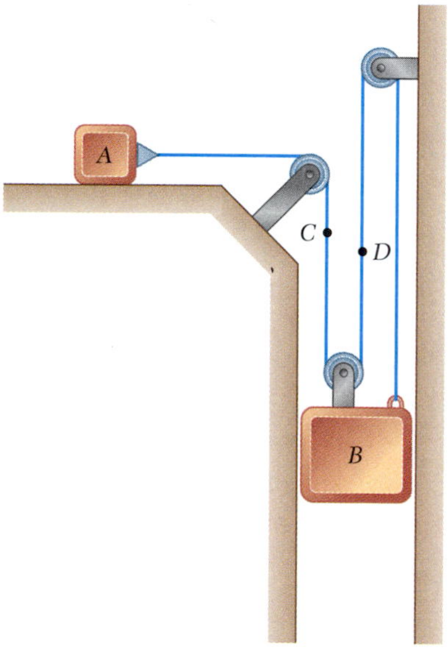

Fig. P11.185

11.186 A particle moves in a straight line with the acceleration shown in the figure. Knowing that the particle starts from the origin with $v_0 = -2$ m/s, (*a*) construct the v–t and x–t curves for $0 < t < 18$ s, (*b*) determine the position and the velocity of the particle and the total distance traveled when $t = 18$ s.

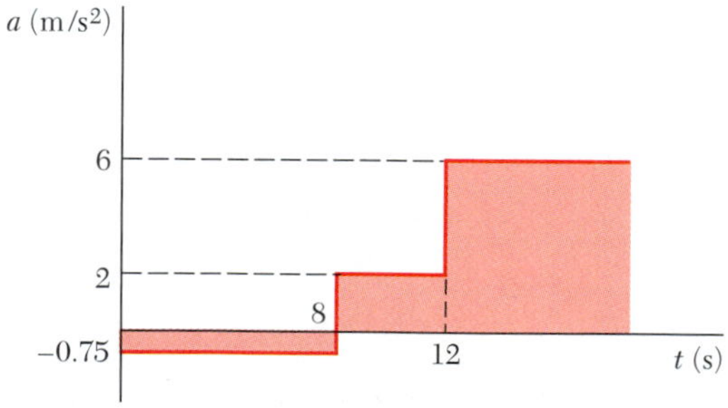

Fig. P11.186

11.187 A temperature sensor is attached to slider AB which moves back and forth through 60 in. The maximum velocities of the slider are 12 in./s to the right and 30 in./s to the left. When the slider is moving to the right, it accelerates and decelerates at a constant rate of 6 in./s^2; when moving to the left, the slider accelerates and decelerates at a constant rate of 20 in./s^2. Determine the time required for the slider to complete a full cycle, and construct the v–t and x–t curves of its motion.

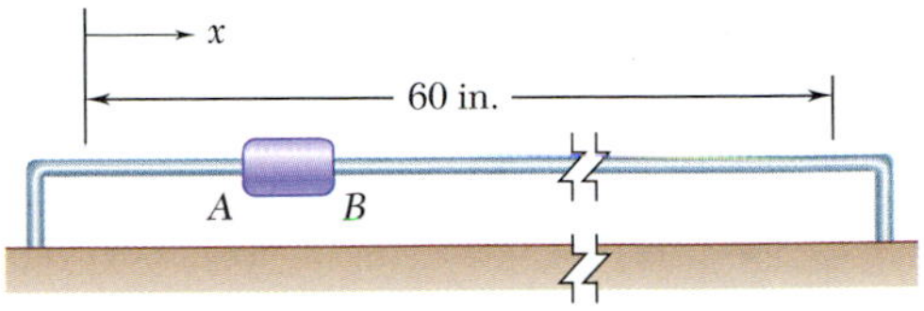

Fig. *P11.187*

11.188 A group of children are throwing balls through a 0.72-m-inner-diameter tire hanging from a tree. A child throws a ball with an initial velocity $\mathbf{v}_0$ at an angle of 3° with the horizontal. Determine the range of values of v_0 for which the ball will go through the tire.

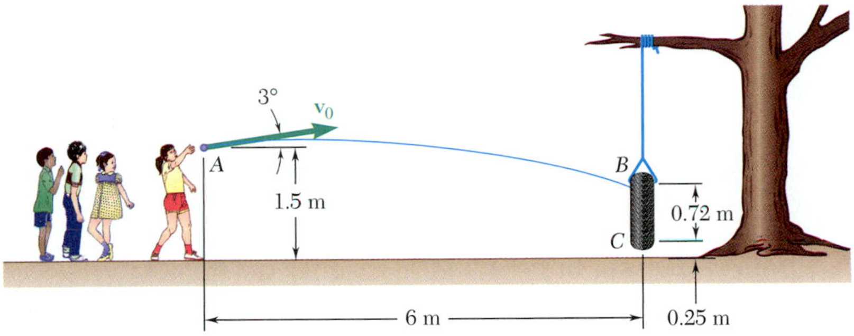

Fig. P11.188

11.189 An oscillating garden sprinkler which discharges water with an initial velocity $\mathbf{v}_0$ of 8 m/s is used to water a vegetable garden. Determine the distance d to the farthest point B that will be watered and the corresponding angle α when (*a*) the vegetables are just beginning to grow, (*b*) the height h of the corn is 1.8 m.

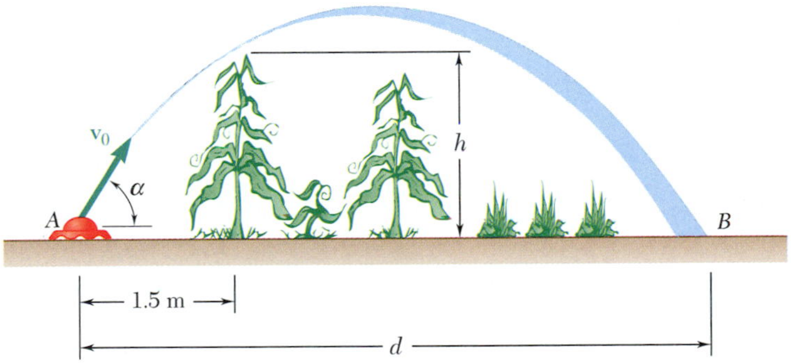

Fig. P11.189

11.190 Knowing that at the instant shown block A has a velocity of 8 in./s and an acceleration of 6 in./s^2 both directed down the incline, determine (*a*) the velocity of block B, (*b*) the acceleration of block B.

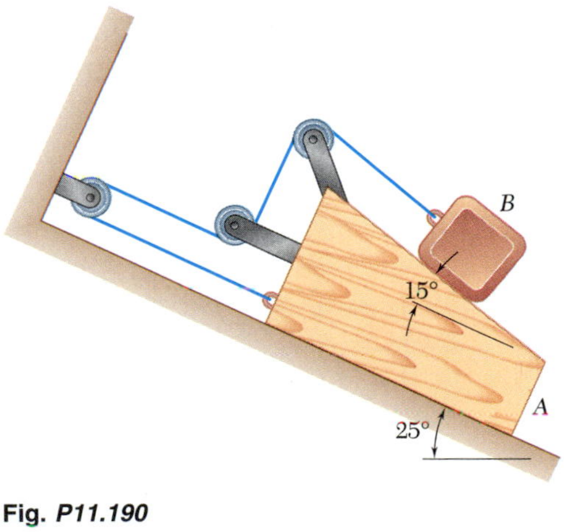

Fig. *P11.190*

11.191 To test its performance, an automobile is driven around a circular test track of diameter d. Determine (*a*) the value of d if when the speed of the automobile is 72 km/h, the normal component of the acceleration is 3.2 m/s^2, (*b*) the speed of the automobile if d = 180 m and the normal component of the acceleration is measured to be 0.6 g.

11.192 A motorist traveling along a straight portion of a highway is decreasing the speed of his automobile at a constant rate before exiting from the highway onto a circular exit ramp with a radius of 560 ft. He continues to decelerate at the same constant rate so that 10 s after entering the ramp, his speed has decreased to 20 mi/h, a speed which he then maintains. Knowing that at this constant speed the total acceleration of the automobile is equal to one-quarter of its value prior to entering the ramp, determine the maximum value of the total acceleration of the automobile.

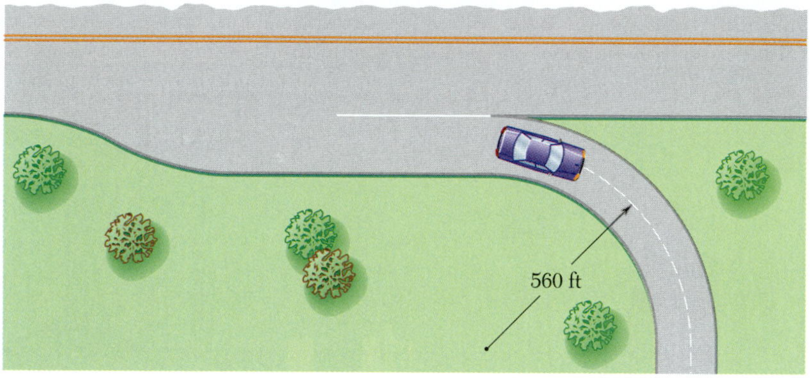

Fig. P11.192

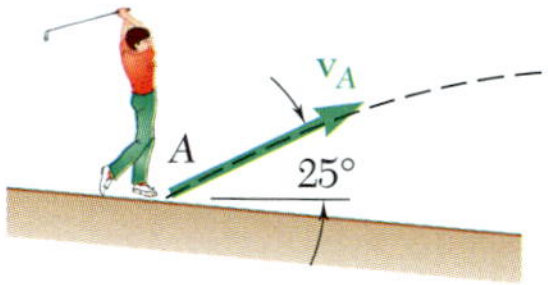

Fig. *P11.193*

11.193 A golfer hits a golf ball from point A with an initial velocity of 50 m/s at an angle of 25° with the horizontal. Determine the radius of curvature of the trajectory described by the ball (*a*) at point A, (*b*) at the highest point of the trajectory.

Computer Problems

11.C1 A projectile enters a resisting medium at $x = 0$ with an initial velocity $v_0 = 300$ m/s and travels a distance d before coming to rest. The velocity of the projectile is defined by the relation $v = v_0 - 2700x$, where v is expressed in meters/second and x in meters. (*a*) Determine the distance d and derive expressions for the position, velocity, and acceleration of the projectile as functions of time. (*b*) Plot the position, velocity, and acceleration of the projectile as functions of time from the time the projectile enters the medium until it penetrates a distance $0.99d$.

11.C2 The magnitude in ft/s^2 of the deceleration due to air resistance of the nose cone of a small experimental rocket is known to be $0.00015v^2$, where v is expressed in ft/s. The nose cone is projected vertically from the ground with an initial velocity of 420 ft/s. Derive expressions for the speed of the cone as a function of height as the cone moves up to its maximum height and as it returns to the ground. Plot the speed of the cone as a function of height for the upward motion and for the downward motion. [*Hint:* The deceleration of the cone as it moves up is $g + 0.00015v^2$ and the acceleration of the cone as it moves down is $g - 0.00015v^2$, where $g = 32.2$ ft/s^2.]

11.C3 The motion of a particle is defined by the equations $x = 30t^2 - 120t$ and $y = 120t^2 - 40t^3$, where x and y are expressed in millimeters and t in seconds. Derive expressions for the velocity and acceleration of the particle as functions of t. Consider the time interval $0 \le t \le 2$ s and plot (*a*) the path of the particle in the xy plane, (*b*) the components of the velocity v_x and v_y and the magnitude of the velocity v, (*c*) the components of the acceleration a_x and a_y and the magnitude of the acceleration a.

11.C4 The rotation of rod OA about O is defined by the relation $\theta = t^3 + 4t$, where θ is expressed in radians and t in seconds. Collar B slides along the rod in such a way that its distance from O is $r = t^3 + 2t^2$, where r is expressed in inches and t in seconds. Derive expressions for the velocity and acceleration of the collar as functions of t. Consider one revolution of the rod starting at $t = 0$ and plot (*a*) the path of the particle in the xy plane, (*b*) the components of the velocity v_r and v_θ and the magnitude of the velocity v, (*c*) the components of the acceleration a_r and a_θ and the magnitude of the acceleration a.

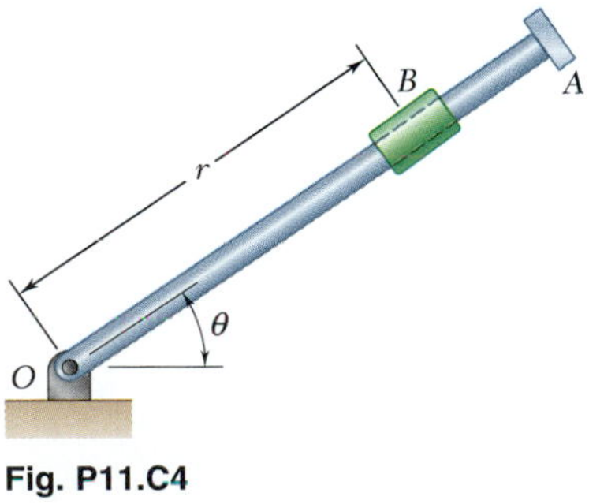

Fig. P11.C4

11.C5 The path of particle P is the ellipse defined by the relations $r = 1.75/(1 - 0.75 \cos \pi t)$ and $\theta = \pi t$, where r is expressed in inches, t in seconds, and θ in radians. Derive expressions for the velocity and acceleration of P as functions of t. Consider the time interval $0 \leq t \leq 2$ s and plot (*a*) the components of the velocity v_r and v_θ and the magnitude of the velocity v, (*b*) the components of the acceleration a_r and a_θ and the magnitude of the acceleration a.

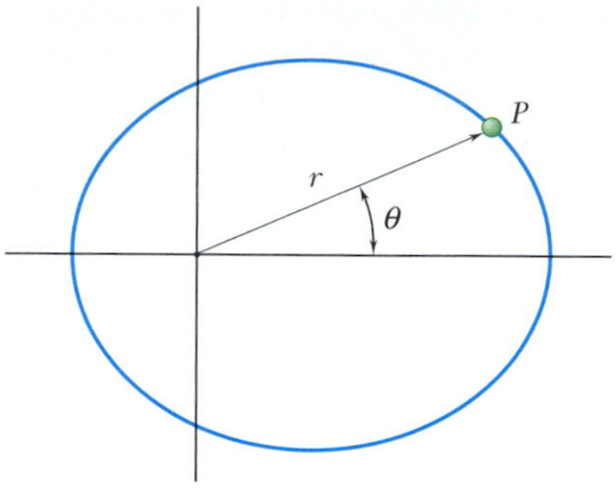

Fig. P11.C5

11.C6 The mechanism shown is known as a Whitworth quick-return mechanism. The input rod AP rotates at a constant rate $\dot{\phi}$, and the pin P is free to slide in the slot of the output rod BD. Using computational software, calculate and plot $\dot{\theta}$ versus ϕ and θ versus ϕ for one revolution of rod AP. Assume $\dot{\phi} = 1$ rad/s, $l = 80$ mm, and (*a*) $b = 50$ mm, (*b*) $b = 60$ mm, (*c*) $b = 70$ mm.

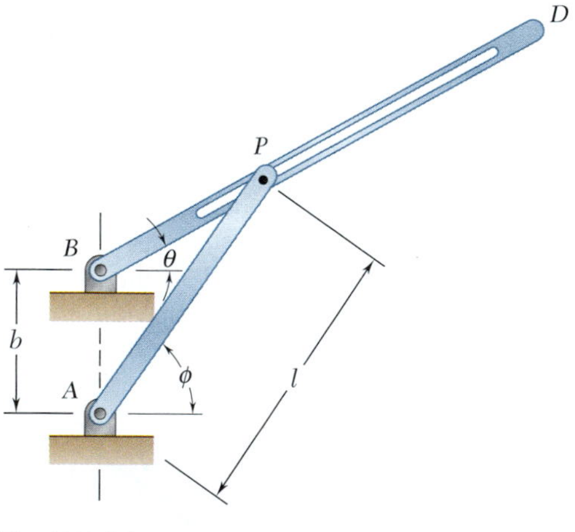

Fig. P11.C6

11.C7 In an amusement park ride, "airplane" A is attached to the 35-ft-long rigid member OB. To operate the ride, the airplane and OB are rotated so that $70° \leq \theta_0 \leq 130°$ and then are allowed to swing freely about O. The airplane is subjected to the acceleration of gravity and to a deceleration due to air resistance, $-kv^2$, which acts in a direction opposite to that of its velocity $\mathbf{v}$. Neglecting the mass and the aerodynamic drag of OB and the friction in the bearing at O, use computational software to determine the speed of the airplane for given values of θ_0 and θ and the value of θ at which the airplane first comes to rest after being released. Use values of θ_0 from 70° to 130° in 30° increments, and determine the maximum speed of the airplane and the first two values of θ at which $v = 0$. For each value of θ_0, let (*a*) $k = 0$, (*b*) $k = 4 \times 10^{-6}\ \text{in}^{-1}$, (*c*) $k = 1 \times 10^{-3}\ \text{in}^{-1}$. (*Hint:* Express the tangential acceleration of the airplane in terms of g, k, and θ. Recall that $v_\theta = r\dot{\theta}$ and use time intervals $\Delta t = 0.008$ s.)

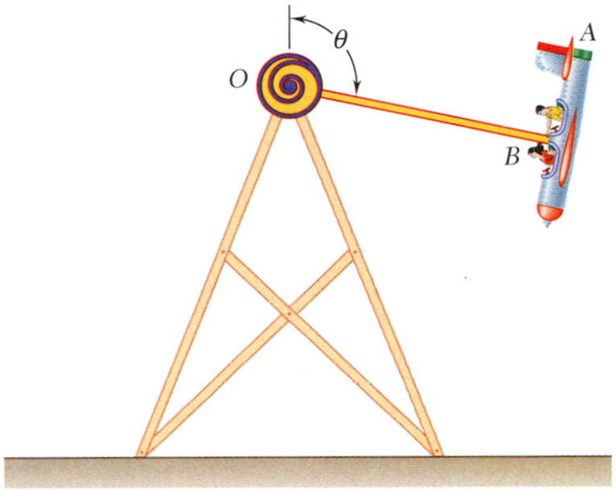

Fig. P11.C7

CHAPTER

12

Kinetics of Particles: Newton's Second Law

A roller-coaster car can travel on a straight path, a curved path in the horizontal plane, or a curved path in the vertical plane. In each case the force of gravity and the forces exerted by the track on the car must be considered as well as the acceleration of the car as studied in the previous chapter. The relation existing among force, mass, and acceleration will be studied in this chapter.

12.1. INTRODUCTION

Newton's first and third laws of motion were used extensively in statics to study bodies at rest and the forces acting upon them. These two laws are also used in dynamics; in fact, they are sufficient for the study of the motion of bodies which have no acceleration. However, when bodies are accelerated, that is, when the magnitude or the direction of their velocity changes, it is necessary to use Newton's second law of motion to relate the motion of the body with the forces acting on it.

In this chapter we will discuss Newton's second law and apply it to the analysis of the motion of particles. As we state in Sec. 12.2, if the resultant of the forces acting on a particle is not zero, the particle will have an acceleration proportional to the magnitude of the resultant and in the direction of this resultant force. Moreover, the ratio of the magnitudes of the resultant force and of the acceleration can be used to define the *mass* of the particle.

In Sec. 12.3, the *linear momentum* of a particle is defined as the product $\mathbf{L} = m\mathbf{v}$ of the mass m and velocity $\mathbf{v}$ of the particle, and it is demonstrated that Newton's second law can be expressed in an alternative form relating the rate of change of the linear momentum with the resultant of the forces acting on that particle.

Section 12.4 stresses the need for consistent units in the solution of dynamics problems and provides a review of the International System of Units (SI units) and the system of U.S. customary units.

In Secs. 12.5 and 12.6 and in the Sample Problems which follow, Newton's second law is applied to the solution of engineering problems, using either rectangular components or tangential and normal components of the forces and accelerations involved. We recall that an actual body—including bodies as large as a car, rocket, or airplane—can be considered as a particle for the purpose of analyzing its motion as long as the effect of a rotation of the body about its mass center can be ignored.

The second part of the chapter is devoted to the solution of problems in terms of radial and transverse components, with particular emphasis on the motion of a particle under a central force. In Sec. 12.7, the *angular momentum* $\mathbf{H}_O$ of a particle about a point O is defined as the moment about O of the linear momentum of the particle: $\mathbf{H}_O = \mathbf{r} \times m\mathbf{v}$. It then follows from Newton's second law that the rate of change of the angular momentum $\mathbf{H}_O$ of a particle is equal to the sum of the moments about O of the forces acting on that particle.

Section 12.9 deals with the motion of a particle under a *central force,* that is, under a force directed toward or away from a fixed point O. Since such a force has zero moment about O, it follows that the angular momentum of the particle about O is conserved. This property greatly simplifies the analysis of the motion of a particle under a central force; in Sec. 12.10 it is applied to the solution of problems involving the orbital motion of bodies under gravitational attraction.

Sections 12.11 through 12.13 are optional. They present a more extensive discussion of orbital motion and contain a number of problems related to space mechanics.

12.2. NEWTON'S SECOND LAW OF MOTION

Newton's second law can be stated as follows:

If the resultant force acting on a particle is not zero, the particle will have an acceleration proportional to the magnitude of the resultant and in the direction of this resultant force.

Newton's second law of motion is best understood by imagining the following experiment: A particle is subjected to a force $\mathbf{F}_1$ of constant direction and constant magnitude F_1. Under the action of that force, the particle is observed to move in a straight line and *in the direction of the force* (Fig. 12.1*a*). By determining the position of the particle at various instants, we find that its acceleration has a constant magnitude a_1. If the experiment is repeated with forces $\mathbf{F}_2$, $\mathbf{F}_3$, . . . , of different magnitude or direction (Fig. 12.1*b* and *c*), we find each time that the particle moves in the direction of the force acting on it and that the magnitudes a_1, a_2, a_3, . . . , of the accelerations are proportional to the magnitudes F_1, F_2, F_3, . . . , of the corresponding forces:

$$\frac{F_1}{a_1} = \frac{F_2}{a_2} = \frac{F_3}{a_3} = \cdots = \text{constant}$$

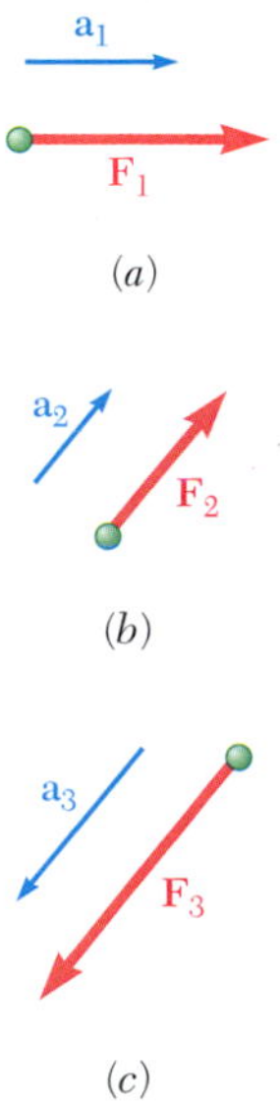

Fig. 12.1

The constant value obtained for the ratio of the magnitudes of the forces and accelerations is a characteristic of the particle under consideration; it is called the *mass* of the particle and is denoted by m. When a particle of mass m is acted upon by a force $\mathbf{F}$, the force $\mathbf{F}$ and the acceleration $\mathbf{a}$ of the particle must therefore satisfy the relation

$$\mathbf{F} = m\mathbf{a} \tag{12.1}$$

This relation provides a complete formulation of Newton's second law; it expresses not only that the magnitudes of $\mathbf{F}$ and $\mathbf{a}$ are proportional but also (since m is a positive scalar) that the vectors $\mathbf{F}$ and $\mathbf{a}$ have the same direction (Fig. 12.2). We should note that Eq. (12.1) still holds when $\mathbf{F}$ is not constant but varies with time in magnitude or direction. The magnitudes of $\mathbf{F}$ and $\mathbf{a}$ remain proportional, and the two vectors have the same direction at any given instant. However, they will not, in general, be tangent to the path of the particle.

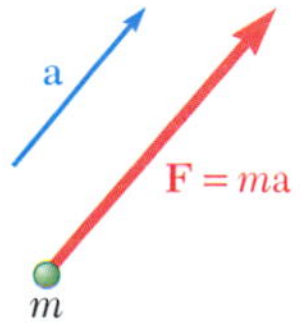

Fig. 12.2

When a particle is subjected simultaneously to several forces, Eq. (12.1) should be replaced by

$$\Sigma\mathbf{F} = m\mathbf{a} \tag{12.2}$$

where $\Sigma\mathbf{F}$ represents the sum, or resultant, of all the forces acting on the particle.

It should be noted that the system of axes with respect to which the acceleration $\mathbf{a}$ is determined is not arbitrary. These axes must have a constant orientation with respect to the stars, and their origin must either be attached to the sun† or move with a constant velocity with

†More accurately, to the mass center of the solar system.

respect to the sun. Such a system of axes is called a *newtonian frame of reference*.† A system of axes attached to the earth does *not* constitute a newtonian frame of reference, since the earth rotates with respect to the stars and is accelerated with respect to the sun. However, in most engineering applications, the acceleration **a** can be determined with respect to axes attached to the earth and Eqs. (12.1) and (12.2) used without any appreciable error. On the other hand, these equations do not hold if **a** represents a relative acceleration measured with respect to moving axes, such as axes attached to an accelerated car or to a rotating piece of machinery.

We observe that if the resultant $\Sigma\mathbf{F}$ of the forces acting on the particle is zero, it follows from Eq. (12.2) that the acceleration **a** of the particle is also zero. If the particle is initially at rest ($\mathbf{v}_0 = 0$) with respect to the newtonian frame of reference used, it will thus remain at rest ($\mathbf{v} = 0$). If originally moving with a velocity $\mathbf{v}_0$, the particle will maintain a constant velocity $\mathbf{v} = \mathbf{v}_0$; that is, it will move with the constant speed v_0 in a straight line. This, we recall, is the statement of Newton's first law (Sec. 2.10). Thus, Newton's first law is a particular case of Newton's second law and can be omitted from the fundamental principles of mechanics.

12.3. LINEAR MOMENTUM OF A PARTICLE. RATE OF CHANGE OF LINEAR MOMENTUM

Replacing the acceleration **a** by the derivative $d\mathbf{v}/dt$ in Eq. (12.2), we write

$$\Sigma\mathbf{F} = m\frac{d\mathbf{v}}{dt}$$

or, since the mass m of the particle is constant,

$$\Sigma\mathbf{F} = \frac{d}{dt}(m\mathbf{v}) \qquad (12.3)$$

The vector $m\mathbf{v}$ is called the *linear momentum,* or simply the *momentum,* of the particle. It has the same direction as the velocity of the particle, and its magnitude is equal to the product of the mass m and the speed v of the particle (Fig. 12.3). Equation (12.3) expresses that *the resultant of the forces acting on the particle is equal to the rate of change of the linear momentum of the particle*. It is in this form that the second law of motion was originally stated by Newton. Denoting by **L** the linear momentum of the particle,

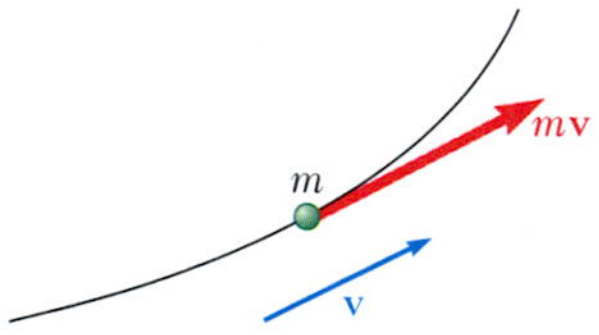

Fig. 12.3

$$\mathbf{L} = m\mathbf{v} \qquad (12.4)$$

and by $\dot{\mathbf{L}}$ its derivative with respect to t, we can write Eq. (12.3) in the alternative form

$$\Sigma\mathbf{F} = \dot{\mathbf{L}} \qquad (12.5)$$

†Since stars are not actually fixed, a more rigorous definition of a newtonian frame of reference (also called an *inertial system*) is *one with respect to which Eq. (12.2) holds*.

It should be noted that the mass m of the particle is assumed to be constant in Eqs. (12.3) to (12.5). Equation (12.3) or (12.5) should therefore not be used to solve problems involving the motion of bodies, such as rockets, which gain or lose mass. Problems of that type will be considered in Sec. 14.12.†

It follows from Eq. (12.3) that the rate of change of the linear momentum $m\mathbf{v}$ is zero when $\Sigma\mathbf{F} = 0$. Thus, *if the resultant force acting on a particle is zero, the linear momentum of the particle remains constant, in both magnitude and direction.* This is the principle of *conservation of linear momentum* for a particle, which can be recognized as an alternative statement of Newton's first law (Sec. 2.10).

12.4. SYSTEMS OF UNITS

In using the fundamental equation $\mathbf{F} = m\mathbf{a}$, the units of force, mass, length, and time cannot be chosen arbitrarily. If they are, the magnitude of the force $\mathbf{F}$ required to give an acceleration $\mathbf{a}$ to the mass m will *not* be numerically equal to the product ma; it will be only proportional to this product. Thus, we can choose three of the four units arbitrarily but must choose the fourth unit so that the equation $\mathbf{F} = m\mathbf{a}$ is satisfied. The units are then said to form a system of consistent kinetic units.

Two systems of consistent kinetic units are currently used by American engineers, the International System of Units (SI units‡) and the system of U.S. customary units. Both systems were discussed in detail in Sec. 1.3 and are described only briefly in this section.

International System of Units (SI Units). In this system, the base units are the units of length, mass, and time, and are called, respectively, the *meter* (m), the *kilogram* (kg), and the *second* (s). All three are arbitrarily defined (Sec. 1.3). The unit of force is a derived unit. It is called the *newton* (N) and is defined as the force which gives an acceleration of 1 m/s^2 to a mass of 1 kg (Fig. 12.4). From Eq. (12.1) we write

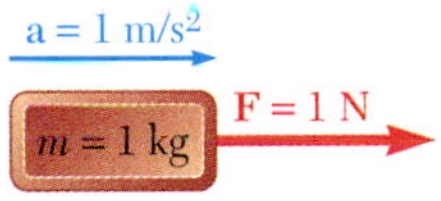

Fig 12.4

$$1\text{ N} = (1\text{ kg})(1\text{ m/s}^2) = 1\text{ kg}\cdot\text{m/s}^2$$

The SI units are said to form an *absolute* system of units. This means that the three base units chosen are independent of the location where measurements are made. The meter, the kilogram, and the second may be used anywhere on the earth; they may even be used on another planet. They will always have the same significance.

The *weight* $\mathbf{W}$ of a body, or *force of gravity* exerted on that body, should, like any other force, be expressed in newtons. Since a body subjected to its own weight acquires an acceleration equal to the acceleration of gravity g, it follows from Newton's second law that the magnitude W of the weight of a body of mass m is

$$W = mg \tag{12.6}$$

†On the other hand, Eqs. (12.3) and (12.5) do hold in *relativistic mechanics*, where the mass m of the particle is assumed to vary with the speed of the particle.

‡SI stands for *Système International d'Unités* (French).

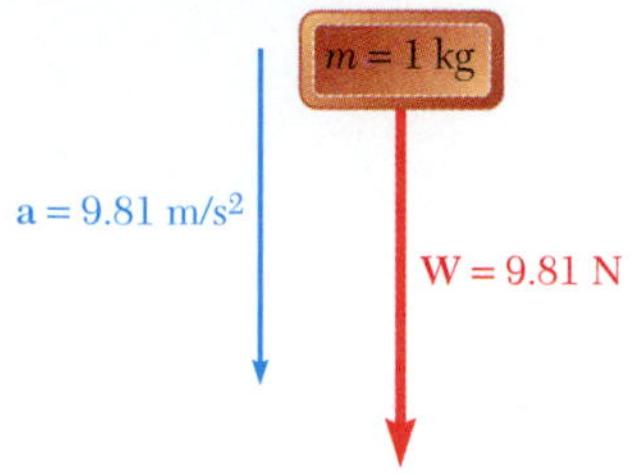

Fig 12.5

Recalling that $g = 9.81\ \text{m/s}^2$, we find that the weight of a body of mass 1 kg (Fig. 12.5) is

$$W = (1\ \text{kg})(9.81\ \text{m/s}^2) = 9.81\ \text{N}$$

Multiples and submultiples of the units of length, mass, and force are frequently used in engineering practice. They are, respectively, the *kilometer* (km) and the *millimeter* (mm); the *megagram*† (Mg) and the *gram* (g); and the *kilonewton* (kN). By definition,

$$1\ \text{km} = 1000\ \text{m} \qquad 1\ \text{mm} = 0.001\ \text{m}$$
$$1\ \text{Mg} = 1000\ \text{kg} \qquad 1\ \text{g} = 0.001\ \text{kg}$$
$$1\ \text{kN} = 1000\ \text{N}$$

The conversion of these units to meters, kilograms, and newtons, respectively, can be effected simply by moving the decimal point three places to the right or to the left.

Units other than the units of mass, length, and time can all be expressed in terms of these three base units. For example, the unit of linear momentum can be obtained by recalling the definition of linear momentum and writing

$$mv = (\text{kg})(\text{m/s}) = \text{kg} \cdot \text{m/s}$$

U.S. Customary Units. Most practicing American engineers still commonly use a system in which the base units are the units of length, force, and time. These units are, respectively, the *foot* (ft), the *pound* (lb), and the *second* (s). The second is the same as the corresponding SI unit. The foot is defined as 0.3048 m. The pound is defined as the *weight* of a platinum standard, called the *standard pound*, which is kept at the National Institute of Standards and Technology outside Washington and the mass of which is 0.453 592 43 kg. Since the weight of a body depends upon the gravitational attraction of the earth, which varies with location, it is specified that the standard pound should be placed at sea level and at a latitude of 45° to properly define a force of 1 lb. Clearly, the U.S. customary units do not form an absolute system of units. Because of their dependence upon the gravitational attraction of the earth, they are said to form a *gravitational* system of units.

While the standard pound also serves as the unit of mass in commercial transactions in the United States, it cannot be so used in engineering computations since such a unit would not be consistent with the base units defined in the preceding paragraph. Indeed, when acted upon by a force of 1 lb, that is, when subjected to its own weight, the standard pound receives the acceleration of gravity, $g = 32.2\ \text{ft/s}^2$ (Fig. 12.6), and not the unit acceleration required by Eq. (12.1). The unit of mass consistent with the foot, the pound, and the second is the mass which receives an acceleration of $1\ \text{ft/s}^2$ when a force of 1 lb is applied to it (Fig. 12.7). This unit, sometimes called a *slug*, can be derived from the equation $F = ma$ after substituting 1 lb and $1\ \text{ft/s}^2$ for F and a, respectively. We write

$$F = ma \qquad 1\ \text{lb} = (1\ \text{slug})(1\ \text{ft/s}^2)$$

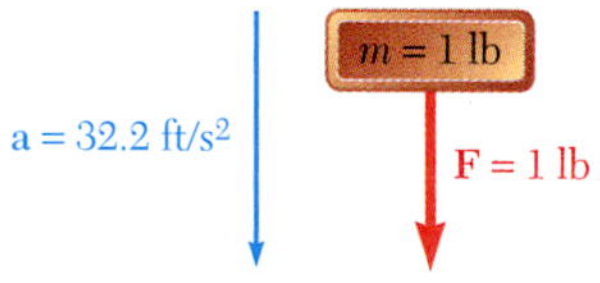

Fig 12.6

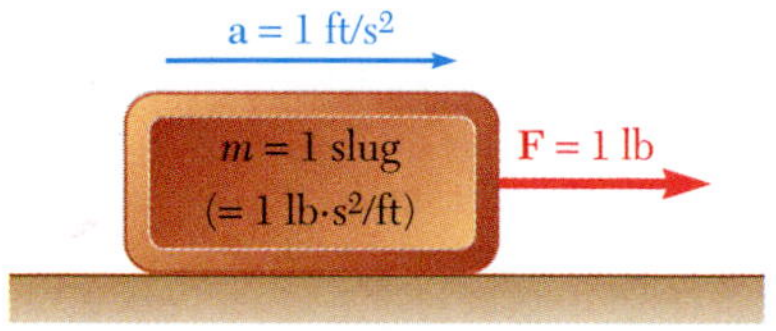

Fig 12.7

†Also known as a *metric ton*.

and obtain

$$1 \text{ slug} = \frac{1 \text{ lb}}{1 \text{ ft/s}^2} = 1 \text{ lb} \cdot \text{s}^2/\text{ft}$$

Comparing Figs. 12.6 and 12.7, we conclude that the slug is a mass 32.2 times larger than the mass of the standard pound.

The fact that bodies are characterized in the U.S. customary system of units by their weight in pounds rather than by their mass in slugs was a convenience in the study of statics, where we were dealing for the most part with weights and other forces and only seldomly with masses. However, in the study of kinetics, which involves forces, masses, and accelerations, it will be repeatedly necessary to express in slugs the mass m of a body, the weight W of which has been given in pounds. Recalling Eq. (12.6), we will write

$$m = \frac{W}{g} \qquad (12.7)$$

where g is the acceleration of gravity ($g = 32.2 \text{ ft/s}^2$).

Units other than the units of force, length, and time can all be expressed in terms of these three base units. For example, the unit of linear momentum can be obtained by using the definition of linear momentum to write

$$mv = (\text{lb} \cdot \text{s}^2/\text{ft})(\text{ft/s}) = \text{lb} \cdot \text{s}$$

Conversion from One System of Units to Another. The conversion from U.S. customary units to SI units, and vice versa, was discussed in Sec. 1.4. You will recall that the conversion factors obtained for the units of length, force, and mass are, respectively,

Length:	$1 \text{ ft} = 0.3048 \text{ m}$
Force:	$1 \text{ lb} = 4.448 \text{ N}$
Mass:	$1 \text{ slug} = 1 \text{ lb} \cdot \text{s}^2/\text{ft} = 14.59 \text{ kg}$

Although it cannot be used as a consistent unit of mass, the mass of the standard pound is, by definition,

$$1 \text{ pound-mass} = 0.4536 \text{ kg}$$

This constant can be used to determine the *mass* in SI units (kilograms) of a body which has been characterized by its *weight* in U.S. customary units (pounds).

12.5. EQUATIONS OF MOTION

Consider a particle of mass m acted upon by several forces. We recall from Sec. 12.2 that Newton's second law can be expressed by the equation

$$\Sigma \mathbf{F} = m\mathbf{a} \qquad (12.2)$$

which relates the forces acting on the particle and the vector $m\mathbf{a}$ (Fig. 12.8). In order to solve problems involving the motion of a particle, however, it will be found more convenient to replace Eq. (12.2) by equivalent equations involving scalar quantities.

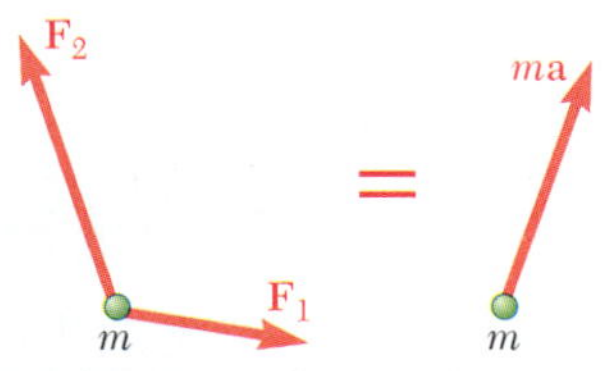

Fig. 12.8

Rectangular Components. Resolving each force **F** and the acceleration **a** into rectangular components, we write

$$\Sigma(F_x\mathbf{i} + F_y\mathbf{j} + F_z\mathbf{k}) = m(a_x\mathbf{i} + a_y\mathbf{j} + a_z\mathbf{k})$$

from which it follows that

$$\Sigma F_x = ma_x \qquad \Sigma F_y = ma_y \qquad \Sigma F_z = ma_z \tag{12.8}$$

Recalling from Sec. 11.11 that the components of the acceleration are equal to the second derivatives of the coordinates of the particle, we have

$$\Sigma F_x = m\ddot{x} \qquad \Sigma F_y = m\ddot{y} \qquad \Sigma F_z = m\ddot{z} \tag{12.8$'$}$$

Consider, as an example, the motion of a projectile. If the resistance of the air is neglected, the only force acting on the projectile after it has been fired is its weight $\mathbf{W} = -W\mathbf{j}$. The equations defining the motion of the projectile are therefore

$$m\ddot{x} = 0 \qquad m\ddot{y} = -W \qquad m\ddot{z} = 0$$

and the components of the acceleration of the projectile are

$$\ddot{x} = 0 \qquad \ddot{y} = -\frac{W}{m} = -g \qquad \ddot{z} = 0$$

where g is 9.81 m/s^2 or 32.2 ft/s^2. The equations obtained can be integrated independently, as shown in Sec. 11.11, to obtain the velocity and displacement of the projectile at any instant.

When a problem involves two or more bodies, equations of motion should be written for each of the bodies (see Sample Probs. 12.3 and 12.4). You will recall from Sec. 12.2 that all accelerations should be measured with respect to a newtonian frame of reference. In most engineering applications, accelerations can be determined with respect to axes attached to the earth, but relative accelerations measured with respect to moving axes, such as axes attached to an accelerated body, cannot be substituted for **a** in the equations of motion.

Photo 12.1 As it travels on the curved portion of a track, the bobsled is subjected to a normal component of acceleration directed toward the center of curvature of its path.

Tangential and Normal Components. Resolving the forces and the acceleration of the particle into components along the tangent to the path (in the direction of motion) and the normal (toward

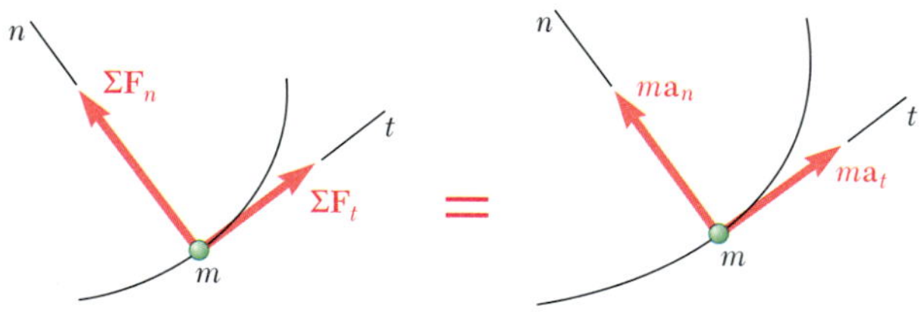

Fig. 12.9

the inside of the path) (Fig. 12.9), and substituting into Eq. (12.2), we obtain the two scalar equations

$$\Sigma F_t = ma_t \qquad \Sigma F_n = ma_n \tag{12.9}$$

Substituting for a_t and a_n from Eqs. (11.40), we have

$$\Sigma F_t = m\frac{dv}{dt} \qquad \Sigma F_n = m\frac{v^2}{\rho} \tag{12.9$'$}$$

The equations obtained may be solved for two unknowns.

12.6. DYNAMIC EQUILIBRIUM

Returning to Eq. (12.2) and transposing the right-hand member, we write Newton's second law in the alternative form

$$\Sigma \mathbf{F} - m\mathbf{a} = 0 \tag{12.10}$$

which expresses that if we add the vector $-m\mathbf{a}$ to the forces acting on the particle, *we obtain a system of vectors equivalent to zero* (Fig. 12.10). The vector $-m\mathbf{a}$, of magnitude ma and of *direction opposite* to that of the acceleration, is called an *inertia vector*. The particle may thus be considered to be in equilibrium under the given forces and the inertia vector. The particle is said to be in *dynamic equilibrium,* and the problem under consideration can be solved by the methods developed earlier in statics.

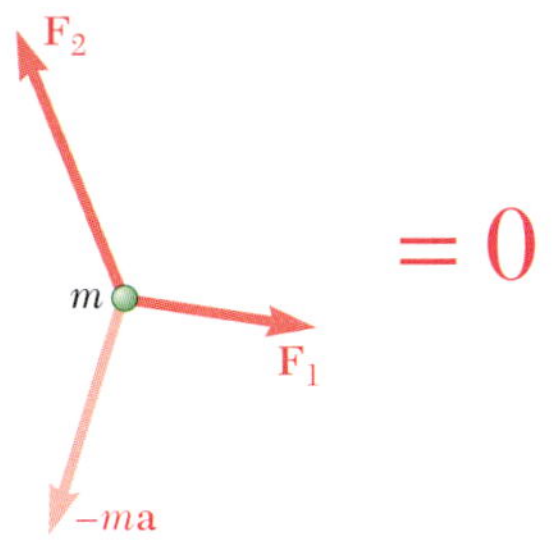

Fig. 12.10

In the case of coplanar forces, all the vectors shown in Fig. 12.10, *including the inertia vector,* can be drawn tip-to-tail to form a closed-vector polygon. Or the sums of the components of all the vectors in Fig. 12.10, again including the inertia vector, can be equated to zero. Using rectangular components, we therefore write

$$\Sigma F_x = 0 \qquad \Sigma F_y = 0 \qquad \textbf{\textit{including inertia vector}} \tag{12.11}$$

When tangential and normal components are used, it is more convenient to represent the inertia vector by its two components $-m\mathbf{a}_t$ and $-m\mathbf{a}_n$ in the sketch itself (Fig. 12.11). The tangential component of the inertia vector provides a measure of the resistance the particle offers to a change in speed, while its normal component (also called *centrifugal force*) represents the tendency of the particle to leave its curved path. We should note that either of these two components may be zero under special conditions: (1) if the particle starts from rest, its initial velocity is zero and the normal component of the inertia vector is zero at $t = 0$; (2) if the particle moves at constant speed along its path, the tangential component of the inertia vector is zero and only its normal component needs to be considered.

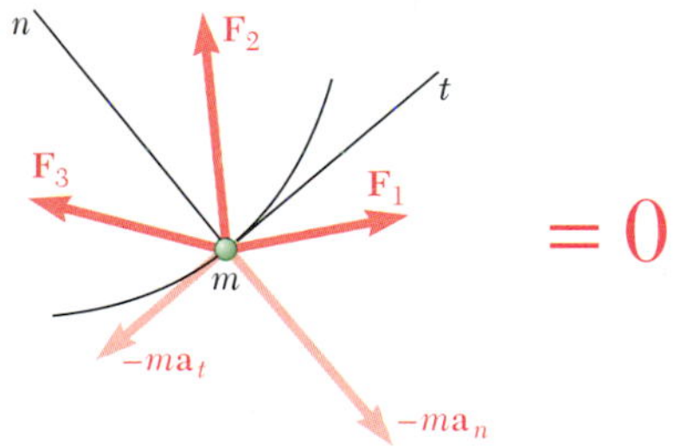

Fig. 12.11

Because they measure the resistance that particles offer when we try to set them in motion or when we try to change the conditions of their motion, inertia vectors are often called *inertia forces*. The inertia forces, however, are not forces like the forces found in statics, which are either contact forces or gravitational forces (weights). Many people, therefore, object to the use of the word "force" when referring to the vector $-m\mathbf{a}$ or even avoid altogether the concept of dynamic equilibrium. Others point out that inertia forces and actual forces, such as gravitational forces, affect our senses in the same way and cannot be distinguished by physical measurements. A man riding in an elevator which is accelerated upward will have the feeling that his weight has suddenly increased; and no measurement made within the elevator could establish whether the elevator is truly accelerated or whether the force of attraction exerted by the earth has suddenly increased.

Sample problems have been solved in this text by the direct application of Newton's second law, as illustrated in Figs. 12.8 and 12.9, rather than by the method of dynamic equilibrium.

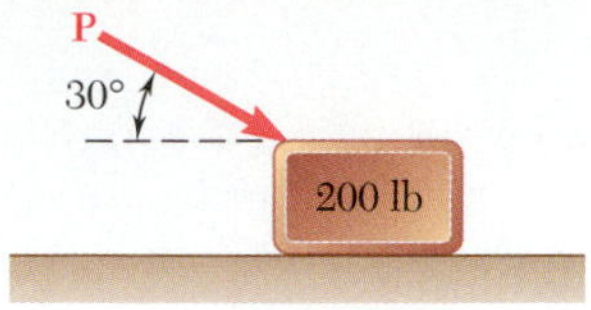

SAMPLE PROBLEM 12.1

A 200-lb block rests on a horizontal plane. Find the magnitude of the force **P** required to give the block an acceleration of 10 ft/s^2 to the right. The coefficient of kinetic friction between the block and the plane is $\mu_k = 0.25$.

SOLUTION

The mass of the block is

$$m = \frac{W}{g} = \frac{200 \text{ lb}}{32.2 \text{ ft/s}^2} = 6.21 \text{ lb} \cdot \text{s}^2\text{/ft}$$

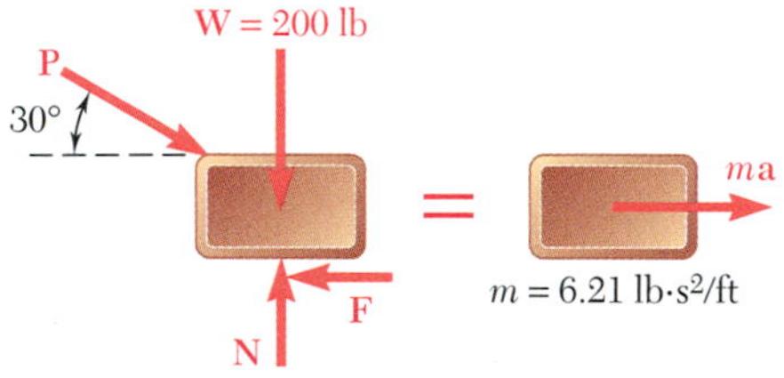

We note that $F = \mu_k N = 0.25N$ and that $a = 10$ ft/s^2. Expressing that the forces acting on the block are equivalent to the vector $m\mathbf{a}$, we write

$\overset{+}{\rightarrow}\Sigma F_x = ma$: $\quad P \cos 30° - 0.25N = (6.21 \text{ lb} \cdot \text{s}^2\text{/ft})(10 \text{ ft/s}^2)$

$P \cos 30° - 0.25N = 62.1 \text{ lb} \qquad (1)$

$+\uparrow\Sigma F_y = 0$: $\quad N - P \sin 30° - 200 \text{ lb} = 0 \qquad (2)$

Solving (2) for N and substituting the result into (1), we obtain

$$N = P \sin 30° + 200 \text{ lb}$$

$$P \cos 30° - 0.25(P \sin 30° + 200 \text{ lb}) = 62.1 \text{ lb} \qquad P = 151 \text{ lb} \blacktriangleleft$$

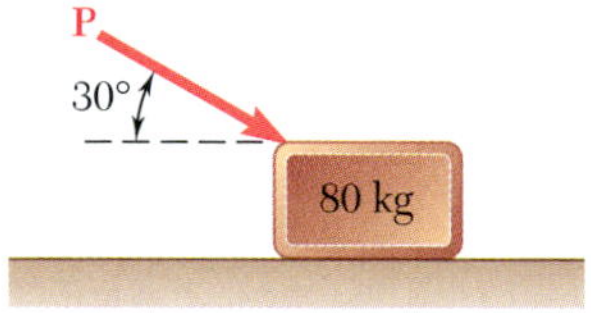

SAMPLE PROBLEM 12.2

An 80-kg block rests on a horizontal plane. Find the magnitude of the force **P** required to give the block an acceleration of 2.5 m/s^2 to the right. The coefficient of kinetic friction between the block and the plane is $\mu_k = 0.25$.

SOLUTION

The weight of the block is

$$W = mg = (80 \text{ kg})(9.81 \text{ m/s}^2) = 785 \text{ N}$$

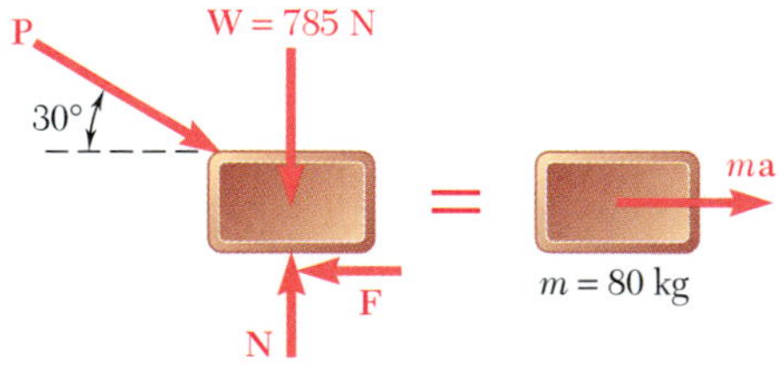

We note that $F = \mu_k N = 0.25N$ and that $a = 2.5$ m/s^2. Expressing that the forces acting on the block are equivalent to the vector $m\mathbf{a}$, we write

$\overset{+}{\rightarrow}\Sigma F_x = ma$: $\quad P \cos 30° - 0.25N = (80 \text{ kg})(2.5 \text{ m/s}^2)$

$P \cos 30° - 0.25N = 200 \text{ N} \qquad (1)$

$+\uparrow\Sigma F_y = 0$: $\quad N - P \sin 30° - 785 \text{ N} = 0 \qquad (2)$

Solving (2) for N and substituting the result into (1), we obtain

$$N = P \sin 30° + 785 \text{ N}$$

$$P \cos 30° - 0.25(P \sin 30° + 785 \text{ N}) = 200 \text{ N} \qquad P = 535 \text{ N} \blacktriangleleft$$

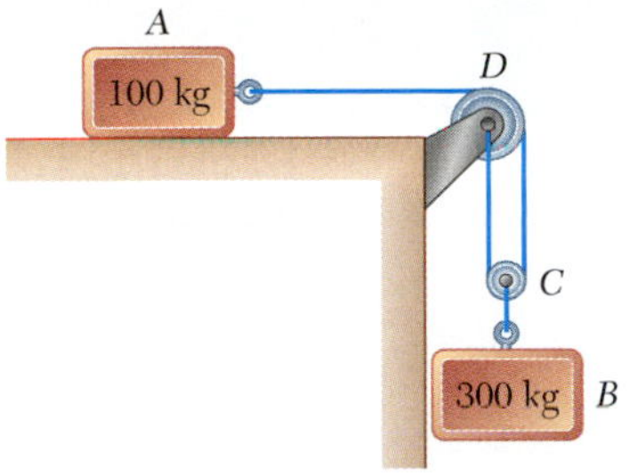

SAMPLE PROBLEM 12.3

The two blocks shown start from rest. The horizontal plane and the pulley are frictionless, and the pulley is assumed to be of negligible mass. Determine the acceleration of each block and the tension in each cord.

SOLUTION

Kinematics. We note that if block A moves through x_A to the right, block B moves down through

$$x_B = \tfrac{1}{2}x_A$$

Differentiating twice with respect to t, we have

$$a_B = \tfrac{1}{2}a_A \qquad (1)$$

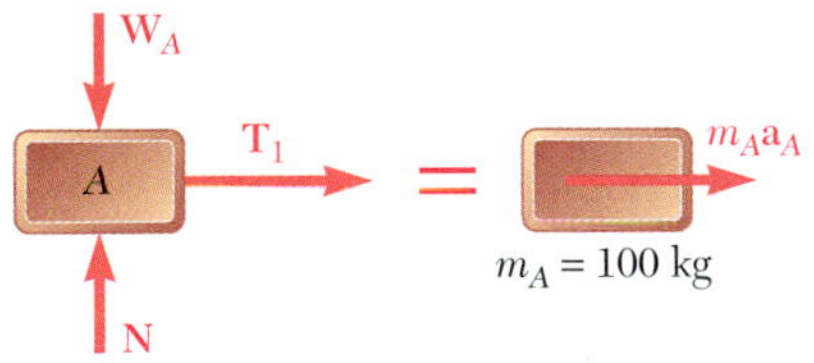

Kinetics. We apply Newton's second law successively to block A, block B, and pulley C.

Block A. Denoting by T_1 the tension in cord ACD, we write

$$\overset{+}{\rightarrow}\Sigma F_x = m_A a_A: \qquad T_1 = 100a_A \qquad (2)$$

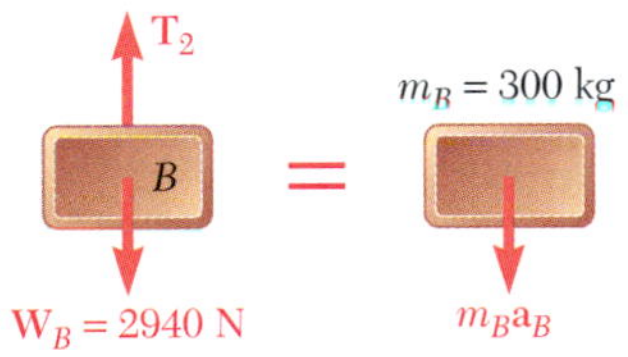

Block B. Observing that the weight of block B is

$$W_B = m_B g = (300 \text{ kg})(9.81 \text{ m/s}^2) = 2940 \text{ N}$$

and denoting by T_2 the tension in cord BC, we write

$$+\downarrow\Sigma F_y = m_B a_B: \qquad 2940 - T_2 = 300a_B$$

or, substituting for a_B from (1),

$$2940 - T_2 = 300(\tfrac{1}{2}a_A)$$
$$T_2 = 2940 - 150a_A \qquad (3)$$

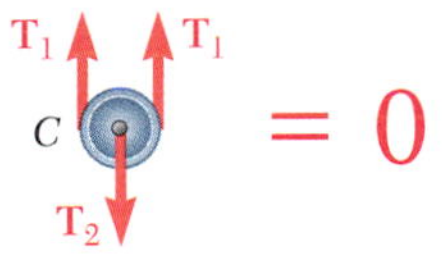

Pulley C. Since m_C is assumed to be zero, we have

$$+\downarrow\Sigma F_y = m_C a_C = 0: \qquad T_2 - 2T_1 = 0 \qquad (4)$$

Substituting for T_1 and T_2 from (2) and (3), respectively, into (4) we write

$$2940 - 150a_A - 2(100a_A) = 0$$
$$2940 - 350a_A = 0 \qquad a_A = 8.40 \text{ m/s}^2 \blacktriangleleft$$

Substituting the value obtained for a_A into (1) and (2), we have

$$a_B = \tfrac{1}{2}a_A = \tfrac{1}{2}(8.40 \text{ m/s}^2) \qquad a_B = 4.20 \text{ m/s}^2 \blacktriangleleft$$
$$T_1 = 100a_A = (100 \text{ kg})(8.40 \text{ m/s}^2) \qquad T_1 = 840 \text{ N} \blacktriangleleft$$

Recalling (4), we write

$$T_2 = 2T_1 \qquad T_2 = 2(840 \text{ N}) \qquad T_2 = 1680 \text{ N} \blacktriangleleft$$

We note that the value obtained for T_2 is *not* equal to the weight of block B.

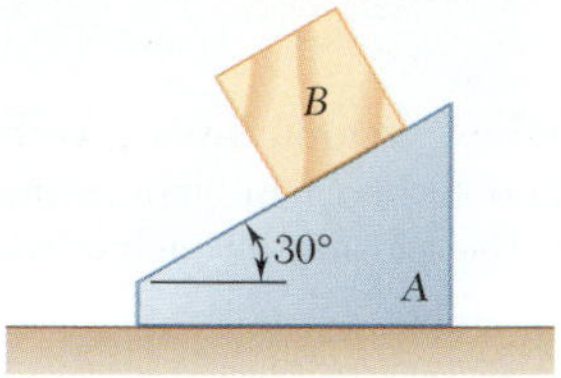

SAMPLE PROBLEM 12.4

The 12-lb block B starts from rest and slides on the 30-lb wedge A, which is supported by a horizontal surface. Neglecting friction, determine (*a*) the acceleration of the wedge, (*b*) the acceleration of the block relative to the wedge.

SOLUTION

Kinematics. We first examine the acceleration of the wedge and the acceleration of the block.

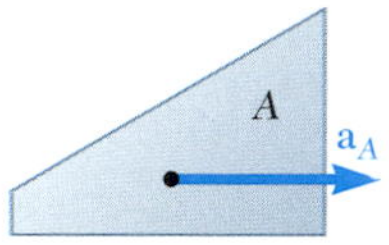

Wedge A. Since the wedge is constrained to move on the horizontal surface, its acceleration $\mathbf{a}_A$ is horizontal. We will assume that it is directed to the right.

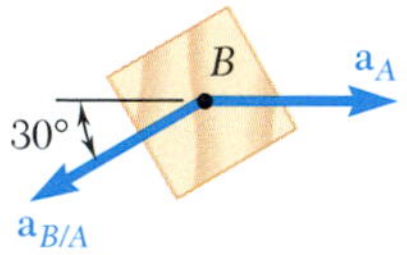

Block B. The acceleration $\mathbf{a}_B$ of block B can be expressed as the sum of the acceleration of A and the acceleration of B relative to A. We have

$$\mathbf{a}_B = \mathbf{a}_A + \mathbf{a}_{B/A}$$

where $\mathbf{a}_{B/A}$ is directed along the inclined surface of the wedge.

Kinetics. We draw the free-body diagrams of the wedge and of the block and apply Newton's second law.

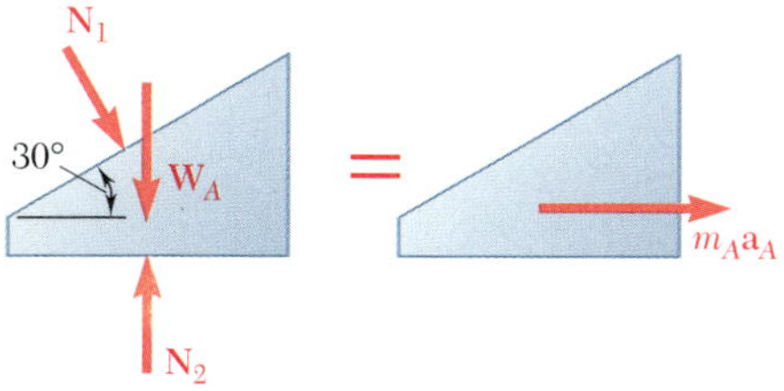

Wedge A. We denote the forces exerted by the block and the horizontal surface on wedge A by $\mathbf{N}_1$ and $\mathbf{N}_2$, respectively.

$$\overset{+}{\rightarrow}\Sigma F_x = m_A a_A: \qquad N_1 \sin 30° = m_A a_A$$
$$0.5N_1 = (W_A/g)a_A \qquad (1)$$

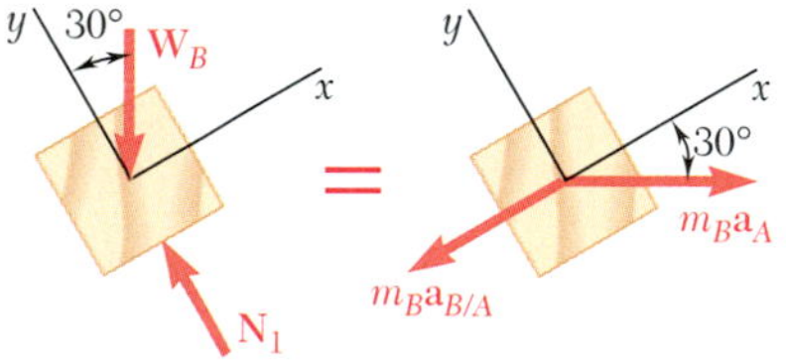

Block B. Using the coordinate axes shown and resolving $\mathbf{a}_B$ into its components $\mathbf{a}_A$ and $\mathbf{a}_{B/A}$, we write

$$+\nearrow\Sigma F_x = m_B a_x: \qquad -W_B \sin 30° = m_B a_A \cos 30° - m_B a_{B/A}$$
$$-W_B \sin 30° = (W_B/g)(a_A \cos 30° - a_{B/A})$$
$$a_{B/A} = a_A \cos 30° + g \sin 30° \qquad (2)$$
$$+\nwarrow\Sigma F_y = m_B a_y: \qquad N_1 - W_B \cos 30° = -m_B a_A \sin 30°$$
$$N_1 - W_B \cos 30° = -(W_B/g)a_A \sin 30° \qquad (3)$$

a. Acceleration of Wedge A. Substituting for N_1 from Eq. (1) into Eq. (3), we have

$$2(W_A/g)a_A - W_B \cos 30° = -(W_B/g)a_A \sin 30°$$

Solving for a_A and substituting the numerical data, we write

$$a_A = \frac{W_B \cos 30°}{2W_A + W_B \sin 30°}g = \frac{(12 \text{ lb}) \cos 30°}{2(30 \text{ lb}) + (12 \text{ lb}) \sin 30°}(32.2 \text{ ft/s}^2)$$
$$a_A = +5.07 \text{ ft/s}^2 \qquad \mathbf{a}_A = 5.07 \text{ ft/s}^2 \rightarrow \blacktriangleleft$$

b. Acceleration of Block B Relative to A. Substituting the value obtained for a_A into Eq. (2), we have

$$a_{B/A} = (5.07 \text{ ft/s}^2) \cos 30° + (32.2 \text{ ft/s}^2) \sin 30°$$
$$a_{B/A} = +20.5 \text{ ft/s}^2 \qquad \mathbf{a}_{B/A} = 20.5 \text{ ft/s}^2 \measuredangle 30° \blacktriangleleft$$

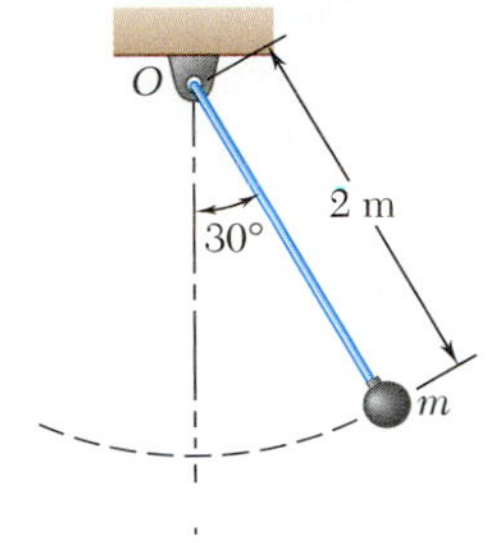

SAMPLE PROBLEM 12.5

The bob of a 2-m pendulum describes an arc of a circle in a vertical plane. If the tension in the cord is 2.5 times the weight of the bob for the position shown, find the velocity and the acceleration of the bob in that position.

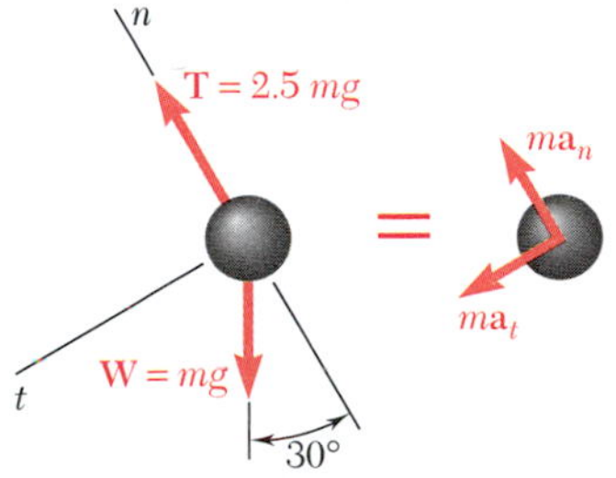

SOLUTION

The weight of the bob is $W = mg$; the tension in the cord is thus 2.5 mg. Recalling that $\mathbf{a}_n$ is directed toward O and assuming $\mathbf{a}_t$ as shown, we apply Newton's second law and obtain

$+\swarrow \Sigma F_t = ma_t$: $\quad mg \sin 30° = ma_t$

$a_t = g \sin 30° = +4.90 \text{ m/s}^2 \qquad \mathbf{a}_t = 4.90 \text{ m/s}^2 \swarrow$ ◀

$+\nwarrow \Sigma F_n = ma_n$: $\quad 2.5\, mg - mg \cos 30° = ma_n$

$a_n = 1.634\, g = +16.03 \text{ m/s}^2 \qquad \mathbf{a}_n = 16.03 \text{ m/s}^2 \nwarrow$ ◀

Since $a_n = v^2/\rho$, we have $v^2 = \rho a_n = (2 \text{ m})(16.03 \text{ m/s}^2)$

$v = \pm 5.66 \text{ m/s} \qquad \mathbf{v} = 5.66 \text{ m/s} \swarrow\!\!\nearrow$ (up or down) ◀

SAMPLE PROBLEM 12.6

Determine the rated speed of a highway curve of radius $\rho = 400$ ft banked through an angle $\theta = 18°$. The *rated speed* of a banked highway curve is the speed at which a car should travel if no lateral friction force is to be exerted on its wheels.

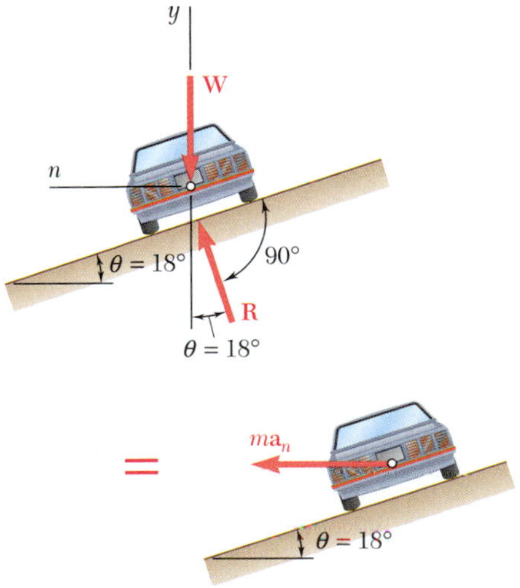

SOLUTION

The car travels in a *horizontal* circular path of radius ρ. The normal component $\mathbf{a}_n$ of the acceleration is directed toward the center of the path; its magnitude is $a_n = v^2/\rho$, where v is the speed of the car in ft/s. The mass m of the car is W/g, where W is the weight of the car. Since no lateral friction force is to be exerted on the car, the reaction **R** of the road is shown perpendicular to the roadway. Applying Newton's second law, we write

$$+\uparrow \Sigma F_y = 0: \qquad R \cos \theta - W = 0 \qquad R = \frac{W}{\cos \theta} \tag{1}$$

$$\overset{+}{\leftarrow} \Sigma F_n = ma_n: \qquad R \sin \theta = \frac{W}{g} a_n \tag{2}$$

Substituting for R from (1) into (2), and recalling that $a_n = v^2/\rho$,

$$\frac{W}{\cos \theta} \sin \theta = \frac{W}{g}\frac{v^2}{\rho} \qquad v^2 = g\rho \tan \theta$$

Substituting $\rho = 400$ ft and $\theta = 18°$ into this equation, we obtain

$$v^2 = (32.2 \text{ ft/s}^2)(400 \text{ ft}) \tan 18°$$

$v = 64.7 \text{ ft/s} \qquad v = 44.1 \text{ mi/h}$ ◀

SOLVING PROBLEMS ON YOUR OWN

In the problems for this lesson, you will apply *Newton's second law of motion,* $\Sigma \mathbf{F} = m\mathbf{a}$, to relate the forces acting on a particle to the motion of the particle.

1. *Writing the equations of motion.* When applying Newton's second law to the types of motion discussed in this lesson, you will find it most convenient to express the vectors **F** and **a** in terms of either their rectangular components or their tangential and normal components.

a. When using rectangular components, and recalling from Sec. 11.11 the expressions found for a_x, a_y, and a_z, you will write

$$\Sigma F_x = m\ddot{x} \qquad \Sigma F_y = m\ddot{y} \qquad \Sigma F_z = m\ddot{z}$$

b. When using tangential and normal components, and recalling from Sec. 11.13 the expressions found for a_t and a_n, you will write

$$\Sigma F_t = m\frac{dv}{dt} \qquad \Sigma F_n = m\frac{v^2}{\rho}$$

2. *Drawing a free-body diagram* showing the applied forces *and an equivalent diagram* showing the vector $m\mathbf{a}$ or its components will provide you with a pictorial representation of Newton's second law [Sample Probs. 12.1 through 12.6]. These diagrams will be of great help to you when writing the equations of motion. Note that when a problem involves two or more bodies, it is usually best to consider each body separately.

3. *Applying Newton's second law.* As we observed in Sec. 12.2, the acceleration used in the equation $\Sigma \mathbf{F} = m\mathbf{a}$ should always be the *absolute acceleration* of the particle (that is, it should be measured with respect to a newtonian frame of reference). Also, *if the sense of the acceleration* **a** *is unknown* or is not easily deduced, assume an arbitrary sense for **a** (usually the positive direction of a coordinate axis) and then let the solution provide the correct sense. Finally, note how the solutions of Sample Probs. 12.3 and 12.4 were divided into a *kinematics* portion and a *kinetics* portion, and how in Sample Prob. 12.4 we used two systems of coordinate axes to simplify the equations of motion.

4. *When a problem involves dry friction,* be sure to review the relevant sections of *Statics* [Secs. 8.1 to 8.3] before attempting to solve that problem. In particular, you should know when each of the equations $F = \mu_s N$ and $F = \mu_k N$ may be used. You should also recognize that if the motion of a system is not specified, it is necessary first to assume a possible motion and then to check the validity of that assumption.

5. *Solving problems involving relative motion.* When a body B moves with respect to a body A, as in Sample Prob. 12.4, it is often convenient to express the acceleration of B as

$$\mathbf{a}_B = \mathbf{a}_A + \mathbf{a}_{B/A}$$

where $\mathbf{a}_{B/A}$ is the acceleration of B relative to A, that is, the acceleration of B as observed from a frame of reference attached to A and in translation. If B is observed to move in a straight line, $\mathbf{a}_{B/A}$ will be directed along that line. On the other hand, if B is observed to move along a circular path, the relative acceleration $\mathbf{a}_{B/A}$ should be resolved into components tangential and normal to that path.

6. *Finally, always consider the implications of any assumption you make.* Thus, in a problem involving two cords, if you assume that the tension in one of the cords is equal to its maximum allowable value, check whether any requirements set for the other cord will then be satisfied. For instance, will the tension T in that cord satisfy the relation $0 \le T \le T_{max}$? That is, will the cord remain taut and will its tension be less than its maximum allowable value?

Problems

12.1 The value of the acceleration of gravity at any latitude ϕ is given by $g = 9.7087(1 + 0.0053 \sin^2 \phi)$ m/s^2, where the effect of the rotation of the earth as well as the fact that the earth is not spherical have been taken into account. Knowing that the mass of a gold bar has been officially designated as 2 kg, determine to four significant figures its mass in kilograms and its weight in newtons at a latitude of (*a*) 0°, (*b*) 45°, (*c*) 60°.

12.2 The acceleration due to gravity on Mars is 3.75 m/s^2. Knowing that the mass of a silver bar has been officially designated as 20 kg, determine, on Mars, its weight in newtons.

12.3 An artificial satellite is in a circular orbit 900 mi above the surface of Venus. The weight of the satellite was determined to be 400 lb before it was launched from earth. Determine the magnitude of the linear momentum of the satellite knowing that its orbital speed is 14.5×10^3 mi/h.

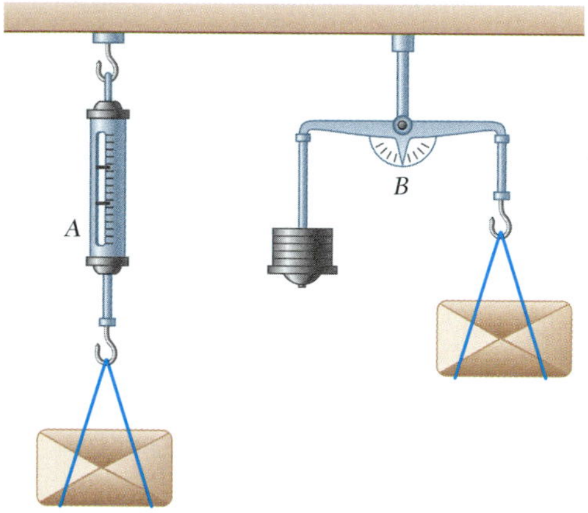

Fig. *P12.4*

12.4 A spring scale A and a lever scale B having equal lever arms are fastened to the roof on an elevator, and identical packages are attached to the scales as shown. Knowing that when the elevator moves downward with an acceleration of 2 ft/s^2 the spring scale indicates a load of 7 lb, determine (*a*) the weight of the packages, (*b*) the load indicated by the spring scale and the mass needed to balance the lever scale when the elevator moves upward with an acceleration of 2 ft/s^2.

12.5 In the braking test of a sports car its velocity is reduced from 110 km/h to zero in a distance of 51 m with slipping impending. Knowing that the coefficient of kinetic friction is 80 percent of the coefficient of static friction, determine (*a*) the coefficient of static friction, (*b*) the stopping distance for the same initial velocity if the car skids. Ignore air resistance and rolling resistance.

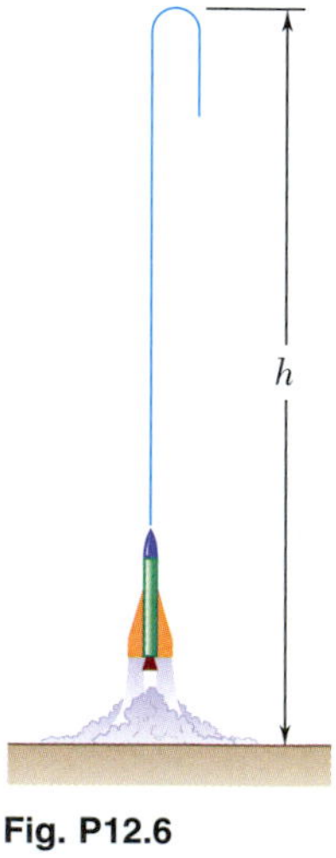

Fig. P12.6

12.6 A 0.1-kg model rocket is launched vertically from rest at time $t = 0$ with a constant thrust of 10 N for one second and no thrust for $t > 1$ s. Neglecting air resistance and the decrease in mass of the rocket, determine (*a*) the maximum height h reached by the rocket, (*b*) the time required to reach this maximum height.

12.7 Determine the maximum theoretical speed that an automobile starting from rest can reach after traveling 1320 ft. Assume that the coefficient of static friction is 0.80 between the tires and the pavement and that (*a*) the automobile has front-wheel drive and the front wheels support 65 percent of the automobile's weight, (*b*) the automobile has rear-wheel drive and the rear wheels support 42 percent of the automobile's weight.

12.8 A hockey player hits a puck so that it comes to rest 4 s after sliding 60 ft on the ice. Determine (*a*) the initial velocity of the puck, (*b*) the coefficient of friction between the puck and the ice.

12.9 A 40-kg package is at rest on an incline when a force **P** is applied to it. Determine the magnitude of **P** if 4 s is required for the package to travel 10 m up the incline. The static and kinetic coefficients of friction between the package and the incline are 0.30 and 0.25, respectively.

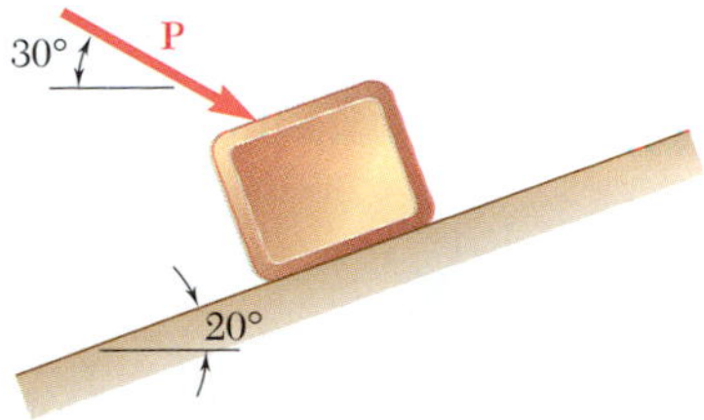

Fig. P12.9

12.10 If an automobile's braking distance from 100 km/h is 60 m on level pavement, determine the automobile's braking distance from 100 km/h when it is (*a*) going up a 6° incline, (*b*) going down a 2-percent incline.

12.11 A tractor-trailer is traveling at 90 km/h when the driver applies his brakes. Knowing that the braking forces of the tractor and the trailer are 16 kN and 60 kN, respectively, determine (*a*) the distance traveled by the tractor-trailer before it comes to a stop, (*b*) the horizontal component of the force in the hitch between the tractor and the trailer while they are slowing down.

Fig. *P12.11* and P12.12

12.12 Solve Prob. 12.11 assuming that a second trailer and dolly, with a combined mass of 11 300 kg, are coupled to the rear of the tractor-trailer. The braking force of the second trailer is 57 kN.

12.13 The two blocks shown are originally at rest. Neglecting the masses of the pulleys and the effect of friction in the pulleys and between the blocks and the incline, determine (*a*) the acceleration of each block, (*b*) the tension in the cable.

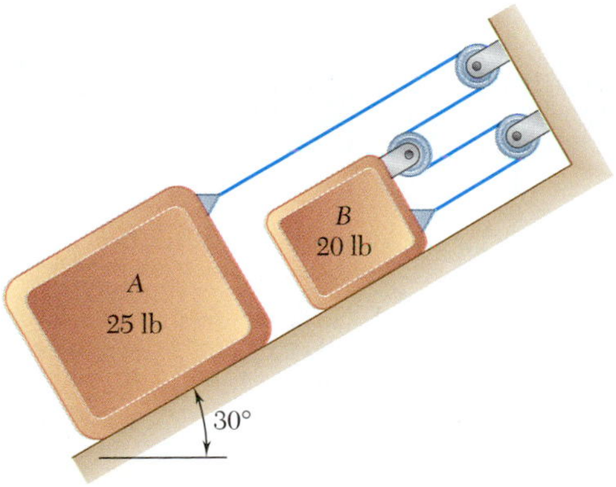

Fig. *P12.13* and *P12.14*

12.14 The two blocks shown are originally at rest. Neglecting the masses of the pulleys and the effect of friction in the pulleys and assuming that the coefficients of friction between both blocks and the incline are $\mu_s = 0.25$ and $\mu_k = 0.20$, determine (*a*) the acceleration of each block, (*b*) the tension in the cable.

12.15 Block A weighs 80 lb, and block B weighs 16 lb. The coefficients of friction between all surfaces of contact are $\mu_s = 0.20$ and $\mu_k = 0.15$. Knowing that $P = 0$, determine (*a*) the acceleration of block B, (*b*) the tension in the cord.

12.16 Block A weighs 80 lb, and block B weighs 16 lb. The coefficients of friction between all surfaces of contact are $\mu_s = 0.20$ and $\mu_k = 0.15$. Knowing that $\mathbf{P} = 10\text{ lb} \rightarrow$, determine (*a*) the acceleration of block B, (*b*) the tension in the cord.

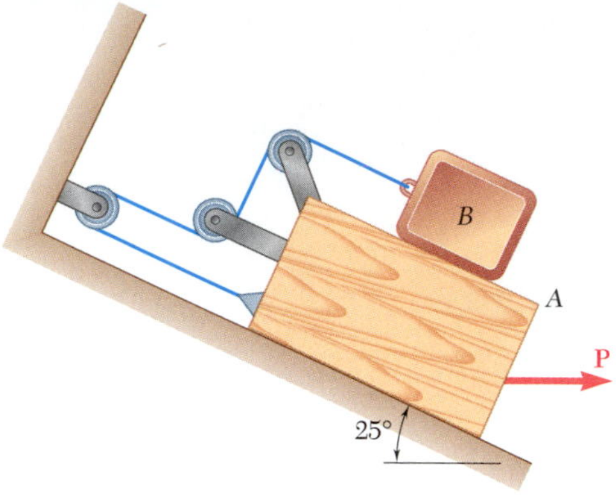

Fig. P12.15 and P12.16

12.17 Boxes A and B are at rest on a conveyor belt that is initially at rest. The belt is suddenly started in an upward direction so that slipping occurs between the belt and the boxes. Knowing that the coefficients of kinetic friction between the belt and the boxes are $(\mu_k)_A = 0.30$ and $(\mu_k)_B = 0.32$, determine the initial acceleration of each box.

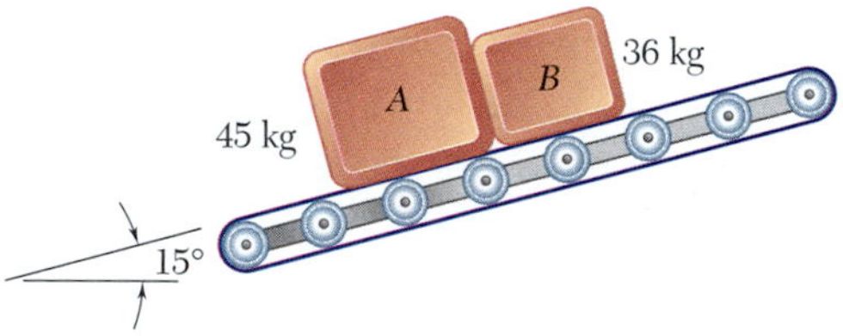

Fig. *P12.17*

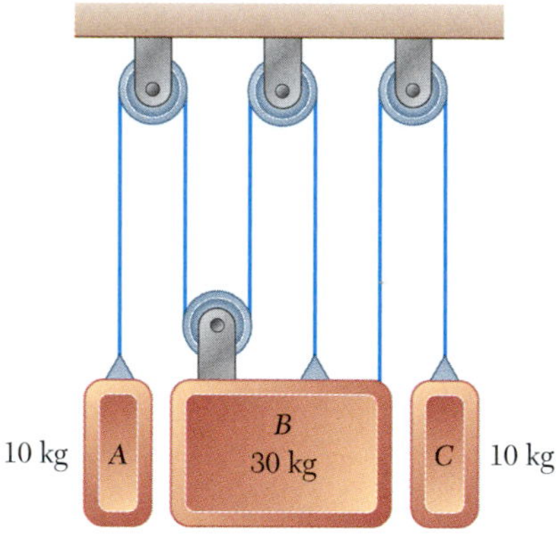

Fig. P12.18

12.18 The system shown is initially at rest. Neglecting the masses of the pulleys and the effect of friction in the pulleys, determine (*a*) the acceleration of each block, (*b*) the tension in each cable.

12.19 Each of the systems shown is initially at rest. Neglecting axle friction and the masses of the pulleys, determine for each system (*a*) the acceleration of block A, (*b*) the velocity of block A after it has moved through 5 ft, (*c*) the time required for block A to reach a velocity of 10 ft/s.

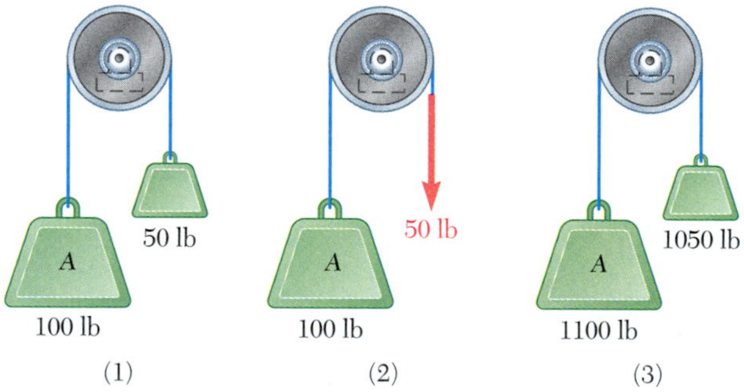

Fig. P12.19

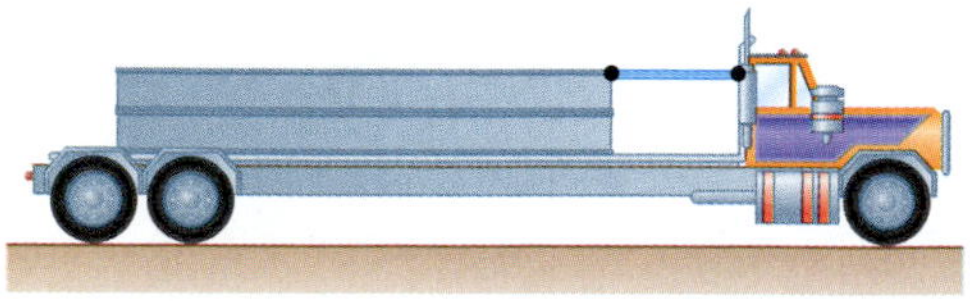

Fig. P12.20

12.20 The flat-bed trailer carries two 1500-kg beams with the upper beam secured by a cable. The coefficients of static friction between the two beams and between the lower beam and the bed of the trailer are 0.25 and 0.30, respectively. Knowing that the load does not shift, determine (*a*) the maximum acceleration of the trailer and the corresponding tension in the cable, (*b*) the maximum deceleration of the trailer.

12.21 A package is at rest on a conveyor belt which is initially at rest. The belt is started and moves to the right for 1.5 s with a constant acceleration of 3.2 m/s^2. The belt then moves with a constant deceleration $\mathbf{a}_2$ and comes to a stop after a total displacement of 4.6 m. Knowing that the coefficients of friction between the package and the belt are $\mu_s = 0.35$ and $\mu_k = 0.25$, determine (*a*) the deceleration $\mathbf{a}_2$ of the belt, (*b*) the displacement of the package relative to the belt as the belt comes to a stop.

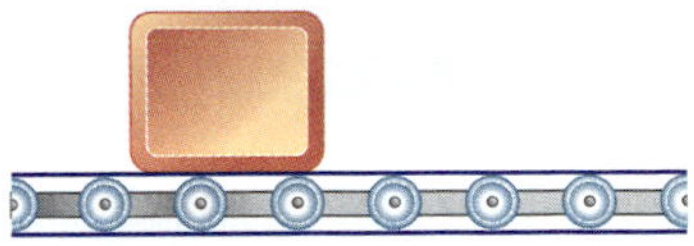

Fig. P12.21

12.22 To transport a series of bundles of shingles *A* to a roof, a contractor uses a motor-driven lift consisting of a horizontal platform *BC* which rides on rails attached to the sides of a ladder. The lift starts from rest and initially moves with a constant acceleration $\mathbf{a}_1$ as shown. The lift then decelerates at a constant rate $\mathbf{a}_2$ and comes to rest at *D*, near the top of the ladder. Knowing that the coefficient of static friction between the bundle of shingles and the horizontal platform is 0.30, determine the largest allowable acceleration $\mathbf{a}_1$ and the largest allowable deceleration $\mathbf{a}_2$ if the bundle is not to slide on the platform.

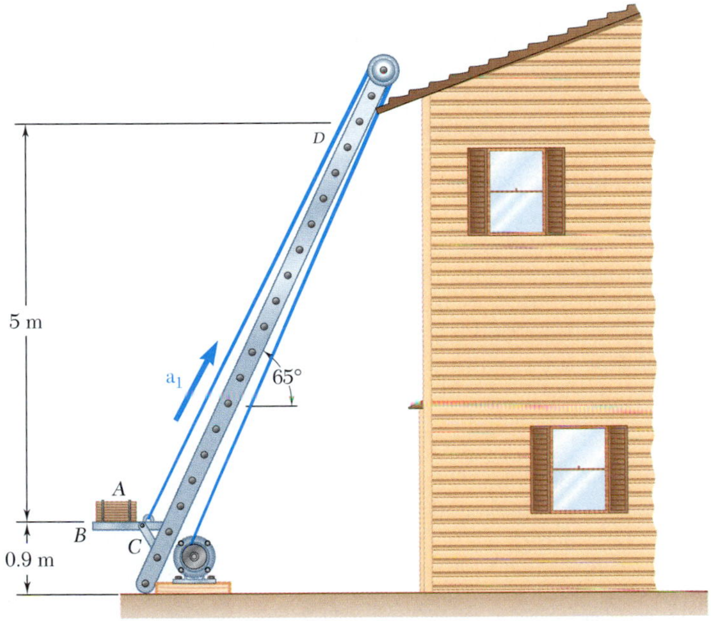

Fig. *P12.22*

12.23 To unload a bound stack of plywood from a truck, the driver first tilts the bed of the truck and then accelerates from rest. Knowing that the coefficients of friction between the bottom sheet of plywood and the bed are $\mu_s = 0.40$ and $\mu_k = 0.30$, determine (*a*) the smallest acceleration of the truck which will cause the stack of plywood to slide, (*b*) the acceleration of the truck which causes corner *A* of the stack of plywood to reach the end of the bed in 0.4 s.

Fig. *P12.23*

12.24 The propellers of a ship of mass m can produce a propulsive force $\mathbf{F}_0$; they produce a force of the same magnitude but opposite direction when the engines are reversed. Knowing that the ship was proceeding forward at its maximum speed v_0 when the engines were put into reverse, determine the distance the ship travels before coming to a stop. Assume that the frictional resistance of the water varies directly with the square of the velocity.

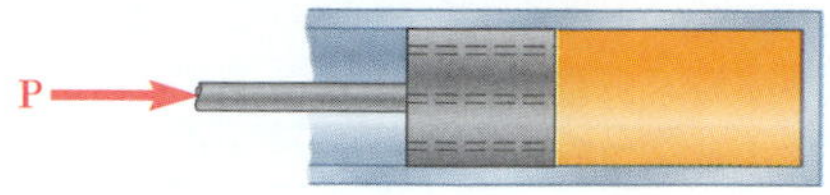

Fig. P12.25

12.25 A constant force **P** is applied to a piston and rod of total mass m to make them move in a cylinder filled with oil. As the piston moves, the oil is forced through orifices in the piston and exerts on the piston a force of magnitude kv in a direction opposite to the motion of the piston. Knowing that the piston starts from rest at $t = 0$ and $x = 0$, show that the equation relating x, v, and t, where x is the distance traveled by the piston and v is the speed of the piston, is linear in each of the variables.

12.26 Determine the maximum theoretical speed that a 1225 kg automobile starting from rest can reach after traveling 400 m if air resistance is considered. Assume that the coefficient of static friction between the tires and the pavement is 0.70, that the automobile has front-wheel drive, that the front wheels support 62 percent of the automobile's weight, and that the aerodynamic drag **D** has a magnitude $D = 0.575v^2$, where D and v are expressed in newtons and m/s, respectively.

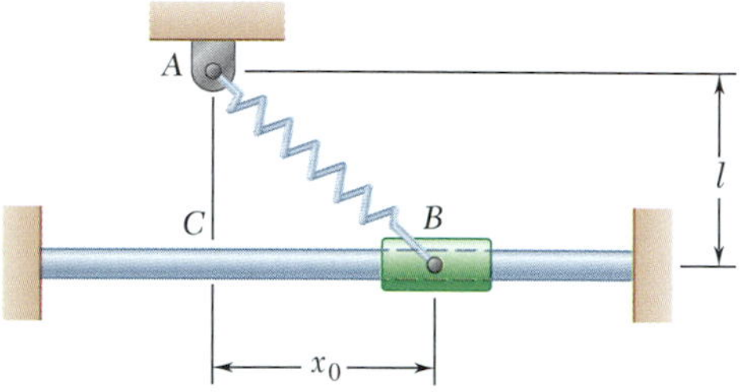

Fig. P12.27

12.27 A spring AB of constant k is attached to a support A and to a collar of mass m. The unstretched length of the spring is l. Knowing that the collar is released from rest at $x = x_0$ and neglecting friction between the collar and the horizontal rod, determine the magnitude of the velocity of the collar as it passes through point C.

12.28 The masses of blocks A, B, and C are $m_A = m_C = 10$ kg, and $m_B = 5$ kg. Knowing that $P = 200$ N and neglecting the masses of the pulleys and the effect of friction, determine (*a*) the acceleration of each block, (*b*) the tension in the cable.

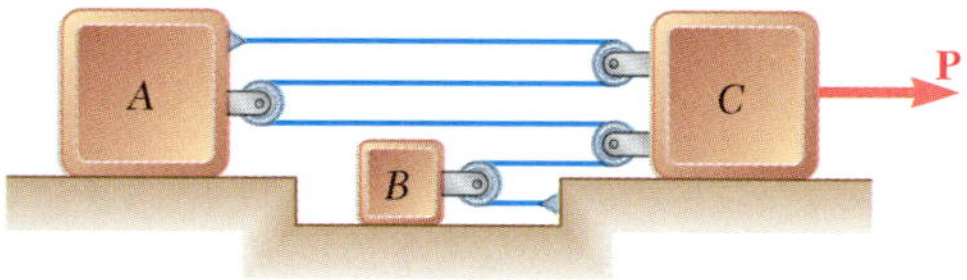

Fig. P12.28 and P12.29

12.29 The coefficients of friction between the three blocks and the horizontal surfaces are $\mu_s = 0.25$ and $\mu_k = 0.20$. The masses of the blocks are $m_A = m_C = 10$ kg, and $m_B = 5$ kg. Knowing that the blocks are initially at rest and that C moves to the right through 0.8 m in 0.4 s, determine (*a*) the acceleration of each block, (*b*) the tension in the cable, (*c*) the force **P**. Neglect axle friction and the masses of the pulleys.

12.30 The 30-lb block B is supported by the 55-lb block A and is attached to a cord to which a 50-lb horizontal force is applied as shown. Neglecting friction, determine (*a*) the acceleration of block A, (*b*) the acceleration of block B relative to A.

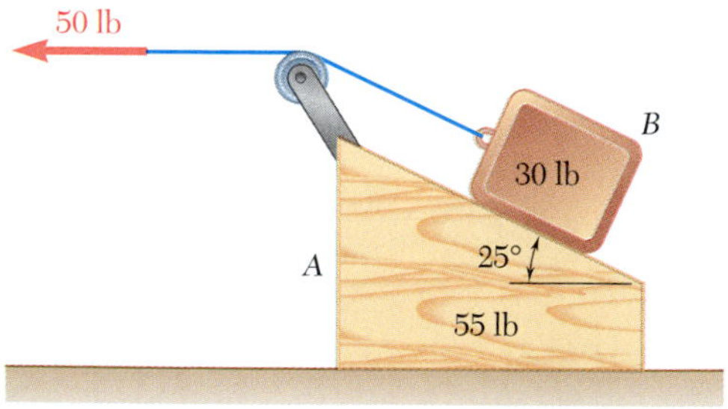

Fig. P12.30

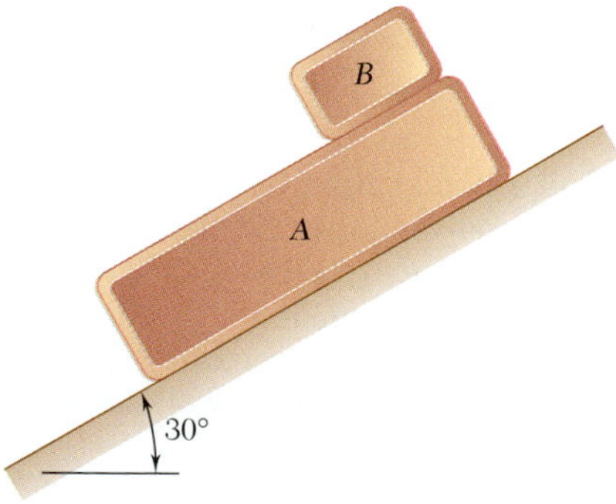

Fig. P12.31

12.31 The coefficients of friction between block B and block A are $\mu_s = 0.12$ and $\mu_k = 0.10$ and the coefficients of friction between block A and the incline are $\mu_s = 0.24$ and $\mu_k = 0.20$. Block A weighs 20 lb and block B weighs 10 lb. Knowing that the system is released from rest in the position shown, determine (*a*) the acceleration of A, (*b*) the velocity of B relative to A at $t = 0.5$ s.

12.32 A 25-kg block A rests on an inclined surface, and a 15-kg counterweight B is attached to a cable as shown. Neglecting friction, determine the acceleration of A and the tension in the cable immediately after the system is released from rest.

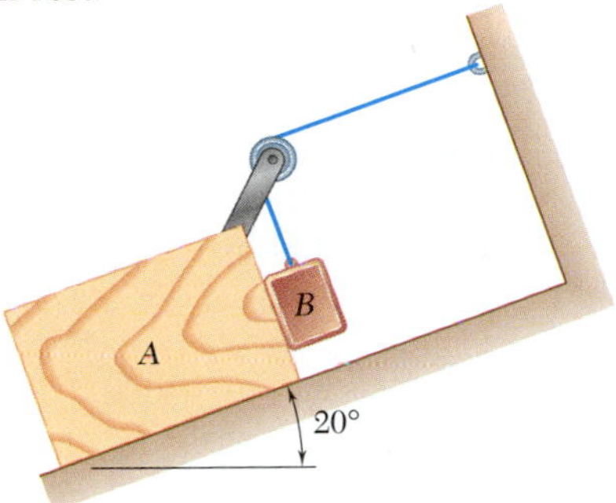

Fig. ***P12.32***

12.33 A 250-kg crate B is suspended from a cable attached to a 20-kg trolley A which rides on an inclined I-beam as shown. Knowing that at the instant shown the trolley has an acceleration of 0.4 m/s^2 up to the right, determine (*a*) the acceleration of B relative to A, (*b*) the tension in cable CD.

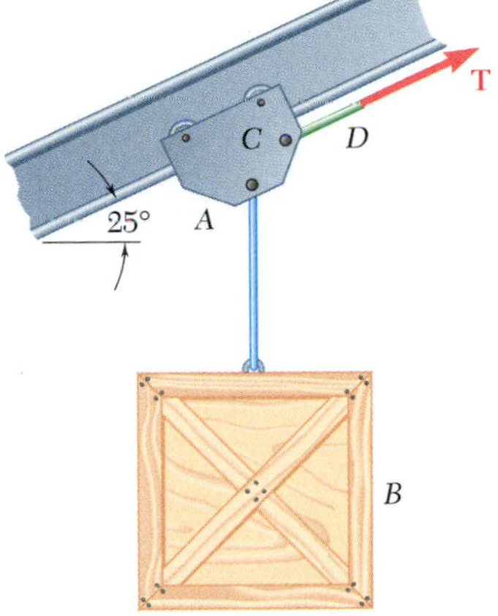

Fig. ***P12.33***

12.34 A single wire ACB of length 2 m passes through a ring at C that is attached to a sphere which revolves at a constant speed v in the horizontal circle shown. Knowing that $\theta_1 = 60°$ and $\theta_2 = 30°$ and that the tension is the same in both portions of the wire, determine the speed v.

12.35 A single wire ACB passes through a ring at C that is attached to a 5-kg sphere which revolves at a constant speed v in the horizontal circle shown. Knowing that $\theta_1 = 50°$ and $d = 0.8$ m and that the tension in both portions of the wire is 34 N, determine (*a*) the angle θ_2, (*b*) the speed v.

12.36 Two wires AC and BC are tied to a 15-lb sphere which revolves at a constant speed v in the horizontal circle shown. Knowing that $\theta_1 = 50°$ and $\theta_2 = 25°$ and that $d = 4$ ft, determine the range of values of v for which both wires are taut.

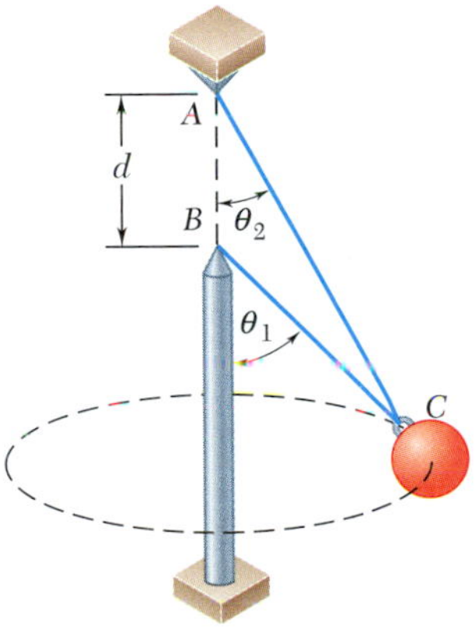

Fig. P12.34, P12.35, and ***P12.36***

12.37 During a hammer thrower's practice swings, the 16-lb head A of the hammer revolves at a constant speed v in a horizontal circle as shown. If $\rho = 3$ ft and $\theta = 60°$, determine (*a*) the tension in wire BC, (*b*) the speed of the hammer's head.

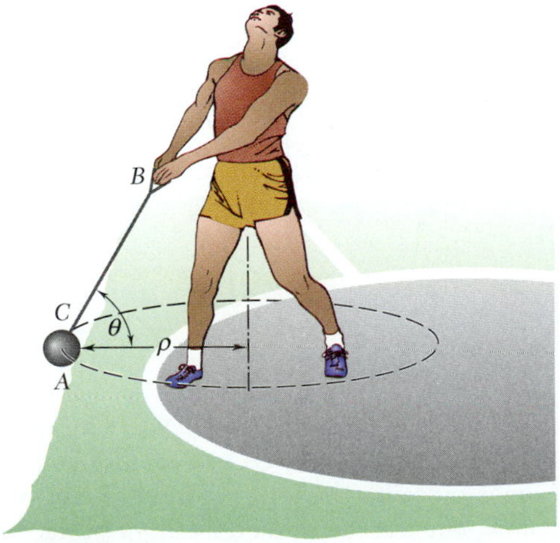

Fig. ***P12.37***

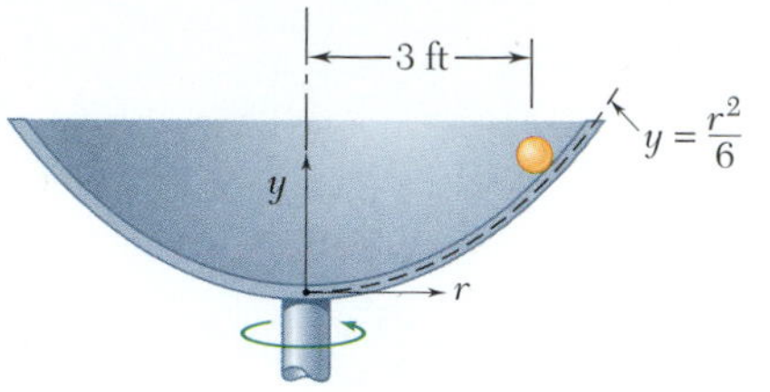

Fig. P12.38

12.38 A 2-lb sphere is at rest relative to a parabolic dish which rotates at a constant rate about a vertical axis. Neglecting friction and knowing that $r = 3$ ft, determine (*a*) the velocity v of the sphere, (*b*) the magnitude of the normal force exerted by the sphere on the inclined surface of the dish.

12.39 A 2-lb collar C slides without friction along the rod OA and is attached to rod BC by a frictionless pin. The rods rotate in the horizontal plane. At the instant shown BC is rotating counterclockwise and the speed of C is 3 ft/s, increasing at a rate of 4 ft/s^2. Determine at this instant, (*a*) the tension in rod BC, (*b*) the force exerted by the collar on rod OA.

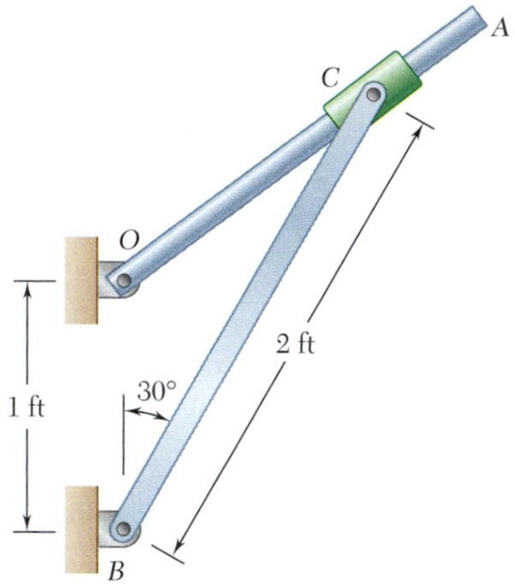

Fig. P12.39

***12.40** The 0.5-kg flyballs of a centrifugal governor revolve at a constant speed v in the horizontal circle of 150-mm radius shown. Neglecting the mass of links AB, BC, AD, and DE and requiring that the links support only tensile forces, determine the range of the allowable values of v so that the magnitudes of the forces in the links do not exceed 75 N.

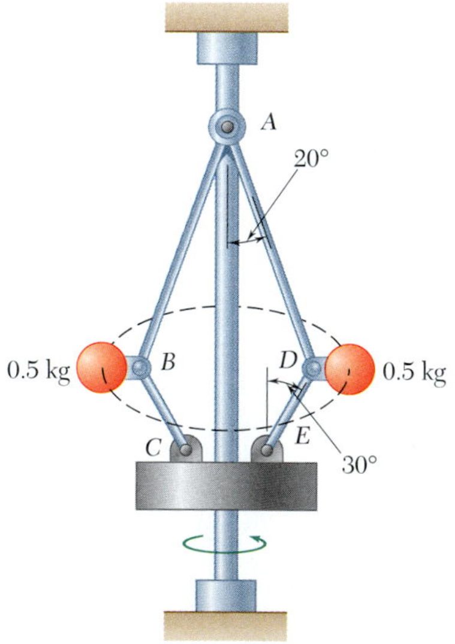

Fig. P12.40

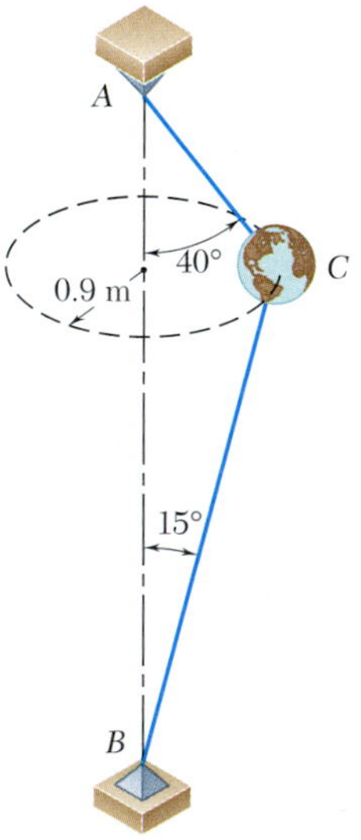

Fig. *P12.41*

****12.41*** As part of an outdoor display, a 5-kg model C of the earth is attached to wires AC and BC and revolves at a constant speed v in the horizontal circle shown. Determine the range of the allowable values of v if both wires are to remain taut and if the tension in either of the wires is not to exceed 116 N.

12.42 An airline pilot climbs to a new flight level along the path shown. Knowing that the speed of the airplane decreases at a constant rate from 180 m/s at point A to 160 m/s at point C, determine the magnitude of the abrupt change in the force exerted on a 90-kg passenger as the airplane passes point B.

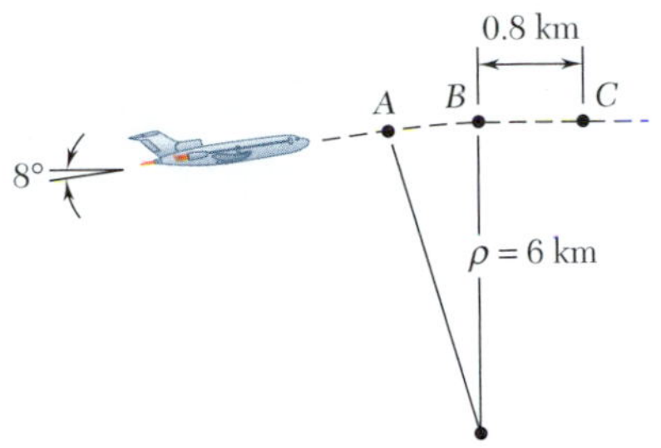

Fig. P12.42 and P12.43

12.43 An airline pilot climbs to a new flight level along the path shown. The motion of the airplane between A and B is defined by the relation $s = t(180 - t)$, where s is the arc length in meters, t is the time in seconds, and $t = 0$ when the airplane is at point A. Determine the force exerted by his seat on a 75-kg passenger (*a*) just after the airplane passes point A, (*b*) just before the airplane reaches point B.

12.44 A 60-lb child sits on a swing and is held in the position shown by a second child. Neglecting the weight of the swing, determine the tension in rope AB (*a*) while the second child holds the swing with his arms outstretched horizontally, (*b*) immediately after the swing is released.

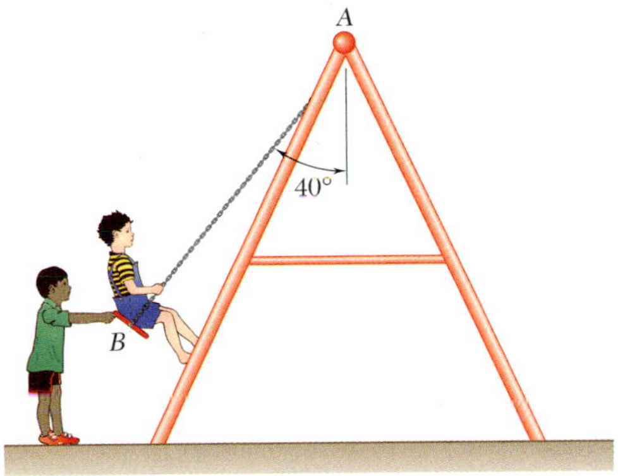

Fig. P12.44

12.45 A 180-lb wrecking ball B is attached to a 40-ft-long steel cable AB and swings in the vertical arc shown. Determine the tension in the cable (*a*) at the top C of the swing, where $\theta = 30°$, (*b*) at the bottom D of the swing, where the speed of B is 18.6 ft/s.

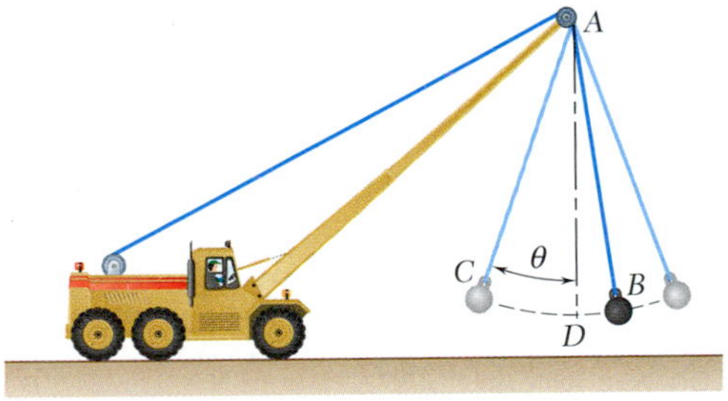

Fig. P12.45

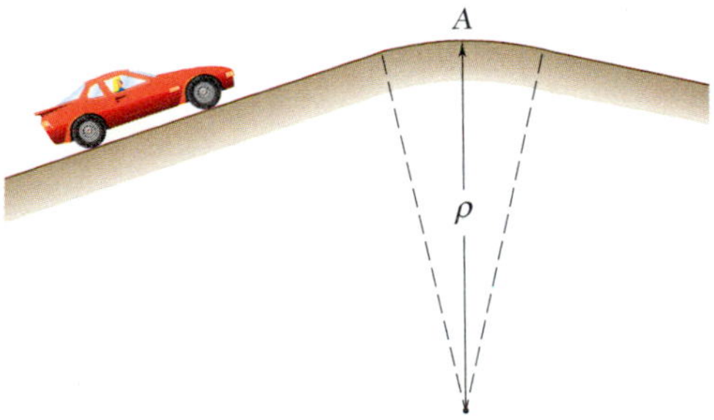

Fig. P12.46

12.46 During a high-speed chase, an 1100-kg sports car traveling at a speed of 160 km/h just loses contact with the road as it reaches the crest A of a hill. (*a*) Determine the radius of curvature ρ of the vertical profile of the road at A. (*b*) Using the value of ρ found in part *a*, determine the force exerted on a 70-kg driver by the seat of his 1400-kg car as the car, traveling at a constant speed of 80 km/h, passes through A.

12.47 A 220-g block B fits inside a small cavity cut in arm OA, which rotates in the vertical plane at a constant rate. When $\theta = 180°$, the spring is stretched to its maximum length and the block exerts a force of 3.5 N on the face of the cavity closest to A. Neglecting friction, determine the range of values of θ for which the block is not in contact with that face of the cavity.

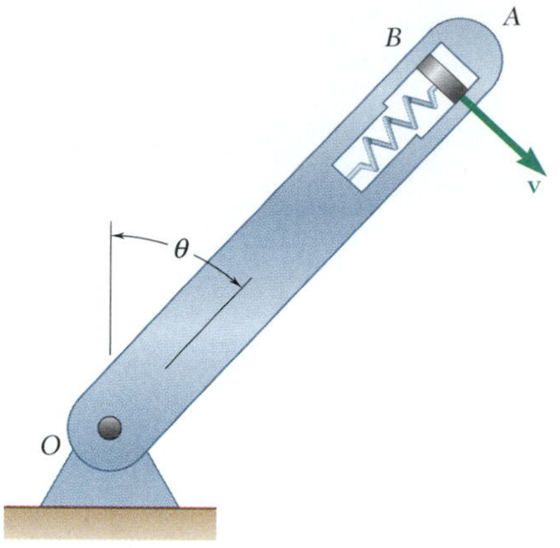

Fig. P12.47

12.48 A small 0.4-lb sphere B is given a downward velocity $\mathbf{v}_0$ and swings freely in the vertical plane, first about O and then about the peg A after the cord comes in contact with the peg. Determine the largest allowable velocity $\mathbf{v}_0$ if the tension in the cord is not to exceed 2.4 lb.

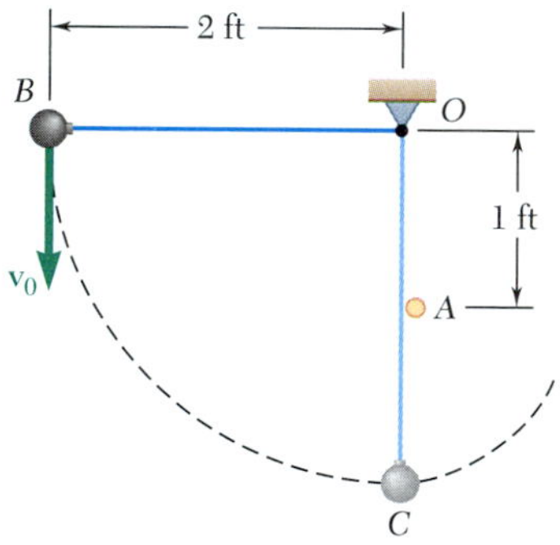

Fig. *P12.48*

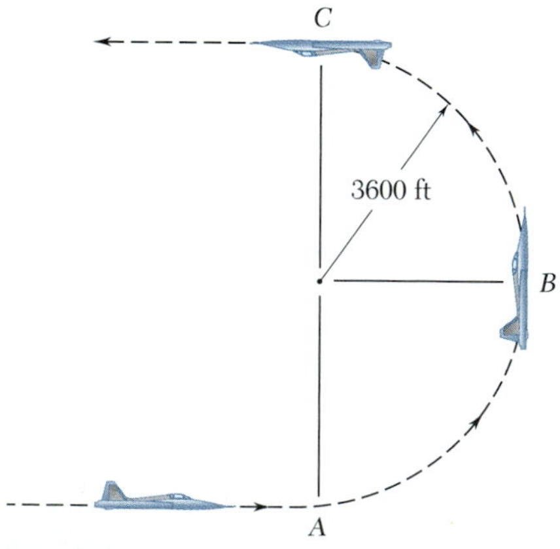

Fig. *P12.49*

12.49 A 120-lb pilot flies a jet trainer in a half vertical loop of 3600-ft radius so that the speed of the trainer decreases at a constant rate. Knowing that the pilot's apparent weights at points A and C are 380 lb and 80 lb, respectively, determine the force exerted on her by the seat of the trainer when the trainer is at point B.

12.50 A car is traveling on a banked road at a constant speed v. Determine the range of values of v for which the car does not skid. Express your answer in terms of the radius r of the curve, the banking angle θ, and the angle of static friction ϕ_s between the tires and the pavement.

12.51 A car traveling at a speed of 110 km/h approaches a curve of radius 50 m. Knowing that the coefficient of static friction between the tires and the road is 0.65, determine by how much the driver must reduce his speed to safely negotiate the curve if the banking angle is (*a*) $\theta = 10°$, (*b*) $\theta = -5°$, because of the presence of a sinkhole under the roadbed.

12.52 Tilting trains such as the *Acela*, which runs from Washington to New York to Boston, are designed to travel safely at high speeds on curved sections of track which were built for slower, conventional trains. As it enters a curve, each car is tilted by hydraulic actuators mounted on its trucks. The tilting feature of the cars also increases passenger comfort by eliminating or greatly reducing the side force $\mathbf{F}_s$ (parallel to the floor of the car) to which passengers feel subjected. For a train traveling at 125 mi/h on a curved section of track banked at an angle $\theta = 8°$ and with a rated speed of 75 mi/h, determine (*a*) the magnitude of the side force felt by a passenger of weight W in a standard car with no tilt ($\phi = 0$), (*b*) the required angle of tilt ϕ if the passenger is to feel no side force. (See Sample Problem 12.6 for the definition of rated speed.)

12.53 Tests carried out with the tilting trains described in Prob. 12.52 revealed that passengers feel queasy when they see through the car windows that the train is rounding a curve at high speed, yet do not feel any side force. Designers, therefore, prefer to reduce, but not eliminate that force. For the train of Prob. 12.52, determine the required angle of tilt ϕ if passengers are to feel side forces equal to 12 percent of their weights.

Fig. P12.52 and P12.53

12.54 A small block B fits inside a slot cut in arm OA which rotates in a vertical plane at a constant rate. The block remains in contact with the end of the slot closest to A and its speed is 1.4 m/s for $0 \leq \theta \leq 150°$. Knowing that the block begins to slide when $\theta = 150°$, determine the coefficient of static friction between the block and the slot.

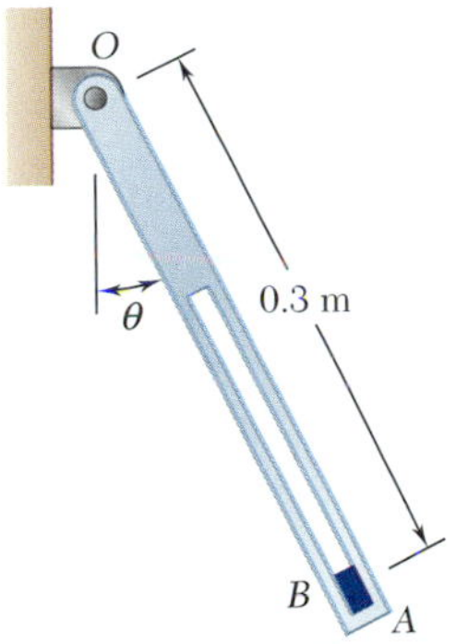

Fig. P12.54

12.55 A 3-kg block is at rest relative to a parabolic dish which rotates at a constant rate about a vertical axis. Knowing that the coefficient of static friction is 0.5 and that $r = 2$ m, determine the maximum allowable velocity v of the block.

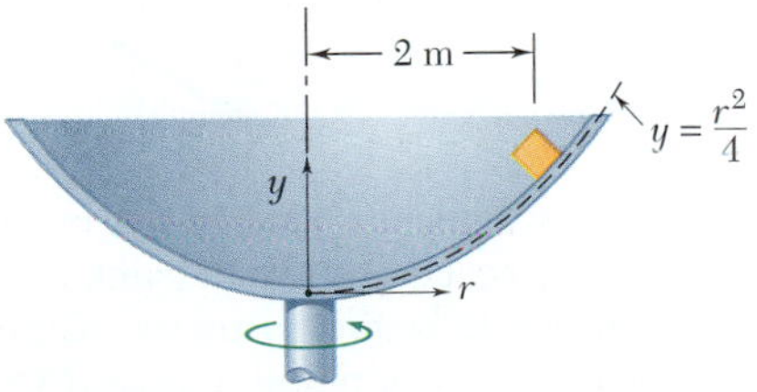

Fig. P12.55

12.56 A small 12-oz collar D can slide on portion AB of a rod which is bent as shown. Knowing that $\alpha = 35°$ and that the rod rotates about the vertical AC at a constant rate of 6 rad/s, determine the value of r for which the collar will not slide on the rod if the effect of friction between the rod and the collar is neglected.

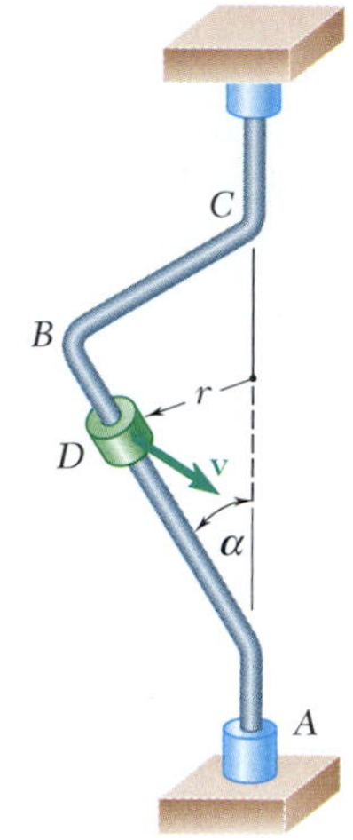

Fig. *P12.56* and *P12.57*

12.57 A small 8-oz collar D can slide on portion AB of a rod which is bent as shown. Knowing that the rod rotates about the vertical AC at a constant rate and that $\alpha = 40°$ and $r = 24$ in., determine the range of values of the speed v for which the collar will not slide on the rod if the coefficient of static friction between the rod and the collar is 0.35.

12.58 Four seconds after a polisher is started from rest, small tufts of fleece from along the circumference of the 10-in.-diameter polishing pad are observed to fly free of the pad. If the polisher is started so that the fleece along the circumference undergoes a constant tangential acceleration of 12 ft/s^2, determine (*a*) the speed v of a tuft as it leaves the pad, (*b*) the magnitude of the force required to free the tuft if the average weight of a tuft is 60×10^{-6} oz.

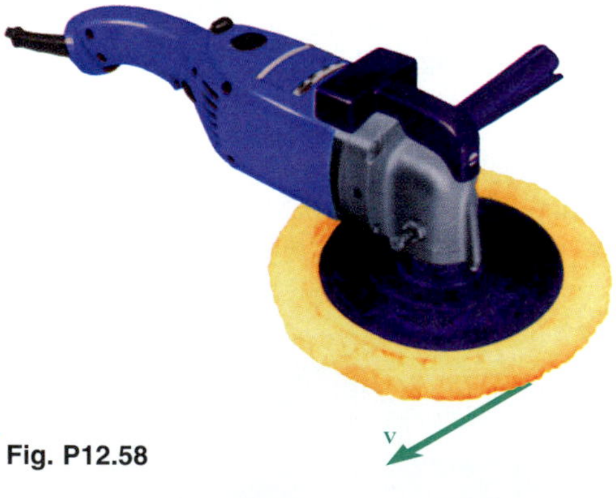

Fig. P12.58

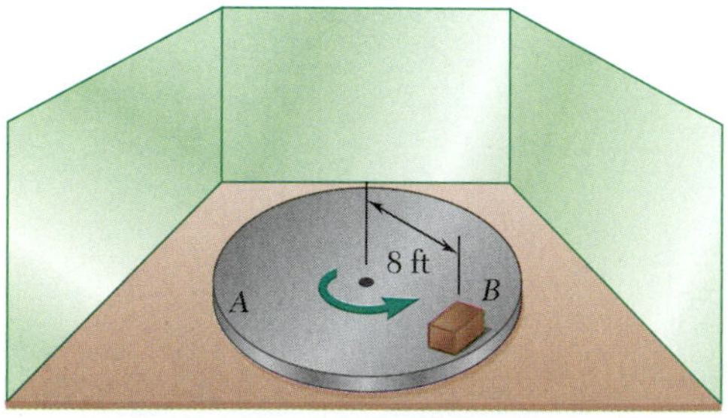

Fig. P12.59

12.59 A turntable A is built into a stage for use in a theatrical production. It is observed during a rehearsal that a trunk B starts to slide on the turntable 12 s after the turntable begins to rotate. Knowing that the trunk undergoes a constant tangential acceleration of 0.75 ft/s^2, determine the coefficient of static friction between the trunk and the turntable.

12.60 The parallel-link mechanism $ABCD$ is used to transport a component I between manufacturing processes at stations E, F, and G by picking it up at a station when $\theta = 0$ and depositing it at the next station when $\theta = 180°$. Knowing that member BC remains horizontal throughout its motion and that links AB and CD rotate at a constant rate in a vertical plane in such a way that $v_B = 0.7$ m/s, determine (*a*) the minimum value of the coefficient of static friction between the component and BC if the component is not to slide on BC while being transferred, (*b*) the values of θ for which sliding is impending.

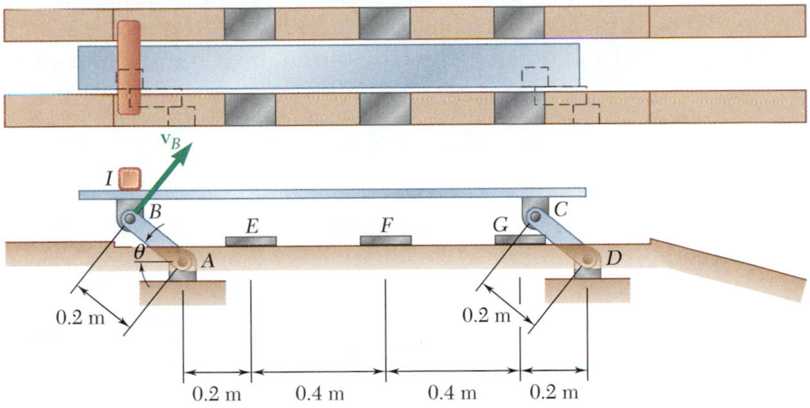

Fig. *P12.60*

12.61 Knowing that the coefficients of friction between the component I and member BC of the mechanism of Prob. 12.60 are $\mu_s = 0.35$ and $\mu_k = 0.25$, determine (*a*) the maximum allowable speed v_B if the component is not to slide on BC while being transferred, (*b*) the values of θ for which sliding is impending.

12.62 In the cathode-ray tube shown, electrons emitted by the cathode and attracted by the anode pass through a small hole in the anode and then travel in a straight line with a speed v_0 until they strike the screen at A. However, if a difference of potential V is established between the two parallel plates, the electrons will be subjected to a force $\mathbf{F}$ perpendicular to the plates while they travel between the plates and will strike the screen at point B, which is at a distance δ from A. The magnitude of the force $\mathbf{F}$ is $F = eV/d$, where $-e$ is the charge of an electron and d is the distance between the plates. Neglecting the effects of gravity, derive an expression for the deflection δ in terms of V, v_0, the charge $-e$ and the mass m of an electron, and the dimensions d, l, and L.

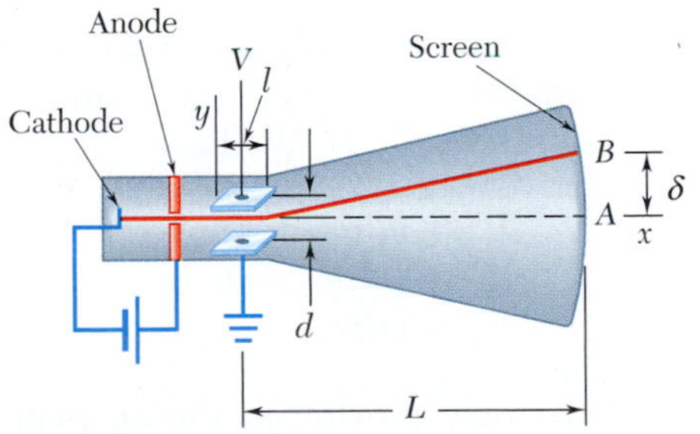

Fig. P12.62 and P12.63

12.63 In Prob. 12.62, determine the smallest allowable value of the ratio d/l in terms of e, m, v_0, and V if at $x = l$ the minimum permissible distance between the path of the electrons and the positive plate is $0.075d$.

12.7. ANGULAR MOMENTUM OF A PARTICLE. RATE OF CHANGE OF ANGULAR MOMENTUM

Consider a particle P of mass m moving with respect to a newtonian frame of reference $Oxyz$. As we saw in Sec. 12.3, the linear momentum of the particle at a given instant is defined as the vector $m\mathbf{v}$ obtained by multiplying the velocity $\mathbf{v}$ of the particle by its mass m. The moment about O of the vector $m\mathbf{v}$ is called the *moment of momentum*, or the *angular momentum*, of the particle about O at that instant and is denoted by $\mathbf{H}_O$. Recalling the definition of the moment of a vector (Sec. 3.6) and denoting by $\mathbf{r}$ the position vector of P, we write

$$\mathbf{H}_O = \mathbf{r} \times m\mathbf{v} \tag{12.12}$$

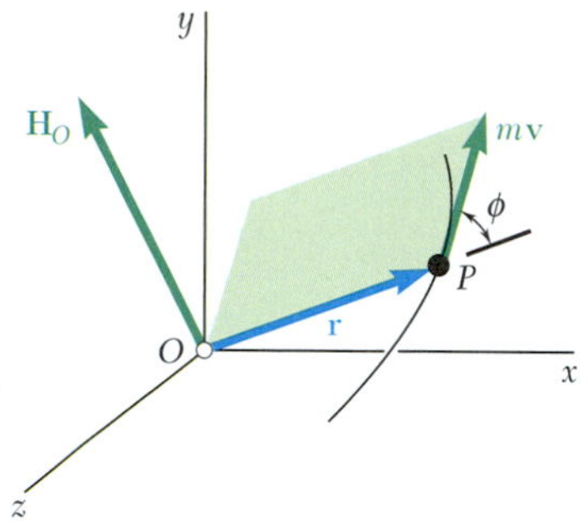

Fig. 12.12

and note that $\mathbf{H}_O$ is a vector perpendicular to the plane containing $\mathbf{r}$ and $m\mathbf{v}$ and of magnitude

$$H_O = rmv \sin \phi \tag{12.13}$$

where ϕ is the angle between $\mathbf{r}$ and $m\mathbf{v}$ (Fig. 12.12). The sense of $\mathbf{H}_O$ can be determined from the sense of $m\mathbf{v}$ by applying the right-hand rule. The unit of angular momentum is obtained by multiplying the units of length and of linear momentum (Sec. 12.4). With SI units, we have

$$(\text{m})(\text{kg} \cdot \text{m/s}) = \text{kg} \cdot \text{m}^2/\text{s}$$

With U.S. customary units, we write

$$(\text{ft})(\text{lb} \cdot \text{s}) = \text{ft} \cdot \text{lb} \cdot \text{s}$$

Resolving the vectors $\mathbf{r}$ and $m\mathbf{v}$ into components and applying formula (3.10), we write

$$\mathbf{H}_O = \begin{vmatrix} \mathbf{i} & \mathbf{j} & \mathbf{k} \\ x & y & z \\ mv_x & mv_y & mv_z \end{vmatrix} \tag{12.14}$$

The components of $\mathbf{H}_O$, which also represent the moments of the linear momentum $m\mathbf{v}$ about the coordinate axes, can be obtained by expanding the determinant in (12.14). We have

$$\begin{aligned} H_x &= m(yv_z - zv_y) \\ H_y &= m(zv_x - xv_z) \\ H_z &= m(xv_y - yv_x) \end{aligned} \tag{12.15}$$

Photo 12.2 During a hammer thrower's practice swings, the hammer acquires angular momentum about a vertical axis at the center of its circular path.

In the case of a particle moving in the xy plane, we have $z = v_z = 0$ and the components H_x and H_y reduce to zero. The angular momentum is thus perpendicular to the xy plane; it is then completely defined by the scalar

$$H_O = H_z = m(xv_y - yv_x) \tag{12.16}$$

which will be positive or negative according to the sense in which the particle is observed to move from O. If polar coordinates are used, we resolve the linear momentum of the particle into radial and transverse components (Fig. 12.13) and write

$$H_O = rmv \sin \phi = rmv_\theta \tag{12.17}$$

or, recalling from (11.45) that $v_\theta = r\dot{\theta}$,

$$H_O = mr^2\dot{\theta} \tag{12.18}$$

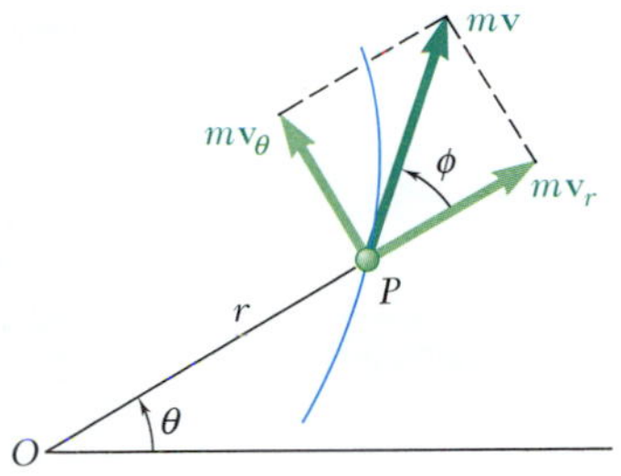

Fig. 12.13

Let us now compute the derivative with respect to t of the angular momentum $\mathbf{H}_O$ of a particle P moving in space. Differentiating both members of Eq. (12.12), and recalling the rule for the differentiation of a vector product (Sec. 11.10), we write

$$\dot{\mathbf{H}}_O = \dot{\mathbf{r}} \times m\mathbf{v} + \mathbf{r} \times m\dot{\mathbf{v}} = \mathbf{v} \times m\mathbf{v} + \mathbf{r} \times m\mathbf{a}$$

Since the vectors $\mathbf{v}$ and $m\mathbf{v}$ are collinear, the first term of the expression obtained is zero; and, by Newton's second law, $m\mathbf{a}$ is equal to the sum $\Sigma\mathbf{F}$ of the forces acting on P. Noting that $\mathbf{r} \times \Sigma\mathbf{F}$ represents the sum $\Sigma\mathbf{M}_O$ of the moments about O of these forces, we write

$$\Sigma\mathbf{M}_O = \dot{\mathbf{H}}_O \tag{12.19}$$

Equation (12.19), which results directly from Newton's second law, states that *the sum of the moments about O of the forces acting on the particle is equal to the rate of change of the moment of momentum, or angular momentum, of the particle about O.*

12.8. EQUATIONS OF MOTION IN TERMS OF RADIAL AND TRANSVERSE COMPONENTS

Consider a particle P, of polar coordinates r and θ, which moves in a plane under the action of several forces. Resolving the forces and the acceleration of the particle into radial and transverse components (Fig. 12.14) and substituting into Eq. (12.2), we obtain the two scalar equations

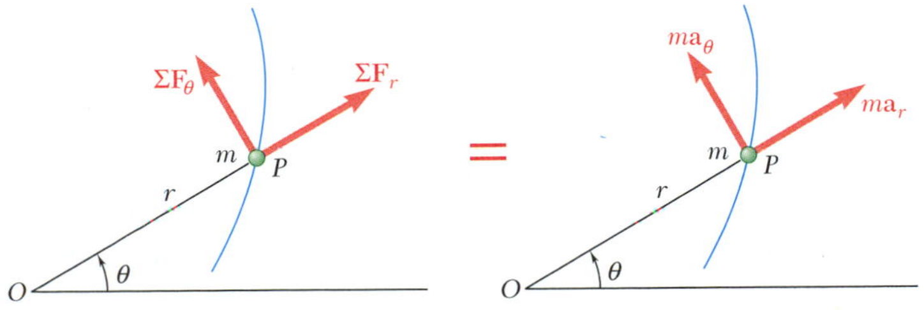

Fig 12.14

$$\Sigma F_r = ma_r \qquad \Sigma F_\theta = ma_\theta \tag{12.20}$$

Substituting for a_r and a_θ from Eqs. (11.46), we have

$$\Sigma F_r = m(\ddot{r} - r\dot{\theta}^2) \tag{12.21}$$

$$\Sigma F_\theta = m(r\ddot{\theta} + 2\dot{r}\dot{\theta}) \tag{12.22}$$

Photo 12.3 The path of specimen being tested in a centrifuge is a horizontal circle. The forces acting on the specimen and its acceleration can be resolved into radial and transverse components with r = constant.

The equations obtained can be solved for two unknowns.

Equation (12.22) could have been derived from Eq. (12.19). Recalling (12.18) and noting that $\Sigma M_O = r\Sigma F_\theta$, Eq. (12.19) yields

$$r\Sigma F_\theta = \frac{d}{dt}(mr^2\dot{\theta}) = m(r^2\ddot{\theta} + 2r\dot{r}\dot{\theta})$$

and, after dividing both members by r,

$$\Sigma F_\theta = m(r\ddot{\theta} + 2\dot{r}\dot{\theta}) \tag{12.22}$$

12.9. MOTION UNDER A CENTRAL FORCE. CONSERVATION OF ANGULAR MOMENTUM

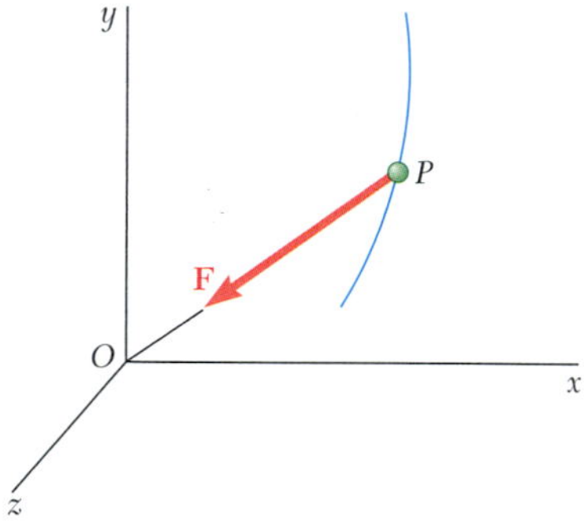

Fig. 12.15

When the only force acting on a particle P is a force $\mathbf{F}$ directed toward or away from a fixed point O, the particle is said to be moving *under a central force,* and the point O is referred to as the *center of force* (Fig. 12.15). Since the line of action of $\mathbf{F}$ passes through O, we must have $\Sigma \mathbf{M}_O = 0$ at any given instant. Substituting into Eq. (12.19), we therefore obtain

$$\dot{\mathbf{H}}_O = 0$$

for all values of t and, integrating in t,

$$\mathbf{H}_O = \text{constant} \tag{12.23}$$

We thus conclude that *the angular momentum of a particle moving under a central force is constant, in both magnitude and direction.*

Recalling the definition of the angular momentum of a particle (Sec. 12.7), we write

$$\mathbf{r} \times m\mathbf{v} = \mathbf{H}_O = \text{constant} \tag{12.24}$$

from which it follows that the position vector $\mathbf{r}$ of the particle P must be perpendicular to the constant vector $\mathbf{H}_O$. Thus, a particle under a central force moves in a fixed plane perpendicular to $\mathbf{H}_O$. The vector $\mathbf{H}_O$ and the fixed plane are defined by the initial position vector $\mathbf{r}_0$

and the initial velocity $\mathbf{v}_0$ of the particle. For convenience, let us assume that the plane of the figure coincides with the fixed plane of motion (Fig. 12.16).

Since the magnitude H_O of the angular momentum of the particle P is constant, the right-hand member in Eq. (12.13) must be constant. We therefore write

$$rmv \sin \phi = r_0 m v_0 \sin \phi_0 \tag{12.25}$$

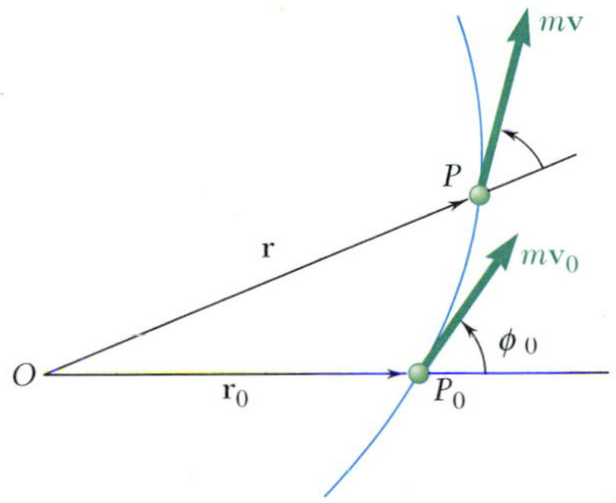

Fig. 12.16

This relation applies to the motion of any particle under a central force. Since the gravitational force exerted by the sun on a planet is a central force directed toward the center of the sun, Eq. (12.25) is fundamental to the study of planetary motion. For a similar reason, it is also fundamental to the study of the motion of space vehicles in orbit about the earth.

Alternatively, recalling Eq. (12.18), we can express the fact that the magnitude H_O of the angular momentum of the particle P is constant by writing

$$mr^2\dot{\theta} = H_O = \text{constant} \tag{12.26}$$

or, dividing by m and denoting by h the angular momentum per unit mass H_O/m,

$$r^2\dot{\theta} = h \tag{12.27}$$

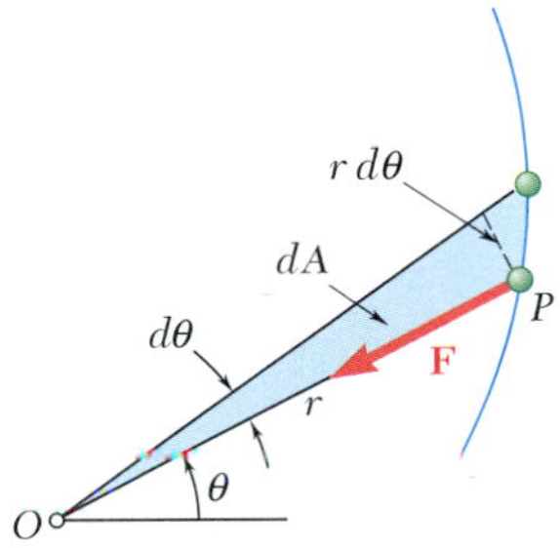

Fig. 12.17

Equation (12.27) can be given an interesting geometric interpretation. Observing from Fig. 12.17 that the radius vector OP sweeps an infinitesimal area $dA = \frac{1}{2}r^2\, d\theta$ as it rotates through an angle $d\theta$, and defining the *areal velocity* of the particle as the quotient dA/dt, we note that the left-hand member of Eq. (12.27) represents twice the areal velocity of the particle. We thus conclude that *when a particle moves under a central force, its areal velocity is constant.*

12.10. NEWTON'S LAW OF GRAVITATION

As you saw in the preceding section, the gravitational force exerted by the sun on a planet or by the earth on an orbiting satellite is an important example of a central force. In this section, you will learn how to determine the magnitude of a gravitational force.

In his *law of universal gravitation,* Newton states that two particles of masses M and m at a distance r from each other attract each other with equal and opposite forces $\mathbf{F}$ and $-\mathbf{F}$ directed along the line joining the particles (Fig. 12.18). The common magnitude F of the two forces is

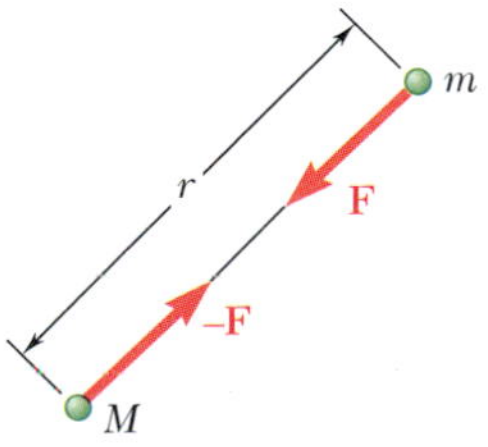

Fig. 12.18

$$F = G\frac{Mm}{r^2} \tag{12.28}$$

where G is a universal constant, called the *constant of gravitation.* Experiments show that the value of G is $(66.73 \pm 0.03) \times 10^{-12}$ m^3/kg $\cdot$ s^2 in SI units or approximately 34.4×10^{-9} ft^4/lb $\cdot$ s^4 in U.S. customary units. Gravitational forces exist between any pair of bodies, but their effect is appreciable only when one of the bodies has a very large mass. The effect of gravitational forces is apparent in the cases of the motion of a planet about the sun, of satellites orbiting about the earth, or of bodies falling on the surface of the earth.

Since the force exerted by the earth on a body of mass m located on or near its surface is defined as the weight $\mathbf{W}$ of the body, we can substitute the magnitude $W = mg$ of the weight for F, and the radius R of the earth for r, in Eq. (12.28). We obtain

$$W = mg = \frac{GM}{R^2}m \qquad \text{or} \qquad g = \frac{GM}{R^2} \tag{12.29}$$

where M is the mass of the earth. Since the earth is not truly spherical, the distance R from the center of the earth depends upon the point selected on its surface, and the values of W and g will thus vary with the altitude and latitude of the point considered. Another reason for the variation of W and g with latitude is that a system of axes attached to the earth does not constitute a newtonian frame of reference (see Sec. 12.2). A more accurate definition of the weight of a body should therefore include a component representing the centrifugal force due to the rotation of the earth. Values of g at sea level vary from 9.781 m/s^2, or 32.09 ft/s^2, at the equator to 9.833 m/s^2, or 32.26 ft/s^2, at the poles.†

The force exerted by the earth on a body of mass m located in space at a distance r from its center can be found from Eq. (12.28). The computations will be somewhat simplified if we note that according to Eq. (12.29), the product of the constant of gravitation G and the mass M of the earth can be expressed as

$$GM = gR^2 \tag{12.30}$$

where g and the radius R of the earth will be given their average values $g = 9.81$ m/s^2 and $R = 6.37 \times 10^6$ m in SI units‡ and $g = 32.2$ ft/s^2 and $R = (3960$ mi$)(5280$ ft/mi$)$ in U.S. customary units.

The discovery of the law of universal gravitation has often been attributed to the belief that, after observing an apple falling from a tree, Newton had reflected that the earth must attract an apple and the moon in much the same way. While it is doubtful that this incident actually took place, it may be said that Newton would not have formulated his law if he had not first perceived that the acceleration of a falling body must have the same cause as the acceleration which keeps the moon in its orbit. This basic concept of the continuity of gravitational attraction is more easily understood today, when the gap between the apple and the moon is being filled with artificial earth satellites.

†A formula expressing g in terms of the latitude ϕ was given in Prob. 12.1.

‡The value of R is easily found if one recalls that the circumference of the earth is $2\pi R = 40 \times 10^6$ m.

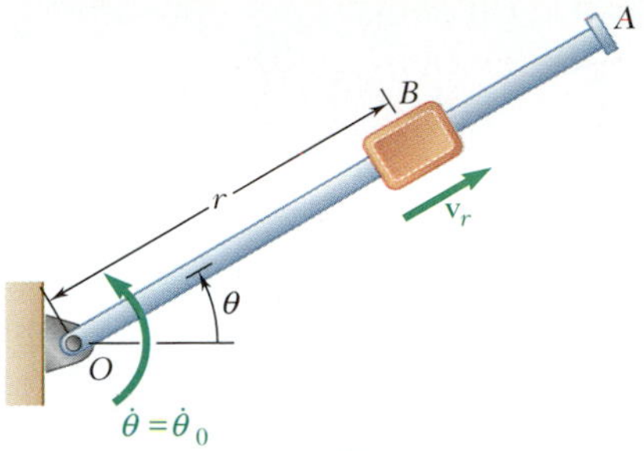

SAMPLE PROBLEM 12.7

A block B of mass m can slide freely on a frictionless arm OA which rotates in a horizontal plane at a constant rate $\dot{\theta}_0$. Knowing that B is released at a distance r_0 from O, express as a function of r, (*a*) the component v_r of the velocity of B along OA, (*b*) the magnitude of the horizontal force $\mathbf{F}$ exerted on B by the arm OA.

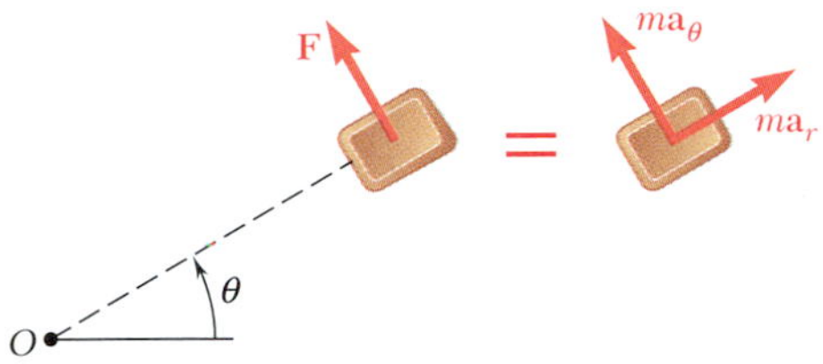

SOLUTION

Since all other forces are perpendicular to the plane of the figure, the only force shown acting on B is the force $\mathbf{F}$ perpendicular to OA.

Equations of Motion. Using radial and transverse components.

$$+\nearrow \Sigma F_r = ma_r: \qquad 0 = m(\ddot{r} - r\dot{\theta}^2) \qquad (1)$$

$$+\nwarrow \Sigma F_\theta = ma_\theta: \qquad F = m(r\ddot{\theta} + 2\dot{r}\dot{\theta}) \qquad (2)$$

***a.* Component v_r of Velocity.** Since $v_r = \dot{r}$, we have

$$\ddot{r} = \dot{v}_r = \frac{dv_r}{dt} = \frac{dv_r}{dr}\frac{dr}{dt} = v_r\frac{dv_r}{dr}$$

Substituting for $\ddot{r}$ in (1), recalling that $\dot{\theta} = \dot{\theta}_0$, and separating the variables,

$$v_r\,dv_r = \dot{\theta}_0^2 r\,dr$$

Multiplying by 2, and integrating from 0 to v_r and from r_0 to r,

$$v_r^2 = \dot{\theta}_0^2(r^2 - r_0^2) \qquad v_r = \dot{\theta}_0(r^2 - r_0^2)^{1/2} \blacktriangleleft$$

***b.* Horizontal Force F.** Setting $\dot{\theta} = \dot{\theta}_0$, $\ddot{\theta} = 0$, $\dot{r} = v_r$ in Eq. (2), and substituting for v_r the expression obtained in part *a*,

$$F = 2m\dot{\theta}_0(r^2 - r_0^2)^{1/2}\dot{\theta}_0 \qquad F = 2m\dot{\theta}_0^2(r^2 - r_0^2)^{1/2} \blacktriangleleft$$

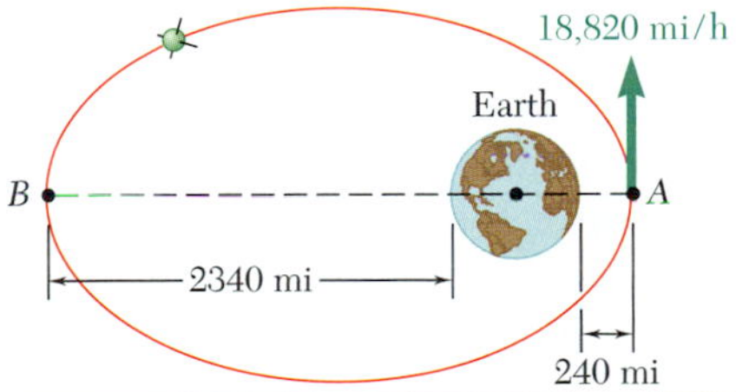

SAMPLE PROBLEM 12.8

A satellite is launched in a direction parallel to the surface of the earth with a velocity of 18,820 mi/h from an altitude of 240 mi. Determine the velocity of the satellite as it reaches its maximum altitude of 2340 mi. It is recalled that the radius of the earth is 3960 mi.

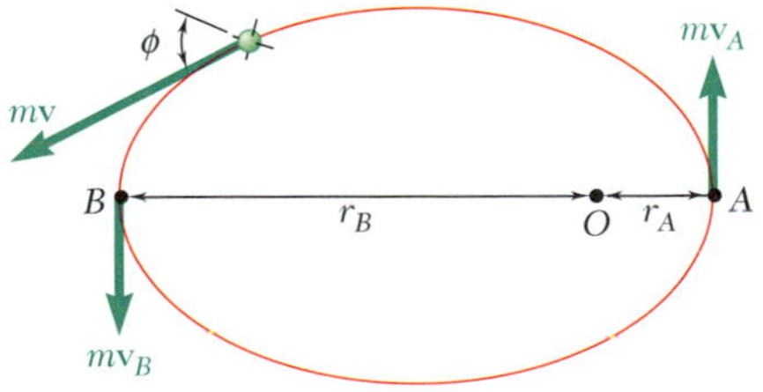

SOLUTION

Since the satellite is moving under a central force directed toward the center O of the earth, its angular momentum $\mathbf{H}_O$ is constant. From Eq. (12.13) we have

$$rmv \sin \phi = H_O = \text{constant}$$

which shows that v is minimum at B, where both r and $\sin \phi$ are maximum. Expressing conservation of angular momentum between A and B.

$$r_A m v_A = r_B m v_B$$

$$v_B = v_A \frac{r_A}{r_B} = (18{,}820 \text{ mi/h})\frac{3960 \text{ mi} + 240 \text{ mi}}{3960 \text{ mi} + 2340 \text{ mi}}$$

$$v_B = 12{,}550 \text{ mi/h} \blacktriangleleft$$

SOLVING PROBLEMS ON YOUR OWN

In this lesson we continued our study of Newton's second law by expressing the force and the acceleration in terms of their *radial and transverse components*, where the corresponding equations of motion are

$$\Sigma F_r = ma_r: \qquad \Sigma F_r = m(\ddot{r} - r\dot{\theta}^2)$$
$$\Sigma F_\theta = ma_\theta: \qquad \Sigma F_\theta = m(r\ddot{\theta} + 2\dot{r}\dot{\theta})$$

We introduced the *moment of the momentum*, or the *angular momentum*, $\mathbf{H}_O$ of a particle about O:

$$\mathbf{H}_O = \mathbf{r} \times m\mathbf{v} \tag{12.12}$$

and found that $\mathbf{H}_O$ is constant when the particle moves under a *central force* with its center located at O.

1. *Using radial and transverse components.* Radial and transverse components were introduced in the last lesson of Chap. 11 [Sec. 11.14]; you should review that material before attempting to solve the following problems. Also, our comments in the preceding lesson regarding the application of Newton's second law (drawing a free-body diagram and a $m\mathbf{a}$ diagram, etc.) still apply [Sample Prob. 12.7]. Finally, note that the solution of that sample problem depends on the application of techniques developed in Chap. 11—you will need to use similar techniques to solve some of the problems of this lesson.

2. *Solving problems involving the motion of a particle under a central force.* In problems of this type, the angular momentum $\mathbf{H}_O$ of the particle about the center of force O is conserved. You will find it convenient to introduce the constant $h = H_O/m$ representing the angular momentum per unit mass. Conservation of the angular momentum of the particle P about O can then be expressed by either of the following equations

$$rv \sin \phi = h \qquad \text{or} \qquad r^2\dot{\theta} = h$$

where r and θ are the polar coordinates of P, and ϕ is the angle that the velocity $\mathbf{v}$ of the particle forms with the line OP (Fig. 12.16). The constant h can be determined from the initial conditions and either of the above equations can be solved for one unknown.

3. *In space-mechanics problems* involving the orbital motion of a planet about the sun, or a satellite about the earth, the moon, or some other planet, the central force **F** is the force of gravitational attraction; it is directed *toward* the center of force O and has the magnitude

$$F = G\frac{Mm}{r^2} \tag{12.28}$$

Note that in the particular case of the gravitational force exerted by the earth, the product GM can be replaced by gR^2, where R is the radius of the earth [Eq. 12.30].

The following two cases of orbital motion are frequently encountered:

a. For a satellite in a circular orbit, the force **F** is normal to the orbit and you can write $F = ma_n$; substituting for F from Eq. (12.28) and observing that $a_n = v^2/\rho = v^2/r$, you will obtain

$$G\frac{Mm}{r^2} = m\frac{v^2}{r} \qquad \text{or} \qquad v^2 = \frac{GM}{r}$$

b. For a satellite in an elliptic orbit, the radius vector **r** and the velocity **v** of the satellite are perpendicular to each other at the points A and B which are, respectively, farthest and closest to the center of force O [Sample Prob. 12.8]. Thus, conservation of angular momentum of the satellite between these two points can be expressed as

$$r_A m v_A = r_B m v_B$$

Problems

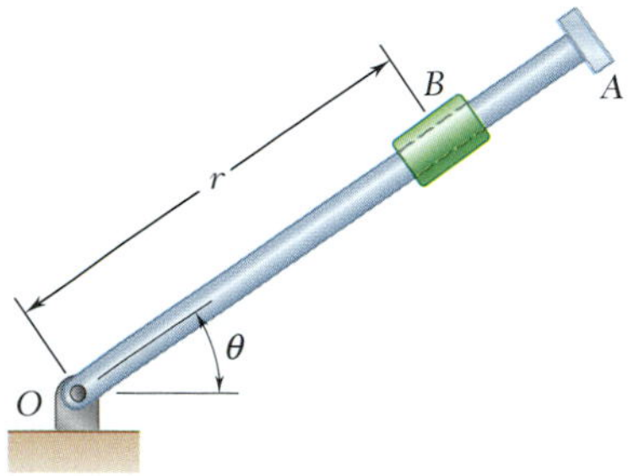

Fig. P12.64 and P12.65

12.64 Rod OA rotates about O in a horizontal plane. The motion of the 400-g collar B is defined by the relations $r = 500 + 300 \sin \pi t$ and $\theta = 2\pi(t^2 - 2t)$, where r is expressed in millimeters, t in seconds, and θ in radians. Determine the radial and transverse components of the force exerted on the collar when (*a*) $t = 0$, (*b*) $t = 0.8$ s.

12.65 Rod OA oscillates about O in a horizontal plane. The motion of the 200-g collar B is defined by the relations $r = 5/(t + 2)$ and $\theta = (4/\pi) \sin \pi t$, where r is expressed in meters, t in seconds, and θ in radians. Determine the radial and transverse components of the force exerted on the collar when (*a*) $t = 2$ s, (*b*) $t = 7$ s.

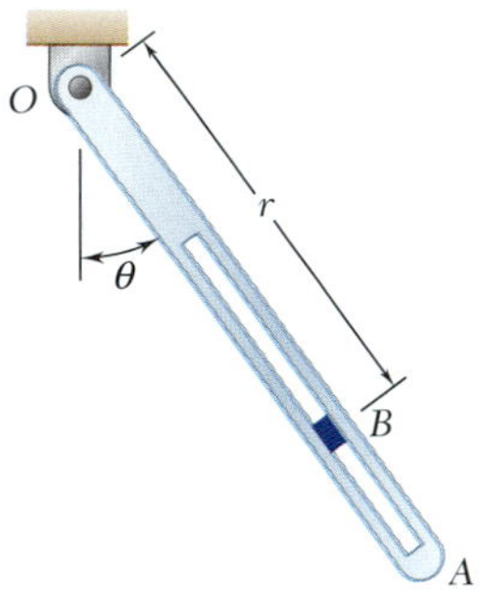

Fig. P12.66 and P12.67

12.66 A 1.2-lb block B slides without friction inside a slot cut in arm OA which rotates in a vertical plane at a constant rate, $\dot{\theta} = 2$ rad/s. At the instant when $\theta = 30°$, $r = 2$ ft and the force exerted on the block by the arm is zero. Determine, at this instant, (*a*) the relative velocity of the block with respect to the arm, (*b*) the relative acceleration of the block with respect to the arm.

12.67 A 1.2-lb block B slides without friction inside a slot cut in arm OA which rotates in a vertical plane. The motion of the rod is defined by the relation $\ddot{\theta} = 10$ rad/s^2, constant. At the instant when $\theta = 45°$, $r = 2.4$ ft and the velocity of the block is zero. Determine, at this instant, (*a*) the force exerted on the block by the arm, (*b*) the relative acceleration of the block with respect to the arm.

12.68 The 6-lb collar B slides on the frictionless arm AA'. The arm is attached to drum D and rotates about O in a horizontal plane at the rate $\dot{\theta} = 0.8t$, where $\dot{\theta}$ and t are expressed in rad/s and seconds, respectively. As the arm-drum assembly rotates, a mechanism within the drum releases cord so that the collar moves outward from O with a constant speed of 1.5 ft/s. Knowing that at $t = 0$, $r = 0$, determine the time at which the tension in the cord is equal to the magnitude of the horizontal force exerted on B by arm AA'.

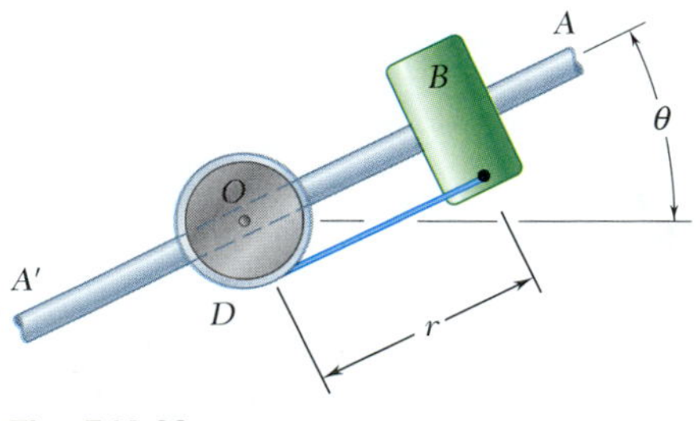

Fig. *P12.68*

12.69 The 4-oz pin B slides along the slot in the rotating arm OC and along the slot DE which is cut in a fixed horizontal plate. Neglecting friction and knowing that arm OC rotates at a constant rate $\dot{\theta}_0 = 10$ rad/s, determine for any given value of θ (*a*) the radial and transverse components of the resultant force **F** exerted on pin B, (*b*) the forces **P** and **Q** exerted on pin B by arm OC and the wall of the slot DE, respectively.

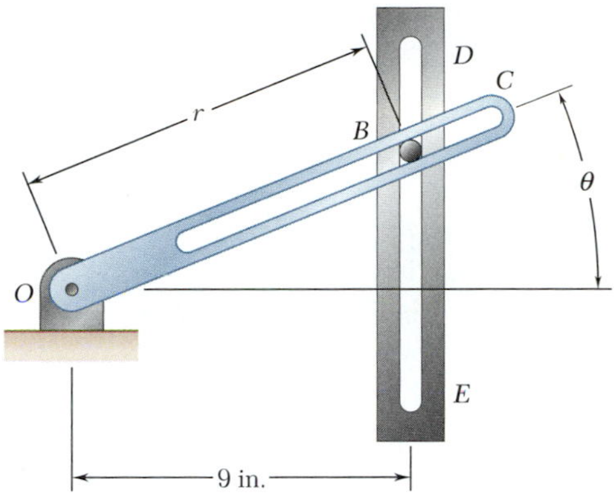

Fig. *P12.69*

12.70 Disk A rotates in a horizontal plane about a vertical axis at the constant rate of $\dot{\theta}_0 = 15$ rad/s. Slider B has a mass of 230 g and moves in a frictionless slot cut in the disk. The slider is attached to a spring of constant $k = 60$ N/m, which is undeformed when $r = 0$. Knowing that at a given instant the acceleration of the slider relative to the disk is $\ddot{r} = -12$ m/s^2 and that the horizontal force exerted on the slider by the disk is 9 N, determine at that instant (*a*) the distance r, (*b*) the radial component of the velocity of the slider.

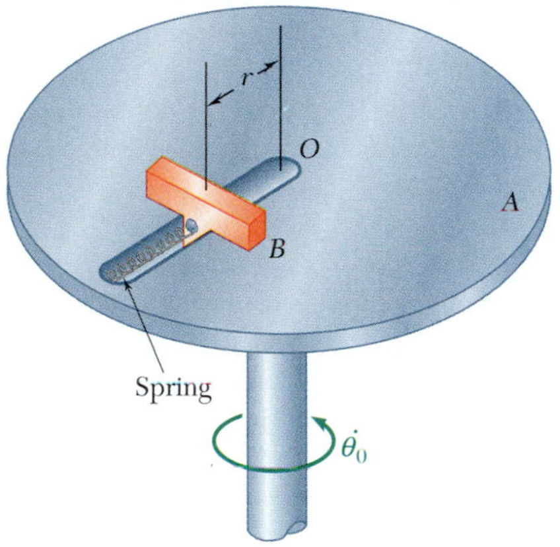

Fig. P12.70

12.71 The horizontal rod OA rotates about a vertical shaft according to the relation $\dot{\theta} = 10t$, where $\dot{\theta}$ and t are expressed in rad/s and seconds, respectively. A 250-g collar B is held by a cord with a breaking strength of 18 N. Neglecting friction, determine, immediately after the cord breaks, (*a*) the relative acceleration of the collar with respect to the rod, (*b*) the magnitude of the horizontal force exerted on the collar by the rod.

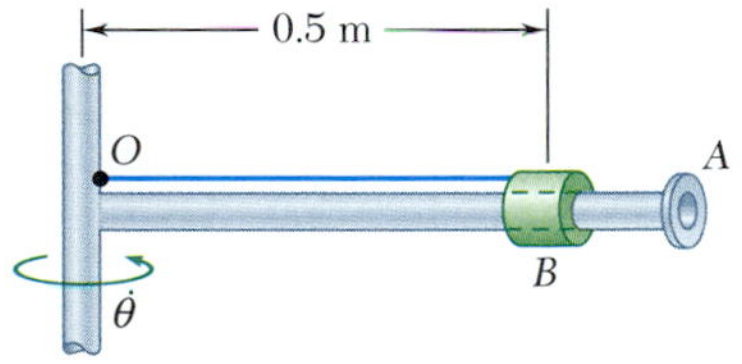

Fig. P12.71

12.72 The two blocks are released from rest when $r = 0.8$ m and $\theta = 30°$. Neglecting the mass of the pulley and the effect of friction in the pulley and between block A and the horizontal surface, determine (*a*) the initial tension in the cable, (*b*) the initial acceleration of block A, (*c*) the initial acceleration of block B.

12.73 The velocity of block A is 2 m/s to the right at the instant when $r = 0.8$ m and $\theta = 30°$. Neglecting the mass of the pulley and the effect of friction in the pulley and between block A and the horizontal surface, determine, at this instant, (*a*) the tension in the cable, (*b*) the acceleration of block A, (*c*) the acceleration of block B.

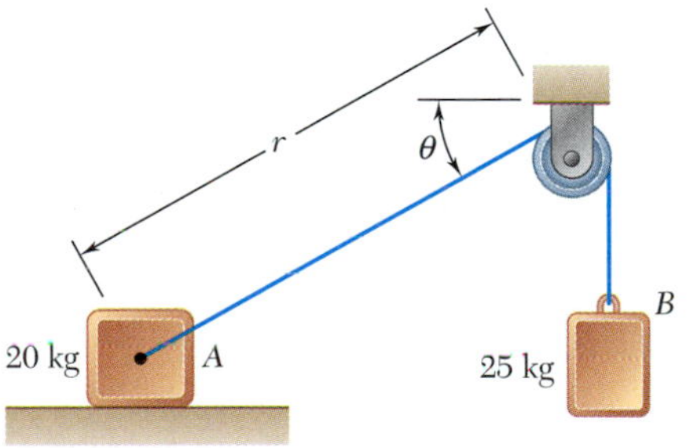

Fig. P12.72 and P12.73

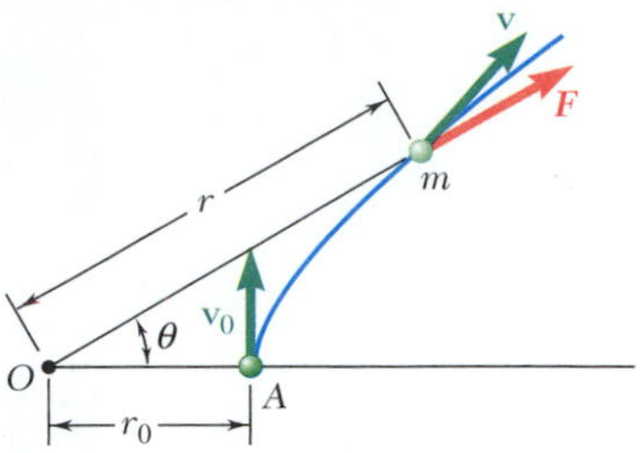

Fig. P12.74 and P12.75

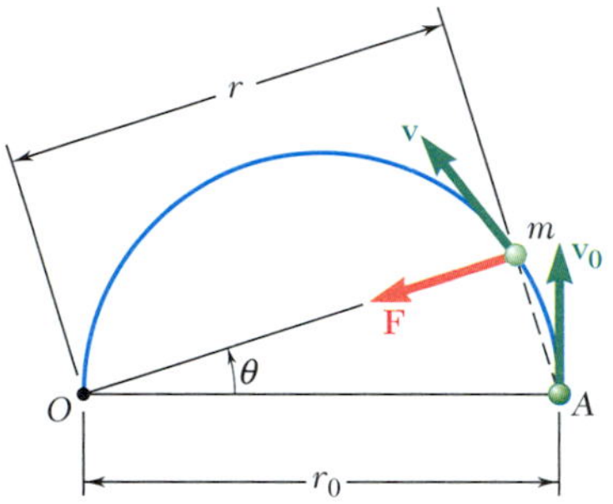

Fig. P12.76 and *P12.77*

12.74 A particle of mass m is projected from point A with an initial velocity $\mathbf{v}_0$ perpendicular to line OA and moves under a central force $\mathbf{F}$ directed away from the center of force O. Knowing that the particle follows a path defined by the equation $r = r_0/\sqrt{\cos 2\theta}$ and using Eq. (12.27), express the radial and transverse components of the velocity $\mathbf{v}$ of the particle as functions of θ.

12.75 For the particle of Prob. 12.74, show (*a*) that the velocity of the particle and the central force $\mathbf{F}$ are proportional to the distance r from the particle to the center of force O, (*b*) that the radius of curvature of the path is proportional to r^3.

12.76 A particle of mass m is projected from point A with an initial velocity $\mathbf{v}_0$ perpendicular to line OA and moves under a central force $\mathbf{F}$ along a semicircular path of diameter OA. Observing that $r = r_0 \cos\theta$ and using Eq. (12.27), show that the speed of the particle is $v = v_0/\cos^2\theta$.

12.77 For the particle of Prob. 12.76, determine the tangential component F_t of the central force $\mathbf{F}$ along the tangent to the path of the particle for (*a*) $\theta = 0$, (*b*) $\theta = 45°$.

12.78 The radius of the orbit of a moon of a given planet is three times as large as the radius of that planet. Denoting by ρ the mean density of the planet, show that the time required by the moon to complete one full revolution about the planet is $9(\pi/G\rho)^{1/2}$, where G is the constant of gravitation.

12.79 Communication satellites are placed in a geosynchronous orbit, that is, in a circular orbit such that they complete one full revolution about the earth in one sidereal day (23.934 hr), and thus appear stationary with respect to the ground. Determine (*a*) the altitude of these satellites above the surface of the earth, (*b*) the velocity with which they describe their orbit. Give the answers in both SI and U.S. customary units.

12.80 Show that the radius r of the orbit of a moon of a given planet can be determined from the radius R of the planet, the acceleration of gravity at the surface of the planet, and the time τ required by the moon to complete one full revolution about the planet. Determine the acceleration of gravity at the surface of the planet Jupiter knowing that $R = 44{,}400$ mi, $\tau = 3.551$ days, and $r = 417{,}000$ mi for its moon Europa.

12.81 A spacecraft is placed into a polar orbit about the planet Mars at an altitude of 230 mi. Knowing that the mean density of Mars is $7.65 \text{ lb} \cdot \text{s}^2/\text{ft}^4$ and that the radius of Mars is 2111 mi, determine (*a*) the time τ required for the spacecraft to complete one full revolution about Mars, (*b*) the velocity with which the spacecraft describes its orbit.

12.82 Determine the mass of the earth knowing that the mean radius of the moon's orbit about the earth is 384.5 Mm and that the moon requires 27.32 days to complete one full revolution about the earth.

12.83 The periodic times of the planet Jupiter's satellites, Ganymede and Callisto, have been observed to be 7.15 days and 16.69 days, respectively. Knowing that the mass of Jupiter is 319 times that of the earth and that the orbits of the two satellites are circular, determine (*a*) the radius of the orbit of Ganymede, (*b*) the velocity with which Callisto describes its orbit. Give the answers in U.S. customary units. (The *periodic time* of a satellite is the time it requires to complete one full revolution about the planet.)

12.84 The periodic time (see Prob. 12.83) of an earth satellite in a circular polar orbit is 120 min. Determine (*a*) the altitude h of the satellite, (*b*) the time during which the satellite is above the horizon for an observer located at the north pole.

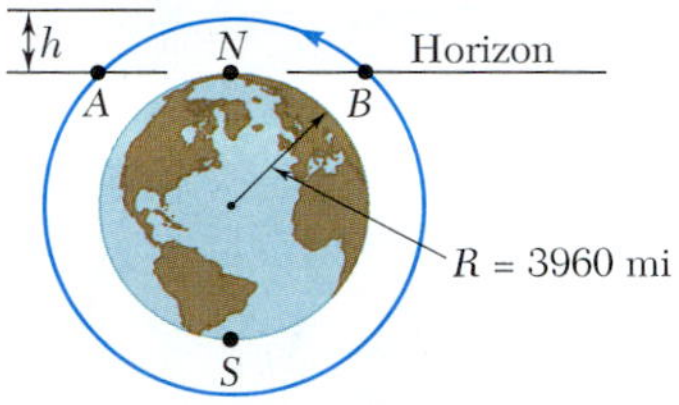

Fig. *P12.84*

12.85 A 540-kg spacecraft first is placed into a circular orbit about the earth at an altitude of 4500 km and then is transferred to a circular orbit about the moon. Knowing that the mass of the moon is 0.01230 times the mass of the earth and that the radius of the moon is 1740 km, determine (*a*) the gravitational force exerted on the spacecraft as it was orbiting the earth, (*b*) the required radius of the orbit of the spacecraft about the moon if the periodic times of the two orbits are to be equal, (*c*) the acceleration of gravity at the surface of the moon. (The *periodic time* of a satellite is the time it requires to complete one full revolution about a planet.)

12.86 As a first approximation to the analysis of a space flight from the earth to Mars, assume the orbits of the earth and Mars are circular and coplanar. The mean distances from the sun to the earth and to Mars are 149.6×10^6 km and 227.8×10^6 km, respectively. To place the spacecraft into an elliptical transfer orbit at point A, its speed is increased over a short interval of time to v_A which is 2.94 km/s faster than the earth's orbital speed. When the spacecraft reaches point B on the elliptical transfer orbit, its speed v_B is increased to the orbital speed of Mars. Knowing that the mass of the sun is 332.8×10^3 times the mass of the earth, determine the increase in speed required at B.

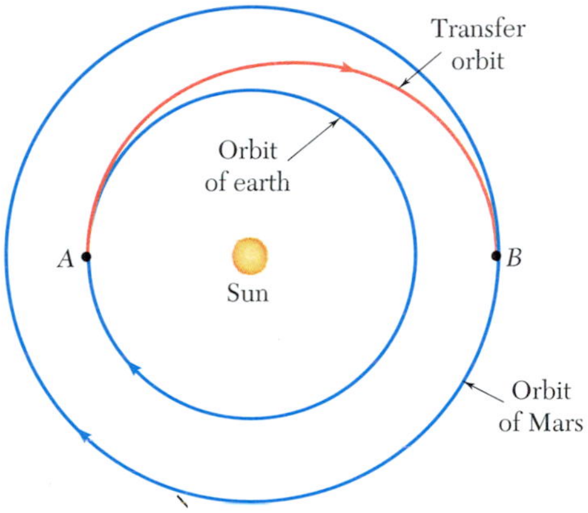

Fig. P12.86

12.87 During a flyby of the earth, the velocity of a spacecraft is 34.2×10^3 ft/s as it reaches its minimum altitude of 600 mi above the surface at point A. At point B the spacecraft is observed to have an altitude of 5160 mi. Assuming that the trajectory of the spacecraft is parabolic, determine its velocity at B.

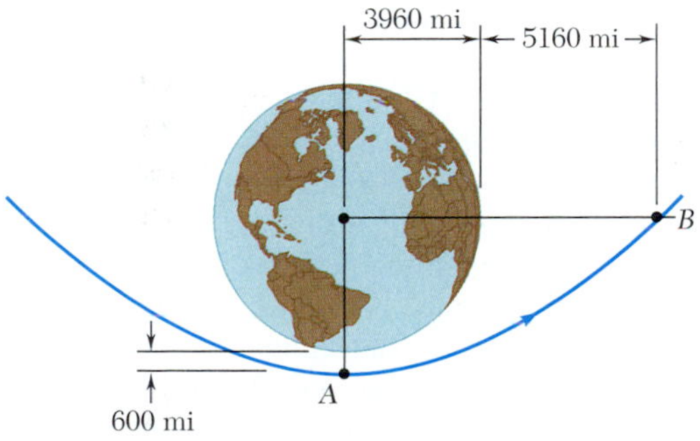

Fig. P12.87

12.88 A space vehicle is in a circular orbit of 1400-mi radius around the moon. To transfer to a smaller orbit of 1300-mi radius, the vehicle is first placed in an elliptic path AB by reducing its speed by 86 ft/s as it passes through A. Knowing that the mass of the moon is 5.03×10^{21} lb $\cdot$ s^2/ft, determine (*a*) the speed of the vehicle as it approaches B on the elliptic path, (*b*) the amount by which its speed should be reduced as it approaches B to insert it into the smaller circular orbit.

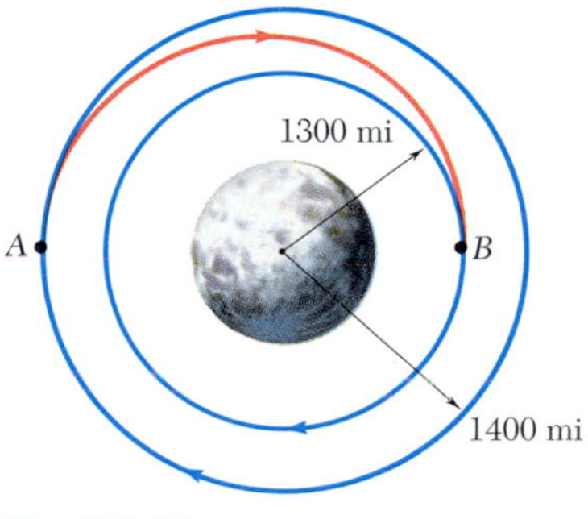

Fig. *P12.88*

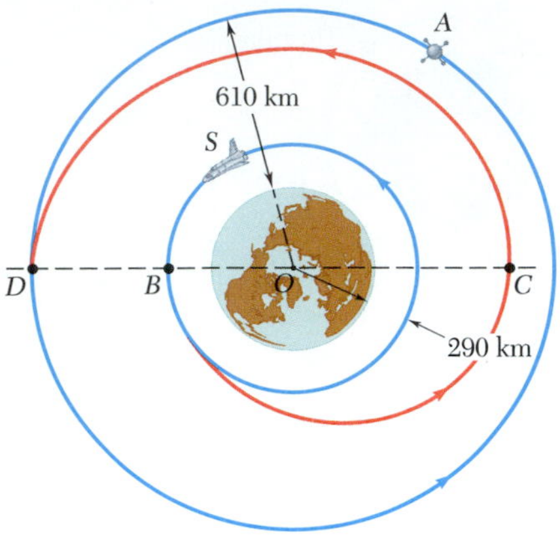

Fig. P12.89

12.89 A space shuttle S and a satellite A are in the circular orbits shown. In order for the shuttle to recover the satellite, the shuttle is first placed in an elliptic path BC by increasing its speed by $\Delta v_B = 85$ m/s as it passes through B. As the shuttle approaches C, its speed is increased by $\Delta v_C = 79$ m/s to insert it into a second elliptic transfer orbit CD. Knowing that the distance from O to C is 6900 km, determine the amount by which the speed of the shuttle should be increased as it approaches D to insert it into the circular orbit of the satellite.

12.90 A 1-lb ball A and a 2-lb ball B are mounted on a horizontal rod which rotates freely about a vertical shaft. The balls are held in the positions shown by pins. The pin holding B is suddenly removed and the ball moves to position C as the rod rotates. Neglecting friction and the mass of the rod and knowing that the initial speed of A is $v_A = 8$ ft/s, determine (*a*) the radial and transverse components of the acceleration of ball B immediately after the pin is removed, (*b*) the acceleration of ball B relative to the rod at that instant, (*c*) the speed of ball A after ball B has reached the stop at C.

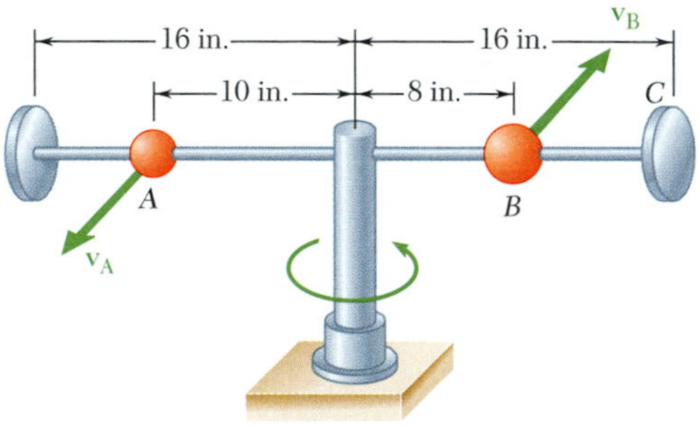

Fig. P12.90

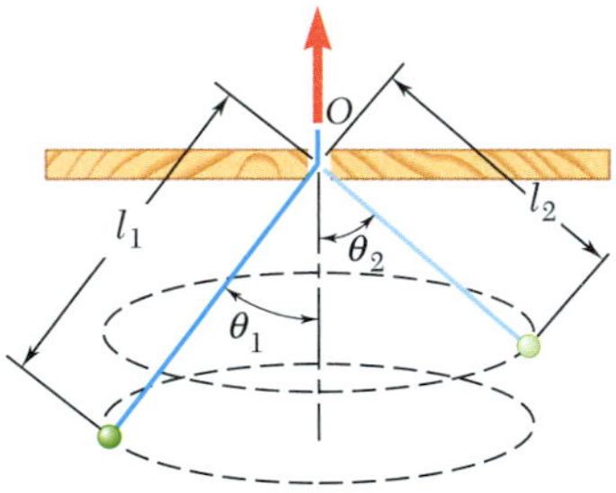

Fig. P12.91

12.91 A small ball swings in a horizontal circle at the end of a cord of length l_1, which forms an angle θ_1 with the vertical. The cord is then slowly drawn through the support at O until the length of the free end is l_2. (*a*) Derive a relation among l_1, l_2, θ_1, and θ_2. (*b*) If the ball is set in motion so that initially $l_1 = 2$ ft and $\theta_1 = 40°$, determine the angle θ_2 when $l_2 = 1.5$ ft.

12.92 Two 1.2-kg collars A and B can slide without friction on a frame, consisting of the horizontal rod OE and the vertical rod CD, which is free to rotate about CD. The two collars are connected by a cord running over a pulley that is attached to the frame at O and a stop prevents collar B from moving. The frame is rotating at the rate $\dot{\theta} = 10$ rad/s and $r = 0.2$ m when the stop is removed allowing collar A to move out along rod OE. Neglecting friction and the mass of the frame, determine (*a*) the tension in the cord and the acceleration of collar A relative to rod OE immediately after the stop is removed, (*b*) the transverse component of the velocity of collar A when $r = 0.3$ m.

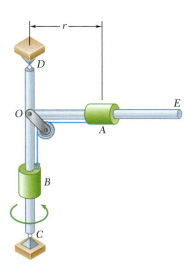

Fig. *P12.92* and *P12.93*

12.93 Two 1.2-kg collars A and B can slide without friction on a frame, consisting of the horizontal rod OE and the vertical rod CD, which is free to rotate about CD. The two collars are connected by a cord running over a pulley that is attached to the frame at O and a stop prevents collar B from moving. The frame is rotating at the rate $\dot{\theta} = 12$ rad/s and $r = 0.2$ m when the stop is removed allowing collar A to move out along rod OE. Neglecting friction and the mass of the frame, determine, for the position $r = 0.4$ m, (*a*) the transverse component of the velocity of collar A, (*b*) the tension in the cord and the acceleration of collar A relative to the rod OE.

*12.11. TRAJECTORY OF A PARTICLE UNDER A CENTRAL FORCE

Consider a particle P moving under a central force $\mathbf{F}$. We propose to obtain the differential equation which defines its trajectory.

Assuming that the force $\mathbf{F}$ is directed toward the center of force O, we note that ΣF_r and ΣF_θ reduce, respectively, to $-F$ and zero in Eqs. (12.21) and (12.22). We therefore write

$$m(\ddot{r} - r\dot{\theta}^2) = -F \tag{12.31}$$

$$m(r\ddot{\theta} + 2\dot{r}\dot{\theta}) = 0 \tag{12.32}$$

These equations define the motion of P. We will, however, replace Eq. (12.32) by Eq. (12.27), which is equivalent to Eq. (12.32), as can easily be checked by differentiating it with respect to t, but which is more convenient to use. We write

$$r^2\dot{\theta} = h \qquad \text{or} \qquad r^2\frac{d\theta}{dt} = h \tag{12.33}$$

Equation (12.33) can be used to eliminate the independent variable t from Eq. (12.31). Solving Eq. (12.33) for $\dot{\theta}$ or $d\theta/dt$, we have

$$\dot{\theta} = \frac{d\theta}{dt} = \frac{h}{r^2} \tag{12.34}$$

from which it follows that

$$\dot{r} = \frac{dr}{dt} = \frac{dr}{d\theta}\frac{d\theta}{dt} = \frac{h}{r^2}\frac{dr}{d\theta} = -h\frac{d}{d\theta}\left(\frac{1}{r}\right) \tag{12.35}$$

$$\ddot{r} = \frac{d\dot{r}}{dt} = \frac{d\dot{r}}{d\theta}\frac{d\theta}{dt} = \frac{h}{r^2}\frac{d\dot{r}}{d\theta}$$

or, substituting for $\dot{r}$ from (12.35),

$$\ddot{r} = \frac{h}{r^2}\frac{d}{d\theta}\left[-h\frac{d}{d\theta}\left(\frac{1}{r}\right)\right]$$

$$\ddot{r} = -\frac{h^2}{r^2}\frac{d^2}{d\theta^2}\left(\frac{1}{r}\right) \tag{12.36}$$

Substituting for θ and $\ddot{r}$ from (12.34) and (12.36), respectively, in Eq. (12.31) and introducing the function $u = 1/r$, we obtain after reductions

$$\frac{d^2u}{d\theta^2} + u = \frac{F}{mh^2u^2} \tag{12.37}$$

In deriving Eq. (12.37), the force $\mathbf{F}$ was assumed directed toward O. The magnitude F should therefore be positive if $\mathbf{F}$ is actually directed toward O (attractive force) and negative if $\mathbf{F}$ is directed away from O (repulsive force). If F is a known function of r and thus of u, Eq. (12.37) is a differential equation in u and θ. This differential equation defines the trajectory followed by the particle under the central force $\mathbf{F}$. The equation of the trajectory can be obtained by solving the differential equation (12.37) for u as a function of θ and determining the constants of integration from the initial conditions.

Photo 12.4 The International Space Station and the space vehicle are subjected to the gravitational pull of the earth and if all other forces are neglected, the trajectory of each will be a circle, an ellipse, a parabola, or a hyperbola.

*12.12. APPLICATION TO SPACE MECHANICS

After the last stages of their launching rockets have burned out, earth satellites and other space vehicles are subjected to only the gravitational pull of the earth. Their motion can therefore be determined from Eqs. (12.33) and (12.37), which govern the motion of a particle under a central force, after F has been replaced by the expression obtained for the force of gravitational attraction.† Setting in Eq. (12.37)

$$F = \frac{GMm}{r^2} = GMmu^2$$

where M = mass of earth
m = mass of space vehicle
r = distance from center of earth to vehicle
$u = 1/r$

we obtain the differential equation

$$\frac{d^2u}{d\theta^2} + u = \frac{GM}{h^2} \qquad (12.38)$$

where the right-hand member is observed to be a constant.

The solution of the differential equation (12.38) is obtained by adding the particular solution $u = GM/h^2$ to the general solution $u = C \cos(\theta - \theta_0)$ of the corresponding homogeneous equation (that is, the equation obtained by setting the right-hand member equal to zero). Choosing the polar axis so that $\theta_0 = 0$, we write

$$\frac{1}{r} = u = \frac{GM}{h^2} + C\cos\theta \qquad (12.39)$$

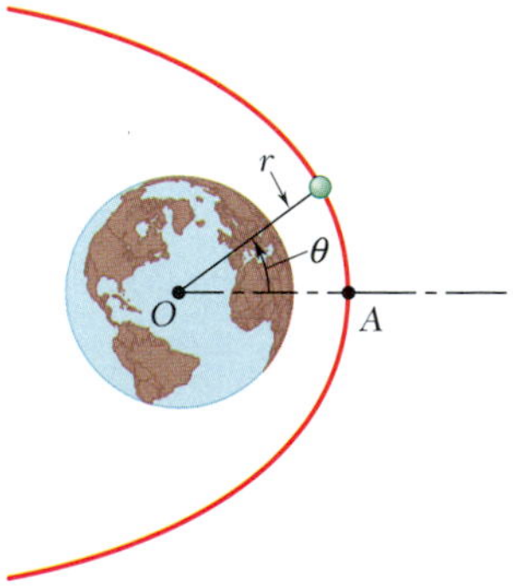

Fig. 12.19

Equation (12.39) is the equation of a *conic section* (ellipse, parabola, or hyperbola) in the polar coordinates r and θ. The origin O of the coordinates, which is located at the center of the earth, is a *focus* of this conic section, and the polar axis is one of its axes of symmetry (Fig. 12.19).

The ratio of the constants C and GM/h^2 defines the *eccentricity* ε of the conic section; letting

$$\varepsilon = \frac{C}{GM/h^2} = \frac{Ch^2}{GM} \qquad (12.40)$$

we can write Eq. (12.39) in the form

$$\frac{1}{r} = \frac{GM}{h^2}(1 + \varepsilon\cos\theta) \qquad (12.39')$$

This equation represents three possible trajectories.

1. $\varepsilon > 1$, or $C > GM/h^2$: There are two values θ_1 and $-\theta_1$ of the polar angle, defined by $\cos\theta_1 = -GM/Ch^2$, for which

†It is assumed that the space vehicles considered here are attracted by the earth only and that their mass is negligible compared with the mass of the earth. If a vehicle moves very far from the earth, its path may be affected by the attraction of the sun, the moon, or another planet.

the right-hand member of Eq. (12.39) becomes zero. For both these values, the radius vector r becomes infinite; the conic section is a *hyperbola* (Fig. 12.20).

2. $\varepsilon = 1$, or $C = GM/h^2$: The radius vector becomes infinite for $\theta = 180°$; the conic section is a *parabola*.
3. $\varepsilon < 1$, or $C < GM/h^2$: The radius vector remains finite for every value of θ; the conic section is an *ellipse*. In the particular case when $\varepsilon = C = 0$, the length of the radius vector is constant; the conic section is a circle.

Let us now see how the constants C and GM/h^2, which characterize the trajectory of a space vehicle, can be determined from the vehicle's position and velocity at the beginning of its free flight. We will assume that, as is generally the case, the powered phase of its flight has been programmed in such a way that as the last stage of the launching rocket burns out, the vehicle has a velocity parallel to the surface of the earth (Fig. 12.21). In other words, we will assume that the space vehicle begins its free flight at the vertex A of its trajectory.†

Denoting the radius vector and speed of the vehicle at the beginning of its free flight by r_0 and v_0, respectively, we observe that the velocity reduces to its transverse component and, thus, that $v_0 = r_0\dot{\theta}_0$. Recalling Eq. (12.27), we express the angular momentum per unit mass h as

$$h = r_0^2\dot{\theta}_0 = r_0 v_0 \qquad (12.41)$$

The value obtained for h can be used to determine the constant GM/h^2. We also note that the computation of this constant will be simplified if we use the relation obtained in Sec. 12.10:

$$GM = gR^2 \qquad (12.30)$$

where R is the radius of the earth ($R = 6.37 \times 10^6$ m or 3960 mi) and g is the acceleration of gravity at the surface of the earth.

The constant C is obtained by setting $\theta = 0$, $r = r_0$ in (12.39):

$$C = \frac{1}{r_0} - \frac{GM}{h^2} \qquad (12.42)$$

Substituting for h from (12.41), we can then easily express C in terms of r_0 and v_0.

Let us now determine the initial conditions corresponding to each of the three fundamental trajectories indicated above. Considering first the parabolic trajectory, we set C equal to GM/h^2 in Eq. (12.42) and eliminate h between Eqs. (12.41) and (12.42). Solving for v_0, we obtain

$$v_0 = \sqrt{\frac{2GM}{r_0}}$$

We can easily check that a larger value of the initial velocity corresponds to a hyperbolic trajectory and a smaller value corresponds to an elliptic orbit. Since the value of v_0 obtained for the parabolic

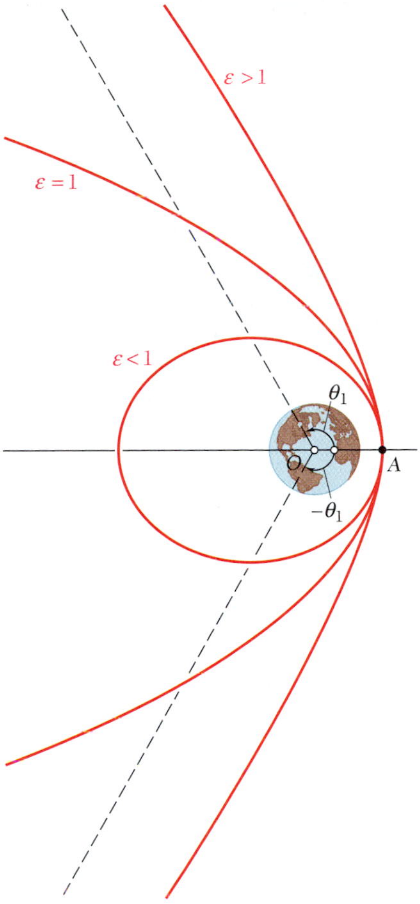

Fig. 12.20

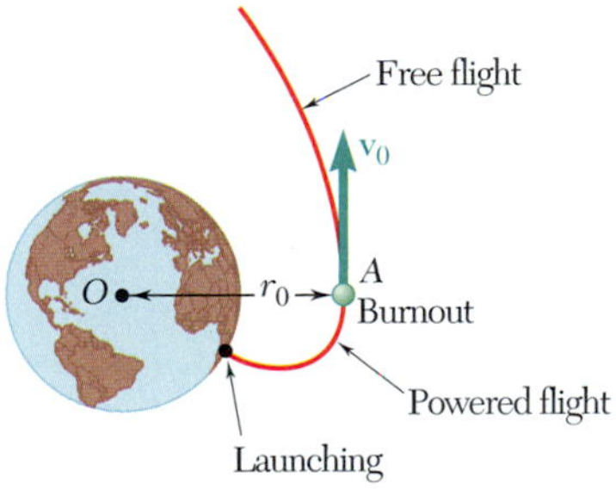

Fig. 12.21

†Problems involving oblique launchings will be considered in Sec. 13.9.

trajectory is the smallest value for which the space vehicle does not return to its starting point, it is called the *escape velocity.* We write therefore

$$v_{esc} = \sqrt{\frac{2GM}{r_0}} \quad \text{or} \quad v_{esc} = \sqrt{\frac{2gR^2}{r_0}} \tag{12.43}$$

if we make use of Eq. (12.30). We note that the trajectory will be (1) hyperbolic if $v_0 > v_{esc}$, (2) parabolic if $v_0 = v_{esc}$, and (3) elliptic if $v_0 < v_{esc}$.

Among the various possible elliptic orbits, the one obtained when $C = 0$, the *circular orbit,* is of special interest. The value of the initial velocity corresponding to a circular orbit is easily found to be

$$v_{circ} = \sqrt{\frac{GM}{r_0}} \quad \text{or} \quad v_{circ} = \sqrt{\frac{gR^2}{r_0}} \tag{12.44}$$

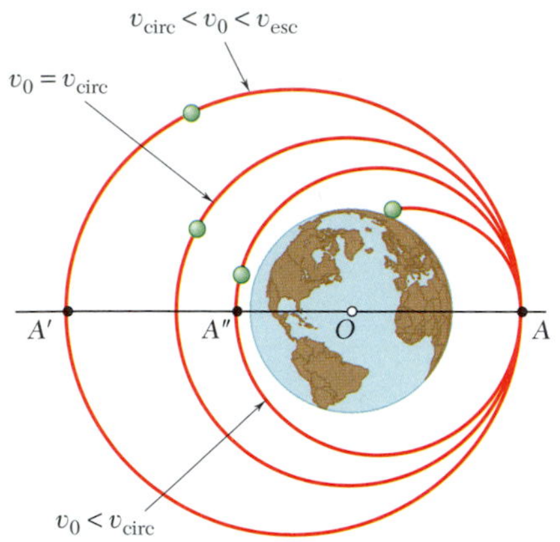

Fig. 12.22

if Eq. (12.30) is taken into account. We note from Fig. 12.22 that for values of v_0 larger than v_{circ} but smaller than v_{esc}, point A where free flight begins is the point of the orbit closest to the earth; this point is called the *perigee,* while point A', which is farthest away from the earth, is known as the *apogee.* For values of v_0 smaller than v_{circ}, point A is the apogee, while point A'', on the other side of the orbit, is the perigee. For values of v_0 much smaller than v_{circ}, the trajectory of the space vehicle intersects the surface of the earth; in such a case, the vehicle does not go into orbit.

Ballistic missiles, which were designed to hit the surface of the earth, also travel along elliptic trajectories. In fact, we should now realize that any object projected in vacuum with an initial velocity v_0 smaller than v_{esc} will move along an elliptic path. It is only when the distances involved are small that the gravitational field of the earth can be assumed uniform and that the elliptic path can be approximated by a parabolic path, as was done earlier (Sec. 11.11) in the case of conventional projectiles.

Periodic Time. An important characteristic of the motion of an earth satellite is the time required by the satellite to describe its orbit. This time, known as the *periodic time* of the satellite, is denoted by τ. We first observe, in view of the definition of areal velocity (Sec. 12.9), that τ can be obtained by dividing the area inside the orbit by the areal velocity. Noting that the area of an ellipse is equal to πab, where a and b denote the semimajor and semiminor axes, respectively, and that the areal velocity is equal to $h/2$, we write

$$\tau = \frac{2\pi ab}{h} \tag{12.45}$$

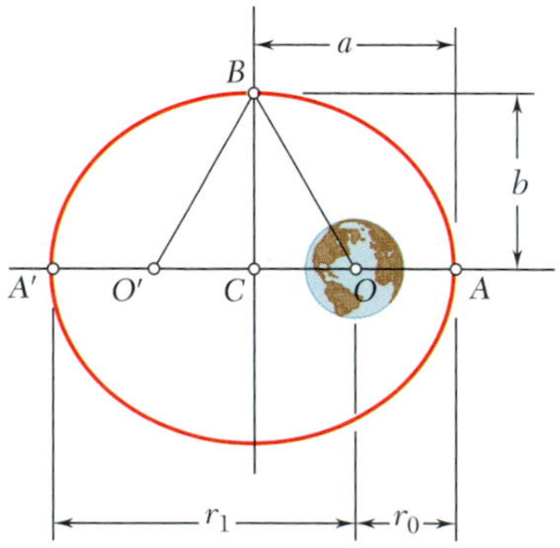

Fig. 12.23

While h can be readily determined from r_0 and v_0 in the case of a satellite launched in a direction parallel to the surface of the earth, the semiaxes a and b are not directly related to the initial conditions. Since, on the other hand, the values r_0 and r_1 of r corresponding to the perigee and apogee of the orbit can easily be determined from Eq. (12.39), we will express the semiaxes a and b in terms of r_0 and r_1.

Consider the elliptic orbit shown in Fig. 12.23. The earth's center is located at O and coincides with one of the two foci of the ellipse, while the points A and A' represent, respectively, the perigee and apogee of the orbit. We easily check that

$$r_0 + r_1 = 2a$$

and thus

$$a = \tfrac{1}{2}(r_0 + r_1) \qquad (12.46)$$

Recalling that the sum of the distances from each of the foci to any point of the ellipse is constant, we write

$$O'B + BO = O'A + OA = 2a \qquad \text{or} \qquad BO = a$$

On the other hand, we have $CO = a - r_0$. We can therefore write

$$b^2 = (BC)^2 = (BO)^2 - (CO)^2 = a^2 - (a - r_0)^2$$
$$b^2 = r_0(2a - r_0) = r_0 r_1$$

and thus

$$b = \sqrt{r_0 r_1} \qquad (12.47)$$

Formulas (12.46) and (12.47) indicate that the semimajor and semiminor axes of the orbit are equal, respectively, to the arithmetic and geometric means of the maximum and minimum values of the radius vector. Once r_0 and r_1 have been determined, the lengths of the semiaxes can be easily computed and substituted for a and b in formula (12.45).

*12.13. KEPLER'S LAWS OF PLANETARY MOTION

The equations governing the motion of an earth satellite can be used to describe the motion of the moon around the earth. In that case, however, the mass of the moon is not negligible compared with the mass of the earth, and the results obtained are not entirely accurate.

The theory developed in the preceding sections can also be applied to the study of the motion of the planets around the sun. Although another error is introduced by neglecting the forces exerted by the planets on one another, the approximation obtained is excellent. Indeed, even before Newton had formulated his fundamental theory, the properties expressed by Eq. (12.39), where M now represents the mass of the sun, and by Eq. (12.33) had been discovered by the German astronomer Johann Kepler (1571–1630) from astronomical observations of the motion of the planets.

Kepler's three *laws of planetary motion* can be stated as follows:

1. Each planet describes an ellipse, with the sun located at one of its foci.
2. The radius vector drawn from the sun to a planet sweeps equal areas in equal times.
3. The squares of the periodic times of the planets are proportional to the cubes of the semimajor axes of their orbits.

The first law states a particular case of the result established in Sec. 12.12, and the second law expresses that the areal velocity of each planet is constant (see Sec. 12.9). Kepler's third law can also be derived from the results obtained in Sec. 12.12.†

†See Prob. 12.120.

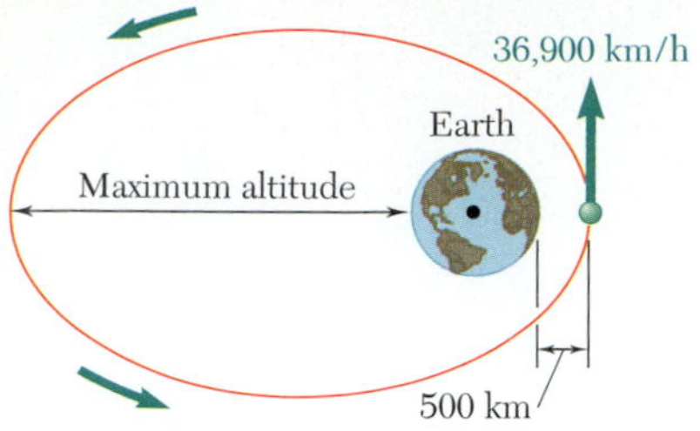

SAMPLE PROBLEM 12.9

A satellite is launched in a direction parallel to the surface of the earth with a velocity of 36 900 km/h from an altitude of 500 km. Determine (*a*) the maximum altitude reached by the satellite, (*b*) the periodic time of the satellite.

SOLUTION

***a.* Maximum Altitude.** After the satellite is launched, it is subjected only to the gravitational attraction of the earth; its motion is thus governed by Eq. (12.39),

$$\frac{1}{r} = \frac{GM}{h^2} + C \cos \theta \qquad (1)$$

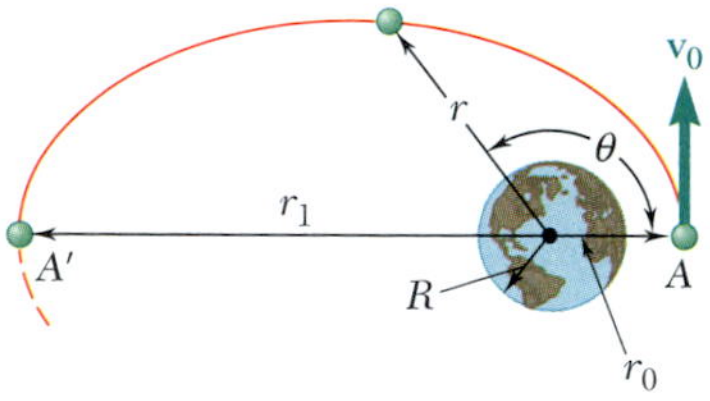

Since the radial component of the velocity is zero at the point of launching A, we have $h = r_0 v_0$. Recalling that for the earth $R = 6370$ km, we compute

$$r_0 = 6370 \text{ km} + 500 \text{ km} = 6870 \text{ km} = 6.87 \times 10^6 \text{ m}$$

$$v_0 = 36\,900 \text{ km/h} = \frac{36.9 \times 10^6 \text{ m}}{3.6 \times 10^3 \text{ s}} = 10.25 \times 10^3 \text{ m/s}$$

$$h = r_0 v_0 = (6.87 \times 10^6 \text{ m})(10.25 \times 10^3 \text{ m/s}) = 70.4 \times 10^9 \text{ m}^2\text{/s}$$

$$h^2 = 4.96 \times 10^{21} \text{ m}^4\text{/s}^2$$

Since $GM = gR^2$, where R is the radius of the earth, we have

$$GM = gR^2 = (9.81 \text{ m/s}^2)(6.37 \times 10^6 \text{ m})^2 = 398 \times 10^{12} \text{ m}^3\text{/s}^2$$

$$\frac{GM}{h^2} = \frac{398 \times 10^{12} \text{ m}^3\text{/s}^2}{4.96 \times 10^{21} \text{ m}^4\text{/s}^2} = 80.3 \times 10^{-9} \text{ m}^{-1}$$

Substituting this value into (1), we obtain

$$\frac{1}{r} = 80.3 \times 10^{-9} \text{ m}^{-1} + C \cos \theta \qquad (2)$$

Noting that at point A we have $\theta = 0$ and $r = r_0 = 6.87 \times 10^6$ m, we compute the constant C:

$$\frac{1}{6.87 \times 10^6 \text{ m}} = 80.3 \times 10^{-9} \text{ m}^{-1} + C \cos 0° \qquad C = 65.3 \times 10^{-9} \text{ m}^{-1}$$

At A', the point on the orbit farthest from the earth, we have $\theta = 180°$. Using (2), we compute the corresponding distance r_1:

$$\frac{1}{r_1} = 80.3 \times 10^{-9} \text{ m}^{-1} + (65.3 \times 10^{-9} \text{ m}^{-1}) \cos 180°$$

$$r_1 = 66.7 \times 10^6 \text{ m} = 66\,700 \text{ km}$$

Maximum altitude = 66 700 km − 6370 km = 60 300 km ◀

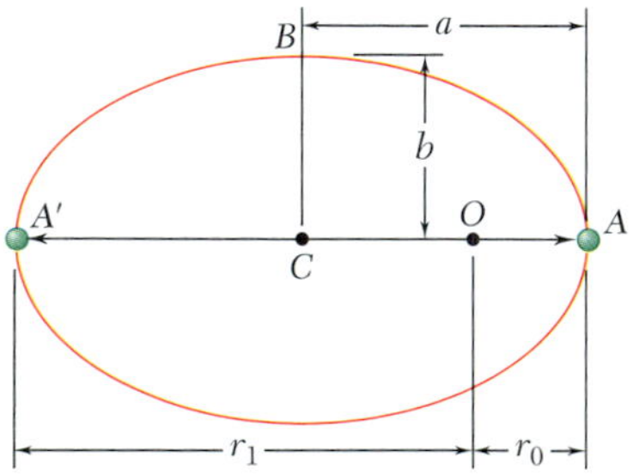

***b.* Periodic Time.** Since A and A' are the perigee and apogee, respectively, of the elliptic orbit, we use Eqs. (12.46) and (12.47) and compute the semimajor and semiminor axes of the orbit:

$$a = \tfrac{1}{2}(r_0 + r_1) = \tfrac{1}{2}(6.87 + 66.7)(10^6) \text{ m} = 36.8 \times 10^6 \text{ m}$$

$$b = \sqrt{r_0 r_1} = \sqrt{(6.87)(66.7)} \times 10^6 \text{ m} = 21.4 \times 10^6 \text{ m}$$

$$\tau = \frac{2\pi ab}{h} = \frac{2\pi(36.8 \times 10^6 \text{ m})(21.4 \times 10^6 \text{ m})}{70.4 \times 10^9 \text{ m}^2\text{/s}}$$

$$\tau = 70.3 \times 10^3 \text{ s} = 1171 \text{ min} = 19 \text{ h } 31 \text{ min} \quad ◀$$

SOLVING PROBLEMS ON YOUR OWN

In this lesson, we continued our study of the motion of a particle under a central force and applied the results to problems in space mechanics. We found that the trajectory of a particle under a central force is defined by the differential equation

$$\frac{d^2u}{d\theta^2} + u = \frac{F}{mh^2u^2} \tag{12.37}$$

where u is the reciprocal of the distance r of the particle to the center of force ($u = 1/r$), F is the magnitude of the central force $\mathbf{F}$, and h is a constant equal to the angular momentum per unit mass of the particle. In space-mechanics problems, $\mathbf{F}$ is the force of gravitational attraction exerted on the satellite or spacecraft by the sun, earth, or other planet about which it travels. Substituting $F = GMm/r^2 = GMmu^2$ into Eq. (12.37), we obtain for that case

$$\frac{d^2u}{d\theta^2} + u = \frac{GM}{h^2} \tag{12.38}$$

where the right-hand member is a constant.

1. *Analyzing the motion of satellites and spacecraft.* The solution of the differential equation (12.38) defines the trajectory of a satellite or spacecraft. It was obtained in Sec. 12.12 and was given in the alternative forms

$$\frac{1}{r} = \frac{GM}{h^2} + C\cos\theta \qquad \text{or} \qquad \frac{1}{r} = \frac{GM}{h^2}(1 + \varepsilon\cos\theta) \qquad (12.39, 12.39')$$

Remember when applying these equations that $\theta = 0$ always corresponds to the perigee (the point of closest approach) of the trajectory (Fig. 12.19) and that h is a constant for a given trajectory. Depending on the value of the eccentricity ε, the trajectory will be a hyperbola, a parabola, or an ellipse.

a. $\varepsilon > 1$: The trajectory is a hyperbola, so that for this case the spacecraft never returns to its starting point.

b. $\varepsilon = 1$: The trajectory is a parabola. This is the limiting case between open (hyperbolic) and closed (elliptic) trajectories. We had observed for this case that the velocity v_0 at the perigee is equal to the escape velocity v_{esc},

$$v_0 = v_{esc} = \sqrt{\frac{2GM}{r_0}} \tag{12.43}$$

Note that the escape velocity is the smallest velocity for which the spacecraft does not return to its starting point.

c. ε < 1: The trajectory is an elliptic orbit. For problems involving elliptic orbits, you may find that the relation derived in Prob. 12.102,

$$\frac{1}{r_0} + \frac{1}{r_1} = \frac{2GM}{h^2}$$

will be useful in the solution of subsequent problems. When you apply this equation, remember that r_0 and r_1 are the distances from the center of force to the perigee ($\theta = 0$) and apogee ($\theta = 180°$), respectively; that $h = r_0 v_0 = r_1 v_1$; and that, for a satellite orbiting the earth, $GM_{\text{earth}} = gR^2$, where R is the radius of the earth. Also recall that the trajectory is a circle when $\varepsilon = 0$.

2. *Determining the point of impact of a descending spacecraft.* For problems of this type, you may assume that the trajectory is elliptic and that the initial point of the descent trajectory is the apogee of the path (Fig. 12.22). Note that at the point of impact, the distance r in Eqs. (12.39) and (12.39′) is equal to the radius R of the body on which the spacecraft lands or crashes. In addition, we have $h = Rv_I \sin \phi_I$, where v_I is the speed of the spacecraft at impact and ϕ_I is the angle that its path forms with the vertical at the point of impact.

3. *Calculating the time to travel between two points on a trajectory.* For central force motion, the time t required for a particle to travel along a portion of its trajectory can be determined by recalling from Sec. 12.9 that the rate at which area is swept per unit time by the position vector $\mathbf{r}$ is equal to one-half of the angular momentum per unit mass h of the particle: $dA/dt = h/2$. It follows, since h is a constant for a given trajectory, that

$$t = \frac{2A}{h}$$

where A is the total area swept in the time t.

a. In the case of an elliptic trajectory, the time required to complete one orbit is called the *periodic time* and is expressed as

$$\tau = \frac{2(\pi ab)}{h} \tag{12.45}$$

where a and b are the semimajor and semiminor axes, respectively, of the ellipse and are related to the distances r_0 and r_1 by

$$a = \frac{1}{2}(r_0 + r_1) \qquad \text{and} \qquad b = \sqrt{r_0 r_1} \qquad (12.46, 12.47)$$

b. Kepler's third law provides a convenient relation between the periodic times of two satellites describing elliptic orbits about the same body [Sec. 12.13]. Denoting the semimajor axes of the two orbits by a_1 and a_2, respectively, and the corresponding periodic times by τ_1 and τ_2, we have

$$\frac{\tau_1^2}{\tau_2^2} = \frac{a_1^3}{a_2^3}$$

c. In the case of a parabolic trajectory, you may be able to use the expression given on the inside of the front cover of the book for a parabolic or a semiparabolic area to calculate the time required to travel between two points of the trajectory.

Problems

12.94 A particle of mass m is projected from point A with an initial velocity $\mathbf{v}_0$ perpendicular to OA and moves under a central force $\mathbf{F}$ along an elliptic path defined by the equation $r = r_0/(2 - \cos\theta)$. Using Eq. (12.37), show that $\mathbf{F}$ is inversely proportional to the square of the distance r from the particle to the center of force O.

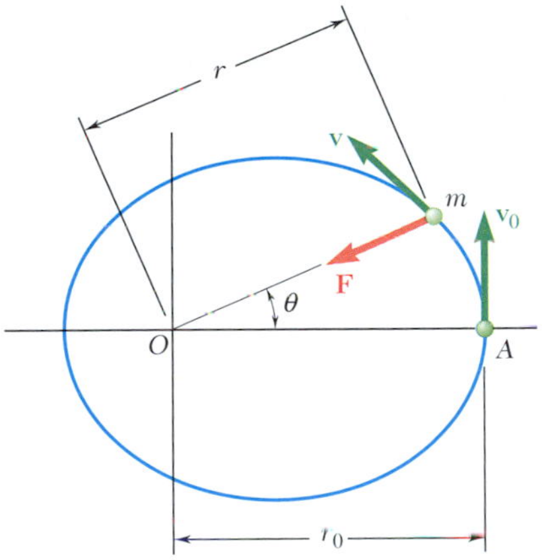

Fig. P12.94

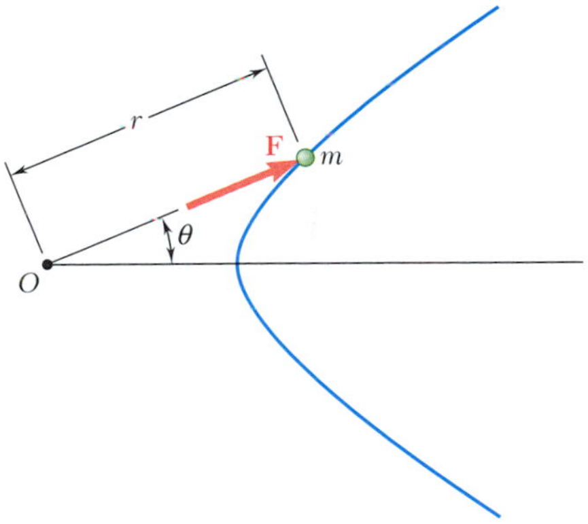

Fig. P12.95

12.95 A particle of mass m describes the path defined by the equation $r = r_0/(6\cos\theta - 5)$ under a central force $\mathbf{F}$ directed away from the center of force O. Using Eq. (12.37), show that $\mathbf{F}$ is inversely proportional to the square of the distance r from the particle to O.

12.96 A particle of mass m describes the parabola $y = x^2/4r_0$ under a central force $\mathbf{F}$ directed toward the center of force C. Using Eq. (12.37) and Eq. (12.39′) with $\varepsilon = 1$, show that $\mathbf{F}$ is inversely proportional to the square of the distance r from the particle to the center of force and that the angular momentum per unit mass $h = \sqrt{2GMr_0}$.

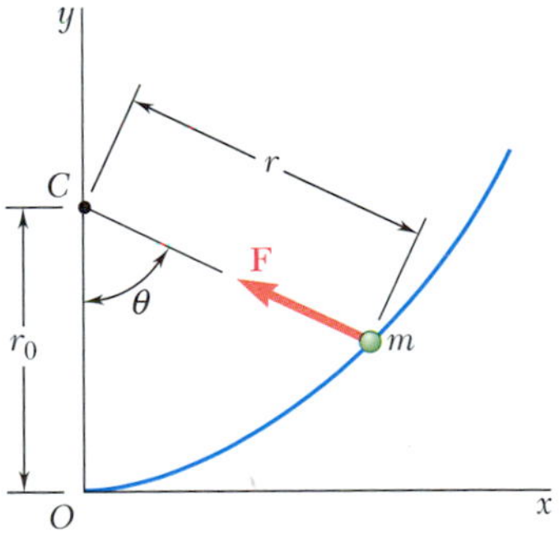

Fig. P12.96

12.97 For the particle of Prob. 12.74, and using Eq. (12.37), show that the central force $\mathbf{F}$ is proportional to the distance r from the particle to the center of force O.

12.98 It was observed that during the Galileo spacecraft's first flyby of the earth, its minimum altitude was 600 mi above the surface of the earth. Assuming that the trajectory of the spacecraft was parabolic, determine the maximum velocity of Galileo during its first flyby of the earth.

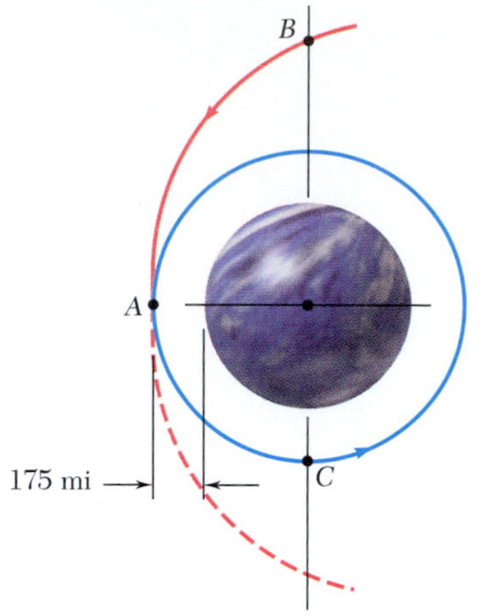

Fig. *P12.99*

12.99 As a space probe approaching the planet Venus on a parabolic trajectory reaches point A closest to the planet, its velocity is decreased to insert it into a circular orbit. Knowing that the mass and the radius of Venus are 334×10^{21} lb · s^2/ft and 3761 mi, respectively, determine (a) the velocity of the probe as it approaches A, (b) the decrease in velocity required to insert it into a circular orbit.

12.100 It was observed that as the Galileo spacecraft reached the point of its trajectory closest to Io, a moon of the planet Jupiter, it was at a distance of 2820 km from the center of Io and had a velocity of 15 km/s. Knowing that the mass of Io is 0.01496 times the mass of the earth, determine the eccentricity of the trajectory of the spacecraft as it approached Io.

12.101 It was observed that during its second flyby of the earth, the Galileo spacecraft had a velocity of 14.1 km/s as it reached its minimum altitude of 303 km above the surface of the earth. Determine the eccentricity of the trajectory of the spacecraft during this portion of its flight.

12.102 A satellite describes an elliptic orbit about a planet of mass M. Denoting by r_0 and r_1, respectively, the minimum and maximum values of the distance r from the satellite to the center of the planet, derive the relation

$$\frac{1}{r_0} + \frac{1}{r_1} = \frac{2GM}{h^2}$$

where h is the angular momentum per unit mass of the satellite.

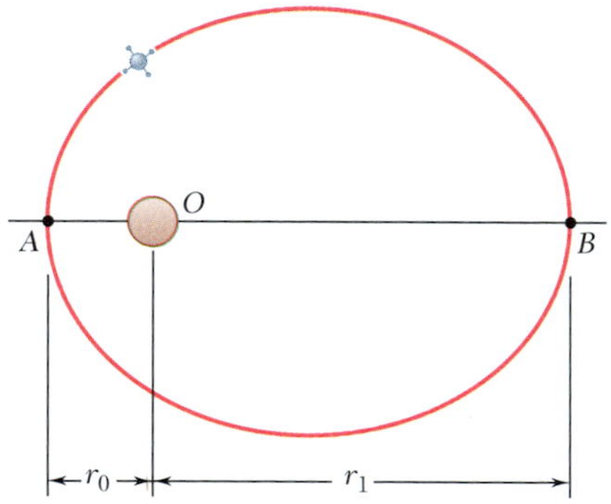

Fig. P12.102

12.103 The Chandra X-ray observatory, launched in 1999, achieved an elliptical orbit of minimum altitude 6200 mi and maximum altitude 86,900 mi above the surface of the earth. Assuming that the observatory was transferred to this orbit from a circular orbit of altitude 6200 mi at point A, determine (a) the increase in speed required at A, (b) the speed of the observatory at B.

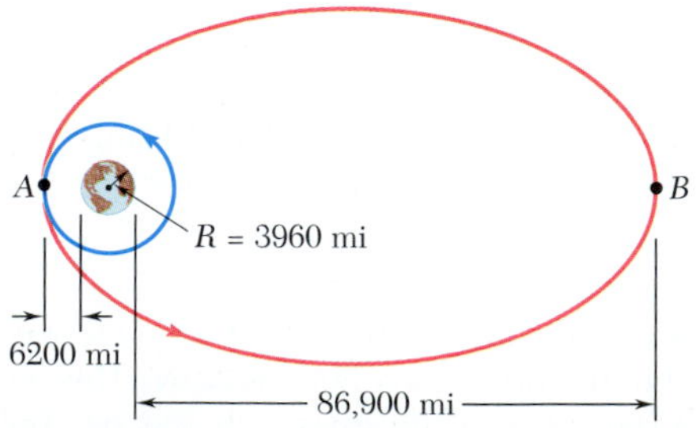

Fig. P12.103

12.104 A satellite describes a circular orbit at an altitude of 19 110 km above the surface of the earth. Determine (*a*) the increase in speed required at point A for the satellite to achieve the escape velocity and enter a parabolic orbit, (*b*) the decrease in speed required at point A for the satellite to enter an elliptic orbit of minimum altitude 6370 km, (*c*) the eccentricity ε of the elliptic orbit.

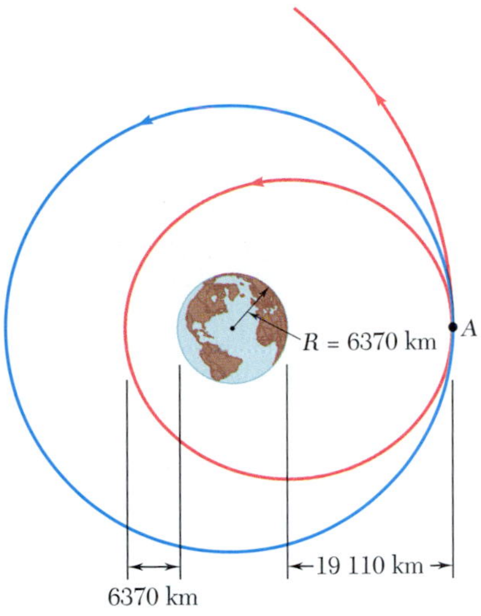

Fig. *P12.104*

12.105 As it describes an elliptic orbit about the sun, a spacecraft reaches a maximum distance of 325×10^6 km from the center of the sun at point A (called the *aphelion*) and a minimum distance of 148×10^6 km at point B (called the *perihelion*). To place the spacecraft in a smaller elliptic orbit with aphelion A' and perihelion B', where A' and B' are located 264.7×10^6 km and 137.6×10^6 km, respectively, from the center of the sun, the speed of the spacecraft is first reduced as it passes through A and then is further reduced as it passes through B'. Knowing that the mass of the sun is 332.8×10^3 times the mass of the earth, determine (*a*) the speed of the spacecraft at A, (*b*) the amounts by which the speed of the spacecraft should be reduced at A and B' to insert it into the desired elliptic orbit.

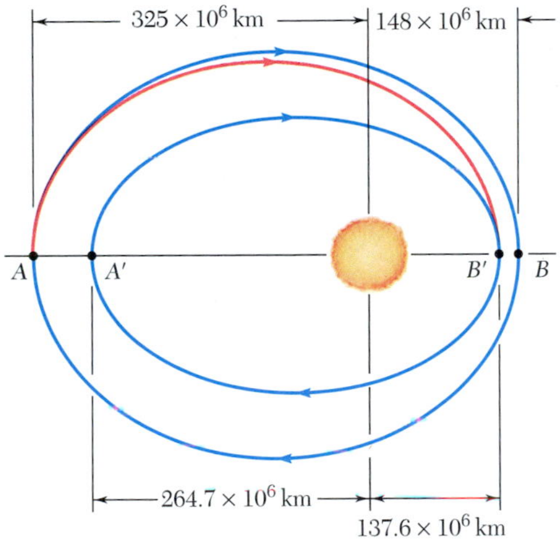

Fig. *P12.105*

12.106 A space probe is in a circular orbit about the planet Mars. The orbit must have a radius of 2500 mi and be located in a specified plane which is different from the plane of the approach trajectory. As the probe reaches A, the point of its original trajectory closest to Mars, it is inserted into a first elliptic transfer orbit by reducing its speed by Δv_A. This orbit brings it to point B with a much reduced velocity. There the probe is inserted into a second transfer orbit located in the specified plane by changing the direction of its velocity and further reducing its speed by Δv_B. Finally, as the probe reaches point C, it is inserted into the desired circular orbit by reducing its speed by Δv_C. Knowing that the mass of Mars is 0.1074 times the mass of the earth, that $r_A = 5625$ mi and $r_B = 112{,}500$ mi, and that the probe approaches A on a parabolic trajectory, determine by how much the speed of the probe should be reduced (*a*) at A, (*b*) at B, (*c*) at C.

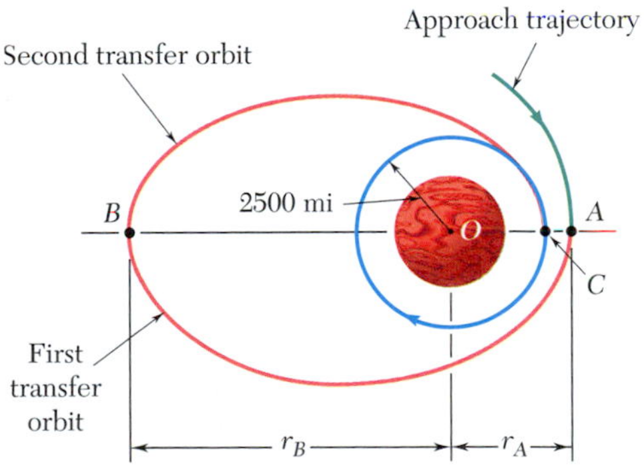

Fig. P12.106

12.107 For the probe of Prob. 12.106, it is known that $r_A = 5625$ mi and that the speed of the probe is reduced by 1300 ft/s as it passes through A. Determine (*a*) the distance from the center of Mars to point B, (*b*) the amounts by which the speed of the probe should be reduced at B and C, respectively.

12.108 Determine the time needed for the space probe of Prob. 12.106 to travel from A to B on its first transfer orbit.

12.109 The Clementine spacecraft described an elliptic orbit of minimum altitude $h_A = 250$ mi and a maximum altitude of $h_B = 1840$ mi above the surface of the moon. Knowing that the radius of the moon is 1080 mi and the mass of the moon is 0.01230 times the mass of the earth, determine the periodic time of the spacecraft.

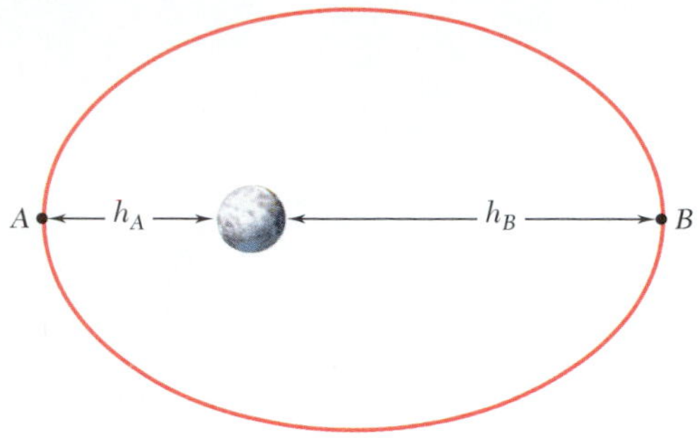

Fig. P12.109

12.110 A spacecraft and a satellite are at diametrically opposite positions in the same circular orbit of altitude 500 km above the earth. As it passes through point A, the spacecraft fires its engine for a short interval of time to increase its speed and enter an elliptic orbit. Knowing that the spacecraft returns to A at the same time the satellite reaches A after completing one and a half orbits, determine (a) the increase in speed required, (b) the periodic time for the elliptic orbit.

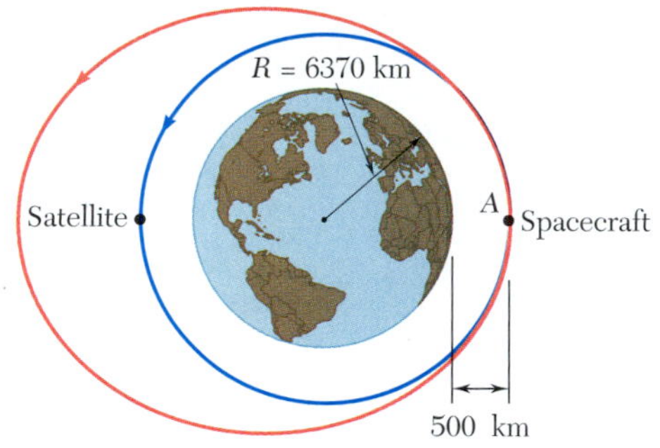

Fig. *P12.110*

12.111 Based on observations made during the 1996 sighting of the comet Hyakutake, it was concluded that the trajectory of the comet is a highly elongated ellipse for which the eccentricity is approximately $\varepsilon = 0.999887$. Knowing that for the 1996 sighting the minimum distance between the comet and the sun was $0.230R_E$, where R_E is the mean distance from the sun to the earth, determine the periodic time of the comet.

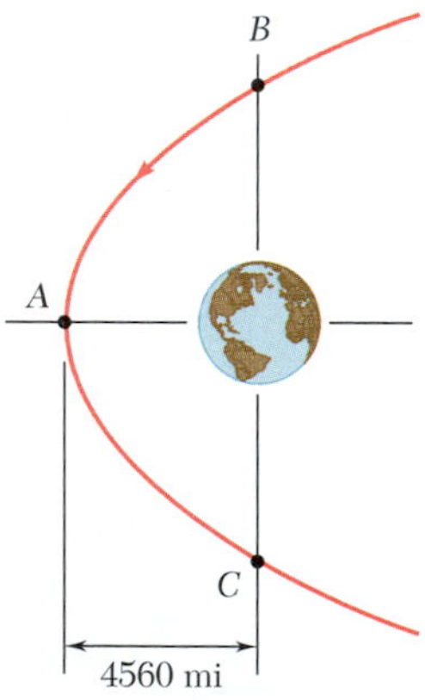

Fig. P12.112

12.112 It was observed that during its first flyby of the earth, the Galileo spacecraft had a velocity of 6.48 mi/s as it reached its minimum distance of 4560 mi from the center of the earth. Assuming that the trajectory of the spacecraft was parabolic, determine the time needed for the spacecraft to travel from B to C on its trajectory.

12.113 Determine the time needed for the space probe of Prob. 12.99 to travel from B to C.

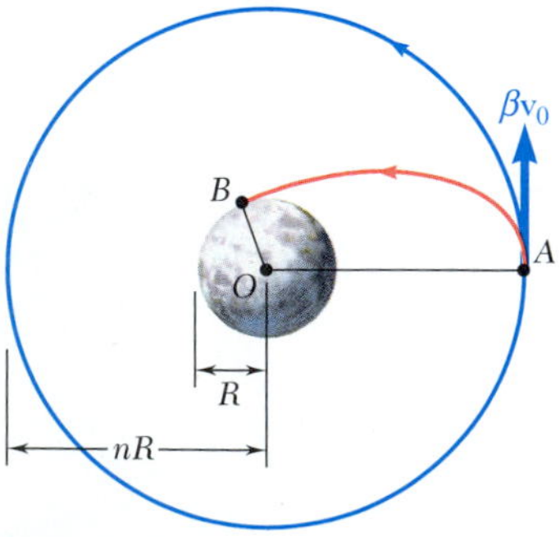

Fig. P12.114

12.114 A space probe is describing a circular orbit of radius nR with a speed v_0 about a planet of radius R and center O. As the probe passes through point A, its speed is reduced from v_0 to βv_0, where $\beta < 1$, to place the probe on a crash trajectory. Express in terms of n and β the angle AOB, where B denotes the point of impact of the probe on the planet.

12.115 Prior to the Apollo missions to the moon, several Lunar Orbiter spacecraft were used to photograph the lunar surface to obtain information regarding possible landing sites. At the conclusion of each mission, the trajectory of the spacecraft was adjusted so that the spacecraft would crash on the moon to further study the characteristics of the lunar surface. Shown below is the elliptic orbit of Lunar Orbiter 2. Knowing that the mass of the moon is 0.01230 times the mass of the earth, determine the amount by which the speed of the orbiter should be reduced at point B so that it impacts the lunar surface at point C. (*Hint.* Point B is the apogee of the elliptic impact trajectory.)

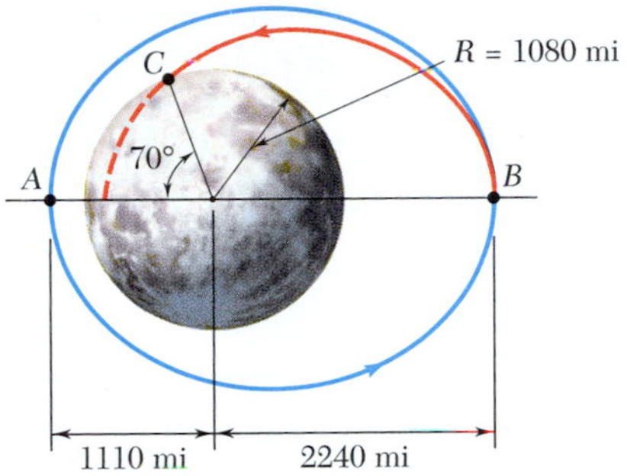

Fig. P12.115

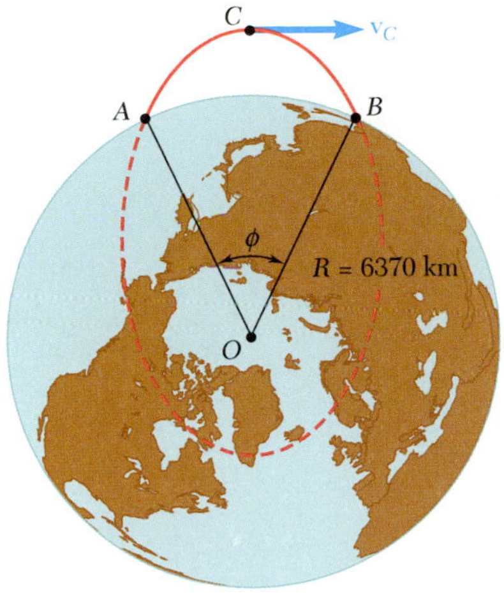

Fig. *P12.116*

12.116 A long-range ballistic trajectory between points A and B on the earth's surface consists of a portion of an ellipse with the apogee at point C. Knowing that point C is 1500 km above the surface of the earth and the range $R\phi$ of the trajectory is 6000 km, determine (a) the velocity of the projectile at C, (b) the eccentricity ε of the trajectory.

12.117 A space shuttle is describing a circular orbit at an altitude of 563 km above the surface of the earth. As it passes through point A, it fires its engine for a short interval of time to reduce its speed by 152 m/s and begin its descent toward the earth. Determine the angle AOB so that the altitude of the shuttle at point B is 121 km. (*Hint.* Point A is the apogee of the elliptic descent orbit.)

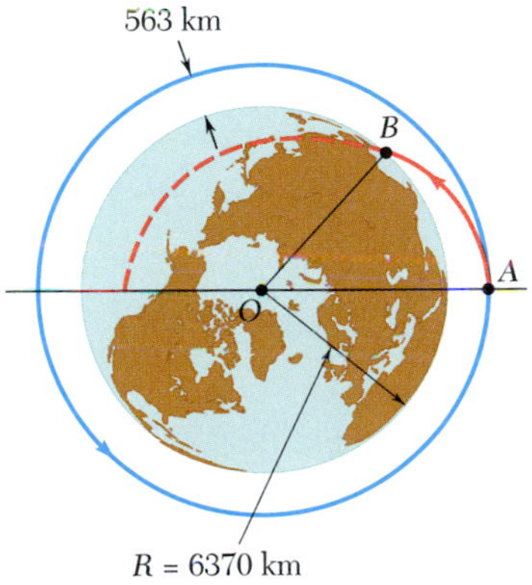

Fig. *P12.117*

12.118 A satellite describes an elliptic orbit about a planet. Denoting by r_0 and r_1 the distances corresponding, respectively, to the perigee and apogee of the orbit, show that the curvature of the orbit at each of these two points can be expressed as

$$\frac{1}{\rho} = \frac{1}{2}\left(\frac{1}{r_0} + \frac{1}{r_1}\right)$$

12.119 (a) Express the eccentricity ε of the elliptic orbit described by a satellite about a planet in terms of the distances r_0 and r_1 corresponding, respectively, to the perigee and apogee of the orbit. (b) Use the result obtained in part a and the data given in Prob. 12.111, where $R_\varepsilon = 93.0 \times 10^6$ mi, to determine the appropriate maximum distance from the sun reached by comet Hyakutake.

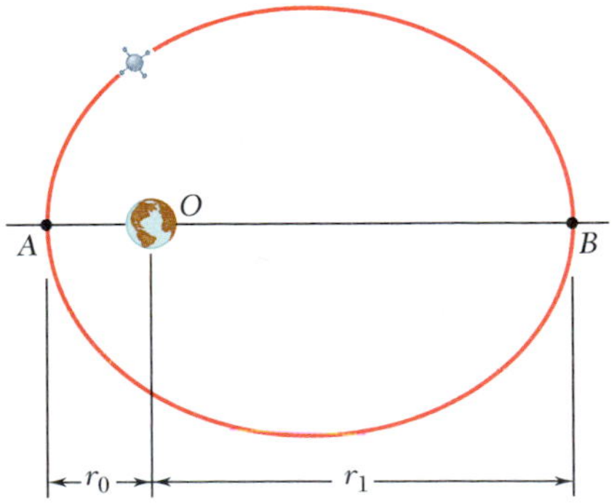

Fig. P12.118 and P12.119

12.120 Derive Kepler's third law of planetary motion from Eqs. (12.39) and (12.45).

12.121 Show that the angular momentum per unit mass h of a satellite describing an elliptic orbit of semimajor axis a and eccentricity ε about a planet of mass M can be expressed as

$$h = \sqrt{GMa(1 - \varepsilon^2)}$$

REVIEW AND SUMMARY FOR CHAPTER 12

This chapter was devoted to Newton's second law and its application to the analysis of the motion of particles.

Newton's second law

Denoting by m the mass of a particle, by $\Sigma\mathbf{F}$ the sum, or resultant, of the forces acting on the particle, and by $\mathbf{a}$ the acceleration of the particle relative to a *newtonian frame of reference* [Sec. 12.2], we wrote

$$\Sigma\mathbf{F} = m\mathbf{a} \tag{12.2}$$

Linear momentum

Introducing the *linear momentum* of a particle, $\mathbf{L} = m\mathbf{v}$ [Sec. 12.3], we saw that Newton's second law can also be written in the form

$$\Sigma\mathbf{F} = \dot{\mathbf{L}} \tag{12.5}$$

which expresses that *the resultant of the forces acting on a particle is equal to the rate of change of the linear momentum of the particle.*

Consistent systems of units

Equation (12.2) holds only if a consistent system of units is used. With SI units, the forces should be expressed in newtons, the masses in kilograms, and the accelerations in m/s^2; with U.S. customary units, the forces should be expressed in pounds, the masses in $lb \cdot s^2/ft$ (also referred to as *slugs*), and the accelerations in ft/s^2 [Sec. 12.4].

Equations of motion for a particle

To solve a problem involving the motion of a particle, Eq. (12.2) should be replaced by equations containing scalar quantities [Sec. 12.5]. Using *rectangular components* of $\mathbf{F}$ and $\mathbf{a}$, we wrote

$$\Sigma F_x = ma_x \qquad \Sigma F_y = ma_y \qquad \Sigma F_z = ma_z \tag{12.8}$$

Using *tangential and normal components,* we had

$$\Sigma F_t = m\frac{dv}{dt} \qquad \Sigma F_n = m\frac{v^2}{\rho} \tag{12.9'}$$

Dynamic equilibrium

We also noted [Sec. 12.6] that the equations of motion of a particle can be replaced by equations similar to the equilibrium equations used in statics if a vector $-m\mathbf{a}$ of magnitude ma but of sense opposite to that of the acceleration is added to the forces applied to the particle; the particle is then said to be in *dynamic equilibrium.* For the sake of uniformity, however, all the Sample Problems were solved by using the equations of motion, first with rectangular components [Sample Probs. 12.1 through 12.4], then with tangential and normal components [Sample Probs. 12.5 and 12.6].

Angular momentum

In the second part of the chapter, we defined the *angular momentum* $\mathbf{H}_O$ of a particle about a point O as the moment about O of the linear momentum $m\mathbf{v}$ of that particle [Sec. 12.7]. We wrote

$$\mathbf{H}_O = \mathbf{r} \times m\mathbf{v} \qquad (12.12)$$

and noted that $\mathbf{H}_O$ is a vector perpendicular to the plane containing $\mathbf{r}$ and $m\mathbf{v}$ (Fig. 12.24) and of magnitude

$$H_O = rmv \sin \phi \qquad (12.13)$$

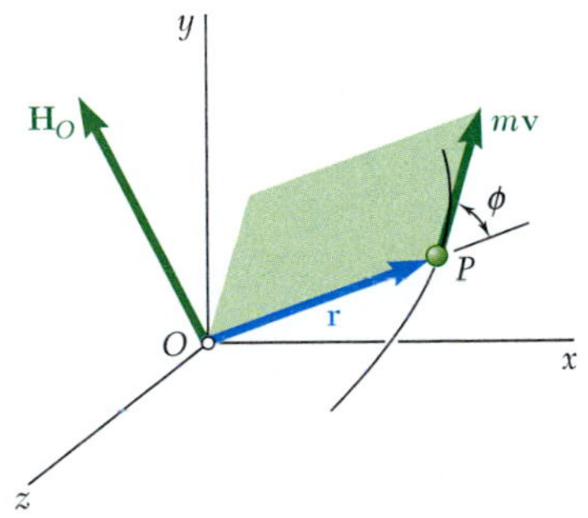

Fig. 12.24

Resolving the vectors $\mathbf{r}$ and $m\mathbf{v}$ into rectangular components, we expressed the angular momentum $\mathbf{H}_O$ in the determinant form

$$\mathbf{H}_O = \begin{vmatrix} \mathbf{i} & \mathbf{j} & \mathbf{k} \\ x & y & z \\ mv_x & mv_y & mv_z \end{vmatrix} \qquad (12.14)$$

In the case of a particle moving in the xy plane, we have $z = v_z = 0$. The angular momentum is perpendicular to the xy plane and is completely defined by its magnitude. We wrote

$$H_O = H_z = m(xv_y - yv_x) \qquad (12.16)$$

Rate of change of angular momentum

Computing the rate of change $\dot{\mathbf{H}}_O$ of the angular momentum $\mathbf{H}_O$, and applying Newton's second law, we wrote the equation

$$\Sigma \mathbf{M}_O = \dot{\mathbf{H}}_O \qquad (12.19)$$

which states that *the sum of the moments about O of the forces acting on a particle is equal to the rate of change of the angular momentum of the particle about O.*

Radial and transverse components

In many problems involving the plane motion of a particle, it is found convenient to use *radial and transverse components* [Sec. 12.8, Sample Prob. 12.7] and to write the equations

$$\Sigma F_r = m(\ddot{r} - r\dot{\theta}^2) \qquad (12.21)$$
$$\Sigma F_\theta = m(r\ddot{\theta} + 2\dot{r}\dot{\theta}) \qquad (12.22)$$

Motion under a central force

When the only force acting on a particle P is a force $\mathbf{F}$ directed toward or away from a fixed point O, the particle is said to be moving *under a central force* [Sec. 12.9]. Since $\Sigma \mathbf{M}_O = 0$ at any given instant, it follows from Eq. (12.19) that $\dot{\mathbf{H}}_O = 0$ for all values of t and, thus, that

$$\mathbf{H}_O = \text{constant} \qquad (12.23)$$

We concluded that *the angular momentum of a particle moving under a central force is constant, both in magnitude and direction,* and that the particle moves in a plane perpendicular to the vector $\mathbf{H}_O$.

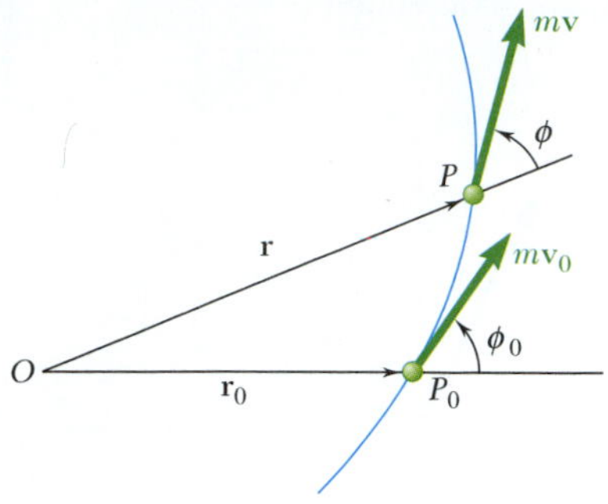

Fig. 12.25

Recalling Eq. (12.13), we wrote the relation

$$rmv \sin \phi = r_0 m v_0 \sin \phi_0 \tag{12.25}$$

for the motion of any particle under a central force (Fig. 12.25). Using polar coordinates and recalling Eq. (12.18), we also had

$$r^2\dot{\theta} = h \tag{12.27}$$

where h is a constant representing the angular momentum per unit mass, H_O/m, of the particle. We observed (Fig. 12.26) that the infinitesimal area dA swept by the radius vector OP as it rotates through $d\theta$ is equal to $\frac{1}{2}r^2\, d\theta$ and, thus, that the left-hand member of Eq. (12.27) represents twice the *areal velocity dA/dt* of the particle. Therefore, *the areal velocity of a particle moving under a central force is constant.*

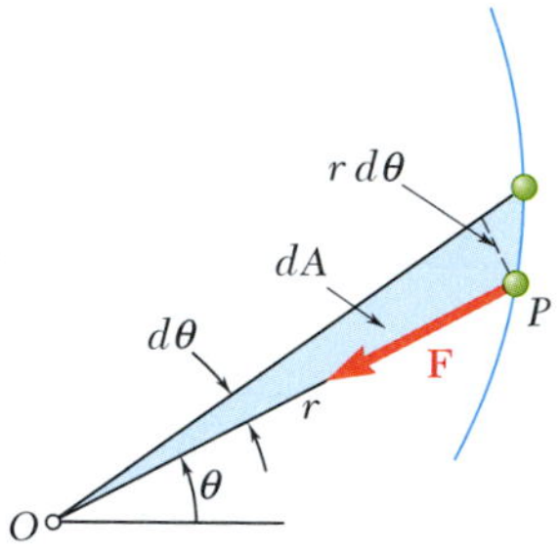

Fig. 12.26

Newton's law of universal gravitation

An important application of the motion under a central force is provided by the orbital motion of bodies under gravitational attraction [Sec. 12.10]. According to *Newton's law of universal gravitation,* two particles at a distance r from each other and of masses M and m, respectively, attract each other with equal and opposite forces $\mathbf{F}$ and $-\mathbf{F}$ directed along the line joining the particles (Fig. 12.27). The common magnitude F of the two forces is

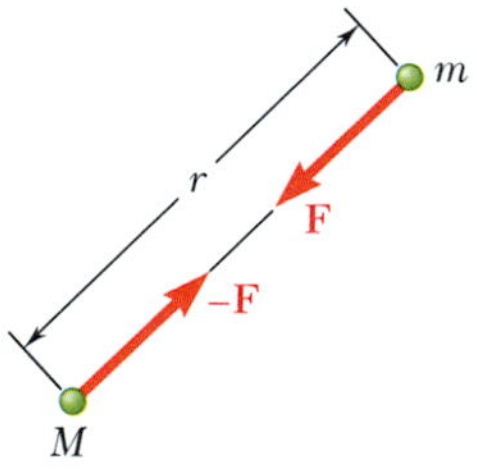

Fig. 12.27

$$F = G\frac{Mm}{r^2} \tag{12.28}$$

where G is the *constant of gravitation.* In the case of a body of mass m subjected to the gravitational attraction of the earth, the product GM, where M is the mass of the earth, can be expressed as

$$GM = gR^2 \tag{12.30}$$

where $g = 9.81$ m/s^2 = 32.2 ft/s^2 and R is the radius of the earth.

Orbital motion

It was shown in Sec. 12.11 that a particle moving under a central force describes a trajectory defined by the differential equation

$$\frac{d^2u}{d\theta^2} + u = \frac{F}{mh^2u^2} \tag{12.37}$$

where $F > 0$ corresponds to an attractive force and $u = 1/r$. In the case of a particle moving under a force of gravitational attraction [Sec. 12.12], we substituted for F the expression given in Eq. (12.28). Measuring θ from the axis OA joining the focus O to the point A of the trajectory closest to O (Fig. 12.28), we found that the solution to Eq. (12.37) was

$$\frac{1}{r} = u = \frac{GM}{h^2} + C \cos \theta \qquad (12.39)$$

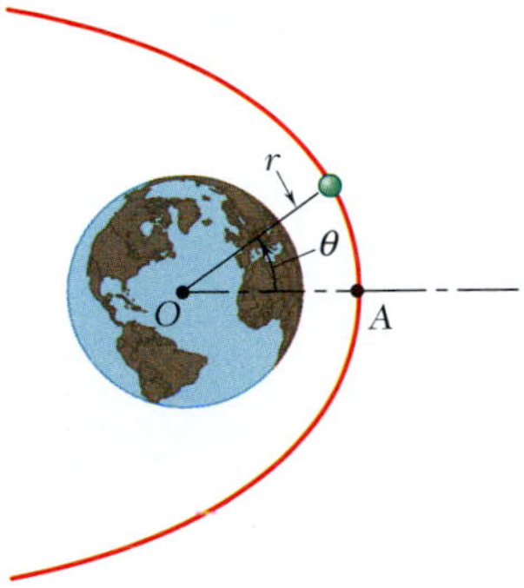

Fig. 12.28

This is the equation of a conic of eccentricity $\varepsilon = Ch^2/GM$. The conic is an *ellipse* if $\varepsilon < 1$, a *parabola* if $\varepsilon = 1$, and a hyperbola if $\varepsilon > 1$. The constants C and h can be determined from the initial conditions; if the particle is projected from point A ($\theta = 0$, $r = r_0$) with an initial velocity $\mathbf{v}_0$ perpendicular to OA, we have $h = r_0 v_0$ [Sample Prob. 12.9].

Escape velocity

It was also shown that the values of the initial velocity corresponding, respectively, to a parabolic and a circular trajectory were

$$v_{esc} = \sqrt{\frac{2GM}{r_0}} \qquad (12.43)$$

$$v_{circ} = \sqrt{\frac{GM}{r_0}} \qquad (12.44)$$

and that the first of these values, called the *escape velocity*, is the smallest value of v_0 for which the particle will not return to its starting point.

Periodic time

The *periodic time* τ of a planet or satellite was defined as the time required by that body to describe its orbit. It was shown that

$$\tau = \frac{2\pi ab}{h} \qquad (12.45)$$

where $h = r_0 v_0$ and where a and b represent the semimajor and semiminor axes of the orbit. It was further shown that these semiaxes are respectively equal to the arithmetic and geometric means of the maximum and minimum values of the radius vector r.

Kepler's laws

The last section of the chapter [Sec. 12.13] presented *Kepler's laws of planetary motion* and showed that these empirical laws, obtained from early astronomical observations, confirm Newton's laws of motion as well as his law of gravitation.

Review Problems

12.122 The acceleration of a package sliding down section AB of incline ABC is 5 m/s^2. Assuming that the coefficient of kinetic friction is the same for each section, determine the acceleration of the package on section BC of the incline.

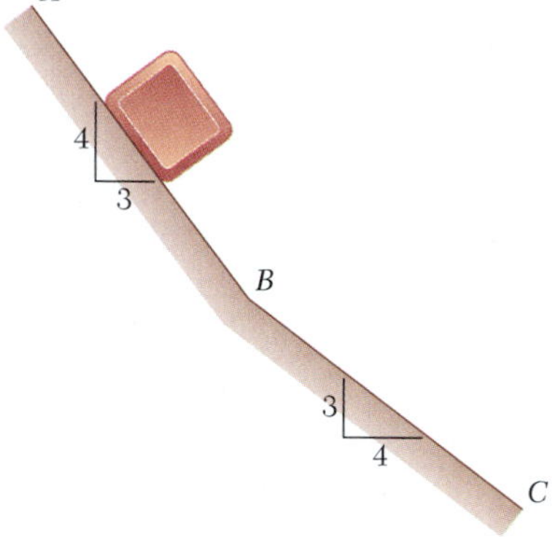

Fig. P12.122

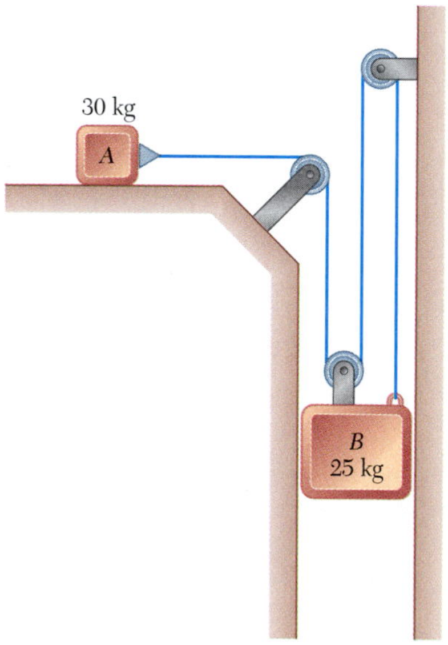

Fig. P12.123

12.123 The two blocks shown are originally at rest. Neglecting the masses of the pulleys and the effect of friction in the pulleys and assuming that the coefficients of friction between block A and the horizontal surface are $\mu_s = 0.25$ and $\mu_k = 0.20$, determine (*a*) the acceleration of each block, (*b*) the tension in the cable.

12.124 The coefficients of friction between package A and the incline are $\mu_s = 0.35$ and $\mu_k = 0.30$. Knowing that the system is initially at rest and that block B comes to rest on block C, determine (*a*) the maximum velocity reached by package A, (*b*) the distance up the incline through which package A will travel.

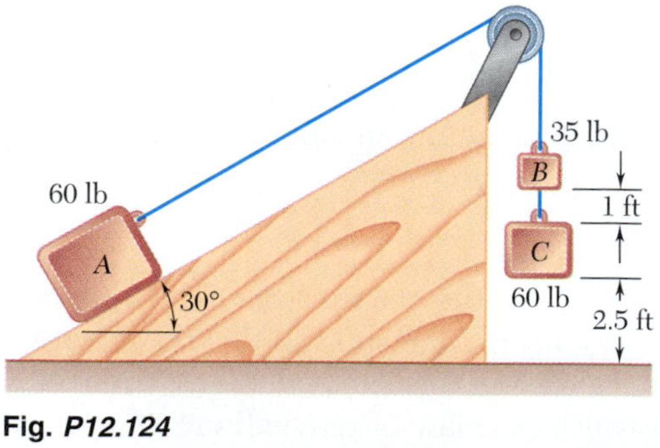

Fig. *P12.124*

12.125 The masses of blocks A, B, and C are $m_A = 4$ kg, $m_B = 10$ kg, and $m_C = 2$ kg. Knowing that $P = 0$ and neglecting the masses of the pulleys and the effect of friction, determine (*a*) the acceleration of each block, (*b*) the tension in the cord.

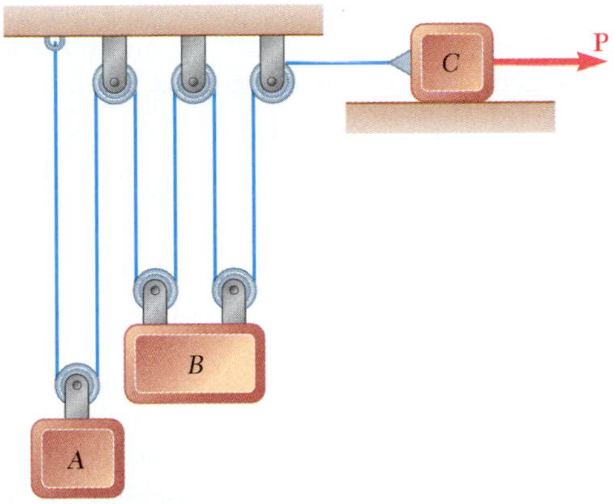

Fig. P12.125

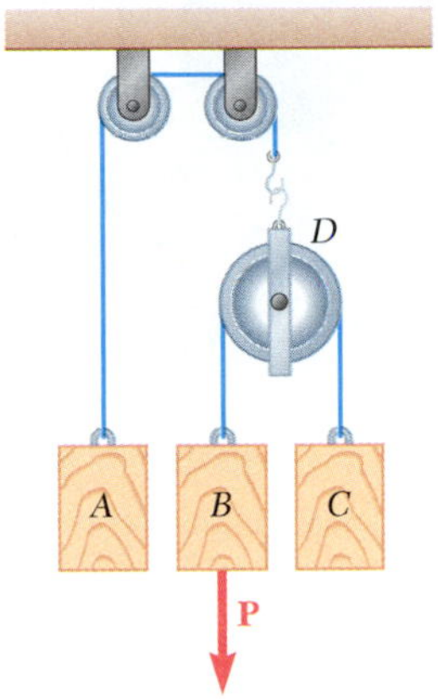

Fig. P12.126

12.126 Block A weighs 20 lb, and blocks B and C weigh 10 lb each. Knowing that the blocks are initially at rest and that B moves through 8 ft in 2 s, determine (*a*) the magnitude of the force **P**, (*b*) the tension in the cord AD. Neglect the masses of the pulleys and axle friction.

12.127 A 12-lb block B rests as shown on the upper surface of a 30-lb wedge A. Neglecting friction, determine immediately after the system is released from rest (*a*) the acceleration of A, (*b*) the acceleration of B relative to A.

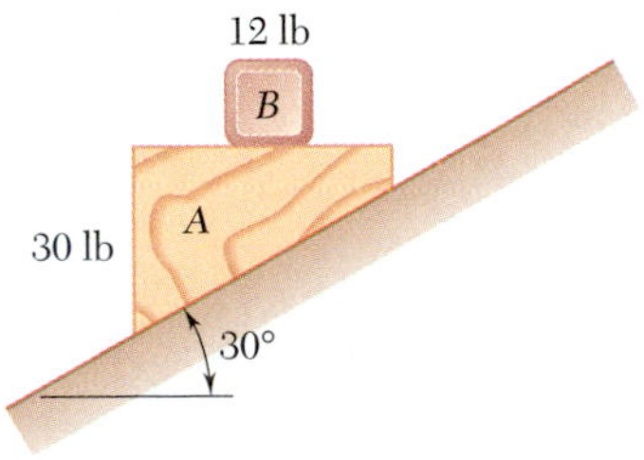

Fig. *P12.127*

12.128 The roller-coaster track shown is contained in a vertical plane. The portion of track between A and B is straight and horizontal, while the portions to the left of A and to the right of B have radii of curvature as indicated. A car is traveling at a speed of 72 km/h when the brakes are suddenly applied, causing the wheels of the car to slide on the track ($\mu_k = 0.25$). Determine the initial deceleration of the car if the brakes are applied as the car (*a*) has almost reached A, (*b*) is traveling between A and B, (*c*) has just passed B.

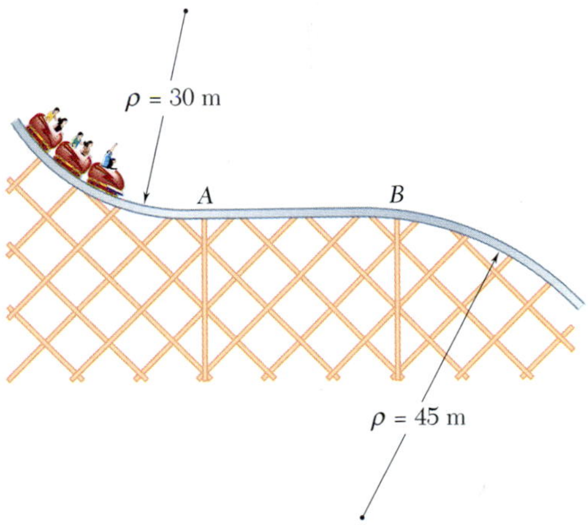

Fig. P12.128

12.129 A satellite is placed into a circular orbit about the planet Saturn at an altitude of 3400 km. The satellite describes its orbit with a velocity of 24.45 km/s. Knowing that the radius of the orbit about Saturn and the periodic time of Atlas, one of Saturn's moons, are 137.64×10^3 km and 0.6019 days, respectively, determine (*a*) the radius of Saturn, (*b*) the mass of Saturn. (The *periodic time* of a satellite is the time it requires to complete one full revolution about the planet.)

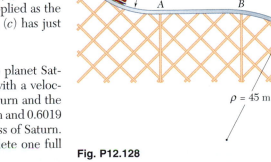

12.130 The periodic times (see Prob. 12.129) of the planet Uranus's moons Juliet and Titania have been observed to be 0.4931 days and 8.706 days, respectively. Knowing that the radius of Juliet's orbit is 40,000 mi, determine (*a*) the mass of Uranus, (*b*) the radius of Titania's orbit.

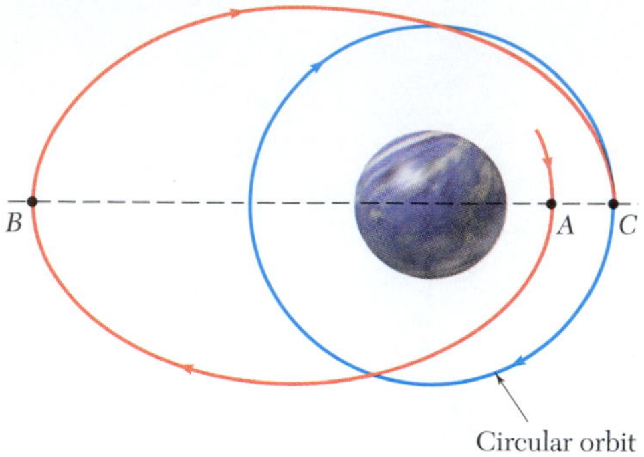

Fig. P12.131

12.131 A space probe is to be placed in a circular orbit of 6420-km radius about the planet Venus. As the probe approaches Venus, its speed is decreased so that, as it reaches point A, its speed and altitude above the surface of the planet are 7420 m/s and 288 km, respectively. The path of the probe from A to B is elliptic, and as the probe approaches B, its speed is increased by $\Delta v_B = 21.4$ m/s to insert it into the elliptic transfer orbit BC. Finally, as the probe passes through C, its speed is decreased by $\Delta v_C = -238$ m/s to insert it into the required circular orbit. Knowing that the mass and the radius of the planet Venus are 4.869×10^{24} kg and 6052 km, respectively, determine (*a*) the speed of the probe as it approaches B on the elliptic path, (*b*) its altitude above the surface of the planet at B.

12.132 To place a communications satellite into a geosynchronous orbit (see Prob. 12.79) at an altitude of 22,240 mi above the surface of the earth, the satellite first is released from a space shuttle, which is in a circular orbit at an altitude of 185 mi, and then is propelled by an upper-stage booster to its final altitude. As the satellite passes through A, the booster's motor is fired to insert the satellite into an elliptic transfer orbit. The booster is again fired at B to insert the satellite into a geosynchronous orbit. Knowing that the second firing increases the speed of the satellite by 4810 ft/s, determine (*a*) the speed of the satellite as it approaches B on the elliptic transfer orbit, (*b*) the increase in speed resulting from the first firing at A.

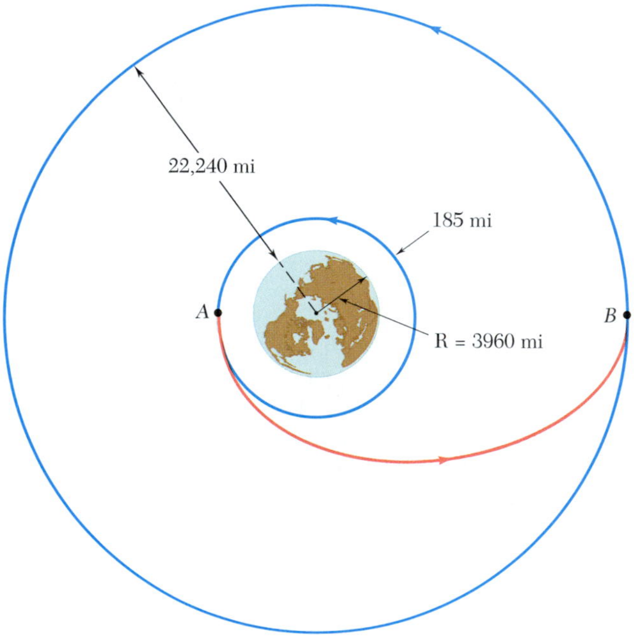

Fig. P12.132

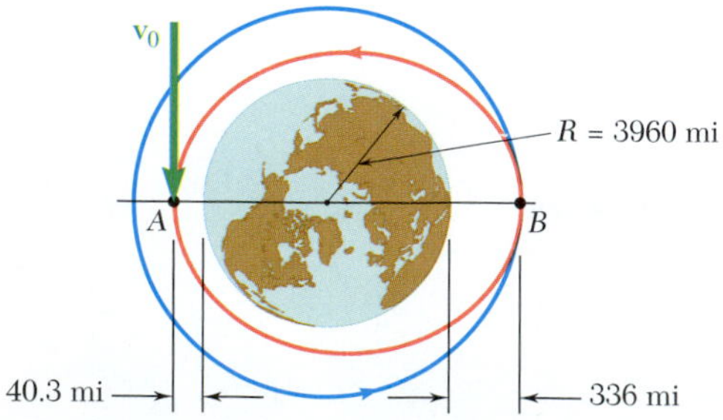

Fig. *P12.133*

12.133 At main engine cutoff of its thirteenth flight, the space shuttle Discovery was in an elliptic orbit of minimum altitude 40.3 mi and maximum altitude 336 mi above the surface of the earth. Knowing that at point A the shuttle had a velocity $\mathbf{v}_0$ parallel to the surface of the earth and that the shuttle was transferred to a circular orbit as it passed through point B, determine (*a*) the speed v_0 of the shuttle at A, (*b*) the increase in speed required at B to insert the shuttle into the circular orbit.

Computer Problems

12.C1 Block B of weight 18 lb is initially at rest as shown on the upper surface of a 45-lb wedge A which is supported by a horizontal surface. A 4-lb block C is connected to block B by a cord, which passes over a pulley of negligible mass. Using computational software and denoting by μ the coefficient of friction at all surfaces, calculate the initial acceleration of the wedge and the initial acceleration of block B relative to the wedge for values of $\mu \geq 0$. Use 0.01 increments for μ until the wedge does not move and then use 0.1 increments until no motion occurs.

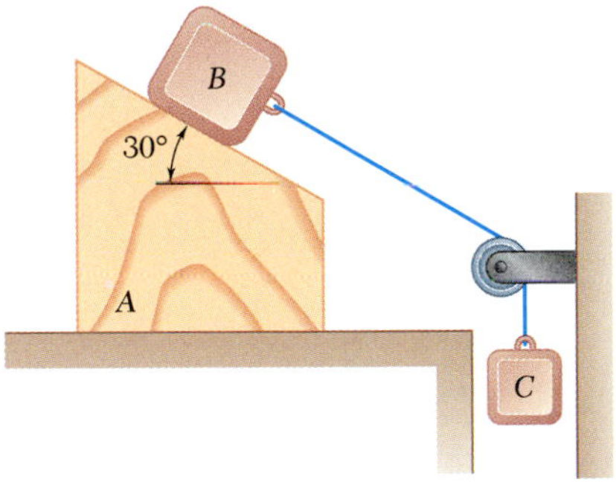

Fig. P12.C1

12.C2 A small 0.50-kg block is at rest at the top of a cylindrical surface. The block is given an initial velocity $\mathbf{v}_0$ to the right of magnitude 3 m/s, which causes it to slide on the cylindrical surface. Using computational software calculate and plot the values of θ at which the block leaves the surface for values of μ_k, the coefficient of kinetic friction between the block and the surface, from 0 to 0.4 using 0.05 increments.

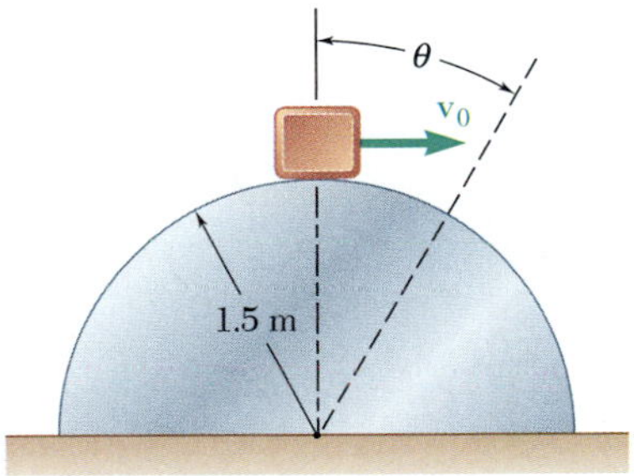

Fig. P12.C2

12.C3 A block of mass m is attached to a spring of constant k. The block is released from rest when the spring is in a horizontal and undeformed position. Knowing that $r_0 = 4$ ft, use computational software to determine (*a*) for $k/m = 15\ \text{s}^{-2}$, 20s^{-2}, and $25\ \text{s}^{-2}$, the length of the spring and the magnitude and direction of the velocity of the block as the block passes directly under the point of suspension of the spring, (*b*) the value of k/m for which that velocity is horizontal.

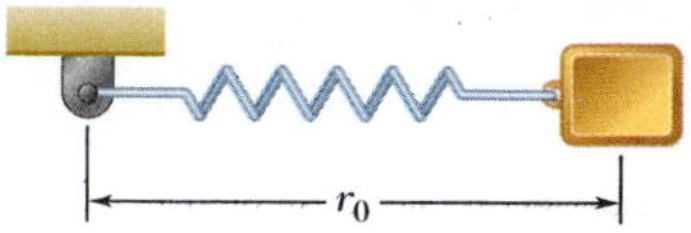

Fig. P12.C3

12.C4 An airplane has a weight of 60,000 lb and its engines develop a constant thrust of 12,500 lb during take-off. The drag **D** exerted on the airplane has a magnitude $D = 0.060v^2$, where D is expressed in lb and v is the speed in ft/s. The airplane starts from rest at the end of the runway and becomes airborne at a speed of 250 ft/s. Determine and plot the position and speed of the airplane as functions of time and the speed as a function of position as the airplane moves down the runway.

12.C5 Two wires AC and BC are tied at point C to a small sphere which revolves at a constant speed v in the horizontal circle shown. Calculate and plot the tensions in the wires as functions of v. Determine the range of values of v for which both wires remain taut.

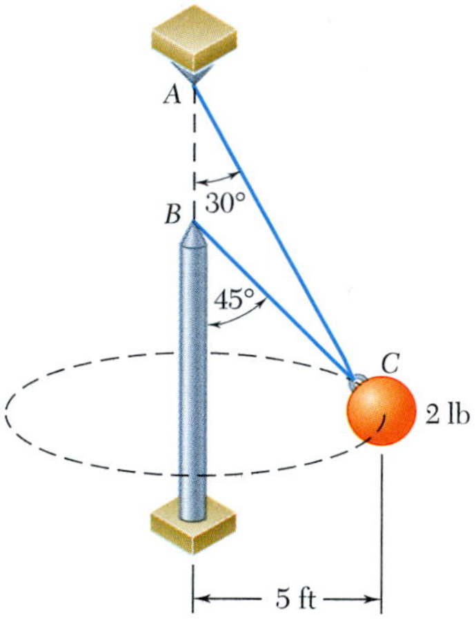

Fig. P12.C5

12.C6 A 10-lb bag is gently pushed off the top of a wall at point A and swings in a vertical plane at the end of a rope of length $l = 5$ ft. Calculate and plot the speed of the bag and the magnitude of the tension in the rope as functions of the angle θ from 0 to 90°.

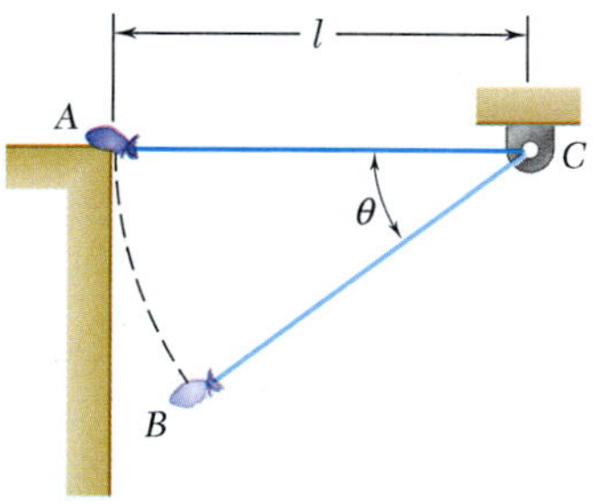

Fig. P12.C6

12.C7 The two-dimensional motion of particle B is defined by the relations $r = t^3 - 2t^2$ and $\theta = t^3 - 4t$, where r is expressed in m, t in seconds, and θ in radians. Knowing that the particle has a mass of 0.25 kg and moves in a horizontal plane, calculate and plot the radial and transverse components and the magnitude of the force acting on the particle as functions of t from 0 to 1.5 s.

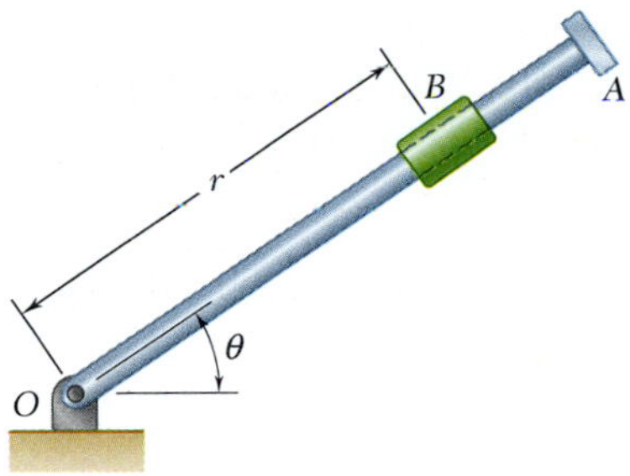

Fig. P12.C7

4250/-

C H A P T E R

Kinetics of Particles: Energy and Momentum Methods

13

As the head of the golf club hits the ball, some of its energy and momentum are transferred to the ball. Considering the energy and/or momentum of a particle is often an effective way of studying its motion.

KINETICS OF PARTICLES: ENERGY AND MOMENTUM METHODS

(4250)(500)(sin 6)

F d

100 d = 13.

13.1. INTRODUCTION

In the preceding chapter, most problems dealing with the motion of particles were solved through the use of the fundamental equation of motion $\mathbf{F} = m\mathbf{a}$. Given a particle acted upon by a force $\mathbf{F}$, we could solve this equation for the acceleration $\mathbf{a}$; then, by applying the principles of kinematics, we could determine from $\mathbf{a}$ the velocity and position of the particle at any time.

Using the equation $\mathbf{F} = m\mathbf{a}$ together with the principles of kinematics allows us to obtain two additional methods of analysis, the *method of work and energy* and the *method of impulse and momentum*. The advantage of these methods lies in the fact that they make the determination of the acceleration unnecessary. Indeed, the method of work and energy directly relates force, mass, velocity, and displacement, while the method of impulse and momentum relates force, mass, velocity, and time.

The method of work and energy will be considered first. In Secs. 13.2 through 13.4, the *work of a force* and the *kinetic energy of a particle* are discussed and the principle of work and energy is applied to the solution of engineering problems. The concepts of *power* and *efficiency* of a machine are introduced in Sec. 13.5.

Sections 13.6 through 13.8 are devoted to the concept of *potential energy* of a conservative force and to the application of the principle of conservation of energy to various problems of practical interest. In Sec. 13.9, the principles of conservation of energy and of conservation of angular momentum are used jointly to solve problems of space mechanics.

The second part of the chapter is devoted to the *principle of impulse and momentum* and to its application to the study of the motion of a particle. As you will see in Sec. 13.11, this principle is particularly effective in the study of the *impulsive motion* of a particle, where very large forces are applied for a very short time interval.

In Secs. 13.12 through 13.14, the *central impact* of two bodies will be considered. It will be shown that a certain relation exists between the relative velocities of the two colliding bodies before and after impact. This relation, together with the fact that the total momentum of the two bodies is conserved, can be used to solve a number of problems of practical interest.

Finally, in Sec. 13.15, you will learn to select from the three fundamental methods presented in Chaps. 12 and 13 the method best suited for the solution of a given problem. You will also see how the principle of conservation of energy and the method of impulse and momentum can be combined to solve problems involving only conservative forces, except for a short impact phase during which impulsive forces must also be taken into consideration.

13.2. WORK OF A FORCE

We will first define the terms *displacement* and *work* as they are used in mechanics.† Consider a particle which moves from a point A to a

†The definition of work was given in Sec. 10.2, and the basic properties of the work of a force were outlined in Secs. 10.2 and 10.6. For convenience, we repeat here the portions of this material which relate to the kinetics of particles.

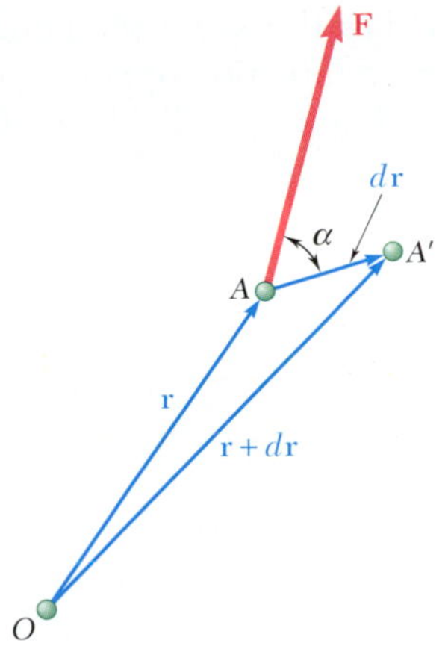

Fig. 13.1

neighboring point A' (Fig. 13.1). If $\mathbf{r}$ denotes the position vector corresponding to point A, the small vector joining A and A' can be denoted by the differential $d\mathbf{r}$; the vector $d\mathbf{r}$ is called the *displacement* of the particle. Now, let us assume that a force $\mathbf{F}$ is acting on the particle. The *work of the force* $\mathbf{F}$ *corresponding to the displacement* $d\mathbf{r}$ is defined as the quantity

$$dU = \mathbf{F} \cdot d\mathbf{r} \tag{13.1}$$

obtained by forming the scalar product of the force $\mathbf{F}$ and the displacement $d\mathbf{r}$. Denoting by F and ds respectively, the magnitudes of the force and of the displacement, and by α the angle formed by $\mathbf{F}$ and $d\mathbf{r}$, and recalling the definition of the scalar product of two vectors (Sec. 3.9), we write

$$dU = F\,ds\cos\alpha \tag{13.1$'$}$$

Using formula (3.30), we can also express the work dU in terms of the rectangular components of the force and of the displacement:

$$dU = F_x dx + F_y dy + F_z dz \tag{13.1$''$}$$

Being a *scalar quantity,* work has a magnitude and a sign but no direction. We also note that work should be expressed in units obtained by multiplying units of length by units of force. Thus, if U.S. customary units are used, work should be expressed in ft · lb or in · lb. If SI units are used, work should be expressed in N · m. The unit of work N · m is called a *joule* (J).† Recalling the conversion factors indicated in Sec. 12.4, we write

$$1 \text{ ft} \cdot \text{lb} = (1 \text{ ft})(1 \text{ lb}) = (0.3048 \text{ m})(4.448 \text{ N}) = 1.356 \text{ J}$$

It follows from (13.1′) that the work dU is positive if the angle α is acute and negative if α is obtuse. Three particular cases are of special

†The joule (J) is the SI unit of *energy,* whether in mechanical form (work, potential energy, kinetic energy) or in chemical, electrical, or thermal form. We should note that even though N · m = J, the moment of a force must be expressed in N · m and not in joules, since the moment of a force is not a form of energy.

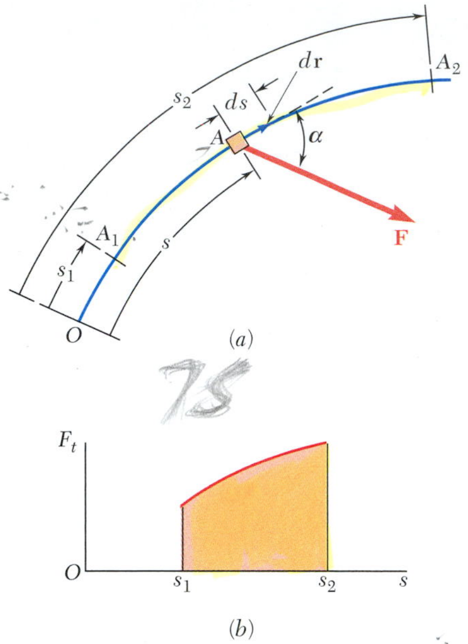

Fig. 13.2

interest. If the force $\mathbf{F}$ has the same direction as $d\mathbf{r}$, the work dU reduces to $F\,ds$. If $\mathbf{F}$ has a direction opposite to that of $d\mathbf{r}$, the work is $dU = -F\,ds$. Finally, if $\mathbf{F}$ is perpendicular to $d\mathbf{r}$, the work dU is zero.

The work of $\mathbf{F}$ during a *finite* displacement of the particle from A_1 to A_2 (Fig. 13.2*a*) is obtained by integrating Eq. (13.1) along the path described by the particle. This work, denoted by $U_{1\to 2}$, is

$$U_{1\to 2} = \int_{A_1}^{A_2} \mathbf{F} \cdot d\mathbf{r} \tag{13.2}$$

Using the alternative expression (13.1′) for the elementary work dU, and observing that $F \cos \alpha$ represents the tangential component F_t of the force, we can also express the work $U_{1\to 2}$ as

$$U_{1\to 2} = \int_{s_1}^{s_2} (F \cos \alpha)\, ds = \int_{s_1}^{s_2} F_t\, ds \tag{13.2′}$$

where the variable of integration s measures the distance traveled by the particle along the path. The work $U_{1\to 2}$ is represented by the area under the curve obtained by plotting $F_t = F \cos \alpha$ against s (Fig. 13.2*b*).

When the force $\mathbf{F}$ is defined by its rectangular components, the expression (13.1″) can be used for the elementary work. We then write

$$U_{1\to 2} = \int_{A_1}^{A_2} (F_x\, dx + F_y\, dy + F_z\, dz) \tag{13.2″}$$

where the integration is to be performed along the path described by the particle.

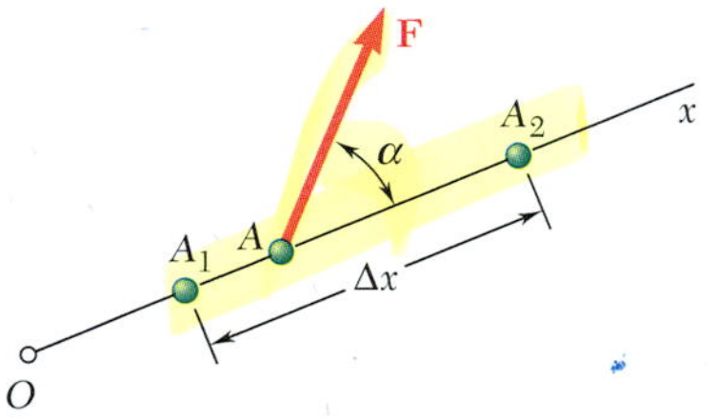

Fig. 13.3

Work of a Constant Force in Rectilinear Motion. When a particle moving in a straight line is acted upon by a force $\mathbf{F}$ of constant magnitude and of constant direction (Fig. 13.3), formula (13.2′) yields

$$U_{1\to 2} = (F \cos \alpha)\, \Delta x \tag{13.3}$$

where α = angle the force forms with direction of motion
Δx = displacement from A_1 to A_2

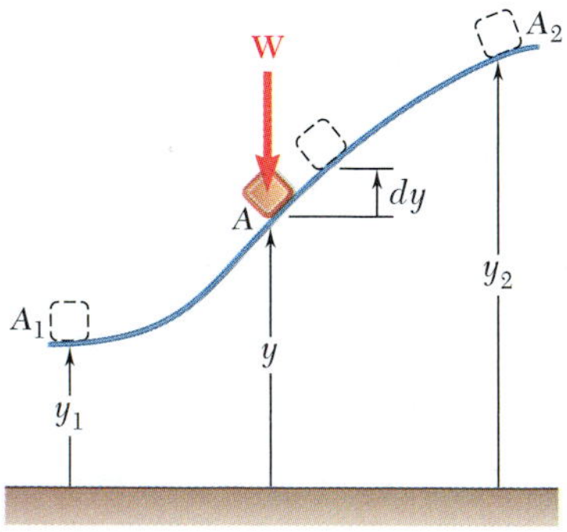

Fig. 13.4

Work of the Force of Gravity. The work of the weight $\mathbf{W}$ of a body, that is, of the force of gravity exerted on that body, is obtained by substituting the components of $\mathbf{W}$ into (13.1″) and (13.2″). With the y axis chosen upward (Fig. 13.4), we have $F_x = 0$, $F_y = -W$, and $F_z = 0$, and we write

$$dU = -W\, dy$$

$$U_{1\to 2} = -\int_{y_1}^{y_2} W\, dy = Wy_1 - Wy_2 \tag{13.4}$$

or

$$U_{1\to 2} = -W(y_2 - y_1) = -W\, \Delta y \tag{13.4′}$$

where Δy is the vertical displacement from A_1 to A_2. The work of the weight **W** is thus equal to *the product of W and the vertical displacement of the center of gravity of the body*. The work is *positive* when $\Delta y < 0$, that is, *when the body moves down*.

Work of the Force Exerted by a Spring. Consider a body A attached to a fixed point B by a spring; it is assumed that the spring is undeformed when the body is at A_0 (Fig. 13.5*a*). Experimental evidence shows that the magnitude of the force **F** exerted by the spring on body A is proportional to the deflection x of the spring measured from the position A_0. We have

$$F = kx \tag{13.5}$$

where k is the *spring constant*, expressed in N/m or kN/m if SI units are used and in lb/ft or lb/in. if U.S. customary units are used.†

The work of the force **F** exerted by the spring during a finite displacement of the body from $A_1(x = x_1)$ to $A_2(x = x_2)$ is obtained by writing

$$dU = -F\,dx = -kx\,dx$$

$$U_{1\rightarrow 2} = -\int_{x_1}^{x_2} kx\,dx = \tfrac{1}{2}kx_1^2 - \tfrac{1}{2}kx_2^2 \tag{13.6}$$

Care should be taken to express k and x in consistent units. For example, if U.S. customary units are used, k should be expressed in lb/ft and x in feet, or k in lb/in. and x in inches; in the first case, the work is obtained in ft · lb, in the second case, in in · lb. We note that the work of the force **F** exerted by the spring on the body is *positive* when $x_2 < x_1$, that is, *when the spring is returning to its undeformed position*.

Since Eq. (13.5) is the equation of a straight line of slope k passing through the origin, the work $U_{1\rightarrow 2}$ of **F** during the displacement from A_1 to A_2 can be obtained by evaluating the area of the trapezoid shown in Fig. 13.5*b*. This is done by computing F_1 and F_2 and multiplying the base Δx of the trapezoid by its mean height $\frac{1}{2}(F_1 + F_2)$. Since the work of the force **F** exerted by the spring is positive for a negative value of Δx, we write

$$U_{1\rightarrow 2} = -\tfrac{1}{2}(F_1 + F_2)\,\Delta x \tag{13.6'}$$

Formula (13.6′) is usually more convenient to use than (13.6) and affords fewer chances of confusing the units involved.

Work of a Gravitational Force. We saw in Sec. 12.10 that two particles of mass M and m at a distance r from each other attract each other with equal and opposite forces **F** and $-$**F**, directed along the line joining the particles and of magnitude

$$F = G\frac{Mm}{r^2}$$

†The relation $F = kx$ is correct under static conditions only. Under dynamic conditions, formula (13.5) should be modified to take the inertia of the spring into account. However, the error introduced by using the relation $F = kx$ in the solution of kinetics problems is small if the mass of the spring is small compared with the other masses in motion.

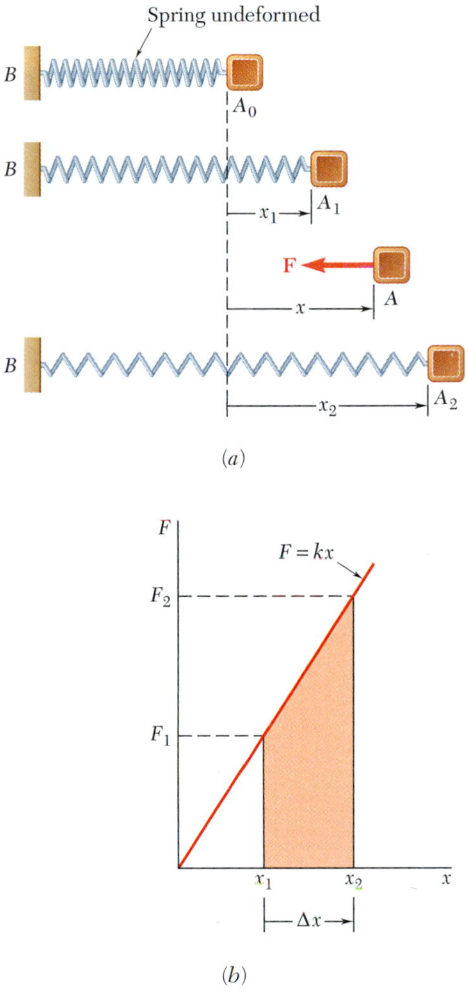

Fig. 13.5

Let us assume that the particle M occupies a fixed position O while the particle m moves along the path shown in Fig. 13.6. The work of the force $\mathbf{F}$ exerted on the particle m during an infinitesimal displacement of the particle from A to A' can be obtained by multiplying the magnitude F of the force by the radial component dr of the displacement. Since $\mathbf{F}$ is directed toward O, the work is negative and we write

$$dU = -F\,dr = -G\frac{Mm}{r^2}\,dr$$

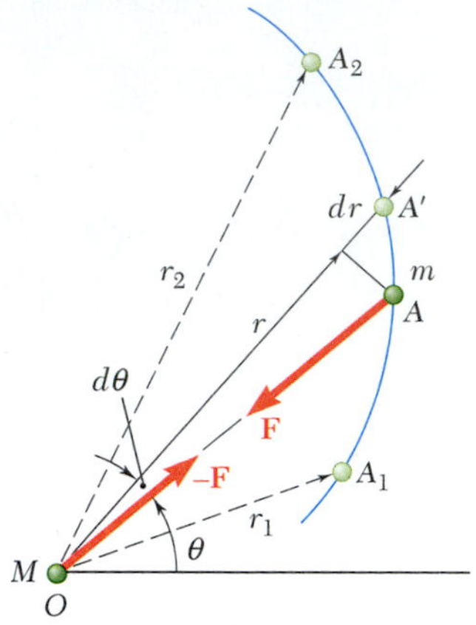

Fig. 13.6

The work of the gravitational force $\mathbf{F}$ during a finite displacement from $A_1(r = r_1)$ to $A_2(r = r_2)$ is therefore

$$U_{1\rightarrow 2} = -\int_{r_1}^{r_2} \frac{GMm}{r^2}\,dr = \frac{GMm}{r_2} - \frac{GMm}{r_1} \qquad (13.7)$$

where M is the mass of the earth. This formula can be used to determine the work of the force exerted by the earth on a body of mass m at a distance r from the center of the earth, when r is larger than the radius R of the earth. Recalling the first of the relations (12.29), we can replace the product GMm in Eq. (13.7) by WR^2, where R is the radius of the earth ($R = 6.37 \times 10^6$ m or 3960 mi) and W is the weight of the body at the surface of the earth.

A number of forces frequently encountered in problems of kinetics *do no work.* They are forces applied to fixed points ($ds = 0$) or acting in a direction perpendicular to the displacement ($\cos\alpha = 0$). Among the forces which do no work are the following: the reaction at a frictionless pin when the body supported rotates about the pin, the reaction at a frictionless surface when the body in contact moves along the surface, the reaction at a roller moving along its track, and the weight of a body when its center of gravity moves horizontally.

13.3. KINETIC ENERGY OF A PARTICLE. PRINCIPLE OF WORK AND ENERGY

Consider a particle of mass m acted upon by a force $\mathbf{F}$ and moving along a path which is either rectilinear or curved (Fig. 13.7). Expressing Newton's second law in terms of the tangential components of the force and of the acceleration (see Sec. 12.5), we write

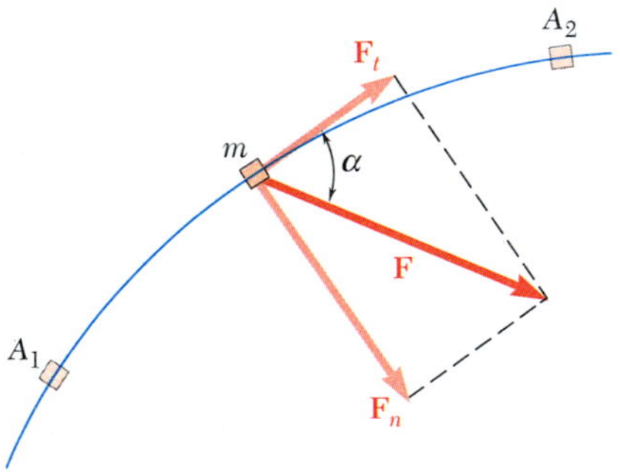

Fig. 13.7

$$F_t = ma_t \qquad \text{or} \qquad F_t = m\frac{dv}{dt}$$

where v is the speed of the particle. Recalling from Sec. 11.9 that $v = ds/dt$, we obtain

$$F_t = m\frac{dv}{ds}\frac{ds}{dt} = mv\frac{dv}{ds}$$

$$F_t\,ds = mv\,dv$$

Integrating from A_1, where $s = s_1$ and $v = v_1$, to A_2, where $s = s_2$ and $v = v_2$, we write

$$\int_{s_1}^{s_2} F_t\,ds = m\int_{v_1}^{v_2} v\,dv = \tfrac{1}{2}mv_2^2 - \tfrac{1}{2}mv_1^2 \qquad (13.8)$$

The left-hand member of Eq. (13.8) represents the work $U_{1\rightarrow 2}$ of the force $\mathbf{F}$ exerted on the particle during the displacement from A_1 to

A_2; as indicated in Sec. 13.2, the work $U_{1\rightarrow2}$ is a scalar quantity. The expression $\frac{1}{2}mv^2$ is also a scalar quantity; it is defined as the kinetic energy of the particle and is denoted by T. We write

$$T = \tfrac{1}{2}mv^2 \tag{13.9}$$

Substituting into (13.8), we have

$$U_{1\rightarrow2} = T_2 - T_1 \tag{13.10}$$

which expresses that, when a particle moves from A_1 to A_2 under the action of a force **F**, *the work of the force* **F** *is equal to the change in kinetic energy of the particle.* This is known as the *principle of work and energy*. Rearranging the terms in (13.10), we write

$$T_1 + U_{1\rightarrow2} = T_2 \tag{13.11}$$

Thus, *the kinetic energy of the particle at A_2 can be obtained by adding to its kinetic energy at A_1 the work done during the displacement from A_1 to A_2 by the force* **F** *exerted on the particle.* Like Newton's second law from which it is derived, the principle of work and energy applies only with respect to a newtonian frame of reference (Sec. 12.2). The speed v used to determine the kinetic energy T should therefore be measured with respect to a newtonian frame of reference.

Since both work and kinetic energy are scalar quantities, their sum can be computed as an ordinary algebraic sum, the work $U_{1\rightarrow2}$ being considered as positive or negative according to the direction of **F**. When several forces act on the particle, the expression $U_{1\rightarrow2}$ represents the total work of the forces acting on the particle; it is obtained by adding algebraically the work of the various forces.

As noted above, the kinetic energy of a particle is a scalar quantity. It further appears from the definition $T = \frac{1}{2}mv^2$ that regardless of the direction of motion of the particle the kinetic energy is always positive. Considering the particular case when $v_1 = 0$ and $v_2 = v$, and substituting $T_1 = 0$ and $T_2 = T$ into (13.10), we observe that the work done by the forces acting on the particle is equal to T. Thus, the kinetic energy of a particle moving with a speed v represents the work which must be done to bring the particle from rest to the speed v. Substituting $T_1 = T$ and $T_2 = 0$ into (13.10), we also note that when a particle moving with a speed v is brought to rest, the work done by the forces acting on the particle is $-T$. Assuming that no energy is dissipated into heat, we conclude that the work done by the forces exerted *by the particle* on the bodies which cause it to come to rest is equal to T. Thus, the kinetic energy of a particle also represents *the capacity to do work associated with the speed of the particle*.

The kinetic energy is measured in the same units as work, that is, in joules if SI units are used and in ft · lb if U.S. customary units are used. We check that, in SI units,

$$T = \tfrac{1}{2}mv^2 = \text{kg}(\text{m/s})^2 = (\text{kg} \cdot \text{m/s}^2)\text{m} = \text{N} \cdot \text{m} = \text{J}$$

while, in customary units,

$$T = \tfrac{1}{2}mv^2 = (\text{lb} \cdot \text{s}^2/\text{ft})(\text{ft/s})^2 = \text{ft} \cdot \text{lb}$$

13.4. APPLICATIONS OF THE PRINCIPLE OF WORK AND ENERGY

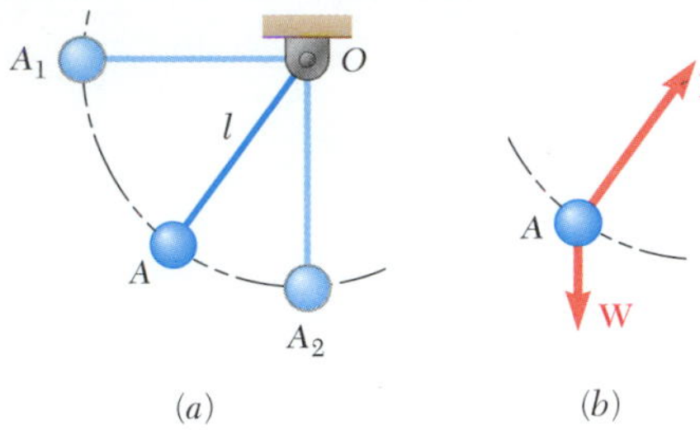

Fig. 13.8

The application of the principle of work and energy greatly simplifies the solution of many problems involving forces, displacements, and velocities. Consider, for example, the pendulum OA consisting of a bob A of weight W attached to a cord of length l (Fig. 13.8*a*). The pendulum is released with no initial velocity from a horizontal position OA_1 and allowed to swing in a vertical plane. We wish to determine the speed of the bob as it passes through A_2, directly under O.

We first determine the work done during the displacement from A_1 to A_2 by the forces acting on the bob. We draw a free-body diagram of the bob, showing all the *actual* forces acting on it, that is, the weight **W** and the force **P** exerted by the cord (Fig. 13.8*b*). (An inertia vector is not an actual force and *should not* be included in the free-body diagram.) We note that the force **P** does no work, since it is normal to the path; the only force which does work is thus the weight **W**. The work of **W** is obtained by multiplying its magnitude W by the vertical displacement l (Sec. 13.2); since the displacement is downward, the work is positive. We therefore write $U_{1\rightarrow 2} = Wl$.

Now considering the kinetic energy of the bob, we find $T_1 = 0$ at A_1 and $T_2 = \frac{1}{2}(W/g)v_2^2$ at A_2. We can now apply the principle of work and energy; recalling formula (13.11), we write:

$$T_1 + U_{1\rightarrow 2} = T_2 \qquad 0 + Wl = \frac{1}{2}\frac{W}{g}v_2^2$$

Solving for v_2, we find $v_2 = \sqrt{2gl}$. We note that the speed obtained is that of a body falling freely from a height l.

The example we have considered illustrates the following advantages of the method of work and energy:

1. In order to find the speed at A_2, there is no need to determine the acceleration in an intermediate position A and to integrate the expression obtained from A_1 to A_2.
2. All quantities involved are scalars and can be added directly, without using x and y components.
3. Forces which do no work are eliminated from the solution of the problem.

What is an advantage in one problem, however, may be a disadvantage in another. It is evident, for instance, that the method of work and energy cannot be used to directly determine an acceleration. It is also evident that in determining a force which is normal to the path of the particle, a force which does no work, the method of work and energy must be supplemented by the direct application of Newton's second law. Suppose, for example, that we wish to determine the tension in the cord of the pendulum of Fig. 13.8*a* as the bob passes through A_2. We draw a free-body diagram of the bob in that position (Fig. 13.9) and express Newton's second law in terms of tangential and normal components. The equations $\Sigma F_t = ma_t$ and $\Sigma F_n = ma_n$ yield, respectively, $a_t = 0$ and

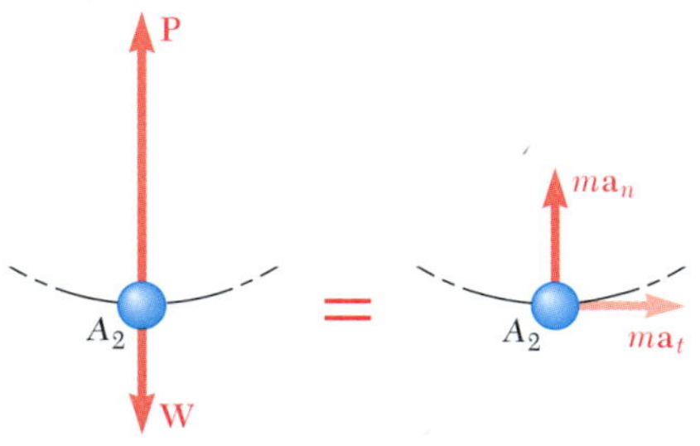

Fig. 13.9

$$P - W = ma_n = \frac{W}{g}\frac{v_2^2}{l}$$

But the speed at A_2 was determined earlier by the method of work and energy. Substituting $v_2^2 = 2gl$ and solving for P, we write

$$P = W + \frac{W}{g}\frac{2gl}{l} = 3W$$

When a problem involves two particles or more, the principle of work and energy can be applied to each particle separately. Adding the kinetic energies of the various particles, and considering the work of all the forces acting on them, we can also write a single equation of work and energy for all the particles involved. We have

$$T_1 + U_{1\rightarrow 2} = T_2 \tag{13.11}$$

where T represents the arithmetic sum of the kinetic energies of the particles involved (all terms are positive) and $U_{1\rightarrow 2}$ is the work of all the forces acting on the particles, *including the forces of action and reaction exerted by the particles on each other.* In problems involving bodies connected by *inextensible cords or links,* however, the work of the forces exerted by a given cord or link on the two bodies it connects cancels out, since the points of application of these forces move through equal distances (see Sample Prob. 13.2).†

Since friction forces have a direction opposite of that of the displacement of the body on which they act, *the work of friction forces is always negative.* This work represents energy dissipated into heat and always results in a decrease in the kinetic energy of the body involved (see Sample Prob. 13.3).

13.5. POWER AND EFFICIENCY

Power is defined as the time rate at which work is done. In the selection of a motor or engine, power is a much more important criterion than is the actual amount of work to be performed. Either a small motor or a large power plant can be used to do a given amount of work; but the small motor may require a month to do the work done by the power plant in a matter of minutes. If ΔU is the work done during the time interval Δt, then the average power during that time interval is

$$\text{Average power} = \frac{\Delta U}{\Delta t}$$

Letting Δt approach zero, we obtain at the limit

$$\text{Power} = \frac{dU}{dt} \tag{13.12}$$

†The application of the method of work and energy to a system of particles is discussed in detail in Chap. 14.

Substituting the scalar product $\mathbf{F} \cdot d\mathbf{r}$ for dU, we can also write

$$\text{Power} = \frac{dU}{dt} = \frac{\mathbf{F} \cdot d\mathbf{r}}{dt}$$

and, recalling that $d\mathbf{r}/dt$ represents the velocity $\mathbf{v}$ of the point of application of $\mathbf{F}$,

$$\text{Power} = \mathbf{F} \cdot \mathbf{v} \tag{13.13}$$

Since power was defined as the time rate at which work is done, it should be expressed in units obtained by dividing units of work by the unit of time. Thus, if SI units are used, power should be expressed in J/s; this unit is called a *watt* (W). We have

$$1 \text{ W} = 1 \text{ J/s} = 1 \text{ N} \cdot \text{m/s}$$

If U.S. customary units are used, power should be expressed in ft · lb/s or in *horsepower* (hp), with the latter defined as

$$1 \text{ hp} = 550 \text{ ft} \cdot \text{lb/s}$$

Recalling from Sec. 13.2 that 1 ft · lb = 1.356 J, we verify that

$$1 \text{ ft} \cdot \text{lb/s} = 1.356 \text{ J/s} = 1.356 \text{ W}$$
$$1 \text{ hp} = 550(1.356 \text{ W}) = 746 \text{ W} = 0.746 \text{ kW}$$

The *mechanical efficiency* of a machine was defined in Sec. 10.5 as the ratio of the output work to the input work:

$$\eta = \frac{\text{output work}}{\text{input work}} \tag{13.14}$$

This definition is based on the assumption that work is done at a constant rate. The ratio of the output to the input work is therefore equal to the ratio of the rates at which output and input work are done, and we have

$$\eta = \frac{\text{power output}}{\text{power input}} \tag{13.15}$$

Because of energy losses due to friction, the output work is always smaller than the input work, and consequently the power output is always smaller than the power input. The mechanical efficiency of a machine is therefore always less than 1.

When a machine is used to transform mechanical energy into electric energy, or thermal energy into mechanical energy, its *overall efficiency* can be obtained from formula (13.15). The overall efficiency of a machine is always less than 1; it provides a measure of all the various energy losses involved (losses of electric or thermal energy as well as frictional losses). Note that it is necessary to express the power output and the power input in the same units before using formula (13.15).

SAMPLE PROBLEM 13.1

An automobile weighing 4000 lb is driven down a 5° incline at a speed of 60 mi/h when the brakes are applied, causing a constant total braking force (applied by the road on the tires) of 1500 lb. Determine the distance traveled by the automobile as it comes to a stop.

SOLUTION

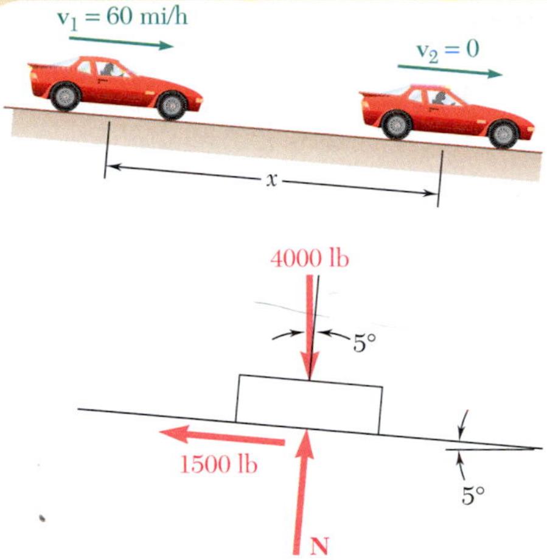

Kinetic Energy

Position 1:

$$v_1 = \left(60 \frac{\text{mi}}{\text{h}}\right)\left(\frac{5280 \text{ ft}}{1 \text{ mi}}\right)\left(\frac{1 \text{ h}}{3600 \text{ s}}\right) = 88 \text{ ft/s}$$

$$T_1 = \tfrac{1}{2}mv_1^2 = \tfrac{1}{2}(4000/32.2)(88)^2 = 481{,}000 \text{ ft} \cdot \text{lb}$$

Position 2: $v_2 = 0 \qquad T_2 = 0$

Work $\quad U_{1\to 2} = -1500x + (4000 \sin 5°)x = -1151x$

Principle of Work and Energy

$$T_1 + U_{1\to 2} = T_2$$
$$481{,}000 - 1151x = 0 \qquad x = 418 \text{ ft} \blacktriangleleft$$

SAMPLE PROBLEM 13.2

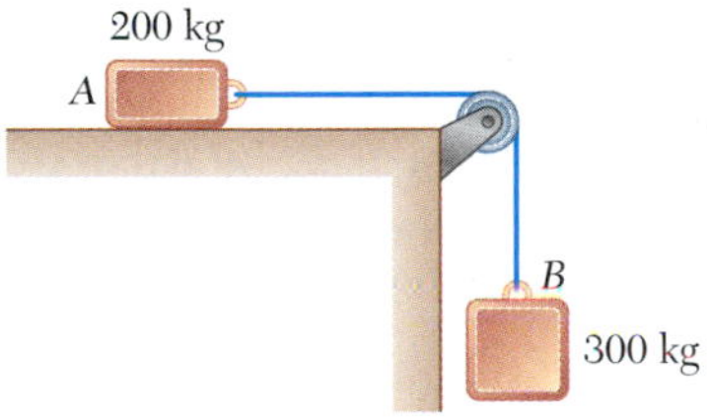

Two blocks are joined by an inextensible cable as shown. If the system is released from rest, determine the velocity of block A after it has moved 2 m. Assume that the coefficient of kinetic friction between block A and the plane is $\mu_k = 0.25$ and that the pulley is weightless and frictionless.

SOLUTION

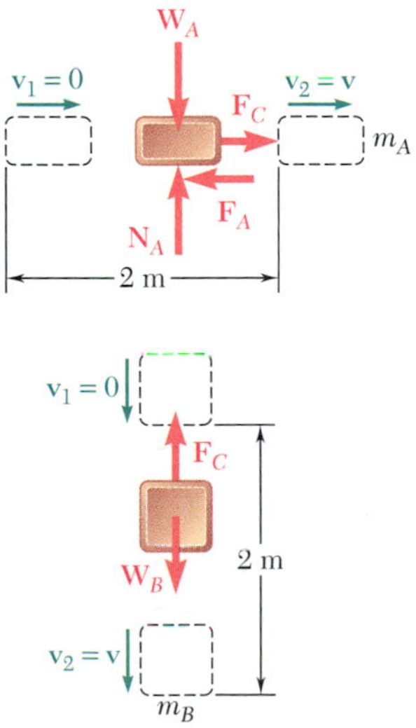

Work and Energy for Block *A*. We denote the friction force by $\mathbf{F}_A$ and the force exerted by the cable by $\mathbf{F}_C$, and write

$$m_A = 200 \text{ kg} \qquad W_A = (200 \text{ kg})(9.81 \text{ m/s}^2) = 1962 \text{ N}$$
$$F_A = \mu_k N_A = \mu_k W_A = 0.25(1962 \text{ N}) = 490 \text{ N}$$
$$T_1 + U_{1\to 2} = T_2: \quad 0 + F_C(2 \text{ m}) - F_A(2 \text{ m}) = \tfrac{1}{2}m_A v^2$$
$$F_C(2 \text{ m}) - (490 \text{ N})(2 \text{ m}) = \tfrac{1}{2}(200 \text{ kg})v^2 \qquad (1)$$

Work and Energy for Block *B*. We write

$$m_B = 300 \text{ kg} \qquad W_B = (300 \text{ kg})(9.81 \text{ m/s}^2) = 2940 \text{ N}$$
$$T_1 + U_{1\to 2} = T_2: \quad 0 + W_B(2 \text{ m}) - F_C(2 \text{ m}) = \tfrac{1}{2}m_B v^2$$
$$(2940 \text{ N})(2 \text{ m}) - F_C(2 \text{ m}) = \tfrac{1}{2}(300 \text{ kg})v^2 \qquad (2)$$

Adding the left-hand and right-hand members of (1) and (2), we observe that the work of the forces exerted by the cable on A and B cancels out:

$$(2940 \text{ N})(2 \text{ m}) - (490 \text{ N})(2 \text{ m}) = \tfrac{1}{2}(200 \text{ kg} + 300 \text{ kg})v^2$$
$$4900 \text{ J} = \tfrac{1}{2}(500 \text{ kg})v^2 \qquad v = 4.43 \text{ m/s} \blacktriangleleft$$

SAMPLE PROBLEM 13.3

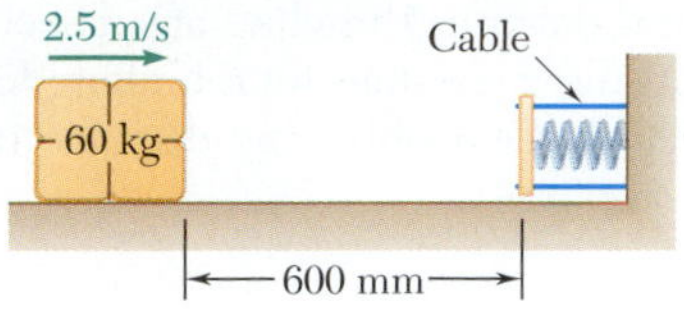

A spring is used to stop a 60-kg package which is sliding on a horizontal surface. The spring has a constant $k = 20$ kN/m and is held by cables so that it is initially compressed 120 mm. Knowing that the package has a velocity of 2.5 m/s in the position shown and that the maximum additional deflection of the spring is 40 mm, determine (*a*) the coefficient of kinetic friction between the package and the surface, (*b*) the velocity of the package as it passes again through the position shown.

SOLUTION

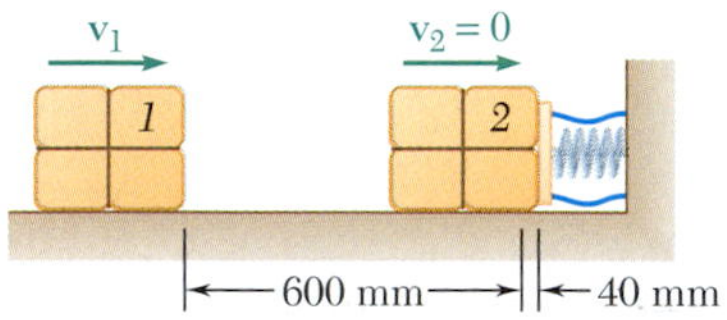

a.* Motion from Position *1* to Position *2

Kinetic Energy. ***Position 1:*** $v_1 = 2.5$ m/s

$$T_1 = \tfrac{1}{2}mv_1^2 = \tfrac{1}{2}(60 \text{ kg})(2.5 \text{ m/s})^2 = 187.5 \text{ N}\cdot\text{m} = 187.5 \text{ J}$$

Position 2: (maximum spring deflection): $v_2 = 0 \qquad T_2 = 0$

Work

Friction Force **F**. We have

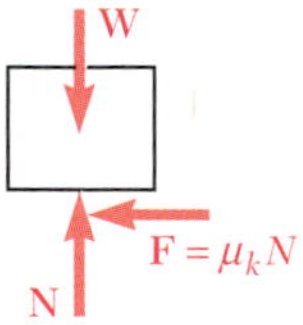

$$F = \mu_k N = \mu_k W = \mu_k mg = \mu_k(60 \text{ kg})(9.81 \text{ m/s}^2) = (588.6 \text{ N})\mu_k$$

The work of **F** is negative and equal to

$$(U_{1\rightarrow 2})_f = -Fx = -(588.6 \text{ N})\mu_k(0.600 \text{ m} + 0.040 \text{ m}) = -(377 \text{ J})\mu_k$$

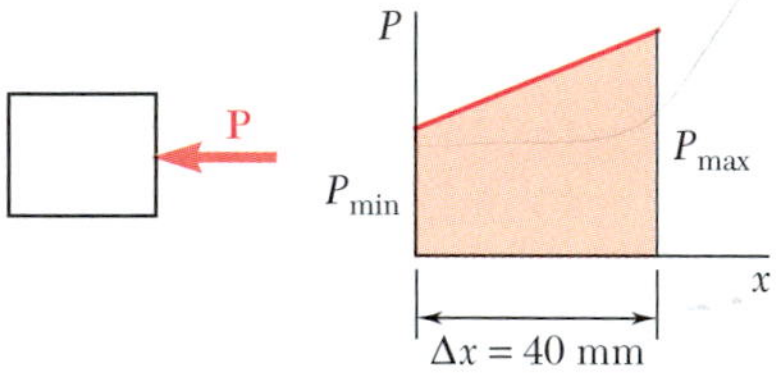

Spring Force **P**. The variable force ***P*** exerted by the spring does an amount of negative work equal to the area under the force-deflection curve of the spring force. We have

$$P_{min} = kx_0 = (20 \text{ kN/m})(120 \text{ mm}) = (20\,000 \text{ N/m})(0.120 \text{ m}) = 2400 \text{ N}$$
$$P_{max} = P_{min} + k\,\Delta x = 2400 \text{ N} + (20 \text{ kN/m})(40 \text{ mm}) = 3200 \text{ N}$$
$$(U_{1\rightarrow 2})_e = -\tfrac{1}{2}(P_{min} + P_{max})\,\Delta x = -\tfrac{1}{2}(2400 \text{ N} + 3200 \text{ N})(0.040 \text{ m}) = -112.0 \text{ J}$$

The total work is thus

$$U_{1\rightarrow 2} = (U_{1\rightarrow 2})_f + (U_{1\rightarrow 2})_e = -(377 \text{ J})\mu_k - 112.0 \text{ J}$$

Principle of Work and Energy

$$T_1 + U_{1\rightarrow 2} = T_2: \qquad 187.5 \text{ J} - (377 \text{ J})\mu_k - 112.0 \text{ J} = 0 \qquad \mu_k = 0.20 \blacktriangleleft$$

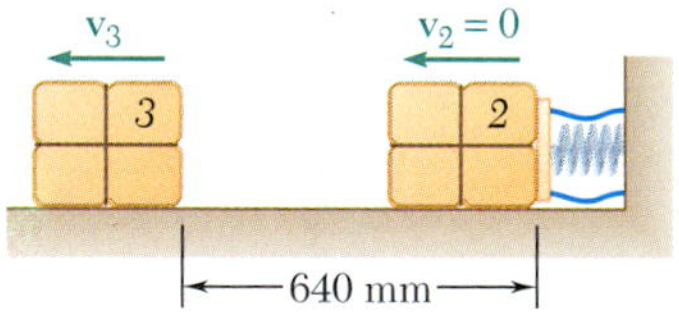

b.* Motion from Position *2* to Position *3

Kinetic Energy. ***Position 2:*** $v_2 = 0 \qquad T_2 = 0$

Position 3: $T_3 = \tfrac{1}{2}mv_3^2 = \tfrac{1}{2}(60 \text{ kg})v_3^2$

Work. Since the distances involved are the same, the numerical values of the work of the friction force **F** and of the spring force **P** are the same as above. However, while the work of **F** is still negative, the work of **P** is now positive.

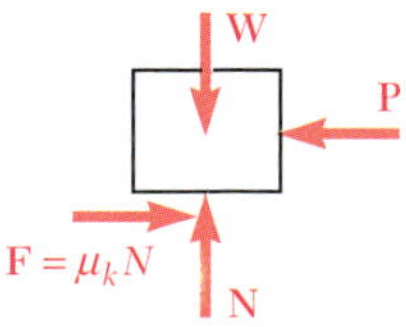

$$U_{2\rightarrow 3} = -(377 \text{ J})\mu_k + 112.0 \text{ J} = -75.5 \text{ J} + 112.0 \text{ J} = +36.5 \text{ J}$$

Principle of Work and Energy

$$T_2 + U_{2\rightarrow 3} = T_3: \qquad 0 + 36.5 \text{ J} = \tfrac{1}{2}(60 \text{ kg})v_3^2$$
$$v_3 = 1.103 \text{ m/s} \qquad \mathbf{v}_3 = 1.103 \text{ m/s} \leftarrow \blacktriangleleft$$

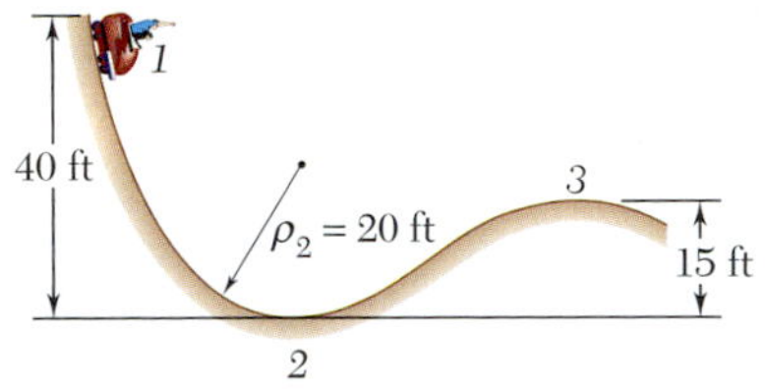

SAMPLE PROBLEM 13.4

A 2000-lb car starts from rest at point *1* and moves without friction down the track shown. (*a*) Determine the force exerted by the track on the car at point *2*, where the radius of curvature of the track is 20 ft. (*b*) Determine the minimum safe value of the radius of curvature at point *3*.

SOLUTION

***a.* Force Exerted by the Track at Point *2*.** The principle of work and energy is used to determine the velocity of the car as it passes through point *2*.

Kinetic Energy. $T_1 = 0 \qquad T_2 = \frac{1}{2}mv_2^2 = \frac{1}{2}\frac{W}{g}v_2^2$

Work. The only force which does work is the weight **W**. Since the vertical displacement from point *1* to point *2* is 40 ft downward, the work of the weight is

$$U_{1\rightarrow 2} = +W(40 \text{ ft})$$

Principle of Work and Energy

$$T_1 + U_{1\rightarrow 2} = T_2 \qquad 0 + W(40 \text{ ft}) = \frac{1}{2}\frac{W}{g}v_2^2$$

$$v_2^2 = 80g = 80(32.2) \qquad v_2 = 50.8 \text{ ft/s}$$

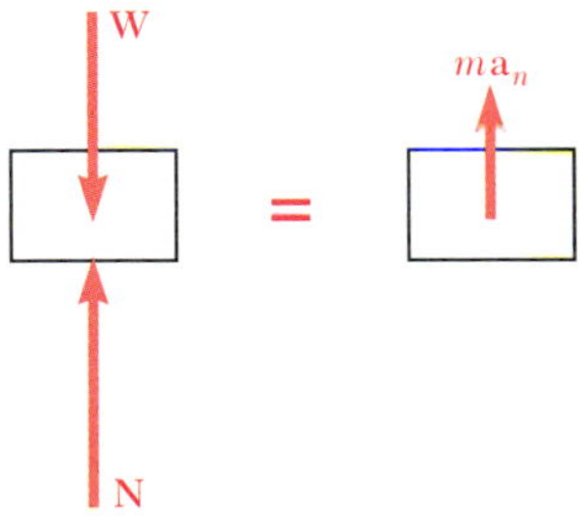

Newton's Second Law at Point 2. The acceleration $\mathbf{a}_n$ of the car at point *2* has a magnitude $a_n = v_2^2/\rho$ and is directed upward. Since the external forces acting on the car are **W** and **N**, we write

$$+\uparrow \Sigma F_n = ma_n: \qquad -W + N = ma_n = \frac{W}{g}\frac{v_2^2}{\rho} = \frac{W}{g}\frac{80g}{20}$$

$$N = 5W \qquad \mathbf{N} = 10{,}000 \text{ lb}\uparrow \blacktriangleleft$$

b.* Minimum Value of ρ at Point *3*.** ***Principle of Work and Energy. Applying the principle of work and energy between point *1* and point *3*, we obtain

$$T_1 + U_{1\rightarrow 3} = T_3 \qquad 0 + W(25 \text{ ft}) = \frac{1}{2}\frac{W}{g}v_3^2$$

$$v_3^2 = 50g = 50(32.2) \qquad v_3 = 40.1 \text{ ft/s}$$

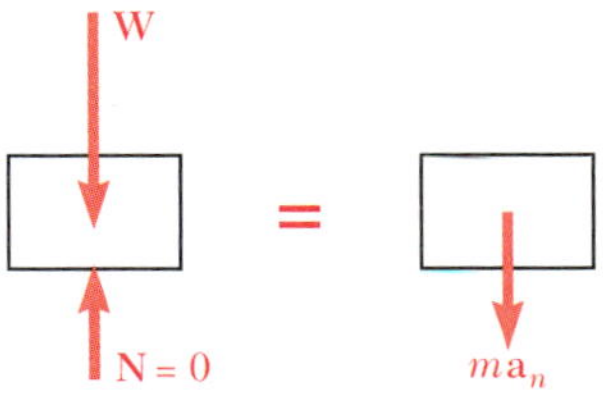

Newton's Second Law at Point 3. The minimum safe value of ρ occurs when **N** = 0. In this case, the acceleration $\mathbf{a}_n$, of magnitude $a_n = v_3^2/\rho$, is directed downward, and we write

$$+\downarrow \Sigma F_n = ma_n: \qquad W = \frac{W}{g}\frac{v_3^2}{\rho} = \frac{W}{g}\frac{50g}{\rho}$$

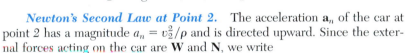

$$\rho = 50 \text{ ft} \blacktriangleleft$$

SAMPLE PROBLEM 13.5

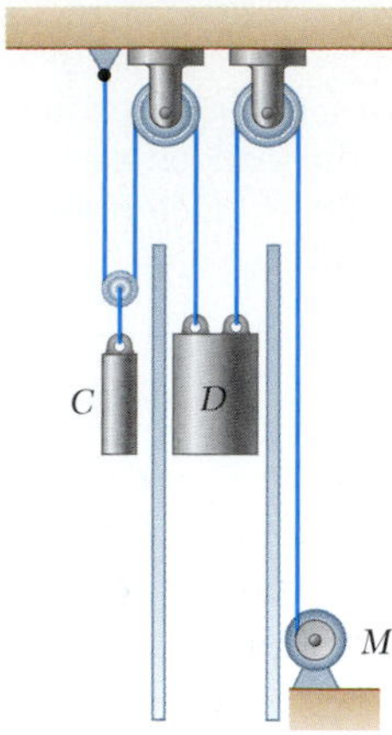

The dumbwaiter D and its load have a combined weight of 600 lb, while the counterweight C weighs 800 lb. Determine the power delivered by the electric motor M when the dumbwaiter (a) is moving up at a constant speed of 8 ft/s, (b) has an instantaneous velocity of 8 ft/s and an acceleration of 2.5 ft/s^2, both directed upward.

SOLUTION

Since the force $\mathbf{F}$ exerted by the motor cable has the same direction as the velocity $\mathbf{v}_D$ of the dumbwaiter, the power is equal to Fv_D, where $v_D = 8$ ft/s. To obtain the power, we must first determine $\mathbf{F}$ in each of the two given situations.

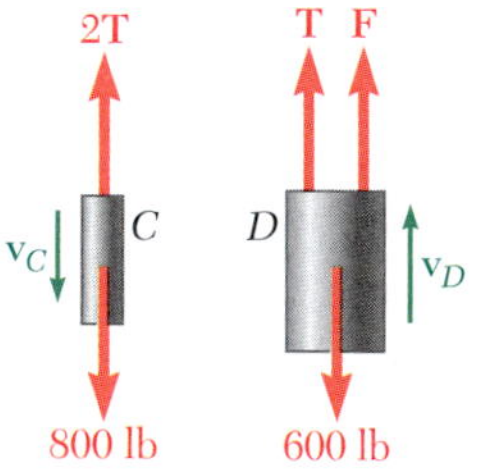

***a.* Uniform Motion.** We have $\mathbf{a}_C = \mathbf{a}_D = 0$; both bodies are in equilibrium.

Free Body C: $\quad +\uparrow \Sigma F_y = 0: \quad 2T - 800 \text{ lb} = 0 \qquad T = 400 \text{ lb}$

Free Body D: $\quad +\uparrow \Sigma F_y = 0: \quad F + T - 600 \text{ lb} = 0$

$$F = 600 \text{ lb} - T = 600 \text{ lb} - 400 \text{ lb} = 200 \text{ lb}$$

$$Fv_D = (200 \text{ lb})(8 \text{ ft/s}) = 1600 \text{ ft}\cdot\text{lb/s}$$

$$\text{Power} = (1600 \text{ ft}\cdot\text{lb/s})\frac{1 \text{ hp}}{550 \text{ ft}\cdot\text{lb/s}} = 2.91 \text{ hp} \quad \blacktriangleleft$$

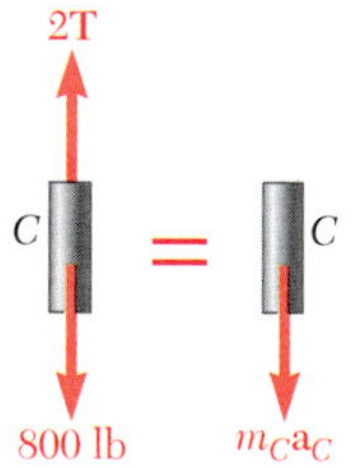

***b.* Accelerated Motion.** We have

$$\mathbf{a}_D = 2.5 \text{ ft/s}^2 \uparrow \qquad \mathbf{a}_C = -\tfrac{1}{2}\mathbf{a}_D = 1.25 \text{ ft/s}^2 \downarrow$$

The equations of motion are

Free Body C: $\quad +\downarrow \Sigma F_y = m_C a_C: \quad 800 - 2T = \dfrac{800}{32.2}(1.25) \quad T = 384.5 \text{ lb}$

Free Body D: $\quad +\uparrow \Sigma F_y = m_D a_D: \quad F + T - 600 = \dfrac{600}{32.2}(2.5)$

$$F + 384.5 - 600 = 46.6 \qquad F = 262.1 \text{ lb}$$

$$Fv_D = (262.1 \text{ lb})(8 \text{ ft/s}) = 2097 \text{ ft}\cdot\text{lb/s}$$

$$\text{Power} = (2097 \text{ ft}\cdot\text{lb/s})\frac{1 \text{ hp}}{550 \text{ ft}\cdot\text{lb/s}} = 3.81 \text{ hp} \quad \blacktriangleleft$$

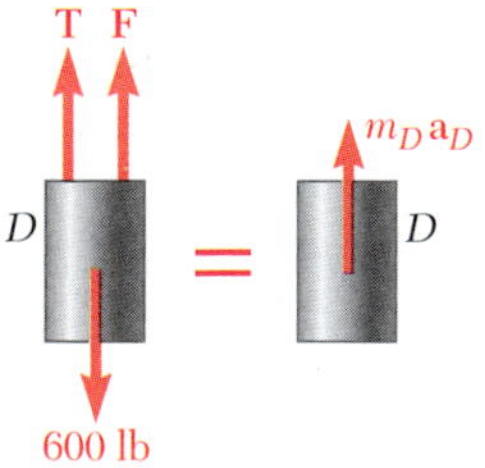

SOLVING PROBLEMS ON YOUR OWN

In the preceding chapter, you solved problems dealing with the motion of a particle by using the fundamental equation $\mathbf{F} = m\mathbf{a}$ to determine the acceleration $\mathbf{a}$. By applying the principles of kinematics you were then able to determine from $\mathbf{a}$ the velocity and displacement of the particle at any time. In this lesson we combined $\mathbf{F} = m\mathbf{a}$ and the principles of kinematics to obtain an additional method of analysis called the *method of work and energy.* This eliminates the need to calculate the acceleration and will enable you to relate the velocities of the particle at two points along its path of motion. To solve a problem by the method of work and energy you will follow these steps:

1. *Computing the work of each of the forces.* The work $U_{1\to2}$ of a given force $\mathbf{F}$ during the finite displacement of the particle from A_1 to A_2 is defined as

$$U_{1\to2} = \int \mathbf{F} \cdot d\mathbf{r} \qquad \text{or} \qquad U_{1\to2} = \int (F \cos \alpha)\, ds \quad (13.2, 13.2')$$

where α is the angle between $\mathbf{F}$ and the displacement $d\mathbf{r}$. The work $U_{1\to2}$ is a scalar quantity and is expressed in ft · lb or in · lb in the U.S. customary system of units and in N · m or joules (J) in the SI system of units. Note that the work done is zero for a force perpendicular to the displacement ($\alpha = 90°$). Negative work is done for $90° < \alpha < 180°$ and in particular for a friction force, which is always opposite in direction to the displacement ($\alpha = 180°$).

The work $U_{1\to2}$ can be easily evaluated in the following cases that you will encounter:

a. Work of a constant force in rectilinear motion

$$U_{1\to2} = (F \cos \alpha)\, \Delta x \qquad (13.3)$$

where α = angle the force forms with the direction of motion
Δx = displacement from A_1 to A_2 (Fig. 13.3)

b. Work of the force of gravity

$$U_{1\to2} = -W\, \Delta y \qquad (13.4')$$

where Δy is the vertical displacement of the center of gravity of the body of weight W. Note that the work is positive when Δy is negative, that is, when the body moves down (Fig. 13.4).

c. Work of the force exerted by a spring

$$U_{1\to2} = \tfrac{1}{2}kx_1^2 - \tfrac{1}{2}kx_2^2 \qquad (13.6)$$

where k is the spring constant and x_1 and x_2 are the elongations of the spring corresponding to the positions A_1 and A_2 (Fig. 13.5). *(continued)*

d. Work of a gravitational force

$$U_{1\rightarrow 2} = \frac{GMm}{r_2} - \frac{GMm}{r_1} \qquad (13.7)$$

for a displacement of the body from $A_1(r = r_1)$ to $A_2(r = r_2)$ (Fig. 13.6).

2. ***Calculate the kinetic energy at A_1 and A_2.*** The kinetic energy T is

$$T = \tfrac{1}{2}mv^2 \qquad (13.9)$$

where m is the mass of the particle and v is the magnitude of its velocity. The units of kinetic energy are the same as the units of work, that is, ft · lb or in · lb if U.S. customary units are used and N · m or joules (J) if SI units are used.

3. ***Substitute the values for the work done $U_{1\rightarrow 2}$ and the kinetic energies T_1 and T_2*** into the equation

$$T_1 + U_{1\rightarrow 2} = T_2 \qquad (13.11)$$

You will now have *one equation* which you can solve for *one unknown.* Note that this equation does not yield the time of travel or the acceleration directly. However, if you know the radius of curvature ρ of the path of the particle at a point where you have obtained the velocity v, you can express the normal component of the acceleration as $a_n = v^2/\rho$ and obtain the normal component of the force exerted on the particle by writing $F_n = mv^2/\rho$.

4. ***Power was introduced in this lesson as the time rate at which work is done, $P = dU/dt$.*** Power is measured in ft · lb/s or *horsepower* (hp) in U.S. customary units and in J/s or *watts* (W) in the SI system of units. To calculate the power, you can use the equivalent formula,

$$P = \mathbf{F} \cdot \mathbf{v} \qquad (13.13)$$

where $\mathbf{F}$ and $\mathbf{v}$ denote the force and the velocity, respectively, at a given time [Sample Prob. 13.5]. In some problems [see, e.g., Prob. 13.50], you will be asked for the *average power,* which can be obtained by dividing the total work by the time interval during which the work is done.

Problems

13.1 A 400-kg satellite is placed in a circular orbit 6394 km above the surface of the earth. At this elevation the acceleration of gravity is 4.09 m/s^2. Knowing that its orbital speed is 20 000 km/h, determine the kinetic energy of the satellite.

13.2 A 1000-lb satellite is placed in a circular orbit 3000 mi above the surface of the earth. At this elevation the acceleration of gravity is 8.03 ft/s^2. Knowing that its orbital speed is 14,000 mi/h, determine the kinetic energy of the satellite.

13.3 An 8-lb stone is dropped from a height h and strikes the ground with a velocity of 85 ft/s. (*a*) Find the kinetic energy of the stone as it strikes the ground and the height h from which it was dropped. (*b*) Solve part *a*, assuming that the same stone is dropped on the moon. (Acceleration of gravity on the moon = 5.31 ft/s^2.)

13.4 A 2-kg stone is dropped from a height h and strikes the ground with a velocity of 24 m/s. (*a*) Find the kinetic energy of the stone as it strikes the ground and the height h from which it was dropped. (*b*) Solve part *a*, assuming that the same stone is dropped on the moon. (Acceleration of gravity on the moon = 1.62 m/s^2.)

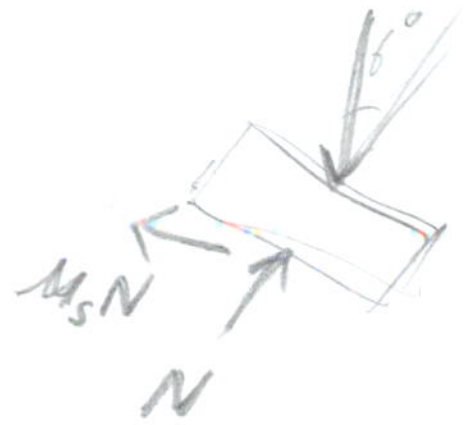

13.5 An automobile weighing 4250 lb starts from rest at point A on a 6° incline and coasts through a distance of 500 ft to point B. The brakes are then applied, causing the automobile to come to a stop at point C, 70 ft from B. Knowing that slipping is impending during the braking period and neglecting air resistance and rolling resistance, determine (*a*) the speed of the automobile at point B, (*b*) the coefficient of static friction between the tires and the road.

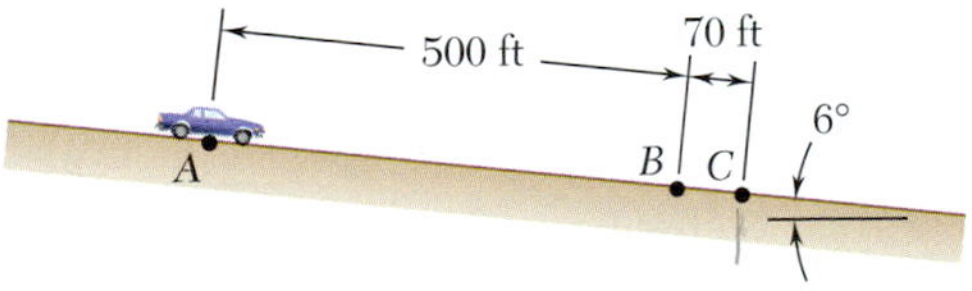

Fig. P13.5 and P13.6

13.6 An automobile weighing 4250 lb starts from rest at point A on a 6° incline and coasts through a distance of 500 ft to point B. The brakes are then fully applied, causing the automobile to skid to a stop at point C, 70 ft from B. Knowing that the coefficient of dynamic friction between the tires and the road is 0.75, determine the work done on the automobile by the combined effects of air resistance and rolling resistance between points A and C.

Fig. P13.7 and P13.8

13.7 Skid marks on a drag race track indicate that the rear (drive) wheels of a car skid for the first 18 m and roll with slipping impending during the remaining 382 m. The front wheels of the car are just off the ground for the first 18 m, and for the remainder of the race 75 percent of the weight of the car is on the rear wheels. Knowing that the speed of the car is 58 km/h at the end of the first 18 m and that the coefficient of kinetic friction is 80 percent of the coefficient of static friction, determine the speed of the car at the end of the 400-m track. Ignore air resistance and rolling resistance.

13.8 In an automobile drag race, the rear drive wheels of a car skid for the first 25 m and roll with slipping impending during the remaining 375 m. The front wheels of the car are just off the ground for the first 25 m, and for the remainder of the race 80 percent of the weight is on the rear wheels. (*a*) What is the required coefficient of static friction if, starting from rest, the car is to reach a peak speed of 275 km/h at the end of the 400-m track? Assume that the coefficient of kinetic friction is 75 percent of the coefficient of static friction and ignore air resistance and rolling resistance. (*b*) How fast is the car going at the end of the first 25 m?

13.9 The 16-lb block *A* is released from rest in the position shown. Neglecting the effect of friction and the masses of the pulleys, determine the velocity of the block after it has moved 2 ft up the incline.

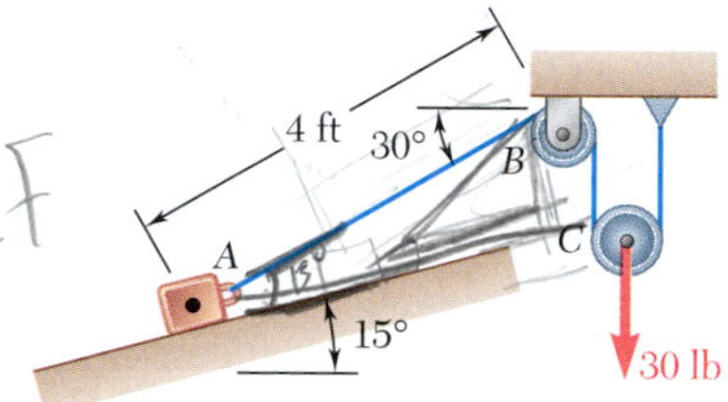

Fig. *P13.9* and *P13.10*

13.10 The 16-lb block *A* is released in the position shown with a velocity of 3 ft/s up the incline. Knowing that the velocity of the block is 9 ft/s after it has moved 2 ft up the incline, determine the work done by the friction force exerted on the block. Neglect the masses of the pulleys.

13.11 Boxes are transported by a conveyor belt with a velocity $\mathbf{v}_0$ to a fixed incline at *A* where they slide and eventually fall off at *B*. Knowing that $\mu_k = 0.40$, determine the velocity of the conveyor belt if the boxes leave the incline at *B* with a velocity of 2 m/s.

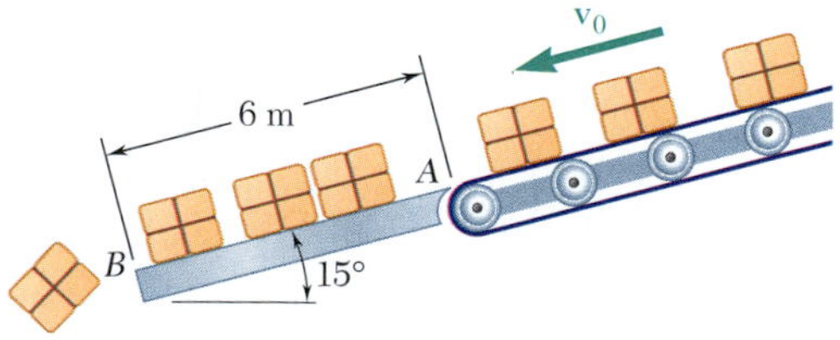

Fig. P13.11 and P13.12

13.12 Boxes are transported by a conveyor belt with a velocity $\mathbf{v}_0$ to a fixed incline at *A* where they slide and eventually fall off at *B*. Knowing that $\mu_k = 0.40$, determine the velocity of the conveyor belt if the boxes are to have a zero velocity at *B*.

13.13 In an ore-mining operation, a bucket full of ore is suspended from a traveling crane which moves slowly along a stationary bridge. The bucket is to swing no more than 12 ft horizontally when the crane is brought to a sudden stop. Determine the maximum allowable horizontal speed v of the crane.

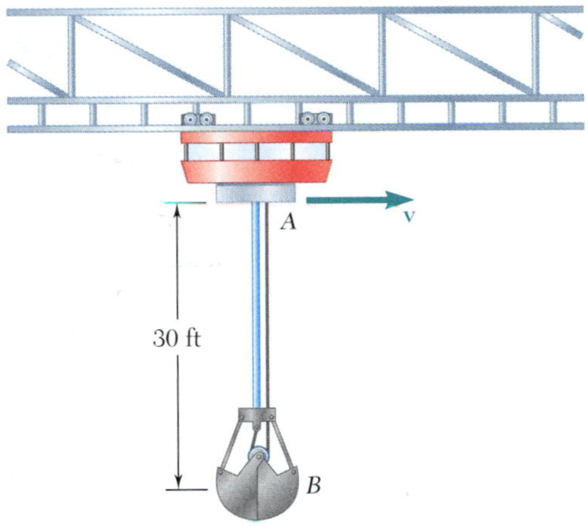

Fig. *P13.13* and *P13.14*

13.14 In an ore-mining operation, a bucket full of ore is suspended from a traveling crane which moves slowly along a stationary bridge. The crane is traveling at a speed of 10 ft/s when it is brought to a sudden stop. Determine the maximum horizontal distance through which the bucket will swing.

13.15 Car B is towing car A with a 5-m cable at a constant speed of 10 m/s on an uphill grade when the brakes of car B are fully applied causing it to skid to a stop. Car A, whose driver had not observed that car B was slowing down, then strikes the rear of car B. Neglecting air resistance and rolling resistance and assuming a coefficient of kinetic friction of 0.9, determine the speed of car A just before the collision.

13.16 Car B is towing car A at a constant speed of 10 m/s on an uphill grade when the brakes of car A are fully applied causing all four wheels to skid. The driver of car B does not change the throttle setting or change gears. The masses of the cars A and B are 1400 kg and 1200 kg, respectively, and the coefficient of kinetic friction is 0.8. Neglecting air resistance and rolling resistance, determine (a) the distance traveled by the cars before they come to a stop, (b) the tension in the cable.

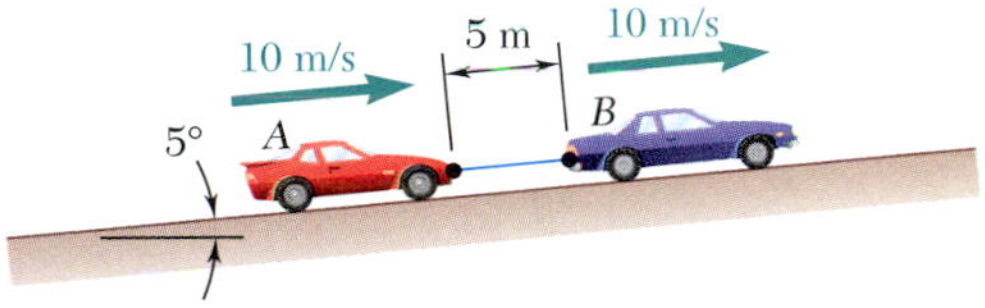

Fig. P13.15 and P13.16

13.17 A trailer truck has a 4400-lb cab and an 18,000-lb trailer. It is traveling on level ground at 60 mi/h and must slow down to a stop in 3000 ft. Determine (a) the average braking force that must be supplied, (b) the average force in the coupling if 60 percent of the braking force is supplied by the trailer and 40 percent by the cab.

13.18 A trailer truck has a 4400-lb cab and an 18,000-lb trailer. Knowing that it is traveling on level ground at 60 mi/h and that an average braking force of 680 lb is applied, determine (a) how far the trailer truck will travel before coming to a stop, (b) the average force in the coupling between the cab and the trailer if the brakes on the trailer fail and all of the braking force is supplied by the cab.

Fig. P13.17 and P13.18

13.19 The system shown, consisting of a 20-kg collar A and a 10-kg counterweight B, is at rest when a constant 500-N force is applied to collar A. (a) Determine the velocity of A just before it hits the support at C. (b) Solve part a assuming that the counterweight B is replaced by a 98.1-N downward force. Ignore friction and the mass of the pulleys.

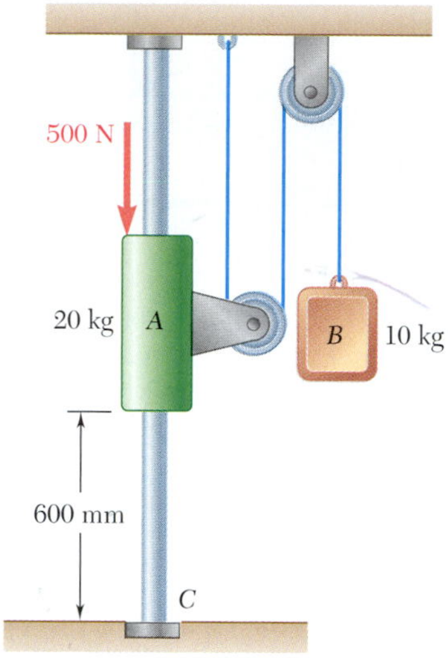

Fig. *P13.19*

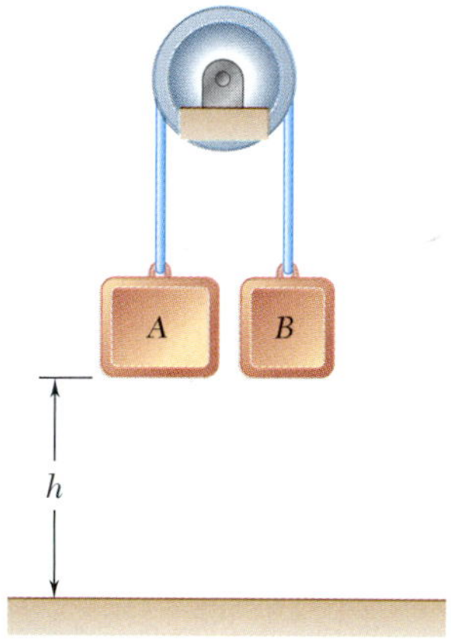

Fig. *P13.20*

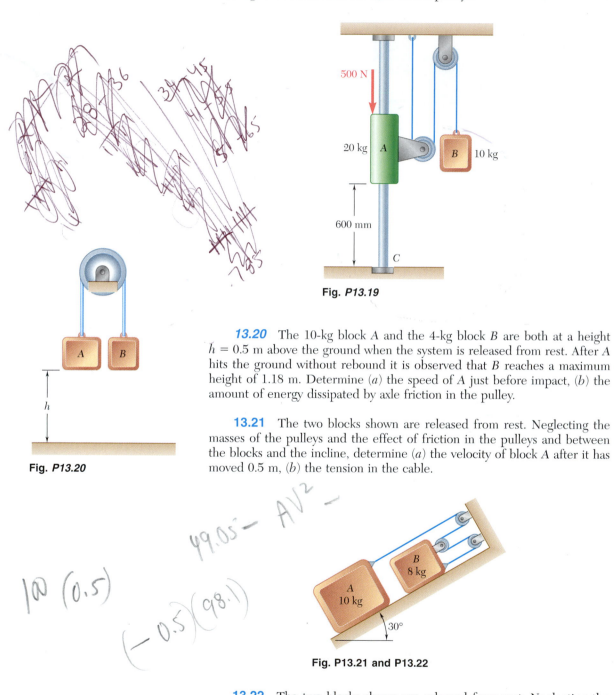

13.20 The 10-kg block A and the 4-kg block B are both at a height $h = 0.5$ m above the ground when the system is released from rest. After A hits the ground without rebound it is observed that B reaches a maximum height of 1.18 m. Determine (a) the speed of A just before impact, (b) the amount of energy dissipated by axle friction in the pulley.

13.21 The two blocks shown are released from rest. Neglecting the masses of the pulleys and the effect of friction in the pulleys and between the blocks and the incline, determine (a) the velocity of block A after it has moved 0.5 m, (b) the tension in the cable.

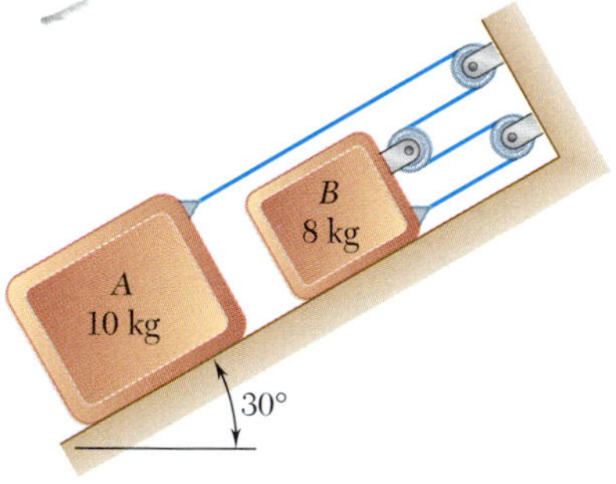

Fig. P13.21 and P13.22

13.22 The two blocks shown are released from rest. Neglecting the masses of the pulleys and the effect of friction in the pulleys and knowing that the coefficients of friction between both blocks and the incline are $\mu_s = 0.25$ and $\mu_k = 0.20$, determine (a) the velocity of block A after it has moved 0.5 m, (b) the tension in the cable.

13.23 Four 3-kg packages are held in place by friction on a conveyor which is disengaged from its drive motor. When the system is released from rest, package *1* leaves the belt at *A* just as package *4* comes onto the inclined portion of the belt at *B*. Determine (*a*) the velocity of package *2* as it leaves the belt at *A*, (*b*) the velocity of package *3* as it leaves the belt at *A*. Neglect the mass of the belt and rollers.

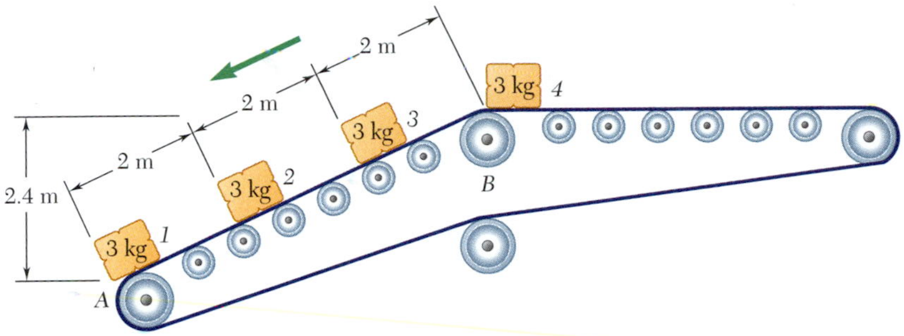

Fig. P13.23

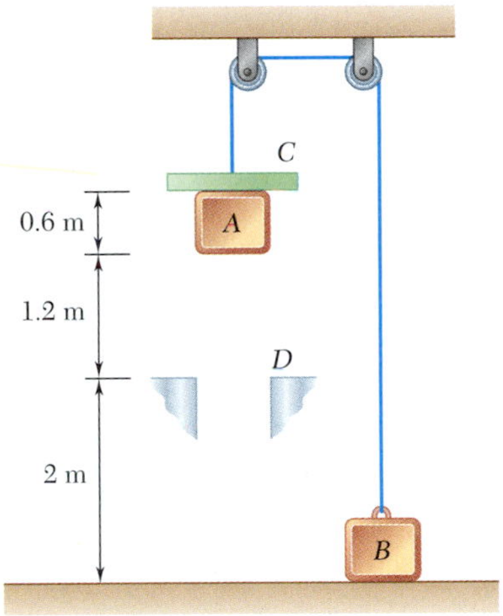

Fig. P13.24

13.24 Two blocks *A* and *B*, of mass 8 kg and 10 kg, respectively, are connected by a cord which passes over pulleys as shown. A 6-kg collar *C* is placed on block *A* and the system is released from rest. After the blocks move 1.8 m, collar *C* is removed and blocks *A* and *B* continue to move. Determine the speed of block *A* just before it strikes the ground.

13.25 A 0.7-lb block rests on top of a 0.5-lb block supported by but not attached to a spring of constant 9 lb/ft. The upper block is suddenly removed. Determine (*a*) the maximum velocity reached by the 0.5-lb block, (*b*) the maximum height reached by the 0.5-lb block.

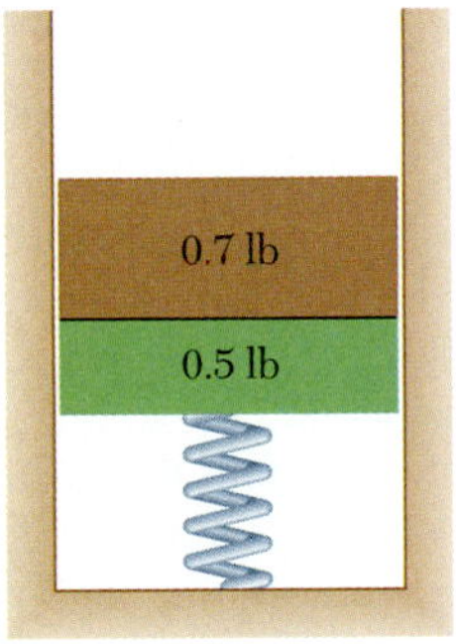

Fig. *P13.25*

13.26 Solve Prob. 13.25, assuming that the 4-lb block is attached to the spring.

13.27 A 4-kg collar *C* slides on a horizontal rod between springs *A* and *B*. If the collar is pushed to the right until spring *B* is compressed 50 mm and released, determine the distance through which the collar will travel, assuming (*a*) no friction between the collar and the rod, (*b*) a coefficient of friction $\mu_k = 0.35$.

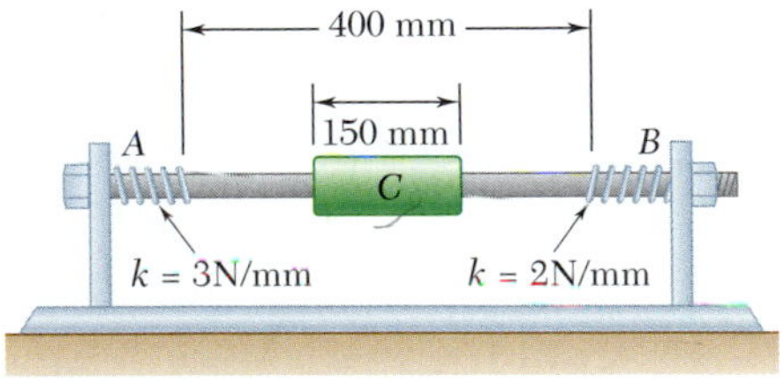

Fig. P13.27

13.28 A 3-kg collar C slides on a frictionless vertical rod. It is pushed up into the position shown, compressing the upper spring by 50 mm and released. Determine (*a*) the maximum deflection of the lower spring, (*b*) the maximum velocity of the collar.

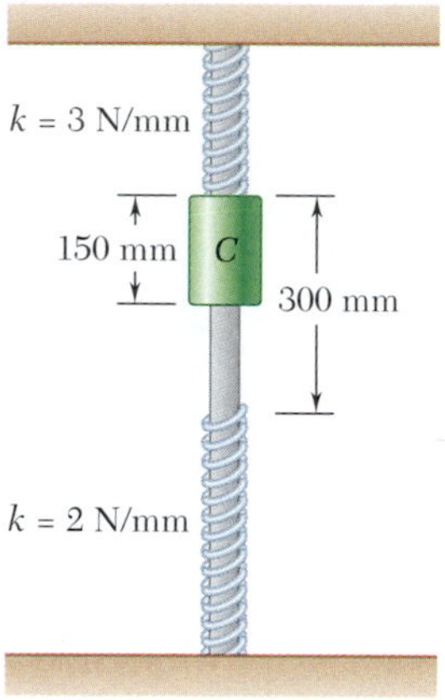

Fig. P13.28

13.29 A 20-lb block is attached to spring A and connected to spring B by a cord and pulley. The block is held in the position shown with both springs unstretched when the support is removed and the block is released with no initial velocity. Knowing that the constant of each spring is 12 lb/in., determine (*a*) the velocity of the block after it has moved down 2 in., (*b*) the maximum velocity achieved by the block.

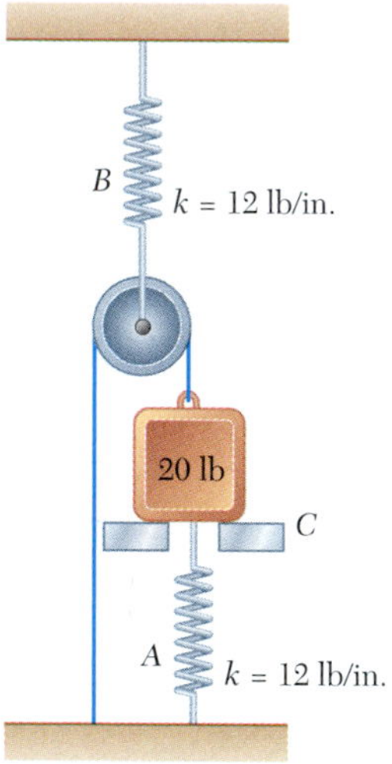

Fig. P13.29 and P13.30

13.30 A 20-lb block is attached to spring A and connected to spring B by a cord and pulley. The block is held in the position shown with spring A stretched 1 in. and spring B unstretched when the support is removed and the block is released with no initial velocity. Knowing that the constant of each spring is 12 lb/in., determine (*a*) the velocity of the block after it has moved down 1 in., (*b*) the maximum velocity achieved by the block.

13.31 An uncontrolled automobile traveling at 100 km/h strikes squarely a highway crash cushion of the type shown in which the automobile is brought to rest by successively crushing steel barrels. The magnitude F of the force required to crush the barrels is shown as a function of the distance x the automobile has moved into the cushion. Knowing that the automobile has a mass of 1000 kg and neglecting the effect of friction, determine (*a*) the distance the automobile will move into the cushion before it comes to rest, (*b*) the maximum deceleration of the automobile.

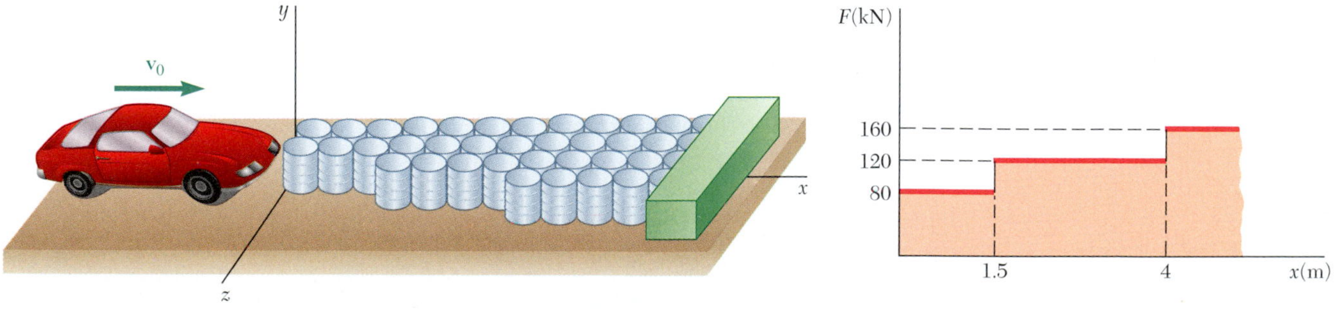

Fig. *P13.31*

13.32 A piston of mass m and cross-sectional area A is in equilibrium under the pressure p at the center of a cylinder closed at both ends. Assuming that the piston is moved to the left a distance $a/2$ and released, and knowing that the pressure on each side of the piston varies inversely with the volume, determine the velocity of the piston as it again reaches the center of the cylinder. Neglect friction between the piston and the cylinder and express your answer in terms of m, a, p, and A.

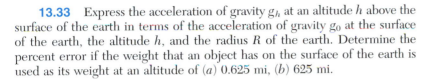

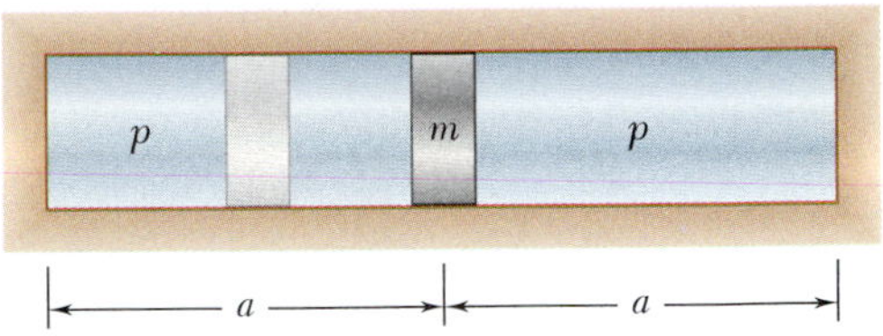

Fig. P13.32

13.33 Express the acceleration of gravity g_h at an altitude h above the surface of the earth in terms of the acceleration of gravity g_0 at the surface of the earth, the altitude h, and the radius R of the earth. Determine the percent error if the weight that an object has on the surface of the earth is used as its weight at an altitude of (*a*) 0.625 mi, (*b*) 625 mi.

13.34 A rocket is fired vertically from the surface of the moon with a velocity v_0. Derive a formula for the ratio h_n/h_u of heights reached with a velocity v, if Newton's law of gravitation is used to calculate h_n and a uniform gravitational field is used to calculate h_u. Express your answer in terms of the acceleration of gravity g_m on the surface of the moon, the radius R_m of the moon, and the velocities v and v_0.

13.35 A meteor starts from rest at a very great distance from the earth. Knowing that the radius of the earth is 6370 km and neglecting all forces except the gravitational attraction of the earth, determine the speed of the meteor (*a*) when it enters the ionosphere at an altitude of 1000 km, (*b*) when it enters the stratosphere at an altitude of 50 km, (*c*) when it strikes the earth's surface.

13.36 During a flyby of the earth, the velocity of a spacecraft is 10.4 km/s as it reaches its minimum altitude of 990 km above the surface at point O. At point B the spacecraft is observed to have an altitude of 8350 km. Assuming that the trajectory of the spacecraft is parabolic, determine its speed at (*a*) point A, (*b*) point B.

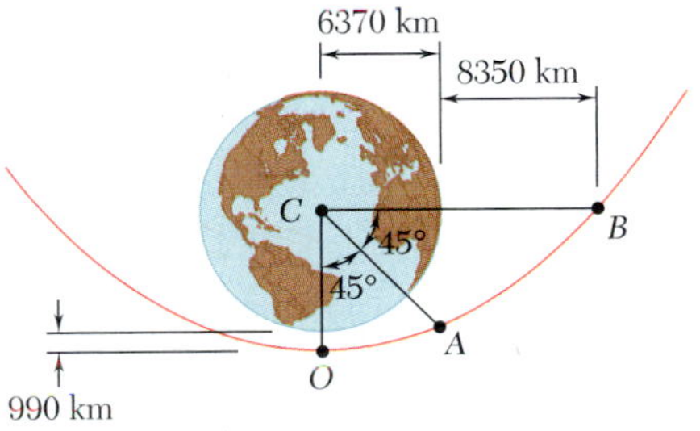

Fig. P13.36

13.37 A 0.75-lb brass (nonmagnetic) block A and an 0.5-lb steel magnet B are in equilibrium in a brass tube under the magnetic repelling force of another steel magnet C located at a distance $x = 0.15$ in. from B. The force is inversely proportional to the square of the distance between B and C. If block A is suddenly removed, determine (*a*) the maximum velocity of B, (*b*) the maximum acceleration of B. Assume that air resistance and friction are negligible.

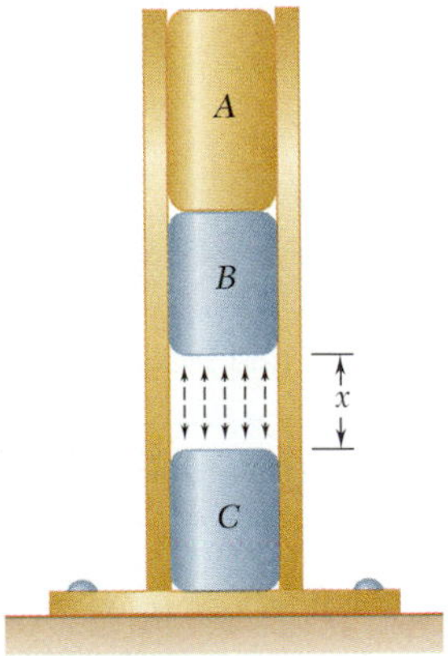

Fig. P13.37

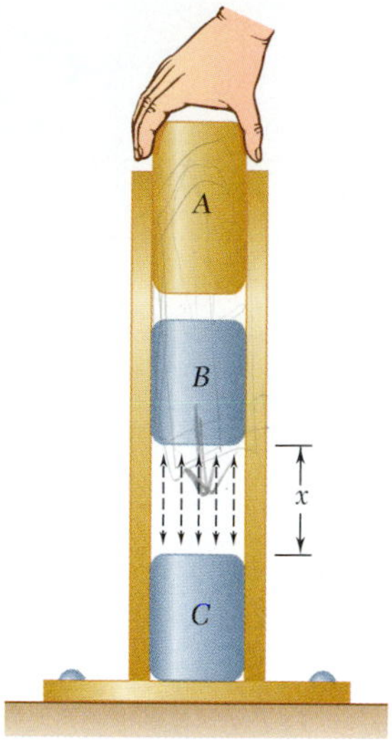

Fig. P13.38

13.38 In a brass tube a 180-g steel magnet B is in equilibrium under the repelling force of another steel magnet C located at a distance $x = 4$ mm from B. The force is inversely proportional to the square of the distance between B and C. If a 270-g brass (nonmagnetic) block A is carefully brought in contact with magnet B and released, determine (*a*) the maximum velocity of A and B, (*b*) the maximum deflection of A and B. Assume that air resistance and friction are negligible.

13.39 The sphere at A is given a downward velocity $\mathbf{v}_0$ and swings in a vertical circle of radius l and center O. Determine the smallest velocity $\mathbf{v}_0$ for which the sphere will reach point B as it swings about point O (*a*) if AO is a rope, (*b*) if AO is a slender rod of negligible mass.

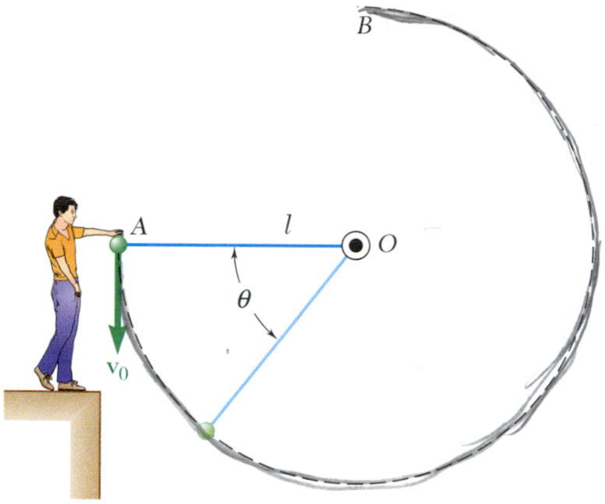

Fig. *P13.39* and *P13.40*

13.40 The sphere at A is given a downward velocity $\mathbf{v}_0$ of magnitude 16 ft/s and swings in a vertical plane at the end of a rope of length $l = 6$ ft attached to a support at O. Determine the angle θ at which the rope will break, knowing that it can withstand a maximum tension equal to twice the weight of the sphere.

13.41 A 3-lb block A rests on a 3-lb block B and is attached to a spring of constant 12 lb/ft. The coefficients of friction between the two blocks are $\mu_s = 0.95$ and $\mu_k = 0.90$ and the coefficients of friction between block B and the horizontal surface are $\mu_s = 0.15$ and $\mu_k = 0.10$. Knowing that the blocks are released from rest when the spring is stretched 4 in., determine (*a*) the velocity of block A as it reaches the position in which the spring is unstretched, (*b*) the maximum velocity of block A.

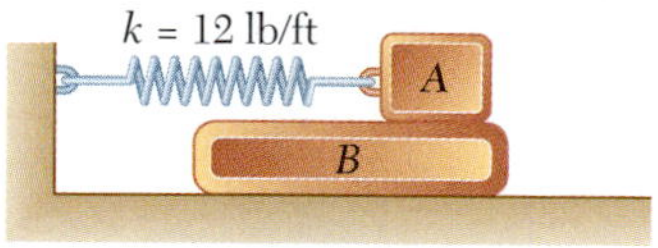

Fig. P13.41 and P13.42

13.42 A 3-lb block A rests on a 3-lb block B and is attached to a spring of constant 12 lb/ft. The coefficients of friction between the two blocks are $\mu_s = 0.35$ and $\mu_k = 0.30$ and the coefficients of friction between block B and the horizontal surface are $\mu_s = 0.15$ and $\mu_k = 0.10$. Knowing that the blocks are released from rest when the spring is stretched 4 in., determine (*a*) the velocity of block A as it reaches the position in which the spring is unstretched, (*b*) the maximum velocity of block A.

13.43 A section of track for a roller coaster consists of two circular arcs AB and CD joined by a straight portion BC. The radius of AB is 27 m and the radius of CD is 72 m. The car and its occupants, of total mass 250 kg, reach point A with practically no velocity and then drop freely along the track. Determine the normal force exerted by the track on the car as the car reaches point B. Ignore air resistance and rolling resistance.

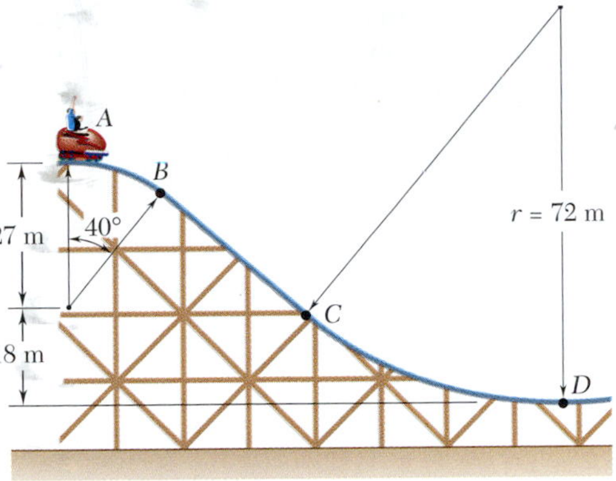

Fig. P13.43 and P13.44

13.44 A section of track for a roller coaster consists of two circular arcs AB and CD joined by a straight portion BC. The radius of AB is 27 m and the radius of CD is 72 m. The car and its occupants, of total mass 250 kg, reach point A with practically no velocity and then drop freely along the track. Determine the maximum and minimum values of the normal force exerted by the track on the car as the car travels from A to D. Ignore air resistance and rolling resistance.

13.45 A small block slides at a speed $v = 3$ m/s on a horizontal surface at a height $h = 1$ m above the ground. Determine (*a*) the angle θ at which it will leave the cylindrical surface BCD, (*b*) the distance x at which it will hit the ground. Neglect friction and air resistance.

13.46 A small block slides at a speed v on a horizontal surface. Knowing that $h = 8$ ft, determine the required speed of the block if it is to leave the cylindrical surface BCD when $\theta = 40°$.

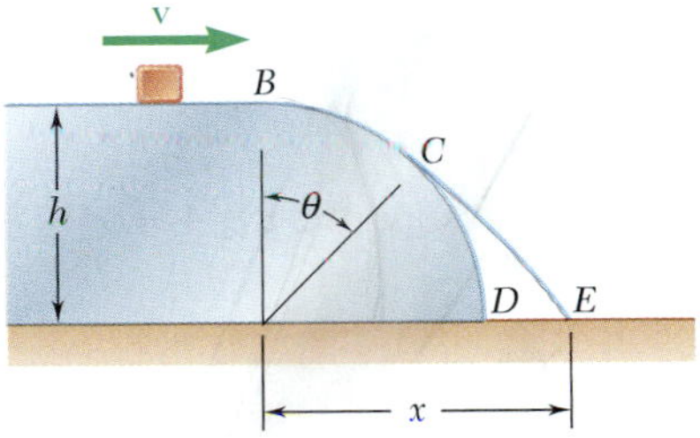

Fig. *P13.45* and *P13.46*

13.47 (*a*) A 60-kg woman rides a 7-kg bicycle up a 3 percent slope at a constant speed of 2 m/s. How much power must be developed by the woman? (*b*) A 90-kg man on a 9-kg bicycle starts down the same slope and maintains a constant speed of 6 m/s by braking. How much power is dissipated by the brakes? Ignore air resistance and rolling resistance.

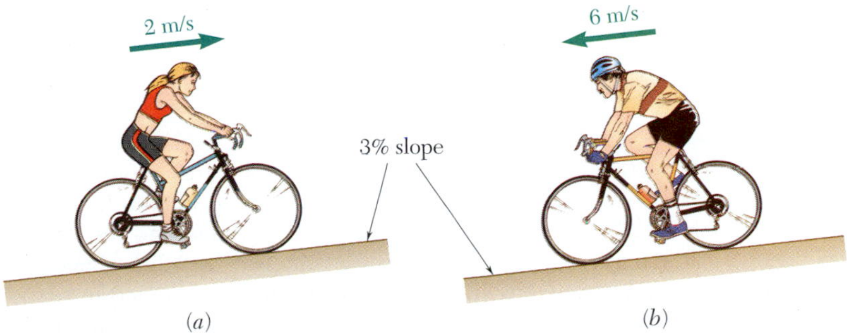

Fig. P13.47

13.48 A 70-kg sprinter starts from rest and accelerates uniformly for 5.4 s over a distance of 35 m. Neglecting air resistance, determine the average power developed by the sprinter.

13.49 A power specification formula is to be derived for electric motors which drive conveyor belts moving solid material at different rates to different heights and distances. Denoting the efficiency of the motors by η and neglecting the power needed to drive the belt itself, derive a formula (*a*) in the SI system of units, for the power P in kW, in terms of the mass flow rate m in kg/h, the height b, and the horizontal distance l in meters, and (*b*) in U.S. customary units, for the power in hp, in terms of the mass flow rate m in tons/h, and the height b and horizontal distance l in feet.

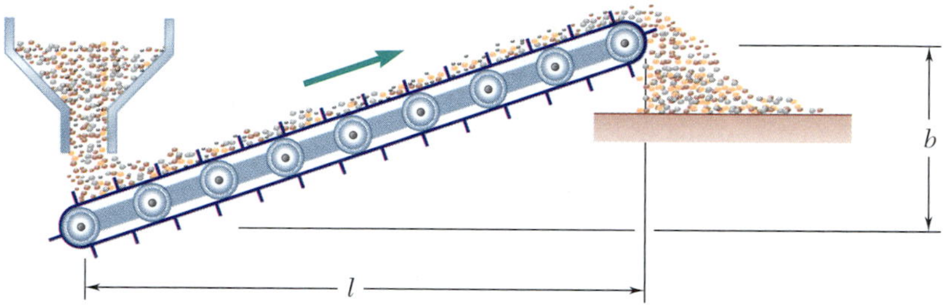

Fig. P13.49

Fig. P13.50 and *P13.51*

13.50 It takes 16 s to raise a 2800-lb car and the supporting 650-lb hydraulic car-lift platform to a height of 6.5 ft. Knowing that the overall conversion efficiency from electric to mechanical power for the system is 82 percent, determine (*a*) the average output power delivered by the hydraulic pump to lift the system, (*b*) the average electric power required.

13.51 The velocity of the lift of Prob. 13.50 increases uniformly from zero to its maximum value in 8 s and then decreases uniformly to zero in 8 s. Knowing that the peak power output of the hydraulic pump is 8 hp when the velocity is maximum, determine the maximum lifting force provided by the pump.

13.52 A 1400-kg automobile starts from rest and travels 400 m during a performance test. The motion of the automobile is defined by the relation $x = 4000 \ln(\cosh 0.03t)$, where x and t are expressed in meters and seconds, respectively. The magnitude of the aerodynamic drag is $D = 0.35v^2$, where D and v are expressed in newtons and m/s, respectively. Determine the power dissipated by the aerodynamic drag when (*a*) $t = 10$ s, (*b*) $t = 15$ s.

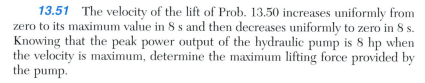

Fig. P13.52 and P13.53

13.53 A 1400-kg automobile starts from rest and travels 400 m during a performance test. The motion of the automobile is defined by the relation $a = 3.6e^{-0.0005x}$, where a and x are expressed in m/s^2 and meters, respectively. The magnitude of the aerodynamic drag is $D = 0.35\,v^2$, where D and v are expressed in newtons and m/s, respectively. Determine the power dissipated by the aerodynamic drag when (*a*) $x = 200$ m, (*b*) $x = 400$ m.

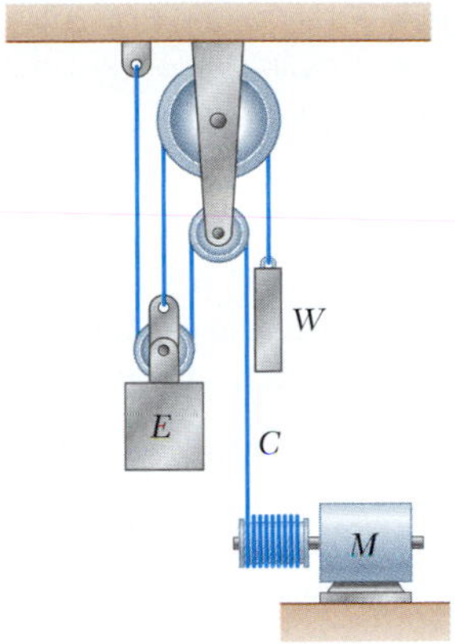

Fig. *P13.54*

13.54 The elevator E has a weight of 6600 lb when fully loaded and is connected as shown to a counterweight W of weight 2200 lb. Determine the power in hp delivered by the motor (*a*) when the elevator is moving down at a constant speed of 1 ft/s, (*b*) when it has an upward velocity of 1 ft/s and a deceleration of 0.18 ft/s^2.

13.6. POTENTIAL ENERGY†

Let us consider again a body of weight **W** which moves along a curved path from a point A_1 of elevation y_1 to a point A_2 of elevation y_2 (Fig. 13.4). We recall from Sec. 13.2 that the work of the force of gravity **W** during this displacement is

$$U_{1\rightarrow 2} = Wy_1 - Wy_2 \qquad (13.4)$$

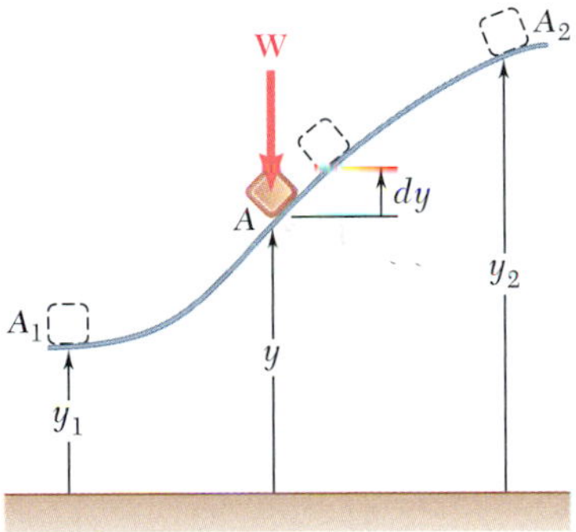

Fig. 13.4 (*repeated*)

The work of **W** may thus be obtained by subtracting the value of the function Wy corresponding to the second position of the body from its value corresponding to the first position. The work of **W** is independent of the actual path followed; it depends only upon the initial and final values of the function Wy. This function is called the *potential energy* of the body with respect to the *force of gravity* **W** and is denoted by V_g. We write

$$U_{1\rightarrow 2} = (V_g)_1 - (V_g)_2 \qquad \text{with } V_g = Wy \qquad (13.16)$$

We note that if $(V_g)_2 > (V_g)_1$, that is, *if the potential energy increases* during the displacement (as in the case considered here), *the work* $U_{1\rightarrow 2}$ *is negative*. If, on the other hand, the work of **W** is positive, the potential energy decreases. Therefore, the potential energy V_g of the body provides a measure of the work which can be done by its weight **W**. Since only the *change* in potential energy, and not the actual value of V_g, is involved in formula (13.16), an arbitrary constant can be added to the expression obtained for V_g. In other words, the level, or datum, from which the elevation y is measured can be chosen arbitrarily. Note that potential energy is expressed in the same units as work, that is, in joules if SI units are used and in ft · lb or in · lb if U.S. customary units are used.

†Some of the material in this section has already been considered in Sec. 10.7.

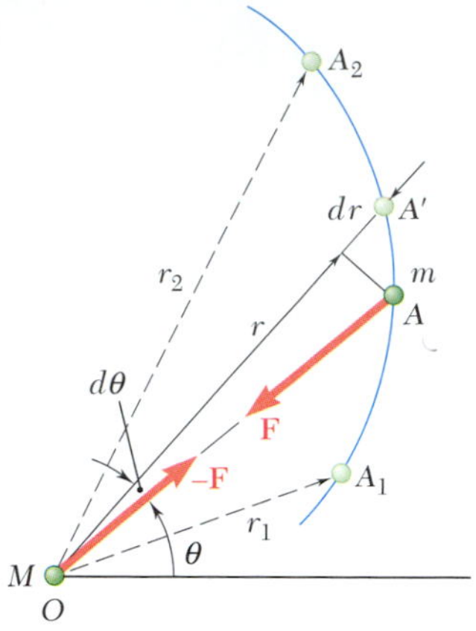

Fig. 13.6 (*repeated*)

It should be noted that the expression just obtained for the potential energy of a body with respect to gravity is valid only as long as the weight **W** of the body can be assumed to remain constant, that is, as long as the displacements of the body are small compared with the radius of the earth. In the case of a space vehicle, however, we should take into consideration the variation of the force of gravity with the distance r from the center of the earth. Using the expression obtained in Sec. 13.2 for the work of a gravitational force, we write (Fig. 13.6).

$$U_{1\rightarrow 2} = \frac{GMm}{r_2} - \frac{GMm}{r_1} \tag{13.7}$$

The work of the force of gravity can therefore be obtained by subtracting the value of the function $-GMm/r$ corresponding to the second position of the body from its value corresponding to the first position. Thus, the expression which should be used for the potential energy V_g when the variation in the force of gravity cannot be neglected is

$$V_g = -\frac{GMm}{r} \tag{13.17}$$

Taking the first of the relations (12.29) into account, we write V_g in the alternative form

$$V_g = -\frac{WR^2}{r} \tag{13.17$'$}$$

where R is the radius of the earth and W is the value of the weight of the body at the surface of the earth. When either of the relations (13.17) or (13.17′) is used to express V_g, the distance r should, of course, be measured from the center of the earth.† Note that V_g is always negative and that it approaches zero for very large values of r.

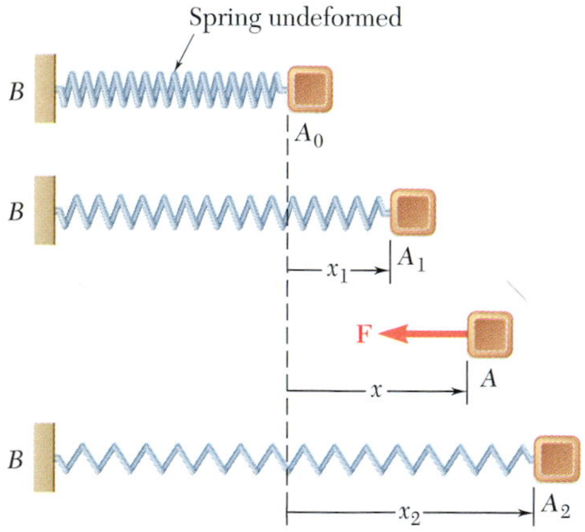

Fig. 13.5 (*repeated*)

Consider now a body attached to a spring and moving from a position A_1, corresponding to a deflection x_1 of the spring, to a position A_2, corresponding to a deflection x_2 of the spring (Fig. 13.5). We recall from Sec. 13.2 that the work of the force **F** exerted by the spring on the body is

$$U_{1\rightarrow 2} = \tfrac{1}{2}kx_1^2 - \tfrac{1}{2}kx_2^2 \tag{13.6}$$

The work of the elastic force is thus obtained by subtracting the value of the function $\frac{1}{2}kx^2$ corresponding to the second position of the body from its value corresponding to the first position. This function is denoted by V_e and is called the *potential energy* of the body with respect to the *elastic force* **F**. We write

$$U_{1\rightarrow 2} = (V_e)_1 - (V_e)_2 \qquad \text{with } V_e = \tfrac{1}{2}kx^2 \tag{13.18}$$

and observe that during the displacement considered, the work of the force **F** exerted by the spring on the body is negative and the

†The expressions given for V_g in (13.17) and (13.17′) are valid only when $r \geq R$, that is, when the body considered is above the surface of the earth.

potential energy V_e increases. You should note that the expression obtained for V_e is valid only if the deflection of the spring is measured from its undeformed position. On the other hand, formula (13.18) can be used even when the spring is rotated about its fixed end (Fig. 13.10*a*). The work of the elastic force depends only upon the initial and final deflections of the spring (Fig. 13.10*b*).

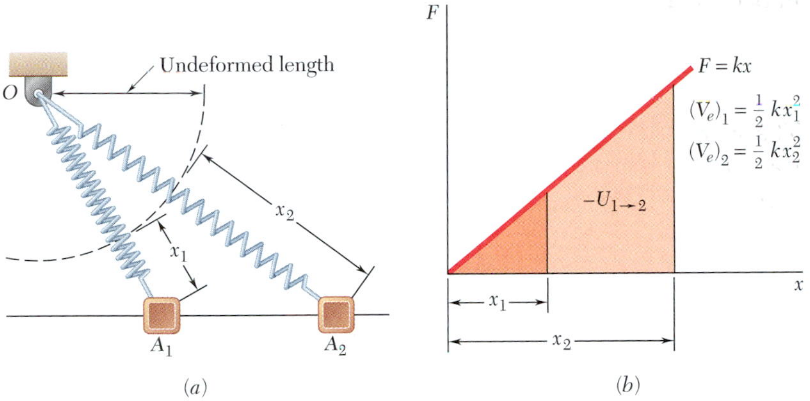

Fig. 13.10

The concept of potential energy can be used when forces other than gravity forces and elastic forces are involved. Indeed, it remains valid as long as the work of the force considered is independent of the path followed by its point of application as this point moves from a given position A_1 to a given position A_2. Such forces are said to be *conservative forces;* the general properties of conservative forces are studied in the following section.

*13.7. CONSERVATIVE FORCES

As indicated in the preceding section, a force **F** acting on a particle A is said to be conservative *if its work $U_{1\to2}$ is independent of the path followed by the particle A as it moves from A_1 to A_2* (Fig. 13.11*a*). We can then write

$$U_{1\to2} = V(x_1, y_1, z_1) - V(x_2, y_2, z_2) \tag{13.19}$$

or, for short,

$$U_{1\to2} = V_1 - V_2 \tag{13.19'}$$

The function $V(x, y, z)$ is called the potential energy, or *potential function,* of **F**.

We note that if A_2 is chosen to coincide with A_1, that is, if the particle describes a closed path (Fig. 13.11*b*), we have $V_1 = V_2$ and the work is zero. Thus for any conservative force **F** we can write

$$\oint \mathbf{F} \cdot d\mathbf{r} = 0 \tag{13.20}$$

where the circle on the integral sign indicates that the path is closed.

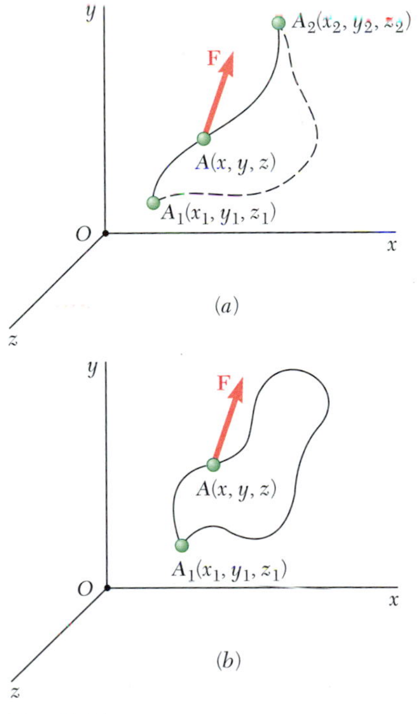

Fig. 13.11

Let us now apply (13.19) between two neighboring points $A(x, y, z)$ and $A'(x + dx, y + dy, z + dz)$. The elementary work dU corresponding to the displacement $d\mathbf{r}$ from A to A' is

$$dU = V(x, y, z) - V(x + dx, y + dy, z + dz)$$

or

$$dU = -dV(x, y, z) \tag{13.21}$$

Thus, the elementary work of a conservative force is an *exact differential.*

Substituting for dU in (13.21) the expression obtained in (13.1″) and recalling the definition of the differential of a function of several variables, we write

$$F_x\,dx + F_y\,dy + F_z\,dz = -\left(\frac{\partial V}{\partial x}dx + \frac{\partial V}{\partial y}dy + \frac{\partial V}{\partial z}dz\right)$$

from which it follows that

$$F_x = -\frac{\partial V}{\partial x} \qquad F_y = -\frac{\partial V}{\partial y} \qquad F_z = -\frac{\partial V}{\partial z} \tag{13.22}$$

It is clear that the components of $\mathbf{F}$ must be functions of the coordinates x, y, and z. Thus, a *necessary* condition for a conservative force is that it depend only upon the position of its point of application. The relations (13.22) can be expressed more concisely if we write

$$\mathbf{F} = F_x\mathbf{i} + F_y\mathbf{j} + F_z\mathbf{k} = -\left(\frac{\partial V}{\partial x}\mathbf{i} + \frac{\partial V}{\partial y}\mathbf{j} + \frac{\partial V}{\partial z}\mathbf{k}\right)$$

The vector in parentheses is known as the *gradient of the scalar function* V and is denoted by **grad** V. We thus write for any conservative force

$$\mathbf{F} = -\mathbf{grad}\,V \tag{13.23}$$

The relations (13.19) to (13.23) were shown to be satisfied by any conservative force. It can also be shown that if a force $\mathbf{F}$ satisfies one of these relations, $\mathbf{F}$ must be a conservative force.

13.8. CONSERVATION OF ENERGY

We saw in the preceding two sections that the work of a conservative force, such as the weight of a particle or the force exerted by a spring, can be expressed as a change in potential energy. When a particle moves under the action of conservative forces, the principle of work and energy stated in Sec. 13.3 can be expressed in a modified form. Substituting for $U_{1\rightarrow 2}$ from (13.19′) into (13.10), we write

$$V_1 - V_2 = T_2 - T_1$$

$$T_1 + V_1 = T_2 + V_2 \tag{13.24}$$

Formula (13.24) indicates that when a particle moves under the action of conservative forces, *the sum of the kinetic energy and of the potential energy of the particle remains constant*. The sum $T + V$ is called the *total mechanical energy* of the particle and is denoted by E.

Consider, for example, the pendulum analyzed in Sec. 13.4, which is released with no velocity from A_1 and allowed to swing in a vertical plane (Fig. 13.12). Measuring the potential energy from the level of A_2, we have, at A_1,

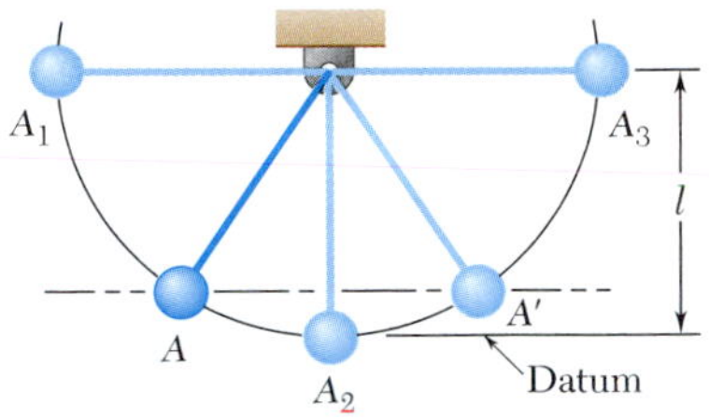

Fig. 13.12

$$T_1 = 0 \qquad V_1 = Wl \qquad T_1 + V_1 = Wl$$

Recalling that at A_2 the speed of the pendulum is $v_2 = \sqrt{2gl}$, we have

$$T_2 = \tfrac{1}{2}mv_2^2 = \frac{1}{2}\frac{W}{g}(2gl) = Wl \qquad V_2 = 0$$
$$T_2 + V_2 = Wl$$

We thus check that the total mechanical energy $E = T + V$ of the pendulum is the same at A_1 and A_2. Whereas the energy is entirely potential at A_1, it becomes entirely kinetic at A_2, and as the pendulum keeps swinging to the right, the kinetic energy is transformed back into potential energy. At A_3, $T_3 = 0$ and $V_3 = Wl$.

Since the total mechanical energy of the pendulum remains constant and since its potential energy depends only upon its elevation, the kinetic energy of the pendulum will have the same value at any two points located on the same level. Thus, the speed of the pendulum is the same at A and at A' (Fig. 13.12). This result can be extended to the case of a particle moving along any given path, regardless of the shape of the path, as long as the only forces acting on the particle are its weight and the normal reaction of the path. The particle of Fig. 13.13, for example, which slides in a vertical plane along a frictionless track, will have the same speed at A, A', and A''.

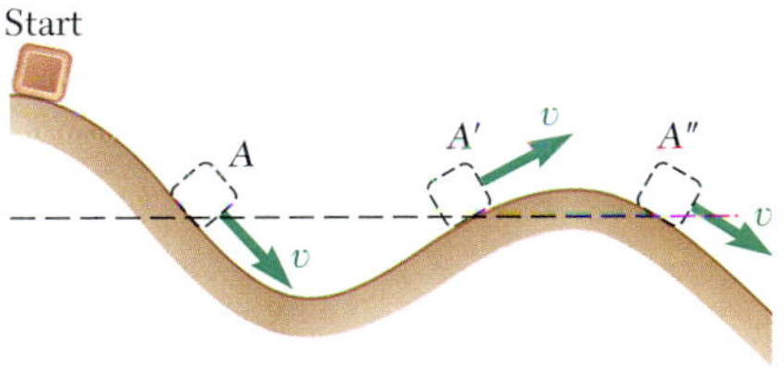

Fig. 13.13

While the weight of a particle and the force exerted by a spring are conservative forces, *friction forces are nonconservative forces*. In other words, *the work of a friction force cannot be expressed as a change in potential energy*. The work of a friction force depends upon the path followed by its point of application; and while the work $U_{1\rightarrow2}$ defined by (13.19) is positive or negative according to the sense of motion, *the work of a friction force*, as we noted in Sec. 13.4, *is always negative*. It follows that when a mechanical system involves friction, its total mechanical energy does not remain constant but decreases. The energy of the system, however, is not lost; it is transformed into heat, and the sum of the *mechanical energy* and of the *thermal energy* of the system remains constant.

Other forms of energy can also be involved in a system. For instance, a generator converts mechanical energy into *electric energy;* a gasoline engine converts *chemical energy* into mechanical energy; a nuclear reactor converts *mass* into thermal energy. If all forms of energy are considered, the energy of any system can be considered as constant and the principle of conservation of energy remains valid under all conditions.

13.9. MOTION UNDER A CONSERVATIVE CENTRAL FORCE. APPLICATION TO SPACE MECHANICS

We saw in Sec. 12.9 that when a particle P moves under a central force $\mathbf{F}$, the angular momentum $\mathbf{H}_O$ of the particle about the center of force O is constant. If the force $\mathbf{F}$ is also conservative, there exists a potential energy V associated with $\mathbf{F}$, and the total energy $E = T + V$ of the particle is constant (Sec. 13.8). Thus, when a particle moves under a conservative central force, both the principle of conservation of angular momentum and the principle of conservation of energy can be used to study its motion.

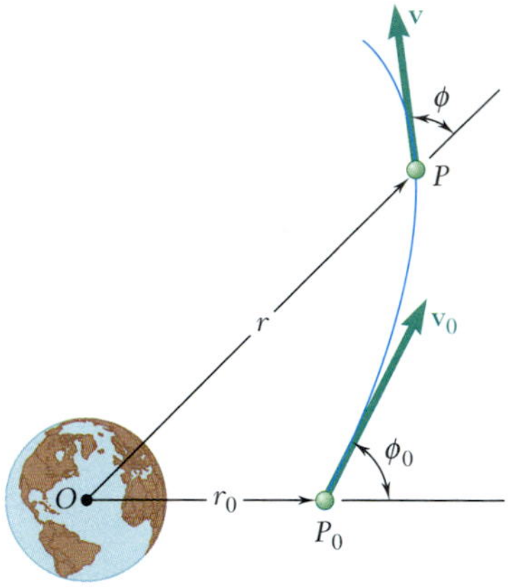

Fig. 13.14

Consider, for example, a space vehicle of mass m moving under the earth's gravitational force. Let us assume that it begins its free flight at point P_0 at a distance r_0 from the center of the earth, with a velocity $\mathbf{v}_0$ forming an angle ϕ_0 with the radius vector OP_0 (Fig. 13.14). Let P be a point of the trajectory described by the vehicle; we denote by r the distance from O to P, by $\mathbf{v}$ the velocity of the vehicle at P, and by ϕ the angle formed by $\mathbf{v}$ and the radius vector OP. Applying the principle of conservation of angular momentum about O between P_0 and P (Sec. 12.9), we write

$$r_0 m v_0 \sin \phi_0 = r m v \sin \phi \qquad (13.25)$$

Recalling the expression (13.17) obtained for the potential energy due to a gravitational force, we apply the principle of conservation of energy between P_0 and P and write

$$T_0 + V_0 = T + V$$

$$\tfrac{1}{2} m v_0^2 - \frac{GMm}{r_0} = \tfrac{1}{2} m v^2 - \frac{GMm}{r} \qquad (13.26)$$

where M is the mass of the earth.

Equation (13.26) can be solved for the magnitude v of the velocity of the vehicle at P when the distance r from O to P is known; Eq. (13.25) can then be used to determine the angle ϕ that the velocity forms with the radius vector OP.

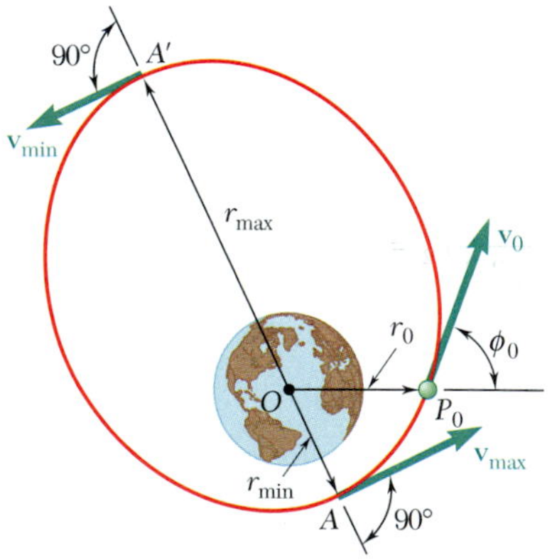

Fig. 13.15

Equations (13.25) and (13.26) can also be used to determine the maximum and minimum values of r in the case of a satellite launched from P_0 in a direction forming an angle ϕ_0 with the vertical OP_0 (Fig. 13.15). The desired values of r are obtained by making $\phi = 90°$ in (13.25) and eliminating v between Eqs. (13.25) and (13.26).

It should be noted that the application of the principles of conservation of energy and of conservation of angular momentum leads to a more fundamental formulation of the problems of space mechanics than does the method indicated in Sec. 12.12. In all cases involving oblique launchings, it will also result in much simpler computations. And while the method of Sec. 12.12 must be used when the actual trajectory or the periodic time of a space vehicle is to be determined, the calculations will be simplified if the conservation principles are first used to compute the maximum and minimum values of the radius vector r.

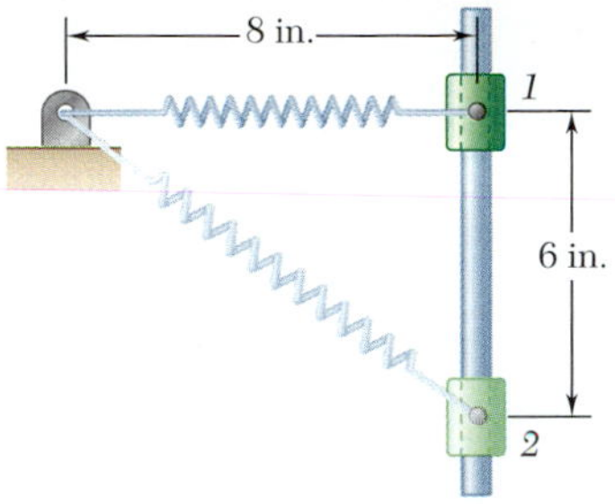

SAMPLE PROBLEM 13.6

A 20-lb collar slides without friction along a vertical rod as shown. The spring attached to the collar has an undeformed length of 4 in. and a constant of 3 lb/in. If the collar is released from rest in position *1*, determine its velocity after it has moved 6 in. to position *2*.

SOLUTION

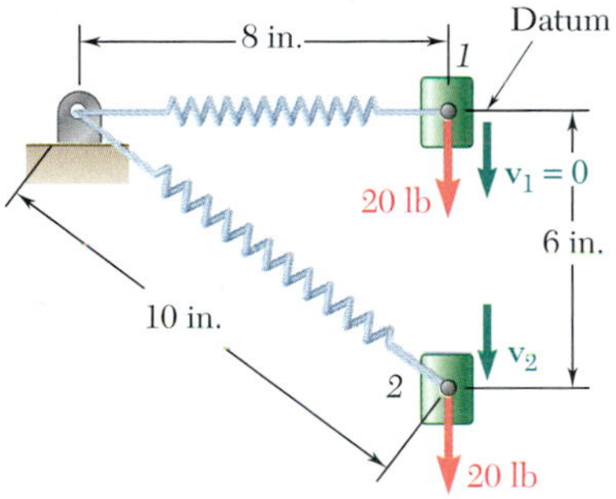

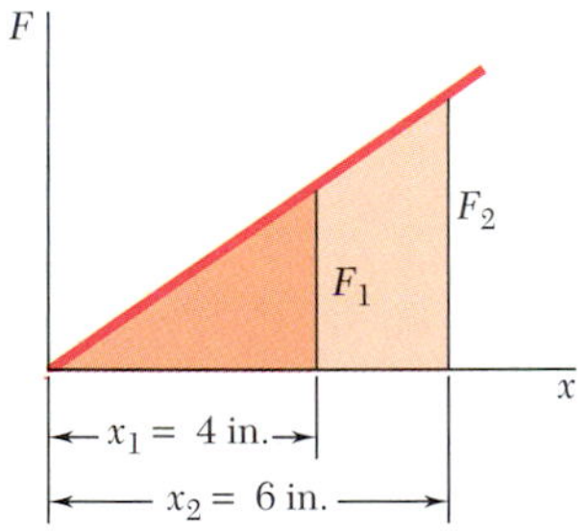

Position *1*. *Potential Energy.* The elongation of the spring is

$$x_1 = 8 \text{ in.} - 4 \text{ in.} = 4 \text{ in.}$$

and we have

$$V_e = \tfrac{1}{2}kx_1^2 = \tfrac{1}{2}(3 \text{ lb/in.})(4 \text{ in.})^2 = 24 \text{ in} \cdot \text{lb}$$

Choosing the datum as shown, we have $V_g = 0$. Therefore,

$$V_1 = V_e + V_g = 24 \text{ in} \cdot \text{lb} = 2 \text{ ft} \cdot \text{lb}$$

Kinetic Energy. Since the velocity in position *1* is zero, $T_1 = 0$.

Position *2*. *Potential Energy.* The elongation of the spring is

$$x_2 = 10 \text{ in.} - 4 \text{ in.} = 6 \text{ in.}$$

and we have

$$V_e = \tfrac{1}{2}kx_2^2 = \tfrac{1}{2}(3 \text{ lb/in.})(6 \text{ in.})^2 = 54 \text{ in} \cdot \text{lb}$$
$$V_g = Wy = (20 \text{ lb})(-6 \text{ in.}) = -120 \text{ in} \cdot \text{lb}$$

Therefore,

$$V_2 = V_e + V_g = 54 - 120 = -66 \text{ in} \cdot \text{lb}$$
$$= -5.5 \text{ ft} \cdot \text{lb}$$

Kinetic Energy

$$T_2 = \tfrac{1}{2}mv_2^2 = \frac{1}{2}\frac{20}{32.2}v_2^2 = 0.311v_2^2$$

Conservation of Energy. Applying the principle of conservation of energy between positions *1* and *2*, we write

$$T_1 + V_1 = T_2 + V_2$$
$$0 + 2 \text{ ft} \cdot \text{lb} = 0.311v_2^2 - 5.5 \text{ ft} \cdot \text{lb}$$
$$v_2 = \pm 4.91 \text{ ft/s}$$

$$\mathbf{v}_2 = 4.91 \text{ ft/s} \downarrow \blacktriangleleft$$

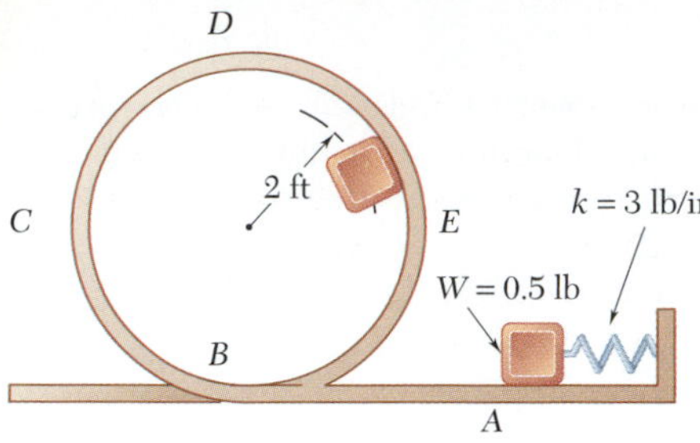

SAMPLE PROBLEM 13.7

The 0.5-lb pellet is pushed against the spring at A and released from rest. Neglecting friction, determine the smallest deflection of the spring for which the pellet will travel around the loop $ABCDE$ and remain at all times in contact with the loop.

SOLUTION

Required Speed at Point *D*. As the pellet passes through the highest point D, its potential energy with respect to gravity is maximum and thus, its kinetic energy and speed are minimum. Since the pellet must remain in contact with the loop, the force $\mathbf{N}$ exerted on the pellet by the loop must be equal to or greater than zero. Setting $\mathbf{N} = 0$, we compute the smallest possible speed v_D.

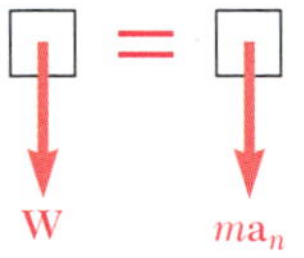

$$+\downarrow \Sigma F_n = ma_n: \qquad W = ma_n \qquad mg = ma_n \qquad a_n = g$$

$$a_n = \frac{v_D^2}{r}: \qquad v_D^2 = ra_n = rg = (2 \text{ ft})(32.2 \text{ ft/s}^2) = 64.4 \text{ ft}^2/\text{s}^2$$

Position *1*. *Potential Energy.* Denoting by x the deflection of the spring and noting that $k = 3$ lb/in. = 36 lb/ft, we write

$$V_e = \tfrac{1}{2}kx^2 = \tfrac{1}{2}(36 \text{ lb/ft})x^2 = 18x^2$$

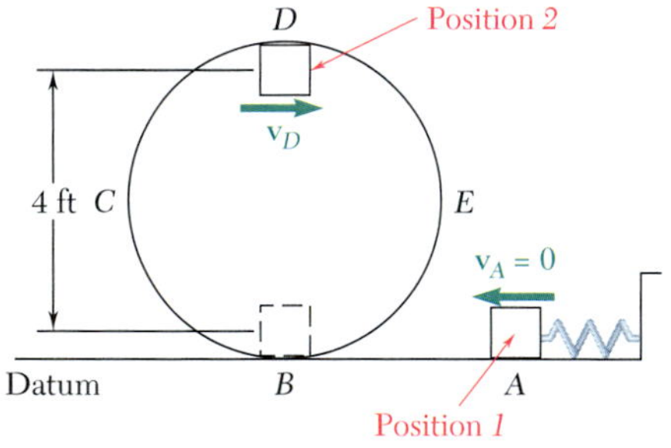

Choosing the datum at A, we have $V_g = 0$; therefore

$$V_1 = V_e + V_g = 18x^2$$

Kinetic Energy. Since the pellet is released from rest, $v_A = 0$ and we have $T_1 = 0$.

Position *2*. *Potential Energy.* The spring is now undeformed; thus $V_e = 0$. Since the pellet is 4 ft above the datum, we have

$$V_g = Wy = (0.5 \text{ lb})(4 \text{ ft}) = 2 \text{ ft} \cdot \text{lb}$$
$$V_2 = V_e + V_g = 2 \text{ ft} \cdot \text{lb}$$

Kinetic Energy. Using the value of v_D^2 obtained above, we write

$$T_2 = \tfrac{1}{2}mv_D^2 = \frac{1}{2}\frac{0.5 \text{ lb}}{32.2 \text{ ft/s}^2}(64.4 \text{ ft}^2/\text{s}^2) = 0.5 \text{ ft} \cdot \text{lb}$$

Conservation of Energy. Applying the principle of conservation of energy between positions *1* and *2*, we write

$$T_1 + V_1 = T_2 + V_2$$
$$0 + 18x^2 = 0.5 \text{ ft} \cdot \text{lb} + 2 \text{ ft} \cdot \text{lb}$$
$$x = 0.3727 \text{ ft} \qquad x = 4.47 \text{ in.} \blacktriangleleft$$

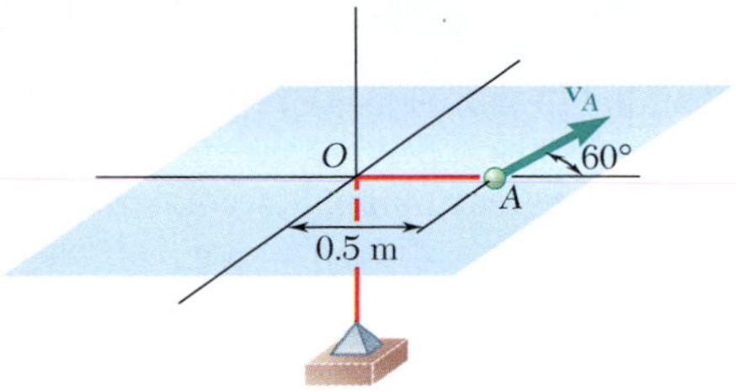

SAMPLE PROBLEM 13.8

A sphere of mass $m = 0.6$ kg is attached to an elastic cord of constant $k =$ 100 N/m, which is undeformed when the sphere is located at the origin O. Knowing that the sphere may slide without friction on the horizontal surface and that in the position shown its velocity $\mathbf{v}_A$ has a magnitude of 20 m/s, determine (*a*) the maximum and minimum distances from the sphere to the origin O, (*b*) the corresponding values of its speed.

SOLUTION

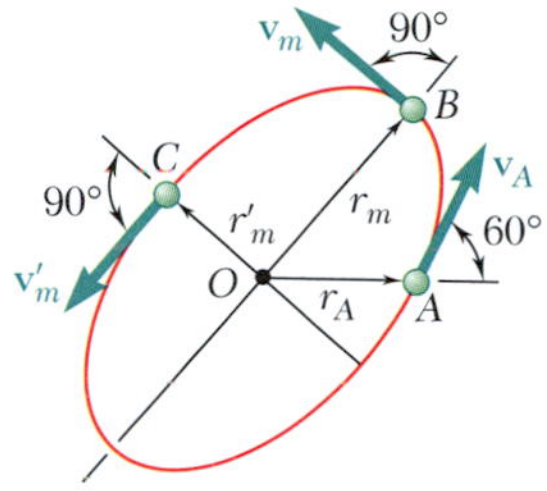

The force exerted by the cord on the sphere passes through the fixed point O, and its work can be expressed as a change in potential energy. It is therefore a conservative central force, and both the total energy of the sphere and its angular momentum about O are conserved.

Conservation of Angular Momentum about *O*. At point B, where the distance from O is maximum, the velocity of the sphere is perpendicular to OB and the angular momentum is $r_m m v_m$. A similar property holds at point C, where the distance from O is minimum. Expressing conservation of angular momentum between A and B, we write

$$r_A m v_A \sin 60° = r_m m v_m$$
$$(0.5 \text{ m})(0.6 \text{ kg})(20 \text{ m/s}) \sin 60° = r_m(0.6 \text{ kg})v_m$$
$$v_m = \frac{8.66}{r_m} \qquad (1)$$

Conservation of Energy

At point A: $T_A = \frac{1}{2}mv_A^2 = \frac{1}{2}(0.6 \text{ kg})(20 \text{ m/s})^2 = 120 \text{ J}$
$V_A = \frac{1}{2}kr_A^2 = \frac{1}{2}(100 \text{ N/m})(0.5 \text{ m})^2 = 12.5 \text{ J}$

At point B: $T_B = \frac{1}{2}mv_m^2 = \frac{1}{2}(0.6 \text{ kg})v_m^2 = 0.3v_m^2$
$V_B = \frac{1}{2}kr_m^2 = \frac{1}{2}(100 \text{ N/m})r_m^2 = 50r_m^2$

Applying the principle of conservation of energy between points A and B, we write

$$T_A + V_A = T_B + V_B$$
$$120 + 12.5 = 0.3v_m^2 + 50r_m^2 \qquad (2)$$

***a.* Maximum and Minimum Values of Distance.** Substituting for v_m from Eq. (1) into Eq. (2) and solving for r_m^2, we obtain

$$r_m^2 = 2.468 \text{ or } 0.1824 \qquad r_m = 1.571 \text{ m}, \; r'_m = 0.427 \text{ m} \blacktriangleleft$$

***b.* Corresponding Values of Speed.** Substituting the values obtained for r_m and r'_m into Eq. (1), we have

$$v_m = \frac{8.66}{1.571} \qquad v_m = 5.51 \text{ m/s} \blacktriangleleft$$

$$v'_m = \frac{8.66}{0.427} \qquad v'_m = 20.3 \text{ m/s} \blacktriangleleft$$

Note. It can be shown that the path of the sphere is an ellipse of center O.

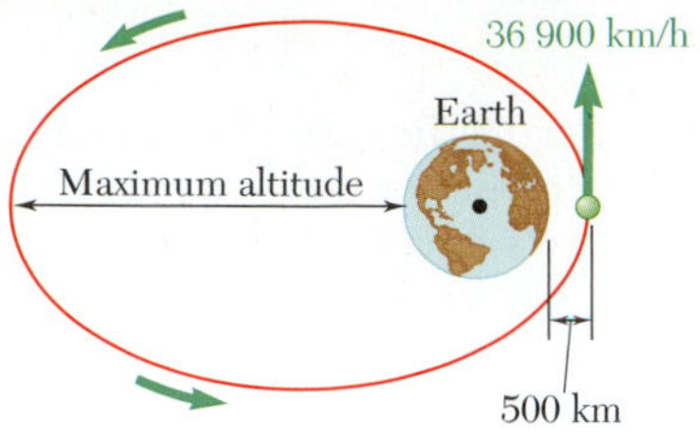

SAMPLE PROBLEM 13.9

A satellite is launched in a direction parallel to the surface of the earth with a velocity of 36 900 km/h from an altitude of 500 km. Determine (*a*) the maximum altitude reached by the satellite, (*b*) the maximum allowable error in the direction of launching if the satellite is to go into orbit and come no closer than 200 km to the surface of the earth.

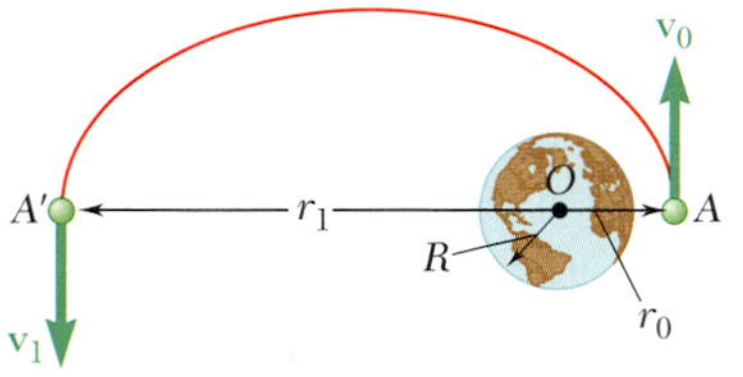

SOLUTION

a. Maximum Altitude. We denote by A' the point of the orbit farthest from the earth and by r_1 the corresponding distance from the center of the earth. Since the satellite is in free flight between A and A', we apply the principle of conservation of energy:

$$T_A + V_A = T_{A'} + V_{A'}$$

$$\tfrac{1}{2}mv_0^2 - \frac{GMm}{r_0} = \tfrac{1}{2}mv_1^2 - \frac{GMm}{r_1} \qquad (1)$$

Since the only force acting on the satellite is the force of gravity, which is a central force, the angular momentum of the satellite about O is conserved. Considering points A and A', we write

$$r_0 m v_0 = r_1 m v_1 \qquad v_1 = v_0 \frac{r_0}{r_1} \qquad (2)$$

Substituting this expression for v_1 into Eq. (1), dividing each term by the mass m, and rearranging the terms, we obtain

$$\tfrac{1}{2}v_0^2\left(1 - \frac{r_0^2}{r_1^2}\right) = \frac{GM}{r_0}\left(1 - \frac{r_0}{r_1}\right) \qquad 1 + \frac{r_0}{r_1} = \frac{2GM}{r_0 v_0^2} \qquad (3)$$

Recalling that the radius of the earth is $R = 6370$ km, we compute

$$r_0 = 6370 \text{ km} + 500 \text{ km} = 6870 \text{ km} = 6.87 \times 10^6 \text{ m}$$
$$v_0 = 36\,900 \text{ km/h} = (36.9 \times 10^6 \text{ m})/(3.6 \times 10^3 \text{ s}) = 10.25 \times 10^3 \text{ m/s}$$
$$GM = gR^2 = (9.81 \text{ m/s}^2)(6.37 \times 10^6 \text{ m})^2 = 398 \times 10^{12} \text{ m}^3/\text{s}^2$$

Substituting these values into (3), we obtain $r_1 = 66.8 \times 10^6$ m.

Maximum altitude $= 66.8 \times 10^6 \text{ m} - 6.37 \times 10^6 \text{ m} = 60.4 \times 10^6 \text{ m} =$ **60 400 km** ◀

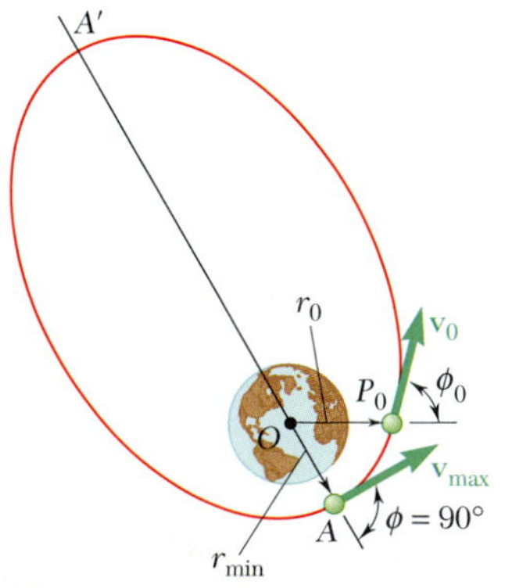

b. Allowable Error in Direction of Launching. The satellite is launched from P_0 in a direction forming an angle ϕ_0 with the vertical OP_0. The value of ϕ_0 corresponding to $r_{min} = 6370$ km $+ 200$ km $= 6570$ km is obtained by applying the principles of conservation of energy and of conservation of angular momentum between P_0 and A:

$$\tfrac{1}{2}mv_0^2 - \frac{GMm}{r_0} = \tfrac{1}{2}mv_{max}^2 - \frac{GMm}{r_{min}} \qquad (4)$$

$$r_0 m v_0 \sin \phi_0 = r_{min} m v_{max} \qquad (5)$$

Solving (5) for v_{max} and then substituting for v_{max} into (4), we can solve (4) for $\sin \phi_0$. Using the values of v_0 and GM computed in part *a* and noting that $r_0/r_{min} = 6870/6570 = 1.0457$, we find

$\sin \phi_0 = 0.9801 \qquad \phi_0 = 90° \pm 11.5° \qquad$ Allowable error $= \pm 11.5°$ ◀

SOLVING PROBLEMS ON YOUR OWN

In this lesson you learned that when the work done by a force **F** acting on a particle A *is independent of the path followed by the particle* as it moves from a given position A_1 to a given position A_2 (Fig. 13.11*a*), then a function V, called *potential energy*, can be defined for the force **F**. Such forces are said to be *conservative forces*, and you can write

$$U_{1\rightarrow 2} = V(x_1, y_1, z_1) - V(x_2, y_2, z_2) \qquad (13.19)$$

or, for short,

$$U_{1\rightarrow 2} = V_1 - V_2 \qquad (13.19')$$

Note that the work is negative when the change in the potential energy is positive, that is, when $V_2 > V_1$.

Substituting the above expression into the equation for work and energy, you can write

$$T_1 + V_1 = T_2 + V_2 \qquad (13.24)$$

which shows that when a particle moves under the action of a conservative force *the sum of the kinetic and potential energies of the particle remains constant.*

Your solution of problems using the above formula will consist of the following steps.

1. *Determine whether all the forces involved are conservative.* If some of the forces are not conservative, for example if friction is involved, you must use the method of work and energy from the previous lesson, since the work done by such forces depends upon the path followed by the particle and a potential function does not exist. If there is no friction and if all the forces are conservative, you can proceed as follows.

2. *Determine the kinetic energy $T = \frac{1}{2}mv^2$ at each end of the path.*

(continued)

3. Compute the potential energy for all the forces involved at each end of the path. You will recall that the following expressions for the potential energy were derived in this lesson.

a. The potential energy of a weight W close to the surface of the earth and at a height y above a given datum,

$$V_g = Wy \tag{13.16}$$

b. The potential energy of a mass m located at a distance r from the center of the earth, large enough so that the variation of the force of gravity must be taken into account,

$$V_g = -\frac{GMm}{r} \tag{13.17}$$

where the distance r is measured from the center of the earth and V_g is equal to zero at $r = \infty$.

c. The potential energy of a body with respect to an elastic force F = kx,

$$V_e = \tfrac{1}{2}kx^2 \tag{13.18}$$

where the distance x is the deflection of the elastic spring measured from its *undeformed* position and k is the spring constant. Note that V_e *depends only upon the deflection* x and not upon the path of the body attached to the spring. Also, V_e is always positive, whether the spring is compressed or elongated.

4. Substitute your expressions for the kinetic and potential energies into Eq. (13.24) above. You will be able to solve this equation for one unknown, for example, for a velocity [Sample Prob. 13.6]. If more than one unknown is involved, you will have to search for another condition or equation, such as the minimum speed [Sample Prob. 13.7] or the minimum potential energy of the particle. For problems involving a central force, a second equation can be obtained by using conservation of angular momentum [Sample Prob. 13.8]. This is especially useful in applications to space mechanics [Sec. 13.9].

Problems

13.55 A force **P** is slowly applied to a plate that is attached to two springs and causes a deflection x_0. In each of the two cases shown, derive an expression for the constant k_e, in terms of k_1 and k_2, of the single spring equivalent to the given system, that is, of the single spring which will undergo the same deflection x_0 when subjected to the same force **P**.

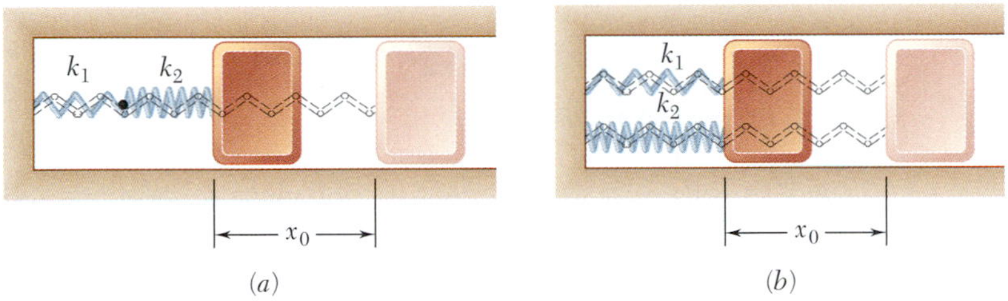

Fig. P13.55

13.56 A 3-kg block can slide without friction in a slot and is attached as shown to three springs of equal length, and of spring constants $k_1 =$ 1 kN/m, $k_2 = 2$ kN/m, $k_3 = 4$ kN/m. The springs are initially unstretched when the block is pushed to the left 45 mm and released. Determine (*a*) the maximum velocity of the block, (*b*) the velocity of the block when it is 18 mm from its initial position.

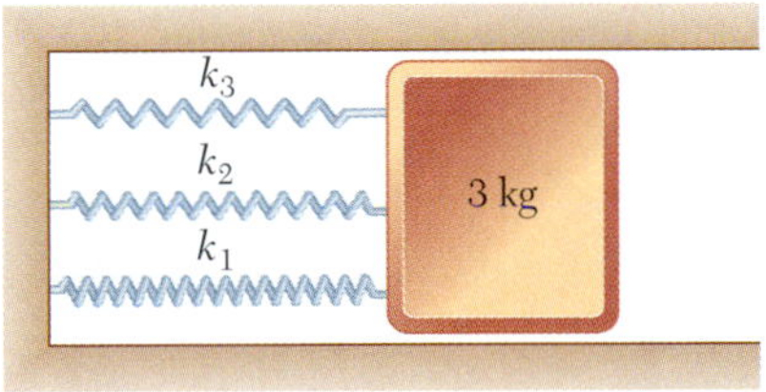

Fig. P13.56

13.57 A 3-lb block can slide without friction in a slot and is connected to two springs of constants $k_1 = 80$ lb/ft and $k_2 = 60$ lb/ft. The springs are initially unstretched when the block is pulled 2 in. to the right and released. Determine (*a*) the maximum velocity of the block, (*b*) the velocity of the block when it is 0.8 in. from its initial position.

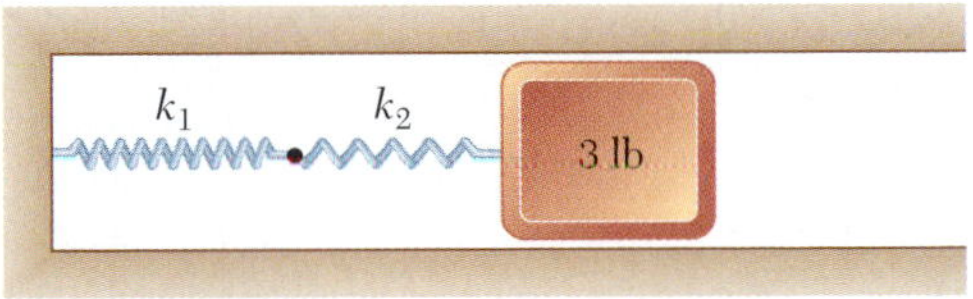

Fig. P13.57

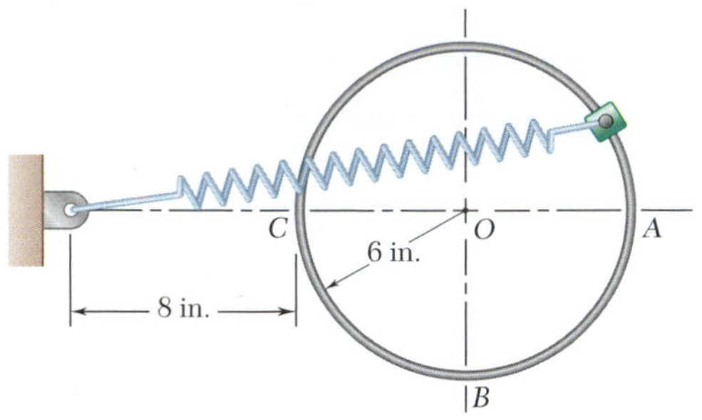

Fig. P13.58

13.58 A 3-lb collar is attached to a spring and slides without friction along a circular rod in a *horizontal* plane. The spring has an undeformed length of 7 in. and a constant $k = 1.5$ lb/in. Knowing that the collar is in equilibrium at A and is given a slight push to get it moving, determine the velocity of the collar (a) as it passes through B, (b) as it passes through C.

13.59 A 2-kg collar can slide without friction along a horizontal rod and is in equilibrium at A when it is pushed 25 mm to the right and released from rest. The springs are undeformed when the collar is at A and the constant of each spring is 500 kN/m. Determine the maximum velocity of the collar.

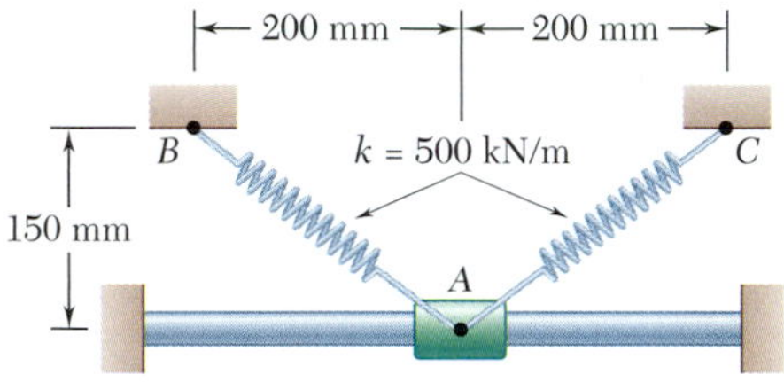

Fig. *P13.59* and *P13.60*

13.60 A 2-kg collar can slide without friction along a horizontal rod and is released from rest at A. The undeformed lengths of springs BA and CA are 250 mm and 225 mm, respectively, and the constant of each spring is 500 kN/m. Determine the velocity of the collar when it has moved 25 mm to the right.

13.61 A thin circular rod is supported in a *vertical plane* by a bracket at A. Attached to the bracket and loosely wound around the rod is a spring of constant $k = 40$ N/m and undeformed length equal to the arc of circle AB. A 200-g collar C, not attached to the spring, can slide without friction along the rod. Knowing that the collar is released from rest when $\theta = 30°$, determine (a) the maximum height above point B reached by the collar, (b) the maximum velocity of the collar.

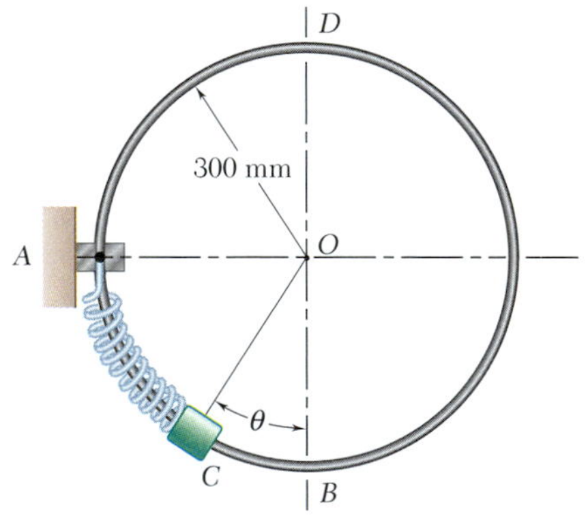

Fig. P13.61 and P13.62

13.62 A thin circular rod is supported in a *vertical plane* by a bracket at A. Attached to the bracket and loosely wound around the rod is a spring of constant $k = 40$ N/m and undeformed length equal to the arc of circle AB. A 200-g collar C, not attached to the spring, can slide without friction along the rod. Knowing that the collar is released from rest at an angle θ with respect to the vertical, determine (a) the smallest value of θ for which the collar will pass through D and reach point A, (b) the velocity of the collar as it reaches point A.

13.63 A 6-lb collar can slide without friction on a vertical rod and is resting in equilibrium on a spring. It is pushed down, compressing the spring 6 in., and released. Knowing that the spring constant is $k = 15$ lb/in., determine (*a*) the maximum height h reached by the collar above its equilibrium position, (*b*) the maximum velocity of the collar.

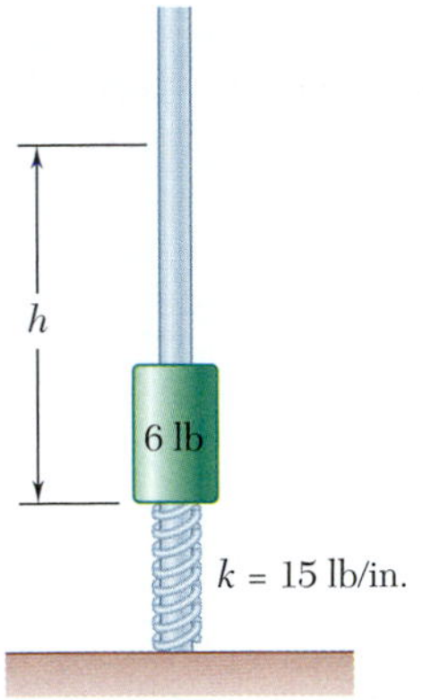

Fig. P13.63 and P13.64

13.64 A 6-lb collar can slide without friction on a vertical rod and is held so it just touches an undeformed spring. Determine the maximum deflection of the spring (*a*) if the collar is slowly released until it reaches an equilibrium position, (*b*) if the collar is suddenly released.

13.65 Blocks A and B weigh 8 lb and 3 lb, respectively, and are connected by a cord-and-pulley system and released from rest in the position shown with the spring undeformed. Knowing that the constant of the spring is 20 lb/ft, determine (*a*) the velocity of block B after it has moved 6 in., (*b*) the maximum velocity of block B, (*c*) the maximum displacement of block B. Ignore friction and the masses of the pulleys and spring.

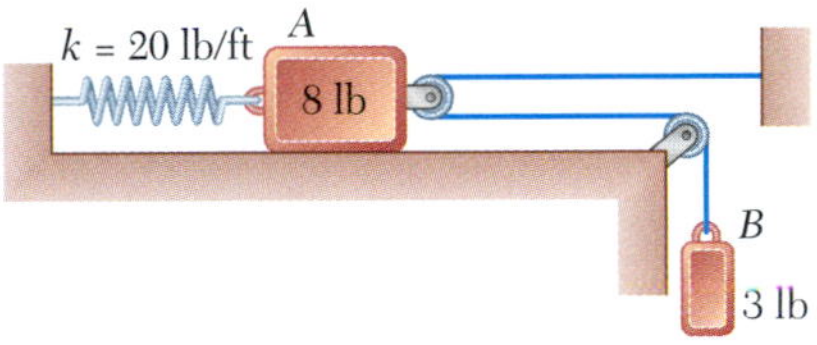

Fig. *P13.65*

13.66 An 8-lb collar A can slide without friction along a vertical rod and is released from rest in the position shown with the springs undeformed. Knowing that the constant of each spring is 20 lb/ft, determine the velocity of the collar after it has moved (*a*) 4 in., (*b*) 7.6 in.

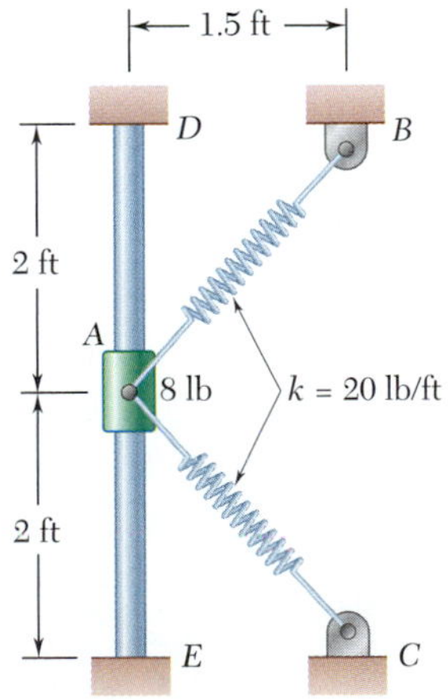

Fig. *P13.66*

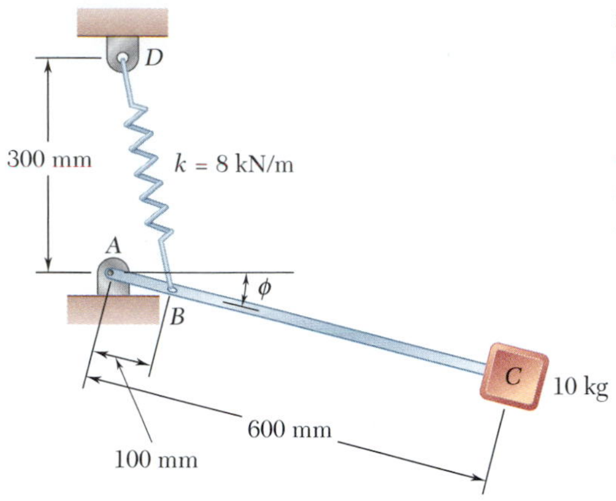

Fig. P13.67

13.67 The system shown is in equilibrium when $\phi = 0$. Knowing that initially $\phi = 90°$ and that block C is given a slight nudge when the system is in that position, determine the velocity of the block as it passes through the equilibrium position $\phi = 0$. Neglect the mass of the rod.

13.68 A 1.2-kg collar can slide along the rod shown. It is attached to an elastic cord anchored at F, which has an undeformed length of 300 mm and a spring constant of 70 N/m. Knowing that the collar is released from rest at A and neglecting friction, determine the speed of the collar (*a*) at B, (*b*) at E.

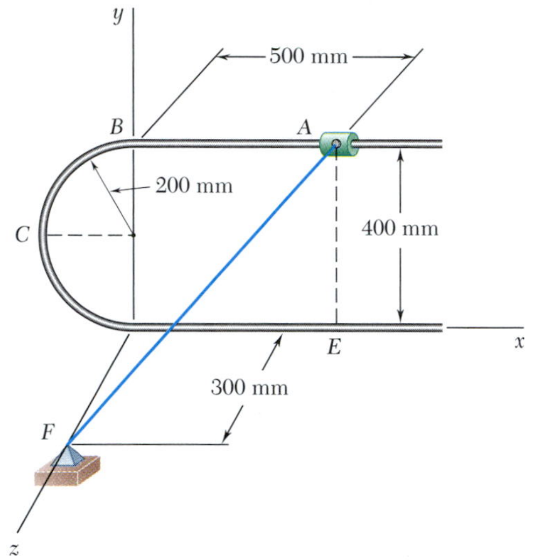

Fig. P13.68

13.69 A 500-g collar can slide without friction along the semicircular rod BCD. The spring is of constant 320 N/m and its undeformed length is 200 mm. Knowing that the collar is released from rest at B, determine (*a*) the speed of the collar as it passes through C, (*b*) the force exerted by the rod on the collar at C.

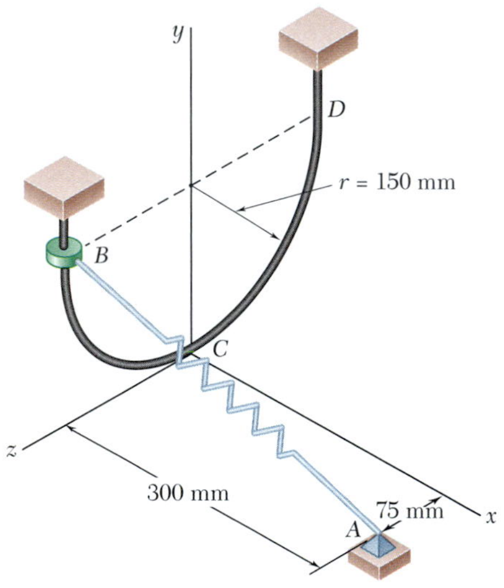

Fig. P13.69

13.70 A thin circular rod is supported in a *vertical plane* by a bracket at A. Attached to the bracket and loosely wound around the rod is a spring of constant $k = 40$ N/m and undeformed length equal to the arc of circle AB. A 200-g collar C is unattached to the spring and can slide without friction along the rod. Knowing that the collar is released from rest when $\theta = 30°$, determine (*a*) the velocity of the collar as it passes through point B, (*b*) the force exerted by the rod on the collar as it passes through B.

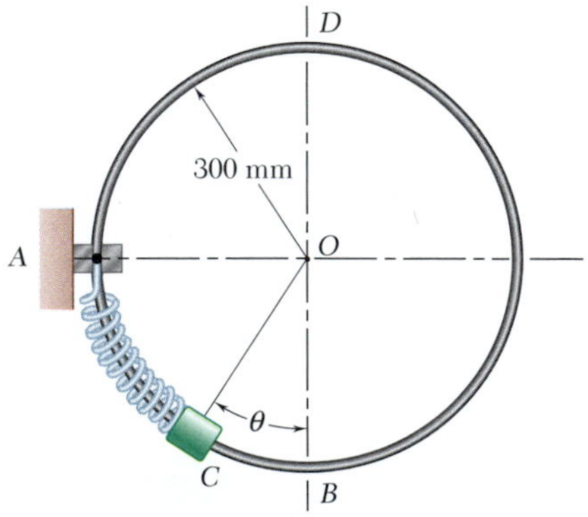

Fig. P13.70

13.71 A 10-oz pellet is released from rest at A and slides without friction along the surface shown. Determine the force exerted on the pellet by the surface (*a*) just before the pellet reaches B, (*b*) immediately after it has passed through B.

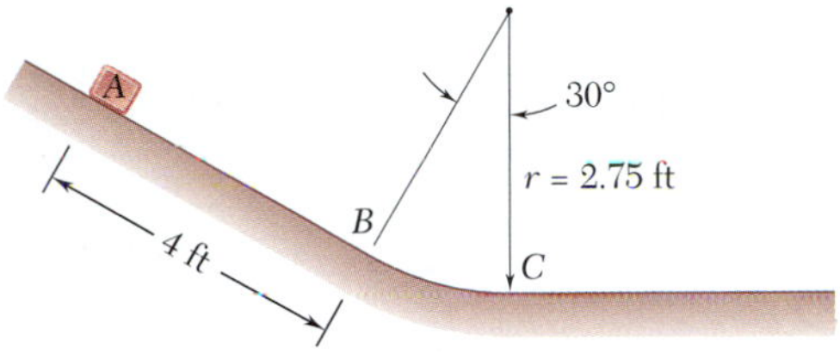

Fig. P13.71 and P13.72

13.72 A 10-oz pellet is released from rest at A and slides without friction along the surface shown. Determine the force exerted on the pellet by the surface (*a*) just before the pellet reaches C, (*b*) immediately after it has passed through C.

13.73 A 1.2-kg collar is attached to a spring and slides without friction along a circular rod in a *vertical* plane. The spring has an undeformed length of 105 mm and a constant $k = 300$ N/m. Knowing that the collar is at rest at C and is given a slight push to get it moving, determine the velocity of the collar and the force exerted by the rod on the collar as it passes through (*a*) point A, (*b*) point B.

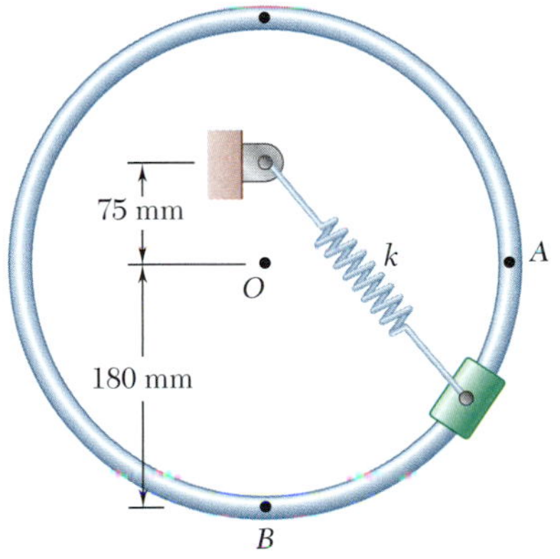

Fig. P13.73 and P13.74

13.74 A 1.2-kg collar is attached to a spring and slides without friction along a circular rod in a *vertical* plane. The spring has an undeformed length of 105 mm and a constant k. The collar is at rest at C and is given a slight push to get it moving. Knowing that the maximum velocity of the collar is achieved as it passes through point A, determine (*a*) the spring constant k, (*b*) the maximum velocity of the collar.

13.75 An 8-oz package is projected upward with a velocity $\mathbf{v}_0$ by a spring at A; it moves around a frictionless loop and is deposited at C. For each of the two loops shown, determine (*a*) the smallest velocity $\mathbf{v}_0$ for which the package will reach C, (*b*) the corresponding force exerted by the package on the loop just before the package leaves the loop at C.

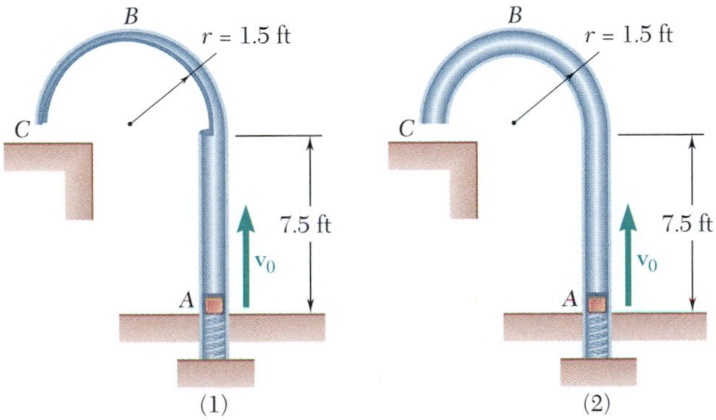

Fig. *P13.75*

13.76 If the package of Prob. 13.75 is not to hit the horizontal surface at C with a speed greater than 10 ft/s, (*a*) show that this requirement can be satisfied only by the second loop, (*b*) determine the largest allowable initial velocity $\mathbf{v}_0$ when the second loop is used.

13.77 Prove that a force $F(x, y, z)$ is conservative if, and only if, the following relations are satisfied:

$$\frac{\partial F_x}{\partial y} = \frac{\partial F_y}{\partial x} \qquad \frac{\partial F_y}{\partial z} = \frac{\partial F_z}{\partial y} \qquad \frac{\partial F_z}{\partial x} = \frac{\partial F_x}{\partial z}$$

13.78 The force $\mathbf{F} = (yz\mathbf{i} + zx\mathbf{j} + xy\mathbf{k})/xyz$ acts on the particle $P(x, y, z)$ which moves in space. (*a*) Using the relation derived in Prob. 13.77, show that this force is a conservative force. (*b*) Determine the potential function associated with $\mathbf{F}$.

13.79 The force $\mathbf{F} = (x\mathbf{i} + y\mathbf{j} + z\mathbf{k})/(x^2 + y^2 + z^2)^{3/2}$ acts on the particle $P(x, y, z)$ which moves in space. (*a*) Using the relation derived in Prob. 13.77, prove that $\mathbf{F}$ is a conservative force. (*b*) Determine the potential function $V(x, y, z)$ associated with $\mathbf{F}$.

***13.80** A force $\mathbf{F}$ acts on a particle $P(x, y)$ which moves in the xy plane. Determine whether $\mathbf{F}$ is a conservative force and compute the work of $\mathbf{F}$ when P describes the path $ABCA$ knowing that (*a*) $\mathbf{F} = (kx + y)\mathbf{i} + (kx + y)\mathbf{j}$, (*b*) $\mathbf{F} = (kx + y)\mathbf{i} + (x + ky)\mathbf{j}$.

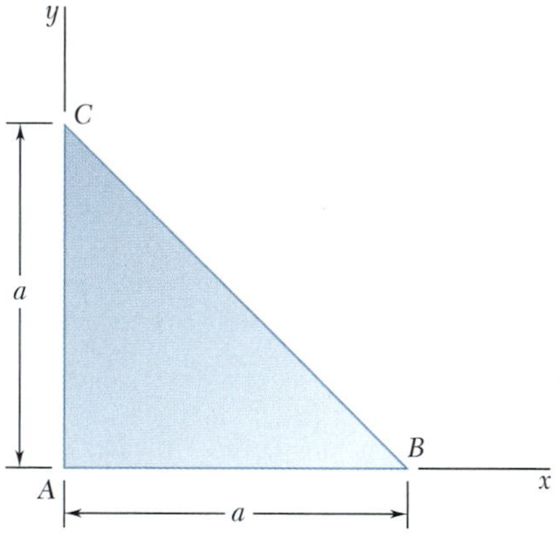

Fig. P13.80

13.81 Certain springs are characterized by increasing stiffness with increasing deformation according to the relation $F = k_1x + k_2x^3$, where F is the force exerted by the spring, k_1 and k_2 are positive constants, and x is the deflection of the spring measured from its undeformed position. Determine (*a*) the potential energy V_e as a function of x, (*b*) the maximum velocity of a particle of mass m attached to the spring and released from rest with $x = x_0$. Neglect friction.

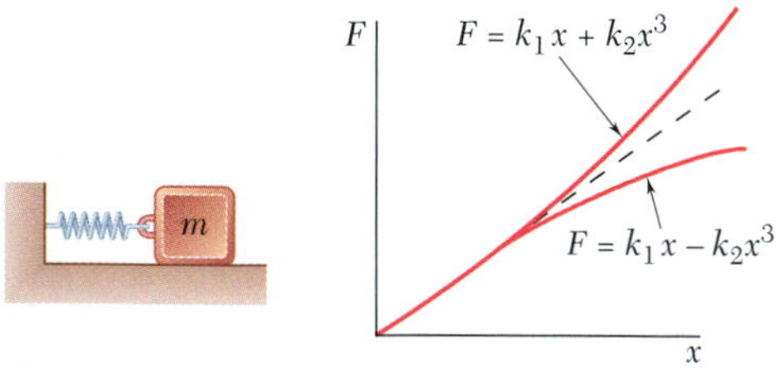

Fig. P13.81 and P13.82

13.82 Certain springs are characterized by decreasing stiffness with increasing deformation according to the relation $F = k_1x - k_2x^3$, where F is the force exerted by the spring, k_1 and k_2 are positive constants, and x is the deflection of the spring measured from its undeformed position. Determine (*a*) the potential energy V_e as a function of x, (*b*) the maximum velocity of a particle of mass m attached to the spring and released from rest with $x = x_0$. Neglect friction.

13.83 Knowing that the velocity of an experimental space probe fired from the earth has a magnitude $v_A = 32.5$ Mm/h at point A, determine the velocity of the probe as it passes through point B.

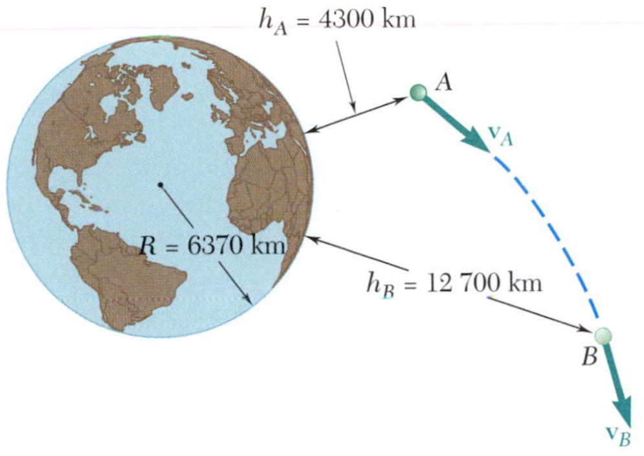

Fig. P13.83

13.84 A lunar excursion module (LEM) was used in the Apollo moon-landing missions to save fuel by making it unnecessary to launch the entire Apollo spacecraft from the moon's surface on its return trip to earth. Check the effectiveness of this approach by computing the energy per kilogram required for a spacecraft to escape the moon's gravitational field if the spacecraft starts from (*a*) the moon's surface, (*b*) a circular orbit 80 km above the moon's surface. Neglect the effect of the earth's gravitational field. (The radius of the moon is 1740 km and its mass is 0.0123 times the mass of the earth.)

13.85 A spacecraft is describing a circular orbit at an altitude of 1500 km above the surface of the earth. As it passes through point A, its speed is reduced by 40 percent and it enters an elliptic crash trajectory with the apogee at point A. Neglecting air resistance, determine the speed of the spacecraft when it reaches the earth's surface at point B.

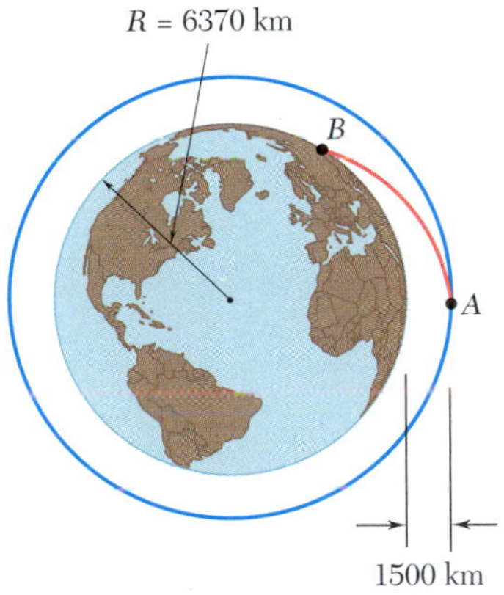

Fig. P13.85

13.86 A satellite describes an elliptic orbit of minimum altitude 606 km above the surface of the earth. The semimajor and semiminor axes are 17 440 km and 13 950 km, respectively. Knowing that the speed of the satellite at point C is 4.78 km/s, determine (*a*) the speed at point A, the perigee, (*b*) the speed at point B, the apogee.

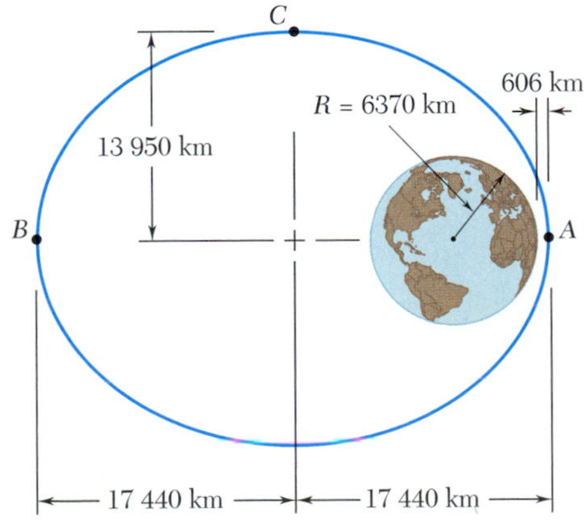

Fig. P13.86

13.87 While describing a circular orbit 200 mi above the earth a space vehicle launches a 6000-lb communications satellite. Determine (*a*) the additional energy required to place the satellite in a geosynchronous orbit at an altitude of 22,000 mi above the surface of the earth, (*b*) the energy required to place the satellite in the same orbit by launching it from the surface of the earth, excluding the energy needed to overcome air resistance. (A *geosynchronous orbit* is a circular orbit in which the satellite appears stationary with respect to the ground.)

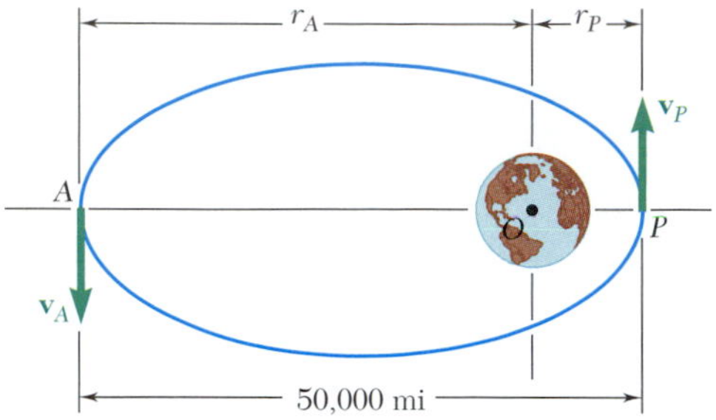

Fig. P13.88

13.88 A satellite is placed into an elliptic orbit about the earth. Knowing that the ratio v_A/v_P of the velocity at the apogee A to the velocity at the perigee P is equal to the ratio r_P/r_A of the distance to the center of the earth at P to that at A, and that the distance between A and P is 50,000 mi, determine the energy per unit weight required to place the satellite in its orbit by launching it from the surface of the earth. Exclude the additional energy needed to overcome the weight of the booster rocket, air resistance, and maneuvering.

13.89 A satellite of mass m describes a circular orbit of radius r about the earth. Express as a function of r (*a*) the potential energy of the satellite, (*b*) its kinetic energy, (*c*) its total energy. Denote the radius of the earth by R and the acceleration of gravity at the surface of the earth by g, and assume that the potential energy of the satellite is zero on its launching pad.

13.90 Observations show that a celestial body traveling at 1900 Mm/h appears to be describing about point B a circle of radius equal to 60 light years. Point B is suspected of being a very dense concentration of mass called a black hole. Determine the ratio M_B/M_S of the mass at B to the mass of the sun. (The mass of the sun is 330 000 times the mass of the earth, and a light year is the distance traveled by light in one year at a speed of 300 Mm/s.

13.91 (*a*) Show that, by setting $r = R + y$ in the right-hand member of Eq. (13.17′) and expanding that member in a power series in y/R, the expression in Eq. (13.16) for the potential energy V_g due to gravity is a first-order approximation for the expression given in Eq. (13.17′). (*b*) Using the same expansion, derive a second-order approximation for V_g.

13.92 How much energy per pound should be imparted to a satellite in order to place it in a circular orbit at an altitude of (*a*) 400 mi, (*b*) 4000 mi?

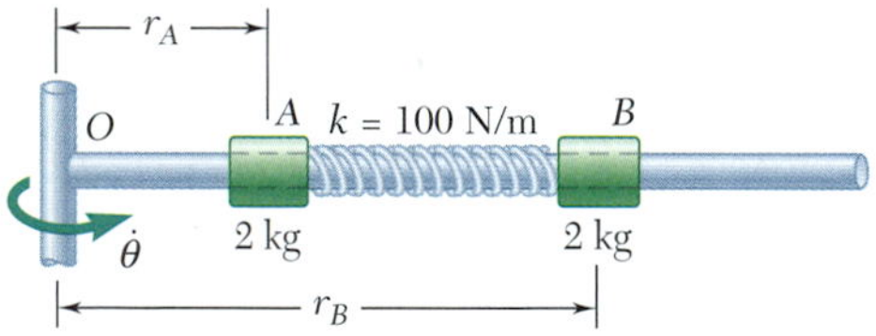

Fig. P13.93 and P13.94

13.93 Two identical 2-kg collars, A and B, are attached to a spring of constant 100 N/m and can slide on a horizontal rod which is free to rotate about a vertical shaft. Collar B is initially prevented from sliding by a stop as the rod rotates at a constant rate $\dot{\theta}_0 = 5$ rad/s and the spring is in compression with $r_A = 1$ m and $r_B = 2.5$ m. After the stop is removed both collars move out along the rod. At the instant when $r_B = 3$ m, determine (*a*) r_A, (*b*) $\dot{\theta}$, (*c*) the total kinetic energy. Neglect friction and the mass of the rod.

13.94 Two identical 2-kg collars, A and B, are attached to a spring of constant 100 N/m and can slide on a horizontal rod which is free to rotate about a vertical shaft. Collar B is initially prevented from sliding by a stop as the rod rotates at a constant rate $\dot{\theta}_0 = 5$ rad/s and the spring is in compression with $r_A = 1$ m and $r_B = 2.5$ m. After the stop is removed both collars move out along the rod. At the instant when the spring is in compression and the total kinetic energy is 185 J, determine (*a*) r_A, (*b*) r_B, (*c*) $\dot{\theta}$. Neglect friction and the mass of the rod.

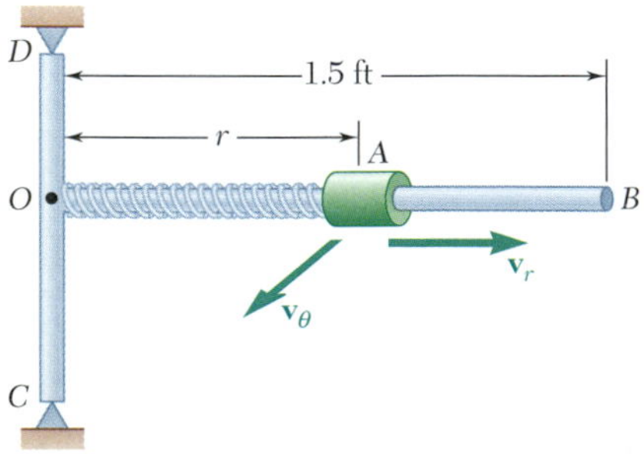

Fig. P13.95

13.95 A 5-lb collar A is attached to a spring of constant 50 lb/ft and undeformed length 0.5 ft. The spring is attached to point O of the frame $DCOB$. The system is set in motion with $r = 0.75$ ft, $v_\theta = 1.5$ ft/s, and $v_r = 0$. Neglecting the mass of the rod and the effect of friction, determine the radial and transverse components of the velocity of the collar when $r = 0.4$ ft.

13.96 A 4-lb collar A and a 1.5-lb collar B can slide without friction on a frame, consisting of the horizontal rod OE and the vertical rod CD, which is free to rotate about CD. The two collars are connected by a cord running over a pulley that is attached to the frame at O. At the instant shown, the velocity $\mathbf{v}_A$ of collar A has a magnitude of 6 ft/s and a stop prevents collar B from moving. If the stop is suddenly removed, determine (a) the velocity of collar A when it is 8 in. from O, (b) the velocity of collar A when collar B comes to rest. (Assume that collar B does not hit O, that collar A does not come off rod OE, and that the mass of the frame is negligible.)

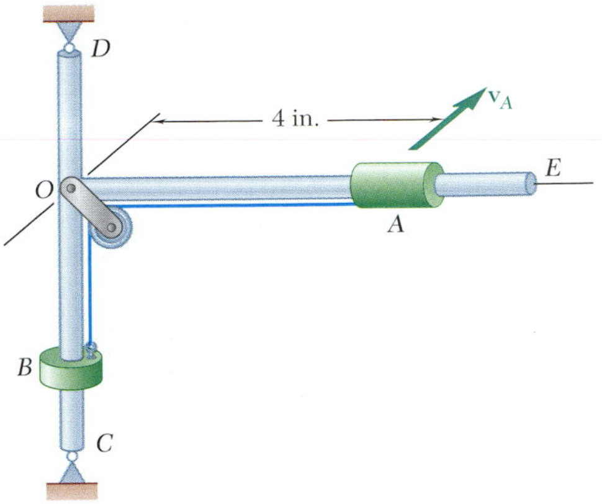

Fig. P13.96

13.97 A 0.7-kg ball that can slide on a *horizontal* frictionless surface is attached to a fixed point O by means of an elastic cord of constant $k = 150$ N/m and undeformed length 600 mm. The ball is placed at point A, 800 mm from O, and given an initial velocity $\mathbf{v}_0$ perpendicular to OA. Determine (a) the smallest allowable value of the initial speed v_0 if the cord is not to become slack, (b) the closest distance d that the ball will come to point O if it is given half the initial speed found in part a.

13.98 A 0.7-kg ball that can slide on a *horizontal* frictionless surface is attached to a fixed point O by means of an elastic cord of constant $k = 150$ N/m and undeformed length 600 mm. The ball is placed at point A, 800 mm from O, and given an initial velocity $\mathbf{v}_0$ perpendicular to OA, allowing the ball to come within a distance $d = 270$ mm of point O after the cord has become slack. Determine (a) the initial speed v_0 of the ball, (b) its maximum speed.

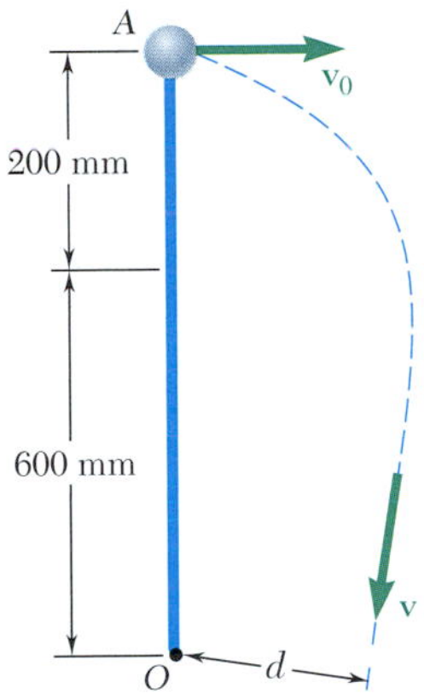

Fig. *P13.97* and *P13.98*

13.99 Using the principles of conservation of energy and conservation of angular momentum, solve part a of Sample Prob. 13.9.

13.100 As a first approximation to the analysis of a space flight from the earth to Mars, it is assumed that the orbits of the earth and Mars are circular and coplanar. The mean distances from the sun to the earth and to Mars are 149.6×10^6 km and 227.8×10^6 km, respectively. To place the spacecraft into an elliptical transfer orbit at point A, its speed is increased over a short interval of time to v_A which is faster than the earth's orbital speed. When the spacecraft reaches point B on the elliptical transfer orbit, its speed v_B is increased to the orbital speed of Mars. Knowing that the mass of the sun is 332.8×10^3 times the mass of the earth, determine the increase in velocity required (a) at A, (b) at B.

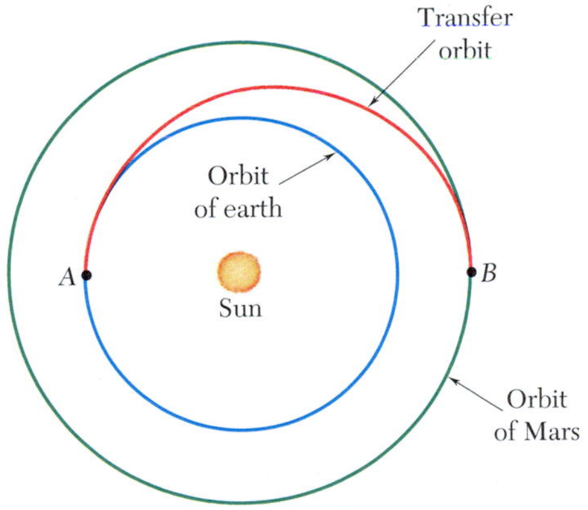

Fig. P13.100

13.101 A spacecraft describing an elliptic orbit about a planet has a maximum speed $v_A = 8$ km/s at its minimum altitude $h_A = 1925$ km above the surface of the planet and a minimum speed $v_B = 2$ km/s at its maximum altitude $h_B = 26\,300$ km. Determine (a) the radius of the planet, (b) the mass of the planet.

13.102 A spacecraft is describing an elliptic orbit of minimum altitude $h_A = 1500$ mi and maximum altitude $h_B = 6000$ mi above the surface of the earth. Determine the speed of the spacecraft at A.

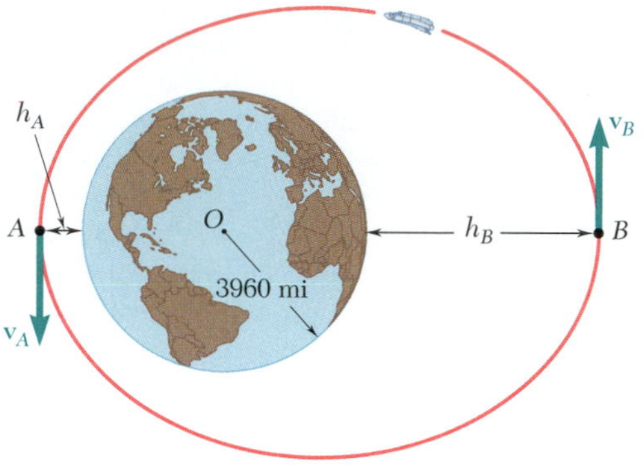

Fig. P13.102

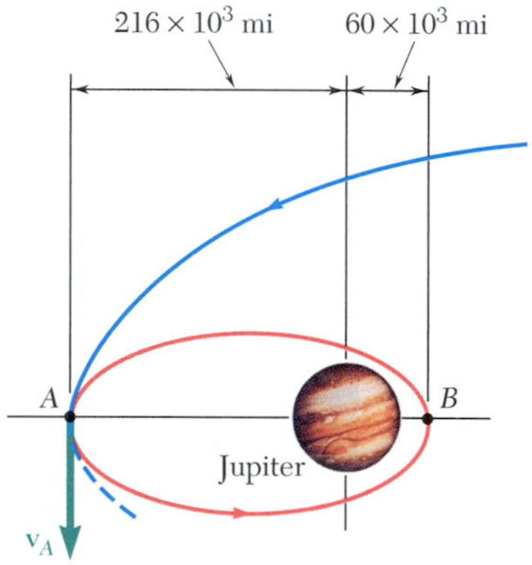

Fig. P13.103

13.103 A spacecraft traveling along a parabolic path toward the planet Jupiter is expected to reach point A with a velocity $\mathbf{v}_A$ of magnitude 18 mi/s. Its engines will then be fired to slow it down, placing it into an elliptic orbit which will bring it to within 60×10^3 mi of Jupiter. Determine the decrease in speed Δv at point A which will place the spacecraft into the required orbit. The mass of Jupiter is 319 times the mass of the earth.

13.104 A spacecraft approaching the planet Saturn reaches point A with a velocity $\mathbf{v}_A$ of magnitude 21 km/s. It is placed in an elliptic orbit about Saturn so that it will be able to periodically examine Tethys, one of Saturn's moons. Tethys is in a circular orbit of radius 295×10^3 km about the center of Saturn, traveling at a speed of 11.3 km/s. Determine (*a*) the decrease in speed required by the spacecraft at A to achieve the desired orbit, (*b*) the speed of the spacecraft when it reaches the orbit of Tethys at B.

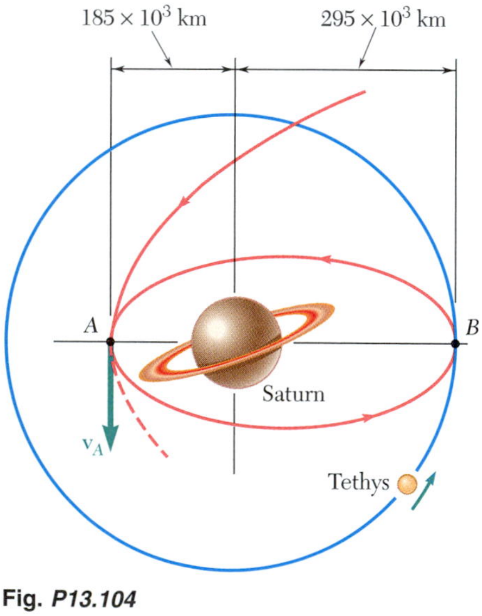

Fig. *P13.104*

13.105 The optimal way of transferring a space vehicle from an inner circular orbit to an outer coplanar orbit is to fire its engines as it passes through A to increase its speed and place it in an elliptic transfer orbit. Another increase in speed as it passes through B will place it in the desired circular orbit. For a vehicle in a circular orbit about the earth at an altitude $h_1 = 320$ km, which is to be transferred to a circular orbit at an altitude $h_2 = 800$ km, determine (*a*) the required increase in speed at A and B, (*b*) the total energy per unit mass required to execute the transfer.

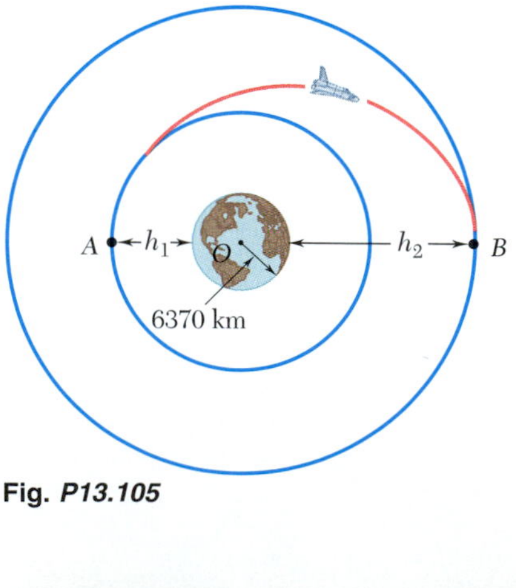

Fig. *P13.105*

13.106 During a flyby of the earth, the velocity of a spacecraft is 10.4 km/s as it reaches its minimum altitude of 990 km above the surface at point A. At point B the spacecraft is observed to have an altitude of 8350 km. Determine (*a*) the magnitude of the velocity at point B, (*b*) the angle ϕ_B.

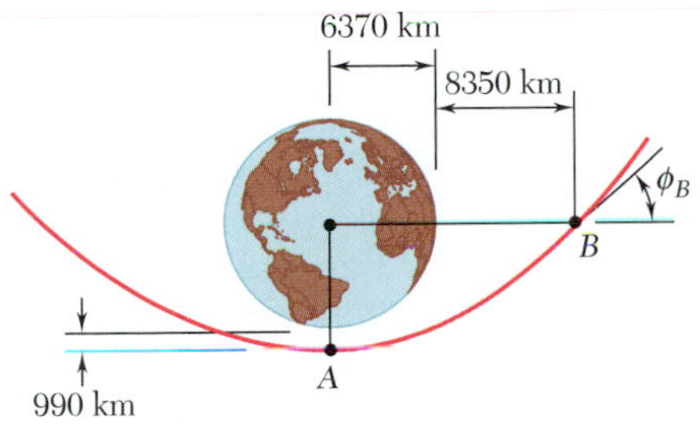

Fig. P13.106

13.107 A satellite describes an elliptic orbit of minimum altitude 606 km above the surface of the earth and semimajor axis 17 440 km. Knowing that the speed of the satellite at point C is 4.78 km/s, apply the principles of conservation of energy and conservation of angular momentum to determine (*a*) the speed at point A, the perigee, (*b*) the semiminor axis b.

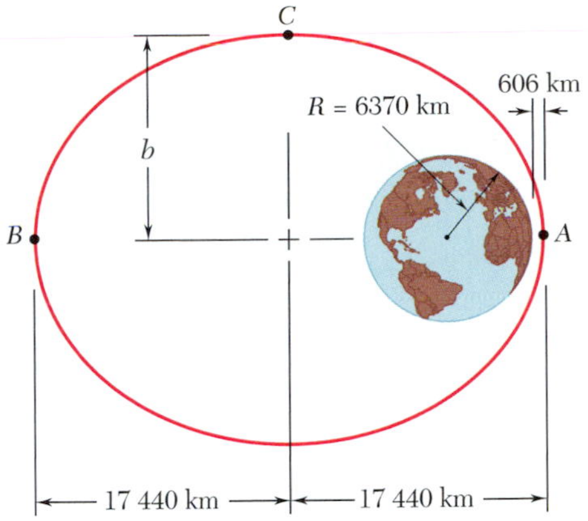

Fig. P13.107

13.108 Upon the LEM's return to the command module, the Apollo spacecraft was turned around so that the LEM faced to the rear. The LEM was then cast adrift with a velocity of 600 ft/s relative to the command module. Determine the magnitude and direction (angle ϕ formed with the vertical OC) of the velocity $\mathbf{v}_C$ of the LEM just before it crashed at C on the moon's surface.

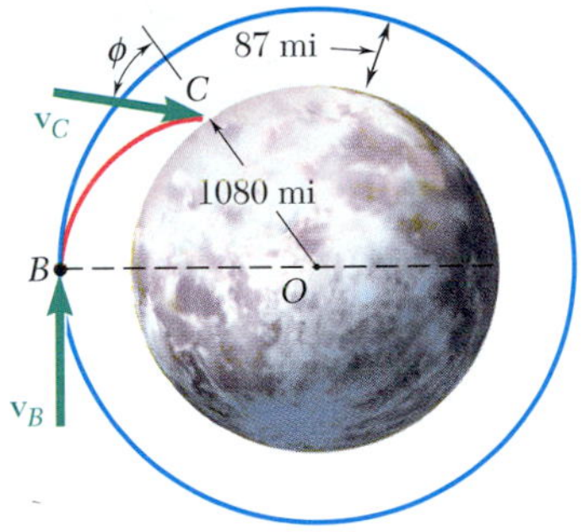

Fig. P13.108

13.109 A space platform is in a circular orbit about the earth at an altitude of 180 mi. As the platform passes through A, a rocket carrying a communications satellite is launched from the platform with a relative velocity of magnitude 2 mi/s in a direction tangent to the orbit of the platform. This was intended to place the rocket in an elliptic transfer orbit bringing it to point B, where the rocket would again be fired to place the satellite in a geosynchronous orbit of radius 26,200 mi. After launching, it was discovered that the relative velocity imparted to the rocket was too large. Determine the angle γ at which the rocket will cross the intended orbit at point C.

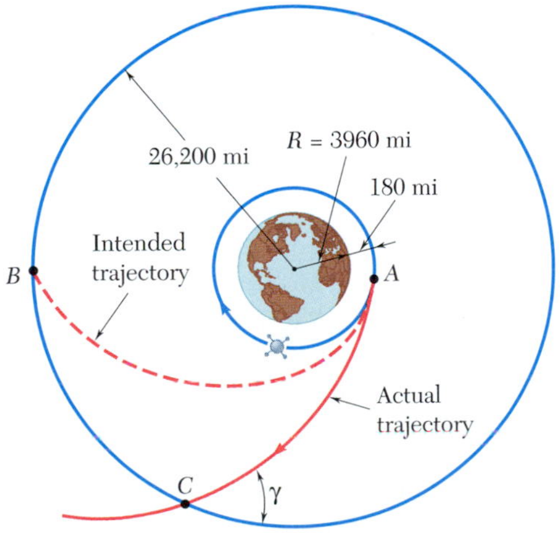

Fig. P13.109

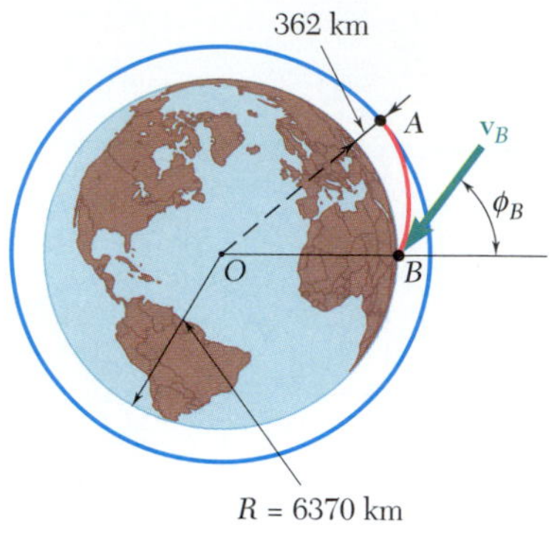

Fig. P13.110

13.110 A space vehicle is in a circular orbit at an altitude of 362 km above the earth. To return to earth, it decreases its speed as it passes through A by firing its engine for a short interval of time in a direction opposite to the direction of its motion. Knowing that the velocity of the space vehicle should form an angle $\phi_B = 60°$ with the vertical as it reaches point B at an altitude of 64.4 km, determine (*a*) the required speed of the vehicle as it leaves its circular orbit at A, (*b*) its speed at point B.

***13.111** In Prob. 13.110, the speed of the space vehicle was decreased as it passed through A by firing its engine in a direction opposite to the direction of motion. An alternative strategy for taking the space vehicle out of its circular orbit would be to turn it around so that its engine would point away from the earth and then give it an incremental velocity $\Delta\mathbf{v}_A$ toward the center O of the earth. This would likely require a smaller expenditure of energy when firing the engine at A, but might result in too fast a descent at B. Assuming this strategy is used with only 50 percent of the energy expenditure used in Prob. 13.110, determine the resulting values of ϕ_B and v_B.

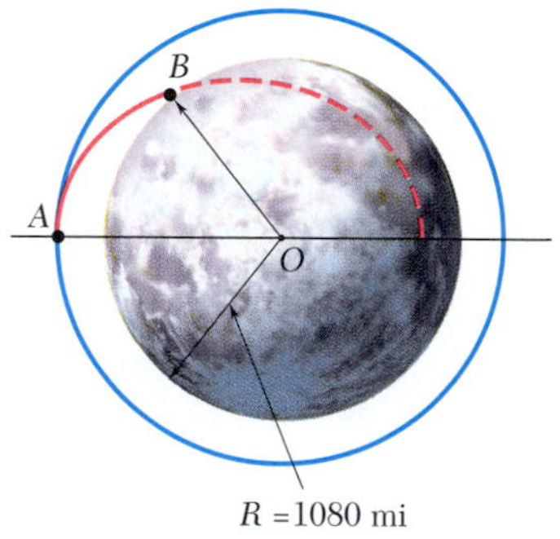

Fig. *P13.112*

13.112 When the lunar excursion module (LEM) was set adrift after returning two of the Apollo astronauts to the command module, which was orbiting the moon at an altitude of 87 mi, its speed was reduced to let it crash on the moon's surface. Determine (*a*) the smallest amount by which the speed of the LEM should have been reduced to make sure that it would crash on the moon's surface, (*b*) the amount by which its speed should have been reduced to cause it to hit the moon's surface at a 45° angle. (*Hint:* Point A is at the apogee of the elliptic crash trajectory. Recall also that the mass of the moon is 0.0123 times the mass of the earth.)

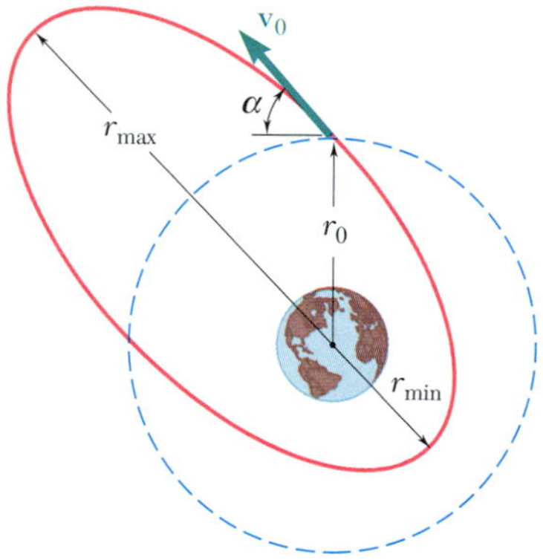

Fig. *P13.113*

13.113 A satellite is projected into space with a velocity $\mathbf{v}_0$ at a distance r_0 from the center of the earth by the last stage of its launching rocket. The velocity $\mathbf{v}_0$ was designed to send the satellite into a circular orbit of radius r_0. However, owing to a malfunction of control, the satellite is not projected horizontally but at an angle α with the horizontal and, as a result, is propelled into an elliptic orbit. Determine the maximum and minimum values of the distance from the center of the earth to the satellite.

13.114 Show that the values v_A and v_P of the speed of an earth satellite at the apogee A and perigee P of an elliptic orbit are defined by the relations

$$v_A^2 = \frac{2GM}{r_A + r_P}\frac{r_P}{r_A} \qquad v_P^2 = \frac{2GM}{r_A + r_P}\frac{r_A}{r_P}$$

where M is the mass of the earth, and r_A and r_P represent, respectively, the maximum and minimum distances of the orbit to the center of the earth.

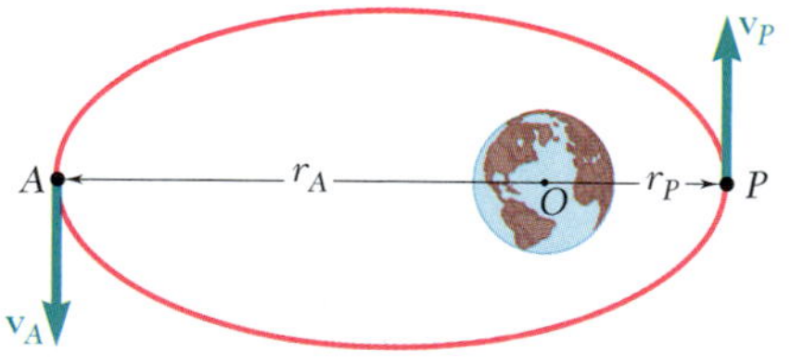

Fig. P13.114 and P13.115

13.115 Show that the total energy E of an earth satellite of mass m describing an elliptic orbit is $E = -GMm/(r_A + r_P)$, where M is the mass of the earth, and r_A and r_P represent, respectively, the maximum and minimum distances of the orbit to the center of the earth. (Recall that the gravitational potential energy of a satellite was defined as being zero at an infinite distance from the earth.)

13.116 A spacecraft of mass m describes a circular orbit of radius r_1 around the earth. (*a*) Show that the additional energy ΔE which must be imparted to the spacecraft to transfer it to a circular orbit of larger radius r_2 is

$$\Delta E = \frac{GMm(r_2 - r_1)}{2r_1 r_2}$$

where M is the mass of the earth. (*b*) Further show that if the transfer from one circular orbit to the other circular orbit is executed by placing the spacecraft on a transitional semielliptic path AB, the amounts of energy ΔE_A and ΔE_B which must be imparted at A and B are, respectively, proportional to r_1 and r_2:

$$\Delta E_A = \frac{r_2}{r_1 + r_2}\Delta E \qquad \Delta E_B = \frac{r_1}{r_1 + r_2}\Delta E$$

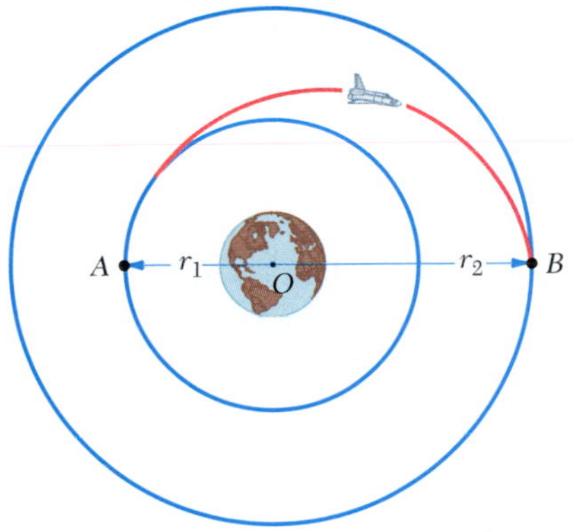

Fig. P13.116

13.117 Using the answers obtained in Prob. 13.113, show that the intended circular orbit and the resulting elliptic orbit intersect at the ends of the minor axis of the elliptic orbit.

13.118 A missile is fired from the ground with an initial velocity $\mathbf{v}_0$ forming an angle ϕ_0 with the vertical. If the missile is to reach a maximum altitude equal to αR, where R is the radius of the earth, (*a*) show that the required angle ϕ_0 is defined by the relation

$$\sin \phi_0 = (1 + \alpha)\sqrt{1 - \frac{\alpha}{1 + \alpha}\left(\frac{v_{esc}}{v_0}\right)^2}$$

where v_{esc} is the escape velocity, (*b*) determine the range of allowable values of v_0.

13.10. PRINCIPLE OF IMPULSE AND MOMENTUM

A third basic method for the solution of problems dealing with the motion of particles will be considered now. This method is based on the principle of impulse and momentum and can be used to solve problems involving force, mass, velocity, and time. It is of particular interest in the solution of problems involving impulsive motion and problems involving impact (Secs. 13.11 and 13.12).

Consider a particle of mass m acted upon by a force $\mathbf{F}$. As we saw in Sec. 12.3, Newton's second law can be expressed in the form

$$\mathbf{F} = \frac{d}{dt}(m\mathbf{v}) \tag{13.27}$$

where $m\mathbf{v}$ is the linear momentum of the particle. Multiplying both sides of Eq. (13.27) by dt and integrating from a time t_1 to a time t_2, we write

$$\mathbf{F}\,dt = d(m\mathbf{v})$$

$$\int_{t_1}^{t_2} \mathbf{F}\,dt = m\mathbf{v}_2 - m\mathbf{v}_1$$

or, transposing the last term,

$$m\mathbf{v}_1 + \int_{t_1}^{t_2} \mathbf{F}\,dt = m\mathbf{v}_2 \tag{13.28}$$

The integral in Eq. (13.28) is a vector known as the *linear impulse*, or simply the *impulse*, of the force $\mathbf{F}$ during the interval of time considered. Resolving $\mathbf{F}$ into rectangular components, we write

$$\mathbf{Imp}_{1\to2} = \int_{t_1}^{t_2} \mathbf{F}\,dt$$
$$= \mathbf{i}\int_{t_1}^{t_2} F_x\,dt + \mathbf{j}\int_{t_1}^{t_2} F_y\,dt + \mathbf{k}\int_{t_1}^{t_2} F_z\,dt \qquad (13.29)$$

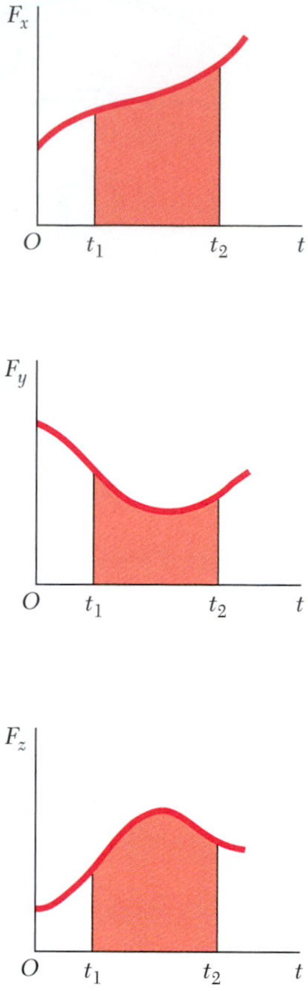

Fig. 13.16

and note that the components of the impulse of the force $\mathbf{F}$ are, respectively, equal to the areas under the curves obtained by plotting the components F_x, F_y, and F_z against t (Fig. 13.16). In the case of a force $\mathbf{F}$ of constant magnitude and direction, the impulse is represented by the vector $\mathbf{F}(t_2 - t_1)$, which has the same direction as $\mathbf{F}$.

If SI units are used, the magnitude of the impulse of a force is expressed in N · s. But, recalling the definition of the newton, we have

$$\text{N}\cdot\text{s} = (\text{kg}\cdot\text{m/s}^2)\cdot\text{s} = \text{kg}\cdot\text{m/s}$$

which is the unit obtained in Sec. 12.4 for the linear momentum of a particle. We thus check that Eq. (13.28) is dimensionally correct. If U.S. customary units are used, the impulse of a force is expressed in lb · s, which is also the unit obtained in Sec. 12.4 for the linear momentum of a particle.

Equation (13.28) expresses that when a particle is acted upon by a force $\mathbf{F}$ during a given time interval, *the final momentum* $m\mathbf{v}_2$ *of the particle can be obtained by adding vectorially its initial momentum* $m\mathbf{v}_1$ *and the impulse of the force* $\mathbf{F}$ *during the time interval considered*

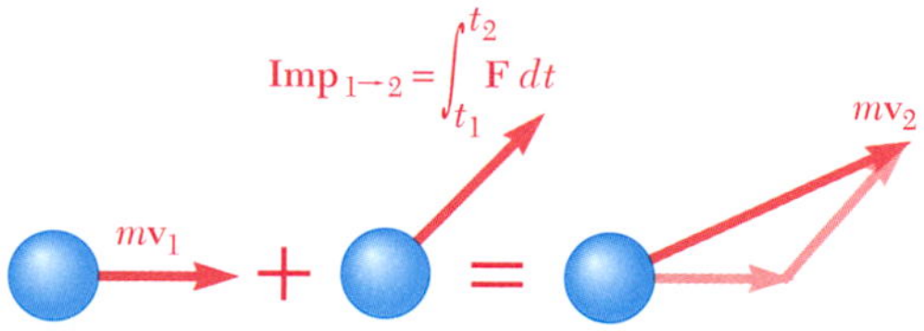

Fig. 13.17

(Fig. 13.17). We write

$$m\mathbf{v}_1 + \mathbf{Imp}_{1\to2} = m\mathbf{v}_2 \qquad (13.30)$$

We note that while kinetic energy and work are scalar quantities, momentum and impulse are vector quantities. To obtain an analytic solution, it is thus necessary to replace Eq. (13.30) by the corresponding component equations

$$(mv_x)_1 + \int_{t_1}^{t_2} F_x\,dt = (mv_x)_2$$
$$(mv_y)_1 + \int_{t_1}^{t_2} F_y\,dt = (mv_y)_2 \qquad (13.31)$$
$$(mv_z)_1 + \int_{t_1}^{t_2} F_z\,dt = (mv_z)_2$$

When several forces act on a particle, the impulse of each of the forces must be considered. We have

$$m\mathbf{v}_1 + \Sigma\ \mathbf{Imp}_{1\to 2} = m\mathbf{v}_2 \tag{13.32}$$

Again, the equation obtained represents a relation between vector quantities; in the actual solution of a problem, it should be replaced by the corresponding component equations.

When a problem involves two particles or more, each particle can be considered separately and Eq. (13.32) can be written for each particle. We can also add vectorially the momenta of all the particles and the impulses of all the forces involved. We write then

$$\Sigma m\mathbf{v}_1 + \Sigma\ \mathbf{Imp}_{1\to 2} = \Sigma m\mathbf{v}_2 \tag{13.33}$$

Since the forces of action and reaction exerted by the particles on each other form pairs of equal and opposite forces, and since the time interval from t_1 to t_2 is common to all the forces involved, the impulses of the forces of action and reaction cancel out and only the impulses of the external forces need be considered.†

If no external force is exerted on the particles or, more generally, if the sum of the external forces is zero, the second term in Eq. (13.33) vanishes and Eq. (13.33) reduces to

$$\Sigma m\mathbf{v}_1 = \Sigma m\mathbf{v}_2 \tag{13.34}$$

which expresses that *the total momentum of the particles is conserved.* Consider, for example, two boats, of mass m_A and m_B, initially at rest, which are being pulled together (Fig. 13.18). If the resistance of the

Fig. 13.18

water is neglected, the only external forces acting on the boats are their weights and the buoyant forces exerted on them. Since these forces are balanced, we write

$$\Sigma m\mathbf{v}_1 = \Sigma m\mathbf{v}_2$$
$$0 = m_A\mathbf{v}'_A + m_B\mathbf{v}'_B$$

where $\mathbf{v}'_A$ and $\mathbf{v}'_B$ represent the velocities of the boats after a finite interval of time. The equation obtained indicates that the boats move in opposite directions (toward each other) with velocities inversely proportional to their masses.‡

†We should note the difference between this statement and the corresponding statement made in Sec. 13.4 regarding the work of the forces of action and reaction between several particles. While the sum of the impulses of these forces is always zero, the sum of their work is zero only under special circumstances, for example, when the various bodies involved are connected by inextensible cords or links and are thus constrained to move through equal distances.

‡Blue equals signs are used in Fig. 13.18 and throughout the remainder of this chapter to express that two systems of vectors are *equipollent,* that is, that they have the same resultant and moment resultant (cf. Sec. 3.19). Red equals signs will continue to be used to indicate that two systems of vectors are *equivalent,* that is, that they have the same effect. This and the concept of conservation of momentum for a system of particles will be discussed in greater detail in Chap. 14.

13.11. IMPULSIVE MOTION

A force acting on a particle during a very short time interval that is large enough to produce a definite change in momentum is called an *impulsive force* and the resulting motion is called an *impulsive motion*. For example, when a baseball is struck, the contact between bat and ball takes place during a very short time interval Δt. But the average value of the force $\mathbf{F}$ exerted by the bat on the ball is very large, and the resulting impulse $\mathbf{F}\,\Delta t$ is large enough to change the sense of motion of the ball (Fig. 13.19).

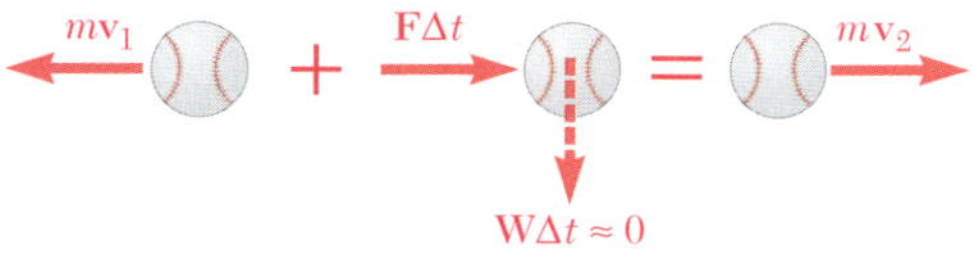

Fig. 13.19

When impulsive forces act on a particle, Eq. (13.32) becomes

$$m\mathbf{v}_1 + \Sigma\mathbf{F}\,\Delta t = m\mathbf{v}_2 \tag{13.35}$$

Any force which is not an impulsive force may be neglected, since the corresponding impulse $\mathbf{F}\,\Delta t$ is very small. *Nonimpulsive forces* include the weight of the body, the force exerted by a spring, or any other force which is *known* to be small compared with an impulsive force. Unknown reactions may or may not be impulsive; their impulses should therefore be included in Eq. (13.35) as long as they have not been proved negligible. The impulse of the weight of the baseball considered above, for example, may be neglected. If the motion of the bat is analyzed, the impulse of the weight of the bat can also be neglected. The impulses of the reactions of the player's hands on the bat, however, should be included; these impulses will not be negligible if the ball is incorrectly hit.

We note that the method of impulse and momentum is particularly effective in the analysis of the impulsive motion of a particle, since it involves only the initial and final velocities of the particle and the impulses of the forces exerted on the particle. The direct application of Newton's second law, on the other hand, would require the determination of the forces as functions of the time and the integration of the equations of motion over the time interval Δt.

In the case of the impulsive motion of several particles, Eq. (13.33) can be used. It reduces to

$$\Sigma m\mathbf{v}_1 + \Sigma\mathbf{F}\,\Delta t = \Sigma m\mathbf{v}_2 \tag{13.36}$$

where the second term involves only impulsive, external forces. If all the external forces acting on the various particles are nonimpulsive, the second term in Eq. (13.36) vanishes and this equation reduces to Eq. (13.34). We write

$$\Sigma m\mathbf{v}_1 = \Sigma m\mathbf{v}_2 \tag{13.34}$$

which expresses that the total momentum of the particles is conserved. This situation occurs, for example, when two particles which are moving freely collide with one another. We should note, however, that while the total momentum of the particles is conserved, their total energy is generally *not* conserved. Problems involving the collision or *impact* of two particles will be discussed in detail in Secs. 13.12 through 13.14.

Photo 13.1 A heavy weight is lifted by the pile driver and then dropped. This delivers a large impulsive force to the pile which is driven into the ground.

SAMPLE PROBLEM 13.10

An automobile weighing 4000 lb is driven down a 5° incline at a speed of 60 mi/h when the brakes are applied, causing a constant total braking force (applied by the road on the tires) of 1500 lb. Determine the time required for the automobile to come to a stop.

SOLUTION

We apply the principle of impulse and momentum. Since each force is constant in magnitude and direction, each corresponding impulse is equal to the product of the force and of the time interval t.

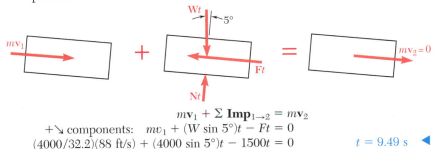

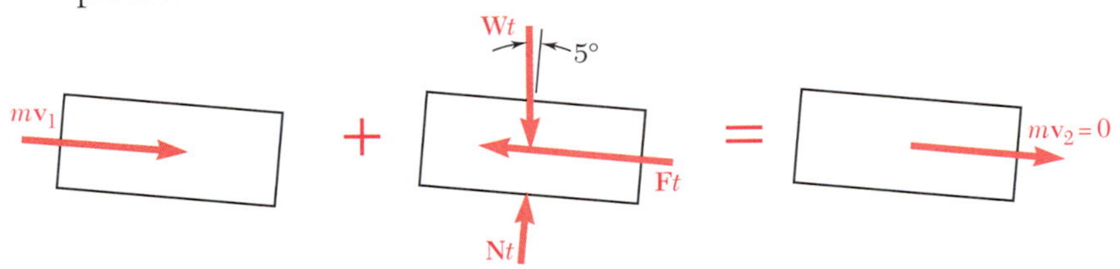

$$m\mathbf{v}_1 + \Sigma\,\mathbf{Imp}_{1\to 2} = m\mathbf{v}_2$$

$+\searrow$ components: $\quad mv_1 + (W \sin 5°)t - Ft = 0$

$$(4000/32.2)(88 \text{ ft/s}) + (4000 \sin 5°)t - 1500t = 0 \qquad t = 9.49 \text{ s} \blacktriangleleft$$

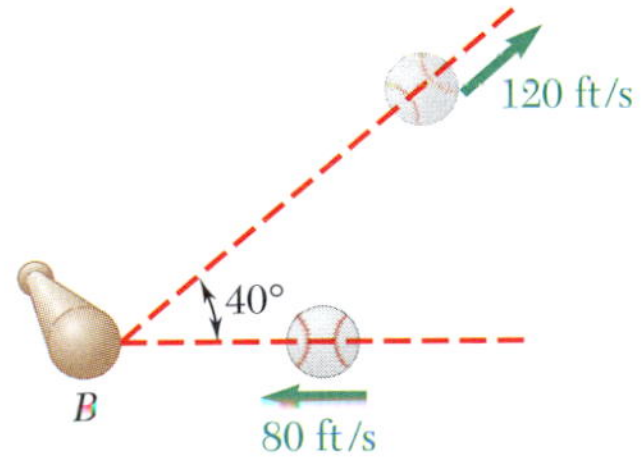

SAMPLE PROBLEM 13.11

A 4-oz baseball is pitched with a velocity of 80 ft/s toward a batter. After the ball is hit by the bat B, it has a velocity of 120 ft/s in the direction shown. If the bat and ball are in contact 0.015 s, determine the average impulsive force exerted on the ball during the impact.

SOLUTION

We apply the principle of impulse and momentum to the ball. Since the weight of the ball is a nonimpulsive force, it can be neglected.

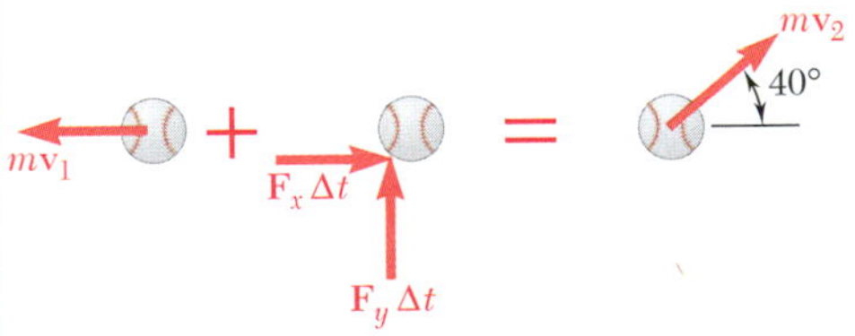

$$m\mathbf{v}_1 + \Sigma\,\mathbf{Imp}_{1\to 2} = m\mathbf{v}_2$$

$\xrightarrow{+}$ x components: $\quad -mv_1 + F_x\,\Delta t = mv_2 \cos 40°$

$$-\frac{\frac{4}{16}}{32.2}(80 \text{ ft/s}) + F_x(0.015 \text{ s}) = \frac{\frac{4}{16}}{32.2}(120 \text{ ft/s}) \cos 40°$$

$$F_x = +89.0 \text{ lb}$$

$+\uparrow y$ components: $\quad 0 + F_y\,\Delta t = mv_2 \sin 40°$

$$F_y(0.015 \text{ s}) = \frac{\frac{4}{16}}{32.2}(120 \text{ ft/s}) \sin 40°$$

$$F_y = +39.9 \text{ lb}$$

From its components F_x and F_y we determine the magnitude and direction of the force $\mathbf{F}$:

$$\mathbf{F} = 97.5 \text{ lb} \measuredangle 24.2° \blacktriangleleft$$

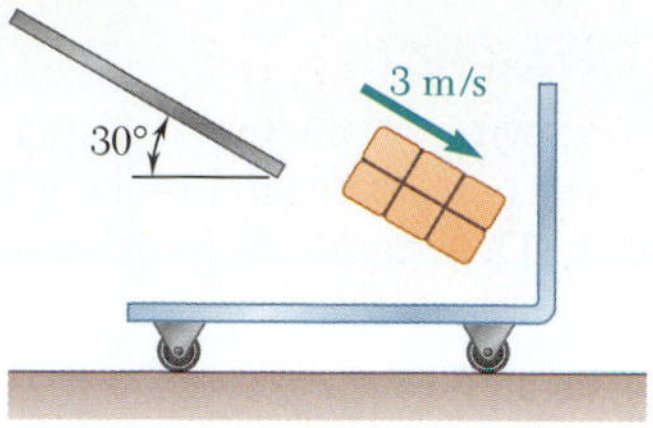

SAMPLE PROBLEM 13.12

A 10-kg package drops from a chute into a 25-kg cart with a velocity of 3 m/s. Knowing that the cart is initially at rest and can roll freely, determine (*a*) the final velocity of the cart, (*b*) the impulse exerted by the cart on the package, (*c*) the fraction of the initial energy lost in the impact.

SOLUTION

We first apply the principle of impulse and momentum to the package-cart system to determine the velocity $\mathbf{v}_2$ of the cart and package. We then apply the same principle to the package alone to determine the impulse $\mathbf{F}\,\Delta t$ exerted on it.

a. Impulse-Momentum Principle: Package and Cart

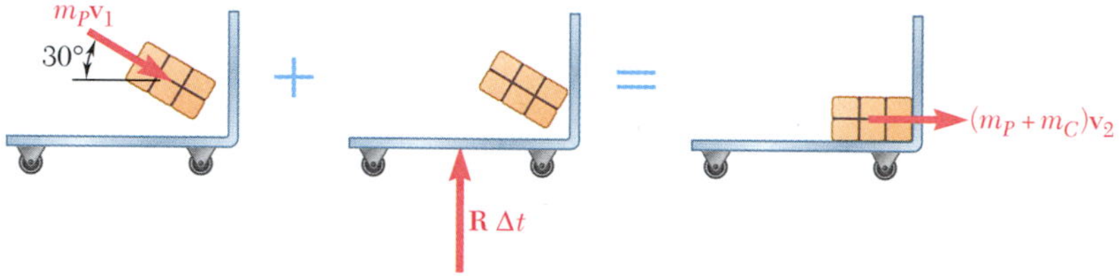

$$m_P\mathbf{v}_1 + \Sigma\,\mathbf{Imp}_{1\to 2} = (m_P + m_C)\mathbf{v}_2$$

$\overset{+}{\rightarrow}$ x components:

$$m_P v_1 \cos 30^\circ + 0 = (m_P + m_C)v_2$$
$$(10\text{ kg})(3\text{ m/s})\cos 30^\circ = (10\text{ kg} + 25\text{ kg})v_2$$
$$\mathbf{v}_2 = 0.742\text{ m/s}\rightarrow \quad ◀$$

We note that the equation used expresses conservation of momentum in the x direction.

b. Impulse-Momentum Principle: Package

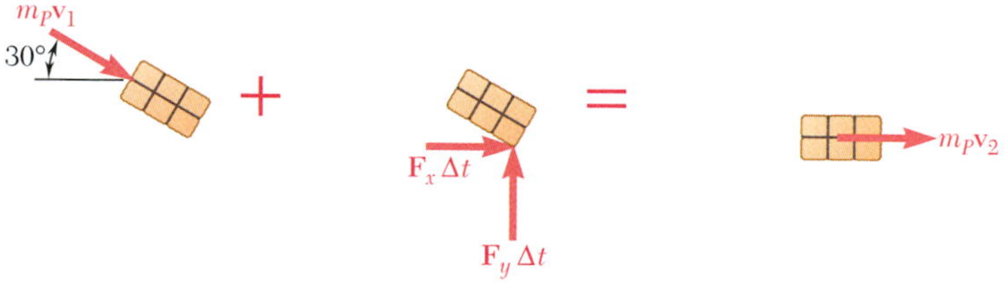

$$m_P\mathbf{v}_1 + \Sigma\,\mathbf{Imp}_{1\to 2} = m_P\mathbf{v}_2$$

$\overset{+}{\rightarrow}$ x components:

$$(10\text{ kg})(3\text{ m/s})\cos 30^\circ + F_x\,\Delta t = (10\text{ kg})(0.742\text{ m/s})$$
$$F_x\,\Delta t = -18.56\text{ N}\cdot\text{s}$$

$+\uparrow y$ components:

$$-m_P v_1 \sin 30^\circ + F_y\,\Delta t = 0$$
$$-(10\text{ kg})(3\text{ m/s})\sin 30^\circ + F_y\,\Delta t = 0$$
$$F_y\,\Delta t = +15\text{ N}\cdot\text{s}$$

The impulse exerted on the package is $\quad \mathbf{F}\,\Delta t = 23.9\text{ N}\cdot\text{s} \measuredangle 38.9^\circ \quad ◀$

***c.* Fraction of Energy Lost.** The initial and final energies are

$$T_1 = \tfrac{1}{2}m_P v_1^2 = \tfrac{1}{2}(10\text{ kg})(3\text{ m/s})^2 = 45\text{ J}$$
$$T_2 = \tfrac{1}{2}(m_P + m_C)v_2^2 = \tfrac{1}{2}(10\text{ kg} + 25\text{ kg})(0.742\text{ m/s})^2 = 9.63\text{ J}$$

The fraction of energy lost is $\quad \dfrac{T_1 - T_2}{T_1} = \dfrac{45\text{ J} - 9.63\text{ J}}{45\text{ J}} = 0.786 \quad ◀$

SOLVING PROBLEMS ON YOUR OWN

In this lesson we integrated Newton's second law to derive the *principle of impulse and momentum* for a particle. Recalling that the *linear momentum* of a particle was defined as the product of its mass m and its velocity $\mathbf{v}$ [Sec. 12.3], we wrote

$$m\mathbf{v}_1 + \Sigma\, \mathbf{Imp}_{1\rightarrow 2} = m\mathbf{v}_2 \tag{13.32}$$

This equation expresses that the linear momentum $m\mathbf{v}_2$ of a particle at time t_2 can be obtained by adding to its linear momentum $m\mathbf{v}_1$ at time t_1 the *impulses* of the forces exerted on the particle during the time interval t_1 to t_2. For computing purposes, the momenta and impulses may be expressed in terms of their rectangular components, and Eq. (13.32) can be replaced by the equivalent scalar equations. The units of momentum and impulse are N · s in the SI system of units and lb · s in U.S. customary units. To solve problems using this equation you can follow these steps:

1. ***Draw a diagram*** showing the particle, its momentum at t_1 and at t_2, and the impulses of the forces exerted on the particle during the time interval t_1 to t_2.

2. ***Calculate the impulse of each force,*** expressing it in terms of its rectangular components if more than one direction is involved. You may encounter the following cases:

a. The time interval is finite and the force is constant.

$$\mathbf{Imp}_{1\rightarrow 2} = \mathbf{F}(t_2 - t_1)$$

b. The time interval is finite and the force is a function of t.

$$\mathbf{Imp}_{1\rightarrow 2} = \int_{t_1}^{t_2} \mathbf{F}(t)\, dt$$

c. The time interval is very small and the force is very large. The force is called an *impulsive force* and its impulse over the time interval $t_2 - t_1 = \Delta t$ is

$$\mathbf{Imp}_{1\rightarrow 2} = \mathbf{F}\, \Delta t$$

Note that this impulse is *zero for a nonimpulsive force* such as the *weight* of a body, the force exerted by a *spring,* or any other force which is known to be small by comparison with the impulsive forces. Unknown reactions, however, *cannot be assumed* to be nonimpulsive and their impulses should be taken into account.

3. ***Substitute the values obtained for the impulses into Eq. (13.32)*** or into the equivalent scalar equations. You will find that the forces and velocities in the problems of this lesson are contained in a plane. You will, therefore, write two scalar equations and solve these equations for *two unknowns.* These unknowns may be a *time* [Sample Prob. 13.10], a *velocity* and an *impulse* [Sample Prob. 13.12], or an *average impulsive force* [Sample Prob. 13.11].

(continued)

4. ***When several particles are involved,*** a separate diagram should be drawn for each particle, showing the initial and final momentum of the particle, as well as the impulses of the forces exerted on the particle.

a. It is usually convenient, however, to first consider a diagram including all the particles. This diagram leads to the equation

$$\Sigma m\mathbf{v}_1 + \Sigma\ \mathbf{Imp}_{1\rightarrow 2} = \Sigma m\mathbf{v}_2 \tag{13.33}$$

where the impulses of *only the forces external to the system* need be considered. Therefore, the two equivalent scalar equations will not contain any of the impulses of the unknown internal forces.

b. If the sum of the impulses of the external forces is zero, Eq. (13.33) reduces to

$$\Sigma m\mathbf{v}_1 = \Sigma m\mathbf{v}_2 \tag{13.34}$$

which expresses that *the total momentum of the particles is conserved.* This occurs either if the resultant of the external forces is zero or, when the time interval Δt is very short (impulsive motion), if all the external forces are nonimpulsive. Keep in mind, however, that the total momentum may be conserved *in one direction,* but not in another (Sample Prob. 13.12].

Problems

13.119 A 2-kg particle is acted upon by a force $\mathbf{F} = -2t^2\mathbf{i} + (3 - t)\mathbf{j}$, where F is expressed in newtons and t in seconds. Knowing that the velocity of the particle is $\mathbf{v}_0 = (3 \text{ m/s})\mathbf{i}$ at $t = 0$, determine (*a*) the time at which the velocity is parallel to the y axis, (*b*) the corresponding velocity of the particle.

13.120 A 3-lb particle is acted upon by a force $\mathbf{F} = (20 \sin 2t)\mathbf{i} + (24 \cos 2t)\mathbf{j}$, where F is expressed in pounds and t in seconds. Determine the magnitude and direction of the velocity of the particle at $t = 6$ s knowing that its velocity is zero at $t = 0$.

13.121 Steep safety ramps are built beside mountain highways to enable vehicles with defective brakes to stop. A 9000-kg truck enters a 15° ramp at a high speed $v_0 = 36$ m/s and travels for 6 s before its speed is reduced to 12 m/s. Assuming constant deceleration, determine (*a*) the magnitude of the braking force, (*b*) the additional time required for the truck to stop. Neglect air resistance and rolling resistance.

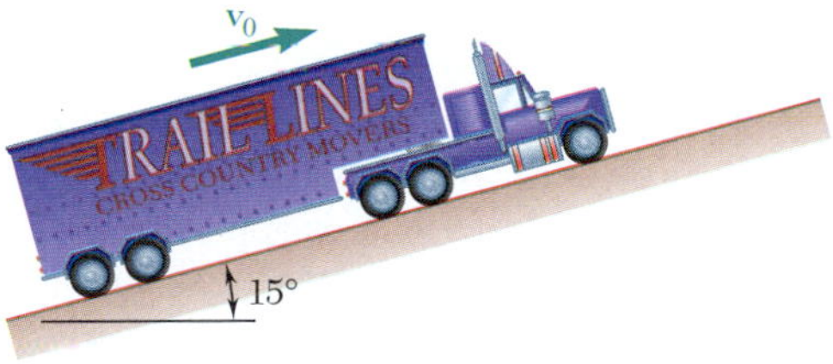

Fig. P13.121 and P13.122

13.122 A 9000-kg truck enters a 15° safety ramp at a high speed v_0 and travels 180 m in 5.5 s before its speed is reduced to $v_0/2$. Assuming constant deceleration, determine (*a*) the initial speed v_0, (*b*) the magnitude of the braking force. Neglect air resistance and rolling resistance.

13.123 Baggage on the floor of the baggage car of a high-speed train is not prevented from moving other than by friction. Determine the smallest allowable value of the coefficient of static friction between a trunk and the floor of the car if the trunk is not to slide when the train decreases its speed at a constant rate from 130 mi/h to 60 mi/h in a time interval of 12 s.

13.124 A truck is traveling on a level road at a speed of 60 mi/h when its brakes are applied to slow it down to 20 mi/h. An antiskid braking system limits the braking force to a value at which the wheels of the truck are just about to slide. Knowing that the coefficient of static friction between the road and the wheels is 0.65, determine the shortest time needed for the truck to slow down.

13.125 The initial velocity of the block in position A is 9 m/s. Knowing that the coefficient of kinetic friction between the block and the plane is $\mu_k = 0.30$, determine the time it takes for the block to reach B with zero velocity, if (*a*) $\theta = 0$, (*b*) $\theta = 20°$.

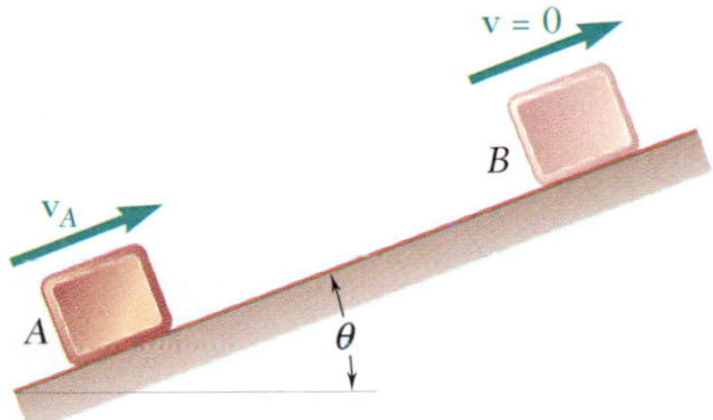

Fig. P13.125

Fig. P13.126

13.126 A 440-kg sailboat with its occupants is running downwind at 12 km/h when its spinnaker is raised to increase its speed. Determine the net force provided by the spinnaker over the 10-s interval that it takes for the boat to reach a speed of 18 km/h.

13.127 In anticipation of a long 6° upgrade, a bus driver accelerates at a constant rate from 50 mi/h to 60 mi/h in 8 s while still on a level section of the highway. Knowing that the speed of the bus is 60 mi/h as it begins to climb the grade at time $t = 0$ and that the driver does not change the setting of the throttle or shift gears, determine (*a*) the speed of the bus when $t = 10$ s, (*b*) the time when the speed is 35 mi/h.

13.128 A 50-lb block is at rest on an incline when a constant horizontal force **P** is applied to it. The static and kinetic coefficients of friction between the block and the incline are 0.4 and 0.3, respectively. Knowing that the speed of the block is 45 ft/s after 6 s, determine the magnitude of **P**.

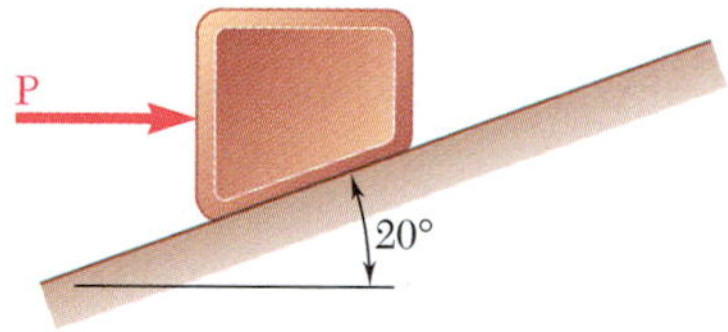

Fig. *P13.128*

13.129 A light train made of two cars travels at 72 km/h. The mass of car *A* is 18 Mg and the mass of car *B* is 13 Mg. When the brakes are suddenly applied, a constant braking force of 19 kN is applied to each car. Determine (*a*) the time required for the train to stop after the brakes are applied, (*b*) the force in the coupling between the cars while the train is slowing down.

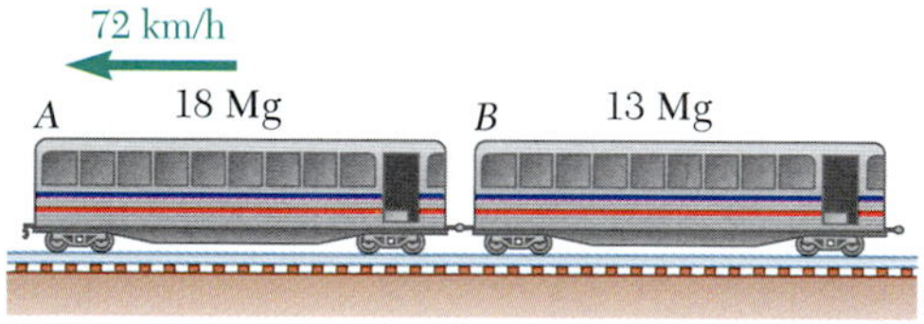

Fig. P13.129

13.130 Solve Prob. 13.129, assuming that a constant braking force of 19 kN is applied to car *B* but the brakes on car *A* are not applied.

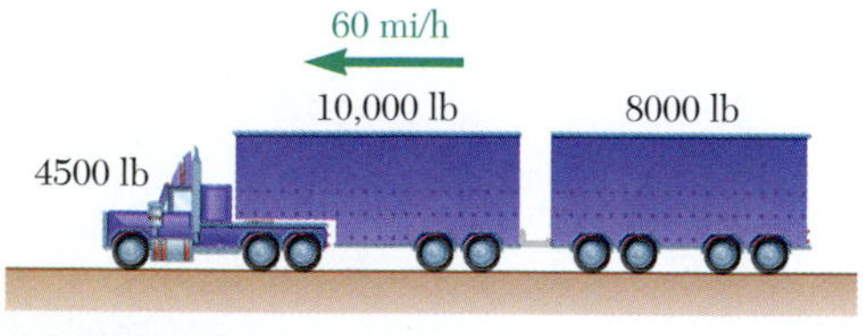

Fig. P13.131

13.131 A tractor-trailer rig with a 4500-lb tractor, a 10,000-lb trailer, and an 8000-lb trailer is traveling on a level road at 60 mi/h. The brakes on the rear trailer fail and the antiskid system of the tractor and front trailer provide the largest possible force which will not cause the wheels to slide. Knowing that the coefficient of static friction is 0.75, determine (*a*) the shortest time for the rig to come to a stop, (*b*) the force in the coupling between the two trailers during that time. Assume that the force exerted by the coupling on each of the two trailers is horizontal.

13.132 A 16-lb cylinder C rests on an 8-lb platform A supported by a cord which passes over the pulleys D and E and is attached to an 8-lb block B. Knowing that the system is released from rest, determine (*a*) the velocity of block B after 0.8 s, (*b*) the force exerted by the cylinder on the platform.

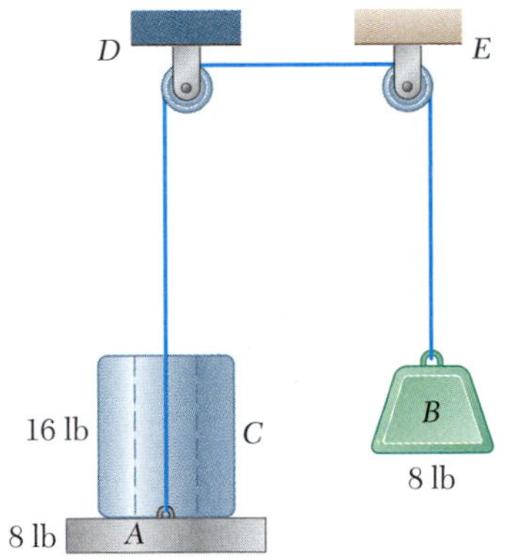

Fig. P13.132

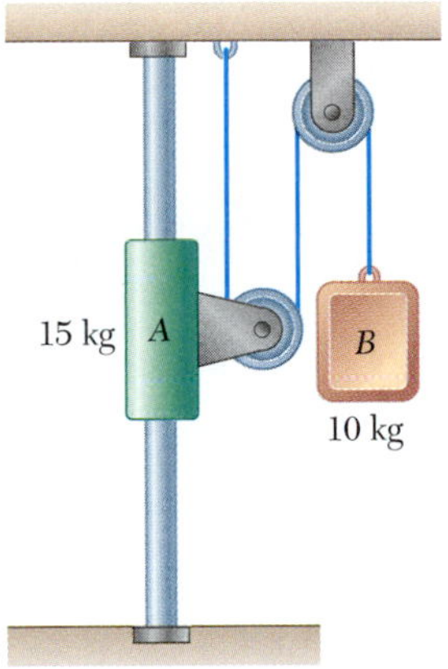

Fig. P13.133

13.133 The system shown is released from rest. Determine the time it takes for the velocity of A to reach 0.6 m/s. Neglect friction and the mass of the pulleys.

13.134 The two blocks shown are released from rest at time $t = 0$. Neglecting the masses of the pulleys and the effect of friction in the pulleys and between the blocks and the incline, determine (*a*) the velocity of block A at $t = 0.5$ s, (*b*) the tension in the cable.

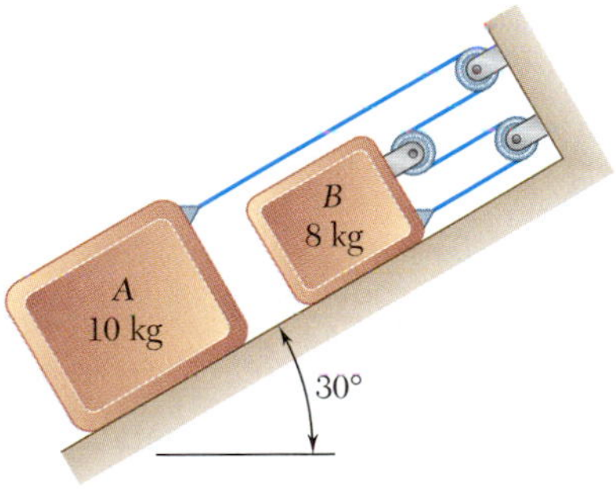

Fig. *P13.134*

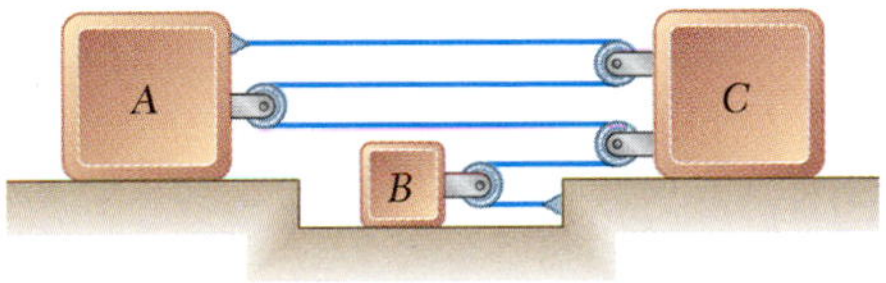

Fig. *P13.135*

13.135 The coefficients of friction between the three blocks and the horizontal surfaces are $\mu_s = 0.25$ and $\mu_k = 0.20$. The masses of the blocks are $m_A = m_C = 10$ kg, and $m_B = 5$ kg. The velocities of blocks A and C at time $t = 0$ are $v_A = 3$ m/s and $v_C = 5$ m/s, both to the right. Determine (*a*) the velocity of each block at $t = 0.5$ s, (*b*) the tension in the cable.

13.136 A 12-lb block which can slide on a frictionless inclined surface is acted upon by a force **P** which varies in magnitude as shown. Knowing that the block is initially at rest, determine (*a*) the velocity of the block at $t = 5$ s, (*b*) the time at which the velocity of the block is zero.

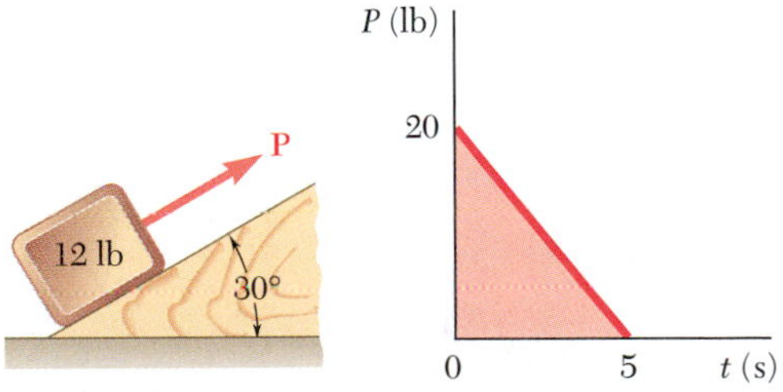

Fig. P13.136

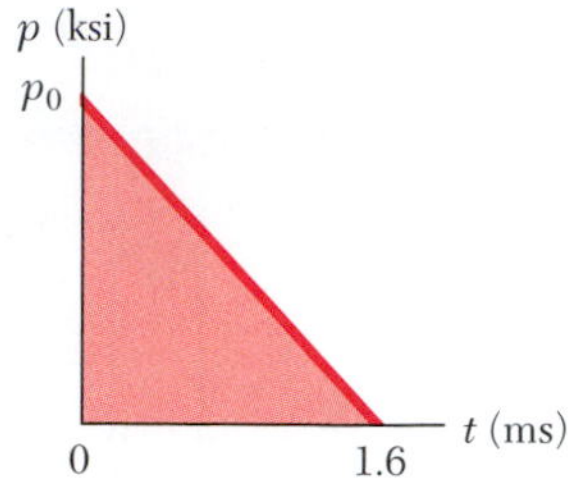

Fig. P13.137

13.137 A simplified model consisting of a single straight line is to be obtained for the variation of pressure inside the 0.4-in.-diameter barrel of a rifle as a 0.7-oz bullet is fired. Knowing that it takes 1.6 ms for the bullet to travel the length of the barrel and that the velocity of the bullet upon exit is 2100 ft/s, determine the value of p_0.

13.138 A 2-kg collar which can slide on a frictionless vertical rod is acted upon by a force **P** which varies in magnitude as shown. Knowing that the collar is initially at rest, determine its velocity at (*a*) $t = 2$ s, (*b*) $t = 3$ s.

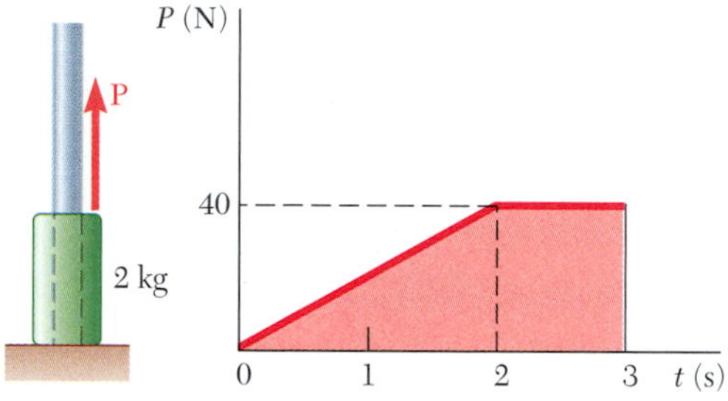

Fig. *P13.138* and P13.139

13.139 A 2-kg collar which can slide on a frictionless vertical rod is acted upon by a force **P** which varies in magnitude as shown. Knowing that the collar is initially at rest, determine (*a*) the maximum velocity of the collar, (*b*) the time when the velocity is zero.

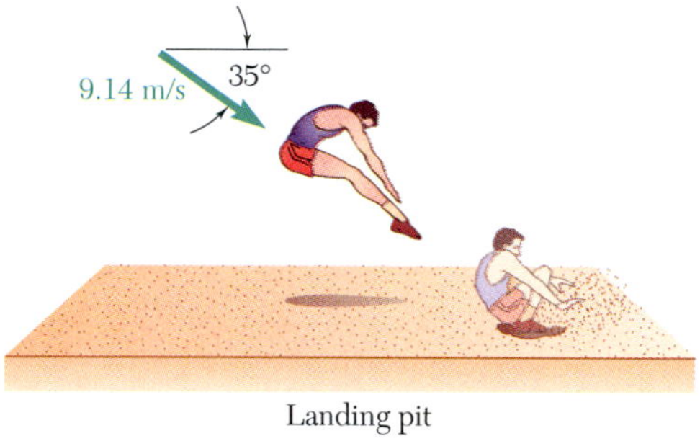

Fig. P13.140

13.140 The last segment of the triple jump track-and-field event is the jump, in which the athlete makes a final leap, landing in a sand-filled pit. Assuming that the velocity of an 84-kg athlete just before landing is 9.14 m/s at an angle of 35° with the horizontal and that the athlete comes to a complete stop in 0.22 s after landing, determine the horizontal component of the average impulsive force exerted on his feet during landing.

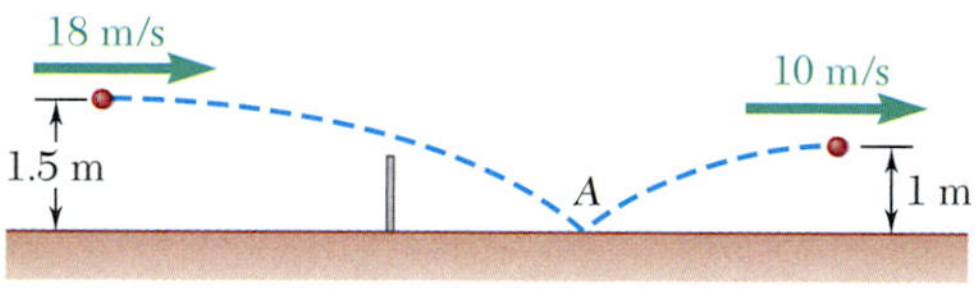

Fig. P13.141

13.141 A player hits a 57-g tennis ball with a horizontal initial velocity of 18 m/s at a height of 1.5 m. The ball bounces at point *A* and rises to a maximum height of 1 m where the velocity is 10 m/s. Knowing that the duration of the impact is 0.004 s, determine the impulsive force exerted on the ball at point *A*.

13.142 A 3-lb collar can slide on a horizontal rod which is free to rotate about a vertical shaft. The collar is initially held at *A* by a cord attached to the shaft. As the rod rotates at a rate $\dot{\theta} = 18$ rad/s, the cord is cut and the collar moves out along the rod and strikes the stop at *B* without rebounding. Neglecting friction and the mass of the rod, determine the magnitude of the impulse of the force exerted by the stop on the collar.

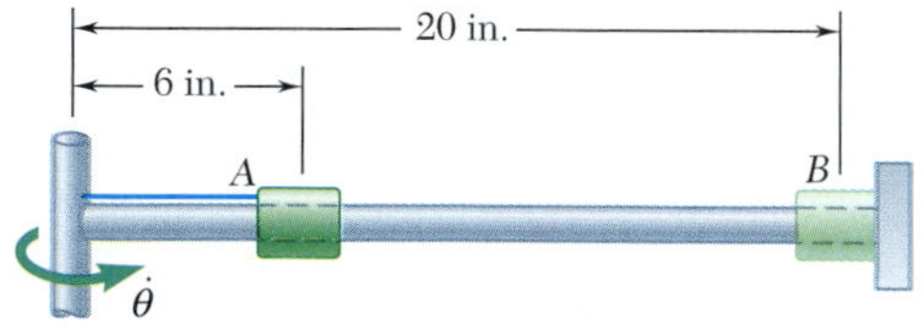

Fig. P13.142

13.143 The triple jump is a track-and-field event in which an athlete gets a running start and tries to leap as far as he can with a hop, step, and jump. Shown in the figure is the initial hop of the athlete. Assuming that he approaches the takeoff line from the left with a horizontal velocity of 30 ft/s, remains in contact with the ground for 0.18 s, and takes off at a 50° angle with a velocity of 36 ft/s, determine the vertical component of the average impulsive force exerted by the ground on his foot. Give your answer in terms of the weight W of the athlete.

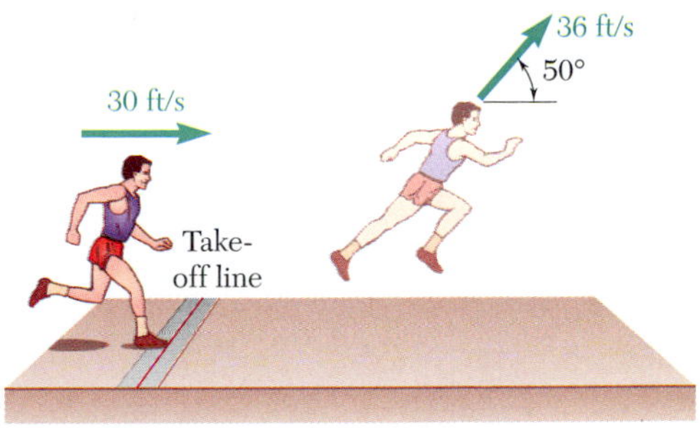

Fig. P13.143

13.144 An estimate of the expected load on over-the-shoulder seat belts is made before designing prototype belts that will be evaluated in automobile crash tests. Assuming that an automobile traveling at 72 km/h is brought to a stop in 110 ms, determine (*a*) the average impulsive force exerted by a 100-kg man on the belt, (*b*) the maximum force F_m exerted on the belt if the force-time diagram has the shape shown.

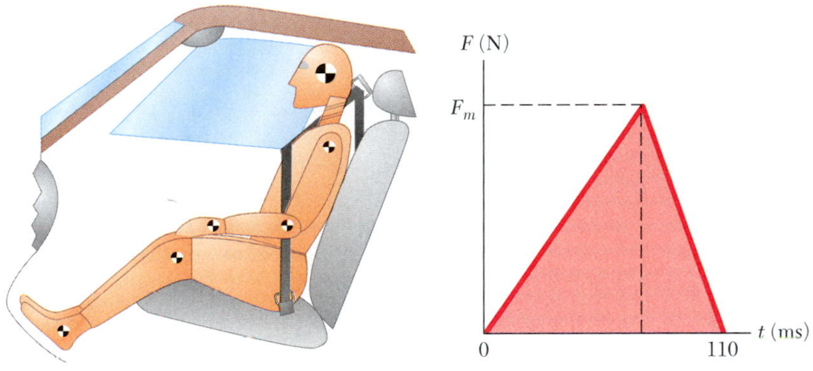

Fig. *P13.144*

13.145 A 45-g golf ball is hit with a golf club and leaves it with a velocity of 38 m/s. We assume that for $0 \leq t \leq t_0$, where t_0 is the duration of the impact, the magnitude F of the force exerted on the ball can be expressed as $F = F_m \sin(\pi t/t_0)$. Knowing that $t_0 = 0.5$ ms, determine the maximum value F_m of the force exerted on the ball.

13.146 The 30-lb suitcase A has been propped up against one end of an 80-lb luggage carrier B and is prevented from sliding down by other luggage. When the luggage is unloaded and the last heavy trunk is removed from the carrier, the suitcase is free to slide down, causing the 80-lb carrier to move to the left with a velocity $\mathbf{v}_B$ of magnitude 2.5 ft/s. Neglecting friction, determine (*a*) the velocity $\mathbf{v}_{A/B}$ of the suitcase relative to the carrier as it rolls on the floor of the carrier, (*b*) the velocity of the carrier after the suitcase hits the right side of the carrier without bouncing back, (*c*) the energy lost in the impact of the suitcase on the floor of the carrier.

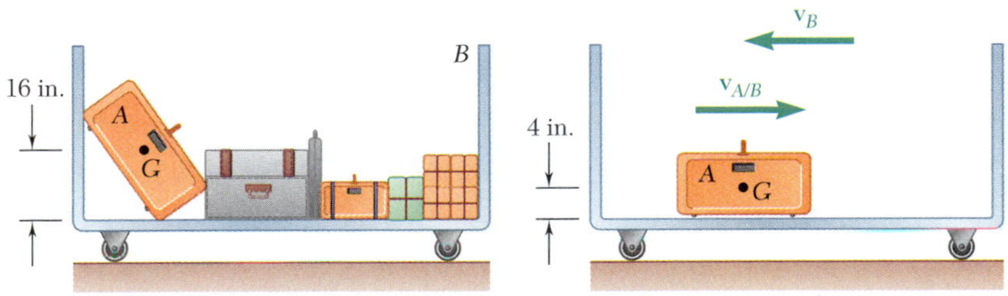

Fig. P13.146

13.147 A 112-Mg tugboat is moving at 2 m/s with a slack towing cable attached to an 88-Mg barge which is at rest. The cable is being unwound from a drum on the tugboat at a constant rate of 1.8 m/s and that rate is maintained after the cable becomes taut. Neglecting the resistance of the water, determine (*a*) the velocity of the tugboat after the cable becomes taut, (*b*) the impulse exerted on the barge as the cable becomes taut.

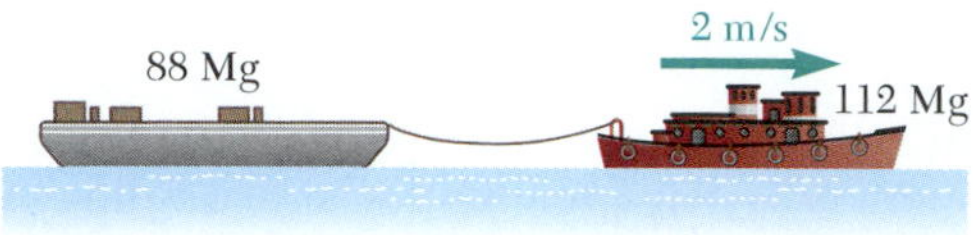

Fig. P13.147

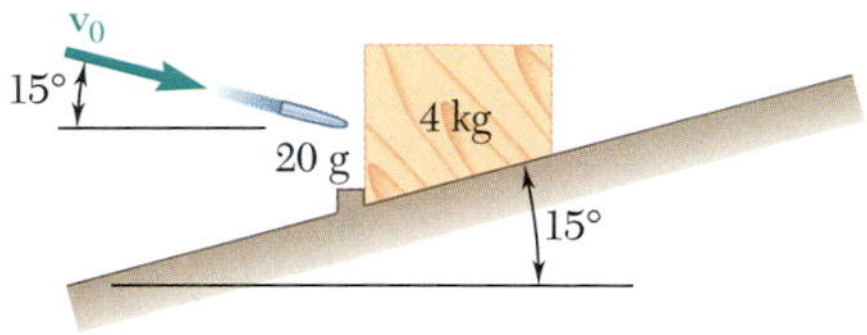

Fig. P13.148

13.148 A 20-g bullet is fired into a 4-kg wooden block and becomes embedded in it. Knowing that the block and bullet then move up the smooth incline for 1.2 s before they come to a stop, determine (*a*) the magnitude of the initial velocity of the bullet, (*b*) the magnitude of the impulse of the force exerted by the bullet on the block.

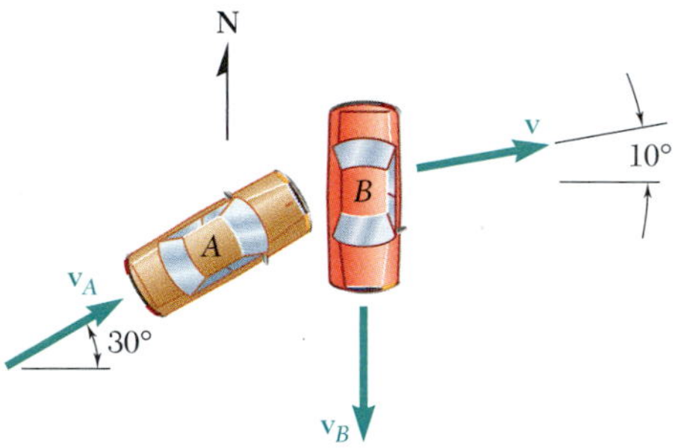

Fig. P13.149

13.149 At an intersection car *B* was traveling south and car *A* was traveling 30° north of east when they slammed into each other. Upon investigation it was found that after the crash the two cars got stuck and skidded off at an angle of 10° north of east. Each driver claimed that he was going at the speed limit of 30 mi/h and that he tried to slow down but couldn't avoid the crash because the other driver was going a lot faster. Knowing that the weights of cars *A* and *B* were 3600 lb and 2800 lb, respectively, determine (*a*) which car was going faster, (*b*) the speed of the faster of the two cars if the slower car was traveling at the speed limit.

13.150 A 4-oz ball moving at a speed of 9 ft/s strikes a 10-oz plate supported by springs. Assuming that no energy is lost in the impact, determine (*a*) the velocity of the ball immediately after impact, (*b*) the impulse of the force exerted by the plate on the ball.

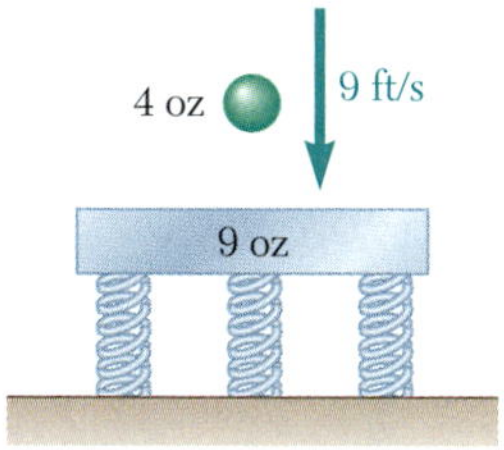

Fig. P13.150

13.151 A 75-g ball is projected from a height of 1.6 m with a horizontal velocity of 2 m/s and bounces from a 400-g smooth plate supported by springs. Knowing that the height of the rebound is 0.6 m, determine (*a*) the velocity of the plate immediately after the impact, (*b*) the energy lost due to the impact.

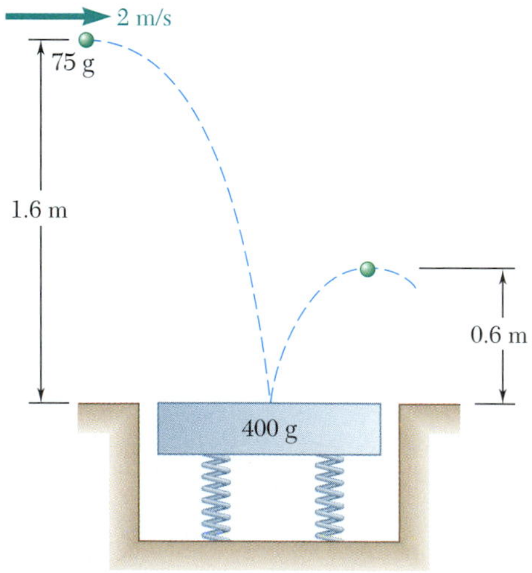

Fig. P13.151

13.152 In order to test the resistance of a chain to impact, the chain is suspended from a 120-kg rigid beam supported by two columns. A rod attached to the last link is then hit by a 30-kg block dropped from a 2-m height. Determine the initial impulse exerted on the chain and the energy absorbed by the chain, assuming that the block does not rebound from the rod and that the columns supporting the beam are (*a*) perfectly rigid, (*b*) equivalent to two perfectly elastic springs.

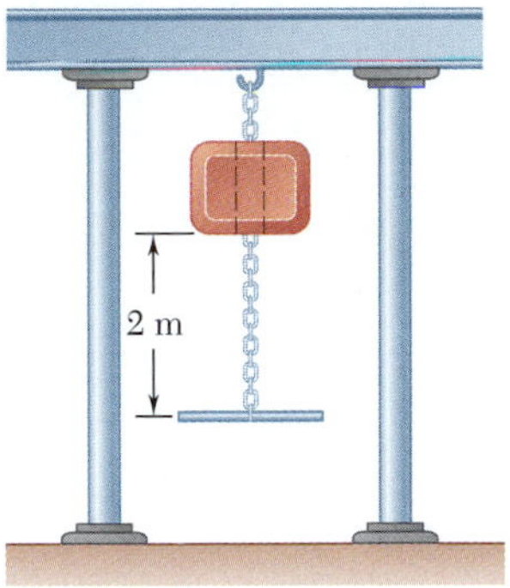

Fig. P13.152

13.153 A 4-lb sphere A is connected to a fixed point O by an inextensible cord of length 3.6 ft. The sphere is resting on a frictionless horizontal surface at a distance of 1.5 ft from O when it is given a velocity $\mathbf{v}_0$ in a direction perpendicular to line OA. It moves freely until it reaches position A', when the cord becomes taut. Determine the maximum allowable velocity $\mathbf{v}_0$ if the impulse of the force exerted on the cord is not to exceed 0.8 lb · s.

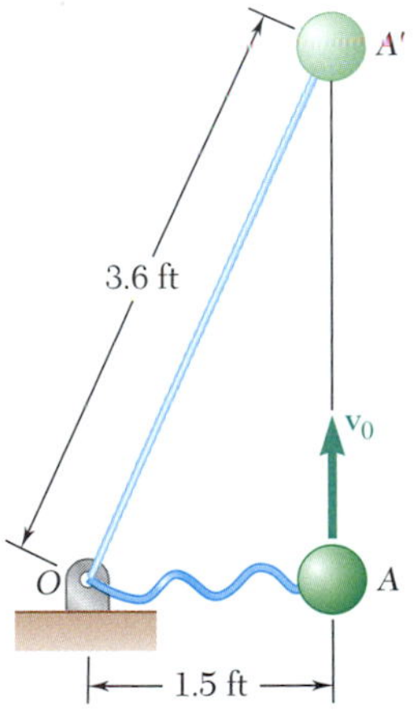

Fig. *P13.153*

13.154 A baseball player catching a ball can soften the impact by pulling his hand back. Assuming that a 5-oz ball reaches his glove at 96 mi/h and that the player pulls his hand back during the impact at an average speed of 25 ft/s over a distance of 8 in., bringing the ball to a stop, determine the average impulsive force exerted on the player's hand.

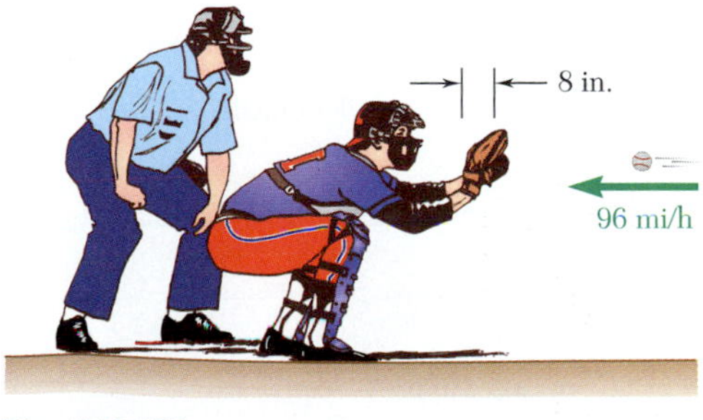

Fig. *P13.154*

13.12. IMPACT

A collision between two bodies which occurs in a very small interval of time and during which the two bodies exert relatively large forces on each other is called an *impact*. The common normal to the surfaces in contact during the impact is called the *line of impact*. If the mass centers on the two colliding bodies are located on this line, the impact is a *central impact*. Otherwise, the impact is said to be *eccentric*. Our present study will be limited to the central impact of two particles. The analysis of the eccentric impact of two rigid bodies will be considered later, in Sec. 17.12.

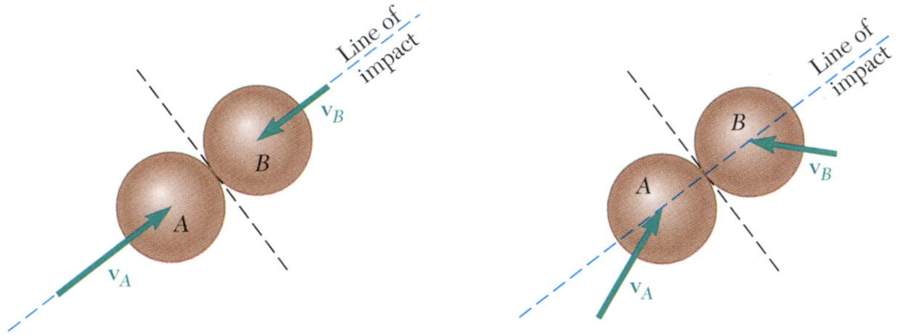

Fig. 13.20 (*a*) Direct central impact (*b*) Oblique central impact

If the velocities of the two particles are directed along the line of impact, the impact is said to be a *direct impact* (Fig. 13.20*a*). If either or both particles move along a line other than the line of impact, the impact is said to be an *oblique impact* (Fig. 13.20*b*).

13.13. DIRECT CENTRAL IMPACT

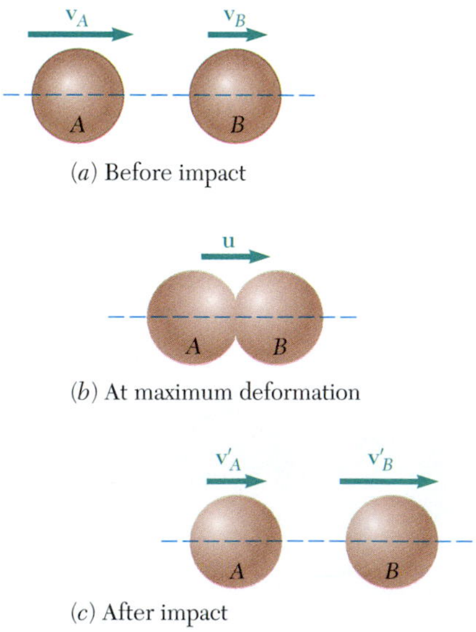

Fig. 13.21

Consider two particles A and B, of mass m_A and m_B, which are moving in the same straight line and to the right with known velocities $\mathbf{v}_A$ and $\mathbf{v}_B$ (Fig. 13.21*a*). If $\mathbf{v}_A$ is larger than $\mathbf{v}_B$, particle A will eventually strike particle B. Under the impact, the two particles, will *deform* and, at the end of the period of deformation, they will have the same velocity $\mathbf{u}$ (Fig. 13.21*b*). A period of *restitution* will then take place, at the end of which, depending upon the magnitude of the impact forces and upon the materials involved, the two particles either will have regained their original shape or will stay permanently deformed. Our purpose here is to determine the velocities $\mathbf{v}'_A$ and $\mathbf{v}'_B$ of the particles at the end of the period of restitution (Fig. 13.21*c*).

Considering first the two particles as a single system, we note that there is no impulsive, external force. Thus, the total momentum of the two particles is conserved, and we write

$$m_A\mathbf{v}_A + m_B\mathbf{v}_B = m_A\mathbf{v}'_A + m_B\mathbf{v}'_B$$

Since all the velocities considered are directed along the same axis, we can replace the equation obtained by the following relation involving only scalar components:

$$m_Av_A + m_Bv_B = m_Av'_A + m_Bv'_B \tag{13.37}$$

A positive value for any of the scalar quantities v_A, v_B, v'_A, or v'_B means that the corresponding vector is directed to the right; a negative value indicates that the corresponding vector is directed to the left.

To obtain the velocities $\mathbf{v}'_A$ and $\mathbf{v}'_B$, it is necessary to establish a second relation between the scalars v'_A and v'_B. For this purpose, let us now consider the motion of particle A during the period of deformation and apply the principle of impulse and momentum. Since the only impulsive force acting on A during this period is the force $\mathbf{P}$ exerted by B (Fig. 13.22*a*), we write, using again scalar components,

$$m_A v_A - \int P\,dt = m_A u \tag{13.38}$$

where the integral extends over the period of deformation. Considering now the motion of A during the period of restitution, and denoting by $\mathbf{R}$ the force exerted by B on A during this period (Fig. 13.22*b*), we write

$$m_A u - \int R\,dt = m_A v'_A \tag{13.39}$$

where the integral extends over the period of restitution.

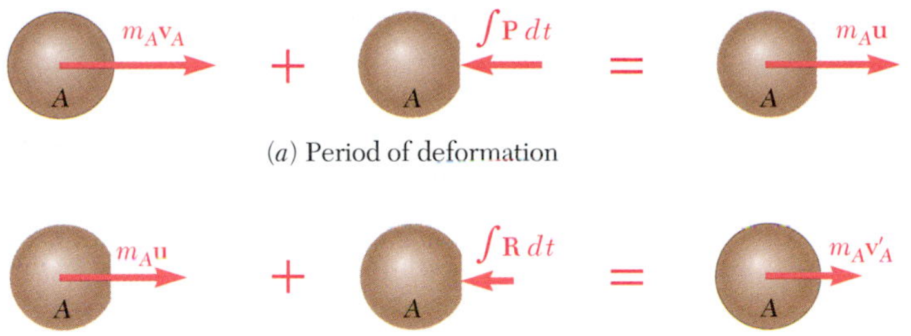

(*a*) Period of deformation

(*b*) Period of restitution

Fig. 13.22

In general, the force $\mathbf{R}$ exerted on A during the period of restitution differs from the force $\mathbf{P}$ exerted during the period of deformation, and the magnitude $\int R\,dt$ of its impulse is smaller than the magnitude $\int P\,dt$ of the impulse of $\mathbf{P}$. The ratio of the magnitudes of the impulses corresponding, respectively, to the period of restitution and to the period of deformation is called the *coefficient of restitution* and is denoted by e. We write

$$e = \frac{\int R\,dt}{\int P\,dt} \tag{13.40}$$

The value of the coefficient e is always between 0 and 1. It depends to a large extent on the two materials involved, but it also varies considerably with the impact velocity and the shape and size of the two colliding bodies.

Solving Eqs. (13.38) and (13.39) for the two impulses and substituting into (13.40), we write

$$e = \frac{u - v'_A}{v_A - u} \tag{13.41}$$

A similar analysis of particle B leads to the relation

$$e = \frac{v'_B - u}{u - v_B} \tag{13.42}$$

Since the quotients in (13.41) and (13.42) are equal, they are also equal to the quotient obtained by adding, respectively, their numerators and their denominators. We have, therefore,

$$e = \frac{(u - v'_A) + (v'_B - u)}{(v_A - u) + (u - v_B)} = \frac{v'_B - v'_A}{v_A - v_B}$$

and

$$v'_B - v'_A = e(v_A - v_B) \tag{13.43}$$

Since $v'_B - v'_A$ represents the relative velocity of the two particles after impact and $v_A - v_B$ represents their relative velocity before impact, formula (13.43) expresses that *the relative velocity of the two particles after impact can be obtained by multiplying their relative velocity before impact by the coefficient of restitution.* This property is used to determine experimentally the value of the coefficient of restitution of two given materials.

The velocities of the two particles after impact can now be obtained by solving Eqs. (13.37) and (13.43) simultaneously for v'_A and v'_B. It is recalled that the derivation of Eqs. (13.37) and (13.43) was based on the assumption that particle B is located to the right of A, and that both particles are initially moving to the right. If particle B is initially moving to the left, the scalar v_B should be considered negative. The same sign convention holds for the velocities after impact: a positive sign for v'_A will indicate that particle A moves to the right after impact, and a negative sign will indicate that it moves to the left.

Two particular cases of impact are of special interest:

1. $e = 0$, *Perfectly Plastic Impact.* When $e = 0$, Eq. (13.43) yields $v'_B = v'_A$. There is no period of restitution, and both particles stay together after impact. Substituting $v'_B = v'_A = v'$ into Eq. (13.37), which expresses that the total momentum of the particles is conserved, we write

$$m_A v_A + m_B v_B = (m_A + m_B)v' \tag{13.44}$$

   This equation can be solved for the common velocity v' of the two particles after impact.

2. $e = 1$, *Perfectly Elastic Impact.* When $e = 1$, Eq. (13.43) reduces to

$$v'_B - v'_A = v_A - v_B \tag{13.45}$$

   which expresses that the relative velocities before and after impact are equal. The impulses received by each particle during the period of deformation and during the period of restitution are equal. The particles move away from each other after impact with the same velocity with which they approached each other before impact. The velocities v'_A and v'_B can be obtained by solving Eqs. (13.37) and (13.45) simultaneously.

It is worth noting that *in the case of a perfectly elastic impact, the total energy of the two particles,* as well as their total momentum, *is conserved.* Equations (13.37) and (13.45) can be written as follows:

$$m_A(v_A - v'_A) = m_B(v'_B - v_B) \tag{13.37'}$$

$$v_A + v'_A = v_B + v'_B \tag{13.45'}$$

Multiplying (13.37′) and (13.45′) member by member, we have

$$m_A(v_A - v'_A)(v_A + v'_A) = m_B(v'_B - v_B)(v'_B + v_B)$$

$$m_A v_A^2 - m_A(v'_A)^2 = m_B(v'_B)^2 - m_B v_B^2$$

Rearranging the terms in the equation obtained and multiplying by $\frac{1}{2}$, we write

$$\tfrac{1}{2}m_A v_A^2 + \tfrac{1}{2}m_B v_B^2 = \tfrac{1}{2}m_A(v'_A)^2 + \tfrac{1}{2}m_B(v'_B)^2 \tag{13.46}$$

which expresses that the kinetic energy of the particles is conserved. It should be noted, however, that *in the general case of impact,* that is, when e is not equal to 1, *the total energy of the particles is not conserved.* This can be shown in any given case by comparing the kinetic energies before and after impact. The lost kinetic energy is in part transformed into heat and in part spent in generating elastic waves within the two colliding bodies.

Photo 13.2 Because the coefficient of restitution is less than one, the tennis ball loses some energy with each bounce.

13.14. OBLIQUE CENTRAL IMPACT

Let us now consider the case when the velocities of the two colliding particles are *not* directed along the line of impact (Fig. 13.23). As indicated in Sec. 13.12, the impact is said to be *oblique.* Since the

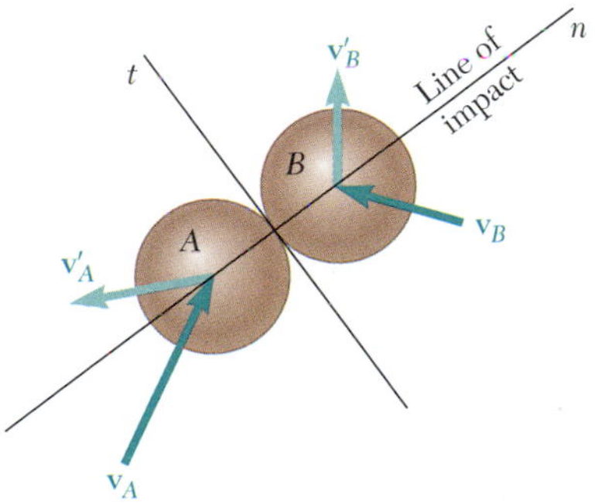

Fig. 13.23

velocities $\mathbf{v}'_A$ and $\mathbf{v}'_B$ of the particles after impact are unknown in direction as well as in magnitude, their determination will require the use of four independent equations.

We choose as coordinate axes the n axis along the line of impact, that is, along the common normal to the surfaces in contact, and the t axis along their common tangent. Assuming that the particles are perfectly *smooth and frictionless,* we observe that the only impulses

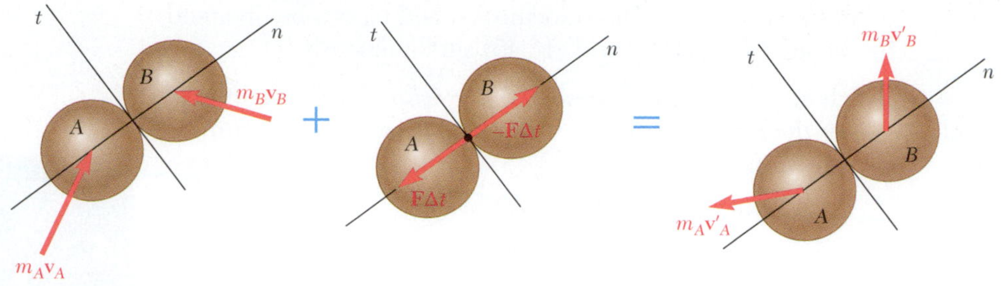

Fig. 13.24

exerted on the particles during the impact are due to internal forces directed along the line of impact, that is, along the n axis (Fig. 13.24). It follows that

1. The component along the t axis of the momentum of each particle, considered separately, is conserved; hence the t component of the velocity of each particle remains unchanged. We write

$$(v_A)_t = (v'_A)_t \qquad (v_B)_t = (v'_B)_t \tag{13.47}$$

2. The component along the n axis of the total momentum of the two particles is conserved. We write

$$m_A(v_A)_n + m_B(v_B)_n = m_A(v'_A)_n + m_B(v'_B)_n \tag{13.48}$$

3. The component along the n axis of the relative velocity of the two particles after impact is obtained by multiplying the n component of their relative velocity before impact by the coefficient of restitution. Indeed, a derivation similar to that given in Sec. 13.13 for direct central impact yields

$$(v'_B)_n - (v'_A)_n = e[(v_A)_n - (v_B)_n] \tag{13.49}$$

Photo 13.3 In a game of pool, the cue ball has just hit another ball, and most of its energy and momentum have been transferred to that ball.

We have thus obtained four independent equations which can be solved for the components of the velocities of A and B after impact. This method of solution is illustrated in Sample Prob. 13.15.

Our analysis of the oblique central impact of two particles has been based so far on the assumption that both particles moved freely before and after the impact. Let us now examine the case when one or both of the colliding particles is constrained in its motion. Consider, for instance, the collision between block A, which is constrained to move on a horizontal surface, and ball B, which is free to move in the plane of the figure (Fig. 13.25). Assuming no friction between the block and the ball, or between the block and the horizontal surface, we note that the impulses exerted on the system consist of the impulses of the internal forces $\mathbf{F}$ and $-\mathbf{F}$ directed along the line of impact, that is, along the n axis, and of the impulse of the external force $\mathbf{F}_{\text{ext}}$ exerted by the horizontal surface on block A and directed along the vertical (Fig. 13.26).

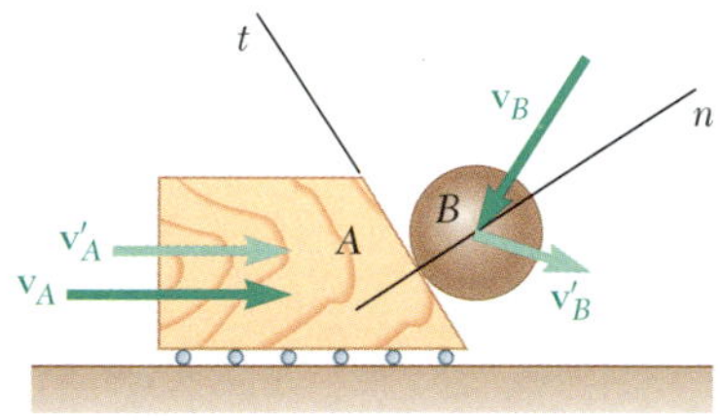

Fig. 13.25

The velocities of block A and ball B immediately after the impact are represented by three unknowns: the magnitude of the velocity $\mathbf{v}'_A$ of block A, which is known to be horizontal, and the magnitude and

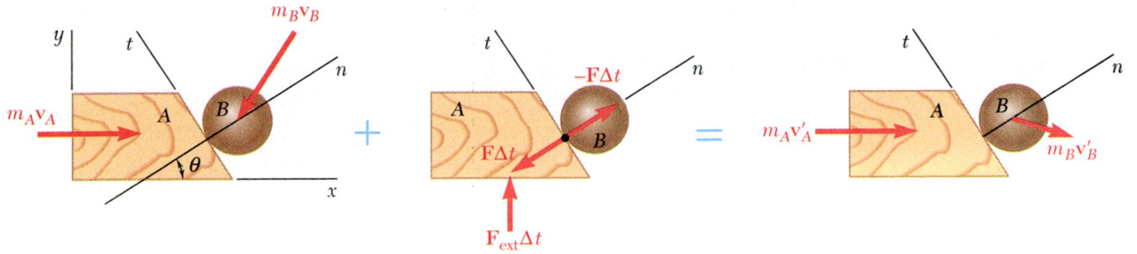

Fig. 13.26

direction of the velocity $\mathbf{v}'_B$ of ball B. We must therefore write three equations by expressing that

1. The component along the t axis of the momentum of ball B is conserved; hence the t component of the velocity of ball B remains unchanged. We write

$$(v_B)_t = (v'_B)_t \qquad (13.50)$$

2. The component along the horizontal x axis of the total momentum of block A and ball B is conserved. We write

$$m_A v_A + m_B(v_B)_x = m_A v'_A + m_B(v'_B)_x \qquad (13.51)$$

3. The component along the n axis of the relative velocity of block A and ball B after impact is obtained by multiplying the n component of their relative velocity before impact by the coefficient of restitution. We write again

$$(v'_B)_n - (v'_A)_n = e[(v_A)_n - (v_B)_n] \qquad (13.49)$$

We should note, however, that in the case considered here, the validity of Eq. (13.49) cannot be established through a mere extension of the derivation given in Sec. 13.13 for the direct central impact of two particles moving in a straight line. Indeed, these particles were not subjected to any external impulse, while block A in the present analysis is subjected to the impulse exerted by the horizontal surface. To prove that Eq. (13.49) is still valid, we will first apply the principle of impulse and momentum to block A over the period of deformation (Fig. 13.27). Considering only the horizontal components, we write

$$m_A v_A - (\textstyle\int P\,dt)\cos\theta = m_A u \qquad (13.52)$$

where the integral extends over the period of deformation and where $\mathbf{u}$ represents the velocity of block A at the end of that period. Considering now the period of restitution, we write in a similar way

$$m_A u - (\textstyle\int R\,dt)\cos\theta = m_A v'_A \qquad (13.53)$$

where the integral extends over the period of restitution.

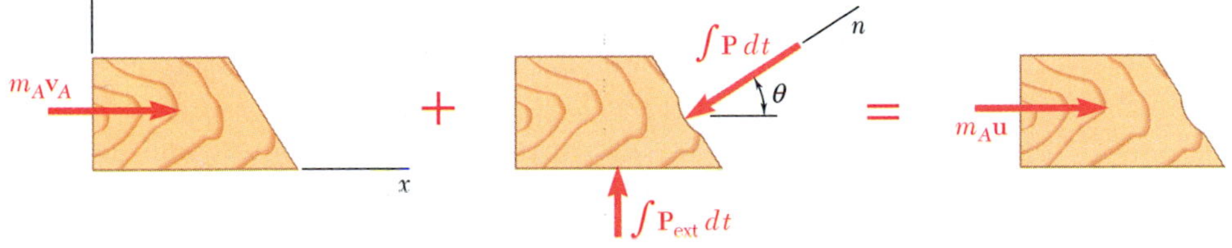

Fig. 13.27

Recalling from Sec. 13.13 the definition of the coefficient of restitution, we write

$$e = \frac{\int R\,dt}{\int P\,dt} \qquad (13.40)$$

Solving Eqs. (13.52) and (13.53) for the integrals $\int P\,dt$ and $\int R\,dt$, and substituting into Eq. (13.40), we have, after reductions,

$$e = \frac{u - v'_A}{v_A - u}$$

or, multiplying all velocities by cos θ to obtain their projections on the line of impact,

$$e = \frac{u_n - (v'_A)_n}{(v_A)_n - u_n} \qquad (13.54)$$

We note that Eq. (13.54) is identical to Eq. (13.41) of Sec. 13.13, except for the subscripts n which are used here to indicate that we are considering velocity components along the line of impact. Since the motion of ball B is unconstrained, the proof of Eq. (13.49) can be completed in the same manner as the derivation of Eq. (13.43) of Sec. 13.13. Thus, we conclude that the relation (13.49) between the components along the line of impact of the relative velocities of two colliding particles remains valid when one of the particles is constrained in its motion. The validity of this relation is easily extended to the case when both particles are constrained in their motion.

13.15. PROBLEMS INVOLVING ENERGY AND MOMENTUM

Photo 13.4 As the automobile crashes into the barrier, considerable energy is dissipated as heat during the permanent deformation of the automobile and the barrier.

You now have at your disposal three different methods for the solution of kinetics problems: the direct application of Newton's second law, $\Sigma\mathbf{F} = m\mathbf{a}$; the method of work and energy; and the method of impulse and momentum. To derive maximum benefit from these three methods, you should be able to choose the method best suited for the solution of a given problem. You should also be prepared to use different methods for solving the various parts of a problem when such a procedure seems advisable.

You have already seen that the method of work and energy is in many cases more expeditious than the direct application of Newton's second law. As indicated in Sec. 13.4, however, the method of work and energy has limitations, and it must sometimes be supplemented by the use of $\Sigma\mathbf{F} = m\mathbf{a}$. This is the case, for example, when you wish to determine an acceleration or a normal force.

For the solution of problems involving no impulsive forces, it will usually be found that the equation $\Sigma\mathbf{F} = m\mathbf{a}$ yields a solution just as fast as the method of impulse and momentum and that the method of work and energy, if it applies, is more rapid and more convenient. However, in problems of impact, the method of impulse and momentum is the only practicable method. A solution based on the direct application of $\Sigma\mathbf{F} = m\mathbf{a}$ would be unwieldy, and the method of work and energy cannot be used since impact (unless perfectly elastic) involves a loss of mechanical energy.

Many problems involve only conservative forces, except for a short impact phase during which impulsive forces act. The solution of such problems can be divided into several parts. The part corresponding to the impact phase calls for the use of the method of impulse and momentum and of the relation between relative velocities, and the other parts can usually be solved by the method of work and energy. If the problem involves the determination of a normal force, however, the use of $\Sigma \mathbf{F} = m\mathbf{a}$ is necessary.

Consider, for example, a pendulum A, of mass m_A and length l, which is released with no velocity from a position A_1 (Fig. 13.28*a*). The pendulum swings freely in a vertical plane and hits a second pendulum B, of mass m_B and same length l, which is initially at rest. After the impact (with coefficient of restitution e), pendulum B swings through an angle θ that we wish to determine.

The solution of the problem can be divided into three parts:

1. *Pendulum A Swings from A_1 to A_2.* The principle of conservation of energy can be used to determine the velocity $(\mathbf{v}_A)_2$ of the pendulum at A_2 (Fig. 13.28*b*).
2. *Pendulum A Hits Pendulum B.* Using the fact that the total momentum of the two pendulums is conserved and the relation between their relative velocities, we determine the velocities $(\mathbf{v}_A)_3$ and $(\mathbf{v}_B)_3$ of the two pendulums after impact (Fig. 13.28*c*).
3. *Pendulum B Swings from B_3 to B_4.* Applying the principle of conservation of energy to pendulum B, we determine the maximum elevation y_4 reached by that pendulum (Fig. 13.28*d*). The angle θ can then be determined by trigonometry.

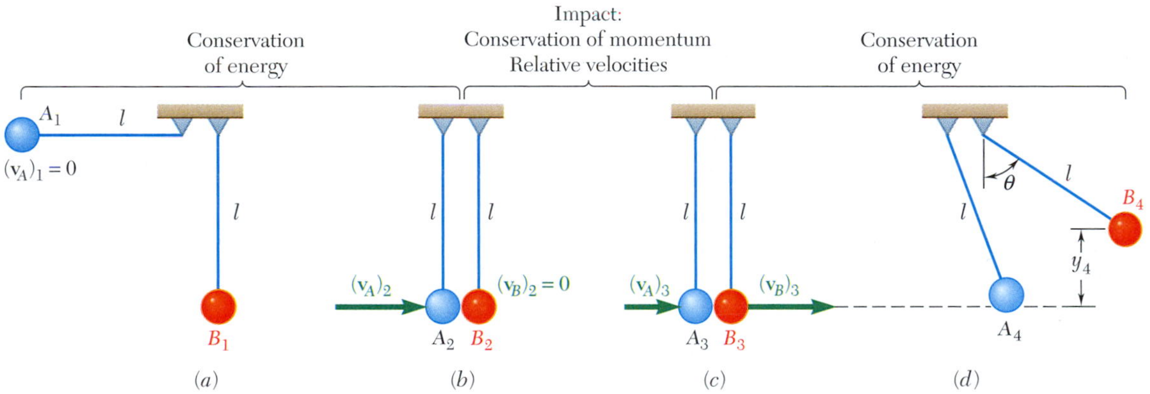

Fig. 13.28

We note that if the tensions in the cords holding the pendulums are to be determined, the method of solution just described should be supplemented by the use of $\Sigma \mathbf{F} = m\mathbf{a}$.

SAMPLE PROBLEM 13.13

A 20-Mg railroad car moving at a speed of 0.5 m/s to the right collides with a 35-Mg car which is at rest. If after the collision the 35-Mg car is observed to move to the right at a speed of 0.3 m/s, determine the coefficient of restitution between the two cars.

SOLUTION

We express that the total momentum of the two cars is conserved.

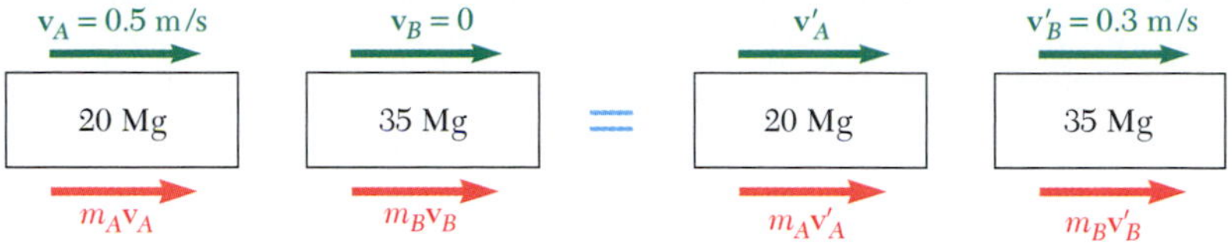

$$m_A\mathbf{v}_A + m_B\mathbf{v}_B = m_A\mathbf{v}'_A + m_B\mathbf{v}'_B$$
$$(20\text{ Mg})(+0.5\text{ m/s}) + (35\text{ Mg})(0) = (20\text{ Mg})v'_A + (35\text{ Mg})(+0.3\text{ m/s})$$
$$v'_A = -0.025\text{ m/s} \qquad \mathbf{v}'_A = 0.025\text{ m/s} \leftarrow$$

The coefficient of restitution is obtained by writing

$$e = \frac{v'_B - v'_A}{v_A - v_B} = \frac{+0.3 - (-0.025)}{+0.5 - 0} = \frac{0.325}{0.5} \qquad e = 0.65 \blacktriangleleft$$

SAMPLE PROBLEM 13.14

A ball is thrown against a frictionless, vertical wall. Immediately before the ball strikes the wall, its velocity has a magnitude v and forms an angle of 30° with the horizontal. Knowing that $e = 0.90$, determine the magnitude and direction of the velocity of the ball as it rebounds from the wall.

SOLUTION

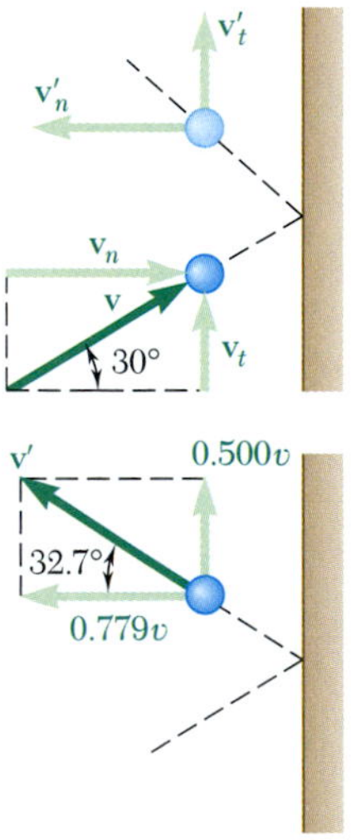

We resolve the initial velocity of the ball into components respectively perpendicular and parallel to the wall:

$$v_n = v\cos 30° = 0.866v \qquad v_t = v\sin 30° = 0.500v$$

Motion Parallel to the Wall. Since the wall is frictionless, the impulse it exerts on the ball is perpendicular to the wall. Thus, the component parallel to the wall of the momentum of the ball is conserved and we have

$$\mathbf{v}'_t = \mathbf{v}_t = 0.500v \uparrow$$

Motion Perpendicular to the Wall. Since the mass of the wall (and earth) is essentially infinite, expressing that the total momentum of the ball and wall is conserved would yield no useful information. Using the relation (13.49) between relative velocities, we write

$$0 - v'_n = e(v_n - 0)$$
$$v'_n = -0.90(0.866v) = -0.779v \qquad \mathbf{v}'_n = 0.779v \leftarrow$$

Resultant Motion. Adding vectorially the components $\mathbf{v}'_n$ and $\mathbf{v}'_t$,

$$\mathbf{v}' = 0.926v \measuredangle 32.7° \blacktriangleleft$$

SAMPLE PROBLEM 13.15

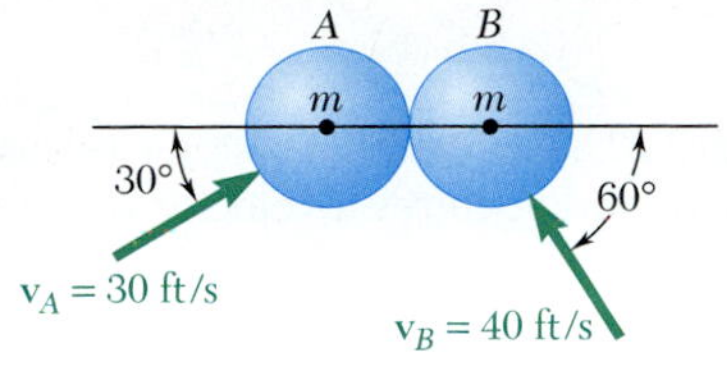

The magnitude and direction of the velocities of two identical frictionless balls before they strike each other are as shown. Assuming $e = 0.90$, determine the magnitude and direction of the velocity of each ball after the impact.

SOLUTION

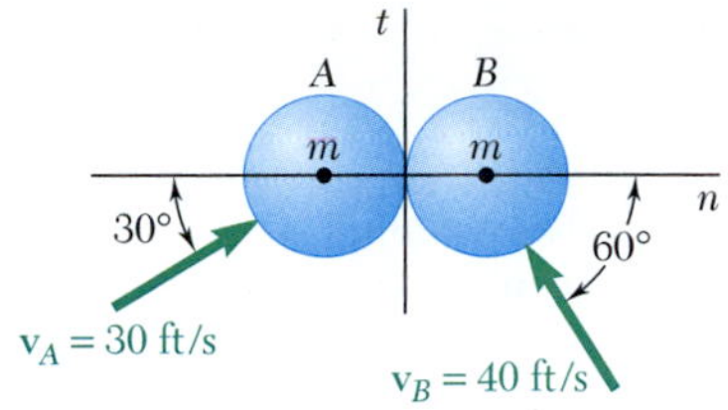

The impulsive forces that the balls exert on each other during the impact are directed along a line joining the centers of the balls called the *line of impact*. Resolving the velocities into components directed, respectively, along the line of impact and along the common tangent to the surfaces in contact, we write

$$(v_A)_n = v_A \cos 30° = +26.0 \text{ ft/s}$$
$$(v_A)_t = v_A \sin 30° = +15.0 \text{ ft/s}$$
$$(v_B)_n = -v_B \cos 60° = -20.0 \text{ ft/s}$$
$$(v_B)_t = v_B \sin 60° = +34.6 \text{ ft/s}$$

Principle of Impulse and Momentum. In the adjoining sketches we show in turn the initial momenta, the impulses, and the final momenta.

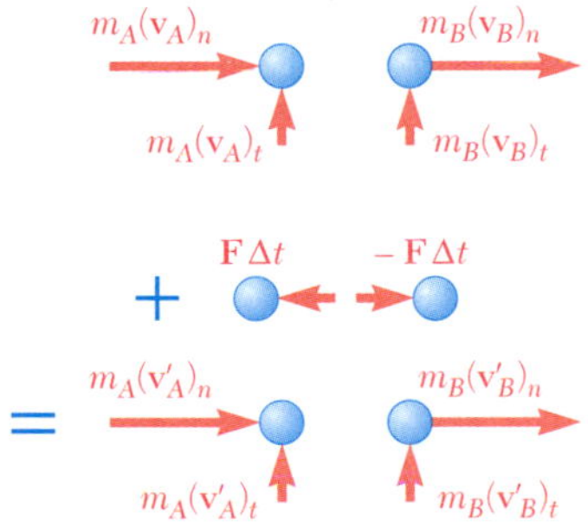

Motion along the Common Tangent. Considering only the t components, we apply the principle of impulse and momentum to each ball *separately*. Since the impulsive forces are directed along the line of impact, the t component of the momentum, and hence the t component of the velocity of each ball, is unchanged. We have

$$(\mathbf{v}'_A)_t = 15.0 \text{ ft/s} \uparrow \qquad (\mathbf{v}'_B)_t = 34.6 \text{ ft/s} \uparrow$$

Motion along the Line of Impact. In the n direction, we consider the two balls as a single system and note that by Newton's third law, the internal impulses are, respectively, $\mathbf{F}\,\Delta t$ and $-\mathbf{F}\,\Delta t$ and cancel. We thus write that the total momentum of the balls is conserved:

$$m_A(v_A)_n + m_B(v_B)_n = m_A(v'_A)_n + m_B(v'_B)_n$$
$$m(26.0) + m(-20.0) = m(v'_A)_n + m(v'_B)_n$$
$$(v_A)_n + (v'_B)_n = 6.0 \quad (1)$$

Using the relation (13.49) between relative velocities, we write

$$(v'_B)_n - (v'_A)_n = e[(v_A)_n - (v_B)_n]$$
$$(v'_B)_n - (v'_A)_n = (0.90)[26.0 - (-20.0)]$$
$$(v'_B)_n - (v'_A)_n = 41.4 \quad (2)$$

Solving Eqs. (1) and (2) simultaneously, we obtain

$$(v'_A)_n = -17.7 \qquad (v'_B)_n = +23.7$$
$$(\mathbf{v}'_A)_n = 17.7 \text{ ft/s} \leftarrow \qquad (\mathbf{v}'_B)_n = 23.7 \text{ ft/s} \rightarrow$$

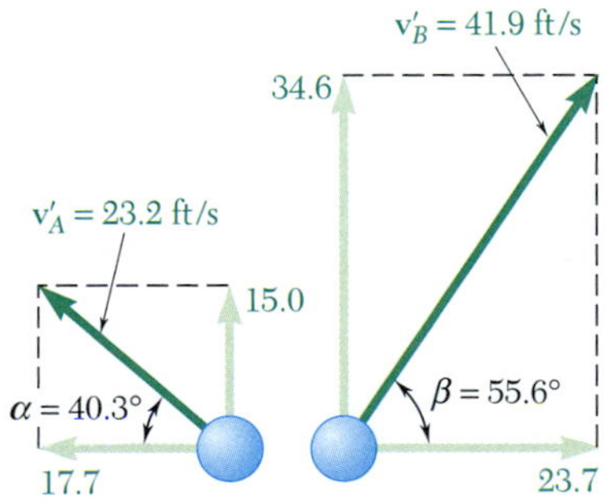

Resultant Motion. Adding vectorially the velocity components of each ball, we obtain

$$\mathbf{v}'_A = 23.2 \text{ ft/s} \measuredangle 40.3° \qquad \mathbf{v}'_B = 41.9 \text{ ft/s} \measuredangle 55.6° \quad \blacktriangleleft$$

SAMPLE PROBLEM 13.16

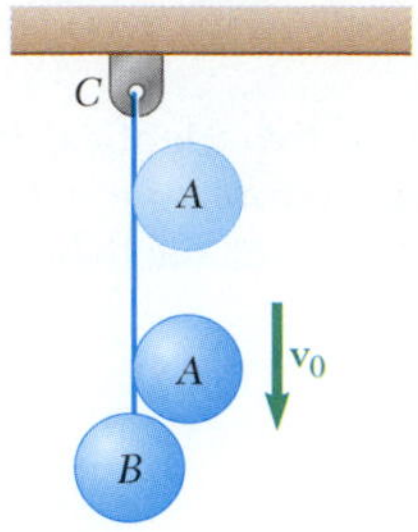

Ball B is hanging from an inextensible cord BC. An identical ball A is released from rest when it is just touching the cord and acquires a velocity $\mathbf{v}_0$ before striking ball B. Assuming perfectly elastic impact ($e = 1$) and no friction, determine the velocity of each ball immediately after impact.

SOLUTION

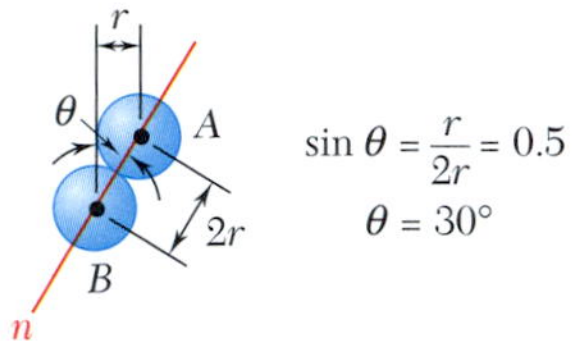

Since ball B is constrained to move in a circle of center C, its velocity $\mathbf{v}_B$ after impact must be horizontal. Thus the problem involves three unknowns: the magnitude v'_B of the velocity of B, and the magnitude and direction of the velocity $\mathbf{v}'_A$ of A after impact.

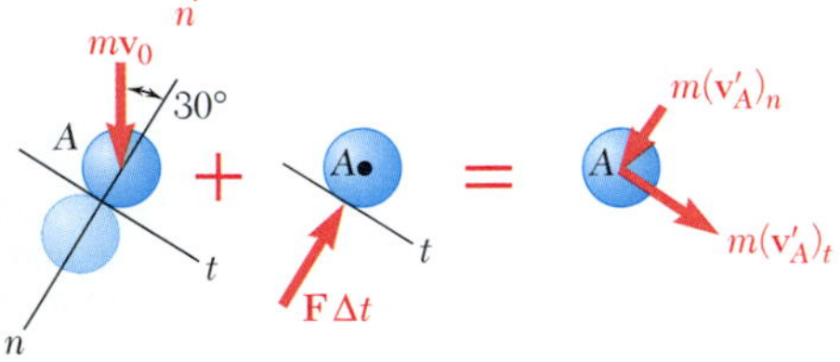

Impulse-Momentum Principle: Ball A

$$m\mathbf{v}_A + \mathbf{F}\,\Delta t = m\mathbf{v}'_A$$

$+\searrow t$ components: $\quad mv_0 \sin 30° + 0 = m(v'_A)_t$

$$(v'_A)_t = 0.5v_0 \tag{1}$$

We note that the equation used expresses conservation of the momentum of ball A along the common tangent to balls A and B.

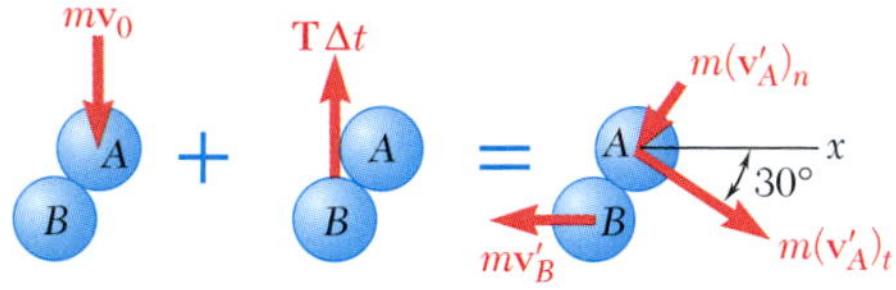

Impulse-Momentum Principle: Balls A and B

$$m\mathbf{v}_A + \mathbf{T}\,\Delta t = m\mathbf{v}'_A + m\mathbf{v}'_B$$

$\overset{+}{\rightarrow} x$ components: $\quad 0 = m(v'_A)_t \cos 30° - m(v'_A)_n \sin 30° - mv'_B$

We note that the equation obtained expresses conservation of the total momentum in the x direction. Substituting for $(v'_A)_t$ from Eq. (1) and rearranging terms, we write

$$0.5(v'_A)_n + v'_B = 0.433v_0 \tag{2}$$

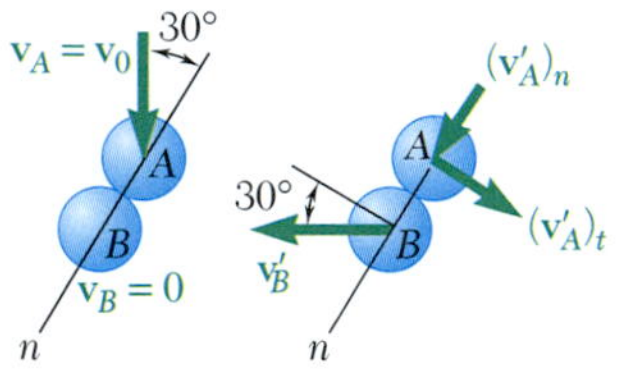

Relative Velocities along the Line of Impact. Since $e = 1$, Eq. (13.49) yields

$$(v'_B)_n - (v'_A)_n = (v_A)_n - (v_B)_n$$
$$v'_B \sin 30° - (v'_A)_n = v_0 \cos 30° - 0$$
$$0.5v'_B - (v'_A)_n = 0.866v_0 \tag{3}$$

Solving Eqs. (2) and (3) simultaneously, we obtain

$$(v'_A)_n = -0.520v_0 \qquad v'_B = 0.693v_0$$

$$\mathbf{v}'_B = 0.693v_0 \leftarrow \quad ◀$$

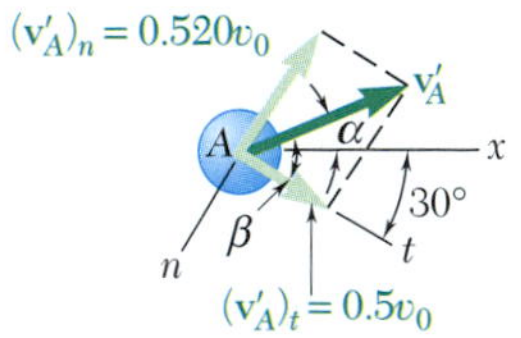

Recalling Eq. (1) we draw the adjoining sketch and obtain by trigonometry

$$v'_A = 0.721v_0 \qquad \beta = 46.1° \qquad \alpha = 46.1° - 30° = 16.1°$$

$$\mathbf{v}'_A = 0.721v_0 \measuredangle 16.1° \quad ◀$$

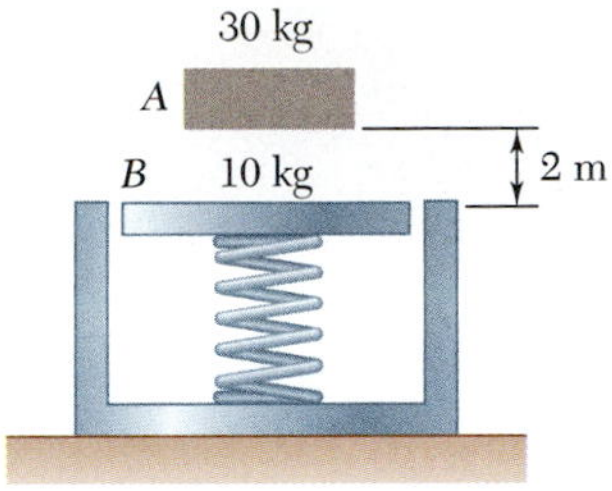

SAMPLE PROBLEM 13.17

A 30-kg block is dropped from a height of 2 m onto the 10-kg pan of a spring scale. Assuming the impact to be perfectly plastic, determine the maximum deflection of the pan. The constant of the spring is $k = 20$ kN/m.

SOLUTION

The impact between the block and the pan *must* be treated separately; therefore we divide the solution into three parts.

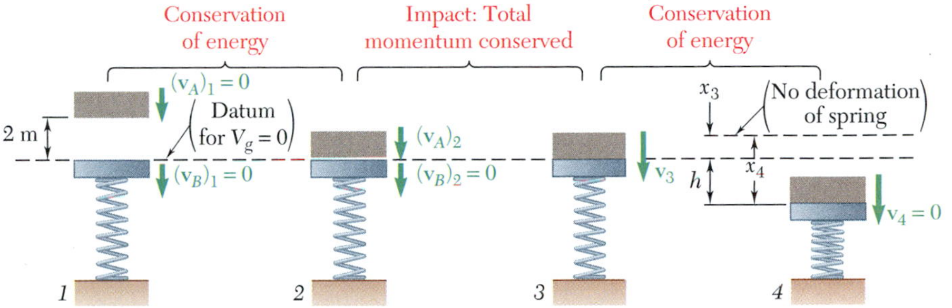

Conservation of Energy. Block: $W_A = (30 \text{ kg})(9.81 \text{ m/s}^2) = 294 \text{ N}$

$$T_1 = \tfrac{1}{2}m_A(v_A)_1^2 = 0 \qquad V_1 = W_A y = (294 \text{ N})(2 \text{ m}) = 588 \text{ J}$$
$$T_2 = \tfrac{1}{2}m_A(v_A)_2^2 = \tfrac{1}{2}(30 \text{ kg})(v_A)_2^2 \qquad V_2 = 0$$
$$T_1 + V_1 = T_2 + V_2: \quad 0 + 588 \text{ J} = \tfrac{1}{2}(30 \text{ kg})(v_A)_2^2 + 0$$
$$(v_A)_2 = +6.26 \text{ m/s} \qquad (\mathbf{v}_A)_2 = 6.26 \text{ m/s}\downarrow$$

Impact: Conservation of Momentum. Since the impact is perfectly plastic, $e = 0$; the block and pan move together after the impact.

$$m_A(v_A)_2 + m_B(v_B)_2 = (m_A + m_B)v_3$$
$$(30 \text{ kg})(6.26 \text{ m/s}) + 0 = (30 \text{ kg} + 10 \text{ kg})v_3$$
$$v_3 = +4.70 \text{ m/s} \qquad \mathbf{v}_3 = 4.70 \text{ m/s}\downarrow$$

Conservation of Energy. Initially the spring supports the weight W_B of the pan; thus the initial deflection of the spring is

$$x_3 = \frac{W_B}{k} = \frac{(10 \text{ kg})(9.81 \text{ m/s}^2)}{20 \times 10^3 \text{ N/m}} = \frac{98.1 \text{ N}}{20 \times 10^3 \text{ N/m}} = 4.91 \times 10^{-3} \text{ m}$$

Denoting by x_4 the total maximum deflection of the spring, we write

$$T_3 = \tfrac{1}{2}(m_A + m_B)v_3^2 = \tfrac{1}{2}(30 \text{ kg} + 10 \text{ kg})(4.70 \text{ m/s})^2 = 442 \text{ J}$$
$$V_3 = V_g + V_e = 0 + \tfrac{1}{2}kx_3^2 = \tfrac{1}{2}(20 \times 10^3)(4.91 \times 10^{-3})^2 = 0.241 \text{ J}$$
$$T_4 = 0$$
$$V_4 = V_g + V_e = (W_A + W_B)(-h) + \tfrac{1}{2}kx_4^2 = -(392)h + \tfrac{1}{2}(20 \times 10^3)x_4^2$$

Noting that the displacement of the pan is $h = x_4 - x_3$, we write

$T_3 + V_3 = T_4 + V_4$:
$$442 + 0.241 = 0 - 392(x_4 - 4.91 \times 10^{-3}) + \tfrac{1}{2}(20 \times 10^3)x_4^2$$
$$x_4 = 0.230 \text{ m} \qquad h = x_4 - x_3 = 0.230 \text{ m} - 4.91 \times 10^{-3} \text{ m}$$
$$h = 0.225 \text{ m} \qquad h = 225 \text{ mm} \blacktriangleleft$$

SOLVING PROBLEMS ON YOUR OWN

This lesson deals with the *impact of two bodies,* that is, with a collision occurring in a very small interval of time. You will solve a number of impact problems by expressing that the total momentum of the two bodies is conserved and noting the relationship which exists between the relative velocities of the two bodies before and after impact.

1. *As a first step in your solution* you should select and draw the following coordinate axes: the t axis, which is tangent to the surfaces of contact of the two colliding bodies, and the n axis, which is normal to the surfaces of contact and defines the *line of impact.* In all the problems of this lesson the line of impact passes through the mass centers of the colliding bodies, and the impact is referred to as a *central impact.*

2. *Next you will draw a diagram* showing the momenta of the bodies before impact, the impulses exerted on the bodies during impact, and the final momenta of the bodies after impact (Fig. 13.24). You will then observe whether the impact is a *direct central impact* or an *oblique central impact.*

3. *Direct central impact.* This occurs when the velocities of bodies A and B before impact are *both directed along the line of impact* (Fig. 13.20a).

a. Conservation of momentum. Since the impulsive forces are internal to the system, you can write that the *total momentum of A and B is conserved,*

$$m_A v_A + m_B v_B = m_A v'_A + m_B v'_B \qquad (13.37)$$

where v_A and v_B denote the velocities of bodies A and B before impact and v'_A and v'_B denote their velocities after impact.

b. Coefficient of restitution. You can also write the following relation between the *relative velocities* of the two bodies before and after impact,

$$v'_B - v'_A = e(v_A - v_B) \qquad (13.43)$$

where e is the coefficient of restitution between the two bodies.

Note that Eqs. (13.37) and (13.43) are scalar equations which can be solved for two unknowns. Also, be careful to adopt a consistent sign convention for all velocities.

4. *Oblique central impact.* This occurs when *one or both* of the initial velocities of the two bodies is *not directed* along the line of impact (Fig. 13.20b). To solve problems of this type, you should *first resolve into components* along the t axis and the n axis the momenta and impulses shown in your diagram.

a. *Conservation of momentum.* Since the impulsive forces act along the line of impact, that is, along the n axis, the component along the t axis of the momentum *of each body* is conserved. Therefore, you can write for each body that the t components of its velocity before and after impact are equal,

$$(v_A)_t = (v'_A)_t \qquad (v_B)_t = (v'_B)_t \tag{13.47}$$

Also, the component along the n axis of the *total momentum* of the system is conserved,

$$m_A(v_A)_n + m_B(v_B)_n = m_A(v'_A)_n + m_B(v'_B)_n \tag{13.48}$$

b. *Coefficient of restitution.* The relation between the relative velocities of the two bodies before and after impact can be written in the n direction only,

$$(v'_B)_n - (v'_A)_n = e[(v_A)_n - (v_B)_n] \tag{13.49}$$

You now have four equations that you can solve for four unknowns. Note that after finding all the velocities, you can determine the impulse exerted by body A on body B by drawing an impulse-momentum diagram for B alone and equating components in the n direction.

c. *When the motion of one of the colliding bodies is constrained,* you must include the impulses of the external forces in your diagram. You will then observe that some of the above relations do not hold. However, in the example shown in Fig. 13.26 the total momentum of the system is conserved in a direction perpendicular to the external impulse. You should also note that when a body A bounces off a fixed surface B, the only conservation of momentum equation which can be used is the first of Eqs. (13.47) [Sample Prob. 13.14].

5. *Remember that energy is lost during most impacts.* The only exception is for *perfectly elastic* impacts ($e = 1$), where energy is conserved. Thus, in the general case of impact, where $e < 1$, the energy is not conserved. Therefore, be careful *not to apply* the principle of conservation of energy through an impact situation. Instead, apply this principle separately to the motions preceding and following the impact [Sample Prob. 13.17].

Problems

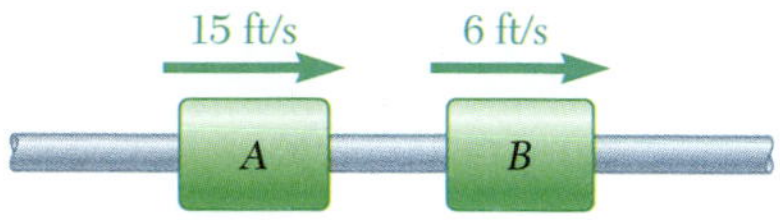

Fig. P13.155

13.155 The identical 2-lb collars A and B are sliding as shown on a frictionless rod. Knowing that the coefficient of restitution is $e = 0.60$, determine (*a*) the velocity of each collar after impact, (*b*) the energy lost during impact.

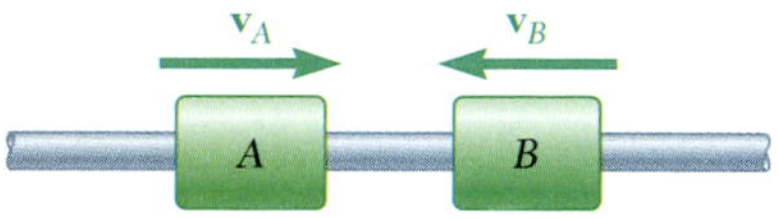

Fig. P13.156

13.156 Collars A and B, of the same mass m, are moving toward each other with the velocities shown. Knowing that the coefficient of restitution between the collars is 0 (plastic impact), show that after impact (*a*) the common velocity of the collars is equal to half the difference in their speeds before impact, (*b*) the loss in kinetic energy is $\frac{1}{4}m(v_A + v_B)^2$.

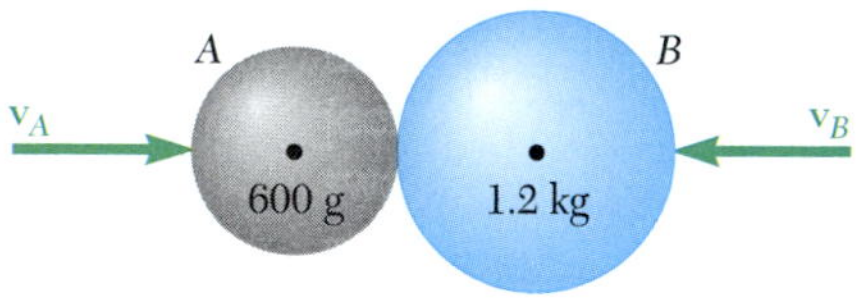

Fig. P13.157 and P13.158

13.157 A 600-g ball A is moving with a velocity $\mathbf{v}_A$ when it is struck by a 1.2-kg ball B which has a velocity $\mathbf{v}_B$ of magnitude $v_B = 6$ m/s. Knowing that the velocity of ball B is zero after impact and that the coefficient of restitution is 0.8, determine the velocity of ball A (*a*) before impact, (*b*) after impact.

13.158 A 600-g ball A is moving with a velocity $\mathbf{v}_A$ of magnitude $v_A = 8$ m/s when it is struck by a 1.2-kg ball B which has a velocity $\mathbf{v}_B$. Knowing that the velocity of ball A is zero after impact and that the coefficient of restitution is 0.2, determine the velocity of ball B (*a*) before impact, (*b*) after impact.

13.159 Two identical cars A and B are at rest on a loading dock with brakes released. Car C, of a slightly different style but of the same weight, has been pushed by dockworkers and hits car B with a velocity of 4.5 ft/s. Knowing that the coefficient of restitution is 0.8 between B and C and 0.5 between A and B, determine the velocity of each car after all collisions have taken place.

Fig. P13.159

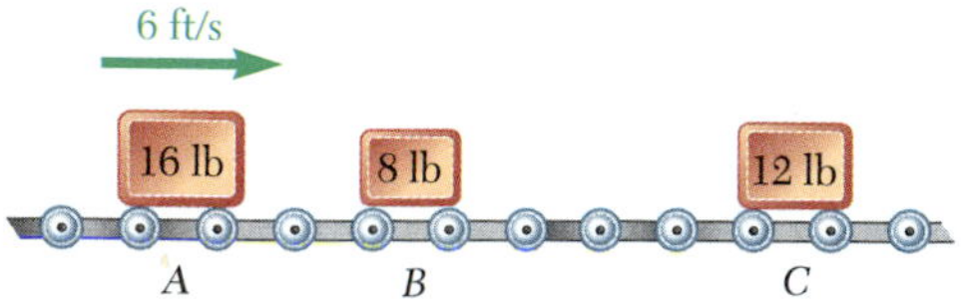

Fig. P13.160

13.160 Packages in an automobile parts supply house are transported to the loading dock by pushing them along a roller track with very little friction. At the instant shown packages B and C are at rest and package A has a velocity of 6 ft/s. Knowing that the coefficient of restitution between the packages is 0.3, determine (*a*) the velocity of package C after A hits B and B hits C, (*b*) the velocity of A after it hits B for the second time.

13.161 Three steel spheres of equal mass are suspended from the ceiling by cords of equal length which are spaced at a distance slightly greater than the diameter of the spheres. After being pulled back and released, sphere A hits sphere B, which then hits sphere C. Denoting by e the coefficient of restitution between the spheres and by $\mathbf{v}_0$ the velocity of A just before it hits B, determine (*a*) the velocities of A and B immediately after the first collision, (*b*) the velocities of B and C immediately after the second collision. (*c*) Assuming now that n spheres are suspended from the ceiling and that the first sphere is pulled back and released as described above, determine the velocity of the last sphere after it is hit for the first time. (*d*) Use the result of part c to obtain the velocity of the last sphere when $n = 8$ and $e = 0.9$.

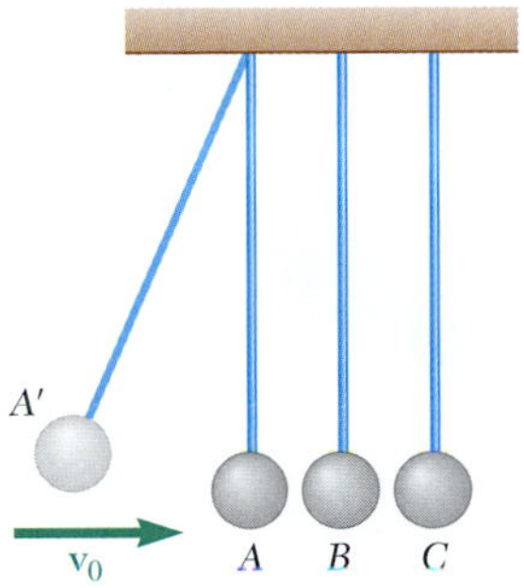

Fig. ***P13.161***

13.162 Two disks sliding on a frictionless horizontal plane with opposite velocities of the same magnitude v_0 hit each other squarely. Disk A is known to have a mass of 6 kg and is observed to have zero velocity after impact. Determine (*a*) the mass of disk B, knowing that the coefficient of restitution between the two disks is 0.5, (*b*) the range of possible values of the mass of disk B if the coefficient of restitution between the two disks is unknown.

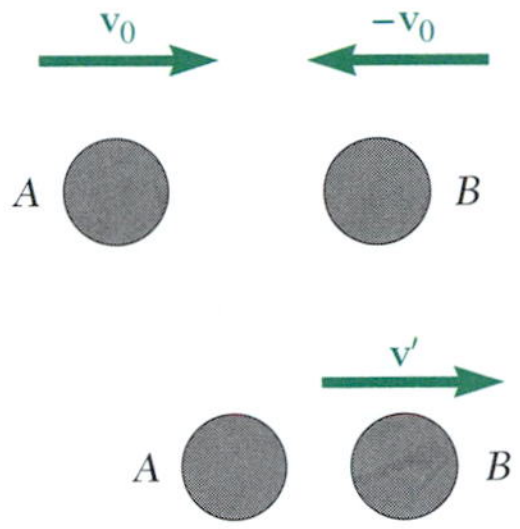

Fig. ***P13.162***

13.163 One of the requirements for tennis balls to be used in official competition is that, when dropped onto a rigid surface from a height of 100 in., the height of the first bounce of the ball must be in the range 53 in. $\leq h \leq$ 58 in. Determine the range of the coefficient of restitution of the tennis balls satisfying this requirement.

13.164 A 7.92-kg sphere A of radius 90 mm moving with a velocity $\mathbf{v}_0$ of magnitude $v_0 = 2$ m/s strikes a 720-g sphere B of radius 40 mm which was at rest. Both spheres are hanging from identical light flexible cords. Knowing that the coefficient of restitution is 0.8, determine the velocity of each sphere immediately after impact.

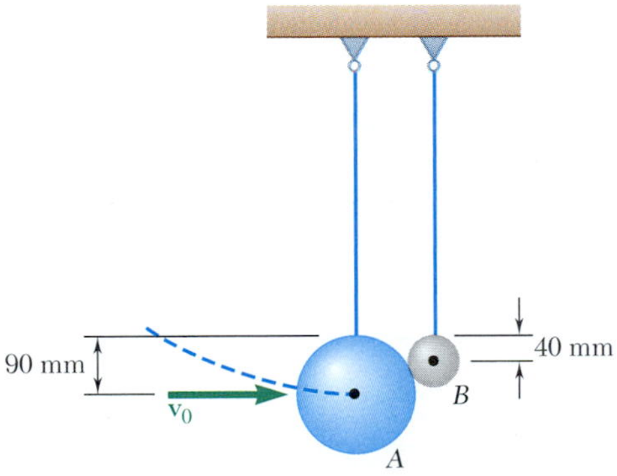

Fig. P13.164

13.165 Solve Prob. 13.164, assuming that the flexible cords from which the spheres are hanging are replaced by light rigid rods.

13.166 Two identical hockey pucks are moving on a hockey rink at the same speed of 10 ft/s in parallel and opposite directions when they strike each other as shown. Assuming a coefficient of restitution $e = 1$, determine the magnitude and direction of the velocity of each puck after impact.

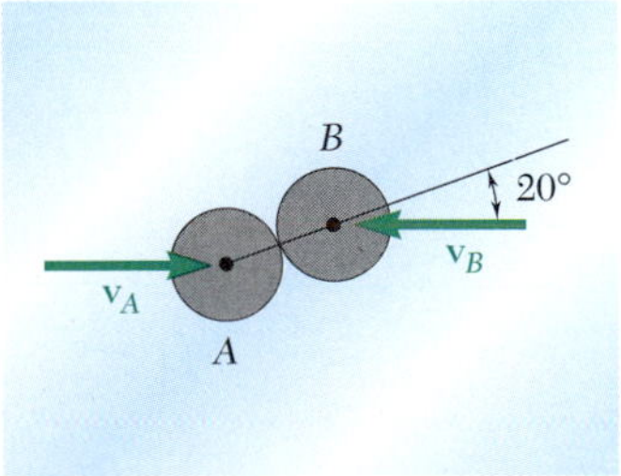

Fig. P13.166

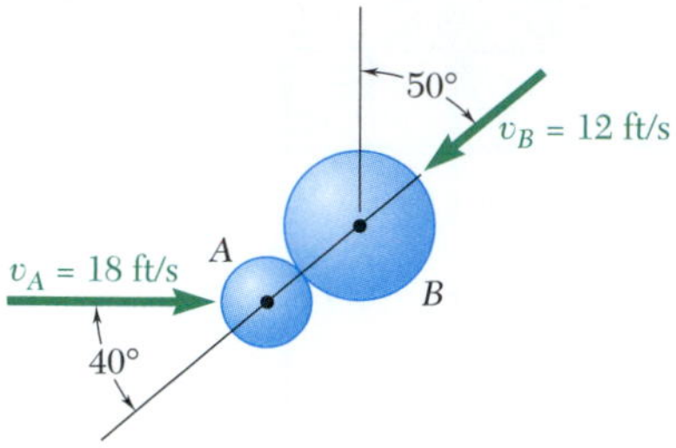

Fig. P13.167

13.167 A 1.5-lb ball A is moving with a velocity of magnitude 18 ft/s when it is hit by a 2.5-lb ball B which has a velocity of magnitude 12 ft/s. Knowing that the coefficient of restitution is 0.8 and assuming no friction, determine the velocity of each ball after impact.

13.168 (*a*) Show that when two identical spheres A and B with coefficient of restitution $e = 1$ collide while moving with velocities $\mathbf{v}_A$ and $\mathbf{v}_B$ which are perpendicular to each other they will rebound with velocities $\mathbf{v}'_A$ and $\mathbf{v}'_B$ which are also perpendicular to each other. (*b*) To verify this property, solve Sample Prob. 13.15, assuming $e = 1$, and determine the angle formed by $\mathbf{v}'_A$ and $\mathbf{v}'_B$.

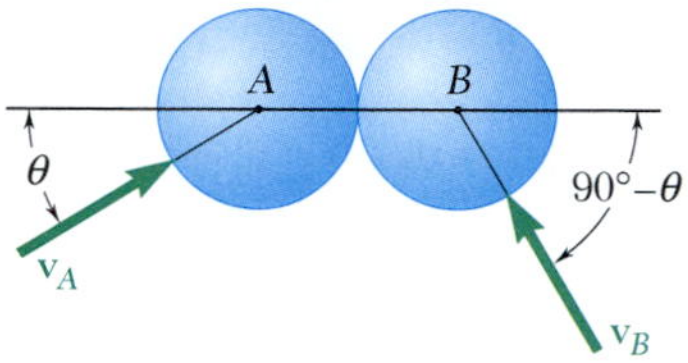

Fig. *P13.168*

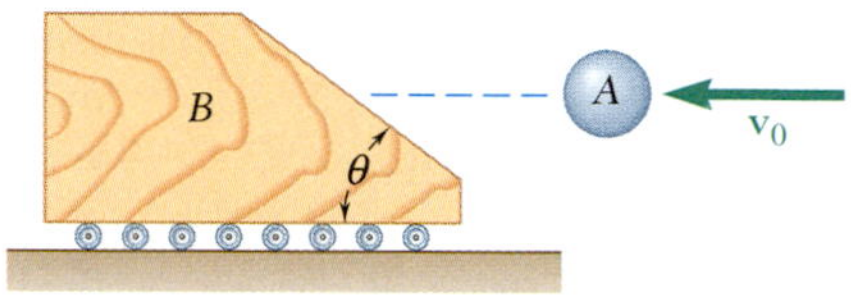

Fig. *P13.169*

13.169 A 3-lb sphere A moving with a velocity $\mathbf{v}_0$ parallel to the ground and of magnitude $v_0 = 6$ ft/s strikes the inclined face of a 12-lb wedge B, which can roll freely on the ground and is initially at rest. Knowing that $\theta = 60°$ and that the coefficient of restitution between the sphere and the wedge is $e = 1$, determine the velocity of the wedge immediately after impact.

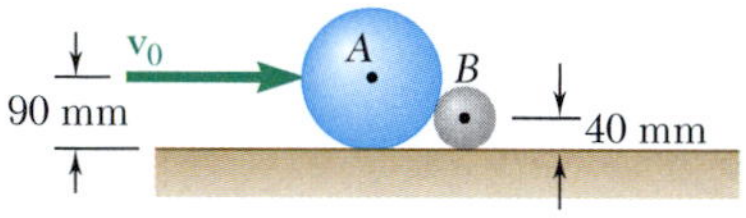

Fig. P13.170

13.170 A 26.4-kg sphere A of radius 90 mm moving with a velocity of magnitude $v_0 = 2$ m/s strikes a 2.4-kg sphere B of radius 40 mm which is initially at rest. All surfaces of contact are frictionless. Assuming perfectly elastic impact ($e = 1$) and no friction, determine the velocity of each sphere immediately after impact.

13.171 The coefficient of restitution is 0.9 between the two 60-mm-diameter billiard balls A and B. Ball A is moving in the direction shown with a velocity of 1 m/s when it strikes ball B, which is at rest. Knowing that after impact B is moving in the x direction, determine (*a*) the angle θ, (*b*) the velocity of B after impact.

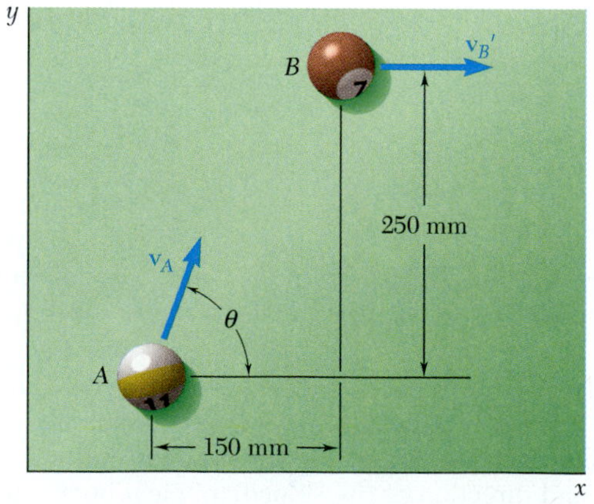

Fig. P13.171

13.172 A ball hits the ground at A with a velocity $\mathbf{v}_0$ of 6 m/s at an angle of 60° with the horizontal. Knowing that $e = 0.6$ between the ball and the ground and that after rebounding the ball reaches point B with a horizontal velocity, determine (a) the distances h and d, (b) the velocity of the ball as it reaches B.

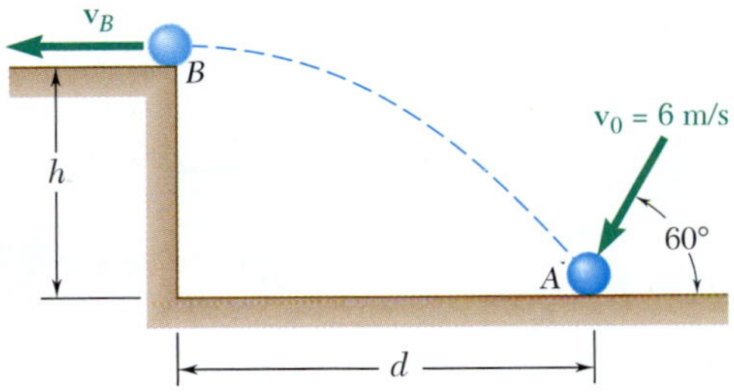

Fig. P13.172

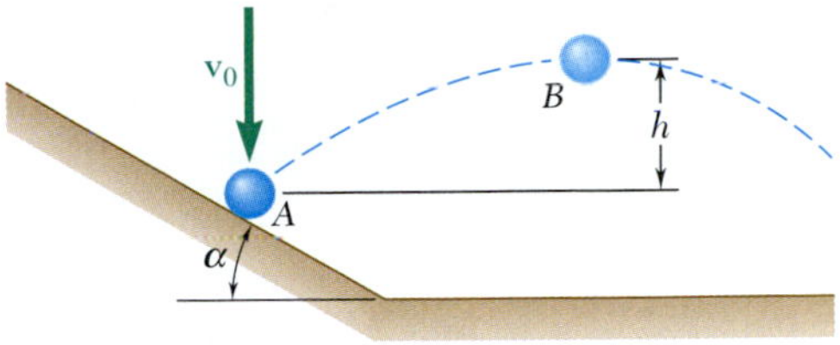

Fig. P13.173

13.173 A sphere rebounds as shown after striking an inclined plane with a vertical velocity $\mathbf{v}_0$ of magnitude $v_0 = 5$ m/s. Knowing that $\alpha = 30°$ and $e = 0.8$ between the sphere and the plane, determine the height h reached by the sphere.

13.174 A boy releases a ball with an initial horizontal velocity at a height of 2 ft. The ball bounces off the ground at point A, bounces off the wall at point B, and hits the ground again at point C. Neglecting friction, determine (a) the coefficient of restitution at point A, (b) the coefficient of restitution at point B.

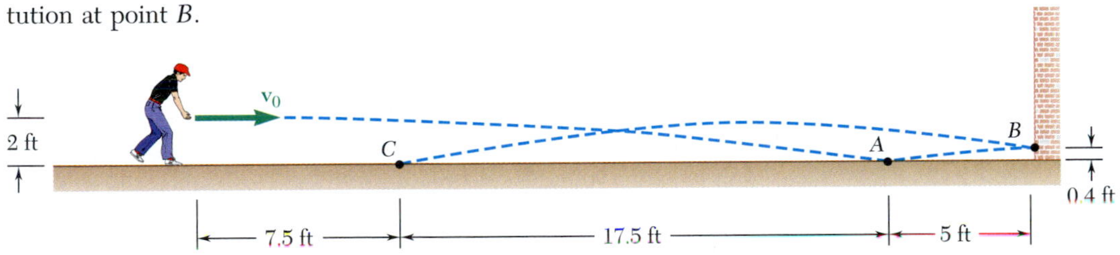

Fig. P13.174

13.175 A girl throws a ball at an inclined wall from a height of 3 ft, hitting the wall at A with a horizontal velocity $\mathbf{v}_0$ of magnitude 25 ft/s. Knowing that the coefficient of restitution between the ball and the wall is 0.9 and neglecting friction, determine the distance d from the foot of the wall to the point B where the ball will hit the ground after bouncing off the wall.

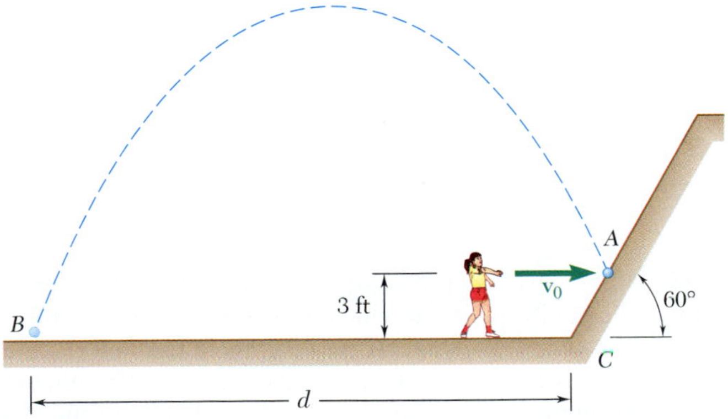

Fig. P13.175

13.176 Two cars of the same mass run head-on into each other at C. After the collision, the cars skid with their brakes locked and come to a stop in the position shown in the lower part of the figure. Knowing that the speed of car A just before impact was 5 km/h and that the coefficient of kinetic friction between the pavement and the tires of both cars is 0.30, determine (*a*) the speed of car B just before impact, (*b*) the effective coefficient of restitution between the two cars.

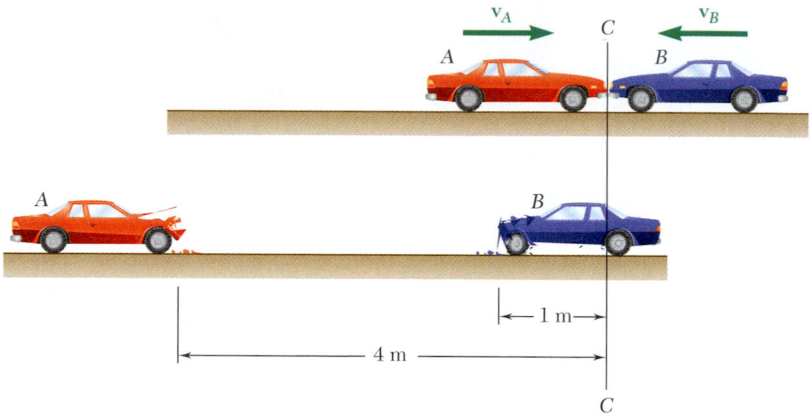

Fig. P13.176

13.177 Blocks A and B each have a mass of 0.4 kg and block C has a mass of 1.2 kg. The coefficient of friction between the blocks and the plane is $\mu_k = 0.30$. Initially block A is moving at a speed $v_0 = 3$ m/s and blocks B and C are at rest (Fig. 1). After A strikes B and B strikes C, all three blocks come to a stop in the positions shown (Fig. 2). Determine (*a*) the coefficients of restitution between A and B and between B and C, (*b*) the displacement x of block C.

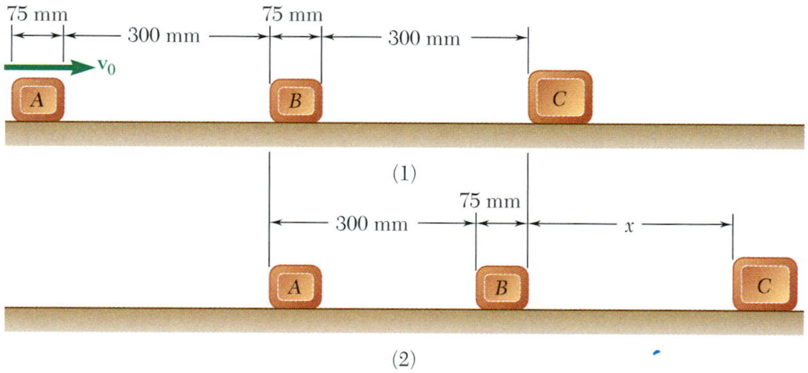

Fig. P13.177

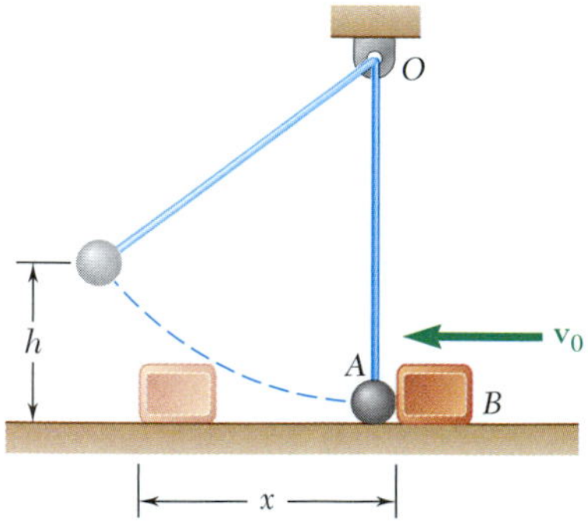

Fig. *P13.178*

13.178 A 2.5-lb block B is moving with a velocity $\mathbf{v}_0$ of magnitude $v_0 =$ 6 ft/s as it hits the 1.5-lb sphere A, which is at rest and hanging from a cord attached at O. Knowing that $\mu_k = 0.6$ between the block and the horizontal surface and $e = 0.8$ between the block and the sphere, determine after impact, (*a*) the maximum height h reached by the sphere, (*b*) the distance x traveled by the block.

13.179 After having been pushed by an airline employee, an empty 80-lb luggage carrier A hits with a velocity of 15 ft/s an identical carrier B containing a 30-lb suitcase equipped with rollers. The impact causes the suitcase to roll into the left wall of carrier B. Knowing that the coefficient of restitution between the two carriers is 0.80 and the coefficient of restitution between the suitcase and the wall of the carrier is 0.30, determine (*a*) the velocity of carrier B after the suitcase hits its wall for the first time, (*b*) the total energy lost in the impact.

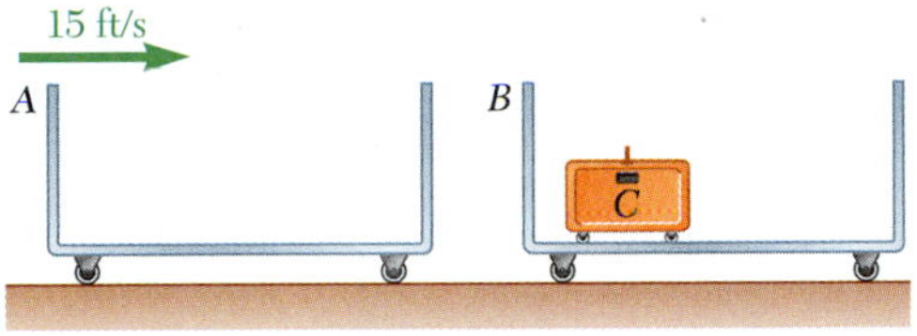

Fig. *P13.179*

13.180 A 300-g collar A is released from rest, slides down a frictionless rod, and strikes a 900-g collar B which is at rest and supported by a spring of constant 500 N/m. Knowing that the coefficient of restitution between the two collars is 0.9, determine (*a*) the maximum distance collar A moves up the rod after impact, (*b*) the maximum distance collar B moves down the rod after impact.

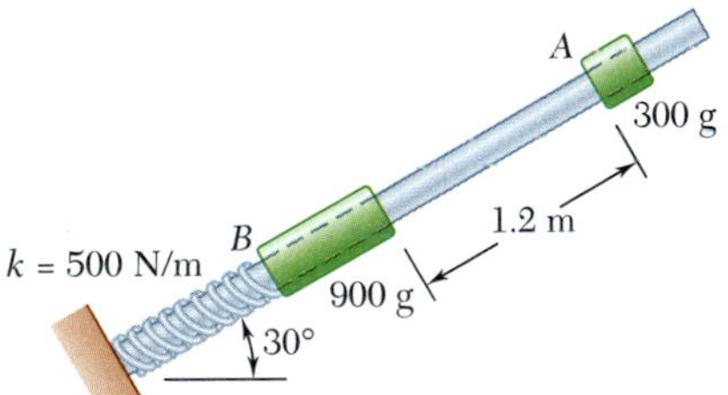

Fig. P13.180 and P13.181

13.181 A 300-g collar A is released from rest, slides down a frictionless rod, and strikes a 900-g collar B which is at rest and supported by a spring of constant 500 N/m. Knowing that the velocity of collar A is zero immediately after impact, determine (*a*) the coefficient of restitution between the two collars, (*b*) the energy lost in the impact, (*c*) the maximum distance collar B moves down the rod after impact.

13.182 A 3-lb block B is attached to an undeformed spring of constant $k = 60$ lb/ft and is resting on a horizontal frictionless surface when it is struck by an identical block A moving at a speed of 16 ft/s. Considering successively the cases when the coefficient of restitution between the two blocks is (1) $e = 1$, (2) $e = 0$, determine (*a*) the maximum deflection of the spring, (*b*) the final velocity of block A.

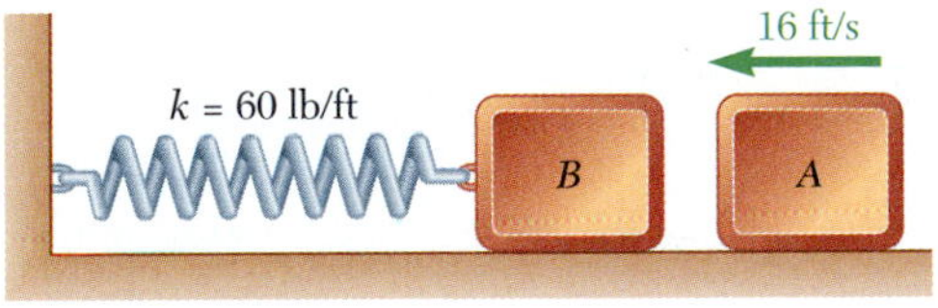

Fig. *P13.182*

13.183 A 0.25-lb ball thrown with a horizontal velocity $\mathbf{v}_0$ strikes a 1.5 lb plate attached to a vertical wall at a height of 36 in. above the ground. It is observed that after rebounding, the ball hits the ground at a distance of 24 in. from the wall when the plate is rigidly attached to the wall (Fig. 1), and at a distance of 10 in. when a foam-rubber mat is placed between the plate and the wall (Fig. 2). Determine (*a*) the coefficient of restitution e between the ball and the plate, (*b*) the initial velocity $\mathbf{v}_0$ of the ball.

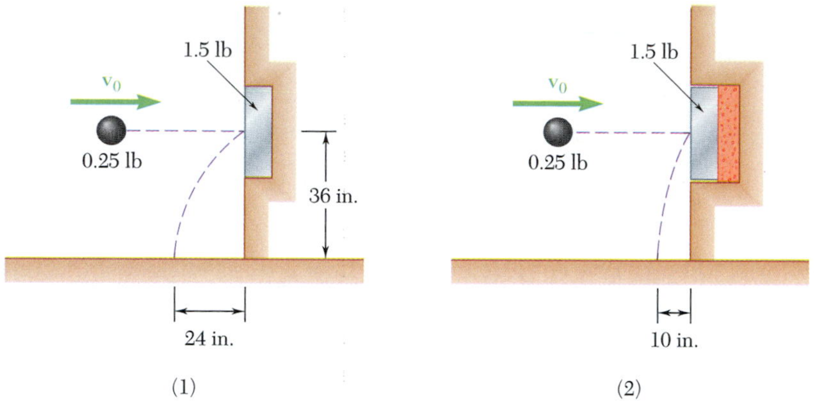

Fig. *P13.183*

13.184 A 23.1-kg sphere A of radius 90 mm moving with a velocity of magnitude $v_0 = 2$ m/s strikes a 2.1-kg sphere B of radius 40 mm which is hanging from an inextensible cord and is initially at rest. Knowing that sphere B swings to a maximum height $h = 0.25$ m, determine the coefficient of restitution between the two spheres.

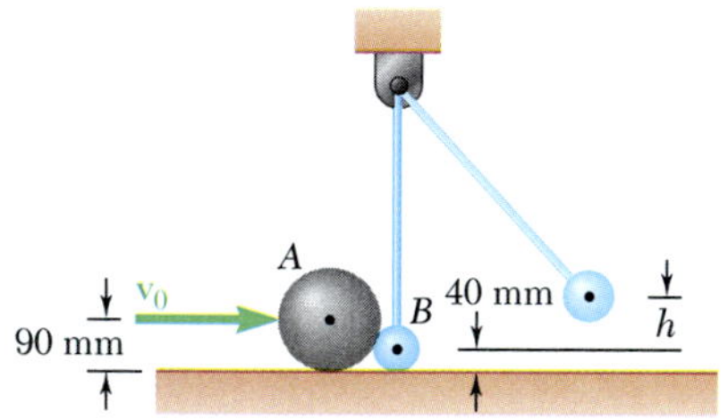

Fig. P13.184 and P13.185

13.185 A 9.13-kg sphere A of radius 90 mm moving with a velocity of magnitude $v_0 = 2$ m/s strikes a 830-g sphere B of radius 40 mm which is hanging from an inextensible cord and is initially at rest. Sphere B swings to a maximum height h after the impact. Determine the range of values of h for values of the coefficient of restitution e between zero and one.

13.186 A 4-oz ball B dropped from a height $h_0 = 4.5$ ft reaches a height $h_2 = 0.75$ ft after bouncing twice from identical 12-oz plates. Plate A rests directly on hard ground, while plate C rests on a foam-rubber mat. Determine (*a*) the coefficient of restitution between the ball and the plates, (*b*) the height h_1 of the ball's first bounce.

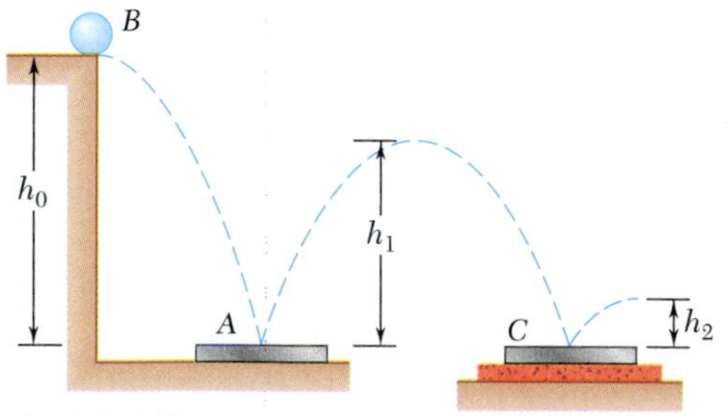

Fig. P13.186

13.187 A 1.5-lb sphere A moving with a velocity $\mathbf{v}_0$ parallel to the ground strikes the inclined face of a 4.5-lb wedge B which can roll freely on the ground and is initially at rest. After impact the sphere is observed from the ground to be moving straight up. Knowing that the coefficient of restitution between the sphere and the wedge is $e = 0.6$, determine (*a*) the angle θ that the inclined face of the wedge makes with the horizontal, (*b*) the energy lost due to the impact.

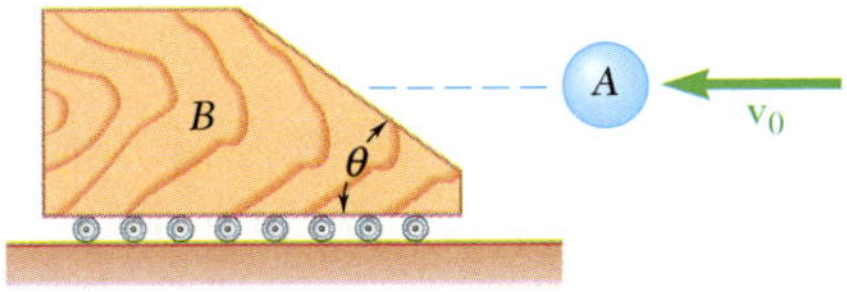

Fig. P13.187

13.188 A 2-kg sphere A strikes the frictionless inclined surface of a 6-kg wedge B at a 90° angle with a velocity of magnitude 4 m/s. The wedge can roll freely on the ground and is initially at rest. Knowing that the coefficient of restitution between the wedge and the sphere is 0.50 and that the inclined surface of the wedge forms an angle $\theta = 40°$ with the horizontal, determine (*a*) the velocities of the sphere and of the wedge immediately after impact, (*b*) the energy lost due to the impact.

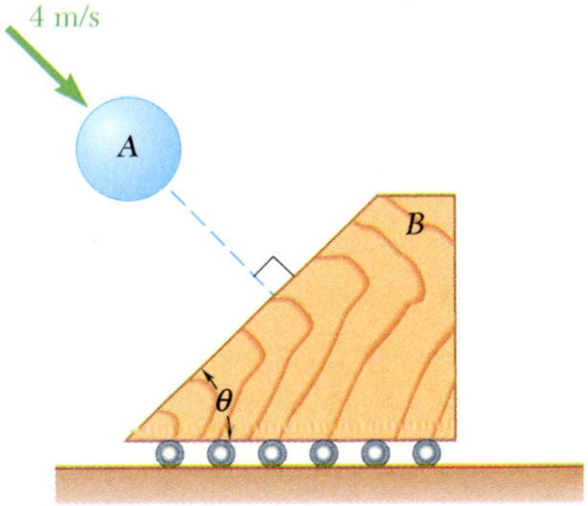

Fig. *P13.188*

13.189 A 340-g ball B is hanging from an inextensible cord attached to a support C. A 170-g ball A strikes B with a velocity $\mathbf{v}_0$ of magnitude 1.5 m/s at an angle of 60° with the vertical. Assuming perfectly elastic impact ($e = 1$) and no friction, determine the height h reached by ball B.

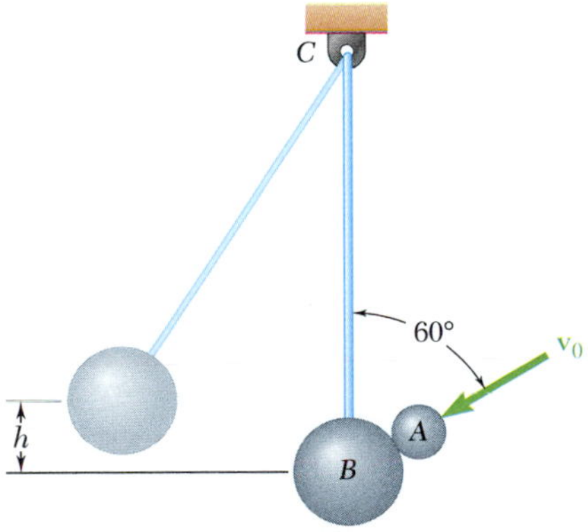

Fig. *P13.189*

REVIEW AND SUMMARY FOR CHAPTER 13

This chapter was devoted to the method of work and energy and to the method of impulse and momentum. In the first half of the chapter we studied the method of work and energy and its application to the analysis of the motion of particles.

Work of a force

We first considered a force **F** acting on a particle A and defined the *work of* **F** *corresponding to the small displacement* $d\mathbf{r}$ [Sec. 13.2] as the quantity

$$dU = \mathbf{F} \cdot d\mathbf{r} \qquad (13.1)$$

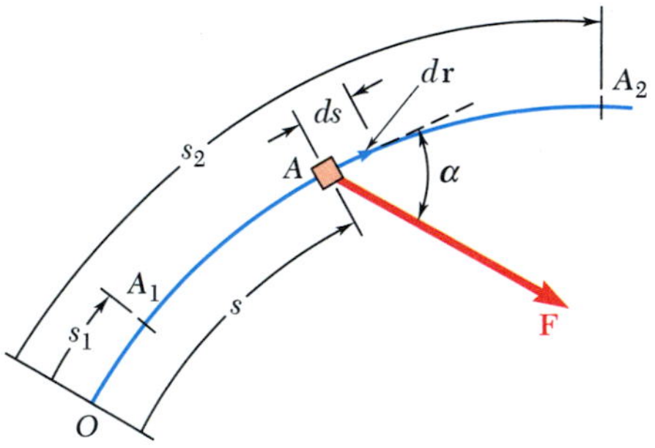

Fig. 13.29

or, recalling the definition of the scalar product of two vectors,

$$dU = F\, ds \cos \alpha \qquad (13.1')$$

where α is the angle between **F** and $d\mathbf{r}$ (Fig. 13.29). The work of **F** during a finite displacement from A_1 to A_2, denoted by $U_{1\rightarrow 2}$, was obtained by integrating Eq. (13.1) along the path described by the particle:

$$U_{1\rightarrow 2} = \int_{A_1}^{A_2} \mathbf{F} \cdot d\mathbf{r} \qquad (13.2)$$

For a force defined by its rectangular components, we wrote

$$U_{1\rightarrow 2} = \int_{A_1}^{A_2} (F_x\, dx + F_y\, dy + F_z\, dz) \qquad (13.2'')$$

Work of a weight

The work of the weight **W** of a body as its center of gravity moves from the elevation y_1 to y_2 (Fig. 13.30) was obtained by substituting $F_x = F_z = 0$ and $F_y = -W$ into Eq. (13.2″) and integrating. We found

$$U_{1\rightarrow 2} = -\int_{y_1}^{y_2} W\, dy = Wy_1 - Wy_2 \qquad (13.4)$$

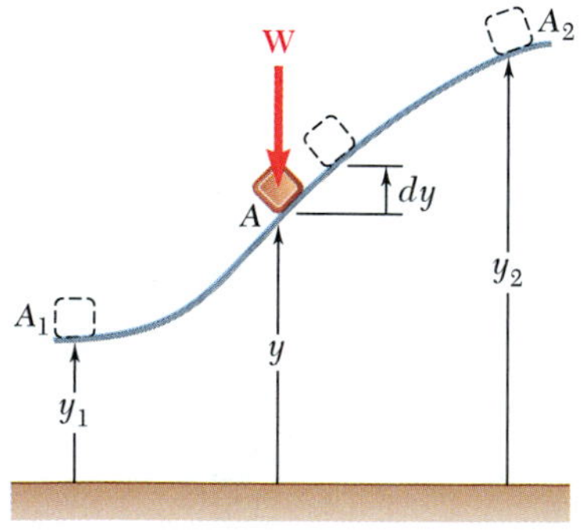

Fig. 13.30

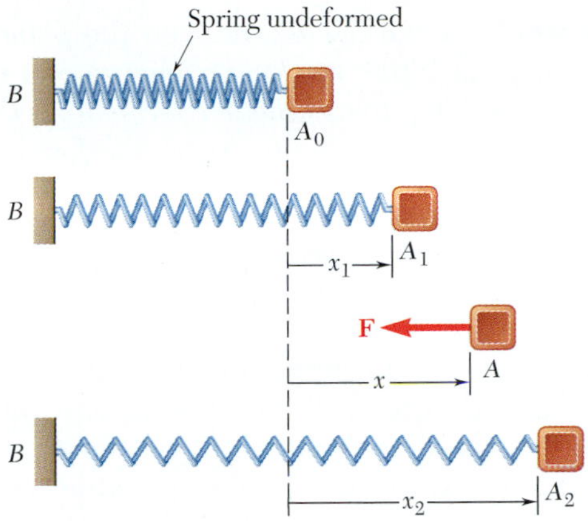

Fig. 13.31

Work of the force exerted by a spring

The work of a force **F** exerted by a spring on a body A during a finite displacement of the body (Fig. 13.31) from $A_1(x = x_1)$ to $A_2(x = x_2)$ was obtained by writing

$$dU = -F\,dx = -kx\,dx$$

$$U_{1\to 2} = -\int_{x_1}^{x_2} kx\,dx = \tfrac{1}{2}kx_1^2 - \tfrac{1}{2}kx_2^2 \qquad (13.6)$$

The work of **F** is therefore positive *when the spring is returning to its undeformed position.*

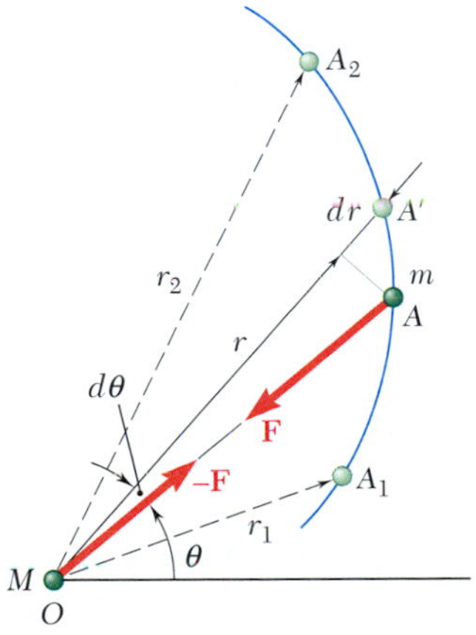

Fig. 13.32

Work of the gravitational force

The *work of the gravitational force* **F** exerted by a particle of mass M located at O on a particle of mass m as the latter moves from A_1 to A_2 (Fig. 13.32) was obtained by recalling from Sec. 12.10 the expression for the magnitude of **F** and writing

$$U_{1\to 2} = -\int_{r_1}^{r_2} \frac{GMm}{r^2}\,dr = \frac{GMm}{r_2} - \frac{GMm}{r_1} \qquad (13.7)$$

Kinetic energy of a particle

The *kinetic energy of a particle* of mass m moving with a velocity **v** [Sec. 13.3] was defined as the scalar quantity

$$T = \tfrac{1}{2}mv^2 \qquad (13.9)$$

Principle of work and energy

From Newton's second law we derived the *principle of work and energy,* which states that *the kinetic energy of a particle at* A_2 *can be obtained by adding to its kinetic energy at* A_1 *the work done during the displacement from* A_1 to A_2 *by the force* **F** *exerted on the particle:*

$$T_1 + U_{1\rightarrow 2} = T_2 \tag{13.11}$$

Method of work and energy

The method of work and energy simplifies the solution of many problems dealing with forces, displacements, and velocities, since it does not require the determination of accelerations [Sec. 13.4]. We also note that it involves only scalar quantities and that forces which do no work need not be considered [Sample Probs. 13.1 and 13.3]. However, this method should be supplemented by the direct application of Newton's second law to determine a force normal to the path of the particle [Sample Prob. 13.4].

Power and mechanical efficiency

The power developed by a machine and its mechanical efficiency were discussed in Sec. 13.5. Power was defined as the time rate at which work is done:

$$\text{Power} = \frac{dU}{dt} = \mathbf{F} \cdot \mathbf{v} \qquad (13.12, 13.13)$$

where **F** is the force exerted on the particle and **v** the velocity of the particle [Sample Prob. 13.5]. The *mechanical efficiency,* denoted by η, was expressed as

$$\eta = \frac{\text{power output}}{\text{power input}} \tag{13.15}$$

Conservative force. Potential energy

When the work of a force **F** is independent of the path followed [Secs. 13.6 and 13.7], the force **F** is said to be a *conservative force,* and its work is equal to *minus the change in the potential energy* V associated with **F**:

$$U_{1\rightarrow 2} = V_1 - V_2 \qquad (13.19')$$

The following expressions were obtained for the potential energy associated with each of the forces considered earlier:

Force of gravity (weight): $\quad V_g = Wy \qquad (13.16)$

Gravitational force: $\quad V_g = -\dfrac{GMm}{r} \qquad (13.17)$

Elastic force exerted by a spring: $\quad V_e = \frac{1}{2}kx^2 \qquad (13.18)$

Substituting for $U_{1\to2}$ from Eq. (13.19′) into Eq. (13.11) and rearranging the terms [Sec. 13.8], we obtained

$$T_1 + V_1 = T_2 + V_2 \qquad (13.24)$$

Principle of conservation of energy

This is the *principle of conservation of energy,* which states that when a particle moves under the action of conservative forces, *the sum of its kinetic and potential energies remains constant*. The application of this principle facilitates the solution of problems involving only conservative forces [Sample Probs. 13.6 and 13.7].

Motion under a gravitational force

Recalling from Sec. 12.9 that, when a particle moves under a central force **F**, its angular momentum about the center of force O remains constant, we observed [Sec. 13.9] that, if the central force **F** is also conservative, the principles of conservation of angular momentum and of conservation of energy can be used jointly to analyze the motion of the particle [Sample Prob. 13.8]. Since the gravitational force exerted by the earth on a space vehicle is both central and conservative, this approach was used to study the motion of such vehicles [Sample Prob. 13.9] and was found particularly effective in the case of an *oblique launching*. Considering the initial position P_0 and an arbitrary position P of the vehicle (Fig. 13.33), we wrote

$$(H_O)_0 = H_O: \qquad r_0 m v_0 \sin \phi_0 = r m v \sin \phi \qquad (13.25)$$

$$T_0 + V_0 = T + V: \qquad \tfrac{1}{2} m v_0^2 - \frac{GMm}{r_0} = \tfrac{1}{2} m v^2 - \frac{GMm}{r} \qquad (13.26)$$

where m was the mass of the vehicle and M the mass of the earth.

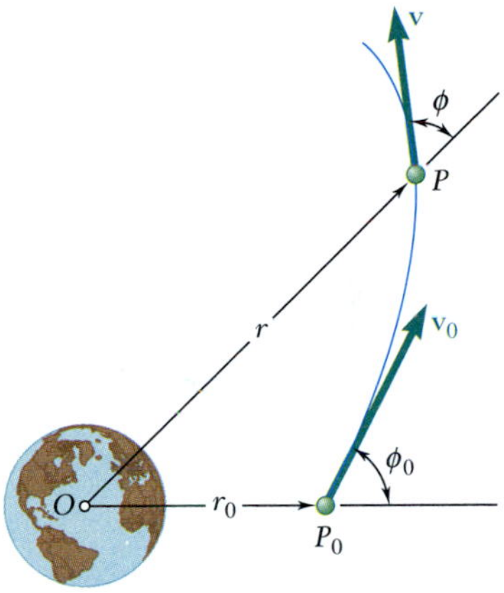

Fig. 13.33

Principle of impulse and momentum for a particle

The second half of the chapter was devoted to the method of impulse and momentum and to its application to the solution of various types of problems involving the motion of particles.

The *linear momentum of a particle* was defined [Sec. 13.10] as the product $m\mathbf{v}$ of the mass m of the particle and its velocity $\mathbf{v}$. From Newton's second law, $\mathbf{F} = m\mathbf{a}$, we derived the relation

$$m\mathbf{v}_1 + \int_{t_1}^{t_2} \mathbf{F}\, dt = m\mathbf{v}_2 \qquad (13.28)$$

where $m\mathbf{v}_1$ and $m\mathbf{v}_2$ represent the momentum of the particle at a time t_1 and a time t_2, respectively, and where the integral defines the *linear impulse of the force* **F** during the corresponding time interval. We wrote therefore

$$m\mathbf{v}_1 + \mathbf{Imp}_{1\to2} = m\mathbf{v}_2 \qquad (13.30)$$

which expresses the principle of impulse and momentum for a particle.

When the particle considered is subjected to several forces, the sum of the impulses of these forces should be used; we had

$$m\mathbf{v}_1 + \Sigma\ \mathbf{Imp}_{1\to 2} = m\mathbf{v}_2 \tag{13.32}$$

Since Eqs. (13.30) and (13.32) involve *vector quantities,* it is necessary to consider their x and y components separately when applying them to the solution of a given problem [Sample Probs. 13.10 and 13.11].

Impulsive motion

The method of impulse and momentum is particularly effective in the study of the *impulsive motion* of a particle, when very large forces, called *impulsive forces,* are applied for a very short interval of time Δt, since this method involves the impulses $\mathbf{F}\ \Delta t$ of the forces, rather than the forces themselves [Sec. 13.11]. Neglecting the impulse of any nonimpulsive force, we wrote

$$m\mathbf{v}_1 + \Sigma\mathbf{F}\ \Delta t = m\mathbf{v}_2 \tag{13.35}$$

In the case of the impulsive motion of several particles, we had

$$\Sigma m\mathbf{v}_1 + \Sigma\mathbf{F}\ \Delta t = \Sigma m\mathbf{v}_2 \tag{13.36}$$

where the second term involves only impulsive, external forces [Sample Prob. 13.12].

In the particular case *when the sum of the impulses of the external forces is zero,* Eq. (13.36) reduces to $\Sigma m\mathbf{v}_1 = \Sigma m\mathbf{v}_2$; that is, *the total momentum of the particles is conserved.*

Direct central impact

In Secs. 13.12 through 13.14, we considered the *central impact* of two colliding bodies. In the case of a *direct central impact* [Sec. 13.13], the two colliding bodies A and B were moving along the *line of impact* with velocities $\mathbf{v}_A$ and $\mathbf{v}_B$, respectively (Fig. 13.34). Two equations could be used to determine their

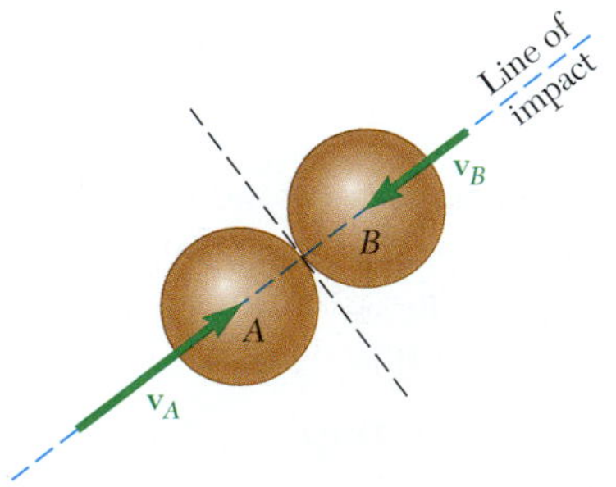

Fig. 13.34

velocities $\mathbf{v}'_A$ and $\mathbf{v}'_B$ after the impact. The first expressed conservation of the total momentum of the two bodies,

$$m_A v_A + m_B v_B = m_A v'_A + m_B v'_B \qquad (13.37)$$

where a positive sign indicates that the corresponding velocity is directed to the right, while the second related the *relative velocities* of the two bodies before and after the impact,

$$v'_B - v'_A = e(v_A - v_B) \qquad (13.43)$$

The constant e is known as the *coefficient of restitution;* its value lies between 0 and 1 and depends in a large measure on the materials involved. When $e = 0$, the impact is said to be *perfectly plastic;* when $e = 1$, it is said to be *perfectly elastic* [Sample Prob. 13.13].

Oblique central impact

In the case of an *oblique central impact* [Sec. 13.14], the velocities of the two colliding bodies before and after the impact were resolved into n components along the line of impact and t components along the common tangent to the surfaces in contact (Fig. 13.35). We observed that the t component of the velocity of each body remained unchanged, while the n components satisfied equations similar to Eqs. (13.37) and (13.43) [Sample Probs. 13.14 and 13.15]. It was shown that although this method was developed for bodies moving freely before and after the impact, it could be extended to the case when one or both of the colliding bodies is constrained in its motion [Sample Prob. 13.16].

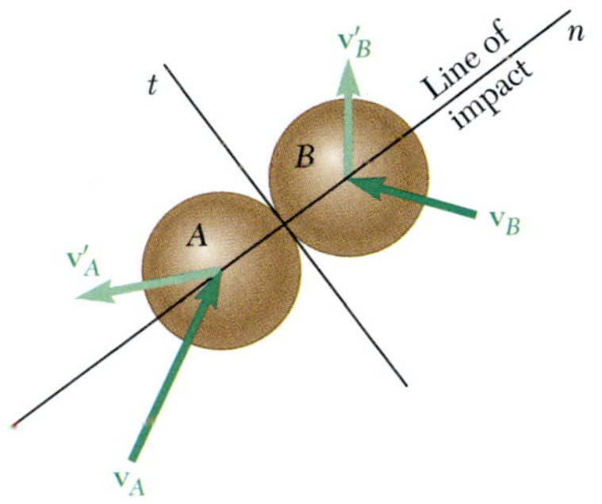

Fig. 13.35

Using the three fundamental methods of kinetic analysis

In Sec. 13.15, we discussed the relative advantages of the three fundamental methods presented in this chapter and the preceding one, namely, Newton's second law, work and energy, and impulse and momentum. We noted that the method of work and energy and the method of impulse and momentum can be combined to solve problems involving a short impact phase during which impulsive forces must be taken into consideration [Sample Prob. 13.17].

Review Problems

Fig. P13.190

13.190 Skid marks on a drag racetrack indicate that the rear (drive) wheels of a car slip for the first 60 ft of the 1320-ft track. (*a*) Knowing that the coefficient of kinetic friction is 0.60, determine the speed of that car at the end of the first 60-ft portion of the track if it starts from rest and the front wheels are just off the ground. (*b*) What is the maximum theoretical speed for the car at the finish line if, after skidding for 60 ft, it is driven without the wheels slipping for the remainder of the race? Assume that while the car is rolling without slipping, 60 percent of the weight of the car is on the rear wheels and the coefficient of static friction is 0.85. Ignore air resistance and rolling resistance.

13.191 A trailer truck enters a 2 percent uphill grade traveling at 40 mi/h and reaches a speed of 65 mi/h in 1000 ft. The cab weighs 4000 lb and the trailer 12,000 lb. Determine (*a*) the average force at the wheels of the cab, (*b*) the average force in the coupling between the cab and the trailer.

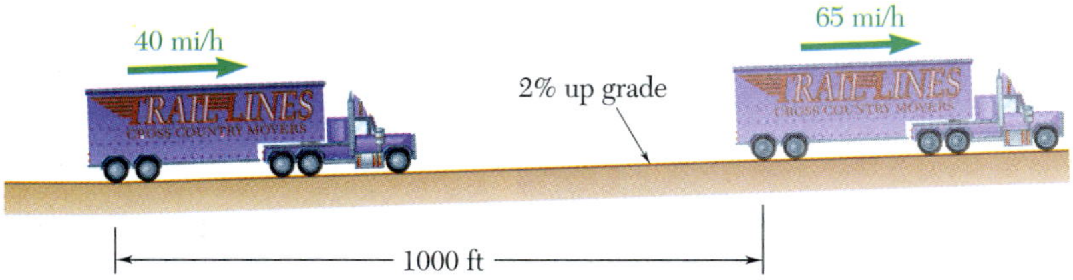

Fig. P13.191

13.192 A small sphere *B* of mass *m* is released from rest in the position shown and swings freely in a vertical plane, first about *O* and then about the peg *A* after the cord comes in contact with the peg. Determine the tension in the cord (*a*) just before the sphere comes in contact with the peg, (*b*) just after it comes in contact with the peg.

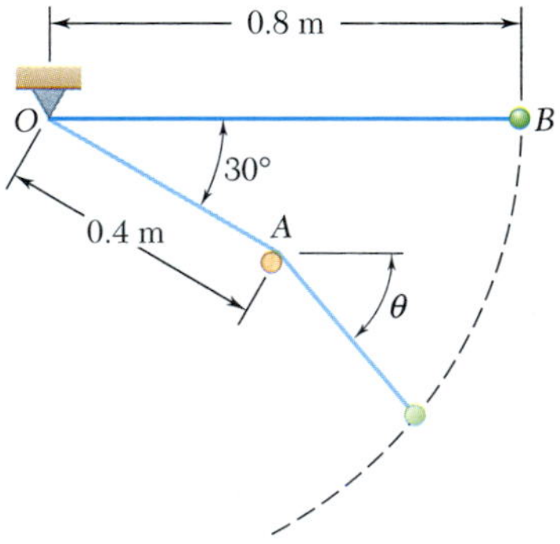

Fig. *P13.192*

13.193 A 500-g collar can slide without friction on the curved rod BC in a *horizontal* plane. Knowing that the undeformed length of the spring is 80 mm and that $k = 400$ kN/m, determine (*a*) the velocity that the collar should be given at A to reach B with zero velocity, (*b*) the velocity of the collar when it eventually reaches C.

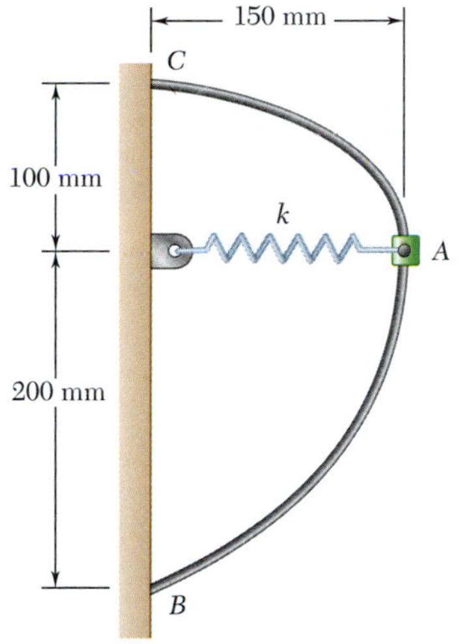

Fig. P13.193

13.194 After completing their moon-exploration mission, the two astronauts forming the crew of an Apollo lunar excursion module (LEM) would prepare to rejoin the command module which was orbiting the moon at an altitude of 140 km. They would fire the LEM's engine, bring it along a curved path to a point A, 8 km above the moon's surface, and shut off the engine. Knowing that the LEM was moving at that time in a direction parallel to the moon's surface and that it then coasted along an elliptic path to rendezvous at B with the command module, determine (*a*) the speed of the LEM at engine shutoff, (*b*) the relative velocity with which the command module approached the LEM at B. (The radius of the moon is 1740 km and its mass is 0.0123 times the mass of the earth.)

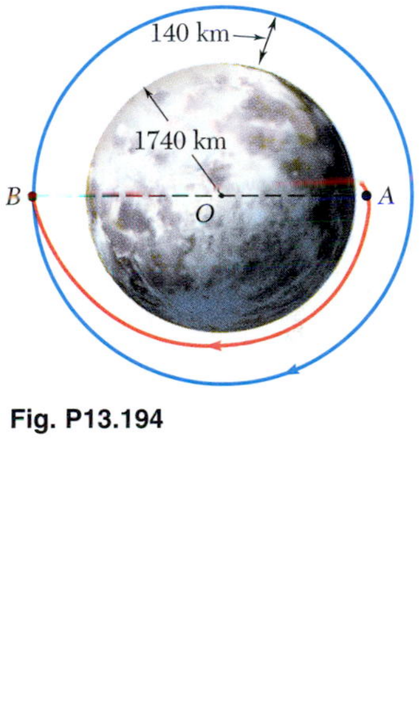

Fig. P13.194

13.195 Skid marks on a drag racetrack indicate that the rear (drive) wheels of a car slip for the first 60 ft of the 1320-ft track. (*a*) Knowing that the coefficient of kinetic friction is 0.60, determine the shortest possible time for the car to travel the initial 60-ft portion of the track if it starts from rest with the front wheels just off the ground. (*b*) Determine the minimum time for the car to run the whole race if, after skidding for 60 ft, the wheels roll without sliding for the remainder of the race. Assume for the rolling portion of the race that 60 percent of the weight is on the rear wheels and the coefficient of static friction is 0.85. Ignore air resistance and rolling resistance.

13.196 A truck is traveling down a road with a 4-percent grade at a speed of 60 mi/h when its brakes are applied to slow it down to 20 mi/h. An antiskid braking system limits the braking force to a value at which the wheels of the truck are just about to slide. Knowing that the coefficient of static friction between the road and the wheels is 0.60, determine the shortest time needed for the truck to slow down.

Fig. *P13.197*

13.197 Two swimmers A and B, of weight 190 lb and 125 lb, respectively, are at diagonally opposite corners of a floating raft when they realize that the raft has broken away from its anchor. Swimmer A immediately starts walking toward B at a speed of 2 ft/s relative to the raft. Knowing that the raft weighs 300 lb, determine (a) the speed of the raft if B does not move, (b) the speed with which B must walk toward A if the raft is not to move.

13.198 Show that for a ball hitting a frictionless surface, $\alpha > \theta$. Also show that the percent loss in kinetic energy due to the impact is $100(1 - e^2)\cos^2\theta$.

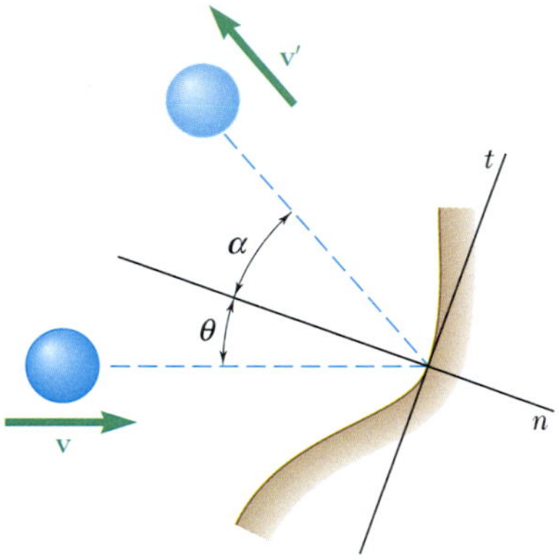

Fig. P13.198

Fig. P13.199

13.199 The three blocks shown are identical. Blocks B and C are at rest when block B is hit by block A, which is moving with a velocity $\mathbf{v}_A$ of 3 ft/s. After the impact, which is assumed to be perfectly plastic ($e = 0$), the velocity of blocks A and B decreases due to friction, while block C picks up speed, until all three blocks are moving with the same velocity $\mathbf{v}$. Knowing that the coefficient of kinetic friction between all surfaces is $\mu_k = 0.20$, determine (a) the time required for the three blocks to reach the same velocity, (b) the total distance traveled by each block during that time.

13.200 A 20-g bullet fired into a 4-kg wooden block suspended from cords AC and BD penetrates the block at point E, halfway between C and D, without hitting cord BD. Determine (a) the maximum height h to which the block and the embedded bullet will swing after impact, (b) the total impulse exerted on the block by the two cords during the impact.

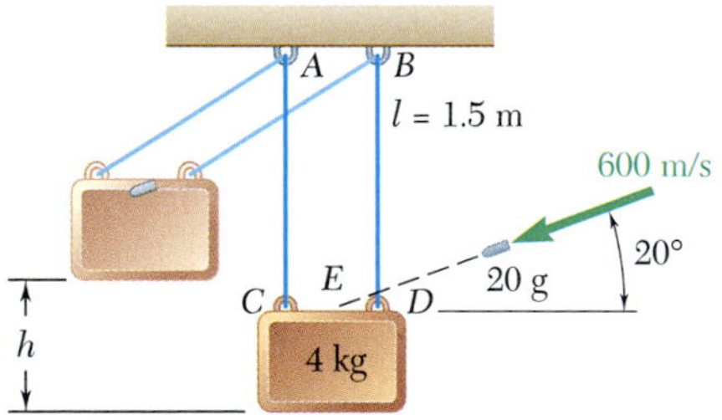

Fig. P13.200

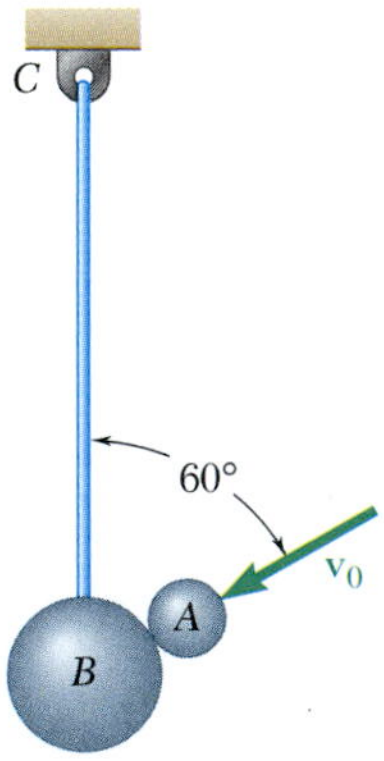

Fig. *P13.201*

13.201 A 700-g ball B is hanging from an inextensible cord attached to a support C. A 350-g ball A strikes B with a velocity $\mathbf{v}_0$ at an angle of 60° with the vertical. Assuming perfectly elastic impact ($e = 1$) and no friction, determine the velocity of each ball immediately after impact. Check that no energy is lost in the impact.

Computer Problems

13.C1 A 6-kg collar is attached to a spring anchored at point C and can slide on a frictionless rod forming an angle of 30° with the vertical. The spring is of constant k and is unstretched when the collar is at A. Knowing that the collar is released from rest at A, use computational software to calculate and plot the speed of the collar at point B for values of k from 20 N/m to 400 N/m.

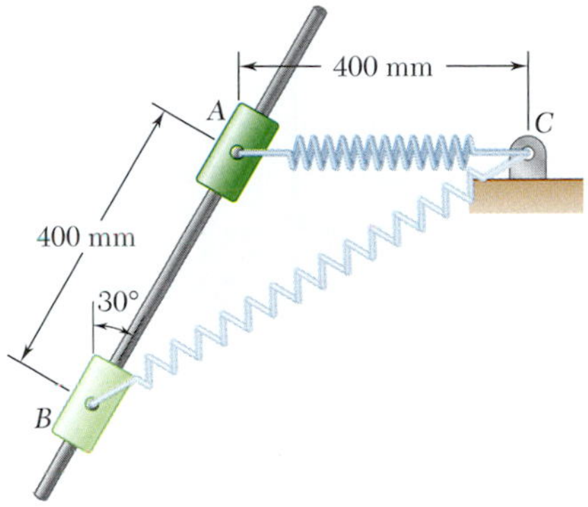

Fig. P13.C1

13.C2 A 10-lb bag is gently pushed off the top of a wall and swings in a vertical plane at the end of a 10-ft rope which can withstand a maximum tension F_m before it breaks. Derive expressions for (*a*) the difference in elevation h between point A and point B where the rope will break, (*b*) the distance d from the vertical wall to the point where the bag will strike the floor. Using computational software, calculate h and d for values of F_m from 10 lb to 25 lb, using 1-lb increments.

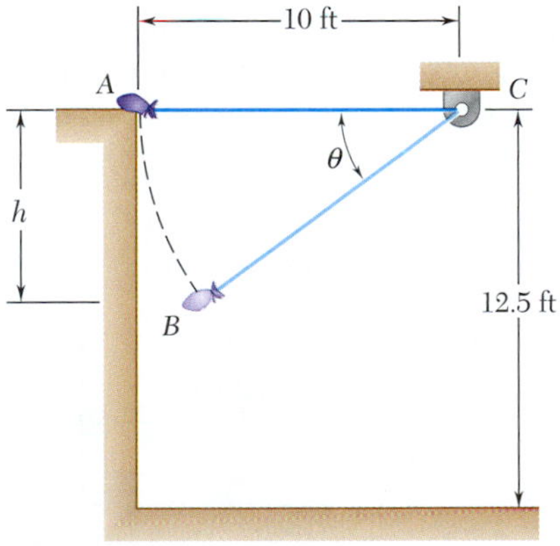

Fig. P13.C2

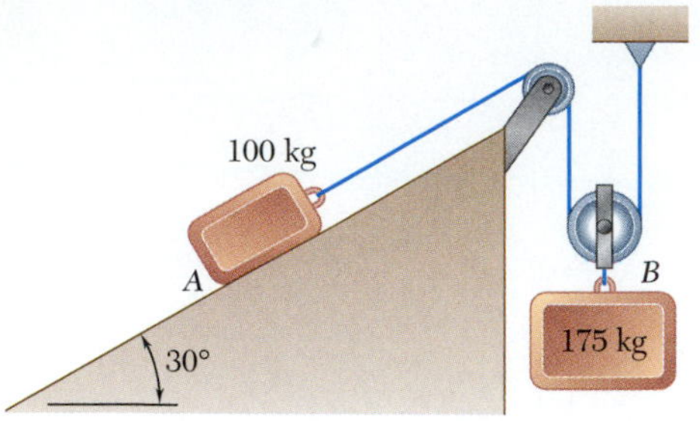

Fig. P13.C3

13.C3 The two blocks shown are released from rest. The coefficients of static and kinetic friction between the block A and the incline are $\mu_s = 0.25$ and $\mu_k = 0.20$. Neglect the masses of the pulleys and the effect of friction in the pulleys. (*a*) Denoting by s the distance block A moves, determine and plot the speed of block A as a function of s for $0 \leq s \leq 1.8$ m. (*b*) Using the expression derived in part (*a*), derive an expression for the distance s as a function of time t and plot this function for values of s from 0 to 1.8 m.

13.C4 A 3-kg block is attached to a cable and to a spring as shown and is initially at rest. The constant of the spring is $k = 2.5$ kN/m and the initial tension in the cable is 18 N. Knowing that the cable is cut, (*a*) derive an expression for the velocity of the block as a function of its displacement s, (*b*) determine the maximum displacement s_m and the maximum speed v_m, (*c*) plot the speed of the block as a function of s for $0 \leq s \leq s_m$.

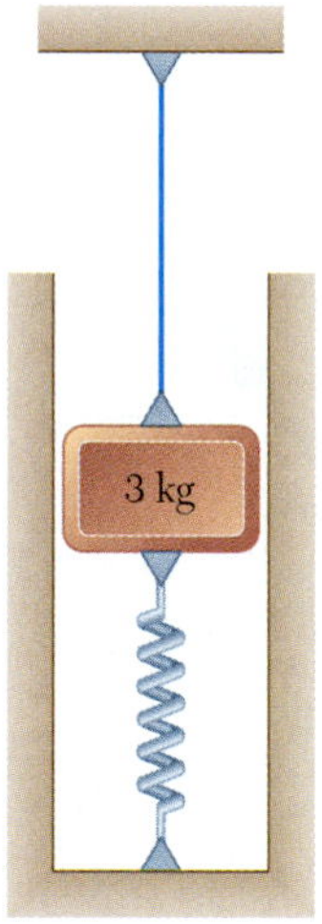

Fig. P13.C4

13.C5 Nonlinear springs are classified as hard or soft, depending upon the curvature of their force-deflection curve (see figure). A delicate instrument having a weight of 10 lb is placed on a spring of length l so that its base is just touching the undeformed spring and then inadvertently released from that position. For each of the following three cases, (*a*) derive an expression for the velocity of the instrument as a function of its displacement x, (*b*) determine the maximum displacement x_m, (*c*) plot the speed of the instrument as a function of x for $0 \leq x \leq x_m$.

(1) Linear spring with a constant $k = 15$ lb/in.
(2) Hard spring for which $F = (15 \text{ lb/in.})x(1 + 0.064x^2)$
(3) Soft spring for which $F = (15 \text{ lb/in.})x(1 - 0.064x^2)$

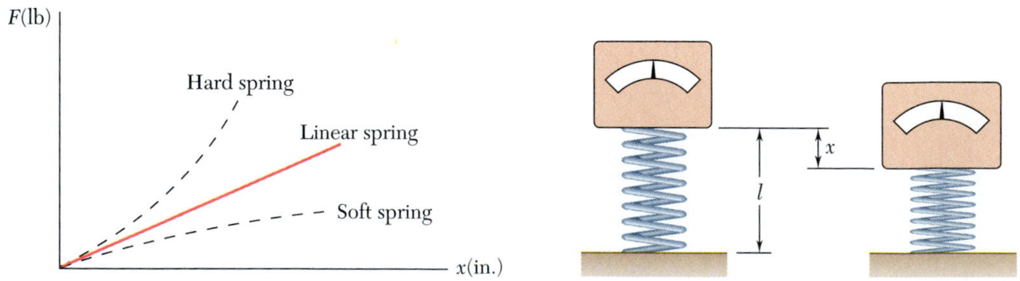

Fig. P13.C5

13.C6 A 1-kg collar is attached to a spring and slides without friction along a circular rod which lies in a vertical plane. The spring has a constant $k = 0.60$ kN/m and is undeformed when the collar is at point B. Knowing that the collar is released from rest at point C, (*a*) determine and plot the speed of the collar as a function of θ for values of θ from 90° to θ_m, where θ_m is the maximum value of θ, (*b*) determine the maximum speed of the collar and the corresponding value of θ.

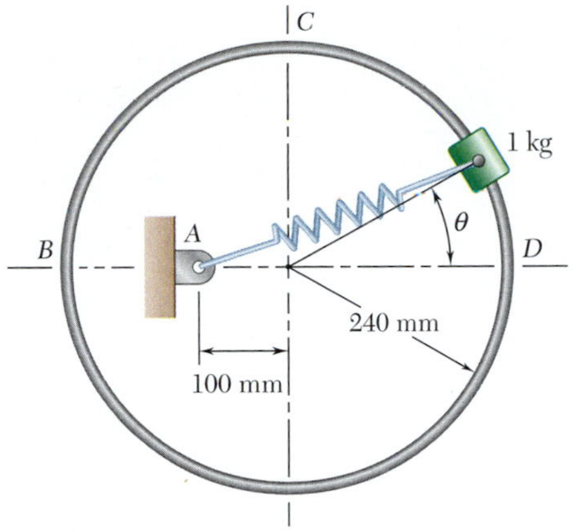

Fig. P13.C6

13.C7 A 25-kg block initially at rest is acted upon by a force **P** which varies in magnitude with time as shown. The coefficients of static and kinetic friction between the block and the horizontal surface are $\mu_s = 0.5$ and $\mu_k = 0.4$. (*a*) Determine and plot the speed of the block as a function of t for $0 \le t \le 16$ s. (*b*) Determine the maximum velocity of the block and the corresponding value of t.

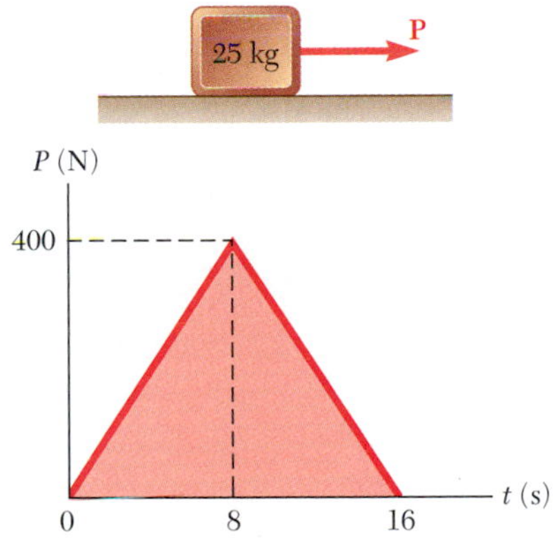

Fig. P13.C7

C H A P T E R

14

Systems of Particles

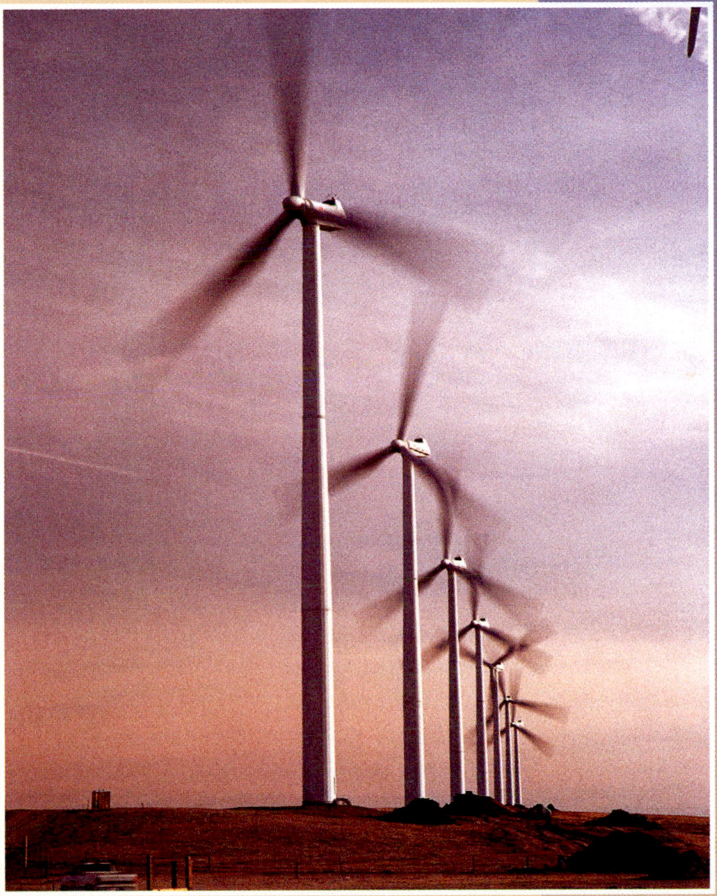

Wind turbine-generators convert some of the kinetic energy of the air particles into electrical energy. The power input is proportional to the cube of the wind speed.

SYSTEMS OF PARTICLES

14.1. INTRODUCTION

In this chapter you will study the motion of *systems of particles*, that is, the motion of a large number of particles considered together. The first part of the chapter is devoted to systems consisting of well-defined particles; the second part considers the motion of variable systems, that is, systems which are continually gaining or losing particles, or doing both at the same time.

In Sec. 14.2, Newton's second law will first be applied to each particle of the system. Defining the *effective force* of a particle as the product $m_i\mathbf{a}_i$ of its mass m_i and its acceleration $\mathbf{a}_i$, we will show that the *external forces* acting on the various particles form a system equipollent to the system of the effective forces, that is, both systems have the same resultant and the same moment resultant about any given point. In Sec. 14.3, it will be further shown that the resultant and moment resultant of the external forces are equal, respectively, to the rate of change of the total linear momentum and of the total angular momentum of the particles of the system.

In Sec. 14.4, the *mass center* of a system of particles is defined and the motion of that point is described, while in Sec. 14.5 the motion of the particles about their mass center is analyzed. The conditions under which the linear momentum and the angular momentum of a system of particles are conserved are discussed in Sec. 14.6, and the results obtained in that section are applied to the solution of various problems.

Sections 14.7 and 14.8 deal with the application of the work-energy principle to a system of particles, and Sec. 14.9 with the application of the impulse-momentum principle. These sections also contain a number of problems of practical interest.

It should be noted that while the derivations given in the first part of this chapter are carried out for a system of independent particles, they remain valid when the particles of the system are rigidly connected, that is, when they form a rigid body. In fact, the results obtained here will form the foundation of our discussion of the kinetics of rigid bodies in Chaps. 16 through 18.

The second part of this chapter is devoted to the study of variable systems of particles. In Sec. 14.11 you will consider steady streams of particles, such as a stream of water diverted by a fixed vane, or the flow of air through a jet engine, and learn to determine the force exerted by the stream on the vane and the thrust developed by the engine. Finally, in Sec. 14.12, you will learn how to analyze systems which gain mass by continually absorbing particles or lose mass by continually expelling particles. Among the various practical applications of this analysis will be the determination of the thrust developed by a rocket engine.

14.2. APPLICATION OF NEWTON'S LAWS TO THE MOTION OF A SYSTEM OF PARTICLES. EFFECTIVE FORCES

In order to derive the equations of motion for a system of n particles, let us begin by writing Newton's second law for each individual particle of the system. Consider the particle P_i, where $1 \leq i \leq n$. Let m_i be the mass of P_i and $\mathbf{a}_i$ its acceleration with respect to the newtonian frame of reference $Oxyz$. The force exerted on P_i by another particle P_j of the system (Fig. 14.1), called an *internal force,* will be denoted by $\mathbf{f}_{ij}$. The resultant of the internal forces exerted on P_i by all the other particles of the system is thus $\sum_{j=1}^{n} \mathbf{f}_{ij}$ (where $\mathbf{f}_{ii}$ has no meaning and is assumed to be equal to zero). Denoting, on the other hand, by $\mathbf{F}_i$ the resultant of all the *external forces* acting on P_i, we write Newton's second law for the particle P_i as follows:

$$\mathbf{F}_i + \sum_{j=1}^{n} \mathbf{f}_{ij} = m_i \mathbf{a}_i \tag{14.1}$$

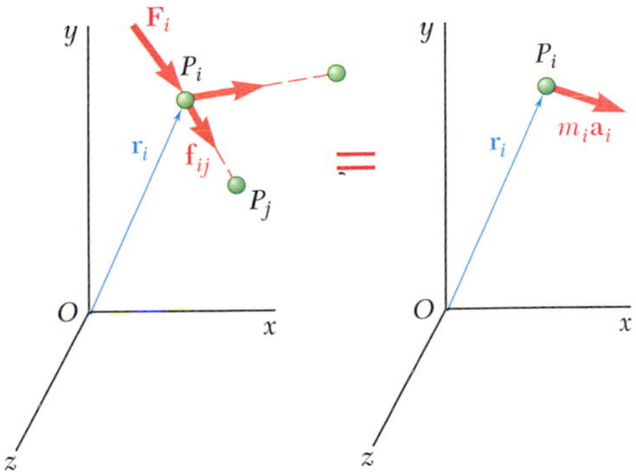

Fig. 14.1

Denoting by $\mathbf{r}_i$ the position vector of P_i and taking the moments about O of the various terms in Eq. (14.1), we also write

$$\mathbf{r}_i \times \mathbf{F}_i + \sum_{j=1}^{n} (\mathbf{r}_i \times \mathbf{f}_{ij}) = \mathbf{r}_i \times m_i \mathbf{a}_i \tag{14.2}$$

Repeating this procedure for each particle P_i of the system, we obtain n equations of the type (14.1) and n equations of the type (14.2), where i takes successively the values $1, 2, \ldots, n$. The vectors $m_i\mathbf{a}_i$ are referred to as the *effective forces* of the particles.† Thus the equations obtained express the fact that the external forces $\mathbf{F}_i$ and the internal forces $\mathbf{f}_{ij}$ acting on the various particles form a system equivalent to the system of the effective forces $m_i\mathbf{a}_i$ (that is, one system may be replaced by the other) (Fig. 14.2).

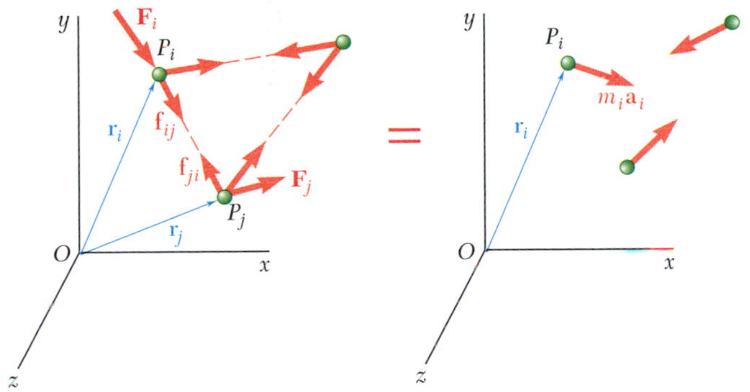

Fig. 14.2

Before proceeding further with our derivation, let us examine the internal forces $\mathbf{f}_{ij}$. We note that these forces occur in pairs $\mathbf{f}_{ij}$, $\mathbf{f}_{ji}$, where $\mathbf{f}_{ij}$ represents the force exerted by the particle P_j on the particle P_i and $\mathbf{f}_{ji}$ represents the force exerted by P_i on P_j (Fig. 14.2). Now, according to Newton's third law (Sec. 6.1), as extended by Newton's law of gravitation to particles acting at a distance (Sec. 12.10), the forces $\mathbf{f}_{ij}$ and $\mathbf{f}_{ji}$ are equal and opposite and have the same line of action. Their sum is therefore $\mathbf{f}_{ij} + \mathbf{f}_{ji} = 0$, and the sum of their moments about O is

$$\mathbf{r}_i \times \mathbf{f}_{ij} + \mathbf{r}_j \times \mathbf{f}_{ji} = \mathbf{r}_i \times (\mathbf{f}_{ij} + \mathbf{f}_{ji}) + (\mathbf{r}_j - \mathbf{r}_i) \times \mathbf{f}_{ji} = 0$$

since the vectors $\mathbf{r}_j - \mathbf{r}_i$ and $\mathbf{f}_{ji}$ in the last term are collinear. Adding

†Since these vectors represent the resultants of the forces acting on the various particles of the system, they can truly be considered as forces.

all the internal forces of the system and summing their moments about O, we obtain the equations

$$\sum_{i=1}^{n}\sum_{j=1}^{n} \mathbf{f}_{ij} = 0 \qquad \sum_{i=1}^{n}\sum_{j=1}^{n} (\mathbf{r}_i \times \mathbf{f}_{ij}) = 0 \tag{14.3}$$

which express the fact that the resultant and the moment resultant of the internal forces of the system are zero.

Returning now to the n equations (14.1), where $i = 1, 2, \ldots, n$, we sum their left-hand members and sum their right-hand members. Taking into account the first of Eqs. (14.3), we obtain

$$\sum_{i=1}^{n} \mathbf{F}_i = \sum_{i=1}^{n} m_i\mathbf{a}_i \tag{14.4}$$

Proceeding similarly with Eqs. (14.2) and taking into account the second of Eqs. (14.3), we have

$$\sum_{i=1}^{n} (\mathbf{r}_i \times \mathbf{F}_i) = \sum_{i=1}^{n} (\mathbf{r}_i \times m_i\mathbf{a}_i) \tag{14.5}$$

Equations (14.4) and (14.5) express the fact that the system of the external forces $\mathbf{F}_i$ and the system of the effective forces $m_i\mathbf{a}_i$ have the same resultant and the same moment resultant. Referring to the definition given in Sec. 3.19 for two equipollent systems of vectors, we can therefore state that *the system of the external forces acting on the particles and the system of the effective forces of the particles are equipollent*† (Fig. 14.3).

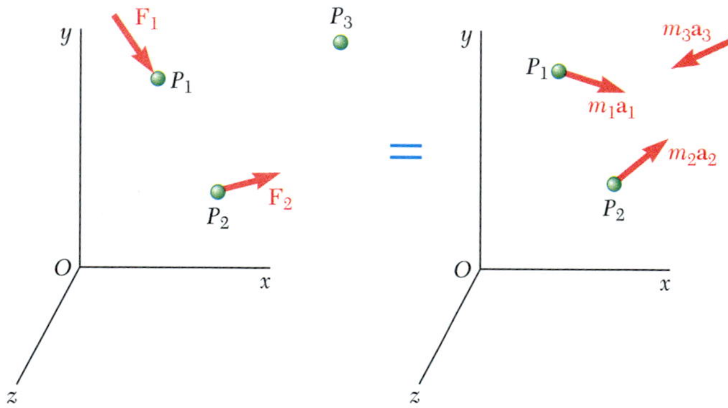

Fig. 14.3

†The result just obtained is often referred to as *d'Alembert's principle*, after the French mathematician Jean le Rond d'Alembert (1717–1783). However, d'Alembert's original statement refers to the motion of a system of connected bodies, with $\mathbf{f}_{ij}$ representing constraint forces which if applied by themselves will not cause the system to move. Since, as it will now be shown, this is in general not the case for the internal forces acting on a system of free particles, the consideration of d'Alembert's principle will be postponed until the motion of rigid bodies is considered (Chap. 16).

Equations (14.3) express the fact that the system of the internal forces $\mathbf{f}_{ij}$ is equipollent to zero. Note, however, that it does *not* follow that the internal forces have no effect on the particles under consideration. Indeed, the gravitational forces that the sun and the planets exert on one another are internal to the solar system and equipollent to zero. Yet these forces are alone responsible for the motion of the planets about the sun.

Similarly, it does not follow from Eqs. (14.4) and (14.5) that two systems of external forces which have the same resultant and the same moment resultant will have the same effect on a given system of particles. Clearly, the systems shown in Figs. 14.4*a* and 14.4*b* have

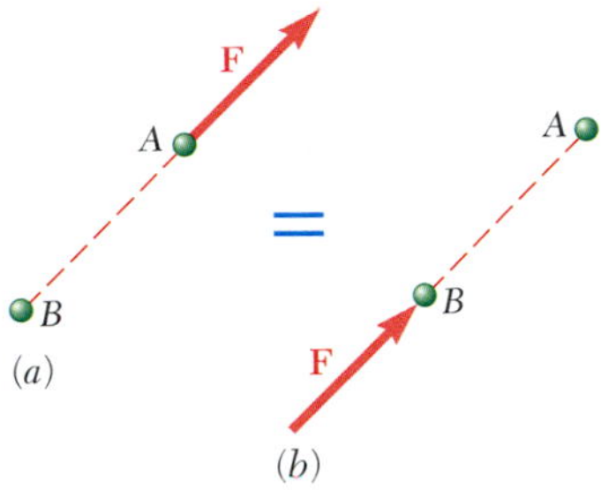

Fig. 14.4

the same resultant and the same moment resultant; yet the first system accelerates particle *A* and leaves particle *B* unaffected, while the second accelerates *B* and does not affect *A*. It is important to recall that when we stated in Sec. 3.19 that two equipollent systems of forces acting on a rigid body are also equivalent, we specifically noted that this property could *not* be extended to a system of forces acting on a set of independent particles such as those considered in this chapter.

In order to avoid any confusion, blue equal signs are used to connect equipollent systems of vectors, such as those shown in Figs. 14.3 and 14.4. These signs indicate that the two systems of vectors have the same resultant and the same moment resultant. Red equals signs will continue to be used to indicate that two systems of vectors are equivalent, that is, that one system can actually be replaced by the other (Fig. 14.2).

14.3. LINEAR AND ANGULAR MOMENTUM OF A SYSTEM OF PARTICLES

Equations (14.4) and (14.5), obtained in the preceding section for the motion of a system of particles, can be expressed in a more condensed form if we introduce the linear and the angular momentum of the system of particles. Defining the linear momentum $\mathbf{L}$ of the system of particles as the sum of the linear momenta of the various particles of the system (Sec. 12.3), we write

$$\mathbf{L} = \sum_{i=1}^{n} m_i \mathbf{v}_i \tag{14.6}$$

Defining the angular momentum $\mathbf{H}_O$ about O of the system of particles in a similar way (Sec. 12.7), we have

$$\mathbf{H}_O = \sum_{i=1}^{n} (\mathbf{r}_i \times m_i\mathbf{v}_i) \tag{14.7}$$

Differentiating both members of Eqs. (14.6) and (14.7) with respect to t, we write

$$\dot{\mathbf{L}} = \sum_{i=1}^{n} m_i\dot{\mathbf{v}}_i = \sum_{i=1}^{n} m_i\mathbf{a}_i \tag{14.8}$$

and

$$\begin{aligned}\dot{\mathbf{H}}_O &= \sum_{i=1}^{n} (\dot{\mathbf{r}}_i \times m_i\mathbf{v}_i) + \sum_{i=1}^{n} (\mathbf{r}_i \times m_i\dot{\mathbf{v}}_i)\\ &= \sum_{i=1}^{n} (\mathbf{v}_i \times m_i\mathbf{v}_i) + \sum_{i=1}^{n} (\mathbf{r}_i \times m_i\mathbf{a}_i)\end{aligned}$$

which reduces to

$$\dot{\mathbf{H}}_O = \sum_{i=1}^{n} (\mathbf{r}_i \times m_i\mathbf{a}_i) \tag{14.9}$$

since the vectors $\mathbf{v}_i$ and $m_i\mathbf{v}_i$ are collinear.

We observe that the right-hand members of Eqs. (14.8) and (14.9) are respectively identical with the right-hand members of Eqs. (14.4) and (14.5). It follows that the left-hand members of these equations are respectively equal. Recalling that the left-hand member of Eq. (14.5) represents the sum of the moments $\mathbf{M}_O$ about O of the external forces acting on the particles of the system, and omitting the subscript i from the sums, we write

$$\Sigma\mathbf{F} = \dot{\mathbf{L}} \tag{14.10}$$

$$\Sigma\mathbf{M}_O = \dot{\mathbf{H}}_O \tag{14.11}$$

These equations express that *the resultant and the moment resultant about the fixed point O of the external forces are respectively equal to the rates of change of the linear momentum and of the angular momentum about O of the system of particles.*

14.4. MOTION OF THE MASS CENTER OF A SYSTEM OF PARTICLES

Equation (14.10) may be written in an alternative form if the *mass center* of the system of particles is considered. The mass center of the system is the point G defined by the position vector $\bar{\mathbf{r}}$, which satisfies the relation

$$m\bar{\mathbf{r}} = \sum_{i=1}^{n} m_i \mathbf{r}_i \tag{14.12}$$

where m represents the total mass $\sum_{i=1}^{n} m_i$ of the particles. Resolving the position vectors $\bar{\mathbf{r}}$ and $\mathbf{r}_i$ into rectangular components, we obtain the following three scalar equations, which can be used to determine the coordinates $\bar{x}, \bar{y}, \bar{z}$ of the mass center:

$$m\bar{x} = \sum_{i=1}^{n} m_i x_i \qquad m\bar{y} = \sum_{i=1}^{n} m_i y_i \qquad m\bar{z} = \sum_{i=1}^{n} m_i z_i \tag{14.12'}$$

Since $m_i g$ represents the weight of the particle P_i, and mg the total weight of the particles, G is also the center of gravity of the system of particles. However, in order to avoid any confusion, G will be referred to as the *mass center* of the system of particles when properties associated with the *mass* of the particles are being discussed, and as the *center of gravity* of the system when properties associated with the *weight* of the particles are being considered. Particles located outside the gravitational field of the earth, for example, have a mass but no weight. We can then properly refer to their mass center, but obviously not to their center of gravity.†

Differentiating both members of Eq. (14.12) with respect to t, we write

$$m\dot{\bar{\mathbf{r}}} = \sum_{i=1}^{n} m_i \dot{\mathbf{r}}_i$$

or

$$m\bar{\mathbf{v}} = \sum_{i=1}^{n} m_i \mathbf{v}_i \tag{14.13}$$

where $\bar{\mathbf{v}}$ represents the velocity of the mass center G of the system of particles. But the right-hand member of Eq. (14.13) is, by definition, the linear momentum $\mathbf{L}$ of the system (Sec. 14.3). We therefore have

$$\mathbf{L} = m\bar{\mathbf{v}} \tag{14.14}$$

and, differentiating both members with respect to t,

$$\dot{\mathbf{L}} = m\bar{\mathbf{a}} \tag{14.15}$$

†It may also be pointed out that the mass center and the center of gravity of a system of particles do not exactly coincide, since the weights of the particles are directed toward the center of the earth and thus do not truly form a system of parallel forces.

where $\bar{\mathbf{a}}$ represents the acceleration of the mass center G. Substituting for $\dot{\mathbf{L}}$ from (14.15) into (14.10), we write the equation

$$\Sigma \mathbf{F} = m\bar{\mathbf{a}} \tag{14.16}$$

which defines the motion of the mass center G of the system of particles.

We note that Eq. (14.16) is identical with the equation we would obtain for a particle of mass m equal to the total mass of the particles of the system, acted upon by all the external forces. We therefore state that *the mass center of a system of particles moves as if the entire mass of the system and all the external forces were concentrated at that point.*

This principle is best illustrated by the motion of an exploding shell. We know that if air resistance is neglected, it can be assumed that a shell will travel along a parabolic path. After the shell has exploded, the mass center G of the fragments of shell will continue to travel along the same path. Indeed, point G must move as if the mass and the weight of all fragments were concentrated at G; it must, therefore, move as if the shell had not exploded.

It should be noted that the preceding derivation does not involve the moments of the external forces. Therefore, *it would be wrong to assume* that the external forces are equipollent to a vector $m\bar{\mathbf{a}}$ attached at the mass center G. This is not in general the case since, as you will see in the next section, the sum of the moments about G of the external forces is not in general equal to zero.

14.5. ANGULAR MOMENTUM OF A SYSTEM OF PARTICLES ABOUT ITS MASS CENTER

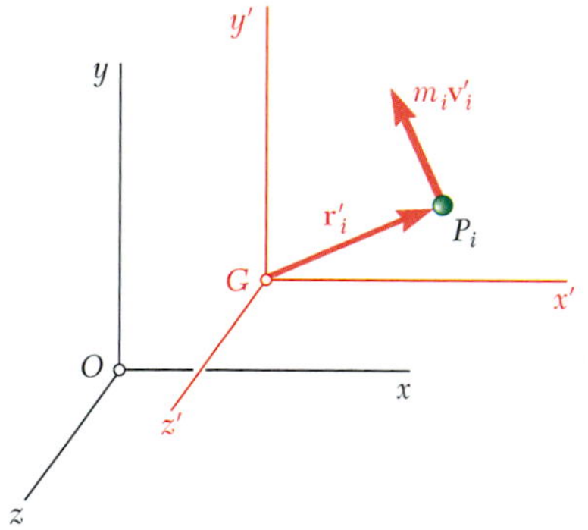

Fig. 14.5

In some applications (for example, in the analysis of the motion of a rigid body) it is convenient to consider the motion of the particles of the system with respect to a centroidal frame of reference $Gx'y'z'$ which translates with respect to the newtonian frame of reference $Oxyz$ (Fig. 14.5). Although a centroidal frame is not, in general, a newtonian frame of reference, it will be seen that the fundamental relation (14.11) holds when the frame $Oxyz$ is replaced by $Gx'y'z'$.

Denoting, respectively, by $\mathbf{r}'_i$ and $\mathbf{v}'_i$ the position vector and the velocity of the particle P_i relative to the moving frame of reference $Gx'y'z'$, we define the *angular momentum* $\mathbf{H}'_G$ of the system of particles *about the mass center* G as follows:

$$\mathbf{H}'_G = \sum_{i=1}^{n} (\mathbf{r}'_i \times m_i\mathbf{v}'_i) \tag{14.17}$$

We now differentiate both members of Eq. (14.17) with respect to t. This operation is similar to that performed in Sec. 14.3 on Eq. (14.7), and so we write immediately

$$\dot{\mathbf{H}}'_G = \sum_{i=1}^{n} (\mathbf{r}'_i \times m_i\mathbf{a}'_i) \tag{14.18}$$

where $\mathbf{a}'_i$ denotes the acceleration of P_i relative to the moving frame of reference. Referring to Sec. 11.12, we write

$$\mathbf{a}_i = \overline{\mathbf{a}} + \mathbf{a}'_i$$

where $\mathbf{a}_i$ and $\overline{\mathbf{a}}$ denote, respectively, the accelerations of P_i and G relative to the frame $Oxyz$. Solving for $\mathbf{a}'_i$ and substituting into (14.18), we have

$$\dot{\mathbf{H}}'_G = \sum_{i=1}^{n} (\mathbf{r}'_i \times m_i\mathbf{a}_i) - \left(\sum_{i=1}^{n} m_i\mathbf{r}'_i\right) \times \overline{\mathbf{a}} \tag{14.19}$$

But, by (14.12), the second sum in Eq. (14.19) is equal to $m\overline{\mathbf{r}}'$ and thus to zero, since the position vector $\overline{\mathbf{r}}'$ of G relative to the frame $Gx'y'z'$ is clearly zero. On the other hand, since $\mathbf{a}_i$ represents the acceleration of P_i relative to a newtonian frame, we can use Eq. (14.1) and replace $m_i\mathbf{a}_i$ by the sum of the internal forces $\mathbf{f}_{ij}$ and of the resultant $\mathbf{F}_i$ of the external forces acting on P_i. But a reasoning similar to that used in Sec. 14.2 shows that the moment resultant about G of the internal forces $\mathbf{f}_{ij}$ of the entire system is zero. The first sum in Eq. (14.19) therefore reduces to the moment resultant about G of the external forces acting on the particles of the system, and we write

$$\Sigma\mathbf{M}_G = \dot{\mathbf{H}}'_G \tag{14.20}$$

which expresses that *the moment resultant about G of the external forces is equal to the rate of change of the angular momentum about G of the system of particles.*

It should be noted that in Eq. (14.17) we defined the angular momentum $\mathbf{H}'_G$ as the sum of the moments about G of the momenta of the particles $m_i\mathbf{v}'_i$ *in their motion relative to the centroidal frame of reference Gx'y'z'*. We may sometimes want to compute the sum $\mathbf{H}_G$ of the moments about G of the momenta of the particles $m_i\mathbf{v}_i$ *in their absolute motion,* that is, in their motion as observed from the newtonian frame of reference $Oxyz$ (Fig. 14.6):

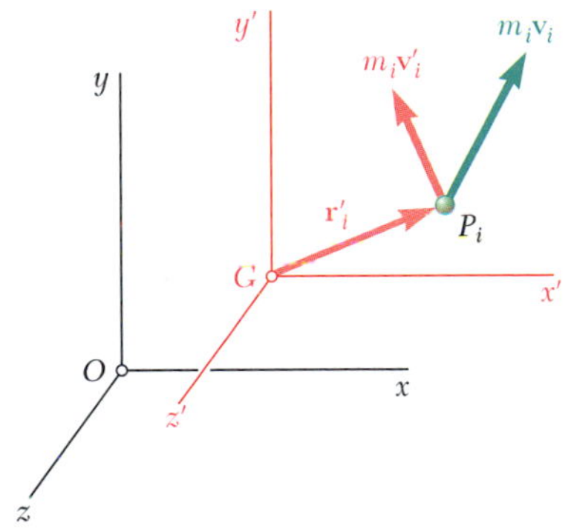

Fig. 14.6

$$\mathbf{H}_G = \sum_{i=1}^{n} (\mathbf{r}'_i \times m_i\mathbf{v}_i) \tag{14.21}$$

Remarkably, the angular momenta $\mathbf{H}'_G$ and $\mathbf{H}_G$ are identically equal. This can be verified by referring to Sec. 11.12 and writing

$$\mathbf{v}_i = \overline{\mathbf{v}} + \mathbf{v}'_i \tag{14.22}$$

Substituting for $\mathbf{v}_i$ from (14.22) into Eq. (14.21), we have

$$\mathbf{H}_G = \left(\sum_{i=1}^{n} m_i\mathbf{r}'_i\right) \times \overline{\mathbf{v}} + \sum_{i=1}^{n} (\mathbf{r}'_i \times m_i\mathbf{v}_i)$$

But, as observed earlier, the first sum is equal to zero. Thus $\mathbf{H}_G$ reduces to the second sum, which, by definition, is equal to $\mathbf{H}'_G$.†

†Note that this property is peculiar to the centroidal frame $Gx'y'z'$ and does not, in general, hold for other frames of reference (see Prob. 14.29).

Taking advantage of the property we have just established, we simplify our notation by dropping the prime (′) from Eq. (14.20) and writing

$$\Sigma \mathbf{M}_G = \dot{\mathbf{H}}_G \tag{14.23}$$

where it is understood that the angular momentum $\mathbf{H}_G$ can be computed by forming the moments about G of the momenta of the particles in their motion with respect to either the newtonian frame $Oxyz$ or the centroidal frame $Gx'y'z'$:

$$\mathbf{H}_G = \sum_{i=1}^{n} (\mathbf{r}'_i \times m_i \mathbf{v}_i) = \sum_{i=1}^{n} (\mathbf{r}'_i \times m_i \mathbf{v}'_i) \tag{14.24}$$

14.6. CONSERVATION OF MOMENTUM FOR A SYSTEM OF PARTICLES

If no external force acts on the particles of a system, the left-hand members of Eqs. (14.10) and (14.11) are equal to zero and these equations reduce to $\dot{\mathbf{L}} = 0$ and $\dot{\mathbf{H}}_O = 0$. We conclude that

$$\mathbf{L} = \text{constant} \qquad \mathbf{H}_O = \text{constant} \tag{14.25}$$

The equations obtained express that the linear momentum of the system of particles and its angular momentum about the fixed point O are conserved.

In some applications, such as problems involving central forces, the moment about a fixed point O of each of the external forces can be zero without any of the forces being zero. In such cases, the second of Eqs. (14.25) still holds; the angular momentum of the system of particles about O is conserved.

Photo 14.1 If the system of particles is defined as the two stages of the space vehicle and no external forces act, the linear momentum of the system and its angular momentum about its mass center are conserved.

The concept of conservation of momentum can also be applied to the analysis of the motion of the mass center G of a system of particles and to the analysis of the motion of the system about G. For example, if the sum of the external forces is zero, the first of Eqs. (14.25) applies. Recalling Eq. (14.14), we write

$$\bar{\mathbf{v}} = \text{constant} \tag{14.26}$$

which expresses that the mass center G of the system moves in a straight line and at a constant speed. On the other hand, if the sum of the moments about G of the external forces is zero, it follows from Eq. (14.23) that the angular momentum of the system about its mass center is conserved:

$$\mathbf{H}_G = \text{constant} \tag{14.27}$$

SAMPLE PROBLEM 14.1

A 200-kg space vehicle is observed at $t = 0$ to pass through the origin of a newtonian reference frame $Oxyz$ with velocity $\mathbf{v}_0 = (150 \text{ m/s})\mathbf{i}$ relative to the frame. Following the detonation of explosive charges, the vehicle separates into three parts A, B, and C, of mass 100 kg, 60 kg, and 40 kg, respectively. Knowing that at $t = 2.5$ s the positions of parts A and B are observed to be $A(555, -180, 240)$ and $B(255, 0, -120)$, where the coordinates are expressed in meters, determine the position of part C at that time.

SOLUTION

Since there is no external force, the mass center G of the system moves with the constant velocity $\mathbf{v}_0 = (150 \text{ m/s})\mathbf{i}$. At $t = 2.5$ s, its position is

$$\bar{\mathbf{r}} = \mathbf{v}_0 t = (150 \text{ m/s})\mathbf{i}(2.5 \text{ s}) = (375 \text{ m})\mathbf{i}$$

Recalling Eq. (14.12), we write

$$m\bar{\mathbf{r}} = m_A\mathbf{r}_A + m_B\mathbf{r}_B + m_C\mathbf{r}_C$$

$$(200 \text{ kg})(375 \text{ m})\mathbf{i} = (100 \text{ kg})[(555 \text{ m})\mathbf{i} - (180 \text{ m})\mathbf{j} + (240 \text{ m})\mathbf{k}] + (60 \text{ kg})[(255 \text{ m})\mathbf{i} - (120 \text{ m})\mathbf{k}] + (40 \text{ kg})\mathbf{r}_C$$

$$\mathbf{r}_C = (105 \text{ m})\mathbf{i} + (450 \text{ m})\mathbf{j} - (420 \text{ m})\mathbf{k} \quad ◀$$

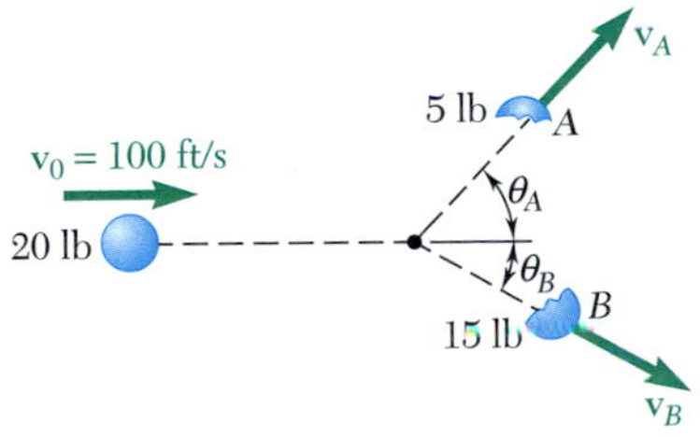

SAMPLE PROBLEM 14.2

A 20-lb projectile is moving with a velocity of 100 ft/s when it explodes into two fragments A and B, weighing 5 lb and 15 lb, respectively. Knowing that immediately after the explosion, fragments A and B travel in directions defined respectively by $\theta_A = 45°$ and $\theta_B = 30°$, determine the velocity of each fragment.

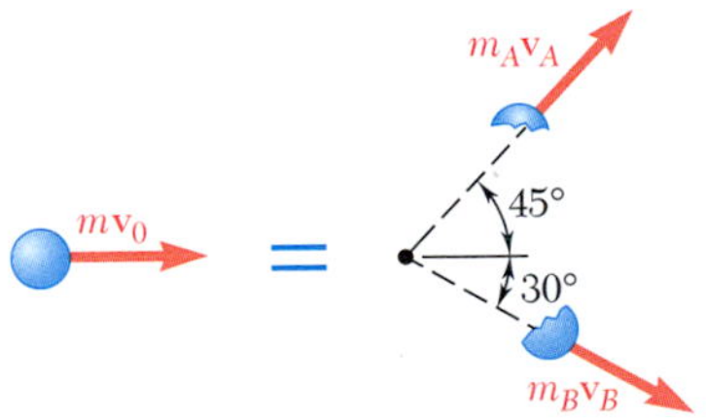

SOLUTION

Since there is no external force, the linear momentum of the system is conserved, and we write

$$m_A\mathbf{v}_A + m_B\mathbf{v}_B = m\mathbf{v}_0$$

$$(5/g)\mathbf{v}_A + (15/g)\mathbf{v}_B = (20/g)\mathbf{v}_0$$

$\xrightarrow{+}$ x components: $\quad 5v_A \cos 45° + 15v_B \cos 30° = 20(100)$

$+\uparrow$ y components: $\quad 5v_A \sin 45° - 15v_B \sin 30° = 0$

Solving simultaneously the two equations for v_A and v_B, we have

$$v_A = 207 \text{ ft/s} \qquad v_B = 97.6 \text{ ft/s}$$

$$\mathbf{v}_A = 207 \text{ ft/s} \angle 45° \qquad \mathbf{v}_B = 97.6 \text{ ft/s} \searrow 30° \quad ◀$$

SOLVING PROBLEMS ON YOUR OWN

This chapter deals with the motion of *systems of particles,* that is, with the motion of a large number of particles considered together, rather than separately. In this first lesson you learned to compute the *linear momentum* and the *angular momentum* of a system of particles. We defined the linear momentum $\mathbf{L}$ of a system of particles as the sum of the linear momenta of the particles and we defined the angular momentum $\mathbf{H}_O$ of the system as the sum of the angular momenta of the particles about O:

$$\mathbf{L} = \sum_{i=1}^{n} m_i\mathbf{v}_i \qquad \mathbf{H}_O = \sum_{i=1}^{n} (\mathbf{r}_i \times m_i\mathbf{v}_i) \qquad (14.6, 14.7)$$

In this lesson, you will solve a number of problems of practical interest, either by observing that the linear momentum of a system of particles is conserved or by considering the motion of the mass center of a system of particles.

1. *Conservation of the linear momentum of a system of particles.* This occurs *when the resultant of the external forces acting on the particles of the system is zero.* You may encounter such a situation in the following types of problems.

a. Problems involving the rectilinear motion of objects such as colliding automobiles and railroad cars. After you have checked that the resultant of the external forces is zero, equate the algebraic sums of the initial momenta and final momenta to obtain an equation which can be solved for one unknown.

b. Problems involving the two-dimensional or three-dimensional motion of objects such as exploding shells, or colliding aircraft, automobiles, or billiard balls. After you have checked that the resultant of the external forces is zero, add vectorially the initial momenta of the objects, add vectorially their final momenta, and equate the two sums to obtain a vector equation expressing that the linear momentum of the system is conserved.

In the case of a two-dimensional motion, this equation can be replaced by two scalar equations which can be solved for two unknowns, while in the case of a three-dimensional motion it can be replaced by three scalar equations which can be solved for three unknowns.

2. *Motion of the mass center of a system of particles.* You saw in Sec. 14.4 that *the mass center of a system of particles moves as if the entire mass of the system and all of the external forces were concentrated at that point.*

a. In the case of a body exploding while in motion, it follows that the mass center of the resulting fragments moves as the body itself would have moved if the explosion had not occurred. Problems of this type can be solved by writing the equation of motion of the mass center of the system in vectorial form and expressing the position vector of the mass center in terms of the position vectors of the various fragments [Eq. (14.12)]. You can then rewrite the vector equation as two or three scalar equations and solve the equations for an equivalent number of unknowns.

b. In the case of the collision of several moving bodies, it follows that the motion of the mass center of the various bodies is unaffected by the collision. Problems of this type can be solved by writing the equation of motion of the mass center of the system in vectorial form and expressing its position vector before and after the collision in terms of the position vectors of the relevant bodies [Eq. (14.12)]. You can then rewrite the vector equation as two or three scalar equations and solve these equations for an equivalent number of unknowns.

Problems

14.1 Car *A* of mass 1800 kg and car *B* of mass 1700 kg are at rest on a 20-Mg flatcar which is also at rest. Cars *A* and *B* then accelerate and quickly reach constant speeds relative to the flatcar of 2.35 m/s and 1.175 m/s, respectively, before decelerating to a stop at the opposite end of the flatcar. Neglecting friction and rolling resistance, determine the velocity of the flatcar when the cars are moving at constant speeds.

Fig. P14.1 and P14.2

14.2 Car *A* of mass 1800 kg and car *B* of mass 1700 kg are at rest on a flatcar which is also at rest. Cars *A* and *B* then accelerate and quickly reach constant speeds relative to the flatcar of 2.55 m/s and 2.50 m/s, respectively, before decelerating to a stop at the opposite end of the flatcar. Knowing that the speed of the flatcar is 0.34 m/s when the cars are moving at constant speeds, determine the mass of the flatcar. Neglect friction and rolling resistance.

14.3 An airline employee tosses two suitcases, of weight 30 lb and 40 lb, respectively, onto a 50-lb baggage carrier in rapid succession. Knowing that the carrier is initially at rest and that the employee imparts a 9-ft/s horizontal velocity to the 30-lb suitcase and a 6-ft/s horizontal velocity to the 40-lb suitcase, determine the final velocity of the baggage carrier if the first suitcase tossed onto the carrier is (*a*) the 30-lb suitcase, (*b*) the 40-lb suitcase.

Fig. P14.3 and P14.4

14.4 An airline employee tosses two suitcases in rapid succession, with a horizontal velocity of 7.2 ft/s, onto a 50-lb baggage carrier which is initially at rest. (*a*) Knowing that the final velocity of the baggage carrier is 3.6 ft/s and that the first suitcase the employee tosses onto the carrier has a weight of 30 lb, determine the weight of the other suitcase. (*b*) What would be the final velocity of the carrier if the employee reversed the order in which he tosses the suitcases?

14.5 A bullet is fired with a horizontal velocity of 500 m/s through a 3-kg block *A* and becomes embedded in a 2.5-kg block *B*. Knowing that blocks *A* and *B* start moving with velocities of 3 m/s and 5 m/s, respectively, determine (*a*) the mass of the bullet, (*b*) its velocity as it travels from block *A* to block *B*.

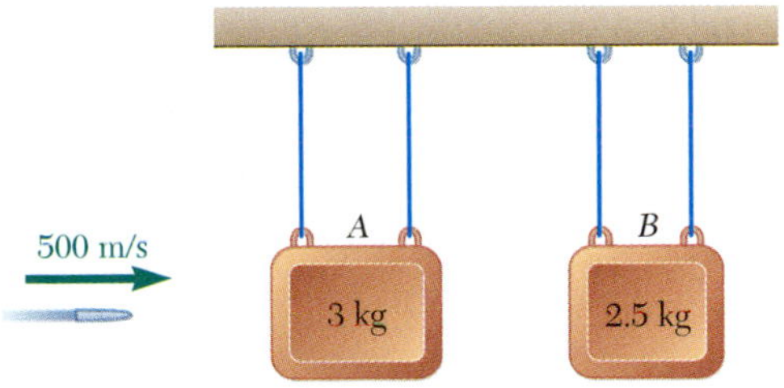

Fig. P14.5

14.6 An 80-Mg railroad engine A coasting at 6.5 km/h strikes a 20-Mg flatcar C carrying a 30-Mg load B which can slide along the floor of the car ($\mu_k = 0.25$). Knowing that the car was at rest with its brakes released and that it automatically coupled with the engine upon impact, determine the velocity of the car (*a*) immediately after impact, (*b*) after the load has slid to a stop relative to the car.

Fig. P14.6

14.7 Three identical cars are being unloaded from an automobile carrier. Cars B and C have just been unloaded and are at rest with their brakes off when car A leaves the unloading ramp with a velocity of 5.76 ft/s and hits car B, which hits car C. Car A then again hits car B. Knowing that the velocity of car B is 5.04 ft/s after the first collision, 0.630 ft/s after the second collision, and 0.709 ft/s after the third collision, determine (*a*) the final velocities of cars A and C, (*b*) the coefficient of restitution for each of the collisions.

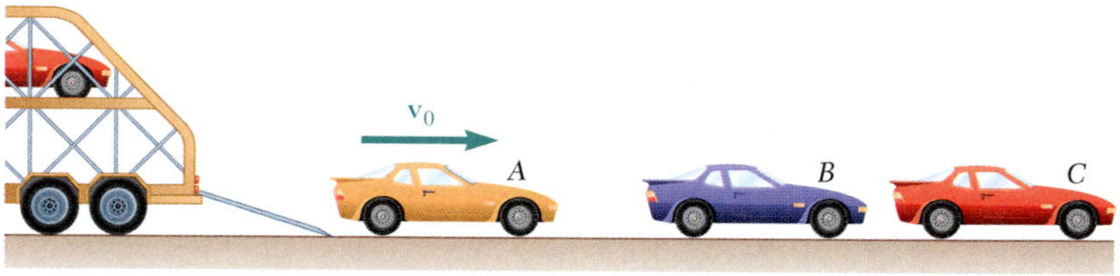

Fig. *P14.7* and *P14.8*

14.8 Three identical cars are being unloaded from an automobile carrier. Cars B and C have just been unloaded and are at rest with their brakes off when car A leaves the unloading ramp with a velocity of 6.00 ft/s and hits car B, which hits car C. Car A then again hits car B. Knowing that the velocity of car A is 1.20 ft/s after its first collision with car B and 1.008 ft/s after its second collision with car B and that the velocity of car C is 3.84 ft/s after it has been hit by car B, determine (*a*) the velocity of car B after each of the three collisions, (*b*) the coefficient of restitution for each of the collisions.

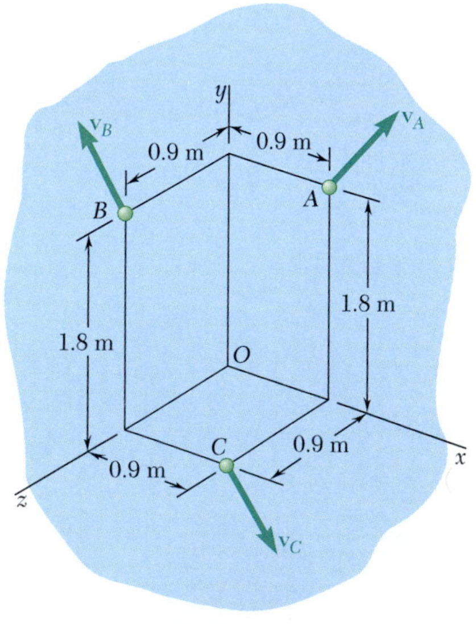

Fig. P14.9 and P14.10

14.9 A system consists of three particles A, B, and C. We know that $m_A = m_B = 2$ kg and $m_C = 14$ kg and that the velocities of the particles, expressed in m/s are, respectively, $\mathbf{v}_A = 14\mathbf{i} + 21\mathbf{j}$, $\mathbf{v}_B = -14\mathbf{i} + 21\mathbf{j}$, and $\mathbf{v}_C = -3\mathbf{j} - 2\mathbf{k}$. Determine the angular momentum $\mathbf{H}_O$ of the system about O.

14.10 For the system of particles of Prob. 14.9, determine (*a*) the position vector $\bar{\mathbf{r}}$ of the mass center G of the system, (*b*) the linear momentum $m\bar{\mathbf{v}}$ of the system, (*c*) the angular momentum $\mathbf{H}_G$ of the system about G. Also verify that the answers to this problem and to Prob. 14.9 satisfy the equation given in Prob. 14.28.

14.11 A system consists of three identical 19.32-lb particles A, B, and C. The velocities of the particles are, respectively, $\mathbf{v}_A = v_A\mathbf{j}$, $\mathbf{v}_B = v_B\mathbf{i}$, and $\mathbf{v}_C = v_C\mathbf{k}$. Knowing that the angular momentum of the system about O, expressed in ft · lb · s is $\mathbf{H}_O = -1.2\mathbf{k}$, determine (*a*) the velocities of the particles, (*b*) the angular momentum of the system about its mass center G.

14.12 A system consists of three identical 19.32-lb particles A, B, and C. The velocities of the particles are, respectively, $\mathbf{v}_A = v_A\mathbf{j}$, $\mathbf{v}_B = v_B\mathbf{i}$, and $\mathbf{v}_C = v_C\mathbf{k}$, and the magnitude of the linear momentum $\mathbf{L}$ of the system is 9 lb · s. Knowing that $\mathbf{H}_G = \mathbf{H}_O$, where $\mathbf{H}_G$ is the angular momentum of the system about its mass center G and $\mathbf{H}_O$ is the angular momentum of the system about O, determine (*a*) the velocities of the particles, (*b*) the angular momentum of the system about O.

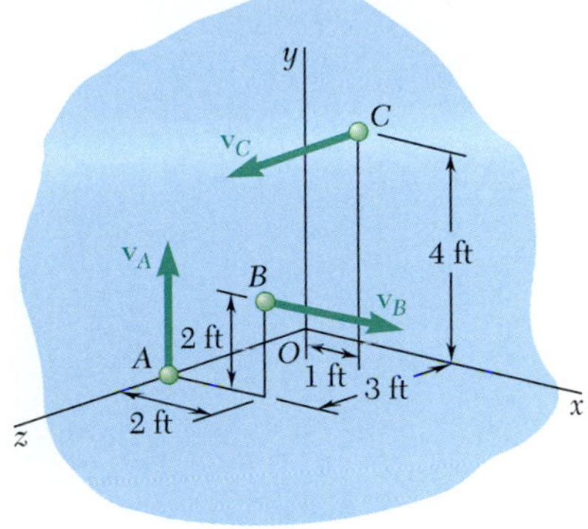

Fig. P14.11 and P14.12

14.13 A system consists of three particles A, B, and C. We know that $m_A = 3$ kg, $m_B = 2$ kg, and $m_C = 4$ kg and that the velocities of the particles expressed in m/s are, respectively, $\mathbf{v}_A = 4\mathbf{i} + 2\mathbf{j} + 2\mathbf{k}$, $\mathbf{v}_B = 4\mathbf{i} + 3\mathbf{j}$, and $\mathbf{v}_C = -2\mathbf{i} + 4\mathbf{j} + 2\mathbf{k}$. Determine the angular momentum $\mathbf{H}_O$ of the system about O.

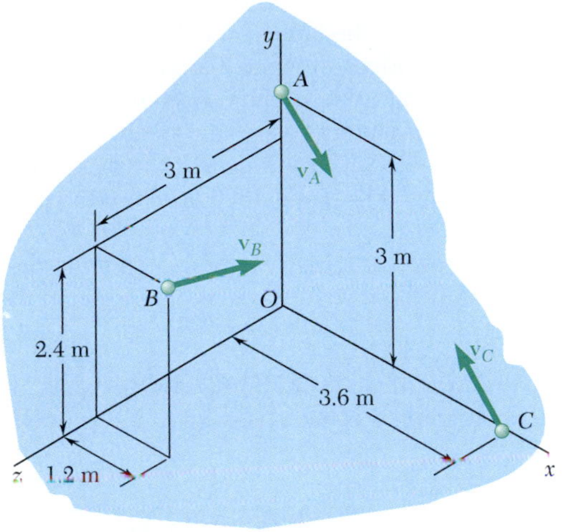

Fig. *P14.13*

14.14 For the system of particles of Prob. 14.13, determine (*a*) the position vector $\bar{\mathbf{r}}$ of the mass center G of the system, (*b*) the linear momentum $m\bar{\mathbf{v}}$ of the system, (*c*) the angular momentum $\mathbf{H}_G$ of the system about G. Also verify that the answers to this problem and to Prob. 14.13 satisfy the equation given in Prob. 14.28.

14.15 A 500-kg space vehicle traveling with a velocity $\mathbf{v}_0 = (450 \text{ m/s})\mathbf{i}$ passes through the origin O at $t = 0$. Explosive charges then separate the vehicle into three parts A, B, and C of 300-kg, 150-kg, and 50-kg, respectively. Knowing that at $t = 4$ s, the positions of parts A and B are observed to be A (1200 m, −350 m, −600 m) and B (2500 m, 450 m, 900 m), determine the corresponding position of part C. Neglect the effect of gravity.

14.16 A 20-kg projectile is passing through the origin O with a velocity $\mathbf{v}_0 = (60 \text{ m/s})\mathbf{i}$ when it explodes into two fragments A and B, of 8-kg and 12-kg, respectively. Knowing that 2 s later the position of fragment A is (120 m, −10 m, −20 m), determine the position of fragment B at the same instant. Assume $a_y = -g = -9.81 \text{ m/s}^2$ and neglect air resistance.

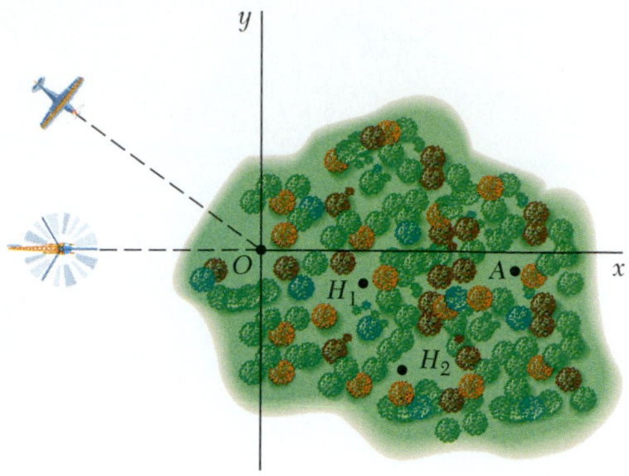

Fig. P14.17

14.17 A small 3000-lb airplane and a 6000-lb helicopter flying at an altitude of 3600 ft are observed to collide directly above a tower located at O in a wooded area. Four minutes earlier the helicopter had been sighted 5.5 mi due west of the tower and the airplane 10 mi west and 7.5 mi north of the tower. As a result of the collision the helicopter was split into two pieces, H_1 and H_2, weighing 2000 lb and 4000 lb, respectively; the airplane remained in one piece as it fell to the ground. Knowing that the two fragments of the helicopter were located at points H_1 (1500 ft, −300 ft) and H_2 (1800 ft, −1500 ft), respectively, and assuming that all pieces hit the ground at the same time, determine the coordinates of the point A where the wreckage of the airplane will be found.

14.18 In Prob. 14.17, knowing that the wreckage of the small airplane was found at A (3600 ft, 240 ft) and the 2000-lb fragment of the helicopter at point H_1 (1200 ft, −600 ft), and assuming that all pieces hit the ground at the same time, determine the coordinates of the point H_2 where the other fragment of the helicopter will be found.

14.19* and *14.20 Car A was at rest 9.28 m south of the point O when it was struck in the rear by car B, which was traveling north at a speed v_B. Car C, which was traveling west at a speed v_C, was 40 m east of point O at the time of the collision. Cars A and B stuck together and, because the pavement was covered with ice, they slid into the intersection and were struck by car C which had not changed its speed. Measurements based on a photograph taken from a traffic helicopter shortly after the second collision indicated that the positions of the cars, expressed in meters, were $\mathbf{r}_A = -10.1\mathbf{i} + 16.9\mathbf{j}$, $\mathbf{r}_B = -10.1\mathbf{i} + 20.4\mathbf{j}$, and $\mathbf{r}_C = -19.8\mathbf{i} - 15.2\mathbf{j}$. Knowing that the masses of cars A, B, and C are, respectively, 1400 kg, 1800 kg, and 1600 kg, solve the problems indicated.

14.19 Knowing that $v_C = 72$ km/h, determine the initial speed of car B and the time elapsed between the first collision and the time the photograph was taken.

14.20 Knowing that the time elapsed between the first collision and the time the photograph was taken was 3.4 s, determine the initial speeds of cars B and C.

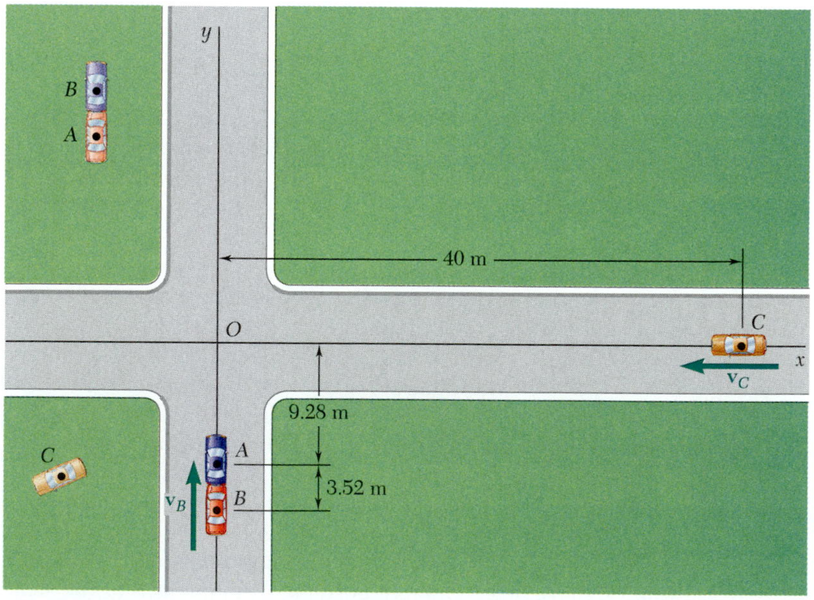

Fig. *P14.19* and *P14.20*

14.21 and 14.22 In a game of pool, ball A is traveling with a velocity $\mathbf{v}_0$ when it strikes balls B and C which are at rest and aligned as shown. Knowing that after the collision the three balls move in the directions indicated and that $v_0 = 4$ m/s and $v_C = 2.1$ m/s, determine the magnitude of the velocity of (a) ball A, (b) ball B.

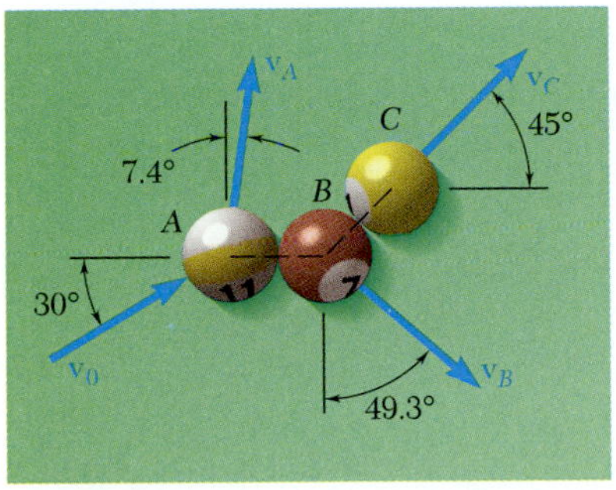

Fig. P14.21

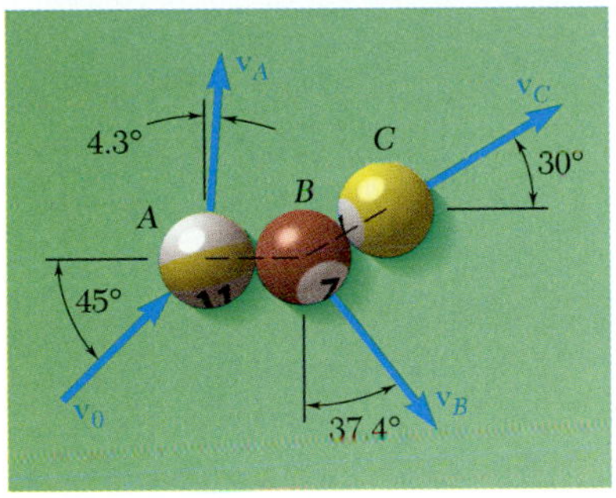

Fig. P14.22

14.23 A 6-lb game bird flying due east 45 ft above the ground with a velocity $\mathbf{v}_B = (30 \text{ ft/s})\mathbf{i}$ is hit by a 2-oz arrow with a velocity $\mathbf{v}_A = (180 \text{ ft/s})\mathbf{j} + (240 \text{ ft/s})\mathbf{k}$, where $\mathbf{j}$ is directed upward. Determine the position of the point P where the bird will hit the ground, relative to the point O located directly under the point of impact.

14.24 Two spheres, each of mass m, can slide freely on a frictionless, horizontal surface. Sphere A is moving at a speed $v_0 = 16$ ft/s when it strikes sphere B which is at rest and the impact causes sphere B to break into two pieces, each of mass $m/2$. Knowing that 0.7 s after the collision one piece reaches point C and 0.9 s after the collision the other piece reaches point D, determine (a) the velocity of sphere A after the collision, (b) the angle θ and the speeds of the two pieces after the collision.

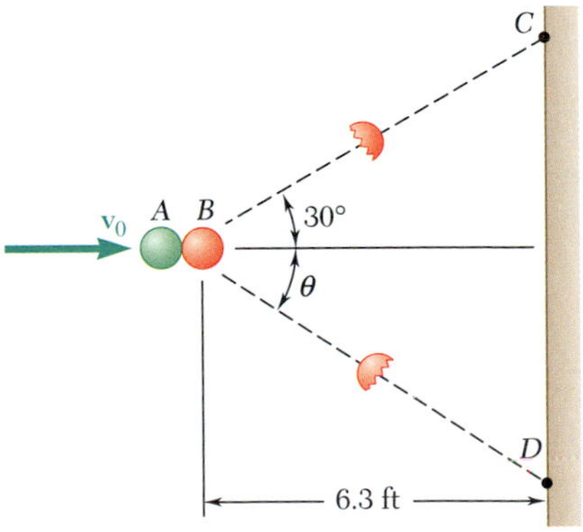

Fig. P14.24

14.25 An 18-lb shell moving with a velocity $\mathbf{v}_0 = (60 \text{ ft/s})\mathbf{i} - (45 \text{ ft/s})\mathbf{j} - (1800 \text{ ft/s})\mathbf{k}$ explodes at point D into three fragments A, B, and C weighing, respectively, 8 lb, 6 lb, and 4 lb. Knowing that the fragments hit the vertical wall at the points indicated, determine the speed of each fragment immediately after the explosion.

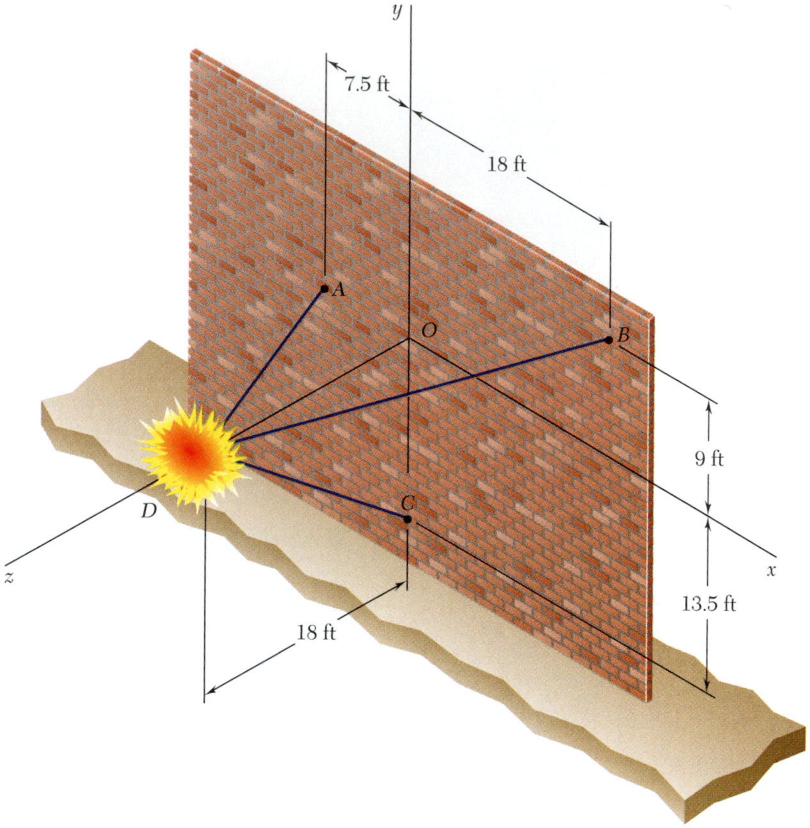

Fig. *P14.25* and *P14.26*

14.26 An 18-lb shell moving with a velocity $\mathbf{v}_0 = (60 \text{ ft/s})\mathbf{i} - (45 \text{ ft/s})\mathbf{j} - (1800 \text{ ft/s})\mathbf{k}$ explodes at point D into three fragments A, B, and C weighing, respectively, 6 lb, 4 lb, and 8 lb. Knowing that the fragments hit the vertical wall at the points indicated, determine the speed of each fragment immediately after the explosion.

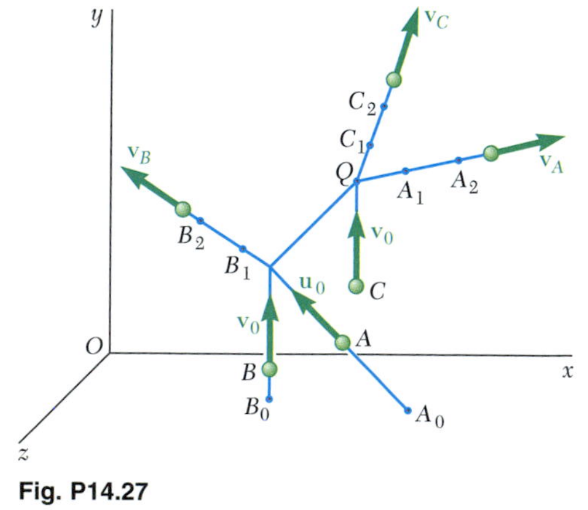

Fig. P14.27

14.27 In a scattering experiment, an alpha particle A is projected with the velocity $\mathbf{u}_0 = -(600 \text{ m/s})\mathbf{i} + (750 \text{ m/s})\mathbf{j} - (800 \text{ m/s})\mathbf{k}$ into a stream of oxygen nuclei moving with a common velocity $\mathbf{v}_0 = (600 \text{ m/s})\mathbf{j}$. After colliding successively with nuclei B and C, particle A is observed to move along the path defined by the points A_1 (280, 240, 120) and A_2 (360, 320, 160), while nuclei B and C are observed to move along paths defined, respectively, by B_1 (147, 220, 130) and B_2 (114, 290, 120), and by C_1 (240, 232, 90) and C_2 (240, 280, 75). All paths are along straight lines and all coordinates are expressed in millimeters. Knowing that the mass of an oxygen nucleus is four times that of an alpha particle, determine the speed of each of the three particles after the collisions.

14.28 Derive the relation

$$\mathbf{H}_O = \bar{\mathbf{r}} \times m\bar{\mathbf{v}} + \mathbf{H}_G$$

between the angular momenta $\mathbf{H}_O$ and $\mathbf{H}_G$ defined in Eqs. (14.7) and (14.24), respectively. The vectors $\bar{\mathbf{r}}$ and $\bar{\mathbf{v}}$ define, respectively, the position and velocity of the mass center G of the system of particles relative to the newtonian frame of reference $Oxyz$, and m represents the total mass of the system.

14.29 Consider the frame of reference $Ax'y'z'$ in translation with respect to the newtonian frame of reference $Oxyz$. We define the angular momentum $\mathbf{H}'_A$ of a system of n particles about A as the sum

$$\mathbf{H}'_A = \sum_{i=1}^{n} \mathbf{r}'_i \times m_i\mathbf{v}'_i \tag{1}$$

of the moments about A of the momenta $m_i\mathbf{v}'_i$ of the particles in their motion relative to the frame $Ax'y'z'$. Denoting by $\mathbf{H}_A$ the sum

$$\mathbf{H}_A = \sum_{i=1}^{n} \mathbf{r}'_i \times m_i\mathbf{v}_i \tag{2}$$

of the moments about A of the momenta $m_i\mathbf{v}_i$ of the particles in their motion relative to the newtonian frame $Oxyz$, show that $\mathbf{H}_A = \mathbf{H}'_A$ at a given instant if, and only if, one of the following conditions is satisfied at that instant: (*a*) A has zero velocity with respect to the frame $Oxyz$, (*b*) A coincides with the mass center G of the system, (*c*) the velocity $\mathbf{v}_A$ relative to $Oxyz$ is directed along the line AG.

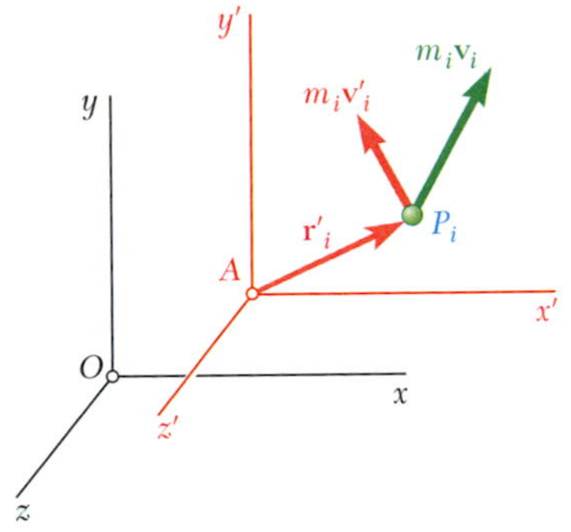

Fig. P14.29

14.30 Show that the relation $\Sigma\mathbf{M}_A = \dot{\mathbf{H}}'_A$, where $\mathbf{H}'_A$ is defined by Eq. (1) of Prob. 14.29 and where $\Sigma\mathbf{M}_A$ represents the sum of the moments about A of the external forces acting on the system of particles, is valid if, and only if, one of the following conditions is satisfied: (*a*) the frame $Ax'y'z'$ is itself a newtonian frame of reference, (*b*) A coincides with the mass center G, (*c*) the acceleration $\mathbf{a}_A$ of A relative to $Oxyz$ is directed along the line AG.

14.7. KINETIC ENERGY OF A SYSTEM OF PARTICLES

The kinetic energy T of a system of particles is defined as the sum of the kinetic energies of the various particles of the system. Referring to Sec. 13.3, we therefore write

$$T = \frac{1}{2}\sum_{i=1}^{n} m_i v_i^2 \tag{14.28}$$

Using a Centroidal Frame of Reference. It is often convenient when computing the kinetic energy of a system comprising a large number of particles (as in the case of a rigid body) to consider separately the motion of the mass center G of the system and the motion of the system relative to a moving frame attached to G.

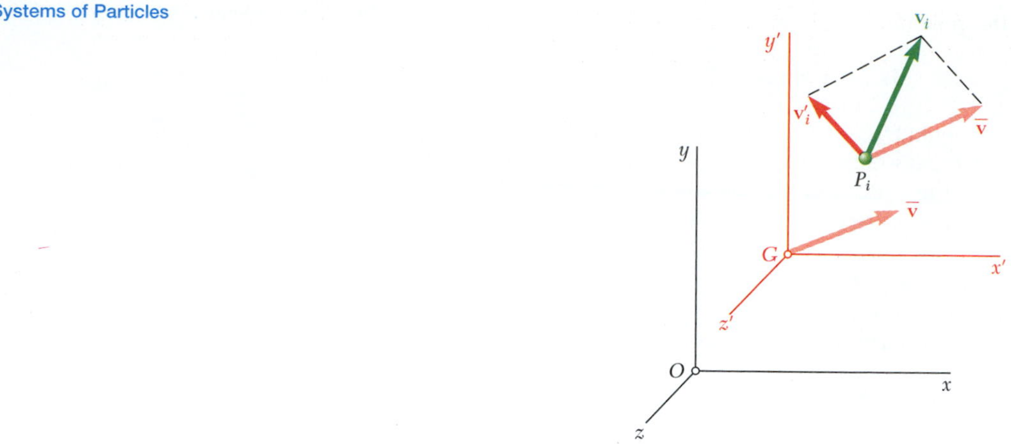

Fig. 14.7

Let P_i be a particle of the system, $\mathbf{v}_i$ its velocity relative to the newtonian frame of reference $Oxyz$, and $\mathbf{v}'_i$ its velocity relative to the moving frame $Gx'y'z'$ which is in translation with respect to $Oxyz$ (Fig. 14.7). We recall from the preceding section that

$$\mathbf{v}_i = \overline{\mathbf{v}} + \mathbf{v}'_i \tag{14.22}$$

where $\overline{\mathbf{v}}$ denotes the velocity of the mass center G relative to the newtonian frame $Oxyz$. Observing that v_i^2 is equal to the scalar product $\mathbf{v}_i \cdot \mathbf{v}_i$, we express the kinetic energy T of the system relative to the newtonian frame $Oxyz$ as follows:

$$T = \frac{1}{2} \sum_{i=1}^{n} m_i v_i^2 = \frac{1}{2} \sum_{i=1}^{n} (m_i \mathbf{v}_i \cdot \mathbf{v}_i)$$

or, substituting for $\mathbf{v}_i$ from (14.22),

$$T = \frac{1}{2} \sum_{i=1}^{n} [m_i(\overline{\mathbf{v}} + \mathbf{v}'_i) \cdot (\overline{\mathbf{v}} + \mathbf{v}'_i)]$$

$$= \frac{1}{2} \left(\sum_{i=1}^{n} m_i \right) \overline{v}^2 + \overline{\mathbf{v}} \cdot \sum_{i=1}^{n} m_i \mathbf{v}'_i + \frac{1}{2} \sum_{i=1}^{n} m_i v_i'^2$$

The first sum represents the total mass m of the system. Recalling Eq. (14.13), we note that the second sum is equal to $m\overline{\mathbf{v}}'$ and thus to zero, since $\overline{\mathbf{v}}'$, which represents the velocity of G relative to the frame $Gx'y'z'$, is clearly zero. We therefore write

$$T = \tfrac{1}{2} m \overline{v}^2 + \frac{1}{2} \sum_{i=1}^{n} m_i v_i'^2 \tag{14.29}$$

This equation shows that the kinetic energy T of a system of particles can be obtained *by adding the kinetic energy of the mass center G* (assuming the entire mass concentrated at G) *and the kinetic energy of the system in its motion relative to the frame $Gx'y'z'$.*

14.8. WORK-ENERGY PRINCIPLE. CONSERVATION OF ENERGY FOR A SYSTEM OF PARTICLES

The principle of work and energy can be applied to each particle P_i of a system of particles. We write

$$T_1 + U_{1\to 2} = T_2 \tag{14.30}$$

for each particle P_i, where $U_{1\to 2}$ represents the work done by the internal forces $\mathbf{f}_{ij}$ and the resultant external force $\mathbf{F}_i$ acting on P_i. Adding the kinetic energies of the various particles of the system and considering the work of all the forces involved, we can apply Eq. (14.30) to the entire system. The quantities T_1 and T_2 now represent the kinetic energy of the entire system and can be computed from either Eq. (14.28) or Eq. (14.29). The quantity $U_{1\to 2}$ represents the work of all the forces acting on the particles of the system. Note that while the internal forces $\mathbf{f}_{ij}$ and $\mathbf{f}_{ji}$ are equal and opposite, the work of these forces will not, in general, cancel out, since the particles P_i and P_j on which they act will, in general, undergo different displacements. Therefore, in computing $U_{1\to 2}$, *we must consider the work of the internal forces* $\mathbf{f}_{ij}$ *as well as the work of the external forces* $\mathbf{F}_i$.

If all the forces acting on the particles of the system are conservative, Eq. (14.30) can be replaced by

$$T_1 + V_1 = T_2 + V_2 \tag{14.31}$$

where V represents the potential energy associated with the internal and external forces acting on the particles of the system. Equation (14.31) expresses the principle of *conservation of energy* for the system of particles.

14.9. PRINCIPLE OF IMPULSE AND MOMENTUM FOR A SYSTEM OF PARTICLES

Integrating Eqs. (14.10) and (14.11) in t from t_1 to t_2, we write

$$\sum \int_{t_1}^{t_2} \mathbf{F}\, dt = \mathbf{L}_2 - \mathbf{L}_1 \tag{14.32}$$

$$\sum \int_{t_1}^{t_2} \mathbf{M}_O\, dt = (\mathbf{H}_O)_2 - (\mathbf{H}_O)_1 \tag{14.33}$$

Recalling the definition of the linear impulse of a force given in Sec. 13.10, we observe that the integrals in Eq. (14.32) represent the linear impulses of the external forces acting on the particles of the system. We shall refer in a similar way to the integrals in Eq. (14.33) as the *angular impulses* about O of the external forces. Thus, Eq. (14.32) expresses that the sum of the linear impulses of the external forces acting on the system is equal to the change in linear momentum of the system. Similarly, Eq. (14.33) expresses that the sum of the angular impulses about O of the external forces is equal to the change in angular momentum about O of the system.

Photo 14.2 Because the concrete block is a system of particles with no initial momentum, the linear momentum $\mathbf{L}_2$ equals the sum of the linear impulses of the external forces acting on the system.

In order to make clear the physical significance of Eqs. (14.32) and (14.33), we will rearrange the terms in these equations and write

$$\mathbf{L}_1 + \sum \int_{t_1}^{t_2} \mathbf{F}\, dt = \mathbf{L}_2 \tag{14.34}$$

$$(\mathbf{H}_O)_1 + \sum \int_{t_1}^{t_2} \mathbf{M}_O\, dt = (\mathbf{H}_O)_2 \tag{14.35}$$

In parts a and c of Fig. 14.8 we have sketched the momenta of the particles of the system at times t_1 and t_2, respectively. In part b we have shown a vector equal to the sum of the linear impulses of the external forces and a couple of moment equal to the sum of the angular impulses about O of the external forces. For simplicity, the particles have

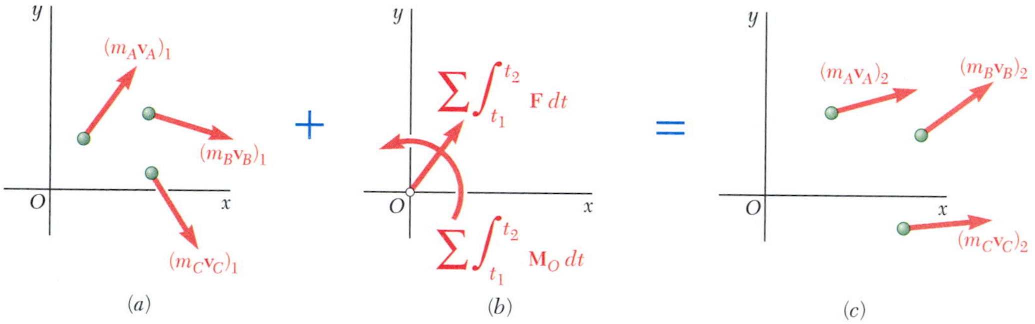

Fig. 14.8

been assumed to move in the plane of the figure, but the present discussion remains valid in the case of particles moving in space. Recalling from Eq. (14.6) that $\mathbf{L}$, by definition, is the resultant of the momenta $m_i\mathbf{v}_i$, we note that Eq. (14.34) expresses that the resultant of the vectors shown in parts a and b of Fig. 14.8 is equal to the resultant of the vectors shown in part c of the same figure. Recalling from Eq. (14.7) that $\mathbf{H}_O$ is the moment resultant of the momenta $m_i\mathbf{v}_i$, we note that Eq. (14.35) similarly expresses that the moment resultant of the vectors in parts a and b of Fig. 14.8 is equal to the moment resultant of the vectors in part c. Together, Eqs. (14.34) and (14.35) thus express that *the momenta of the particles at time t_1 and the impulses of the external forces from t_1 to t_2 form a system of vectors equipollent to the system of the momenta of the particles at time t_2*. This has been indicated in Fig. 14.8 by the use of blue plus and equals signs.

If no external force acts on the particles of the system, the integrals in Eqs. (14.34) and (14.35) are zero, and these equations yield

$$\mathbf{L}_1 = \mathbf{L}_2 \tag{14.36}$$

$$(\mathbf{H}_O)_1 = (\mathbf{H}_O)_2 \tag{14.37}$$

We thus check the result obtained in Sec. 14.6: If no external force acts on the particles of a system, the linear momentum and the angular momentum about O of the system of particles are conserved. The system of the initial momenta is equipollent to the system of the final momenta, and it follows that the angular momentum of the system of particles about *any* fixed point is conserved.

SAMPLE PROBLEM 14.3

For the 200-kg space vehicle of Sample Prob. 14.1, it is known that at t = 2.5 s, the velocity of part A is $\mathbf{v}_A = (270 \text{ m/s})\mathbf{i} - (120 \text{ m/s})\mathbf{j} + (160 \text{ m/s})\mathbf{k}$ and the velocity of part B is parallel to the xz plane. Determine the velocity of part C.

SOLUTION

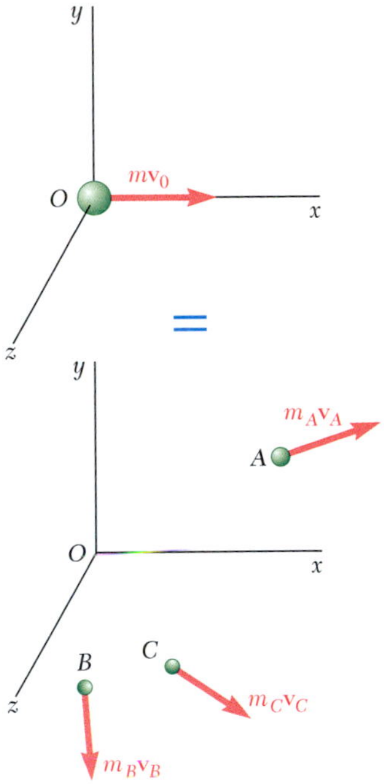

Since there is no external force, the initial momentum $m\mathbf{v}_0$ is equipollent to the system of the final momenta. Equating first the sums of the vectors in both parts of the adjoining sketch, and then the sums of their moments about O, we write

$$\mathbf{L}_1 = \mathbf{L}_2: \qquad m\mathbf{v}_0 = m_A\mathbf{v}_A + m_B\mathbf{v}_B + m_C\mathbf{v}_C \tag{1}$$

$$(\mathbf{H}_O)_1 = (\mathbf{H}_O)_2: \qquad 0 = \mathbf{r}_A \times m_A\mathbf{v}_A + \mathbf{r}_B \times m_B\mathbf{v}_B + \mathbf{r}_C \times m_C\mathbf{v}_C \tag{2}$$

Recalling from Sample Prob. 14.1 that $\mathbf{v}_0 = (150 \text{ m/s})\mathbf{i}$,

$$m_A = 100 \text{ kg} \qquad m_B = 60 \text{ kg} \qquad m_C = 40 \text{ kg}$$
$$\mathbf{r}_A = (555 \text{ m})\mathbf{i} - (180 \text{ m})\mathbf{j} + (240 \text{ m})\mathbf{k}$$
$$\mathbf{r}_B = (255 \text{ m})\mathbf{i} - (120 \text{ m})\mathbf{k}$$
$$\mathbf{r}_C = (105 \text{ m})\mathbf{i} + (450 \text{ m})\mathbf{j} - (420 \text{ m})\mathbf{k}$$

and using the information given in the statement of this problem, we rewrite Eqs. (1) and (2) as follows:

$$200(150\mathbf{i}) = 100(270\mathbf{i} - 120\mathbf{j} + 160\mathbf{k}) + 60[(v_B)_x\mathbf{i} + (v_B)_z\mathbf{k}] + 40[(v_C)_x\mathbf{i} + (v_C)_y\mathbf{j} + (v_C)_z\mathbf{k}] \tag{1$'$}$$

$$0 = 100\begin{vmatrix} \mathbf{i} & \mathbf{j} & \mathbf{k} \\ 555 & -180 & 240 \\ 270 & -120 & 160 \end{vmatrix} + 60\begin{vmatrix} \mathbf{i} & \mathbf{j} & \mathbf{k} \\ 255 & 0 & -120 \\ (v_B)_x & 0 & (v_B)_z \end{vmatrix} + 40\begin{vmatrix} \mathbf{i} & \mathbf{j} & \mathbf{k} \\ 105 & 450 & -420 \\ (v_C)_x & (v_C)_y & (v_C)_z \end{vmatrix} \tag{2$'$}$$

Equating to zero the coefficient of $\mathbf{j}$ in (1′) and the coefficients of $\mathbf{i}$ and $\mathbf{k}$ in (2′), we write, after reductions, the three scalar equations

$$(v_C)_y - 300 = 0$$
$$450(v_C)_z + 420(v_C)_y = 0$$
$$105(v_C)_y - 450(v_C)_x - 45\,000 = 0$$

which yield, respectively,

$$(v_C)_y = 300 \qquad (v_C)_z = -280 \qquad (v_C)_x = -30$$

The velocity of part C is thus

$$\mathbf{v}_C = -(30 \text{ m/s})\mathbf{i} + (300 \text{ m/s})\mathbf{j} - (280 \text{ m/s})\mathbf{k} \quad ◀$$

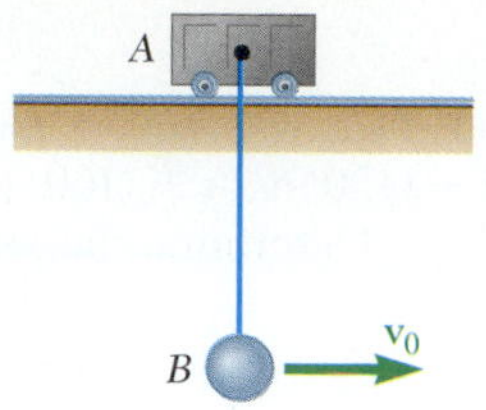

SAMPLE PROBLEM 14.4

Ball B, of mass m_B, is suspended from a cord of length l attached to cart A, of mass m_A, which can roll freely on a frictionless horizontal track. If the ball is given an initial horizontal velocity $\mathbf{v}_0$ while the cart is at rest, determine (*a*) the velocity of B as it reaches its maximum elevation, (*b*) the maximum vertical distance h through which B will rise. (It is assumed that $v_0^2 < 2gl$.)

SOLUTION

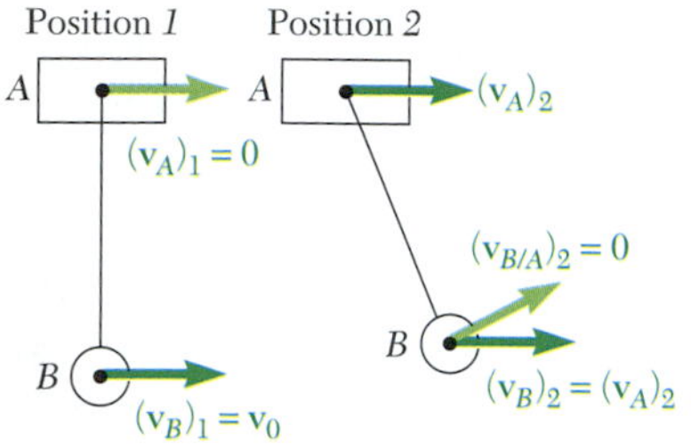

The impulse-momentum principle and the principle of conservation of energy will be applied to the cart-ball system between its initial position *1* and position *2*, when B reaches its maximum elevation.

Velocities ***Position 1:*** $(\mathbf{v}_A)_1 = 0 \qquad (\mathbf{v}_B)_1 = \mathbf{v}_0$ (1)

Position 2: When ball B reaches its maximum elevation, its velocity $(\mathbf{v}_{B/A})_2$ relative to its support A is zero. Thus, at that instant, its absolute velocity is

$$(\mathbf{v}_B)_2 = (\mathbf{v}_A)_2 + (\mathbf{v}_{B/A})_2 = (\mathbf{v}_A)_2 \tag{2}$$

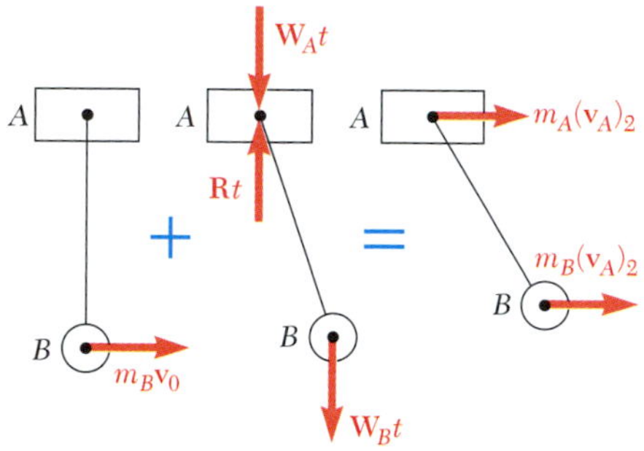

Impulse-Momentum Principle. Noting that the external impulses consist of $\mathbf{W}_At$, $\mathbf{W}_Bt$, and $\mathbf{R}t$, where $\mathbf{R}$ is the reaction of the track on the cart, and recalling (1) and (2), we draw the impulse-momentum diagram and write

$$\Sigma m\mathbf{v}_1 + \Sigma \text{ Ext Imp}_{1\to 2} = \Sigma m\mathbf{v}_2$$

$\xrightarrow{+}$ x components: $\qquad m_Bv_0 = (m_A + m_B)(v_A)_2$

which expresses that the linear momentum of the system is conserved in the horizontal direction. Solving for $(v_A)_2$:

$$(v_A)_2 = \frac{m_B}{m_A + m_B}v_0 \qquad (\mathbf{v}_B)_2 = (\mathbf{v}_A)_2 = \frac{m_B}{m_A + m_B}v_0 \rightarrow \blacktriangleleft$$

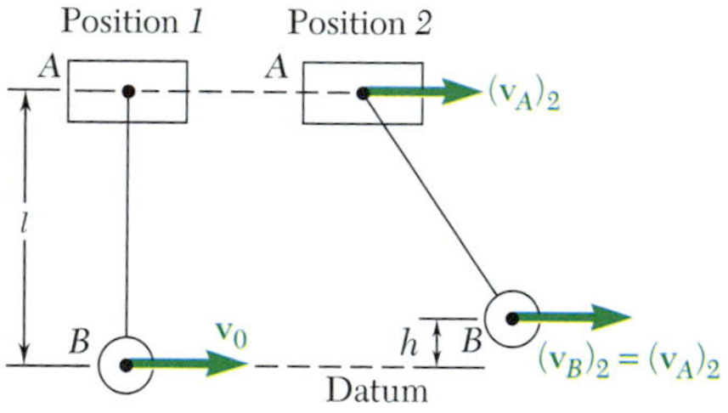

Conservation of Energy

Position 1.	***Potential Energy:***	$V_1 = m_Agl$
	Kinetic Energy:	$T_1 = \frac{1}{2}m_Bv_0^2$
Position 2.	***Potential Energy:***	$V_2 = m_Agl + m_Bgh$
	Kinetic Energy:	$T_2 = \frac{1}{2}(m_A + m_B)(v_A)_2^2$

$T_1 + V_1 = T_2 + V_2$: $\quad \frac{1}{2}m_Bv_0^2 + m_Agl = \frac{1}{2}(m_A + m_B)(v_A)_2^2 + m_Agl + m_Bgh$

Solving for h, we have

$$h = \frac{v_0^2}{2g} - \frac{m_A + m_B}{m_B}\frac{(v_A)_2^2}{2g}$$

or, substituting for $(v_A)_2$ the expression found above,

$$h = \frac{v_0^2}{2g} - \frac{m_B}{m_A + m_B}\frac{v_0^2}{2g} \qquad h = \frac{m_A}{m_A + m_B}\frac{v_0^2}{2g} \blacktriangleleft$$

Remarks. (1) Recalling that $v_0^2 < 2gl$, it follows from the last equation that $h < l$; we thus check that B stays below A as assumed in our solution.

(2) For $m_A \gg m_B$, the answers obtained reduce to $(\mathbf{v}_B)_2 = (\mathbf{v}_A)_2 = 0$ and $h = v_0^2/2g$; B oscillates as a simple pendulum with A fixed. For $m_A \ll m_B$, they reduce to $(\mathbf{v}_B)_2 = (\mathbf{v}_A)_2 = \mathbf{v}_0$ and $h = 0$; A and B move with the same constant velocity $\mathbf{v}_0$.

SAMPLE PROBLEM 14.5

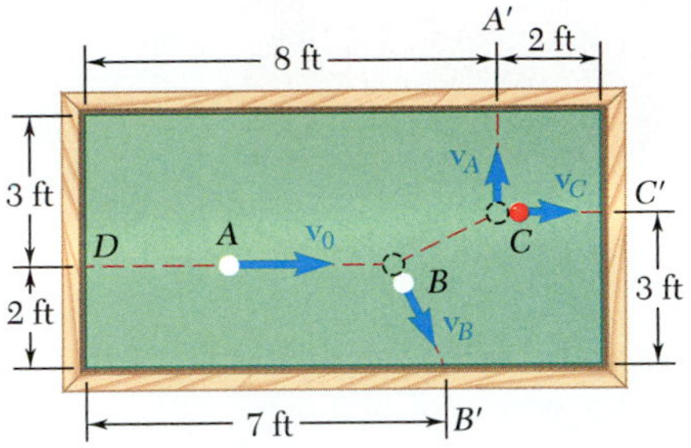

In a game of billiards, ball A is given an initial velocity $\mathbf{v}_0$ of magnitude $v_0 = 10$ ft/s along line DA parallel to the axis of the table. It hits ball B and then ball C, which are both at rest. Knowing that A and C hit the sides of the table squarely at points A' and C', respectively, that B hits the side obliquely at B', and assuming frictionless surfaces and perfectly elastic impacts, determine the velocities $\mathbf{v}_A$, $\mathbf{v}_B$, and $\mathbf{v}_C$ with which the balls hit the sides of the table. (*Remark.* In this sample problem and in several of the problems which follow, the billiard balls are assumed to be particles moving freely in a horizontal plane, rather than the rolling and sliding spheres they actually are.)

SOLUTION

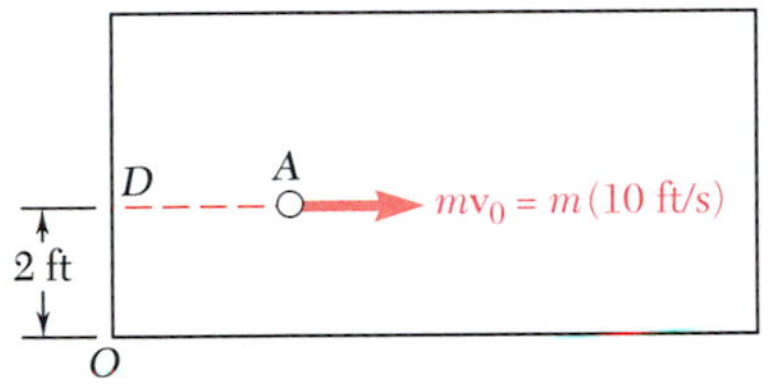

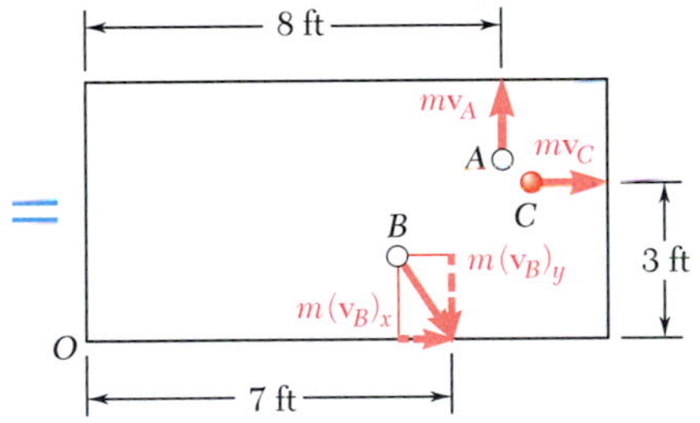

Conservation of Momentum. Since there is no external force, the initial momentum $m\mathbf{v}_0$ is equipollent to the system of momenta after the two collisions (and before any of the balls hits the side of the table). Referring to the adjoining sketch, we write

$$\xrightarrow{+} x \text{ components:} \qquad m(10 \text{ ft/s}) = m(v_B)_x + mv_C \tag{1}$$

$$+\uparrow y \text{ components:} \qquad 0 = mv_A - m(v_B)_y \tag{2}$$

$$+\circlearrowleft \text{ moments about } O\text{:} \quad -(2 \text{ ft})m(10 \text{ ft/s}) = (8 \text{ ft})mv_A - (7 \text{ ft})m(v_B)_y - (3 \text{ ft})mv_C \tag{3}$$

Solving the three equations for v_A, $(v_B)_x$, and $(v_B)_y$ in terms of v_C,

$$v_A = (v_B)_y = 3v_C - 20 \qquad (v_B)_x = 10 - v_C \tag{4}$$

Conservation of Energy. Since the surfaces are frictionless and the impacts are perfectly elastic, the initial kinetic energy $\frac{1}{2}mv_0^2$ is equal to the final kinetic energy of the system:

$$\tfrac{1}{2}mv_0^2 = \tfrac{1}{2}m_A v_A^2 + \tfrac{1}{2}m_B v_B^2 + \tfrac{1}{2}m_C v_C^2$$

$$v_A^2 + (v_B)_x^2 + (v_B)_y^2 + v_C^2 = (10 \text{ ft/s})^2 \tag{5}$$

Substituting for v_A, $(v_B)_x$, and $(v_B)_y$ from (4) into (5), we have

$$2(3v_C - 20)^2 + (10 - v_C)^2 + v_C^2 = 100$$

$$20v_C^2 - 260v_C + 800 = 0$$

Solving for v_C, we find $v_C = 5$ ft/s and $v_C = 8$ ft/s. Since only the second root yields a positive value for v_A after substitution into Eqs. (4), we conclude that $v_C = 8$ ft/s and

$$v_A = (v_B)_y = 3(8) - 20 = 4 \text{ ft/s} \qquad (v_B)_x = 10 - 8 = 2 \text{ ft/s}$$

$\mathbf{v}_A = 4$ ft/s $\uparrow$ $\qquad \mathbf{v}_B = 4.47$ ft/s $\measuredangle$ 63.4° $\qquad \mathbf{v}_C = 8$ ft/s $\rightarrow$ ◀

SOLVING PROBLEMS ON YOUR OWN

In the preceding lesson we defined the linear momentum and the angular momentum of a system of particles. In this lesson we defined the *kinetic energy T* of a system of particles:

$$T = \frac{1}{2}\sum_{i=1}^{n} m_i v_i^2 \qquad (14.28)$$

The solutions of the problems in the preceding lesson were based on the conservation of the linear momentum of a system of particles or on the observation of the motion of the mass center of a system of particles. In this lesson you will solve problems involving the following:

1. Computation of the kinetic energy lost in collisions. The kinetic energy T_1 of the system of particles before the collisions and its kinetic energy T_2 after the collisions are computed from Eq. (14.28) and are subtracted from each other. Keep in mind that, while linear momentum and angular momentum are vector quantities, kinetic energy is a *scalar* quantity.

2. Conservation of linear momentum and conservation of energy. As you saw in the preceding lesson, when the resultant of the external forces acting on a system of particles is zero, the linear momentum of the system is conserved. In problems involving two-dimensional motion, expressing that the initial linear momentum and the final linear momentum of the system are equipollent yields two algebraic equations. Equating the initial total energy of the system of particles (including potential energy as well as kinetic energy) to its final total energy yields an additional equation. Thus, you can write three equations which can be solved for three unknowns [Sample Prob. 14.5]. Note that if the resultant of the external forces is not zero but has a fixed direction, the component of the linear momentum in a direction perpendicular to the resultant is still conserved; the number of equations which can be used is then reduced to two [Sample Prob. 14.4].

3. Conservation of linear and angular momentum. When no external forces act on a system of particles, both the linear momentum of the system and its angular momentum about some arbitrary point are conserved. In the case of three-dimensional motion, this will enable you to write as many as six equations, although you may need to solve only some of them to obtain the desired answers [Sample Prob. 14.3]. In the case of two-dimensional motion, you will be able to write three equations which can be solved for three unknowns.

4. Conservation of linear and angular momentum and conservation of energy. In the case of the two-dimensional motion of a system of particles which are not subjected to any external forces, you will obtain two algebraic equations by expressing that the linear momentum of the system is conserved, one equation by writing that the angular momentum of the system about some arbitrary point is conserved, and a fourth equation by expressing that the total energy of the system is conserved. These equations can be solved for four unknowns.

Problems

14.31 In Prob. 14.1, determine the total work done by the engines of cars A and B while the cars are accelerating to reach constant speeds.

14.32 In Prob. 14.5, determine the energy lost as the bullet (*a*) passes through block A, (*b*) becomes embedded in block B.

14.33 Assuming that the airline employee of Prob. 14.3 first tosses the 30-lb suitcase on the baggage carrier, determine the energy lost (*a*) as the first suitcase hits the carrier, (*b*) as the second suitcase hits the carrier.

14.34 Determine the energy lost as a result of the series of collisions described in Prob. 14.7, knowing that each car weighs 3000 lb.

14.35 Two automobiles A and B, of mass m_A and m_B, respectively, are traveling in opposite directions when they collide head on. The impact is assumed perfectly plastic, and it is further assumed that the energy absorbed by each automobile is equal to its loss of kinetic energy with respect to a moving frame of reference attached to the mass center of the two-vehicle system. Denoting by E_A and E_B, respectively, the energy absorbed by automobile A and by automobile B, (*a*) show that $E_A/E_B = m_B/m_A$, that is, the amount of energy absorbed by each vehicle is inversely proportional to its mass, (*b*) compute E_A and E_B, knowing that $m_A = 2400$ kg and $m_B = 1350$ kg and that the speeds of A and B are, respectively, 135 km/h and 90 km/h.

Fig. *P14.35*

14.36 It is assumed that each of the two automobiles involved in the collision described in Prob. 14.35 had been designed to safely withstand a test in which it crashed into a solid, immovable wall at the speed v_0. The severity of the collision of Prob. 14.35 may then be measured for each vehicle by the ratio of the energy absorbed in the collision to the energy it absorbed in the test. On that basis, show that the collision described in Prob. 14.35 is $(m_A/m_B)^2$ times more severe for automobile B than for automobile A.

14.37 Solve Sample Prob. 14.4, assuming that cart A is given an initial horizontal velocity $\mathbf{v}_0$ while ball B is at rest.

14.38 In a game of pool, ball A is moving with a velocity $\mathbf{v}_0 = v_0\mathbf{i}$ when it strikes balls B and C, which are at rest side by side. Assuming frictionless surfaces and perfectly elastic impact (that is, conservation of energy), determine the final velocity of each ball, assuming that the path of A is (*a*) perfectly centered and that A strikes B and C simultaneously, (*b*) not perfectly centered and that A strikes B slightly before it strikes C.

Fig. P14.38

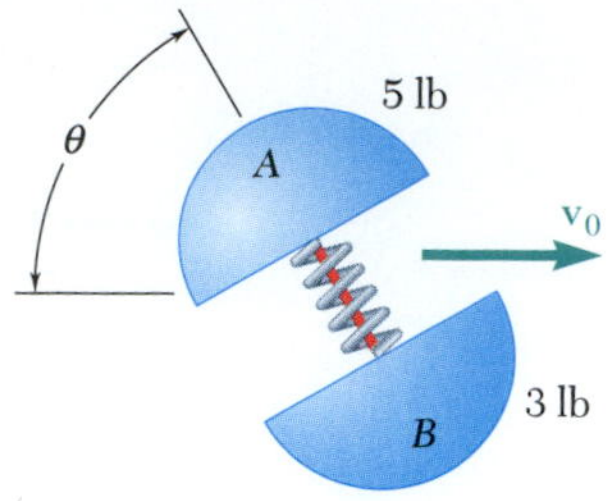

Fig. P14.39

14.39 Two hemispheres are held together by a cord which maintains a spring under compression (the spring is not attached to the hemispheres). The potential energy of the compressed spring is 90 ft · lb and the assembly has an initial velocity $\mathbf{v}_0$ of magnitude $v_0 = 24$ ft/s. Knowing that the cord is severed when $\theta = 30°$, causing the hemispheres to fly apart, determine the resulting velocity of each hemisphere.

14.40 Solve Prob. 14.39, knowing that the cord is severed when $\theta = 120°$.

14.41 and 14.42 In a game of pool, ball A is moving with a velocity $\mathbf{v}_0$ of magnitude $v_0 = 5$ m/s when it strikes balls B and C, which are at rest and aligned as shown. Knowing that after the collision the three balls move in the directions indicated and assuming frictionless surfaces and perfectly elastic impact (that is, conservation of energy), determine the magnitudes of the velocities $\mathbf{v}_A$, $\mathbf{v}_B$, and $\mathbf{v}_C$.

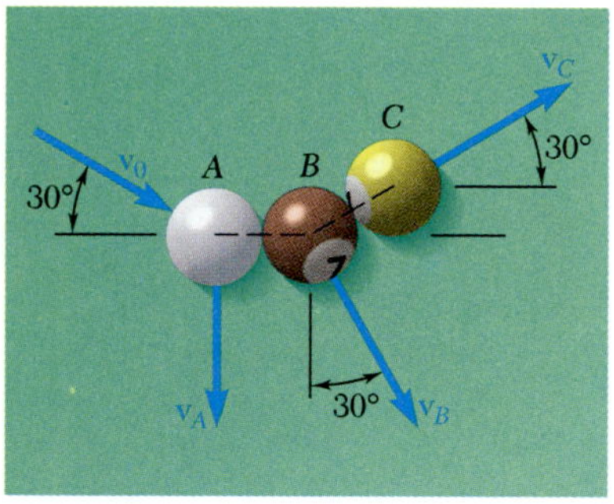

Fig. P14.41

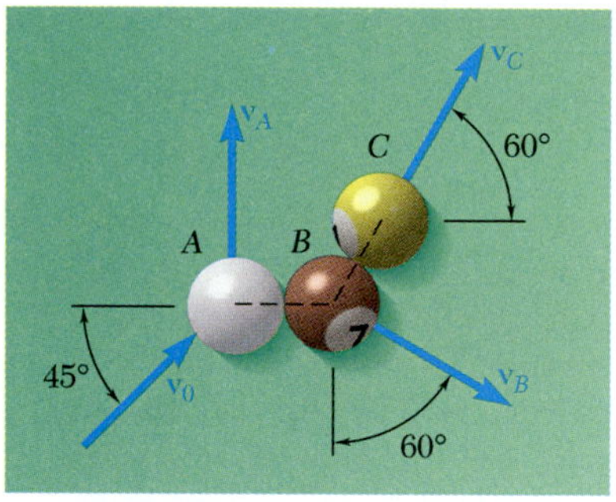

Fig. P14.42

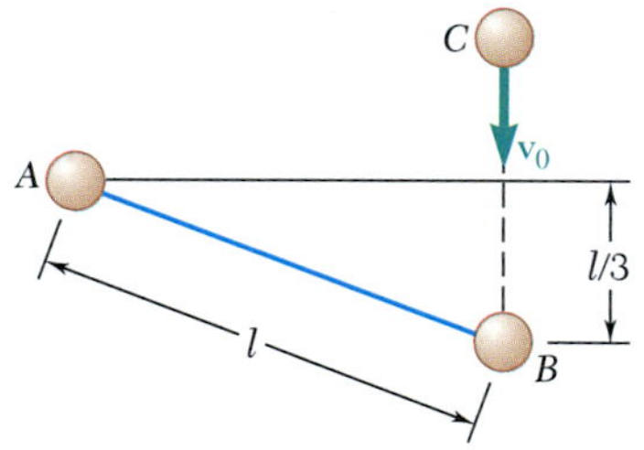

Fig. *P14.43*

14.43 Three spheres, each of mass m, can slide freely on a frictionless, horizontal surface. Spheres A and B are attached to an inextensible, inelastic cord of length l and are at rest in the position shown when sphere B is struck squarely by sphere C which is moving with a velocity $\mathbf{v}_0$. Knowing that the cord is taut when sphere B is struck by sphere C and assuming perfectly elastic impact between B and C, and thus conservation of energy for the entire system, determine the velocity of each sphere immediately after impact.

14.44 Three spheres, each of mass m, can slide freely on a frictionless, horizontal surface. Spheres A and B are attached to an inextensible, inelastic cord of length l and are at rest in the position shown when sphere B is struck squarely by sphere C which is moving with a velocity $\mathbf{v}_0$. Knowing that the cord is slack when sphere B is struck by sphere C and assuming perfectly elastic impact between B and C, determine (a) the velocity of each sphere immediately after the cord becomes taut, (b) the fraction of the initial kinetic energy of the system which is dissipated when the cord becomes taut.

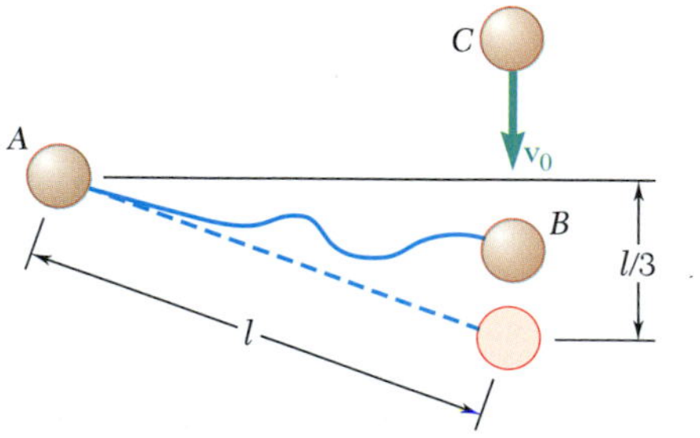

Fig. *P14.44*

14.45 In the scattering experiment of Prob. 14.27, it is known that the alpha particle is projected from A_0 (300, 0, 300) and that it collides with the oxygen nucleus C at Q (240, 200, 100), where all coordinates are expressed in millimeters. Determine the coordinates of point B_0 where the original path of nucleus B intersects the xz plane. (*Hint:* Express that the angular momentum of the three particles about Q is conserved.)

14.46 A 240-kg space vehicle traveling with a velocity $\mathbf{v}_0 = (300 \text{ m/s})\mathbf{k}$ passes through the origin O. Explosive charges then separate the vehicle into three parts A, B, and C, with masses of 40 kg, 80 kg, and 120 kg, respectively. Knowing that shortly thereafter the positions of the three parts are A (48, 48, 432), B (120, 264, 648), and C (−96, −192, 384), where the coordinates are expressed in meters, that the velocity of B is $\mathbf{v}_B = (100 \text{ m/s})\mathbf{i} + (220 \text{ m/s})\mathbf{j} + (440 \text{ m/s})\mathbf{k}$, and that the x component of the velocity of C is −80 m/s, determine the velocity of part A.

14.47 Four small disks A, B, C, and D can slide freely on a frictionless horizontal surface. Disks B, C, and D are connected by light rods and are at rest in the position shown when disk B is struck squarely by disk A which is moving to the right with a velocity $\mathbf{v}_0 = (38.5 \text{ ft/s})\mathbf{i}$. The weights of the disks are $W_A = W_B = W_C = 15$ lb, and $W_D = 30$ lb. Knowing that the velocities of the disks immediately after the impact are $\mathbf{v}_A = \mathbf{v}_B = (8.25 \text{ ft/s})\mathbf{i}$, $\mathbf{v}_C = v_C\mathbf{i}$, and $\mathbf{v}_D = v_D\mathbf{i}$, determine (*a*) the speeds v_C and v_D, (*b*) the fraction of the initial kinetic energy of the system which is dissipated during the collision.

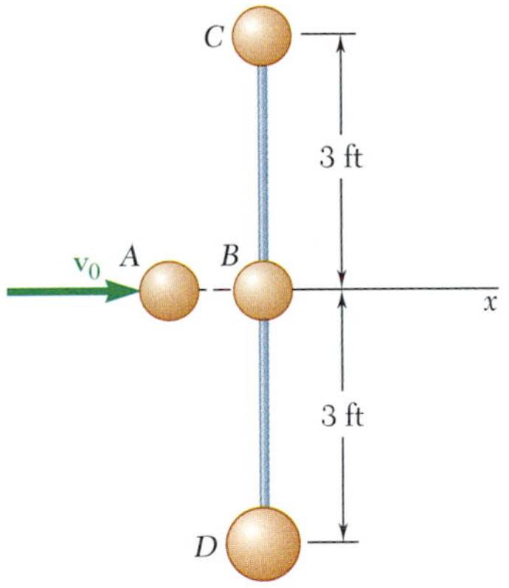

Fig. P14.47 and P14.48

14.48 Four small disks A, B, C, and D can slide freely on a frictionless horizontal surface. Disks B, C, and D are connected by light rods and are at rest in the position shown when disk B is struck squarely by disk A which is moving to the right with a velocity $\mathbf{v}_0 = (38.5 \text{ ft/s})\mathbf{i}$. The weights of the disks are $W_A = W_B = W_C = 15$ lb, and $W_D = 30$ lb. Knowing that the velocities of the disks immediately after the impact are $\mathbf{v}_A = 0$, $\mathbf{v}_B = (10.5 \text{ ft/s})\mathbf{i}$, $\mathbf{v}_C = v_C\mathbf{i}$, and $\mathbf{v}_D = v_D\mathbf{i}$, determine (*a*) the speeds v_C and v_D, (*b*) the fraction of the initial kinetic energy of the system which is dissipated during the collision.

14.49 Two small spheres A and B, with masses of 2.5 kg and 1 kg, respectively, are connected by a rigid rod of negligible mass. The two spheres are resting on a horizontal, frictionless surface when A is suddenly given the velocity $\mathbf{v}_0 = (3.5 \text{ m/s})\mathbf{i}$. Determine (*a*) the linear momentum of the system and its angular momentum about its mass center G, (*b*) the velocities of A and B after the rod AB has rotated through 180°.

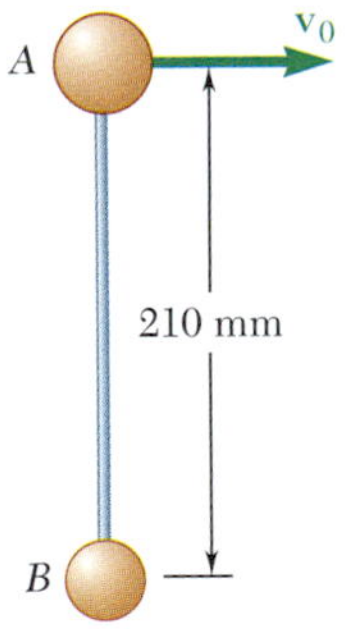

Fig. P14.49

14.50 Solve Prob. 14.49, assuming that it is B which is suddenly given the velocity $\mathbf{v}_0 = (3.5 \text{ m/s})\mathbf{i}$.

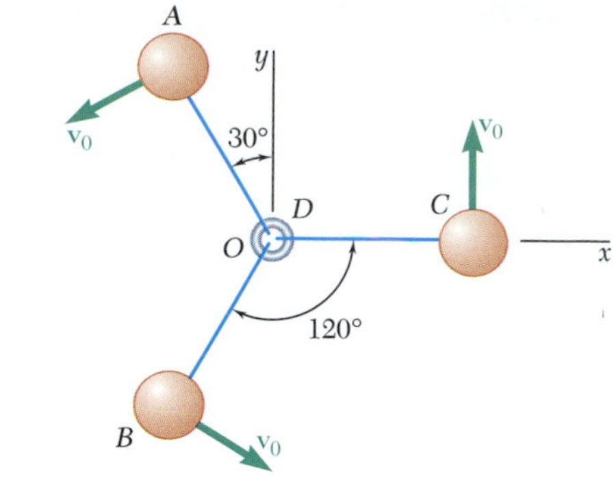

Fig. *P14.51* and *P14.52*

14.51 Three small spheres A, B, and C, each of mass m, are connected to a small ring D of negligible mass by means of three inextensible, inelastic cords of length l which are equally spaced. The spheres can slide freely on a frictionless horizontal surface and are rotating initially at a speed v_0 about ring D which is at rest. Suddenly the cord CD breaks. After the other two cords have again become taut, determine (*a*) the speed of ring D, (*b*) the relative speed at which spheres A and B rotate about D, (*c*) the fraction of the original energy of spheres A and B which is dissipated when cords AD and BD again became taut.

14.52 Three small spheres A, B, and C, each of mass m, are connected to a small ring D of mass $2m$ by means of three inextensible, inelastic cords of length l which are equally spaced. The spheres can slide freely on a frictionless horizontal surface and are rotating initially at a speed v_0 about ring D which is at rest. The cord CD breaks at time $t = 0$ when the spheres and ring are in the position shown. Considering the system which consists of ring D and spheres A and B, determine for $t > 0$ (*a*) the position vector $\bar{\mathbf{r}}$ of the mass center G of the system, (*b*) the angular momentum $\mathbf{H}_G$ of the system about G, (*c*) the angular momentum $\mathbf{H}_O$ of the system about the origin O, (*d*) the kinetic energy T of the system.

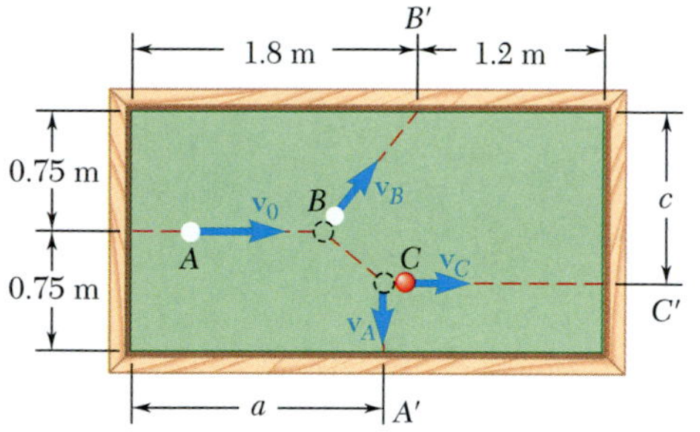

Fig. P14.53

14.53 In a game of billiards, ball A is given an initial velocity $\mathbf{v}_0$ along the longitudinal axis of the table. It hits ball B and then ball C, which are both at rest. Balls A and C are observed to hit the sides of the table squarely at A' and C', respectively, and ball B is observed to hit the side obliquely at B'. Knowing that $v_0 = 4$ m/s, $v_A = 1.92$ m/s, and $a = 1.65$ m, determine (*a*) the velocities $\mathbf{v}_B$ and $\mathbf{v}_C$ of balls B and C, (*b*) the point C' where ball C hits the side of the table. Assume frictionless surfaces and perfectly elastic impacts (that is, conservation of energy).

14.54 For the game of billiards of Prob. 14.53, it is now assumed that $v_0 = 5$ m/s, $v_C = 3.2$ m/s, and $c = 1.22$ m. Determine (*a*) the velocities $\mathbf{v}_A$ and $\mathbf{v}_B$ of balls A and B, (*b*) the point A' where ball A hits the side of the table.

14.55 Three small identical spheres A, B, and C, which can slide on a horizontal, frictionless surface, are attached to three strings of length l which are tied to a ring G. Initially the spheres rotate about the ring which moves along the x axis with a velocity $\mathbf{v}_0$. Suddenly the ring breaks and the three spheres move freely in the xy plane. Knowing that $\mathbf{v}_A = (8.66 \text{ ft/s})\mathbf{j}$, $\mathbf{v}_C = (15 \text{ ft/s})\mathbf{i}$, $a = 0.866$ ft, and $d = 0.5$ ft, determine (*a*) the initial velocity of the ring, (*b*) the length l of the strings, (*c*) the rate in rad/s at which the spheres were rotating about G.

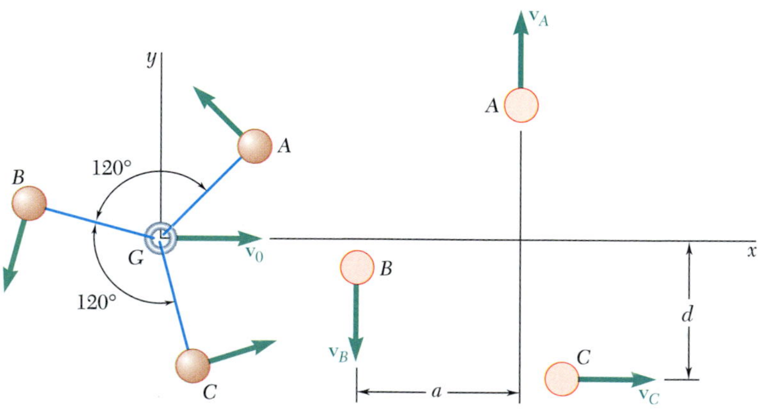

Fig. P14.55 and P14.56

14.56 Three small identical spheres A, B, and C, which can slide on a horizontal, frictionless surface, are attached to three strings, 0.25 ft long, which are tied to a ring G. Initially the spheres rotate counterclockwise about the ring with a relative velocity of 2.5 ft/s and the ring moves along the x axis with a velocity $\mathbf{v}_0 = (1.25 \text{ ft/s})\mathbf{i}$. Suddenly the ring breaks and the three spheres move freely in the xy plane with A and B following paths parallel to the y axis at a distance $a = 0.433$ ft from each other and C following a path parallel to the x axis. Determine (*a*) the velocity of each sphere, (*b*) the distance d.

14.57 Two small disks, A and B, of mass 2.4 kg and 1.2 kg, respectively, can slide on a horizontal, frictionless surface. They are connected by a cord, 900 mm long, and spin counterclockwise about their mass center G at a rate of 8 rad/s. At $t = 0$, the coordinates of G are $\bar{x}_0 = 0$, $\bar{y}_0 = 2.48$ m, and its velocity is $\bar{\mathbf{v}}_0 = (1.92 \text{ m/s})\mathbf{i} + (0.48 \text{ m/s})\mathbf{j}$. Shortly thereafter the cord breaks; disk A is then observed to move along a path parallel to the y axis and disk B along a path which intersects the x axis at a distance $b = 8$ m from O. Determine (*a*) the velocities of A and B after the cord breaks, (*b*) the distance a from the y axis to the path of A.

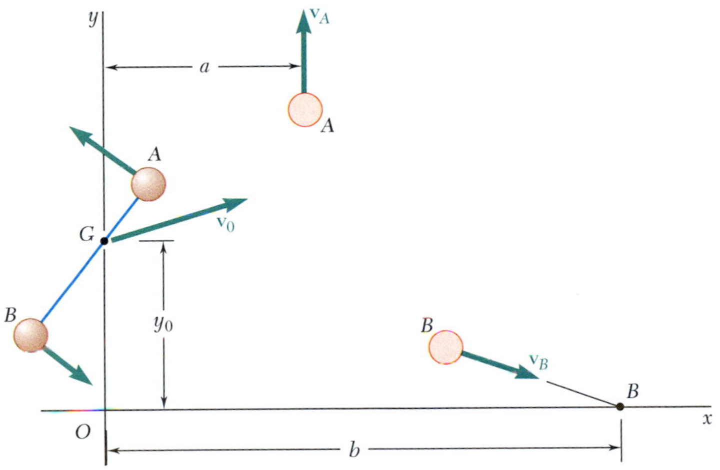

Fig. ***P14.57*** **and** ***P14.58***

14.58 Two small disks, A and B, of mass 3 kg and 1.5 kg, respectively, can slide on a horizontal and frictionless surface. They are connected by a cord of negligible mass and spin about their mass center G. At $t = 0$, G is moving with the velocity $\bar{\mathbf{v}}_0$ and its coordinates are $\bar{x}_0 = 0$, $\bar{y}_0 = 2.5$ m. Shortly thereafter, the cord breaks and disk A is observed to move with a velocity $\bar{\mathbf{v}}_A = (2.56 \text{ m/s})\mathbf{j}$ in a straight line and at a distance $a = 1.861$ m from the y axis, while B moves with a velocity $\bar{\mathbf{v}}_B = (3.6 \text{ m/s})\mathbf{i} - (2.24 \text{ m/s})\mathbf{j}$ along a path intersecting the x axis at a distance $b = 7.2$ m from the origin O. Determine (*a*) the initial velocity $\bar{\mathbf{v}}_0$ of the mass center G of the two disks, (*b*) the length of the cord initially connecting the two disks, (*c*) the rate in rad/s at which the disks were spinning about G.

*14.10. VARIABLE SYSTEMS OF PARTICLES

All the systems of particles considered so far consisted of well-defined particles. These systems did not gain or lose any particles during their motion. In a large number of engineering applications, however, it is necessary to consider *variable systems of particles,* that is, systems which are continually gaining or losing particles, or doing both at the same time. Consider, for example, a hydraulic turbine. Its analysis involves the determination of the forces exerted by a stream of water on rotating blades, and we note that the particles of water in contact with the blades form an everchanging system which continually acquires and loses particles. Rockets furnish another example of variable systems, since their propulsion depends upon the continual ejection of fuel particles.

We recall that all the kinetics principles established so far were derived for constant systems of particles, which neither gain nor lose particles. We must therefore find a way to reduce the analysis of a variable system of particles to that of an auxiliary constant system. The procedure to follow is indicated in Secs. 14.11 and 14.12 for two broad categories of applications: a steady stream of particles and a system that is gaining or losing mass.

*14.11. STEADY STREAM OF PARTICLES

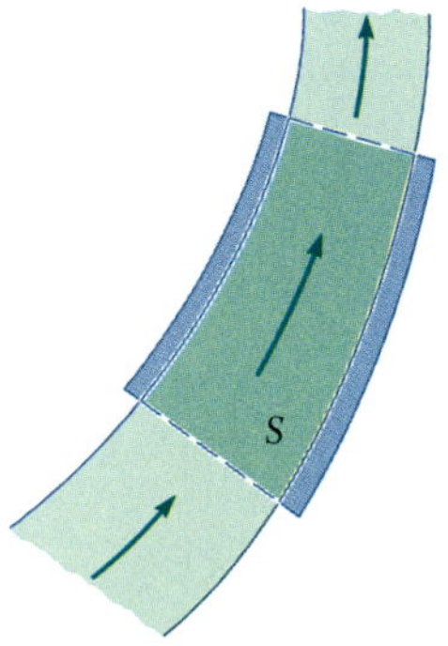

Fig. 14.9

Consider a steady stream of particles, such as a stream of water diverted by a fixed vane or a flow of air through a duct or through a blower. In order to determine the resultant of the forces exerted on the particles in contact with the vane, duct, or blower, we isolate these particles and denote by S the system thus defined (Fig. 14.9). We observe that S is a variable system of particles, since it continually gains particles flowing in and loses an equal number of particles flowing out. Therefore, the kinetics principles that have been established so far cannot be directly applied to S.

However, we can easily define an auxiliary system of particles which does remain constant for a short interval of time Δt. Consider at time t the system S *plus* the particles which will enter S during the interval at time Δt (Fig. 14.10a). Next, consider at time $t + \Delta t$ the system S *plus* the particles which have left S during the interval Δt (Fig. 14.10c). Clearly, *the same particles are involved in both cases,* and we can apply to those particles the principle of impulse and momentum. Since the total mass m of the system S remains constant, the particles entering the system and those leaving the system in the time Δt must have the same mass Δm. Denoting by $\mathbf{v}_A$ and $\mathbf{v}_B$, respectively, the velocities of the particles entering S at A and leaving S at B, we represent the momentum of the particles entering S by $(\Delta m)\mathbf{v}_A$ (Fig. 14.10a) and the momentum of the particles leaving S by $(\Delta m)\mathbf{v}_B$ (Fig. 14.10c). We also represent by appropriate vectors the momenta $m_i\mathbf{v}_i$ of the particles forming S and the impulses of the forces exerted on S and indicate by blue plus and equals signs that the system of the momenta and impulses in parts a and b of Fig. 14.10 is equipollent to the system of the momenta in part c of the same figure.

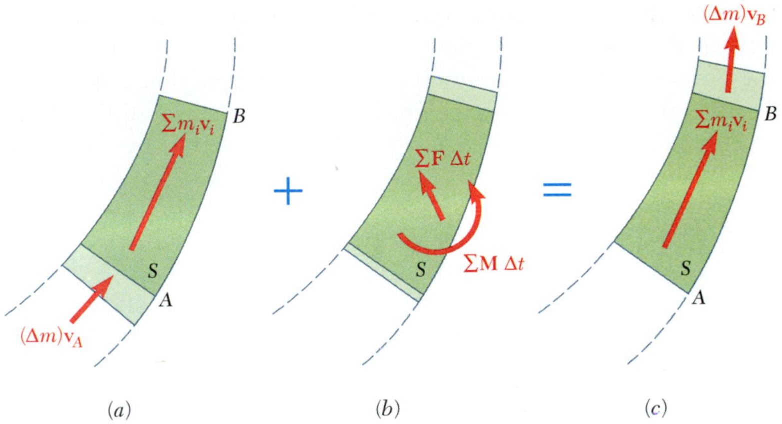

Fig. 14.10

The resultant $\Sigma m_i\mathbf{v}_i$ of the momenta of the particles of S is found on both sides of the equals sign and can thus be omitted. We conclude that *the system formed by the momentum $(\Delta m)\mathbf{v}_A$ of the particles entering S in the time Δt and the impulses of the forces exerted on S during that time is equipollent to the momentum $(\Delta m)\mathbf{v}_B$ of the particles leaving S in the same time Δt*. We can therefore write

$$(\Delta m)\mathbf{v}_A + \Sigma\mathbf{F}\,\Delta t = (\Delta m)\mathbf{v}_B \tag{14.38}$$

A similar equation can be obtained by taking the moments of the vectors involved (see Sample Prob. 14.6). Dividing all terms of Eq. (14.38) by Δt and letting Δt approach zero, we obtain at the limit

$$\Sigma\mathbf{F} = \frac{dm}{dt}(\mathbf{v}_B - \mathbf{v}_A) \tag{14.39}$$

where $\mathbf{v}_B - \mathbf{v}_A$ represents the difference between the *vector* $\mathbf{v}_B$ and the *vector* $\mathbf{v}_A$.

If SI units are used, dm/dt is expressed in kg/s and the velocities in m/s; we check that both members of Eq. (14.39) are expressed in the same units (newtons). If U.S. customary units are used, dm/dt must be expressed in slugs/s and the velocities in ft/s; we check again that both members of the equation are expressed in the same units (pounds).†

The principle we have established can be used to analyze a large number of engineering applications. Some of the more common of these applications will be considered next.

†It is often convenient to express the mass rate of flow dm/dt as the product ρQ, where ρ is the density of the stream (mass per unit volume) and Q its volume rate of flow (volume per unit time). If SI units are used, ρ is expressed in kg/m^3 (for instance, $\rho = 1000$ kg/m^3 for water) and Q in m^3/s. However, if U.S. customary units are used, ρ will generally have to be computed from the corresponding specific weight γ (weight per unit volume), $\rho = \gamma/g$. Since γ is expressed in lb/ft^3 (for instance, $\gamma = 62.4$ lb/ft^3 for water), ρ is obtained in slugs/ft^3. The volume rate of flow Q is expressed in ft^3/s.

Fluid Stream Diverted by a Vane. If the vane is fixed, the method of analysis given above can be applied directly to find the force $\mathbf{F}$ exerted by the vane on the stream. We note that $\mathbf{F}$ is the only force which needs to be considered since the pressure in the stream is constant (atmospheric pressure). The force exerted by the stream on the vane will be equal and opposite to $\mathbf{F}$. If the vane moves with a constant velocity, the stream is not steady. However, it will appear steady to an observer moving with the vane. We should therefore choose a system of axes moving with the vane. Since this system of axes is not accelerated, Eq. (14.38) can still be used, but $\mathbf{v}_A$ and $\mathbf{v}_B$ must be replaced by the *relative velocities* of the stream with respect to the vane (see Sample Prob. 14.7).

Fluid Flowing through a Pipe. The force exerted by the fluid on a pipe transition such as a bend or a contraction can be determined by considering the system of particles S in contact with the transition. Since, in general, the pressure in the flow will vary, the forces exerted on S by the adjoining portions of the fluid should also be considered.

Photo 14.3 The analysis of a jet engine reduces to that of a steady stream of air particles with all velocities measured with respect to the airplane.

Jet Engine. In a jet engine, air enters with no velocity through the front of the engine and leaves through the rear with a high velocity. The energy required to accelerate the air particles is obtained by burning fuel. The mass of the burned fuel in the exhaust gases will usually be small enough compared with the mass of the air flowing through the engine that it can be neglected. Thus, the analysis of a jet engine reduces to that of an airstream. This stream can be considered as a steady stream if all velocities are measured with respect to the airplane. It will be assumed, therefore, that the airstream enters the engine with a velocity $\mathbf{v}$ of magnitude equal to the speed of the airplane and leaves with a velocity $\mathbf{u}$ equal to the relative velocity of the exhaust gases (Fig. 14.11). Since the intake and exhaust pressures are nearly atmospheric, the only external force which needs to be considered is the force exerted by the engine on the airstream. This force is equal and opposite to the thrust.†

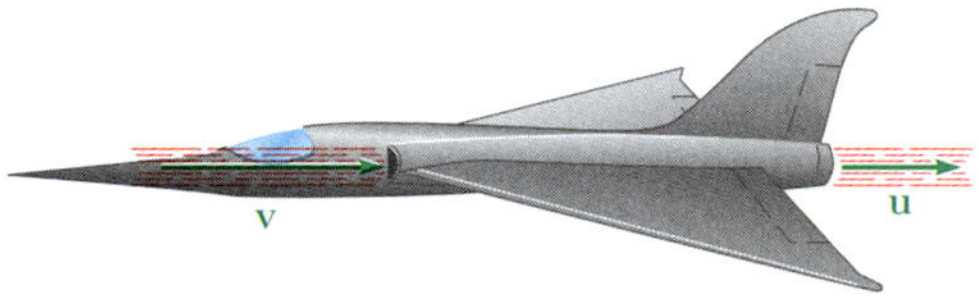

Fig. 14.11

†Note that if the airplane is accelerated, it cannot be used as a newtonian frame of reference. The same result will be obtained for the thrust, however, by using a reference frame at rest with respect to the atmosphere, since the air particles will then be observed to enter the engine with no velocity and to leave it with a velocity of magnitude $u - v$.

Fan. We consider the system of particles S shown in Fig. 14.12. The velocity $\mathbf{v}_A$ of the particles entering the system is assumed equal to zero, and the velocity $\mathbf{v}_B$ of the particles leaving the system is the velocity of the *slipstream*. The rate of flow can be obtained by multiplying v_B by the cross-sectional area of the slipstream. Since the pressure all around S is atmospheric, the only external force acting on S is the thrust of the fan.

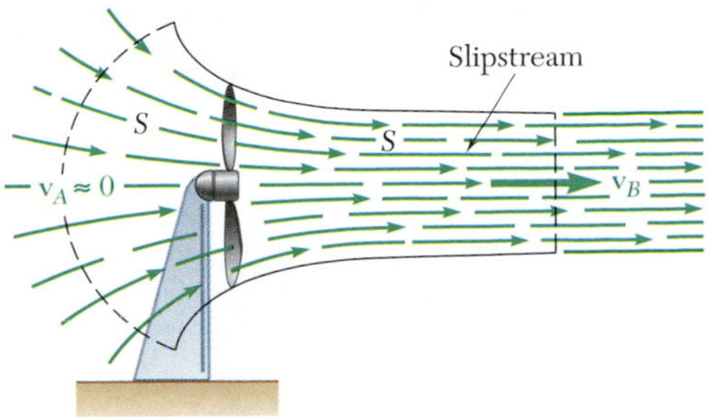

Fig. 14.12

Helicopter. The determination of the thrust created by the rotating blades of a hovering helicopter is similar to the determination of the thrust of a fan. The velocity $\mathbf{v}_A$ of the air particles as they approach the blades is assumed to be zero, and the rate of flow is obtained by multiplying the magnitude of the velocity $\mathbf{v}_B$ of the slipstream by its cross-sectional area.

*14.12. SYSTEMS GAINING OR LOSING MASS

Let us now analyze a different type of variable system of particles, namely, a system which gains mass by continually absorbing particles or loses mass by continually expelling particles. Consider the system S shown in Fig. 14.13. Its mass, equal to m at the instant t, increases

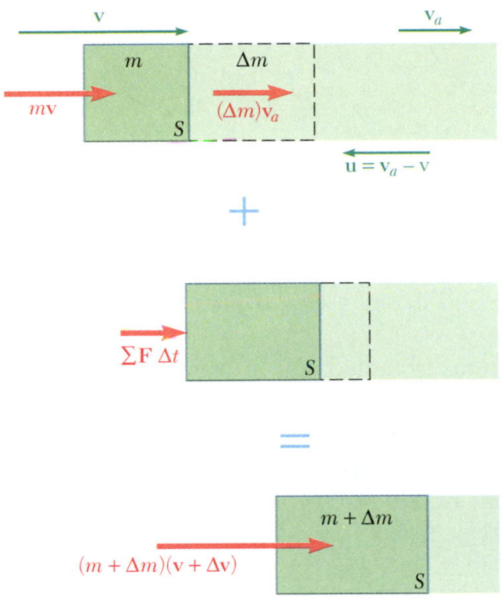

Fig. 14.13

by Δm in the interval of time Δt. In order to apply the principle of impulse and momentum to the analysis of this system, we must consider at time t the system S *plus* the particles of mass Δm which S absorbs during the time interval Δt. The velocity of S at time t is denoted by $\mathbf{v}$, the velocity of S at time $t + \Delta t$ is denoted by $\mathbf{v} + \Delta\mathbf{v}$, and the absolute velocity of the particles absorbed is denoted by $\mathbf{v}_a$. Applying the principle of impulse and momentum, we write

$$m\mathbf{v} + (\Delta m)\mathbf{v}_a + \Sigma\mathbf{F}\,\Delta t = (m + \Delta m)(\mathbf{v} + \Delta\mathbf{v}) \qquad (14.40)$$

Solving for the sum $\Sigma\mathbf{F}\,\Delta t$ of the impulses of the external forces acting on S (excluding the forces exerted by the particles being absorbed), we have

$$\Sigma\mathbf{F}\,\Delta t = m\Delta\mathbf{v} + \Delta m(\mathbf{v} - \mathbf{v}_a) + (\Delta m)(\Delta\mathbf{v}) \qquad (14.41)$$

Introducing the *relative velocity* $\mathbf{u}$ with respect to S of the particles which are absorbed, we write $\mathbf{u} = \mathbf{v}_a - \mathbf{v}$ and note, since $v_a < v$, that the relative velocity $\mathbf{u}$ is directed to the left, as shown in Fig. 14.13. Neglecting the last term in Eq. (14.41), which is of the second order, we write

$$\Sigma\mathbf{F}\,\Delta t = m\,\Delta\mathbf{v} - (\Delta m)\mathbf{u}$$

Dividing through by Δt and letting Δt approach zero, we have at the limit†

$$\Sigma\mathbf{F} = m\frac{d\mathbf{v}}{dt} - \frac{dm}{dt}\mathbf{u} \qquad (14.42)$$

Rearranging the terms and recalling that $d\mathbf{v}/dt = \mathbf{a}$, where $\mathbf{a}$ is the acceleration of the system S, we write

$$\Sigma\mathbf{F} + \frac{dm}{dt}\mathbf{u} = m\mathbf{a} \qquad (14.43)$$

which shows that the action on S of the particles being absorbed is equivalent to a thrust

$$\mathbf{P} = \frac{dm}{dt}\mathbf{u} \qquad (14.44)$$

which tends to slow down the motion of S, since the relative velocity $\mathbf{u}$ of the particles is directed to the left. If SI units are used, dm/dt is expressed in kg/s, the relative velocity u in m/s, and the corresponding thrust in newtons. If U.S. customary units are used, dm/dt must be expressed in slugs/s, u in ft/s, and the corresponding thrust in pounds.‡

The equations obtained can also be used to determine the motion of a system S losing mass. In this case, the rate of change of mass is negative, and the action on S of the particles being expelled is equivalent to a thrust in the direction of $-\mathbf{u}$, that is, in the direction opposite to that in which the particles are being expelled. A *rocket* represents a typical case of a system continually losing mass (see Sample Prob. 14.8).

Photo 14.4 As the shuttle's booster rockets are fired, the gas particles they eject provide the thrust required for liftoff.

†When the absolute velocity $\mathbf{v}_a$ of the particles absorbed is zero, $\mathbf{u} = -\mathbf{v}$, and formula (14.42) becomes

$$\Sigma\mathbf{F} = \frac{d}{dt}(m\mathbf{v})$$

Comparing the formula obtained to Eq. (12.3) of Sec. 12.3, we observe that Newton's second law can be applied to a system gaining mass, *provided that the particles absorbed are initially at rest.* It may also be applied to a system losing mass, *provided that the velocity of the particles expelled is zero* with respect to the frame of reference selected.

‡See footnote on page 887.

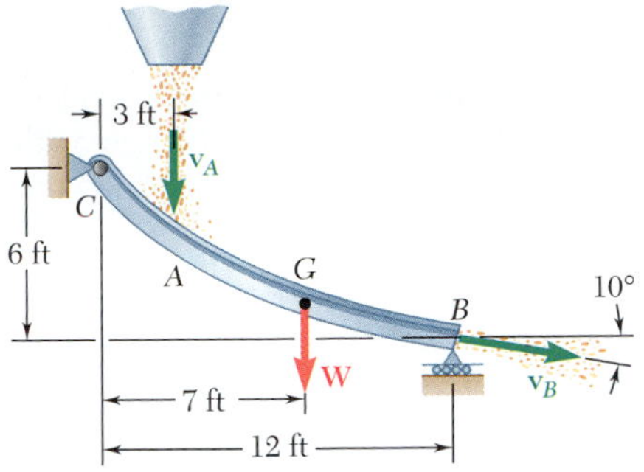

SAMPLE PROBLEM 14.6

Grain falls from a hopper onto a chute CB at the rate of 240 lb/s. It hits the chute at A with a velocity of 20 ft/s and leaves at B with a velocity of 15 ft/s, forming an angle of 10° with the horizontal. Knowing that the combined weight of the chute and of the grain it supports is a force **W** of magnitude 600 lb applied at G, determine the reaction at the roller support B and the components of the reaction at the hinge C.

SOLUTION

We apply the principle of impulse and momentum for the time interval Δt to the system consisting of the chute, the grain it supports, and the amount of grain which hits the chute in the interval Δt. Since the chute does not move, it has no momentum. We also note that the sum $\Sigma m_i \mathbf{v}_i$ of the momenta of the particles supported by the chute is the same at t and $t + \Delta t$ and can thus be omitted.

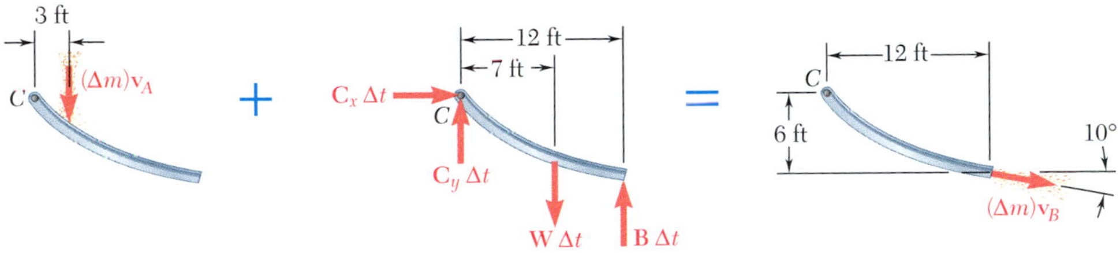

Since the system formed by the momentum $(\Delta m)\mathbf{v}_A$ and the impulses is equipollent to the momentum $(\Delta m)\mathbf{v}_B$, we write

$\xrightarrow{+}$ x components:
$$C_x\,\Delta t = (\Delta m)v_B \cos 10° \qquad (1)$$

$+\uparrow$ y components:
$$-(\Delta m)v_A + C_y\,\Delta t - W\,\Delta t + B\,\Delta t = -(\Delta m)v_B \sin 10° \qquad (2)$$

$+\circlearrowleft$ moments about C:
$$-3(\Delta m)v_A - 7(W\,\Delta t) + 12(B\,\Delta t) = 6(\Delta m)v_B \cos 10° - 12(\Delta m)v_B \sin 10° \qquad (3)$$

Using the given data, $W = 600$ lb, $v_A = 20$ ft/s, $v_B = 15$ ft/s, and $\Delta m/\Delta t = 240/32.2 = 7.45$ slugs/s, and solving Eq. (3) for B and Eq. (1) for C_x,

$$12B = 7(600) + 3(7.45)(20) + 6(7.45)(15)(\cos 10° - 2 \sin 10°)$$
$$12B = 5075 \qquad B = 423 \text{ lb} \qquad \mathbf{B} = 423 \text{ lb} \uparrow \blacktriangleleft$$
$$C_x = (7.45)(15) \cos 10° = 110.1 \text{ lb} \qquad \mathbf{C}_x = 110.1 \text{ lb} \rightarrow \blacktriangleleft$$

Substituting for B and solving Eq. (2) for C_y,

$$C_y = 600 - 423 + (7.45)(20 - 15 \sin 10°) = 307 \text{ lb}$$
$$\mathbf{C}_y = 307 \text{ lb} \uparrow \blacktriangleleft$$

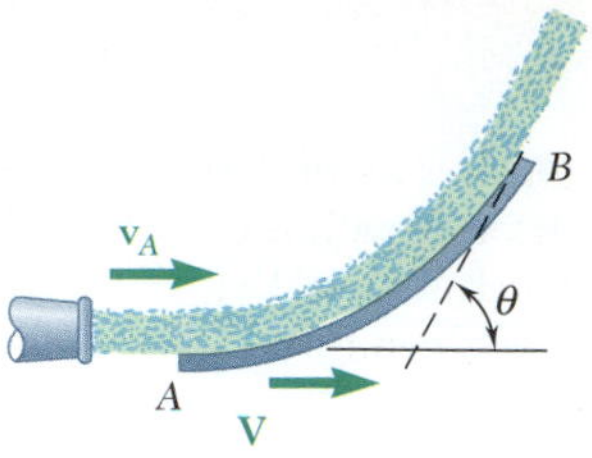

SAMPLE PROBLEM 14.7

A nozzle discharges a stream of water of cross-sectional area A with a velocity $\mathbf{v}_A$. The stream is deflected by a *single* blade which moves to the right with a constant velocity $\mathbf{V}$. Assuming that the water moves along the blade at constant speed, determine (*a*) the components of the force $\mathbf{F}$ exerted by the blade on the stream, (*b*) the velocity $\mathbf{V}$ for which maximum power is developed.

SOLUTION

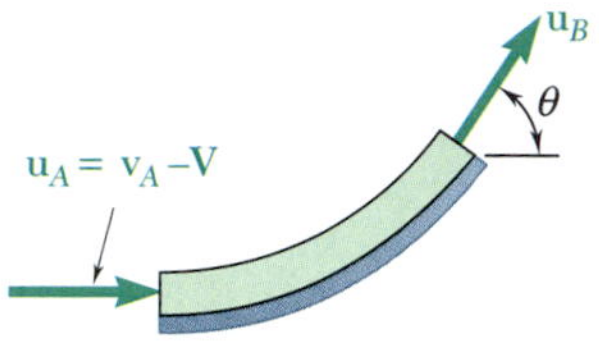

***a.* Components of Force Exerted on Stream.** We choose a coordinate system which moves with the blade at a constant velocity $\mathbf{V}$. The particles of water strike the blade with a relative velocity $\mathbf{u}_A = \mathbf{v}_A - \mathbf{V}$ and leave the blade with a relative velocity $\mathbf{u}_B$. Since the particles move along the blade at a constant speed, the relative velocities $\mathbf{u}_A$ and $\mathbf{u}_B$ have the same magnitude u. Denoting the density of water by ρ, the mass of the particles striking the blade during the time interval Δt is $\Delta m = A\rho(v_A - V)\,\Delta t$; an equal mass of particles leaves the blade during Δt. We apply the principle of impulse and momentum to the system formed by the particles in contact with the blade and the particles striking the blade in the time Δt.

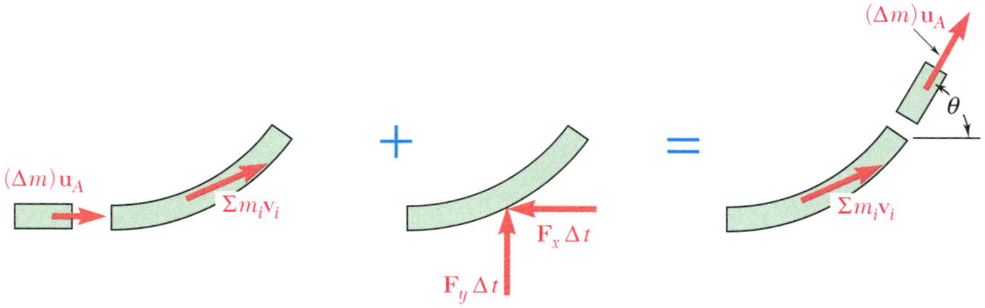

Recalling that $\mathbf{u}_A$ and $\mathbf{u}_B$ have the same magnitude u, and omitting the momentum $\Sigma m_i \mathbf{v}_i$ which appears on both sides, we write

$\xrightarrow{+}$ x components: $\quad (\Delta m)u - F_x\,\Delta t = (\Delta m)u \cos\theta$

$+\uparrow y$ components: $\quad +F_y\,\Delta t = (\Delta m)u \sin\theta$

Substituting $\Delta m = A\rho(v_A - V)\,\Delta t$ and $u = v_A - V$, we obtain

$$\mathbf{F}_x = A\rho(v_A - V)^2(1 - \cos\theta) \leftarrow \qquad \mathbf{F}_y = A\rho(v_A - V)^2 \sin\theta \uparrow \quad ◀$$

***b.* Velocity of Blade for Maximum Power.** The power is obtained by multiplying the velocity V of the blade by the component F_x of the force exerted by the stream on the blade.

$$\text{Power} = F_x V = A\rho(v_A - V)^2(1 - \cos\theta)V$$

Differentiating the power with respect to V and setting the derivative equal to zero, we obtain

$$\frac{d(\text{power})}{dV} = A\rho(v_A^2 - 4v_A V + 3V^2)(1 - \cos\theta) = 0$$

$$V = v_A \qquad V = \tfrac{1}{3}v_A \qquad \text{For maximum power } \mathbf{V} = \tfrac{1}{3}v_A \rightarrow \quad ◀$$

Note. These results are valid only when a *single* blade deflects the stream. Different results are obtained when a series of blades deflects the stream, as in a Pelton-wheel turbine. (See Prob. 14.81.)

SAMPLE PROBLEM 14.8

A rocket of initial mass m_0 (including shell and fuel) is fired vertically at time $t = 0$. The fuel is consumed at a constant rate $q = dm/dt$ and is expelled at a constant speed u relative to the rocket. Derive an expression for the magnitude of the velocity of the rocket at time t, neglecting the resistance of the air.

SOLUTION

At time t, the mass of the rocket shell and remaining fuel is $m = m_0 - qt$, and the velocity is $\mathbf{v}$. During the time interval Δt, a mass of fuel $\Delta m = q\ \Delta t$ is expelled with a speed u relative to the rocket. Denoting by $\mathbf{v}_e$ the absolute velocity of the expelled fuel, we apply the principle of impulse and momentum between time t and time $t + \Delta t$.

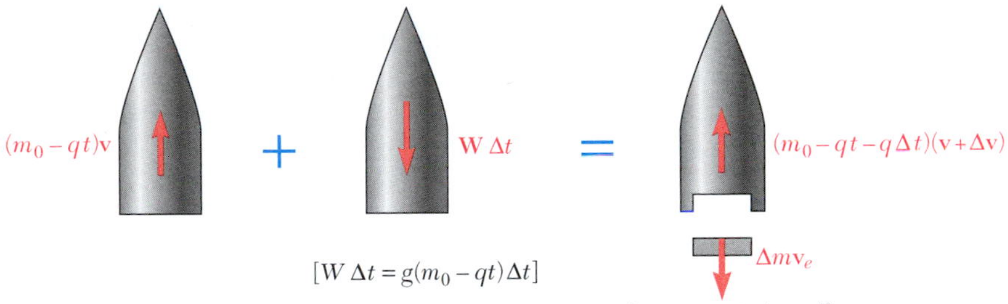

$[W\,\Delta t = g(m_0 - qt)\Delta t]$

$[\Delta m v_e = q\,\Delta t(u - v)]$

We write

$$(m_0 - qt)v - g(m_0 - qt)\ \Delta t = (m_0 - qt - q\ \Delta t)(v + \Delta v) - q\ \Delta t(u - v)$$

Dividing through by Δt and letting Δt approach zero, we obtain

$$-g(m_0 - qt) = (m_0 - qt)\frac{dv}{dt} - qu$$

Separating variables and integrating from $t = 0$, $v = 0$ to $t = t$, $v = v$,

$$dv = \left(\frac{qu}{m_0 - qt} - g\right) dt \qquad \int_0^v dv = \int_0^t \left(\frac{qu}{m_0 - qt} - g\right) dt$$

$$v = [-u \ln (m_0 - qt) - gt]_0^t \qquad v = u \ln \frac{m_0}{m_0 - qt} - gt \quad ◀$$

Remark. The mass remaining at time t_f, after all the fuel has been expended, is equal to the mass of the rocket shell $m_s = m_0 - qt_f$, and the maximum velocity attained by the rocket is $v_m = u \ln (m_0/m_s) - gt_f$. Assuming that the fuel is expelled in a relatively short period of time, the term gt_f is small and we have $v_m \approx u \ln (m_0/m_s)$. In order to escape the gravitational field of the earth, a rocket must reach a velocity of 11.18 km/s. Assuming $u = 2200$ m/s and $v_m = 11.18$ km/s, we obtain $m_0/m_s = 161$. Thus, to project each kilogram of the rocket shell into space, it is necessary to consume more than 161 kg of fuel if a propellant yielding $u = 2200$ m/s is used.

SOLVING PROBLEMS ON YOUR OWN

This lesson is devoted to the study of the motion of *variable systems of particles*, that is, systems which are continually *gaining or losing particles* or doing both at the same time. The problems you will be asked to solve will involve (1) *steady streams of particles* and (2) *systems gaining or losing mass.*

1. To solve problems involving a steady stream of particles, you will consider a portion S of the stream and express that the system formed by the momentum of the particles entering S at A in the time Δt and the impulses of the forces exerted on S during that time is equipollent to the momentum of the particles leaving S at B in the same time Δt (Fig. 14.10). Considering only the resultants of the vector systems involved, you can write the vector equation

$$(\Delta m)\mathbf{v}_A + \Sigma\mathbf{F}\,\Delta t = (\Delta m)\mathbf{v}_B \tag{14.38}$$

You may want to consider as well the moments about a given point of the vector systems involved to obtain an additional equation [Sample Prob. 14.6], but many problems can be solved using Eq. (14.38) or the equation obtained by dividing all terms by Δt and letting Δt approach zero,

$$\Sigma\mathbf{F} = \frac{dm}{dt}(\mathbf{v}_B - \mathbf{v}_A) \tag{14.39}$$

where $\mathbf{v}_B - \mathbf{v}_A$ represents a *vector subtraction* and where the mass rate of flow dm/dt can be expressed as the product ρQ of the density ρ of the stream (mass per unit volume) and the volume rate of flow Q (volume per unit time). If U.S. customary units are used, ρ is expressed as the ratio γ/g, where γ is the specific weight of the stream and g is the acceleration of gravity.

Typical problems involving a steady stream of particles have been described in Sec. 14.11. You may be asked to determine the following:

a. Thrust caused by a diverted flow. Equation (14.39) is applicable, but you will get a better understanding of the problem if you use a solution based on Eq. (14.38).

b. Reactions at supports of vanes or conveyor belts. First draw a diagram showing on one side of the equals sign the momentum $(\Delta m)\mathbf{v}_A$ of the particles impacting the vane or belt in the time Δt, as well as the impulses of the loads and reactions at the supports during that time, and showing on the other side the momentum $(\Delta m)\mathbf{v}_B$ of the particles leaving the vane or belt in the time Δt [Sample Prob. 14.6]. Equating the x components, y components, and moments of the quantities on both sides of the equals sign will yield three scalar equations which can be solved for three unknowns.

c. Thrust developed by a jet engine, a propeller, or a fan. In most cases, a single unknown is involved, and that unknown can be obtained by solving the scalar equation derived from Eq. (14.38) or Eq. (14.39).

2. ***To solve problems involving systems gaining mass,*** you will consider the system S, which has a mass m and is moving with a velocity $\mathbf{v}$ at time t, and the particles of mass Δm with velocity $\mathbf{v}_a$ that S will absorb in the time interval Δt (Fig. 14.13). You will then express that the total momentum of S and of the particles that will be absorbed, *plus* the impulse of the external forces exerted on S, are equipollent to the momentum of S at time $t + \Delta t$. Noting that the mass of S and its velocity at that time are, respectively, $m + \Delta m$ and $\mathbf{v} + \Delta\mathbf{v}$, you will write the vector equation

$$m\mathbf{v} + (\Delta m)\mathbf{v}_a + \Sigma\mathbf{F}\,\Delta t = (m + \Delta m)(\mathbf{v} + \Delta\mathbf{v}) \qquad (14.40)$$

As was shown in Sec. 14.12, if you introduce the relative velocity $\mathbf{u} = \mathbf{v}_a - \mathbf{v}$ of the particles being absorbed, you obtain the following expression for the resultant of the external forces applied to S:

$$\Sigma\mathbf{F} = m\frac{d\mathbf{v}}{dt} - \frac{dm}{dt}\mathbf{u} \qquad (14.42)$$

Furthermore, it was shown that the action on S of the particles being absorbed is equivalent to a thrust

$$\mathbf{P} = \frac{dm}{dt}\mathbf{u} \qquad (14.44)$$

exerted in the direction of the relative velocity of the particles being absorbed.

Examples of systems gaining mass are conveyor belts and moving railroad cars being loaded with gravel or sand, and chains being pulled out of a pile.

3. ***To solve problems involving systems losing mass,*** such as rockets and rocket engines, you can use Eqs. (14.40) through (14.44), provided that you give negative values to the increment of mass Δm and to the rate of change of mass dm/dt. It follows that the thrust defined by Eq. (14.44) will be exerted in a direction opposite to the direction of the relative velocity of the particles being ejected.

Problems

14.59 The nozzle shown discharges a stream of water at a flow rate $Q = 1.8 \text{ m}^3/\text{min}$ with a velocity $\mathbf{v}$ of magnitude 20 m/s. The stream is split into two streams with equal flow rates by a wedge which is kept in a fixed position. Determine the components (drag and lift) of the force exerted by the stream on the wedge.

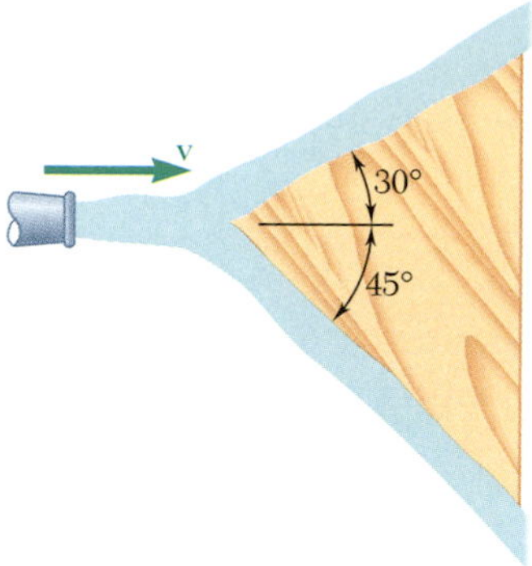

Fig. P14.59 and P14.60

14.60 The nozzle shown discharges a stream of water at a flow rate $Q = 1.92 \text{ m}^3/\text{min}$ with a velocity $\mathbf{v}$ of magnitude 16 m/s. The stream is split into two streams of equal flow rates by a wedge which is moving to the left at a constant speed of 4 m/s. Determine the components (drag and lift) of the force exerted by the stream on the wedge.

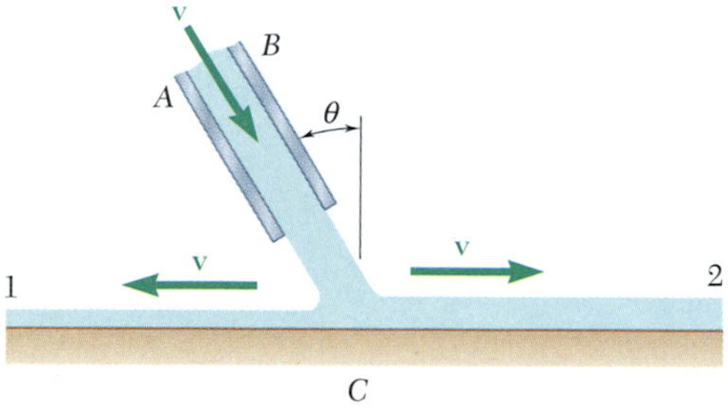

Fig. P14.61 and P14.62

14.61 Water flows in a continuous sheet from between two plates A and B with a velocity $\mathbf{v}$ of magnitude 90 ft/s. The stream is split into two parts by a smooth horizontal plate C. Knowing that the rates of flow in each of the two resulting streams are, respectively, $Q_1 = 26$ gal/min and $Q_2 = 130$ gal/min, determine (*a*) the angle θ, (*b*) the total force exerted by the stream on the horizontal plate. (*Note:* $1 \text{ ft}^3 = 7.48$ gal.)

14.62 Water flows in a continuous sheet from between two plates A and B with a velocity $\mathbf{v}$ of magnitude 120 ft/s. The stream is split into two parts by a smooth horizontal plate C. Determine the rates of flow Q_1 and Q_2 in each of the two resulting streams, knowing that $\theta = 30°$ and that the total force exerted by the stream on the horizontal plate is a 125-lb vertical force.

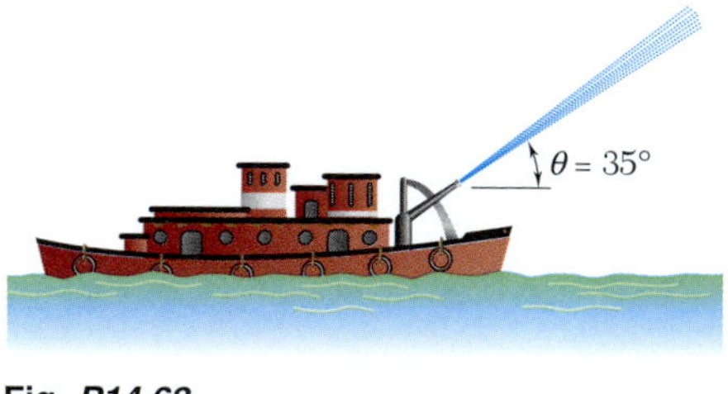

Fig. ***P14.63***

14.63 A hose discharges water at a rate of $8 \text{ m}^3/\text{min}$ with a velocity of 50 m/s from the bow of a fireboat. Determine the thrust developed by the engine to keep the fireboat in a stationary position.

14.64 Tree limbs and branches are being fed at A at the rate of 10 lb/s into a shredder which spews the resulting wood chips at C with a velocity of 60 ft/s. Determine the horizontal component of the force exerted by the shredder on the truck hitch at D.

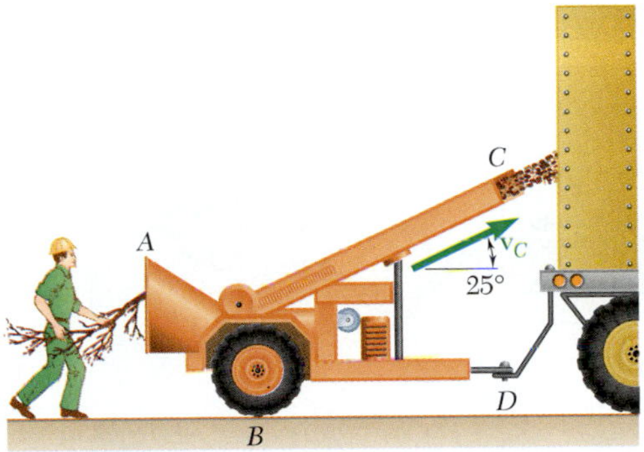

Fig. P14.64

14.65 A stream of water flowing at a rate of 1.2 m^3/min and moving with a velocity of magnitude 30 m/s at both A and B is deflected by a vane welded to a hinged plate. Knowing that the combined mass of the vane and plate is 20 kg with the mass center at point G, determine (*a*) the angle θ, (*b*) the reaction at C.

14.66 A stream of water flowing at a rate of 1.2 m^3/min and moving with a velocity of magnitude v at both A and B is deflected by a vane welded to a hinged plate. The combined mass of the vane and plate is 20 kg with the mass center at point G. Knowing that $\theta = 45°$, determine (*a*) the speed v of the flow, (*b*) the reaction at C.

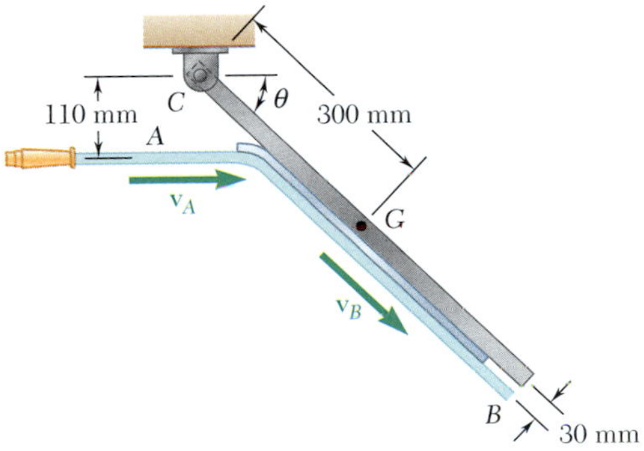

Fig. P14.65 and P14.66

14.67 The nozzle shown discharges water at the rate of 40 ft^3/min. Knowing that at both A and B the stream of water moves with a velocity of magnitude 75 ft/s and neglecting the weight of the vane, determine the components of the reactions at C and D.

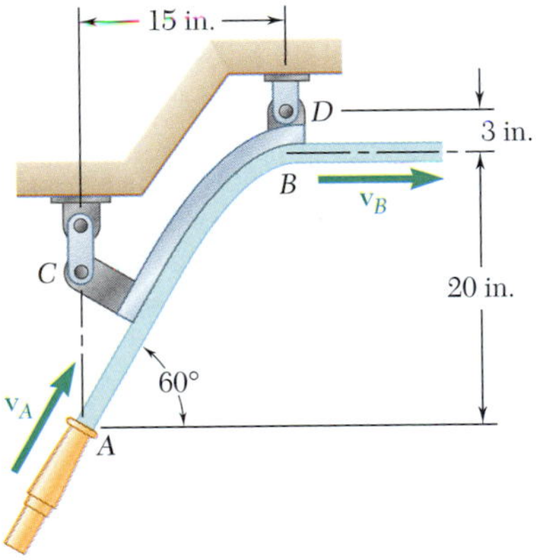

Fig. P14.67

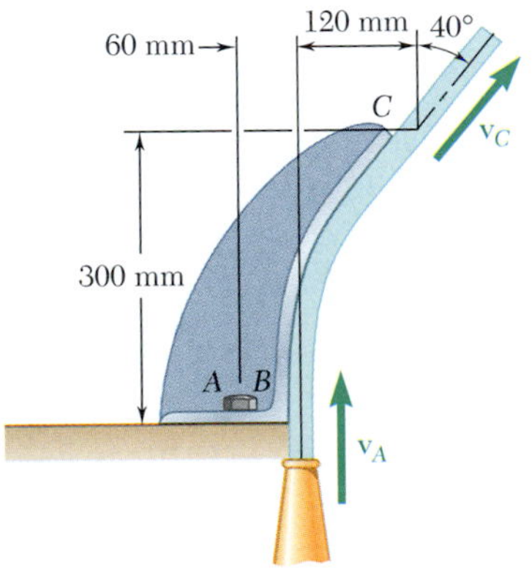

Fig. P14.68

14.68 The nozzle shown discharges water at the rate of 800 L/min. Knowing that at both B and C the stream of water moves with a velocity of magnitude 30 m/s, and neglecting the weight of the vane, determine the force-couple system which must be applied at A to hold the vane in place.

14.69 Coal is being discharged from a first conveyor belt at the rate of 240 lb/s. It is received at *A* by a second belt which discharges it again at *B*. Knowing that $v_1 = 9$ ft/s and $v_2 = 12.25$ ft/s and that the second belt assembly and the coal it supports have a total weight of 944 lb, determine the components of the reactions at *C* and *D*.

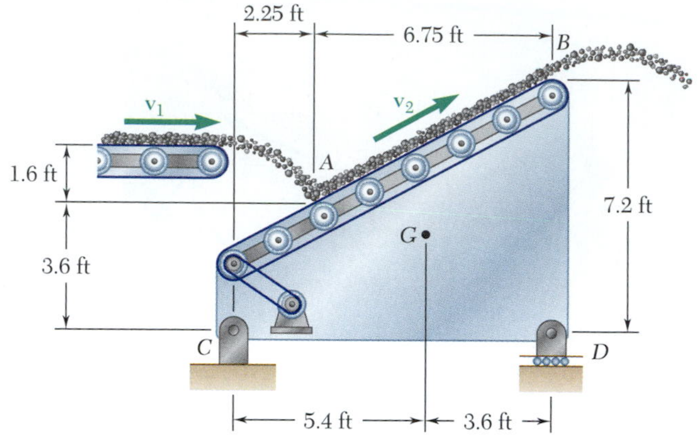

Fig. *P14.69*

14.70 A stream of water having a cross-sectional area of 1.5 in^2 and moving with a velocity of magnitude 60 ft/s at both *A* and *B* is deflected by two vanes which are welded as shown to a vertical plate. Knowing that the combined weight of the plate and vanes is 10 lb, determine the reactions at *C* and *D*.

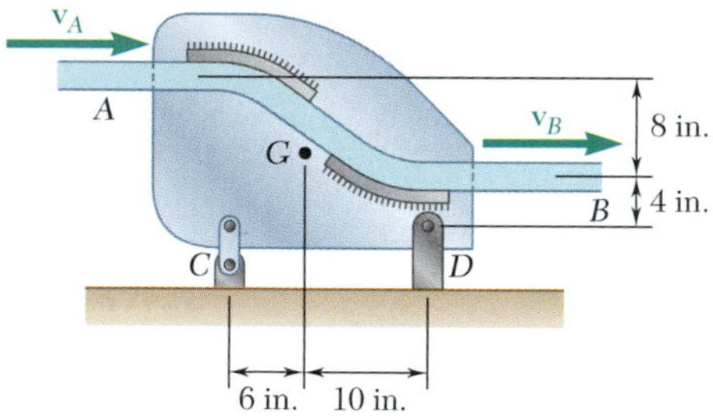

Fig. *P14.70*

14.71 The total drag due to air friction on a jet airplane cruising in level flight at a speed of 760 mi/h is 10 kips. Knowing that the exhaust velocity is 2400 ft/s relative to the airplane, determine the rate in lb/s at which the air must pass through the engine.

14.72 While cruising in level flight at a speed of 560 mi/h, a jet plane scoops in air at a rate of 200 lb/s and discharges it with a velocity of 1980 ft/s relative to the airplane. Determine the total drag due to air friction on the airplane.

14.73 The jet engine shown scoops in air at *A* at a rate of 90 kg/s and discharges it at *B* with a velocity of 600 m/s relative to the airplane. Determine the magnitude and line of action of the propulsive thrust developed by the engine when the speed of the airplane is (*a*) 480 km/h, (*b*) 960 km/h.

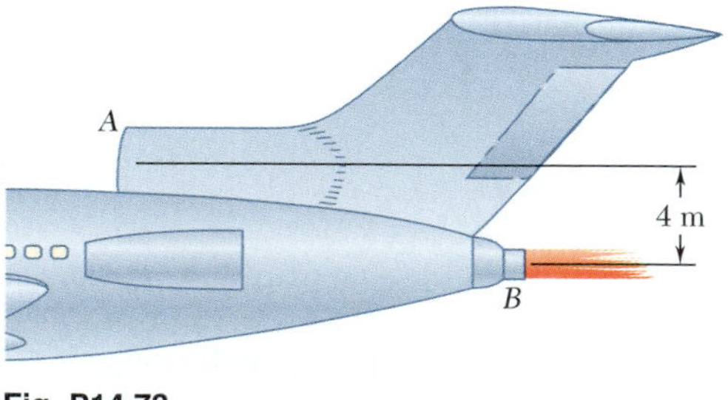

Fig. P14.73

14.74 The helicopter shown can produce a maximum downward air speed of 24 m/s in a 9-m-diameter slipstream. Knowing that the weight of the helicopter and crew is 15 kN and assuming $\rho = 1.21 \text{ kg/m}^3$ for air, determine the maximum load that the helicopter can lift while hovering in midair.

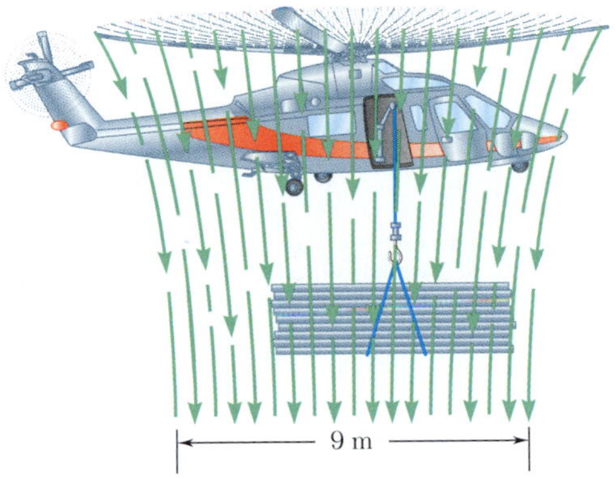

Fig. P14.74

14.75 Prior to take-off the pilot of a 3000-kg twin-engine airplane tests the reversible-pitch propellers by increasing the reverse thrust with the brakes at point B locked. Knowing that point G is the center of gravity of the airplane, determine the velocity of the air in the two 2.2-m-diameter slipstreams when the nose wheel A begins to lift off the ground. Assume $\rho = 1.21 \text{ kg/m}^3$ and neglect the approach velocity of the air.

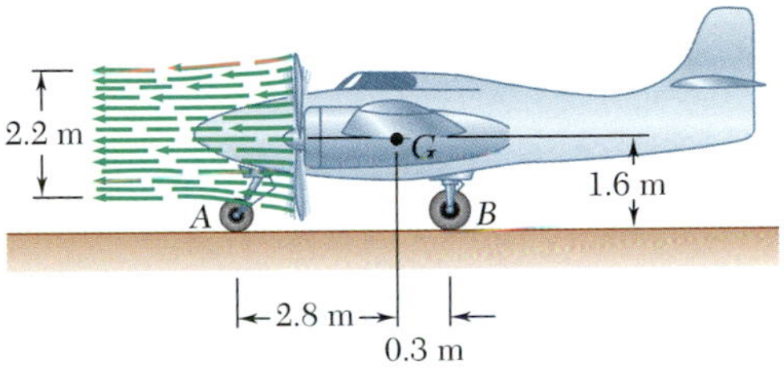

Fig. *P14.75* and *P14.76*

14.76 Prior to take-off the pilot of a 3000-kg twin-engine airplane tests the reversible-pitch propellers with the brakes at point B locked. Knowing that the velocity of the air in the two 2.2-m-diameter slipstreams is 20 m/s and that point G is the center of gravity of the airplane, determine the reactions at points A and B. Assume $\rho = 1.21 \text{ kg/m}^3$ and neglect the approach velocity of the air.

14.77 A 32-kip jet airplane maintains a constant speed of 480 mi/h while climbing at an angle $\alpha = 18°$. The airplane scoops in air at a rate of 600 lb/s and discharges it with a velocity of 2000 ft/s relative to the airplane. If the pilot changes to a horizontal flight while maintaining the same engine setting, determine (a) the initial acceleration of the plane, (b) the maximum horizontal speed that will be attained. Assume that the drag due to air friction is proportional to the square of the speed.

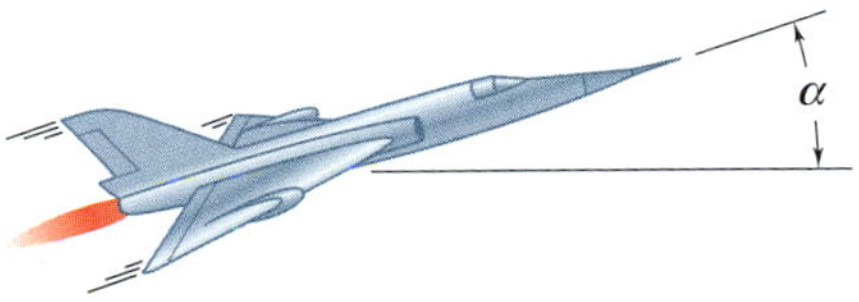

Fig. P14.77

14.78 A jet airliner is cruising at a speed of 900 km/h with its engines scooping in air at the rate of 360 kg/s and discharging it with a velocity of 620 m/s relative to the plane when a control surface malfunction suddenly causes a 20 percent increase in drag. Knowing that the pilot maintains level flight with the same mass flow rate and discharge relative velocity, determine the new cruising speed. Assume that the drag due to air friction is proportional to the square of the speed of the plane.

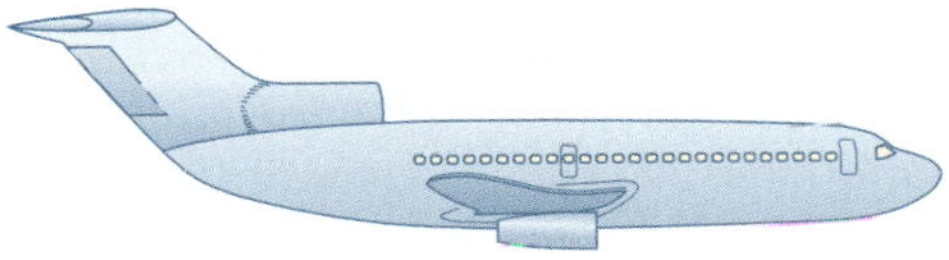

Fig. P14.78

14.79 The wind-turbine-generator shown operates at a wind speed of 18 mi/h with an efficiency of 0.4. Knowing that the area swept out by the blades is a circle of diameter $d = 22$ ft and that $\gamma = 0.076$ lb/ft^3, determine (*a*) the kinetic energy of the air particles entering the 22-ft-diameter circle per second, (*b*) the output power.

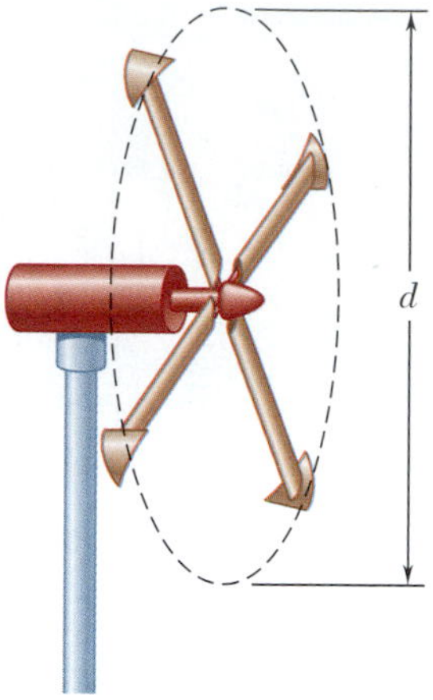

Fig. P14.79 and P14.80

14.80 The wind-turbine-generator shown has an output-power rating of 4.7 hp for a wind speed of 21 mi/h and operates at an efficiency of 0.35. Knowing that $\gamma = 0.076$ lb/ft^3, determine (*a*) the diameter d of the circular area swept out by the blades, (*b*) the kinetic energy of the air particles entering the circular area per second.

14.81 The propeller of a small airplane has a 2-m-diameter slipstream and produces a thrust of 3.6 kN when the airplane is at rest on the ground. Assuming $\rho = 1.21$ kg/m^3 for air, determine (*a*) the speed of the air in the slipstream, (*b*) the volume of air passing through the propeller per second, (*c*) the kinetic energy imparted per second to the air in the slipstream.

14.82 While cruising in level flight at a speed of 900 km/h, a jet airplane scoops in air at the rate of 91 kg/s and discharges it with a velocity of 670 m/s relative to the airplane. Determine (*a*) the power actually used to propel the airplane, (*b*) the total power developed by the engine, (*c*) the mechanical efficiency of the airplane.

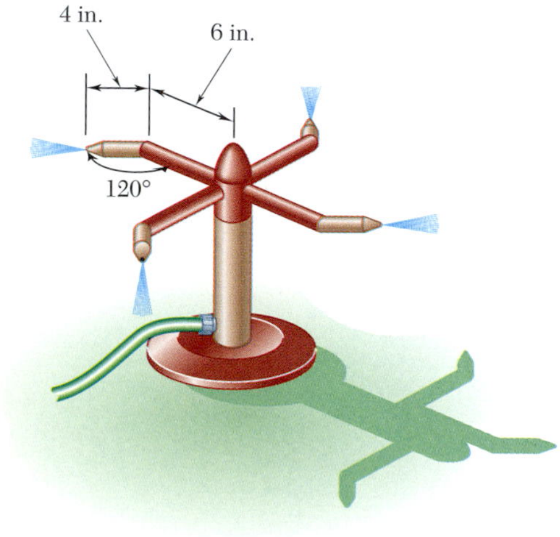

Fig. P14.83

14.83 A garden sprinkler has four rotating arms, each of which consists of two horizontal straight sections of pipe forming an angle of 120°. Each arm discharges water at a rate of 5 gal/min with a velocity of 60 ft/s relative to the arm. Knowing that the friction between the moving and stationary parts of the sprinkler is equivalent to a couple of magnitude $M = 0.275$ lb · ft, determine the constant rate at which the sprinkler rotates (1 ft^3 = 7.48 gal).

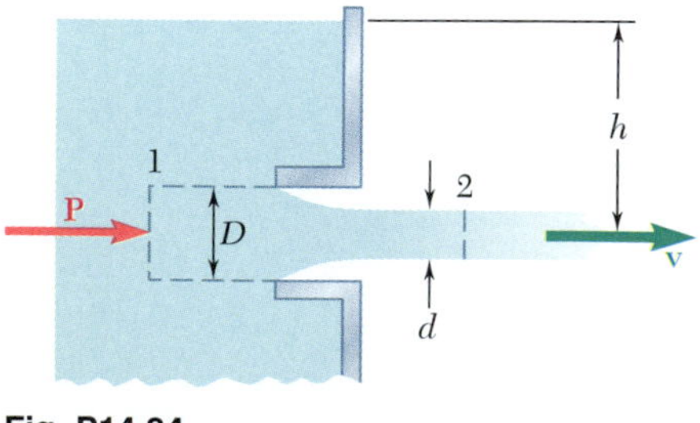

Fig. P14.84

14.84 A circular reentrant orifice (also called Borda's mouthpiece) of diameter D is placed at a depth h below the surface of a tank. Knowing that the speed of the issuing stream is $v = \sqrt{2gh}$ and assuming that the speed of approach v_1 is zero, show that the diameter of the stream is $d = D/\sqrt{2}$. (*Hint:* Consider the section of water indicated, and note that P is equal to the pressure at a depth h multiplied by the area of the orifice.)

*14.85 The depth of water flowing in a rectangular channel of width b at a speed v_1 and a depth d_1 increases to a depth d_2 at a *hydraulic jump*. Express the rate of flow Q in terms of b, d_1, and d_2.

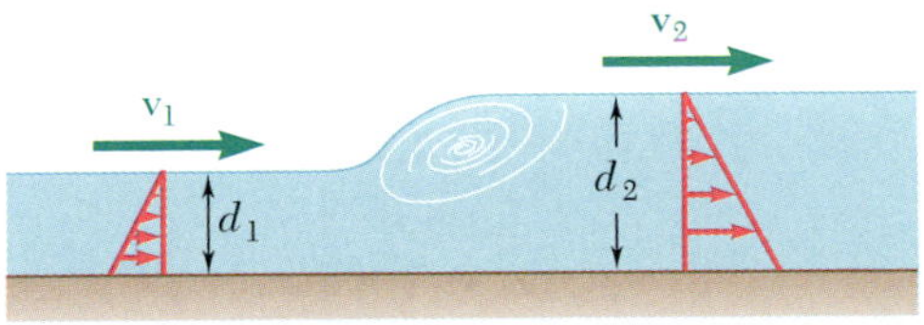

Fig. P14.85

*14.86 Determine the rate of flow in the channel of Prob. 14.85, knowing that $b = 3$ m, $d_1 = 1.25$ m, and $d_2 = 1.5$ m.

14.87 The final component of a conveyor system receives sand at a rate of 100 kg/s at A and discharges it at B. The sand is moving horizontally at A and B with a velocity of magnitude $v_A = v_B = 4.5$ m/s. Knowing that the combined weight of the component and of the sand it supports is $W = 4$ kN, determine the reactions at C and D.

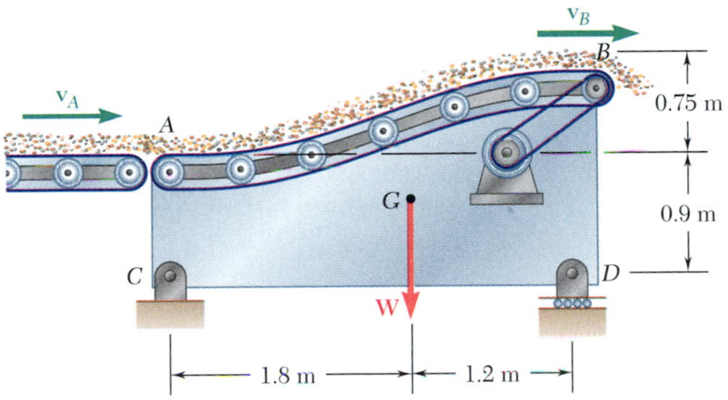

Fig. P14.87

14.88 A railroad car of length L and mass m_0 when empty is moving freely on a horizontal track while being loaded with sand from a stationary chute at a rate $dm/dt = q$. Knowing that the car was approaching the chute at a speed v_0, determine (*a*) the mass of the car and its load after the car has cleared the chute, (*b*) the speed of the car at that time.

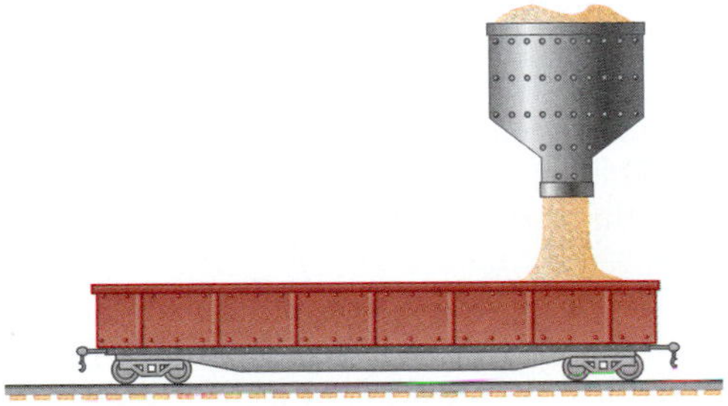

Fig. P14.88

14.89 The ends of a chain lie in piles at A and C. When given an initial speed v, the chain keeps moving freely at that speed over the pulley at B. Neglecting friction, determine the required value of h.

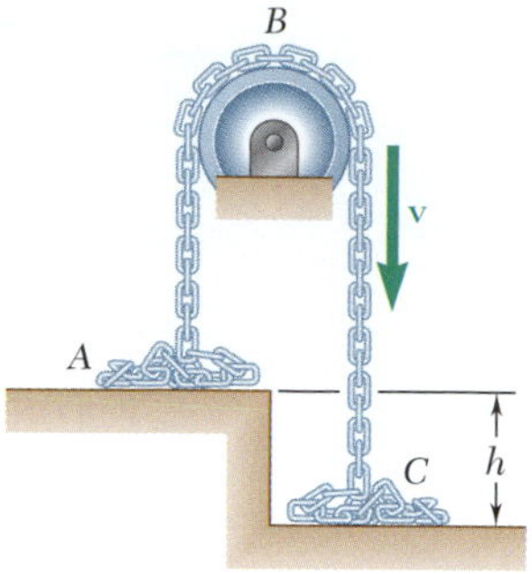

Fig. P14.89

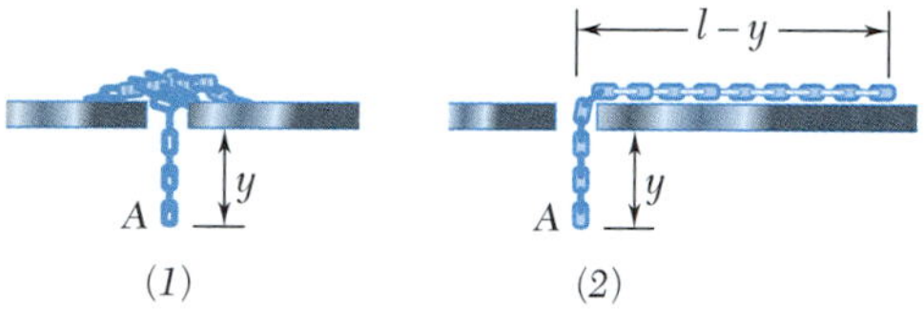

Fig. P14.90

14.90 A chain of length l and mass m falls through a small hole in a plate. Initially, when y is very small, the chain is at rest. In each case shown, determine (*a*) the acceleration of the first link A as a function of y, (*b*) the velocity of the chain as the last link passes through the hole. In case *1* assume that the individual links are at rest until they fall through the hole; in case *2* assume that at any instant all links have the same speed. Ignore the effect of friction.

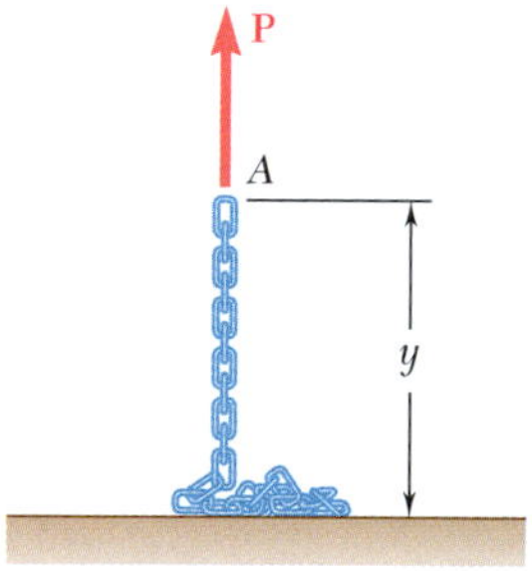

Fig. *P14.91*

14.91 A chain of length l and mass m lies in a pile on the floor. If its end A is raised vertically at a constant speed v, express in terms of the length y of chain which is off the floor at any given instant (*a*) the magnitude of the force **P** applied at A, (*b*) the reaction of the floor.

14.92 Solve Prob. 14.91, assuming that the chain is being lowered to the floor at a constant speed v.

14.93 A rocket has a mass of 960 kg, including 800 kg of fuel, which is consumed at the rate of 10 kg/s and ejected with a relative velocity of 3600 m/s. Knowing that the rocket is fired vertically from the ground, determine its acceleration (*a*) as it is fired, (*b*) as the last particle of fuel is being consumed.

14.94 A rocket has a mass of 1500 kg, including 1200 kg of fuel, which is consumed at the rate of 15 kg/s. Knowing that the rocket is fired vertically from the ground and that its acceleration increases by 220 m/s^2 from the time it is fired to the time the last particle of fuel has been consumed, determine the relative velocity with which the fuel is being ejected.

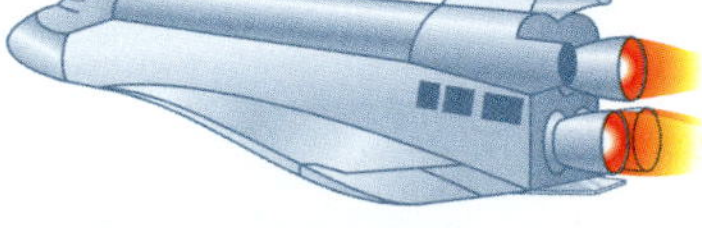

Fig. P14.95 and P14.96

14.95 The main propulsion system of a space shuttle consists of three identical rocket engines, each of which burns the hydrogen-oxygen propellant at the rate of 750 lb/s and ejects it with a relative velocity of 12,500 ft/s. Determine the total thrust provided by the three engines.

14.96 The main propulsion system of a space shuttle consists of three identical rocket engines which provide a total thrust of 1200 kips. Determine the rate at which the hydrogen-oxygen propellant is burned by each of the three engines, knowing that it is ejected with a relative velocity of 12,500 ft/s.

14.97 The mass of a spacecraft, including fuel, is 5800 kg when the rocket engines are fired to increase its velocity by 120 m/s. Knowing that 500 kg of fuel is consumed, determine the relative velocity of the fuel ejected.

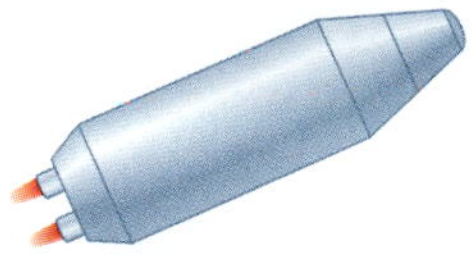

Fig. P14.97 and P14.98

14.98 The rocket engines of a spacecraft are fired to increase its velocity by 150 m/s. Knowing that 600 kg of fuel is ejected at a relative velocity of 1800 m/s, determine the mass of the spacecraft after the firing.

14.99 A 1200-lb spacecraft is mounted on top of a rocket weighing 42,000 lb, including 40,000 lb of fuel. Knowing that the fuel is consumed at a rate of 500 lb/s and ejected with a relative velocity of 12,000 ft/s, determine the maximum speed imparted to the spacecraft when the rocket is fired vertically from the ground.

14.100 The rocket used to launch the 1200-lb spacecraft of Prob. 14.99 is redesigned to include two stages A and B, each weighing 21,300 lb, including 20,000 lb of fuel. The fuel is again consumed at a rate of 500 lb/s and ejected with a relative velocity of 12,000 ft/s. Knowing that when stage A expels its last particle of fuel, its casing is released and jettisoned, determine (a) the speed of the rocket at that instant, (b) the maximum speed imparted to the spacecraft.

Fig. *P14.99*

Fig. *P14.100*

14.101 Determine the altitude reached by the spacecraft of Prob. 14.99 when all the fuel of its launching rocket has been consumed.

14.102 For the spacecraft and the two-stage launching rocket of Prob. 14.100, determine the altitude at which (a) stage A of the rocket is released, (b) the fuel of both stages has been consumed.

14.103 Determine the distance traveled by the spacecraft of Prob. 14.97 during the rocket engine firing, knowing that its initial speed was 2500 m/s and the duration of the firing was 60 s.

14.104 For the rocket of Prob. 14.93, determine (a) the altitude at which all fuel has been consumed, (b) the velocity of the rocket at that time.

14.105 In a rocket, the kinetic energy imparted to the consumed and ejected fuel is wasted as far as propelling the rocket is concerned. The useful power is equal to the product of the force available to propel the rocket and the speed of the rocket. If v is the speed of the rocket and u is the relative speed of the expelled fuel, show that the mechanical efficiency of the rocket is $\eta = 2uv/(u^2 + v^2)$. Explain why $\eta = 1$ when $u = v$.

14.106 In a jet airplane, the kinetic energy imparted to exhaust gases is wasted as far as propelling the airplane is concerned. The useful power is equal to the product of the force available to propel the airplane and the speed of the airplane. If v is the speed of the airplane and u is the relative speed of the expelled gases, show that the mechanical efficiency of the airplane is $\eta = 2v/(u + v)$. Explain why $\eta = 1$ when $u = v$.

REVIEW AND SUMMARY FOR CHAPTER 14

In this chapter we analyzed the motion of *systems of particles,* that is, the motion of a large number of particles considered together. In the first part of the chapter we considered systems consisting of well-defined particles, while in the second part we analyzed systems which are continually gaining or losing particles, or doing both at the same time.

Effective forces

We first defined the *effective force* of a particle P_i of a given system as the product $m_i\mathbf{a}_i$ of its mass m_i and its acceleration $\mathbf{a}_i$ with respect to a newtonian frame of reference centered at O [Sec. 14.2]. We then showed that *the system of the external forces acting on the particles and the system of the effective forces of the particles are equipollent;* that is, both systems have the *same resultant* and the *same moment resultant* about O:

$$\sum_{i=1}^{n} \mathbf{F}_i = \sum_{i=1}^{n} m_i\mathbf{a}_i \tag{14.4}$$

$$\sum_{i=1}^{n} (\mathbf{r}_i \times \mathbf{F}_i) = \sum_{i=1}^{n} (\mathbf{r}_i \times m_i\mathbf{a}_i) \tag{14.5}$$

Linear and angular momentum of a system of particles

Defining the *linear momentum* $\mathbf{L}$ and the *angular momentum* $\mathbf{H}_O$ *about point* O of the system of particles [Sec. 14.3] as

$$\mathbf{L} = \sum_{i=1}^{n} m_i\mathbf{v}_i \qquad \mathbf{H}_O = \sum_{i=1}^{n} (\mathbf{r}_i \times m_i\mathbf{v}_i) \tag{14.6, 14.7}$$

we showed that Eqs. (14.4) and (14.5) can be replaced by the equations

$$\Sigma\mathbf{F} = \dot{\mathbf{L}} \qquad \Sigma\mathbf{M}_O = \dot{\mathbf{H}}_O \tag{14.10, 14.11}$$

which express that *the resultant and the moment resultant about O of the external forces are, respectively, equal to the rates of change of the linear momentum and of the angular momentum about O of the system of particles.*

Motion of the mass center of a system of particles

In Sec. 14.4, we defined the mass center of a system of particles as the point G whose position vector $\bar{\mathbf{r}}$ satisfies the equation

$$m\bar{\mathbf{r}} = \sum_{i=1}^{n} m_i\mathbf{r}_i \tag{14.12}$$

where m represents the total mass $\sum_{i=1}^{n} m_i$ of the particles. Differentiating both members of Eq. (14.12) twice with respect to t, we obtained the relations

$$\mathbf{L} = m\overline{\mathbf{v}} \qquad \dot{\mathbf{L}} = m\overline{\mathbf{a}} \qquad (14.14, 14.15)$$

where $\overline{\mathbf{v}}$ and $\overline{\mathbf{a}}$ represent, respectively, the velocity and the acceleration of the mass center G. Substituting for $\dot{\mathbf{L}}$ from (14.15) into (14.10), we obtained the equation

$$\Sigma\mathbf{F} = m\overline{\mathbf{a}} \qquad (14.16)$$

from which we concluded that *the mass center of a system of particles moves as if the entire mass of the system and all the external forces were concentrated at that point* [Sample Prob. 14.1].

Angular momentum of a system of particles about its mass center

In Sec. 14.5 we considered the motion of the particles of a system with respect to a centroidal frame $Gx'y'z'$ attached to the mass center G of the system and in translation with respect to the newtonian frame $Oxyz$ (Fig. 14.14). We defined the *angular momentum* of the system *about its mass center G* as the sum of the moments about G of the momenta $m_i\mathbf{v}_i'$ of the particles in their motion relative to the frame $Gx'y'z'$. We also noted that the same result can be obtained by considering the moments about G of the momenta $m_i\mathbf{v}_i$ of the particles in their absolute motion. We therefore wrote

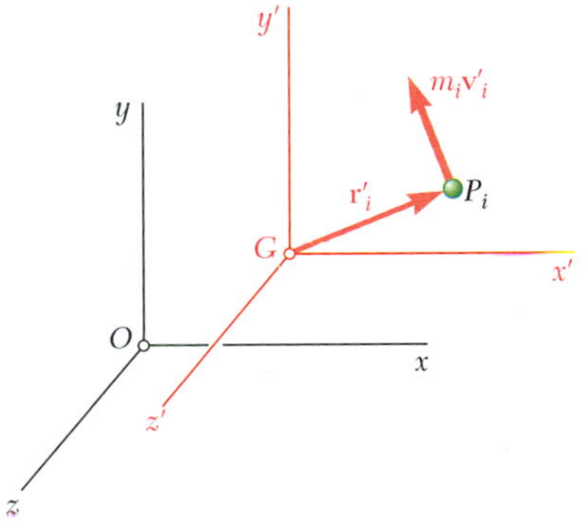

Fig. 14.14

$$\mathbf{H}_G = \sum_{i=1}^{n} (\mathbf{r}_i' \times m_i\mathbf{v}_i) = \sum_{i=1}^{n} (\mathbf{r}_i' \times m_i\mathbf{v}_i') \qquad (14.24)$$

and derived the relation

$$\Sigma\mathbf{M}_G = \dot{\mathbf{H}}_G \qquad (14.23)$$

which expresses that *the moment resultant about G of the external forces is equal to the rate of change of the angular momentum about G of the system of particles.* As will be seen later, this relation is fundamental to the study of the motion of rigid bodies.

Conservation of momentum

When no external force acts on a system of particles [Sec. 14.6], it follows from Eqs. (14.10) and (14.11) that the linear momentum $\mathbf{L}$ and the angular momentum $\mathbf{H}_O$ of the system are conserved [Sample Probs. 14.2 and 14.3]. In problems involving central forces, the angular momentum of the system about the center of force O will also be conserved.

Kinetic energy of a system of particles

The kinetic energy T of a system of particles was defined as the sum of the kinetic energies of the particles [Sec. 14.7]:

$$T = \frac{1}{2}\sum_{i=1}^{n} m_i v_i^2 \qquad (14.28)$$

Using the centroidal frame of reference $Gx'y'z'$ of Fig. 14.14, we noted that the kinetic energy of the system can also be obtained by adding the kinetic energy $\frac{1}{2}m\bar{v}^2$ associated with the motion of the mass center G and the kinetic energy of the system in its motion relative to the frame $Gx'y'z'$:

$$T = \frac{1}{2}m\bar{v}^2 + \frac{1}{2}\sum_{i=1}^{n} m_i v_i'^2 \qquad (14.29)$$

Principle of work and energy

The *principle of work and energy* can be applied to a system of particles as well as to individual particles [Sec. 14.8]. We wrote

$$T_1 + U_{1\rightarrow 2} = T_2 \qquad (14.30)$$

and noted that $U_{1\rightarrow 2}$ represents the work of *all* the forces acting on the particles of the system, internal as well as external.

Conservation of energy

If all the forces acting on the particles of the system are *conservative,* we can determine the potential energy V of the system and write

$$T_1 + V_1 = T_2 + V_2 \qquad (14.31)$$

which expresses the *principle of conservation of energy* for a system of particles.

Principle of impulse and momentum

We saw in Sec. 14.9 that the *principle of impulse and momentum* for a system of particles can be expressed graphically as shown in Fig. 14.15. It states that the momenta of the particles at time t_1 and the impulses of the external forces from t_1 to t_2 form a system of vectors equipollent to the system of the momenta of the particles at time t_2.

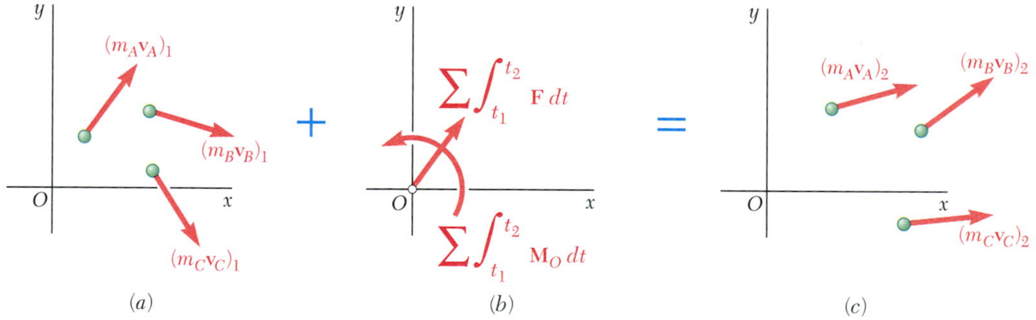

Fig. 14.15

If no external force acts on the particles of the system, the systems of momenta shown in parts *a* and *c* of Fig. 14.15 are equipollent and we have

$$\mathbf{L}_1 = \mathbf{L}_2 \qquad (\mathbf{H}_O)_1 = (\mathbf{H}_O)_2 \qquad (14.36, 14.37)$$

Use of conservation principles in the solution of problems involving systems of particles

Many problems involving the motion of systems of particles can be solved by applying simultaneously the principle of impulse and momentum and the principle of conservation of energy [Sample Prob. 14.4] or by expressing that the linear momentum, angular momentum, and energy of the system are conserved [Sample Prob. 14.5].

Variable systems of particles

Steady stream of particles

In the second part of the chapter, we considered *variable systems of particles.* First we considered a *steady stream of particles,* such as a stream of water diverted by a fixed vane or the flow of air through a jet engine [Sec. 14.11]. Applying the principle of impulse and momentum to a system S of particles during a time interval Δt, and including the particles which enter the system at A during that time interval and those (of the same mass Δm) which leave the system at B, we concluded that *the system formed by the momentum $(\Delta m)\mathbf{v}_A$ of the particles entering S in the time Δt and the impulses of the forces exerted on S during that time is equipollent to the momentum $(\Delta m)\mathbf{v}_B$ of the particles leaving S in the same time* Δt (Fig. 14.16). Equating the x components, y components, and moments about a fixed point of the vectors involved, we could obtain as many as three equations, which could be solved for the desired unknowns [Sample Probs. 14.6 and 14.7]. From this result, we could also derive the following expression for the resultant $\Sigma\mathbf{F}$ of the forces exerted on S,

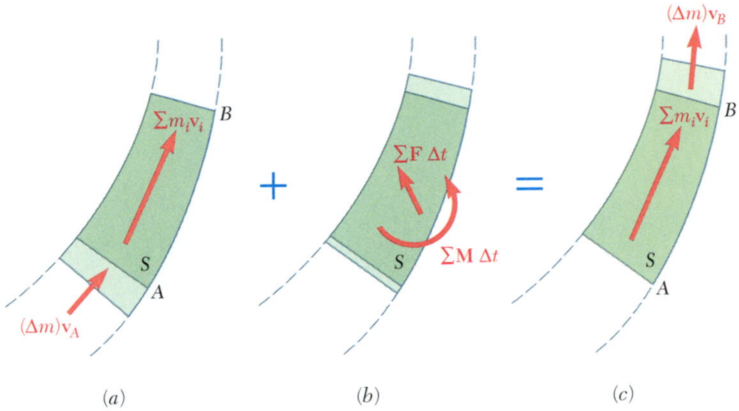

Fig. 14.16

$$\Sigma\mathbf{F} = \frac{dm}{dt}(\mathbf{v}_B - \mathbf{v}_A) \tag{14.39}$$

where $\mathbf{v}_B - \mathbf{v}_A$ represents the difference between the *vectors* $\mathbf{v}_B$ and $\mathbf{v}_A$ and where dm/dt is the mass rate of flow of the stream (see footnote, page 887).

Systems gaining or losing mass

Considering next a system of particles gaining mass by continually absorbing particles or losing mass by continually expelling particles [Sec. 14.12], as in the case of a rocket, we applied the principle of impulse and momentum to the system during a time interval Δt, being careful to include the particles gained or lost during that time interval [Sample Prob. 14.8]. We also noted that the action on a system S of the particles being *absorbed* by S was equivalent to a thrust

$$\mathbf{P} = \frac{dm}{dt}\mathbf{u} \tag{14.44}$$

where dm/dt is the rate at which mass is being absorbed, and $\mathbf{u}$ is the velocity of the particles *relative to S.* In the case of particles being *expelled* by S, the rate dm/dt is negative and the thrust $\mathbf{P}$ is exerted in a direction opposite to that in which the particles are being expelled.

Review Problems

14.107 A 180-lb man and a 120-lb woman stand side by side at the same end of a 300-lb boat, ready to dive, each with a 16-ft/s velocity relative to the boat. Determine the velocity of the boat after they have both dived, if (*a*) the woman dives first, (*b*) the man dives first.

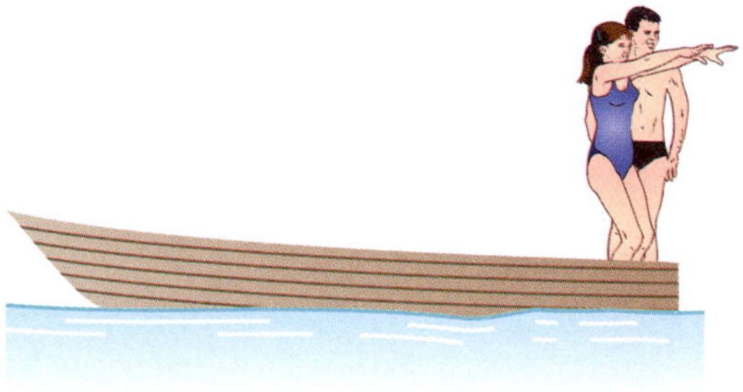

Fig. P14.107

14.108 A 180-lb man and a 120-lb woman stand at opposite ends of a 300-lb boat, ready to dive, each with a 16-ft/s velocity relative to the boat. Determine the velocity of the boat after they have both dived, if (*a*) the woman dives first, (*b*) the man dives first.

Fig. P14.108

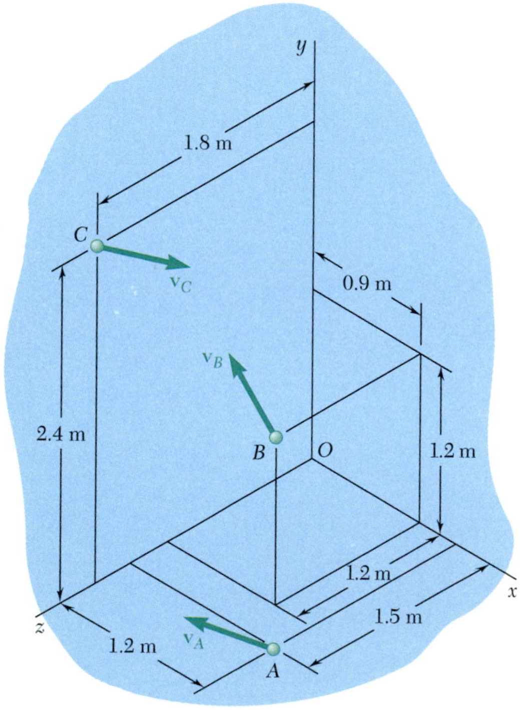

Fig. *P14.109*

14.109 A system consists of three particles A, B, and C. We know that $m_A = 3$ kg, $m_B = 4$ kg, and $m_C = 5$ kg and that the velocities of the particles expressed in m/s are, respectively, $\mathbf{v}_A = -4\mathbf{i} + 4\mathbf{j} + 6\mathbf{k}$, $\mathbf{v}_B = -6\mathbf{i} + 8\mathbf{j} + 4\mathbf{k}$, and $\mathbf{v}_C = 2\mathbf{i} - 6\mathbf{j} - 4\mathbf{k}$. Determine the angular momentum $\mathbf{H}_O$ of the system about O.

14.110 For the system of particles in Prob. 14.109, determine (*a*) the position vector $\bar{\mathbf{r}}$ of the mass center G of the system, (*b*) the linear momentum $m\bar{\mathbf{v}}$ of the system, (*c*) the angular momentum $\mathbf{H}_G$ of the system about G. Also verify that the answers to this problem and to Prob. 14.109 satisfy the equation given in Prob. 14.28.

14.111 and 14.112 Car A was traveling east at high speed when it collided at point O with car B, which was traveling north at 72 km/h. Car C, which was traveling west at 90 km/h, was 10 m east and 3 m north of point O at the time of the collision. Because the pavement was wet, the driver of car C could not prevent his car from sliding into the other two cars, and the three cars, stuck together, kept sliding until they hit the utility pole P. Knowing that the masses of cars A, B, and C are, respectively, 1500 kg, 1300 kg, and 1200 kg, and neglecting the forces exerted on the cars by the wet pavement, solve the problems indicated.

14.111 Knowing that the coordinates of the utility pole are $x_P = 18$ m and $y_P = 13.9$ m, determine (*a*) the time elapsed from the first collision to the stop at P, (*b*) the speed of car A.

14.112 Knowing that the speed of car A was 129.6 km/h and that the time elapsed from the first collision to the stop at P was 2.4 s, determine the coordinates of the utility pole P.

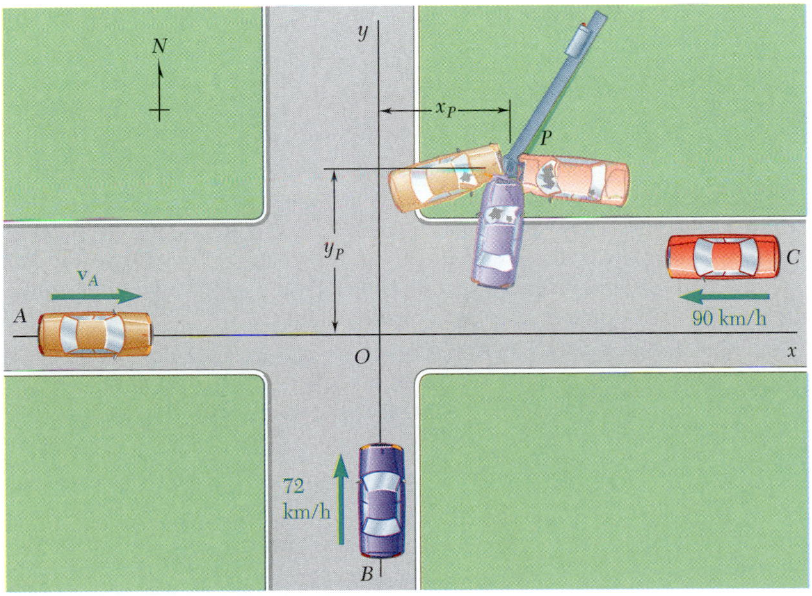

Fig. P14.111 and *P14.112*

14.113 In Prob. 14.107, determine the work done by the woman and by the man as each dives from the boat, assuming that the woman dives first.

14.114 A rotary power plow is used to remove snow from a level section of railroad track. The plow car is placed ahead of an engine which propels it at a constant speed of 12 mi/h. The plow car clears 180 tons of snow per minute, projecting it in the direction shown with a velocity of 40 ft/s relative to the plow car. Neglecting friction, determine (*a*) the force exerted by the engine on the plow car, (*b*) the lateral force exerted by the track on the plow car.

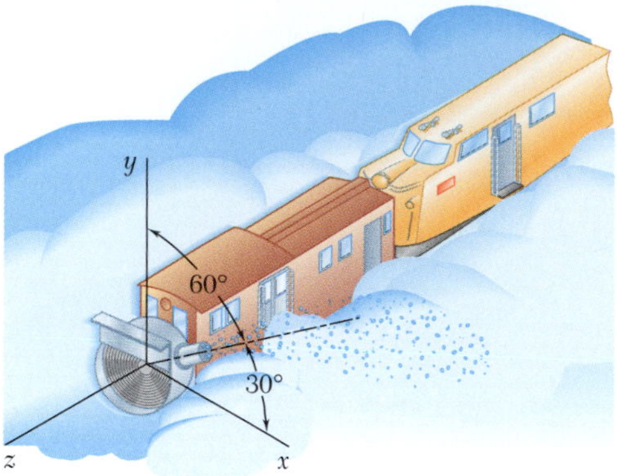

Fig. P14.114

14.115 The stream of water shown flows at a rate of 150 gal/min and moves with a velocity of magnitude 60 ft/s at both A and B. The vane is supported by a pin and bracket at C and by a load cell at D which can exert only a horizontal force. Neglecting the weight of the vane, determine the components of the reactions at C and D (1 ft^3 = 7.48 gal).

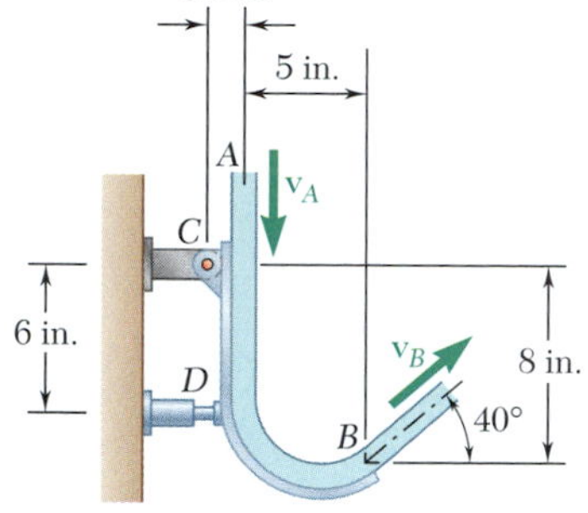

Fig. *P14.115*

14.116 In order to shorten the distance required for landing, a jet airplane is equipped with moveable vanes which partially reverse the direction of the air discharged by each of its engines. Each engine scoops in the air at a rate of 120 kg/s and discharges it with a velocity of 600 m/s relative to the engine. At an instant when the speed of the airplane is 270 km/h, determine the reverse thrust provided by each of the engines.

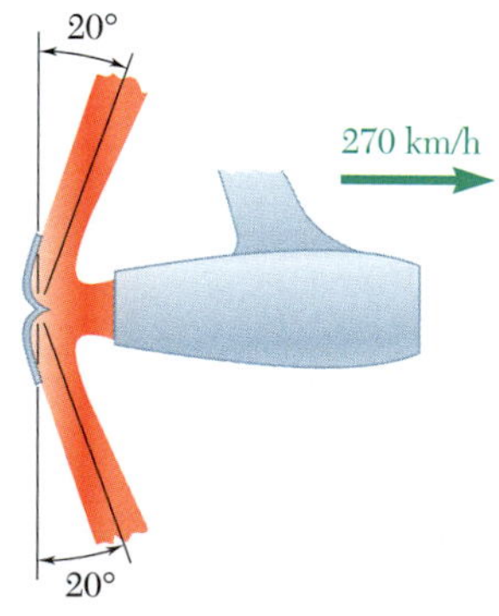

Fig. P14.116

14.117 A space vehicle describing a circular orbit about the earth at a speed of 24 Mm/h releases its front end capsule which has a gross mass of 600 kg, including 400 kg of fuel. Knowing that the fuel is consumed at the rate of 20 kg/s and ejected with a relative velocity of 3000 m/s, determine (*a*) the tangential acceleration of the capsule as its engine is fired, (*b*) the maximum speed attained by the capsule.

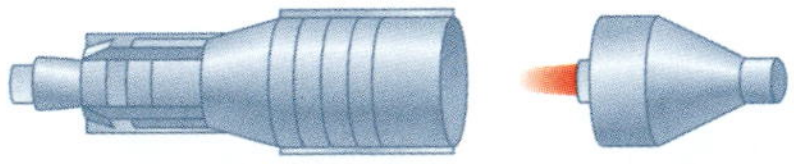

Fig. P14.117

14.118 A jet airliner is cruising at a speed of 600 mi/h with each of its three engines discharging air with a velocity of 2000 ft/s relative to the plane. Determine the speed of the airliner after it has lost the use of (*a*) one of its engines, (*b*) two of its engines. Assume that the drag due to air friction is proportional to the square of the speed and that the remaining engines keep operating at the same rate.

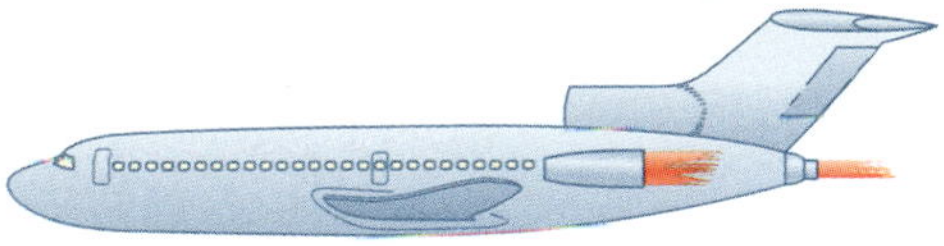

Fig. *P14.118*

Computer Problems

14.C1 A 9-kg block B starts from rest and slides down the inclined surface of a 15-kg wedge A which is supported by a horizontal surface. Neglecting friction and denoting by s the distance traveled by block B down the surface of the wedge, derive expressions for (*a*) the speed of block B relative to wedge A, (*b*) the speed of wedge A. (*c*) Plot the speed of B relative to A and the speed of A as functions of s for $0 \leq s \leq 0.6$ m.

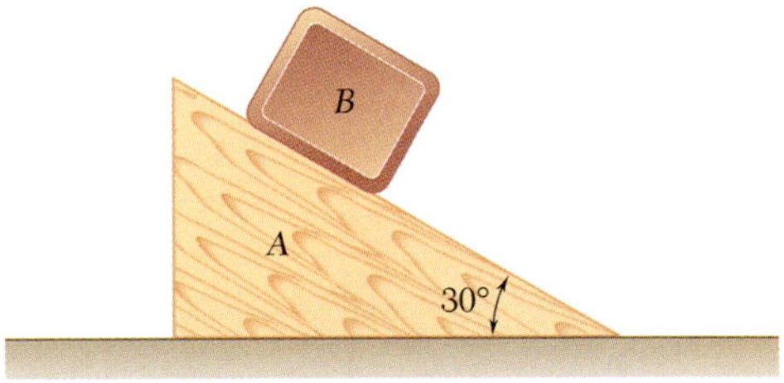

Fig. P14.C1

14.C2 A system of particles consists of n particles A_i of weight W_i and coordinates x_i, y_i, and z_i, having velocities of components $(v_x)_i$, $(v_y)_i$, and $(v_z)_i$. Derive expressions for the components of the angular momentum of the system about the origin O of the coordinates. Use computational software to solve (*a*) Prob. 14.9, (*b*) Prob. 14.13.

14.C3 A shell moving with a velocity of known components v_x, v_y, and v_z explodes into three fragments of masses m_1, m_2, and m_3 at point A_0 at a distance d from a vertical wall. Derive expressions for the speeds of the three fragments immediately after the explosion, knowing the coordinates x_i and y_i of the points A_i (i = 1, 2, 3) where the fragments hit the wall. Use computational software to solve (*a*) Prob.14.25, (*b*) Prob. 14.26.

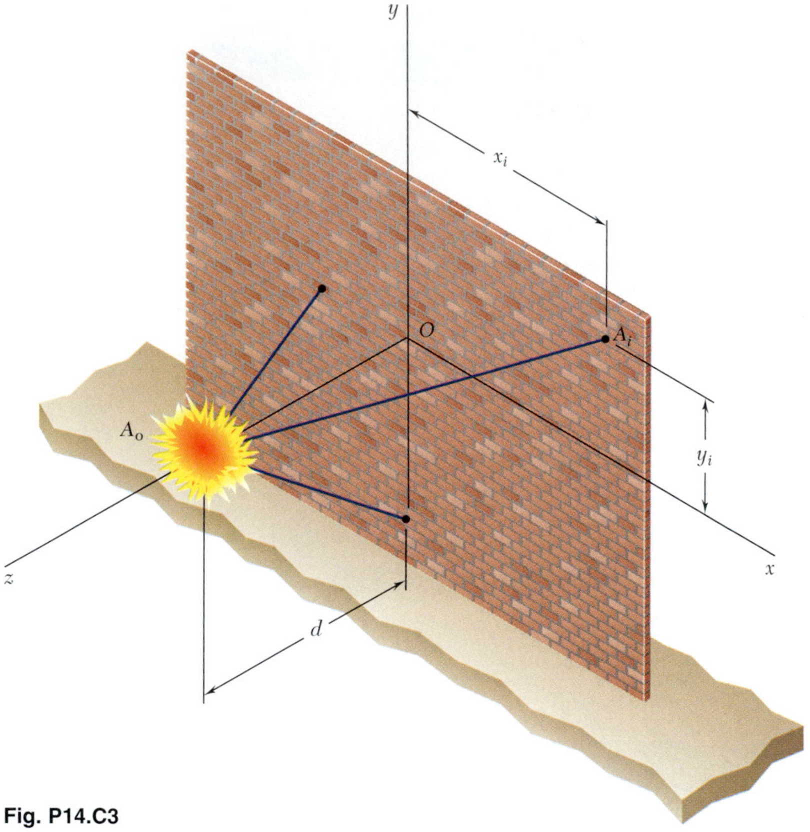

Fig. P14.C3

14.C4 As a 12,000-lb training plane lands on an aircraft carrier at a speed of 150 mi/h, its tail hooks into the end of a 260-ft long chain which lies in a pile below deck. The chain has a weight per unit length of 24 lb/ft and there is no other retarding force. Denoting by x the distance traveled by the plane while the chain is being pulled out, use computational software to calculate and plot as functions of x (*a*) the elapsed time, (*b*) the speed of the plane, (*c*) the deceleration of the plane.

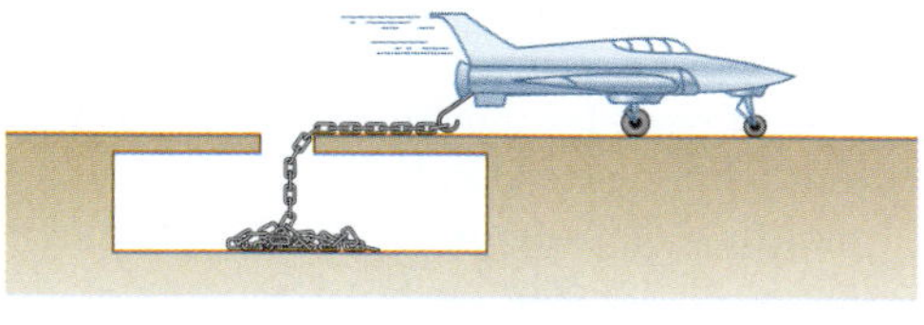

Fig. P14.C4

14.C5 A 32,000-lb jet airplane maintains a constant speed of 645 mi/h while climbing at an angle $\alpha = 18°$. The airplane scoops in air at a rate of 600 lb/s and discharges it with a velocity of 2000 mi/h relative to the airplane. Knowing that the pilot changes the angle of climb α while maintaining the same engine setting, calculate and plot as functions of α for $0 \le \alpha \le 20°$ (*a*) the initial acceleration of the plane, (*b*) the maximum speed that will be attained. Assume that the drag due to air friction is proportional to the square of the speed.

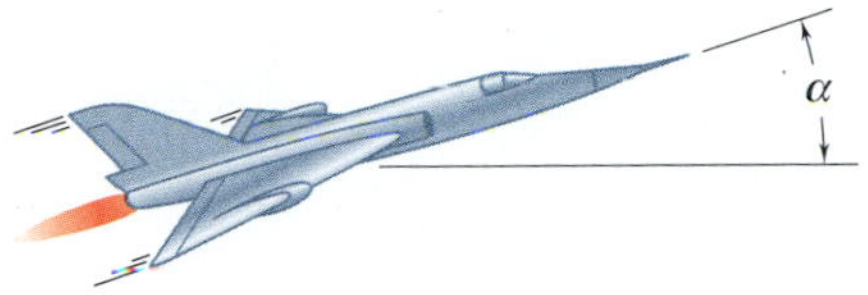

Fig. P14.C5

14.C6 A rocket has a mass of 1200 kg, including 1000 kg of fuel, which is consumed at the rate of 12.5 kg/s and ejected with a relative velocity of 3.6 km/s. Knowing that the rocket is fired vertically from the ground and assuming a constant value for the acceleration of gravity, use computational software to calculate and plot from the time of ignition to the time when the last particle of fuel is being consumed (*a*) the acceleration a of the rocket in m/s^2, (*b*) its velocity v in m/s, (*c*) its elevation h above the ground in kilometers. (*Hint:* Use for v the expression derived in Sample Prob. 14.8, and integrate this expression analytically to obtain h.)

14.C7 Two hemispheres are held together by a cord which maintains a spring under compression (the spring is not attached to the hemispheres). The potential energy of the compressed spring is 90 ft · lb and the assembly has an initial velocity $\mathbf{v}_0$ of magnitude $v_0 = 24$ ft/s. Knowing that the cord is severed causing the hemispheres to fly apart, calculate and plot the magnitude of the resulting velocity for each hemisphere as a function of θ for $30° \le \theta \le 120°$.

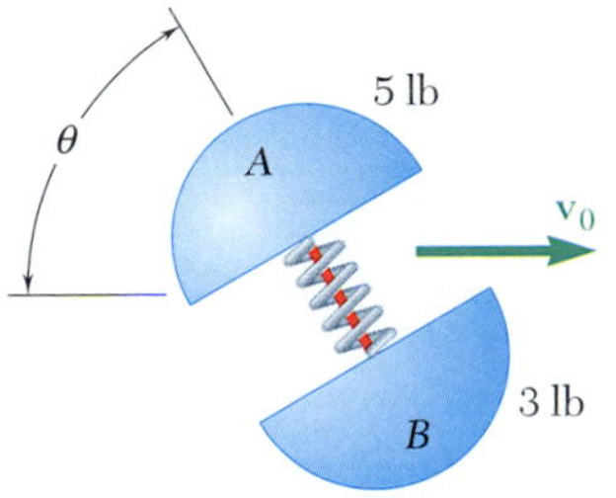

Fig. P14.C7

CHAPTER

15

Kinematics of Rigid Bodies

In this chapter you will learn to analyze the motion of mechanical systems such as the gear system shown. You may be asked, for example, to determine the angular velocity and angular acceleration of one gear, knowing the angular velocity and angular acceleration of another gear.

15.1. INTRODUCTION

In this chapter, the kinematics of *rigid bodies* will be considered. You will investigate the relations existing between the time, the positions, the velocities, and the accelerations of the various particles forming a rigid body. As you will see, the various types of rigid-body motion can be conveniently grouped as follows:

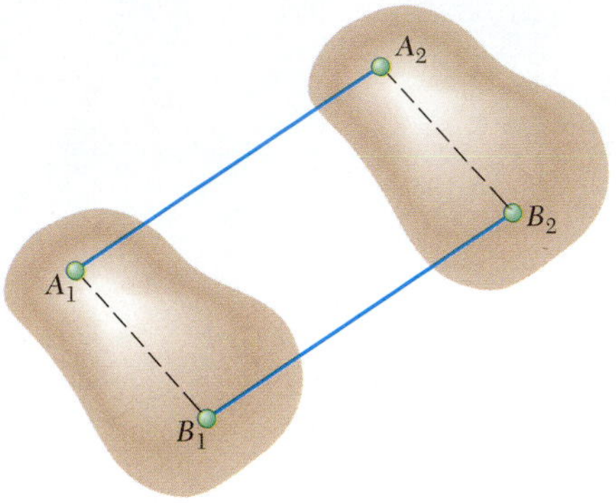

Fig. 15.1

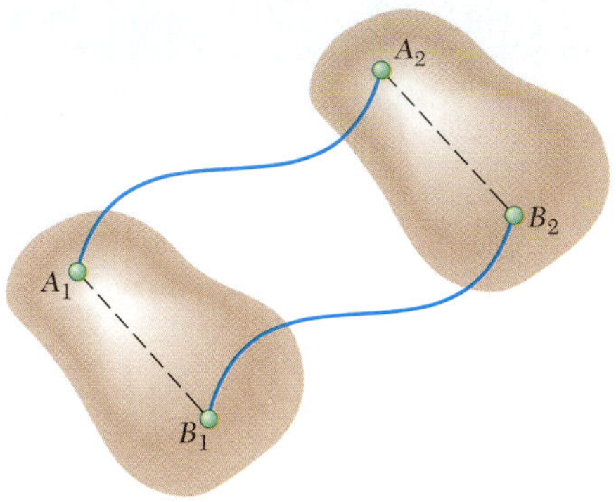

Fig. 15.2

1. *Translation.* A motion is said to be a translation if any straight line inside the body keeps the same direction during the motion. It can also be observed that in a translation all the particles forming the body move along parallel paths. If these paths are straight lines, the motion is said to be a *rectilinear translation* (Fig. 15.1); if the paths are curved lines, the motion is a *curvilinear translation* (Fig. 15.2).

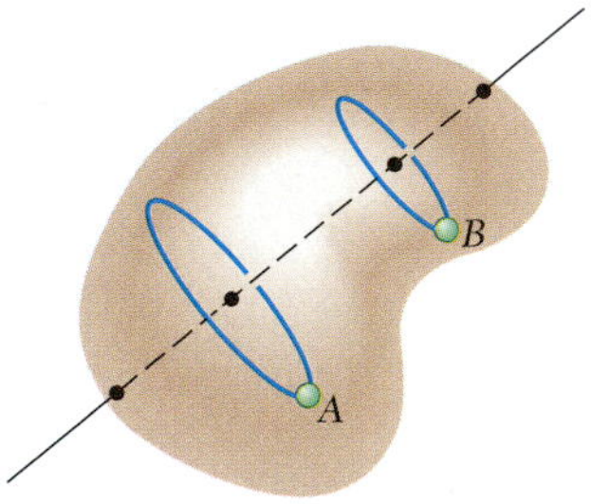

Fig. 15.3

2. *Rotation about a Fixed Axis.* In this motion, the particles forming the rigid body move in parallel planes along circles centered on the same fixed axis (Fig. 15.3). If this axis, called the *axis of rotation,* intersects the rigid body, the particles located on the axis have zero velocity and zero acceleration.

   Rotation should not be confused with certain types of curvilinear translation. For example, the plate shown in Fig. 15.4*a* is in curvilinear translation, with all its particles moving along *parallel* circles, while the plate shown in Fig. 15.4*b* is in rotation, with all its particles moving along *concentric* circles.

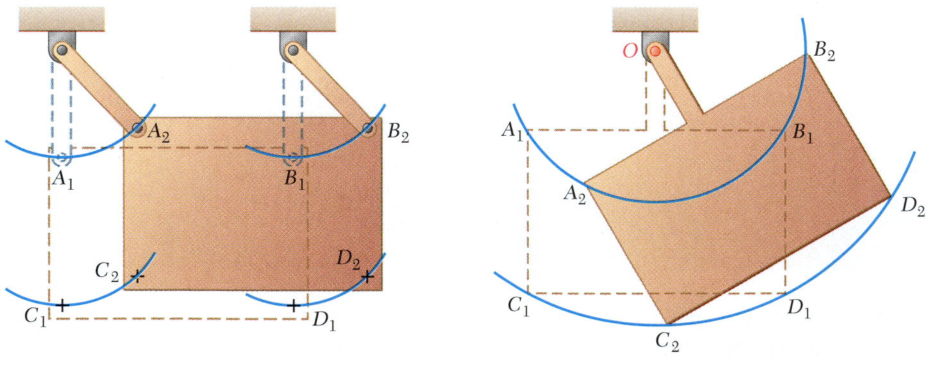

(*a*) Curvilinear translation

(*b*) Rotation

Fig. 15.4

In the first case, any given straight line drawn on the plate will maintain the same direction, whereas in the second case, point O remains fixed.

Because each particle moves in a given plane, the rotation of a body about a fixed axis is said to be a *plane motion*.

3. *General Plane Motion.* There are many other types of plane motion, that is, motions in which all the particles of the body move in parallel planes. Any plane motion which is neither a rotation nor a translation is referred to as a general plane motion. Two examples of general plane motion are given in Fig. 15.5.

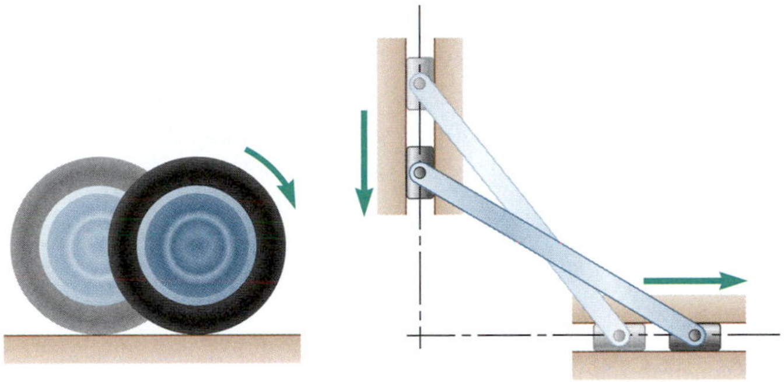

(*a*) Rolling wheel (*b*) Sliding rod

Fig. 15.5

4. *Motion about a Fixed Point.* The three-dimensional motion of a rigid body attached at a fixed point O, for example, the motion of a top on a rough floor (Fig. 15.6), is known as motion about a fixed point.

5. *General Motion.* Any motion of a rigid body which does not fall in any of the categories above is referred to as a general motion.

After a brief discussion in Sec. 15.2 of the motion of translation, the rotation of a rigid body about a fixed axis is considered in Sec. 15.3. The *angular velocity* and the *angular acceleration* of a rigid body about a fixed axis will be defined, and you will learn to express the velocity and the acceleration of a given point of the body in terms of its position vector and the angular velocity and angular acceleration of the body.

The following sections are devoted to the study of the general plane motion of a rigid body and to its application to the analysis of mechanisms such as gears, connecting rods, and pin-connected linkages. Resolving the plane motion of a slab into a translation and a rotation (Secs. 15.5 and 15.6), we will then express the velocity of a point B of the slab as the sum of the velocity of a reference point A and of the velocity of B relative to a frame of reference translating with A (that is, moving with A but not rotating). The same approach is used later in Sec. 15.8 to express the acceleration of B in terms of the acceleration of A and of the acceleration of B relative to a frame translating with A.

KINEMATICS OF RIGID BODIES

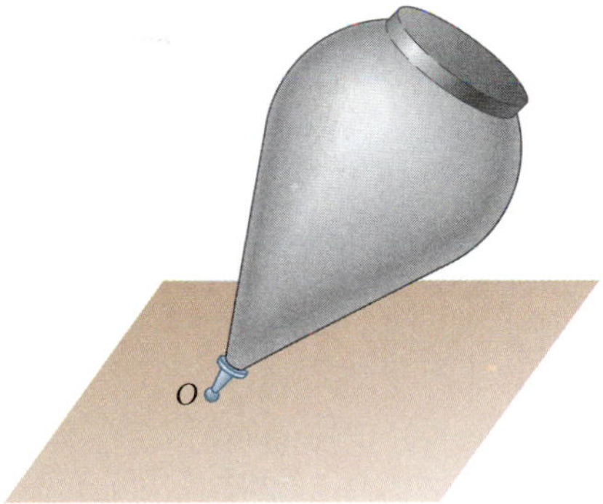

Fig. 15.6

An alternative method for the analysis of velocities in plane motion, based on the concept of *instantaneous center of rotation,* is given in Sec. 15.7; and still another method of analysis, based on the use of parametric expressions for the coordinates of a given point, is presented in Sec. 15.9.

The motion of a particle relative to a rotating frame of reference and the concept of *Coriolis acceleration* are discussed in Secs. 15.10 and 15.11, and the results obtained are applied to the analysis of the plane motion of mechanisms containing parts which slide on each other.

The remaining part of the chapter is devoted to the analysis of the three-dimensional motion of a rigid body, namely, the motion of a rigid body with a fixed point and the general motion of a rigid body. In Secs. 15.12 and 15.13, a fixed frame of reference or a frame of reference in translation will be used to carry out this analysis; in Secs. 15.14 and 15.15, the motion of the body relative to a rotating frame or to a frame in general motion will be considered, and the concept of Coriolis acceleration will again be used.

15.2. TRANSLATION

Consider a rigid body in translation (either rectilinear or curvilinear translation), and let A and B be any two of its particles (Fig. 15.7*a*). Denoting, respectively, by $\mathbf{r}_A$ and $\mathbf{r}_B$ the position vectors of A and B with respect to a fixed frame of reference and by $\mathbf{r}_{B/A}$ the vector joining A and B, we write

$$\mathbf{r}_B = \mathbf{r}_A + \mathbf{r}_{B/A} \tag{15.1}$$

Let us differentiate this relation with respect to t. We note that from the very definition of a translation, the vector $\mathbf{r}_{B/A}$ must maintain a constant direction; its magnitude must also be constant, since A and

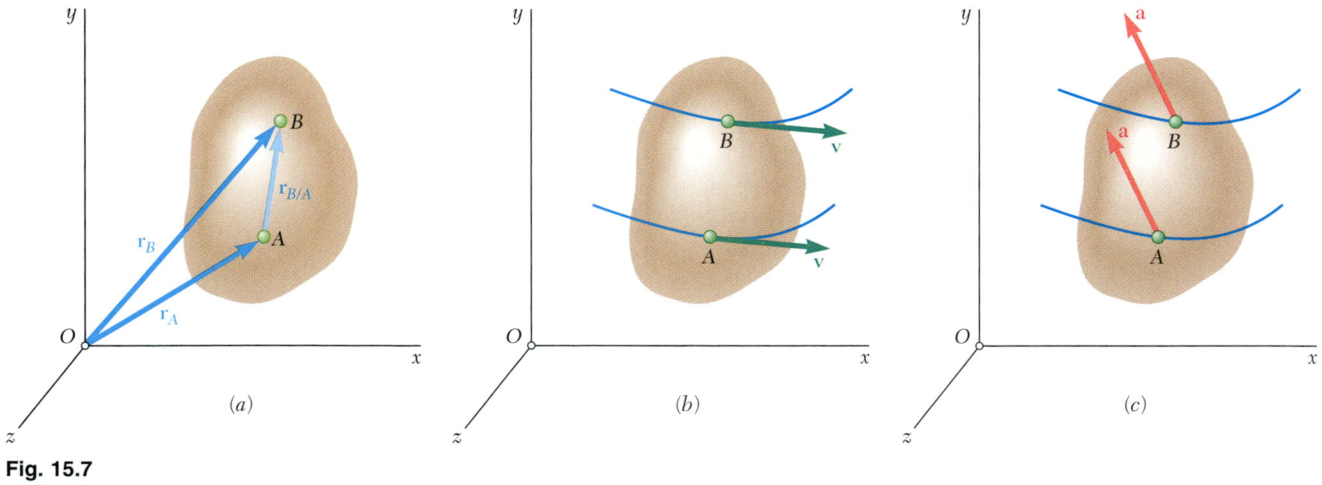

Fig. 15.7

B belong to the same rigid body. Thus, the derivative of $\mathbf{r}_{B/A}$ is zero and we have

$$\mathbf{v}_B = \mathbf{v}_A \tag{15.2}$$

Differentiating once more, we write

$$\mathbf{a}_B = \mathbf{a}_A \tag{15.3}$$

Thus, *when a rigid body is in translation, all the points of the body have the same velocity and the same acceleration at any given instant* (Fig. 15.7*b* and *c*). In the case of curvilinear translation, the velocity and acceleration change in direction as well as in magnitude at every instant. In the case of rectilinear translation, all particles of the body move along parallel straight lines, and their velocity and acceleration keep the same direction during the entire motion.

15.3. ROTATION ABOUT A FIXED AXIS

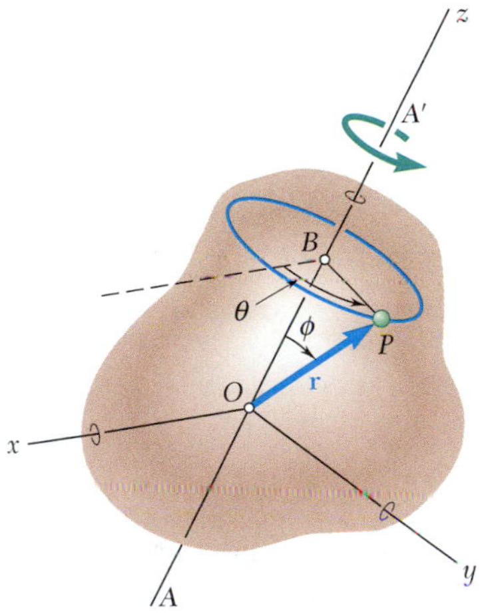

Fig. 15.8

Consider a rigid body which rotates about a fixed axis AA'. Let P be a point of the body and $\mathbf{r}$ its position vector with respect to a fixed frame of reference. For convenience, let us assume that the frame is centered at point O on AA' and that the z axis coincides with AA' (Fig. 15.8). Let B be the projection of P on AA'; since P must remain at a constant distance from B, it will describe a circle of center B and of radius $r \sin \phi$, where ϕ denotes the angle formed by $\mathbf{r}$ and AA'.

The position of P and of the entire body is completely defined by the angle θ the line BP forms with the zx plane. The angle θ is known as the *angular coordinate* of the body and is defined as positive when viewed as counterclockwise from A'. The angular coordinate will be expressed in radians (rad) or, occasionally, in degrees (°) or revolutions (rev). We recall that

$$1 \text{ rev} = 2\pi \text{ rad} = 360°$$

We recall from Sec. 11.9 that the velocity $\mathbf{v} = d\mathbf{r}/dt$ of a particle P is a vector tangent to the path of P and of magnitude $v = ds/dt$. Observing that the length Δs of the arc described by P when the body rotates through $\Delta\theta$ is

$$\Delta s = (BP)\,\Delta\theta = (r \sin \phi)\,\Delta\theta$$

and dividing both members by Δt, we obtain at the limit, as Δt approaches zero,

$$v = \frac{ds}{dt} = r\dot{\theta} \sin \phi \tag{15.4}$$

where $\dot{\theta}$ denotes the time derivative of θ. (Note that the angle θ depends on the position of P within the body, but the rate of change $\dot{\theta}$ is itself independent of P.) We conclude that the velocity $\mathbf{v}$ of P is

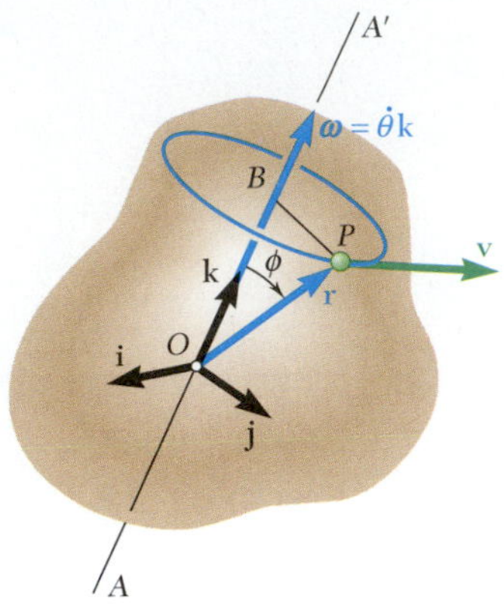

Fig. 15.9

a vector perpendicular to the plane containing AA' and $\mathbf{r}$, and of magnitude v defined by (15.4). But this is precisely the result we would obtain if we drew along AA' a vector $\boldsymbol{\omega} = \dot{\theta}\mathbf{k}$ and formed the vector product $\boldsymbol{\omega} \times \mathbf{r}$ (Fig. 15.9). We thus write

$$\mathbf{v} = \frac{d\mathbf{r}}{dt} = \boldsymbol{\omega} \times \mathbf{r} \tag{15.5}$$

The vector

$$\boldsymbol{\omega} = \omega\mathbf{k} = \dot{\theta}\mathbf{k} \tag{15.6}$$

which is directed along the axis of rotation, is called the *angular velocity* of the body and is equal in magnitude to the rate of change $\dot{\theta}$ of the angular coordinate; its sense may be obtained by the right-hand rule (Sec. 3.6) from the sense of rotation of the body.†

The acceleration $\mathbf{a}$ of the particle P will now be determined. Differentiating (15.5) and recalling the rule for the differentiation of a vector product (Sec. 11.10), we write

$$\begin{aligned}\mathbf{a} &= \frac{d\mathbf{v}}{dt} = \frac{d}{dt}(\boldsymbol{\omega} \times \mathbf{r}) \\ &= \frac{d\boldsymbol{\omega}}{dt} \times \mathbf{r} + \boldsymbol{\omega} \times \frac{d\mathbf{r}}{dt} \\ &= \frac{d\boldsymbol{\omega}}{dt} \times \mathbf{r} + \boldsymbol{\omega} \times \mathbf{v} \end{aligned} \tag{15.7}$$

Photo 15.1 For the central gear rotating about a fixed axis, the angular velocity and angular acceleration of that gear are vectors directed along the vertical axis of rotation.

The vector $d\boldsymbol{\omega}/dt$ is denoted by $\boldsymbol{\alpha}$ and is called the *angular acceleration* of the body. Substituting also for $\mathbf{v}$ from (15.5), we have

$$\mathbf{a} = \boldsymbol{\alpha} \times \mathbf{r} + \boldsymbol{\omega} \times (\boldsymbol{\omega} \times \mathbf{r}) \tag{15.8}$$

Differentiating (15.6) and recalling that $\mathbf{k}$ is constant in magnitude and direction, we have

$$\boldsymbol{\alpha} = \alpha\mathbf{k} = \dot{\omega}\mathbf{k} = \ddot{\theta}\mathbf{k} \tag{15.9}$$

Thus, the angular acceleration of a body rotating about a fixed axis is a vector directed along the axis of rotation, and is equal in magnitude to the rate of change $\dot{\omega}$ of the angular velocity. Returning to (15.8), we note that the acceleration of P is the sum of two vectors. The first vector is equal to the vector product $\boldsymbol{\alpha} \times \mathbf{r}$; it is tangent to the circle described by P and therefore represents the tangential component of the acceleration. The second vector is equal to the *vector triple product* $\boldsymbol{\omega} \times (\boldsymbol{\omega} \times \mathbf{r})$ obtained by forming the vector product of $\boldsymbol{\omega}$ and $\boldsymbol{\omega} \times \mathbf{r}$; since $\boldsymbol{\omega} \times \mathbf{r}$ is tangent to the circle described by P, the vector triple product is directed toward the center B of the circle and therefore represents the normal component of the acceleration.

†It will be shown in Sec. 15.12 in the more general case of a rigid body rotating simultaneously about axes having different directions that angular velocities obey the parallelogram law of addition and thus are actually vector quantities.

Rotation of a Representative Slab. The rotation of a rigid body about a fixed axis can be defined by the motion of a representative slab in a reference plane perpendicular to the axis of rotation. Let us choose the xy plane as the reference plane and assume that it coincides with the plane of the figure, with the z axis pointing out of the paper (Fig. 15.10). Recalling from (15.6) that $\boldsymbol{\omega} = \omega\mathbf{k}$, we note

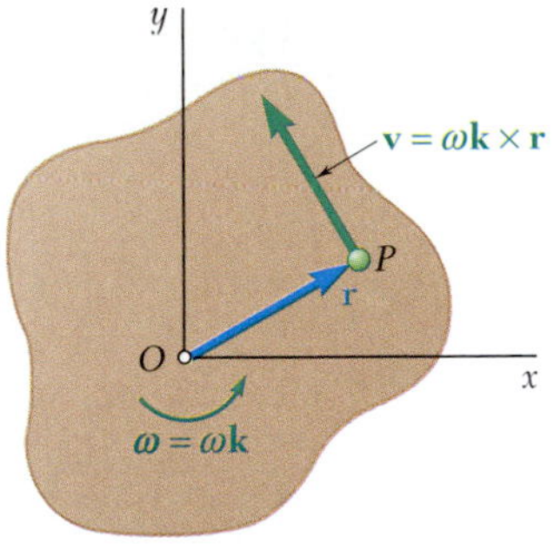

Fig. 15.10

that a positive value of the scalar ω corresponds to a counterclockwise rotation of the representative slab, and a negative value to a clockwise rotation. Substituting $\omega\mathbf{k}$ for $\boldsymbol{\omega}$ into Eq. (15.5), we express the velocity of any given point P of the slab as

$$\mathbf{v} = \omega\mathbf{k} \times \mathbf{r} \tag{15.10}$$

Since the vectors $\mathbf{k}$ and $\mathbf{r}$ are mutually perpendicular, the magnitude of the velocity $\mathbf{v}$ is

$$v = r\omega \tag{15.10'}$$

and its direction can be obtained by rotating $\mathbf{r}$ through 90° in the sense of rotation of the slab.

Substituting $\boldsymbol{\omega} = \omega\mathbf{k}$ and $\boldsymbol{\alpha} = \alpha\mathbf{k}$ into Eq. (15.8), and observing that cross-multiplying $\mathbf{r}$ twice by $\mathbf{k}$ results in a 180° rotation of the vector $\mathbf{r}$, we express the acceleration of point P as

$$\mathbf{a} = \alpha\mathbf{k} \times \mathbf{r} - \omega^2\mathbf{r} \tag{15.11}$$

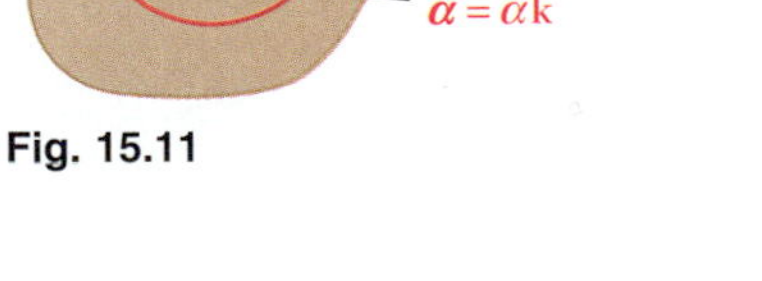

Fig. 15.11

Resolving $\mathbf{a}$ into tangential and normal components (Fig. 15.11), we write

$$\begin{aligned} \mathbf{a}_t &= \alpha\mathbf{k} \times \mathbf{r} \qquad & a_t &= r\alpha \\ \mathbf{a}_n &= -\omega^2\mathbf{r} \qquad & a_n &= r\omega^2 \end{aligned} \tag{15.11'}$$

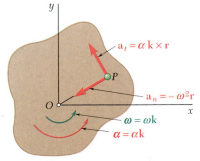

The tangential component $\mathbf{a}_t$ points in the counterclockwise direction if the scalar α is positive, and in the clockwise direction if α is negative. The normal component $\mathbf{a}_n$ always points in the direction opposite to that of $\mathbf{r}$, that is, toward O.

15.4. EQUATIONS DEFINING THE ROTATION OF A RIGID BODY ABOUT A FIXED AXIS

The motion of a rigid body rotating about a fixed axis AA' is said to be *known* when its angular coordinate θ can be expressed as a known function of t. In practice, however, the rotation of a rigid body is seldom defined by a relation between θ and t. More often, the conditions of motion will be specified by the type of angular acceleration that the body possesses. For example, α may be given as a function of t, as a function of θ, or as a function of ω. Recalling the relations (15.6) and (15.9), we write

$$\omega = \frac{d\theta}{dt} \tag{15.12}$$

$$\alpha = \frac{d\omega}{dt} = \frac{d^2\theta}{dt^2} \tag{15.13}$$

or, solving (15.12) for dt and substituting into (15.13),

$$\alpha = \omega\frac{d\omega}{d\theta} \tag{15.14}$$

Since these equations are similar to those obtained in Chap. 11 for the rectilinear motion of a particle, their integration can be performed by following the procedure outlined in Sec. 11.3.

Two particular cases of rotation are frequently encountered:

1. *Uniform Rotation.* This case is characterized by the fact that the angular acceleration is zero. The angular velocity is thus constant, and the angular coordinate is given by the formula

$$\theta = \theta_0 + \omega t \tag{15.15}$$

2. *Uniformly Accelerated Rotation.* In this case, the angular acceleration is constant. The following formulas relating angular velocity, angular coordinate, and time can then be derived in a manner similar to that described in Sec. 11.5. The similarity between the formulas derived here and those obtained for the rectilinear uniformly accelerated motion of a particle is apparent.

$$\begin{aligned} \omega &= \omega_0 + \alpha t \\ \theta &= \theta_0 + \omega_0 t + \tfrac{1}{2}\alpha t^2 \\ \omega^2 &= \omega_0^2 + 2\alpha(\theta - \theta_0) \end{aligned} \tag{15.16}$$

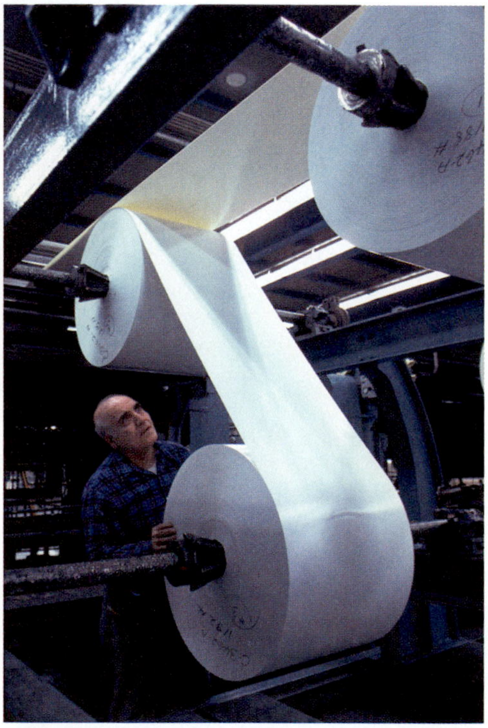

Photo 15.2 If the lower roll has a constant angular velocity, the speed of the paper being wound onto it increases as the radius of the roll increases.

It should be emphasized that formula (15.15) can be used only when $\alpha = 0$, and formulas (15.16) can be used only when $\alpha =$ constant. In any other case, the general formulas (15.12) to (15.14) should be used.

SAMPLE PROBLEM 15.1

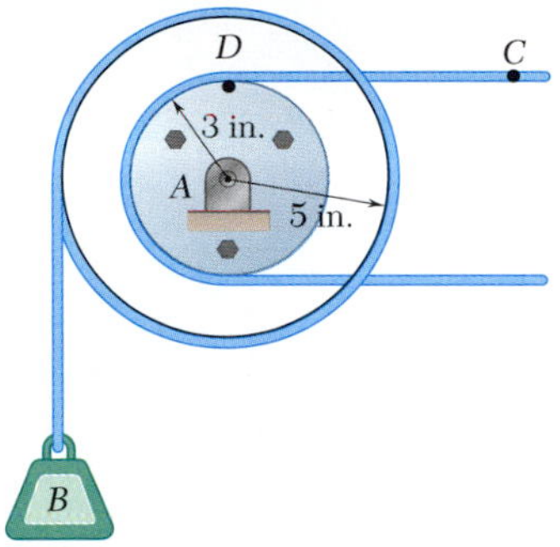

Load B is connected to a double pulley by one of the two inextensible cables shown. The motion of the pulley is controlled by cable C, which has a constant acceleration of 9 in./s^2 and an initial velocity of 12 in./s, both directed to the right. Determine (*a*) the number of revolutions executed by the pulley in 2 s, (*b*) the velocity and change in position of the load B after 2 s, and (*c*) the acceleration of point D on the rim of the inner pulley at $t = 0$.

SOLUTION

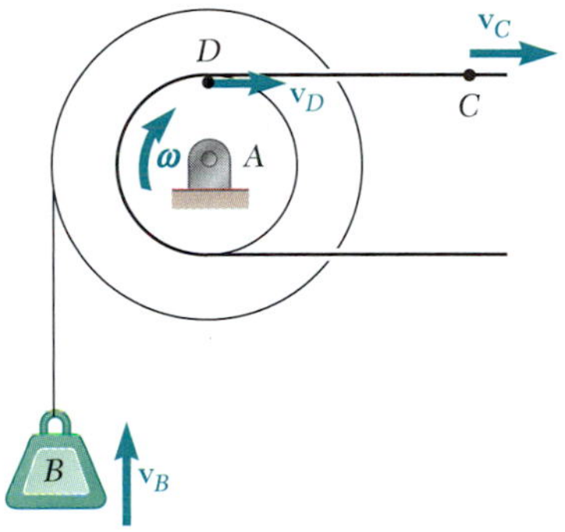

***a.* Motion of Pulley.** Since the cable is inextensible, the velocity of point D is equal to the velocity of point C and the tangential component of the acceleration of D is equal to the acceleration of C.

$$(\mathbf{v}_D)_0 = (\mathbf{v}_C)_0 = 12 \text{ in./s} \rightarrow \qquad (\mathbf{a}_D)_t = \mathbf{a}_C = 9 \text{ in./s}^2 \rightarrow$$

Noting that the distance from D to the center of the pulley is 3 in., we write

$$(v_D)_0 = r\omega_0 \qquad 12 \text{ in./s} = (3 \text{ in.})\omega_0 \qquad \boldsymbol{\omega}_0 = 4 \text{ rad/s} \downarrow$$
$$(a_D)_t = r\alpha \qquad 9 \text{ in./s}^2 = (3 \text{ in.})\alpha \qquad \boldsymbol{\alpha} = 3 \text{ rad/s}^2 \downarrow$$

Using the equations of uniformly accelerated motion, we obtain, for $t = 2$ s,

$$\omega = \omega_0 + \alpha t = 4 \text{ rad/s} + (3 \text{ rad/s}^2)(2 \text{ s}) = 10 \text{ rad/s}$$
$$\boldsymbol{\omega} = 10 \text{ rad/s} \downarrow$$
$$\theta = \omega_0 t + \tfrac{1}{2}\alpha t^2 = (4 \text{ rad/s})(2 \text{ s}) + \tfrac{1}{2}(3 \text{ rad/s}^2)(2 \text{ s})^2 = 14 \text{ rad}$$
$$\theta = 14 \text{ rad} \downarrow$$

$$\text{Number of revolutions} = (14 \text{ rad})\left(\frac{1 \text{ rev}}{2\pi \text{ rad}}\right) = 2.23 \text{ rev} \quad \blacktriangleleft$$

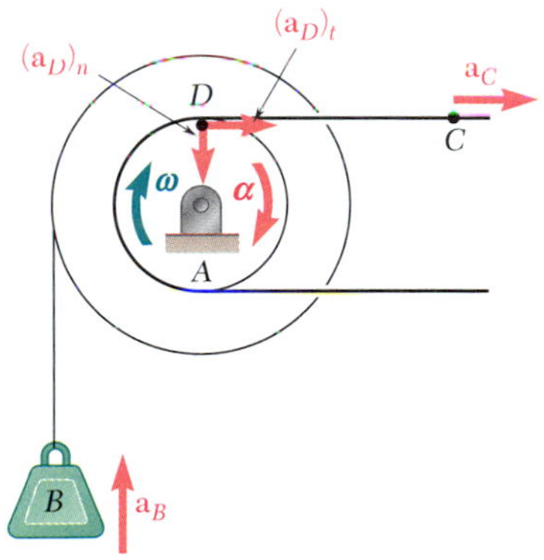

***b.* Motion of Load *B*.** Using the following relations between linear and angular motion, with $r = 5$ in., we write

$$v_B = r\omega = (5 \text{ in.})(10 \text{ rad/s}) = 50 \text{ in./s} \qquad \mathbf{v}_B = 50 \text{ in./s} \uparrow \quad \blacktriangleleft$$
$$\Delta y_B = r\theta = (5 \text{ in.})(14 \text{ rad}) = 70 \text{ in.} \qquad \Delta y_B = 70 \text{ in. upward} \quad \blacktriangleleft$$

***c.* Acceleration of Point *D* at *t* = 0.** The tangential component of the acceleration is

$$(\mathbf{a}_D)_t = \mathbf{a}_C = 9 \text{ in./s}^2 \rightarrow$$

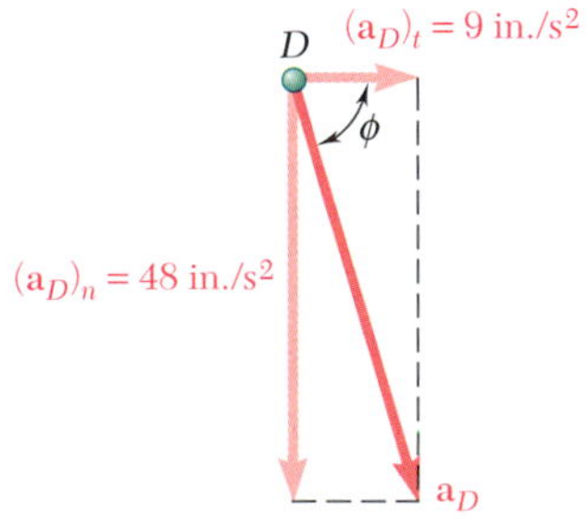

Since, at $t = 0$, $\omega_0 = 4$ rad/s, the normal component of the acceleration is

$$(a_D)_n = r_D\omega_0^2 = (3 \text{ in.})(4 \text{ rad/s})^2 = 48 \text{ in./s}^2 \qquad (\mathbf{a}_D)_n = 48 \text{ in./s}^2 \downarrow$$

The magnitude and direction of the total acceleration can be obtained by writing

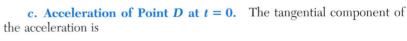

$$\tan\phi = (48 \text{ in./s}^2)/(9 \text{ in./s}^2) \qquad \phi = 79.4°$$
$$a_D \sin 79.4° = 48 \text{ in./s}^2 \qquad a_D = 48.8 \text{ in./s}^2$$
$$\mathbf{a}_D = 48.8 \text{ in./s}^2 \measuredangle 79.4° \quad \blacktriangleleft$$

SOLVING PROBLEMS ON YOUR OWN

In this lesson we began the study of the motion of rigid bodies by considering two particular types of motion of rigid bodies: *translation* and *rotation* about a *fixed axis*.

1. *Rigid body in translation.* At any given instant, all the points of a rigid body in translation have the *same velocity* and the *same acceleration* (Fig. 15.7).

2. *Rigid body rotating about a fixed axis.* The position of a rigid body rotating about a fixed axis was defined at any given instant by the *angular coordinate* θ, which is usually measured in *radians*. Selecting the unit vector $\mathbf{k}$ along the fixed axis and in such a way that the rotation of the body appears counterclockwise as seen from the tip of $\mathbf{k}$, we defined the *angular velocity* $\boldsymbol{\omega}$ and the *angular acceleration* $\boldsymbol{\alpha}$ of the body:

$$\boldsymbol{\omega} = \dot{\theta}\mathbf{k} \qquad \boldsymbol{\alpha} = \ddot{\theta}\mathbf{k} \qquad (15.6,\ 15.9)$$

In solving problems, keep in mind that the vectors $\boldsymbol{\omega}$ and $\boldsymbol{\alpha}$ are both directed along the fixed axis of rotation and that their sense can be obtained by the right-hand rule.

a. *The velocity of a point P* of a body rotating about a fixed axis was found to be

$$\mathbf{v} = \boldsymbol{\omega} \times \mathbf{r} \qquad (15.5)$$

where $\boldsymbol{\omega}$ is the angular velocity of the body and $\mathbf{r}$ is the position vector drawn from any point on the axis of rotation to point P (Fig. 15.9).

b. *The acceleration of point P* was found to be

$$\mathbf{a} = \boldsymbol{\alpha} \times \mathbf{r} + \boldsymbol{\omega} \times (\boldsymbol{\omega} \times \mathbf{r}) \qquad (15.8)$$

Since vector products are not commutative, *be sure to write the vectors in the order shown* when using either of the above two equations.

3. *Rotation of a representative slab.* In many problems, you will be able to reduce the analysis of the rotation of a three-dimensional body about a fixed axis to the study of the rotation of a representative slab in a plane perpendicular to the fixed axis. The z axis should be directed along the axis of rotation and point out of the paper. Thus, the representative slab will be rotating in the xy plane about the origin O of the coordinate system (Fig. 15.10).

To solve problems of this type you should do the following:

a. *Draw a diagram of the representative slab,* showing its dimensions, its angular velocity and angular acceleration, as well as the vectors representing the velocities and accelerations of the points of the slab for which you have or seek information.

b. Relate the rotation of the slab and the motion of points of the slab by writing the equations

$$v = r\omega \qquad (15.10')$$

$$a_t = r\alpha \qquad a_n = r\omega^2 \qquad (15.11')$$

Remember that the velocity $\mathbf{v}$ and the component $\mathbf{a}_t$ of the acceleration of a point P of the slab are tangent to the circular path described by P. The directions of $\mathbf{v}$ and $\mathbf{a}_t$ are found by rotating the position vector $\mathbf{r}$ through 90° in the sense indicated by $\boldsymbol{\omega}$ and $\boldsymbol{\alpha}$, respectively. The normal component $\mathbf{a}_n$ of the acceleration of P is always directed toward the axis of rotation.

4. Equations defining the rotation of a rigid body. You must have been pleased to note the similarity existing between the equations defining the rotation of a rigid body about a fixed axis [Eqs. (15.12) through (15.16)] and those in Chap. 11 defining the rectilinear motion of a particle [Eqs. (11.1) through (11.8)]. All you have to do to obtain the new set of equations is to substitute θ, ω, and α for x, v, and a in the equations of Chap. 11.

Problems

15.1 The motion of a cam is defined by the relation $\theta = 4t^3 - 12t^2 + 15$, where θ is expressed in radians and t in seconds. Determine the angular coordinate, the angular velocity, and the angular acceleration of the cam when (*a*) $t = 0$, (*b*) $t = 6$ s.

15.2 For the cam in Prob. 15.1, determine the time, angular coordinate, and angular acceleration when the angular velocity is zero.

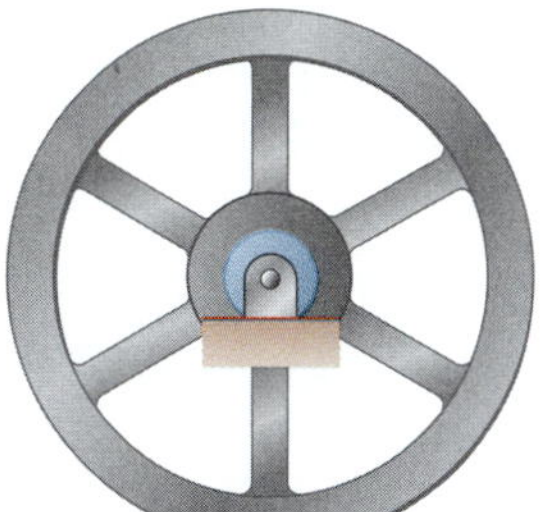

Fig. P15.3 and P15.4

15.3 The motion of an oscillating flywheel is defined by the relation $\theta = \theta_0 e^{-3\pi t} \cos 4\pi t$, where θ is expressed in radians and t in seconds. Knowing that $\theta_0 = 0.5$ rad, determine the angular coordinate, the angular velocity, and the angular acceleration of the flywheel when (*a*) $t = 0$, (*b*) $t = 0.125$ s.

15.4 The motion of an oscillating flywheel is defined by the relation $\theta = \theta_0 e^{-7\pi t/6} \sin 4\pi t$, where θ is expressed in radians and t in seconds. Knowing that $\theta_0 = 0.4$ rad, determine the angular coordinate, the angular velocity, and the angular acceleration of the flywheel when (*a*) $t = 0.125$ s, (*b*) $t = \infty$.

15.5 The rotor of a gas turbine is rotating at a speed of 7200 rpm when the turbine is shut down. It is observed that 5 min is required for the rotor to coast to rest. Assuming uniformly accelerated motion, determine (*a*) the angular acceleration, (*b*) the number of revolutions that the rotor executes before coming to rest.

15.6 When the power to an electric motor is turned on the motor reaches its rated speed of 2400 rpm in 4 s, and when the power is turned off the motor coasts to rest in 40 s. Assuming uniformly accelerated motion, determine the number of revolutions that the motor executes (*a*) in reaching its rated speed, (*b*) in coasting to rest.

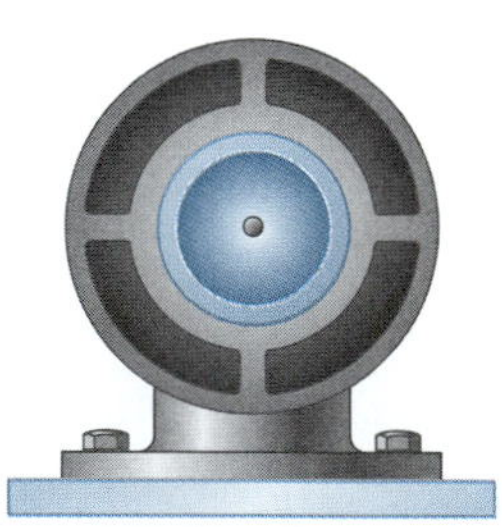

Fig. P15.6

15.7 The angular acceleration of an oscillating disk is defined by the relation $\alpha = -k\theta$. Determine (*a*) the value of k for which $\omega = 12$ rad/s when $\theta = 0$ and $\theta = 6$ rad when $\omega = 0$, (*b*) the angular velocity of the disk when $\theta = 3$ rad.

15.8 The angular acceleration of a shaft is defined by the relation $\alpha = -0.5\omega$, where α is expressed in rad/s^2 and ω in rad/s. Knowing that at $t = 0$ the angular velocity of the shaft is 30 rad/s, determine (*a*) the number of revolutions the shaft will execute before coming to rest, (*b*) the time required for the shaft to come to rest, (*c*) the time required for the angular velocity of the shaft to reduce to 2 percent of its initial value.

15.9 The assembly shown consists of two rods and a rectangular plate *BCDE* which are welded together. The assembly rotates about the axis *AB* with a constant angular velocity of 10 rad/s. Knowing that the rotation is counterclockwise as viewed from *B*, determine the velocity and acceleration of corner *E*.

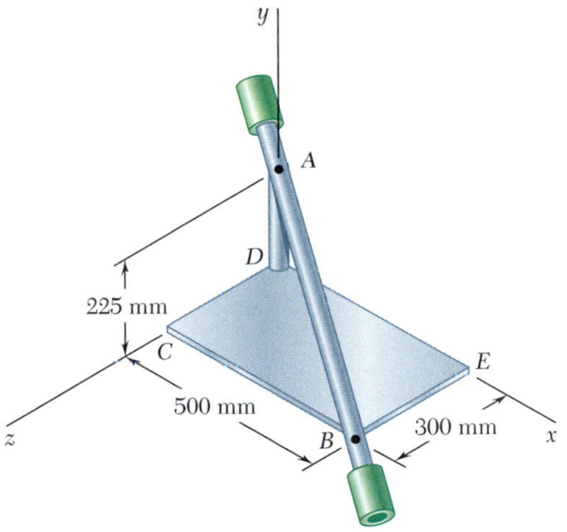

Fig. P15.9

15.10 In Prob. 15.9, determine the velocity and acceleration of corner *C*, assuming that the angular velocity is 10 rad/s and decreases at the rate of 20 rad/s^2.

15.11 The rectangular block shown rotates about the diagonal *OA* with a constant angular velocity of 6.76 rad/s. Knowing that the rotation is counterclockwise as viewed from *A*, determine the velocity and acceleration of point *B* at the instant shown.

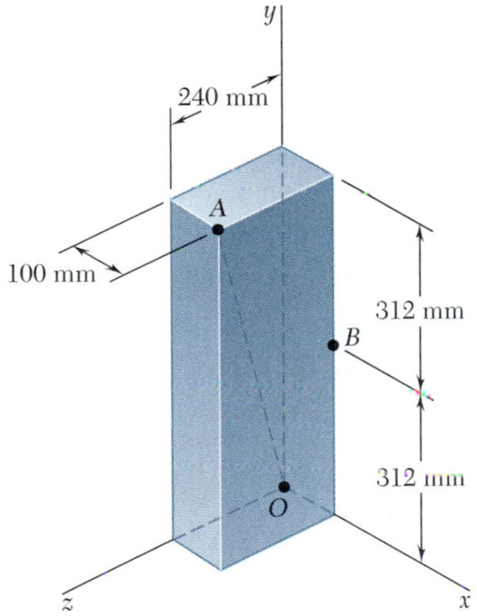

Fig. *P15.11*

15.12 In Prob. 15.11, determine the velocity and acceleration of point *B* at the instant shown, assuming that the angular velocity is 3.38 rad/s and decreases at the rate of 5.07 rad/s^2.

15.13 The bent rod *ABCDE* rotates about a line joining points *A* and *E* with a constant angular velocity of 12 rad/s. Knowing that the rotation is clockwise as viewed from *E*, determine the velocity and acceleration of corner *C*.

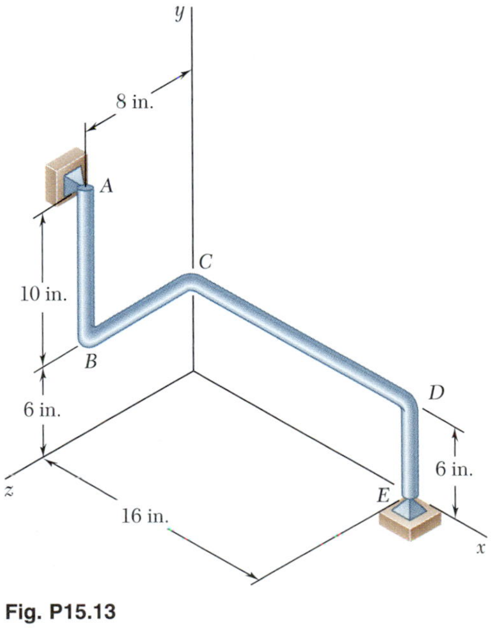

Fig. P15.13

15.14 In Prob. 15.13, determine the velocity and acceleration of corner *B*, assuming that the angular velocity is 12 rad/s and increases at the rate of 60 rad/s^2.

15.15 The earth makes one complete revolution on its axis in 23 h 56 min. Knowing that the mean radius of the earth is 6370 km, determine the linear velocity and acceleration of a point on the surface of the earth (*a*) at the equator, (*b*) at Philadelphia, latitude 40° north, (*c*) at the North Pole.

15.16 The earth makes one complete revolution around the sun in 365.24 days. Assuming that the orbit of the earth is circular and has a radius of 93,000,000 mi, determine the velocity and acceleration of the earth.

15.17 The belt shown moves over two pulleys without slipping. At the instant shown the pulleys are rotating clockwise and the speed of point B on the belt is 4 m/s, increasing at the rate of 32 m/s^2. Determine, at this instant, (a) the angular velocity and angular acceleration of each pulley, (b) the acceleration of point P on pulley C.

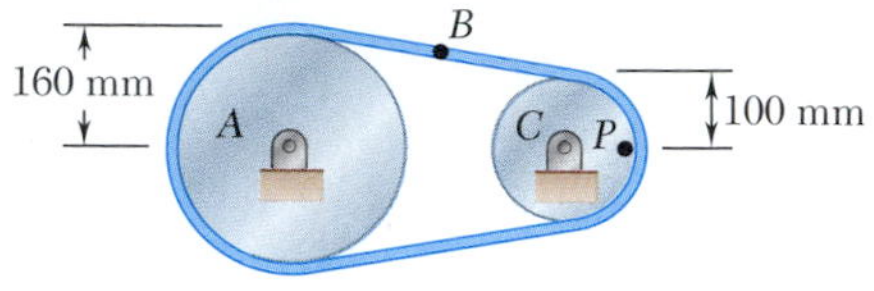

Fig. P15.17 and P15.18

15.18 The belt shown moves over two pulleys without slipping. Pulley A starts from rest with a clockwise angular acceleration defined by the relation $\alpha = 120 - 0.002\omega^2$, where α is expressed in rad/s^2 and ω is expressed in rad/s. Determine, after one-half revolution of pulley A, (a) the magnitude of the acceleration of point B on the belt, (b) the acceleration of point P on pulley C.

15.19 The belt sander shown is initially at rest. If the driving drum B has a constant angular acceleration of 120 rad/s^2 counterclockwise, determine the magnitude of the acceleration of the belt at point C when (a) $t = 0.5$ s, (b) $t = 2$ s.

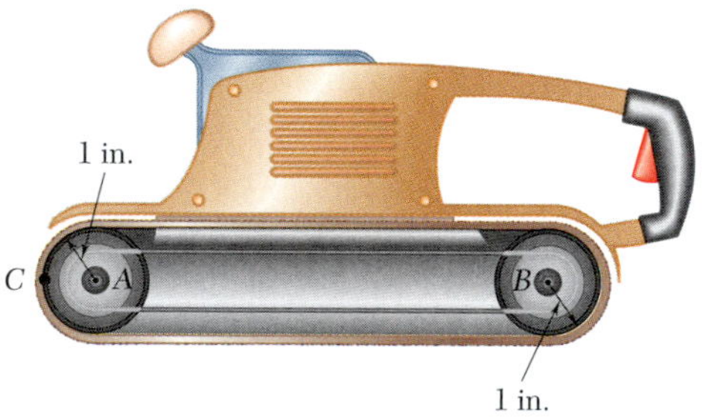

Fig. P15.19 and P15.20

15.20 The rated speed of drum B of the belt sander shown is 2400 rpm. When the power is turned off, it is observed that the sander coasts from its rated speed to rest in 10 s. Assuming uniformly decelerated motion, determine the velocity and acceleration of point C of the belt, (a) immediately before the power is turned off, (b) 9 s later.

15.21 A series of small machine components being moved by a conveyor belt passes over a 120-mm-radius idler pulley. At the instant shown, the velocity of point A is 300 mm/s to the left and its acceleration is 180 mm/s^2 to the right. Determine (a) the angular velocity and angular acceleration of the idler pulley, (b) the total acceleration of the machine component at B.

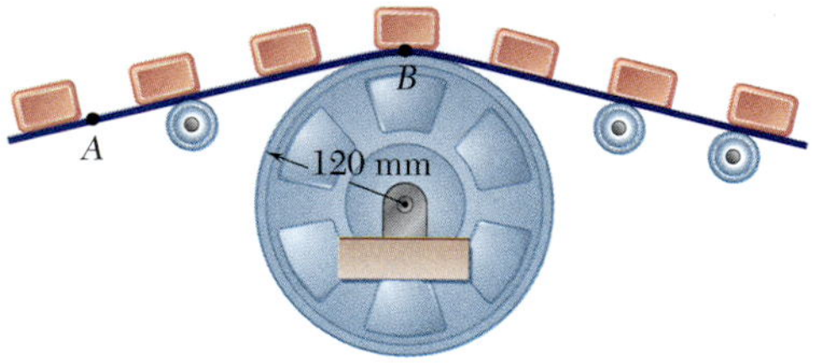

Fig. *P15.21* and *P15.22*

15.22 A series of small machine components being moved by a conveyor belt passes over a 120-mm-radius idler pulley. At the instant shown, the angular velocity of the idler pulley is 4 rad/s clockwise. Determine the angular acceleration of the pulley for which the magnitude of the total acceleration of the machine component at B is 2400 mm/s^2.

15.23 A gear reduction system consists of three gears A, B, and C. Knowing that gear A rotates clockwise with a constant angular velocity ω_A = 600 rpm, determine (a) the angular velocities of gears B and C, (b) the accelerations of the points on gears B and C which are in contact.

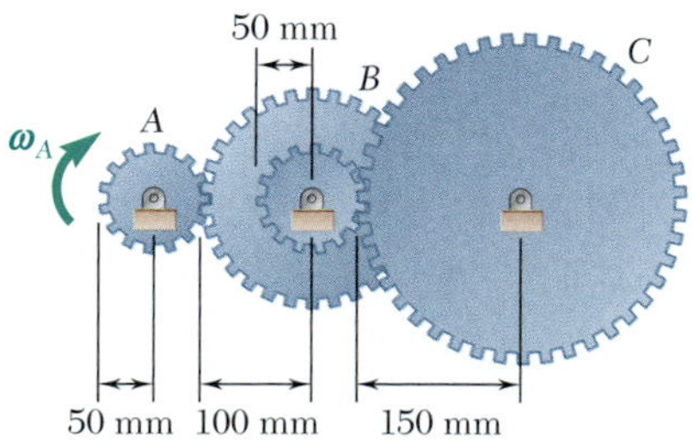

Fig. P15.23 and P15.24

15.24 A gear reduction system consists of three gears A, B, and C. Gear A starts from rest at time $t = 0$ and rotates clockwise with constant angular acceleration. Knowing that the angular velocity of gear A is 600 rpm at time $t = 2$ s, determine (a) the angular accelerations of gears B and C, (b) the accelerations of the points on gears B and C which are in contact when $t = 0.5$ s.

15.25 Ring B has an inner radius r_2 and hangs from the horizontal shaft A as shown. Knowing that shaft A rotates with a constant angular velocity ω_A and that no slipping occurs, derive a relation in terms of r_1, r_2, r_3, and ω_A for (a) the angular velocity of ring B, (b) the acceleration of the points of shaft A and ring B which are in contact.

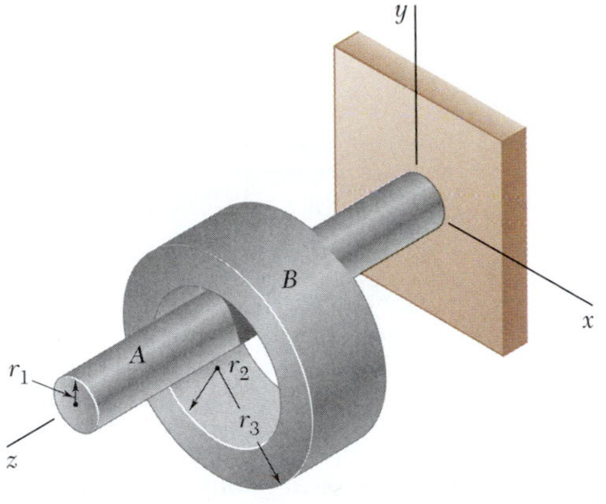

Fig. *P15.25* and *P15.26*

15.26 Ring B has an inner radius r_2 and hangs from the horizontal shaft A as shown. Shaft A rotates with a constant angular velocity of 25 rad/s and no slipping occurs. Knowing that $r_1 = 0.5$ in., $r_2 = 2.5$ in., and $r_3 = 3.5$ in., determine (a) the angular velocity of ring B, (b) the acceleration of the points of shaft A and ring B which are in contact, (c) the magnitude of the acceleration of a point on the outside surface of ring B.

15.27 Cylinder A is moving downward with a velocity of 3 m/s when the brake is suddenly applied to the drum. Knowing that the cylinder moves 6 m downward before coming to rest and assuming uniformly accelerated motion, determine (a) the angular acceleration of the drum, (b) the time required for the cylinder to come to rest.

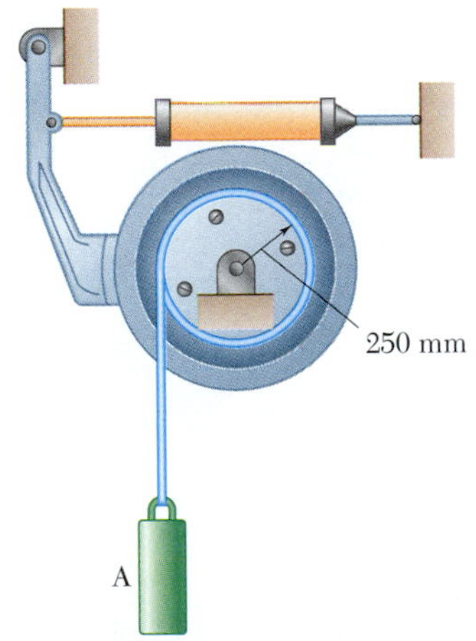

Fig. P15.27 and P15.28

15.28 The system shown is held at rest by the brake-and-drum system shown. After the brake is partially released at $t = 0$, it is observed that the cylinder moves 5 m in 4.5 s. Assuming uniformly accelerated motion, determine (a) the angular acceleration of the drum, (b) the angular velocity of the drum at $t = 3.5$ s.

15.29 Two blocks and a pulley are connected by inextensible cords as shown. The relative velocity of block A with respect to block B is 2.5 ft/s to the left at time $t = 0$ and 1.25 ft/s to the left when $t = 0.25$ s. Knowing that the angular acceleration of the pulley is constant, find (*a*) the relative acceleration of block A with respect to block B, (*b*) the distance block A moves relative to block B during the interval $0 \leq t \leq 0.25$ s.

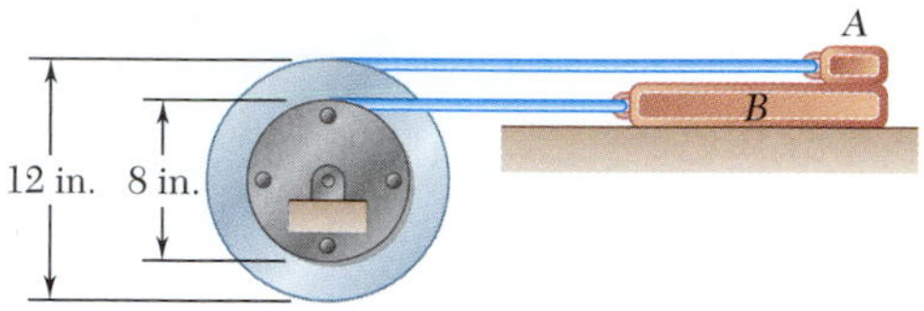

Fig. P15.29 and P15.30

15.30 Two blocks and a pulley are connected by inextensible cords as shown. The relative velocity of block A with respect to block B is 3 ft/s to the left at time $t = 0$ and 1.5 ft/s to the left after one-half revolution of the pulley. Knowing that the angular acceleration of the pulley is constant, find (*a*) the relative acceleration of block A with respect to block B, (*b*) the distance block A moves relative to block B during the interval $0 \leq t \leq 0.3$ s.

15.31 Disk B is at rest when it is brought into contact with disk A which is rotating freely at 450 rpm clockwise. After 6 s of slippage, during which each disk has a constant angular acceleration, disk A reaches a final angular velocity of 140 rpm clockwise. Determine the angular acceleration of each disk during the period of slippage.

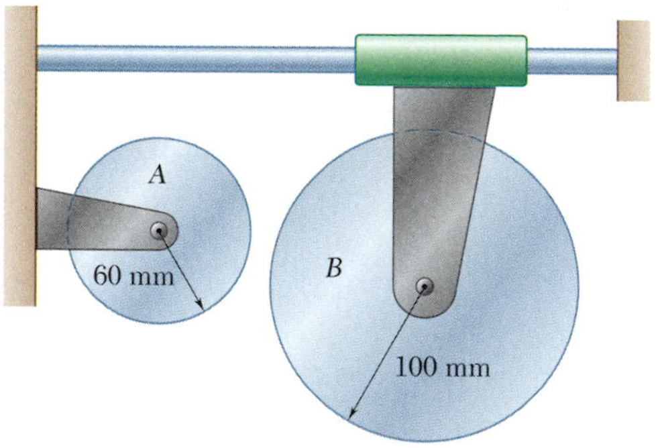

Fig. P15.31 and P15.32

15.32 A simple friction drive consists of two disks A and B. Initially, disk A has a clockwise angular velocity of 500 rpm and disk B is at rest. It is known that disk A will coast to rest in 60 s with constant angular deceleration. However, rather than waiting until both disks are at rest to bring them together, disk B is given a constant angular acceleration of 2.5 rad/s^2 counterclockwise. Determine (*a*) at what time the disks can be brought together if they are not to slip, (*b*) the angular velocity of each disk as contact is made.

15.33 Two friction disks A and B are to be brought into contact without slipping when the angular velocity of disk A is 240 rpm counterclockwise. Disk A starts from rest at time $t = 0$ and is given a constant angular acceleration of magnitude α. Disk B starts from rest at time $t = 2$ s and is given a constant clockwise angular acceleration, also of magnitude α. Determine (*a*) the required angular acceleration magnitude α, (*b*) the time at which the contact occurs.

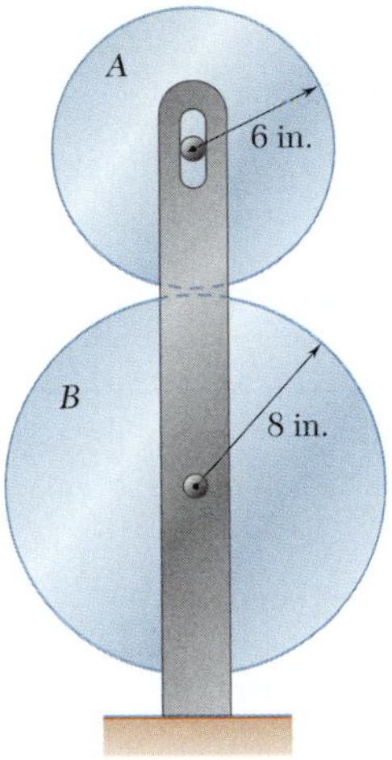

Fig. *P15.33* and *P15.34*

15.34 Two friction disks A and B are brought into contact when the angular velocity of disk A is 240 rpm counterclockwise and disk B is at rest. A period of slipping follows and disk B makes 2 revolutions before reaching its final angular velocity. Assuming that the angular acceleration of each disk is constant and inversely proportional to the cube of its radius, determine (*a*) the angular acceleration of each disk, (*b*) the time during which the disks slip.

***15.35** In a continuous printing process, paper is drawn into the presses at a constant speed v. Denoting by r the radius of the paper roll at any given time and by b the thickness of the paper, derive an expression for the angular acceleration of the paper roll.

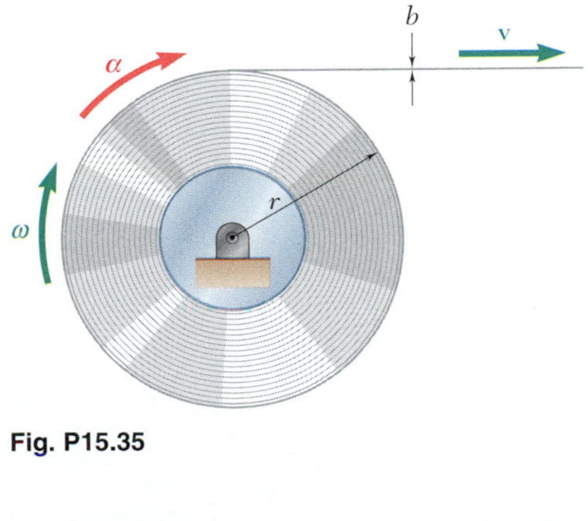

Fig. P15.35

***15.36** Television recording tape is being rewound on a VCR reel which rotates with a constant angular velocity ω_0. Denoting by r the radius of the reel at any given time and by b the thickness of the tape, derive an expression for the acceleration of the tape as it approaches the reel.

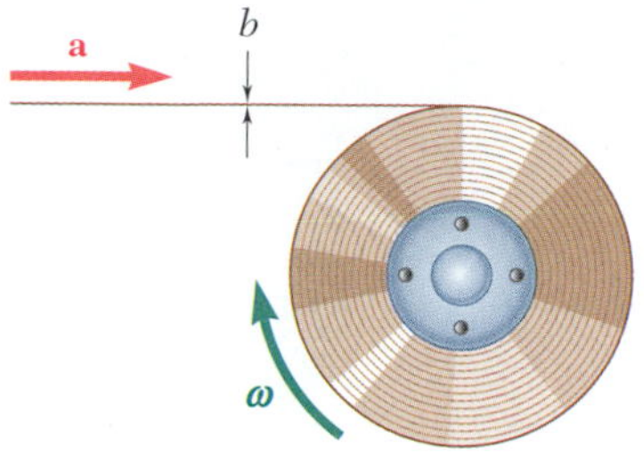

Fig. P15.36

15.5. GENERAL PLANE MOTION

As indicated in Sec. 15.1, we understand by general plane motion a plane motion which is neither a translation nor a rotation. As you will presently see, however, *a general plane motion can always be considered as the sum of a translation and a rotation.*

Consider, for example, a wheel rolling on a straight track (Fig. 15.12). Over a certain interval of time, two given points A and B will have moved, respectively, from A_1 to A_2 and from B_1 to B_2. The same result could be obtained through a translation which would bring A and B into A_2 and B_1' (the line AB remaining vertical), followed by a rotation about A bringing B into B_2. Although the original rolling motion differs from the combination of translation and rotation when these motions are taken in succession, the original motion can be exactly duplicated by a combination of simultaneous translation and rotation.

Another example of plane motion is given in Fig. 15.13, which represents a rod whose extremities slide along a horizontal and a vertical track, respectively. This motion can be replaced by a translation in a horizontal direction and a rotation about A (Fig. 15.13a)

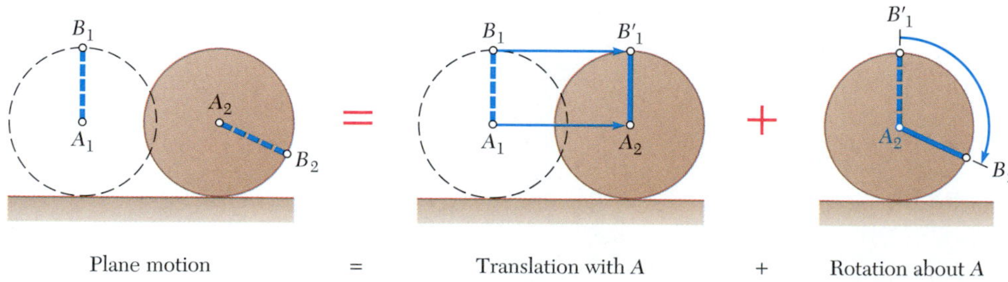

Fig. 15.12

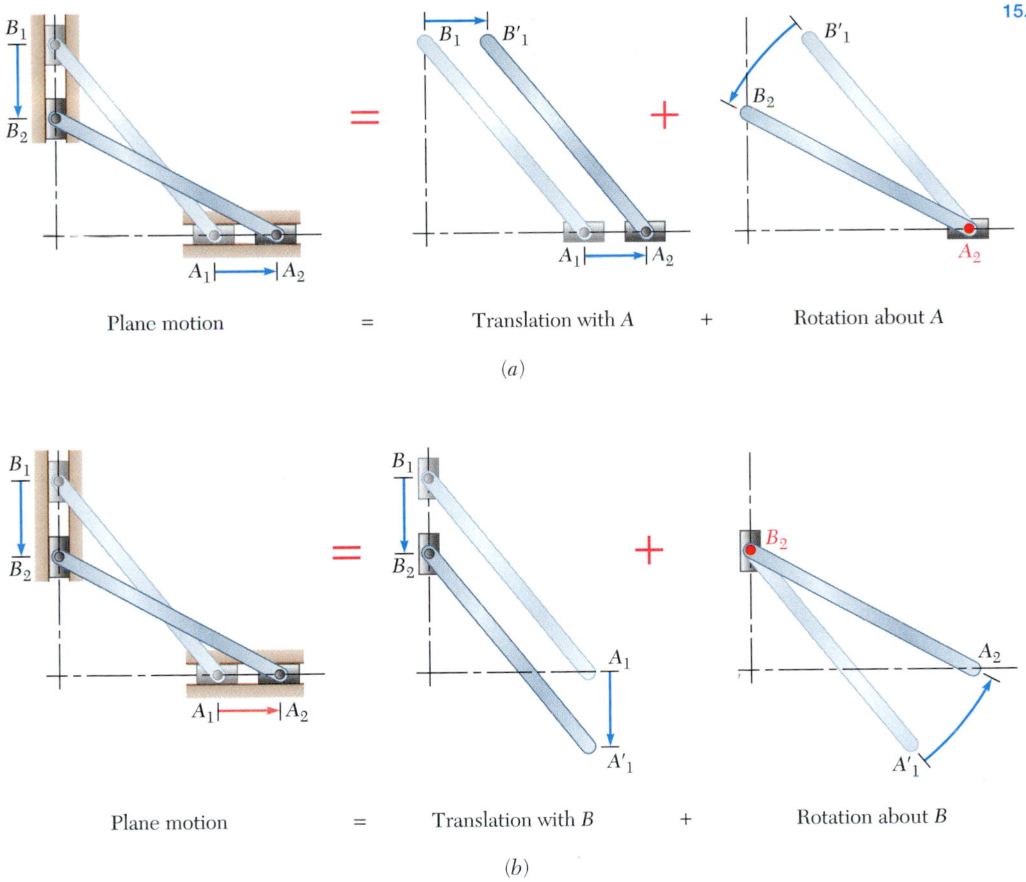

Fig. 15.13

or by a translation in a vertical direction and a rotation about B (Fig. 15.13*b*).

In the general case of plane motion, we will consider a small displacement which brings two particles A and B of a representative slab, respectively, from A_1 and B_1 into A_2 and B_2 (Fig. 15.14). This displacement can be divided into two parts: in one, the particles move into A_2 and B_1' while the line AB maintains the same direction; in the other, B moves into B_2 while A remains fixed. The first part of the motion is clearly a translation and the second part a rotation about A.

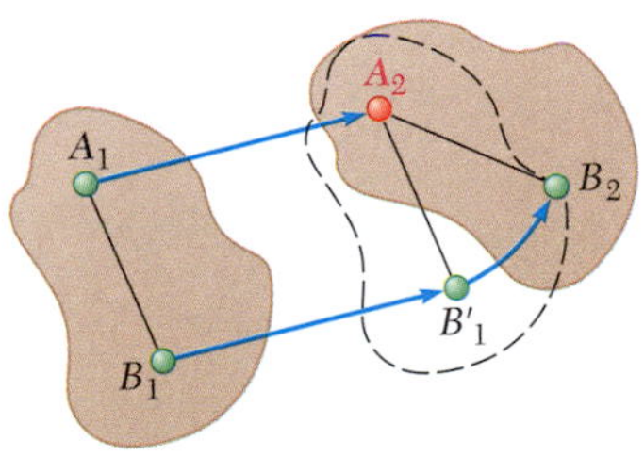

Fig. 15.14

Recalling from Sec. 11.12 the definition of the relative motion of a particle with respect to a moving frame of reference—as opposed to its absolute motion with respect to a fixed frame of reference—we can restate as follows the result obtained above: Given two particles A and B of a rigid slab in plane motion, the relative motion of B with respect to a frame attached to A and of fixed orientation is a rotation. To an observer moving with A but not rotating, particle B will appear to describe an arc of a circle centered at A.

15.6. ABSOLUTE AND RELATIVE VELOCITY IN PLANE MOTION

We saw in the preceding section that any plane motion of a slab can be replaced by a translation defined by the motion of an arbitrary reference point A and a simultaneous rotation about A. The absolute velocity $\mathbf{v}_B$ of a particle B of the slab is obtained from the relative-velocity formula derived in Sec. 11.12,

$$\mathbf{v}_B = \mathbf{v}_A + \mathbf{v}_{B/A} \tag{15.17}$$

where the right-hand member represents a vector sum. The velocity $\mathbf{v}_A$ corresponds to the translation of the slab with A, while the relative velocity $\mathbf{v}_{B/A}$ is associated with the rotation of the slab about A and is measured with respect to axes centered at A and of fixed orientation (Fig. 15.15). Denoting by $\mathbf{r}_{B/A}$ the position vector of B relative to A, and by $\omega\mathbf{k}$ the angular velocity of the slab with respect to axes of fixed orientation, we have from (15.10) and (15.10′)

$$\mathbf{v}_{B/A} = \omega\mathbf{k} \times \mathbf{r}_{B/A} \qquad v_{B/A} = r\omega \tag{15.18}$$

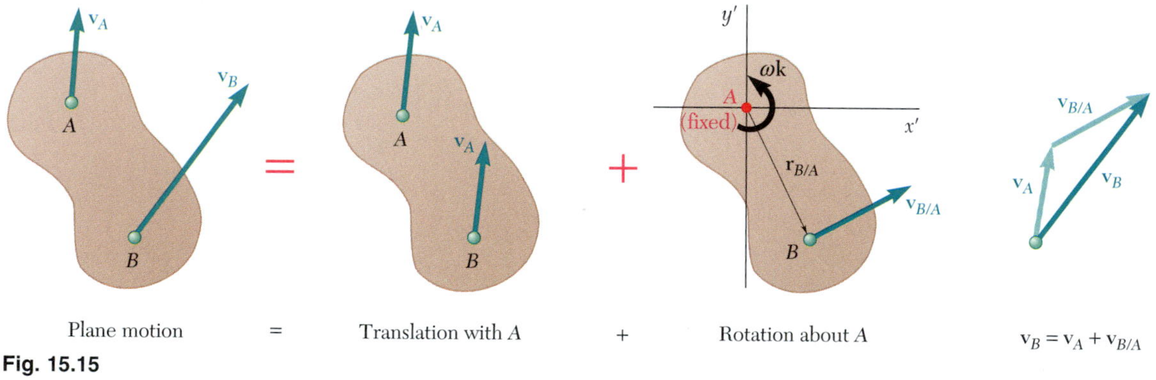

Fig. 15.15

where r is the distance from A to B. Substituting for $\mathbf{v}_{B/A}$ from (15.18) into (15.17), we can also write

$$\mathbf{v}_B = \mathbf{v}_A + \omega\mathbf{k} \times \mathbf{r}_{B/A} \tag{15.17′}$$

As an example, let us again consider the rod AB of Fig. 15.13. Assuming that the velocity $\mathbf{v}_A$ of end A is known, we propose to find the velocity $\mathbf{v}_B$ of end B and the angular velocity $\boldsymbol{\omega}$ of the rod, in terms of the velocity $\mathbf{v}_A$, the length l, and the angle θ. Choosing A as a reference point, we express that the given motion is equivalent to a translation with A and simultaneous rotation about A (Fig. 15.16). The absolute velocity of B must therefore be equal to the vector sum

$$\mathbf{v}_B = \mathbf{v}_A + \mathbf{v}_{B/A} \tag{15.17}$$

We note that while the direction of $\mathbf{v}_{B/A}$ is known, its magnitude $l\omega$ is unknown. However, this is compensated for by the fact that the direction of $\mathbf{v}_B$ is known. We can therefore complete the diagram of Fig. 15.16. Solving for the magnitudes v_B and ω, we write

$$v_B = v_A \tan\theta \qquad \omega = \frac{v_{B/A}}{l} = \frac{v_A}{l\cos\theta} \tag{15.19}$$

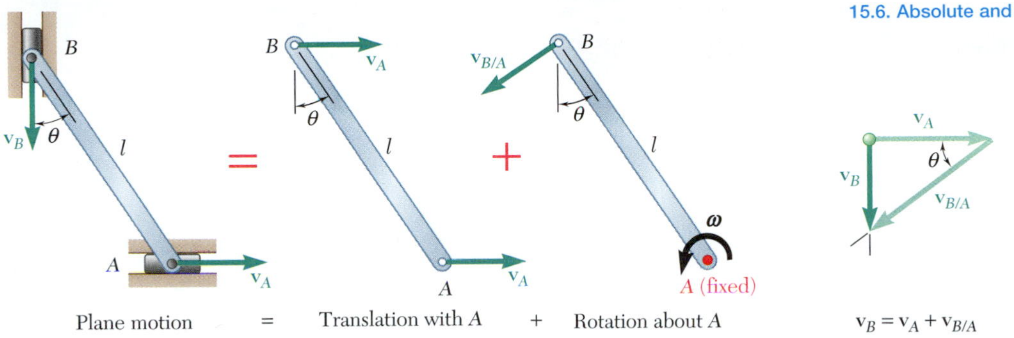

Fig. 15.16

The same result can be obtained by using B as a point of reference. Resolving the given motion into a translation with B and a simultaneous rotation about B (Fig. 15.17), we write the equation

$$\mathbf{v}_A = \mathbf{v}_B + \mathbf{v}_{A/B} \tag{15.20}$$

which is represented graphically in Fig. 15.17. We note that $\mathbf{v}_{A/B}$ and $\mathbf{v}_{B/A}$ have the same magnitude $l\omega$ but opposite sense. The sense of the relative velocity depends, therefore, upon the point of reference which has been selected and should be carefully ascertained from the appropriate diagram (Fig. 15.16 or 15.17).

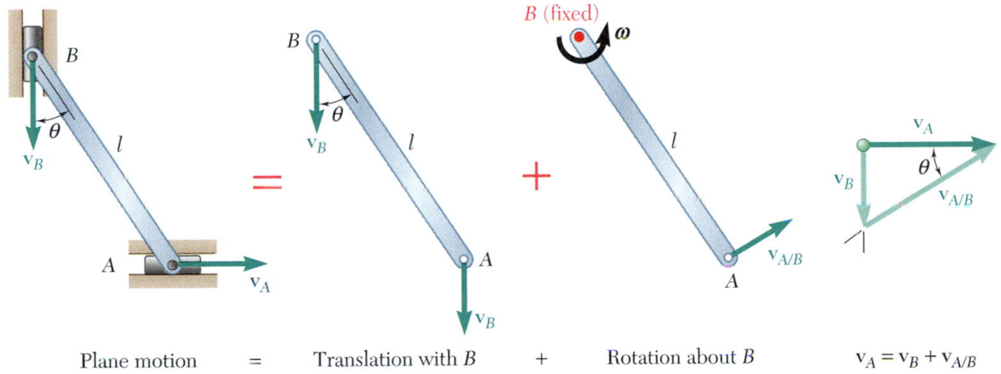

Fig. 15.17

Finally, we observe that the angular velocity $\boldsymbol{\omega}$ of the rod in its rotation about B is the same as in its rotation about A. It is measured in both cases by the rate of change of the angle θ. This result is quite general; we should therefore bear in mind that *the angular velocity* $\boldsymbol{\omega}$ *of a rigid body in plane motion is independent of the reference point.*

Most mechanisms consist not of one but of *several* moving parts. When the various parts of a mechanism are pin-connected, the analysis of the mechanism can be carried out by considering each part as a rigid body, keeping in mind that the points where two parts are connected must have the same absolute velocity (see Sample Prob. 15.3). A similar analysis can be used when gears are involved, since the teeth in contact must also have the same absolute velocity. However, when a mechanism contains parts which slide on each other, the relative velocity of the parts in contact must be taken into account (see Secs. 15.10 and 15.11).

SAMPLE PROBLEM 15.2

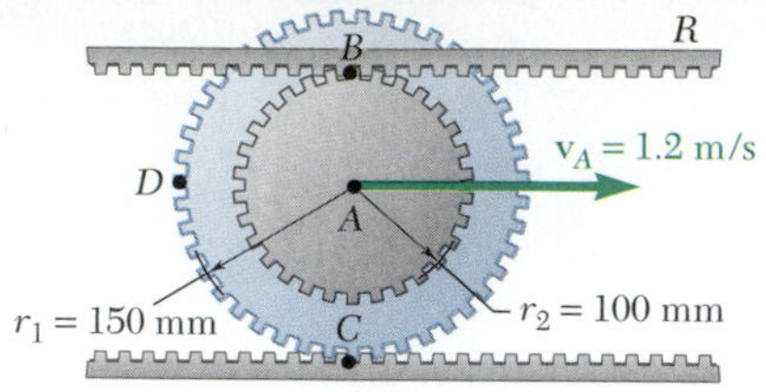

The double gear shown rolls on the stationary lower rack; the velocity of its center A is 1.2 m/s directed to the right. Determine (a) the angular velocity of the gear, (b) the velocities of the upper rack R and of point D of the gear.

SOLUTION

***a.* Angular Velocity of the Gear.** Since the gear rolls on the lower rack, its center A moves through a distance equal to the outer circumference $2\pi r_1$ for each full revolution of the gear. Noting that 1 rev $= 2\pi$ rad, and that when A moves to the right ($x_A > 0$) the gear rotates clockwise ($\theta < 0$), we write

$$\frac{x_A}{2\pi r_1} = -\frac{\theta}{2\pi} \qquad x_A = -r_1\theta$$

Differentiating with respect to the time t and substituting the known values $v_A = 1.2$ m/s and $r_1 = 150$ mm $= 0.150$ m, we obtain

$$v_A = -r_1\omega \qquad 1.2 \text{ m/s} = -(0.150 \text{ m})\omega \qquad \omega = -8 \text{ rad/s}$$

$$\boldsymbol{\omega} = \omega\mathbf{k} = -(8 \text{ rad/s})\mathbf{k} \quad ◀$$

where $\mathbf{k}$ is a unit vector pointing out of the paper.

***b.* Velocities.** The rolling motion is resolved into two component motions: a translation with the center A and a rotation about the center A. In the translation, all points of the gear move with the same velocity $\mathbf{v}_A$. In the rotation, each point P of the gear moves about A with a relative velocity $\mathbf{v}_{P/A} = \omega\mathbf{k} \times \mathbf{r}_{P/A}$, where $\mathbf{r}_{P/A}$ is the position vector of P relative to A.

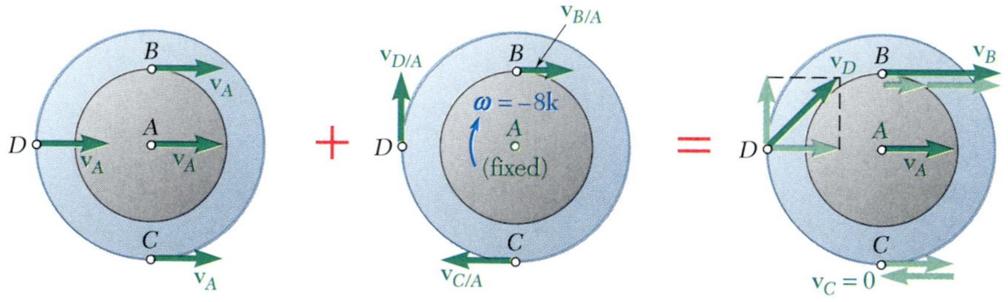

Translation + Rotation = Rolling Motion

Velocity of Upper Rack. The velocity of the upper rack is equal to the velocity of point B; we write

$$\begin{aligned}\mathbf{v}_R = \mathbf{v}_B &= \mathbf{v}_A + \mathbf{v}_{B/A} = \mathbf{v}_A + \omega\mathbf{k} \times \mathbf{r}_{B/A}\\ &= (1.2 \text{ m/s})\mathbf{i} - (8 \text{ rad/s})\mathbf{k} \times (0.100 \text{ m})\mathbf{j}\\ &= (1.2 \text{ m/s})\mathbf{i} + (0.8 \text{ m/s})\mathbf{i} = (2 \text{ m/s})\mathbf{i}\end{aligned}$$

$$\mathbf{v}_R = 2 \text{ m/s} \rightarrow \quad ◀$$

Velocity of Point D

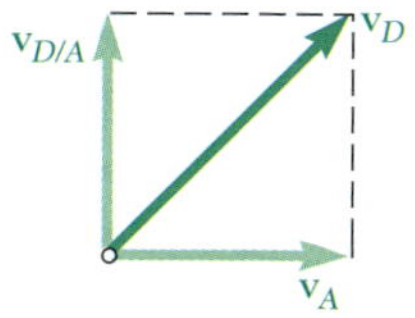

$$\begin{aligned}\mathbf{v}_D &= \mathbf{v}_A + \mathbf{v}_{D/A} = \mathbf{v}_A + \omega\mathbf{k} \times \mathbf{r}_{D/A}\\ &= (1.2 \text{ m/s})\mathbf{i} - (8 \text{ rad/s})\mathbf{k} \times (-0.150 \text{ m})\mathbf{i}\\ &= (1.2 \text{ m/s})\mathbf{i} + (1.2 \text{ m/s})\mathbf{j}\end{aligned}$$

$$\mathbf{v}_D = 1.697 \text{ m/s} \measuredangle 45° \quad ◀$$

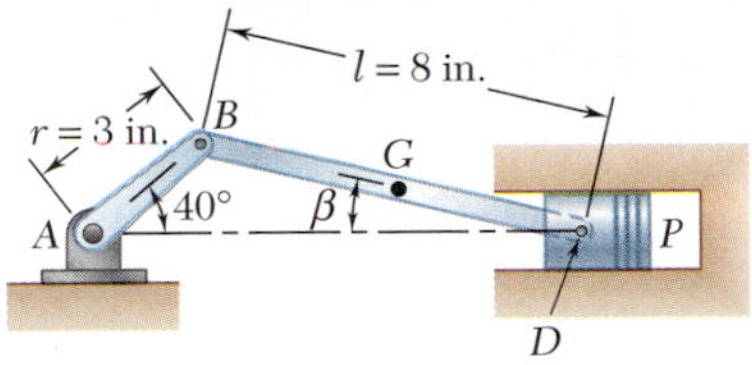

SAMPLE PROBLEM 15.3

In the engine system shown, the crank AB has a constant clockwise angular velocity of 2000 rpm. For the crank position indicated, determine (*a*) the angular velocity of the connecting rod BD, (*b*) the velocity of the piston P.

SOLUTION

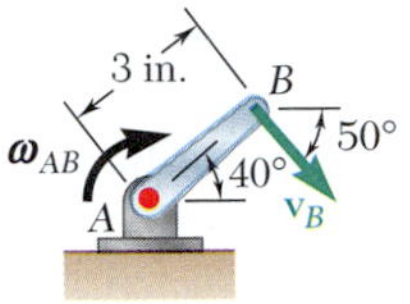

Motion of Crank *AB*. The crank AB rotates about point A. Expressing ω_{AB} in rad/s and writing $v_B = r\omega_{AB}$, we obtain

$$\omega_{AB} = \left(2000\frac{\text{rev}}{\text{min}}\right)\left(\frac{1\text{ min}}{60\text{ s}}\right)\left(\frac{2\pi\text{ rad}}{1\text{ rev}}\right) = 209.4\text{ rad/s}$$

$$v_B = (AB)\omega_{AB} = (3\text{ in.})(209.4\text{ rad/s}) = 628.3\text{ in./s}$$
$$\mathbf{v}_B = 628.3\text{ in./s} \searrow 50°$$

Motion of Connecting Rod *BD*. We consider this motion as a general plane motion. Using the law of sines, we compute the angle β between the connecting rod and the horizontal:

$$\frac{\sin 40°}{8\text{ in.}} = \frac{\sin\beta}{3\text{ in.}} \qquad \beta = 13.95°$$

The velocity $\mathbf{v}_D$ of the point D where the rod is attached to the piston must be horizontal, while the velocity of point B is equal to the velocity $\mathbf{v}_B$ obtained above. Resolving the motion of BD into a translation with B and a rotation about B, we obtain

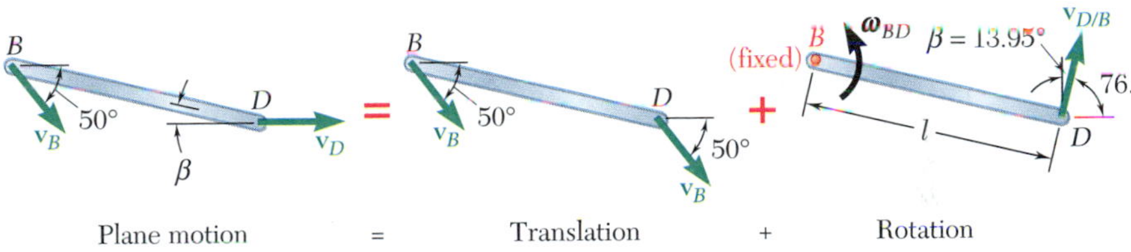

Expressing the relation between the velocities $\mathbf{v}_D$, $\mathbf{v}_B$, and $\mathbf{v}_{D/B}$, we write

$$\mathbf{v}_D = \mathbf{v}_B + \mathbf{v}_{D/B}$$

We draw the vector diagram corresponding to this equation. Recalling that $\beta = 13.95°$, we determine the angles of the triangle and write

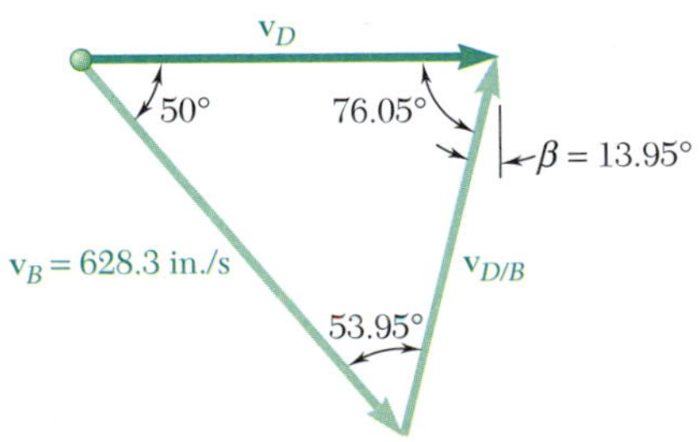

$$\frac{v_D}{\sin 53.95°} = \frac{v_{D/B}}{\sin 50°} = \frac{628.3\text{ in./s}}{\sin 76.05°}$$

$$v_{D/B} = 495.9\text{ in./s} \qquad \mathbf{v}_{D/B} = 495.9\text{ in./s} \measuredangle 76.05°$$
$$v_D = 523.4\text{ in./s} = 43.6\text{ ft/s} \qquad \mathbf{v}_D = 43.6\text{ ft/s} \rightarrow$$

$$\mathbf{v}_P = \mathbf{v}_D = 43.6\text{ ft/s} \rightarrow \quad ◀$$

Since $v_{D/B} = l\omega_{BD}$, we have

$$495.9\text{ in./s} = (8\text{ in.})\omega_{BD} \qquad \boldsymbol{\omega}_{BD} = 62.0\text{ rad/s} \circlearrowleft \quad ◀$$

SOLVING PROBLEMS ON YOUR OWN

In this lesson you learned to analyze the velocity of bodies in *general plane motion.* You found that a general plane motion can always be considered as the sum of the two motions you studied in the last lesson, namely, *a translation and a rotation.*

To solve a problem involving the velocity of a body in plane motion you should take the following steps.

1. Whenever possible determine the velocity of the points of the body where the body is connected to another body whose motion is known. That other body may be an arm or crank rotating with a given angular velocity [Sample Prob. 15.3].

2. Next start drawing a "diagram equation" to use in your solution (Figs. 15.15 and 15.16). This "equation" will consist of the following diagrams.

a. Plane motion diagram: Draw a diagram of the body including all dimensions and showing those points for which you know or seek the velocity.

b. Translation diagram: Select a reference point A for which you know the direction and/or the magnitude of the velocity $\mathbf{v}_A$, and draw a second diagram showing the body in translation with all of its points having the same velocity $\mathbf{v}_A$.

c. Rotation diagram: Consider point A as a fixed point and draw a diagram showing the body in rotation about A. Show the angular velocity $\boldsymbol{\omega} = \omega\mathbf{k}$ of the body and the relative velocities with respect to A of the other points, such as the velocity $\mathbf{v}_{B/A}$ of B relative to A.

3. Write the relative-velocity formula

$$\mathbf{v}_B = \mathbf{v}_A + \mathbf{v}_{B/A}$$

While you can solve this vector equation analytically by writing the corresponding scalar equations, you will usually find it easier to solve it by using a vector triangle (Fig. 15.16).

4. A different reference point can be used to obtain an equivalent solution. For example, if point B is selected as the reference point, the velocity of point A is expressed as

$$\mathbf{v}_A = \mathbf{v}_B + \mathbf{v}_{A/B}$$

Note that the relative velocities $\mathbf{v}_{B/A}$ and $\mathbf{v}_{A/B}$ have the same magnitude but opposite sense. Relative velocities, therefore, depend upon the reference point that has been selected. The angular velocity, however, is independent of the choice of reference point.

Problems

15.37 Rod AB can slide freely along the floor and the inclined plane. At the instant shown, the velocity of end A is 1.4 m/s to the left. Determine (*a*) the angular velocity of the rod, (*b*) the velocity of end B of the rod.

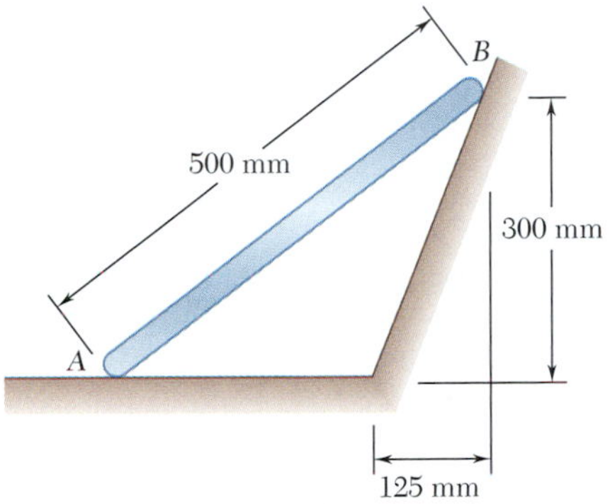

Fig. P15.37 and P15.38

15.38 Rod AB can slide freely along the floor and the inclined plane. At the instant shown the angular velocity of the rod is 4.2 rad/s counterclockwise. Determine (*a*) the velocity of end A of the rod, (*b*) the velocity of end B of the rod.

15.39 Collar A moves up with a velocity of 3.6 ft/s. At the instant shown when $\theta = 25°$, determine (*a*) the angular velocity of rod AB, (*b*) the velocity of collar B.

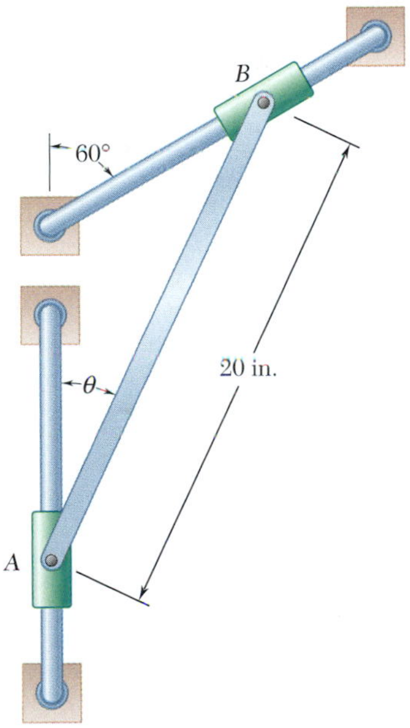

Fig. P15.39

15.40 Rod AB moves over a small wheel at C while end A moves to the right with a constant velocity of 25 in./s. At the instant shown, determine (*a*) the angular velocity of the rod, (*b*) the velocity of end B of the rod.

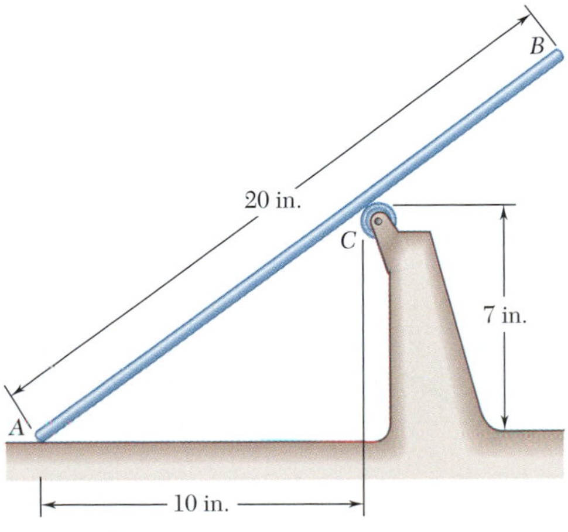

Fig. P15.40

15.41 The motion of rod AB is guided by pins attached at A and B which slide in the slots shown. At the instant shown, $\theta = 40°$ and the pin at B moves upward to the left with a constant velocity of 150 mm/s. Determine (*a*) the angular velocity of the rod, (*b*) the velocity of the pin at end A.

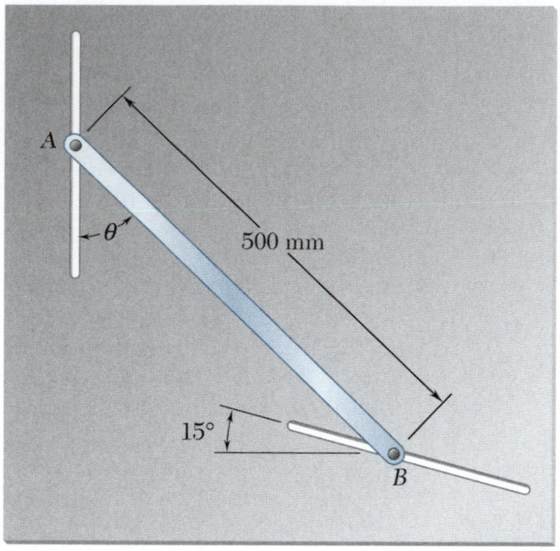

Fig. *P15.41* and *P15.42*

15.42 The motion of rod AB is guided by pins attached at A and B which slide in the slots shown. At the instant shown, $\theta = 30°$ and the pin at A moves downward with a constant velocity of 225 mm/s. Determine (*a*) the angular velocity of the rod, (*b*) the velocity of the pin at end B.

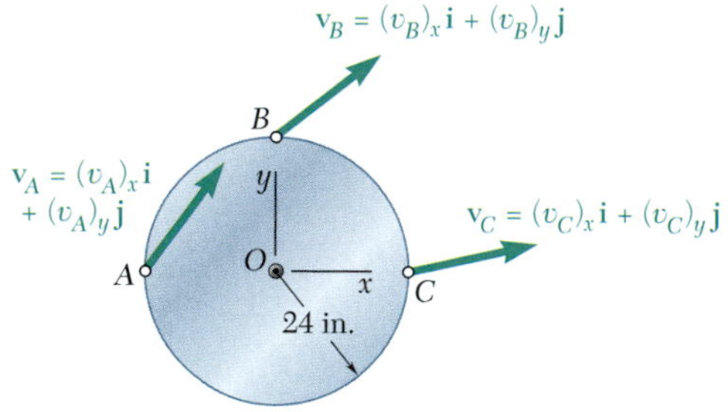

Fig. P15.43

15.43 The disk shown moves in the xy plane. Knowing that $(v_A)_y = -282$ in./s, $(v_B)_x = -296$ in./s, and $(v_C)_x = -56$ in./s, determine (*a*) the angular velocity of the disk, (*b*) the velocity of point B.

15.44 In Prob. 15.43, determine (*a*) the velocity of point O, (*b*) the point of the disk with zero velocity.

15.45 The sheet metal form shown moves in the xy plane. Knowing that $(v_A)_x = 100$ mm/s, $(v_B)_y = -75$ mm/s, and $(v_C)_x = 400$ mm/s, determine (*a*) the angular velocity of the plate, (*b*) the velocity of point A.

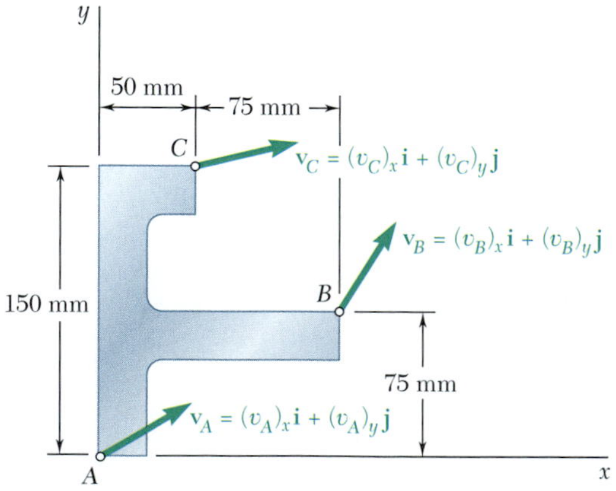

Fig. P15.45

15.46 In Prob. 15.45, determine the locus of points of the sheet metal form for which the magnitude of the velocity is 200 mm/s.

15.47 In the planetary gear system shown, the radius of gears A, B, C, and D is 60 mm and the radius of the outer gear E is 180 mm. Knowing that gear E has an angular velocity of 120 rpm clockwise and that the central gear has an angular velocity of 150 rpm clockwise, determine (*a*) the angular velocity of each planetary gear, (*b*) the angular velocity of the spider connecting the planetary gears.

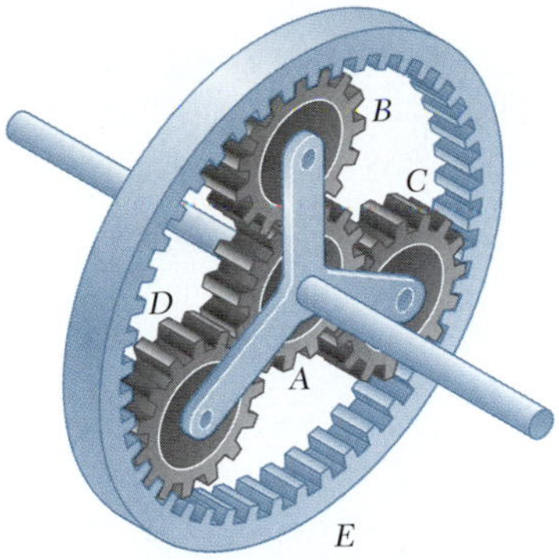

Fig. P15.47 and P15.48

15.48 In the planetary gear system shown, the radius of the central gear A is a, the radius of the planetary gears is b, and the radius of the outer gear E is $a + 2b$. The angular velocity of gear A is ω_A clockwise, and the outer gear is stationary. If the angular velocity of the spider BCD is to be $\omega_A/5$ clockwise, determine (*a*) the required value of the ratio b/a, (*b*) the corresponding angular velocity of each planetary gear.

15.49 The outer gear C rotates with an angular velocity of 5 rad/s clockwise. Knowing that the inner gear A is stationary, determine (*a*) the angular velocity of the intermediate gear B, (*b*) the angular velocity of the arm AB.

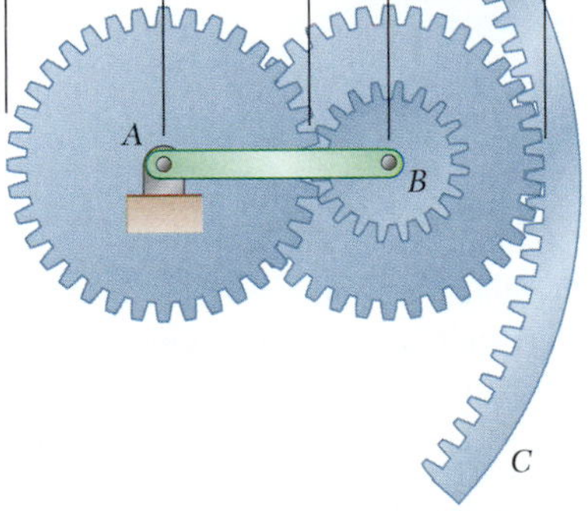

Fig. P15.49 and P15.50

15.50 The intermediate gear B rotates with an angular velocity of 20 rad/s clockwise. Knowing that the outer gear C is stationary, determine (*a*) the angular velocity of the inner gear A, (*b*) the angular velocity of the arm AB.

15.51 and 15.52 Arm ACB rotates about point C with an angular velocity of 40 rad/s counterclockwise. Two friction disks A and B are pinned at their centers to arm ACB as shown. Knowing that the disks roll without slipping at surfaces of contact, determine the angular velocity of (*a*) disk A, (*b*) disk B.

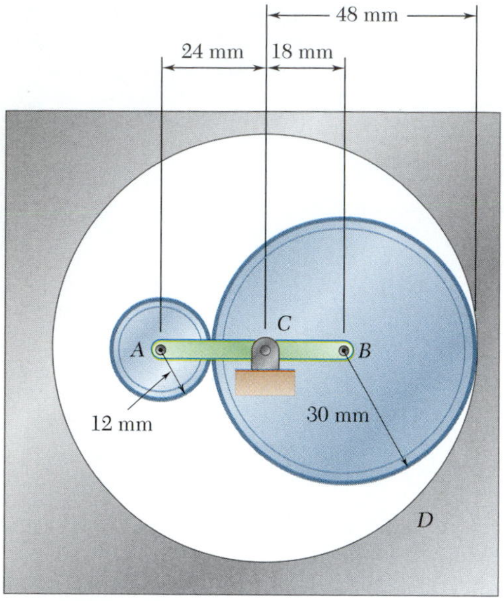

Fig. P15.51

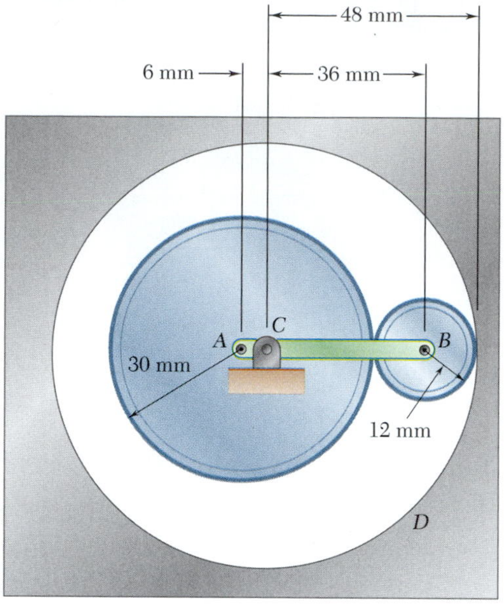

Fig. P15.52

15.53 Gear A rotates with an angular velocity of 120 rpm clockwise. Knowing that the angular velocity of arm AB is 90 rpm clockwise, determine the corresponding angular velocity of gear B.

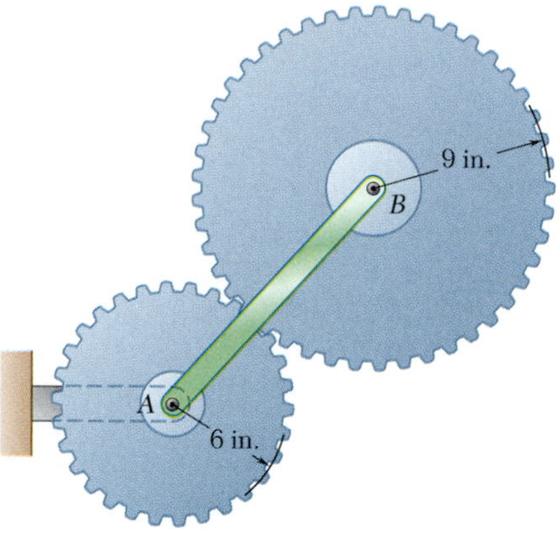

Fig. *P15.53* and *P15.54*

15.54 Arm AB rotates with an angular velocity of 42 rpm clockwise. Determine the required angular velocity of gear A for which (*a*) the angular velocity of gear B is 20 rpm counterclockwise, (*b*) the motion of gear B is a curvilinear translation.

15.55 Knowing that the disk has a constant angular velocity of 15 rad/s clockwise, determine the angular velocity of bar BD and the velocity of collar D when (*a*) $\theta = 0$, (*b*) $\theta = 90°$, (*c*) $\theta = 180°$.

15.56 The disk has a constant angular velocity of 20 rad/s clockwise. (*a*) Determine the two values of the angle θ for which the velocity of collar D is zero. (*b*) For each of these values of θ, determine the corresponding value of the angular velocity of bar BD.

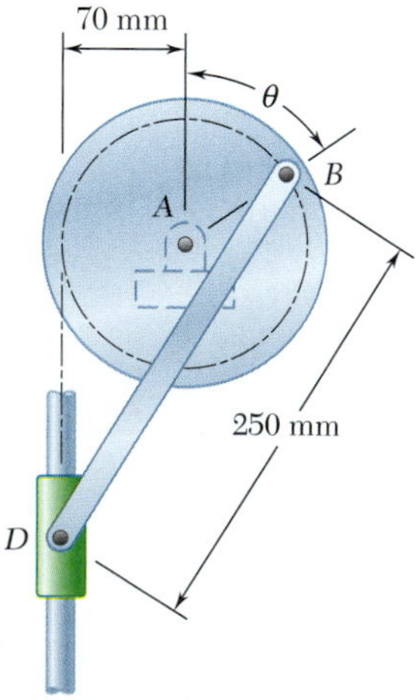

Fig. P15.55 and P15.56

15.57 In the engine system shown, $l = 8$ in. and $b = 3$ in. Knowing that the crank AB rotates with a constant angular velocity of 1000 rpm clockwise, determine the velocity of piston P and the angular velocity of the connecting rod when (*a*) $\theta = 0$, (*b*) $\theta = 90°$.

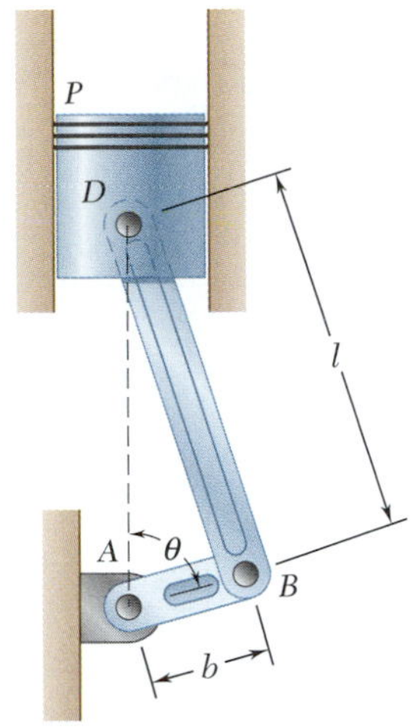

Fig. P15.57 and P15.58

15.58 In the engine system shown in Fig. P15.57 and P15.58, $l = 8$ in. and $b = 3$ in. Knowing that the crank AB rotates with a constant angular velocity of 1000 rpm clockwise, determine the velocity of piston P and the angular velocity of the connecting rod when $\theta = 60°$.

15.59 In the eccentric shown, a disk of 40-mm-radius revolves about shaft O that is located 10 mm from the center A of the disk. The distance between the center A of the disk and the pin at B is 160 mm. Knowing that the angular velocity of the disk is 900 rpm clockwise, determine the velocity of the block when $\theta = 30°$.

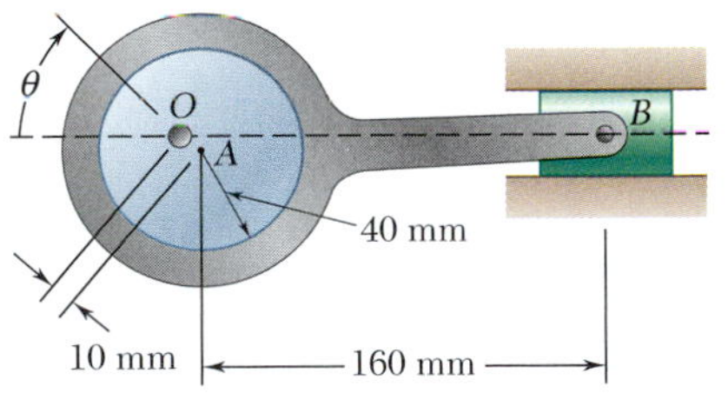

Fig. P15.59

15.60 Determine the velocity of the block of Prob. 15.59 when $\theta = 120°$.

15.61 A straight rack rests on a gear of radius r and is attached to a block B as shown. Denoting by ω_D the clockwise angular velocity of gear D and by θ the angle formed by the rack and the horizontal, derive expressions for the velocity of block B and the angular velocity of the rack in terms of r, θ, and ω_D.

15.62 A straight rack rests on a gear of radius $r = 3$ in. and is attached to a block B as shown. Knowing that at the instant shown the angular velocity of gear D is 15 rpm counterclockwise and $\theta = 20°$, determine (*a*) the velocity of block B, (*b*) the angular velocity of the rack.

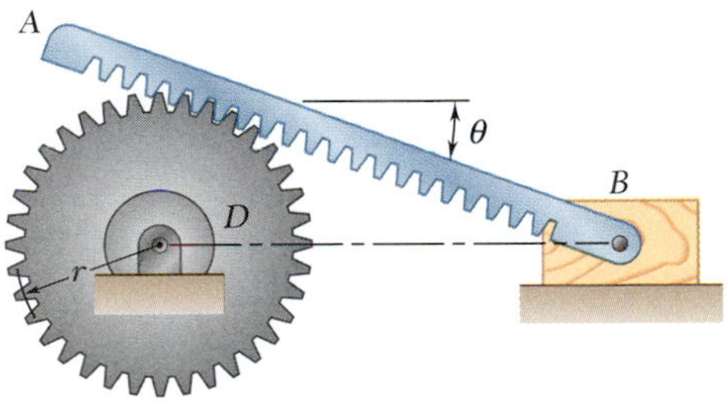

Fig. *P15.61* and *P15.62*

15.63 Bar AB is rotating clockwise and, at the instant shown, the magnitude of the velocity of point G is 3.6 m/s. Determine the angular velocity of each of the three bars at that instant.

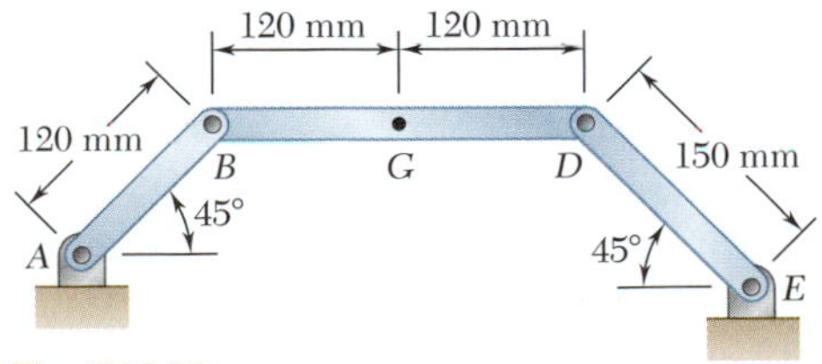

Fig. P15.63

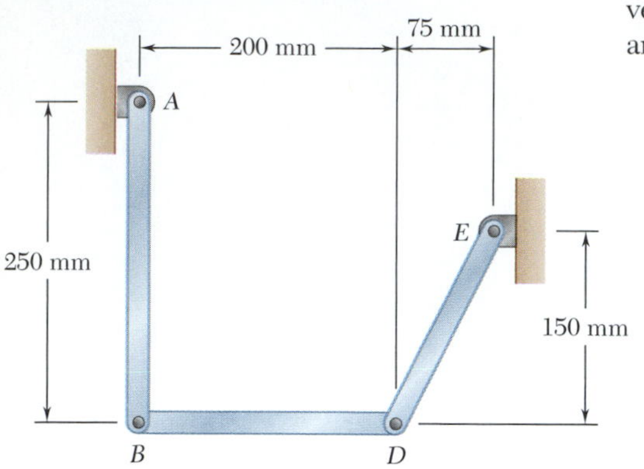

Fig. P15.64

15.64 through 15.66 In the position shown, bar AB has an angular velocity of 4 rad/s clockwise. Determine the angular velocity of bars BD and DE.

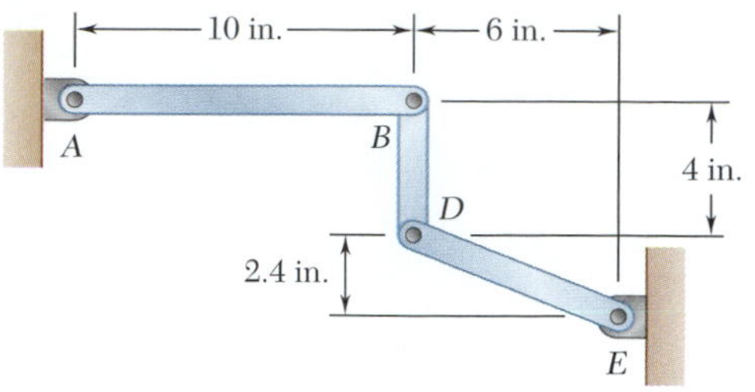

Fig. P15.65

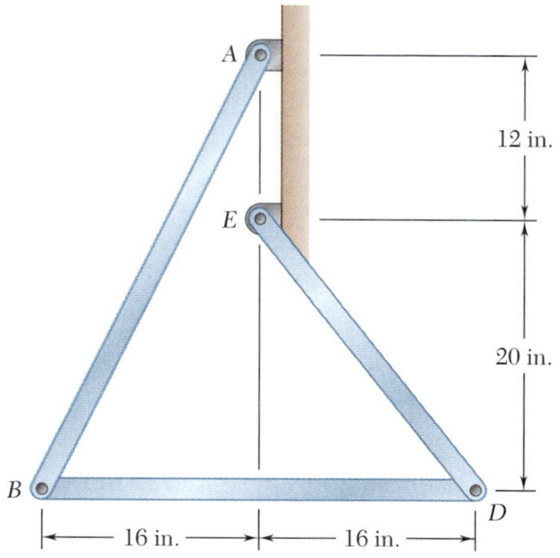

Fig. P15.66

15.67 At the instant shown, bar AB has a constant angular velocity of 25 rad/s counterclockwise. Determine at that instant (*a*) the angular velocity of the rectangular plate $FBDH$, (*b*) the velocity of point F.

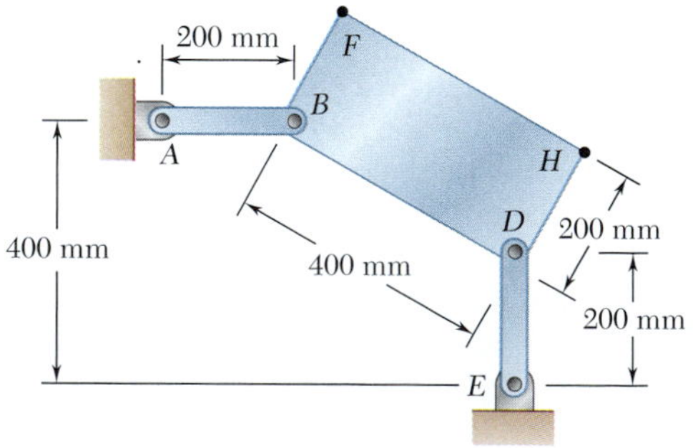

Fig. *P15.67* and *P15.68*

15.68 At the instant shown, bar DE has a constant angular velocity of 35 rad/s clockwise. Determine at that instant (*a*) the angular velocity of the rectangular plate $FBDH$, (*b*) the point on the plate $FBDH$ with zero velocity.

15.69 The 4-in.-radius wheel shown rolls to the left with a velocity of 45 in./s. Knowing that the distance AD is 2.5 in., determine the velocity of the collar and the angular velocity of rod AB when (*a*) $\beta = 0$, (*b*) $\beta = 90°$.

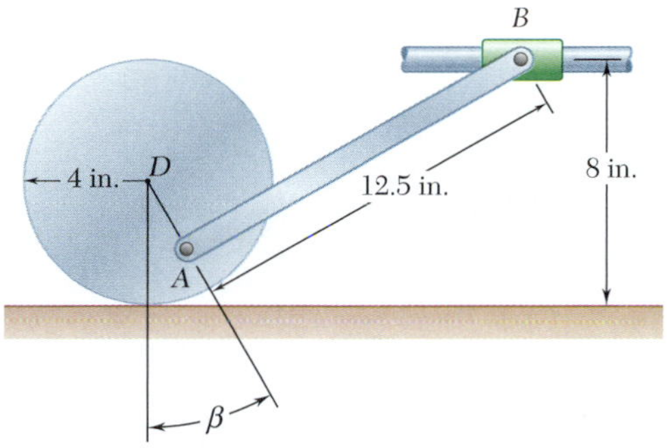

Fig. P15.69

15.70 An automobile travels to the right at a constant speed of 80 km/h. If the diameter of the wheel is 560 mm, determine the velocities of points B, C, D, and E on the rim of the wheel.

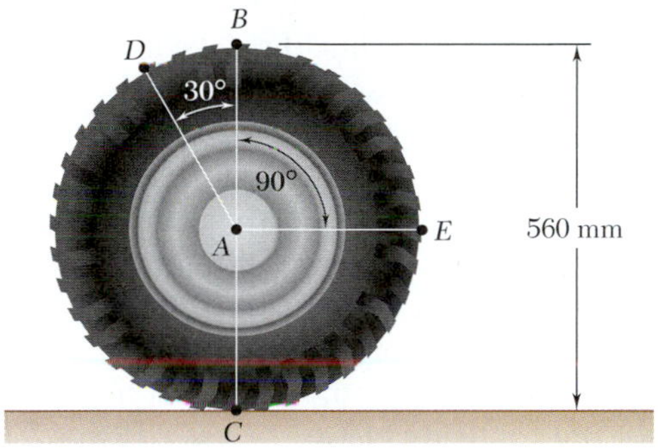

Fig. P15.70

15.7. INSTANTANEOUS CENTER OF ROTATION IN PLANE MOTION

Consider the general plane motion of a slab. We propose to show that at any given instant the velocities of the various particles of the slab are the same as if the slab were rotating about a certain axis perpendicular to the plane of the slab, called the *instantaneous axis of rotation.* This axis intersects the plane of the slab at a point C, called the *instantaneous center of rotation* of the slab.

We first recall that the plane motion of a slab can always be replaced by a translation defined by the motion of an arbitrary reference point A and by a rotation about A. As far as the velocities are concerned, the translation is characterized by the velocity $\mathbf{v}_A$ of the reference point A and the rotation is characterized by the angular velocity $\boldsymbol{\omega}$ of the slab (which is independent of the choice of A). Thus, the velocity $\mathbf{v}_A$ of point A and the angular velocity $\boldsymbol{\omega}$ of the slab define completely the

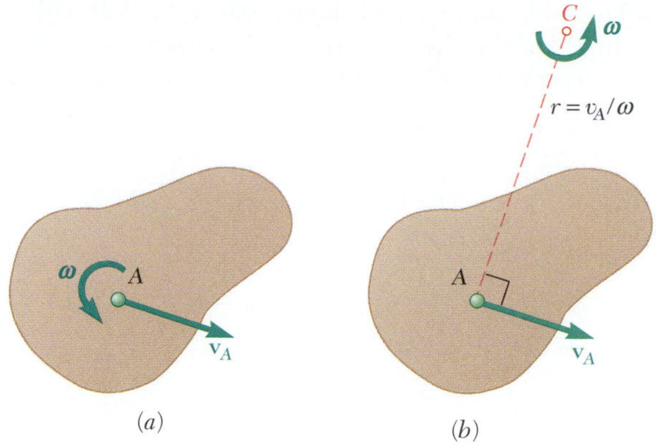

Fig. 15.18

velocities of all the other particles of the slab (Fig. 15.18*a*). Now let us assume that $\mathbf{v}_A$ and $\boldsymbol{\omega}$ are known and that they are both different from zero. (If $\mathbf{v}_A = 0$, point A is itself the instantaneous center of rotation, and if $\boldsymbol{\omega} = 0$, all the particles have the same velocity $\mathbf{v}_A$.) These velocities could be obtained by letting the slab rotate with the angular velocity $\boldsymbol{\omega}$ about a point C located on the perpendicular to $\mathbf{v}_A$ at a distance $r = v_A/\omega$ from A as shown in Fig. 15.18*b*. We check that the velocity of A would be perpendicular to AC and that its magnitude would be $r\omega = (v_A/\omega)\omega = v_A$. Thus the velocities of all the other particles of the slab would be the same as originally defined. Therefore, *as far as the velocities are concerned, the slab seems to rotate about the instantaneous center* C at the instant considered.

The position of the instantaneous center can be defined in two other ways. If the directions of the velocities of two particles A and B of the slab are known and if they are different, the instantaneous center C is obtained by drawing the perpendicular to $\mathbf{v}_A$ through A and the perpendicular to $\mathbf{v}_B$ through B and determining the point in which these two lines intersect (Fig. 15.19*a*). If the velocities $\mathbf{v}_A$ and $\mathbf{v}_B$ of two particles A and B are perpendicular to the line AB and if their magnitudes are known, the instantaneous center can be found by intersecting the line AB with the line joining the extremities of the vectors $\mathbf{v}_A$ and $\mathbf{v}_B$ (Fig. 15.19*b*). Note that if $\mathbf{v}_A$ and $\mathbf{v}_B$ were parallel in Fig. 15.19*a* or if $\mathbf{v}_A$ and $\mathbf{v}_B$ had the same magnitude in Fig. 15.19*b*,

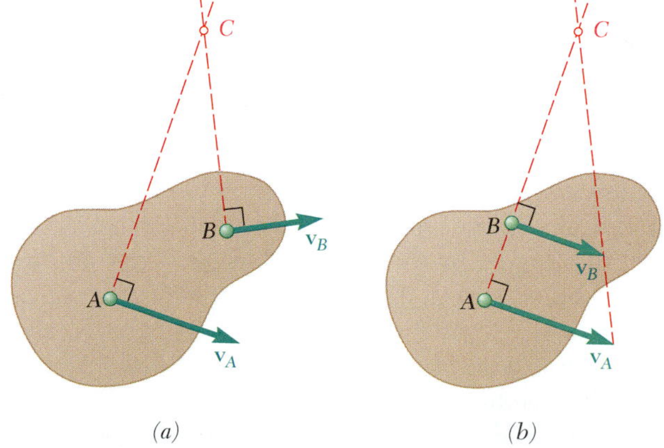

Fig. 15.19

the instantaneous center C would be at an infinite distance and $\boldsymbol{\omega}$ would be zero; all points of the slab would have the same velocity.

To see how the concept of instantaneous center of rotation can be put to use, let us consider again the rod of Sec. 15.6. Drawing the perpendicular to $\mathbf{v}_A$ through A and the perpendicular to $\mathbf{v}_B$ through B (Fig. 15.20), we obtain the instantaneous center C. At the instant

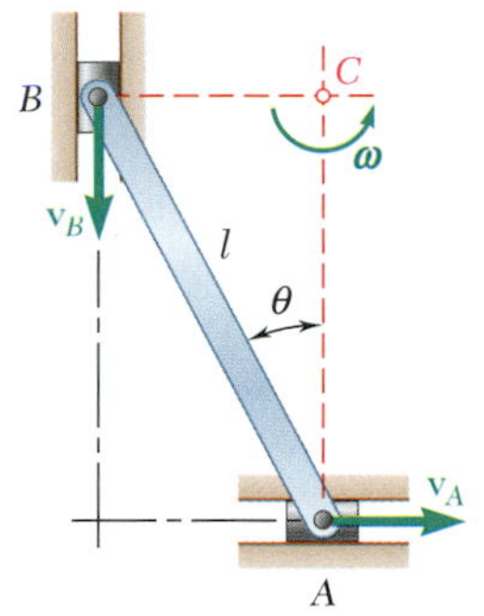

Fig. 15.20

considered, the velocities of all the particles of the rod are thus the same as if the rod rotated about C. Now, if the magnitude v_A of the velocity of A is known, the magnitude ω of the angular velocity of the rod can be obtained by writing

$$\omega = \frac{v_A}{AC} = \frac{v_A}{l \cos \theta}$$

The magnitude of the velocity of B can then be obtained by writing

$$v_B = (BC)\omega = l \sin \theta \frac{v_A}{l \cos \theta} = v_A \tan \theta$$

Note that only *absolute* velocities are involved in the computation.

The instantaneous center of a slab in plane motion can be located either on the slab or outside the slab. If it is located on the slab, the particle C coinciding with the instantaneous center at a given instant t must have zero velocity at that instant. However, it should be noted that the instantaneous center of rotation is valid only at a given instant. Thus, the particle C of the slab which coincides with the instantaneous center at time t will generally not coincide with the instantaneous center at time $t + \Delta t$; while its velocity is zero at time t, it will probably be different from zero at time $t + \Delta t$. This means that, in general, the particle C *does not have zero acceleration* and, therefore, that the *accelerations* of the various particles of the slab *cannot* be determined as if the slab were rotating about C.

As the motion of the slab proceeds, the instantaneous center moves in space. But it was just pointed out that the position of the instantaneous center on the slab keeps changing. Thus, the instantaneous center describes one curve in space, called the *space centrode*, and another curve on the slab, called the *body centrode* (Fig. 15.21). It can be shown that at any instant, these two curves are tangent at C and that as the slab moves, the body centrode appears to *roll* on the space centrode.

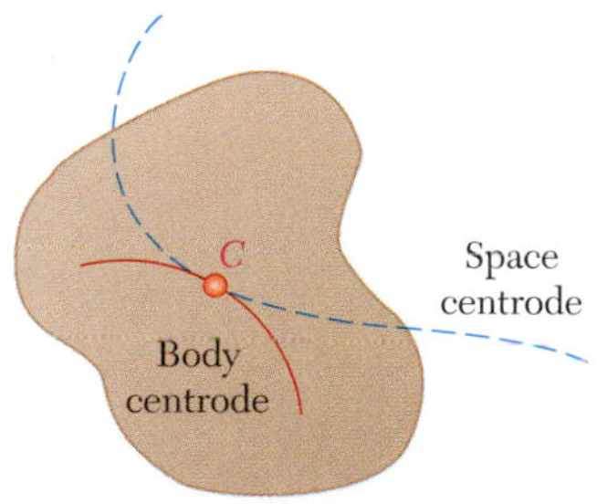

Fig. 15.21

SAMPLE PROBLEM 15.4

Solve Sample Prob. 15.2, using the method of the instantaneous center of rotation.

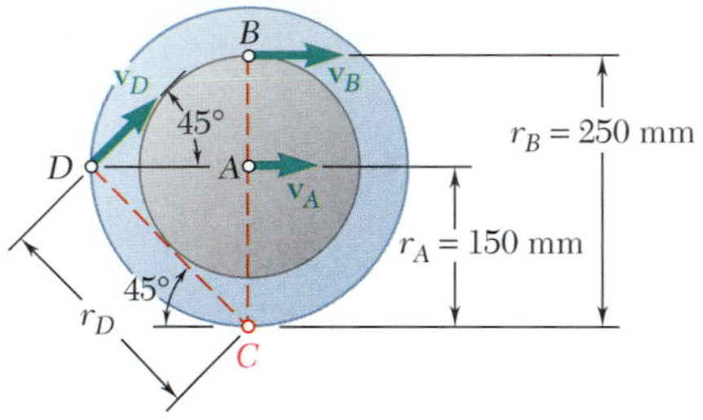

SOLUTION

***a.* Angular Velocity of the Gear.** Since the gear rolls on the stationary lower rack, the point of contact C of the gear with the rack has no velocity; point C is therefore the instantaneous center of rotation. We write

$$v_A = r_A\omega \qquad 1.2 \text{ m/s} = (0.150 \text{ m})\omega$$

$$\boldsymbol{\omega} = 8 \text{ rad/s} \circlearrowright \quad ◀$$

***b.* Velocities.** As far as velocities are concerned, all points of the gear seem to rotate about the instantaneous center.

Velocity of Upper Rack. Recalling that $v_R = v_B$, we write

$$v_R = v_B = r_B\omega \qquad v_R = (0.250 \text{ m})(8 \text{ rad/s}) = 2 \text{ m/s}$$

$$\mathbf{v}_R = 2 \text{ m/s} \rightarrow \quad ◀$$

Velocity of Point D. Since $r_D = (0.150 \text{ m})\sqrt{2} = 0.2121$ m, we write

$$v_D = r_D\omega \qquad v_D = (0.2121 \text{ m})(8 \text{ rad/s}) = 1.697 \text{ m/s}$$

$$\mathbf{v}_D = 1.697 \text{ m/s} \measuredangle 45° \quad ◀$$

SAMPLE PROBLEM 15.5

Solve Sample Prob. 15.3, using the method of the instantaneous center of rotation.

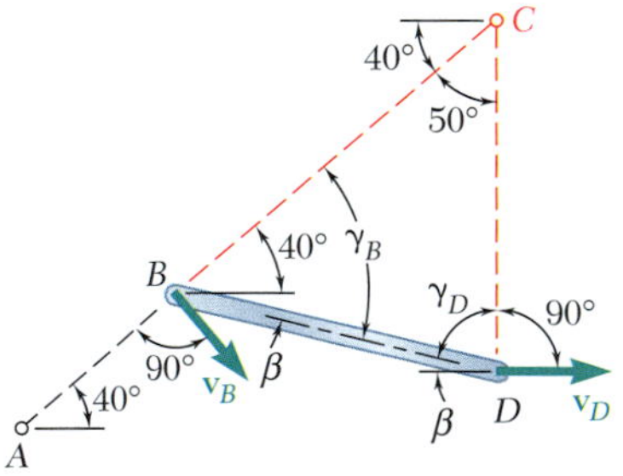

SOLUTION

Motion of Crank *AB*. Referring to Sample Prob. 15.3, we obtain the velocity of point B; $\mathbf{v}_B = 628.3$ in./s $\measuredangle$ 50°.

Motion of the Connecting Rod *BD*. We first locate the instantaneous center C by drawing lines perpendicular to the absolute velocities $\mathbf{v}_B$ and $\mathbf{v}_D$. Recalling from Sample Prob. 15.3 that $\beta = 13.95°$ and that $BD = 8$ in., we solve the triangle BCD.

$$\gamma_B = 40° + \beta = 53.95° \qquad \gamma_D = 90° - \beta = 76.05°$$

$$\frac{BC}{\sin 76.05°} = \frac{CD}{\sin 53.95°} = \frac{8 \text{ in.}}{\sin 50°}$$

$$BC = 10.14 \text{ in.} \qquad CD = 8.44 \text{ in.}$$

Since the connecting rod BD seems to rotate about point C, we write

$$v_B = (BC)\omega_{BD}$$
$$628.3 \text{ in./s} = (10.14 \text{ in.})\omega_{BD}$$

$$\boldsymbol{\omega}_{BD} = 62.0 \text{ rad/s} \circlearrowleft \quad ◀$$

$$v_D = (CD)\omega_{BD} = (8.44 \text{ in.})(62.0 \text{ rad/s})$$
$$= 523 \text{ in./s} = 43.6 \text{ ft/s}$$

$$\mathbf{v}_P = \mathbf{v}_D = 43.6 \text{ ft/s} \rightarrow \quad ◀$$

SOLVING PROBLEMS ON YOUR OWN

In this lesson we introduced the *instantaneous center of rotation* in plane motion. This provides us with an alternative way for solving problems involving the *velocities* of the various points of a body in plane motion.

As its name suggests, the *instantaneous center of rotation* is the point about which you can assume a body is rotating at a given instant, as you determine the velocities of the points of the body at that instant.

A. To determine the instantaneous center of rotation of a body in plane motion, you should use one of the following procedures.

1. If the velocity $\mathbf{v}_A$ of a point A and the angular velocity ω of the body are both known (Fig. 15.18):

a. Draw a sketch of the body, showing point A, its velocity $\mathbf{v}_A$, and the angular velocity $\boldsymbol{\omega}$ of the body.

b. From A draw a line perpendicular to $\mathbf{v}_A$ on the side of $\mathbf{v}_A$ from which this velocity is viewed as having *the same sense as* $\boldsymbol{\omega}$.

c. Locate the instantaneous center C on this line, at a distance $r = v_A/\omega$ from point A.

2. If the directions of the velocities of two points A and B are known and are different (Fig. 15.19*a*):

a. Draw a sketch of the body, showing points A and B and their velocities $\mathbf{v}_A$ and $\mathbf{v}_B$.

b. From A and B draw lines perpendicular to $\mathbf{v}_A$ and $\mathbf{v}_B$, respectively. The instantaneous center C is located at the point where the two lines intersect.

c. If the velocity of one of the two points is known, you can determine the angular velocity of the body. For example, if you know v_A, you can write $\omega = v_A/AC$, where AC is the distance from point A to the instantaneous center C.

3. If the velocities of two points A and B are known and are both perpendicular to the line AB (Fig. 15.19*b*):

a. Draw a sketch of the body, showing points A and B with their velocities $\mathbf{v}_A$ and $\mathbf{v}_B$ *drawn to scale.*

b. Draw a line through points A and B, and another line through the tips of the vectors $\mathbf{v}_A$ and $\mathbf{v}_B$. The instantaneous center C is located at the point where the two lines intersect.

(*continued*)

c. The angular velocity of the body is obtained by either dividing $\mathbf{v}_A$ by AC or $\mathbf{v}_B$ by BC.

d. If the velocities $\mathbf{v}_A$ and $\mathbf{v}_B$ have the same magnitude, the two lines drawn in part *b* do not intersect; the instantaneous center C is at an infinite distance. The angular velocity $\boldsymbol{\omega}$ is zero and *the body is in translation.*

B. Once you have determined the instantaneous center and the angular velocity of a body, you can determine the velocity $\mathbf{v}_P$ of any point P of the body in the following way.

1. Draw a sketch of the body, showing point P, the instantaneous center of rotation C, and the angular velocity $\boldsymbol{\omega}$.

2. Draw a line from P to the instantaneous center C and measure or calculate the distance from P to C.

3. The velocity $\mathbf{v}_P$ is a vector perpendicular to the line PC, of the same sense as $\boldsymbol{\omega}$, and of magnitude $v_P = (PC)\omega$.

Finally, keep in mind that the instantaneous center of rotation can be used *only* to determine velocities. *It cannot be used to determine accelerations.*

Problems

15.71 A 5-m beam *AE* is being lowered by means of two overhead cranes. At the instant shown it is known that the velocity of point *D* is 1 m/s downward and the velocity of point *E* is 1.5 m/s downward. Determine (*a*) the instantaneous center of rotation of the beam, (*b*) the velocity of point *A*.

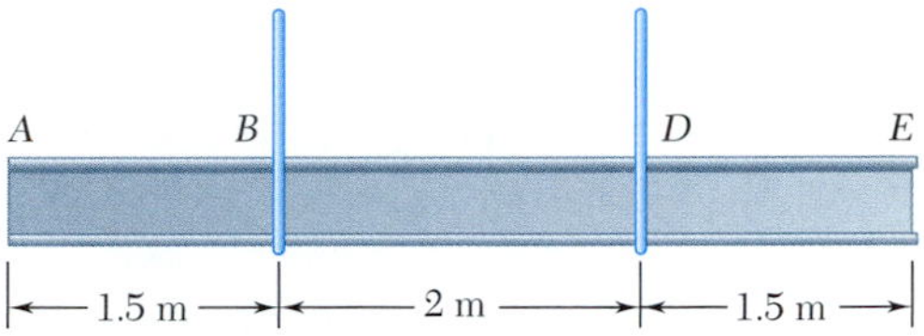

Fig. P15.71 and P15.72

15.72 A 5-m beam *AE* is being lowered by means of two overhead cranes. At the instant shown it is known that the velocity of point *A* is 542 mm/s downward and the velocity of point *E* is 292 mm/s upward. Determine (*a*) the instantaneous center of rotation of the beam, (*b*) the velocity of point *D*.

15.73 At the instant shown during deceleration, the velocity of an automobile is 40 ft/s to the right. Knowing that the velocity of the contact point *A* of the wheel with the ground is 5 ft/s to the right, determine (*a*) the instantaneous center of rotation of the wheel, (*b*) the velocity of point *B*, (*c*) the velocity of point *D*.

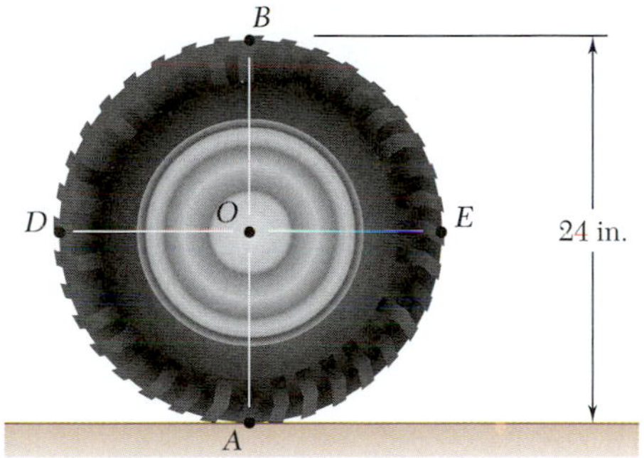

Fig. P15.73 and P15.74

15.74 At the instant shown during acceleration, the velocity of an automobile is 40 ft/s to the right. Knowing that the velocity of the contact point *A* of the wheel with the ground is 5 ft/s to the left, determine (*a*) the instantaneous center of rotation of the wheel, (*b*) the velocity of point *B*, (*c*) the velocity of point *E*.

15.75 and 15.76 A 60-mm-radius drum is rigidly attached to a 100-mm-radius drum as shown. One of the drums rolls without sliding on the surface shown, and a cord is wound around the other drum. Knowing that end *E* of the cord is pulled to the left with a velocity of 120 mm/s, determine (*a*) the angular velocity of the drums, (*b*) the velocity of the center of the drums, (*c*) the length of cord wound or unwound per second.

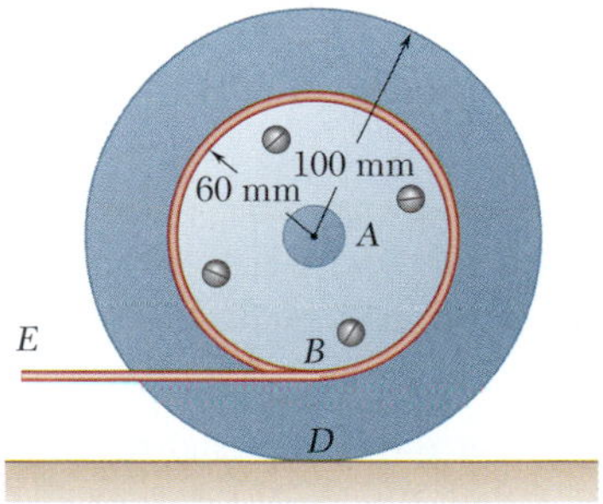

Fig. P15.75

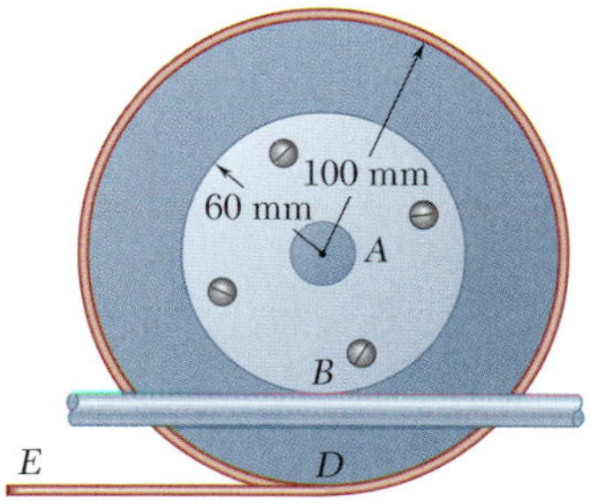

Fig. P15.76

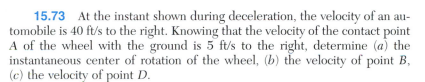

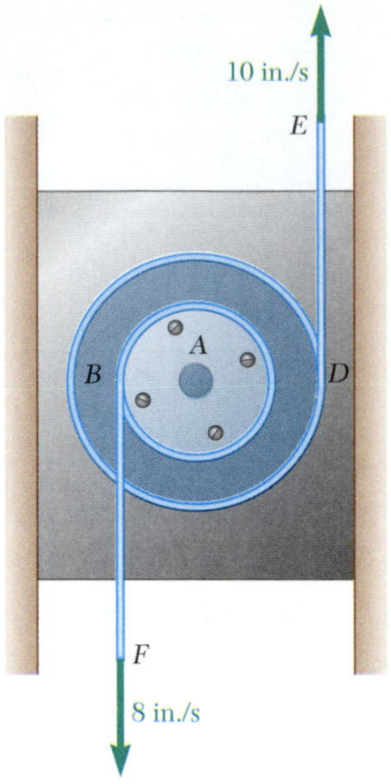

Fig. *P15.77*

15.77 A double pulley is attached to a slider block by a pin at *A*. The 1.5-in.-radius inner pulley is rigidly attached to the 3-in.-radius outer pulley. Knowing that each of the two cords is pulled at a constant speed as shown, determine (*a*) the instantaneous center of rotation of the double pulley, (*b*) the velocity of the slider block, (*c*) the number of inches of cord wrapped or unwrapped on each pulley per second.

15.78 Solve Prob. 15.77, assuming that cord *E* is pulled upward at a speed of 8 in./s and cord *F* is pulled downward at a speed of 10 in./s.

15.79 Knowing that at the instant shown the angular velocity of bar *DC* is 18 rad/s counterclockwise, determine (*a*) the angular velocity of bar *AB*, (*b*) the angular velocity of bar *BC*, (*c*) the velocity of the midpoint of bar *BC*.

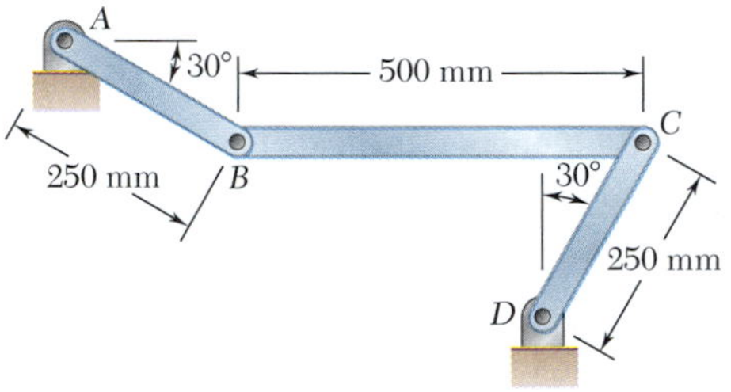

Fig. P15.79 and P15.80

15.80 Knowing that at the instant shown bar *AB* is rotating counterclockwise and that the magnitude of the velocity of the midpoint of bar *BC* is 2.6 m/s, determine (*a*) the angular velocity of bar *AB*, (*b*) the angular velocity of bar *BC*, (*c*) the angular velocity of bar *DC*.

15.81 Knowing that at the instant shown the velocity of collar *A* is 45 in./s to the left, determine (*a*) the angular velocity of rod *ADB*, (*b*) the velocity of point *B*.

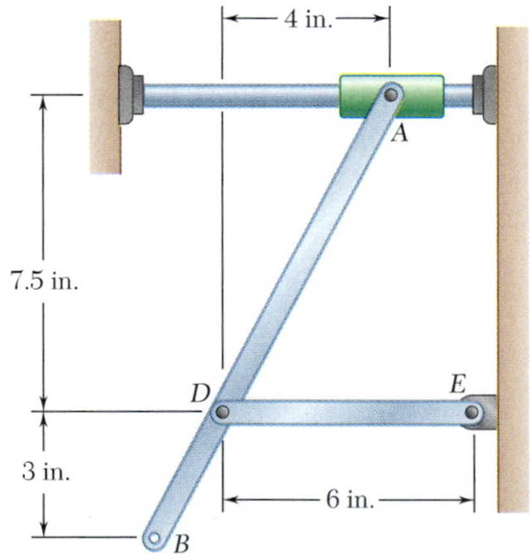

Fig. P15.81 and P15.82

15.82 Knowing that at the instant shown the angular velocity of rod *DE* is 2.4 rad/s clockwise, determine (*a*) the velocity of collar *A*, (*b*) the velocity of point *B*.

15.83 An overhead door is guided by wheels at A and B that roll in horizontal and vertical tracks. Knowing that when $\theta = 40°$ the velocity of wheel B is 0.6 m/s upward, determine (*a*) the angular velocity of the door, (*b*) the velocity of end D of the door.

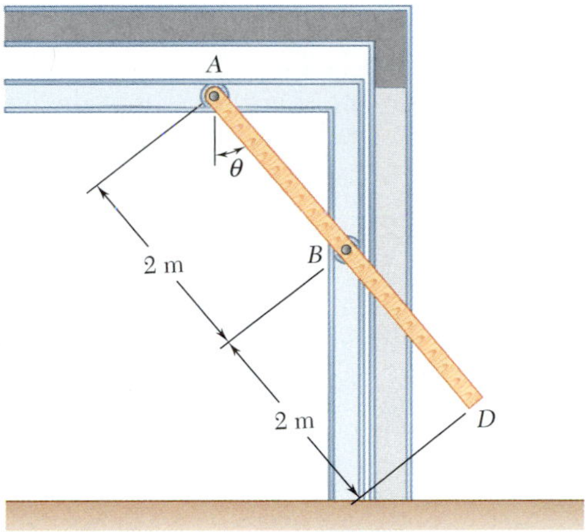

Fig. *P15.83*

15.84 Rod ABD is guided by wheels at A and B that roll in horizontal and vertical tracks. Knowing that at the instant shown $\beta = 60°$ and the velocity of wheel B is 800 mm/s downward, determine (*a*) the angular velocity of the rod, (*b*) the velocity of point D.

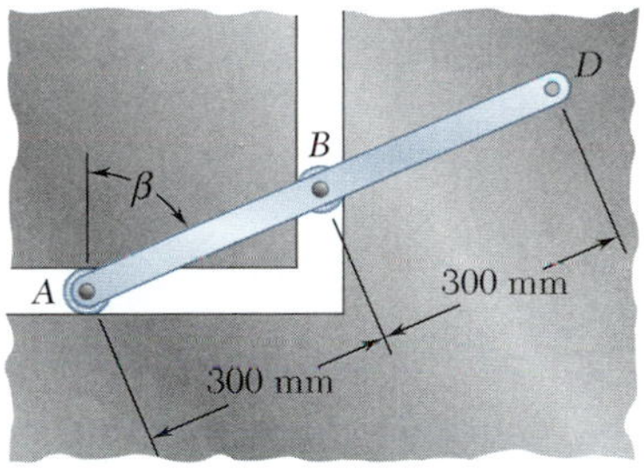

Fig. *P15.84*

15.85 Small wheels have been attached to the ends of bar AB and roll freely along the surfaces shown. Knowing that the velocity of wheel B is 2.5 m/s to the right at the instant shown, determine (*a*) the velocity of end A of the bar, (*b*) the angular velocity of the bar, (*c*) the velocity of the midpoint of the bar.

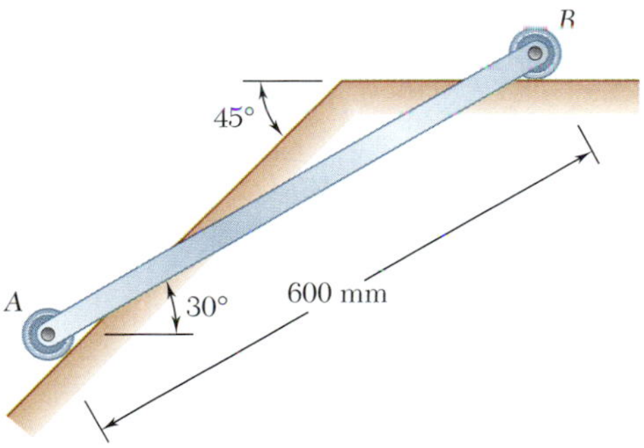

Fig. P15.85

15.86 At the instant shown, the angular velocity of bar DE is 8 rad/s counterclockwise. Determine (*a*) the angular velocity of bar BD, (*b*) the angular velocity of bar AB, (*c*) the velocity of the midpoint of bar BD.

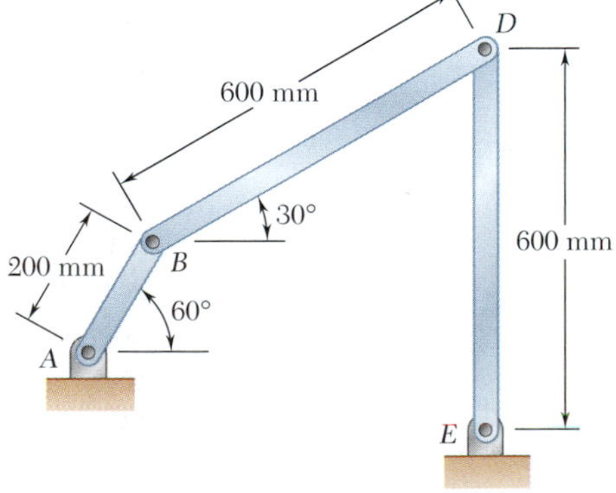

Fig. P15.86

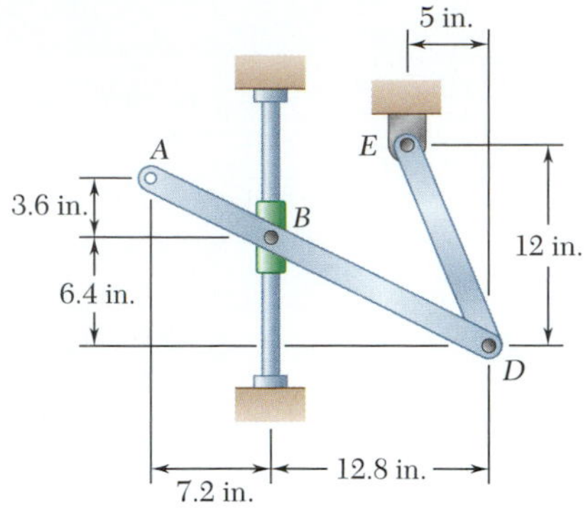

Fig. P15.87 and P15.88

15.87 Arm *ABD* is connected by pins to a collar at *B* and to crank *DE*. Knowing that the velocity of collar *B* is 16 in./s upward, determine (*a*) the angular velocity of arm *ABD*, (*b*) the velocity of point *A*.

15.88 Arm *ABD* is connected by pins to a collar at *B* and to crank *DE*. Knowing that the angular velocity of crank *DE* is 1.2 rad/s counterclockwise, determine (*a*) the angular velocity of arm *ABD*, (*b*) the velocity of point *A*.

15.89 The pin at *B* is attached to member *ABD* and can slide freely along the slot cut in the fixed plate. Knowing that at the instant shown the angular velocity of arm *DE* is 3 rad/s clockwise, determine (*a*) the angular velocity of member *ABD*, (*b*) the velocity of point *A*.

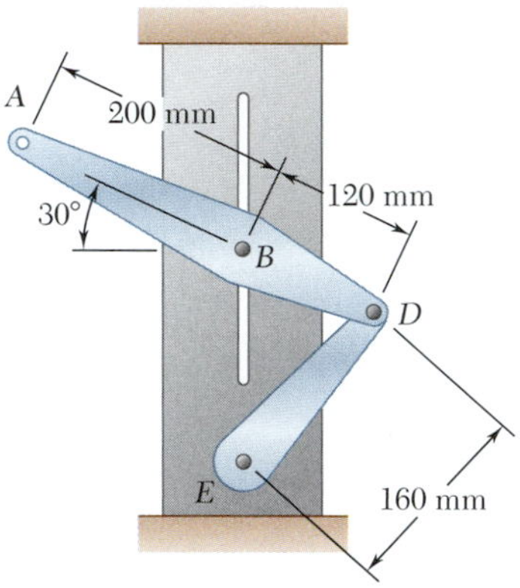

Fig. *P15.89*

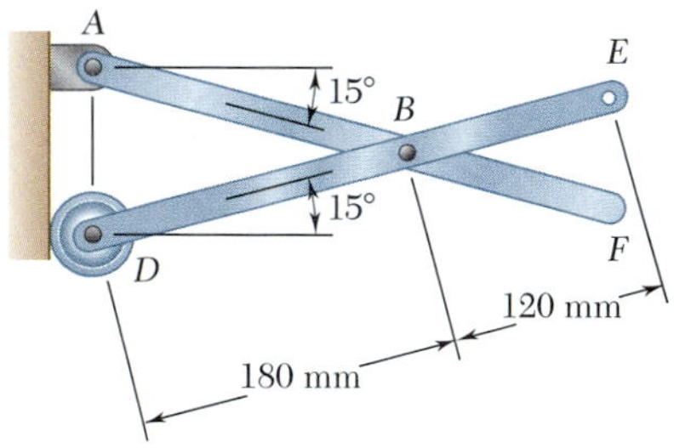

Fig. *P15.90*

15.90 Two identical rods *ABF* and *DBE* are connected by a pin at *B*. Knowing that at the instant shown the velocity of point *D* is 200 mm/s upward, determine the velocity of (*a*) point *E*, (*b*) point *F*.

15.91 Two rods *AB* and *BD* are connected to three collars as shown. Knowing that collar *A* moves downward with a velocity of 6 in./s, determine at the instant shown (*a*) the angular velocity of each rod, (*b*) the velocity of collar *D*.

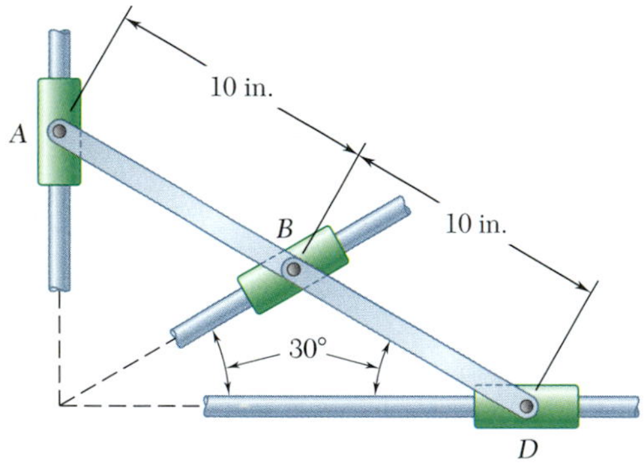

Fig. P15.91

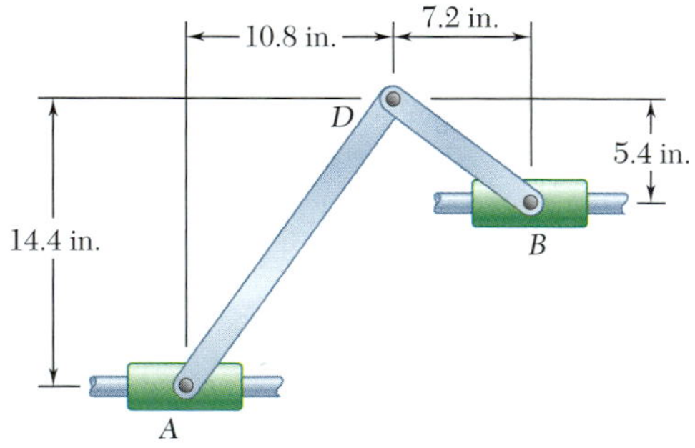

Fig. P15.92

15.92 At the instant shown, the velocity of collar *A* is 1.4 ft/s to the right and the velocity of collar *B* is 3.6 ft/s to the left. Determine (*a*) the angular velocity of bar *AD*, (*b*) the angular velocity of bar *BD*, (*c*) the velocity of point *D*.

15.93 Two rods AB and DE are connected as shown. Knowing that point D moves to the left with a velocity of 800 mm/s, determine at the instant shown (*a*) the angular velocity of each rod, (*b*) the velocity of point A.

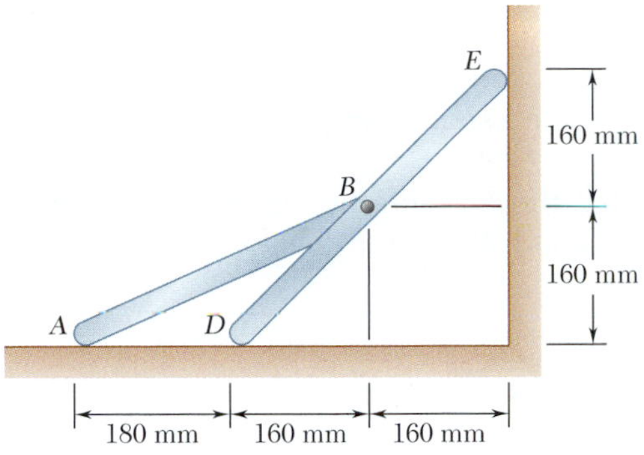

Fig. P15.93

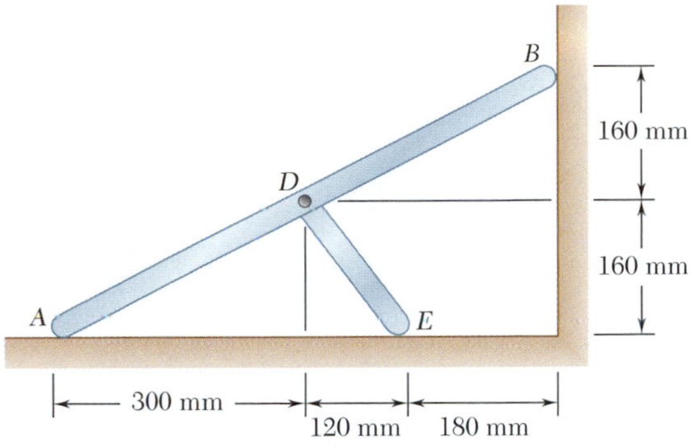

Fig. P15.94

15.94 Two rods AB and DE are connected as shown. Knowing that point B moves downward with a velocity of 1.2 m/s, determine at the instant shown (*a*) the angular velocity of each rod, (*b*) the velocity of point E.

15.95 Two 20-in. rods AB and DE are connected as shown. Point D is the midpoint of rod AB, and at the instant shown rod DE is horizontal. Knowing that the velocity of point A is 1 ft/s downward, determine (*a*) the angular velocity of rod DE, (*b*) the velocity of point E.

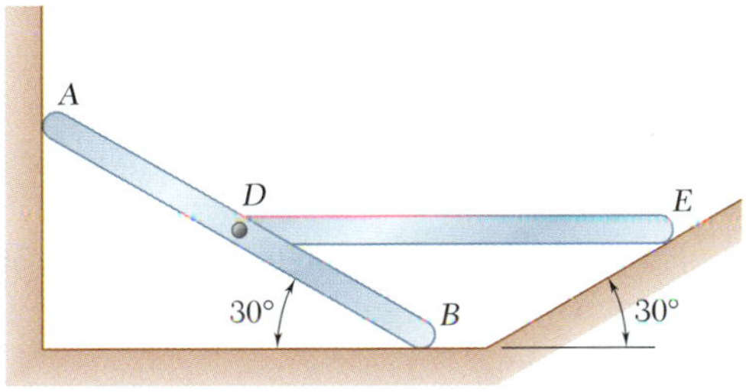

Fig. *P15.95*

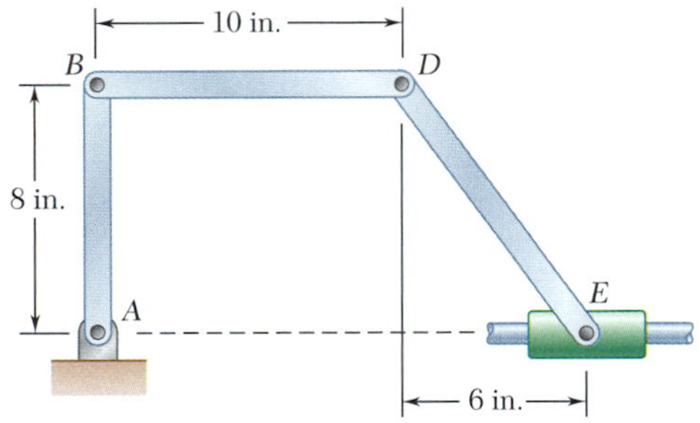

Fig. *P15.96*

15.96 At the instant shown, the angular velocity of bar AB is 4.5 rad/s counterclockwise and the velocity of collar E is 28 in./s to the right. Determine (*a*) the angular velocity of bar BD, (*b*) the angular velocity of bar DE, (*c*) the velocity of point D.

15.97 Describe the space centrode and the body centrode of rod ABD of Prob. 15.84. (*Hint:* The body centrode does not have to lie on a physical portion of the rod.)

15.98 Describe the space centrode and the body centrode of the gear of Sample Prob. 15.2 as the gear rolls on the stationary horizontal track.

15.99 Using the method of Sec. 15.7, solve Prob. 15.59.

15.100 Using the method of Sec. 15.7, solve Prob. 15.70.

15.101 Using the method of Sec. 15.7, solve Prob. 15.65.

15.102 Using the method of Sec. 15.7, solve Prob. 15.66.

15.8. ABSOLUTE AND RELATIVE ACCELERATION IN PLANE MOTION

We saw in Sec. 15.5 that any plane motion can be replaced by a translation defined by the motion of an arbitrary reference point A and a simultaneous rotation about A. This property was used in Sec. 15.6 to determine the velocity of the various points of a moving slab. The same property will now be used to determine the acceleration of the points of the slab.

We first recall that the absolute acceleration $\mathbf{a}_B$ of a particle of the slab can be obtained from the relative-acceleration formula derived in Sec. 11.12,

$$\mathbf{a}_B = \mathbf{a}_A + \mathbf{a}_{B/A} \tag{15.21}$$

where the right-hand member represents a vector sum. The acceleration $\mathbf{a}_A$ corresponds to the translation of the slab with A, while the relative acceleration $\mathbf{a}_{B/A}$ is associated with the rotation of the slab about A and is measured with respect to axes centered at A and of fixed orientation. We recall from Sec. 15.3 that the relative acceleration $\mathbf{a}_{B/A}$ can be resolved into two components, a *tangential component* $(\mathbf{a}_{B/A})_t$ perpendicular to the line AB, and a *normal component* $(\mathbf{a}_{B/A})_n$ directed toward A (Fig. 15.22). Denoting by $\mathbf{r}_{B/A}$ the position vector of B relative to A and, respectively, by $\omega\mathbf{k}$ and $\alpha\mathbf{k}$ the angular velocity and angular acceleration of the slab with respect to axes of fixed orientation, we have

$$\begin{aligned} (\mathbf{a}_{B/A})_t &= \alpha\mathbf{k} \times \mathbf{r}_{B/A} \qquad & (a_{B/A})_t &= r\alpha \\ (\mathbf{a}_{B/A})_n &= -\omega^2\mathbf{r}_{B/A} \qquad & (a_{B/A})_n &= r\omega^2 \end{aligned} \tag{15.22}$$

where r is the distance from A to B. Substituting into (15.21) the expressions obtained for the tangential and normal components of $\mathbf{a}_{B/A}$, we can also write

$$\mathbf{a}_B = \mathbf{a}_A + \alpha\mathbf{k} \times \mathbf{r}_{B/A} - \omega^2\mathbf{r}_{B/A} \tag{15.21'}$$

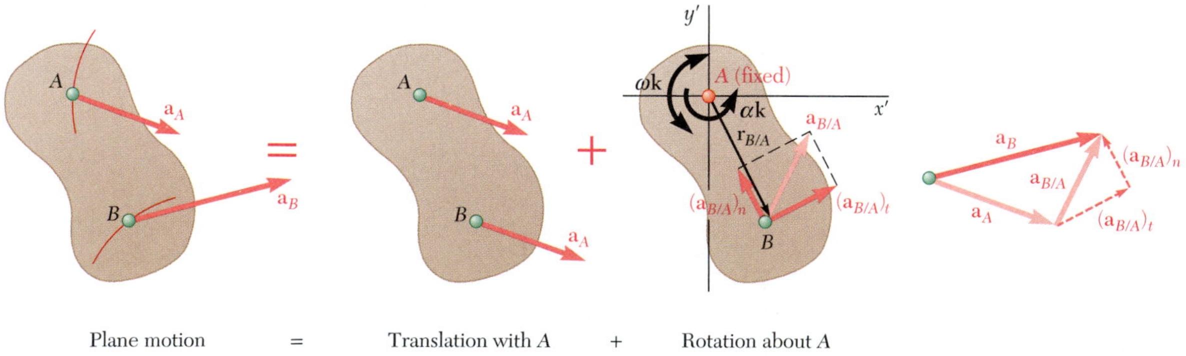

Fig. 15.22

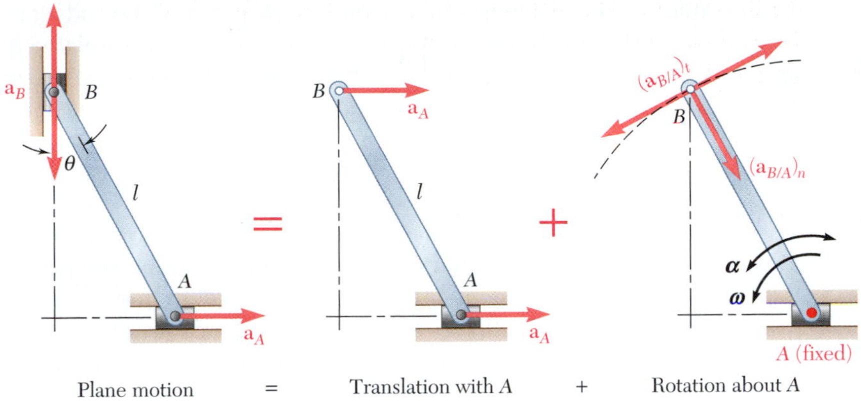

Fig. 15.23

As an example, let us again consider the rod AB whose extremities slide, respectively, along a horizontal and a vertical track (Fig. 15.23). Assuming that the velocity $\mathbf{v}_A$ and the acceleration $\mathbf{a}_A$ of A are known, we propose to determine the acceleration $\mathbf{a}_B$ of B and the angular acceleration $\boldsymbol{\alpha}$ of the rod. Choosing A as a reference point, we express that the given motion is equivalent to a translation with A and a rotation about A. The absolute acceleration of B must be equal to the sum

$$\begin{aligned} \mathbf{a}_B &= \mathbf{a}_A + \mathbf{a}_{B/A} \\ &= \mathbf{a}_A + (\mathbf{a}_{B/A})_n + (\mathbf{a}_{B/A})_t \end{aligned} \tag{15.23}$$

where $(\mathbf{a}_{B/A})_n$ has the magnitude $l\omega^2$ and is *directed toward* A, while $(\mathbf{a}_{B/A})_t$ has the magnitude $l\alpha$ and is perpendicular to AB. Students should note that there is no way to tell whether the tangential component $(\mathbf{a}_{B/A})_t$ is directed to the left or to the right, and therefore both possible directions for this component are indicated in Fig. 15.23. Similarly, both possible senses for $\mathbf{a}_B$ are indicated, since it is not known whether point B is accelerated upward or downward.

Equation (15.23) has been expressed geometrically in Fig. 15.24. Four different vector polygons can be obtained, depending upon the sense of $\mathbf{a}_A$ and the relative magnitude of a_A and $(a_{B/A})_n$. If we are to determine a_B and α from one of these diagrams, we must know not only a_A and θ but also ω. The angular velocity of the rod should therefore be separately determined by one of the methods indicated in Secs. 15.6 and 15.7. The values of a_B and α can then be obtained by considering successively the x and y components of the vectors shown in Fig. 15.24. In the case of polygon a, for example, we write

$\overset{+}{\rightarrow} x$ components: $\qquad 0 = a_A + l\omega^2 \sin\theta - l\alpha \cos\theta$

$+\uparrow y$ components: $\qquad -a_B = -l\omega^2 \cos\theta - l\alpha \sin\theta$

and solve for a_B and α. The two unknowns can also be obtained by direct measurement on the vector polygon. In that case, care should be taken to draw first the known vectors $\mathbf{a}_A$ and $(\mathbf{a}_{B/A})_n$.

It is quite evident that the determination of accelerations is considerably more involved than the determination of velocities. Yet in

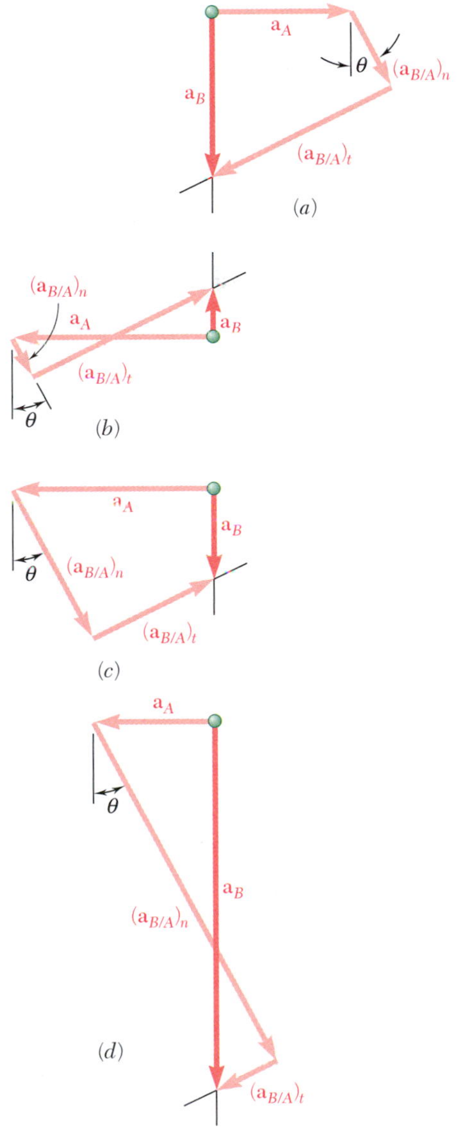

Fig. 15.24

Photo 15.3 The central gear rotates about a fixed axis and is pin-connected to three bars which are in general plane motion.

the example considered here, the extremities A and B of the rod were moving along straight tracks, and the diagrams drawn were relatively simple. If A and B had moved along curved tracks, it would have been necessary to resolve the accelerations $\mathbf{a}_A$ and $\mathbf{a}_B$ into normal and tangential components and the solution of the problem would have involved six different vectors.

When a mechanism consists of several moving parts which are pin-connected, the analysis of the mechanism can be carried out by considering each part as a rigid body, keeping in mind that the points at which two parts are connected must have the same absolute acceleration (see Sample Prob. 15.7). In the case of meshed gears, the tangential components of the accelerations of the teeth in contact are equal, but their normal components are different.

*15.9. ANALYSIS OF PLANE MOTION IN TERMS OF A PARAMETER

In the case of certain mechanisms, it is possible to express the coordinates x and y of all the significant points of the mechanism by means of simple analytic expressions containing a single parameter. It is sometimes advantageous in such a case to determine the absolute velocity and the absolute acceleration of the various points of the mechanism directly, since the components of the velocity and of the acceleration of a given point can be obtained by differentiating the coordinates x and y of that point.

Let us consider again the rod AB whose extremities slide, respectively, in a horizontal and a vertical track (Fig. 15.25). The coordinates x_A and y_B of the extremities of the rod can be expressed in terms of the angle θ the rod forms with the vertical:

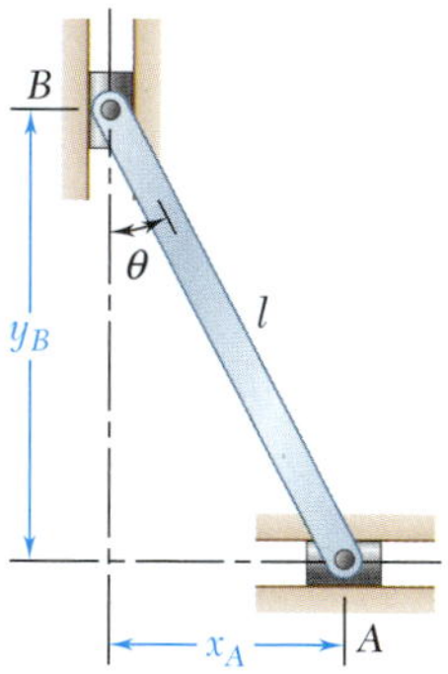

Fig. 15.25

$$x_A = l \sin\theta \qquad y_B = l \cos\theta \tag{15.24}$$

Differentiating Eqs. (15.24) twice with respect to t, we write

$$\begin{aligned} v_A &= \dot{x}_A = l\dot{\theta}\cos\theta \\ a_A &= \ddot{x}_A = -l\dot{\theta}^2 \sin\theta + l\ddot{\theta}\cos\theta \end{aligned}$$

$$\begin{aligned} v_B &= \dot{y}_B = -l\dot{\theta}\sin\theta \\ a_B &= \ddot{y}_B = -l\dot{\theta}^2 \cos\theta - l\ddot{\theta}\sin\theta \end{aligned}$$

Recalling that $\dot{\theta} = \omega$ and $\ddot{\theta} = \alpha$, we obtain

$$v_A = l\omega\cos\theta \qquad v_B = -l\omega\sin\theta \tag{15.25}$$

$$a_A = -l\omega^2 \sin\theta + l\alpha\cos\theta \qquad a_B = -l\omega^2\cos\theta - l\alpha\sin\theta \tag{15.26}$$

We note that a positive sign for v_A or a_A indicates that the velocity $\mathbf{v}_A$ or the acceleration $\mathbf{a}_A$ is directed to the right; a positive sign for v_B or a_B indicates that $\mathbf{v}_B$ or $\mathbf{a}_B$ is directed upward. Equations (15.25) can be used, for example to determine v_B and ω when v_A and θ are known. Substituting for ω in (15.26), we can then determine a_B and α if a_A is known.

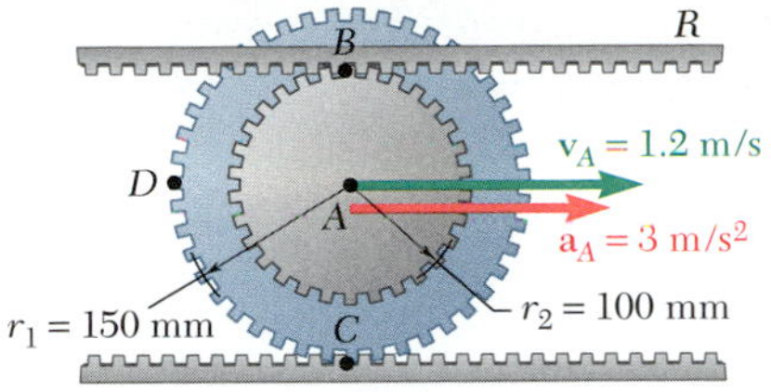

SAMPLE PROBLEM 15.6

The center of the double gear of Sample Prob. 15.2 has a velocity of 1.2 m/s to the right and an acceleration of 3 m/s^2 to the right. Recalling that the lower rack is stationary, determine (*a*) the angular acceleration of the gear, (*b*) the acceleration of points *B*, *C*, and *D* of the gear.

SOLUTION

a. Angular Acceleration of the Gear. In Sample Prob. 15.2, we found that $x_A = -r_1\theta$ and $v_A = -r_1\omega$. Differentiating the latter with respect to time, we obtain $a_A = -r_1\alpha$.

$$v_A = -r_1\omega \qquad 1.2 \text{ m/s} = -(0.150 \text{ m})\omega \qquad \omega = -8 \text{ rad/s}$$
$$a_A = -r_1\alpha \qquad 3 \text{ m/s}^2 = -(0.150 \text{ m})\alpha \qquad \alpha = -20 \text{ rad/s}^2$$
$$\boldsymbol{\alpha} = \alpha\mathbf{k} = -(20 \text{ rad/s}^2)\mathbf{k} \quad ◀$$

b. Accelerations. The rolling motion of the gear is resolved into a translation with *A* and a rotation about *A*.

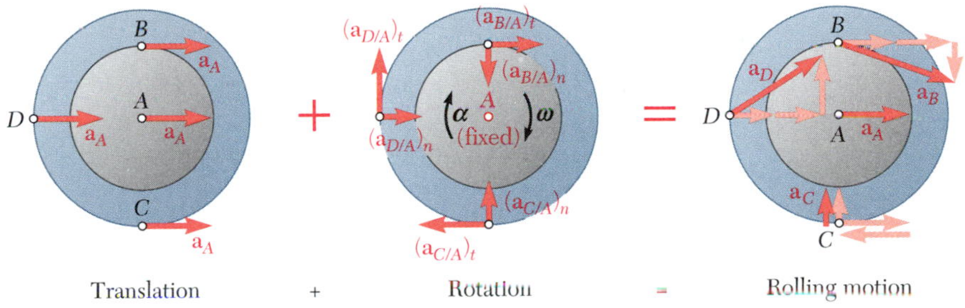

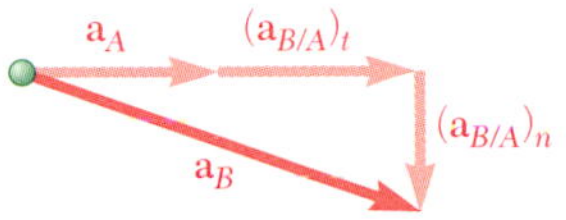

Acceleration of Point B. Adding vectorially the accelerations corresponding to the translation and to the rotation, we obtain

$$\begin{aligned}
\mathbf{a}_B &= \mathbf{a}_A + \mathbf{a}_{B/A} = \mathbf{a}_A + (\mathbf{a}_{B/A})_t + (\mathbf{a}_{B/A})_n \\
&= \mathbf{a}_A + \alpha\mathbf{k} \times \mathbf{r}_{B/A} - \omega^2\mathbf{r}_{B/A} \\
&= (3 \text{ m/s}^2)\mathbf{i} - (20 \text{ rad/s}^2)\mathbf{k} \times (0.100 \text{ m})\mathbf{j} - (8 \text{ rad/s})^2(0.100 \text{ m})\mathbf{j} \\
&= (3 \text{ m/s}^2)\mathbf{i} + (2 \text{ m/s}^2)\mathbf{i} - (6.40 \text{ m/s}^2)\mathbf{j}
\end{aligned}$$
$$\mathbf{a}_B = 8.12 \text{ m/s}^2 \measuredangle 52.0° \quad ◀$$

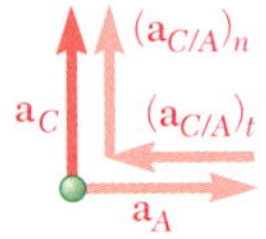

Acceleration of Point C

$$\begin{aligned}
\mathbf{a}_C &= \mathbf{a}_A + \mathbf{a}_{C/A} = \mathbf{a}_A + \alpha\mathbf{k} \times \mathbf{r}_{C/A} - \omega^2\mathbf{r}_{C/A} \\
&= (3 \text{ m/s}^2)\mathbf{i} - (20 \text{ rad/s}^2)\mathbf{k} \times (-0.150 \text{ m})\mathbf{j} - (8 \text{ rad/s})^2(-0.150 \text{ m})\mathbf{j} \\
&= (3 \text{ m/s}^2)\mathbf{i} - (3 \text{ m/s}^2)\mathbf{i} + (9.60 \text{ m/s}^2)\mathbf{j}
\end{aligned}$$
$$\mathbf{a}_C = 9.60 \text{ m/s}^2 \uparrow \quad ◀$$

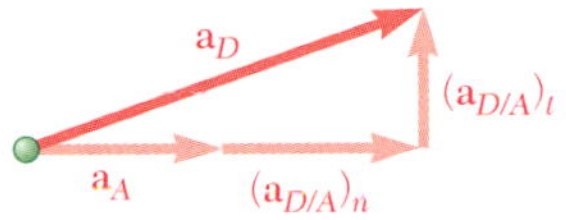

Acceleration of Point D

$$\begin{aligned}
\mathbf{a}_D &= \mathbf{a}_A + \mathbf{a}_{D/A} = \mathbf{a}_A + \alpha\mathbf{k} \times \mathbf{r}_{D/A} - \omega^2\mathbf{r}_{D/A} \\
&= (3 \text{ m/s}^2)\mathbf{i} - (20 \text{ rad/s}^2)\mathbf{k} \times (-0.150 \text{ m})\mathbf{i} - (8 \text{ rad/s})^2(-0.150 \text{ m})\mathbf{i} \\
&= (3 \text{ m/s}^2)\mathbf{i} + (3 \text{ m/s}^2)\mathbf{j} + (9.60 \text{ m/s}^2)\mathbf{i}
\end{aligned}$$
$$\mathbf{a}_D = 12.95 \text{ m/s}^2 \measuredangle 13.4° \quad ◀$$

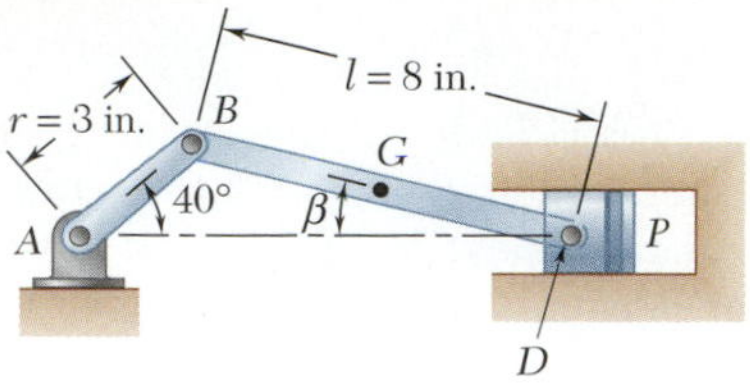

SAMPLE PROBLEM 15.7

Crank AB of the engine system of Sample Prob. 15.3 has a constant clockwise angular velocity of 2000 rpm. For the crank position shown, determine the angular acceleration of the connecting rod BD and the acceleration of point D.

SOLUTION

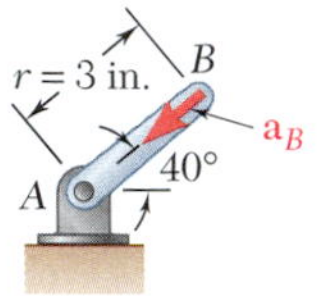

Motion of Crank *AB*. Since the crank rotates about A with constant $\omega_{AB} = 2000$ rpm $= 209.4$ rad/s, we have $\alpha_{AB} = 0$. The acceleration of B is therefore directed toward A and has a magnitude

$$a_B = r\omega_{AB}^2 = (\tfrac{3}{12}\text{ ft})(209.4\text{ rad/s})^2 = 10{,}962\text{ ft/s}^2$$
$$\mathbf{a}_B = 10{,}962\text{ ft/s}^2 \measuredangle 40^\circ$$

Motion of the Connecting Rod *BD*. The angular velocity $\boldsymbol{\omega}_{BD}$ and the value of β were obtained in Sample Prob. 15.3:

$$\boldsymbol{\omega}_{BD} = 62.0\text{ rad/s} \circlearrowleft \qquad \beta = 13.95^\circ$$

The motion of BD is resolved into a translation with B and a rotation about B. The relative acceleration $\mathbf{a}_{D/B}$ is resolved into normal and tangential components:

$$(a_{D/B})_n = (BD)\omega_{BD}^2 = (\tfrac{8}{12}\text{ ft})(62.0\text{ rad/s})^2 = 2563\text{ ft/s}^2$$
$$(\mathbf{a}_{D/B})_n = 2563\text{ ft/s}^2 \measuredangle 13.95^\circ$$
$$(a_{D/B})_t = (BD)\alpha_{BD} = (\tfrac{8}{12})\alpha_{BD} = 0.6667\alpha_{BD}$$
$$(\mathbf{a}_{D/B})_t = 0.6667\alpha_{BD} \measuredangle 76.05^\circ$$

While $(\mathbf{a}_{D/B})_t$ must be perpendicular to BD, its sense is not known.

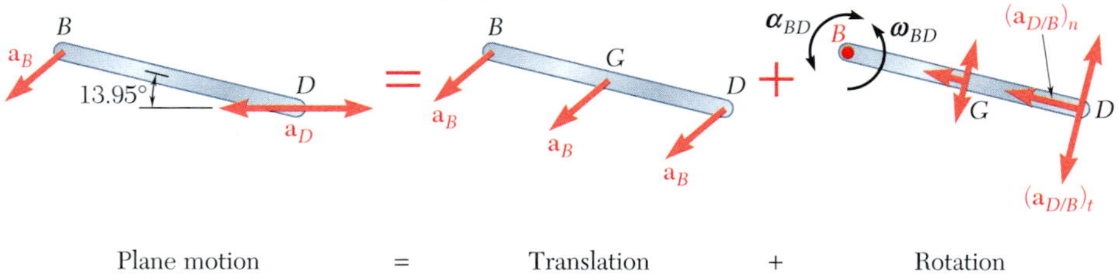

Noting that the acceleration $\mathbf{a}_D$ must be horizontal, we write

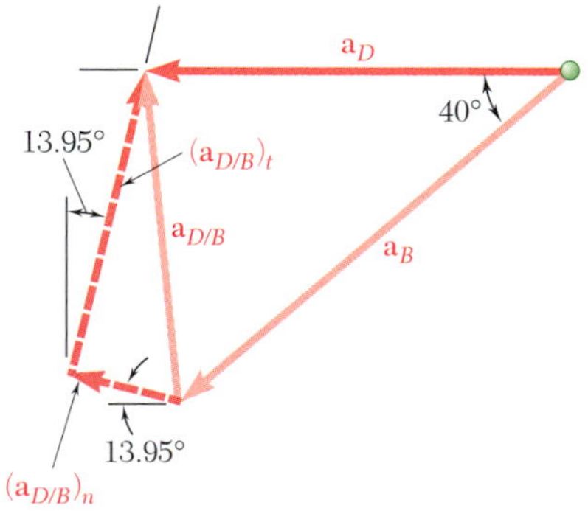

$$\mathbf{a}_D = \mathbf{a}_B + \mathbf{a}_{D/B} = \mathbf{a}_B + (\mathbf{a}_{D/B})_n + (\mathbf{a}_{D/B})_t$$
$$[a_D \leftrightarrow] = [10{,}962 \measuredangle 40^\circ] + [2563 \measuredangle 13.95^\circ] + [0.6667\alpha_{BD} \measuredangle 76.05^\circ]$$

Equating x and y components, we obtain the following scalar equations:

$\xrightarrow{+} x$ components:

$$-a_D = -10{,}962\cos 40^\circ - 2563\cos 13.95^\circ + 0.6667\alpha_{BD}\sin 13.95^\circ$$

$+\uparrow y$ components:

$$0 = -10{,}962\sin 40^\circ + 2563\sin 13.95^\circ + 0.6667\alpha_{BD}\cos 13.95^\circ$$

Solving the equations simultaneously, we obtain $\alpha_{BD} = +9940$ rad/s^2 and $a_D = +9290$ ft/s^2. The positive signs indicate that the senses shown on the vector polygon are correct; we write

$$\boldsymbol{\alpha}_{BD} = 9940\text{ rad/s}^2 \circlearrowleft \quad ◀$$
$$\mathbf{a}_D = 9290\text{ ft/s}^2 \leftarrow \quad ◀$$

SAMPLE PROBLEM 15.8

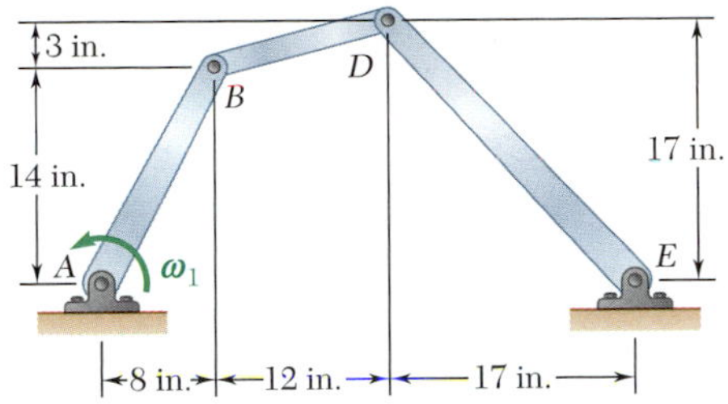

The linkage *ABDE* moves in the vertical plane. Knowing that in the position shown crank *AB* has a constant angular velocity $\boldsymbol{\omega}_1$ of 20 rad/s counterclockwise, determine the angular velocities and angular accelerations of the connecting rod *BD* and of the crank *DE*.

SOLUTION

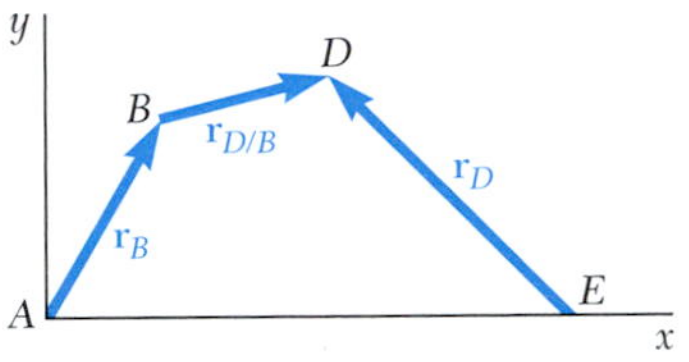

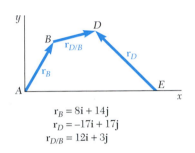

$\mathbf{r}_B = 8\mathbf{i} + 14\mathbf{j}$
$\mathbf{r}_D = -17\mathbf{i} + 17\mathbf{j}$
$\mathbf{r}_{D/B} = 12\mathbf{i} + 3\mathbf{j}$

This problem could be solved by the method used in Sample Prob. 15.7. In this case, however, the vector approach will be used. The position vectors $\mathbf{r}_B$, $\mathbf{r}_D$, and $\mathbf{r}_{D/B}$ are chosen as shown in the sketch.

Velocities. Since the motion of each element of the linkage is contained in the plane of the figure, we have

$$\boldsymbol{\omega}_{AB} = \omega_{AB}\mathbf{k} = (20\text{ rad/s})\mathbf{k} \qquad \boldsymbol{\omega}_{BD} = \omega_{BD}\mathbf{k} \qquad \boldsymbol{\omega}_{DE} = \omega_{DE}\mathbf{k}$$

where $\mathbf{k}$ is a unit vector pointing out of the paper. We now write

$$\mathbf{v}_D = \mathbf{v}_B + \mathbf{v}_{D/B}$$
$$\omega_{DE}\mathbf{k} \times \mathbf{r}_D = \omega_{AB}\mathbf{k} \times \mathbf{r}_B + \omega_{BD}\mathbf{k} \times \mathbf{r}_{D/B}$$
$$\omega_{DE}\mathbf{k} \times (-17\mathbf{i} + 17\mathbf{j}) = 20\mathbf{k} \times (8\mathbf{i} + 14\mathbf{j}) + \omega_{BD}\mathbf{k} \times (12\mathbf{i} + 3\mathbf{j})$$
$$-17\omega_{DE}\mathbf{j} - 17\omega_{DE}\mathbf{i} = 160\mathbf{j} - 280\mathbf{i} + 12\omega_{BD}\mathbf{j} - 3\omega_{BD}\mathbf{i}$$

Equating the coefficients of the unit vectors $\mathbf{i}$ and $\mathbf{j}$, we obtain the following two scalar equations:

$$-17\omega_{DE} = -280 - 3\omega_{BD}$$
$$-17\omega_{DE} = +160 + 12\omega_{BD}$$
$$\boldsymbol{\omega}_{BD} = -(29.33\text{ rad/s})\mathbf{k} \qquad \boldsymbol{\omega}_{DE} = (11.29\text{ rad/s})\mathbf{k} \quad ◀$$

Accelerations. Noting that at the instant considered crank *AB* has a constant angular velocity, we write

$$\boldsymbol{\alpha}_{AB} = 0 \qquad \boldsymbol{\alpha}_{BD} = \alpha_{BD}\mathbf{k} \qquad \boldsymbol{\alpha}_{DE} = \alpha_{DE}\mathbf{k}$$
$$\mathbf{a}_D = \mathbf{a}_B + \mathbf{a}_{D/B} \tag{1}$$

Each term of Eq. (1) is evaluated separately:

$$\begin{aligned}
\mathbf{a}_D &= \alpha_{DE}\mathbf{k} \times \mathbf{r}_D - \omega_{DE}^2\mathbf{r}_D \\
&= \alpha_{DE}\mathbf{k} \times (-17\mathbf{i} + 17\mathbf{j}) - (11.29)^2(-17\mathbf{i} + 17\mathbf{j}) \\
&= -17\alpha_{DE}\mathbf{j} - 17\alpha_{DE}\mathbf{i} + 2170\mathbf{i} - 2170\mathbf{j} \\
\mathbf{a}_B &= \alpha_{AB}\mathbf{k} \times \mathbf{r}_B - \omega_{AB}^2\mathbf{r}_B = 0 - (20)^2(8\mathbf{i} + 14\mathbf{j}) \\
&= -3200\mathbf{i} - 5600\mathbf{j} \\
\mathbf{a}_{D/B} &= \alpha_{BD}\mathbf{k} \times \mathbf{r}_{D/B} - \omega_{BD}^2\mathbf{r}_{D/B} \\
&= \alpha_{BD}\mathbf{k} \times (12\mathbf{i} + 3\mathbf{j}) - (29.33)^2(12\mathbf{i} + 3\mathbf{j}) \\
&= 12\alpha_{BD}\mathbf{j} - 3\alpha_{BD}\mathbf{i} - 10{,}320\mathbf{i} - 2580\mathbf{j}
\end{aligned}$$

Substituting into Eq. (1) and equating the coefficients of $\mathbf{i}$ and $\mathbf{j}$, we obtain

$$-17\alpha_{DE} + 3\alpha_{BD} = -15{,}690$$
$$-17\alpha_{DE} - 12\alpha_{BD} = -6010$$
$$\boldsymbol{\alpha}_{BD} = -(645\text{ rad/s}^2)\mathbf{k} \qquad \boldsymbol{\alpha}_{DE} = (809\text{ rad/s}^2)\mathbf{k} \quad ◀$$

SOLVING PROBLEMS ON YOUR OWN

This lesson was devoted to the determination of the *accelerations* of the points of a *rigid body in plane motion*. As you did previously for velocities, you will again consider the plane motion of a rigid body as the sum of two motions, namely, *a translation and a rotation*.

To solve a problem involving accelerations in plane motion you should use the following steps:

1. *Determine the angular velocity of the body.* To find $\boldsymbol{\omega}$ you can either

a. Consider the motion of the body as the sum of a translation and a rotation as you did in See. 15.6, or

b. Use the instantaneous center of rotation of the body as you did in Sec. 15.7. However, *keep in mind that you cannot use the instantaneous center to determine accelerations*.

2. *Start drawing a "diagram equation"* to use in your solution. This "equation" will involve the following diagrams (Fig. 15.44).

a. Plane motion diagram. Draw a sketch of the body including all dimensions, as well as the angular velocity $\boldsymbol{\omega}$. Show the angular acceleration $\boldsymbol{\alpha}$ with its magnitude and sense if you know them. Also show those points for which you know or seek the accelerations, indicating all that you know about these accelerations.

b. Translation diagram. Select a reference point A for which you know the direction, the magnitude, or a component of the acceleration $\mathbf{a}_A$. Draw a second diagram showing the body in translation with each point having the same acceleration as point A.

c. Rotation diagram. Considering point A as a fixed reference point, draw a third diagram showing the body in rotation about A. Indicate the normal and tangential components of the relative accelerations of other points, such as the components $(\mathbf{a}_{B/A})_n$ and $(\mathbf{a}_{B/A})_t$ of the acceleration of point B with respect to point A.

3. *Write the relative-acceleration formula*

$$\mathbf{a}_B = \mathbf{a}_A + \mathbf{a}_{B/A} \qquad \text{or} \qquad \mathbf{a}_B = \mathbf{a}_A + (\mathbf{a}_{B/A})_n + (\mathbf{a}_{B/A})_t$$

The sample problems illustrate three different ways to use this vector equation:

a. If α is given or can easily be determined, you can use this equation to determine the accelerations of various points of the body [Sample Prob. 15.6].

b. If α cannot easily be determined, select for point B a point for which you know the direction, the magnitude, or a component of the acceleration $\mathbf{a}_B$ and draw a vector diagram of the equation. Starting at the same point, draw all known acceleration components in tip-to-tail fashion for each member of the equation. Complete the diagram by drawing the two remaining vectors in appropriate directions and in such a way that the two sums of vectors end at a common point.

The magnitudes of the two remaining vectors can be found either graphically or analytically. Usually an analytic solution will require the solution of two simultaneous equations [Sample Prob. 15.7]. However, by first considering the components of the various vectors in a direction perpendicular to one of the unknown vectors, you may be able to obtain an equation in a single unknown.

One of the two vectors obtained by the method just described will be $(\mathbf{a}_{B/A})_t$, from which you can compute α. Once α has been found, the vector equation can be used to determine the acceleration of any other point of the body.

c. A full vector approach can also be used to solve the vector equation. This is illustrated in Sample Prob. 15.8.

4. The analysis of plane motion in terms of a parameter completed this lesson. This method should be used *only if it is possible* to express the coordinates x and y of all significant points of the body in terms of a single parameter (Sec. 15.9). By differentiating twice with respect to t the coordinates x and y of a given point, you can determine the rectangular components of the absolute velocity and absolute acceleration of that point.

Problems

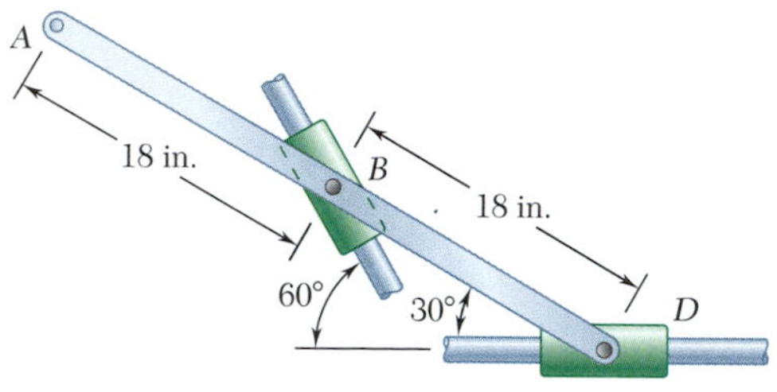

Fig. P15.103 and P15.104

15.103 Collars B and D are pin-connected to bar ABD and can slide along fixed rods. At the instant shown the angular velocity of bar ABD is zero and the acceleration of point D is 24 ft/s^2 to the right. Determine (*a*) the angular acceleration of the bar, (*b*) the acceleration of point B, (*c*) the acceleration of point A.

15.104 Collars B and D are pin-connected to bar ABD and can slide along fixed rods. At the instant shown the angular velocity of bar ABD is zero and its angular acceleration is 12 rad/s^2 clockwise. Determine (*a*) the acceleration of point D, (*b*) the acceleration of point B, (*c*) the acceleration of point A.

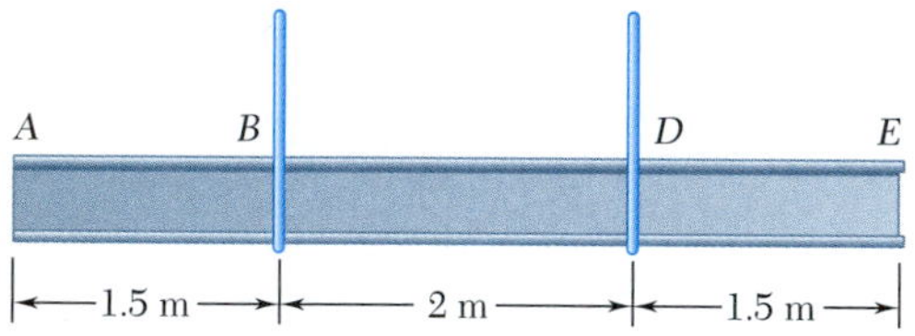

Fig. P15.105 and P15.106

15.105 A 5-m steel beam is lowered by means of two cables unwinding at the same speed from overhead cranes. As the beam approaches the ground, the crane operators apply brakes to slow the unwinding motion. At the instant considered the deceleration of the cable attached at B is 2.5 m/s^2, while that of the cable attached at D is 1.5 m/s^2. Determine (*a*) the angular acceleration of the beam, (*b*) the acceleration of points A and E.

15.106 For a 5-m steel beam AE the acceleration of point A is 2 m/s^2 downward and the angular acceleration of the beam is 1.2 rad/s^2 counterclockwise. Knowing that at the instant considered the angular velocity of the beam is zero, determine the acceleration (*a*) of cable B, (*b*) of cable D.

15.107* and *15.108 Bar BDE is attached to two links AB and CD. Knowing that at the instant shown link AB rotates with a constant angular velocity of 3 rad/s clockwise, determine the acceleration (*a*) of point D, (*b*) of point E.

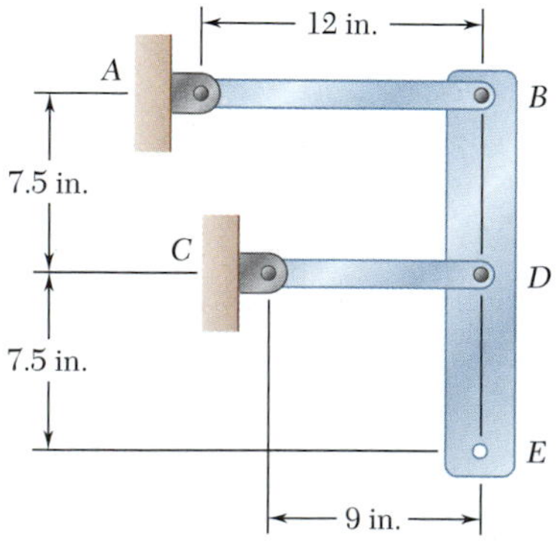

Fig. *P15.107*

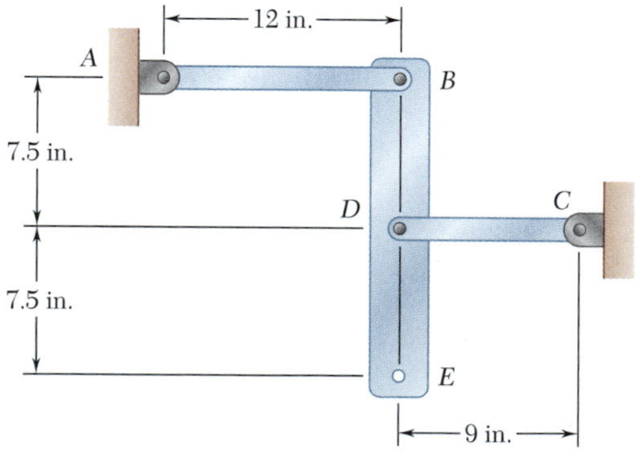

Fig. *P15.108*

15.109 At the instant shown the angular velocity of the wheel is 2 rad/s clockwise and its angular acceleration is 3 rad/s^2 counterclockwise. Knowing that the wheel rolls without slipping, determine the location of the point on the wheel with zero acceleration at this instant.

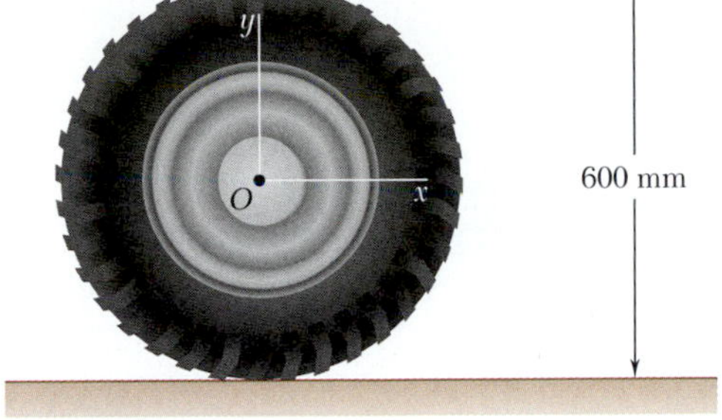

Fig. P15.109

15.110 The motion of the 3-in.-radius cylinder is controlled by the cord shown. Knowing that end E of the cord has a velocity of 12 in./s and an acceleration of 19.2 in./s^2, both directed upward, determine the acceleration (*a*) of point A, (*b*) of point B.

15.111 The motion of the 3-in.-radius cylinder is controlled by the cord shown. Knowing that end E of the cord has a velocity of 12 in./s and an acceleration of 19.2 in./s^2, both directed upward, determine the accelerations of points C and D of the cylinder.

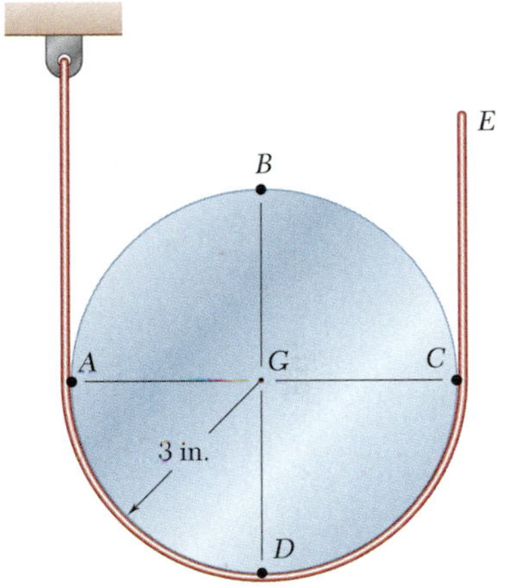

Fig. P15.110 and P15.111

15.112 A wheel rolls without slipping on a fixed cylinder. Knowing that at the instant shown the angular velocity of the wheel is 10 rad/s clockwise and its angular acceleration is 30 rad/s^2 counterclockwise, determine the acceleration of (*a*) point A, (*b*) point B, (*c*) point C.

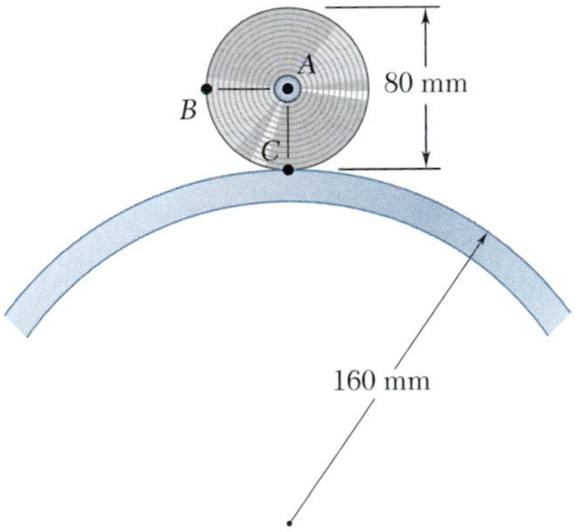

Fig. P15.112

15.113 The 360-mm-radius flywheel is rigidly attached to a 30-mm-radius shaft that can roll along parallel rails. Knowing that at the instant shown the center of the shaft has a velocity of 24 mm/s and an acceleration of 10 mm/s^2, both directed down to the left, determine the acceleration (*a*) of point A, (*b*) of point B.

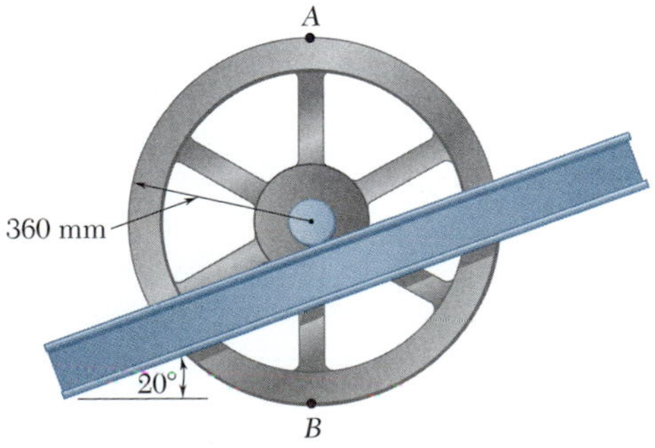

Fig. *P15.113*

15.114 The 6-in.-radius drum rolls without slipping on a belt that moves to the left with a constant velocity of 12 in./s. At an instant when the velocity and acceleration of the center D of the drum are as shown, determine the accelerations of points A, B, and C of the drum.

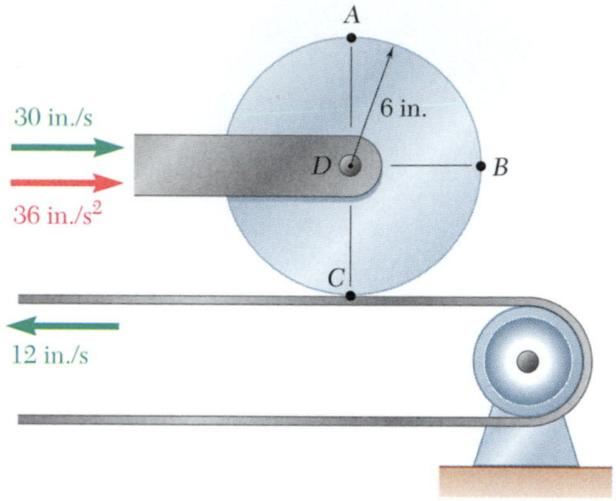

Fig. P15.114

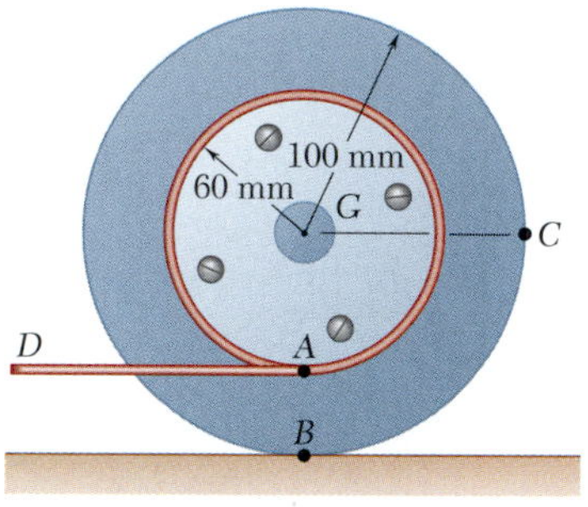

Fig. *P15.115*

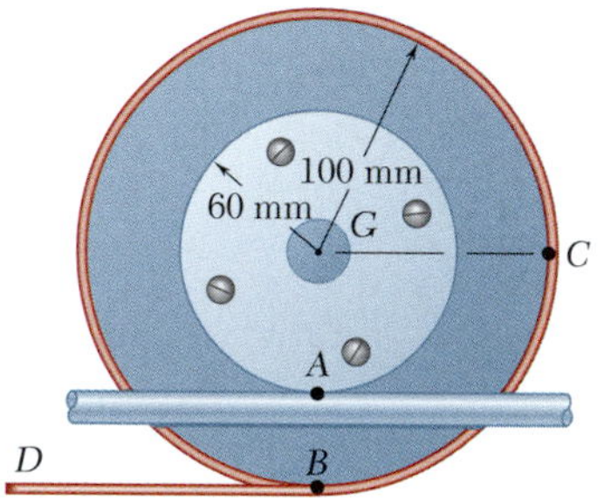

Fig. *P15.116*

15.115 and 15.116 A 60-mm-radius drum is rigidly attached to a 100-mm-radius drum as shown. One of the drums rolls without sliding on the surface shown, and a cord is wound around the other drum. Knowing that at the instant shown end D of the cord has a velocity of 160 mm/s and an acceleration of 600 mm/s^2, both directed to the left, determine the accelerations of points A, B, and C of the drums.

15.117 Arm AB has a constant angular velocity of 16 rad/s counterclockwise. At the instant when $\theta = 0$, determine the acceleration (*a*) of collar D, (*b*) of the midpoint G of bar BD.

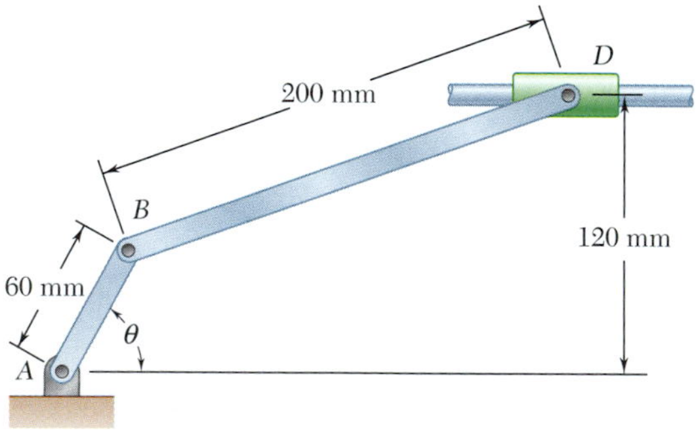

Fig. P15.117 and P15.118

15.118 Arm AB has a constant angular velocity of 16 rad/s counterclockwise. At the instant when $\theta = 90°$, determine the acceleration (*a*) of collar D, (*b*) of the midpoint G of bar BD.

15.119 Collar D slides on a fixed horizontal rod with a constant velocity of 4 ft/s to the right. Knowing that at the instant shown $x = 8$ in., determine (*a*) the angular acceleration of bar BD, (*b*) the angular acceleration of bar AB.

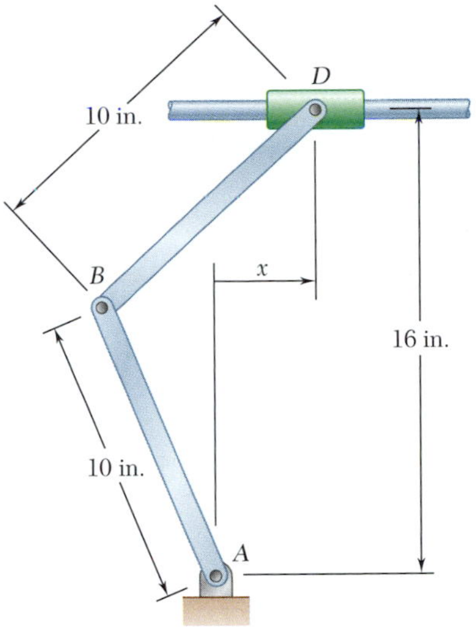

Fig. P15.119 and P15.120

15.120 Collar D slides on a fixed horizontal rod with a constant velocity of 2 ft/s to the right. Knowing that at the instant shown $x = 0$, determine (*a*) the angular acceleration of bar BD, (*b*) the angular acceleration of bar AB.

15.121 Collar D slides on a fixed vertical rod. Knowing that the disk has a constant angular velocity of 15 rad/s clockwise, determine the angular acceleration of bar BD and the acceleration of collar D when (*a*) $\theta = 0$, (*b*) $\theta = 90°$, (*c*) $\theta = 180°$.

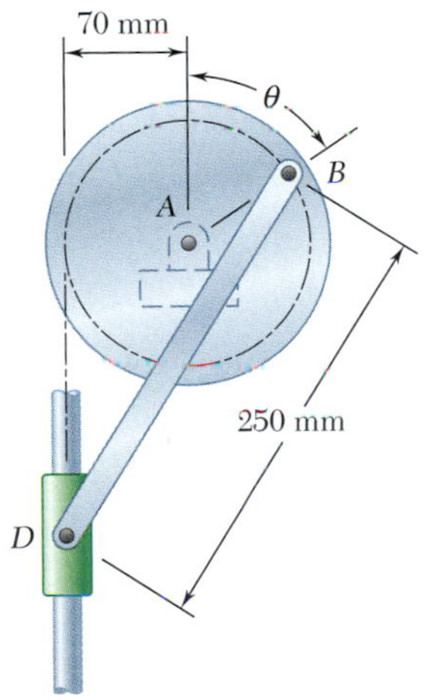

Fig. P15.121

15.122 In the planetary gear system shown the radius of gears A, B, C, and D is 60 mm and the radius of the outer gear E is 180 mm. Knowing that gear A has a constant angular velocity of 150 rpm clockwise and that the outer gear E is stationary, determine the magnitude of the acceleration of the tooth of gear D that is in contact with (*a*) gear A, (*b*) gear E.

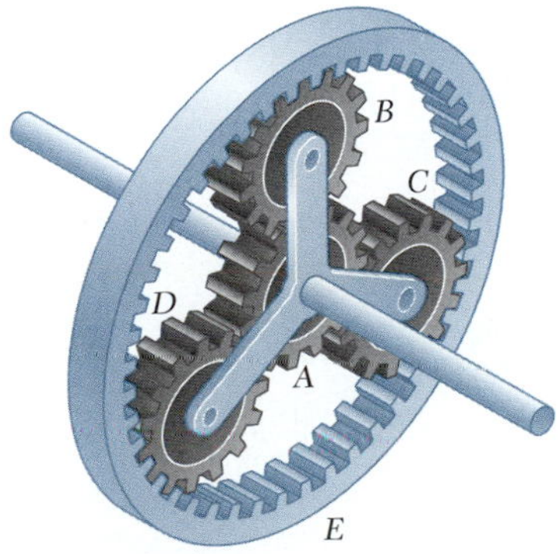

Fig. P15.122

15.123 Knowing that crank AB rotates about point A with a constant angular velocity of 900 rpm clockwise, determine the acceleration of the piston P when $\theta = 60°$.

15.124 Knowing that crank AB rotates about point A with a constant angular velocity of 900 rpm clockwise, determine the acceleration of the piston P when $\theta = 120°$.

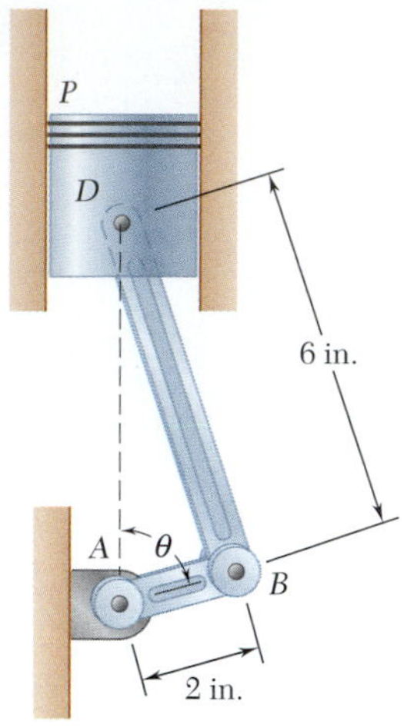

Fig. *P15.123* and *P15.124*

15.125 Knowing that at the instant shown bar AB has a constant angular velocity of 19 rad/s clockwise, determine (*a*) the angular acceleration of bar BGD, (*b*) the angular acceleration of bar DE.

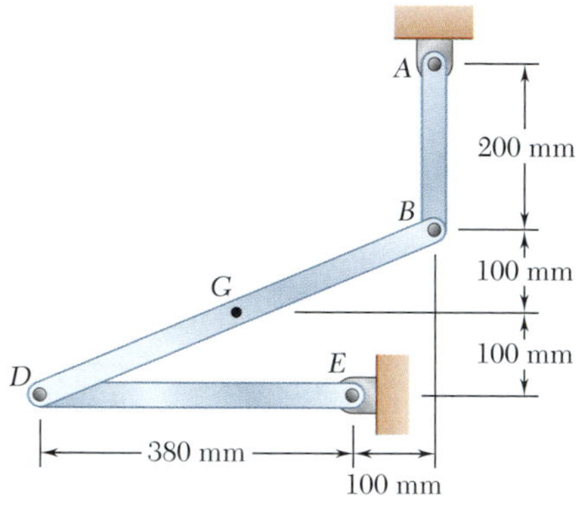

Fig. P15.125 and P15.126

15.126 Knowing that at the instant shown bar DE has a constant angular velocity of 18 rad/s clockwise, determine (*a*) the acceleration of point B, (*b*) the acceleration of point G.

15.127 Knowing that at the instant shown rod AB has a constant angular velocity of 6 rad/s clockwise, determine the acceleration of point D.

15.128 Knowing that at the instant shown rod AB has a constant angular velocity of 6 rad/s clockwise, determine (*a*) the angular acceleration of member BDE, (*b*) the acceleration of point E.

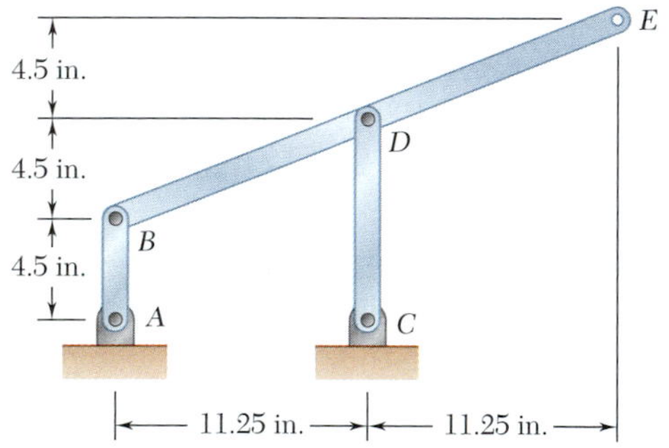

Fig. P15.127 and P15.128

15.129 At the instant shown the angular velocity of bar DE is 4 rad/s clockwise and its angular acceleration is 10 rad/s^2 counterclockwise. Determine (*a*) the angular acceleration of bar BGD, (*b*) the acceleration of point G.

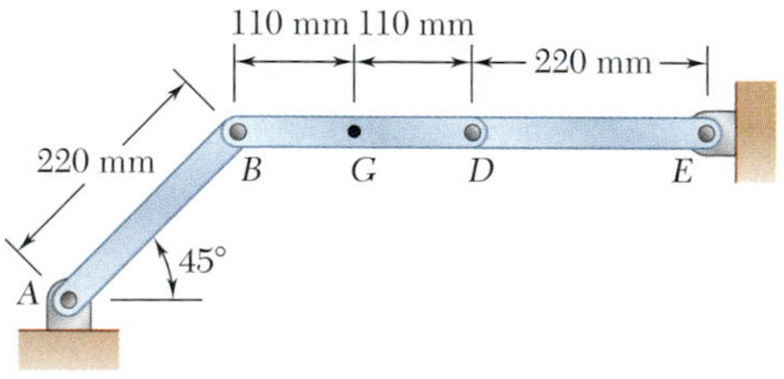

Fig. P15.129

15.130 At the instant shown rod AB has a constant angular velocity of 8 rad/s clockwise. Knowing that $l = 12$ in., determine the acceleration of the midpoint C of member BD.

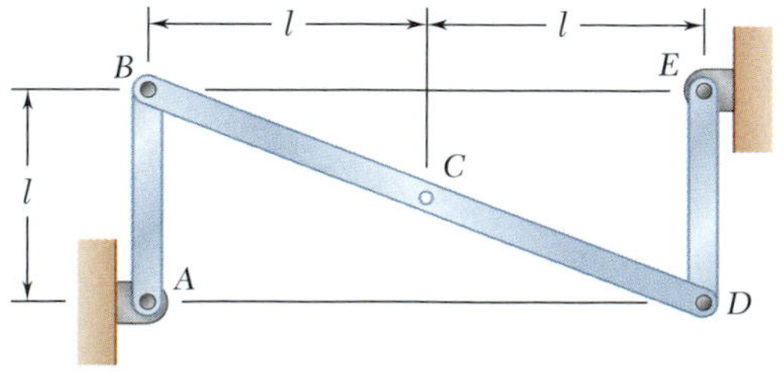

Fig. *P15.130*

15.131 and *15.132* Knowing that at the instant shown bar AB has a constant angular velocity of 4 rad/s clockwise, determine the angular acceleration (*a*) of bar BD, (*b*) of bar DE.

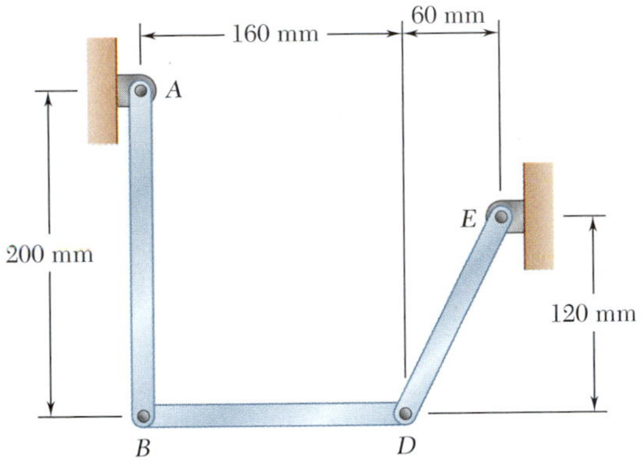

Fig. P15.131

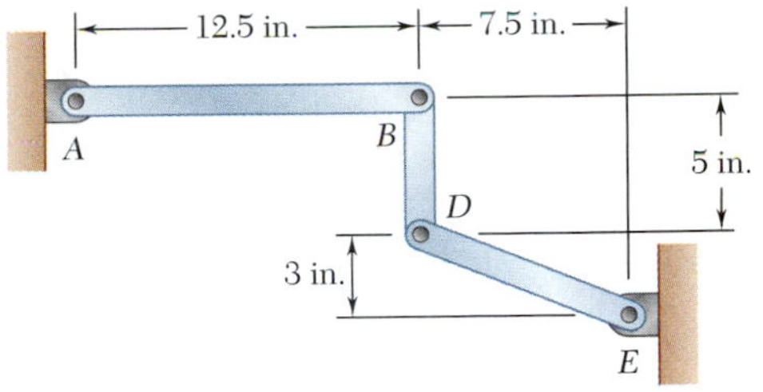

Fig. *P15.132*

15.133 and *15.134* Solve the problem indicated making full use of the vector approach as done in Sample Prob. 15.8.

15.133 Prob. 15.131.

15.134 Prob. 15.132.

15.135 Denoting by $\mathbf{r}_A$ the position vector of point A of a rigid slab that is in plane motion, show that (*a*) the position vector $\mathbf{r}_C$ of the instantaneous center of rotation is

$$\mathbf{r}_C = \mathbf{r}_A + \frac{\boldsymbol{\omega} \times \mathbf{v}_A}{\omega^2}$$

where $\boldsymbol{\omega}$ is the angular velocity of the slab and $\mathbf{v}_A$ is the velocity of point A, (*b*) the acceleration of the instantaneous center of rotation is zero if, and only if,

$$\mathbf{a}_A = \frac{\alpha}{\omega}\mathbf{v}_A + \boldsymbol{\omega} \times \mathbf{v}_A$$

where $\boldsymbol{\alpha} = \alpha\mathbf{k}$ is the angular acceleration of the slab.

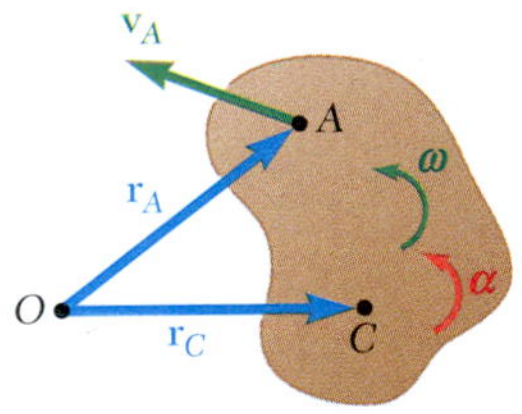

Fig. P15.135

***15.136** Collar D slides on a fixed vertical rod. The disk rotates with a constant clockwise angular velocity $\boldsymbol{\omega}$. Using the method of Sec. 15.9, derive an expression for the velocity of point D in terms of θ, ω, b, and l.

***15.137** Collar D slides on a fixed vertical rod. The disk rotates with a constant clockwise angular velocity $\boldsymbol{\omega}$. Using the method of Sec. 15.9, derive an expression for the angular acceleration of rod BD in terms of θ, ω, b, and l.

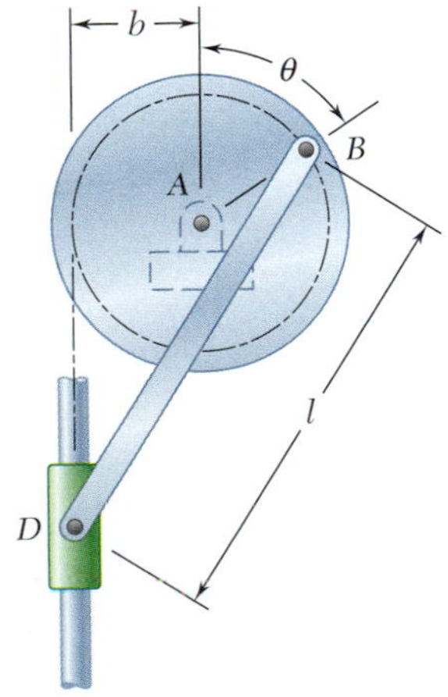

Fig. P15.136 and P15.137

***15.138** The drive disk of the Scotch crosshead mechanism shown has an angular velocity $\boldsymbol{\omega}$ and an angular acceleration $\boldsymbol{\alpha}$, both directed counterclockwise. Using the method of Sec. 15.9, derive expressions for the velocity and acceleration of point B.

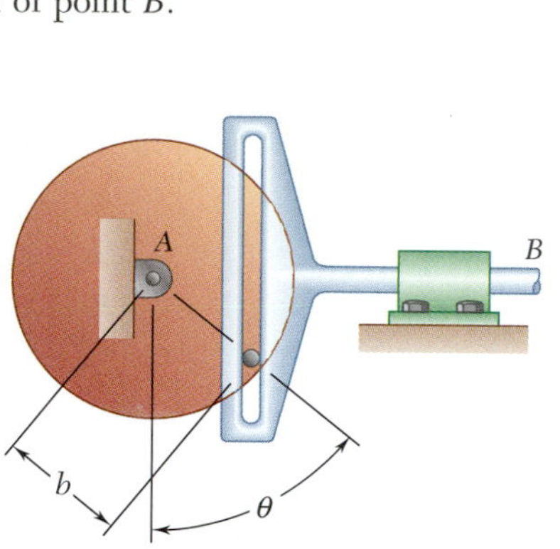

Fig. P15.138

***15.139** Rod AB moves over a small wheel at C while end A moves to the right with a constant velocity $\mathbf{v}_A$. Using the method of Sec. 15.9, derive expressions for the angular velocity and angular acceleration of the rod.

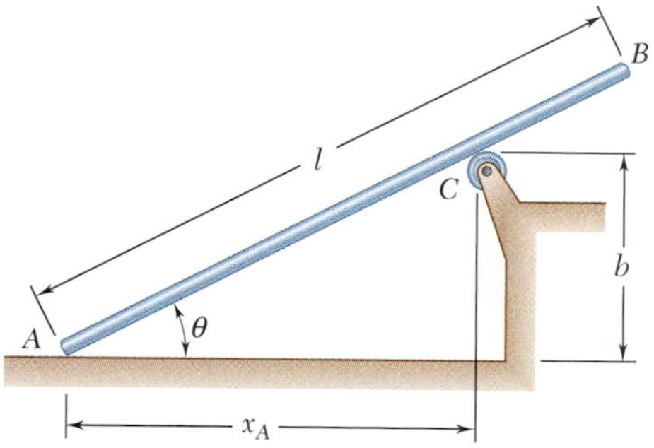

Fig. P15.139 and P15.140

***15.140** Rod AB moves over a small wheel at C while end A moves to the right with a constant velocity $\mathbf{v}_A$. Using the method of Sec. 15.9, derive expressions for the horizontal and vertical components of the velocity of point B.

***15.141** A disk of radius r rolls to the right with a constant velocity $\mathbf{v}$. Denoting by P the point of the rim in contact with the ground at $t = 0$, derive expressions for the horizontal and vertical components of the velocity of P at any time t.

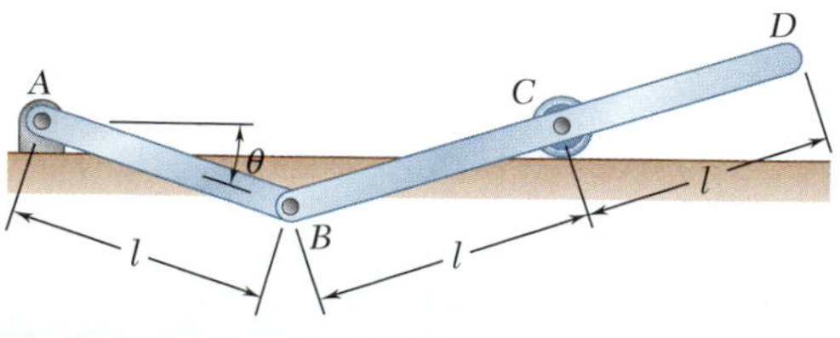

Fig. P15.142 and P15.143

***15.142** At the instant shown, rod AB rotates with a constant angular velocity $\boldsymbol{\omega}$ and an angular acceleration $\boldsymbol{\alpha}$, both clockwise. Using the method of Sec. 15.9, derive expressions for the velocity and acceleration of point C.

***15.143** At the instant shown, rod AB rotates with a constant angular velocity $\boldsymbol{\omega}$ and an angular acceleration $\boldsymbol{\alpha}$, both clockwise. Using the method of Sec. 15.9, derive expressions for the horizontal and vertical components of the velocity and acceleration of point D.

15.144 Crank AB rotates with a constant clockwise angular velocity $\boldsymbol{\omega}$. Using the method of Sec. 15.9, derive expressions for the angular velocity of rod BD and the velocity of the point on the rod coinciding with point E in terms of θ, ω, b, and l.

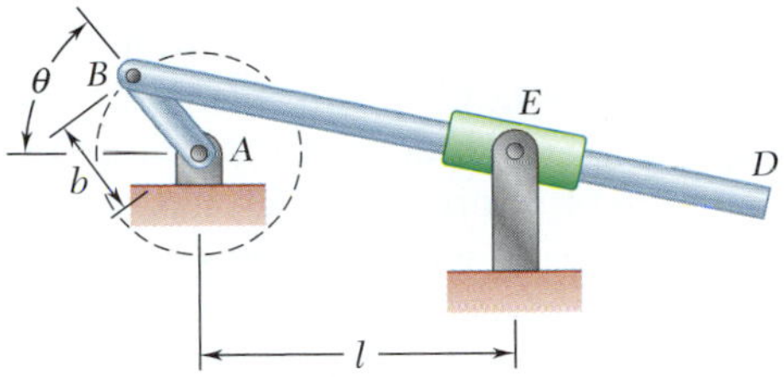

Fig. *P15.144* and *P15.145*

15.145 Crank AB rotates with a constant clockwise angular velocity $\boldsymbol{\omega}$. Using the method of Sec. 15.9, derive an expression for the angular acceleration of rod BD in terms of θ, ω, b, and l.

***15.146** A wheel of radius r rolls without slipping along the inside of a fixed cylinder of radius R with a constant angular velocity $\boldsymbol{\omega}$. Denoting by P the point of the wheel in contact with the cylinder at $t = 0$, derive expressions for the horizontal and vertical components of the velocity of P at any time t. (The curve described by point P is a *hypocycloid*.)

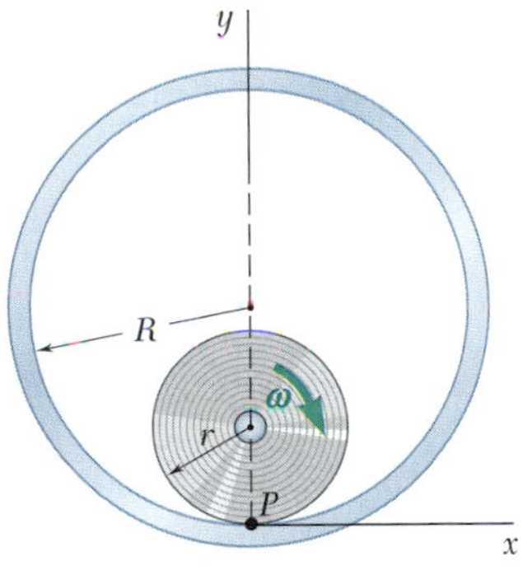

Fig. P15.146

***15.147** In Prob. 15.146, show that the path of P is a vertical straight line when $r = R/2$. Derive expressions for the corresponding velocity and acceleration of P at any time t.

15.10. RATE OF CHANGE OF A VECTOR WITH RESPECT TO A ROTATING FRAME

We saw in Sec. 11.10 that the rate of change of a vector is the same with respect to a fixed frame and with respect to a frame in translation. In this section, the rates of change of a vector $\mathbf{Q}$ with respect to a fixed frame and with respect to a rotating frame of reference will be considered.† You will learn to determine the rate of change of $\mathbf{Q}$ with respect to one frame of reference when $\mathbf{Q}$ is defined by its components in another frame.

†It is recalled that the selection of a fixed frame of reference is arbitrary. Any frame may be designated as "fixed"; all others will then be considered as moving.

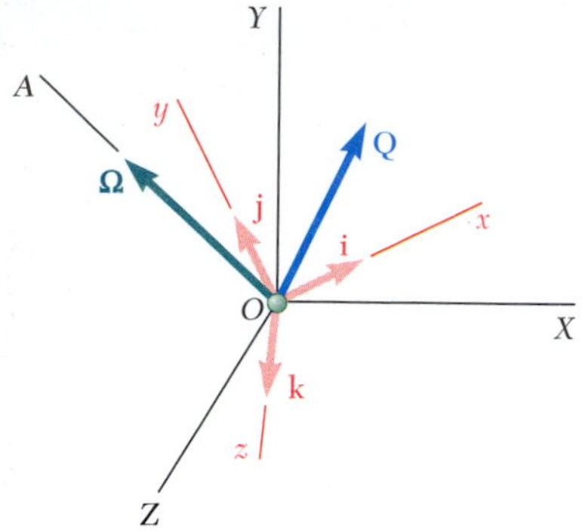

Fig. 15.26

Consider two frames of reference centered at O, a fixed frame $OXYZ$ and a frame $Oxyz$ which rotates about the fixed axis OA; let $\mathbf{\Omega}$ denote the angular velocity of the frame $Oxyz$ at a given instant (Fig. 15.26). Consider now a vector function $\mathbf{Q}(t)$ represented by the vector $\mathbf{Q}$ attached at O; as the time t varies, both the direction and the magnitude of $\mathbf{Q}$ change. Since the variation of $\mathbf{Q}$ is viewed differently by an observer using $OXYZ$ as a frame of reference and by an observer using $Oxyz$, we should expect the rate of change of $\mathbf{Q}$ to depend upon the frame of reference which has been selected. Therefore, the rate of change of $\mathbf{Q}$ with respect to the fixed frame $OXYZ$ will be denoted by $(\dot{\mathbf{Q}})_{OXYZ}$, and the rate of change of $\mathbf{Q}$ with respect to the rotating frame $Oxyz$ will be denoted by $(\dot{\mathbf{Q}})_{Oxyz}$. We propose to determine the relation existing between these two rates of change.

Let us first resolve the vector $\mathbf{Q}$ into components along the x, y, and z axes of the rotating frame. Denoting by $\mathbf{i}$, $\mathbf{j}$, and $\mathbf{k}$ the corresponding unit vectors, we write

$$\mathbf{Q} = Q_x\mathbf{i} + Q_y\mathbf{j} + Q_z\mathbf{k} \qquad (15.27)$$

Differentiating (15.27) with respect to t and considering the unit vectors $\mathbf{i}$, $\mathbf{j}$, $\mathbf{k}$ as fixed, we obtain the rate of change of $\mathbf{Q}$ *with respect to the rotating frame Oxyz:*

$$(\dot{\mathbf{Q}})_{Oxyz} = \dot{Q}_x\mathbf{i} + \dot{Q}_y\mathbf{j} + \dot{Q}_z\mathbf{k} \qquad (15.28)$$

To obtain the rate of change of $\mathbf{Q}$ *with respect to the fixed frame OXYZ*, we must consider the unit vectors $\mathbf{i}$, $\mathbf{j}$, $\mathbf{k}$ as variable when differentiating (15.27). We therefore write

$$(\dot{\mathbf{Q}})_{OXYZ} = \dot{Q}_x\mathbf{i} + \dot{Q}_y\mathbf{j} + \dot{Q}_z\mathbf{k} + Q_x\frac{d\mathbf{i}}{dt} + Q_y\frac{d\mathbf{j}}{dt} + Q_z\frac{d\mathbf{k}}{dt} \qquad (15.29)$$

Recalling (15.28), we observe that the sum of the first three terms in the right-hand member of (15.29) represents the rate of change $(\dot{\mathbf{Q}})_{Oxyz}$. We note, on the other hand, that the rate of change $(\dot{\mathbf{Q}})_{OXYZ}$ would reduce to the last three terms in (15.29) if the vector $\mathbf{Q}$ were fixed within the frame $Oxyz$, since $(\dot{\mathbf{Q}})_{Oxyz}$ would then be zero. But in that case, $(\dot{\mathbf{Q}})_{OXYZ}$ would represent the velocity of a particle located at the tip of $\mathbf{Q}$ and belonging to a body rigidly attached to the frame $Oxyz$. Thus, the last three terms in (15.29) represent the velocity of that particle; since the frame $Oxyz$ has an angular velocity $\mathbf{\Omega}$ with respect to $OXYZ$ at the instant considered, we write, by (15.5),

$$Q_x\frac{d\mathbf{i}}{dt} + Q_y\frac{d\mathbf{j}}{dt} + Q_z\frac{d\mathbf{k}}{dt} = \mathbf{\Omega} \times \mathbf{Q} \qquad (15.30)$$

Substituting from (15.28) and (15.30) into (15.29), we obtain the fundamental relation

$$(\dot{\mathbf{Q}})_{OXYZ} = (\dot{\mathbf{Q}})_{Oxyz} + \mathbf{\Omega} \times \mathbf{Q} \qquad (15.31)$$

We conclude that the rate of change of the vector $\mathbf{Q}$ with respect to the fixed frame $OXYZ$ is made of two parts: The first part represents the rate of change of $\mathbf{Q}$ with respect to the rotating frame $Oxyz$; the second part, $\mathbf{\Omega} \times \mathbf{Q}$, is induced by the rotation of the frame $Oxyz$.

The use of relation (15.31) simplifies the determination of the rate of change of a vector $\mathbf{Q}$ with respect to a fixed frame of reference $OXYZ$ when the vector $\mathbf{Q}$ is defined by its components along the axes of a rotating frame $Oxyz$, since this relation does not require the separate computation of the derivatives of the unit vectors defining the orientation of the rotating frame.

15.11. PLANE MOTION OF A PARTICLE RELATIVE TO A ROTATING FRAME. CORIOLIS ACCELERATION

Consider two frames of reference, both centered at O and both in the plane of the figure, a fixed frame OXY and a rotating frame Oxy (Fig. 15.27). Let P be a particle moving in the plane of the figure. The position vector $\mathbf{r}$ of P is the same in both frames, but its rate of change depends upon the frame of reference which has been selected.

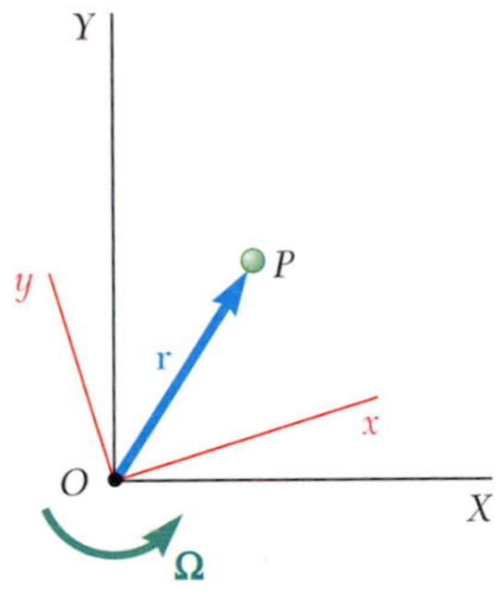

Fig. 15.27

The absolute velocity $\mathbf{v}_P$ of the particle is defined as the velocity observed from the fixed frame OXY and is equal to the rate of change $(\dot{\mathbf{r}})_{OXY}$ of $\mathbf{r}$ with respect to that frame. We can, however, express $\mathbf{v}_P$ in terms of the rate of change $(\dot{\mathbf{r}})_{Oxy}$ observed from the rotating frame if we make use of Eq. (15.31). Denoting by $\boldsymbol{\Omega}$ the angular velocity of the frame Oxy with respect to OXY at the instant considered, we write

$$\mathbf{v}_P = (\dot{\mathbf{r}})_{OXY} = \boldsymbol{\Omega} \times \mathbf{r} + (\dot{\mathbf{r}})_{Oxy} \tag{15.32}$$

But $(\dot{\mathbf{r}})_{Oxy}$ defines the velocity of the particle P *relative to the rotating* frame Oxy. Denoting the rotating frame by $\mathcal{F}$ for short, we represent the velocity $(\dot{\mathbf{r}})_{Oxy}$ of P relative to the rotating frame by $\mathbf{v}_{P/\mathcal{F}}$. Let us imagine that a rigid slab has been attached to the rotating frame. Then $v_{P/\mathcal{F}}$ represents the velocity of P along the path that it describes on that slab (Fig. 15.28), and the term $\boldsymbol{\Omega} \times \mathbf{r}$ in (15.32) represents the velocity $\mathbf{v}_{P'}$ of the point P' of the slab—or rotating frame—which coincides with P at the instant considered. Thus, we have

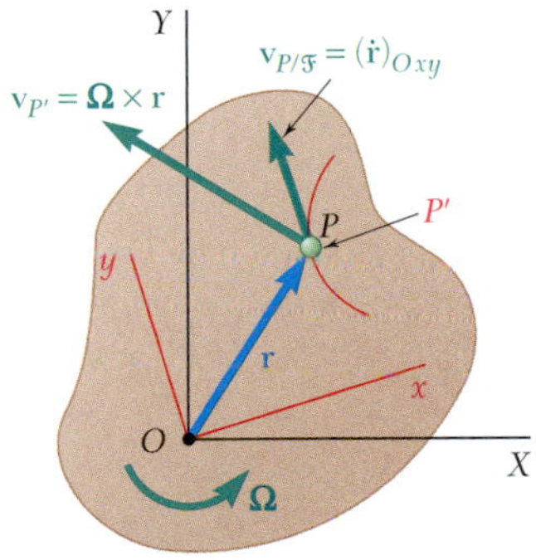

Fig. 15.28

$$\mathbf{v}_P = \mathbf{v}_{P'} + \mathbf{v}_{P/\mathcal{F}} \tag{15.33}$$

where $\mathbf{v}_P$ = absolute velocity of particle P
$\mathbf{v}_{P'}$ = velocity of point P' of moving frame $\mathcal{F}$ coinciding with P
$\mathbf{v}_{P/\mathcal{F}}$ = velocity of P relative to moving frame $\mathcal{F}$

The absolute acceleration $\mathbf{a}_P$ of the particle is defined as the rate of change of $\mathbf{v}_P$ with respect to the fixed frame OXY. Computing the rates of change with respect to OXY of the terms in (15.32), we write

$$\mathbf{a}_P = \dot{\mathbf{v}}_P = \dot{\boldsymbol{\Omega}} \times \mathbf{r} + \boldsymbol{\Omega} \times \dot{\mathbf{r}} + \frac{d}{dt}[(\dot{\mathbf{r}})_{Oxy}] \tag{15.34}$$

where all derivatives are defined with respect to OXY, except where indicated otherwise. Referring to Eq. (15.31), we note that the last term in (15.34) can be expressed as

$$\frac{d}{dt}[(\dot{\mathbf{r}})_{Oxy}] = (\ddot{\mathbf{r}})_{Oxy} + \boldsymbol{\Omega} \times (\dot{\mathbf{r}})_{Oxy}$$

On the other hand, $\dot{\mathbf{r}}$ represents the velocity $\mathbf{v}_P$ and can be replaced by the right-hand member of Eq. (15.32). After completing these two substitutions into (15.34), we write

$$\mathbf{a}_P = \dot{\mathbf{\Omega}} \times \mathbf{r} + \mathbf{\Omega} \times (\mathbf{\Omega} \times \mathbf{r}) + 2\mathbf{\Omega} \times (\dot{\mathbf{r}})_{Oxy} + (\ddot{\mathbf{r}})_{Oxy} \qquad (15.35)$$

Referring to the expression (15.8) obtained in Sec. 15.3 for the acceleration of a particle in a rigid body rotating about a fixed axis, we note that the sum of the first two terms represents the acceleration $\mathbf{a}_{P'}$ of the point P' of the rotating frame which coincides with P at the instant considered. On the other hand, the last term defines the acceleration $\mathbf{a}_{P/\mathcal{F}}$ of P relative to the rotating frame. If it were not for the third term, which has not been accounted for, a relation similar to (15.33) could be written for the accelerations, and $\mathbf{a}_P$ could be expressed as the sum of $\mathbf{a}_{P'}$ and $\mathbf{a}_{P/\mathcal{F}}$. However, it is clear that *such a relation would be incorrect* and that we must include the additional term. This term, which will be denoted by $\mathbf{a}_c$, is called the *complementary acceleration*, or *Coriolis acceleration*, after the French mathematician de Coriolis (1792–1843). We write

$$\mathbf{a}_P = \mathbf{a}_{P'} + \mathbf{a}_{P/\mathcal{F}} + \mathbf{a}_c \qquad (15.36)$$

where $\mathbf{a}_P$ = absolute acceleration of particle P
$\mathbf{a}_{P'}$ = acceleration of point P' of moving frame $\mathcal{F}$ coinciding with P
$\mathbf{a}_{P/\mathcal{F}}$ = acceleration of P relative to moving frame $\mathcal{F}$
$\mathbf{a}_c = 2\mathbf{\Omega} \times (\dot{\mathbf{r}})_{Oxy} = 2\mathbf{\Omega} \times \mathbf{v}_{P/\mathcal{F}}$
= complementary, or Coriolis, acceleration†

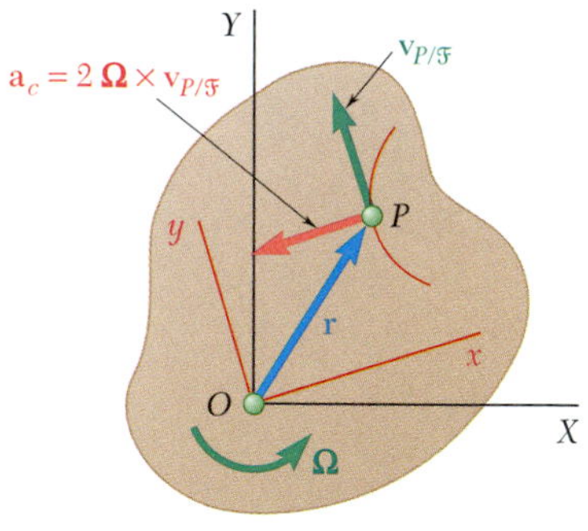

Fig. 15.29

We note that since point P' moves in a circle about the origin O, its acceleration $\mathbf{a}_{P'}$ has, in general, two components: a component $(\mathbf{a}_{P'})_t$ tangent to the circle, and a component $(\mathbf{a}_{P'})_n$ directed toward O. Similarly, the acceleration $\mathbf{a}_{P/\mathcal{F}}$ generally has two components: a component $(\mathbf{a}_{P/\mathcal{F}})_t$ tangent to the path that P describes on the rotating slab, and a component $(\mathbf{a}_{P/\mathcal{F}})_n$ directed toward the center of curvature of that path. We further note that since the vector $\mathbf{\Omega}$ is perpendicular to the plane of motion, and thus to $\mathbf{v}_{P/\mathcal{F}}$, the magnitude of the Coriolis acceleration $\mathbf{a}_c = 2\mathbf{\Omega} \times \mathbf{v}_{P/\mathcal{F}}$ is equal to $2\Omega v_{P/\mathcal{F}}$, and its direction can be obtained by rotating the vector $\mathbf{v}_{P/\mathcal{F}}$ through 90° in the sense of rotation of the moving frame (Fig. 15.29). The Coriolis acceleration reduces to zero when either $\mathbf{\Omega}$ or $\mathbf{v}_{P/\mathcal{F}}$ is zero.

The following example will help in understanding the physical meaning of the Coriolis acceleration. Consider a collar P which is

†It is important to note the difference between Eq. (15.36) and Eq. (15.21) of Sec. 15.8. When we wrote

$$\mathbf{a}_B = \mathbf{a}_A + \mathbf{a}_{B/A} \qquad (15.21)$$

in Sec. 15.8, we were expressing the absolute acceleration of point B as the sum of its acceleration $\mathbf{a}_{B/A}$ relative to a *frame in translation* and of the acceleration $\mathbf{a}_A$ of a point of that frame. We are now trying to relate the absolute acceleration of point P to its acceleration $\mathbf{a}_{P/\mathcal{F}}$ relative to a *rotating frame* $\mathcal{F}$ and to the acceleration $\mathbf{a}_{P'}$ of the point P' of that frame which coincides with P; Eq. (15.36) shows that because the frame is rotating, it is necessary to include an additional term representing the Coriolis acceleration $\mathbf{a}_c$.

made to slide at a constant relative speed u along a rod OB rotating at a constant angular velocity $\boldsymbol{\omega}$ about O (Fig. 15.30*a*). According to formula (15.36), the absolute acceleration of P can be obtained by adding vectorially the acceleration $\mathbf{a}_A$ of the point A of the rod coinciding with P, the relative acceleration $\mathbf{a}_{P/OB}$ of P with respect to the rod, and the Coriolis acceleration $\mathbf{a}_c$. Since the angular velocity $\boldsymbol{\omega}$ of the rod is constant, $\mathbf{a}_A$ reduces to its normal component $(\mathbf{a}_A)_n$ of magnitude $r\omega^2$; and since u is constant, the relative acceleration $\mathbf{a}_{P/OB}$ is zero. According to the definition given above, the Coriolis acceleration is a vector perpendicular to OB, of magnitude $2\omega u$, and directed as shown in the figure. The acceleration of the collar P consists, therefore, of the two vectors shown in Fig. 15.30*a*. Note that the result obtained can be checked by applying the relation (11.44).

To understand better the significance of the Coriolis acceleration, let us consider the absolute velocity of P at time t and at time $t + \Delta t$ (Fig. 15.30*b*). The velocity at time t can be resolved into its components $\mathbf{u}$ and $\mathbf{v}_A$; the velocity at time $t + \Delta t$ can be resolved into its components $\mathbf{u}'$ and $\mathbf{v}_{A'}$. Drawing these components from the same origin (Fig. 15.30*c*), we note that the change in velocity during the time Δt can be represented by the sum of three vectors, $\overrightarrow{RR'}$, $\overrightarrow{TT''}$, and $\overrightarrow{T''T'}$. The vector $\overrightarrow{TT''}$ measures the change in direction of the velocity $\mathbf{v}_A$, and the quotient $\overrightarrow{TT''}/\Delta t$ represents the acceleration $\mathbf{a}_A$ when Δt approaches zero. We check that the direction of $\overrightarrow{TT''}$ is that of $\mathbf{a}_A$ when Δt approaches zero and that

$$\lim_{\Delta t \to 0} \frac{TT''}{\Delta t} = \lim_{\Delta t \to 0} v_A \frac{\Delta\theta}{\Delta t} = r\omega\omega = r\omega^2 = a_A$$

The vector $\overrightarrow{RR'}$ measures the change in direction of $\mathbf{u}$ due to the rotation of the rod; the vector $\overrightarrow{T''T'}$ measures the change in magnitude of $\mathbf{v}_A$ due to the motion of P on the rod. The vectors $\overrightarrow{RR'}$ and $\overrightarrow{T''T'}$ result from the *combined effect* of the relative motion of P and of the rotation of the rod; they would vanish if *either* of these two motions stopped. It is easily verified that the sum of these two vectors defines the Coriolis acceleration. Their direction is that of $\mathbf{a}_c$ when Δt approaches zero, and since $RR' = u\,\Delta\theta$ and $T''T' = v_{A'} - v_A = (r + \Delta r)\omega - r\omega = \omega\,\Delta r$, we check that a_c is equal to

$$\lim_{\Delta t \to 0} \left(\frac{RR'}{\Delta t} + \frac{T''T'}{\Delta t} \right) = \lim_{\Delta t \to 0} \left(u \frac{\Delta\theta}{\Delta t} + \omega \frac{\Delta r}{\Delta t} \right) = u\omega + \omega u = 2\omega u$$

Formulas (15.33) and (15.36) can be used to analyze the motion of mechanisms which contain parts sliding on each other. They make it possible, for example, to relate the absolute and relative motions of sliding pins and collars (see Sample Probs. 15.9 and 15.10). The concept of Coriolis acceleration is also very useful in the study of long-range projectiles and of other bodies whose motions are appreciably affected by the rotation of the earth. As was pointed out in Sec. 12.2, a system of axes attached to the earth does not truly constitute a newtonian frame of reference; such a system of axes should actually be considered as rotating. The formulas derived in this section will therefore facilitate the study of the motion of bodies with respect to axes attached to the earth.

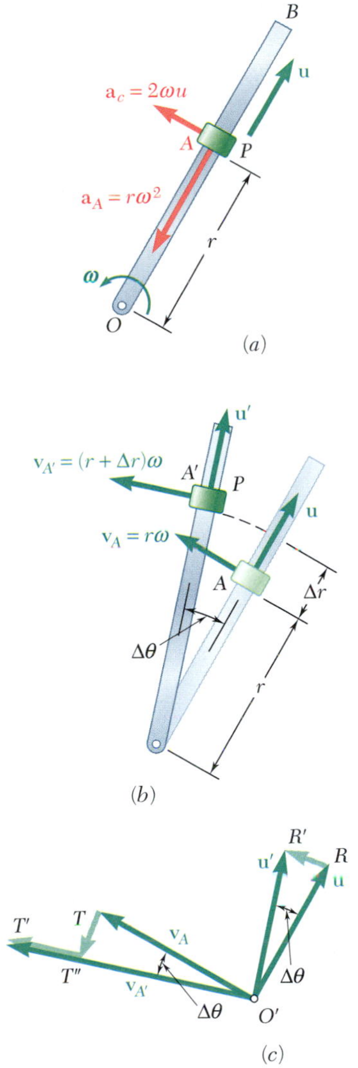

Fig. 15.30

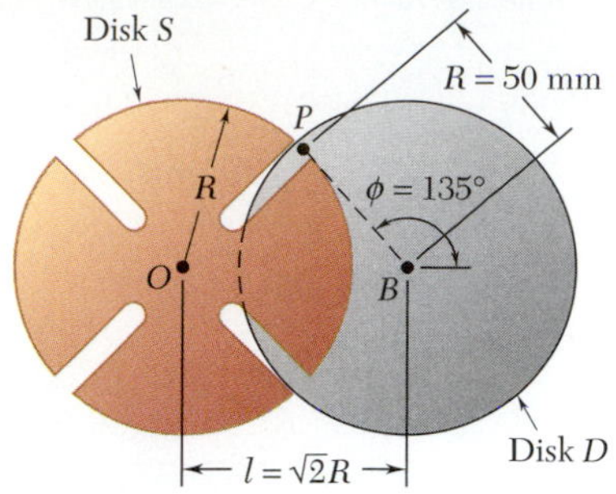

SAMPLE PROBLEM 15.9

The Geneva mechanism shown is used in many counting instruments and in other applications where an intermittent rotary motion is required. Disk D rotates with a constant counterclockwise angular velocity $\boldsymbol{\omega}_D$ of 10 rad/s. A pin P is attached to disk D and slides along one of several slots cut in disk S. It is desirable that the angular velocity of disk S be zero as the pin enters and leaves each slot; in the case of four slots, this will occur if the distance between the centers of the disks is $l = \sqrt{2}\,R$.

At the instant when $\phi = 150°$, determine (*a*) the angular velocity of disk S, (*b*) the velocity of pin P relative to disk S.

SOLUTION

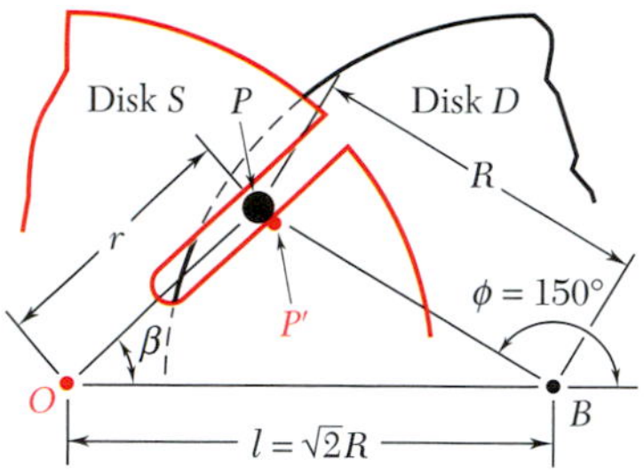

We solve triangle OPB, which corresponds to the position $\phi = 150°$. Using the law of cosines, we write

$$r^2 = R^2 + l^2 - 2Rl\cos 30° = 0.551R^2 \qquad r = 0.742R = 37.1 \text{ mm}$$

From the law of sines,

$$\frac{\sin\beta}{R} = \frac{\sin 30°}{r} \qquad \sin\beta = \frac{\sin 30°}{0.742} \qquad \beta = 42.4°$$

Since pin P is attached to disk D, and since disk D rotates about point B, the magnitude of the absolute velocity of P is

$$v_P = R\omega_D = (50 \text{ mm})(10 \text{ rad/s}) = 500 \text{ mm/s}$$
$$\mathbf{v}_P = 500 \text{ mm/s} \measuredangle 60°$$

We consider now the motion of pin P along the slot in disk S. Denoting by P' the point of disk S which coincides with P at the instant considered and selecting a rotating frame $\mathcal{S}$ attached to disk S, we write

$$\mathbf{v}_P = \mathbf{v}_{P'} + \mathbf{v}_{P/\mathcal{S}}$$

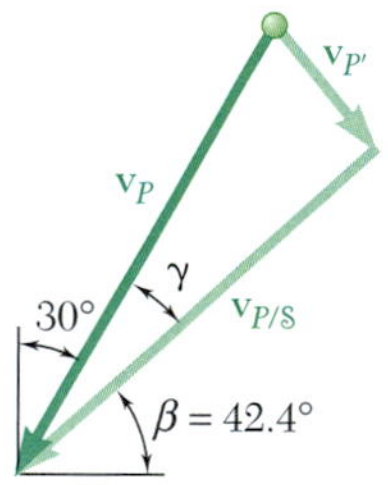

Noting that $\mathbf{v}_{P'}$ is perpendicular to the radius OP and that $\mathbf{v}_{P/\mathcal{S}}$ is directed along the slot, we draw the velocity triangle corresponding to the equation above. From the triangle, we compute

$$\gamma = 90° - 42.4° - 30° = 17.6°$$
$$v_{P'} = v_P \sin\gamma = (500 \text{ mm/s})\sin 17.6°$$
$$\mathbf{v}_{P'} = 151.2 \text{ mm/s} \measuredangle 42.4°$$
$$v_{P/\mathcal{S}} = v_P \cos\gamma = (500 \text{ mm/s})\cos 17.6°$$
$$\mathbf{v}_{P/S} = \mathbf{v}_{P/\mathcal{S}} = 477 \text{ mm/s} \measuredangle 42.4° \quad ◀$$

Since $\mathbf{v}_{P'}$ is perpendicular to the radius OP, we write

$$v_{P'} = r\omega_{\mathcal{S}} \qquad 151.2 \text{ mm/s} = (37.1 \text{ mm})\omega_{\mathcal{S}}$$
$$\boldsymbol{\omega}_S = \boldsymbol{\omega}_{\mathcal{S}} = 4.08 \text{ rad/s} \circlearrowright \quad ◀$$

SAMPLE PROBLEM 15.10

In the Geneva mechanism of Sample Prob. 15.9, disk D rotates with a constant counterclockwise angular velocity $\boldsymbol{\omega}_D$ of magnitude 10 rad/s. At the instant when $\phi = 150°$, determine the angular acceleration of disk S.

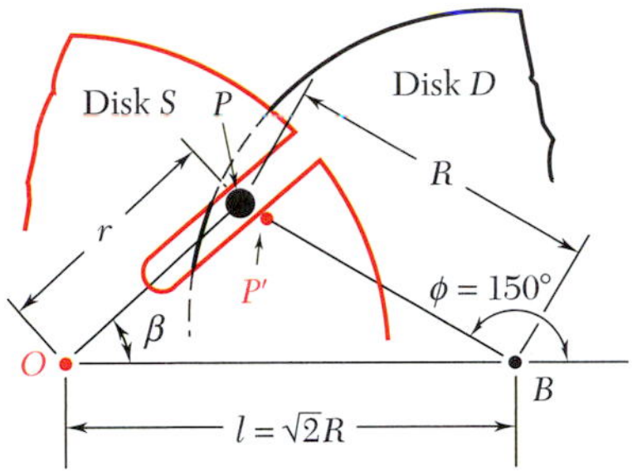

SOLUTION

Referring to Sample Prob. 15.9, we obtain the angular velocity of the frame $\mathcal{S}$ attached to disk S and the velocity of the pin relative to $\mathcal{S}$:

$$\omega_{\mathcal{S}} = 4.08 \text{ rad/s} \circlearrowright$$
$$\beta = 42.4° \qquad \mathbf{v}_{P/\mathcal{S}} = 477 \text{ mm/s} \nearrow 42.4°$$

Since pin P moves with respect to the rotating frame $\mathcal{S}$, we write

$$\mathbf{a}_P = \mathbf{a}_{P'} + \mathbf{a}_{P/\mathcal{S}} + \mathbf{a}_c \qquad (1)$$

Each term of this vector equation is investigated separately.

Absolute Acceleration $\mathbf{a}_P$. Since disk D rotates with a constant angular velocity, the absolute acceleration $\mathbf{a}_P$ is directed toward B. We have

$$a_P = R\omega_D^2 = (500 \text{ mm})(10 \text{ rad/s})^2 = 5000 \text{ mm/s}^2$$
$$\mathbf{a}_P = 5000 \text{ mm/s}^2 \searrow 30°$$

Acceleration $\mathbf{a}_{P'}$ of the Coinciding Point P'. The acceleration $\mathbf{a}_{P'}$ of the point P' of the frame $\mathcal{S}$ which coincides with P at the instant considered is resolved into normal and tangential components. (We recall from Sample Prob. 15.9 that $r = 37.1$ mm.)

$$(a_{P'})_n = r\omega_{\mathcal{S}}^2 = (37.1 \text{ mm})(4.08 \text{ rad/s})^2 = 618 \text{ mm/s}^2$$
$$(\mathbf{a}_{P'})_n = 618 \text{ mm/s}^2 \nearrow 42.4°$$
$$(a_{P'})_t = r\alpha_{\mathcal{S}} = 37.1\alpha_{\mathcal{S}} \qquad (\mathbf{a}_{P'})_t = 37.1\alpha_{\mathcal{S}} \nwarrow 42.4°$$

Relative Acceleration $\mathbf{a}_{P/\mathcal{S}}$. Since the pin P moves in a straight slot cut in disk S, the relative acceleration $\mathbf{a}_{P/\mathcal{S}}$ must be parallel to the slot; that is, its direction must be $\swarrow$ 42.4°.

Coriolis Acceleration $\mathbf{a}_c$. Rotating the relative velocity $\mathbf{v}_{P/\mathcal{S}}$ through 90° in the sense of $\boldsymbol{\omega}_{\mathcal{S}}$, we obtain the direction of the Coriolis component of the acceleration: $\nwarrow$ 42.4°. We write

$$a_c = 2\omega_{\mathcal{S}} v_{P/\mathcal{S}} = 2(4.08 \text{ rad/s})(477 \text{ mm/s}) = 3890 \text{ mm/s}^2$$
$$\mathbf{a}_c = 3890 \text{ mm/s}^2 \nwarrow 42.4°$$

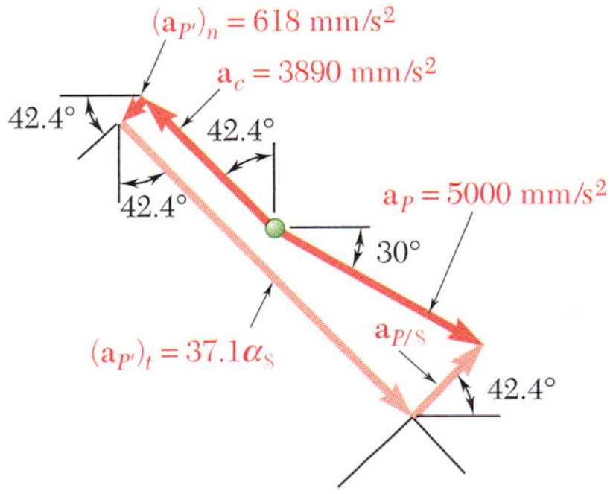

We rewrite Eq. (1) and substitute the accelerations found above:

$$\mathbf{a}_P = (\mathbf{a}_{P'})_n + (\mathbf{a}_{P'})_t + \mathbf{a}_{P/\mathcal{S}} + \mathbf{a}_c$$
$$[5000 \searrow 30°] = [618 \nearrow 42.4°] + [37.1\alpha_{\mathcal{S}} \nwarrow 42.4°] + [a_{P/\mathcal{S}} \swarrow 42.4°] + [3890 \nwarrow 42.4°]$$

Equating components in a direction perpendicular to the slot,

$$5000 \cos 17.6° = 37.1\alpha_{\mathcal{S}} - 3890$$
$$\boldsymbol{\alpha}_S = \boldsymbol{\alpha}_{\mathcal{S}} = 233 \text{ rad/s}^2 \circlearrowright \quad \blacktriangleleft$$

SOLVING PROBLEMS ON YOUR OWN

In this lesson you studied the rate of change of a vector with respect to a rotating frame and then applied your knowledge to the analysis of the plane motion of a particle relative to a rotating frame.

1. *Rate of change of a vector with respect to a fixed frame and with respect to a rotating frame.* Denoting by $(\dot{\mathbf{Q}})_{OXYZ}$ the rate of change of a vector **Q** with respect to a fixed frame $OXYZ$ and by $(\dot{\mathbf{Q}})_{Oxyz}$ its rate of change with respect to a rotating frame $Oxyz$, we obtained the fundamental relation

$$(\dot{\mathbf{Q}})_{OXYZ} = (\dot{\mathbf{Q}})_{Oxyz} + \boldsymbol{\Omega} \times \mathbf{Q} \tag{15.31}$$

where $\boldsymbol{\Omega}$ is the angular velocity of the rotating frame.

This fundamental relation will now be applied to the solution of two-dimensional problems.

2. *Plane motion of a particle relative to a rotating frame.* Using the above fundamental relation and designating by $\mathcal{F}$ the rotating frame, we obtained the following expressions for the velocity and the acceleration of a particle P:

$$\mathbf{v}_P = \mathbf{v}_{P'} + \mathbf{v}_{P/\mathcal{F}} \tag{15.33}$$
$$\mathbf{a}_P = \mathbf{a}_{P'} + \mathbf{a}_{P/\mathcal{F}} + \mathbf{a}_c \tag{15.36}$$

In these equations:

a. The subscript P refers to the absolute motion of the particle P, that is, to its motion with respect to a fixed frame of reference OXY.

b. The subscript P′ refers to the motion of the point P' of the rotating frame $\mathcal{F}$ which coincides with P at the instant considered.

c. The subscript P/ℱ refers to the motion of the particle P relative to the rotating frame $\mathcal{F}$.

d. The term $\mathbf{a}_c$ represents the Coriolis acceleration of point P. Its magnitude is $2\Omega v_{P/\mathcal{F}}$, and its direction is found by rotating $\mathbf{v}_{P/\mathcal{F}}$ through 90° in the sense of rotation of the frame $\mathcal{F}$.

You should keep in mind that the Coriolis acceleration should be taken into account whenever a part of the mechanism you are analyzing is moving with respect to another part that is rotating. The problems you will encounter in this lesson involve collars that slide on rotating rods, booms that extend from cranes rotating in a vertical plane, etc.

When solving a problem involving a rotating frame, you will find it convenient to draw vector diagrams representing Eqs. (15.33) and (15.36), respectively, and use these diagrams to obtain either an analytical or a graphical solution.

Problems

15.148 and 15.149 Pin P is attached to the wheel shown and slides in a slot cut in bar BD. The wheel rolls to the right without slipping with a constant angular velocity of 20 rad/s. Knowing that $x = 480$ mm when $\theta = 0$, determine the angular velocity of the bar and the relative velocity of pin P with respect to the rod for the given data.

15.148 (*a*) $\theta = 0$, (*b*) $\theta = 90°$.

15.149 $\theta = 30°$.

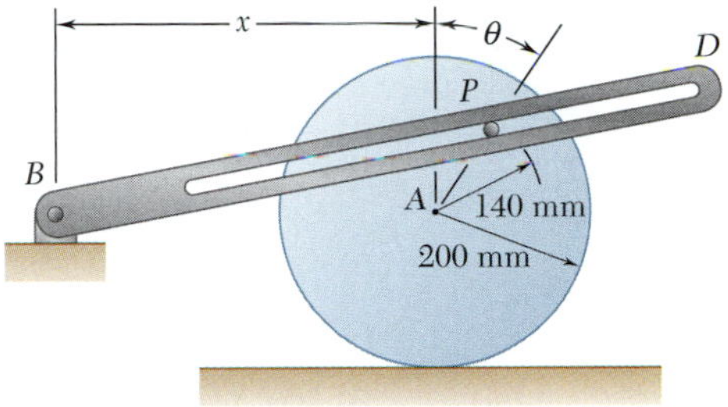

Fig. P15.148 and P15.149

15.150 and 15.151 Two rotating rods are connected by a slider block P. The velocity $\mathbf{v}_0$ of the slider block relative to the rod on which it slides has a constant velocity of 30 in./s and is directed outward. Determine the angular velocity of each rod for the position shown.

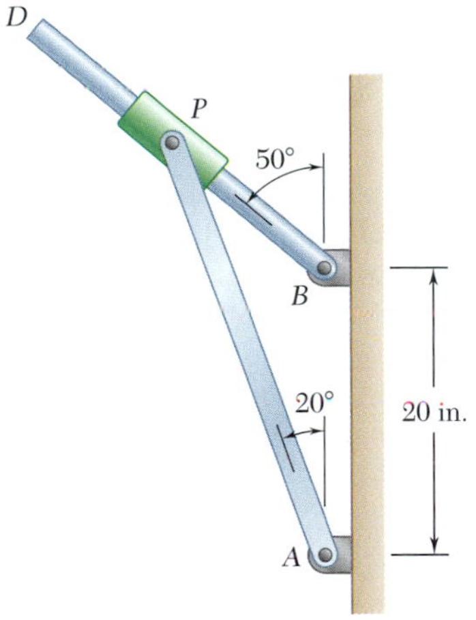

Fig. P15.150

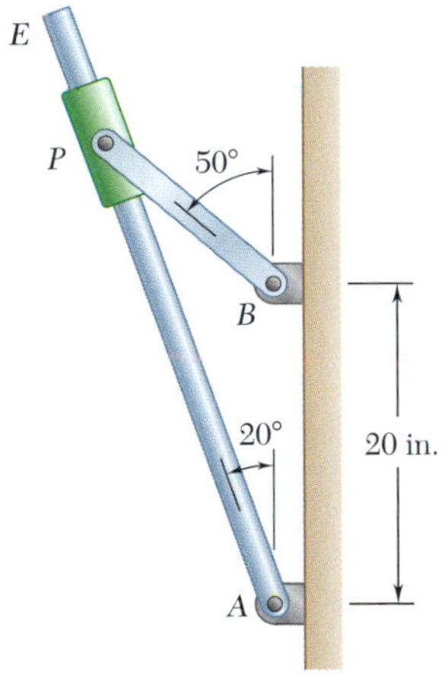

Fig. P15.151

15.152 and 15.153 Pin P is attached to the collar shown; the motion of the pin is guided by a slot cut in rod BD and by the collar that slides on rod AE. Knowing that at the instant considered the rods rotate clockwise with constant angular velocities, determine for the given data the velocity of pin P.

15.152 $\omega_{AE} = 8$ rad/s, $\omega_{BD} = 3$ rad/s

15.153 $\omega_{AE} = 7$ rad/s, $\omega_{BD} = 4.8$ rad/s

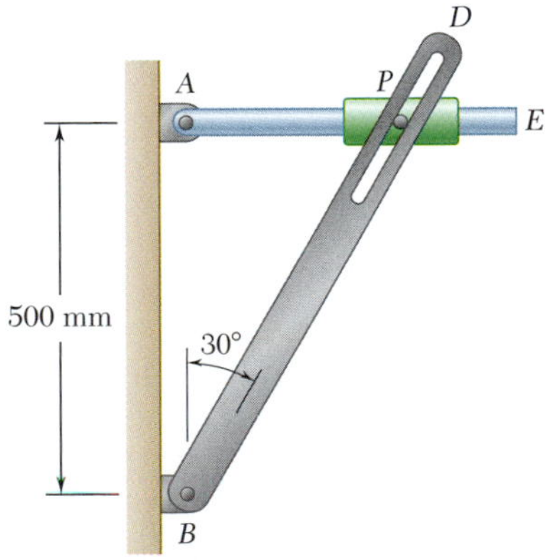

Fig. P15.152 and P15.153

15.154 Knowing that at the instant shown the angular velocity of bar AB is 15 rad/s clockwise and the angular velocity of bar EF is 10 rad/s clockwise, determine (*a*) the angular velocity of rod DE, (*b*) the relative velocity of collar B with respect to rod DE.

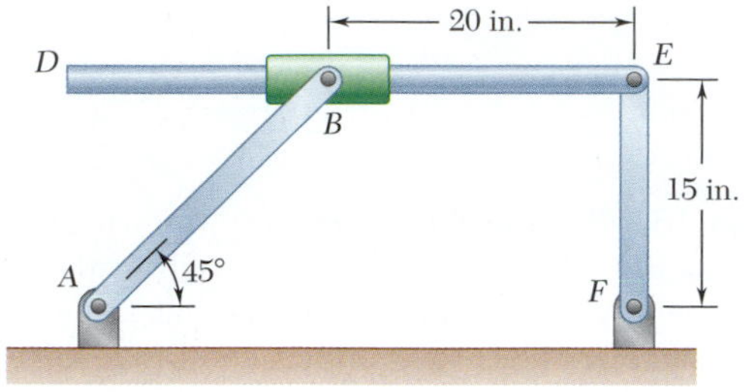

Fig. *P15.154* and *P15.155*

15.155 Knowing that at the instant shown the angular velocity of rod DE is 10 rad/s clockwise and the angular velocity of bar EF is 15 rad/s counterclockwise, determine (*a*) the angular velocity of bar AB, (*b*) the relative velocity of collar B with respect to rod DE.

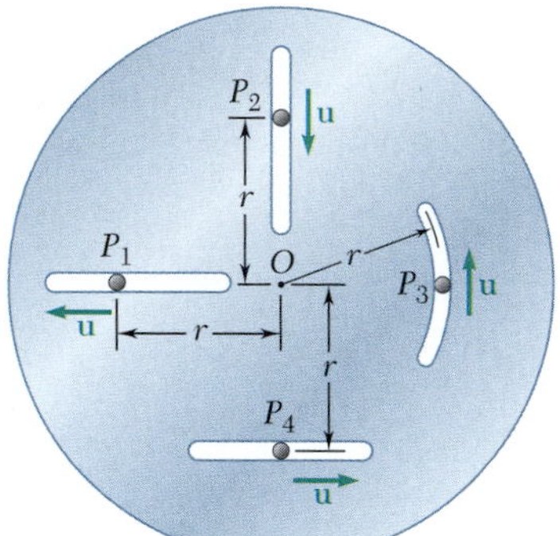

Fig. *P15.156*

15.156 Four pins slide in four separate slots cut in a circular plate as shown. When the plate is at rest, each pin has a velocity directed as shown and of the same constant magnitude u. If each pin maintains the same velocity relative to the plate when the plate rotates about O with a constant counterclockwise angular velocity $\boldsymbol{\omega}$, determine the acceleration of each pin.

15.157 Solve Prob. 15.156, assuming that the plate rotates about O with a constant clockwise angular velocity $\boldsymbol{\omega}$.

15.158 Pin P is attached to the collar shown; the motion of the pin is guided by a slot cut in bar BD and by the collar that slides on rod AE. Rod AE rotates with a constant angular velocity of 5 rad/s clockwise and the distance from A to P increases at a constant rate of 2 m/s. Determine at the instant shown (*a*) the angular acceleration of bar BD, (*b*) the relative acceleration of pin P with respect to bar BD.

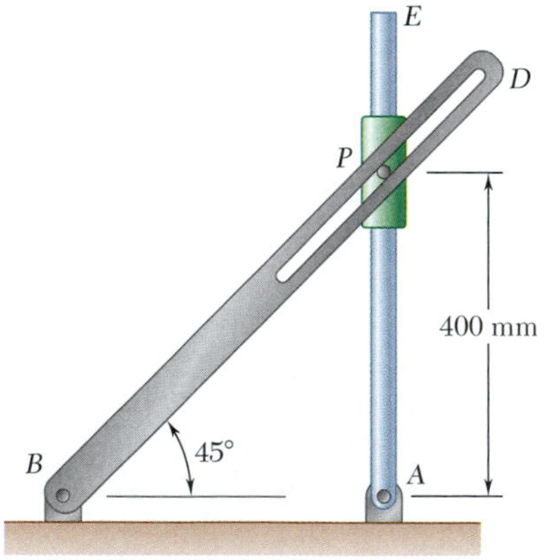

Fig. P15.158 and P15.159

15.159 Pin P is attached to the collar shown; the motion of the pin is guided by a slot cut in bar BD and by the collar that slides on rod AE. Bar BD rotates with a constant angular velocity of 5 rad/s counterclockwise and the distance from B to P decreases at a constant rate of 3.5 m/s. Determine at the instant shown (*a*) the angular acceleration of rod AE, (*b*) the relative acceleration of pin P with respect to rod AE.

15.160 At the instant shown the length of the boom AB is being *decreased* at the constant rate of 0.6 ft/s and the boom is being lowered at the constant rate of 0.08 rad/s. Determine (*a*) the velocity of point B, (*b*) the acceleration of point B.

15.161 At the instant shown the length of the boom AB is being *increased* at the constant rate of 0.6 ft/s and the boom is being lowered at the constant rate of 0.08 rad/s. Determine (*a*) the velocity of point B, (*b*) the acceleration of point B.

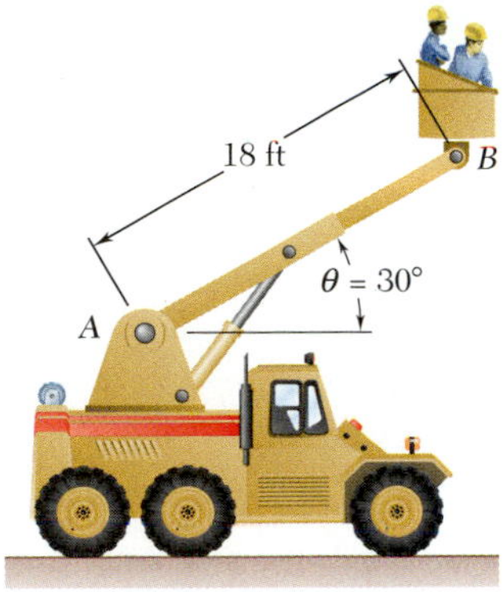

Fig. P15.160 and P15.161

15.162 The cage of a mine elevator moves downward at a constant speed of 12.2 m/s. Determine the magnitude and direction of the Coriolis acceleration of the cage if the elevator is located (*a*) at the equator, (*b*) at latitude 40° north, (*c*) at latitude 40° south.

15.163 A rocket sled is tested on a straight track that is built along a meridian. Knowing that the track is located at latitude 40° north, determine the Coriolis acceleration of the sled when it is moving north at a speed of 600 mi/h.

15.164 The motion of nozzle D is controlled by arm AB. At the instant shown the arm is rotating counterclockwise at the constant rate $\omega = 2.4$ rad/s and portion BC is being extended at the constant rate $u = 250$ mm/s with respect to the arm. For each of the arrangements shown, determine the acceleration of nozzle D.

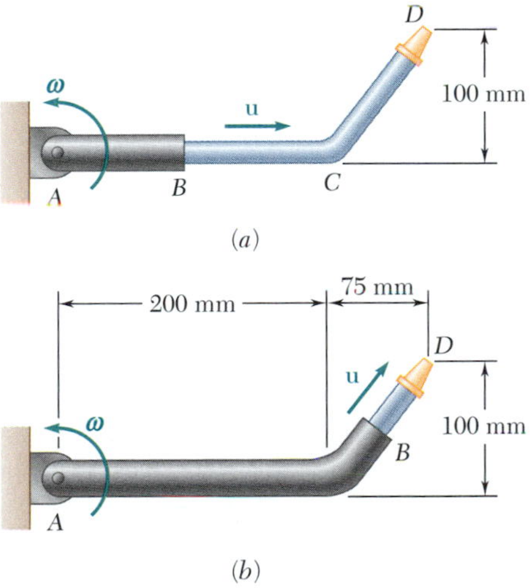

Fig. P15.164

15.165 Solve Prob. 15.164, assuming that the direction of the relative velocity **u** is reversed so that portion BD is being retracted.

15.166 Collar P slides toward point A at a constant relative speed of 2 m/s along rod AB which rotates counterclockwise with a constant angular velocity of 5 rad/s. At the instant shown determine (*a*) the velocity and acceleration of point P, (*b*) the velocity and acceleration of point D.

15.167 Collar P slides toward point A at a constant relative speed of 3.2 m/s along rod AB which rotates clockwise with a constant angular velocity of 4 rad/s. At the instant shown determine (*a*) the angular velocities of bars PD and DE, (*b*) the angular accelerations of bars PD and DE.

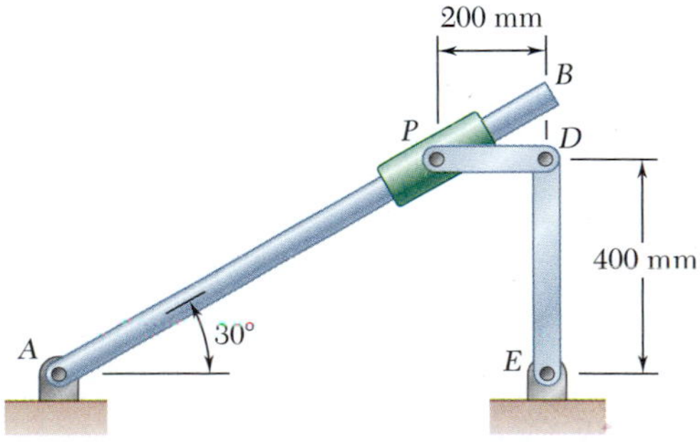

Fig. *P15.166* and *P15.167*

15.168 and 15.169 A chain is looped around two gears of radius 2 in. that can rotate freely with respect to the 16-in. arm AB. The chain moves about arm AB in a clockwise direction at the constant rate of 4 in./s relative to the arm. Knowing that in the position shown arm AB rotates clockwise about A at the constant rate $\omega = 0.75$ rad/s, determine the acceleration of each of the chain links indicated.

15.168 Links *1* and *2*.

15.169 Links *3* and *4*.

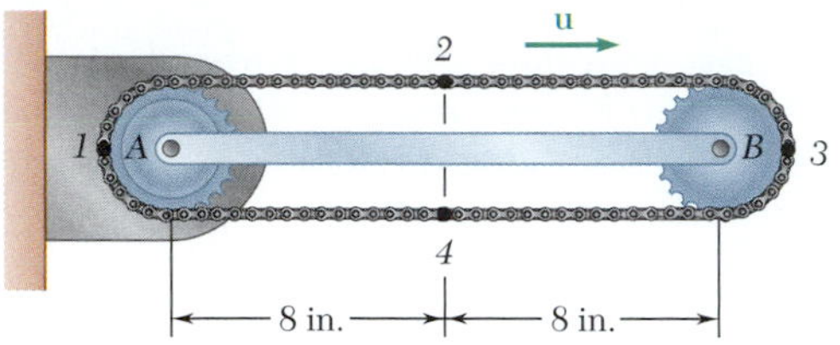

Fig. P15.168 and P15.169

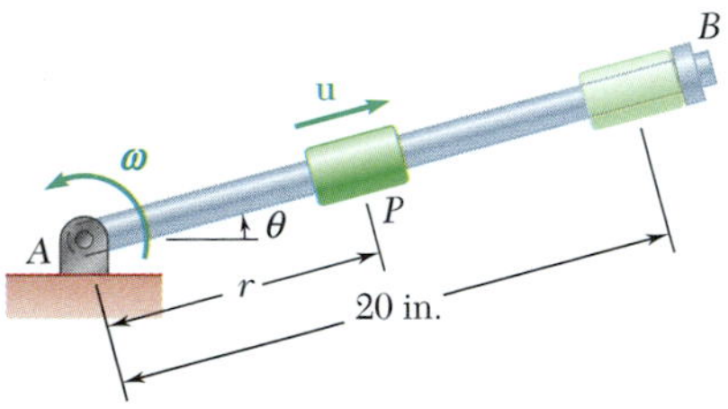

Fig. *P15.170*

15.170 The collar P slides outward at a constant relative speed u along rod AB, which rotates counterclockwise with a constant angular velocity of 20 rpm. Knowing that $r = 10$ in. when $\theta = 0$ and that the collar reaches B when $\theta = 90°$, determine the magnitude of the acceleration of the collar P just as it reaches B.

15.171 Pin P slides in a circular slot cut in the plate shown at a constant relative speed $u = 180$ mm/s. Knowing that at the instant shown the plate rotates clockwise about A at the constant rate $\omega = 6$ rad/s, determine the acceleration of the pin if it is located at (*a*) point A, (*b*) point B, (*c*) point C.

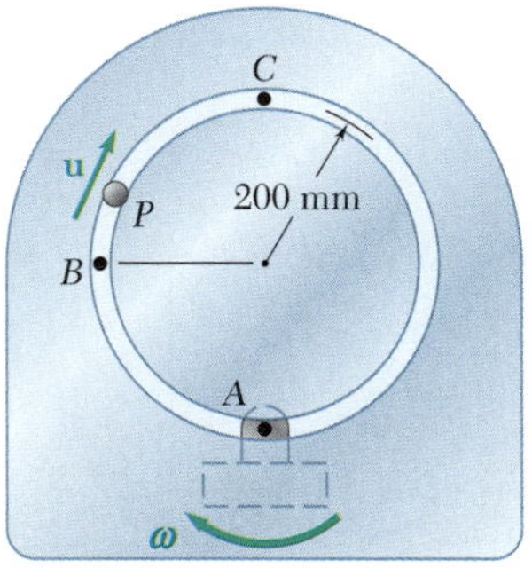

Fig. P15.171 and P15.172

15.172 Pin P slides in a circular slot cut in the plate shown at a constant relative speed $u = 180$ mm/s. Knowing that at the instant shown the angular velocity $\boldsymbol{\omega}$ of the plate is 6 rad/s clockwise and is decreasing at the rate of 10 rad/s^2, determine the acceleration of the pin if it is located at (*a*) point A, (*b*) point B, (*c*) point C.

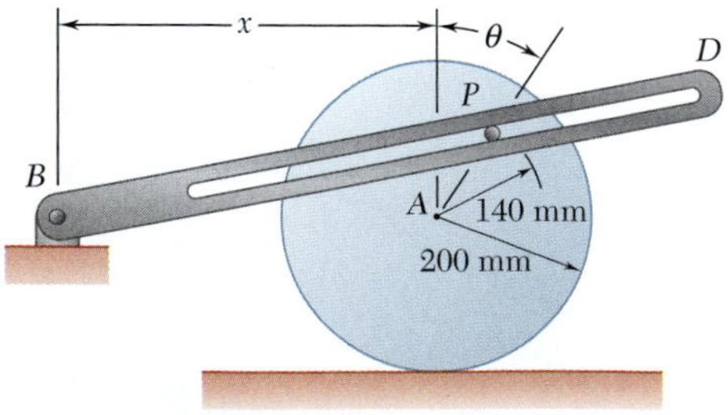

Fig. P15.173 and P15.174

15.173 and 15.174 Pin P is attached to the wheel shown and slides in a slot cut in bar BD. The wheel rolls to the right without slipping with a constant angular velocity of 20 rad/s. Knowing that $x = 480$ mm when $\theta = 0$, determine (*a*) the angular acceleration of the bar, (*b*) the relative acceleration of pin P with respect to the bar for the given data.

15.173 $\theta = 0$.

15.174 $\theta = 90°$.

15.175 and 15.176 Knowing that at the instant shown the rod attached at B rotates with a constant counterclockwise angular velocity $\boldsymbol{\omega}_B$ of 6 rad/s, determine the angular velocity and angular acceleration of the rod attached at A.

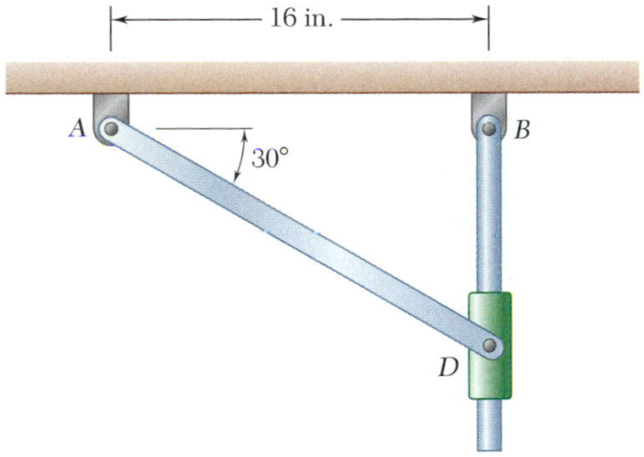

Fig. P15.175

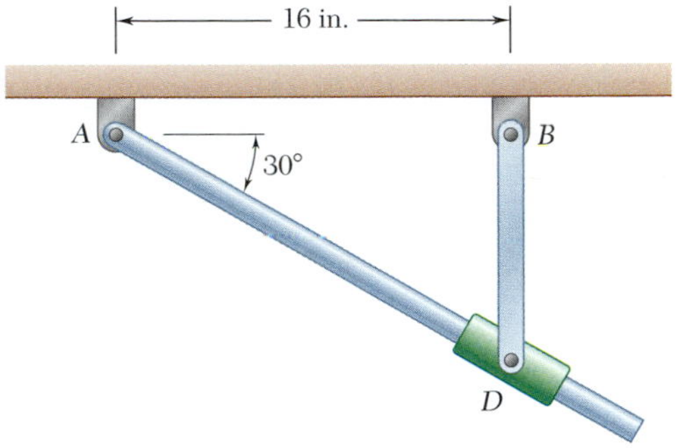

Fig. P15.176

15.177 The Geneva mechanism shown is used to provide an intermittent rotary motion of disk S. Disk D rotates with a constant counterclockwise angular velocity $\boldsymbol{\omega}_D$ of 8 rad/s. A pin P is attached to disk D and can slide in one of six equally spaced slots cut in disk S. It is desirable that the angular velocity of disk S be zero as the pin enters and leaves each of the six slots; this will occur if the distance between the centers of the disks and the radii of the disks are related as shown. Determine the angular velocity and angular acceleration of disk S at the instant when $\phi = 150°$.

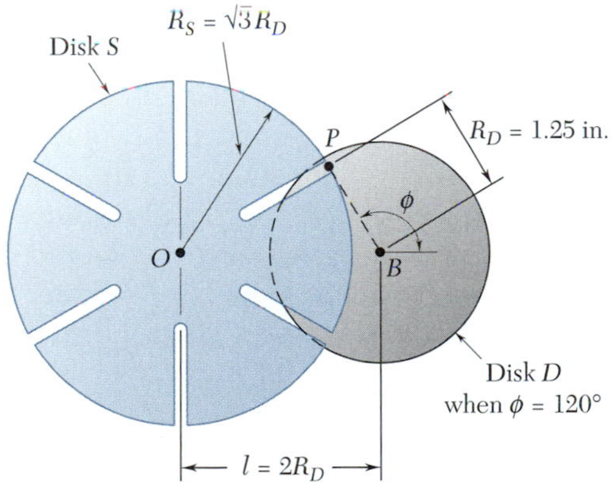

Fig. P15.177

15.178 In Prob. 15.177, determine the angular velocity and angular acceleration of disk S at the instant when $\phi = 135°$.

15.179 The disk shown rotates with a constant clockwise angular velocity of 12 rad/s. At the instant shown, determine (*a*) the angular velocity and angular acceleration of rod BD, (*b*) the velocity and acceleration of the point of the rod coinciding with E.

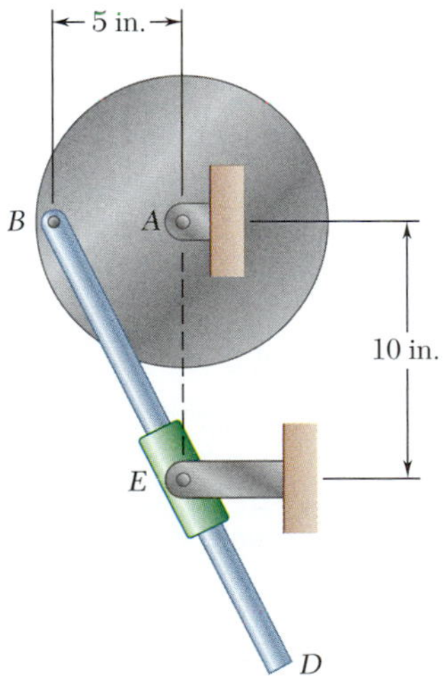

Fig. *P15.179*

15.180 Collar B slides along rod AC and is attached to a block that moves in a vertical slot. Knowing that $R = 18$ in., $\theta = 30°$, $\omega = 6$ rad/s, and $\alpha = 4$ rad/s^2, determine the velocity and acceleration of collar B.

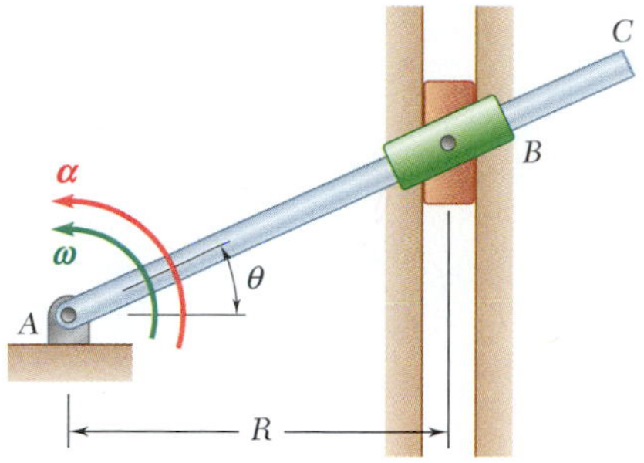

Fig. ***P15.180***

*15.12. MOTION ABOUT A FIXED POINT

In Sec. 15.3 the motion of a rigid body constrained to rotate about a fixed axis was considered. The more general case of the motion of a rigid body which has a fixed point O will now be examined.

First, it will be proved that *the most general displacement of a rigid body with a fixed point O is equivalent to a rotation of the body about an axis through O.*† Instead of considering the rigid body itself, we can detach a sphere of center O from the body and analyze the motion of that sphere. Clearly, the motion of the sphere completely characterizes the motion of the given body. Since three points define the position of a solid in space, the center O and two points A and B on the surface of the sphere will define the position of the sphere and thus the position of the body. Let A_1 and B_1 characterize the position of the sphere at one instant, and let A_2 and B_2 characterize its position at a later instant (Fig. 15.31*a*). Since the sphere is rigid, the lengths of the arcs of great circle A_1B_1 and A_2B_2 must be equal, but except for this requirement, the positions of A_1, A_2, B_1, and B_2 are arbitrary. We propose to prove that the points A and B can be brought, respectively, from A_1 and B_1 into A_2 and B_2 by a single rotation of the sphere about an axis.

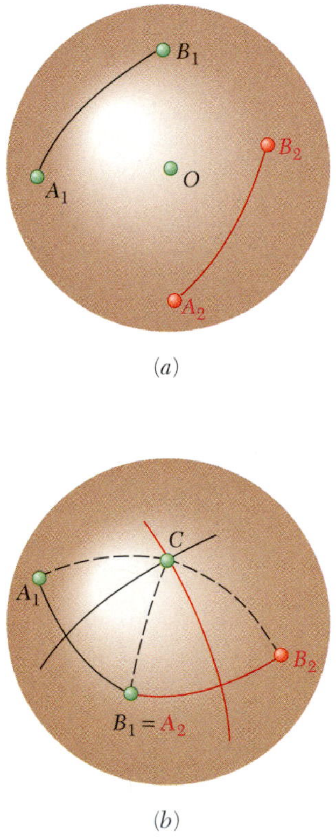

Fig. 15.31

For convenience, and without loss of generality, we select point B so that its initial position coincides with the final position of A; thus, $B_1 = A_2$ (Fig. 15.31*b*). We draw the arcs of great circle A_1A_2, A_2B_2 and the arcs bisecting, respectively, A_1A_2 and A_2B_2. Let C be the point of intersection of these last two arcs; we complete the construction by drawing A_1C, A_2C, and B_2C. As pointed out above, because of the rigidity of the sphere, $A_1B_1 = A_2B_2$. Since C is by construction equidistant from A_1, A_2, and B_2, we also have $A_1C = A_2C = B_2C$. As a result, the spherical triangles A_1CA_2 and B_1CB_2 are congruent and the

†This is known as *Euler's theorem.*

angles A_1CA_2 and B_1CB_2 are equal. Denoting by θ the common value of these angles, we conclude that the sphere can be brought from its initial position into its final position by a single rotation through θ about the axis OC.

It follows that the motion during a time interval Δt of a rigid body with a fixed point O can be considered as a rotation through $\Delta\theta$ about a certain axis. Drawing along that axis a vector of magnitude $\Delta\theta/\Delta t$ and letting Δt approach zero, we obtain at the limit the *instantaneous axis of rotation* and the angular velocity $\boldsymbol{\omega}$ of the body at the instant considered (Fig. 15.32). The velocity of a particle P of the body can then be obtained, as in Sec. 15.3, by forming the vector product of $\boldsymbol{\omega}$ and of the position vector $\mathbf{r}$ of the particle:

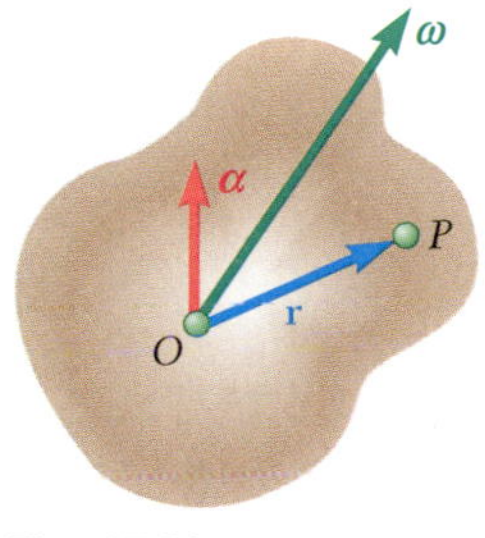

Fig. 15.32

$$\mathbf{v} = \frac{d\mathbf{r}}{dt} = \boldsymbol{\omega} \times \mathbf{r} \tag{15.37}$$

The acceleration of the particle is obtained by differentiating (15.37) with respect to t. As in Sec. 15.3 we have

$$\mathbf{a} = \boldsymbol{\alpha} \times \mathbf{r} + \boldsymbol{\omega} \times (\boldsymbol{\omega} \times \mathbf{r}) \tag{15.38}$$

where the angular acceleration $\boldsymbol{\alpha}$ is defined as the derivative

$$\boldsymbol{\alpha} = \frac{d\boldsymbol{\omega}}{dt} \tag{15.39}$$

of the angular velocity $\boldsymbol{\omega}$.

In the case of the motion of a rigid body with a fixed point, the direction of $\boldsymbol{\omega}$ and of the instantaneous axis of rotation changes from one instant to the next. The angular acceleration $\boldsymbol{\alpha}$ therefore reflects the change in direction of $\boldsymbol{\omega}$ as well as its change in magnitude and, in general, *is not directed along the instantaneous axis of rotation*. While the particles of the body located on the instantaneous axis of rotation have zero velocity at the instant considered, they do not have zero acceleration. Also, the accelerations of the various particles of the body *cannot* be determined as if the body were rotating permanently about the instantaneous axis.

Recalling the definition of the velocity of a particle with position vector $\mathbf{r}$, we note that the angular acceleration $\boldsymbol{\alpha}$, as expressed in (15.39), represents the velocity of the tip of the vector $\boldsymbol{\omega}$. This property may be useful in the determination of the angular acceleration of a rigid body. For example, it follows that the vector $\boldsymbol{\alpha}$ is tangent to the curve described in space by the tip of the vector $\boldsymbol{\omega}$.

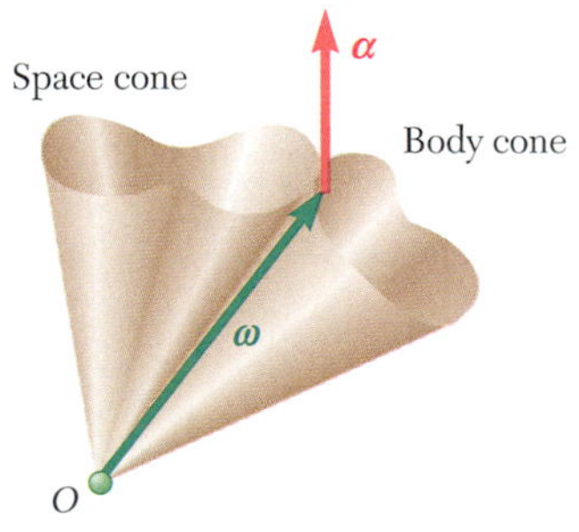

Fig. 15.33

We should note that the vector $\boldsymbol{\omega}$ moves within the body, as well as in space. It thus generates two cones called, respectively, the *body cone* and the *space cone* (Fig. 15.33).† It can be shown that at any given instant, the two cones are tangent along the instantaneous axis of rotation and that as the body moves, the body cone appears to *roll* on the space cone.

†It is recalled that a *cone* is, by definition, a surface generated by a straight line passing through a fixed point. In general, the cones considered here *will not be circular cones*.

Before concluding our analysis of the motion of a rigid body with a fixed point, we should prove that angular velocities are actually vectors. As indicated in Sec. 2.3, some quantities, such as the *finite rotations* of a rigid body, have magnitude and direction but do not obey the parallelogram law of addition; these quantities cannot be considered as vectors. In contrast, angular velocities (and also *infinitesimal rotations*), as will be demonstrated presently, *do obey* the parallelogram law and thus are truly vector quantities.

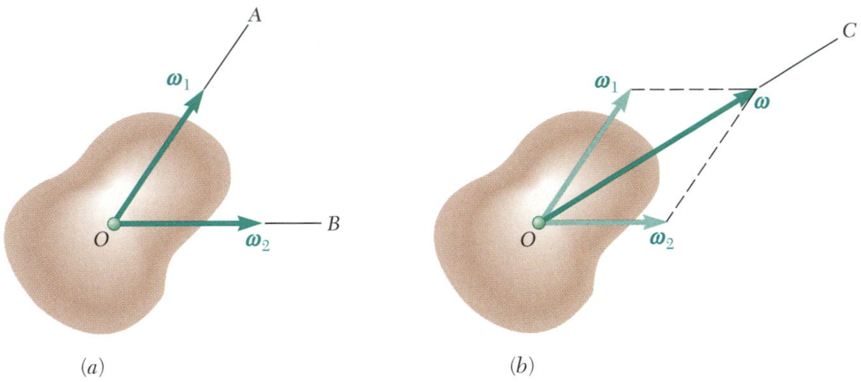

Fig. 15.34

Photo 15.4 When the ladder rotates about its fixed base, its angular velocity can be obtained by adding the angular velocities which correspond to simultaneous rotations about two different axes.

Consider a rigid body with a fixed point O which at a given instant rotates simultaneously about the axes OA and OB with angular velocities $\boldsymbol{\omega}_1$ and $\boldsymbol{\omega}_2$ (Fig. 15.34*a*). We know that this motion must be equivalent at the instant considered to a single rotation of angular velocity $\boldsymbol{\omega}$. We propose to show that

$$\boldsymbol{\omega} = \boldsymbol{\omega}_1 + \boldsymbol{\omega}_2 \tag{15.40}$$

that is, that the resulting angular velocity can be obtained by adding $\boldsymbol{\omega}_1$ and $\boldsymbol{\omega}_2$ by the parallelogram law (Fig. 15.34*b*).

Consider a particle P of the body, defined by the position vector $\mathbf{r}$. Denoting, respectively, by $\mathbf{v}_1$, $\mathbf{v}_2$, and $\mathbf{v}$ the velocity of P when the body rotates about OA only, about OB only, and about both axes simultaneously, we write

$$\mathbf{v} = \boldsymbol{\omega} \times \mathbf{r} \qquad \mathbf{v}_1 = \boldsymbol{\omega}_1 \times \mathbf{r} \qquad \mathbf{v}_2 = \boldsymbol{\omega}_2 \times \mathbf{r} \tag{15.41}$$

But the vectorial character of *linear* velocities is well established (since they represent the derivatives of position vectors). We therefore have

$$\mathbf{v} = \mathbf{v}_1 + \mathbf{v}_2$$

where the plus sign indicates vector addition. Substituting from (15.41), we write

$$\boldsymbol{\omega} \times \mathbf{r} = \boldsymbol{\omega}_1 \times \mathbf{r} + \boldsymbol{\omega}_2 \times \mathbf{r}$$
$$\boldsymbol{\omega} \times \mathbf{r} = (\boldsymbol{\omega}_1 + \boldsymbol{\omega}_2) \times \mathbf{r}$$

where the plus sign still indicates vector addition. Since the relation obtained holds for an arbitrary $\mathbf{r}$, we conclude that (15.40) must be true.

*15.13. GENERAL MOTION

The most general motion of a rigid body in space will now be considered. Let A and B be two particles of the body. We recall from Sec. 11.12 that the velocity of B with respect to the fixed frame of reference $OXYZ$ can be expressed as

$$\mathbf{v}_B = \mathbf{v}_A + \mathbf{v}_{B/A} \tag{15.42}$$

where $\mathbf{v}_{B/A}$ is the velocity of B relative to a frame $AX'Y'Z'$ attached to A and of fixed orientation (Fig. 15.35). Since A is fixed in this frame, the motion of the body relative to $AX'Y'Z'$ is the motion of a body with a fixed point. The relative velocity $\mathbf{v}_{B/A}$ can therefore be obtained from (15.37) after $\mathbf{r}$ has been replaced by the position vector $\mathbf{r}_{B/A}$ of B relative to A. Substituting for $\mathbf{v}_{B/A}$ into (15.42), we write

$$\mathbf{v}_B = \mathbf{v}_A + \boldsymbol{\omega} \times \mathbf{r}_{B/A} \tag{15.43}$$

where $\boldsymbol{\omega}$ is the angular velocity of the body at the instant considered.

The acceleration of B is obtained by a similar reasoning. We first write

$$\mathbf{a}_B = \mathbf{a}_A + \mathbf{a}_{B/A}$$

and, recalling Eq. (15.38),

$$\mathbf{a}_B = \mathbf{a}_A + \boldsymbol{\alpha} \times \mathbf{r}_{B/A} + \boldsymbol{\omega} \times (\boldsymbol{\omega} \times \mathbf{r}_{B/A}) \tag{15.44}$$

where $\boldsymbol{\alpha}$ is the angular acceleration of the body at the instant considered.

Equations (15.43) and (15.44) show that *the most general motion of a rigid body is equivalent, at any given instant, to the sum of a translation,* in which all the particles of the body have the same velocity and acceleration as a reference particle A, *and of a motion in which particle A is assumed to be fixed.*†

It is easily shown, by solving (15.43) and (15.44) for $\mathbf{v}_A$ and $\mathbf{a}_A$, that the motion of the body with respect to a frame attached to B would be characterized by the same vectors $\boldsymbol{\omega}$ and $\boldsymbol{\alpha}$ as its motion relative to $AX'Y'Z'$. The angular velocity and angular acceleration of a rigid body at a given instant are thus independent of the choice of reference point. On the other hand, one should keep in mind that whether the moving frame is attached to A or to B, it should maintain a fixed orientation; that is, it should remain parallel to the fixed reference frame $OXYZ$ throughout the motion of the rigid body. In many problems it will be more convenient to use a moving frame which is allowed to rotate as well as to translate. The use of such moving frames will be discussed in Secs. 15.14 and 15.15.

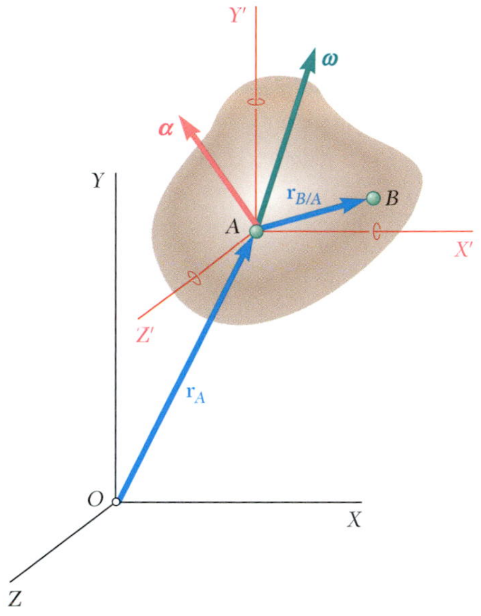

Fig. 15.35

†It is recalled from Sec. 15.12 that, in general, the vectors $\boldsymbol{\omega}$ and $\boldsymbol{\alpha}$ are not collinear, and that the accelerations of the particles of the body in their motion relative to the frame $AX'Y'Z'$ cannot be determined as if the body were rotating permanently about the instantaneous axis through A.

SAMPLE PROBLEM 15.11

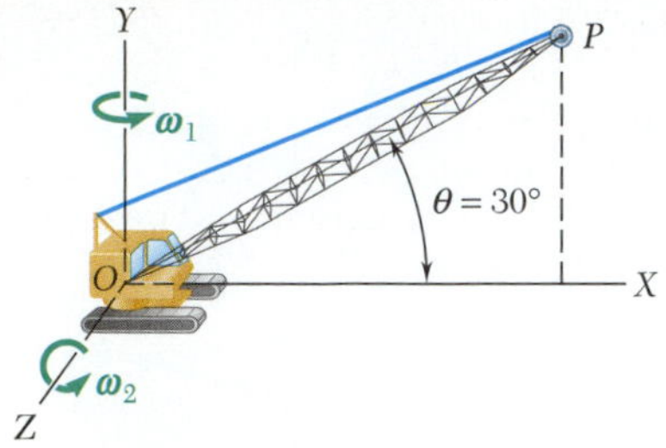

The crane shown rotates with a constant angular velocity $\boldsymbol{\omega}_1$ of 0.30 rad/s. Simultaneously, the boom is being raised with a constant angular velocity $\boldsymbol{\omega}_2$ of 0.50 rad/s relative to the cab. Knowing that the length of the boom OP is $l = 12$ m, determine (*a*) the angular velocity $\boldsymbol{\omega}$ of the boom, (*b*) the angular acceleration $\boldsymbol{\alpha}$ of the boom, (*c*) the velocity $\mathbf{v}$ of the tip of the boom, (*d*) the acceleration $\mathbf{a}$ of the tip of the boom.

SOLUTION

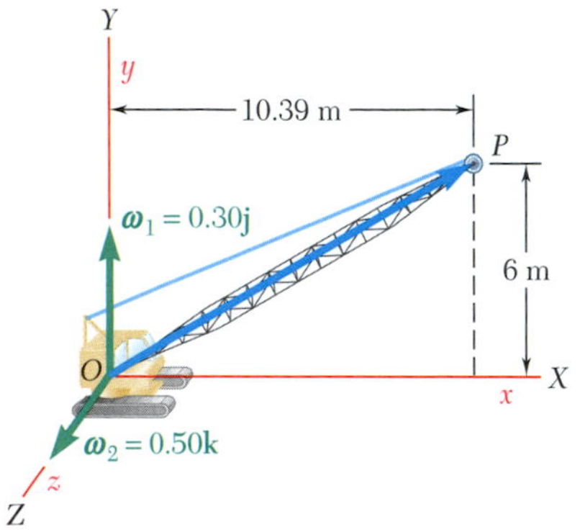

***a.* Angular Velocity of Boom.** Adding the angular velocity $\boldsymbol{\omega}_1$ of the cab and the angular velocity $\boldsymbol{\omega}_2$ of the boom relative to the cab, we obtain the angular velocity $\boldsymbol{\omega}$ of the boom at the instant considered:

$$\boldsymbol{\omega} = \boldsymbol{\omega}_1 + \boldsymbol{\omega}_2 \qquad \boldsymbol{\omega} = (0.30 \text{ rad/s})\mathbf{j} + (0.50 \text{ rad/s})\mathbf{k} \quad \blacktriangleleft$$

***b.* Angular Acceleration of Boom.** The angular acceleration $\boldsymbol{\alpha}$ of the boom is obtained by differentiating $\boldsymbol{\omega}$. Since the vector $\boldsymbol{\omega}_1$ is constant in magnitude and direction, we have

$$\boldsymbol{\alpha} = \dot{\boldsymbol{\omega}} = \dot{\boldsymbol{\omega}}_1 + \dot{\boldsymbol{\omega}}_2 = 0 + \dot{\boldsymbol{\omega}}_2$$

where the rate of change $\dot{\boldsymbol{\omega}}_2$ is to be computed with respect to the fixed frame $OXYZ$. However, it is more convenient to use a frame $Oxyz$ attached to the cab and rotating with it, since the vector $\boldsymbol{\omega}_2$ also rotates with the cab and therefore has zero rate of change with respect to that frame. Using Eq. (15.31) with $\mathbf{Q} = \boldsymbol{\omega}_2$ and $\boldsymbol{\Omega} = \boldsymbol{\omega}_1$, we write

$$(\dot{\mathbf{Q}})_{OXYZ} = (\dot{\mathbf{Q}})_{Oxyz} + \boldsymbol{\Omega} \times \mathbf{Q}$$
$$(\dot{\boldsymbol{\omega}}_2)_{OXYZ} = (\dot{\boldsymbol{\omega}}_2)_{Oxyz} + \boldsymbol{\omega}_1 \times \boldsymbol{\omega}_2$$
$$\boldsymbol{\alpha} = (\dot{\boldsymbol{\omega}}_2)_{OXYZ} = 0 + (0.30 \text{ rad/s})\mathbf{j} \times (0.50 \text{ rad/s})\mathbf{k}$$

$$\boldsymbol{\alpha} = (0.15 \text{ rad/s}^2)\mathbf{i} \quad \blacktriangleleft$$

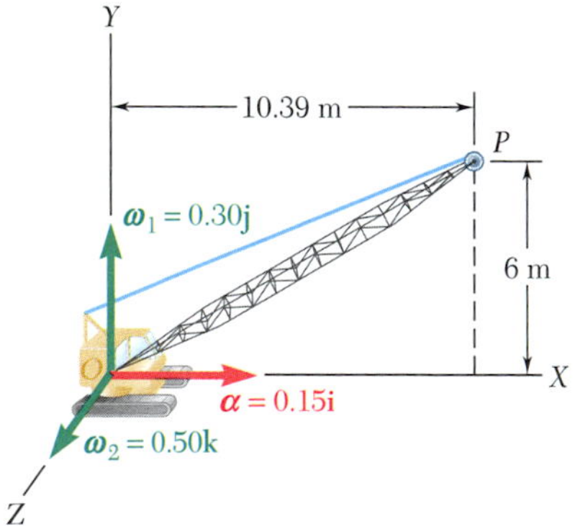

***c.* Velocity of Tip of Boom.** Noting that the position vector of point P is $\mathbf{r} = (10.39 \text{ m})\mathbf{i} + (6 \text{ m})\mathbf{j}$ and using the expression found for $\boldsymbol{\omega}$ in part *a*, we write

$$\mathbf{v} = \boldsymbol{\omega} \times \mathbf{r} = \begin{vmatrix} \mathbf{i} & \mathbf{j} & \mathbf{k} \\ 0 & 0.30 \text{ rad/s} & 0.50 \text{ rad/s} \\ 10.39 \text{ m} & 6 \text{ m} & 0 \end{vmatrix}$$

$$\mathbf{v} = -(3 \text{ m/s})\mathbf{i} + (5.20 \text{ m/s})\mathbf{j} - (3.12 \text{ m/s})\mathbf{k} \quad \blacktriangleleft$$

***d.* Acceleration of Tip of Boom.** Recalling that $\mathbf{v} = \boldsymbol{\omega} \times \mathbf{r}$, we write

$$\mathbf{a} = \boldsymbol{\alpha} \times \mathbf{r} + \boldsymbol{\omega} \times (\boldsymbol{\omega} \times \mathbf{r}) = \boldsymbol{\alpha} \times \mathbf{r} + \boldsymbol{\omega} \times \mathbf{v}$$

$$\mathbf{a} = \begin{vmatrix} \mathbf{i} & \mathbf{j} & \mathbf{k} \\ 0.15 & 0 & 0 \\ 10.39 & 6 & 0 \end{vmatrix} + \begin{vmatrix} \mathbf{i} & \mathbf{j} & \mathbf{k} \\ 0 & 0.30 & 0.50 \\ -3 & 5.20 & -3.12 \end{vmatrix}$$

$$= 0.90\mathbf{k} - 0.94\mathbf{i} - 2.60\mathbf{i} - 1.50\mathbf{j} + 0.90\mathbf{k}$$

$$\mathbf{a} = -(3.54 \text{ m/s}^2)\mathbf{i} - (1.50 \text{ m/s}^2)\mathbf{j} + (1.80 \text{ m/s}^2)\mathbf{k} \quad \blacktriangleleft$$

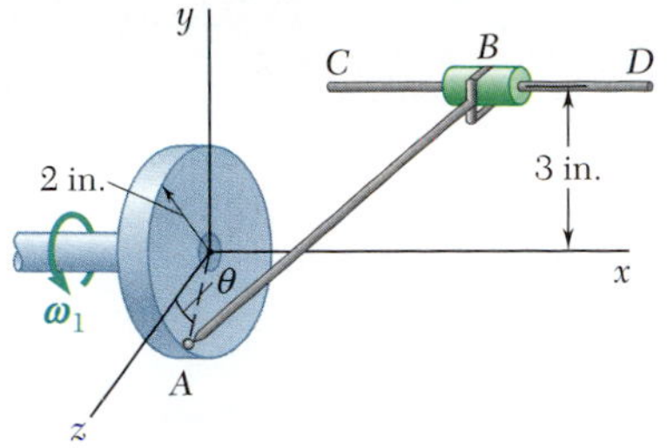

SAMPLE PROBLEM 15.12

The rod AB, of length 7 in., is attached to the disk by a ball-and-socket connection and to the collar B by a clevis. The disk rotates in the yz plane at a constant rate $\omega_1 = 12$ rad/s, while the collar is free to slide along the horizontal rod CD. For the position $\theta = 0$, determine (*a*) the velocity of the collar, (*b*) the angular velocity of the rod.

SOLUTION

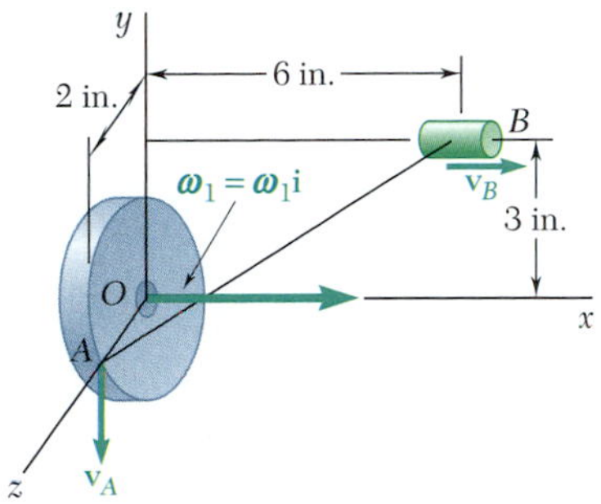

$\boldsymbol{\omega}_1 = 12\,\mathbf{i}$
$\mathbf{r}_A = 2\mathbf{k}$
$\mathbf{r}_B = 6\mathbf{i} + 3\mathbf{j}$
$\mathbf{r}_{B/A} = 6\mathbf{i} + 3\mathbf{j} - 2\mathbf{k}$

***a*. Velocity of Collar.** Since point A is attached to the disk and since collar B moves in a direction parallel to the x axis, we have

$$\mathbf{v}_A = \boldsymbol{\omega}_1 \times \mathbf{r}_A = 12\mathbf{i} \times 2\mathbf{k} = -24\mathbf{j} \qquad \mathbf{v}_B = v_B\mathbf{i}$$

Denoting by $\boldsymbol{\omega}$ the angular velocity of the rod, we write

$$\mathbf{v}_B = \mathbf{v}_A + \mathbf{v}_{B/A} = \mathbf{v}_A + \boldsymbol{\omega} \times \mathbf{r}_{B/A}$$

$$v_B\mathbf{i} = -24\mathbf{j} + \begin{vmatrix} \mathbf{i} & \mathbf{j} & \mathbf{k} \\ \omega_x & \omega_y & \omega_z \\ 6 & 3 & -2 \end{vmatrix}$$

$$v_B\mathbf{i} = -24\mathbf{j} + (-2\omega_y - 3\omega_z)\mathbf{i} + (6\omega_z + 2\omega_x)\mathbf{j} + (3\omega_x - 6\omega_y)\mathbf{k}$$

Equating the coefficients of the unit vectors, we obtain

$$v_B = \qquad -2\omega_y - 3\omega_z \qquad (1)$$
$$24 = 2\omega_x \qquad + 6\omega_z \qquad (2)$$
$$0 = 3\omega_x - 6\omega_y \qquad (3)$$

Clearly, the three equations obtained cannot be solved for the four unknowns v_B, ω_x, ω_y, and ω_z. An additional equation reflecting the type of connection at B will be obtained in part *b*. However, since the velocity of B does not depend upon that connection, we should be able to obtain v_B by eliminating the other unknowns from Eqs. (1), (2), and (3). We note that ω_x can be eliminated by multiplying Eq. (2) by 3, Eq. (3) by −2, and adding, and that ω_y and ω_z can be eliminated in a similar way. We verify that multiplying Eqs. (1), (2), and (3), respectively, by 6, 3, and −2 and adding eliminates all components of ω and yields an equation that can be solved for $\mathbf{v}_B$:

$$6v_B + 72 = 0 \qquad v_B = -12 \qquad \mathbf{v}_B = -(12 \text{ in./s})\mathbf{i} \quad ◀$$

***b*. Angular Velocity of Rod *AB*.** We noted that the angular velocity cannot be determined from Eqs. (1), (2), and (3) alone. An additional equation is obtained by considering the constraint imposed by the clevis at B.

The collar-clevis connection at B permits rotation of AB about the rod CD and also about an axis perpendicular to the plane containing AB and CD. It prevents rotation of AB about the axis EB, which is perpendicular to CD and lies in the plane containing AB and CD. Thus the projection of $\boldsymbol{\omega}$ on $\mathbf{r}_{E/B}$ must be zero and we write†

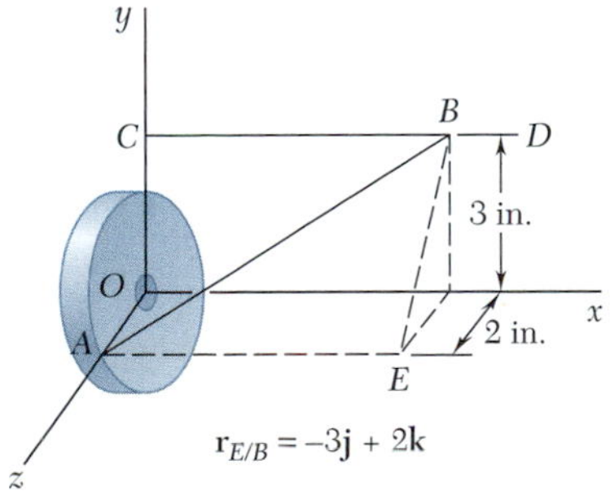

$\mathbf{r}_{E/B} = -3\mathbf{j} + 2\mathbf{k}$

$$\boldsymbol{\omega} \cdot \mathbf{r}_{E/B} = 0 \qquad (\omega_x\mathbf{i} + \omega_y\mathbf{j} + \omega_z\mathbf{k}) \cdot (-3\mathbf{j} + 2\mathbf{k}) = 0$$
$$-3\omega_y + 2\omega_z = 0 \qquad (4)$$

Solving Eqs. (1) through (4) simultaneously, we obtain

$$v_B = -12 \qquad \omega_x = 3.69 \qquad \omega_y = 1.846 \qquad \omega_z = 2.77$$

$$\boldsymbol{\omega} = (3.69 \text{ rad/s})\mathbf{i} + (1.846 \text{ rad/s})\mathbf{j} - (2.77 \text{ rad/s})\mathbf{k} \quad ◀$$

†We could also note that the direction of EB is that of the vector triple product $\mathbf{r}_{B/C} \times (\mathbf{r}_{B/C} \times \mathbf{r}_{B/A})$ and write $\boldsymbol{\omega} \cdot [\mathbf{r}_{B/C} \times (\mathbf{r}_{B/C} \times \mathbf{r}_{B/A})] = 0$. This formulation would be particularly useful if the rod CD were skew.

SOLVING PROBLEMS ON YOUR OWN

In this lesson you started the study of the *kinematics of rigid bodies in three dimensions*. You first studied the *motion of a rigid body about a fixed point* and then the *general motion of a rigid body*.

A. *Motion of a rigid body about a fixed point.* To analyze the motion of a point B of a body rotating about a fixed point O you may have to take some or all of the following steps.

1. *Determine the position vector* r connecting the fixed point O to point B.

2. *Determine the angular velocity* $\boldsymbol{\omega}$ *of the body* with respect to a fixed frame of reference. The angular velocity $\boldsymbol{\omega}$ will often be obtained by adding two component angular velocities $\boldsymbol{\omega}_1$ and $\boldsymbol{\omega}_2$ [Sample Prob. 15.11].

3. *Compute the velocity of B* by using the equation

$$\mathbf{v} = \boldsymbol{\omega} \times \mathbf{r} \tag{15.37}$$

Your computation will usually be facilitated if you express the vector product as a determinant.

4. *Determine the angular acceleration* $\boldsymbol{\alpha}$ *of the body.* The angular acceleration $\boldsymbol{\alpha}$ represents the rate of change $(\dot{\boldsymbol{\omega}})_{OXYZ}$ of the vector $\boldsymbol{\omega}$ *with respect to a fixed frame of reference* $OXYZ$ and reflects both *a change in magnitude and a change in direction* of the angular velocity. However, when computing $\boldsymbol{\alpha}$ you may find it convenient to first compute the rate of change $(\dot{\boldsymbol{\omega}})_{Oxyz}$ of $\boldsymbol{\omega}$ with respect to a rotating frame of reference $Oxyz$ of your choice and use Eq. (15.31) of the preceding lesson to obtain $\boldsymbol{\alpha}$. You will write

$$\boldsymbol{\alpha} = (\dot{\boldsymbol{\omega}})_{OXYZ} = (\dot{\boldsymbol{\omega}})_{Oxyz} + \boldsymbol{\Omega} \times \boldsymbol{\omega}$$

where $\boldsymbol{\Omega}$ is the angular velocity of the rotating frame $Oxyz$ [Sample Prob. 15.11].

5. *Compute the acceleration of B* by using the equation

$$\mathbf{a} = \boldsymbol{\alpha} \times \mathbf{r} + \boldsymbol{\omega} \times (\boldsymbol{\omega} \times \mathbf{r}) \tag{15.38}$$

Note that the vector product $(\boldsymbol{\omega} \times \mathbf{r})$ represents the velocity of point B and was computed in step 3. Also, the computation of the first vector product in (15.38) will be facilitated if you express this product in determinant form. Remember that, as was the case with the plane motion of a rigid body, the instantaneous axis of rotation *cannot* be used to determine accelerations.

B. *General motion of a rigid body.* The general motion of a rigid body may be considered as *the sum of a translation and a rotation*. Keep the following in mind:

a. *In the translation part of the motion,* all the points of the body have the *same velocity* $\mathbf{v}_A$ *and the same acceleration* $\mathbf{a}_A$ as the point A of the body that has been selected as the reference point.

b. *In the rotation part of the motion,* the same reference point A is assumed to be a *fixed point*.

1. *To determine the velocity of a point B* of the rigid body when you know the velocity $\mathbf{v}_A$ of the reference point A and the angular velocity $\boldsymbol{\omega}$ of the body, you simply add $\mathbf{v}_A$ to the velocity $\mathbf{v}_{B/A} = \boldsymbol{\omega} \times \mathbf{r}_{B/A}$ of B in its rotation about A:

$$\mathbf{v}_B = \mathbf{v}_A + \boldsymbol{\omega} \times \mathbf{r}_{B/A} \tag{15.43}$$

As indicated earlier, the computation of the vector product will usually be facilitated if you express this product in determinant form.

Equation (15.43) can also be used to determine the magnitude of $\mathbf{v}_B$ when its direction is known, even if $\boldsymbol{\omega}$ is not known. While the corresponding three scalar equations are linearly dependent and the components of $\boldsymbol{\omega}$ are indeterminate, these components can be eliminated and $\mathbf{v}_A$ can be found by using an appropriate linear combination of the three equations [Sample Prob. 15.12, part *a*]. Alternatively, you can assign an arbitrary value to one of the components of $\boldsymbol{\omega}$ and solve the equations for $\mathbf{v}_A$. However, an additional equation must be sought in order to determine the true values of the components of $\boldsymbol{\omega}$ [Sample Prob. 15.12, part *b*].

2. *To determine the acceleration of a point B* of the rigid body when you know the acceleration $\mathbf{a}_A$ of the reference point A and the angular acceleration $\boldsymbol{\alpha}$ of the body, you simply add $\mathbf{a}_A$ to the acceleration of B in its rotation about A, as expressed by Eq. (15.38):

$$\mathbf{a}_B = \mathbf{a}_A + \boldsymbol{\alpha} \times \mathbf{r}_{B/A} + \boldsymbol{\omega} \times (\boldsymbol{\omega} \times \mathbf{r}_{B/A}) \tag{15.44}$$

Note that the vector product $(\boldsymbol{\omega} \times \mathbf{r}_{B/A})$ represents the velocity $\mathbf{v}_{B/A}$ of B relative to A and may already have been computed as part of your calculation of $\mathbf{v}_B$. We also remind you that the computation of the other two vector products will be facilitated if you express these products in determinant form.

The three scalar equations associated with Eq. (15.44) can also be used to determine the magnitude of $\mathbf{a}_B$ when its direction is known, even if $\boldsymbol{\omega}$ and $\boldsymbol{\alpha}$ are not known. While the components of $\boldsymbol{\omega}$ and $\boldsymbol{\alpha}$ are indeterminate, you can assign arbitrary values to one of the components of $\boldsymbol{\omega}$ and to one of the components of $\boldsymbol{\alpha}$ and solve the equations for $\mathbf{a}_B$.

Problems

15.181 The bowling ball shown rolls without slipping on the horizontal xz plane with an angular velocity $\boldsymbol{\omega} = \omega_x\mathbf{i} + \omega_y\mathbf{j} + \omega_z\mathbf{k}$. Knowing that $\mathbf{v}_A = (4.8\text{ m/s})\mathbf{i} - (4.8\text{ m/s})\mathbf{j} + (3.6\text{ m/s})\mathbf{k}$ and $\mathbf{v}_D = (9.6\text{ m/s})\mathbf{i} + (7.2\text{ m/s})\mathbf{k}$, determine (*a*) the angular velocity of the bowling ball, (*b*) the velocity of its center C.

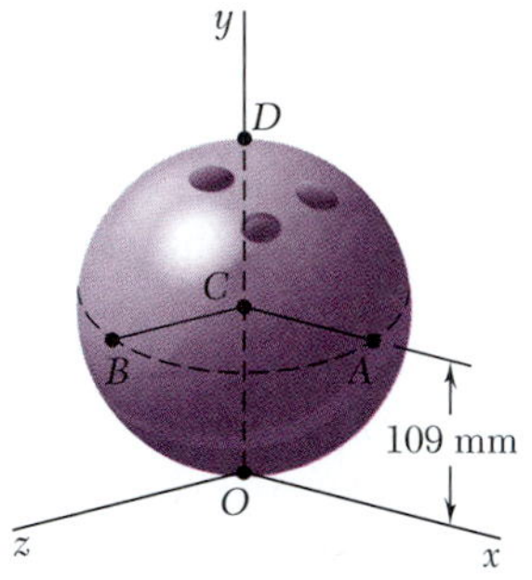

Fig. P15.181 and P15.182

15.182 The bowling ball shown rolls without slipping on the horizontal xz plane with an angular velocity $\boldsymbol{\omega} = \omega_x\mathbf{i} + \omega_y\mathbf{j} + \omega_z\mathbf{k}$. Knowing that $\mathbf{v}_B = (3.6\text{ m/s})\mathbf{i} - (4.8\text{ m/s})\mathbf{j} + (4.8\text{ m/s})\mathbf{k}$ and $\mathbf{v}_D = (7.2\text{ m/s})\mathbf{i} + (9.6\text{ m/s})\mathbf{k}$, determine (*a*) the angular velocity of the bowling ball, (*b*) the velocity of its center C.

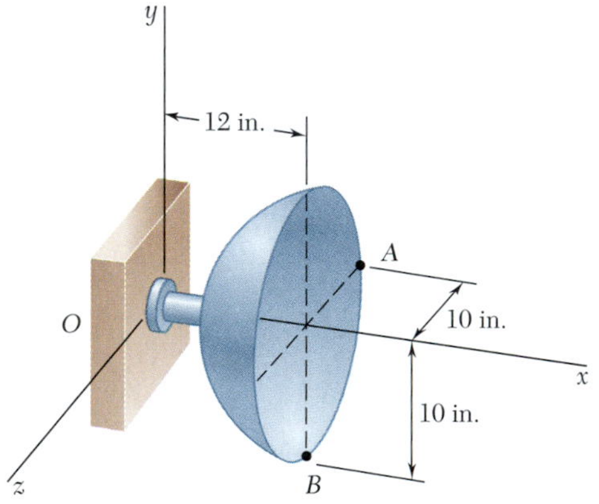

Fig. P15.183

15.183 At the instant considered the radar antenna shown rotates about the origin of coordinates with an angular velocity $\boldsymbol{\omega} = \omega_x\mathbf{i} + \omega_y\mathbf{j} + \omega_z\mathbf{k}$. Knowing that $(v_A)_y = 15$ in./s, $(v_B)_y = 9$ in./s, and $(v_B)_z = 18$ in./s, determine (*a*) the angular velocity of the antenna, (*b*) the velocity of point A.

15.184 The blade assembly of an oscillating fan rotates with a constant angular velocity $\boldsymbol{\omega}_1 = -(450\text{ rpm})\mathbf{i}$ with respect to the motor housing. Determine the angular acceleration of the blade assembly, knowing that at the instant shown the angular velocity and the angular acceleration of the motor housing are, respectively, $\boldsymbol{\omega}_2 = -(3\text{ rpm})\mathbf{j}$ and $\alpha_2 = 0$.

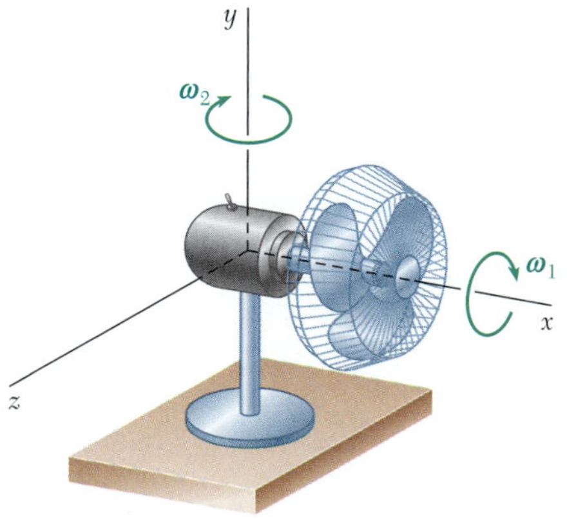

Fig. P15.184

15.185 Gear *A* is constrained to roll on the fixed gear *B* but is free to rotate about axle *AD*. Axle *AD* is connected by a clevis to the vertical shaft *DE* which rotates as shown with a constant angular velocity $\boldsymbol{\omega}_1$. Determine (*a*) the angular velocity of gear *A*, (*b*) the angular acceleration of gear *A*.

15.186 Gear *A* is constrained to roll on gear *B* but is free to rotate about axle *AD*. Axle *AD* is connected by a clevis to the vertical shaft *DE* which rotates as shown with a constant angular velocity $\boldsymbol{\omega}_1$. Knowing that gear *B* rotates with a constant angular velocity $\boldsymbol{\omega}_2$, determine (*a*) the angular velocity of gear *A*, (*b*) the angular acceleration of gear *A*.

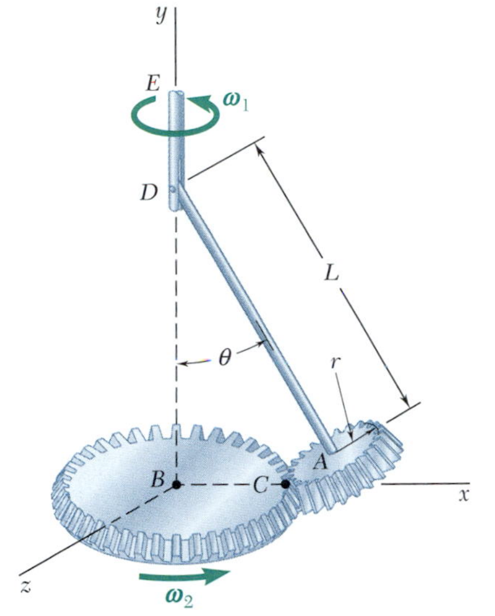

Fig. ***P15.185*** **and** ***P15.186***

15.187 The L-shaped arm *BCD* rotates about the *z* axis with a constant angular velocity $\boldsymbol{\omega}_1$ of 5 rad/s. Knowing that the 7.5-in.-radius disk rotates about *BC* with a constant angular velocity $\boldsymbol{\omega}_2$ of 4 rad/s, determine the angular acceleration of the disk.

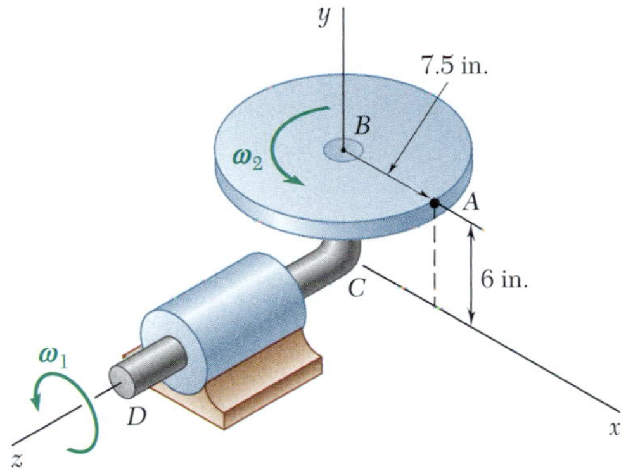

Fig. P15.187

15.188 In Prob. 15.187, determine (*a*) the velocity of point *A*, (*b*) the acceleration of point *A*.

15.189 A gun barrel of length $OP = 12$ ft is mounted on a turret as shown. To keep the gun aimed at a moving target the azimuth angle β is being increased at the rate $d\beta/dt = 30°/\text{s}$ and the elevation angle γ is being increased at the rate $d\gamma/dt = 10°/\text{s}$. For the position $\beta = 90°$ and $\gamma = 30°$, determine (*a*) the angular velocity of the barrel, (*b*) the angular acceleration of the barrel, (*c*) the velocity and acceleration of point *P*.

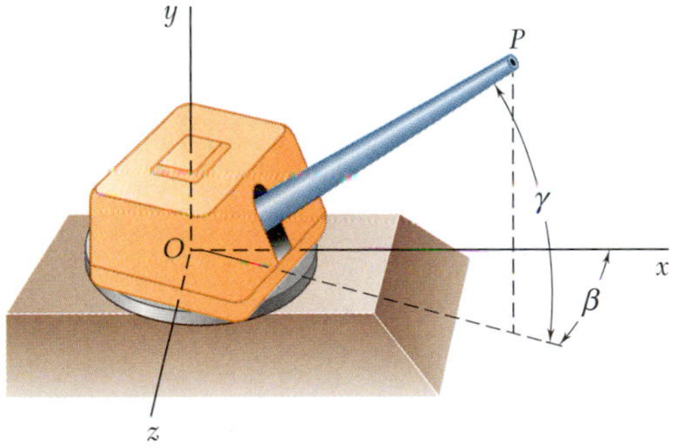

Fig. P15.189

15.190 A 60-mm-radius disk spins at the constant rate $\omega_2 = 4$ rad/s about an axis held by a housing attached to a horizontal rod that rotates at the constant rate $\omega_1 = 5$ rad/s. For the position shown, determine (*a*) the angular acceleration of the disk, (*b*) the acceleration of point *P* on the rim of the disk if $\theta = 0$, (*c*) the acceleration of point *P* on the rim of the disk if $\theta = 90°$.

15.191 A 60-mm-radius disk spins at the constant rate $\omega_2 = 4$ rad/s about an axis held by a housing attached to a horizontal rod that rotates at the constant rate $\omega_1 = 5$ rad/s. Knowing that $\theta = 30°$, determine the acceleration of point *P* on the rim of the disk.

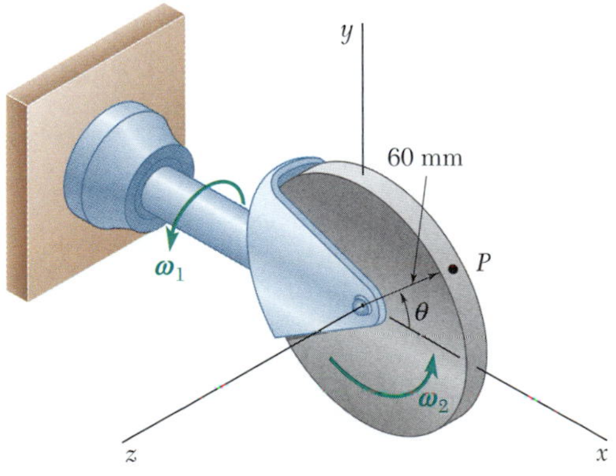

Fig. P15.190 and P15.191

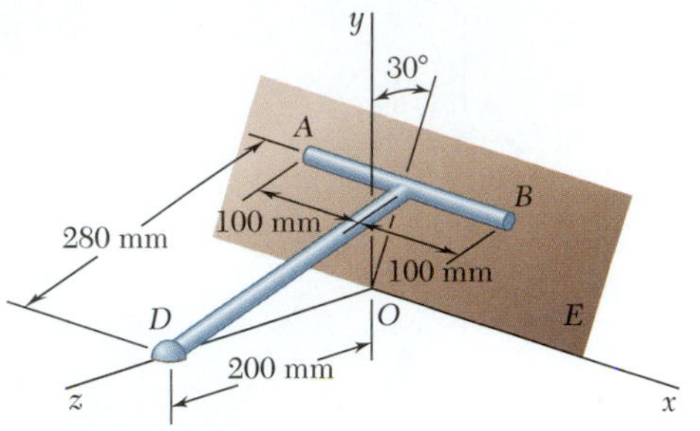

Fig. P15.192

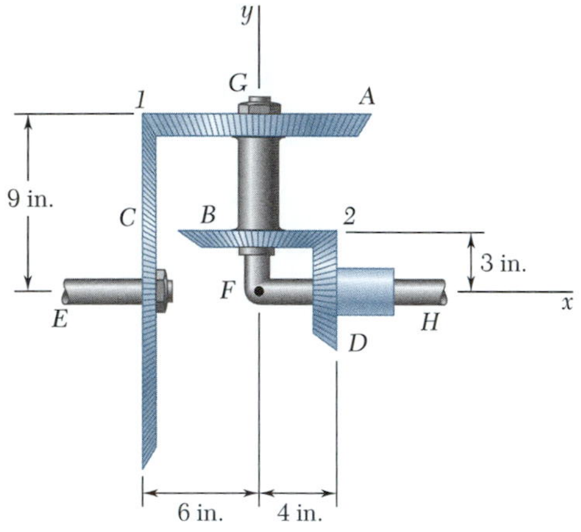

Fig. *P15.194*

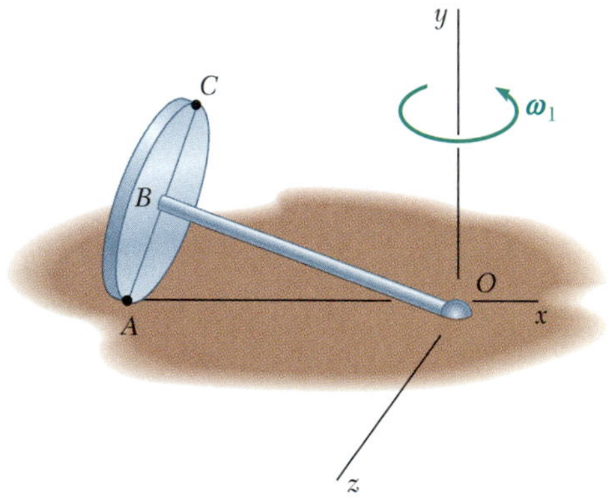

Fig. *P15.195*

15.192 Two rods are welded together to form the assembly shown that is attached to a fixed ball-and-socket joint at D. Rod AB moves on the inclined plane E that is perpendicular to the yz plane. Knowing that at the instant shown the speed of point B is 100 mm/s and $\omega_y < 0$, determine (*a*) the angular velocity of the assembly, (*b*) the velocity of point A.

15.193 In Prob. 15.192 the speed of point B is known to be increasing at the rate of 200 mm/s^2. For the position shown, determine (*a*) the angular acceleration of the assembly, (*b*) the radius of curvature of the path of point B.

15.194 In the planetary gear system shown, gears A and B are rigidly connected to each other and rotate as a unit about shaft FG. Gears C and D rotate with constant angular velocities of 15 rad/s and 30 rad/s, respectively, both counterclockwise when viewed from the right. Choosing the z axis pointing out of the plane of the figure, determine (*a*) the common angular velocity of gears A and B, (*b*) the common angular acceleration of gears A and B, (*c*) the acceleration of the tooth of gear B that is in contact with gear D at point 2.

15.195 A 1.5-in.-radius wheel is mounted on an axle OB of length 5 in. The wheel rolls without sliding on the horizontal floor, and the axle is perpendicular to the plane of the wheel. Knowing that the system rotates about the y axis at a constant rate $\omega_1 = 2.4$ rad/s, determine (*a*) the angular velocity of the wheel, (*b*) the angular acceleration of the wheel, (*c*) the acceleration of point C located at the highest point on the rim of the wheel.

15.196 Rod AB is connected by ball-and-socket joints to collar A and to the 320-mm-diameter disk C. Knowing that disk C rotates counterclockwise at the constant rate $\omega_0 = 3$ rad/s in the zx plane, determine the velocity of collar A for the position shown.

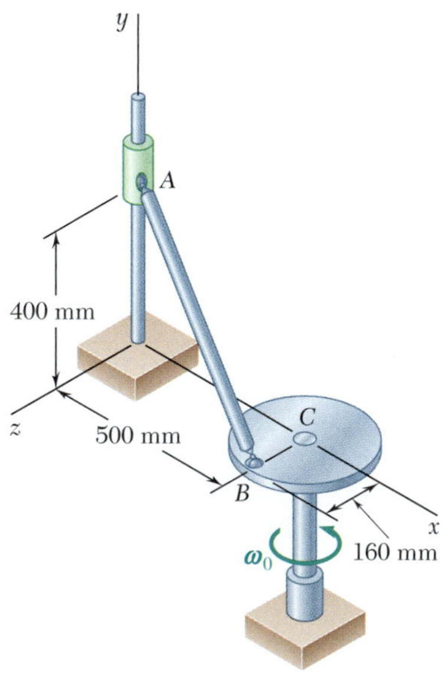

Fig. P15.196

15.197 Rod AB of length 580 mm is connected by ball-and-socket joints to the rotating crank BC and to the collar A. Crank BC is of length 160 mm and rotates in the horizontal xz plane at the constant rate $\omega_0 = 10$ rad/s. At the instant shown, when crank BC is parallel to the z axis, determine the velocity of collar A.

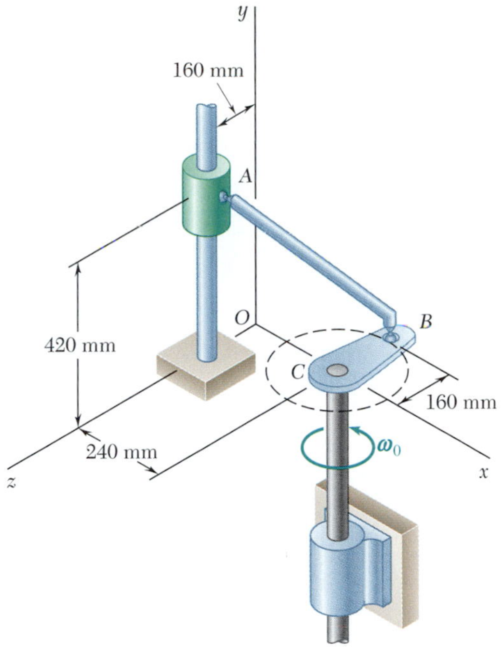

Fig. P15.197

15.198 Rod AB of length 13 in. is connected by ball-and-socket joints to collars A and B, which slide along the two rods shown. Knowing that collar B moves toward point D at a constant speed of 36 in./s, determine the velocity of collar A when $b = 4$ in.

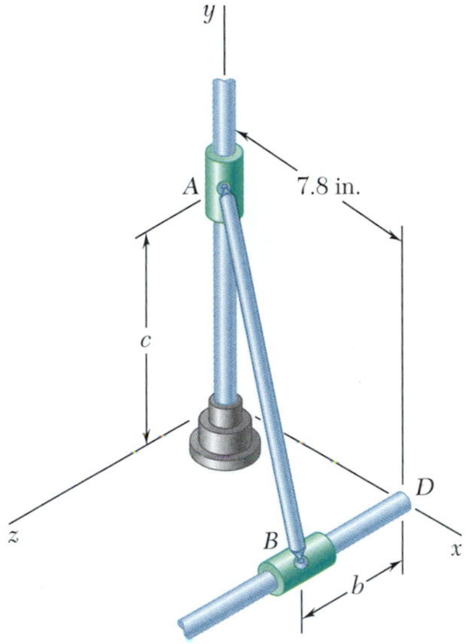

Fig. P15.198 and P15.199

15.199 Rod AB of length 13 in. is connected by ball-and-socket joints to collars A and B, which slide along the two rods shown. Knowing that collar B moves away from point D at a constant speed of 64 in./s, determine the velocity of collar A when $b = 6.24$ in.

15.200 Rod AB of length 500 mm is connected by ball-and-socket joints to collars A and B, which slide along the two rods shown. Knowing that collar B moves toward point E at a constant speed of 200 mm/s, determine the velocity of collar A as collar B passes through point D.

15.201 Rod AB of length 500 mm is connected by ball-and-socket joints to collars A and B, which slide along the two rods shown. Knowing that collar B moves toward point E at a constant speed of 200 mm/s, determine the velocity of collar A as collar B passes through point C.

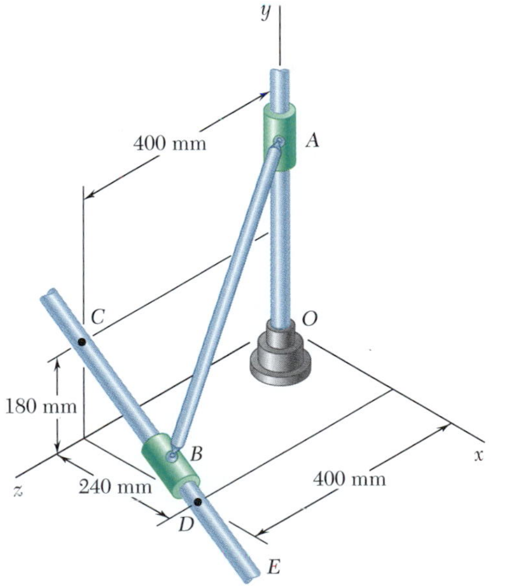

Fig. P15.200 and P15.201

15.202 Rod AB of length 15 in. is connected by ball-and-socket joints to collars A and B, which slide along the two rods shown. Knowing that collar B moves toward point D at a constant speed of 2.5 in./s, determine the velocity of collar A when $c = 4$ in.

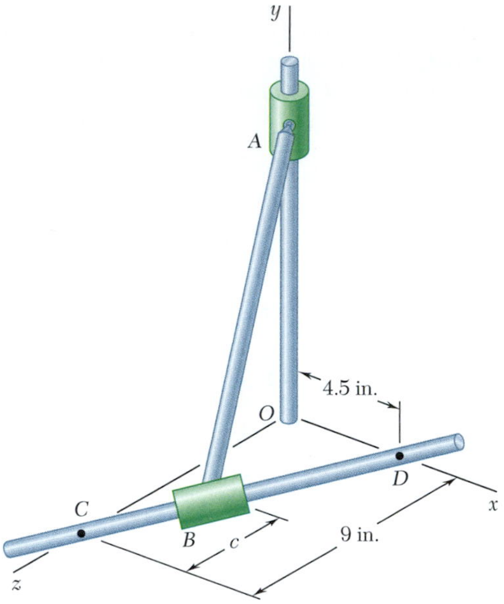

Fig. P15.202 and P15.203

15.203 Rod AB of length 15 in. is connected by ball-and-socket joints to collars A and B, which slide along the two rods shown. Knowing that collar B moves toward point D at a constant speed of 2.5 in./s, determine the velocity of collar A when $c = 6$ in.

15.204 Two shafts AC and EG, which lie in the vertical yz plane, are connected by a universal joint at D. Shaft AC rotates with a constant angular velocity $\boldsymbol{\omega}_1$ as shown. At a time when the arm of the crosspiece attached to shaft AC is vertical, determine the angular velocity of shaft EG.

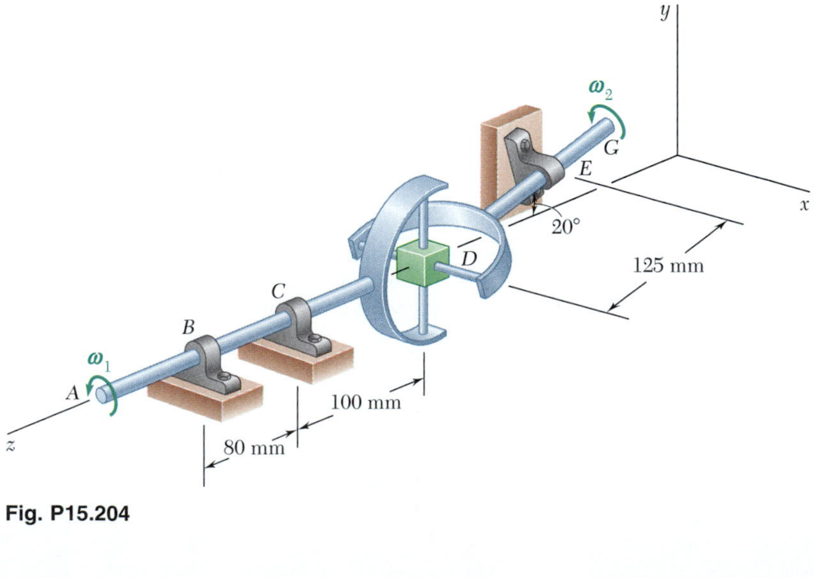

Fig. P15.204

15.205 Solve Prob. 15.204 assuming that the arm of the crosspiece attached to shaft AC is horizontal.

15.206 Rod AB of length 275 mm is connected by a ball-and-socket joint to collar A and by a clevis connection to collar B. Knowing that collar B moves down at a constant speed of 1.35 m/s, determine at the instant shown (a) the angular velocity of the rod, (b) the velocity of collar A.

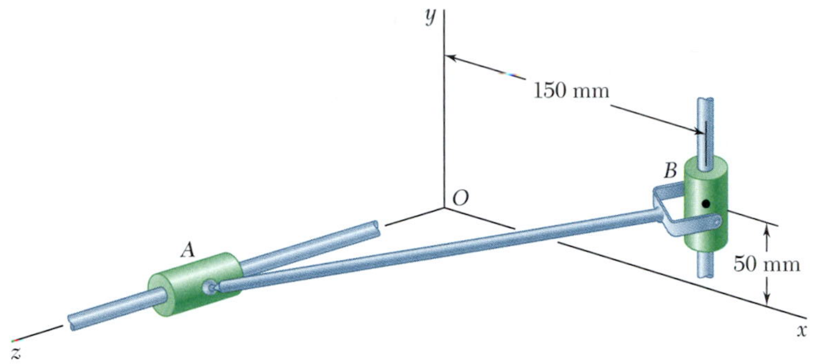

Fig. P15.206

15.207 Rod BC of length 840 mm is connected by a ball-and-socket joint to collar B and by a clevis connection to collar C. Knowing that collar B moves toward A at a constant speed of 390 mm/s, determine at the instant shown (a) the angular velocity of the rod, (b) the velocity of collar C.

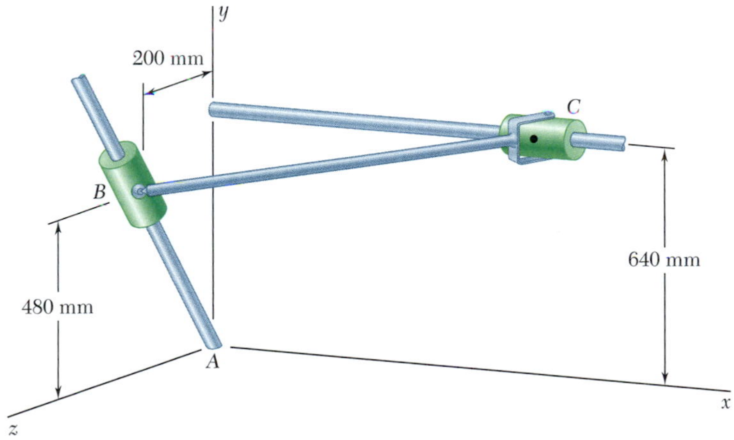

Fig. *P15.207*

15.208 through 15.213 For the mechanism of the problem indicated, determine the acceleration of collar A.

15.208 Mechanism of Prob. 15.202.
15.209 Mechanism of Prob. 15.203.
15.210 Mechanism of Prob. 15.198.
15.211 Mechanism of Prob. 15.199.
15.212 Mechanism of Prob. 15.196.
15.213 Mechanism of Prob. 15.197.

*15.14. THREE-DIMENSIONAL MOTION OF A PARTICLE RELATIVE TO A ROTATING FRAME. CORIOLIS ACCELERATION

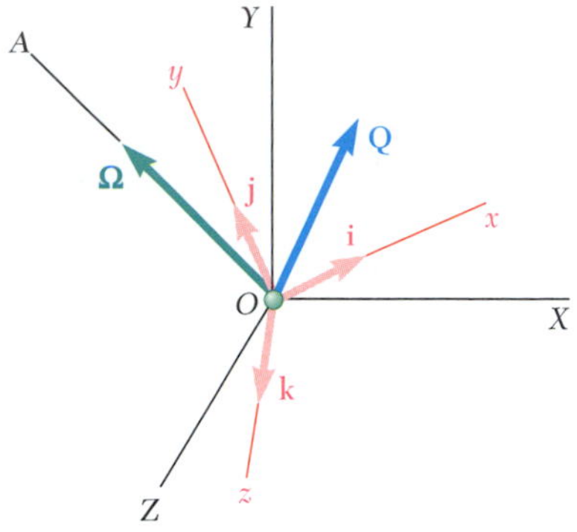

Fig. 15.36

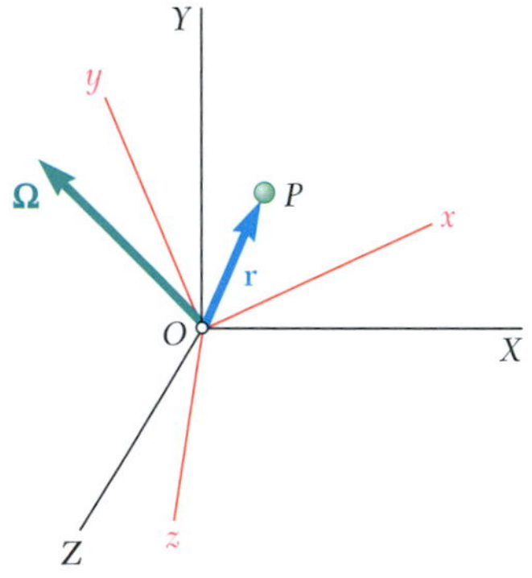

Fig. 15.37

We saw in Sec. 15.10 that given a vector function $\mathbf{Q}(t)$ and two frames of reference centered at O—a fixed frame $OXYZ$ and a rotating frame $Oxyz$—the rates of change of $\mathbf{Q}$ with respect to the two frames satisfy the relation

$$(\dot{\mathbf{Q}})_{OXYZ} = (\dot{\mathbf{Q}})_{Oxyz} + \boldsymbol{\Omega} \times \mathbf{Q} \qquad (15.31)$$

We had assumed at the time that the frame $Oxyz$ was constrained to rotate about a fixed axis OA. However, the derivation given in Sec. 15.10 remains valid when the frame $Oxyz$ is constrained only to have a fixed point O. Under this more general assumption, the axis OA represents the *instantaneous* axis of rotation of the frame $Oxyz$ (Sec. 15.12) and the vector $\boldsymbol{\Omega}$, its angular velocity at the instant considered (Fig. 15.36).

Let us now consider the three-dimensional motion of a particle P relative to a rotating frame $Oxyz$ constrained to have a fixed origin O. Let $\mathbf{r}$ be the position vector of P at a given instant and $\boldsymbol{\Omega}$ be the angular velocity of the frame $Oxyz$ with respect to the fixed frame $OXYZ$ at the same instant (Fig. 15.37). The derivations given in Sec. 15.11 for the two-dimensional motion of a particle can be readily extended to the three-dimensional case, and the absolute velocity $\mathbf{v}_P$ of P (that is, its velocity with respect to the fixed frame $OXYZ$) can be expressed as

$$\mathbf{v}_P = \boldsymbol{\Omega} \times \mathbf{r} + (\dot{\mathbf{r}})_{Oxyz} \qquad (15.45)$$

Denoting by $\mathcal{F}$ the rotating frame $Oxyz$, we write this relation in the alternative form

$$\mathbf{v}_P = \mathbf{v}_{P'} + \mathbf{v}_{P/\mathcal{F}} \qquad (15.46)$$

where $\mathbf{v}_P$ = absolute velocity of particle P
$\mathbf{v}_{P'}$ = velocity of point P' of moving frame $\mathcal{F}$ coinciding with P
$\mathbf{v}_{P/\mathcal{F}}$ = velocity of P relative to moving frame $\mathcal{F}$

The absolute acceleration $\mathbf{a}_P$ of P can be expressed as

$$\mathbf{a}_P = \dot{\boldsymbol{\Omega}} \times \mathbf{r} + \boldsymbol{\Omega} \times (\boldsymbol{\Omega} \times \mathbf{r}) + 2\boldsymbol{\Omega} \times (\dot{\mathbf{r}})_{Oxyz} + (\ddot{\mathbf{r}})_{Oxyz} \qquad (15.47)$$

An alternative form is

$$\mathbf{a}_P = \mathbf{a}_{P'} + \mathbf{a}_{P/\mathcal{F}} + \mathbf{a}_c \qquad (15.48)$$

where $\mathbf{a}_P$ = absolute acceleration of particle P
$\mathbf{a}_{P'}$ = acceleration of point P' of moving frame $\mathcal{F}$ coinciding with P

$\mathbf{a}_{P/\mathcal{F}}$ = acceleration of P relative to moving frame $\mathcal{F}$
$\mathbf{a}_c = 2\boldsymbol{\Omega} \times (\dot{\mathbf{r}})_{Oxyz} = 2\boldsymbol{\Omega} \times \mathbf{v}_{P/\mathcal{F}}$
= complementary, or Coriolis, acceleration†

We note that the Coriolis acceleration is perpendicular to the vectors $\boldsymbol{\Omega}$ and $\mathbf{v}_{P/\mathcal{F}}$. However, since these vectors are usually not perpendicular to each other, the magnitude of $\mathbf{a}_c$ is in general *not* equal to $2\Omega v_{P/\mathcal{F}}$, as was the case for the plane motion of a particle. We further note that the Coriolis acceleration reduces to zero when the vectors $\boldsymbol{\Omega}$ and $\mathbf{v}_{P/\mathcal{F}}$ are parallel, or when either of them is zero.

Rotating frames of reference are particularly useful in the study of the three-dimensional motion of rigid bodies. If a rigid body has a fixed point O, as was the case for the crane of Sample Prob. 15.11, we can use a frame $Oxyz$ which is neither fixed nor rigidly attached to the rigid body. Denoting by $\boldsymbol{\Omega}$ the angular velocity of the frame $Oxyz$, we then resolve the angular velocity $\boldsymbol{\omega}$ of the body into the components $\boldsymbol{\Omega}$ and $\boldsymbol{\omega}_{B/\mathcal{F}}$, where the second component represents the angular velocity of the body relative to the frame $Oxyz$ (see Sample Prob. 15.14). An appropriate choice of the rotating frame often leads to a simpler analysis of the motion of the rigid body than would be possible with axes of fixed orientation. This is especially true in the case of the general three-dimensional motion of a rigid body, that is, when the rigid body under consideration has no fixed point (see Sample Prob. 15.15).

*15.15. FRAME OF REFERENCE IN GENERAL MOTION

Consider a fixed frame of reference $OXYZ$ and a frame $Axyz$ which moves in a known, but arbitrary, fashion with respect to $OXYZ$ (Fig. 15.38). Let P be a particle moving in space. The position of P is defined at any instant by the vector $\mathbf{r}_P$ in the fixed frame, and by the vector $\mathbf{r}_{P/A}$ in the moving frame. Denoting by $\mathbf{r}_A$ the position vector of A in the fixed frame, we have

$$\mathbf{r}_P = \mathbf{r}_A + \mathbf{r}_{P/A} \qquad (15.49)$$

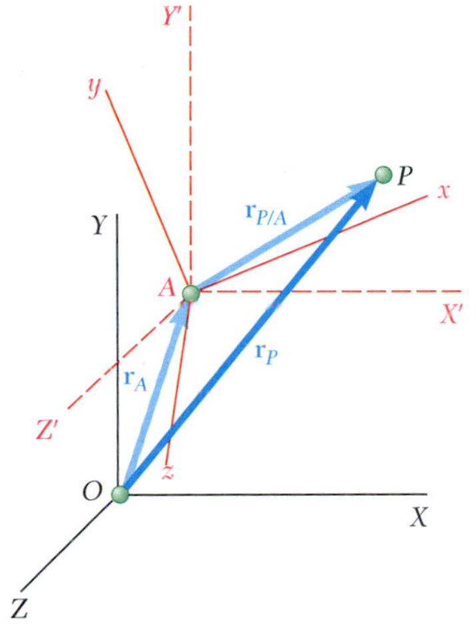

Fig. 15.38

The absolute velocity $\mathbf{v}_P$ of the particle is obtained by writing

$$\mathbf{v}_P = \dot{\mathbf{r}}_P = \dot{\mathbf{r}}_A + \dot{\mathbf{r}}_{P/A} \qquad (15.50)$$

where the derivatives are defined with respect to the fixed frame $OXYZ$. The first term in the right-hand member of (15.50) thus represents the velocity $\mathbf{v}_A$ of the origin A of the moving axes. On the other hand, since the rate of change of a vector is the same with respect to a fixed frame and with respect to a frame in translation (Sec. 11.10), the second term can be regarded as the velocity $\mathbf{v}_{P/A}$ of P relative to the frame $AX'Y'Z'$ of the same orientation as $OXYZ$ and the same origin as $Axyz$. We therefore have

$$\mathbf{v}_P = \mathbf{v}_A + \mathbf{v}_{P/A} \qquad (15.51)$$

†It is important to note the difference between Eq. (15.48) and Eq. (15.21) of Sec. 15.8. See the footnote on page 974.

But the velocity $\mathbf{v}_{P/A}$ of P relative to $AX'Y'Z'$ can be obtained from (15.45) by substituting $\mathbf{r}_{P/A}$ for $\mathbf{r}$ in that equation. We write

$$\mathbf{v}_P = \mathbf{v}_A + \boldsymbol{\Omega} \times \mathbf{r}_{P/A} + (\dot{\mathbf{r}}_{P/A})_{Axyz} \tag{15.52}$$

where $\boldsymbol{\Omega}$ is the angular velocity of the frame $Axyz$ at the instant considered.

The absolute acceleration $\mathbf{a}_P$ of the particle is obtained by differentiating (15.51) and writing

$$\mathbf{a}_P = \dot{\mathbf{v}}_P = \dot{\mathbf{v}}_A + \dot{\mathbf{v}}_{P/A} \tag{15.53}$$

where the derivatives are defined with respect to either of the frames $OXYZ$ or $AX'Y'Z'$. Thus, the first term in the right-hand member of (15.53) represents the acceleration $\mathbf{a}_A$ of the origin A of the moving axes and the second term represents the acceleration $\mathbf{a}_{P/A}$ of P relative to the frame $AX'Y'Z'$. This acceleration can be obtained from (15.47) by substituting $\mathbf{r}_{P/A}$ for $\mathbf{r}$. We therefore write

$$\begin{aligned}\mathbf{a}_P = \mathbf{a}_A + \dot{\boldsymbol{\Omega}} \times \mathbf{r}_{P/A} + \boldsymbol{\Omega} \times (\boldsymbol{\Omega} \times \mathbf{r}_{P/A}) \\ + 2\boldsymbol{\Omega} \times (\dot{\mathbf{r}}_{P/A})_{Axyz} + (\ddot{\mathbf{r}}_{P/A})_{Axyz}\end{aligned} \tag{15.54}$$

Photo 15.5 The motion of air particles in a hurricane can be considered as motion relative to a frame of reference attached to the Earth and rotating with it.

Formulas (15.52) and (15.54) make it possible to determine the velocity and acceleration of a given particle with respect to a fixed frame of reference, when the motion of the particle is known with respect to a moving frame. These formulas become more significant, and considerably easier to remember, if we note that the sum of the first two terms in (15.52) represents the velocity of the point P' of the moving frame which coincides with P at the instant considered, and that the sum of the first three terms in (15.54) represents the acceleration of the same point. Thus, the relations (15.46) and (15.48) of the preceding section are still valid in the case of a reference frame in general motion, and we can write

$$\mathbf{v}_P = \mathbf{v}_{P'} + \mathbf{v}_{P/\mathcal{F}} \tag{15.46}$$
$$\mathbf{a}_P = \mathbf{a}_{P'} + \mathbf{a}_{P/\mathcal{F}} + \mathbf{a}_c \tag{15.48}$$

where the various vectors involved have been defined in Sec. 15.14.

It should be noted that if the moving reference frame $\mathcal{F}$ (or $Axyz$) is in translation, the velocity and acceleration of the point P' of the frame which coincides with P become, respectively, equal to the velocity and acceleration of the origin A of the frame. On the other hand, since the frame maintains a fixed orientation, $\mathbf{a}_c$ is zero, and the relations (15.46) and (15.48) reduce, respectively, to the relations (11.33) and (11.34) derived in Sec. 11.12.

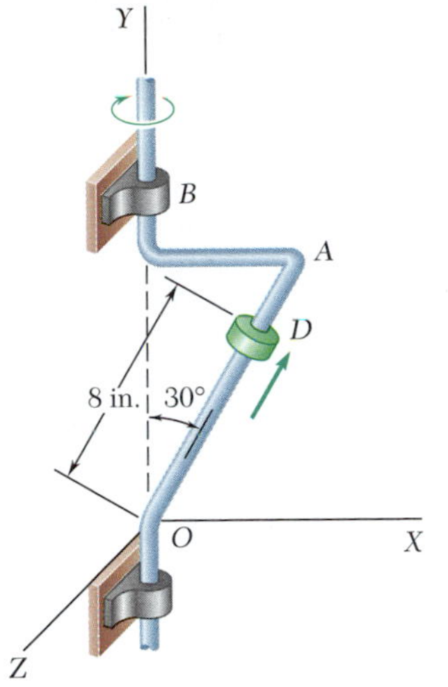

SAMPLE PROBLEM 15.13

The bent rod OAB rotates about the vertical OB. At the instant considered, its angular velocity and angular acceleration are, respectively, 20 rad/s and 200 rad/s^2, both clockwise when viewed from the positive Y axis. The collar D moves along the rod, and at the instant considered, $OD = 8$ in. The velocity and acceleration of the collar relative to the rod are, respectively, 50 in./s and 600 in./s^2, both upward. Determine (*a*) the velocity of the collar, (*b*) the acceleration of the collar.

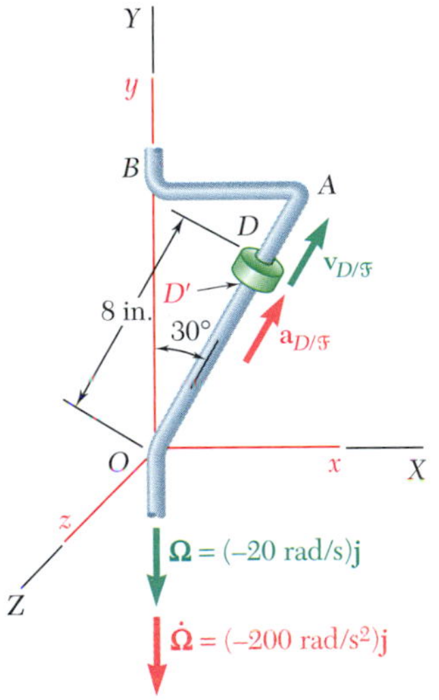

SOLUTION

Frames of Reference. The frame $OXYZ$ is fixed. We attach the rotating frame $Oxyz$ to the bent rod. Its angular velocity and angular acceleration relative to $OXYZ$ are therefore $\mathbf{\Omega} = (-20 \text{ rad/s})\mathbf{j}$ and $\dot{\mathbf{\Omega}} = (-200 \text{ rad/s}^2)\mathbf{j}$, respectively. The position vector of D is

$$\mathbf{r} = (8 \text{ in.})(\sin 30°\mathbf{i} + \cos 30°\mathbf{j}) = (4 \text{ in.})\mathbf{i} + (6.93 \text{ in.})\mathbf{j}$$

***a.* Velocity $\mathbf{v}_D$.** Denoting by D' the point of the rod which coincides with D and by $\mathcal{F}$ the rotating frame $Oxyz$, we write from Eq. (15.46)

$$\mathbf{v}_D = \mathbf{v}_{D'} + \mathbf{v}_{D/\mathcal{F}} \tag{1}$$

where

$$\mathbf{v}_{D'} = \mathbf{\Omega} \times \mathbf{r} = (-20 \text{ rad/s})\mathbf{j} \times [(4 \text{ in.})\mathbf{i} + (6.93 \text{ in.})\mathbf{j}] = (80 \text{ in./s})\mathbf{k}$$
$$\mathbf{v}_{D/\mathcal{F}} = (50 \text{ in./s})(\sin 30°\mathbf{i} + \cos 30°\mathbf{j}) = (25 \text{ in./s})\mathbf{i} + (43.3 \text{ in./s})\mathbf{j}$$

Substituting the values obtained for $\mathbf{v}_{D'}$ and $\mathbf{v}_{D/\mathcal{F}}$ into (1), we find

$$\mathbf{v}_D = (25 \text{ in./s})\mathbf{i} + (43.3 \text{ in./s})\mathbf{j} + (80 \text{ in./s})\mathbf{k} \quad ◀$$

***b.* Acceleration $\mathbf{a}_D$.** From Eq. (15.48) we write

$$\mathbf{a}_D = \mathbf{a}_{D'} + \mathbf{a}_{D/\mathcal{F}} + \mathbf{a}_c \tag{2}$$

where

$$\begin{aligned}
\mathbf{a}_{D'} &= \dot{\mathbf{\Omega}} \times \mathbf{r} + \mathbf{\Omega} \times (\mathbf{\Omega} \times \mathbf{r}) \\
&= (-200 \text{ rad/s}^2)\mathbf{j} \times [(4 \text{ in.})\mathbf{i} + (6.93 \text{ in.})\mathbf{j}] - (20 \text{ rad/s})\mathbf{j} \times (80 \text{ in./s})\mathbf{k} \\
&= +(800 \text{ in./s}^2)\mathbf{k} - (1600 \text{ in./s}^2)\mathbf{i} \\
\mathbf{a}_{D/\mathcal{F}} &= (600 \text{ in./s}^2)(\sin 30°\mathbf{i} + \cos 30°\mathbf{j}) = (300 \text{ in./s}^2)\mathbf{i} + (520 \text{ in./s}^2)\mathbf{j} \\
\mathbf{a}_c &= 2\mathbf{\Omega} \times \mathbf{v}_{D/\mathcal{F}} \\
&= 2(-20 \text{ rad/s})\mathbf{j} \times [(25 \text{ in./s})\mathbf{i} + (43.3 \text{ in./s})\mathbf{j}] = (1000 \text{ in./s}^2)\mathbf{k}
\end{aligned}$$

Substituting the values obtained for $\mathbf{a}_{D'}$, $\mathbf{a}_{D/\mathcal{F}}$, and $\mathbf{a}_c$ into (2),

$$\mathbf{a}_D = -(1300 \text{ in./s}^2)\mathbf{i} + (520 \text{ in./s}^2)\mathbf{j} + (1800 \text{ in./s}^2)\mathbf{k} \quad ◀$$

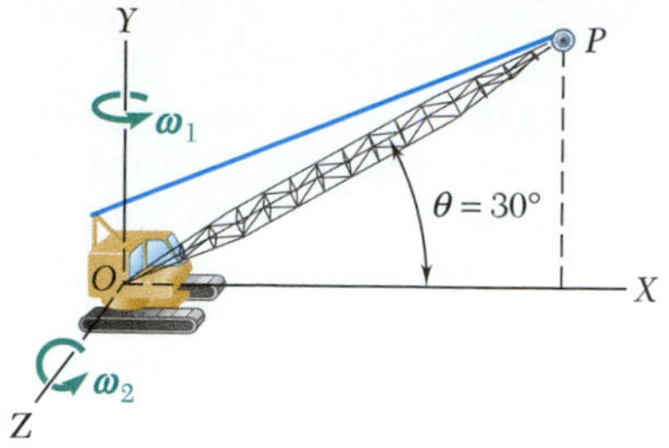

SAMPLE PROBLEM 15.14

The crane shown rotates with a constant angular velocity $\boldsymbol{\omega}_1$ of 0.30 rad/s. Simultaneously, the boom is being raised with a constant angular velocity $\boldsymbol{\omega}_2$ of 0.50 rad/s relative to the cab. Knowing that the length of the boom OP is $l = 12$ m, determine (*a*) the velocity of the tip of the boom, (*b*) the acceleration of the tip of the boom.

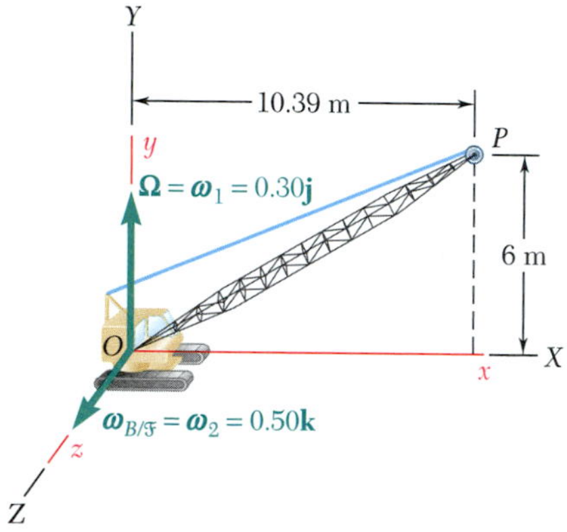

SOLUTION

Frames of Reference. The frame $OXYZ$ is fixed. We attach the rotating frame $Oxyz$ to the cab. Its angular velocity with respect to the frame $OXYZ$ is therefore $\boldsymbol{\Omega} = \boldsymbol{\omega}_1 = (0.30 \text{ rad/s})\mathbf{j}$. The angular velocity of the boom relative to the cab and the rotating frame $Oxyz$ (or $\mathcal{F}$, for short) is $\boldsymbol{\omega}_{B/\mathcal{F}} = \boldsymbol{\omega}_2 = (0.50 \text{ rad/s})\mathbf{k}$.

***a.* Velocity $\mathbf{v}_P$.** From Eq. (15.46) we write

$$\mathbf{v}_P = \mathbf{v}_{P'} + \mathbf{v}_{P/\mathcal{F}} \tag{1}$$

where $\mathbf{v}_{P'}$ is the velocity of the point P' of the rotating frame which coincides with P:

$$\mathbf{v}_{P'} = \boldsymbol{\Omega} \times \mathbf{r} = (0.30 \text{ rad/s})\mathbf{j} \times [(10.39 \text{ m})\mathbf{i} + (6 \text{ m})\mathbf{j}] = -(3.12 \text{ m/s})\mathbf{k}$$

and where $\mathbf{v}_{P/\mathcal{F}}$ is the velocity of P relative to the rotating frame $Oxyz$. But the angular velocity of the boom relative to $Oxyz$ was found to be $\boldsymbol{\omega}_{B/\mathcal{F}} = (0.50 \text{ rad/s})\mathbf{k}$. The velocity of its tip P relative to $Oxyz$ is therefore

$$\begin{aligned}\mathbf{v}_{P/\mathcal{F}} = \boldsymbol{\omega}_{B/\mathcal{F}} \times \mathbf{r} &= (0.50 \text{ rad/s})\mathbf{k} \times [(10.39 \text{ m})\mathbf{i} + (6 \text{ m})\mathbf{j}] \\ &= -(3 \text{ m/s})\mathbf{i} + (5.20 \text{ m/s})\mathbf{j}\end{aligned}$$

Substituting the values obtained for $\mathbf{v}_{P'}$ and $\mathbf{v}_{P/\mathcal{F}}$ into (1), we find

$$\mathbf{v}_P = -(3 \text{ m/s})\mathbf{i} + (5.20 \text{ m/s})\mathbf{j} - (3.12 \text{ m/s})\mathbf{k} \quad ◀$$

***b.* Acceleration $\mathbf{a}_P$.** From Eq. (15.48) we write

$$\mathbf{a}_P = \mathbf{a}_{P'} + \mathbf{a}_{P/\mathcal{F}} + \mathbf{a}_c \tag{2}$$

Since $\boldsymbol{\Omega}$ and $\boldsymbol{\omega}_{B/\mathcal{F}}$ are both constant, we have

$$\begin{aligned}\mathbf{a}_{P'} &= \boldsymbol{\Omega} \times (\boldsymbol{\Omega} \times \mathbf{r}) = (0.30 \text{ rad/s})\mathbf{j} \times (-3.12 \text{ m/s})\mathbf{k} = -(0.94 \text{ m/s}^2)\mathbf{i} \\ \mathbf{a}_{P/\mathcal{F}} &= \boldsymbol{\omega}_{B/\mathcal{F}} \times (\boldsymbol{\omega}_{B/\mathcal{F}} \times \mathbf{r}) \\ &= (0.50 \text{ rad/s})\mathbf{k} \times [-(3 \text{ m/s})\mathbf{i} + (5.20 \text{ m/s})\mathbf{j}] \\ &= -(1.50 \text{ m/s}^2)\mathbf{j} - (2.60 \text{ m/s}^2)\mathbf{i} \\ \mathbf{a}_c &= 2\boldsymbol{\Omega} \times \mathbf{v}_{P/\mathcal{F}} \\ &= 2(0.30 \text{ rad/s})\mathbf{j} \times [-(3 \text{ m/s})\mathbf{i} + (5.20 \text{ m/s})\mathbf{j}] = (1.80 \text{ m/s}^2)\mathbf{k}\end{aligned}$$

Substituting for $\mathbf{a}_{P'}$, $\mathbf{a}_{P/\mathcal{F}}$, and $\mathbf{a}_c$ into (2), we find

$$\mathbf{a}_P = -(3.54 \text{ m/s}^2)\mathbf{i} - (1.50 \text{ m/s}^2)\mathbf{j} + (1.80 \text{ m/s}^2)\mathbf{k} \quad ◀$$

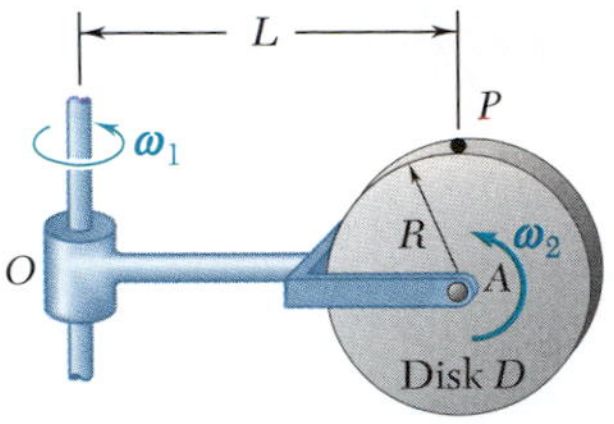

SAMPLE PROBLEM 15.15

Disk D, of radius R, is pinned to end A of the arm OA of length L located in the plane of the disk. The arm rotates about a vertical axis through O at the constant rate ω_1, and the disk rotates about A at the constant rate ω_2. Determine (*a*) the velocity of point P located directly above A, (*b*) the acceleration of P, (*c*) the angular velocity and angular acceleration of the disk.

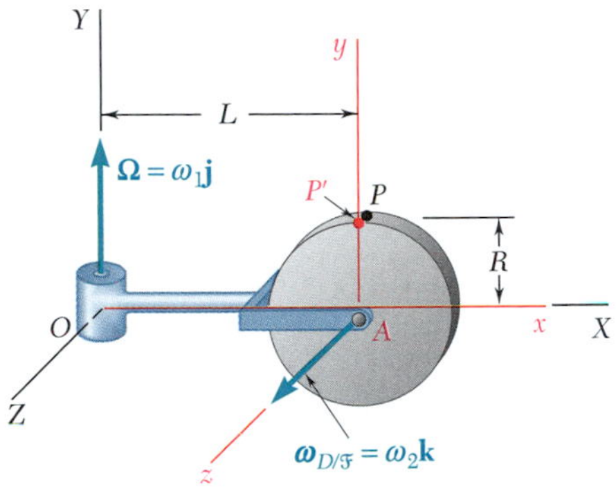

SOLUTION

Frames of Reference. The frame $OXYZ$ is fixed. We attach the moving frame $Axyz$ to the arm OA. Its angular velocity with respect to the frame $OXYZ$ is therefore $\boldsymbol{\Omega} = \omega_1\mathbf{j}$. The angular velocity of disk D relative to the moving frame $Axyz$ (or $\mathcal{F}$, for short) is $\boldsymbol{\omega}_{D/\mathcal{F}} = \omega_2\mathbf{k}$. The position vector of P relative to O is $\mathbf{r} = L\mathbf{i} + R\mathbf{j}$, and its position vector relative to A is $\mathbf{r}_{P/A} = R\mathbf{j}$.

***a.* Velocity $\mathbf{v}_P$.** Denoting by P' the point of the moving frame which coincides with P, we write from Eq. (15.46)

$$\mathbf{v}_P = \mathbf{v}_{P'} + \mathbf{v}_{P/\mathcal{F}} \tag{1}$$

where $\mathbf{v}_{P'} = \boldsymbol{\Omega} \times \mathbf{r} = \omega_1\mathbf{j} \times (L\mathbf{i} + R\mathbf{j}) = -\omega_1 L\mathbf{k}$

$$\mathbf{v}_{P/\mathcal{F}} = \boldsymbol{\omega}_{D/\mathcal{F}} \times \mathbf{r}_{P/A} = \omega_2\mathbf{k} \times R\mathbf{j} = -\omega_2 R\mathbf{i}$$

Substituting the values obtained for $\mathbf{v}_{P'}$ and $\mathbf{v}_{P/\mathcal{F}}$ into (1), we find

$$\mathbf{v}_P = -\omega_2 R\mathbf{i} - \omega_1 L\mathbf{k} \quad ◀$$

***b.* Acceleration $\mathbf{a}_P$.** From Eq. (15.48) we write

$$\mathbf{a}_P = \mathbf{a}_{P'} + \mathbf{a}_{P/\mathcal{F}} + \mathbf{a}_c \tag{2}$$

Since $\boldsymbol{\Omega}$ and $\boldsymbol{\omega}_{D/\mathcal{F}}$ are both constant, we have

$$\mathbf{a}_{P'} = \boldsymbol{\Omega} \times (\boldsymbol{\Omega} \times \mathbf{r}) = \omega_1\mathbf{j} \times (-\omega_1 L\mathbf{k}) = -\omega_1^2 L\mathbf{i}$$
$$\mathbf{a}_{P/\mathcal{F}} = \boldsymbol{\omega}_{D/\mathcal{F}} \times (\boldsymbol{\omega}_{D/\mathcal{F}} \times \mathbf{r}_{P/A}) = \omega_2\mathbf{k} \times (-\omega_2 R\mathbf{i}) = -\omega_2^2 R\mathbf{j}$$
$$\mathbf{a}_c = 2\boldsymbol{\Omega} \times \mathbf{v}_{P/\mathcal{F}} = 2\omega_1\mathbf{j} \times (-\omega_2 R\mathbf{i}) = 2\omega_1\omega_2 R\mathbf{k}$$

Substituting the values obtained into (2), we find

$$\mathbf{a}_P = -\omega_1^2 L\mathbf{i} - \omega_2^2 R\mathbf{j} + 2\omega_1\omega_2 R\mathbf{k} \quad ◀$$

***c.* Angular Velocity and Angular Acceleration of Disk.**

$$\boldsymbol{\omega} = \boldsymbol{\Omega} + \boldsymbol{\omega}_{D/\mathcal{F}} \qquad \boldsymbol{\omega} = \omega_1\mathbf{j} + \omega_2\mathbf{k} \quad ◀$$

Using Eq. (15.31) with $\mathbf{Q} = \boldsymbol{\omega}$, we write

$$\boldsymbol{\alpha} = (\dot{\boldsymbol{\omega}})_{OXYZ} = (\dot{\boldsymbol{\omega}})_{Axyz} + \boldsymbol{\Omega} \times \boldsymbol{\omega}$$
$$= 0 + \omega_1\mathbf{j} \times (\omega_1\mathbf{j} + \omega_2\mathbf{k})$$

$$\boldsymbol{\alpha} = \omega_1\omega_2\mathbf{i} \quad ◀$$

SOLVING PROBLEMS ON YOUR OWN

In this lesson you concluded your study of the kinematics of rigid bodies by learning how to use an auxiliary frame of reference $\mathcal{F}$ to analyze the three-dimensional motion of a rigid body. This auxiliary frame may be a *rotating frame* with a fixed origin O, or it may be a *frame in general motion.*

A. *Using a rotating frame of reference.* As you approach a problem involving the use of a rotating frame $\mathcal{F}$ you should take the following steps.

1. *Select the rotating frame $\mathcal{F}$ that you wish to use* and draw the corresponding coordinate axes x, y, and z from the fixed point O.

2. *Determine the angular velocity $\boldsymbol{\Omega}$ of the frame $\mathcal{F}$* with respect to a fixed frame $OXYZ$. In most cases, you will have selected a frame which is attached to some rotating element of the system; $\boldsymbol{\Omega}$ will then be the angular velocity of that element.

3. *Designate as P' the point of the rotating frame $\mathcal{F}$* that coincides with the point P of interest at the instant you are considering. Determine the velocity $\mathbf{v}_{P'}$ and the acceleration $\mathbf{a}_{P'}$ of point P'. Since P' is part of $\mathcal{F}$ and has the same position vector $\mathbf{r}$ as P, you will find that

$$\mathbf{v}_{P'} = \boldsymbol{\Omega} \times \mathbf{r} \qquad \text{and} \qquad \mathbf{a}_{P'} = \boldsymbol{\alpha} \times \mathbf{r} + \boldsymbol{\Omega} \times (\boldsymbol{\Omega} \times \mathbf{r})$$

where $\boldsymbol{\alpha}$ is the angular acceleration of $\mathcal{F}$. However, in many of the problems that you will encounter, the angular velocity of $\mathcal{F}$ is constant in both magnitude and direction, and $\boldsymbol{\alpha} = 0$.

4. *Determine the velocity and acceleration of point P with respect to the frame $\mathcal{F}$.* As you are trying to determine $\mathbf{v}_{P/\mathcal{F}}$ and $\mathbf{a}_{P/\mathcal{F}}$ you will find it useful to visualize the motion of P on frame $\mathcal{F}$ when the frame is not rotating. If P is a point of a rigid body $\mathcal{B}$ which has an angular velocity $\boldsymbol{\omega}_{\mathcal{B}}$ and an angular acceleration $\boldsymbol{\alpha}_{\mathcal{B}}$ relative to $\mathcal{F}$ [Sample Prob. 15.14], you will find that

$$\mathbf{v}_{P/\mathcal{F}} = \boldsymbol{\omega}_{\mathcal{B}} \times \mathbf{r} \qquad \text{and} \qquad \mathbf{a}_{P/\mathcal{F}} = \boldsymbol{\alpha}_{\mathcal{B}} \times \mathbf{r} + \boldsymbol{\omega}_{\mathcal{B}} \times (\boldsymbol{\omega}_{\mathcal{B}} \times \mathbf{r})$$

In many of the problems that you will encounter, the angular velocity of body $\mathcal{B}$ relative to frame $\mathcal{F}$ is constant in both magnitude and direction, and $\boldsymbol{\alpha}_{\mathcal{B}} = 0$.

5. *Determine the Coriolis acceleration.* Considering the angular velocity $\boldsymbol{\Omega}$ of frame $\mathcal{F}$ and the velocity $\mathbf{v}_{P/\mathcal{F}}$ of point P relative to that frame, which was computed in the previous step, you write

$$\mathbf{a}_c = 2\boldsymbol{\Omega} \times \mathbf{v}_{P/\mathcal{F}}$$

6. The velocity and the acceleration of P with respect to the fixed frame OXYZ can now be obtained by adding the expressions you have determined:

$$\mathbf{v}_P = \mathbf{v}_{P'} + \mathbf{v}_{P/\mathcal{F}} \tag{15.46}$$

$$\mathbf{a}_P = \mathbf{a}_{P'} + \mathbf{a}_{P/\mathcal{F}} + \mathbf{a}_c \tag{15.48}$$

B. Using a frame of reference in general motion. The steps that you will take differ only slightly from those listed under A. They consist of the following:

1. Select the frame $\mathcal{F}$ that you wish to use and a reference point A in that frame, from which you will draw the coordinate axes, x, y, and z defining that frame. You will consider the motion of the frame as the sum of a *translation with A and a rotation about A.*

2. Determine the velocity $\mathbf{v}_A$ of point A and the angular velocity $\mathbf{\Omega}$ of the frame. In most cases, you will have selected a frame which is attached to some element of the system; $\mathbf{\Omega}$ will then be the angular velocity of that element.

3. Designate as P′ the point of frame $\mathcal{F}$ that coincides with the point P of interest at the instant you are considering, and determine the velocity $\mathbf{v}_{P'}$ and the acceleration $\mathbf{a}_{P'}$ of that point. In some cases, this can be done by visualizing the motion of P if that point were prevented from moving with respect to $\mathcal{F}$ [Sample Prob. 15.15]. A more general approach is to recall that the motion of P' is the sum of a translation with the reference point A and a rotation about A. The velocity $\mathbf{v}_{P'}$ and the acceleration $\mathbf{a}_{P'}$ of P', therefore, can be obtained by adding $\mathbf{v}_A$ and $\mathbf{a}_A$, respectively, to the expressions found in paragraph A3 and replacing the position vector $\mathbf{r}$ by the vector $\mathbf{r}_{P/A}$ drawn from A to P:

$$\mathbf{v}_{P'} = \mathbf{v}_A + \mathbf{\Omega} \times \mathbf{r}_{P/A} \qquad \mathbf{a}_{P'} = \mathbf{a}_A + \boldsymbol{\alpha} \times \mathbf{r}_{P/A} + \mathbf{\Omega} \times (\mathbf{\Omega} \times \mathbf{r}_{P/A})$$

Steps 4, 5, and 6 are the same as in Part A, except that the vector $\mathbf{r}$ should again be replaced by $\mathbf{r}_{P/A}$. Thus, Eqs. (15.46) and (15.48) can still be used to obtain the velocity and the acceleration of P with respect to the fixed frame of reference $OXYZ$.

Problems

15.214 The bent pipe shown rotates at the constant rate $\omega_1 = 10$ rad/s. Knowing that a ball bearing D moves in portion BC of the pipe toward end C at a constant relative speed $u = 2$ ft/s, determine at the instant shown (*a*) the velocity of D, (*b*) the acceleration of D.

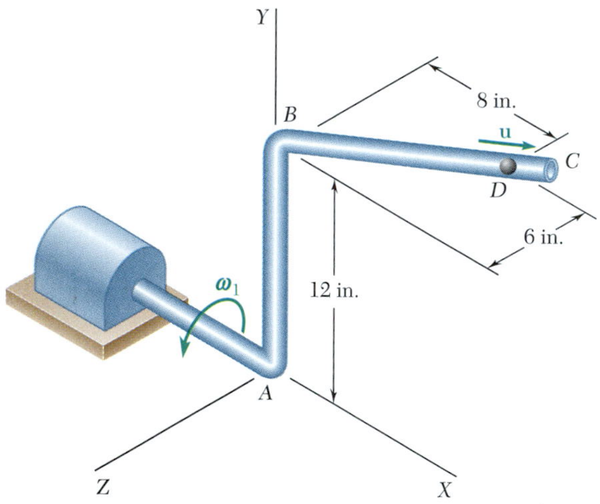

Fig. P15.214

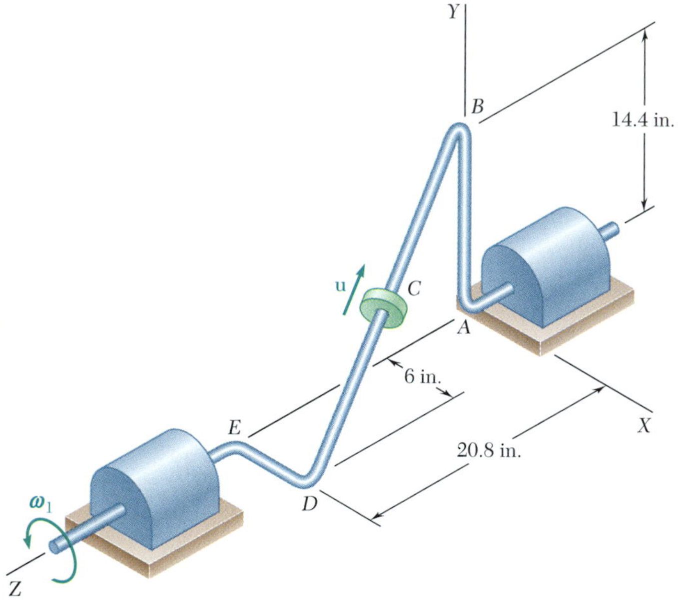

Fig. P15.215

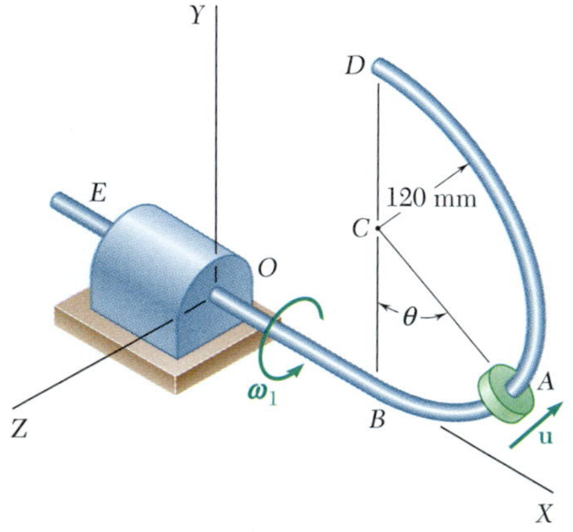

Fig. P15.216 and P15.217

15.215 The bent rod shown rotates at the constant rate $\omega_1 = 5$ rad/s and collar C moves toward point B at a constant relative speed $u = 39$ in./s. Knowing that collar C is halfway between points B and D at the instant shown, determine its velocity and acceleration.

15.216 The bent rod EBD rotates at the constant rate $\omega_1 = 8$ rad/s. Knowing that collar A moves upward along the rod at a constant relative speed $u = 600$ mm/s and that $\theta = 60°$, determine (*a*) the velocity of A, (*b*) the acceleration of A.

15.217 The bent rod EBD rotates at the constant rate $\omega_1 = 8$ rad/s. Knowing that collar A moves upward along the rod at a constant relative speed $u = 600$ mm/s and that $\theta = 120°$, determine (*a*) the velocity of A, (*b*) the acceleration of A.

15.218 A square plate of side 360 mm is hinged at A and B to a clevis. The plate rotates at the constant rate $\omega_2 = 4$ rad/s with respect to the clevis, which itself rotates at the constant rate $\omega_1 = 3$ rad/s about the Y axis. For the position shown, determine (*a*) the velocity of point C, (*b*) the acceleration of point C.

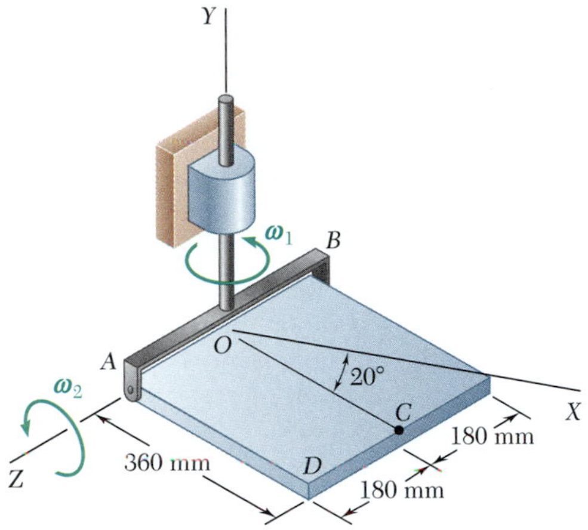

Fig. P15.218 and P15.219

15.219 A square plate of side 360 mm is hinged at A and B to a clevis. The plate rotates at the constant rate $\omega_2 = 4$ rad/s with respect to the clevis, which itself rotates at the constant rate $\omega_1 = 3$ rad/s about the Y axis. For the position shown, determine (*a*) the velocity of corner D, (*b*) the acceleration of corner D.

15.220 The rectangular plate shown rotates at the constant rate $\omega_2 = 12$ rad/s with respect to arm AE, which itself rotates at the constant rate $\omega_1 = 9$ rad/s about the Z axis. For the position shown, determine (*a*) the velocity of corner B, (*b*) the acceleration of corner B.

15.221 The rectangular plate shown rotates at the constant rate $\omega_2 = 12$ rad/s with respect to arm AE, which itself rotates at the constant rate $\omega_1 = 9$ rad/s about the Z axis. For the position shown, determine (*a*) the velocity of corner C, (*b*) the acceleration of corner C.

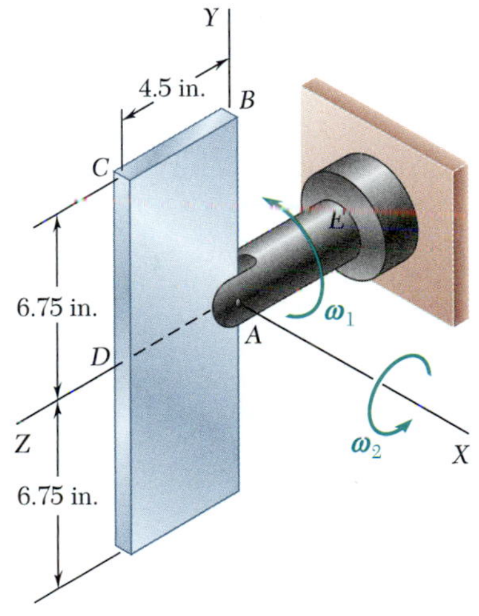

Fig. *P15.220* and *P15.221*

15.222 Solve Prob. 15.221 assuming that at the instant shown the angular velocity $\boldsymbol{\omega}_2$ of the plate with respect to arm AE is 12 rad/s and is decreasing at the rate of 60 rad/s^2, while the angular velocity $\boldsymbol{\omega}_1$ of the arm about the Z axis is 9 rad/s and is decreasing at the rate of 45 rad/s^2.

15.223 Solve Prob. 15.214 assuming that at the instant shown the angular velocity ω_1 of the pipe is 10 rad/s and is decreasing at the rate of 15 rad/s^2, while the relative speed u of the ball bearing is 2 ft/s and is increasing at the rate of 10 ft/s^2.

15.224 Solve Prob. 15.215 assuming that at the instant shown the angular velocity ω_1 of the rod is 5 rad/s and is increasing at the rate of 10 rad/s^2, while the relative speed u of the collar C is 39 in./s and is decreasing at the rate of 260 in./s^2.

15.225 Using the method of Sec. 15.14, solve Prob. 15.190.

15.226 Using the method of Sec. 15.14, solve Prob. 15.191.

15.227 Using the method of Sec. 15.14, solve Prob. 15.185.

15.228 Using the method of Sec. 15.14, solve Prob. 15.186.

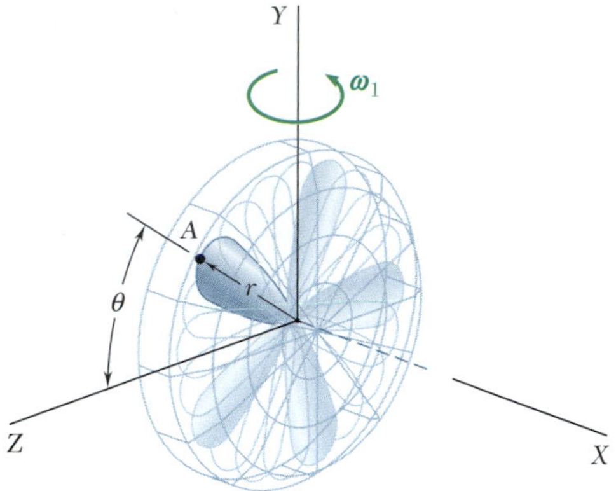

Fig. P15.229

15.229 The rotor of a fan rotates about the Y axis at a constant rate $\omega_1 = 0.8$ rad/s. At the same time, the blades rotate at a constant rate $\omega_2 = d\theta/dt = 300$ rpm. Knowing that the distance between the rotor and point A on the tip of the blade is $r = 160$ mm and that $\theta = 45°$ at the instant shown, determine the velocity and acceleration of point A.

15.230 The crane shown rotates at the constant rate $\omega_1 = 0.25$ rad/s; simultaneously, the telescoping boom is being lowered at the constant rate $\omega_2 = 0.40$ rad/s. Knowing that at the instant shown the length of the boom is 6 m and is increasing at the constant rate $u = 0.45$ m/s, determine the velocity and acceleration of point B.

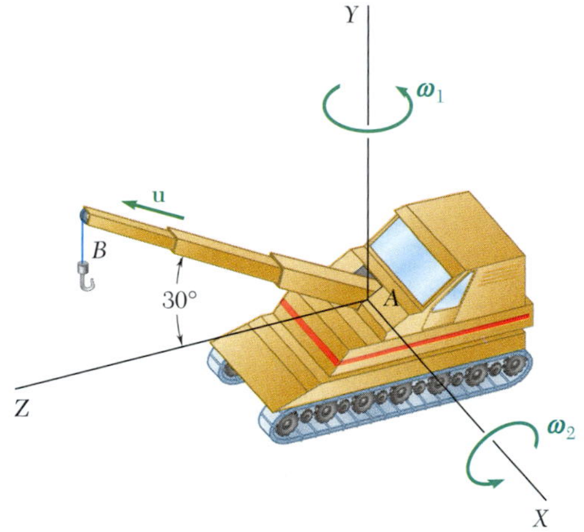

Fig. P15.230

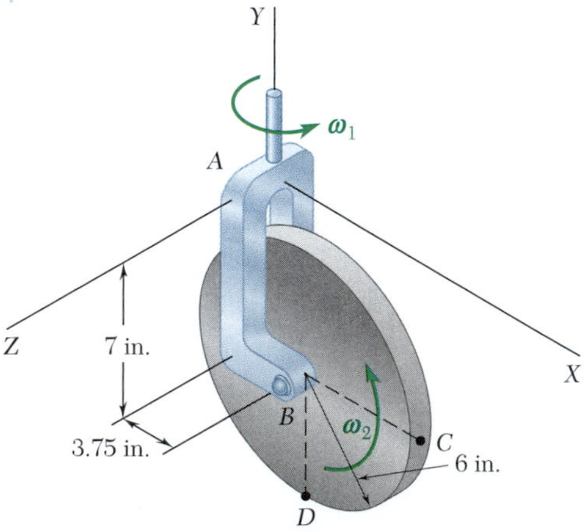

Fig. P15.231 and P15.232

15.231 A disk of radius 6 in. rotates at the constant rate $\omega_2 = 5$ rad/s with respect to the arm AB, which itself rotates at the constant rate $\omega_1 = 3$ rad/s. For the position shown, determine the velocity and acceleration of point C.

15.232 A disk of radius 6 in. rotates at the constant rate $\omega_2 = 5$ rad/s with respect to the arm AB, which itself rotates at the constant rate $\omega_1 = 3$ rad/s. For the position shown, determine the velocity and acceleration of point D.

15.233 The 400-mm bar AB is made to rotate at the constant rate $\omega_2 = d\theta/dt = 8$ rad/s with respect to the frame CD which itself rotates at the constant rate $\omega_1 = 12$ rad/s about the Y axis. Knowing that $\theta = 60°$ at the instant shown, determine the velocity and acceleration of point A.

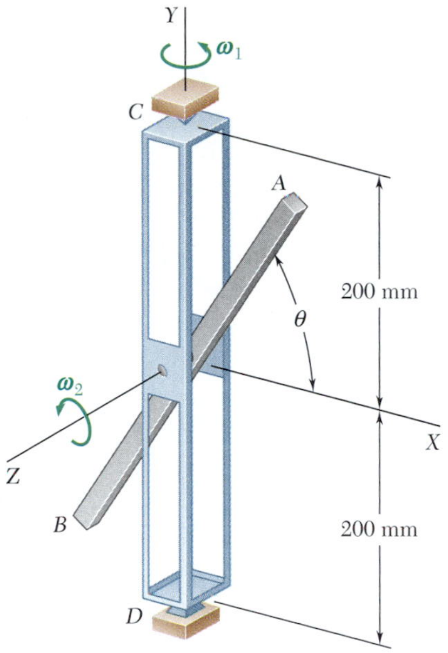

Fig. P15.233 and P15.234

15.234 The 400-mm bar AB is made to rotate at the rate $\omega_2 = d\theta/dt$ with respect to the frame CD which itself rotates at the rate ω_1 about the Y axis. At the instant shown $\omega_1 = 12$ rad/s, $d\omega_1/dt = -16$ rad/s^2, $\omega_2 = 8$ rad/s, $d\omega_2/dt = 10$ rad/s^2, and $\theta = 60°$. Determine the velocity and acceleration of point A at this instant.

15.235 and 15.236 In the position shown the thin rod moves at a constant speed $u = 60$ mm/s out of the tube BC. At the same time tube BC rotates at the constant rate $\omega_2 = 1.5$ rad/s with respect to arm CD. Knowing that the entire assembly rotates about the X axis at the constant rate $\omega_1 = 1.2$ rad/s, determine the velocity and acceleration of end A of the rod.

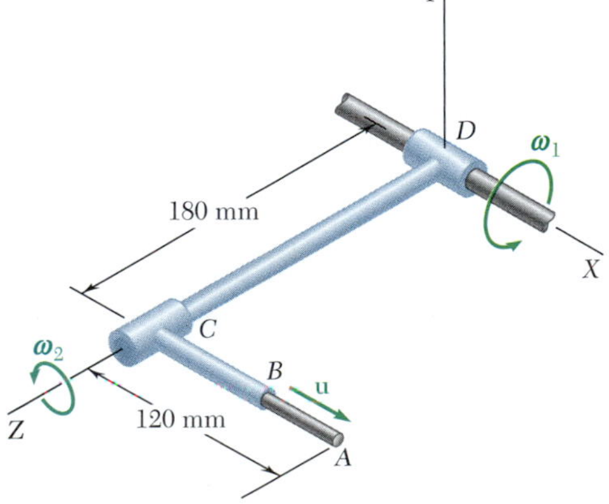

Fig. P15.235

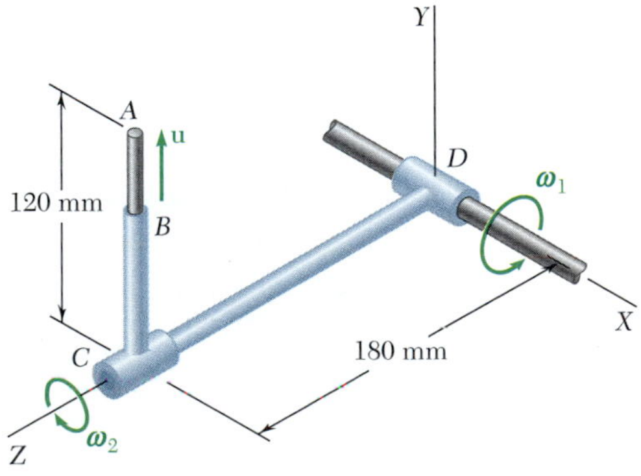

Fig. P15.236

15.237 A disk of 9-in. radius rotates at the constant rate $\omega_2 = 12$ rad/s with respect to arm CD, which itself rotates at the constant rate $\omega_1 = 8$ rad/s about the Y axis. Determine at the instant shown the velocity and acceleration of point A on the rim of the disk.

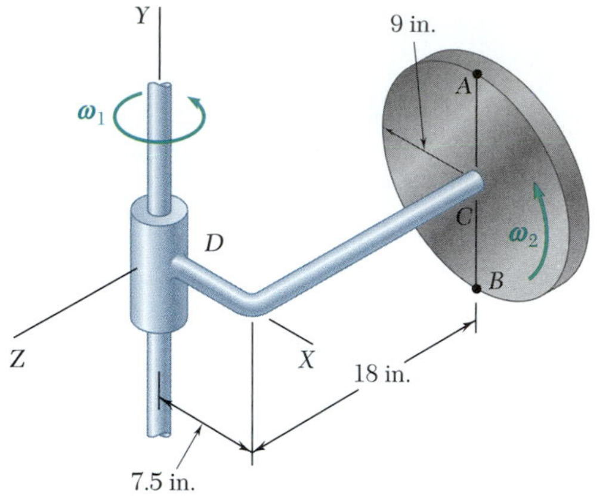

Fig. P15.237 and P15.238

15.238 A disk of 9-in. radius rotates at the constant rate $\omega_2 = 12$ rad/s with respect to arm CD, which itself rotates at the constant rate $\omega_1 = 8$ rad/s about the Y axis. Determine at the instant shown the velocity and acceleration of point B on the rim of the disk.

15.239 The vertical plate shown is welded to arm EFG, and the entire unit rotates at the constant rate $\omega_1 = 1.6$ rad/s about the Y axis. At the same time, a continuous link belt moves around the perimeter of the plate at a constant speed $u = 90$ mm/s. For the position shown, determine the acceleration of the link of the belt located (*a*) at point A, (*b*) at point B.

15.240 The vertical plate shown is welded to arm EFG, and the entire unit rotates at the constant rate $\omega_1 = 1.6$ rad/s about the Y axis. At the same time, a continuous link belt moves around the perimeter of the plate at a constant speed $u = 90$ mm/s. For the position shown, determine the acceleration of the link of the belt located (*a*) at point C, (*b*) at point D.

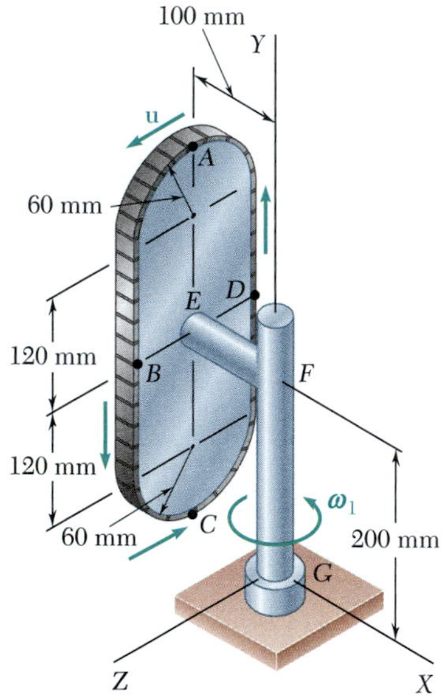

Fig. P15.239 and P15.240

15.241 The cylinder shown rotates at the constant rate $\omega_2 = 8$ rad/s with respect to rod CD, which itself rotates at the constant rate $\omega_1 = 6$ rad/s about the X axis. For the position shown, determine the velocity and acceleration of point A on the edge of the cylinder.

15.242 The cylinder shown rotates at the constant rate $\omega_2 = 8$ rad/s with respect to rod CD, which itself rotates at the constant rate $\omega_1 = 6$ rad/s about the X axis. For the position shown, determine the velocity and acceleration of point B on the edge of the cylinder.

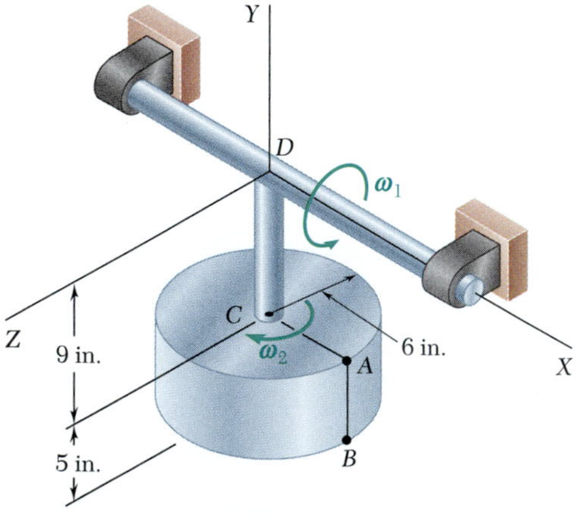

Fig. *P15.241* and *P15.242*

REVIEW AND SUMMARY FOR CHAPTER 15

This chapter was devoted to the study of the kinematics of rigid bodies.

Rigid body in translation

We first considered the *translation* of a rigid body [Sec. 15.2] and observed that in such a motion, *all points of the body have the same velocity and the same acceleration at any given instant.*

Rigid body in rotation about a fixed axis

We next considered the *rotation* of a rigid body about a fixed axis [Sec. 15.3]. The position of the body is defined by the angle θ that the line BP, drawn from the axis of rotation to a point P of the body, forms with a fixed plane (Fig. 15.39). We found that the magnitude of the velocity of P is

$$v = \frac{ds}{dt} = r\dot{\theta} \sin \phi \qquad (15.4)$$

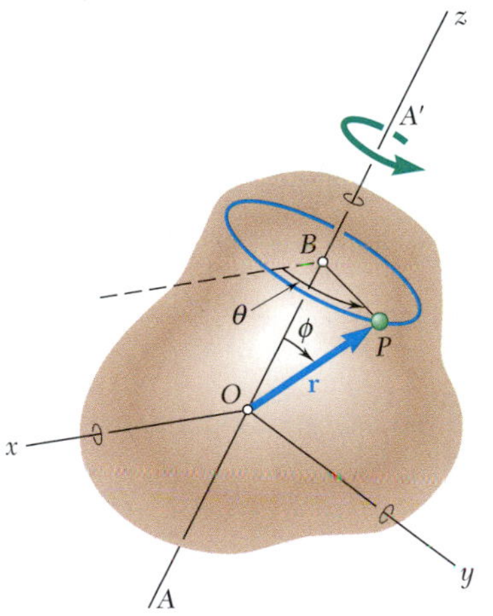

Fig. 15.39

where $\dot{\theta}$ is the time derivative of θ. We then expressed the velocity of P as

$$\mathbf{v} = \frac{d\mathbf{r}}{dt} = \boldsymbol{\omega} \times \mathbf{r} \qquad (15.5)$$

where the vector

$$\boldsymbol{\omega} = \omega\mathbf{k} = \dot{\theta}\mathbf{k} \qquad (15.6)$$

is directed along the fixed axis of rotation and represents the *angular velocity* of the body.

Denoting by $\boldsymbol{\alpha}$ the derivative $d\boldsymbol{\omega}/dt$ of the angular velocity, we expressed the acceleration of P as

$$\mathbf{a} = \boldsymbol{\alpha} \times \mathbf{r} + \boldsymbol{\omega} \times (\boldsymbol{\omega} \times \mathbf{r}) \qquad (15.8)$$

Differentiating (15.6), and recalling that $\mathbf{k}$ is constant in magnitude and direction, we found that

$$\boldsymbol{\alpha} = \alpha\mathbf{k} = \dot{\omega}\mathbf{k} = \ddot{\theta}\mathbf{k} \qquad (15.9)$$

The vector $\boldsymbol{\alpha}$ represents the *angular acceleration* of the body and is directed along the fixed axis of rotation.

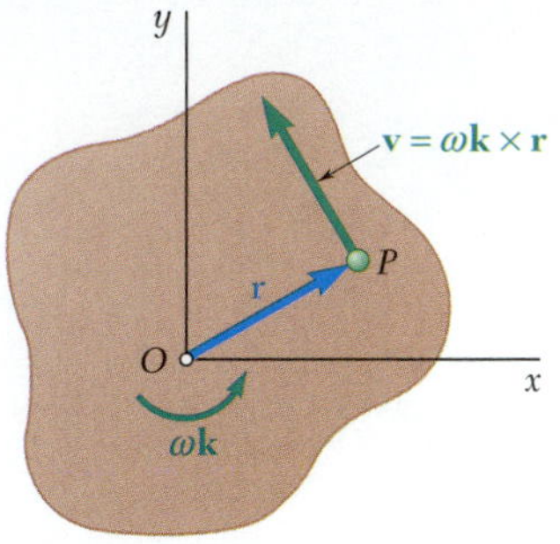

Fig. 15.40

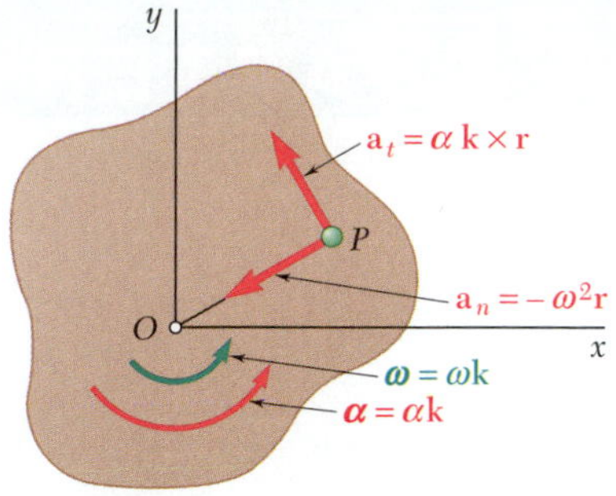

Fig. 15.41

Rotation of a representative slab

Next we considered the motion of a representative slab located in a plane perpendicular to the axis of rotation of the body (Fig. 15.40). Since the angular velocity is perpendicular to the slab, the velocity of a point P of the slab was expressed as

$$\mathbf{v} = \omega\mathbf{k} \times \mathbf{r} \tag{15.10}$$

where $\mathbf{v}$ is contained in the plane of the slab. Substituting $\boldsymbol{\omega} = \omega\mathbf{k}$ and $\boldsymbol{\alpha} = \alpha\mathbf{k}$ into (15.8), we found that the acceleration of P could be resolved into tangential and normal components (Fig. 15.41) respectively equal to

Tangential and normal components

$$\begin{aligned} \mathbf{a}_t &= \alpha\mathbf{k} \times \mathbf{r} & a_t &= r\alpha \\ \mathbf{a}_n &= -\omega^2\mathbf{r} & a_n &= r\omega^2 \end{aligned} \tag{15.11$'$}$$

Angular velocity and angular acceleration of rotating slab

Recalling Eqs. (15.6) and (15.9), we obtained the following expressions for the *angular velocity* and the *angular acceleration* of the slab [Sec. 15.4]:

$$\omega = \frac{d\theta}{dt} \tag{15.12}$$

$$\alpha = \frac{d\omega}{dt} = \frac{d^2\theta}{dt^2} \tag{15.13}$$

or

$$\alpha = \omega\frac{d\omega}{d\theta} \tag{15.14}$$

We noted that these expressions are similar to those obtained in Chap. 11 for the rectilinear motion of a particle.

Two particular cases of rotation are frequently encountered: *uniform rotation* and *uniformly accelerated rotation*. Problems involving either of these motions can be solved by using equations similar to those used in Secs. 11.4 and 11.5 for the uniform rectilinear motion and the uniformly accelerated rectilinear motion of a particle, but where x, v, and a are replaced by θ, ω, and α, respectively [Sample Prob. 15.1].

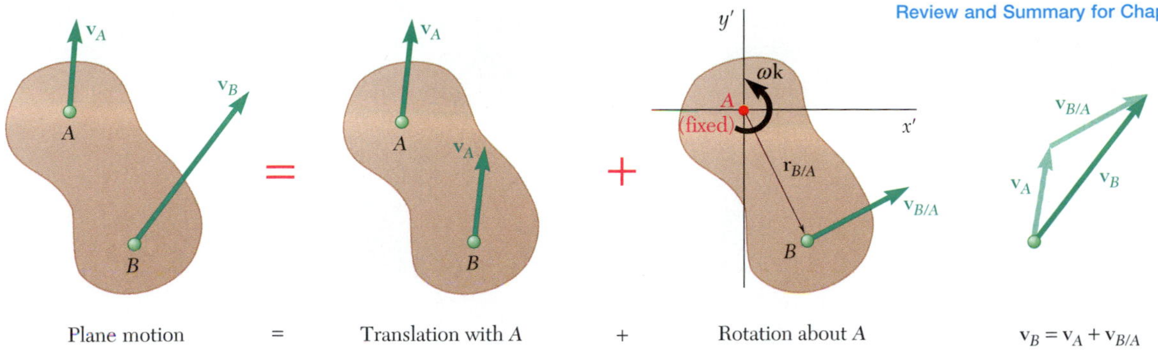

Fig. 15.42

Velocities in plane motion

The *most general plane motion* of a rigid slab can be considered as the *sum of a translation and a rotation* [Sec. 15.5]. For example, the slab shown in Fig. 15.42 can be assumed to translate with point A, while simultaneously rotating about A. It follows [Sec. 15.6] that the velocity of any point B of the slab can be expressed as

$$\mathbf{v}_B = \mathbf{v}_A + \mathbf{v}_{B/A} \tag{15.17}$$

where $\mathbf{v}_A$ is the velocity of A and $\mathbf{v}_{B/A}$ the relative velocity of B with respect to A or, more precisely, with respect to axes $x'y'$ translating with A. Denoting by $\mathbf{r}_{B/A}$ the position vector of B relative to A, we found that

$$\mathbf{v}_{B/A} = \omega\mathbf{k} \times \mathbf{r}_{B/A} \qquad v_{B/A} = r\omega \tag{15.18}$$

The fundamental equation (15.17) relating the absolute velocities of points A and B and the relative velocity of B with respect to A was expressed in the form of a vector diagram and used to solve problems involving the motion of various types of mechanisms [Sample Probs. 15.2 and 15.3].

Instantaneous center of rotation

Another approach to the solution of problems involving the velocities of the points of a rigid slab in plane motion was presented in Sec. 15.7 and used in Sample Probs. 15.4 and 15.5. It is based on the determination of the *instantaneous center of rotation* C of the slab (Fig. 15.43).

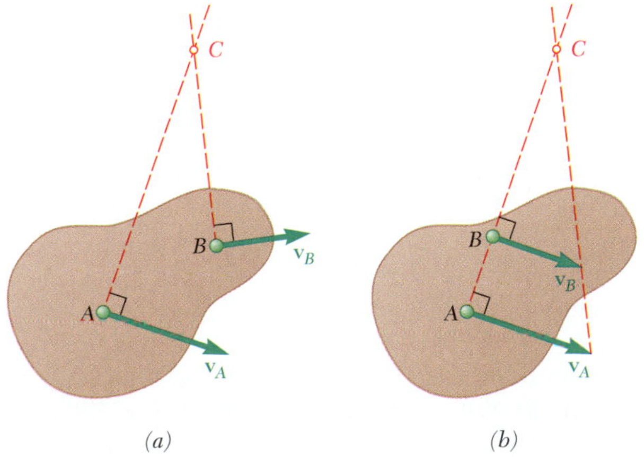

Fig. 15.43

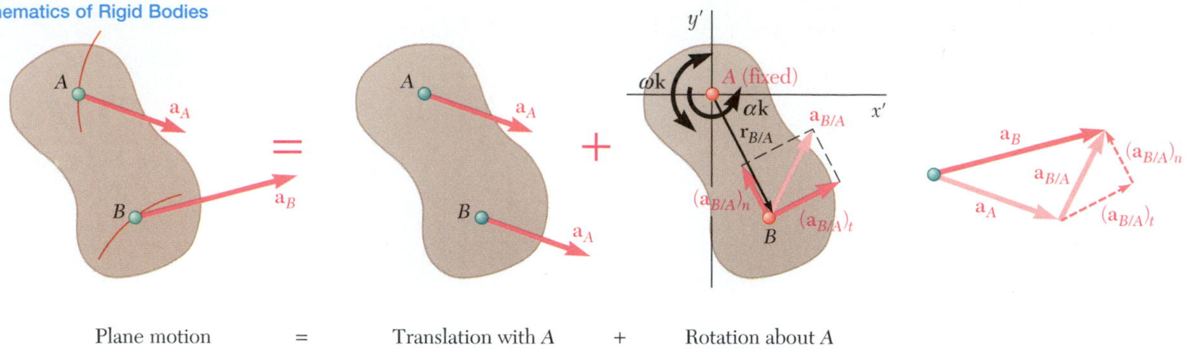

Fig. 15.44

Accelerations in plane motion

The fact that any plane motion of a rigid slab can be considered as the sum of a translation of the slab with a reference point A and a rotation about A was used in Sec. 15.8 to relate the absolute accelerations of any two points A and B of the slab and the relative acceleration of B with respect to A. We had

$$\mathbf{a}_B = \mathbf{a}_A + \mathbf{a}_{B/A} \tag{15.21}$$

where $\mathbf{a}_{B/A}$ consisted of a *normal component* $(\mathbf{a}_{B/A})_n$ of magnitude $r\omega^2$ directed toward A, and a *tangential component* $(\mathbf{a}_{B/A})_t$ of magnitude $r\alpha$ perpendicular to the line AB (Fig. 15.44). The fundamental relation (15.21) was expressed in terms of vector diagrams or vector equations and used to determine the accelerations of given points of various mechanisms [Sample Probs. 15.6 through 15.8]. It should be noted that the instantaneous center of rotation C considered in Sec. 15.7 cannot be used for the determination of accelerations, since point C, in general, does *not* have zero acceleration.

Coordinates expressed in terms of a parameter

In the case of certain mechanisms, it is possible to express the coordinates x and y of all significant points of the mechanism by means of simple analytic expressions containing a *single parameter*. The components of the absolute velocity and acceleration of a given point are then obtained by differentiating twice with respect to the time t the coordinates x and y of that point [Sec. 15.9].

Rate of change of a vector with respect to a rotating frame

While the rate of change of a vector is the same with respect to a fixed frame of reference and with respect to a frame in translation, the rate of change of a vector with respect to a rotating frame is different. Therefore, in order to study the motion of a particle relative to a rotating frame we first had to compare the rates of change of a general vector $\mathbf{Q}$ with respect to a fixed frame $OXYZ$ and with respect to a frame $Oxyz$ rotating with an angular velocity $\boldsymbol{\Omega}$ [Sec. 15.10] (Fig. 15.45). We obtained the fundamental relation

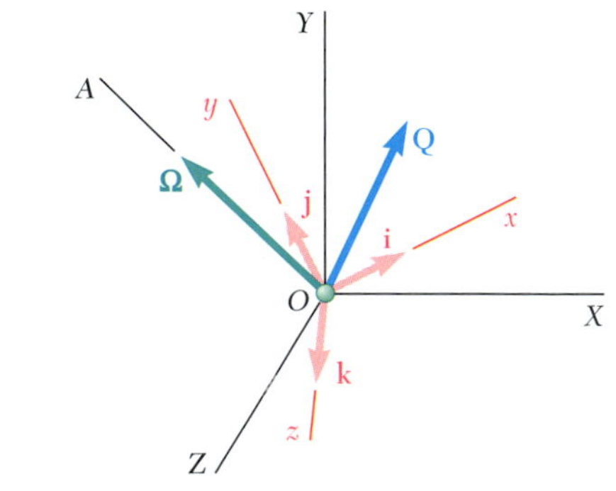

Fig. 15.45

$$(\dot{\mathbf{Q}})_{OXYZ} = (\dot{\mathbf{Q}})_{Oxyz} + \boldsymbol{\Omega} \times \mathbf{Q} \tag{15.31}$$

and we concluded that the rate of change of the vector $\mathbf{Q}$ with respect to the fixed frame $OXYZ$ is made of two parts: The first part represents the rate of change of $\mathbf{Q}$ with respect to the rotating frame $Oxyz$; the second part, $\boldsymbol{\Omega} \times \mathbf{Q}$, is induced by the rotation of the frame $Oxyz$.

The next part of the chapter [Sec. 15.11] was devoted to the two-dimensional kinematic analysis of a particle P moving with respect to a frame $\mathcal{F}$ rotating with an angular velocity $\boldsymbol{\Omega}$ about a fixed axis (Fig. 15.46). We found that the absolute velocity of P could be expressed as

Plane motion of a particle relative to a rotating frame

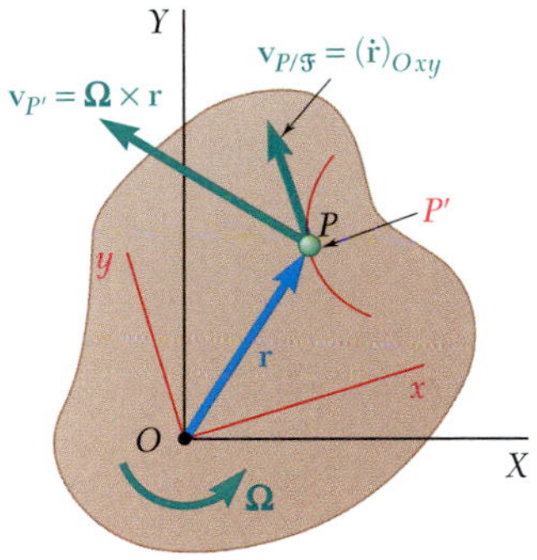

Fig. 15.46

$$\mathbf{v}_P = \mathbf{v}_{P'} + \mathbf{v}_{P/\mathcal{F}} \tag{15.33}$$

where $\mathbf{v}_P$ = absolute velocity of particle P
$\mathbf{v}_{P'}$ = velocity of point P' of moving frame $\mathcal{F}$ coinciding with P
$\mathbf{v}_{P/\mathcal{F}}$ = velocity of P relative to moving frame $\mathcal{F}$

We noted that the same expression for $\mathbf{v}_P$ is obtained if the frame is in translation rather than in rotation. However, when the frame is in rotation, the expression for the acceleration of P is found to contain an additional term $\mathbf{a}_c$ called the *complementary acceleration* or *Coriolis acceleration*. We wrote

$$\mathbf{a}_P = \mathbf{a}_{P'} + \mathbf{a}_{P/\mathcal{F}} + \mathbf{a}_c \tag{15.36}$$

where $\mathbf{a}_P$ = absolute acceleration of particle P
$\mathbf{a}_{P'}$ = acceleration of point P' of moving frame $\mathcal{F}$ coinciding with P
$\mathbf{a}_{P/\mathcal{F}}$ = acceleration of P relative to moving frame $\mathcal{F}$
$\mathbf{a}_c = 2\boldsymbol{\Omega} \times (\dot{\mathbf{r}})_{Oxy} = 2\boldsymbol{\Omega} \times \mathbf{v}_{P/\mathcal{F}}$
= complementary, or Coriolis, acceleration

Since $\boldsymbol{\Omega}$ and $\mathbf{v}_{P/\mathcal{F}}$ are perpendicular to each other in the case of plane motion, the Coriolis acceleration was found to have a magnitude $a_c = 2\Omega v_{P/\mathcal{F}}$ and to point in the direction obtained by rotating the vector $\mathbf{v}_{P/\mathcal{F}}$ through 90° in the sense of rotation of the moving frame. Formulas (15.33) and (15.36) can be used to analyze the motion of mechanisms which contain parts sliding on each other [Sample Probs. 15.9 and 15.10].

Motion of a rigid body with a fixed point

The last part of the chapter was devoted to the study of the kinematics of rigid bodies in three dimensions. We first considered the motion of a rigid body with a fixed point [Sec. 15.12]. After proving that the most general displacement of a rigid body with a fixed point O is equivalent to a rotation of the body about an axis through O, we were able to define the angular velocity $\boldsymbol{\omega}$ and the *instantaneous axis of rotation* of the body at a given instant. The velocity of a point P of the body (Fig. 15.47) could again be expressed as

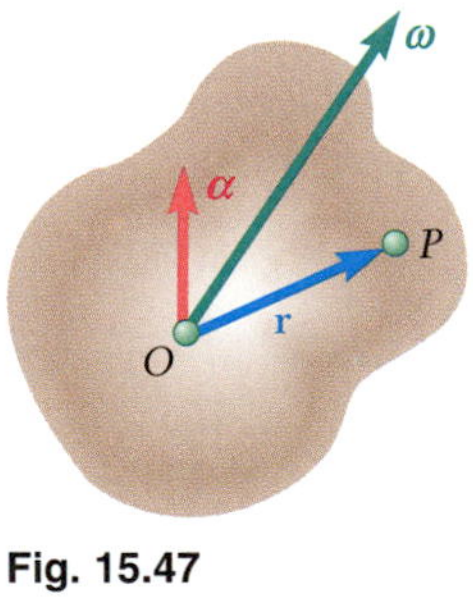

Fig. 15.47

$$\mathbf{v} = \frac{d\mathbf{r}}{dt} = \boldsymbol{\omega} \times \mathbf{r} \tag{15.37}$$

Differentiating this expression, we also wrote

$$\mathbf{a} = \boldsymbol{\alpha} \times \mathbf{r} + \boldsymbol{\omega} \times (\boldsymbol{\omega} \times \mathbf{r}) \tag{15.38}$$

However, since the direction of $\boldsymbol{\omega}$ changes from one instant to the next, the angular acceleration $\boldsymbol{\alpha}$ is, in general, not directed along the instantaneous axis of rotation [Sample Prob. 15.11].

General motion in space

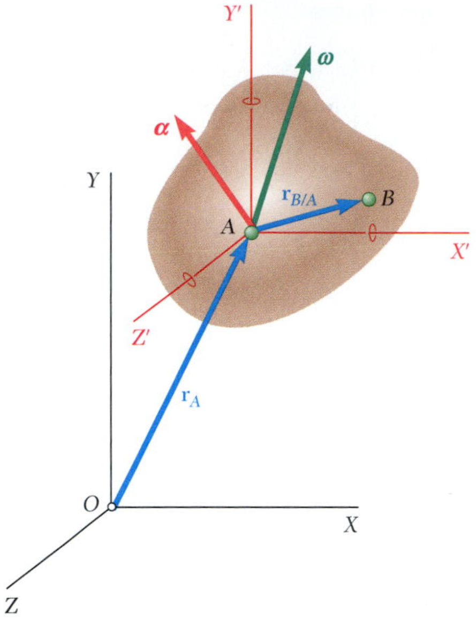

Fig. 15.48

It was shown in Sec. 15.13 that *the most general motion of a rigid body in space is equivalent, at any given instant, to the sum of a translation and a rotation.* Considering two particles A and B of the body, we found that

$$\mathbf{v}_B = \mathbf{v}_A + \mathbf{v}_{B/A} \tag{15.42}$$

where $\mathbf{v}_{B/A}$ is the velocity of B relative to a frame $AX'Y'Z'$ attached to A and of fixed orientation (Fig. 15.48). Denoting by $\mathbf{r}_{B/A}$ the position vector of B relative to A, we wrote

$$\mathbf{v}_B = \mathbf{v}_A + \boldsymbol{\omega} \times \mathbf{r}_{B/A} \tag{15.43}$$

where $\boldsymbol{\omega}$ is the angular velocity of the body at the instant considered [Sample Prob. 15.12]. The acceleration of B was obtained by a similar reasoning. We first wrote

$$\mathbf{a}_B = \mathbf{a}_A + \mathbf{a}_{B/A}$$

and, recalling Eq. (15.38),

$$\mathbf{a}_B = \mathbf{a}_A + \boldsymbol{\alpha} \times \mathbf{r}_{B/A} + \boldsymbol{\omega} \times (\boldsymbol{\omega} \times \mathbf{r}_{B/A}) \tag{15.44}$$

Three-dimensional motion of a particle relative to a rotating frame

In the final two sections of the chapter we considered the three-dimensional motion of a particle P relative to a frame $Oxyz$ rotating with an angular velocity $\boldsymbol{\Omega}$ with respect to a fixed frame $OXYZ$ (Fig. 15.49). In Sec. 15.14 we expressed the absolute velocity $\mathbf{v}_P$ of P as

$$\mathbf{v}_P = \mathbf{v}_{P'} + \mathbf{v}_{P/\mathcal{F}} \tag{15.46}$$

where $\mathbf{v}_P$ = absolute velocity of a particle P

$\mathbf{v}_{P'}$ = velocity of point P' of moving frame $\mathcal{F}$ coinciding with P

$\mathbf{v}_{P/\mathcal{F}}$ = velocity of P relative to moving frame $\mathcal{F}$

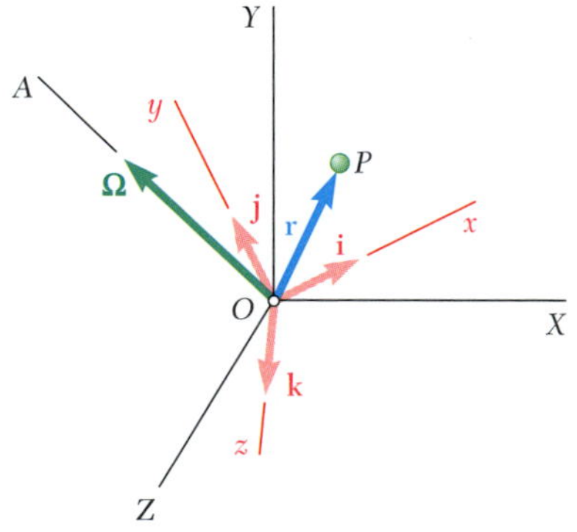

Fig. 15.49

The absolute acceleration $\mathbf{a}_P$ of P was then expressed as

$$\mathbf{a}_P = \mathbf{a}_{P'} + \mathbf{a}_{P/\mathcal{F}} + \mathbf{a}_c \qquad (15.48)$$

where $\mathbf{a}_P$ = absolute acceleration of particle P
$\mathbf{a}_{P'}$ = acceleration of point P' of moving frame $\mathcal{F}$ coinciding with P
$\mathbf{a}_{P/\mathcal{F}}$ = acceleration of P relative to moving frame $\mathcal{F}$
$\mathbf{a}_c = 2\boldsymbol{\Omega} \times (\dot{\mathbf{r}})_{Oxyz} = 2\boldsymbol{\Omega} \times \mathbf{v}_{P/\mathcal{F}}$
= complementary, or Coriolis, acceleration

It was noted that the magnitude a_c of the Coriolis acceleration is not equal to $2\Omega v_{P/\mathcal{F}}$ [Sample Prob. 15.13] except in the special case when $\boldsymbol{\Omega}$ and $\mathbf{v}_{P/\mathcal{F}}$ are perpendicular to each other.

Frame of reference in general motion

We also observed [Sec. 15.15] that Eqs. (15.46) and (15.48) remain valid when the frame $Axyz$ moves in a known, but arbitrary, fashion with respect to the fixed frame $OXYZ$ (Fig. 15.50), provided that the motion of A is included in the terms $\mathbf{v}_{P'}$ and $\mathbf{a}_{P'}$ representing the absolute velocity and acceleration of the coinciding point P'.

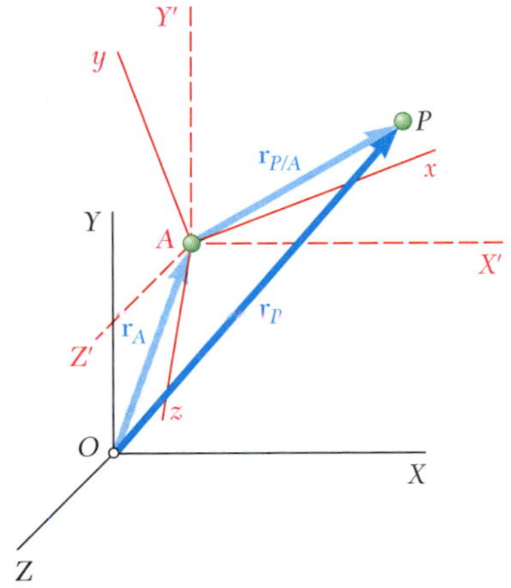

Fig. 15.50

Rotating frames of reference are particularly useful in the study of the three-dimensional motion of rigid bodies. Indeed, there are many cases where an appropriate choice of the rotating frame will lead to a simpler analysis of the motion of the rigid body than would be possible with axes of fixed orientation [Sample Probs. 15.14 and 15.15].

Review Problems

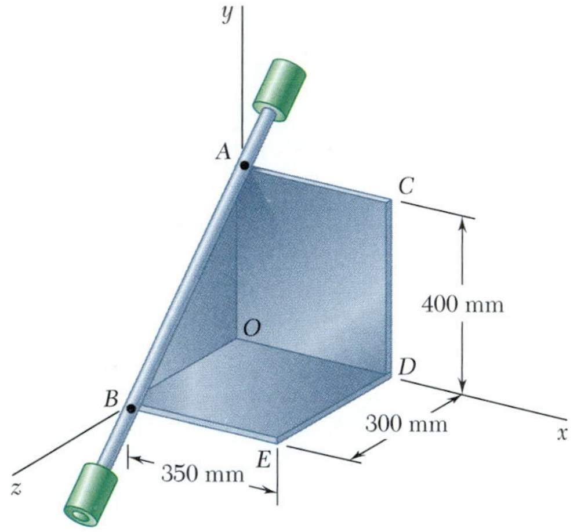

Fig. P15.243

15.243 A triangular plate and two rectangular plates are welded to each other and to the straight rod *AB*. The entire welded unit rotates about axis *AB* with a constant angular velocity of 5 rad/s. Knowing that at the instant considered the velocity of corner *E* is directed downward, determine the velocity and acceleration of corner *D*.

15.244 Two blocks and a pulley are connected by inextensible cords as shown. The pulley has an initial angular velocity of 0.8 rad/s counterclockwise and a constant angular acceleration of 1.8 rad/s^2 clockwise. After 5 s of motion, determine the velocity and position of (*a*) block *A*, (*b*) block *B*.

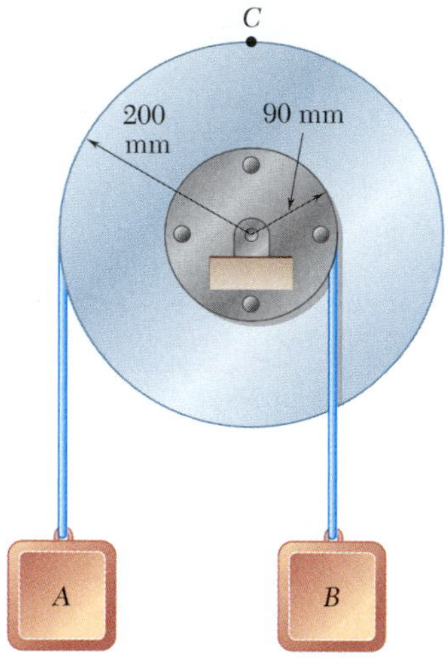

Fig. P15.244

15.245 Knowing that crank *AB* has a constant angular velocity of 160 rpm counterclockwise, determine the angular velocity of rod *BD* and the velocity of collar *D* when (*a*) $\theta = 0$, (*b*) $\theta = 90°$.

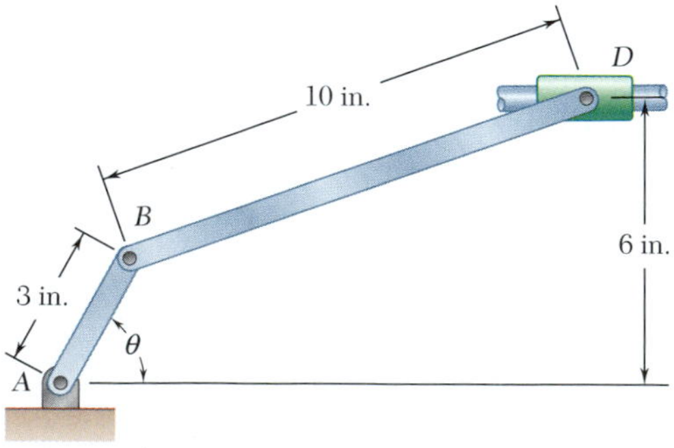

Fig. *P15.245*

15.246 In the position shown, bar DE has a constant angular velocity of 15 rad/s clockwise. Determine (a) the distance b for which the velocity of point F is vertical, (b) the corresponding velocity of point F.

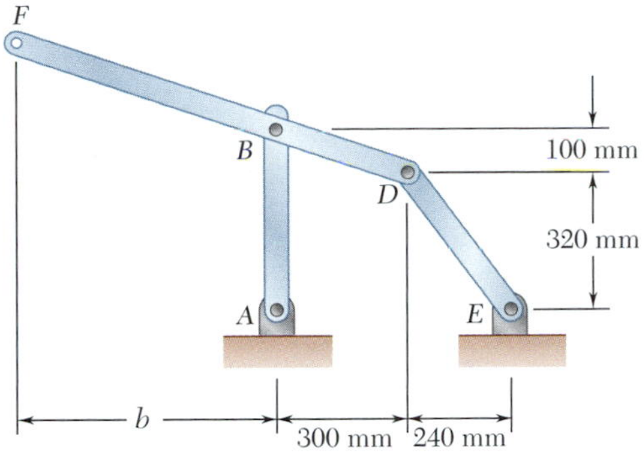

Fig. P15.246

15.247 The spool of tape shown and its frame assembly are pulled upward at a speed v_A = 750 mm/s. Knowing that the 80-mm-radius spool has an angular velocity of 15 rad/s clockwise and that at the instant shown the total thickness of the tape on the spool is 20 mm, determine (a) the instantaneous center of rotation of the spool, (b) the velocities of points B and D.

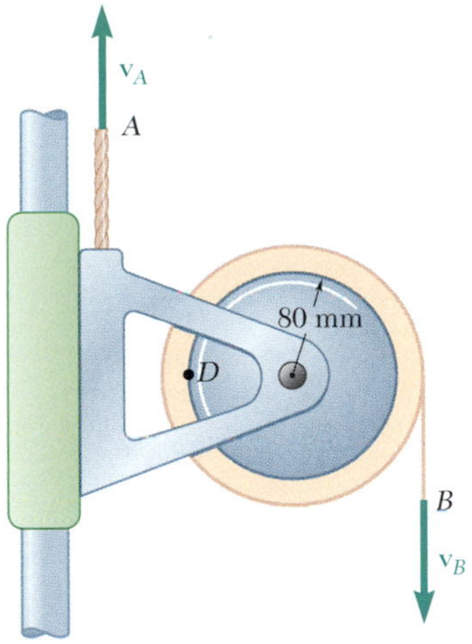

Fig. P15.247

15.248 Knowing that at the instant shown the angular velocity of rod AB is 15 rad/s clockwise, determine (a) the angular velocity of rod BD, (b) the velocity of the midpoint of rod BD.

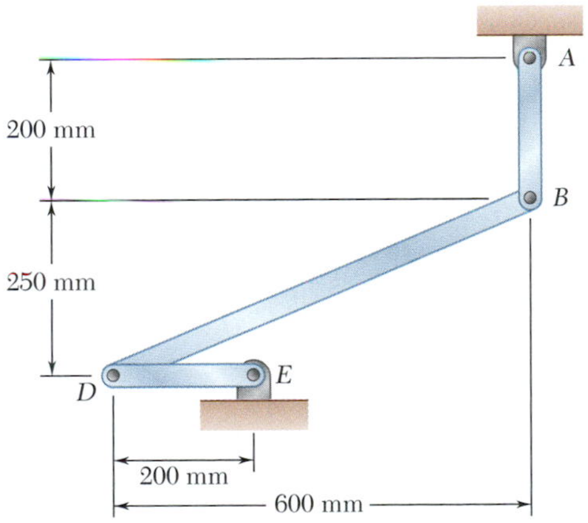

Fig. *P15.248*

15.249 An automobile travels to the left at a constant speed of 48 mi/h. Knowing that the diameter of the wheel is 22 in., determine the acceleration (a) of point B, (b) of point C, (c) of point D.

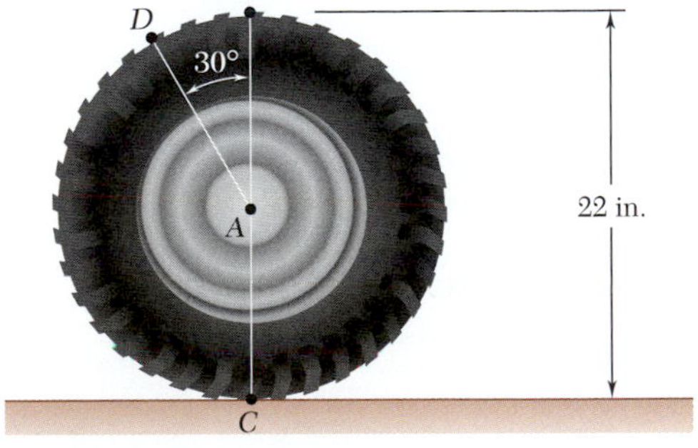

Fig. P15.249

15.250 Knowing that at the instant shown rod AB has a constant angular velocity of 15 rad/s counterclockwise, determine (*a*) the angular acceleration of rod DE, (*b*) the acceleration of point D.

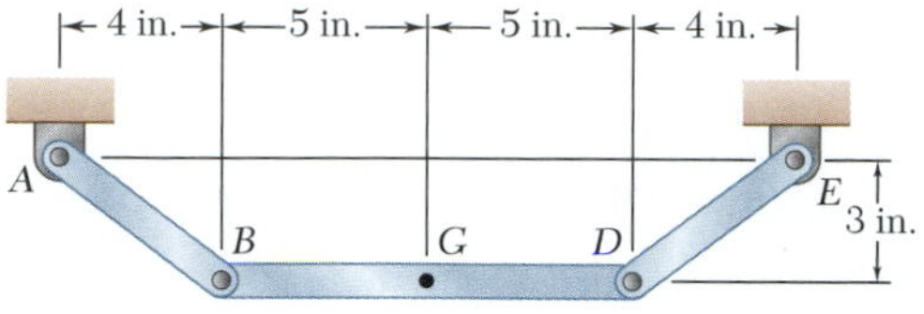

Fig. P15.250

15.251 Two rods AE and BD pass through holes drilled into a hexagonal block. (The holes are drilled in different planes so that the rods will not touch each other.) Knowing that at the instant considered rod AE rotates counterclockwise with a constant angular velocity $\boldsymbol{\omega}$, determine the relative velocity of the block with respect to each rod when $\theta = 45°$.

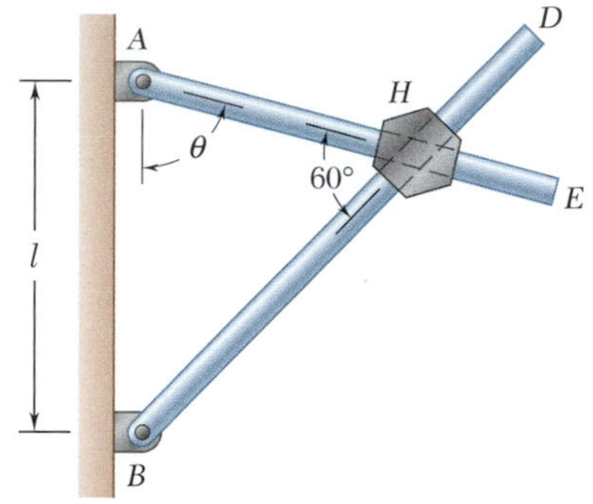

Fig. *P15.251*

15.252 The motion of pin D is guided by a slot cut in rod AB and a slot cut in the fixed plate. Knowing that at the instant shown rod AB rotates with an angular velocity of 3 rad/s and an angular acceleration of 5 rad/s^2, both clockwise, determine the acceleration of pin D.

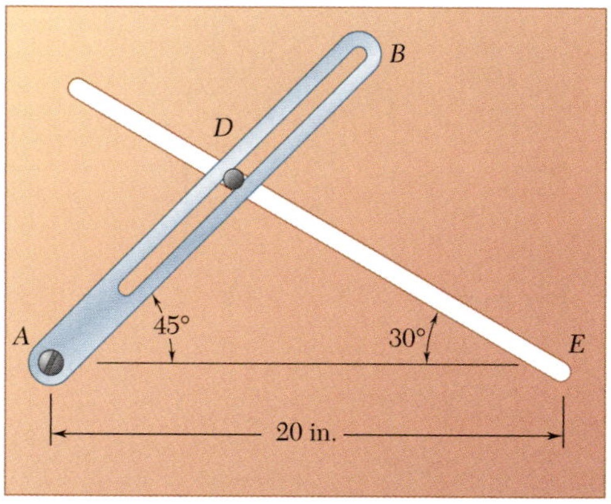

Fig. P15.252

15.253 Three rods are welded together to form the corner assembly shown that is attached to a fixed ball-and-socket joint at O. The end of rod OA moves on the inclined plane D that is perpendicular to the xy plane. The end of rod OB moves on the horizontal plane E. Knowing that at the instant shown $\mathbf{v}_B = -(15 \text{ in./s})\mathbf{k}$, determine (*a*) the angular velocity of the assembly, (*b*) the velocity of point C.

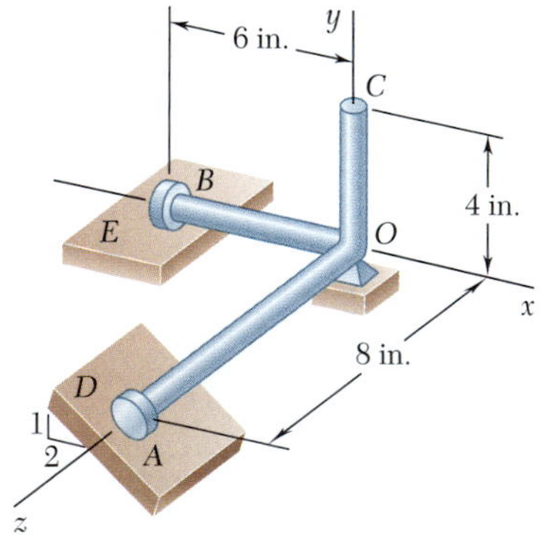

Fig. P15.253

15.254 The arm AB of length 5 m is used to provide an elevated platform for construction workers. In the position shown, arm AB is being raised at the constant rate $d\theta/dt = 0.25$ rad/s; simultaneously, the unit is being rotated counterclockwise about the Y axis at the constant rate $\omega_1 = 0.15$ rad/s. Knowing that $\theta = 20°$, determine the velocity and acceleration of point B.

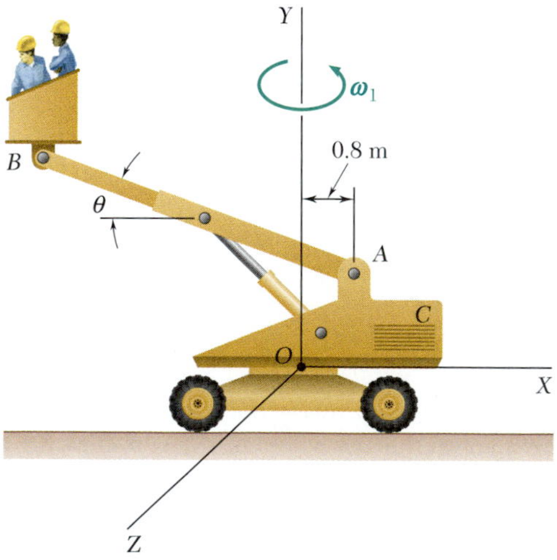

Fig. *P15.254*

Computer Problems

15.C1 The disk shown has a constant angular velocity of 500 rpm counterclockwise. Knowing that rod BD is 10 in. long, use computational software to determine and plot, for values of θ from 0 to 360°, (*a*) the velocity of collar D, (*b*) the angular velocity of rod BD. (*c*) Determine the two values of θ for which the speed of collar D is zero.

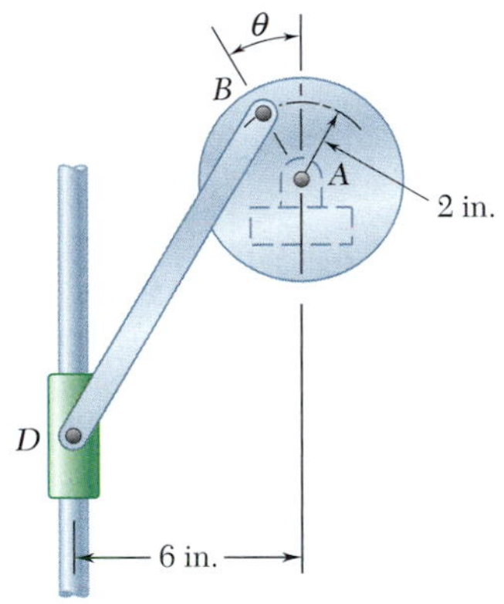

Fig. P15.C1

15.C2 The 4-in.-radius wheel shown rolls to the left with a constant velocity of 45 in./s. Knowing that the distance AD is 2.5 in., determine and plot, for values of β from 0 to 360°, (*a*) the velocity of collar B, (*b*) the angular velocity of rod AB.

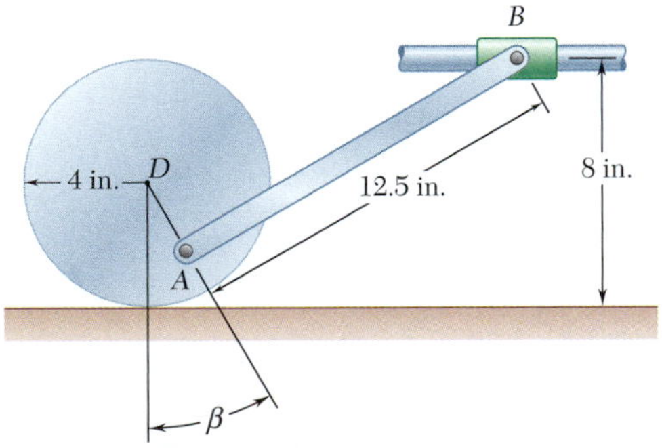

Fig. P15.C2

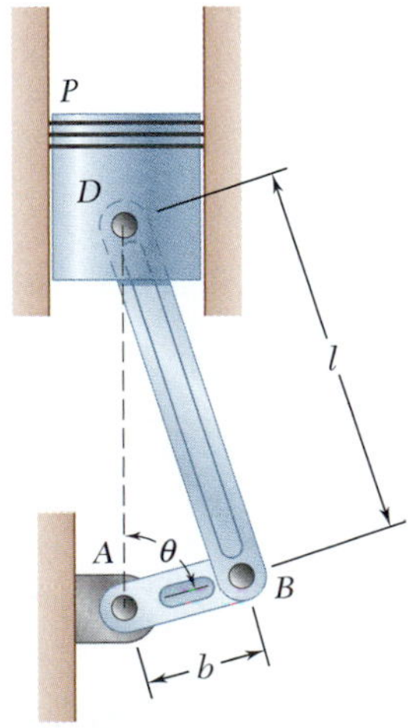

Fig. P15.C3

15.C3 In the engine system shown, $l = 8$ in. and $b = 3$ in. Knowing that crank AB rotates with a constant angular velocity of 1000 rpm clockwise, use computational software to determine and plot, for values of θ from 0 to 180°, (*a*) the angular velocity and angular acceleration of rod BD, (*b*) the velocity and acceleration of the piston P.

15.C4 Knowing that crank AB has a constant angular velocity of 160 rpm counterclockwise, determine and plot, for values of θ from 0 to 360°, (*a*) the angular velocity and angular acceleration of rod BD, (*b*) the velocity and acceleration of collar D.

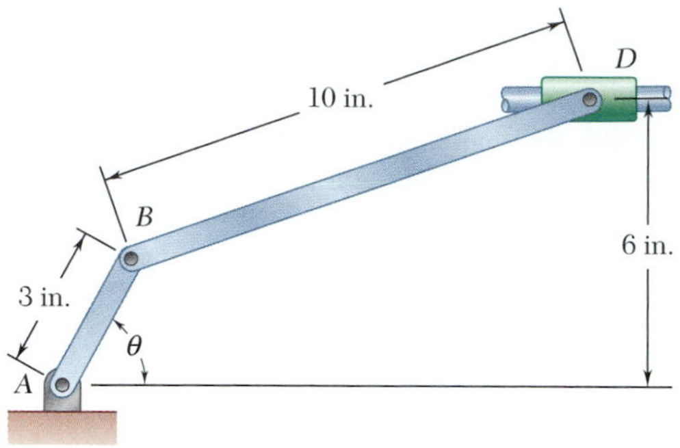

Fig. P15.C4

15.C5 Two rotating rods are connected by a slider block P as shown. Knowing that rod BP rotates with a constant angular velocity of 6 rad/s counterclockwise, use computational software to determine and plot, for values of θ from 0 to 180°, (*a*) the angular velocity and angular acceleration of rod AE. (*b*) Determine the value of θ for which the angular acceleration α_{AE} of rod AE is maximum and the corresponding value of α_{AE}.

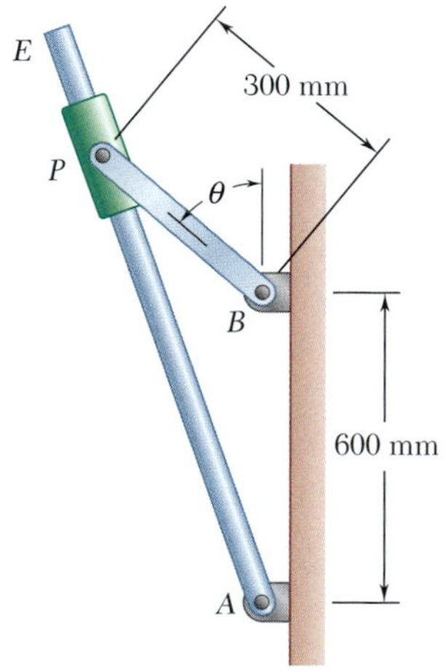

Fig. P15.C5

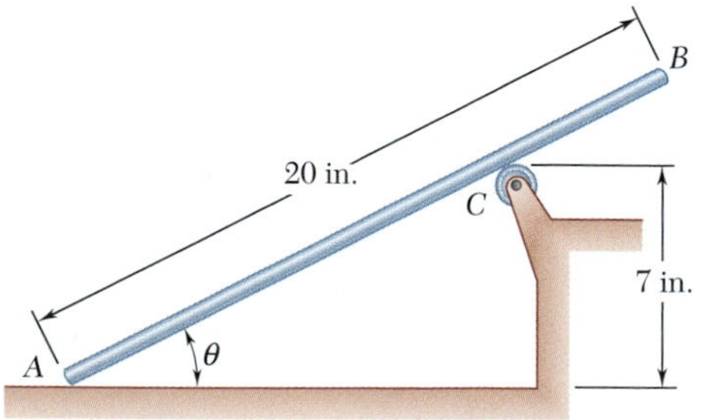

Fig. P15.C6

15.C6 Rod AB moves over a small wheel at C while end A moves to the right with a constant velocity of 9 in./s. Use computational software to determine and plot, for values of θ from 20° to 90°, (*a*) the velocity of point B, (*b*) the angular acceleration of the rod. (*c*) Determine the value of θ for which the angular acceleration α of the rod is maximum and the corresponding value of α.

15.C7 Rod AB of length 500 mm is connected by ball-and-socket joints to collars A and B, which slide along the two rods shown. Collar B moves towards support E at a constant speed of 400 mm/s. Denoting by d the distance from point C to collar B, use computational software to determine and plot the velocity of collar A for values of d from 0 to 300 mm.

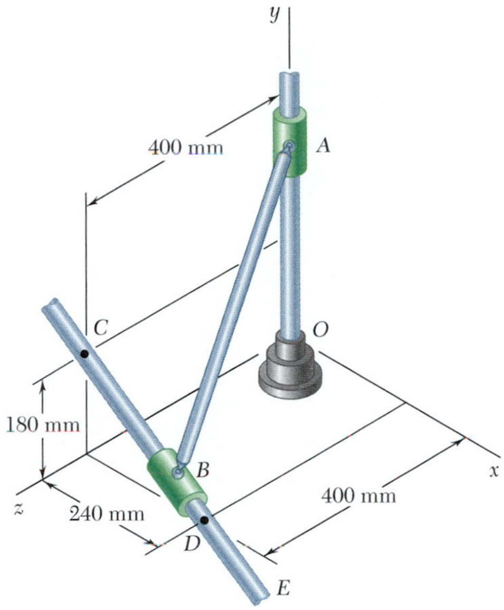

Fig. P15.C7

C H A P T E R

16

Plane Motion of Rigid Bodies: Forces and Accelerations

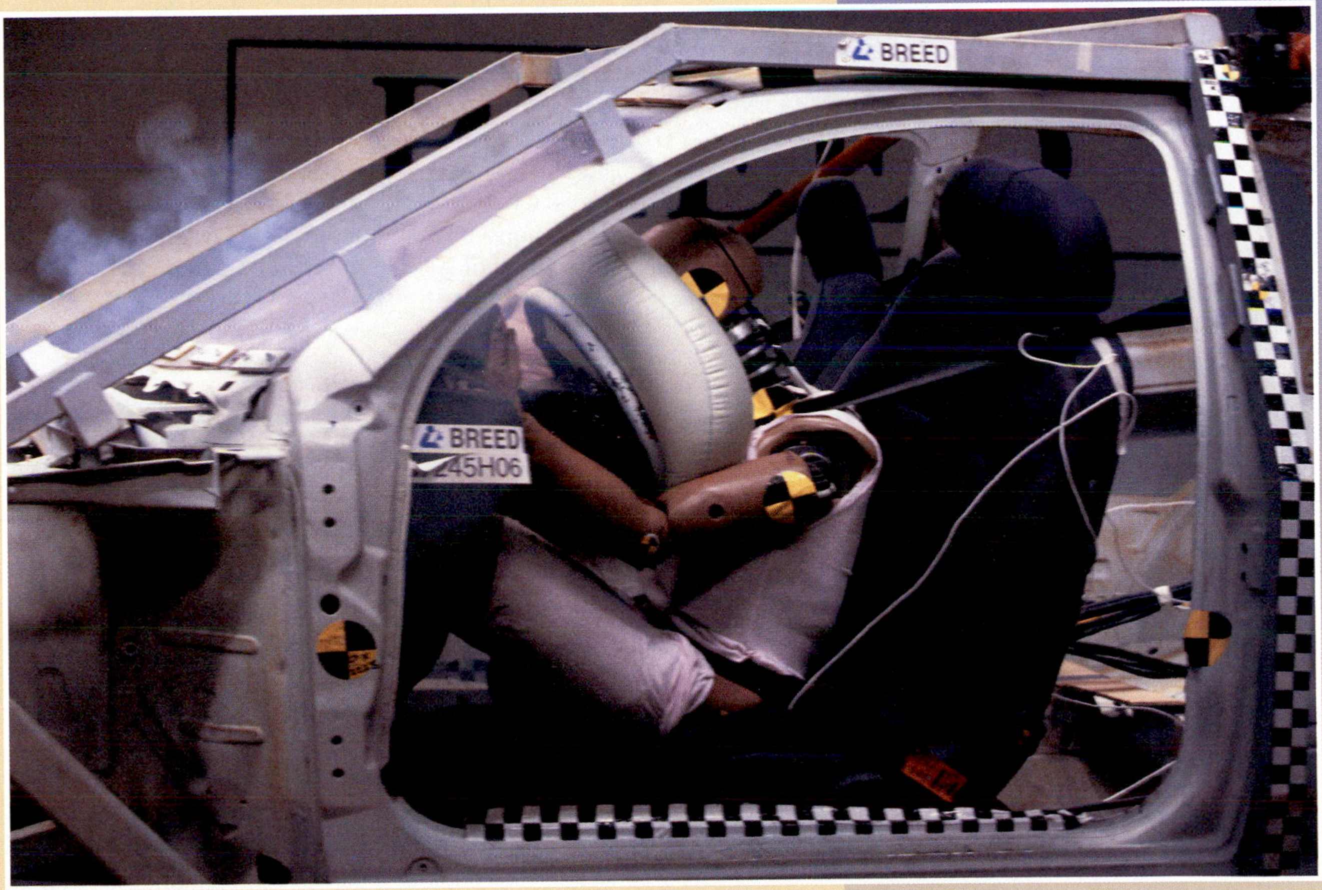

Accelerometers attached at several points on the crash test dummy provide acceleration data which engineers can use to estimate the forces generated during the motion. In this chapter and in Chaps. 17 and 18 you will learn to analyze the motion of a rigid body by considering the motion of its mass center, the motion relative to its mass center, and the external forces acting on it.

16.1. INTRODUCTION

In this chapter and in Chaps. 17 and 18, you will study the *kinetics of rigid bodies,* that is, the relations existing between the forces acting on a rigid body, the shape and mass of the body, and the motion produced. In Chaps. 12 and 13, you studied similar relations, assuming then that the body could be considered as a particle, that is, that its mass could be concentrated in one point and that all forces acted at that point. The shape of the body, as well as the exact location of the points of application of the forces, will now be taken into account. You will also be concerned not only with the motion of the body as a whole but also with the motion of the body about its mass center.

Our approach will be to consider rigid bodies as made of large numbers of particles and to use the results obtained in Chap. 14 for the motion of systems of particles. Specifically, two equations from Chap. 14 will be used: Eq. (14.16), $\Sigma\mathbf{F} = m\bar{\mathbf{a}}$, which relates the resultant of the external forces and the acceleration of the mass center G of the system of particles, and Eq. (14.23), $\Sigma\mathbf{M}_G = \dot{\mathbf{H}}_G$, which relates the moment resultant of the external forces and the angular momentum of the system of particles about G.

Except for Sec. 16.2, which applies to the most general case of the motion of a rigid body, the results derived in this chapter will be limited in two ways: (1) They will be restricted to the *plane motion* of rigid bodies, that is, to a motion in which each particle of the body remains at a constant distance from a fixed reference plane. (2) The rigid bodies considered will consist only of plane slabs and of bodies which are symmetrical with respect to the reference plane.† The study of the plane motion of nonsymmetrical three-dimensional bodies and, more generally, the motion of rigid bodies in three-dimensional space will be postponed until Chap. 18.

In Sec. 16.3, we define the angular momentum of a rigid body in plane motion and show that the rate of change of the angular momentum $\dot{\mathbf{H}}_G$ about the mass center is equal to the product $\bar{I}\boldsymbol{\alpha}$ of the centroidal mass moment of inertia $\bar{I}$ and the angular acceleration $\boldsymbol{\alpha}$ of the body. D'Alembert's principle, introduced in Sec. 16.4, is used to prove that the external forces acting on a rigid body are equivalent to a vector $m\bar{\mathbf{a}}$ attached at the mass center and a couple of moment $\bar{I}\boldsymbol{\alpha}$.

In Sec. 16.5, we derive the principle of transmissibility using only the parallelogram law and Newton's laws of motion, allowing us to remove this principle from the list of axioms (Sec. 1.2) required for the study of the statics and dynamics of rigid bodies.

Free-body-diagram equations are introduced in Sec. 16.6 and will be used in the solution of all problems involving the plane motion of rigid bodies.

After considering the plane motion of connected rigid bodies in Sec. 16.7, you will be prepared to solve a variety of problems involving the translation, centroidal rotation, and unconstrained motion of rigid bodies. In Sec. 16.8 and in the remaining part of the chapter, the solution of problems involving noncentroidal rotation, rolling motion, and other partially constrained plane motions of rigid bodies will be considered.

†Or, more generally, bodies which have a principal centroidal axis of inertia perpendicular to the reference plane.

16.2. EQUATIONS OF MOTION FOR A RIGID BODY

Consider a rigid body acted upon by several external forces $\mathbf{F}_1$, $\mathbf{F}_2$, $\mathbf{F}_3$, . . . (Fig. 16.1). We can assume that the body is made of a large number n of particles of mass Δm_i ($i = 1, 2, \ldots, n$) and apply the results obtained in Chap. 14 for a system of particles (Fig. 16.2). Considering first the motion of the mass center G of the body with respect to the newtonian frame of reference $Oxyz$, we recall Eq. (14.16) and write

$$\Sigma \mathbf{F} = m\overline{\mathbf{a}} \tag{16.1}$$

where m is the mass of the body and $\overline{\mathbf{a}}$ is the acceleration of the mass center G. Turning now to the motion of the body relative to the centroidal frame of reference $Gx'y'z'$, we recall Eq. (14.23) and write

$$\Sigma \mathbf{M}_G = \dot{\mathbf{H}}_G \tag{16.2}$$

where $\dot{\mathbf{H}}_G$ represents the rate of change of $\mathbf{H}_G$, the angular momentum about G of the system of particles forming the rigid body. In the following, $\mathbf{H}_G$ will simply be referred to as the *angular momentum of the rigid body about its mass center G*. Together Eqs. (16.1) and (16.2) express that *the system of the external forces is equipollent to the system consisting of the vector $m\overline{\mathbf{a}}$ attached at G and the couple of moment $\dot{\mathbf{H}}_G$* (Fig. 16.3).†

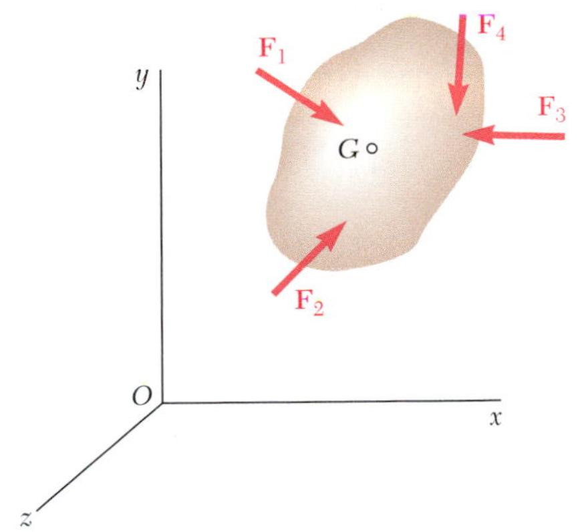

Fig. 16.1

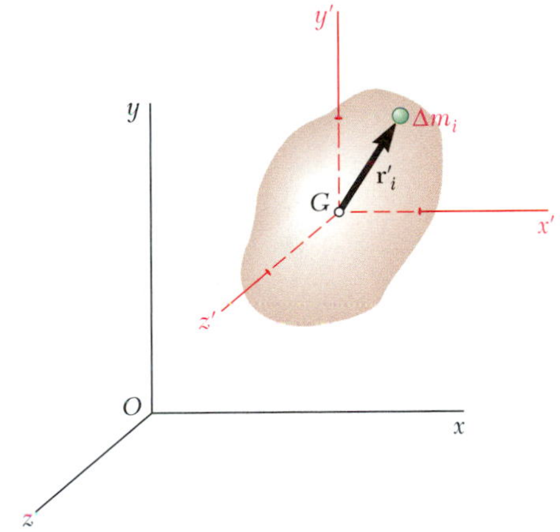

Fig. 16.2

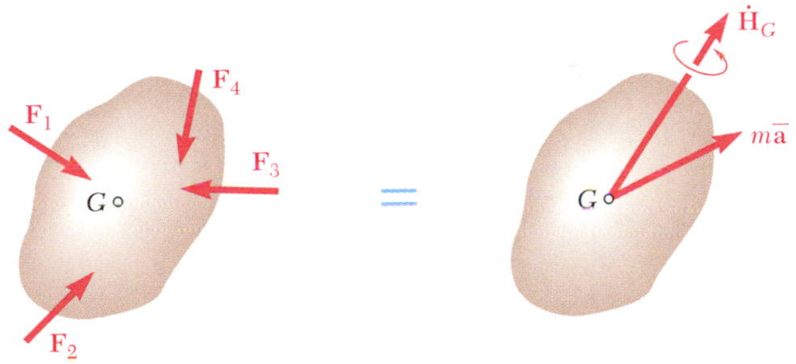

Fig. 16.3

Equations (16.1) and (16.2) apply in the most general case of the motion of a rigid body. In the rest of this chapter, however, our analysis will be limited to the *plane motion* of rigid bodies, that is, to a motion in which each particle remains at a constant distance from a fixed reference plane, and it will be assumed that the rigid bodies considered consist only of plane slabs and of bodies which are symmetrical with respect to the reference plane. Further study of the plane motion of nonsymmetrical three-dimensional bodies and of the motion of rigid bodies in three-dimensional space will be postponed until Chap. 18.

Photo 16.1 The system of external forces acting on the man and wakeboard includes the weights, the tension in the tow rope, and the forces exerted by the water and the air.

†Since the systems involved act on a rigid body, we could conclude at this point, by referring to Sec. 3.19, that the two systems are *equivalent* as well as equipollent and use red rather than blue equals signs in Fig. 16.3. However, by postponing this conclusion, we will be able to arrive at it independently (Secs. 16.4 and 18.5), thereby eliminating the necessity of including the principle of transmissibility among the axioms of mechanics (Sec. 16.5).

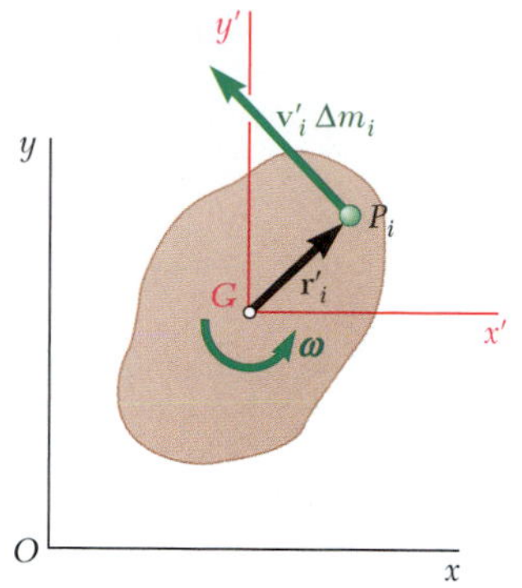

Fig. 16.4

16.3. ANGULAR MOMENTUM OF A RIGID BODY IN PLANE MOTION

Consider a rigid slab in plane motion. Assuming that the slab is made of a large number n of particles P_i of mass Δm_i and recalling Eq. (14.24) of Sec. 14.5, we note that the angular momentum $\mathbf{H}_G$ of the slab about its mass center G can be computed by taking the moments about G of the momenta of the particles of the slab in their motion with respect to either of the frames Oxy or $Gx'y'$ (Fig. 16.4). Choosing the latter course, we write

$$\mathbf{H}_G = \sum_{i=1}^{n} (\mathbf{r}'_i \times \mathbf{v}'_i\,\Delta m_i) \qquad (16.3)$$

where $\mathbf{r}'_i$ and $\mathbf{v}'_i\,\Delta m_i$ denote, respectively, the position vector and the linear momentum of the particle P_i relative to the centroidal frame of reference $Gx'y'$. But since the particle belongs to the slab, we have $\mathbf{v}'_i = \boldsymbol{\omega} \times \mathbf{r}'_i$, where $\boldsymbol{\omega}$ is the angular velocity of the slab at the instant considered. We write

$$\mathbf{H}_G = \sum_{i=1}^{n} [\mathbf{r}'_i \times (\boldsymbol{\omega} \times \mathbf{r}'_i)\,\Delta m_i]$$

Referring to Fig. 16.4, we easily verify that the expression obtained represents a vector of the same direction as $\boldsymbol{\omega}$ (that is, perpendicular to the slab) and of magnitude equal to $\omega \Sigma r_i'^2\,\Delta m_i$. Recalling that the sum $\Sigma r_i'^2\,\Delta m_i$ represents the moment of inertia $\bar{I}$ of the slab about a centroidal axis perpendicular to the slab, we conclude that the angular momentum $\mathbf{H}_G$ of the slab about its mass center is

$$\mathbf{H}_G = \bar{I}\boldsymbol{\omega} \qquad (16.4)$$

Differentiating both members of Eq. (16.4) we obtain

$$\dot{\mathbf{H}}_G = \bar{I}\dot{\boldsymbol{\omega}} = \bar{I}\boldsymbol{\alpha} \qquad (16.5)$$

Thus the rate of change of the angular momentum of the slab is represented by a vector of the same direction as $\boldsymbol{\alpha}$ (that is, perpendicular to the slab) and of magnitude $\bar{I}\alpha$.

It should be kept in mind that the results obtained in this section have been derived for a rigid slab in plane motion. As you will see in Chap. 18, they remain valid in the case of the plane motion of rigid bodies which are symmetrical with respect to the reference plane.† However, they do not apply in the case of nonsymmetrical bodies or in the case of three-dimensional motion.

†Or, more generally, bodies which have a principal centroidal axis of inertia perpendicular to the reference plane.

16.4. PLANE MOTION OF A RIGID BODY. D'ALEMBERT'S PRINCIPLE

Consider a rigid slab of mass m moving under the action of several external forces $\mathbf{F}_1$, $\mathbf{F}_2$, $\mathbf{F}_3$, . . . , contained in the plane of the slab (Fig. 16.5). Substituting for $\dot{\mathbf{H}}_G$ from Eq. (16.5) into Eq. (16.2) and writing the fundamental equations of motion (16.1) and (16.2) in scalar form, we have

$$\Sigma F_x = m\bar{a}_x \qquad \Sigma F_y = m\bar{a}_y \qquad \Sigma M_G = \bar{I}\alpha \tag{16.6}$$

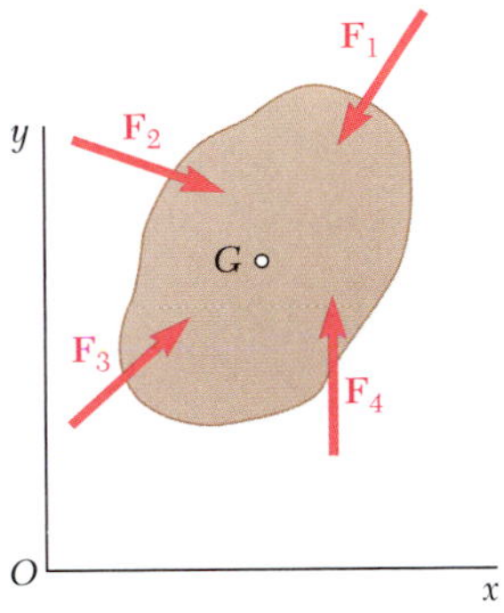

Fig. 16.5

Equations (16.6) show that the acceleration of the mass center G of the slab and its angular acceleration $\boldsymbol{\alpha}$ are easily obtained once the resultant of the external forces acting on the slab and their moment resultant about G have been determined. Given appropriate initial conditions, the coordinates $\bar{x}$ and $\bar{y}$ of the mass center and the angular coordinate θ of the slab can then be obtained by integration at any instant t. Thus *the motion of the slab is completely defined by the resultant and moment resultant about G of the external forces acting on it.*

This property, which will be extended in Chap. 18 to the case of the three-dimensional motion of a rigid body, is characteristic of the motion of a rigid body. Indeed, as we saw in Chap. 14, the motion of a system of particles which are not rigidly connected will in general depend upon the specific external forces acting on the various particles, as well as upon the internal forces.

Since the motion of a rigid body depends only upon the resultant and moment resultant of the external forces acting on it, it follows that *two systems of forces which are equipollent,* that is, which have the same resultant and the same moment resultant, *are also equivalent;* that is, they have exactly the same effect on a given rigid body.†

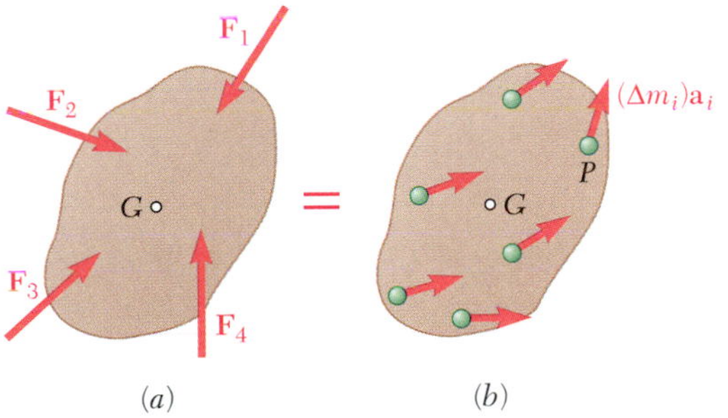

Fig. 16.6

Consider in particular the system of the external forces acting on a rigid body (Fig. 16.6*a*) and the system of the effective forces associated with the particles forming the rigid body (Fig. 16.6*b*). It was shown in Sec. 14.2 that the two systems thus defined are equipollent. But since the particles considered now form a rigid body, it follows from the discussion above that the two systems are also equivalent. We can thus state that *the external forces acting on a rigid body are equivalent to the effective forces of the various particles forming the body.* This statement is referred to as *d'Alembert's principle* after the French mathematician Jean le Rond d'Alembert (1717–1783), even though d'Alembert's original statement was written in a somewhat different form.

The fact that the system of external forces is *equivalent* to the system of the effective forces has been emphasized by the use of a red equals sign in Fig. 16.6 and also in Fig. 16.7, where using results obtained earlier in this section, we have replaced the effective forces by a vector $m\bar{\mathbf{a}}$ attached at the mass center G of the slab and a couple of moment $\bar{I}\boldsymbol{\alpha}$.

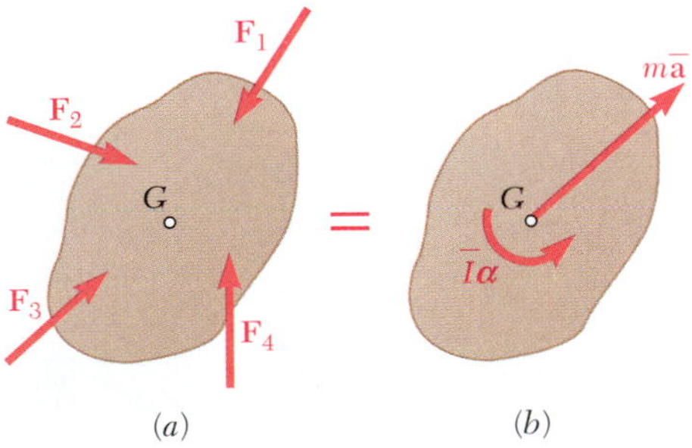

Fig. 16.7

†This result has already been derived in Sec. 3.19 from the principle of transmissibility (Sec. 3.3). The present derivation is independent of that principle, however, and will make possible its elimination from the axioms of mechanics (Sec. 16.5).

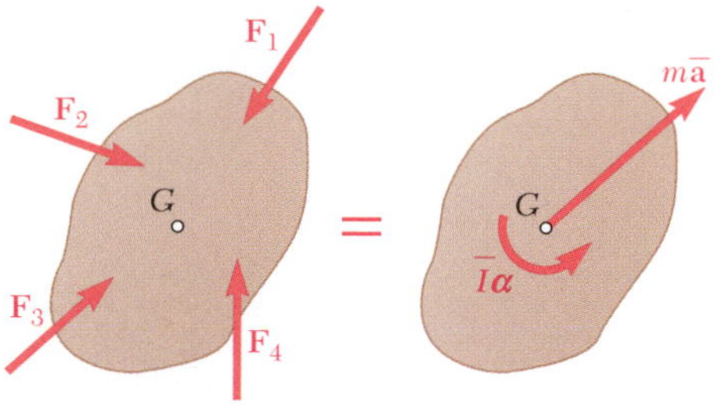

Fig. 16.7 (***repeated***)

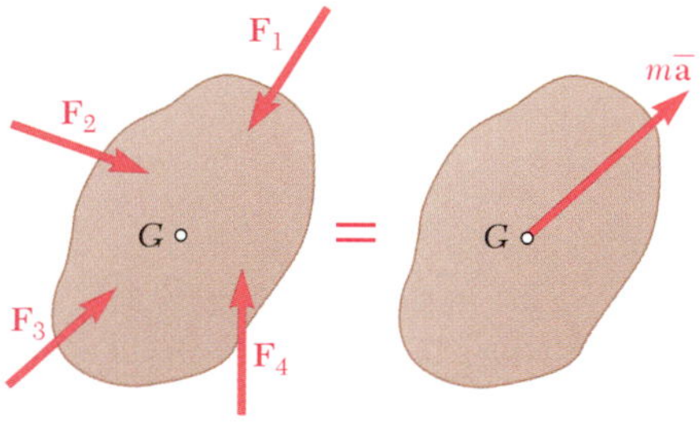

Fig. 16.8 Translation.

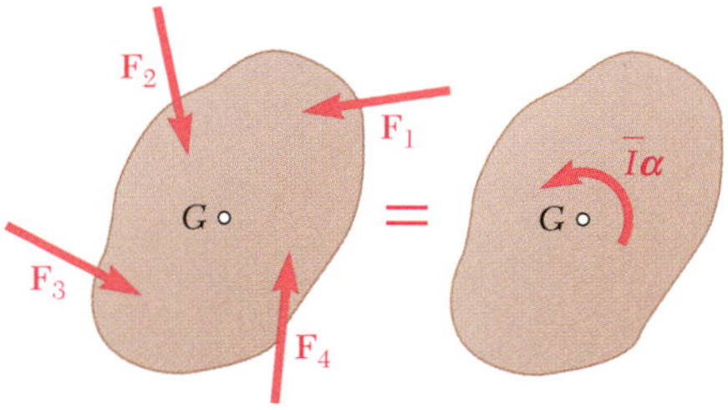

Fig. 16.9 Centroidal rotation.

Photo 16.2 When the vehicle is at rest with the engine running, the motion of each gear is an example of centroidal rotation.

Translation. In the case of a body in translation, the angular acceleration of the body is identically equal to zero and its effective forces reduce to the vector $m\bar{\mathbf{a}}$ attached at G (Fig. 16.8). Thus, the resultant of the external forces acting on a rigid body in translation passes through the mass center of the body and is equal to $m\bar{\mathbf{a}}$.

Centroidal Rotation. When a slab, or, more generally, a body symmetrical with respect to the reference plane, rotates about a fixed axis perpendicular to the reference plane and passing through its mass center G, we say that the body is in *centroidal rotation*. Since the acceleration $\bar{\mathbf{a}}$ is identically equal to zero, the effective forces of the body reduce to the couple $\bar{I}\boldsymbol{\alpha}$ (Fig. 16.9). Thus, the external forces acting on a body in centroidal rotation are equivalent to a couple of moment $\bar{I}\boldsymbol{\alpha}$.

General Plane Motion. Comparing Fig. 16.7 with Figs. 16.8 and 16.9, we observe that from the point of view of *kinetics*, the most general plane motion of a rigid body symmetrical with respect to the reference plane can be replaced by the sum of a translation and a centroidal rotation. We should note that this statement is more restrictive than the similar statement made earlier from the point of view of *kinematics* (Sec. 15.5), since we now require that the mass center of the body be selected as the reference point.

Referring to Eqs. (16.6), we observe that the first two equations are identical with the equations of motion of a particle of mass m acted upon by the given forces $\mathbf{F}_1$, $\mathbf{F}_2$, $\mathbf{F}_3$, . . . We thus check that *the mass center G of a rigid body in plane motion moves as if the entire mass of the body were concentrated at that point, and as if all the external forces acted on it.* We recall that this result has already been obtained in Sec. 14.4 in the general case of a system of particles, the particles being not necessarily rigidly connected. We also note, as we did in Sec. 14.4, that the system of the external forces does not, in general, reduce to a single vector $m\bar{\mathbf{a}}$ attached at G. Therefore, in the general case of the plane motion of a rigid body, *the resultant of the external forces acting on the body does not pass through the mass center of the body.*

Finally, it should be observed that the last of Eqs. (16.6) would still be valid if the rigid body, while subjected to the same applied forces, were constrained to rotate about a fixed axis through G. Thus, *a rigid body in plane motion rotates about its mass center as if this point were fixed.*

*16.5. A REMARK ON THE AXIOMS OF THE MECHANICS OF RIGID BODIES

The fact that two equipollent systems of external forces acting on a rigid body are also equivalent, that is, have the same effect on that rigid body, has already been established in Sec. 3.19. But there it was derived from the *principle of transmissibility,* one of the axioms used in our study of the statics of rigid bodies. It should be noted that this axiom has not been used in the present chapter because Newton's second and third laws of motion make its use unnecessary in the study of the dynamics of rigid bodies.

In fact, the principle of transmissibility may now be *derived* from the other axioms used in the study of mechanics. This principle stated, without proof (Sec. 3.3), that the conditions of equilibrium or motion of a rigid body remain unchanged if a force **F** acting at a given point of the rigid body is replaced by a force **F**′ of the same magnitude and same direction, but acting at a different point, provided that the two forces have the same line of action. But since **F** and **F**′ have the same moment about any given point, it is clear that they form two equipollent systems of external forces. Thus, we may now *prove,* as a result of what we established in the preceding section, that **F** and **F**′ have the same effect on the rigid body (Fig. 3.3).

The principle of transmissibility can therefore be removed from the list of axioms required for the study of the mechanics of rigid bodies. These axioms are reduced to the parallelogram law of addition of vectors and to Newton's laws of motion.

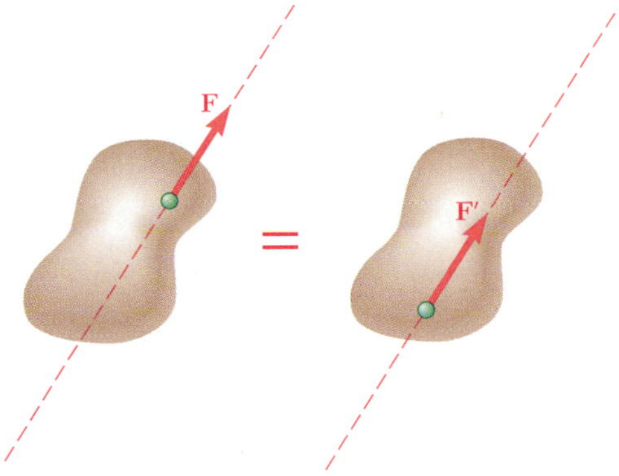

Fig. 3.3 (*repeated*)

16.6. SOLUTION OF PROBLEMS INVOLVING THE MOTION OF A RIGID BODY

We saw in Sec. 16.4 that when a rigid body is in plane motion, there exists a fundamental relation between the forces $\mathbf{F}_1$, $\mathbf{F}_2$, $\mathbf{F}_3$, . . . , acting on the body, the acceleration $\bar{\mathbf{a}}$ of its mass center, and the angular acceleration $\boldsymbol{\alpha}$ of the body. This relation, which is represented in Fig. 16.7 in the form of a *free-body-diagram equation,* can be used to determine the acceleration $\bar{\mathbf{a}}$ and the angular acceleration $\boldsymbol{\alpha}$ produced by a given system of forces acting on a rigid body or, conversely, to determine the forces which produce a given motion of the rigid body.

The three algebraic equations (16.6) can be used to solve problems of plane motion.† However, our experience in statics suggests that the solution of many problems involving rigid bodies could be simplified by an appropriate choice of the point about which the moments of the forces are computed. It is therefore preferable to remember the relation existing between the forces and the accelerations in the pictorial form shown in Fig. 16.7 and to derive from this fundamental relation the component or moment equations which fit best the solution of the problem under consideration.

The fundamental relation shown in Fig. 16.7 can be presented in an alternative form if we add to the external forces an inertia vector $-m\bar{\mathbf{a}}$ of sense opposite to that of $\bar{\mathbf{a}}$, attached at G, and an inertia couple $-\bar{I}\boldsymbol{\alpha}$ of moment equal in magnitude to $\bar{I}\alpha$ and of sense opposite to that of $\boldsymbol{\alpha}$ (Fig. 16.10). The system obtained is equivalent to zero, and the rigid body is said to be in *dynamic equilibrium.*

Whether the principle of equivalence of external and effective forces is directly applied, as in Fig. 16.7, or whether the concept of dynamic equilibrium is introduced, as in Fig. 16.10, the use of free-body-diagram equations showing vectorially the relationship existing between the forces applied on the rigid body and the resulting linear and angular accelerations presents considerable advantages over the blind application of formulas (16.6). These advantages can be summarized as follows:

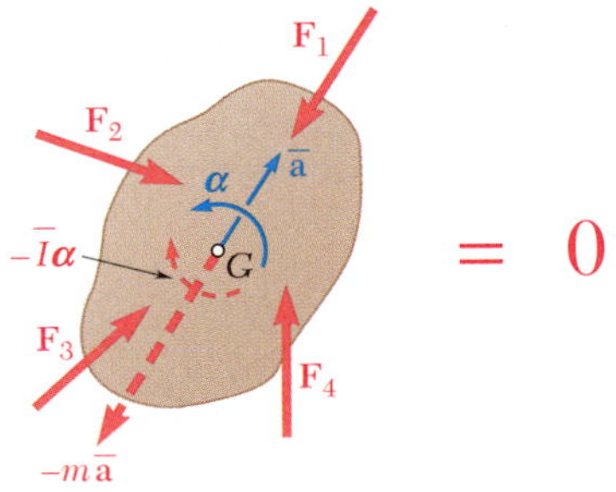

Fig. 16.10

†We recall that the last of Eqs. (16.6) is valid only in the case of the plane motion of a rigid body symmetrical with respect to the reference plane. In all other cases, the methods of Chap. 18 should be used.

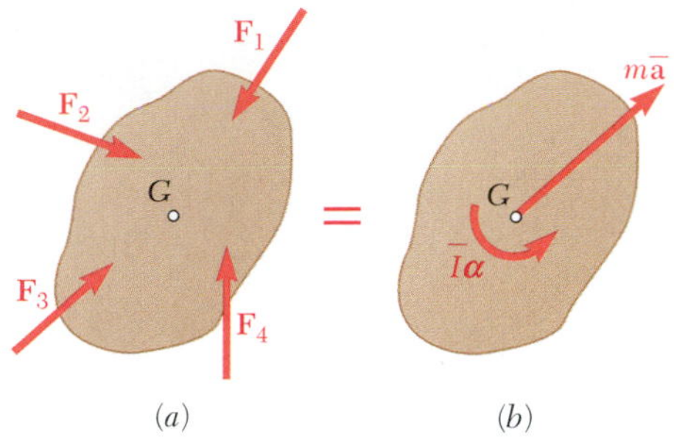

Fig. 16.7 (*repeated*)

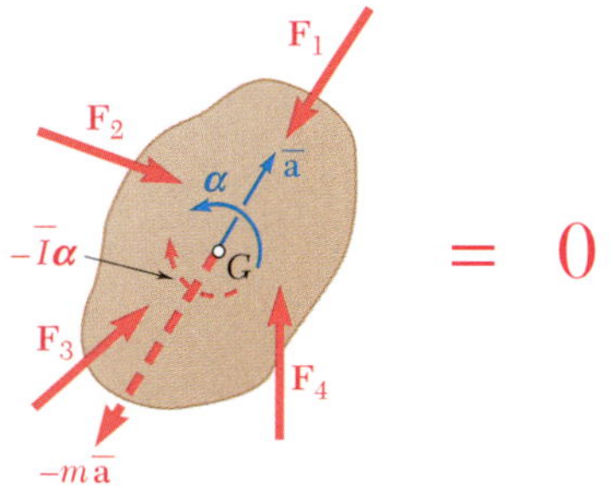

Fig. 16.10 (*repeated*)

Photo 16.3 The forklift and moving load can be analyzed as a system of two connected rigid bodies in plane motion.

1. The use of a pictorial representation provides a much clearer understanding of the effect of the forces on the motion of the body.
2. This approach makes it possible to divide the solution of a dynamics problem into two parts: In the first part, the analysis of the kinematic and kinetic characteristics of the problem leads to the free-body diagrams of Fig. 16.7 or 16.10; in the second part, the diagram obtained is used to analyze the various forces and vectors involved by the methods of Chap. 3.
3. A unified approach is provided for the analysis of the plane motion of a rigid body, regardless of the particular type of motion involved. While the kinematics of the various motions considered may vary from one case to the other, the approach to the kinetics of the motion is consistently the same. In every case a diagram will be drawn showing the external forces, the vector $m\bar{\mathbf{a}}$ associated with the motion of G, and the couple $\bar{I}\boldsymbol{\alpha}$ associated with the rotation of the body about G.
4. The resolution of the plane motion of a rigid body into a translation and a centroidal rotation, which is used here, is a basic concept which can be applied effectively throughout the study of mechanics. It will be used again in Chap. 17 with the method of work and energy and the method of impulse and momentum.
5. As you will see in Chap. 18, this approach can be extended to the study of the general three-dimensional motion of a rigid body. The motion of the body will again be resolved into a translation and a rotation about the mass center, and free-body-diagram equations will be used to indicate the relationship existing between the external forces and the rates of change of the linear and angular momentum of the body.

16.7. SYSTEMS OF RIGID BODIES

The method described in the preceding section can also be used in problems involving the plane motion of several connected rigid bodies. For each part of the system, a diagram similar to Fig. 16.7 or Fig. 16.10 can be drawn. The equations of motion obtained from these diagrams are solved simultaneously.

In some cases, as in Sample Prob. 16.3, a single diagram can be drawn for the entire system. This diagram should include all the external forces, as well as the vectors $m\bar{\mathbf{a}}$ and the couples $\bar{I}\boldsymbol{\alpha}$ associated with the various parts of the system. However, internal forces such as the forces exerted by connecting cables, can be omitted since they occur in pairs of equal and opposite forces and are thus equipollent to zero. The equations obtained by expressing that the system of the external forces is equipollent to the system of the effective forces can be solved for the remaining unknowns.†

It is not possible to use this second approach in problems involving more than three unknowns, since only three equations of motion are available when a single diagram is used. We need not elaborate upon this point, since the discussion involved would be completely similar to that given in Sec. 6.11 in the case of the equilibrium of a system of rigid bodies.

†Note that we cannot speak of *equivalent* systems since we are not dealing with a single rigid body.

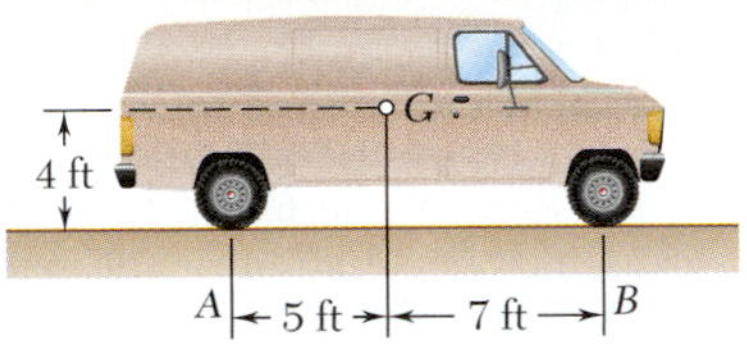

SAMPLE PROBLEM 16.1

When the forward speed of the truck shown was 30 ft/s, the brakes were suddenly applied, causing all four wheels to stop rotating. It was observed that the truck skidded to rest in 20 ft. Determine the magnitude of the normal reaction and of the friction force at each wheel as the truck skidded to rest.

SOLUTION

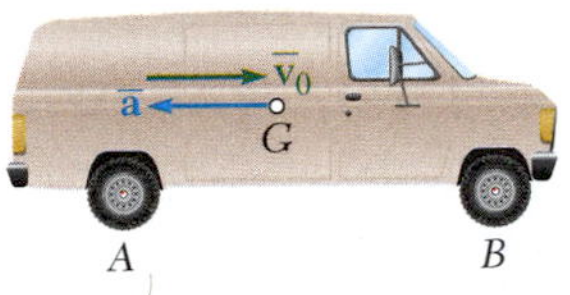

Kinematics of Motion. Choosing the positive sense to the right and using the equations of uniformly accelerated motion, we write

$$\bar{v}_0 = +30 \text{ ft/s} \qquad \bar{v}^2 = \bar{v}_0^2 + 2\bar{a}\bar{x} \qquad 0 = (30)^2 + 2\bar{a}(20)$$
$$\bar{a} = -22.5 \text{ ft/s}^2 \qquad \bar{\mathbf{a}} = 22.5 \text{ ft/s}^2 \leftarrow$$

Equations of Motion. The external forces consist of the weight **W** of the truck and of the normal reactions and friction forces at the wheels. (The vectors $\mathbf{N}_A$ and $\mathbf{F}_A$ represent the sum of the reactions at the rear wheels, while $\mathbf{N}_B$ and $\mathbf{F}_B$ represent the sum of the reactions at the front wheels.) Since the truck is in translation, the effective forces reduce to the vector $m\bar{\mathbf{a}}$ attached at G. Three equations of motion are obtained by expressing that the system of the external forces is equivalent to the system of the effective forces.

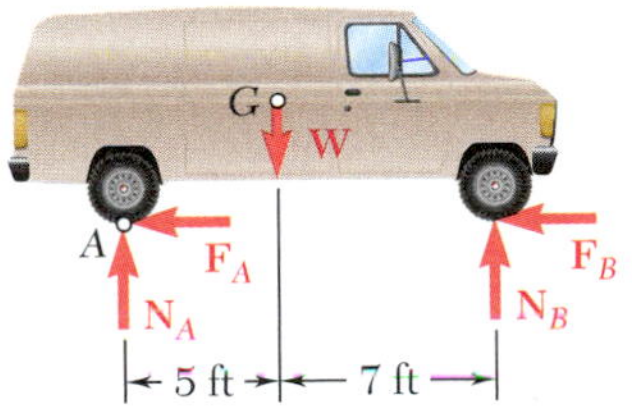

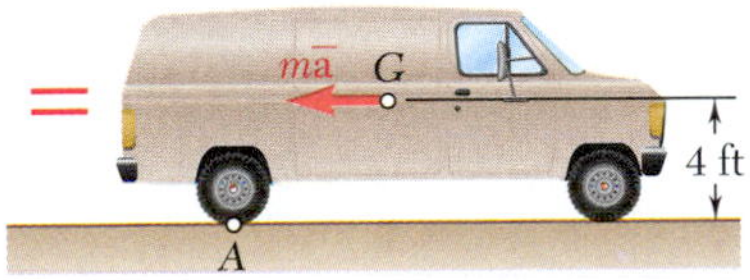

$$+\uparrow \Sigma F_y = \Sigma(F_y)_{\text{eff}}: \qquad N_A + N_B - W = 0$$

Since $F_A = \mu_k N_A$ and $F_B = \mu_k N_B$, where μ_k is the coefficient of kinetic friction, we find that

$$F_A + F_B = \mu_k(N_A + N_B) = \mu_k W$$

$$\overset{+}{\rightarrow} \Sigma F_x = \Sigma(F_x)_{\text{eff}}: \qquad -(F_A + F_B) = -m\bar{a}$$
$$-\mu_k W = -\frac{W}{32.2 \text{ ft/s}^2}(22.5 \text{ ft/s}^2)$$
$$\mu_k = 0.699$$

$$+\circlearrowleft \Sigma M_A = \Sigma(M_A)_{\text{eff}}: \qquad -W(5 \text{ ft}) + N_B(12 \text{ ft}) = m\bar{a}(4 \text{ ft})$$
$$-W(5 \text{ ft}) + N_B(12 \text{ ft}) = \frac{W}{32.2 \text{ ft/s}^2}(22.5 \text{ ft/s}^2)(4 \text{ ft})$$
$$N_B = 0.650W$$
$$F_B = \mu_k N_B = (0.699)(0.650W) \qquad F_B = 0.454W$$

$$+\uparrow \Sigma F_y = \Sigma(F_y)_{\text{eff}}: \qquad N_A + N_B - W = 0$$
$$N_A + 0.650W - W = 0$$
$$N_A = 0.350W$$
$$F_A = \mu_k N_A = (0.699)(0.350W) \qquad F_A = 0.245W$$

Reactions at Each Wheel. Recalling that the values computed above represent the sum of the reactions at the two front wheels or the two rear wheels, we obtain the magnitude of the reactions at each wheel by writing

$$N_{\text{front}} = \tfrac{1}{2}N_B = 0.325W \qquad N_{\text{rear}} = \tfrac{1}{2}N_A = 0.175W \quad ◀$$
$$F_{\text{front}} = \tfrac{1}{2}F_B = 0.227W \qquad F_{\text{rear}} = \tfrac{1}{2}F_A = 0.122W \quad ◀$$

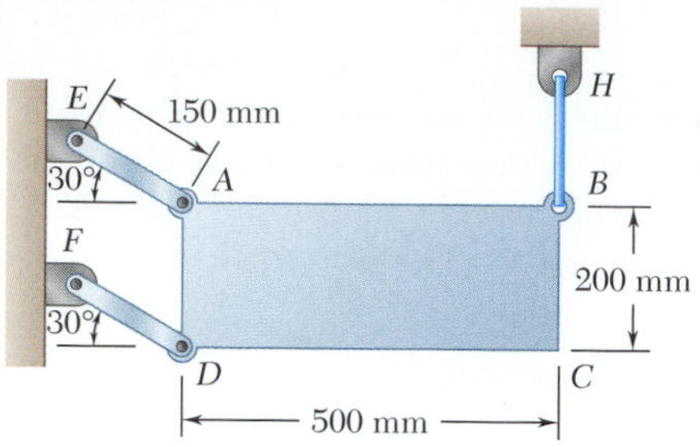

SAMPLE PROBLEM 16.2

The thin plate *ABCD* of mass 8 kg is held in the position shown by the wire *BH* and two links *AE* and *DF*. Neglecting the mass of the links, determine immediately after wire *BH* has been cut (*a*) the acceleration of the plate, (*b*) the force in each link.

SOLUTION

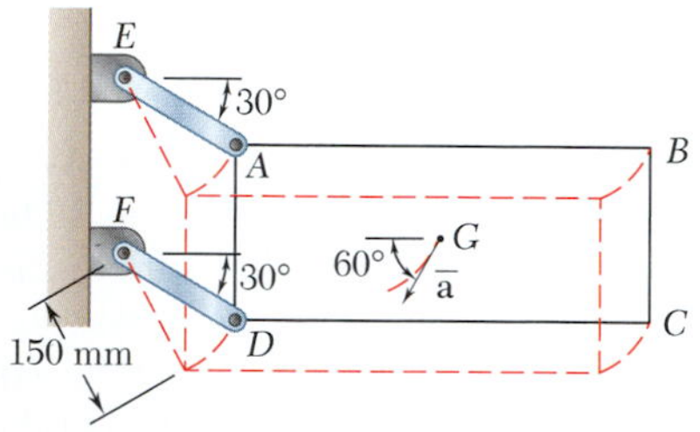

Kinematics of Motion. After wire *BH* has been cut, we observe that corners *A* and *D* move along parallel circles of radius 150 mm centered, respectively, at *E* and *F*. The motion of the plate is thus a curvilinear translation; the particles forming the plate move along parallel circles of radius 150 mm.

At the instant wire *BH* is cut, the velocity of the plate is zero. Thus the acceleration $\bar{\mathbf{a}}$ of the mass center *G* of the plate is tangent to the circular path which will be described by *G*.

Equations of Motion. The external forces consist of the weight **W** and the forces $\mathbf{F}_{AE}$ and $\mathbf{F}_{DF}$ exerted by the links. Since the plate is in translation, the effective forces reduce to the vector $m\bar{\mathbf{a}}$ attached at *G* and directed along the *t* axis. A free-body-diagram equation is drawn to show that the system of the external forces is equivalent to the system of the effective forces.

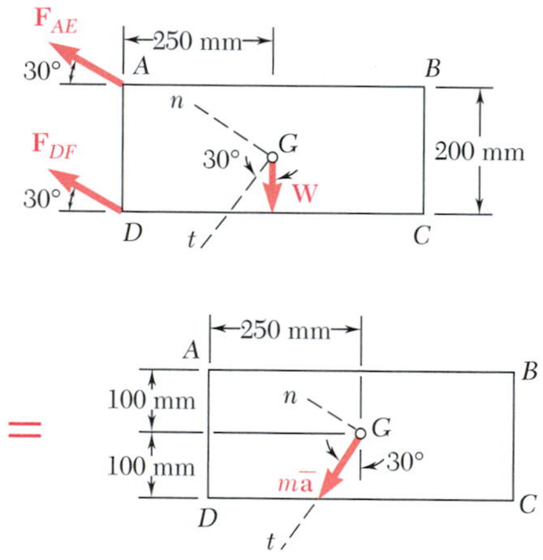

a. Acceleration of the Plate.

$+\swarrow \Sigma F_t = \Sigma(F_t)_{\text{eff}}$:

$$W \cos 30° = m\bar{a}$$
$$mg \cos 30° = m\bar{a}$$
$$\bar{a} = g \cos 30° = (9.81\ \text{m/s}^2) \cos 30° \qquad (1)$$

$$\bar{\mathbf{a}} = 8.50\ \text{m/s}^2 \measuredangle 60° \quad ◀$$

b. Forces in Links AE and DF.

$+\nwarrow \Sigma F_n = \Sigma(F_n)_{\text{eff}}$: $\quad F_{AE} + F_{DF} - W \sin 30° = 0 \qquad (2)$

$+\circlearrowright \Sigma M_G = \Sigma(M_G)_{\text{eff}}$:

$$(F_{AE} \sin 30°)(250\ \text{mm}) - (F_{AE} \cos 30°)(100\ \text{mm}) + (F_{DF} \sin 30°)(250\ \text{mm}) + (F_{DF} \cos 30°)(100\ \text{mm}) = 0$$
$$38.4 F_{AE} + 211.6 F_{DF} = 0$$
$$F_{DF} = -0.1815 F_{AE} \qquad (3)$$

Substituting for F_{DF} from (3) into (2), we write

$$F_{AE} - 0.1815 F_{AE} - W \sin 30° = 0$$
$$F_{AE} = 0.6109 W$$
$$F_{DF} = -0.1815(0.6109 W) = -0.1109 W$$

Noting that $W = mg = (8\ \text{kg})(9.81\ \text{m/s}^2) = 78.48$ N, we have

$$F_{AE} = 0.6109(78.48\ \text{N}) \qquad F_{AE} = 47.9\ \text{N}\ T \quad ◀$$
$$F_{DF} = -0.1109(78.48\ \text{N}) \qquad F_{DF} = 8.70\ \text{N}\ C \quad ◀$$

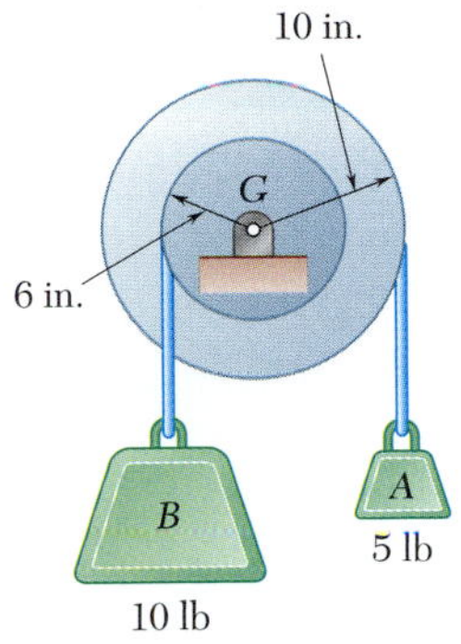

SAMPLE PROBLEM 16.3

A pulley weighing 12 lb and having a radius of gyration of 8 in. is connected to two blocks as shown. Assuming no axle friction, determine the angular acceleration of the pulley and the acceleration of each block.

SOLUTION

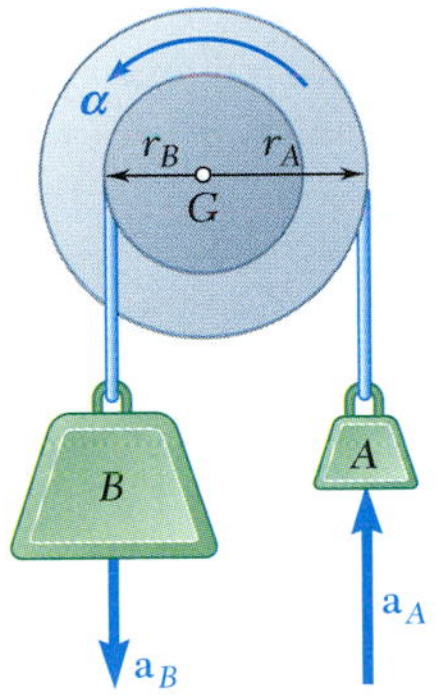

Sense of Motion. Although an arbitrary sense of motion can be assumed (since no friction forces are involved) and later checked by the sign of the answer, we may prefer to determine the actual sense of rotation of the pulley first. The weight of block B required to maintain the equilibrium of the pulley when it is acted upon by the 5-lb block A is first determined. We write

$$+\circlearrowleft \Sigma M_G = 0: \qquad W_B(6 \text{ in.}) - (5 \text{ lb})(10 \text{ in.}) = 0 \qquad W_B = 8.33 \text{ lb}$$

Since block B actually weighs 10 lb, the pulley will rotate counterclockwise.

Kinematics of Motion. Assuming $\boldsymbol{\alpha}$ counterclockwise and noting that $a_A = r_A\alpha$ and $a_B = r_B\alpha$, we obtain

$$\mathbf{a}_A = (\tfrac{10}{12} \text{ ft})\alpha \uparrow \qquad \mathbf{a}_B = (\tfrac{6}{12} \text{ ft})\alpha \downarrow$$

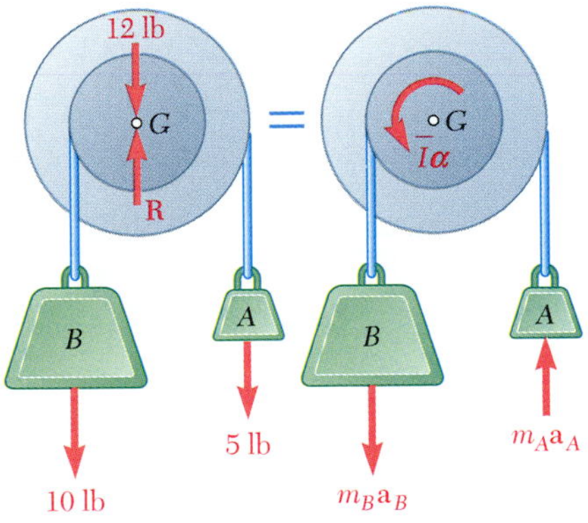

Equations of Motion. A single system consisting of the pulley and the two blocks is considered. Forces external to this system consist of the weights of the pulley and the two blocks and of the reaction at G. (The forces exerted by the cables on the pulley and on the blocks are internal to the system considered and cancel out.) Since the motion of the pulley is a centroidal rotation and the motion of each block is a translation, the effective forces reduce to the couple $\bar{I}\boldsymbol{\alpha}$ and the two vectors $m\mathbf{a}_A$ and $m\mathbf{a}_B$. The centroidal moment of inertia of the pulley is

$$\bar{I} = m\bar{k}^2 = \frac{W}{g}\bar{k}^2 = \frac{12 \text{ lb}}{32.2 \text{ ft/s}^2}(\tfrac{8}{12} \text{ ft})^2 = 0.1656 \text{ lb} \cdot \text{ft} \cdot \text{s}^2$$

Since the system of the external forces is equipollent to the system of the effective forces, we write

$$+\circlearrowleft \Sigma M_G = \Sigma(M_G)_{\text{eff}}:$$

$$(10 \text{ lb})(\tfrac{6}{12} \text{ ft}) - (5 \text{ lb})(\tfrac{10}{12} \text{ ft}) = +\bar{I}\alpha + m_B a_B(\tfrac{6}{12} \text{ ft}) + m_A a_A(\tfrac{10}{12} \text{ ft})$$

$$(10)(\tfrac{6}{12}) - (5)(\tfrac{10}{12}) = 0.1656\alpha + \tfrac{10}{32.2}(\tfrac{6}{12}\alpha)(\tfrac{6}{12}) + \tfrac{5}{32.2}(\tfrac{10}{12}\alpha)(\tfrac{10}{12})$$

$$\alpha = +2.374 \text{ rad/s}^2 \qquad \boldsymbol{\alpha} = 2.37 \text{ rad/s}^2 \circlearrowleft \blacktriangleleft$$

$$a_A = r_A\alpha = (\tfrac{10}{12} \text{ ft})(2.374 \text{ rad/s}^2) \qquad \mathbf{a}_A = 1.978 \text{ ft/s}^2 \uparrow \blacktriangleleft$$

$$a_B = r_B\alpha = (\tfrac{6}{12} \text{ ft})(2.374 \text{ rad/s}^2) \qquad \mathbf{a}_B = 1.187 \text{ ft/s}^2 \downarrow \blacktriangleleft$$

SAMPLE PROBLEM 16.4

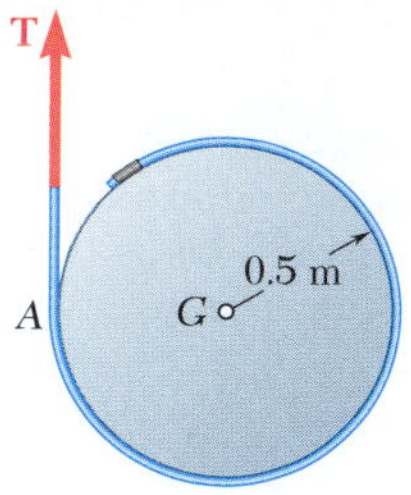

A cord is wrapped around a homogeneous disk of radius $r = 0.5$ m and mass $m = 15$ kg. If the cord is pulled upward with a force **T** of magnitude 180 N, determine (*a*) the acceleration of the center of the disk, (*b*) the angular acceleration of the disk, (*c*) the acceleration of the cord.

SOLUTION

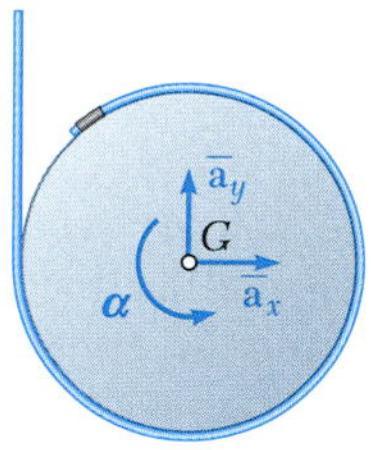

Equations of Motion. We assume that the components $\bar{\mathbf{a}}_x$ and $\bar{\mathbf{a}}_y$ of the acceleration of the center are directed, respectively, to the right and upward and that the angular acceleration of the disk is counterclockwise. The external forces acting on the disk consist of the weight **W** and the force **T** exerted by the cord. This system is equivalent to the system of the effective forces, which consists of a vector of components $m\bar{\mathbf{a}}_x$ and $m\bar{\mathbf{a}}_y$ attached at G and a couple $\bar{I}\boldsymbol{\alpha}$. We write

$\overset{+}{\rightarrow}\Sigma F_x = \Sigma(F_x)_{\text{eff}}$: $\quad 0 = m\bar{a}_x \qquad \bar{\mathbf{a}}_x = 0$ ◀

$+\uparrow\Sigma F_y = \Sigma(F_y)_{\text{eff}}$: $\quad T - W = m\bar{a}_y$

$$\bar{a}_y = \frac{T - W}{m}$$

Since $T = 180$ N, $m = 15$ kg, and $W = (15 \text{ kg})(9.81 \text{ m/s}^2) = 147.1$ N, we have

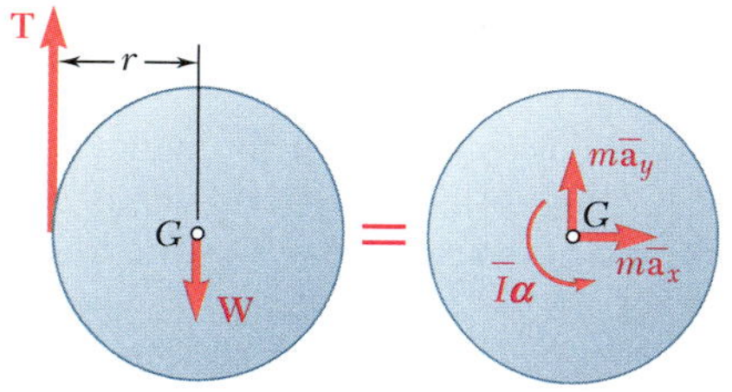

$$\bar{a}_y = \frac{180 \text{ N} - 147.1 \text{ N}}{15 \text{ kg}} = +2.19 \text{ m/s}^2 \qquad \bar{\mathbf{a}}_y = 2.19 \text{ m/s}^2 \uparrow \blacktriangleleft$$

$+\circlearrowleft\Sigma M_G = \Sigma(M_G)_{\text{eff}}$: $\quad -Tr = \bar{I}\alpha$

$$-Tr = (\tfrac{1}{2}mr^2)\alpha$$

$$\alpha = -\frac{2T}{mr} = -\frac{2(180 \text{ N})}{(15 \text{ kg})(0.5 \text{ m})} = -48.0 \text{ rad/s}^2$$

$$\boldsymbol{\alpha} = 48.0 \text{ rad/s}^2 \circlearrowright \blacktriangleleft$$

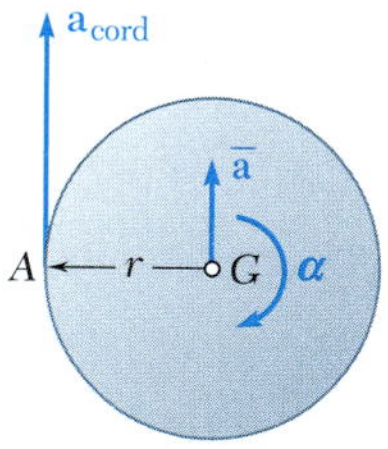

Acceleration of Cord. Since the acceleration of the cord is equal to the tangential component of the acceleration of point A on the disk, we write

$$\mathbf{a}_{\text{cord}} = (\mathbf{a}_A)_t = \bar{\mathbf{a}} + (\mathbf{a}_{A/G})_t$$
$$= [2.19 \text{ m/s}^2 \uparrow] + [(0.5 \text{ m})(48 \text{ rad/s}^2)\uparrow]$$

$$\mathbf{a}_{\text{cord}} = 26.2 \text{ m/s}^2 \uparrow \blacktriangleleft$$

SAMPLE PROBLEM 16.5

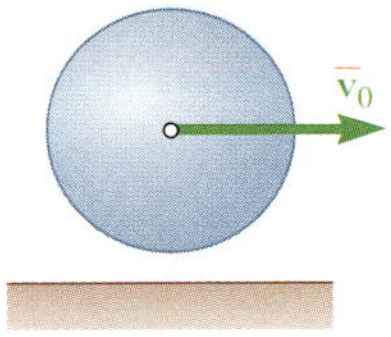

A uniform sphere of mass m and radius r is projected along a rough horizontal surface with a linear velocity $\overline{\mathbf{v}}_0$ and no angular velocity. Denoting by μ_k the coefficient of kinetic friction between the sphere and the floor, determine (*a*) the time t_1 at which the sphere will start rolling without sliding, (*b*) the linear velocity and angular velocity of the sphere at time t_1.

SOLUTION

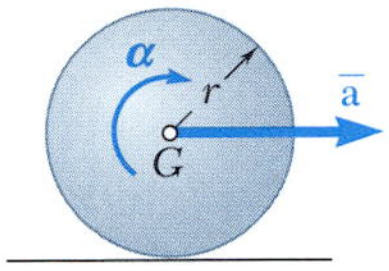

Equations of Motion. The positive sense is chosen to the right for $\overline{\mathbf{a}}$ and clockwise for $\boldsymbol{\alpha}$. The external forces acting on the sphere consist of the weight **W**, the normal reaction **N**, and the friction force **F**. Since the point of the sphere in contact with the surface is sliding to the right, the friction force **F** is directed to the left. While the sphere is sliding, the magnitude of the friction force is $F = \mu_k N$. The effective forces consist of the vector $m\overline{\mathbf{a}}$ attached at G and the couple $\overline{I}\boldsymbol{\alpha}$. Expressing that the system of the external forces is equivalent to the system of the effective forces, we write

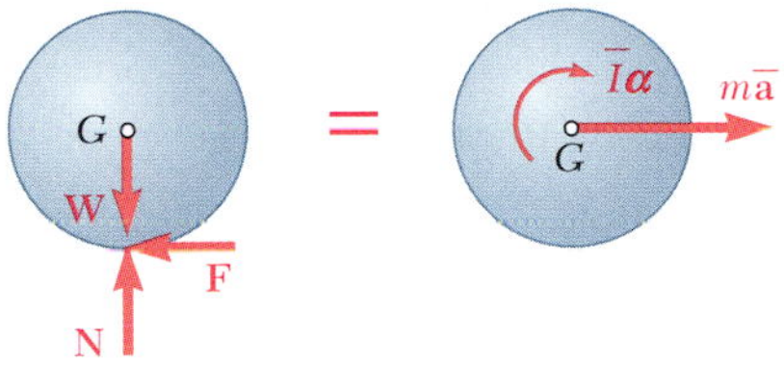

$+\uparrow\Sigma F_y = \Sigma(F_y)_{\text{eff}}$: $\quad N - W = 0$

$$N = W = mg \qquad F = \mu_k N = \mu_k mg$$

$\xrightarrow{+}\Sigma F_x = \Sigma(F_x)_{\text{eff}}$: $\quad -F = m\overline{a} \qquad -\mu_k mg = m\overline{a} \qquad \overline{a} = -\mu_k g$

$+\circlearrowright\Sigma M_G = \Sigma(M_G)_{\text{eff}}$: $\quad Fr = \overline{I}\alpha$

Noting that $\overline{I} = \frac{2}{5}mr^2$ and substituting the value obtained for F, we write

$$(\mu_k mg)r = \frac{2}{5}mr^2\alpha \qquad \alpha = \frac{5}{2}\frac{\mu_k g}{r}$$

Kinematics of Motion. As long as the sphere both rotates and slides, its linear and angular motions are uniformly accelerated.

$$t = 0,\ \overline{v} = \overline{v}_0 \qquad \overline{v} = \overline{v}_0 + \overline{a}t = \overline{v}_0 - \mu_k g t \tag{1}$$

$$t = 0,\ \omega_0 = 0 \qquad \omega = \omega_0 + \alpha t = 0 + \left(\frac{5}{2}\frac{\mu_k g}{r}\right)t \tag{2}$$

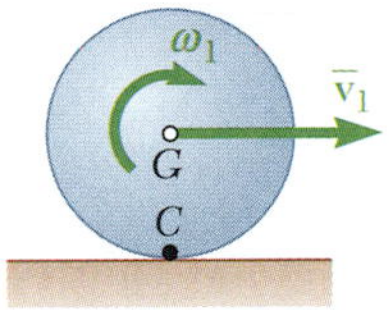

The sphere will start rolling without sliding when the velocity $\mathbf{v}_C$ of the point of contact C is zero. At that time, $t = t_1$, point C becomes the instantaneous center of rotation, and we have

$$\overline{v}_1 = r\omega_1 \tag{3}$$

Substituting in (3) the values obtained for $\overline{v}_1$ and ω_1 by making $t = t_1$ in (1) and (2), respectively, we write

$$\overline{v}_0 - \mu_k g t_1 = r\left(\frac{5}{2}\frac{\mu_k g}{r}t_1\right) \qquad t_1 = \frac{2}{7}\frac{\overline{v}_0}{\mu_k g} \quad ◀$$

Substituting for t_1 into (2), we have

$$\omega_1 = \frac{5}{2}\frac{\mu_k g}{r}t_1 = \frac{5}{2}\frac{\mu_k g}{r}\left(\frac{2}{7}\frac{\overline{v}_0}{\mu_k g}\right) \qquad \omega_1 = \frac{5}{7}\frac{\overline{v}_0}{r} \qquad \boldsymbol{\omega}_1 = \frac{5}{7}\frac{\overline{v}_0}{r}\circlearrowright \quad ◀$$

$$\overline{v}_1 = r\omega_1 = r\left(\frac{5}{7}\frac{\overline{v}_0}{r}\right) \qquad \overline{v}_1 = \frac{5}{7}\overline{v}_0 \qquad \overline{\mathbf{v}}_1 = \frac{5}{7}\overline{v}_0 \rightarrow \quad ◀$$

SOLVING PROBLEMS ON YOUR OWN

This chapter deals with the *plane motion* of rigid bodies, and in this first lesson we considered rigid bodies that are free to move under the action of applied forces.

1. Effective forces. We first recalled that a rigid body consists of a large number of particles. The effective forces of the particles forming the body were found to be equivalent to a vector $m\bar{\mathbf{a}}$ attached at the mass center G of the body and a couple of moment $\bar{I}\boldsymbol{\alpha}$ [Fig. 16.7]. Noting that the applied forces are equivalent to the effective forces, we wrote

$$\Sigma F_x = m\bar{a}_x \qquad \Sigma F_y = m\bar{a}_y \qquad \Sigma M_G = \bar{I}\alpha \tag{16.5}$$

where $\bar{a}_x$ and $\bar{a}_y$ are the x and y components of the acceleration of the mass center G of the body and α is the angular acceleration of the body. It is important to note that when these equations are used, *the moments of the applied forces must be computed with respect to the mass center of the body.* However, you learned a more efficient method of solution based on the use of a free-body-diagram equation.

2. Free-body-diagram equation. Your first step in the solution of a problem should be to draw a *free-body-diagram equation.*

a. A free-body-diagram equation consists of two diagrams representing two equivalent systems of vectors. *In the first diagram* you should show *the forces exerted on the body,* including the applied forces, the reactions at the supports, and the weight of the body. *In the second diagram* you should show the vector $m\bar{\mathbf{a}}$ and the couple $\bar{I}\boldsymbol{\alpha}$ representing *the effective forces.*

b. Using a free-body-diagram equation allows you to *sum components in any direction and to sum moments about any point.* When writing the three equations of motion needed to solve a given problem, you can therefore select one or more equations involving a single unknown. Solving these equations first and substituting the values obtained for the unknowns into the remaining equation(s) will yield a simpler solution.

3. Plane motion of a rigid body. The problems that you will be asked to solve will fall into one of the following categories.

a. Rigid body in translation. For a body in translation, the angular acceleration is zero. The effective forces reduce to *the vector* $m\bar{\mathbf{a}}$ applied at the mass center [Sample Probs. 16.1 and 16.2].

b. Rigid body in centroidal rotation. For a body in centroidal rotation, the acceleration of the mass center is zero. The effective forces reduce to *the couple* $\bar{I}\boldsymbol{\alpha}$ [Sample Prob. 16.3].

c. Rigid body in general plane motion. You can consider the general plane motion of a rigid body as the sum of a translation and a centroidal rotation. The effective forces are equivalent to the vector $m\bar{\mathbf{a}}$ and the couple $\bar{I}\boldsymbol{\alpha}$ [Sample Probs. 16.4 and 16.5].

4. Plane motion of a system of rigid bodies. You first should draw a free-body-diagram equation that includes all the rigid bodies of the system. A vector $m\bar{\mathbf{a}}$ and a couple $\bar{I}\boldsymbol{\alpha}$ are attached to each body. However, the forces exerted on each other by the various bodies of the system can be omitted, since they occur in pairs of equal and opposite forces.

a. If no more than three unknowns are involved, you can use this free-body-diagram equation and sum components in any direction and sum moments about any point to obtain equations that can be solved for the desired unknowns [Sample Prob. 16.3].

b. If more than three unknowns are involved, you must draw a separate free-body-diagram equation for each of the rigid bodies of the system. Both internal forces and external forces should be included in each of the free-body-diagram equations, and care should be taken to represent with equal and opposite vectors the forces that two bodies exert on each other.

Problems

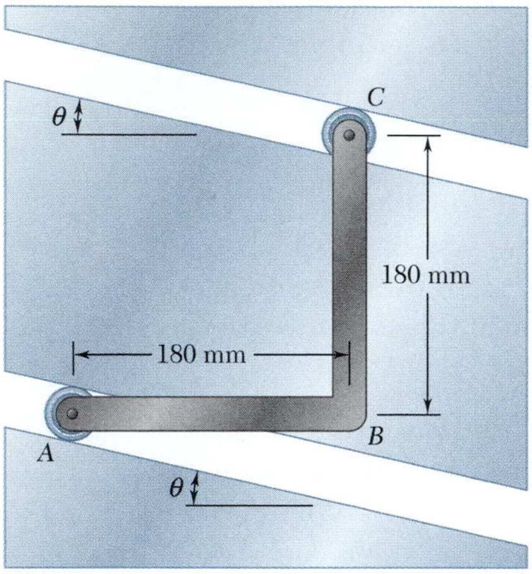

Fig. P16.1 and P16.2

16.1 Two identical 0.4-kg slender rods *AB* and *BC* are welded together to form an L-shaped assembly. The assembly is guided by two small wheels that roll freely in inclined parallel slots cut in a vertical plate. Knowing that $\theta = 30°$, determine (*a*) the acceleration of the assembly, (*b*) the reactions at *A* and *C*.

16.2 Two identical 0.4-kg slender rods *AB* and *BC* are welded together to form an L-shaped assembly. The assembly is guided by two small wheels that roll freely in inclined parallel slots cut in a vertical plate. Determine (*a*) the angle of inclination θ for which the reaction at *A* is zero, (*b*) the corresponding acceleration of the assembly.

16.3 A 60-lb uniform thin panel is placed in a truck with end *A* resting on a rough horizontal surface and end *B* supported by a smooth vertical surface. Knowing that the deceleration of the truck is 12 ft/s^2, determine (*a*) the reactions at ends *A* and *B*, (*b*) the minimum required coefficient of static friction at end *A*.

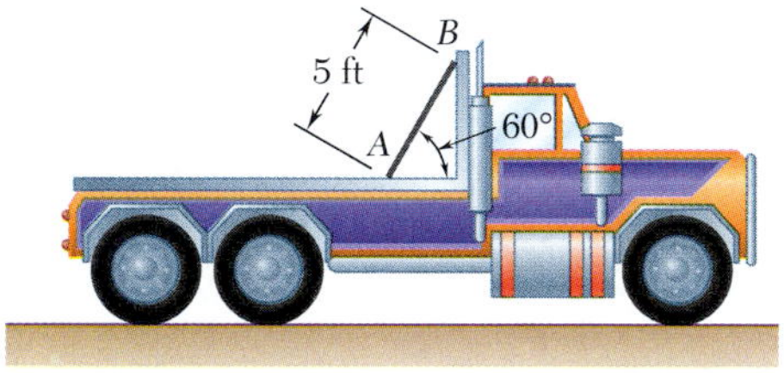

Fig. P16.3 and P16.4

16.4 A 60-lb uniform thin panel is placed in a truck with end *A* resting on a rough horizontal surface and end *B* supported by a smooth vertical surface. Knowing that the panel remains in the position shown, determine (*a*) the maximum allowable acceleration of the truck, (*b*) the corresponding minimum required coefficient of static friction at end *A*.

16.5 Knowing that the coefficient of static friction between the tires and the road is 0.80 for the automobile shown, determine the maximum possible acceleration on a level road, assuming (*a*) four-wheel drive, (*b*) rear-wheel drive, (*c*) front-wheel drive.

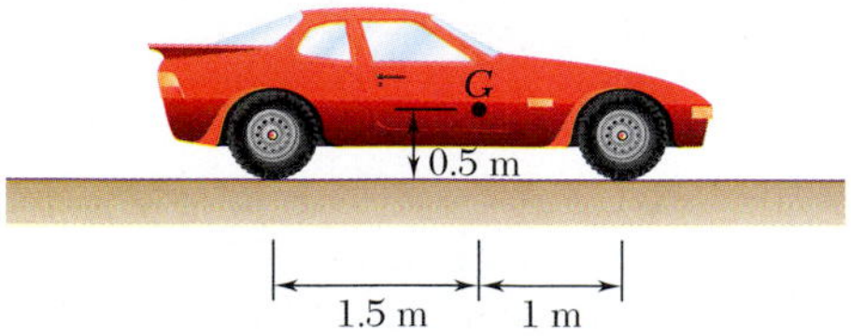

Fig. P16.5

16.6 For the truck of Sample Prob. 16.1, determine the distance through which the truck will skid if (*a*) the rear-wheel brakes fail to operate, (*b*) the front-wheel brakes fail to operate.

16.7 A 50-lb cabinet is mounted on casters that allow it to move freely ($\mu = 0$) on the floor. If a 25-lb force is applied as shown, determine (*a*) the acceleration of the cabinet, (*b*) the range of values of h for which the cabinet will not tip.

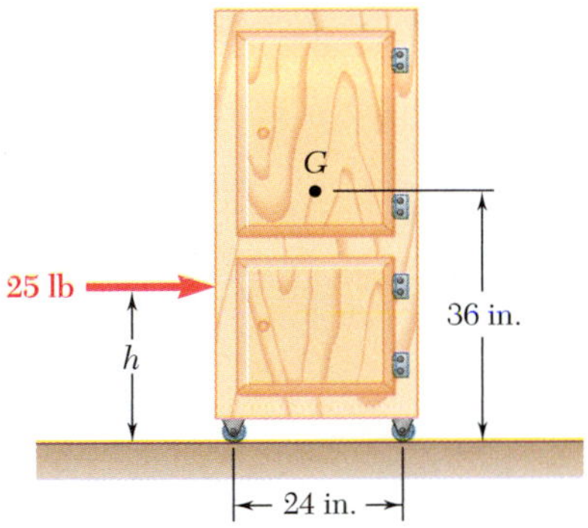

Fig. *P16.7*

16.8 Solve Prob. 16.7 assuming that the casters are locked and slide on the rough floor ($\mu_s = 0.25$).

16.9 The support bracket shown is used to transport a cylindrical can from one elevation to another. Knowing that $\mu_s = 0.30$ between the can and the bracket, determine (*a*) the magnitude of the upward acceleration **a** for which the can will slide on the bracket, (*b*) the smallest ratio h/d for which the can will tip before it slides.

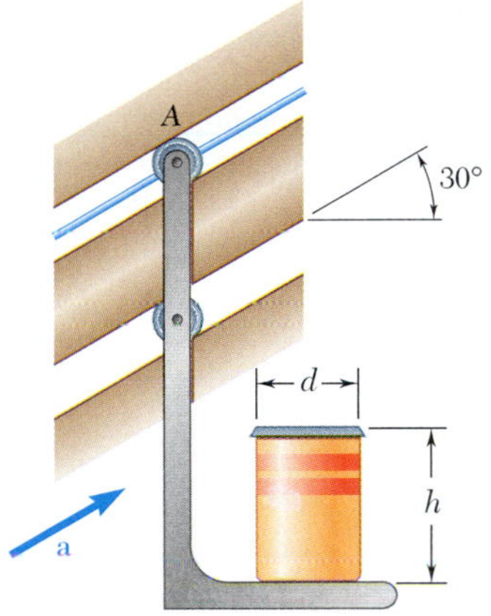

Fig. P16.9

16.10 Solve Prob. 16.9 assuming that the acceleration **a** of the bracket is downward.

16.11 An 8-lb uniform slender rod AB is held in position by two ropes and the link CA which has a negligible weight. After rope BD is cut the assembly rotates in a vertical plane under the combined effect of gravity and a 4 lb · ft couple **M** applied to link CA as shown. Determine, immediately after rope BD has been cut, (*a*) the acceleration of rod AB, (*b*) the tension in rope EB.

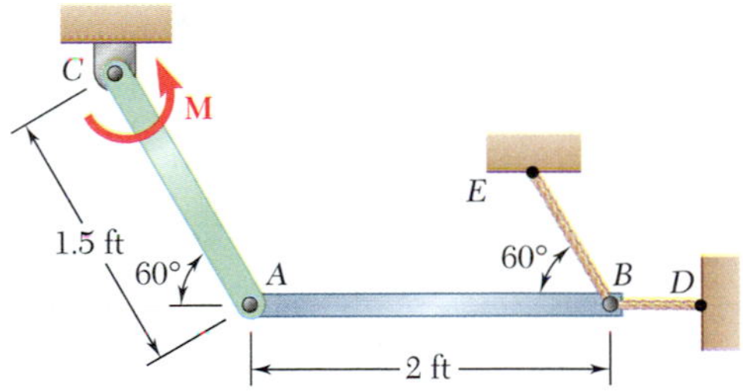

Fig. P16.11 and *P16.12*

16.12 An 8-lb uniform slender rod AB is held in position by two ropes and the link CA which has a negligible weight. After rope BD is cut the assembly rotates in a vertical plane under the combined effect of gravity and a couple **M** applied to link CA as shown. Knowing that the magnitude of the acceleration of rod AB is 12 ft/s^2 immediately after rope BD has been cut, determine (*a*) the magnitude of the couple **M**, (*b*) the tension in rope EB.

16.13 A uniform circular plate of mass 6 kg is attached to two links AC and BD of the same length. Knowing that the plate is released from rest in the position shown, in which lines joining G to A and B are, respectively, horizontal and vertical, determine (*a*) the acceleration of the plate, (*b*) the tension in each link.

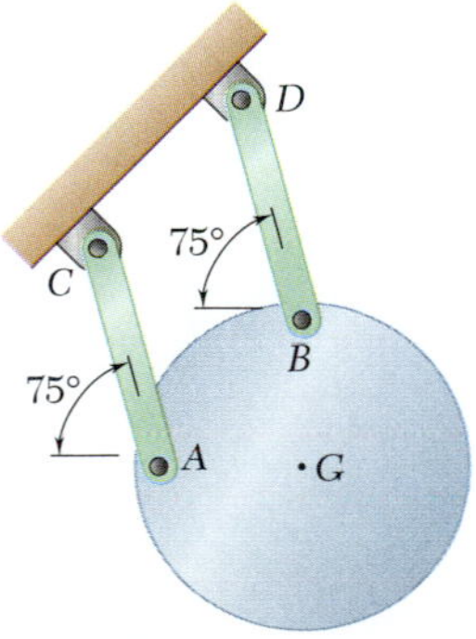

Fig. P16.13

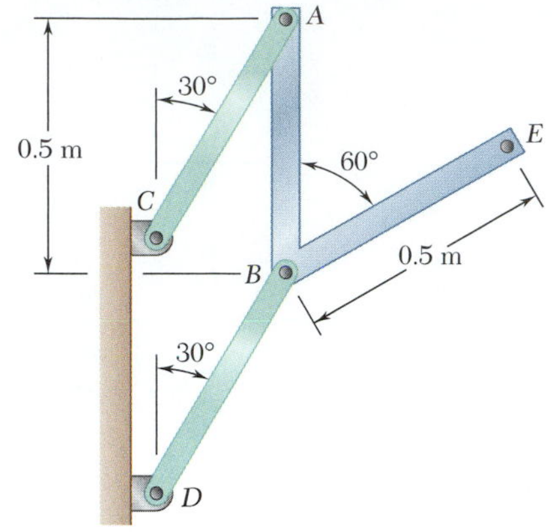

Fig. P16.14

16.14 Bars AB and BE, each of mass 4 kg, are welded together and are pin-connected to two links AC and BD. Knowing that the assembly is released from rest in the position shown and neglecting the masses of the links, determine (*a*) the acceleration of the assembly, (*b*) the forces in the links.

16.15 Cranks BE and CF, each of length 15 in., are made to rotate at a constant speed of 90 rpm counterclockwise. For the position shown, and knowing that $\mathbf{P} = 0$, determine the vertical components of the forces exerted on the 15-lb uniform rod $ABCD$ by pins B and C.

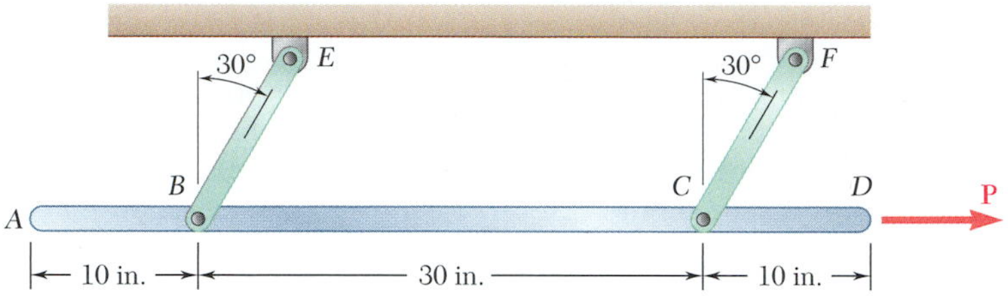

Fig. P16.15 and *P16.16*

16.16 At the instant shown the angular velocity of links BE and CF is 6 rad/s counterclockwise and is decreasing at the rate of 12 rad/s^2. Knowing that the length of each link is 15 in. and neglecting the weight of the links, determine (*a*) the force $\mathbf{P}$, (*b*) the corresponding force in each link. The weight of rod AD is 15 lb.

***16.17** Draw the shear and bending-moment diagrams for each of the bars AB and BE of Prob. 16.14.

***16.18** Draw the shear and bending-moment diagrams for the horizontal rod $ABCD$ of Prob. 16.15.

16.19 For a rigid slab in translation, show that the system of the effective forces consists of vectors $(\Delta m_i)\bar{\mathbf{a}}$ attached to the various particles of the slab, where $\bar{\mathbf{a}}$ is the acceleration of the mass center G of the slab. Further show, by computing their sum and the sum of their moments about G, that the effective forces reduce to a single vector $m\bar{\mathbf{a}}$ attached at G.

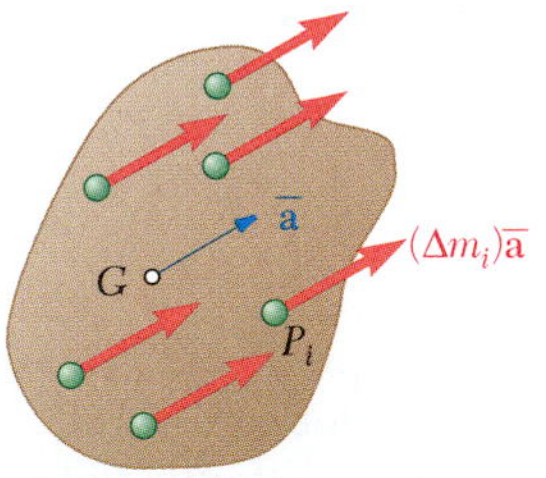

Fig. P16.19

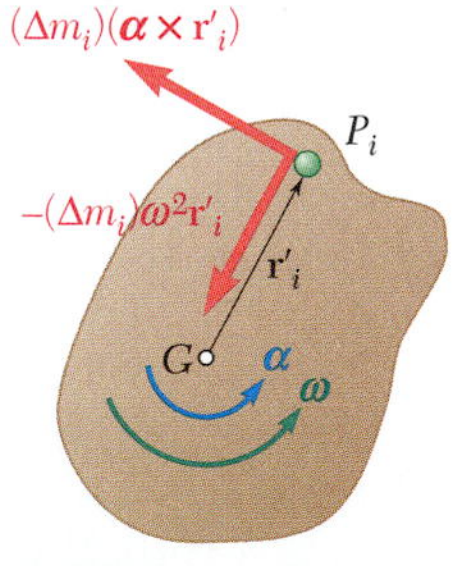

Fig. P16.20

16.20 For a rigid slab in centroidal rotation, show that the system of the effective forces consists of vectors $-(\Delta m_i)\omega^2\mathbf{r}'_i$ and $(\Delta m_i)(\boldsymbol{\alpha} \times \mathbf{r}'_i)$ attached to the various particles P_i of the slab, where $\boldsymbol{\omega}$ and $\boldsymbol{\alpha}$ are the angular velocity and angular acceleration of the slab, and where $\mathbf{r}'_i$ denotes the position of the particle P_i relative to the mass center G of the slab. Further show, by computing their sum and the sum of their moments about G, that the effective forces reduce to a couple $\bar{I}\boldsymbol{\alpha}$.

16.21 It takes 10 min for a 2.4-Mg flywheel to coast to rest from an angular velocity of 300 rpm. Knowing that the radius of gyration of the flywheel is 1 m, determine the average magnitude of the couple due to kinetic friction in the bearing.

16.22 The rotor of an electric motor has an angular velocity of 3600 rpm when the load and power are cut off. The 120-lb rotor, which has a centroidal radius of gyration of 9 in., then coasts to rest. Knowing that kinetic friction results in a couple of magnitude 2.5 lb · ft exerted on the rotor, determine the number of revolutions that the rotor executes before coming to rest.

16.23 A 20-lb uniform disk is placed in contact with an inclined surface and a constant 7.5 lb · ft couple **M** is applied as shown. The weight of the link *AB* is negligible. Knowing that the coefficient of kinetic friction at *D* is 0.4, determine (*a*) the angular acceleration of the disk, (*b*) the force in the link *AB*.

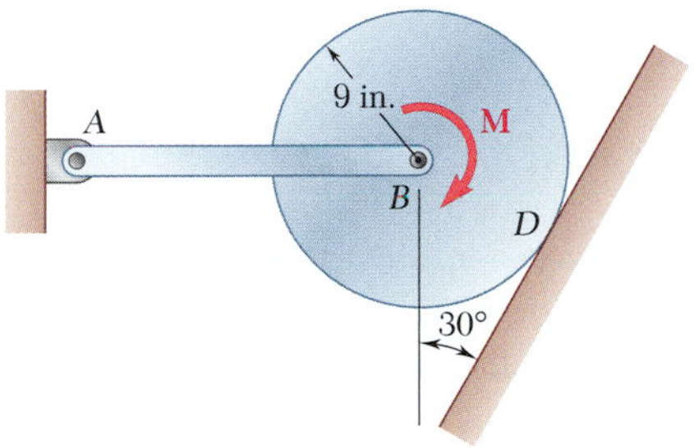

Fig. P16.23 and P16.24

16.24 A 20-lb uniform disk is placed in contact with an inclined surface and a constant couple **M** is applied as shown. The weight of the uniform link *AB* is 10 lb and the coefficient of kinetic friction at *D* is 0.25. Knowing that the angular acceleration of the disk is 16 rad/s^2 clockwise, determine (*a*) the magnitude of the couple **M**, (*b*) the force exerted on the disk at *D*.

16.25 The 160-mm-radius brake drum is attached to a larger flywheel that is not shown. The total mass moment of inertia of the drum and the flywheel is 18 kg · m^2 and the coefficient of kinetic friction between the drum and the brake shoe is 0.35. Knowing that the angular velocity of the flywheel is 360 rpm counterclockwise when a force **P** of magnitude 300 N is applied to the pedal *C*, determine the number of revolutions executed by the flywheel before it comes to rest.

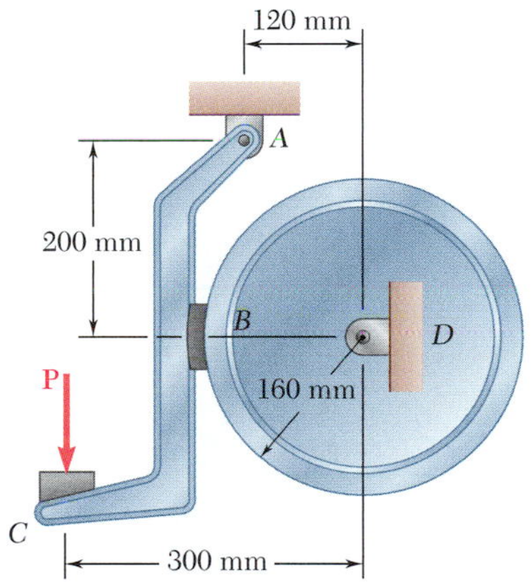

Fig. P16.25

16.26 Solve Prob. 16.25 assuming that the initial angular velocity of the flywheel is 360 rpm clockwise.

16.27 The flywheel shown has a radius of 600 mm, a mass of 144 kg, and a radius of gyration of 450 mm. An 18-kg block *A* is attached to a wire that is wrapped around the flywheel, and the system is released from rest. Neglecting the effect of friction, determine (*a*) the acceleration of block *A*, (*b*) the speed of block *A* after it has moved 1.8 m.

16.28 In order to determine the mass moment of inertia of a flywheel of radius 1.5 ft, a 20-lb block is attached to a wire that is wrapped around the flywheel. The block is released and is observed to fall 12 ft in 4.5 s. To eliminate bearing friction from the computations, a second block of weight 40 lb is used and is observed to fall 12 ft in 2.8 s. Assuming that the moment of the couple due to friction remains constant, determine the mass moment of inertia of the flywheel.

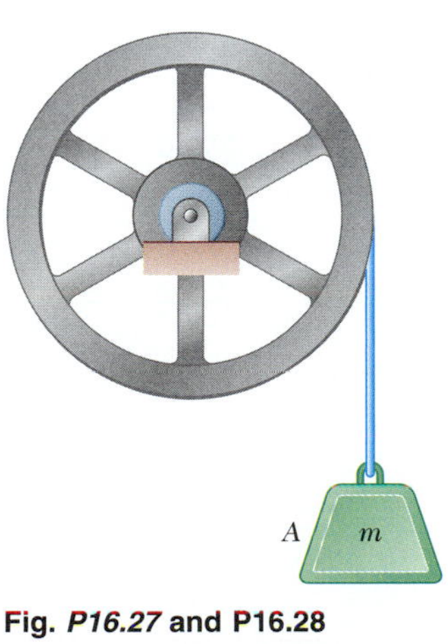

Fig. *P16.27* and P16.28

16.29 Each of the double pulleys shown has a mass moment of inertia of 20 kg · m^2 and is initially at rest. The outside radius is 400 mm, and the inner radius is 200 mm. Determine (*a*) the angular acceleration of each pulley, (*b*) the angular velocity of each pulley after point *A* on the cord has moved 3 m.

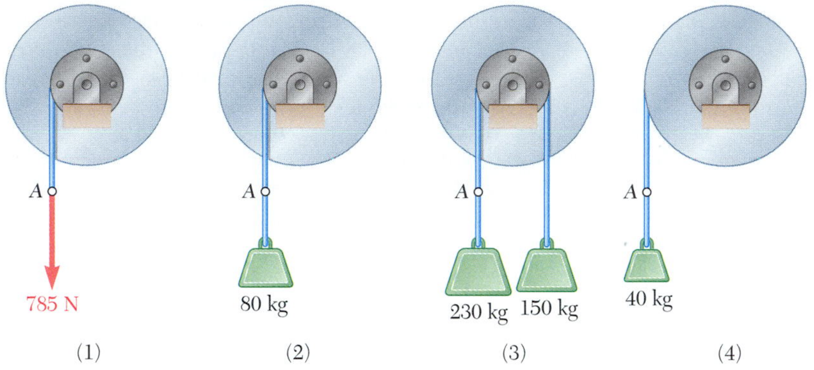

Fig. P16.29

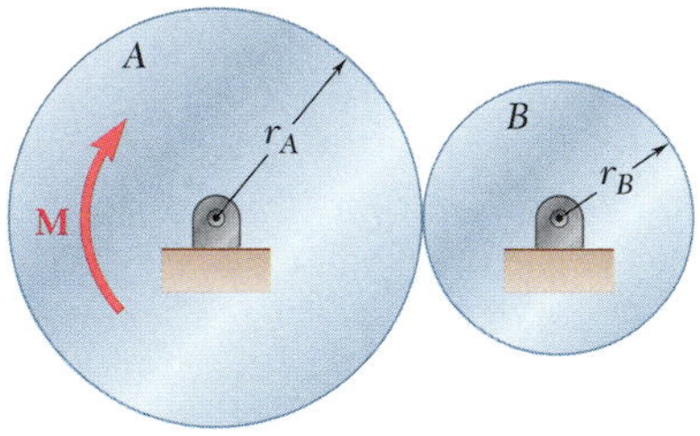

Fig. P16.30

16.30 The weight and radius of friction disk *A* are W_A = 12 lb and r_A = 6 in.; the weight and radius of friction disk *B* are W_B = 6 lb and r_B = 4 in. The disks are at rest when a couple **M** of moment 7.5 lb · in. is applied to disk *A*. Assuming that no slipping occurs between the disks, determine (*a*) the angular acceleration of each disk, (*b*) the friction force that disk *A* exerts on disk *B*.

16.31 Solve Prob. 16.30 assuming that the couple **M** is applied to disk *B*.

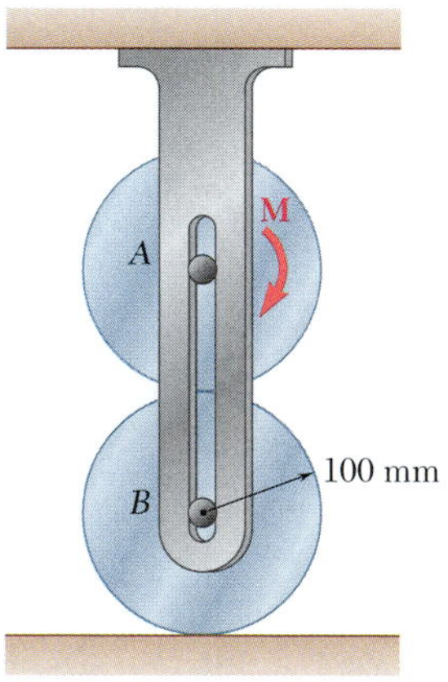

Fig. P16.32 and P16.33

16.32 Two identical 8-kg uniform cylinders are at rest when a constant couple **M** of magnitude 5 N · m is applied to cylinder *A*. Knowing that the coefficient of kinetic friction between cylinder *B* and the horizontal surface is 0.2 and that no slipping occurs between the cylinders, determine (*a*) the angular acceleration of each cylinder, (*b*) the minimum allowable value of the coefficient of static friction between the cylinders.

16.33 Two identical 8-kg uniform cylinders are at rest when a constant couple **M** of magnitude 5 N · m is applied to cylinder *A*. Knowing that the coefficient of kinetic friction between the cylinders is 0.45 and the coefficient of kinetic friction between cylinder *B* and the horizontal surface is 0.15, determine the angular acceleration of each cylinder.

16.34 Disk *A* has a mass of 8 kg and an initial angular velocity of 480 rpm clockwise; disk *B* has a mass of 4 kg and is initially at rest. The disks are brought together by applying a horizontal force of magnitude 30 N to the axle of disk *A*. Knowing that μ_k = 0.15 between the disks and neglecting bearing friction, determine (*a*) the angular acceleration of each disk, (*b*) the final angular velocity of each disk.

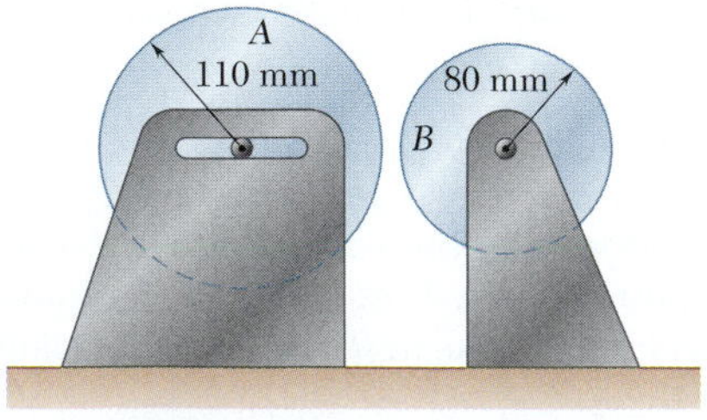

Fig. P16.34

16.35 Solve Prob. 16.34 assuming that initially disk A is at rest and disk B has an angular velocity of 480 rpm clockwise.

16.36 Disk B has an angular velocity $\boldsymbol{\omega}_0$ when it is brought into contact with disk A, which is at rest. Show that (*a*) the final angular velocities of the disks are independent of the coefficient of friction μ_k between the disks as long as $\mu_k \neq 0$, (*b*) the final angular velocity of disk B depends only upon $\boldsymbol{\omega}_0$ and the ratio of the masses m_A and m_B of the two disks.

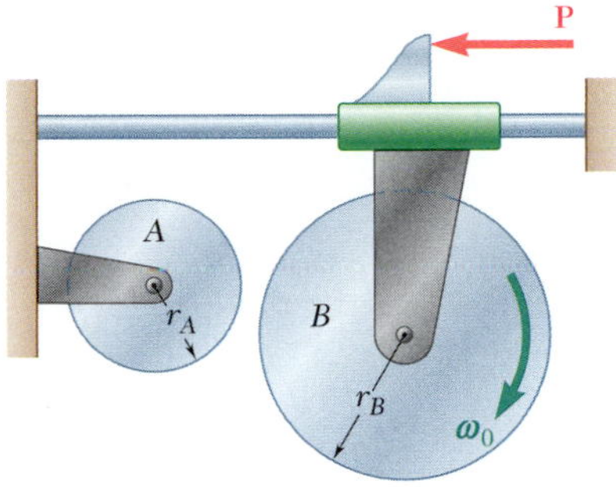

Fig. P16.36, P16.37, and *P16.38*

16.37 The 12-lb disk A has a radius $r_A = 6$ in. and an initial angular velocity $\boldsymbol{\omega}_0 = 750$ rpm clockwise. The 30-lb disk B has a radius $r_B = 10$ in. and is at rest. A force **P** of magnitude 5 lb is then applied to bring the disks into contact. Knowing that $\mu_k = 0.25$ between the disks and neglecting bearing friction, determine (*a*) the angular acceleration of each disk, (*b*) the final angular velocity of each disk.

16.38 Solve Prob. 16.37 assuming that disk A is initially at rest and that disk B has an angular velocity of 750 rpm clockwise.

16.39 A cylinder of radius r and mass m rests on two small casters A and B as shown. Initially, the cylinder is at rest and is set in motion by rotating caster B clockwise at high speed so that slipping occurs between the cylinder and caster B. Denoting by μ_k the coefficient of kinetic friction and neglecting the moment of inertia of the free caster A, derive an expression for the angular acceleration of the cylinder.

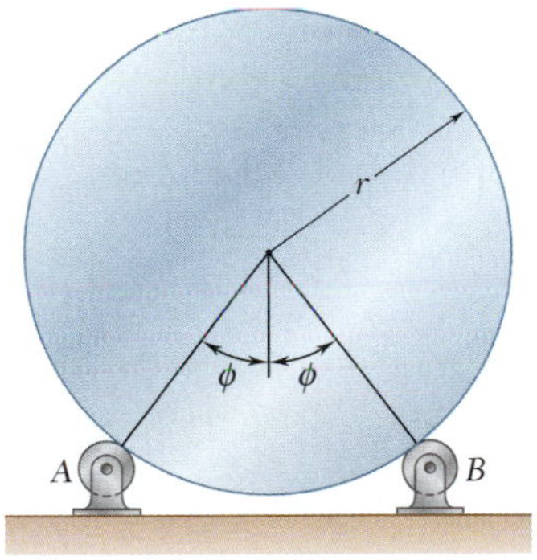

Fig. P16.39

16.40 In Prob. 16.39 assume that no slipping can occur between caster B and the cylinder (such a case would exist if the cylinder and caster had gear teeth along their rims). Derive an expression for the maximum allowable counterclockwise acceleration $\boldsymbol{\alpha}$ of the cylinder if it is not to lose contact with the caster at A.

16.41 Show that the system of the effective forces for a rigid slab in plane motion reduces to a single vector, and express the distance from the mass center G of the slab to the line of action of this vector in terms of the centroidal radius of gyration $\bar{k}$ of the slab, the magnitude $\bar{a}$ of the acceleration of G, and the angular acceleration α.

16.42 For a rigid slab in plane motion, show that the system of the effective forces consists of vectors $(\Delta m_i)\bar{\mathbf{a}}$, $-(\Delta m_i)\omega^2\mathbf{r}'_i$, and $(\Delta m_i)(\boldsymbol{\alpha} \times \mathbf{r}'_i)$ attached to the various particles P_i of the slab, where $\bar{\mathbf{a}}$ is the acceleration of the mass center G of the slab, $\boldsymbol{\omega}$ is the angular velocity of the slab, $\boldsymbol{\alpha}$ is its angular acceleration, and $\mathbf{r}'_i$ denotes the position vector of the particle P_i relative to G. Further show, by computing their sum and the sum of their moments about G, that the effective forces reduce to a vector $m\bar{\mathbf{a}}$ attached at G and a couple $\bar{I}\boldsymbol{\alpha}$.

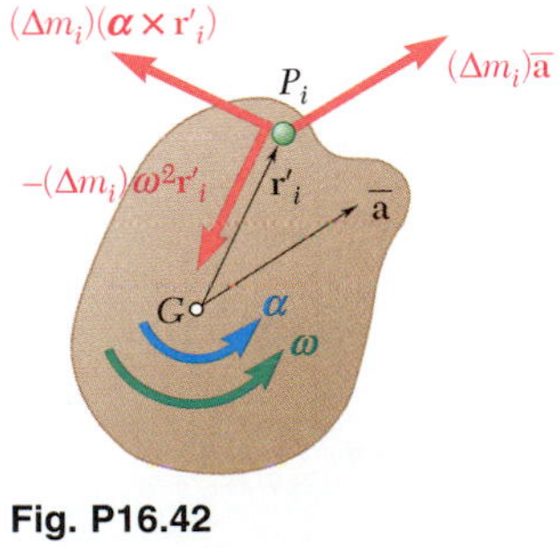

Fig. P16.42

16.43 A uniform slender rod AB rests on a frictionless horizontal surface, and a force $\mathbf{P}$ of magnitude 2 N is applied at A in a direction perpendicular to the rod. Knowing that the rod has a mass of 3 kg, determine the acceleration of (*a*) point A, (*b*) point B.

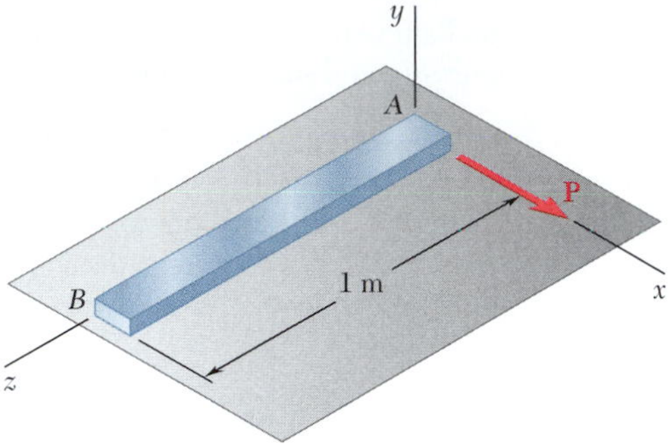

Fig. P16.43

16.44 (*a*) In Prob. 16.43, determine the point of the rod AB at which the force $\mathbf{P}$ should be applied if the acceleration of point B is to be zero. (*b*) Knowing that $P = 2$ N, determine the corresponding acceleration of point A.

16.45 and 16.46 A force $\mathbf{P}$ of magnitude 0.75 lb is applied to a tape wrapped around the body indicated. Knowing that the body rests on a frictionless horizontal surface, determine the acceleration of (*a*) point A, (*b*) point B.

16.45 A thin hoop of weight 6 lb.
16.46 A uniform disk of weight 6 lb.

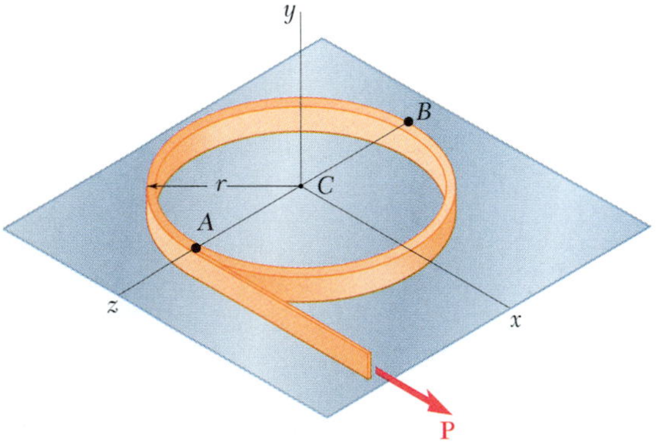

Fig. P16.45

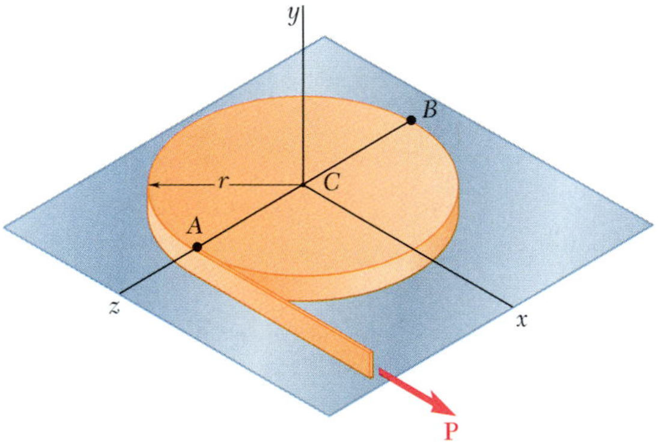

Fig. P16.46 and P16.47

16.47 A force $\mathbf{P}$ is applied to a tape wrapped around a uniform disk that rests on a frictionless horizontal surface. Show that for each 360° rotation of the disk the center of the disk will move a distance πr.

16.48 A uniform semicircular plate of mass 6 kg is suspended from three vertical wires at points A, B, and C, and a force $\mathbf{P}$ of magnitude 5 N is applied to point B. Immediately after $\mathbf{P}$ is applied, determine the acceleration of (*a*) the mass center of the plate, (*b*) point C.

16.49 Immediately after the force $\mathbf{P}$ is applied to the plate of Prob. 16.48, determine the acceleration of (*a*) point A, (*b*) point B.

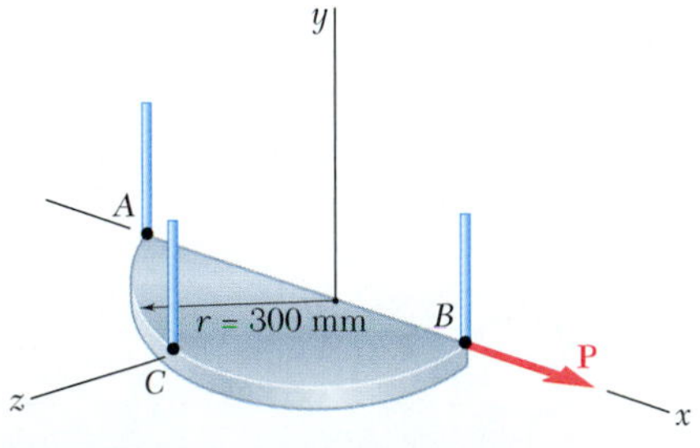

Fig. P16.48 and *P16.49*

16.50 A drum of 10-in. radius is attached to a disk of radius $r_A = 7.5$ in. The disk and drum have a combined weight of 12 lb and a combined radius of gyration of 6 in. and are suspended by two cords. Knowing that $T_A = 9$ lb and $T_B = 6$ lb, determine the accelerations of points A and B on the cords.

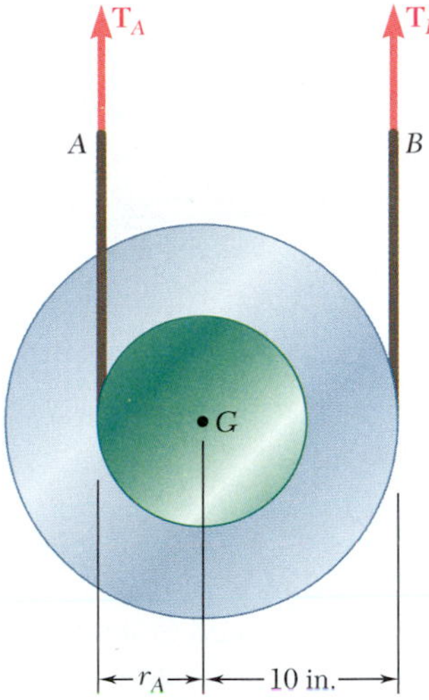

Fig. P16.50 and *P16.51*

16.51 A drum of 10-in. radius is attached to a disk of radius $r_A =$ 6.92 in. The disk and drum have a combined weight of 12 lb and are suspended by two cords. Knowing that the acceleration of point B on the cord is zero, $T_A = 10$ lb, and $T_B = 5$ lb, determine the combined radius of gyration of the disk and drum.

16.52 and 16.53 A 5-m beam of mass 200 kg is lowered by means of two cables unwinding from overhead cranes. As the beam approaches the ground, the crane operators apply brakes to slow the unwinding motion. Knowing that the deceleration of cable A is 5 m/s^2 and the deceleration of cable B is 0.5 m/s^2, determine the tension in each cable.

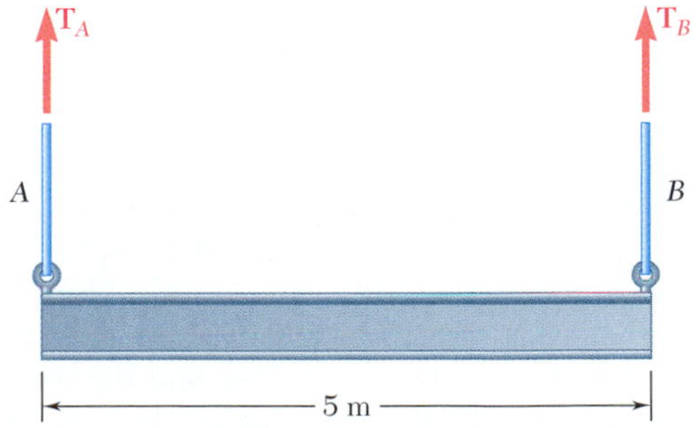

Fig. P16.52

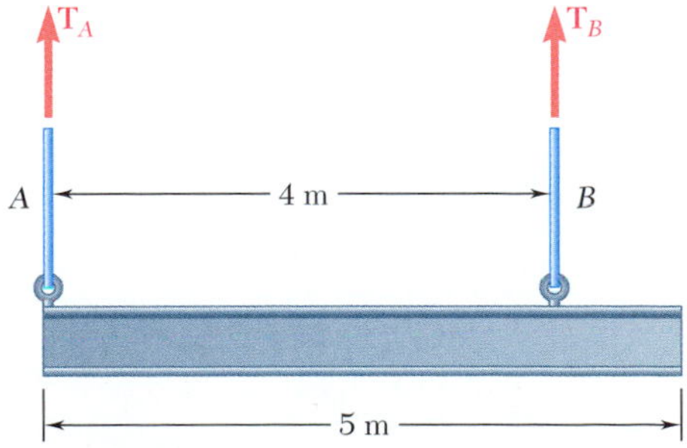

Fig. P16.53

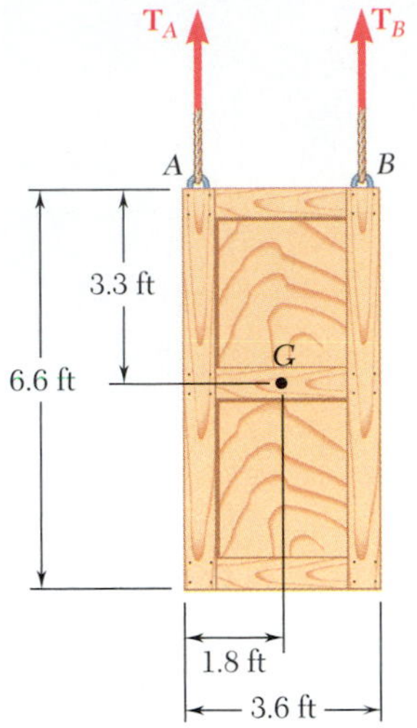

Fig. P16.54

16.54 and *16.55* The 400-lb crate shown is lowered by means of two overhead cranes. Knowing that at the instant shown the deceleration of cable A is 3 ft/s^2 and that of cable B is 1 ft/s^2, determine the tension in each cable.

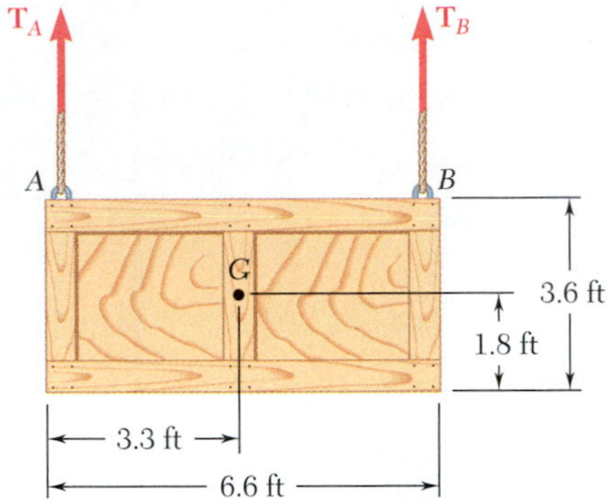

Fig. *P16.55*

16.56 The uniform disk shown, of mass m and radius r, rotates counterclockwise. Its center C is constrained to move in a slot cut in the vertical member AB and a horizontal force **P** is applied at B to maintain contact at D between the disk and the vertical wall. The disk moves downward under the influence of gravity and the friction at D. Denoting by μ_k the coefficient of kinetic friction between the disk and the wall and neglecting friction in the vertical slot, determine (*a*) the angular acceleration of the disk, (*b*) the acceleration of the center C of the disk.

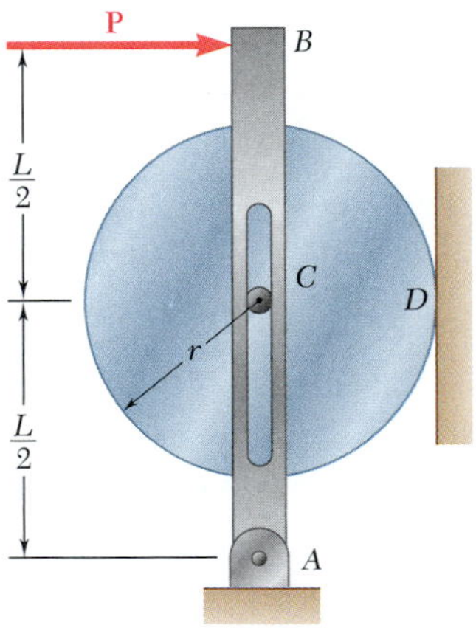

Fig. P16.56 and P16.57

16.57 The 6-kg uniform disk shown, of radius $r = 80$ mm, rotates counterclockwise. Its center C is constrained to move in a slot cut in the vertical member AB and a 50-N horizontal force **P** is applied at B to maintain contact at D between the disk and the vertical wall. The disk moves downward under the influence of gravity and the friction at D. Knowing that the coefficient of kinetic friction between the disk and the wall is 0.12 and neglecting friction in the vertical slot, determine (*a*) the angular acceleration of the disk, (*b*) the acceleration of the center C of the disk.

16.58 through 16.60 A beam AB of mass m and of uniform cross section is suspended from two springs as shown. If spring 2 breaks, determine at that instant (*a*) the angular acceleration of the beam, (*b*) the acceleration of point A, (*c*) the acceleration of point B.

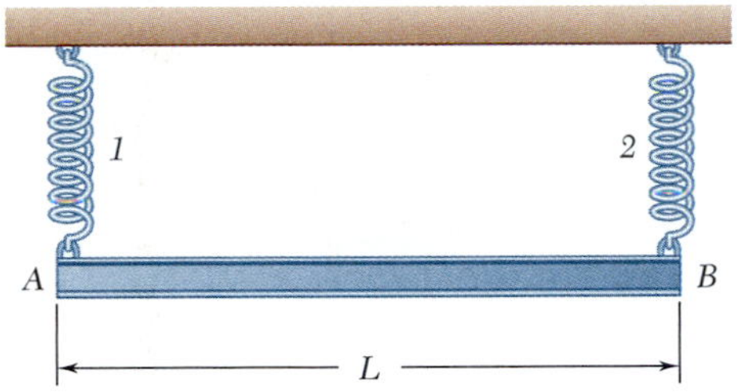

Fig. P16.58

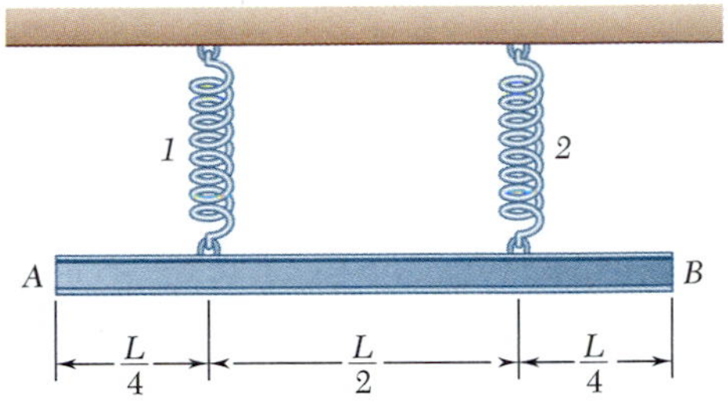

Fig. P16.59

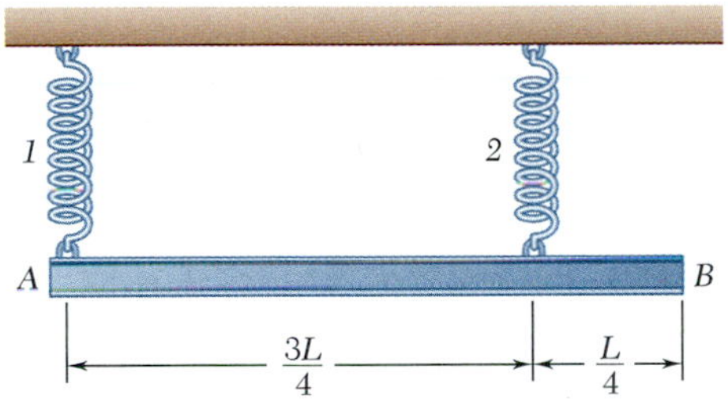

Fig. P16.60

16.61 through 16.63 A thin plate of the shape indicated and of mass m is suspended from two springs as shown. If spring 2 breaks, determine the acceleration at that instant (*a*) of point A, (*b*) of point B.

16.61 A circular plate of radius b.
16.62 A thin hoop of radius b.
16.63 A square plate of side b.

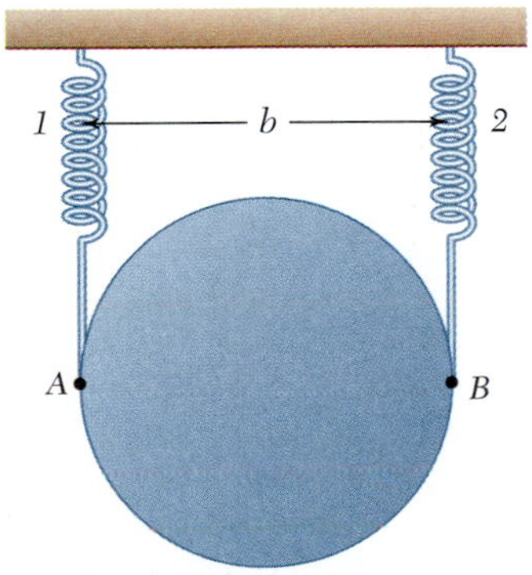

Fig. P16.61

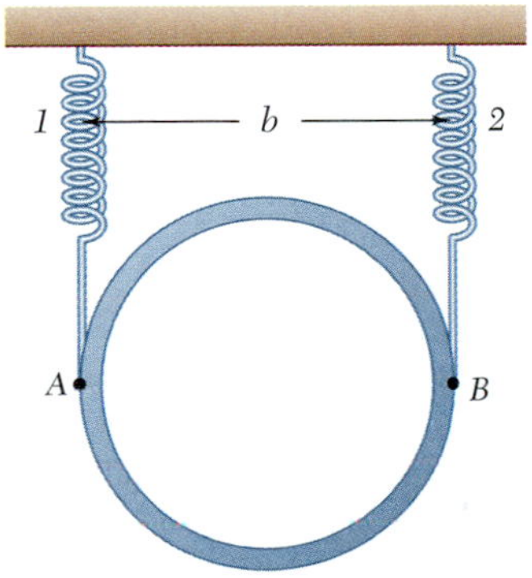

Fig. *P16.62*

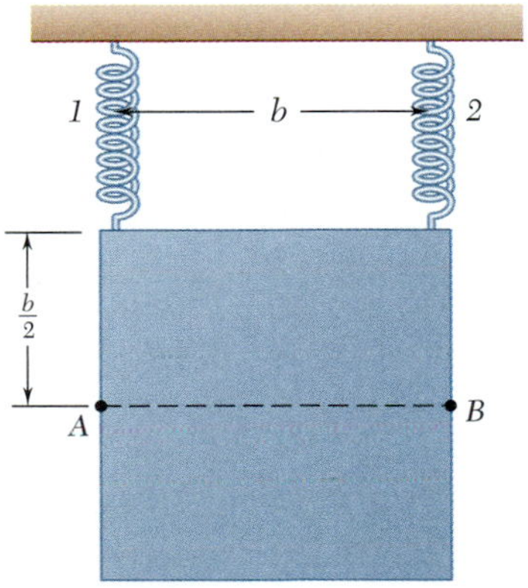

Fig. P16.63

16.64 A bowler projects a 200-mm-diameter ball of mass 5 kg along an alley with a forward velocity $\mathbf{v}_0$ of 5 m/s and a backspin $\boldsymbol{\omega}_0$ of 9 rad/s. Knowing that the coefficient of kinetic friction between the ball and the alley is 0.10, determine (*a*) the time t_1 at which the ball will start rolling without sliding, (*b*) the speed of the ball at time t_1, (*c*) the distance the ball will have traveled at time t_1.

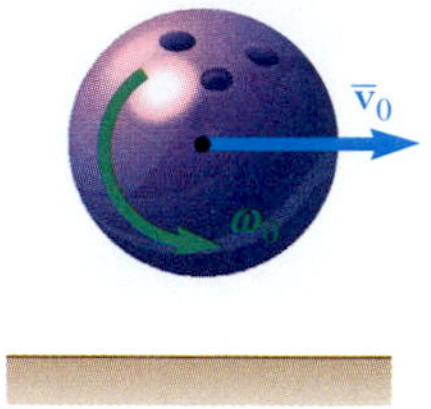

Fig. P16.64

16.65 Solve Prob. 16.64 assuming that the bowler projects the ball with the same forward velocity but with a backspin of 18 rad/s.

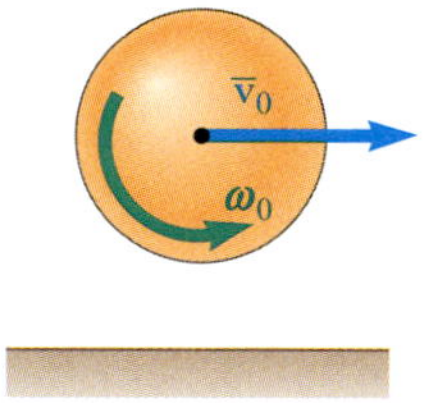

Fig. P16.66

16.66 A sphere of radius r and mass m is projected along a rough horizontal surface with the initial velocities indicated. If the final velocity of the sphere is to be zero, express, in terms of v_0, r, and μ_k, (*a*) the required magnitude of $\boldsymbol{\omega}_0$, (*b*) the time t_1 for the sphere to come to rest, (*c*) the distance the sphere will move before coming to rest.

16.67 Solve Prob. 16.66 assuming that the sphere is replaced by a uniform thin hoop of radius r and mass m.

16.68 A uniform sphere of radius r and mass m is placed with no initial velocity on a belt that moves to the right with a constant velocity $\mathbf{v}_1$. Denoting by μ_k the coefficient of kinetic friction between the sphere and the belt, determine (*a*) the time t_1 at which the sphere will start rolling without sliding, (*b*) the linear and angular velocities of the sphere at time t_1.

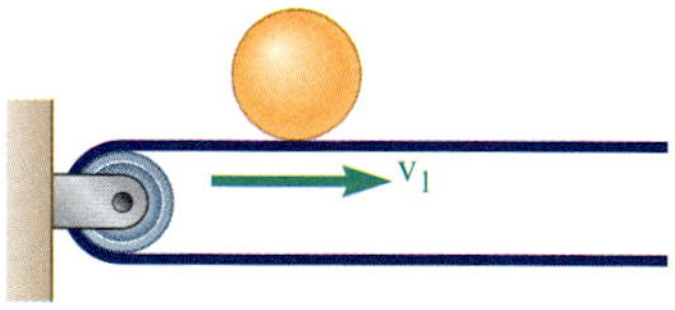

Fig. P16.68

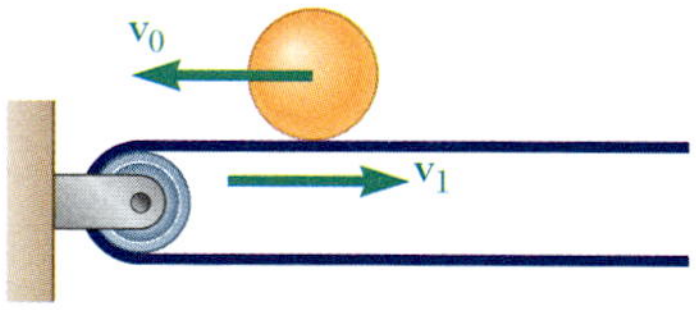

Fig. *P16.69*

16.69 A sphere of radius r and mass m has a linear velocity $\mathbf{v}_0$ directed to the left and no angular velocity as it is placed on a belt moving to the right with a constant placed velocity $\mathbf{v}_1$. If after first sliding on the belt the sphere is to have no linear velocity relative to the ground as it starts rolling on the belt without sliding, determine in terms of v_1 and the coefficient of kinetic friction μ_k between the sphere and the belt (*a*) the required value of v_0, (*b*) time t_1 at which the sphere will start rolling on the belt, (*c*) the distance the sphere will have moved relative to the ground at time t_1.

16.8. CONSTRAINED PLANE MOTION

Most engineering applications deal with rigid bodies which are moving under given constraints. For example, cranks must rotate about a fixed axis, wheels must roll without sliding, and connecting rods must describe certain prescribed motions. In all such cases, definite relations exist between the components of the acceleration $\bar{\mathbf{a}}$ of the mass center G of the body considered and its angular acceleration $\boldsymbol{\alpha}$; the corresponding motion is said to be a *constrained motion*.

The solution of a problem involving a constrained plane motion calls first for a *kinematic analysis* of the problem. Consider, for example, a slender rod AB of length l and mass m whose extremities are connected to blocks of negligible mass which slide along horizontal and vertical frictionless tracks. The rod is pulled by a force $\mathbf{P}$ applied at A (Fig. 16.11). We know from Sec. 15.8 that the acceleration $\bar{\mathbf{a}}$ of the mass center G of the rod can be determined at any given instant from the position of the rod, its angular velocity, and its angular acceleration at that instant. Suppose, for example, that the values of θ, ω, and α are known at a given instant and that we wish to determine the corresponding value of the force $\mathbf{P}$, as well as the reactions at A and B. We should first *determine the components $\bar{a}_x$ and $\bar{a}_y$ of the acceleration of the mass center* G by the method of Sec. 15.8. We next apply d'Alembert's principle (Fig. 16.12), using the expressions obtained for $\bar{a}_x$ and $\bar{a}_y$. The unknown forces $\mathbf{P}$, $\mathbf{N}_A$, and $\mathbf{N}_B$ can then be determined by writing and solving the appropriate equations.

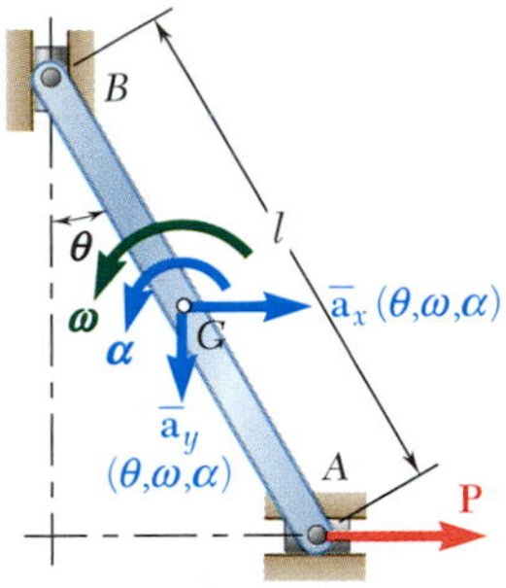

Fig. 16.11

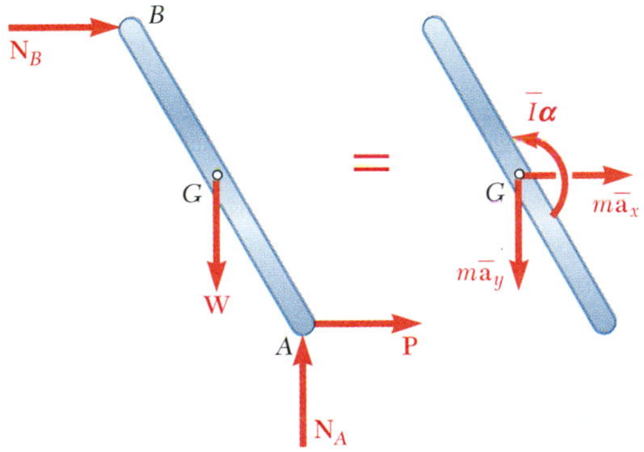

Fig. 16.12

Suppose now that the applied force $\mathbf{P}$, the angle θ, and the angular velocity ω of the rod are known at a given instant and that we wish to find the angular acceleration α of the rod and the components $\bar{a}_x$ and $\bar{a}_y$ of the acceleration of its mass center at that instant, as well as the reactions at A and B. The preliminary kinematic study of the problem will have for its object *to express the components $\bar{a}_x$ and $\bar{a}_y$ of the acceleration of G in terms of the angular acceleration α of the rod*. This will be done by first expressing the acceleration of a suitable reference point such as A in terms of the angular acceleration α. The components $\bar{a}_x$ and $\bar{a}_y$ of the acceleration of G can then be determined in terms of α, and the expressions obtained carried into Fig. 16.12. Three equations can then be derived in terms of α, N_A, and N_B and solved for the three unknowns (see Sample Prob. 16.10). Note that the

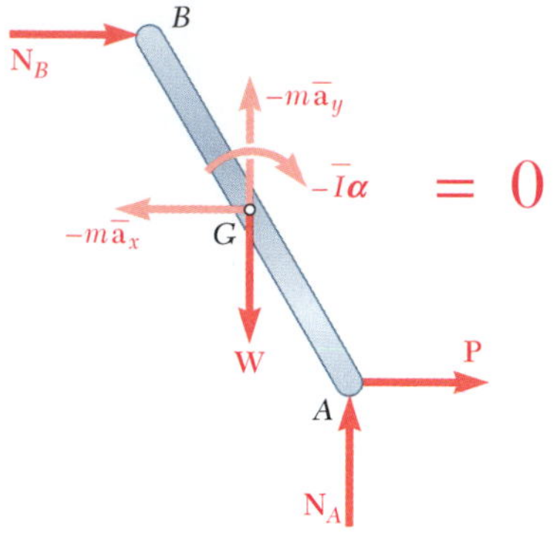

Fig. 16.13

method of dynamic equilibrium can also be used to carry out the solution of the two types of problems we have considered (Fig. 16.13).

When a mechanism consists of *several moving parts*, the approach just described can be used with each part of the mechanism. The procedure required to determine the various unknowns is then similar to the procedure followed in the case of the equilibrium of a system of connected rigid bodies (Sec. 6.11).

Earlier, we analyzed two particular cases of constrained plane motion: the translation of a rigid body, in which the angular acceleration of the body is constrained to be zero, and the centroidal rotation, in which the acceleration $\bar{\mathbf{a}}$ of the mass center of the body is constrained to be zero. Two other particular cases of constrained plane motion are of special interest: the *noncentroidal rotation* of a rigid body and the *rolling motion* of a disk or wheel. These two cases can be analyzed by one of the general methods described above. However, in view of the range of their applications, they deserve a few special comments.

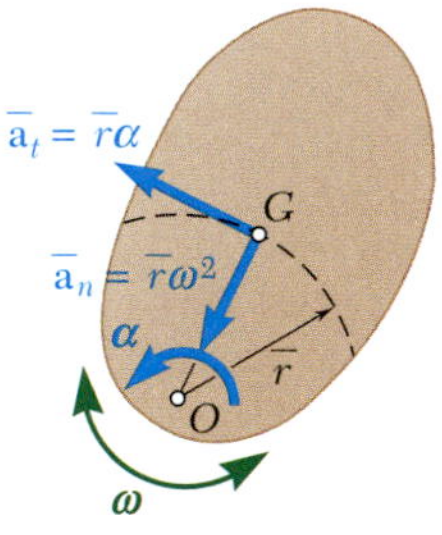

Fig. 16.14

Noncentroidal Rotation. The motion of a rigid body constrained to rotate about a fixed axis which does not pass through its mass center is called *noncentroidal rotation*. The mass center G of the body moves along a circle of radius $\bar{r}$ centered at the point O, where the axis of rotation intersects the plane of reference (Fig. 16.14). Denoting, respectively, by $\boldsymbol{\omega}$ and $\boldsymbol{\alpha}$ the angular velocity and the angular acceleration of the line OG, we obtain the following expressions for the tangential and normal components of the acceleration of G:

$$\bar{a}_t = \bar{r}\alpha \qquad \bar{a}_n = \bar{r}\omega^2 \tag{16.7}$$

Since line OG belongs to the body, its angular velocity $\boldsymbol{\omega}$ and its angular acceleration $\boldsymbol{\alpha}$ also represent the angular velocity and the angular acceleration of the body in its motion relative to G. Equations (16.7) define, therefore, the kinematic relation existing between the motion of the mass center G and the motion of the body about G. They should be used to eliminate $\bar{a}_t$ and $\bar{a}_n$ from the equations obtained by applying d'Alembert's principle (Fig. 16.15) or the method of dynamic equilibrium (Fig. 16.16).

An interesting relation is obtained by equating the moments about the fixed point O of the forces and vectors shown, respectively, in parts a and b of Fig. 16.15. We write

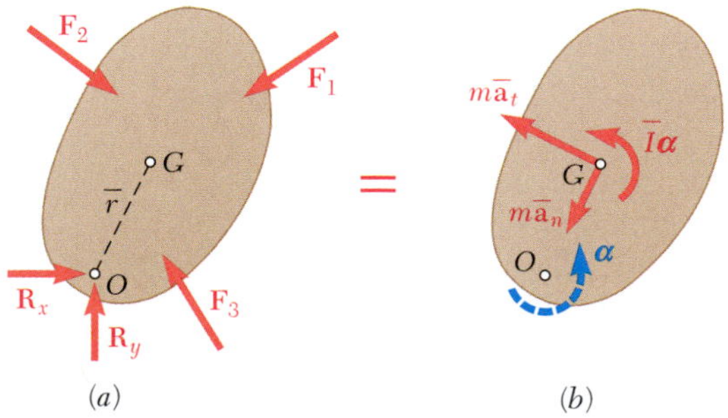

Fig. 16.15

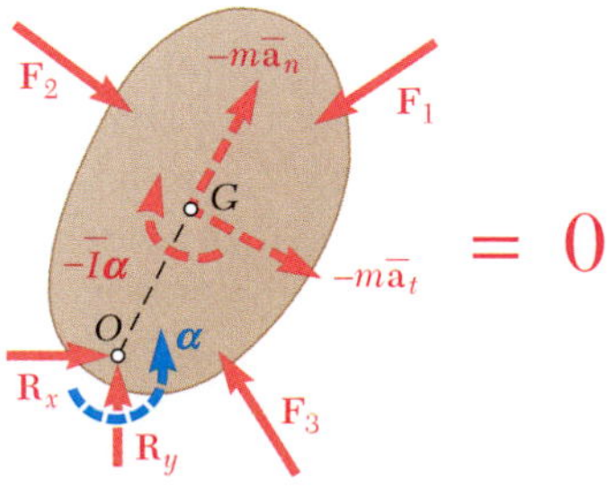

Fig. 16.16

$$+\circlearrowleft \Sigma M_O = \bar{I}\alpha + (m\bar{r}\alpha)\bar{r} = (\bar{I} + m\bar{r}^2)\alpha$$

But according to the parallel-axis theorem, we have $\bar{I} + m\bar{r}^2 = I_O$, where I_O denotes the moment of inertia of the rigid body about the fixed axis. We therefore write

$$\Sigma M_O = I_O\alpha \tag{16.8}$$

Although formula (16.8) expresses an important relation between the sum of the moments of the external forces about the fixed point O and the product $I_O\alpha$, it should be clearly understood that this formula does not mean that the system of the external forces is equivalent to a couple of moment $I_O\alpha$. The system of the effective forces, and thus the system of the external forces, reduces to a couple only when O coincides with G—that is, *only when the rotation is centroidal* (Sec. 16.4). In the more general case of noncentroidal rotation, the system of the external forces does not reduce to a couple.

A particular case of noncentroidal rotation is of special interest—the case of *uniform rotation,* in which the angular velocity $\boldsymbol{\omega}$ is constant. Since $\boldsymbol{\alpha}$ is zero, the inertia couple in Fig. 16.16 vanishes and the inertia vector reduces to its normal component. This component (also called *centrifugal force*) represents the tendency of the rigid body to break away from the axis of rotation.

Rolling Motion. Another important case of plane motion is the motion of a disk or wheel rolling on a plane surface. If the disk is constrained to roll without sliding, the acceleration $\bar{\mathbf{a}}$ of its mass center G and its angular acceleration $\boldsymbol{\alpha}$ are not independent. Assuming that the disk is balanced, so that its mass center and its geometric center coincide, we first write that the distance $\bar{x}$ traveled by G during a rotation θ of the disk is $\bar{x} = r\theta$, where r is the radius of the disk. Differentiating this relation twice, we write

$$\bar{a} = r\alpha \tag{16.9}$$

Recalling that the system of the effective forces in plane motion reduces to a vector $m\bar{\mathbf{a}}$ and a couple $\bar{I}\boldsymbol{\alpha}$, we find that in the particular case of the rolling motion of a balanced disk, the effective forces reduce to a vector of magnitude $mr\alpha$ attached at G and to a couple of magnitude $\bar{I}\alpha$. We may thus express that the external forces are equivalent to the vector and couple shown in Fig. 16.17.

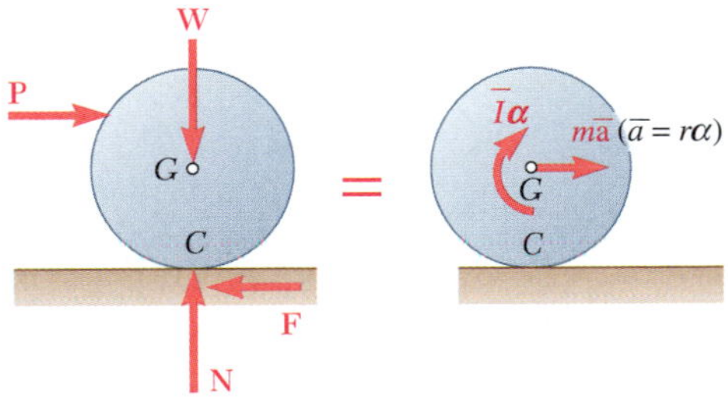

Fig. 16.17

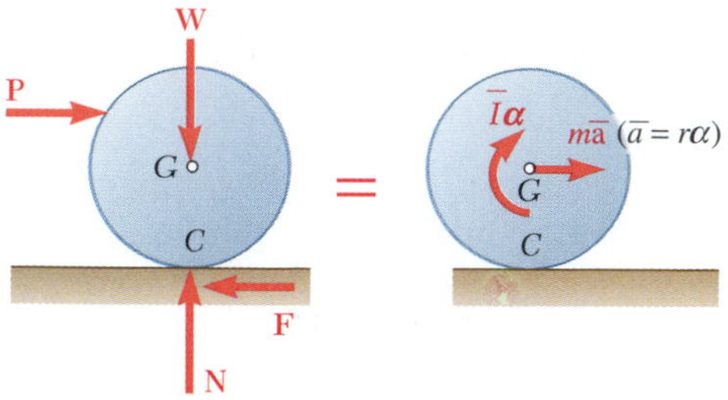

Fig. 16.17 (*repeated*)

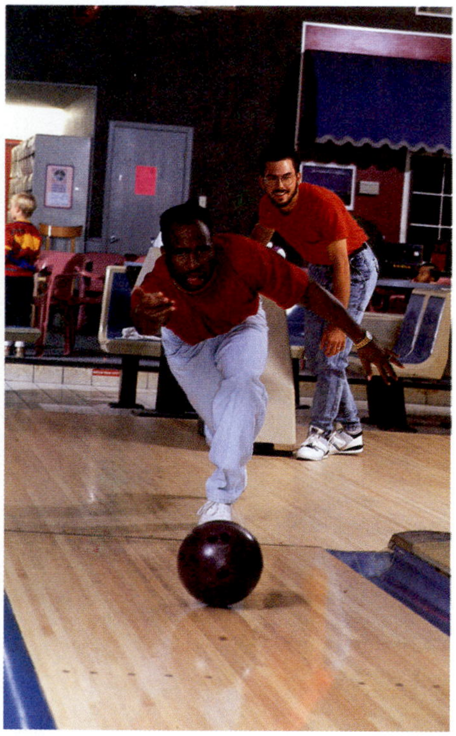

Photo 16.4 As the ball hits the bowling alley, it first spins and slides, then rolls without sliding.

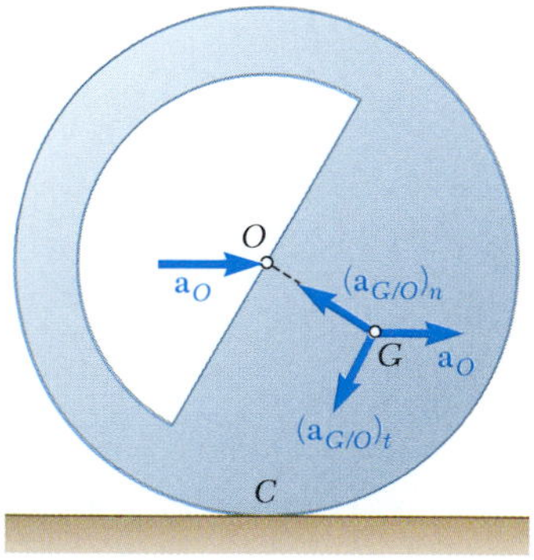

Fig. 16.18

When a disk *rolls without sliding,* there is no relative motion between the point of the disk in contact with the ground and the ground itself. Thus as far as the computation of the friction force **F** is concerned, a rolling disk can be compared with a block at rest on a surface. The magnitude F of the friction force can have any value, as long as this value does not exceed the maximum value $F_m = \mu_s N$, where μ_s is the coefficient of static friction and N is the magnitude of the normal force. In the case of a rolling disk, the magnitude F of the friction force should therefore be determined independently of N by solving the equation obtained from Fig. 16.17.

When *sliding is impending,* the friction force reaches its maximum value $F_m = \mu_s N$ and can be obtained from N.

When the disk *rotates and slides* at the same time, a relative motion exists between the point of the disk which is in contact with the ground and the ground itself, and the force of friction has the magnitude $F_k = \mu_k N$, where μ_k is the coefficient of kinetic friction. In this case, however, the motion of the mass center G of the disk and the rotation of the disk about G are independent, and $\bar{a}$ is not equal to $r\alpha$.

These three different cases can be summarized as follows:

Rolling, no sliding:	$F \le \mu_s N$	$\bar{a} = r\alpha$
Rolling, sliding impending:	$F = \mu_s N$	$\bar{a} = r\alpha$
Rotating and sliding:	$F = \mu_k N$	$\bar{a}$ *and* α *independent*

When it is not known whether or not a disk slides, it should first be assumed that the disk rolls without sliding. If F is found smaller than or equal to $\mu_s N$, the assumption is proved correct. If F is found larger than $\mu_s N$, the assumption is incorrect and the problem should be started again, assuming rotating and sliding.

When a disk is *unbalanced,* that is, when its mass center G does not coincide with its geometric center O, the relation (16.9) does not hold between $\bar{a}$ and α. However, a similar relation holds between the magnitude a_O of the acceleration of the geometric center and the angular acceleration α of an unbalanced disk which rolls without sliding. We have

$$a_O = r\alpha \tag{16.10}$$

To determine $\bar{a}$ in terms of the angular acceleration α and the angular velocity ω of the disk, we can use the relative-acceleration formula

$$\begin{aligned}\bar{\mathbf{a}} = \bar{\mathbf{a}}_G &= \mathbf{a}_O + \mathbf{a}_{G/O} \\ &= \mathbf{a}_O + (\mathbf{a}_{G/O})_t + (\mathbf{a}_{G/O})_n\end{aligned} \tag{16.11}$$

where the three component accelerations obtained have the directions indicated in Fig. 16.18 and the magnitudes $a_O = r\alpha$, $(a_{G/O})_t = (OG)\alpha$, and $(a_{G/O})_n = (OG)\omega^2$.

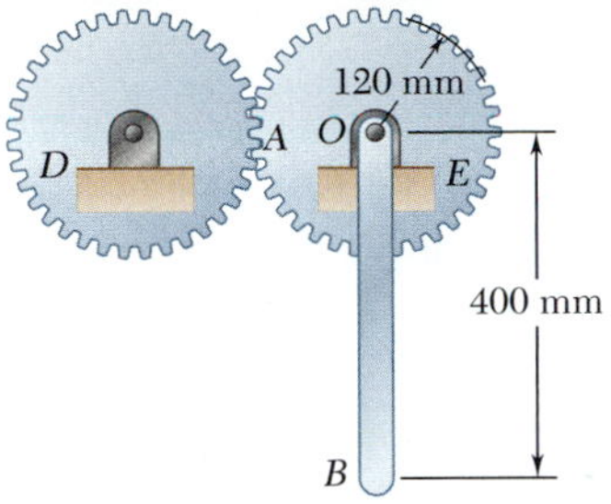

SAMPLE PROBLEM 16.6

The portion AOB of a mechanism consists of a 400-mm steel rod OB welded to a gear E of radius 120 mm which can rotate about a horizontal shaft O. It is actuated by a gear D and, at the instant shown, has a clockwise angular velocity of 8 rad/s and a counterclockwise angular acceleration of 40 rad/s^2. Knowing that rod OB has a mass of 3 kg and gear E a mass of 4 kg and a radius of gyration of 85 mm, determine (*a*) the tangential force exerted by gear D on gear E, (*b*) the components of the reaction at shaft O.

SOLUTION

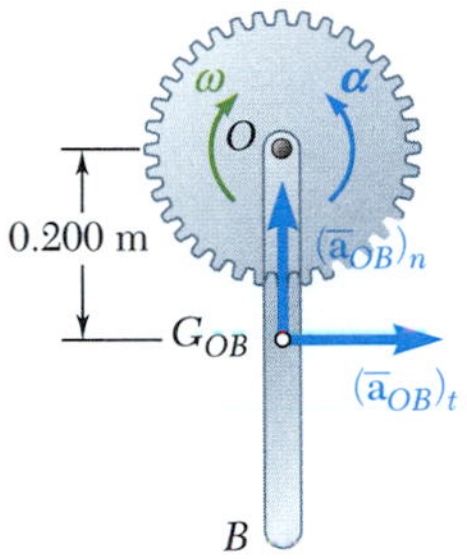

In determining the effective forces of the rigid body AOB, gear E and rod OB will be considered separately. Therefore, the components of the acceleration of the mass center G_{OB} of the rod will be determined first:

$$(\bar{a}_{OB})_t = \bar{r}\alpha = (0.200 \text{ m})(40 \text{ rad/s}^2) = 8 \text{ m/s}^2$$
$$(\bar{a}_{OB})_n = \bar{r}\omega^2 = (0.200 \text{ m})(8 \text{ rad/s})^2 = 12.8 \text{ m/s}^2$$

Equations of Motion. Two sketches of the rigid body AOB have been drawn. The first shows the external forces consisting of the weight $\mathbf{W}_E$ of gear E, the weight $\mathbf{W}_{OB}$ of the rod OB, the force $\mathbf{F}$ exerted by gear D, and the components $\mathbf{R}_x$ and $\mathbf{R}_y$ of the reaction at O. The magnitudes of the weights are, respectively,

$$W_E = m_E g = (4 \text{ kg})(9.81 \text{ m/s}^2) = 39.2 \text{ N}$$
$$W_{OB} = m_{OB} g = (3 \text{ kg})(9.81 \text{ m/s}^2) = 29.4 \text{ N}$$

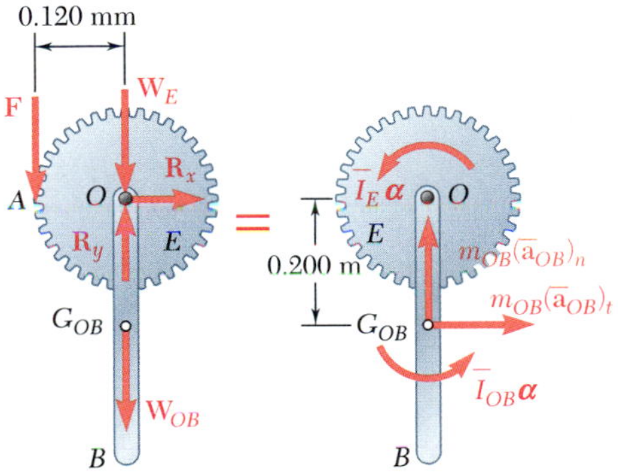

The second sketch shows the effective forces, which consist of a couple $\bar{I}_E\boldsymbol{\alpha}$ (since gear E is in centroidal rotation) and of a couple and two vector components at the mass center of OB. Since the accelerations are known, we compute the magnitudes of these components and couples:

$$\bar{I}_E\alpha = m_E\bar{k}_E^2\alpha = (4 \text{ kg})(0.085 \text{ m})^2(40 \text{ rad/s}^2) = 1.156 \text{ N}\cdot\text{m}$$
$$m_{OB}(\bar{a}_{OB})_t = (3 \text{ kg})(8 \text{ m/s}^2) = 24.0 \text{ N}$$
$$m_{OB}(\bar{a}_{OB})_n = (3 \text{ kg})(12.8 \text{ m/s}^2) = 38.4 \text{ N}$$
$$\bar{I}_{OB}\alpha = (\tfrac{1}{12}m_{OB}L^2)\alpha = \tfrac{1}{12}(3 \text{ kg})(0.400 \text{ m})^2(40 \text{ rad/s}^2) = 1.600 \text{ N}\cdot\text{m}$$

Expressing that the system of the external forces is equivalent to the system of the effective forces, we write the following equations:

$+\circlearrowleft\Sigma M_O = \Sigma(M_O)_{\text{eff}}$:

$$F(0.120 \text{ m}) = \bar{I}_E\alpha + m_{OB}(\bar{a}_{OB})_t(0.200 \text{ m}) + \bar{I}_{OB}\alpha$$
$$F(0.120 \text{ m}) = 1.156 \text{ N}\cdot\text{m} + (24.0 \text{ N})(0.200 \text{ m}) + 1.600 \text{ N}\cdot\text{m}$$

$$F = 63.0 \text{ N} \qquad \mathbf{F} = 63.0 \text{ N}\downarrow \blacktriangleleft$$

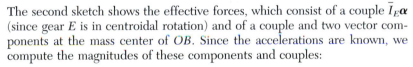

$\xrightarrow{+}\Sigma F_x = \Sigma(F_x)_{\text{eff}}$: $\qquad R_x = m_{OB}(\bar{a}_{OB})_t$

$$R_x = 24.0 \text{ N} \qquad \mathbf{R}_x = 24.0 \text{ N} \rightarrow \blacktriangleleft$$

$+\uparrow\Sigma F_y = \Sigma(F_y)_{\text{eff}}$: $\qquad R_y - F - W_E - W_{OB} = m_{OB}(\bar{a}_{OB})_n$

$$R_y - 63.0 \text{ N} - 39.2 \text{ N} - 29.4 \text{ N} = 38.4 \text{ N}$$

$$R_y = 170.0 \text{ N} \qquad \mathbf{R}_y\ 170.0 \text{ N}\uparrow \blacktriangleleft$$

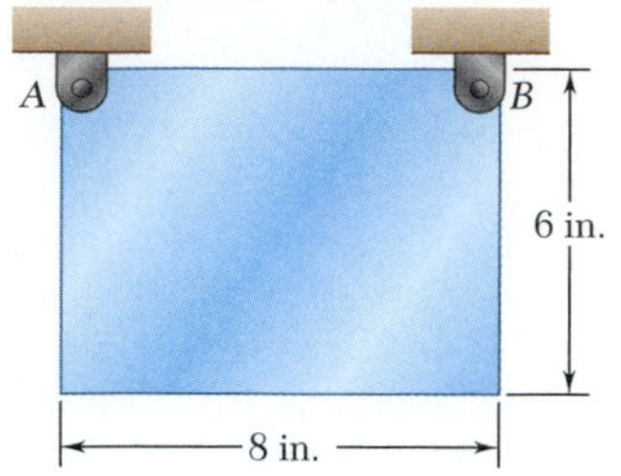

SAMPLE PROBLEM 16.7

A 6 × 8 in. rectangular plate weighing 60 lb is suspended from two pins A and B. If pin B is suddenly removed, determine (a) the angular acceleration of the plate, (b) the components of the reaction at pin A, immediately after pin B has been removed.

SOLUTION

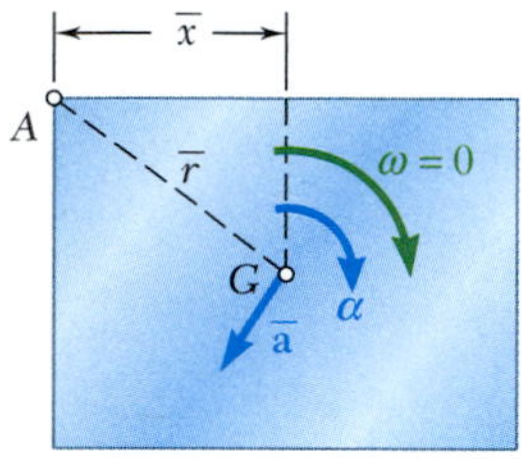

***a.* Angular Acceleration.** We observe that as the plate rotates about point A, its mass center G describes a circle of radius $\bar{r}$ with center at A.

Since the plate is released from rest ($\omega = 0$), the normal component of the acceleration of G is zero. The magnitude of the acceleration $\bar{\mathbf{a}}$ of the mass center G is thus $\bar{a} = \bar{r}\alpha$. We draw the diagram shown to express that the external forces are equivalent to the effective forces:

$+\circlearrowright\Sigma M_A = \Sigma(M_A)_{\text{eff}}$: $\qquad W\bar{x} = (m\bar{a})\bar{r} + \bar{I}\alpha$

Since $\bar{a} = \bar{r}\alpha$, we have

$$W\bar{x} = m(\bar{r}\alpha)\bar{r} + \bar{I}\alpha \qquad \alpha = \frac{W\bar{x}}{\dfrac{W}{g}\bar{r}^2 + \bar{I}} \tag{1}$$

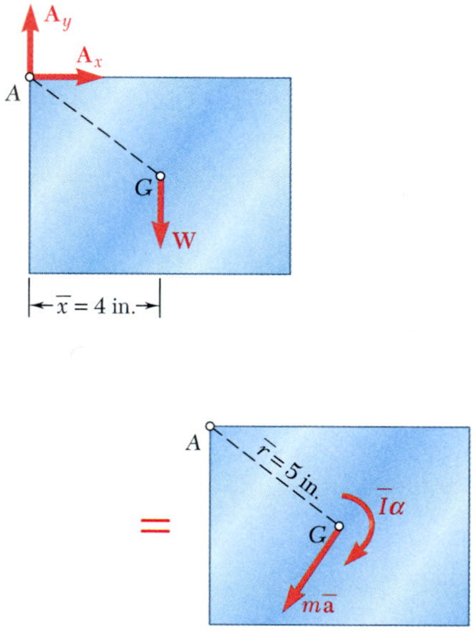

The centroidal moment of inertia of the plate is

$$\bar{I} = \frac{m}{12}(a^2 + b^2) = \frac{60 \text{ lb}}{12(32.2 \text{ ft/s}^2)}[(\tfrac{8}{12} \text{ ft})^2 + (\tfrac{6}{12} \text{ ft})^2]$$
$$= 0.1078 \text{ lb} \cdot \text{ft} \cdot \text{s}^2$$

Substituting this value of $\bar{I}$ together with $W = 60$ lb, $\bar{r} = \frac{5}{12}$ ft, and $\bar{x} = \frac{4}{12}$ ft into Eq. (1), we obtain

$$\alpha = +46.4 \text{ rad/s}^2 \qquad \boldsymbol{\alpha} = 46.4 \text{ rad/s}^2 \circlearrowright \blacktriangleleft$$

***b.* Reaction at A.** Using the computed value of α, we determine the magnitude of the vector $m\bar{\mathbf{a}}$ attached at G.

$$m\bar{a} = m\bar{r}\alpha = \frac{60 \text{ lb}}{32.2 \text{ ft/s}^2}(\tfrac{5}{12} \text{ ft})(46.4 \text{ rad/s}^2) = 36.0 \text{ lb}$$

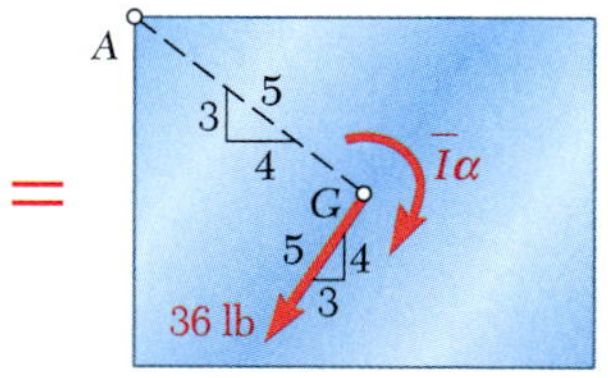

Showing this result on the diagram, we write the equations of motion

$\xrightarrow{+}\Sigma F_x = \Sigma(F_x)_{\text{eff}}$: $\qquad A_x = -\frac{3}{5}(36 \text{ lb})$
$\qquad = -21.6 \text{ lb} \qquad \mathbf{A}_x = 21.6 \text{ lb} \leftarrow \blacktriangleleft$

$+\uparrow\Sigma F_y = \Sigma(F_y)_{\text{eff}}$: $\qquad A_y - 60 \text{ lb} = -\frac{4}{5}(36 \text{ lb})$
$\qquad A_y = +31.2 \text{ lb} \qquad \mathbf{A}_y = 31.2 \text{ lb} \uparrow \blacktriangleleft$

The couple $\bar{I}\boldsymbol{\alpha}$ is not involved in the last two equations; nevertheless, it should be indicated on the diagram.

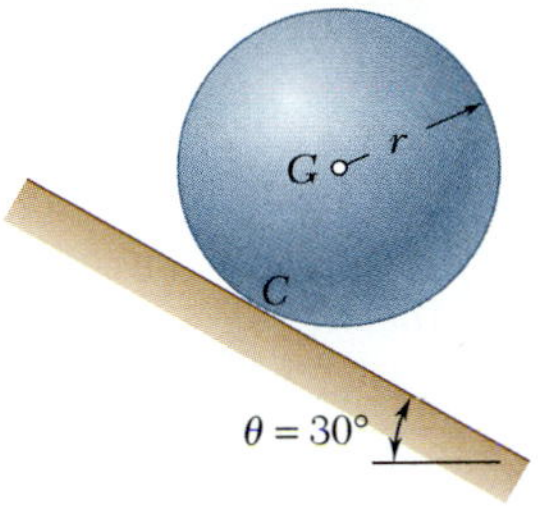

SAMPLE PROBLEM 16.8

A sphere of radius r and weight W is released with no initial velocity on the incline and rolls without slipping. Determine (*a*) the minimum value of the coefficient of static friction compatible with the rolling motion, (*b*) the velocity of the center G of the sphere after the sphere has rolled 10 ft, (*c*) the velocity of G if the sphere were to move 10 ft down a frictionless 30° incline.

SOLUTION

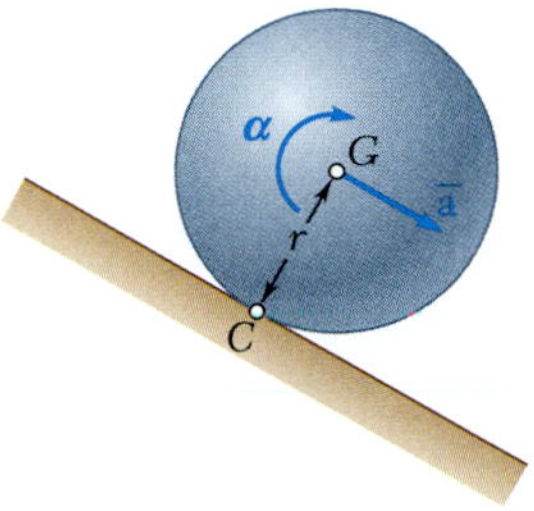

a. Minimum μ_s for Rolling Motion. The external forces **W**, **N**, and **F** form a system equivalent to the system of effective forces represented by the vector $m\bar{\mathbf{a}}$ and the couple $\bar{I}\boldsymbol{\alpha}$. Since the sphere rolls without sliding, we have $\bar{a} = r\alpha$.

$$+\circlearrowright \Sigma M_C = \Sigma(M_C)_{\text{eff}}: \qquad (W \sin\theta)r = (m\bar{a})r + \bar{I}\alpha$$
$$(W \sin\theta)r = (mr\alpha)r + \bar{I}\alpha$$

Noting that $m = W/g$ and $\bar{I} = \frac{2}{5}mr^2$, we write

$$(W\sin\theta)r = \left(\frac{W}{g}r\alpha\right)r + \frac{2}{5}\frac{W}{g}r^2\alpha \qquad \alpha = +\frac{5g\sin\theta}{7r}$$

$$\bar{a} = r\alpha = \frac{5g\sin\theta}{7} = \frac{5(32.2\text{ ft/s}^2)\sin 30°}{7} = 11.50\text{ ft/s}^2$$

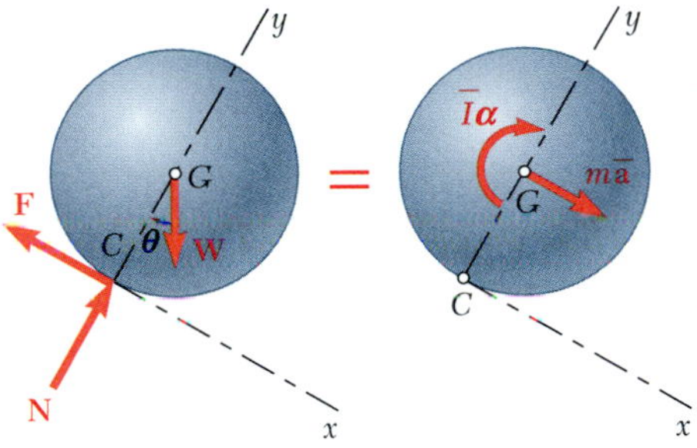

$$+\searrow \Sigma F_x = \Sigma(F_x)_{\text{eff}}: \qquad W\sin\theta - F = m\bar{a}$$
$$W\sin\theta - F = \frac{W}{g}\frac{5g\sin\theta}{7}$$
$$F = +\tfrac{2}{7}W\sin\theta = \tfrac{2}{7}W\sin 30° \qquad \mathbf{F} = 0.143W \;\searrow 30°$$

$$+\nearrow \Sigma F_y = \Sigma(F_y)_{\text{eff}}: \qquad N - W\cos\theta = 0$$
$$N = W\cos\theta = 0.866W \qquad \mathbf{N} = 0.866W \;\angle 60°$$

$$\mu_s = \frac{F}{N} = \frac{0.143W}{0.866W} \qquad \mu_s = 0.165 \blacktriangleleft$$

b. Velocity of Rolling Sphere. We have uniformly accelerated motion:

$$\bar{v}_0 = 0 \qquad \bar{a} = 11.50\text{ ft/s}^2 \qquad \bar{x} = 10\text{ ft} \qquad \bar{x}_0 = 0$$
$$\bar{v}^2 = \bar{v}_0^2 + 2\bar{a}(\bar{x} - \bar{x}_0) \qquad \bar{v}^2 = 0 + 2(11.50\text{ ft/s}^2)(10\text{ ft})$$
$$\bar{v} = 15.17\text{ ft/s} \qquad \bar{\mathbf{v}} = 15.17\text{ ft/s} \;\searrow 30° \blacktriangleleft$$

c. Velocity of Sliding Sphere. Assuming now no friction, we have $F = 0$ and obtain

$$+\circlearrowright \Sigma M_G = \Sigma(M_G)_{\text{eff}}: \qquad 0 = \bar{I}\alpha \qquad \alpha = 0$$

$$+\searrow \Sigma F_x = \Sigma(F_x)_{\text{eff}}: \qquad W\sin 30° = m\bar{a} \qquad 0.50W = \frac{W}{g}\bar{a}$$

$$\bar{a} = +16.1\text{ ft/s}^2 \qquad \bar{\mathbf{a}} = 16.1\text{ ft/s}^2 \;\searrow 30°$$

Substituting $\bar{a} = 16.1\text{ ft/s}^2$ into the equations for uniformly accelerated motion, we obtain

$$\bar{v}^2 = \bar{v}_0^2 + 2\bar{a}(\bar{x} - \bar{x}_0) \qquad \bar{v}^2 = 0 + 2(16.1\text{ ft/s}^2)(10\text{ ft})$$
$$\bar{v} = 17.94\text{ ft/s} \qquad \bar{\mathbf{v}} = 17.94\text{ ft/s} \;\searrow 30° \blacktriangleleft$$

SAMPLE PROBLEM 16.9

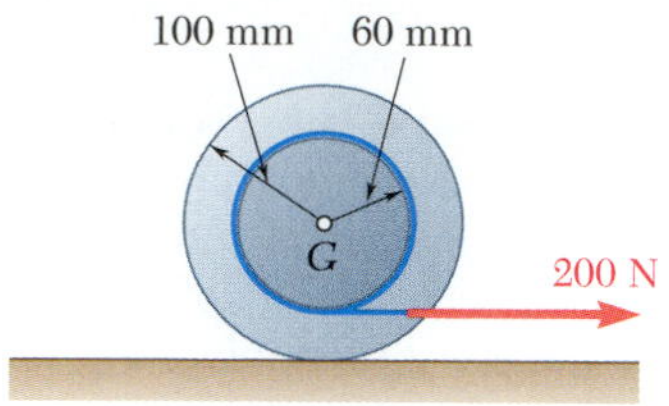

A cord is wrapped around the inner drum of a wheel and pulled horizontally with a force of 200 N. The wheel has a mass of 50 kg and a radius of gyration of 70 mm. Knowing that $\mu_s = 0.20$ and $\mu_k = 0.15$, determine the acceleration of G and the angular acceleration of the wheel.

SOLUTION

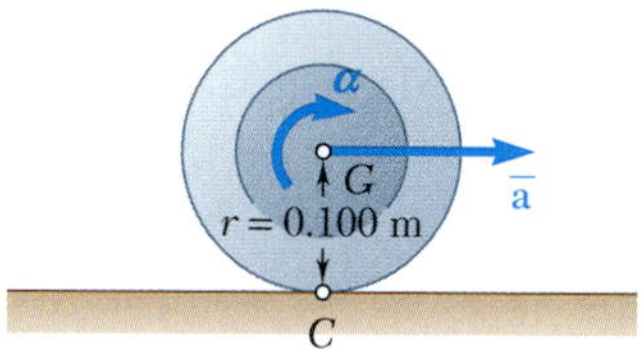

***a.* Assume Rolling without Sliding.** In this case, we have

$$\bar{a} = r\alpha = (0.100 \text{ m})\alpha$$

We can determine whether this assumption is justified by comparing the friction force obtained with the maximum available friction force. The moment of inertia of the wheel is

$$\bar{I} = m\bar{k}^2 = (50 \text{ kg})(0.070 \text{ m})^2 = 0.245 \text{ kg}\cdot\text{m}^2$$

Equations of Motion

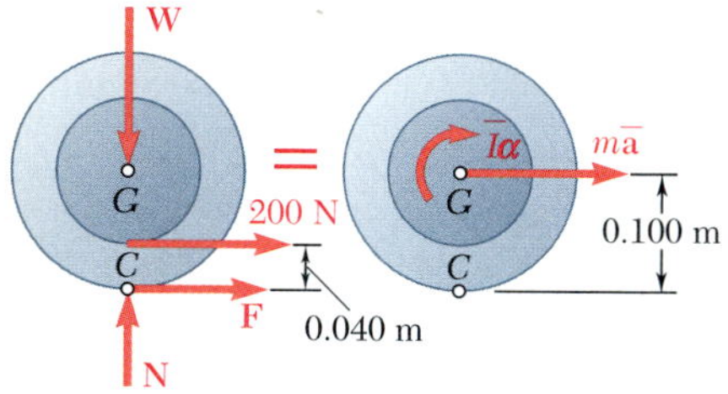

$$+\circlearrowright\Sigma M_C = \Sigma(M_C)_{\text{eff}}: \quad (200 \text{ N})(0.040 \text{ m}) = m\bar{a}(0.100 \text{ m}) + \bar{I}\alpha$$
$$8.00 \text{ N}\cdot\text{m} = (50 \text{ kg})(0.100 \text{ m})\alpha(0.100 \text{ m}) + (0.245 \text{ kg}\cdot\text{m}^2)\alpha$$
$$\alpha = +10.74 \text{ rad/s}^2$$
$$\bar{a} = r\alpha = (0.100 \text{ m})(10.74 \text{ rad/s}^2) = 1.074 \text{ m/s}^2$$

$$\overset{+}{\rightarrow}\Sigma F_x = \Sigma(F_x)_{\text{eff}}: \quad F + 200 \text{ N} = m\bar{a}$$
$$F + 200 \text{ N} = (50 \text{ kg})(1.074 \text{ m/s}^2)$$
$$F = -146.3 \text{ N} \qquad \mathbf{F} = 146.3 \text{ N} \leftarrow$$

$$+\uparrow\Sigma F_y = \Sigma(F_y)_{\text{eff}}:$$
$$N - W = 0 \qquad N = W = mg = (50 \text{ kg})(9.81 \text{ m/s}^2) = 490.5 \text{ N}$$
$$\mathbf{N} = 490.5 \text{ N} \uparrow$$

Maximum Available Friction Force

$$F_{\text{max}} = \mu_s N = 0.20(490.5 \text{ N}) = 98.1 \text{ N}$$

Since $F > F_{\text{max}}$, the assumed motion is impossible.

***b.* Rotating and Sliding.** Since the wheel must rotate and slide at the same time, we draw a new diagram, where $\bar{\mathbf{a}}$ and $\boldsymbol{\alpha}$ are independent and where

$$F = F_k = \mu_k N = 0.15(490.5 \text{ N}) = 73.6 \text{ N}$$

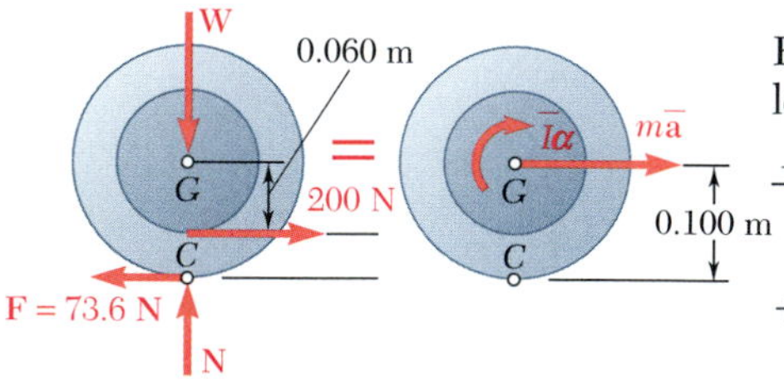

From the computation of part *a*, it appears that **F** should be directed to the left. We write the following equations of motion:

$$\overset{+}{\rightarrow}\Sigma F_x = \Sigma(F_x)_{\text{eff}}: \quad 200 \text{ N} - 73.6 \text{ N} = (50 \text{ kg})\bar{a}$$
$$\bar{a} = +2.53 \text{ m/s}^2 \qquad \bar{\mathbf{a}} = 2.53 \text{ m/s}^2 \rightarrow \blacktriangleleft$$

$$+\circlearrowright\Sigma M_G = \Sigma(M_G)_{\text{eff}}:$$
$$(73.6 \text{ N})(0.100 \text{ m}) - (200 \text{ N})(0.060 \text{ m}) = (0.245 \text{ kg}\cdot\text{m}^2)\alpha$$
$$\alpha = -18.94 \text{ rad/s}^2 \qquad \boldsymbol{\alpha} = 18.94 \text{ rad/s}^2 \circlearrowleft \blacktriangleleft$$

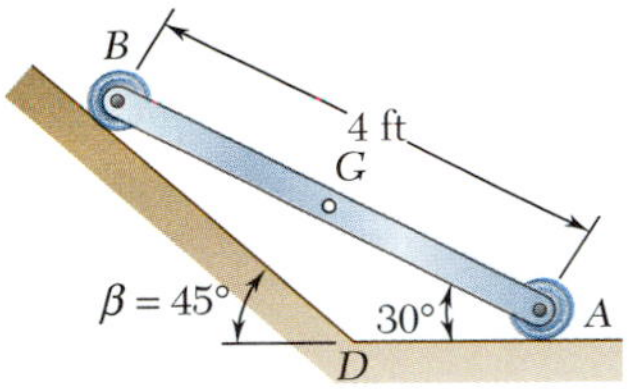

SAMPLE PROBLEM 16.10

The extremities of a 4-ft rod weighing 50 lb can move freely and with no friction along two straight tracks as shown. If the rod is released with no velocity from the position shown, determine (*a*) the angular acceleration of the rod, (*b*) the reactions at *A* and *B*.

SOLUTION

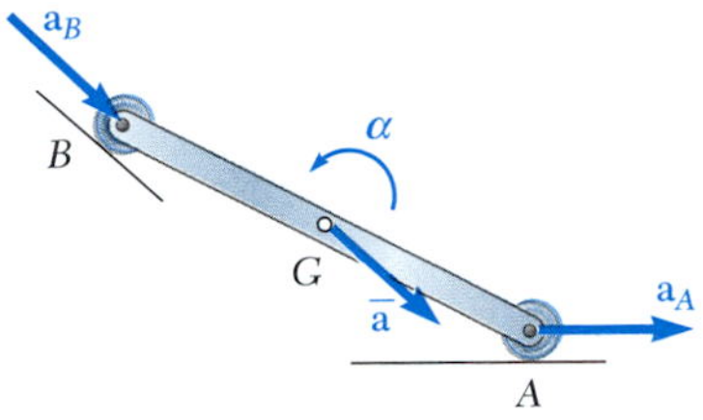

Kinematics of Motion. Since the motion is constrained, the acceleration of *G* must be related to the angular acceleration $\boldsymbol{\alpha}$. To obtain this relation, we first determine the magnitude of the acceleration $\mathbf{a}_A$ of point *A* in terms of α. Assuming that $\boldsymbol{\alpha}$ is directed counterclockwise and noting that $a_{B/A} = 4\alpha$, we write

$$\mathbf{a}_B = \mathbf{a}_A + \mathbf{a}_{B/A}$$
$$[a_B \searrow 45°] = [a_A \rightarrow] + [4\alpha \nearrow 60°]$$

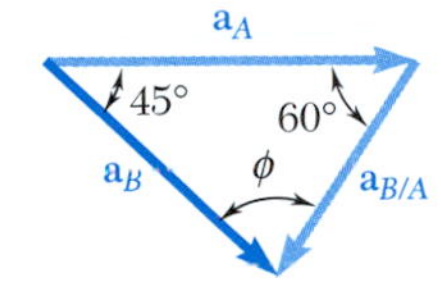

Noting that $\phi = 75°$ and using the law of sines, we obtain

$$a_A = 5.46\alpha \qquad a_B = 4.90\alpha$$

The acceleration of *G* is now obtained by writing

$$\bar{\mathbf{a}} = \mathbf{a}_G = \mathbf{a}_A + \mathbf{a}_{G/A}$$
$$\bar{\mathbf{a}} = [5.46\alpha \rightarrow] + [2\alpha \nearrow 60°]$$

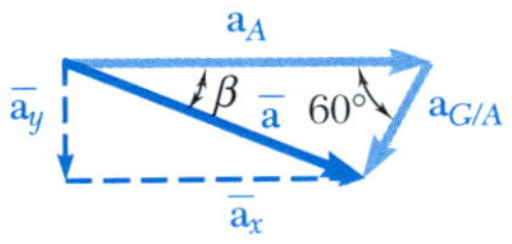

Resolving $\bar{\mathbf{a}}$ into *x* and *y* components, we obtain

$$\bar{a}_x = 5.46\alpha - 2\alpha \cos 60° = 4.46\alpha \qquad \bar{\mathbf{a}}_x = 4.46\alpha \rightarrow$$
$$\bar{a}_y = -2\alpha \sin 60° = -1.732\alpha \qquad \bar{\mathbf{a}}_y = 1.732\alpha \downarrow$$

Kinetics of Motion. We draw a free-body-diagram equation expressing that the system of the external forces is equivalent to the system of the effective forces represented by the vector of components $m\bar{\mathbf{a}}_x$ and $m\bar{\mathbf{a}}_y$ attached at *G* and the couple $\bar{I}\boldsymbol{\alpha}$. We compute the following magnitudes:

$$\bar{I} = \tfrac{1}{12}ml^2 = \frac{1}{12}\frac{50 \text{ lb}}{32.2 \text{ ft/s}^2}(4 \text{ ft})^2 = 2.07 \text{ lb} \cdot \text{ft} \cdot \text{s}^2 \qquad \bar{I}\alpha = 2.07\alpha$$

$$m\bar{a}_x = \frac{50}{32.2}(4.46\alpha) = 6.93\alpha \qquad m\bar{a}_y = -\frac{50}{32.2}(1.732\alpha) = -2.69\alpha$$

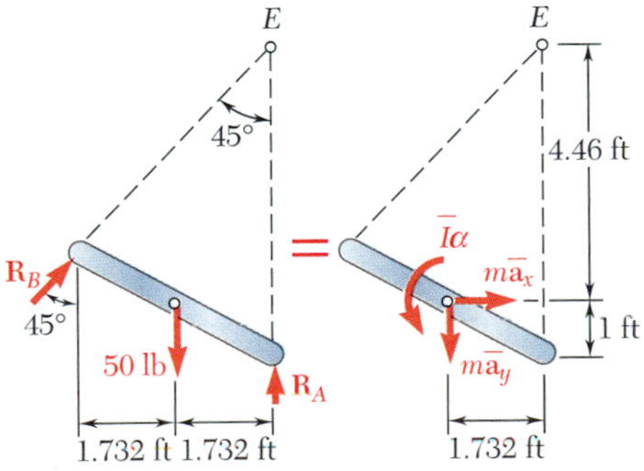

Equations of Motion

$+\circlearrowleft\Sigma M_E = \Sigma(M_E)_{\text{eff}}$:

$$(50)(1.732) = (6.93\alpha)(4.46) + (2.69\alpha)(1.732) + 2.07\alpha$$
$$\alpha = +2.30 \text{ rad/s}^2 \qquad \boldsymbol{\alpha} = 2.30 \text{ rad/s}^2 \circlearrowleft \blacktriangleleft$$

$\xrightarrow{+}\Sigma F_x = \Sigma(F_x)_{\text{eff}}$: $\quad R_B \sin 45° = (6.93)(2.30) = 15.94$

$$R_B = 22.5 \text{ lb} \qquad \mathbf{R}_B = 22.5 \text{ lb} \measuredangle 45° \blacktriangleleft$$

$+\uparrow\Sigma F_y = \Sigma(F_y)_{\text{eff}}$: $\quad R_A + R_B \cos 45° - 50 = -(2.69)(2.30)$

$$R_A = -6.19 - 15.94 + 50 = 27.9 \text{ lb} \qquad \mathbf{R}_A = 27.9 \text{ lb} \uparrow \blacktriangleleft$$

SOLVING PROBLEMS ON YOUR OWN

In this lesson we considered the *plane motion of rigid bodies under constraints*. We found that the types of constraints involved in engineering problems vary widely. For example, a rigid body may be constrained to rotate about a fixed axis or to roll on a given surface, or it may be pin-connected to collars or to other bodies.

1. *Your solution of a problem involving the constrained motion of a rigid body,* will, in general, consist of two steps. First, you will consider the *kinematics of the motion,* and then you will solve the *kinetics portion of the problem.*

2. *The kinematic analysis of the motion* is done by using the methods you learned in Chap. 15. Due to the constraints, linear and angular accelerations will be related. (They will *not* be independent, as they were in the last lesson.) You should establish *relationships among the accelerations* (angular as well as linear), and your goal should be to express all accelerations in terms of a *single unknown acceleration*. This is the first step taken in the solution of each of the sample problems in this lesson.

a. For a body in noncentroidal rotation, the components of the acceleration of the mass center are $\bar{a}_t = \bar{r}\alpha$ and $\bar{a}_n = \bar{r}\omega^2$, where ω will generally be known [Sample Probs. 16.6 and 16.7].

b. For a rolling disk or wheel, the acceleration of the mass center is $\bar{a} = r\alpha$ [Sample Prob. 16.8].

c. For a body in general plane motion, your best course of action, if neither $\bar{a}$ nor α is known or readily obtainable, is to express $\bar{a}$ in terms of α [Sample Prob. 16.10].

3. *The kinetic analysis of the motion* is carried out as follows.

a. Start by drawing a free-body-diagram equation. This was done in all the sample problems of this lesson. In each case the left-hand diagram shows the external forces, including the applied forces, the reactions, and the weight of the body. The right-hand diagram shows the vector $m\bar{\mathbf{a}}$ and the couple $\bar{I}\boldsymbol{\alpha}$.

b. Next, reduce the number of unknowns in the free-body-diagram equation by using the relationships among the accelerations that you found in your kinematic analysis. You will then be ready to consider equations that can be written by summing components or moments. Choose first an equation that involves a single unknown. After solving for that unknown, substitute the value obtained into the other equations, which you will then solve for the remaining unknowns.

4. *When solving problems involving rolling disks or wheels,* keep in mind the following.

a. *If sliding is impending,* the friction force exerted on the rolling body has reached its maximum value, $F_m = \mu_s N$, where N is the normal force exerted on the body and μ_s is the coefficient of *static friction* between the surfaces of contact.

b. *If sliding is not impending,* the friction force F can have *any value* smaller than F_m and should, therefore, be considered as an independent unknown. After you have determined F, be sure to check that it is smaller than F_m; if it is not, *the body does not roll,* but rotates and slides as described in the next paragraph.

c. *If the body rotates and slides at the same time,* then the body is *not rolling* and the acceleration $\bar{a}$ of the mass center is *independent* of the angular acceleration α of the body: $\bar{a} \neq r\alpha$. On the other hand, the friction force has a well-defined value, $F = \mu_k N$, where μ_k is the coefficient of *kinetic friction* between the surfaces of contact.

d. *For an unbalanced rolling disk or wheel,* the relation $\bar{a} = r\alpha$ between the acceleration $\bar{a}$ of the mass center G and the angular acceleration α of the disk or wheel *does not hold anymore*. However, a similar relation holds between the acceleration a_O of the *geometric center* O and the angular acceleration α of the disk or wheel: $a_O = r\alpha$. This relation can be used to express $\bar{a}$ in terms of α and ω (Fig. 16.18).

5. *For a system of connected rigid bodies,* the goal of your *kinematic analysis* should be to determine all the accelerations from the given data, or to express them all in terms of a single unknown. (For systems with several degrees of freedom, you will need to use as many unknowns as there are degrees of freedom.)

Your *kinetic analysis* will generally be carried out by drawing a free-body-diagram equation for the entire system, as well as for one or several of the rigid bodies involved. In the latter case, both internal and external forces should be included, and care should be taken to represent with equal and opposite vectors the forces that two bodies exert on each other.

Problems

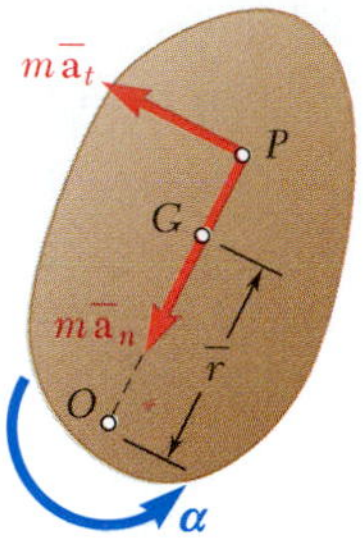

Fig. P16.70

16.70 Show that the couple $\bar{I}\alpha$ of Fig. 16.15 can be eliminated by attaching the vectors $m\bar{\mathbf{a}}_t$ and $m\bar{\mathbf{a}}_n$ at a point P called the *center of percussion*, located on line OG at a distance $GP = \bar{k}^2/\bar{r}$ from the mass center of the body.

16.71 A uniform slender rod of length $L = 900$ mm and mass $m = 1$ kg hangs freely from a hinge at A. If a force $\mathbf{P}$ of magnitude 3.5 N is applied at B horizontally to the left ($h = L$), determine (a) the angular acceleration of the rod, (b) the components of the reaction at A.

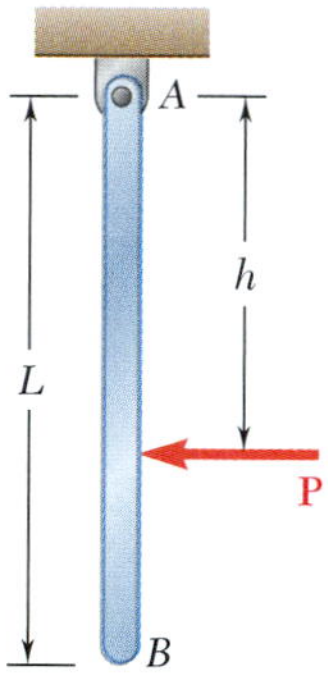

Fig. P16.71

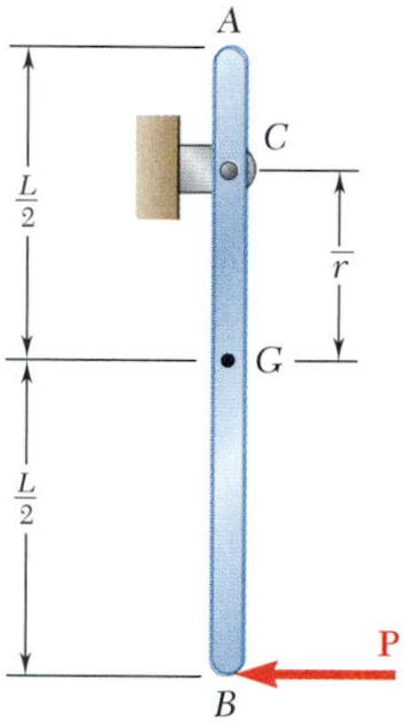

Fig. P16.72

16.72 A uniform slender rod of length $L = 36$ in. and weight $W = 10$ lb hangs freely from a hinge at C. A horizontal force $\mathbf{P}$ of magnitude 15 lb is applied at end B. Knowing that $\bar{r} = 9$ in., determine (a) the angular acceleration of the rod, (b) the components of the reaction at C.

16.73 In Prob. 16.71, determine (a) the distance h for which the horizontal component of the reaction at A is zero, (b) the corresponding angular acceleration of the rod.

16.74 In Prob. 16.72, determine (a) the distance $\bar{r}$ for which the horizontal component of the reaction at C is zero, (b) the corresponding angular acceleration of the rod.

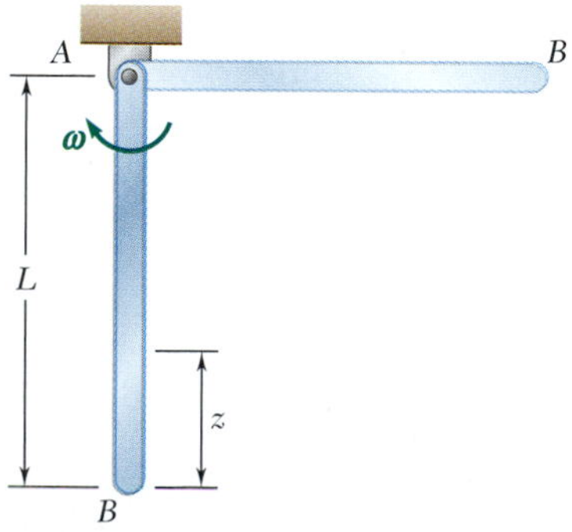

Fig. P16.75

16.75 A uniform slender rod AB of length L and mass m is pivoted at end A and released from a horizontal position. The angular velocity of the rod as it passes through the vertical position is known to be $\omega = \sqrt{3g/L}$. (a) Express the tension in the rod at a distance z from end B in terms of z, m, g, and L for the vertical position. (b) Knowing that the weight of the rod is 10 N, determine the maximum tension in the rod for the vertical position.

16.76 A large flywheel is mounted on a horizontal shaft and rotates at a constant rate of 1200 rpm. Experimental data indicate that the total force exerted by the flywheel on the shaft varies from 12 kip upward to 18 kip downward. Determine (a) the weight of the flywheel, (b) the distance from the center of the shaft to the mass center of the flywheel.

16.77 A uniform slender rod of weight 0.24 lb/ft is used to form the assembly shown. The assembly rotates clockwise at a constant rate of 120 rpm under the combined effect of gravity and the couple **M** which varies in magnitude and sense. Determine the magnitude and sense of the couple **M** and the reaction at point A for (*a*) $\theta = 90°$, (*b*) $\theta = 180°$.

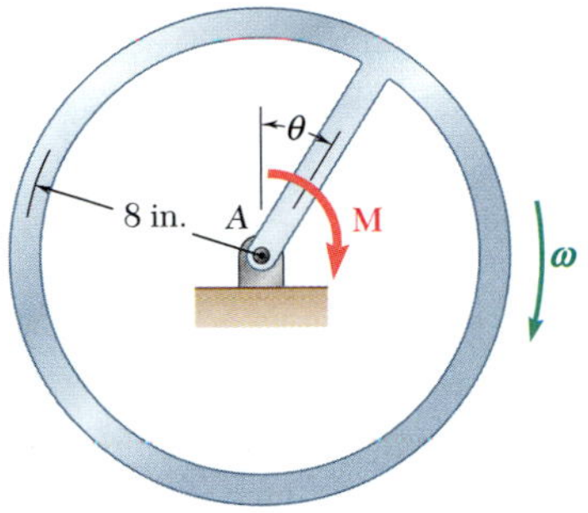

Fig. P16.77

16.78 The shutter shown was formed by removing one quarter of a disk of 15-mm radius and is used to interrupt a beam of light emitting from a lens at C. Knowing that the shutter has a mass of 50 g and rotates at the constant rate of 24 cycles per second, determine the magnitude of the force exerted by the shutter on the shaft at A.

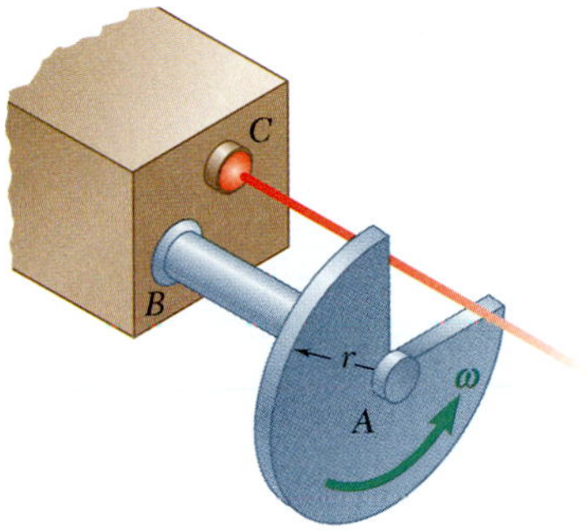

Fig. P16.78

16.79 A uniform rod of length L and mass m is supported as shown. If the cable attached at B suddenly breaks, determine (*a*) the acceleration of end B, (*b*) the reaction at the pin support.

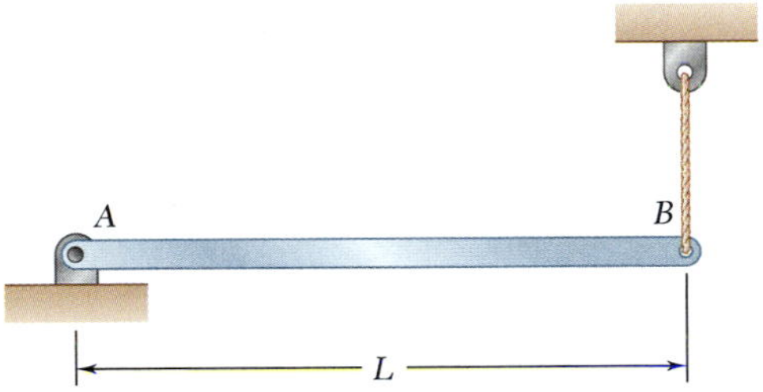

Fig. *P16.79*

16.80 A uniform rod of length L and mass m is supported as shown with $b = 0.2L$ when the cable attached to end B suddenly breaks. Determine at this instant (*a*) the acceleration of end B, (*b*) the reaction at the pin support.

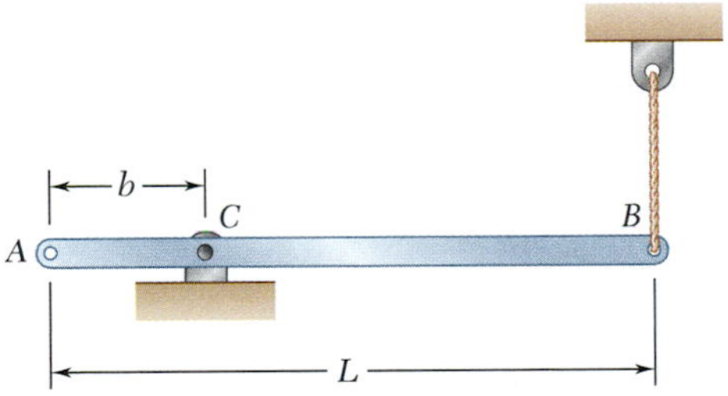

Fig. P16.80 and P16.81

16.81 A uniform rod of length L and mass m is supported as shown. If the cable attached at B suddenly breaks, determine (*a*) the distance b for which the acceleration of end A is maximum, (*b*) the corresponding acceleration of end A and the reaction at C.

16.82 A half cylinder of mass m is released from rest in the position shown and swings freely about the horizontal diameter AB. Determine, at this instant, (*a*) the acceleration of point D, (*b*) the acceleration of the mass center of the half cylinder.

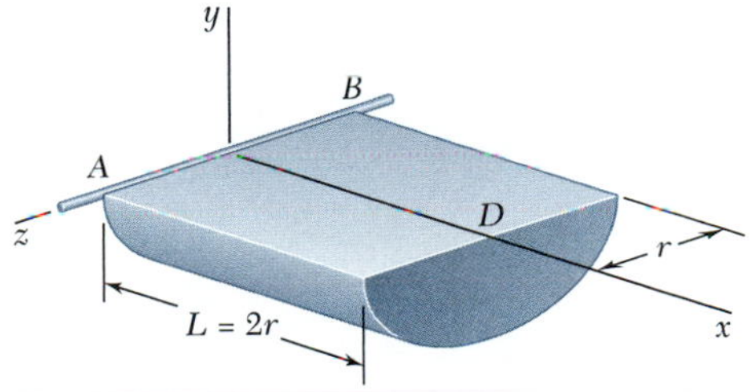

Fig. P16.82

16.83 Two identical 2-kg slender rods *AB* and *BC* are connected by a pin at *B* and by the cord *AC*. The assembly rotates in a vertical plane under the combined effect of gravity and an 8 N · m couple **M** applied to rod *AB*. Knowing that in the position shown the angular velocity of the assembly is zero, determine (*a*) the angular acceleration of the assembly, (*b*) the tension in cord *AC*.

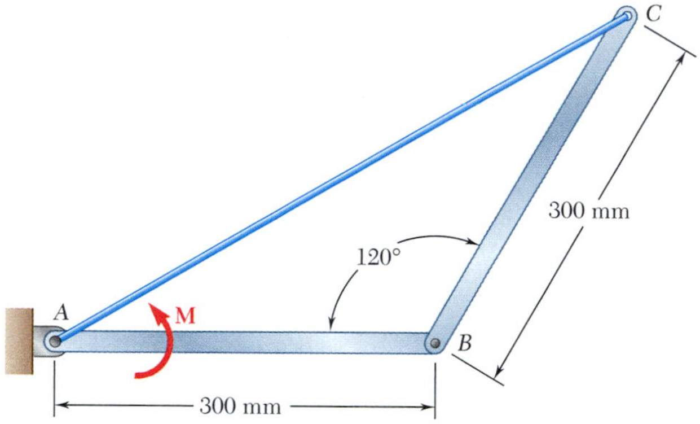

Fig. P16.83 and *P16.84*

16.84 Two identical 2-kg slender rods *AB* and *BC* are connected by a pin at *B* and by the cord *AC*. The assembly rotates in a vertical plane under the combined effect of gravity and a couple **M** applied to rod *AB*. Knowing that in the position shown the angular velocity of the assembly is zero and the tension in cord *AC* is 2 N, determine (*a*) the angular acceleration of the assembly, (*b*) the magnitude of the couple **M**.

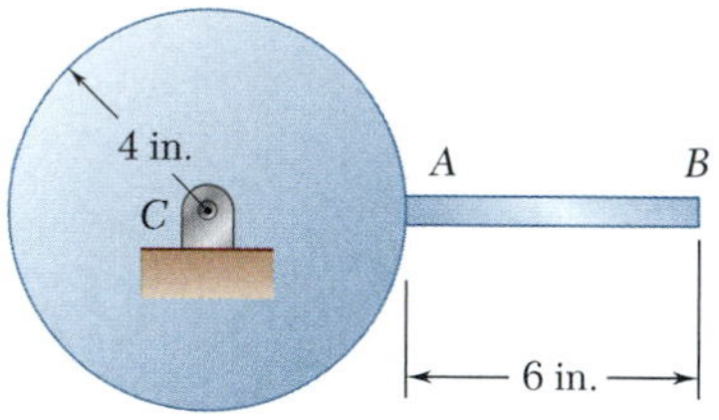

Fig. P16.85

16.85 A 3-lb slender rod is welded to a 10-lb uniform disk as shown. The assembly swings freely about *C* in a vertical plane. Knowing that in the position shown the assembly has an angular velocity of 10 rad/s clockwise, determine (*a*) the angular acceleration of the assembly, (*b*) the components of the reaction at *C*.

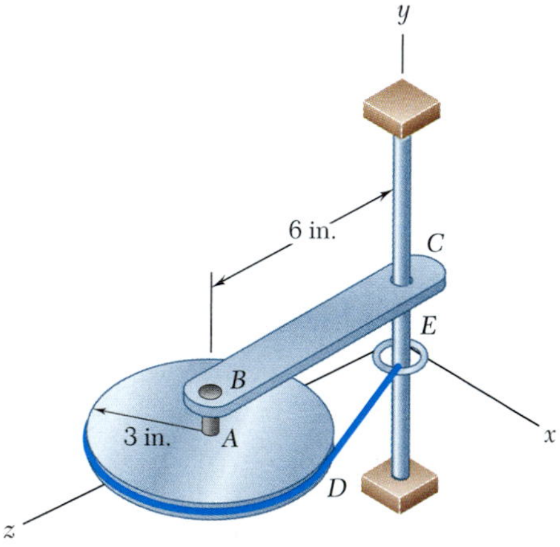

Fig. P16.86

16.86 A 10-lb uniform disk is attached to the 6-lb uniform rod *BC* by means of a frictionless pin *AB*. An elastic cord is wound around the edge of the disk and is attached to a ring at *E*. Both ring *E* and rod *BC* can rotate freely about the vertical shaft. Knowing that the system is released from rest when the tension in the elastic cord is 3 lb, determine (*a*) the angular acceleration of the disk, (*b*) the acceleration of the center of the disk.

16.87 Derive the equation $\Sigma M_C = I_C\alpha$ for the rolling disk of Fig. 16.17, where ΣM_C represents the sum of the moments of the external forces about the instantaneous center *C*, and I_C is the moment of inertia of the disk about *C*.

16.88 Show that in the case of an unbalanced disk, the equation derived in Prob. 16.87 is valid only when the mass center *G*, the geometric center *O*, and the instantaneous center *C* happen to lie in a straight line.

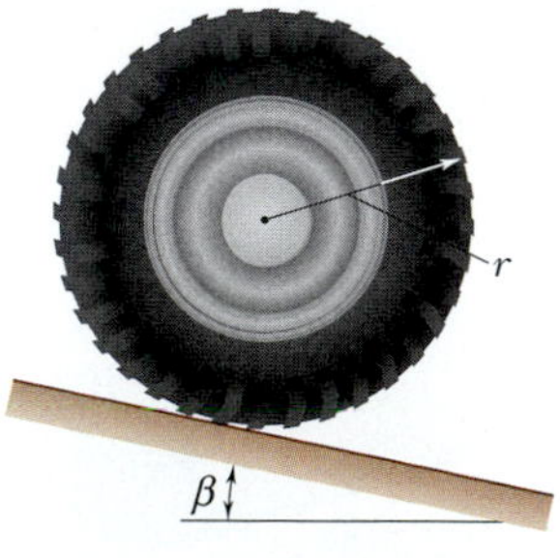

Fig. P16.89

16.89 A wheel of radius r and centroidal radius of gyration $\bar{k}$ is released from rest on the incline and rolls without sliding. Derive an expression for the acceleration of the center of the wheel in terms of r, $\bar{k}$, β, and g.

16.90 A flywheel is rigidly attached to a shaft of 30-mm radius that can roll along parallel rails as shown. When released from rest, the system rolls 5 m in 40 s. Determine the centroidal radius of gyration of the system.

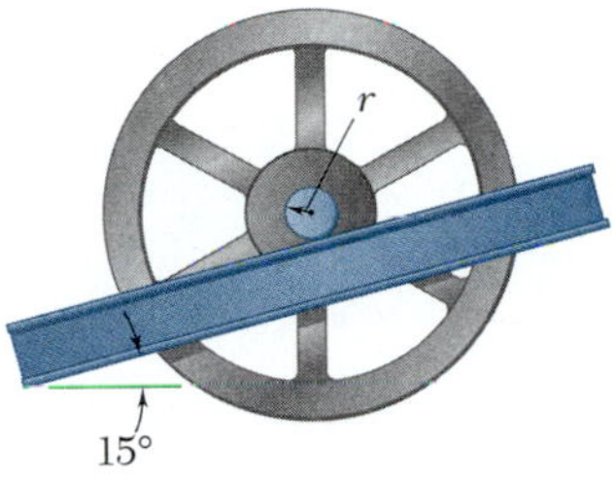

Fig. P16.90 and P16.91

16.91 A flywheel of centroidal radius of gyration $\bar{k}$ is rigidly attached to a shaft that can roll along parallel rails. Denoting by μ_k the coefficient of static friction between the shaft and the rails, derive an expression for the largest angle of inclination β for which no slipping will occur.

16.92 A homogeneous sphere S, a uniform cylinder C, and a thin pipe P are in contact when they are released from rest on the incline shown. Knowing that all three objects roll without slipping, determine, after 6 s of motion, the clear distance between (*a*) the pipe and the cylinder, (*b*) the cylinder and the sphere. Give the answers in both U.S. customary and SI units.

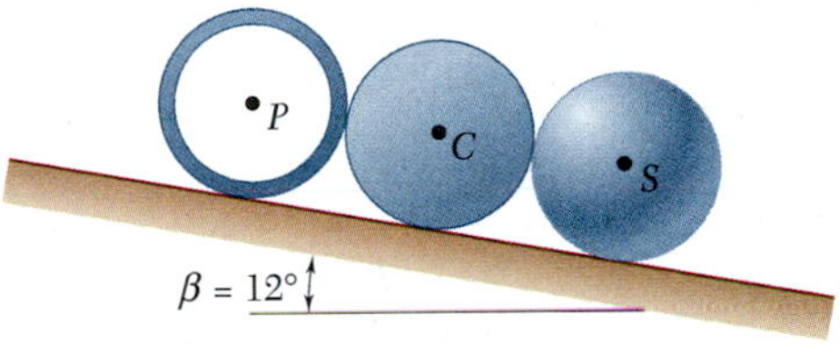

Fig. P16.92

16.93 through 16.96 A drum of 80-mm radius is attached to a disk of 160-mm radius. The disk and drum have a combined mass of 5 kg and combined radius of gyration of 120 mm. A cord is attached as shown and pulled with a force **P** of magnitude 20 N. Knowing that the coefficients of static and kinetic friction are $\mu_s = 0.25$ and $\mu_k = 0.20$, respectively, determine (*a*) whether or not the disk slides, (*b*) the angular acceleration of the disk and the acceleration of G.

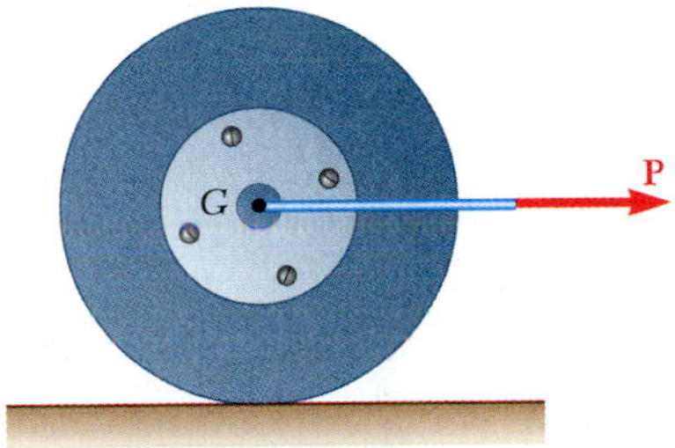

Fig. P16.93 and P16.97

16.97 through 16.100 A drum of 4-in. radius is attached to a disk of 8-in. radius. The disk and drum have a total weight of 10 lb and combined radius of gyration of 6 in. A cord is attached as shown and pulled with a force **P** of magnitude 5 lb. Knowing that the disk rolls without sliding, determine (*a*) the angular acceleration of the disk and the acceleration of G, (*b*) the minimum value of the coefficient of static friction compatible with this motion.

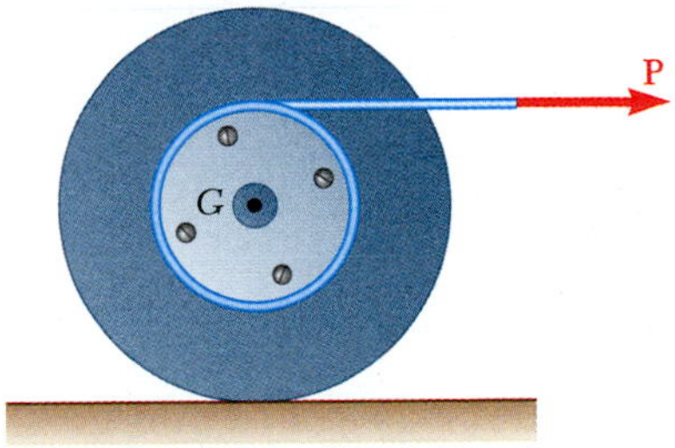

Fig. P16.94 and P16.98

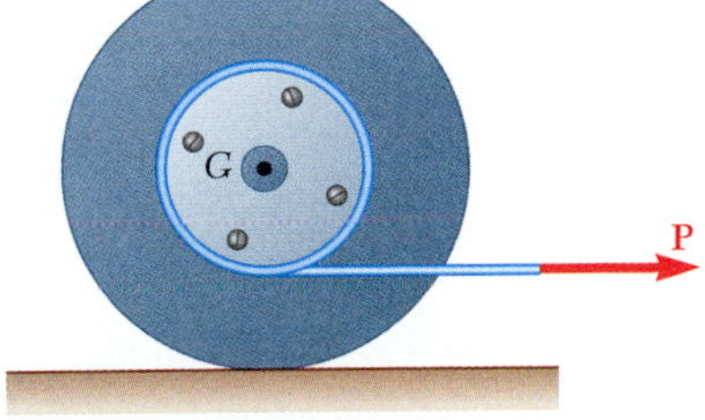

Fig. P16.95 and P16.99

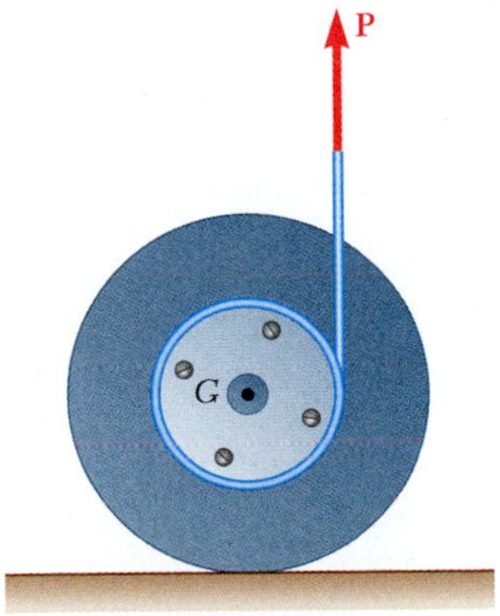

Fig. P16.96 and P16.100

16.101 and 16.102 Two uniform disks A and B, each of mass 2 kg, are connected by a 2.5-kg rod CD as shown. A counterclockwise couple $\mathbf{M}$ of moment 2.25 N · m is applied to disk A. Knowing that the disks roll without sliding, determine (*a*) the acceleration of the center of each disk, (*b*) the horizontal component of the force exerted on disk B by pin D.

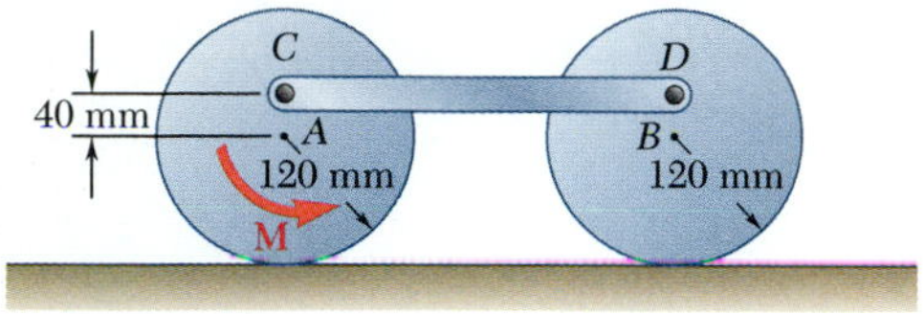

Fig. P16.101

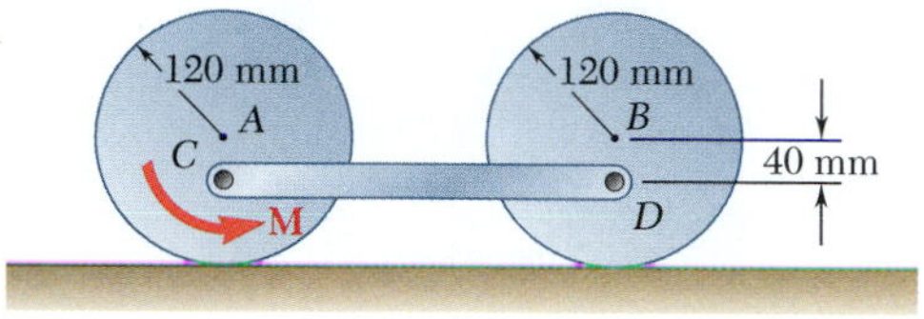

Fig. P16.102

16.103 A 5-kg uniform square plate is supported by two identical 1.5-kg uniform slender rods AD and BE. It is held in the position shown by rope CF. Determine, immediately after rope CF has been cut, (*a*) the acceleration of the plate, (*b*) the force exerted on the plate at point B.

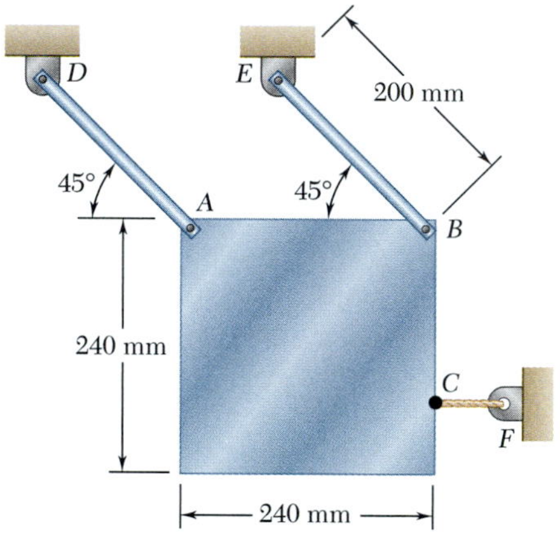

Fig. P16.103

16.104 Solve Prob. 16.103 assuming that rod AD is replaced by a cable of negligible mass.

16.105 A half section of a uniform thin pipe of mass m is at rest when a force $\mathbf{P}$ is applied as shown. Assuming that the section rolls without sliding, determine (*a*) its initial angular acceleration, (*b*) the minimum value of the coefficient of static friction consistent with the motion.

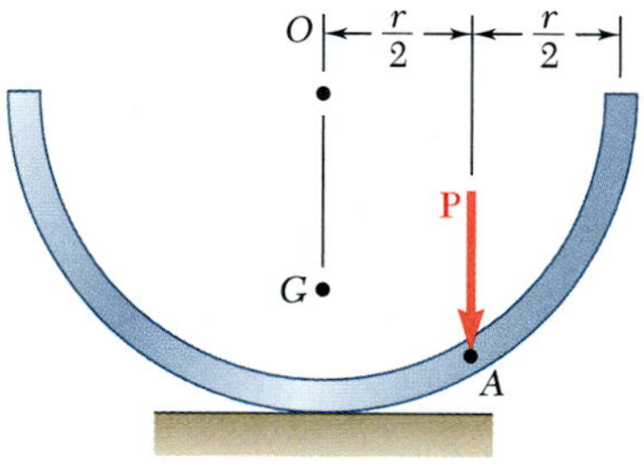

Fig. P16.105

16.106 Solve Prob. 16.105 assuming that the force $\mathbf{P}$ applied at point A is directed horizontally to the right.

16.107 A small clamp of mass m_B is attached at B to a hoop of mass m_h. The system is released from rest when $\theta = 90°$ and rolls without sliding. Knowing that $m_h = 5m_B$, determine (*a*) the angular acceleration of the hoop, (*b*) the horizontal and vertical components of the acceleration of B.

16.108 A small clamp of mass m_B is attached at B to a hoop of mass m_h. Knowing that the system is released from rest and rolls without sliding, derive an expression for the angular acceleration of the hoop in terms of m_B, m_h, r, and θ.

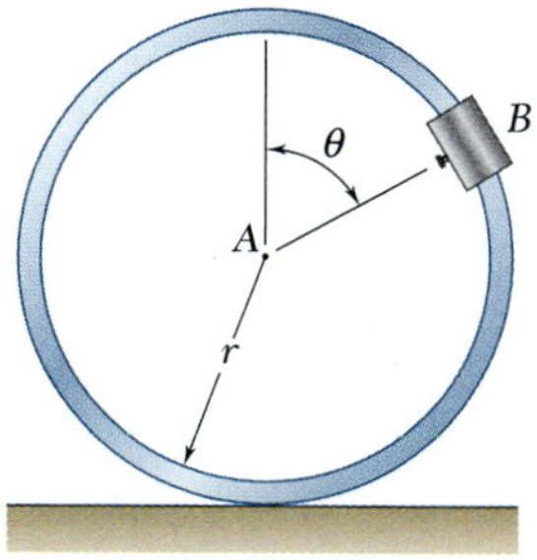

Fig. P16.107 and *P16.108*

16.109 The center of gravity G of a 3.5-lb unbalanced tracking wheel is located at a distance $r = 0.9$ in. from its geometric center B. The radius of the wheel is $R = 3$ in. and its centroidal radius of gyration is 2.2 in. At the instant shown the center B of the wheel has a velocity of 1.05 ft/s and an acceleration of 3.6 ft/s^2, both directed to the left. Knowing that the wheel rolls without sliding and neglecting the mass of the driving yoke AB, determine the horizontal force **P** applied to the yoke.

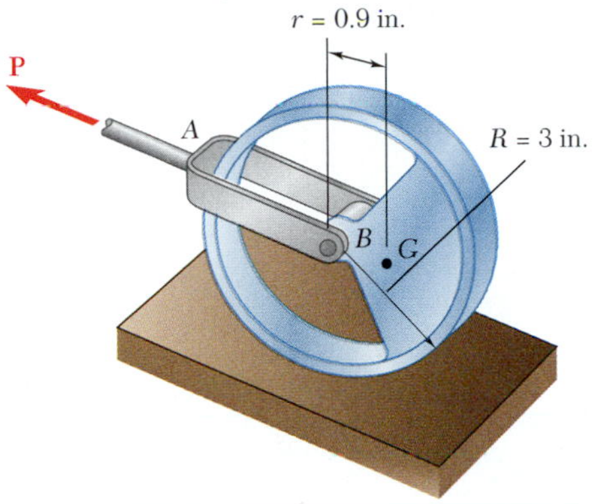

Fig. P16.109

16.110 For the wheel of Prob. 16.109, determine the horizontal force **P** applied to the yoke, knowing that at the instant shown the center B has a velocity of 1.05 ft/s and an acceleration of 3.6 ft/s^2, both directed to the right.

16.111 The 4-lb uniform rod AB is in the shape of a quarter of a circle and its ends are attached to collars of negligible weight that slide without friction along fixed rods. Knowing that rod AB is released from rest in the position shown, determine immediately after release (*a*) the angular acceleration of the rod, (*b*) the reaction at end B.

16.112 The 4-lb uniform rod AB is in the shape of a quarter of a circle and its ends are attached to collars of negligible weight that slide along fixed rods. Rod AB is released from rest in the position shown. Assuming no friction at A and a friction force of magnitude 0.2 lb at B, determine the angular acceleration of the rod immediately after release.

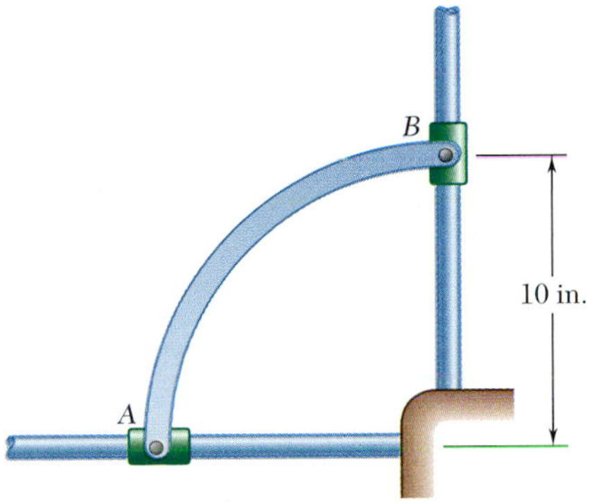

Fig. P16.111 and P16.112

16.113 The uniform rod AB of mass m and length $2L$ is attached to collars of negligible mass that slide without friction along fixed rods. If the rod is released from rest in the position shown, derive an expression for (*a*) the angular acceleration of the rod, (*b*) the reaction at A.

16.114 The uniform rod AB of mass $m = 6$ kg and total length $2L = 600$ mm is attached to collars of negligible mass that slide without friction along fixed rods. If rod AB is released from rest when $\theta = 30°$, determine immediately after release (*a*) the angular acceleration of the rod, (*b*) the reaction at A.

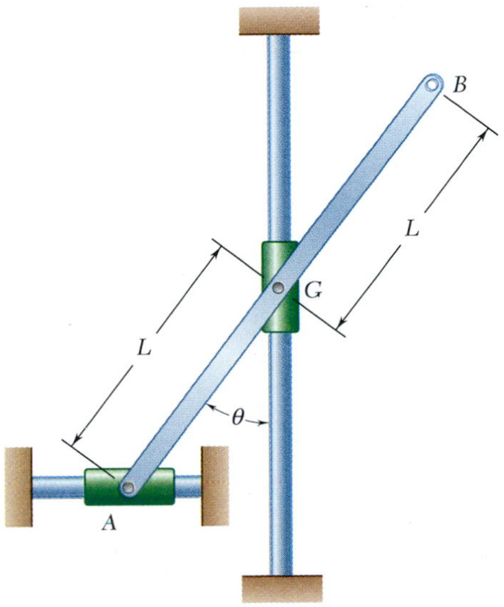

Fig. P16.113 and *P16.114*

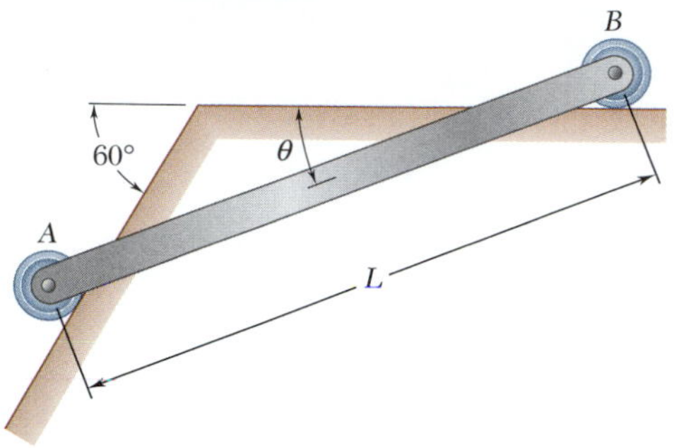

Fig. P16.115

16.115 The motion of the uniform rod AB of mass 8 kg and length $L = 900$ mm is guided by small wheels of negligible mass that roll on the surface shown. If the rod is released from rest when $\theta = 20°$, determine (*a*) the angular acceleration of the rod, (*b*) the components of the reaction at A.

16.116 The 8-kg uniform slender rod AB is supported by a small wheel at C and end A can slide on the horizontal surface. Knowing that the rod is released from rest in the position shown and neglecting friction at A and C, determine immediately after release (*a*) the angular acceleration of the rod, (*b*) the reaction at point A.

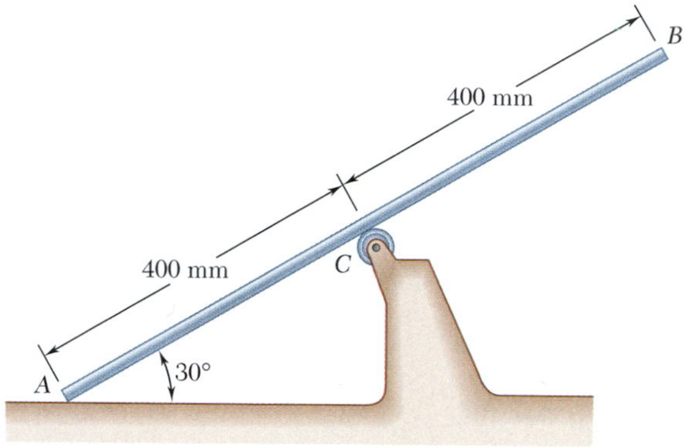

Fig. P16.116

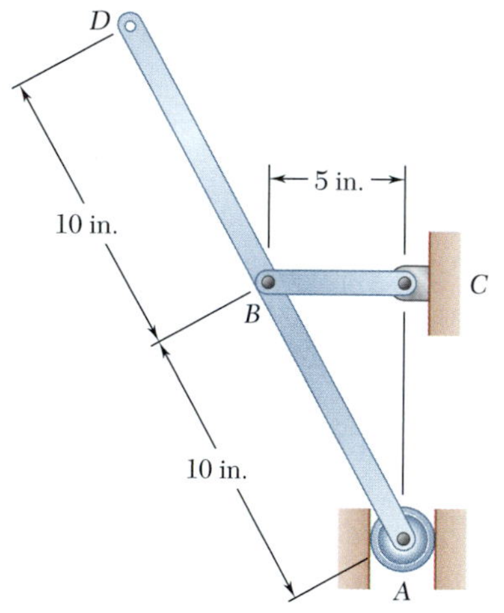

Fig. P16.117

16.117 The 10-lb uniform rod ABD is attached to the crank BC and is fitted with a small wheel that can roll without friction along a vertical slot. Knowing that at the instant shown crank BC rotates with an angular velocity of 6 rad/s clockwise and an angular acceleration of 15 rad/s^2 counterclockwise, determine the reaction at A.

16.118 The 8.05-lb uniform slender bar BD is attached to bar AB and a wheel of negligible mass which rolls on a circular surface. Knowing that at the instant shown bar AB has an angular velocity of 6 rad/s and no angular acceleration, determine the reaction at point D.

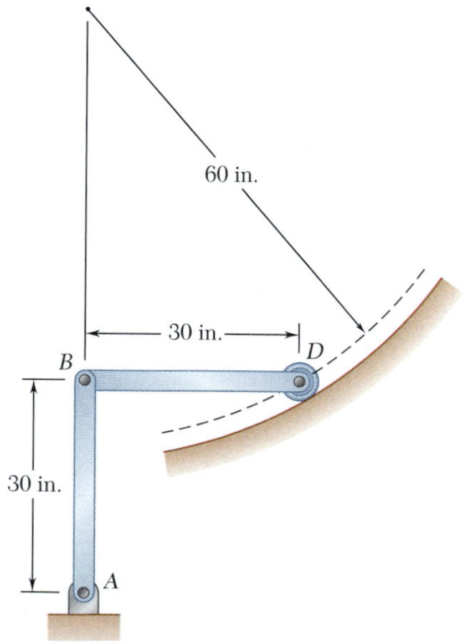

Fig. P16.118

16.119 The 300-mm uniform rod BD of mass 3 kg is connected as shown to crank AB and to collar D of negligible mass, which can slide freely along a horizontal rod. Knowing that crank AB rotates counterclockwise at the constant rate of 300 rpm, determine the reaction at D when $\theta = 0$.

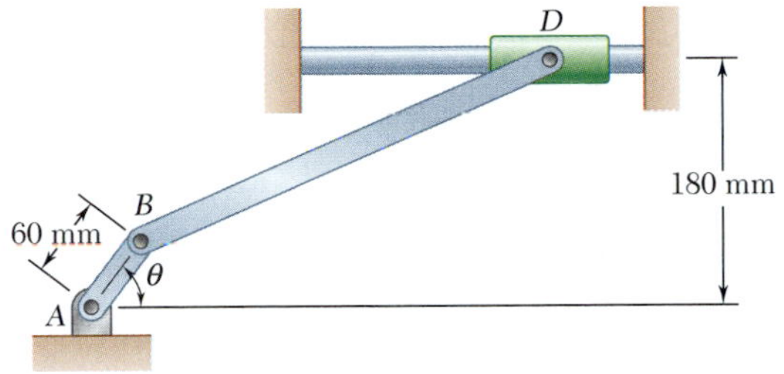

Fig. P16.119

16.120 Solve Prob. 16.119 when $\theta = 180°$.

16.121 Solve Prob. 16.119 when $\theta = 90°$.

16.122 The collar B of negligible weight can slide freely on the 8-lb uniform rod CD. Knowing that in the position shown crank AB rotates with an angular velocity of 5 rad/s and an angular acceleration of 60 rad/s^2, both clockwise, determine the force exerted on rod CD by collar B.

16.123 The collar B of negligible weight can slide freely on the 8-lb uniform rod CD. Knowing that in the position shown crank AB rotates with an angular velocity of 5 rad/s and an angular acceleration of 60 rad/s^2, both counterclockwise, determine the force exerted on rod CD by collar B.

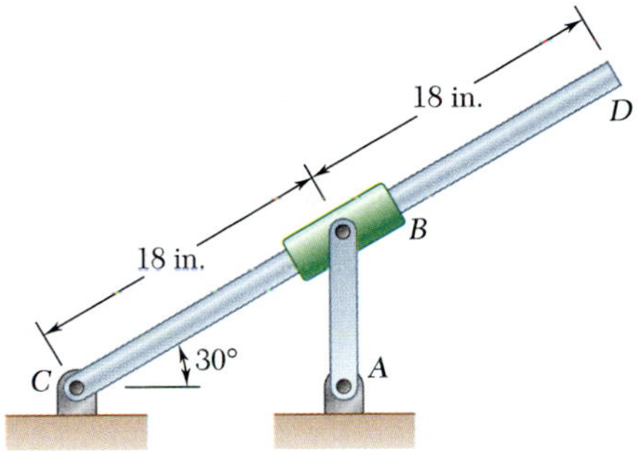

Fig. P16.122 and *P16.123*

16.124 A driver starts his car with the door on the passenger's side wide open ($\theta = 0$). The 36-kg door has a centroidal radius of gyration $\bar{k} =$ 250 mm, and its mass center is located at a distance $r = 440$ mm from its vertical axis of rotation. Knowing that the driver maintains a constant acceleration of 2 m/s^2, determine the angular velocity of the door as it slams shut ($\theta = 90°$).

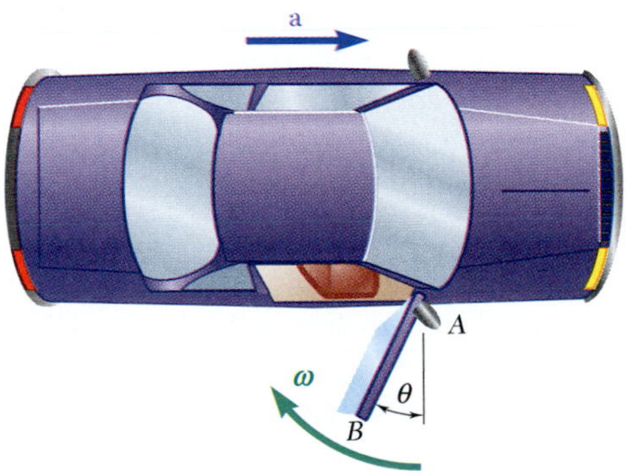

Fig. P16.124

16.125 For the car of Prob. 16.124, determine the smallest constant acceleration that the driver can maintain if the door is to close and latch, knowing that as the door hits the frame its angular velocity must be at least 2 rad/s for the latching mechanism to operate.

16.126 Two 3-kg uniform bars are connected to form the linkage shown. Neglecting the effect of friction, determine the reaction at D immediately after the linkage is released from rest in the position shown.

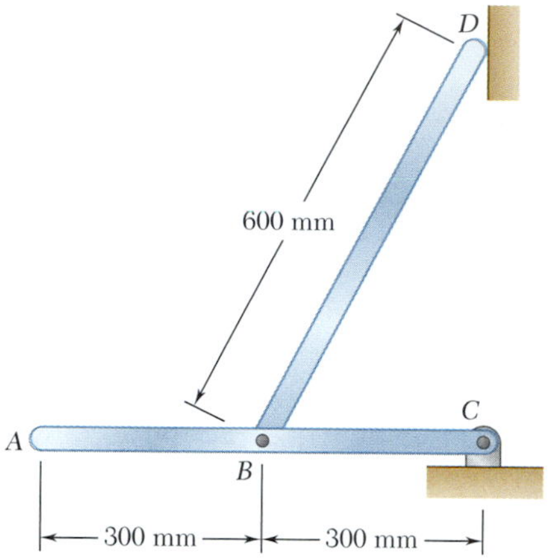

Fig. P16.126

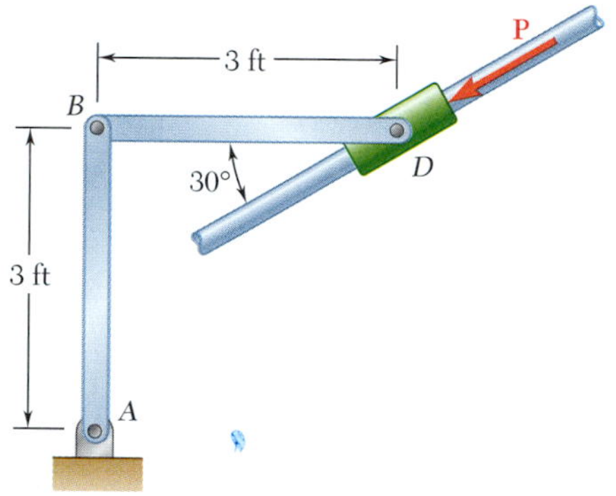

Fig. P16.127

16.127 The linkage ABD is formed by connecting two 8.05-lb bars and a collar of negligible weight. The motion of the linkage is controlled by the force **P** applied to the collar. Knowing that at the instant shown the angular velocity and angular acceleration of bar AB are zero and 10 rad/s^2 counterclockwise, respectively, determine the force **P**.

16.128 Solve Prob. 16.127 assuming that at the instant shown the angular velocity and angular acceleration of bar AB are 5 rad/s clockwise and zero, respectively.

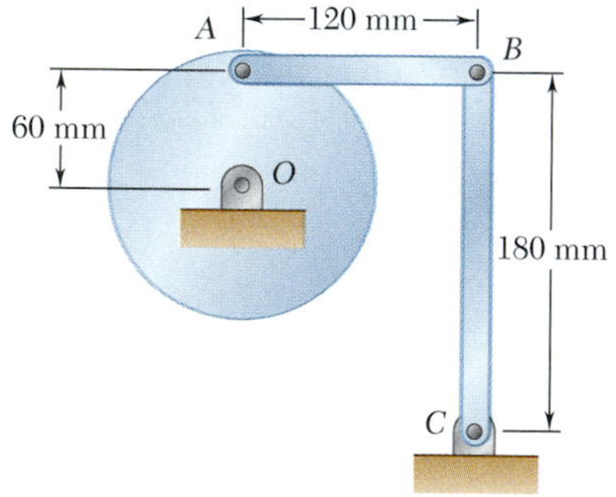

Fig. P16.129 and P16.130

16.129 The 2-kg rod AB and the 3-kg rod BC are connected as shown to a disk that is made to rotate in a vertical plane at a constant angular velocity of 6 rad/s clockwise. For the position shown, determine the forces exerted at A and B on rod AB.

16.130 The 2-kg rod AB and the 3-kg rod BC are connected as shown to a disk that is made to rotate in a vertical plane. Knowing that at the instant shown the disk has an angular acceleration of 18 rad/s^2 clockwise and no angular velocity, determine the components of the forces exerted at A and B on rod AB.

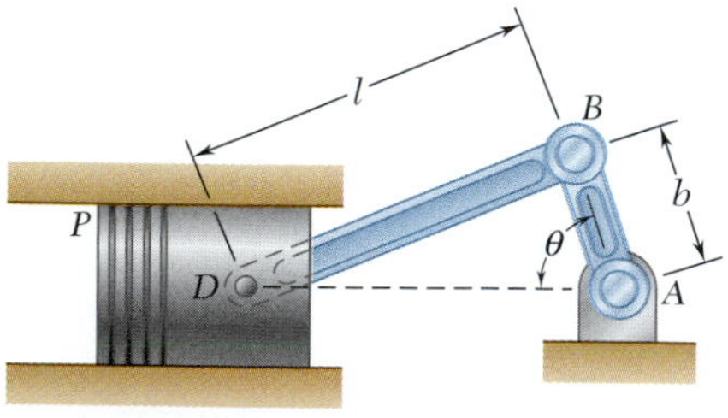

Fig. P16.131

16.131 In the engine system shown $l = 10$ in. and $b = 4$ in. The connecting rod BD is assumed to be a 3-lb uniform slender rod and is attached to the 4.5-lb piston P. During a test of the system, crank AB is made to rotate with a constant angular velocity of 600 rpm clockwise with no force applied to the face of the piston. Determine the forces exerted on the connecting rod at B and D when $\theta = 180°$. (Neglect the effect of the weight of the rod.)

16.132 Solve Prob. 16.131 when $\theta = 0$.

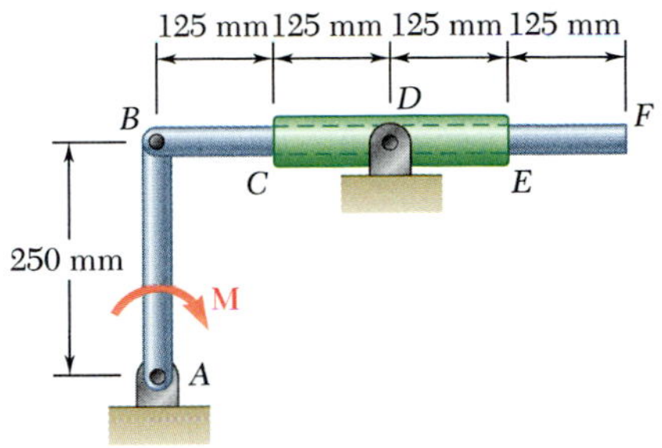

Fig. P16.133 and P16.134

16.133 The 2-kg uniform slender rod *AB*, the 4-kg uniform slender rod *BF*, and the 2-kg uniform thin sleeve *CE* are connected as shown and move without friction in a vertical plane. The motion of the linkage is controlled by the couple **M** applied to rod *AB*. Knowing that at the instant shown the angular velocity of rod *AB* is 15 rad/s and the magnitude of the couple **M** is 7 N · m, determine (*a*) the angular acceleration of rod *AB*, (*b*) the reaction at point *D*.

16.134 The 2-kg uniform slender rod *AB*, the 4-kg uniform slender rod *BF*, and the 2-kg uniform thin sleeve *CE* are connected as shown and move without friction in a vertical plane. The motion of the linkage is controlled by the couple **M** applied to rod *AB*. Knowing that at the instant shown the angular velocity of rod *AB* is 30 rad/s and the angular acceleration of rod *AB* is 96 rad/s^2 clockwise, determine (*a*) the magnitude of the couple **M**, (*b*) the reaction at point *D*.

***16.135** The uniform slender 4.1-lb bar *BD* is attached to the uniform 12.3-lb uniform disk by a pin at *B* and released from rest in the position shown. Assuming that the disk rolls without slipping, determine (*a*) the initial reaction at the contact point *A*, (*b*) the corresponding smallest allowable value of the coefficient of static friction.

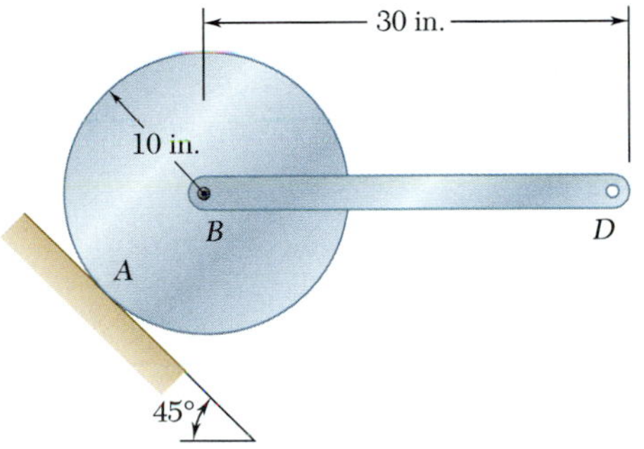

Fig. P16.135

****16.136*** Two disks, each of mass *m* and radius *r* are connected as shown by a continuous chain belt of negligible mass. If a pin at point *C* of the chain belt is suddenly removed, determine (*a*) the angular acceleration of each disk, (*b*) the tension in the left-hand portion of the belt, (*c*) the acceleration of the center of disk *B*.

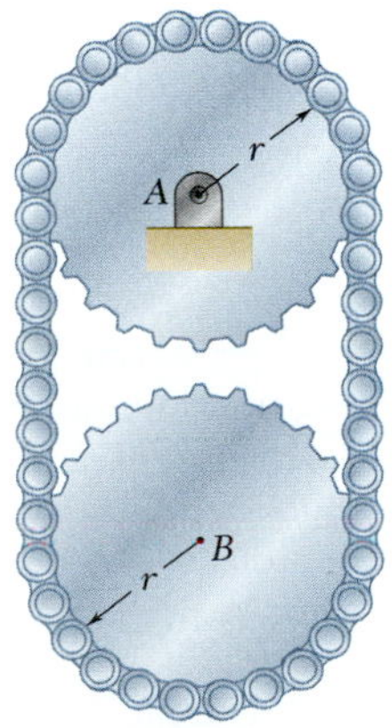

Fig. *P16.136*

***16.137** A uniform slender bar AB of mass m is suspended as shown from a uniform disk of the same mass m. Determine the accelerations of points A and B immediately after a horizontal force **P** has been applied at B.

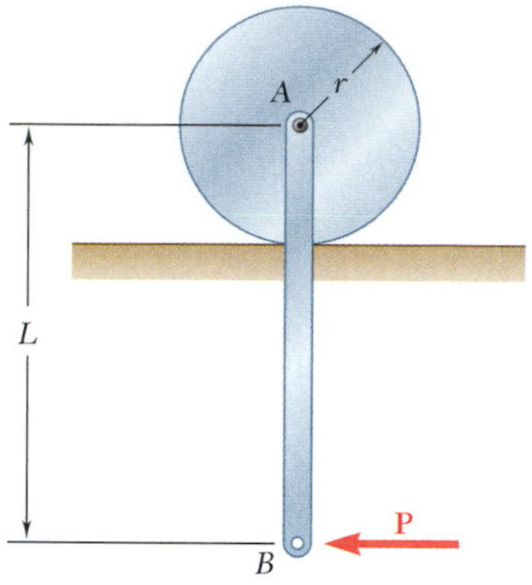

Fig. *P16.137*

***16.138 and *16.139** The 3-kg cylinder B and the 2-kg wedge A are held at rest in the position shown by cord C. Assuming that the cylinder rolls without sliding on the wedge and neglecting friction between the wedge and the ground, determine, immediately after cord C has been cut, (*a*) the acceleration of the wedge, (*b*) the angular acceleration of the cylinder.

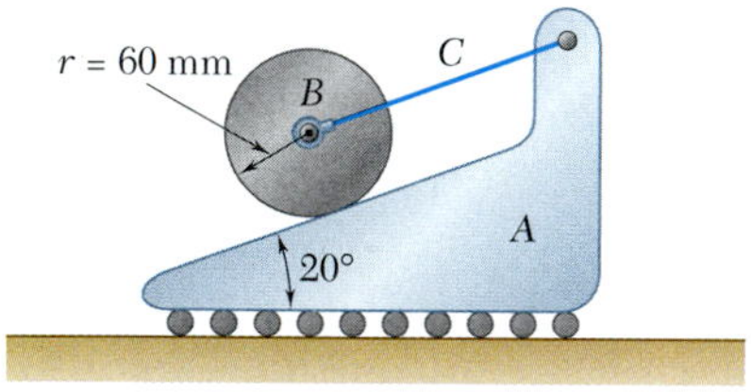

Fig. P16.138

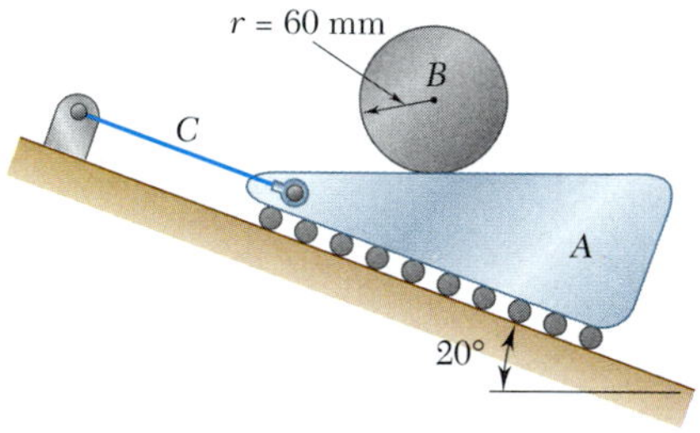

Fig. P16.139

***16.140** Each of the 6-lb bars AB and BC is of length $L = 25$ in. A horizontal force **P** of magnitude 5 lb is applied to bar BC as shown. Knowing that $b = L$ (**P** is applied at C), determine the angular acceleration of each bar.

***16.141** Each of the 6-lb bars AB and BC is of length $L = 25$ in. A horizontal force **P** of magnitude 5 lb is applied to bar BC. For the position shown, determine (*a*) the distance b for which the bars move as if they formed a single rigid body, (*b*) the corresponding angular acceleration of the bars.

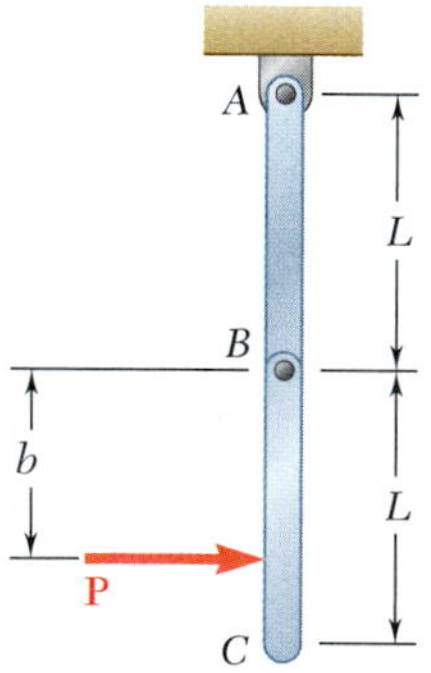

Fig. P16.140 and P16.141

***16.142** (*a*) Determine the magnitude and the location of the maximum bending moment in the rod of Prob. 16.71. (*b*) Show that the answer to part *a* is independent of the mass of the rod.

***16.143** Draw the shear and bending-moment diagrams for the rod of Prob. 16.79 immediately after the cable at B breaks.

REVIEW AND SUMMARY FOR CHAPTER 16

In this chapter, we studied the *kinetics of rigid bodies,* that is, the relations existing between the forces acting on a rigid body, the shape and mass of the body, and the motion produced. Except for the first two sections, which apply to the most general case of the motion of a rigid body, our analysis was restricted to the *plane motion of rigid slabs* and rigid bodies symmetrical with respect to the reference plane. The study of the plane motion of non-symmetrical rigid bodies and of the motion of rigid bodies in three-dimensional space will be considered in Chap. 18.

Fundamental equations of motion for a rigid body

We first recalled [Sec. 16.2] the two fundamental equations derived in Chap. 14 for the motion of a system of particles and observed that they apply in the most general case of the motion of a rigid body. The first equation defines the motion of the mass center G of the body; we have

$$\Sigma \mathbf{F} = m\overline{\mathbf{a}} \tag{16.1}$$

where m is the mass of the body and $\overline{\mathbf{a}}$ the acceleration of G. The second is related to the motion of the body relative to a centroidal frame of reference; we wrote

$$\Sigma \mathbf{M}_G = \dot{\mathbf{H}}_G \tag{16.2}$$

where $\dot{\mathbf{H}}_G$ is the rate of change of the angular momentum $\mathbf{H}_G$ of the body about its mass center G. Together, Eqs. (16.1) and (16.2) express that *the system of the external forces is equipollent to the system consisting of the vector $m\overline{\mathbf{a}}$ attached at G and the couple of moment $\dot{\mathbf{H}}_G$* (Fig. 16.19).

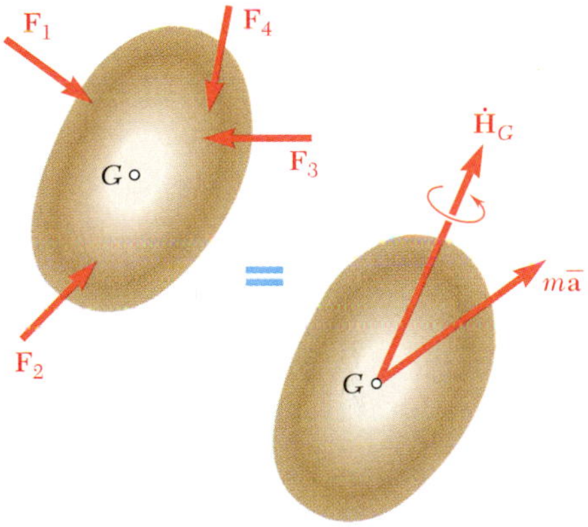

Fig. 16.19

Angular momentum in plane motion

Restricting our analysis at this point and for the rest of the chapter to the plane motion of rigid slabs and rigid bodies symmetrical with respect to the reference plane, we showed [Sec. 16.3] that the angular momentum of the body could be expressed as

$$\mathbf{H}_G = \overline{I}\boldsymbol{\omega} \tag{16.4}$$

where $\overline{I}$ is the moment of inertia of the body about a centroidal axis perpendicular to the reference plane and $\boldsymbol{\omega}$ is the angular velocity of the body. Differentiating both members of Eq. (16.4), we obtained

$$\dot{\mathbf{H}}_G = \overline{I}\dot{\boldsymbol{\omega}} = \overline{I}\boldsymbol{\alpha} \tag{16.5}$$

which shows that in the restricted case considered here, the rate of change of the angular momentum of the rigid body can be represented by a vector of the same direction as $\boldsymbol{\alpha}$ (that is, perpendicular to the plane of reference) and of magnitude $\bar{I}\alpha$.

Equations for the plane motion of a rigid body

It follows from the above [Sec. 16.4] that the plane motion of a rigid slab or of a rigid body symmetrical with respect to the reference plane is defined by the three scalar equations

$$\Sigma F_x = m\bar{a}_x \qquad \Sigma F_y = m\bar{a}_y \qquad \Sigma M_G = \bar{I}\alpha \qquad (16.6)$$

d'Alembert's principle

It further follows that *the external forces acting on the rigid body are actually* ***equivalent*** *to the effective forces of the various particles forming the body*. This statement, known as *d'Alembert's principle,* can be expressed in the form of the vector diagram shown in Fig. 16.20, where the effective forces have been represented by a vector $m\bar{\mathbf{a}}$ attached at G and a couple $\bar{I}\boldsymbol{\alpha}$. In the particular case of a slab in *translation,* the effective forces shown in part b of this figure reduce to the single vector $m\bar{\mathbf{a}}$, while in the particular case of a slab in *centroidal rotation,* they reduce to the single couple $\bar{I}\boldsymbol{\alpha}$; in any other case of plane motion, both the vector $m\bar{\mathbf{a}}$ and the couple $\bar{I}\boldsymbol{\alpha}$ should be included.

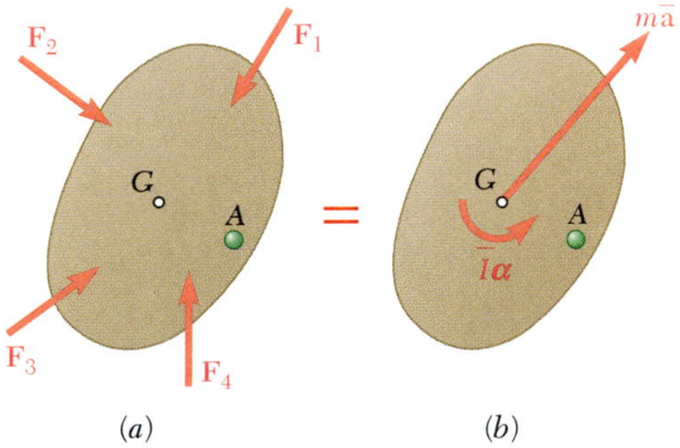

Fig. 16.20

Free-body-diagram equation

Any problem involving the plane motion of a rigid slab may be solved by drawing a *free-body-diagram equation* similar to that of Fig. 16.20 [Sec. 16.6]. Three equations of motion can then be obtained by equating the x components, y components, and moments about an arbitrary point A, of the forces and vectors involved [Sample Probs. 16.1, 16.2, 16.4, and 16.5]. An alternative solution can be obtained by adding to the external forces an *inertia vector* $-m\bar{\mathbf{a}}$ of sense opposite to that of $\bar{\mathbf{a}}$, attached at G, and an *inertia couple* $-\bar{I}\boldsymbol{\alpha}$ of sense opposite to that of $\boldsymbol{\alpha}$. The system obtained in this way is equivalent to zero, and the slab is said to be in *dynamic equilibrium.*

Connected rigid bodies

The method described above can also be used to solve problems involving the plane motion of several connected rigid bodies [Sec. 16.7]. A free-body-diagram equation is drawn for each part of the system and the equations of motion obtained are solved simultaneously. In some cases, however, a single diagram can be drawn for the entire system, including all the external forces as well as the vectors $m\bar{\mathbf{a}}$ and the couples $\bar{I}\boldsymbol{\alpha}$ associated with the various parts of the system [Sample Prob. 16.3].

Constrained plane motion

In the second part of the chapter, we were concerned with rigid bodies *moving under given constraints* [Sec. 16.8]. While the kinetic analysis of the constrained plane motion of a rigid slab is the same as above, it must be supplemented by a *kinematic analysis* which has for its object to express the components $\bar{a}_x$ and $\bar{a}_y$ of the acceleration of the mass center G of the slab in terms of its angular acceleration α. Problems solved in this way included the *noncentroidal rotation* of rods and plates [Sample Probs. 16.6 and 16.7], the *rolling motion* of spheres and wheels [Sample Probs. 16.8 and 16.9], and the plane motion of *various types of linkages* [Sample Prob. 16.10].

Review Problems

16.144 The uniform slender rod *AB* is welded to the hub *D*, and the system rotates about the vertical axis *DE* with a constant angular velocity $\boldsymbol{\omega}$. (*a*) Denoting by w the weight per unit length of the rod, express the tension in the rod at a distance z from end *A* in terms of w, l, z, and $\boldsymbol{\omega}$. (*b*) Determine the tension in the rod for $w = 0.25$ lb/ft, $l = 1.2$ ft, $z = 0.9$ ft, and $\omega = 150$ rpm.

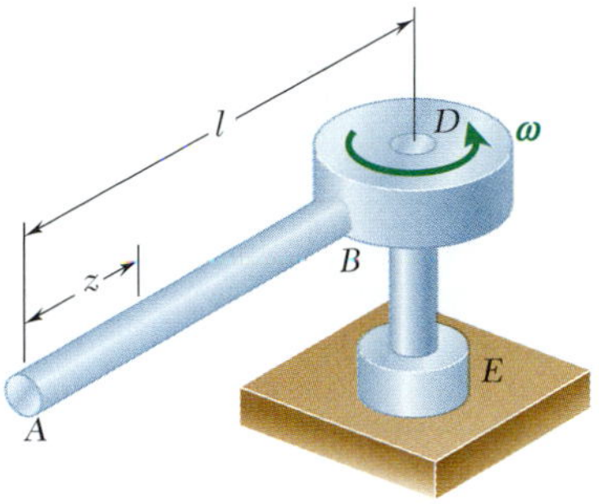

Fig. P16.144

16.145 Two uniform disks, each of mass m and radius r, are connected by an inextensible cable and roll without sliding on the surfaces shown. Knowing that the system is released from rest when $\beta = 15°$, determine the acceleration of the center (*a*) of disk *A*, (*b*) of disk *B*.

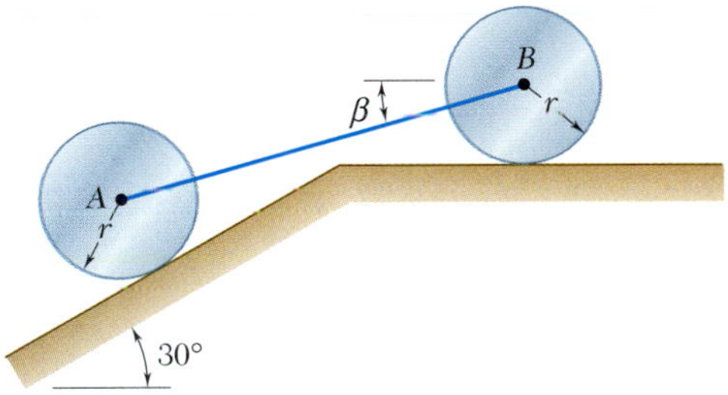

Fig. P16.145

16.146 Solve Prob. 16.145 assuming that $\beta = 5°$.

16.147 A rectangular plate of mass 5 kg is suspended from four vertical wires, and a force **P** of magnitude 6 N is applied to corner *C* as shown. Immediately after **P** is applied, determine the acceleration of (*a*) the midpoint of edge *BC*, (*b*) corner *B*.

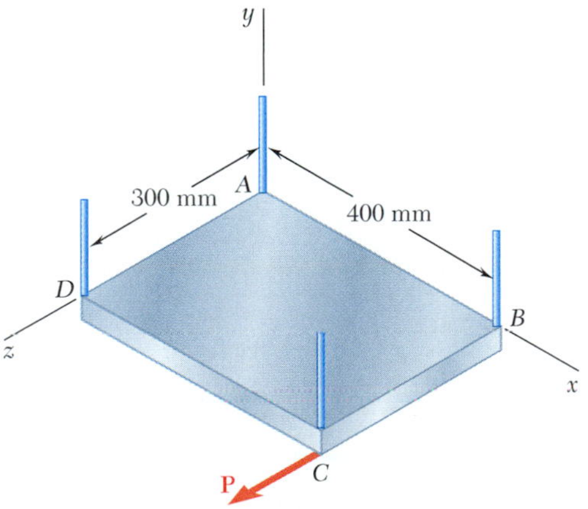

Fig. *P16.147*

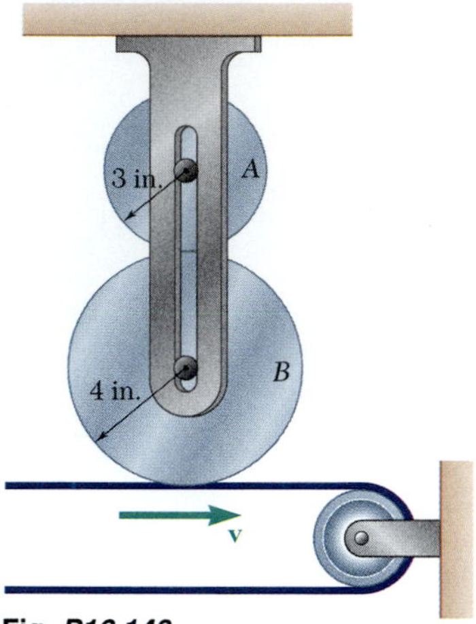

Fig. ***P16.148***

16.148 Cylinders A and B, weighing 4 lb and 9 lb, respectively, are initially at rest. Knowing that the coefficient of kinetic friction is 0.20 between the cylinders and between cylinder B and the belt, determine the angular acceleration of each cylinder immediately after cylinder B is placed in contact with the belt.

16.149 The ends of the 10-kg uniform rod AB are attached to collars of negligible weight that slide without friction along fixed rods. If the rod is released from rest when $\theta = 25°$, determine immediately after release (*a*) the angular acceleration of the rod, (*b*) the reaction at A, (*c*) the reaction at B.

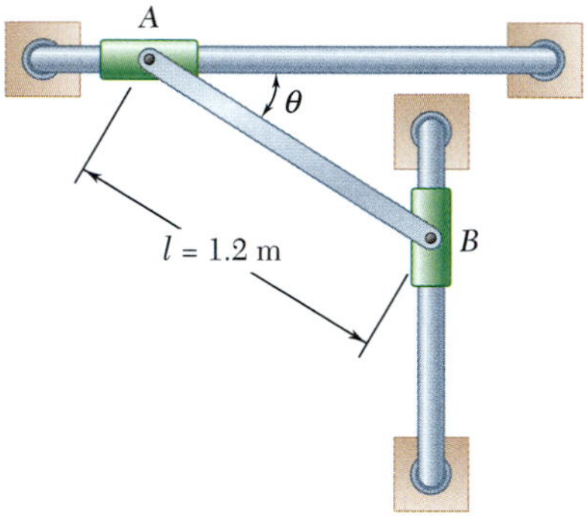

Fig. P16.149 and P16.150

16.150 The ends of the 10-kg uniform rod AB are attached to collars of negligible weight that slide without friction along fixed rods. A vertical force **P** is applied to collar B when $\theta = 25°$, causing the collar to start from rest with an upward acceleration of 12 m/s^2. Determine (*a*) the force **P**, (*b*) the reaction at A.

16.151 The axis of a 5-in.-radius disk is fitted into a slot that forms an angle of 30° with the vertical. The disk is at rest when it is placed in contact with a conveyor belt moving at constant speed. Knowing that the coefficient of kinetic friction between the disk and the belt is 0.20 and neglecting bearing friction, determine the angular acceleration of the disk while slipping occurs.

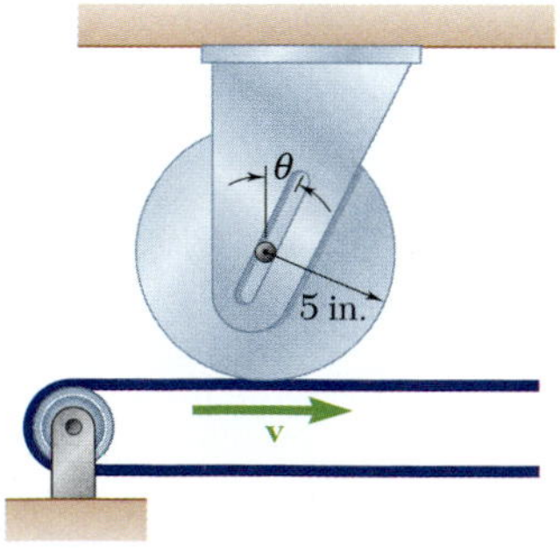

Fig. P16.151

16.152 Solve Prob. 16.151 assuming that the direction of motion of the conveyor belt is reversed.

16.153 The 5-kg uniform bar *BDE* is attached to the two rods *AB* and *CD*. Knowing that at the instant shown rod *AB* has an angular velocity of 6 rad/s clockwise and no angular acceleration, determine the horizontal component of the reaction (*a*) at *B*, (*b*) at *D*.

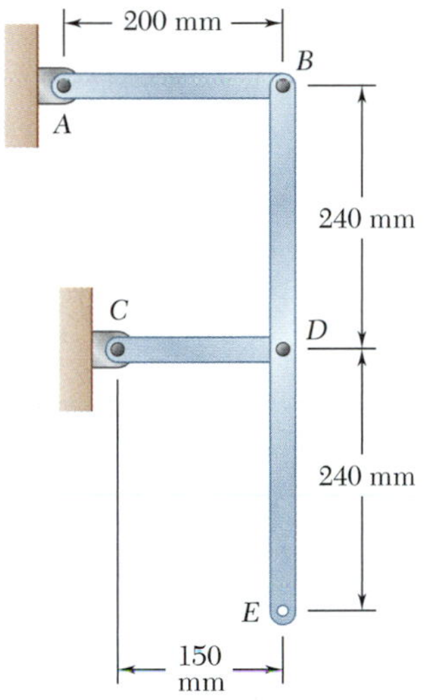

Fig. *P16.153*

16.154 The motion of the 3-lb rod *AB* is guided by two small wheels that roll freely in a horizontal slot cut in a vertical plate. If a force **P** of magnitude 5 lb is applied at *B*, determine (*a*) the acceleration of the rod, (*b*) the reactions at *A* and *B*.

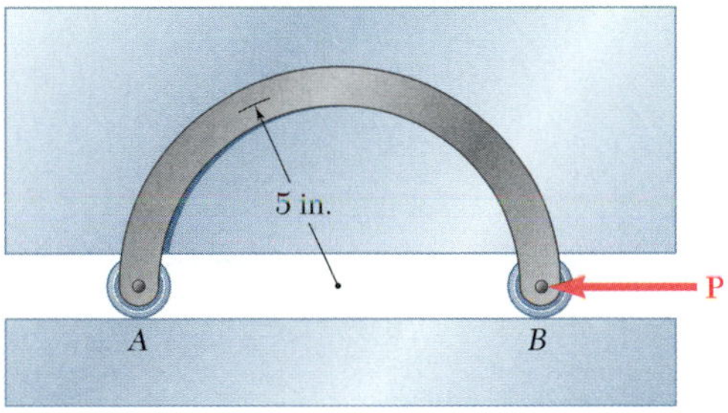

Fig. *P16.154*

16.155 The linkage *ABCD* is formed by connecting the 3-kg bar *BC* to the 1.5-kg bars *AB* and *CD*. The motion of the linkage is controlled by the couple **M** applied to bar *AB*. Knowing that at the instant shown bar *AB* has an angular velocity of 24 rad/s clockwise and no angular acceleration, determine (*a*) the couple **M**, (*b*) the components of the force exerted at *B* on rod *BC*.

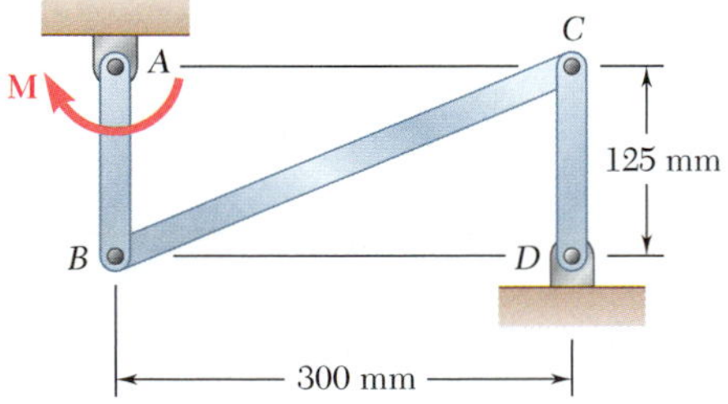

Fig. P16.155

Computer Problems

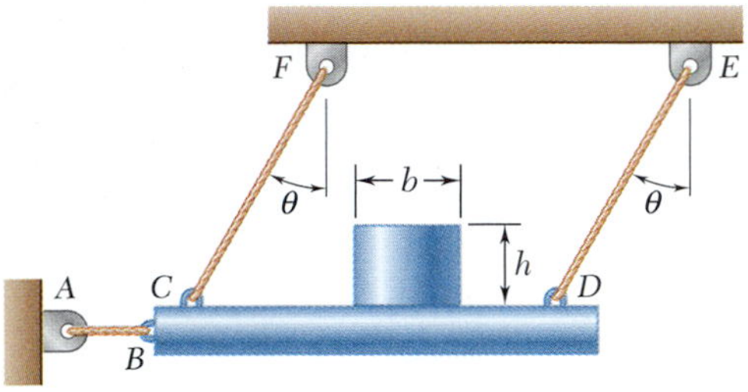

Fig. P16.C1

16.C1 A 15-kg cylinder of diameter $b = 160$ mm and height $h = 140$ mm is placed on a 5-kg platform CD that is held in the position shown by three cables. It is desired to determine the minimum value of μ_s between the cylinder and the platform for which the cylinder does not slip on the platform, immediately after cable AB is cut. Using computational software calculate and plot the minimum allowable value of μ_s for values of θ from 0 to 40°. Knowing that the actual value of μ_s is 0.75, determine the value of θ at which slipping impends.

16.C2 The cranks BE and CF are made to rotate at a constant speed of 90 rpm counterclockwise. Determine and plot the vertical components of the forces exerted on the 12-lb uniform rod $ABCD$ by pins B and C as functions of the angle θ from 0 to 360°.

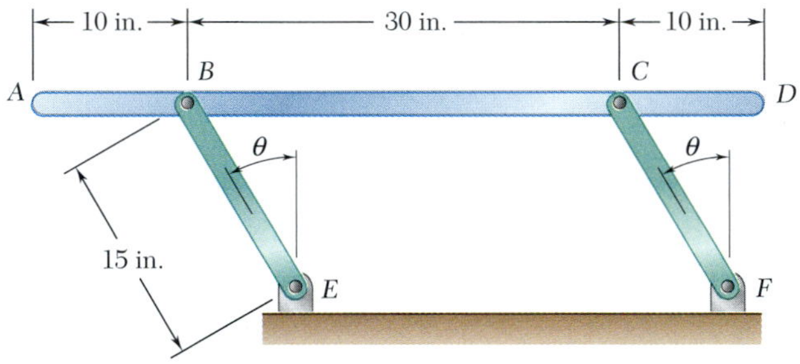

Fig. P16.C2

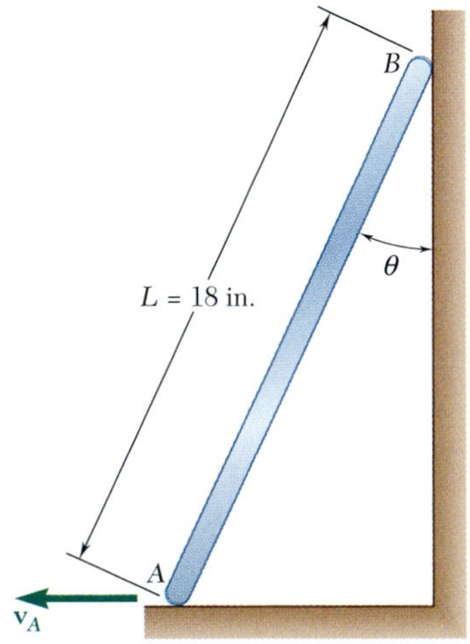

Fig. P16.C3

16.C3 End A of the 10-lb rod AB is moved to the left at a constant speed $v_A = 60$ in./s. Using computational software calculate and plot the normal reactions at ends A and B of the rod for values of θ from 0 to 40°. Determine the value of θ at which end B of the rod loses contact with the wall.

16.C4 A uniform slender bar AB of weight W is suspended from springs AC and BD as shown. Using computational software calculate and plot the accelerations of ends A and B, immediately after spring AC has broken, for values of θ from 0 to 90°.

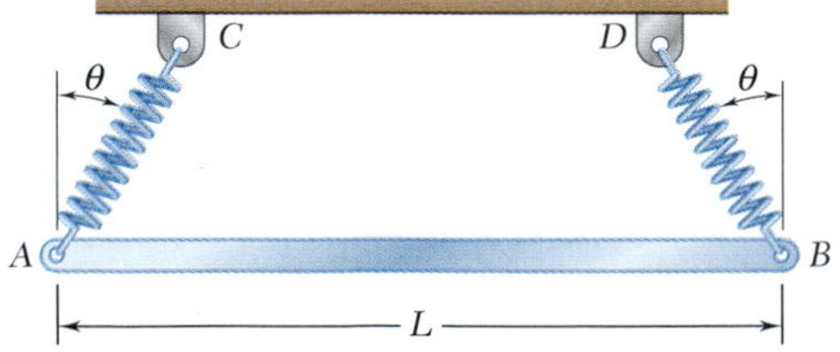

Fig. P16.C4

16.C5 A 60-lb plank rests on two horizontal pipes AB and CD of a scaffolding. The pipes are 10 ft apart, and the plank overhangs 2 ft at each end. When pipe CD suddenly breaks, a 140-lb worker is standing on the plank at a distance x from pipe AB. Calculate and plot the initial acceleration of the worker for values of x from 0 to 10 ft.

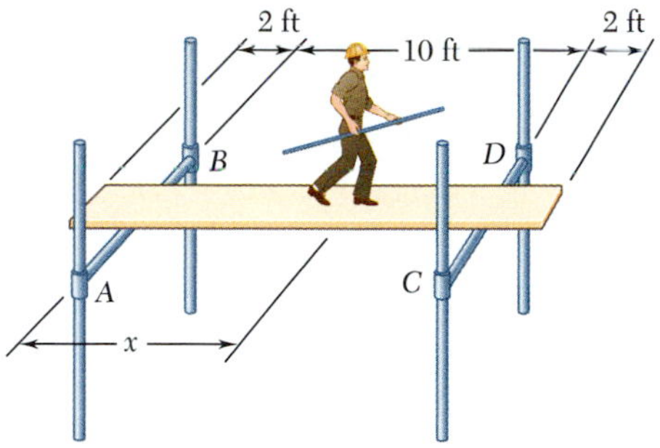

Fig. P16.C5

16.C6 In the engine system of Prob. 15.C3 of Chap. 15, the connecting rod BD is assumed to be a 5-lb slender rod and is attached to the 6-lb piston P. Knowing that during a test of the system crank AB is made to rotate with a constant angular velocity of 1000 rpm clockwise with no force applied to the face of the piston, use computational software to calculate and plot the horizontal and vertical components of the dynamic reactions exerted on the connecting rod at B and D for values of θ from 0 to 180°.

16.C7 The uniform slender bar AB, of mass 10 kg and length 720 mm, is suspended as shown from the 15-kg cart C. Knowing that the system is released from rest and neglecting the effect of friction, calculate and plot the initial acceleration of the cart and the initial angular acceleration of the bar as functions of θ from 0 to 90°.

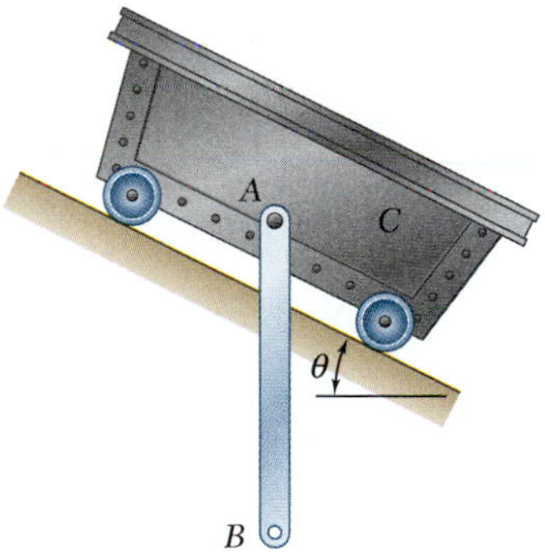

Fig. P16.C7

C H A P T E R

Plane Motion of Rigid Bodies: Energy and Momentum Methods

17

In this chapter the energy and momentum methods will be added to the tools available for your study of the motion of rigid bodies. For example, when the rotating bat contacts the ball, it applies an impulsive force to the ball, and the momentum and kinetic energy of both the bat and ball change abruptly.

PLANE MOTION OF RIGID BODIES: ENERGY AND MOMENTUM METHODS

17.1. INTRODUCTION

In this chapter the method of work and energy and the method of impulse and momentum will be used to analyze the plane motion of rigid bodies and of systems of rigid bodies.

The method of work and energy will be considered first. In Secs. 17.2 through 17.5, the work of a force and of a couple will be defined, and an expression for the kinetic energy of a rigid body in plane motion will be obtained. The principle of work and energy will then be used to solve problems involving displacements and velocities. In Sec. 17.6, the principle of conservation of energy will be applied to the solution of a variety of engineering problems.

In the second part of the chapter, the principle of impulse and momentum will be applied to the solution of problems involving velocities and time (Secs. 17.8 and 17.9) and the concept of conservation of angular momentum will be introduced and discussed (Sec. 17.10).

In the last part of the chapter (Secs. 17.11 and 17.12), problems involving the eccentric impact of rigid bodies will be considered. As was done in Chap. 13, where we analyzed the impact of particles, the coefficient of restitution between the colliding bodies will be used together with the principle of impulse and momentum in the solution of impact problems. It will also be shown that the method used is applicable not only when the colliding bodies move freely after the impact but also when the bodies are partially constrained in their motion.

17.2. PRINCIPLE OF WORK AND ENERGY FOR A RIGID BODY

The principle of work and energy will now be used to analyze the plane motion of rigid bodies. As was pointed out in Chap. 13, the method of work and energy is particularly well adapted to the solution of problems involving velocities and displacements. Its main advantage resides in the fact that the work of forces and the kinetic energy of particles are scalar quantities.

In order to apply the principle of work and energy to the analysis of the motion of a rigid body, it will again be assumed that the rigid body is made of a large number n of particles of mass Δm_i. Recalling Eq. (14.30) of Sec. 14.8, we write

$$T_1 + U_{1\rightarrow 2} = T_2 \tag{17.1}$$

where T_1, T_2 = initial and final values of total kinetic energy of particles forming the rigid body

$U_{1\rightarrow 2}$ = work of all forces acting on various particles of body

The total kinetic energy

$$T = \frac{1}{2} \sum_{i=1}^{n} \Delta m_i v_i^2 \tag{17.2}$$

is obtained by adding positive scalar quantities and is itself a positive scalar quantity. You will see later how T can be determined for various types of motion of a rigid body.

The expression $U_{1\to2}$ in (17.1) represents the work of all the forces acting on the various particles of the body, whether these forces are internal or external. However, as you will see presently, the total work of the internal forces holding together the particles of a rigid body is zero. Consider two particles A and B of a rigid body and the two equal and opposite forces $\mathbf{F}$ and $-\mathbf{F}$ they exert on each other (Fig. 17.1). While, in general, small displacements $d\mathbf{r}$ and $d\mathbf{r}'$ of the two particles are different, the components of these displacements along AB must be equal; otherwise, the particles would not remain at the same distance from each other and the body would not be rigid. Therefore, the work of $\mathbf{F}$ is equal in magnitude and opposite in sign to the work of $-\mathbf{F}$, and their sum is zero. Thus, the total work of the internal forces acting on the particles of a rigid body is zero, and *the expression $U_{1\to2}$ in Eq.* (17.1) *reduces to the work of the external forces* acting on the body during the displacement considered.

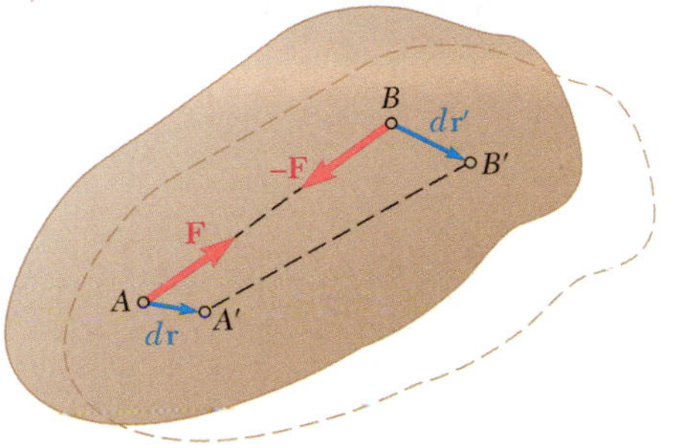

Fig. 17.1

17.3. WORK OF FORCES ACTING ON A RIGID BODY

We saw in Sec. 13.2 that the work of a force $\mathbf{F}$ during a displacement of its point of application from A_1 to A_2 is

$$U_{1\to2} = \int_{A_1}^{A_2} \mathbf{F} \cdot d\mathbf{r} \tag{17.3}$$

or

$$U_{1\to2} = \int_{s_1}^{s_2} (F \cos \alpha)\, ds \tag{17.3'}$$

where F is the magnitude of the force, α is the angle it forms with the direction of motion of its point of application A, and s is the variable of integration which measures the distance traveled by A along its path.

In computing the work of the external forces acting on a rigid body, it is often convenient to determine the work of a couple without considering separately the work of each of the two forces forming the couple. Consider the two forces $\mathbf{F}$ and $-\mathbf{F}$ forming a couple of moment $\mathbf{M}$ and acting on a rigid body (Fig. 17.2). Any small displacement of the rigid body bringing A and B, respectively, into A' and B'' can be divided into two parts: in one part points A and B undergo equal displacements $d\mathbf{r}_1$; in the other part A' remains fixed while B' moves into B'' through a displacement $d\mathbf{r}_2$ of magnitude $ds_2 = r\, d\theta$. In the first part of the motion, the work of $\mathbf{F}$ is equal in magnitude and opposite in sign to the work of $-\mathbf{F}$ and their sum is zero. In the second part of the motion, only force $\mathbf{F}$ works, and its work is $dU = F\, ds_2 = Fr\, d\theta$. But the product Fr is equal to the magnitude M of the moment of the couple. Thus, the work of a couple of moment $\mathbf{M}$ acting on a rigid body is

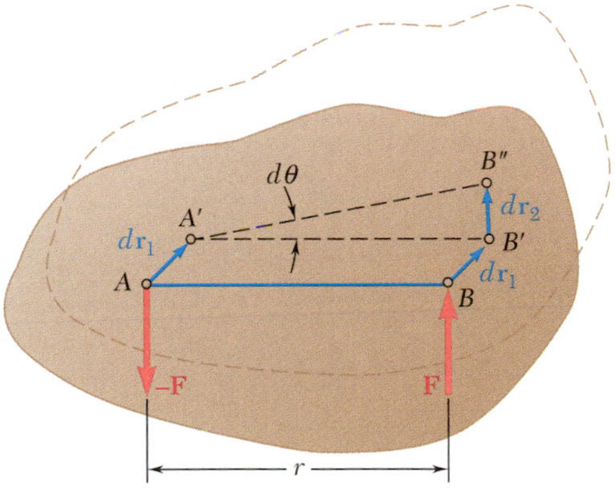

Fig. 17.2

$$dU = M\, d\theta \tag{17.4}$$

where $d\theta$ is the small angle expressed in radians through which the body rotates. We again note that work should be expressed in units obtained by multiplying units of force by units of length. The work of the couple during a finite rotation of the rigid body is obtained by integrating both members of (17.4) from the initial value θ_1 of the angle θ to its final value θ_2. We write

$$U_{1\rightarrow 2} = \int_{\theta_1}^{\theta_2} M\, d\theta \qquad (17.5)$$

When the moment **M** *of the couple is constant,* formula (17.5) reduces to

$$U_{1\rightarrow 2} = M(\theta_2 - \theta_1) \qquad (17.6)$$

It was pointed out in Sec. 13.2 that a number of forces encountered in problems of kinetics *do no work*. They are forces applied to fixed points or acting in a direction perpendicular to the displacement of their point of application. Among the forces which do no work the following have been listed: the reaction at a frictionless pin when the body supported rotates about the pin, the reaction at a frictionless surface when the body in contact moves along the surface, and the weight of a body when its center of gravity moves horizontally. We can add now that *when a rigid body rolls without sliding on a fixed surface, the friction force* **F** *at the point of contact C does no work.* The velocity $\mathbf{v}_C$ of the point of contact C is zero, and the work of the friction force **F** during a small displacement of the rigid body is

$$dU = F\, ds_C = F(v_C\, dt) = 0$$

17.4. KINETIC ENERGY OF A RIGID BODY IN PLANE MOTION

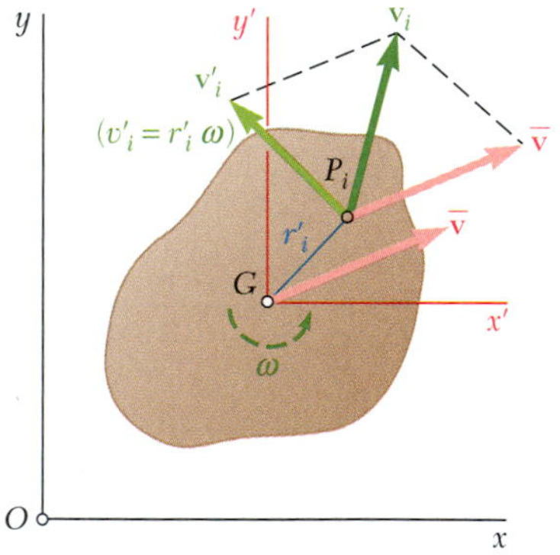

Fig. 17.3

Consider a rigid body of mass m in plane motion. We recall from Sec. 14.7 that, if the absolute velocity $\mathbf{v}_i$ of each particle P_i of the body is expressed as the sum of the velocity $\overline{\mathbf{v}}$ of the mass center G of the body and of the velocity $\mathbf{v}'_i$ of the particle relative to a frame $Gx'y'$ attached to G and of fixed orientation (Fig. 17.3), the kinetic energy of the system of particles forming the rigid body can be written in the form

$$T = \tfrac{1}{2}m\bar{v}^2 + \frac{1}{2}\sum_{i=1}^{n} \Delta m_i\, v_i'^2 \qquad (17.7)$$

But the magnitude v'_i of the relative velocity of P_i is equal to the product $r'_i\omega$ of the distance r'_i of P_i from the axis through G perpendicular to the plane of motion and of the magnitude ω of the angular velocity of the body at the instant considered. Substituting into (17.7), we have

$$T = \tfrac{1}{2}m\bar{v}^2 + \frac{1}{2}\left(\sum_{i=1}^{n} r_i'^2\, \Delta m_i\right)\omega^2 \qquad (17.8)$$

or, since the sum represents the moment of inertia $\bar{I}$ of the body about the axis through G,

$$T = \tfrac{1}{2}m\bar{v}^2 + \tfrac{1}{2}\bar{I}\omega^2 \qquad (17.9)$$

We note that in the particular case of a body in translation ($\omega = 0$), the expression obtained reduces to $\frac{1}{2}m\bar{v}^2$, while in the case of a centroidal rotation ($\bar{v} = 0$), it reduces to $\frac{1}{2}\bar{I}\omega^2$. We conclude that the kinetic energy of a rigid body in plane motion can be separated into two parts: (1) the kinetic energy $\frac{1}{2}m\bar{v}^2$ associated with the motion of the mass center G of the body, and (2) the kinetic energy $\frac{1}{2}\bar{I}\omega^2$ associated with the rotation of the body about G.

Noncentroidal Rotation. The relation (17.9) is valid for any type of plane motion and can therefore be used to express the kinetic energy of a rigid body rotating with an angular velocity $\boldsymbol{\omega}$ about a fixed axis through O (Fig. 17.4). In that case, however, the kinetic energy of the body can be expressed more directly by noting that the speed v_i of the particle P_i is equal to the product $r_i\omega$ of the distance r_i of P_i from the fixed axis and the magnitude ω of the angular velocity of the body at the instant considered. Substituting into (17.2), we write

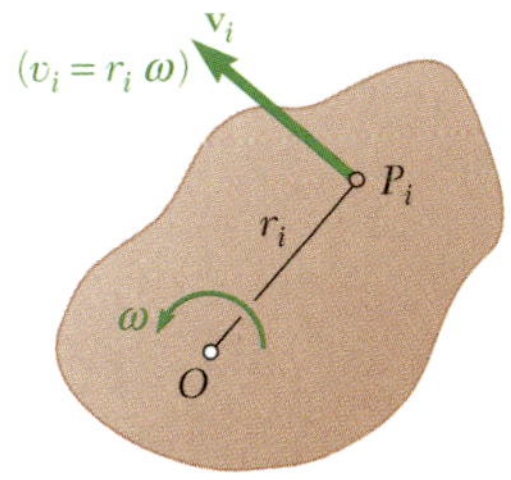

Fig. 17.4

$$T = \frac{1}{2}\sum_{i=1}^{n} \Delta m_i (r_i\omega)^2 = \frac{1}{2}\left(\sum_{i=1}^{n} r_i^2\, \Delta m_i\right)\omega^2$$

or, since the last sum represents the moment of inertia I_O of the body about the fixed axis through O,

$$T = \tfrac{1}{2}I_O\omega^2 \qquad (17.10)$$

We note that the results obtained are not limited to the motion of plane slabs or to the motion of bodies which are symmetrical with respect to the reference plane, and can be applied to the study of the plane motion of any rigid body, regardless of its shape. However, since Eq. (17.9) is applicable to any plane motion while Eq. (17.10) is applicable only in cases involving noncentroidal rotation, Eq. (17.9) will be used in the solution of all the sample problems.

17.5. SYSTEMS OF RIGID BODIES

When a problem involves several rigid bodies, each rigid body can be considered separately and the principle of work and energy can be applied to each body. Adding the kinetic energies of all the particles and considering the work of all the forces involved, we can also write the equation of work and energy for the entire system. We have

$$T_1 + U_{1\rightarrow 2} = T_2 \qquad (17.11)$$

where T represents the arithmetic sum of the kinetic energies of the rigid bodies forming the system (all terms are positive) and $U_{1\rightarrow 2}$ represents the work of all the forces acting on the various bodies, whether these forces are *internal* or *external* from the point of view of the system as a whole.

The method of work and energy is particularly useful in solving problems involving pin-connected members, blocks and pulleys connected by inextensible cords, and meshed gears. In all these cases, the internal forces occur by pairs of equal and opposite forces, and the points of application of the forces in each pair *move through equal distances* during a small displacement of the system. As a result, the work of the internal forces is zero and $U_{1\to2}$ reduces to the work of the *forces external to the system*.

17.6. CONSERVATION OF ENERGY

We saw in Sec. 13.6 that the work of conservative forces, such as the weight of a body or the force exerted by a spring, can be expressed as a change in potential energy. When a rigid body, or a system of rigid bodies, moves under the action of conservative forces, the principle of work and energy stated in Sec. 17.2 can be expressed in a modified form. Substituting for $U_{1\to2}$ from (13.19′) into (17.1), we write

$$T_1 + V_1 = T_2 + V_2 \tag{17.12}$$

Formula (17.12) indicates that when a rigid body, or a system of rigid bodies, moves under the action of conservative forces, *the sum of the kinetic energy and of the potential energy of the system remains constant*. It should be noted that in the case of the plane motion of a rigid body, the kinetic energy of the body should include both the *translational* term $\frac{1}{2}m\bar{v}^2$ and the *rotational* term $\frac{1}{2}\bar{I}\omega^2$.

As an example of application of the principle of conservation of energy, let us consider a slender rod AB, of length l and mass m, whose extremities are connected to blocks of negligible mass sliding along horizontal and vertical tracks. We assume that the rod is released with no initial velocity from a horizontal position (Fig. 17.5*a*), and we wish to determine its angular velocity after it has rotated through an angle θ (Fig. 17.5*b*).

Since the initial velocity is zero, we have $T_1 = 0$. Measuring the potential energy from the level of the horizontal track, we write $V_1 = 0$. After the rod has rotated through θ, the center of gravity G of the

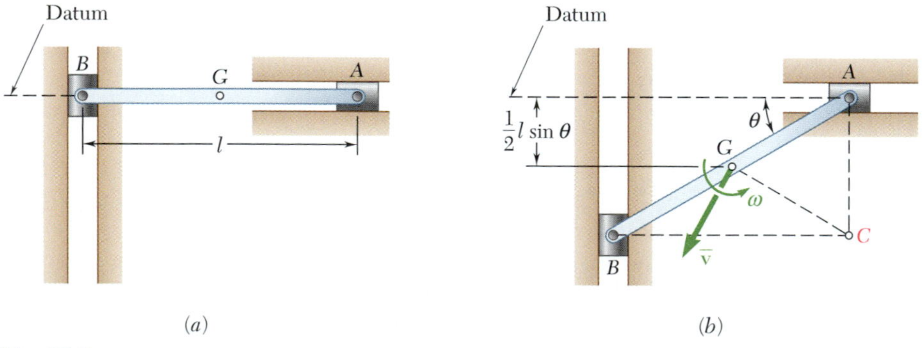

Fig. 17.5

rod is at a distance $\frac{1}{2}l \sin\theta$ below the reference level and we have

$$V_2 = -\tfrac{1}{2}Wl \sin\theta = -\tfrac{1}{2}mgl \sin\theta$$

Observing that in this position the instantaneous center of the rod is located at C and that $CG = \frac{1}{2}l$, we write $\bar{v}_2 = \frac{1}{2}l\omega$ and obtain

$$T_2 = \tfrac{1}{2}m\bar{v}_2^2 + \tfrac{1}{2}\bar{I}\omega_2^2 = \tfrac{1}{2}m(\tfrac{1}{2}l\omega)^2 + \tfrac{1}{2}(\tfrac{1}{12}ml^2)\omega^2$$
$$= \frac{1}{2}\frac{ml^2}{3}\omega^2$$

Applying the principle of conservation of energy, we write

$$T_1 + V_1 = T_2 + V_2$$
$$0 = \frac{1}{2}\frac{ml^2}{3}\omega^2 - \tfrac{1}{2}mgl \sin\theta$$
$$\omega = \left(\frac{3g}{l}\sin\theta\right)^{1/2}$$

The advantages of the method of work and energy, as well as its shortcomings, were indicated in Sec. 13.4. Here we should add that the method of work and energy must be supplemented by the application of d'Alembert's principle when reactions at fixed axles, rollers, or sliding blocks are to be determined. For example, in order to compute the reactions at the extremities A and B of the rod of Fig. 17.5*b*, a diagram should be drawn to express that the system of the external forces applied to the rod is equivalent to the vector $m\bar{\mathbf{a}}$ and the couple $\bar{I}\boldsymbol{\alpha}$. The angular velocity $\boldsymbol{\omega}$ of the rod, however, is determined by the method of work and energy before the equations of motion are solved for the reactions. The complete analysis of the motion of the rod and of the forces exerted on the rod requires, therefore, the combined use of the method of work and energy and of the principle of equivalence of the external and effective forces.

Photo 17.1 By using the principle of conservation of energy and a free-body-diagram equation, you will find that a force larger than four times his own weight will be exerted on the hands of the gymnast if, keeping his body straight, he swings through 180° about the horizontal bar.

17.7. POWER

Power was defined in Sec. 13.5 as the time rate at which work is done. In the case of a body acted upon by a force $\mathbf{F}$, and moving with a velocity $\mathbf{v}$, the power was expressed as follows:

$$\text{Power} = \frac{dU}{dt} = \mathbf{F}\cdot\mathbf{v} \qquad (13.13)$$

In the case of a rigid body rotating with an angular velocity $\boldsymbol{\omega}$ and acted upon by a couple of moment $\mathbf{M}$ parallel to the axis of rotation, we have, by (17.4),

$$\text{Power} = \frac{dU}{dt} = \frac{M\,d\theta}{dt} = M\omega \qquad (17.13)$$

The various units used to measure power, such as the watt and the horsepower, were defined in Sec. 13.5.

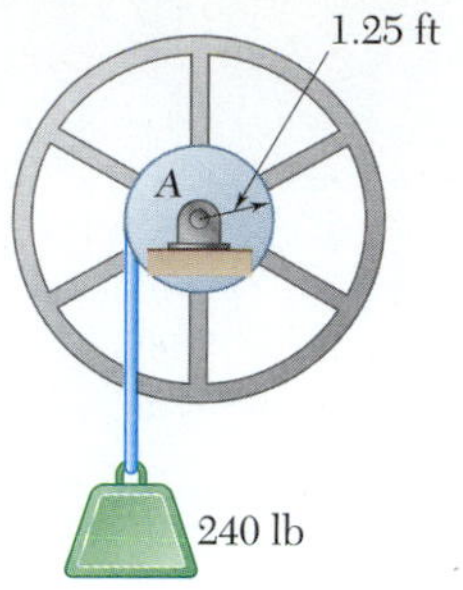

SAMPLE PROBLEM 17.1

A 240-lb block is suspended from an inextensible cable which is wrapped around a drum of 1.25-ft radius rigidly attached to a flywheel. The drum and flywheel have a combined centroidal moment of inertia $\bar{I} = 10.5 \text{ lb} \cdot \text{ft} \cdot \text{s}^2$. At the instant shown, the velocity of the block is 6 ft/s directed downward. Knowing that the bearing at A is poorly lubricated and that the bearing friction is equivalent to a couple **M** of magnitude 60 lb · ft, determine the velocity of the block after it has moved 4 ft downward.

SOLUTION

We consider the system formed by the flywheel and the block. Since the cable is inextensible, the work done by the internal forces exerted by the cable cancels. The initial and final positions of the system and the external forces acting on the system are as shown.

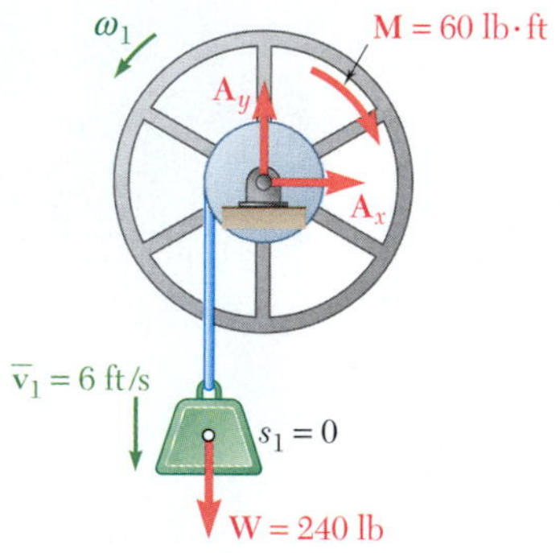

Kinetic Energy. ***Position 1.***

Block: $\bar{v}_1 = 6 \text{ ft/s}$

Flywheel: $\omega_1 = \dfrac{\bar{v}_1}{r} = \dfrac{6 \text{ ft/s}}{1.25 \text{ ft}} = 4.80 \text{ rad/s}$

$$
\begin{aligned}
T_1 &= \tfrac{1}{2} m \bar{v}_1^2 + \tfrac{1}{2} \bar{I} \omega_1^2 \\
&= \frac{1}{2} \frac{240 \text{ lb}}{32.2 \text{ ft/s}^2} (6 \text{ ft/s})^2 + \frac{1}{2}(10.5 \text{ lb} \cdot \text{ft} \cdot \text{s}^2)(4.80 \text{ rad/s})^2 \\
&= 255 \text{ ft} \cdot \text{lb}
\end{aligned}
$$

Position 2. Noting that $\omega_2 = \bar{v}_2/1.25$, we write

$$
\begin{aligned}
T_2 &= \tfrac{1}{2} m \bar{v}_2^2 + \tfrac{1}{2} \bar{I} \omega_2^2 \\
&= \frac{1}{2} \frac{240}{32.2} (\bar{v}_2)^2 + \left(\frac{1}{2}\right)(10.5)\left(\frac{\bar{v}_2}{1.25}\right)^2 = 7.09 \bar{v}_2^2
\end{aligned}
$$

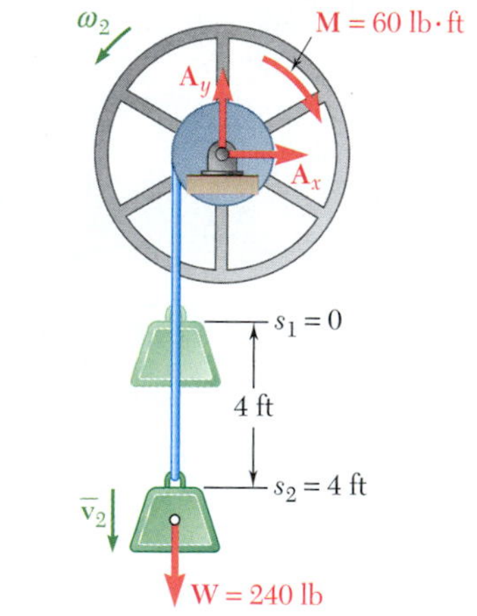

Work. During the motion, only the weight **W** of the block and the friction couple **M** do work. Noting that **W** does positive work and that the friction couple **M** does negative work, we write

$$
s_1 = 0 \qquad s_2 = 4 \text{ ft}
$$

$$
\theta_1 = 0 \qquad \theta_2 = \frac{s_2}{r} = \frac{4 \text{ ft}}{1.25 \text{ ft}} = 3.20 \text{ rad}
$$

$$
\begin{aligned}
U_{1 \to 2} &= W(s_2 - s_1) - M(\theta_2 - \theta_1) \\
&= (240 \text{ lb})(4 \text{ ft}) - (60 \text{ lb} \cdot \text{ft})(3.20 \text{ rad}) \\
&= 768 \text{ ft} \cdot \text{lb}
\end{aligned}
$$

Principle of Work and Energy

$$
\begin{aligned}
T_1 + U_{1 \to 2} &= T_2 \\
255 \text{ ft} \cdot \text{lb} + 768 \text{ ft} \cdot \text{lb} &= 7.09 \bar{v}_2^2 \\
\bar{v}_2 &= 12.01 \text{ ft/s}
\end{aligned}
$$

$\bar{\mathbf{v}}_2 = 12.01 \text{ ft/s} \downarrow$ ◀

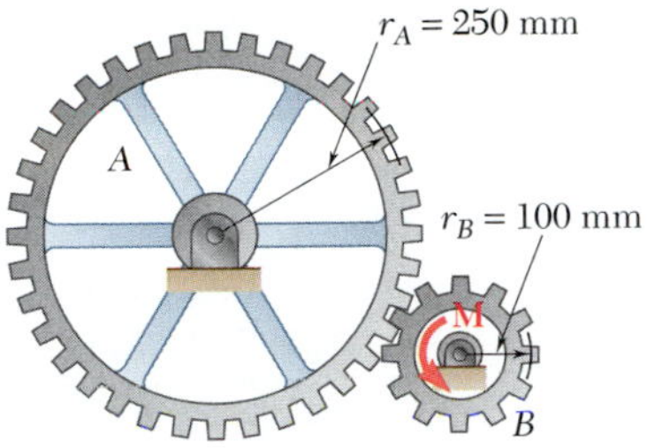

SAMPLE PROBLEM 17.2

Gear A has a mass of 10 kg and a radius of gyration of 200 mm; gear B has a mass of 3 kg and a radius of gyration of 80 mm. The system is at rest when a couple **M** of magnitude 6 N · m is applied to gear B. Neglecting friction, determine (a) the number of revolutions executed by gear B before its angular velocity reaches 600 rpm, (b) the tangential force which gear B exerts on gear A.

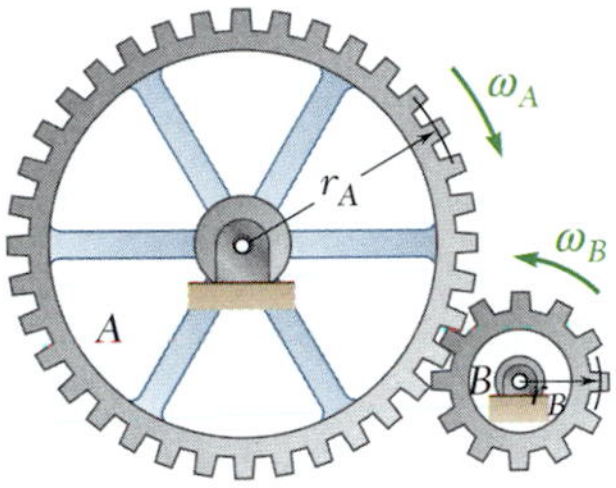

SOLUTION

Motion of Entire System. Noting that the peripheral speeds of the gears are equal, we write

$$r_A\omega_A = r_B\omega_B \qquad \omega_A = \omega_B\frac{r_B}{r_A} = \omega_B\frac{100\text{ mm}}{250\text{ mm}} = 0.40\omega_B$$

For $\omega_B = 600$ rpm, we have

$$\omega_B = 62.8\text{ rad/s} \qquad \omega_A = 0.40\omega_B = 25.1\text{ rad/s}$$
$$\bar{I}_A = m_A\bar{k}_A^2 = (10\text{ kg})(0.200\text{ m})^2 = 0.400\text{ kg}\cdot\text{m}^2$$
$$\bar{I}_B = m_B\bar{k}_B^2 = (3\text{ kg})(0.080\text{ m})^2 = 0.0192\text{ kg}\cdot\text{m}^2$$

Kinetic Energy. Since the system is initially at rest, $T_1 = 0$. Adding the kinetic energies of the two gears when $\omega_B = 600$ rpm, we obtain

$$T_2 = \tfrac{1}{2}\bar{I}_A\omega_A^2 + \tfrac{1}{2}\bar{I}_B\omega_B^2$$
$$= \tfrac{1}{2}(0.400\text{ kg}\cdot\text{m}^2)(25.1\text{ rad/s})^2 + \tfrac{1}{2}(0.0192\text{ kg}\cdot\text{m}^2)(62.8\text{ rad/s})^2$$
$$= 163.9\text{ J}$$

Work. Denoting by θ_B the angular displacement of gear B, we have

$$U_{1\rightarrow 2} = M\theta_B = (6\text{ N}\cdot\text{m})(\theta_B\text{ rad}) = (6\theta_B)\text{ J}$$

Principle of Work and Energy

$$T_1 + U_{1\rightarrow 2} = T_2$$
$$0 + (6\theta_B)\text{ J} = 163.9\text{ J}$$
$$\theta_B = 27.32\text{ rad} \qquad \theta_B = 4.35\text{ rev} \blacktriangleleft$$

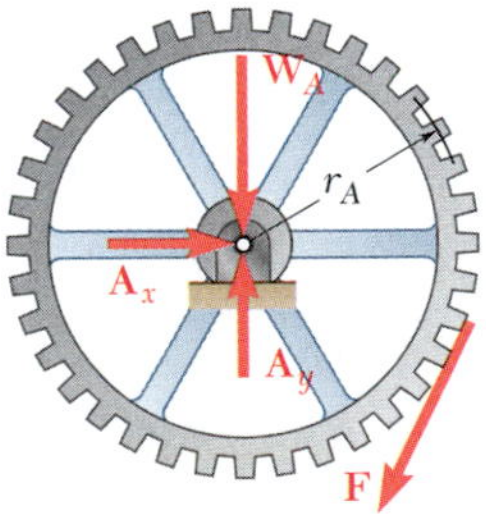

Motion of Gear A. ***Kinetic Energy.*** Initially, gear A is at rest, so $T_1 = 0$. When $\omega_B = 600$ rpm, the kinetic energy of gear A is

$$T_2 = \tfrac{1}{2}\bar{I}_A\omega_A^2 = \tfrac{1}{2}(0.400\text{ kg}\cdot\text{m}^2)(25.1\text{ rad/s})^2 = 126.0\text{ J}$$

Work. The forces acting on gear A are as shown. The tangential force **F** does work equal to the product of its magnitude and of the length $\theta_A r_A$ of the arc described by the point of contact. Since $\theta_A r_A = \theta_B r_B$, we have

$$U_{1\rightarrow 2} = F(\theta_B r_B) = F(27.3\text{ rad})(0.100\text{ m}) = F(2.73\text{ m})$$

Principle of Work and Energy

$$T_1 + U_{1\rightarrow 2} = T_2$$
$$0 + F(2.73\text{ m}) = 126.0\text{ J}$$
$$F = +46.2\text{ N} \qquad \mathbf{F} = 46.2\text{ N}\swarrow \blacktriangleleft$$

SAMPLE PROBLEM 17.3

A sphere, a cylinder, and a hoop, each having the same mass and the same radius, are released from rest on an incline. Determine the velocity of each body after it has rolled through a distance corresponding to a change in elevation h.

SOLUTION

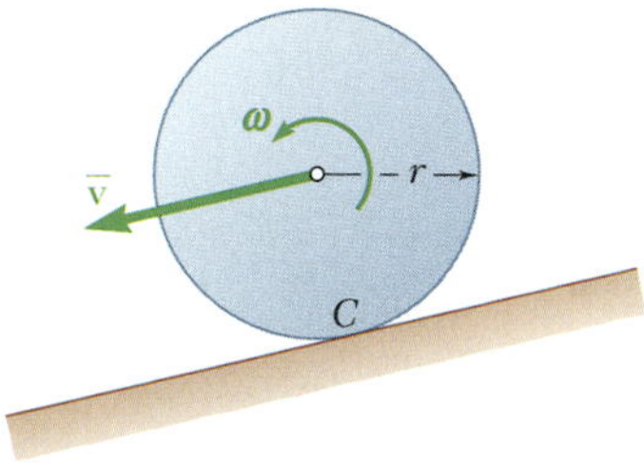

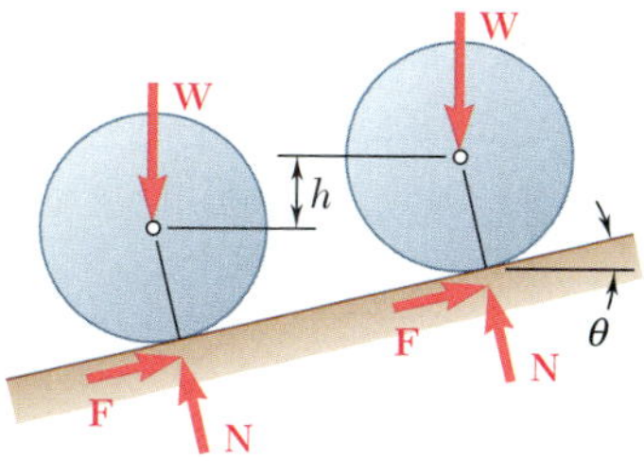

The problem will first be solved in general terms, and then results for each body will be found. We denote the mass by m, the centroidal moment of inertia by $\bar{I}$, the weight by W, and the radius by r.

Kinematics. Since each body rolls, the instantaneous center of rotation is located at C and we write

$$\omega = \frac{\bar{v}}{r}$$

Kinetic Energy

$$T_1 = 0$$

$$T_2 = \tfrac{1}{2}m\bar{v}^2 + \tfrac{1}{2}\bar{I}\omega^2$$

$$= \tfrac{1}{2}m\bar{v}^2 + \tfrac{1}{2}\bar{I}\left(\frac{\bar{v}}{r}\right)^2 = \tfrac{1}{2}\left(m + \frac{\bar{I}}{r^2}\right)\bar{v}^2$$

Work. Since the friction force **F** in rolling motion does no work,

$$U_{1\rightarrow 2} = Wh$$

Principle of Work and Energy

$$T_1 + U_{1\rightarrow 2} = T_2$$

$$0 + Wh = \tfrac{1}{2}\left(m + \frac{\bar{I}}{r^2}\right)\bar{v}^2 \qquad \bar{v}^2 = \frac{2Wh}{m + \bar{I}/r^2}$$

Noting that $W = mg$, we rearrange the result and obtain

$$\bar{v}^2 = \frac{2gh}{1 + \bar{I}/mr^2}$$

Velocities of Sphere, Cylinder, and Hoop. Introducing successively the particular expression for $\bar{I}$, we obtain

Sphere:	$\bar{I} = \tfrac{2}{5}mr^2$	$\bar{v} = 0.845\sqrt{2gh}$ ◀
Cylinder:	$\bar{I} = \tfrac{1}{2}mr^2$	$\bar{v} = 0.816\sqrt{2gh}$ ◀
Hoop:	$\bar{I} = mr^2$	$\bar{v} = 0.707\sqrt{2gh}$ ◀

Remark. Let us compare the results with the velocity attained by a frictionless block sliding through the same distance. The solution is identical to the above solution except that $\omega = 0$; we find $\bar{v} = \sqrt{2gh}$.

Comparing the results, we note that the velocity of the body is independent of both its mass and radius. However, the velocity does depend upon the quotient $\bar{I}/mr^2 = \bar{k}^2/r^2$, which measures the ratio of the rotational kinetic energy to the translational kinetic energy. Thus the hoop, which has the largest $\bar{k}$ for a given radius r, attains the smallest velocity, while the sliding block, which does not rotate, attains the largest velocity.

SAMPLE PROBLEM 17.4

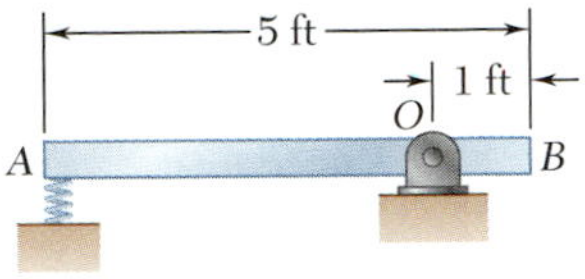

A 30-lb slender rod AB is 5 ft long and is pivoted about a point O which is 1 ft from end B. The other end is pressed against a spring of constant k = 1800 lb/in. until the spring is compressed 1 in. The rod is then in a horizontal position. If the rod is released from this position, determine its angular velocity and the reaction at the pivot O as the rod passes through a vertical position.

SOLUTION

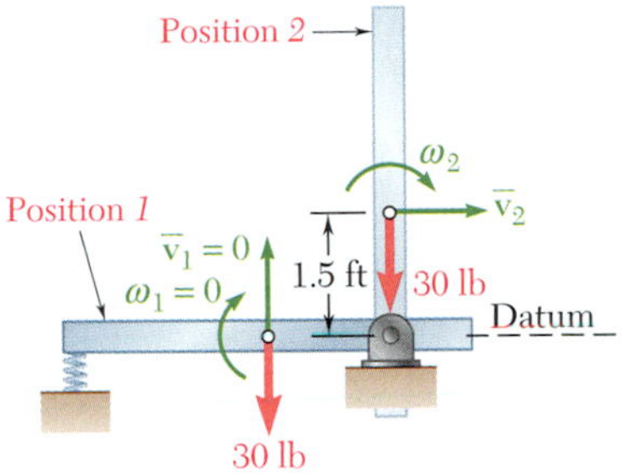

Position 1. ***Potential Energy.*** Since the spring is compressed 1 in., we have x_1 = 1 in.

$$V_e = \tfrac{1}{2}kx_1^2 = \tfrac{1}{2}(1800 \text{ lb/in.})(1 \text{ in.})^2 = 900 \text{ in} \cdot \text{lb}$$

Choosing the datum as shown, we have $V_g = 0$; therefore,

$$V_1 = V_e + V_g = 900 \text{ in} \cdot \text{lb} = 75 \text{ ft} \cdot \text{lb}$$

Kinetic Energy. Since the velocity in position *1* is zero, we have $T_1 = 0$.

Position 2. ***Potential Energy.*** The elongation of the spring is zero, and we have $V_e = 0$. Since the center of gravity of the rod is now 1.5 ft above the datum,

$$V_g = (30 \text{ lb})(+1.5 \text{ ft}) = 45 \text{ ft} \cdot \text{lb}$$
$$V_2 = V_e + V_g = 45 \text{ ft} \cdot \text{lb}$$

Kinetic Energy. Denoting by $\boldsymbol{\omega}_2$ the angular velocity of the rod in position *2*, we note that the rod rotates about O and write $\bar{v}_2 = \bar{r}\omega_2 = 1.5\omega_2$.

$$\bar{I} = \tfrac{1}{12}ml^2 = \frac{1}{12}\frac{30 \text{ lb}}{32.2 \text{ ft/s}^2}(5 \text{ ft})^2 = 1.941 \text{ lb} \cdot \text{ft} \cdot \text{s}^2$$

$$T_2 = \tfrac{1}{2}m\bar{v}_2^2 + \tfrac{1}{2}\bar{I}\omega_2^2 = \frac{1}{2}\frac{30}{32.2}(1.5\omega_2)^2 + \tfrac{1}{2}(1.941)\omega_2^2 = 2.019\omega_2^2$$

Conservation of Energy

$$T_1 + V_1 = T_2 + V_2$$
$$0 + 75 \text{ ft} \cdot \text{lb} = 2.019\omega_2^2 + 45 \text{ ft} \cdot \text{lb}$$

$$\boldsymbol{\omega}_2 = 3.86 \text{ rad/s} \circlearrowright \quad ◀$$

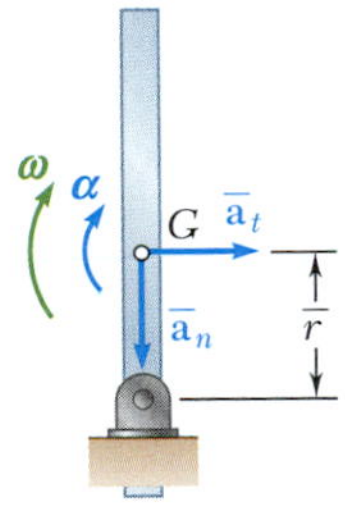

Reaction in Position 2. Since ω_2 = 3.86 rad/s, the components of the acceleration of G as the rod passes through position *2* are

$$\bar{a}_n = \bar{r}\omega_2^2 = (1.5 \text{ ft})(3.86 \text{ rad/s})^2 = 22.3 \text{ ft/s}^2 \qquad \bar{\mathbf{a}}_n = 22.3 \text{ ft/s}^2 \downarrow$$
$$a_t = \bar{r}\alpha \qquad \bar{\mathbf{a}}_t = \bar{r}\alpha \rightarrow$$

We express that the system of external forces is equivalent to the system of effective forces represented by the vector of components $m\bar{\mathbf{a}}_t$ and $m\bar{\mathbf{a}}_n$ attached at G and the couple $\bar{I}\boldsymbol{\alpha}$.

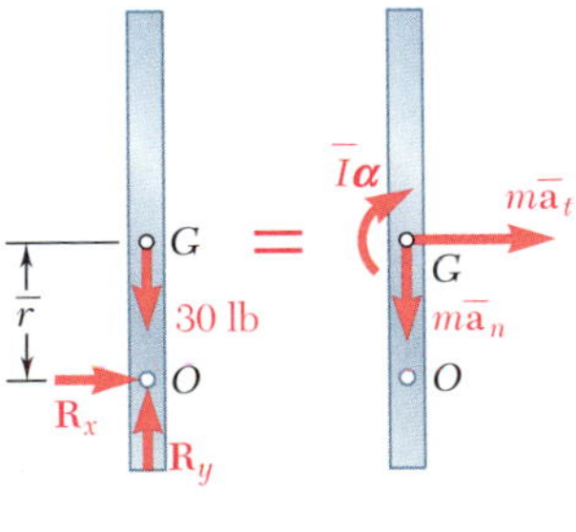

$+\circlearrowright\Sigma M_O = \Sigma(M_O)_{\text{eff}}$:	$0 = \bar{I}\alpha + m(\bar{r}\alpha)\bar{r}$	$\alpha = 0$
$\overset{+}{\rightarrow}\Sigma F_x = \Sigma(F_x)_{\text{eff}}$:	$R_x = m(\bar{r}\alpha)$	$R_x = 0$
$+\uparrow\Sigma F_y = \Sigma(F_y)_{\text{eff}}$:	$R_y - 30 \text{ lb} = -m\bar{a}_n$	

$$R_y - 30 \text{ lb} = -\frac{30 \text{ lb}}{32.2 \text{ ft/s}^2}(22.3 \text{ ft/s}^2)$$

$$R_y = +9.22 \text{ lb} \qquad \mathbf{R} = 9.22 \text{ lb} \uparrow \quad ◀$$

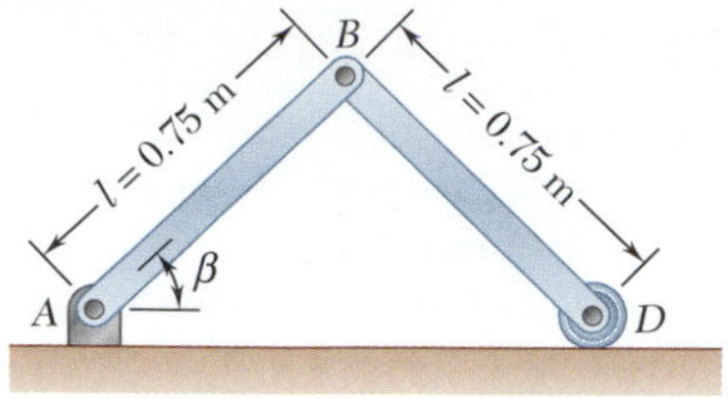

SAMPLE PROBLEM 17.5

Each of the two slender rods shown is 0.75 m long and has a mass of 6 kg. If the system is released from rest with $\beta = 60°$, determine (*a*) the angular velocity of rod *AB* when $\beta = 20°$, (*b*) the velocity of point *D* at the same instant.

SOLUTION

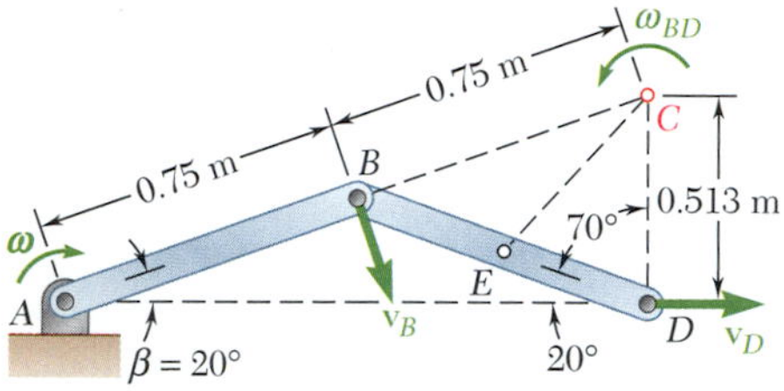

Kinematics of Motion When $\beta = 20°$. Since $\mathbf{v}_B$ is perpendicular to the rod *AB* and $\mathbf{v}_D$ is horizontal, the instantaneous center of rotation of rod *BD* is located at *C*. Considering the geometry of the figure, we obtain

$$BC = 0.75 \text{ m} \qquad CD = 2(0.75 \text{ m}) \sin 20° = 0.513 \text{ m}$$

Applying the law of cosines to triangle *CDE*, where *E* is located at the mass center of rod *BD*, we find $EC = 0.522$ m. Denoting by ω the angular velocity of rod *AB*, we have

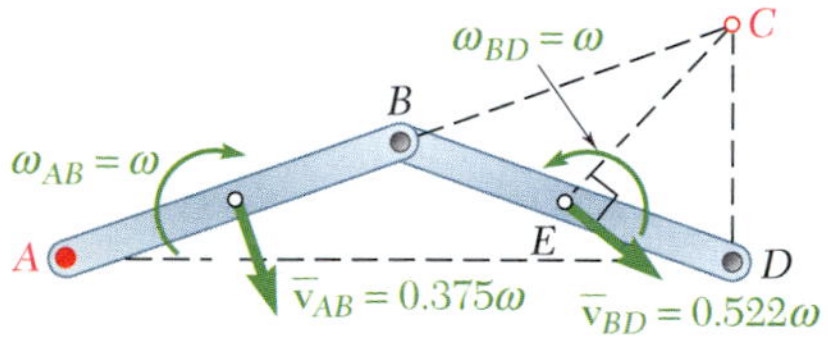

$$\overline{v}_{AB} = (0.375 \text{ m})\omega \qquad \overline{\mathbf{v}}_{AB} = 0.375\omega \searrow$$
$$v_B = (0.75 \text{ m})\omega \qquad \overline{\mathbf{v}}_B = 0.75\omega \searrow$$

Since rod *BD* seems to rotate about point *C*, we write

$$v_B = (BC)\omega_{BD} \qquad (0.75 \text{ m})\omega = (0.75 \text{ m})\omega_{BD} \qquad \boldsymbol{\omega}_{BD} = \omega \circlearrowleft$$
$$\overline{v}_{BD} = (EC)\omega_{BD} = (0.522 \text{ m})\omega \qquad \overline{\mathbf{v}}_{BD} = 0.522\omega \searrow$$

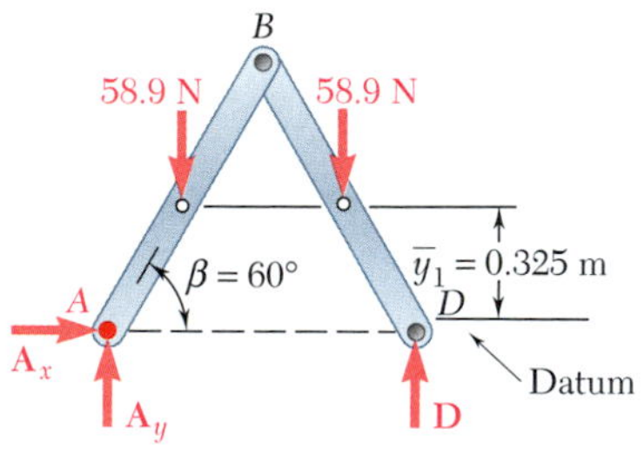

Position *1*

Position *1*. *Potential Energy.* Choosing the datum as shown, and observing that $W = (6 \text{ kg})(9.81 \text{ m/s}^2) = 58.86$ N, we have

$$V_1 = 2W\overline{y}_1 = 2(58.86 \text{ N})(0.325 \text{ m}) = 38.26 \text{ J}$$

Kinetic Energy. Since the system is at rest, $T_1 = 0$.

Position *2*. *Potential Energy*

$$V_2 = 2W\overline{y}_2 = 2(58.86 \text{ N})(0.1283 \text{ m}) = 15.10 \text{ J}$$

Kinetic Energy

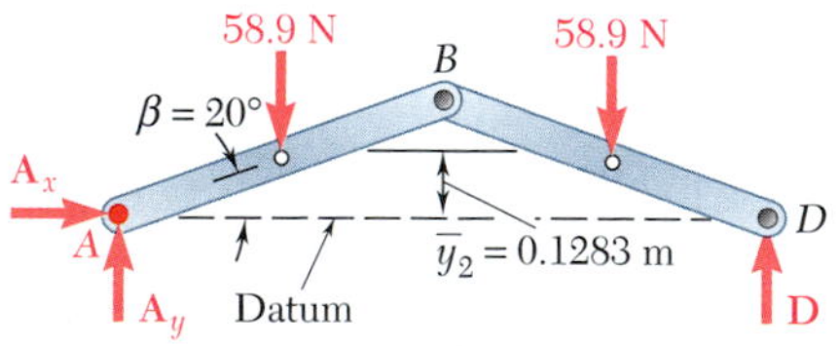

Position *2*

$$\overline{I}_{AB} = \overline{I}_{BD} = \tfrac{1}{12}ml^2 = \tfrac{1}{12}(6 \text{ kg})(0.75 \text{ m})^2 = 0.281 \text{ kg}\cdot\text{m}^2$$
$$T_2 = \tfrac{1}{2}m\overline{v}_{AB}^2 + \tfrac{1}{2}\overline{I}_{AB}\omega_{AB}^2 + \tfrac{1}{2}m\overline{v}_{BD}^2 + \tfrac{1}{2}\overline{I}_{BD}\omega_{BD}^2$$
$$= \tfrac{1}{2}(6)(0.375\omega)^2 + \tfrac{1}{2}(0.281)\omega^2 + \tfrac{1}{2}(6)(0.522\omega)^2 + \tfrac{1}{2}(0.281)\omega^2$$
$$= 1.520\omega^2$$

Conservation of Energy

$$T_1 + V_1 = T_2 + V_2$$
$$0 + 38.26 \text{ J} = 1.520\omega^2 + 15.10 \text{ J}$$
$$\omega = 3.90 \text{ rad/s} \qquad \boldsymbol{\omega}_{AB} = 3.90 \text{ rad/s} \circlearrowright \quad \blacktriangleleft$$

Velocity of Point *D*

$$v_D = (CD)\omega = (0.513 \text{ m})(3.90 \text{ rad/s}) = 2.00 \text{ m/s}$$
$$\mathbf{v}_D = 2.00 \text{ m/s} \rightarrow \quad \blacktriangleleft$$

SOLVING PROBLEMS ON YOUR OWN

In this lesson we introduced energy methods to determine the velocity of rigid bodies for various positions during their motion. As you found out previously in Chap. 13, energy methods should be considered for problems involving displacements and velocities.

1. ***The method of work and energy,*** when applied to all of the particles forming a rigid body, yields the equation

$$T_1 + U_{1\rightarrow 2} = T_2 \tag{17.1}$$

where T_1 and T_2 are, respectively, the initial and final values of the total kinetic energy of the particles forming the body and $U_{1\rightarrow 2}$ is the *work done by the external forces* exerted on the rigid body.

a. Work of forces and couples. To the expression for the work of a force (Chap. 13), we added the expression for the work of a couple and wrote

$$U_{1\rightarrow 2} = \int_{A_1}^{A_2} \mathbf{F} \cdot d\mathbf{r} \qquad U_{1\rightarrow 2} = \int_{\theta_1}^{\theta_2} M\, d\theta \qquad (17.3,\ 17.5)$$

When the moment of a couple is constant, the work of the couple is

$$U_{1\rightarrow 2} = M(\theta_2 - \theta_1) \tag{17.6}$$

where θ_1 and θ_2 are expressed in radians [Sample Probs. 17.1 and 17.2].

b. The kinetic energy of a rigid body in plane motion was found by considering the motion of the body as the sum of a translation with its mass center and a rotation about the mass center.

$$T = \tfrac{1}{2}m\bar{v}^2 + \tfrac{1}{2}\bar{I}\omega^2 \tag{17.9}$$

where $\bar{v}$ is the velocity of the mass center and ω is the angular velocity of the body [Sample Probs. 17.3 and 17.4].

2. ***For a system of rigid bodies*** we again used the equation

$$T_1 + U_{1\rightarrow 2} = T_2 \tag{17.1}$$

where T is the sum of the kinetic energies of the bodies forming the system and U is the work done by *all the forces acting on the bodies,* internal as well as external. Your computations will be simplified if you keep the following in mind.

a. The forces exerted on each other by pin-connected members or by meshed gears are equal and opposite, and, since they have the same point of application, they undergo equal small displacements. Therefore, *their total work is zero* and can be omitted from your calculations [Sample Prob. 17.2].

(continued)

b. The forces exerted by an inextensible cord on the two bodies it connects have the same magnitude and their points of application move through equal distances, but the work of one force is positive and the work of the other is negative. Therefore, *their total work is zero* and can again be omitted from your calculations [Sample Prob. 17.1].

c. The forces exerted by a spring on the two bodies it connects also have the same magnitude, but their points of application will generally move through different distances. Therefore, *their total work is usually not zero* and should be taken into account in your calculations.

3. *The principle of conservation of energy* can be expressed as

$$T_1 + V_1 = T_2 + V_2 \tag{17.12}$$

where V represents the potential energy of the system. This principle can be used when a body or a system of bodies is acted upon by conservative forces, such as the force exerted by a spring or the force of gravity [Sample Probs. 17.4 and 17.5].

4. *The last section of this lesson was devoted to power,* which is the time rate at which work is done. For a body acted upon by a couple of moment **M**, the power can be expressed as

$$\text{Power} = M\omega \tag{17.13}$$

where ω is the angular velocity of the body expressed in rad/s. As you did in Chap. 13, you should express power either in watts or in horsepower (1 hp = 550 ft · lb/s).

Problems

17.1 It is known that 1500 revolutions are required for the 2720-kg flywheel to coast to rest from an angular velocity of 300 rpm. Knowing that the radius of gyration of the flywheel is 914 mm, determine the average magnitude of the couple due to kinetic friction in the bearings.

17.2 The rotor of an electric motor has an angular velocity of 3600 rpm when the load and power are cut off. The 110-lb rotor, which has a centroidal radius of gyration of 9 in., then coasts to rest. Knowing that the kinetic friction of the rotor produces a couple of magnitude 2.5 lb · ft, determine the number of revolutions that the rotor executes before coming to rest.

17.3 The flywheel of a punching machine has a weight of 650 lb and a radius of gyration of 30 in. Each punching operation requires 1800 ft · lb of work. (*a*) Knowing that the speed of the flywheel is 300 rpm just before a punching, determine the speed immediately after the punching. (*b*) If a constant 15 lb · ft couple is applied to the shaft of the flywheel, determine the number of revolutions executed before the speed is again 300 rpm.

17.4 The flywheel of a small punching machine rotates at 360 rpm. Each punching operation requires 2034 J of work and it is desired that the speed of the flywheel after each punching be not less that 95 percent of the original speed. (*a*) Determine the required moment of inertia of the flywheel. (*b*) If a constant 24.4 N · m couple is applied to the shaft of the flywheel, determine the number of revolutions that must occur between two successive punchings, knowing that the initial velocity is to be 360 rpm at the start of each punching.

17.5 Two uniform disks of the same material are attached to a shaft as shown. Disk A has a mass of 20 kg and a radius r = 150 mm. Disk B is twice as thick as disk A. Knowing that a couple **M** of magnitude 30 N · m is applied to disk A when the system is at rest, determine the radius nr of disk B if the angular velocity of the system is to be 480 rpm after 5 revolutions.

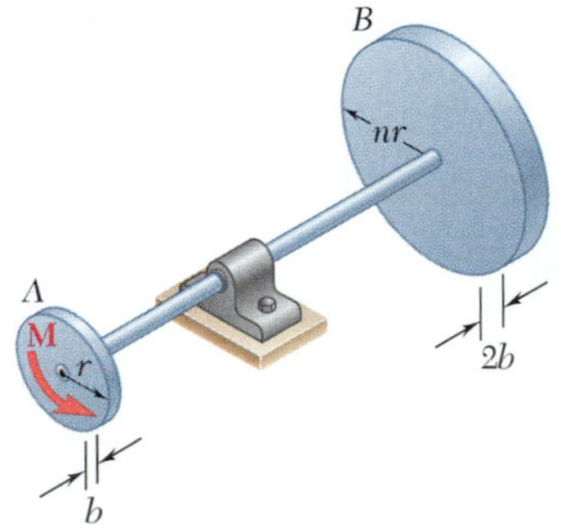

Fig. P17.5

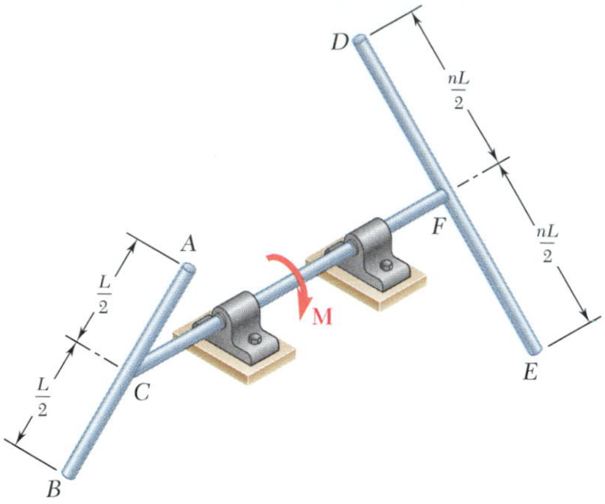

Fig. P17.6

17.6 Two uniform slender rods AB and DE of weight w per unit length are attached to a shaft CF as shown. The length of rod AB is L and the length of rod DE is nL. A couple **M** of constant moment is applied to shaft CF when the system is at rest and removed after the system has executed one complete revolution. Neglecting the mass of the shaft, determine the length of rod DE which results in the largest final speed of point D.

17.7 The uniform cylinder A, of mass m and radius r, has an angular velocity ω_0 when it is brought into contact with an identical cylinder B which is at rest. The coefficient of kinetic friction at the contact point D is μ_k. After a period of slipping, the cylinders attain constant angular velocities of equal magnitude and opposite direction at the same time. Knowing that the final kinetic energy of the system is one-half the initial kinetic energy of cylinder A, derive expressions for the number of revolutions executed by each cylinder before it attains a constant angular velocity.

17.8 The uniform 8-lb cylinder A, of radius $r = 6$ in., has an angular velocity $\omega_0 = 50$ rad/s when it is brought into contact with an identical cylinder B which is at rest. The coefficient of kinetic friction at the contact point D is μ_k. After a period of slipping, the cylinders attain constant angular velocities of equal magnitude and opposite direction at the same time. Knowing that cylinder A executes three revolutions before it attains a constant angular velocity and cylinder B executes one revolution before it attains a constant angular velocity, determine (*a*) the final angular velocity of each cylinder, (*b*) the coefficient of kinetic friction μ_k.

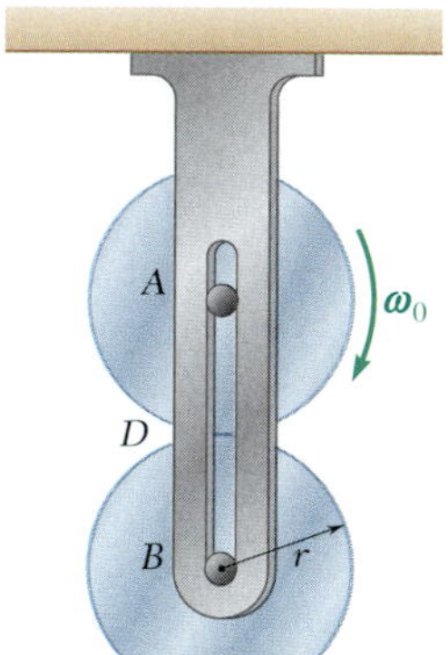

Fig. *P17.7* and P17.8

17.9 Each of the gears A and B has a weight of 5 lb and a radius of gyration of 4 in., while gear C has a weight of 25 lb and a radius of gyration of 7.5 in. A couple **M** of magnitude 6.75 lb · ft is applied to gear C. Determine (*a*) the number of revolutions of gear C required for its angular velocity to increase from 100 to 450 rpm, (*b*) the corresponding tangential force acting on gear A.

17.10 Solve Prob. 17.9 assuming that the 6.75 lb · ft couple is applied to gear B.

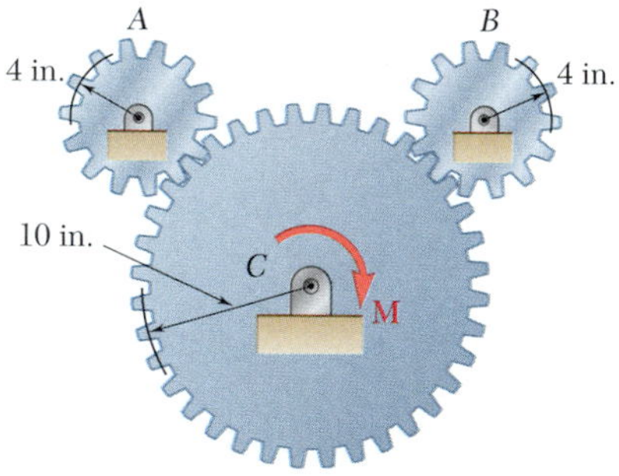

Fig. P17.9

17.11 The double pulley shown has a mass of 14 kg and a centroidal radius of gyration of 165 mm. Cylinder A and block B are attached to cords that are wrapped on the pulleys as shown. The coefficient of kinetic friction between block B and the surface is 0.25. Knowing that the system is released from rest in the position shown, determine (*a*) the velocity of cylinder A as it strikes the ground, (*b*) the total distance that block B moves before coming to rest.

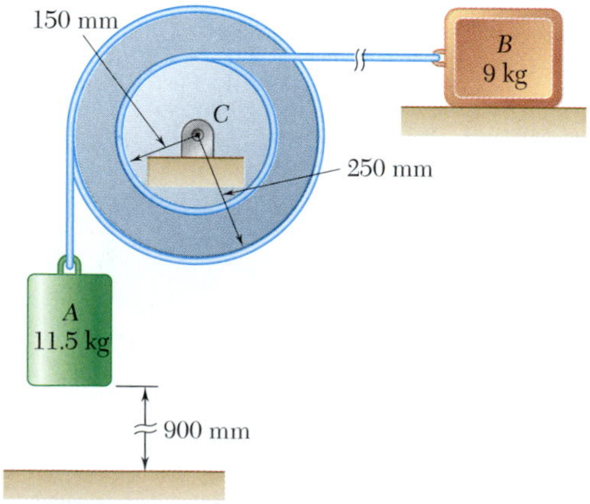

Fig. P17.11

17.12 The 200-mm-radius brake drum is attached to a larger flywheel that is not shown. The total mass moment of inertia of the flywheel and drum is 19 kg · m^2 and the coefficient of kinetic friction between the drum and the brake shoe is 0.35. Knowing that the initial angular velocity of the flywheel is 360 rpm counterclockwise, determine the vertical force **P** that must be applied to the pedal C if the system is to stop in 100 revolutions.

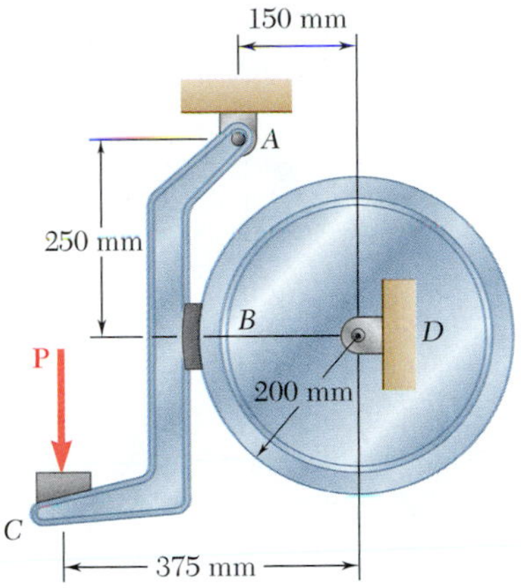

Fig. P17.12

17.13 Solve Prob. 17.12 assuming that the initial angular velocity of the flywheel is 360 rpm clockwise.

***17.14* and 17.15** A slender 6-kg rod can rotate in a vertical plane about a pivot at B. A spring of constant $k = 600$ N/m and an unstretched length of 225 mm is attached to the rod as shown. Knowing that the rod is released from rest in the position shown, determine its angular velocity after it has rotated through 90°.

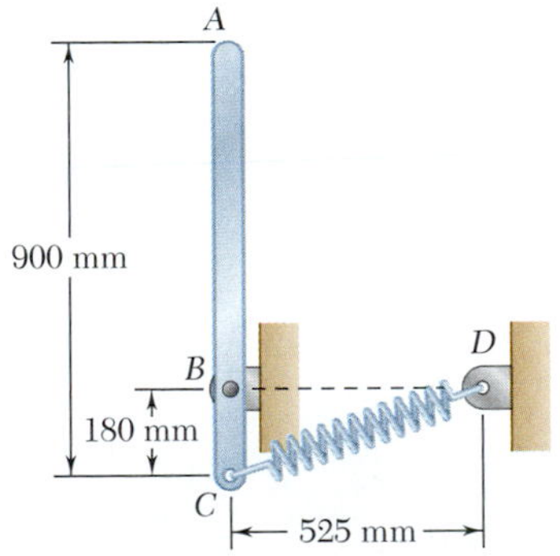

Fig. *P17.14*

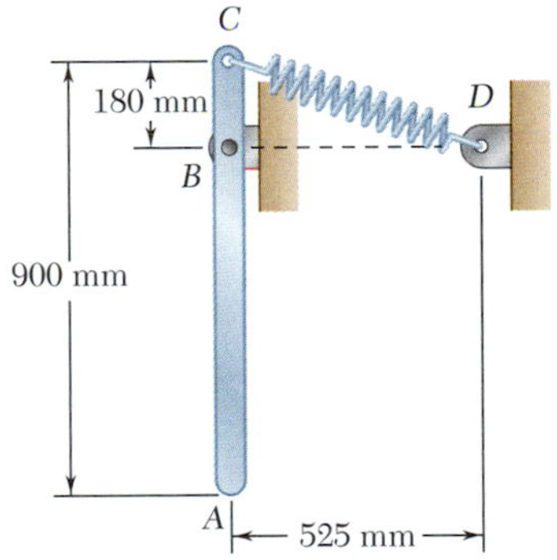

Fig. P17.15

17.16 A slender rod of length l is pivoted about a point C located at a distance b from its center G. It is released from rest in a horizontal position and swings freely. Determine (a) the distance b for which the angular velocity of the rod as it passes through a vertical position is maximum, (b) the corresponding values of its angular velocity and of the reaction at C.

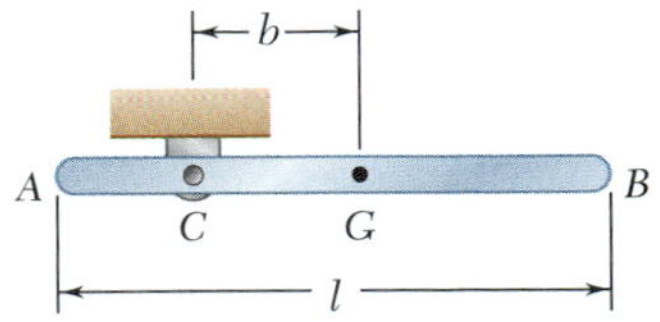

Fig. *P17.16*

17.17 A 73-kg gymnast is executing a series of full-circle swings on the horizontal bar. In the position shown he has a small and negligible clockwise angular velocity and will maintain his body straight and rigid as he swings downward. Assuming that during the swing the centroidal radius of gyration of his body is 457 mm, determine his angular velocity and the force exerted on his hands after he has rotated through (*a*) 90°, (*b*) 180°.

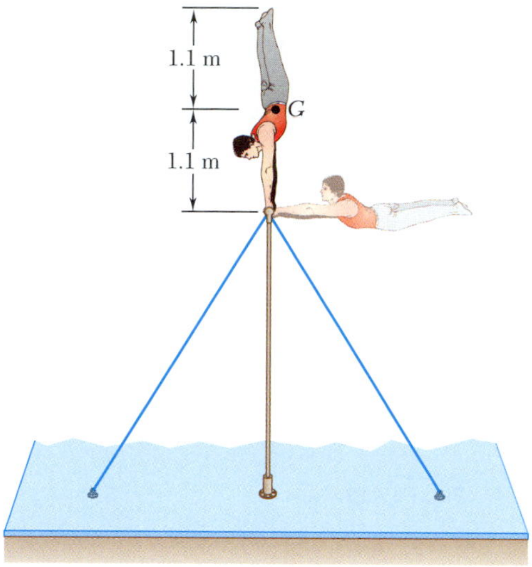

Fig. P17.17

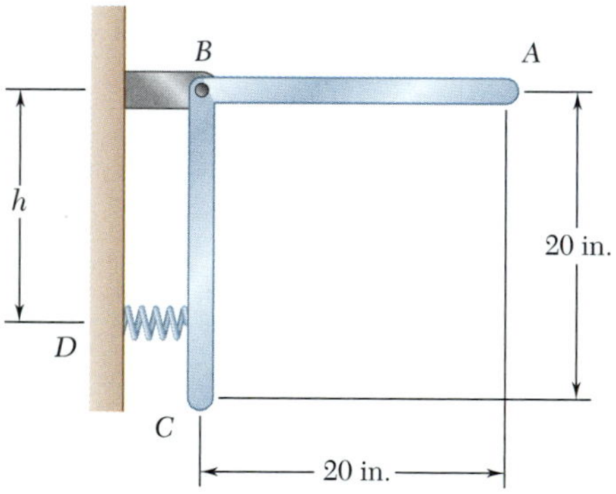

Fig. P17.18

17.18 Two identical slender rods *AB* and *BC* are welded together to form an L-shaped assembly. The assembly is pressed against a spring at *D* and released from the position shown. Knowing that the maximum angle of rotation of the assembly in its subsequent motion is 90° counterclockwise, determine the magnitude of the angular velocity of the assembly as it passes through the position where rod *AB* forms an angle of 30° with the horizontal.

17.19 The 60-lb turbine disk has a centroidal radius of gyration of 7 in. and is rotating clockwise at a constant rate of 60 rpm when a small blade of weight 0.1 lb at point *A* becomes loose and is thrown off. Neglecting friction, determine the change in the angular velocity of the turbine disk after it has rotated through (*a*) 90°, (*b*) 270°.

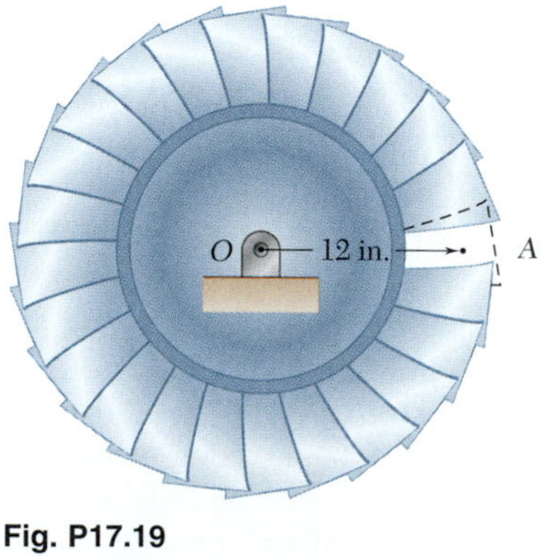

Fig. P17.19

17.20 A 24-kg uniform cylindrical roller, initially at rest, is acted upon by a 100 N force as shown. Knowing that the body rolls without slipping, determine (*a*) the velocity of its center *G* after it has moved 1.8 m, (*b*) the friction force required to prevent slipping.

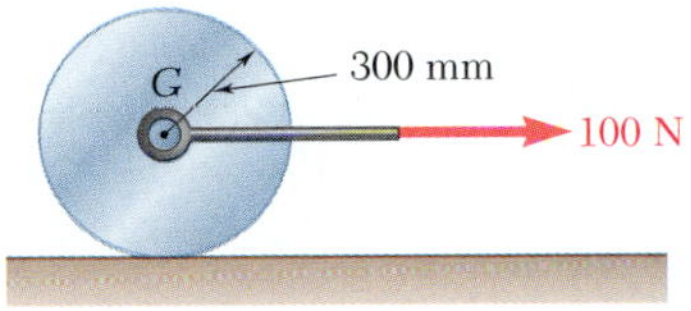

Fig. *P17.20*

17.21 A rope is wrapped around a cylinder of radius *r* and mass *m* as shown. Knowing that the cylinder is released from rest, determine the velocity of the center of the cylinder after it has moved downward a distance *s*.

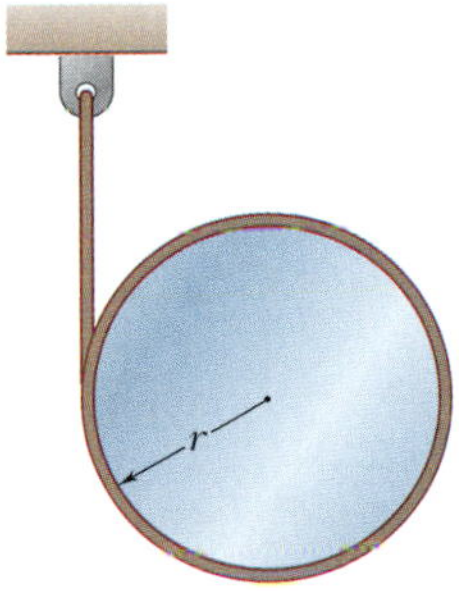

Fig. P17.21

17.22 A collar *B*, of mass *m* and negligible dimensions, is attached to the rim of a hoop of the same mass *m* and of radius *r* that rolls without sliding on a horizontal surface. Determine the angular velocity $\boldsymbol{\omega}_1$ of the hoop in terms of *g* and *r* when *B* is directly above the center *A*, knowing that the angular velocity of the hoop is $4\boldsymbol{\omega}_1$ when *B* is directly below *A*.

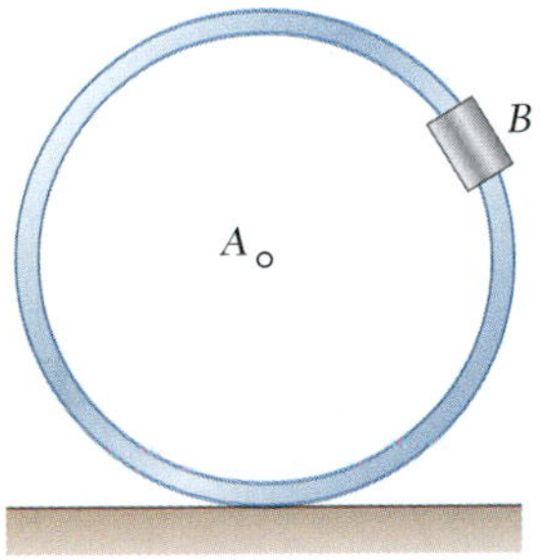

Fig. *P17.22*

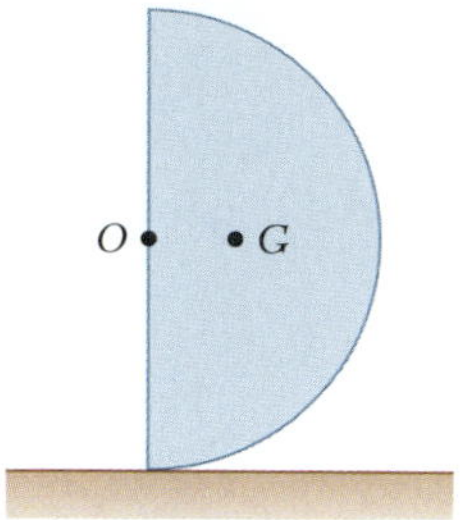

Fig. *P17.23*

17.23 A half-cylinder of mass m and radius r is released from rest in the position shown. Knowing that the half-cylinder rolls without sliding, determine (*a*) its angular velocity after it has rolled through 90°, (*b*) the reaction at the horizontal surface at the same instant. [*Hint:* Note that $GO = 4r/3\pi$ and that, by the parallel-axis theorem, $\bar{I} = \frac{1}{2}mr^2 - m(GO)^2$.]

17.24 The 1.5-kg uniform slender bar AB is connected to the 3-kg gear B which meshes with the stationary outer gear C. The centroidal radius of gyration of gear B is 30 mm. Knowing that the system is released from rest in the position shown, determine (*a*) the angular velocity of the bar as it passes through the vertical position, (*b*) the corresponding angular velocity of gear B.

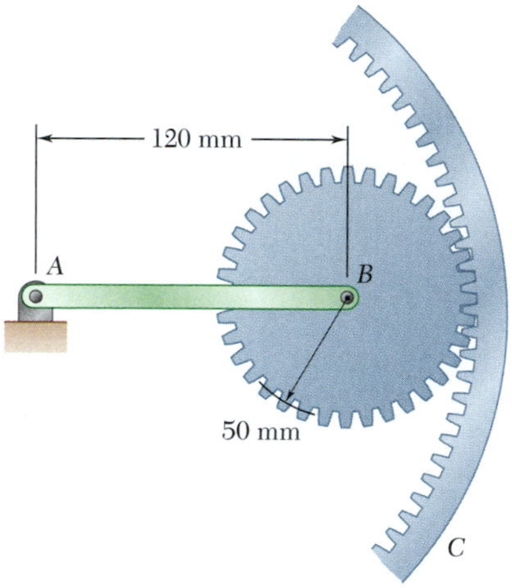

Fig. *P17.24* and P17.25

17.25 The 1.5-kg uniform slender bar AB is connected to the 3-kg gear B which meshes with the stationary outer gear C. The centroidal radius of gyration of gear B is 30 mm. The system is released in the position shown when the angular velocity of the bar is 4 rad/s counterclockwise. Knowing that the magnitude of the angular velocity of the bar as it passes through the vertical position is 12 rad/s, determine the amount of energy dissipated in friction.

17.26 through 17.28 The 18-lb cradle is supported as shown by two uniform disks that roll without sliding at all surfaces of contact. The weight of each disk is $W = 12$ lb and the radius of each disk is $r = 4$ in. Knowing that the system is initially at rest, determine the velocity of the cradle after it has moved 15 in.

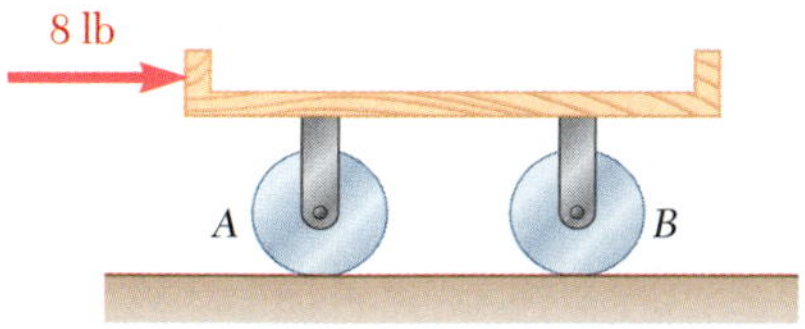

Fig. P17.26

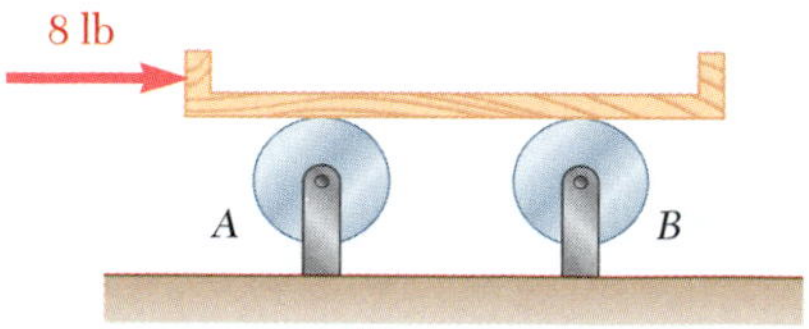

Fig. P17.27

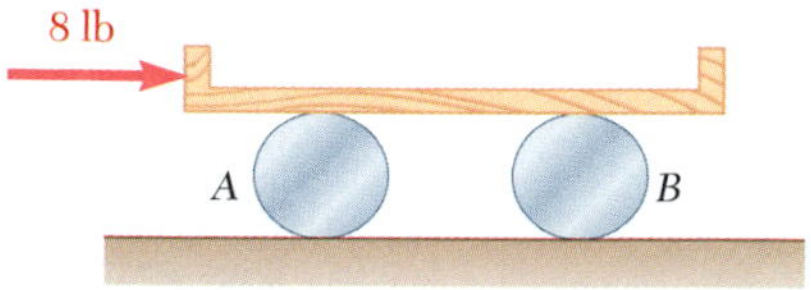

Fig. P17.28

17.29 The mechanism shown is one of two identical mechanisms attached to the two sides of a 200-lb uniform rectangular door. Edge *ABC* of the door is guided by wheels of negligible mass that roll in horizontal and vertical tracks. A spring of constant $k = 40$ lb/ft is attached to wheel *B*. Knowing that the door is released from rest in the position $\theta = 30°$ with the spring unstretched, determine the velocity of wheel *A* just as the door reaches the vertical position.

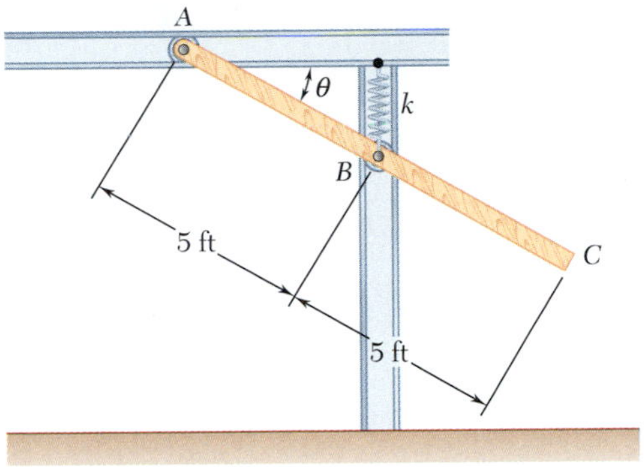

Fig. P17.29 and P17.30

17.30 The mechanism shown is one of two identical mechanisms attached to the two sides of a 200-lb uniform rectangular door. Edge *ABC* of the door is guided by wheels of negligible mass that roll in horizontal and vertical tracks. A spring of constant *k* is attached to wheel *B* in such a way that its tension is zero when $\theta = 30°$. Knowing that the door is released from rest in the position $\theta = 45°$ and reaches the vertical position with an angular velocity of 0.6 rad/s, determine the spring constant *k*.

17.31 The motion of the uniform slender 2.4-kg rod *AB* is guided at *A* and *C* by collars of negligible mass. The system is released from rest in the position $\theta = 45°$. Knowing that the applied force **P** is zero, determine the angular velocity of rod *AB* when $\theta = 30°$.

17.32 The motion of the uniform slender 2.4-kg rod *AB* is guided at *A* and *C* by collars of negligible mass. The system is released from rest in the position $\theta = 30°$. Knowing that the magnitude of the force **P** applied to collar *A* is 10 N, determine the angular velocity of rod *AB* when $\theta = 45°$.

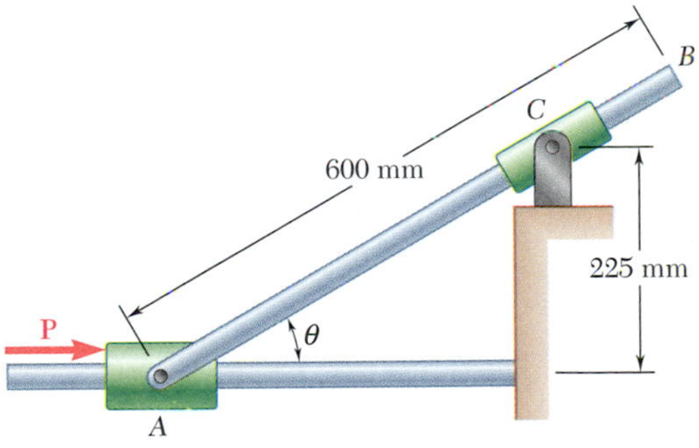

Fig. P17.31 and P17.32

17.33 The motion of the uniform rod *AB* is guided by small wheels of negligible mass that roll on the surface shown. If the rod is released from rest when $\theta = 0$, determine the velocities of *A* and *B* when $\theta = 30°$.

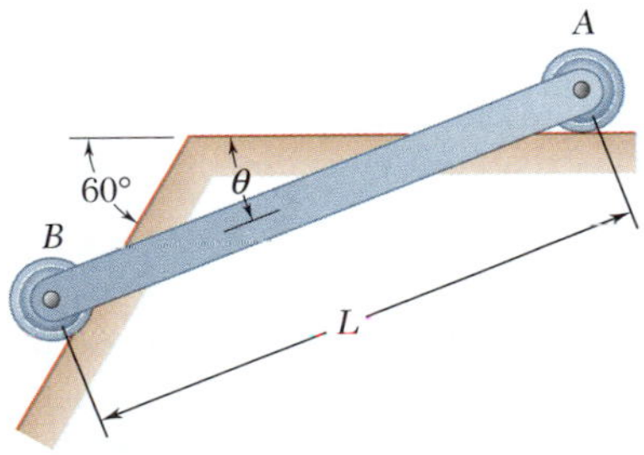

Fig. *P17.33*

17.34 Two uniform rods, each of mass m and length L, are connected to form the linkage shown. End D of rod BD can slide freely in the horizontal slot, while end A of rod AB is supported by a pin and bracket. If end D is moved slightly to the left and then released, determine its velocity (*a*) when it is directly below A, (*b*) when rod AB is vertical.

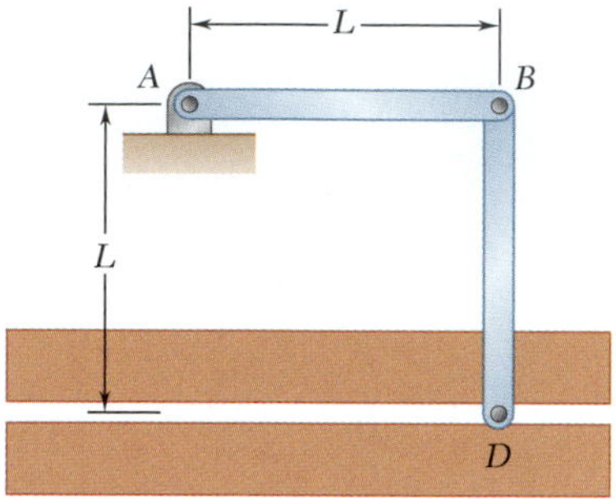

Fig. *P17.34*

17.35 The uniform rods AB and BC are of mass 2.4 kg and 4 kg, respectively, and the small wheel at C is of negligible mass. If the wheel is moved slightly to the right and then released, determine the velocity of pin B after rod AB has rotated through 90°.

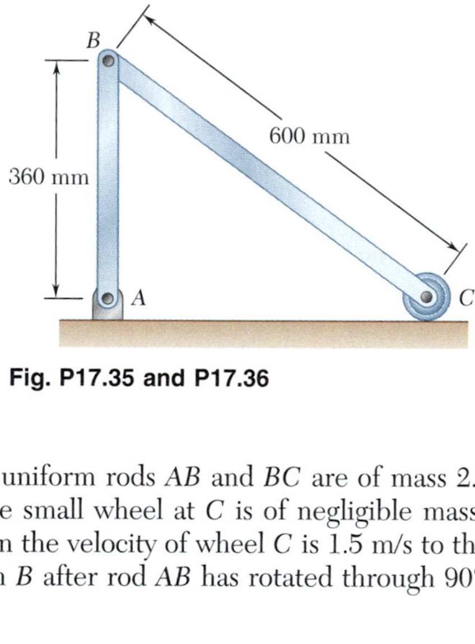

Fig. P17.35 and P17.36

17.36 The uniform rods AB and BC are of mass 2.4 kg and 4 kg, respectively, and the small wheel at C is of negligible mass. Knowing that in the position shown the velocity of wheel C is 1.5 m/s to the right, determine the velocity of pin B after rod AB has rotated through 90°.

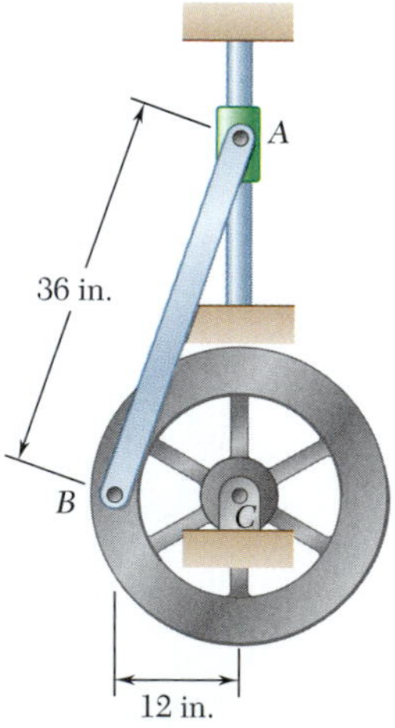

Fig. P17.37 and P17.38

17.37 The 8-lb rod AB is attached to a collar of negligible weight at A and to a flywheel at B. The flywheel has a weight of 32 lb and a radius of gyration of 9 in. Knowing that in the position shown the angular velocity of the flywheel is 60 rpm clockwise, determine the velocity of the flywheel when point B is directly below C.

17.38 If in Prob. 17.37 the angular velocity of the flywheel is to be the same in the position shown and when point B is directly above C, determine the required value of its angular velocity in the position shown.

17.39 The flywheel shown has a radius of 20 in., a weight of 250 lb and a radius of gyration of 15 in. A 44-lb block A is attached to a wire that is wrapped around the flywheel. Determine the power delivered by the electric motor attached to the shaft of the flywheel at the instant when the velocity of block A is 25 ft/s up and its acceleration is (a) zero, (b) 3 ft/s^2 up.

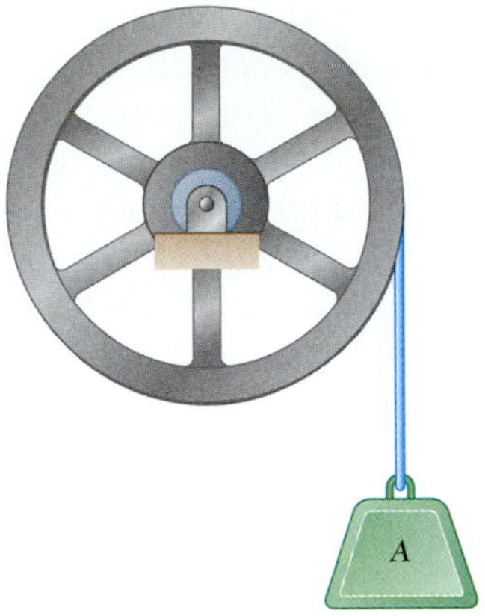

Fig. P17.39

17.40 Knowing that the maximum allowable couple that can be applied to a shaft is 1.75 kN · m, determine the maximum power that can be transmitted by the shaft at (a) 180 rpm, (b) 480 rpm.

17.41 A motor attached to shaft AB develops 3.36 kW while running at a constant speed of 720 rpm. Determine the magnitude of the couple exerted (a) on shaft AB, (b) on shaft CD.

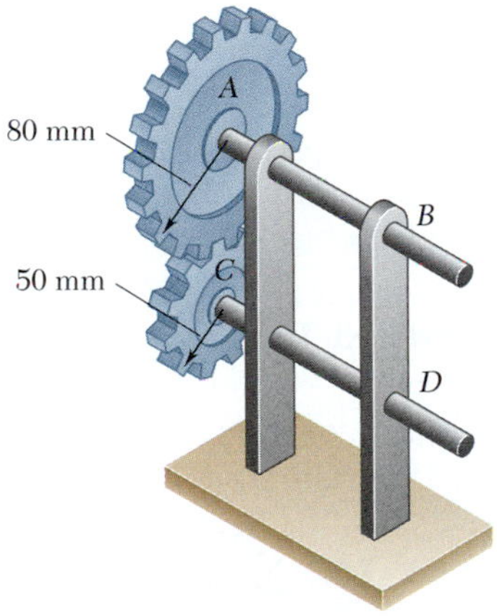

Fig. P17.41

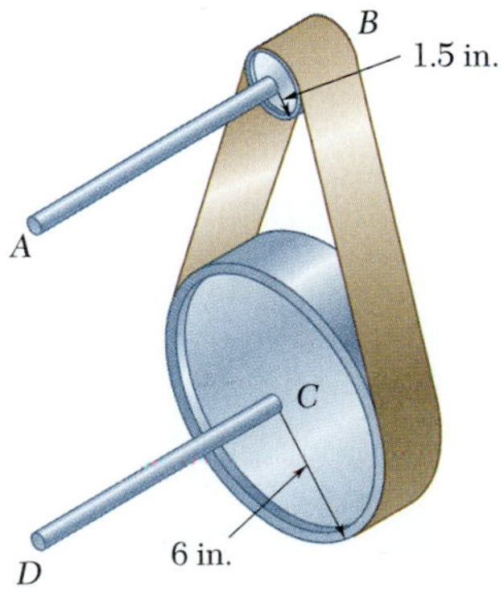

Fig. P17.42

17.42 The shaft-disk-belt arrangement shown is used to transmit 3.2 hp from point A to point D. Knowing that the maximum allowable couples that can be applied to shafts AB and CD are 18 lb · ft and 58 lb · ft, respectively, determine the required minimum speed of shaft AB.

17.8. PRINCIPLE OF IMPULSE AND MOMENTUM FOR THE PLANE MOTION OF A RIGID BODY

The principle of impulse and momentum will now be applied to the analysis of the plane motion of rigid bodies and of systems of rigid bodies. As was pointed out in Chap. 13, the method of impulse and momentum is particularly well adapted to the solution of problems involving time and velocities. Moreover, the principle of impulse and momentum provides the only practicable method for the solution of problems involving impulsive motion or impact (Secs. 17.11 and 17.12).

Considering again a rigid body as made of a large number of particles P_i, we recall from Sec. 14.9 that the system formed by the momenta of the particles at time t_1 and the system of the impulses of the external forces applied from t_1 to t_2 are together equipollent to the system formed by the momenta of the particles at time t_2. Since the vectors associated with a rigid body can be considered as sliding vectors, it follows (Sec. 3.19) that the systems of vectors shown in Fig. 17.6 are not only equipollent but truly *equivalent* in the sense

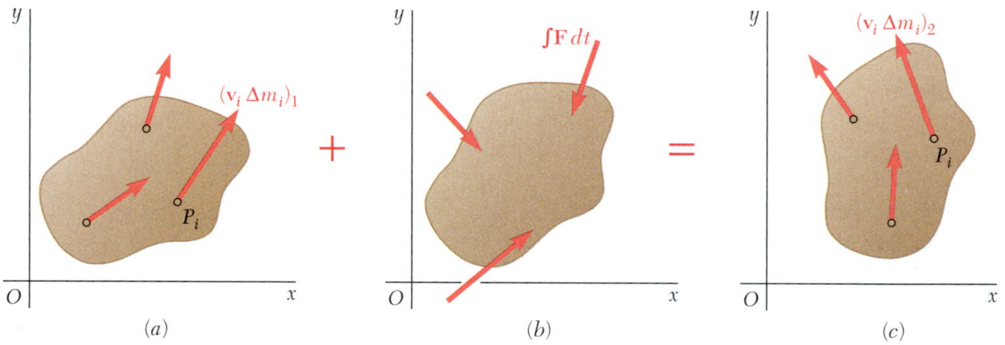

Fig. 17.6

that the vectors on the left-hand side of the equals sign can be transformed into the vectors on the right-hand side through the use of the fundamental operations listed in Sec. 3.13. We therefore write

$$\textbf{Syst Momenta}_1 + \textbf{Syst Ext Imp}_{1\rightarrow 2} = \textbf{Syst Momenta}_2 \tag{17.14}$$

But the momenta $\mathbf{v}_i\,\Delta m_i$ of the particles can be reduced to a vector attached at G, equal to their sum

$$\mathbf{L} = \sum_{i=1}^{n} \mathbf{v}_i\,\Delta m_i$$

and a couple of moment equal to the sum of their moments about G

$$\mathbf{H}_G = \sum_{i=1}^{n} \mathbf{r}'_i \times \mathbf{v}_i\,\Delta m_i$$

We recall from Sec. 14.3 that $\mathbf{L}$ and $\mathbf{H}_G$ define, respectively, the linear momentum and the angular momentum about G of the system of particles forming the rigid body. We also note from Eq. (14.14) that $\mathbf{L} = m\overline{\mathbf{v}}$. On the other hand, restricting the present analysis to the plane motion of a rigid slab or of a rigid body symmetrical with respect to the reference plane, we recall from Eq. (16.4) that $\mathbf{H}_G = \overline{I}\boldsymbol{\omega}$. We thus conclude that the system of the momenta $\mathbf{v}_i\,\Delta m_i$ is equivalent to the *linear momentum vector* $m\overline{\mathbf{v}}$ attached at G and to the *angular momentum couple* $\overline{I}\boldsymbol{\omega}$ (Fig. 17.7). Observing that the system of

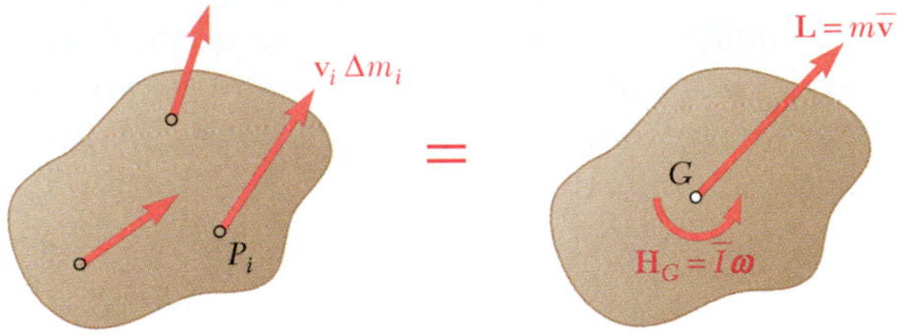

Fig. 17.7

momenta reduces to the vector $m\overline{\mathbf{v}}$ in the particular case of a translation ($\boldsymbol{\omega} = 0$) and to the couple $\overline{I}\boldsymbol{\omega}$ in the particular case of a centroidal rotation ($\overline{\mathbf{v}} = 0$), we verify once more that the plane motion of a rigid body symmetrical with respect to the reference plane can be resolved into a translation with the mass center G and a rotation about G.

Replacing the system of momenta in parts *a* and *c* of Fig. 17.6 by the equivalent linear momentum vector and angular momentum couple, we obtain the three diagrams shown in Fig. 17.8. This figure

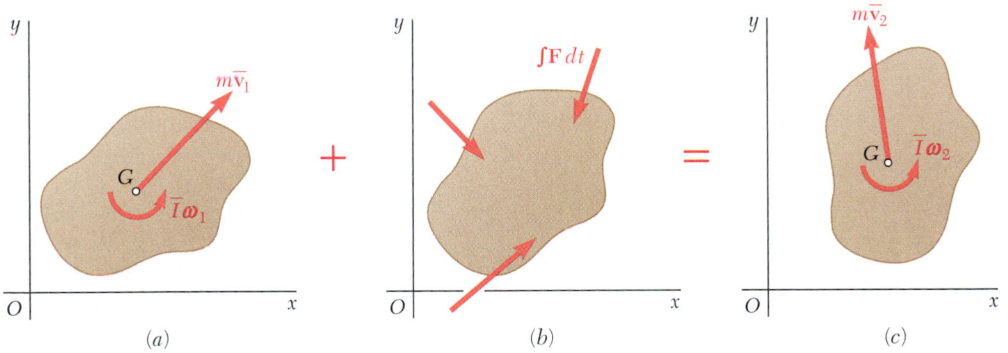

Fig. 17.8

expresses as a free-body-diagram equation the fundamental relation (17.14) in the case of the plane motion of a rigid slab or of a rigid body symmetrical with respect to the reference plane.

Three equations of motion can be derived from Fig. 17.8. Two equations are obtained by summing and equating the x and y *components* of the momenta and impulses, and the third equation is obtained by summing and equating the *moments* of these vectors *about any given point*. The coordinate axes can be chosen fixed in space, or allowed to move with the mass center of the body while maintaining a fixed direction. In either case, the point about which moments are taken should keep the same position relative to the coordinate axes during the interval of time considered.

In deriving the three equations of motion for a rigid body, care should be taken not to add linear and angular momenta indiscriminately. Confusion can be avoided by remembering that $m\bar{v}_x$ and $m\bar{v}_y$ represent the *components of a vector,* namely, the linear momentum vector $m\bar{\mathbf{v}}$, while $\bar{I}\omega$ represents the *magnitude of a couple,* namely, the angular momentum couple $\bar{I}\boldsymbol{\omega}$. Thus the quantity $\bar{I}\omega$ should be added only to the *moment* of the linear momentum $m\bar{\mathbf{v}}$, never to this vector itself nor to its components. All quantities involved will then be expressed in the same units, namely $\mathrm{N \cdot m \cdot s}$ or $\mathrm{lb \cdot ft \cdot s}$.

Noncentroidal Rotation. In this particular case of plane motion, the magnitude of the velocity of the mass center of the body is $\bar{v} = \bar{r}\omega$, where $\bar{r}$ represents the distance from the mass center to the fixed axis of rotation and $\boldsymbol{\omega}$ represents the angular velocity of the body at the instant considered; the magnitude of the momentum vector attached at G is thus $m\bar{v} = m\bar{r}\omega$. Summing the moments about O of the momentum vector and momentum couple (Fig. 17.9) and using

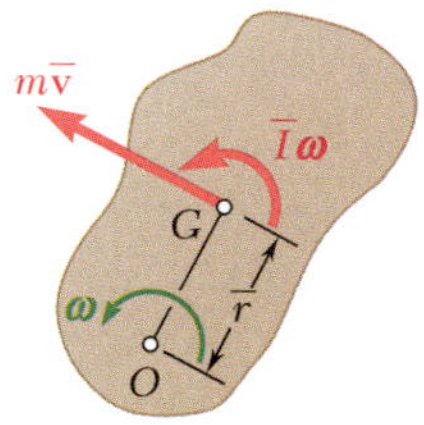

Fig. 17.9

the parallel-axis theorem for moments of inertia, we find that the angular momentum $\mathbf{H}_O$ of the body about O has the magnitude†

$$\bar{I}\omega + (m\bar{r}\omega)\bar{r} = (\bar{I} + m\bar{r}^2)\omega = I_O\omega \tag{17.15}$$

Equating the moments about O of the momenta and impulses in (17.14), we write

$$I_O\omega_1 + \sum \int_{t_1}^{t_2} M_O\, dt = I_O\omega_2 \tag{17.16}$$

In the general case of plane motion of a rigid body symmetrical with respect to the reference plane, Eq. (17.16) can be used with respect to the instantaneous axis of rotation under certain conditions. It is recommended, however, that all problems of plane motion be solved by the general method described earlier in this section.

†Note that the sum $\mathbf{H}_A$ of the moments about an arbitrary point A of the momenta of the particles of a rigid slab is, in general, *not* equal to $I_A\boldsymbol{\omega}$. (See Prob. 17.67.)

17.9. SYSTEMS OF RIGID BODIES

The motion of several rigid bodies can be analyzed by applying the principle of impulse and momentum to each body separately (Sample Prob. 17.6). However, in solving problems involving no more than three unknowns (including the impulses of unknown reactions), it is often convenient to apply the principle of impulse and momentum to the system as a whole. The momentum and impulse diagrams are drawn for the entire system of bodies. For each moving part of the system, the diagrams of momenta should include a momentum vector, a momentum couple, or both. Impulses of forces internal to the system can be omitted from the impulse diagram, since they occur in pairs of equal and opposite vectors. Summing and equating successively the x components, y components, and moments of all vectors involved, one obtains three relations which express that the momenta at time t_1 and the impulses of the external forces form a system equipollent to the system of the momenta at time t_2.† Again, care should be taken not to add linear and angular momenta indiscriminately; each equation should be checked to make sure that consistent units have been used. This approach has been used in Sample Prob. 17.8 and, further on, in Sample Probs. 17.9 and 17.10.

17.10. CONSERVATION OF ANGULAR MOMENTUM

When no external force acts on a rigid body or a system of rigid bodies, the impulses of the external forces are zero and the system of the momenta at time t_1 is equipollent to the system of the momenta at time t_2. Summing and equating successively the x components, y components, and moments of the momenta at times t_1 and t_2, we conclude that the total linear momentum of the system is conserved in any direction and that its total angular momentum is conserved about any point.

There are many engineering applications, however, in which *the linear momentum is not conserved* yet *the angular momentum* $\mathbf{H}_O$ of the system about a given point O *is conserved* that is, in which

$$(\mathbf{H}_O)_1 = (\mathbf{H}_O)_2 \qquad (17.17)$$

Such cases occur when the lines of action of all external forces pass through O or, more generally, when the sum of the angular impulses of the external forces about O is zero.

Problems involving *conservation of angular momentum* about a point O can be solved by the general method of impulse and momentum, that is, by drawing momentum and impulse diagrams as described in Secs. 17.8 and 17.9. Equation (17.17) is then obtained by summing and equating moments about O (Sample Prob. 17.8). As you will see later in Sample Prob. 17.9, two additional equations can be written by summing and equating x and y components and these equations can be used to determine two unknown linear impulses, such as the impulses of the reaction components at a fixed point.

Photo 17.2 A figure skater at the beginning and at the end of a spin. By using the principle of conservation of angular momentum you will find that her angular velocity is much higher at the end of the spin.

†Note that as in Sec. 16.7, we cannot speak of *equivalent* systems since we are not dealing with a single rigid body.

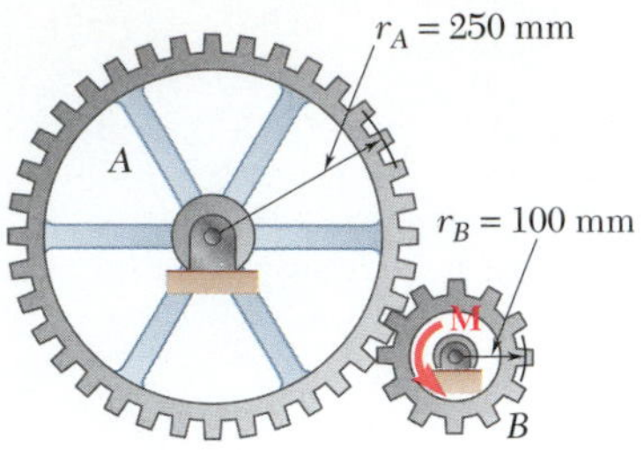

SAMPLE PROBLEM 17.6

Gear A has a mass of 10 kg and a radius of gyration of 200 mm, and gear B has a mass of 3 kg and a radius of gyration of 80 mm. The system is at rest when a couple $\mathbf{M}$ of magnitude 6 N · m is applied to gear B. (These gears were considered in Sample Prob. 17.2.) Neglecting friction, determine (a) the time required for the angular velocity of gear B to reach 600 rpm, (b) the tangential force which gear B exerts on gear A.

SOLUTION

We apply the principle of impulse and momentum to each gear separately. Since all forces and the couple are constant, their impulses are obtained by multiplying them by the unknown time t. We recall from Sample Prob. 17.2 that the centroidal moments of inertia and the final angular velocities are

$$\bar{I}_A = 0.400 \text{ kg} \cdot \text{m}^2 \qquad \bar{I}_B = 0.0192 \text{ kg} \cdot \text{m}^2$$
$$(\omega_A)_2 = 25.1 \text{ rad/s} \qquad (\omega_B)_2 = 62.8 \text{ rad/s}$$

Principle of Impulse and Momentum for Gear *A*. The systems of initial momenta, impulses, and final momenta are shown in three separate sketches.

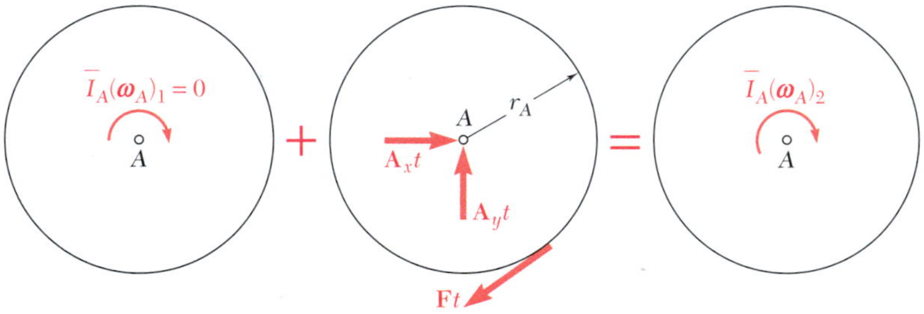

$$\textbf{Syst Momenta}_1 + \textbf{Syst Ext Imp}_{1\to 2} = \textbf{Syst Momenta}_2$$

+↺moments about A: $\quad 0 - Ftr_A = -\bar{I}_A(\omega_A)_2$

$$Ft(0.250 \text{ m}) = (0.400 \text{ kg} \cdot \text{m}^2)(25.1 \text{ rad/s})$$
$$Ft = 40.2 \text{ N} \cdot \text{s}$$

Principle of Impulse and Momentum for Gear *B*.

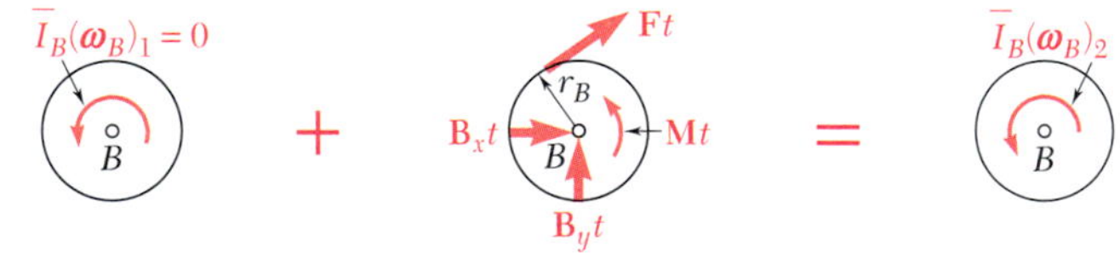

$$\textbf{Syst Momenta}_1 + \textbf{Syst Ext Imp}_{1\to 2} = \textbf{Syst Momenta}_2$$

+↺moments about B: $\quad 0 + Mt - Ftr_B = \bar{I}_B(\omega_B)_2$

$$+(6 \text{ N} \cdot \text{m})t - (40.2 \text{ N} \cdot \text{s})(0.100 \text{ m}) = (0.0192 \text{ kg} \cdot \text{m}^2)(62.8 \text{ rad/s})$$

$t = 0.871$ s ◀

Recalling that $Ft = 40.2$ N · s, we write

$$F(0.871 \text{ s}) = 40.2 \text{ N} \cdot \text{s} \qquad F = +46.2 \text{ N}$$

Thus, the force exerted by gear B on gear A is $\quad \mathbf{F} = 46.2 \text{ N} \swarrow$ ◀

SAMPLE PROBLEM 17.7

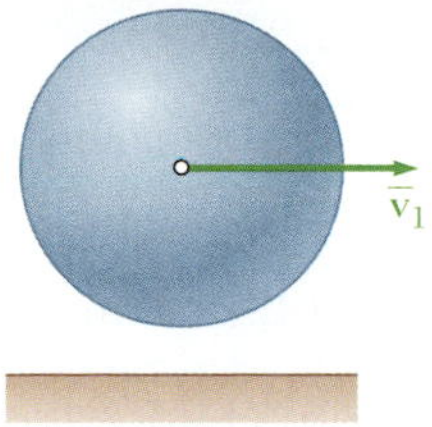

A uniform sphere of mass m and radius r is projected along a rough horizontal surface with a linear velocity $\bar{\mathbf{v}}_1$ and no angular velocity. Denoting by μ_k the coefficient of kinetic friction between the sphere and the surface, determine (*a*) the time t_2 at which the sphere will start rolling without sliding, (*b*) the linear and angular velocities of the sphere at time t_2.

SOLUTION

While the sphere is sliding relative to the surface, it is acted upon by the normal force **N**, the friction force **F**, and its weight **W** of magnitude $W = mg$.

Principle of Impulse and Momentum. We apply the principle of impulse and momentum to the sphere from the time $t_1 = 0$ when it is placed on the surface until the time $t_2 = t$ when it starts rolling without sliding.

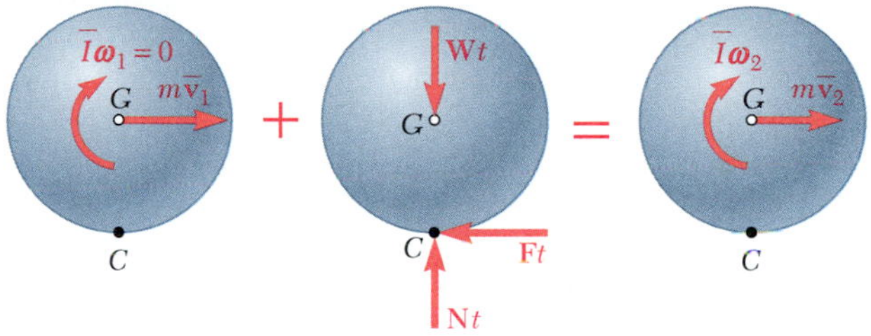

Syst Momenta$_1$ + Syst Ext Imp$_{1\to2}$ = Syst Momenta$_2$

$+\uparrow y$ components:	$Nt - Wt = 0$	(1)
$\xrightarrow{+} x$ components:	$m\bar{v}_1 - Ft = m\bar{v}_2$	(2)
$+\downarrow$ moments about G:	$Ftr = \bar{I}\omega_2$	(3)

From (1) we obtain $N = W = mg$. During the entire time interval considered, sliding occurs at point C and we have $F = \mu_k N = \mu_k mg$. Substituting $\mu_k mg$ for F into (2), we write

$$m\bar{v}_1 - \mu_k mgt = m\bar{v}_2 \qquad \bar{v}_2 = \bar{v}_1 - \mu_k gt \tag{4}$$

Substituting $F = \mu_k mg$ and $\bar{I} = \frac{2}{5}mr^2$ into (3),

$$\mu_k mgtr = \tfrac{2}{5}mr^2\omega_2 \qquad \omega_2 = \frac{5}{2}\frac{\mu_k g}{r}t \tag{5}$$

The sphere will start rolling without sliding when the velocity $\mathbf{v}_C$ of the point of contact is zero. At that time, point C becomes the instantaneous center of rotation, and we have $\bar{v}_2 = r\omega_2$. Substituting from (4) and (5), we write

$$\bar{v}_2 = r\omega_2 \qquad \bar{v}_1 - \mu_k gt = r\left(\frac{5}{2}\frac{\mu_k g}{r}t\right) \qquad t = \frac{2}{7}\frac{\bar{v}_1}{\mu_k g} \quad \blacktriangleleft$$

Substituting this expression for t into (5),

$$\omega_2 = \frac{5}{2}\frac{\mu_k g}{r}\left(\frac{2}{7}\frac{\bar{v}_1}{\mu_k g}\right) \qquad \omega_2 = \frac{5}{7}\frac{\bar{v}_1}{r} \qquad \boldsymbol{\omega}_2 = \frac{5}{7}\frac{\bar{v}_1}{r}\downarrow \quad \blacktriangleleft$$

$$\bar{v}_2 = r\omega_2 \qquad \bar{v}_2 = r\left(\frac{5}{7}\frac{\bar{v}_1}{r}\right) \qquad \bar{\mathbf{v}}_2 = \tfrac{5}{7}\bar{v}_1 \rightarrow \quad \blacktriangleleft$$

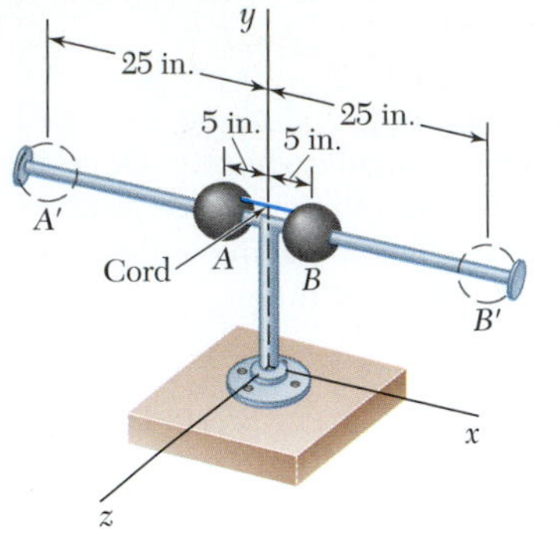

SAMPLE PROBLEM 17.8

Two solid spheres of radius 3 in., weighing 2 lb each, are mounted at A and B on the horizontal rod $A'B'$, which rotates freely about the vertical with a counterclockwise angular velocity of 6 rad/s. The spheres are held in position by a cord which is suddenly cut. Knowing that the centroidal moment of inertia of the rod and pivot is $\bar{I}_R = 0.25 \text{ lb} \cdot \text{ft} \cdot \text{s}^2$, determine (*a*) the angular velocity of the rod after the spheres have moved to positions A' and B', (*b*) the energy lost due to the plastic impact of the spheres and the stops at A' and B'.

SOLUTION

***a*. Principle of Impulse and Momentum.** In order to determine the final angular velocity of the rod, we will express that the initial momenta of the various parts of the system and the impulses of the external forces are together equipollent to the final momenta of the system.

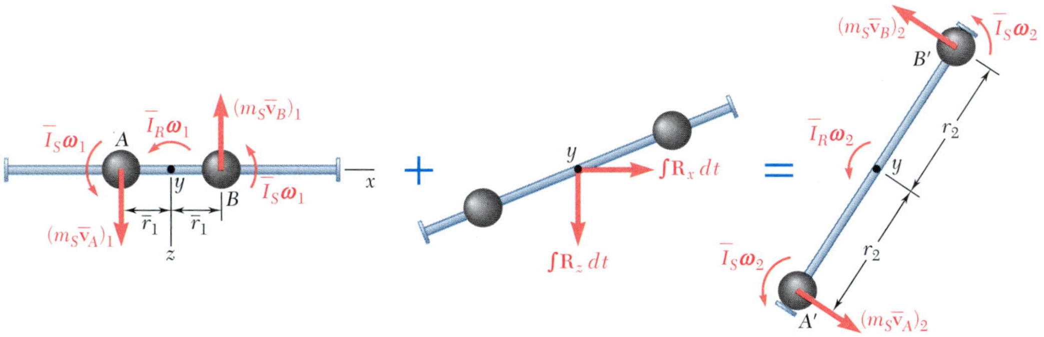

Syst Momenta$_1$ + Syst Ext Imp$_{1\to2}$ = Syst Momenta$_2$

Observing that the external forces consist of the weights and the reaction at the pivot, which have no moment about the y axis, and noting that $\bar{v}_A = \bar{v}_B = \bar{r}\omega$, we equate moments about the y axis:

$$2(m_S\bar{r}_1\omega_1)\bar{r}_1 + 2\bar{I}_S\omega_1 + \bar{I}_R\omega_1 = 2(m_S\bar{r}_2\omega_2)\bar{r}_2 + 2\bar{I}_S\omega_2 + \bar{I}_R\omega_2$$

$$(2m_S\bar{r}_1^2 + 2\bar{I}_S + \bar{I}_R)\omega_1 = (2m_S\bar{r}_2^2 + 2\bar{I}_S + \bar{I}_R)\omega_2 \tag{1}$$

which expresses that *the angular momentum of the system about the y axis is conserved*. We now compute

$$\bar{I}_S = \tfrac{2}{5}m_S a^2 = \tfrac{2}{5}(2 \text{ lb}/32.2 \text{ ft/s}^2)(\tfrac{3}{12} \text{ ft})^2 = 0.00155 \text{ lb} \cdot \text{ft} \cdot \text{s}^2$$

$$m_S\bar{r}_1^2 = (2/32.2)(\tfrac{5}{12})^2 = 0.0108 \qquad m_S\bar{r}_2^2 = (2/32.2)(\tfrac{25}{12})^2 = 0.2696$$

Substituting these values, and $\bar{I}_R = 0.25$ and $\omega_1 = 6$ rad/s into (1):

$$0.275(6 \text{ rad/s}) = 0.792\omega_2 \qquad \boldsymbol{\omega}_2 = 2.08 \text{ rad/s} \circlearrowleft \blacktriangleleft$$

***b*. Energy Lost.** The kinetic energy of the system at any instant is

$$T = 2(\tfrac{1}{2}m_S\bar{v}^2 + \tfrac{1}{2}\bar{I}_S\omega^2) + \tfrac{1}{2}\bar{I}_R\omega^2 = \tfrac{1}{2}(2m_S\bar{r}^2 + 2\bar{I}_S + \bar{I}_R)\omega^2$$

Recalling the numerical values found above, we have

$$T_1 = \tfrac{1}{2}(0.275)(6)^2 = 4.95 \text{ ft} \cdot \text{lb} \qquad T_2 = \tfrac{1}{2}(0.792)(2.08)^2 = 1.713 \text{ ft} \cdot \text{lb}$$

$$\Delta T = T_2 - T_1 = 1.71 - 4.95 \qquad \Delta T = -3.24 \text{ ft} \cdot \text{lb} \blacktriangleleft$$

SOLVING PROBLEMS ON YOUR OWN

In this lesson you learned to use the method of impulse and momentum to solve problems involving the plane motion of rigid bodies. As you found out previously in Chap. 13, this method is most effective when used in the solution of problems involving velocities and time.

1. *The principle of impulse and momentum for the plane motion of a rigid body* is expressed by the following vector equation:

$$\textbf{Syst Momenta}_1 + \textbf{Syst Ext Imp}_{1\rightarrow 2} = \textbf{Syst Momenta}_2 \qquad (17.14)$$

where **Syst Momenta** represents the system of the momenta of the particles forming the rigid body, and **Syst Ext Imp** represents the system of all the external impulses exerted during the motion.

a. The system of the momenta of a rigid body is equivalent to a linear momentum vector $m\overline{\mathbf{v}}$ attached at the mass center of the body and an angular momentum couple $\bar{I}\boldsymbol{\omega}$ (Fig. 17.7).

b. You should draw a free-body-diagram equation for the rigid body to express graphically the above vector equation. Your diagram equation will consist of three sketches of the body, representing respectively the initial momenta, the impulses of the external forces, and the final momenta. It will show that the system of the initial momenta and the system of the impulses of the external forces are together equivalent to the system of the final momenta (Fig. 17.8).

c. By using the free-body-diagram equation, you can sum components in any direction and sum moments about any point. When summing moments about a point, remember to include the *angular momentum* $\bar{I}\omega$ of the body, as well as the *moments* of the components of its *linear momentum*. In most cases you will be able to select and solve an equation that involves only one unknown. This was done in all the sample problems of this lesson.

2. *In problems involving a system of rigid bodies,* you can apply the principle of impulse and momentum to the system as a whole. Since internal forces occur in equal and opposite pairs, they will not be part of your solution [Sample Prob. 17.8].

3. *Conservation of angular momentum about a given axis* occurs when, for a system of rigid bodies, *the sum of the moments of the external impulses about that axis is zero.* You can indeed easily observe from the free-body-diagram equation that the initial and final angular momenta of the system about that axis are equal and, thus, that *the angular momentum of the system about the given axis is conserved.* You can then sum the angular momenta of the various bodies of the system and the moments of their linear momenta about that axis to obtain an equation which can be solved for one unknown [Sample Prob. 17.8].

Problems

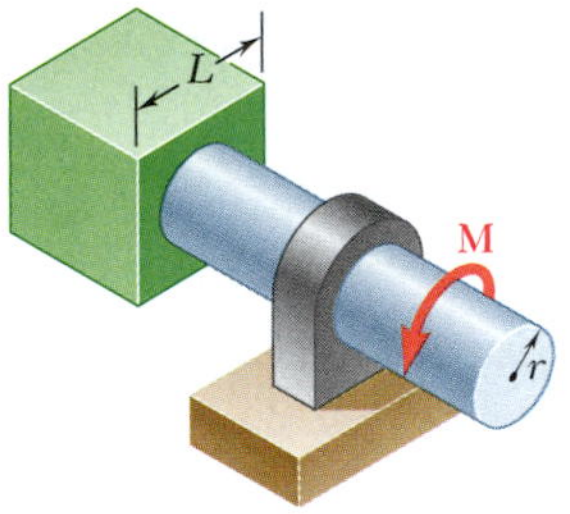

Fig. *P17.45* and P17.46

17.43 The rotor of an electric motor has a weight of 70 lb and a radius of gyration of 10 in. It is observed that 5.3 min is required for the rotor to coast to rest from an angular velocity of 3600 rpm. Determine the average magnitude of the couple due to kinetic friction in the bearings of the rotor.

17.44 A 1.815-Mg flywheel with a radius of gyration of 686 mm is allowed to coast to rest from an angular velocity of 450 rpm. Knowing that the kinetic friction produces a couple of magnitude 14.1 N · m, determine the time required for the flywheel to coast to rest.

17.45 A uniform 72-kg cube is attached to a uniform 68-kg circular shaft as shown and a couple **M** of constant magnitude is applied to the shaft when the system is at rest. Knowing that $r = 100$ mm, $L = 300$ mm, and the angular velocity of the system is 960 rpm after 4 s, determine the magnitude of the couple **M**.

17.46 A uniform 168-lb cube is attached to a uniform 156-lb circular shaft as shown and a couple **M** of constant magnitude 180 lb · in. is applied to the shaft. Knowing that $r = 4$ in. and $L = 12$ in., determine the time required for the angular velocity of the system to increase from 1000 rpm to 2000 rpm.

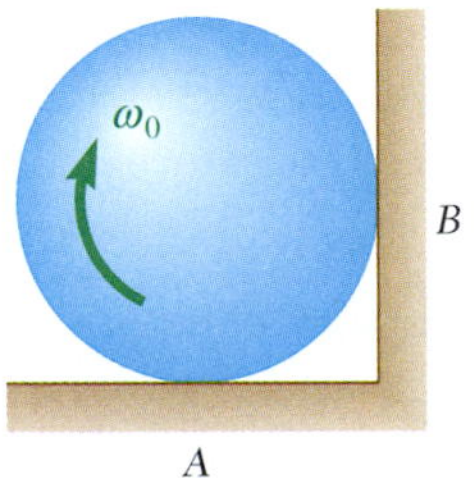

Fig. P17.47 and P17.48

17.47 A sphere of radius r and weight W with an initial clockwise angular velocity $\boldsymbol{\omega}_0$ is placed in the corner formed by the floor and a vertical wall. Denoting by μ_k the coefficient of kinetic friction at A and B, derive an expression for the time required for the sphere to come to rest.

17.48 A 6-lb sphere of radius $r = 5$ in. with an initial clockwise angular velocity $\boldsymbol{\omega}_0 = 90$ rad/s is placed in the corner formed by the floor and a vertical wall. Knowing that the coefficient of kinetic friction is 0.10 at A and B, determine the time required for the sphere to come to rest.

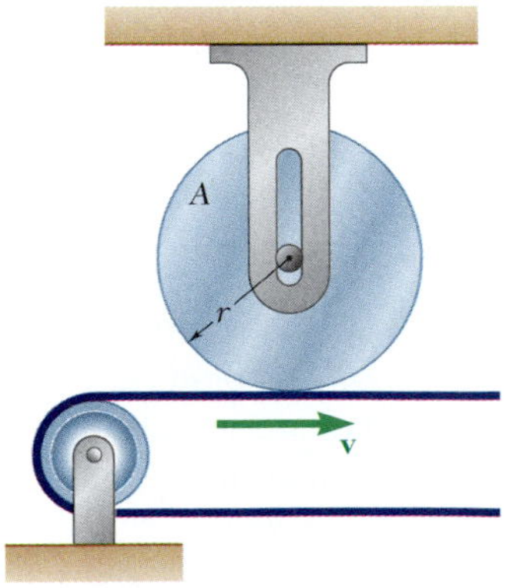

Fig. P17.49 and P17.50

17.49 A disk of constant thickness, initially at rest, is placed in contact with a belt that moves with a constant velocity **v**. Denoting by μ_k the coefficient of kinetic friction between the disk and the belt, derive an expression for the time required for the disk to reach a constant angular velocity.

17.50 Disk A, of mass 2 kg and radius $r = 60$ mm, is at rest when it is placed in contact with a belt which moves at a constant speed $v = 15$ m/s. Knowing that $\mu_k = 0.20$ between the disk and the belt, determine the time required for the disk to reach a constant angular velocity.

17.51 The 700-lb flywheel of a small hoisting engine has a radius of gyration of 24 in. If the power is cut off when the angular velocity of the flywheel is 100 rpm clockwise, determine the time required for the system to come to rest.

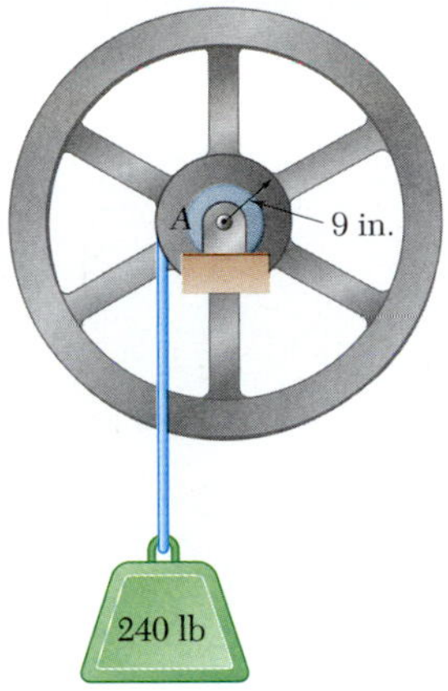

Fig. P17.51

17.52 In Prob. 17.51, determine the time required for the angular velocity of the flywheel to be reduced to 40 rpm clockwise.

17.53 Each of the gears A and B weighs 1.5 lb and has a radius of gyration of 1.6 in., while gear C weighs 8 lb and has a radius of gyration of 4 in. Assume that kinetic friction in the bearings of gears A, B, and C produces couples of constant magnitude 0.1 lb · ft, 0.1 lb · ft, and 0.2 lb · ft, respectively. Knowing that the initial angular velocity of gear C is 2000 rpm, determine the time required for the system to come to rest.

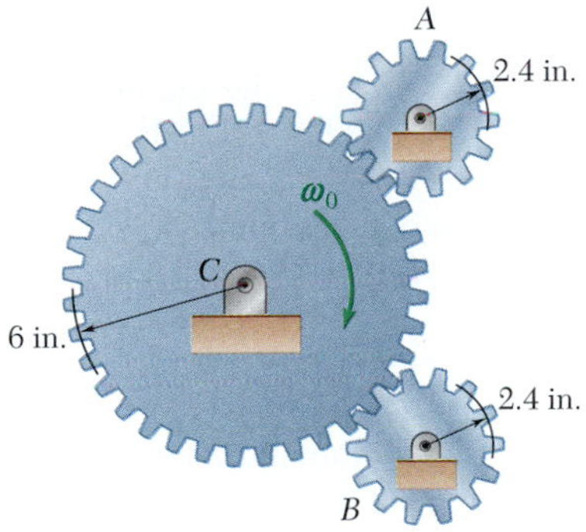

Fig. P17.53

17.54 Two identical uniform cylinders of mass m and radius r are at rest at time $t = 0$ when a couple **M** of constant magnitude $M < mgr$ is applied to cylinder A. Knowing that the coefficient of kinetic friction between cylinder B and the horizontal surface is $\mu_k < \dfrac{M}{2mgr}$ and that no slipping occurs between the two cylinders, derive an expression for the angular velocity of cylinder B at time t.

17.55 Two identical 8-kg uniform cylinders of radius $r = 100$ mm are at rest when a couple **M** of constant magnitude 5 N · m is applied to cylinder A. Slipping occurs between the two cylinders and between cylinder B and the horizontal surface. Knowing that the coefficient of kinetic friction is 0.5 between the two cylinders and 0.2 between cylinder B and the horizontal surface, determine the angular velocity of each cylinder after 5 s.

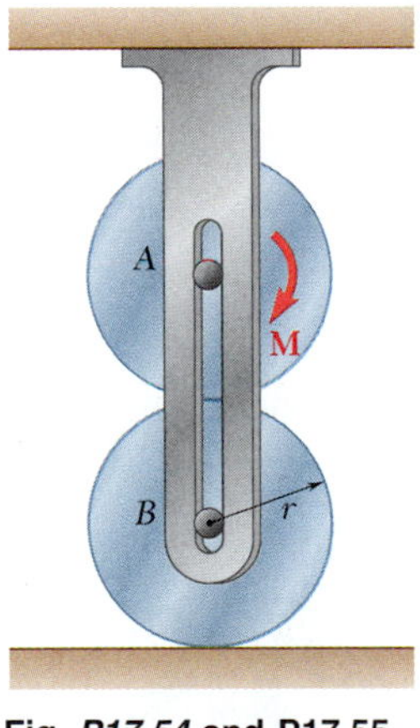

Fig. *P17.54* and P17.55

17.56 Show that the system of momenta for a rigid slab in plane motion reduces to a single vector, and express the distance from the mass center G to the line of action of this vector in terms of the centroidal radius of gyration $\bar{k}$ of the slab, the magnitude $\bar{v}$ of the velocity of G, and the angular velocity ω.

17.57 Show that, when a rigid slab rotates about a fixed axis through O perpendicular to the slab, the system of the momenta of its particles is equivalent to a single vector of magnitude $m\bar{r}\omega$, perpendicular to the line OG, and applied to a point P on this line, called the *center of percussion,* at a distance $GP = \bar{k}^2/\bar{r}$ from the mass center of the slab.

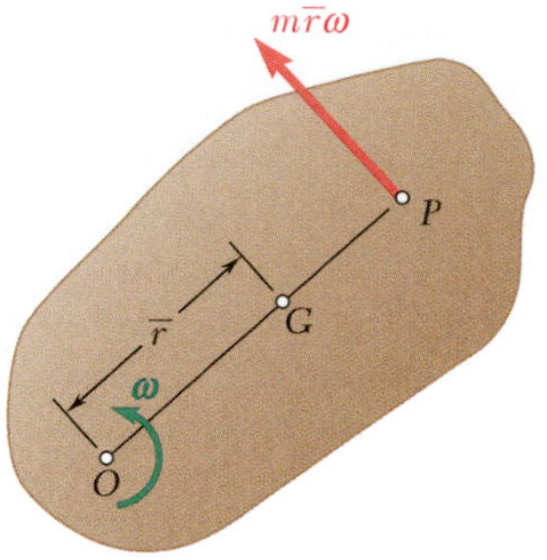

Fig. P17.57

17.58 Show that the sum $\mathbf{H}_A$ of the moments about a point A of the momenta of the particles of a rigid slab in plane motion is equal to $I_A\boldsymbol{\omega}$, where $\boldsymbol{\omega}$ is the angular velocity of the slab at the instant considered and I_A is the moment of inertia of the slab about A, if and only if one of the following conditions is satisfied: (*a*) A is the mass center of the slab, (*b*) A is the instantaneous center of rotation, (*c*) the velocity of A is directed along a line joining point A and the mass center G.

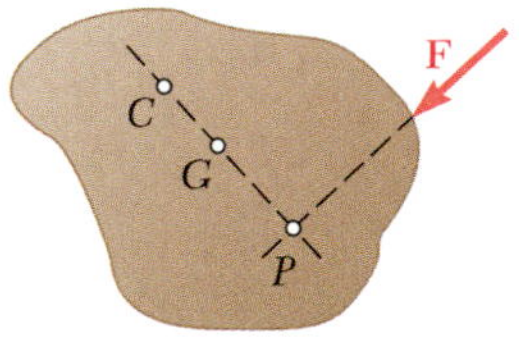

Fig. P17.59

17.59 Consider a rigid slab initially at rest and subjected to an impulsive force **F** contained in the plane of the slab. We define the *center of percussion* P as the point of intersection of the line of action of **F** with the perpendicular drawn from G. (*a*) Show that the instantaneous center of rotation C of the slab is located on line GP at a distance $GC = \bar{k}^2/GP$ on the opposite side of G. (*b*) Show that if the center of percussion were located at C the instantaneous center of rotation would be located at P.

17.60 A wheel of radius r and centroidal radius of gyration $\bar{k}$ is released from rest on the incline shown at time $t = 0$. Assuming that the wheel rolls without sliding, determine (*a*) the velocity of its center at time t, (*b*) the coefficient of static friction required to prevent slipping.

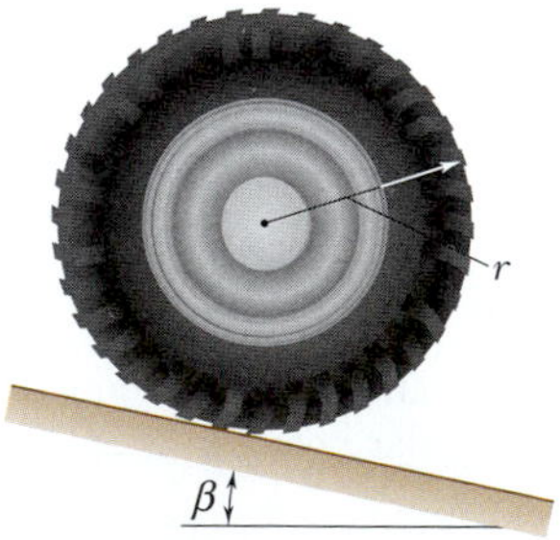

Fig. P17.60

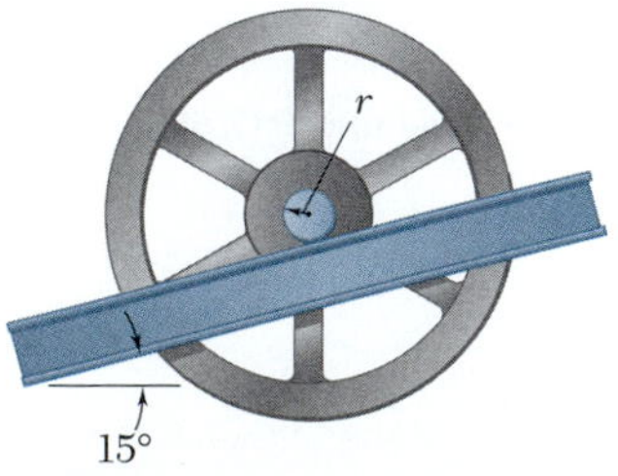

Fig. P17.61

17.61 A flywheel is rigidly attached to a 38-mm-radius shaft that rolls without sliding along parallel rails. Knowing that after being released from rest the system attains a speed of 152 mm/s in 30 s, determine the centroidal radius of gyration of the system.

17.62 A drum of 4-in. radius is attached to a disk of 8-in. radius. The disk and drum have a combined weight of 10 lb and a combined radius of gyration of 6 in. A cord is attached to the drum at *A* and pulled with a constant force **P** of magnitude 5 lb. Knowing that the disk rolls without sliding and that its initial angular velocity is 10 rad/s clockwise, determine (*a*) the angular velocity of the disk after 5 s, (*b*) the corresponding total impulse of the friction force exerted on the disk at *B*.

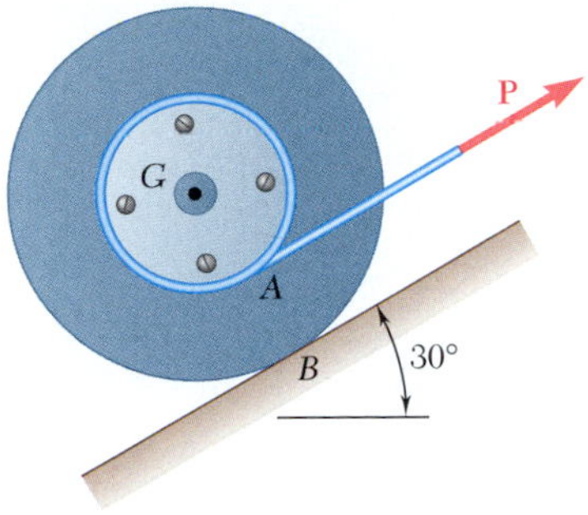

Fig. P17.62

17.63 The bar *AB* of negligible mass is attached by pins to two disks and the system is released from rest in the position shown. Disk *A* has a mass of 6 kg and a radius of gyration of 90 mm. Disk *B* has a mass of 3 kg and a radius of gyration of 80 mm. Assuming that disk *A* does not reach the corner *D*, determine (*a*) the velocity of the bar *AB* after 0.6 s, (*b*) the corresponding total impulse of the force exerted by disk *A* on the bar.

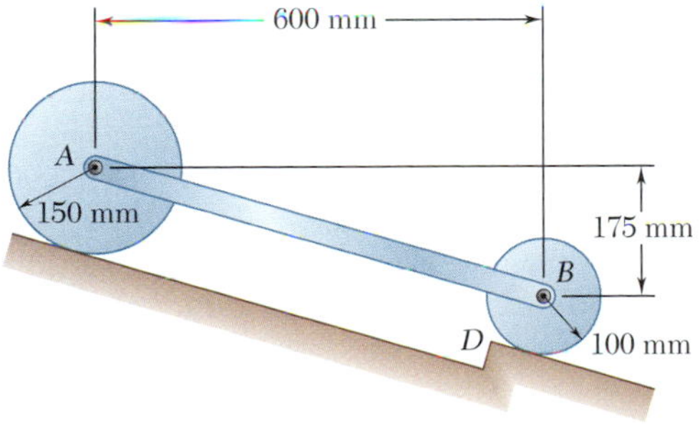

Fig. P17.63 and P17.64

17.64 The 5-kg uniform bar *AB* is attached by pins to two uniform disks and the system is released from rest in the position shown. The mass of disk *A* is 6 kg and that of disk *B* is 3 kg. Assuming that disk *A* does not reach the corner *D*, determine the velocity of bar *AB* after 0.6 s.

17.65* and *17.66 A 12-in.-radius cylinder of weight 16 lb rests on a 6-lb carriage. The system is at rest when a force **P** of magnitude 2.5 lb is applied as shown for 1.2 s. Knowing that the cylinder rolls without sliding on the carriage and neglecting the mass of the wheels of the carriage, determine the resulting velocity of (*a*) the carriage, (*b*) the center of the cylinder.

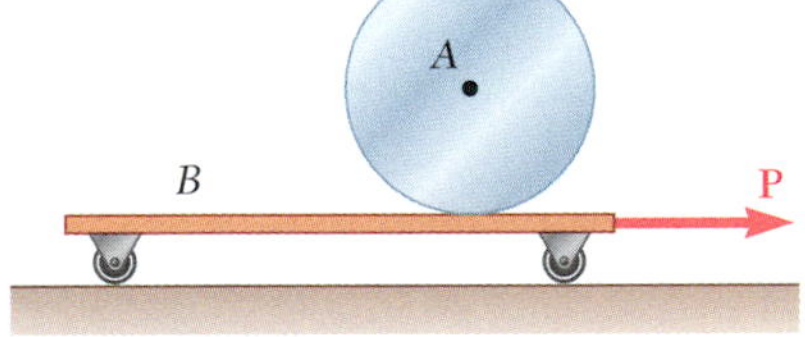

Fig. *P17.65*

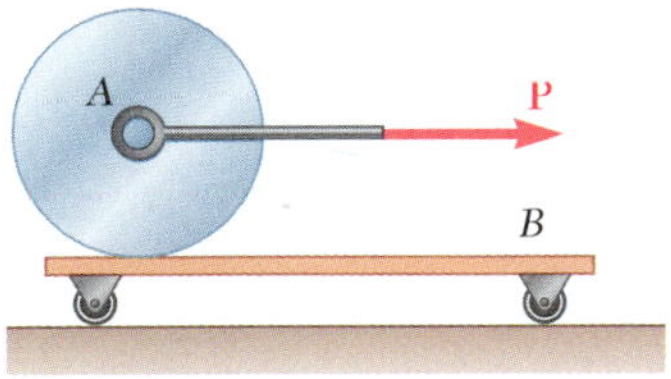

Fig. *P17.66*

17.67 A 30-lb double pulley has a radius of gyration of 5 in. and is attached to a 20-lb slider block by a pin at point G. The system is at rest when constant forces $\mathbf{P}_A$ and $\mathbf{P}_B$ are applied to the cords as shown. Knowing that after 2 s the velocity of point A is 11 ft/s to the left and the velocity of point B is 19 ft/s to the right, determine the magnitudes of the forces $\mathbf{P}_A$ and $\mathbf{P}_B$.

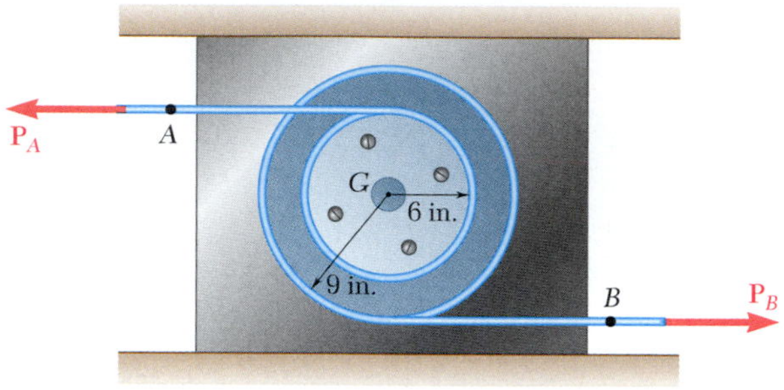

Fig. P17.67

Fig. *P17.68*

17.68 A sphere of radius r and mass m is placed on a horizontal floor with no linear velocity but with a clockwise angular velocity $\boldsymbol{\omega}_0$. Denoting by μ_k the coefficient of kinetic friction between the sphere and the floor, determine (*a*) the time t_1 at which the sphere will start rolling without sliding, (*b*) the linear and angular velocities of the sphere at time t_1.

Fig. P17.69

17.69 A sphere of radius r and mass m is projected along a rough horizontal surface with the initial velocities shown. If the final velocity of the sphere is to be zero, express (*a*) the required magnitude of $\boldsymbol{\omega}_0$ in terms of v_0 and r, (*b*) the time required for the sphere to come to rest in terms of v_0 and the kinetic coefficient of friction μ_k.

17.70 A 1.134-kg disk of radius 100 mm is attached to the yoke BCD by means of short shafts fitted in bearings at B and D. The 0.68-kg yoke has a radius of gyration of 75 mm about the x axis. Initially the assembly is rotating at 120 rpm with the disk in the plane of the yoke ($\theta = 0$). If the disk is slightly disturbed and rotates with respect to the yoke until $\theta = 90°$, where it is stopped by a small bar at D, determine the final angular velocity of the assembly.

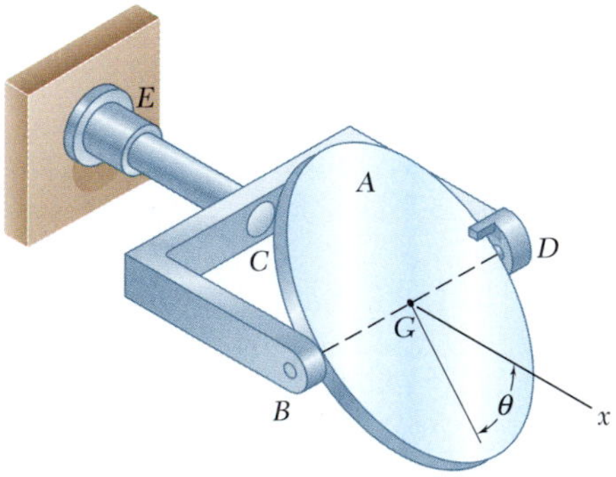

Fig. P17.70

17.71 A semicircular panel of radius r is attached with hinges to a circular plate of radius r and initially held in the vertical position as shown. The plate and the panel are made of the same material and have the same thickness. Knowing that the entire assembly is rotating freely with an initial angular velocity $\boldsymbol{\omega}_0$, determine the angular velocity of the assembly after the panel has been released and comes to rest against the plate.

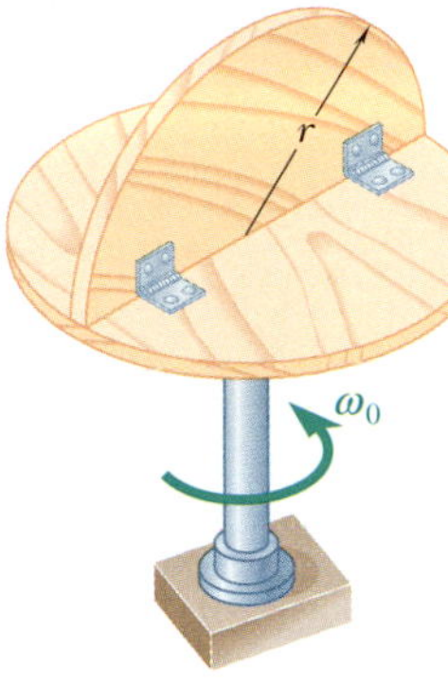

Fig. P17.71

17.72 Two 0.36-kg balls are put successively into the center C of the slender 1.8-kg tube AB. Knowing that when the first ball is put into the tube the initial angular velocity of the tube is 8 rad/s and neglecting the effect of friction, determine the angular velocity of the tube just after (a) the first ball has left the tube, (b) the second ball has left the tube.

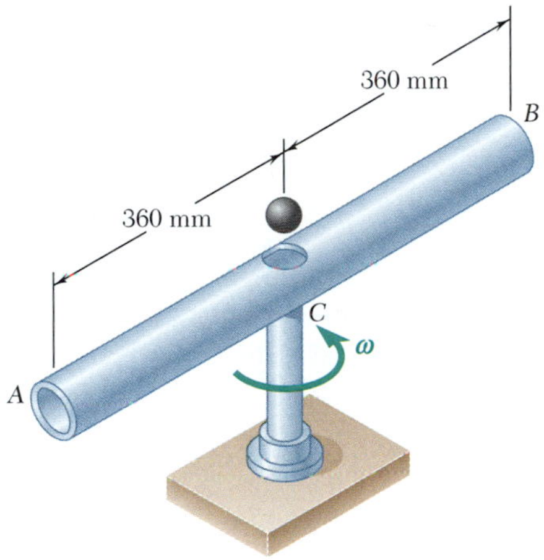

Fig. P17.72

17.73 Determine the final angular velocity of the tube of Prob. 17.72, assuming that a single ball of mass 0.72 kg is introduced into the tube.

17.74 The 60-lb uniform disk A and the bar BC are at rest and the 10-lb uniform disk D has an initial angular velocity $\boldsymbol{\omega}_1$ of magnitude 440 rpm when the compressed spring is released and disk D contacts disk A. The system rotates freely about the vertical spindle BE. After a period of slippage, disk D rolls without slipping. Knowing that the magnitude of the final angular velocity of disk D is 176 rpm, determine the final angular velocities of bar BC and disk A. Neglect the mass of bar BC.

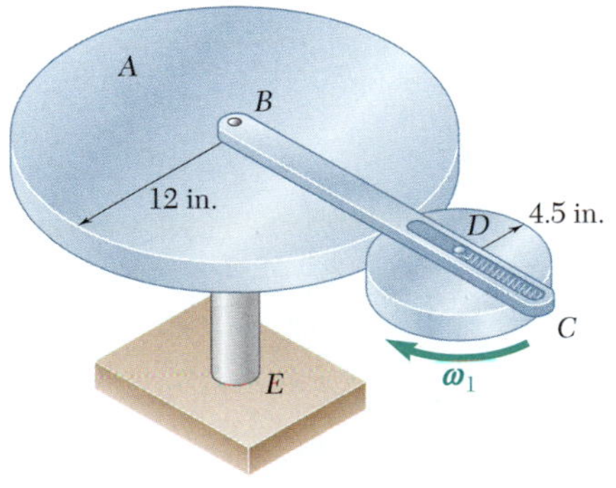

Fig. P17.74

17.75 In the helicopter shown, a vertical tail rotor is used to prevent rotation of the cab as the speed of the main blades is changed. Assuming that the tail rotor is not operating, determine the final angular velocity of the cab after the speed of the main blades has been changed from 180 to 240 rpm. (The speed of the main blades is measured relative to the cab, and the cab has a centroidal moment of inertia of 881 $kg \cdot m^2$. Each of the four main blades is assumed to be a slender 4.27-m rod of mass 24.9 kg.)

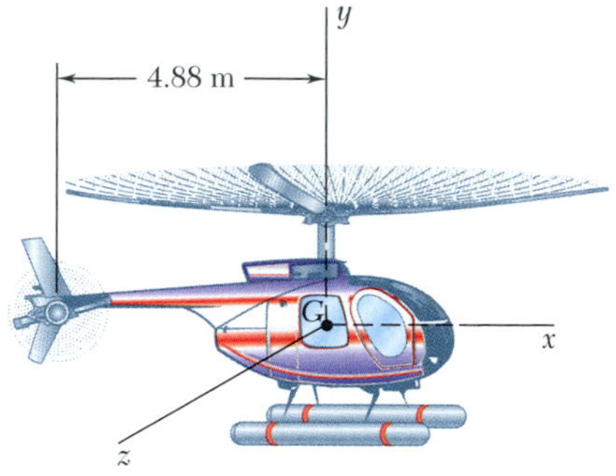

Fig. P17.75

17.76 Assuming that the tail rotor in Prob. 17.75 is operating and that the angular velocity of the cab remains zero, determine the final horizontal velocity of the cab when the speed of the main blades is changed from 180 to 240 rpm. The cab has a mass of 567 kg and is initially at rest. Also determine the force exerted by the tail rotor if the change in speed takes place uniformly in 12 s.

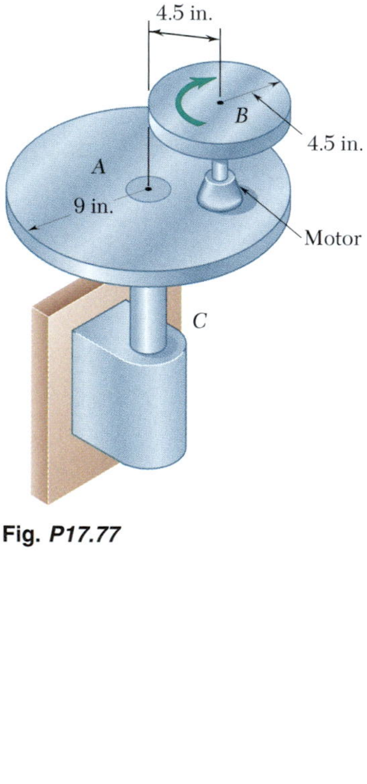

Fig. *P17.77*

17.77 The 8-lb disk B is attached to the shaft of a motor mounted on plate A, which can rotate freely about the vertical shaft C. The motor-plate-shaft unit has a moment of inertia of 0.14 $lb \cdot ft \cdot s^2$ with respect to the axis of the shaft. If the motor is started when the system is at rest, determine the angular velocities of the disk and of the plate after the motor has attained its normal operating speed of 360 rpm.

17.78 A small 3-kg collar C can slide freely on a thin ring of mass 4.5 kg and radius 325 mm. The ring is welded to a short vertical shaft, which can rotate freely in a fixed bearing. Initially the ring has an angular velocity of 35 rad/s and the collar is at the top of the ring ($\theta = 0$) when it is given a slight nudge. Neglecting the effect of friction, determine (a) the angular velocity of the ring as the collar passes through the position $\theta = 90°$, (b) the corresponding velocity of the collar relative to the ring.

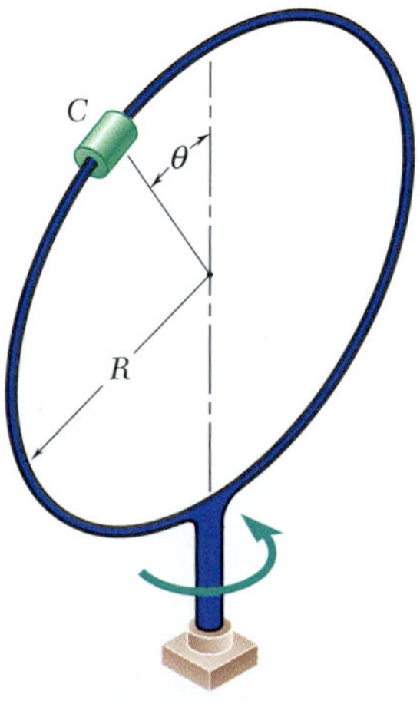

Fig. *P17.78*

17.79 Collar C has a weight of 18 lb and can slide freely on rod AB, which in turn can rotate freely in a horizontal plane. The assembly is rotating with an angular velocity $\boldsymbol{\omega}$ of 1.5 rad/s when a spring located between A and C is released, projecting the collar along the rod with an initial relative speed of $v_r = 4.5$ ft/s. Knowing that the combined mass moment of inertia about B of the rod and spring is $0.84 \text{ lb} \cdot \text{ft} \cdot \text{s}^2$, determine (*a*) the minimum distance between the collar and point B in the ensuing motion, (*b*) the corresponding angular velocity of the assembly.

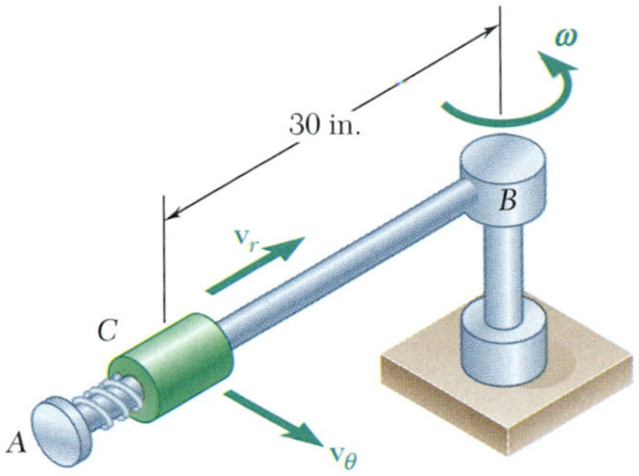

Fig. P17.79

17.80 In Prob. 17.79, determine the required magnitude of the initial relative speed v_r if during the ensuing motion the minimum distance between collar C and point B is to be 15 in.

17.81 A 1.8-kg collar A and a 0.7-kg collar B can slide without friction on a frame, consisting of the horizontal rod OE and the vertical rod CD, which is free to rotate about its vertical axis of symmetry. The two collars are connected by a cord running over a pulley that is attached to the frame at O. At the instant shown, the velocity $\mathbf{v}_A$ of collar A has a magnitude of 2.1 m/s and a stop prevents collar B from moving. The stop is suddenly removed and collar A moves toward E. As it reaches a distance of 0.12 m from O, the magnitude of its velocity is observed to be 2.5 m/s. Determine at that instant the magnitude of the angular velocity of the frame and the moment of inertia of the frame and pulley system about CD.

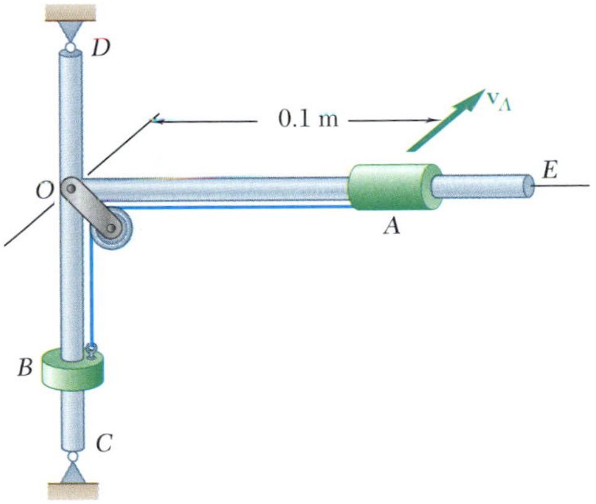

Fig. P17.81

17.82 In Prob. 17.72, determine the velocity of each ball relative to the tube as it leaves the tube.

17.83 In Prob. 17.73, determine the velocity of the ball relative to the tube as it leaves the tube.

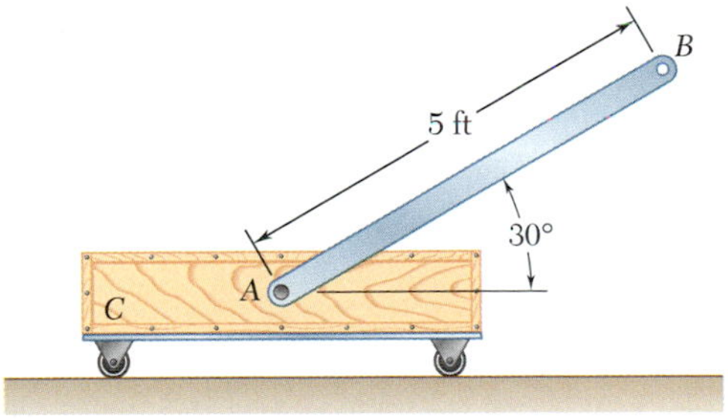

Fig. P17.84

17.84 Rod AB has a weight of 6 lb and is attached to a 10-lb cart C. Knowing that the system is released from rest in the position shown and neglecting friction, determine (*a*) the velocity of point B as rod AB passes through a vertical position, (*b*) the corresponding velocity of the cart C.

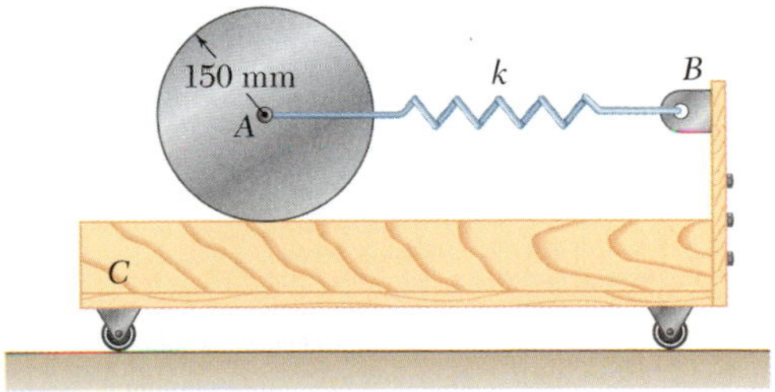

Fig. P17.85

17.85 A 3-kg uniform cylinder A can roll without sliding on a 5-kg cart C and is attached to a spring AB of constant $k = 100$ N/m as shown. The system is released from rest when the spring is stretched 20 mm. Neglecting wheel friction, determine the velocity of the cart and the angular velocity of the cylinder when the spring first reaches its undeformed state.

17.11. IMPULSIVE MOTION

You saw in Chap. 13 that the method of impulse and momentum is the only practicable method for the solution of problems involving the impulsive motion of a particle. Now you will find that problems involving the impulsive motion of a rigid body are particularly well suited to a solution by the method of impulse and momentum. Since the time interval considered in the computation of linear impulses and angular impulses is very short, the bodies involved can be assumed to occupy the same position during that time interval, making the computation quite simple.

17.12. ECCENTRIC IMPACT

In Secs. 13.13 and 13.14, you learned to solve problems of *central impact,* that is, problems in which the mass centers of the two colliding bodies are located on the line of impact. You will now analyze the *eccentric impact* of two rigid bodies. Consider two bodies which collide, and denote by $\mathbf{v}_A$ and $\mathbf{v}_B$ the velocities before impact *of the two points of contact* A and B (Fig. 17.10*a*). Under the impact, the two bodies

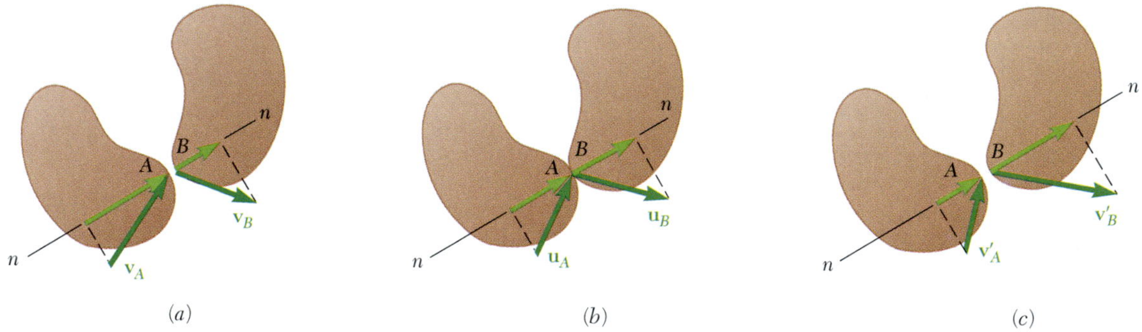

Fig. 17.10

will *deform,* and at the end of the period of deformation, the velocities $\mathbf{u}_A$ and $\mathbf{u}_B$ of A and B will have equal components along the line of impact nn (Fig. 17.10*b*). A period of *restitution* will then take place, at the end of which A and B will have velocities $\mathbf{v}'_A$ and $\mathbf{v}'_B$ (Fig. 17.10*c*). Assuming that the bodies are frictionless, we find that the forces they exert on each other are directed along the line of impact. Denoting the magnitude of the impulse of one of these forces during the period of deformation by $\int P\,dt$ and the magnitude of its impulse during the period of restitution by $\int R\,dt$, we recall that the coefficient of restitution e is defined as the ratio

$$e = \frac{\int R\,dt}{\int P\,dt} \tag{17.18}$$

We propose to show that the relation established in Sec. 13.13 between the relative velocities of two particles before and after impact also holds between the components along the line of impact of the relative velocities of the two points of contact A and B. We propose to show, therefore, that

$$(v'_B)_n - (v'_A)_n = e[(v_A)_n - (v_B)_n] \tag{17.19}$$

It will first be assumed that the motion of each of the two colliding bodies of Fig. 17.10 is unconstrained. Thus the only impulsive forces exerted on the bodies during the impact are applied at A and B, respectively. Consider the body to which point A belongs and draw the three momentum and impulse diagrams corresponding to the period of deformation (Fig. 17.11). We denote by $\overline{\mathbf{v}}$ and $\overline{\mathbf{u}}$, respectively,

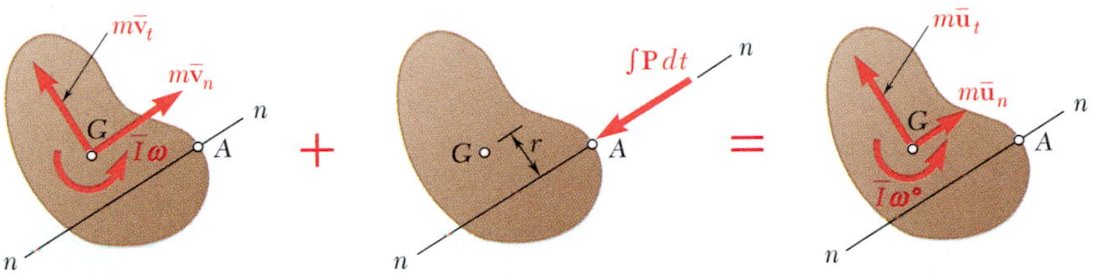

Fig. 17.11

the velocity of the mass center at the beginning and at the end of the period of deformation, and we denote by $\boldsymbol{\omega}$ and $\boldsymbol{\omega}^*$ the angular velocity of the body at the same instants. Summing and equating the components of the momenta and impulses along the line of impact nn, we write

$$m\bar{v}_n - \int P\,dt = m\bar{u}_n \tag{17.20}$$

Summing and equating the moments about G of the momenta and impulses, we also write

$$\bar{I}\omega - r\int P\,dt = \bar{I}\omega^* \tag{17.21}$$

where r represents the perpendicular distance from G to the line of impact. Considering now the period of restitution, we obtain in a similar way

$$m\bar{u}_n - \int R\,dt = m\bar{v}'_n \tag{17.22}$$
$$\bar{I}\omega^* - r\int R\,dt = \bar{I}\omega' \tag{17.23}$$

where $\overline{\mathbf{v}}'$ and $\boldsymbol{\omega}'$ represent, respectively, the velocity of the mass center and the angular velocity of the body after impact. Solving (17.20) and (17.22) for the two impulses and substituting into (17.18), and then solving (17.21) and (17.23) for the same two impulses and substituting again into (17.18), we obtain the following two alternative expressions for the coefficient of restitution:

$$e = \frac{\bar{u}_n - \bar{v}'_n}{\bar{v}_n - \bar{u}_n} \qquad e = \frac{\omega^* - \omega'}{\omega - \omega^*} \tag{17.24}$$

Multiplying by r the numerator and denominator of the second expression obtained for e, and adding respectively to the numerator and denominator of the first expression, we have

$$e = \frac{\overline{u}_n + r\omega^* - (\overline{v}'_n + r\omega')}{\overline{v}_n + r\omega - (\overline{u}_n + r\omega^*)} \tag{17.25}$$

Observing that $\overline{v}_n + r\omega$ represents the component $(v_A)_n$ along nn of the velocity of the point of contact A and that, similarly, $\overline{u}_n + r\omega^*$ and $\overline{v}'_n + r\omega'$ represent, respectively, the components $(u_A)_n$ and $(v'_A)_n$, we write

$$e = \frac{(u_A)_n - (v'_A)_n}{(v_A)_n - (u_A)_n} \tag{17.26}$$

The analysis of the motion of the second body leads to a similar expression for e in terms of the components along nn of the successive velocities of point B. Recalling that $(u_A)_n = (u_B)_n$, and eliminating these two velocity components by a manipulation similar to the one used in Sec. 13.13, we obtain relation (17.19).

If one or both of the colliding bodies is constrained to rotate about a fixed point O, as in the case of a compound pendulum (Fig. 17.12*a*), an impulsive reaction will be exerted at O (Fig. 17.12*b*). Let us verify that while their derivation must be modified, Eqs. (17.26) and (17.19) remain valid. Applying formula (17.16) to the period of deformation and to the period of restitution, we write

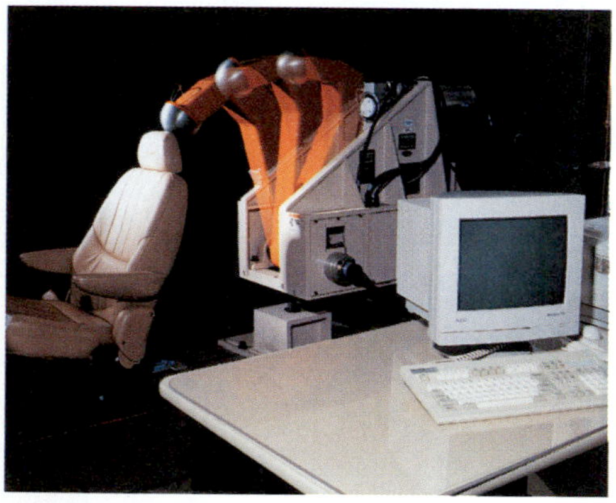
Photo 17.3 The method of impulse and momentum is used in the analysis of data from a pendulum impact test. The system shown was developed to test for compliance with safety standards for seat backs, instrument panels, and other vehicle interior components.

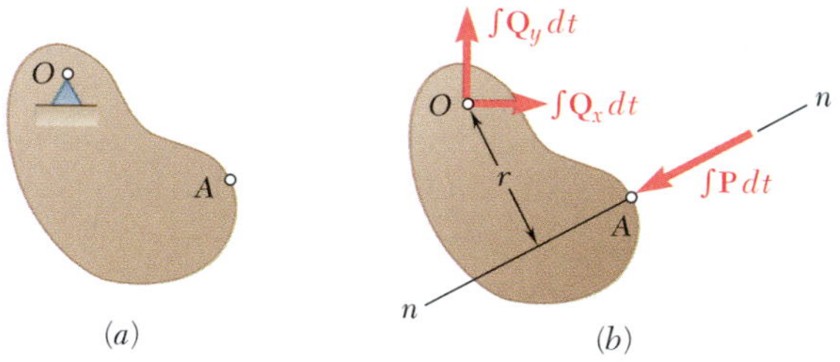

Fig. 17.12

$$I_O\omega - r\int P\,dt = I_O\omega^* \tag{17.27}$$

$$I_O\omega^* - r\int R\,dt = I_O\omega' \tag{17.28}$$

where r represents the perpendicular distance from the fixed point O to the line of impact. Solving (17.27) and (17.28) for the two impulses and substituting into (17.18), and then observing that $r\omega$, $r\omega^*$, and $r\omega'$ represent the components along nn of the successive velocities of point A, we write

$$e = \frac{\omega^* - \omega'}{\omega - \omega^*} = \frac{r\omega^* - r\omega'}{r\omega - r\omega^*} = \frac{(u_A)_n - (v'_A)_n}{(v_A)_n - (u_A)_n}$$

and check that Eq. (17.26) still holds. Thus Eq. (17.19) remains valid when one or both of the colliding bodies is constrained to rotate about a fixed point O.

In order to determine the velocities of the two colliding bodies after impact, relation (17.19) should be used in conjunction with one or several other equations obtained by applying the principle of impulse and momentum (Sample Prob. 17.10).

SAMPLE PROBLEM 17.9

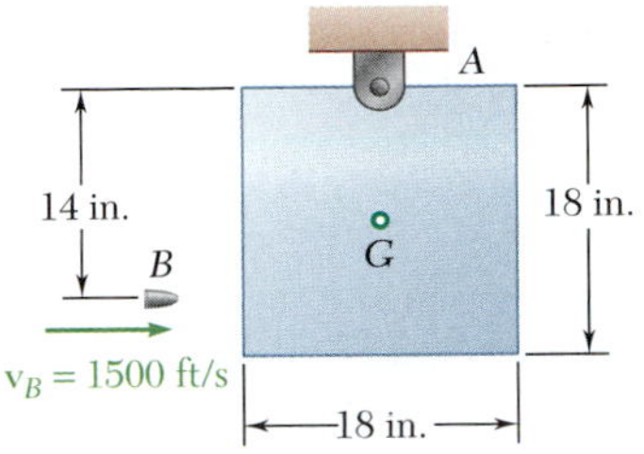

A 0.05-lb bullet B is fired with a horizontal velocity of 1500 ft/s into the side of a 20-lb square panel suspended from a hinge at A. Knowing that the panel is initially at rest, determine (*a*) the angular velocity of the panel immediately after the bullet becomes embedded, (*b*) the impulsive reaction at A, assuming that the bullet becomes embedded in 0.0006 s.

SOLUTION

Principle of Impulse and Momentum. We consider the bullet and the panel as a single system and express that the initial momenta of the bullet and panel and the impulses of the external forces are together equipollent to the final momenta of the system. Since the time interval $\Delta t = 0.0006$ s is very short, we neglect all nonimpulsive forces and consider only the external impulses $\mathbf{A}_x\,\Delta t$ and $\mathbf{A}_y\,\Delta t$.

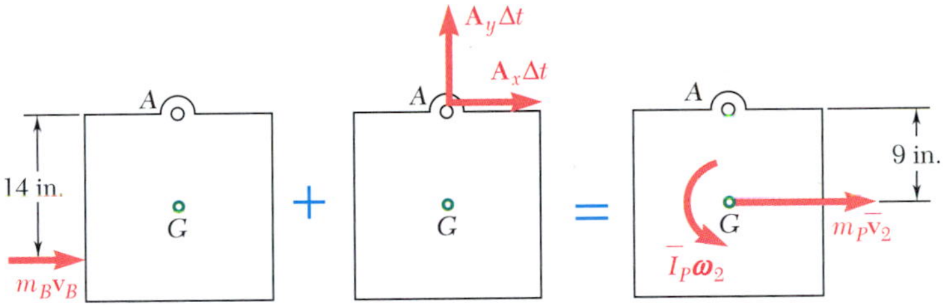

Syst Momenta$_1$ + Syst Ext Imp$_{1\to2}$ = Syst Momenta$_2$

+↺moments about A: $\quad m_Bv_B(\frac{14}{12}\text{ ft}) + 0 = m_P\bar{v}_2(\frac{9}{12}\text{ ft}) + \bar{I}_P\omega_2 \qquad (1)$

$\xrightarrow{+}x$ components: $\quad m_Bv_B + A_x\,\Delta t = m_P\bar{v}_2 \qquad (2)$

$+\uparrow y$ components: $\quad 0 + A_y\,\Delta t = 0 \qquad (3)$

The centroidal mass moment of inertia of the square panel is

$$\bar{I}_P = \tfrac{1}{6}m_Pb^2 = \frac{1}{6}\left(\frac{20\text{ lb}}{32.2}\right)(\tfrac{18}{12}\text{ ft})^2 = 0.2329\text{ lb}\cdot\text{ft}\cdot\text{s}^2$$

Substituting this value as well as the given data into (1) and noting that

$$\bar{v}_2 = (\tfrac{9}{12}\text{ ft})\omega_2$$

we write

$$\left(\frac{0.05}{32.2}\right)(1500)(\tfrac{14}{12}) = 0.2329\omega_2 + \left(\frac{20}{32.2}\right)(\tfrac{9}{12}\omega_2)(\tfrac{9}{12})$$

$$\omega_2 = 4.67\text{ rad/s} \qquad \boldsymbol{\omega}_2 = 4.67\text{ rad/s}\circlearrowleft \blacktriangleleft$$

$$\bar{v}_2 = (\tfrac{9}{12}\text{ ft})\omega_2 = (\tfrac{9}{12}\text{ ft})(4.67\text{ rad/s}) = 3.50\text{ ft/s}$$

Substituting $\bar{v}_2 = 3.50$ ft/s, $\Delta t = 0.0006$ s, and the given data into Eq. (2), we have

$$\left(\frac{0.05}{32.2}\right)(1500) + A_x(0.0006) = \left(\frac{20}{32.2}\right)(3.50)$$

$$A_x = -259\text{ lb} \qquad \mathbf{A}_x = 259\text{ lb}\leftarrow \blacktriangleleft$$

From Eq. (3), we find $\quad A_y = 0 \qquad \mathbf{A}_y = 0 \blacktriangleleft$

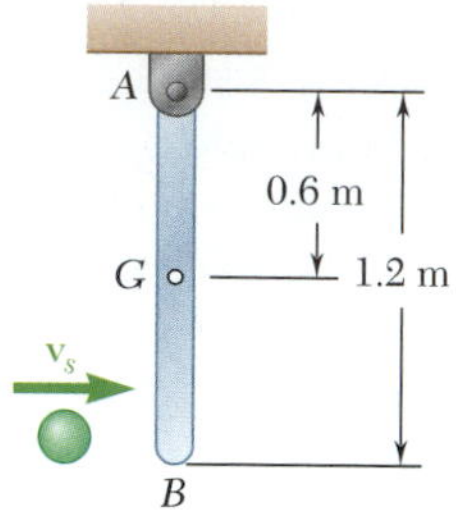

SAMPLE PROBLEM 17.10

A 2-kg sphere moving horizontally to the right with an initial velocity of 5 m/s strikes the lower end of an 8-kg rigid rod AB. The rod is suspended from a hinge at A and is initially at rest. Knowing that the coefficient of restitution between the rod and the sphere is 0.80, determine the angular velocity of the rod and the velocity of the sphere immediately after the impact.

SOLUTION

Principle of Impulse and Momentum. We consider the rod and sphere as a single system and express that the initial momenta of the rod and sphere and the impulses of the external forces are together equipollent to the final momenta of the system. We note that the only impulsive force external to the system is the impulsive reaction at A.

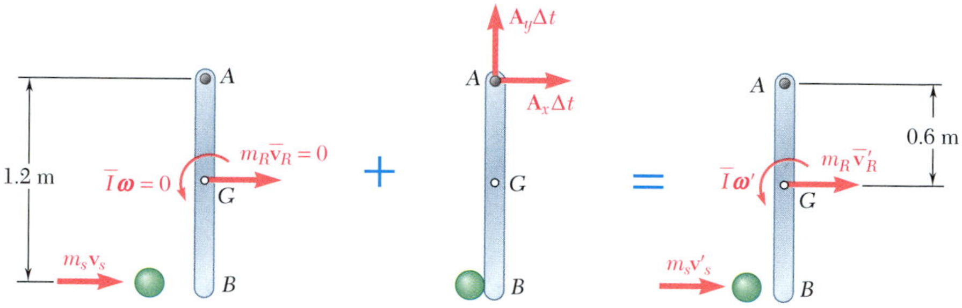

Syst Momenta$_1$ + Syst Ext Imp$_{1\rightarrow2}$ = Syst Momenta$_2$

+↺moments about A:

$$m_s v_s(1.2\text{ m}) = m_s v'_s(1.2\text{ m}) + m_R\bar{v}'_R\,(0.6\text{ m}) + \bar{I}\omega' \tag{1}$$

Since the rod rotates about A, we have $\bar{v}'_R = \bar{r}\omega' = (0.6\text{ m})\omega'$. Also,

$$\bar{I} = \tfrac{1}{12}mL^2 = \tfrac{1}{12}(8\text{ kg})(1.2\text{ m})^2 = 0.96\text{ kg}\cdot\text{m}^2$$

Substituting these values and the given data into Eq. (1), we have

$$(2\text{ kg})(5\text{ m/s})(1.2\text{ m}) = (2\text{ kg})v'_s(1.2\text{ m}) + (8\text{ kg})(0.6\text{ m})\omega'(0.6\text{ m}) + (0.96\text{ kg}\cdot\text{m}^2)\omega'$$

$$12 = 2.4v'_s + 3.84\omega' \tag{2}$$

Relative Velocities. Choosing positive to the right, we write

$$v'_B - v'_s = e(v_s - v_B)$$

Substituting $v_s = 5$ m/s, $v_B = 0$, and $e = 0.80$, we obtain

$$v'_B - v'_s = 0.80(5\text{ m/s}) \tag{3}$$

Again noting that the rod rotates about A, we write

$$v'_B = (1.2\text{ m})\omega' \tag{4}$$

Solving Eqs. (2) to (4) simultaneously, we obtain

$\omega' = 3.21$ rad/s $\boldsymbol{\omega}' = 3.21$ rad/s ↺ ◀

$v'_s = -0.143$ m/s $\mathbf{v}'_s = 0.143$ m/s ← ◀

SAMPLE PROBLEM 17.11

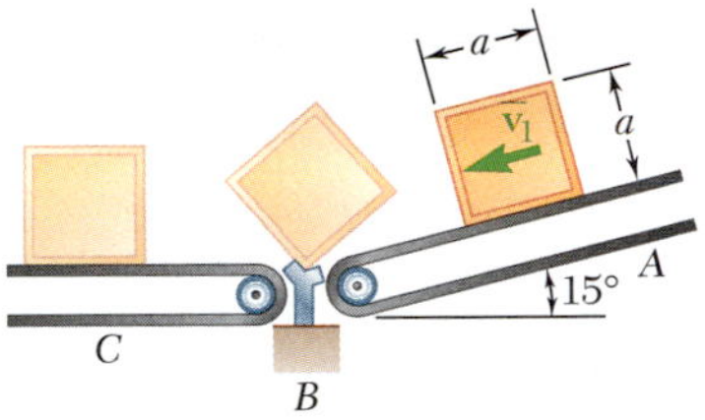

A square package of side a and mass m moves down a conveyor belt A with a constant velocity $\bar{\mathbf{v}}_1$. At the end of the conveyor belt, the corner of the package strikes a rigid support at B. Assuming that the impact at B is perfectly plastic, derive an expression for the smallest magnitude of the velocity $\bar{\mathbf{v}}_1$ for which the package will rotate about B and reach conveyor belt C.

SOLUTION

Principle of Impulse and Momentum. Since the impact between the package and the support is perfectly plastic, the package rotates about B during the impact. We apply the principle of impulse and momentum to the package and note that the only impulsive force external to the package is the impulsive reaction at B.

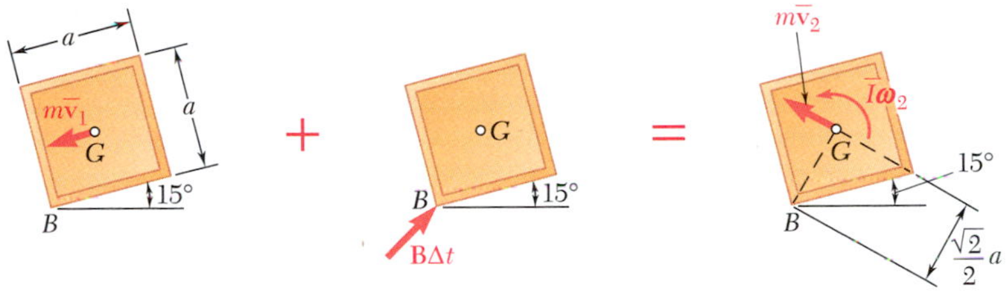

Syst Momenta$_1$ + Syst Ext Imp$_{1\to2}$ = Syst Momenta$_2$

$+\circlearrowleft$moments about B: $\quad (m\bar{v}_1)(\frac{1}{2}a) + 0 = (m\bar{v}_2)(\frac{1}{2}\sqrt{2}a) + \bar{I}\omega_2 \qquad (1)$

Since the package rotates about B, we have $\bar{v}_2 = (GB)\omega_2 = \frac{1}{2}\sqrt{2}a\omega_2$. We substitute this expression, together with $\bar{I} = \frac{1}{6}ma^2$, into Eq. (1):

$$(m\bar{v}_1)(\tfrac{1}{2}a) = m(\tfrac{1}{2}\sqrt{2}a\omega_2)(\tfrac{1}{2}\sqrt{2}a) + \tfrac{1}{6}ma^2\omega_2 \qquad \bar{v}_1 = \tfrac{4}{3}a\omega_2 \qquad (2)$$

Principle of Conservation of Energy. We apply the principle of conservation of energy between position 2 and position 3.

Position 2

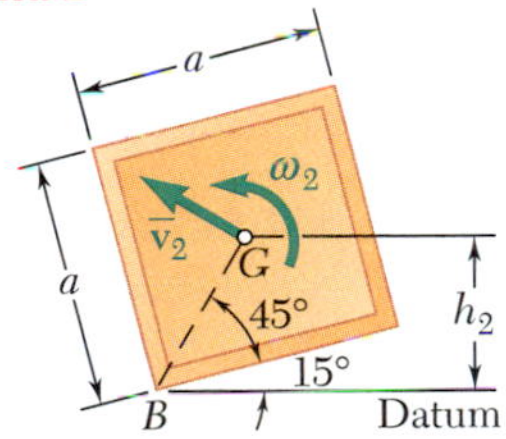

$GB = \frac{1}{2}\sqrt{2}a = 0.707a$

$h_2 = GB \sin(45° + 15°)$

$= 0.612a$

Position 2. $V_2 = Wh_2$. Recalling that $\bar{v}_2 = \frac{1}{2}\sqrt{2}a\omega_2$, we write

$$T_2 = \tfrac{1}{2}m\bar{v}_2^2 + \tfrac{1}{2}\bar{I}\omega_2^2 = \tfrac{1}{2}m(\tfrac{1}{2}\sqrt{2}a\omega_2)^2 + \tfrac{1}{2}(\tfrac{1}{6}ma^2)\omega_2^2 = \tfrac{1}{3}ma^2\omega_2^2$$

Position 3. Since the package must reach conveyor belt C, it must pass through position 3 where G is directly above B. Also, since we wish to determine the smallest velocity for which the package will reach this position, we choose $\bar{v}_3 = \omega_3 = 0$. Therefore $T_3 = 0$ and $V_3 = Wh_3$.

Position 3

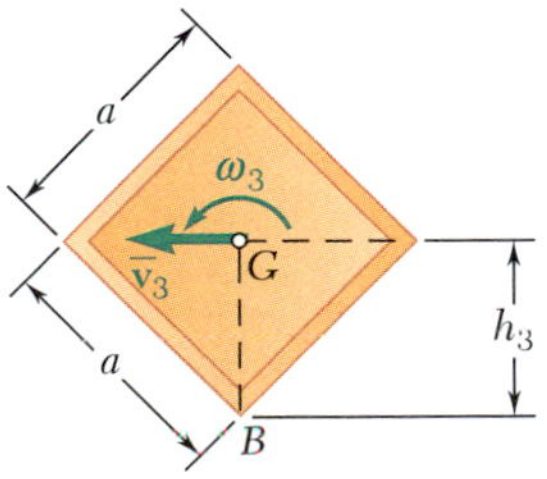

$h_3 = GB = 0.707a$

Conservation of Energy

$$T_2 + V_2 = T_3 + V_3$$
$$\tfrac{1}{3}ma^2\omega_2^2 + Wh_2 = 0 + Wh_3$$
$$\omega_2^2 = \frac{3W}{ma^2}(h_3 - h_2) = \frac{3g}{a^2}(h_3 - h_2) \qquad (3)$$

Substituting the computed values of h_2 and h_3 into Eq. (3), we obtain

$$\omega_2^2 = \frac{3g}{a^2}(0.707a - 0.612a) = \frac{3g}{a^2}(0.095a) \qquad \omega_2 = \sqrt{0.285\, g/a}$$

$$\bar{v}_1 = \tfrac{4}{3}a\omega_2 = \tfrac{4}{3}a\sqrt{0.285g/a} \qquad \bar{v}_1 = 0.712\sqrt{ga} \blacktriangleleft$$

SOLVING PROBLEMS ON YOUR OWN

This lesson was devoted to the *impulsive motion* and to the *eccentric impact of rigid bodies.*

1. *Impulsive motion* occurs when a rigid body is subjected to a very large force **F** for a very short interval of time Δt; the resulting impulse $\mathbf{F}\,\Delta t$ is both finite and different from zero. Such forces are referred to as *impulsive forces* and are encountered whenever there is an impact between two rigid bodies. Forces for which the impulse is zero are referred to as *nonimpulsive forces.* As you saw in Chap. 13, the following forces can be assumed to be nonimpulsive: the *weight* of a body, the force exerted by a *spring,* and any other force which is *known* to be small by comparison with the impulsive forces. Unknown reactions, however, *cannot be assumed* to be nonimpulsive.

2. *Eccentric impact of rigid bodies.* You saw that when two bodies collide, the velocity components along the line of impact of the *points of contact A and B* before and after impact satisfy the following equation:

$$(v'_B)_n - (v'_A)_n = e[(v_A)_n - (v_B)_n] \qquad (17.19)$$

where the left-hand member is the *relative velocity after the impact,* and the right-hand member is the product of the coefficient of restitution and the *relative velocity before the impact.*

This equation expresses the same relation between the velocity components of the points of contact before and after an impact that you used for particles in Chap. 13.

3. *To solve a problem involving an impact* you should use the *method of impulse and momentum* and take the following steps.

a. Draw a free-body-diagram equation of the body that will express that the system consisting of the momenta immediately before impact and of the impulses of the external forces is equivalent to the system of the momenta immediately after impact.

b. The free-body-diagram equation will relate the velocities before and after impact and the impulsive forces and reactions. In some cases, you will be able to determine the unknown velocities and impulsive reactions by solving equations obtained by summing components and moments [Sample Prob. 17.9].

c. In the case of an impact in which $e > 0$, the number of unknowns will be greater than the number of equations that you can write by summing components and moments, and you should supplement the equations obtained from the free-body-diagram equation with Eq. (17.19), which relates the relative velocities of the points of contact before and after impact [Sample Prob. 17.10].

d. During an impact you must use the method of impulse and momentum. However, *before and after the impact* you can, if necessary, use some of the other methods of solution that you have learned, such as the method of work and energy [Sample Prob. 17.11].

Problems

17.86 A 65-lb uniform circular plate of radius r is supported by a ball-and-socket joint at point A and is at rest in the vertical xy plane when a bullet weighing 0.03 lb is fired with the velocity $\mathbf{v}_0 = -(700 \text{ ft/s})\mathbf{k}$ and hits the plate at point C. Knowing that $r = 16$ in. and $h = 28$ in., determine (*a*) the angular velocity of the plate immediately after the bullet becomes embedded, (*b*) the impulsive reaction at point A, assuming that the bullet becomes embedded in 1.1 ms.

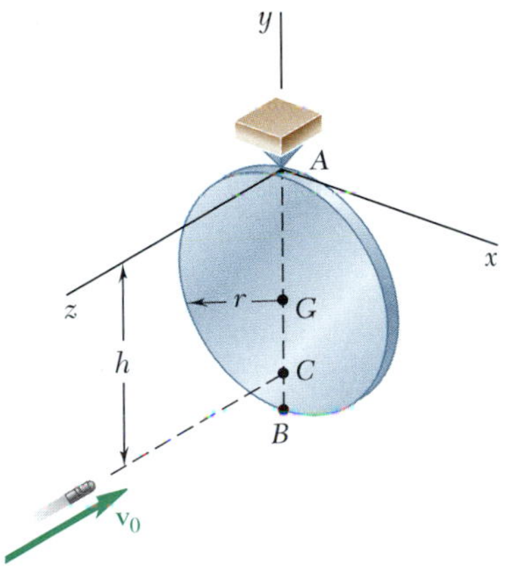

Fig. P17.86

17.87 In Prob. 17.86, determine (*a*) the required distance h if the impulsive reaction at point A is to be zero, (*b*) the corresponding velocity of the mass center G of the plate immediately after the bullet becomes embedded at C.

17.88 A 40-g bullet is fired with a horizontal velocity of 600 m/s into the lower end of a slender 7-kg bar of length $L = 600$ mm. Knowing that $h = 240$ mm and that the bar is initially at rest, determine (*a*) the angular velocity of the bar immediately after the bullet becomes embedded, (*b*) the impulsive reaction at C, assuming that the bullet becomes embedded in 0.001 s.

17.89 In Prob. 17.88, determine (*a*) the required distance h if the impulsive reaction at C is to be zero, (*b*) the corresponding angular velocity of the bar immediately after the bullet becomes embedded.

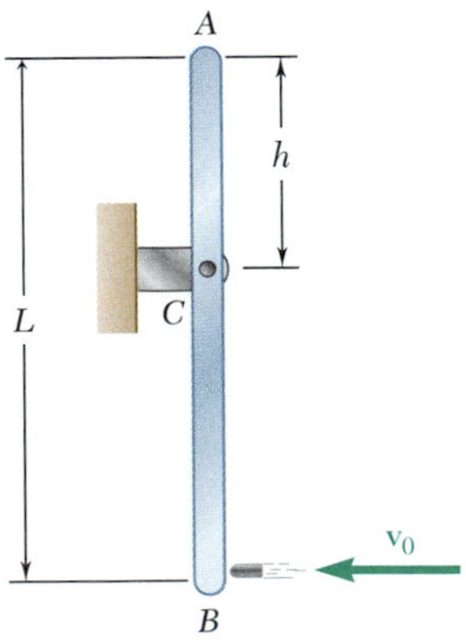

Fig. P17.88

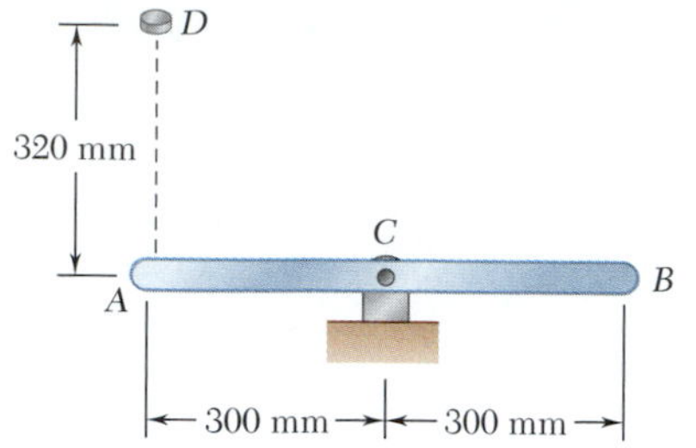

Fig. P17.90

17.90 A 15-g magnet D is released from rest in the position shown, falls a distance of 320 mm, and becomes attached at A to the 200-g steel bar AB. Assuming that the impact is perfectly plastic, determine the angular velocity of the bar and the velocity of the magnet immediately after the impact.

17.91 The uniform slender rod AB of weight 5 lb and length 30 in. forms an angle $\beta = 30°$ with the vertical as it strikes the smooth corner shown with a vertical velocity $\mathbf{v}_1$ of magnitude 8 ft/s and no angular velocity. Assuming that the impact is perfectly plastic, determine the angular velocity of the rod immediately after the impact.

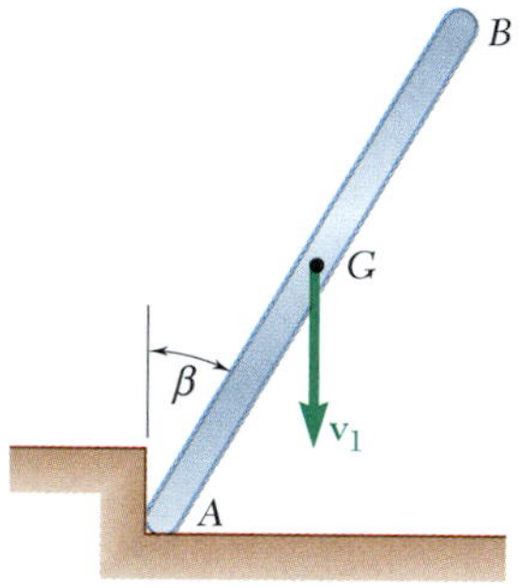

Fig. P17.91 and P17.92

17.92 The uniform slender rod AB of weight 5 lb and length 30 in. forms an angle β with the vertical as it strikes the smooth corner shown with a vertical velocity $\mathbf{v}_1$ of magnitude 8 ft/s and no angular velocity. Knowing that the magnitude of the velocity of the mass center G of the rod immediately after the impact is 2 ft/s and assuming that the impact is perfectly plastic, determine (*a*) the angle β, (*b*) the impulse exerted on the rod at point A.

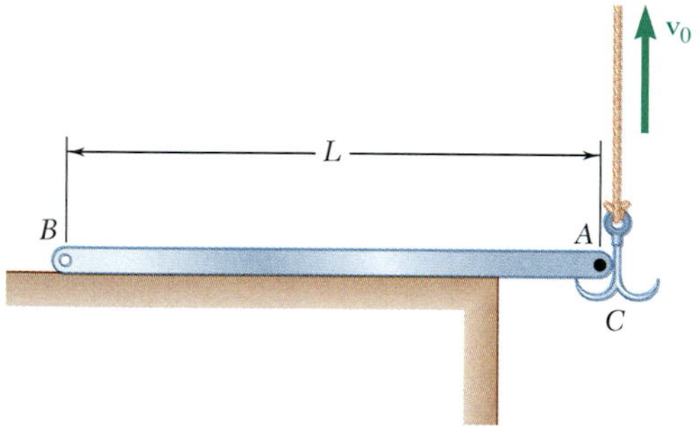

Fig. *P17.93*

17.93 A uniform slender rod AB of mass m is at rest on a frictionless horizontal surface when hook C engages a small pin at A. Knowing that the hook is pulled upward with a constant velocity $\mathbf{v}_0$, determine the impulse exerted on the rod (*a*) at A, (*b*) at B. Assume that the velocity of the hook is unchanged and that the impact is perfectly plastic.

17.94 A uniform slender rod AB of mass m and length L has a vertical velocity of magnitude v_1 and no angular velocity ($\omega_1 = 0$) when it strikes a rigid frictionless support at point C. Knowing that $h = L/4$ and assuming perfectly plastic impact, determine (*a*) the angular velocity of the rod and the velocity of its mass center immediately after the impact, (*b*) the impulse exerted on the rod at point C.

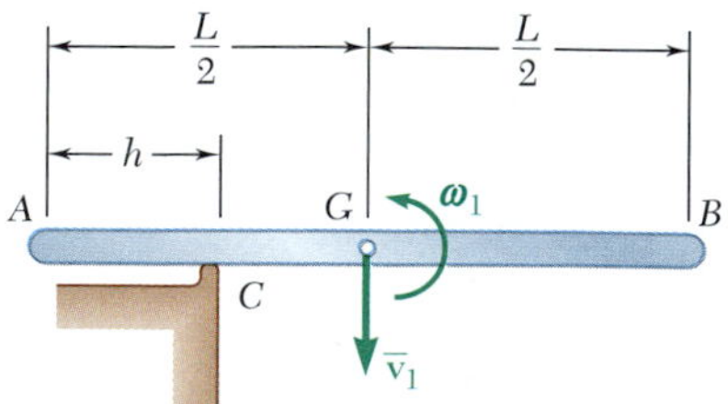

Fig. P17.94 and *P17.95*

17.95 A uniform slender rod of mass m and length L strikes a rigid frictionless support at point C with an angular velocity of magnitude ω_1 when the velocity of its mass center G is zero ($\bar{v}_1 = 0$). Knowing that the angular velocity of the rod immediately after the impact is $\omega_1/2$, counterclockwise, and assuming perfectly elastic impact, determine (*a*) the ratio h/L, (*b*) the velocity of the mass center of the rod immediately after the impact, (*c*) the impulse exerted on the rod at point C.

17.96 A uniform slender rod AB is at rest on a frictionless horizontal table when end A of the rod is struck by a hammer which delivers an impulse that is perpendicular to the rod. In the subsequent motion, determine the distance b through which the rod will move each time it completes a full revolution.

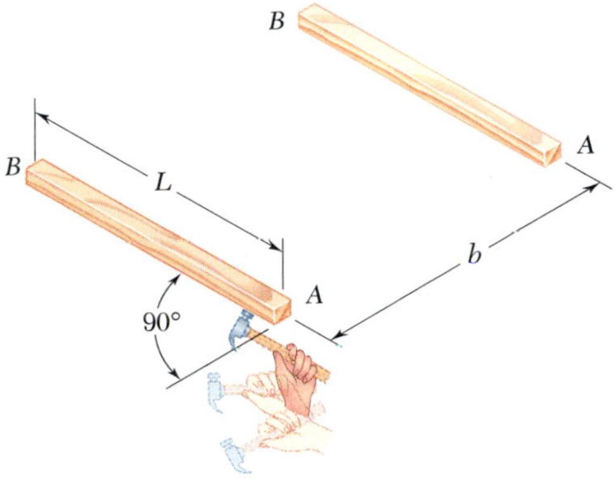

Fig. P17.96

17.97 A uniform sphere of radius r rolls down the incline shown without slipping. It hits a horizontal surface and, after slipping for a while, it starts rolling again. Assuming that the sphere does not bounce as it hits the horizontal surface, determine its angular velocity and the velocity of its mass center after it has resumed rolling.

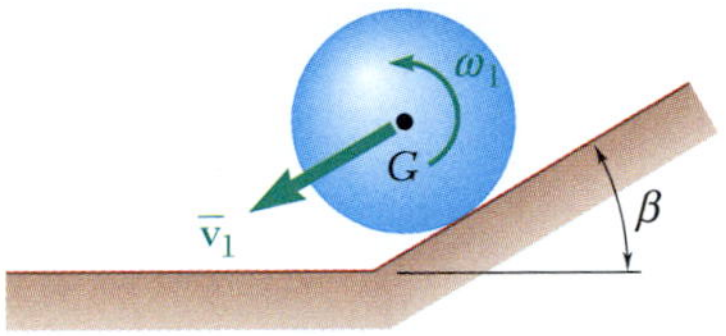

Fig. P17.97

17.98 The slender rod AB of length L forms an angle β with the vertical axis as it strikes the frictionless surface shown with a vertical velocity $\bar{\mathbf{v}}_1$ and no angular velocity. Assuming that the impact is perfectly elastic, derive an expression for the angular velocity of the rod immediately after the impact.

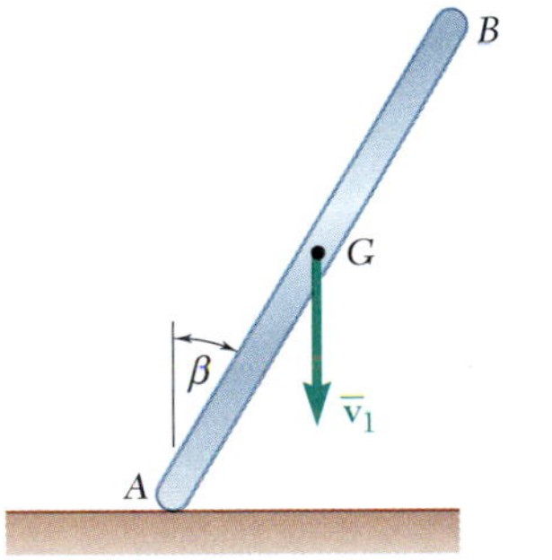

Fig. P17.98

17.99 Solve Prob. 17.98, assuming that the impact between rod AB and the frictionless surface is perfectly plastic.

17.100 A uniformly loaded square crate is falling freely with a velocity $\mathbf{v}_0$ when cable AB suddenly becomes taut. Assuming that the impact is perfectly plastic, determine the angular velocity of the crate and the velocity of its mass center immediately after the cable becomes taut.

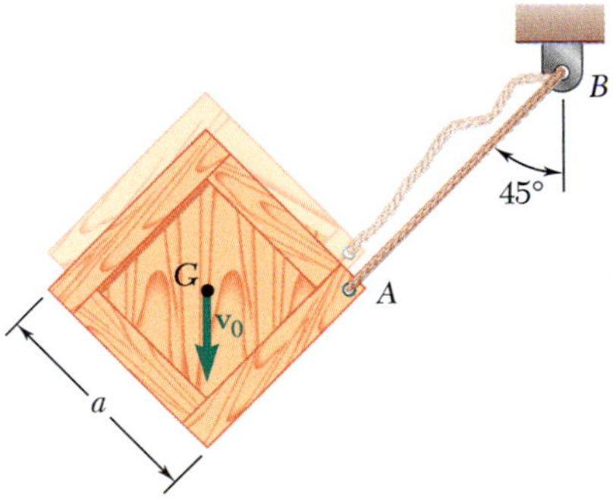

Fig. P17.100

17.101 A uniform slender rod AB of mass m and length L is falling freely with a velocity $\mathbf{v}_0$ when end B strikes a smooth inclined surface as shown. Assuming that the impact is perfectly elastic, determine the angular velocity of the rod and the velocity of its mass center immediately after the impact.

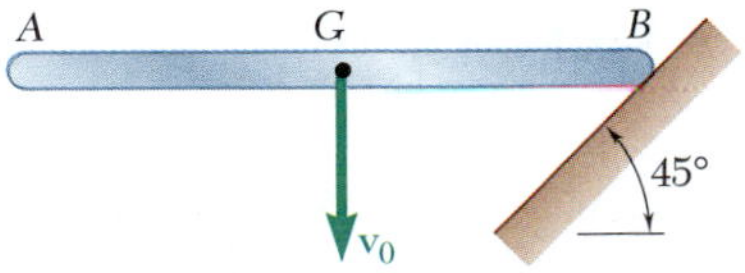

Fig. P17.101

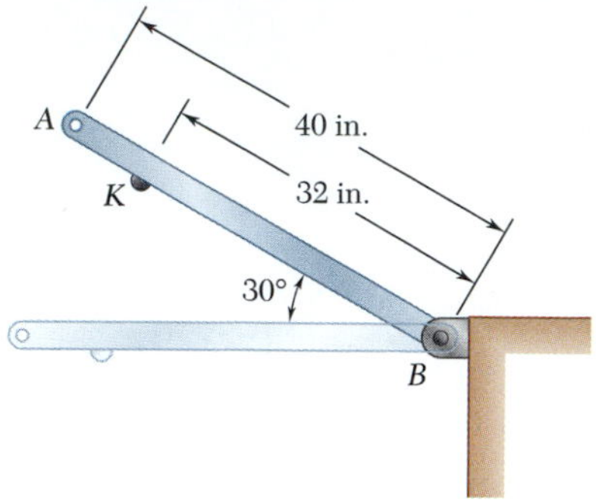

Fig. *P17.102*

17.102 A slender 10-lb rod is released from rest in the position shown. It is observed that after the rod strikes the vertical surface it rebounds to a horizontal position. (*a*) Determine the coefficient of restitution between knob *K* and the surface. (*b*) Show that the same rebound can be expected for any position of knob *K*. (Neglect the mass of knob *K*.)

17.103 A uniform slender rod *AB* of mass *m* and length *L* is released from rest in the position shown. Knowing that the impact between knob *B* and the horizontal surface is perfectly elastic, determine (*a*) the angular velocity of the rod immediately after the impact, (*b*) the impulses exerted on the rod at points *A* and *B*. (Neglect the masses of the knobs at points *A* and *B* and the friction between the knobs and the horizontal surface.)

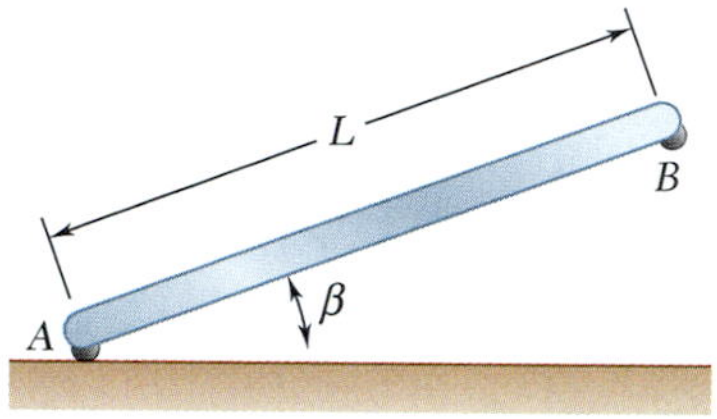

Fig. P17.103

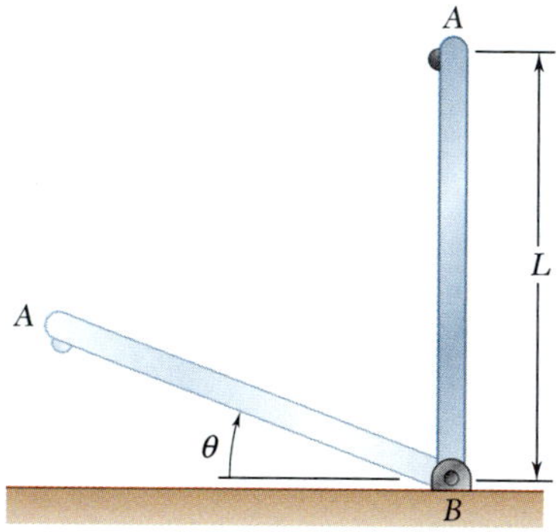

Fig. *P17.104*

17.104 The uniform slender rod *AB* is in equilibrium in the position shown when end *A* is given a slight nudge, causing the rod to rotate counterclockwise and hit the horizontal surface. Knowing that the coefficient of restitution between the knob at *A* and the horizontal surface is 0.50, determine the maximum angle of rebound θ of the rod.

17.105 A 1.25-oz bullet is fired with a horizontal velocity of 950 ft/s into the 18-lb wooden beam *AB*. The beam is suspended from a collar of negligible weight that can slide along a horizontal rod. Neglecting friction between the collar and the rod, determine the maximum angle of rotation of the beam during its subsequent motion.

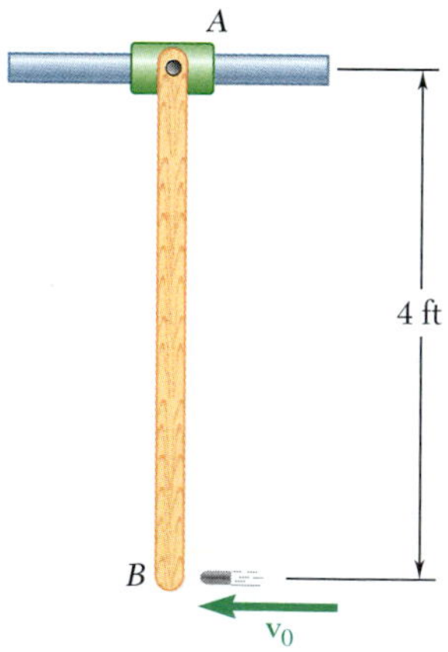

Fig. P17.105

17.106 For the beam of Prob. 17.105, determine the velocity of the 1.25-oz bullet for which the maximum angle of rotation of the beam will be 90°.

17.107 A uniformly loaded square crate is released from rest with its corner D directly above A; it rotates about A until its corner B strikes the floor, and then rotates about B. The floor is sufficiently rough to prevent slipping and the impact at B is perfectly plastic. Denoting by $\boldsymbol{\omega}_0$ the angular velocity of the crate immediately after B strikes the floor, determine (*a*) the angular velocity of the crate immediately before B strikes the floor, (*b*) the fraction of the kinetic energy of the crate lost during the impact, (*c*) the angle θ through which the crate will rotate after B strikes the floor.

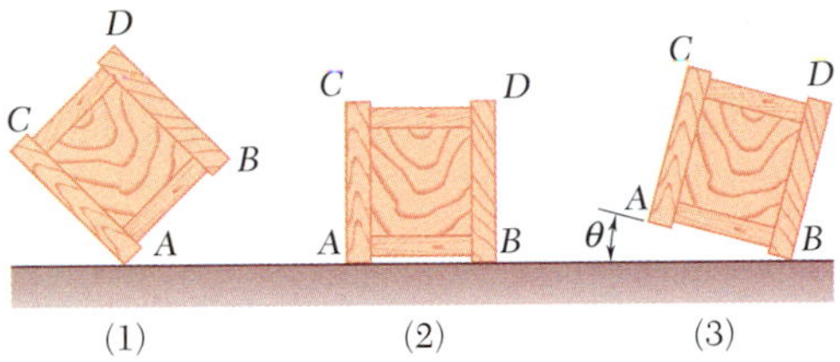

Fig. P17.107

17.108 A uniform slender rod AB of length $L = 600$ mm is placed with its center equidistant from two supports that are located at a distance $b = 100$ mm from each other. End B of the rod is raised a distance $h_0 = 80$ mm and released; the rod then rocks on the supports as shown. Assuming that the impact at each support is perfectly plastic and that no slipping occurs between the rod and the supports, determine (*a*) the height h_1 reached by end A after the first impact, (*b*) the height h_2 reached by end B after the second impact.

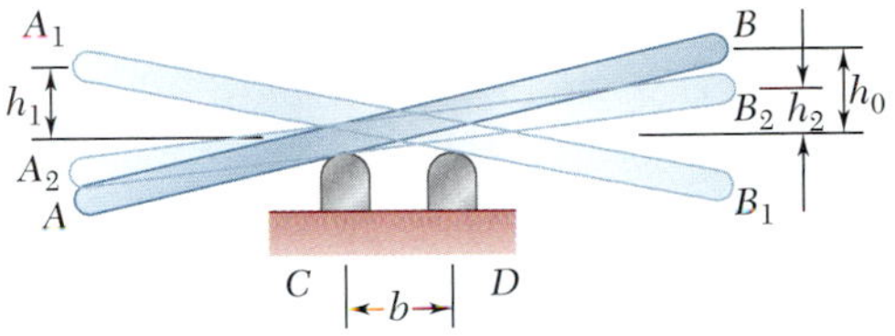

Fig. P17.108

17.109 The 9-kg rigid body BD consists of two identical 60-mm-radius spheres and the rod which connects them and has a centroidal radius of gyration of 250 mm. The body is at rest on a horizontal frictionless surface when it is struck by the 3-kg sphere A which has a radius of 60 mm and is moving as shown with a velocity $\mathbf{v}_1$ of magnitude 4 m/s. Assuming a perfectly plastic impact, determine immediately after the impact, (*a*) the angular velocity of body BD, (*b*) the velocity of point G.

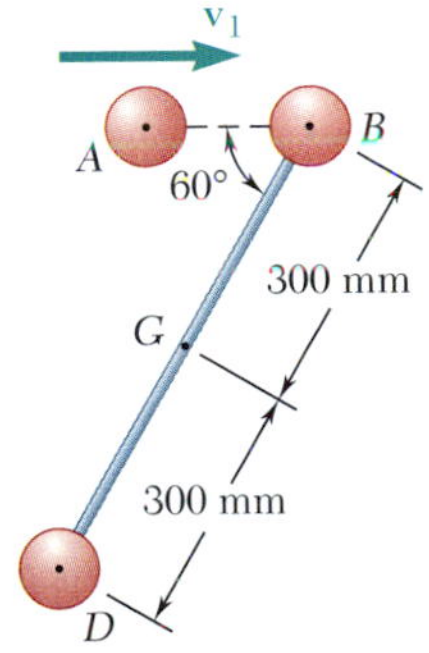

Fig. P17.109

17.110 Solve Prob. 17.109, assuming that the impact is perfectly elastic.

17.111 A slender rod AB is released from rest in the position shown. It swings down to a vertical position and strikes a second and identical rod CD which is resting on a frictionless surface. Assuming that the coefficient of restitution between the rods is 0.4, determine the velocity of rod CD immediately after the impact.

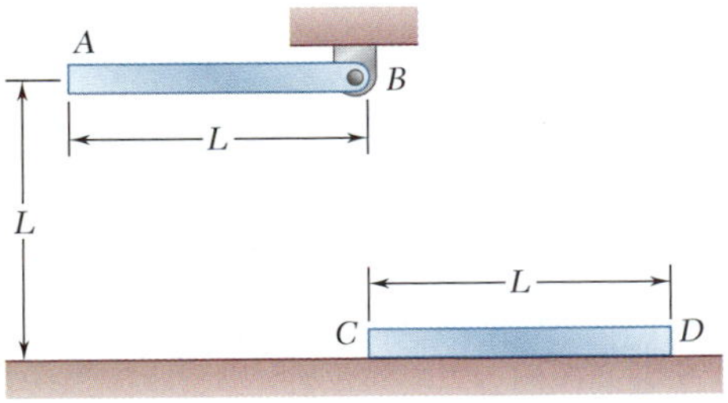

Fig. *P17.111*

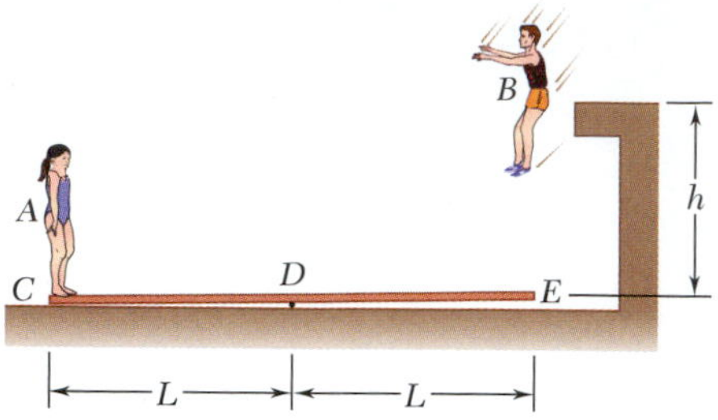

Fig. P17.113

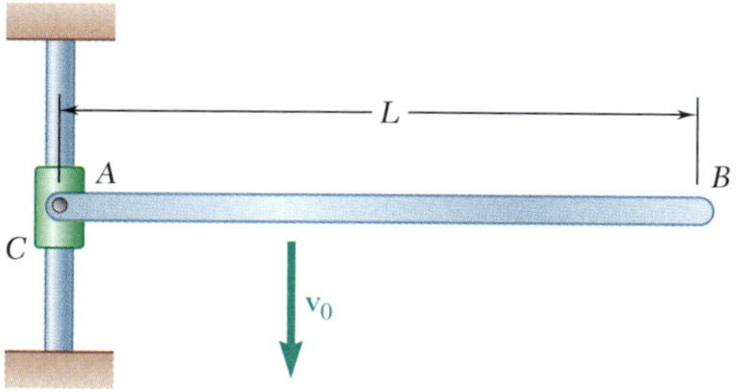

Fig. P17.116

17.112 Solve Prob. 17.111 assuming that the impact between the rods is perfectly elastic.

17.113 The plank *CDE* has a weight of 30 lb and rests on a small pivot at *D*. The 110-lb gymnast *A* is standing on the plank at *C* when the 140-lb gymnast *B* jumps from a height of 7.5 ft and strikes the plank at *E*. Assuming perfectly plastic impact and that gymnast *A* is standing absolutely straight, determine the height to which gymnast *A* will rise.

17.114 Solve Prob. 17.113 assuming that the gymnasts change places so that gymnast *A* jumps onto the plank while gymnast *B* stands at *C*.

17.115 Member *ABC* has a weight of 5 lb and is attached to a pin support at *B*. A 1.5-lb sphere *D* strikes end *C* of member *ABC* with a vertical velocity $\mathbf{v}_1$ of 9 ft/s. Knowing that $L = 30$ in. and that the coefficient of restitution between the sphere and member *ABC* is 0.5, determine immediately after the impact (*a*) the angular velocity of member *ABC*, (*b*) the velocity of the sphere.

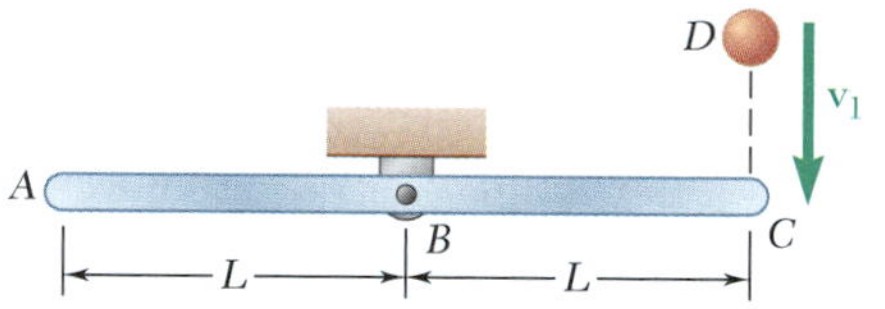

Fig. P17.115

17.116 The uniform slender rod *AB* of mass m_{AB} is attached by a pin to collar *C* of mass m_C and the system is falling freely with a velocity $\mathbf{v}_0$ when collar *C* strikes a horizontal surface as shown. Denoting by *e* the coefficient of restitution between the collar and the surface, determine (*a*) the angular velocity of the rod immediately after the impact, (*b*) the impulse exerted on collar *C* by the surface.

17.117 The 2.5-kg slender rod *AB* is released from rest in the position shown and swings to a vertical position where it strikes the 1.5-kg slender rod *CD*. Knowing that the coefficient of restitution between the knob *K* attached to rod *AB* and rod *CD* is 0.8, determine the maximum angle θ_m through which rod *CD* will rotate after the impact.

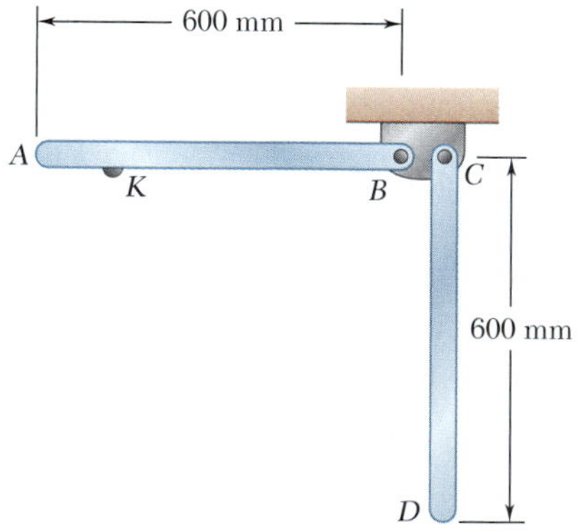

Fig. P17.117

17.118 Sphere A of mass m and radius r rolls without slipping with a velocity $\overline{\mathbf{v}}_1$ on a horizontal surface when it hits squarely an identical sphere B that is at rest. Denoting by μ_k the coefficient of kinetic friction between the spheres and the surface, neglecting friction between the spheres, and assuming perfectly elastic impact, determine (*a*) the linear and angular velocities of each sphere immediately after the impact, (*b*) the velocity of each sphere after it has started rolling uniformly.

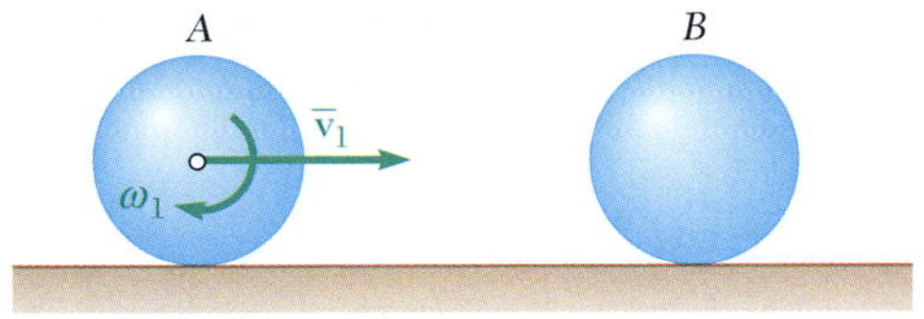

Fig. P17.118

17.119 A small rubber ball of radius r is thrown against a rough floor with a velocity $\overline{\mathbf{v}}_A$ of magnitude $\overline{v}_0$ and a backspin $\boldsymbol{\omega}_A$ of magnitude ω_0. It is observed that the ball bounced from A to B, then from B to A, then from A to B, etc. Assuming perfectly elastic impact, determine the required magnitude ω_0 of the backspin in terms of $\overline{v}_0$ and r.

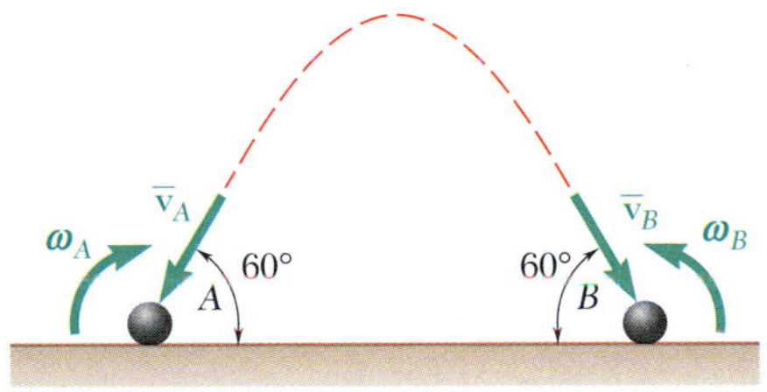

Fig. P17.119

17.120 In a game of pool, ball A is rolling without slipping with a velocity $\overline{\mathbf{v}}_0$ as it hits obliquely ball B, which is at rest. Denoting by r the radius of each ball and by μ_k the coefficient of kinetic friction between the balls, and assuming perfectly elastic impact, determine (*a*) the linear and angular velocity of each ball immediately after the impact, (*b*) the velocity of ball B after it has started rolling uniformly.

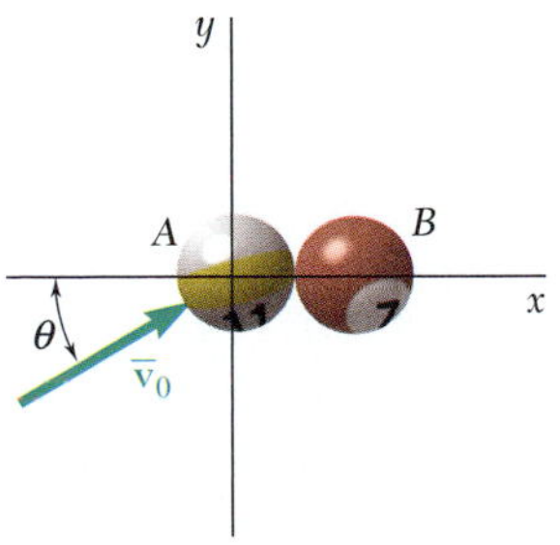

Fig. P17.120

REVIEW AND SUMMARY FOR CHAPTER 17

In this chapter we again considered the method of work and energy and the method of impulse and momentum. In the first part of the chapter we studied the method of work and energy and its application to the analysis of the motion of rigid bodies and systems of rigid bodies.

Principle of work and energy for a rigid body

In Sec. 17.2, we first expressed the principle of work and energy for a rigid body in the form

$$T_1 + U_{1\rightarrow 2} = T_2 \qquad (17.1)$$

where T_1 and T_2 represent the initial and final values of the kinetic energy of the rigid body and $U_{1\rightarrow 2}$ represents the work of the *external forces* acting on the rigid body.

Work of a force or a couple

In Sec. 17.3, we recalled the expression found in Chap. 13 for the work of a force **F** applied at a point A, namely

$$U_{1\rightarrow 2} = \int_{s_1}^{s_2} (F \cos \alpha)\, ds \qquad (17.3')$$

where F was the magnitude of the force, α the angle it formed with the direction of motion of A, and s the variable of integration measuring the distance traveled by A along its path. We also derived the expression for the *work of a couple of moment* **M** applied to a rigid body during a rotation in θ of the rigid body:

$$U_{1\rightarrow 2} = \int_{\theta_1}^{\theta_2} M\, d\theta \qquad (17.5)$$

Kinetic energy in plane motion

We then derived an expression for the kinetic energy of a rigid body in plane motion [Sec. 17.4]. We wrote

$$T = \tfrac{1}{2}m\bar{v}^2 + \tfrac{1}{2}\bar{I}\omega^2 \qquad (17.9)$$

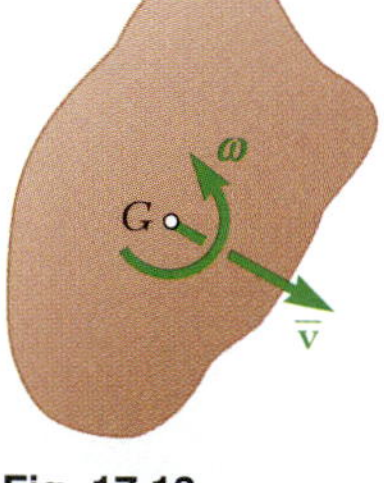

Fig. 17.13

where $\bar{v}$ is the velocity of the mass center G of the body, ω is the angular velocity of the body, and $\bar{I}$ is its moment of inertia about an axis through G perpendicular to the plane of reference (Fig. 17.13) [Sample Prob. 17.3]. We noted that the kinetic energy of a rigid body in plane motion can be separated into two parts: (1) the kinetic energy $\frac{1}{2}m\bar{v}^2$ associated with the motion of the mass center G of the body, and (2) the kinetic energy $\frac{1}{2}\bar{I}\omega^2$ associated with the rotation of the body about G.

Kinetic energy in rotation

For a rigid body rotating about a fixed axis through O with an angular velocity $\boldsymbol{\omega}$, we had

$$T = \tfrac{1}{2} I_O \omega^2 \tag{17.10}$$

where I_O was the moment of inertia of the body about the fixed axis. We noted that the result obtained is not limited to the rotation of plane slabs or of bodies symmetrical with respect to the reference plane, but is valid regardless of the shape of the body or of the location of the axis of rotation.

Systems of rigid bodies

Equation (17.1) can be applied to the motion of systems of rigid bodies [Sec. 17.5] as long as all the forces acting on the various bodies involved—internal as well as external to the system—are included in the computation of $U_{1\rightarrow2}$. However, in the case of systems consisting of pin-connected members, or blocks and pulleys connected by inextensible cords, or meshed gears, the points of application of the internal forces move through equal distances and the work of these forces cancels out [Sample Probs. 17.1 and 17.2].

Conservation of energy

When a rigid body, or a system of rigid bodies, moves under the action of conservative forces, the principle of work and energy can be expressed in the form

$$T_1 + V_1 = T_2 + V_2 \tag{17.12}$$

which is referred to as the *principle of conservation of energy* [Sec. 17.6]. This principle can be used to solve problems involving conservative forces such as the force of gravity or the force exerted by a spring [Sample Probs. 17.4 and 17.5]. However, when a reaction is to be determined, the principle of conservation of energy must be supplemented by the application of d'Alembert's principle [Sample Prob. 17.4].

Power

In Sec. 17.7, we extended the concept of power to a rotating body subjected to a couple, writing

$$\text{Power} = \frac{dU}{dt} = \frac{M\,d\theta}{dt} = M\omega \tag{17.13}$$

where M is the magnitude of the couple and ω the angular velocity of the body.

The middle part of the chapter was devoted to the method of impulse and momentum and its application to the solution of various types of problems involving the plane motion of rigid slabs and rigid bodies symmetrical with respect to the reference plane.

Principle of impulse and momentum for a rigid body

We first recalled the *principle of impulse and momentum* as it was derived in Sec. 14.9 for a system of particles and applied it to the *motion of a rigid body* [Sec. 17.8]. We wrote

$$\textbf{Syst Momenta}_1 + \textbf{Syst Ext Imp}_{1\rightarrow2} = \textbf{Syst Momenta}_2 \tag{17.14}$$

Next we showed that for a rigid slab or a rigid body symmetrical with respect to the reference plane, the system of the momenta of the particles forming the body is equivalent to a vector $m\overline{\mathbf{v}}$ attached at the mass center G of the body and a couple $\bar{I}\boldsymbol{\omega}$ (Fig. 17.14). The

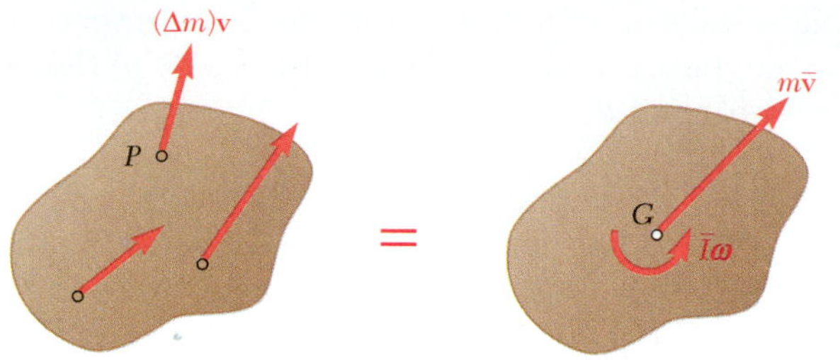

Fig. 17.14

vector $m\overline{\mathbf{v}}$ is associated with the translation of the body with G and represents the *linear momentum* of the body, while the couple $\bar{I}\boldsymbol{\omega}$ corresponds to the rotation of the body about G and represents the *angular momentum* of the body about an axis through G.

Equation (17.14) can be expressed graphically as shown in Fig. 17.15 by drawing three diagrams representing respectively the system of the initial momenta of the body, the impulses of the external forces acting on the body, and the system of the final momenta of the body.

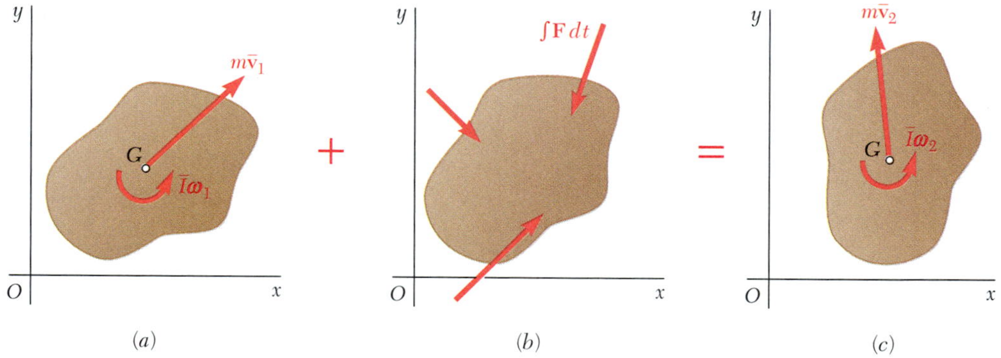

Fig. 17.15

Summing and equating respectively the *x components*, the *y components*, and the *moments about any given point* of the vectors shown in that figure, we obtain three equations of motion which can be solved for the desired unknowns [Sample Probs. 17.6 and 17.7].

In problems dealing with several connected rigid bodies [Sec. 17.9], each body can be considered separately [Sample Prob. 17.6], or, if no more than three unknowns are involved, the principle of impulse and momentum can be applied to the entire system, considering the impulses of the external forces only [Sample Prob. 17.8].

Conservation of angular momentum

When the lines of action of all the external forces acting on a system of rigid bodies pass through a given point O, the angular momentum of the system about O is conserved [Sec. 17.10]. It was suggested that problems involving conservation of angular momentum be solved by the general method described above [Sample Prob. 17.8].

Impulsive motion

The last part of the chapter was devoted to the *impulsive motion* and the *eccentric impact* of rigid bodies. In Sec. 17.11, we recalled that the method of impulse and momentum is the only practicable method for the solution of problems involving impulsive motion and that the computation of impulses in such problems is particularly simple [Sample Prob. 17.9].

Eccentric impact

In Sec. 17.12, we recalled that the eccentric impact of two rigid bodies is defined as an impact in which the mass centers of the colliding bodies are *not* located on the line of impact. It was shown that in such a situation a relation similar to that derived in Chap. 13 for the central impact of two particles and involving the coefficient of restitution e still holds, but that *the velocities of points A and B where contact occurs during the impact should be used.* We have

$$(v'_B)_n - (v'_A)_n = e[(v_A)_n - (v_B)_n] \qquad (17.19)$$

where $(v_A)_n$ and $(v_B)_n$ are the components along the line of impact of the velocities of A and B before the impact, and $(v'_A)_n$ and $(v'_B)_n$ are their components after the impact (Fig. 17.16). Equation

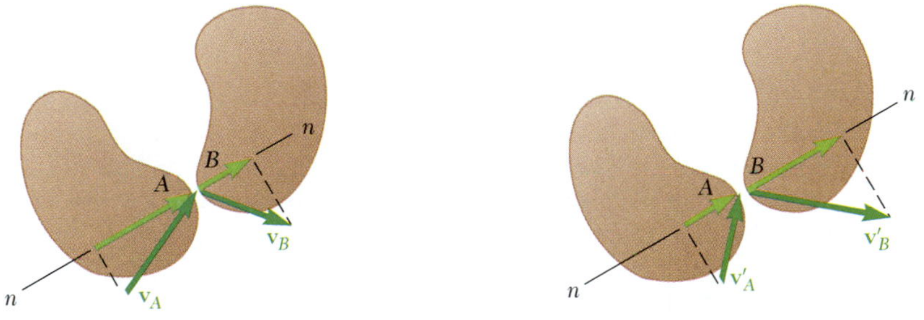

(*a*) Before impact

(*b*) After impact

Fig. 17.16

(17.19) is applicable not only when the colliding bodies move freely after the impact but also when the bodies are partially constrained in their motion. It should be used in conjunction with one or several other equations obtained by applying the principle of impulse and momentum [Sample Prob. 17.10]. We also considered problems where the method of impulse and momentum and the method of work and energy can be combined [Sample Prob. 17.11].

Review Problems

17.121 Disk B has an initial angular velocity $\boldsymbol{\omega}_0$ when it is brought into contact with disk A which is at rest. Show that the final angular velocity of disk B depends only on ω_0 and the ratio of the masses m_A and m_B of the two disks.

17.122 The 7.5-lb disk A has a radius $r_A = 6$ in. and is initially at rest. The 10-lb disk B has a radius $r_B = 8$ in. and an angular velocity $\boldsymbol{\omega}_0$ of 900 rpm when it is brought into contact with disk A. Neglecting friction in the bearings, determine (*a*) the final angular velocity of each disk, (*b*) the total impulse of the friction force exerted on disk A.

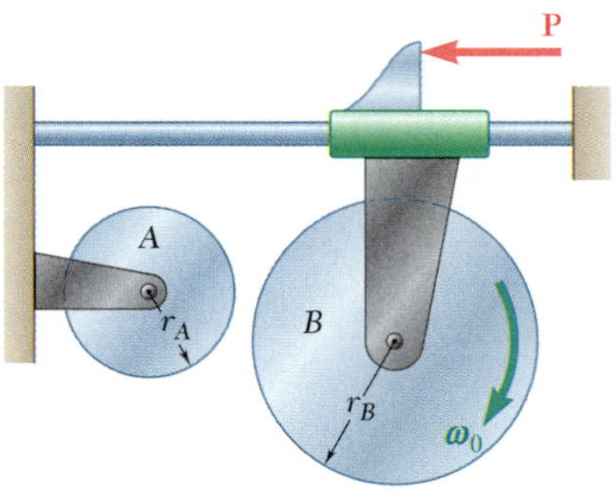

Fig. P17.121 and *P17.122*

17.123 A uniform disk of constant thickness and initially at rest is placed in contact with the belt shown, which moves at a constant speed v = 25 m/s. Knowing that the coefficient of kinetic friction between the disk and the belt is 0.15, determine (*a*) the number of revolutions executed by the disk before it reaches a constant angular velocity, (*b*) the time required for the disk to reach that constant angular velocity.

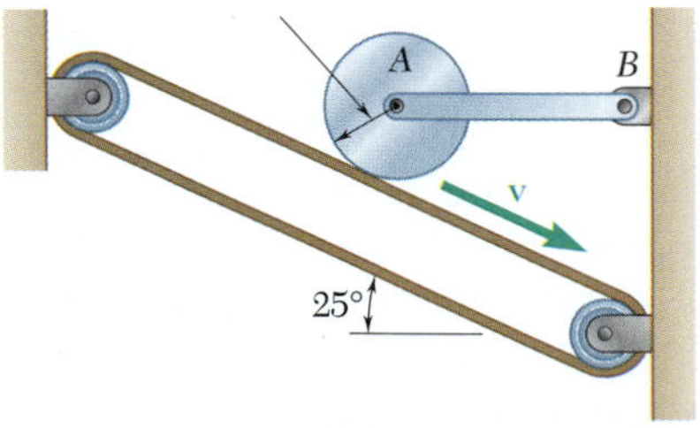

Fig. P17.123

17.124 A uniform slender rod of length L is dropped onto rigid supports at A and B. Since support B is slightly lower than support A, the rod strikes A with a velocity $\overline{\mathbf{v}}_1$ before it strikes B. Assuming perfectly elastic impact at both A and B, determine the angular velocity of the rod and the velocity of its mass center immediately after the rod (*a*) strikes support A, (*b*) strikes support B, (*c*) again strikes support A.

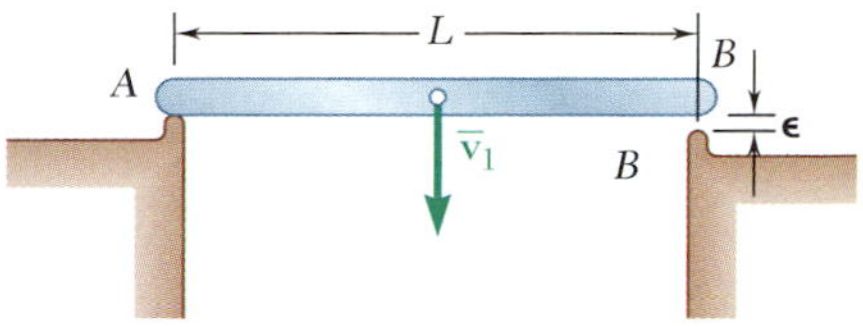

Fig. P17.124

17.125 Two uniform cylinders, each of weight $W = 14$ lb and radius $r = 5$ in., are connected by a belt as shown. Knowing that the initial angular velocity of cylinder B is 30 rad/s counterclockwise, determine (*a*) the distance through which cylinder A will rise before the angular velocity of cylinder B is reduced to 5 rad/s, (*b*) the tension in the portion of belt connecting the two cylinders.

17.126 Two uniform cylinders, each of weight $W = 14$ lb and radius $r = 5$ in., are connected by a belt as shown. Knowing that at the instant shown the angular velocity of cylinder B is 30 rad/s counterclockwise, determine (*a*) the time required for the angular velocity of cylinder B to be reduced to 5 rad/s, (*b*) the tension in the portion of belt connecting the two cylinders.

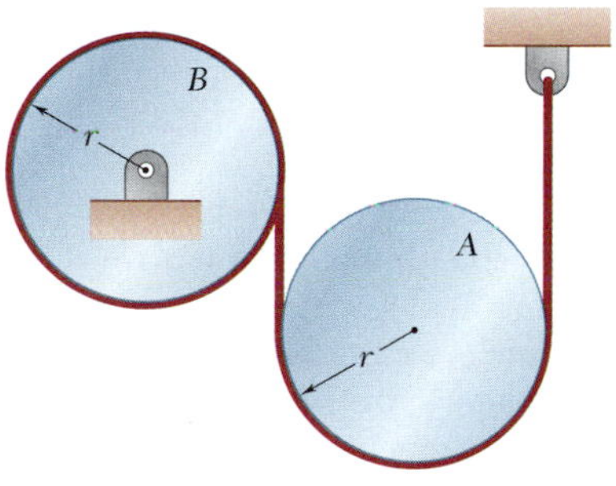

Fig. P17.125 and P17.126

17.127 A 3-kg bar AB is attached by a pin at D to a 4-kg square plate, which can rotate freely about a vertical axis. Knowing that the angular velocity of the plate is 120 rpm when the bar is vertical, determine (*a*) the angular velocity of the plate after the bar has swung into a horizontal position and has come to rest against pin C, (*b*) the energy lost during the plastic impact at C.

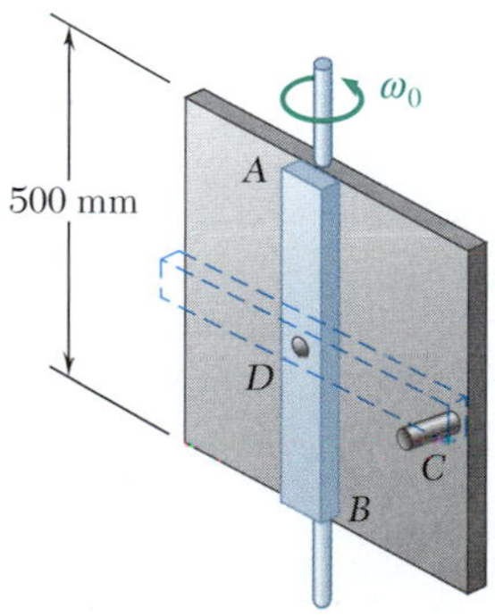

Fig. *P17.127*

17.128 The double pulley shown has a mass of 3 kg and a radius of gyration of 100 mm. Knowing that when the pulley is at rest, a force **P** of magnitude 24 N is applied to cord B, determine (a) the velocity of the center of the pulley after 1.5 s, (b) the tension in cord C.

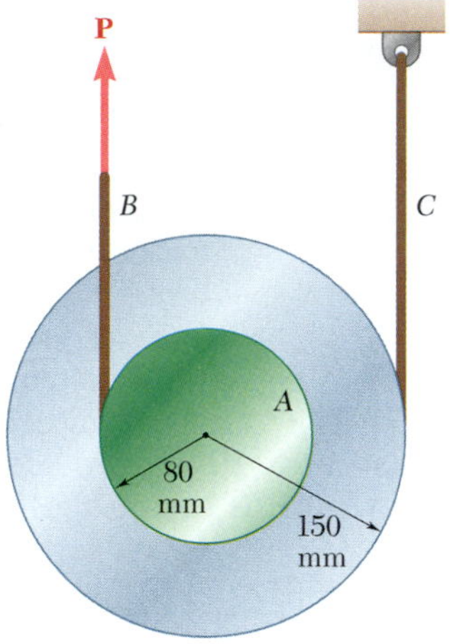

Fig. P17.128

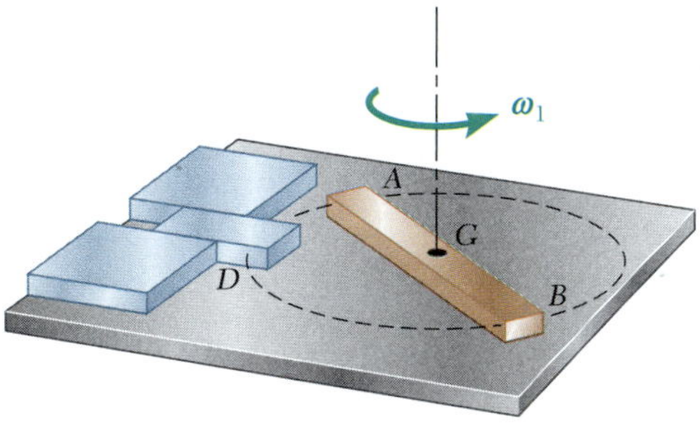

Fig. P17.129

17.129 A uniform slender bar of length L and mass m is supported by a frictionless horizontal table. Initially the bar is spinning about its mass center G with a constant angular velocity $\boldsymbol{\omega}_1$. Suddenly latch D is moved to the right and is struck by end A of the bar. Assuming that the impact of A and D is perfectly plastic, determine the angular velocity of the bar and the velocity of its mass center immediately after the impact.

17.130 Solve Prob. 17.129 assuming that the impact of A and D is perfectly elastic.

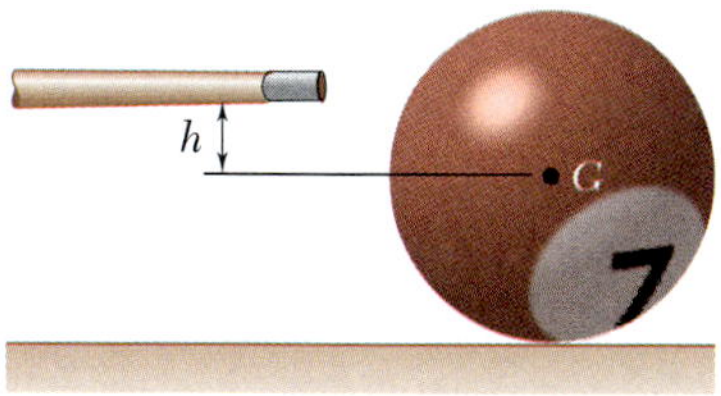

Fig. P17.131

17.131 At what height h above its mass center G should a billiard ball of radius r be struck horizontally by a cue if the ball is to start rolling without sliding?

17.132 The ends of a 9-lb rod AB are constrained to move along slots cut in a vertical plane as shown. A spring of constant $k = 3$ lb/in. is attached to end A in such a way that its tension is zero when $\theta = 0$. If the rod is released from rest when $\theta = 0$, determine the angular velocity of the rod and the velocity of end B when $\theta = 30°$.

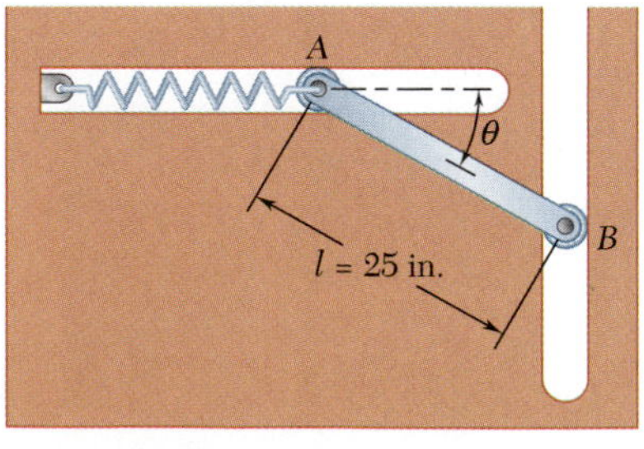

Fig. *P17.132*

Computer Problems

17.C1 The 10-lb slender rod AB is welded to the 6-lb uniform disk, which rotates about a pivot at A. A spring of constant $k = 4.5$ lb/ft is attached to the disk and is unstretched when rod AB is horizontal. The assembly is released from rest in the position shown. Calculate and plot the angular velocity of the assembly after it has rotated through an angle θ, for values of θ from 0 to θ_m, the angle of maximum rotation. Determine the maximum angular velocity of the assembly and the corresponding value of θ.

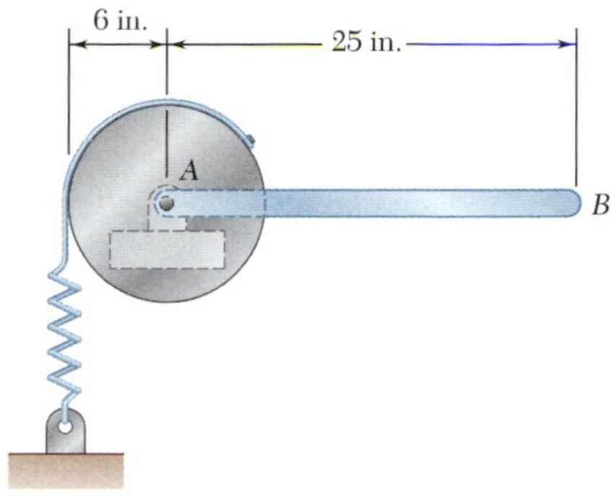

Fig. P17.C1

17.C2 Two 3-kg slender rods are welded to the edge of a 4-kg uniform disk as shown. The assembly is released from rest in the position shown and swings freely about the pivot C. Calculate and plot the angular velocity of the assembly after it has rotated through an angle θ, for values of θ from 0 to θ_m, the angle of maximum rotation. Determine the maximum angular velocity of the assembly and the corresponding value of θ.

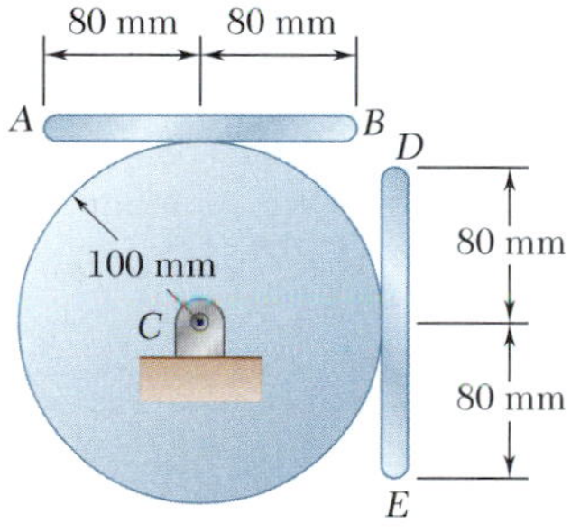

Fig. P17.C2

17.C3 The uniform slender rod AB of length $L = 40$ in. and weight 10 lb rests on a small wheel at D and is attached to a collar of negligible mass that can slide freely on the vertical rod EF. Knowing that $a = 10$ in. and that the rod is released from rest when $\theta = 0$, use computational software to calculate and plot the angular velocity of the rod and the velocity of end A for values of θ from 0 to 50°. Determine the maximum angular velocity of the rod and the corresponding value of θ.

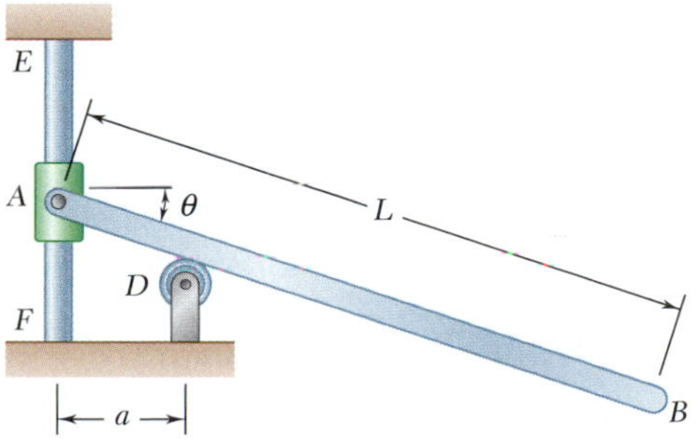

Fig. P17.C3

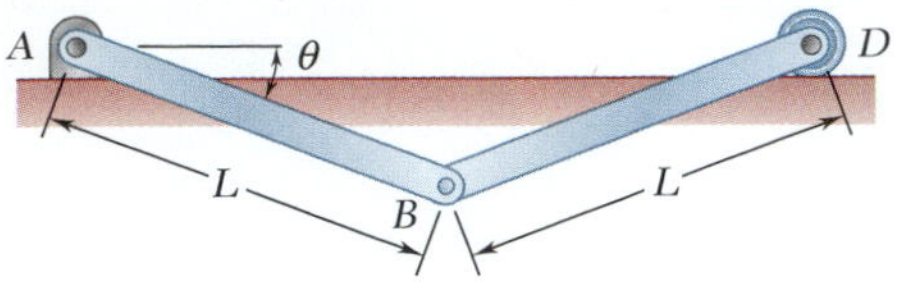

Fig. P17.C4

17.C4 Each of the two identical slender bars shown has a length $L =$ 600 mm. Knowing that the system is released from rest when the bars are horizontal, use computational software to calculate and plot the angular velocity of rod AB and the velocity of point D for values of θ from 0 to 90°.

17.C5 Collar C has a weight of 5 lb and can slide without friction on rod AB. A spring of constant 3.75 lb/in. and an unstretched length $r_0 = 25$ in. is attached as shown to the collar and to the hub B. The total mass moment of inertia of the rod, hub, and spring is known to be 0.22 lb · ft · s^2 about B. Initially the collar is held at a distance of 25 in. from the axis of rotation by a small pin protruding from the rod. The pin is suddenly removed as the assembly is rotating in a horizontal plane with angular velocity ω_0 of 10 rad/s. Denoting by r the distance of the collar from the axis of rotation, use computational software to calculate and plot the angular velocity of the assembly and the velocity of the collar relative to the rod for values of r from 25 in. to 35 in. Determine the maximum value of r in the ensuing motion.

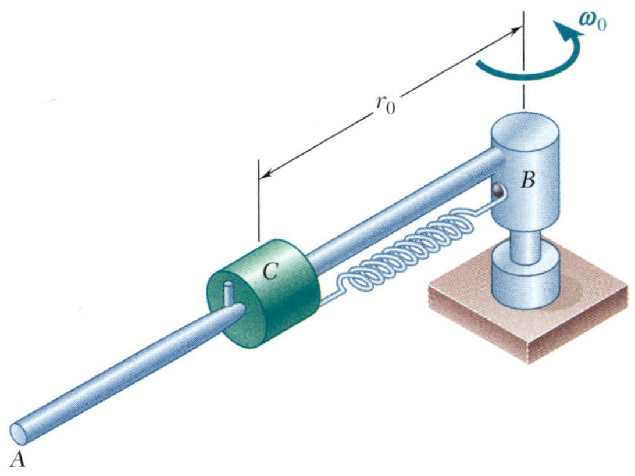

Fig. P17.C5

17.C6 Rod AB has a weight of 6 lb and is attached at A to a 10-lb cart C. Knowing that the system is released from rest when $\theta = 45°$ and neglecting friction, use computational software to calculate the velocity of the cart and the velocity of end B of the rod for values of θ from +45° to −90° using 15° decrements. Using appropriate smaller decrements, determine the value of θ for which the velocity of the cart to the left is maximum and the corresponding value of the velocity.

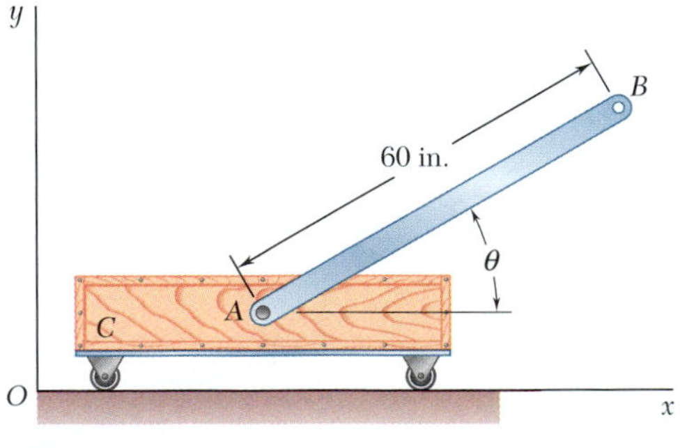

Fig. P17.C6

17.C7 A uniform 200-mm-radius sphere rolls over a series of parallel horizontal bars equally spaced at a distance d. As it rotates without slipping about a given bar, the sphere strikes the next bar and starts rotating about that bar without slipping, until it strikes the next bar, and so on. Assuming perfectly plastic impact and knowing that the sphere has an angular velocity ω_0 of 1.5 rad/s as its mass center G is directly above bar A, use computational software to calculate, for values of the spacing d from 20 to 120 mm and using 10-mm increments, (a) the angular velocity ω_1 of the sphere as G passes directly above bar B, (b) the number of bars over which the sphere will roll after leaving bar A.

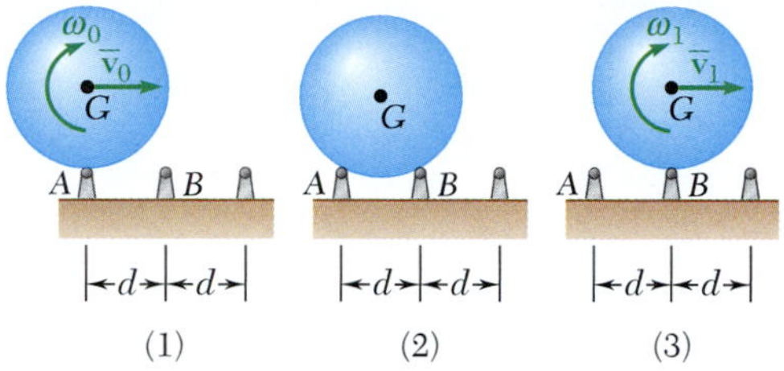

Fig. P17.C7

CHAPTER

Kinetics of Rigid Bodies in Three Dimensions

18

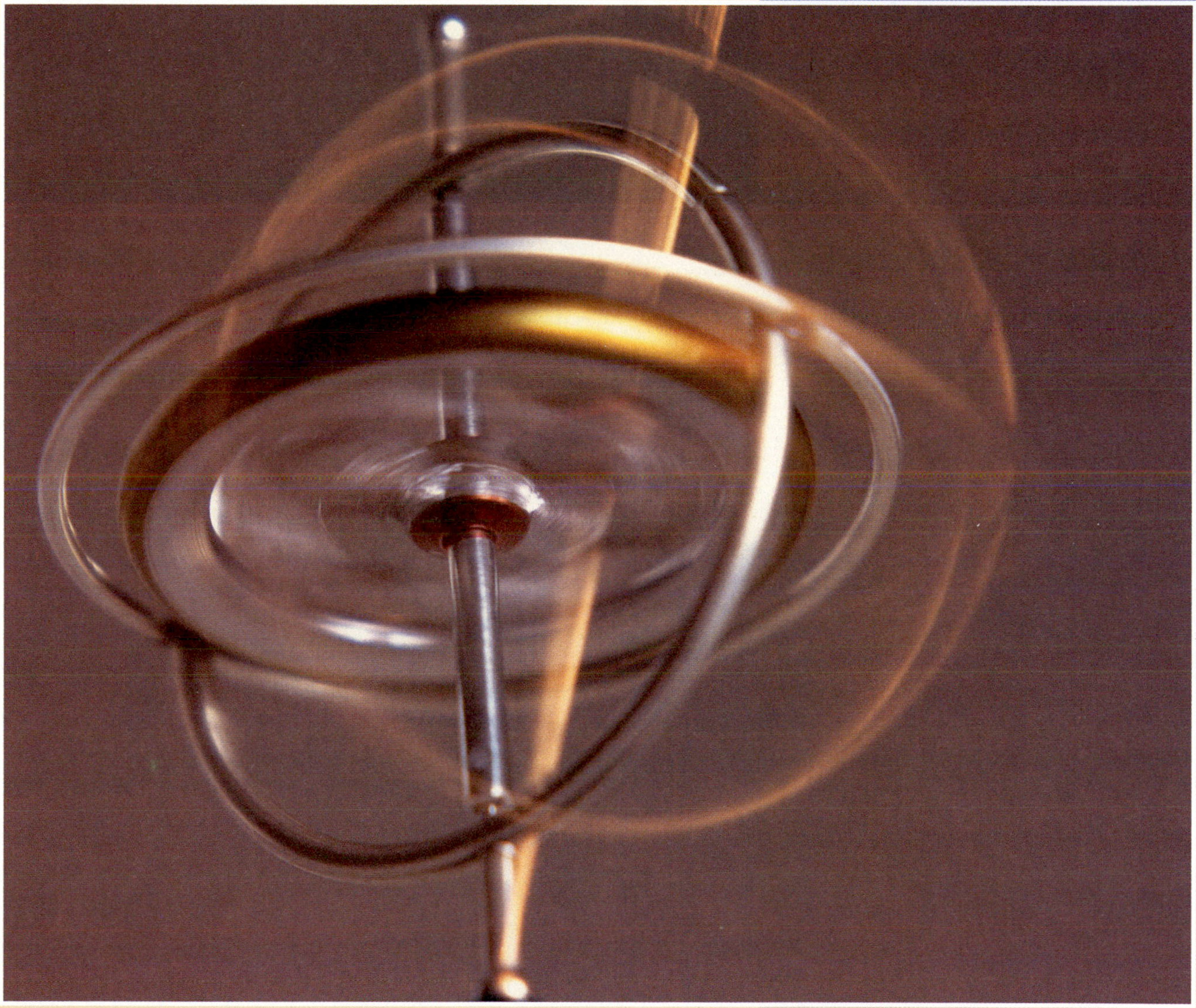

While the general principles that you learned in earlier chapters can be used again to solve problems involving the three-dimensional motion of rigid bodies, the solution of these problems requires a new approach and is considerably more involved than the solution of two-dimensional problems. One example is the study of the motion of a spinning rotor supported by a suspension which allows the axis of the rotor to assume any orientation in space.

*18.1. INTRODUCTION

In Chaps. 16 and 17 we were concerned with the plane motion of rigid bodies and of systems of rigid bodies. In Chap. 16 and in the second half of Chap. 17 (momentum method), our study was further restricted to that of plane slabs and of bodies symmetrical with respect to the reference plane. However, many of the fundamental results obtained in these two chapters remain valid in the case of the motion of a rigid body in three dimensions.

For example, the two fundamental equations

$$\Sigma \mathbf{F} = m\overline{\mathbf{a}} \tag{18.1}$$

$$\Sigma \mathbf{M}_G = \dot{\mathbf{H}}_G \tag{18.2}$$

on which the analysis of the plane motion of a rigid body was based, remain valid in the most general case of motion of a rigid body. As indicated in Sec. 16.2, these equations express that the system of the external forces is equipollent to the system consisting of the vector $m\overline{\mathbf{a}}$ attached at G and the couple of moment $\dot{\mathbf{H}}_G$ (Fig. 18.1). However, the relation $\mathbf{H}_G = \bar{I}\boldsymbol{\omega}$, which enabled us to determine the angular momentum of a rigid slab and which played an important part in the solution of problems involving the plane motion of slabs and bodies symmetrical with respect to the reference plane, ceases to be valid in the case of nonsymmetrical bodies or three-dimensional motion. Thus in the first part of the chapter, in Sec. 18.2, a more general method for computing the angular momentum $\mathbf{H}_G$ of a rigid body in three dimensions will be developed.

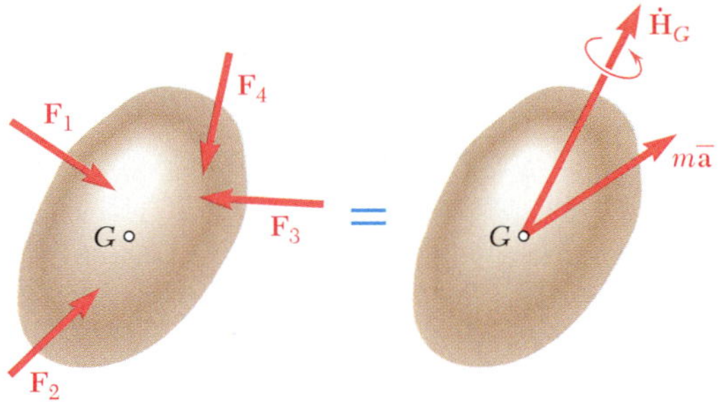

Fig. 18.1

Similarly, although the main feature of the impulse-momentum method discussed in Sec. 17.7, namely, the reduction of the momenta of the particles of a rigid body to a linear momentum vector $m\overline{\mathbf{v}}$ attached at the mass center G of the body and an angular momentum couple $\mathbf{H}_G$, remains valid, the relation $\mathbf{H}_G = \bar{I}\boldsymbol{\omega}$ must be discarded and replaced by the more general relation developed in Sec. 18.2 before this method can be applied to the three-dimensional motion of a rigid body (Sec. 18.3).

We also note that the work-energy principle (Sec. 17.2) and the principle of conservation of energy (Sec. 17.6) still apply in the case of the motion of a rigid body in three dimensions. However, the expression obtained in Sec. 17.4 for the kinetic energy of a rigid body in plane motion will be replaced by a new expression developed in Sec. 18.4 for a rigid body in three-dimensional motion.

In the second part of the chapter, you will first learn to determine the rate of change $\dot{\mathbf{H}}_G$ of the angular momentum $\mathbf{H}_G$ of a three-dimensional rigid body, using a rotating frame of reference with respect to which the moments and products of inertia of the body remain constant (Sec. 18.5). Equations (18.1) and (18.2) will then be expressed in the form of free-body-diagram equations, which can be used to solve various problems involving the three-dimensional motion of rigid bodies (Secs. 18.6 through 18.8).

The last part of the chapter (Secs. 18.9 through 18.11) is devoted to the study of the motion of the gyroscope or, more generally, of an axisymmetrical body with a fixed point located on its axis of symmetry. In Sec. 18.10, the particular case of the steady precession of a gyroscope will be considered, and, in Sec. 18.11, the motion of an axisymmetrical body subjected to no force, except its own weight, will be analyzed.

*18.2. ANGULAR MOMENTUM OF A RIGID BODY IN THREE DIMENSIONS

In this section you will see how the angular momentum $\mathbf{H}_G$ of a body about its mass center G can be determined from the angular velocity $\boldsymbol{\omega}$ of the body in the case of three-dimensional motion.

According to Eq. (14.24), the angular momentum of the body about G can be expressed as

$$\mathbf{H}_G = \sum_{i=1}^{n} (\mathbf{r}'_i \times \mathbf{v}'_i \, \Delta m_i) \tag{18.3}$$

where $\mathbf{r}'_i$ and $\mathbf{v}'_i$ denote, respectively, the position vector and the velocity of the particle P_i, of mass Δm_i, relative to the centroidal frame $Gxyz$ (Fig. 18.2). But $\mathbf{v}'_i = \boldsymbol{\omega} \times \mathbf{r}'_i$, where $\boldsymbol{\omega}$ is the angular

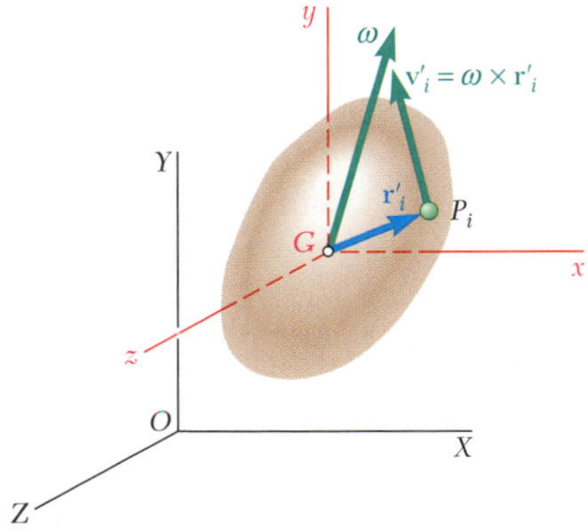

Fig. 18.2

velocity of the body at the instant considered. Substituting into (18.3), we have

$$\mathbf{H}_G = \sum_{i=1}^{n} [\mathbf{r}'_i \times (\boldsymbol{\omega} \times \mathbf{r}'_i)\, \Delta m_i]$$

Recalling the rule for determining the rectangular components of a vector product (Sec. 3.5), we obtain the following expression for the x component of the angular momentum:

$$\begin{aligned} H_x &= \sum_{i=1}^{n} [y_i(\boldsymbol{\omega} \times \mathbf{r}'_i)_z - z_i(\boldsymbol{\omega} \times \mathbf{r}'_i)_y]\, \Delta m_i \\ &= \sum_{i=1}^{n} [y_i(\omega_x y_i - \omega_y x_i) - z_i(\omega_z x_i - \omega_x z_i)]\, \Delta m_i \\ &= \omega_x \sum_i (y_i^2 + z_i^2)\, \Delta m_i - \omega_y \sum_i x_i y_i\, \Delta m_i - \omega_z \sum_i z_i x_i\, \Delta m_i \end{aligned}$$

Replacing the sums by integrals in this expression and in the two similar expressions which are obtained for H_y and H_z, we have

$$\begin{aligned} H_x &= \omega_x \int (y^2 + z^2)\, dm - \omega_y \int xy\, dm - \omega_z \int zx\, dm \\ H_y &= -\omega_x \int xy\, dm + \omega_y \int (z^2 + x^2)\, dm - \omega_z \int yz\, dm \\ H_z &= -\omega_x \int zx\, dm - \omega_y \int yz\, dm + \omega_z \int (x^2 + y^2)\, dm \end{aligned} \tag{18.4}$$

We note that the integrals containing squares represent the *centroidal mass moments of inertia* of the body about the x, y, and z axes, respectively (Sec. 9.11); we have

$$\bar{I}_x = \int (y^2 + z^2)\, dm \qquad \bar{I}_y = \int (z^2 + x^2)\, dm \qquad \bar{I}_z = \int (x^2 + y^2)\, dm \tag{18.5}$$

Similarly, the integrals containing products of coordinates represent the *centroidal mass products of inertia* of the body (Sec. 9.16); we have

$$\bar{I}_{xy} = \int xy\, dm \qquad \bar{I}_{yz} = \int yz\, dm \qquad \bar{I}_{zx} = \int zx\, dm \tag{18.6}$$

Substituting from (18.5) and (18.6) into (18.4), we obtain the components of the angular momentum $\mathbf{H}_G$ of the body about its mass center:

$$\begin{aligned} H_x &= +\bar{I}_x \omega_x - \bar{I}_{xy} \omega_y - \bar{I}_{xz} \omega_z \\ H_y &= -\bar{I}_{yx} \omega_x + \bar{I}_y \omega_y - \bar{I}_{yz} \omega_z \\ H_z &= -\bar{I}_{zx} \omega_x - \bar{I}_{zy} \omega_y + \bar{I}_z \omega_z \end{aligned} \tag{18.7}$$

The relations (18.7) show that the operation which transforms the vector $\boldsymbol{\omega}$ into the vector $\mathbf{H}_G$ (Fig. 18.3) is characterized by the array of moments and products of inertia

$$\begin{pmatrix} \bar{I}_x & -\bar{I}_{xy} & -\bar{I}_{xz} \\ -\bar{I}_{yx} & \bar{I}_y & -\bar{I}_{yz} \\ -\bar{I}_{zx} & -\bar{I}_{zy} & \bar{I}_z \end{pmatrix} \tag{18.8}$$

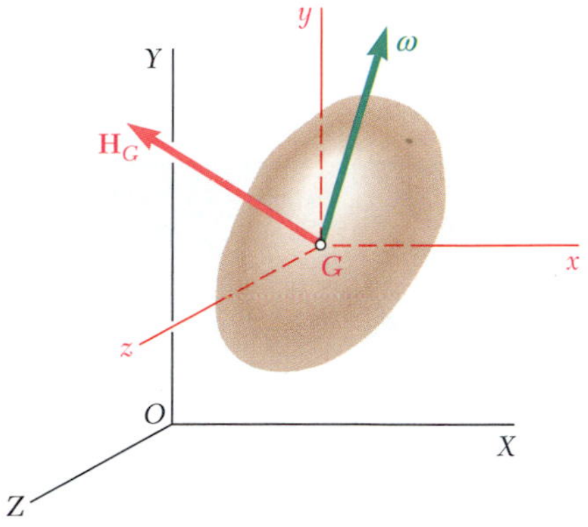

Fig. 18.3

The array (18.8) defines the *inertia tensor* of the body at its mass center G.† A new array of moments and products of inertia would be obtained if a different system of axes were used. The transformation characterized by this new array, however, would still be the same. Clearly, the angular momentum $\mathbf{H}_G$ corresponding to a given angular velocity $\boldsymbol{\omega}$ is independent of the choice of the coordinate axes. As was shown in Secs. 9.17 and 9.18, it is always possible to select a system of axes $Gx'y'z'$, called *principal axes of inertia,* with respect to which all the products of inertia of a given body are zero. The array (18.8) takes then the diagonalized form

$$\begin{pmatrix} \bar{I}_{x'} & 0 & 0 \\ 0 & \bar{I}_{y'} & 0 \\ 0 & 0 & \bar{I}_{z'} \end{pmatrix} \tag{18.9}$$

where $\bar{I}_{x'}, \bar{I}_{y'}, \bar{I}_{z'}$ represent the *principal centroidal moments of inertia* of the body, and the relations (18.7) reduce to

$$H_{x'} = \bar{I}_{x'}\omega_{x'} \qquad H_{y'} = \bar{I}_{y'}\omega_{y'} \qquad H_{z'} = \bar{I}_{z'}\omega_{z'} \tag{18.10}$$

We note that if the three principal centroidal moments of inertia $\bar{I}_{x'}, \bar{I}_{y'}, \bar{I}_{z'}$ are equal, the components $H_{x'}, H_{y'}, H_{z'}$ of the angular momentum about G are proportional to the components $\omega_{x'}, \omega_{y'}, \omega_{z'}$ of the angular velocity, and the vectors $\mathbf{H}_G$ and $\boldsymbol{\omega}$ are collinear. In general, however, the principal moments of inertia will be different, and the vectors $\mathbf{H}_G$ and $\boldsymbol{\omega}$ *will have different directions,* except when two of the three components of $\boldsymbol{\omega}$ happen to be zero, that is, when $\boldsymbol{\omega}$ is directed along one of the coordinate axes. Thus, *the angular momentum* $\mathbf{H}_G$ *of a rigid body and its angular velocity* $\boldsymbol{\omega}$ *have the same direction if, and only if,* $\boldsymbol{\omega}$ *is directed along a principal axis of inertia.*‡

†Setting $\bar{I}_x = I_{11}$, $\bar{I}_y = I_{22}$, $\bar{I}_z = I_{33}$, and $-\bar{I}_{xy} = I_{12}$, $-\bar{I}_{xz} = I_{13}$, etc., we may write the inertia tensor (18.8) in the standard form

$$\begin{pmatrix} I_{11} & I_{12} & I_{13} \\ I_{21} & I_{22} & I_{23} \\ I_{31} & I_{32} & I_{33} \end{pmatrix}$$

Denoting by H_1, H_2, H_3 the components of the angular momentum $\mathbf{H}_G$ and by $\omega_1, \omega_2, \omega_3$ the components of the angular velocity $\boldsymbol{\omega}$, we can write the relations (18.7) in the form

$$H_i = \sum_j I_{ij}\omega_j$$

where i and j take the values 1, 2, 3. The quantities I_{ij} are said to be the *components* of the inertia tensor. Since $I_{ij} = I_{ji}$, the inertia tensor is a *symmetric tensor of the second order.*

‡In the particular case when $\bar{I}_{x'} = \bar{I}_{y'} = \bar{I}_{z'}$, any line through G can be considered as a principal axis of inertia, and the vectors $\mathbf{H}_G$ and $\boldsymbol{\omega}$ are always collinear.

Since this condition is satisfied in the case of the plane motion of a rigid body symmetrical with respect to the reference plane, we were able in Secs. 16.3 and 17.8 to represent the angular momentum $\mathbf{H}_G$ of such a body by the vector $\bar{I}\boldsymbol{\omega}$. We must realize, however, that this result cannot be extended to the case of the plane motion of a nonsymmetrical body, or to the case of the three-dimensional motion of a rigid body. Except when $\boldsymbol{\omega}$ happens to be directed along a principal axis of inertia, the angular momentum and angular velocity of a rigid body have different directions, and the relation (18.7) or (18.10) must be used to determine $\mathbf{H}_G$ from $\boldsymbol{\omega}$.

Photo 18.1 The design of a robotic welder for an automobile assembly line requires a three-dimensional study of both kinematics and kinetics.

Reduction of the Momenta of the Particles of a Rigid Body to a Momentum Vector and a Couple at *G*. We saw in Sec. 17.8 that the system formed by the momenta of the various particles of a rigid body can be reduced to a vector $\mathbf{L}$ attached at the mass center G of the body, representing the linear momentum of the body, and to a couple $\mathbf{H}_G$, representing the angular momentum of the body about G (Fig. 18.4). We are now in a position to determine the vector $\mathbf{L}$ and the couple $\mathbf{H}_G$ in the most general case of three-dimensional motion of a rigid body. As in the case of the two-dimensional motion considered in Sec. 17.8, the linear momentum $\mathbf{L}$ of the body is equal to the product $m\bar{\mathbf{v}}$ of its mass m and the velocity $\bar{\mathbf{v}}$ of its mass center G. The angular momentum $\mathbf{H}_G$, however, can no longer be obtained by simply multiplying the angular velocity $\boldsymbol{\omega}$ of the body by the scalar $\bar{I}$; it must now be obtained from the components of $\boldsymbol{\omega}$ and from the centroidal moments and products of inertia of the body through the use of Eq. (18.7) or (18.10).

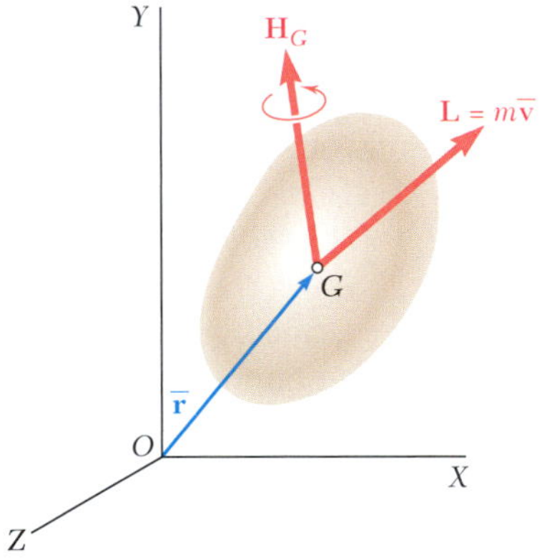

Fig. 18.4

We should also note that once the linear momentum $m\bar{\mathbf{v}}$ and the angular momentum $\mathbf{H}_G$ of a rigid body have been determined, its angular momentum $\mathbf{H}_O$ about any given point O can be obtained by adding the moments about O of the vector $m\bar{\mathbf{v}}$ and of the couple $\mathbf{H}_G$. We write

$$\mathbf{H}_O = \bar{\mathbf{r}} \times m\bar{\mathbf{v}} + \mathbf{H}_G \qquad (18.11)$$

Angular Momentum of a Rigid Body Constrained to Rotate about a Fixed Point. In the particular case of a rigid body constrained to rotate in three-dimensional space about a fixed point O (Fig. 18.5*a*), it is sometimes convenient to determine the angular

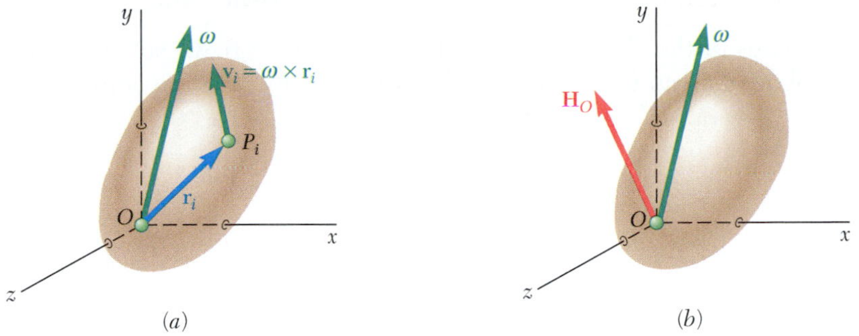

Fig. 18.5

momentum $\mathbf{H}_O$ of the body about the fixed point O. While $\mathbf{H}_O$ could be obtained by first computing $\mathbf{H}_G$ as indicated above and then using Eq. (18.11), it is often advantageous to determine $\mathbf{H}_O$ directly from the angular velocity $\boldsymbol{\omega}$ of the body and its moments and products of inertia with respect to a frame $Oxyz$ centered at the fixed point O. Recalling Eq. (14.7), we write

$$\mathbf{H}_O = \sum_{i=1}^{n} (\mathbf{r}_i \times \mathbf{v}_i \, \Delta m_i) \qquad (18.12)$$

where $\mathbf{r}_i$ and $\mathbf{v}_i$ denote, respectively, the position vector and the velocity of the particle P_i with respect to the fixed frame $Oxyz$. Substituting $\mathbf{v}_i = \boldsymbol{\omega} \times \mathbf{r}_i$, and after manipulations similar to those used in the earlier part of this section, we find that the components of the angular momentum $\mathbf{H}_O$ (Fig. 18.5*b*) are given by the relations

$$\begin{aligned} H_x &= +I_x\omega_x - I_{xy}\omega_y - I_{xz}\omega_z \\ H_y &= -I_{yx}\omega_x + I_y\omega_y - I_{yz}\omega_z \\ H_z &= -I_{zx}\omega_x - I_{zy}\omega_y + I_z\omega_z \end{aligned} \qquad (18.13)$$

where the moments of inertia I_x, I_y, I_z and the products of inertia I_{xy}, I_{yz}, I_{zx} are computed with respect to the frame $Oxyz$ centered at the fixed point O.

*18.3. APPLICATION OF THE PRINCIPLE OF IMPULSE AND MOMENTUM TO THE THREE-DIMENSIONAL MOTION OF A RIGID BODY

Before we can apply the fundamental equation (18.2) to the solution of problems involving the three-dimensional motion of a rigid body, we must learn to compute the derivative of the vector $\mathbf{H}_G$. This will be done in Sec. 18.5. The results obtained in the preceding section can, however, be used right away to solve problems by the impulse-momentum method.

Photo 18.2 As a result of the impulsive force applied by the bowling ball, a pin acquires both linear momentum and angular momentum.

Recalling that the system formed by the momenta of the particles of a rigid body reduces to a linear momentum vector $m\bar{\mathbf{v}}$ attached at the mass center G of the body and an angular momentum couple $\mathbf{H}_G$, we represent graphically the fundamental relation

$$\textbf{Syst Momenta}_1 + \textbf{Syst Ext Imp}_{1\rightarrow 2} = \textbf{Syst Momenta}_2 \qquad (17.4)$$

by means of the three sketches shown in Fig. 18.6. To solve a given problem, we can use these sketches to write appropriate component

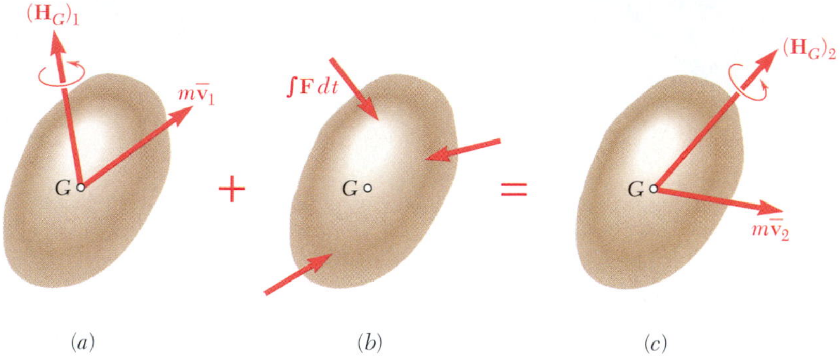

Fig. 18.6

and moment equations, keeping in mind that the components of the angular momentum $\mathbf{H}_G$ are related to the components of the angular velocity $\boldsymbol{\omega}$ by Eqs. (18.7) of the preceding section.

In solving problems dealing with the motion of a body rotating about a fixed point O, it will be convenient to eliminate the impulse of the reaction at O by writing an equation involving the moments of the momenta and impulses about O. We recall that the angular momentum $\mathbf{H}_O$ of the body about the fixed point O can be obtained either directly from Eqs. (18.13) or by first computing its linear momentum $m\bar{\mathbf{v}}$ and its angular momentum $\mathbf{H}_G$ and then using Eq. (18.11).

*18.4. KINETIC ENERGY OF A RIGID BODY IN THREE DIMENSIONS

Consider a rigid body of mass m in three-dimensional motion. We recall from Sec. 14.6 that if the absolute velocity $\mathbf{v}_i$ of each particle P_i of the body is expressed as the sum of the velocity $\bar{\mathbf{v}}$ of the mass center G of the body and of the velocity $\mathbf{v}'_i$ of the particle relative to a frame $Gxyz$ attached to G and of fixed orientation (Fig. 18.7), the kinetic energy of the system of particles forming the rigid body can be written in the form

$$T = \tfrac{1}{2}m\bar{v}^2 + \frac{1}{2}\sum_{i=1}^{n} \Delta m_i v_i'^2 \qquad (18.14)$$

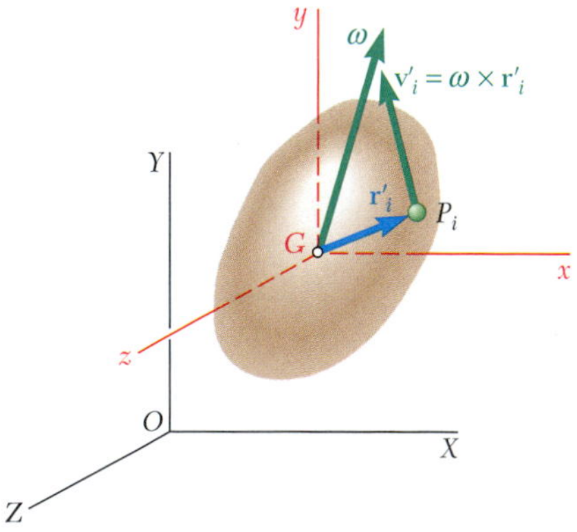

Fig. 18.7

where the last term represents the kinetic energy T' of the body relative to the centroidal frame $Gxyz$. Since $v'_i = |\mathbf{v}'_i| = |\boldsymbol{\omega} \times \mathbf{r}'_i|$, we write

$$T' = \frac{1}{2}\sum_{i=1}^{n} \Delta m_i v_i'^2 = \frac{1}{2}\sum_{i=1}^{n} |\boldsymbol{\omega} \times \mathbf{r}'_i|^2\, \Delta m_i$$

Expressing the square in terms of the rectangular components of the vector product, and replacing the sums by integrals, we have

$$\begin{aligned} T' &= \tfrac{1}{2}\int[(\omega_x y - \omega_y x)^2 + (\omega_y z - \omega_z y)^2 + (\omega_z x - \omega_x z)^2]\, dm \\ &= \tfrac{1}{2}[\omega_x^2 \int(y^2 + z^2)\, dm + \omega_y^2 \int(z^2 + x^2)\, dm + \omega_z^2 \int(x^2 + y^2)\, dm \\ &\qquad - 2\omega_x\omega_y \int xy\, dm - 2\omega_y\omega_z \int yz\, dm - 2\omega_z\omega_x \int zx\, dm] \end{aligned}$$

or, recalling the relations (18.5) and (18.6),

$$T' = \tfrac{1}{2}(\bar{I}_x\omega_x^2 + \bar{I}_y\omega_y^2 + \bar{I}_z\omega_z^2 - 2\bar{I}_{xy}\omega_x\omega_y - 2\bar{I}_{yz}\omega_y\omega_z - 2\bar{I}_{zx}\omega_z\omega_x) \tag{18.15}$$

Substituting into (18.14) the expression (18.15) we have just obtained for the kinetic energy of the body relative to centroidal axes, we write

$$\begin{aligned} T = \tfrac{1}{2}m\bar{v}^2 + \tfrac{1}{2}(\bar{I}_x\omega_x^2 + \bar{I}_y\omega_y^2 + \bar{I}_z\omega_z^2 - 2\bar{I}_{xy}\omega_x\omega_y \\ - 2\bar{I}_{yz}\omega_y\omega_z - 2\bar{I}_{zx}\omega_z\omega_x) \end{aligned} \tag{18.16}$$

If the axes of coordinates are chosen so that they coincide at the instant considered with the principal axes x', y', z' of the body, the relation obtained reduces to

$$T = \tfrac{1}{2}m\bar{v}^2 + \tfrac{1}{2}(\bar{I}_{x'}\omega_{x'}^2 + \bar{I}_{y'}\omega_{y'}^2 + \bar{I}_{z'}\omega_{z'}^2) \tag{18.17}$$

where $\bar{\mathbf{v}}$ = velocity of mass center
$\boldsymbol{\omega}$ = angular velocity
m = mass of rigid body
$\bar{I}_{x'}, \bar{I}_{y'}, \bar{I}_{z'}$ = principal centroidal moments of inertia

The results we have obtained enable us to apply to the three-dimensional motion of a rigid body the principles of work and energy (Sec. 17.2) and conservation of energy (Sec. 17.6).

Kinetic Energy of a Rigid Body with a Fixed Point. In the particular case of a rigid body rotating in three-dimensional space about a fixed point O, the kinetic energy of the body can be expressed in terms of its moments and products of inertia with respect to axes attached at O (Fig. 18.8). Recalling the definition of the kinetic energy of a system of particles, and substituting $v_i = |\mathbf{v}_i| = |\boldsymbol{\omega} \times \mathbf{r}_i|$, we write

$$T = \frac{1}{2}\sum_{i=1}^{n} \Delta m_i v_i^2 = \frac{1}{2}\sum_{i=1}^{n} |\boldsymbol{\omega} \times \mathbf{r}_i|^2\, \Delta m_i \tag{18.18}$$

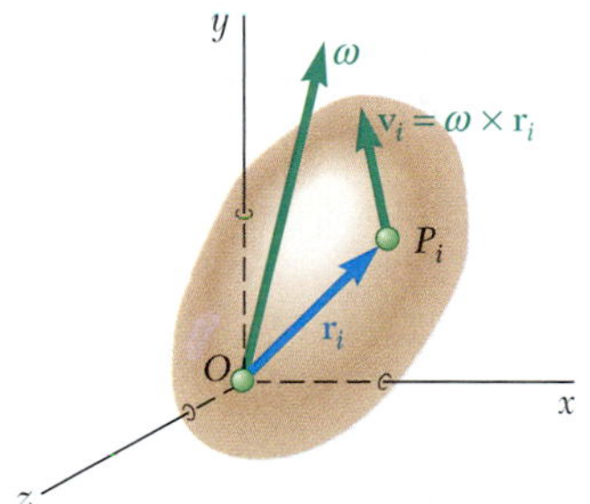

Fig. 18.8

Manipulations similar to those used to derive Eq. (18.15) yield

$$T = \tfrac{1}{2}(I_x\omega_x^2 + I_y\omega_y^2 + I_z\omega_z^2 - 2I_{xy}\omega_x\omega_y - 2I_{yz}\omega_y\omega_z - 2I_{zx}\omega_z\omega_x) \tag{18.19}$$

or, if the principal axes x', y', z' of the body at the origin O are chosen as coordinate axes,

$$T = \tfrac{1}{2}(I_{x'}\omega_{x'}^2 + I_{y'}\omega_{y'}^2 + I_{z'}\omega_{z'}^2) \tag{18.20}$$

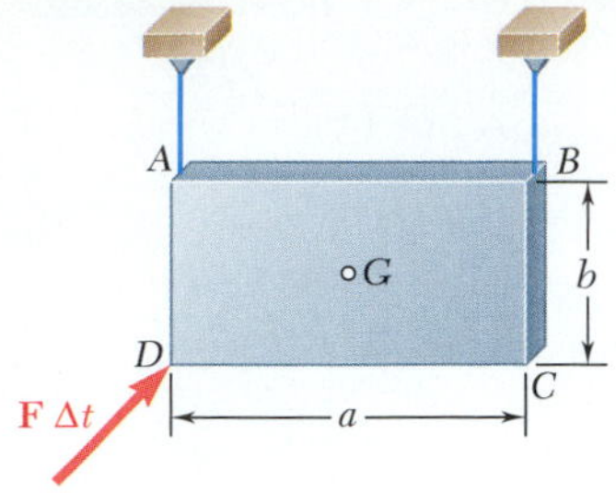

SAMPLE PROBLEM 18.1

A rectangular plate of mass m suspended from two wires at A and B is hit at D in a direction perpendicular to the plate. Denoting by $\mathbf{F}\,\Delta t$ the impulse applied at D, determine immediately after the impact (*a*) the velocity of the mass center G, (*b*) the angular velocity of the plate.

SOLUTION

Assuming that the wires remain taut and thus that the components $\bar{v}_y$ of $\bar{\mathbf{v}}$ and ω_z of $\boldsymbol{\omega}$ are zero after the impact, we have

$$\bar{\mathbf{v}} = \bar{v}_x\mathbf{i} + \bar{v}_z\mathbf{k} \qquad \boldsymbol{\omega} = \omega_x\mathbf{i} + \omega_y\mathbf{j}$$

and since the x, y, z axes are principal axes of inertia,

$$\mathbf{H}_G = \bar{I}_x\omega_x\mathbf{i} + \bar{I}_y\omega_y\mathbf{j} \qquad \mathbf{H}_G = \tfrac{1}{12}mb^2\omega_x\mathbf{i} + \tfrac{1}{12}ma^2\omega_y\mathbf{j} \qquad (1)$$

Principle of Impulse and Momentum. Since the initial momenta are zero, the system of the impulses must be equivalent to the system of the final momenta:

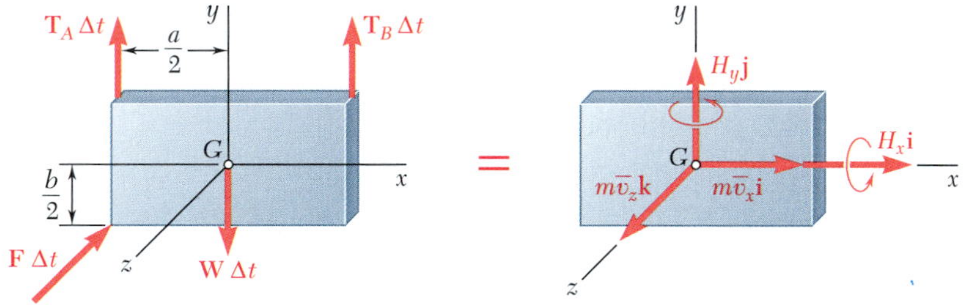

***a.* Velocity of Mass Center.** Equating the components of the impulses and momenta in the x and z directions:

x components: $\quad 0 = m\bar{v}_x \qquad \bar{v}_x = 0$

z components: $\quad -F\,\Delta t = m\bar{v}_z \qquad \bar{v}_z = -F\,\Delta t/m$

$$\bar{\mathbf{v}} = \bar{v}_x\mathbf{i} + \bar{v}_z\mathbf{k} \qquad \bar{\mathbf{v}} = -(F\,\Delta t/m)\mathbf{k} \quad ◀$$

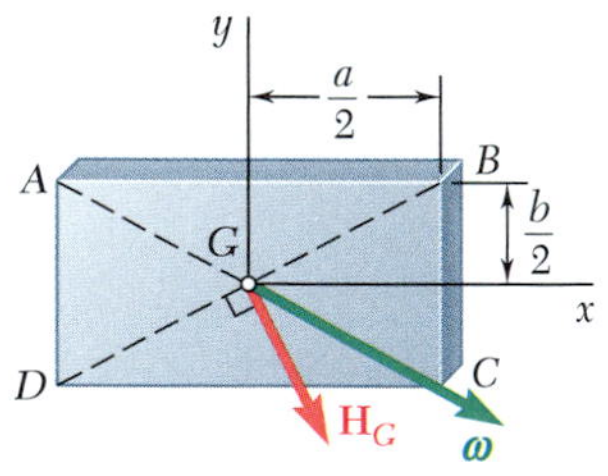

***b.* Angular Velocity.** Equating the moments of the impulses and momenta about the x and y axes:

About x axis: $\quad \tfrac{1}{2}bF\,\Delta t = H_x$

About y axis: $\quad -\tfrac{1}{2}aF\,\Delta t = H_y$

$$\mathbf{H}_G = H_x\mathbf{i} + H_y\mathbf{j} \qquad \mathbf{H}_G = \tfrac{1}{2}bF\,\Delta t\mathbf{i} - \tfrac{1}{2}aF\,\Delta t\mathbf{j} \qquad (2)$$

Comparing Eqs. (1) and (2), we conclude that

$$\omega_x = 6F\,\Delta t/mb \qquad \omega_y = -6F\,\Delta t/ma$$

$$\boldsymbol{\omega} = \omega_x\mathbf{i} + \omega_y\mathbf{j} \qquad \boldsymbol{\omega} = (6F\,\Delta t/mab)(a\mathbf{i} - b\mathbf{j}) \quad ◀$$

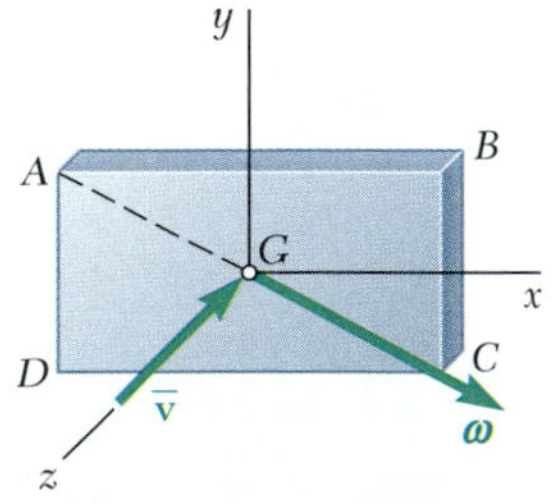

We note that $\boldsymbol{\omega}$ is directed along the diagonal AC.

Remark: Equating the y components of the impulses and momenta, and their moments about the z axis, we obtain two additional equations which yield $T_A = T_B = \tfrac{1}{2}W$. We thus verify that the wires remain taut and that our assumption was correct.

SAMPLE PROBLEM 18.2

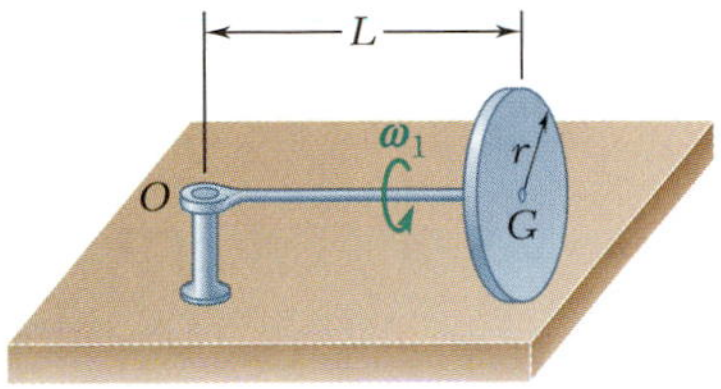

A homogeneous disk of radius r and mass m is mounted on an axle OG of length L and negligible mass. The axle is pivoted at the fixed point O, and the disk is constrained to roll on a horizontal floor. Knowing that the disk rotates counterclockwise at the rate ω_1 about the axle OG, determine (*a*) the angular velocity of the disk, (*b*) its angular momentum about O, (*c*) its kinetic energy, (*d*) the vector and couple at G equivalent to the momenta of the particles of the disk.

SOLUTION

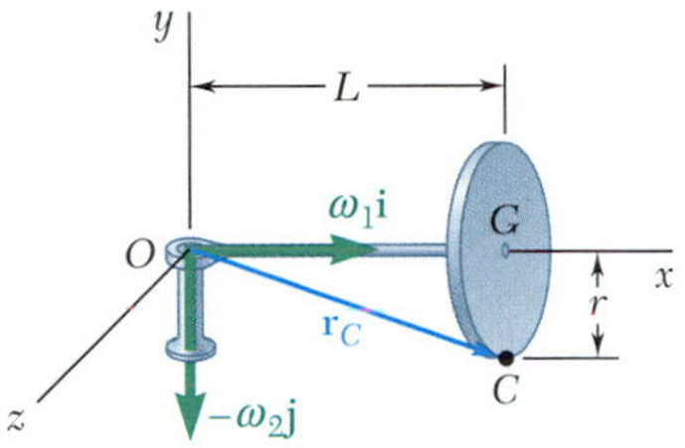

***a.* Angular Velocity.** As the disk rotates about the axle OG it also rotates with the axle about the y axis at a rate ω_2 clockwise. The total angular velocity of the disk is therefore

$$\boldsymbol{\omega} = \omega_1\mathbf{i} - \omega_2\mathbf{j} \tag{1}$$

To determine ω_2 we write that the velocity of C is zero:

$$\mathbf{v}_C = \boldsymbol{\omega} \times \mathbf{r}_C = 0$$
$$(\omega_1\mathbf{i} - \omega_2\mathbf{j}) \times (L\mathbf{i} - r\mathbf{j}) = 0$$
$$(L\omega_2 - r\omega_1)\mathbf{k} = 0 \qquad \omega_2 = r\omega_1/L$$

Substituting into (1) for ω_2: $\boldsymbol{\omega} = \omega_1\mathbf{i} = (r\omega_1/L)\mathbf{j}$ ◀

***b.* Angular Momentum about *O*.** Assuming the axle to be part of the disk, we can consider the disk to have a fixed point at O. Since the x, y, and z axes are principal axes of inertia for the disk,

$$H_x = I_x\omega_x = (\tfrac{1}{2}mr^2)\omega_1$$
$$H_y = I_y\omega_y = (mL^2 + \tfrac{1}{4}mr^2)(-r\omega_1/L)$$
$$H_z = I_z\omega_z = (mL^2 + \tfrac{1}{4}mr^2)0 = 0$$
$$\mathbf{H}_O = \tfrac{1}{2}mr^2\omega_1\mathbf{i} - m(L^2 + \tfrac{1}{4}r^2)(r\omega_1/L)\mathbf{j} \quad ◀$$

***c.* Kinetic Energy.** Using the values obtained for the moments of inertia and the components of $\boldsymbol{\omega}$, we have

$$T = \tfrac{1}{2}(I_x\omega_x^2 + I_y\omega_y^2 + I_z\omega_z^2) = \tfrac{1}{2}[\tfrac{1}{2}mr^2\omega_1^2 + m(L^2 + \tfrac{1}{4}r^2)(-r\omega_1/L)^2]$$
$$T = \tfrac{1}{8}mr^2\left(6 + \frac{r^2}{L^2}\right)\omega_1^2 \quad ◀$$

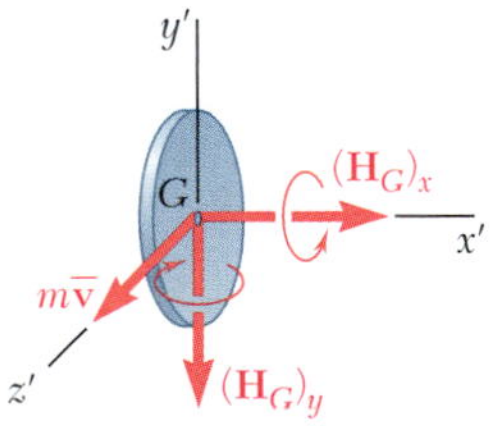

***d.* Momentum Vector and Couple at *G*.** The linear momentum vector $m\overline{\mathbf{v}}$ and the angular momentum couple $\mathbf{H}_G$ are

$$m\overline{\mathbf{v}} = mr\omega_1\mathbf{k} \quad ◀$$

and

$$\mathbf{H}_G = \overline{I}_{x'}\omega_x\mathbf{i} + \overline{I}_{y'}\omega_y\mathbf{j} + \overline{I}_{z'}\omega_z\mathbf{k} = \tfrac{1}{2}mr^2\omega_1\mathbf{i} + \tfrac{1}{4}mr^2(-r\omega_1/L)\mathbf{j}$$
$$\mathbf{H}_G = \tfrac{1}{2}mr^2\omega_1\left(\mathbf{i} - \frac{r}{2L}\mathbf{j}\right) \quad ◀$$

SOLVING PROBLEMS ON YOUR OWN

In this lesson you learned to compute the *angular momentum of a rigid body in three dimensions* and to apply the principle of impulse and momentum to the three-dimensional motion of a rigid body. You also learned to compute the *kinetic energy of a rigid body in three dimensions*. It is important for you to keep in mind that, except for very special situations, the angular momentum of a rigid body in three dimensions *cannot* be expressed as the product $\bar{I}\boldsymbol{\omega}$ and, therefore, *will not have the same direction as the angular velocity* $\boldsymbol{\omega}$ (Fig. 18.3).

1. To compute the angular momentum $\mathbf{H}_G$ of a rigid body about its mass center G, you must first determine the angular velocity $\boldsymbol{\omega}$ of the body with respect to a system of axes *centered at G and of fixed orientation*. Since you will be asked in this lesson to determine the angular momentum of the body *at a given instant only*, select the system of axes which will be most convenient for your computations.

a. If the principal axes of inertia of the body at G are known, use these axes as coordinate axes x', y', and z', since the corresponding products of inertia of the body will be equal to zero. Resolve $\boldsymbol{\omega}$ into components $\omega_{x'}$, $\omega_{y'}$, and $\omega_{z'}$ along these axes and compute the principal moments of inertia $\bar{I}_{x'}$, $\bar{I}_{y'}$, and $\bar{I}_{z'}$. The corresponding components of the angular momentum $\mathbf{H}_G$ are

$$H_{x'} = \bar{I}_{x'}\omega_{x'} \qquad H_{y'} = \bar{I}_{y'}\omega_{y'} \qquad H_{z'} = \bar{I}_{z'}\omega_{z'} \tag{18.10}$$

b. If the principal axes of inertia of the body at G are not known, you must use Eqs. (18.7) to determine the components of the angular momentum $\mathbf{H}_G$. These equations require prior computation of the *product of inertia* of the body as well as prior computation of its moments of inertia with respect to the selected axes.

c. The magnitude and direction cosines of $\mathbf{H}_G$ are obtained from formulas similar to those used in Statics [Sec. 2.12]. We have

$$H_G = \sqrt{H_x^2 + H_y^2 + H_z^2}$$

$$\cos\theta_x = \frac{H_x}{H_G} \qquad \cos\theta_y = \frac{H_y}{H_G} \qquad \cos\theta_z = \frac{H_z}{H_G}$$

d. Once $\mathbf{H}_G$ has been determined, you can obtain the angular momentum of the body *about any given point O* by observing from Fig. 18.4 that

$$\mathbf{H}_O = \bar{\mathbf{r}} \times m\bar{\mathbf{v}} + \mathbf{H}_G \tag{18.11}$$

where $\bar{\mathbf{r}}$ is the position vector of G relative to O, and $m\bar{\mathbf{v}}$ is the linear momentum of the body.

2. To compute the angular momentum $\mathbf{H}_O$ of a rigid body with a fixed point O, follow the procedure described in paragraph 1, except that you should now use axes centered at the fixed point O.

a. If the principal axes of inertia of the body at O are known, resolve $\boldsymbol{\omega}$ into components along these axes [Sample Prob. 18.2]. The corresponding components of the angular momentum $\mathbf{H}_G$ are obtained from equations similar to Eqs. (18.10).

b. If the principal axes of inertia of the body at O are not known, you must compute the products as well as the moments of inertia of the body with respect to the axes that you have selected and use Eqs. (18.13) to determine the components of the angular momentum $\mathbf{H}_O$.

3. ***To apply the principle of impulse and momentum*** to the solution of a problem involving the three-dimensional motion of a rigid body, you will use the same vector equation that you used for plane motion in Chap. 17,

$$\textbf{Syst Momenta}_1 + \textbf{Syst Ext Imp}_{1\rightarrow 2} = \textbf{Syst Momenta}_2 \qquad (17.4)$$

where the initial and final systems of momenta are each represented by a *linear-momentum vector* $m\overline{\mathbf{v}}$ and an *angular-momentum couple* $\mathbf{H}_G$. Now, however, these vector-and-couple systems should be represented in three dimensions as shown in Fig. 18.6, and $\mathbf{H}_G$ should be determined as explained in paragraph 1.

a. In problems involving the application of a known impulse to a rigid body, draw the free-body-diagram equation corresponding to Eq. (17.4). Equating the components of the vectors involved, you will determine the final linear momentum $m\overline{\mathbf{v}}$ of the body and, thus, the corresponding velocity $\overline{\mathbf{v}}$ of its mass center. Equating moments about G, you will determine the final angular momentum $\mathbf{H}_G$ of the body. You will then substitute the values obtained for the components of $\mathbf{H}_G$ into Eqs. (18.10) or (18.7) and solve these equations for the corresponding values of the components of the angular velocity $\boldsymbol{\omega}$ of the body [Sample Prob. 18.1].

b. In problems involving unknown impulses, draw the free-body-diagram equation corresponding to Eq. (17.4) and write equations which do not involve the unknown impulses. Such equations can be obtained by equating moments about the point or line of impact.

4. ***To compute the kinetic energy of a rigid body with a fixed point O,*** resolve the angular velocity $\boldsymbol{\omega}$ into components along axes of your choice and compute the moments and products of inertia of the body with respect to these axes. As was the case for the computation of the angular momentum, use the principal axes of inertia x', y', and z' if you can easily determine them. The products of inertia will then be zero [Sample Prob. 18.2], and the expression for the kinetic energy will reduce to

$$T = \tfrac{1}{2}(I_{x'}\omega_{x'}^2 + I_{y'}\omega_{y'}^2 + I_{z'}\omega_{z'}^2) \qquad (18.20)$$

If you must use axes other than the principal axes of inertia, the kinetic energy of the body should be expressed as shown in Eq. (18.19).

5. ***To compute the kinetic energy of a rigid body in general motion,*** consider the motion as the sum of a *translation with the mass center G and a rotation about G*. The kinetic energy associated with the translation is $\frac{1}{2}m\overline{v}^2$. If principal axes of inertia can be used, the kinetic energy associated with the rotation about G can be expressed in the form used in Eq. (18.20). The total kinetic energy of the rigid body is then

$$T = \tfrac{1}{2}m\overline{v}^2 + \tfrac{1}{2}(\overline{I}_{x'}\omega_{x'}^2 + \overline{I}_{y'}\omega_{y'}^2 + \overline{I}_{z'}\omega_{z'}^2) \qquad (18.17)$$

If you must use axes other than the principal axes of inertia to determine the kinetic energy associated with the rotation about G, the total kinetic energy of the body should be expressed as shown in Eq. (18.16).

Problems

18.1 A rectangular plate of mass 8 kg is attached to a shaft as shown. Knowing that the plate has an angular velocity of constant magnitude $\omega = 5$ rad/s, determine the angular momentum $\mathbf{H}_G$ of the plate about its mass center G.

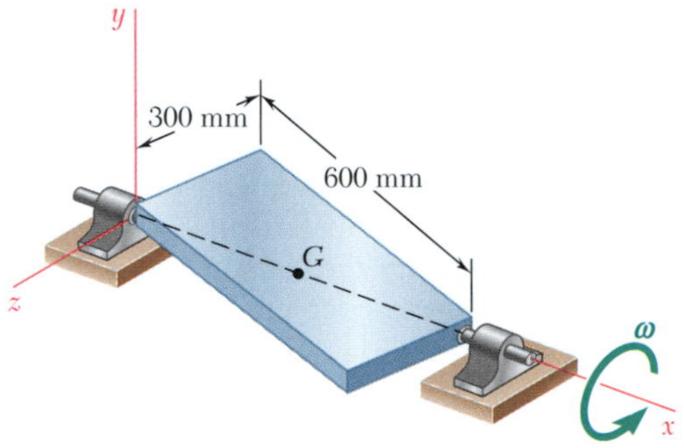

Fig. P18.1

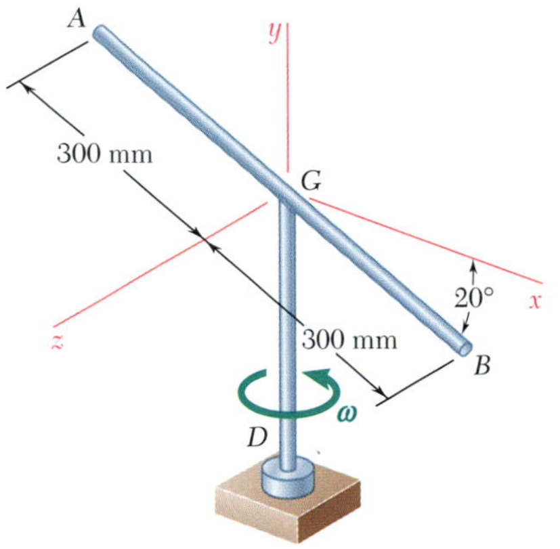

Fig. P18.2

18.2 A uniform 1.8-kg rod AB is welded at its midpoint G to a vertical shaft GD. Knowing that the shaft rotates with an angular velocity of constant magnitude $\omega = 1200$ rpm, determine the angular momentum $\mathbf{H}_G$ of the rod about G.

18.3 Two uniform rods AB and CE, each of weight 3 lb and length 24 in., are welded to each other at their midpoints. Knowing that this assembly has an angular velocity of constant magnitude $\omega = 12$ rad/s, determine the magnitude and direction of the angular momentum $\mathbf{H}_D$ of the assembly about D.

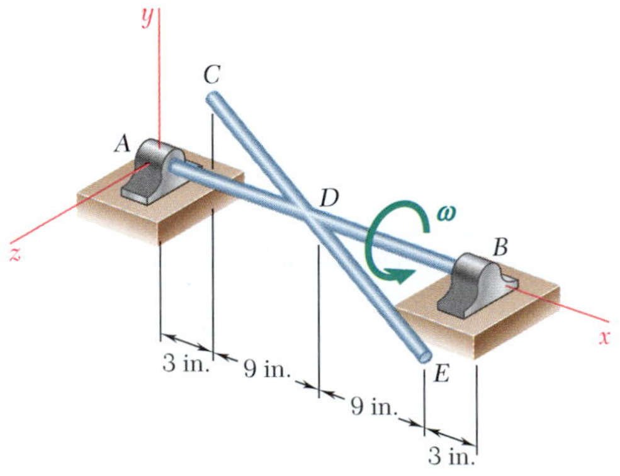

Fig. P18.3

18.4 A thin homogeneous square plate of mass m and side a is welded to a vertical shaft AB with which it forms an angle of 45°. Knowing that the shaft rotates with a constant angular velocity $\boldsymbol{\omega}$, determine the angular momentum $\mathbf{H}_A$ of the plate about point A.

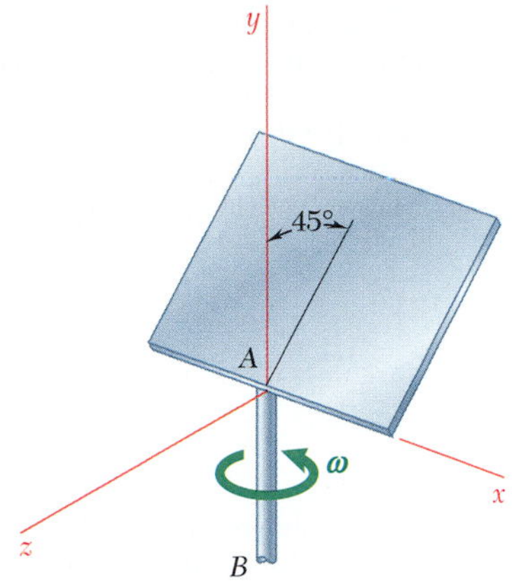

Fig. P18.4

18.5 A homogeneous disk of mass $m = 6$ kg rotates at the constant rate $\omega_1 = 16$ rad/s with respect to arm ABC, which is welded to a shaft DCE rotating at the constant rate $\omega_2 = 8$ rad/s. Determine the angular momentum $\mathbf{H}_A$ of the disk about its center A.

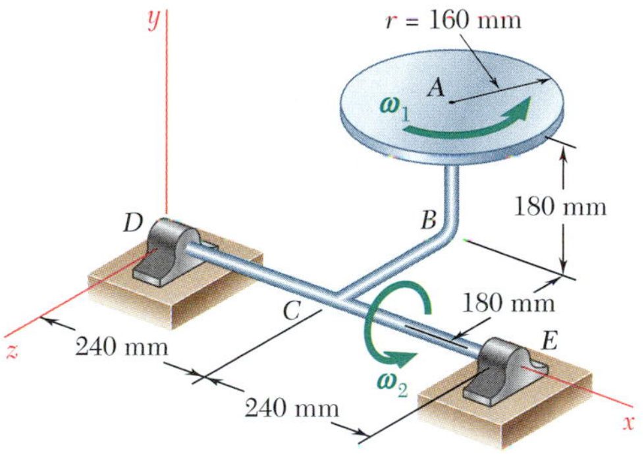

Fig. P18.5

18.6 A homogeneous disk of mass $m = 8$ kg rotates at the constant rate $\omega_1 = 12$ rad/s with respect to arm OA, which itself rotates at the constant rate $\omega_2 = 4$ rad/s about the y axis. Determine the angular momentum $\mathbf{H}_A$ of the disk about its center A.

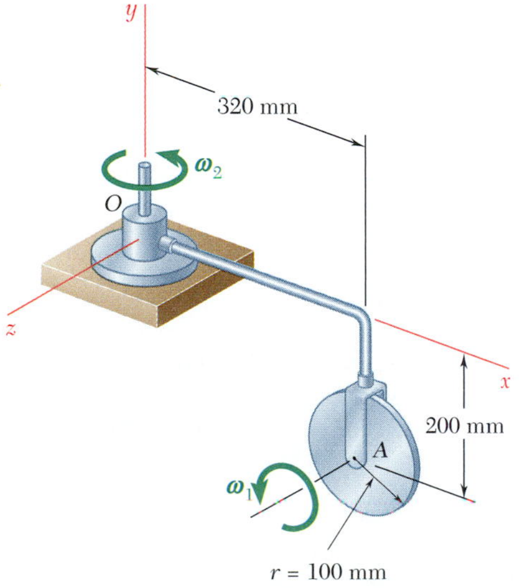

Fig. P18.6

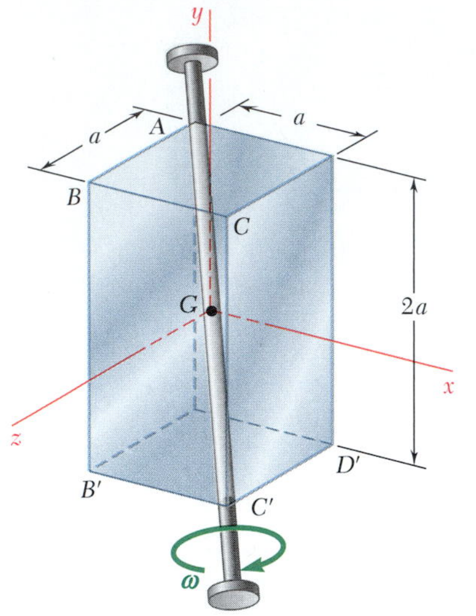

Fig. *P18.7*

18.7 A solid rectangular parallelepiped of mass m has a square base of side a and a length $2a$. Knowing that it rotates at the constant rate ω about its diagonal AC' and that its rotation is observed from A as counterclockwise, determine (*a*) the magnitude of the angular momentum $\mathbf{H}_G$ of the parallelepiped about its mass center G, (*b*) the angle that $\mathbf{H}_G$ forms with the diagonal AC'.

18.8 Solve Prob. 18.7, assuming that the solid rectangular parallelepiped has been replaced by a hollow one consisting of six thin metal plates welded together.

18.9 Determine the angular momentum $\mathbf{H}_D$ of the disk of Prob. 18.5 about point D.

18.10 Determine the angular momentum $\mathbf{H}_O$ of the disk of Prob. 18.6 about the fixed point O.

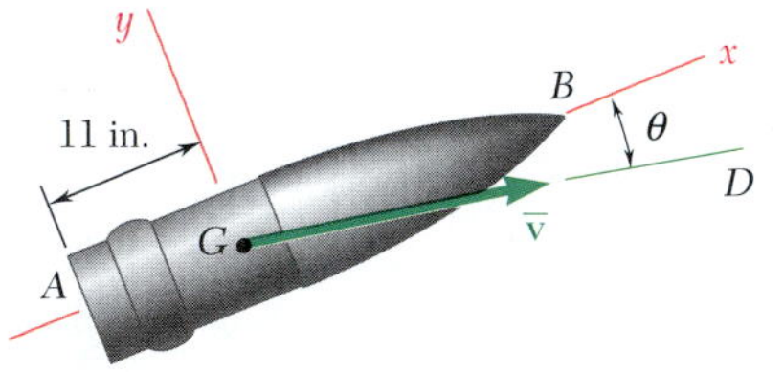

Fig. P18.11

18.11 The 60-lb projectile shown has a radius of gyration of 2.4 in. about its axis of symmetry Gx and a radius of gyration of 10 in. about the transverse axis Gy. Its angular velocity $\boldsymbol{\omega}$ can be resolved into two components: one component, directed along Gx, measures the *rate of spin* of the projectile, while the other component, directed along GD, measures its *rate of precession*. Knowing that $\theta = 5°$ and that the angular momentum of the projectile about its mass center G is $\mathbf{H}_G = (0.640 \text{ lb}\cdot\text{ft}\cdot\text{s})\mathbf{i} - (0.018 \text{ lb}\cdot\text{ft}\cdot\text{s})\mathbf{j}$, determine (*a*) the rate of spin, (*b*) the rate of precession.

18.12 Determine the angular momentum $\mathbf{H}_A$ of the projectile of Prob. 18.11 about the center A of its base, knowing that its mass center G has a velocity $\overline{\mathbf{v}}$ of 1950 ft/s. Give your answer in terms of components respectively parallel to the x and y axes shown and to a third axis z pointing toward you.

18.13 Determine the angular momentum $\mathbf{H}_O$ of the disk of Sample Prob. 18.2 from the expressions obtained for its linear momentum $m\overline{\mathbf{v}}$ and its angular momentum $\mathbf{H}_G$, using Eqs. (18.11). Verify that the result obtained is the same as that obtained by direct computation.

18.14 (*a*) Show that the angular momentum $\mathbf{H}_B$ of a rigid body about point B can be obtained by adding to the angular momentum $\mathbf{H}_A$ of that body about point A the vector product of the vector $\mathbf{r}_{A/B}$ drawn from B to A and the linear momentum $m\overline{\mathbf{v}}$ of the body:

$$\mathbf{H}_B = \mathbf{H}_A + \mathbf{r}_{A/B} \times m\overline{\mathbf{v}}$$

(*b*) Further show that when a rigid body rotates about a fixed axis, its angular momentum is the same about any two points A and B located on the fixed axis ($\mathbf{H}_A = \mathbf{H}_B$) if, and only if, the mass center G of the body is located on the fixed axis.

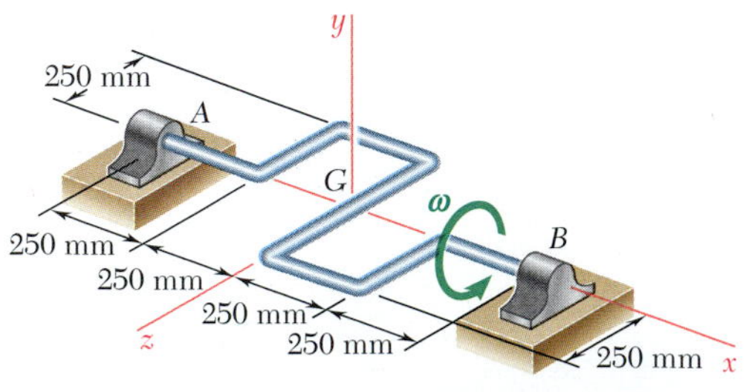

Fig. P18.15

18.15 A 5-kg rod of uniform cross section is used to form the shaft shown. Knowing that the shaft rotates with a constant angular velocity $\boldsymbol{\omega}$ of magnitude 12 rad/s, determine (*a*) the angular momentum $\mathbf{H}_G$ of the shaft about its mass center G, (*b*) the angle formed by $\mathbf{H}_G$ and the axis AB.

18.16 Determine the angular momentum of the shaft of Prob. 18.15 about (*a*) point *A*, (*b*) point *B*.

18.17 The triangular plate shown has a weight of 16 lb and is welded to a vertical shaft *AB*. Knowing that the plate rotates at the constant rate ω = 12 rad/s, determine its angular momentum about (*a*) point *C*, (*b*) point *A*. (*Hint:* To solve part *b* find $\bar{\mathbf{v}}$ and use the property indicated in part *a* of Prob. 18.14.)

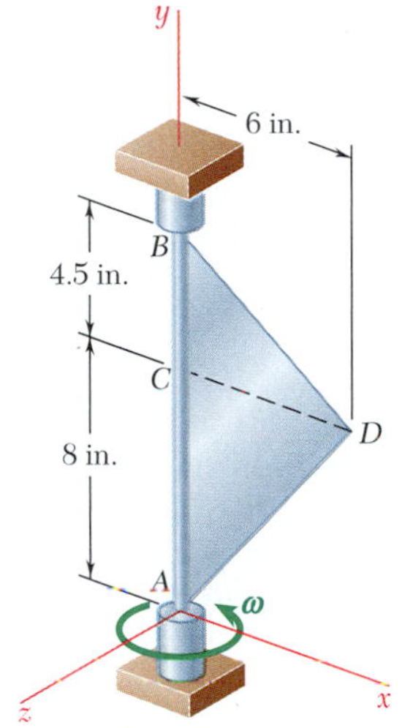

Fig. P18.17 and P18.18

18.18 The triangular plate shown has a weight of 16 lb and is welded to a vertical shaft *AB*. Knowing that the plate rotates at the constant rate ω = 12 rad/s, determine its angular momentum about (*a*) point *C*, (*b*) point *B*. (See hint of Prob. 18.17.)

18.19 Two L-shaped arms, each of mass 5 kg, are welded at the one-third points of the 600 mm shaft *AB* to form the assembly shown. Knowing that the assembly rotates at the constant rate of 360 rpm, determine (*a*) the angular momentum $\mathbf{H}_A$ of the assembly about point *A*, (*b*) the angle formed by $\mathbf{H}_A$ and *AB*.

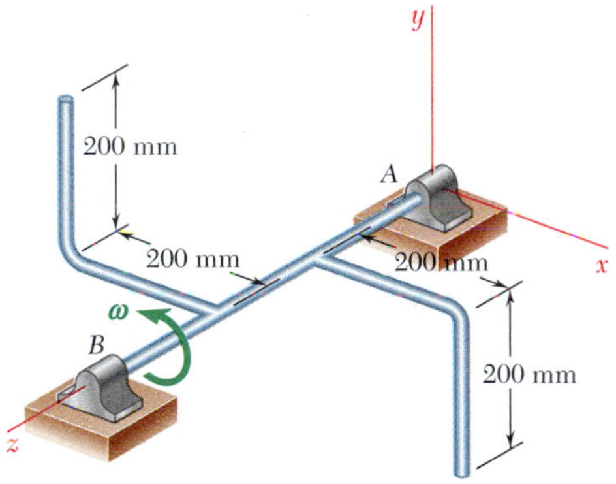

Fig. *P18.19*

18.20 For the assembly of Prob. 18.19, determine (*a*) the angular momentum $\mathbf{H}_B$ of the assembly about point *B*, (*b*) the angle formed by $\mathbf{H}_B$ and *BA*.

18.21 One of the sculptures displayed on a university campus consists of a hollow cube made of six aluminum sheets, each 1.5 × 1.5 m, welded together and reinforced with internal braces of negligible mass. The cube is mounted on a fixed base *A* and can rotate freely about its vertical diagonal *AB*. As she passes by this display on the way to a class in mechanics, an engineering student grabs corner *C* of the cube and pushes it for 1.2 s in a direction perpendicular to the plane *ABC* with an average force of 50 N. Having observed that it takes 5 s for the cube to complete one full revolution, she flips out her calculator and proceeds to determine the mass of the cube. What is the result of her calculation? (*Hint:* The perpendicular distance from the diagonal joining two vertices of a cube to any of its other six vertices can be obtained by multiplying the side of the cube by $\sqrt{2/3}$.)

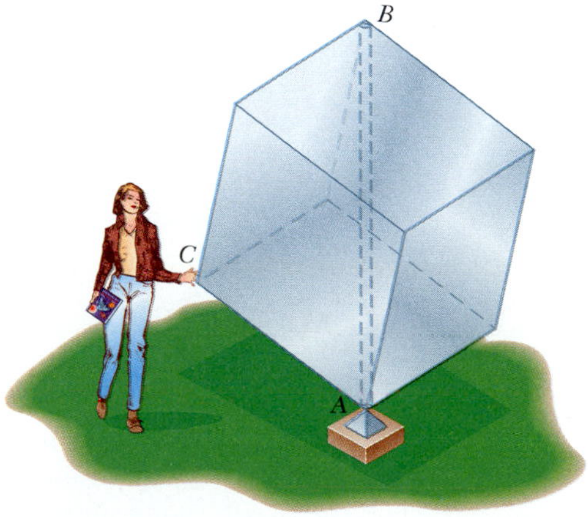

Fig. P18.21

18.22 If the aluminum cube of Prob. 18.21 were replaced by a cube of the same size, made of six plywood sheets of mass 8 kg each, how long would it take for the cube to complete one full revolution if the student pushed its corner C in the same way that she pushed the corner of the aluminum cube?

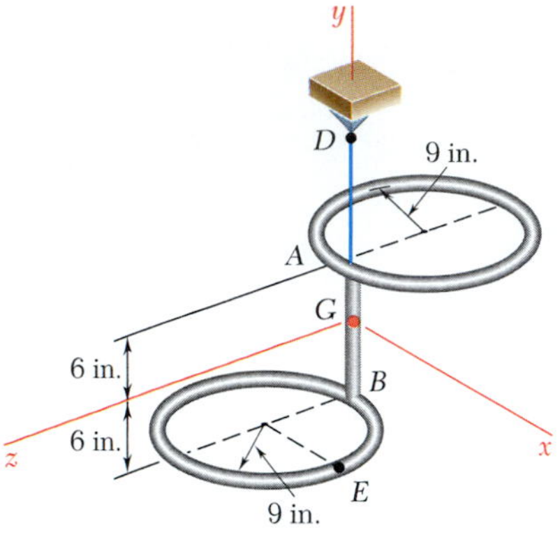

Fig. P18.23 and P18.24

18.23 A wire of uniform cross section weighing 0.8 lb/ft is used to form the wire figure shown, which is suspended from cord AD. Knowing that an impulse $\mathbf{F}\Delta t = (1.2 \text{ lb} \cdot \text{s})\mathbf{j}$ is applied to the wire figure at point E, determine (*a*) the velocity of the mass center G of the wire figure, (*b*) the angular velocity of the wire figure.

18.24 A wire of uniform cross section weighing 0.8 lb/ft is used to form the wire figure shown, which is suspended from cord AD. Knowing that an impulse $\mathbf{F}\Delta t = -(1.2 \text{ lb} \cdot \text{s})\mathbf{i}$ is applied to the wire figure at point E, determine (*a*) the velocity of the mass center G of the wire figure, (*b*) the angular velocity of the wire figure.

18.25 A uniform rod of mass m is bent into the shape shown and is suspended from a wire attached to its mass center G. The bent rod is hit at A in a direction perpendicular to the plane containing the rod (in the positive x direction). Denoting the corresponding impulse by $\mathbf{F}\Delta t$, determine immediately after the impact (*a*) the velocity of the mass center G, (*b*) the angular velocity of the rod.

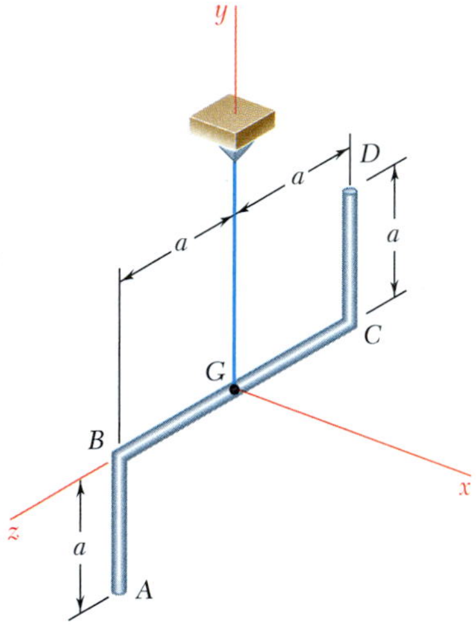

Fig. P18.25

18.26 Solve Prob. 18.25 assuming that the bent rod is hit at B.

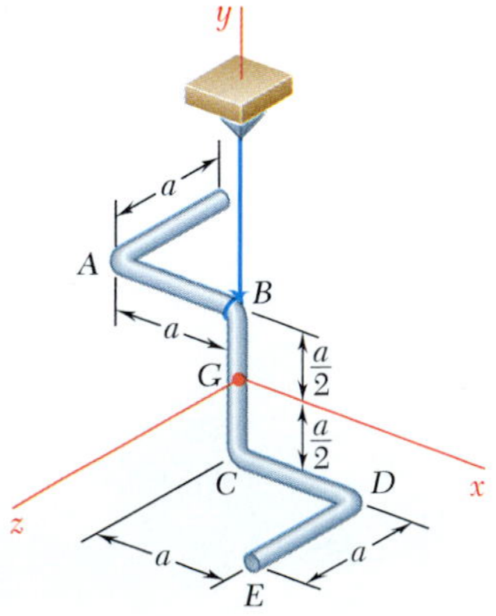

Fig. *P18.27*

18.27 A uniform rod of mass m and length $5a$ is bent into the shape shown and is suspended from a wire attached at point B. Knowing that the rod is hit at point A in the negative y direction and denoting the corresponding impulse by $-(F\Delta t)\mathbf{j}$, determine immediately after the impact (*a*) the velocity of the mass center G, (*b*) the angular velocity of the rod.

18.28 Solve Prob. 18.27, assuming that the rod is hit at point C in the negative z direction.

18.29 A circular plate of radius a and mass m supported by a ball-and-socket joint at point A is rotating about the y axis with a constant angular velocity $\boldsymbol{\omega} = \omega_0 \mathbf{j}$ when an obstruction is suddenly introduced at point B in the xy plane. Assuming the impact at point B to be perfectly plastic ($e = 0$), determine immediately after the impact (a) the angular velocity of the plate, (b) the velocity of its mass center G.

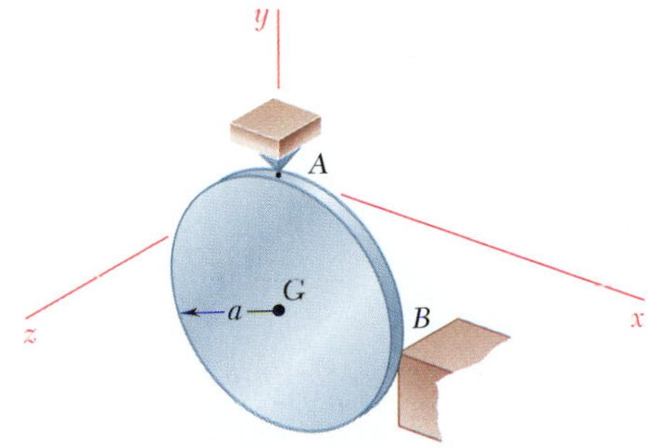

Fig. P18.29

18.30 Determine the impulse exerted on the plate of Prob. 18.29 during the impact by (a) the obstruction at point B, (b) the support at point A.

18.31 A rectangular plate of mass m is falling with a velocity $\bar{\mathbf{v}}_0$ and no angular velocity when its corner C strikes an obstruction. Assuming the impact to be perfectly plastic ($e = 0$), determine the angular velocity of the plate immediately after the impact.

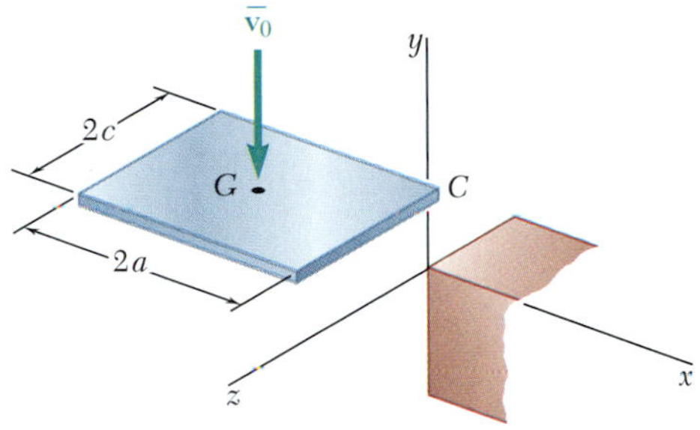

Fig. P18.31

18.32 For the plate of Prob. 18.31, determine (a) the velocity of its mass center G immediately after the impact, (b) the impulse exerted on the plate by the obstruction during the impact.

18.33 The coordinate axes shown represent the principal centroidal axes of inertia of a 1500-kg space probe whose radii of gyration are $k_x = 0.55$ m, $k_y = 0.57$ m, and $k_z = 0.50$ m. The probe has no angular velocity when a 0.14-kg meteorite strikes one of its solar panels at A with a velocity $\mathbf{v}_0 = (720 \text{ m/s})\mathbf{i} - (900 \text{ m/s})\mathbf{j} + (960 \text{ m/s})\mathbf{k}$ relative to the probe. Knowing that the meteorite emerges on the other side of the panel with no change in the direction of its velocity, but with a speed reduced by 20 percent, determine the final angular velocity of the probe.

18.34 The coordinate axes shown represent the principal centroidal axes of inertia of a 1500-kg space probe whose radii of gyration are $k_x = 0.55$ m, $k_y = 0.57$ m, and $k_z = 0.50$ m. The probe has no angular velocity when a 0.14-kg meteorite strikes one of its solar panels at A and emerges on the other side of the panel with no change in the direction of its velocity, but with a speed reduced by 25 percent. Knowing that the final angular velocity of the probe is $\boldsymbol{\omega} = (0.05 \text{ rad/s})\mathbf{i} - (0.12 \text{ rad/s})\mathbf{j} + \omega_z\mathbf{k}$ and that the x component of the resulting change in the velocity of the mass center of the probe is -17 mm/s, determine (a) the component ω_z of the final angular velocity of the probe, (b) the relative velocity $\mathbf{v}_0$ with which the meteorite strikes the panel.

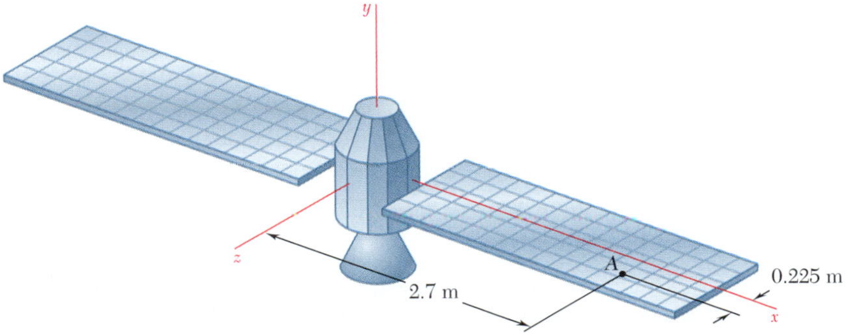

Fig. P18.33 and P18.34

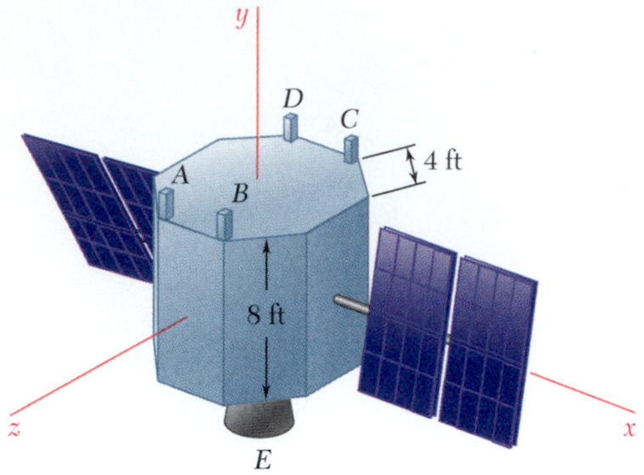

Fig. *P18.35*

18.35 A 5000-lb probe in orbit about the moon is 8 ft high and has octagonal bases of sides 4 ft. The coordinate axes shown are the principal centroidal axes of inertia of the probe, and its radii of gyration are $k_x = 2.45$ ft, $k_y = 2.65$ ft, and $k_z = 2.55$ ft. The probe is equipped with a main 125-lb thruster E and with four 5-lb thrusters A, B, C, and D which can expel fuel in the positive y direction. The probe has an angular velocity $\boldsymbol{\omega} = (0.040 \text{ rad/s})\mathbf{i} + (0.060 \text{ rad/s})\mathbf{k}$ when two of the 5-lb thrusters are used to reduce the angular velocity to zero. Determine (*a*) which of the thrusters should be used, (*b*) the operating time of each of these thrusters, (*c*) for how long the main thruster E should be activated if the velocity of the mass center of the probe is to remain unchanged.

18.36 Solve Prob. 18.35 assuming that the angular velocity of the probe is $\boldsymbol{\omega} = (0.060 \text{ rad/s})\mathbf{i} - (0.040 \text{ rad/s})\mathbf{k}$.

18.37 Denoting, respectively, by $\boldsymbol{\omega}$, $\mathbf{H}_O$, and T the angular velocity, the angular momentum, and the kinetic energy of a rigid body with a fixed point O, (*a*) prove that $\mathbf{H}_O \cdot \boldsymbol{\omega} = 2T$; (*b*) show that the angle θ between $\boldsymbol{\omega}$ and $\mathbf{H}_O$ will always be acute.

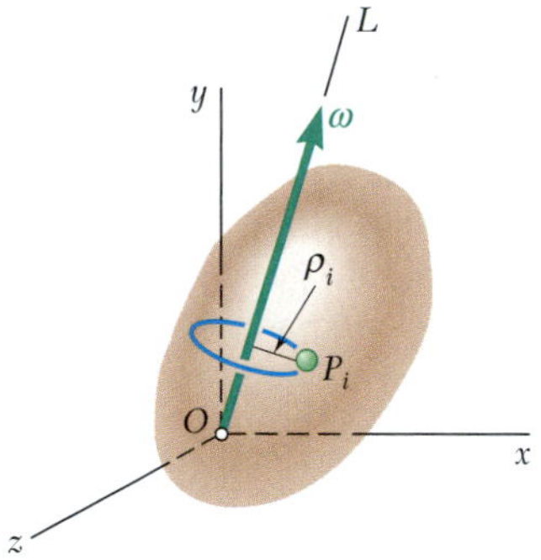

Fig. P18.38

18.38 Show that the kinetic energy of a rigid body with a fixed point O can be expressed as $T = \frac{1}{2}I_{OL}\omega^2$, where ω is the instantaneous angular velocity of the body and I_{OL} is its moment of inertia about the line of action OL of ω. Derive this expression (*a*) from Eqs. (9.46) and (18.19), (*b*) by considering T as the sum of the kinetic energies of particles P_i describing circles of radius ρ_i about line OL.

18.39 Determine the kinetic energy of the rectangular plate of Prob. 18.1.

18.40 Determine the kinetic energy of rod AB of Prob. 18.2.

18.41 Determine the kinetic energy of the assembly of Prob. 18.3.

18.42 Determine the kinetic energy of the square plate of Prob. 18.4.

18.43 Determine the kinetic energy of the disk of Prob. 18.5.

18.44 Determine the kinetic energy of the disk of Prob. 18.6.

18.45 Determine the kinetic energy of the solid parallelepiped of Prob. 18.7.

18.46 Determine the kinetic energy of the hollow parallelepiped of Prob. 18.8.

18.47 Determine the kinetic energy of the shaft of Prob. 18.15.

18.48 Determine the kinetic energy of the assembly of Prob. 18.19.

18.49 Determine the kinetic energy of the triangular plate of Prob. 18.17.

18.50 Determine the kinetic energy imparted to the cube of Prob. 18.21.

18.51 Determine the kinetic energy lost when the plate of Prob. 18.29 hits the obstruction at point B.

18.52 Determine the kinetic energy lost when corner C of the plate of Prob. 18.31 hits the obstruction.

18.53 Determine the kinetic energy of the space probe of Prob. 18.33 in its motion about its mass center after its collision with the meteorite.

18.54 Determine the kinetic energy of the space probe of Prob. 18.34 in its motion about its mass center after its collision with the meteorite.

*18.5. MOTION OF A RIGID BODY IN THREE DIMENSIONS

As indicated in Sec. 18.2, the fundamental equations

$$\Sigma \mathbf{F} = m\bar{\mathbf{a}} \tag{18.1}$$

$$\Sigma \mathbf{M}_G = \dot{\mathbf{H}}_G \tag{18.2}$$

remain valid in the most general case of the motion of a rigid body. Before Eq. (18.2) could be applied to the three-dimensional motion of a rigid body, however, it was necessary to derive Eqs. (18.7), which relate the components of the angular momentum $\mathbf{H}_G$ and those of the angular velocity $\boldsymbol{\omega}$. It still remains for us to find an effective and convenient way for computing the components of the derivative $\dot{\mathbf{H}}_G$ of the angular momentum.

Since $\mathbf{H}_G$ represents the angular momentum of the body in its motion relative to centroidal axes $GX'Y'Z'$ of fixed orientation (Fig. 18.9), and since $\dot{\mathbf{H}}_G$ represents the rate of change of $\mathbf{H}_G$ with respect to the same axes, it would seem natural to use components of $\boldsymbol{\omega}$ and $\mathbf{H}_G$ along the axes X', Y', Z' in writing the relations (18.7). But since the body rotates, its moments and products of inertia would change continually, and it would be necessary to determine their values as functions of the time. It is therefore more convenient to use axes x, y, z attached to the body, ensuring that its moments and products of inertia will maintain the same values during the motion. This is permissible since, as indicated earlier, the transformation of $\boldsymbol{\omega}$ into $\mathbf{H}_G$ is independent of the system of coordinate axes selected. The angular velocity $\boldsymbol{\omega}$, however, should still be *defined* with respect to the frame $GX'Y'Z'$ of fixed orientation. The vector $\boldsymbol{\omega}$ may then be *resolved* into components along the rotating x, y, and z axes. Applying the relations (18.7), we obtain the *components* of the vector $\mathbf{H}_G$ along the rotating axes. The vector $\mathbf{H}_G$, however, represents the angular momentum about G of the body *in its motion relative to the frame* $GX'Y'Z'$.

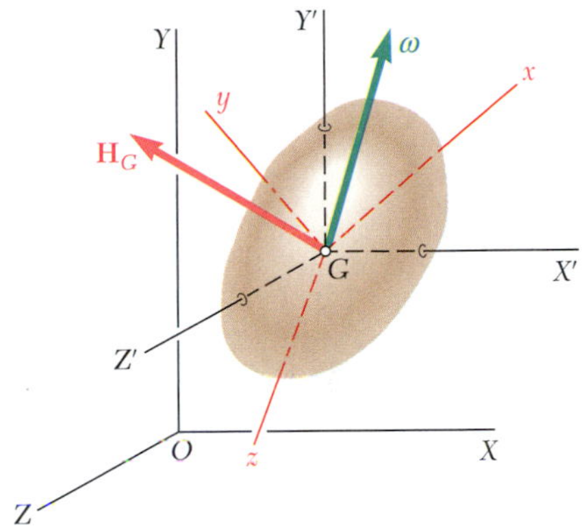

Fig. 18.9

Differentiating with respect to t the components of the angular momentum in (18.7), we define the rate of change of the vector $\mathbf{H}_G$ with respect to the rotating frame $Gxyz$:

$$(\dot{\mathbf{H}}_G)_{Gxyz} = \dot{H}_x\mathbf{i} + \dot{H}_y\mathbf{j} + \dot{H}_z\mathbf{k} \tag{18.21}$$

where $\mathbf{i}$, $\mathbf{j}$, $\mathbf{k}$ are the unit vectors along the rotating axes. Recalling from Sec. 15.10 that the rate of change $\dot{\mathbf{H}}_G$ of the vector $\mathbf{H}_G$ with respect to the frame $GX'Y'Z'$ is found by adding to $(\dot{\mathbf{H}}_G)_{Gxyz}$ the vector product $\boldsymbol{\Omega} \times \mathbf{H}_G$, where $\boldsymbol{\Omega}$ denotes the angular velocity of the rotating frame, we write

$$\dot{\mathbf{H}}_G = (\dot{\mathbf{H}}_G)_{Gxyz} + \boldsymbol{\Omega} \times \mathbf{H}_G \tag{18.22}$$

where $\mathbf{H}_G$ = angular momentum of body with respect to frame $GX'Y'Z'$ of fixed orientation

$(\dot{\mathbf{H}}_G)_{Gxyz}$ = rate of change of $\mathbf{H}_G$ with respect to rotating frame $Gxyz$, to be computed from the relations (18.7) and (18.21)

$\boldsymbol{\Omega}$ = angular velocity of rotating frame $Gxyz$

Substituting for $\dot{\mathbf{H}}_G$ from (18.22) into (18.2), we have

$$\Sigma \mathbf{M}_G = (\dot{\mathbf{H}}_G)_{Gxyz} + \boldsymbol{\Omega} \times \mathbf{H}_G \tag{18.23}$$

If the rotating frame is attached to the body, as has been assumed in this discussion, its angular velocity $\boldsymbol{\Omega}$ is identically equal to the angular velocity $\boldsymbol{\omega}$ of the body. There are many applications, however, where it is advantageous to use a frame of reference which is not actually attached to the body but rotates in an independent manner. For example, if the body considered is axisymmetrical, as in Sample Prob. 18.5 or Sec. 18.9, it is possible to select a frame of reference with respect to which the moments and products of inertia of the body remain constant, but which rotates less than the body itself.† As a result, it is possible to obtain simpler expressions for the angular velocity $\boldsymbol{\omega}$ and the angular momentum $\mathbf{H}_G$ of the body than could have been obtained if the frame of reference had actually been attached to the body. It is clear that in such cases the angular velocity $\boldsymbol{\Omega}$ of the rotating frame and the angular velocity $\boldsymbol{\omega}$ of the body are different.

*18.6. EULER'S EQUATIONS OF MOTION. EXTENSION OF D'ALEMBERT'S PRINCIPLE TO THE MOTION OF A RIGID BODY IN THREE DIMENSIONS

If the x, y, and z axes are chosen to coincide with the principal axes of inertia of the body, the simplified relations (18.10) can be used to determine the components of the angular momentum $\mathbf{H}_G$. Omitting the primes from the subscripts, we write

$$\mathbf{H}_G = \bar{I}_x\omega_x\mathbf{i} + \bar{I}_y\omega_y\mathbf{j} + \bar{I}_z\omega_z\mathbf{k} \tag{18.24}$$

where $\bar{I}_x$, $\bar{I}_y$, and $\bar{I}_z$ denote the principal centroidal moments of inertia of the body. Substituting for $\mathbf{H}_G$ from (18.24) into (18.23) and setting $\boldsymbol{\Omega} = \boldsymbol{\omega}$, we obtain the three scalar equations

$$\begin{aligned}\Sigma M_x &= \bar{I}_x\,\dot{\omega}_x - (\bar{I}_y - \bar{I}_z)\omega_y\omega_z \\ \Sigma M_y &= \bar{I}_y\,\dot{\omega}_y - (\bar{I}_z - \bar{I}_x)\omega_z\omega_x \\ \Sigma M_z &= \bar{I}_z\,\dot{\omega}_z - (\bar{I}_x - \bar{I}_y)\omega_x\omega_y\end{aligned} \tag{18.25}$$

These equations, called *Euler's equations of motion* after the Swiss mathematician Leonhard Euler (1707–1783), can be used to analyze the motion of a rigid body about its mass center. In the following sections, however, Eq. (18.23) will be used in preference to Eqs. (18.25), since the former is more general and the compact vectorial form in which it is expressed is easier to remember.

Writing Eq. (18.1) in scalar form, we obtain the three additional equations

$$\Sigma F_x = m\bar{a}_x \qquad \Sigma F_y = m\bar{a}_y \qquad \Sigma F_z = m\bar{a}_z \tag{18.26}$$

which, together with Euler's equations, form a system of six differential equations. Given appropriate initial conditions, these differential

†More specifically, the frame of reference will have no spin (see Sec. 18.9).

equations have a unique solution. Thus, the motion of a rigid body in three dimensions is completely defined by the resultant and the moment resultant of the external forces acting on it. This result will be recognized as a generalization of a similar result obtained in Sec. 16.4 in the case of the plane motion of a rigid slab. It follows that in three as well as two dimensions, two systems of forces which are equipollent are also equivalent; that is, they have the same effect on a given rigid body.

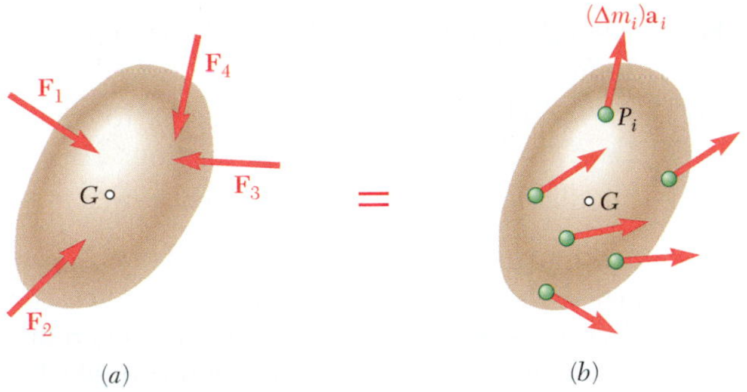

Fig. 18.10

Considering in particular the system of the external forces acting on a rigid body (Fig. 18.10*a*) and the system of the effective forces associated with the particles forming the rigid body (Fig. 18.10*b*), we can state that the two systems—which were shown in Sec. 14.2 to be equipollent—are also equivalent. This is the extension of d'Alembert's principle to the three-dimensional motion of a rigid body. Replacing the effective forces in Fig. 18.10*b* by an equivalent force-couple system, we verify that the system of the external forces acting on a rigid body in three-dimensional motion is equivalent to the system consisting of the vector $m\bar{\mathbf{a}}$ attached at the mass center G of the body and the couple of moment $\dot{\mathbf{H}}_G$ (Fig. 18.11), where $\dot{\mathbf{H}}_G$ is obtained from the relations (18.7) and (18.22). Note that the equivalence of the systems of vectors shown in Fig. 18.10 and in Fig. 18.11 has been indicated by *red* equals signs. Problems involving the three-dimensional motion of a rigid body can be solved by considering the free-body-diagram equation represented in Fig. 18.11 and writing appropriate scalar equations relating the components or moments of the external and effective forces (see Sample Prob. 18.3).

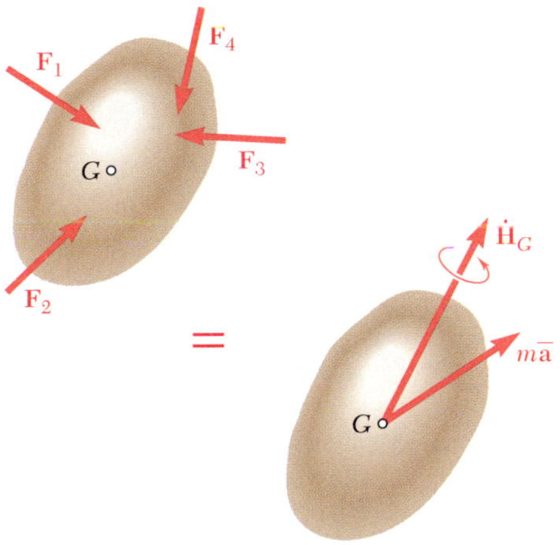

Fig. 18.11

*18.7. MOTION OF A RIGID BODY ABOUT A FIXED POINT

When a rigid body is constrained to rotate about a fixed point O, it is desirable to write an equation involving the moments about O of the external and effective forces, since this equation will not contain the unknown reaction at O. While such an equation can be obtained from Fig. 18.11, it may be more convenient to write it by considering the rate of change of the angular momentum $\mathbf{H}_O$ of the body about the fixed point O (Fig. 18.12). Recalling Eq. (14.11), we write

$$\Sigma \mathbf{M}_O = \dot{\mathbf{H}}_O \tag{18.27}$$

where $\dot{\mathbf{H}}_O$ denotes the rate of change of the vector $\mathbf{H}_O$ with respect to the fixed frame $OXYZ$. A derivation similar to that used in Sec. 18.5

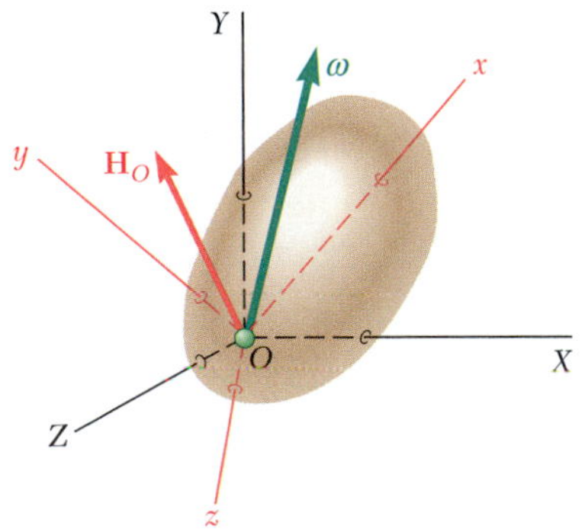

Fig. 18.12

Photo 18.3 The revolving radio telescope is an example of a structure constrained to rotate about a fixed point.

enables us to relate $\dot{\mathbf{H}}_O$ to the rate of change $(\dot{\mathbf{H}}_O)_{Oxyz}$ of $\mathbf{H}_O$ with respect to the rotating frame $Oxyz$. Substitution into (18.27) leads to the equation

$$\Sigma \mathbf{M}_O = (\dot{\mathbf{H}}_O)_{Oxyz} + \mathbf{\Omega} \times \mathbf{H}_O \tag{18.28}$$

where $\Sigma \mathbf{M}_O$ = sum of moments about O of forces applied to rigid body

$\mathbf{H}_O$ = angular momentum of body with respect to fixed frame $OXYZ$

$(\dot{\mathbf{H}}_O)_{Oxyz}$ = rate of change of $\mathbf{H}_O$ with respect to rotating frame $Oxyz$, to be computed from relations (18.13)

$\mathbf{\Omega}$ = angular velocity of rotating frame $Oxyz$

If the rotating frame is attached to the body, its angular velocity $\mathbf{\Omega}$ is identically equal to the angular velocity $\boldsymbol{\omega}$ of the body. However, as indicated in the last paragraph of Sec. 18.5, there are many applications where it is advantageous to use a frame of reference which is not actually attached to the body but rotates in an independent manner.

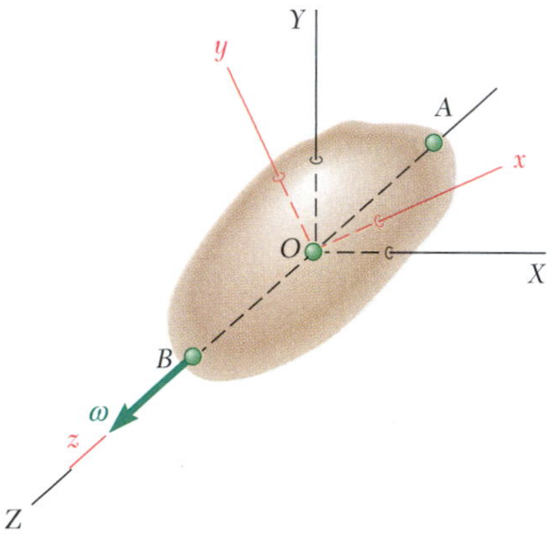

Fig. 18.13

*18.8. ROTATION OF A RIGID BODY ABOUT A FIXED AXIS

Equation (18.28), which was derived in the preceding section, will be used to analyze the motion of a rigid body constrained to rotate about a fixed axis AB (Fig. 18.13). First, we note that the angular velocity of the body with respect to the fixed frame $OXYZ$ is represented by the vector $\boldsymbol{\omega}$ directed along the axis of rotation. Attaching the moving frame of reference $Oxyz$ to the body, with the z axis along AB, we have $\boldsymbol{\omega} = \omega\mathbf{k}$. Substituting $\omega_x = 0$, $\omega_y = 0$, $\omega_z = \omega$ into the relations (18.13), we obtain the components along the rotating axes of the angular momentum $\mathbf{H}_O$ of the body about O:

$$H_x = -I_{xz}\omega \qquad H_y = -I_{yz}\omega \qquad H_z = I_z\omega$$

Since the frame $Oxyz$ is attached to the body, we have $\mathbf{\Omega} = \boldsymbol{\omega}$ and Eq. (18.28) yields

$$\begin{aligned}\Sigma \mathbf{M}_O &= (\dot{\mathbf{H}}_O)_{Oxyz} + \boldsymbol{\omega} \times \mathbf{H}_O \\ &= (-I_{xz}\mathbf{i} - I_{yz}\mathbf{j} + I_z\mathbf{k})\dot{\omega} + \omega\mathbf{k} \times (-I_{xz}\mathbf{i} - I_{yz}\mathbf{j} + I_z\mathbf{k})\omega \\ &= (-I_{xz}\mathbf{i} - I_{yz}\mathbf{j} + I_z\mathbf{k})\alpha + (-I_{xz}\mathbf{j} + I_{yz}\mathbf{i})\omega^2\end{aligned}$$

The result obtained can be expressed by the three scalar equations

$$\begin{aligned}\Sigma M_x &= -I_{xz}\alpha + I_{yz}\omega^2 \\ \Sigma M_y &= -I_{yz}\alpha - I_{xz}\omega^2 \\ \Sigma M_z &= I_z\alpha\end{aligned} \tag{18.29}$$

When the forces applied to the body are known, the angular acceleration α can be obtained from the last of Eqs. (18.29). The angular velocity ω is then determined by integration and the values obtained for α and ω substituted into the first two equations (18.29). These equations plus the three equations (18.26) which define the motion of the mass center of the body can then be used to determine the reactions at the bearings A and B.

It is possible to select axes other than the ones shown in Fig. 18.13 to analyze the rotation of a rigid body about a fixed axis. In many cases, the principal axes of inertia of the body will be found more advantageous. It is therefore wise to revert to Eq. (18.28) and to select the system of axes which best fits the problem under consideration.

If the rotating body is symmetrical with respect to the xy plane, the products of inertia I_{xz} and I_{yz} are equal to zero and Eqs. (18.29) reduce to

$$\Sigma M_x = 0 \qquad \Sigma M_y = 0 \qquad \Sigma M_z = I_z\alpha \tag{18.30}$$

which is in accord with the results obtained in Chap. 16. If, on the other hand, the products of inertia I_{xz} and I_{yz} are different from zero, the sum of the moments of the external forces about the x and y axes will also be different from zero, even when the body rotates at a constant rate ω. Indeed, in the latter case, Eqs. (18.29) yield

$$\Sigma M_x = I_{yz}\omega^2 \qquad \Sigma M_y = -I_{xz}\omega^2 \qquad \Sigma M_z = 0 \tag{18.31}$$

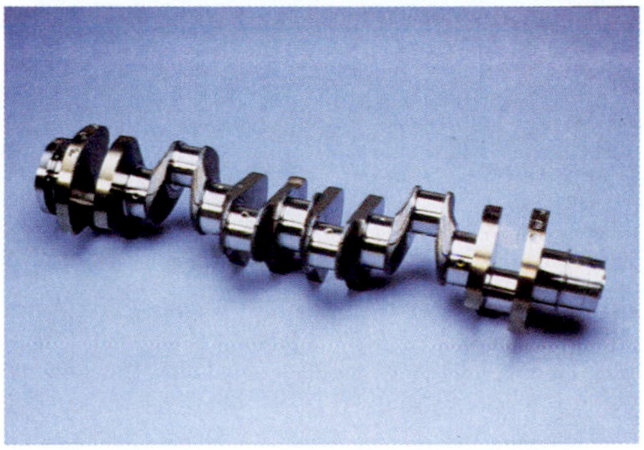

Photo 18.4 The forces exerted by a rotating automobile crankshaft on its bearings are the static and dynamic reactions. The crankshaft can be designed to be dynamically as well as statically balanced.

This last observation leads us to discuss the *balancing of rotating shafts*. Consider, for instance, the crankshaft shown in Fig. 18.14*a*, which is symmetrical about its mass center G. We first observe that when the crankshaft is at rest, it exerts no lateral thrust on its supports, since its center of gravity G is located directly above A. The shaft is said to be *statically balanced*. The reaction at A, often referred to as a *static reaction*, is vertical and its magnitude is equal to the weight W of the shaft. Let us now assume that the shaft rotates with a constant angular velocity $\boldsymbol{\omega}$. Attaching our frame of reference to the shaft, with its origin at G, the z axis along AB, and the y axis in the plane of symmetry of the shaft (Fig. 18.14*b*), we note that I_{xz} is zero and that I_{yz} is positive. According to Eqs. (18.31), the external forces include a couple of moment $I_{yz}\omega^2\mathbf{i}$. Since this couple is formed by the reaction at B and the horizontal component of the reaction at A, we have

$$\mathbf{A}_y = \frac{I_{yz}\omega^2}{l}\mathbf{j} \qquad \mathbf{B} = -\frac{I_{yz}\omega^2}{l}\mathbf{j} \tag{18.32}$$

Since the bearing reactions are proportional to ω^2, the shaft will have a tendency to tear away from its bearings when rotating at high speeds. Moreover, since the bearing reactions $\mathbf{A}_y$ and $\mathbf{B}$, called *dynamic reactions*, are contained in the yz plane, they rotate with the shaft and cause the structure supporting it to vibrate. These undesirable effects will be avoided if, by rearranging the distribution of mass around the shaft or by adding corrective masses, we let I_{yz} become equal to zero. The dynamic reactions $\mathbf{A}_y$ and $\mathbf{B}$ will vanish and the reactions at the bearings will reduce to the static reaction $\mathbf{A}_z$, the direction of which is fixed. The shaft will then be *dynamically as well as statically balanced.*

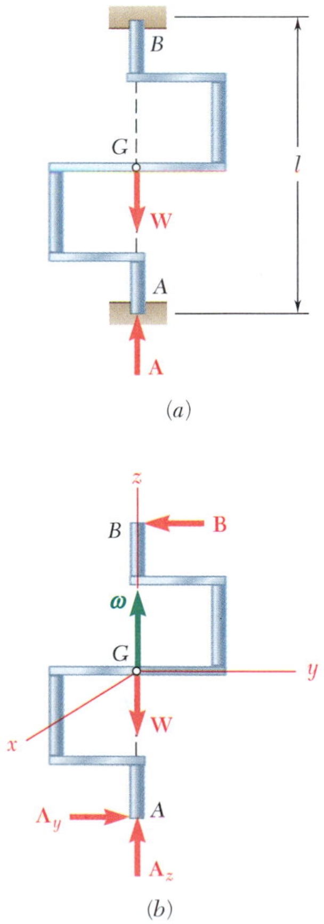

Fig. 18.14

SAMPLE PROBLEM 18.3

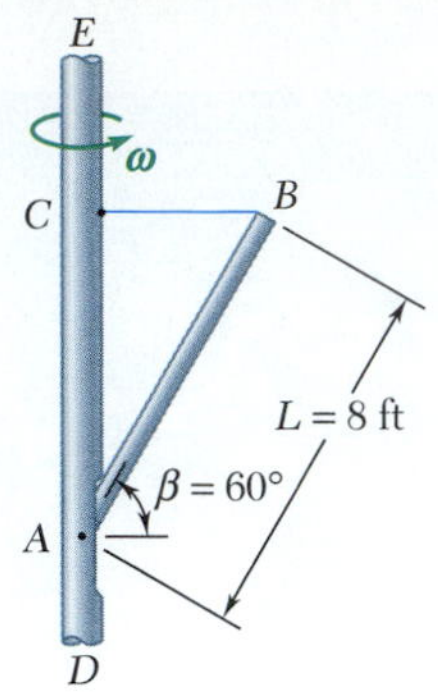

A slender rod AB of length $L = 8$ ft and weight $W = 40$ lb is pinned at A to a vertical axle DE which rotates with a constant angular velocity $\boldsymbol{\omega}$ of 15 rad/s. The rod is maintained in position by means of a horizontal wire BC attached to the axle and to the end B of the rod. Determine the tension in the wire and the reaction at A.

SOLUTION

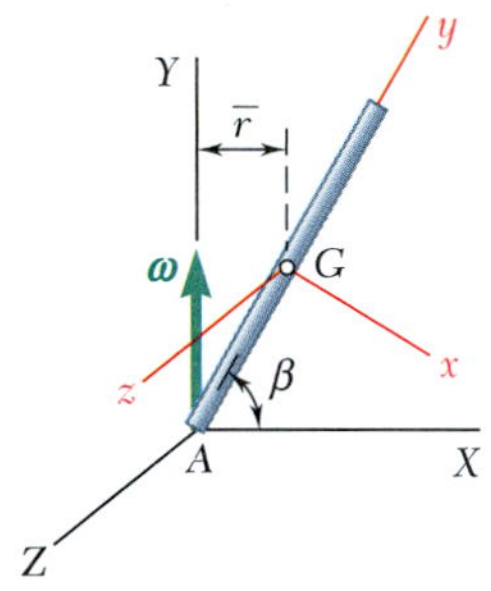

The effective forces reduce to the vector $m\bar{\mathbf{a}}$ attached at G and the couple $\dot{\mathbf{H}}_G$. Since G describes a horizontal circle of radius $\bar{r} = \frac{1}{2}L \cos \beta$ at the constant rate ω, we have

$$\bar{\mathbf{a}} = \mathbf{a}_n = -\bar{r}\omega^2\mathbf{I} = -(\tfrac{1}{2}L \cos \beta)\omega^2\mathbf{I} = -(450 \text{ ft/s}^2)\mathbf{I}$$

$$m\bar{\mathbf{a}} = \frac{40}{g}(-450\mathbf{I}) = -(559 \text{ lb})\mathbf{I}$$

Determination of $\dot{\mathbf{H}}_G$. We first compute the angular momentum $\mathbf{H}_G$. Using the principal centroidal axes of inertia x, y, z, we write

$$\bar{I}_x = \tfrac{1}{12}mL^2 \qquad \bar{I}_y = 0 \qquad \bar{I}_z = \tfrac{1}{12}mL^2$$

$$\omega_x = -\omega \cos \beta \qquad \omega_y = \omega \sin \beta \qquad \omega_z = 0$$

$$\mathbf{H}_G = \bar{I}_x\omega_x\mathbf{i} + \bar{I}_y\omega_y\mathbf{j} + \bar{I}_z\omega_z\mathbf{k}$$

$$\mathbf{H}_G = -\tfrac{1}{12}mL^2\omega \cos \beta \, \mathbf{i}$$

The rate of change $\dot{\mathbf{H}}_G$ of $\mathbf{H}_G$ with respect to axes of fixed orientation is obtained from Eq. (18.22). Observing that the rate of change $(\dot{\mathbf{H}}_G)_{Gxyz}$ of $\mathbf{H}_G$ with respect to the rotating frame $Gxyz$ is zero, and that the angular velocity $\boldsymbol{\Omega}$ of that frame is equal to the angular velocity $\boldsymbol{\omega}$ of the rod, we have

$$\dot{\mathbf{H}}_G = (\dot{\mathbf{H}}_G)_{Gxyz} + \boldsymbol{\omega} \times \mathbf{H}_G$$

$$\dot{\mathbf{H}}_G = 0 + (-\omega \cos \beta \, \mathbf{i} + \omega \sin \beta \, \mathbf{j}) \times (-\tfrac{1}{12}mL^2\omega \cos \beta \, \mathbf{i})$$

$$\dot{\mathbf{H}}_G = \tfrac{1}{12}mL^2\omega^2 \sin \beta \cos \beta \, \mathbf{k} = (645 \text{ lb} \cdot \text{ft})\mathbf{k}$$

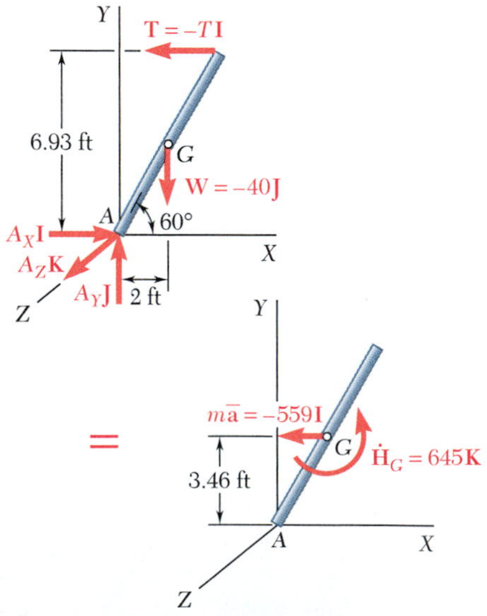

Equations of Motion. Expressing that the system of the external forces is equivalent to the system of the effective forces, we write

$\Sigma\mathbf{M}_A = \Sigma(\mathbf{M}_A)_{\text{eff}}$:

$$6.93\mathbf{J} \times (-T\mathbf{I}) + 2\mathbf{I} \times (-40\mathbf{J}) = 3.46\mathbf{J} \times (-559\mathbf{I}) + 645\mathbf{K}$$

$$(6.93T - 80)\mathbf{K} = (1934 + 645)\mathbf{K} \qquad T = 384 \text{ lb} \quad ◀$$

$\Sigma\mathbf{F} = \Sigma\mathbf{F}_{\text{eff}}$: $\quad A_X\mathbf{I} + A_Y\mathbf{J} + A_Z\mathbf{K} - 384\mathbf{I} - 40\mathbf{J} = -559\mathbf{I}$

$$\mathbf{A} = -(175 \text{ lb})\mathbf{I} + (40 \text{ lb})\mathbf{J} \quad ◀$$

Remark. The value of T could have been obtained from $\mathbf{H}_A$ and Eq. (18.28). However, the method used here also yields the reaction at A. Moreover, it draws attention to the effect of the asymmetry of the rod on the solution of the problem by clearly showing that both the vector $m\bar{\mathbf{a}}$ and the couple $\dot{\mathbf{H}}_G$ must be used to represent the effective forces.

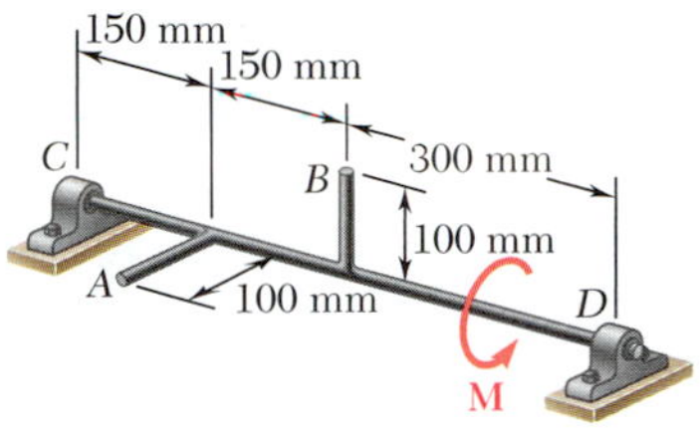

SAMPLE PROBLEM 18.4

Two 100-mm rods A and B, each of mass 300 g, are welded to shaft CD which is supported by bearings at C and D. If a couple $\mathbf{M}$ of magnitude equal to 6 N · m is applied to the shaft, determine the components of the dynamic reactions at C and D at the instant when the shaft has reached an angular velocity at 1200 rpm. Neglect the moment of inertia of the shaft itself.

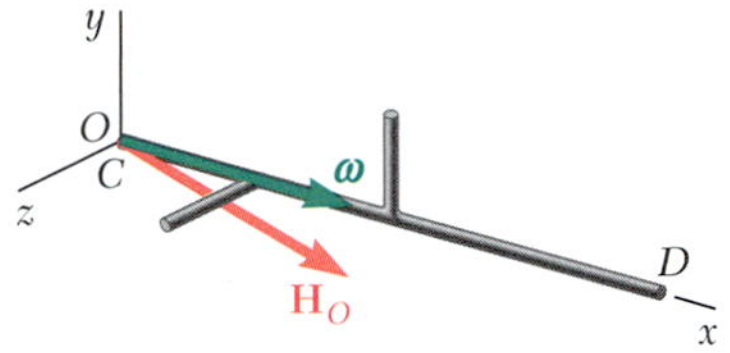

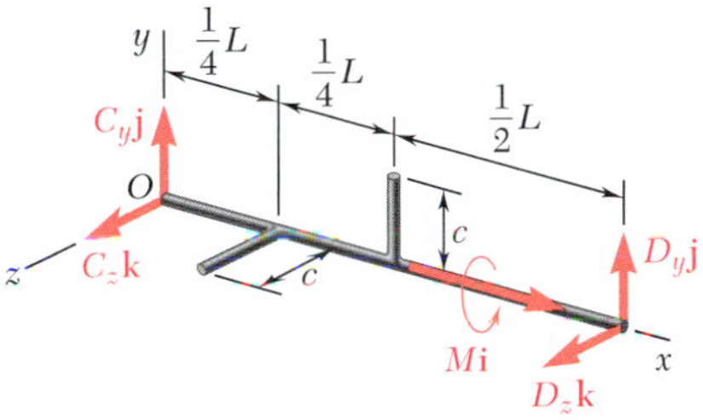

SOLUTION

Angular Momentum about *O*. We attach to the body the frame of reference $Oxyz$ and note that the axes chosen are not principal axes of inertia for the body. Since the body rotates about the x axis, we have $\omega_x = \omega$ and $\omega_y = \omega_z = 0$. Substituting into Eqs. (18.13),

$$H_x = I_x\omega \qquad H_y = -I_{xy}\omega \qquad H_z = -I_{xz}\omega$$
$$\mathbf{H}_O = (I_x\mathbf{i} - I_{xy}\mathbf{j} - I_{xz}\mathbf{k})\omega$$

Moments of the External Forces about *O*. Since the frame of reference rotates with the angular velocity $\boldsymbol{\omega}$, Eq. (18.28) yields

$$\begin{aligned}\Sigma\mathbf{M}_O &= (\dot{\mathbf{H}}_O)_{Oxyz} + \boldsymbol{\omega} \times \mathbf{H}_O \\ &= (I_x\mathbf{i} - I_{xy}\mathbf{j} - I_{xz}\mathbf{k})\alpha + \omega\mathbf{i} \times (I_x\mathbf{i} - I_{xy}\mathbf{j} - I_{xz}\mathbf{k})\omega \\ &= I_x\alpha\mathbf{i} - (I_{xy}\alpha - I_{xz}\omega^2)\mathbf{j} - (I_{xz}\alpha + I_{xy}\omega^2)\mathbf{k}\end{aligned} \qquad (1)$$

Dynamic Reaction at *D*. The external forces consist of the weights of the shaft and rods, the couple $\mathbf{M}$, the static reactions at C and D, and the dynamic reactions at C and D. Since the weights and static reactions are balanced, the external forces reduce to the couple $\mathbf{M}$ and the dynamic reactions $\mathbf{C}$ and $\mathbf{D}$ as shown in the figure. Taking moments about O, we have

$$\Sigma\mathbf{M}_O = L\mathbf{i} \times (D_y\mathbf{j} + D_z\mathbf{k}) + M\mathbf{i} = M\mathbf{i} - D_zL\mathbf{j} + D_yL\mathbf{k} \qquad (2)$$

Equating the coefficients of the unit vector $\mathbf{i}$ in (1) and (2):

$$M = I_x\alpha \qquad M = 2(\tfrac{1}{3}mc^2)\alpha \qquad \alpha = 3M/2mc^2$$

Equating the coefficients of $\mathbf{k}$ and $\mathbf{j}$ in (1) and (2):

$$D_y = -(I_{xz}\alpha + I_{xy}\omega^2)/L \qquad D_z = (I_{xy}\alpha - I_{xz}\omega^2)/L \qquad (3)$$

Using the parallel-axis theorem, and noting that the product of inertia of each rod is zero with respect to centroidal axes, we have

$$I_{xy} = \Sigma m\bar{x}\bar{y} = m(\tfrac{1}{2}L)(\tfrac{1}{2}c) = \tfrac{1}{4}mLc$$
$$I_{xz} = \Sigma m\bar{x}\bar{z} = m(\tfrac{1}{4}L)(\tfrac{1}{2}c) = \tfrac{1}{8}mLc$$

Substituting into (3) the values found for I_{xy}, I_{xz}, and α:

$$D_y = -\tfrac{3}{16}(M/c) - \tfrac{1}{4}mc\omega^2 \qquad D_z = \tfrac{3}{8}(M/c) - \tfrac{1}{8}mc\omega^2$$

Substituting $\omega = 1200$ rpm $= 125.7$ rad/s, $c = 0.100$ m, $M = 6$ N · m, and $m = 0.300$ kg, we have

$$D_y = -129.8 \text{ N} \qquad D_z = -36.8 \text{ N} \quad ◀$$

Dynamic Reaction at *C*. Using a frame of reference attached at D, we obtain equations similar to Eqs. (3), which yield

$$C_y = -152.2 \text{ N} \qquad C_z = -155.2 \text{ N} \quad ◀$$

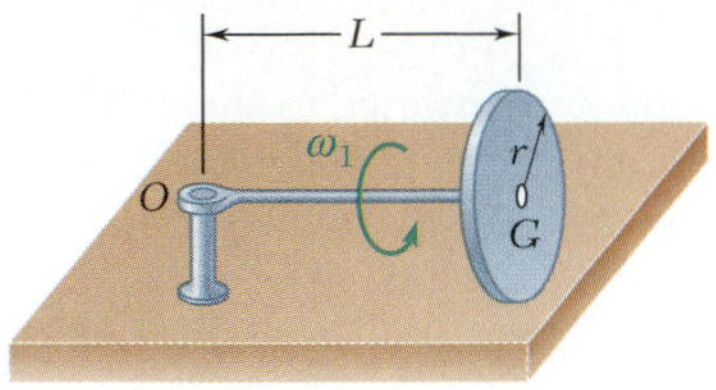

SAMPLE PROBLEM 18.5

A homogeneous disk of radius r and mass m is mounted on an axle OG of length L and negligible mass. The axle is pivoted at the fixed point O and the disk is constrained to roll on a horizontal floor. Knowing that the disk rotates counterclockwise at the constant rate ω_1 about the axle, determine (*a*) the force (assumed vertical) exerted by the floor on the disk, (*b*) the reaction at the pivot O.

SOLUTION

The effective forces reduce to the vector $m\bar{\mathbf{a}}$ attached at G and the couple $\dot{\mathbf{H}}_G$. Recalling from Sample Prob. 18.2 that the axle rotates about the y axis at the rate $\omega_2 = r\omega_1/L$, we write

$$m\bar{\mathbf{a}} = -mL\omega_2^2\mathbf{i} = -mL(r\omega_1/L)^2\mathbf{i} = -(mr^2\omega_1^2/L)\mathbf{i} \qquad (1)$$

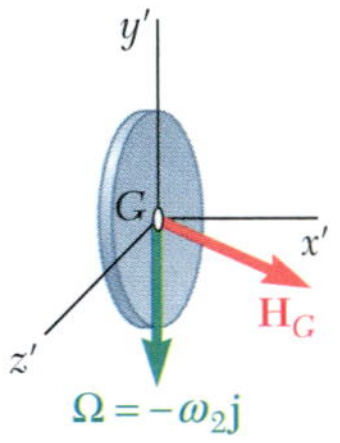

Determination of $\dot{\mathbf{H}}_G$. We recall from Sample Prob. 18.2 that the angular momentum of the disk about G is

$$\mathbf{H}_G = \tfrac{1}{2}mr^2\omega_1\left(\mathbf{i} - \frac{r}{2L}\mathbf{j}\right)$$

where $\mathbf{H}_G$ is resolved into components along the rotating axes x', y', z', with x' along OG and y' vertical. The rate of change $\dot{\mathbf{H}}_G$ of $\mathbf{H}_G$ with respect to axes of fixed orientation is obtained from Eq. (18.22). Noting that the rate of change $(\dot{\mathbf{H}}_G)_{Gx'y'z'}$ of $\mathbf{H}_G$ with respect to the rotating frame is zero, and that the angular velocity $\boldsymbol{\Omega}$ of that frame is

$$\boldsymbol{\Omega} = -\omega_2\mathbf{j} = -\frac{r\omega_1}{L}\mathbf{j}$$

we have

$$\begin{aligned}\dot{\mathbf{H}}_G &= (\dot{\mathbf{H}}_G)_{Gx'y'z'} + \boldsymbol{\Omega} \times \mathbf{H}_G \\ &= 0 - \frac{r\omega_1}{L}\mathbf{j} \times \tfrac{1}{2}mr^2\omega_1\left(\mathbf{i} - \frac{r}{2L}\mathbf{j}\right) \\ &= \tfrac{1}{2}mr^2(r/L)\omega_1^2\mathbf{k} \qquad (2)\end{aligned}$$

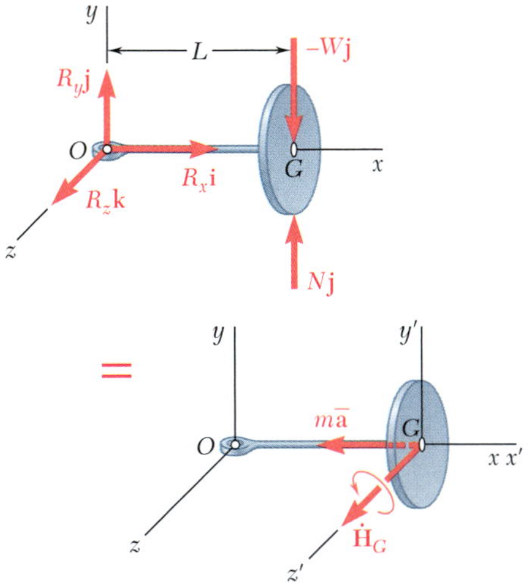

Equations of Motion. Expressing that the system of the external forces is equivalent to the system of the effective forces, we write

$\Sigma\mathbf{M}_O = \Sigma(\mathbf{M}_O)_{\text{eff}}$: $\quad L\mathbf{i} \times (N\mathbf{j} - W\mathbf{j}) = \dot{\mathbf{H}}_G$

$$(N - W)L\mathbf{k} = \tfrac{1}{2}mr^2(r/L)\omega_1^2\mathbf{k}$$

$$N = W + \tfrac{1}{2}mr(r/L)^2\omega_1^2 \qquad \mathbf{N} = [W + \tfrac{1}{2}mr(r/L)^2\omega_1^2]\mathbf{j} \quad (3) \blacktriangleleft$$

$\Sigma\mathbf{F} = \Sigma\mathbf{F}_{\text{eff}}$: $\quad \mathbf{R} + N\mathbf{j} - W\mathbf{j} = m\bar{\mathbf{a}}$

Substituting for N from (3), for $m\bar{\mathbf{a}}$ from (1), and solving for $\mathbf{R}$, we have

$$\mathbf{R} = -(mr^2\omega_1^2/L)\mathbf{i} - \tfrac{1}{2}mr(r/L)^2\omega_1^2\mathbf{j}$$

$$\mathbf{R} = -\frac{mr^2\omega_1^2}{L}\left(\mathbf{i} + \frac{r}{2L}\mathbf{j}\right) \blacktriangleleft$$

SOLVING PROBLEMS ON YOUR OWN

In this lesson you will be asked to solve problems involving the *three-dimensional motion of rigid bodies*. The method you will use is basically the same that you used in Chap. 16 in your study of the plane motion of rigid bodies. You will draw a free-body-diagram equation showing that the system of the external forces is equivalent to the system of the effective forces, and you will equate sums of components and sums of moments on both sides of this equation. Now, however, the system of the effective forces will be represented by the vector $m\bar{\mathbf{a}}$ and a couple vector $\dot{\mathbf{H}}_G$, the determination of which will be explained in paragraphs 1 and 2 below.

To solve a problem involving the three-dimensional motion of a rigid body, you should take the following steps:

1. *Determine the angular momentum $\mathbf{H}_G$ of the body about its mass center G* from its angular velocity $\boldsymbol{\omega}$ with respect to a frame of reference $GX'Y'Z'$ of fixed orientation. This is an operation you learned to perform in the preceding lesson. However, since the configuration of the body will be changing with time, it will now be necessary for you to use an auxiliary system of axes $Gx'y'z'$ (Fig. 18.9) to compute the components of $\boldsymbol{\omega}$ and the moments and products of inertia of the body. These axes may be rigidly attached to the body, in which case their angular velocity is equal to $\boldsymbol{\omega}$ [Sample Probs. 18.3 and 18.4], or they may have an angular velocity $\boldsymbol{\Omega}$ of their own [Sample Prob. 18.5].

Recall the following from the preceding lesson:

a. If the principal axes of inertia of the body at G are known, use these axes as coordinate axes x', y', and z', since the corresponding products of inertia of the body will be equal to zero. (Note that if the body is axisymmetric, these axes do not need to be rigidly attached to the body.) Resolve $\boldsymbol{\omega}$ into components $\omega_{x'}$, $\omega_{y'}$, and $\omega_{z'}$ along these axes and compute the principal moments of inertia $\bar{I}_{x'}$, $\bar{I}_{y'}$, and $\bar{I}_{z'}$. The corresponding components of the angular momentum $\mathbf{H}_G$ are

$$H_{x'} = \bar{I}_{x'}\omega_{x'} \qquad H_{y'} = \bar{I}_{y'}\omega_{y'} \qquad H_{z'} = \bar{I}_{z'}\omega_{z'} \tag{18.10}$$

b. If the principal axes of inertia of the body at G are not known, you must use Eqs. (18.7) to determine the components of the angular momentum $\mathbf{H}_G$. These equations require your prior computation of the *products of inertia* of the body, as well as of its moments of inertia, with respect to the selected axes.

2. *Compute the rate of change $\dot{\mathbf{H}}_G$ of the angular momentum $\mathbf{H}_G$ with respect to the frame $GX'Y'Z'$.* Note that this frame has a *fixed orientation,* while the frame $Gx'y'z'$ you used when you calculated the components of the vector $\boldsymbol{\omega}$ was a *rotating frame*. We refer you to our discussion in Sec. 15.10 of the rate of change of a vector with respect to a rotating frame. Recalling Eq. (15.31), you will express the rate of change $\dot{\mathbf{H}}_G$ as follows:

$$\dot{\mathbf{H}}_G = (\dot{\mathbf{H}}_G)_{Gx'y'z'} + \boldsymbol{\Omega} \times \mathbf{H}_G \tag{18.22}$$

The first term in the right-hand member of Eq. (18.22) represents the rate of change of $\mathbf{H}_G$ with respect to the rotating frame $Gx'y'z'$. This term will drop out if $\boldsymbol{\omega}$—and thus, $\mathbf{H}_G$—

(continued)

remain constant in both magnitude and direction when viewed from that frame. On the other hand, if any of the time derivatives $\dot{\omega}_{x'}$, $\dot{\omega}_{y'}$, and $\dot{\omega}_{z'}$ is different from zero, $(\dot{\mathbf{H}}_G)_{Gx'y'z'}$ will also be different from zero, and its components should be determined by differentiating Eqs. (18.10) with respect to t. Finally, we remind you that if the rotating frame is rigidly attached to the body, its angular velocity will be the same as that of the body, and $\mathbf{\Omega}$ can be replaced by $\boldsymbol{\omega}$.

3. *Draw the free-body-diagram equation for the rigid body,* showing that the system of the external forces exerted on the body is equivalent to the vector $m\bar{\mathbf{a}}$ applied at G and the couple vector $\dot{\mathbf{H}}_G$ (Fig. 18.11). By equating components in any direction and moments about any point, you can write as many as six independent scalar equations of motion [Sample Probs. 18.3 and 18.5].

4. *When solving problems involving the motion of a rigid body about a fixed point O,* you may find it convenient to use the following equation, derived in Sec. 18.7, which eliminates the components of the reaction at the support O,

$$\Sigma\mathbf{M}_O = (\dot{\mathbf{H}}_O)_{Oxyz} + \mathbf{\Omega} \times \mathbf{H}_O \qquad (18.28)$$

where the first term in the right-hand member represents the rate of change of $\mathbf{H}_O$ with respect to the rotating frame $Oxyz$, and where $\mathbf{\Omega}$ is the angular velocity of that frame.

5. *When determining the reactions at the bearings of a rotating shaft,* use Eq. (18.28) and take the following steps:

a. Place the fixed point O at one of the two bearings supporting the shaft and attach the rotating frame $Oxyz$ to the shaft, with one of the axes directed along it. Assuming, for instance, that the x axis has been aligned with the shaft, you will have $\mathbf{\Omega} = \boldsymbol{\omega} = \omega\mathbf{i}$ [Sample Prob. 18.4].

b. Since the selected axes, usually, will not be the principal axes of inertia at O, you must compute the *products of inertia* of the shaft, as well as its moments of inertia, with respect to these axes, and use Eqs. (18.13) to determine $\mathbf{H}_O$. Assuming again that the x axis has been aligned with the shaft, Eqs. (18.13) reduce to

$$H_x = I_x\omega \qquad H_y = -I_{yz}\omega \qquad H_z = -I_{zx}\omega \qquad (18.13')$$

which shows that $\mathbf{H}_O$ *will not be directed along the shaft.*

c. To obtain $\dot{\mathbf{H}}_O$, substitute the expressions obtained into Eq. (18.28), and let $\mathbf{\Omega} = \boldsymbol{\omega} = \omega\mathbf{i}$. If the angular velocity of the shaft is constant, the first term in the right-hand member of the equation will drop out. However, if the shaft has an angular acceleration $\boldsymbol{\alpha} = \alpha\mathbf{i}$, the first term will not be zero and must be determined by differentiating with respect to t the expressions in (18.13′). The result will be equations similar to Eqs. (18.13′), with ω replaced by α.

d. Since point O coincides with one of the bearings, the three scalar equations corresponding to Eq. (18.28) can be solved for the components of the dynamic reaction at the other bearing. If the mass center G of the shaft is located on the line joining the two bearings, the effective force $m\bar{\mathbf{a}}$ will be zero. Drawing the free-body-diagram equation of the shaft, you will then observe that the components of the dynamic reaction at the first bearing must be equal and opposite to those you have just determined. If G is not located on the line joining the two bearings, you can determine the reaction at the first bearing by placing the fixed point O at the second bearing and repeating the earlier procedure [Sample Prob. 18.4]; or you can obtain additional equations of motion from the free-body-diagram equation of the shaft, making sure to first determine and include the effective force $m\bar{\mathbf{a}}$ applied at G.

e. Most problems call for the determination of the "dynamic reactions" at the bearings, that is, for the *additional forces* exerted by the bearings on the shaft when the shaft is rotating. When determining dynamic reactions, ignore the effect of static loads, such as the weight of the shaft.

Problems

18.55 Determine the rate of change $\dot{\mathbf{H}}_G$ of the angular momentum $\mathbf{H}_G$ of the plate of Prob. 18.1.

18.56 Determine the rate of change $\dot{\mathbf{H}}_G$ of the angular momentum $\mathbf{H}_G$ of rod *AB* of Prob. 18.2.

18.57 Determine the rate of change $\dot{\mathbf{H}}_D$ of the angular momentum $\mathbf{H}_D$ of the assembly of Prob. 18.3.

18.58 Determine the rate of change $\dot{\mathbf{H}}_A$ of the angular momentum $\mathbf{H}_A$ of the square plate of Prob. 18.4.

18.59 Determine the rate of change $\dot{\mathbf{H}}_A$ of the angular momentum $\mathbf{H}_A$ of the disk of Prob. 18.5.

18.60 Determine the rate of change $\dot{\mathbf{H}}_A$ of the angular momentum $\mathbf{H}_A$ of the disk of Prob. 18.6.

18.61 Determine the rate of change $\dot{\mathbf{H}}_D$ of the angular momentum $\mathbf{H}_D$ of the assembly of Prob. 18.3, assuming that at the instant considered the assembly has an angular velocity $\boldsymbol{\omega} = (12 \text{ rad/s})\mathbf{i}$ and an angular acceleration $\boldsymbol{\alpha} = (96 \text{ rad/s}^2)\mathbf{i}$.

18.62 Determine the rate of change $\dot{\mathbf{H}}_D$ of the angular momentum $\mathbf{H}_D$ of the assembly of Prob. 18.3, assuming that at the instant considered the assembly has an angular velocity $\boldsymbol{\omega} = (12 \text{ rad/s})\mathbf{i}$ and an angular acceleration $\boldsymbol{\alpha} = -(96 \text{ rad/s}^2)\mathbf{i}$.

18.63 A homogeneous 4-kg disk is mounted on the horizontal shaft *AB*. The plane of the disk forms a 20° angle with the *yz* plane as shown. Knowing that the shaft rotates with a constant angular velocity $\boldsymbol{\omega}$ of magnitude 10 rad/s, determine the dynamic reactions at points *A* and *B*.

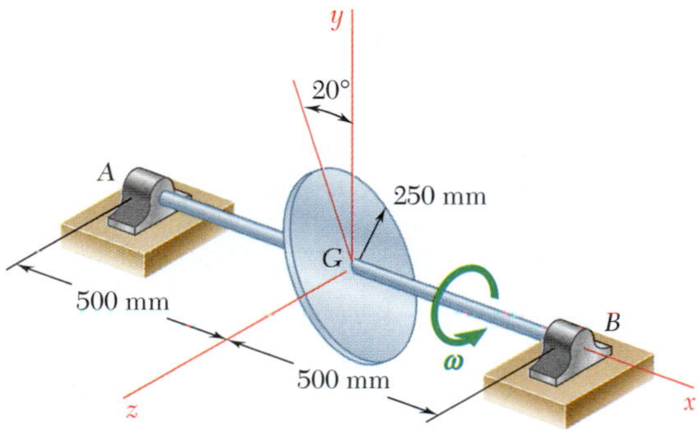

Fig. P18.63

18.64 A thin homogeneous 0.8-kg rod is used to form the shaft shown. Knowing that the shaft rotates with a constant angular velocity $\boldsymbol{\omega}$ of magnitude 20 rad/s, determine the dynamic reactions at points A and B.

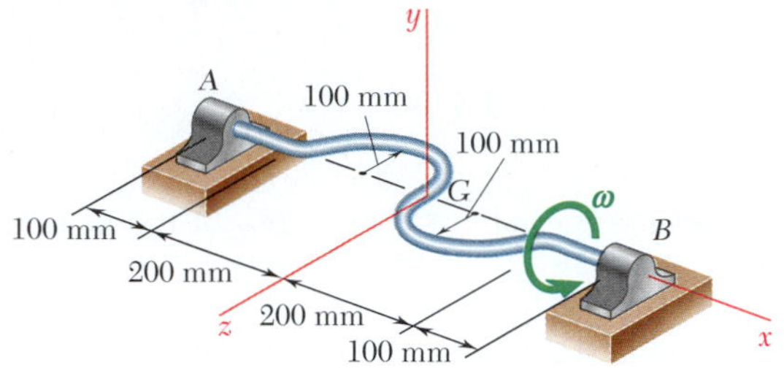

Fig. P18.64

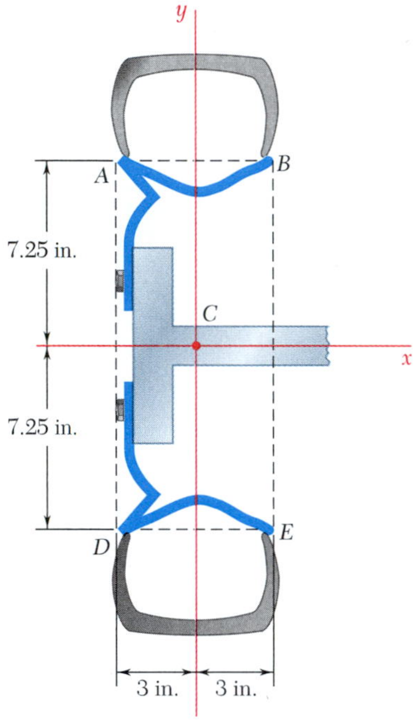

Fig. P18.65 and P18.66

18.65 When the 40-lb wheel shown is attached to a balancing machine and made to spin at a rate of 750 rpm, it is found that the forces exerted by the wheel on the machine are equivalent to a force-couple system consisting of a force $\mathbf{F} = (36.2 \text{ lb})\mathbf{j}$ applied at C and a couple $\mathbf{M}_C = (10.85 \text{ lb} \cdot \text{ft})\mathbf{k}$, where the unit vectors form a triad which rotates with the wheel. (*a*) Determine the distance from the axis of rotation to the mass center of the wheel and the products of inertia I_{xy} and I_{zx}, (*b*) If only two corrective weights are to be used to balance the wheel statically and dynamically, what should these weights be and at which of the points A, B, D, or E should they be placed?

18.66 After attaching the 40-lb wheel shown to a balancing machine and making it spin at the rate of 900 rpm, a mechanic has found that to balance the wheel both statically and dynamically, he should use two corrective weights, a 6-oz weight placed at B and a 2-oz weight placed at D. Using a right-handed frame of reference rotating with the wheel (with the z axis perpendicular to the plane of the figure), determine before the corrective weights have been attached (*a*) the distance from the axis of rotation to the mass center of the wheel and the products of inertia I_{xy} and I_{zx}, (*b*) the force-couple system at C equivalent to the forces exerted by the wheel on the machine.

18.67 The assembly shown consists of pieces of sheet aluminum of uniform thickness and of total mass 1.5 kg welded to a light axle supported by bearings A and B. Knowing that the assembly rotates at the constant rate $\omega = 240$ rpm, determine the dynamic reactions at A and B.

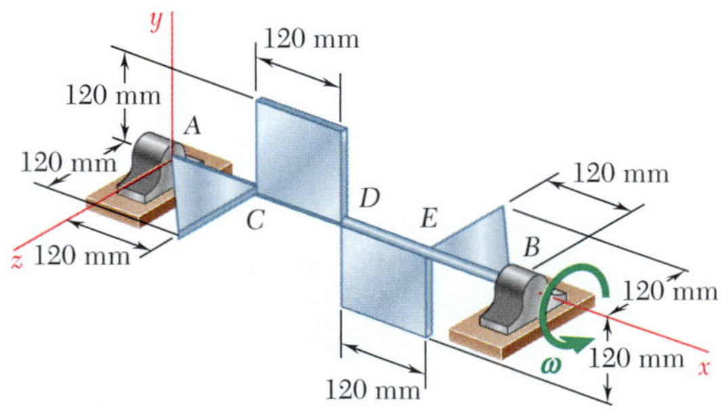

Fig. *P18.67*

18.68 Solve Prob. 18.67 assuming $\omega = 360$ rpm.

18.69 Knowing that the assembly of Prob. 18.63 is initially at rest ($\omega = 0$) when a couple of moment $\mathbf{M}_0 = (25\ \text{N} \cdot \text{m})\mathbf{i}$ is applied to the shaft, determine (*a*) the resulting angular acceleration of the assembly, (*b*) the dynamic reactions at points *A* and *B* immediately after the couple has been applied.

18.70 Knowing that the shaft of Prob. 18.64 is initially at rest ($\omega = 0$), determine (*a*) the magnitude of the couple $\mathbf{M}_0 = M_0\mathbf{i}$ required to impart to the shaft an angular acceleration $\boldsymbol{\alpha} = (150\ \text{rad/s}^2)\mathbf{i}$, (*b*) the dynamic reactions at points *A* and *B* immediately after the couple has been applied.

18.71 A 5-lb piece of sheet metal with dimensions 8 × 32 in. was bent to form the component shown. The component is at rest ($\omega = 0$) when a couple $\mathbf{M}_0 = (0.6\ \text{lb} \cdot \text{ft})\mathbf{k}$ is applied to it. Determine (*a*) the angular acceleration of the component, (*b*) the dynamic reactions at *A* and *B* immediately after the couple is applied.

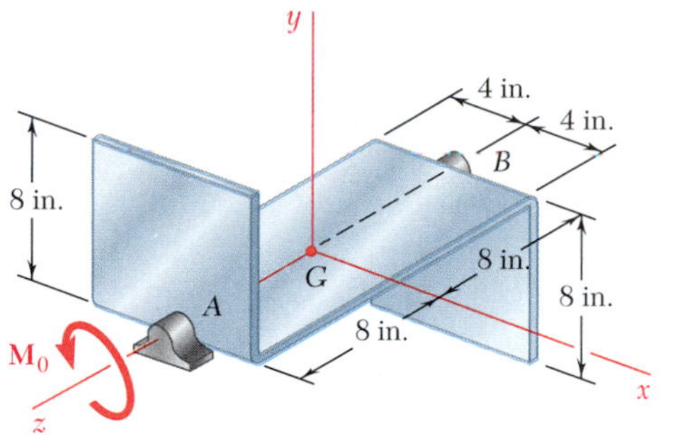

Fig. P18.71

18.72 For the component of Prob. 18.71, determine the dynamic reactions at *A* and *B* after one full revolution of the component.

18.73 The shaft of Prob. 18.64 is initially at rest ($\omega = 0$) when a couple of moment $\mathbf{M}_0 = M_0\mathbf{i}$ is applied. Knowing that the dynamic reaction at point *B* is $(1.5\ \text{N})\mathbf{j}$ immediately after the couple has been applied, determine (*a*) the initial angular acceleration of the shaft, (*b*) the applied couple $\mathbf{M}_0$.

18.74 The assembly of Prob. 18.67 is initially at rest ($\omega = 0$) when a couple $\mathbf{M}_0$ is applied to axle *AB*. Knowing that the resulting angular acceleration of the assembly is $\boldsymbol{\alpha} = (150\ \text{rad/s}^2)\mathbf{i}$, determine (*a*) the couple $\mathbf{M}_0$, (*b*) the dynamic reactions at *A* and *B* immediately after the couple is applied.

18.75 The assembly shown has a mass of 6 kg and consists of four thin 400-mm-diameter semicircular aluminum plates welded to a light 1-m-long shaft *AB*. The assembly is at rest ($\omega = 0$) at time $t = 0$ when a couple $\mathbf{M}_0$ is applied to it as shown, causing the assembly to complete one full revolution in 2 s. Determine (*a*) the couple $\mathbf{M}_0$, (*b*) the dynamic reactions at *A* and *B* at $t = 0$.

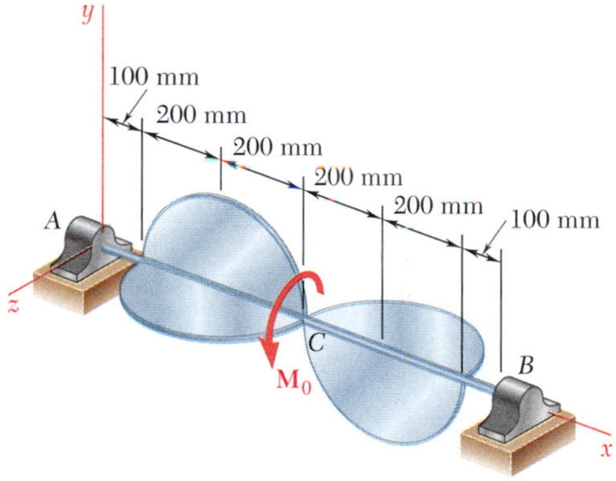

Fig. P18.75

18.76 For the assembly of Prob. 18.75, determine the dynamic reactions at *A* and *B* at $t = 2$ s.

18.77 The flywheel of an automobile engine, which is rigidly attached to the crankshaft, is equivalent to a 20-in.-diameter, 0.75-in.-thick steel plate. Determine the magnitude of the couple exerted by the flywheel on the horizontal crankshaft as the automobile travels around an unbanked curve of 600-ft radius at a speed of 60 mi/h, with the flywheel rotating at 2700 rpm. Assume the automobile to have (*a*) a rear-wheel drive with the engine mounted longitudinally, (*b*) a front-wheel drive with the engine mounted transversely. (Specific weight of steel = 0.284 lb/in^3.)

18.78 An airplane propeller weighs 360 lb and has a radius of gyration of 32 in. Knowing that the propeller rotates at 1600 rpm as the airplane is traveling on a vertical circular path of 2000-ft radius at 340 mi/h, determine the magnitude of the couple exerted by the propeller on its shaft due to the rotation of the airplane.

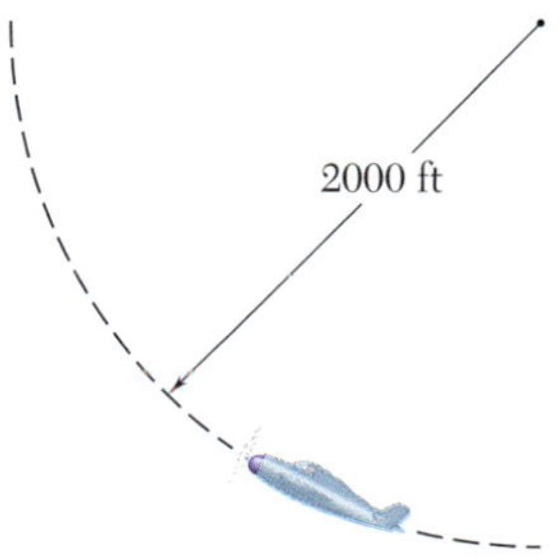

Fig. P18.78

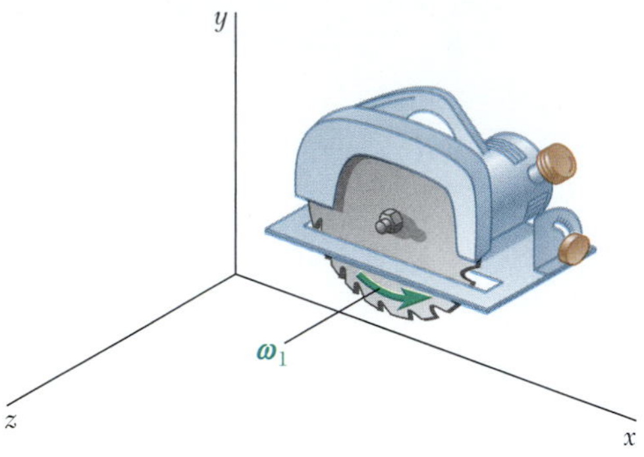

Fig. P18.79

18.79 The blade of a portable saw and the rotor of its motor have a total mass of 1.75 kg and a combined radius of gyration of 30 mm. Knowing that the blade rotates as shown at the rate $\omega_1 = 2500$ rpm, determine the magnitude and direction of the couple **M** that a worker must exert on the handle of the saw to rotate it with a constant angular velocity $\boldsymbol{\omega}_2 = -(2.4 \text{ rad/s})\mathbf{j}$.

18.80 The blade of an oscillating fan and the rotor of its motor have a total mass of 300 g and a combined radius of gyration of 75 mm. They are supported by bearings at A and B, 125 mm apart, and rotate at the rate $\omega_1 = 1800$ rpm. Determine the dynamic reactions at A and B when the motor casing has an angular velocity $\boldsymbol{\omega}_2 = (0.6 \text{ rad/s})\mathbf{j}$.

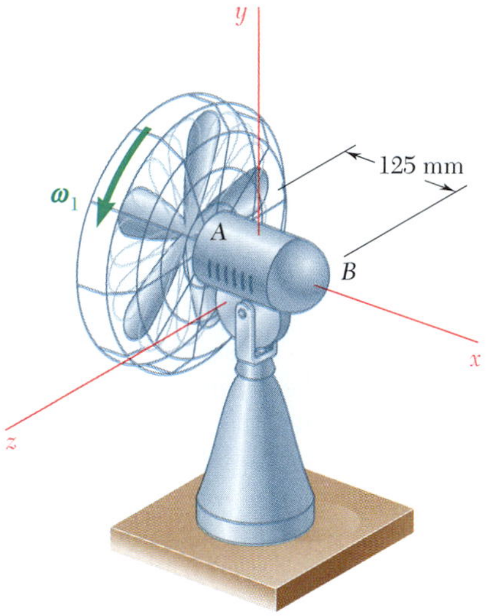

Fig. P18.80

18.81 Each wheel of an automobile has a weight of 48 lb, a diameter of 23 in., and a radius of gyration of 9 in. The automobile travels around an unbanked curve of radius 450 ft at a speed of 65 mi/h. Knowing that the transverse distance between the wheels is 4.5 ft, determine the additional normal force exerted by the ground on each outside wheel due to the motion of the car.

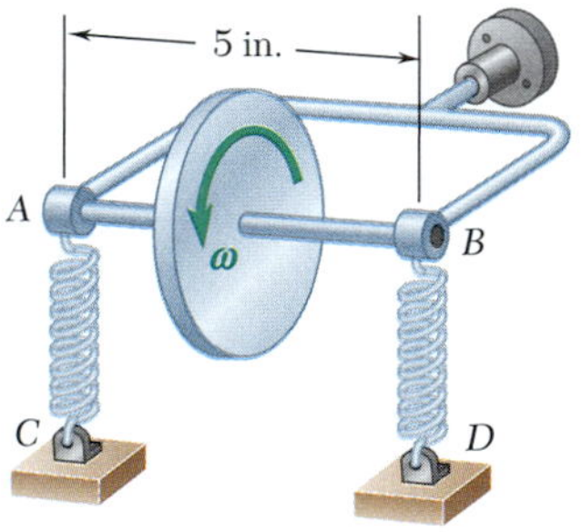

Fig. *P18.82*

18.82 The essential structure of a certain type of aircraft turn indicator is shown. Each spring has a constant of 40 lb/ft, and the 7-oz uniform disk of 2-in. radius spins at the rate of 10,000 rpm. The springs are stretched and exert equal vertical forces on yoke AB when the airplane is traveling in a straight path. Determine the angle through which the yoke will rotate when the pilot executes a horizontal turn of 2250-ft radius to the right at a speed of 500 mi/h. Indicate whether point A will move up or down.

18.83 A thin ring of radius $r = 200$ mm is attached by a collar at point A to a vertical shaft which rotates with a constant angular velocity $\boldsymbol{\omega}$. Determine (*a*) the constant angle β that the plane of the ring forms with the vertical when $\omega = 10$ rad/s, (*b*) the largest value of ω for which the ring will remain vertical ($\beta = 0$).

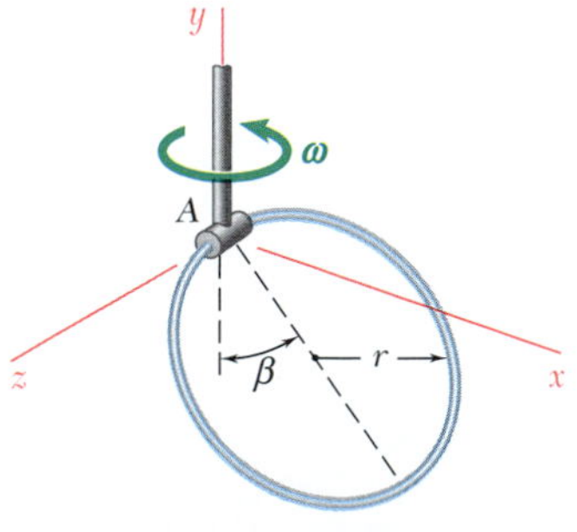

Fig. P18.83 and P18.84

18.84 A thin ring of radius $r = 300$ mm is attached by a collar at point A to a vertical shaft which rotates with a constant angular velocity $\boldsymbol{\omega}$. Determine (*a*) the value of ω for which the ring forms a constant angle $\beta = 45°$ with the vertical y axis, (*b*) the largest value of ω for which the ring will remain vertical ($\beta = 0$).

18.85 A uniform square plate of side $a = 9$ in. is hinged at points A and B to a clevis which rotates with a constant angular velocity $\boldsymbol{\omega}$ about a vertical axis. Determine (*a*) the constant angle β that the plate forms with the horizontal x axis when $\omega = 12$ rad/s, (*b*) the largest value of ω for which the plate remains vertical ($\beta = 90°$).

18.86 A uniform square plate of side $a = 12$ in. is hinged at points A and B to a clevis which rotates with a constant angular velocity $\boldsymbol{\omega}$ about a vertical axis. Determine (*a*) the value of ω for which the plate forms a constant angle $\beta = 60°$ with the horizontal x axis, (*b*) the largest value of ω for which the plate remains vertical ($\beta = 90°$).

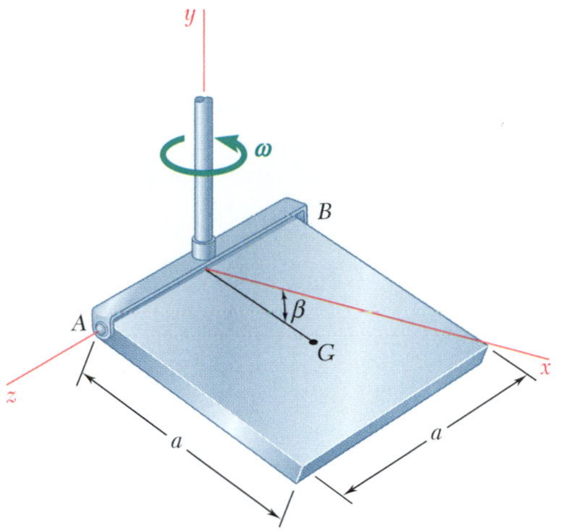

Fig. P18.85 and P18.86

18.87 and 18.88 The slender rod AB is attached by a clevis to arm BCD which rotates with a constant angular velocity $\boldsymbol{\omega}$ about the centerline of its vertical portion CD. Determine the magnitude of the angular velocity $\boldsymbol{\omega}$.

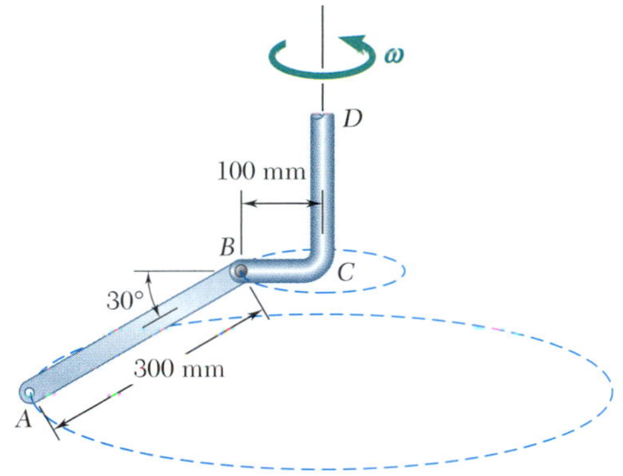

Fig. P18.87

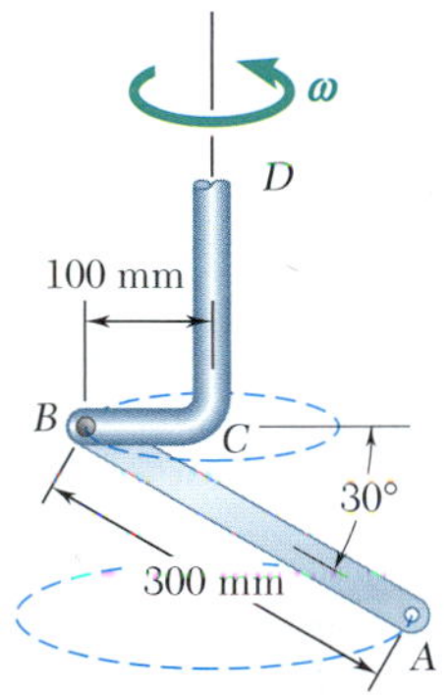

Fig. P18.88

18.89 The 2-lb gear A is constrained to roll on the fixed gear B, but is free to rotate about axle AD. Axle AD, of length 20 in. and negligible weight, is connected by a clevis to the vertical shaft DE which rotates as shown with a constant angular velocity $\boldsymbol{\omega}_1$. Assuming that gear A can be approximated by a thin disk of radius 4 in., determine the largest allowable value of ω_1 if gear A is not to lose contact with gear B.

18.90 Determine the force $\mathbf{F}$ exerted by gear B on gear A of Prob. 18.89 when shaft DE rotates with the constant angular velocity $\boldsymbol{\omega}_1 = 4$ rad/s. (*Hint:* The force $\mathbf{F}$ must be perpendicular to the line drawn from D to C.)

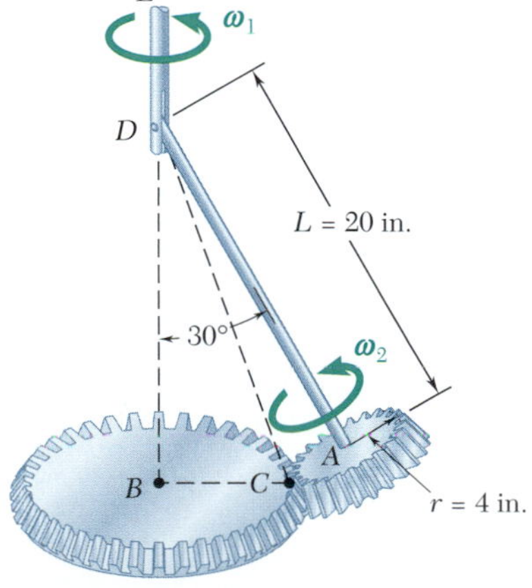

Fig. *P18.89*

18.91 The 300-g disk shown spins at the rate $\omega_1 = 750$ rpm, while axle AB rotates as shown with an angular velocity $\boldsymbol{\omega}_2$ of 6 rad/s. Determine the dynamic reactions at A and B.

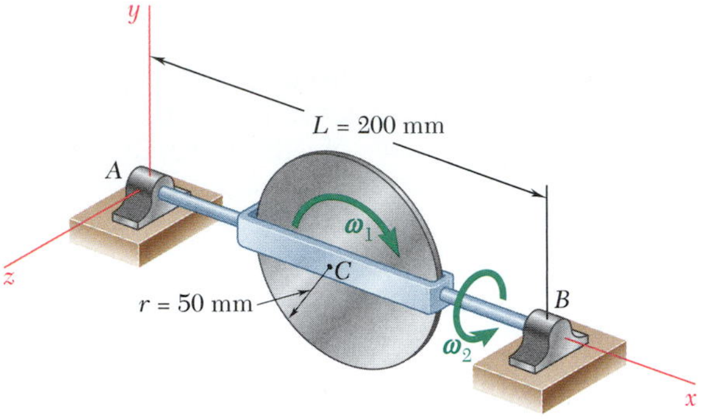

Fig. P18.91 and P18.92

18.92 The 300-g disk shown spins at the rate $\omega_1 = 750$ rpm, while axle AB rotates as shown with an angular velocity $\boldsymbol{\omega}_2$. Determine the maximum allowable magnitude of $\boldsymbol{\omega}_2$ if the dynamic reactions at A and B are not to exceed 1 N each.

18.93 Two disks, each of weight 10 lb and radius 12 in., spin as shown at the rate $\omega_1 = 1200$ rpm about a rod AB of negligible mass which rotates about the horizontal z axis at the rate $\omega_2 = 60$ rpm. (*a*) Determine the dynamic reactions at points C and D. (*b*) Solve part *a* assuming that the direction of spin of disk A is reversed.

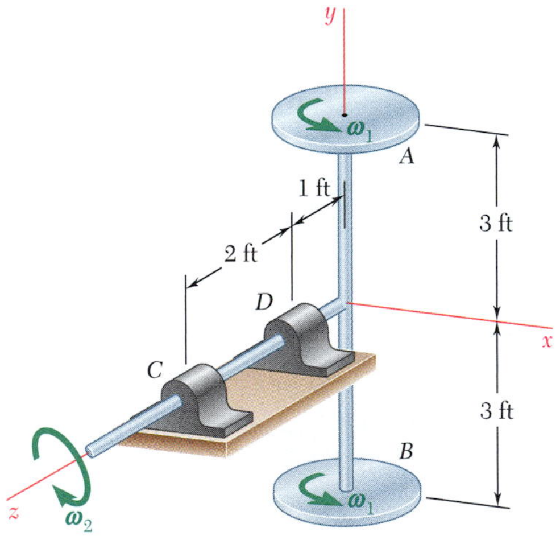

Fig. P18.93 and P18.94

18.94 Two disks, each of weight 10 lb and radius 12 in., spin as shown at the rate $\omega_1 = 1200$ rpm about a rod AB of negligible mass which rotates about the horizontal z axis at the rate ω_2. Determine the maximum allowable value of ω_2 if the magnitudes of the dynamic reactions at points C and D are not to exceed 80 lb each.

18.95 A homogeneous disk of mass $m = 4$ kg rotates at the constant rate $\omega_1 = 12$ rad/s with respect to arm OA, which itself rotates at the constant rate $\omega_2 = 4$ rad/s about the y axis. Determine the force-couple system representing the dynamic reactions at the support O.

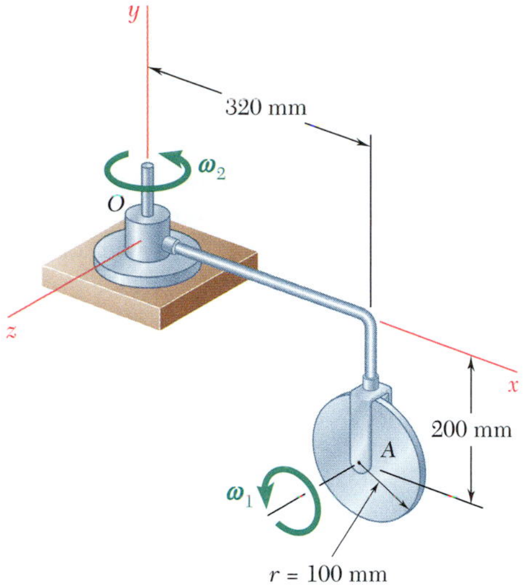

Fig. P18.95

18.96 A homogeneous disk of mass $m = 3$ kg rotates at the constant rate $\omega_1 = 16$ rad/s with respect to arm ABC, which is welded to shaft DCE rotating at the constant rate $\omega_2 = 8$ rad/s. Determine the dynamic reactions at D and E.

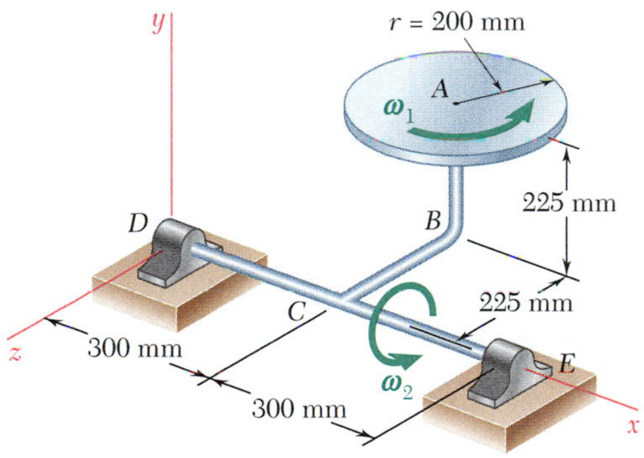

Fig. P18.96

***18.97** A 100-lb advertising panel of length $2a = 7.2$ ft and width $2b = 4.8$ ft is kept rotating at a constant rate ω_1 about its horizontal axis by a small electric motor attached at A to frame ACB. This frame itself is kept rotating at a constant rate ω_2 about a vertical axis by a second motor attached at C to the column CD. Knowing that the panel and the frame complete a full revolution in 6 s and 12 s, respectively, express, as a function of the angle θ, the dynamic reaction exerted on column CD by its support at D.

***18.98** For the system of Prob. 18.97, show that (a) the dynamic reaction at D is independent of the length $2a$ of the panel, (b) the ratio M_1/M_2 of the magnitudes of the couples exerted by the motors at A and C, respectively, is independent of the dimensions and mass of the panel and is equal to $\omega_2/2\omega_1$ at any given instant.

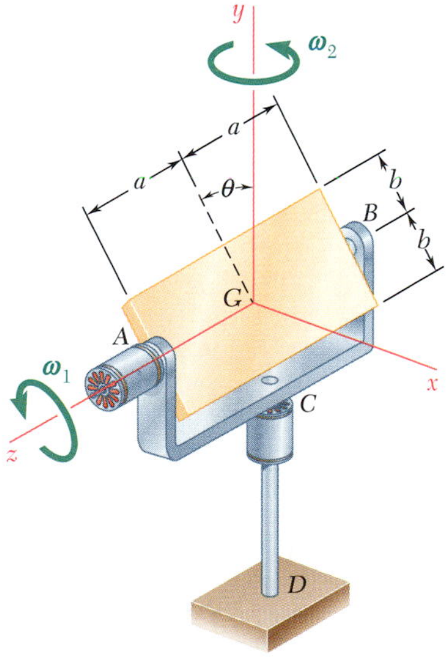

Fig. P18.97

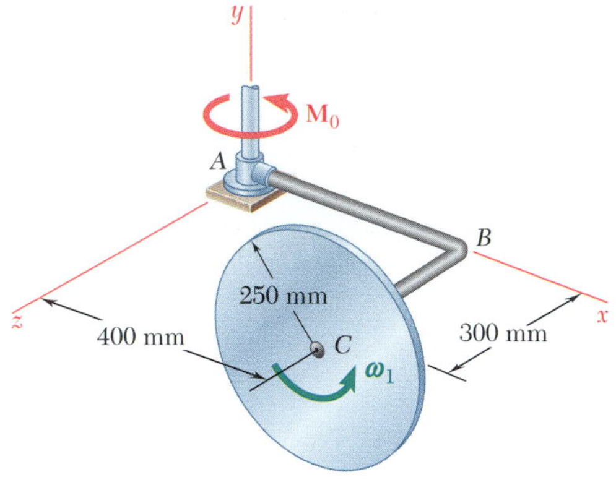

Fig. P18.99 and P18.100

18.99 A 40-kg disk of radius 250 mm spins as shown at the constant rate $\omega_1 = 80$ rad/s with respect to the bent axle ABC. The system is at rest when a couple $\mathbf{M}_0 = (2 \text{ N} \cdot \text{m})\mathbf{j}$ is applied for 5 s and then removed. Determine the force-couple system representing the dynamic reaction at the support at point A after the couple was removed. Assume that the bent axle ABC has a negligible mass.

18.100 A 40-kg disk of radius 250 mm spins as shown at the constant rate $\omega_1 = 80$ rad/s with respect to the bent axle ABC. The system is at rest when a couple $\mathbf{M}_0$ is applied as shown for 3 s and then removed. Knowing that the maximum angular velocity reached by the bent axle ABC is 2.4 rad/s, determine (*a*) the couple $\mathbf{M}_0$, (*b*) the force-couple system representing the dynamic reaction at the support at point A just before the couple was removed. Assume that the bent axle ABC has a negligible mass.

18.101 A 5-lb homogeneous disk of radius 4 in. rotates with an angular velocity $\boldsymbol{\omega}_1$ with respect to arm ABC, which is welded to a shaft DCE rotating as shown at the constant rate $\omega_2 = 12$ rad/s. Friction in the bearing at A causes ω_1 to decrease at the rate of 15 rad/s^2. Determine the dynamic reactions at D and E at a time when ω_1 has decreased to 50 rad/s.

18.102 A 5-lb homogeneous disk of radius 4 in. rotates at the constant rate $\omega_1 = 50$ rad/s with respect to arm ABC, which is welded to a shaft DCE. Knowing that at the instant shown shaft DCE has an angular velocity of $\boldsymbol{\omega}_2 = (12 \text{ rad/s})\mathbf{k}$ and an angular acceleration $\boldsymbol{\alpha}_2 = (8 \text{ rad/s}^2)\mathbf{k}$, determine (*a*) the couple which must be applied to shaft DCE to produce that angular acceleration, (*b*) the corresponding dynamic reactions at D and E.

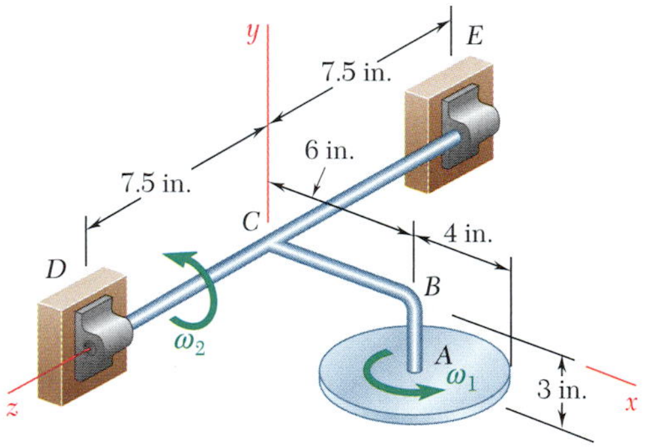

Fig. P18.101 and P18.102

18.103 It is assumed that at the instant shown the disk of Prob. 18.95 has an angular velocity $\boldsymbol{\omega}_1 = (12 \text{ rad/s})\mathbf{k}$, which is decreasing at the rate of 4 rad/s^2 due to friction in the bearing at A. Recalling that arm OA rotates with a constant angular velocity $\boldsymbol{\omega}_2 = (4 \text{ rad/s})\mathbf{j}$, determine the force-couple system representing the dynamic reaction at the support at O.

18.104 It is assumed that at the instant shown shaft DCE of Prob. 18.96 has an angular velocity $\boldsymbol{\omega}_2 = (8 \text{ rad/s})\mathbf{i}$ and an angular acceleration $\boldsymbol{\alpha}_2 = (6 \text{ rad/s}^2)\mathbf{i}$. Recalling that the disk rotates with a constant angular velocity $\boldsymbol{\omega}_1 = (16 \text{ rad/s})\mathbf{j}$, determine (*a*) the couple that must be applied to shaft DCE to produce the given angular acceleration, (*b*) the corresponding dynamic reactions at D and E.

*18.9. MOTION OF A GYROSCOPE. EULERIAN ANGLES

A *gyroscope* consists essentially of a rotor which can spin freely about its geometric axis. When mounted in a Cardan's suspension (Fig. 18.15), a gyroscope can assume any orientation, but its mass center must remain fixed in space. In order to define the position of a gyroscope at a given instant, let us select a fixed frame of reference $OXYZ$, with the origin O located at the mass center of the gyroscope and the Z axis directed along the line defined by the bearings A and A' of the outer gimbal. We will consider a reference position of the gyroscope in which the two gimbals and a given diameter DD' of the rotor are located in the fixed YZ plane (Fig. 18.15*a*). The gyroscope can be brought from this reference position into any arbitrary position (Fig. 18.15*b*) by means of the following steps: (1) a rotation of the outer gimbal through an angle ϕ about the axis AA', (2) a rotation of the inner gimbal through θ about BB', and (3) a rotation of the rotor through ψ about CC'. The angles ϕ, θ, and ψ are called the *Eulerian angles;* they completely characterize the position of the gyroscope at any given instant. Their derivatives $\dot{\phi}$, $\dot{\theta}$, and $\dot{\psi}$ define, respectively, the rate of *precession*, the rate of *nutation*, and the rate of *spin* of the gyroscope at the instant considered.

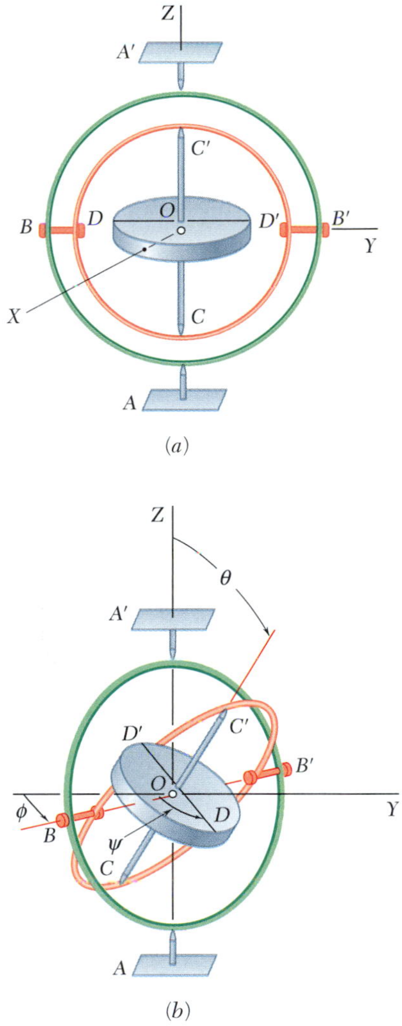

Fig. 18.15

In order to compute the components of the angular velocity and of the angular momentum of the gyroscope, we will use a rotating system of axes $Oxyz$ *attached to the inner gimbal,* with the y axis along BB' and the z axis along CC' (Fig. 18.16). These axes are principal axes of inertia for the gyroscope. While they follow it in its precession and nutation, however, they do not spin; for that reason, they are more convenient to use than axes actually attached to the gyroscope. The angular velocity $\boldsymbol{\omega}$ of the gyroscope with respect to the fixed frame of reference $OXYZ$ will now be expressed as the sum of three partial angular velocities corresponding, respectively, to the precession, the nutation, and the spin of the gyroscope. Denoting by $\mathbf{i}$, $\mathbf{j}$, and $\mathbf{k}$ the unit vectors along the rotating axes, and by $\mathbf{K}$ the unit vector along the fixed Z axis, we have

$$\boldsymbol{\omega} = \dot{\phi}\mathbf{K} + \dot{\theta}\mathbf{j} + \dot{\psi}\mathbf{k} \tag{18.33}$$

Since the vector components obtained for $\boldsymbol{\omega}$ in (18.33) are not orthogonal (Fig. 18.16), the unit vector $\mathbf{K}$ will be resolved into components along the x and z axes; we write

$$\mathbf{K} = -\sin\theta\,\mathbf{i} + \cos\theta\,\mathbf{k} \tag{18.34}$$

and, substituting for $\mathbf{K}$ into (18.33),

$$\boldsymbol{\omega} = -\dot{\phi}\sin\theta\,\mathbf{i} + \dot{\theta}\mathbf{j} + (\dot{\psi} + \dot{\phi}\cos\theta)\mathbf{k} \tag{18.35}$$

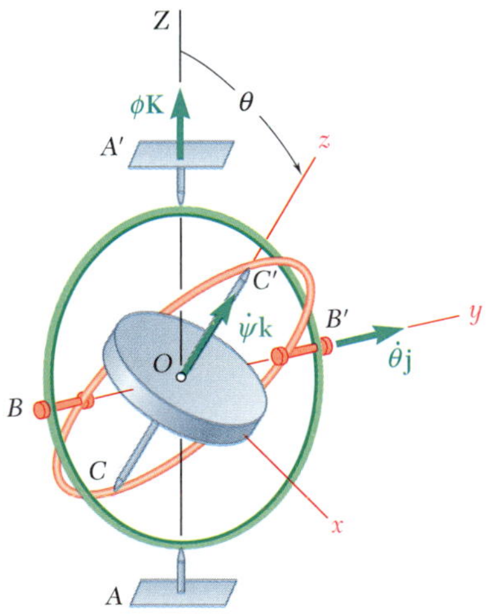

Fig. 18.16

Since the coordinate axes are principal axes of inertia, the components of the angular momentum $\mathbf{H}_O$ can be obtained by multiplying the components of $\boldsymbol{\omega}$ by the moments of inertia of the rotor about the x, y, and z axes, respectively. Denoting by I the moment of inertia of the rotor about its spin axis, by I' its moment of inertia about a transverse axis through O, and neglecting the mass of the gimbals, we write

$$\mathbf{H}_O = -I'\dot{\phi} \sin \theta \, \mathbf{i} + I'\dot{\theta}\mathbf{j} + I(\dot{\psi} + \dot{\phi} \cos \theta)\mathbf{k} \tag{18.36}$$

Recalling that the rotating axes are attached to the inner gimbal and thus do not spin, we express their angular velocity as the sum

$$\boldsymbol{\Omega} = \dot{\phi}\mathbf{K} + \dot{\theta}\mathbf{j} \tag{18.37}$$

or, substituting for $\mathbf{K}$ from (18.34),

$$\boldsymbol{\Omega} = -\dot{\phi} \sin \theta \, \mathbf{i} + \dot{\theta}\mathbf{j} + \dot{\phi} \cos \theta \, \mathbf{k} \tag{18.38}$$

Substituting for $\mathbf{H}_O$ and $\boldsymbol{\Omega}$ from (18.36) and (18.38) into the equation

$$\Sigma \mathbf{M}_O = (\dot{\mathbf{H}}_O)_{Oxyz} + \boldsymbol{\Omega} \times \mathbf{H}_O \tag{18.28}$$

we obtain the three differential equations

$$\begin{aligned} \Sigma M_x &= -I'(\ddot{\phi} \sin \theta + 2\dot{\theta}\dot{\phi} \cos \theta) + I\dot{\theta} \, (\dot{\psi} + \dot{\phi} \cos \theta) \\ \Sigma M_y &= I'(\ddot{\theta} - \dot{\phi}^2 \sin \theta \cos \theta) + I\dot{\phi} \sin \theta \, (\dot{\psi} + \dot{\phi} \cos \theta) \\ \Sigma M_z &= I \frac{d}{dt} (\dot{\psi} + \dot{\phi} \cos \theta) \end{aligned} \tag{18.39}$$

The equations (18.39) define the motion of a gyroscope subjected to a given system of forces when the mass of its gimbals is neglected. They can also be used to define the motion of an *axisymmetrical body* (or body of revolution) attached at a point on its axis of symmetry, and to define the motion of an axisymmetrical body about its mass center. While the gimbals of the gyroscope helped us visualize the Eulerian angles, it is clear that these angles can be used to define the position of any rigid body with respect to axes centered at a point of the body, regardless of the way in which the body is actually supported.

Since the equations (18.39) are nonlinear, it will not be possible, in general, to express the Eulerian angles ϕ, θ, and ψ as analytical functions of the time t, and numerical methods of solution may have to be used. However, as you will see in the following sections, there are several particular cases of interest which can be analyzed easily.

*18.10. STEADY PRECESSION OF A GYROSCOPE

Let us now investigate the particular case of gyroscopic motion in which the angle θ, the rate of precession $\dot{\phi}$, and the rate of spin $\dot{\psi}$ remain constant. We propose to determine the forces which must be applied to the gyroscope to maintain this motion, known as the *steady precession* of a gyroscope.

Instead of applying the general equations (18.39), we will determine the sum of the moments of the required forces by computing the rate of change of the angular momentum of the gyroscope in the particular case considered. We first note that the angular velocity $\boldsymbol{\omega}$ of the gyroscope, its angular momentum $\mathbf{H}_O$, and the angular velocity $\boldsymbol{\Omega}$ of the rotating frame of reference (Fig. 18.17) reduce, respectively, to

$$\boldsymbol{\omega} = -\dot{\phi} \sin \theta \, \mathbf{i} + \omega_z \mathbf{k} \tag{18.40}$$

$$\mathbf{H}_O = -I'\dot{\phi} \sin \theta \, \mathbf{i} + I\omega_z \mathbf{k} \tag{18.41}$$

$$\boldsymbol{\Omega} = -\dot{\phi} \sin \theta \, \mathbf{i} + \dot{\phi} \cos \theta \, \mathbf{k} \tag{18.42}$$

where $\omega_z = \dot{\psi} + \dot{\phi} \cos \theta$ = rectangular component along spin axis of total angular velocity of gyroscope

Since θ, $\dot{\phi}$, and $\dot{\psi}$ are constant, the vector $\mathbf{H}_O$ is constant in magnitude and direction with respect to the rotating frame of reference and its rate of change $(\dot{\mathbf{H}}_O)_{Oxyz}$ with respect to that frame is zero. Thus Eq. (18.28) reduces to

$$\Sigma \mathbf{M}_O = \boldsymbol{\Omega} \times \mathbf{H}_O \tag{18.43}$$

which yields, after substitutions from (18.41) and (18.42),

$$\Sigma \mathbf{M}_O = (I\omega_z - I'\dot{\phi} \cos \theta)\dot{\phi} \sin \theta \, \mathbf{j} \tag{18.44}$$

Since the mass center of the gyroscope is fixed in space, we have, by (18.1), $\Sigma \mathbf{F} = 0$; thus, the forces which must be applied to the gyroscope to maintain its steady precession reduce to a couple of moment equal to the right-hand member of Eq. (18.44). We note that *this couple should be applied about an axis perpendicular to the precession axis and to the spin axis of the gyroscope* (Fig. 18.18).

In the particular case when the precession axis and the spin axis are at a right angle to each other, we have $\theta = 90°$ and Eq. (18.44) reduces to

$$\Sigma \mathbf{M}_O = I\dot{\psi}\dot{\phi}\mathbf{j} \tag{18.45}$$

Thus, if we apply to the gyroscope a couple $\mathbf{M}_O$ about an axis perpendicular to its axis of spin, the gyroscope will precess about an axis perpendicular to both the spin axis and the couple axis, in a sense such that the vectors representing the spin, the couple, and the precession, respectively, form a right-handed triad (Fig. 18.19).

Because of the relatively large couples required to change the orientation of their axles, gyroscopes are used as stabilizers in torpedoes

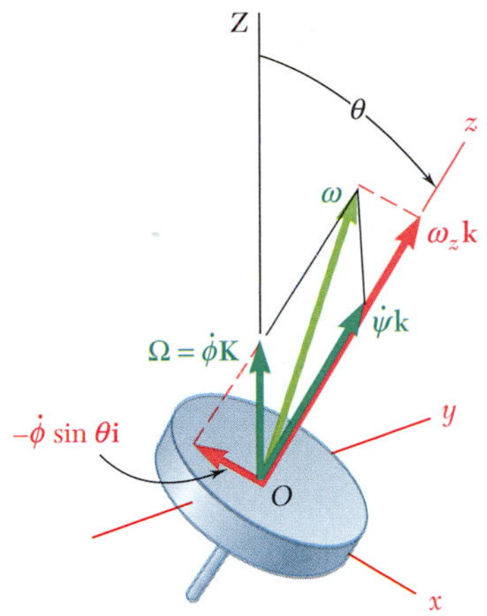

Fig. 18.17

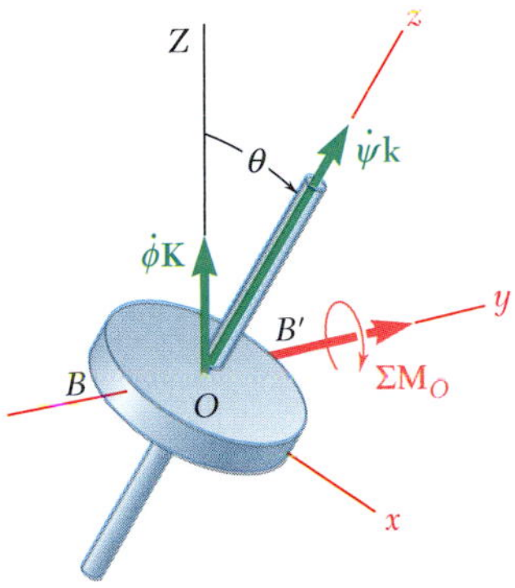

Fig. 18.18

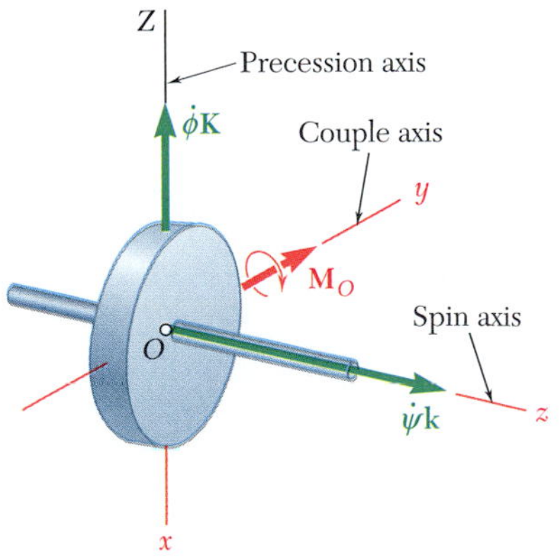

Fig. 18.19

and ships. Spinning bullets and shells remain tangent to their trajectory because of gyroscopic action. And a bicycle is easier to keep balanced at high speeds because of the stabilizing effect of its spinning wheels. However, gyroscopic action is not always welcome and must be taken into account in the design of bearings supporting rotating shafts subjected to forced precession. The reactions exerted by its propellers on an airplane which changes its direction of flight must also be taken into consideration and compensated for whenever possible.

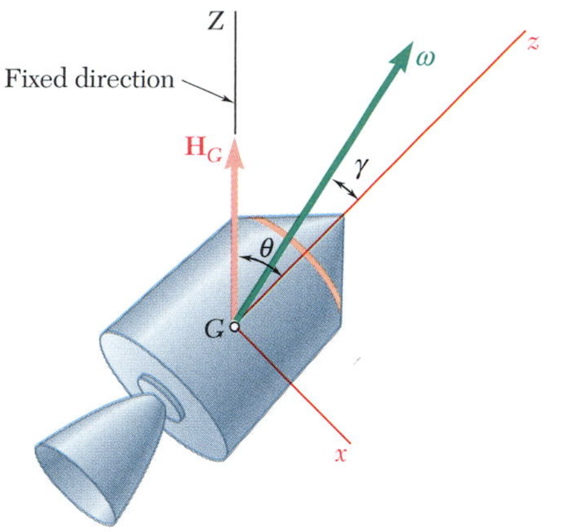

Fig. 18.20

*18.11. MOTION OF AN AXISYMMETRICAL BODY UNDER NO FORCE

In this section you will analyze the motion about its mass center of an axisymmetrical body under no force except its own weight. Examples of such a motion are furnished by projectiles, if air resistance is neglected, and by artificial satellites and space vehicles after burnout of their launching rockets.

Since the sum of the moments of the external forces about the mass center G of the body is zero, Eq. (18.2) yields $\dot{\mathbf{H}}_G = 0$. It follows that the angular momentum $\mathbf{H}_G$ of the body about G is constant. Thus, the direction of $\mathbf{H}_G$ is fixed in space and can be used to define the Z axis, or axis of precession (Fig. 18.20). Selecting a rotating system of axes $Gxyz$ with the z axis along the axis of symmetry of the body, the x axis in the plane defined by the Z and z axes, and the y axis pointing away from you, we have

$$H_x = -H_G \sin\theta \qquad H_y = 0 \qquad H_z = H_G \cos\theta \qquad (18.46)$$

where θ represents the angle formed by the Z and z axes, and H_G denotes the constant magnitude of the angular momentum of the body about G. Since the x, y, and z axes are principal axes of inertia for the body considered, we can write

$$H_x = I'\omega_x \qquad H_y = I'\omega_y \qquad H_z = I\omega_z \qquad (18.47)$$

where I denotes the moment of inertia of the body about its axis of symmetry and I' denotes its moment of inertia about a transverse axis through G. It follows from Eqs. (18.46) and (18.47) that

$$\omega_x = -\frac{H_G \sin\theta}{I'} \qquad \omega_y = 0 \qquad \omega_z = \frac{H_G\cos\theta}{I} \qquad (18.48)$$

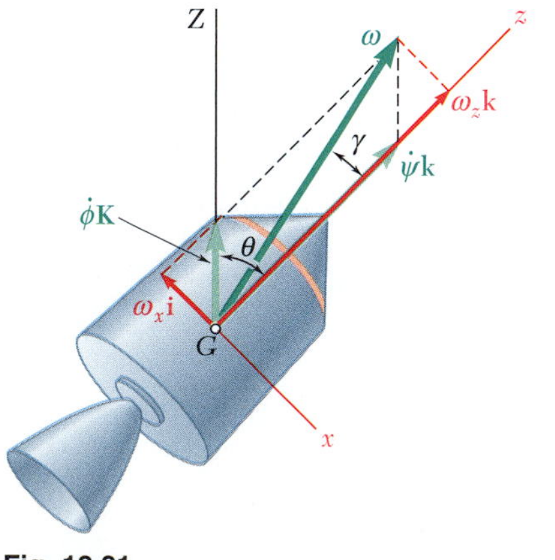

Fig. 18.21

The second of the relations obtained shows that the angular velocity $\boldsymbol{\omega}$ has no component along the y axis, that is, along an axis perpendicular to the Zz plane. Thus, the angle θ formed by the Z and z axes remains constant and *the body is in steady precession about the Z axis.*

Dividing the first and third of the relations (18.48) member by member, and observing from Fig. 18.21 that $-\omega_x/\omega_z = \tan\gamma$, we obtain the following relation between the angles γ and θ that the vectors $\boldsymbol{\omega}$ and $\mathbf{H}_G$, respectively, form with the axis of symmetry of the body:

$$\tan\gamma = \frac{I}{I'}\tan\theta \qquad (18.49)$$

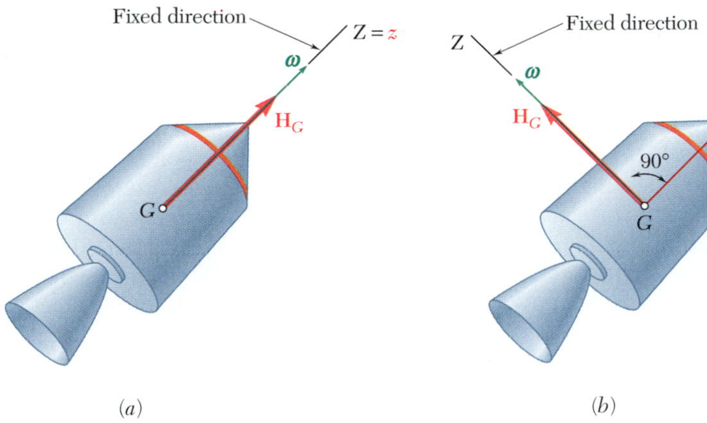

Fig. 18.22

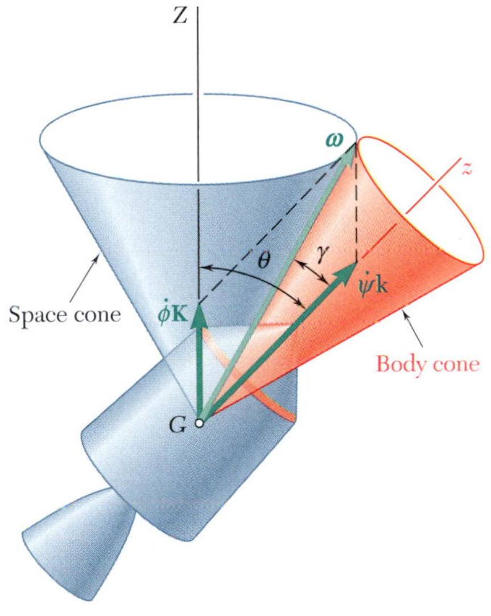

Fig. 18.23

There are two particular cases of motion of an axisymmetrical body under no force which involve no precession: (1) If the body is set to spin about its axis of symmetry, we have $\omega_x = 0$ and, by (18.47), $H_x = 0$; the vectors $\boldsymbol{\omega}$ and $\mathbf{H}_G$ have the same orientation and the body keeps spinning about its axis of symmetry (Fig. 18.22*a*). (2) If the body is set to spin about a transverse axis, we have $\omega_z = 0$ and, by (18.47), $H_z = 0$; again $\boldsymbol{\omega}$ and $\mathbf{H}_G$ have the same orientation and the body keeps spinning about the given transverse axis (Fig. 18.22*b*).

Considering now the general case represented in Fig. 18.21, we recall from Sec. 15.12 that the motion of a body about a fixed point—or about its mass center—can be represented by the motion of a body cone rolling on a space cone. In the case of steady precession, the two cones are circular, since the angles γ and $\theta - \gamma$ that the angular velocity $\boldsymbol{\omega}$ forms, respectively, with the axis of symmetry of the body and with the precession axis are constant. Two cases should be distinguished:

1. $I < I'$. This is the case of an elongated body, such as the space vehicle of Fig. 18.23. By (18.49) we have $\gamma < \theta$; the vector $\boldsymbol{\omega}$ lies inside the angle ZGz; the space cone and the body cone are tangent externally; the spin and the precession are both observed as counterclockwise from the positive z axis. The precession is said to be *direct*.

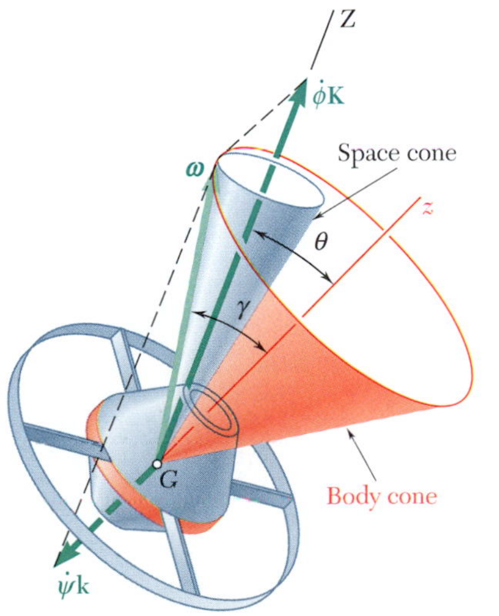

Fig. 18.24

2. $I > I'$. This is the case of a flattened body, such as the satellite of Fig. 18.24. By (18.49) we have $\gamma > \theta$; since the vector $\boldsymbol{\omega}$ must lie outside the angle ZGz, the vector $\dot{\psi}\mathbf{k}$ has a sense opposite to that of the z axis; the space cone is inside the body cone; the precession and the spin have opposite senses. The precession is said to be *retrograde*.

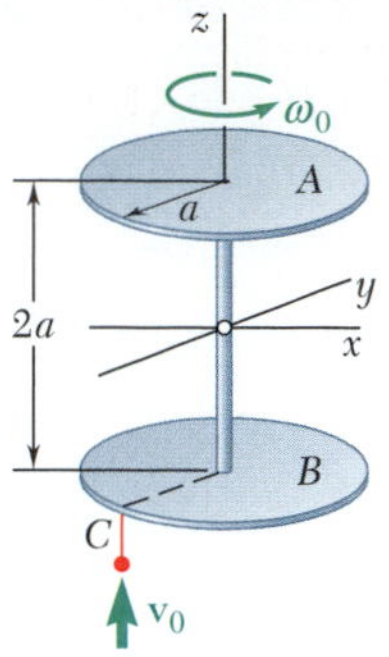

SAMPLE PROBLEM 18.6

A space satellite of mass m is known to be dynamically equivalent to two thin disks of equal mass. The disks are of radius $a = 800$ mm and are rigidly connected by a light rod of length $2a$. Initially the satellite is spinning freely about its axis of symmetry at the rate $\omega_0 = 60$ rpm. A meteorite, of mass $m_0 = m/1000$ and traveling with a velocity $\mathbf{v}_0$ of 2000 m/s relative to the satellite, strikes the satellite and becomes embedded at C. Determine (*a*) the angular velocity of the statellite immediately after impact, (*b*) the precession axis of the ensuing motion, (*c*) the rates of precession and spin of the ensuing motion.

SOLUTION

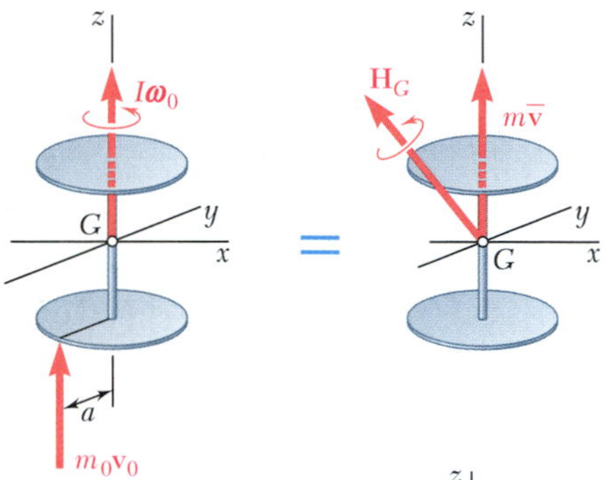

Moments of Inertia. We note that the axes shown are principal axes of inertia for the satellite and write

$$I = I_z = \tfrac{1}{2}ma^2 \qquad I' = I_x = I_y = 2[\tfrac{1}{4}(\tfrac{1}{2}m)a^2 + (\tfrac{1}{2}m)a^2] = \tfrac{5}{4}ma^2$$

Principle of Impulse and Momentum. We consider the satellite and the meteorite as a single system. Since no external force acts on this system, the momenta before and after impact are equipollent. Taking moments about G, we write

$$-a\mathbf{j} \times m_0v_0\mathbf{k} + I\omega_0\mathbf{k} = \mathbf{H}_G$$

$$\mathbf{H}_G = -m_0v_0a\mathbf{i} + I\omega_0\mathbf{k} \qquad (1)$$

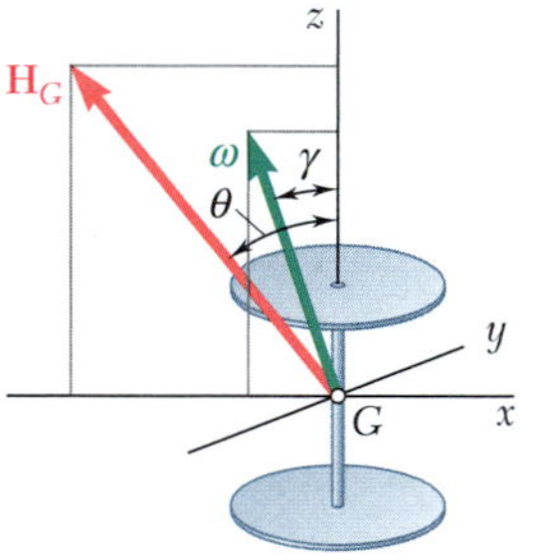

Angular Velocity after Impact. Substituting the values obtained for the components of $\mathbf{H}_G$ and for the moments of inertia into

$$H_x = I_x\omega_x \qquad H_y = I_y\omega_y \qquad H_z = I_z\omega_z$$

we write

$$-m_0v_0a = I'\omega_x = \tfrac{5}{4}ma^2\omega_x \qquad 0 = I'\omega_y \qquad I\omega_0 = I\omega_z$$

$$\omega_x = -\frac{4}{5}\frac{m_0v_0}{ma} \qquad \omega_y = 0 \qquad \omega_z = \omega_0 \qquad (2)$$

For the satellite considered we have $\omega_0 = 60$ rpm $= 6.283$ rad/s, $m_0/m = \frac{1}{1000}$, $a = 0.800$ m, and $v_0 = 2000$ m/s; we find

$$\omega_x = -2 \text{ rad/s} \qquad \omega_y = 0 \qquad \omega_z = 6.283 \text{ rad/s}$$

$$\omega = \sqrt{\omega_x^2 + \omega_z^2} = 6.594 \text{ rad/s} \qquad \tan\gamma = \frac{-\omega_x}{\omega_z} = +0.3183$$

$\omega = 63.0$ rpm $\quad \gamma = 17.7°$ ◀

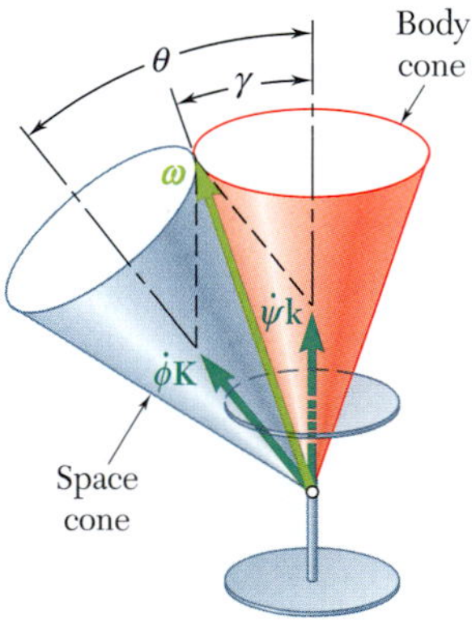

Precession Axis. Since in free motion the direction of the angular momentum $\mathbf{H}_G$ is fixed in space, the satellite will precess about this direction. The angle θ formed by the precession axis and the z axis is

$$\tan\theta = \frac{-H_x}{H_z} = \frac{m_0v_0a}{I\omega_0} = \frac{2m_0v_0}{ma\omega_0} = 0.796 \qquad \theta = 38.5° \blacktriangleleft$$

Rates of Precession and Spin. We sketch the space and body cones for the free motion of the satellite. Using the law of sines, we compute the rates of precession and spin.

$$\frac{\omega}{\sin\theta} = \frac{\dot{\phi}}{\sin\gamma} = \frac{\dot{\psi}}{\sin(\theta - \gamma)}$$

$\dot{\phi} = 30.8$ rpm $\quad \dot{\psi} = 35.9$ rpm ◀

SOLVING PROBLEMS ON YOUR OWN

In this lesson we analyzed the motion of *gyroscopes* and of other *axisymmetrical bodies* with a fixed point O. In order to define the position of these bodies at any given instant, we introduced the three *Eulerian angles* ϕ, θ, and ψ (Fig. 18.15), and noted that their time derivatives define, respectively, the rate of *precession*, the rate of *nutation* and the rate of *spin* (Fig. 18.16). The problems you will encounter fall into one of the following categories.

1. Steady precession. This is the motion of a gyroscope or other axisymmetrical body with a fixed point located on its axis of symmetry, in which the angle θ, the rate of precession $\dot{\phi}$, and the rate of spin $\dot{\psi}$ all remain constant.

a. Using the rotating frame of reference Oxyz shown in Fig. 18.17, which *precesses* with the body, *but does not spin* with it, we obtained the following expressions for the angular velocity $\boldsymbol{\omega}$ of the body, its angular momentum $\mathbf{H}_O$, and the angular velocity $\boldsymbol{\Omega}$ of the frame $Oxyz$:

$$\boldsymbol{\omega} = -\dot{\phi} \sin\theta\, \mathbf{i} + \omega_z \mathbf{k} \qquad (18.40)$$

$$\mathbf{H}_O = -I'\dot{\phi} \sin\theta\, \mathbf{i} + I\,\omega_z \mathbf{k} \qquad (18.41)$$

$$\boldsymbol{\Omega} = -\dot{\phi} \sin\theta\, \mathbf{i} + \dot{\phi} \cos\theta\, \mathbf{k} \qquad (18.42)$$

where I = moment of inertia of body about its axis of symmetry
I' = moment of inertia of body about a transverse axis through O
ω_z = *rectangular* component of $\boldsymbol{\omega}$ along z axis = $\dot{\psi} + \dot{\phi}\cos\theta$

b. The sum of the moments about O of the forces applied to the body is equal to the rate of change of its angular momentum, as expressed by Eq. (18.28). But, since θ and the rates of change $\dot{\phi}$ and $\dot{\psi}$ are constant, it follows from Eq. (18.41) that $\mathbf{H}_O$ remains constant in magnitude and direction when viewed from the frame $Oxyz$. Thus, its rate of change is zero with respect to that frame and you can write

$$\Sigma\mathbf{M}_O = \boldsymbol{\Omega} \times \mathbf{H}_O \qquad (18.43)$$

where $\boldsymbol{\Omega}$ and $\mathbf{H}_O$ are defined, respectively by Eq. (18.42) and Eq. (18.41). The equation obtained shows that the moment resultant at O of the forces applied to the body is perpendicular to both the axis of precession and the axis of spin (Fig. 18.18).

c. Keep in mind that the method described applies, not only to gyroscopes, where the fixed point O coincides with the mass center G, but also *to any axisymmetrical body with a fixed point O located on its axis of symmetry*. This method, therefore, can be used to analyze the *steady precession of a top* on a rough floor.

d. When an axisymmetrical body has no fixed point, but is in steady precession about its mass center G, you should draw a *free-body-diagram equation* showing that the system of the external forces exerted on the body (including the body's weight) is equivalent to the vector $m\bar{\mathbf{a}}$ applied at G and the couple vector $\dot{\mathbf{H}}_G$. You can use Eqs. (18.40) through (18.42), replacing $\mathbf{H}_O$ with $\mathbf{H}_G$, and express the moment of the couple as

$$\dot{\mathbf{H}}_G = \boldsymbol{\Omega} \times \mathbf{H}_G$$

You can then use the free-body-diagram equation to write as many as six independent scalar equations.

(continued)

2. *Motion of an axisymmetrical body under no force, except its own weight.* We have $\Sigma \mathbf{M}_G = 0$ and, thus, $\dot{\mathbf{H}}_G = 0$; it follows that *the angular momentum* $\mathbf{H}_G$ is constant in magnitude and direction (Sec. 18.11). The body is in *steady precession* with the precession axis GZ directed along $\mathbf{H}_G$ (Fig. 18.20). Using the rotating frame $Gxyz$ and denoting by γ the angle that $\boldsymbol{\omega}$ forms with the spin axis Gz (Fig. 18.21), we obtained the following relation between γ and the angle θ formed by the precession and spin axes:

$$\tan \gamma = \frac{I}{I'} \tan \theta \qquad (18.49)$$

The precession is said to be *direct* if $I < I'$ (Fig. 18.23) and *retrograde* if $I > I'$ (Fig. 18.24).

a. *In many of the problems* dealing with the motion of an axisymmetrical body under no force, you will be asked to determine the *precession axis* and the *rates of precession and spin* of the body, knowing the magnitude of its *angular velocity* $\boldsymbol{\omega}$ and the angle γ that it forms with the axis of symmetry Gz (Fig. 18.21). From Eq. (18.49) you will determine the angle θ that the precession axis GZ forms with Gz and resolve $\boldsymbol{\omega}$ into its two *oblique components* $\dot{\phi}\mathbf{K}$ and $\dot{\psi}\mathbf{k}$. Using the law of sines, you will then determine the rate of precession $\dot{\phi}$ and the rate of spin $\dot{\psi}$.

b. *In other problems,* the body will be subjected to *a given impulse* and you will first determine the resulting *angular momentum* $\mathbf{H}_G$. Using Eqs. (18.10), you will calculate the rectangular components of the angular velocity $\boldsymbol{\omega}$, its magnitude ω, and the angle γ that it forms with the axis of symmetry. You will then determine the *precession axis* and the *rates of precession and spin* as described above [Sample Prob. 18.6].

3. *General motion of an axisymmetric body with a fixed point O located on its axis of symmetry, and subjected only to its own weight.* This is a motion in which the angle θ is allowed to vary. At any given instant you should take into account the rate of precession $\dot{\phi}$, the rate of spin $\dot{\psi}$, *and the rate of nutation* $\dot{\theta}$, none of which will remain constant. An example of such a motion is the motion of a top, which is discussed in Probs. 18.139 and 18.140. The rotating frame of reference $Oxyz$ that you will use is still the one shown in Fig. 18.18, but this frame will now rotate about the y axis at the rate $\dot{\theta}$. Equations (18.40), (18.41), and (18.42), therefore, should be replaced by the following equations:

$$\boldsymbol{\omega} = -\dot{\phi} \sin \theta\, \mathbf{i} + \dot{\theta}\mathbf{j} + (\dot{\psi} + \dot{\phi} \cos \theta)\, \mathbf{k} \qquad (18.40')$$

$$\mathbf{H}_O = -I'\dot{\phi} \sin \theta\, \mathbf{i} + I'\dot{\theta}\mathbf{j} + I(\dot{\psi} + \dot{\phi} \cos \theta)\, \mathbf{k} \qquad (18.41')$$

$$\boldsymbol{\Omega} = -\dot{\phi} \sin \theta\, \mathbf{i} + \dot{\theta}\mathbf{j} + \dot{\phi} \cos \theta\, \mathbf{k} \qquad (18.42')$$

Since substituting these expressions into Eq. (18.44) would lead to nonlinear differential equations, it is preferable, whenever feasible, to apply the following conservation principles.

a. *Conservation of energy.* Denoting by c the distance between the fixed point O and the mass center G of the body, and by E the total energy, you will write

$$T + V = E: \qquad \tfrac{1}{2}(I'\omega_x^2 + I'\omega_y^2 + I\omega_z^2) + mgc \cos \theta = E$$

and substitute for the components of $\boldsymbol{\omega}$ the expressions obtained in Eq. (18.40′). Note that c will be positive or negative, depending upon the position of G relative to O. Also, $c = 0$ if G coincides with O; the *kinetic energy* is then conserved.

b. *Conservation of the angular momentum about the axis of precession.* Since the support at O is located on the Z axis, and since the weight of the body and the Z axis are both vertical and, thus, parallel to each other, it follows that $\Sigma M_Z = 0$ and, thus, that H_Z remains constant. This can be expressed by writing that the scalar product $\mathbf{K} \cdot \mathbf{H}_O$ is constant, where $\mathbf{K}$ is the unit vector along the Z axis.

c. *Conservation of the angular momentum about the axis of spin.* Since the support at O and the center of gravity G are both located on the z axis, it follows that $\Sigma M_z = 0$ and, thus, that H_z remains constant. This is expressed by writing that the coefficient of the unit vector $\mathbf{k}$ in Eq. (18.41′) is constant. Note that this last conservation principle cannot be applied when the body is restrained from spinning about its axis of symmetry, but in that case the only variables are θ and ϕ.

Problems

18.105 A uniform thin disk of 150-mm diameter is attached to the end of a rod AB of negligible mass which is supported by a ball-and-socket joint at point A. Knowing that the disk is observed to precess about the vertical axis AC at the constant rate of 36 rpm in the sense indicated and that its axis of symmetry AB forms an angle $\beta = 60°$ with AC, determine the rate at which the disk spins about rod AB.

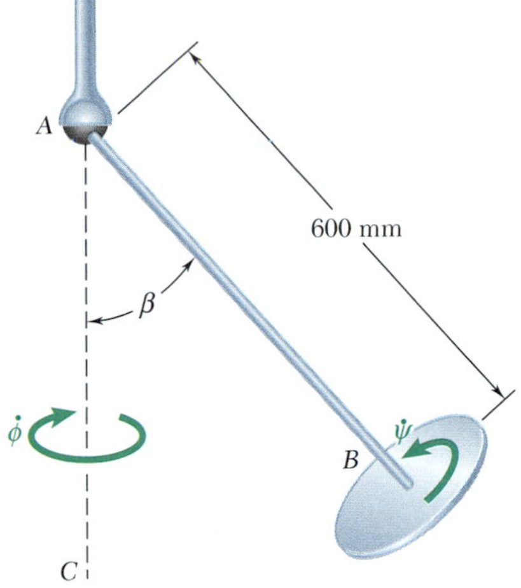

Fig. P18.105 and P18.106

18.106 A uniform thin disk of 150-mm diameter is attached to the end of a rod AB of negligible mass which is supported by a ball-and-socket joint at point A. Knowing that the disk is spinning about its axis of symmetry AB at the rate of 2100 rpm in the sense indicated and that AB forms an angle $\beta = 45°$ with the vertical axis AC, determine the two possible rates of steady precession of the disk about the axis AC.

18.107 A solid aluminum sphere of radius 4 in. is welded to the end of an 8-in.-long rod AB of negligible weight which is supported by a ball-and-socket joint at A. Knowing that the sphere is observed to precess about a vertical axis at the constant rate of 50 rpm in the sense indicated and that rod AB forms an angle $\beta = 25°$ with the vertical, determine the rate of spin of the sphere about line AB.

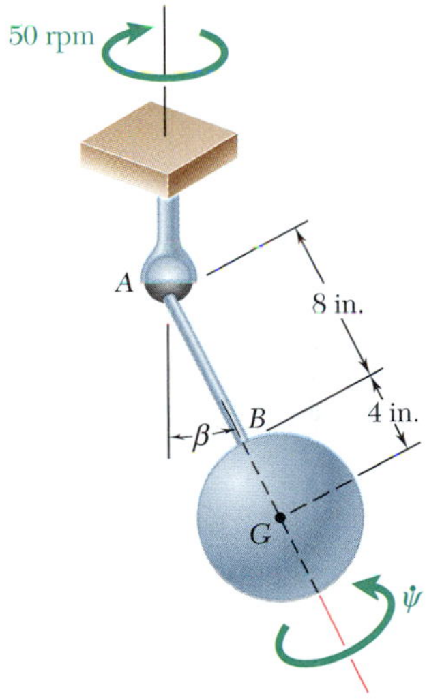

Fig. P18.107 and *P18.108*

18.108 A solid aluminum sphere of radius 4 in. is welded to the end of an 8-in.-long rod AB of negligible weight which is supported by a ball-and-socket joint at A. Knowing that the sphere spins as shown about line AB at the rate of 800 rpm, determine the angle β for which the sphere will precess about the vertical axis at the constant rate of 50 rpm in the sense indicated.

18.109 The 3-oz top shown is supported at the fixed point O. The radii of gyration of the top with respect to its axis of symmetry and with respect to a transverse axis through O are 1.05 in. and 2.25 in., respectively. Knowing that $c = 1.875$ in. and that the rate of spin of the top about its axis of symmetry is 1800 rpm, determine the two possible rates of steady precession corresponding to $\theta = 30°$.

18.110 The top shown is supported at the fixed point O and its moments of inertia about its axis of symmetry and about a transverse axis through O are denoted, respectively, by I and I'. (*a*) Show that the condition for steady precession of the top is

$$(I\omega_z - I'\dot\phi\cos\theta)\dot\phi = Wc$$

where $\dot\phi$ is the rate of precession and ω_z is the rectangular component of the angular velocity along the axis of symmetry of the top. (*b*) Show that if the rate of spin $\dot\psi$ of the top is very large compared with its rate of precession $\dot\phi$, the condition for steady precession is $I\dot\psi\dot\phi \approx Wc$. (*c*) Determine the percentage error introduced when this last relation is used to approximate the slower of the two rates of precession obtained for the top of Prob. 18.109.

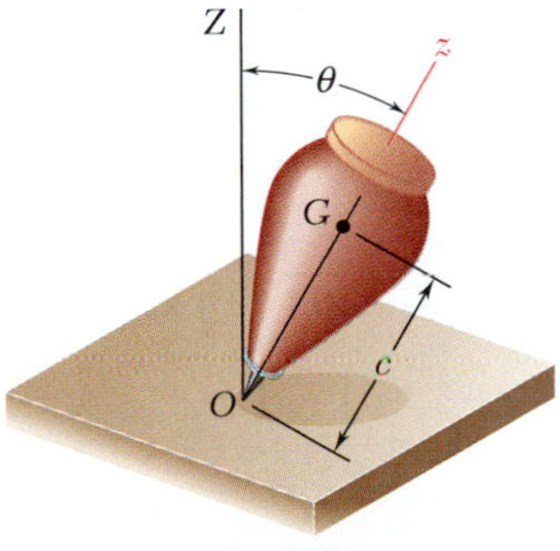

Fig. *P18.109* and P18.110

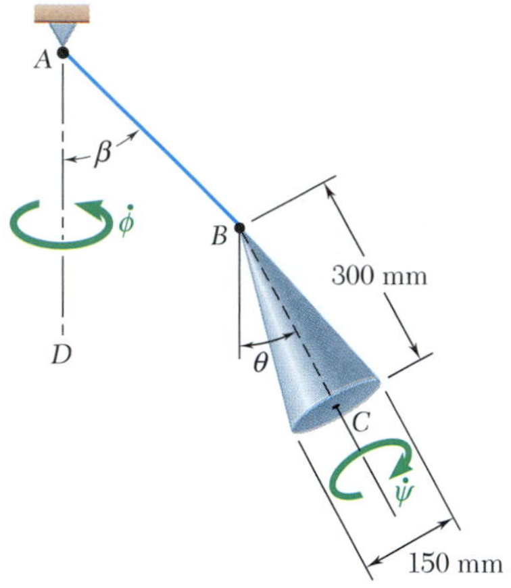

Fig. P18.111 and P18.112

18.111 A solid cone is attached as shown to cord AB. The cone spins about its axis BC at the constant rate $\dot{\psi} = 1200$ rpm and precesses about the vertical axis through point A at the constant rate $\dot{\phi} = 1000$ rpm. Determine the angle β for which the axis BC of the cone is aligned with the cord AB $(\theta = \beta)$.

18.112 A solid cone is attached as shown to cord AB. Knowing that the angles that cord AB and the axis BC of the cone form with the vertical axis are, respectively, $\beta = 45°$ and $\theta = 30°$, and that the cone precesses at the constant rate $\dot{\phi} = 8$ rad/s in the sense indicated, determine (*a*) the rate of spin $\dot{\psi}$ of the cone about its axis BC, (*b*) the length of cord AB.

18.113 A solid sphere of radius $c = 60$ mm is attached as shown to cord AB. The sphere is observed to precess at the constant rate $\dot{\phi} = 6$ rad/s about the vertical axis AD. Knowing that $\beta = 40°$, determine the angle θ that the diameter BC forms with the vertical when the sphere (*a*) has no spin, (*b*) spins about its diameter BC at the rate $\dot{\psi} = 50$ rad/s, (*c*) spins about BC at the rate $\dot{\psi} = -50$ rad/s.

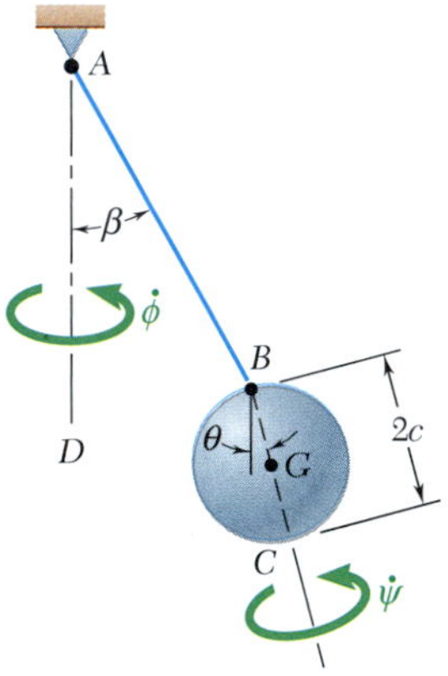

Fig. P18.113 and P18.114

18.114 A solid sphere of radius $c = 60$ mm is attached as shown to a cord AB of length 300 mm. The sphere spins about its diameter BC and precesses about the vertical axis AD. Knowing that $\theta = 20°$ and $\beta = 35°$, determine (*a*) the rate of spin of the sphere, (*b*) its rate of precession.

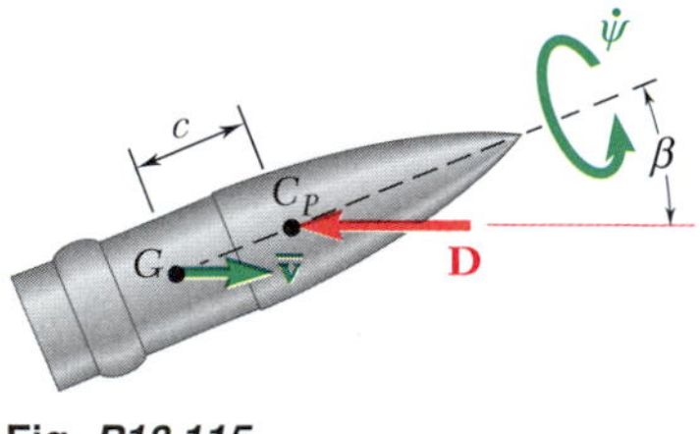

Fig. *P18.115*

18.115 A high-speed photographic record shows that a certain projectile was fired with a horizontal velocity $\bar{\mathbf{v}}$ of 1800 ft/s and with its axis of symmetry forming an angle $\beta = 3°$ with the horizontal. The rate of spin $\dot{\psi}$ of the projectile was 6000 rpm, and the atmospheric drag was equivalent to a force $\mathbf{D}$ of 25 lb acting at the center of pressure C_P located at a distance $c = 7.5$ in. from G. (*a*) Knowing that the projectile has a weight of 40 lb and a radius of gyration of 2.5 in. with respect to its axis of symmetry, determine its approximate rate of steady precession. (*b*) If it is further known that the radius of gyration of the projectile with respect to a transverse axis through G is 10 in., determine the exact values of the two possible rates of precession.

18.116 If the earth were a sphere, the gravitational attraction of the sun, moon, and planets would at all times be equivalent to a single force **R** acting at the mass center of the earth. However, the earth is actually an oblate spheroid and the gravitational system acting on the earth is equivalent to a force **R** and a couple **M**. Knowing that the effect of the couple **M** is to cause the axis of the earth to precess about the axis *GA* at the rate of one revolution in 25,800 years, determine the average magnitude of the couple **M** applied to the earth. Assume that the average specific weight of the earth is 344 lb/ft^3, that the average radius of the earth is 3960 mi, and that $\bar{I} = \frac{2}{5}mR^2$. (*Note:* This forced precession is known as the precession of the equinoxes and is not to be confused with the free precession discussed in Prob. 18.121.)

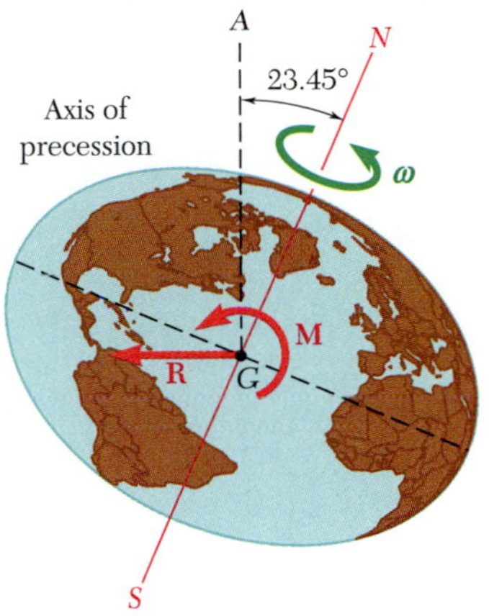

Fig. *P18.116*

18.117 Show that for an axisymmetrical body under no force, the rates of precession and spin can be expressed, respectively, as

$$\dot{\phi} = \frac{H_G}{I'}$$

and

$$\dot{\psi} = \frac{H_G \cos\theta(I' - I)}{II'}$$

where H_G is the constant value of the angular momentum of the body.

18.118 (*a*) Show that for an axisymmetrical body under no force, the rate of precession can be expressed as

$$\dot{\phi} = \frac{I\omega_2}{I' \cos\theta}$$

where ω_2 is the rectangular component of $\boldsymbol{\omega}$ along the axis of symmetry of the body. (*b*) Use this result to check that the condition (18.44) for steady precession is satisfied by an axisymmetrical body under no force.

18.119 Show that the angular velocity vector $\boldsymbol{\omega}$ of an axisymmetrical body under no force is observed from the body itself to rotate about the axis of symmetry at the constant rate

$$n = \frac{I' - I}{I'}\omega_2$$

where ω_2 is the rectangular component of $\boldsymbol{\omega}$ along the axis of symmetry of the body.

18.120 For an axisymmetrical body under no force, prove (*a*) that the rate of retrograde precession can never be less than twice the rate of spin of the body about its axis of symmetry, (*b*) that in Fig. 18.24 the axis of symmetry of the body can never lie within the space cone.

18.121 Using the relation given in Prob. 18.119, determine the period of precession of the north pole of the earth about the axis of symmetry of the earth. The earth may be approximated by an oblate spheroid of axial moment of inertia I and of transverse moment of inertia $I' = 0.99671I$. (*Note:* Actual observations show a period of precession of the north pole of about 432.5 mean solar days; the difference between the observed and computed periods is due to the fact that the earth is not a perfectly rigid body. The free precession considered here should not be confused with the much slower precession of the equinoxes, which is a forced precession. See Prob. 18.116.)

18.122 A coin is tossed into the air. It is observed to spin at the rate of 600 rpm about an axis GC perpendicular to the coin and to precess about the vertical direction GD. Knowing that GC forms an angle of 15° with GD, determine (*a*) the angle that the angular velocity $\boldsymbol{\omega}$ of the coin forms with GD, (*b*) the rate of precession of the coin about GD.

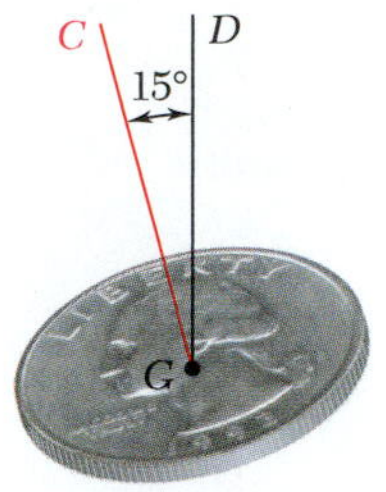

Fig. P18.122

18.123 The angular velocity vector of a football which has just been kicked is horizontal, and its axis of symmetry OC is oriented as shown. Knowing that the magnitude of the angular velocity is 200 rpm and that the ratio of the axial and transverse moments of inertia is $I/I' = \frac{1}{3}$, determine (*a*) the orientation of the axis of precession OA, (*b*) the rates of precession and spin.

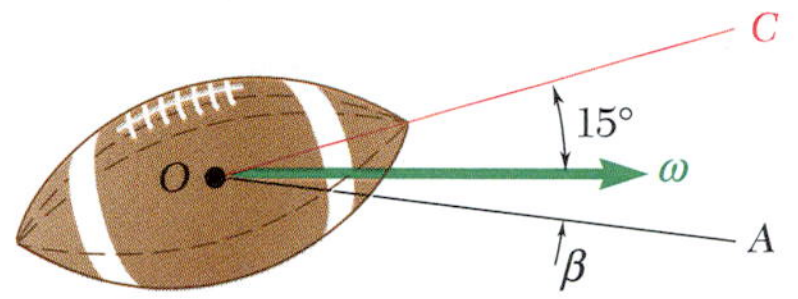

Fig. P18.123

18.124 A space station consists of two sections A and B of equal masses, which are rigidly connected. Each section is dynamically equivalent to a homogeneous cylinder of length 15 m and radius 3 m. Knowing that the station is precessing about the fixed direction GD at the constant rate of 2 rev/h, determine the rate of spin of the station about its axis of symmetry CC'.

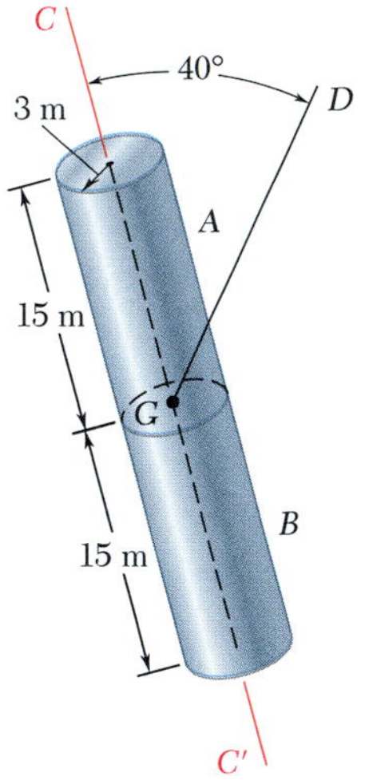

Fig. *P18.124*

18.125 If the connection between sections A and B of the space station of Prob. 18.124 is severed when the station is oriented as shown in Fig. P18.124 and if the two sections are gently pushed apart along their common axis of symmetry, determine (*a*) the angle between the spin axis and the new precession axis of section A, (*b*) the rate of precession of section A, (*c*) its rate of spin.

18.126 A 5-kip satellite is 7.2 ft high and has octagonal bases of sides 3.6 ft. The coordinate axes shown are the principal centroidal axes of inertia of the satellite, and its radii of gyration are $k_x = k_z = 2.7$ ft and $k_y = 2.94$ ft. The satellite is equipped with a main 125-lb thruster E and four 5-lb thrusters A, B, C, and D, which can expel fuel in the positive y direction. The satellite is spinning at the rate of 36 rev/h about its axis of symmetry Gy, which maintains a fixed direction in space, when thrusters A and B are activated for 2 s. Determine (*a*) the precession axis of the satellite, (*b*) its rate of precession, (*c*) its rate of spin.

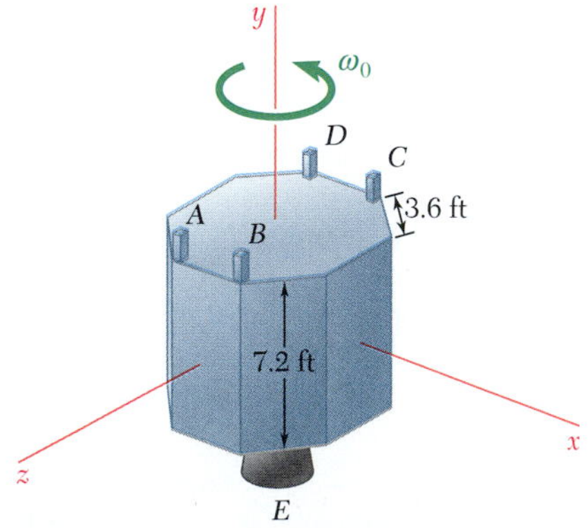

Fig. P18.126

18.127 Solve Prob. 18.126 assuming that thrusters *A* and *D* (instead of *A* and *B*) are activated for 2 s.

18.128 Solve Sample Prob. 18.6 assuming that the meteorite strikes the satellite at *C* with a velocity $\mathbf{v}_0 = (2000 \text{ m/s})\mathbf{i}$.

18.129 A homogeneous square plate of mass m and side c is held at points *A* and *B* by a frame of negligible mass which is supported by bearings at points *C* and *D*. The plate is free to rotate about *AB*, and the frame is free to rotate about the vertical *CD*. Knowing that, initially, $\theta_0 = 45°$, $\dot{\theta}_0 = 0$, and $\dot{\phi}_0 = 8$ rad/s, determine for the ensuing motion (*a*) the range of values of θ, (*b*) the minimum value of $\dot{\phi}$, (*c*) the maximum value of $\dot{\theta}$.

18.130 A homogeneous square plate of mass m and side c is held at points *A* and *B* by a frame of negligible mass which is supported by bearings at points *C* and *D*. The plate is free to rotate about *AB*, and the frame is free to rotate about the vertical *CD*. Initially the plate lies in the plane of the frame ($\theta_0 = 90°$) and the frame has an angular velocity $\dot{\phi}_0 = 8$ rad/s. If the plate is slightly disturbed, determine for the ensuing motion (*a*) the minimum value of $\dot{\phi}$, (*b*) the maximum value of $\dot{\theta}$.

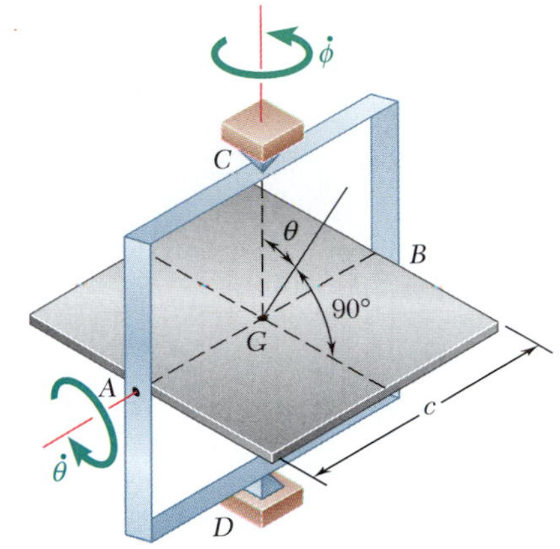

Fig. P18.129 and P18.130

18.131 A homogeneous disk of mass m is connected at *A* and *B* to a fork-ended shaft of negligible mass which is supported by a bearing at *C*. The disk is free to rotate about its horizontal diameter *AB* and the shaft is free to rotate about a vertical axis through *C*. Initially the disk lies in a vertical plane ($\theta_0 = 90°$) and the shaft has an angular velocity $\dot{\phi}_0 = 16$ rad/s. If the disk is slightly disturbed, determine for the ensuing motion (*a*) the minimum value of $\dot{\phi}$, (*b*) the maximum value of $\dot{\theta}$.

18.132 A homogeneous disk of mass m is connected at *A* and *B* to a fork-ended shaft of negligible mass which is supported by a bearing at *C*. The disk is free to rotate about its horizontal diameter *AB* and the shaft is free to rotate about a vertical axis through *C*. Knowing that, initially, $\theta_0 = 40°$, $\dot{\theta}_0 = 0$, and $\dot{\phi}_0 = 16$ rad/s, determine for the ensuing motion (*a*) the range of values of θ, (*b*) the minimum value of $\dot{\phi}$, (*c*) the maximum value of $\dot{\theta}$.

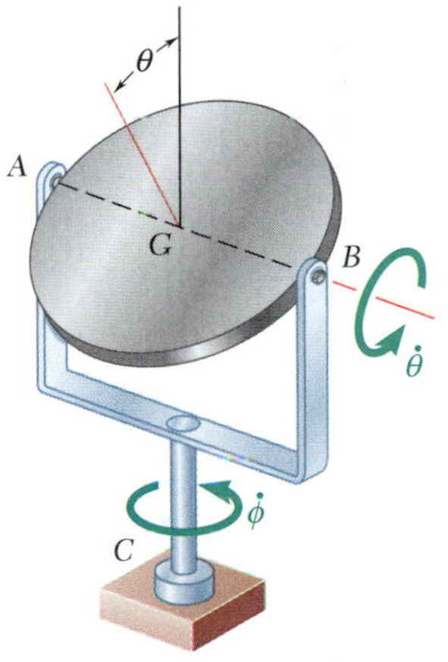

Fig. P18.131 and P18.132

18.133 A homogeneous disk of radius 9 in. is welded to a rod *AG* of length 18 in. and of negligible weight which is connected by a clevis to a vertical shaft *AB*. The rod and disk can rotate freely about a horizontal axis *AC*, and shaft *AB* can rotate freely about a vertical axis. Initially rod *AG* is horizontal ($\theta_0 = 90°$) and has no angular velocity about *AC*. Knowing that the maximum value $\dot{\phi}_m$ of the angular velocity of shaft *AB* in the ensuing motion is twice its initial value $\dot{\phi}_0$, determine (*a*) the minimum value of θ, (*b*) the initial angular velocity $\dot{\phi}_0$ of shaft *AB*.

18.134 A homogeneous disk of radius 9 in. is welded to a rod *AG* of length 18 in. and of negligible weight which is connected by a clevis to a vertical shaft *AB*. The rod and disk can rotate freely about a horizontal axis *AC*, and shaft *AB* can rotate freely about a vertical axis. Initially rod *AG* is horizontal ($\theta_0 = 90°$) and has no angular velocity about *AC*. Knowing that the smallest value of θ in the ensuing motion is 30°, determine (*a*) the initial angular velocity of shaft *AB*, (*b*) its maximum angular velocity.

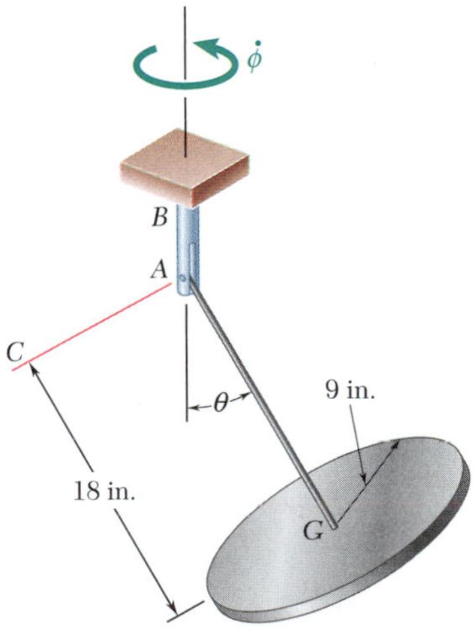

Fig. *P18.133* and *P18.134*

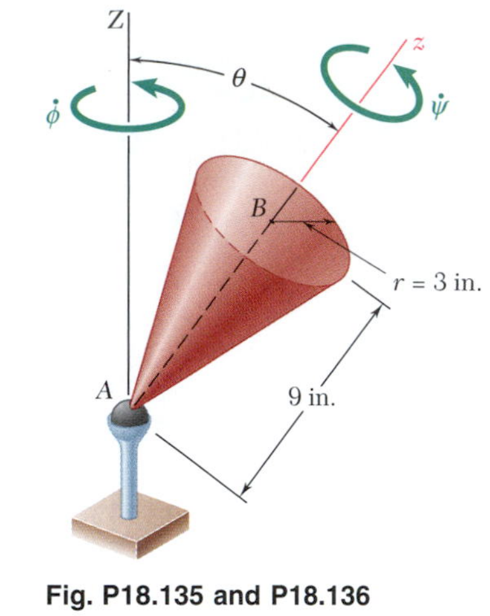

Fig. P18.135 and P18.136

***18.135** The top shown is supported at the fixed point O. Denoting by ϕ, θ, and ψ the Eulerian angles defining the position of the top with respect to a fixed frame of reference, consider the general motion of the top in which all Eulerian angles vary.

(*a*) Observing that $\Sigma M_Z = 0$ and $\Sigma M_z = 0$, and denoting by I and I', respectively, the moments of inertia of the top about its axis of symmetry and about a transverse axis through O, derive the two first-order differential equations of motion

$$I'\dot{\phi}\sin^2\theta + I(\dot{\psi} + \dot{\phi}\cos\theta)\cos\theta = \alpha \tag{1}$$

$$I(\dot{\psi} + \dot{\phi}\cos\theta) = \beta \tag{2}$$

where α and β are constants depending upon the initial conditions. These equations express that the angular momentum of the top is conserved about both the Z and z axes, that is, that the rectangular component of $\mathbf{H}_O$ along each of these axes is constant.

(*b*) Use Eqs. (1) and (2) to show that the rectangular component ω_z of the angular velocity of the top is constant and that the rate of precession $\dot{\phi}$ depends upon the value of the angle of nutation θ.

***18.136** (*a*) Applying the principle of conservation of energy, derive a third differential equation for the general motion of the top of Prob. 18.135.

(*b*) Eliminating the derivatives $\dot{\phi}$ and $\dot{\psi}$ from the equation obtained and from the two equations of Prob. 18.135, show that the rate of nutation $\dot{\theta}$ is defined by the differential equation $\dot{\theta}^2 = f(\theta)$, where

$$f(\theta) = \frac{1}{I'}\left(2E - \frac{\beta^2}{I} - 2mgc\cos\theta\right) - \left(\frac{\alpha - \beta\cos\theta}{I'\sin\theta}\right)^2 \tag{1}$$

(*c*) Further show, by introducing the auxiliary variable $x = \cos\theta$, that the maximum and minimum values of θ can be obtained by solving for x the cubic equation

$$\left(2E - \frac{\beta^2}{I} - 2mgcx\right)(1 - x^2) - \frac{1}{I'}(\alpha - \beta x)^2 = 0 \tag{2}$$

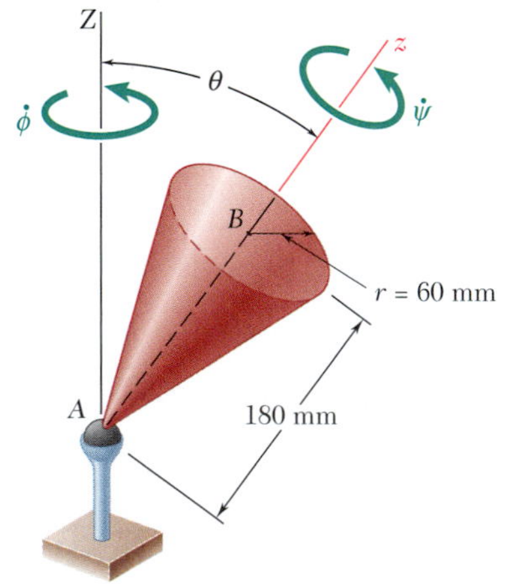

Fig. P18.137 and P18.138

***18.137** A solid cone of height 180 mm with a circular base of radius 60 mm is supported by a ball and socket at A. The cone is released from the position $\theta_0 = 30°$ with a rate of spin $\dot{\psi}_0 = 300$ rad/s, a rate of precession $\dot{\phi}_0 = 20$ rad/s, and a zero rate of nutation. Determine (*a*) the maximum value of θ in the ensuing motion, (*b*) the corresponding values of the rates of spin and precession. [*Hint:* Use Eq. (2) of Prob. 18.136; you can either solve this equation numerically or reduce it to a quadratic equation, since one of its roots is known.]

***18.138** A solid cone of height 180 mm with a circular base of radius 60 mm is supported by a ball and socket at A. The cone is released from the position $\theta_0 = 30°$ with a rate of spin $\dot{\psi}_0 = 300$ rad/s, a rate of precession $\dot{\phi}_0 = -4$ rad/s, and a zero rate of nutation. Determine (*a*) the maximum value of θ in the ensuing motion, (*b*) the corresponding values of the rates of spin and precession, (*c*) the value of θ for which the sense of the precession is reversed. (See hint of Prob. 18.137.)

***18.139** A homogeneous disk of radius 9 in. is welded to a rod AG of length 18 in. and of negligible weight which is supported by a ball and socket at A. The disk is released with a rate of spin $\dot{\psi}_0 = 50$ rad/s, with zero rates of precession and nutation, and with rod AG horizontal ($\theta_0 = 90°$). Determine (*a*) the smallest value of θ in the ensuing motion, (*b*) the rates of precession and spin as the disk passes through its lowest position.

***18.140** A homogeneous disk of radius 9 in. is welded to a rod AG of length 18 in. and of negligible weight which is supported by a ball and socket at A. The disk is released with a rate of spin $\dot{\psi}_0$, counterclockwise as seen from A, with zero rates of precession and nutation, and with rod AG horizontal ($\theta_0 = 90°$). Knowing that the smallest value of θ in the ensuing motion is 30°, determine (*a*) the rate of spin $\dot{\psi}_0$ of the disk in its initial position, (*b*) the rates of precession and spin as the disk passes through its lowest position.

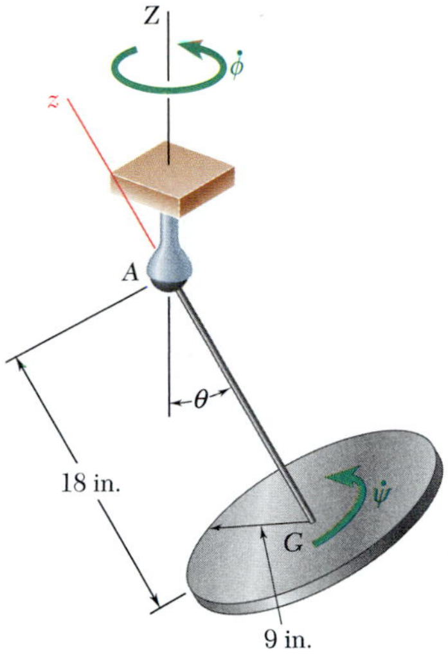

Fig. *P18.139* and *P18.140*

***18.141** Consider a rigid body of arbitrary shape which is attached at its mass center O and subjected to no force other than its weight and the reaction of the support at O.

(*a*) Prove that the angular momentum $\mathbf{H}_O$ of the body about the fixed point O is constant in magnitude and direction, that the kinetic energy T of the body is constant, and that the projection along $\mathbf{H}_O$ of the angular velocity $\boldsymbol{\omega}$ of the body is constant.

(*b*) Show that the tip of the vector $\boldsymbol{\omega}$ describes a curve on a fixed plane in space (called the *invariable plane*), which is perpendicular to $\mathbf{H}_O$ and at a distance $2T/H_O$ from O.

(*c*) Show that with respect to a frame of reference attached to the body and coinciding with its principal axes of inertia, the tip of the vector $\boldsymbol{\omega}$ appears to describe a curve on an ellipsoid of equation

$$I_x\omega_x^2 + I_y\omega_y^2 + I_z\omega_z^2 = 2T = \text{constant}$$

The ellipsoid (called the *Poinsot ellipsoid*) is rigidly attached to the body and is of the same shape as the ellipsoid of inertia, but of a different size.

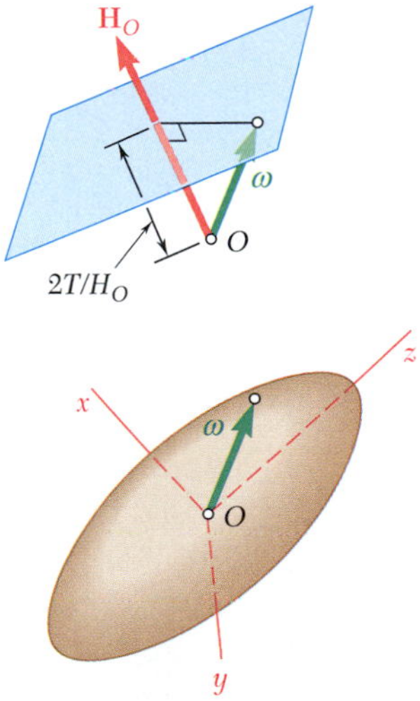

Fig. P18.141

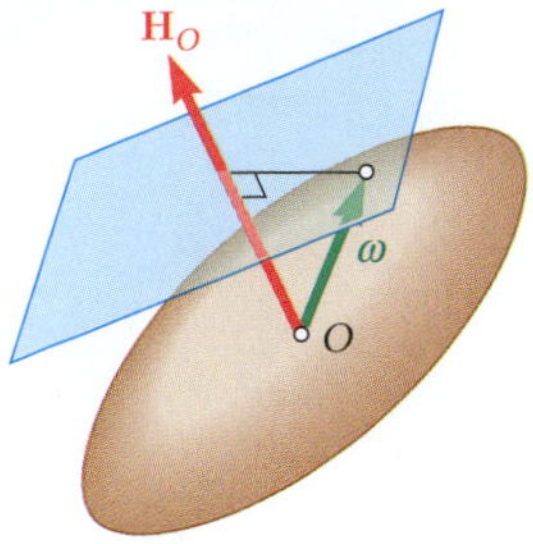

Fig. P18.142

***18.142** Referring to Prob. 18.141, (*a*) prove that the Poinsot ellipsoid is tangent to the invariable plane, (*b*) show that the motion of the rigid body must be such that the Poinsot ellipsoid appears to roll on the invariable plane. [*Hint:* In part *a*, show that the normal to the Poinsot ellipsoid at the tip of $\boldsymbol{\omega}$ is parallel to $\mathbf{H}_O$. It is recalled that the direction of the normal to a surface of equation $F(x, y, z) =$ constant at a point P is the same as that of **grad** F at point P.]

***18.143** Using the results obtained in Probs. 18.141 and 18.142, show that for an axisymmetrical body attached at its mass center O and under no force other than its weight and the reaction at O, the Poinsot ellipsoid is an ellipsoid of revolution and the space and body cones are both circular and are tangent to each other. Further show that (*a*) the two cones are tangent externally, and the precession is direct, when $I < I'$, where I and I' denote, respectively, the axial and transverse moment of inertia of the body, (*b*) the space cone is inside the body cone, and the precession is retrograde, when $I > I'$.

***18.144** Refer to Probs. 18.141 and 18.142.

(*a*) Show that the curve (called *polhode*) described by the tip of the vector $\boldsymbol{\omega}$ with respect to a frame of reference coinciding with the principal axes of inertia of the rigid body is defined by the equations

$$I_x\omega_x^2 + I_y\omega_y^2 + I_z\omega_z^2 = 2T = \text{constant} \qquad (1)$$
$$I_x^2\omega_x^2 + I_y^2\omega_y^2 + I_z^2\omega_z^2 = H_O^2 = \text{constant} \qquad (2)$$

and that the curve can, therefore, be obtained by intersecting the Poinsot ellipsoid with the ellipsoid defined by Eq. (2).

(*b*) Further show, assuming $I_x > I_y > I_z$, that the polhodes obtained for various values of H_O have the shapes indicated in the figure.

(*c*) Using the result obtained in part *b*, show that a rigid body under no force can rotate about a fixed centroidal axis if, and only if, that axis coincides with one of the principal axes of inertia of the body, and that the motion will be stable if the axis of rotation coincides with the major or minor axis of the Poinsot ellipsoid (z or x axis in the figure) and unstable if it coincides with the intermediate axis (y axis).

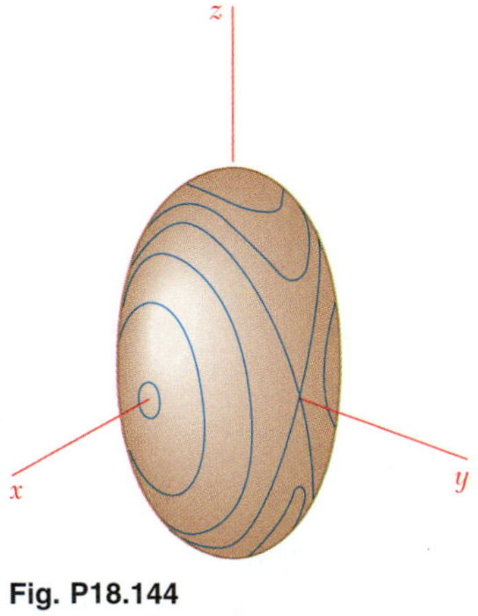

Fig. P18.144

REVIEW AND SUMMARY FOR CHAPTER 18

This chapter was devoted to the kinetic analysis of the motion of rigid bodies in three dimensions.

Fundamental equations of motion for a rigid body

We first noted [Sec. 18.1] that the two fundamental equations derived in Chap. 14 for the motion of a system of particles

$$\Sigma \mathbf{F} = m\overline{\mathbf{a}} \tag{18.1}$$

$$\Sigma \mathbf{M}_G = \dot{\mathbf{H}}_G \tag{18.2}$$

provide the foundation of our analysis, just as they did in Chap. 16 in the case of the plane motion of rigid bodies. The computation of the angular momentum $\mathbf{H}_G$ of the body and of its derivative $\dot{\mathbf{H}}_G$, however, are now considerably more involved.

Angular momentum of a rigid body in three dimensions

In Sec. 18.2, we saw that the rectangular components of the angular momentum $\mathbf{H}_G$ of a rigid body can be expressed as follows in terms of the components of its angular velocity $\boldsymbol{\omega}$ and of its centroidal moments and products of inertia:

$$\begin{aligned} H_x &= +\bar{I}_x\omega_x - \bar{I}_{xy}\omega_y - \bar{I}_{xz}\omega_z \\ H_y &= -\bar{I}_{yx}\omega_x + \bar{I}_y\omega_y - \bar{I}_{yz}\omega_z \\ H_z &= -\bar{I}_{zx}\omega_x - \bar{I}_{zy}\omega_y + \bar{I}_z\omega_z \end{aligned} \tag{18.7}$$

If *principal axes of inertia* $Gx'y'z'$ are used, these relations reduce to

$$H_{x'} = \bar{I}_{x'}\omega_{x'} \qquad H_{y'} = \bar{I}_{y'}\omega_{y'} \qquad H_{z'} = \bar{I}_{z'}\omega_{z'} \tag{18.10}$$

We observed that, in general, *the angular momentum* $\mathbf{H}_G$ *and the angular velocity* $\boldsymbol{\omega}$ *do not have the same direction* (Fig. 18.25). They will, however, have the same direction if $\boldsymbol{\omega}$ is directed along one of the principal axes of inertia of the body.

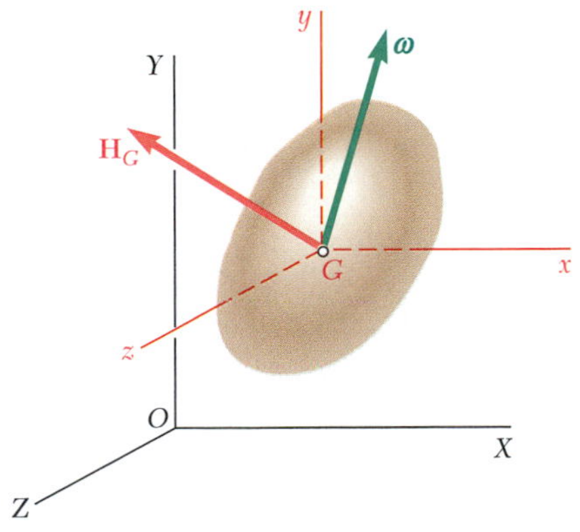

Fig. 18.25

Angular momentum about a given point

Recalling that the system of the momenta of the particles forming a rigid body can be reduced to the vector $m\overline{\mathbf{v}}$ attached at G and the couple $\mathbf{H}_G$ (Fig. 18.26), we noted that, once the linear momentum $m\overline{\mathbf{v}}$ and the angular momentum $\mathbf{H}_G$ of a rigid body have been determined, the angular momentum $\mathbf{H}_O$ of the body about any given point O can be obtained by writing

$$\mathbf{H}_O = \overline{\mathbf{r}} \times m\overline{\mathbf{v}} + \mathbf{H}_G \tag{18.11}$$

Rigid body with a fixed point

In the particular case of a rigid body *constrained to rotate about a fixed point* O, the components of the angular momentum $\mathbf{H}_O$ of the body about O can be obtained directly from the components of its angular velocity and from its moments and products of inertia with respect to axes through O. We wrote

$$\begin{aligned} H_x &= +I_x\omega_x - I_{xy}\omega_y - I_{xz}\omega_z \\ H_y &= -I_{yx}\omega_x + I_y\omega_y - I_{yz}\omega_z \\ H_z &= -I_{zx}\omega_x - I_{zy}\omega_y + I_z\omega_z \end{aligned} \tag{18.13}$$

Principle of impulse and momentum

The *principle of impulse and momentum* for a rigid body in three-dimensional motion [Sec. 18.3] is expressed by the same fundamental formula that was used in Chap. 17 for a rigid body in plane motion,

$$\textbf{Syst Momenta}_1 + \textbf{Syst Ext Imp}_{1\rightarrow 2} = \textbf{Syst Momenta}_2 \tag{17.4}$$

but the systems of the initial and final momenta should now be represented as shown in Fig. 18.26, and $\mathbf{H}_G$ should be computed from the relations (18.7) or (18.10) [Sample Probs. 18.1 and 18.2].

Kinetic energy of a rigid body in three dimensions

The *kinetic energy* of a rigid body in three-dimensional motion can be divided into two parts [Sec. 18.4], one associated with the motion of its mass center G and the other with its motion about G. Using principal centroidal axes x', y', z', we wrote

$$T = \tfrac{1}{2}m\overline{v}^2 + \tfrac{1}{2}(\overline{I}_{x'}\omega_{x'}^2 + \overline{I}_{y'}\omega_{y'}^2 + \overline{I}_{z'}\omega_{z'}^2) \tag{18.17}$$

where $\overline{\mathbf{v}}$ = velocity of mass center
$\boldsymbol{\omega}$ = angular velocity
m = mass of rigid body
$\overline{I}_{x'}, \overline{I}_{y'}, \overline{I}_{z'}$ = principal centroidal moments of inertia

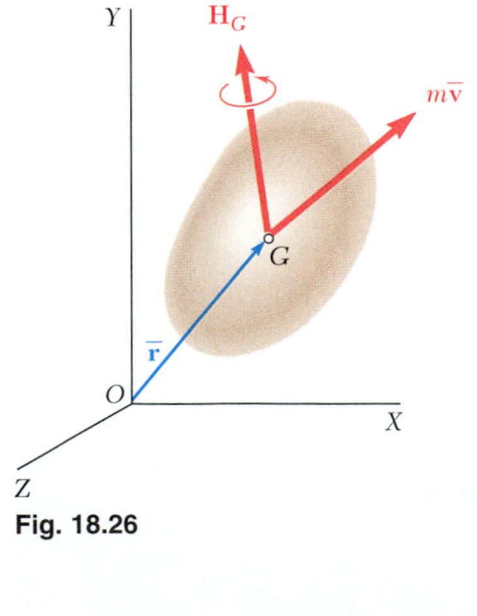

Fig. 18.26

We also noted that, in the case of a rigid body *constrained to rotate about a fixed point O*, the kinetic energy of the body can be expressed as

$$T = \tfrac{1}{2}(I_{x'}\omega_{x'}^2 + I_{y'}\omega_{y'}^2 + I_{z'}\omega_{z'}^2) \qquad (18.20)$$

where the x', y', and z' axes are the principal axes of inertia of the body at O. The results obtained in Sec. 18.4 make it possible to extend to the three-dimensional motion of a rigid body the application of the *principle of work and energy* and of the *principle of conservation of energy*.

Using a rotating frame to write the equations of motion of a rigid body in space

The second part of the chapter was devoted to the application of the fundamental equations

$$\Sigma \mathbf{F} = m\bar{\mathbf{a}} \qquad (18.1)$$

$$\Sigma \mathbf{M}_G = \dot{\mathbf{H}}_G \qquad (18.2)$$

to the motion of a rigid body in three dimensions. We first recalled [Sec. 18.5] that $\mathbf{H}_G$ represents the angular momentum of the body relative to a centroidal frame $GX'Y'Z'$ of fixed orientation (Fig. 18.27)

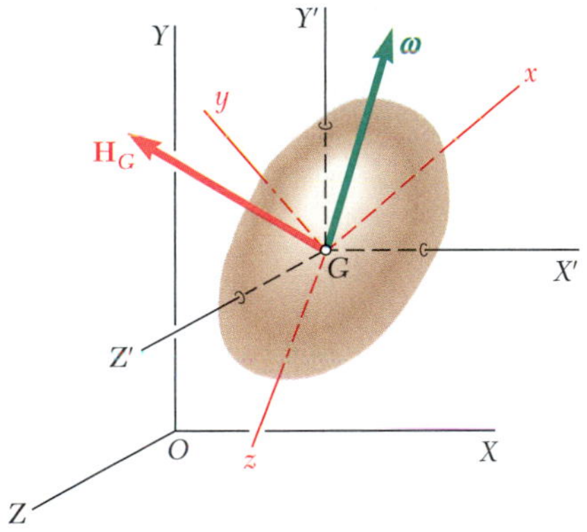

Fig. 18.27

and that $\dot{\mathbf{H}}_G$ in Eq. (18.2) represents the rate of change of $\mathbf{H}_G$ with respect to that frame. We noted that, as the body rotates, its moments and products of inertia with respect to the frame $GX'Y'Z'$ change continually. Therefore, it is more convenient to use a rotating frame $Gxyz$ when resolving $\boldsymbol{\omega}$ into components and computing the moments and products of inertia that will be used to determine $\mathbf{H}_G$ from Eqs. (18.7) or (18.10). However, since $\dot{\mathbf{H}}_G$ in Eq. (18.2) represents the rate of change of $\mathbf{H}_G$ with respect to the frame $GX'Y'Z'$ of fixed orientation, we must use the method of Sec. 15.10 to determine its value. Recalling Eq. (15.31), we wrote

$$\dot{\mathbf{H}}_G = (\dot{\mathbf{H}}_G)_{Gxyz} + \boldsymbol{\Omega} \times \mathbf{H}_G \qquad (18.22)$$

where $\mathbf{H}_G$ = angular momentum of body with respect to frame $GX'Y'Z'$ of fixed orientation

$(\dot{\mathbf{H}}_G)_{Gxyz}$ = rate of change of $\mathbf{H}_G$ with respect to rotating frame $Gxyz$, to be computed from relations (18.7)

$\boldsymbol{\Omega}$ = angular velocity of the rotating frame $Gxyz$

Substituting for $\dot{\mathbf{H}}_G$ from (18.22) into (18.2), we obtained

$$\Sigma \mathbf{M}_G = (\dot{\mathbf{H}}_G)_{Gxyz} + \boldsymbol{\Omega} \times \mathbf{H}_G \tag{18.23}$$

If the rotating frame is actually attached to the body, its angular velocity $\boldsymbol{\Omega}$ is identically equal to the angular velocity $\boldsymbol{\omega}$ of the body. There are many applications, however, where it is advantageous to use a frame of reference which is not attached to the body but rotates in an independent manner [Sample Prob. 18.5].

Euler's equations of motion. d'Alembert's principle

Setting $\boldsymbol{\Omega} = \boldsymbol{\omega}$ in Eq. (18.23), using principal axes, and writing this equation in scalar form, we obtained *Euler's equations of motion* [Sec. 18.6]. A discussion of the solution of these equations and of the scalar equations corresponding to Eq. (18.1) led us to extend d'Alembert's principle to the three-dimensional motion of a rigid body and to conclude that the system of the external forces acting on the rigid body is not only equipollent, but actually *equivalent* to the effective forces of the body represented by the vector $m\bar{\mathbf{a}}$ and the couple $\dot{\mathbf{H}}_G$ (Fig. 18.28). Problems involving the three-dimensional motion of a rigid body can be solved by considering the free-body-diagram equation represented in Fig. 18.28 and writing appropriate scalar equations relating the components or moments of the external and effective forces [Sample Probs. 18.3 and 18.5].

Free-body-diagram equation

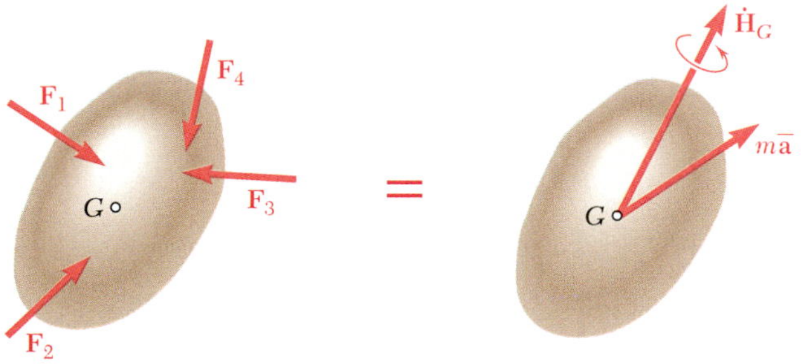

Fig. 18.28

Rigid body with a fixed point

In the case of a rigid body *constrained to rotate about a fixed point O*, an alternative method of solution, involving the moments of the forces and the rate of change of the angular momentum about point O, can be used. We wrote [Sec. 18.7]:

$$\Sigma \mathbf{M}_O = (\dot{\mathbf{H}}_O)_{Oxyz} + \boldsymbol{\Omega} \times \mathbf{H}_O \tag{18.28}$$

where $\Sigma \mathbf{M}_O$ = sum of moments about O of forces applied to rigid body

$\mathbf{H}_O$ = angular momentum of body with respect to fixed frame $OXYZ$

$(\dot{\mathbf{H}}_O)_{Oxyz}$ = rate of change of $\mathbf{H}_O$ with respect to rotating frame $Oxyz$, to be computed from relations (18.13)

$\boldsymbol{\Omega}$ = angular velocity of rotating frame $Oxyz$

This approach can be used to solve certain problems involving the rotation of a rigid body about a fixed axis [Sec. 18.8], for example, an unbalanced rotating shaft [Sample Prob. 18.4].

Motion of a gyroscope

In the last part of the chapter, we considered the motion of *gyroscopes* and other *axisymmetrical bodies*. Introducing the *Eulerian angles* ϕ, θ, and ψ to define the position of a gyroscope (Fig. 18.29), we observed that their derivatives $\dot{\phi}$, $\dot{\theta}$, and $\dot{\psi}$ represent, respectively, the rates of *precession, nutation,* and *spin* of the gyroscope [Sec. 18.9]. Expressing the angular velocity $\boldsymbol{\omega}$ in terms of these derivatives, we wrote

$$\boldsymbol{\omega} = -\dot{\phi} \sin \theta \, \mathbf{i} + \dot{\theta}\mathbf{j} + (\dot{\psi} + \dot{\phi} \cos \theta)\mathbf{k} \qquad (18.35)$$

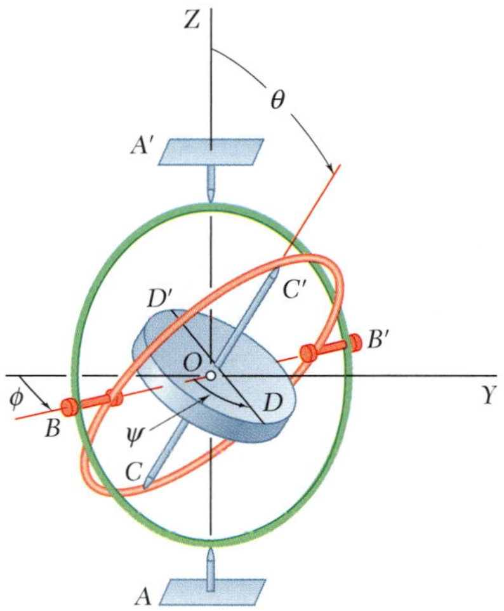

Fig. 18.29

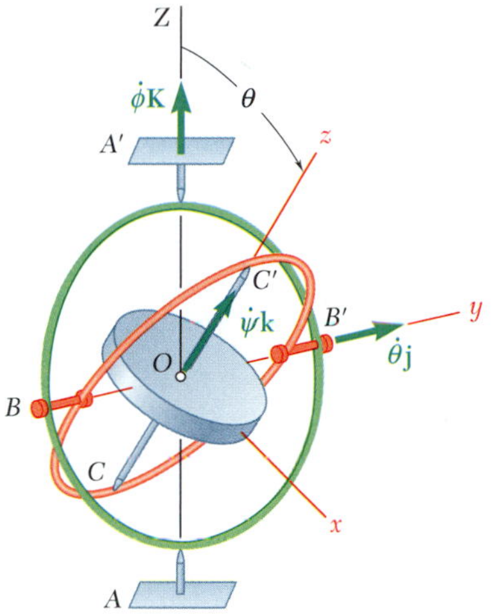

Fig. 18.30

where the unit vectors are associated with a frame $Oxyz$ attached to the inner gimbal of the gyroscope (Fig. 18.30) and rotate, therefore, with the angular velocity

$$\boldsymbol{\Omega} = -\dot{\phi} \sin \theta \, \mathbf{i} + \dot{\theta}\mathbf{j} + \dot{\phi} \cos \theta \, \mathbf{k} \qquad (18.38)$$

Denoting by I the moment of inertia of the gyroscope with respect to its spin axis z and by I' its moment of inertia with respect to a transverse axis through O, we wrote

$$\mathbf{H}_O = -I'\dot{\phi} \sin \theta \, \mathbf{i} + I'\dot{\theta}\mathbf{j} + I(\dot{\psi} + \dot{\phi} \cos \theta)\mathbf{k} \qquad (18.36)$$

Substituting for $\mathbf{H}_O$ and $\boldsymbol{\Omega}$ into Eq. (18.28) led us to the differential equations defining the motion of the gyroscope.

Steady precession

In the particular case of the *steady precession* of a gyroscope [Sec. 18.10], the angle θ, the rate of precession $\dot{\phi}$, and the rate of spin $\dot{\psi}$ remain constant. We saw that such a motion is possible only if the moments of the external forces about O satisfy the relation

$$\Sigma\mathbf{M}_O = (I\omega_z - I'\dot{\phi} \cos \theta)\dot{\phi} \sin \theta\mathbf{j} \qquad (18.44)$$

that is, if the external forces reduce to a couple of moment equal to the right-hand member of Eq. (18.44) and applied *about an axis perpendicular to the precession axis and to the spin axis* (Fig. 18.31). The chapter ended with a discussion of the motion of an axisymmetrical body spinning and precessing *under no force* [Sec. 18.11; Sample Prob. 18.6].

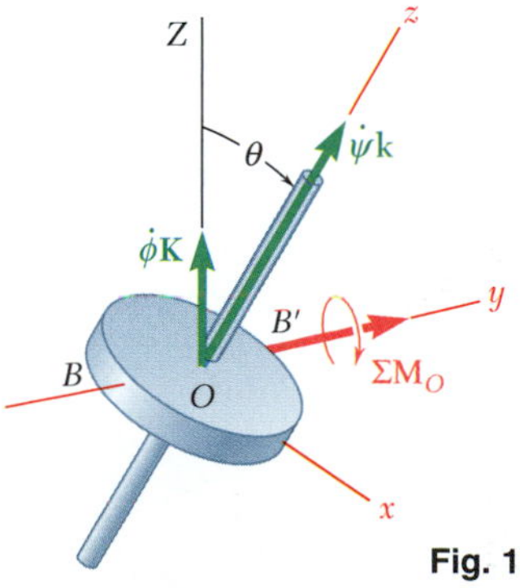

Fig. 18.31

Review Problems

18.145 A thin, homogeneous disk of mass m and radius r spins at the constant rate ω_1 about an axle held by a fork-ended vertical rod which rotates at the constant rate ω_2. Determine the angular momentum $\mathbf{H}_G$ of the disk about its mass center G.

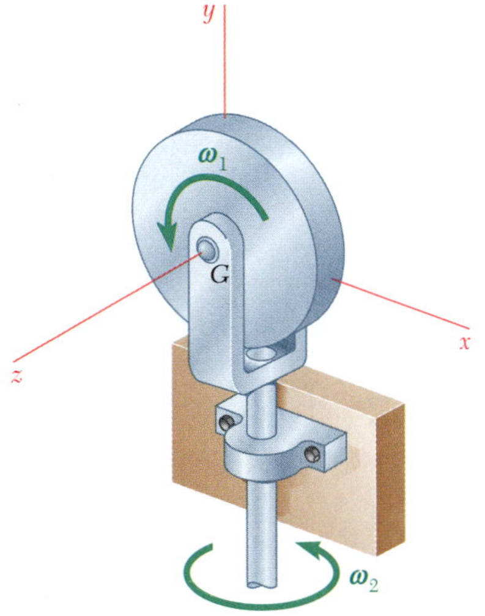

Fig. P18.145

18.146 Determine the kinetic energy of the disk of Prob. 18.145.

18.147 Two circular plates, each of mass 4 kg, are rigidly connected by a rod AB of negligible mass and are suspended from point A as shown. Knowing that an impulse $\mathbf{F}\,\Delta t = -(2.4\ \text{N} \cdot \text{s})\mathbf{k}$ is applied at point D, determine (*a*) the velocity of the mass center G of the assembly, (*b*) the angular velocity of the assembly.

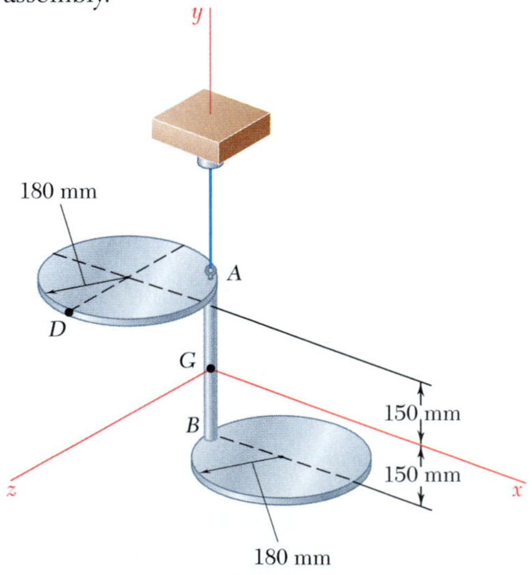

Fig. P18.147

18.148 Two L-shaped arms, each weighing 5 lb, are welded to the one-third points of the 24-in. shaft AB. Knowing that shaft AB rotates at the constant rate $\omega = 180$ rpm, determine (*a*) the angular momentum $\mathbf{H}_A$ of the body about A, (*b*) the angle $\mathbf{H}_A$ forms with the shaft.

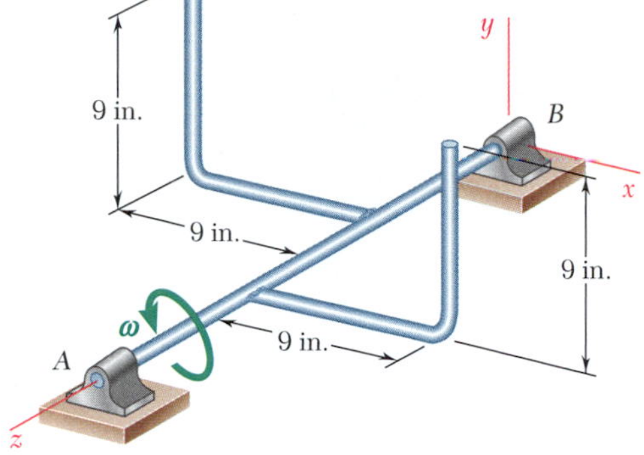

Fig. *P18.148*

18.149 A square plate of side a and mass m supported by a ball-and-socket joint at A is rotating about the y axis with a constant angular velocity $\boldsymbol{\omega} = \omega_0\mathbf{j}$ when an obstruction is suddenly introduced at B in the xy plane. Assuming the impact at B is perfectly plastic ($e = 0$), determine immediately after impact (*a*) the angular velocity of the plate, (*b*) the velocity of its mass center G.

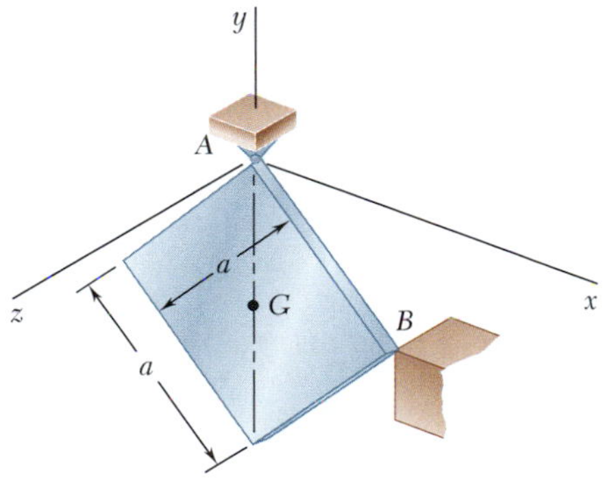

Fig. P18.149

18.150 Determine the impulse exerted on the plate of Prob. 18.149 during the impact by (*a*) the obstruction at B, (*b*) the support at A.

18.151 The 16-lb shaft shown has a uniform cross section. Knowing that the shaft rotates at the constant rate $\omega = 12$ rad/s, determine the dynamic reactions at A and B.

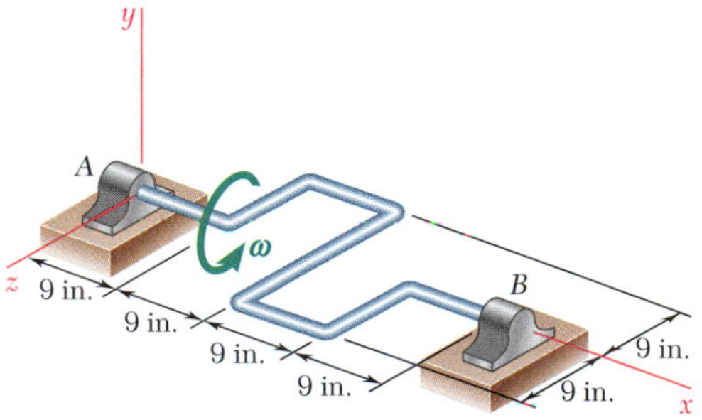

Fig. P18.151

18.152 Each of the two triangular plates shown has a mass of 5 kg and is welded to a vertical shaft AB. Knowing that the assembly rotates at the constant rate $\omega = 8$ rpm, determine the dynamic reactions at A and B.

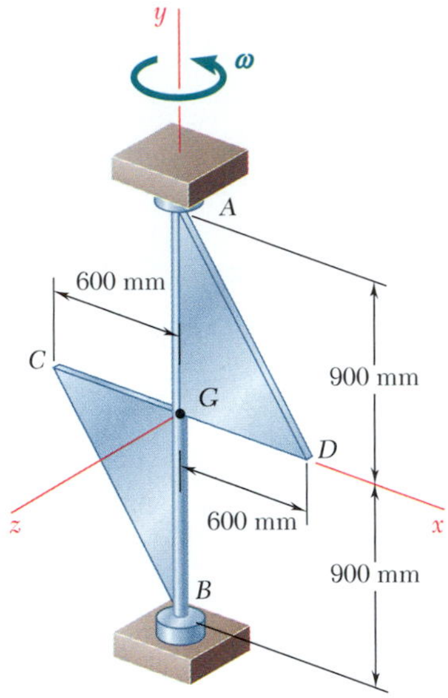

Fig. *P18.152*

18.153 Knowing that the assembly of Prob. 18.152 is initially at rest ($\omega = 0$) when a couple of moment $\mathbf{M}_0 = (36\text{ N}\cdot\text{m})\mathbf{j}$ is applied to the shaft, determine (*a*) the resulting angular acceleration of the assembly, (*b*) the dynamic reactions at A and B immediately after the couple is applied.

18.154 A slender rod is bent to form a square frame of side 6 in. The frame is attached by a collar at A to a vertical shaft which rotates with a constant angular velocity $\boldsymbol{\omega}$. Determine (*a*) the angle β that line AB forms with the horizontal x axis when $\omega = 9.8$ rad/s, (*b*) the largest value of ω for which $\beta = 90°$.

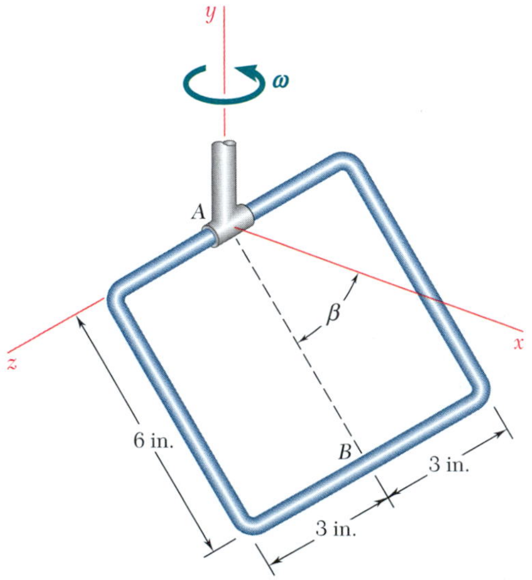

Fig. P18.154

18.155 A 3-kg homogeneous disk of radius 60 mm spins as shown at the constant rate $\omega_1 = 60$ rad/s. The disk is supported by the fork-ended rod AB, which is welded to the vertical shaft CBD. The system is at rest when a couple $\mathbf{M}_0 = (0.4\ \text{N} \cdot \text{m})\mathbf{j}$ is applied to the shaft for 2 s and then removed. Determine the dynamic reactions at C and D after the couple has been removed.

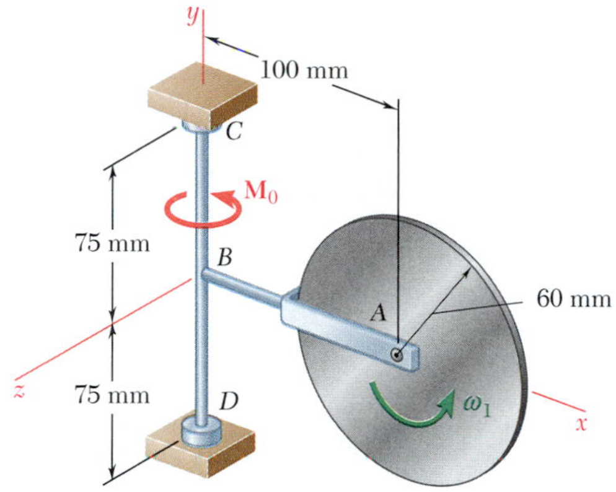

Fig. P18.155

18.156 A solid cone of height 9 in. with a circular base of radius 3 in. is supported by a ball-and-socket joint at A. Knowing that the cone is observed to precess about the vertical axis AC at the constant rate of 40 rpm in the sense indicated and that its axis of summetry AB forms an angle $\beta = 40°$ with AC, determine the rate at which the cone spins about the axis AB.

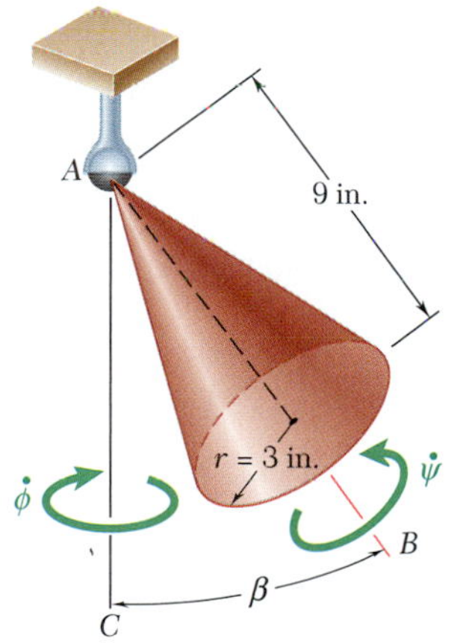

Fig. *P18.156*

Computer Problems

18.C1 A thin homogeneous disk of weight W and radius r is mounted on the horizontal axle AB. The plane of the disk forms an angle β with the plane normal to the axle. The axle rotates with a constant angular velocity ω. For values of β from 0 to 90°, determine and plot (*a*) the angle θ formed by the axle and the angular momentum of the disk about G, (*b*) the kinetic energy of the disk, and (*c*) the magnitude of the rate of change of the angular momentum of the disk about G.

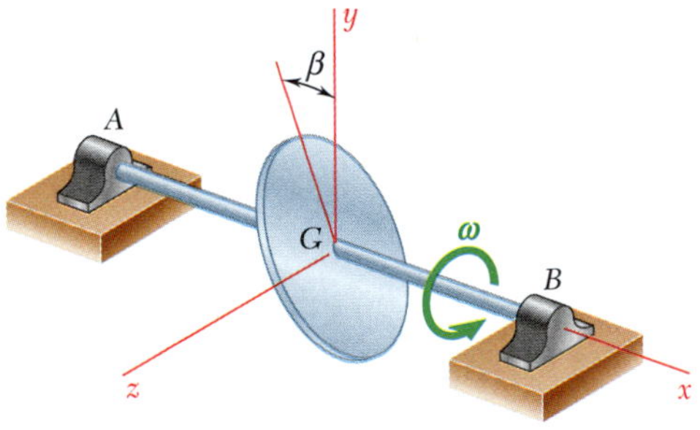

Fig. P18.C1

18.C2 A wire of uniform cross section and mass per unit length $m' = 0.065$ kg/m is used to form the wire figure shown, which is suspended from cord AD. An impulse $\mathbf{F}\,\Delta t = (2\text{ N}\cdot\text{s})\mathbf{j}$ is applied to the wire figure at point E. Use computational software to calculate and plot, immediately after the impact, for values of θ from 0 to 180°, (*a*) the velocity of the mass center of the wire figure, (*b*) the angular velocity of the figure.

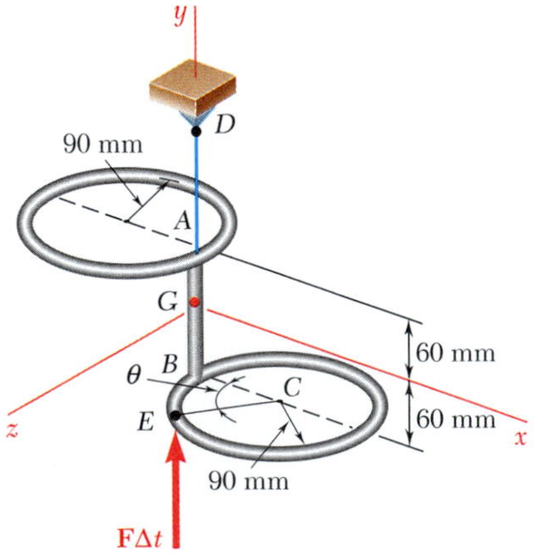

Fig. P18.C2

18.C3 Two L-shaped arms, each of mass 2 kg, are welded to the one-third points of the 600-mm shaft *AB* of negligible mass. The assembly is at rest ($\omega = 0$) at time $t = 0$ when a couple of moment $\mathbf{M}_0 = (1.6 \text{ N} \cdot \text{m})\mathbf{k}$ is applied to the shaft. Determine the dynamic reactions exerted on the shaft by the bearings at *A* and *B* at any time *t*. Resolve these reactions into components directed along *x* and *y* axes rotating with the assembly. Plot the components of the reactions and the magnitudes of the reactions from $t = 0$ to $t = 0.5$ s.

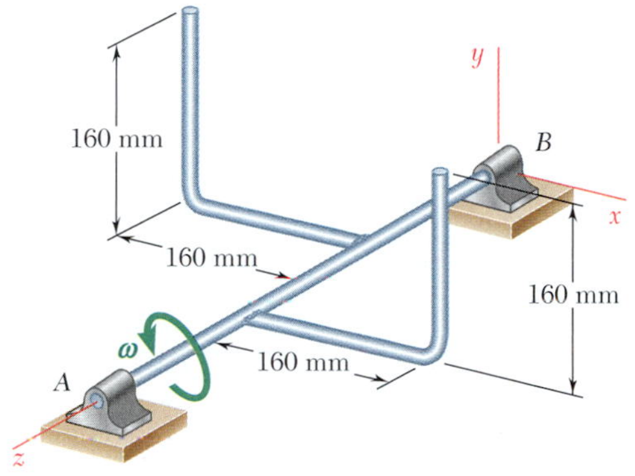

Fig. P18.C3

18.C4 A couple $\mathbf{M}_0 = (0.036 \text{ N} \cdot \text{m})\mathbf{i}$ is applied to an assembly consisting of pieces of sheet aluminum of uniform thickness and of total mass 1.35 kg, which are welded to a light axle supported by bearings at *A* and *B*. Determine the dynamic reactions exerted by the bearings on the axle at any time *t* after the couple has been applied. Resolve these reactions into components directed along *y* and *z* axes rotating with the assembly. Use computational software (*a*) to calculate and plot the components of the reactions from $t = 0$ to $t = 2$ s, (*b*) to determine the time at which the *z* components of the reactions at *A* and *B* are equal to zero.

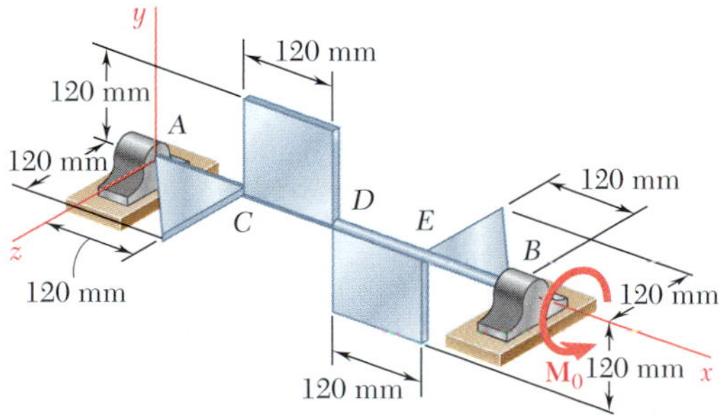

Fig. P18.C4

18.C5 A 5-lb homogeneous disk of radius 4 in. can rotate with respect to arm ABC, which is welded to a shaft DCE supported by bearings at D and E. Both the arm and the shaft are of negligible mass. At time $t = 0$ a couple $\mathbf{M}_0 = (5 \text{ lb} \cdot \text{in.})\mathbf{k}$ is applied to shaft DCE. Knowing that at $t = 0$ the angular velocity of the disk is $\boldsymbol{\omega}_1 = (60 \text{ rad/s})\mathbf{j}$ and that friction in the bearing at A causes the magnitude of $\boldsymbol{\omega}_1$ to decrease at the rate of 15 rad/s^2, determine the dynamic reactions exerted on the shaft by the bearings at D and E at any time t. Resolve these reactions into components directed along x and y axes rotating with the shaft. Use computational software (*a*) to calculate and plot the components of the reactions from $t = 0$ to $t = 4$ s, (*b*) to determine the times t_1 and t_2 at which the x and y components of the reactions at E are respectively equal to zero.

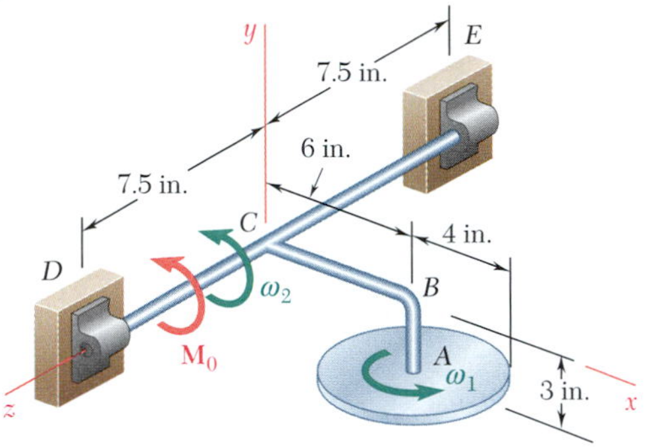

Fig. P18.C5

18.C6 A homogeneous disk of radius 9 in. is welded to a rod AG of length 18 in. and of negligible mass which is connected by a clevis to a vertical shaft AB. The rod and disk can rotate freely about a horizontal axis AC, and shaft AB can rotate freely about a vertical axis. Initially rod AG forms a given angle θ_0 with the downward vertical and its angular velocity $\dot{\theta}_0$ about AC is zero. Shaft AB is then given an angular velocity $\dot{\phi}_0$ about the vertical. Use computational software (*a*) to calculate the minimum value θ_m of the angle θ in the ensuing motion and the period of oscillation in θ, that is, the time required for θ to regain its initial value θ_0, (*b*) to compute and plot the angular velocity $\dot{\phi}$ of shaft AB for values of θ from θ_0 to θ_m. Apply this software with the initial conditions (*i*) $\theta_0 = 90°$, $\dot{\phi}_0 = 5$ rad/s, (*ii*) $\theta_0 = 90°$, $\dot{\phi}_0 = 10$ rad/s, (*iii*) $\theta_0 = 60°$, $\dot{\phi}_0 = 5$ rad/s. [*Hint:* Use the principle of conservation of energy and the fact that the angular momentum of the body about the vertical through A is conserved to obtain an equation of the form $\dot{\theta}^2 = f(\theta)$. This equation can be integrated by a numerical method.]

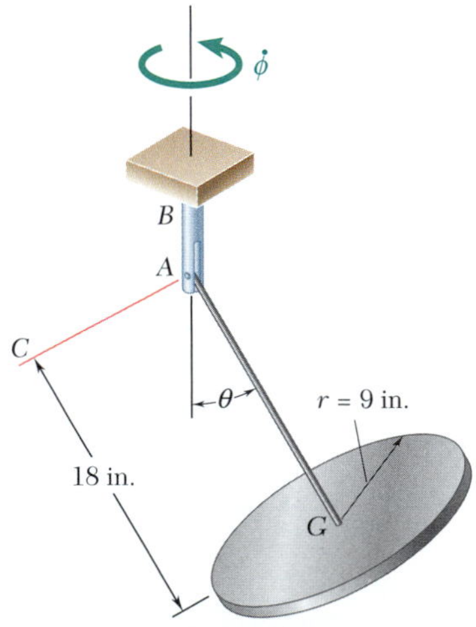

Fig. P18.C6

18.C7 A homogeneous disk of radius 9 in. is welded to a rod AG of length 18 in. and of negligible mass which is supported by a ball-and-socket joint at A. The disk is released in the position $\theta = \theta_0$ with a rate of spin $\dot{\psi}_0$, a rate of precession $\dot{\phi}_0$, and a zero rate of nutation. Use computational software (*a*) to calculate the minimum value θ_m of the angle θ in the ensuing motion and the period of oscillation in θ, that is, the time required for θ to regain its initial value θ_0, (*b*) to compute and plot the rate of spin $\dot{\psi}$ and the rate of precession $\dot{\phi}$ for values of θ from θ_0 to θ_m. Apply this software with the initial conditions (*i*) $\theta_0 = 90°$, $\dot{\psi}_0 = 50$ rad/s, $\dot{\phi}_0 = 0$, (*ii*) $\theta_0 = 90°$, $\dot{\psi}_0 = 0$, $\dot{\phi}_0 = 5$ rad/s, (*iii*) $\theta_0 = 90°$, $\dot{\psi}_0 = 50$ rad/s, $\dot{\phi}_0 = 5$ rad/s, (*iv*) $\theta_0 = 90°$, $\dot{\psi}_0 = 10$ rad/s, $\dot{\phi}_0 = 5$ rad/s, (*v*) $\theta_0 = 60°$, $\dot{\psi}_0 = 0$, $\dot{\phi}_0 = 5$ rad/s, (*vi*) $\theta_0 = 60°$, $\dot{\psi}_0 = 50$ rad/s, $\dot{\phi}_0 = 5$ rad/s. [*Hint:* Use the principle of conservation of energy and the fact that the angular momentum of the body is conserved about both the Z and z axes to obtain an equation of the form $\dot{\theta}^2 = f(\theta)$. This equation can be integrated by a numerical method.]

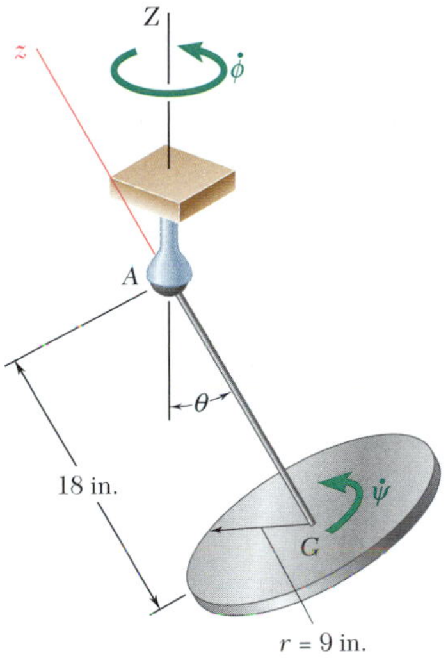

Fig. P18.C7

C H A P T E R

19

Mechanical Vibrations

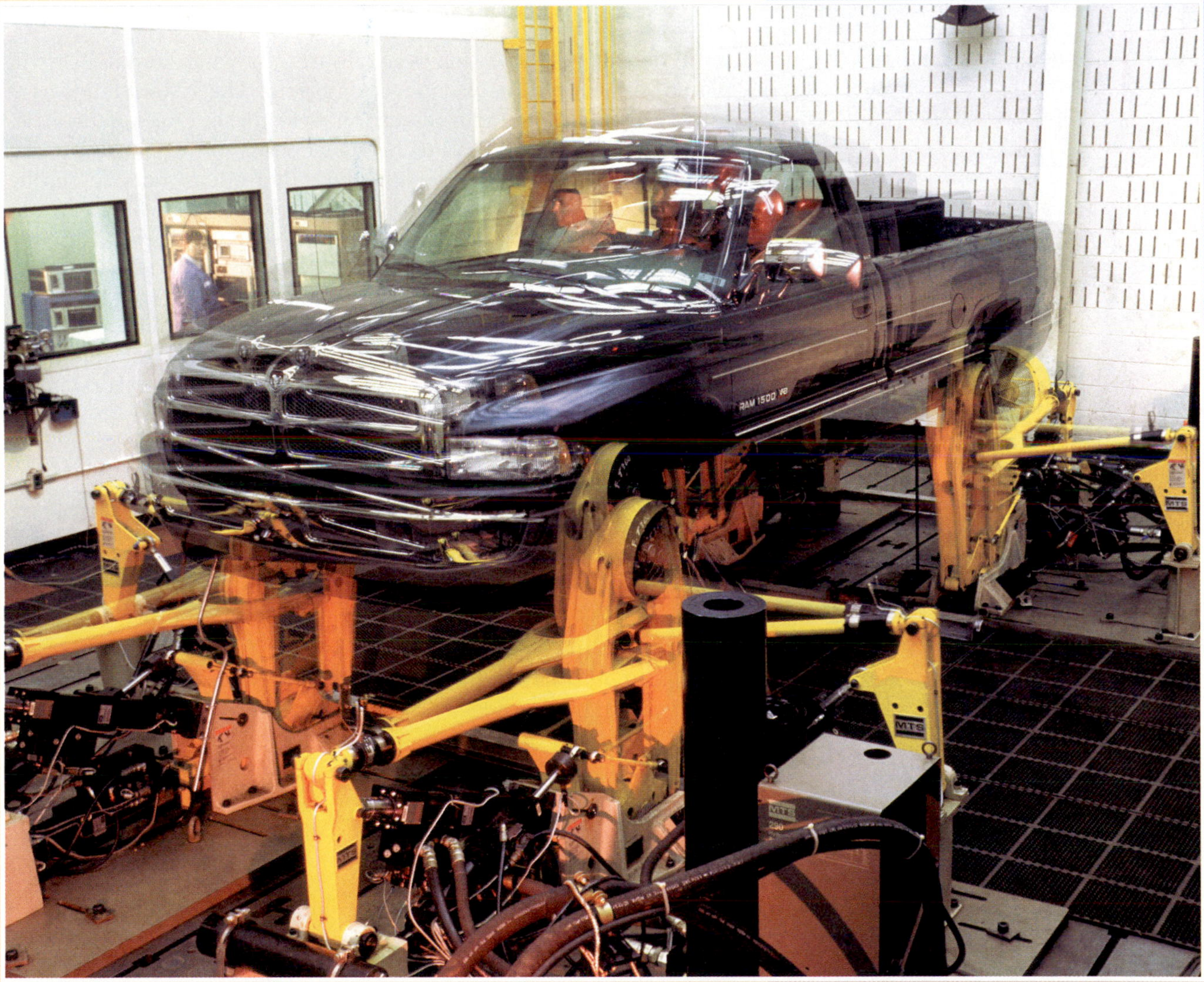

Mechanical systems may undergo *free vibrations* or they may be subjected to *forced vibrations.* The vibrations are *damped* when friction forces are present and *undamped* otherwise. The truck suspension consists essentially of springs and shock absorbers which cause the body of the truck to undergo *damped forced vibrations* when the wheels are subjected to periodic forces in the vehicle dynamics test shown.

19.1. INTRODUCTION

A *mechanical vibration* is the motion of a particle or a body which oscillates about a position of equilibrium. Most vibrations in machines and structures are undesirable because of the increased stresses and energy losses which accompany them. They should therefore be eliminated or reduced as much as possible by appropriate design. The analysis of vibrations has become increasingly important in recent years owing to the current trend toward higher-speed machines and lighter structures. There is every reason to expect that this trend will continue and that an even greater need for vibration analysis will develop in the future.

The analysis of vibrations is a very extensive subject to which entire texts have been devoted. Our present study will therefore be limited to the simpler types of vibrations, namely, the vibrations of a body or a system of bodies with one degree of freedom.

A mechanical vibration generally results when a system is displaced from a position of stable equilibrium. The system tends to return to this position under the action of restoring forces (either elastic forces, as in the case of a mass attached to a spring, or gravitational forces, as in the case of a pendulum). But the system generally reaches its original position with a certain acquired velocity which carries it beyond that position. Since the process can be repeated indefinitely, the system keeps moving back and forth across its position of equilibrium. The time interval required for the system to complete a full cycle of motion is called the *period* of the vibration. The number of cycles per unit time defines the *frequency*, and the maximum displacement of the system from its position of equilibrium is called the *amplitude* of the vibration.

When the motion is maintained by the restoring forces only, the vibration is said to be a *free vibration* (Secs. 19.2 to 19.6). When a periodic force is applied to the system, the resulting motion is described as a *forced vibration* (Sec. 19.7). When the effects of friction can be neglected, the vibrations are said to be *undamped*. However, all vibrations are actually *damped* to some degree. If a free vibration is only slightly damped, its amplitude slowly decreases until, after a certain time, the motion comes to a stop. But if damping is large enough to prevent any true vibration, the system then slowly regains its original position (Sec. 19.8). A damped forced vibration is maintained as long as the periodic force which produces the vibration is applied. The amplitude of the vibration, however, is affected by the magnitude of the damping forces (Sec. 19.9).

VIBRATIONS WITHOUT DAMPING

19.2. FREE VIBRATIONS OF PARTICLES. SIMPLE HARMONIC MOTION

Consider a body of mass m attached to a spring of constant k (Fig. 19.1*a*). Since at the present time we are concerned only with the motion of its mass center, we will refer to this body as a particle. When the particle is in static equilibrium, the forces acting on it are its weight **W** and the force **T** exerted by the spring, of magnitude

$T = k\delta_{st}$, where δ_{st} denotes the elongation of the spring. We have, therefore,

$$W = k\delta_{st}$$

Suppose now that the particle is displaced through a distance x_m from its equilibrium position and released with no initial velocity. If x_m has been chosen smaller than δ_{st}, the particle will move back and forth through its equilibrium position; a vibration of amplitude x_m has been generated. Note that the vibration can also be produced by imparting a certain initial velocity to the particle when it is in its equilibrium position $x = 0$ or, more generally, by starting the particle from any given position $x = x_0$ with a given initial velocity $\mathbf{v}_0$.

To analyze the vibration, let us consider the particle in a position P at some arbitrary time t (Fig. 19.1*b*). Denoting by x the displacement OP measured from the equilibrium position O (positive downward), we note that the forces acting on the particle are its weight $\mathbf{W}$ and the force $\mathbf{T}$ exerted by the spring which, in this position, has a magnitude $T = k(\delta_{st} + x)$. Recalling that $W = k\delta_{st}$, we find that the magnitude of the resultant $\mathbf{F}$ of the two forces (positive downward) is

$$F = W - k(\delta_{st} + x) = -kx \tag{19.1}$$

Thus the *resultant* of the forces exerted on the particle is proportional to the displacement OP *measured from the equilibrium position.* Recalling the sign convention, we note that $\mathbf{F}$ is always directed *toward* the equilibrium position O. Substituting for F into the fundamental equation $F = ma$ and recalling that a is the second derivative $\ddot{x}$ of x with respect to t, we write

$$m\ddot{x} + kx = 0 \tag{19.2}$$

Note that the same sign convention should be used for the acceleration $\ddot{x}$ and for the displacement x, namely, positive downward.

The motion defined by Eq. (19.2) is called a *simple harmonic motion.* It is characterized by the fact that *the acceleration is proportional to the displacement and of opposite direction.* We can verify that each of the functions $x_1 = \sin(\sqrt{k/m}\,t)$ and $x_2 = \cos(\sqrt{k/m}\,t)$ satisfies Eq. (19.2). These functions, therefore, constitute two *particular solutions* of the differential equation (19.2). The *general solution* of Eq. (19.2) is obtained by multiplying each of the particular solutions by an arbitrary constant and adding. Thus, the general solution is expressed as

$$x = C_1x_1 + C_2x_2 = C_1 \sin\left(\sqrt{\frac{k}{m}}t\right) + C_2 \cos\left(\sqrt{\frac{k}{m}}t\right) \tag{19.3}$$

We note that x is a *periodic function* of the time t and does, therefore, represent a vibration of the particle P. The coefficient of t in the expression we have obtained is referred to as the *natural circular frequency* of the vibration and is denoted by ω_n. We have

$$\text{Natural circular frequency} = \omega_n = \sqrt{\frac{k}{m}} \tag{19.4}$$

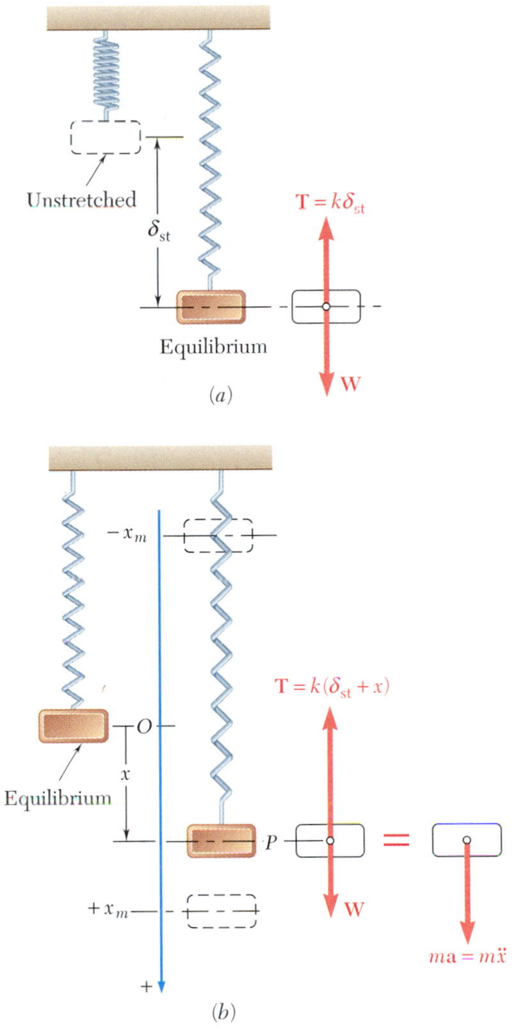

Fig. 19.1

Substituting for $\sqrt{k/m}$ into Eq. (19.3), we write

$$x = C_1 \sin \omega_n t + C_2 \cos \omega_n t \tag{19.5}$$

This is the general solution of the differential equation

$$\ddot{x} + \omega_n^2 x = 0 \tag{19.6}$$

which can be obtained from Eq. (19.2) by dividing both terms by m and observing that $k/m = \omega_n^2$. Differentiating twice both members of Eq. (19.5) with respect to t, we obtain the following expressions for the velocity and the acceleration at time t:

$$v = \dot{x} = C_1\omega_n \cos \omega_n t - C_2\omega_n \sin \omega_n t \tag{19.7}$$

$$a = \ddot{x} = -C_1\omega_n^2 \sin \omega_n t - C_2\omega_n^2 \cos \omega_n t \tag{19.8}$$

The values of the constants C_1 and C_2 depend upon the *initial conditions* of the motion. For example, we have $C_1 = 0$ if the particle is displaced from its equilibrium position and released at $t = 0$ with no initial velocity, and we have $C_2 = 0$ if the particle is started from O at $t = 0$ with a certain initial velocity. In general, substituting $t = 0$ and the initial values x_0 and v_0 of the displacement and the velocity into Eqs. (19.5) and (19.7), we find that $C_1 = v_0/\omega_n$ and $C_2 = x_0$.

The expressions obtained for the displacement, velocity, and acceleration of a particle can be written in a more compact form if we observe that Eq. (19.5) expresses that the displacement $x = OP$ is the sum of the x components of two vectors $\mathbf{C}_1$ and $\mathbf{C}_2$, respectively, of magnitude C_1 and C_2, directed as shown in Fig. 19.2*a*. As t varies, both vectors rotate clockwise; we also note that the magnitude of their resultant $\overrightarrow{OQ}$ is equal to the maximum displacement x_m. The simple harmonic motion of P along the x axis can thus be obtained by projecting on this axis the motion of a point Q describing an *auxiliary circle* of radius x_m *with a constant angular velocity* ω_n (which explains the name of natural *circular* frequency given to ω_n). Denoting by ϕ the angle formed by the vectors $\overrightarrow{OQ}$ and $\mathbf{C}_1$, we write

$$OP = OQ \sin (\omega_n t + \phi) \tag{19.9}$$

which leads to new expressions for the displacement, velocity, and acceleration of P:

$$x = x_m \sin (\omega_n t + \phi) \tag{19.10}$$

$$v = \dot{x} = x_m\omega_n \cos (\omega_n t + \phi) \tag{19.11}$$

$$a = \ddot{x} = -x_m\omega_n^2 \sin (\omega_n t + \phi) \tag{19.12}$$

The displacement-time curve is represented by a sine curve (Fig. 19.2*b*); the maximum value x_m of the displacement is called the *amplitude* of the vibration, and the angle ϕ which defines the initial position of Q on the circle is called the *phase angle.* We note from Fig. 19.2 that a full *cycle* is described as the angle $\omega_n t$ increases by 2π rad. The corresponding value of t, denoted by τ_n, is called the *period* of the free vibration and is measured in seconds. We have

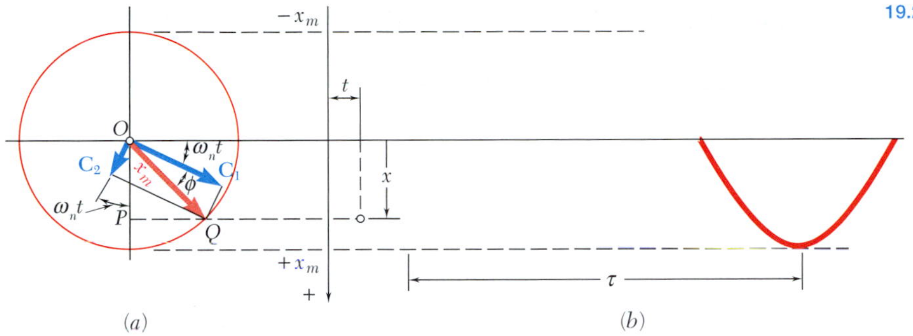

Fig. 19.2

$$\text{Period} = \tau_n = \frac{2\pi}{\omega_n} \tag{19.13}$$

The number of cycles described per unit of time is denoted by f_n and is known as the *natural frequency* of the vibration. We write

$$\text{Natural frequency} = f_n = \frac{1}{\tau_n} = \frac{\omega_n}{2\pi} \tag{19.14}$$

The unit of frequency is a frequency of 1 cycle per second, corresponding to a period of 1 s. In terms of base units the unit of frequency is thus 1/s or s^{-1}. It is called a *hertz* (Hz) in the SI system of units. It also follows from Eq. (19.14) that a frequency of 1 s^{-1} or 1 Hz corresponds to a circular frequency of 2π rad/s. In problems involving angular velocities expressed in revolutions per minute (rpm), we have 1 rpm $= \frac{1}{60}\ s^{-1} = \frac{1}{60}$ Hz, or 1 rpm $= (2\pi/60)$ rad/s.

Recalling that ω_n was defined in (19.4) in terms of the constant k of the spring and the mass m of the particle, we observe that the period and the frequency are independent of the initial conditions and of the amplitude of the vibration. Note that τ_n and f_n depend on the *mass* rather than on the *weight* of the particle and thus are independent of the value of g.

The velocity-time and acceleration-time curves can be represented by sine curves of the same period as the displacement-time curve, but with different phase angles. From Eqs. (19.11) and (19.12), we note that the maximum values of the magnitudes of the velocity and acceleration are

$$v_m = x_m\omega_n \qquad a_m = x_m\omega_n^2 \tag{19.15}$$

Since the point Q describes the auxiliary circle, of radius x_m, at the constant angular velocity ω_n, its velocity and acceleration are equal, respectively, to the expressions (19.15). Recalling Eqs. (19.11) and (19.12), we find, therefore, that the velocity and acceleration of P can be obtained at any instant by projecting on the x axis vectors of magnitudes $v_m = x_m\omega_n$ and $a_m = x_m\omega_n^2$ representing, respectively, the velocity and acceleration of Q at the same instant (Fig. 19.3).

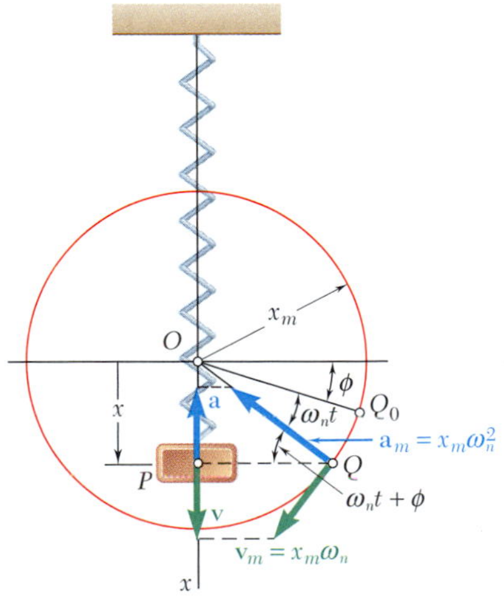

Fig. 19.3

The results obtained are not limited to the solution of the problem of a mass attached to a spring. They can be used to analyze the rectilinear motion of a particle *whenever the resultant* **F** *of the forces acting on the particle is proportional to the displacement x and directed toward O.* The fundamental equation of motion $F = ma$ can then be written in the form of Eq. (19.6), which is characteristic of a simple harmonic motion. Observing that the coefficient of x must be equal to ω_n^2, we can easily determine the natural circular frequency ω_n of the motion. Substituting the value obtained for ω_n into Eqs. (19.13) and (19.14), we then obtain the period τ_n and the natural frequency f_n of the motion.

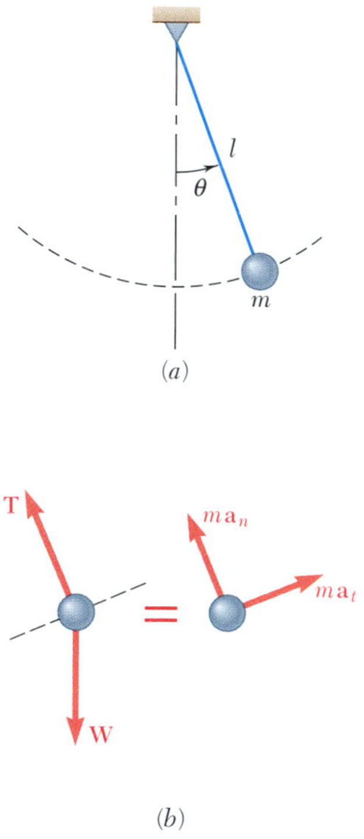

Fig. 19.4

19.3. SIMPLE PENDULUM (APPROXIMATE SOLUTION)

Most of the vibrations encountered in engineering applications can be represented by a simple harmonic motion. Many others, although of a different type, can be *approximated* by a simple harmonic motion, provided that their amplitude remains small. Consider, for example, a *simple pendulum*, consisting of a bob of mass m attached to a cord of length l, which can oscillate in a vertical plane (Fig. 19.4*a*). At a given time t, the cord forms an angle θ with the vertical. The forces acting on the bob are its weight **W** and the force **T** exerted by the cord (Fig. 19.4*b*). Resolving the vector $m\mathbf{a}$ into tangential and normal components, with $m\mathbf{a}_t$ directed to the right, that is, in the direction corresponding to increasing values of θ, and observing that $a_t = l\alpha = l\ddot{\theta}$, we write

$$\Sigma F_t = ma_t: \qquad -W \sin\theta = ml\ddot{\theta}$$

Noting that $W = mg$ and dividing through by ml, we obtain

$$\ddot{\theta} + \frac{g}{l}\sin\theta = 0 \tag{19.16}$$

For oscillations of small amplitude, we can replace $\sin\theta$ by θ, expressed in radians, and write

$$\ddot{\theta} + \frac{g}{l}\theta = 0 \tag{19.17}$$

Comparison with Eq. (19.6) shows that the differential equation (19.17) is that of a simple harmonic motion with a natural circular frequency ω_n equal to $(g/l)^{1/2}$. The general solution of Eq. (19.17) can, therefore, be expressed as

$$\theta = \theta_m \sin(\omega_n t + \phi)$$

where θ_m is the amplitude of the oscillations and ϕ is a phase angle. Substituting into Eq. (19.13) the value obtained for ω_n, we get the following expression for the period of the small oscillations of a pendulum of length l:

$$\tau_n = \frac{2\pi}{\omega_n} = 2\pi\sqrt{\frac{l}{g}} \tag{19.18}$$

*19.4. SIMPLE PENDULUM (EXACT SOLUTION)

Formula (19.18) is only approximate. To obtain an exact expression for the period of the oscillations of a simple pendulum, we must return to Eq. (19.16). Multiplying both terms by $2\dot{\theta}$ and integrating from an initial position corresponding to the maximum deflection, that is, $\theta = \theta_m$ and $\dot{\theta} = 0$, we write

$$\left(\frac{d\theta}{dt}\right)^2 = \frac{2g}{l}(\cos\theta - \cos\theta_m)$$

Replacing $\cos\theta$ by $1 - 2\sin^2(\theta/2)$ and $\cos\theta_m$ by a similar expression, solving for dt, and integrating over a quarter period from $t = 0$, $\theta = 0$ to $t = \tau_n/4$, $\theta = \theta_m$, we have

$$\tau_n = 2\sqrt{\frac{l}{g}}\int_0^{\theta_m}\frac{d\theta}{\sqrt{\sin^2(\theta_m/2) - \sin^2(\theta/2)}}$$

The integral in the right-hand member is known as an *elliptic integral;* it cannot be expressed in terms of the usual algebraic or trigonometric functions. However, setting

$$\sin(\theta/2) = \sin(\theta_m/2)\sin\phi$$

we can write

$$\tau_n = 4\sqrt{\frac{l}{g}}\int_0^{\pi/2}\frac{d\phi}{\sqrt{1 - \sin^2(\theta_m/2)\sin^2\phi}} \qquad (19.19)$$

where the integral obtained, commonly denoted by K, can be calculated by using a numerical method of integration. It can also be found in *tables of elliptic integrals* for various values of $\theta_m/2$.† In order to compare the result just obtained with that of the preceding section, we write Eq. (19.19) in the form

$$\tau_n = \frac{2K}{\pi}\left(2\pi\sqrt{\frac{l}{g}}\right) \qquad (19.20)$$

Formula (19.20) shows that the actual value of the period of a simple pendulum can be obtained by multiplying the approximate value given in Eq. (19.18) by the correction factor $2K/\pi$. Values of the correction factor are given in Table 19.1 for various values of the amplitude θ_m. We note that for ordinary engineering computations the correction factor can be omitted as long as the amplitude does not exceed 10°.

Table 19.1. Correction Factor for the Period of a Simple Pendulum

θ_m	0°	10°	20°	30°	60°	90°	120°	150°	180°
K	1.571	1.574	1.583	1.598	1.686	1.854	2.157	2.768	∞
$2K/\pi$	1.000	1.002	1.008	1.017	1.073	1.180	1.373	1.762	∞

†See, for example, *Standard Mathematical Tables,* Chemical Rubber Publishing Company, Cleveland, Ohio.

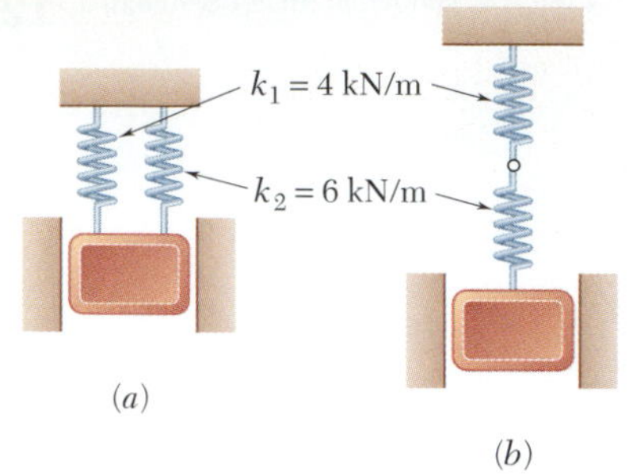

SAMPLE PROBLEM 19.1

A 50-kg block moves between vertical guides as shown. The block is pulled 40 mm down from its equilibrium position and released. For each spring arrangement, determine the period of the vibration, the maximum velocity of the block, and the maximum acceleration of the block.

SOLUTION

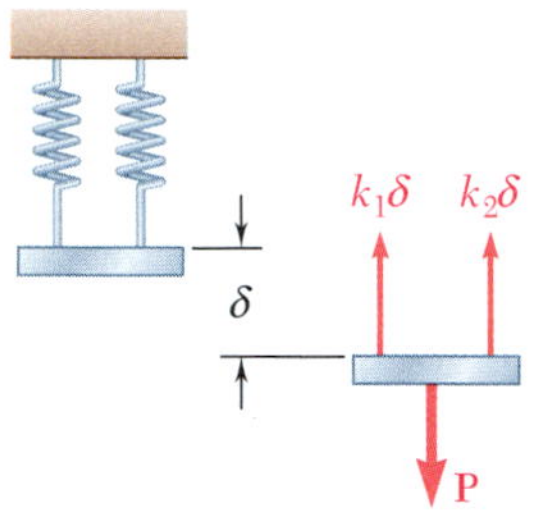

a. Springs Attached in Parallel. We first determine the constant k of a single spring equivalent to the two springs *by finding the magnitude of the force* **P** required to cause a given deflection δ. Since for a deflection δ the magnitudes of the forces exerted by the springs are, respectively, $k_1\delta$ and $k_2\delta$, we have

$$P = k_1\delta + k_2\delta = (k_1 + k_2)\delta$$

The constant k of the single equivalent spring is

$$k = \frac{P}{\delta} = k_1 + k_2 = 4\text{ kN/m} + 6\text{ kN/m} = 10\text{ kN/m} = 10^4\text{ N/m}$$

Period of Vibration: Since $m = 50$ kg, Eq. (19.4) yields

$$\omega_n^2 = \frac{k}{m} = \frac{10^4\text{ N/m}}{50\text{ kg}} \qquad \omega_n = 14.14\text{ rad/s}$$

$$\tau_n = 2\pi/\omega_n \qquad \tau_n = 0.444\text{ s} \blacktriangleleft$$

Maximum Velocity: $v_m = x_m\omega_n = (0.040\text{ m})(14.14\text{ rad/s})$

$$v_m = 0.566\text{ m/s} \qquad \mathbf{v}_m = 0.566\text{ m/s} \updownarrow \blacktriangleleft$$

Maximum Acceleration: $a_m = x_m\omega_n^2 = (0.040\text{ m})(14.14\text{ rad/s})^2$

$$a_m = 8.00\text{ m/s}^2 \qquad \mathbf{a}_m = 8.00\text{ m/s}^2 \updownarrow \blacktriangleleft$$

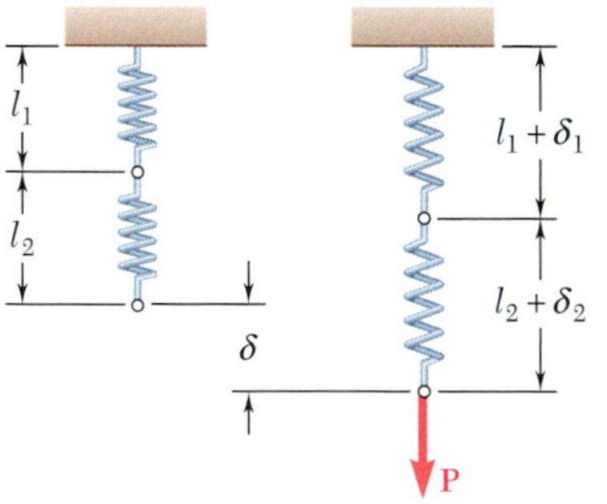

b. Springs Attached in Series. We first determine the constant k of a single spring equivalent to the two springs *by finding the total elongation* δ of the springs under a given static load **P**. To facilitate the computation, a static load of magnitude $P = 12$ kN is used.

$$\delta = \delta_1 + \delta_2 = \frac{P}{k_1} + \frac{P}{k_2} = \frac{12\text{ kN}}{4\text{ kN/m}} + \frac{12\text{ kN}}{6\text{ kN/m}} = 5\text{ m}$$

$$k = \frac{P}{\delta} = \frac{12\text{ kN}}{5\text{ m}} = 2.4\text{ kN/m} = 2400\text{ N/m}$$

Period of Vibration: $\omega_n^2 = \frac{k}{m} = \frac{2400\text{ N/m}}{50\text{ kg}} \qquad \omega_n = 6.93\text{ rad/s}$

$$\tau_n = \frac{2\pi}{\omega_n} \qquad \tau_n = 0.907\text{ s} \blacktriangleleft$$

Maximum Velocity: $v_m = x_m\omega_n = (0.040\text{ m})(6.93\text{ rad/s})$

$$v_m = 0.277\text{ m/s} \qquad \mathbf{v}_m = 0.277\text{ m/s} \updownarrow \blacktriangleleft$$

Maximum Acceleration: $a_m = x_m\omega_n^2 = (0.040\text{ m})(6.93\text{ rad/s})^2$

$$a_m = 1.920\text{ m/s}^2 \qquad \mathbf{a}_m = 1.920\text{ m/s}^2 \updownarrow \blacktriangleleft$$

SOLVING PROBLEMS ON YOUR OWN

This chapter deals with *mechanical vibrations,* that is, with the motion of a particle or body oscillating about a position of equilibrium.

In this first lesson, we saw that a *free vibration* of a particle occurs when the particle is subjected to a force proportional to its displacement and of opposite direction, such as the force exerted by a spring (Fig. 19.1). The resulting motion, called a *simple harmonic motion,* is characterized by the differential equation

$$m\ddot{x} + kx = 0 \tag{19.2}$$

where x is the displacement of the particle, $\ddot{x}$ is its acceleration, m is its mass, and k is the constant of the spring. The solution of this differential equation was found to be

$$x = x_m \sin (\omega_n t + \phi) \tag{19.10}$$

where x_m = amplitude of the vibration
$\omega_n = \sqrt{k/m}$ = natural circular frequency (rad/s)
ϕ = phase angle (rad)

We also defined the *period* of the vibration as the time $\tau_n = 2\pi/\omega_n$ needed for the particle to complete one cycle, and the *natural frequency* as the number of cycles per second, $f_n = 1/\tau_n = \omega_n/2\pi$, expressed in Hz or s^{-1}. Differentiating Eq. (19.10) twice yields the velocity and the acceleration of the particle at any time. The maximum values of the velocity and acceleration were found to be

$$v_m = x_m\omega_n \qquad a_m = x_m\omega_n^2 \tag{19.15}$$

To determine the parameters in Eq. (19.10) you can follow these steps.

1. *Draw a free-body diagram showing the forces exerted on the particle* when the particle is at a distance x from its position of equilibrium. The resultant of these forces will be proportional to x and its direction will be opposite to the positive direction of x [Eq. (19.1)].

2. *Write the differential equation of motion* by equating to $m\ddot{x}$ the resultant of the forces found in step 1. Note that once a positive direction for x has been chosen, the same sign convention should be used for the acceleration $\ddot{x}$. After transposition, you will obtain an equation of the form of Eq. (19.2).

3. *Determine the natural circular frequency ω_n* by dividing the coefficient of x by the coefficient of $\ddot{x}$ in this equation and taking the square root of the result obtained. Make sure that ω_n is expressed in rad/s.

(continued)

4. *Determine the amplitude x_m and the phase angle ϕ* by substituting the value obtained for ω_n and the initial values of x and $\dot{x}$ into Eq. (19.10) and the equation obtained by differentiating Eq. (19.10) with respect to t.

Equation (19.10) and the two equations obtained by differentiating Eq. (19.10) twice with respect to t can now be used to find the displacement, velocity, and acceleration of the particle at any time. Equations (19.15) yield the maximum velocity v_m and the maximum acceleration a_m.

5. *You also saw that for the small oscillations of a simple pendulum,* the angle θ that the cord of the pendulum forms with the vertical satisfies the differential equation

$$\ddot{\theta} + \frac{g}{l}\,\theta = 0 \tag{19.17}$$

where l is the length of the cord and where θ is expressed in radians [Sec. 19.3]. This equation defines again a *simple harmonic motion,* and its solution is of the same form as Eq. (19.10),

$$\theta = \theta_m \sin(\omega_n t + \phi)$$

where the natural circular frequency $\omega_n = \sqrt{g/l}$ is expressed in rad/s. The determination of the various constants in this expression is carried out in a manner similar to that described above. Remember that the velocity of the bob is tangent to the path and that its magnitude is $v = l\dot{\theta}$, while the acceleration of the bob has a tangential component $\mathbf{a}_t$, of magnitude $a_t = l\ddot{\theta}$, and a normal component $\mathbf{a}_n$ directed toward the center of the path and of magnitude $a_n = l\dot{\theta}^2$.

Problems

19.1 Determine the maximum velocity and maximum acceleration of a particle which moves in simple harmonic motion with an amplitude of 400 mm and a period of 1.4 s.

19.2 Determine the amplitude and maximum velocity of a particle which moves in simple harmonic motion with a maximum acceleration of 21.6 ft/s^2 and a frequency of 8 Hz.

19.3 A particle moves in simple harmonic motion. Knowing that the amplitude is 15 in. and the maximum acceleration is 15 ft/s^2, determine the maximum velocity of the particle and the frequency of its motion.

19.4 A simple pendulum consisting of a bob attached to a cord oscillates in a vertical plane with a period of 1.3 s. Assuming simple harmonic motion and knowing that the maximum velocity of the bob is 0.4 m/s, determine (*a*) the amplitude of the motion in degrees, (*b*) the maximum tangential acceleration of the bob.

19.5 A simple pendulum consisting of a bob attached to a cord of length $l = 40$ in. oscillates in a vertical plane. Assuming simple harmonic motion and knowing that the bob is released from rest when $\theta = 6°$, determine (*a*) the frequency of oscillation, (*b*) the maximum velocity of the bob.

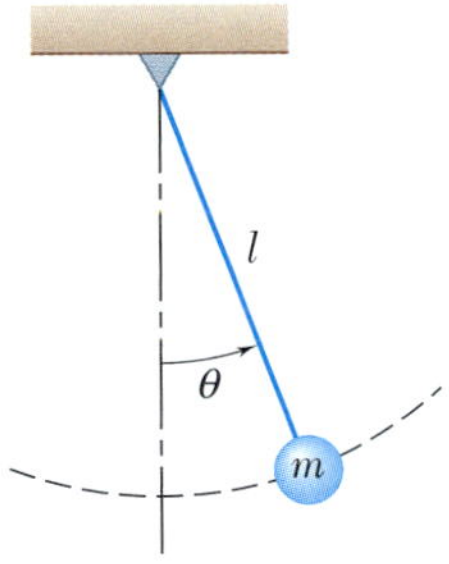

Fig. P19.4 and P19.5

19.6 A 10-kg block is initially held so that the vertical spring attached as shown is undeformed. Knowing that the block is suddenly released from rest, determine (*a*) the amplitude and frequency of the resulting motion, (*b*) the maximum velocity and maximum acceleration of the block.

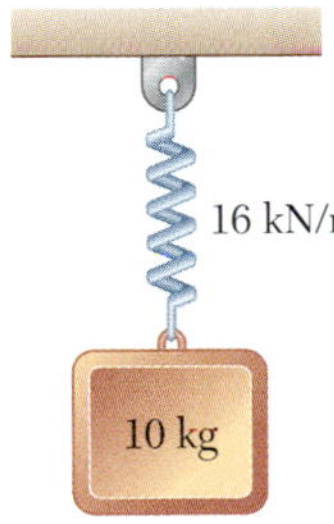

Fig. P19.6

19.7 A 70-lb block is attached to a spring and can move without friction in a slot as shown. The block is in its equilibrium position when it is struck by a hammer which imparts to the block an initial velocity of 10 ft/s. Determine (*a*) the period and frequency of the resulting motion, (*b*) the amplitude of the motion and the maximum acceleration of the block.

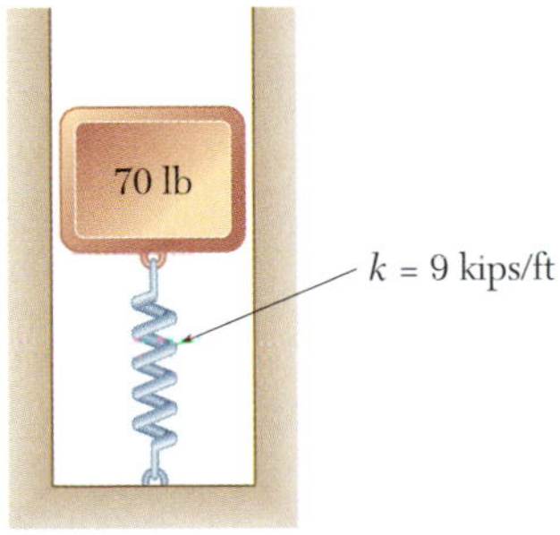

Fig. *P19.7*

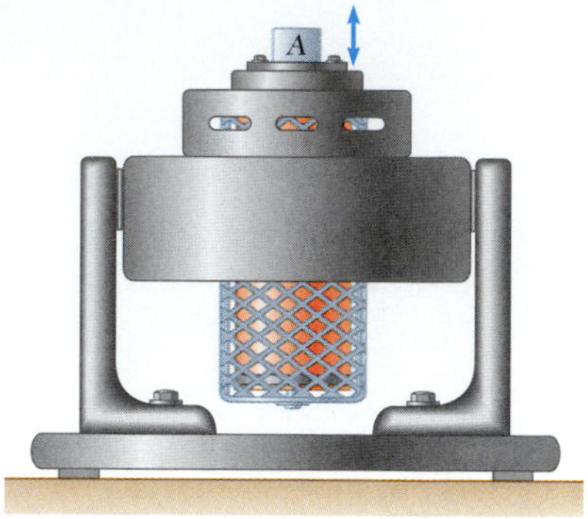

Fig. P19.8

19.8 An instrument package A is bolted to a shaker table as shown. The table moves vertically in simple harmonic motion at the same frequency as the variable-speed motor which drives it. The package is to be tested at a peak acceleration of 50 m/s^2. Knowing that the amplitude of the shaker table is 58 mm, determine (*a*) the required speed of the motor in rpm, (*b*) the maximum velocity of the table.

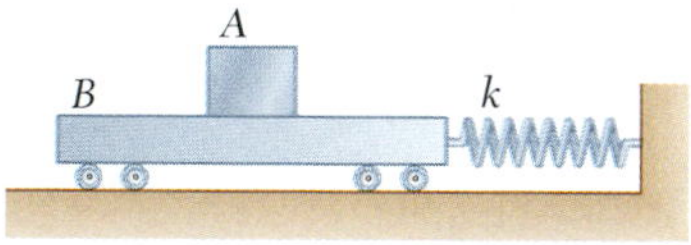

Fig. P19.9

19.9 A 5-kg block A rests on a 20-kg plate B which is attached to an unstretched spring of constant k = 900 N/m. Plate B is slowly moved 60 mm to the left and released from rest. Assuming that block A does not slip on the plate, determine (*a*) the amplitude and frequency of the resulting motion, (*b*) the corresponding smallest allowable value of the coefficient of static friction.

19.10 The motion of a particle is described by the equation $x = 10 \sin 2t + 8 \cos 2t$, where x is expressed in feet and t in seconds. Determine (*a*) the period of the motion, (*b*) its amplitude, (*c*) its phase angle.

19.11 The motion of a particle is described by the equation $x = 0.2 \cos 10\pi t + 0.15 \sin(10\pi t - \pi/3)$, where x is expressed in feet and t in seconds. Determine (*a*) the period of the resulting motion, (*b*) its amplitude, (*c*) its phase angle.

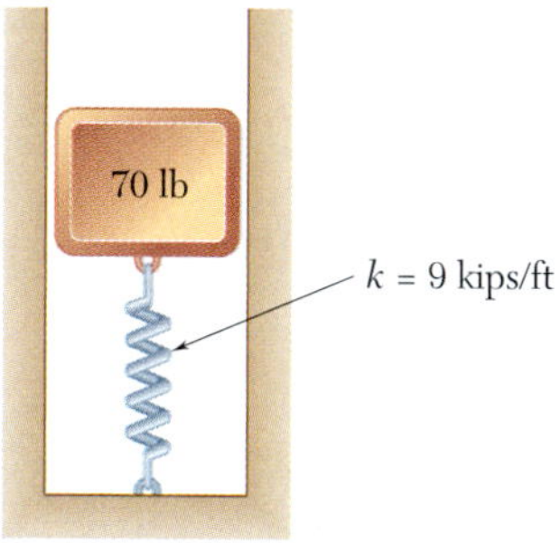

Fig. P19.12

19.12 A 70-lb block attached to a spring of constant k = 9 kips/ft can move without friction in a slot as shown. The block is given an initial 15-in. displacement downward from its equilibrium position and released. Determine 1.5 s after the block has been released (*a*) the total distance traveled by the block, (*b*) the acceleration of the block.

19.13 A 1.4-kg block is supported as shown by a spring of constant k = 400 N/m which can act in tension or compression. The block is in its equilibrium position when it is struck from below by a hammer which imparts to the block an upward velocity of 2.5 m/s. Determine (*a*) the time required for the block to move 60 mm upward, (*b*) the corresponding velocity and acceleration of the block.

Fig. P19.13

19.14 In Prob. 19.13, determine the position, velocity, and acceleration of the block 0.90 s after it has been struck by the hammer.

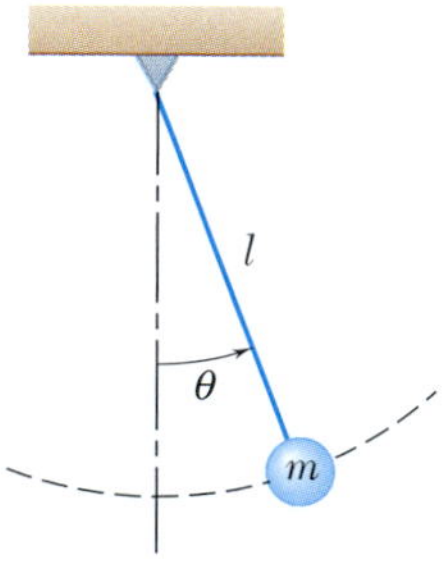

Fig. P19.15

19.15 The bob of a simple pendulum of length l = 40 in. is released from rest when $\theta = +5°$. Assuming simple harmonic motion, determine 1.6 s after release (*a*) the angle θ, (*b*) the magnitudes of the velocity and acceleration of the bob.

19.16 A 10-lb collar rests on but is not attached to the spring shown. It is observed that when the collar is pushed down 9 in. or more and released, it loses contact with the spring. Determine (*a*) the spring constant, (*b*) the position, velocity, and acceleration of the collar 0.16 s after it has been pushed down 9 in. and released.

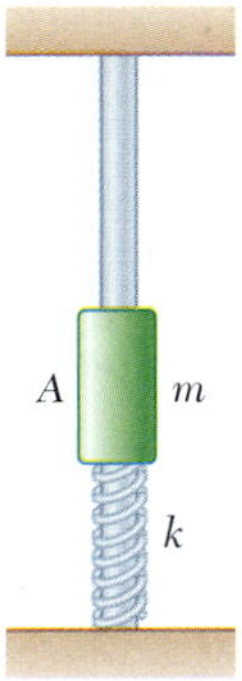

Fig. *P19.16*

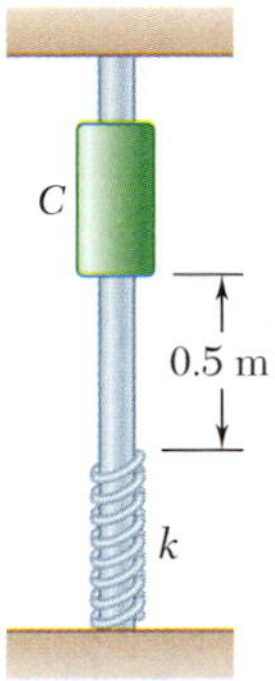

Fig. P19.17

19.17 A 5-kg collar *C* is released from rest in the position shown and slides without friction on a vertical rod until it hits a spring of constant k = 720 N/m which it compresses. The velocity of the collar is reduced to zero, and the collar reverses the direction of its motion and returns to its initial position. The cycle is then repeated. Determine (*a*) the period of the motion of the collar, (*b*) the velocity of the collar 0.4 s after it was released. (*Note:* This is a periodic motion, but not simple harmonic motion.)

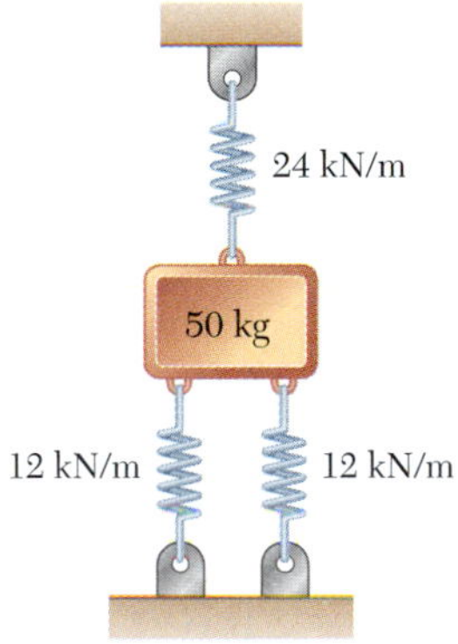

Fig. P19.18

19.18 and *19.19* A 50-kg block is supported by the spring arrangement shown. The block is moved vertically downward from its equilibrium position and released. Knowing that the amplitude of the resulting motion is 60 mm, determine (*a*) the period and frequency of the motion, (*b*) the maximum velocity and maximum acceleration of the block.

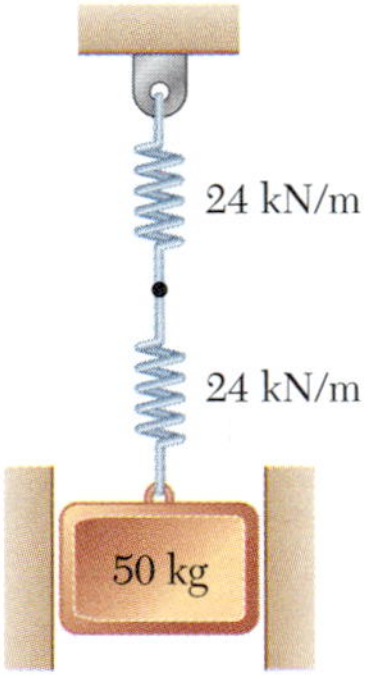

Fig. *P19.19*

19.20 A 13.6-kg block is supported by the spring arrangement shown. If the block is moved from its equilibrium 44 mm vertically downward and released, determine (*a*) the period and frequency of the resulting motion, (*b*) the maximum velocity and acceleration of the block.

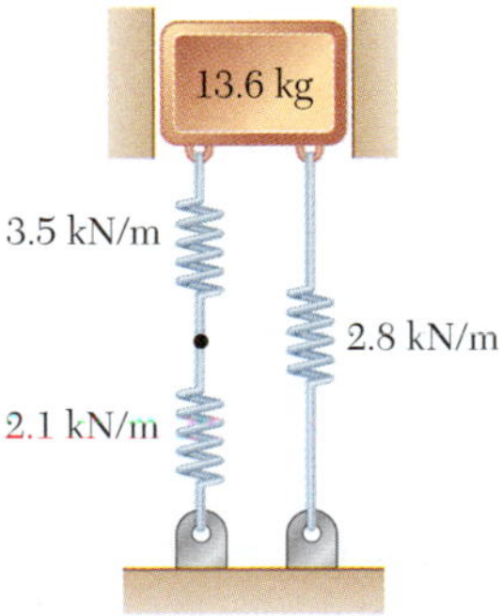

Fig. P19.20

19.21 A 10-lb block, attached to the lower end of a spring whose upper end is fixed, vibrates with a period of 6.8 s. Knowing that the constant k of the spring is inversely proportional to its length, determine the period of a 6-lb block which is attached to the center of the same spring if the upper and lower ends of the spring are fixed.

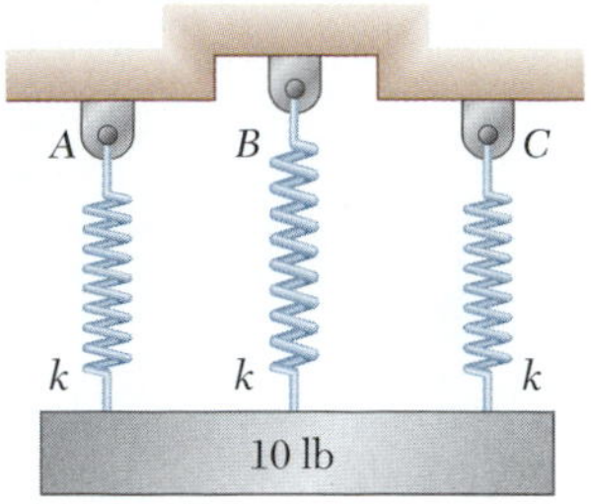

Fig. P19.22

19.22 A 10-lb block is supported as shown by three springs, each of which has a constant k. In the equilibrium position the tensions in springs A, B, and C are 3 lb, 4 lb, and 3 lb, respectively. The block is moved vertically downward 0.5 in. from its equilibrium position and released from rest. Knowing that in the resulting motion the minimum tension in spring B is zero, determine (a) the spring constant k, (b) the frequency of the motion, (c) the maximum compression in spring A.

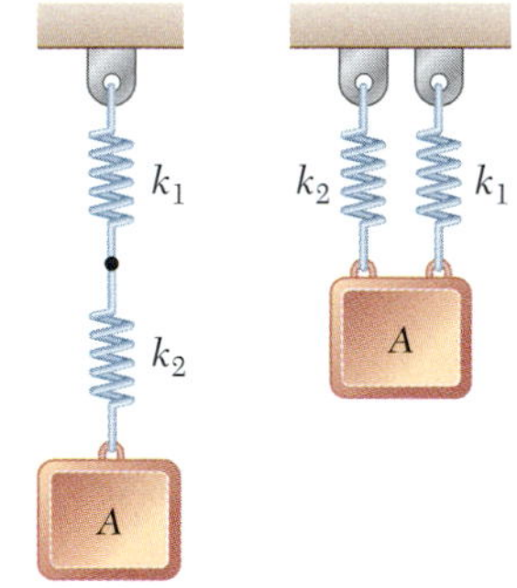

Fig. *P19.23*

19.23 Two springs of constants k_1 and k_2 are connected in series to a block A that vibrates in simple harmonic motion with a period of 5 s. When the same two springs are connected in parallel to the same block, the block vibrates with a period of 2 s. Determine the ratio k_1/k_2 of the two spring constants.

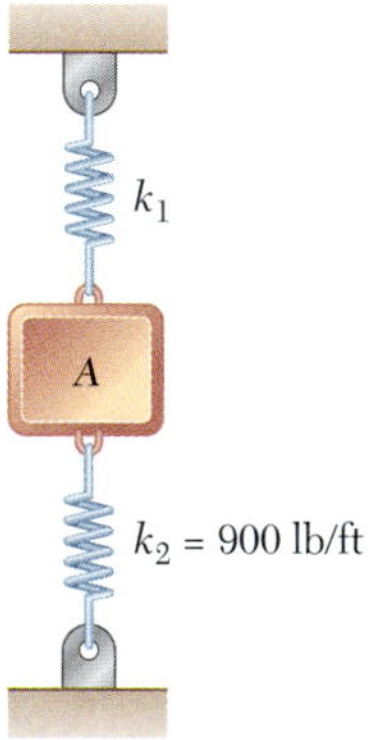

Fig. P19.24

19.24 The period of vibration of the system shown is observed to be 0.7 s. After the spring of constant $k_2 = 900$ lb/ft is removed from the system, the period is observed to be 0.9 s. Determine (a) the constant k_1, (b) the weight of block A.

19.25 The period of small oscillations of the system shown is observed to be 1.6 s. After a 14-lb collar is placed on top of collar A, the period of oscillation is observed to be 2.1 s. Determine (a) the weight of collar A, (b) the spring constant k.

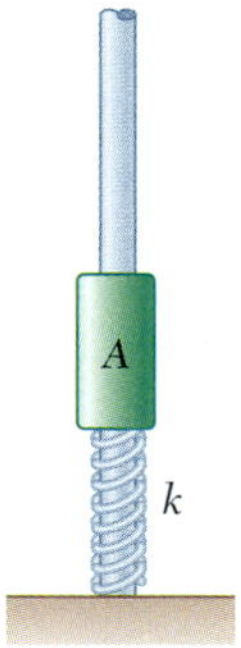

Fig. P19.25

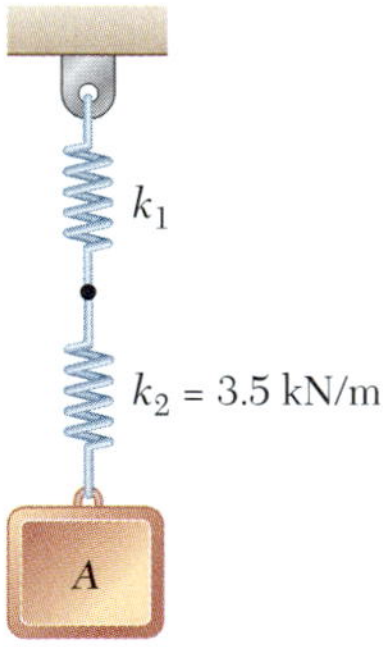

Fig. P19.26

19.26 The period of vibration of the system shown is observed to be 0.2 s. After the spring of constant $k_2 = 3.5$ kN/m is removed and block A is connected to the spring of constant k_1, the period is observed to be 0.12 s. Determine (a) the constant k_1 of the remaining spring, (b) the mass of block A.

19.27 The 50-kg block D is supported as shown by three springs, each of which has a constant k. The period of vertical vibration is to remain unchanged when the block is replaced by a 60-kg block, spring A is replaced by a spring of constant k_A, and the other two springs are unchanged. Knowing that the period of vibration is 0.4 s, determine the values of k and k_A.

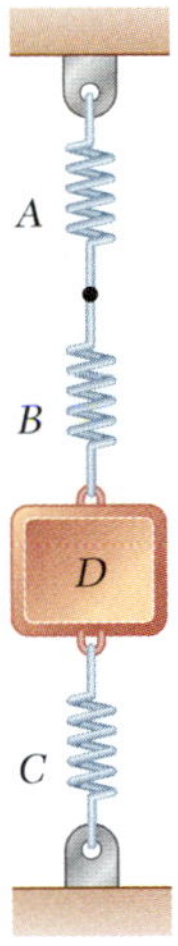

Fig. P19.27

19.28 From mechanics of materials it is known that when a static load **P** is applied at the end B of a uniform metal rod fixed at end A, the length of the rod will increase by an amount $\delta = PL/AE$, where L is the length of the undeformed rod, A is its cross-sectional area, and E is the modulus of elasticity of the material. Knowing that $L = 18$ in. and $E = 29 \times 10^6$ lb/in^2 and that the diameter of the rod is 0.32 in., and neglecting the weight of the rod, determine (*a*) the equivalent spring constant of the rod, (*b*) the frequency of the vertical vibrations of a block of weight $W = 16$ lb attached to end B of the same rod.

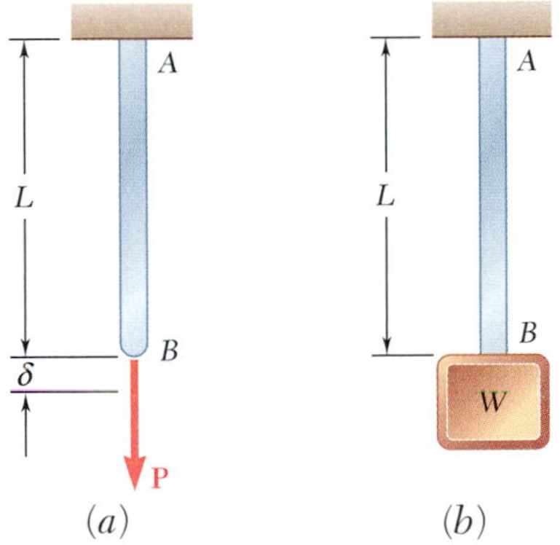

Fig. *P19.28*

19.29 From mechanics of materials it is known that for a simply supported beam of uniform cross section a static load **P** applied at the center will cause a deflection $\delta_A = PL^3/48EI$, where L is the length of the beam, E is the modulus of elasticity, and I is the moment of inertia of the cross-sectional area of the beam. Knowing that $L = 5$ m, $E = 200$ GPa, and $I = 20 \times 10^{-6}$m^4, determine (*a*) the equivalent spring constant of the beam, (*b*) the frequency of vibration of a 750 kg block attached to the center of the beam. Neglect the mass of the beam and assume that the load remains in contact with the beam.

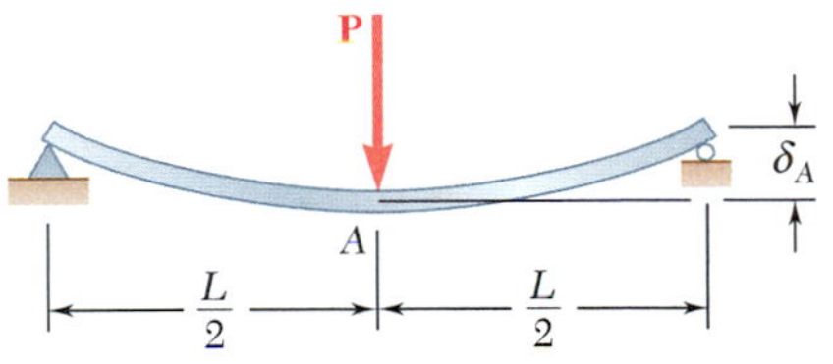

Fig. P19.29

19.30 A 2.4-in. deflection of the second floor of a building is measured directly under a newly installed 12.3-kips piece of rotating machinery which has a slightly unbalanced rotor. Assuming that the deflection of the floor is proportional to the load it supports, determine (*a*) the equivalent spring constant of the floor system, (*b*) the speed in rpm of the rotating machinery that should be avoided if it is not to coincide with the natural frequency of the floor-machinery system.

***19.31** The force-deflection equation for a nonlinear spring fixed at one end is $F = 4x^{1/2}$ where F is the force, expressed in newtons, applied at the other end and x is the deflection expressed in meters. (*a*) Determine the deflection x_0 if a 100-g block is suspended from the spring and is at rest. (*b*) Assuming that the slope of the force-deflection curve at the point corresponding to this loading can be used as an equivalent spring constant, determine the frequency of vibration of the block if it is given a very small downward displacement from its equilibrium position and released.

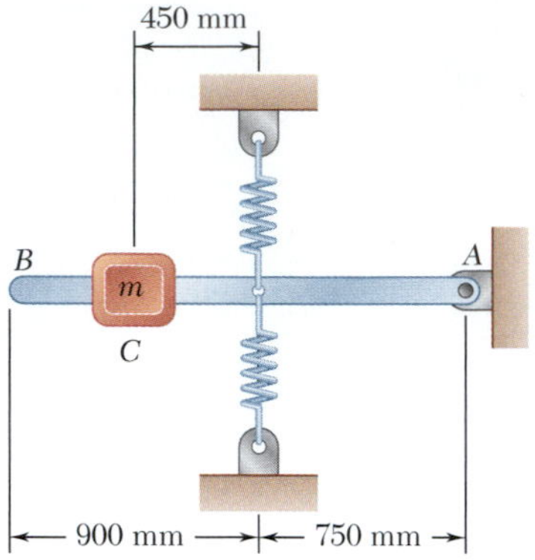

Fig. P19.32

19.32 Rod AB is attached to a hinge at A and to two springs, each of constant $k = 1.35$ kN/m. (*a*) Determine the mass m of the block C for which the period of small oscillations is 4 s. (*b*) If end B is depressed 60 mm and released, determine the maximum velocity of block C. Neglect the mass of the rod and assume that each spring can act in either tension or compression.

19.33 Denoting by δ_{st} the static deflection of a beam under a given load, show that the frequency of vibration of the load is

$$f = \frac{1}{2\pi}\sqrt{\frac{g}{\delta_{st}}}$$

Neglect the mass of the beam, and assume that the load remains in contact with the beam.

***19.34** Expanding the integrand in Eq. (19.19) of Sec. 19.4 into a series of even powers of sin ϕ and integrating, show that the period of a simple pendulum of length l may be approximated by the formula

$$\tau = 2\pi\sqrt{\frac{l}{g}}\left(1 + \frac{1}{4}\sin^2\frac{\theta_m}{2}\right)$$

where θ_m is the amplitude of the oscillations.

***19.35** Using the formula given in Prob. 19.34, determine the amplitude θ_m for which the period of a simple pendulum is $\frac{3}{4}$ percent longer than the period of the same pendulum for small oscillations.

***19.36** Using the data of Table 19.1, determine the period of a simple pendulum of length $l = 600$ mm (*a*) for small oscillations, (*b*) for oscillations of amplitude $\theta_m = 30°$, (*c*) for oscillations of amplitude $\theta_m = 120°$.

***19.37** Using the data of Table 19.1, determine the length in inches of a simple pendulum which oscillates with a period of 3 s and an amplitude of $\theta_m = 60°$.

19.5. FREE VIBRATIONS OF RIGID BODIES

The analysis of the vibrations of a rigid body or of a system of rigid bodies possessing a single degree of freedom is similar to the analysis of the vibrations of a particle. An appropriate variable, such as a distance x or an angle θ, is chosen to define the position of the body or system of bodies, and an equation relating this variable and its second derivative with respect to t is written. If the equation obtained is of the same form as (19.6), that is, if we have

$$\ddot{x} + \omega_n^2 x = 0 \qquad \text{or} \qquad \ddot{\theta} + \omega_n^2\theta = 0 \tag{19.21}$$

the vibration considered is a simple harmonic motion. The period and natural frequency of the vibration can then be obtained by identifying ω_n and substituting its value into Eqs. (19.13) and (19.14).

In general, a simple way to obtain one of Eqs. (19.21) is to express that the system of the external forces is equivalent to the system of the effective forces by drawing a free-body-diagram equation for an arbitrary value of the variable and writing the appropriate equation of motion. We recall that our goal should be *the determination of the coefficient* of the variable x or θ, *not* the determination of the variable itself or of the derivative $\ddot{x}$ or $\ddot{\theta}$. Setting this coefficient equal to ω_n^2, we obtain the natural circular frequency ω_n, from which τ_n and f_n can be determined.

The method we have outlined can be used to analyze vibrations which are truly represented by a simple harmonic motion, or vibrations of small amplitude which can be *approximated* by a simple harmonic motion. As an example, let us determine the period of the small oscillations of a square plate of side $2b$ which is suspended from the midpoint O of one of its sides (Fig. 19.5*a*). We consider the plate in an arbitrary position defined by the angle θ that the line OG forms with the vertical and draw a free-body-diagram equation to express that the weight $\mathbf{W}$ of the plate and the components $\mathbf{R}_x$ and $\mathbf{R}_y$ of the reaction at O are equivalent to the vectors $m\overline{\mathbf{a}}_t$ and $m\overline{\mathbf{a}}_n$ and to the couple $\overline{I}\boldsymbol{\alpha}$ (Fig. 19.5*b*). Since the angular velocity and angular acceleration of the plate are equal, respectively, to $\dot{\theta}$ and $\ddot{\theta}$, the magnitudes of the two vectors are, respectively, $mb\ddot{\theta}$ and $mb\dot{\theta}^2$, while the moment of the couple is $\overline{I}\ddot{\theta}$. In previous applications of this method (Chap. 16), we tried whenever possible to assume the correct sense for the acceleration. Here, however, we must assume the same positive sense for θ and $\ddot{\theta}$ in order to obtain an equation of the form (19.21). Consequently, the angular acceleration $\ddot{\theta}$ will be assumed positive counterclockwise, even though this assumption is obviously unrealistic. Equating moments about O, we write

$$+\circlearrowleft \qquad -W(b \sin \theta) = (mb\ddot{\theta})b + \overline{I}\ddot{\theta}$$

Noting that $\overline{I} = \frac{1}{12}m[(2b)^2 + (2b)^2] = \frac{2}{3}mb^2$ and $W = mg$, we obtain

$$\ddot{\theta} + \frac{3}{5}\frac{g}{b}\sin\theta = 0 \tag{19.22}$$

For oscillations of small amplitude, we can replace $\sin\theta$ by θ, expressed in radians, and write

$$\ddot{\theta} + \frac{3}{5}\frac{g}{b}\theta = 0 \tag{19.23}$$

Comparison with (19.21) shows that the equation obtained is that of a simple harmonic motion and that the natural circular frequency ω_n of the oscillations is equal to $(3g/5b)^{1/2}$. Substituting into (19.13), we find that the period of the oscillations is

$$\tau_n = \frac{2\pi}{\omega_n} = 2\pi\sqrt{\frac{5b}{3g}} \tag{19.24}$$

The result obtained is valid only for oscillations of small amplitude. A more accurate description of the motion of the plate is obtained by comparing Eqs. (19.16) and (19.22). We note that the two equations are identical if we choose l equal to $5b/3$. This means that the plate will oscillate as a simple pendulum of length $l = 5b/3$, and the results of Sec. 19.4 can be used to correct the value of the period given in (19.24). The point A of the plate located on line OG at a distance $l = 5b/3$ from O is defined as the *center of oscillation* corresponding to O (Fig. 19.5*a*).

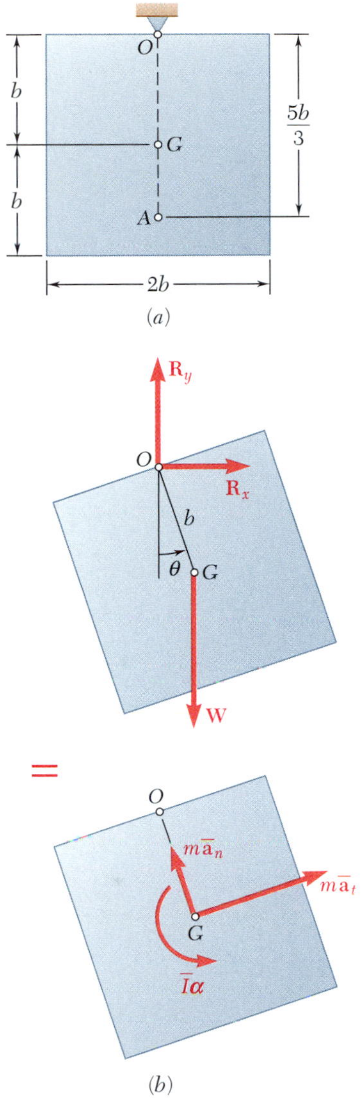

Fig. 19.5

SAMPLE PROBLEM 19.2

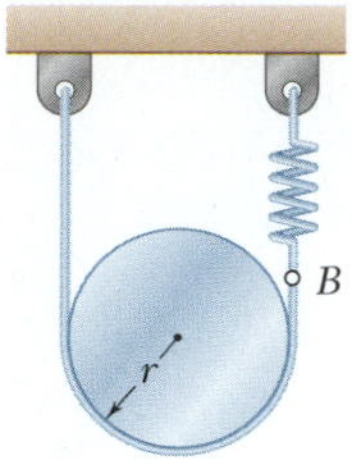

A cylinder of weight W and radius r is suspended from a looped cord as shown. One end of the cord is attached directly to a rigid support, while the other end is attached to a spring of constant k. Determine the period and natural frequency of the vibrations of the cylinder.

SOLUTION

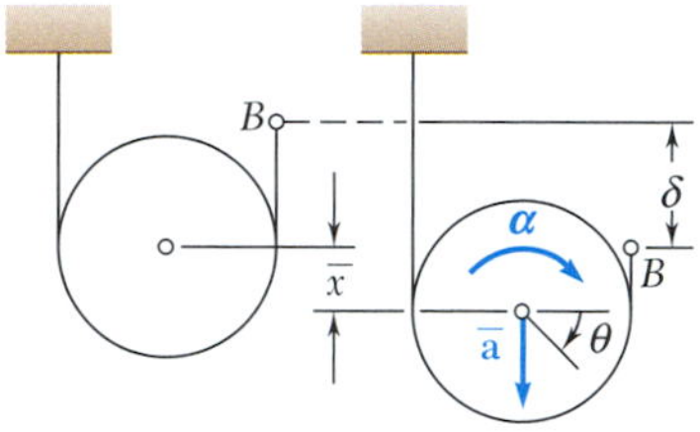

Kinematics of Motion. We express the linear displacement and the acceleration of the cylinder in terms of the angular displacement θ. Choosing the positive sense clockwise and measuring the displacements from the equilibrium position, we write

$$\bar{x} = r\theta \qquad \delta = 2\bar{x} = 2r\theta$$

$$\boldsymbol{\alpha} = \ddot{\theta} \circlearrowright \qquad \bar{a} = r\alpha = r\ddot{\theta} \qquad \bar{\mathbf{a}} = r\ddot{\theta} \downarrow \tag{1}$$

Equations of Motion. The system of external forces acting on the cylinder consists of the weight $\mathbf{W}$ and of the forces $\mathbf{T}_1$ and $\mathbf{T}_2$ exerted by the cord. We express that this system is equivalent to the system of effective forces represented by the vector $m\bar{\mathbf{a}}$ attached at G and the couple $\bar{I}\boldsymbol{\alpha}$.

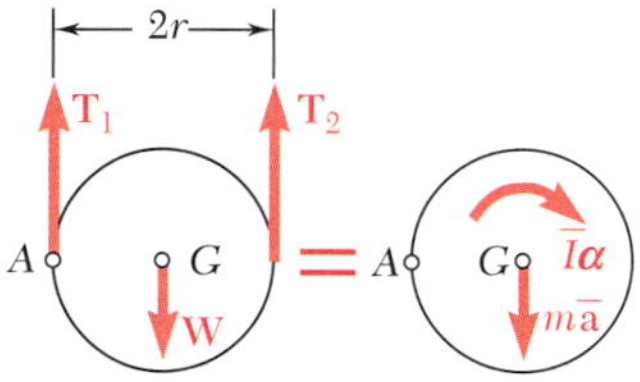

$$+\circlearrowright \Sigma M_A = \Sigma(M_A)_{\text{eff}}: \qquad Wr - T_2(2r) = m\bar{a}r + \bar{I}\alpha \tag{2}$$

When the cylinder is in its position of equilibrium, the tension in the cord is $T_0 = \frac{1}{2}W$. We note that for an angular displacement θ, the magnitude of $\mathbf{T}_2$ is

$$T_2 = T_0 + k\delta = \tfrac{1}{2}W + k\delta = \tfrac{1}{2}W + k(2r\theta) \tag{3}$$

Substituting from (1) and (3) into (2), and recalling that $\bar{I} = \frac{1}{2}mr^2$, we write

$$Wr - (\tfrac{1}{2}W + 2kr\theta)(2r) = m(r\ddot{\theta})r + \tfrac{1}{2}mr^2\ddot{\theta}$$

$$\ddot{\theta} + \frac{8}{3}\frac{k}{m}\theta = 0$$

The motion is seen to be simple harmonic, and we have

$$\omega_n^2 = \frac{8}{3}\frac{k}{m} \qquad \omega_n = \sqrt{\frac{8}{3}\frac{k}{m}}$$

$$\tau_n = \frac{2\pi}{\omega_n} \qquad \tau_n = 2\pi\sqrt{\frac{3}{8}\frac{m}{k}} \quad \blacktriangleleft$$

$$f_n = \frac{\omega_n}{2\pi} \qquad f_n = \frac{1}{2\pi}\sqrt{\frac{8}{3}\frac{k}{m}} \quad \blacktriangleleft$$

SAMPLE PROBLEM 19.3

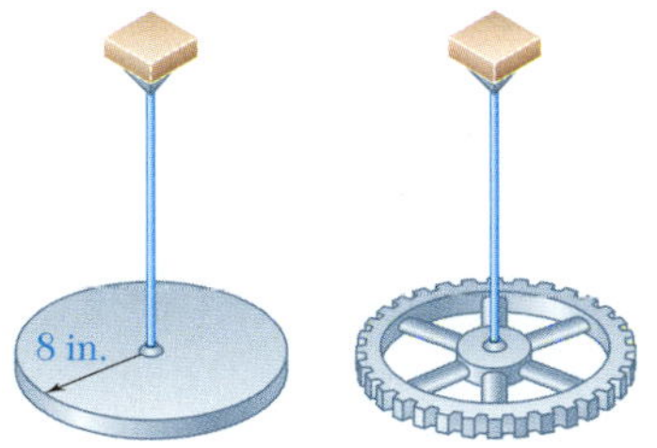

A circular disk, weighing 20 lb and of radius 8 in., is suspended from a wire as shown. The disk is rotated (thus twisting the wire) and then released; the period of the torsional vibration is observed to be 1.13 s. A gear is then suspended from the same wire, and the period of torsional vibration for the gear is observed to be 1.93 s. Assuming that the moment of the couple exerted by the wire is proportional to the angle of twist, determine (*a*) the torsional spring constant of the wire, (*b*) the centroidal moment of inertia of the gear, (*c*) the maximum angular velocity reached by the gear if it is rotated through 90° and released.

SOLUTION

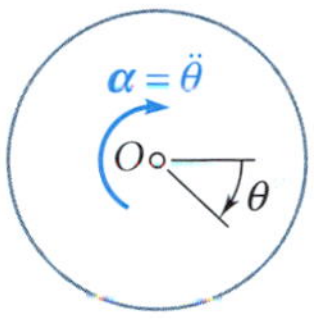

a. Vibration of Disk. Denoting by θ the angular displacement of the disk, we express that the magnitude of the couple exerted by the wire is $M = K\theta$, where K is the torsional spring constant of the wire. Since this couple must be equivalent to the couple $\bar{I}\boldsymbol{\alpha}$ representing the effective forces of the disk, we write

$$+\circlearrowleft \Sigma M_O = \Sigma(M_O)_{\text{eff}}: \qquad +K\theta = -\bar{I}\ddot{\theta}$$

$$\ddot{\theta} + \frac{K}{\bar{I}}\theta = 0$$

The motion is seen to be simple harmonic, and we have

$$\omega_n^2 = \frac{K}{\bar{I}} \qquad \tau_n = \frac{2\pi}{\omega_n} \qquad \tau_n = 2\pi\sqrt{\frac{\bar{I}}{K}} \qquad (1)$$

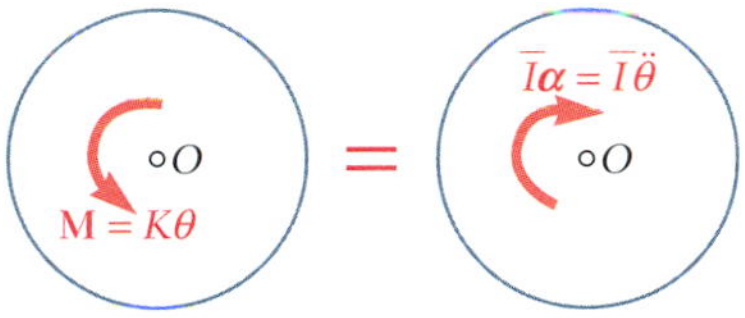

For the disk, we have

$$\tau_n = 1.13 \text{ s} \qquad \bar{I} = \tfrac{1}{2}mr^2 = \frac{1}{2}\left(\frac{20 \text{ lb}}{32.2 \text{ ft/s}^2}\right)\left(\frac{8}{12}\text{ ft}\right)^2 = 0.138 \text{ lb} \cdot \text{ft} \cdot \text{s}^2$$

Substituting into (1), we obtain

$$1.13 = 2\pi\sqrt{\frac{0.138}{K}} \qquad K = 4.27 \text{ lb} \cdot \text{ft/rad} \quad \blacktriangleleft$$

b. Vibration of Gear. Since the period of vibration of the gear is 1.93 s and $K = 4.27$ lb · ft/rad, Eq. (1) yields

$$1.93 = 2\pi\sqrt{\frac{\bar{I}}{4.27}} \qquad \bar{I}_{\text{gear}} = 0.403 \text{ lb} \cdot \text{ft} \cdot \text{s}^2 \quad \blacktriangleleft$$

c. Maximum Angular Velocity of Gear. Since the motion is simple harmonic, we have

$$\theta = \theta_m \sin \omega_n t \qquad \omega = \theta_m \omega_n \cos \omega_n t \qquad \omega_m = \theta_m \omega_n$$

Recalling that $\theta_m = 90° = 1.571$ rad and $\tau = 1.93$ s, we write

$$\omega_m = \theta_m \omega_n = \theta_m\left(\frac{2\pi}{\tau}\right) = (1.571 \text{ rad})\left(\frac{2\pi}{1.93 \text{ s}}\right)$$

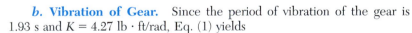

$$\omega_m = 5.11 \text{ rad/s} \quad \blacktriangleleft$$

SOLVING PROBLEMS ON YOUR OWN

In this lesson you saw that a rigid body, or a system of rigid bodies, whose position can be defined by a single coordinate x or θ, will execute a simple harmonic motion if the differential equation obtained by applying Newton's second law is of the form

$$\ddot{x} + \omega_n^2 x = 0 \qquad \text{or} \qquad \ddot{\theta} + \omega_n^2 \theta = 0 \tag{19.21}$$

Your goal should be to determine ω_n, from which you can obtain the period τ_n and the natural frequency f_n. Taking into account the initial conditions, you can then write an equation of the form

$$x = x_m \sin (\omega_n t + \phi) \tag{19.10}$$

where x should be replaced by θ if a rotation is involved. To solve the problems in this lesson, you will follow these steps:

1. ***Choose a coordinate which will measure the displacement of the body*** from its equilibrium position. You will find that many of the problems in this lesson involve the rotation of a body about a fixed axis and that the angle measuring the rotation of the body from its equilibrium position is the most convenient coordinate to use. In problems involving the general plane motion of a body, where a coordinate x (and possibly a coordinate y) is used to define the position of the mass center G of the body, and a coordinate θ is used to measure its rotation about G, find kinematic relations which will allow you to express x (and y) in terms of θ [Sample Prob. 19.2].

2. ***Draw a free-body-diagram equation*** to express that the system of the external forces is equivalent to the system of the effective forces, which consists of the vector $m\bar{\mathbf{a}}$ and the couple $\bar{I}\boldsymbol{\alpha}$, where $\bar{a} = \ddot{x}$ and $\alpha = \ddot{\theta}$. Be sure that each applied force or couple is drawn in a direction consistent with the assumed displacement and that the senses of $\bar{\mathbf{a}}$ and $\boldsymbol{\alpha}$ are, respectively, those in which the coordinates x and θ are increasing.

3. ***Write the differential equations of motion*** by equating the sums of the components of the external and effective forces in the x and y directions and the sums of their moments about a given point. If necessary, use the kinematic relations developed in step 1 to obtain equations involving only the coordinate θ. If θ is a small angle, replace $\sin \theta$ by θ and $\cos \theta$ by 1, if these functions appear in your equations. Eliminating any unknown reactions, you will obtain an equation of the type of Eqs. (19.21). Note that in problems involving a body rotating about a fixed axis, you can immediately obtain such an equation by equating the moments of the external and effective forces about the fixed axis.

4. Comparing the equation you have obtained with one of Eqs. (19.21), you can identify ω_n^2 and, thus, determine the natural circular frequency ω_n. Remember that the object of your analysis is *not to solve* the differential equation you have obtained, *but to identify* ω_n^2.

5. Determine the amplitude and the phase angle ϕ by substituting the value obtained for ω_n and the initial values of the coordinate and its first derivative into Eq. (19.10) and the equation obtained by differentiating (19.10) with respect to t. From Eq. (19.10) and the two equations obtained by differentiating (19.10) twice with respect to t, and using the kinematic relations developed in step 1, you will be able to determine the position, velocity, and acceleration of any point of the body at any given time.

6. In problems involving torsional vibrations, the torsional spring constant K is expressed in N · m/rad or lb · ft/rad. The product of K and the angle of twist θ, expressed in radians, yields the moment of the restoring couple, which should be equated to the sum of the moments of the effective forces or couples about the axis of rotation [Sample Prob. 19.3].

Problems

19.38 The 10-lb uniform rod *AC* is attached to springs of constant $k = 50$ lb/ft at *B* and $k = 62$ lb/ft at *C*, which can act in tension or compression. If the end *C* is depressed slightly and released, determine (*a*) the frequency of vibration, (*b*) the amplitude of the motion of point *C*, knowing that the maximum velocity of that point is 2.7 ft/s.

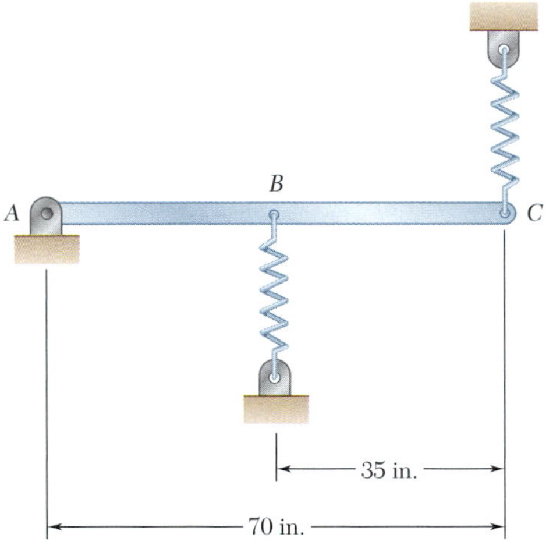

Fig. P19.38

19.39 The 9-kg uniform rod *AB* is attached to springs at *A* and *B*, each of constant 850 N/m, which can act in tension or compression. If the end *A* of the rod is depressed slightly and released, determine (*a*) the frequency of vibration, (*b*) the amplitude of the angular motion of the rod, knowing that the maximum velocity of point *A* is 1.1 mm/s.

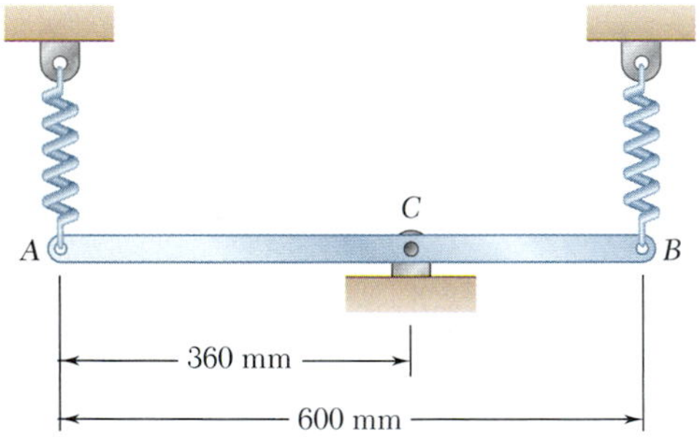

Fig. *P19.39*

19.40 Two identical 3-kg uniform rods are hinged at point *B* and attached to a spring of constant $k = 4$ kN/m. The rods are guided by small wheels of negligible mass and the system is in equilibrium when the rods are horizontal as shown. Knowing that point *B* is depressed 20 mm and released, determine (*a*) the frequency of the resulting vibration, (*b*) the magnitude of the maximum angular acceleration of rod *AB*.

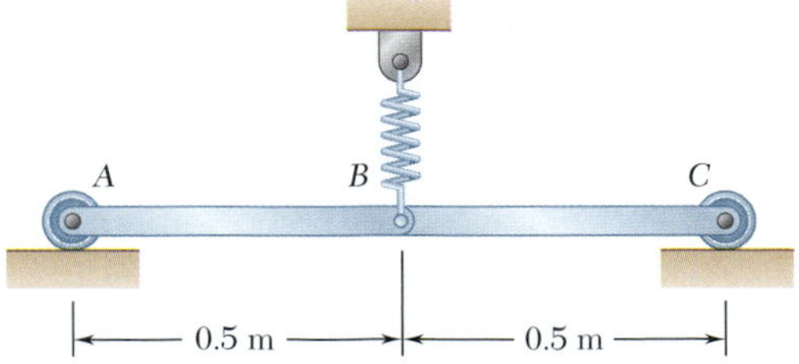

Fig. P19.40

19.41 A 1.2-kg uniform rod AB is attached to a hinge at A and to two springs, each of constant $k = 450$ N/m. (*a*) Determine the mass m of block C for which the period of small oscillations is 0.6 s. (*b*) If end B is depressed 60 mm and released, determine the maximum velocity of block C.

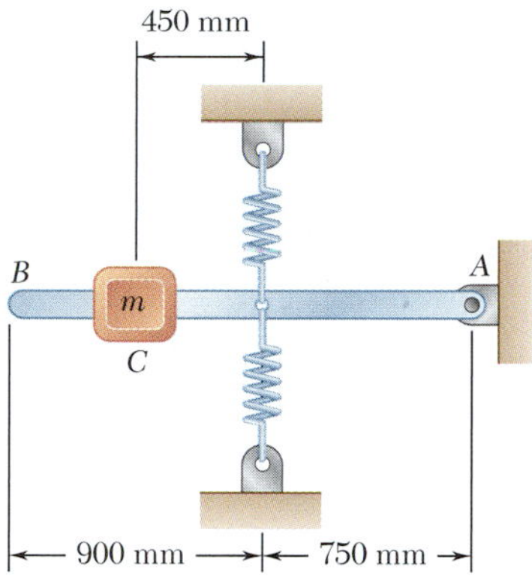

Fig. *P19.41*

19.42 A 12-lb uniform cylinder can roll without sliding on a horizontal surface and is attached by a pin at point C to the 8-lb horizontal bar AB. The bar is attached to two springs, each of constant $k = 30$ lb/in. as shown. Knowing that the bar is moved 0.5 in. to the right of the equilibrium position and released, determine (*a*) the period of vibration of the system, (*b*) the magnitude of the maximum velocity of bar AB.

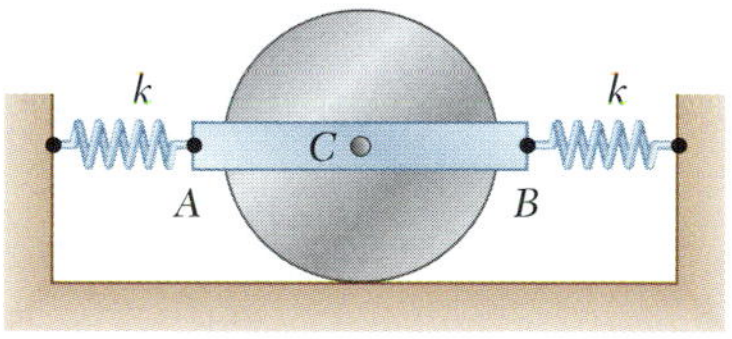

Fig. P19.42 and P19.43

19.43 A 12-lb uniform cylinder is assumed to roll without sliding on a horizontal surface and is attached by a pin at point C to the 8-lb horizontal bar AB. The bar is attached to two springs, each of constant $k = 20$ lb/in. as shown. Knowing that the coefficient of static friction between the cylinder and the surface is 0.5, determine the maximum amplitude of the motion of point C which is compatible with the assumption of rolling.

19.44 A 270-kg flywheel has a diameter of 1.2 m and a radius of gyration of 0.5 m. A belt is placed over the rim and attached to two springs, each of constant $k = 13$ kN/m. The initial tension in the belt is sufficient to prevent slipping. If the end C of the belt is pulled 25 mm to the right and released, determine (*a*) the period of vibration, (*b*) the maximum angular velocity of the flywheel.

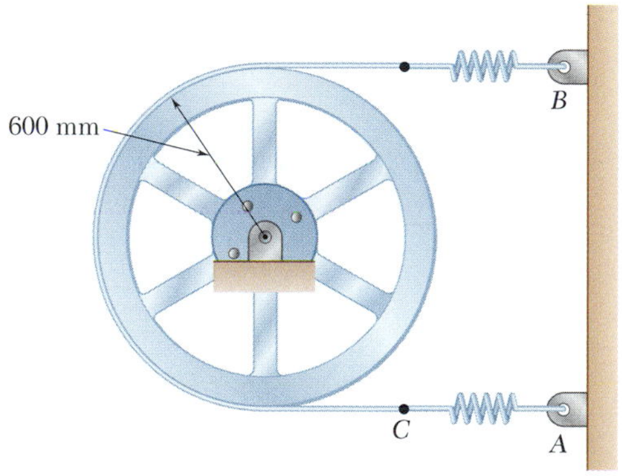

Fig. P19.44

19.45 For the uniform semicircular plate of radius r, determine the period of small oscillations if the plate (*a*) is suspended from point A as shown, (*b*) is suspended from a pin located at point B.

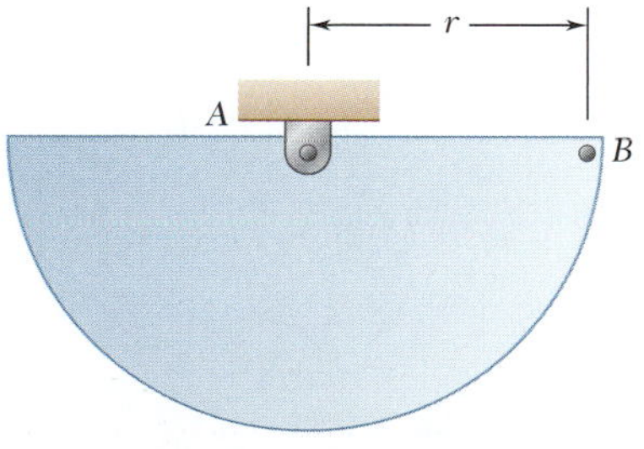

Fig. P19.45

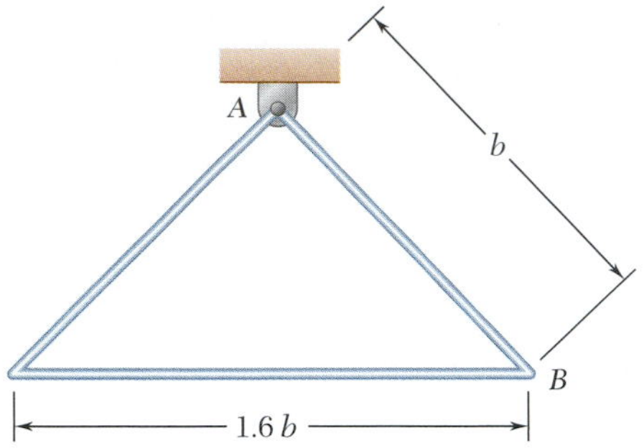

Fig. P19.46

19.46 A thin homogeneous wire is bent into the shape of an isosceles triangle of sides b, b, and $1.6b$. Determine the period of small oscillations if the wire (a) is suspended from point A as shown, (b) is suspended from point B.

19.47 A semicircular hole is cut in a uniform square plate which is attached to a frictionless pin at its geometric center O. Determine (a) the period of small oscillations of the plate, (b) the length of a simple pendulum which has the same period.

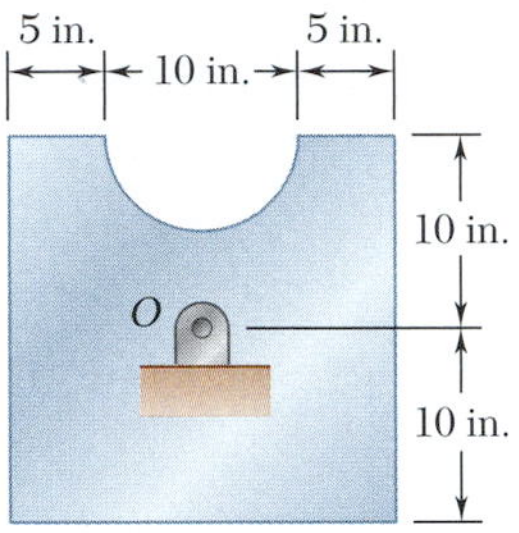

Fig. P19.47

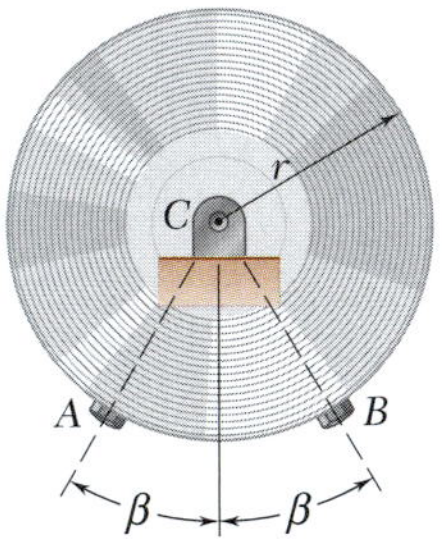

Fig. P19.48 and P19.49

19.48 Two small weights w are attached at A and B to the rim of a uniform disk of radius r and weight W. Denoting by τ_0 the period of small oscillations when $\beta = 0$, determine the angle β for which the period of small oscillations is $2\tau_0$.

19.49 Two 50-g weights are attached at A and B to the rim of a 1.5-kg uniform disk of radius $r = 80$ mm. Determine the frequency of small oscillations when $\beta = 45°$.

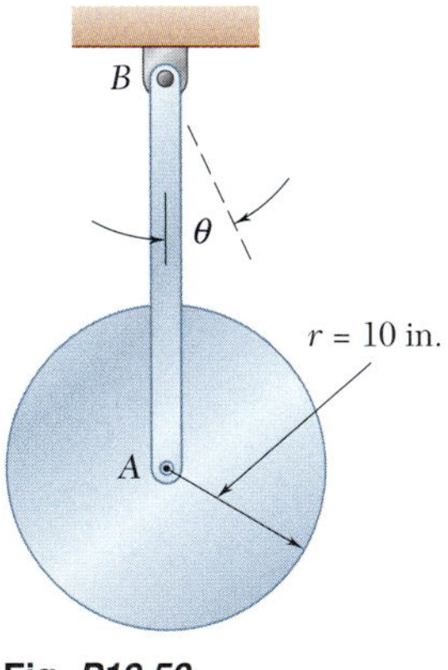

Fig. *P19.50*

19.50 A uniform disk of radius $r = 10$ in. is attached at A to a 26-in. rod AB of negligible weight which can rotate freely in a vertical plane about B. Determine the period of small oscillations (a) if the disk is free to rotate in a bearing at A, (b) if the rod is riveted to the disk at A.

19.51 A small collar of mass 1-kg is rigidly attached to a 3-kg uniform rod of length $L = 1$ m. Determine the period of small oscillations of the rod when (a) $d = 1$ m, (b) $d = 0.7$ m.

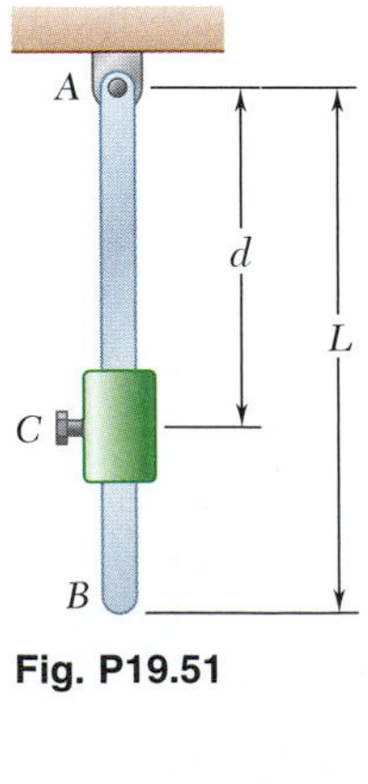

Fig. P19.51

19.52 A *compound pendulum* is defined as a rigid slab which oscillates about a fixed point O, called the center of suspension. Show that the period of oscillation of a compound pendulum is equal to the period of a simple pendulum of length OA, where the distance from A to the mass center G is $GA = \bar{k}^2\sqrt{r}$. Point A is defined as the center of oscillation and coincides with the center of percussion defined in Prob. 17.57.

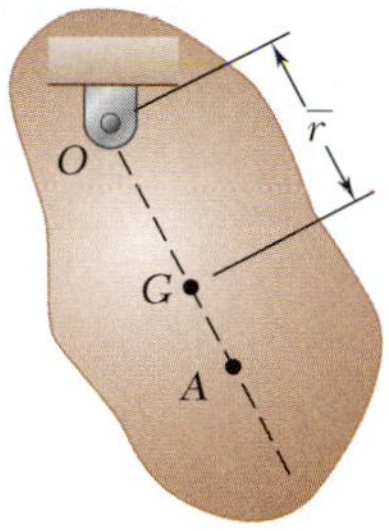

Fig. P19.52 and P19.53

19.53 A rigid slab oscillates about a fixed point O. Show that the smallest period of oscillation occurs when the distance $\bar{r}$ from point O to the mass center G is equal to $\bar{k}$.

19.54 Show that if the compound pendulum of Prob. 19.52 is suspended from A instead of O, the period of oscillation is the same as before and the new center of oscillation is located at O.

19.55 The 16-lb uniform bar AB is hinged at C and is attached at A to a spring of constant $k = 50$ lb/ft. If end A is given a small displacement and released, determine (*a*) the frequency of small oscillations, (*b*) the smallest value of the spring constant k for which oscillations will occur.

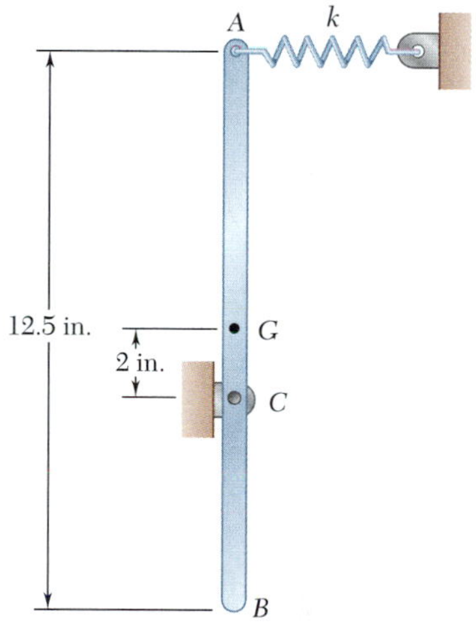

Fig. P19.55

19.56 A 10-kg uniform equilateral triangular plate is suspended from a pin located at one of its vertices and is attached to two springs, each of constant $k = 200$ N/m. Knowing that the plate is given a small angular displacement and released, determine the frequency of the resulting vibration.

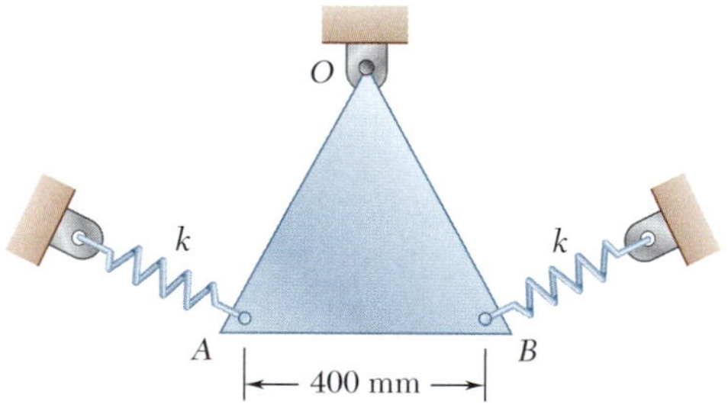

Fig. P19.56

19.57 Two uniform rods, each of weight $W = 24$ lb and length $L = 40$ in., are welded together to form the assembly shown. Knowing that the constant of each spring is $k = 50$ lb/ft and that end A is given a small displacement and released, determine the frequency of the resulting motion.

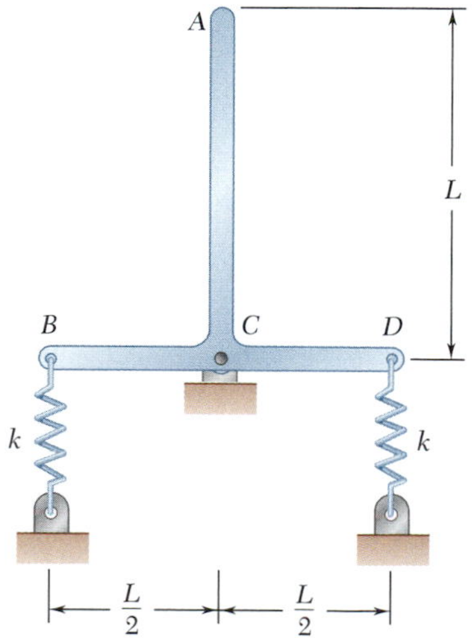

Fig. *P19.57*

19.58 A uniform disk of radius r and mass m can roll without slipping on a cylindrical surface and is attached to bar ABC of length L and negligible mass. The bar is attached at point A to a spring of constant k and can rotate freely about point B in the vertical plane. Knowing that end A is given a small displacement and released, determine the frequency of the resulting vibration in terms of m, L, k, and g.

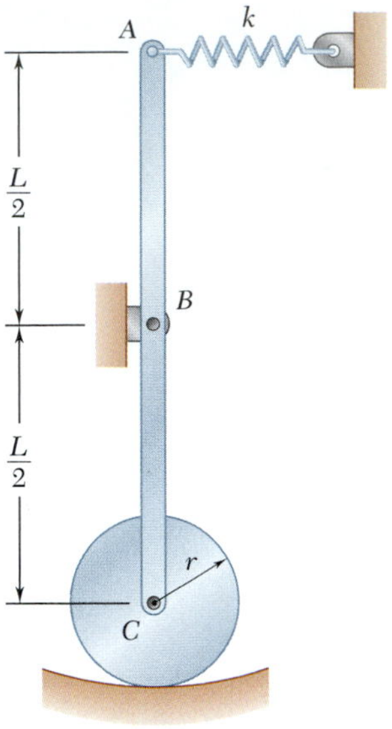

Fig. P19.58

19.59 A uniform disk of radius $r = 10$ in. is attached at A to a 26-in. rod AB of negligible weight which can rotate freely in a vertical plane about B. If the rod is displaced 2° from the position shown and released, determine the magnitude of the maximum velocity of point A, assuming that the disk (*a*) is free to rotate in a bearing at A, (*b*) is riveted to the rod at A.

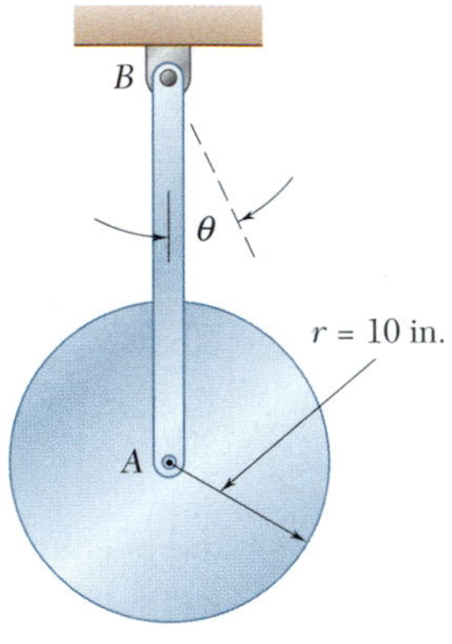

Fig. P19.59

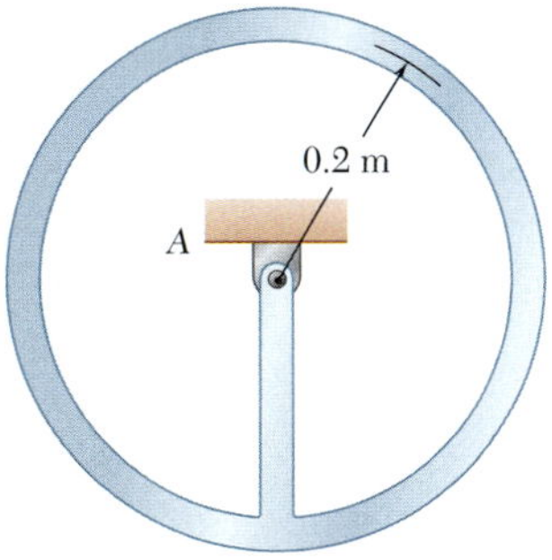

Fig. P19.60

19.60 A homogeneous rod of mass per unit length equal to 0.4 kg/m is used to form the assembly shown, which rotates freely about pivot A in a vertical plane. Knowing that the assembly is displaced 2° clockwise from its equilibrium position and released, determine its angular velocity and angular acceleration 5 s later.

19.61 A homogeneous wire bent to form the figure shown is attached to a pin support at A. Knowing that $r = 12$ in. and that point B is pushed down 1 in. and released, determine the magnitude of the velocity of B, 8 s later.

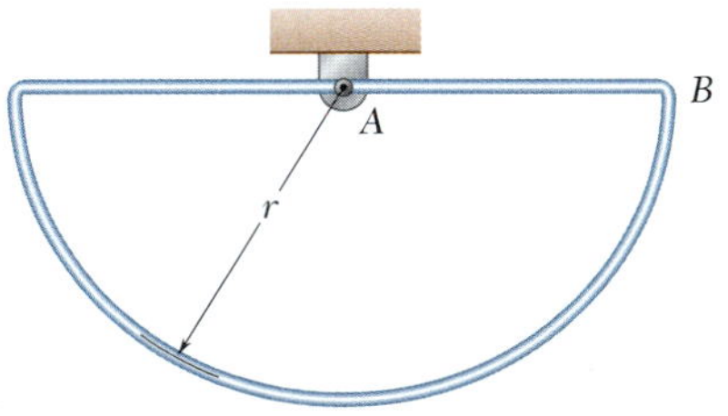

Fig. *P19.61* and P19.62

19.62 A homogeneous wire bent to form the figure shown is attached to a pin support at A. Knowing that $r = 320$ mm and that point B is pushed down 30 mm and released, determine the magnitude of the acceleration of B, 10 s later.

19.63 A uniform disk of radius $r = 6$ in. is welded at its center to two elastic rods of equal length with fixed ends A and B. Knowing that the disk rotates through an 8° angle when a 0.375 lb · ft couple is applied to the disk and that it oscillates with a period of 1.3 s when the couple is removed, determine (*a*) the weight of the disk, (*b*) the period of vibration if one of the rods is removed.

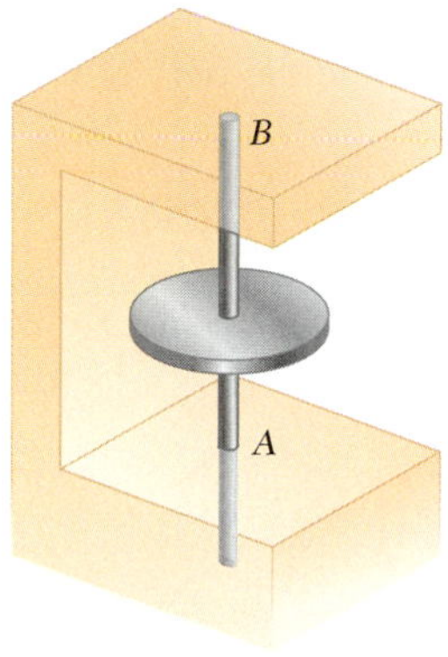

Fig. P19.63

19.64 A 5-kg uniform rod CD of length $l = 0.7$ m is welded at C to two elastic rods, which have fixed ends at A and B and are known to have a combined torsional spring constant $K = 24$ N · m/rad. Determine the period of small oscillation, if the equilibrium position of CD is (*a*) vertical as shown, (*b*) horizontal.

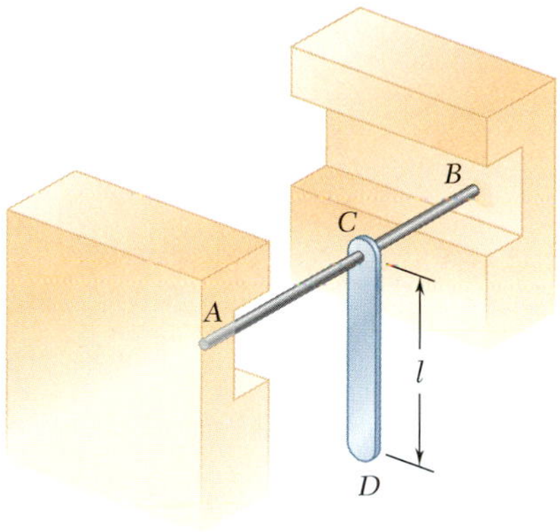

Fig. *P19.64*

19.65 A 60-kg uniform circular plate is welded to two elastic rods which have fixed ends at supports A and B as shown. The torsional spring constant of each rod is 200 N · m/rad, and the system is in equilibrium when the plate is vertical. Knowing that the plate is rotated 2° about axis AB and released, determine (*a*) the period of oscillation, (*b*) the magnitude of the maximum velocity of the mass center G of the plate.

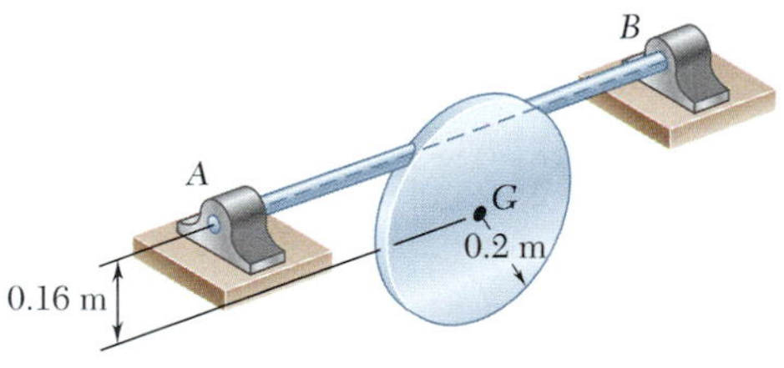

Fig. P19.65

19.66 A horizontal platform P is held by several rigid bars which are connected to a vertical wire. The period of oscillation of the platform is found to be 2.2 s when the platform is empty and 3.8 s when an object A of uniform moment of inertia is placed on the platform with its mass center directly above the center of the plate. Knowing that the wire has a torsional constant $K = 27$ N · m/rad, determine the centroidal moment of inertia of object A.

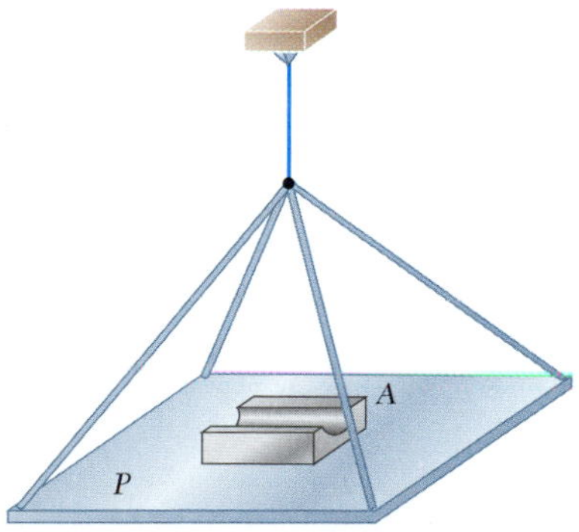

Fig. *P19.66*

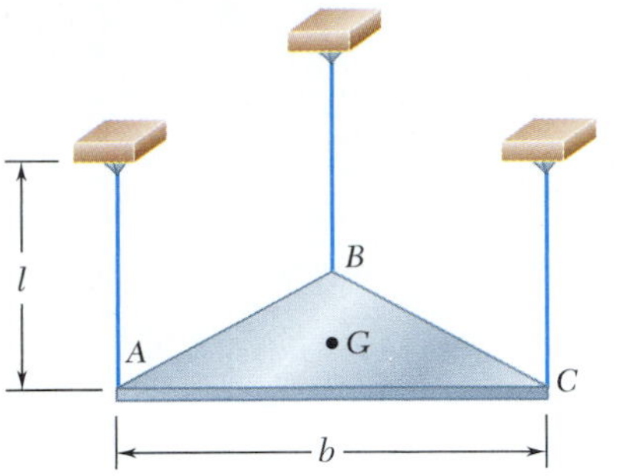

Fig. P19.67

19.67 A uniform equilateral triangular plate of side b is suspended from three vertical wires of the same length l. Determine the period of small oscillations of the plate when (a) it is rotated through a small angle about a vertical axis through its mass center G, (b) it is given a small horizontal displacement in a direction perpendicular to AB.

19.68 A 4-lb circular disk of radius $r = 40$ in. is suspended at its center C from wires AB and BC soldered together at B. The torsional spring constants of the wires are $K_1 = 3$ lb · ft/rad for AB and $K_2 = 1.5$ lb · ft/rad for BC. Determine the period of oscillation of the disk about the axis AC.

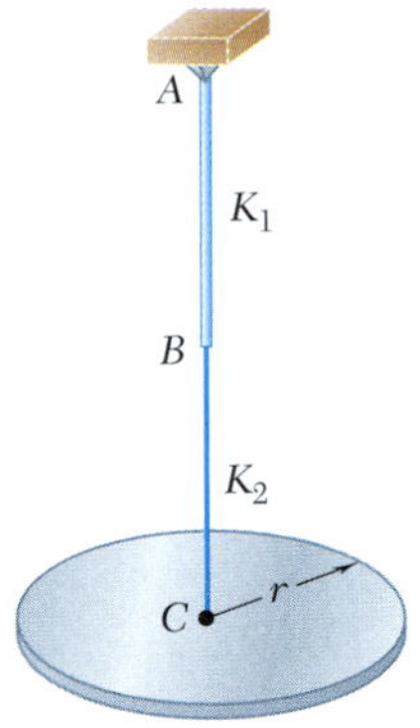

Fig. P19.68

19.6. APPLICATION OF THE PRINCIPLE OF CONSERVATION OF ENERGY

We saw in Sec. 19.2 that when a particle of mass m is in simple harmonic motion, the resultant **F** of the forces exerted on the particle has a magnitude proportional to the displacement x measured from the position of equilibrium O and is directed toward O; we write $F = -kx$. Referring to Sec. 13.6, we note that **F** is a *conservative force* and that the corresponding potential energy is $V = \frac{1}{2}kx^2$, where V is assumed equal to zero in the equilibrium position $x = 0$. Since the velocity of the particle is equal to $\dot{x}$, its kinetic energy is $T = \frac{1}{2}m\dot{x}^2$ and we can express that the total energy of the particle is conserved by writing

$$T + V = \text{constant} \qquad \tfrac{1}{2}m\dot{x}^2 + \tfrac{1}{2}kx^2 = \text{constant}$$

Dividing through by $m/2$ and recalling from Sec. 19.2 that $k/m = \omega_n^2$, where ω_n is the natural circular frequency of the vibration, we have

$$\dot{x}^2 + \omega_n^2 x^2 = \text{constant} \tag{19.25}$$

Equation (19.25) is characteristic of a simple harmonic motion, since it can be obtained from Eq. (19.6) by multiplying both terms by $2\dot{x}$ and integrating.

The principle of conservation of energy provides a convenient way for determining the period of vibration of a rigid body or of a system of rigid bodies possessing a single degree of freedom, once it has been established that the motion of the system is a simple harmonic motion or that it can be approximated by a simple harmonic motion. Choosing an appropriate variable, such as a distance x or an angle θ, we consider two particular positions of the system:

1. *The displacement of the system is maximum;* we have $T_1 = 0$, and V_1 can be expressed in terms of the amplitude x_m or θ_m (choosing $V = 0$ in the equilibrium position).
2. *The system passes through its equilibrium position;* we have $V_2 = 0$, and T_2 can be expressed in terms of the maximum velocity $\dot{x}_m$ or the maximum angular velocity $\dot{\theta}_m$.

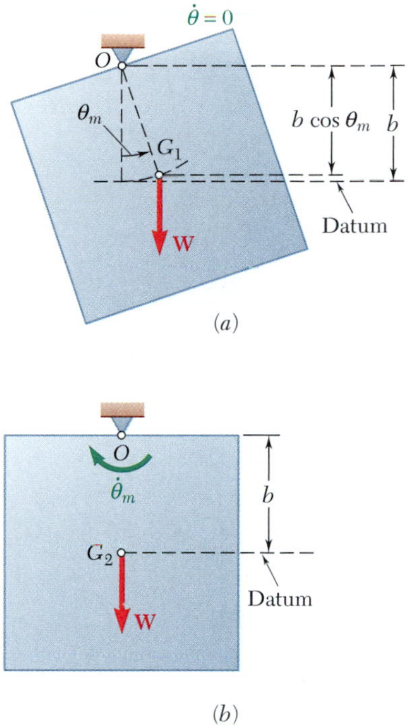

Fig. 19.6

We then express that the total energy of the system is conserved and write $T_1 + V_1 = T_2 + V_2$. Recalling from (19.15) that for simple harmonic motion the maximum velocity is equal to the product of the amplitude and of the natural circular frequency ω_n, we find that the equation obtained can be solved for ω_n.

As an example, let us consider again the square plate of Sec. 19.5. In the position of maximum displacement (Fig. 19.6*a*), we have

$$T_1 = 0 \qquad V_1 = W(b - b \cos \theta_m) = Wb(1 - \cos \theta_m)$$

or, since $1 - \cos \theta_m = 2 \sin^2 (\theta_m/2) \approx 2(\theta_m/2)^2 = \theta_m^2/2$ for oscillations of small amplitude,

$$T_1 = 0 \qquad V_1 = \tfrac{1}{2}Wb\theta_m^2 \tag{19.26}$$

As the plate passes through its position of equilibrium (Fig. 19.6*b*), its velocity is maximum and we have

$$T_2 = \tfrac{1}{2}m\bar{v}_m^2 + \tfrac{1}{2}\bar{I}\omega_m^2 = \tfrac{1}{2}mb^2\dot{\theta}_m^2 + \tfrac{1}{2}\bar{I}\dot{\theta}_m^2 \qquad V_2 = 0$$

or, recalling from Sec. 19.5 that $\bar{I} = \tfrac{2}{3}mb^2$,

$$T_2 = \tfrac{1}{2}(\tfrac{5}{3}mb^2)\dot{\theta}_m^2 \qquad V_2 = 0 \tag{19.27}$$

Substituting from (19.26) and (19.27) into $T_1 + V_1 = T_2 + V_2$, and noting that the maximum velocity $\dot{\theta}_m$ is equal to the product $\theta_m\omega_n$, we write

$$\tfrac{1}{2}Wb\theta_m^2 = \tfrac{1}{2}(\tfrac{5}{3}mb^2)\theta_m^2\omega_n^2 \tag{19.28}$$

which yields $\omega_n^2 = 3g/5b$ and

$$\tau_n = \frac{2\pi}{\omega_n} = 2\pi\sqrt{\frac{5b}{3g}} \tag{19.29}$$

as previously obtained.

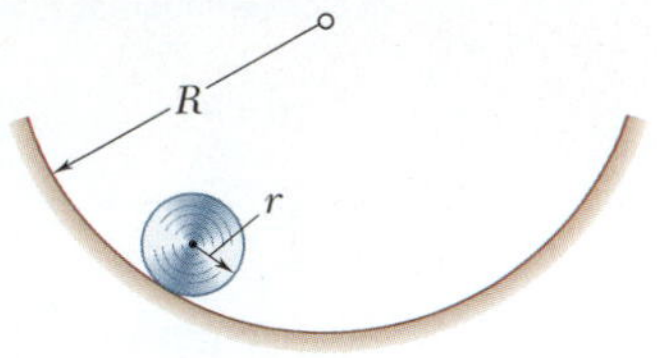

SAMPLE PROBLEM 19.4

Determine the period of small oscillations of a cylinder of radius r which rolls without slipping inside a curved surface of radius R.

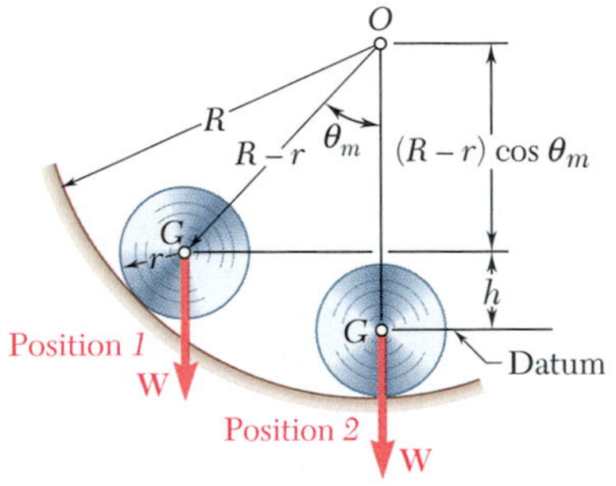

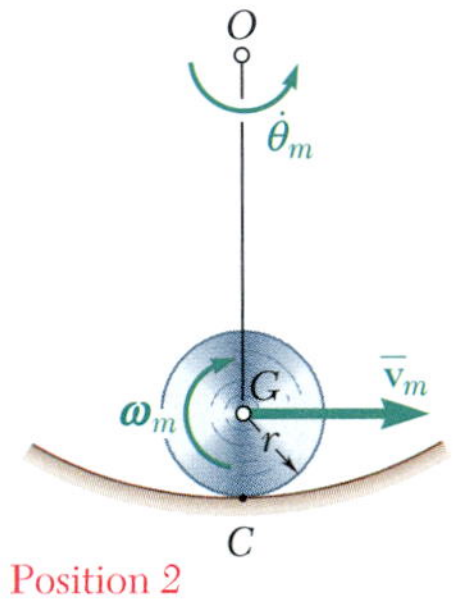

SOLUTION

We denote by θ the angle which line OG forms with the vertical. Since the cylinder rolls without slipping, we may apply the principle of conservation of energy between position *1*, where $\theta = \theta_m$, and position *2*, where $\theta = 0$.

Position *1*

Kinetic Energy. Since the velocity of the cylinder is zero, $T_1 = 0$.

Potential Energy. Choosing a datum as shown and denoting by W the weight of the cylinder, we have

$$V_1 = Wh = W(R - r)(1 - \cos\theta)$$

Noting that for small oscillations $(1 - \cos\theta) = 2\sin^2(\theta/2) \approx \theta^2/2$, we have

$$V_1 = W(R - r)\frac{\theta_m^2}{2}$$

Position 2. Denoting by $\dot{\theta}_m$ the angular velocity of line OG as the cylinder passes through position *2*, and observing that point C is the instantaneous center of rotation of the cylinder, we write

$$\bar{v}_m = (R - r)\dot{\theta}_m \qquad \omega_m = \frac{\bar{v}_m}{r} = \frac{R - r}{r}\dot{\theta}_m$$

Kinetic Energy

$$\begin{aligned} T_2 &= \tfrac{1}{2}m\bar{v}_m^2 + \tfrac{1}{2}\bar{I}\omega_m^2 \\ &= \tfrac{1}{2}m(R - r)^2\dot{\theta}_m^2 + \tfrac{1}{2}(\tfrac{1}{2}mr^2)\left(\frac{R - r}{r}\right)^2\dot{\theta}_m^2 \\ &= \tfrac{3}{4}m(R - r)^2\dot{\theta}_m^2 \end{aligned}$$

Potential Energy

$$V_2 = 0$$

Conservation of Energy

$$T_1 + V_1 = T_2 + V_2$$

$$0 + W(R - r)\frac{\theta_m^2}{2} = \tfrac{3}{4}m(R - r)^2\dot{\theta}_m^2 + 0$$

Since $\dot{\theta}_m = \omega_n\theta_m$ and $W = mg$, we write

$$mg(R - r)\frac{\theta_m^2}{2} = \tfrac{3}{4}m(R - r)^2(\omega_n\theta_m)^2 \qquad \omega_n^2 = \frac{2}{3}\frac{g}{R - r}$$

$$\tau_n = \frac{2\pi}{\omega_n} \qquad \tau_n = 2\pi\sqrt{\frac{3}{2}\frac{R - r}{g}} \blacktriangleleft$$

SOLVING PROBLEMS ON YOUR OWN

In the problems which follow you will be asked to use the *principle of conservation of energy* to determine the period or natural frequency of the simple harmonic motion of a particle or rigid body. Assuming that you choose an angle θ to define the position of the system (with $\theta = 0$ in the equilibrium position), as you will in most of the problems in this lesson, you will express that the total energy of the system is conserved, $T_1 + V_1 = T_2 + V_2$, between the position 1 of maximum displacement ($\theta_1 = \theta_m$, $\dot{\theta}_1 = 0$) and the position 2 of maximum velocity ($\dot{\theta}_2 = \dot{\theta}_m$, $\theta_2 = 0$). It follows that T_1 and V_2 will both be zero, and the energy equation will reduce to $V_1 = T_2$, where V_1 and T_2 are homogeneous quadratic expressions in θ_m and $\dot{\theta}_m$, respectively. Recalling that, for a simple harmonic motion, $\dot{\theta}_m = \theta_m \omega_n$ and substituting this product into the energy equation, you will obtain, after reduction, an equation that you can solve for ω_n^2. Once you have determined the natural circular frequency ω_n, you can obtain the period τ_n and the natural frequency f_n of the vibration.

The steps that you should take are as follows:

1. *Calculate the potential energy V_1 of the system in its position of maximum displacement.* Draw a sketch of the system in its position of maximum displacement and express the potential energy of all the forces involved (internal as well as external) in terms of the maximum displacement x_m or θ_m.

a. *The potential energy associated with the weight W of a body is $V_g = Wy$,* where y is the elevation of the center of gravity G of the body above its equilibrium position. If the problem you are solving involves the oscillation of a rigid body about a horizontal axis through a point O located at a distance b from G (Fig. 19.6), express y in terms of the angle θ that the line OG forms with the vertical: $y = b(1 - \cos\theta)$. But, for small values of θ, you can replace this expression with $y = \frac{1}{2}b\theta^2$ [Sample Prob. 19.4]. Therefore, when θ reaches its maximum value θ_m, and for oscillations of small amplitude, you can express V_g as

$$V_g = \tfrac{1}{2}Wb\theta_m^2$$

Note that *if G is located above O* in its equilibrium position (instead of below O, as we have assumed), the vertical displacement y will be negative and should be approximated as $y = -\frac{1}{2}b\theta^2$, which will result in a negative value for V_g. In the absence of other forces, the equilibrium position will be unstable, and the system will not oscillate. (See, for instance, Prob. 19.91.)

b. *The potential energy associated with the elastic force exerted by a spring is $V_e = \frac{1}{2}kx^2$,* where k is the constant of the spring and x its deflection. In problems involving the rotation of a body about an axis, you will generally have $x = a\theta$, where a is the distance from the axis of rotation to the point of the body

(continued)

where the spring is attached, and where θ is the angle of rotation. Therefore, when x reaches its maximum value x_m and θ reaches its maximum value θ_m, you can express V_e as

$$V_e = \tfrac{1}{2}kx_m^2 = \tfrac{1}{2}ka^2\theta_m^2$$

c. The potential energy V_1 of the system in its position of maximum displacement is obtained by adding the various potential energies that you have computed. It will be equal to the product of a constant and θ_m^2.

2. *Calculate the kinetic energy T_2 of the system in its position of maximum velocity.* Note that this position is also the equilibrium position of the system.

a. If the system consists of a single rigid body, the kinetic energy T_2 of the system will be the sum of the kinetic energy associated with the motion of the mass center G of the body and the kinetic energy associated with the rotation of the body about G. You will write, therefore,

$$T_2 = \tfrac{1}{2}m\bar{v}_m^2 + \tfrac{1}{2}\bar{I}\omega_m^2$$

Assuming that the position of the body has been defined by an angle θ, express $\bar{v}_m$ and ω_m in terms of the rate of change $\dot{\theta}_m$ of θ as the body passes through its equilibrium position. The kinetic energy of the body will thus be expressed as the product of a constant and $\dot{\theta}_m^2$. Note that if θ measures the rotation of the body about its mass center, as was the case for the plate of Fig. 19.6, then $\omega_m = \dot{\theta}_m$. In other cases, however, the kinematics of the motion should be used to derive a relation between ω_m and $\dot{\theta}_m$ [Sample Prob. 19.4].

b. If the system consists of several rigid bodies, repeat the above computation for each of the bodies, using the same coordinate θ, and add the results obtained.

3. *Equate the potential energy V_1 of the system to its kinetic energy T_2,*

$$V_1 = T_2$$

and, recalling the first of Eqs. (19.15), replace $\dot{\theta}_m$ in the right-hand term by the product of the amplitude θ_m and the circular frequency ω_n. Since both terms now contain the factor θ_m^2, this factor can be canceled and the resulting equation can be solved for the circular frequency ω_n.

Problems

19.69 Two small spheres, A and C, each of mass m, are attached to rod AB, which is supported by a pin and bracket at B and by a spring CD of constant k. Knowing that the mass of the rod is negligible and that the system is in equilibrium when the rod is horizontal, determine the frequency of the small oscillations of the system.

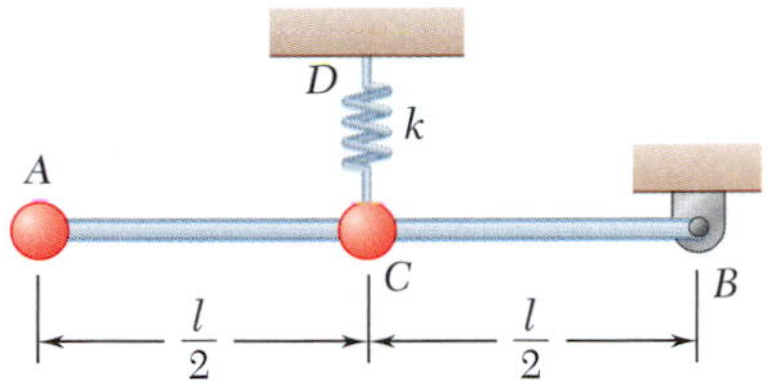

Fig. P19.69

19.70 A 0.7-kg sphere A and a 0.5-kg sphere C are attached to the ends of a rod AC of negligible mass which can rotate in a vertical plane about an axis at B. Determine the period of small oscillations of the rod.

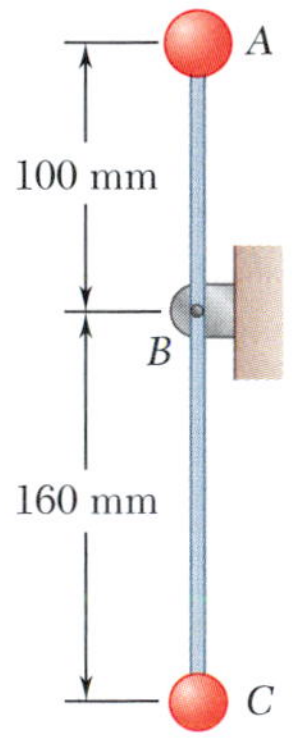

Fig. P19.70

19.71 Two blocks, each of weight 3 lb, are attached to links which are pin-connected to bar BC as shown. The weights of the links and bar are negligible, and the blocks can slide without friction. Block D is attached to a spring of constant $k = 4$ lb/in. Knowing that block A is moved 0.5 in. from its equilibrium position and released, determine the magnitude of the maximum velocity of block D during the resulting motion.

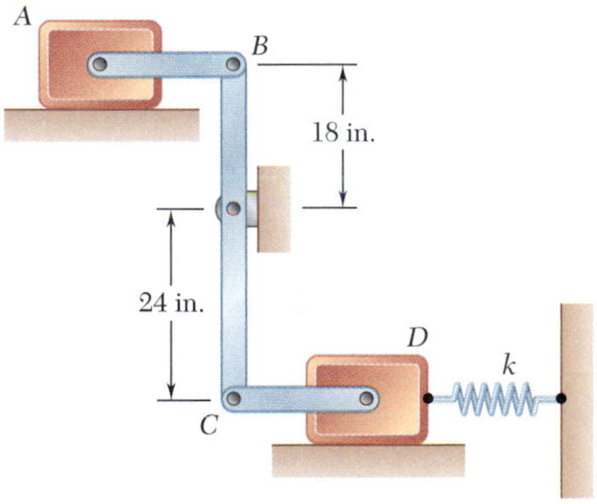

Fig. P19.71 and *P19.72*

19.72 Two blocks, each of weight 3 lb, are attached to links which are pin-connected to bar BC as shown. The weights of the links and bar are negligible, and the blocks can slide without friction. Block D is attached to a spring of constant $k = 4$ lb/in. Knowing that block A is at rest when it is struck horizontally with a mallet and given an initial velocity of 10 in./s, determine the magnitude of the maximum displacement of block D during the resulting motion.

19.73 A uniform rod AB can rotate in a vertical plane about a horizontal axis at C located at a distance c above the mass center G of the rod. For small oscillations determine the value of c for which the frequency of the motion will be maximum.

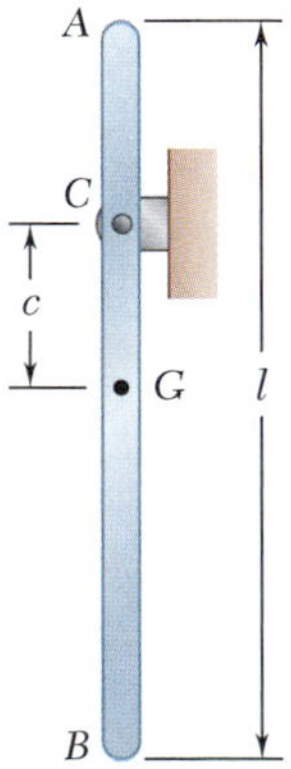

Fig. *P19.73*

19.74 A thin uniform plate cut into the shape of a quarter circle can rotate in a vertical plane about a horizontal axis at point O. Determine the period of small oscillations of the plate.

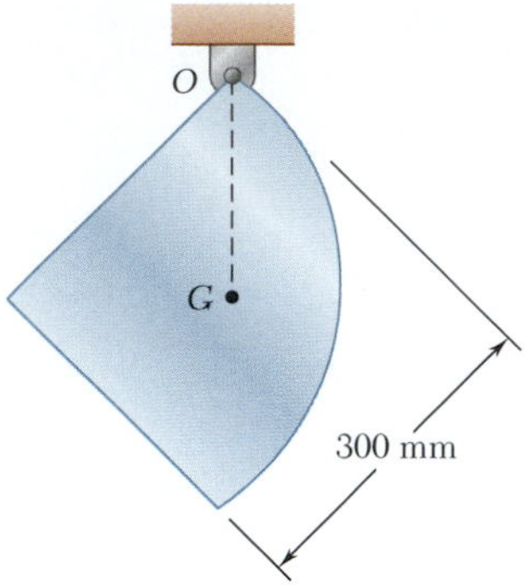

Fig. P19.74

19.75 The inner rim of a 38-kg flywheel is placed on a knife edge, and the period of its small oscillations is found to be 1.26 s. Determine the centroidal moment of inertia of the flywheel.

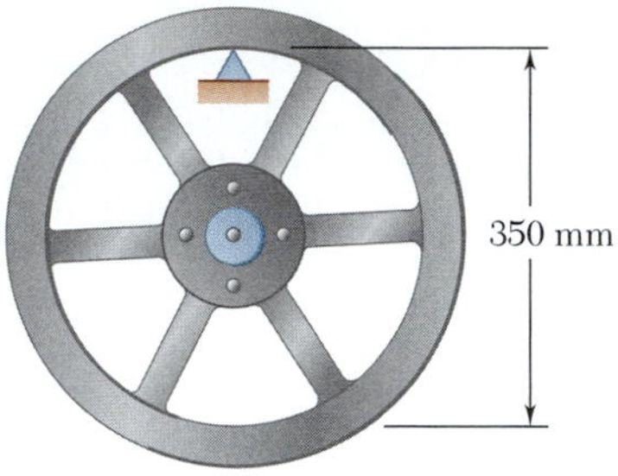

Fig. P19.75

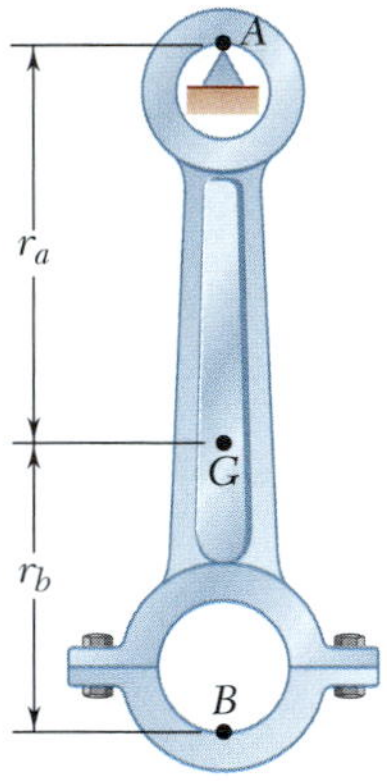

Fig. P19.76

19.76 A connecting rod is supported by a knife edge at point A; the period of its small oscillations is observed to be 0.895 s. The rod is then inverted and supported by a knife edge at point B and the period of its small oscillations is observed to be 0.805 s. Knowing that $r_a + r_b = 10.5$ in., determine (*a*) the location of the mass center G, (*b*) the centroidal radius of gyration $\bar{k}$.

19.77 A uniform disk of radius r and mass m can roll without slipping on a cylindrical surface and is attached to bar ABC of length L and negligible mass. The bar is attached to a spring of constant k and can rotate freely in the vertical plane about point B. Knowing that end A is given a small displacement and released, determine the frequency of the resulting oscillations in terms of m, L, k, and g.

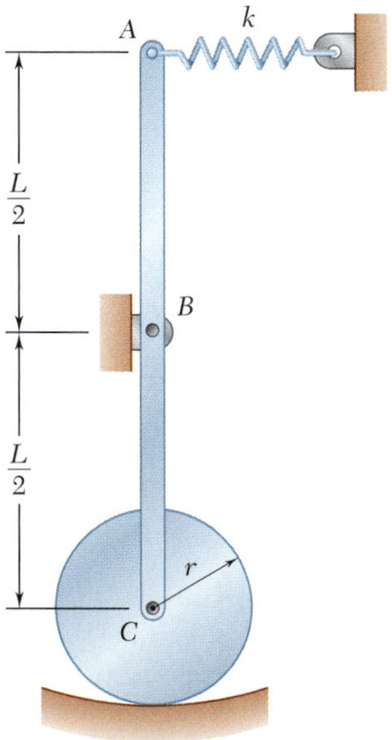

Fig. P19.77

19.78 A 7-kg uniform cylinder can roll without sliding on an incline and is attached to a spring AB as shown. If the center of the cylinder is moved 10 mm down the incline and released, determine (*a*) the period of vibration, (*b*) the maximum velocity of the center of the cylinder.

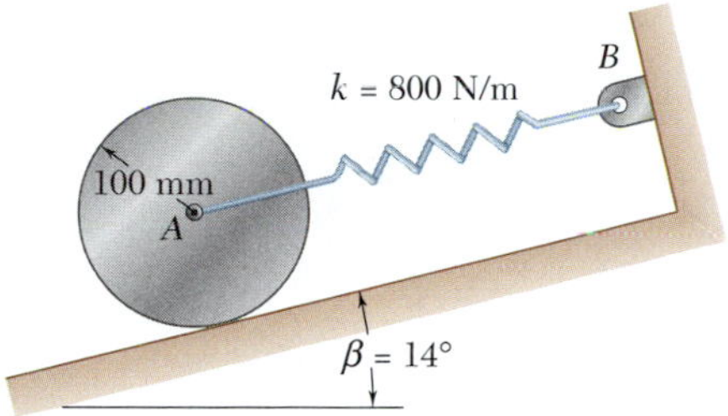

Fig. *P19.78*

19.79 Two uniform rods, each of mass $m = 0.6$ kg and length $l = 160$ mm, are welded together to form the assembly shown. Knowing that the constant of each spring is $k = 120$ N/m and that end A is given a small displacement and released, determine the frequency of the resulting motion.

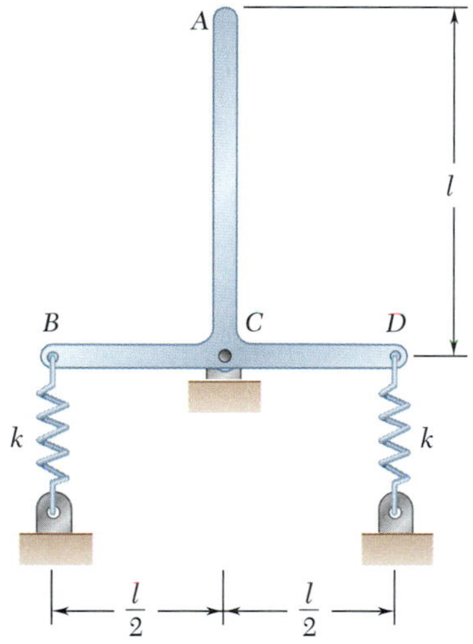

Fig. *P19.79*

19.80 A slender 20-lb bar AB of length $l = 2$ ft is connected to two collars of negligible mass. Collar A is attached to a spring of constant $k = 100$ lb/ft and can slide on a horizontal rod, while collar B can slide freely on a vertical rod. Knowing that the system is in equilibrium when bar AB is vertical and that collar A is given a small displacement and released, determine the period of the resulting vibrations.

19.81 A slender 10-lb bar AB of length $l = 2$ ft is connected to two collars, each of weight 5 lb. Collar A is attached to a spring of constant $k = 100$ lb/ft and can slide on a horizontal rod, while collar B can slide freely on a vertical rod. Knowing that the system is in equilibrium when bar AB is vertical and that collar A is given a small displacement and released, determine the period of the resulting vibrations.

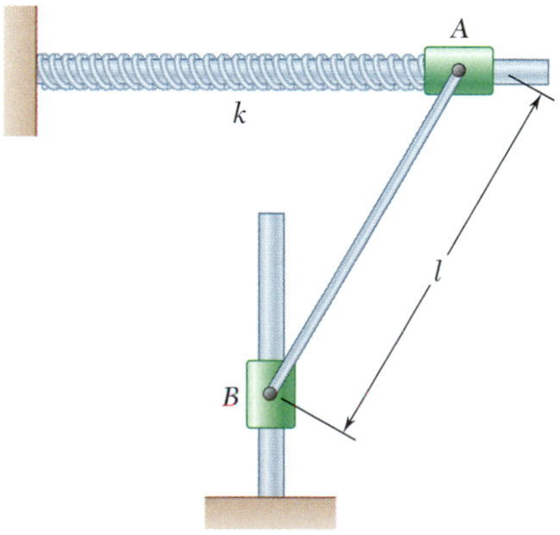

Fig. P19.80 and P19.81

19.82 A 6-lb slender rod AB is bolted to a 10-lb uniform disk. A spring of constant 25 lb/ft is attached to the disk and is unstretched in the position shown. If end B of the rod is given a small displacement and released, determine the period of vibration of the system.

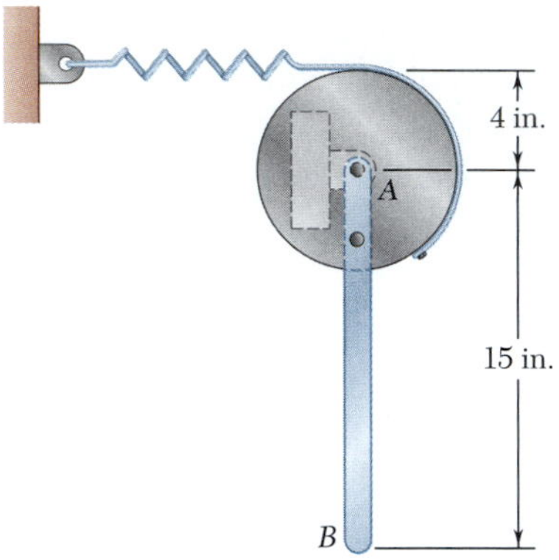

Fig. P19.82

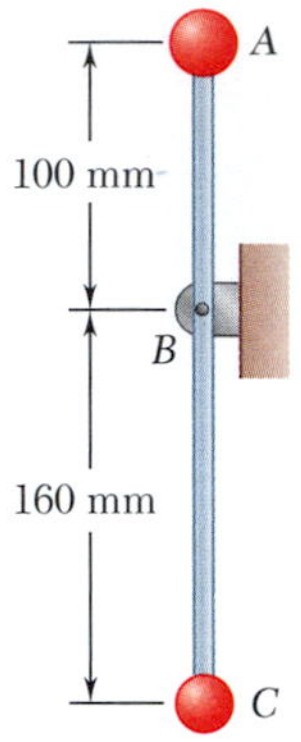

Fig. *P19.83*

19.83 A 0.7-kg sphere A and a 0.5-kg sphere C are attached to the ends of a 1-kg rod AC which can rotate in a vertical plane about an axis at B. Determine the period of small oscillations of the rod.

19.84 Spheres A and C, each of weight W, are attached to the ends of a homogeneous rod of the same weight W and of length $2l$ which is bent as shown. The system is allowed to oscillate about a frictionless pin at B. Knowing that $\beta = 40°$ and $l = 25$ in., determine the frequency of small oscillations.

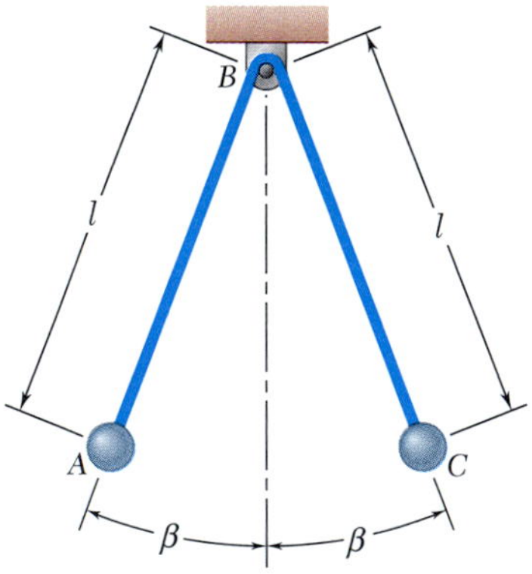

Fig. P19.84

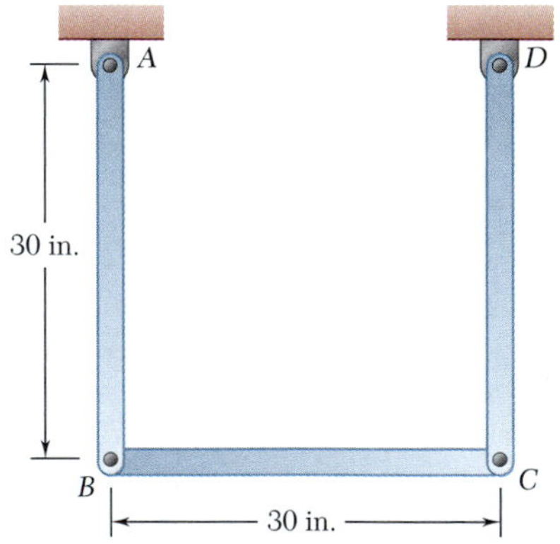

Fig. P19.85

19.85 Three identical 8-lb uniform slender bars are connected by pins as shown and can move in a vertical plane. Knowing that bar BC is given a small displacement and released, determine the period of vibration of the system.

19.86 and 19.87 Two uniform rods AB and CD, each of length l and mass m, are attached to gears as shown. Knowing that the mass of gear C is m and that the mass of gear A is $4m$, determine the period of small oscillations of the system.

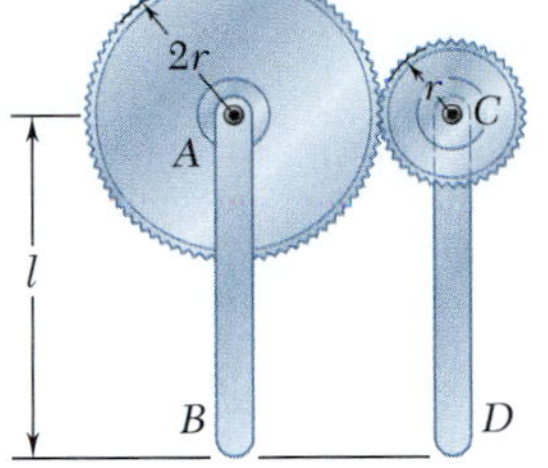

Fig. P19.86

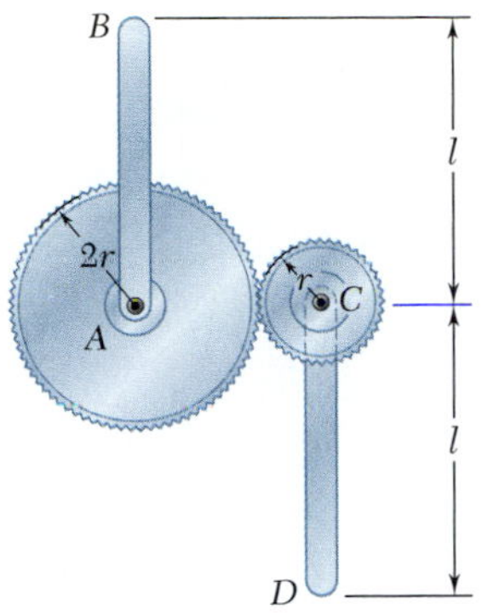

Fig. P19.87

19.88 A 5-kg uniform rod CD is welded at C to a shaft of negligible mass which is welded to the centers of two 10-kg uniform disks A and B. Knowing that the disks roll without sliding, determine the period of small oscillations of the system.

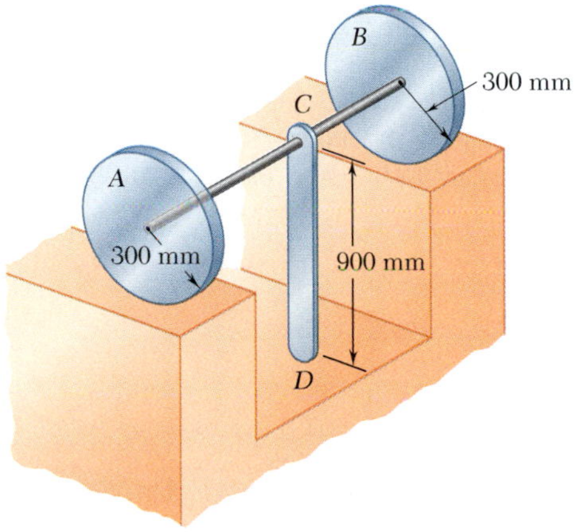

Fig. *P19.88*

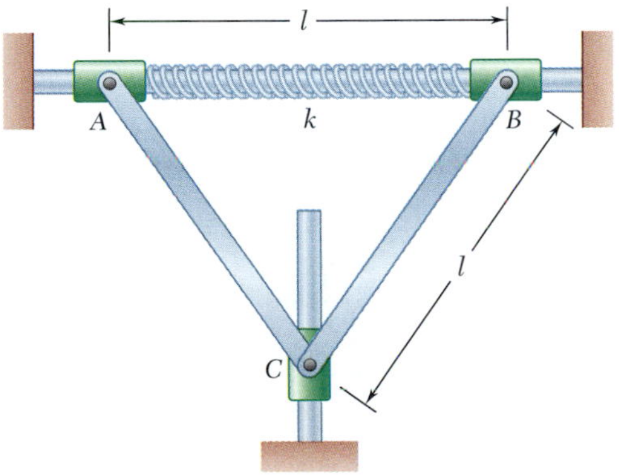

Fig. P19.89

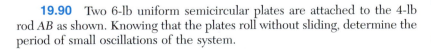

19.89 Three collars, each of mass m, are connected by pins to bars AC and BC, each of length l and negligible mass. Collars A and B can slide without friction on a horizontal rod and are connected by a spring of constant k. Collar C can slide without friction on a vertical rod, and the system is in equilibrium in the position shown. Knowing that collar C is given a small displacement and released, determine the frequency of the resulting motion of the system.

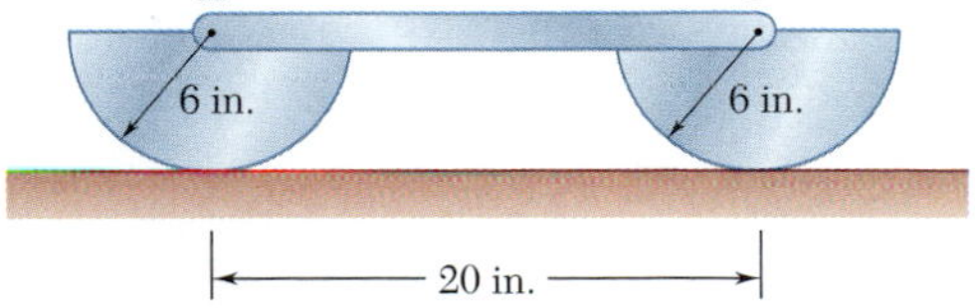

Fig. P19.90

19.90 Two 6-lb uniform semicircular plates are attached to the 4-lb rod AB as shown. Knowing that the plates roll without sliding, determine the period of small oscillations of the system.

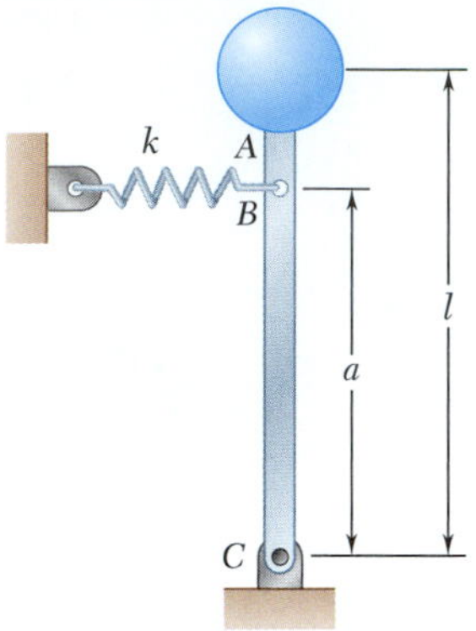

Fig. *P19.91* and P19.92

19.91 An inverted pendulum consisting of a sphere of weight W and a bar ABC of length l and negligible weight is supported by a pin and bracket at C. A spring of constant k is attached to the bar at B and is undeformed when the bar is in the vertical position shown. Determine (*a*) the frequency of small oscillations, (*b*) the smallest value of a for which the oscillations will occur.

19.92 For the inverted pendulum of Prob. 19.91 and for given values of k, a, and l, it is observed that $f = 1.5$ Hz when $m = 0.9$ kg and that $f = 0.8$ Hz when $m = 1.8$ kg. Determine the largest value of m for which small oscillations will occur.

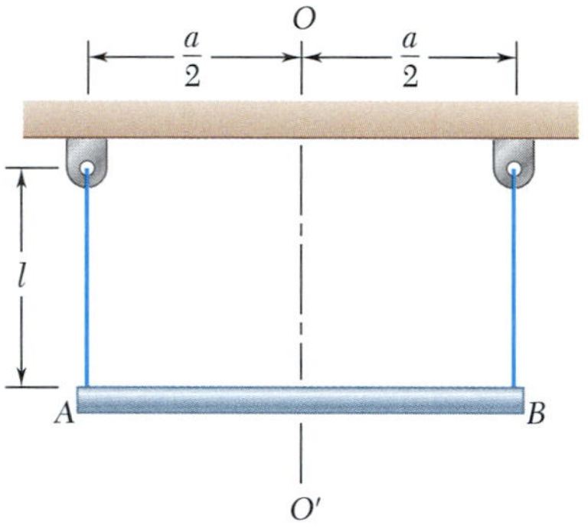

Fig. *P19.93*

19.93 A section of uniform pipe is suspended from two vertical cables attached at A and B. Determine the frequency of oscillation when the pipe is given a small rotation about the centroidal axis OO' and released.

19.94 A 4-lb uniform rod ABC is supported by a pin at B and is attached to a spring at C. It is connected at A to a 4-lb block DE which is attached to a spring and can roll without friction. Knowing that each spring can act in tension or compression, determine the frequency of small oscillations of the system when the rod is rotated through a small angle and released.

19.95 A uniform 3-kg disk can roll without slipping on a cylindrical surface and is attached to a 2-kg uniform slender bar AB. The bar is attached to a spring of constant $k = 300$ N/m and can rotate freely in the vertical plane about point A. Knowing that end B is given a small displacement and released, determine the period of the resulting vibration.

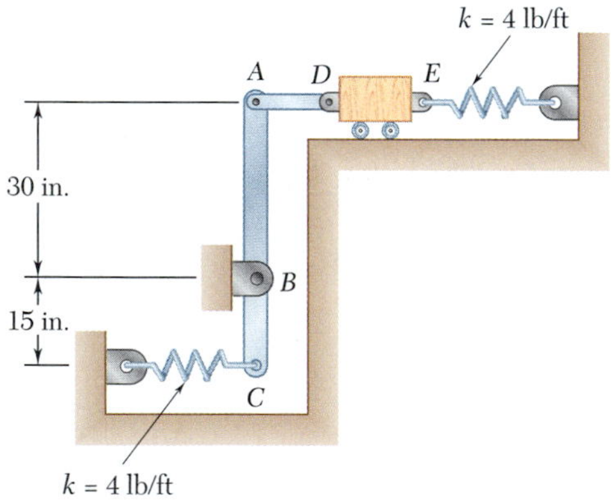

Fig. P19.94

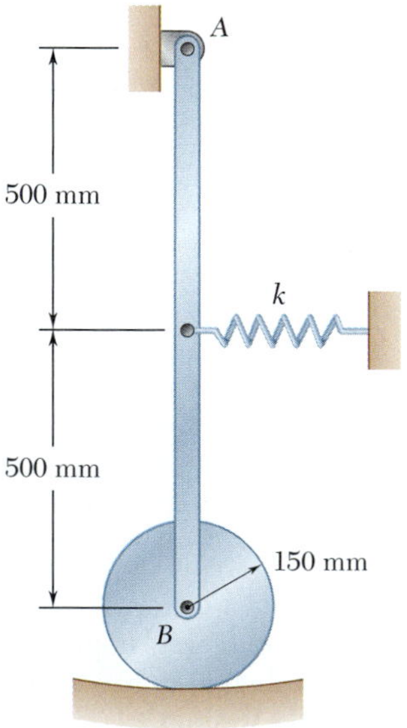

Fig. P19.95

***19.96** As a submerged body moves through a fluid, the particles of the fluid flow around the body and thus acquire kinetic energy. In the case of a sphere moving in an ideal fluid, the total kinetic energy acquired by the fluid is $\frac{1}{4}(\gamma/g)Vv^2$, where γ is the specific weight of the fluid, V is the volume of the sphere, and v is the velocity of the sphere. Consider a 1-lb hollow spherical shell of radius 4 in. which is held submerged in a tank of water by a spring of constant 40 lb/ft. (*a*) Neglecting fluid friction, determine the period of vibration of the shell when it is displaced vertically and then released. (*b*) Solve part *a*, assuming that the tank is accelerated upward at the constant rate of 24 ft/s^2.

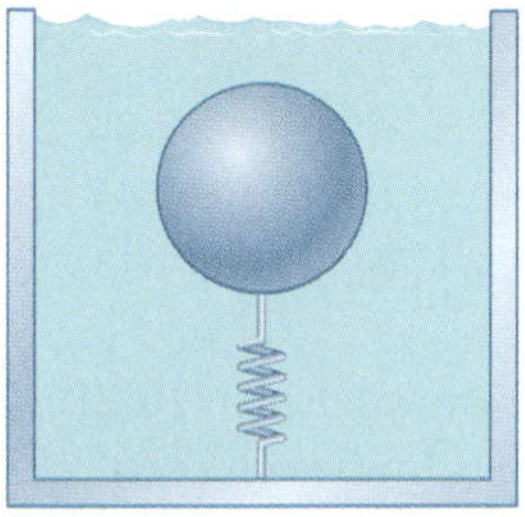

Fig. P19.96

***19.97** A thin plate of length l rests on a half cylinder of radius r. Derive an expression for the period of small oscillations of the plate.

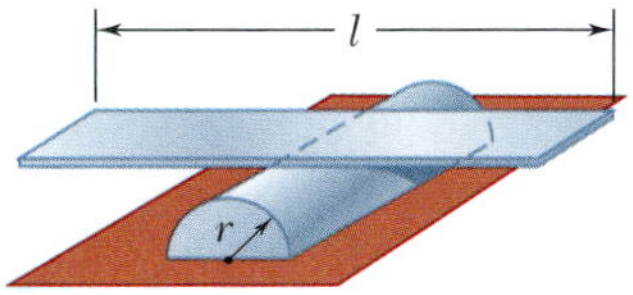

Fig. P19.97

19.7. FORCED VIBRATIONS

The most important vibrations from the point of view of engineering applications are the *forced vibrations* of a system. These vibrations occur when a system is subjected to a periodic force or when it is elastically connected to a support which has an alternating motion.

Consider first the case of a body of mass m suspended from a spring and subjected to a periodic force **P** of magnitude $P = P_m \sin \omega_f t$, where ω_f is the circular frequency of **P** and is referred to as the *forced circular frequency* of the motion (Fig. 19.7). This force may be an actual external force applied to the body, or it may be a centrifugal force produced by the rotation of some unbalanced part of the body (see Sample Prob. 19.5). Denoting by x the displacement of the body measured from its equilibrium position, we write the equation of motion,

$$+\downarrow \Sigma F = ma: \qquad P_m \sin \omega_f t + W - k(\delta_{st} + x) = m\ddot{x}$$

Recalling that $W = k\delta_{st}$, we have

$$m\ddot{x} + kx = P_m \sin \omega_f t \qquad (19.30)$$

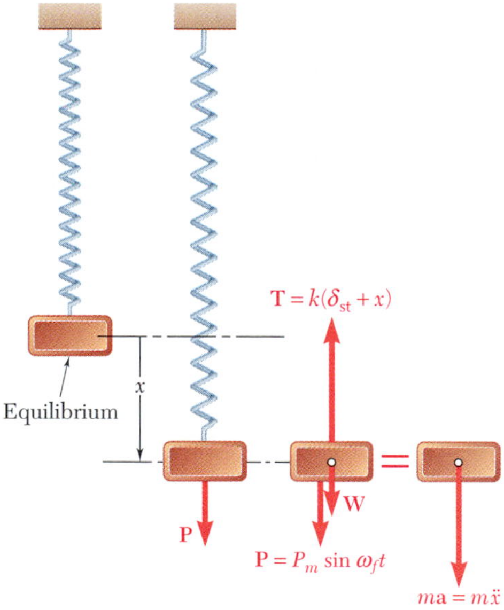

Fig. 19.7

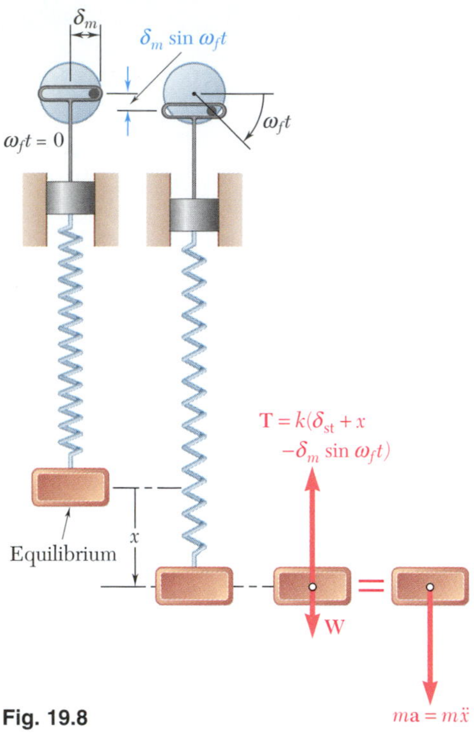

Fig. 19.8

Next we consider the case of a body of mass m suspended from a spring attached to a moving support whose displacement δ is equal to $\delta_m \sin \omega_f t$ (Fig. 19.8). Measuring the displacement x of the body from the position of static equilibrium corresponding to $\omega_f t = 0$, we find that the total elongation of the spring at time t is $\delta_{st} + x - \delta_m \sin \omega_f t$. The equation of motion is thus

$$+\downarrow \Sigma F = ma: \qquad W - k(\delta_{st} + x - \delta_m \sin \omega_f t) = m\ddot{x}$$

Recalling that $W = k\delta_{st}$, we have

$$m\ddot{x} + kx = k\delta_m \sin \omega_f t \tag{19.31}$$

We note that Eqs. (19.30) and (19.31) are of the same form and that a solution of the first equation will satisfy the second if we set $P_m = k\delta_m$.

A differential equation such as (19.30) or (19.31), possessing a right-hand member different from zero, is said to be *nonhomogeneous*. Its general solution is obtained by adding a particular solution of the given equation to the general solution of the corresponding *homogeneous* equation (with right-hand member equal to zero). A *particular solution* of (19.30) or (19.31) can be obtained by trying a solution of the form

$$x_{part} = x_m \sin \omega_f t \tag{19.32}$$

Substituting x_{part} for x into (19.30), we find

$$-m\omega_f^2 x_m \sin \omega_f t + kx_m \sin \omega_f t = P_m \sin \omega_f t$$

which can be solved for the amplitude,

$$x_m = \frac{P_m}{k - m\omega_f^2}$$

Recalling from (19.4) that $k/m = \omega_n^2$, where ω_n is the natural circular frequency of the system, we write

$$x_m = \frac{P_m/k}{1 - (\omega_f/\omega_n)^2} \tag{19.33}$$

Substituting from (19.32) into (19.31), we obtain in a similar way

$$x_m = \frac{\delta_m}{1 - (\omega_f/\omega_n)^2} \tag{19.33'}$$

The homogeneous equation corresponding to (19.30) or (19.31) is Eq. (19.2), which defines the free vibration of the body. Its general solution, called the *complementary function*, was found in Sec. 19.2:

$$x_{comp} = C_1 \sin \omega_n t + C_2 \cos \omega_n t \tag{19.34}$$

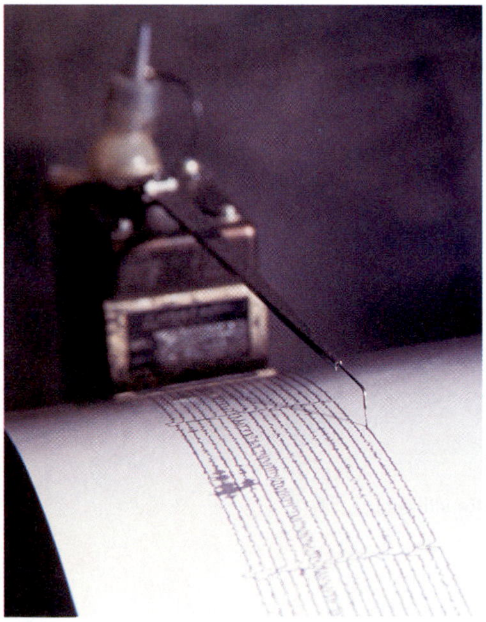

Photo 19.1 A seismometer operates by measuring the amount of electrical energy needed to keep a mass centered in the housing in the presence of strong ground shaking.

Adding the particular solution (19.32) to the complementary function (19.34), we obtain the *general solution* of Eqs. (19.30) and (19.31):

$$x = C_1 \sin \omega_n t + C_2 \cos \omega_n t + x_m \sin \omega_f t \qquad (19.35)$$

We note that the vibration obtained consists of two superposed vibrations. The first two terms in Eq. (19.35) represent a free vibration of the system. The frequency of this vibration is the *natural frequency* of the system, which depends only upon the constant k of the spring and the mass m of the body, and the constants C_1 and C_2 can be determined from the initial conditions. This free vibration is also called a *transient* vibration, since in actual practice it will soon be damped out by friction forces (Sec. 19.9).

The last term in (19.35) represents the *steady-state* vibration produced and maintained by the impressed force or impressed support movement. Its frequency is the *forced frequency* imposed by this force or movement, and its amplitude x_m, defined by (19.33) or (19.33′), depends upon the *frequency ratio* ω_f/ω_n. The ratio of the amplitude x_m of the steady-state vibration to the static deflection P_m/k caused by a force P_m, or to the amplitude δ_m of the support movement, is called the *magnification factor*. From (19.33) and (19.33′), we obtain

$$\text{Magnification factor} = \frac{x_m}{P_m/k} = \frac{x_m}{\delta_m} = \frac{1}{1 - (\omega_f/\omega_n)^2} \qquad (19.36)$$

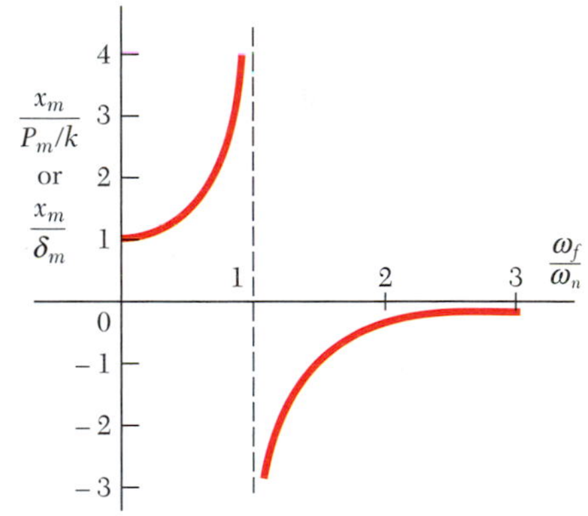

Fig. 19.9

The magnification factor has been plotted in Fig. 19.9 against the frequency ratio ω_f/ω_n. We note that when $\omega_f = \omega_n$, the amplitude of the forced vibration becomes infinite. The impressed force or impressed support movement is said to be in *resonance* with the given system. Actually, the amplitude of the vibration remains finite because of damping forces (Sec. 19.9); nevertheless, such a situation should be avoided, and the forced frequency should not be chosen too close to the natural frequency of the system. We also note that for $\omega_f < \omega_n$ the coefficient of $\sin \omega_f t$ in (19.35) is positive, while for $\omega_f > \omega_n$ this coefficient is negative. In the first case the forced vibration is *in phase* with the impressed force or impressed support movement, while in the second case it is 180° *out of phase*.

Finally, let us observe that the velocity and the acceleration in the steady-state vibration can be obtained by differentiating twice with respect to t the last term of Eq. (19.35). Their maximum values are given by expressions similar to those of Eqs. (19.15) of Sec. 19.2, except that these expressions now involve the amplitude and the circular frequency of the forced vibration:

$$v_m = x_m \omega_f \qquad a_m = x_m \omega_f^2 \qquad (19.37)$$

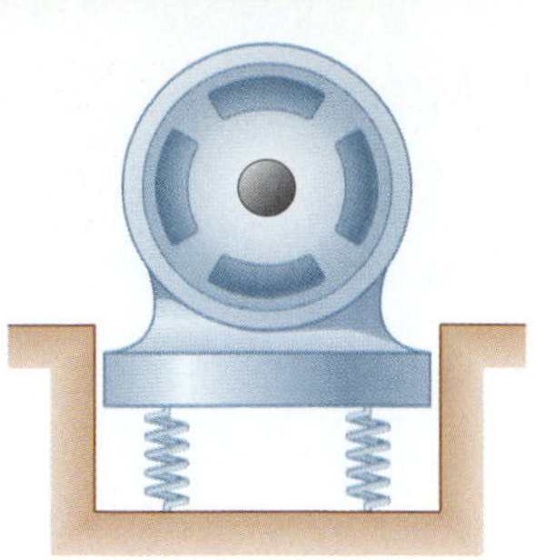

SAMPLE PROBLEM 19.5

A motor weighing 350 lb is supported by four springs, each having a constant of 750 lb/in. The unbalance of the rotor is equivalent to a weight of 1 oz located 6 in. from the axis of rotation. Knowing that the motor is constrained to move vertically, determine (*a*) the speed in rpm at which resonance will occur, (*b*) the amplitude of the vibration of the motor at a speed of 1200 rpm.

SOLUTION

***a*. Resonance Speed.** The resonance speed is equal to the natural circular frequency ω_n (in rpm) of the free vibration of the motor. The mass of the motor and the equivalent constant of the supporting springs are

$$m = \frac{350 \text{ lb}}{32.2 \text{ ft/s}^2} = 10.87 \text{ lb} \cdot \text{s}^2\text{/ft}$$

$$k = 4(750 \text{ lb/in.}) = 3000 \text{ lb/in.} = 36{,}000 \text{ lb/ft}$$

$$\omega_n = \sqrt{\frac{k}{m}} = \sqrt{\frac{36{,}000}{10.87}} = 57.5 \text{ rad/s} = 549 \text{ rpm}$$

Resonance speed = 549 rpm ◀

***b*. Amplitude of Vibration at 1200 rpm.** The angular velocity of the motor and the mass of the equivalent 1-oz weight are

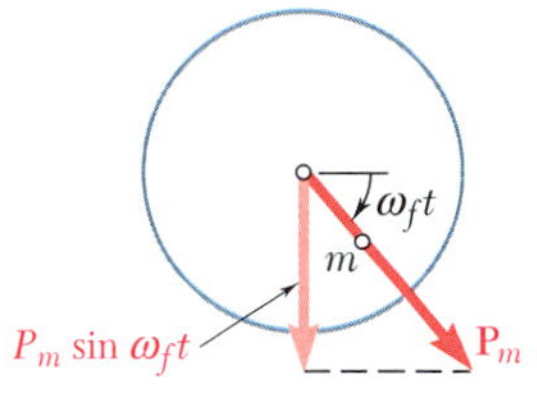

$$\omega = 1200 \text{ rpm} = 125.7 \text{ rad/s}$$

$$m = (1 \text{ oz})\frac{1 \text{ lb}}{16 \text{ oz}}\frac{1}{32.2 \text{ ft/s}^2} = 0.001941 \text{ lb} \cdot \text{s}^2\text{/ft}$$

The magnitude of the centrifugal force due to the unbalance of the rotor is

$$P_m = ma_n = mr\omega^2 = (0.001941 \text{ lb} \cdot \text{s}^2\text{/ft})(\tfrac{6}{12} \text{ ft})(125.7 \text{ rad/s})^2 = 15.33 \text{ lb}$$

The static deflection that would be caused by a constant load P_m is

$$\frac{P_m}{k} = \frac{15.33 \text{ lb}}{3000 \text{ lb/in.}} = 0.00511 \text{ in.}$$

The forced circular frequency ω_f of the motion is the angular velocity of the motor,

$$\omega_f = \omega = 125.7 \text{ rad/s}$$

Substituting the values of P_m/k, ω_f, and ω_n into Eq. (19.33), we obtain

$$x_m = \frac{P_m/k}{1 - (\omega_f/\omega_n)^2} = \frac{0.00511 \text{ in.}}{1 - (125.7/57.5)^2} = -0.001352 \text{ in.}$$

$x_m = 0.001352$ in. (out of phase) ◀

Note. Since $\omega_f > \omega_n$, the vibration is 180° out of phase with the centrifugal force due to the unbalance of the rotor. For example, when the unbalanced mass is directly below the axis of rotation, the position of the motor is $x_m = 0.001352$ in. above the position of equilibrium.

SOLVING PROBLEMS ON YOUR OWN

This lesson was devoted to the analysis of the *forced vibrations* of a mechanical system. These vibrations occur either when the system is subjected to a periodic force **P** (Fig. 19.7), or when it is elastically connected to a support which has an alternating motion (Fig. 19.8). In the first case, the motion of the system is defined by the differential equation

$$m\ddot{x} + kx = P_m \sin \omega_f t \tag{19.30}$$

where the right-hand member represents the magnitude of the force **P** at a given instant. In the second case, the motion is defined by the differential equation

$$m\ddot{x} + kx = k\delta_m \sin \omega_f t \tag{19.31}$$

where the right-hand member is the product of the spring constant k and the displacement of the support at a given instant. You will be concerned only with the *steady-state* motion of the system, which is defined by a *particular solution* of these equations, of the form

$$x_{\text{part}} = x_m \sin \omega_f t \tag{19.32}$$

***1. If the forced vibration is caused by a periodic force* P,** of amplitude P_m and circular frequency ω_f, the amplitude of the vibration is

$$x_m = \frac{P_m/k}{1 - (\omega_f/\omega_n)^2} \tag{19.33}$$

where ω_n is the *natural circular frequency* of the system $\omega_n = \sqrt{k/m}$, and k is the spring constant. Note that the circular frequency of the vibration is ω_f and that the amplitude x_m does not depend upon the initial conditions. For $\omega_f = \omega_n$, the denominator in Eq. (19.33) is zero and x_m is infinite (Fig. 19.9); the impressed force **P** is said to be in *resonance* with the system. Also, for $\omega_f < \omega_n$, x_m is positive and the vibration is *in phase* with **P**, while, for $\omega_f > \omega_n$, x_m is negative and the vibration is *out of phase.*

a. In the problems which follow, you may be asked to determine one of the parameters in Eq. (19.33) when the others are known. We suggest that you keep Fig. 19.9 in front of you when solving these problems. For example, if you are asked to find the frequency at which the amplitude of a forced vibration has a given value, but you do not know whether the vibration is in or out of phase with respect to the impressed force, you should note from Fig. 19.9 that there can be two frequencies satisfying this requirement, one corresponding to a positive value of x_m and to a vibration in phase with the impressed force, and the other corresponding to a negative value of x_m and to a vibration out of phase with the impressed force.

(continued)

b. Once you have obtained the amplitude x_m of the motion of a component of the system from Eq. (19.33), you can use Eqs. (19.37) to determine the maximum values of the velocity and acceleration of that component:

$$v_m = x_m \omega_f \qquad a_m = x_m \omega_f^2 \tag{19.37}$$

c. When the impressed force P is due to the unbalance of the rotor of a motor, its maximum value is $P_m = mr\omega_f^2$, where m is the mass of the rotor, r is the distance between its mass center and the axis of rotation, and ω_f is equal to the angular velocity ω of the rotor expressed in rad/s [Sample Prob. 19.5].

2. *If the forced vibration is caused by the simple harmonic motion of a support,* of amplitude δ_m and circular frequency ω_f, the amplitude of the vibration is

$$x_m = \frac{\delta_m}{1 - (\omega_f/\omega_n)^2} \tag{19.33'}$$

where ω_n is the *natural circular frequency* of the system, $\omega_n = \sqrt{k/m}$. Again, note that the circular frequency of the vibration is ω_f and that the amplitude x_m does not depend upon the initial conditions.

a. Be sure to read our comments in paragraphs 1, 1a, and 1b, since they apply equally well to a vibration caused by the motion of a support.

b. If the maximum acceleration a_m of the support is specified, rather than its maximum displacement δ_m, remember that, since the motion of the support is a simple harmonic motion, you can use the relation $a_m = \delta_m \omega_f^2$ to determine δ_m; the value obtained is then substituted into Eq. (19.33′).

Problems

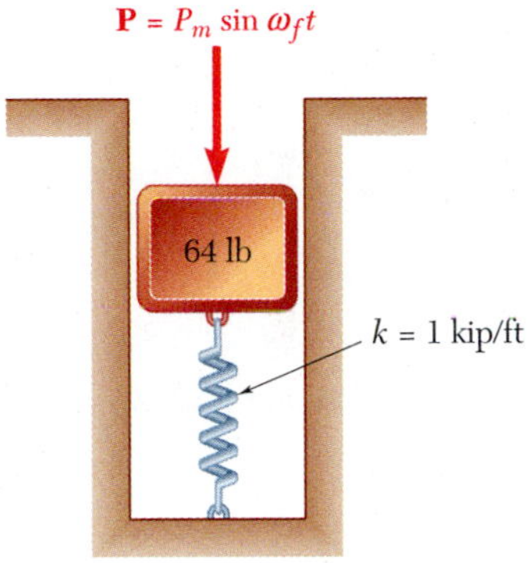

Fig. P19.98

19.98 A 64-lb block is attached to a spring of constant $k = 1$ kip/ft and can move without friction in a vertical slot as shown. It is acted upon by a periodic force of magnitude $P = P_m \sin \omega_f t$, where $\omega_f = 10$ rad/s. Knowing that the amplitude of the motion is 0.75 in., determine P_m.

19.99 A 4-kg collar can slide on a frictionless horizontal rod and is attached to a spring of constant 450 N/m. It is acted upon by a periodic force of magnitude $P = P_m \sin \omega_f t$, where $P_m = 13$ N. Determine the amplitude of the motion of the collar if (*a*) $\omega_f = 5$ rad/s, (*b*) $\omega_f = 10$ rad/s.

19.100 A 4-kg collar can slide on a frictionless horizontal rod and is attached to a spring of constant k. It is acted upon by a periodic force of magnitude $P = P_m \sin \omega_f t$, where $P_m = 9$ N and $\omega_f = 5$ rad/s. Determine the value of the spring constant k knowing that the motion of the collar has an amplitude of 150 mm and is (*a*) in phase with the applied force, (*b*) out of phase with the applied force.

19.101 A collar of mass m which slides on a frictionless horizontal rod is attached to a spring of constant k and is acted upon by a periodic force of magnitude $P = P_m \sin \omega_f t$. Determine the range of values of ω_f for which the amplitude of the vibration exceeds three times the static deflection caused by a constant force of magnitude P_m.

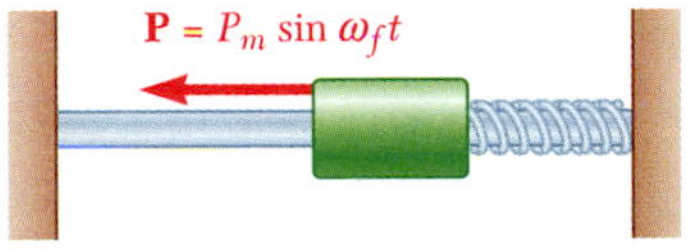

Fig. P19.99, *P19.100*, and P19.101

19.102 A small 20-kg block A is attached to the rod BC of negligible mass which is supported at B by a pin and bracket and at C by a spring of constant $k = 2$ kN/m. The system can move in a vertical plane and is in equilibrium when the rod is horizontal. The rod is acted upon at C by a periodic force **P** of magnitude $P = P_m \sin \omega_f t$, where $P_m = 6$ N. Knowing that $b = 200$ mm, determine the range of values of ω_f for which the amplitude of vibration of block A exceeds 3.5 mm.

19.103 A small 20-kg block A is attached to the rod BC of negligible mass which is supported at B by a pin and bracket and at C by a spring of constant $k = 2$ kN/m. The system can move in a vertical plane and is in equilibrium when the rod is horizontal. The rod is acted upon at C by a periodic force **P** of magnitude $P = P_m \sin \omega_f t$, where $P_m = 6$ N. Knowing that $\omega_f = 15$ rad/s, determine the range of values of b for which the amplitude of vibration of block A exceeds 3.5 mm.

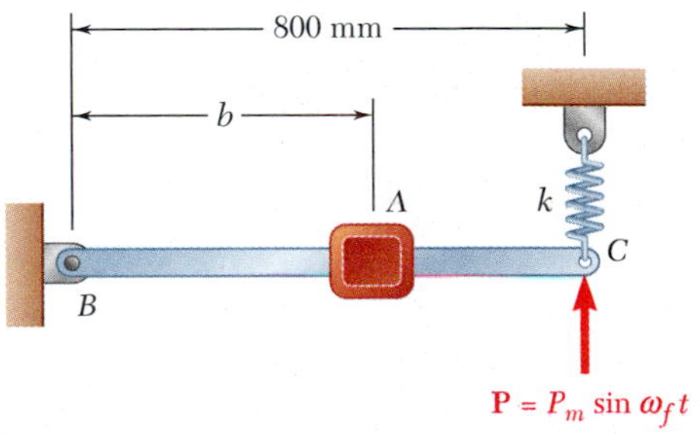

Fig. P19.102 and *P19.103*

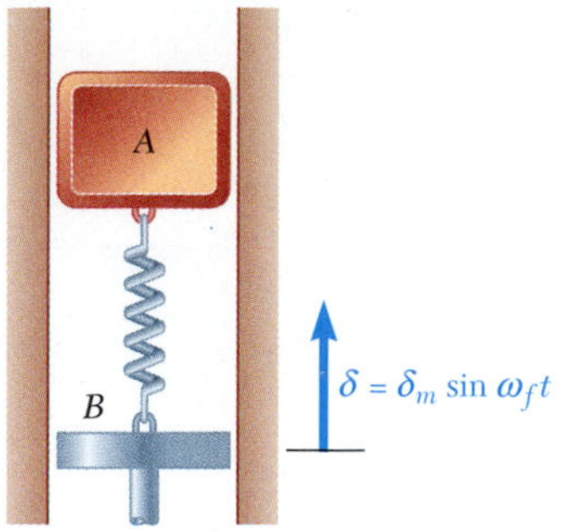

Fig. P19.104 and *P19.105*

19.104 A 2-kg block A slides in a vertical frictionless slot and is connected to a moving support B by means of a spring AB of constant $k =$ 117 N/m. Knowing that the displacement of the support is $\delta = \delta_m \sin \omega_f t$, where $\delta_m = 100$ mm and $\omega_f = 5$ rad/s, determine (*a*) the amplitude of the motion of the block, (*b*) the amplitude of the fluctuating force exerted by the spring on the block.

19.105 A 16-lb block A slides in a vertical frictionless slot and is connected to a moving support B by means of a spring AB of constant $k =$ 130 lb/ft. Knowing that the displacement of the support is $\delta = \delta_m \sin \omega_f t$, where $\delta_m = 6$ in., determine the range of values of ω_f for which the amplitude of the fluctuating force exerted by the spring on the block is less than 30 lb.

19.106 A cantilever beam AB supports a block which causes a static deflection of 40 mm at B. Assuming that the support at A undergoes a vertical periodic displacement $\delta = \delta_m \sin \omega_f t$, where $\delta_m = 10$ mm, determine the range of values of ω_f for which the amplitude of the motion of the block will be less than 20 mm. Neglect the mass of the beam and assume that the block does not leave the beam.

Fig. P19.106

19.107 A small 4-lb sphere B is attached to the bar AB of negligible weight which is supported at A by a pin and bracket and connected at C to a moving support D by means of a spring of constant $k = 20$ lb/in. Knowing that support D undergoes a vertical displacement $\delta = \delta_m \sin \omega_f t$, where $\delta_m =$ 0.1 in. and $\omega_f = 15$ rad/s, determine (*a*) the magnitude of the maximum angular velocity of bar AB, (*b*) the magnitude of the maximum acceleration of sphere B.

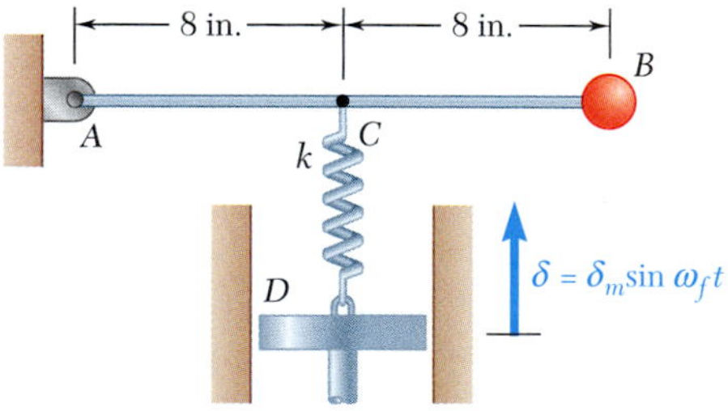

Fig. P19.107

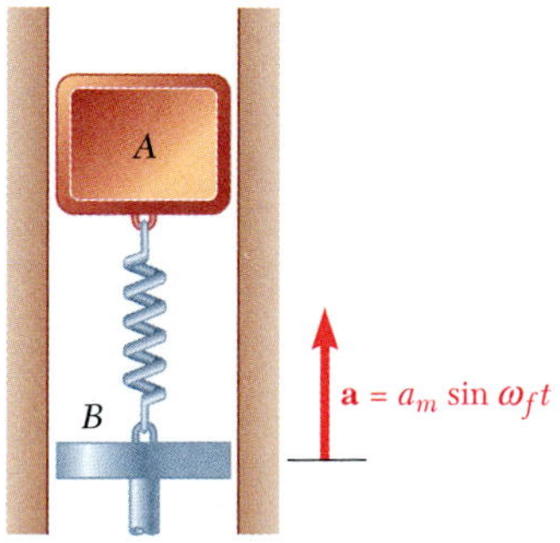

Fig. *P19.108*

19.108 A 16-lb block A slides in a vertical frictionless slot and is connected to a moving fixed support B by means of a spring of constant $k =$ 10 lb/ft. Knowing that the acceleration of the support is $a = a_m \sin \omega_f t$, where $a_m = 4.5$ ft/s^2 and $\omega_f = 5$ rad/s, determine (*a*) the maximum displacement of block A, (*b*) the amplitude of the fluctuating force exerted by the spring on the block.

19.109 A 360-g ball is connected to a paddle by means of an elastic cord AB of constant $k = 70$ N/m. Knowing that the paddle is moved vertically according to the relation $\delta = \delta_m \sin \omega_f t$, with an amplitude $\delta_m = 200$ mm and a frequency $f_f = 0.5$ Hz, determine the steady-state amplitude of the motion of the ball.

19.110 A 360-g ball is connected to a paddle by means of an elastic cord AB of constant $k = 70$ N/m. Knowing that the paddle is moved vertically according to the relation $\delta = \delta_m \sin \omega_f t$, where $\delta_m = 200$ mm, determine the maximum allowable circular frequency ω_f if the cord is not to become slack.

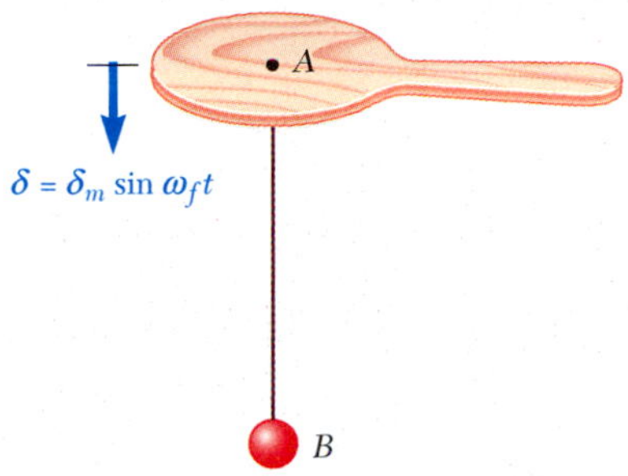

Fig. P19.109 and P19.110

19.111 The 2.4-lb bob of a simple pendulum of length $l = 30$ in. is suspended from a 2.8-lb collar C. The collar is forced to move according to the relation $x_C = \delta_m \sin \omega_f t$, with an amplitude $\delta_m = 0.5$ in. and a frequency $f_f = 0.5$ Hz. Determine (*a*) the amplitude of the motion of the bob, (*b*) the force that must be applied to collar C to maintain the motion.

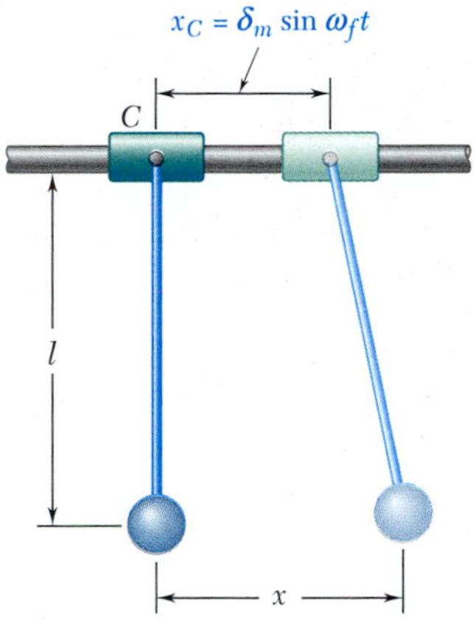

Fig. P19.111

19.112 A motor of mass M is supported by springs with an equivalent spring constant k. The unbalance of its rotor is equivalent to a mass m located at a distance r from the axis of rotation. Show that when the angular velocity of the rotor is ω_f, the amplitude x_m of the motion of the motor is

$$x_m = \frac{r(m/M)(\omega_f/\omega_n)^2}{1 - (\omega_f/\omega_n)^2}$$

where $\omega_n = \sqrt{k/M}$.

19.113 As the rotational speed of a spring-supported 200-lb motor is increased, the amplitude of the vibration due to the unbalance of its 30-lb rotor first increases and then decreases. It is observed that as very high speeds are reached, the amplitude of the vibration approaches 0.165 in. Determine the distance between the mass center of the rotor and its axis of rotation. (*Hint:* Use the formula derived in Prob. 19.112.)

19.114 As the rotational speed of a spring-supported motor is slowly increased from 200 to 300 rpm, the amplitude of the vibration due to the unbalance of its rotor is observed to increase continuously from 0.125 to 0.4 in. Determine the speed at which resonance will occur.

19.115 A 200-kg motor is supported by springs having a total constant of 215 kN/m. The unbalance of the rotor is equivalent to a 30-g mass located 200 mm from the axis of rotation. Determine the range of allowable values of the motor speed if the amplitude of the vibration is not to exceed 1.5 mm.

19.116 A 100-kg motor is bolted to a light horizontal beam. The unbalance of its rotor is equivalent to a 57-g mass located 100 mm from the axis of rotation. Knowing that resonance occurs at a motor speed of 400 rpm, determine the amplitude of the steady-state vibration at (*a*) 800 rpm, (*b*) 200 rpm, (*c*) 425 rpm.

Fig. P19.116 and P19.117

19.117 A 360-lb motor is bolted to a light horizontal beam. The unbalance of its rotor is equivalent to a 0.9-oz weight located 7.5 in. from the axis of rotation, and the static deflection of the beam due to the weight of the motor is 0.6 in. The amplitude of the vibration due to the unbalance can be decreased by adding a plate to the base of the motor. If the amplitude of vibration is to be less than 2.16×10^{-3} in. for motor speeds above 300 rpm, determine the required weight of the plate.

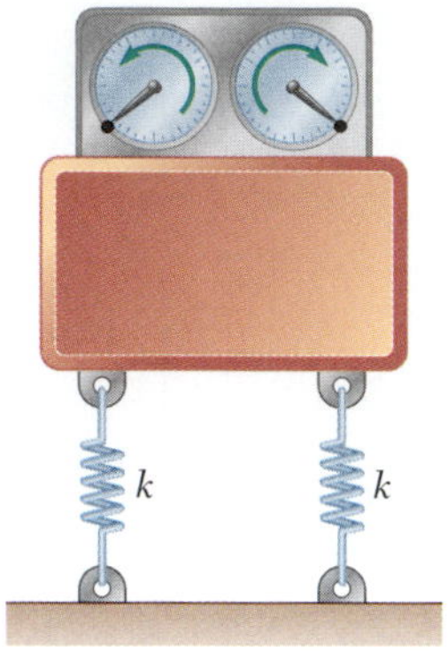

Fig. P19.118

19.118 A counter-rotating eccentric mass exciter consisting of two rotating 100-g masses describing circles of radius r at the same speed but in opposite senses is placed on a machine element to induce a steady-state vibration of the element. The total mass of the system is 300 kg, the constant of each spring is $k = 600$ kN/m, and the rotational speed of the exciter is 1200 rpm. Knowing that the amplitude of the total fluctuating force exerted on the foundation is 160 N, determine the radius r.

19.119 A 360-lb motor is supported by springs of total constant 12.5 kips/ft. The unbalance of the rotor is equivalent to a 0.9-oz weight located 7.5 in. from the axis of rotation. Determine the range of speeds of the motor for which the amplitude of the fluctuating force exerted on the foundation is less than 5 lb.

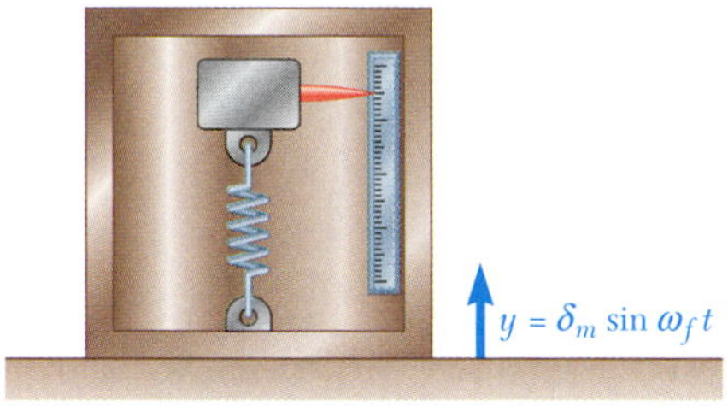

Fig. P19.120 and *P19.121*

19.120 A vibrometer used to measure the amplitude of vibrations consists of a box containing a mass-spring system with a known natural frequency of 150 Hz. The box is rigidly attached to a surface which is moving according to the equation $y = \delta_m \sin \omega_f t$. If the amplitude z_m of the motion of the mass relative to the box is used as a measure of the amplitude δ_m of the vibration of the surface, determine (*a*) the percent error when the frequency of the vibration is 750 Hz, (*b*) the frequency at which the error is zero.

19.121 A certain accelerometer consists essentially of a box containing a mass-spring system with a known natural frequency of 1760 Hz. The box is rigidly attached to a surface which is moving according to the equation $y = \delta_m \sin \omega_f t$. If the amplitude z_m of the motion of the mass relative to the box times a scale factor ω_n^2 is used as a measure of the maximum acceleration $a_m = \delta_m \omega_f^2$ of the vibrating surface, determine the percent error when the frequency of the vibration is 480 Hz.

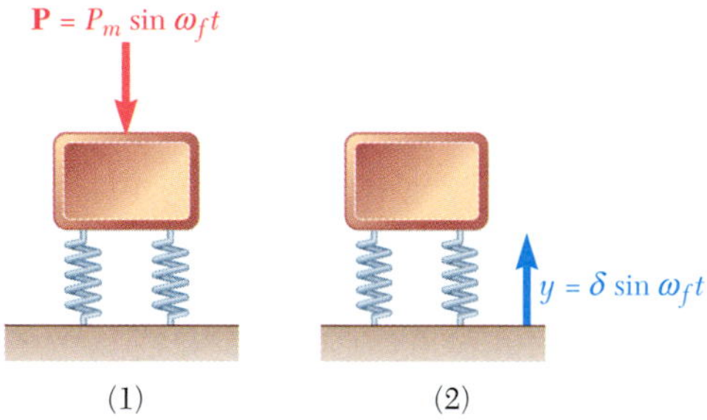

Fig. *P19.122*

19.122 Figures (1) and (2) show how springs can be used to support a block in two different situations. In Fig. (1) they help decrease the amplitude of the fluctuating force transmitted by the block to the foundation. In Fig. (2) they help decrease the amplitude of the fluctuating displacement transmitted by the foundation to the block. The ratio of the transmitted force to the impressed force or the ratio of the transmitted displacement to the impressed displacement is called the *transmissibility*. Derive an equation for the transmissibility for each situation. Give your answer in terms of the ratio ω_f/ω_n of the frequency ω_f of the impressed force or impressed displacement to the natural frequency ω_n of the spring-mass system. Show that in order to cause any reduction in transmissibility, the ratio ω_f/ω_n must be greater than $\sqrt{2}$.

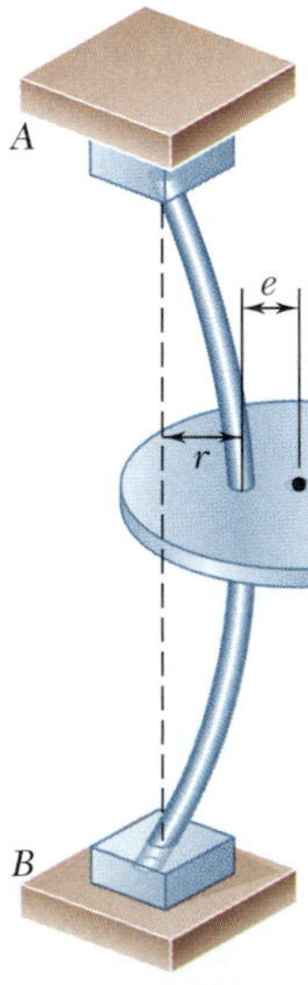

Fig. P19.123

19.123 A 27-kg disk is attached with an eccentricity $e = 150\ \mu$m to the midpoint of a vertical shaft AB which revolves at a constant angular velocity ω_f. Knowing that the spring constant k for horizontal movement of the disk is 580 kN/m, determine (*a*) the angular velocity ω_f at which resonance will occur, (*b*) the deflection r of the shaft when $\omega_f = 1200$ rpm.

19.124 A small trailer and its load have a total weight of 500-lb. The trailer is supported by two springs, each of constant 350 lb/ft, and is pulled over a road, the surface of which can be approximated by a sine curve with an amplitude of 2 in. and a wavelength of 15 ft (that is, the distance between successive crests is 15 ft and the vertical distance from crest to trough is 4 in.). Determine (*a*) the speed at which resonance will occur, (*b*) the amplitude of the vibration of the trailer at a speed of 30 mi/h.

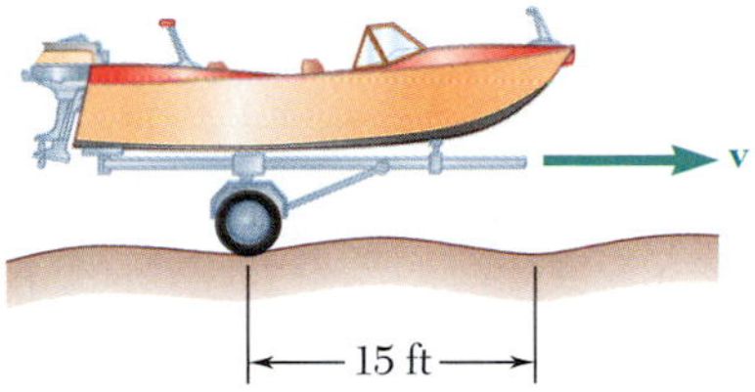

Fig. P19.124

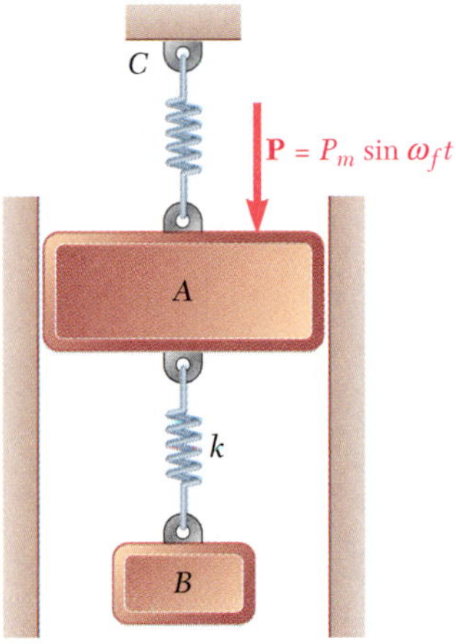

Fig. P19.125

19.125 Block A can move without friction in the slot as shown and is acted upon by a vertical periodic force of magnitude $P = P_m \sin \omega_f t$, where $\omega_f = 2$ rad/s and $P_m = 5$ lb. A spring of constant k is attached to the bottom of block A and to a 44-lb block B. Determine (*a*) the value of the constant k which will prevent a steady-state vibration of block A, (*b*) the corresponding amplitude of the vibration of block B.

DAMPED VIBRATIONS

*19.8. DAMPED FREE VIBRATIONS

The vibrating systems considered in the first part of this chapter were assumed free of damping. Actually all vibrations are damped to some degree by friction forces. These forces can be caused by *dry friction,* or *Coulomb friction,* between rigid bodies, by *fluid friction* when a rigid body moves in a fluid, or by *internal friction* between the molecules of a seemingly elastic body.

A type of damping of special interest is the *viscous damping* caused by fluid friction at low and moderate speeds. Viscous damping is characterized by the fact that the friction force is *directly proportional and opposite to the velocity* of the moving body. As an example, let us again consider a body of mass m suspended from a spring of constant k, assuming that the body is attached to the plunger of a dashpot (Fig. 19.10). The magnitude of the friction force exerted on the plunger by the surrounding fluid is equal to $c\dot{x}$, where the constant c, expressed in N · s/m or lb · s/ft and known as the *coefficient of viscous damping,* depends upon the physical properties of the fluid and the construction of the dashpot. The equation of motion is

$$+\downarrow \Sigma F = ma: \qquad W - k(\delta_{st} + x) - c\dot{x} = m\ddot{x}$$

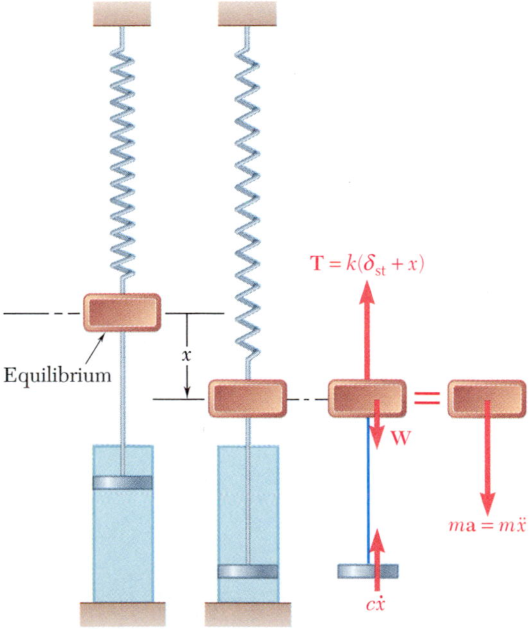

Fig. 19.10

Recalling that $W = k\delta_{st}$, we write

$$m\ddot{x} + c\dot{x} + kx = 0 \tag{19.38}$$

Substituting $x = e^{\lambda t}$ into (19.38) and dividing through by $e^{\lambda t}$ we write the *characteristic equation*

$$m\lambda^2 + c\lambda + k = 0 \tag{19.39}$$

and obtain the roots

$$\lambda = -\frac{c}{2m} \pm \sqrt{\left(\frac{c}{2m}\right)^2 - \frac{k}{m}} \tag{19.40}$$

Defining the *critical damping coefficient* c_c as the value of c which makes the radical in Eq. (19.40) equal to zero, we write

$$\left(\frac{c_c}{2m}\right)^2 - \frac{k}{m} = 0 \qquad c_c = 2m\sqrt{\frac{k}{m}} = 2m\omega_n \tag{19.41}$$

where ω_n is the natural circular frequency of the system in the absence of damping. We can distinguish three different cases of damping, depending upon the value of the coefficient c.

1. *Heavy damping:* $c > c_c$. The roots λ_1 and λ_2 of the characteristic equation (19.39) are real and distinct, and the general solution of the differential equation (19.38) is

$$x = C_1 e^{\lambda_1 t} + C_2 e^{\lambda_2 t} \tag{19.42}$$

This solution corresponds to a nonvibratory motion. Since λ_1 and λ_2 are both negative, x approaches zero as t increases indefinitely. However, the system actually regains its equilibrium position after a finite time.

2. *Critical damping:* $c = c_c$. The characteristic equation has a double root $\lambda = -c_c/2m = -\omega_n$, and the general solution of (19.38) is

$$x = (C_1 + C_2 t)e^{-\omega_n t} \tag{19.43}$$

The motion obtained is again nonvibratory. Critically damped systems are of special interest in engineering applications since they regain their equilibrium position in the shortest possible time without oscillation.

3. *Light damping:* $c < c_c$. The roots of Eq. (19.39) are complex and conjugate, and the general solution of (19.38) is of the form

$$x = e^{-(c/2m)t}(C_1 \sin \omega_d t + C_2 \cos \omega_d t) \tag{19.44}$$

where ω_d is defined by the relation

$$\omega_d^2 = \frac{k}{m} - \left(\frac{c}{2m}\right)^2$$

Substituting $k/m = \omega_n^2$ and recalling (19.41), we write

$$\omega_d = \omega_n \sqrt{1 - \left(\frac{c}{c_c}\right)^2} \qquad (19.45)$$

where the constant c/c_c is known as the *damping factor.* Even though the motion does not actually repeat itself, the constant ω_d is commonly referred to as the *circular frequency* of the damped vibration. A substitution similar to the one used in Sec. 19.2 enables us to write the general solution of Eq. (19.38) in the form

$$x = x_0 e^{-(c/2m)t} \sin (\omega_d t + \phi) \qquad (19.46)$$

The motion defined by Eq. (19.46) is vibratory with diminishing amplitude (Fig. 19.11), and the time interval $\tau_d = 2\pi/\omega_d$ separating two successive points where the curve defined by Eq. (19.46) touches one of the limiting curves shown in Fig. 19.11 is commonly referred to as the *period of the damped vibration.* Recalling Eq. (19.45), we observe that $\omega_d < \omega_n$ and, thus, that τ_d is larger than the period of vibration τ_n of the corresponding undamped system.

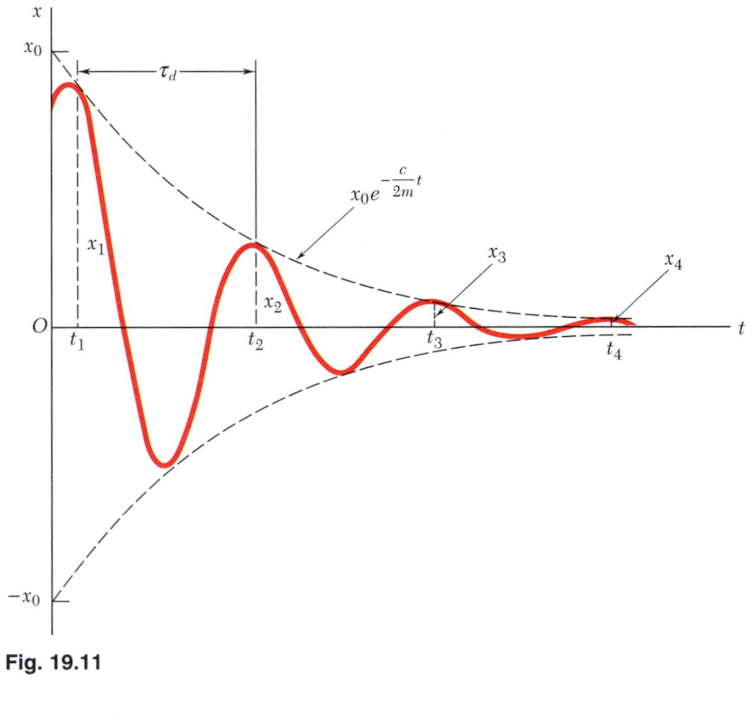

Fig. 19.11

*19.9. DAMPED FORCED VIBRATIONS

If the system considered in the preceding section is subjected to a periodic force **P** of magnitude $P = P_m \sin \omega_f t$, the equation of motion becomes

$$m\ddot{x} + c\dot{x} + kx = P_m \sin \omega_f t \tag{19.47}$$

Photo 19.2 The automobile suspension shown consists essentially of a spring and a shock absorber, which will cause the body of the car to undergo *damped forced vibrations* when the car is driven over an uneven road.

The general solution of (19.47) is obtained by adding a particular solution of (19.47) to the complementary function or general solution of the homogeneous equation (19.38). The complementary function is given by (19.42), (19.43), or (19.44), depending upon the type of damping considered. It represents a *transient* motion which is eventually damped out.

Our interest in this section is centered on the steady-state vibration represented by a particular solution of (19.47) of the form

$$x_{\text{part}} = x_m \sin (\omega_f t - \varphi) \tag{19.48}$$

Substituting x_{part} for x into (19.47), we obtain

$$-m\omega_f^2 x_m \sin (\omega_f t - \varphi) + c\omega_f x_m \cos (\omega_f t - \varphi) + kx_m \sin (\omega_f t - \varphi) = P_m \sin \omega_f t$$

Making $\omega_f t - \varphi$ successively equal to 0 and to $\pi/2$, we write

$$c\omega_f x_m = P_m \sin \varphi \tag{19.49}$$

$$(k - m\omega_f^2)\, x_m = P_m \cos \varphi \tag{19.50}$$

Squaring both members of (19.49) and (19.50) and adding, we have

$$[(k - m\omega_f^2)^2 + (c\omega_f)^2]\, x_m^2 = P_m^2 \tag{19.51}$$

Solving (19.51) for x_m and dividing (19.49) and (19.50) member by member, we obtain, respectively,

$$x_m = \frac{P_m}{\sqrt{(k - m\omega_f^2)^2 + (c\omega_f)^2}} \qquad \tan \varphi = \frac{c\omega_f}{k - m\omega_f^2} \tag{19.52}$$

Recalling from (19.4) that $k/m = \omega_n^2$, where ω_n is the circular frequency of the undamped free vibration, and from (19.41) that $2m\omega_n = c_c$, where c_c is the critical damping coefficient of the system, we write

$$\frac{x_m}{P_m/k} = \frac{x_m}{\delta_m} = \frac{1}{\sqrt{[1 - (\omega_f/\omega_n)^2]^2 + [2(c/c_c)(\omega_f/\omega_n)]^2}} \tag{19.53}$$

$$\tan \varphi = \frac{2(c/c_c)(\omega_f/\omega_n)}{1 - (\omega_f/\omega_n)^2} \tag{19.54}$$

Formula (19.53) expresses the magnification factor in terms of the frequency ratio ω_f/ω_n and damping factor c/c_c. It can be used to determine the amplitude of the steady-state vibration produced by an impressed force of magnitude $P = P_m \sin \omega_f t$ or by an impressed support movement $\delta = \delta_m \sin \omega_f t$. Formula (19.54) defines in terms of the same parameters the *phase difference* φ between the impressed force or impressed support movement and the resulting steady-state vibration of the damped system. The magnification factor has been plotted against the frequency ratio in Fig. 19.12 for various values of the damping factor. We observe that the amplitude of a forced vibration can be kept small by choosing a large coefficient of viscous damping c or by keeping the natural and forced frequencies far apart.

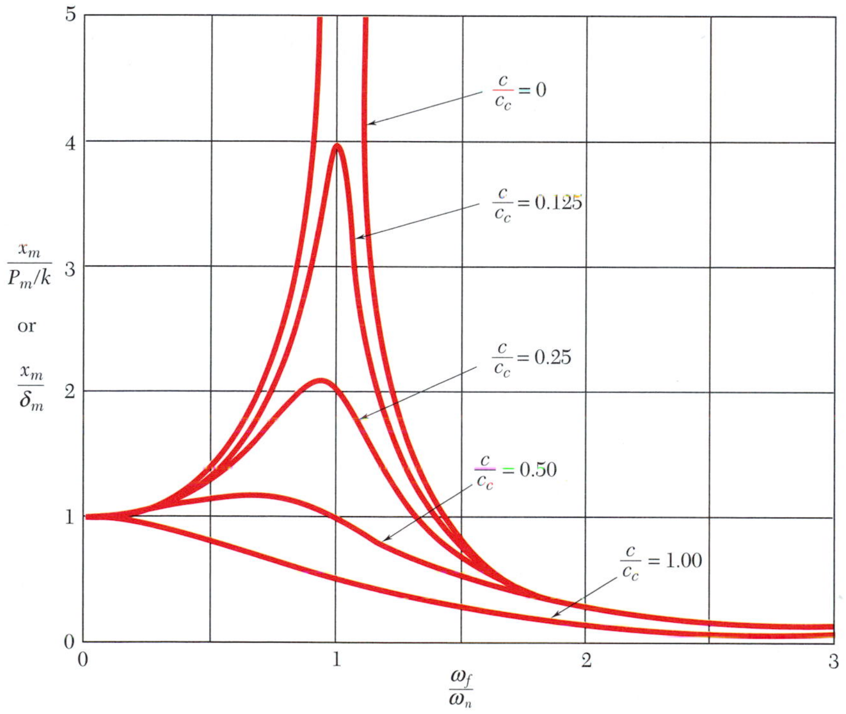

Fig. 19.12

*19.10. ELECTRICAL ANALOGUES

Oscillating electrical circuits are characterized by differential equations of the same type as those obtained in the preceding sections. Their analysis is therefore similar to that of a mechanical system, and the results obtained for a given vibrating system can be readily extended to the equivalent circuit. Conversely, any result obtained for an electrical circuit will also apply to the corresponding mechanical system.

Consider an electrical circuit consisting of an inductor of inductance L, a resistor of resistance R, and a capacitor of capacitance C, connected in series with a source of alternating voltage $E = E_m \sin \omega_f t$ (Fig. 19.13). It is recalled from elementary circuit theory† that if i denotes the current in the circuit and q denotes the electric charge on the capacitor, the drop in potential is $L(di/dt)$ across the inductor, Ri across the resistor, and q/C across the capacitor. Expressing that the algebraic sum of the applied voltage and of the drops in potential around the circuit loop is zero, we write

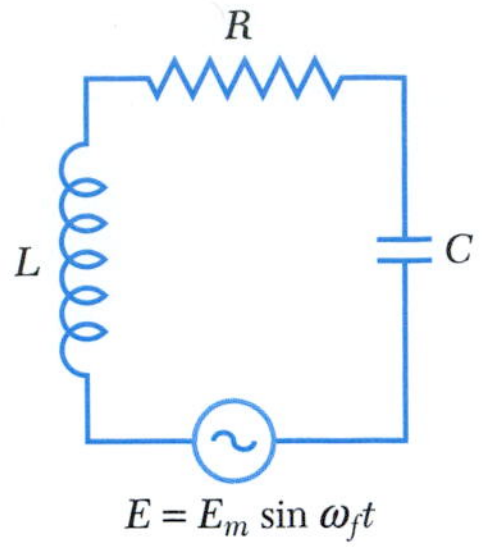

Fig. 19.13

$$E_m \sin \omega_f t - L\frac{di}{dt} - Ri - \frac{q}{C} = 0 \tag{19.55}$$

Rearranging the terms and recalling that at any instant the current i is equal to the rate of change $\dot{q}$ of the charge q, we have

$$L\ddot{q} + R\dot{q} + \frac{1}{C}q = E_m \sin \omega_f t \tag{19.56}$$

We verify that Eq. (19.56), which defines the oscillations of the electrical circuit of Fig. 19.13, is of the same type as Eq. (19.47), which characterizes the damped forced vibrations of the mechanical system of Fig. 19.10. By comparing the two equations, we can construct a table of the analogous mechanical and electrical expressions.

Table 19.2. Characteristics of a Mechanical System and of Its Electrical Analogue

Mechanical System		Electrical Circuit	
m	Mass	L	Inductance
c	Coefficient of viscous damping	R	Resistance
k	Spring constant	$1/C$	Reciprocal of capacitance
x	Displacement	q	Charge
v	Velocity	i	Current
P	Applied force	E	Applied voltage

†See C. R. Paul, S. A. Nasar, and L. E. Unnewehr, *Introduction to Electrical Engineering*, 2nd ed., McGraw-Hill, New York, 1992.

Table 19.2 can be used to extend the results obtained in the preceding sections for various mechanical systems to their electrical analogues. For instance, the amplitude i_m of the current in the circuit of Fig. 19.13 can be obtained by noting that it corresponds to the maximum value v_m of the velocity in the analogous mechanical system. Recalling from the first of Eqs. (19.37) that $v_m = x_m\omega_f$, substituting for x_m from Eq. (19.52), and replacing the constants of the mechanical system by the corresponding electrical expressions, we have

$$i_m = \frac{\omega_f E_m}{\sqrt{\left(\frac{1}{C} - L\omega_f^2\right)^2 + (R\omega_f)^2}}$$

$$i_m = \frac{E_m}{\sqrt{R^2 + \left(L\omega_f - \frac{1}{C\omega_f}\right)^2}} \qquad (19.57)$$

The radical in the expression obtained is known as the *impedance* of the electrical circuit.

The analogy between mechanical systems and electrical circuits holds for transient as well as steady-state oscillations. The oscillations of the circuit shown in Fig. 19.14, for instance, are analogous to the damped free vibrations of the system of Fig. 19.10. As far as the initial conditions are concerned, we should note that closing the switch S when the charge on the capacitor is $q = q_0$ is equivalent to releasing the mass of the mechanical system with no initial velocity from the position $x = x_0$. We should also observe that if a battery of constant voltage E is introduced in the electrical circuit of Fig. 19.14, closing the switch S will be equivalent to suddenly applying a force of constant magnitude P to the mass of the mechanical system of Fig. 19.10.

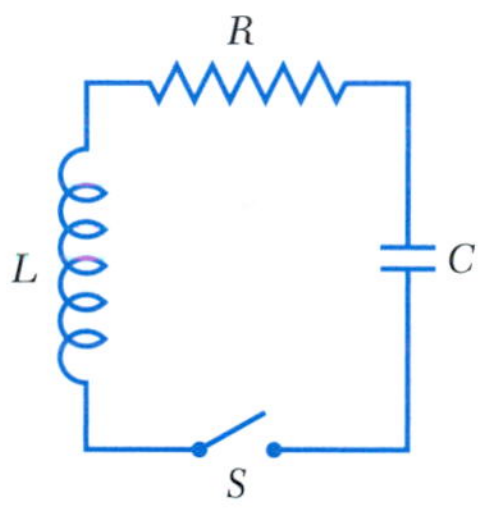

Fig. 19.14

The discussion above would be of questionable value if its only result were to make it possible for mechanics students to analyze electrical circuits without learning the elements of circuit theory. It is hoped that this discussion will instead encourage students to apply to the solution of problems in mechanical vibrations the mathematical techniques they may learn in later courses in circuit theory. The chief value of the concept of electrical analogue, however, resides in its application to *experimental methods* for the determination of the characteristics of a given mechanical system. Indeed, an electrical circuit is much more easily constructed than is a mechanical model, and the fact that its characteristics can be modified by varying the inductance, resistance, or capacitance of its various components makes the use of the electrical analogue particularly convenient.

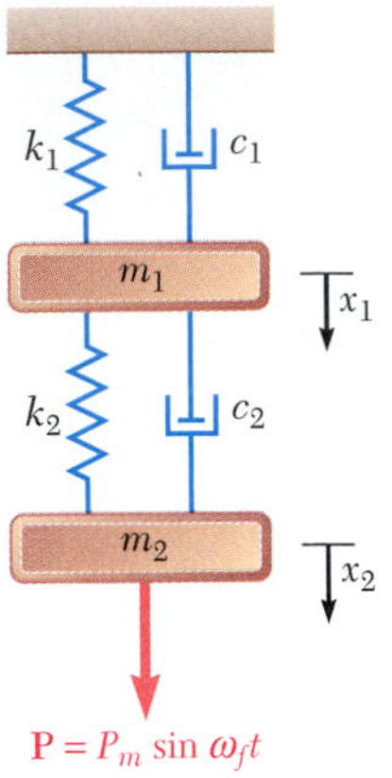

Fig. 19.15

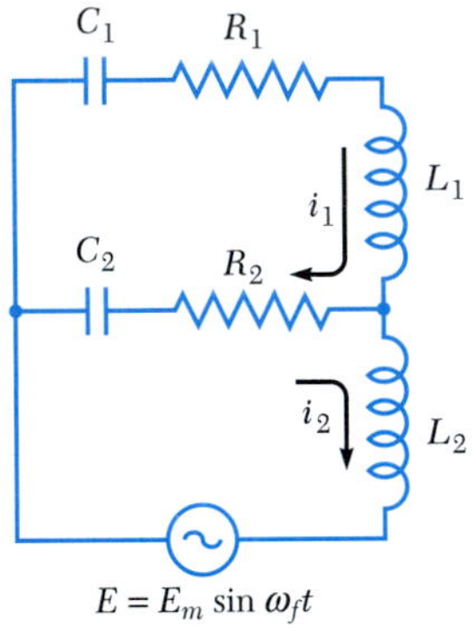

Fig. 19.16

To determine the electrical analogue of a given mechanical system, we focus our attention on each moving mass in the system and observe which springs, dashpots, or external forces are applied directly to it. An equivalent electrical loop can then be constructed to match each of the mechanical units thus defined; the various loops obtained in that way will together form the desired circuit. Consider, for instance, the mechanical system of Fig. 19.15. We observe that the mass m_1 is acted upon by two springs of constants k_1 and k_2 and by two dashpots characterized by the coefficients of viscous damping c_1 and c_2. The electrical circuit should therefore include a loop consisting of an inductor of inductance L_1 proportional to m_1, of two capacitors of capacitance C_1 and C_2 inversely proportional to k_1 and k_2, respectively, and of two resistors of resistance R_1 and R_2, proportional to c_1 and c_2, respectively. Since the mass m_2 is acted upon by the spring k_2 and the dashpot c_2, as well as the force $P = P_m \sin \omega_f t$, the circuit should also include a loop containing the capacitor C_2, the resistor R_2, the new inductor L_2, and the voltage source $E = E_m \sin \omega_f t$ (Fig. 19.16).

To check that the mechanical system of Fig. 19.15 and the electrical circuit of Fig. 19.16 actually satisfy the same differential equations, the equations of motion for m_1 and m_2 will first be derived. Denoting, respectively, by x_1 and x_2 the displacements of m_1 and m_2 from their equilibrium positions, we observe that the elongation of the spring k_1 (measured from the equilibrium position) is equal to x_1, while the elongation of the spring k_2 is equal to the relative displacement $x_2 - x_1$ of m_2 with respect to m_1. The equations of motion for m_1 and m_2 are therefore

$$m_1\ddot{x}_1 + c_1\dot{x}_1 + c_2(\dot{x}_1 - \dot{x}_2) + k_1x_1 + k_2(x_1 - x_2) = 0 \tag{19.58}$$

$$m_2\ddot{x}_2 + c_2(\dot{x}_2 - \dot{x}_1) + k_2(x_2 - x_1) = P_m \sin \omega_f t \tag{19.59}$$

Consider now the electrical circuit of Fig. 19.16; we denote, respectively, by i_1 and i_2 the current in the first and second loops, and by q_1 and q_2 the integrals $\int i_2\, dt$ and $\int i_2\, dt$. Noting that the charge on the capacitor C_1 is q_1, while the charge on C_2 is $q_1 - q_2$, we express that the sum of the potential differences in each loop is zero and obtain the following equations

$$L_1\ddot{q}_1 + R_1\dot{q}_1 + R_2(\dot{q}_1 - \dot{q}_2) + \frac{q_1}{C_1} + \frac{q_1 - q_2}{C_2} = 0 \tag{19.60}$$

$$L_2\ddot{q}_2 + R_2(\dot{q}_2 - \dot{q}_1) + \frac{q_1 - q_2}{C_2} = E_m \sin \omega_f t \tag{19.61}$$

We easily check that Eqs. (19.60) and (19.61) reduce to (19.58) and (19.59), respectively, when the substitutions indicated in Table 19.2 are performed.

SOLVING PROBLEMS ON YOUR OWN

In this lesson a more realistic model of a vibrating system was developed by including the effect of the *viscous damping* caused by fluid friction. Viscous damping was represented in Fig. 19.10 by the force exerted on the moving body by a plunger moving in a dashpot. This force is equal in magnitude to $c\dot{x}$, where the constant c, expressed in N · s/m or lb · s/ft, is known as the *coefficient of viscous damping*. Keep in mind that the same sign convention should be used for x, $\dot{x}$, and $\ddot{x}$.

1. *Damped free vibrations.* The differential equation defining this motion was found to be

$$m\ddot{x} + c\dot{x} + kx = 0 \tag{19.38}$$

To obtain the solution of this equation, calculate the *critical damping coefficient* c_c, using the formula

$$c_c = 2m\sqrt{k/m} = 2m\omega_n \tag{19.41}$$

where ω_n is the natural circular frequency of the *undamped* system.

a. If c > c_c (heavy damping), the solution of Eq. (19.38) is

$$x = C_1 e^{\lambda_1 t} + C_2 e^{\lambda_2 t} \tag{19.42}$$

where

$$\lambda_{1,2} = -\frac{c}{2m} \pm \sqrt{\left(\frac{c}{2m}\right)^2 - \frac{k}{m}} \tag{19.40}$$

and where the constants C_1 and C_2 can be determined from the initial conditions $x(0)$ and $\dot{x}(0)$. This solution corresponds to a nonvibratory motion.

b. If c = c_c (critical damping), the solution of Eq. (19.38) is

$$x = (C_1 + C_2 t)e^{-\omega_n t} \tag{19.43}$$

which also corresponds to a nonvibratory motion.

c. If c < c_c (light damping), the solution of Eq. (19.38) is

$$x = x_0 e^{-(c/2m)t} \sin(\omega_d t + \phi) \tag{19.46}$$

where

$$\omega_d = \omega_n \sqrt{1 - \left(\frac{c}{c_c}\right)^2} \tag{19.45}$$

and where x_0 and ϕ can be determined from the initial conditions $x(0)$ *and* $\dot{x}(0)$. This solution corresponds to oscillations of decreasing amplitude and of period $\tau_d = 2\pi/\omega_d$ (Fig. 19.11).

(continued)

2. ***Damped forced vibrations.*** These vibrations occur when a system with viscous damping is subjected to a periodic force **P** of magnitude $P = P_m \sin \omega_f t$ or when it is elastically connected to a support with an alternating motion $\delta = \delta_m \sin \omega_f t$. In the first case the motion is defined by the differential equation

$$m\ddot{x} + c\dot{x} + kx = P_m \sin \omega_f t \qquad (19.47)$$

and in the second case by a similar equation obtained by replacing P_m with $k\delta_m$. You will be concerned only with the *steady-state* motion of the system, which is defined by a *particular solution* of these equations, of the form

$$x_{\text{part}} = x_m \sin (\omega_f t - \varphi) \qquad (19.48)$$

where

$$\frac{x_m}{P_m/k} = \frac{x_m}{\delta_m} = \frac{1}{\sqrt{[1 - (\omega_f/\omega_n)^2]^2 + [2(c/c_c)(\omega_f/\omega_n)]^2}} \qquad (19.53)$$

and

$$\tan \varphi = \frac{2(c/c_c)(\omega_f/\omega_n)}{1 - (\omega_f/\omega_n)^2} \qquad (19.54)$$

The expression given in Eq. (19.53) is referred to as the *magnification factor* and has been plotted against the frequency ratio ω_f/ω_n in Fig. 19.12 for various values of the damping factor c/c_c. In the problems which follow you may be asked to determine one of the parameters in Eqs. (19.53) and (19.54) when the others are known.

Problems

19.126 Show that in the case of heavy damping ($c > c_c$), a body never passes through its position of equilibrium O (*a*) if it is released with no initial velocity from an arbitrary position, or (*b*) if it is started from O with an arbitrary initial velocity.

19.127 Show that in the case of heavy damping ($c > c_c$), a body released from an arbitrary position with an arbitrary initial velocity cannot pass more than once through its equilibrium position.

19.128 In the case of light damping, the displacements x_1, x_2, x_3, shown in Fig. 19.11 may be assumed equal to the maximum displacements. Show that the ratio of any two successive maximum displacements x_n and x_{n+1} is constant and that the natural logarithm of this ratio, called the *logarithmic decrement*, is

$$\ln \frac{x_n}{x_{n+1}} = \frac{2\pi(c/c_c)}{\sqrt{1-(c/c_c)^2}}$$

19.129 In practice, it is often difficult to determine the logarithmic decrement of a system with light damping defined in Prob. 19.128 by measuring two successive maximum displacements. Show that the logarithmic decrement can also be expressed as $(1/k)\ \ln\ (x_n/x_{n+k})$, where k is the number of cycles between readings of the maximum displacement.

19.130 In a system with light damping ($c < c_c$), the period of vibration is commonly defined as the time interval $\tau_d = 2\pi/\omega_d$ corresponding to two successive points where the displacement-time curve touches one of the limiting curves shown in Fig. 19.11. Show that the interval of time (*a*) between a maximum positive displacement and the following negative displacement is $\frac{1}{2}\tau_d$, (*b*) between two successive zero displacements is $\frac{1}{2}\tau_d$, (*c*) between a maximum positive displacement and the following zero displacement is greater than $\frac{1}{4}\tau_d$.

19.131 The block shown is depressed 24 mm from its equilibrium position and released. Knowing that after 10 cycles the maximum displacement of the block is 10 mm, determine (*a*) the damping factor c/c_c, (*b*) the value of the coefficient of viscous damping. (*Hint:* See Probs. 19.128 and 19.129.)

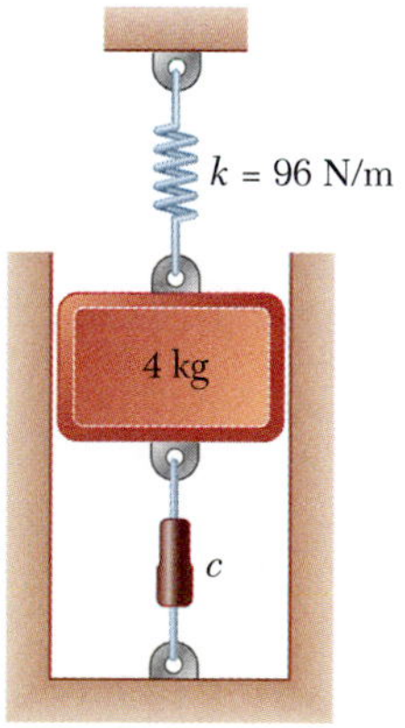

Fig. P19.131

19.132 Successive maximum displacements of a spring-mass-dashpot system similar to that shown in Fig. 19.10 are 1.25, 0.75, and 0.45 in. Knowing that $W = 36$ lb and $k = 175$ lb/ft, determine (*a*) the damping factor c/c_c, (*b*) the value of the coefficient of viscous damping c. (*Hint:* See Probs. 19.128 and 19.129.)

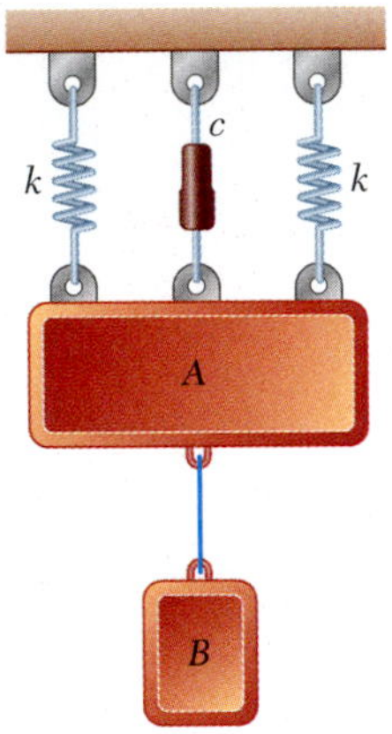

Fig. P19.133 and P19.134

19.133 A 2-lb block B is connected by a cord to a 5-lb block A which is suspended as shown from two springs, each of constant $k = 12.5$ lb/ft, and a dashpot of damping coefficient $c = 0.5$ lb · s/ft. Knowing that the system is at rest when the cord connecting A and B is cut, determine the minimum tension that will occur in each spring during the resulting motion.

19.134 A 2-lb block B is connected by a cord to a 5-lb block A which is suspended as shown from two springs, each of constant $k = 12.5$ lb/ft, and a dashpot of damping coefficient $c = 4$ lb · s/ft. Knowing that the system is at rest when the cord connecting A and B is cut, determine the velocity of block A after 0.1 s.

19.135 The barrel of a field gun weighs 1800 lb and is returned into firing position after recoil by a recuperator of constant $c = 1320$ lb · s/ft. Determine (*a*) the constant k which should be used for the recuperator to return the barrel into firing position in the shortest possible time without any oscillation, (*b*) the time needed for the barrel to move back two-thirds of the way from its maximum-recoil position to its firing position.

19.136 A uniform rod of mass m is supported by a pin at A and a spring of constant k at B and is connected at D to a dashpot of damping coefficient c. Determine in terms of m, k, and c, for small oscillations, (*a*) the differential equation of motion, (*b*) the critical damping coefficient c_c.

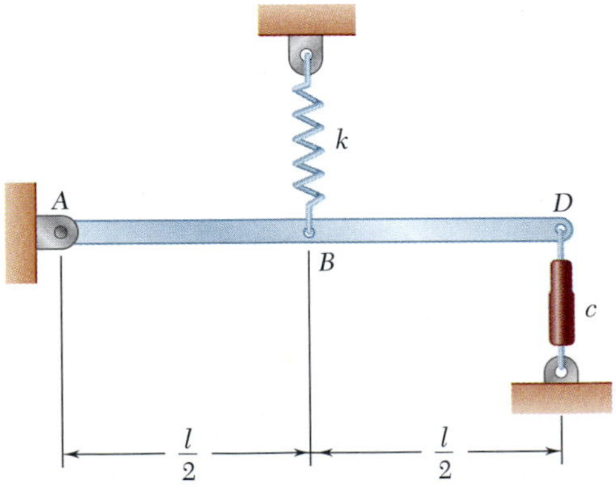

Fig. P19.136

19.137 A 1.8-kg uniform rod is supported by a pin at O and a spring at A and is connected to a dashpot at B. Determine (a) the differential equation of motion for small oscillations, (b) the angle that the rod will form with the horizontal 2.5 s after end B has been pushed 23 mm down and released.

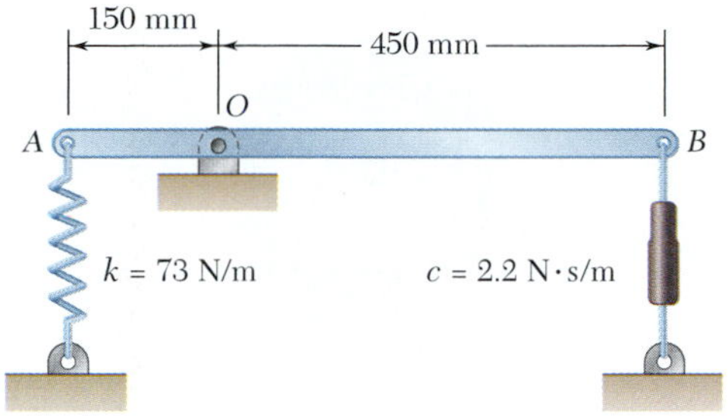

Fig. P19.137

19.138 A 500-kg machine element is supported by two springs, each of constant 4.4 kN/m. A periodic force of 135-N amplitude is applied to the element with a frequency of 2.8 Hz. Knowing that the coefficient of damping is 1.6 kN · s/m, determine the amplitude of the steady-state vibration of the element.

19.139 In Prob. 19.138, determine the required value of the constant of each spring if the amplitude of the steady-state vibration is to be 1.3 mm.

19.140 In the case of the forced vibration of a system, determine the range of values of the damping factor c/c_c for which the magnification factor will always decrease as the frequency ratio ω_f/ω_n increases.

19.141 Show that for a small value of the damping factor c/c_c, the maximum amplitude of a forced vibration occurs when $\omega_f \approx \omega_n$ and that the corresponding value of the magnification factor is $\frac{1}{2}(c/c_c)$.

19.142 A 30-lb motor is supported by four springs, each of constant 3.75 kips/ft. The unbalance of the motor is equivalent to a weight of 0.64 oz located 5 in. from the axis of rotation. Knowing that the motor is constrained to move vertically, determine the amplitude of the steady-state vibration of the motor at a speed of 1500 rpm, assuming (a) that no damping is present, (b) that the damping factor c/c_c is equal to 1.3.

19.143 A 36-lb motor is bolted to a light horizontal beam which has a static deflection of 0.075 in. due to the weight of the motor. Knowing that the unbalance of the rotor is equivalent to a weight of 0.64 oz located 6.25 in. from the axis of rotation, determine the amplitude of the vibration of the motor at a speed of 900 rpm, assuming (a) that no damping is present, (b) that the damping factor c/c_c is equal to 0.055.

19.144 A 45-kg motor is bolted to a light horizontal beam which has a static deflection of 6 mm due to the weight of the motor. The unbalance of the motor is equivalent to a mass of 110 g located 75 mm from the axis of rotation. Knowing that the amplitude of the vibration of the motor is 0.25 mm at a speed of 300 rpm, determine (a) the damping factor c/c_c, (b) the coefficient of damping c.

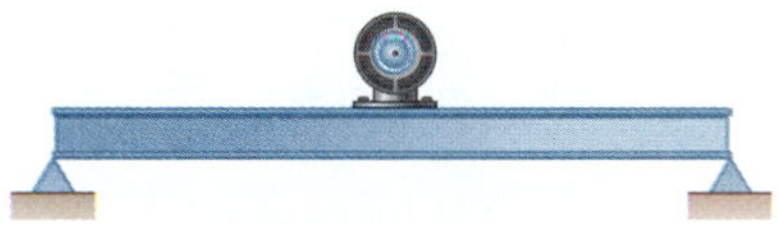

Fig. P19.143 and P19.144

19.145 A 200-lb motor is supported by four springs, each of constant 7.5 kips/ft, and is connected to the ground by a dashpot having a coefficient of damping $c = 490$ lb · s/ft. The motor is constrained to move vertically, and the amplitude of its motion is observed to be 0.10 in. at a speed of 1200 rpm. Knowing that the weight of the rotor is 30 lb, determine the distance between the mass center of the rotor and the axis of the shaft.

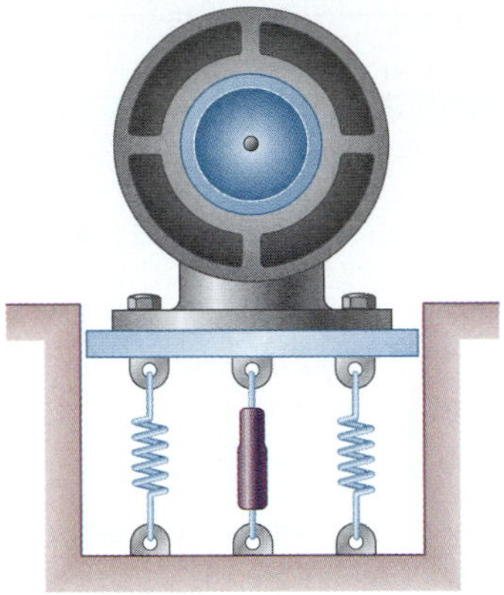

Fig. *P19.145*

19.146 The unbalance of the rotor of a 400-lb motor is equivalent to a 3-oz weight located 6 in. from the axis of rotation. The pad which is placed between the motor and the foundation is equivalent to a spring of constant $k = 500$ lb/ft in parallel with a dashpot of constant c. Knowing that the magnitude of the maximum acceleration of the motor is 0.03 ft/s^2 at a speed of 100 rpm, determine the damping factor c/c_c.

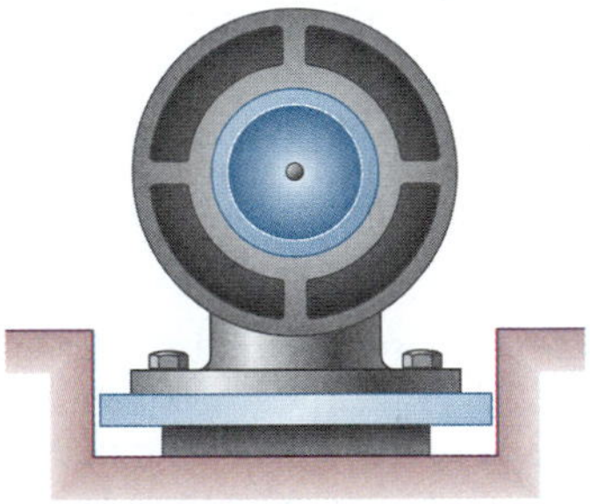

Fig. P19.146

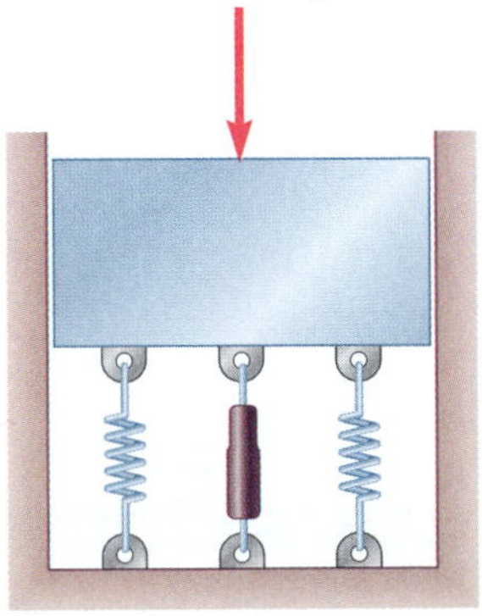

Fig. P19.147 and *P19.148*

19.147 A machine element is supported by springs and is connected to a dashpot as shown. Show that if a periodic force of magnitude $P = P_m \sin \omega_f t$ is applied to the element, the amplitude of the fluctuating force transmitted to the foundation is

$$F_m = P_m \sqrt{\frac{1 + [2(c/c_c)(\omega_f/\omega_n)]^2}{[1 - (\omega_f/\omega_n)^2]^2 + [2(c/c_c)(\omega_f/\omega_n)]^2}}$$

19.148 A 91-kg machine element supported by four springs, each of constant 175 N/m, is subjected to a periodic force of frequency 0.8 Hz and amplitude 89 N. Determine the amplitude of the fluctuating force transmitted to the foundation if (*a*) a dashpot with a coefficient of damping $c = 365$ N · s/m is connected to the machine and to the ground, (*b*) the dashpot is removed.

***19.149** For a steady-state vibration with damping under a harmonic force, show that the mechanical energy dissipated per cycle by the dashpot is $E = \pi c x_m^2 \omega_f$, where c is the coefficient of damping, x_m is the amplitude of the motion, and ω_f is the circular frequency of the harmonic force.

****19.150*** The suspension of an automobile can be approximated by the simplified spring-dashpot system shown. (*a*) Write the differential equation defining the vertical displacement of the mass m when the system moves at a speed v over a road with a sinusoidal cross section of amplitude δ_m and wavelength L. (*b*) Derive an expression for the amplitude of the vertical displacement of the mass m.

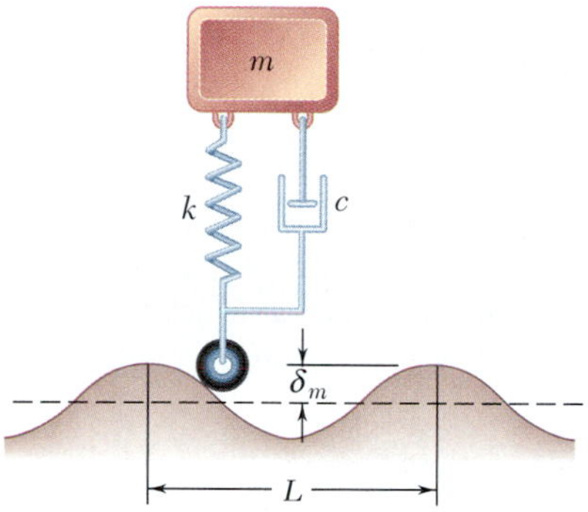

Fig. *P19.150*

***19.151** Two blocks A and B, each of mass m, are supported as shown by three springs of the same constant k. Blocks A and B are connected by a dashpot and block B is connected to the ground by two dashpots, each dashpot having the same coefficient of damping c. Block A is subjected to a force of magnitude $P = P_m \sin \omega_f t$. Write the differential equations defining the displacements x_A and x_B of the two blocks from their equilibrium positions.

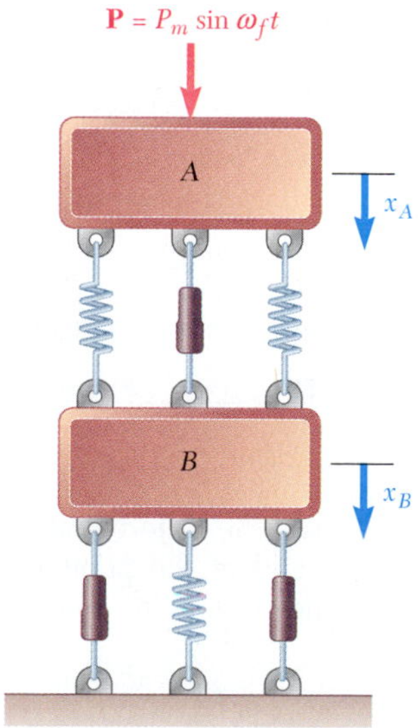

Fig. P19.151

19.152 Express in terms of L, C, and E the range of values of the resistance R for which oscillations will take place in the circuit shown when switch S is closed.

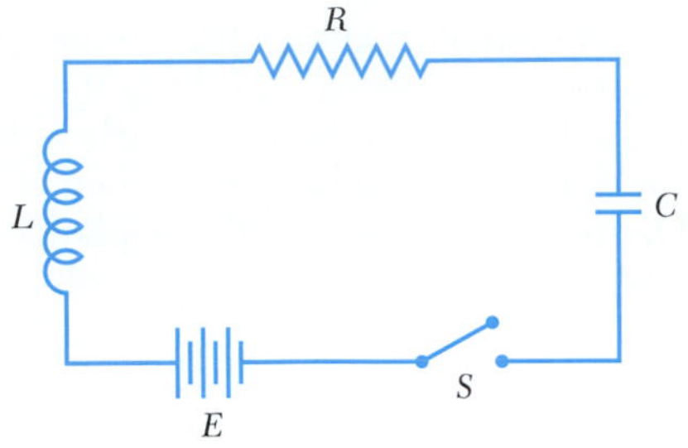

Fig. P19.152

19.153 Consider the circuit of Prob. 19.152 when the capacitor C is removed. If switch S is closed at time $t = 0$, determine (*a*) the final value of the current in the circuit, (*b*) the time t at which the current will have reached $(1 - 1/e)$ times its final value. (The desired value of t is known as the time constant of the circuit.)

19.154 and 19.155 Draw the electrical analogue of the mechanical system shown. (*Hint:* Draw the loops corresponding to the free bodies m and A.)

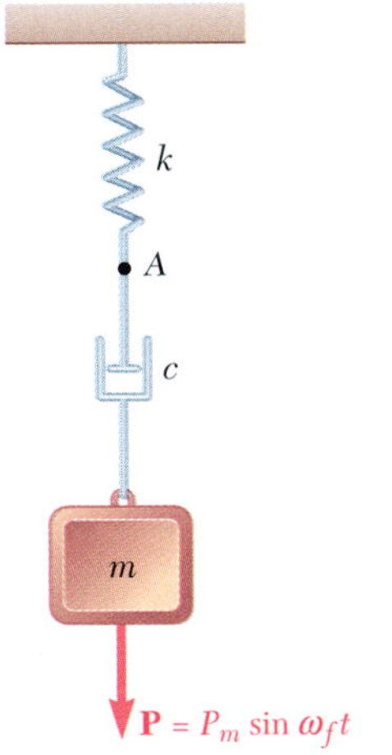

Fig. P19.154 and P19.156

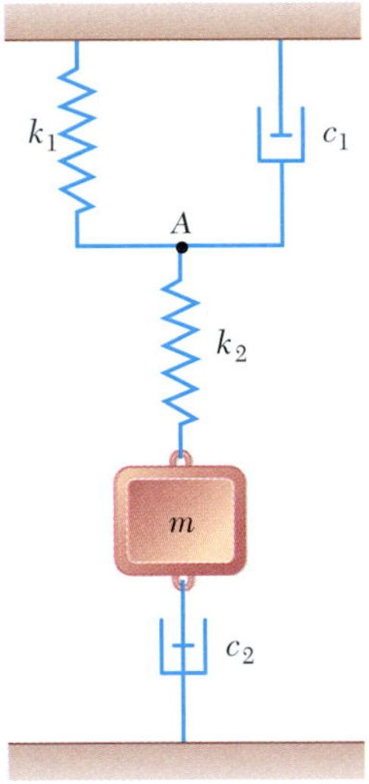

Fig. P19.155 and P19.157

19.156 and 19.157 Write the differential equations defining (*a*) the displacements of the mass m and of the point A, (*b*) the charges on the capacitors of the electrical analogue.

REVIEW AND SUMMARY FOR CHAPTER 19

This chapter was devoted to the study of *mechanical vibrations*, that is, to the analysis of the motion of particles and rigid bodies oscillating about a position of equilibrium. In the first part of the chapter [Secs. 19.2 through 19.7], we considered *vibrations without damping*, while the second part was devoted to *damped vibrations* [Secs. 19.8 through 19.10].

Free vibrations of a particle

In Sec. 19.2, we considered the *free vibrations of a particle*, that is, the motion of a particle P subjected to a restoring force proportional to the displacement of the particle—such as the force exerted by a spring. If the displacement x of the particle P is measured from its equilibrium position O (Fig. 19.17), the resultant $\mathbf{F}$ of the forces acting on P (including its weight) has a magnitude kx and is directed toward O. Applying Newton's second law $F = ma$ and recalling that $a = \ddot{x}$, we wrote the differential equation

$$m\ddot{x} + kx = 0 \qquad (19.2)$$

or, setting $\omega_n^2 = k/m$,

$$\ddot{x} + \omega_n^2 x = 0 \qquad (19.6)$$

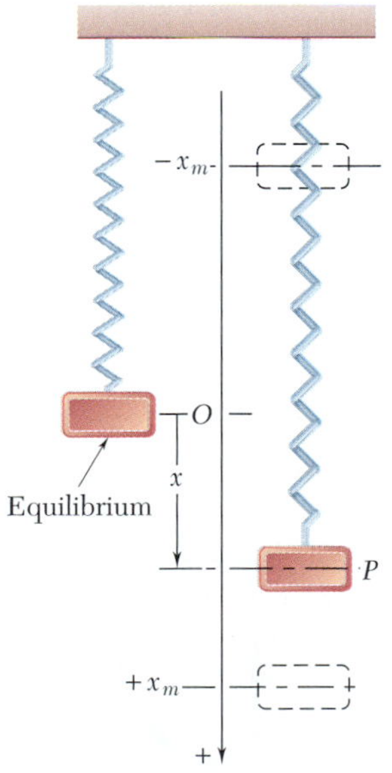

Fig. 19.17

The motion defined by this equation is called a *simple harmonic motion.*

The solution of Eq. (19.6), which represents the displacement of the particle P, was expressed as

$$x = x_m \sin (\omega_n t + \phi) \tag{19.10}$$

where x_m = amplitude of the vibration
$\omega_n = \sqrt{k/m}$ = natural circular frequency
ϕ = phase angle

The *period of the vibration* (that is, the time required for a full cycle) and its *natural frequency* (that is, the number of cycles per second) were expressed as

$$\text{Period} = \tau_n = \frac{2\pi}{\omega_n} \tag{19.13}$$

$$\text{Natural frequency} = f_n = \frac{1}{\tau_n} = \frac{\omega_n}{2\pi} \tag{19.14}$$

The velocity and acceleration of the particle were obtained by differentiating Eq. (19.10), and their maximum values were found to be

$$v_m = x_m\omega_n \qquad a_m = x_m\omega_n^2 \tag{19.15}$$

Since all the above parameters depend directly upon the natural circular frequency ω_n and thus upon the ratio k/m, it is essential in any given problem to calculate the value of the constant k; this can be done by determining the relation between the restoring force and the corresponding displacement of the particle [Sample Prob. 19.1].

It was also shown that the oscillatory motion of the particle P can be represented by the projection on the x axis of the motion of a point Q describing an auxiliary circle of radius x_m with the constant angular velocity ω_n (Fig. 19.18). The instantaneous values of the velocity and acceleration of P can then be obtained by projecting on the x axis the vectors $\mathbf{v}_m$ and $\mathbf{a}_m$ representing, respectively, the velocity and acceleration of Q.

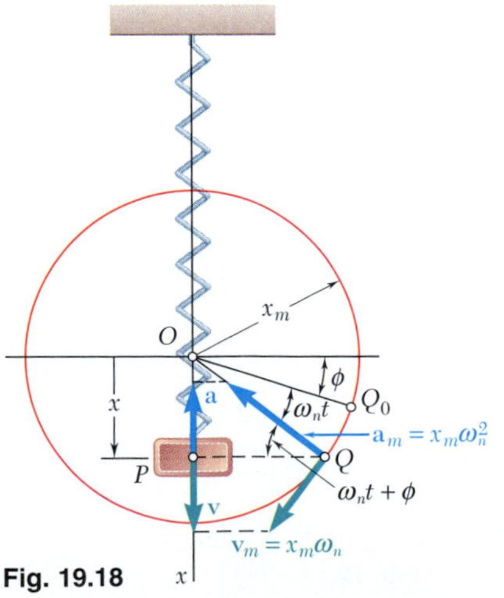

Fig. 19.18

Simple pendulum

While the motion of a *simple pendulum* is not truly a simple harmonic motion, the formulas given above can be used with $\omega_n^2 = g/l$ to calculate the period and natural frequency of the *small oscillations* of a simple pendulum [Sec. 19.3]. Large-amplitude oscillations of a simple pendulum were discussed in Sec. 19.4.

Free vibrations of a rigid body

The *free vibrations of a rigid body* can be analyzed by choosing an appropriate variable, such as a distance x or an angle θ, to define the position of the body, drawing a free-body-diagram equation to express the equivalence of the external and effective forces, and writing an equation relating the selected variable and its second derivative [Sec. 19.5]. If the equation obtained is of the form

$$\ddot{x} + \omega_n^2 x = 0 \qquad \text{or} \qquad \ddot{\theta} + \omega_n^2 \theta = 0 \tag{19.21}$$

the vibration considered is a simple harmonic motion and its period and natural frequency can be obtained *by identifying* ω_n and substituting its value into Eqs. (19.13) and (19.14) [Sample Probs. 19.2 and 19.3].

Using the principle of conservation of energy

The *principle of conservation of energy* can be used as an alternative method for the determination of the period and natural frequency of the simple harmonic motion of a particle or rigid body [Sec. 19.6]. Choosing again an appropriate variable, such as θ, to define the position of the system, we express that the total energy of the system is conserved, $T_1 + V_1 = T_2 + V_2$, between the position of maximum displacement ($\theta_1 = \theta_m$) and the position of maximum velocity ($\dot{\theta}_2 = \dot{\theta}_m$). If the motion considered is simple harmonic, the two members of the equation obtained consist of homogeneous quadratic expressions in θ_m and $\dot{\theta}_m$, respectively.† Substituting $\dot{\theta}_m = \theta_m \omega_n$ in this equation, we can factor out θ_m^2 and solve for the circular frequency ω_n [Sample Prob. 19.4].

Forced vibrations

In Sec. 19.7, we considered the *forced vibrations* of a mechanical system. These vibrations occur when the system is subjected to a periodic force (Fig. 19.19) or when it is elastically connected to a support which has an alternating motion (Fig. 19.20). Denoting by ω_f the forced circular frequency, we found that in the first case, the motion of the system was defined by the differential equation

$$m\ddot{x} + kx = P_m \sin \omega_f t \tag{19.30}$$

and that in the second case it was defined by the differential equation

$$m\ddot{x} + kx = k\delta_m \sin \omega_f t \tag{19.31}$$

The general solution of these equations is obtained by adding a particular solution of the form

$$x_{\text{part}} = x_m \sin \omega_f t \tag{19.32}$$

†If the motion considered can only be *approximated* by a simple harmonic motion, such as for the small oscillations of a body under gravity, the potential energy must be approximated by a quadratic expression in θ_m.

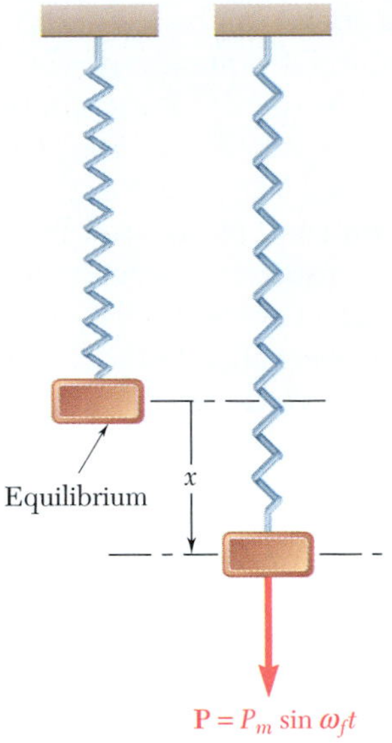

Fig. 19.19

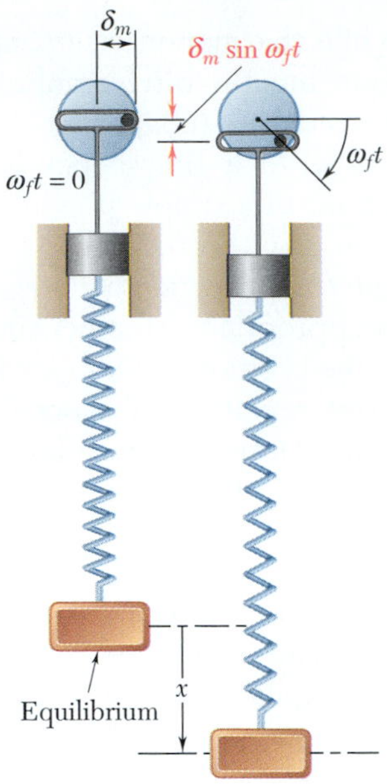

Fig. 19.20

to the general solution of the corresponding homogeneous equation. The particular solution (19.32) represents a *steady-state vibration* of the system, while the solution of the homogeneous equation represents a *transient free vibration* which can generally be neglected.

Dividing the amplitude x_m of the steady-state vibration by P_m/k in the case of a periodic force, or by δ_m in the case of an oscillating support, we defined the *magnification factor* of the vibration and found that

$$\text{Magnification factor} = \frac{x_m}{P_m/k} = \frac{x_m}{\delta_m} = \frac{1}{1 - (\omega_f/\omega_n)^2} \qquad (19.36)$$

According to Eq. (19.36), the amplitude x_m of the forced vibration *becomes infinite when* $\omega_f = \omega_n$, that is, *when the forced frequency is equal to the natural frequency of the system.* The impressed force or impressed support movement is then said to be in *resonance* with the system [Sample Prob. 19.5]. Actually the amplitude of the vibration remains finite, due to damping forces.

Damped free vibrations

In the last part of the chapter, we considered the *damped vibrations* of a mechanical system. First, we analyzed the *damped free vibrations* of a system with *viscous damping* [Sec. 19.8]. We found that the motion of such a system was defined by the differential equation

$$m\ddot{x} + c\dot{x} + kx = 0 \qquad (19.38)$$

where c is a constant called the *coefficient of viscous damping.* Defining the *critical damping coefficient* c_c as

$$c_c = 2m\sqrt{\frac{k}{m}} = 2m\omega_n \qquad (19.41)$$

where ω_n is the natural circular frequency of the system in the absence of damping, we distinguished three different cases of damping, namely, (1) *heavy damping,* when $c > c_c$; (2) *critical damping,* when $c = c_c$; and (3) *light damping,* when $c < c_c$. In the first two cases, the system when disturbed tends to regain its equilibrium position without any oscillation. In the third case, the motion is vibratory with diminishing amplitude.

Damped forced vibrations

In Sec. 19.9, we considered the *damped forced vibrations* of a mechanical system. These vibrations occur when a system with viscous damping is subjected to a periodic force **P** of magnitude $P = P_m \sin \omega_f t$ or when it is elastically connected to a support with an alternating motion $\delta = \delta_m \sin \omega_f t$. In the first case, the motion of the system was defined by the differential equation

$$m\ddot{x} + c\dot{x} + kx = P_m \sin \omega_f t \qquad (19.47)$$

and in the second case by a similar equation obtained by replacing P_m by $k\delta_m$ in (19.47).

The *steady-state vibration* of the system is represented by a particular solution of Eq. (19.47) of the form

$$x_{\text{part}} = x_m \sin (\omega_f t - \varphi) \qquad (19.48)$$

Dividing the amplitude x_m of the steady-state vibration by P_m/k in the case of a periodic force, or by δ_m in the case of an oscillating support, we obtained the following expression for the magnification factor:

$$\frac{x_m}{P_m/k} = \frac{x_m}{\delta_m} = \frac{1}{\sqrt{[1 - (\omega_f/\omega_n)^2]^2 + [2(c/c_c)(\omega_f/\omega_n)]^2}} \qquad (19.53)$$

where $\omega_n = \sqrt{k/m}$ = natural circular frequency of undamped system
$c_c = 2m\omega_n$ = critical damping coefficient
c/c_c = damping factor

We also found that the *phase difference* φ between the impressed force or support movement and the resulting steady-state vibration of the damped system was defined by the relation

$$\tan \varphi = \frac{2(c/c_c)(\omega_f/\omega_n)}{1 - (\omega_f/\omega_n)^2} \qquad (19.54)$$

Electrical analogues

The chapter ended with a discussion of *electrical analogues* [Sec. 19.10], in which it was shown that the vibrations of mechanical systems and the oscillations of electrical circuits are defined by the same differential equations. Electrical analogues of mechanical systems can therefore be used to study or predict the behavior of these systems.

Review Problems

19.158 A 75-mm-radius hole is cut in a 200-mm-radius uniform disk which is attached to a frictionless pin at its geometric center A. Determine (*a*) the period of small oscillations of the disk, (*b*) the length of a simple pendulum which has the same period.

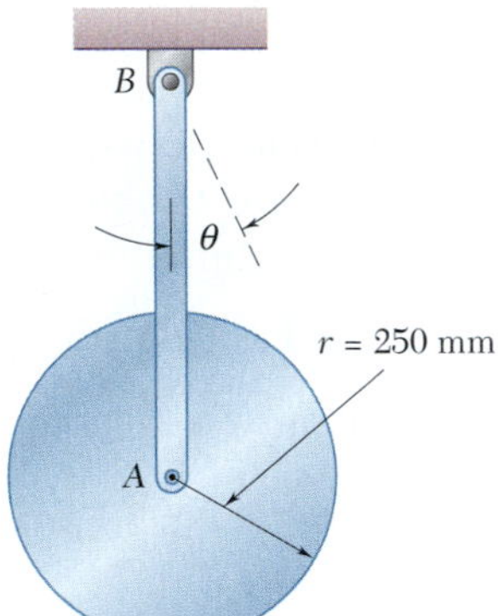

Fig. P19.158

19.159 An instrument package B is placed on table C as shown. The table is made to move horizontally in simple harmonic motion with an amplitude of 3 in. Knowing that the coefficient of static friction is 0.65, determine the largest allowable frequency of the motion if the package is not to slide on the table.

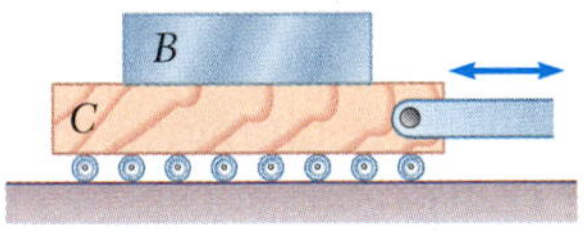

Fig. P19.159

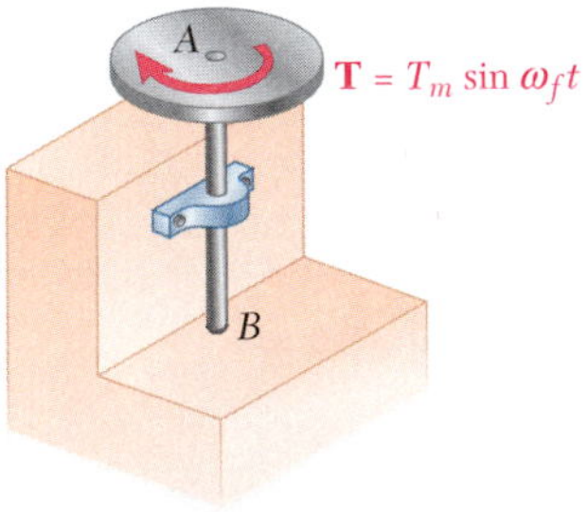

Fig. P19.160 and P19.161

19.160 A 16-lb uniform disk of radius 10 in. is welded to a vertical shaft with a fixed end at B. The disk rotates through an angle of 3° when a static couple of magnitude 3.5 lb · ft is applied to it. If the disk is acted upon by a periodic torsional couple of magnitude $T = T_m \sin \omega_f t$, where $T_m =$ 4.2 lb · ft, determine the range of values of ω_f for which the amplitude of the vibration is less than the angle of rotation caused by a static couple of magnitude T_m.

19.161 For the disk of Prob. 19.160 determine the range of values of ω_f for which the amplitude of the vibration will be less than 3.5°.

19.162 A beam ABC is supported by a pin connection at A and by rollers at B. A 120-kg block placed on the end of the beam causes a static deflection of 15 mm at C. Assuming that the support at A undergoes a vertical periodic displacement $\delta = \delta_m \sin \omega_f t$, where $\delta_m = 10$ mm and $\omega_f = 18$ rad/s, and the support at B does not move, determine the maximum acceleration of the block at C. Neglect the weight of the beam and assume that the block does not leave the beam.

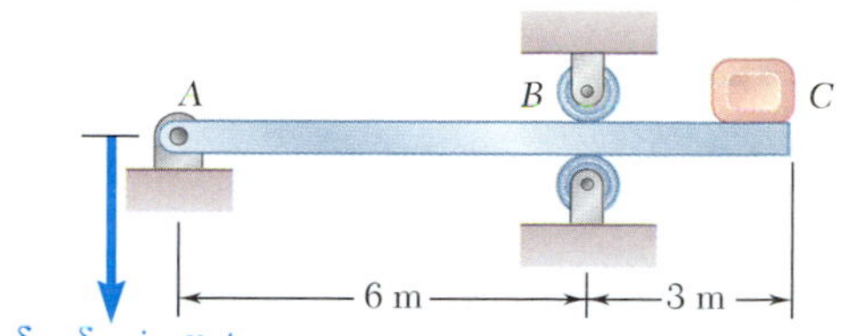

Fig. P19.162

19.163 A rod of mass m and length L rests on two pulleys A and B which rotate in opposite directions as shown. Denoting by μ_k the coefficient of kinetic friction between the rod and the pulleys, determine the frequency of vibration if the rod is given a small displacement to the right and released.

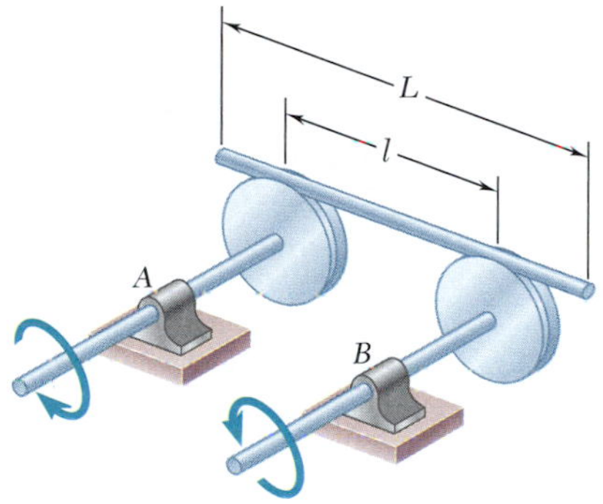

Fig. P19.163

19.164 A 4-kg block A is dropped from a height of 800 mm onto a 9-kg block B which is at rest. Block B is supported by a spring of constant $k = 1500$ N/m and is attached to a dashpot of damping coefficient $c = 230$ N · s/m. Knowing that there is no rebound, determine the maximum distance the blocks will move after the impact.

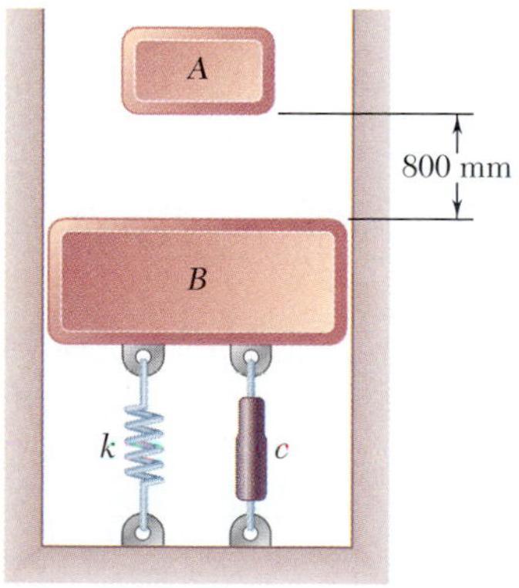

Fig. P19.164

19.165 Solve Prob. 19.164 assuming that the damping coefficient of the dashpot is $c = 300 \text{ N} \cdot \text{s/m}$.

19.166 A 30-lb uniform cylinder can roll without sliding on a 15° incline. A belt is attached to the rim of the cylinder, and a spring holds the cylinder at rest in the position shown. If the center of the cylinder is moved 2 in. down the incline and released, determine (*a*) the period of vibration, (*b*) the maximum acceleration of the center of the cylinder.

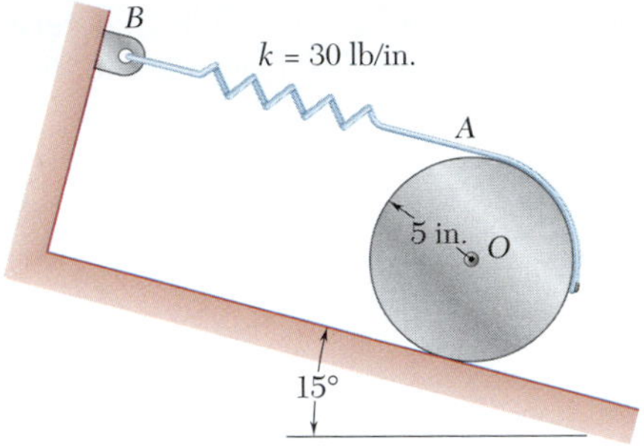

Fig. P19.166

19.167 An 800-g rod *AB* is bolted to a 1.2-kg disk. A spring of constant $k = 12$ N/m is attached to the center of the disk at *A* and to the wall at *C*. Knowing that the disk rolls without sliding, determine the period of small oscillations of the system.

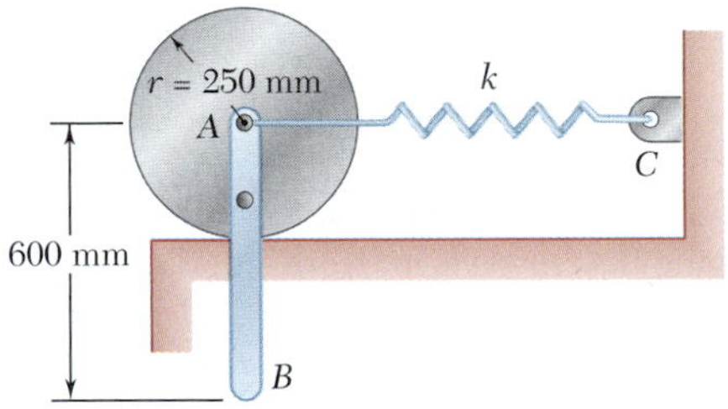

Fig. P19.167

19.168 A 1.8-kg collar *A* is attached to a spring of constant 800 N/m and can slide without friction on a horizontal rod. If the collar is moved 70 mm to the left from its equilibrium position and released, determine the maximum velocity and maximum acceleration of the collar during the resulting motion.

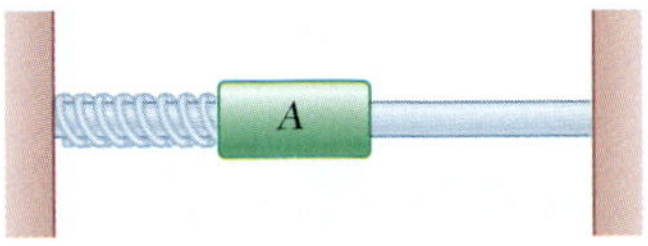

Fig. P19.168

19.169 A 30-lb block is supported by the spring arrangement shown. The block is moved from its equilibrium position 0.8 in. vertically downward and released. Knowing that the period of the resulting motion is 1.5 s, determine (*a*) the constant k, (*b*) the maximum velocity and maximum acceleration of the block.

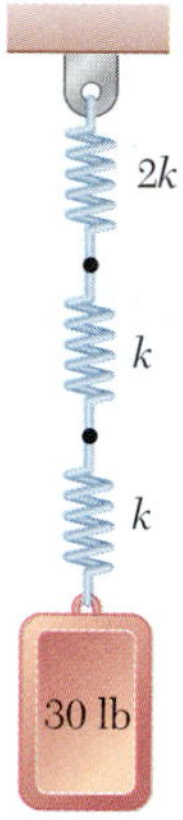

Fig. *P19.169*

Computer Problems

19.C1 The force-deflection equation for a class of nonlinear springs fixed at one end is $F = 0.5x^{1/n}$ where F is the magnitude, expressed in pounds, of the force applied at the other end of the spring and x is the deflection expressed in feet. Knowing that a block of weight W is suspended from the spring and is given a small downward displacement from its equilibrium position, use computational software to calculate and plot the frequency of vibration of the block for values of W equal to 0.5, 0.75, and 1.0 lb and values of n from 1 to 2. Assume that the slope of the force deflection curve at the point corresponding to $F = W$ can be used as an equivalent spring constant.

19.C2 The slender rod AB of weight W_{AB} is attached to two identical collars, each of weight W. The system lies in a horizontal plane and is in equilibrium in the position shown when the collar A is given a small displacement and released. Calculate and plot the period of vibration for values of the angle β from 10° to 80° for weight ratios W/W_{AB} of 0, 2, and 4.

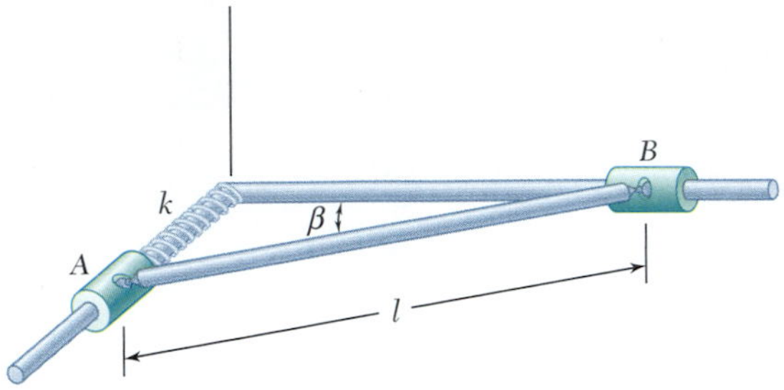

Fig. P19.C2

19.C3 A 30-lb motor is supported by four springs, each of constant 350 lb/in. The unbalance of the motor is equivalent to a weight of 0.04 lb located 5 in. from the axis of rotation. Knowing that the motor is constrained to move vertically, use computational software to calculate and plot the amplitude of the vibration and the maximum acceleration of the motor for motor speeds from 0 to 1250 rpm and from 1300 to 2500 rpm.

19.C4 Solve Prob. 19.C3, assuming that a dashpot having a coefficient of damping $c = 160$ lb · s/ft has been connected to the motor base and to the ground. Consider motor speeds from 0 to 2500 rpm.

19.C5 A machine element supported by springs and connected to a dashpot is subjected to a periodic force of magnitude $P = P_m \sin w_f t$. The *transmissibility* T_m of the system is defined as the ratio F_m/P_m of the maximum value F_m of the fluctuating periodic force transmitted to the foundation to the maximum value P_m of the periodic force applied to the machine element. Use computational software to calculate and plot the value of T_m for frequency ratio w_f/w_n from 0 to 5 and for damping factors c/c_c equal to 0.1, 0.4, 0.7, and 1. (*Hint:* Use the formula given in Prob. 19.147.)

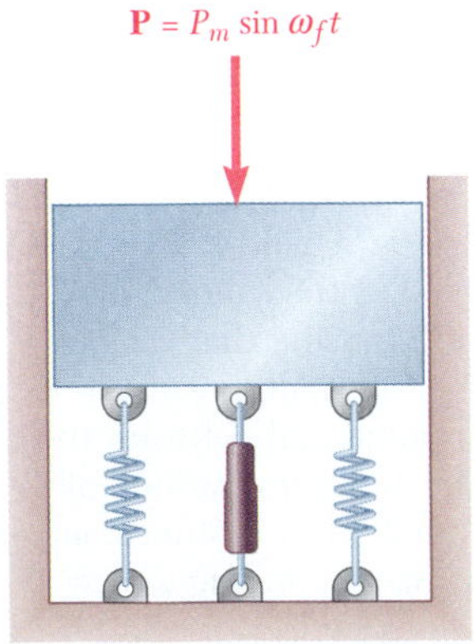

Fig. P19.C5

19.C6 The suspension of an automobile can be approximated by the simplified spring-and-dashpot system shown. Write the differential equation defining the vertical motion of the mass relative to the road when the system moves at a speed v over a road with a sinusoidal cross section of amplitude δ_m and wave length L. Determine and plot the amplitude of the displacement of the mass relative to the road as a function of the speed v for damping factors $c/c_c = 0.1$, 0.5, and 0.9.

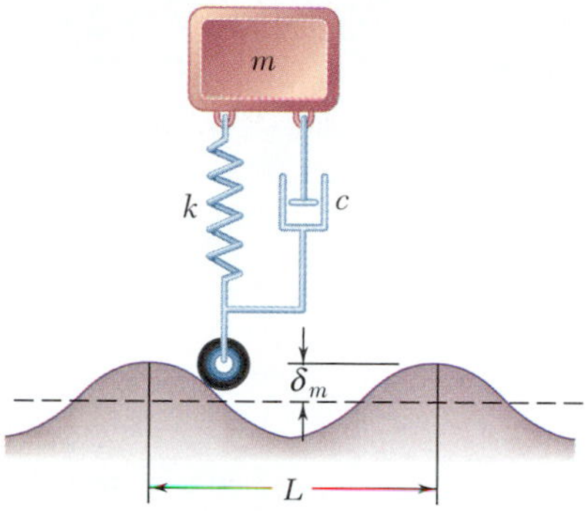

Fig. P19.C6

19.C7 A block of mass $m = 90$ kg, supported by two springs of constants $k_1 = k_2 = 85$ kN/m, is subjected to a periodic force of maximum magnitude $P_m = 500$ N. The coefficient of damping is $c = 2500$ N · s/m. (*a*) Calculate and plot the amplitude of the steady-state vibration of the block for values of ω_f from 0 to 100 rad/s. (*b*) Determine the maximum amplitude and the corresponding frequency.

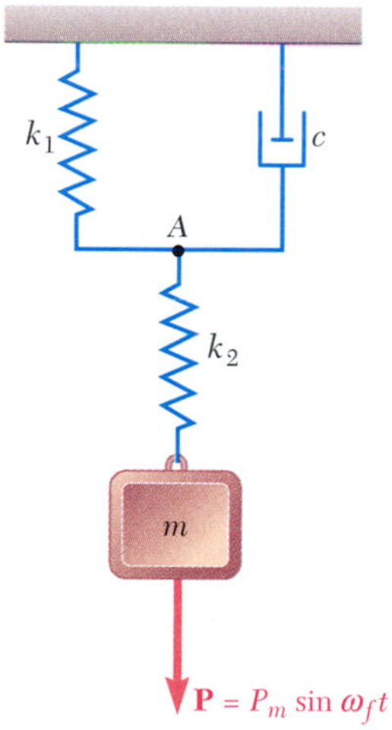

Fig. P19.C7

APPENDIX

Fundamentals of Engineering Examination

Engineers are required to be licensed when their work directly affects the public health, safety, and welfare. The intent is to ensure that engineers have met minimum qualifications involving competence, ability, experience, and character. The licensing process involves an initial exam, called the *Fundamentals of Engineering Examination*, professional experience, and a second exam, called the *Principles and Practice of Engineering*. Those who successfully complete these requirements are licensed as a *Professional Engineer*. The exams are developed under the auspices of the *National Council of Examiners for Engineering and Surveying*.

The first exam, the *Fundamentals of Engineering Examination*, can be taken just before or after graduation from a four-year accredited engineering program. The exam stresses subject material in a typical undergraduate engineering program, including statics and dynamics. The topics included in the exam cover much of the material in this book. The following is a list of the main topic areas, with references to the appropriate sections in this book. Also included are problems that can be solved to review this material.

Concurrent Force Systems (2.2–2.9; 2.12–2.14)
Problems: 2.33, 2.35, 2.36, 2.38, 2.77, 2.86, 2.93, 2.94, 2.98

Vector Forces (3.4–3.11)
Problems: 3.16, 3.17, 3.25, 3.31, 3.37, 3.40

Equilibrium in Two Dimensions (2.11; 4.1–4.7)
Problems: 4.3, 4.11, 4.14, 4.21, 4.34, 4.37, 4.70, 4.81

Equilibrium in Three Dimensions (2.15; 4.8–4.9)
Problems: 4.102, 4.104, 4.111, 4.114, 4.122, 4.126, 4.138, 4.141, 4.146

Centroid of an Area (5.2–5.7)
Problems: 5.6, 5.28, 5.38, 5.51, 5.53, 5.90, 5.96, 5.97, 5.119, 5.131, 5.133

Analysis of Trusses (6.2–6.7)
Problems: 6.2, 6.4, 6.33, 6.42, 6.43, 6.57

Equilibrium of Two-Dimensional Frames (6.9–6.11)
Problems: 6.77, 6.82, 6.84, 6.92, 6.93

Shear and Bending Moment (7.3–7.6)
Problems: 7.22, 7.25, 7.29, 7.33, 7.43, 7.47, 7.65, 7.75

Fiction (8.2–8.5; 8.10)
Problems: 8.11, 8.15, 8.21, 8.32, 8.50, 8.52, 8.101, 8.104, 8.105, 8.106

Moments of Inertia (9.2–9.10)
Problems: 9.3, 9.31, 9.32, 9.33, 9.75, 9.78, 9.84, 9.90, 9.103, 9.105

Kinematics (11.1–11.6; 11.9–11.14; 15.2–15.8)
Problems: 11.7, 11.35, 11.61, 11.66, 11.102, 15.6, 15.39, 15.57, 15.65, 15.81, 15.110, 15.138

Force, Mass, and Acceleration (12.1–12.6; 16.2–16.8)
Problems: 12.5, 12.13, 12.30, 12.31, 12.38, 12.45, 12.55, 16.5, 16.6, 16.9, 16.15, 16.22, 16.30, 16.54, 16.58, 16.72, 16.76, 16.131

Work and Energy (13.1–13.6; 13.8; 17.1–17.7)
Problems: 13.5, 13.7, 13.13, 13.39, 13.41, 13.50, 13.57, 13.58, 13.63, 17.2, 17.3, 17.18, 17.21

Impulse and Momentum (13.10–13.15; 17.8–17.12)
Problems: 13.123, 13.131, 13.138, 13.149, 13.155, 13.159, 17.43, 17.48, 17.51, 17.60, 17.88, 17.98

Vibrations (19.1–19.3; 19.5–19.7)
Problems: 19.3, 19.12, 19.21, 19.24, 19.30, 19.51, 19.63, 19.78, 19.98, 19.106, 19.114

PHOTO CREDITS

CHAPTER 1

Opener (both): © Bettmann/CORBIS; **p. 4:** © NASA; **p. 9:** © Robert Rathe Photography; **p. 10:** Photo copyright 1998, Lockheed Martin.

CHAPTER 2

Opener: © David Parker, 1997/Science Photo Library; **p. 37:** © H. David Seawell/CORBIS; **p. 57:** © D'Arazien/Getty Images.

CHAPTER 3

Opener: © Steve Kaufman/CORBIS; **p. 107:** © Elliot Eisenberg; **p. 112:** © Steve Hix/Getty Images; **p. 124:** © Steve Raymer/CORBIS; **p. 127:** © Elliot Eisenberg; **p. 128:** © Elliot Eisenberg.

CHAPTER 4

Opener: Courtesy Central Artery/Tunnel Project, Boston MA; **p. 159 (both):** © Elliot Eisenberg; **p. 160 (top):** © Lucinda Dowell; **p. 160 (middle and bottom):** Courtesy National Information Services for Earthquake Engineering, University of California, Berkeley; **p. 192 (both):** © Elliot Eisenberg.

CHAPTER 5

Opener: Courtesy Central Artery/Tunnel Project, Boston MA; **p. 220:** © Christies Images/SuperStock; **p. 238:** © Annie Griffiths Belt/CORBIS; **p. 248:** © Ghislain & Marie David de Lossy/Getty Images; **p. 249:** © Bruce Hands/Stock, Boston; **p. 259:** © NASA.

CHAPTER 6

Opener: Natalie Fobes/CORBIS; **p. 286:** Courtesy National Information Services for Earthquake Engineering, University of California, Berkeley; **p. 288:** © Ferdinand Beer; **p. 289:** © Sabina Dowell; **p. 293:** © EyeWire Collection/Getty Images; **p. 331:** Courtesy of Luxo ASA, Oslo, Norway.

CHAPTER 7

Opener (both): © Graeme-Peacock; **p. 355:** © Sabina Dowell; **p. 362:** © Elliot Eisenberg; **p. 383:** © Michael S. Yamashita/CORBIS; **p. 386:** © Brian Yarvin/Photo Researchers; **p. 395:** Steve Kahn/Getty Images.

CHAPTER 8

Opener: Richard Cummins/CORBIS; **p. 416:** © Chuck Savage/CORBIS; **p. 431:** © Ted Spiegel/CORBIS; **p. 451:** © Adam Woolfitt/CORBIS; **p. 452:** © Elliot Eisenberg.

CHAPTER 9

Opener: © Steve Dunwell/Getty Images; **p. 484:** © Ed Eckstein/CORBIS; **p. 513:** Courtesy of Caterpillar Engine Division.

CHAPTER 10

Opener: Courtesy of Stamm Manufacturing, Fort Pierce, FL; **p. 559:** Courtesy of Altec Industries; **p. 563:** Courtesy of DE-STA-CO Industries.

CHAPTER 11

Opener: © Tom Twomey/Check Six; **p. 605:** © Jonathan Nourok/PhotoEdit; **p. 646:** © Duomo/CORBIS; **p. 647:** Navy; **p. 664:** © Phyllis Picardi/Index Stock.

CHAPTER 12

Opener: © Lester Lefkowitz/CORBIS; **p. 698:** © Mike Powell/Getty Images; **p. 718:** © Bohemian Nomad Picturemakers/CORBIS; **p. 720:** © J. Von Rummelhoff, Mediaworks, University of California, Davis; **p. 732:** NASA.

CHAPTER 13

Opener: © James Noble/CORBIS; **p. 808:** © Banut AB, Sweden; **p. 823:** © Andrew Davidhazy/RIT; **p. 824:** © Tom McCarthy/Index Stock; p. **826:** © Sabina Dowell.

CHAPTER 14

Opener: Warren Gretz/National Renewable Energy Laboratory; **p. 864:** NASA; **p. 875:** © Peter M. Fisher/CORBIS; **p. 888:** © The Boeing Company; **p. 890:** NASA.

CHAPTER 15

Opener: Courtesy of The Gear Works Inc., Seattle; **p. 920:** © Zefa-Faltner/CORBIS; **p. 922:** © Joseph Nettis/Stock Boston Inc./PictureQuest; **p. 958:** Royalty-Free/CORBIS; **p. 986:** © Northrop Grumman/Index Stock Imagery/PictureQuest; **p. 1000:** © Stocktrek/CORBIS.

CHAPTER 16

Opener: © David Woods/CORBIS; **p. 1027:** © Neil Rabinowitz/CORBIS; **p. 1030:** © Larry Bray/Getty Images; **p. 1032:** © Tony Arruza/CORBIS; **p. 1054:** © Robert E. Daemmrich.

CHAPTER 17

Opener: © Chuck Savage/CORBIS; **p. 1087:** © David Madison Sports Images; **p. 1107:** (both) © Photography by Leah; **p. 1122:** Courtesy MTS Systems Corporation.

CHAPTER 18

Opener: © Rick Bostick/Index Stock; **p. 1150:** © SuperStock; **p. 1152:** © Matthias Kulka/CORBIS; **p. 1168:** © Roger Ressmeyer/CORBIS; **p. 1169:** Courtesy of Caterpillar Engine Division.

CHAPTER 19

Opener: Courtesy of MTS Systems Corporation; **p. 1252:** © Tony Freeman/Index Stock; **p. 1264:** © George White Location Photography.

Index

Answers to Problems

Answers to problems with a number set in straight type are given on this and the following pages. Answers to problems set in italic are not listed. Answers to computer problems are given at www.mhhe.com/beerjohnston7.

CHAPTER 2

2.1 8.40 kN ⦪ 19.0°.
2.2 575 N ⦫ 67.0°.
2.3 37 lb ⦪ 76.0°.
2.5 *(a)* 107.6 N. *(b)* 75.0 N.
2.6 *(a)* 89.2 N. *(b)* 55.8 N.
2.7 *(a)* 113.3°. *(b)* 71.5 lb.
2.8 *(a)* 22.7°. *(b)* 69.1 lb.
2.9 *(a)* 489 N. *(b)* 738 N.
2.10 *(a)* 43.5°. *(b)* 513 N.
2.13 *(a)* 10.0 lb ↓. *(b)* 17.32 lb.
2.14 *(a)* 206 N →. *(b)* 295 N.
2.15 25.2 lb ⦬ 8.13°.
2.16 8.38 kN ⦪ 18.76°.
2.17 576 N ⦫ 66.4°.
2.18 37.2 lb ⦪ 76.6°.
2.21 (20 kN) 15.32 kN, 12.86 kN; (30 kN) −10.26 kN, 28.2 kN; (42 kN) −39.5 kN, 14.36 kN.
2.22 (40 lb) −30.6 lb, −25.7 lb; (60 lb) 30.0 lb, −52.0 lb; (80 lb) 72.5 lb, 33.8 lb.
2.23 (204 lb) −96.0 lb, 180.0 lb; (212 lb) 112.0 lb, 180.0 lb; (400 lb) −320 lb, −240 lb.
2.25 *(a)* 176.2 N. *(b)* 113.3 N.
2.26 *(a)* 1674 N. *(b)* 1371 N.
2.27 *(a)* 339 lb. *(b)* 218 lb ↓.
2.28 *(a)* 25.9 lb. *(b)* 6.70 lb.
2.31 530 N ∡ 38.5°.
2.32 65.2 kN ⦬ 58.2°.
2.33 84.2 lb ⦪ 31.4°.
2.34 327 lb ⦬ 21.5°.
2.35 1007 N ∡ 4.91°.
2.36 1019 N ∡ 26.1°.
2.39 *(a)* 40.3°. *(b)* 252 lb.
2.40 *(a)* 116.0 lb. *(b)* 60.0 lb.
2.41 *(a)* 7.29 kN. *(b)* 9.03 kN.
2.43 *(a)* 441 lb. *(b)* 487 lb.
2.44 *(a)* 5.23 kN. *(b)* 0.503 kN.
2.45 *(a)* 169.7 lb. *(b)* 348 lb.
2.46 *(a)* 2.81 kN. *(b)* 2.66 kN.
2.49 $F_C = 332$ lb; $F_D = 368$ lb.
2.50 $F_B = 672$ lb; $F_D = 232$ lb.
2.53 *(a)* 750 N. *(b)* 1190 N.
2.54 $0 \leq W \leq 609$ N.
2.55 *(a)* 218 kN. *(b)* 15.11 kN.
2.56 *(a)* 33.1 kN. *(b)* 141.7 kN.
2.57 *(a)* 66.2 N. *(b)* 208 N.
2.58 *(a)* 62.8 N. *(b)* 758 mm.
2.59 *(a)* 55.0°. *(b)* 2.41 kN.
2.60 4.14 m.
2.61 *(a)* 270 lb. *(b)* 77.5°.
2.64 *(a)* 90.0°. *(b)* 68.9 lb.
2.65 40.0 N.
2.66 415 mm.
2.68 *(b)* 916 N. *(d)* 687 N.
2.69 149.1 lb ⦬ 32.3° or 274 lb ⦫ 32.3°.
2.70 *(a)* 24.2°. *(b)* 146.0 lb.
2.71 *(a)* 2.30 kN. *(b)* 3.53 kN.
2.73 *(a)* +157.0 lb, +100.0 lb, −73.2 lb. *(b)* 38.3°, 60.0°, 111.5°.
2.75 *(a)* +2.46 kN, −2.70 kN, +2.07 kN. *(b)* 54.1°, 130.0°, 60.5°.
2.76 *(a)* −1.076 kN, −2.55 kN, +2.31 kN. *(b)* 107.4°, 135.0°, 50.1°.
2.77 *(a)* 576 N. *(b)* 67.5°, 30.0°, 108.7°.
2.78 *(a)* 200 N. *(b)* 112.5°, 30.0°, 108.7°.
2.79 *(a)* −30.0 lb, +103.9 lb, +52.0 lb. *(b)* 104.5°, 30.0°, 64.3°.
2.81 $F = 900$ lb; $\theta_x = 27.3°$, $\theta_y = 73.2°$, $\theta_z = 110.8°$.
2.82 $F = 1300$ N; $\theta_x = 72.1°$, $\theta_y = 157.4°$, $\theta_z = 76.7°$.
2.83 *(a)* 135.0°. *(b)* $F_x = 121.7$ N, $F_z = 158.5$ N; $F = 283$ N.
2.86 *(a)* $F_y = -359$ lb, $F_z = -437$ lb. *(b)* $\theta_x = 70.5°$, $\theta_y = 126.7°$.
2.87 +400 N, +2000 N, −500 N.
2.89 +96.0 lb, +144.0 lb, −108.0 lb.
2.90 −75.0 lb, +144.0 lb, +108.0 lb.
2.91 +120.0 N, −140.0 N, −120.0 N.
2.94 701 lb; $\theta_x = 77.4°$, $\theta_y = 37.4°$, $\theta_z = 55.5°$.
2.95 1826 N; $\theta_x = 48.2°$, $\theta_y = 116.6°$, $\theta_z = 53.4°$.
2.96 1826 N; $\theta_x = 50.6°$, $\theta_y = 117.6°$, $\theta_z = 51.8°$.
2.97 379 N; $\theta_x = 129.3°$, $\theta_y = 40.0°$, $\theta_z = 96.1°$.
2.98 *(a)* 1807 lb. *(b)* 2.47 kips; $\theta_x = 90.0°$, $\theta_y = 139.2°$, $\theta_z = 49.2°$.
2.100 $T_{AB} = 245$ lb; $T_{AD} = 257$ lb.
2.101 279 N.
2.103 239 lb ↑.
2.104 215 lb ↑.
2.105 8.41 kN.
2.107 4.63 kN.
2.108 $T_{AB} = 2.63$ kN; $T_{AC} = 3.82$ kN; $T_{AD} = 2.43$ kN.
2.109 1.603 lb.
2.111 6.66 kN ↑.
2.112 8.81 kN ↑.
2.113 211 lb.
2.115 $T_{AD} = T_{CD} = 135.1$ N; $T_{BD} = 46.9$ N.
2.116 $T_{AB} = 4.33$ kN; $T_{AC} = 2.36$ kN; $T_{AD} = 2.37$ kN.
2.117 $T_{AB} = 115.9$ lb; $T_{AC} = 12.78$ lb; $T_{AD} = 121.9$ lb.
2.119 $T_{BE} = 0.299$ lb; $T_{CF} = 1.002$ lb; $T_{DG} = 1.117$ lb.
2.121 $T_{AB} = 551$ N; $T_{AC} = 503$ N.
2.122 $T_{AB} = 306$ N; $T_{AC} = 756$ N.
2.123 $T_{AB} = 46.5$ lb; $T_{AC} = 34.2$ lb; $T_{ADE} = 110.8$ lb.

2.124 (*a*) 50.0 lb. (*b*) 162.0 lb. (*c*) 468 lb.
2.125 $P = 160.0$ N; $Q = 240$ N.
2.127 (*a*) 2.27 kN. (*b*) 1.963 kN.
2.129 (*a*) 523 lb. (*b*) 428 lb.
2.131 (*a*) 454 N. (*b*) 1202 N.
2.132 (*a*) 52.0 lb. (*b*) 45.0 lb.
2.134 $F_C = 6.40$ kN; $F_D = 4.80$ kN.
2.135 $F_B = 15.00$ kN; $F_C = 8.00$ kN.
2.137 68.6 in.
2.139 748 N; $\theta_x = 120.1°$, $\theta_y = 52.5°$, $\theta_z = 128.0°$.
2.140 (*a*) 368 lb →. (*b*) 213 lb.

CHAPTER 3

3.1 1.788 N · m ↺.
3.2 14.74 N ⦢ 35.2°.
3.3 50.6° or 59.1°.
3.4 12.21 lb · in. ↺.
3.6 (*a*) 80.0 N · m ↻. (*b*) 205 N. (*c*) 177.8 N ⦣ 20.0°.
3.7 (*a*) 283 lb · ft ↺. (*b*) 36.0 lb ⦣ 55.9°.
3.9 27.9 N · m ↺.
3.10 30.8 N · m ↺.
3.11 (*a*) 7600 lb · in. ↺. (*b*) 7600 lb · in. ↺.
3.12 250 lb.
3.16 84.0 mm.
3.17 (*a*) $(-4\mathbf{i} + 7\mathbf{j} + 10\mathbf{k})/\sqrt{165}$.
(*b*) $(-2\mathbf{i} + 2\mathbf{j} - 3\mathbf{k})/\sqrt{17}$.
3.19 (*a*) $-(8\text{ N}\cdot\text{m})\mathbf{i} - (7\text{ N}\cdot\text{m})\mathbf{j} - (18\text{ N}\cdot\text{m})\mathbf{k}$.
(*b*) $(17\text{ N}\cdot\text{m})\mathbf{i} + (4\text{ N}\cdot\text{m})\mathbf{j} + (31\text{ N}\cdot\text{m})\mathbf{k}$.
(*c*) 0.
3.21 $(5.24\text{ kN}\cdot\text{m})\mathbf{i} - (3.75\text{ kN}\cdot\text{m})\mathbf{k}$.
3.23 $(42.2\text{ lb}\cdot\text{in.})\mathbf{i} + (40.6\text{ lb}\cdot\text{in.})\mathbf{j} - (50.4\text{ lb}\cdot\text{in.})\mathbf{k}$.
3.25 (*a*) $-(1840\text{ lb}\cdot\text{ft})\mathbf{i} - (368\text{ lb}\cdot\text{ft})\mathbf{k}$.
(*b*) $-(1840\text{ lb}\cdot\text{ft})\mathbf{i} + (2160\text{ lb}\cdot\text{ft})\mathbf{j} + (2740\text{ lb}\cdot\text{ft})\mathbf{k}$.
3.26 $(2.98\text{ N}\cdot\text{m})\mathbf{i} - (3.16\text{ N}\cdot\text{m})\mathbf{j} - (2.28\text{ N}\cdot\text{m})\mathbf{k}$.
3.27 5.49 m.
3.28 4.45 m.
3.30 1.141 m.
3.32 10.96 ft.
3.33 10.96 ft.
3.34 2.08 m.
3.35 $\mathbf{P}\cdot\mathbf{Q} = 17$; $\mathbf{P}\cdot\mathbf{S} = 9$; $\mathbf{Q}\cdot\mathbf{S} = -82$.
3.37 43.6°.
3.39 (*a*) 134.1°. (*b*) −230 N.
3.40 (*a*) 65.0°. (*b*) 188.3 N.
3.41 (*a*) 71.1°. (*b*) 9.73 N.
3.43 (*a*) 187 in^3. (*b*) 46 in^3.
3.44 1.700.
3.45 $M_x = -30.7$ N · m; $M_y = 12.96$ N · m;
$M_z = -2.38$ N · m.
3.46 $M_x = -25.1$ N · m; $M_y = 10.60$ N · m; $M_z = 39.9$ N · m.
3.47 63.0 N.
3.48 49.0 N.
3.49 4.00 ft.
3.51 $P = 40.6$ lb; $\phi = 9.93°$; $\theta = 48.9°$.
3.53 365 lb · in.
3.55 172.9 N · m.
3.56 −187.3 N · m.
3.57 $aP/\sqrt{2}$.
3.58 (*b*) $a/\sqrt{2}$.
3.59 −18.46 kip · ft.
3.62 4.79 in.
3.63 17.50 in.
3.64 2.75 m.
3.66 295 mm.
3.68 (*a*) 8.40 N · m ↺. (*b*) 0.700 m. (*c*) 31.6°.
3.69 (*a*) 12.00 lb. (*b*) 10.29 lb. (*c*) 6.36 lb.
3.70 (*a*) 8.62 N · m ↺. (*b*) 53.1° ⦣. (*c*) 16.11 N.
3.71 (*a*) 48.0 N · m ↻. (*b*) 243 mm.
3.72 $M = 10.00$ N · m; $\theta_x = 90.0°$, $\theta_y = 143.1°$, $\theta_z = 126.9°$.
3.73 $M = 162.3$ N · m; $\theta_x = 67.5°$, $\theta_y = 23.6°$, $\theta_z = 83.2°$.
3.74 $M = 68.8$ lb · in.; $\theta_x = 76.4°$, $\theta_y = 13.56°$, $\theta_z = 90.0°$.
3.76 $M = 350$ N · m; $\theta_x = 33.1°$, $\theta_y = 64.8°$, $\theta_z = 69.9°$.
3.78 (*a*) 1724 N · m ↻. (*b*) 958 N · m ↻.
3.79 (*a*) $\mathbf{F}_B = 20$ lb ←; 100 lb · in. ↺. (*b*) $\mathbf{F}_C = 25$ lb ↓;
$\mathbf{F}_D = 25$ lb ↑.
3.81 $\mathbf{F}_B = 100.0$ N ⦥ 30.0°; $\mathbf{F}_C = 140.0$ N ⦥ 30.0°.
3.83 $\mathbf{F}_A = 93.8$ lb ⦤ 60.0°; $\mathbf{F}_C = 156.3$ lb ⦤ 60.0°.
3.84 (*a*) $\mathbf{F} = (60.0\text{ lb})\mathbf{k}$; $\mathbf{M} = (70.0\text{ lb}\cdot\text{ft})\mathbf{j}$.
(*b*) $\mathbf{F} = (60.0\text{ lb})\mathbf{k}$; 1.167 ft from *A* along *AB*.
(*c*) 23.3 lb.
3.85 (*a*) $\mathbf{F} = 800$ N ↓; $d = 3.00$ m.
(*b*) $\mathbf{F} = 800$ N ↓; $d = 0$.
3.86 (*a*) $\mathbf{F}_A = 110.0$ N ⦣ 20.0°; $\mathbf{M}_A = 3.25$ N · m ↺.
(*b*) $\mathbf{F}_E = 110.0$ N ⦣ 20.0°; 31.5 mm below *A*.
3.88 (*a*) 10.00 lb; 5.61 in. to the left of *B*. (*b*) 77.0° or −13.03°.
3.90 $\mathbf{F} = -(270\text{ lb})\mathbf{i}$; $\mathbf{M} = (648\text{ lb}\cdot\text{in.})\mathbf{j} - (1080\text{ lb}\cdot\text{in.})\mathbf{k}$.
3.91 $\mathbf{F} = (36.0\text{ N})\mathbf{i} - (288\text{ N})\mathbf{j} + (144.0\text{ N})\mathbf{k}$;
$\mathbf{M} = (1080\text{ N}\cdot\text{m})\mathbf{i} - (270\text{ N}\cdot\text{m})\mathbf{k}$.
3.92 $\mathbf{F} = -(18.00\text{ N})\mathbf{i} - (336\text{ N})\mathbf{j} + (144.0\text{ N})\mathbf{k}$;
$\mathbf{M} = (1080\text{ N}\cdot\text{m})\mathbf{i} + (135.0\text{ N}\cdot\text{m})\mathbf{k}$.
3.93 $\mathbf{F} = (2.00\text{ lb})\mathbf{i} + (38.0\text{ lb})\mathbf{j} - (24.0\text{ lb})\mathbf{k}$;
$\mathbf{M} = (792\text{ lb}\cdot\text{in.})\mathbf{i} + (708\text{ lb}\cdot\text{in.})\mathbf{j} + (1187\text{ lb}\cdot\text{in.})\mathbf{k}$.
3.95 $\mathbf{F} = (270\text{ N})\mathbf{i} - (90.0\text{ N})\mathbf{j} + (135.0\text{ N})\mathbf{k}$;
$\mathbf{M} = (88.4\text{ N}\cdot\text{m})\mathbf{i} + (97.2\text{ N}\cdot\text{m})\mathbf{j} - (42.0\text{ N}\cdot\text{m})\mathbf{k}$.
3.96 (*a*) $-(2.40\text{ N})\mathbf{j}$. (*b*) $x = -16.89$ mm, $z = -24.5$ mm.
3.98 (*a*) *Loading a*: $\mathbf{R} = 300$ lb ↓; $\mathbf{M} = 500$ lb · ft ↺.
Loading b: $\mathbf{R} = 300$ lb ↓; $\mathbf{M} = 450$ lb · ft ↻.
Loading c: $\mathbf{R} = 300$ lb ↓; $\mathbf{M} = 450$ lb · ft ↺.
Loading d: $\mathbf{R} = 200$ lb ↑; $\mathbf{M} = 450$ lb · ft ↺.
Loading e: $\mathbf{R} = 300$ lb ↓; $\mathbf{M} = 100$ lb · ft ↻.
Loading f: $\mathbf{R} = 300$ lb ↓; $\mathbf{M} = 400$ lb · ft ↺.
Loading g: $\mathbf{R} = 500$ lb ↓; $\mathbf{M} = 500$ lb · ft ↺.
Loading h: $\mathbf{R} = 300$ lb ↓; $\mathbf{M} = 450$ lb · ft ↺.
(*b*) Loadings *c* and *h*.
3.99 Loading *f*.
3.100 (*a*) $\mathbf{R} = 300$ lb ↓; 1.500 ft. (*b*) $\mathbf{R} = 200$ lb ↑; 2.25 ft.
(*c*) $\mathbf{R} = 300$ lb ↓; 0.333 ft.
3.101 Force-couple system at *E*.
3.104 (*a*) 2.61 m to the right of *A*. (*b*) 1.493 m.
3.105 $\mathbf{R} = 362$ N ⦤ 81.9°; $M = 327$ N · m.
3.106 (*a*) $\mathbf{R} = 45.4$ lb ⦤ 65.4°.
(*b*) 4.63 in. to the left of *B* and 10.09 in. below *B*.
3.108 (*a*) $\mathbf{R} = 18.61$ kN ⦢ 55.2°; 1.014 mm to the left of *EF*.
(*b*) 55.9°.
3.109 (*a*) 22.2°.
(*b*) $\mathbf{R} = 18.98$ kN ⦢ 67.8°; 3.28 mm to the left of *EF*.
3.110 773 N ⦢ 79.0°; 9.54 m to the right of *A*.
3.111 (*a*) 1.471 lb ⦣ 55.6°. (*b*) 19.12 in. to the left of *B*.
(*c*) 1.389 in. above and to the left of *A*.
3.113 350 N ⦤ 21.4°; 92.6 mm to the left of *B* and 27.4 mm to the right of *F*.
3.114 (*a*) $\mathbf{R} = F$ ⦢ $\tan^{-1}(a^2/2bx)$;
$\mathbf{M} = 2Fb^2(x - x^3/a^2)/\sqrt{a^4 + 4b^2x^2}$ ↺. (*b*) 369 mm.

3.115 **R** = −(7 lb)**i** − (8 lb)**j** − (4 lb)**k**; **M** = −(12 lb · in.)**i** + (10.5 lb · in.)**j** + (7.5 lb · in.)**k**.
3.116 **R** = −(840 N)**j** − (679 N)**k**; **M** = (4.50 N · m)**i** + (656 N · m)**j** − (440 N · m)**k**.
3.117 **A** = (8 N)**i** − (160 N)**j** + (10 N)**k**; **B** = −(48 N)**i** + (160 N)**j** + (10 N)**k**.
3.119 (*a*) 53.1°. (*b*) **R** = (8 lb)**j** + (1 lb)**k**; **M** = −(192 lb · in.)**i** + (360 lb · in.)**k**.
3.120 (*a*) **R** = (8.66 lb)**j**; **M** = (55.0 lb · in.)**i** + (45.0 lb · in.)**j** + (77.9 lb · in.)**k**. (*b*) Clockwise.
3.121 **R** = (13.64 N)**i** − (41.3 N)**j** − (6.36 N)**k**; **M** = (10.82 N · m)**i** − (2.55 N · m)**j** − (10.03 N · m)**k**.
3.122 **R** = −(40 N)**i** + (280 N)**j** − (100 N)**k**; **M** = −(135 N · m)**i** + (205 N · m)**j** + (741 N · m)**k**.
3.123 235 lb; x = 7.69 ft, z = 9.10 ft.
3.124 x = 7.03 ft, z = 3.55 ft.
3.125 3400 kN ↓; x = 30.4 m, z = 14.39 m.
3.126 a = 19.48 m; b = 17.47 m.
3.129 (*a*) P; $\theta_x = 90°$, $\theta_y = 90°$, $\theta_z = -180°$. (*b*) $3a$. (*c*) Axis of wrench is parallel to the z axis at $x = 0$, $y = a$.
3.131 (*a*) −(21 N)**j**. (*b*) 0.571 m. (*c*) Axis of wrench is parallel to the y axis at $x = 0$, z = 41.7 mm.
3.133 (*a*) −(21 lb)**j** − (20 lb)**k**. (*b*) 19.10 in. (*c*) x = 21.1 in., z = −7.43 in.
3.135 (*a*) 3P(2**i** − 20**j** − **k**)/25. (*b*) −0.0988a. (*c*) x = 2.00a, z = −1.990a.
3.137 **R** = (4 lb)**i** + (6 lb)**j** − (2 lb)**k**; p = 1.226 ft; y = −3.68 ft, z = −1.031 ft.
3.138 **A** = (M/b)**i** + $R(1 + a/b)$**k**; **B** = $-(M/b)$**i** − (aR/b)**k**.
3.142 (*a*) **F** = 360 N ⦬ 50.0°; **M** = 230 N · m ↺. (*b*) $\mathbf{F}_A$ = 253 N ↓; $\mathbf{F}_D$ = 232 N ⦬ 5.52°.
3.144 (*a*) −(600 N)**k**; d = 90 mm below ED. (*b*) −(600 N)**k**; d = 90 mm above ED.
3.145 (*a*) 196.2 N · m ↻. (*b*) 199.0 N ⦪ 59.5°.
3.147 283 lb.
3.149 **R** = (8.40 lb)**i** − (19.20 lb)**j** − (3.20 lb)**k**; **M** = (71.6 lb · ft)**i** + (56.8 lb · ft)**j** − (65.2 lb · ft)**k**.
3.150 (*a*) 12.39 N · m ↻. (*b*) 12.39 N · m ↻. (*c*) 12.39 N · m ↻.
3.151 185.2 lb ∠ 11.84°; 23.3 in. to the left of the vertical centerline of the motor.
3.152 ϕ = 24.6°; d = 34.6 in.

CHAPTER 4

4.1 (*a*) 12.13 kN ↑. (*b*) 16.81 kN ↑.
4.2 (*a*) 1.182 kN ↓. (*b*) 2.49 kN ↑.
4.3 (*a*) 834 lb ↑. (*b*) 916 lb ↑.
4.4 (*a*) 751 lb ↑. (*b*) 874 lb ↑.
4.5 (*a*) **A** = 20.0 N ↓; **B** = 150.0 N ↑. (*b*) **A** = 10.0 N ↓; **B** = 140.0 N ↑.
4.6 40.0 mm.
4.9 2.32 kg ≤ m_A ≤ 266 kg.
4.10 140.0 N ≤ T_C ≤ 180.0 N.
4.13 6.00 kips ≤ P ≤ 73.5 kips.
4.14 1.00 in. ≤ a ≤ 5.00 in.
4.15 (*a*) 28.0 N ∠ 60.0°. (*b*) 28.2 N ⦬ 6.62°.
4.16 (*a*) 300 N. (*b*) 584 N ∠ 77.5°.
4.17 (*a*) **A** = 60.0 lb ↑; **C** = 170.9 lb ⦬ 20.6°. (*b*) **A** = 97.4 lb ∠ 60.0°; **C** = 225 lb ⦬ 22.0°.
4.18 (*a*) **A** = 40.0 lb ∠ 30°; **B** = 40.0 lb ⦪ 30°. (*b*) **A** = 115.5 lb ⦫ 12.55°; **B** = 130.2 lb ⦪ 30.0°.
4.21 (*a*) 312 N ↓. (*b*) 859 N ∠ 21.3°.
4.22 932 N.
4.23 (*a*) **A** = 47.5 lb →; **B** = 42.5 lb ⦪ 28.1°. (*b*) **A** = 67.2 lb ∠ 45.0°; **B** = 46.5 lb ⦬ 36.3°.
4.24 (*a*) **A** = 51.5 lb ∠ 22.8°; **B** = 37.5 lb ←. (*b*) **A** = 50.6 lb ⦫ 20.2°; **B** = 53.0 lb ⦪ 45.0°.
4.27 (*a*) **A** = 37.3 N ⦫ 58.0°; **E** = 140.5 N ∠ 60.0°. (*b*) **A** = 0; **E** = 127.3 N ∠ 45.0°.
4.28 (*a*) 477 N. (*b*) 715 N ⦪ 60.5°.
4.29 T = 128.4 N; **A** = 145.8 N ⦫ 81.0°.
4.30 T = 404 N; **A** = 304 N ⦫ 86.9°.
4.31 T = 20.0 lb; **C** = 22.4 lb ∠ 26.6°.
4.33 (*a*) 1.305P ⦪ 40.0°. (*b*) 1.013P ∠ 9.14°.
4.34 (*a*) 32.5 lb. (*b*) 56.0 lb ⦬ 2.05°.
4.35 $T = 2P/3$; **C** = 0.577P →.
4.37 T_{BE} = 646 lb; T_{CF} = 192.1 lb; **D** = 750 lb ←.
4.38 (*a*) 17.68 N. (*b*) $\mathbf{N}_A$ = 24.1 N ⦪ 45.0°; $\mathbf{N}_D$ = 24.1 N ∠ 45.0°.
4.39 (*a*) 0°. (*b*) 12.50 N. (*c*) $\mathbf{N}_A$ = 15.31 N ⦪ 45.0°; $\mathbf{N}_D$ = 15.31 N ∠ 45.0°.
4.41 (*a*) $\mathbf{N}_B$ = 2.77 N ↑; $\mathbf{N}_C$ = 3.70 N ⦪ 39.8°. (*b*) 5.26 N ∠ 77.3°.
4.42 (*a*) 6.46 N. (*b*) $\mathbf{N}_B$ = 4.47 N ↑; $\mathbf{N}_C$ = 5.98 N ⦪ 39.8°.
4.43 **A** = **D** = 0; **B** = 192.9 lb ←; **C** = 28.0 lb →.
4.45 (*a*) **A** = 20.0 lb ↑; $\mathbf{M}_A$ = 30.0 lb · ft ↺. (*b*) **A** = 28.3 lb ∠ 45.0°; $\mathbf{M}_A$ = 30.0 lb · ft ↺. (*c*) **A** = 40.0 lb ↑; $\mathbf{M}_A$ = 60.0 lb · ft ↺.
4.46 **C** = 33.9 N ∠ 45.0°; $\mathbf{M}_C$ = 0.120 N · m ↺.
4.47 **C** = 35.8 N ∠ 26.6°; $\mathbf{M}_C$ = 0.800 N · m ↺.
4.48 T_{max} = 444 lb; T_{min} = 347 lb.
4.49 (*a*) **A** = 14.72 N ↑; $\mathbf{M}_A$ = 3.43 N · m ↺. (*b*) **A** = 8.72 N ↑; $\mathbf{M}_A$ = 1.034 N · m ↺.
4.50 2.33 N ≤ F_E ≤ 14.83 N.
4.53 (*a*) $\sin^{-1}(2M/Wl)$. (*b*) 22.0°.
4.54 (*a*) $W/2(1 - \tan\theta)$. (*b*) 39.8°.
4.55 (*a*) $\tan^{-1}(2P/mg)$. (*b*) 63.9°.
4.56 (*a*) $\sin\theta + \cos\theta = M/Pl$. (*b*) 17.11°; 72.9°.
4.57 (*a*) $2\sin^{-1}[kl/\sqrt{2}(kl - P)]$. (*b*) 141.1°.
4.59 (*a*) $\tan\theta - \sin\theta = W/kl$. (*b*) 58.0°.
4.61 (1) Completely constrained; determinate; equilibrium; **A** = 116.6 N ∠ 59.0°, **B** = 60.0 N ←. (2) Improperly constrained; indeterminate; no equilibrium. (3) Partially constrained; determinate; equilibrium; **A** = **C** = 50 N ↑. (4) Completely constrained; determinate; equilibrium; **A** = 50 N ↑, **B** = 78.1 N ⦪ 39.8°, **C** = 60 N →. (5) Completely constrained; indeterminate; equilibrium; $\mathbf{A}_y$ = 50 N ↑. (6) Completely constrained; indeterminate; equilibrium; $\mathbf{A}_x$ = 60 N →, $\mathbf{B}_x$ = 60 N ←. (7) Completely constrained; determinate; equilibrium; **A** = **C** = 50 N ↑. (8) Improperly constrained; indeterminate; no equilibrium.
4.63 P = 28.0 lb; **A** = 35.0 lb ⦪ 36.9°.
4.64 a ≥ 2.31 in.
4.65 (*a*) 312 N ↓. (*b*) 859 N ∠ 21.3°.
4.66 932 N.
4.69 **C** = 270 N ⦪ 56.3°; **D** = 167.7 N ⦫ 26.6°.
4.70 **A** = 276 N ⦫ 45.0°; **C** = 379 N ⦪ 59.0°.
4.71 215 N ⦪ 80.8°.
4.72 189.2 N ∠ 80.9°.
4.73 (*a*) 183.1 lb. (*b*) 177.5 lb ⦪ 32.3°.
4.74 (*a*) 366 lb. (*b*) 300 lb ⦬ 1.921°.

4.76 77.5 N ∠ 30.0°.
4.77 (*a*) 11.82 kN ∠ 75.0°. (*b*) 10.68 kN ∠ 73.4°.
4.79 (*a*) 45.4 N ∠ 45.0°. (*b*) 387 N ∠ 85.2°.
4.81 T = 1200 lb; **B** = 1500 lb ∠ 36.9°.
4.82 T = 544 N; **C** = 865 N ∠ 33.7°.
4.83 (*a*) **A** = 40.0 lb ∠ 30°; **B** = 40.0 lb ∠ 30°.
(*b*) **A** = 115.5 lb ∠ 12.55°; **B** = 130.2 lb ∠ 30.0°.
4.84 (*a*) 477 N. (*b*) 715 N ∠ 60.5°.
4.85 (*a*) 1.743P ∠ 55.0°. (*b*) 1.088P ∠ 23.2°.
4.86 (*a*) 1.305P ∠ 40.0°. (*b*) 1.013P ∠ 9.14°.
4.87 $\sqrt{(S^2 - L^2)/3}$
4.88 (*a*) 129.1 mm. (*b*) 17.10 N. (*c*) 8.71 N ←.
4.90 (*a*) 43.0°. (*b*) **A** = 45.7 N ←; **B** = 108.2 N ∠ 65.0°.
4.91 34.5°.
4.94 (*a*) 18.43°. (*b*) $mg/5$.
4.95 (*a*) 23.5°. (*b*) 136.3 mm.
4.96 **C** = (964 N)**j** − (843 N)**k**; **D** = −(3130 N)**j** + (482 N)**k**.
4.97 **C** = (189.3 N)**j** − (482 N)**k**; **D** = −(4160 N)**j** + (964 N)**k**.
4.99 (*a*) 34.0 lb. (*b*) **C** = −(4.20 lb)**j** − (27.2 lb)**k**;
D = (18.20 lb)**j** − (40.8 lb)**k**.
4.100 (*a*) 20.6 lb. (*b*) **C** = −(2.30 lb)**j** − (21.2 lb)**k**;
D = (9.98 lb)**j** − (35.6 lb)**k**.
4.102 T_A = 92.0 N; $T_B = T_C$ = 76.6 N.
4.103 m = 1.786 kg; x = 200 mm, z = 100 mm.
4.104 (*a*) $\mathbf{A}_y$ = 2.95 N ↑; $\mathbf{B}_y = \mathbf{C}_y$ = 0.684 N ↑. (*b*) 54.1°.
4.105 T_A = 5.00 lb; T_C = 3.18 lb; T_D = 21.8 lb.
4.106 (*a*) 1.600 ft. (*b*) T_A = 5.00 lb; T_C = 0; T_D = 25.0 lb.
4.107 $T_A = 0.211W$; $T_B = 0.366W$; $T_C = 0.423W$.
4.110 (*a*) 6.00 lb. (*b*) 0. (*c*) 6.00 lb.
4.111 (*a*) T_A = 24.5 N; $T_B = T_C$ = 36.8 N. (*b*) 150.0 mm.
4.112 $T_{BD} = T_{BE}$ = 707 N; **C** = −(200 N)**j** + (866 N)**k**.
4.113 $T_{BD} = T_{BE}$ = 5.24 kN; **A** = (5.71 kN)**i** − (2.67 kN)**j**.
4.114 T_{AD} = 486 lb; T_{BE} = 672 lb;
C = (1008 lb)**i** − (48.0 lb)**j** + (72.0 lb)**k**.
4.116 T_{BD} = 360 lb; T_{BE} = 180.0 lb;
A = −(90.0 lb)**i** + (540 lb)**j** + (60.0 lb)**k**.
4.118 T_{EG} = 1.758 kN; T_{ICFH} = 2.06 kN;
D = (4610 N)**i** − (529 N)**j** − (36.4 N)**k**.
4.119 T_{EG} = 1.507 kN; T_{ICFH} = 1.767 kN;
D = (3950 N)**i** − (373 N)**j** − (31.2 N)**k**.
4.120 (*a*) 374 N. (*b*) **E** = (104.9 N)**i** − (152.3 N)**k**;
F = (115.1 N)**i** + (176.0 N)**j** + (482 N)**k**.
4.122 (*a*) 97.1 N. (*b*) **A** = −(23.5 N)**i** + (63.8 N)**j** − (7.85 N)**k**;
B = (9.81 N)**j** + (66.7 N)**k**.
4.124 (*a*) 9.19 lb. (*b*) **A** = (6.93 lb)**j** − (1.106 lb)**k**;
B = −(1.778 lb)**i** + (1.067 lb)**j** + (5.25 lb)**k**.
4.125 (*a*) 22.8 lb. (*b*) **A** = (11.73 lb)**j** − (1.106 lb)**k**;
B = −(1.244 lb)**i** − (18.13 lb)**j** + (5.25 lb)**k**.
4.126 (*a*) 100.1 lb. (*b*) **A** = (33.3 lb)**i** + (109.6 lb)**j** + (41.1 lb)**k**;
B = (32.9 lb)**j** + (53.9 lb)**k**.
4.127 (*a*) 180.5 lb. (*b*) **A** = −(57.0 lb)**i** + (109.6 lb)**j** − (10.96 lb)**k**;
B = (32.9 lb)**j** + (106.0 lb)**k**.
4.128 (*a*) 22.4 N. (*b*) **D** = (6.00 N)**k**;
$\mathbf{M}_D$ = (0.1320 N · m)**i** + (1.008 N · m)**j** + (0.0896 N · m)**k**.
4.129 T_{CF} = 333 N; T_{DE} = 750 N; **A** = (267 N)**i** + (450 N)**k**;
$\mathbf{M}_A$ = −(54.0 N · m)**i**.
4.130 (*a*) 374 N. (*b*) **E** = (220 N)**i** + (176.0 N)**j** + (330 N)**k**;
$\mathbf{M}_E$ = −(62.7 N · m)**i** + (14.96 N · m)**k**.
4.131 (*a*) 9.19 lb. (*b*) **B** = −(1.778 lb)**i** + (8.00 lb)**j** + (4.15 lb)**k**;
$\mathbf{M}_B$ = −(33.2 lb · in.)**j** − (208 lb · in.)**k**.
4.132 T_{DI} = 164.8 N; T_{EH} = 933 N; T_{FG} = 187.8 N;
A = (1094 N)**i** + (98.7 N)**j** − (21.3 N)**k**.
4.133 T_{DI} = 180.6 N; T_{EH} = 960 N; T_{FG} = 100.9 N;
A = (1067 N)**i** + (110.4 N)**j** − (11.45 N)**k**.
4.136 **B** = (75.0 N)**k**; **C** = +(36.0 N)**j** − (20.0 N)**k**;
D = −(36.0 N)**j** + (5.00 N)**k**.
4.138 **A** = (120.0 N)**j** − (150.0 N)**k**; **B** = (180.0 N)**i** + (150 N)**k**;
C = −(180.0 N)**i** + (120.0 N)**j**.
4.140 (*a*) (3.20 lb)**i** + (4.27 lb)**j**.
(*b*) **A** = −(3.20 lb)**i** + (5.73 lb)**j** − (3.00 lb)**k**;
B = (3.00 lb)**k**.
4.141 (*a*) 1.575 lb. (*b*) **A** = −(0.975 lb)**i** + (4.40 lb)**j** − (0.300 lb)**k**;
$\mathbf{N}_B$ = (0.800 lb)**j** + (0.600 lb)**k**.
4.142 29.1 lb.
4.144 8.73 lb.
4.145 9.05 lb.
4.146 853 N.
4.148 (112.5 N)**j**.
4.149 (*a*) x = 2.00 m, y = 4.00 m. (*b*) 131.6 N.
4.151 (*a*) x = 0, y = 29.6 in., z = 25.0 in. (*b*) $\mathbf{N}_D$ = (8.98 lb)**i**.
4.152 (*a*) **D** = 20.0 lb ↓; $\mathbf{M}_D$ = 20.0 lb · ft ↺.
(*b*) **D** = 10.00 lb ↓; $\mathbf{M}_D$ = 30.0 lb · ft. ↻.
4.154 **A** = 244 N →; **D** = 344 N ∠ 22.2°.
4.156 (*a*) 325 lb ↑. (*b*) 1175 lb ↑.
4.157 **C** = 7.07 lb ∠ 45.0°; $\mathbf{M}_C$ = 43.0 lb · in. ↻.
4.158 **C** = 7.07 lb ∠ 45.0°; $\mathbf{M}_C$ = 45.0 lb · in. ↻.
4.159 (*a*) 462 N. (*b*) **C** = −(336 N)**j** + (467 N)**k**;
D = (505 N)**j** − (66.7 N)**k**.
4.161 T_{BE} = 975 N; T_{CF} = 600 N; T_{DG} = 625 N;
A = (2100 N)**i** + (175.0 N)**j** − (375 N)**k**.
4.163 (*a*) 16.00 in. (*b*) 267 mm.

CHAPTER 5

5.1 $\overline{X}$ = 5.60 in., $\overline{Y}$ = 6.60 in.
5.2 $\overline{X}$ = 96.4 mm, $\overline{Y}$ = 34.7 mm.
5.4 $\overline{X}$ = 1.421 mm, $\overline{Y}$ = 12.42 mm.
5.5 $\overline{X}$ = 49.4 mm, $\overline{Y}$ = 93.8 mm.
5.6 $\overline{X}$ = 0.721 in., $\overline{Y}$ = 3.40 in.
5.7 $\overline{X} = \overline{Y}$ = 31.1 mm.
5.9 1.340.
5.10 $\dfrac{r_1 + r_2}{\pi - 2\alpha}\cos\alpha$.
5.11 $\overline{X}$ = 0, $\overline{Y}$ = 0.632 in.
5.12 $\overline{X}$ = −15.83 mm, $\overline{Y}$ = 3.34 mm.
5.15 $\overline{X}$ = 5.95 in., $\overline{Y}$ = 14.41 in.
5.16 $\overline{X}$ = 3.21 in., $\overline{Y}$ = 3.31 in.
5.17 $(Q_x)_1 = 25.0 \times 10^3$ mm³; $(Q_x)_2 = -25.0 \times 10^3$ mm³.
5.18 $(Q_x)_1 = 1.393 \times 10^6$ mm³; $(Q_x)_2 = -1.393 \times 10^6$ mm³.
5.20 115.3 lb.
5.21 $\overline{X}$ = 4.67 in., $\overline{Y}$ = 6.67 in.
5.23 $\overline{X}$ = 1.441 mm, $\overline{Y}$ = 12.72 mm.
5.25 (*a*) 1.470 N. (*b*) 6.13 N ∠ 83.1°.
5.26 63.6°.
5.28 12.00 in.
5.29 (*a*) 0.472a. (*b*) 0.387a.
5.31 $\bar{x} = a/3$, $\bar{y} = 2h/3$.
5.32 $\bar{x} = 2a/5$, $\bar{y} = 4h/7$.
5.33 $\bar{x} = 16a/35$, $\bar{y} = 16h/35$.
5.35 $\bar{x} = 0$, $\bar{y} = \dfrac{2}{3}R\,\dfrac{3 - \sin^2\alpha}{2\alpha + \sin 2\alpha}\sin\alpha$.
5.37 $\bar{x} = 39a/50$, $\bar{y} = -39b/175$.
5.38 $\bar{x} = a/4$, $\bar{y} = 3b/10$.
5.39 $\bar{x} = 5L/4$, $\bar{y} = 33a/40$.

5.41 $\bar{x} = 0.546a, \bar{y} = 0.423b$.
5.43 $-2r\sqrt{2}/3\pi$.
5.44 $2a/5$.
5.45 $\bar{x} = -9.27a, \bar{y} = 3.09a$.
5.46 $\bar{x} = 0.363L, \bar{y} = 0.1653L$.
5.47 (*a*) $V = 2.21 \times 10^6$ mm^3; $A = 89.8 \times 10^3$ mm^2. (*b*) $V = 4.36 \times 10^6$ mm^3; $A = 202 \times 10^3$ mm^2.
5.48 (*a*) $V = 24.0 \times 10^3$ mm^3; $A = 4.50 \times 10^3$ mm^2. (*b*) $V = 26.5 \times 10^3$ mm^3; $A = 5.12 \times 10^3$ mm^2.
5.50 (*a*) π^2a^2b. (*b*) $2\pi^2a^2b$. (*c*) $2\pi a^2b/3$.
5.51 $V = 122.0$ in^3; $A = 488$ in^2.
5.53 720 mm^3.
5.54 (*a*) 3.24×10^3 mm^2. (*b*) 2.74×10^3 mm^2. (*c*) 2.80×10^3 mm^2.
5.56 0.214 lb.
5.57 2.21 kN.
5.59 0.443 lb.
5.60 1013 mm^2.
5.61 (*a*) $R = 1800$ lb, 10.80 ft to the right of A. (*b*) $\mathbf{A} = 1800$ lb ↑; $\mathbf{M}_A = 19.44$ kip · ft ↺.
5.62 (*a*) $R = 3600$ N, 3.75 m to the right of A. (*b*) $\mathbf{A} = 1350$ N ↑; $\mathbf{B} = 2250$ N ↑.
5.63 $\mathbf{A} = 1000$ lb ↑; $\mathbf{B} = 800$ lb ↑.
5.65 $\mathbf{A} = 34.0$ N ↑; $\mathbf{M}_A = 2.92$ N · m ↺.
5.66 $\mathbf{A} = 270$ N ↑; $\mathbf{B} = 720$ N ↓.
5.67 $\mathbf{A} = 3.33$ kN ↑; $\mathbf{M}_A = 6.33$ kN · m ↺.
5.69 (*a*) 1.608 ft. (*b*) $\mathbf{A} = \mathbf{B} = 152.2$ lb ↑.
5.71 $\mathbf{B} = 7.04$ kN ↑; $\mathbf{C} = 15.46$ kN ↑.
5.72 (*a*) 3.19 kN/m ↑. (*b*) 1.375 kN ↑.
5.73 $w_A = 6.67$ kN/m; $R_B = 29.0$ kN.
5.75 (*a*) $\mathbf{H} = 1589$ kN →, $\mathbf{V} = 5.93$ MN ↑. (*b*) 10.48 m to the right of A. (*c*) $\mathbf{R} = 1675$ kN ◸ 18.43°.
5.76 (*a*) $\mathbf{H} = 10.11$ kips →, $\mathbf{V} = 79.8$ kips ↑. (*b*) 15.90 ft to the right of A. (*c*) $\mathbf{R} = 13.53$ kips ◸ 41.6°.
5.77 8.32 ft.
5.78 $\mathbf{T} = 9.10$ kips ←; $\mathbf{A} = 21.6$ kips ←.
5.80 4.75%.
5.82 43.5 kN ◸ 30.0°.
5.83 $\mathbf{A} = 563$ lb ◸ 37.1°; $\mathbf{B} = 1014$ lb ↑.
5.85 1.110 m.
5.86 1.152 m.
5.87 $\mathbf{A} = 800$ lb ◹ 41.8°; $\mathbf{D} = 11.33$ lb →.
5.89 (*a*) 42.1 N ◺ 63.4°. (*b*) $\mathbf{R}_H = 37.7$ N ↓; $\mathbf{M}_H = 0.452$ N · m ↻.
5.90 (*a*) $-1.118R$. (*b*) 0.884.
5.91 $-(2h^2 - 3b^2)/2(4h - 3b)$.
5.92 $-2a(2h - b)/\pi(4h - 3b)$.
5.93 (*a*) $0.548L$. (*b*) $2\sqrt{3}$.
5.94 2.44 in.
5.95 0.749 in.
5.96 64.2 mm.
5.97 19.51 mm.
5.100 $\bar{X} = 48.1$ mm, $\bar{Y} = 68.3$ mm, $\bar{Z} = 75.8$ mm.
5.101 $\bar{X} = 20.4$ mm, $\bar{Y} = -4.55$ mm, $\bar{Z} = 29.0$ mm.
5.103 $\bar{X} = 4.54$ in., $\bar{Y} = 6.59$ in., $\bar{Z} = 6.00$ in.
5.104 $\bar{X} = 70.2$ mm, $\bar{Y} = 198.7$ mm, $\bar{Z} = 0$.
5.106 $\bar{X} = 340$ mm, $\bar{Y} = 314$ mm, $\bar{Z} = 283$ mm.
5.107 $\bar{X} = 150.0$ mm, $\bar{Y} = 11.42$ mm, $\bar{Z} = 119.4$ mm.
5.108 $\bar{X} = 5.83$ in., $\bar{Y} = 5.15$ in., $\bar{Z} = 3.94$ in.
5.109 $\bar{X} = 3.06$ ft, $\bar{Y} = 0.539$ ft, $\bar{Z} = 2.99$ ft.
5.110 $\bar{X} = 0.525$ m, $\bar{Y} = 1.241$ m, $\bar{Z} = 0.406$ m.
5.111 $\bar{X} = 0.610$ m, $\bar{Y} = 2.70$ m, $\bar{Z} = 0$.
5.113 $\bar{X} = \bar{Z} = 0$, $\bar{Y} = 10.51$ mm above the base.
5.115 $\bar{X} = 274$ mm, $\bar{Y} = -103.6$ mm.
5.116 $\bar{x}_1 = 21a/88$; $\bar{x}_2 = 27a/40$.
5.117 $\bar{x}_1 = 21h/88$; $\bar{x}_2 = 27h/40$.
5.119 $\bar{x} = 5a/16, \bar{y} = \bar{z} = 0$.
5.121 $\bar{x} = h, \bar{y} = 0.869a, \bar{z} = 0$.
5.122 $\bar{x} = 1.297a, \bar{y} = \bar{z} = 0$.
5.123 $\bar{x} = \bar{z} = 0, \bar{y} = 0.374b$.
5.125 $\bar{x} = 0, \bar{y} = \bar{z} = -R/2$.
5.127 $V = 9.50$ m^3; $\bar{x} = 3.82$ m.
5.129 $\bar{x} = 0, \bar{y} = 5h/16, \bar{z} = -a/4$.
5.130 $\bar{X} = 7.22$ in., $\bar{Y} = 9.56$ in.
5.132 $\bar{X} = 19.27$ mm, $\bar{Y} = 26.6$ mm.
5.134 0.739 m.
5.135 $V = \pi L^3/6$.
5.137 $\mathbf{B} = 1360$ lb ↑; $\mathbf{C} = 2360$ lb ↑.
5.139 (*a*) $-0.402a$. (*b*) 2/5 and 2/3.
5.140 $\bar{X} = 0.1452$ m, $\bar{Y} = 0.396$ m, $\bar{Z} = 0.370$ m.
5.141 $\bar{x} = h/6, \bar{y} = \bar{z} = 0$.

CHAPTER 6

6.1 $F_{AB} = 125.0$ N C; $F_{AC} = 260$ N C; $F_{BC} = 100.0$ N T.
6.2 $F_{AB} = 3.40$ kips T; $F_{AC} = 4.00$ kips T; $F_{BC} = 5.00$ kips C.
6.4 $F_{AB} = 1.200$ kN T; $F_{AC} = 3.36$ kN C; $F_{AD} = 3.48$ kN T; $F_{BC} = F_{CD} = 2.52$ kN C.
6.5 $F_{AB} = F_{AE} = 6.71$ kN T; $F_{AC} = F_{AD} = 10.00$ kN C; $F_{BC} = F_{DE} = 6.00$ kN C; $F_{CD} = 2.00$ kN T.
6.6 $F_{AB} = 4.38$ kips C; $F_{AD} = 8.13$ kips T; $F_{BC} = 3.70$ kips C; $F_{BD} = 6.08$ kips C; $F_{CD} = 3.70$ kips T.
6.7 $F_{AB} = F_{DE} = 200$ N C; $F_{AC} = F_{BC} = F_{CD} = F_{CE} = 520$ N T; $F_{BE} = 480$ N C.
6.9 $F_{AB} = F_{FH} = 15.00$ kN C; $F_{AC} = F_{CE} = F_{EG} = F_{GH} = 12.00$ kN T; $F_{BC} = F_{FG} = 0$; $F_{BD} = F_{DF} = 11.93$ kN C; $F_{BE} = F_{EF} = 0.714$ kN C; $F_{DE} = 0.856$ kN T.
6.10 $F_{AB} = F_{FH} = 15.00$ kN C; $F_{AC} = F_{CE} = F_{EG} = F_{GH} = 12.00$ kN T; $F_{BC} = F_{FG} = 0$; $F_{BD} = F_{DF} = 10.00$ kN C; $F_{BE} = F_{EF} = 5.00$ kN C; $F_{DE} = 6.00$ kN T.
6.11 $F_{AB} = F_{FG} = 44.3$ kips C; $F_{AC} = F_{EG} = 40.5$ kips T; $F_{BC} = F_{EF} = 11.25$ kips C; $F_{BD} = F_{DF} = 36.9$ kips C; $F_{CD} = F_{DE} = 11.25$ kips T; $F_{CE} = 27.0$ kips T.
6.13 $F_{AB} = 3.10$ kips C; $F_{AC} = F_{CE} = 2.93$ kips T; $F_{BC} = F_{FG} = 0$; $F_{BD} = F_{DF} = 2.02$ kips C; $F_{BE} = 1.000$ kip C; $F_{DE} = 1.100$ kips T; $F_{EF} = 0.750$ kip C; $F_{EG} = F_{GH} = 2.68$ kips T; $F_{FH} = 2.83$ kips C.
6.15 $F_{AB} = F_{FH} = 11.25$ kN C; $F_{AC} = F_{CE} = F_{EG} = F_{GH} = 6.75$ kN T; $F_{BC} = F_{FG} = 6.00$ kN T; $F_{BD} = F_{DF} = 9.00$ kN C; $F_{BE} = F_{EF} = 3.75$ kN T; $F_{DE} = 0$.
6.16 $F_{AB} = 9.38$ kN C; $F_{AC} = F_{CE} = 5.63$ kN T; $F_{BC} = 6.00$ kN T; $F_{BD} = F_{DF} = 6.75$ kN C; $F_{BE} = 1.875$ kN T; $F_{DE} = F_{FG} = 0$; $F_{EF} = 5.63$ kN T; $F_{EG} = F_{GH} = 3.38$ kN T; $F_{FH} = 5.63$ kN C.
6.17 $F_{AB} = 5.42$ kips C; $F_{AC} = 6.17$ kips T; $F_{BC} = 1.152$ kips C; $F_{BD} = 5.76$ kips C; $F_{CD} = 2.06$ kips T; $F_{CE} = 4.11$ kips T; $F_{DE} = 2.30$ kips C.

6.18 $F_{DF} = 6.09$ kips C; $F_{DG} = 2.06$ kips T; $F_{EG} = 4.11$ kips T;
$F_{FG} = 1.152$ kips C; $F_{FH} = 6.43$ kips C;
$F_{GH} = 6.17$ kips T.

6.19 $F_{AB} = 4.95$ kips C; $F_{AC} = F_{CE} = 3.91$ kips T; $F_{BC} = 0$;
$F_{BD} = 3.54$ kips C; $F_{BE} = 1.000$ kip C;
$F_{DE} = 0.500$ kip T; $F_{DF} = 2.52$ kips C;
$F_{DG} = 0.280$ kip C; $F_{EG} = 2.80$ kips T.

6.20 $F_{FG} = 1.750$ kips T; $F_{FH} = 2.52$ kips C;
$F_{GH} = 0.838$ kip T; $F_{GI} = F_{IK} = F_{KL} = 1.677$ kips T;
$F_{HI} = F_{IJ} = F_{JK} = 0$; $F_{HJ} = F_{JL} = 2.12$ kips C.

6.21 $F_{AB} = F_{DF} = 2.86$ kN T; $F_{AC} = F_{EF} = 2.86$ kN C;
$F_{BC} = F_{DE} = 0.750$ kN C; $F_{BD} = 2.76$ kN T;
$F_{BE} = F_{EH} = 0$;
$F_{CE} = 2.76$ kN C; $F_{CH} = F_{EJ} = 1.500$ kN C.

6.24 $F_{AB} = F_{BD} = 41.0$ kN T; $F_{AC} = 40.0$ kN C;
$F_{BC} = 3.00$ kN C;
$F_{CD} = 6.71$ kN T; $F_{CE} = 46.0$ kN C; $F_{DE} = 1.650$ kN C;
$F_{DG} = 47.2$ kN T; $F_{EF} = 50.0$ kN C; $F_{EG} = 4.29$ kN T;
$F_{FG} = 40.0$ kN C; $F_{FH} = 30.0$ kN C.

6.27 Trusses of Probs. 6.13 and 6.25 are simple trusses.

6.28 Trusses of Probs. 6.19, 6.21, and 6.23 are simple trusses.

6.30 *BC, BE, DE, EF, FG, IJ, KN, MN.*

6.32 *BC, LM.*

6.33 *BF, BG, DH, EH, GJ, HJ.*

6.34 *BC, BE, DE, FH, HI, IJ, OQ, QR.*

6.35 (*a*) *BC, FG.* (*b*) *BC, HI, IJ, JK.*

6.36 $F_{AB} = F_{AD} = 2240$ lb C; $F_{AC} = 1755$ lb C;
$F_{BC} = F_{CD} = 422$ lb T; $F_{BD} = 633$ lb T.

6.37 $F_{AB} = F_{AD} = 7.02$ kips T; $F_{AC} = 13.78$ kips C;
$F_{BC} = F_{CD} = 3.31$ kips T; $F_{BD} = 4.77$ kips C.

6.38 $F_{AB} = F_{AD} = 488$ lb C; $F_{AC} = 2080$ lb T;
$F_{BC} = F_{CD} = 1000$ lb C; $F_{BD} = 560$ lb T.

6.39 $F_{AB} = 1342$ N C; $F_{AC} = 0$; $F_{AD} = 1500$ N T;
$F_{AE} = 300$ N T; $F_{BC} = 400$ N T; $F_{BD} = 600$ N C;
$F_{CD} = 566$ N C; $F_{CE} = 200$ N C; $F_{DE} = 100.0$ N C.

6.40 (*b*) $F_{AE} = 504$ N T; $F_{BE} = 746$ N C; $F_{DE} = F_{EF} = 0$;
$F_{EG} = 730$ N T; $F_{EH} = 480$ N C.

6.41 (*b*) $F_{BG} = 696$ N C; $F_{CG} = F_{DG} = F_{FG} = 0$;
$F_{EG} = 730$ N T; $F_{GH} = 550$ N C.

6.42 $F_{CE} = 4.00$ kN T; $F_{DE} = 1.300$ kN T; $F_{DF} = 4.50$ kN C.

6.44 $F_{CE} = 2.87$ kips T; $F_{DE} = 2.28$ kips T; $F_{DF} = 5.13$ kips C.

6.45 $F_{GI} = 5.13$ kips T; $F_{GJ} = 1.235$ kips T; $F_{HJ} = 6.15$ kips C.

6.46 $F_{CF} = 13.00$ kN T; $F_{EF} = 0.559$ kN T; $F_{EG} = 13.50$ kN C.

6.48 $F_{CE} = 14.00$ kips T; $F_{DE} = 1.692$ kips C;
$F_{DF} = 12.73$ kips C.

6.50 $F_{DF} = 13.00$ kN C; $F_{DG} = 4.22$ kN C; $F_{EG} = 16.22$ kN T.

6.51 $F_{GI} = 16.22$ kN T; $F_{HI} = 1.000$ kN T; $F_{HJ} = 17.33$ kN C.

6.52 $F_{BD} = 5.95$ kips C; $F_{CD} = 1.250$ kips T;
$F_{CE} = 4.50$ kips T.

6.53 $F_{EG} = 3.38$ kips T; $F_{FG} = 1.602$ kips T; $F_{FH} = 4.46$ kips C.

6.55 $F_{FH} = 5.10$ kips C; $F_{GI} = 3.89$ kips T; $F_{GJ} = 2.40$ kips T.

6.56 $F_{AB} = 36.4$ kN T; $F_{AG} = 20.0$ kN T; $F_{FG} = 51.6$ kN C.

6.57 $F_{AE} = 77.6$ kN T; $F_{EF} = 51.6$ kN C; $F_{FJ} = 82.0$ kN C.

6.58 $F_{BE} = 0.783$ kN C; $F_{CE} = 3.70$ kN T; $F_{DF} = 3.35$ kN C.

6.59 $F_{GI} = 4.79$ kN T; $F_{HJ} = 5.35$ kN C; $F_{IJ} = 2.97$ kN C.

6.60 $F_{AF} = 2.50$ kips T; $F_{EJ} = 1.500$ kips T.

6.62 $F_{DG} = 3.60$ kN C; $F_{FH} = 3.60$ kN T.

6.63 $F_{GJ} = 0.833$ kN C; $F_{HK} = 3.60$ kN T; $F_{IL} = 3.60$ kN C.

6.64 $F_{BG} = 27.4$ kN T; $F_{DG} = 9.13$ kN T.

6.65 $F_{CF} = 18.25$ kN T; $F_{CH} = 36.5$ kN T.

6.66 $F_{BD} = 3.83$ kips C; $F_{CE} = 3.31$ kips T;
$F_{CD} = 0.606$ kip T.

6.67 $F_{BD} = 2.63$ kips C; $F_{CE} = 2.37$ kips T; $F_{BE} = 0.303$ kip T.

6.68 $F_{CE} = 3.00$ kips T; $F_{DF} = 6.00$ kips C; $F_{DE} = 5.00$ kips T.

6.70 (*a*) Completely constrained, determinate.
(*b*) Partially constrained.
(*c*) Improperly constrained.

6.72 (*a*) Completely constrained, determinate.
(*b*) Completely constrained, indeterminate.
(*c*) Improperly constrained.

6.74 (*a*) Completely constrained, indeterminate.
(*b*) Completely constrained, determinate.
(*c*) Improperly constrained.

6.75 (*a*) Completely constrained, determinate.
(*b*) Partially constrained.
(*c*) Completely constrained, indeterminate.

6.76 (*a*) 112.5 lb $\measuredangle$ 36.9°. (*b*) 73.9 lb $\measuredangle$ 66.0°.

6.78 (*a*) 29.3 N C. (*b*) $\mathbf{A} = 25.1$ N $\measuredangle$ 16.69°.

6.79 (*a*) 25.5 N C. (*b*) $\mathbf{A} = 40.6$ N $\measuredangle$ 53.8°.

6.80 $\mathbf{A}_x = 100.0$ kN $\leftarrow$, $\mathbf{A}_y = 80.0$ kN $\uparrow$; $\mathbf{B}_x = 100.0$ kN $\leftarrow$,
$\mathbf{B}_y = 40.0$ kN $\downarrow$; $\mathbf{C}_x = 200$ kN $\rightarrow$, $\mathbf{C}_y = 40.0$ kN $\downarrow$.

6.81 $\mathbf{A} = 100.0$ kN $\leftarrow$; $\mathbf{B}_x = 100.0$ kN $\rightarrow$, $\mathbf{B}_y = 40.0$ kN $\uparrow$;
$\mathbf{C} = 40.0$ kN $\downarrow$.

6.82 $\mathbf{A}_x = 4.50$ kips $\leftarrow$, $\mathbf{A}_y = 5.00$ kips $\downarrow$; $\mathbf{B} = 2.25$ kips $\rightarrow$;
$\mathbf{C}_x = 2.25$ kips $\rightarrow$, $\mathbf{C}_y = 5.00$ kips $\uparrow$.

6.84 (*a*) $\mathbf{A}_x = 96.0$ lb $\leftarrow$, $\mathbf{A}_y = 112.0$ lb $\uparrow$; $\mathbf{E}_x = 96.0$ lb $\rightarrow$,
$\mathbf{E}_y = 48.0$ lb $\uparrow$. (*b*) $\mathbf{A}_x = 96.0$ lb $\leftarrow$, $\mathbf{A}_y = 32.0$ lb $\uparrow$;
$\mathbf{E}_x = 96.0$ lb $\rightarrow$, $\mathbf{E}_y = 128.0$ lb $\uparrow$.

6.85 (*a*) $\mathbf{A}_x = 72.0$ N $\leftarrow$, $\mathbf{A}_y = 48.0$ N $\uparrow$; $\mathbf{E}_x = 72.0$ N $\rightarrow$,
$\mathbf{E}_y = 72.0$ N $\uparrow$. (*b*) $\mathbf{A}_x = 32.0$ N $\leftarrow$, $\mathbf{A}_y = 8.00$ N $\uparrow$;
$\mathbf{E}_x = 32.0$ N $\rightarrow$, $\mathbf{E}_y = 112.0$ N $\uparrow$.

6.86 (*a*) $\mathbf{A}_x = 36.0$ lb $\leftarrow$, $\mathbf{A}_y = 18.00$ lb $\downarrow$; $\mathbf{E}_x = 36.0$ lb $\rightarrow$,
$\mathbf{E}_y = 18.00$ lb $\uparrow$. (*b*) $\mathbf{A}_x = 36.0$ lb $\leftarrow$, $\mathbf{A}_y = 12.00$ lb $\uparrow$;
$\mathbf{E}_x = 36.0$ lb $\rightarrow$, $\mathbf{E}_y = 12.00$ lb $\downarrow$.

6.88 (*a*) $\mathbf{A} = 192.0$ N $\downarrow$; $\mathbf{B} = 432$ N $\uparrow$. (*b*) $\mathbf{A}_x = 320$ N $\rightarrow$,
$\mathbf{A}_y = 192.0$ N $\downarrow$; $\mathbf{B}_x = 320$ N $\leftarrow$, $\mathbf{B}_y = 432$ N $\uparrow$.

6.89 (*a*) $\mathbf{B}_x = 128.0$ N $\rightarrow$, $\mathbf{B}_y = 40.0$ N $\uparrow$; $\mathbf{F}_x = 128.0$ N $\leftarrow$,
$\mathbf{F}_y = 152.0$ N $\uparrow$. (*b*) $\mathbf{B}_x = 128.0$ N $\rightarrow$, $\mathbf{B}_y = 136.0$ N $\uparrow$;
$\mathbf{F}_x = 128.0$ N $\leftarrow$, $\mathbf{F}_y = 56.0$ N $\uparrow$. (*c*) $\mathbf{B}_x = 128.0$ N $\rightarrow$,
$\mathbf{B}_y = 40.0$ N $\uparrow$; $\mathbf{F}_x = 128.0$ N $\leftarrow$, $\mathbf{F}_y = 152.0$ N $\uparrow$.

6.92 $\mathbf{A}_x = 45.0$ lb $\leftarrow$, $\mathbf{A}_y = 75.0$ lb $\uparrow$;
$\mathbf{E}_x = 45.0$ lb $\rightarrow$, $\mathbf{E}_y = 135.0$ lb $\uparrow$.

6.93 $\mathbf{B}_x = 175.0$ lb $\leftarrow$, $\mathbf{B}_y = 50.0$ lb $\downarrow$;
$\mathbf{E}_x = 175.0$ lb $\rightarrow$, $\mathbf{E}_y = 125.0$ lb $\uparrow$.

6.94 $\mathbf{A}_x = 1710$ N $\rightarrow$, $\mathbf{A}_y = 785$ N $\uparrow$;
$\mathbf{E}_x = 1710$ N $\leftarrow$, $\mathbf{E}_y = 1099$ N $\uparrow$.

6.95 $\mathbf{A}_x = 1318$ N $\rightarrow$, $\mathbf{A}_y = 361$ N $\uparrow$;
$\mathbf{E}_x = 1318$ N $\leftarrow$, $\mathbf{E}_y = 1523$ N $\uparrow$.

6.96 (*a*) $\mathbf{A} = 31.8$ kips $\uparrow$; $\mathbf{B} = 31.2$ kips $\uparrow$.
(*b*) $\mathbf{C} = 8.71$ kips $\leftarrow$; $\mathbf{D} = 43.3$ kips $\measuredangle$ 78.4°.

6.97 (*a*) $\mathbf{A} = 1.778$ kips $\uparrow$; $\mathbf{B} = 16.22$ kips $\uparrow$.
(*b*) $\mathbf{C} = 32.7$ kips $\leftarrow$; $\mathbf{D} = 35.0$ kips $\measuredangle$ 20.8°.

6.98 $\mathbf{A}_x = 3.04$ kN $\leftarrow$, $\mathbf{A}_y = 2.70$ kN $\downarrow$; $\mathbf{B}_x = 6.08$ kN $\rightarrow$,
$\mathbf{B}_y = 1.800$ kN $\uparrow$; $\mathbf{E}_x = 3.04$ kN $\leftarrow$, $\mathbf{E}_y = 0.900$ kN $\uparrow$.

6.99 $\mathbf{A}_x = 6.5$ kips $\leftarrow$, $\mathbf{A}_y = 2$ kips $\downarrow$; $\mathbf{B}_x = 18$ kips $\rightarrow$,
$\mathbf{B}_y = 3$ kips $\uparrow$; $\mathbf{E}_x = 11.5$ kips $\leftarrow$, $\mathbf{E}_y = 1$ kip $\downarrow$.

6.100 $\mathbf{B}_x = 2120$ N $\rightarrow$, $\mathbf{B}_y = 1540$ N $\downarrow$;
$\mathbf{C}_x = 1195$ N $\leftarrow$, $\mathbf{C}_y = 1540$ N $\uparrow$.

6.102 (*a*) $\mathbf{C}_x = 100$ N $\leftarrow$, $\mathbf{C}_y = 100$ N $\uparrow$; $\mathbf{D}_x = 100$ N $\rightarrow$,
$\mathbf{D}_y = 20$ N $\downarrow$. (*b*) $\mathbf{E}_x = 100$ N $\leftarrow$, $\mathbf{E}_y = 180$ N $\uparrow$.

6.104 (*a*) 2.02 kN $\measuredangle$ 18.18°. (*b*) 714 N T.

6.105 (*a*) 3.17 kN $\measuredangle$ 6.52°. (*b*) 2.74 kN T.

6.106 (*a*) $\mathbf{A}_x = 50.0$ kips $\rightarrow$, $\mathbf{A}_y = 30.5$ kips $\uparrow$.
(*b*) $\mathbf{B}_x = 50.0$ kips $\leftarrow$, $\mathbf{B}_y = 2.50$ kips $\downarrow$.

6.108 **A** = 327 N →; **B** = 827 N ←; **D** = 621 N ↑; **E** = 246 N ↑.
6.109 $F_{AF} = P/4\ C$; $F_{BG} = F_{DG} = P/\sqrt{2}\ C$; $F_{EH} = P/4\ T$.
6.111 $F_{AF} = P/\sqrt{2}\ C$; $F_{BG} = 0$; $F_{DG} = P\sqrt{2}\ C$; $F_{EH} = P/\sqrt{2}\ T$.
6.112 $F_{AF} = M_0/4a\ C$; $F_{BG} = F_{DG} = M_0/a\sqrt{2}\ T$; $F_{EH} = 3M_0/4a\ C$.
6.113 **A** = $P/15$ ↑; **D** = $2P/15$ ↑; **E** = $8P/15$ ↑; **H** = $4P/15$ ↑.
6.114 **A** = $3P/13$ ↑; **D** = $P/13$ ↑; **F** = $9P/13$ ↑.
6.115 (*a*) **A** = 2.24*P* ∠ 26.6°; **B** = 2*P* →; frame is rigid. (*b*) Frame is not rigid. (*c*) **A** = *P* ↑; **B** = *P* ↓; **C** = *P* ↑; frame is rigid.
6.116 (*a*) Frame is not rigid. (*b*) Reactions can be found for an arbitrary value of B_x; frame is rigid. (*c*) **A** = 2.09*P* ⦩ 16.70°; **B** = 2.04*P* ∠ 11.31°; frame is rigid.
6.117 (*a*) **A** = 2.06*P* ∠ 14.04°; **B** = 2.06*P* ⦩ 14.04°; frame is rigid. (*b*) Frame is not rigid. (*c*) **A** = 1.250*P* ⦩ 36.9°; **B** = 1.031*P* ∠ 14.04°; frame is rigid.
6.118 282 N →.
6.120 (*a*) 179.0 lb ↓. (*b*) 135.7 lb ⦪ 61.3°.
6.121 (*a*) 80.5 lb ↓. (*b*) 181.9 lb ⦪ 61.3°.
6.122 455 N ⦪ 16.17°.
6.125 (*a*) 7.00 kips ←. (*b*) 17.50 kips ←.
6.127 93.8 lb · in. ↻.
6.128 463 lb · in. ↻.
6.129 110.6 N · m ↻.
6.130 14.83°.
6.131 15.22 N · m ↺.
6.132 12.50 N · m ↺.
6.133 3.33 kip · in. ↻.
6.135 **D** = 2.52 kips ←; **F** = 1.694 kips ⦪ 16.46°.
6.137 **F** = 3.97 kN →; **H** = 3.31 kN ⦫ 20.9°.
6.138 **B** = 1117 N ⦫ 18.43°; **D** = 1117 N ⦪ 18.43°.
6.139 1680 N.
6.141 1.408 kips.
6.142 156.6 N.
6.143 50.9 N ⦫ 30.0°.
6.145 60.0 lb.
6.146 133.3 lb.
6.147 (*a*) 12.15 kN ⦩ 44.7°. (*b*) 10.63 kN ⦩ 35.7°.
6.148 (*a*) 14.29 kN ⦩ 0.261°. (*b*) 14.47 kN ⦩ 9.10°.
6.149 (*a*) 10.72 kN ∠ 21.8°. (*b*) 35.4 kN ⦩ 45.0°.
6.151 (*a*) 30.0 mm. (*b*) 80.0 N · m ↻.
6.153 (*a*) 57.7 N · m. (*b*) **B** = 0; **D** = −(48.1 N)**k**; **E** = (48.1 N)**k**.
6.154 (*a*) 43.3 N · m. (*b*) **B** = −(50.0 N)**k**; **D** = (83.3 N)**k**; **E** = −(33.3 N)**k**.
6.155 $\mathbf{E}_x$ = 2450 lb →, $\mathbf{E}_y$ = 3790 lb ↑; $\mathbf{F}_x$ = 648 lb →, $\mathbf{F}_y$ = 2890 lb ↓; $\mathbf{H}_x$ = 3100 lb ←, $\mathbf{H}_y$ = 900 lb ↓.
6.157 $F_{AB} = F_{BD} = F_{DF} = F_{FH}$ = 32.0 kips *T*; F_{AC} = 34.2 kips *C*; $F_{BC} = F_{BE} = F_{DE} = F_{DG} = F_{FG} = 0$; $F_{CE} = F_{EG}$ = 34.2 kips *C*; F_{GH} = 48.2 kips *C*.
6.159 F_{BC} = 4.10 kips *T*; F_{BH} = 4.39 kips *T*; F_{GH} = 8.00 kips *C*.
6.161 $\mathbf{B}_x$ = 10.00 kN →, $\mathbf{B}_y$ = 13.75 kN ↑; $\mathbf{D}_x$ = 22.0 kN ←, $\mathbf{D}_y$ = 13.75 kN ↓.
6.162 F_{AB} = 15.90 kN *C*; F_{AC} = 13.50 kN *T*; F_{BC} = 16.80 kN *C*; F_{BD} = 13.50 kN *C*; F_{CD} = 15.90 kN *T*.
6.164 F_{DF} = 5.45 kN *C*; F_{DG} = 1.000 kN *T*; F_{EG} = 4.65 kN *T*.
6.165 F_{GI} = 4.65 kN *T*; F_{HI} = 1.800 kN *C*; F_{HJ} = 4.65 kN *C*.
6.166 (*a*) 80.0 lb *T*. (*b*) 72.1 lb ⦫ 16.10°.
6.167 **E** = 970 lb →; **F** = 633 lb ⦫ 39.2°.

CHAPTER 7

7.1 (On *JD*) **F** = 0; **V** = 40.0 lb ↑; **M** = 160.0 lb · in. ↺.
7.2 (On *AJ*) **F** = 0; **V** = 60.0 lb ←; **M** = 60.0 lb · in. ↺.
7.4 (On *AJ*) **F** = 129.9 N ↘; **V** = 25.0 N ↗; **M** = 4.00 N · m ↻.
7.5 (On *AJ*) **F** = 624 N →; **V** = 260 N ↓; **M** = 52.0 N · m ↺.
7.7 (On *AJ*) **F** = 26.0 lb ↖; **V** = 15.00 lb ↗; **M** = 234 lb · in. ↻.
7.9 (On *CJ*) **F** = 110.4 N ↘; **V** = 135.8 N ↙; **M** = 50.4 N · m ↺.
7.10 (*a*) 140.0 N at *C*. (*b*) 156.5 N at *B* and *D*. (*c*) 89.6 N · m at *C*.
7.11 (On *AJ*) **F** = 48.7 lb ⦪ 60°; **V** = 64.3 lb ∠ 30°; **M** = 309 lb · in. ↻.
7.13 (On *CJ*) **F** = 5.93 N ↗; **V** = 412 N ↘; **M** = 34.8 N · m ↻.
7.14 (On *KD*) **F** = 48.0 N ↙; **V** = 83.1 N ↘; **M** = 17.86 N · m ↻.
7.15 (On *BJ*) **F** = 62.5 lb ↘; **V** = 30.0 lb ↗; **M** = 90.0 lb · ft ↺.
7.16 (On *AK*) **F** = 140.0 lb ←; **V** = 22.5 lb ↓; **M** = 54.0 lb · ft ↻.
7.19 (On *CJ*) **F** = 680 N ↘; **V** = 87.2 N ↙; **M** = 18.97 N · m ↺.
7.20 (On *AK*) **F** = 514 N ↙; **V** = 87.2 N ↖; **M** = 18.97 N · m ↻.
7.21 (*a*) (On *AJ*) **F** = 0; **V** = $P/2$ →; **M** = $Pa/2$ ↺. (*b*) (On *AJ*) **F** = 2*P* ↓; **V** = $3P/2$ ←; **M** = $3Pa/2$ ↻. (*c*) (On *AJ*) **F** = $2P/7$ ↓; **V** = $3P/14$ →; **M** = $3Pa/14$ ↺.
7.22 (*a*) (On *AJ*) **F** = $P/2$ ↓; **V** = 0; **M** = 0. (*b*) (On *AJ*) **F** = $11P/14$ ↓; **V** = $2P/7$ ←; **M** = $2Pa/7$ ↻. (*c*) (On *AJ*) **F** = $5P/2$ ↓; **V** = 2*P* ←; **M** = 2*Pa* ↻.
7.23 (On *CJ*) 0.0666*Wr* ↻.
7.24 (On *AJ*) 0.1788*Wr* ↻.
7.25 (On *AJ*) 0.0557*Wr* ↺.
7.27 0.357*Wr* for $\theta = 49.3°$.
7.29 (*b*) $wL/4$; $3wL^2/32$.
7.30 (*b*) $wL/2$; $3wL^2/8$.
7.33 (*b*) 12.63 kips; 29.8 kip · ft.
7.34 (*b*) 9.50 kN; 4.80 kN · m.
7.35 (*b*) 8.00 kips; 33.0 kip · ft.
7.36 (*b*) 810 N; 162.0 N · m.
7.37 (*b*) 14.00 kips; 114.0 kip · ft.
7.38 (*b*) 14.40 kips; 57.6 kip · ft.
7.40 (*b*) 22.0 kN; 22.0 kN · m.
7.41 (*b*) 4.50 kN; 13.50 kN · m.
7.43 (*b*) 540 lb; 203 lb · ft.
7.44 (*b*) 600 lb; 450 lb · ft.
7.45 (*b*) 1.833 kN; 799 N · m.
7.47 (*a*) +800 N; +640 N · m. (*b*) −400 N; +160 N · m.
7.49 7.50 kN; 7.20 kN · m.
7.50 (*b*) 768 N; 624 N · m.
7.51 (*a*) 0.932 ft. (*b*) 173.7 lb · ft.
7.52 (*a*) 478 mm. (*b*) 580 N · m.
7.53 (*a*) 40.0 kN. (*b*) 40.0 kN · m.
7.54 (*a*) 2.13 ft. (*b*) 4.25 kip · ft.
7.55 (*a*) 6.93 ft. (*b*) 402 lb · ft.
7.58 (*b*) $wL/4$; $3wL^2/32$.
7.59 (*b*) $wL/2$; $3wL^2/8$.
7.62 (*b*) 12.63 kips; 29.8 kip · ft.
7.63 (*b*) 810 N; 162.0 N · m.
7.64 (*b*) 6.60 kN; 5.82 kN · m.
7.65 (*b*) 10.50 kips; 30.0 kip · ft.
7.66 (*b*) 14.00 kips; 114.0 kip · ft.
7.67 (*b*) 14.40 kips; 57.6 kip · ft.
7.69 (*b*) 22.0 kN; 22.0 kN · m.
7.71 (*b*) 7.00 kN; 30.0 kN · m.
7.72 (*b*) 8.44 kip · ft, 4.25 ft from *A*.
7.73 (*b*) 42.2 kip · ft, 8.20 ft from *A*.
7.74 (*a*) 35.0 kN · m. (*b*) 57.1 kN · m.

7.75 (*a*) 22.5 kN · m, 1.500 m from *A*.
(*b*) 30.2 kN · m, 1.350 m from *A*.
7.78 (*b*) 25.0 kN · m, 2.50 m from *A*.
7.79 (*b*) 13.44 kN · m, 1.833 m from *A*.
7.80 (*a*) $V = (w_0/6L)(3x^2 - 6Lx + 2L^2)$; $M = (w_0/6L)(x^3 - 3Lx^2 + 2L^2x)$. (*c*) $0.0642w_0L^2$, at $x = 0.423L$.
7.81 $V = (w_0/L)(-2x^2 + Lx)$; $M = (w_0/6L)(-4x^3 + 3Lx^2)$. (*c*) $w_0L^2/6$, at $x = L$.
7.82 (*b*) $1.359w_0a^2$, at $x = 2.10a$.
7.84 (*a*) **P** = 850 lb ↓; **Q** = 1275 lb ↓. (*b*) $M_C = -1000$ lb · ft.
7.86 (*a*) **P** = 6.60 kN ↓; **Q** = 0.600 kN ↓.
(*b*) $V_{max} = 7.40$ kN, at *A*;
$M_{max} = 7.13$ kN · m, 1.333 m from *A*.
7.87 (*a*) **P** = 0.660 kN ↓; **Q** = 2.28 kN ↓.
(*b*) $|V|_{max} = 3.81$ kN, at *B*;
$M_{max} = 4.23$ kN · m, 3.45 m from *A*.
7.88 (*a*) 2.25 ft. (*b*) $\mathbf{A}_x = 800$ lb ←, $\mathbf{A}_y = 450$ lb ↑.
(*c*) 1000 lb.
7.90 (*a*) 3.35 kN ⦨ 17.35°. (*b*) 3.88 kN ⦢ 34.5°.
7.91 (*a*) 10.67 kN ⦨ 2.15°. (*b*) 11.26 kN ⦢ 18.65°.
7.92 (*a*) 929 lb ⦢ 23.8°. (*b*) $d_B = 2.22$ ft; $d_D = 3.28$ ft.
7.94 (*a*) 12.29 kN. (*b*) 11.00 m.
7.95 (*a*) 15.62 kN. (*b*) 8.00 m.
7.96 (*a*) 113.9 N. (*b*) 1.626 m.
7.98 32.0 lb.
7.99 34.9 lb.
7.100 $P = 108.0$ kN; $Q = 45.0$ kN.
7.102 (*a*) 3.69 kN. (*b*) 60.1 m.
7.104 (*a*) 50,200 kips. (*b*) 3575 ft.
7.105 (*a*) 56,400 kips. (*b*) 4284 ft.
7.106 (*a*) 313 mm. (*b*) 1.433°.
7.108 3.75 ft.
7.109 (*a*) $\sqrt{3L\Delta/8}$. (*b*) 3.67 m.
7.110 (*a*) 58,900 kips. (*b*) 29.2°.
7.111 (*a*) 24.0 ft to the left of *B*. (*b*) 3450 lb.
7.112 (*a*) 1.713 m. (*b*) 6.19 m.
7.113 (*a*) 2.22 m. (*b*) 5.36 kN.
7.117 11.00 m.
7.118 8.00 m.
7.120 $y = h[1 - \cos(\pi x/L)]$; $T_0 = w_0L^2/h\pi^2$; $T_{max} = (w_0L/\pi)\sqrt{(L^2/h^2\pi^2) + 1}$.
7.122 (*a*) 7.92 ft. (*b*) 1.164 lb.
7.123 (*a*) 29.7 ft. (*b*) 61.5 lb.
7.124 199.5 ft.
7.126 (*a*) 9.93 N →. (*b*) 14.83 m.
7.128 (*a*) 8.83 m. (*b*) 24.5 N →.
7.130 18.69 ft.
7.131 (*a*) 4.22 m. (*b*) 80.3°.
7.132 5.23 m.
7.133 2.68 m and 5.99 m.
7.134 12.14 lb.
7.136 27.3 lb →.
7.137 40.7 lb →.
7.138 (*a*) $a = 26.8$ m; $b = 20.4$ m. (*b*) 35.3 m.
7.139 (*a*) $a = 20.1$ m; $b = 15.29$ m. (*b*) 26.5 m.
7.140 (*a*) Point *C* is 23.9 ft to the left of *B*. (*b*) 14.42 lb.
7.141 (*a*) Point *C* is 20.7 ft to the left of *B*. (*b*) 9.55 lb.
7.143 (*a*) $1.326T_m/w$. (*b*) 12.72 km.
7.146 (*a*) 0.338. (*b*) $\theta_B = 56.5°$; $T_m = 0.755wL$.
7.147 (*b*) 11.50 kips; 14.00 kip · ft.
7.149 (*a*) 2.28 m. (*b*) $\mathbf{D}_x = 13.67$ kN →, $\mathbf{D}_y = 7.80$ kN ↑.
(*c*) 15.94 kN.
7.151 (On *BJ*) 0.289*Wr* ↰.
7.152 (On *BJ*) 0.417*Wr* ↰.
7.153 (On *CJ*) **F** = 625 N ↓; **V** = 120.0 N ←; **M** = 27.0 N · m ↰.
7.155 900 N; 675 N · m.
7.156 3040 lb · ft, 4.50 ft from *A*.
7.157 (*a*) 48.0 lb. (*b*) 10.00 ft.

CHAPTER 8

8.1 Block moves; **F** = 193.2 N ↗.
8.2 Equilibrium; **F** = 246 N ↗.
8.3 Equilibrium; **F** = 0.677 lb ↙.
8.5 $1.542 \text{ lb} \le P \le 12.08$ lb.
8.6 (*a*) 41.7 lb. (*b*) 14.04°.
8.7 (*a*) 50.5°. (*b*) 66.5°.
8.9 (*a*) $0 \le \theta \le 54.9°$. (*b*) $168.7° \le \theta \le 180.0°$.
8.11 (*a*) 72.0 lb ←. (*b*) 40.0 lb ←.
8.12 (*a*) 56.0 lb ←. (*b*) 40.0 lb ←.
8.13 43.0°.
8.15 (*a*) 141.3 N. (*b*) 117.7 N. (*c*) 50.5 N.
8.17 $Wr\mu_s(1 + \mu_s)/(1 + \mu_s^2)$.
8.18 (*a*) 0.360*Wr*. (*b*) 0.422*Wr*.
8.19 1374 lb · in.
8.21 0.200.
8.22 0.1835.
8.23 (*a*) 4.62° and 48.2°. (*b*) 0.526*W* and 0.374*W*.
8.25 0.1865.
8.27 0.283.
8.28 (*a*) 28.1 lb ←. (*b*) 0.440 in.
8.30 $(\mu_s)_A = 0.1364$; $(\mu_s)_C = 0.1512$.
8.31 $(\mu_s)_A = 0.1875$; $(\mu_s)_C = 0.1977$.
8.32 (*a*) Plate is in equilibrium.
(*b*) Plate moves downward.
8.34 (*a*) $W \le 34.6$ N. (*b*) $17.82 \text{ N} \le W \le 98.2$ N.
8.35 (*a*) $W \le 4.07$ N and $W \ge 86.4$ N. (*b*) $W \ge 246$ N.
8.36 $-11.71 \text{ lb} \le P \le 8.59$ lb.
8.37 $-12.31 \text{ N} \le P \le 105.3$ N.
8.39 37.9 N.
8.40 24.5 lb · ft ↰.
8.42 0.1757.
8.43 35.8°.
8.45 1.268*W*.
8.46 (*a*) 4.92 lb. (*b*) $\mathbf{C}_x = 1.082$ lb ←; $\mathbf{C}_y = 104.7$ lb ↑.
8.47 (*a*) 10.17 lb. (*b*) $\mathbf{C}_x = 6.52$ lb →; $\mathbf{C}_y = 105.4$ lb ↑.
8.48 1.957 kN.
8.49 2.13 kN.
8.52 2.46 kips ←.
8.54 (*a*) 1.647 kN. (*b*) 1.647 kN.
8.55 (*a*) 789 N. (*b*) 1100 N.
8.57 1.509 lb.
8.58 (*a*) Wedge is forced up and out from between plates.
(*b*) Wedge will become self-locked at $\theta = 4.20°$.
8.59 305 N.
8.60 (*b*) 57.7 lb ←.
8.62 (*a*) 821 N →. (*b*) Plank moves.
8.63 (*a*) 818 N ←. (*b*) Plank does not move.
8.64 0.384.
8.67 14.45 N · m.
8.70 52.2 lb · in.

8.71 36.1 lb · in.
8.72 15.31 N · m.
8.73 4.14 N · m.
8.74 14.48 N · m.
8.75 0.0787.
8.76 90.8 lb.
8.77 83.6 lb.
8.80 (*a*) 0.344. (*b*) 120.2 N.
8.81 T_{AB} = 290 N; T_{CD} = 310 N; T_{EF} = 331 N.
8.82 T_{AB} = 310 N; T_{CD} = 290 N; T_{EF} = 272 N.
8.83 (*a*) 1.177 kips. (*b*) 1.349°.
8.84 253 N.
8.85 250 N.
8.88 (*a*) 219 lb. (*b*) 164.0 lb.
8.89 9.95 in.
8.90 18.39 N.
8.91 15.25 lb · in.
8.95 13.80 N.
8.96 35.0 lb.
8.97 1.200 mm.
8.98 42.7 N.
8.99 (*a*) 306 lb. (*b*) 252 lb.
8.101 (*a*) 0.329. (*b*) 2.67 turns.
8.102 (*a*) 0.213. (*b*) 25.0 lb.
8.103 11.85 lb $\leq W_B \leq$ 33.8 lb.
8.104 286 N $\leq P \leq$ 4840 N.
8.107 40.1 N · m.
8.108 28.6 N · m.
8.109 (*a*) 338 lb · in. (*b*) 168.8 lb.
8.111 (*a*) 11.98 lb. (*b*) 1.900.
8.113 (*a*) 432 N · m. (*b*) 0.219.
8.114 (*a*) 646 N. (*b*) 47.0 N.
8.115 0.361.
8.116 (*a*) 23.3 lb. (*b*) 95.1 lb. (*c*) 68.8 lb.
8.118 (*a*) 1.847 kg $\leq m_A \leq$ 34.7 kg. (*b*) 2.81 kg $\leq m_A \leq$ 22.8 kg.
8.119 (*a*) 5.55 kg. (*b*) 8.00 kg.
8.120 (*a*) 2.28 kg $\leq m_A \leq$ 28.1 kg. (*b*) 3.46 kg $\leq m_A \leq$ 18.49 kg.
8.121 (*a*) 7.79 kg. (*b*) 5.40 kg.
8.122 1.074 lb.
8.124 (*a*) 538 N · m ↻. (*b*) 1.142 kN ↓.
8.125 (*a*) 538 N · m ↺. (*b*) 879 N ↓.
8.126 0.350.
8.127 0.266.
8.130 81.7 N · m.
8.131 (*a*) 638 lb · in. (*b*) 219 lb.
8.133 48.3° $\leq \theta \leq$ 78.7°.
8.135 31.0°.
8.137 14.23 kg $\leq m \leq$ 175.7 kg.
8.139 665 lb.
8.140 35.1 N · m.
8.141 (*b*) 2.69 lb.
8.142 143.0 lb $\leq P \leq$ 483 lb.
8.143 (*a*) 11.43 lb. (*b*) Motion impends at *C*.

CHAPTER 9

9.1 $5a^3b/33$.
9.2 $3a^3b/10$.
9.3 $b^3h/4$.
9.4 $47a^3b/60$.
9.5 $5ab^3/17$.
9.8 $0.0430ab^3$.
9.9 $4ab^3/15$.
9.10 $ab^3/4$.
9.13 $2.16a^3b$.
9.14 $0.1606a^3b$.
9.15 $ab^3/10$; $b/\sqrt{5}$.
9.17 $a^3b/6$; $a/\sqrt{3}$.
9.18 $15a^3b/91$; $0.524a$.
9.20 $1.523a^3h$; $1.404a$.
9.21 $12a^4$; $1.414a$.
9.22 $4ab(a^2 + 5b^2)/3$; $\sqrt{(a^2 + 5b^2)/3}$.
9.23 $ab(7a^2 + 52b^2)/210$; $\sqrt{(7a^2 + 52b^2)/70}$.
9.26 (*b*) −10.56%; −0.772%; −0.0488%.
9.27 $31\pi a^4/20$; $1.153a$.
9.28 $bh(12h^2 + b^2)/48$; $\sqrt{(12h^2 + b^2)/24}$.
9.31 2.44 in^4; 1.093 in.
9.32 7.36×10^6 mm^4; 32.0 mm.
9.33 0.402 in^4; 0.444 in.
9.35 $I_x = 60.9a^4$; $I_y = 11.86a^4$.
9.37 $\bar{I}$ = 20.6 in^4; d_2 = 2.50 in.
9.38 A = 5.00 in^2; $\bar{I}$ = 12.19 in^4.
9.39 $J_B = 608 \times 10^6$ mm^4; $J_D = 1215 \times 10^6$ mm^4.
9.40 4.33×10^6 mm^4.
9.41 $\bar{I}_x$ = 11.71 in^4; $\bar{I}_y$ = 36.4 in^4.
9.43 $\bar{I}_x = 30.6 \times 10^6$ mm^4; $\bar{I}_y = 12.03 \times 10^6$ mm^4.
9.44 $\bar{I}_x = 433 \times 10^6$ mm^4; $\bar{I}_y = 733 \times 10^6$ mm^4.
9.46 (*a*) 96.4×10^3 in^4. (*b*) 96.2×10^3 in^4.
9.47 (*a*) 122.4×10^6 mm^4. (*b*) 64.4×10^6 mm^4.
9.48 (*a*) 1392×10^6 mm^4. (*b*) 729×10^6 mm^4.
9.49 $\bar{I}_x$ = 982 in^4, $\bar{I}_y$ = 101.0 in^4;
$\bar{k}_x$ = 5.66 in., $\bar{k}_y$ = 1.814 in.
9.51 $\bar{I}_x = 126.4 \times 10^6$ mm^4, $\bar{I}_y = 108.8 \times 10^6$ mm^4;
$\bar{k}_x$ = 94.2 mm, $\bar{k}_y$ = 87.5 mm.
9.52 12.29 mm.
9.53 $\bar{I}_x$ = 9.34 in^4; $\bar{I}_y$ = 131.2 in^4.
9.55 b = 91.2 mm; $\bar{I}_x = 11.34 \times 10^6$ mm^4,
$\bar{I}_y = 34.0 \times 10^6$ mm^4.
9.56 (*a*) 364 mm.
(*b*) $\bar{I}_x = 34.1 \times 10^6$ mm^4, $\bar{I}_y = 110.2 \times 10^6$ mm^4.
9.57 $3\pi r/16$.
9.60 $1.224r$.
9.61 $F_A = F_B$ = 217 lb; $F_C = F_D$ = 222 lb.
9.62 2.66 m.
9.63 225 mm.
9.64 4.38 in.
9.67 $a^4/2$.
9.68 $a^2b^2/12$.
9.69 $-b^2h^2/8$.
9.70 $a^2b^2(4 + \pi^2)/8\pi^2$.
9.72 501 in^4.
9.73 138.2×10^6 mm^4.
9.74 -159.6×10^3 mm^4.
9.75 1.573×10^6 mm^4.
9.77 -262×10^6 mm^4.
9.78 2.81 in^4.
9.80 $\bar{I}_{x'} = 3.31 \times 10^3$ in^4; $\bar{I}_{y'} = 2.31 \times 10^3$ in^4;
$\bar{I}_{x'y'} = 1.947 \times 10^3$ in^4.
9.82 $\bar{I}_{x'} = 4.61 \times 10^6$ mm^4; $\bar{I}_{y'} = 3.82 \times 10^6$ mm^4;
$\bar{I}_{x'y'} = -3.83 \times 10^6$ mm^4.
9.83 $\bar{I}_{x'} = 149.9 \times 10^3$ mm^4; $\bar{I}_{y'} = 469 \times 10^3$ mm^4;
$\bar{I}_{x'y'} = 143.5 \times 10^3$ mm^4.

9.84 $\bar{I}_{x'} = 5.30 \text{ in}^4$; $\bar{I}_{y'} = 6.73 \text{ in}^4$; $\bar{I}_{x'y'} = 4.38 \text{ in}^4$.
9.86 7.22°; 4820 in^4, 802 in^4.
9.88 12.06°; $8.06 \times 10^6 \text{ mm}^4$, $0.365 \times 10^6 \text{ mm}^4$.
9.89 −24.0°; $524 \times 10^3 \text{ mm}^4$, $94.9 \times 10^3 \text{ mm}^4$.
9.90 −19.64°; 10.45 in^4, 1.577 in^4.
9.92 $\bar{I}_{x'} = 3.31 \times 10^3 \text{ in}^4$; $\bar{I}_{y'} = 2.31 \times 10^3 \text{ in}^4$;
$\bar{I}_{x'y'} = 1.947 \times 10^3 \text{ in}^4$.
9.94 $\bar{I}_{x'} = 4.61 \times 10^6 \text{ mm}^4$; $\bar{I}_{y'} = 3.82 \times 10^6 \text{ mm}^4$;
$\bar{I}_{x'y'} = -3.83 \times 10^6 \text{ mm}^4$.
9.95 $\bar{I}_{x'} = 149.9 \times 10^3 \text{ mm}^4$; $\bar{I}_{y'} = 469 \times 10^3 \text{ mm}^4$;
$\bar{I}_{x'y'} = 143.5 \times 10^3 \text{ mm}^4$.
9.96 $\bar{I}_{x'} = 5.30 \text{ in}^4$; $\bar{I}_{y'} = 6.73 \text{ in}^4$; $\bar{I}_{x'y'} = 4.38 \text{ in}^4$.
9.97 20.2°; $1.754a^4$, $0.209a^4$.
9.98 7.22°; 4820 in^4, 802 in^4.
9.100 29.7°; $405 \times 10^6 \text{ mm}^4$, $83.9 \times 10^6 \text{ mm}^4$.
9.101 −24.0°; $524 \times 10^3 \text{ mm}^4$, $94.9 \times 10^3 \text{ mm}^4$.
9.103 −20.0°; 39.6 in^4, 5.37 in^4.
9.104 −30.1°; $884 \times 10^6 \text{ mm}^4$, $281 \times 10^6 \text{ mm}^4$.
9.105 (*a*) $-215 \times 10^3 \text{ mm}^4$. (*b*) −28.1°. (*c*) $568 \times 10^3 \text{ mm}^4$.
9.106 19.64°; 10.45 in^4, 1.577 in^4.
9.107 −24.0°; $8.33 \times 10^6 \text{ mm}^4$, $1.509 \times 10^6 \text{ mm}^4$.
9.108 (*a*) −7.50°. (*b*) 646 in^4, 274 in^4.
9.112 $\pm 159.4 \times 10^3 \text{ mm}^4$.
9.113 (*a*) $5mr_2^2/16$. (*b*) $0.1347mr_2^2$.
9.114 (*a*) $0.0699mb^2$. (*b*) $m(a^2 + 0.279b^2)/4$.
9.116 (*a*) $5ma^2/3$. (*b*) $11ma^2/18$.
9.117 (*a*) $ma^2/6$. (*b*) $m(a^2 + 7b^2)/6$.
9.119 (*a*) $5ma^2/18$. (*b*) $3.61ma^2$.
9.120 (*a*) $0.994ma^2$. (*b*) $2.33ma^2$.
9.121 $5mb^2/18$.
9.123 (*a*) $1.085ma^2$. (*b*) $3.67ma^2$.
9.124 $m(b^2 + h^2)/10$.
9.126 $m(a^2 + b^2)/5$.
9.127 $I_x = I_y = ma^2/4$; $I_z = ma^2/2$.
9.128 (*a*) $mh^2/6$. (*b*) $m(a^2 + 4h^2 \sin^2 \theta)/24$.
(*c*) $m(a^2 + 4h^2 \cos^2 \theta)/24$.
9.129 $1.964 \times 10^{-6} \text{ lb} \cdot \text{ft} \cdot \text{s}^2$; 0.345 in.
9.130 $1.184 \times 10^{-6} \text{ kg} \cdot \text{m}^2$; 8.19 mm.
9.131 $ma^2/2$; $a/\sqrt{2}$.
9.132 2.00 in.
9.133 $281 \times 10^{-3} \text{ kg} \cdot \text{m}^2$.
9.136 (*a*) 46.0 mm. (*b*) $8.54 \times 10^{-3} \text{ kg} \cdot \text{m}^2$; 45.4 mm.
9.137 $I_x = 2.14 \text{ lb} \cdot \text{ft} \cdot \text{s}^2$; $I_y = 4.83 \text{ lb} \cdot \text{ft} \cdot \text{s}^2$; $I_z = 4.03 \text{ lb} \cdot \text{ft} \cdot \text{s}^2$.
9.139 $I_x = 612 \times 10^{-6} \text{ kg} \cdot \text{m}^2$; $I_y = 737 \times 10^{-6} \text{ kg} \cdot \text{m}^2$;
$I_z = 250 \times 10^{-6} \text{ kg} \cdot \text{m}^2$.
9.141 $I_x = 37.4 \times 10^{-3} \text{ kg} \cdot \text{m}^2$; $I_y = 90.9 \times 10^{-3} \text{ kg} \cdot \text{m}^2$;
$I_z = 76.7 \times 10^{-3} \text{ kg} \cdot \text{m}^2$.
9.142 $I_x = 15.40 \times 10^{-3} \text{ lb} \cdot \text{ft} \cdot \text{s}^2$; $I_y = 999 \times 10^{-3} \text{ lb} \cdot \text{ft} \cdot \text{s}^2$;
$I_z = 988 \times 10^{-3} \text{ lb} \cdot \text{ft} \cdot \text{s}^2$.
9.143 (*a*) $34.1 \times 10^{-3} \text{ lb} \cdot \text{ft} \cdot \text{s}^2$. (*b*) $50.1 \times 10^{-3} \text{ lb} \cdot \text{ft} \cdot \text{s}^2$.
(*c*) $34.9 \times 10^{-3} \text{ lb} \cdot \text{ft} \cdot \text{s}^2$.
9.144 $21.2 \text{ kg} \cdot \text{m}^2$.
9.145 $3.98 \text{ kg} \cdot \text{m}^2$.
9.147 (*a*) $9.88 \times 10^{-3} \text{ lb} \cdot \text{ft} \cdot \text{s}^2$. (*b*) $11.53 \times 10^{-3} \text{ lb} \cdot \text{ft} \cdot \text{s}^2$.
(*c*) $2.19 \times 10^{-3} \text{ lb} \cdot \text{ft} \cdot \text{s}^2$.
9.148 $I_x = I_z = 6.85 \times 10^{-3} \text{ kg} \cdot \text{m}^2$; $I_y = 12.63 \times 10^{-3} \text{ kg} \cdot \text{m}^2$.
9.149 $I_x = 23.2 \times 10^{-3} \text{ kg} \cdot \text{m}^2$; $I_y = 21.4 \times 10^{-3} \text{ kg} \cdot \text{m}^2$;
$I_z = 17.99 \times 10^{-3} \text{ kg} \cdot \text{m}^2$.
9.151 $I_{xy} = 0.488 \times 10^{-3} \text{ lb} \cdot \text{ft} \cdot \text{s}^2$; $I_{yz} = 1.184 \times 10^{-3} \text{ lb} \cdot \text{ft} \cdot \text{s}^2$;
$I_{zx} = 2.70 \times 10^{-3} \text{ lb} \cdot \text{ft} \cdot \text{s}^2$.
9.152 $I_{xy} = 5.94 \times 10^{-3} \text{ lb} \cdot \text{ft} \cdot \text{s}^2$; $I_{yz} = 3.45 \times 10^{-3} \text{ lb} \cdot \text{ft} \cdot \text{s}^2$;
$I_{zx} = 11.19 \times 10^{-3} \text{ lb} \cdot \text{ft} \cdot \text{s}^2$.
9.153 $I_{xy} = 36.2 \times 10^{-3} \text{ kg} \cdot \text{m}^2$; $I_{yz} = 8.49 \times 10^{-3} \text{ kg} \cdot \text{m}^2$;
$I_{zx} = 20.6 \times 10^{-3} \text{ kg} \cdot \text{m}^2$.
9.154 $I_{xy} = -691 \times 10^{-6} \text{ kg} \cdot \text{m}^2$; $I_{yz} = 203 \times 10^{-6} \text{ kg} \cdot \text{m}^2$;
$I_{zx} = -848 \times 10^{-6} \text{ kg} \cdot \text{m}^2$.
9.157 $I_{xy} = -0.1931 \text{ kg} \cdot \text{m}^2$; $I_{yz} = 0.310 \text{ kg} \cdot \text{m}^2$;
$I_{zx} = 2.26 \text{ kg} \cdot \text{m}^2$.
9.158 $I_{xy} = -187.5 \times 10^{-6} \text{ lb} \cdot \text{ft} \cdot \text{s}^2$; $I_{yz} = 0$;
$I_{zx} = -102.7 \times 10^{-6} \text{ lb} \cdot \text{ft} \cdot \text{s}^2$.
9.159 $I_{xy} = -11wa^3/g$; $I_{yz} = wa^3(\pi + 6)/2g$; $I_{zx} = -wa^3/4g$.
9.162 $I_{xy} = -m'R_1^3/2$; $I_{yz} = m'R_1^3/2$; $I_{zx} = -m'R_2^3/2$.
9.164 (*a*) $mac/20$. (*b*) $I_{xy} = mab/20$; $I_{yz} = mbc/20$.
9.165 $ma^2(10h^2 + 3a^2)/12(h^2 + a^2)$.
9.166 $3.22ma^2$.
9.167 $44.3 \times 10^{-3} \text{ lb} \cdot \text{ft} \cdot \text{s}^2$.
9.168 $4.96 \times 10^{-3} \text{ lb} \cdot \text{ft} \cdot \text{s}^2$.
9.169 $5Wa^2/18g$.
9.170 $5\rho ta^4/12$.
9.171 $6.74 \text{ kg} \cdot \text{m}^2$.
9.173 $25.3 \times 10^{-3} \text{ kg} \cdot \text{m}^2$.
9.175 (*a*) $b/a = 2.00$; $c/a = 2.00$. (*b*) $b/a = 1.000$; $c/a = 0.500$.
9.176 (*a*) 2.00. (*b*) $\sqrt{2/3}$.
9.177 (*a*) $1/\sqrt{3}$. (*b*) $\sqrt{7/12}$.
9.179 (*a*) $ma^2/6$. (*b*) $I_{x'} = ma^2/6$; $I_{y'} = I_{z'} = 11ma^2/12$.
9.181 (*a*) $K_1 = 0.363ma^2$; $K_2 = 1.583ma^2$; $K_3 = 1.720ma^2$.
(*b*) $(\theta_x)_1 = (\theta_z)_1 = 49.7°$, $(\theta_y)_1 = 113.7°$;
$(\theta_x)_2 = 45.0°$, $(\theta_y)_2 = 90.0°$, $(\theta_z)_2 = 135.0°$;
$(\theta_x)_3 = (\theta_z)_3 = 73.5°$, $(\theta_y)_3 = 23.7°$.
9.182 (*a*) $K_1 = 34.9 \times 10^{-3} \text{ lb} \cdot \text{ft} \cdot \text{s}^2$; $K_2 = 34.0 \times 10^{-3} \text{ lb} \cdot \text{ft} \cdot \text{s}^2$;
$K_3 = 50.2 \times 10^{-3} \text{ lb} \cdot \text{ft} \cdot \text{s}^2$.
(*b*) $(\theta_x)_1 = (\theta_y)_1 = 90.0°$, $(\theta_z)_1 = 0$;
$(\theta_x)_2 = 3.43°$, $(\theta_y)_2 = 86.6°$, $(\theta_z)_2 = 90.0°$;
$(\theta_x)_3 = 93.4°$, $(\theta_y)_3 = 3.41°$, $(\theta_z)_3 = 90.0°$.
9.183 (*a*) $K_1 = 1.180 \times 10^{-3} \text{ lb} \cdot \text{ft} \cdot \text{s}^2$;
$K_2 = 10.72 \times 10^{-3} \text{ lb} \cdot \text{ft} \cdot \text{s}^2$;
$K_3 = 11.70 \times 10^{-3} \text{ lb} \cdot \text{ft} \cdot \text{s}^2$.
(*b*) $(\theta_x)_1 = 72.5°$, $(\theta_y)_1 = 83.0°$, $(\theta_z)_1 = 18.89°$;
$(\theta_x)_2 = 19.38°$, $(\theta_y)_2 = 83.7°$, $(\theta_z)_2 = 108.3°$;
$(\theta_x)_3 = 98.2°$, $(\theta_y)_3 = 9.46°$, $(\theta_z)_3 = 94.7°$.
9.184 (*a*) $K_1 = 0.1639Wa^2/g$; $K_2 = 1.054Wa^2/g$; $K_3 = 1.115Wa^2/g$.
(*b*) $(\theta_x)_1 = 36.7°$, $(\theta_y)_1 = 71.6°$, $(\theta_z)_1 = 59.5°$;
$(\theta_x)_2 = 74.9°$, $(\theta_y)_2 = 54.5°$, $(\theta_z)_2 = 140.5°$;
$(\theta_x)_3 = 57.5°$, $(\theta_y)_3 = 138.8°$, $(\theta_z)_3 = 112.4°$.
9.185 (*a*) $K_1 = 0.203\rho ta^4$; $K_2 = 0.698\rho ta^4$; $K_3 = 0.765\rho ta^4$.
(*b*) $(\theta_x)_1 = 40.2°$, $(\theta_y)_1 = 50.0°$, $(\theta_z)_1 = 86.7°$;
$(\theta_x)_2 = 56.2°$, $(\theta_y)_2 = 134.5°$, $(\theta_z)_2 = 63.4°$;
$(\theta_x)_3 = 70.8°$, $(\theta_y)_3 = 108.0°$, $(\theta_z)_3 = 153.2°$.
9.186 (*a*) $K_1 = 10.02 \times 10^{-3} \text{ lb} \cdot \text{ft} \cdot \text{s}^2$;
$K_2 = 18.62 \times 10^{-3} \text{ lb} \cdot \text{ft} \cdot \text{s}^2$;
$K_3 = 22.9 \times 10^{-3} \text{ lb} \cdot \text{ft} \cdot \text{s}^2$.
(*b*) $(\theta_x)_1 = 35.2°$, $(\theta_y)_1 = (\theta_z)_1 = 65.9°$;
$(\theta_x)_2 = 90.0°$, $(\theta_y)_2 = 45.0°$, $(\theta_z)_2 = 135.0°$;
$(\theta_x)_3 = 54.7°$, $(\theta_y)_3 = (\theta_z)_3 = 125.3°$.
9.187 $a^3b/6$.
9.188 $\bar{I}_x = 18.13 \text{ in}^4$; $\bar{I}_y = 4.51 \text{ in}^4$.
9.190 (*a*) $7ma^2/18$. (*b*) $0.819ma^2$.
9.192 (*a*) $\bar{I}_{x'} = 8.83 \text{ in}^4$; $\bar{I}_{y'} = 3.20 \text{ in}^4$; $\bar{I}_{x'y'} = 3.43 \text{ in}^4$.
(*b*) 57.7° clockwise.
9.194 (*a*) 176.9 mm. (*b*) 87.8 mm.
9.196 $I_x = 28.3 \times 10^{-3} \text{ kg} \cdot \text{m}^2$, $I_y = 183.8 \times 10^{-3} \text{ kg} \cdot \text{m}^2$;
$k_x = 42.9$ mm, $k_y = 109.3$ mm.
9.197 $I_x = 4a^4$; $I_y = 16a^4/3$.
9.198 17.19 mm.

CHAPTER 10

10.1 60 N $\downarrow$.
10.2 40 lb $\downarrow$.
10.4 6.00 N · m ↻.
10.5 (*a*) 12.00 lb *T*; 2.40 in. $\rightarrow$. (*b*) 28.0 lb *T*; 5.60 in. $\rightarrow$.
10.7 (*a*) 75.0 N $\uparrow$. (*b*) 225 N $\downarrow$.
10.8 500 N $\downarrow$.
10.9 $(3P/2)\tan\theta$.
10.12 $P[(l/a)\cos^3\theta - 1]$.
10.13 1803 lb.
10.14 5.42°.
10.15 $3Pa/2$.
10.18 (*a*) $Pl\sin 2\theta$. (*b*) $3Pl\cos\theta$. (*c*) $Pl\sin\theta$.
10.19 (*a*) 126.9 lb · ft ↺. (*b*) 81.4 lb · ft ↺.
10.20 (*a*) 369 lb $\rightarrow$. (*b*) 477 lb $\rightarrow$.
10.21 7.70 lb ⦪ 30.0°.
10.22 71.5 lb · ft ↺.
10.25 39.2°.
10.26 32.3°.
10.27 40.8°.
10.28 19.81° and 51.9°.
10.30 25.0°.
10.31 39.7° and 69.0°.
10.32 15.03° and 36.9°.
10.33 40.2°.
10.36 31.0°.
10.37 60.4°.
10.39 61.2°.
10.40 78.7°; 324°; 379°.
10.41 397 N ⦩ 44.7°.
10.42 22.9° and 70.7°.
10.43 31.8 N ⦫ 35.0°.
10.44 15.27°.
10.45 9.47 kN $\nwarrow$.
10.46 10.20 kN $\nwarrow$.
10.47 $\eta = 1/(1 + \mu\cot\alpha)$.
10.49 5.00 kip · ft; 1.363 kip · ft.
10.50 $\eta = \tan\theta/\tan(\theta + \phi_s)$.
10.52 150.6 lb; 126.6 lb.
10.53 $\mathbf{A}$ = 15.00 lb $\downarrow$; $\mathbf{M}_A$ = 195.0 lb · ft ↺.
10.54 $\mathbf{D}$ = 315 lb $\downarrow$.
10.55 21.8 mm (shortened).
10.57 41.7 mm $\downarrow$.
10.60 25.0°.
10.61 39.7° and 69.0°.
10.62 40.2°.
10.65 31.0°.
10.69 θ = 14.48°, unstable; θ = 90.0°, stable; θ = 165.5°, unstable; θ = 270°, stable.
10.70 θ = 0, unstable; θ = 137.8°, stable.
10.73 θ = 61.2°, stable.
10.74 θ = 78.7°, stable; θ = 324°, unstable; θ = 379°, stable.
10.75 1.118 kg, stable.
10.76 109.4°, stable.
10.78 301 mm.
10.80 θ = 9.69°, stable; θ = 33.8°, unstable; θ = 90.0°, stable.
10.81 (*a*) $0 \le W \le 90.0$ lb. (*b*) $0 \le \theta \le 45.0°$.
10.82 θ = 13.19°, stable; θ = 76.8°, unstable.
10.83 49.1°.
10.84 70.9°.
10.85 54.9°.
10.87 61.4°.
10.89 $k > 6.94$ lb/in.
10.90 12.25 in.
10.91 $0 \le P < ka$.
10.94 $0 \le P < ka/10$.
10.95 $0 \le P < 320$ N.
10.96 $Q > 432$ N.
10.99 $0 \le P < 0.219ka$.
10.100 $0 \le P < 30.7$ lb.
10.102 17.90 kips $\nearrow$.
10.104 (*a*) and (*b*) $P < ka^2/2l$.
10.105 270 N $\uparrow$.
10.107 θ = 0 and θ = 180.0°, unstable; θ = 75.5° and θ = 285°, stable.
10.108 $\mathbf{A}$ = 250 N $\uparrow$; $\mathbf{M}_A$ = 450 N · m ↺.
10.109 1050 N $\uparrow$.
10.110 $(Pl/a)\sin\theta\cos^2\theta$.
10.112 113.8°.

CHAPTER 11

11.1 x = 992 mm, v = 506 mm/s, a = 240 mm/s^2.
11.2 t = 1.686 s, x = −14.85 m, a = 28.7 m/s^2.
11.3 x = 9.59 in., v = 175.5 in./s, a = −959 in./s^2.
11.4 x = 9.59 in., a = −17.55 in./s^2.
11.7 (*a*) 6 s. (*b*) v = 288 m/s, a = 96 m/s^2, 944 m.
11.8 (*a*) 2 s, 4 s. (*b*) x = 20 mm, 44 mm.
11.9 v = 42.4 mm/s, x = 199.5 mm.
11.10 v = −453 mm/s, x = 131.0 mm.
11.11 $x = 0.00686t^4 + 7t + 36$, $v = 0.0274t^3 + 7$.
11.12 (*a*) 35.2 ft, 17.6 ft. (*b*) 18.4 ft.
11.15 281 m^{-2}.
11.16 v = 3.25 m/s, x = 0.1366 m.
11.17 (*a*) 12 in./s. (*b*) 19 in., 21 in.
11.18 x = 20.7 in., v = 8.11 in./s.
11.21 (*a*) 5.74 ft/s. (*b*) 1.079 s.
11.22 (*a*) 22.5 ft. (*b*) 38.4 ft/s.
11.23 1.609 m/s.
11.24 (*a*) 0.3 m. (*b*) 0.461 m.
11.27 −0.0216 ft/s^2, 21.1 s.
11.28 (*a*) 7.15 km. (*b*) -52.1×10^{-6} m/s^2. (*c*) 49.9 min.
11.32 (*a*) $x = 2.36v_0T$, $a = \pi v_0/T$. (*b*) 0.363 v_0.
11.33 (*a*) 2 m/s^2. (*b*) 60 m/s.
11.34 (*a*) 6 s. (*b*) 60 m.
11.35 (*a*) 73.5 ft/s $\rightarrow$. (*b*) 61.5 ft/s $\rightarrow$. (*c*) 43.8 ft.
11.36 (*a*) 7.33 ft/s^2. (*b*) 8 s.
11.39 (*a*) 325 m. (*b*) 2.95 m $\rightarrow$.
11.40 (*a*) 44.2 m/s $\uparrow$. (*b*) 39.2 m/s $\uparrow$.
11.41 (*a*) 0.472 mi. (*b*) 40.5 mi/h.
11.42 (*a*) a_A = −6.91 ft/s^2, a_B = 6.66 ft/s^2. (*b*) 2.60 s before *A* reaches the exchange zone.
11.43 (*a*) 2.08 m/s^2. (*b*) 3.27 m/s^2.
11.44 (*a*) 3.59 m/s^2. (*b*) 7.08 s. (*c*) 278 m.
11.47 (*a*) 8 ft/s $\uparrow$. (*b*) 2 ft/s $\downarrow$. (*c*) 6 ft/s $\downarrow$.
11.48 (*a*) $\mathbf{a}_A$ = 0.12 ft/s^2 $\uparrow$; $\mathbf{a}_C$ = 0.48 ft/s^2 $\downarrow$. (*b*) $\mathbf{v}_B$ = 0.48 ft/s $\uparrow$; 0.48 ft $\uparrow$.
11.49 (*a*) $\mathbf{a}_A$ = 38 mm/s^2 $\downarrow$; $\mathbf{a}_B$ = 25.3 mm/s^2 $\uparrow$; $\mathbf{a}_C$ = 12.67 mm/s^2 $\downarrow$. (*b*) $\mathbf{v}_B$ = 203 mm/s $\uparrow$; 811 mm $\uparrow$.

11.50 (*a*) 915 mm/s ↑. (*b*) 305 mm/s ↑. (*c*) 1830 mm/s ↓. (*d*) 1220 mm/s ↓.
11.51 (*a*) 600 mm/s →. (*b*) 1200 mm/s ←. (*c*) 900 mm/s ←.
11.52 (*a*) $\mathbf{a}_A = 50.8$ mm/s² →; $\mathbf{a}_B = 25.4$ mm/s² ←. (*b*) $\mathbf{v}_B = 152.5$ mm/s ←; 458 mm ←.
11.55 (*a*) $(6.25 - 1.5\,t^2)$ mm/s². (*b*) 18 mm.
11.56 (*a*) $\mathbf{a}_B = 50$ mm/s² ↑; $\mathbf{a}_C = 77.5$ mm/s² ↓. (*b*) 0.645 s. (*c*) 16.13 mm.
11.57 (*a*) 2 in./s →. (*b*) 0. (*c*) $\mathbf{a}_A = 1.667$ in./s² ←; $\mathbf{a}_C = 1.25$ in./s² ←.
11.58 (*a*) $\mathbf{a}_B = 3.5$ in./s² ←; $\mathbf{a}_C = 5$ in./s² →. (*b*) $\mathbf{v}_A = 4$ in./s ←; $\mathbf{v}_B = 6$ in./s →. (*c*) 12 in./s ←.
11.61 (*b*) $x = -8$ ft, $v = 8$ ft/s, 88 ft.
11.62 (*b*) 5.83 s.
11.63 (*b*) 1383 m. (*c*) 9 s, 49.5 s.
11.64 (*b*) 420 m. (*c*) 10.69 s, 40 s.
11.65 (*a*) 44.1 s. (*b*) 152.3 ft/s² ↑.
11.66 7.45 in./s.
11.69 6.77 ft/s².
11.70 (*a*) 0.609 s. (*b*) $v = 1.252$ ft/s, $x = 9.73$ ft.
11.71 (*a*) $t = 576$ s, $x = 11.2$ km. (*b*) 0.262 m/s².
11.72 (*a*) −0.0395 m/s². (*b*) 0.444 m/s.
11.75 (*a*) 13.17 m. (*b*) 17.91 s.
11.76 Car *B* is 43.1 m in front of car *A*.
11.77 (*a*) 1.973 s. (*b*) 0.365 m/s, 0.1825 m/s.
11.78 (*a*) 9.58 min. (*b*) 31.3 km/h.
11.81 (*a*) 1.913 ft/s. (*b*) 0.836 ft.
11.82 (*a*) 66.6 mi/h. (*b*) 1980 ft.
11.83 (*a*) 3.19 s. (*b*) 62.6 m.
11.84 (*a*) 5.6 m/s². (*b*) 3.33 m/s².
11.86 −16 ft.
11.89 (*a*) 0, 6 ft/s² ↘ 36.9°. (*b*) 3 ft/s ↘ 36.9°, 0. (*c*) 0, 6 ft/s² ↖ 36.9°.
11.90 (*a*) 9.42 ft/s ↑. (*b*) 7.26 ft/s ←. (*c*) 3.14 ft/s ↓.
11.91 (*a*) 6.14 m/s. (*b*) $t = 1.757$ s, $x = 1.206$ m, $y = 0.761$ m, ↗ 10.9°.
11.94 (*a*) $\mathbf{r} = 20$ in. ↑; $\mathbf{v} = 43.4$ in./s ↘ 46.3°; $\mathbf{a} = 743$ in./s² ↙ 85.4°. (*b*) $\mathbf{r} = 18.10$ in. ↘ 6.0°; $\mathbf{v} = 5.65$ in./s ↗ 31.8°; $\mathbf{a} = 70.3$ in./s² ↖ 86.9°.
11.95 $v = \sqrt{R^2(1 + \omega_n{}^2t^2) + c^2}$, $a = \omega_n R\sqrt{4 + \omega_n{}^2t^2}$.
11.96 (*a*) $v = 3$ ft/s, $a = 3.61$ ft/s². (*b*) 3.82 s.
11.97 (*a*) 115.3 km/h $\leq v_0 \leq$ 148.0 km/h. (*b*) 6.66°, 4.05°.
11.98 4.87 m/s $\leq v_0 \leq$ 11.44 m/s.
11.99 (*a*) 2.87 s. (*b*) 265 ft. (*c*) 33.1 ft.
11.100 10.14 ft $< h <$ 17.05 ft.
11.101 215 m.
11.102 $0 \leq d \leq 0.524$ m.
11.105 6.98 m/s.
11.106 (*a*) 9.34 m/s. (*b*) 9.27 m/s.
11.107 2.27 ft/s $\leq v_0 \leq$ 4.00 ft/s.
11.108 8.84 ft/s.
11.111 (*a*) 3.96°. (*b*) 375 m. (*c*) 17.51 s.
11.112 (*a*) 14.87°. (*b*) 0.1052 s.
11.115 (*a*) 4.13 m. (*b*) 59.5°.
11.116 (*a*) 108.7 m. (*b*) 30°. (*c*) 27.2 m.
11.117 78.1 km/h ↗ 56.3°.
11.118 (*a*) 1.545 m/s ↙ 62.8°. (*b*) 1.586 m/s ↘ 60°.
11.119 (*a*) 4.93 ft/s ↗ 38.6°. (*b*) 4.81 ft/s ↗ 58.3°.
11.120 1.822 knots ↘ 17.37°.
11.123 (*a*) 400 mm/s² ↑. (*b*) 894 mm/s ↗ 63.4°.
11.124 (*a*) 6.97 in./s ↖ 66.6°. (*b*) 5.72 in./s ↘ 68.6°.
11.125 (*a*) 3.21 ft/s² ↘ 22.4°. (*b*) 6.43 ft/s ↘ 22.4°.
11.126 (*a*) 3.11 ft. (*b*) 40.2 ft/s ↘ 86.4°.
11.127 (*a*) 1.175 m/s ↖ 10°. (*b*) 0.1675 m/s ↖ 10°.
11.128 3.34 m/s ↙ 81.9°.
11.129 14.30 km/h ↗ 36.4°.
11.130 24 km/h ↙ 58.6°.
11.133 (*a*) 0.407 ft/s². (*b*) 0.0333 ft/s². (*c*) 0.00593 ft/s².
11.134 (*a*) 90 ft/s. (*b*) 87 ft/s.
11.137 (*a*) 0.8 in./s². (*b*) 1.084 in./s².
11.138 8.56 s.
11.139 0.680 m/s².
11.140 (*a*) 83.8 km/h. (*b*) 3.71 m/s².
11.143 (*a*) 326 m. (*b*) 62.8 m.
11.144 (*a*) 0.75 m. (*b*) 1.155 m. (*c*) 0.75 m.
11.145 (*a*) 27.2 ft/s ↗ 40°. (*b*) 13.48 ft.
11.146 54.5 ft/s ↗ 4.0°; 54.5 ft/s ↘ 4.0°.
11.147 (*a*) 0.634 m. (*b*) 9.07 m.
11.148 v_B^3/gv_A.
11.151 $(R^2 + c^2)/2\omega_n R$.
11.152 2.50 ft.
11.153 149.8 Gm.
11.154 1425 Gm.
11.155 16,200 mi/h.
11.156 7700 mi/h.
11.159 528 km.
11.160 50.7 h.
11.163 (*a*) $\mathbf{v} = (12 \text{ ft/s})\mathbf{e}_r - (144 \text{ ft/s})\mathbf{e}_\theta$; $\mathbf{a} = -(1740 \text{ ft/s}^2)\mathbf{e}_r + (96 \text{ ft/s}^2)\mathbf{e}_\theta$. (*b*) $\mathbf{v} = (96 \text{ ft/s})\mathbf{e}_\theta$; $\mathbf{a} = -(384 \text{ ft/s}^2)\mathbf{e}_r$; Path is circle of radius 24 ft.
11.164 (*a*) $\mathbf{v} = 3\pi b\ \mathbf{e}_\theta$; $\mathbf{a} = -4\pi^2 b\ \mathbf{e}_r$. (*b*) $\theta = 2n\pi$, $n = 0, 1, 2. \ldots$
11.165 (*a*) $\mathbf{v} = (2 \text{ m/s})\mathbf{e}_r$; $\mathbf{a} = (8 \text{ m/s}^2)\mathbf{e}_\theta$. (*b*) $\mathbf{v} = (4.24 \text{ m/s})\mathbf{e}_r + (1.414 \text{ m/s})\mathbf{e}_\theta$; $\mathbf{a} = (5.66 \text{ m/s}^2)\mathbf{e}_r + (5.66 \text{ m/s}^2)\mathbf{e}_\theta$.
11.166 (*a*) $v = 2.83t$ m/s, $a = 2.83$ m/s². (*b*) ∞; Path is a straight line.
11.169 $\dfrac{\dot{\theta} d \sin\beta}{\sin^2(\beta - \theta)}$.
11.170 (*a*) $\dot{r} = -\omega d/2$, $\dot{\theta} = \omega/2$. (*b*) $\ddot{r} = -\sqrt{3}\omega^2 d/4$, $\ddot{\theta} = 0$.
11.171 140.7 mi/h.
11.172 $v = 97.9$ km/h; $\beta = 49.7°$.
11.175 $b\theta\omega^2 e^{\frac{1}{2}\theta^2}\sqrt{4 + \theta^2}$.
11.176 $\dfrac{b\omega^2}{\theta^4}\sqrt{36 + 4\theta^2 + \theta^4}$.
11.180 $\tan^{-1}\left[R(2 + \omega_n^2 t^2)/c\sqrt{4 + \omega_n^2 t^2}\right]$.
11.181 (*a*) $\theta_x = 90°$, $\theta_y = 123.7°$, $\theta_z = 33.7°$. (*b*) $\theta_x = 103.4°$, $\theta_y = 134.3°$, $\theta_z = 47.4°$.
11.182 (*a*) 0, 4 s. (*b*) 168.5 m.
11.183 (*a*) ±4 ft/s. (*b*) 4.8 s. (*c*) 7.8 ft, −7.67 ft.
11.185 (*a*) $\mathbf{a}_A = 20$ m/s² →; $\mathbf{a}_B = 6.67$ m/s² ↓. (*b*) 13.33 m/s ↓; 13.33 m ↓.
11.186 (*b*) $x = 52$ m, $v = 36$ m/s, 164 m.
11.188 10.64 m/s $\leq v_0 \leq$ 14.48 m/s.
11.189 (*a*) $d = 6.52$ m, $\alpha = 45°$. (*b*) $d = 5.84$ m, $\alpha = 58.2°$.
11.191 (*a*) 250 m. (*b*) 82.9 km/h.
11.192 15.95 ft/s².

CHAPTER 12

12.1 $m = 2.000$ kg at all latitudes.
(*a*) 19.56 N. (*b*) 19.61 N. (*c*) 19.64 N.
12.2 75 N.
12.5 (*a*) 0.933. (*b*) 63.8 m.
12.6 (*a*) 460 m. (*b*) 10.19 s.
12.7 (*a*) 143.2 mi/h. (*b*) 115.2 mi/h.
12.8 (*a*) 30 ft/s. (*b*) 0.233.
12.9 612 N.
12.10 (*a*) 51.7 m. (*b*) 61.9 m.
12.12 (*a*) 61.1 m. (*b*) 18.78 kN tension.
12.15 (*a*) 3.24 ft/s^2 $\searrow$ 25°. (*b*) 10.54 lb.
12.16 (*a*) 6.49 ft/s^2 $\searrow$ 25°. (*b*) 12.16 lb.
12.18 (*a*) $\mathbf{a}_A = 2.26$ m/s^2 $\downarrow$; $\mathbf{a}_B = 0.755$ m/s^2 $\uparrow$;
$\mathbf{a}_C = 0.755$ m/s^2 $\downarrow$.
(*b*) $T_A = 75.5$ N, $T_C = 90.6$ N.
12.19 *System 1:* (*a*) 10.73 ft/s^2 $\downarrow$. (*b*) 10.36 ft/s $\downarrow$. (*c*) 0.932 s.
System 2: (*a*) 16.10 ft/s^2 $\downarrow$. (*b*) 12.69 ft/s $\downarrow$. (*c*) 0.621 s.
System 3: (*a*) 0.749 ft/s^2 $\downarrow$. (*b*) 2.74 ft/s $\downarrow$. (*c*) 13.35 s.
12.20 (*a*) $\mathbf{a} = 8.34$ m/s^2 $\rightarrow$; $T = 16.19$ kN. (*b*) 2.45 m/s^2.
12.21 (*a*) 11.52 m/s^2 $\leftarrow$. (*b*) 0.787 m $\rightarrow$.
12.24 $0.347\ mv_0^2/F_0$.
12.27 $\sqrt{k/m}\,(\sqrt{l^2 + x_0^2} - l)$.
12.28 (*a*) $\mathbf{a}_A = 7.27$ m/s^2 $\rightarrow$; $\mathbf{a}_B = 9.70$ m/s^2 $\rightarrow$;
$\mathbf{a}_C = 10.30$ m/s^2 $\rightarrow$.
(*b*) 24.2 N.
12.29 (*a*) $\mathbf{a}_A = 6.83$ m/s^2 $\rightarrow$; $\mathbf{a}_B = 9.76$ m/s^2 $\rightarrow$; $\mathbf{a}_C = 10$ m/s^2 $\rightarrow$.
(*b*) 29.3 N. (*c*) 237 N.
12.30 (*a*) 8.63 ft/s^2 $\leftarrow$. (*b*) 32.2 ft/s^2 $\searrow$ 25°.
12.31 (*a*) 9.13 ft/s^2 $\nearrow$ 30°. (*b*) 2.09 ft/s $\nearrow$ 30°.
12.34 2.49 m/s.
12.35 (*a*) 36.9°. (*b*) 3.88 m/s.
12.38 (*a*) 9.83 ft/s. (*b*) 2.83 lb.
12.39 (*a*) 0.389 lb. (*b*) 0.272 lb $\swarrow$ 53.8°.
12.40 $0.732\text{ m/s} \le v \le 4.34\text{ m/s}$.
12.42 434 N.
12.43 (*a*) 327 N $\searrow$ 73.6°. (*b*) 402 N $\searrow$ 68.1°.
12.44 (*a*) 39.2 lb. (*b*) 23.0 lb.
12.45 (*a*) 155.9 lb. (*b*) 228 lb.
12.46 (*a*) 201 m. (*b*) 515 N $\uparrow$.
12.47 $-51.6° \le \theta \le 51.6°$.
12.50 $\sqrt{gr\tan(\theta - \phi_s)} \le v \le \sqrt{gr\tan(\theta + \phi_s)}$.
12.51 (*a*) 33.0 km/h. (*b*) 51.8 km/h.
12.52 (*a*) 0.247*W*. (*b*) 13.33°.
12.53 6.91°.
12.54 0.400.
12.55 7.67 m/s.
12.58 (*a*) 48 ft/s. (*b*) 0.644×10^{-3} lb.
12.59 0.315.
12.62 $elVL/mdv_0^2$.
12.63 $\dfrac{1.085}{v_0}\sqrt{\dfrac{eV}{m}}$.
12.64 (*a*) $F_r = -31.6$ N, $F_\theta = -6.96$ N.
(*b*) $F_r = -2.41$ N, $F_\theta = 4.93$ N.
12.65 (*a*) $F_r = -3.97$ N, $F_\theta = 0.5$ N.
(*b*) $F_r = -1.775$ N, $F_\theta = 0.0988$ N.
12.66 (*a*) 4.03 ft/s $\searrow$ 60°. (*b*) 35.9 ft/s^2 $\swarrow$ 60°.
12.67 (*a*) 1.743 lb $\nearrow$ 45°. (*b*) 22.8 ft/s^2 $\swarrow$ 45°.
12.70 (*a*) 335 mm. (*b*) 1.304 m/s.
12.71 (*a*) $\ddot{r} = 72$ m/s^2 . (*b*) 1.25 N.
12.72 (*a*) 126.6 N. (*b*) 5.48 m/s^2 $\rightarrow$. (*c*) 4.75 m/s^2 $\downarrow$.
12.73 (*a*) 142.7 N. (*b*) 6.18 m/s^2 $\rightarrow$. (*c*) 4.10 m/s^2 $\downarrow$.
12.74 $v_r = v_0 \sin 2\theta/\sqrt{\cos 2\theta}$, $v_\theta = v_0\sqrt{\cos 2\theta}$.
12.79 (*a*) 35.8×10^3 km, 22,200 mi.
(*b*) 3070 m/s, 10,090 ft/s.
12.80 81.5 ft/s^2.
12.81 (*a*) 1.941 h. (*b*) 7580 mi/h.
12.82 6.04×10^{24} kg.
12.85 (*a*) 1.819 kN. (*b*) 2510 km. (*c*) 1.617 m/s^2.
12.86 2.64 km/s.
12.87 24.2×10^3 ft/s $\nearrow$ 45°.
12.89 15.94 m/s.
12.90 (*a*) $a_r = a_\theta = 0$. (*b*) 61.4 ft/s^2. (*c*) 2.98 ft/s.
12.91 (*a*) $l_1^3 \sin^3\theta_1 \tan\theta_1 = l_2^3 \sin^3\theta_2 \tan\theta_2$. (*b*) 49.8°.
12.100 105.5.
12.101 2.33.
12.103 (*a*) 5530 ft/s. (*b*) 2430 ft/s.
12.106 (*a*) 243 ft/s. (*b*) 159.7 ft/s. (*c*) 4270 ft/s.
12.107 (*a*) 17,710 mi. (*b*) $|\Delta v_B| = 792$ ft/s; $|\Delta v_C| = 3470$ ft/s.
12.108 123.6 h.
12.109 4.98 h.
12.112 1.043 h.
12.113 50.9 min.
12.114 $\cos^{-1}[(1 - n\beta^2)/(1 - \beta^2)]$.
12.115 263 ft/s.
12.119 (*a*) $(r_1 - r_0)/(r_1 + r_0)$. (*b*) 3.79×10^{11} mi.
12.122 2.09 m/s^2 $\swarrow$ 36.9°.
12.123 (*a*) $\mathbf{a}_A = 0.698$ m/s^2 $\rightarrow$; $\mathbf{a}_B = 0.233$ m/s^2 $\downarrow$.
(*b*) 79.8 N.
12.125 (*a*) $\mathbf{a}_A = 0.316$ m/s^2 $\downarrow$; $\mathbf{a}_B = 2.22$ m/s^2 $\downarrow$;
$\mathbf{a}_C = 9.49$ m/s^2 $\leftarrow$.
(*b*) 18.99 N.
12.126 (*a*) 1.656 lb. (*b*) 20.8 lb.
12.128 (*a*) 5.79 m/s^2. (*b*) 2.45 m/s^2. (*c*) 0.230 m/s^2.
12.129 (*a*) 60.3×10^3 km. (*b*) 570×10^{24} kg.
12.131 (*a*) 6410 m/s. (*b*) 1287 km.
12.132 (*a*) 5280 ft/s. (*b*) 8000 ft/s.

CHAPTER 13

13.1 6.17 GJ.
13.2 6.55×10^9 ft · lb.
13.5 (*a*) 58.0 ft/s. (*b*) 0.856.
13.6 -31.3×10^3 ft · lb.
13.7 265 km/h.
13.8 (*a*) 0.933. (*b*) 66.7 km/h.
13.11 4.36 m/s $\nearrow$.
13.12 3.87 m/s $\nearrow$.
13.15 9.09 m/s.
13.16 (*a*) 11.88 m. (*b*) 5.05 kN.
13.17 (*a*) 898 lb. (*b*) 182.8 lb *C*.
13.18 (*a*) 3960 ft. (*b*) 546 lb *C*.
13.21 (*a*) 1.818 m/s $\nearrow$ 30°. (*b*) 16.02 N.
13.22 (*a*) 1.152 m/s $\nearrow$ 30°. (*b*) 18.79 N.
13.23 (*a*) 3.96 m/s. (*b*) 5.60 m/s.
13.24 1.683 m/s.
13.27 (*a*) 341 mm. (*b*) 182.0 mm.
13.28 (*a*) 127.8 mm. (*b*) 2.95 m/s.
13.29 (*a*) 1.638 ft/s $\downarrow$. (*b*) 1.892 ft/s $\downarrow$.
13.30 (*a*) 2.56 ft/s $\downarrow$. (*b*) 3.03 ft/s $\downarrow$.
13.33 $g_h = \dfrac{g_0}{[(h/R) + 1]^2}$. (*a*) 0.0316%. (*b*) 25.4%.

13.34 $1/[1 - (v_0^2 - v^2)/2g_mR_m]$.

13.37 (*a*) 0.521 ft/s. (*b*) 48.3 ft/s^2.

13.38 (*a*) 0.1030 m/s. (*b*) 240 mm.

13.41 (*a*) 2.24 ft/s ←. (*b*) 2.27 ft/s ←.

13.42 (*a*) 2.81 ft/s ←. (*b*) 2.93 ft/s ←.

13.43 731 N.

13.44 N_{min} = 731 N at *B*; N_{max} = 5520 N at *D*.

13.47 (*a*) 39.4 W. (*b*) 174.7 W.

13.48 1.089 kW.

13.49 (*a*) $0.278 \times 10^{-6}\ mgb/\eta$. (*b*) $1.010 \times 10^{-3}\ Wb/\eta$.

13.50 (*a*) 2.55 hp. (*b*) 3.11 hp.

13.52 (*a*) 14.95 kW. (*b*) 45.4 kW.

13.53 (*a*) 17.75 kW. (*b*) 46.7 kW.

13.55 (*a*) $k_e = k_1k_2/(k_1 + k_2)$. (*b*) $k_e = k_1 + k_2$.

13.56 (*a*) 2.17 m/s. (*b*) 1.992 m/s.

13.57 (*a*) 3.20 ft/s. (*b*) 2.93 ft/s.

13.58 (*a*) 11.66 ft/s. (*b*) 15.01 ft/s.

13.61 (*a*) 0.292 m. (*b*) 2.39 m/s.

13.62 (*a*) 43.1°. (*b*) 2.43 m/s.

13.63 (*a*) 45.2 in. (*b*) 15.54 ft/s.

13.64 (*a*) 0.4 in. (*b*) 0.8 in.

13.67 2.06 m/s.

13.68 (*a*) 2.71 m/s. (*b*) 3.60 m/s.

13.69 (*a*) 2.94 m/s. (*b*) $-(33.9\ \text{N})\mathbf{i} + (33.3\ \text{N})\mathbf{j}$.

13.70 (*a*) 2.39 m/s. (*b*) 5.78 N ↑.

13.71 (*a*) 0.541 lb. (*b*) 1.450 lb.

13.72 (*a*) 1.702 lb. (*b*) 0.625 lb.

13.73 (*a*) 14.88 N →. (*b*) 23.6 N ↓.

13.74 (*a*) 340 N/m. (*b*) 1.112 m/s ↓.

13.78 (*b*) $V = -\ln xyz$.

13.79 (*b*) $V = \dfrac{1}{\sqrt{x^2 + y^2 + z^2}}$.

13.80 (*a*) Not conservative, $U = (k - 1)a^2/2$.
(*b*) Conservative, $U = 0$.

13.81 (*a*) $k_1x^2/2 + k_2x^4/4$. (*b*) $\sqrt{(k_1/m)x_0^2 + (k_2/2m)x_0^4}$.

13.82 (*a*) $k_1x^2/2 - k_2x^4/4$. (*b*) $\sqrt{(k_1/m)x_0^2 - (k_2/2m)x_0^4}$.

13.85 6.48 km/s.

13.86 (*a*) 9.56 km/s. (*b*) 2.39 km/s.

13.87 (*a*) 50.1×10^9 ft · lb. (*b*) 115.9×10^9 ft · lb.

13.88 19.25×10^6 ft · lb/lb.

13.89 (*a*) $V = mgR[1 - (R/r)]$. (*b*) $T = \frac{1}{2}mgR^2/r$.
(*c*) $E = mgR[1 - (R/2r)]$.

13.90 1.205×10^9.

13.93 (*a*) 1.2 m. (*b*) 3.47 rad/s. (*c*) 191.8 J.

13.94 (*a*) 1.054 m. (*b*) 2.64 m. (*c*) 4.50 rad/s.

13.95 v_r = 3.35 ft/s; v_θ = 2.81 ft/s.

13.96 (*a*) $|v|$ = 4.77 ft/s. (*b*) 0.552 ft/s.

13.100 (*a*) 2.94 km/s. (*b*) 2.64 km/s.

13.101 (*a*) 6200 km. (*b*) 4.87×10^{24} kg.

13.102 25.1×10^3 ft/s.

13.103 10.16 mi/s.

13.106 (*a*) 7.35 km/s. (*b*) 45.0°.

13.107 (*a*) 9.56 km/s. (*b*) 13,950 km.

13.108 5160 ft/s; 79.9°.

13.109 66.8°.

13.110 (*a*) 3.45 km/s. (*b*) 4.17 km/s.

13.111 9.34 km/s; 59.4°.

13.118 (*b*) $v_{esc}\sqrt{\alpha/(1 + \alpha)} \le v_0 \le v_{esc}\sqrt{(1 + \alpha)/(2 + \alpha)}$.

13.119 (*a*) 2.08 s. (*b*) 2.04 m/s.

13.121 (*a*) 13.15 kN. (*b*) 3 s.

13.122 (*a*) 157.1 km/h. (*b*) 12.85 kN.

13.123 0.266.

13.124 2.80 s.

13.125 (*a*) 3.06 s. (*b*) 0.963 s.

13.126 73.3 N.

13.129 (*a*) 16.32 s. (*b*) 3.06 kN *T*.

13.130 (*a*) 32.6 s. (*b*) 11.03 kN *T*.

13.131 (*a*) 5.65 s. (*b*) 3870 lb *C*.

13.132 (*a*) 12.88 ft/s. (*b*) 8 lb ↓.

13.133 0.673 s.

13.136 (*a*) 53.7 ft/s ⦣ 30°. (*b*) 8.34 s.

13.137 28.4 ksi.

13.139 (*a*) 15.38 m/s ↑. (*b*) 4.57 s.

13.140 2.86 kN.

13.141 180.9 N ⦥ 50.9°.

13.142 0.800 lb · s.

13.143 5.76*W*.

13.146 (*a*) 9.17 ft/s →. (*b*) zero. (*c*) 1.532 ft · lb.

13.147 (*a*) 1.912 m/s →. (*b*) 9.86 kN · s →.

13.148 (*a*) 707 m/s. (*b*) 14.09 N · s.

13.149 (*a*) Car A. (*b*) 67.2 mi/h.

13.150 (*a*) 3.46 ft/s ↑. (*b*) 0.0967 lb · s ↑.

13.151 (*a*) 1.694 m/s ↓. (*b*) 0.1619 J.

13.152 (*a*) 187.9 N · s, 587 J. (*b*) 150.3 N · s, 471 J.

13.155 (*a*) $\mathbf{v}_A$ = 7.8 ft/s →; $\mathbf{v}_B$ = 13.2 ft/s →. (*b*) 0.805 ft · lb.

13.157 (*a*) 4 m/s →. (*b*) 8 m/s ←.

13.158 (*a*) 2 m/s ←. (*b*) 2 m/s →.

13.159 $\mathbf{v}_A$ = 3.04 ft/s ←; $\mathbf{v}_B$ = 1.013 ft/s ←; $\mathbf{v}_C$ = 0.45 ft/s ←.

13.160 (*a*) 2.70 ft/s →. (*b*) 2.42 ft/s →.

13.163 $0.728 \le e \le 0.762$.

13.164 $\mathbf{v}_A$ = 1.741 m/s →; $\mathbf{v}_B$ = 3.08 m/s ⦣ 22.6°.

13.165 $\mathbf{v}_A$ = 1.7 m/s →; $\mathbf{v}_B$ = 3.3 m/s →.

13.166 $\mathbf{v}_A$ = 10 ft/s ⦦ 40°; $\mathbf{v}_B$ = 10 ft/s ⦣ 40°.

13.167 $\mathbf{v}_A$ = 19.12 ft/s ⦦ 72.2°; $\mathbf{v}_B$ = 5.41 ft/s ⦣ 40°.

13.170 $\mathbf{v}_A$ = 1.677 m/s ⦣ 4.68°; $\mathbf{v}_B$ = 3.61 m/s →.

13.171 (*a*) 70.2°. (*b*) 0.322 m/s →.

13.172 (*a*) *h* = 0.496 m, *d* = 0.953 m. (*b*) 3 m/s ←.

13.173 156.1 mm.

13.174 (*a*) 0.6. (*b*) 0.9.

13.175 13.25 ft.

13.176 (*a*) 31.2 km/h. (*b*) 0.241.

13.177 (*a*) e_{AB} = 0.506, e_{BC} = 0.333. (*b*) 44.0 mm.

13.180 (*a*) 217 mm. (*b*) 69.1 mm.

13.181 (*a*) 1/3. (*b*) 1.177 J. (*c*) 48.5 mm.

13.184 0.226.

13.185 166.5 mm $\le h \le$ 666 mm.

13.186 (*a*) 0.923. (*b*) 3.83 ft.

13.187 (*a*) 62.7°. (*b*) 0.00932 v_0^2 ft · lb (for v_0 in ft/s).

13.190 (*a*) 32.8 mi/h. (*b*) 142.5 mi/h.

13.191 (*a*) 1723 lb. (*b*) 1292 lb *T*.

13.193 (*a*) 87.2 m/s. (*b*) 105.8 m/s.

13.194 (*a*) 1704 m/s. (*b*) 29.8 m/s.

13.196 3.26 s.

13.199 (*a*) 0.0776 s. (*b*) d_C = 0.01941 ft;
$d_A = d_B$ = 0.0766 ft.

13.200 (*a*) 401 mm. (*b*) 4.10 N · s.

CHAPTER 14

14.1 0.265 m/s →.

14.2 22.5 Mg.

14.3 (*a*) 4.25 ft/s →. (*b*) 4.25 ft/s →.

14.4 (*a*) 20 lb. (*b*) 3.6 ft/s →.
14.5 (*a*) 43.4 g. (*b*) 293 m/s.
14.6 (*a*) 5.2 km/h →. (*b*) 4 km/h →.
14.9 zero.
14.10 (*a*) (0.8 m)**i** + (0.4 m)**j** + (0.8 m)**k**.
(*b*) (42 kg · m/s)**i** − (28 kg · m/s)**k**.
(*c*) (44.8 kg · m²/s)**i** − (22.4 kg · m²/s)**j** − (33.6 kg · m²/s)**k**.
14.11 (*a*) (4.00 ft/s)**j**; (1.000 ft/s)**i**; (3.00 ft/s)**k**.
(*b*) (1.200 ft · lb · s)**i** + (0.600 ft · lb · s)**j** − (2.40 ft · lb · s)**k**.
14.12 (*a*) (10 ft/s)**j**; (5 ft/s)**i**; (10 ft/s)**k**.
(*b*) (6 ft · lb · s)**i** + (3 ft · lb · s)**j** − (6 ft · lb · s)**k**.
14.15 (3300 m)**i** + (750 m)**j** + (900 m)**k**.
14.16 (120 m)**i** − (26.0 m)**j** + (13.33 m)**k**.
14.17 3510 ft, −267 ft.
14.18 1881 ft, −1730 ft.
14.21 (*a*) 2.01 m/s. (*b*) 2.27 m/s.
14.22 (*a*) 2.83 m/s. (*b*) 1.313 m/s.
14.23 (52.6 ft)**i** + (8.77 ft)**k**.
14.24 (*a*) 8 ft/s →. (*b*) $\theta = 36.6°$, $v_C = 10.39$ ft/s, $v_D = 8.72$ ft/s.
14.27 $v_A = 919$ m/s; $v_B = 717$ m/s; $v_C = 619$ m/s.
14.31 5320 J.
14.32 (*a*) 3550 J. (*b*) 1830 J.
14.33 (*a*) 23.6 ft · lb. (*b*) 2.85 ft · lb.
14.34 597 ft · lb.
14.37 (*a*) $\dfrac{m_A}{m_A + m_B} v_0$ →. (*b*) $\dfrac{m_A}{m_A + m_B} \dfrac{v_0^2}{2g}$.
14.38 (*a*) $\mathbf{v}_A = 0.2\, v_0$ ←; $\mathbf{v}_B = 0.693\, v_0$ ⦣ 30°;
$\mathbf{v}_C = 0.693\, v_0$ ⦤ 30°.
(*b*) $\mathbf{v}_A = 0.25\, v_0$ ⦥ 60°; $\mathbf{v}_B = 0.866\, v_0$ ⦣ 30°;
$\mathbf{v}_C = 0.433\, v_0$ ⦤ 30°.
14.39 $\mathbf{v}_A = 12.00$ ft/s ⦣ 60.3°; $\mathbf{v}_B = 56.8$ ft/s ⦤ 17.8°.
14.40 $\mathbf{v}_A = 38.9$ ft/s ⦣ 27.7°; $\mathbf{v}_B = 30.8$ ft/s ⦤ 77.6°.
14.41 $v_A = 2.5$ m/s; $v_B = 2.17$ m/s; $v_C = 3.75$ m/s.
14.42 $v_A = 3.54$ m/s; $v_B = 3.06$ m/s; $v_C = 1.768$ m/s.
14.45 $x = 181.7$ mm, $y = 0$, $z = 139.4$ mm.
14.46 (40 m/s)**i** + (40 m/s)**j** + (260 m/s)**k**.
14.47 (*a*) $v_C = 11$ ft/s; $v_D = 5.5$ ft/s. (*b*) 0.786.
14.48 (*a*) $v_C = 14$ ft/s; $v_D = 7$ ft/s. (*b*) 0.727.
14.49 (*a*) $\mathbf{L} = 8.75$ kg · m/s →; $\mathbf{H}_G = 0.525$ kg · m²/s ↻.
(*b*) $\mathbf{v}_A = 1.5$ m/s →; $\mathbf{v}_B = 5$ m/s →.
14.50 (*a*) $\mathbf{L} = 3.5$ kg · m/s →; $\mathbf{H}_G = 0.525$ kg · m²/s ↺.
(*b*) $\mathbf{v}_A = 2$ m/s →; $\mathbf{v}_B = 1.5$ m/s ←.
14.53 (*a*) $\mathbf{v}_B = 2.4$ m/s ⦣ 53.1°; $\mathbf{v}_C = 2.56$ m/s →.
(*b*) $c = 1.059$ m.
14.54 (*a*) $\mathbf{v}_A = 2.4$ m/s ↓; $\mathbf{v}_B = 3$ m/s ⦣ 53.1°. (*b*) $a = 1.864$ m.
14.55 (*a*) 5 ft/s →. (*b*) 0.5 ft. (*c*) 20 rad/s ↺.
14.56 (*a*) $v_A = 2.17$ ft/s ↑; $v_B = 2.17$ ft/s ↓; $v_C = 3.75$ ft/s →.
(*b*) 0.25 ft.
14.59 128.1 N →; 62.1 N ↑.
14.60 170.7 N →; 82.8 N ↑.
14.61 (*a*) 41.8°. (*b*) 45.2 lb ↓.
14.62 $Q_1 = 0.1552$ ft³/s, $Q_2 = 0.466$ ft³/s.
14.65 (*a*) 35.4°. (*b*) 187.3 N ⦥ 53.8°.
14.66 (*a*) 26.0 m/s. (*b*) 230 N ⦥ 48.4°.
14.67 $C_x = 0$, $C_y = 29.0$ lb ↓; $D_x = 48.5$ lb →, $D_y = 54.9$ lb ↓.
14.68 $\mathbf{A} = 274$ N ⦤ 20.0°; $M_A = 46.0$ N · m ↻.
14.71 251 lb/s.
14.72 7200 lb.
14.73 (*a*) 42.0 kN, 1.143 m below *B*. (*b*) 30.0 kN,
3.20 m below *B*.
14.74 29.3 kN.
14.77 (*a*) 9.95 ft/s². (*b*) 586 mi/h.
14.78 840 km/h.
14.79 (*a*) 8250 ft · lb/s. (*b*) 6.00 hp.
14.80 (*a*) 16.51 ft. (*b*) 7390 ft · lb/s.
14.83 412 rpm.
14.87 $\mathbf{C} = 1.712$ kN ↑; $\mathbf{D} = 2.29$ kN ↑.
14.88 (*a*) $m_0 e^{qL/m_0 v_0}$. (*b*) $v_0 e^{-qL/m_0 v_0}$.
14.89 v^2/g.
14.90 *Case 1:* (*a*) $g/3$. (*b*) $\sqrt{2gl/3}$.
Case 2: (*a*) gy/l. (*b*) $\sqrt{gl}$.
14.93 (*a*) 27.7 m/s² ↑. (*b*) 215 m/s² ↑.
14.94 5500 m/s.
14.95 873 kips.
14.96 1030 lb/s.
14.97 1331 m/s.
14.98 6900 kg.
14.101 120.1 mi.
14.102 (*a*) 20.0 mi. (*b*) 126.8 mi.
14.103 153.5 km.
14.104 (*a*) 153.4 km. (*b*) 5670 m/s.
14.107 (*a*) 9.2 ft/s ←. (*b*) 9.37 ft/s ←.
14.108 (*a*) 2.8 ft/s ←. (*b*) 0.229 ft/s ←.
14.110 (*a*) (0.6 m)**i** + (1.4 m)**j** + (1.525 m)**k**.
(*b*) −(26 kg · m/s)**i** + (14 kg · m/s)**j** + (14 kg · m/s)**k**.
(*c*) −(29.5 kg · m²/s)**i** − (16.75 kg · m²/s)**j** +
(3.2 kg · m²/s)**k**.
14.111 (*a*) 2 s. (*b*) 144 km/h.
14.113 *Woman:* 382 ft · lb; *man:* 447 ft · lb.
14.114 (*a*) (3280 lb)**k**. (*b*) (6450 lb)**i**.
14.116 33.6 kN.
14.117 (*a*) 100 m/s². (*b*) 35.9 Mm/h.

CHAPTER 15

15.1 (*a*) 15 rad; 0; −24 rad/s². (*b*) 447 rad;
288 rad/s; 120 rad/s².
15.2 For $t = 0$: $\theta = 15$ rad, $\alpha = -24$ rad/s²;
For $t = 2$ s: $\theta = -1$ rad, $\alpha = 24$ rad/s².
15.3 (*a*) 0.5 rad; −4.71 rad/s; −34.5 rad/s².
(*b*) 0; −1.934 rad/s; 36.5 rad/s².
15.4 (*a*) 0.253 rad; −0.927 rad/s; −36.6 rad/s².
(*b*) 0; 0; 0.
15.5 (*a*) −2.51 rad/s². (*b*) 18,000 rev.
15.6 (*a*) 80 rev. (*b*) 800 rev.
15.9 $\mathbf{v}_E$ = (1.08 m/s)**i** + (2.4 m/s)**j**;
$\mathbf{a}_E$ = −(11.52 m/s²)**i** + (5.18 m/s²)**j** + (23.1 m/s²)**k**.
15.10 $\mathbf{v}_C$ = −(2.4 m/s)**j** − (1.8 m/s)**k**;
$\mathbf{a}_C$ = (18 m/s²)**i** + (19.2 m/s²)**j** − (15.6 m/s²)**k**.
15.13 $\mathbf{v}_C$ = −(24 in./s)**i** − (64 in./s)**j** + (80 in./s)**k**.
$\mathbf{a}_C$ = (896 in./s²)**i** + (544 in./s²)**j** + (704 in./s²)**k**.
15.14 $\mathbf{v}_B$ = (40 in./s)**i** + (80 in./s)**k**;
$\mathbf{a}_B$ = (840 in./s²)**i** + (800 in./s²)**j** + (80 in./s²)**k**.
15.16 66,700 mi/h, 19.47×10^{-3} ft/s².
15.17 (*a*) $\omega_A = 25$ rad/s ↻; $\alpha_A = 200$ rad/s² ↻;
$\omega_C = 40$ rad/s ↻; $\alpha_C = 320$ rad/s² ↻.
(*b*) 163.2 m/s² ⦥ 11.31°.
15.18 (*a*) 18.96 m/s². (*b*) 192.7 m/s² ⦥ 5.65°.
15.19 (*a*) 300 ft/s². (*b*) 4800 ft/s².
15.20 (*a*) $v_C = 20.9$ ft/s, $a_C = 5260$ ft/s².
(*b*) $v_C = 2.09$ ft/s, $a_C = 52.7$ ft/s².

15.23 (*a*) $\omega_B = 300$ rpm ↺; $\omega_C = 100$ rpm ↻.
(*b*) $\mathbf{a}_B = 49.3\ \text{m/s}^2$ ←; $\mathbf{a}_C = 16.45\ \text{m/s}^2$ →.
15.24 (*a*) $\alpha_B = 15.71\ \text{rad/s}^2$ ↺; $\alpha_C = 5.24\ \text{rad/s}^2$ ↻.
(*b*) $\mathbf{a}_B = 3.18\ \text{m/s}^2$ ⦪ 14.3°; $\mathbf{a}_C = 1.294\ \text{m/s}^2$ ∠ 37.4°.
15.27 (*a*) 3 rad/s^2 ↻. (*b*) 4 s.
15.28 (*a*) 1.975 rad/s^2 ↺. (*b*) 6.91 rad/s ↺.
15.29 (*a*) 5 ft/s^2 →. (*b*) 5.63 in. ←.
15.30 (*a*) 6.45 ft/s^2 →. (*b*) 7.32 in. ←.
15.31 $\alpha_A = 5.41\ \text{rad/s}^2$ ↺; $\alpha_B = 1.466\ \text{rad/s}^2$ ↺.
15.32 (*a*) 10.39 s. (*b*) $\omega_A = 413$ rpm ↻; $\omega_B = 248$ rpm ↺.
15.35 $\alpha = bv^2/2\pi r^3$.
15.36 $a = b\omega_0^2/2\pi$.
15.37 (*a*) 3 rad/s ↻. (*b*) 1.3 m/s ⦩ 67.4°.
15.38 (*a*) 1.96 m/s →. (*b*) 1.82 m/s ∠ 67.4°.
15.39 (*a*) 2.28 rad/s ↻. (*b*) 3.98 ft/s ∠ 30°.
15.40 (*a*) 1.175 rad/s ↺. (*b*) 1.869 ft/s ∠ 59.1°.
15.43 (*a*) 10 rad/s ↺. (*b*) $-(296\ \text{in./s})\mathbf{i} - (42\ \text{in./s})\mathbf{j}$.
15.44 (*a*) $-(56\ \text{in./s})\mathbf{i} - (42\ \text{in./s})\mathbf{j}$. (*b*) $x = 4.2$ in., $y = -5.6$ in.
15.45 (*a*) $-(2\ \text{rad/s})\mathbf{k}$. (*b*) $(100\ \text{mm/s})\mathbf{i} + (175\ \text{mm/s})\mathbf{j}$.
15.47 (*a*) 105 rpm ↻. (*b*) 127.5 rpm ↻.
15.48 (*a*) 1.5. (*b*) $\omega_A/3$ ↺.
15.49 (*a*) 8.33 rad/s ↻. (*b*) 2.78 rad/s ↻.
15.50 (*a*) 30 rad/s ↺. (*b*) 13.33 rad/s ↺.
15.51 (*a*) 200 rad/s ↺. (*b*) 24 rad/s ↻.
15.52 (*a*) 104 rad/s ↺. (*b*) 120 rad/s ↻.
15.55 (*a*) $\boldsymbol{\omega}_{BD} = 4.38$ rad/s ↻; $\mathbf{v}_D = 0.306$ m/s ↑.
(*b*) $\boldsymbol{\omega}_{BD} = 0$; $\mathbf{v}_D = 1.05$ m/s ↓.
(*c*) $\boldsymbol{\omega}_{BD} = 4.38$ rad/s ↺; $\mathbf{v}_D = 0.306$ m/s ↓.
15.56 (*b*) For $\theta = 22.9°$: $\omega_{BD} = 5.6$ rad/s ↻;
For $\theta = 192.6°$: $\omega_{BD} = 5.6$ rad/s ↺.
15.57 (*a*) $\mathbf{v}_P = 0$; $\omega_{BD} = 39.3$ rad/s ↺.
(*b*) $\mathbf{v}_P = 26.2$ ft/s ↓; $\omega_{BD} = 0$.
15.58 $\mathbf{v}_P = 27.2$ ft/s ↓; $\omega_{BD} = 20.8$ rad/s ↺.
15.59 497 mm/s ←.
15.60 791 mm/s ←.
15.63 $\omega_{AB} = 42.4$ rad/s ↻; $\omega_{BD} = 30$ rad/s ↺;
$\omega_{DE} = 33.9$ rad/s ↻.
15.64 $\omega_{BD} = 2.5$ rad/s ↺; $\omega_{DE} = 6.67$ rad/s ↻.
15.65 $\omega_{BD} = 4$ rad/s ↻; $\omega_{DE} = 6.67$ rad/s ↺.
15.66 $\omega_{BD} = 5.2$ rad/s ↻; $\omega_{DE} = 6.4$ rad/s ↻.
15.69 (*a*) $\mathbf{v}_B = 16.88$ in./s ←; $\omega_{AB} = 0$.
(*b*) $\mathbf{v}_B = 35.5$ in./s ←; $\omega_{AB} = 2.37$ rad/s ↻.
15.70 $\mathbf{v}_B = 44.4$ m/s →; $\mathbf{v}_C = 0$; $\mathbf{v}_D = 42.9$ m/s ∠ 15°;
$\mathbf{v}_E = 31.4$ m/s ⦫ 45°.
15.71 (*a*) 0.5 m to the right of *A*. (*b*) 0.1667 m/s ↑.
15.72 (*a*) 0.251 m to the left of *D*. (*b*) 41.8 mm/s ↑.
15.73 (*a*) 1.714 in. below *A*. (*b*) 75 ft/s →. (*c*) 53.2 ft/s ∠ 41.2°.
15.74 (*a*) 1.333 in. above *A*. (*b*) 85 ft/s →. (*c*) 60.2 ft/s ⦫ 48.4°.
15.75 (*a*) 3 rad/s ↺. (*b*) 300 mm/s ←. (*c*) wound 180 mm/s.
15.76 (*a*) 3 rad/s ↻. (*b*) 180 mm/s →. (*c*) unwound 300 mm/s.
15.79 (*a*) 31.2 rad/s ↻. (*b*) 18 rad/s ↺. (*c*) 4.5 m/s ⦩ 30°.
15.80 (*a*) 18.01 rad/s ↺. (*b*) 10.4 rad/s ↻. (*c*) 10.4 rad/s ↻.
15.81 (*a*) 6 rad/s ↺. (*b*) 38.1 in./s ⦫ 61.8°.
15.82 (*a*) 27 in./s →. (*b*) 22.9 in./s ⦪ 61.8°.
15.85 (*a*) 2.24 m/s ∠ 45°. (*b*) 3.05 rad/s ↻.
(*c*) 2.19 m/s ∠ 21.2°.
15.86 (*a*) 8 rad/s ↻. (*b*) 41.6 rad/s ↺. (*c*) 6.35 m/s ⦪ 19.1°.
15.87 (*a*) 1.579 rad/s ↻. (*b*) 28.0 in./s ∠ 78.3°.
15.88 (*a*) 2.25 rad/s ↺. (*b*) 39.8 in./s ⦩ 78.3°.
15.91 (*a*) $\omega_{AB} = 1.039$ rad/s ↺; $\omega_{BD} = 0.346$ rad/s ↻.
(*b*) 3.46 in./s →.
15.92 (*a*) 2.67 rad/s ↺. (*b*) 4 rad/s ↻. (*c*) 3 ft/s ⦪ 53.1°.
15.93 (*a*) $\omega_{AB} = 1.176$ rad/s ↻; $\omega_{DE} = 2.5$ rad/s ↻.
(*b*) 588 mm/s ←.
15.94 (*a*) $\omega_{AB} = 2$ rad/s ↻; $\omega_{DE} = 5$ rad/s ↺. (*b*) 480 mm/s →.
15.97 *Space centrode:* Quarter circle of 300-mm. radius centered at intersection of tracks.
Body centrode: Semicircle of 150-mm radius with center on rod *AB* equidistant from *A* and *B*.
15.98 *Space centrode:* Lower rack.
Body centrode: Circumference of gear.
15.103 (*a*) 16 rad/s^2 ↺. (*b*) 24 ft/s^2 ⦫ 60°. (*c*) 41.6 ft/s^2 ↓.
15.104 (*a*) 18 ft/s^2 ←. (*b*) 18 ft/s^2 ⦪ 60°. (*c*) 31.2 ft/s^2 ↑.
15.105 (*a*) 0.5 rad/s^2 ↻. (*b*) $\mathbf{a}_A = 3.25\ \text{m/s}^2$ ↑; $\mathbf{a}_E = 0.75\ \text{m/s}^2$ ↑.
15.106 (*a*) 0.2 m/s^2 ↓. (*b*) 2.2 m/s^2 ↑.
15.109 $x = -144$ mm, $y = -108$ mm.
15.110 (*a*) 12 in./s^2 →. (*b*) 9.90 in./s^2 ⦩ 14.0°.
15.111 $\mathbf{a}_C = 22.6\ \text{in./s}^2$ ⦪ 58.0°; $\mathbf{a}_D = 23.6\ \text{in./s}^2$ ∠ 66.0°.
15.112 (*a*) 1.442 m/s^2 ⦩ 33.7°. (*b*) 3.44 m/s^2 ⦫ 35.5°.
(*c*) 3.2 m/s^2 ↑.
15.114 $\mathbf{a}_A = 25.2\ \text{ft/s}^2$ ⦫ 76.2°; $\mathbf{a}_B = 21.7\ \text{ft/s}^2$ ⦩ 7.9°;
$\mathbf{a}_C = 24.5\ \text{ft/s}^2$ ↑.
15.117 (*a*) 24.4 m/s^2 ←. (*b*) 19.86 m/s^2 ←.
15.118 (*a*) 4.83 m/s^2 ←. (*b*) 8.05 m/s^2 ⦩ 72.5°.
15.119 (*a*) 28.8 rad/s^2 ↺. (*b*) 17.28 rad/s^2 ↻.
15.120 (*a*) 3 rad/s^2 ↺. (*b*) 3 rad/s^2 ↻.
15.121 (*a*) $\boldsymbol{\alpha}_{BD} = 5.58\ \text{rad/s}^2$ ↻; $\mathbf{a}_D = 10.77\ \text{m/s}^2$ ↓.
(*b*) $\boldsymbol{\alpha}_{BD} = 76.0\ \text{rad/s}^2$ ↺; $\mathbf{a}_D = 10.65\ \text{m/s}^2$ ↓.
(*c*) $\boldsymbol{\alpha}_{BD} = 5.58\ \text{rad/s}^2$ ↻; $\mathbf{a}_D = 20.7\ \text{m/s}^2$ ↑.
15.122 (*a*) 1.851 m/s^2. (*b*) 5.55 m/s^2.
15.125 (*a*) 228 rad/s^2 ↺. (*b*) 92 rad/s^2 ↻.
15.126 (*a*) 41.4 m/s^2 ⦪ 78.6°. (*b*) 60.9 m/s^2 ∠ 19.5°.
15.127 87.2 in./s^2 ⦩ 68.2°.
15.128 (*a*) 7.20 rad/s^2 ↺. (*b*) 64.8 in./s^2 ←.
15.129 (*a*) 22 rad/s^2 ↺. (*b*) 7.02 m/s^2 ⦫ 41.2°.
15.131 (*a*) 24.8 rad/s^2 ↺. (*b*) 30.6 rad/s^2 ↻.
15.136 $\dfrac{b^2\omega\cos\theta(1+\sin\theta)}{\sqrt{l^2 - b^2(1+\sin\theta)^2}} - b\omega\sin\theta$ ↑.
15.137 $\dfrac{b^3\omega^2\cos^2\theta(1+\sin\theta)}{[l^2 - b^2(1+\sin\theta)^2]^{3/2}} - \dfrac{b\omega^2\sin\theta}{\sqrt{l^2 - b^2(1+\sin\theta)^2}}$ ↻.
15.138 $v_B = b\omega\cos\theta$; $a_B = b\alpha\cos\theta - b\omega^2\sin\theta$.
15.141 $v_x = v[1 - \cos(vt/r)]$, $v_y = v\sin(vt/r)$.
15.142 $\mathbf{v}_C = 2l\omega\sin\theta$ ←; $\mathbf{a}_C = 2l(\alpha\sin\theta + \omega^2\cos\theta)$ ←.
15.143 $\mathbf{v}_D = -(3l\omega\sin\theta)\mathbf{i} + (l\omega\cos\theta)\mathbf{j}$;
$\mathbf{a}_D = -(3l\alpha\sin\theta + 3l\omega^2\cos\theta)\mathbf{i} + (l\alpha\cos\theta - l\omega^2\sin\theta)\mathbf{j}$.
15.146 $\mathbf{v}_P = r\omega\left[\cos\dfrac{r\omega t}{R-r} - \cos\omega t\right]\mathbf{i} +$
$r\omega\left[\sin\dfrac{r\omega t}{R-r} + \sin\omega t\right]\mathbf{j}$.
15.147 $\mathbf{v} = (R\omega\sin\omega t)\mathbf{j}$; $\mathbf{a} = (R\omega^2\cos\omega t)\mathbf{j}$.
15.148 (*a*) $\boldsymbol{\omega}_{BD} = 3.81$ rad/s ↻; $\mathbf{v}_{P/BD} = 6.53$ m/s ∠ 16.26°.
(*b*) $\boldsymbol{\omega}_{BD} = 3.00$ rad/s ↻; $\mathbf{v}_{P/BD} = 4$ m/s →.
15.149 $\boldsymbol{\omega}_{BD} = 3.82$ rad/s ↻; $\mathbf{v}_{P/BD} = 6.06$ m/s ∠ 10.49°.
15.150 $\boldsymbol{\omega}_{BD} = 3.80$ rad/s ↺; $\boldsymbol{\omega}_{AP} = 1.958$ rad/s ↺.
15.151 $\boldsymbol{\omega}_{AE} = 1.696$ rad/s ↻; $\boldsymbol{\omega}_{BP} = 4.39$ rad/s ↻.
15.152 2.40 m/s ⦫ 73.9°.
15.153 2.87 m/s ⦫ 44.8°.
15.158 (*a*) 37.5 rad/s^2 ↻. (*b*) 7.07 m/s^2 ∠ 45°.
15.159 (*a*) 10.31 rad/s^2 ↻. (*b*) 15.31 m/s^2 ↑.
15.160 (*a*) 1.560 ft/s ⦫ 82.6°. (*b*) 0.1500 ft/s^2 ⦪ 9.8°.
15.161 (*a*) 1.560 ft/s ⦫ 37.4°. (*b*) 0.1500 ft/s^2 ⦩ 69.8°.

15.163 82.5×10^{-3} ft/s² to left of sled.
15.164 (*a*) 1.702 m/s² ↘ 21.5°. (*b*) 2.55 m/s² ↘ 3.2°.
15.165 (*a*) 2.38 m/s² ↙ 48.3°. (*b*) 1.438 m/s² ↙ 64.3°.
15.168 $\mathbf{a}_1$ = 15.13 in./s² →; $\mathbf{a}_2$ = 8.43 in./s² ↙ 57.7°.
15.169 $\mathbf{a}_3$ = 24.1 in./s² ←; $\mathbf{a}_4$ = 8.43 in./s² ↘ 57.7°.
15.171 (*a*) 2.32 m/s² ↑. (*b*) 11.94 m/s² ↖ 37.1°. (*c*) 16.72 m/s² ↓.
15.172 (*a*) 2.32 m/s² ↑. (*b*) 11.88 m/s² ↖ 50.7°.
(*c*) 17.19 m/s² ↙ 76.5°.
15.173 (*a*) 8.09 rad/s² ↻. (*b*) 8.43 m/s² ↙ 16.3°.
15.174 (*a*) 25.7 rad/s² ↺. (*b*) 47.6 m/s² ←.
15.175 ω_A = 6 rad/s ↺; α_A = 62.4 rad/s² ↻.
15.176 ω_A = 1.5 rad/s ↺; α_A = 7.79 rad/s² ↺.
15.177 ω_S = 3.81 rad/s ↻; α_S = 81.4 rad/s² ↻.
15.178 ω_S = 1.526 rad/s ↻; α_S = 57.6 rad/s² ↻.
15.181 (*a*) (33.0 rad/s)**i** − (44.0 rad/s)**k**.
(*b*) (4.8 m/s)**i** + (3.6 m/s)**k**.
15.182 (*a*) (44.0 rad/s)**i** − (33.0 rad/s)**k**.
(*b*) (3.6 m/s)**i** + (4.8 m/s)**k**.
15.183 (*a*) (0.6 rad/s)**i** − (2 rad/s)**j** + (0.75 rad/s)**k**.
(*b*) (20 in./s)**i** + (15 in./s)**j** + (24 in./s)**k**.
15.184 −(14.80 rad/s²)**k**.
15.187 −(20 rad/s²)**i**.
15.188 (*a*) −(2.5 ft/s)**i** + (3.13 ft/s)**j** − (2.5 ft/s)**k**.
(*b*) −(25.6 ft/s²)**i** − (12.5 ft/s²)**j**.
15.189 (*a*) −(0.1745 rad/s)**i** − (0.524 rad/s)**j**.
(*b*) −(0.0914 rad/s²)**k**.
(*c*) $\mathbf{v}_P$ = −(5.44 ft/s)**i** + (1.814 ft/s)**j** − (1.047 ft/s)**k**;
$\mathbf{a}_P$ = (1.097 ft/s²)**i** − (0.1828 ft/s²)**j** − (3.17 ft/s²)**k**.
15.190 (*a*) −(20 rad/s²)**j**. (*b*) −(0.96 m/s²)**i** + (2.4 m/s²)**k**.
(*c*) −(2.46 ft/s²)**j**.
15.191 −(0.831 ft/s²)**i** − (1.230 ft/s²)**j** + (2.08 ft/s²)**k**.
15.192 (*a*) −(0.207 rad/s)**j** − (0.358 rad/s)**k**.
(*b*) (91.0 mm/s)**i** + (35.8 mm/s)**j** − (20.7 mm/s)**k**.
15.193 (*a*) −(0.414 rad/s²)**j** − (0.717 rad/s²)**k**. (*b*) 242 mm.
15.196 −(0.6 m/s)**j**.
15.197 (0.914 m/s)**j**.
15.198 (15 in./s)**j**.
15.199 −(48 in./s)**j**.
15.200 −(0.333 m/s)**j**.
15.201 −(0.12 m/s)**j**.
15.204 $\omega_1/\cos 20°$.
15.205 $\omega_1 \cos 20°$.
15.206 (*a*) −(4.15 rad/s)**i** + (0.615 rad/s)**j** − (2.77 rad/s)**k**.
(*b*) (0.3 m/s)**k**.
15.208 −(0.476 m/s²)**j**.
15.209 −(0.438 m/s²)**j**.
15.210 −(158.4 in./s²)**j**.
15.211 −(769 in./s²)**j**.
15.214 (*a*) (1.6 ft/s)**i** + (5 ft/s)**j** + (8.8 ft/s)**k**.
(*b*) −(76 ft/s²)**j** + (50 ft/s²)**k**.
15.215 $\mathbf{v}_C$ = −(45 in./s)**i** + (36.6 in./s)**j** − (31.2 in./s)**k**;
$\mathbf{a}_C$ = −(291 in./s²)**i** − (270 in./s²)**j**.
15.216 (*a*) (0.3 m/s)**i** + (0.520 m/s)**j** + (0.48 m/s)**k**.
(*b*) −(2.60 m/s²)**i** − (2.34 m/s²)**j** + (8.31 m/s²)**k**.
15.217 (*a*) −(0.3 m/s)**i** + (0.520 m/s)**j** + (1.44 m/s)**k**.
(*b*) −(2.60 m/s²)**i** − (13.02 m/s²)**j** + (8.31 m/s²)**k**.
15.218 (*a*) (0.493 m/s)**i** + (1.353 m/s)**j** − (1.015 m/s)**k**.
(*b*) −(8.46 m/s²)**i** + (1.970 m/s²)**j** − (2.96 m/s²)**k**.
15.219 (*a*) (1.033 m/s)**i** + (1.353 m/s)**j** − (1.015 m/s)**k**.
(*b*) −(8.46 m/s²)**i** + (1.970 m/s²)**j** − (4.58 m/s²)**k**.
15.222 (*a*) −(60.8 in./s)**i** − (54 in./s)**j** + (81 in./s)**k**.
(*b*) (106.3 ft/s²)**i** − (104.1 ft/s²)**j** − (87.8 ft/s²)**k**.
15.223 (*a*) (1.6 ft/s)**i** + (5 ft/s)**j** + (8.8 ft/s)**k**.
(*b*) (8 ft/s²)**i** − (83.5 ft/s²)**j** + (29 ft/s²)**k**.
15.224 $\mathbf{v}_C$ = −(45 in./s)**i** + (36.6 in./s)**j** − (31.2 in./s)**k**;
$\mathbf{a}_C$ = −(303 in./s²)**i** − (384 in./s²)**j** + (208 in./s²)**k**.
15.229 $\mathbf{v}_A$ = (0.09 m/s)**i** + (3.55 m/s)**j** − (3.55 m/s)**k**;
$\mathbf{a}_A$ = −(5.7 m/s²)**i** − (111.7 m/s²)**j** − (111.7 m/s²)**k**.
15.230 $\mathbf{v}_B$ = (1.299 m/s)**i** − (1.853 m/s)**j** + (1.590 m/s)**k**;
$\mathbf{a}_B$ = (0.795 m/s²)**i** − (0.792 m/s²)**j** − (0.976 m/s²)**k**.
15.231 $\mathbf{v}_C$ = (30 in./s)**j** − (29.3 in./s)**k**; $\mathbf{a}_C$ = −(238 in./s²)**i**.
15.232 $\mathbf{v}_D$ = (30 in./s)**i** − (11.25 in./s)**k**;
$\mathbf{a}_D$ = −(33.8 in./s²)**i** + (150 in./s²)**j** − (180 in./s²)**k**.
15.235 $\mathbf{v}_A$ = (60 mm/s)**i** − (36 mm/s)**j**;
$\mathbf{a}_A$ = −(270 mm/s²)**i** + (180 mm/s²)**j** + (172.8 mm/s²)**k**.
15.236 $\mathbf{v}_A$ = (180 mm/s)**i** − (156 mm/s)**j** + (144 mm/s)**k**;
$\mathbf{a}_A$ = (180 mm/s²)**i** − (443 mm/s²)**j** − (115.2 mm/s²)**k**.
15.237 $\mathbf{v}_A$ = −(21 ft/s)**i** − (5 ft/s)**k**;
$\mathbf{a}_A$ = −(40 ft/s²)**i** − (108 ft/s²)**j** + (240 ft/s²)**k**.
15.238 $\mathbf{v}_B$ = −(3 ft/s)**i** − (5 ft/s)**k**;
$\mathbf{a}_B$ = −(40 ft/s²)**i** + (108 ft/s²)**j** − (48 ft/s²)**k**.
15.239 (*a*) (544 mm/s²)**i** − (135 mm/s²)**j**.
(*b*) (256 mm/s²)**i** − (153.6 mm/s²)**k**.
15.240 (*a*) −(32 mm/s²)**i** + (135 mm/s²)**j**.
(*b*) (256 mm/s²)**i** + (153.6 mm/s²)**k**.
15.243 $\mathbf{v}_D$ = −(1.2 m/s)**i** − (1.05 m/s)**j** − (1.4 m/s)**k**;
$\mathbf{a}_D$ = −(8.75 m/s²)**i** + (3.6 m/s²)**j** + (4.8 m/s²)**k**.
15.244 (*a*) 1.640 m/s ↑; 3.70 m ↑. (*b*) 0.738 m/s ↓; 1.665 m ↓.
15.246 (*a*) 0.900 m. (*b*) 10.80 m/s ↓.
15.247 (*a*) 50 mm to the right of center of spool.
(*b*) $\mathbf{v}_B$ = 750 mm/s ↓; $\mathbf{v}_D$ = 1.950 m/s ↑.
15.249 (*a*) 5410 ft/s² ↓. (*b*) 5410 ft/s² ↑. (*c*) 5410 ft/s² ↖ 60°.
15.250 (*a*) 1080 rad/s² ↻. (*b*) 460 ft/s² ↘ 64.9°.
15.252 105.3 in./s² ↖ 30°.
15.253 (*a*) −(1.25 rad/s)**i** − (2.5 rad/s)**j**. (*b*) −(5 in./s)**k**.

CHAPTER 16

16.1 (*a*) 4.91 m/s² ↖ 30°.
(*b*) **A** = 2.94 N ↙ 60°; **C** = 9.74 N ∠ 60°.
16.2 (*a*) 18.43°. (*b*) 3.10 m/s² ↖ 18.43°.
16.3 (*a*) **A** = 60.3 lb ∠ 84.2°; **B** = 28.5 lb ←. (*b*) 0.1023.
16.4 (*a*) 18.59 ft/s² →. (*b*) 0.577.
16.5 (*a*) 7.85 m/s². (*b*) 3.74 m/s². (*c*) 4.06 m/s².
16.6 (*a*) 36.8 ft. (*b*) 42.3 ft.
16.9 (*a*) 0.419*g*. (*b*) 3.33.
16.10 (*a*) 0.295*g*. (*b*) 3.33.
16.11 (*a*) 5.37 ft/s² ↙ 30°. (*b*) 4.23 lb.
16.13 (*a*) 2.54 m/s² ↙ 15°. (*b*) F_{AC} = 12.01 N *T*;
F_{BD} = 44.8 N *T*.
16.14 (*a*) 4.91 m/s² ↖ 30°. (*b*) F_{AC} = 0; F_{BD} = 68.0 N *C*.
16.15 $\mathbf{B}_y = \mathbf{C}_y$ = 29.9 lb ↑.
16.17 *Just above B:* $|V|$ = 16.99 N, $|M|$ = 4.25 N · m.
16.18 *Just to the right of B:* V = 17.94 lb; *At center,*
M = 74.7 lb · in.
16.21 125.7 N · m.
16.22 9480 rev.
16.23 (*a*) 2.35 rad/s² ↻. (*b*) 15.74 lb *C*.
16.24 (*a*) 9.34 lb · ft. (*b*) 36.0 lb ↘ 44.0°.
16.25 125.2 rev.
16.28 20.3 lb · ft · s².

16.29 *Case 1:* (*a*) 7.85 rad/s^2 ↺. (*b*) 15.35 rad/s ↺.
Case 2: (*a*) 6.77 rad/s^2 ↺. (*b*) 14.25 rad/s ↺.
Case 3: (*a*) 4.46 rad/s^2 ↺. (*b*) 11.57 rad/s ↺.
Case 4: (*a*) 5.95 rad/s^2 ↺. (*b*) 9.44 rad/s ↺.
16.30 (*a*) $\alpha_A = 8.94$ rad/s^2 ↻; $\alpha_B = 13.42$ rad/s^2 ↺.
(*b*) 0.417 lb ↓.
16.31 (*a*) $\alpha_A = 13.42$ rad/s^2 ↺; $\alpha_B = 20.1$ rad/s^2 ↻.
(*b*) 1.25 lb ↓.
16.32 (*a*) $\alpha_A = 23.3$ rad/s^2 ↻; $\alpha_B = 23.3$ rad/s^2 ↺. (*b*) 0.519.
16.33 $\alpha_A = 36.7$ rad/s^2 ↻; $\alpha_B = 29.4$ rad/s^2 ↺.
16.34 (*a*) $\alpha_A = 10.23$ rad/s^2 ↺; $\alpha_B = 28.1$ rad/s^2 ↺.
(*b*) $\omega_A = 320$ rpm ↻; $\omega_B = 440$ rpm ↺.
16.35 (*a*) $\alpha_A = 10.23$ rad/s^2 ↺; $\alpha_B = 28.1$ rad/s^2 ↺.
(*b*) $\omega_A = 116.4$ rpm ↺; $\omega_B = 160$ rpm ↻.
16.37 (*a*) $\alpha_A = 13.42$ rad/s^2 ↺; $\alpha_B = 3.22$ rad/s^2 ↺.
(*b*) $\omega_A = 214$ rpm ↻; $\omega_B = 128.6$ rpm ↺.
16.39 $\dfrac{2\mu_k g \sin\phi}{r(\sin 2\phi - \mu_k \cos 2\phi)}$ ↺.
16.43 (*a*) 2.67 m/s^2 →. (*b*) 1.333 m/s^2 ←.
16.45 (*a*) (8.05 ft/s^2)**i**. (*b*) 0.
16.46 (*a*) (12.08 ft/s^2)**i**. (*b*) −(4.03 ft/s^2)**i**.
16.48 (*a*) (0.833 m/s^2)**i**. (*b*) (0.1969 m/s^2)**i**.
16.50 $\mathbf{a}_A = 12.24$ ft/s^2 ↑; $\mathbf{a}_B = 2.46$ ft/s^2 ↑.
16.52 $T_A = 1331$ N; $T_B = 1181$ N.
16.53 $T_A = 1017$ N; $T_B = 1383$ N.
16.54 $T_A = 221$ lb; $T_B = 203$ lb.
16.56 (*a*) $4\mu_k P/mr$ ↻. (*b*) $g + 2\mu_k P/m$ ↓.
16.57 (*a*) 50 rad/s^2 ↻. (*b*) 11.81 m/s^2 ↓.
16.58 (*a*) $3g/L$ ↻. (*b*) g ↑. (*c*) $2g$ ↓.
16.59 (*a*) $3g/2L$ ↻. (*b*) $g/4$ ↑. (*c*) $5g/4$ ↓.
16.60 (*a*) $2g/L$ ↻. (*b*) $g/3$ ↑. (*c*) $5g/3$ ↓.
16.61 (*a*) $g/2$ ↑. (*b*) $3g/2$ ↓.
16.63 (*a*) $g/4$ ↑. (*b*) $5g/4$ ↓.
16.64 (*a*) 1.718 s. (*b*) 3.31 m/s. (*c*) 7.14 m.
16.65 (*a*) 1.980 s. (*b*) 3.06 m/s. (*c*) 7.98 m.
16.66 (*a*) $5v_0/2r$. (*b*) $v_0/\mu_k g$. (*c*) $v_0^2/2\mu_k g$.
16.68 (*a*) $2v_1/7\mu_k g$. (*b*) $\mathbf{v} = 2v_1/7$ →; $\omega = 5v_1/7r$ ↺.
16.71 (*a*) 11.67 rad/s^2 ↻. (*b*) $\mathbf{A}_x = 1.75$ N ←; $\mathbf{A}_y = 9.81$ N ↑.
16.72 (*a*) 82.8 rad/s^2 ↻. (*b*) $\mathbf{C}_x = 4.29$ lb ←; $\mathbf{C}_y = 10$ lb ↑.
16.73 (*a*) 600 mm. (*b*) 7.78 rad/s^2 ↻.
16.75 (*a*) $(mgz/L)(4 - 3\,z/2L)$. (*b*) 25 N.
16.77 (*a*) $\mathbf{M} = 0.64$ lb · in ↺; $\mathbf{R} = 1.194$ lb ∠ 77.3°.
(*b*) $\mathbf{M} = 0$; $\mathbf{R} = 1.427$ lb ↑.
16.78 3.41 N.
16.80 (*a*) $18g/13$ ↓. (*b*) $25mg/52$ ↑.
16.81 (*a*) $L/3$. (*b*) $g/2$ ↑; $3mg/4$ ↑.
16.82 (*a*) $-(24g/19)\mathbf{j}$. (*b*) $-(4g/19)[(4/\pi)\mathbf{i} + 3\mathbf{j}]$.
16.83 (*a*) 5.90 rad/s^2 ↻. (*b*) 5.68 N.
16.85 (*a*) 34.4 rad/s^2 ↻. (*b*) $\mathbf{C}_x = 5.43$ lb ←; $\mathbf{C}_y = 11.13$ lb ↑.
16.86 (*a*) (77.3 rad/s^2)**j**. (*b*) −(4.03 ft/s^2)**i**.
16.89 $\dfrac{r^2 g \sin\beta}{(r^2 + \bar{k}^2)}$.
16.90 604 mm.
16.91 $\tan\beta = \mu_s[1 + (r/\bar{k})^2]$.
16.92 (*a*) 6.12 m or 20.1 ft. (*b*) 1.748 m or 5.74 ft.
16.93 (*a*) Does not slide. (*b*) 16 rad/s^2 ↻; 2.56 m/s^2 →.
16.94 (*a*) Does not slide. (*b*) 24 rad/s^2 ↻; 3.84 m/s^2 →.
16.97 (*a*) 15.46 rad/s^2 ↻; 10.30 ft/s^2 →. (*b*) 0.18.
16.98 (*a*) 23.2 rad/s^2 ↻; 15.46 ft/s^2 →. (*b*) 0.02.
16.99 (*a*) 7.73 rad/s^2 ↻; 5.15 ft/s^2 →. (*b*) 0.34.
16.100 (*a*) 7.73 rad/s^2 ↺; 5.15 ft/s^2 ←. (*b*) 0.32.
16.101 (*a*) $\mathbf{a}_A = \mathbf{a}_B = 1.795$ m/s^2 ←. (*b*) 4.04 N ←.
16.102 (*a*) $\mathbf{a}_A = \mathbf{a}_B = 2.64$ m/s^2 ←. (*b*) 11.87 N ←.
16.103 (*a*) 7.51 m/s^2 ∠ 45°. (*b*) 2.04 N ↓.
16.104 (*a*) 7.25 m/s^2 ∠ 45°. (*b*) 1.577 ∠ 45°.
16.105 (*a*) $0.688\,P/mr$. (*b*) $0.25P/(mg + P)$.
16.107 (*a*) $g/12r$ ↻. (*b*) $g/12$ →; $g/12$ ↓.
16.109 1.831 lb ←.
16.111 (*a*) 31.0 rad/s^2 ↻. (*b*) 1.165 lb ←.
16.112 26.7 rad/s^2 ↻.
16.113 (*a*) $\dfrac{3g\sin\theta}{L(1 + 3\sin^2\theta)}$ ↻. (*b*) $\dfrac{mg}{1 + 3\sin^2\theta}$ ↑.
16.116 (*a*) 6.37 rad/s^2 ↻. (*b*) 7.85 N ↑.
16.118 5.32 lb ∠ 60°.
16.119 98.7 N ↑.
16.120 34.6 N ↓.
16.121 3.61 N ↓.
16.122 3.21 lb ∠ 60°.
16.124 2.62 rad/s ↻.
16.125 1.164 m/s^2 →.
16.126 1.214 N ←.
16.127 7.37 lb ∠ 30°.
16.129 $\mathbf{A} = 8.13$ N ↑; $\mathbf{B} = 8.61$ N ↑.
16.130 $\mathbf{A}_x = 3.24$ N →; $\mathbf{A}_y = 9.81$ N ↑;
$\mathbf{B}_x = 1.08$ N ←; $\mathbf{B}_y = 9.81$ N ↑.
16.131 $\mathbf{B} = 208$ lb ←; $\mathbf{D} = 110.3$ →.
16.133 (*a*) 24 rad/s^2 ↻. (*b*) 143.2 N ↑.
16.134 (*a*) 28 N · m. (*b*) 396 N ↑.
16.135 (*a*) 10.48 lb ∠ 60.3°. (*b*) 0.273.
16.138 (*a*) 1.950 m/s^2 →. (*b*) 57.6 rad/s^2 ↺.
16.139 (*a*) 5.19 m/s^2 ∠ 20°. (*b*) 54.2 rad/s^2 ↺.
16.140 $\alpha_{AB} = 11.04$ rad/s^2 ↻; $\alpha_{BC} = 55.2$ rad/s^2 ↺.
16.141 (*a*) 0.947 ft. (*b*) 7.03 rad/s^2 ↺.
16.142 0.606 N · m, 520 mm below *A*.
16.143 Just to the right of *A*, $V = mg/4$; $L/3$ to the right of *A*, $M = mgL/27$.
16.144 (*a*) $\dfrac{\omega^2 wz}{g}(l - z/2)$. (*b*) 1.293 lb.
16.145 (*a*) $g/6$ ∠ 30°. (*b*) $g/6$ ←.
16.146 (*a*) $0.1824g$ ∠ 30°. (*b*) $0.1659g$ ←.
16.149 (*a*) 11.11 rad/s^2 ↻. (*b*) 37.7 N ↑. (*c*) 28.2 N →.
16.150 (*a*) 97.8 N ↑. (*b*) 60.3 N ↑.
16.151 27.7 rad/s^2 ↺.
16.152 34.9 rad/s^2 ↻.
16.155 (*a*) 15 N · m ↺. (*b*) $\mathbf{B}_x = 120$ N →; $\mathbf{B}_y = 88.2$ N ↑.

CHAPTER 17

17.1 119.0 N · m.
17.2 8690 rev.
17.3 (*a*) 296 rpm. (*b*) 19.10 rev.
17.5 155.6 mm.
17.6 1.260 *L*.
17.8 (*a*) $\omega_A = 25$ rad/s ↻; $\omega_B = 25$ rad/s ↺. (*b*) 0.386.
17.9 (*a*) 10.57 rev. (*b*) 1.157 lb ∠.
17.11 (*a*) 2.93 m/s ↓. (*b*) 2.36 m.
17.12 317 N ↓.
17.13 365 N ↓.
17.15 5.44 rad/s ↻.
17.17 (*a*) 3.90 rad/s ↻; 1.226 kN ∠ 4.93°.
(*b*) 5.52 rad/s ↻; 3160 N ↑.

17.18 6.29 rad/s.
17.19 (*a*) −0.241 rpm. (*b*) +0.240 rpm.
17.21 $\sqrt{4gs/3}$ ↓.
17.25 0.1936 J.
17.26 3.45 ft/s →.
17.27 4.63 ft/s →.
17.28 4.88 ft/s →.
17.29 15.54 ft/s →.
17.30 54.8 lb/ft.
17.31 1.359 rad/s ↻.
17.32 1.080 rad/s ↺.
17.35 3.25 m/s ↓.
17.36 3.96 m/s ↓.
17.37 81.3 rpm ↻.
17.38 97.8 rpm ↻.
17.39 (*a*) 2 hp. (*b*) 2.78 hp.
17.40 (*a*) 33.0 kW. (*b*) 88.0 kW.
17.41 (*a*) 44.6 N · m. (*b*) 27.9 N · m.
17.42 1159 rpm.
17.43 1.790 lb · ft.
17.44 47.6 min.
17.46 7.95 s.
17.47 $\dfrac{2r\omega_0(1+\mu_k^2)}{5g\mu_k(1+\mu_k)}$.
17.48 4.28 s.
17.49 $\dfrac{\bar{v}}{2g\mu_k}$.
17.50 3.82 s.
17.51 5.30 s.
17.52 3.18 s.
17.53 11.36 s.
17.55 $\omega_A = 134.5$ rad/s ↻; $\omega_B = 98.1$ rad/s ↺.
17.56 $\bar{k}^2\omega/\bar{v}$.
17.60 (*a*) $\dfrac{r^2 gt \sin\beta}{r^2+\bar{k}^2}$ ↖ β. (*b*) $\dfrac{\bar{k}^2 \tan\beta}{r^2+\bar{k}^2}$.
17.61 0.850 m.
17.62 (*a*) 28.6 rad/s ↺. (*b*) 8 lb · s ↗ 30°.
17.63 (*a*) 1.134 m/s ↖ 16.3°. (*b*) 0.635 N · s ↖ 16.3°.
17.64 1.247 m/s ↖ 16.3°.
17.67 $P_A = 1.087$ lb, $P_B = 1.863$ lb.
17.69 (*a*) $5\bar{v}_0/2r$. (*b*) $\bar{v}_0/\mu_k g$.
17.70 84.2 rpm.
17.71 $5\omega_0/6$.
17.72 (*a*) 5 rad/s. (*b*) 3.13 rad/s.
17.73 3.64 rad/s.
17.74 $\omega_{BC} = 36$ rpm ↻; $\omega_A = 16.5$ rpm ↺.
17.75 24.4 rpm.
17.76 1.169 m/s, 64.9 N.
17.79 (*a*) 17.33 in. (*b*) 3.24 rad/s.
17.80 5.16 ft/s.
17.81 $\omega = 18.83$ rad/s, $I_{CD} = 0.0508$ kg · m².
17.83 1.942 m/s.
17.84 (*a*) 25.8 ft/s ←. (*b*) 5.95 ft/s →.
17.85 $\mathbf{v} = 40.8$ mm/s ←; $\omega = 0.726$ rad/s ↻.
17.86 (*a*) (0.339 rad/s)**i**. (*b*) −(237 lb)**k**.
17.88 (*a*) 35.9 rad/s ↻. (*b*) 8.39 kN →.
17.89 (*a*) 0.2 m. (*b*) 33.5 rad/s ↻.
17.90 $\omega = 1.534$ rad/s ↺; $\mathbf{v}_A = 0.460$ m/s ↓.
17.91 2.4 rad/s ↻.
17.92 (*a*) 19.47°. (*b*) 1.176 lb · s ↙ 75.6°.
17.94 (*a*) $\omega = 12v_1/7L$ ↻; $\bar{\mathbf{v}} = 3v_1/7$ ↓. (*b*) $4mv_1/7$ ↑.
17.96 $\pi L/3$.
17.97 $\omega = \frac{1}{7}(2 + 5\cos\beta)\omega_1$ ↺; $\bar{\mathbf{v}} = \frac{1}{7}(2 + 5\cos\beta)\bar{v}_1$ ←.
17.98 $\dfrac{12\bar{v}_1 \sin\beta}{L(3\sin^2\beta + 1)}$.
17.100 $\omega = 3\sqrt{2}v_0/5a$ ↺; $\bar{\mathbf{v}} = 0.825\ v_0$ ↖ 76.0°.
17.101 $\omega = 2.4v_0/L$ ↺; $\bar{\mathbf{v}} = 0.721\ v_0$ ↗ 56.3°.
17.103 (*a*) $\sqrt{3g\sin\beta/L}$ ↺.
(*b*) $\mathbf{A}\Delta t = m\sqrt{gL\sin\beta/3}$ ↑; $\mathbf{B}\Delta t = 2m\sqrt{gL\sin\beta/3}$ ↑.
17.105 52.8°
17.106 1510 ft/s.
17.107 (*a*) $\omega_0/4$ ↻. (*b*) 15/16. (*c*) 1.50°.
17.108 (*a*) 57.3 mm. (*b*) 41.0 mm.
17.109 (*a*) 3.27 rad/s ↻. (*b*) 0.787 m/s →.
17.110 (*a*) 6.55 rad/s ↻. (*b*) 1.575 m/s →.
17.112 $\sqrt{3gL}/2$.
17.113 2.17 ft.
17.115 (*a*) 2.56 rad/s ↻. (*b*) 1.895 ft/s ↓.
17.116 (*a*) $3v_0(1+e)/2L$ ↻. (*b*) $v_0(1+e)(m_C + \frac{m_{AB}}{4})$ ↑.
17.117 105.4°.
17.118 (*a*) $\bar{\mathbf{v}}_A = 0$, $\omega_A = \omega_1$ ↻; $\bar{\mathbf{v}}_B = v_1$ →, $\omega_B = 0$.
(*b*) $\bar{\mathbf{v}}_A = 2\bar{v}_1/7$ →; $\bar{\mathbf{v}}_B = 5\bar{v}_1/7$ →.
17.119 $5v_0/4r$.
17.120 (*a*) $\bar{\mathbf{v}}_A = \bar{v}_0 \sin\theta\ \mathbf{j}$, $\bar{\mathbf{v}}_B = \bar{v}_0\cos\theta\ \mathbf{i}$;
$\omega_A = (\bar{v}_0/r)(-\sin\theta\mathbf{i} + \cos\theta\mathbf{j})$, $\omega_B = 0$. (*b*) $\frac{5}{7}(v_0\cos\theta)\mathbf{i}$.
17.123 (*a*) 118.7 rev. (*b*) 7.16 s.
17.124 (*a*) $\omega = 3\bar{v}_1/L$ ↻; $\bar{\mathbf{v}} = \bar{v}_1/2$ ↓.
(*b*) $\omega = 3\bar{v}_1/L$ ↺; $\bar{\mathbf{v}} = \bar{v}_1/2$ ↑. (*c*) $\omega = 0$; $\bar{\mathbf{v}} = \bar{v}_1$ ↑.
17.125 (*a*) 2.06 ft. (*b*) 4 lb.
17.126 (*a*) 0.566 s. (*b*) 4 lb.
17.128 (*a*) 2.55 m/s ↑. (*b*) 10.53 N.
17.129 $\omega = \omega_1/4$ ↺; $\bar{\mathbf{v}} = \omega_1 L/8$ ↑.
17.131 $2r/5$.

CHAPTER 18

18.1 (0.480 kg · m²/s)**i** − (0.360 kg · m²/s)**k**.
18.2 (2.18 kg · m²/s)**i** + (5.99 kg · m²/s)**j**.
18.3 0.246 lb · ft · s; $\theta_x = 48.6°$, $\theta_y = 41.4°$, $\theta_z = 90°$.
18.4 $\dfrac{ma^2\omega}{12}[3\mathbf{j} + 2\mathbf{k}]$.
18.5 (0.307 kg · m²/s)**i** + (1.229 kg · m²/s)**j**.
18.6 (0.08 kg · m²/s)**j** + (0.48 kg · m²/s)**k**.
18.9 (3.42 kg · m²/s)**i** − (0.845 kg · m²/s)**j** + (2.07 kg · m²/s)**k**.
18.10 (2.05 kg · m²/s)**i** + (3.36 kg · m²/s)**j** + (0.48 kg · m²/s)**k**.
18.11 (*a*) 8.43 rad/s. (*b*) 0.1596 rad/s.
18.12 (0.640 lb · ft · s)**i** − (0.018 lb · ft · s)**j** − (290 lb · ft · s)**k**.
18.15 (*a*) (1.563 kg · m²/s)**i** − (0.938 kg · m²/s)**k**. (*b*) 31.0°.
18.16 (*a*) (1.563 kg · m²/s)**i** − (0.938 kg · m²/s)**k**.
(*b*) (1.563 kg · m²/s)**i** − (0.938 kg · m²/s)**k**.
18.17 (*a*) (0.0725 lb · ft · s)**i** + (0.248 lb · ft · s)**j**.
(*b*) −(0.590 lb · ft · s)**i** + (0.248 lb · ft · s)**j**.
18.18 (*a*) (0.0725 lb · ft · s)**i** + (0.248 lb · ft · s)**j**.
(*b*) (0.445 lb · ft · s)**i** + (0.248 lb · ft · s)**j**.
18.21 93.6 kg.
18.22 2.57 s.
18.23 (*a*) (4.63 ft/s)**j**.
(*b*) −(3.49 rad/s)**i** − (3.09 rad/s)**j** + (9.26 rad/s)**k**.
18.24 (*a*) −(4.63 ft/s)**i**. (*b*) −(2.39 rad/s)**j** − (3.09 rad/s)**k**.

18.25 (*a*) $(F\Delta t/m)\mathbf{i}$. (*b*) $(12F\Delta t/7ma)(-\mathbf{j} + 5\mathbf{k})$.
18.26 (*a*) $(F\Delta t/m)\mathbf{i}$. (*b*) $(12F\Delta t/7ma)(2\mathbf{j} - 3\mathbf{k})$.
18.29 (*a*) $(\omega_0/6)(-\mathbf{i} + \mathbf{j})$. (*b*) $(a\omega_0/6)\mathbf{k}$.
18.30 (*a*) $(5ma\omega_0/24)\mathbf{k}$. (*b*) $-(ma\omega_0/24)\mathbf{k}$.
18.31 $\frac{3}{7}\bar{v}_0\left(\frac{1}{c}\mathbf{i} + \frac{1}{a}\mathbf{k}\right)$.
18.32 (*a*) $-\frac{6}{7}\bar{v}_0\mathbf{j}$. (*b*) $\frac{1}{7}m\bar{v}_0\mathbf{j}$.
18.33 (0.0125 rad/s)**i** − (0.1396 rad/s)**j** − (0.1814 rad/s)**k**.
18.34 (*a*) −0.726 rad/s.
(*b*) −(729 m/s)**i** − (2880 m/s)**j** + (558 m/s)**k**.
18.39 1.2 J.
18.40 376 J.
18.41 0.978 ft · lb.
18.42 $ma^2\omega^2/8$.
18.43 23.5 J.
18.44 9.59 J.
18.47 9.38 J.
18.48 237 J.
18.49 1.491 ft · lb.
18.51 $5ma^2\omega_0^2/48$.
18.52 $m\bar{v}_0^2/14$.
18.55 (1.8 N · m)**j**.
18.56 −(274 N · m)**k**.
18.57 (2.22 lb · ft)**k**.
18.58 $(ma^2\omega^2/6)\mathbf{i}$.
18.59 (9.83 N · m)**k**.
18.60 (1.92 N · m)**i**.
18.63 **A** = −(2.01 N)**j**; **B** = (2.01 N)**j**.
18.64 **A** = (2.58 N)**k**; **B** = −(2.58 N)**k**.
18.65 (*a*) 0.0567 in. above the axis,
$I_{xy} = 1.759 \times 10^{-3}$ lb · ft · s², $I_{zx} = 0$.
(*b*) 0.498 oz at *A*; 5.50 oz at *E*.
18.66 (*a*) 0.0453 in. below the axis,
$I_{xy} = -2.35 \times 10^{-3}$ lb · ft · s², $I_{zx} = 0$.
(*b*) $\mathbf{F}_C$ = −(41.7 lb)**j**; $\mathbf{M}_C$ = −(20.8 lb · ft)**k**.
18.69 (*a*) 212 rad/s². (*b*) **A** = (4.27 N)**k**; **B** = −(4.27 N)**k**.
18.70 (*a*) 0.455 N · m. (*b*) **A** = (0.966 N)**j**; **B** = −(0.966 N)**j**.
18.71 (*a*) 34.8. rad/s². (*b*) **A** = −(0.450 lb)**i**; **B** = (0.450 lb)**i**.
18.72 **A** = −(0.450 lb)**i** − (5.65 lb)**j**; **B** = (0.450 lb)**i** + (5.65 lb)**j**.
18.75 (*a*) (0.1885 N · m)**i**.
(*b*) **A** = −(0.16 N)**j** + (0.16 N)**k**;
B = (0.16 N)**j** − (0.16 N)**k**.
18.76 **A** = −(2.17 N)**j** − (1.851 N)**k**;
B = (2.17 N)**j** + (1.851 N)**k**.
18.77 (*a*) 29.9 lb · ft. (*b*) 29.9 lb · ft.
18.78 3320 lb · ft.
18.79 −(0.990 N · m)**i**.
18.80 **A** = (1.527 N)**j**; **B** = −(1.527 N)**j**.
18.83 (*a*) 70.9°. (*b*) 5.72 rad/s.
18.84 (*a*) 5.55 rad/s. (*b*) 4.67 rad/s.
18.85 (*a*) 26.6°. (*b*) 8.02 rad/s.
18.86 (*a*) 7.47 rad/s. (*b*) 6.95 rad/s.
18.87 7.89 rad/s.
18.88 15.24 rad/s.
18.91 **A** = (0.884 N)**k**; **B** = −(0.884 N)**k**.
18.92 6.79 rad/s.
18.93 (*a*) **C** = −(122.6 lb)**j**; **D** = (122.6 lb)**j**. (*b*) **C** = **D** = 0.
18.94 39.2 rpm.
18.95 **R** = −(20.5 N)**i**; $\mathbf{M}_O^R$ = (0.960 N · m)**i** − (4.10 N · m)**k**.
18.96 **D** = −(34.4 N)**j** + (21.6 N)**k**; **E** = −(8.8 N)**j** + (21.6 N)**k**.
18.99 **R** = −(14.17 N)**i** − (10.63 N)**k**; $\mathbf{M}_A^R$ = (94.1 N · m)**i**.
18.100 (*a*) (8.5 N · m)**j**.
(*b*) **R** = −(82.6 N)**i** − (81.9 N)**k**; $\mathbf{M}_A^R$ = (240 N · m)**i**.
18.101 **D** = −(5.69 lb)**i** + (6.94 lb)**j**;
E = −(5.49 lb)**i** − (1.346 lb)**j**.
18.102 (*a*) (0.423 lb · ft)**k**.
(*b*) **D** = −(5.43 lb)**i** + (7.25 lb)**j**;
E = −(5.43 lb)**i** − (1.035 lb)**j**.
18.105 $\dot{\psi}$ = 3010 rpm.
18.106 $\dot{\varphi}$ = −59.1 rpm, 35.8 rpm.
18.107 302 rpm.
18.110 (*c*) −6.13%.
18.111 36.9°.
18.112 (*a*) 130.1 rad/s. (*b*) 57.7 mm.
18.113 (*a*) 40°. (*b*) 25.8°. (*c*) 72.4°.
18.114 (*a*) 63.2 rad/s. (*b*) 5.97 rad/s.
18.122 (*a*) 13.19°. (*b*) 1242 rpm (retrograde).
18.123 (*a*) β = 23.8°. (*b*) Precession, 82.6 rpm; spin, 128.8 rpm.
18.126 (*a*) θ_x = 44.1°, θ_y = 45.9°, θ_z = 90°.
(*b*) 61.3 rev/h. (*c*) 6.68 rev/h.
18.127 (*a*) θ_x = 90°, θ_y = 23.1°, θ_z = 66.9°.
(*b*) 46.4 rev/h. (*c*) 6.68 rev/h.
18.128 (*a*) 109.4 rpm; γ_x = 90°, γ_y = 100.05°, γ_z = 10.05°.
(*b*) θ_x = 90°, θ_y = 113.9°, θ_z = 23.9°.
(*c*) Precession, 47.1 rpm; spin, 64.6 rpm.
18.129 (*a*) $-45° \le \theta \le 45°$. (*b*) 6 rad/s. (*c*) 4.90 rad/s.
18.130 (*a*) 4 rad/s. (*b*) 5.66 rad/s.
18.131 (*a*) 8 rad/s. (*b*) 11.31 rad/s.
18.132 (*a*) $-40° \le \theta \le 40°$. (*b*) 12.69 rad/s. (*c*) 9.16 rad/s.
18.137 (*a*) 47.0°. (*b*) Spin, 307 rad/s; precession, 15.25 rad/s.
18.138 (*a*) 76.3°. (*b*) Spin, 294 rad/s; precession, 9.62 rad/s.
(*c*) 36.5°.
18.145 $\frac{1}{4}mr^2(\omega_2\mathbf{j} + 2\omega_1\mathbf{k})$.
18.146 $\frac{1}{8}mr^2(\omega_2^2 + 2\omega_1^2)$.
18.147 (*a*) −(0.3 m/s)**k**. (*b*) −(0.962 rad/s)**i** − (0.577 rad/s)**j**.
18.149 (*a*) $\frac{1}{8}\omega_0(-\mathbf{i} + \mathbf{j})$. (*b*) $0.0884\omega_0 a\mathbf{k}$.
18.150 (*a*) $0.1031\, m\omega_0 a\mathbf{k}$. (*b*) $-0.01473\, m\omega_0 a\mathbf{k}$.
18.151 **A** = (3.35 lb)**k**; **B** = −(3.35 lb)**k**.
18.153 (*a*) 60 rad/s². (*b*) **A** = −(15 N)**k**; **B** = +(15 N)**k**.
18.154 (*a*) 53.6°. (*b*) 8.79 rad/s.
18.155 **C** = −(89.8 N)**i** + (52.8 N)**k**;
D = −(89.8 N)**i** − (52.8 N)**k**.

CHAPTER 19

19.1 1.795 m/s; 8.06 m/s².
19.2 0.1026 in.; 5.16 in./s.
19.3 4.33 ft/s; 0.551 Hz.
19.4 (*a*) 11.29°. (*b*) 1.933 m/s².
19.5 (*a*) 0.495 Hz. (*b*) 13.02 in./s.
19.6 (*a*) 6.13 mm; 6.37 Hz. (*b*) 245 mm/s; 9.81 m/s².
19.8 (*a*) 280 rpm. (*b*) 1.703 m/s.
19.9 (*a*) 60 mm; 0.955 Hz. (*b*) 0.220.
19.10 (*a*) 3.14 s. (*b*) 12.81 ft. (*c*) 38.7°.
19.12 (*a*) 77.1 ft. (*b*) 3320 ft/s².
19.13 (*a*) 0.0247 s. (*b*) 2.29 m/s ↑; 17.14 m/s² ↓.

19.15 (*a*) 1.288°. (*b*) 0.874 ft/s; 0.759 ft/s^2.
19.17 (*a*) 0.943 s. (*b*) 2.45 m/s ↓.
19.18 (*a*) 0.203 s; 4.93 Hz. (*b*) 1.859 m/s; 57.6 m/s^2.
19.20 (*a*) 0.361 s; 2.77 Hz. (*b*) 0.765 m/s; 13.31 m/s^2.
19.21 2.63 s.
19.22 (*a*) 96 lb/ft. (*b*) 4.85 Hz. (*c*) 1 lb.
19.24 (*a*) 1378 lb/ft. (*b*) 910 lb.
19.25 (*a*) 19.37 lb. (*b*) 9.28 lb/ft.
19.26 (*a*) 6.22 kN/m. (*b*) 2.27 kg.
19.27 $k = 8.22$ kN/m; $k_A = 32.9$ kN/m.
19.29 (*a*) 1536 kN/m. (*b*) 7.20 Hz.
19.30 (*a*) 5130 lb/in. (*b*) 121.2 rpm.
19.31 (*a*) 60.1 mm. (*b*) 1.437 Hz.
19.32 (*a*) 427 kg. (*b*) 68.5 mm/s.
19.35 19.95°.
19.36 (*a*) 1.554 s. (*b*) 1.581 s. (*c*) 2.13 s.
19.37 76.5 in.
19.38 (*a*) 4.27 Hz. (*b*) 0.1006 ft.
19.40 (*a*) 7.12 Hz. (*b*) 80 rad/s^2.
19.42 (*a*) 0.210 s. (*b*) 14.93 in./s.
19.43 1.083 in.
19.44 (*a*) 0.534 s. (*b*) 0.491 rad/s.
19.45 (*a*) $6.82\sqrt{r/g}$. (*b*) $7.38\sqrt{r/g}$.
19.46 (*a*) $6.33\sqrt{b/g}$. (*b*) $6.67\sqrt{b/g}$.
19.47 (*a*) 2.81 s. (*b*) 6.44 ft.
19.48 75.5°.
19.49 0.508 Hz.
19.51 (*a*) 1.794 s. (*b*) 1.651 s.
19.55 (*a*) 3.01 Hz. (*b*) 5.64 lb/ft.
19.56 1.815 Hz.
19.58 $f_n = (1/2\pi)\sqrt{\dfrac{2k}{3m} + \dfrac{4g}{3L}}$.
19.59 (*a*) 3.50 in./s. (*b*) 3.38 in./s.
19.60 0.01345 rad/s ↺; 0.1268 rad/s^2 ↺.
19.62 0.365 m/s^2.
19.63 (*a*) 29.6 lb. (*b*) 1.838 s.
19.65 (*a*) 0.413 s. (*b*) 85.0 mm/s.
19.67 (*a*) $\pi\sqrt{l/g}$. (*b*) $2\pi\sqrt{l/g}$.
19.68 5.22 s.
19.69 $f_n = (1/2\pi)\sqrt{\dfrac{k}{5m}}$.
19.70 2.82 s.
19.71 12.11 in./s.
19.74 1.048 s.
19.75 1.460 kg · m^2.
19.76 (*a*) $r_a = 6.40$ in. (*b*) 3.03 in.
19.77 $f_n = (1/2\pi)\sqrt{\dfrac{2k}{3m} + \dfrac{4g}{3L}}$.
19.80 0.279 s.
19.81 0.312 s.
19.82 0.831 s.
19.84 0.567 Hz.
19.85 1.598 s.
19.86 $\tau_n = 2\pi\sqrt{\dfrac{12r^2 + 2l^2}{3gl}}$.
19.87 $\tau_n = 2\pi\sqrt{\dfrac{60r^2 + 10l^2}{9gl}}$.
19.89 $f_n = (1/2\pi)\sqrt{\dfrac{12k}{7m} + \dfrac{8g}{7\sqrt{3}l}}$.
19.90 1.192 s.
19.92 2.99 kg.
19.94 0.866 Hz.
19.95 1.336 s.
19.96 (*a*) 0.423 s. (*b*) 0.423 s.
19.97 $\dfrac{\pi l}{\sqrt{3gr}}$.
19.98 50.1 lb.
19.99 (*a*) 37.1 mm (in phase). (*b*) 260 mm (in phase).
19.101 $\sqrt{2k/3m} < \omega_f < \sqrt{4k/3m}$.
19.102 35.5 rad/s $< \omega_f <$ 44.1 rad/s.
19.104 (*a*) 174.6 mm. (*b*) 8.73 N.
19.106 $\omega_f <$ 11.07 rad/s; $\omega_f >$ 19.18 rad/s.
19.107 (*a*) 0.351 rad/s. (*b*) 7.02 ft/s^2.
19.109 211 mm.
19.110 6.26 rad/s.
19.111 (*a*) 2.14 in. (*b*) $-0.1669 \sin \pi t$ (lb).
19.114 457 rpm.
19.115 $\omega \leq$ 310 rpm; $\omega \geq$ 316 rpm.
19.116 (*a*) 0.076 mm (out of phase). (*b*) 0.019 mm (in phase). (*c*) 0.499 mm (out of phase).
19.117 70.2 lb.
19.118 149.3 mm.
19.119 $\omega <$ 286 rpm; $\omega >$ 367 rpm.
19.120 (*a*) 4.17%. (*b*) 106.1 Hz.
19.123 (*a*) 1400 rpm. (*b*) 0.416 mm.
19.124 (*a*) 10.93 mi/h. (*b*) 0.0255 ft (out of phase).
19.125 (*a*) 5.47 lb/ft. (*b*) 0.915 ft (out of phase).
19.131 (*a*) 0.01393. (*b*) 0.546 N · s/m.
19.133 1.831 lb.
19.134 0.358 ft/s ↑.
19.135 (*a*) 7790 lb/ft. (*b*) 0.1939 s.
19.136 (*a*) $\dfrac{d^2\theta}{dt^2} + (3c/m)\dfrac{d\theta}{dt} + (3k/4m)\theta = 0$. (*b*) $\sqrt{km/3}$.
19.137 (*a*) $0.0945\dfrac{d^2\theta}{dt^2} + 0.446\dfrac{d\theta}{dt} + 1.643\theta = 0$.
(*b*) 24.2×10^{-6} rad (above the horizontal).
19.139 127.4 kN/m.
19.140 $c/c_c \geq \sqrt{1/2}$.
19.143 (*a*) 0.01653 in. (out of phase). (*b*) 0.01622 in.
19.144 (*a*) 0.1269. (*b*) 462 N · s/m.
19.146 0.477.
19.151 $m\dfrac{d^2x_A}{dt^2} + c\left(\dfrac{dx_A}{dt} - \dfrac{dx_B}{dt}\right) + 2k(x_A - x_B) = P_m \sin \omega_f t$.

$$m\frac{d^2x_B}{dt^2} + 3c\frac{dx_B}{dt} - c\frac{dx_A}{dt} + 3kx_B - 2kx_A = 0.$$

19.152 $R < 2\sqrt{L/C}$.
19.153 (*a*) E/R. (*b*) L/R.
19.156 (*a*) $c\dfrac{d}{dt}(x_A - x_m) + kx_A = 0$.

$$m\frac{d^2x_m}{dt^2} + c\frac{d}{dt}(x_m - x_A) = P_m \sin \omega_f t.$$

(*b*) $R\dfrac{d}{dt}(q_A - q_m) + \dfrac{1}{C}q_A = 0$.

$$L\frac{d^2q_m}{dt^2} + R\frac{d}{dt}(q_m - q_A) = E_m \sin \omega_f t.$$

19.157 (*a*) $c_1\frac{dx_A}{dt} + (k_1 + k_2)x_A - k_2x_m = 0.$

$m\frac{d^2x_m}{dt^2} + c_2\frac{dx_m}{dt} + k_2(x_m - x_A) = 0.$

(*b*) $R_1\frac{dq_A}{dt} + \left(\frac{1}{C_1} + \frac{1}{C_2}\right)q_A - \frac{1}{C_2}q_m = 0.$

$L\frac{d^2q_m}{dt^2} + R_2\frac{dq_m}{dt} + \frac{1}{C_2}(q_m - q_A) = 0.$

19.158 (*a*) 2.28 s. (*b*) 1.294 m.
19.159 1.456 Hz.
19.160 $\omega_f > 27.8$ rad/s.
19.161 $\omega_f > 28.0$ rad/s.
19.162 3.21 m/s^2.
19.163 $(1/2\pi)\sqrt{2\mu_k g/l}$.
19.164 56.9 mm.
19.167 1.327 s.

SI Prefixes

Multiplication Factor	Prefix†	Symbol
1 000 000 000 000 = 10^{12}	tera	T
1 000 000 000 = 10^{9}	giga	G
1 000 000 = 10^{6}	mega	M
1 000 = 10^{3}	kilo	k
100 = 10^{2}	hecto‡	h
10 = 10^{1}	deka‡	da
0.1 = 10^{-1}	deci‡	d
0.01 = 10^{-2}	centi‡	c
0.001 = 10^{-3}	milli	m
0.000 001 = 10^{-6}	micro	μ
0. 000 000 001 = 10^{-9}	nano	n
0.000 000 000 001 = 10^{-12}	pico	p
0.000 000 000 000 001 = 10^{-15}	femto	f
0.000 000 000 000 000 001 = 10^{-18}	atto	a

† The first syllable of every prefix is accented so that the prefix will retain its identity. Thus, the preferred pronunciation of kilometer places the accent on the first syllable, not the second.

‡ The use of these prefixes should be avoided, except for the measurement of areas and volumes and for the nontechnical use of centimeter, as for body and clothing measurements.

Principal SI Units Used in Mechanics

Quantity	Unit	Symbol	Formula
Acceleration	Meter per second squared	. . .	m/s^2
Angle	Radian	rad	†
Angular acceleration	Radian per second squared	. . .	rad/s^2
Angular velocity	Radian per second	. . .	rad/s
Area	Square meter	. . .	m^2
Density	Kilogram per cubic meter	. . .	kg/m^3
Energy	Joule	J	$\text{N} \cdot \text{m}$
Force	Newton	N	$\text{kg} \cdot \text{m/s}^2$
Frequency	Hertz	Hz	s^{-1}
Impulse	Newton-second	. . .	$\text{kg} \cdot \text{m/s}$
Length	Meter	m	‡
Mass	Kilogram	kg	‡
Moment of a force	Newton-meter	. . .	$\text{N} \cdot \text{m}$
Power	Watt	W	J/s
Pressure	Pascal	Pa	N/m^2
Time	Second	s	‡
Velocity	Meter per second	. . .	m/s
Volume, solids	Cubic meter	. . .	m^3
Liquids	Liter	L	$10^{-3}\ \text{m}^3$
Work	Joule	J	$\text{N} \cdot \text{m}$

† Supplementary unit (1 revolution = 2π rad = 360°).

‡ Base unit.

U.S. Customary Units and Their SI Equivalents

Quantity	U.S. Customary Units	SI Equivalent
Acceleration	ft/s^2	$0.3048\ m/s^2$
	$in./s^2$	$0.0254\ m/s^2$
Area	ft^2	$0.0929\ m^2$
	in^2	$645.2\ mm^2$
Energy	ft · lb	1.356 J
Force	kip	4.448 kN
	lb	4.448 N
	oz	0.2780 N
Impulse	lb · s	4.448N · s
Length	ft	0.3048 m
	in.	25.40 mm
	mi	1.609 km
Mass	oz mass	28.35 g
	lb mass	0.4536 kg
	slug	14.59 kg
	ton	907.2 kg
Moment of a force	lb · ft	1.356 N · m
	lb · in.	0.1130 N · m
Moment of inertia		
Of an area	in^4	$0.4162 \times 10^6\ mm^4$
Of a mass	$lb \cdot ft \cdot s^2$	$1.356\ kg \cdot m^2$
Momentum	lb · s	4.448 kg · m/s
Power	ft · lb/s	1.356 W
	hp	745.7 W
Pressure	lb/ft^2	47.88 Pa
	lb/in^2 (psi)	6.895 kPa
Velocity	ft/s	0.3048 m/s
	in./s	0.0254 m/s
	mi/h (mph)	0.4470 m/s
	mi/h (mph)	1.609 km/h
Volume	ft^3	$0.02832\ m^3$
	in^3	$16.39\ cm^3$
Liquids	gal	3.785 L
	qt	0.9464 L
Work	ft · lb	1.356 J

Centroids of Common Shapes of Areas and Lines

Shape		$\overline{x}$	$\overline{y}$	Area
Triangular area	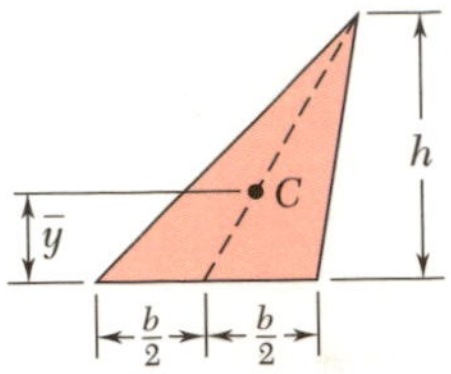		$\frac{h}{3}$	$\frac{bh}{2}$
Quarter-circular area	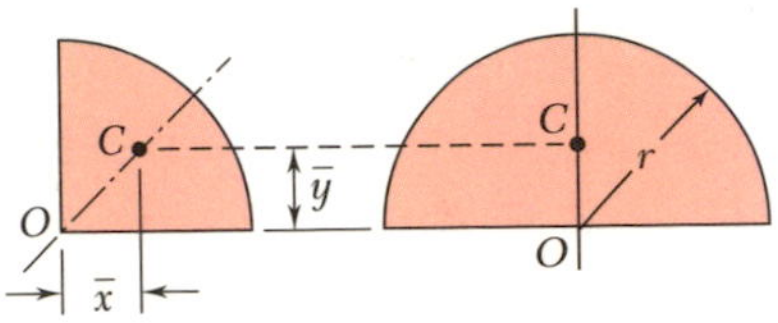	$\frac{4r}{3\pi}$	$\frac{4r}{3\pi}$	$\frac{\pi r^2}{4}$
Semicircular area		0	$\frac{4r}{3\pi}$	$\frac{\pi r^2}{2}$
Semiparabolic area	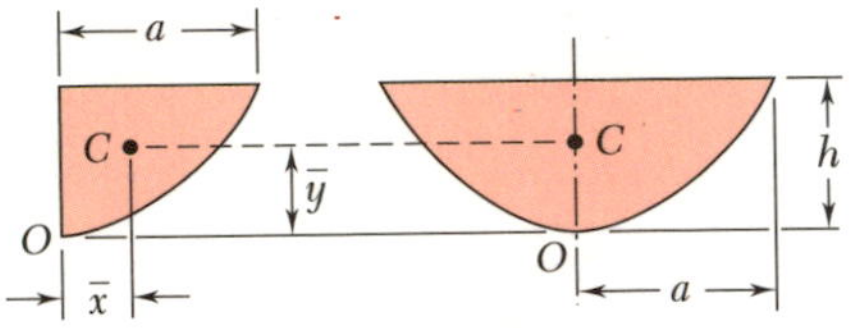	$\frac{3a}{8}$	$\frac{3h}{5}$	$\frac{2ah}{3}$
Parabolic area		0	$\frac{3h}{5}$	$\frac{4ah}{3}$
Parabolic spandrel	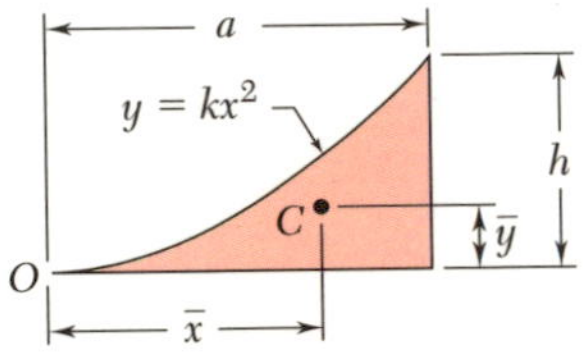	$\frac{3a}{4}$	$\frac{3h}{10}$	$\frac{ah}{3}$
Circular sector	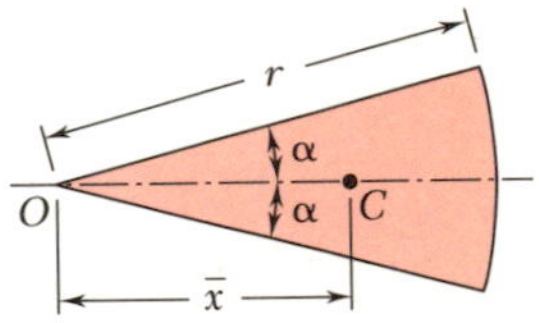	$\frac{2r\sin\alpha}{3\alpha}$	0	αr^2
Quarter-circular arc	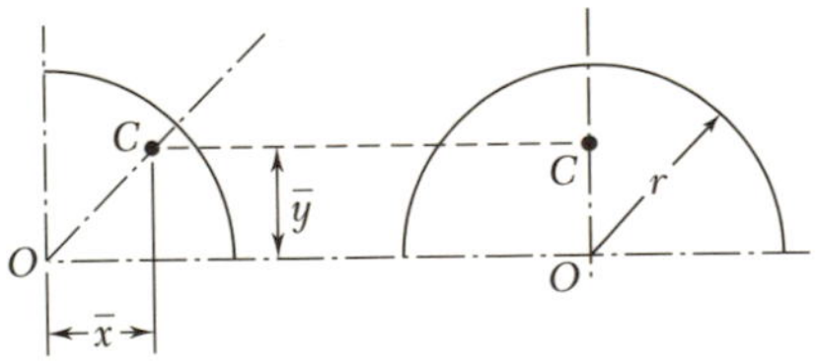	$\frac{2r}{\pi}$	$\frac{2r}{\pi}$	$\frac{\pi r}{2}$
Semicircular arc		0	$\frac{2r}{\pi}$	πr
Arc of circle	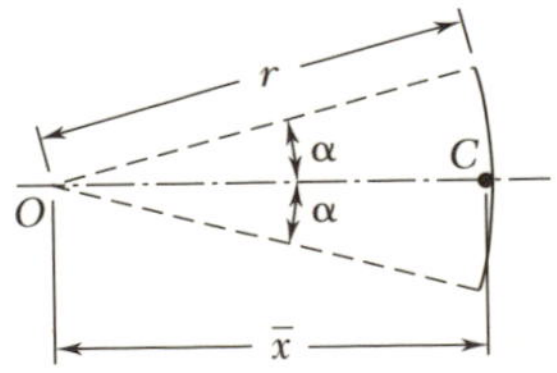	$\frac{r\sin\alpha}{\alpha}$	0	$2\alpha r$